Theory and Applications of Computational Chemistry
The First Forty Years

Theory and Applications of Computational Chemistry The First Forty Years

Editors

CLIFFORD E. DYKSTRA

Department of Chemistry
Indiana University–Purdue University Indianapolis (IUPUI)
Indianapolis, IN, U.S.A.

GERNOT FRENKING

Fachbereich Chemie
Phillips-Universität Marburg
Marburg, Germany

KWANG S. KIM

Department of Chemistry
Pohang University of Science and Technology
Pohang, Republic of Korea

GUSTAVO E. SCUSERIA

Department of Chemistry
Rice University
Houston, TX, U.S.A.

ELSEVIER

2005

Amsterdam – Boston – Heidelberg – London – New York – Oxford
Paris – San Diego – San Francisco – Singapore – Sydney – Tokyo

ELSEVIER B.V.
Radarweg 29
P.O. Box 211, 1000 AE
Amsterdam, The Netherlands

ELSEVIER Inc.
525 B Street
Suite 1900, San Diego
CA 92101-4495, USA

ELSEVIER Ltd.
The Boulevard
Langford Lane, Kidlington
Oxford OX5 1GB, UK

ELSEVIER Ltd
84 Theobalds Road
London WC1X 8RR
UK

First edition 2005

Library of Congress Cataloging in Publication Data
A catalog record is available from the Library of Congress.

British Library Cataloguing in Publication Data
A catalogue record is available from the British Library.

ISBN: 0-444-51719-7 (hardback)
ISBN: 0-444-51904-1 (flexi-cover)

⊗ The paper used in this publication meets the requirements of ANSI/NISO Z39.48-1992 (Permanence of Paper).
Printed in The Netherlands.

Dedication

"The Editors dedicate this volume to those investigators in theoretical and computational chemistry (TACC) who celebrated their 60th birthday in 2004, or thereabouts, and to all those on whom TACC's rapid development over the last 40 years has depended. In particular, we extend our heartfelt thanks and appreciation to one member of that select group who has been a mentor for all four us, Professor H.F. Schaefer. His efforts to put computers to work solving real chemical problems, starting from fundamental principles, have advanced the chemical sciences in countless ways."

Contents

Chapter 1. Computing technologies, theories, and algorithms. The making of 40 years and more of theoretical and computational chemistry
Clifford E. Dykstra, Gernot Frenking, Kwang S. Kim and Gustavo E. Scuseria 1
1.1 Introduction 1
1.2 Technology and methodology 2
1.3 Outlook 6
1.4 Acknowledgements 7
1.5 References 7

Chapter 2. Dynamical, time-dependent view of molecular theory
Yngve Öhrn and Erik Deumens 9
2.1 Introduction 9
2.2 Molecular Hamiltonian 12
2.3 The time-dependent variational principle in quantum mechanics 18
2.4 Coherent states 21
2.4.1 Gaussian wave packet as a coherent state 21
2.4.1.1 Gaussian wave packet with evolving width 25
2.4.2 The determinantal coherent state for *N* electrons 29
2.5 Minimal electron nuclear dynamics (END) 32
2.6 Rendering of dynamics 37
2.7 Acknowledgements 39
2.8 References 39

Chapter 3. Computation of non-covalent binding affinities
J. Andrew McCammon 41
3.1 Introduction 41
3.2 Current methods 42

3.3 Future prospects 43
3.4 Concluding perspective: molecular dynamics simulations and drug discovery 44
3.5 Acknowledgements 45
3.6 References 45

Chapter 4. Electrodynamics in computational chemistry
Linlin Zhao, Shengli Zou, Encai Hao and George C. Schatz 47
4.1 Introduction 47
4.2 Electrodynamics of metal nanoparticles 49
4.2.1 Methods 49
4.2.2 Dielectric constants 50
4.2.3 Spherical particles 51
4.2.4 Effects of particle shape 51
4.2.5 Effects of solvent and of surrounding layers 52
4.2.6 Local electric fields and SERS 55
4.3 Electronic structure studies of surface enhanced Raman spectra 59
4.3.1 Surface models and electronic structure methods 59
4.3.2 Applications 60
4.4 Acknowledgements 63
4.5 References 64

Chapter 5. Variational transition state theory
Bruce C. Garrett and Donald G. Truhlar 67
5.1 Introduction 67
5.2 Gas phase reactions 68
5.2.1 Classical mechanical theory 68
5.2.2 Inclusion of quantum mechanical effects 72
5.2.3 Improved prescriptions for the reaction coordinate and dividing surface 75
5.2.4 Spectroscopy of the transition state 77
5.2.5 Applications 77
5.3 Reactions in condensed phases 77
5.3.1 Reactions in rigid environments and application to reactions in crystals or at crystal–vapor interfaces 78
5.3.2 Reactions in fluid environments with a single reaction coordinate 79
5.3.3 Reactions in fluid environments with an ensemble of reaction coordinates 82
5.4 Summary and conclusions 83
5.5 Acknowledgements 84
5.6 References 84

Chapter 6. Computational chemistry: attempting to simulate large molecular systems
Enrico Clementi 89
6.1 Introduction 89
6.2 The long preparation and the seeding time: 1930–1960 90
6.3 Quantum Chemistry and the Laboratory of Molecular Structure and Spectra, Chicago, 1960 92
6.4 My Hartree–Fock, MC-SCF and density functional period: the 1960 decade 95
6.5 From Schrödinger to Newton; my second simulation period 103
6.6 Statistical and fluid dynamic simulations, and also computers hardware development in the Hudson valley 105
6.7 Back to the beginning: a new approach to an old problem 108
6.8 Conclusions 111
6.9 Acknowledgements 111
6.10 References 111

Chapter 7. The beginnings of coupled-cluster theory: an eyewitness account
Josef Paldus 115
7.1 'Prehistory' 116
7.2 Gestation. 119
7.3 Birth 124
7.4 Growing pains 129
7.5 Maturation 132
7.6 Quo vadis?. 134
7.7 Acknowledgements 140
7.8 References 140

Chapter 8. Controlling quantum phenomena with photonic reagents
Herschel Rabitz. 149
8.1 How can control of quantum dynamics phenomena be achieved?. . . . 149
8.2 Why does quantum control with photonic reagents appear to be so easy? 156
8.3 What is occurring during the process of controlling quantum dynamics phenomena? 158
8.4 Conclusion. 162
8.5 References 162

Chapter 9. First-principles calculations of anharmonic vibrational spectroscopy of large molecules
R.B. Gerber, G.M. Chaban, B. Brauer and Y. Miller 165
9.1 Introduction 165

9.2 Anharmonic vibrational spectroscopy methods . . . 167
9.2.1 Perturbation theory . . . 168
9.2.2 The vibrational self-consistent field approach . . . 169
9.2.3 Grid methods . . . 172
9.2.4 Diffusion quantum Monte Carlo . . . 172
9.2.5 Semiclassical methods . . . 173
9.3 *Ab initio* vibrational spectroscopy . . . 173
9.3.1 Fitting *ab initio* potentials *versus* direct *ab initio* spectroscopy calculations . . . 173
9.3.2 *Ab initio* VSCF and CC-VSCF . . . 174
9.3.2.1 VSCF equations . . . 174
9.3.2.2 Representations of the potential . . . 175
9.3.2.3 CC-VSCF equations . . . 177
9.3.2.4 Anharmonic infrared intensities . . . 178
9.3.2.5 Electronic structure methods used with VSCF . . . 178
9.3.2.6 Improvements and extensions of VSCF and CC-VSCF . . . 179
9.3.3 *Ab initio* anharmonic calculations using perturbation theory . . . 180
9.4 Applications and performance . . . 180
9.4.1 Performance for large molecules . . . 180
9.4.2 Spectroscopy calculations as a test of *ab initio* and DFT force fields . . . 182
9.4.3 Vibrational spectroscopy of hydrogen-bonded clusters . . . 182
9.4.4 *Ab initio* spectroscopy and the identification of new molecular species . . . 185
9.4.5 *Ab initio* spectroscopy and the elucidation of complex spectra . . . 186
9.4.6 Overtones and combination mode transitions . . . 186
9.4.7 Open-shell systems . . . 187
9.5 Future directions . . . 188
9.5.1 Larger systems . . . 188
9.5.2 Quest for increased accuracy . . . 189
9.5.3 Time-domain spectroscopy . . . 189
9.6 Acknowledgements . . . 189
9.7 References . . . 190

Chapter 10. Finding minima, transition states, and following reaction pathways on *ab initio* potential energy surfaces
Hrant P. Hratchian and H. Bernhard Schlegel . . . 195
10.1 Introduction . . . 195
10.2 Background . . . 196

10.2.1 Potential energy surfaces 196
10.2.2 Analytic PES derivatives 198
10.2.3 Coordinate systems 201
10.3 Minimization 202
10.3.1 Newton methods 203
10.3.2 GDIIS 207
10.3.3 QM/MM optimizations 209
10.3.4 Finding surface intersections and points of closest approach 210
10.3.5 Practical considerations. 212
10.3.5.1 Starting structure 212
10.3.5.2 Coordinate system 213
10.3.5.3 Minimization algorithm 215
10.3.5.4 Hessian quality 215
10.3.5.5 Tips for difficult minimizations 217
10.4 Transition state optimization. 218
10.4.1 Local methods 219
10.4.2 Climbing, bracketing, and interpolation methods 220
10.4.3 Path optimization methods 224
10.4.4 Practical considerations. 227
10.4.4.1 Building an initial structure 227
10.4.4.2 Coordinate system 228
10.4.4.3 Algorithm choice 229
10.4.4.4 Hessian quality 230
10.4.4.5 Verifying TSs 230
10.5 Reaction path following 230
10.5.1 First-order methods 232
10.5.2 Second-order methods 234
10.5.3 Higher order integrators 236
10.5.4 Dynamic reaction path 237
10.5.5 Practical considerations. 238
10.5.5.1 Algorithm choice 238
10.5.5.2 Projected frequencies and coupling matrix elements 241
10.5.5.3 Bifurcation 242
10.5.5.4 Tips for difficult reaction path calculations 242
10.6 Summary and outlook 243
10.7 References 243

Chapter 11. Progress in the quantum description of vibrational motion of polyatomic molecules
Joel M. Bowman, Stuart Carter and Nicholas C. Handy. . . . 251
11.1 Introduction 251
11.2 Beyond the harmonic approximation 252

11.3 Vibrational CI theory 254
11.3.1 The *n*-mode representation of the potential 255
11.3.2 Results of selected calculations 257
11.3.3 The 'Reaction Path' version of MULTIMODE 260
11.4 Current bottlenecks and future progress 263
11.5 Acknowledgements 265
11.6 References 265

Chapter 12. Toward accurate computations in photobiology
Adalgisa Sinicropi and Massimo Olivucci 269
12.1 Introduction 269
12.2 *Ab initio* quantum chemical methods for excited states 272
12.3 Fate of light energy in photobiology 276
12.3.1 GFP spectroscopy 278
12.3.2 Rh spectroscopy 279
12.3.3 The photoisomerization path of Rh 282
12.3.4 Nature of the energy storage 284
12.4 From photobiology to biomimetic molecular switches 285
12.5 Conclusions 287
12.6 Acknowledgements 288
12.7 References 288

Chapter 13. The nature of the chemical bond in the light of an energy decomposition analysis
Matthias Lein and Gernot Frenking 291
13.1 Introduction 291
13.2 Energy decomposition analysis 295
13.3 Bonding in main-group compounds 296
13.3.1 Diatomic molecules H_2, N_2, CO, BF 297
13.3.2 Dipnicogens N_2–Bi_2 302
13.3.3 Dihalogens F_2–I_2 303
13.3.4 Nonpolar single bonds of the first octal row H_nE–EH_n ($E = Li$–F; $n = 0$–3) 305
13.3.5 Nonpolar multiple bonds HB=BH, $H_2C{=}CH_2$, HN=NH and HC≡CH 308
13.3.6 Nonpolar group-14 bonds H_3E–EH_3 ($E = C$–Pb) 310
13.3.7 Donor–acceptor bonds Y_3B–NX_3 and Y_3B–PX_3 (X, Y = H, Me, Cl) 311
13.3.8 Main group metallocenes ECp_2 (E = Be–Ba, Zn, Si–Pb) and ECp (E = Li–Cs, B–Tl) 314
13.3.9 Bonding in SF_6 and XeF_6 and a comparison with WF_6 322
13.4 Bonding in transition metal compounds 326
13.4.1 Carbonyl complexes $TM(CO)_6^q$ ($TM^q = Hf^{2-}$, Ta^-, W, Re^+, Os^{2+}, Ir^{3+}) 326

13.4.2	Group-13 heteroleptic diyl complexes $(CO)_4Fe{-}ER$ ($E = B{-}Tl$; R = Cp, Ph, Me) and homolytic diyl complexes $Fe(EMe)_5$ ($E = B{-}Tl$) and $TM(EMe)_4$ (TM = Ni, Pd, Pt; $E = B{-}Tl$)	332
13.4.3	Carbene and carbyne complexes and heavier homologues $(CO)_5W{-}CH_2$, $(CO)_5W{-}E(OH)_2$, $Cl_4W{-}EH_2$, $Cl(CO)_4W{-}EH$ and $Cl_3W{-}EH$ (E = C, Si, Ge, Sn, Pb)	342
13.4.4	Ethylene and acetylene complexes $(CO)_5TM{-}C_2H_x$ and $Cl_4TM{-}C_2H_x$ (TM = Cr, Mo, W), $(CO)_4TM{-}C_2H_x$ (TM = Fe, Ru, Os) and $TM^+{-}C_2H_x$ (TM = Cu, Ag, Au)	347
13.4.5	Phosphane complexes $(CO)_5TM{-}PX_3$ (TM = Cr, Mo, W; X = H, Me, F, Cl)	354
13.4.6	Dihydrogen complexes $TM(CO)_5(\eta^2\text{-}H_2)$ (TM = Cr, Mo, W) and $W(CO)_3X_2(\eta^2\text{-}H_2)$ (X = CO, PH_3, PCl_3, PMe_3)	357
13.4.7	Metallocene complexes $Fe(\eta^5\text{-}E_5)_2$ and $Ti(\eta^5\text{-}E_5)_2^{2-}$ (E = CH, N, P, As, Sb) and bis(benzene)chromium	360
13.5	Conclusion	366
13.6	Acknowledgements	367
13.7	References	367

Chapter 14. Superoperator many-body theory of molecular currents: non-equilibrium Green functions in real time
Upendra Harbola and Shaul Mukamel . . . 373

14.1	Introduction	373
14.2	Dyson equations for superoperator Green functions	375
14.3	The calculation of molecular currents	382
14.4	Discussion	384
14.5	Acknowledgements	384
	Appendix 14A: Superoperator expressions for the Keldysh Green functions	385
	Appendix 14B: Superoperator Green function expression for the current	387
	Appendix 14C: Self-energies for superoperator Green functions	389
	Appendix 14D: Dyson equations in the +/− representation	393
	Appendix 14E: Wick's theorem for superoperators	394
14.6	References	395

Chapter 15. Role of computational chemistry in the theory of unimolecular reaction rates
William L. Hase and Reinhard Schinke . . . 397

15.1	Introduction	398

15.2 Role of computational chemistry . . . 400
15.2.1 The lifetime distribution . . . 400
15.2.2 Intrinsic and apparent non-RRKM behavior . . . 403
15.2.3 Phase space structures . . . 405
15.2.4 Resonance states . . . 409
15.2.5 Steps in unimolecular reaction rates . . . 413
15.2.6 Impact of direct dynamics simulations . . . 415
15.3 The future . . . 419
15.4 Acknowledgements . . . 420
15.5 References . . . 420

Chapter 16. Molecular dynamics: an account of its evolution
Raymond Kapral and Giovanni Ciccotti . . . 425
16.1 Introduction . . . 425
16.2 Early days . . . 426
16.3 Classical period of classical molecular dynamics . . . 428
16.4 Quantum mechanics and molecular dynamics . . . 432
16.5 Coarse grained and mesoscopic dynamics . . . 435
16.6 Conclusion . . . 437
16.7 Acknowledgements . . . 437
16.8 References . . . 438

Chapter 17. Equations of motion methods for computing electron affinities and ionization potentials
Jack Simons . . . 443
17.1 Introduction . . . 443
17.2 Basics of EOM theory as applied to EAs and IPs . . . 445
17.2.1 The EA equations of motion . . . 445
17.2.2 The analogous equations of motion for ionization potentials . . . 447
17.2.3 The rank of the operators . . . 448
17.2.4 Equations of lower rank for both EAs and IPs . . . 449
17.2.5 Summary . . . 449
17.3 Practical implementations of EOM theories for EAs and IPs . . . 450
17.3.1 The Møller–Plesset based approximations . . . 450
17.3.2 Relationship to Greens functions/propagators . . . 453
17.3.3 The natural orbital or extended Koopmans' theorem approach . . . 454
17.3.4 Multiconfiguration-based approximations . . . 454
17.3.5 Coupled-cluster based EOM . . . 455
17.4 Some special cases . . . 457
17.4.1 Calculating EAs as IPs . . . 457
17.4.2 Metastable anion states . . . 457

17.5 Summary 461
17.6 Acknowledgements 461
17.7 References 461

Chapter 18. Multireference coupled cluster method based on the Brillouin–Wigner perturbation theory
Petr Čársky, Jiří Pittner and Ivan Hubač 465
18.1 Introduction 465
18.2 Single-reference *versus* multireference methods 466
18.3 Overview of multireference CC methods 468
18.4 Multireference Brillouin–Wigner coupled cluster method 470
18.5 Intruder states and size extensivity 472
18.6 Performance of the multireference Brillouin–Wigner CC method and applications 476
18.7 Summary 479
18.8 Acknowledgements 479
18.9 References 479

Chapter 19. Electronic structure: the momentum perspective
Ajit J. Thakkar 483
19.1 Introduction 483
19.2 Momentum–space wave functions 484
19.3 Densities and density matrices 487
19.4 Properties of the momentum density 490
19.5 Experimental determination of momentum densities 491
19.6 *Ab initio* computations 494
19.7 Illustrative calculations 494
19.8 Concluding remarks 502
19.9 Acknowledgements 502
19.10 References 502

Chapter 20. Recent advances in *ab initio,* density functional theory, and relativistic electronic structure theory
Haruyuki Nakano, Takahito Nakajima, Takao Tsuneda and Kimihiko Hirao 507
20.1 Introduction 507
20.2 Multireference perturbation theory and valence bond description of electronic structures of molecules 508
20.2.1 Multireference perturbation theory 508
20.2.1.1 Multireference Møller–Plesset perturbation method 509
20.2.1.2 Multiconfigurational quasi-degenerate perturbation theory (MC-QDPT) 512

20.2.1.3 Application of multireference perturbation theory to singlet–triplet splitting of CH_2 and CF_2 513
20.2.1.4 Extension of reference wavefunctions—quasi-degenerate perturbation theory with quasi-complete active space self-consistent field reference functions (QCAS-QDPT) 514
20.2.1.5 Further extension of reference wavefunctions—quasi-degenerate perturbation theory with general-multiconfiguration space self-consistent field reference functions (GMC-QDPT) 516
20.2.1.6 Application of QCAS- and GMC-QDPT 517
20.2.1.6.1 Transition state barrier height for the unimolecular dissociation reaction of formaldehyde $H_2CO \rightarrow H_2 + CO$ 517
20.2.1.6.2 Valence excitation energies for formaldehyde 518
20.2.1.6.3 The most stable structure of SiC_3 521
20.2.2 Valence bond description of complete active space self-consistent field function 523
20.2.2.1 The CASVB method 523
20.2.2.2 Description of electronic structure of benzene 525
20.2.2.3 Description of chemical reaction—hydrogen exchange reactions $H_2 + X \rightarrow H + HX$ ($X = F$, Cl, Br, and I) 526
20.3 Long-range and other corrections for density functionals 529
20.3.1 Conventional correction schemes in density functional theory 529
20.3.2 Long-range correction schemes for exchange functionals 532
20.3.3 Applicabilities of long-range correction scheme. 533
20.3.3.1 Van der Waals calculations 534
20.3.3.2 Time-dependent density functional calculations 534
20.3.3.3 Transition metal dimer calculations. 538
20.3.3.4 Other calculations 540
20.4 Relativistic molecular theory 540
20.4.1 Introduction 540
20.4.2 Four-component relativistic molecular theory 542
20.4.2.1 Dirac–Hartree–Fock and Dirac–Kohn–Sham methods 542
20.4.2.2 Generally contracted Gaussian-type spinors and kinetic balance 543

20.4.2.3 Efficient evaluation of electron repulsion integrals 544
20.4.2.4 Relativistic pseudospectral approach 545
20.4.3 Two-component relativistic molecular theory 548
20.4.3.1 Approximate relativistic Hamiltonians 548
20.4.3.2 RESC method 548
20.4.3.3 Douglas–Kroll method 549
20.4.3.4 Extended Douglas–Kroll transformations applied to the relativistic many-electron Hamiltonian . . . 550
20.5 Summary 553
20.6 References 554

Chapter 21. Semiempirical quantum-chemical methods in computational chemistry
Walter Thiel 559
21.1 Introduction 559
21.2 Historical overview 560
21.3 Established methods 563
21.3.1 Basic concepts 563
21.3.2 MNDO and related methods 564
21.4 Selected recent developments 566
21.4.1 Beyond the MNDO model: orthogonalization corrections 566
21.4.2 Implementation of d orbitals in MNDO-type methods 567
21.4.3 Modified general-purpose methods. 568
21.4.4 Special-purpose parametrizations 569
21.4.5 Computational aspects 570
21.4.6 Linear scaling methods 571
21.4.7 Hybrid methods. 572
21.5 Selected recent applications 573
21.6 Summary and outlook 576
21.7 Acknowledgements 577
21.8 References 577

Chapter 22. Size-consistent state-specific multi-reference methods: a survey of some recent developments
Dola Pahari, Sudip Chattopadhyay, Sanghamitra Das, Debashis Mukherjee and Uttam Sinha Mahapatra 581
22.1 Introduction 582
22.2 The SS-MRCC formalism 589
22.2.1 General developments for the complete model space 589
22.2.2 The use of anonymous parentage for inactive excitations in SS-MRCC method: API-SSMRCC theory 593

22.2.3 Proof of the connectivity of the API-SSMRCC formalism . . . 598
22.3 Emergence of state-specific multi-reference perturbation theory SS-MRPT from SS-MRCC theory . . . 599
22.3.1 Choice of the zeroth order Hamiltonians . . . 601
22.4 Emergence of the SS-MRCEPA(I) methods from SS-MRCC . . . 602
22.5 The size-extensive state-specific MRCC formalism using an IMS . . . 606
22.6 Results and discussion . . . 611
22.6.1 H4 model . . . 613
22.6.2 Insertion of Be into H_2: BeH_2 model . . . 615
22.6.3 LiH molecule . . . 619
22.6.4 BH molecule . . . 624
22.6.5 First and second order electrical property: LiH molecule . . . 627
22.7 Summary and conclusions . . . 629
22.8 Acknowledgements . . . 630
22.9 References . . . 631

Chapter 23. The valence bond diagram approach: a paradigm for chemical reactivity
Sason Shaik and Philippe C. Hiberty . . . 635
23.1 Introduction . . . 635
23.2 VB diagrams for chemical reactivity . . . 638
23.3 VBSCD—the origins of barriers in chemical reactions . . . 638
23.3.1 Bridges, causes, and causality: a VBSCD perspective . . . 640
23.3.2 Comments on quantitative applications of VBSCDs . . . 642
23.3.3 Comments on and some qualitative applications of VBSCDs . . . 643
23.3.3.1 Radical exchange reactions . . . 644
23.3.3.2 Electrocyclic and transition metal catalyzed bond activation reactions . . . 646
23.3.3.3 Reactions between nucleophiles and electrophiles . . . 647
23.3.4 Making stereochemical predictions with the VBSCD model . . . 650
23.4 Valence bond configuration mixing diagrams . . . 652
23.4.1 General features of the VBCMD . . . 652
23.4.2 VBCMD with ionic intermediate curves . . . 652
23.4.2.1 Proton transfer processes . . . 653
23.4.2.2 Nucleophilic substitution on silicon—stable hypercoordinated species . . . 654
23.4.3 VBCMD with intermediates nascent from 'foreign states' . . . 656
23.4.3.1 The $S_{RN}2$ and $S_{RN}2^c$ mechanisms . . . 656

23.5 Additional applications of VB diagrams 658
23.5.1 VBSCD: A general model for electronic delocalization. . . . 658
23.5.2 VBSCD: The twin-state concept and its link to photochemical reactivity 659
23.6 Prospective 663
23.7 Acknowledgements 664
Appendix 23A: Computing mono-determinant VB wave functions with standard *ab initio* programs. 664
23.8 References 665

Chapter 24. Progress in the development of exchange-correlation functionals
Gustavo E. Scuseria and Viktor N. Staroverov 669
24.1 Introduction 669
24.2 Kohn–Sham density functional theory. 671
24.2.1 Motivation for density functional theory 671
24.2.2 Kohn–Sham scheme 673
24.3 Exchange and correlation density functionals 675
24.3.1 Exchange-correlation energy 675
24.3.2 Ingredients of density functional approximations 677
24.3.3 Analytic properties of exchange-correlation functionals . . . 679
24.4 Strategies for designing density functionals 680
24.5 Local density approximations 682
24.5.1 Local density approximation for exchange 682
24.5.2 Local density approximation for correlation 684
24.6 Density-gradient expansion. 686
24.7 Constraint satisfaction 688
24.7.1 Corrections on the asymptotic behavior 688
24.7.2 Normalization of the exchange-correlation hole 692
24.7.3 Systematic constraint satisfaction. 695
24.8 Modeling the exchange-correlation hole 699
24.8.1 Exchange functionals based on a model hole 699
24.8.2 Functionals based on a correlated wave function 701
24.8.3 Functionals based on a model pair correlation function . . . 703
24.8.4 Functionals based on a density matrix expansion 704
24.9 Empirical fits 706
24.10 Mixing exact and approximate exchange 708
24.10.1 Global hybrids 708
24.10.2 Local hybrids. 711
24.10.3 Screened hybrids. 712
24.11 Implementation and performance 714
24.12 Conclusion 716
24.13 Acknowledgements. 717
24.14 References 717

Chapter 25. Multiconfigurational quantum chemistry
Björn O. Roos 725
25.1 Introduction 725
25.2 The density matrix and the natural orbitals 727
25.3 The hydrogen molecule. 730
25.4 Degeneracy and near degeneracy 734
25.4.1 Static and dynamic electron correlation 736
25.5 Multiconfigurational wave functions 738
25.5.1 A brief historical expose 738
25.5.2 The MCSCF wave function 739
25.5.3 The complete active space SCF method 740
25.5.4 Choosing the active space 740
25.5.4.1 Atoms and atomic ions 741
25.5.4.2 Small molecules 741
25.5.4.3 Electronic spectroscopy for organic molecules 742
25.5.4.4 Transition metal compounds 742
25.5.4.5 Lanthanide and actinide chemistry 743
25.6 Dynamic correlation and the CASPT2 method 744
25.7 The relativistic regime 747
25.8 Three examples 748
25.8.1 The ozone molecule 749
25.8.2 The allyl radical 751
25.8.2.1 The ground state 752
25.8.2.2 The electronic spectrum 754
25.8.3 The PbF molecule 756
25.9 Conclusions 760
25.10 Acknowledgements 761
25.11 References 761

Chapter 26. Concepts of perturbation, orbital interaction, orbital mixing and orbital occupation
Myung-Hwan Whangbo 765
26.1 Introduction 765
26.2 Orbital interaction on the basis of effective one-electron Hamiltonian 766
26.2.1 Exact relationship between two sets of molecular orbitals 766
26.2.2 Perturbation analysis and orbital interaction 767
26.2.2.1 Non-degenerate perturbation 768
26.2.2.2 Degenerate perturbation 769
26.2.3 Normal *versus* counterintuitive orbital interaction 770
26.3 Effect of electron–electron repulsion 772

26.3.1 Configuration interaction 773
26.3.2 States with different orbital occupancy 774
26.3.3 Mapping between electronic and spin Hamiltonians 775
26.3.4 Spin polarization 777
26.3.5 Non-equivalent orbital interactions in an open-shell system 778
26.3.6 Orbital ordering in magnetic solids 780
26.4 Spin–orbit coupling and orbital mixing 782
26.5 Concluding remarks 783
26.6 Acknowledgements 783
26.7 References 783

Chapter 27. G2, G3 and associated quantum chemical models for accurate theoretical thermochemistry
Krishnan Raghavachari and Larry A. Curtiss 785
27.1 Introduction 785
27.2 Thermochemical test sets 787
27.3 G2 theory 789
27.3.1 Assessment of G2 theory 793
27.4 G3 theory 794
27.4.1 Assessment of G3 theory 797
27.5 G3X theory 799
27.5.1 Assessment of G3X theory 801
27.6 G3S theory 803
27.7 G3 theory for third-row elements 807
27.8 Applications 808
27.9 Summary and concluding remarks 809
27.10 Acknowledgement 810
27.11 References 810

Chapter 28. Factors that affect conductance at the molecular level
Charles W. Bauschlicher Jr. and Alessandra Ricca 813
28.1 Introduction 813
28.2 Molecular electronics 814
28.3 Carbon nanotubes as molecular sensors 820
28.4 Conclusions and outlook 827
28.5 Acknowledgements 829
28.6 References 829

Chapter 29. The CH···O hydrogen bond: a historical account
Steve Scheiner 831
29.1 Introduction 831

29.2 Early thinking . . . 832
29.2.1 1970s: the beginning of quantum chemical study . . . 833
29.2.2 1980s: more accurate calculations . . . 834
29.2.3 1990s: proliferation and diversification . . . 835
29.3 A surprising observation . . . 837
29.3.1 Status at the end of the 20th century . . . 838
29.4 The 21st century . . . 839
29.4.1 H-bond: to be or not to be? . . . 840
29.4.2 Underlying reasons for blue shift . . . 843
29.4.3 Other approaches—similar findings . . . 846
29.4.4 Another interpretation . . . 847
29.5 Future perspectives . . . 851
29.6 References . . . 852

Chapter 30. *Ab initio* and DFT calculations on the Cope rearrangement, a reaction with a chameleonic transition state
Weston Thatcher Borden . . . 859
30.1 Introduction . . . 859
30.2 Information from experiments about the Cope TS . . . 861
30.3 MINDO/3, AM1, and CASSCF calculations on the Cope TS . . . 862
30.4 Inclusion of dynamic electron correlation . . . 863
30.5 Substituent effects on the Cope rearrangement . . . 865
30.6 Summary . . . 870
30.7 Acknowledgements . . . 872
30.8 References . . . 872

Chapter 31. High-temperature quantum chemical molecular dynamics simulations of carbon nanostructure self-assembly processes
Stephan Irle, Guishan Zheng, Marcus Elstner and Keiji Morokuma . . . 875
31.1 Introduction . . . 876
31.2 Previous theoretical investigations toward fullerene formation mechanisms . . . 877
31.3 Computational methodology . . . 879
31.4 Self-assembly capping process of open-ended carbon nanotubes . . . 880
31.5 Self-assembly of fullerene molecules from ensembles of randomly oriented C_2 molecules . . . 883
31.6 Conclusions . . . 887
31.7 Acknowledgements . . . 887
31.8 References . . . 887

Chapter 32. Computational chemistry of isomeric fullerenes and endofullerenes
Zdeněk Slanina and Shigeru Nagase 891
32.1 Introduction 891
32.2 Relative stabilities of isomers 892
32.3 Energetics and thermodynamics of carbon clusters 893
32.4 Small carbon clusters 896
32.5 Generation of cages 897
32.6 Smaller fullerenes 897
32.7 Higher fullerenes 898
32.8 Endohedral metallofullerenes 901
32.9 Concluding remarks 907
32.10 Acknowledgements 907
32.11 References 908

Chapter 33. On the importance of many-body forces in clusters and condensed phase
Krzysztof Szalewicz, Robert Bukowski and
Bogumil Jeziorski 919
33.1 Introduction 919
33.2 Definitions 921
33.3 Historical perspective 923
33.4 Perturbation theory of intermolecular interactions 927
33.5 Overview of pair contributions 928
33.6 Perturbation theory of nonadditive forces 930
33.7 Comparison of nonadditive effects for selected systems 933
33.8 Physical interpretation of nonadditive components 937
33.8.1 Third-order induction energy 939
33.9 Case studies of nonadditive effects in clusters 942
33.9.1 Helium trimer 942
33.9.2 Argon trimer and condensed phase 943
33.9.3 Ar–Ar–HF trimer 946
33.9.4 $(H_2O)_2$ HCl trimer 947
33.10 Three-body effects in open-shell clusters 948
33.10.1 $Ar_2NO^-(^3\Sigma^-)$ trimer 948
33.10.2 High-spin sodium trimer 948
33.10.3 Ar_2O^- ionic trimer—the case of orbital degeneracy 949
33.11 Water clusters and condensed phase 951
33.11.1 Two-body potentials for water 951
33.11.2 Three-body potentials for water 953
33.11.3 Simulations of liquid water 954
13.12 Acknowledgements 958
13.13 References 958

Chapter 34. Clusters to functional molecules, nanomaterials, and molecular devices: theoretical exploration
Kwang S. Kim, P. Tarakeshwar and Han Myoung Lee 963
34.1 Introduction 963
34.2 Theoretical background 966
34.3 Clusters 967
34.3.1 Aqueous clusters 967
34.3.2 Metallic clusters 974
34.3.3 Weakly bound clusters 976
34.4 Ionophores, receptors, and chemical sensors 980
34.5 Nanomaterials 983
34.6 Molecular devices 987
34.7 Concluding remarks 989
34.8 Acknowledgements 989
34.9 References 989

Chapter 35. Monte Carlo simulations of the finite temperature properties of $(H_2O)_6$
R.A. Christie and K.D. Jordan 995
35.1 Introduction 995
35.2 Methodology 998
35.3 Results 1001
35.3.1 Energetics of $(H_2O)_6$; basis set and thermal effects 1001
35.3.2 Error analysis of the truncated *n*-body approximation for *E*. 1001
35.3.3 Inherent structures 1002
35.3.4 Radial distribution function 1002
35.3.5 Temperature dependence of the energy and heat capacity of $(H_2O)_6$ 1003
35.4 Conclusions 1005
35.5 Acknowledgements 1006
35.6 References 1006

Chapter 36. Computational quantum chemistry on polymer chains: aspects of the last half century
Jean-Marie André 1011
36.1 Introduction 1011
36.2 Electronic structure of polymers: methodology (1965–till date). . . . 1012
36.3 Band structure calculations and photoelectron spectra 1016
36.4 Band structure calculations and (semi)conducting properties (1978–till date) 1020
36.5 Band structure calculations and non-linear optical properties 1025
36.6 Band structure calculations and electron transfer Marcus theory. . . . 1033

36.7 Conclusions 1041
36.8 Acknowledgements 1042
36.9 References 1042

Chapter 37. Forty years of *ab initio* calculations on intermolecular forces
Paul E.S. Wormer and Ad van der Avoird 1047
37.1 Introduction 1047
37.2 Prehistory: before computers 1048
37.3 Antiquity: the sixties 1049
37.3.1 Supermolecular methods 1049
37.3.2 Perturbation methods 1051
37.4 The middle ages: era of mainframes 1053
37.4.1 Unexpanded dispersion 1054
37.4.2 Multipole-expanded dispersion 1056
37.4.3 Applications 1058
37.5 Modern times: revolution and democracy 1059
37.5.1 The SAPT method 1060
37.5.2 The coupled cluster method 1063
37.5.3 Latest developments 1064
Appendix 37A: Relationship between dispersion and $\boldsymbol{E}_{\mathrm{MP2}}^{\mathrm{AB}}$ 1069
37.6 References 1072

Chapter 38. Applied density functional theory and the deMon codes 1964–2004
D.R. Salahub, A. Goursot, J. Weber, A.M. Köster and A. Vela 1079
38.1 Introduction. From the 1920s to the 1960s 1079
38.2 The 1970s 1081
38.3 The 1980s 1083
38.4 The 1990s 1086
38.5 The 2000s 1089
38.6 Résumé 1091
38.7 Acknowledgements 1091
38.8 References 1092

Chapter 39. SAC–CI method applied to molecular spectroscopy
M. Ehara, J. Hasegawa and H. Nakatsuji 1099
39.1 Introduction 1099
39.2 SAC–CI method 1102
39.3 Excited and ionized states of π-conjugated organic compounds 1106

39.3.1 Excitation and ionization spectra of furan and thiophene 1106
39.3.2 *p*-Benzoquinone and its anion radical 1108
39.3.3 Aniline: Effect of the amino-group conformation to the excitation spectrum 1111
39.4 Collision-induced absorption spectra of CsXe SYSTEM 1112
39.5 Transition metal complexes 1115
39.5.1 CrO_2Cl_2 1115
39.5.2 Tetraoxo complexes: CrO_4^{2-}, MoO_4^{2-}, MnO_4^{2-}, TcO_4^{2-}, RuO_4^{2-} and OsO_4^{2-} 1116
39.5.3 Excited states and ^{95}Mo NMR chemical shift of $MoO_{4-n}S_n^{2-}$ $(n = 0–4)$ and $MoSe_4^{2-}$ 1118
39.6 Photochemistry of transition metal complex, $Ni(CO)_4$ 1120
39.7 Porphyrins and related compounds. 1121
39.7.1 Excited states of free-base phthalocyanine 1122
39.7.2 Bacterial photosynthetic reaction center 1124
39.8 Inner-shell ionization spectroscopy 1125
39.8.1 Core-electron binding energy 1125
39.8.2 Inner-shell satellite spectrum 1126
39.8.3 Vibrational spectrum of inner-shell ionization 1127
39.9 Geometries of molecular excited states 1128
39.9.1 Malonaldehyde 1128
39.9.2 Multi-electron processes; C_2 and CO^+ 1129
39.9.3 Acetylene and CNC 1132
39.10 Hyperfine splitting constants 1133
39.11 Summary 1136
39.12 Acknowledgements 1137
39.13 References 1137

Chapter 40. Forty years of Fenske–Hall molecular orbital theory
Charles Edwin Webster and Michael B. Hall 1143
40.1 Introduction 1143
40.2 Illustrative example 1144
40.3 Theory 1146
40.4 Transition metal clusters 1150
40.5 Conclusions 1163
40.6 Acknowledgements 1163
40.7 References 1163

Chapter 41. Advances in electronic structure theory: GAMESS a decade later
Mark S. Gordon and Michael W. Schmidt 1167
41.1 Introduction 1167

41.2	QM methods	1168
	41.2.1 Variational methods	1168
	41.2.2 Many-body methods	1172
	41.2.3 Excited states, non-adiabatic and relativistic methods	1173
	41.2.4 Properties related to nuclear energy derivatives	1175
	41.2.5 Other properties	1176
41.3	Scalable electronic structure theory	1177
41.4	QM/MM Methods	1181
	41.4.1 Discrete solvent approaches	1181
	41.4.2 Surface chemistry	1183
	41.4.3 Continuum solvent methods	1184
41.5	Summary and prognosis	1184
41.6	Acknowledgements	1185
41.7	References	1185

Chapter 42. How and why coupled-cluster theory became the pre-eminent method in an *ab initio* quantum chemistry
Rodney J. Bartlett . . . 1191

42.1	Introduction	1191
42.2	Origins: $\exp(T_2)|0\rangle$	1192
42.3	Higher excitations in CC theory: $\exp(T_1 + T_2 + T_3 + \cdots)|0\rangle$	1197
42.4	Analytical gradients and the CC functional: $E = \langle 0|(1 + \Lambda)\bar{H}|0\rangle$, $E^k = \langle 0|(1 + \Lambda)\bar{H}^k|0\rangle$	1202
42.5	Excited states: $\bar{H}R_k = R_k\omega_k$	1207
42.6	Developments for large molecules and polymers	1213
42.7	Acknowledgements	1216
42.8	References	1216

Biographical sketches of contributors . . . 1223

Subject Index . . . 1267

Theory and Applications of Computational Chemistry: The First Forty Years
Edited by C. Dykstra et al.

CHAPTER 1

Computing technologies, theories, and algorithms. The making of 40 years and more of theoretical and computational chemistry

Clifford E. Dykstra[1], Gernot Frenking[2], Kwang S. Kim[3] and Gustavo E. Scuseria[4]

[1] *Department of Chemistry, Indiana University–Purdue University Indianapolis, Indianapolis, IN 46202, USA*
[2] *Fachbereich Chemie, Philipps-Universität Marburg, Hans-Meerwein-Str., D-35032 Marburg, Germany*
[3] *Department of Chemistry, Pohang University of Science and Technology, San 31, Hyojadong, Namgu, Pohang 790-784, South Korea*
[4] *Department of Chemistry, Rice University, 6100 Main Street, Houston, Texas 77005-1892, USA*

Abstract

In their earliest stages, theoretical chemistry and computational chemistry were often distinct, with the latter having its developments and capabilities directly tied to the state of computer technology. This was especially true for the part of computational chemistry with some of the highest computing demands, *ab initio* electronic structure theory. The point in time where *ab initio* electronic structure calculations were first beginning to be carried out is roughly 40 years ago, and since that time, all parts of theoretical chemistry have developed links to computation. The evolution in such links and connections has had a powerful impact on science, something of historical, theoretical, and practical interest.

1.1 INTRODUCTION

Computational chemists have often given close attention to the specific technologies associated with computer memory, data storage, processor speed, and program development software. In more recent times, the list of technologies can be extended to include processor architecture, data transmission, graphical displays, networking, and interfaces. The state of technology, 40 years ago, provides a fascinating contrast with

References p. 7

contemporary computing capabilities. For instance, 40 years ago, compilers for a powerful high-level language known as FORTRAN were rather new. Prior to that, programs were developed with more root-level instructions *via* languages such as ASSEMBLER. A notable piece of early *ab initio* quantum chemistry work came from the IBM Research Laboratory in California which produced calculational results for a number of linear molecules as complex as HCCCN with a program in the 'FAP Assembly Language for an IBM 7094' [1].

FORTRAN, whose name encodes 'formula translation,' was certainly the most algebraic/number-crunching language when it was introduced. Its usefulness led to a sequence of versions, with FORTRAN IV being a mid-1960s form that was widely available and persisted as a standard version for some time. From the standpoint of writing code, FORTRAN was very well suited to linear algebra algorithms. This had an influence on computational chemistry developments. Theoretical work that cast problems in a matrix form could be quickly exploited for computer calculations. In some cases, presentation of a new method in a form that brought out a matrix structure helped popularize the method. It proved important that interested theoretical chemists were able to see how something could be coded. Without question, compiler technology has at times impacted computational chemistry methodology.

Forty years after the heyday of FORTRAN IV, everything about computing is orders of magnitude better. One could argue that theoretical and computational chemistry simply marched along with the computer developments, but that would not be a fair characterization. In applying theoretical ideas using computers, chemistry consistently pushed the limits of computing capability. As a result, investigators were keen on exploiting any variation in equipment and architecture that became available, and this had the effect of drawing attention to the computational practicability of any new theoretical idea.

This chapter introduces a volume of work that shows the enormous breadth of theoretical and computational chemistry today, a volume which gives insights and historical perspectives on certain crucial developments and establishes how theory and computation have become more and more linked as methodologies and technologies have advanced. This first chapter recalls certain themes or trends in computing technology's influence on computational chemistry, these having been selected as samples, with no claim of being comprehensive. A reverse effect, the role of computational chemistry in shaping computer technology, becomes clear in certain places of the overall story covered by the entire volume. Now, to begin a discussion of historical aspects of computing in chemistry, it is worth noting what was not in existence and not normally available for the first investigators: communication between computers, high-level languages for programming other than FORTRAN, fast and sizable data storage, fast and extensive memory, and even in some cases, the terminals to access computers conveniently.

1.2 TECHNOLOGY AND METHODOLOGY

Through almost two of the first four decades of computational chemistry, a crucial hands-on tool for investigators was the punch card, commonly called an 'IBM card' (Fig. 1.1). These held source codes and input data in a matrix of 80 columns by 12 rows. Editing

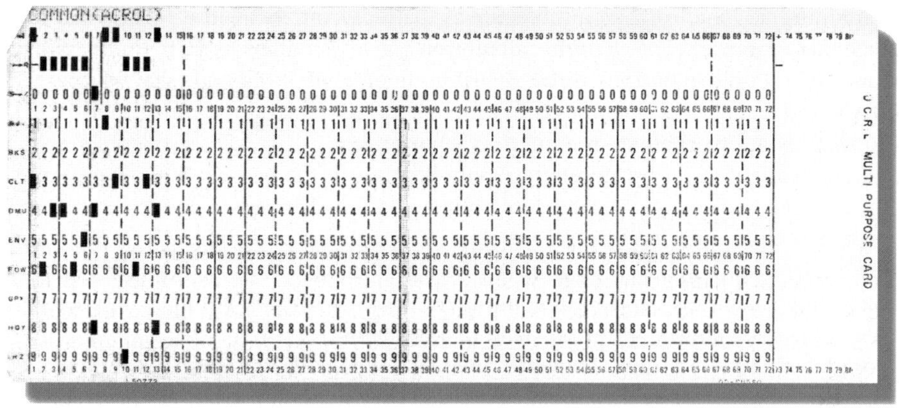

Fig. 1.1. Punch cards for entering data or holding program source code had 80 by 12 spots to be punched or left filled. In other words, there were 960 bits per card. The maximum information in a 15–20 kg box of 2000 cards was a mere 240 kB, though rarely were cards used with formats that allowed for maximum data storage.

a FORTRAN program required keypunching cards and inserting them in proper sequence in a stack of cards. Code could not be generated anywhere near as quickly as is possible with contemporary screen editors and graphical tools. Editing code was a laborious process, and accidentally dropping a box of cards could be a disaster. Yet, a lot of codes were produced and a lot of semi-empirical calculations, *ab initio* calculations, calculations of classical and semi-classical reaction trajectories, statistical mechanical modeling, and quantum dynamical calculations were performed—with punch card communication to the processor.

An interesting marker in time for the start of computational chemistry was the founding of the Quantum Chemistry Program Exchange (QCPE) in 1962. QCPE set out to be a distributor of software that chemists had written, becoming more than that as programming efforts got more involved [2]. The utility of semi-empirical electronic structure methods, when first induced to the QCPE collection, sharply increased QCPE's distribution activity [2]. In many respects, the spread of semi-empirical codes in the earliest days of computational chemistry did more than anything else to show that there would be strong demand for debugged, working code to carry out calculational tasks of interest to chemists. It was the start of what would spark the creation of many of today's for-profit companies that market computational chemistry software.

Electronic structure was the heavyweight among fields of theoretical chemistry when it came to computation in the early days, and QCPE's catalog certainly reflected that. Semi-empirical electronic structure methods, which remain important today, approximate or eliminate many of the electron–electron repulsion integrals of an *ab initio* approach and thereby reduce calculational costs enormously. They were the first methods that could be applied to the electronic structure of real molecular problems, though the approximations and empiricism would present unexpected problems from time to time. By the mid-1970's, *ab initio* SCF calculations for small organic molecules done with modest Gaussian basis sets could be called routine. In fact, as of December 31, 1969, the closing date for the first compilation of *ab initio* calculations by Richards et al. [3], there were already about 300

References p. 7

species (counting an ionic form of a molecule as distinct from its neutral forms) that had been subjected to some level of *ab initio* calculation, mostly small basis SCF. Future Nobel prize winner John Pople undertook direct distribution of an *ab initio* code that was named GAUSSIAN, and even one of the first versions, GAUSSIAN70 (dating from around 1970), proved to be a lasting workhorse in numerous laboratories.

Computers available to most computational chemists in the earliest days were large, central systems (mainframes) where users collected to read-in sets of punch cards and recover some printed output. A small revolution, which we can rightfully dub a mini revolution got its start in the early-to-mid 1970's with the entrepreneurial development of small computing machines called minicomputers, along with low-volume, inexpensive card readers, printers, simple terminals, and disk systems.

Minicomputers did two things. They provided less expensive computing, which meant more computing cycles per year in most cases, and they put control and operation in the hands of the users, which is part of the reason costs were reduced. The Miller–Schaefer experiment with a Datacraft 6024 minicomputer (Fig. 1.2) at Berkeley (University of California) was probably the initial volley in this revolution against expensive, central computer centers. The experiment turned out to be more than finding and demonstrating cost-effectiveness. It also affected methods development, especially in electronic structure. The Final Report to the (US) National Science Foundation on the Miller–Schaefer project titled 'Large Scale Scientific Computation via Minicomputer' indicated that classical trajectory calculations for simple chemical reactions could be handled 'without major alteration of presently existing codes,' but that problems such as MCSCF called for 'algorithms designed to optimally utilize the relatively small memory of the minicomputer.' Minicomputers had substantially less memory than mainframes of the day, and algorithms or even whole new methods were needed. The reformulation

Fig. 1.2. The minicomputer revolution in computational chemistry began as machines like this Datacraft 6024/4 at the University of California (1973) came into use. This was a 24-bit machine (48 bits per word in extended precision). Note the two rows of switches. These could be used to directly set the bit strings for one word of memory at a time, a low-level feature that at least proved helpful for certain crashes.

of the electron correlation problem in terms of matrices, the same size as used in SCF was one example [4,5].

Digital Equipment Corporation (DEC) introduced a sophisticated minicomputer, the VAX 11/780, in the mid- to late-1970s. At prices of 300 k USD (and up), machines could be acquired that were as fast as 1960s multi-million dollar mainframes. Double precision meant 64-bit words, and multiple external disk drives could be added to provide hundreds of MB of storage. DEC was not the only producer of minicomputers, and before long, with different vendors and models, their use for computational chemistry became fairly widespread.

There was another mini revolution following fast on the spread of minicomputers. Parallel and vector computing architectures were emerging by the early 1980's (available to a few even earlier), and computational chemists were adapting algorithms as soon as they could get machine time. However, doing so-called for consideration of many aspects of computer technology. It was not simply moving a box of cards from the central mainframe to the minicomputer site. As pointed out in a 1984 report by J. S. Binkley [6]:

> Modern, high-speed vector processing machines have some idiosyncrasies that most developers of scalar algorithms have not previously encountered... [T]heir I/O capabilities employ peripheral devices that are comparable to those available on super minicomputers. Thus, algorithms that were developed on older machines where CP-time and I/O times were well balanced become terribly I/O bound.

One of the specific types of solutions for *ab initio* electronic structure was direct methods wherein intermediate quantities (two-electron integrals) normally stored on disk were recomputed when needed [7]. Binkley's report went on to say that the effort to adapt to the special features of vector and parallel architectures led to 'the production of better scalar algorithms.' In other words, the basic ideas behind algorithms were influenced by the technology, in this case, computer architecture, and this is really a very significant and constant theme in the evolution of theoretical and computational chemistry.

There has probably been a third mini revolution, though one more diffuse in time. It could be called the workstation or even laptop revolution. Computing technology has advanced so far and so fast that high-level calculations can be performed on battery-powered devices of 100 cubic inches (1.6 L) or so. Memory is inexpensive and external storage can seem limitless on these systems. Commercial software from a half-dozen vendors, if not more, is available for most standard types of computational chemistry. This state of technology has also had an impact on the evolution of theoretical and computational chemistry, and from one view, that impact has been integration of approaches. For instance, molecular mechanics calculations, originally a stand-alone type of computation, are frequently carried out on laptops and workstations to provide an initial geometry for an *ab initio* optimization, and they are directly tied in the emerging QM/MM treatments. *Ab initio* treatments are combined with dynamical and statistical understanding so that calculations often give thermodynamic values, not simply single molecule energies. Molecular mechanics, molecular dynamics, Monte Carlo, and quantum Monte Carlo calculations, newcomers compared to SCF calculations are widespread, and increasingly, they are being incorporated into broad-based computational work on biomolecular problems. The typical capability of these powerful computers, the laptops and workstations, facilitates interfacing and combining what are

References p. 7

otherwise large, stand-alone blocks of code, or at least it provides for a wide variety of calculational tools being available in one computer system.

These days, graphical control (input) and graphical analysis of the results are highly sophisticated, and the user does not always need the detailed theoretical understanding to use the software. This points to a newer theme. It is that in many problem areas, theory has been so effectively translated into computational form that computations are commonplace tools for everyday chemists as much as an IR or NMR spectrometer. It is fair to argue that in some, maybe equal, proportion, hardware advances and algorithmic/methodological advances have yielded the remarkable capabilities for computational chemistry that exist now. Certain chapters in this volume provide a glimpse of the evolution to genuine problem solving *via* computation in areas such as electrical conductance at the molecular level, photobiology, fullerenes, carbon nanostructures, and nanotechnology. There will be more in the next 40 years!

1.3 OUTLOOK

A clear point to this introduction to the volume about the first 40 years of the theory and applications of computational chemistry (TACC) is that the connections between technology and methodology, and between pure theoretical efforts and application *via* computation are firmly in place. As well, computational chemistry has been driven by the ever-evolving (ever-changing) technology of electronic computing both in terms of the applications that can be approached and in the methodology. Advances in hardware technology and in algorithms have been intertwined throughout. Of the many TACC pioneers, a number of whom are pictured here (Figs. 1.3 and 1.4), had there been none who pursued the use of computers to solve chemical problems until the current state of

Fig. 1.3. Birthday cakes for some of the speakers in the Theory and Applications of Computational Chemistry Conference held at Gyeongju, Korea during February 15–20, 2004. Celebrating their 60th birthday in 2004 are, from left to right, Tom Ziegler, Helmut Schwarz, M. H. Whangbo, Ad van der Avoird, Jack Simons, Henry F. Schaefer, Per E. M. Siegbahn, Michael A. Robb, Leo Radom, Hiroshi Nakatsuji, Poul Jørgensen, Kimihiko Hirao, William L. Hase, Michael B. Hall, Mark S. Gordon, R. Benny Gerber, Petr Cársky, Rodney J. Bartlett, Yitzhak Apeloig, Jean-Marie André and Herschel A. Rabitz.

Fig. 1.4. A number of the speakers in the Theory and Applications of Computational Chemistry Conference (Gyeongju, Korea) whose pioneering efforts in the field cover all or most of the last 40 years. From left to right: Michael L. Klein, Björn Roos, Jean-Louis Rivail, Pekka Pyykkö, Keiji Morokuma, Rudolph A. Marcus, Enrico Clementi, Nicholas C. Handy, William H. Miller and Peter Pulay.

computing technology had been achieved, there would likely be differences in the methods, techniques, and algorithms that are in use. Hence, the link with technology is not only a part of the history of computational chemistry, but it is also something that provides insights to where the field stands, why some things are done the way they are done, and hopefully guidance of how forthcoming computing technology might best be used in chemistry. That guidance, of course, is subject to how well we can predict where computing technology goes from here. The pioneers in TACC managed that just right, and most likely the ones to follow will, too.

1.4 ACKNOWLEDGEMENTS

Two of us wish to thank the National Science Foundation for support of work in computational and theoretical chemistry via Grant CHE-9982156 (CED) and Grant CHE-9982156 (GES).

1.5 REFERENCES

1 A.D. McLean and M. Yoshimine, Computation of molecular properties and structures, IBM J. Res. Dev., 12 (1968) 206.
2 R. Counts, The Changing Role of QCPE (www.qcpe.indiana.edu/history/changing_role_1989.html).
3 W.G. Richards, T.E.H. Walker and R.K. Hinkley, A bibliography of ab initio molecular wave functions, Clarendon Press, Oxford, 1971.
4 R. Ahlrichs and F. Driessler, Determination of pair natural orbitals. A new method to solve the multi-configuration Hartree–Fock problem for two-electron wave functions, Theor. Chim. Acta, 36 (1975) 275.
5 W. Meyer, Theory of self-consistent electron pairs. An iterative method for correlated many-electron wavefunctions, J. Chem. Phys., 64 (1976) 2901; C.E. Dykstra, H.F. Schaefer and W. Meyer, A theory of self-consistent electron pairs. Computational methods and preliminary applications, J. Chem. Phys., 65 (1976) 2740.
6 J.S. Binkley, Algorithmic considerations in large mainframe computers, NATO ASI Ser. C, 133 (1984) 209.
7 J. Almlöf and P.R. Taylor, Computational aspects of direct SCF and MCSCF methods, NATO ASI Ser. C, 133 (1984) 107.

Theory and Applications of Computational Chemistry: The First Forty Years
Edited by C. Dykstra et al.

CHAPTER 2

Dynamical, time-dependent view of molecular theory

Yngve Öhrn and Erik Deumens

Quantum Theory Project, Departments of Chemistry and Physics, University of Florida PO BOX 118435, Gainesville, FL 32611-8435, USA

Abstract

In this chapter we present a time-dependent, direct, nonadiabatic theory of molecular processes. We put this approach in contrast to the current theory paradigm of approximate separation of electronic and nuclear dynamics, which proceeds *via* approximate electronic stationary states and corresponding potential energy surfaces (PESs). This established approach in all its variants has provided a basis for qualitative understanding of rate processes and, for systems with few nuclear degrees of freedom; it has produced quantitative data that can be used to guide experiments. In this picture the dynamics of the reacting system takes place on a stationary electronic state potential surface and may under the influence of nonadiabatic coupling terms 'jump' to another potential surface the probability of such transitions often viewed as a statistical 'surface hopping' [J. Chem. Phys., 55 (1971) 562] event. The time-dependent, direct, and nonadiabatic theory presented here is fully dynamical in that the evolving state, which describes the simultaneous dynamics of electrons and nuclei of a reacting system [Rev. Mod. Phys., 66 (3) (1994) 917] changes in time under the mutual instantaneous forces in a Cartesian laboratory system of coordinates. This approach, which has been applied to reactive collisions involving polyatomic molecules over a large range of energies, proceeds without predetermined PESs, uses total molecular wave functions that are parameterized as generalized coherent states, and imposes no constraints on molecular geometries.

2.1 INTRODUCTION

Chemistry may be characterized as the science of transformation of matter at low to moderate energies. The study of chemical reactions, i.e. how products are formed from reactants, is central to chemistry [1]. Since the beginning of modern chemistry one has studied kinetics and attempted to infer reaction mechanisms by varying controllable parameters, such as concentrations, temperature, substituents, etc. and determined the effect on product yields. However, not until recent times has the actual microscopic action of the reacting species been accessible to the chemist and the early theoretical models developed to describe reactions were hence of a phenomenological nature. For instance, the early work of Lindemann, Hinshelwood, Rice, Ramsparger, and Kassel on unimolecular rate theory makes reasonable assumptions about the details of the dynamics of the reacting molecular system without actually knowing whether they are true or not. Later the seminal work of Eyring [2] introducing absolute reaction rate theory, also known as activated complex theory or transition state theory, can be labeled as the beginning of modern theoretical studies of chemical reaction rates by providing interpretations of the Arrhenius rate parameters in terms of molecular structure and properties. Common to this early theory and to most current work on chemical reactions is the notion that the atomic nuclei of participating species move subject to forces derived from a potential.

Chemical reactions in bulk are analyzed in terms of simple steps, called elementary chemical reactions. Such elementary steps can be characterized as encounters of reactant molecules to form product molecules. Such encounters take place in various media such as a solvent or at a metal surface. The perhaps simplest and purest form of elementary reactions take place in gas phase allowing experimental control over initial and final states of participating species with the use of modern laser technology, and thus obtaining detailed information from which to draw conclusions about reaction mechanisms. This is often done in intricate molecular beam experiments involving ultrafast-pulsed lasers to monitor the reacting molecular system on picosecond to femtosecond time scales. In this way the experiment is not limited to viewing only the 'opening act', the reactants, and the 'final act' the separated products, but can actually enjoy the entire 'play' as the reaction proceeds from reactants to products.

The importance of the time parameter in the study of reactions is clear already in the early kinetic studies, where it takes the form of the inverse rate constant. From the point of view of fundamental chemical theory, elementary chemical reactions are simply the detailed dynamics of electrons and atomic nuclei that constitute the total molecular system of reacting species, which is governed by the time-dependent Schrödinger equation

$$H\Psi = \mathrm{i}\hbar \frac{\partial \Psi}{\partial t} \tag{1}$$

It is now a generally accepted view that electrons and atomic nuclei are 'the fundamental particles' of chemistry and that the time-dependent Schördinger equation is the central equation for the study of molecular structure and dynamics, and thus also for chemical reactions in general. In spite of the tremendous advances in the power and speed of electronic computers and in generally available sophisticated software for finding adequate approximate solutions to the Schrödinger equation, accurate treatments are still limited to rather simple systems, for which predictive results can be obtained from theory alone. Nevertheless much can be achieved by approximations to the time-dependent Schrödinger equation and even with approximate solutions to approximate equations.

For time-independent Hamiltonians H one often equivalently studies the time-independent Schrödinger equation

$$H\Phi = E\Phi \tag{2}$$

not as a boundary value problem but seeking solutions for general energy values E. The explicitly time-dependent formulation is more readily treated as an initial value problem.

The disparate masses of even the lightest atomic nucleus and an electron have led to the widely accepted view that an adequate description of low energy processes can be achieved by assuming an effective separation of electronic and nuclear degrees of freedom. Considering the electron dynamics to take place in the field of stationary nuclei leads to the introduction of an electronic Hamiltonian operator H_{el} consisting of the kinetic energy operator of electrons, their mutual Coulombic interactions, and the Coulombic attraction terms to each of the atomic nuclei. The corresponding Schrödinger eigenvalue problem

$$H_{\mathrm{el}}(\vec{R})|n\rangle = E_n(\vec{R})|n\rangle \tag{3}$$

has the solutions $|n\rangle$, which are electronic stationary states with characteristic electronic energies $E_n(\vec{R})$. The lowest eigenvalue $E_0(\vec{R})$ corresponds to the electronic ground state $|0\rangle$ can be obtained at some level of approximation for various chosen nuclear geometries $\vec{R}$ in an internal coordinate system attached to the nuclear framework. For $\bar{N}$ nuclei such a function in $3\bar{N} - 6$ dimensions is commonly interpolated and fitted to some analytical representation to also yield the energy at intermediate nuclear geometries. As a practical matter most electronic structure codes use Cartesian coordinates to solve Eq. (3). Adding the Coulomb repulsion terms of the nuclei to this function one obtains the ground state potential energy surface (PES). Commonly this PES becomes the potential energy for the nuclear dynamics, which can be treated classically, semi-classically, or fully quantum mechanically.

Obviously, the electronic energies $E_n(\vec{R})$ for $n \neq 0$ corresponds in a similar manner to potential surfaces for electronically excited states. Each PES usually exhibits considerable structure for a polyatomic system and will provide useful pictures with reactant and product valleys, local minima corresponding to stable species, and transition states serving as gateways for the system to travel from one valley to another. However, for the number of nuclear degrees of freedom beyond six, i.e. for more than four-atom systems it becomes extremely cumbersome to produce the PES's and quite complicated to visualize the topology. Furthermore, when more than one PES is needed, which is not unusual, there is a need for nonadiabatic coupling terms, which also may need interpolation in order to provide useful information.

For those few systems for which one or more accurate PES have been determined this strategy, of proceeding *via* precalculated potentials, works quite well [3]. Detailed quantum dynamics obtains accurate differential and integral state–state cross sections and rate coefficients in agreement with the best experiments for some small systems. However, as the complexity of the reacting system increases it becomes increasingly difficult to proceed *via* full PESs. One way out of this problem is to identify some active modes and eliminate or discretize degrees of freedom that are either changing slowly throughout the critical part of the dynamics or which are not directly involved. This reduced dimensionality dynamics [4] has been successful in some cases, but introduces some arbitrariness or bias and hard to control errors. Another procedure that has gained recent prominence is so called direct classical dynamics (see, e.g. [5]). This approach calculates the forces on the nuclei during a classical trajectory making it necessary to compute the PES only in those points where the dynamics take the nuclei. Since the reacting system can tumble one commonly performs the calculations in $3\bar{N}-3$ dimensions also including the rotational degrees of freedom but no coupling terms.

In the following sections we consider the Coulombic Hamiltonian of a general molecular system and comment upon the difference between using an internal set of coordinates with axes fixed in the molecular system and employing a Cartesian laboratory system of axes. The study of reactive molecular systems in terms of stationary molecular electronic states and their PESs as described above is put in contrast to a direct, nonadiabatic, time-dependent treatment, which is fully dynamical in that the evolving state, which describes the simultaneous dynamics of participating electrons and nuclei, changes in time under the mutual instantaneous forces. The wave function parameters carry the time-dependence and in the choice of parameters it is useful to consider generalized coherent states. This approach proceeds without predetermined PES, and the dynamical equations that describe the time evolution of the total system state vector are derived using the time-dependent variational principle. A minimal form of this dynamical approach to molecular processes is discussed in some detail.

2.2 MOLECULAR HAMILTONIAN

The molecular Hamiltonian contains a variety of terms. If we limit the description to Coulombic interactions we can write (we are using the subscripts i and j for electron labels and k and l for nuclear labels)

$$H = -\frac{\hbar^2}{2m}\sum_i \nabla^2_{x_i} - \frac{\hbar^2}{2}\sum_k \frac{1}{m_k}\nabla^2_{X_k} - \sum_i\sum_k \frac{Z_k e^2}{|\vec{x}_i - \vec{X}_k|} + \frac{1}{2}\sum_i\sum_{j\neq i}\frac{e^2}{|\vec{x}_i - \vec{x}_j|} + \frac{1}{2}\sum_k\sum_l \frac{Z_k Z_l e^2}{|\vec{X}_k - \vec{X}_l|} \quad (4)$$

where the terms in order are the operators of kinetic energy of the electrons, kinetic energy of the nuclei, the electron–nuclear attraction energy, the electron–electron repulsion energy, and the nuclear–nuclear repulsion energy. Obviously, if it were not for the electron–nuclear attraction terms the electronic and the nuclear energetics and dynamics would be decoupled. Although this term is not small much of molecular quantum mechanics can be performed with an effective decoupling of electronic and nuclear degrees of freedom or rather a discretization of the nuclear coordinates.

One normally proceeds by eliminating the translational motion and chooses the origin of a molecule-fixed coordinate system. A suitable choice is the center of mass of the nuclei. This choice introduces no additional coupling terms between the nuclear and the electronic degrees of freedom. However, it introduces reduced masses and so-called mass polarization terms. Such terms are of the form

$$\frac{\hbar^2}{2M_0}\sum_i \sum_{j\neq i} \nabla_i \nabla_j \tag{5}$$

for the electrons and a similar term for the nuclei, where M_0 is the total mass of all the nuclei. Because of this small factor these terms are small and often neglected. Nevertheless, these omissions must be recognized as a source of errors. The potential energy terms (collectively given the symbol U) all depend on the interparticle distances, which are unaffected by the transformation to internal coordinates. Obviously, the choice of an internal origin of coordinates leads to the elimination of three degrees of freedom, which means, say, that the position of one of the nuclei is dependent on the position of all the others.

Example. We consider a general molecular system in the laboratory frame with the center of mass

$$\vec{\rho} = \frac{1}{M}\left[\sum_k m_k \vec{X}_k + m\sum_i \vec{x}_i\right] \tag{6}$$

The internal coordinates relative to the center of mass of the nuclei are

$$\vec{r}_i = \vec{x}_i - \frac{1}{M_0}\sum_k m_k \vec{X}_k \tag{7}$$

for the electrons, and

$$\vec{R}_l = \vec{X}_l - \frac{1}{M_0}\sum_k m_k \vec{X}_k \tag{8}$$

for the nuclei. The position of one nucleus, say p is then obtained as

$$\vec{R}_p = -\frac{1}{m_p}\sum_{k\neq p} m_k \vec{R}_k \tag{9}$$

from the fact that the center of mass of the nuclei is the origin. In the above expressions we have used the notations $M_0 = \sum_k m_k$ and $M = M_0 + mN$, where N is the number of electrons in the system.

The kinetic energy terms are now altered and we can see how they change by using the chain rule of differentiation. Note that the Cartesian components of the position coordinates are such that

$$\vec{\rho} = (\xi, \eta, \zeta), \qquad \vec{x} = (a, b, c), \qquad \vec{X} = (A, B, C), \qquad \vec{r} = (x, y, z), \qquad \vec{R} = (X, Y, Z)$$

and for example

$$\frac{\partial}{\partial a} = \frac{\partial}{\partial x}\frac{\partial x}{\partial a} + \frac{\partial}{\partial \xi}\frac{\partial \xi}{\partial a} = \frac{\partial}{\partial x} + \frac{m}{M}\frac{\partial}{\partial \xi} \tag{10}$$

yielding

$$\nabla_{x_i} = \nabla_i + \frac{m}{M}\nabla_\rho \tag{11}$$

Similarly we obtain for $k \neq p$

$$\begin{aligned}\frac{\partial}{\partial A_k} &= \frac{\partial}{\partial X_k}\frac{\partial X_k}{\partial A_k} + \frac{\partial}{\partial \xi}\frac{\partial \xi}{\partial A_k} + \sum_l \frac{\partial}{\partial X_l}\frac{\partial X_l}{\partial A_k} + \sum_i \frac{\partial}{\partial x_i}\frac{\partial x_i}{\partial A_k} \\ &= \frac{\partial}{\partial X_k} + \frac{m_k}{M}\frac{\partial}{\partial \xi} - \frac{m_k}{M_0}\sum_{l\neq p}\frac{\partial}{\partial X_l} - \frac{m_k}{M_0}\sum_i \frac{\partial}{\partial x_i}\end{aligned}$$

leading to

$$\nabla_{X_k} = \nabla_k + \frac{m_k}{M}\nabla_\rho - \frac{m_k}{M_0}\sum_{l\neq p}\nabla_l - \frac{m_k}{M_0}\sum_i \nabla_i \tag{12}$$

(for $k \neq p$) and for the particular nucleus p

$$\nabla_{X_p} = \frac{m_p}{M}\nabla_\rho - \frac{m_p}{M_0}\sum_{l\neq p}\nabla_l - \frac{m_p}{M_0}\sum_i \nabla_i \tag{13}$$

Insertion of these expressions in the molecular Hamiltonian yields [6]

$$\begin{aligned}H(p) = &-\frac{\hbar^2}{2M}\nabla_\rho^2 - \frac{\hbar^2}{2}\left[\frac{1}{m} + \frac{1}{M_0}\right]\sum_i \nabla_i^2 - \frac{\hbar^2}{2M_0}\sum_{i\neq j}\nabla_i\nabla_j \\ &-\frac{\hbar^2}{2}\sum_{k\neq p}\left[\frac{1}{m_k} - \frac{1}{M_0}\right]\nabla_k^2 + \frac{\hbar^2}{2M_0}\sum_{\substack{k\neq l \\ k,l\neq p}}\nabla_k\nabla_l + U\end{aligned} \tag{14}$$

In the above example the choice of molecule fixed origin is the center of mass of the nuclei, but any other point could have been chosen, and a different choice will give a different internal Hamiltonian. For instance, if we choose the origin to be centered on a particular nucleus p, which might be the preferred choice if that nucleus has a much greater mass than all the others, then the corresponding Hamiltonian is

$$\begin{aligned}\bar{H}(p) = &-\frac{\hbar^2}{2M}\nabla_\rho^2 - \frac{\hbar^2}{2}\left[\frac{1}{m} + \frac{1}{m_p}\right]\sum_i \nabla_i^2 - \frac{\hbar^2}{2m_p}\sum_{i\neq j}\nabla_i\nabla_j - \frac{\hbar^2}{m_p}\sum_{k\neq p}\sum_i \nabla_k\nabla_i \\ &-\frac{\hbar^2}{2}\sum_{k\neq p}\left[\frac{1}{m_k} + \frac{1}{m_p}\right]\nabla_k^2 - \frac{\hbar^2}{2m_p}\sum_{\substack{k\neq l \\ k,l\neq p}}\nabla_k\nabla_l + U\end{aligned} \tag{15}$$

where coupling terms between the nuclear and the electronic degrees of freedom appears in the kinetic energy.

In general one can use a product wave function with one factor depending only on the center of mass coordinates $\vec{\rho}$ and a second factor depending on the rest, i.e. all the internal coordinates and the three rotational degrees of freedom.

One normally proceeds by defining the electronic Hamiltonian

$$H_{\mathrm{el}} = -\frac{\hbar^2}{2}\left[\frac{1}{m} + \frac{1}{M_0}\right]\sum_i \nabla_i^2 - \frac{\hbar^2}{2M_0}\sum_{i\neq j}\nabla_i\nabla_j - \sum_i\sum_k \frac{Z_k e^2}{|\vec{r}_i - \vec{R}_k|}$$
$$+\frac{1}{2}\sum_i\sum_{j\neq i}\frac{e^2}{|\vec{r}_i - \vec{r}_j|} + \frac{1}{2}\sum_k\sum_l \frac{Z_k Z_l e^2}{|\vec{R}_k - \vec{R}_l|} \tag{16}$$

where the nuclear coordinates are now representing a fixed nuclear framework (classical nuclei held fixed). The solution of the eigenvalue problem of the electronic Schrödinger equation

$$H_{\mathrm{el}}\Phi_k(\vec{r};\vec{R}) = V_k(\vec{R})\Phi_k(\vec{r};\vec{R}) \tag{17}$$

for a fixed nuclear geometry yields the (approximate) electronic stationary states and the associated energy eigenvalues for that single nuclear configuration. This is then commonly repeated for a number of different nuclear geometries usually chosen by some rationale so as to cover a particular minimum in the energy or some barrier, etc. If a sufficient number of nuclear geometry points are used and they are chosen to cover all possible distortions of the molecule we get one or more PES $V_k(\vec{R})$. The (approximate) eigenfunctions and the associated PESs are known only in a number of discrete points in $3\bar{N} - 6$ dimensions (for $\bar{N}$ nuclei), since for their determination also the rotational motion is assumed to have been eliminated ($V_k(\vec{R})$ is rotation invariant) and the electronic stationary states and PESs are constructed in a molecule-fixed coordinate system. Further use of these quantities for the study of molecular processes involving nuclear motion must then involve interpolation of some form. Note that the common procedure of just neglecting the nuclear kinetic energy terms in the total Hamiltonian Eq. (4) will not produce the reduced masses and the mass polarization terms.

The Schrödinger equation for the system when the center of mass kinetic energy term has been eliminated can be expressed as

$$[T_{\mathrm{n}} + H_{\mathrm{el}}]\Psi(\vec{r},\vec{R}) = E\Psi(\vec{r},\vec{R}) \tag{18}$$

where

$$T_{\mathrm{n}} = -\frac{\hbar^2}{2}\sum_{k\neq p}\left[\frac{1}{m_k} - \frac{1}{M_0}\right]\nabla_k^2 + \frac{\hbar^2}{2M_0}\sum_{\substack{k\neq l\\ k,l\neq p}}\nabla_k\nabla_l \tag{19}$$

It is then common to introduce the stationary electronic states as a basis, such that

$$\Psi(\vec{r},\vec{R}) = \sum_k \Phi_k(\vec{r};\vec{R})\chi_k(\vec{R}) \tag{20}$$

and by inserting this into Eq. (18), multiplying from the left by Φ_l^*, integrating over the electronic degrees of freedom, and using the fact that the stationary electronic states at a fixed nuclear geometry are orthogonal, one obtains

$$[(\Phi_l|T_\mathrm{n}\Phi_l)_r + V_l(\vec{R})]\chi_l(\vec{R}) - E\chi_l(\vec{R}) = -\sum_{k\neq l}(\Phi_l|T_\mathrm{n}\Phi_k)_r\chi_k(\vec{R}) \tag{21}$$

This is a set of coupled partial differential equations with the terms $(\Phi_l|T_\mathrm{n}\Phi_k)_r$ on the right-hand side being the so called nonadiabatic coupling terms, with the subscript r indicating integration over electronic coordinates. These equations form the basis of the close-coupling approach to atomic and molecular scattering [7]. When these terms are neglected we obtain what is called the adiabatic approximation. The nuclear dynamics is then described by the Schrödinger equation

$$[(\Phi_l|T_\mathrm{n}\Phi_l)_r + V_l(\vec{R})]\chi_l(\vec{R}) = E\chi_l(\vec{R}) \tag{22}$$

The term $(\Phi_l|T_\mathrm{n}\Phi_l)_r$ can be expressed in more detail using the fact that

$$-\frac{\hbar^2}{2}\sum_{k\neq p}\left[\frac{1}{m_k} - \frac{1}{M_0}\right](\Phi_l|\nabla_k^2\Phi_l)_r + \frac{\hbar^2}{2M_0}\sum_{\substack{k\neq m\\ k,m\neq p}}(\Phi_l|\nabla_k\nabla_m\Phi_l)_r \tag{23}$$

is an operator that acts on the nuclear wave function $\chi_l(\vec{R})$, and, for instance,

$$(\Phi_l|\nabla_k^2\Phi_l)_r\chi_l(\vec{R}) = \nabla_k^2\chi_l(\vec{R}) + 2(\Phi_l|\nabla_k\Phi_l)_r\cdot\nabla_k\chi_l(\vec{R}) + \chi_l(\vec{R})(\Phi_l|\nabla_k^2\Phi_l)_r \tag{24}$$

Now, the integral $(\Phi_l|\nabla_k\Phi_l)_r$ over electronic degrees of freedom involves differentiation with respect to parameters under the integral sign and since the electronic stationary states are assumed to be normalized and orthogonal, we get for the case of real wave functions

$$(\Phi_l|\nabla_k\Phi_l)_r = \frac{1}{2}\nabla_k(\Phi_l|\Phi_l)_r = 0 \tag{25}$$

So, we can define the potential energy term

$$B_l(\vec{R}) = -\frac{\hbar^2}{2}\sum_k\left(\frac{1}{m_k} - \frac{1}{M_0}\right)(\Phi_l|\nabla_k^2\Phi_l)_r + \frac{\hbar^2}{2M_0}\sum_{\substack{k\neq m\\ k,m\neq p}}(\Phi_l|\nabla_k\nabla_m\Phi_l)_r \tag{26}$$

which is usually small but not necessarily unimportant. The Schrödinger equation for the nuclear motion then becomes

$$[T_\mathrm{n} + V_l(\vec{R}) + B_l(\vec{R})]\chi_l(\vec{R}) = E\chi_l(\vec{R}) \tag{27}$$

and when the mass polarization terms in the kinetic energy operator are neglected we can write

$$\left[-\frac{\hbar^2}{2}\sum_{k\neq p}\left(\frac{1}{m_k} - \frac{1}{M_0}\right)\nabla_k^2 + V_l(\vec{R}) + B_l(\vec{R})\right]\chi_l(\vec{R}) = E\chi_l(\vec{R}) \tag{28}$$

which is the normal result of the so called adiabatic approximation. When also the $B_l(\vec{R})$, is neglected one calls the result the Born–Oppenheimer approximation [8].

Example. We study the case of a diatomic molecule. The kinetic energy operator can for this case be expressed as

$$-\frac{\hbar^2}{2}\left(\frac{1}{m_1}-\frac{1}{M_0}\right)\nabla_1^2 \tag{29}$$

and when $\vec{R}=\vec{R}_1-\vec{R}_2$, and the center of mass of the nuclei is the origin, so $\vec{R}_2=-(m_1/m_2)\vec{R}_1$, which yields $\vec{R}=(m_1+m_2)\vec{R}_1/m_2$, and the kinetic energy can be expressed in terms of $\vec{R}$ as

$$-\frac{\hbar^2}{2}\left(\frac{1}{m_1}-\frac{1}{M_0}\right)\nabla_1^2=-\frac{\hbar^2}{2}\left(\frac{1}{m_1}+\frac{1}{m_2}\right)\nabla^2=-\frac{\hbar^2}{2\mu}\nabla^2 \tag{30}$$

The above equation then becomes

$$\left[-\frac{\hbar^2}{2\mu}\nabla^2+V_l(R)+B_l(R)\right]\chi_l(\vec{R})=E\chi_l(\vec{R}) \tag{31}$$

where

$$-\frac{\hbar^2}{2\mu}\nabla^2=-\frac{\hbar^2}{2\mu}\left[\frac{\partial^2}{\partial R^2}+\frac{2}{R}\frac{\partial}{\partial R}\right]+\frac{\vec{J}^2}{2\mu R^2} \tag{32}$$

with $\vec{J}^2$, the total angular momentum operator square of rotational motion and $\mu=m_1m_2/(m_1+m_2)$. Since for a diatomic the potential energy depends only on the bond distance R, the wave functions can be expressed as

$$\chi(\vec{R})=\upsilon(R)Y_{JM}(\theta,\varphi) \tag{33}$$

where $Y_{JM}(\theta,\varphi)$ is a spherical harmonic and the Schrödinger equation for vibrational motion is

$$\left[-\frac{\hbar^2}{2\mu}\left(\frac{\partial^2}{\partial R^2}+\frac{2}{R}\frac{\partial}{\partial R}\right)+\frac{\hbar^2J(J+1)}{2\mu R^2}+V(R)+B(R)\right]\upsilon(R)=E\upsilon(R) \tag{34}$$

showing the so-called centrifugal term. This illustrates that the vibrational and rotational motions of a molecule are indeed coupled.

References pp. 39–40

2.3 THE TIME-DEPENDENT VARIATIONAL PRINCIPLE IN QUANTUM MECHANICS

The time-dependent variational principle in quantum mechanics [9] starts from the quantum mechanical action [10,11]

$$A = \int_{t_1}^{t_2} L(\psi^*, \psi)\, \mathrm{d}t \tag{35}$$

where the quantum mechanical Lagrangian is

$$L(\psi^*, \psi) = \left\langle \psi | \mathrm{i}\hbar \frac{\partial}{\partial t} - H | \psi \right\rangle / \langle \psi | \psi \rangle \tag{36}$$

and H is the quantum mechanical Hamiltonian of the system. When the wave function ψ is completely general and allowed to vary in the entire Hilbert space then the TDVP yields the time-dependent Schrödinger equation. However, if the possible wave function variations are restricted in any way, such as is the case for a wave function represented in a finite basis and being of a particular functional form, then the corresponding Lagrangian will generate an approximation to the Schrödinger time evolution.

We consider a wave function expressed in terms of a set of (in general complex) parameters $\mathbf{z}$ (e.g. molecular orbital coefficients, average nuclear positions and momenta, etc.). These parameters are time-dependent and can be expressed as $z_\alpha \equiv z_\alpha(t)$ and thought of as arranged in a column or row array. We write

$$\psi = \psi(\mathbf{z}) = |\mathbf{z}\rangle \tag{37}$$

and employ the principle of least action

$$\delta A = \int_{t_1}^{t_2} \delta L(\psi^*, \psi)\, \mathrm{d}t = 0 \tag{38}$$

with the Lagrangian

$$L = \left[\frac{\mathrm{i}}{2} \langle \mathbf{z} | \dot{\mathbf{z}} \rangle - \frac{\mathrm{i}}{2} \langle \dot{\mathbf{z}} | \mathbf{z} \rangle - \langle \mathbf{z} | H | \mathbf{z} \rangle \right] / \langle \mathbf{z} | \mathbf{z} \rangle \tag{39}$$

where we have put $\hbar = 1$, and write the symmetric form of the time derivative term. One way to see how this can come about is to consider

$$\int_{t_1}^{t_2} \frac{(\partial / \partial t) \langle \mathbf{z} | \mathbf{z} \rangle}{\langle \mathbf{z} | \mathbf{z} \rangle}\, \mathrm{d}t = 0 \tag{40}$$

which holds if we require

$$\langle \mathbf{z}(t_2) | \mathbf{z}(t_2) \rangle = \langle \mathbf{z}(t_1) | \mathbf{z}(t_1) \rangle \tag{41}$$

as our boundary condition.

The variation of the Lagrangian can be expressed in more detail as

$$\delta L = \frac{i}{2}[\langle \delta \mathbf{z}|\dot{\mathbf{z}}\rangle - \langle \delta\dot{\mathbf{z}}|\mathbf{z}\rangle - \langle \delta\mathbf{z}|H|\mathbf{z}\rangle]/\langle \mathbf{z}|\mathbf{z}\rangle - \left[\frac{i}{2}\langle \mathbf{z}|\dot{\mathbf{z}}\rangle - \frac{i}{2}\langle \dot{\mathbf{z}}|\mathbf{z}\rangle - \langle \mathbf{z}|H|\mathbf{z}\rangle\right] \times \langle \delta\mathbf{z}|\mathbf{z}\rangle/\langle \mathbf{z}|\mathbf{z}\rangle^2 + \text{complex conjugate} \tag{42}$$

We would like to get rid of all the terms that contain the variation $\delta\dot{\mathbf{z}}$. To this end we add and subtract the total time derivative

$$\frac{d}{dt}\frac{\langle \delta\mathbf{z}|\mathbf{z}\rangle}{\langle \mathbf{z}|\mathbf{z}\rangle} = \frac{\langle \delta\dot{\mathbf{z}}|\mathbf{z}\rangle + \langle \delta\mathbf{z}|\dot{\mathbf{z}}\rangle}{\langle \mathbf{z}|\mathbf{z}\rangle} - \frac{\langle \delta\mathbf{z}|\mathbf{z}\rangle}{\langle \mathbf{z}|\mathbf{z}\rangle^2}\frac{d}{dt}\langle \mathbf{z}|\mathbf{z}\rangle \tag{43}$$

and its complex conjugate to write

$$\delta L = \frac{i}{2}\frac{\langle \delta\mathbf{z}|\dot{\mathbf{z}}\rangle}{\langle \mathbf{z}|\mathbf{z}\rangle} + \frac{i}{2}\frac{\langle \delta\mathbf{z}|\dot{\mathbf{z}}\rangle}{\langle \mathbf{z}|\mathbf{z}\rangle} - \frac{i}{2}\frac{\langle \delta\mathbf{z}|\mathbf{z}\rangle}{\langle \mathbf{z}|\mathbf{z}\rangle^2}\frac{d}{dt}\langle \mathbf{z}|\mathbf{z}\rangle - \frac{i}{2}\frac{d}{dt}\frac{\langle \delta\mathbf{z}|\mathbf{z}\rangle}{\langle \mathbf{z}|\mathbf{z}\rangle} - \frac{\langle \delta\mathbf{z}|H|\mathbf{z}\rangle}{\langle \mathbf{z}|\mathbf{z}\rangle} - \frac{\langle \delta\mathbf{z}|\mathbf{z}\rangle}{\langle \mathbf{z}|\mathbf{z}\rangle^2}\left[\frac{i}{2}\langle \mathbf{z}|\dot{\mathbf{z}}\rangle - \frac{i}{2}\langle \dot{\mathbf{z}}|\mathbf{z}\rangle - \langle \mathbf{z}|H|\mathbf{z}\rangle\right] + \text{complex conjugate} \tag{44}$$

The time integration involved in $\delta A = 0$ eliminates the total derivative terms since due to the boundary conditions they are zero, i.e.

$$\langle \delta\mathbf{z}(t_2)|\mathbf{z}(t_2)\rangle - \langle \delta\mathbf{z}(t_1)|\mathbf{z}(t_1)\rangle = 0 \tag{45}$$

which follows from Eq. (41) and the fact that $|\delta\mathbf{z}\rangle$ and $\langle\delta\mathbf{z}|$ are independent variations.

The surviving terms of δL can be expressed as

$$i\frac{\langle \delta\mathbf{z}|\dot{\mathbf{z}}\rangle}{\langle \mathbf{z}|\mathbf{z}\rangle} - \frac{\langle \delta\mathbf{z}|H|\mathbf{z}\rangle}{\langle \mathbf{z}|\mathbf{z}\rangle} - \frac{\langle \delta\mathbf{z}|\mathbf{z}\rangle}{\langle \mathbf{z}|\mathbf{z}\rangle^2}[i\langle \mathbf{z}|\dot{\mathbf{z}}\rangle - \langle \mathbf{z}|H|\mathbf{z}\rangle] + \text{complex conjugate} \tag{46}$$

Since $\delta\mathbf{z}$ and $\delta\mathbf{z}^*$ can be considered as independent variations one can conclude that

$$\left(i\frac{\partial}{\partial t} - H\right)|\mathbf{z}\rangle = \frac{\langle \mathbf{z}|i\partial/\partial t - H|\mathbf{z}\rangle}{\langle \mathbf{z}|\mathbf{z}\rangle}|\mathbf{z}\rangle \tag{47}$$

which is the Schrödinger equation if the right-hand side is zero. By explicitly considering the overall wave function phase we can eliminate the right-hand side. We write

$$|\mathbf{z}\rangle \rightarrow e^{-i\gamma}|\mathbf{z}\rangle \tag{48}$$

with γ only a function of time and obtain

$$\langle \mathbf{z}|i\partial/\partial t - H|\mathbf{z}\rangle \rightarrow \langle \mathbf{z}|e^{i\gamma}(i\partial/\partial t - H)e^{-i\gamma}|\mathbf{z}\rangle = \dot{\gamma}\langle \mathbf{z}|\mathbf{z}\rangle + \langle \mathbf{z}|i\partial/\partial t - H|\mathbf{z}\rangle = 0 \tag{49}$$

which means that the time derivative of the overall phase must be

$$-\dot{\gamma} = \frac{\langle \mathbf{z}|i\partial/\partial t - H|\mathbf{z}\rangle}{\langle \mathbf{z}|\mathbf{z}\rangle} \tag{50}$$

We introduce the notations $S(\mathbf{z},\mathbf{z}^*) = \langle \mathbf{z}|\mathbf{z}\rangle$, and $E(\mathbf{z},\mathbf{z}^*) = \langle \mathbf{z}|H|\mathbf{z}\rangle/\langle \mathbf{z}|\mathbf{z}\rangle$, which leads to the equation

$$-\dot{\gamma} = \frac{\mathrm{i}}{2}\sum_\alpha\left[\dot{z}_\alpha\frac{\partial}{\partial z_\alpha} - \dot{z}_\alpha^*\frac{\partial}{\partial z_\alpha^*}\right]\ln S(\mathbf{z},\mathbf{z}^*) - E(\mathbf{z},\mathbf{z}^*) \tag{51}$$

where we have used the chain rule of differentiation. Note that for a stationary state all $\dot{z} = 0$, and $E(\mathbf{z},\mathbf{z}^*) = E$ yielding $\gamma = Et$ and the phase factor $\mathrm{e}^{-\mathrm{i}Et/\hbar}$. The above expression for δA can be similarly written as

$$0 = \delta A = \int_{t_1}^{t_2}\delta L\,\mathrm{d}t \tag{52}$$

$$= \int_{t_1}^{t_2}\left\{\sum_\alpha\left[\sum_\beta \mathrm{i}\frac{\partial^2 \ln S}{\partial z_\alpha^*\partial z_\beta}\dot{z}_\beta - \frac{\partial E}{\partial z_\alpha^*}\right]\delta z_\alpha^*\right. \tag{53}$$

$$\left. + \sum_\alpha\left[\sum_\beta -\mathrm{i}\frac{\partial^2 \ln S}{\partial z_\alpha\partial z_\beta^*}\dot{z}_\beta^* - \frac{\partial E}{\partial z_\alpha}\right]\delta z_\alpha\right\}\mathrm{d}t \tag{54}$$

where say $\delta\mathbf{z} = \sum_\alpha(\partial/\partial z_\alpha)\delta z_\alpha$, and since δz_α and δz_α^* are independent variations we can write

$$\mathrm{i}\sum_\beta C_{\alpha\beta}\dot{z}_\beta = \frac{\partial E}{\partial z_\alpha^*} \tag{55}$$

$$-\mathrm{i}\sum_\beta C_{\alpha\beta}^*\dot{z}_\beta^* = \frac{\partial E}{\partial z_\alpha} \tag{56}$$

where $C_{\alpha\beta} = \partial^2 \ln S/\partial z_\alpha^*\partial z_\beta$. We introduce the matrix $\mathbf{C} = \{C_{\alpha\beta}\}$ and assume it to be invertible to write

$$\begin{bmatrix}\dot{\mathbf{z}}\\ \dot{\mathbf{z}}^*\end{bmatrix} = \begin{bmatrix}-\mathrm{i}\mathbf{C}^{-1} & 0\\ 0 & \mathrm{i}\mathbf{C}^{*-1}\end{bmatrix}\begin{bmatrix}\partial E/\partial\mathbf{z}^*\\ \partial E/\partial\mathbf{z}\end{bmatrix} \tag{57}$$

which is a matrix equation in block form. One may introduce a generalized Poisson bracket by considering two general differentiable functions $f(\mathbf{z},\mathbf{z}^*)$, and $g(\mathbf{z},\mathbf{z}^*)$ and write

$$\{f,g\} = [\,\partial f/\partial\mathbf{z}^T \quad \partial f/\mathbf{z}^\dagger]\begin{bmatrix}-\mathrm{i}\mathbf{C}^{-1} & 0\\ 0 & -\mathrm{i}\mathbf{C}^{*-1}\end{bmatrix}\begin{bmatrix}\partial g/\partial\mathbf{z}^*\\ \partial g/\partial\mathbf{z}\end{bmatrix}$$

$$= -\mathrm{i}\sum_{\alpha,\beta}\left[\frac{\partial f}{\partial z_\alpha}(\mathbf{C}^{-1})_{\alpha\beta}\frac{\partial g}{\partial z_\beta^*} - \frac{\partial g}{\partial z_\alpha}(\mathbf{C}^{-1})_{\alpha\beta}\frac{\partial f}{\partial z_\beta^*}\right] \tag{58}$$

It follows that

$$\dot{\mathbf{z}} = \{\mathbf{z},E\}, \qquad \dot{\mathbf{z}}^* = \{\mathbf{z}^*,E\} \tag{59}$$

which shows that the time evolution of the wave function parameters, and thus, of the wave function, is governed by Hamilton-like equations. Such a set of coupled first-order differential equations in time plus the equation for the evolution of the overall phase can be integrated by a great variety of methods. Schematically we write $\dot{\mathbf{z}}(t) = F(\mathbf{z}(t))$ and proceed by finite steps such that $\Delta\mathbf{z}_i = F(\mathbf{z}_i)\Delta t$ and $\mathbf{z}_{i+1} = \mathbf{z}_i + \Delta\mathbf{z}_i$; $\mathbf{z}(0) = \mathbf{z}_0$.

2.4 COHERENT STATES

The discussion of the TDVP in the previous chapter exploits a family of state vectors $|\mathbf{z}\rangle$ labeled by a set of time-dependent, and complex parameters $\mathbf{z} = \{z_1, z_2, \ldots, z_M\}$ [12]. Such parameter spaces should be continuous and *complete* in the sense that as the state vector evolves in time and the complex parameters assume all possible values throughout their range, all states of the particular form $|\mathbf{z}\rangle$ are obtained. Such demands on parameter spaces are satisfied by (generalized) *coherent states* [13] which relates the parameters to a particular Lie group G. Typically one chooses a unitary irreducible representation of G and a corresponding lowest (or highest) weight state $|0\rangle$ of such a representation. A maximal subgroup H of G that leaves $|0\rangle$ invariant is called the stability group and the cosets of G by H provide a suitable nonredundant set of parameters to label the coherent state. In general one would also require the existence of a positive measure $\mathrm{d}\mathbf{z}$ on this parameter space such that when the integral

$$\int |\mathbf{z}\rangle\langle\mathbf{z}|\mathrm{d}\mathbf{z} = I \tag{60}$$

is taken over the range of the parameter space one obtains the identity.

Already the notion of continuity of the labels rules out as coherent states some familiar sets of states used in quantum mechanics. For instance, a set of discrete orthogonal states, such as a set of orthonormal basis functions $\{|n\rangle\}$ cannot be coherent states.

2.4.1 Gaussian wave packet as a coherent state

A Gaussian wave packet in one dimension can be expressed as ($\hbar = 1$)

$$\psi(x) \propto \exp\left[-\frac{1}{2}\left(\frac{x-q}{b}\right)^2 + \mathrm{i}px\right] \tag{61}$$

where we take the view that the parameters p, and q are time dependent, while the width parameter b is time-independent. The interpretation of these parameters is evident from the definition of quantum mechanical averages, i.e.

$$\langle x\rangle = \int_{-\infty}^{\infty} x\,\mathrm{e}^{-((x-q)/b)^2}\,\mathrm{d}x \Big/ \int_{-\infty}^{\infty} \mathrm{e}^{-((x-q)/b)^2}\,\mathrm{d}x = \langle\psi|x\psi\rangle/\langle\psi|\psi\rangle \tag{62}$$

Since the average value $\langle x - q\rangle = 0$, it follows that $q = \langle x\rangle$, which is the average position of the wave packet. The average momentum of the wave packet

$$\langle\hat{p}\rangle = \left\langle -\mathrm{i}\frac{\partial}{\partial x}\right\rangle = p - \langle x - q\rangle/b^2 = p \tag{63}$$

defines the parameter p. The square of the width of the wave packet

$$\langle(x-q)^2\rangle = -\partial\langle\psi|\psi\rangle/\partial(1/b^2)/\langle\psi|\psi\rangle = \sqrt{\pi}\frac{b^3}{2}/\sqrt{\pi}b \tag{64}$$

making the width $\Delta x = \langle(x-q)^2\rangle^{1/2} = b/\sqrt{2}$.

The Gaussian wave packet has a number of interesting properties. For instance, it has a minimal uncertainty product $\Delta x\Delta p = 1/2$ in units of $\hbar$. This follows from

$$\langle\hat{p}^2\rangle = \left\langle -\hbar^2\frac{\mathrm{d}^2}{\mathrm{d}x^2}\right\rangle = p^2 + \frac{\hbar^2}{2b^2} \tag{65}$$

and $\Delta p = \langle\hat{p}^2 - p^2\rangle^{1/2} = 1/b\sqrt{2}$, $(\hbar = 1)$.

In addition the Gaussian wave packet can be written as a displaced harmonic oscillator ground state, such that

$$\psi(x) = \exp[\mathrm{i}px]\exp[-\mathrm{i}q\hat{p}]\exp\left[-\frac{1}{2}\left(\frac{x}{b}\right)^2\right] \tag{66}$$

i.e. the oscillator ground state is displaced, $x \to x - q$, and boosted $0 \to p$. This can be seen from

$$\begin{aligned} \mathrm{e}^{-\mathrm{i}q\hat{p}}\exp\left[-\frac{1}{2}\left(\frac{x}{b}\right)^2\right] &= \left[1 - q\frac{\partial}{\partial x} + \frac{q^2}{2}\frac{\partial^2}{\partial x^2} - \cdots\right]\exp\left[-\frac{1}{2}\left(\frac{x}{b}\right)^2\right] \\ &= \exp\left[-\frac{1}{2}\left(\frac{x-q}{b}\right)^2\right] \end{aligned} \tag{67}$$

The Gaussian wave packet is a coherent state and can be expressed as a superposition of oscillator states. This means that

$$\psi(x) = \sum_n c_n|n\rangle = \exp[za^\dagger - z^*a]|0\rangle \tag{68}$$

where $|n\rangle$ is a harmonic oscillator eigenstate a and $a^\dagger$ are harmonic oscillator field operators and z is a suitable complex combination of wave function parameters.

This can be seen from the result

$$\psi(x) \propto \mathrm{e}^{\mathrm{i}px}\mathrm{e}^{-\mathrm{i}q\hat{p}}\mathrm{e}^{-(1/2)(x/b)^2} \propto \mathrm{e}^{-\mathrm{i}(q\hat{p}-px)}\mathrm{e}^{-(1/2)(x/b)^2} \tag{69}$$

where, since x and $\hat{p}$ do not commute, the last step is nontrivial. Introducing the complex parameter $z = (q/b + \mathrm{i}bp)/\sqrt{2}$, and observing that the harmonic oscillator field operators can be expressed as

$$a^\dagger = -\mathrm{i}((b\hat{p} + \mathrm{i}x)/b)/\sqrt{2} \tag{70}$$

$$a = \mathrm{i}((b\hat{p} - \mathrm{i}x)/b)/\sqrt{2} \tag{71}$$

we can write $za^{\dagger} - z^{*}a = -\mathrm{i}(q\hat{p} - px)$, and

$$\psi(x) \propto \mathrm{e}^{za^{\dagger} - z^{*}a}\, \mathrm{e}^{-(1/2)(x/b)^{2}} \propto \mathrm{e}^{za^{\dagger} - z^{*}a}|0\rangle \tag{72}$$

The last expression is the 'classical' or canonical *coherent state* $|z\rangle$. The Baker–Campbell–Hausdorff (BCH) formula, yields

$$\mathrm{e}^{A+B} = \exp\left[-\frac{1}{2}[A,B]_{-}\right]\mathrm{e}^{A}\,\mathrm{e}^{B} \tag{73}$$

which is true for the case when the commutator $[A,B]_{-}$ commutes with A and B. When applied to

$$|z\rangle = \mathrm{e}^{za^{\dagger} - z^{*}a}|0\rangle \tag{74}$$

the BCH formula yields

$$|z\rangle = \mathrm{e}^{-(1/2)|z|^{2}}\,\mathrm{e}^{za^{\dagger}}\,e^{-z^{*}a}|0\rangle = \mathrm{e}^{-(1/2)|z|^{2}}\,\mathrm{e}^{za^{\dagger}}|0\rangle = \mathrm{e}^{-(1/2)|z|^{2}}\sum_{n=0}^{\infty}(n!)^{-1}(za^{\dagger})^{n}|0\rangle$$
$$= \mathrm{e}^{-(1/2)|z|^{2}}\sum_{n=0}^{\infty}(n!)^{-1/2}(z)^{n}|n\rangle \tag{75}$$

The Gaussian wave packet in this form is the original 'coherent state'. Generalizations of this concept have been made, in particular the work of Perelomov [14] has introduced so-called group-related coherent states. Such a state is formed by the action of a Lie group operator $\exp\{\sum_{m} z_{m}F_{m}\}$ acting on a reference state $|0\rangle$. The $\{z_{m}\}$ are the, in general complex, Lie group parameters, and $\{F_{m}\}$ are the generators of the corresponding Lie algebra. The reference state is usually a lowest weight state and called the fiducial state. It is commonly invariant to some of the group elements, thus defining a so-called stability group of the fiducial state. The parameters labeling the coherent state are then associated with the left coset of the Lie group with respect to the stability group. This assures nonredundancy of parameters. The canonical coherent state has this form in terms of the so-called Weyl group, whose Lie algebra generators are $\{1, a, a^{\dagger}\}$. The one parameter stability group is just the phase factor $\mathrm{e}^{\mathrm{i}\alpha}$ and the coset representative is $\mathrm{e}^{za^{\dagger} - z^{*}a}$.

The scalar product of two coherent states

$$\langle z_{1}|z_{2}\rangle = \exp\{-(|z_{1}|^{2} + |z_{2}|^{2})/2\}\sum_{n}\frac{(z_{1}^{*}z_{2})^{n}}{n!} = \exp\left\{-\frac{1}{2}|z_{1}|^{2} + z_{1}^{*}z_{2} - \frac{1}{2}|z_{2}|^{2}\right\} \tag{76}$$

i.e. a nowhere vanishing continuous function of the parameters. The canonical coherent state has the property that

$$a|z\rangle = z|z\rangle \tag{77}$$

which can easily be seen from

$$e^{-za^\dagger} a e^{za^\dagger} = a + z[a, a^\dagger]_- + \frac{z^2}{2}\Big[[a, a^\dagger]_-, a^\dagger\Big]_- + \cdots = a + z \tag{78}$$

which means that $e^{-za^\dagger} a e^{za^\dagger}|0\rangle = z|0\rangle$. Furthermore, the coherent state for all the values of the complex parameter z is a set of states satisfying the resolution of the identity

$$\begin{aligned}\pi^{-1}\int |z\rangle\langle z|\mathrm{d}^2 z &= \frac{1}{\pi}\sum_{n,m}(n!m!)^{-1/2}\int \mathrm{e}^{-|z|^2}(z^*)^n(z)^m|m\rangle\langle n|\mathrm{d}^2 z \\ &= \frac{1}{\pi}\sum_{n,m}(n!m!)^{-1/2}\int_0^\infty \mathrm{e}^{-|z|^2}|z|^{n+m+1}\mathrm{d}|z|\int_0^{2\pi}\mathrm{e}^{\mathrm{i}(m-n)\phi}\,\mathrm{d}\phi|m\rangle\langle n| \\ &= \sum_n (n!)^{-1}\int_0^\infty \mathrm{e}^{-|z|^2}|z|^{2n}\,\mathrm{d}|z|^2|n\rangle\langle n| = \sum_n |n\rangle\langle n| = 1\end{aligned} \tag{79}$$

This result permits us to write

$$|z'\rangle = \pi^{-1}\int |z\rangle\langle z|z'\rangle\mathrm{d}^2 z \tag{80}$$

which illustrates the overcompleteness of the set $\{|z\rangle\}$. Thus, as a set of functions labeled by the continuous complex parameter z the coherent state satisfies the resolution of the identity and is inherently linearly dependent.

Considering the time evolution of a harmonic oscillator with $|z\rangle$ as the initial state, we obtain

$$\mathrm{e}^{-\mathrm{i}Ht}|z\rangle = \mathrm{e}^{-\mathrm{i}t\omega(n+(1/2))}|z\rangle = \mathrm{e}^{-(1/2)|z|^2}\sum_{n=0}^{\infty}(n!)^{-1/2}(z\mathrm{e}^{-\mathrm{i}t\omega})^n|n\rangle\,\mathrm{e}^{-\mathrm{i}t\omega/2} \propto |\mathrm{e}^{-\mathrm{i}t\omega}z\rangle \tag{81}$$

This shows that the coherent state evolves into other coherent states by a time-dependent label change that follows the classical oscillator solution.

Application of the TDVP to the wave packet dynamics with the coherent state $|z\rangle$ is straightforward. We note that

$$S(z^*, z') = \langle z|z'\rangle = \exp\left(-\frac{1}{2}|z|^2 + z^* z' - \frac{1}{2}|z'|^2\right) \tag{82}$$

$$E(z^*, z) = \langle z|H|z\rangle/\langle z|z\rangle = \omega\langle z|\left(a^\dagger a + \frac{1}{2}\right)|z\rangle/\langle z|z\rangle = \omega\left(z^* z + \frac{1}{2}\right) \tag{83}$$

and that the dynamical equations become

$$\begin{bmatrix}\mathrm{i} & 0\\ 0 & -\mathrm{i}\end{bmatrix}\begin{bmatrix}\dot{z}\\ \dot{z}^*\end{bmatrix} = \begin{bmatrix}\omega z\\ \omega z^*\end{bmatrix} \tag{84}$$

since $C = \partial^2 \ln S/\partial z^* \partial z'|_{z'=z} = 1$. The equation $i\dot{z} = \omega z$ becomes in more detail

$$\frac{i}{\sqrt{2}}(\dot{q}/b + ib\dot{p}) = \omega\frac{1}{\sqrt{2}}(q/b + ibp) \tag{85}$$

assuming a constant width wave packet. One easily deduces that

$$\ddot{p} = -\omega^2 p \tag{86}$$

$$\ddot{q} = -\omega^2 q \tag{87}$$

i.e. in an oscillator field with $b = 1/\sqrt{m\omega}$ the Gaussian wave packet is coherent and that its average position has a harmonic motion $q(t) = q_0 \cos \omega t + (p_0/m\omega)\sin \omega t$, while $p(t) = p_0 \cos \omega t - m\omega \sin \omega t$.

2.4.1.1 Gaussian wave packet with evolving width

A more general coherent state description of a Gaussian wave packet is required when we allow the width parameter to evolve in time. The corresponding Lie group is then Sp(2, *R*), which is isomorphic to SU(1, 1) or SO(2, 1). The generators of the Sp(2, *R*) Lie algebra are

$$t_1 = \frac{i}{2}\begin{bmatrix} -1 & 0 \\ 0 & 1 \end{bmatrix}, \qquad t_2 = \frac{i}{2}\begin{bmatrix} 0 & 1 \\ 1 & 0 \end{bmatrix}, \qquad t_3 = \frac{i}{2}\begin{bmatrix} 0 & 1 \\ -1 & 0 \end{bmatrix} \tag{88}$$

satisfying the relations

$$[t_1, t_2] = -it_3, \qquad [t_2, t_3] = it_1, \qquad [t_3, t_1] = it_2 \tag{89}$$

where the different signs on the right indicate that we are dealing with a noncompact group.

Realization of the generators in terms of a Cartesian coordinate x and its conjugate momentum p, such that $[x, p] = i\hbar$, are

$$\begin{aligned} &t_1 \to T_1 = -(xp + px)/4\hbar, \qquad t_2 \to T_2 = (p^2/2\mu - \mu\omega^2 x^2/2)/2\hbar\omega, \\ &t_3 \to T_3 = (p^2/2\mu + \mu\omega^2 x^2/2)/2\hbar\omega \end{aligned} \tag{90}$$

Another useful realization obtains in terms of the oscillator field operator

$$a = \left(ip/\sqrt{\mu\omega\hbar} + \sqrt{\frac{\mu\omega}{\hbar}}x\right)/\sqrt{2} \tag{91}$$

and its adjoint. We write

$$\begin{aligned} &T_+ = -T_2 + iT_1 = \frac{1}{2}a^\dagger a^\dagger, \qquad T_- = -T_2 - iT_1 = \frac{1}{2}aa, \\ &T_0 = T_3 = \frac{1}{4}(a^\dagger a + aa^\dagger) \end{aligned} \tag{92}$$

A common parameterization of the Sp(2, R) group is in terms of Euler angles, such that a group element would be

$$g(\alpha, \beta, \gamma) = \mathrm{e}^{\mathrm{i}\alpha t_3}\,\mathrm{e}^{\mathrm{i}\beta t_1}\,\mathrm{e}^{\mathrm{i}\gamma t_3} \tag{93}$$

One can readily show that

$$\mathrm{e}^{\mathrm{i}\beta t_1} = \begin{bmatrix} \mathrm{e}^{\beta/2} & 0 \\ 0 & \mathrm{e}^{\beta/2} \end{bmatrix} \tag{94}$$

and

$$\mathrm{e}^{\mathrm{i}\gamma t_3} = \begin{bmatrix} \cos\gamma/2 & -\sin\gamma/2 \\ \sin\gamma/2 & \cos\gamma/2 \end{bmatrix} \tag{95}$$

The operator $\exp[\mathrm{i}\beta T_1]$ is a scale transformation as can be seen from the relation

$$e\mathrm{i}\beta T_1 f(x) = \exp\left[-\frac{\beta}{2}x\frac{\partial}{\partial x} - \frac{\beta}{4}\right]f(x) = \mathrm{e}^{-\beta/4}f(\mathrm{e}^{-\beta/2}x) \tag{96}$$

This result is readily shown by considering the transformation of powers of the coordinate. For instance, by using the power series representation

$$f(x) = \sum_{n=0}^{\infty}\frac{f^{(n)}(0)}{n!}x^n \tag{97}$$

and the defining expansion of an exponential operator

$$\exp\left[-\frac{\beta}{2}\left(x\frac{\partial}{\partial x} + \frac{1}{2}\right)\right] = \sum_{k=0}^{\infty}\frac{\beta^k}{k!}\left[-\frac{\beta}{2}\left(x\frac{\partial}{\partial x} + \frac{1}{2}\right)\right]^k \tag{98}$$

noting that

$$\left[-\frac{1}{2}\left(x\frac{\partial}{\partial x} + \frac{1}{2}\right)\right]^k x^n = (-1)^k\left(\frac{2n+1}{4}\right)^k x^n \tag{99}$$

and

$$\begin{aligned}
&\left[\sum_{k=0}^{\infty}\left(\frac{-\beta}{4}\right)^k\frac{1}{k!}\right]\times\left[\sum_{l=0}^{\infty}\left(\frac{-\beta}{2}\right)^l\frac{1}{l!}x\right]^n \\
&= \left[\sum_{k=0}^{\infty}\left(\frac{-\beta}{4}\right)^k\frac{1}{k!}\right]\times\left[\sum_{l=0}^{\infty}\left(\frac{-n\beta}{2}\right)^l\frac{1}{l!}\right]x^n \\
&= x^n\sum_{j=0}^{\infty}\left(\frac{-\beta}{4}\right)^j\frac{1}{j!}\sum_{k=0}^{j}\binom{j}{k}(2n)^{j-k} \\
&= \sum_{j=0}^{\infty}\left(\frac{-(2n+1)\beta}{4}\right)^j\frac{x^n}{j!}
\end{aligned} \tag{100}$$

the scaling property is shown.

As discussed above, defining a coherent state involves choosing a Lie group G, a unitary irreducible representation, the lowest (or highest) weight state of such a representation, a stability subgroup, and corresponding cosets. This will yield a useful parameter space in which to describe the coherent state. In analogy with the compact Lie group SO(3), the irreducible representations of $\mathrm{Sp}(2, R)$ are labeled by the eigenvalues of T_3 corresponding to the lowest weight state. The harmonic oscillator ground state

$$|0\rangle \propto \exp\left[-\frac{1}{2}\left(\frac{x}{b}\right)^2\right] \tag{101}$$

with $b^2 = \hbar/\mu\omega$ is the lowest weight state of an irreducible representation labeled by $k = 1/8$, where

$$T_3|0\rangle = 2k|0\rangle \tag{102}$$

We then identify the stability group H as the set

$$H = h|T(h)|0\rangle = \mathrm{e}^{\mathrm{i}\sigma_h}|0\rangle \tag{103}$$

Each element of the coset space G/H then corresponds to a coherent state. The decomposition of the group into cosets, taking advantage of the stability group properties, reduces the parameter space of the coherent state to a nonredundant set. In our case the stability group is SO(2) and we can write

$$\mathrm{e}^{\mathrm{i}\omega T_3}|0\rangle = \mathrm{e}^{\mathrm{i}\omega(1/4)(a^\dagger a + aa^\dagger)}|0\rangle \mathrm{e}^{\mathrm{i}\omega 2k}|0\rangle = \mathrm{e}^{\mathrm{i}\omega/4}|0\rangle \tag{104}$$

A new choice of parameters is r, s, ω, i.e.

$$g(\alpha, \beta, \gamma) \rightarrow g(r, s, \omega) = \begin{bmatrix} \sqrt{r} & 0 \\ 0 & 1/\sqrt{r} \end{bmatrix} \begin{bmatrix} 1 & 0 \\ s & 1 \end{bmatrix} \begin{bmatrix} \cos\omega/2 & -\sin\omega/2 \\ \sin\omega/2 & \cos\omega/2 \end{bmatrix} \tag{105}$$

where the new parameters are identified as

$$r = \cosh\beta + \sinh\beta\cos\alpha, \qquad s = \sinh\beta\sin\alpha \tag{106}$$

$$\omega = \alpha + \gamma - \arctan\left(\frac{\sin\alpha\sinh\beta/2}{\cosh\beta/2 + \cos\alpha\sinh\beta/2}\right) \tag{107}$$

An element of G can now be expressed as

$$g(r, s, \omega) = \mathrm{e}^{\mathrm{i}t_1\ln r}\,\mathrm{e}^{\mathrm{i}s(t_3 - t_2)}\mathrm{e}^{\mathrm{i}\omega t_3} \tag{108}$$

and going to the unitary irreducible representation carried by even functions the coherent state becomes

$$|r, s\rangle = T(r, s, 0)|0\rangle = \mathrm{e}^{\mathrm{i}T_1\ln r}\,\mathrm{e}^{\mathrm{i}s(T_3 - T_2)}\,\mathrm{e}^{-(1/2)(x/b)^2} = \exp\left[(\mathrm{i}s - 1)\frac{x^2}{r2b^2}\right] \tag{109}$$

where Eq. (90) is used and $b = 1/\sqrt{\mu\omega}$. The parameters are related to average values of the generators, such that

$$\langle r, s|T_3 - T_2|r, s\rangle = \frac{1}{b\sqrt{\pi r}}\int_{-\infty}^{\infty}\frac{x^2}{2b^2}\exp(-x^2/rb^2)\,\mathrm{d}x = \frac{b^3 r\sqrt{\pi r}}{2b^2 2b\sqrt{\pi r}} = \frac{r}{4} \tag{110}$$

and

$$\begin{aligned}\langle r, s|T_3|r, s\rangle &= \left\langle\left(\frac{\mathrm{i}}{2}x\frac{\partial}{\partial x} + \frac{\mathrm{i}}{4}\right)\right\rangle \\ &= \frac{1}{b\sqrt{\pi r}}\int_{-\infty}^{\infty}\left[\mathrm{i}\frac{x^2}{2b^2 r}(\mathrm{i}s - 1) + \frac{\mathrm{i}}{4}\right]\exp(-x^2/rb^2)\,\mathrm{d}x = -\frac{s}{4}\end{aligned} \tag{111}$$

where s is real and r is real and positive (see [15]).

A convenient reparameterization of the wave packet in terms of u and w can be accomplished as

$$b^2 r = 2w^2, \qquad s = 2uw \tag{112}$$

so that the Gaussian wave packet becomes

$$\psi(x) \propto \exp\left[-\left(\frac{1 - 2\mathrm{i}uw}{4w^2}\right)(x - q)^2 + \mathrm{i}px\right] = |p, q, u, w\rangle \tag{113}$$

The Hamiltonian

$$H = -\frac{\hbar^2}{2\mu}\frac{\partial^2}{\partial x^2} + V(x) \tag{114}$$

yields the wave packet average energy

$$E(p, q, u, w) = \frac{p^2}{2\mu} + \frac{u^2}{2\mu} + U(q, w) \tag{115}$$

with $\hbar = 1$ and

$$U(q, w) = \frac{1}{8\mu w^2} + (2\pi)^{-1/2}\int_{-\infty}^{\infty}\mathrm{e}^{-(1/2)y^2}V(wy + q)\,\mathrm{d}y \tag{116}$$

We can now apply the TDVP equations to study the propagation of this wave packet. The elements of the dynamical metric are

$$\eta_{rs} = \mathrm{i}\left[\frac{\partial^2}{\partial r'\partial s} - \frac{\partial^2}{\partial r\partial s'}\right]\ln S \tag{117}$$

with

$$S = \langle p', q', u', w' | p, q, u, w \rangle = \int_{-\infty}^{\infty} \exp[-ax^2 + bx + c] \mathrm{d}x$$

$$= \mathrm{e}^c \, \mathrm{e}^{b^2/4a} \int_{-\infty}^{\infty} \exp[-(\sqrt{a}x - b/2\sqrt{a})^2] \mathrm{d}x = \mathrm{e}^c \, \mathrm{e}^{b^2/4a} \sqrt{\pi/a} \tag{118}$$

and where

$$a = \left[\frac{1 + 2\mathrm{i}u'w'}{4w'^2} + \frac{1 - 2\mathrm{i}uw}{4w^2} \right]$$

$$b = 2\left[\frac{1 + 2\mathrm{i}u'w'}{4w'^2} q' + \frac{1 - 2\mathrm{i}uw}{4w^2} q - \mathrm{i}p' + \mathrm{i}p \right] \tag{119}$$

$$c = -\left[\frac{1 + 2\mathrm{i}u'w'}{4w'^2} q'^2 + \frac{1 - 2\mathrm{i}uw}{4w^2} q^2 \right]$$

Differentiation of

$$\ln S = c + b^2/4b - \frac{1}{2} \ln a + \frac{1}{2} \ln \pi \tag{120}$$

yields the elements of the upper triangle

$$\eta_{pq} = 1, \quad \eta_{pu} = 0, \quad \eta_{pw} = 0, \quad \eta_{qu} = 0, \quad \eta_{qw} = 0, \quad \eta_{uw} = 1 \tag{121}$$

in the antisymmetric metric matrix $\{\eta_{rs}\}$. The TDVP equations then become

$$\begin{bmatrix} 0 & 1 & 0 & 0 \\ -1 & 0 & 0 & 0 \\ 0 & 0 & 0 & 1 \\ 0 & 0 & -1 & 0 \end{bmatrix} \begin{bmatrix} \dot{p} \\ \dot{q} \\ \dot{u} \\ \dot{w} \end{bmatrix} = \begin{bmatrix} \partial E/\partial p \\ \partial E/\partial q \\ \partial E/\partial u \\ \partial E/\partial w \end{bmatrix} \tag{122}$$

or in more detail

$$\dot{q} = \frac{p}{\mu}, \qquad \dot{p} = -\frac{\partial U}{\partial q}, \qquad \dot{w} = \frac{u}{\mu}, \qquad \dot{u} = -\frac{\partial U}{\partial w} \tag{123}$$

which look very much like the classical Hamilton's equations.

2.4.2 The determinantal coherent state for *N* electrons

For an N-electron system we choose a set of N spin orbitals $\mathbf{u}^{\cdot} = \{u_1, u_2, \cdots, u_N\}$ and form a determinantal wave function

$$\det\{u_1(x_1)u_2(x_2)\cdots u_N(x_N)\} \tag{124}$$

or in second quantization the state vector

$$|0\rangle = \prod_{i=1}^{N} a_i^{\dagger}|\text{vac}\rangle \tag{125}$$

with $|\text{vac}\rangle$ the true vacuum state. We call this the reference state. The basis (of rank K) and the associated field operators are divided into two sets, those that refer to the reference state denoted by $\mathbf{a}^{\cdot}$ and the rest denoted by $\mathbf{a}^{\text{o}}$, i.e.

$$(\mathbf{a}^{\cdot}, \mathbf{a}^{\text{o}}) \tag{126}$$

and of course a similar partition of the creators, so that the reference state is

$$|0\rangle = \prod_{k=1}^{N} a_k^{\cdot\dagger}|\text{vac}\rangle \tag{127}$$

The creation operators and the basis transform in the same manner, so when we apply a general unitary transformation to the basis, the creation operators suffer the same transformation. We can write in matrix form

$$(\mathbf{b}^{\cdot\dagger}, \mathbf{b}^{\text{o}\dagger}) = (\mathbf{a}^{\cdot\dagger}, \mathbf{a}^{\text{o}\dagger})\begin{pmatrix} \mathbf{U}^{\cdot} & \mathbf{U}' \\ \mathbf{U}'' & \mathbf{U}^{\text{o}} \end{pmatrix} \tag{128}$$

and conclude that the reference state becomes

$$\begin{aligned}
\prod_{i=1}^{N} b_i^{\cdot\dagger}|\text{vac}\rangle &= \prod_{i=1}^{N}\left[\sum_{l=1}^{N} a_l^{\cdot\dagger} U_{li}^{\cdot} + \sum_{j=N+1}^{K} a_j^{\text{o}\dagger} U_{ji}''\right]|\text{vac}\rangle \\
&= \prod_{i=1}^{N}\left[\sum_{l=1}^{N}\left\{a_l^{\cdot\dagger} + \sum_{j=N+1}^{K}\sum_{k=1}^{N} a_j^{\text{o}\dagger} U_{jk}''\left(\mathbf{U}^{\cdot -1}\right)_{kl}\right\}U_{li}^{\cdot}\right]|\text{vac}\rangle \\
&= \alpha\prod_{i=1}^{N}\left[a_i^{\cdot\dagger} + \sum_{j=N+1}^{K}\sum_{k=1}^{N} a_j^{\text{o}\dagger} U_{jk}''\left(\mathbf{U}^{\cdot -1}\right)_{ki}\right]|\text{vac}\rangle \\
&= \alpha\prod_{i=1}^{N}\left[1 + \sum_{j=N+1}^{K}\sum_{k=1}^{N} a_j^{\text{o}\dagger} U_{jk}''\left(\mathbf{U}^{\cdot -1}\right)_{ki} a_i^{\cdot}\right] a_i^{\cdot\dagger}|\text{vac}\rangle \\
&= \alpha\prod_{i=1}^{N}\left[1 + \sum_{j=N+1}^{K}\sum_{k=1}^{N} a_j^{\text{o}\dagger} U_{jk}''\left(\mathbf{U}^{\cdot -1}\right)_{ki} a_i^{\cdot}\right]\prod_{l=1}^{N} a_l^{\cdot\dagger}|\text{vac}\rangle
\end{aligned} \tag{129}$$

If we introduce the complex parameters

$$z_{ji} = \sum_{k=1}^{N} U_{jk}''\left(\mathbf{U}^{\cdot -1}\right)_{ki} \tag{130}$$

and write the unnormalized state vector

$$\prod_{i=1}^{N} b_i^{\dagger} |\mathrm{vac}\rangle = |\mathbf{z}\rangle \tag{131}$$

with the parameters being time dependent. In terms of the orthonormal spin orbital basis we write

$$\begin{aligned}
|\mathbf{z}\rangle &= \prod_{i=1}^{N} \left[1 + \sum_{j=N+1}^{K} z_{ji} a_j^{\mathrm{o}\dagger} a_i^{\cdot} \right] |0\rangle \\
&= \prod_{i=1}^{N} \prod_{j=N+1}^{K} \left[1 + z_{ji} a_j^{\mathrm{o}\dagger} a_i^{\cdot} \right] |0\rangle \\
&= \prod_{i=1}^{N} \prod_{j=N+1}^{K} \exp\left[z_{ji} a_j^{\mathrm{o}\dagger} a_i^{\cdot} \right] |0\rangle \\
&= \exp\left[\sum_{i=1}^{N} \sum_{j=N+1}^{K} z_{ji} a_j^{\mathrm{o}\dagger} a_i^{\cdot} \right] |0\rangle
\end{aligned} \tag{132}$$

In going from the second to the third line in the above equation we have used the fact that the electron field operators are nilpotent and this also is the reason that the exponentiation in the fourth line works. The end result is true because all the operators $a_j^{\mathrm{o}\dagger} a_i^{\cdot}$ commute.

From these equations it follows straightforwardly that the wave function representative of this state is

$$\det\left[\chi_i(x_j) \right] \tag{133}$$

with the 'dynamical spin orbitals'

$$\chi_i = u_i + \sum_{j=N+1}^{K} u_j z_{ji} \tag{134}$$

and where the parameters z_{ji} are complex and are considered to be functions of the time parameter t. As they change during a process involving the electronic system the determinantal state vector can in principle become any determinantal wave function possible to express in the spin orbital basis. The spin orbitals χ_i are not orthonormal even if the basis $\{u_k\}$ is and in actual application one will often use a raw basis of atomic spin orbitals (often built from Gaussian type orbitals), which are not orthonormal. In such a case the exponential form of the determinantal state is not applicable and one has to deal with the full complications of the nonunit metric of the basis (see [16]). This is indeed possible and has been coded into the ENDyne program system that uses narrow wave packet nuclei and single determinantal electron states in an explicitly time-dependent, nonadiabatic treatment of molecular processes.

A determinantal wave function expressed in this form is a coherent state. The associated Lie group is the unitary group $U(K)$ and the reference state Eq. (127) is the lowest weight state of the irreducible representation $[1^N 0^{K-N}]$ of U(K). The stability group is $U(N) \times U(K-N)$. The norm in an orthonormal basis of spin orbitals is

$$\langle \mathbf{z} | \mathbf{z} \rangle = \det[1 + \mathbf{z}^\dagger \mathbf{z}] \tag{135}$$

which means that we can define an appropriate measure $\mathbf{dz}$ in the parameter space such that the resolution of the identity

$$\int |\mathbf{z}\rangle \langle \mathbf{z}| \mathbf{dz} = 1 \tag{136}$$

holds. The derivation of the form of the measure $\mathbf{dz}$ is nontrivial [17].

2.5 MINIMAL ELECTRON NUCLEAR DYNAMICS (END)

The time-dependent variational principle introduced above with the quantum mechanical action

$$A = \int_{t_1}^{t_2} L \, \mathrm{d}t \tag{137}$$

and the quantum mechanical Lagrangian ($\hbar = 1$)

$$L = \langle \psi | \frac{\mathrm{i}}{2} \left(\frac{\partial}{\partial t} - \frac{\tilde{\partial}}{\partial t} \right) - H | \psi \rangle / \langle \psi | \psi \rangle \tag{138}$$

proceeds by making the action stationary, which leads to the Euler–Lagrange equations

$$\frac{\mathrm{d}}{\mathrm{d}t} \frac{\partial L}{\partial \dot{q}} = \frac{\partial L}{\partial q} \tag{139}$$

once the dynamical variables q have been chosen. The END theory starts with this principle of least action, chooses a particular family of wave functions ψ and a basis set for its description and derives a set of Euler–Lagrange equations as the dynamical equations that for that choice of wave function form and basis represent the time-dependent Schrödinger equation and can be integrated in time to yield the time-evolving state vector or the system under consideration.

A simple choice of wave function family for a molecular system that makes sense is a product state vector

$$|\psi\rangle = |z, R, P\rangle |R, P\rangle = |z\rangle |\phi\rangle \tag{140}$$

with $|z, R, P\rangle = |z\rangle$ an electronic state vector, and $|\phi\rangle = |R, P\rangle$ a nuclear state vector. It should be noted that this product form is not an adiabatic type wave function since the electronic part is parametrically dependent on the (time-dependent) average nuclear positions R (and average momenta P), not their actual position (and momentum) coordinates. This corresponds rather to a Born–Huang type product and is more like adiabatic wave function and could in principle be resolved into a number of adiabatic states.

This lowest of END approximations aims at a description in terms of classical nuclei, so a starting nuclear wave function is

$$|\phi\rangle = \prod_{j,k} \exp\left[-\frac{1}{2}\left(\frac{X_{jk} - R_{jk}}{b}\right)^2 + \mathrm{i}P_{jk}(X_{jk} - R_{jk})\right] \tag{141}$$

i.e. of product form, where the parameters R_{jk} and P_{jk} are the Cartesian components of the average nuclear positions and momenta, respectively. The simplest electronic wave function that makes any sense is a complex spin unrestricted Thouless determinant [16,18]

$$|z, R, P\rangle \equiv |z\rangle = \det\{\chi_h(x_p)\} \tag{142}$$

with

$$\chi_h = u_h + \sum_p u_p z_{ph},\ 1 \le h \le N \tag{143}$$

where $\{u_i\}_1^K$ is a set of atomic spin orbitals (GTO's) centered on the average nuclear positions

$$\vec{R}_k = (\, R_{1k} \quad R_{2k} \quad R_{3k}\,) \tag{144}$$

In particular for high collision energies it is advantageous to include in this atomic basis so-called electronic translation factors $\exp\{\mathrm{i}\vec{k}\cdot\vec{r}\}$, with $\vec{k} = m\vec{P}_k/M_k$ for nucleus k to produce traveling atomic orbitals. The Lagrangian can now be written as

$$\begin{aligned} L = \frac{\mathrm{i}}{2}\Big[& \langle\phi|\frac{\partial}{\partial t}|\phi\rangle/\langle\phi|\phi\rangle - \langle\phi|\frac{\tilde{\partial}}{\partial t}|\phi\rangle/\langle\phi|\phi\rangle + \langle z|\frac{\partial}{\partial t}|z\rangle/\langle z|z\rangle \\ & - \langle z|\frac{\tilde{\partial}}{\partial t}|z\rangle/\langle z|z\rangle\Big] - \langle\psi H|\psi\rangle/\langle\psi|\psi\rangle \end{aligned} \tag{145}$$

We then use the fact that the parameters z, R, and P are considered to be time-dependent and employ the chain rule of differentiation to write

$$\begin{aligned} L = \frac{\mathrm{i}}{2}\sum_{i,l}\Bigg\{ & \Bigg[\langle\phi| - \frac{\partial}{\partial X_{il}}|\phi\rangle/\langle\phi|\phi\rangle + \langle\phi| - \frac{\partial}{\partial X_{il}}|\phi\rangle/\langle\phi|\phi\rangle + \frac{\partial \ln S}{\partial R_{il}} - \frac{\partial \ln S}{\partial R'_{il}}\Bigg]\dot{R}_{il} \\ & + \Bigg[\frac{\partial \ln S}{\partial P_{il}} - \frac{\partial \ln S}{\partial P'_{il}}\Bigg]\dot{P}_{il}\Bigg\} + \frac{\mathrm{i}}{2}\sum_{p,h}\left(\frac{\partial \ln S}{\partial z_{ph}}\dot{z}_{ph} - \frac{\partial \ln S}{\partial z^*_{ph}}\dot{z}^*_{ph}\right) - \langle\psi|H|\psi\rangle/\langle\psi|\psi\rangle \end{aligned} \tag{146}$$

where we have used the form of the nuclear wave function, which is such that the differentiation with respect to R_{il} is the negative of the differentiation with respect to the nuclear coordinate X_{il}. We have also introduced the overlap between two electronic wave functions $S = \langle z|z\rangle = S(z^*, R', P'; z, R, P)$.

The narrow wave packet limit ($b \to 0, \hbar \to 0, \hbar/b^2 \to 1$) leads to the Lagrangian

$$L = \sum_{j,l} \left\{ \left(P_{jl} + \frac{i}{2} \left[\frac{\partial \ln S}{\partial R_{jl}} - \frac{\partial \ln S}{\partial R'_{jl}} \right] \right) \dot{R}_{jl} + \frac{i}{2} \left[\frac{\partial \ln S}{\partial P_{jl}} - \frac{\partial \ln S}{\partial P'_{jl}} \right] \dot{P}_{jl} \right\} + \frac{i}{2} \sum_{p,h} \left(\frac{\partial \ln S}{\partial z_{ph}} \dot{z}_{ph} - \frac{\partial \ln S}{\partial z^*_{ph}} \dot{z}^*_{ph} \right) - E \tag{147}$$

where

$$E = \sum_{j,l} \frac{P_{jl}^2}{2M_l} - \frac{\langle z|H_{\mathrm{el}}|z\rangle}{\langle z|z\rangle} \tag{148}$$

and contains the nuclear–nuclear repulsion terms, and where in the derivatives R', P' are put equal to R and P, respectively, after differentiation. When the total Hamiltonian is written

$$H = T_{\mathrm{n}} + T_{\mathrm{e}} + V_{\mathrm{ne}} + V_{\mathrm{ee}} + V_{\mathrm{nn}} \tag{149}$$

with the nuclear kinetic energy terms T_{n}, the electron kinetic energy terms T_{e}, the nuclear–electron interaction terms V_{ne}, the electron–electron interaction terms V_{ee}, and the nuclear–nuclear interaction terms V_{nn} one can discern the electronic Hamiltonian

$$H_{\mathrm{el}} = T_{\mathrm{e}} + V_{\mathrm{ne}} + V_{\mathrm{ee}} + V_{\mathrm{nn}} \tag{150}$$

Using this Lagrangian for quantum electrons and classical nuclei and choosing as the dynamical variable $q = R_{ik}$ we get

$$\frac{\partial L}{\partial \dot{R}_{ik}} = P_{ik} + \frac{i}{2} \left[\frac{\partial \ln S}{\partial R_{ik}} - \frac{\partial \ln S}{\partial R'_{ik}} \right] \tag{151}$$

and

$$\begin{aligned} \frac{d}{dt} \frac{\partial L}{\partial \dot{R}_{ik}} = \dot{P}_{ik} &+ \frac{i}{2} \sum_{j,l} \left[\frac{\partial^2 \ln S}{\partial R_{ik} \partial R_{jl}} + \frac{\partial^2 \ln S}{\partial R_{ik} \partial R'_{jl}} - \frac{\partial^2 \ln S}{\partial R'_{ik} \partial R_{jl}} - \frac{\partial^2 \ln S}{\partial R'_{ik} \partial R'_{jl}} \right] \dot{R}_{jl} \\ &+ \frac{i}{2} \sum_{j,l} \left[\frac{\partial^2 \ln S}{\partial R_{ik} \partial P_{jl}} + \frac{\partial^2 \ln S}{\partial R_{ik} \partial P'_{jl}} - \frac{\partial^2 \ln S}{\partial R'_{ik} \partial P_{jl}} - \frac{\partial^2 \ln S}{\partial R'_{ik} \partial P'_{jl}} \right] \dot{P}_{jl} \\ &+ \frac{i}{2} \sum_{p,h} \left[\frac{\partial^2 \ln S}{\partial R_{ik} \partial z_{ph}} \dot{z}_{ph} - \frac{\partial^2 \ln S}{\partial R'_{ik} \partial z_{ph}} \dot{z}_{ph} + \frac{\partial^2 \ln S}{\partial R_{ik} \partial z^*_{ph}} \dot{z}^*_{ph} - \frac{\partial^2 \ln S}{\partial R'_{ik} \partial z^*_{ph}} \dot{z}^*_{ph} \right] \end{aligned} \tag{152}$$

This expression should equal

$$\frac{\partial L}{\partial R_{ik}} = \frac{\mathrm{i}}{2}\sum_{p,h}\left[\frac{\partial^2 \ln S}{\partial R_{ik}\partial z_{ph}}\dot{z}_{ph} + \frac{\partial^2 \ln S}{\partial R'_{ik}\partial z_{ph}}\dot{z}_{ph} - \frac{\partial^2 \ln S}{\partial R_{ik}\partial z^*_{ph}}\dot{z}^*_{p} - \frac{\partial^2 \ln S}{\partial R'_{ik}\partial z^*_{ph}}\dot{z}^*_{ph}\right]$$
$$+ \frac{\mathrm{i}}{2}\sum_{j,l}\left[\frac{\partial^2 \ln S}{\partial R_{ik}\partial R_{jl}} - \frac{\partial^2 \ln S}{\partial R_{ik}\partial R'_{jl}} + \frac{\partial^2 \ln S}{\partial R'_{ik}\partial R_{jl}} - \frac{\partial^2 \ln S}{\partial R'_{ik}\partial R'_{jl}}\right]\dot{R}_{jl}$$
$$+ \frac{\mathrm{i}}{2}\sum_{j,l}\left[\frac{\partial^2 \ln S}{\partial R_{ik}\partial P_{jl}} - \frac{\partial^2 \ln S}{\partial R_{ik}\partial P'_{jl}} + \frac{\partial^2 \ln S}{\partial R'_{ik}\partial P_{jl}} - \frac{\partial^2 \ln S}{\partial R'_{ik}\partial P'_{jl}}\right]\dot{P}_{jl} - \frac{\partial E}{\partial R_{ik}} \tag{153}$$

and cancellation of terms leads to the following result

$$\dot{P}_{ik} + 2\sum_{j,l}\left[\Im\left\{\frac{\partial^2 \ln S}{\partial R'_{ik}\partial R_{jl}}\right\}\dot{R}_{jl} + \Im\left\{\frac{\partial^2 \ln S}{\partial R'_{ik}\partial P_{jl}}\dot{P}_{jl}\right\}\right]$$
$$- \mathrm{i}\sum_{p,h}\left[\frac{\partial^2 \ln S}{\partial R_{ik}\partial z_{ph}}\dot{z}_{ph} - \frac{\partial^2 \ln S}{\partial R_{ik}\partial z^*_{ph}}\dot{z}^*_{ph}\right] = -\frac{\partial E}{\partial R_{ik}} \tag{154}$$

or by collecting the three Cartesian components for each nucleus (such as $\mathbf{R}_k = R_{1k}, R_{2k}, R_{3k}$) and the z parameters in a rectangular matrix $\mathbf{z} = \{z_{ph}\}$ we can write the more compact form[1]

$$-\dot{\mathbf{P}}_k + \sum_l\left[\mathbf{C}_{\mathbf{R}_k\mathbf{R}_l}\dot{\mathbf{R}}_l + \mathbf{C}_{\mathbf{R}_k\mathbf{P}_l}\dot{\mathbf{P}}_l\right] + \mathrm{i}\mathbf{C}^\dagger_{\mathbf{R}_k}\dot{\mathbf{z}} - \mathrm{i}\mathbf{C}^T_{\mathbf{R}_k}\dot{\mathbf{z}}^* = \vec{\nabla}_{\mathbf{R}_k}E \tag{155}$$

where

$$(\mathbf{C}_{\mathbf{RR}})_{ik;jl} = (\mathbf{C}_{R_kR_l})_{ij} = -2\Im\frac{\partial^2 \ln S}{\partial R'_{ik}\partial R_{jl}}\Big|_{R'=R,\,P'=P} \tag{156}$$

and

$$(\mathbf{C}_{\mathbf{R}})_{ph;ik} = (\mathbf{C}_{\mathbf{R}_k})_{ph;i} = (C_{R_{ik}})_{ph} = \frac{\partial^2 \ln S}{\partial z^*_{ph}\partial R_{ik}}\Big|_{R'=R,\,P'=P} \tag{157}$$

Similarly, for $q = P_k$ we obtain

$$\dot{\mathbf{R}}_k + \sum_l\left[\mathbf{C}_{\mathbf{P}_k\mathbf{R}_l}\dot{\mathbf{R}}_l + \mathbf{C}_{\mathbf{P}_k\mathbf{P}_l}\dot{\mathbf{P}}_l\right] + \mathrm{i}\mathbf{C}^\dagger_{\mathbf{P}_k}\dot{\mathbf{z}} - \mathrm{i}\mathbf{C}^T_{\mathbf{P}_k}\dot{\mathbf{z}}^* = \vec{\nabla}_{\mathbf{P}_k}E \tag{158}$$

and for $q = z^*_p$ the equation is

$$\mathrm{i}\mathbf{C}\dot{\mathbf{z}} + \mathrm{i}\mathbf{C}_{\mathbf{R}}\dot{\mathbf{R}} + \mathrm{i}\mathbf{C}_{\mathbf{P}}\dot{\mathbf{P}} = \frac{\partial E}{\partial \mathbf{z}^*} \tag{159}$$

[1] One should note that $(\mathbf{C}^\dagger_{\mathbf{R}})_{ph;ik} = (\mathbf{C}^*_{\mathbf{R}})_{ik;ph} = \partial^2 \ln S/\partial R'_{ik}\partial z_{ph}$ and that $(\mathbf{C}^T_{\mathbf{R}})_{ph;ik} = (\mathbf{C}_{\mathbf{R}})_{ik;ph} = \partial^2 \ln S/\,\partial R''_{ik}\partial z^*_{ph}$.

The complex conjugate of this equation obtains when we choose $q = z_p$. All these equations of motion can be put together in matrix form

$$\begin{bmatrix} i\mathbf{C} & 0 & i\mathbf{C}_{\mathbf{R}} & i\mathbf{C}_{\mathbf{P}} \\ 0 & -i\mathbf{C}^* & -i\mathbf{C}^*_{\mathbf{R}} & -i\mathbf{C}^*_{\mathbf{P}} \\ i\mathbf{C}^\dagger_{\mathbf{R}} & -i\mathbf{C}^T_{\mathbf{R}} & \mathbf{C}_{\mathbf{RR}} & -\mathbf{I}+\mathbf{C}_{\mathbf{RP}} \\ i\mathbf{C}^\dagger_{\mathbf{P}} & -i\mathbf{C}^T_{\mathbf{P}} & \mathbf{I}+\mathbf{C}_{\mathbf{PR}} & \mathbf{C}_{\mathbf{PP}} \end{bmatrix} \begin{bmatrix} \dot{\mathbf{z}} \\ \dot{\mathbf{z}}^* \\ \dot{\mathbf{R}} \\ \dot{\mathbf{P}} \end{bmatrix} = \begin{bmatrix} \partial E/\partial \mathbf{z}^* \\ \partial E/\partial \mathbf{z} \\ \partial E/\partial \mathbf{R} \\ \partial E/\partial \mathbf{P} \end{bmatrix} \tag{160}$$

The integration of this set of coupled first-order differential equation can be done in a number of ways. Care must be taken since there are basically rather two different time scales involved, i.e. that of the nuclear dynamics and that of the normally considerably faster electron dynamics. It should be observed that this END takes place in a Cartesian laboratory reference frame, which means that the overall translation as well as overall rotation of the molecular system is included. This offers no complications since the equations of motion satisfy basic conservation laws and, thus, total momentum and angular momentum are conserved. At any time in the evolution of the molecular system can the overall translation be isolated and eliminated if so should be deemed necessary. This level of theory [16,19] is implemented in the program system ENDyne [20], and has been applied to atomic and molecular reactive collisions. Calculations of cross sections, differential as well as integral, yield results in excellent agreement with the best experiments.

END theory at this level of approximation can be characterized as full nonlinear time-dependent Hartree–Fock with moving classical nuclei. It is a direct, nonadiabatic approach to molecular processes. One might surmise that such a method will do well at hyperthermal collision energies where many potential surfaces and associated nonadiabatic coupling terms will come into play. In particular, ion–atom and ion–molecule reactions should be well described, since the direct as well as the charge transfer processes can be treated on an equal footing. This is indeed true, and comparisons of calculated and measured direct *absolute* cross sections for a number of systems at keV energies, such as H^+, H, and He on He and Ne (see [21–23]) have been published. Also, charge exchange, energy loss, and differential absolute cross sections for H^+ and H on H and H_2 in excellent agreement with the best experiments have been obtained with minimal END [24,25]. The same is true for H^+ on atomic nitrogen, oxygen, and fluorine [26], and N_2 [27].

Differential cross sections and state–state processes are more sensitive to the level of treatment than are integral cross sections, for which minimal END can be shown to give results in good agreement with experiment also for lower energies, in some cases down to a fraction of an eV. Results for H_2^+ on H_2 [28], and H^+ on C_2H_2 [29], and on C_2H_6 [30].

Cross sections and possible mechanisms have also been studied for the reactions [31–33]

$$D_2 + NH_3^+ \rightarrow NH_3D^+ + D \tag{161}$$

$$D_2 + NH_3^+ \rightarrow NH_2D^+ + D + H \tag{162}$$

with discovery of a two-step process for $D \rightarrow H$ exchange.

That the simple minimal END wave function for the reacting system can get such good results is puzzling to some and deserves a comment. Obviously, a single determinantal electronic wave function, even if it is complex and has nonorthogonal spin orbitals, in a 'static' time-independent calculation, can only yield results of SCF quality. However, in a time-dependent 'dynamic' calculation with the wave function adjusting at each time step to the moving nuclei such a simple representation of the electrons provides a surprisingly flexible description of the electron dynamics.

The classical description of the nuclei means that a product molecular species vibrates and rotates as a classical object. Using the notion that coherent states bridges the classical and quantum descriptions, one can view this classical motion as an evolving state, which can be resolved into rovibrational states. In the case of low excitation energies, when to a good approximation each vibrational mode and the rotations can be considered to be decoupled, each normal vibrational mode can be represented by the evolving state in Eq. (75). The energy of such a state is $\hbar\omega(|z|^2 + 1/2)$, yielding the vibrational excitation energy for such a mode to be $|z|^2 = E_{\text{vib}}/\hbar\omega$, where the vibrational energy E_{vib} is obtained from a so-called generalized Prony analysis [25,34–36] of the END trajectories. In this manner vibrationally resolved cross sections can be calculated *a posteriori*. A similar treatment can be given the rotational motion *via* another choice of coherent states [37].

The END theory is a general approach [38] to approximate the time-dependent Schrödinger equation and is not limited to treatments of molecular reaction dynamics. Other problems that have been studied include intramolecular electron transfer [16], the effect of intense laser light on the vibrational dynamics of small molecules [39], and solitonic charge transport in polyenes [40].

2.6 RENDERING OF DYNAMICS

When the forces between reactants are derived from a precalculated PES it is possible to produce informative pictures with reactant valleys and product valleys perhaps connected by saddles indicating transition states. Time-laps photography or movies of dynamical events may show probabilities in terms of nuclear wave functions evolving on one surface and then transfer to another surface if nonadiabatic coupling terms are present.

In the case of direct nonadiabatic dynamics the rendering of dynamical events needs some rethinking. When the nuclei are treated as classical particles or by narrow wave packets movies of dynamically changing ball and stick models can be quite effective and informative. It is possible to use the rendering of such trajectories for finding errors in the dynamics and to illustrate mechanisms.

When the participating electronic degrees of freedom are treated dynamically, rather than being integrated out to provide the average forces, then one can augment the rendering with an evolving electronic charge density of the reacting system. Sometimes a simple picture, such as is provided by a spherical electron cloud around each atom, the size of which changes in time with the electron population, is useful and provides the crucial information about the studied process. This manner of depicting the reacting system is shown in Fig. 2.1.

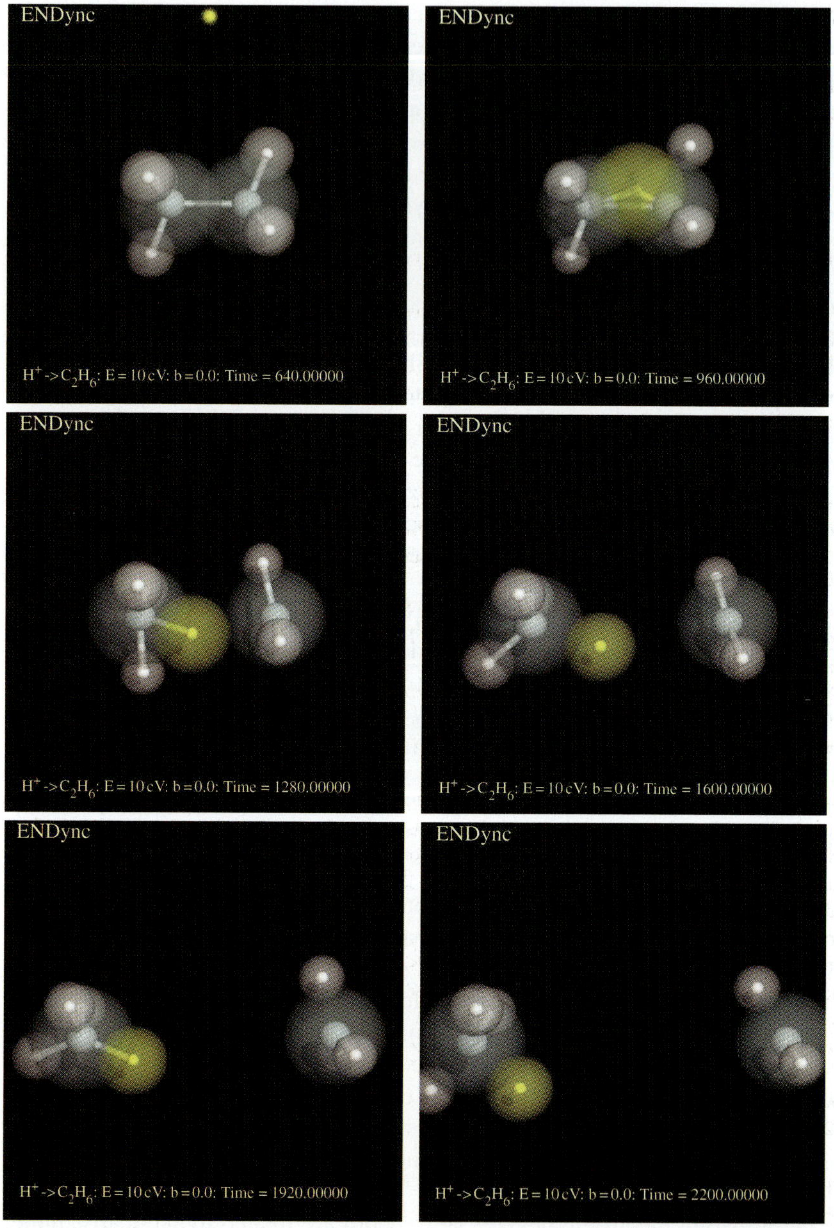

Fig. 2.1. Six snap shots of $H^+ + C_2H_6 \rightarrow CH_4 + CH_3^+$ with the dynamical electrons represented by a sphere around each nucleus with the size proportional to the electronic population on each atom. The H^+ approaches from above in the first frame, and polarizes the C–C bond in the second frame. The third through sixth frames show the products departing rovibrationally excited.

2.7 ACKNOWLEDGEMENTS

We would like to thank Remigio Cabrera-Trujillo for providing the figures, and we acknowledge support from the Office of Naval Research.

2.8 REFERENCES

1 R.D. Levine and R.B. Bernstein, Molecular reaction dynamics and chemical reactivity, Oxford University, New York, 1987.
2 J.C. Giddings and H. Eyring, Equilibrium theory of unimolecular reactions, J. Chem. Phys., 22 (1954) 538.
3 W.H. Miller, Recent advances in quantum mechanical reactive scattering theory, Annu. Rev. Phys. Chem., 41 (1990) 245–281.
4 J.M. Bowman, Overview of reduced dimensionality quantum approaches to reactive scattering, Theor. Chem. Acc., 108 (2002) 125–133.
5 J.B. Liu, K.Y. Song, W.L. Hase and S.L. Anderson, Direct dynamics trajectory study of vibrational effects: can polyani rules be generalized to a polyatomic system? JACS, 126 (2004) 8602.
6 A. Fröman, Isotope effects and electronic energy in molecules, J. Chem. Phys, 36 (1962) 1490.
7 M. Kimura and N.F. Lane, The low-energy, heavy-particle collisions—a close-coupling treatment, in: D. Bates and B. Bederson (Eds.), Advances in atomic, molecular and optical physics, Academic, New York, 1990, p. 79.
8 M. Born and J.R. Oppenheimer, Zur Ouantentheorie der Molekeln, Ann. Physik, 84 (1927) 457.
9 J. Broeckhove, L. Lathouwers, E. Kesteloot and P. Van Leuven, On the equivalence of time dependent variational principles, Chem. Phys. Lett., 149 (1988) 547.
10 P.A.M. Dirac, The principles of quantum mechanics, The international series of monographs on physics, Vol. 27, Oxford University Press, New York, 1930.
11 P. Kramer and M. Saraceno, Geometry of the time-dependent variational principle, Quantum mechanics, Springer, New York, 1981.
12 Y. Öhrn and E. Deumens, Electron nuclear dynamics with coherent states, in: D.H. Feng and J.R. Klauder (Eds.), Proceedings of the International Symposium on Coherent States: Past, Present, and Future, Oak Ridge Associated Universities, World Scientific, Singapore, 1993.
13 J.R. Klauder and B.-S. Skagerstam, Coherent states, Applications in physics and mathematical physics, World Scientific, Singapore, 1985.
14 A.M. Perelomov, Coherent states for arbitrary Lie group, Commun. Math. Phys., 26 (1972) 222–236.
15 M. Sugiura, Unitary representations and harmonic analysis, John Wiley & Sons, New York, 1975).
16 E. Deumens, A. Diz, R. Longo and Y. Öhrn, Time-dependent theoretical treatments of the dynamics of electrons and nuclei in molecular systems, Rev. Mod. Phys., 66 (3) (1994) 917–983.
17 J.P. Blaizot and H. Orland, Path integrals for the nuclear many-body problem, Phys. Rev. C, 24 (1981) 1740.
18 D.J. Thouless, Stability conditions and nuclear rotations in the Hartree–Fock theory, Nucl. Phys., 21 (1960) 225.
19 Y. Öhrn, E. Deumens, A. Diz, R. Longo, J. Oreiro and H. Taylor, Time evolution of electrons and nuclei in molecular systems, in: J. Broeckhove and L. Lathouwers (Eds.), Time-dependent quantum molecular dynamics, Plenum, New York, 1992, pp. 279–292.
20 E. Deumens, T. Helgaker, A. Diz, H. Taylor, J. Oreiro, B. Mogensen, J.A. Morales, M.C. Neto and R. Cabrera-Trujillo and D. Jacquemin, ENDyne version 5 Software for Electron Nuclear Dynamics, Quantum Theory Project, University of Florida, Gainesville, FL, 2002, http://www.qtp.ufl.edu/endyne.html
21 R. Cabrera-Trujillo, J.R. Sabin, Y. Öhrn and E. Deumens, Direct differential cross section calculations for ion–atom and atom–atom collisions in the keV range, Phys. Rev. A, 61 (2000) 032719.
22 R. Cabrera-Trujillo, J.R. Sabin, Y. Öhrn and E. Deumens, Charge exchange and threshold effects in the energy loss of slow projectiles, Phys. Rev. Lett., 84 (2000) 5300.

References pp. 39–40

23 R. Cabrera-Trujillo, E. Deumens, Y. Öhrn and J.R. Sabin, Impact parameter dependence of electronic and nuclear energy loss of swift ions: $H^+ \rightarrow He$ and $H^+ \rightarrow H$, Nucl. Instrum. Methods, B168 (2000) 484.
24 R. Cabrera-Trujillo, Y. Öhrn, E. Deumens and J.R. Sabin, Trajectory and molecular binding effects in stopping cross section for hydrogen beams on H_2, J. Chem. Phys., 116 (2002) 2783.
25 J.A. Morales, A.C. Diz, E. Deumens and Y. Öhrn, Electron nuclear dynamics of $H^+ + H_2$ collisions at $E_{lab} = 30$ eV, J. Chem. Phys., 103 (1995) 9968–9980.
26 R. Cabrera-Trujillo, Y. Öhrn, E. Deumens and J.R. Sabin, Stopping cross section in the low to intermediate energy range: Study of proton and hydrogen atom collisions with atomic N, O, and F, Phys. Rev. A, 62 (2000) 052714.
27 R. Cabrera-Trujillo, Y. Öhrn, E. Deumens, J.R. Sabin and B.G. Lindsay, Theoretical and experimental studies of the $H^+ + N_2$ system: Differential cross sections for direct and charge-transfer scattering at keV energies, Phys. Rev. A, 66 (2002) 042712.
28 Y. Öhrn, J. Oreiro and E. Deumens, Bond making and bond breaking in molecular dynamics, Int. J. Quantum Chem., 58 (1996) 583.
29 S.A. Malinovskaya, R. Cabrera-Trujillo, J.R. Sabin, E. Deumens and Y. Öhrn, Dynamics of collisions of protons with acetylene molecules at 30 ev, J. Chem. Phys., 117 (2002) 1103.
30 R. Cabrera-Trujillo, J.R. Sabin, Y. Öhrn and E. Deumens, Energy loss studies of protons colliding with ethane: Preliminary results, J. Electron Spectrosc., 129 (2003) 303–308.
31 M. Coutinho-Neto, E. Deumens and Y. Öhrn, Abstraction and exchange mechanisms for the $D_2 + NH_3^+$: Reaction at hyperthermal energies, J. Chem. Phys., 116 (2002) 2794.
32 R.J.S. Morrison, W.E. Conway, T. Ebata and R.N. Zare, Vibrationally state-selected reactions of ammonia ions. I. $NH_3^+(\nu) + D_2$, J. Chem. Phys., 84 (1986) 5527.
33 J.C. Poutsma, M.A. Everest, J.E. Flad, G.C. Jones, Jr. and R.N. Zare, State-selected studies of the reaction of $NH_3^+(\nu_1,\nu_2)$ with D_2, Chem. Phys. Lett., 305 (1999) 343.
34 B. de Prony, J.E. Polytech, 1 (2) (1795) 24–76.
35 A. Blass, E. Deumens and Y. Öhrn, Rovibrational analysis of molecular collisions using coherent states, J. Chem. Phys., 115 (2001) 8366.
36 J.A. Morales, A.C. Diz, E. Deumens and Y. Öhrn, Molecular vibrational state distributions in collisions, Chem. Phys. Lett., 233 (1995) 392.
37 J.A. Morales, E. Deumens and Y. Öhrn, On rotational coherent states in molecular quantum dynamics, J. Math. Phys., 40 (1999) 766.
38 Y. Öhrn and E. Deumens, Towards an ab initio treatment of the time-dependent Schrödinger equation of molecular systems, J. Phys. Chem., 103 (1999) 9545.
39 J. Broeckhove, M.D. Coutinho-Neto, E. Deumens and Y. Öhrn, Electron nuclear dynamics of LiH and HF in an intense laser field, Phys. Rev. A, 56 (1997) 4996.
40 B. Champagne, E. Deumens and Y. Öhrn, Vibrations and soliton dynamics of positively charged polyacetylene chains, J. Chem. Phys., 107 (1997) 5433.

Theory and Applications of Computational Chemistry: The First Forty Years
Edited by C. Dykstra et al.

CHAPTER 3

Computation of non-covalent binding affinities

J. Andrew McCammon

Department of Chemistry and Biochemistry and Department of Pharmacology, NSF Center for Theoretical Biological Physics, Howard Hughes Medical Institute, University of California at San Diego, La Jolla, CA 92093-0365, USA

Abstract

The ability to accurately predict and analyze molecular recognition is being achieved by advances in several areas of theoretical chemistry and related fields such as applied mathematics and computational science. This chapter provides an overview of the history, current state and future prospects for computational studies of molecular recognition.

3.1 INTRODUCTION

Non-covalent binding is central to many processes in the chemical sciences. To name only a few, examples include the formation of host–guest complexes in physical organic chemistry, the self-assembly of supramolecular structures in nanotechnology, and the binding of drugs to their receptors in medicinal chemistry. The examples just listed, and many others, involve binding in solution. Thus, an understanding of the strength of binding requires consideration of desolvation (or solvent reorganization) in addition to the direct interactions between the binding partners. The present account describes the evolution of the computational methods for treating such problems, and a perspective on the current state of this area of computational chemistry. In what follows, I will focus on examples related to drug discovery, but it is clear that the methods have much more general applicability.

In virtually all chemistry courses from the elementary level upward, it is shown that the equilibrium constant for the binding of molecules A and B to form the complex AB depends exponentially on the standard free energy change associated with complexation. It has long been recognized that if one could compute the standard free energy

References pp. 45–46

of complexation of solute molecules, it would be possible both to gain a deeper understanding of the origins of molecular recognition, and to contemplate the 'first principles' design of pharmaceuticals and other compounds. Such calculations were attempted, for example, by the Scheraga group as early as 1972 [1], although limitations in computer power did not allow inclusion of solvation or entropic effects in this work.

In 1986, Wong and McCammon [2] combined the statistical mechanical theory of free energy with atomistic simulations of solvent and solutes to calculate, for the first time, the relative standard free energy of binding of different small inhibitor molecules to an enzyme. The necessary statistical mechanical theory had been available for many years. But several other factors and advances in computational and theoretical chemistry during the 14 years between the two calculations were required to enable the 1986 work to succeed. One important advance was the introduction of molecular dynamics simulations into biochemistry in 1977 [3]. In such simulations, as is now well known, Newton's equations of motion are used with a detailed model of the forces within a molecular system to generate trajectories of the atomic fluctuations. Another factor that made the 1986 work possible was the growth of computer power, by roughly a factor of 600 over the 14-year period according to Moore's law. Although the 1977 molecular dynamics simulation was limited to a very small protein with no explicit solvent, the 1986 work utilized a molecular dynamics simulation of the large enzyme trypsin in a bath of explicitly represented water molecules. But, as with many advances in computational chemistry, the key factor leading to the 1986 breakthrough was a new theoretical element, in this case the concept of using thermodynamic cycles to relate the desired relative free energy to that of two non-physical processes: computational 'alchemical' transformations of one inhibitor into another one, in solution and in the binding site [4].

Subsequent work has shown that free energy calculations that involve systems as large as proteins or other macromolecules can provide usefully accurate results in favorable cases. But, in general, there are difficulties in achieving precise and accurate results with reasonable amounts of computer time, even using current state-of-the-art machines. These difficulties arise primarily from the incomplete sampling of the rough, many-dimensional potential energy surfaces of such systems. Below, I mention several lines of work that hold promise for making free energy calculations faster and more accurate for biomolecular systems.

3.2 CURRENT METHODS

For calculations of relative free energies of binding, the theoretical framework outlined by Tembe and McCammon [4] has often been used essentially without change. This framework recognizes that brute force calculations of standard free energies of binding will encounter convergence problems related to the dramatic changes in solvation of the binding partners, conformational changes that require physical times longer that those that can be explored by simulation, etc. Tembe and McCammon [4] introduced the use of thermodynamic cycle analyses that allow the desired relative free energies to be computed in terms of 'alchemical' transformations, as described above. The advantage is that only relatively localized changes occur in the simulated system, at least in favorable cases.

Calculation of the standard free energy of binding itself can be viewed as a special case of the above, in which one of the pair of ligands contains no atoms [5]. Some care is required to be sure that such calculations yield answers that actually correspond to the desired standard state [6,7]. Unfortunately, many calculations of free energies of binding have not made appropriate contact with a standard state, so that results in the literature must be interpreted with caution.

It has been mentioned that perhaps the greatest limitation to the precision of free energy calculations to date has been the often-inadequate sampling of a representative set of configurations of the system. Increases in computer power of course increase the 'radius of convergence' of such calculations. Such increases come not only from the Moore's Law improvements in hardware, but also from algorithmic advances for parallelization and for increasing time steps in molecular dynamics [8]. New methods on the physical/theoretical side have also been developed to speed convergence. One such method is the use of soft-core solute models, so that one simulation can generate an adequate reference ensemble for a family of alchemical changes [9,10]. Hamelberg et al. [11] have recently introduced an 'accelerated molecular dynamics' method that substantially improves sampling, while preserving the ability to recover thermodynamic data. The 'lambda dynamics' method of Kong and Brooks [12] increases the efficiency of free energy calculations by treating the coupling parameter as a dynamic variable.

More rapid convergence of free energy calculations can also be obtained by replacing part of the system with a simpler model, such as a continuum model for the solvent. This has the advantage of obviating the need for sampling the configurations of this part of the system, and it also reduces the computation time so that longer simulations are possible for the rest of the system. In view of the important role that specific hydrogen bonds may play, the combination of fully atomistic simulations with subsequent continuum analyses is probably a more reliable procedure than using a continuum solvent model exclusively. The Kollman group demonstrated impressive success with this approach to calculations of free energies of binding [13]. Below, I will outline how this approach can be further developed, starting with a rigorous basis in statistical mechanics, to calculate standard free energies of binding.

Calculations of relative free energies of binding often involve the alteration of bond lengths in the course of an alchemical simulation. When the bond lengths are subject to constraints, a correction is needed for variation of the Jacobian factor in the expression for the free energy. Although a number of expressions for the correction formula have been described in the literature, the correct expressions are those presented by Boresch and Karplus [14].

3.3 FUTURE PROSPECTS

It was noted above that a continuum treatment of the solvent can be helpful, although representing certain solvent molecules explicitly may be necessary. The expressions for handling the free energy contributions in such hybrid models have been derived by Gilson et al. [6].

Two remaining problems relating to the continuum treatment of solvation include the slowness of Poisson–Boltzmann calculations, when these are used to treat electrostatic effects, and the difficulty of keeping buried, explicit solvent in equilibrium with the external solvent when, e.g. there are changes in nearby solute groups in an alchemical simulation. Faster methods for solving the Poisson–Boltzmann equation by means of adaptive mesh techniques [15] and other techniques [16] are becoming available, however. For buried solvent molecules, open ensemble methods should be helpful, although extension of the existing methods to allow for solute flexibility is needed [17].

It is not uncommon for protons to be taken up or released upon formation of a biomolecular complex. There is a need for methods that automatically probe for the correct protonation state in free energy calculations. This problem is complicated by the fact that proteins adapt to and stabilize whatever protonation state is assigned to them during the course of a molecular dynamics simulation [18]. Recently, a method has been described that allows for titration state adjustments for a protein that is undergoing molecular dynamics simulation in a continuum solvent model [19]. It should be possible to extend this approach to include ligand-binding processes. When the change in protonation state during ligand binding is known, equations are available to account for the addition or removal of protons from the solvent in the overall calculation of the free energy change [6].

Finally, it is sometimes true that the binding of a ligand to a protein is associated with large changes in conformation of one or both molecules. Our group has recently been developing a 'relaxed complex' approach to the particularly challenging case in which the protein undergoes conformational changes [20,21]. The basic idea is simple. One selects a large number of 'snapshot' conformations from a molecular dynamics simulation of the unliganded protein. Methods such as accelerated molecular dynamics can be used to generate snapshots of a more diverse set of protein conformations [11]. One or a set of ligand molecules can then be docked to these snapshots, using any convenient rapid docking algorithm. The most tightly bound complexes can then be rescored using higher accuracy methods, ideally based on rigorous statistical mechanical foundations [22].

3.4 CONCLUDING PERSPECTIVE: MOLECULAR DYNAMICS SIMULATIONS AND DRUG DISCOVERY

It is worth noting that molecular dynamics simulations have already contributed greatly to drug discovery and thereby to the improvement of human health. The most dramatic contributions have been in the introduction of the HIV protease inhibitors that have extended the lives of thousands of patients with HIV infections. AIDS was first recognized as an infectious disease caused by HIV viruses in the early 1980s. Data from the US Centers for Disease control show a disheartening increase in the US death toll from HIV infections that had reached 50,000 per year by 1995. The addition of the HIV protease inhibitors to the treatment regimen starting in late 1995 led to an abrupt drop in this death rate, to fewer than 20,000 per year by 1998. The first HIV protease inhibitors were, in part, products of structure-based discovery [23,24]. Molecular dynamics plays a role in any structure-based discovery program even in the earliest stages, as a tool for refining crystallographic or NMR-derived structural data. Beyond this, molecular dynamics has been used to generate samples of conformations of hypothetical lead compounds in the binding sites of rigid models of target receptors. Molecular dynamics has also been used to explore induced-fit effects, the conformational changes that a receptor may exhibit upon the binding of a ligand. All of these approaches, and even free energy calculations with molecular dynamics, contributed to the successful introduction of HIV protease inhibitors as life-saving drugs [23,24].

Although challenges remain, and provide fruitful grounds for basic research, it is clear that the types of methods described here will play an increasing role in drug discovery research. Indeed, several drugs in addition to the HIV protease inhibitors have also been shaped in part by such methods, including drugs for cancer and influenza. As the methods for sampling conformations of protein molecules improve, we can hope that the simulations will reach the state that they have for the more moderately sized host–guest systems. In this realm, there are cases in which experimental results have been redone and corrected in response to free energy calculations [25,26].

3.5 ACKNOWLEDGEMENTS

This work has been supported in part by grants from NIH, NSF, the NSF Center for Theoretical Biological Physics, the National Biomedical Computing Resource, the NSF Supercomputer Centers, the W.M. Keck Foundation, and Accelrys, Inc.

3.6 REFERENCES

1 K.E.B. Platzer, F.A. Momany and H.A. Scheraga, Int. J. Peptide Protein Res., 4 (1972) 201.
2 C.F. Wong and J.A. McCammon, J. Am. Chem. Soc., 108 (1986) 3830.
3 J.A. McCammon, B.R. Gelin and M. Karplus, Nature, 267 (1977) 585.
4 B.L. Tembe and J.A. McCammon, Comput. Chem., 8 (1984) 281.
5 W.L. Jorgensen, J.K. Buckner, S. Boudon and J. Tirado-Rives, J. Chem. Phys., 89 (1988) 3742.
6 M.K. Gilson, J.A. Given, B.L. Bush and J.A. McCammon, Biophys. J., 72 (1997) 1047.
7 J. Hermans and L. Wang, J. Am. Chem. Soc., 119 (1997) 2707.
8 T. Schlick, R.D. Skeel, A.T. Brunger, L.V. Kale, J.A. Board, J. Hermans and K. Schulten, J. Comput. Phys., 151 (1999) 9.
9 H. Liu, A.E. Mark and W.F. van Gunsteren, J. Phys. Chem., 100 (1996) 9485.
10 T.Z. Mordasini and J.A. McCammon, J. Phys. Chem. B, 104 (2000) 360.
11 D. Hamelberg, J. Mongan and J.A. McCammon, J. Chem. Phys., 120 (2004) 11919.
12 X. Kong and C.L. Brooks, J. Chem. Phys., 105 (1996) 2414.
13 I. Massova and P.A. Kollman, Perspect. Drug Discov. Des., 18 (2000) 113.
14 S. Boresch and M. Karplus, J. Chem. Phys., 105 (1996) 5145.
15 N.A. Baker, D. Sept, S. Joseph, M.J. Holst and J.A. McCammon, Proc. Natl Acad. Sci. USA, 98 (2001) 10037.
16 R. Luo, L. David and M.K. Gilson, J. Comput. Chem., 23 (2002) 1244.
17 H. Resat, T.J. Marrone and J.A. McCammon, Biophys. J., 72 (1997) 522.

18 S.T. Wlodek, J. Antosiewicz and J.A. McCammon, Protein Sci., 6 (1997) 373.
19 J. Mongan, D.A. Case and J.A. McCammon, J. Comput. Chem., 25 (2004) 2038.
20 J.H. Lin, A.L. Perryman, J.R. Schames and J.A. McCammon, J. Am. Chem. Soc., 124 (2002) 5632.
21 J.H. Lin, A.L. Perryman, J.R. Schames and J.A. McCammon, Biopolymers, 68 (2003) 47.
22 J.M.J. Swanson, R.H. Henchman and J.A. McCammon, Biophys. J., 86 (2004) 67.
23 C.N. Hodge, T.P. Straatsma, J.A. McCammon, A. Wlodawer, in: W. Chiu, R.M. Burnett and R. Garcea (Eds.), Structural biology of viruses, Oxford University Press, New York, 1997.
24 M.R. Reddy, J.M.D. Varney, V. Kalish, V.N. Viswanadhan and K. Appelt, J. Med. Chem., 37 (1994) 1145.
25 W.R. Cannon, J.D. Madura, R.P. Thummel and J.A. McCammon, J. Am. Chem. Soc., 115 (1993) 879.
26 M.L. Lamb and W.L. Jorgensen, Curr. Opin. Chem. Biol., 1 (1997) 449.

Theory and Applications of Computational Chemistry: The First Forty Years
Edited by C. Dykstra et al.

CHAPTER 4

Electrodynamics in computational chemistry

Linlin Zhao, Shengli Zou, Encai Hao and George C. Schatz

Department of Chemistry, Northwestern University, Evanston, IL 60208-3113, USA

Abstract

This chapter is concerned with the interaction of light with particles that are sufficiently large in size that classical electrodynamics can be used to describe the optical properties of the particle. Most of the review is concerned with metal nanoparticles (silver and gold), where the linear optical properties (extinction, absorption, scattering) are strongly dependent on particle size, shape, and local dielectric environment. To describe such problems, there has been much recent progress in the use of computational electrodynamics methods including the discrete dipole approximation, the finite difference time domain method, and other methods, and we review recent applications of these methods that have been used to interpret the experiments. The review also describes the development of electronic structure methods for describing the Raman spectra of molecules adsorbed on the surface of the metal particles, including methods that explicitly include the electric fields from classical electrodynamics calculations in the calculation of Raman intensities.

4.1 INTRODUCTION

An important topic in computational chemistry is the interaction of molecules with electromagnetic fields. This interaction leads to a number of familiar processes, including light absorption, emission, fluorescence, and scattering, all of which refer to the linear interaction of a molecule with the field, and there are many important non-linear interactions as well. In essentially all computational chemistry studies of these processes, the electromagnetic field is assumed to be periodic in time, and constant in space, meaning that the molecule is assumed to be so small compared to the wavelength of light that spatial variation of the field can be ignored. While this is an appropriate assumption

References pp. 64–65

for isolated small molecules, the spatial variation of the field becomes crucial in studies of nanoscale materials, and under this circumstance, classical electrodynamics plays an important role.

Although classical electrodynamics is not a subject that one normally associates with computational chemistry, the subject has been of long-term interest to the chemistry of nanoscale materials, even predating the discovery of quantum mechanics, and especially in the last 10 years it has emerged as a major tool for describing both the linear and non-linear optical properties of materials. Another important change in the last 10 years is that electrodynamics has made the transition from a primarily analytical theory to a primarily computational theory (using computational algorithms that are in many cases familiar to chemists in other contexts). This has proven to be extremely useful, as advances in lithography and supra-molecular assembly have increasingly provided materials with dimensions up to 1000 nm where issues of nanoscale size and shape play an important role in determining optical response.

This review is concerned with the advances in our understanding of chemical problems that have occurred as a result of developments in computational electrodynamics, with an emphasis on problems involving the optical properties of nanoscale metal particles. In addition, in part of the review we describe theoretical methods that mix classical electrodynamics with molecular quantum mechanics, and which thereby enable one to describe the optical properties of molecules that interact with nanoparticles. Our focus will be on linear optical properties, and on the interaction of electromagnetic fields with materials that are large enough in size that the size of the wavelength matters. We will not consider intense laser fields, or the interaction of fields with atoms or small molecules.

Electrodynamics was one of the great triumphs of 19th century science, and especially of J.C. Maxwell [1]. Although the importance of Maxwell's theory in describing materials at the nanoscale was not immediately evident from his work, Maxwell's theory eventually saw significant applications, most notably as a result of Mie theory [2], developed in 1908, in which the scattering of light by a spherical particle was solved analytically. Mie theory applies to particles of any size, but its application to nanoscale metal particles led to the resolution of a long-standing problem (going back to Faraday [3]) as to the origin of the size dependence of the absorption spectra of metal particles.

Many metal particles, especially silver and gold, show a strong absorption band in the visible portion of the spectrum that arises from collective excitation of the conduction electrons. Fig. 4.1 shows figuratively how the electron cloud can oscillate relative to the positions of the nuclei, leading to a characteristic oscillation frequency that is associated

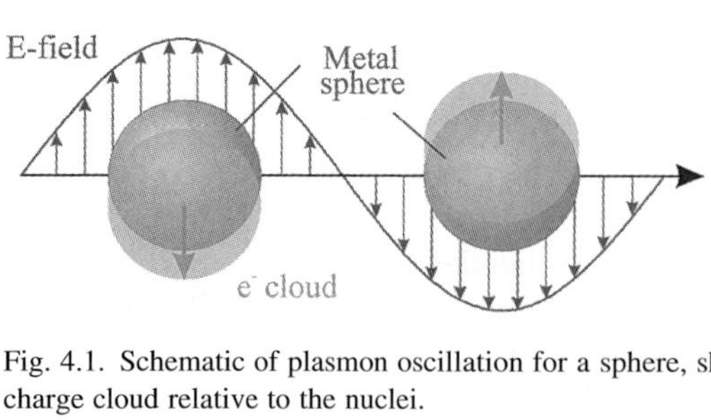

Fig. 4.1. Schematic of plasmon oscillation for a sphere, showing the displacement of the conduction electron charge cloud relative to the nuclei.

with what is called plasmon excitation. For a spherical particle, the plasmon frequency depends on the density of conduction electrons, and therefore on the chemical identity of the elements. In addition, it depends on other aspects of the electronic structure of the metal, such as the location of the filled d-orbitals relative to the Fermi energy, as d-orbital excitation provides a mechanism for localized excitation rather than the delocalized excitation that arises from conduction band excitation. The plasmon frequency also depends on the size and shape of the particle as we show below. For the noble metals, copper, silver, and gold, the plasmon wavelength is typically in the visible region of the electromagnetic spectrum. Given this, as well as the favorable chemistry of silver and gold which allows for the easy preparation of nanoparticles both in air and in solution, there have been very extensive studies of plasmon resonance spectra of particles made from these elements, as reviewed by Kreibig [4,5].

In the past few years, silver and gold particles have been developed for use in chemical and biological sensing applications [6–9] in which the intense absorption associated with the plasmon resonances has been found to be superior to fluorescence for some applications, so the practical significance of plasmon resonance excitation is now well established. This work has been accompanied by substantial theory work [10–39] in which the optical properties have been studied by computational electrodynamics methods for problems which range from isolated metal nanoparticles and nanoholes to arrays and aggregates of nanoparticles. In addition, although much of the interest in optical properties has centered on extinction and scattering, there is also interest in the surface enhanced Raman spectroscopy (SERS) of molecules adsorbed on the surfaces of the particles [38], and in non-linear optical properties of metal particles such as hyper-Rayleigh scattering [26].

This article is divided into two sections. In the first section, we overview the recent computational electrodynamics studies that have been performed on metallic (silver or gold) particles with an emphasis on problems of more interest to chemistry, such as the detection of molecules through adsorption-induced shifts of the plasmon resonance wavelength. In the second section, we turn our attention to a subject of more direct interest to theoretical chemistry, namely the calculation of SERS intensities using electronic structure methods. The challenge to the electronic structure community here is how to treat the interaction of an electronically localized system like a molecule with an electronically delocalized structure like a metal particle that is tens of nanometer in dimension. There have been attempts at dealing with this problem that we will describe, but this is a field that is still in a relatively primitive state, so our review will also consider new developments in the field that are likely to be important in the future.

4.2 ELECTRODYNAMICS OF METAL NANOPARTICLES

4.2.1 Methods

In this section we review the many recent studies that have been performed using computational electrodynamics methods with gold and silver nanoparticles, often with molecular adsorbates that one wishes to detect. Until about 10 years ago, almost all studies of this type were done using Mie theory, which is specifically for spherical particles, and we will use Mie theory for the first applications that we present. While this is useful for qualitative insight, and is often adequate for treating colloidal particles where there is always a distribution of particle shapes and sizes, the modern generation of research in this area has largely shifted to particles that are synthesized using methods that yield more uniform particle shapes (often as plates, rods, or triangles) and size distributions as this produces plasmon resonance spectra that are reproducible and controllable. To model these spectra, a number of computational electrodynamics methods have been developed, including the discrete dipole approximation (DDA), [40–42] the finite difference time domain (FDTD) method, [43,44] and the multiple multipole method (MMP) [45–47]. These methods provide the capability of describing light scattering from particles or assemblies of particles in which the total size of the system is less than 200 nm in each dimension. Thus the bulk of the results that we discuss will be based on computational electrodynamics calculations, particularly the DDA method.

4.2.2 Dielectric constants

One limitation of both Mie theory and the computational electrodynamics methods is that the results are only as good as the dielectric constants that are used. Ideally, one would be able to calculate such information directly from electronic structure calculations, but in reality this is not practical for metals like silver and gold so the information is derived from experimental data that is obtained for bulk metal (or more typically for films). Here we use experimental dielectric constants from Hunter and Lynch (HL) [48] (with some smoothing [12] as the HL compilation combines data from different sources that do not overlap perfectly). Other compilations of dielectric constant information are also available, but the HL compilation is relatively recent, and it provides a careful analysis of data from many sources that attempt to reduce problems from void formation in the film structure.

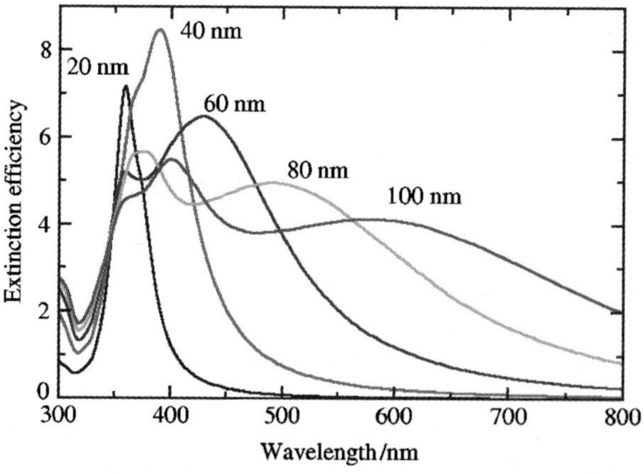

Fig. 4.2. Extinction spectra of spherical silver particles with radii 20–100 nm.

This use of bulk dielectric constants for nanoparticle calculations is appropriate for particles that are large enough (larger than the conduction electron mean free path), such as particles having radii $\geq$20 nm considered in Fig. 4.2. For smaller particles, one needs to correct the dielectric constant for the effect of scattering of the conduction electrons from the particle surfaces. Procedures for doing this have been studied in several places, as recently reviewed by Coronado and Schatz [37].

4.2.3 Spherical particles

To begin our presentation, we consider silver spheres using Mie theory. Fig. 4.2 shows the spectra that we obtain from standard Mie expressions [49] for particle radii that range from 20 to 100 nm. For small particles (20 nm), one sees a single extinction maximum at about 360 nm that corresponds to the localized surface plasmon resonance (LSPR) in which all of the electrons in the particle are oscillating in-phase with the applied radiation (dipolar excitation). The figure shows that this resonance red-shifts and broadens as the wavelength is increased. This occurs because of the finite size of the particle relative to the wavelength of light, with the increased width arising because of radiative damping (i.e. the particle emits light at a rate that is fast enough to limit the ability of the particle to be excited) and the increased wavelength due to dynamic depolarization (i.e. the incident field oscillates as function of position such that it is not possible to excite all the electrons in-phase).

We also see in Fig. 4.2 that new peaks appear as the particle radius increases. These correspond to higher multipole resonances associated with the sphere. For example, the first new peak that appears after the dipole resonance red-shifts sufficiently corresponds to quadrupole excitation. Such a higher resonance is not important for particles close to the size of molecules, but becomes increasingly important as the particle size increases.

4.2.4 Effects of particle shape

Particle shape plays an important role in determining the plasmon resonance wavelength of a silver nanoparticle. Fig. 4.3 shows typical results for silver spheroids (a three-dimensional object that is obtained by rotating an ellipse about its long axis (prolate spheroid) or a short axis (oblate spheroid) as a function of the ratio of major to minor axis. Spheroids are one of the few particle shapes for which one can solve Maxwell's equations exactly, so in this case we used a solution due to Voshchinnikov and Farafonov (VF) [50] to generate these results. The spheroids in this case are oblate in shape, and they show that the dipole plasmon resonance peak shifts to the red (from 500 to over 1000 nm) as the object goes from spherical to more and more oblate. This is a typical result for an anisotropic nanoparticle. In addition, a strong quadrupole plasmon peak occurs at 370 nm, which is most intense for a sphere shape, and is strongly suppressed in oblate particles.

Fig. 4.4 shows extinction spectra from many different shaped silver particles (spheres, cylinder, cube, prism, pyramid), all for an effective radius (radius of a sphere of equal volume) of 50 nm. This shows very clearly that the plasmon maximum is strongly shape

References pp. 64–65

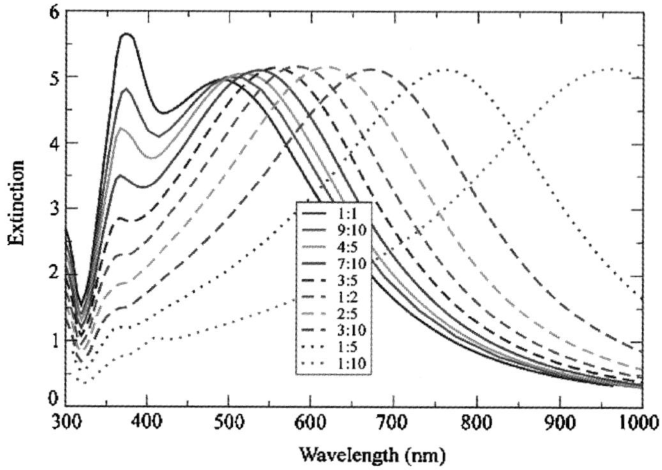

Fig. 4.3. Extinction spectra of oblate spheroids for minor:major axis ratios ranging from 1:1 (a sphere) to 1:10. Each spheroid has the same volume, taken to be that for a sphere whose radius is 80 nm.

dependent, with the sphere shape giving the bluest wavelength of all the particles, and particles with sharp points giving the reddest plasmons (pyramids and prisms). In this picture there are weak secondary peaks on the blue side of the dipole resonance peak, some of which may be due to quadrupole and other higher multipole states.

4.2.5 Effects of solvent and of surrounding layers

Fig. 4.5 shows the effect of a solvent on the plasmon lineshape. Here we see the expected red shifting of all resonance structures as the refractive index is increased from 1.0 to 2.0.

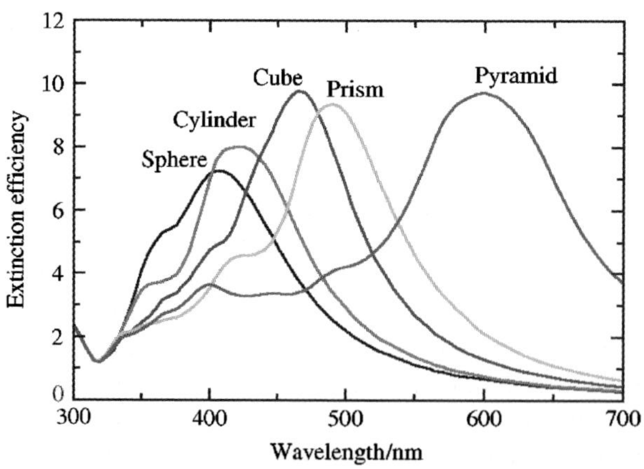

Fig. 4.4. Extinction spectra of silver nanoparticles having the following shapes: sphere, cylinder, cube, triangular prism, and tetrahedron (pyramid). Each particle has the same volume, taken to be that of a sphere whose radius is 50 nm.

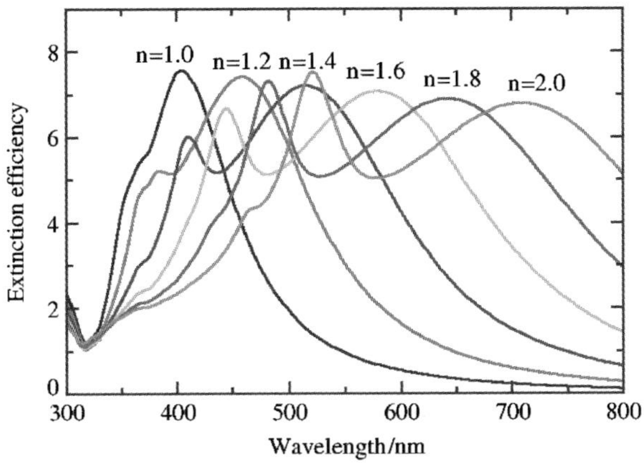

Fig. 4.5. Effect of the surrounding medium (*n* is the index of refraction of the medium) on the extinction spectrum of a 50 nm spherical silver nanoparticle.

In addition, we see that the quadrupole peak becomes more and more noticeable as index is increased. This is due to the fact that the dielectric response of silver leads to strong damping for wavelengths below 400 nm where the quadrupole resonance is located for the particle in vacuum, but not for higher refractive indices when the plasmon resonance is shifted to wavelengths longer than 400 nm.

Fig. 4.6a considers a spherical core–shell particle in which the core is taken to be vacuum and the shell is silver. The particle radius is 50 nm, so when the shell thickness is 50 nm we recover the solid particle result. As the shell becomes thinner, the plasmon resonance red-shifts considerably, very much like we see for highly oblate spheroids. Fig. 4.6a assumes that the dielectric constant of silver is independent of shell thickness, so the resonance width does not change much when the shell becomes thin. However, the correct dielectric response needs to include for finite size effects (as noted above) when the shell thickness is smaller than the conduction electron mean free path. Fig. 4.6b shows what happens to the spectrum in Fig. 4.6a when the finite size effect is incorporated, and we see that it has a significant effect for shells below 10 nm thickness, leading to much broader plasmon lineshapes.

Recently, a new class of chemical and biological sensors has been developed based on the shift of the plasmon resonance wavelength that arises when the analyte of interest binds to the surface of the nanoparticle [9,38]. The effect is very much like that considered in Fig. 4.5, except that the particles considered are anisotropic (truncated tetrahedron shape), the particles are on a surface (glass or mica, typically) and the analyte layer thickness can be varied. Fig. 4.7 shows an example of the type of information studied, here showing the plasmon wavelength shift associated with binding many layers of molecules on the surface of the particle as a function of the layer thickness. The curves labeled B and D show results obtained in experimental studies for particles with dimensions 100 nm (in-plane) and 30 or 50 nm (out of plane), while curves A and C show calculated results for the same two structures. The molecules on the particle surface were

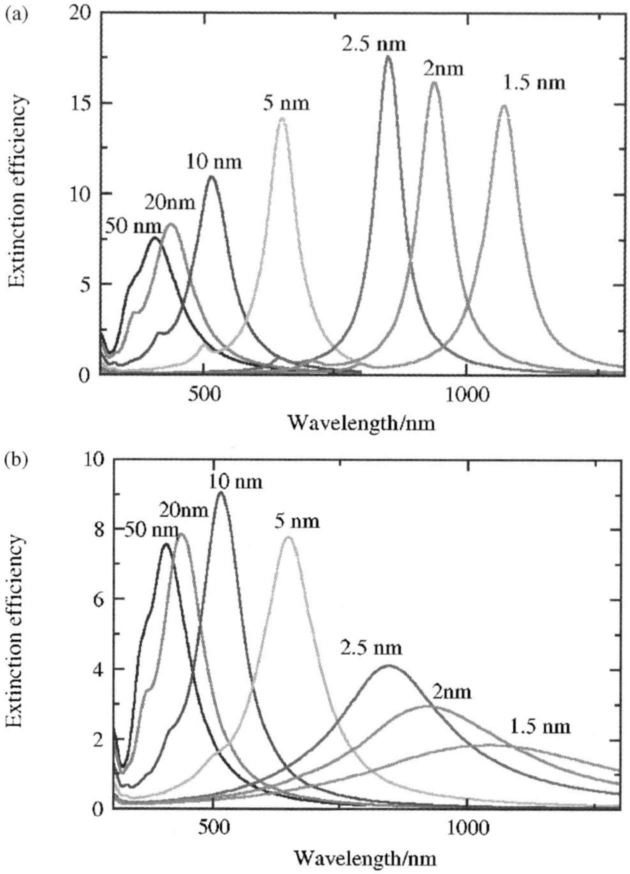

Fig. 4.6. (a) Extinction spectra of a 50 nm core–shell particle, with vacuum core and silver shell thickness varying from 1.5 to 40 nm. (b) Same as in (a) except the dielectric constant is corrected for size-dependent (surface scattering) effects.

carboxy-terminated alkane thioate self-assembled monolayers, with successive layers 'glued' together using Cu^{2+} ions. The index of refraction of the layers was found by fitting to be about 1.65 (i.e. larger than the expected bulk phase result of about 1.50), which is understandable given that the layer is ordered.

The results in Fig. 4.7 show that the plasmon wavelength increases rapidly with layer thickness for small thickness, but then it saturates for large thickness. This behavior is very much like the dependence of the externally induced electric field on distance from the surface. Theory and experiment in Fig. 4.7 are in good agreement, and the theory indicates that the field is largely asymptotic for layer thicknesses larger than 30 nm. This indicates that a 30–50 nm particle has a field whose range is comparable to the size of the particle. This distance is much smaller than the range associated with surface plasmon resonances for flat surfaces (where 200 nm distances are common).

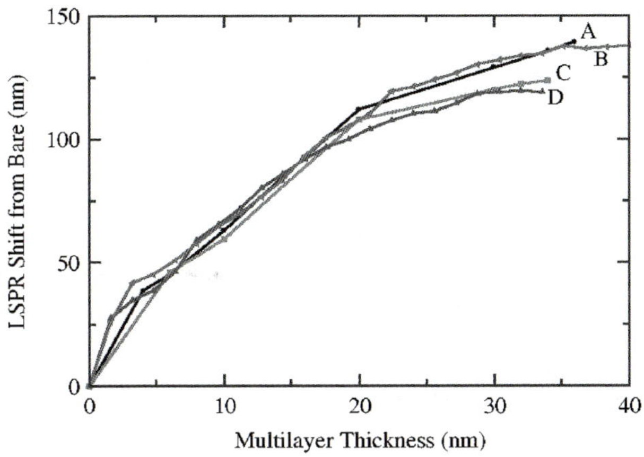

Fig. 4.7. Shift of the plasmon resonance wavelength for particles coated with molecules whose refractive index is 1.65 from that for a bare silver nanoparticle taken to have a truncated tetrahedral shape. (A) Calculation for a particle with dimensions: bisector, $a = 100$ nm, height $b = 30$ nm. (B) Experiment results for $a = 100$ nm, $b = 30$ nm. (C) Calculation for $a = 100$ nm, $b = 50$ nm. (D) Experiment for $a = 100$ nm, $b = 50$ nm.

4.2.6 Local electric fields and SERS

One reason for the behavior in Fig. 4.7 can be inferred from Figs. 4.8 and 4.9, in which we examine the electric fields near the particle surfaces. Fig. 4.8 shows contours of the field for the truncated tetrahedron particle, for light polarized along the z-axis in the figure. We see that the bottom tips of the particle are hot compared to the top and sides, with $|\mathbf{E}|^2$ being 12,000 at the tips, and only 6 on the cool sidewalls. Since the plasmon wavelength

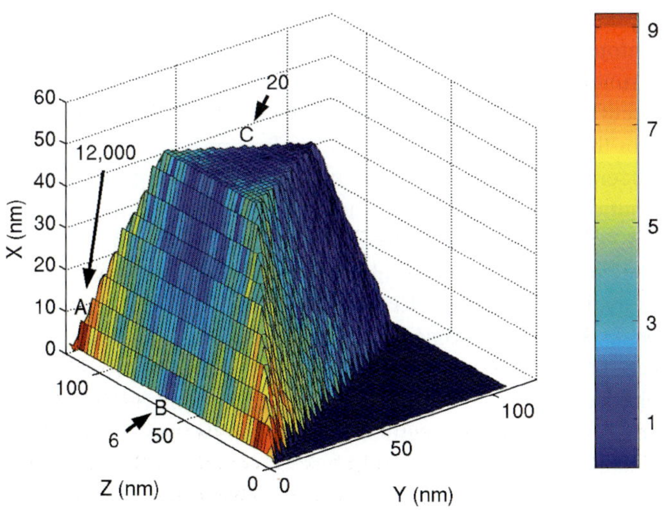

Fig. 4.8. Plots of the logarithm of the square of the electric field around the surface of a truncated tetrahedral silver nanoparticle having $a = 100$ nm, $b = 50$ nm. The field is averaged over distances from the surface between 0 and 2.0 nm from the surface.

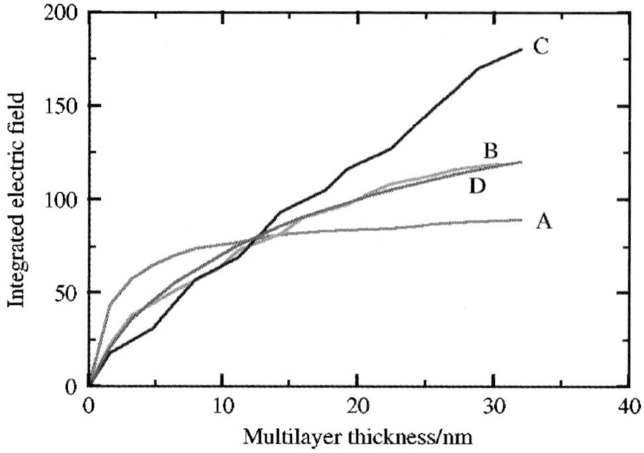

Fig. 4.9. Integrated electric field as a function of adsorbate layer thickness. The points labeled A, B, and C refer to specific locations on the surface of the particle as indicated in Fig. 4.8, while D is the average over the particle surface.

shift is approximately proportional to field (as demonstrated explicitly in Fig. 4.9), we see that the wavelength shift is primarily determined by the hot spots for those molecules which are at the particle surfaces. However, Fig. 4.9 shows that the intensity of the hot spot diminishes rapidly with distance from the particle surface, while intensity for the cool regions has much longer range dependence. The results in Fig. 4.7 are the average over hot and cool regions, so they average over the long-range and short-range effects.

Electromagnetic hot spots play an important role in many spectroscopic observations. Fig. 4.8 shows how much the local field varies for truncated tetrahedron particles, but an important question is how the field changes when other particle shapes are considered. Fig. 4.10 shows local field contours for several other particle shapes including a sphere, cylinder, cube, prism, and untruncated tetrahedron. This shows the peak field can vary a lot, with the smallest $|\mathbf{E}|^2$ occurring for a sphere (less than 10^2), and the largest for particles with sharp tips (10^4 for the tetrahedron). These $|\mathbf{E}|^2$ values provide an important estimate of the intensity enhancement expected in single molecule SERS enhancements, as the electromagnetic contribution to this intensity is approximately equal to the peak value of the product $|\mathbf{E}(\omega)|^2|\mathbf{E}(\omega')|^2$ where ω is the incident frequency and ω' is the Stokes frequency. (This is to be contrasted with the enhancement factor for conventional SERS, which depends on the average of $|\mathbf{E}(\omega)|^2|\mathbf{E}(\omega')|^2$ over all the molecules on the surface.) Since the plasmon peaks are usually somewhat broader than the shift between ω and ω', the intensity enhancement is approximately $|\mathbf{E}|^4$. Thus if $|\mathbf{E}|^2$ is approximately 10^4, the Raman intensity enhancement is close to 10^8. This may seem to be a large number, but in fact estimates of this enhancement based on single molecule SERS measurements are much larger (10^{12}–10^{15}) [9,38,51–55].

One issue in the comparison of theory and experiment for single molecule SERS is that the experiments have mostly been done for aggregates of particles. It is well known that fields can be enhanced at the junction between two closely spaced particles, so a key question is whether the additional field associated with small clusters of particles is

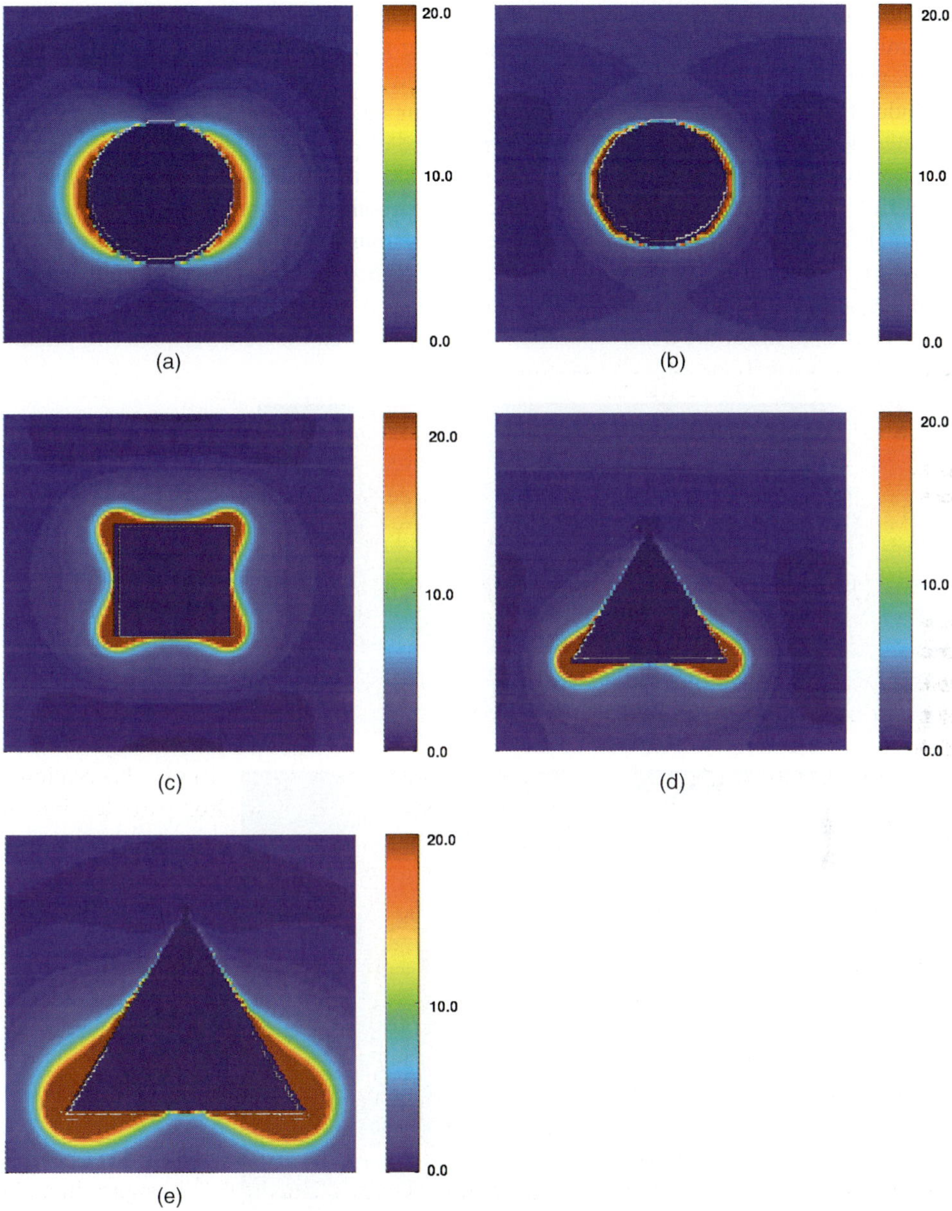

Fig. 4.10. Electric field ($|\mathbf{E}|^2$) around silver nanoparticles with the following shape: (a) sphere; (b) cylinder; (c) cube; (d) prism; (e) tetrahedron. All particles have the same effective radius of 50 nm as considered in Fig. 4.4. The wavelengths and peak $|\mathbf{E}|^2$ values are: (a) 407.5 nm and 54; (b) 421.5 nm and 96; (c) 466 nm and 745; (d) 489.4 nm and 1330; (e) 599.0 nm and 9770.

sufficient to explain the observed enhancements. Fig. 4.11 shows examples of electrodynamics results for dimers of particles, including spheres and triangular prisms. This shows that the peak $|\mathbf{E}|^2$ for a dimer of spheres is about 10^4 for a particle spacing of 2 nm, so indeed it is possible to obtain greatly enhanced fields for dimers. Even larger enhancements are possible if the spheres are spaced closer than 2 nm [29]; however, the plasmon wavelength where this occurs is often shifted too far to the red (>1000 nm) to be useful for SERS measurements. There are also concerns about the applicability of classical electrodynamics with local dielectric constants to such problems [43]. However, another approach to generating large fields is to consider dimers of anisotropic particles such as triangular prisms. These are included in Fig. 4.11 for two orientations of the prisms, both for the 2 nm separation. Here we see that the field between two prisms

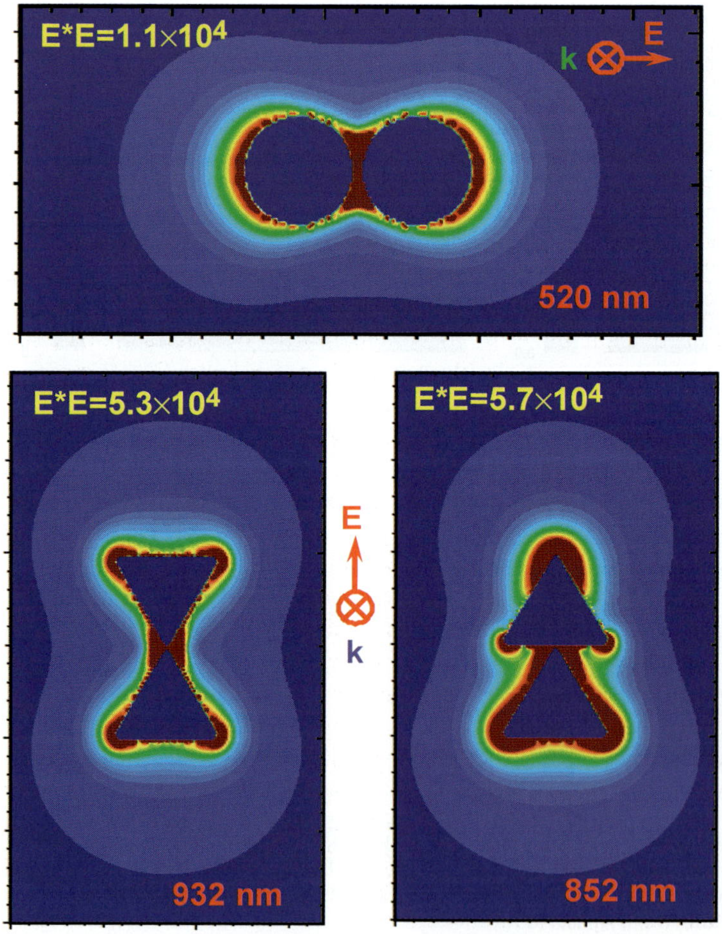

Fig. 4.11. Contours of $|\mathbf{E}|^2$ for dimers of silver particles, including two spheres (radius 19 nm, separation 2 nm) and two triangular prisms placed head-to-head and head-to-tail (edge length 60 nm, width 12 nm, snip 2 nm and separation 2 nm).

is larger than that between two spheres, although not dramatically so. Based on these and other results, we find that $|\mathbf{E}|^4$ enhancements up to 10^{10} can be obtained for dimers of particles, which provide a rationalization of the origin of most of the enhancement factor observed in single molecule SERS. This indicates that there must be some important non-electromagnetic contributions to SERS, i.e. the 'chemical enhancement', and in Section 4.3, we present some recent results regarding this mechanism.

4.3 ELECTRONIC STRUCTURE STUDIES OF SURFACE ENHANCED RAMAN SPECTRA

4.3.1 Surface models and electronic structure methods

To study the 'chemical' contribution to SERS, it is necessary to use electronic structure methods to model the interaction of molecules with metal particles. However, as we describe below, this is a complicated procedure, and as a result, there have not been many studies of this sort, and the available studies are not very sophisticated. Indeed, the description of SERS intensities associated with molecules adsorbed onto nanoparticle surfaces is one of the most difficult challenges in modern quantum chemistry. The problem here is that one needs to treat the electronic properties of both the (microscopic) adsorbed molecule and the (macroscopic) metal surface consistently in both ground and excited states in order to properly model the SERS experiment. Two simplifying models can be imagined [56]: (1) a single molecule (or a regular array of molecules) on a flat perfect crystal surface and (2) a single molecule on a small metal cluster. The first model is in principle amenable to treatment by band structure methods, and the closest work to this have been jellium/RPA studies of induced dipoles (usually point dipoles) near semi-infinite flat surfaces [57–59]. These studies were very useful in sorting out contributions to SERS and other processes that can arise on flat surfaces, but they do not treat the molecular electronic properties consistently, and they miss the influence of roughness of the surface that is known to play a very important role in the SERS intensity.

The second model, that of a molecule on a cluster, is more easily related to experiment, especially to SERS observations for colloidal systems and for metal island films, as the molecular electronic structure properties can be consistently included in the calculation. The basic idea here is that the molecule–cluster complex is more or less analogous to a metal–ligand or perhaps cluster–ligand coordination complex, and as a result, its Raman intensity might be increased in much the same way as that of ligand complexes [60]. Since the Fermi energy of most metals lies in a region intermediate between the energy of the HOMO and the LUMO of many molecules, the creation of this complex might lead to resonances in the visible region of the spectrum, even for colorless adsorbates, due to possible metal-to-molecule or molecule-to-metal transitions. These states can contribute resonantly to the Raman intensity of the molecule–cluster complex, thus increasing its magnitude further. The biggest problem for this second model is that it is impossible to choose the cluster used in calculations large enough to directly match what is studied in the experiments. Even the smallest metal particles for which SERS has been observed so far contain thousands of atoms, and it is likely that smaller clusters are less SERS active due to broadening of the plasmon resonance due to surface scattering.

Different approaches to using this model on small clusters in a way that provides potentially meaningful conclusions about large clusters have been proposed by Arenas [61,62], Corni [63–65], Nakatsuji [66,67], and Schatz [68–70]. Arenas's approach assumes that interaction with the surface leads to transfer of an electron to the molecule to produce an excited state that is the same as the ground state of the negative ion of the molecule. In this case the SERS spectrum is described as a resonance Raman effect that couples the neutral molecule ground state to the anionic state. The relative intensities of the different vibrational modes are then determined by Franck–Condon factors from *ab initio* calculations. Corni proposed a two-step procedure, which is to treat the metal cluster as a continuous body characterized by an electrodynamic response that is described using the DDA method, and then to treat the molecule at the *ab initio* level with an effective Hamiltonian that includes the molecule–metal interaction in the quasi-static approximation (long wavelength limit). In both Nakatsuji and Schatz's approaches, *ab initio*, mostly time-dependent Hartree–Fock (TDHF), methods have been used to calculate Raman intensities for small molecules interacting with simple metal clusters in which all atoms in the molecule and metal are described with Gaussian-orbital basis functions. In these calculations, the excited molecular orbitals are given width factors (or damping constants) that are characteristic of macroscopic metal clusters, thus forcing the clusters to have polarization properties that are like the macroscopic cluster. Also, in the TDHF version of the theory, the SERS intensities are evaluated at frequencies that correspond to resonant excitations of the metal clusters (the small cluster analog of plasmon resonances). These models are thus able to describe the influence of charge transfer states on the Raman spectra without making any *ad hoc* assumptions about charge transfer, and they also include dynamic coupling of the induced polarization in the molecule with that in the metal. Since the molecule and metal cluster are described with standard quantum chemistry methods, it is possible to determine the complete Raman spectrum (vibrational frequencies and relative intensities), including the effects of molecule–metal interactions. The primary flaw with this model is that it does not include the metal electrodynamic response other than is contained in the small cluster electrodynamics, and as a result the enhancement factor cannot be quantitatively determined.

4.3.2 Applications

Recently, we have proposed a methodology [71] similar to the last approach, but using semiempirical molecular orbital methods instead of TDHF methods, to calculate the frequency-dependent polarizability properties of the molecule–surface complex. Although this is a lower level description of the electronic structure, the use of semiempirical methods allows us to describe more complex molecules than has been considered in earlier studies. In the following, we present some recent results for the pyridine–copper tetramer system, and we examine the influence of molecule–metal distance on the Raman intensities for this model. There are two components to the calculation of the Raman intensity: (1) calculation of the ground state structure and normal coordinates and (2) calculation of the derivative of the frequency-dependent polarizability with respect to these normal coordinates. For all the results to be presented, we have used DFT calculations with the B3LYP exchange-correlation functional, 6-311G basis set for the pyridine molecule, and the SBKJC basis set and effective core potential for the copper tetramer, based on GAMESS [72], to calculate the normal coordinates. The molecular polarizabilities were calculated using two different methods: zero frequency DFT calculations based on GAMESS, and a sum-over-states (SOS) [68,73,74] evaluation of the frequency-dependent polarizabilities that uses excited state energies and dipole matrix elements from CIS (singles CI) calculations using the INDO/S semiempirical method in the program CNDO [75]. The GAMESS DFT calculation directly determines the zero frequency polarizabilities and Raman intensities, but we had to use finite differencing to determine zero and non-zero frequency polarizability derivatives and Raman spectra in the INDO/S-SOS calculations. For the GAMESS calculations we have determined the minimum energy structure of pyridine, the Cu tetramer, and the pyridine/Cu complex, along with the vibrational frequencies, normal coordinates and Raman intensities. INDO/S does not provide meaningful minimum energy structures, so the GAMESS geometry and Hessian have been used in all of the semiempirical calculations. To estimate the SERS intensity, we assume that the surface orients the C_{2v} axis of pyridine perpendicular to it. According to surface selection rules [76,77], since the field near the surface is dominated by its normal component, only the components of polarizability derivatives perpendicular to the surface contribute to spectra. Denoting this perpendicular direction as z, the intensities are proportional to the zz components of polarizability derivatives.

Fig. 4.12 presents the static SER spectra of pyridine, a Cu_4 cluster, and the Cu_4–pyridine complex. The structure of the Cu_4–pyridine complex is also presented, while Fig. 4.13 gives the experimental and the calculated frequency-dependent SER spectra of pyridine only. Note that we have used a Y-shaped Cu_4 cluster structure, rather than the global minimum rhombus shape, as this leads to better behaved DFT calculations. As we can see from Fig. 4.12, the static INDO/S and DFT spectra are in reasonable agreement for all except the mode at 1253 cm^{-1} for pyridine, and 1253 and 1500 cm^{-1} for the Cu_4–pyridine complex. The intensities of these modes are suppressed in the INDO/S results relative to DFT, but as shown in Fig. 4.13, the frequency-dependent INDO/S-SOS results match experiment reasonably well, with the four major SERS-active vibrational modes at around 630, 993, 1050, and 1600 cm^{-1} having about the right relative intensities. Moreover, both the DFT and INDO/S results in Fig. 4.12 suggest that for the Cu_4–pyridine complex orientation considered here, the spectrum of Cu_4–pyridine complex is approximately a combination of the spectrum of pyridine and that of Cu_4.

Table 4.1 summarizes the variation of the enhancement factor of the Cu_4–pyridine complex relative to an isolated pyridine molecule as a function of closest N–Cu distance. The applied frequency for both the Cu_4–pyridine complex and the pyridine molecule here is 2.81 eV (same as in Fig. 4.13), which corresponds to the resonant peak associated with the lowest energy state of the Cu_4 cluster. This frequency is higher than would be appropriate for experimental studies of SERS on Cu nanoparticles, but this is one of the problems associated with using a small cluster instead of a nanoparticle. The results show that for a fixed closest N–Cu distance, the enhancement factor is different for different vibrational modes, and for the same vibrational mode, the enhancement factor

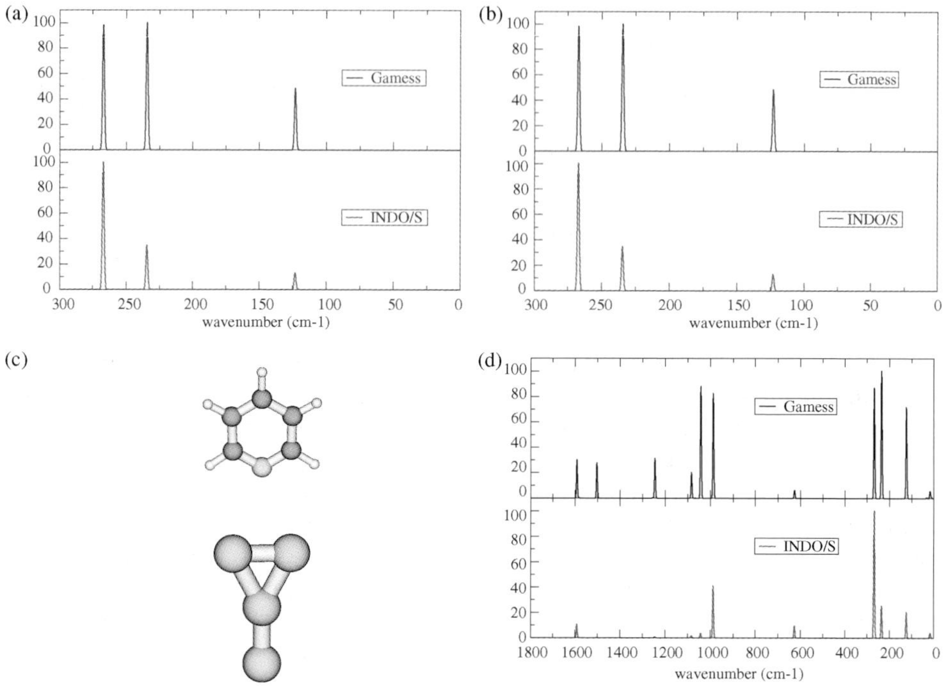

Fig. 4.12. Static SERS spectra of pyridine (a), Cu_4 (b), and Cu_4–pyridine complex (d). GAMESS results *vs* INDO/S. Geometry obtained from GAMESS DFT/B3LYP calculation, with 6-311G basis set for pyridine and SBKJC for Cu_4. Orientation of Cu_4–pyridine complex is shown in (c), with the closest N–Cu distance being 3.87 Å.

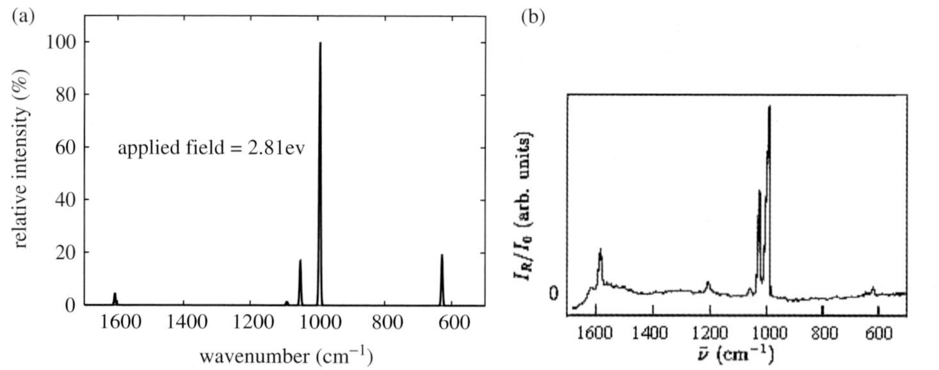

Fig. 4.13. (a) Frequency-dependent SER spectrum of pyridine, INDO/S-SOS, with applied field being 2.81 eV. (b) Experimental SER spectrum of pyridine absorbed on a rough silver electrode in water at −0.25 V *vs* a saturated Ag/AgCl/KCl reference electrode.

Table 4.1 Calculated enhancement factor of Cu_4–pyridine complex relative to pyridine molecule only, for individual SERS-active vibrational mode, as a function of closest N–Cu distance

	$D(Cu_4\text{–py}) = 3.87$ (Å)	$D(Cu_4\text{–py}) = 4.84$ (Å)	$D(Cu_4\text{–py}) = 5.81$ (Å)	$D(Cu_4\text{–py}) = 6.80$ (Å)	$D(Cu_4\text{–py}) = 7.78$ (Å)
628.89 (cm^{-1})	4.83	115.16	119.47	1.33	1.08
992.33 (cm^{-1})	1.62	29.64	28.84	1.12	1.18
1049.73 (cm^{-1})	4.82	87.80	85.80	1.33	0.82
1607.83 (cm^{-1})	16.20	458.51	452.76	0.68	0.73

The applied field for both Cu_4–pyridine complex and pyridine molecule here is 2.81 eV.

first increases then decreases as the closest N–Cu distance increases. The largest enhancement is found when the closest N–Cu distance is in the range 4.84–5.81 Å, with a value of almost 200. The N–Cu distance here is large enough that chemical interactions between the molecule and cluster are not large. The large enhancement for these geometries is therefore determined by electrostatic and electrodynamic interactions. This is a similar conclusion to that of earlier work with cluster models [70], in which it was demonstrated that local field enhancement (the analog of the enhancement effect described by the electrodynamics calculations) is the dominant effect. This is a short-ranged effect both here and in Ref. [70] because the metal cluster is extremely small. It is also possible that a mechanism termed ‘Raman reflection’ in which the oscillating charge distribution in the molecule produces oscillating polarization in the cluster also plays a role (as noted earlier [70]), but this has not been verified in the present studies. The drop in intensity at shorter distances is likely due to detuning of the cluster resonance from the chosen frequency due to chemical interactions between the molecule and cluster. This is an effect that would not occur if the cluster were replaced by a nanoparticle (as the conduction band is always on resonance in the latter case); however, what we learn from the present calculations is that no other enhancement mechanism plays an important role when the molecule and metal cluster have strong chemical interactions.

The results in Figs. 4.12 and 4.13, and in Table 4.1 show that the current level of description of SERS using electronic structure methods is quite primitive. Thus there are significant opportunities for improved work in this area, especially through the introduction of TD-DFT methods. However, a problem that must be dealt with in future work is the direct (frequency-dependent) coupling of the molecule to a realistic size metal nanoparticle that is described using electrodynamics. Such QM/ED (quantum mechanics/electrodynamics) calculations represent a major challenge for future advances in electronic structure theory.

4.4 ACKNOWLEDGEMENTS

We acknowledge support of the Northwestern Material Research Center (EH and GCS), the Nanoscale Science and Engineering Initiative of the National Science Foundation under NSF Award Number EEC-0118025 (SZ and GCS) and AFOSR MURI Grant F49620-01-1-0381 (LZ and GCS). Any opinions, findings, and conclusions or recommendations expressed in this material are those of the author(s) and do not necessarily reflect those of the National Science Foundation.

4.5 REFERENCES

1 J.C. Maxwell, A treatise on electricity and magnetism, Clarendon Press, Oxford, 1873.
2 G. Mie, Ann. Phys., 25 (1908) 377.
3 M. Faraday, Philos. Trans., 147 (1857) 145.
4 U. Kreibig and M. Vollmer, Optical properties of metal clusters, Springer, Berlin, 1995.
5 U. Kreibig, Handbook of optical properties, CRC press, Boca Raton, FL, 1997, p. 145.
6 C.A. Mirkin, R.L. Letsinger, R.C. Mucic and J.J. Storhoff, Nature, 382 (1996) 607.
7 R. Elghanian, J.J. Storhoff, R.C. Mucic, R.L. Letsinger and C.A. Mirkin, Science, 277 (1997) 1078.
8 C.L. Haynes, C.R. Yonzon, X. Zhang, J.T. Walsh and R.P. Van Duyne, Anal. Chem., 76 (2004) 78.
9 A.J. Haes, L. Chang, W.L. Klein and R.P. Van Duyne, J. Amer. Chem. Soc., 127 (2005) 2264.
10 T.R. Jensen, G.C. Schatz and R.P. Van Duyne, J. Phys. Chem. B, 103 (1999) 2394.
11 T.R. Jensen, M.L. Duval, K.L. Kelly, A. Lazarides, G.C. Schatz and R.P. Van Duyne, J. Phys. Chem., 103 (1999) 9846.
12 T. Jensen, K.L. Kelly, A. Lazarides and G.C. Schatz, J. Cluster Sci., 10 (1999) 295.
13 K.L. Kelly, A.A. Lazarides and G.C. Schatz, Comput. Sci. Eng., 3 (2001) 67.
14 K.L. Kelly, T.R. Jensen, A.A. Lazarides and G.C. Schatz, in: D. Feldheim, C. Foss (Eds.), Metal nanoparticles: Synthesis, characterization and applications, Marcel-Dekker, New York, 2002, p. 89.
15 K.L. Kelly, E. Coronado, L. Zhao and G.C. Schatz, J. Phys. Chem. B, 107 (2003) 668.
16 A.A. Lazarides and G.C. Schatz, J. Phys. Chem., 104 (2000) 460.
17 A.A. Lazarides and G.C. Schatz, J. Chem. Phys., 112 (2000) 2987.
18 A.A. Lazarides, K.L. Kelly, T.R. Jensen and G.C. Schatz, J. Mol. Struct. (THEOCHEM), 529 (2000) 59.
19 A.A. Lazarides, K.L. Kelly and G.C. Schatz, Mater. Res. Soc. Symp. Proc., 635 (2001) 1.
20 M.D. Malinsky, K.L. Kelly, G.C. Schatz and R.P. Van Duyne, J. Phys. Chem., 105 (2001) 2343.
21 M.D. Malinsky, K.L. Kelly, G.C. Schatz and R.P. Van Duyne, J. Am. Chem. Soc., 123 (2001) 1471.
22 G.C. Schatz, J. Mol. Struct. (THEOCHEM), 573 (2001) 73.
23 R. Jin, Y. Cao, C.A. Mirkin, K.L. Kelly, G.C. Schatz and J.-G. Zheng, Science, 294 (2001) 1901.
24 R.C. Johnson, J. Li, J.T. Hupp and G.C. Schatz, Chem. Phys. Lett., 356 (2002) 534.
25 R. Jin, Y.C. Cao, E. Hao, G.S. Métraux, G.C. Schatz and C.A. Mirkin, Nature, 425 (2003) 478.
26 E. Hao, G.C. Schatz, R.C. Johnson and J.T. Hupp, J. Chem. Phys., 117 (2002) 5963.
27 E. Hao, K.L. Kelly, J.T. Hupp and G.C. Schatz, J. Am. Chem. Soc., 124 (2002) 15182.
28 E. Hao and G.C. Schatz, J. Chem. Phys., 120 (2004) 357.
29 E. Hao, S. Li, R.C. Bailey, S. Zou, G.C. Schatz and J.T. Hupp, J. Phys. Chem. B, 108 (2004) 1224.
30 E. Hao, R.C. Bailey, G.C. Schatz, J.T. Hupp and S. Li, Nano Lett., 4 (2004) 327.
31 E. Hao, G.C. Schatz and J.T. Hupp, J. Fluorescence, 14 (2004) 331.
32 L. Zhao, K.L. Kelly and G.C. Schatz, J. Phys. Chem. B, 107 (2003) 7343.
33 S. Zou, L. Zhao and G.C. Schatz, SPIE Proc., Plasmonics: Metallic nanostructures and their optical properties, 5221 (2003) 174.
34 S. Zou, N. Janel and G.C. Schatz, J. Chem. Phys., 120 (2004) 10871.
35 L. Yin, V.K. Vlasko-Vlasov, A. Rydh, J. Pearson, U. Welp, S.-H. Chang, S.K. Gray, G.C. Schatz, D.E. Brown and C.W. Kimball, Appl. Phys. Lett., 85 (2004) 467.
36 C.L. Haynes, A.D. McFarland, L. Zhao, G.C. Schatz, R.P. Van Duyne, L. Gunnarsson, J. Prikulis, B. Kasemo and M. Käll, J. Phys. Chem. B, 107 (2003) 7337.
37 E.A. Coronado and G.C. Schatz, J. Chem. Phys., 119 (2003) 3926.
38 G.C. Schatz and R.P. Van Duyne, in: J.M. Chalmers, P.R. Griffiths (Eds.), Handbook of vibrational spectroscopy, Wiley, New York, 2002, p. 759.
39 A.J. Haes, S. Zou, G.C. Schatz and R.P. Van Duyne, J. Phys. Chem. B, 108 (2004) 109.
40 B.T. Draine and J.J. Goodman, Astrophys. J., 405 (1993) 685.
41 B.T. Draine and P.J. Flatau, J. Opt. Soc. Am. A, 11 (1994) 1491.

42 W.H. Yang, G.C. Schatz and R.P. Van Duyne, J. Chem. Phys., 103 (1995) 869.
43 A. Taflove and S.C. Hagness, Computational electrodynamics: the finite-difference time-domain method, 2000.
44 R.X. Bian, R.C. Dunn, X. Xie and P.T. Leung, Phys. Rev. Lett., 75 (1995) 4772.
45 E. Moreno, D. Erni, C. Hafner and R. Vahldieck, Opt. Quantum Electron., 34 (2002) 1051.
46 L. Novotny, D.W. Pohl and B. Hecht, Opt. Lett., 20 (1995) 970.
47 L. Novotny, R.X. Bian and X.S. Xie, Phys. Rev. Lett., 79 (1997) 645.
48 D.W. Lynch and W.R. Hunter, in: E.D. Palik (Ed.), Handbook of optical constants of solids, Academic Press, New York, 1985, p. 350.
49 C.F. Bohren and D.R. Huffman, Absorption and scattering of light by small particles, Wiley-Interscience, New York, 1983.
50 N.V. Voshchinnikov and V.G. Farafonov, Astrophys. Space Sci., 204 (1993) 19.
51 S. Nie and S.R. Emory, Science, 275 (1997) 1102.
52 K. Kneipp, H. Kneipp, G. Deinum, I. Itzkan, R.R. Dasari and M.S. Feld, Appl. Spectrosc., 52 (1998) 175.
53 K. Kneipp, H. Kneipp, I. Itzkan and R.R. Dasari, Chem. Rev., 99 (1999) 2957.
54 A.M. Michaels, M. Nirmal and L.E. Brus, J. Am. Chem. Soc., 121 (1999) 9932.
55 H. Xu, E.J. Bjerneld and M. Käll, Phys. Rev. Lett., 83 (1999) 4357.
56 G.C. Schatz, Chem. Res., 17 (1984) 370.
57 P.J. Feibelman, Phys. Rev. B: Condens. Matter, 22 (1980) 3654.
58 T. Maniv and H. Metiu, Surf. Sci., 101 (1980) 399.
59 S.L. McCall and P.M. Platzman, Phys. Rev. B: Condens. Matter, 22 (1980) 1660.
60 M. Moskovits, L.-L. Tay, J. Yang and T. Haslett, in: V.M. Shalaev (Ed.), Optical properties of nanostructured random media, Topics Appl. Phys., Springer, Berlin, 2002, p. 215.
61 J.F. Arenas, I.L. Tocon, J.C. Otero and J.I. Marcos, J. Phys. Chem., 100 (1996) 9254.
62 J.F. Arenas, J. Soto, I.L. Tocon, D.J. Fernandez, J.C. Otero and J.I. Marcos, J. Chem. Phys., 116 (2002) 7207.
63 S. Corni and J. Tomasi, Chem. Phys. Lett., 342 (2001) 135.
64 S. Corni and J. Tomasi, J. Chem. Phys., 114 (2001) 3739.
65 S. Corni and J. Tomasi, J. Chem. Phys., 116 (2002) 1156.
66 H. Nakai and H. Nakatsuji, J. Chem. Phys., 103 (1995) 2386.
67 O. Kitao and H. Nakatsuji, J. Chem. Phys., 88 (1988) 4913.
68 W.-H. Yang and G.C. Schatz, J. Chem. Phys., 97 (1992) 3831.
69 P.K.K. Pandey and G.C. Schatz, Chem. Phys. Lett., 88 (1982) 193.
70 P.K.K. Pandey and G.C. Schatz, J. Chem. Phys., 80 (1984) 2959.
71 L. Zhao and G.C. Schatz, in: N.J. Halas and T.R. Huser (Eds.), N.J. Proc. SPIE, Plasmonics: Metallic nanostructures and their optical properties, 5512, 10–19, II.
72 M. Schmidt, K.K. Baldridge, J.A. Boatz, S.T. Elbert, M.S. Gordon, J.H. Jensen, S. Koseki, N. Matsunaga, K.A. Nguyen, S.J. Su, T.L. Windus, M. Dupuis and J.A. Montgometry, GAMESS 2004, development version, J. Comput. Chem., 14 (2004) 1347.
73 D.R. Kanis, M.A. Ratner, T.J. Marks and M.C. Zerner, Chem. Mater., 3 (1991) 19.
74 D.R. Kanis, M.A. Ratner and T.J. Marks, J. Am. Chem. Soc., 112 (1990) 8203.
75 J.R. Reimers, CNDO/INDO, 1997.
76 M. Moskovits, J. Chem. Phys., 77 (1982) 4408.
77 R.G. Greenler and T.L. Slager, Spectrochim. Acta Part A, 29 (1973) 193.

Theory and Applications of Computational Chemistry: The First Forty Years
Edited by C. Dykstra et al.

CHAPTER 5

Variational transition state theory

Bruce C. Garrett[1] and Donald G. Truhlar[2]

[1] *Chemical Sciences Division, Pacific Northwest National Laboratory, Richland, WA 99352, USA*
[2] *Department of Chemistry and Supercomputing Institute, University of Minnesota, Minneapolis, MN 55455-0431, USA*

Abstract

We present an overview of variational transition state theory from the perspective of the dynamical formulation of the theory. This formulation provides a firm classical mechanical foundation for a quantitative theory of reaction rate constants, and it provides a sturdy framework for the consistent inclusion of corrections for quantum mechanical effects and the effects of condensed phases. A central construct of the theory is the dividing surface separating reaction and product regions of phase space. We focus on the robust nature of the method offered by the flexibility of the dividing surface, which allows the accurate treatment of a variety of systems from activated and barrierless reactions in the gas phase, reactions in rigid environments, and reactions in liquids and enzymes.

5.1 INTRODUCTION

Transition state theory (TST) is a theoretical framework for calculating and interpreting rate constants of chemical reactions. TST was formulated nearly 70 years ago [1–3] with the goal of providing a computational tool for predicting rate constants from a knowledge of the potential energy surface (PES) controlling the reaction. However, this goal was not attained until much later because theoretical chemists were unable to calculate the PES with sufficient accuracy. Thus, for many years, TST was used primarily for correlating and interpreting rate constants, including kinetic isotope effects, and that aspect of its use continues and will be discussed briefly. The major focus of this article, however, is on the more recent use of TST as a computational tool for predicting rate constants. It is a testament to the utility of TST that its development has continued over an unusually long period of time. For historical perspectives on TST, we refer the reader to previous reviews and perspectives [4–6]. In this work, we review variational transition state theory

References pp. 84–87

(VTST) with multidimensional tunneling (MT) contributions and its implementation for calculating rate constants for chemical reactions in gaseous and condensed phases.

5.2 GAS PHASE REACTIONS

TST provides a means to evaluate the equilibrium rate constant, that is the rate constant for an equilibrium ensemble of reactants described by a Boltzmann distribution of states [7]. The method also assumes that the reaction is electronically adiabatic so that the reaction dynamics of the system can be described by motion of the atoms on a single PES [8]. TST is most easily described in its classical mechanical form, because when classical mechanics is valid, TST can be derived by making a single approximation, the so-called fundamental dynamical assumption [3]. We first review classical TST and the justification for the variational form [9] of the theory before discussing approximate methods for including quantum mechanical effects into the theory.

Our starting point for describing the computational approach to rate constant calculations is to define the Hamiltonian H for a reaction of N atoms with masses m_i, $i = 1, \ldots, N$, and Cartesian coordinates $R_{i\zeta}$, $i = 1, \ldots, N$, $\zeta = x, y, z$, and conjugate momentum $P_{i\zeta}$, $i = 1, \ldots, N$, $\zeta = x, y, z$, which can be reduced to the compact form

$$H(\mathbf{q}, \mathbf{p}) = \frac{\mathbf{p}^2}{2\mu} + V(\mathbf{q}) \tag{1}$$

by introducing the mass-scaled coordinates

$$q_{i\zeta} = \sqrt{\frac{m_i}{\mu}} R_{i\zeta} \tag{2}$$

and momenta

$$p_{i\zeta} = \sqrt{\frac{\mu}{m_i}} P_{i\zeta} \tag{3}$$

where μ is a reduced mass that is chosen for convenience, and $V(\mathbf{q})$ is the PES as a function of the mass-scaled coordinates.

5.2.1 Classical mechanical theory

Exact classical equilibrium rate constants for electronically adiabatic reactions are obtained by evaluating the flux of reactive trajectories through a dividing surface separating the reactant and product regions of phase space $(\mathbf{q}, \mathbf{p})$. The dividing surface plays an important role in TST and its variational formulation, and it is worthwhile being explicit about how it is defined and used. For a system of N atoms, the total phase space, excluding overall translation and rotation, has $6(N - 2)$ degrees of freedom. The dividing surface is then a $6N - 13$ degree of freedom hypersurface that is defined by the constraint $Z(\mathbf{q}, \mathbf{p}) = 0$. We consider only those dividing surface that separate reactants from products, where by convention $Z(\mathbf{q}, \mathbf{p})$ is a function that is negative for reactant-side regions and positive for product-side regions of phase space. The flux of trajectories through the dividing surface at phase point $(\mathbf{q}, \mathbf{p})$ on the dividing surface is given by

$$\left.\frac{\mathrm{d}Z}{\mathrm{d}t}\right|_{Z=0} = \left.\left(\frac{\partial Z}{\partial \mathbf{q}}\frac{\mathrm{d}\mathbf{q}}{\mathrm{d}t} - \frac{\partial Z}{\partial \mathbf{p}}\frac{\mathrm{d}\mathbf{p}}{\mathrm{d}t}\right)\right|_{Z=0} = \left.\left(\frac{\mathbf{p}}{\mu}\frac{\partial Z}{\partial \mathbf{q}} - \frac{\partial V}{\partial \mathbf{q}}\frac{\partial Z}{\partial \mathbf{p}}\right)\right|_{Z=0} \tag{4}$$

The reactant rate constant is then obtained by the proper averaging of this flux through a dividing surface. The dividing surface can be defined to be in the interaction region (e.g. near the saddle point of the reaction) as well as in the reactant region. The question that arises is what weighting of phase points to use in this averaging for arbitrary choices of dividing surface, and there are two parts to this question. One is the correct statistical weight and the other is the weight that should be used to select only reactive trajectories. For the choice of dividing surface in the reactant region, the correct statistical weight to obtain an equilibrium rate constant is an equilibrium distribution of states. For example, the correct statistical weight for a canonical rate constant is just the Boltzmann factor, $\exp(-\beta H)$ [7]. Liouville's theorem of classical mechanics [10] shows that an equilibrium distribution of reactant states will evolve into an equilibrium distribution at the dividing surface and products [11]. Thus, the correct statistical weight for any choice of dividing surface is an equilibrium distribution.

Now we turn to the question of the proper weight to select reactive trajectories. First, note that the flux given by Eq. (4) can be both positive and negative. Averaging the flux over an equilibrium distribution with no additional weighting will include all trajectories, that is, those that cross the dividing surface in both directions, including nonreactive as well as reactive trajectories, so that the equilibrium average is zero. Stated another way, the principle of detailed balance insures that the equilibrium average of the total flux at the dividing surface vanishes. The *total forward flux* is obtained by giving nonzero weight to those phase space points on the dividing surface for which $\mathrm{d}Z/\mathrm{d}t$ is positive. However, some of these phase space points may actually lie on trajectories that recross the dividing surface or are nonreactive or both (this includes trajectories that originate in the products region and recross the dividing surface one or more times, reactive trajectories that originate in the reactants region and cross an odd number of times, and nonreactive trajectories that originate in the reactants region and cross an even number of times). The *total one-way reactive flux* of reactants toward products is obtained by projecting out only those phase space points that lie on trajectories originating in reactants and ending up in products. For this choice of weighting, a reactive trajectory that recrosses the dividing surface multiple times will have multiple contributions to the averaged flux. The *net one-way reactive flux* of reactants to products is obtained by counting each of these trajectories only at its first crossing, even if it crosses more than once. The latter two are much harder to calculate because they require following trajectories to make sure they are reactive. (The fundamental assumption of TST, as we shall see, is that the latter *net one-way reactive flux* equals the former *forward flux*.) A compact expression for the *net one-way reactive flux* through the dividing surface is written as

$$F_Z(\mathbf{q}, \mathbf{p}) = \delta(Z)\left(\frac{\mathbf{p}}{\mu}\frac{\partial Z}{\partial \mathbf{q}} - \frac{\partial V}{\partial \mathbf{q}}\frac{\partial Z}{\partial \mathbf{p}}\right)\chi(\mathbf{q}, \mathbf{p}) \tag{5}$$

where the Dirac delta function $\delta(Z)$ restricts the phase space to the hypersurface dividing reactants from products (i.e. $Z = 0$). The characteristic function $\chi(\mathbf{q}, \mathbf{p})$ projects out the net contribution from reactive trajectories and therefore requires following trajectories from phase points on the dividing surface forward and backward in time to make sure they originated in reactants and finished in products and that only the first crossing is counted. With these definitions, the exact classical rate constant for a canonical ensemble at temperature T is obtained from a phase-space average of the reactive flux

$$k_{\mathrm{C}}(T) = \frac{1}{\Theta(T)(2\pi\hbar)^{3N}} \int \mathbf{dqdp} \exp(-\beta H) F_Z \tag{6}$$

where $\hbar = h/2\pi$, h is Planck's constant, $\beta = 1/k_{\mathrm{B}}T$, k_{B} is Boltzmann's constant, and $\Theta(T)$ is a normalization factor appropriate for the type of reaction (unimolecular or bimolecular). For a unimolecular reaction the normalization factor is given by

$$\Theta(T) = \frac{1}{(2\pi\hbar)^{3N}} \int \mathbf{dqdp} \exp(-\beta H)\theta(-Z) = q_{\mathrm{CM}}^{\mathrm{tr}}(T) Q_{\mathrm{C}}^{\mathrm{R}}(T) \tag{7}$$

where the Heaviside step function [$\theta(x)$ is zero for $x < 0$ and 1 for $x > 0$] restricts the phase-space average to the reactant region, $Q_{\mathrm{C}}^{\mathrm{R}}(T)$ is the vibrational–rotational partition function of the reactant species, and $q_{\mathrm{CM}}^{\mathrm{tr}}(T)$ is the translational partition function for the center-of-mass motion of the system. For bimolecular reactions $\Theta(T)$ takes the form

$$\Theta(T) = q_{\mathrm{CM}}^{\mathrm{tr}}(T)\Phi_{\mathrm{C}}^{\mathrm{R}}(T) \tag{8}$$

where $\Phi_{\mathrm{C}}^{\mathrm{R}}(T)$ is the reactant partition function per unit volume and includes the translational partition function per unit volume for the relative motion of the two reaction species. The center-of-mass translational partition function in the normalization term cancels an identical factor in the phase space average in the numerator of Eq. (6).

The fundamental dynamical assumption of TST [3] approximates the reactive flux through the dividing surface by the forward flux of all trajectories (reactive and nonreactive). If the reaction coordinate were truly separable, then there would be no reversals in the reaction coordinate, as long as the potential energy has a negative second derivative along the reaction coordinate, so the fundamental assumption may be restated as an assumption that the reaction coordinate is separable and the effective potential along the reaction coordinate is convex. In this case the characteristic function is replaced by the TST approximation

$$\chi^{\mathrm{TST}}(\mathbf{q}, \mathbf{p}; Z) = \theta\left(\frac{\mathbf{p}}{\mu}\frac{\partial Z}{\partial \mathbf{q}} - \frac{\partial V}{\partial \mathbf{q}}\frac{\partial Z}{\partial \mathbf{p}}\right) \tag{9}$$

and the reactive flux is approximated by

$$F_Z^{\mathrm{TST}}(\mathbf{q}, \mathbf{p}) = \delta(Z)\left(\frac{\mathbf{p}}{\mu}\frac{\partial Z}{\partial \mathbf{q}} - \frac{\partial V}{\partial \mathbf{q}}\frac{\partial Z}{\partial \mathbf{p}}\right)\theta\left(\frac{\mathbf{p}}{\mu}\frac{\partial Z}{\partial \mathbf{q}} - \frac{\partial V}{\partial \mathbf{q}}\frac{\partial Z}{\partial \mathbf{p}}\right) \tag{10}$$

In this approximation it is assumed that trajectories with positive flux at the dividing surface (i.e. those with $\chi^{\mathrm{TST}}(\mathbf{q}, \mathbf{p}; Z) > 0$) are reactive, and there is no need to follow trajectories of phase points forward and backward in time. The exact classical rate constant is independent of the definition of the dividing surface, since every reactive trajectory must cross any dividing surface separating reactants from products. The generalized TST expression is not independent of the choice of arbitrary dividing surface and is written as a explicit function of Z

$$k_{\mathrm{C}}^{\mathrm{GT}}(T,Z) = \frac{1}{(2\pi\hbar)^{3N}\Theta(T)} \int \mathrm{d}\mathbf{q}\mathrm{d}\mathbf{p}\, \exp(-\beta H) F_{Z}^{\mathrm{TST}}(\mathbf{q},\mathbf{p}) \tag{11}$$

TST counts all forward crossings as reactive (each reactive trajectory must cross the dividing surface with positive flux at least once), and errs by counting as reactive nonreactive trajectories that cross the dividing surface an even number of times and overcounting reactive trajectories that recross the dividing surface. Therefore, the classical TST rate constant is always greater than or equal to the exact classical equilibrium rate constant, and this fact is the basis for variational TST (VTST) in which the dividing surface is optimized to minimize the rate constant and thereby give the best upper bound to it [6,9].

In our most straightforward implementation of VTST for gas-phase reactions, rather than allow arbitrary orientations of the dividing surface, we consider a one-parameter sequence of dividing surfaces that are defined in terms of a reaction path [12,13]. This procedure is applicable to complex problems, and it immediately provides a practical improvement over the conventional choice of placing the dividing surface at the saddle point. A robust choice for the reaction path is the minimum energy path (MEP), that is, the path of steepest descent in the mass-scaled coordinates [14]. The coordinates on this path are denoted $\mathbf{q}^{\mathrm{MEP}}(s)$ as a function of a progress variable s, and the path is defined by

$$\mathbf{q}'(s) \equiv \frac{\mathrm{d}}{\mathrm{d}s}\mathbf{q}^{\mathrm{MEP}}(s) = -\left.\frac{\partial_{\mathbf{q}} V}{|\partial_{\mathbf{q}} V|}\right|_{\mathbf{q}=\mathbf{q}^{\mathrm{MEP}}(s)} \tag{12}$$

This equation is integrated along the reaction coordinate s from the saddle point at $s = 0$ with an initial step along the eigenvector for the unbound mode into the product region ($s < 0$) and reactant ($s > 0$) region. Generalized transition-state dividing surfaces are constrained to be hyperplanes that are orthogonal to the reaction path and are defined by their location s along the reaction coordinate. Given the vector, $\mathbf{q}'(s)$, tangent to the MEP at point s, the dividing surface orthogonal to this tangent vector is defined by

$$Z_s(\mathbf{q},\mathbf{p}) = \mathbf{q}'(s)[\mathbf{q} - \mathbf{q}^{\mathrm{MEP}}(s)] = 0 \tag{13}$$

For s not equal to 0, a dividing surface is called a generalized transition state (to distinguish it from the conventional transition state where the gradient is zero). With this choice of dividing surface the generalized TST rate expression given by Eq. (11) reduces for a bimolecular reaction to

$$k_{\mathrm{C}}^{\mathrm{GT}}(T,s) = k_{\mathrm{C}}^{\mathrm{GT}}(T;Z_s) = \sigma\frac{k_{\mathrm{B}}T}{h}\frac{Q_{\mathrm{C}}^{\mathrm{GT}}(T,s)}{\Phi_{\mathrm{C}}^{\mathrm{R}}(T)}\exp[-\beta V^{\mathrm{MEP}}(s)] \tag{14}$$

where $V^{\mathrm{MEP}}(s)$ is the potential evaluated on the MEP at s, the zero of energy of the potential is defined to be the reactant equilibrium geometry (i.e. $V^{\mathrm{MEP}}(s)$ at reactants is zero), and $Q_{\mathrm{C}}^{\mathrm{GT}}(T,s)$ is the classical mechanical generalized transition state partition function for the bound modes orthogonal to the reaction path at s. Note that this partition function is defined with its zero of energy at the local minimum of the potentials for the bound modes orthogonal to the reaction coordinate. The symmetry factor σ accounts for the fact that the generalized transition state partition function is computed for one reaction path, and for reactions with equivalent reaction paths, this partition function needs to be multiplied by the number of equivalent ways the reaction can proceed. For example, in the reaction $D + H_2 \rightarrow DH + H$, D can react with either H atom, so there are two equivalent pathways and $\sigma = 2$. The symmetry factor does not occur in the more general expression for the rate constant given in Eq. (11) because the phase space average includes all equivalent configurations that lie in the dividing surface. The interested reader is directed to a more detailed discussion of symmetry factors in TST by Pechukas [15]. The canonical variational theory (CVT) rate constant is obtained by minimizing Eq. (14) with respect to s

$$k_{\mathrm{C}}^{\mathrm{CVT}}(T) = \min_{s} k_{\mathrm{C}}^{\mathrm{GT}}(T, s) = k_{\mathrm{C}}^{\mathrm{GT}}[T, s_{\mathrm{C}}^{\mathrm{CVT}}(T)] \tag{15}$$

where $s_{\mathrm{C}}^{\mathrm{CVT}}(T)$ is the location of the dividing surface that minimizes Eq. (14) at temperature T. Eq. (14) provides a quantitative framework for discussing activation energy and steric effects, with the former originating mainly in the exponential term and the latter mainly in the partition function of the transition state.

The expression in Eq. (14) can be recast into a thermodynamic formulation as [13]

$$k_{\mathrm{C}}^{\mathrm{GT}}(T, s) = \frac{k_{\mathrm{B}}T}{h} K^{0} \exp[-\Delta G_{\mathrm{C}}^{\mathrm{GT},0}(T, s)/RT] \tag{16}$$

in which K^0 defines the standard state and $\Delta G_{\mathrm{C}}^{\mathrm{GT},0}(T, s)$ is the standard state free energy of formation of the generalized transition state. Minimizing the canonical rate constant as in Eq. (15) is equivalent to maximizing the free energy of activation with respect to the location of the dividing surface [13,16].

Throughout this section, all rate constant expressions have had a subscript C to denote classical mechanics. Consistent with notation used in previous work, no subscript is used when quantum mechanical effects are included.

5.2.2 Inclusion of quantum mechanical effects

For many reactions, especially those including light atoms, such as reactions involving the transfer of a hydrogen atom, proton, or hydride ion, classical mechanics is not sufficient, and quantum mechanical effects on the motion of the atoms must be included. The inclusion of quantum mechanical effects can also be important when zero-point energies for modes with high vibrational frequencies change along the reaction coordinate. High-frequencies can result from heavy-atom vibrations with large force constants, as well as vibrational modes dominated by light atom motion. In addition quantum effects may be smaller, but not necessarily negligible, for rearrangements dominated by motions of heavier atoms. A rigorous quantum mechanical formulation of TST that employs the fundamental assumption as its only approximation has been discussed [17] with the conclusion that no rigorous quantum version of TST exists that does not require a solution of the full multidimensional reaction dynamics [18]. Therefore, additional approximations are needed to include quantum mechanical effects into TST. The standard approach is an *ad hoc* procedure [1,19] that replaces classical partition functions by approximate quantum mechanical ones and then includes correction factors for quantum mechanical effects (such as tunneling) on the reaction coordinate motion. Replacing the classical partition functions in Eq. (14) by quantum mechanical ones results in a quantized generalized transition state rate constant given by

$$k^{\mathrm{GT}}(T,s) = \sigma \frac{k_{\mathrm{B}}T}{h} \frac{Q^{\mathrm{GT}}(T,s)}{\Phi^{\mathrm{R}}(T)} \exp[-\beta V^{\mathrm{MEP}}(s)] \tag{17}$$

The classical phase-space averages for bound modes in Eq. (11) are replaced by quantum mechanical sums over states. If one assumes separable rotation and uses an independent normal mode approximation, the potential becomes decoupled, and one-dimensional energy levels for the bound modes may be conveniently computed. In this case, the quantized partition function is given by the product of partition functions for each mode. Within the harmonic approximation the independent-mode partition functions are given by an analytical expression, and the vibrational generalized transition state partition function reduces to

$$Q^{\mathrm{GT}}_{\mathrm{HO}}(T,s) = \prod_m \frac{1}{2\sinh[\hbar\omega_m(s)\beta/2]} \tag{18}$$

where $\omega_m(s)$ is the harmonic frequency for mode m at location s along the MEP. When low frequency modes are present, the harmonic approximation is very often not valid, and methods for including anharmonicity must be considered [20].

Harmonic frequencies for bound modes at stationary points on the PES (i.e. locations where $\partial V/\partial \mathbf{q} = \mathbf{0}$) are obtained by diagonalizing the matrix of second derivatives (properly mass-weighted or mass-scaled), which is denoted the Hessian matrix. Frequencies for modes at other locations along the reaction coordinate require special attention since the first derivatives are not zero, and diagonalization of the Hessian would mix the reaction coordinate mode with bound vibrations. Determination of the bound modes orthogonal to the reaction coordinate can be accomplished in a straightforward manner by working in a space normal to the tangent to the reaction path [12] or by projecting out the reaction coordinate motion from the Hessian matrix [21]. The latter method often has the unsatisfactory characteristic that transitional modes (those that correspond to translations or rotations of the reactants or products in an asymptotic region but that evolve into or from a hindered rotation or bending vibration in the interaction region) can have unphysical imaginary frequencies (corresponding to negative eigenvalues of the Hessian matrix) along the reaction coordinate [22]. The use of curvilinear internal coordinates is an attractive alternative that helps with this problem [23].

The quantized generalized transition state rate constant equation (17) is a hybrid expression in which the bound modes are treated quantum mechanically but the reaction coordinate motion is treated classically. Consistent quantum mechanical treatments of reaction coordinate motion rely on the fact that the adiabatic theory of reactions [24,25] is equivalent to one form of VTST (microcanonical VTST) when the reaction coordinate is treated classically [13,26]. In the adiabatic theory of reactions, coordinates orthogonal to the reaction coordinate are treated as 'fast' variables. Reaction probabilities $P^{\mathrm{A}}(\mathbf{n}, E)$ for a state defined by quantum numbers $\mathbf{n}$ at total scattering energy E are obtained by considering the dynamics on the one-mathematical-dimensional vibrationally adiabatic potential. In the harmonic approximation the vibrationally adiabatic potential is written

$$V^{\mathrm{A}}(s, \mathbf{n}) = V^{\mathrm{MEP}}(s) + \sum_m \hbar\omega_m(s)\left(n_m + \frac{1}{2}\right) \tag{19}$$

where the sum is over the bound modes of the generalized transition state at s, and the energy level for state n_m of mode m at location s along the reaction coordinate is given by the harmonic approximation. The reaction probabilities $P^{\mathrm{A}}(\mathbf{n}, E)$ are then thermally averaged to yield the rate constant. When reaction coordinate motion is treated classically, the adiabatic theory of reactions yields an expression for the thermal rate constant which is equivalent to that obtained from microcanonical variational theory even though the approximations in the two theories are very different [13,26]. Since the one-dimensional scattering problem can be treated quantum mechanically, a multiplicative tunneling correction factor for the adiabatic theory of reactions can be obtained, and the equivalency of microcanonical VTST and adiabatic theory makes it consistent to use the same correction factor to account for the quantization of reaction coordinate motion in the variational theory.

The adiabatic approximation is made in a curvilinear coordinate system, and although the potential energy term is simple, the kinetic energy term is complicated by factors dependent upon the curvature of the reaction path [21,25,27]. As shown by Skodje et al. [28], the most successful methods for including the multidimensional effect of the reaction path curvature in the adiabatic calculations of the reaction probabilities specify a tunneling path that 'cuts the corner' and shortens the tunneling length. Marcus and Coltrin [29] found the optimum tunneling path for the collinear $H + H_2$ reaction by finding the path that gave the least exponential damping. General multidimensional tunneling (MT) methods, applicable to polyatomic reactions, have been developed that are appropriate for systems with both small [28,30] and large [12,31,32] reaction path curvature, as well as more general methods that optimize tunneling paths by a least-imaginary-action principle [31,33]. In practice it is usually sufficient to optimize the imaginary action from among a small set of choices by choosing either the small-curvature tunneling approximation, or the large-curvature tunneling approximation, which gives more tunneling at a given tunneling energy; this is called microcanonical optimized multidimensional tunneling (μOMT), or, for short, optimized multidimensional tunneling (OMT) [32,34].

The quantum mechanical CVT rate constant with the tunneling correction factor, $\kappa^{\mathrm{MT}}(T)$, included is given by [35]

$$k^{\mathrm{CVT/MT}}(T) = \kappa^{\mathrm{MT}}(T)k^{\mathrm{CVT}}(T) \tag{20}$$

Rather than compute the tunneling through all adiabatic potentials that contribute significantly to the tunneling correction factor, the tunneling correction factor is approximated by tunneling through just the ground-state potential [35]

$$\kappa^{\mathrm{MT}}(T) = \frac{\int_0^{\infty} \mathrm{d}E\, \mathrm{e}^{-\beta E} P^{\mathrm{MT}}(\mathbf{n}=0, E)}{\int_0^{\infty} \mathrm{d}E\, \mathrm{e}^{-\beta E} P_{\mathrm{C}}(\mathbf{n}=0, E)} \tag{21}$$

where $P^{\mathrm{MT}}(\mathbf{n}=0, E)$ is the multidimensional tunneling approximation to the probability for tunneling through the ground-state adiabatic potential at total energy E and $P_C(\mathbf{n}= 0, E)$ is the classical analog, which is zero for total energy E below the maximum of the adiabatic barrier and one for energy above it. At low temperatures where tunneling corrections are most important, quantized systems tend to be in the ground state, and this approach provides a good approximation. As temperature increases, tunneling through excited-state adiabatic potentials would contribute relatively more, but tunneling becomes less important and the correction factor decreases until at sufficiently high temperatures it tends to unity. The ground-state method gives the correct high temperature limit, and for intermediate temperatures, the tunneling through excited-state adiabatic potentials is approximated (implicitly) by the tunneling probabilities for the ground-state potential with the energy scale shifted by the difference in the excited-state and ground-state energies at the ground-state maximum.

After including variational and quantum effects, the quasithermodynamic variables of TST, like entropy of activation and energy of activation, may be decomposed into 'substantial' contributions that derive from formulas analogous to those of statistical thermodynamics [36] and 'nonsubstantial' contributions deriving from the transmission coefficient [37].

5.2.3 Improved prescriptions for the reaction coordinate and dividing surface

The formalism summarized above is well suited for bimolecular reactions with tight transition states and simple barrier potentials. In such cases we have found that the variational transition state can be found by optimization of a one-parameter sequence of dividing surfaces orthogonal to the reaction path, where the reaction path is defined as the MEP. However, although dividing surfaces defined as hyperplanes perpendicular to the tangent to the MEP (as described in Section 5.2.2) are very serviceable, a number of improvements have been put forth.

It is important to emphasize, in describing these improvements, that the generalized transition-state dividing surface is defined by the MEP only on the reaction path itself. The full definition of the transition state dividing surface for VTST calculations is usually specified by starting with a global definition of a reaction path, after which a one-parameter sequence of dividing surfaces is defined by the location at which the dividing surface crosses the path. In our original work [12,38] we took the dividing surface to locally be a plane orthogonal to the MEP in isoinertial coordinates, but we approximated the partition functions in a way [13,26] that provided physical results even when this surface is unphysical beyond a certain distance from the MEP. The first important improvement is to use curved dividing surfaces defined in terms of nonredundant [22,23] or redundant [39] internal coordinates. Because the neglect of anharmonic bend–stretch interactions is less serious in curvilinear internal coordinates than in rectilinear coordinates [15,40], a dividing surface defined in curvilinear internal coordinates is much more physical, and the harmonic frequencies calculated this way provide much better harmonic approximations to the bound motions in the dividing surface and hence to $Q^{\mathrm{GT}}(T, s)$.

A next higher level of refinement is to *optimize* the orientation of the dividing surface rather than assume that it is normal (in either rectilinear or curvilinear coordinates) to the MEP [40–42]. This variational optimization should in principle improve the rate, just as optimizing the location at which the dividing surface intersects the MEP improves the rate. Even more significantly though it allows one [40,42] to carry out VTST calculations without calculating (or without converging) an MEP, which is often expensive (because it requires small step sizes to follow it) and/or hard to converge.

The use of curvilinear coordinates and optimization of the orientation of the dividing surface are important for quantitative calculations on simple barrier reactions, but even more flexibility in the dividing surfaces is required to obtain quantitative results for very loose variational transition states such as those for barrier-less association reactions or their reverse (dissociation reactions without an intrinsic barrier).

In the context of association reactions, an algorithm in which the reaction coordinate definition is optimized along with the dividing surface along a one-parameter sequence of paths is called 'variable reaction coordinate' (VRC) variational transition state theory [43,44]. In the last few years there has been considerable progress in optimizing VRCs for barrier-less association reactions with strictly loose transition states. A strictly loose transition state is defined as one in which the conserved vibrational modes are uncoupled from the transitional modes and have the same frequencies in the variational transition state as in the associating reagents [45,46]. Conserved vibrational modes are modes that occur in both the associating fragments and the association complex, whereas transition modes (already mentioned above) include overall rotation of the complex and vibrations of the complex that transform into fragment rotations and relative translations upon dissociation of the complex. Progress has included successively refined treatments of the definition of the dividing surface and of the definition of the reaction coordinate (the coordinate that is missing in the transition state) [43,46–48], and elegant derivations of rate expression for these successively improved reaction coordinates [48,49]. A guiding principle in the choice of reaction coordinate in all these methods is to make the reaction coordinate correspond to a physical motion that is relatively uncoupled from the motions orthogonal to it, because the assumption of a separable reaction coordinate is a key aspect of the fundamental assumption of TST, as mentioned above. The recent variational implementation of the multifaceted dividing surface (MDS) VRC version of VTST seems to have brought the theory to a flexible enough state that it is suitable for application to a wide variety of practical applications to complex barrier-less association reactions of polyatomic molecules [50].

5.2.4 Spectroscopy of the transition state

The quantization of transition state energy levels is not simply a mathematical device to add quantum effects to the partition functions. The quantized levels actually show up as structure in the exact quantum mechanical rate constants as functions of total energy [51]. The interpretation of this structure provides clear evidence for quantized dynamical bottlenecks, both near to and distant from the saddle points, as reviewed elsewhere [52]. Quantized variational transition states have also been observed in molecular beam scattering experiments [53]. Analysis of the reactive flux in state-to-state terms from reactant states to transition state levels to product states provides the ultimate limit of resolution allowed by quantum mechanics [53,54]. Quantized energy levels of the variational transition state have been used to rederive TST using the language of quantum mechanical resonance scattering theory [55].

5.2.5 Applications

Recent applications of VTST/MT to polyatomic gas-phase reactions that illustrate the power of the theory include the reactions of chlorine atoms with hydrogen molecules [56] and the reactions of hydrogen atoms with ethylene [57,58] and methane [59], including kinetic isotope effects.

5.3 REACTIONS IN CONDENSED PHASES

Compared to gas-phase reactions, those in condensed-phases systems are more difficult to treat because of the close proximity of other atoms and molecules to the reacting species; interactions with the surroundings affect both the energetics and dynamics of a reaction. Short-range effects, due to atoms and molecules in the immediately vicinity of the reacting species, can be treated using an embedded cluster approach in which the reacting species and a finite number of surrounding atoms are included explicitly in calculations of reaction energies and dynamics using standard approaches used in gas-phase calculations. A major challenge for condensed-phase reactions arises from the fact that long-range interactions of the reacting species with atoms or molecules, which are not directly participating in the reaction, can affect both the energies and dynamics of reactions thus requiring calculations on much larger systems than the embedded cluster models. Sometimes such effects can be included by continuum solvent models [60] or by the use of collective reaction coordinates [61,62], or both.

We first consider the case of a condensed-phase environment in which the atoms have well defined equilibrium geometries and the fluctuations around these geometries are small compared to the interatomic distances (e.g. solids). In this type of rigid environment the long-range effects of the environment are restricted to the reaction energetics and the dynamics, to the extent that dynamics may be separated from energetics, can be treated accurately using an embedded cluster model. We then discuss two approaches to treating reactions in an environment in which the atoms and molecules have many local equilibrium structures that are interconnected by small barriers (i.e. ones that can be easily surmounted by thermal fluctuations). In this type of fluid environment both the energetic and dynamic effects of anharmonicity and perhaps also long-range interactions need to be considered.

The average environment that the reacting species encounter in gas phase and condensed fluid environments is isotropic and translationally invariant. This is not true for rigid environments with well-defined lattice sites, e.g. the average environment that a reacting species sees near a lattice site is very different from that near an interstitial site.

5.3.1 Reactions in rigid environments and application to reactions in crystals or at crystal–vapor interfaces

Solid-state reactions, at least in a crystal, are characterized by the need to consider only one or a few environments for the reacting subsystem. The reacting subsystem may interact with this environment by a variety of interactions, including electrostatic and dispersion forces, hydrogen bonding, and dielectric screening of intramolecular Coulomb interactions due to the electric field lines passing through the environmental medium. These effects must be included in the Hamiltonian. One also needs to consider effects such as relaxation of the lattice around the reacting subsystem or reconstruction of a surface in absence and presence of the reacting subsystem. These considerations also apply to the calculation of equilibrium properties in the solid state, but a question that arises only when one considers dynamics is the inclusion of medium degrees of freedom in the dividing surface, or medium participation in the reaction coordinate.

As mentioned above, these effects can all be included by an embedded cluster approach [63]. In this method one starts with a large but finite rigid lattice representing the crystal and adds the reacting substance as a substitutional impurity, defect, or interstitial in the solid (absorbate) or on its surface (adsorbate). The origin is defined in some convenient way to be at or near the center of the reacting substance. All lattice atoms within a distance R_0 of the origin are fully included in the Hamiltonian, that is, they are treated dynamically on an equal footing with the atoms of the absorbate or adsorbate.

The other atoms in the lattice are held rigid. As such they enter the Hamiltonian and they help to enforce the macroscopic habit on the environment of the reacting substance, but they do not participate dynamically. In principle, both the size of the lattice and the value of R_0 are increased until the results converge.

This kind of treatment requires only minor changes in a VTST computer program. The most significant changes are that there are no translational or rotational partition functions or coordinates to project out of the Hessians. Reactants have $3N$ vibrations, rather than $3N - 6$, and transition states have $3N - 1$ vibrations rather than $3N - 7$.

Examples of applications that have been studied by this method are surface diffusion of hydrogen atoms on metal surfaces [63–66], bulk diffusion [64,66], subsurface-to-surface transport [66], and dissociative chemisorption [67].

5.3.2 Reactions in fluid environments with a single reaction coordinate

The prescription for using VTST for gas-phase and solid-state reactions relies on locating saddle points and reaction paths. For reactions in fluids, there are many saddle points that are close in energy and that differ significantly only in the configuration of solvent molecules [68]. Similarly, the PESs of large clusters often have multiple minima, often separately by low barriers [69], and when a molecule reacts on or with a large cluster, these minima lead to multiple saddle points. The multiple saddle points may be considered to result from the large anharmonicity of the solvent (i.e. of the liquid-phase solvent or the large-cluster microsolvent), and their treatment requires special consideration. As in the case of solid-state reactions, VTST including quantum mechanical effects can be applied to an embedded cluster model, which contains solvent molecules as well as the reacting subsystem; however, unlike the solid-state case where the remainder of the system is fixed in a rigid configuration, it is important to sample over configurations of the remainder of the solvent or to use a mean-field representation that includes an average over solvent configurations.

To generalize the procedures of Sections 5.2 and 5.3.1 to the liquid phase, one can start from the full microscopic description of the system. The Hamiltonian for the whole system is partitioned into a gas-phase component, as given in Eq. (1), for a reactive embedded molecule or embedded cluster (note: an embedded cluster is often called a supermolecule) in the absence of the solvent, and a solvent component that includes coupling between the solvent and reactive subsystem:

$$H(\mathbf{q}, \mathbf{p}, \mathbf{x}, \mathbf{p}_x) = \frac{\mathbf{p}^2}{2\mu} + V_{\mathrm{C}}(\mathbf{q}) + \frac{\mathbf{p}_x^2}{2\mu} + V_{\mathrm{S}}(\mathbf{x}, \mathbf{q}) \quad (22)$$

where $\mathbf{q}$ and $\mathbf{p}$ are coordinates and conjugate momentum of the embedded molecule or cluster (henceforth called the solute), $\mathbf{x}$ and $\mathbf{p}_x$ are the solvent coordinates and conjugate momentum, $V_{\mathrm{C}}(\mathbf{q})$ is the PES for the isolated solute, and $V_{\mathrm{S}}(\mathbf{x}, \mathbf{q})$ is the PES for the solvent including the solvent interactions with the solute. We first consider the classical mechanical expression for the rate constant for this model

$$k_{\mathrm{C}}^{\mathrm{GT}}(T; Z) = \frac{1}{(2\pi\hbar)^{3N+3N_{\mathrm{S}}}\Theta(T)} \int \mathrm{d}\mathbf{q}\mathrm{d}\mathbf{p} \int \mathrm{d}\mathbf{x}\mathrm{d}\mathbf{p}_x \exp(-\beta H) F_{\mathrm{R}}^{\mathrm{TST}}(\mathbf{q}, \mathbf{p}, \mathbf{x}, \mathbf{p}_x) \quad (23)$$

where $3N_{\mathrm{S}}$ is the number of degrees of freedom of the solvent, and the normalization factor $\Theta(T)$ includes the partition function for both the solute and solvent. The reactive flux is explicitly written as a function of solvent coordinates as well as those of the solute. However, if the dividing surface is only a function of the solute coordinates and momentum, Eq. (23) takes on a simpler form

$$k_{\mathrm{C,ES}}^{\mathrm{GT}}(T; Z) = \frac{1}{(2\pi\hbar)^{3N}\Theta(T)} \int \mathrm{d}\mathbf{q}\mathrm{d}\mathbf{p} \exp\left[-\beta\left(\frac{\mathbf{p}^2}{2\mu} + W(\mathbf{q})\right)\right] F_{\mathrm{R}}^{\mathrm{TST}}(\mathbf{q}, \mathbf{p}) \quad (24)$$

where the potential of mean force (PMF) [36] $W(\mathbf{q})$ is defined by

$$\exp[-\beta W(\mathbf{q})] = \frac{1}{(2\pi\hbar)^{3N_S}} \int \mathrm{d}\mathbf{x}\mathrm{d}\mathbf{p}_x \exp\left[-\beta\left(\frac{\mathbf{p}_x^2}{2\mu} + V_C(\mathbf{q}) + V_S(\mathbf{q},\mathbf{x})\right)\right] \tag{25}$$

The normalization factor can also be rewritten in terms of the PMF

$$\Theta(T) = \frac{1}{(2\pi\hbar)^{3N}} \int \mathrm{d}\mathbf{q}\mathrm{d}\mathbf{p} \exp\left[-\beta\left(\frac{\mathbf{p}^2}{2\mu} + W(\mathbf{q})\right)\right]\theta(-Z) \tag{26}$$

so that the rate constant takes the form of the gas-phase expression with an effective, solvent-averaged potential. Classically, the only approximation is the fundamental dynamical assumption of TST, with the restricted choice of dividing surface to not include any of the solvent coordinates or momentum. We use the subscript ES on the rate constant to denote an equilibrium solvation model, in which the only influence of the solvent is to modify the interaction potential. It has long been realized that the effect of equilibrium solvation can be included in TST using the PMF [5,70–74]; however, most previous calculations of equilibrium solvation contributions to the TST rate constant express the PMF as a function of a single coordinate—the reaction coordinate. Eq. (24) differs by the fact that the rate constant is written as a function of a multidimensional equilibrium solvation PES.

Although $W(\mathbf{q})$ may be defined entirely in terms of condensed-phase averages, as in Eq. (25), it can also be defined with reference to gas-phase calculations, in which case it equals the gas-phase free energy of activation plus the free energy of solvation [60, 74–77]. Thus, the equilibrium solvation rate constant given in Eq. (24) takes the thermodynamic form for the rate constant (as given in Eq. (16) for the gas phase)

$$k_{\mathrm{C,ES}}^{\mathrm{GT}}(T,s) = \frac{k_{\mathrm{B}}T}{h} K^0 \exp[-\Delta G_{\mathrm{C,ES}}^{\mathrm{GT},0}(T,s)/RT] \tag{27}$$

where the equilibrium-solvation standard-state free energy of activation is expressed as

$$\Delta G_{\mathrm{C,ES}}^{\mathrm{GT},0}(T,s) = \Delta G_{\mathrm{C}}^{\mathrm{GT},0}(T,s) + \Delta G_{\mathrm{solv}}^{\mathrm{GT},0}(T,s) - \Delta G_{\mathrm{solv}}^{\mathrm{R},0}(T) \tag{28}$$

and $\Delta G_{\mathrm{solv}}^{\mathrm{GT},0}(T,s)$ and $\Delta G_{\mathrm{solv}}^{\mathrm{R},0}(T)$ are the free energies for solvating the generalized transition state at s and reactants, respectively.

If the reaction path and dividing surface are optimized in the gas phase, but the rate constant is calculated with the equilibrium solvation Hamiltonian, the resulting rate constant is called separable equilibrium solvation (SES) [57]. However, if the reaction path and dividing surface are optimized with the equilibrium solvation potential, the result is labeled equilibrium solvation path (ESP) [57,78].

The assumption of equilibration of the solvent at all geometries of the solute neglects the coupling of solvent dynamics with solute dynamics. For example, if the solute dynamics are rapid with respect to the solvent motion, the solvent will not have time to equilibrate as the solute rearranges. Fluctuations of solvent molecules can induce recrossings of the dividing surface and a breakdown of the fundamental dynamical assumption of TST. This type of influence of the liquid on the reaction rate is often referred to as a nonequilibrium or dynamical solvent effect [5,70,71,73,79].

Formulating TST in terms of an approximate flux through a dividing surface reveals a means to mitigate the effects of this breakdown. Variational optimization of a dividing surface that is a function of solvent coordinates as well as the coordinates of the embedded cluster will minimize the recrossings induced by solvent reorganization. However, this procedure leads to more complicated calculations of the ensemble averages in Eq. (23) because they are constrained to configurations lying on the dividing surface. Evaluations of averages over solvent degrees of freedom for fixed configurations of the solute, as required when the dividing surface is a function of only solute coordinates, are standard in PMF calculations [75], whereas evaluations of ensemble averages for more general dividing surfaces are much more complicated. (The formulation of classical TST for arbitrary dividing surfaces [80] is related to the formulation of potentials of mean force for constrained molecular dynamics [81] and to PMF calculations for more complicated reaction coordinates [82].)

This use of arbitrary reaction coordinates also allows [62] inclusion of nonequilibrium solvation effects in VTST by using the solvent energy-gap reaction coordinate that first arose [83] in weak-overlap charge-transfer theory. In electron transfer kinetics, nonequilibrium effects are often studied by a spin boson model [84] in which two diabatic states of the electronic wave function are linearly coupled to a harmonic bath. The coupling constant is a parameter of the model. A more general model, the generalized Langevin treatment [71,85], also involves a linearly coupled harmonic bath, but it has the advantage that the friction coefficient is related by the fluctuation-description theorem [86] to the time autocorrelation function of the force exerted on the system by the bath. In classical mechanics, the time autocorrelation function can be obtained from a molecular dynamics simulation, and one can use this kind of friction estimate to model nonequilibrium solvation [87,88]. This is accomplished by describing solvent frictional effects by a collection of harmonic oscillators that are linearly coupled to the solute degrees of freedom. In the limit of a continuum of oscillators, the classical dynamics for this model are equivalent to the generalized Langevin equation for solute dynamics [89]. When the reaction is treated as a reaction coordinate coupled to a harmonic bath, harmonic TST with a dividing surface at the saddle point that includes dependence on the harmonic solvent coordinates yields the Kramers [90] and Grote–Hynes [91] theories, as reviewed elsewhere [92,93]. This simple model of a reaction in solution is surprisingly robust for describing dynamical solvent affects, even for systems for which the reaction dynamics appear to be controlled by anharmonic solvent reorganization [94]. In an attempt to develop a widely applicable practical scheme for estimating the conditions under which nonequilibrium solvent effects are important and to increase physical insight, the friction has also been further approximated in terms of effective diffusion constants [78,95]. This has the advantage that it neither assumes the validity of classical mechanics nor requires large-scale simulations. We have presented VTST treatments of multidimensional embedded cluster models linearly coupled to harmonic solvent coordinates, including variationally optimizing the dividing surface (including both harmonic solvent and embedded cluster coordinates) and quantum mechanical effects [73,87,95]; and these treatments can be viewed as a generalization of

Grote–Hynes theory to include multidimensional, anharmonic (in the embedded cluster coordinates), and quantum mechanical effects.

The treatment of quantum mechanical effects, particularly tunneling, is especially challenging for reactions in liquids. Comparisons of calculations using VTST methods including multidimensional tunneling corrections for gas-phase reactions, as outlined in Section 5.2.2, with benchmark calculations for a model reaction in solution (a reaction coordinate linearly coupled to a harmonic bath) have shown that the methods are capable of quantitative accuracy for these types of systems and that the largest deficiencies arise from the approximate treatment of anhamonicity, not tunneling [96]. A general prescription for treating tunneling in the presence of a bath, based upon a multidimensional tunneling approximation using the PMF, has been given [97] and applied [76,95,98]. VTST has been used to study the effect of nonequilibrium solvation on quantum mechanical tunneling in models of hydrogen addition to benzene [78] and hydrogen abstraction from methanol [95] and to examine the importance of multidimensional tunneling for a model of a proton transfer reaction in explicit solvent [88].

5.3.3 Reactions in fluid environments with an ensemble of reaction coordinates

For some reactions in solution, it may be necessary to include a large number of saddle points and reaction paths. These reaction paths might differ primarily in the conformation of the bath. One can always attempt to model this situation, as in Section 5.3.2, by using a single reaction coordinate and a mean field representation of the bath, but this will not always be valid, and it requires considerable physical insight to properly include solvent motion in the reaction coordinate, when that is necessary.

In order to treat this kind of system more reliably, ensemble-averaged VTST [99–101] has been developed. In this method, the calculation is divided into two stages. In stage one, one uses a predefined chemical reaction coordinate z to calculate a one-dimensional PMF. The maximum value of this PMF defines the stage-1 free energy of activation,

$$\Delta G^{(1)} = \Delta G^{\ddagger,o}(z_*) \tag{29}$$

at location z_*. In carrying out this calculation, we quantized the vibrations perpendicular to z by a new method developed for this purpose [102]. Because of this quantization, the resulting free energy is not completely classical and we call it quasiclassical. In the process of calculating this stage-1 quasiclassical free energy of activation we sample a large number of systems with $z_* - \delta_z \leq z \leq z_* + \delta_z$, where δ_z is a small numerical tolerance (in principle δ_z should be zero). The equilibrium ensemble corresponding to this small range of z is called the stage-1 transition state ensemble.

In stage 2, one calculates a transmission coefficient for the stage-1 rate constant. This transmission coefficient is calculated by an ensemble average over CVT/μOMT calculations for various reaction paths (labeled $\alpha = 1, 2, ..., N_\alpha$) that pass through N_α samples from stage-1 transition state ensemble. In particular, the final ensemble-averaged

VTST with multidimensional tunneling (EA-VTST/MT) rate constant is given by

$$k(T) = \gamma(T)\frac{k_B T}{h}K^0 \exp(-\Delta G^{(1)}/RT) \tag{30}$$

where

$$\gamma(T) = \frac{\sum_{\alpha=1}^{N_\alpha} \kappa_\alpha(T)\Gamma_\alpha(T)}{N_\alpha} \tag{31}$$

In stage 1, all atoms are treated on an equal footing. However in stage 2, the system is divided into N_1 primary-zone atoms and N_2 secondary-zone atoms. For each α, the N_2 secondary-zone atoms are frozen and the N_1 primary-zone atoms are optimized to the nearest saddle point, then a MEP is calculated, again with N_2 atoms frozen. In both steps, the secondary-zone atoms are not neglected; they provided an effective potential field that is included in the Hamiltonian. Continuing in this fashion, we calculate a free energy of activation profile $\Delta G_\alpha(T)$ for the primary subsystem in the effective field of the secondary subsystem; this is reminiscent of the method in Section 5.3.1. Then

$$\Gamma_\alpha(T) = \exp\{-[\Delta G_\alpha(T) - \Delta G^{(1)}(T)]/RT\} \tag{32}$$

Finally one calculates a transmission coefficient $\kappa_\alpha(T)$ that accounts for tunneling and nonclassical reflection, and we use this in Eq. (31) to calculate the overall transmission coefficient $\gamma(T)$.

Some points should be noted about this treatment. The first is that the transmission coefficient and the quasithermodynamic free energy of activation are not independent. If we choose a poor reaction coordinate for stage 1, then the Γ_α values in stage 2 may be very small. The second is that the procedure used for step 2 allows the secondary subsystem to participate in the reaction coordinate. In other words, since each conformation of the secondary subsystem has its own reaction path, the reaction path does depend on the coordinates of the secondary subsystem.

It is also possible to include a third stage in which the secondary zone is relaxed as a function of s for each α [99]. Although this is more expensive, it is not necessarily more accurate because the transition state passage might be well modeled by an ensemble average of essentially fixed secondary-zone structures [93].

EA-VTST/OMT has been applied successfully to several enzyme reactions, as reviewed elsewhere [100,103].

5.4 SUMMARY AND CONCLUSIONS

"It is no criticism of a chemical theory to call it 'approximate' or 'limited'. The value of a theory is measured by the strength of its predictions within its restricted range of applicability" [104]. TST is an approximate theory with a very broad range of applicability, covering elementary reaction rate constants for virtually all kinds of chemical reactions, provided that the reactants are in local thermal or microcanonical equilibrium, and it can even be extended to certain questions in state-selected chemistry. It provides a language for discussing and analyzing activation energy, steric effects, and solvent and other environmental effects. When the transition state is variationally optimized and quantum effects, especially vibrational zero point energy and tunneling, are included, it provides a quantitative theory as well. It seems unimaginable that it can ever become obsolete, and we expect that research on its application to more and more complex processes will continue to abound.

5.5 ACKNOWLEDGEMENTS

This work was supported in part at both Pacific Northwest National Laboratory (PNNL) and the University of Minnesota (UM) by the Division of Chemical Sciences, Office of Basic Energy Sciences, US Department of Energy (DOE) and by the National Science Foundation at UM. Battelle operates PNNL for DOE.

5.6 REFERENCES

1 H. Eyring, J. Chem. Phys., 3 (1935) 107.
2 H. Eyring, Trans. Faraday Soc., 34 (1938) 41; M.G. Evans and M. Polanyi, Trans. Faraday Soc., 31 (1935) 875; M.G. Evans, Trans. Faraday Soc., 34 (1938) 49.
3 E. Wigner, Trans. Faraday Soc., 34 (1938) 29.
4 D.G. Truhlar and R.E. Wyatt, Ann. Rev. Phys. Chem., 27 (1976) 1; K.J. Laidler and M.C. King, J. Phys. Chem., 87 (1983) 2657; D.G. Truhlar, B.C. Garrett and S.J. Klippenstein, J. Phys. Chem., 100 (1996) 12771; W.H. Miller, Faraday Disc., 110 (1998) 1; B.C. Garrett, Theor. Chem. Acc., 103 (2000) 200.
5 D.G. Truhlar, W.L. Hase and J.T. Hynes, J. Phys. Chem., 87 (1983) 2664.
6 D.G. Truhlar and B.C. Garrett, Ann. Rev. Phys. Chem., 35 (1984) 159.
7 M.A. Eliason and J.O. Hirschfelder, J. Chem. Phys., 30 (1959) 1426; R.K. Boyd, Chem. Rev., 77 (1977) 93.
8 J.N. Murrell, S. Carter, S.C. Farantos, P. Huxley and A.J.C. Varandas, Molecular potential energy surfaces, Wiley, New York, 1984; D.G. Truhlar, in: R.A. Myers (Ed.), The encyclopedia of physical sciences and technology, 3rd ed., Vol. 13, Academic Press, New York, 2001, p. 9.
9 E. Wigner, J. Chem. Phys., 5 (1937) 720.
10 H. Goldstein, Classical mechanics, Addison-Wesley, Reading, MA, 1965.
11 J.B. Anderson, J. Chem. Phys., 58 (1973) 4684; J.B. Anderson, Adv. chem. phys., 91 (1995) 381; I. Mayer, J. Chem. Phys., 60 (1974) 2564.
12 B.C. Garrett and D.G. Truhlar, J. Chem. Phys., 70 (1979) 1593.
13 B.C. Garrett and D.G. Truhlar, J. Phys. Chem., 83 (1979) 1052.
14 R.E.J. Weston, J. Chem. Phys., 31 (1959) 892; I. Shavitt, J. Chem. Phys., 49 (1968) 4048; R.A. Marcus, J. Chem. Phys., 49 (1968) 2617; D.G. Truhlar and A. Kuppermann, J. Am. Chem. Soc., 93 (1971) 1840; K. Fukui, in: R. Daudel and B. Pullman (Eds.), The world of quantum chemistry, Reidel, Dordrecht (1974); H.F. Schaefer, III, Chem. Brit., 11 (1975) 227.
15 M.A. Pariseau, I. Suzuki and J. Overend, J. Chem. Phys., 42 (1965) 2335.
16 A. Tweedale and K.J. Laidler, J. Chem. Phys., 53 (1970) 2045.
17 F.J. McLafferty, Chem. Phys. Lett., 27 (1974) 511; W.H. Miller, J. Chem. Phys., 61 (1974) 1823.
18 W.H. Miller, Acc. Chem. Res., 26 (1993) 174.
19 S. Glasstone, K.J. Laidler and H. Eyring, The theory of rate processes, McGraw-Hill, New York, 1941.
20 B.C. Garrett and D.G. Truhlar, J. Phys. Chem., 83 (1979) 1915; D.G. Truhlar, A.D. Isaacson, R.T. Skodje and B.C. Garrett, J. Phys. Chem., 86 (1982) 2252; B.C. Garrett and D.G. Truhlar, J. Chem. Phys., 81 (1984) 309; D.G. Truhlar, J. Comput. Chem., 12 (1991) 266.

21 W.H. Miller, N.C. Handy and J.E. Adams, J. Chem. Phys., 72 (1980) 99.
22 G.A. Natanson, B.C. Garrett, T.N. Truong, T. Joseph and D.G. Truhlar, J. Chem. Phys., 94 (1991) 7875.
23 K.A. Nguyen, C.F. Jackels and D.G. Truhlar, J. Chem. Phys., 104 (1996) 6491.
24 R.A. Marcus, J. Chem. Phys., 45 (1966) 4450; R.A. Marcus, J. Chem. Phys., 46 (1967) 959; R.A. Marcus, J. Chem. Phys., 49 (1968) 2610; D.G. Truhlar, J. Chem. Phys., 53 (1970) 2041.
25 R.A. Marcus, J. Chem. Phys., 45 (1966) 4493.
26 B.C. Garrett and D.G. Truhlar, J. Phys. Chem., 83 (1979) 1079.
27 R.A. Marcus, J. Chem. Phys., 41 (1964) 2614; G.L. Hofacker, Z. Naturforsch. A, 18 (1963) 607; G.L. Hofacker and N. Rösch, Ber. Bunsen Ges. Phys. Chem., 77 (1973) 661.
28 R.T. Skodje, D.G. Truhlar and B.C. Garrett, J. Chem. Phys., 77 (1982) 5955.
29 R.A. Marcus and M.E. Coltrin, J. Chem. Phys., 67 (1977) 2609.
30 R.T. Skodje, D.G. Truhlar and B.C. Garrett, J. Phys. Chem., 85 (1981) 3019; D.-h. Lu, T.N. Truong, V.S. Melissas, G.C. Lynch, Y.-P. Liu, B.C. Garrett, R. Steckler, A.D. Isaacson, S.N. Rai, G.C. Hancock, J.G. Lauderdale, T. Joseph and D.G. Truhlar, Comp. Phys. Comm., 71 (1992) 235; Y.-P. Liu, G.C. Lynch, T.N. Truong, D.-h. Lu, D.G. Truhlar and B.C. Garrett, J. Am. Chem. Soc., 115 (1993) 2408.
31 B.C. Garrett and D.G. Truhlar, J. Chem. Phys., 79 (1983) 4931.
32 Y.-P. Liu, D.-h. Lu, A. Gonzalez-Lafont, D.G. Truhlar and B.C. Garrett, J. Am. Chem. Soc., 115 (1993) 7806.
33 D.G. Truhlar, A.D. Isaacson, B.C. Garrett, in: M. Baer (Ed.), Theory of chemical reaction dynamics, Vol. IV, CRC Press, Boca Raton, FL, 1985, p. 65.
34 A. Fernandez-Ramos and D.G. Truhlar, J. Chem. Phys., 114 (2001) 191.
35 B.C. Garrett, D.G. Truhlar, R.S. Grev and A.W. Magnuson, J. Phys. Chem., 84 (1980) 1730.
36 D.A. McQuarrie, Statistical mechanics, Harper and Row, New York, 1976.
37 D.G. Truhlar and B.C. Garrett, J. Am. Chem. Soc., 111 (1989) 1232; S.C. Tucker, D.G. Truhlar, in: J. Bertran, I.G. Csizmadia (Eds.), New theoretical concepts for understanding organic reactions, Kluwer, Dordrecht (1989) 291; M. Garcia-Viloca, J. Gao, M. Karplus and D.G. Truhlar, Science, 303 (2004) 186.
38 A.D. Isaacson and D.G. Truhlar, J. Chem. Phys., 76 (1982) 1380.
39 Y.-Y. Chuang and D.G. Truhlar, J. Phys. Chem. A, 102 (1998) 242.
40 J. Villa and D.G. Truhlar, Theor. Chem. Acc., 97 (1997) 317.
41 A. Gonzalez-Lafont, J. Villa, J.M. Lluch, J. Bertran, R. Steckler and D.G. Truhlar, J. Phys. Chem. A, 102 (1998) 3420.
42 P.L. Fast and D.G. Truhlar, J. Chem. Phys., 109 (1998) 3721.
43 S.J. Klippenstein, J. Chem. Phys., 96 (1992) 367.
44 S.J. Klippenstein, J. Phys. Chem., 98 (1994) 11459.
45 D.M. Wardlaw and R.A. Marcus, J. Chem. Phys., 83 (1985) 3462; M. Pesa, M.J. Pilling, S.H. Robertson and D.M. Wardlaw, J. Phys. Chem. A, 102 (1998) 8526.
46 D.M. Wardlaw and R.A. Marcus, Adv. Chem. Phys., 70 (1988) 231.
47 S. Robertson, A.F. Wagner and D.M. Wardlaw, J. Chem. Phys., 113 (2000) 2648.
48 S. Robertson, A.F. Wagner and D.M. Wardlaw, J. Phys. Chem. A, 106 (2002) 2598.
49 S.C. Smith, J. Chem. Phys., 111 (1999) 1830; Y. Georgievskii and S.J. Klippenstein, J. Chem. Phys., 118 (2003) 5442.
50 Y. Georgievskii and S.J. Klippenstein, J. Phys. Chem. A, 107 (2003) 9776.
51 D.C. Chatfield, R.S. Friedman, D.G. Truhlar, B.C. Garrett and D.W. Schwenke, J. Am. Chem. Soc., 113 (1991) 486.
52 D.C. Chatfield, R.S. Friedman, D.W. Schwenke and D.G. Truhlar, J. Phys. Chem., 96 (1992) 2414; D.C. Chatfield, R.S. Friedman, S.L. Mielke, G.C. Lynch, T.C. Allison, D.G. Truhlar and D.W. Schwenke, in: R.E. Wyatt, J.Z.H. Zhang (Eds.), Dynamics of molecules and chemical reactions, Marcel Dekker, New York (1996) 323.
53 D.C. Chatfield, R.S. Friedman, D.G. Truhlar and D.W. Schwenke, Faraday Disc., 91 (1991) 289.
54 D.C. Chatfield, R.S. Friedman, G.C. Lynch, D.G. Truhlar and D.W. Schwenke, J. Chem. Phys., 98 (1993) 342; D.C. Chatfield, S.L. Mielke, T.C. Allison and D.G. Truhlar, J. Chem. Phys., 112 (2000) 8387.
55 D.G. Truhlar and B.C. Garrett, J. Phys. Chem., 96 (1992) 6515.

56 S.L. Mielke, T.C. Allison, D.G. Truhlar and D.W. Schwenke, J. Phys. Chem., 100 (1996) 13588.
57 J. Villa, A. Gonzalez-Lafont, J.M. Lluch and D.G. Truhlar, J. Am. Chem. Soc., 120 (1998) 5559.
58 J. Villa, J.C. Corchado, A. Gonzalez-Lafont, J.M. Lluch and D.G. Truhlar, J. Am. Chem. Soc., 120 (1998) 12141; J. Villa, J.C. Corchado, A. Gonzalez-Lafont, J.M. Lluch and D.G. Truhlar, J. Phys. Chem. A, 103 (1999) 5061.
59 J. Pu, J.C. Corchado and D.G. Truhlar, J. Chem. Phys., 115 (2001) 6266; J. Pu and D.G. Truhlar, J. Chem. Phys., 117 (2002) 1479; J. Pu and D.G. Truhlar, J. Chem. Phys., 117 (2002) 10675.
60 C.J. Cramer and D.G. Truhlar, Chem. Rev., 99 (1999) 2161.
61 D.G. Truhlar, G.K. Schenter and B.C. Garrett, J. Chem. Phys., 98 (1993) 5756.
62 G.K. Schenter, B.C. Garrett and D.G. Truhlar, J. Phys. Chem. B, 105 (2001) 9672.
63 J.G. Lauderdale and D.G. Truhlar, Surf. Sci., 164 (1985) 558; J.G. Lauderdale and D.G. Truhlar, J. Chem. Phys., 84 (1986) 1843; S.E. Wonchoba, W.P. Hu and D.G. Truhlar, in: H.L. Sellers, J.T. Golab (Eds.), Theoretical and computational approaches to interface phenomena, Plenum, New York (1994) 1.
64 B.M. Rice, B.C. Garrett, M.L. Koszykowski, S.M. Foiles and M.S. Daw, J. Chem. Phys., 92 (1990) 775.
65 S.E. Wonchoba and D.G. Truhlar, J. Chem. Phys., 99 (1993) 9637; S.E. Wonchoba, W.P. Hu and D.G. Truhlar, Phys. Rev. B, 51 (1995) 9985.
66 S.E. Wonchoba and D.G. Truhlar, Phys. Rev. B, 53 (1996) 11222.
67 T.N. Truong, G. Hancock and D.G. Truhlar, Surf. Sci., 214 (1989) 523.
68 A. Warshel, J. Phys. Chem., 83 (1979) 1640; J.G. Harris and F.H. Stillinger, Chem. Phys., 149 (1990) 63.
69 C.J. Tsai and K.D. Jordan, J. Phys. Chem., 97 (1993) 1227.
70 D. Chandler, J. Chem. Phys., 68 (1978) 2959; B.J. Berne, M. Borkovec and J.E. Straub, J. Phys. Chem., 92 (1988) 3711.
71 J.T. Hynes, in: M. Baer (Ed.), Theory of chemical reaction dynamics, Vol. IV, CRC Press, Boca Raton, FL, 1985, p. 171.
72 J.T. Hynes, Ann. Rev. Phys. Chem., 36 (1985) 573.
73 B.C. Garrett and G.K. Schenter, Int. Rev. Phys. Chem., 13 (1994) 263.
74 B.C. Garrett, D.G. Truhlar, in: P.v.R. Schleyer, N.L. Allinger, T. Clark, J. Gasteiger, P.A. Kollman and H.F. Schaefer, III (Eds.), Encyclopedia of computational chemistry, Vol. 5, Wiley, Chichester, 1998, p. 3094.
75 W.L. Jorgensen, Adv. Chem. Phys., 2 (1988) 469; W.L. Jorgensen, Acc. Chem. Res., 22 (1989) 184.
76 Y.-Y. Chuang, C.J. Cramer and D.G. Truhlar, Int. J. Quant. Chem., 70 (1998) 887.
77 Y.-Y. Chuang, M.L. Radhakrishnan, P.L. Fast, C.J. Cramer and D.G. Truhlar, J. Phys. Chem. A, 103 (1999) 4893.
78 B.C. Garrett and G.K. Schenter, in: C.J. Cramer, D.G. Truhlar (Eds.), Structure and reactivity in aqueous solution, Vol. 568, American Chemical Society, Washington, DC, 1994, p. 122.
79 S.C. Tucker and D.G. Truhlar, J. Am. Chem. Soc., 112 (1990) 3347.
80 G.K. Schenter, B.C. Garrett and D.G. Truhlar, J. Chem. Phys., 119 (2003) 5828.
81 E.A. Carter, G. Ciccotti, J.T. Hynes and R. Kapral, Chem. Phys. Lett., 156 (1989) 472.
82 M. Sprik and G. Ciccotti, J. Chem. Phys., 109 (1998) 7737; E. Darve and A. Pohorille, J. Chem. Phys., 115 (2001) 9169.
83 R.A. Marcus, J. Chem. Phys., 24 (1956) 966; R.A. Marcus, Disc. Faraday Soc., 29 (1960) 21.
84 H. Wang, X.Y. Song, D. Chandler and W.H. Miller, J. Chem. Phys., 110 (1999) 4828; M. Thoss, H.B. Wang and W.H. Miller, J. Chem. Phys., 115 (2001) 2991.
85 S.A. Adelman and J.D. Doll, J. Chem. Phys., 64 (1976) 2375; T.F. George, K.-T. Lee, W.C. Murphy, M. Hutchinson and H.-W. Lee, in: M. Baer (Ed.), Theory of chemical reaction dynamics, Vol. IV CRC Press, Boca Raton, FL (1985) 139.
86 R. Kubo, Rept. Prog. Theor. Phys., 29 (1966) 255.
87 G.K. Schenter, R.P. McRae and B.C. Garrett, J. Chem. Phys., 97 (1992) 9116; R.P. McRae, G.K. Schenter and B.C. Garrett, Faraday Trans., 93 (1997) 997.
88 R.P. McRae, G.K. Schenter, B.C. Garrett, Z. Svetlicic and D.G. Truhlar, J. Chem. Phys., 115 (2001) 8460.
89 R. Zwanzig, J. Stat. Phys., 9 (1973) 215.
90 H.A. Kramers, Physica (Utrecht), 7 (1940) 284.

91 R.F. Grote and J.T. Hynes, J. Chem. Phys., 73 (1980) 2715.

92 P. Hanggi, P. Talkner and M. Borkovec, Rev. Mod. Phys., 62 (1990) 251; S.C. Tucker, in: P. Talkner and P. Hänggi (Eds.), New trends in Kramers' reaction rate theory, Kluwer, Dordrecht (1995) p. 5.

93 J.T. Hynes, in: O. Tapia, J. Bertrán (Eds.), Solvent effects and chemical reactivity, Kluwer, Dordrecht, 1996, p. 231.

94 D.G. Truhlar and B.C. Garrett, J. Phys. Chem. B, 104 (2000) 1069.

95 Y.-Y. Chuang and D.G. Truhlar, J. Am. Chem. Soc., 121 (1999) 10157.

96 R.P. McRae, G.K. Schenter, B.C. Garrett, G.R. Haynes, G.A. Voth and G.C. Schatz, J. Chem. Phys., 97 (1992) 7392; R.P. McRae and B.C. Garrett, J. Chem. Phys., 98 (1993) 6929.

97 D.G. Truhlar, Y.-P. Liu, G.K. Schenter and B.C. Garrett, J. Phys. Chem., 98 (1994) 8396.

98 R.M. Nicoll, I.H. Hillier and D.G. Truhlar, J. Am. Chem. Soc., 123 (2001) 1459.

99 C. Alhambra, J. Corchado, M.L. Sanchez, M. Garcia-Viloca, J. Gao and D.G. Truhlar, J. Phys. Chem. B, 105 (2001) 11326; T.D. Poulsen, M. Garcia-Viloca, J. Gao and D.G. Truhlar, J. Phys. Chem. B, 107 (2003) 9567.

100 D.G. Truhlar, J. Gao, C. Alhambra, M. Garcia-Viloca, J. Corchado, M.L. Sanchez and J. Villa, Acc. Chem. Res., 35 (2002) 341; D.G. Truhlar, J. Gao, M. Garcia-Viloca, C. Alhambra, J. Corchado, M.L. Sanchez, and T.D. Poulsen, Int. J. Quant. Chem. 100 (2004) 1136.

101 M. Garcia-Viloca, C. Alhambra, D.G. Truhlar and J. Gao, J. Comp. Chem., 24 (2003) 177.

102 M. Garcia-Viloca, C. Alhambra, D.G. Truhlar and J. Gao, J. Chem. Phys., 114 (2001) 9953.

103 J. Gao and D.G. Truhlar, Ann. Rev. Phys. Chem., 53 (2002) 467.

104 H.S. Johnston, Gas phase reaction rate theory, Ronald, New York, 1966.

Theory and Applications of Computational Chemistry: The First Forty Years
Edited by C. Dykstra et al.

CHAPTER 6

Computational chemistry: attempting to simulate large molecular systems

Enrico Clementi

Via Carloni 38, 22100 Como, Italy

Abstract

Computational chemistry is a very vast field dealing with atomic and molecular systems, considered at different complexity levels either as discretized quantum mechanical systems, or as statistical ensembles, amenable to Monte Carlo and Molecular Dynamic treatments, or as continuous matter fluid-dynamical distributions, modeled with Navier-Stokes equations. At the upper limits of complexity we encounter the mechanics of living matter, a most fascinating area still highly unexplored.

6.1 INTRODUCTION

A flashback from a 1960 routine: the University of Chicago, Department of Physics, Ryerson building, third floor. At about 5:00pm Prof. Robert S. Mulliken knocks at my office door, opens it, and whispers some greetings, keeping unaltered his unique and characteristic smile. The routine continues: he sits at my desk—on the extra chair for visitors—then after having emptied my over-filled ashtray, he looks at me, by that time already standing at the office large blackboard, ready to fill him in on the details of the last round of computations performed at the Wright Patterson Air Force Base in Dayton, Ohio. Well,...perhaps I should insert a few footnotes and start from the beginning for the benefit of those who have been around less than a half of a century.

The Second World War was fought not only with soldiers on the battlefront but also with the civilians, particularly scientists and engineers. The electronic computer was one among the many innovations of that period, needed in the USA to build atomic weapons, and in Germany to design missiles. After the war, however, it was soon most evident that the computer could open doors to many new dimensions, affecting any aspect of human life, including military supremacy.

References pp. 111–114

Since prehistoric times, mankind's physical limitations and handicaps have been mitigated with the invention of tools and related technologies; indeed the evolution and the history of tool-making parallels the evolution and the history of humanity. However, the traditional tools were all devised and then used to enhance the physical capabilities of man, whereas the computer constitutes a unique and different kind of tool, devised and used to enhance and complement our 'brain', our intellectual capacity.

It follows as no surprise that a few scientists, coming mainly from theoretical physics, mathematics and also a few from chemistry, did look at the computers as 'the' way to bring together the relatively new field of quantum mechanics and the prevailing hypotheses on the nature of the forces which held together the atoms in a molecule. The meeting in 1951 at Shelter Island in Long Island, USA (reported in the Proc. Nat. Acad. Sci. USA, by Parr and Crawford) was an occasion for an important discussion on the topic; among the participants I would like to mention—in addition to Parr and Crawford—Mulliken and Roothaan from the University of Chicago, Slater from MIT, Boston, Löwdin from Uppsala University, Matsen from the University of Austin, and Hirschfelder from the University of Wisconsin. The access to computers was obviously a practical bottleneck, but the formulation of a general theory on the chemical-bond and the solution of related mathematical algorithms were formidable intellectual barriers.

6.2 THE LONG PREPARATION AND THE SEEDING TIME: 1930–1960

The computation of the potential energy curve of the simplest molecules, H_2 in particular, became the obvious test case, not only because there were good spectroscopic and thermodynamic data for eventual comparison between laboratory data with theoretical results [1], but mainly because at the time it was self evident that to test a theory one should start by testing the simplest cases, to avoid the complexity of larger systems, where accidental agreement by cancellation of errors could likely occur, invalidating the test. Parenthetically, one could see that this very simple bit of good sense, a pillar of the scientific method, was later on rather ignored by DFT proponents [2].

As we all know, since the early 1930s, there were two main approaches to explain (rationalize?) molecular binding, one due to Heitler and London, HL for short [3], the second due to Herzberg, Hund and Mulliken, namely the molecular orbital approximation, MO for short [1]. For a comparison between HL and MO results on the ground state of H_2 we refer to a review on this topic [4]. In the HL approximation, the electrons in a molecule are considered as perturbed electrons of the constituting atoms, whereas in the MO approximation the electrons are shared among the nuclei of the constituting atoms, and therefore one does not have to explicitly make reference to the original atomic orbital; of course this concept becomes somewhat fuzzy for inner shell electrons and for all electrons near the molecular dissociation limit.

With evident oversimplification, one can say that the MO picture can provide a reasonably good description for electronic excitation processes, including ground state binding, less so if one wants to know the energy of the molecule as it dissociates into its component atoms, whereas the HL approach, particularly after the conceptual extension due for example to Slater, Pauling, and Eyring leading to the Valence Bond, VB, approach provided a good general language for the formation and stability of molecules [5–8]. Reasonably well approximated (and manual) computations for both methods on the hydrogen molecule ground state potential surface were performed [4], and it was clear that neither of the two approaches was quantitatively exact: more accurate computations and improved models were needed, a difficult task especially for molecules more complex than H_2. The chemistry community overall did favor the VB approach (until around 1960), partly enchanted by the towering figure of Linus Pauling, partly because the VB method was the one which allowed a description of chemical formation and reactions. The physics community, however, knowledgeable of the atomic Hartree model, was leaning toward the MO approach. It is of interest to read Slater's comment [9] 'More and more is becoming clear that the method of molecular orbital represents an approach to chemical problems which is more satisfactory than any other, and that as it becomes more and more quantitative, it can be hoped to lay a really exact foundation for chemical theory. We must remember one reservation, however: the method of molecular orbital does not lead to correct behavior at infinite internuclear distances.'

With the advent of the electronic computers in the early 1950s it was finally possible to reexamine the two main approaches, and to take a closer look at the error implicit in these formulations. Thus we witness a three-way effort: first, to establish a solid theoretical method, second, to develop all the needed mathematical algorithms for its implementation and third to be able to effectively use the new tool, the electronic computer, to obtain routinely numerical solutions. Incidentally, at the time there were relatively few quantum chemical textbooks 'computationally relevant', computers were still extremely expensive and rare; in short computational chemistry hardly existed for lack of teachers, students, hardware and software.

In the early 1950s, still a chemistry student at the University of Pavia, in Italy, I recall a lecture by Prof. C. Coulson; as Paul on the road to Damascus, I became sure that his type of approach, quantum chemistry, was the one to be followed! Thus I enrolled in a few classes given to theoretical physics students, since nothing relevant to modern molecular theories was offered in the curriculum of my chemistry department. Later, in my first postdoctoral year, 1955, at the Polytechnic of Milan, Prof. G. Natta gave me the task to explain why a newly synthesized shining polymer, polyacetylene, had a metallic-like reflectivity and to relate this property to the still experimentally unknown structure of the polymer. Well,...despite being the only chemist in Natta staff to have taken a theoretical physics class, and despite my hurried reading of Coulson's volume [7], I did not solve the assigned problem, and that fact gave me a chip on my shoulder. Incidentally, this is one of the motivations at the origin of my interest for C_2 and large carbon molecules [10,11] started in 1958–1959 on a semi-empirical level with Kenneth S. Pitzer in Berkeley, and later continued in Chicago [12,13] and in Kingston, New York [14].

Another flashback: 1957, Tallahassee, Florida State University, Department of Chemistry, Prof. M. Kasha's laboratory on a typical Saturday afternoon. Kasha and all his group are in a self-study program going through Eyring, Walter and Kimball [8], page after page, with enthusiasm and great care. The chemistry departments at the universities, particularly in the USA, were preparing the future computational chemists.

6.3 QUANTUM CHEMISTRY AND THE LABORATORY OF MOLECULAR STRUCTURE AND SPECTRA, CHICAGO, 1960

Let me return to 1960 in Chicago, a period I personally remember well. Professors Robert S. Mulliken and Clemens C. Roothaan Laboratory of Molecular Structure and Spectra, LMSS, was the Mecca of theoretical chemistry. By the time I started my research appointment in 1960, LMSS had Prof. Clemens C. Roothaan as its second faculty member, Research Associated from USA, Japan, England, India, Spain, Italy, a number of graduate students, later well-known theoreticians, and a most valid group of 'associated non-contract personnel' which in 1960 included Dr Wlodzimierz Kolos. Anticipating by decades other theoretical laboratories, LMSS was structured along three well-defined lines: computer program development, computational applications, and an experimental laboratory in high resolution ultraviolet spectroscopy. Prof. Roothaan (nicknamed C^2, 'C squared') had the full responsibility for computer program development, programming assistance and consultation for the 'computational application group', development of mathematical algorithms and new methods in theoretical chemistry. Prof. Mulliken (nicknamed RSM) had the responsibility to follow closely the 'application people' who had a task either defined by RSM or mutually agreed upon, for the more senior members of LMSS, and to supervise the spectroscopic laboratory, located in the basement of the building. The third task of Prof. Mulliken was one of public relations both with US government agencies and the international scientific community. The atmosphere at LMSS was one of true excitement, since we were fully aware of being members of an exceptional laboratory recognized worldwide; we felt part of 'the' best team in the world.

In the decade 1950–1960 much progress was made and it is fair to say that computational chemistry was conceived in that period. The paper by Roothaan in 1951 [15] proposed to obtain, for closed shell molecules, wave functions expressed by orbitals obtained with the expansion method, using analytical basis set variationally optimized with the self consistent field technique, expanding the previous proposal by Hartree for atomic systems. The original work, made available to the scientific community prior to publication as a technical report, contained a set of examples, including the benzene molecule, omitted in the final publication [15] since it was 'a trivial example', according to the author. That paper essentially marks the end of Mulliken LCAO-MO concepts, since the expansion method does not have to rely on the specification of atomic orbitals. Indeed Roothaan's work starts the era of the analytical (to distinguish it from the numerical) molecular Hartree–Fock technique. The extension of the theory and related computer programs for systems with open shell electrons [16] completed that effort, a landmark and a basic contribution to theoretical and computational chemistry.

Incidentally, we note that at that time it was common for large research centers to collect, publish and distribute their yearly scientific results either as reprints but often also as preprints in order to quickly diffuse new ideas and computational results. In particular the technical reports from LMSS had a very large and free dissemination; its distribution list included research centers for over two hundred 'domestic' locations (USA), over one hundred 'foreign' locations, particularly in England, in addition to a very extended list of US government agencies. I still treasure today a few of the red covered volumes from the early 1960s.

The Chicago group was very careful in its use of semantics, not a surprise, for those familiar with Mulliken's way of thinking, writing and expressing himself (nearly an obsession). The term 'Hartree–Fock' would be used only for sufficiently large and optimized basis sets to ensure insignificant deviation from an eventual numerical solution, the true Hartree–Fock limit. For all other cases one would simply talk of SCF-MO. Today widespread misuse of the term Hartree–Fock definitively represents a step backward, especially when, with exceedingly truncated basis sets, one magnifies the superposition error thus obtaining 'apparently good' binding energies. Once I wrote a paper on the subject, hoping to wake up some of my colleagues, a rather naive idea [17].

To obtain a Hartree–Fock or a SCF-MO solution one needs to compute, store and retrieve a list of integrals over the basis set. That was a most difficult task first tackled for diatomic and for linear molecules using the so-called Slater-type-orbital, STO basis set (however, Slater would call the STO the 'exponential' type orbital). A general solution for four center integrals for non-linear molecules was very hard to find, short of paying a horrible price for computer time, implied, for example by performing a three-dimensional numerical integration for electron 2 in the analytical potential of electron 1. I should mention that the slow and difficult task of obtaining computer solutions for molecular integrals, was preceded by the use of tables, where for given orbital exponents, or related variable, tabular entries and simple formulas with interpolation methods were available; I recall, for example the tables by Kotani from Japan, by Preuss from Germany, by Matsen and by Roothaan from the USA. The first tabulations were those for the overlap integrals, a particularly simple task also in tabular form. Later, with the advent of computers, one would essentially compute the tabular entry rather than look it up in a printed set of tables. I shall return later to other aspects of these computations.

But by the time I arrived in Chicago, Roothaan and his collaborators' effort was not only to secure standard computer programs to deal with Hartree–Fock and SCF-MO functions, but to go beyond the Hartree–Fock approximation toward a fully correlated wave function both with CI techniques and with Hylleraas-type correlation methods. Wlodzimierz Kolos' deservingly most celebrated work [18] is one among the output of this period; as usual Roothaan was selecting the hardest way to make progress. At the same time in England S.F. Boys and G.B. Cook [19] were also considering linear combinations of determinants, a direct, nearly unelegant, cumbersome, brute force but pragmatic, realistic and feasible algorithm. The notion of taking a linear combination of determinants goes back to the mid-1930s, as one can see by reading J.C. Slater's excellent volumes in atomic, molecular and solid state quantum mechanics [20–22]. Concerning S.F. Boys, I must mention his series of papers 'Electronic wave functions' started in 1950 [23].

The main stream of the chemistry community either 'de facto' ignored most of this theoretical effort, or considered it with the sympathetic eye of the wise father looking at children's play: indeed, everybody knows that chemistry is to be carried out in a laboratory, and in a laboratory only.

There is an excellent summary of the status of computational chemistry in 1959, in a report of the Boulder, Colorado, 'Conference on Molecular Quantum Mechanics',

June 1959, published in the *Reviews of Modern Physics*, Volume 32 (April 1960 issue). Even if some of the papers presented at the meeting, were not included, when already submitted to other Journals [10], this volume not only summarizes most of the field, but reports the view of a highly recognized leader in applied quantum mechanics, Prof. Coulson, from the Oxford Mathematical Institute. Coulson in effect divided 'computational chemistry' into two opposing efforts, the 'semi-empirical' one and the '*ab initio*'. Concerning the '*ab initio*', Coulson's opinion is sharp: 'I see little chances—and even less desirability—of dealing in this accurate manner with systems containing more than 20 electrons. For I cannot help recalling Hartree's remark, that if we were to print the wave function values for the ground state of the iron atom with sufficiently small interval in all the electronic coordinates, we should require a whole library to house the books in which they were printed; and that there are not enough atoms in the solar system to make the paper and ink necessary to do the same thing for the uranium atom.' For Coulson the future of computational chemistry manifestly belonged to semi-empirical methods. Clearly for the *ab initio* minority Coulson's opinion was rejected as unimaginative and even uninformed. There was an apocryphal rumor circulating in Boulder: Coulson did realize that only the American industry would be capable of manufacturing and marketing the computer hardware of the future, and England's theoretical chemistry would not keep in step with the USA, an unbearable thought!

For reasons apparent later on in this paper, I hurriedly mention that E. Wigner in 1934 [24] introduced the term 'correlation energy' to identify the error in the Hartree–Fock method, and proposed a simple functional of the electronic density to account for the error; the approach was called 'statistical method' (since it was based on the density rather than on the wave function). Wigner's ideas were used in the computation of the binding energy in metals, and later extended to atoms, particularly by Gombas and the Hungarian school [25a] and even later (1972) to molecules by myself [25b,c]. Related and rather parallel to Wigner innovation, there is Slater's brilliant (even if not fully original) exchange energy approximation [9] also expressed as a density functional (obtained from the free electron gas density).

I do realize that these introductory pages are very incomplete. To present a proper analysis of quantum chemistry in the period 1930–1960 would require a special volume, which, if written by a critical and competent mind could show that the progress from 1960 to 2000 would have been more notable and rapid if in 1960 we would have had a full grasp of the many accomplishments of the time and a mind free from prejudices.

Let me mention, and only as an example among many, the work of R. Daudel and his school concerning aspects of chemical reactivity [26]. It is also of interest to recall that relationships between electronic density and molecular properties, for example and biological properties were the subject of publications by Otto Schmidt in Germany, since the late 1930s. Mademoiselle Alberte Bucher, from the Daudel group, later continued in this direction; as Madame Alberte Pullman, she also summarized the early ideas in a first volume [27], the basis of a subsequent and more general volume [28] and of a very vast literature. It is understandable, but nevertheless unfortunate, that at the time it was assumed that far frontiers for quantum chemistry, like biological sciences, could be tackled with rough approximations and rudimentary methods; there was no understanding that the enormous complexity of biological systems requires the maximum of rigor and realism to avoid explanation based on results plagued with error cancellations, neither suspected and even less understood. Unfortunately this negative trend, from time to time, reappears even today. A fully phenomenological approach can be superior to a highly approximated version of a rigorous and exact method.

The organic chemists also considered quantum chemistry and from an early time; we recall in this context the well-accepted and popular volume by Streitwieser [29]. In the inorganic chemistry area we cannot avoid recalling the crystal field theory advances, and refer for example to C. Ballhausen [30]. For additional references on this period we refer also to the volumes published in memory of Per-Olov Löwdin [31]. I take this occasion to stress the exceptional educational effort by Per-Olov in forming today's computational chemists; his Florida meetings and his summer school in Sweden had a truly catalytic effect and introduced chemistry to the necessary level of mathematics rigor. As stressed above, there should be many and many more references, omitted only because I must cover the period 1960–2000, which 'all in all' mainly harvested what was seeded from 1930 to 1960.

6.4 MY HARTREE–FOCK, MC-SCF AND DENSITY FUNCTIONAL PERIOD: THE 1960 DECADE

In Chicago, B. Ransil computations [32] for the ground state of hydrogen fluoride created a puzzle, when it appeared that the MO picture seemingly suggested a charge transfer opposite to the expected one. Was the Hartree–Fock basically untrustworthy The fluorine atom Slater Type Orbital, STO, basis set was minimal, but the hydrogen atom basis was extended with a 2p polarization function attempting to improve the description; at the time the basis set superposition, BSS, error was still an unknown disease and therefore the puzzle. RSM asked me to look into the problem and I decided to push the MO-SCF computation as much as feasible, namely, to the limits (a) of the computer code, written by D. McLean, M. Yoshimine, and A. Weiss, under the close supervision of 'C^2', (b) of the computer time availability in Dayton, Ohio, and (c) of my enthusiasm and physical endurance.

All orbital exponents were optimized, assuming a parabolic behavior (but often exploring a larger range of values) for a double-zeta type STO basis set with $2p_\sigma$ and $2p_\pi$ on hydrogen and $3d_\pi$ polarization functions on fluorine. Notice the distinction in the orbital exponents between $2p_\sigma$ and $2p_\pi$, later on generally neglected. The importance of polarization functions was a well-known fact, documented in Linnett's public notes, written during a sabbatical in Berkeley, California in the late 1950s. The ground state potential energy curve for the HF molecule was computed for six internuclear distances, and for each point and each choice of orbital exponents the dipole moment was also computed, attempting to learn more on the mysterious charge transfer. An energy minimum was obtained with -100.05604 hartrees at a distance of 1.7326 bohr with a binding energy of 0.147 hartrees and with computed dipole moment of 1.984 debyes; the HF total energy limit was estimated be about 0.004 hartrees deeper. The corresponding spectroscopic laboratory data are: internuclear separation of 1.74 bohr, dipole moment of 1.74 debyes, and binding of 0.2235 hartrees. In addition, a few excited states, the ionization potential and an analysis of the correlation error were presented, stressing for the first time the importance to separate the total molecular correlation into atomic and molecular components.

The hydrogen fluoride riddle was solved and the era of molecular Hartree–Fock computations was open. The need to have adequately large, polarized and balanced basis sets, is tantamount to require near to Hartree–Fock basis sets. The work was quickly written, submitted and printed [33]; it was the first Hartree–Fock type computation for a 'many electron' molecule with the clear message not to waste computer time and human effort with minimal basis sets. Well,...this was a rather strong message! Indeed, minimal basis sets computations littered (embellished?) the computational chemistry literature for decades to follow, with the more or less acceptable excuse that computer time can be too expensive or not available.

Today we would assign the task to obtain an Hartree–Fock type function for the HF molecule as a weekend home-work to computational chemistry students. In the 1960s it took few hours per computation on the Dayton 'supercomputer', the Scientific Univac 1103 A. There were 16 orbital exponents to be optimized, thus a minimum of $16 \times 4 = 64$ computations, yielding a total of a few hundred computations, when multiplied by the number of internuclear distances. Actually the computations were more numerous, since I made a detailed analysis of the energy and dipole moment variations related to the orbital exponent optimization. In practice, it implied many trips from Chicago to Dayton, Ohio for few months, hundreds and hundreds of computer output pages to be analyzed for the input and output details and for energetic information (at each SCF iteration). The orbital exponent optimization was carried out explicitly for the molecule in consideration, rather than accepting pre-fabricated and often over truncated basis sets, a typical situation in the 1960–2000 period and today. Shortly after my work a very similar computation by R.K. Nesbet appeared in Journal of Chemical Physics, with a nearly insignificantly worse energy but an improved dipole moment and, somewhat later, there was a paper by M. Karplus attempting to discriminate which of the two (nearly equivalent) computations was the 'best one'. These details are reported to provide a feeling for the situation of computational chemistry in 1960. Incidentally, today we know that both computations fell short of a near to Hartree–Fock energy; for example G. Corongiu has obtained for the HF molecule an energy at its minimum of -100.071137 hartrees [34], obtained with a Gaussian basis (10,5,4/6,5,4) for the hydrogen and (18,13,5,4/12,6,5,4) for the fluorine atom. Perhaps the relevant observation is not which of the above quoted computations (Clementi *versus* Nesbet) was the best one, but how far are these computations from the correct laboratory data!

Wigner [24] and afterward with more rigor Löwdin [35] defined the correlation energy as the difference between the exact non-relativistic energy and the Hartree–Fock energy (leaving most of us in suspense for a more general and accurate definition that recognizes overlapping areas between relativistic effects and correlation energy). In the 1960s the general idea (mainly a prevailing guess) was that 'grosso modo' there is a contribution to the correlation energy between one or one and one half eV per electron pair; this notion was badly in need of quantitative clarification, since clearly we are talking of a large fraction of the molecular binding energy. For example, for the HF molecule, the molecular correlation energy is about 2 eV [38], thus larger than generally guessed. This clarification was one of my first tasks in California; there is little point in performing Hartree–Fock computations unless the magnitude of the implied error is known and well understood.

Having accepted in the summer of 1961 a staff position with the IBM Research Laboratory in San Jose, California (after much advice and discussions with RSM, since I was torn between an offer from Prof. Herzberg, in Ottawa, a second one by the University of Genoa, in Italy, at a center interested in neuron chemical physics (Mike Kasha's influence) and a third one by IBM) and being at liberty to decide on my research line, as long as it would deal with the use of computers in science in general, chemistry included, I felt finally free and ready to start a long range and very systematic research plan to explain 'living matter mechanisms' *via* the law of physics, within the logical rigor of mathematics and expressed in the form of unambiguous and reproducible computer outputs. Verified laboratory experiments would decide on the correctness of the theoretical frame, and *vice versa*, theory and computer simulations would expand the limits of the laboratory. However, having only a feeling, rather than a knowledge, of what is a correct definition of 'living matter', I generally use the surrogate and simpler expression 'large molecular systems', when thinking of 'living matter'. Clearly, the obvious operational way to start to tackle 'large molecule computational chemistry' was to systematically perform simulations in the Hartree–Fock approximation and to quantitatively calibrate the magnitude of the correlation energy error. *Gradus ad Parnassum* [36]: I shall start with atoms, then pass to simple and later to more complex and large molecular systems. It is a zoom-out methodology, and the resolution of the simulated description will change each time it is felt to have reached the upper complexity limit for a given methodology. From the quantum chemistry of atoms and molecules, to the statistical description of N, P, T or N, V, T ensembles, to the fluid dynamical description of continuous matter, to the organization of continuous matter into specific self generating macro systems, cells in particular. It was 1960, the world was still recovering from the horrors of the war; there had to be a better, just and without wars world and a rational future!

My first selected task was to set a new standard in atomic computations—not a few Hartree–Fock functions for published paper but dozens and dozens for atoms and ions, positive and negative, ground and excited states, to make it clear that the time had come for a new era in simulations, the Computational Chemistry Era. The functions were supposed to help as the eventual starting point in the optimization of molecular basis sets, assuming, incorrectly, that the four center general geometry STO two electron integrals bottleneck would soon be broken. At the same time I wanted to put IBM San Jose on the world map of computational chemistry. Several hundred wave functions were published quickly [37–41], a very hard work at that time when one had to punch input cards for each orbital exponent to be optimized, and repeat the process thousands of times! The work was collected in a volume the 'Table of Atomic Functions' [42]. Few years later, I returned to the atomic work, this time with Carla Roetti, extending the original work limited to atoms and ions from $Z = 2$ to $Z = 36$ to atoms and ions from $Z = 2$ to $Z = 54$ [43]. The tables [42,43] have been requested by hundreds of scientists. Incidentally, the most heavy users of these functions at first were the solid state physicists; the chemists were still far behind.

The next obvious step was to obtain reliable estimate of the atomic correlation energy, but to do this I was in need of computing a reasonable accuracy the relativistic correction at least for the 'non-relativistic atoms' from $Z = 1$ to $Z = 36$, the back bone of the chemical world; the task was accomplished quickly [44]. At this point, using the laboratory available atomic ionization potentials (or suitable extrapolations obtained using the atomic data), it was simple to estimate rather accurately (within 0.001 hartrees) the atomic correlation correction, but more important, it was possible to learn of the notable regularity of the correction [45–47] and from this to extrapolate from atoms to simple molecules [48]. The correlation energy data were revised a few times but it took about 20 years before someone would produce a relatively final set of data [49].

In the mean time IBM San Jose, LMSS in Chicago, and few other centers, were busy in computing systematically at the Hartree–Fock level, mostly diatomic or small molecules. Thus the Hartree–Fock Era was advancing. The main obstacle—lack of computer time and therefore a justification for minimal basis set computations and for rough semi-empirical work in polyatomic molecules. A related problem—still no progress in the generation of a general four-center electron–electron integral package with STO.

The atomic work brought up as very evident the atomic Z-effect, namely the near degeneracy effect first pointed out by Hartree [50], namely the non-dynamical correlation energy discussed by Sinanoglu [51]. Incidentally, at that time Sinanoglu was proposing to compute with reasonable accuracy the correlation energy without the need to compute a Hartree–Fock energy (or density); it is interesting to recall that this idea was praised by leading quantum chemists [52].

I recall that the main stream computational chemistry was bent to fully solve the correlation problem with a single 'technology'. At the time various kinds of configuration interaction algorithms were considered, including configurations with non-orthogonal orbitals. P.-O. Löwdin had written the equations needed to compute the interaction energy for determinants containing non-orthogonal orbital. That work is essential for the 'different orbital for different spins' methods, a concept very dear to Löwdin and, in my opinion, still insufficiently explored in computational chemistry.

Contrary to main stream computational chemistry, I felt that it would be worthy to learn if one could substantially, but systematically, reduce the correlation energy error [53] with some simple algorithm, being clear of the difficulties one would encounter in a full elimination of the correlation error.

I confess that I had serious misgivings about the practicality in performing configuration interaction computations, if the goal was to compute 'large molecules'. It all started with a work, unpublished, on LiO and MgO, using Nesbet's CI code, shortly after his arrival in San Jose. For LiO I managed to select a few different subsets each of 100 configurations (adopting Boys' criteria) and to diagonalise the secular equation; at the time, handling 100×100 matrices was a very serious task. To my knowledge that was the largest CI computation ever attempted, but the energy gain was totally disappointing, around 1.5 eV. for LiO! For MgO I never managed to complete one run after a few days of computer time! After much thinking and mental extrapolations on eventual computer advances even in the far future, I came to the conclusion that CI would never be the method for large molecules. Of course I, like LMSS people, had still much hope for the emerging Multi-Configuration-SCF techniques, but I felt that it was not a task for me, since for large molecules it would take computers of enormous power, perhaps available only after my lifetime. In addition the task was clearly more adapted to someone very skilled and much interested in programming, and willing to spend many years on that single task. Today, considering for example the work by Shavitt or by Schaefer and by others excellent quantum chemists dealing with CI type techniques, I am somewhat less confident on the validity of my early extrapolation on the future of the CI method for large molecules.

It was this type of consideration that brought me in 1963 and 1964 to the coulomb hole method [55]. Merely the simple sight of the electron–electron interaction integral formula, where for a given space location of electron 1 it is possible for electron 2 to occupy the same space position, tells you that something must be done to correct this unphysical behavior. If in addition one has read the Wigner work [24], then one would immediately 'carve' around electron 1 a variable size region, an approximated Coulomb hole, impermeable to electron 2. Clearly the carving has to be done when one computes the average field, thus it is a part of the SCF process. The equation for the coulomb integrals with a coulomb hole were easily derived, and the corresponding code was added to an atomic SCF program. The size of the carving was made proportional—*via* a rather complicated expression—to the product density of the STO for the electron 1 with the STO for electron 2, with an empirical proportionality constant. An atomic computer code was made available to any interested user. Soon thereafter, I extended the Coulomb hole correction method by computing atoms with a Gombas-type density functional [25a]; these density functional computations, were later extended to molecular density functional computations for CH_4 [25b] and C_2H_2, C_2H_4, and C_2H_6 [25c], the 'first density functional molecular computation in quantum chemistry'. Unknown to most US computational chemists, since the early 1940s, the Wigner correction was carefully considered in Europe, particularly by collaborators of Wigner, like Gombas and Gaspar in Hungary. It was during a Löwdin meeting in Hungary that I had the pleasure to meet and discuss this with the two Hungarian scientists.

The importance of team work learned particularly at Chicago LMSS, and the need to create a laboratory with critical mass, lead in fast sequence to my promotion from staff member, to group leader, to department manager (the name of the department was 'Large-Scale Scientific Computations'). From this expansion the concomitant need to hire promising scientists; among them A.D. McLean, M. Yoshimine, P. Bagus and later B. Liu, all from LMSS, R.K. Nesbet from J.C. Slater group, W.A. Lester, Jr. from the Bernstein-Hirschfelder group in Madison, Wisconsin, and even later J. Barker and D. Henderson, at that time from the Waterloo University in Canada. A small group in fluid dynamics was strengthened with the arrival of J. Fromm, from the famous T2 group in Los Alamos. I did consider all of them as colleagues, and, thanks to the liberal support from IBM, I provided as much computer time as they requested, travels to meetings, and a good number of postdoctoral collaborators, chosen from all over the world.

It was not too easy to induce IBM management to start a postdoctoral program at the IBM Research division, but I was determined, since I considered this addition as the only avenue to ensure that my department, and therefore also the IBM Research

Division, could be on equal footing with the best USA Universities. With the 'explicit support and complicity' of J.C. Slater, L.H. Thomas, C.A. Coulson, M. Kotani, R.S. Mulliken, M. Simonetta, R. Daudel, and P.-O. Löwdin and a bit of 'savoir faire' and plenty of good luck, I managed by 1965 to get the program approved and financed. By now hundreds and hundreds of scientists have participated in the IBM Research postdoctoral program, open first in my department at IBM San Jose (later in the Almaden location), then in Yorktown Heights, New York, and in the Zurich laboratory, namely the three IBM Research Division locations. The postdoctoral program (or visiting scientist, to use IBM proper terminology) provided IBM with a stream of new blood, with ideas, with a network of relations, useful also for marketing, and at the same time opened to computational chemistry many Universities, particularly where computer time was still scarce or not available; note that for many years our scientific computer codes were given away for free to the visitors or any requesting scientist. This two-way stream was the case—to mention only a very few examples from the early period—of B. Roos, P. Siegbahn, J. Almlöf, and U. Walgreen from Sweden, A. Veillard, E. Kochanski, and C. Salez from France, G. Dierksen and W. Kramer from the Max Plank in Garching, and W. von Niessen, H. Kistenmaker and W. Mayer also from Germany, H. Popkie from Canada and G.C. Lie from Chicago. IBM management did overall fully appreciate these initiatives and in 1968 honored me with the nomination to IBM Fellow, thus I was free to do any research I wanted, at any IBM location, and with a good research budget (I decided to remain in San Jose, to continue my work, but I asked to be relieved of managerial duties).

The first assault to atomic systems was therefore successfully carried out, but the progress in the molecular field was in my view much too slow; McLean and Yoshimine were once more rewriting the linear molecule computer package, somewhat unable to deal with the very difficult task of creating a general four center integral program; R.K. Nesbet was getting more and more interested in scattering, thus I decided to take on me the burden to produce a general molecular computer code. Two additional motivations: firstly, the MIT group was working on a general molecular code, later known as POLYATOM, and secondly, I had the pleasure to host, as a visiting scientist, S. Huzinaga previously met at LMSS in Chicago. With Sigeru's help I decided that a Gaussian basis set package could be somewhat faster than a corresponding STO package, despite the 3-fold (or likely more) increase in the basis set size. In addition the integral equations, available from MIT were carefully rechecked by Huzinaga, lastly I had hired from the IBM San Jose site a programmer, Mr. D. Davis, willing to experiment the scientific computing field. After 1 year, by December 1965, the documentation manual of IBMOL (an obvious abbreviation for 'IBM molecular package'), with table of test molecules and timing was ready to be used and both manual and the corresponding code were distributed free of charge to quantum chemists [56]. The code was open ended, limited to closed-shell molecules, the basis set could be either primitive or contracted, up to and including *f* functions [57]. I spent the Christmas holidays in a computer room, to obtain the barrier to internal rotation in ethane [58]. By June 1966 a second version of the code was available, much improved, for closed and open shells, with a rudimental inclusion of point group symmetry features [59] with Alain Veillard as first author; I take occasion of this work to express all my admiration to Alain, a most competent, efficient and constructive computational chemist from France. Subsequently, several versions of IBMOL were written.

The Hartree–Fock model was known to fail in the case of near degeneracy, as pointed out by Hartree [50], and the correct approach was to use a linear combination of the few nearly degenerate configurations, optimized self-consistently; this is the well known 'generalized Hartree–Fock method' or the Multi-Configuration-SCF method. The four electron Be ground state and especially the corresponding iso-electronic series of ions, exhibit very clearly the degeneracy problem, with a correlation energy sharply increasing with the atomic number Z.

With Veillard we decided to extend the SCF atomic code and to include more than one determinant, obtaining variationally at the same time both the weight of the determinants and the orbital expansion coefficients [54]. This was our beginning in the Multi-Configuration-SCF work. It was natural to think to adapt the IBMOL package as to include Multi-Configuration. There was no reason not to consider all the configurations which can be generated from the ground state one, as it was done in the Complete CI methodology, thus we proposed the Complete Multi-Configuration Self Consistent Field Theory [60]. In San Jose, A. Veillard met G. Dierksen, and this provided the base of the Strasbourg series of colloquia to implement and extend the Complete Multi-Configuration Self-Consistent-Field Theory; several years later B. Roos also entered the MC-SCF arena developing the well known CASSCF set of programs.

In the mean time McLean and Yoshimine had completely rewritten the linear molecule code 'McL-Yosh Linear Molecular Program. 1.' available also from QCPE. At the end of 1967, 'The Tables of Linear Molecule Wave Functions' were available [61], with the introductory comment that these were a prelude to 'forthcoming tables at the Multi-Configuration-SCF level', following Das and Wahl [62], Clementi and Veillard [54,60], and J. Hinze [63] (at LMSS) works. At that time the San Jose collaboration on MC-SCF with B. Roos group had not yet started; the idea to invest on the MC-SCF path was widely accepted in quantum chemistry.

1967, back in Chicago for a 1 year visiting professorial lecturer (during my San Jose period—1961 to 1974—Mulliken and Roothaan were our very frequent visitors), this time with a general geometry molecular code and ample machine time, even if in San Jose and at Yorktown Heights, and not in Chicago. Mulliken was eager to see the computation of a relatively large molecule exhibiting charge transfer, but had no idea of the required computer time. The choice to study either NH_3 or $N(CH_3)_3$ at two different distances from HCl was influenced by recent studies by RSM in molecular complexes, carried out not in the MO approximation but in the Valence Bond approximation, since RSM felt that VB was superior to MO in describing chemical processes. I rather carefully estimated the required computer time for a full potential with HCl approaching NH_3, the target, relaxing the molecular geometry for each choice of the Carbon-Chlorine distance. I immediately dismissed the need to use VB methodology, after all I was addressing two interacting closed shell systems, thus an overall single closed shell system. For computational reason I also dismissed to consider $N(CH_3)_3$. It was an exciting and hard project, very near to the limit of the computational resources at my disposal. Physically it was a very demanding year, each Saturday and Sunday in the computer room in Yorktown, New York (joined by A. Veillard, at the time visiting Moskowitz in New York and busy at extending IBMOL in the MC-SCF direction), and once a month in San Jose, to supervise my department progress. The rest of the week in Chicago to analyze the outputs and prepare the inputs, to briefly talk with LMSS people, including Prof. Kolos, who would give me his inputs for the H_2 molecule, also to be computed at IBM Yorktown. I badly needed some moment to relax a bit, and for a short period, once a week, I would spend an afternoon oil painting in a painter's studio in Evanston [64].

To map the potential surface of NH_3 reacting with HCl [65,66] was simply a thrill! Before I always did associate these displaces to VB computations, now it was a MO one, it was an all electron computer simulation with a robust method and a double zeta type plus polarization basis set. The scientific community was also excited, and the work eventually made its way also to the *New York Times* daily newspaper [67] and to *Time* magazine [68]. The work was baptized by the French community 'the supermolecule approach' a term I never did much appreciate. It was an important moment for computational chemistry and a clear cut victory for the Hartree–Fock method. In this period in San Jose, as a consequence of my invitation, arrived Jean-Marie and Marie-Claude André; Jean-Marie wanted to continue his pioneering work on polymer-MO, initiated in Belgium in Leroy group. I advised him to 'go *ab initio*', after all that was the right tide, and I had the proper facilities. He accepted and in no time started his well-recognized research plan in polymer quantum chemistry, the base for his well-known Namur group, while Marie-Claude was working with me at the first all-electron *ab initio* computation on molecules of biological interest in molecular genetics. I confess I was reacting to the trend prevailing at the time that insisted that for molecular biology one could be computationally sloppy! After all that to me was 'a kind of personal affront', given my intense interest in 'large molecular systems'. Thus, not surprisingly, we started with computations on the four DNA bases and on a base pair, to learn more on the base pair hydrogen bonds [69,70]. Incidentally, many more molecules were considered in that period and the series of papers 'Study of the Electronic Structure of Molecules' grew from the first paper [71] to paper number 23 [72].

The time was ripe to look more into the interaction between molecules, and this is the topic of the series of papers 'Study of the Structure of Molecular Complexes' from a first paper [73] dealing with a water molecule interacting with a lithium ion, to the paper number 14 [74], all concerned with water–water, water interacting with positive and negative ions and water interacting with molecules of eventual biological interest. The one or two readers of this paper are referred to my publication list printed in the Int. J. Quantum Chem. in a 1992 volume, kindly dedicated to me [75].

By the beginning of the 1970s, I could rather clearly foresee what could be obtained from computational chemistry, even if it was necessary to wait for 'operational details', like the level of efficiency of Multi-Configuration and geometry optimization codes, the rate of increase in the interest from Research and Development centers to daily use computational techniques and, most of all, the willingness of the electronic industry to market computers of drastically increased performance and lower price; the technology was available, but the market was not too clearly defined.

In 1972 I wrote a paper [76] to summarize my overall conclusion, a forecast and blue print for computational quantum chemistry: use a combination of *ab initio* computations, when feasible, and density functional SCF techniques, by that time already successfully tested [25b,c], for example with Wigner type functionals in the SCF contest, for larger systems, and focus on simplifications to reduce the computational time dependency on the size of basis set, by approximations of the integrals of small magnitude. In that paper I pointed out that the computer time dependence was most clearly not a fourth power of the basis set number, but somewhat less than a square power, and this limit could be decreased further. Considering the DFT explosion after 1985, I conclude I was correct on the first point. Looking at the work for example championed by Almlöf, Scuseria, and others, it is evident I was correct also on the second point. However, these horizons were not sufficiently evident to many colleagues in the field. For example at that time J. Pople was also presenting his forecast for computational chemistry dismissing *ab initio* chemistry in favor of semi-empirical approximations. Beside the IBM San Jose center, LMSS, Löwdin's groups in Uppsala, Sweden, and in Gainesville, Florida, Slater's group at MIT and the new 'affiliations' related to these laboratories, and the 'traditional' theoretical centers, particularly in Europe and Japan, there were relatively few other centers for *ab initio* computational chemistry; the prevailing theoretical and computational approach was semi-empirical.

Let me open a parenthesis on computer development, the life blood of computational sciences. The computer industry was moving fast, but still very much far from the needs of computational demanding jobs. The computer era started de facto in 1946 with the ENIAC computer, followed by the IBM-650, and other first generation computers (ferrite memory, vacuum tubes, no operating systems, no compilers, etc.). In the second computer generation (1955–1965) we find, for example the IBM-7090, IBM-7094, Univac, and the CDC computers (transistors, magnetized core memory, tapes, punched cards, Fortran, etc.). The situation changed with the third computer generation, 1965–1975 (integrated circuits, channels, magnetic disks, system software, virtual memory, time-sharing), typified, for example by the computers IBM System 360, but particularly by the appearance of the mini-computers, especially the VAX systems, which were entering in many chemistry departments, starting the mass-computational chemistry approach we know today. This was also the time for early hardware experimentations with special purpose computers with parallel architecture and of my interest in this new technique.

6.5 FROM SCHRÖDINGER TO NEWTON; MY SECOND SIMULATION PERIOD

In any mechanics, there is an implicit complexity limit which is more efficiently dealt with by renormalizing the objects of motion, and thus neglecting explicit account of constant background interactions. Thus in dealing in quantum mechanics with the electronic structure of atoms we do not have to explicitly deal with the details of subnuclear particle interactions, which are normalized to a single charged, somewhat spherical body of a given average radius, or, more simply, to a point charge with a given mass. Equivalently, if we consider an ensemble of a given number of atoms and/or molecules, at a given temperature and pressure (or volume) we do not have to explicitly deal with the detailed nuclear–electron and electron–electron interactions within the atoms and/or molecules constituting the ensemble. Finally, in considering the mechanics of even larger systems we assume a continuous distribution of matter and the renormalized particle in fluid dynamics is the control volume element with conserved mass and angular momentum.

In moving toward large molecular systems, it was for me the time to advance from computational quantum chemistry to Monte Carlo and Molecular Dynamic simulations, and to make use of the molecular interactions obtained from quantum chemical computations during the last few years to define force fields. From here the first work on liquid water [77]. I was 'organizationally ready' with the hiring of excellent scientists as J. Barker and D. Henderson, and of a fluid dynamicist, J. Fromm (kept in reserve for the next and forthcoming 'renormalization' effort). To ensure a positive transition, I also invited R. O. Watts, from Canberra, Australia. One could wonder why it was needed to increase the San Jose team, rather than move into the new area with the existing collaborators. The simple answer is that most scientists like to stay in a 'given well-defined problem area' for a life time, especially when recognized as area leaders. My interest was in 'large molecular systems', but that implied a succession of 'areas'.

The use of Wigner type correlation correction to Hartree–Fock energies [78] and/or the inclusion of dispersion forces [79] and/or the use of CI energies [80] to define different potentials in Monte Carlo simulations of liquid water, underscores the problem on the reliability of *ab initio* potentials for force fields. Note that at the time the force fields were obtained only semi-empirically, but I was championing the *ab initio* banner.

Soon the excitement to see for the first time graphical representations of computed solvation shells for solvated ions from Monte Carlo simulations [81]. The next step was to go to even more complex systems, like enzymes, proteins, and particularly nucleic acids, A-DNA [82], B-DNA [83], without and with counterions [84] and in solution. The quantum biology community was taken by surprise, but soon accepted the new path as a new but necessary computational standard. I was proud to have forcefully recalled that the correct dictionary of quantum biology must contain terms like temperature, volume and free energy; eventually, I was elected president of the International Society of Quantum Biology.

Personally, on my private side, this period was 'a bit confused and even turbulent'. In 1972 Hildegard Cornelius, my first wife, and I decided to divorce by mutual consent. In an ensuing possibly somewhat not too lucid state of mind, this is my current opinion, I left in 1974 IBM San Jose to accept the offer to build a computational department in Novara, Italy, at the Donegani research center, with the chemical company Montedison. It turns out that, unknown to me [85], the company had serious financial and political problems; indeed, by 1977 Montedison essentially disbanded much of the research effort initiated only a few years before. However, thanks to my old director of the San Jose laboratory, Dr. Art Anderson, I was given the opportunity to return to IBM, this time in Poughkeepsie, and later in Kingston, in the state of New York. Dr. Anderson at the time, after having been promoted to vice director of the IBM Research Division, and later site manager of San Jose plant, was the vice-president in charge of all IBM manufacturing divisions.

One of the pleasant memories from my Novara period is the collaboration with Octavio Novaro from the department of Physics at UNAM, in Mexico city. His notable experience in industrial catalysis and his competence in quantum mechanics and quantum chemistry coupled to a marvelous human character made him and his collaborators unique partners to study the mechanism of the Ziegler-Natta reaction. That simulation [86] marks the beginning of computer simulations in chemical catalysis, as he himself has later recalled [87]; about 10 years later, once more computational chemists considered catalysis as an area of computational interest [88]. In the Novara time I renew the contacts with Pavia, and in one occasion I was awarded the gold medal 'Teresiana' (the Empress Maria Teresa did foster the University system in North Italy). In Novara there were other visitors for example W. Kolos, J.M. André, G.C. Lie from IBM San Jose, J. Fernandez Rico, from Madrid, B. Jonssonn from Lund, Sweden and a number of Italian young collaborators entering computational chemistry for the first time.

Novara also marks the beginning of my collaboration with G. Corongiu [89] started with the writing of her doctoral dissertation, a joint study between the quantum chemistry group in Pisa and my department in Novara ('Dipartimento di calcolo chimico'). Since then, her collaboration has been continuous and very close as documented, for example in the over 100 papers listed in Ref. [73].

6.6 STATISTICAL AND FLUID DYNAMIC SIMULATIONS, AND ALSO COMPUTERS HARDWARE DEVELOPMENT IN THE HUDSON VALLEY

The graphics for the water molecules solvating DNA [82–84] was hand drawn; I and Dr. Corongiu were starting a new research group in IBM Poughkeepsie, in a manufacturing environment with no scientific graphics facilities. Dr. Corongiu would point the compass—on a large millimeter paper—at the hydrogen or oxygen atom computer coordinates, which I would read from the computer output and then she would draw the atom circumference; an extenuating effort, done simply because in a very new location with no scientific computation experience and related facilities one had to adapt himself in the expectation the situation would soon improve, as indeed it did.

Let me open a second parenthesis on computer development. The fourth computer generation, 1975–1990, was characterized by very large system integration, VLSI, vector processing, large memories up to Giga bytes, and a broad spectrum of products from high speed supercomputers to PC, workstations, UNIX became popular, CRAY, IBM-3090, Fujitsu, Hitachi and NEC and also the beginning of parallelism. Prices came down, performance and easy use increased and communication became important. Computational chemistry became a must for more and more chemists, even if the computer users had less and less awareness of the computational details of computer programs, and hardly understood that the computed answer could be incorrect, because of limitations of the selected method. In this computer generation and even more in the following years, internet, communications, commercial computer programs, computer servers, personal computers, desktop, graphics, Window, and Linux were common words, memory and disk space seemed unlimited, price/performance improved yearly, but faith in the computer replaced knowledge of the instrument and its software. Computational chemistry was becoming a part of the global economy.

Back to Poughkeepsie in 1978. The computer time was available in large amount, assuming one was capable and willing to use the leftover computer time in a ten thousand IBM employee site, where the electronic components were assembled for most of the IBM mainframe computers worldwide. At the beginning, during day time the computer was mainly for debugging and testing, in night time and especially during weekends the only problem for us was to physically endure in the enormous computer rooms of the IBM plant, keeping fully busy two or, at times, four dedicated computers from evening to early morning. Corongiu quickly become an expert in mounting disks—with our dedicated operating system—and tapes and in supervising the night crews of computer operators. The programming, input preparation, and all the standard scientific activities were left for the day time.

Molecular dynamics and its extension to fluid dynamics were the two main computational efforts, but a new version of IBMOL and of other codes had to be written once more. The IBM postdoctoral program was eventually extended also to Poughkeepsie site and later to the Kingston site, with 'full industrial IBM efficiency', namely, I had to prepare a yearly detailed research plan with inclusion of both permanent and temporary (the postdoctorals and the visitors) personnel, and since IBM was in a growth period, the expected yearly manpower increase! My work was supposed to contribute to IBM understanding of the future needs in the scientific market for computer hardware and software; this did imply for me close contacts with IBM Headquarters in White Plains and in Armonk, with Dr. Anderson and his staff, and a very active participation for me and for all my collaborators to many IBM scientific meetings, in USA and abroad, where IBM would announce and market his products.

In 1981 at a dinner for my birthday, Dr. Weeler, my direct manager, presented me his birthday gift, an IBM 4341, a small but flexible mainframe computer which could be linked channel to channel to other mainframe computers. We had grown to a good size group with permanent members and visitors, carrying out essentially the same research life as in San Jose, but in a different environment, that made us 'white flies'. Contacts with the San Jose group were reestablished, but the interest were notably different, they still working at MC-SCF and relatively small systems, with B. Liu as department manager, we moving to more complex systems both in computational chemistry and in computer hardware.

Time to time I am invited to present a seminar in locations of my previous collaborators, some of them by now internationally well known; it is a most pleasant feeling to renew contacts with my 'extended family'. The highly successful meeting in Gyeongju, Korea, 2004, organized by Prof. Kwang S. Kim was one of these occasions. Others were with Andrés group in Namur, where the University of Notre Dame de la Paix awarded me with an 'Honoris causa' degree or celebrated my 65th anniversary.

The relativistic correction was for a long time, since early 1960s [44] high on my list. An atomic relativistic code was implemented [90], and a molecular relativistic code was started, and eventually ultimated [91] at the L. Pasteur University, in Strasbourg, France, my last (last?) working place, following an invitation from Jean-Marie Lehn.

Computer graphics was also an important topic, essential to visualize the complexity of large molecular systems performed at the molecular dynamic and fluid dynamics level. Water as a liquid [92] and as a solvent or as interstitial medium continued to be investigated, with Corongiu more and more behind this effort and with excellent and highly motivated visitors dealing with molecular complex systems like proflavine-DNA or the gramicidin channel [93]. A rather strong effort was mounted to obtain Hyllaraas CI functions for very small systems, using Gaussian type functions. That, in turn, required the derivation and subsequent coding of the complicated 4 electron integrals, tested for several states of H_2 and H_3^+ [94]. Artificial intelligence was experimented to solve conformational problems and a collaboration in artificial intelligence was established with a specialized center in Finland. The preparation of *ab initio* derived force field for proteins was managed by Corongiu, who was most validly assisted by a young scientist from Japan, Dr. Misako Aida [95]. Fluid dynamics was carried out in the traditional way, for example to study tidal behavior and atmospheric pollution, but the specific interest was to link micro systems to macro systems. For simulating very large systems we also wrote and use a cellular automata program. A Benard cell simulation was a first step toward simulations with a cellular type organization [96]. In this contest it is a pleasure to recall the collaboration with I. Prigogine and his group from Bruxelles, in particular the visits by M. Mareshal, E. Kestenmont, F. Baras, and M. Mansour; I had previously contacts with I. Prigogine, during a few month stay in Bruxelles as a Chair Francqui recipient. Talking of visiting quantum chemists, let me mention once more the Andrés, D. Vercauteren, and L. Leherte from Namur, working mostly on zeolite chemistry [97] and 'newcomers' for example V. Carravetta from Pisa, R. Buenker from Wupperthal, J. Ball from St. Andrews and many more in the software area. Again, W. Kolos did visit us for a short time and so did C.C. Roothaan. M. Dupuis joined the Kingston staff, eventually supported by a valid group of postdoctorals and so did G.C. Lie.

It is rather difficult to give a correct account of the many directions of the Poughkeepsie-Kingston effort; a complete, long and exhausting documentation is given by the three large volumes of MOTECC, below mentioned, but a rather short account is available in Phil. Trans. R. Soc. Lond. [98]. An overall account is reported in section 6.8.

The computer industry in the early 1980 was in ferment; the vector machines were too expensive and did not fulfill the user needs, particularly the fast growing number of small and medium users, the new market. Parallelism was a possible new trend, started mainly in non-traditional industrial hardware centers, particularly at Universities. The development of a parallel computer in my department in those Hudson valley years was prompted by the feeling that I had obligations to IBM. In addition, an IBM strong in the scientific area was directly beneficial to our work. Finally, several non-IBM computational chemistry groups had hardware of definitively higher performance and I was determined to have equivalent facilities. To design, assemble, and test a parallel high performance computer and its system software it was a nearly impossible challenge, that demanded a large attention from early 1980 to 1991, when I left IBM Kingston to move to IBM Italy. I shall not summarize the hardware and software system work and the parallelization of applications in computational chemistry: molecular quantum mechanics, Monte Carlo, Molecular Dynamics, Car-Parrinello, several Fluid dynamic computer programs. All this effort is documented in a relatively recent review paper [99]. IBM Kingston was the place were the first parallelization in computational chemistry were successfully carried out and were a handful of quantum chemists, a few physicists and a few system programmers build, tested and heavily used a self-made parallel computer, faster than CRAY, 'the unquestioned supercomputer' of the time.

Computational chemistry was becoming in my mind more and more simply a part of computational sciences, with blurred boundaries of no essential value. Indeed, we prepared an extended animated movie 'The unity of Science' linking quantum chemistry, molecular dynamics and fluid dynamics computer simulations with parallelism; the audio comments were given in English, French, German, Italian and Chinese. This was the spirit also behind the volumes Modern Techniques in Computational Chemistry, MOTECC [100], and Methods and Techniques in Computational Chemistry, METECC [101], inspired from of the eighteen century Encyclopedia tradition; these volumes are a contribution to move computational chemistry toward the 'global simulations' at first and hopefully later toward the 'cognizant simulation' level [98,100,101].

I was also slowly realizing that even if this evolution was evolving along a path ineluctable in the long range, it was unlikely to materialize in the next decades, because 'most scientists like to stay in a given well defined problem area for a life time'. Still I hope that true progress might come particularly from updating university curricula and from some young computational chemists, perhaps in one of the new countries, entering now the club of the 'affluent and technically most advanced societies'. I often recall that the spirit of the sixties was essential to the development and advances of computational chemistry and of the computer industry.

Perhaps true new progress with breakthroughs, rather than the expected steady growth, needs a right level of preparation, enthusiasm and also some naivety.

In the late eighties IBM was entering a critical period, following a time of un-controlled growth; I felt it was advisable to leave USA and I asked to be transferred to IBM-Italy. A new laboratory in Sardinia, CRS4, was for a short time the new work location [75, 100a]. Soon thereafter, I left for the University L. Pasteur in Strasbourg [100b].

6.7 BACK TO THE BEGINNING: A NEW APPROACH TO AN OLD PROBLEM

Let me briefly return on old concepts. The advantage to deal with the correlation energy problem *via* corrections to the Hartree–Fock average potential is the backbone of the density functional approach from Wigner seminal ideas, to our work, and to popular DFT parameterizations. It was clear that the corrections cannot be limited to the electron-electron interaction, but affect all interactions of the Hamiltonian, namely also the kinetic and the nuclear-electron interactions. This view is the starting point of the series of papers on the HFCC method (started at the L. Pasteur University in Strasbourg [102] and continued with a A. von Humboldt-Stiftung in Bonn and at the Max-Plank in Garching), which provides not only a new method successfully tested on hundreds of molecules, but also an explanation of the origin of the chemical bond, with a partitioning of the atomization energy into different interactions [103] leading to the conclusion that the binding is due about 50% to exchange and 50% to the remaining interactions (classical mechanics type interaction).

In the HFCC semi-empirical method there is a first correlation correction for the individual atoms of the molecule, a second for the molecular binding and a third for van der Waals interactions. In the HFCC work we have always stressed that the use of a HF single determinant is simply a temporary approximation to test different parameterization schemes.

However, the point I wish to address in this section concerns what should be considered as a proper reference function, namely the function on which to add the correlation correction. Note that generally it is assumed that the HF is the proper reference function.

In the early seventies with G.C. Lie we felt that the proper reference function would be a limited MC-SCF to ensure proper dissociation of the molecule. Therefore, we performed density functional computations starting not with a single determinant but with this short linear combination, yielding correct dissociation limits [103,104]; 10 years later, the idea was revisited by Sabin [105].

We now propose the feasibility to use as reference function a linear combination – in the MC-SCF spirit – of a Hartree–Fock, HF, function, Ψ_{HF}, and of a Heitler–London, HL, function, Ψ_{HF}. This combination, $\Psi_{\mathrm{HF-HL}} = a\Psi_{\mathrm{HF}} + b\Psi_{\mathrm{HL}}$, is designed to ensure a reasonable behavior both in the binding and in the dissociation internuclear distances (recall J.C. Slater's comment [9]). The short-hand notation HF-HL will be used to refer to the MC-SCF HF-HL reference function. Note that one could obtain seemingly equivalent results selecting RHF, restricted HF, and UHF, unrestricted HF, but this would lead to the well-known quantum mechanical problem implicit in the UHF use.

In Fig. 6.1 we report very preliminary results [106] on H_2, LiH and Li_2. The RHF, UHF (in the figure we use dashed lines for both potentials) and the HL (long dashed lines) are compared with the MC-SCF HF-HL potential energy (solid line), and with the experimental values (dotted line). The improvement of the MC-SCF HF-HL energy relative to the HF energies is notable, nearly irrelevant with respect to the HL energy; appreciable also the quality of the overall shape of the MC-SCF HF-HL potential curves, from the energy minimum to dissociation. For the case of H_2, we report in addition *ab initio* MC-SCF function results obtained with the six configurations $1\sigma_g$, $1\sigma_u$, $2\sigma_g$, $2\sigma_u$, $1\pi_u$, $1\pi_g$; the resulting binding energy is 107.47 kcal/mol *versus* 109.48 kcal/mol from Kolos [8]; delta and higher configurations are needed to reach the experimental value. The basis sets used are large gaussian basis with extended polarization; the optimization process is likely nearly full, but some marginal improvement can be expected.

A main advantage of the $\Psi_{\mathrm{HF-HL}}$ is that the correlation error is now a smooth decreasing function for all the internuclear distances, and this is the case both for covalent and ionic bonds, thus the ideal starting point for the HFCC corrections. A second appeal of the $\Psi_{\mathrm{HF\text{-}HL}}$ reference function is that one can either proceed semi-empirically as already experimented [102,103], using however the $\Psi_{\mathrm{HF\text{-}HL}}$ starting point, or one can introduce as much *ab initio* correlation as wished, but with a somewhat non-conventional approach in the MC-SCF use. Indeed in the Ψ_{HL} part of $\Psi_{\mathrm{HF\text{-}HL}}$ one can either use the traditional HL approach or substitute it with an appropriate pre-fabricated linear combination of determinants designed to introduce correlation in specific atoms, $\Psi'_{\mathrm{HL}} = \Sigma a_i \Psi_{\mathrm{HL}(i)}$; note that in this simplified notation the index i refers to atoms.

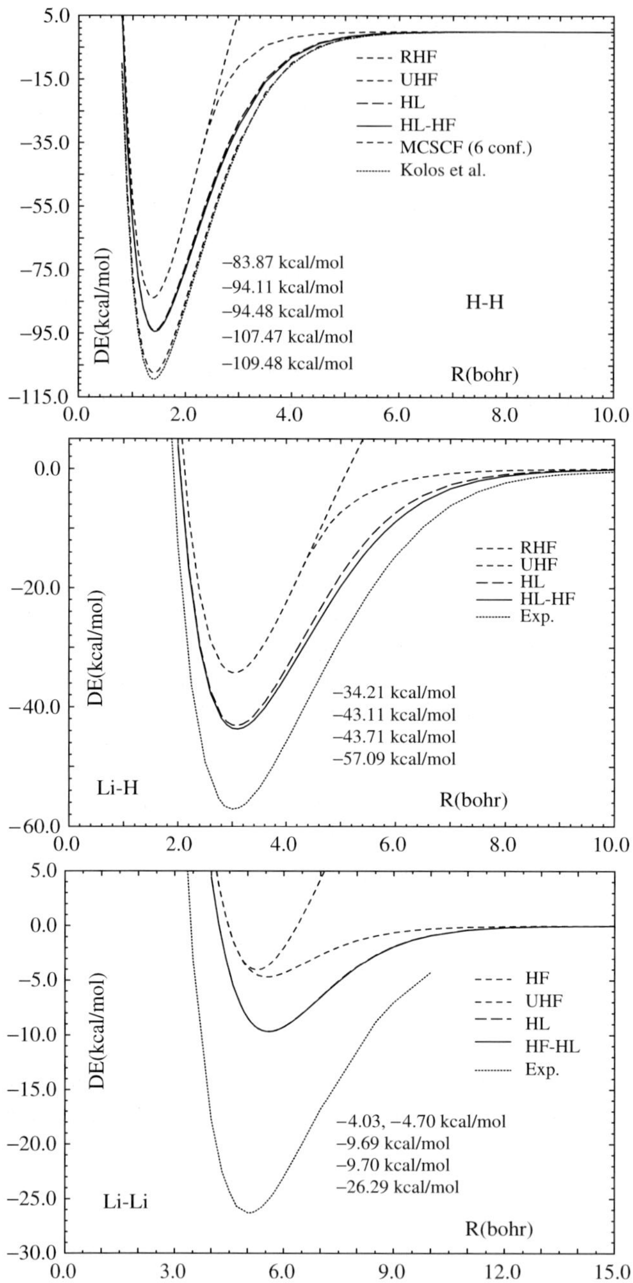

Fig. 6.1. Potential energy computed in the RHF, UHF, HL and MC-SCF HF-HL approximations and experimental data. For H_2 we include also the potential for a six configuration MC-SCF (see text).

This extension does not have to be applied to all the atoms of the molecule, but to as many as chemically relevant; for example in the study of interacting molecules one can select the relevant part of the atoms in the super-molecule. In equivalent fashion one can add configurations to the Ψ_{HF} part of the $\Psi_{HF\text{-}HL}$ function, $\Psi'_{HF} = \Sigma b_j \Psi_{HF(j)}$ introducing mainly 'molecular correlation'; note that in this simplified notation the index j refers to electronic configurations for the molecule. We recall that the presence of non-orthogonal orbitals does not constitute a problem since long ago considered and resolved by P.-O. Löwdin [35].

In conclusion we propose two approaches to the HFCC method [106]. The *ab initio* HFCC, with $\Psi_{HFCC} = a' \Psi'_{HL} + b' \Psi'_{HF}$, allows one to selectively correlate both the atoms of the molecular system (essentially atomic correlation) and that part of the electronic density which is shared among atoms (molecular correlation). The semi-empirical HFCC is designed to do the same but at much lower computational cost. Details on the HFHHL method are given in Ref. [106].

6.8 CONCLUSIONS

We should not consider the evolution of computational chemistry as a 'single rather linear' road, but rather as a network of approaches, with many bifurcations, overlaps, and detours; in general, this is the way toward progress. I confess, I am looking with some anxiety to the unbalance between the scarce development of new methods and the abundance of applications, more and more in 'standard areas'. We should likely move with more rigor toward 'large molecules', namely toward simulations, on living cells and later tissues and even more later toward full living organisms, developing new approaches, new algorithms, new 'renormalization' levels, essential to extend our frontiers. An extended discussion on computer simulations for cellular studies is given in [106]. A bit more thinking and a bit less computing.

6.9 ACKNOWLEDGEMENTS

The work described in this review has been made possible thanks to the collaboration of many, only a few—for space limitations—explicitly mentioned, but all of them, 78 in total, essential; let me recall that the strength of the river flow is due to the contributions from many creeks and torrents. A special mention to J.M. André and M. Kasha for proofreading this manuscript.

6.10 REFERENCES

1 G. Herzberg, Spectra of diatomic molecules, D. Van Nostrand Company, Inc, Princeton, NJ, 1951.
2 R.G. Par and W. Yang, Density functional theory of atoms and molecules, Oxford University Press, Oxford, 1985.
3 W. Heitler and F. London, Z. Physik, 44 (1927) 355, Note: the computed value for H_2 is $D_e = 3.20$ eV.
4 The simplest LCAO-MO yields $D_e = 3.4850$ eV; for a review on the computations of H_2 see A.D. McLean, A. Weiss and M. Yoshimine, Revs. Mod. Phys. 32 (1960) 211.

5 J.H. Van Vleck and A. Sherman, The quantum theory of valence, Revs. Mod. Phys., 7 (1935) 167.
6 L. Pauling, The nature of the chemical bond, Cornell University Press, Ithaca, 1939.
7 C.A. Coulson, Valence, Oxford University Press, New York, 1952.
8 H. Eyring, J. Walter and G.E. Kimball, Quantum chemistry, McGraw-Hill Book Company, Inc, New York, 1951.
9 J.C. Slater, Electronic structure of atoms and molecules, Technical Report No. 3. Solid-State and Molecular Theory Group. MIT, Cambridge, Mass., 1953.
10 E. Clementi and K.S. Pitzer, J. Chem. Phys., 32 (1960) 656.
11 K.S. Pitzer and E. Clementi, J. Am. Chem. Soc., 81 (1959) 4477.
12 E. Clementi and A.D. McLean, J. Chem. Phys., 36 (1962) 45.
13 E. Clementi, J. Am. Chem. Soc., 83 (1961) 4501.
14 G. Corongiu and E. Clementi, Int. J. Quantum Chem., 42 (1992) 1185.
15 C.C. Roothaan, Revs. Mod. Phys., 23 (1951) 69.
16 C.C. Roothaan, Revs. Mod. Phys., 32 (1960) 179.
17 E. Clementi, J. Chem. Phys., 84 (1980) 2122.
18 W. Kolos and C.C. Roothaan, Revs. Mod. Phys., 32 (1960) 219; W. Kolos and L. Wolniewicz, J. Chem. Phys., 41 (1964) 3663; W. Kolos and L. Wolniewicz, J. Chem. Phys., 49 (1968) 404.
19 S.F. Boys and G.B. Cook, Revs. Mod. Phys., 32 (1960) 285.
20 J.C. Slater, Quantum theory of atomic structures, McGraw-Hill Book Company, New York, 1960.
21 J.C. Slater, Quantum theory of matter, 2nd edn., McGraw-Hill Book Company, New York, 1968.
22 J.C. Slater, Electronic structure of molecules, McGraw-Hill Book Company, New York, 1965.
23 S.F. Boys, Proc. R. Soc. (Lond.) A, 200 (1950).
24 E. Wigner, Phys. Rev., 46 (1934) 1002, Note: there is no reference to this work in Ref. [2].
25 (a) Gombas, Pseudopotentiale, Springer, Berlin, 1967; (b) E. Clementi and H. Popkie, J. Am. Chem. Soc., 94, 1972, pp. 4057; (c) E. Clementi and H. Popkie, J. Chem. Phys., 57, 1972, pp. 4870.
26 R. Daudel, R. Lefebvre and C. Moser, Quantum chemistry, Interscience, New York, 1959.
27 A. Pullman and B. Pullman, Cancerisation par les substances chemique et structure moleculaire, Masson, Paris, 1955.
28 B. Pullman and A. Pullman, Quantum biochemistry, Interscience, New York, 1963.
29 A. Streitwieser, Molecular orbital theory for organic chemists, Wiley, New York, 1951.
30 C.J. Ballhausen, Introduction to ligand field theory, McGraw-Hill, New York, 1962.
31 E. Brändas and E.S. Kryachko (Eds.), Fundamental world of quantum chemistry, Kluwer Academic Publishers, Boston, 2003.
32 B. Ransil, Rews. Mod. Phys., 32 (1960) 245.
33 E. Clementi, J. Chem. Phys., 36 (1962) 33.
34 G. Corongiu, The HFHHL method: a combination of Hartree–Fock and Heitler–London approximations. Invited paper, Sanibel Symposium 2005.
35 P.-O. Löwdin, Adv. Chem. Phys., 2 (1959) 207.
36 Muzio Clementi, Gradus ad Parnassum ou l'art de toucher le piano démontré par exercices dans le style sévère et dans le style élégant. A well known collection of piano exercises written to master the musical instrument via a systematic increase in "complexity".
37 E. Clementi, J. Chem. Phys., 38 (1963) 996.
38 E. Clementi, J. Chem. Phys., 38 (1963) 1106.
39 E. Clementi and D.L. Raimondi, J. Chem. Phys., 38 (1963) 2686.
40 E. Clementi, J. Chem. Phys., 40 (1964) 1944.
41 E. Clementi, J. Chem. Phys., 41 (1964) 295.
42 E. Clementi, Tables of atomic functions, IBM J. Res. Dev. Suppl., 9 (1965) 2.
43 E. Clementi and C. Roetti, Roothaan-Hartree–Fock wave functions, Special issue of atomic data and nuclear data tables, Vol. 14, Academic Press, New York, 1974.
44 H. Hartmann and E. Clementi, Phys. Rev., 133A (1964) 1295; E. Clementi, J. Mol. Spectr., 12 (1964) 18.
45 E. Clementi, J. Chem. Phys., 36 (1963) 2248.
46 E. Clementi, J. Chem. Phys., 39 (1963) 175.
47 E. Clementi, J. Chem. Phys., 42 (1965) 2783.

48 E. Clementi, J. Chem. Phys., 39 (1963) 487.
49 S. Chackravorty and E. Clementi, Phys. Rev. A, 39 (1989) 2290.
50 D.R. Hartree, W. Hartree and B. Swirles, Tran. R. Soc. A, 236 (1939) 299.
51 O. Sinanoglu, J. Chem. Phys., 36 (1962) 706.
52 R. Parr, Quantum theory of molecules, Benjamin, New York, 1963.
53 E. Clementi, W. Kraemer and C. Salez, J. Chem. Phys., 53 (1970) 125.
54 E. Clementi and A. Veillard, J. Chem. Phys., 43 (1968) 2415; E. Clementi and A. Veillard, J. Chem. Phys., 44 (1966) 3050.
55 E. Clementi, IBM J. Res. Dev., 9 (1965) 2.
56 D.R. Davies and E. Clementi, IBMOL: Computation of wave functions for molecules of general geometry, IBM Research Laboratory, San Jose, California, 1965.
57 E. Clementi and R.D. Davies, J. Comp. Chem., 1 (1966) 223.
58 E. Clementi and R.D. Davies, J. Chem. Phys., 45 (1966) 2593.
59 A. Veillard and E. Clementi, IBMOL, Version 2. IBM Research Laboratory Technical Report, August, 1966.
60 A. Veillard and E. Clementi, Theor. Chim. Acta, 7 (1967) 133.
61 A.D. McLean and M. Yoshimine, Table of Linear Molecular Functions, Supplement to IBM J. Res. Dev. November 1967.
62 G. Das and A.C. Wahl, J. Chem. Phys., 44 (1966) 3050.
63 J. Hinze, J. Chem. Phys., 50 (1973) 6424.
64 Note. I am presently a regular student at the 'Accademia delle Belle Arti' in Como; I can hardly describe the feeling in finding myself in the drawing class room, with the model and my new colleagues, all of them one half century younger than I, but with a few years of painting experience gained at the 'Licei artistici'.
65 E. Clementi, J. Chem. Phys., 46 (1967) 3842.
66 E. Clementi, J. Chem. Phys., (1967) 3851; E. Clementi, J. Chem. Phys., 47 (1967) 2223.
67 The New York Time, October 10, 1968.
68 Time magazine, November 8, 1968.
69 E. Clementi, J.M. André, M.C. André, D. Klint and D. Hahn, Acta Physica Hungaricae, 27 (1969) 493.
70 E. Clementi, J. Mehl and W. von Niessen, J. Chem. Phys., 54 (1971) 508.
71 E. Clementi, J. Chem. Phys., 46 (1967) 3842.
72 E. Clementi and G. Corongiu, Theor. Chim. Acta, 543 (2001) 39.
73 E. Clementi and H. Popkie, J. Chem. Phys., 57 (1972) 1077.
74 E. Clementi and R. Barsotti, Theor. Chim. Acta (Berl.), 43 (1976) 101.
75 (a) Int. J. Quantum Chem. 42–43 (1992); (b) Int. J. Quantum Chem. 42 (1992) 529 for E. Clementi publication list, (c) and Int. J. Quantum Chem. 42 (1992) 547 for an extended curriculum vitae, which complements the present review.
76 E. Clementi, Proc. Natl Acad. Sci. USA, 69 (1972) 2942.
77 H. Popkie, H. Kisstenmacher and E. Clementi, J. Chem. Phys., 59 (1973) 1325.
78 H. Popkie, H. Kisstenmacher, E. Clementi and R.O. Watts, J. Chem. Phys., 60 (1974) 4455.
79 G.C. Lie and E. Clementi, J. Chem. Phys., 62 (1975) 2195.
80 O. Matsuoka, E. Clementi and M. Yoshimine, J. Chem. Phys. 64 (1976) 1351; G.C. Lie, E. Clementi and M. Yoshimine, J. Chem. Phys. 64 (1976) 2314.
81 E. Clementi, Determination of liquid water structure, coordination numbers for ions and solvation for biological molecules, Lecture notes in chemistry, Vol. 2, Springer, Berlin, 1976.
82 E. Clementi and G. Corongiu, Biopolymers, 18 (1979) 2431.
83 E. Clementi and G. Corongiu, Int. J. Quantum Chem., 16 (1979) 897.
84 E. Clementi and G. Corongiu, in: R.H. Sharma (Ed.), Biomolecular stereodynamics, Vol. 1, Adenine Press, New York, 1981.
85 Note. Before accepting Montedison invitation, I received a very confidential report on Montedison status, written by a large and most reputable French chemical industry. The report was so negative that unfortunately I dismissed it as 'obviously incorrect'.
86 O. Novaro, E. Blaisten-Barojas, E. Clementi, G. Giunchi and M.E. Ruiz-Viscaya, J. Chem. Phys., 68 (1978) 2337.

87 O. Novaro, Int. J. Quantum Chem., 42 (1992) 1047.
88 N. Koga, S.Q. Jin and K. Morokuma, J. Am. Chem. Soc., 110 (1988) 3417.
89 L. Carozzo, G. Corongiu, C. Petrongolo and E. Clementi, J. Chem. Phys., 68 (1978) 787.
90 (a) A.K. Mohanti and E. Clementi, Chem. Phys. Lett. 175, (1989) 348; (b) A.K. Mohanti and E. Clementi, J. Chem. Phys., 93 (1990) 1829.
91 L. Pisani and E. Clementi, J. Chem. Phys., 101 (1994) 3079.
92 E. Clementi and G. Corongiu, Int. J. Quantum Chem., 10 (1983) 31; G.C. Lie and E. Clementi, Phys. Rev., 33 (1986) 2679; J.H. Detrich, G. Corongiu and E. Clementi, Chem. Phys. Lett., 112 (1984) 426; U. Niesar, G. Corongiu, E. Clementi, G.R. Kneller and D.K. Bhattacharya, J. Phys. Chem., 94 (1990) 7949.
93 K.S. Kim and E. Clementi, J. Phys. Chem., 89 (1985) 3655; K.S. Kim and E. Clementi, J. Am. Chem. Soc., 107 (1985) 227; K.S. Kim and E. Clementi, J. Am. Chem. Soc., 107 (1985) 5504; K.S. Kim, H.L. Nguyen, P.K. Swaminatham and E. Clementi, J. Phys. Chem., 89 (1985) 2870.
94 A. Largo-Cabrerizo and E. Clementi, J. Comp. Chem., 8 (1987) 1191; A. Largo-Cabrerizo, C. Urdaneta, G.C. Lie and E. Clementi, Int. J. Quantum Chem., 21 (1987) 677; D. Frey, G.C. Lie and E. Clementi, J. Chem. Phys., 91 (1989) 2369; A. Preiskorn, G.C. Lie and E. Clementi, J. Chem. Phys., 92 (1990) 4941.
95 M. Aida, G. Corongiu and E. Clementi, Int. J. Quantum Chem., 42 (1992) 1353.
96 L. Hannon, G.C. Lie and E. Clementi, J. Stat. Phys., 51 (1988) 965.
97 L. Leherte, D.P. Vercauteren, E.G. Derouane, G.C. Lie, E. Clementi, J.M. André, in: P.A. Jacobs and R.A. Van Santen (Eds.), Zeolite: Facts, figures, future, Elsevier, Amsterdam, 1989.
98 E. Clementi, Phil. Trans. R. Soc. Lond. A, 326 (1988) 445.
99 E. Clementi (Ed.), MOTECC 89, 90, 91 Modern Methods in Computational Chemistry, Science Publishers BV, Leiden, The Netherlands, 1989, 1990, 1991.
100 E. Clementi and G. Corongiu (Eds.), METECC 93, Methods and techniques in computational chemistry, STEF, Cagliari, Italy, 1993; E. Clementi and G. Corongiu (Eds.), METECC 95, Methods and techniques in computational chemistry, STEF, Cagliari, Italy, 1995.
101 E. Clementi and G. Corongiu, Int. J. Quantum Chem., 62 (1997) 571.
102 E. Clementi and G. Corongiu, The origin of molecular atomization energy with the HF and the HFCC models, in: E. Brändas and E.S. Kriachko (Eds.), Fundamental word of quantum chemistry, Kluwer Academic Publishers, Boston, 2003.
103 G.C. Lie and E. Clementi, J. Chem. Phys., 60 (1974) 1275.
104 G.C. Lie and E. Clementi, J. Chem. Phys., 60 (1974) 1288.
105 A. Sabin, Int. J. Quantum Chem., 22 (1988) 59.
106 G. Corongiu and E. Clementi, to be submitted. Work partially presented at the Gyeongju, Korea, TACC meeting, 2004 and in more detail at the Atlanta symposium in honor of J.-M. André, April 2004, and at the 2005 Sanibel Symposium. See E. Clementi, and G. Corongiu, Comments on computational chemistry: From diatomic molecules to large biochemical systems, Int. J. Quantum Chem., in press, 2005.

Theory and Applications of Computational Chemistry: The First Forty Years
Edited by C. Dykstra et al.

CHAPTER 7

The beginnings of coupled-cluster theory: an eyewitness account

Josef Paldus

Department of Applied Mathematics, Department of Chemistry, and Guelph-Waterloo Center for Graduate Work in Chemistry—Waterloo Campus, University of Waterloo, Waterloo, Ont., Canada N2L 3G1

Les chiffres autrefois hameçons de zéros
infiniment variés mijotaient en l'atome
indéfiniment nus indéfiniment beaux
mais leur compte était bon et les voici vaillants
chevauchant l'explosion ô jeunesse ô jeunesse
que le graphe griffait de son zig en zaguant…
(Raymond Queneau: Petite Cosmogonie Portative, Ier chant)[1]

Abstract

This is a quasi-historical outline of the origins of coupled-cluster (CC) theory, its implementation, and its exploitation in the computation of atomic and molecular electronic structure. It is primarily intended to be about the making of CC theory rather than about the theory itself. The references to the literature are handled accordingly. The CC methods are nowadays extensively exploited in a great variety of problems, yet the initial steps were not at all obvious and took considerable time to germinate, develop and, eventually, flourish. This highly personal—and thus biased—viewpoint attempts to describe the sources that stimulated the derivation of the explicit form of CC theories, their working equations and algorithms, as well as their first exploitations. It barely touches on the subsequent

[1] I am not competent to translate Queneau's surrealistic poems that are full of neologisms and hapax legomena, often employing words that are—in his own estimation—'petit-laroussables'. Nonetheless, responding to the urging of the referee and the editors, here is a rough attempt: "Numbers once fish-hooks of zeros/infinitely varied simmer in an atom/endlessly bare forever beautiful/but settling their affair they are here brave/riding the explosion oh youthfulness oh youthfulness/let the graph (diagram) engrave its zig in zag…" (R. Queneau: Concise Portable Cosmogony).

developments, which brought the CC methodology to its present status, and which would require a much more extensive treatment than is the scope of the present article. These accomplishments will be undoubtedly covered by other authors.

Jécris $\vec{F} = m\vec{\gamma}$
et la fléche prend son essor
Jécris $E_2 - E_1 = h\nu$
et la feuille boit le soleil
....
Jécris $\Box A_\mu = j_\mu, F_{\mu\nu} = \partial_\mu A_\nu - \partial_\nu A_\mu$
et la lumière EST
Jécris $i\hbar \frac{\partial \Psi}{\partial t} = (-\hbar^2 \sum_k \frac{\Delta_k}{2m_k} + \sum_i \sum_j \frac{e_i e_j}{|\vec{r}_i - \vec{r}_j|})\Psi$
et le sel se fait cristal
....
Jécris $(\frac{\partial^2}{\partial t^2} - c^2 \vec{\Delta}^2)p(\vec{r}, t) = 0$
et j'entends ta voix.
J'écris, je·décris, je décréte.
L' Univers se plie á la lettre.
Nul atome n'est censé ignorer la loi.
(Jean-Marc Lévy-Leblond: Lois de la Nature)[2]

7.1 'PREHISTORY'

natura non facit saltum[3]

In contrast to established scientific disciplines, such as physics and chemistry, whose roots reach back several centuries, the field of quantum chemistry is a relative newcomer. When, in 1937, the very first textbook bearing this name in its title[4] was published [2], its author, Hans Hellmann, who was later executed by Stalin's henchmen, wrote in the preface: "Quantum Chemistry is a young science that came into existence only a decade ago" [3]. Yet, already at that time, the author was compelled to limit the relevant material, lest the size of the book—comprising well over 300 pages—turn out to be excessive [3].

Indeed, the pioneering work of Heitler and London [4] on hydrogen molecule is generally recognized as the milestone demarcating the birth of the discipline of quantum chemistry. Since that time, the utility and need to elucidate the phenomenon of chemical bonding from first principles became recognized by molecular physicists and chemists of all sorts alike.

[2] "I write...and the arrow takes flight/I write...and the leaf drinks the sun/.../I write...and there IS light/I write...and salt forms a crystal/.../I write...and I hear your voice./I write, I describe, I decree./The Univers bows to the letter./No atom can ignore the law." (J.-M. Lévy-Leblond: The Laws of Nature).

[3] Nature makes no leaps.

[4] Note that the textbook by Pauling and Wilson [1] appeared 2 years earlier than Ref. [2]. However, it bears the title: 'Introduction to Quantum Mechanics'.

The work of Heitler and London gave rise to the *valence bond* (VB) theory, developed primarily by Pauling and collaborators [1], whose qualitative version in the form of the so-called *resonance theory* was enthusiastically embraced by organic chemists. My own first introduction to the subject, while still a high-school student (1950–1953), came through the very nice university textbooks on general and inorganic [5] and organic [6] chemistry by Otto Wichterle. I was particularly fascinated by his brief introduction to the old quantum mechanics, including the Bohr model of the hydrogen atom, and its astonishing simplicity in explaining the observed spectral series. His textbooks rely heavily on the concepts of 'resonance', in spite of the fact that this theory, as advanced by Wheland [7] in the West and by Syrkin and Djatkina [8] in the East, was severely attacked and, in fact, even labelled as 'bourgeois pseudo-science' by Soviet philosophers. When Professor Wichterle was criticized for what he was teaching his students, he smartly countered his critics by pointing out that resonance theory can perhaps mislead a poor philosopher, but not his students, who in couple of weeks learn how to exploit the theory to their advantage without any danger. Fortunately, thanks to such efforts, Czech and Slovak quantum chemistry were not delayed in their development, as was the case in the Soviet Union. In fact, the same is true for both Hungary and Poland, where strong schools in this field arose and continue to thrive.

It is well known that, in view of the sheer mathematical complexity that we face when we try to solve Schrödinger's equation for the simplest molecular systems (already for the one-electron H_2^+ cation [9]), we have to focus on the design of computationally manageable, yet reliable, approximation schemes, based on various model Hamiltonians. Even when ignoring the relativistic effects and freezing the nuclear motion *via* the Born–Oppenheimer approximation [10], the problem is still too formidable for any system having more than two electrons. For this very reason, almost all quantum-mechanical treatments of the molecular electronic structure are based on finite dimensional models.

Yet, even the finite dimensional standard VB approach runs into a number of difficulties, such as the 'nightmare' of the inner shells [11], neglect of overlap integrals, and the so-called $N!$ 'catastrophe' (see, e.g. Ref. [12]). For this very reason, sometime during the second World War, VB theory was eclipsed by the computationally much more amenable molecular orbital (MO) method, relying on the independent-particle model (IPM), which reduces the N-electron problem to effectively a one-electron, though highly non-linear, one. A very important conceptual advance was achieved by the exploitation of the variation principle, which led to the formulation of Hartree–Fock (HF) equations [13,14] (*cf.* also Refs. [15,16]). These integro-differential equations are solved iteratively by generating a suitable self-consistent field (SCF). The numerical solution of these equations for one-center atomic problems became a reality in the 1950s, primarily owing to the earlier efforts by Hartree and Hartree [17] (see Ref. [18] for a historical account).

The availability of numerical solutions of HF equations is still restricted to at most two-center (or linear) systems. Nonetheless, the so-called 'analytic' approach, using suitable basis sets, enabled the computation of SCF solutions within the Roothaan linear combination of atomic orbitals (LCAO) SCF formalism [19]. Generation of such solutions, even for systems with several hundreds of electrons, is nowadays routine, although the handling of general open-shell states can still be frustrating at times, due to the possible multiplicity of various SCF solutions.

References pp. 140–147

The fact that the HF approximation yields well over 99% of the total energy led initially to a general belief that SCF wave functions are sufficiently accurate for the computation of various properties of chemical interest. Unfortunately, once SCF solutions became available for molecular systems, this hope was shattered. Moreover, during the late 1960s, it became abundantly clear that, in spite of the conceptual importance of HF solutions and their utility in supplying quantitative, or at least semiquantitative, information about some molecular properties, their ability to provide a reliable description of the electronic structure was, in general, rather limited. Indeed, the first dramatic indication of the inadequacy of the SCF solutions emerged from Wahl's study of the F_2 molecule [20]. Using a carefully optimized basis set, Wahl showed that at the SCF level of approximation, F_2 is not bound: the negative SCF dissociation energy he obtained was as large as 1.6 eV!

The shortcomings of IPM are nowadays well recognized. Even though the HF total energies are very accurate, say within 0.1% of their exact value, they are not accurate enough to describe many chemical phenomena or properties of interest. For example, the total energies of the first-row diatomics are of the order of 10^2 hartree, so we need at least two orders of magnitude higher precision (i.e. $10^{-3}\%$) to achieve the so-called chemical accuracy of 1 kcal/mol or ~ 1 millihartree. The situation is even more critical when considering non-energetic properties. For this reason, all present day quantitative studies account, in one way or another, for the many-electron correlation effects that are lacking in IPM descriptions.

Presently, the widely used post-Hartree–Fock approaches to the correlation problem in molecular electronic structure calculations are basically of two kinds, namely, those of variational and those of perturbative nature. The former are typified by various configuration interaction (CI) or shell-model methods, and employ the linear Ansatz for the wave function in the spirit of Ritz variation principle (*cf.*, e.g. Ref. [21]). However, since the dimension of the CI problem rapidly increases with increasing size of the system and size of the atomic orbital (AO) basis set employed (see, e.g. the so-called Paldus–Weyl dimension formula [22,23]), one has to rely in actual applications on truncated CI expansions (referred to as a limited CI), despite the fact that these expansions are slowly convergent, even when based on the optimal natural orbitals (NOs). Unfortunately, such limited CI expansions (usually truncated at the doubly excited level relative to the IPM reference, resulting in the CISD method) are unable to properly describe the so-called *dynamic* correlation, which requires that higher than doubly excited configurations be taken into account. Moreover, the energies obtained with the limited CI method are not size-extensive.

Both the above shortcomings can be largely avoided by relying on multireference (MR) CI approaches, which account for configurations up to a chosen excitation level relative to those spanning the model space (usually singles and doubles, yielding the MR CISD method). This approach is particularly useful when handling a manifold of closely lying states, a situation that is invariably encountered when exploring the entire potential energy surfaces (PESs) or curves (PECs) near the dissociation limit, since it is able to properly account for the so-called *static* and *non-dynamic* correlations arising in degenerate and quasi-degenerate situations, respectively. Yet, even the MR CISD methods cannot properly handle dynamic correlation, since this would require a large reference space and thus N-electron spaces of too large a dimension. For this very reason, even the MR CISD results are subsequently corrected for both their lack of dynamic correlation (usually *via* low-order perturbation theory) and size extensivity (*via* various versions of semiempirical Davidson-type corrections).

The approaches of a perturbative nature exploit, in one way or another, the many-body perturbation theory (MBPT), whether of a finite or infinite order. Those of the Rayleigh–Schrödinger (RS) type are generally preferred for their size-extensive character at any level of truncation, since those of the Brillouin–Wigner (BW) type do not possess this property. The RS PT that employs a Hartree–Fock reference is then usually referred to as the Møller–Plesset PT or MPx, with $x = 2, 3, \ldots$ designating the order of PT terms that are retained [24]. Already MP2 can provide a very useful estimate of the correlation effects and is very often relied upon when handling relatively large systems. The general nature of PT expansions prefers the truncation at the even orders of PT [25,26]. Since, however, the fourth-order MBPT already requires triply excited intermediates, its practical exploitation is computationally demanding.

Another possibility is to increase the order of PT while restricting the excitation level of the intervening intermediate states. This is most easily done when the excitations are limited to doubles, yielding the nth order MBPT with doubles, or DMBPT(n). In fact, in this case, the summation can be carried out to infinite order, yielding the DMBPT(∞) method [27]. However, this result can be easily seen to be equivalent to the linear version of coupled-cluster (CC) theory that is restricted to two-body amplitudes, namely, to the L-CPMET (linear coupled-pair many-electron theory) or, more succinctly, to L-CCD (linear CC with doubles) in current terminology.

The need for the inclusion of higher-order effects increases with the degree of quasi-degeneracy of the state considered. For this reason, much effort has been devoted to the formulation of the so-called MR MBPT [28–30]. Here, however, a number of ambiguities arises, which often limits the development of practical algorithms (*cf.*, e.g. attempts to extend the so-called CAS-PT2 method, which is based on the complete active space self-consistent field (CAS SCF) reference, to higher than the second order). In fact, we shall see that the same problem manifests itself, even when extending the standard SR CC theory to the MR case.

In any case, the general formulation of the MBPT—primarily due to Brueckner [31], Goldstone [32], Hugenholtz [33], and Hubbard [34]—has shed much light on the structure of fully correlated, exact N-fermion wave functions, and was essential to the development of perturbative-type methods, including CC theory. For this reason, the next section is devoted to this topic.

7.2 GESTATION

Siehe! nahe blinkt
Ein Wasserquell;der ist auch unser. Nimm
Dein Trinkgefäß, die hohle Kürbis, daß der Trank

Die Seele mir erfrische.
(Friedrich Hölderlin: Der Tod des Empedokles)[5]

A clear delineation of correlation effects was formulated in the early 1960s by Löwdin [35], who also introduced the concept of the 'symmetry dilemma' [36] and the related notions of various broken-spin IPM solutions, namely, the unrestricted-projected, projected-unrestricted, and generalized HF solutions [37]. Furthermore, his pioneering work on density matrices, which led to the concept of *natural orbitals* [38], and especially his series of papers on perturbation theory [39], provided the major impetus for the study of correlation effects in the context of atomic and molecular electronic structure. At that time, computing technology hardly enabled computations at the *ab initio* level, so that he and his collaborators developed the alternant molecular orbital (AMO) method [40,41], in order to account, at least approximately, for the correlation effects within the context of semiempirical Hamiltonians.

The field of atomic and molecular electronic structure was not the only one in which the correlation problem turned out to be of key importance. Indeed, this problem arises in any many-body quantum-mechanical system, particularly for many-fermion systems. For this very reason, much attention was devoted to it by solid-state and nuclear physicists alike. The important model systems that were intensely studied at that time were an electron gas in the former case and the model of infinite nuclear matter in the later case.

Since both these models represent extended systems, the exploitation of the shell-model or CI-type variational methods was *a priori* excluded. This placed emphasis on the development of perturbative approaches for this type of problem. It was soon realized that the most efficient methodological approach must be based on a mathematical formalism that was originally developed in quantum field theory. Moreover, in view of the extended character of the studied systems, it was absolutely essential that the method employed yields energies that are linearly proportional to the particle number N in the system or, in today's parlance, that it must be size extensive, so that the limiting procedure when $N \to \infty$ makes sense. In terms of MBPT, this implies that only the connected or linked energy terms be present in the perturbation series, a requirement that automatically leads to the Rayleigh–Schrödinger PT.

The conjecture that the unlinked terms exactly cancel out in every order of the RS MBPT, as implied by Brueckner's observation [31] that this is the case up to and including the fourth order, was soon proved by Goldstone [32]. At almost the same time, independent derivations were also provided by Hugenholtz [33] and Hubbard [34].

Goldstone's approach exploited a time-dependent PT in the interaction picture and was based on the Gell-Mann and Low adiabatic theorem [42], as was the work of Hubbard [34], while Hugenholtz [33] employed a time-independent PT. The time-dependent version is thus based on the evolution operator or propagator $U(t)$,

$$U(t) = \exp(\mathrm{i}H_0 t)\exp(-\mathrm{i}Ht) = \exp(\mathrm{i}H_0 t)\sum_n |n\rangle \exp(-\mathrm{i}E_n t)\langle n| \qquad (1)$$

[5] "See! Near sparkles/a spring of water; which also belongs to us. Take/your drinking bowl, a hollow gourd, so that the drink/will refresh my soul." (E. Hölderlin: The Death of Empedocles).

while the time-independent approach relies on the Green function or the resolvent operator $G(\epsilon)$,

$$G(\epsilon) = \frac{1}{\epsilon - H} = \sum_n |n\rangle \frac{1}{\epsilon - E_n} \langle n| \tag{2}$$

where H is the Hamiltonian of the system,

$$H = H_0 + V \tag{3}$$

with H_0 designating the chosen IPM Hamiltonian that is associated with the system. Clearly, $|n\rangle \equiv |\Psi_n\rangle$ and E_n are the eigenstates and eigenvalues of H, respectively.

The relationship between the time-dependent and time-independent approaches, given by a Fourier transform between $G(\epsilon)$ and $U(t)$, was elucidated by Bloch [43]. Of course, in either approach, the final result is time-independent, since, after all, we are solving a time-independent or stationary Schrödinger equation. It would thus seem that the time-independent approach—as followed by Hugenholtz—would turn out to be more natural and simpler. This is indeed the case when we are primarily interested in the energy. However, in order to elucidate the PT structure of the exact wave function, the time-dependent approach is beneficial (*cf.* Ref. [34]).

Needless to say, it was the second quantization formalism of quantum field theory, enabling the exploitation of Wick's theorem together with a representation *via* Feynman-like graphs or diagrams—the mathematical techniques relied upon by all the above authors [32–34]—that made it possible to carry out the general proof of the extensive nature of RSPT and to unscramble the general structure of MBPT wave functions and energies. The principal results of these efforts are usually referred to as the *linked cluster* and *connected cluster* theorems (see below).

There has been much confusion in the literature concerning the concept of linked *vs* connected clusters or corresponding diagrams. In fact, already all three pioneers of MBPT [32–34] differ in their definition of these concepts. In our view, and in most present day works, the following distinction is most useful:

- A *linked diagram* has no disconnected vacuum component(s).
- A *connected diagram* is a connected graph in the sense of the standard graph theory.
- A *vacuum diagram* is a diagram having only internal fermion lines (i.e. oriented lines connecting two vertices).

Thus, roughly speaking, a connected diagram consists of a single 'piece', while a disconnected diagram has several distinct subgraphs. Further, an unlinked diagram involves one or more disconnected vacuum diagrams. Connected diagrams are automatically linked, but linked diagrams can either be connected or disconnected, as long as no disconnected component is a vacuum diagram. Clearly, the energy is represented in terms of vacuum diagrams, while diagrams involving external or free fermion lines describe wave functions.

The two main theorems mentioned above may then be formulated as follows (for a brief outline, see, e.g. Refs. [44,45]):

Linked Cluster Theorem. Only linked (or connected) *vacuum* diagrams contribute to the energy, while all unlinked (or disconnected) vacuum diagrams mutually cancel out, so that

$$\Delta E = E - \langle \Phi_0 | H | \Phi_0 \rangle = \sum_{n=0}^{\infty} \langle \Phi_0 | V(RV)^n | \Phi_0 \rangle_{\mathrm{L}} \tag{4}$$

where R is the RS resolvent

$$R = \sum_{i \neq 0} \frac{|\Phi_i\rangle\langle\Phi_i|}{\varepsilon_0 - \varepsilon_i} \tag{5}$$

Here, ε_i and $|\Phi_i\rangle$ are the eigenvalues and eigenstates of the unperturbed Hamiltonian H_0 and the subscript $i = 0$ corresponds to the non-degenerate ground state.
Similarly, the exact wave function $|\Psi\rangle \equiv |\Psi_0\rangle$ is given by all non-vacuum linked diagrams, namely,

$$|\Psi\rangle = \sum_{n=0}^{\infty} \{(RV)^n |\Phi_0\rangle\}_{\mathrm{L}} \tag{6}$$

In each case, the subscript 'L' indicates the linked component.

The connected cluster theorem, which is the basis of the CC theory as will be seen below, then fine-tunes the linked-cluster theorem for the wave function in the following sense:

Connected Cluster Theorem. Define the cluster operator T that generates all the *connected* wave function diagrams by

$$T|\Phi_0\rangle = \sum_{n=1}^{\infty} \{(RV)^n |\Phi_0\rangle\}_{\mathrm{C}} \tag{7}$$

the subscript 'C' now indicating that only contributions from connected diagrams are to be included. Then

$$|\Psi\rangle = \exp(T)|\Phi_0\rangle \tag{8}$$

This theorem was first clearly stated by Hubbard [34].

Going into the finer details of the diagrammatic representation, it is important to note the distinction between the graphs employed by Goldstone [32] (and Hubbard [34]) and those used by Hugenholtz [33]. Basically, the Goldstone diagrams employ non-symmetrized interaction vertices that are associated with the two-body Coulomb integrals $\langle ab|v|cd\rangle$,

$$\langle ab|v|cd\rangle = \int \phi_a^*(1)\phi_b^*(2)v(1,2)\phi_c(1)\phi_d(2)\mathrm{d}\tau_1\, \mathrm{d}\tau_2 \tag{9}$$

with ϕ_i designating the one-electron IPM (spin) orbitals, while Hugenholtz diagrams use vertices representing the anti-symmetrized two-electron integrals, namely,

$$\langle ab|v|cd\rangle_A = \langle ab|v|cd\rangle - \langle ab|v|dc\rangle \tag{10}$$

Goldstone's interaction vertices are in fact 'supervertices' involving two simple vertices—each associated with single ingoing and outgoing, oriented, fermion lines representing the corresponding annihilation and creation operators of the two-body terms in the Hamiltonian—that are interconnected by a non-oriented interaction (usually dashed or dotted) line, while the Hugenholtz vertices result from merging simple vertices of a Goldstone vertex into a single point-like vertex, while eliminating the interaction line. Both of these representations have their advantages and disadvantages (see, e.g. Refs. [46,47] for details).

The main benefit of the Hugenholtz representation is a greatly reduced number of resulting diagrams that need to be considered (by roughly a factor of 2^n for the nth order PT diagrams), while its weakness is the fact that it does not allow an unambiguous determination of the phase of the corresponding algebraic expressions. This shortcoming is not present when Goldstone diagrams are employed, the phase being given by the simple rule $(-1)^{h+l}$, where h is the number of internal hole lines and l designates the number of closed loops of oriented (fermion) lines. In fact, both desirable features may be conveniently combined by exploiting the so-called Brandow diagrams [29] (see also Čížek [48]), namely, by representing each Hugenholtz diagram by one Goldstone version (regardless of which one) and by accordingly adapting the rules for their algebraic evaluation. The assignment of the weights, given by the reciprocal value of the number of automorphisms of a graph (stripped of the labels), is easy for both kinds of diagrams.

Another advantage of Goldstone diagrams is the possibility to eliminate the explicit spin-dependence for spin-free Hamiltonians by simply associating a factor of 2 with each closed loop (i.e. a factor of 2^l with the entire diagram). Of course, in this case, all the Goldstone diagrams are required. Moreover, it is not difficult to show that the corresponding spin-adapted configurations represent non-orthogonal valence bond (VB) types of states [49]. One can still use the reduced number of Hugenholtz-type diagrams and carry out the spin-adaptation by relying on the graphical techniques of spin algebras (*cf.*, e.g. Refs. [50,51]; for a short introduction see the Appendix of Ref. [52]). Clearly, in this case, any spin coupling can be employed, including the preferable particle–particle hole–hole (pp–hh) one, which leads to orthonormal states with desirable transformation properties (*cf.*, Refs. [52,53]).

Let me add, finally, a few personal remarks. The reading of Goldstone's paper by someone who is not intimately familiar with mathematical techniques of quantum field theory may be a rather frustrating task, since the paper sketches out only the main steps of the proof, while leaving out all the technical details. To fill in all these details is far from being trivial, especially if mathematical rigor is not to be sacrificed, as witnessed by the fact that even some textbooks present an incorrect or at least misleading exposition of these details.

My colleague and friend Jiří Čížek and myself have been greatly helped in this regard by Volodja Tolmachev, whose little known monograph [54] of 150 pages essentially represents a detailed and mathematically rigorous version of the original Goldstone paper of about a dozen pages. While in the 1960s no textbooks presenting the MBPT in detail were available, several such texts are available today (see, e.g. Refs. [55–59]; for a brief outline of the modern version of the time-independent approach, see also Refs. [44–47]).

Clearly, the main ideas and results of the MBPT were developed in the context of studies of extended systems. In the development of electron gas theory (for the challenging high-density region), much progress had already been made during the 1950s, starting with the important work of Macke [60], and culminating with the classical paper by Gell-Mann and Brueckner [61]. These developments showed that at high densities, an electron gas has a logarithmic singularity, which can be described by MBPT if the ring diagrams are properly summed to infinite order. Similarly, the nuclear many-body problem was vigorously pursued in the 1950s by relying on the reaction matrix theory of Brueckner (see Ref. [62] for references to earlier work), which was then reformulated in terms of Feynman-like diagrams by Goldstone [32]. In this case, the ladder diagrams play the decisive role. These results were later built upon by Bethe, Brandow and others.

In the context of atomic and molecular electronic structure, it was Hugh Kelly [63] who first applied Goldston's MBPT to closed-shell-type atoms by developing several very successful approximation schemes. In the molecular context, the handling of the many-electron correlation problem was very much influenced by the work of Sinanoğlu [64,65] and Nesbet [66], who emphasized the importance of pair (i.e. two-body) clusters and developed techniques based on Bethe–Goldstone equations for their approximate treatment. These approaches treat more or less independent pair clusters, while neglecting the inter-cluster interactions.

Another important development was the realization of the importance of the 'size-consistency' and 'size-extensivity' in the studies of associative or dissociative chemical processes by Primas [67], as well as his clear delineation of the relationship between the configuration interaction and the exponential coupled-cluster Ansätze.

Unlike the electron gas and nuclear matter, where a major role is played by ring and ladder diagrams, respectively, the atomic and molecular correlation problem—or, in fact, the finite nuclei problem—requires a proper account of both types of diagrams, as well as of their combinations. This was certainly the main motivation for the development of CC methodology, which we shall discuss in the next section.

7.3 BIRTH

...rem acu tetigisti!
(Titus Maccius Plautus, Rudens)[6]

The exponential cluster expansion has a long history in statistical physics, where it is known as the Ursell and Mayer linked-cluster expansion for the partition function [68]. A very general argument for the exponential form of the exact wave function can be found in a paper on the origins of the CC method by Kümmel [69]. Nonetheless,

[6] Literally: "You have touched the thing with a needle"; meaning: "You have hit the nail on the head" (Plautus, The Rope).

historically, it was the study of the MBPT that led to the CC method. Although its exponential structure is already implicit in the Gell-Mann and Low representation [42] of a bound-state wave function in quantum field theory, it was really Hubbard's paper [34] where it appears in a clear and explicit form, and which led Coester and Kümmel [70] to the idea of directly computing the relevant *amplitudes* rather than the energy terms in the MBPT expansion (for the first-hand account, see Ref. [69]).

As already alluded to above, MBPT originated from attempts to solve the many-fermion problem for electron gas and nuclear matter models. In the electron gas case, already the second-order PT contribution to the energy diverges. The correlation problem in nuclei is even more demanding, due to the hard-core or nearly hard-core behavior of the internucleon potential, which already invalidates the Hartree–Fock approximation. Although much has been achieved in this direction by Brueckner [31], Goldstone [32] and others [71], the need for a general approach, which would simultaneously account for both ladder and ring diagrams, as well as their combinations, was very much in demand.

Thus, the crucial step in this direction was Coester's realization that one should focus on the amplitudes rather than the energies themselves. However, the equations for these amplitudes [70] were given in a very general, implicit form and could not be used in actual applications. Indeed, to derive the working equations and to gain sufficient insight into the structure of the theory—which may appear to be a very straightforward exercise these days—was no small task at the time. It was a heroic effort of my cherished colleague Jiří Čížek that led to the formulation of a general procedure of deriving the exact explicit form of the desired CC equations (see, e.g. Bartlett's 'perspective' paper [72]). His task was magnified by the fact that he was very much on his own in this effort, at least initially, and could only rely on the available literary sources. It took considerable time before his highly mathematical 1966 paper [73] was fully appreciated by quantum chemists.

To illustrate this point, let me quote one of the leading theoreticians in our field, Roy McWeeny, who in his 1967 Inaugural Lecture at the University of Sheffield [74] showed a page from Čížek's 1966 paper (page 4262 of Ref. [73]) displaying the L-CCD diagrams and CCD equations, which he commented by the statement:

> ...believe it or not, it is taken from one of the Chemistry Journals!

In fact, the main thrust of McWeeny's lecture was to emphasize "the indivisibility of Science and the artificiality of barriers and boundaries" between various disciplines, as well as his deep "conviction that Mathematics provides the common thread on which all the exact sciences could be strung".

A more recent witness to this effect may be found in a paper on 'A Biography of the CC Method' by Kümmel himself, presented in a special session at the 11th International Conference on Recent Progress in Many-Body Theories in Manchester [75], honoring his 80th birthday. In a section entitled 'The Slow Start and the Complexity of the CC Method', we read [76]

> Considering the fact that the CC method was well understood around the late fifties it looks strange that nothing happened with it until 1966, as Jiří Čížek published his first paper on a quantum chemistry problem [73]. He had looked into the 1957 and 1960 [70] papers published in Nuclear Physics by Fritz and myself. I always found it quite

> remarkable that a quantum chemist would open an issue of a nuclear physics journal. **I myself at that time had almost given up the CC method as not tractable**[7] and, of course, I never looked into the quantum chemistry journals. The result was that I learnt about Jiří's work as late as in the early seventies, when he sent me a big parcel with reprints of the many papers he and Joe Paldus had written until then.

It is certainly very gratifying that great progress has been achieved in this direction, although another danger may be lurking in the offing, thanks to the ever-increasing build up and availability not only of various 'black box' codes, but also of various symbol manipulation packages, such as MAPLE or MATHEMATICA. In spite of their indisputable utility to any researcher, the undeniable fact remains that these tools are more and more often employed by users lacking an adequate background in the underlying theory or mathematics (see, e.g. the introductory part of Robb's 'perspective' paper [77] on the unitary group approach (UGA). For an outline of UGA, see Refs. [22,23,78,79] and references therein; for a brief introduction see Ref. [80]). Undoubtedly, this could lead to very undesirable consequences in a long run. Perhaps the best compromise would be to strive for modular-type codes, with a well-defined input and output, which could be assembled or exploited as needed, not unlike the standard linear algebra or MAPLE and MATHEMATICA packages.

At the time when Jiří was struggling with MBPT and the related mathematical methods of quantum field theory, I was in Ottawa as a post-doctoral fellow learning the art of spectroscopy in the Larger Molecules Section, directed by Don Ramsay, which was a part of the spectroscopy laboratory under the wings of the famous GH (Gerhard Herzberg). Soon after my return to Prague in 1964, I became very interested in Jiří's work and very much encouraged him to pursue this line of research in his Ph.D. (called at that time C.Sc. according to Soviet system) thesis, supervised by our common department head, Jaroslav Koutecký, who had also been my thesis supervisor 3 years earlier. This was very much facilitated by the fact that Jiří and myself shared a small office in the attic of the Institute of Physical Chemistry of the Czechoslovak Academy of Sciences.

Soon afterwards, we were able to meet V.V. [Vladimir (Volodja) Venjaminovich] Tolmachev in Moscow, who taught us much about Goldstone's MBPT, which is very nicely explained in great detail in his little known lecture notes already quoted above [54] (this text was later expanded into a full monograph [81]). However, Volodja was very much 'hooked' on the time-dependent approach and did not much appreciate the time-independent route taken by Jiří and later by myself as well. He focused at the time on the extension of the theory to the multireference case and atomic MBPT applications, which culminated in his Frascati Lectures [28].

Čížek's 1966 paper [73] and the subsequent paper from the Frascati Volume [48], both resulting from his Ph.D. thesis, are often portrayed as a formulation of what is today characterized as the CCD (CC with doubles), which he referred to as the CPMET (coupled-pair many-electron theory) method. It is true that the actual explicit equations that are given in his paper are the CCD equations. It must be emphasized, however, that the essence of the paper is an entirely general formulation of the CC method, and the CCD

[7] Emphasis mine.

equations—which, after all, represent the most important approximation in the CC theory—are given only as an illustrative example. The paper thus provides a completely general algorithm enabling the derivation of explicit CC equations at any level of truncation of the cluster amplitudes using the diagrammatic approach. Once this procedure is mastered, it is a relatively simple—though often laborious—task to write down the relevant CC equations for the higher-than-pair cluster amplitudes (their actual implementation and exploitation is, of course, another story).

One puzzling aspect in the diagrammatic approach concerns the determination of correct weight factors that are associated with each resulting diagram. This was the reason why Jiří introduced the concept of 'skeletons' in addition to actual diagrams. A skeleton results when we strip a given diagram of all free (i.e. summation) spin orbital or orbital labels. The desired weight factor is then given as the reciprocal value of the number of all automorphisms (i.e. of the order of the group of automorphisms) of the skeleton [46]. Since for the linked Goldstone-type MBPT energy diagrams these factors can only be 1 or 1/2, the usefulness of this concept for the more general situations was not fully appreciated, as witnessed by several errata in the literature correcting these factors.

The diagrammatic technique has certainly played an important role in the development of both MBPT and CC theories, even though an algebraic approach is equally expedient, particularly in the CC case. A purely algebraic formalism is certainly more suitable for an automatic computerized implementation of the formalism and, in fact, the only feasible one for the symmetry-adapted CC approaches based on UGA [22, 23,82–84] (see also below). The reason for a partiality to graphical representation is perhaps due to the fact that most human brains are geometric in nature (i.e. image oriented), rather than algebraic, particularly when dealing with such concepts as connectivity or linkage.

In the 1960s and early 1970s, even the formalism of second quantization was shunned by most quantum chemists, even deemed as an unnecessary extravagance. For this reason, we wrote a paper [85] in which we derived the CCD equations using the standard first quantization wave-function formalism, without any diagrams (for a similar CCSD version, see Ref. [86], written in connection with the appearance of the quadratic CI [87]). Yet, even here, we tried for too much brevity and compact mathematical notation. For example, the key expression for the disconnected quadruples in terms of doubles, later on usually written in its full form listing all 18 terms (see, e.g. the last Eq. (33) of Ref. [88]; note already a much more compact form used in Eq. (11) of Ref. [89]),

$$\begin{aligned} c_0^{-1} c_{ijkl}^{abcd} = {} & d_{ij}^{ab} d_{kl}^{cd} - d_{ik}^{ab} d_{jl}^{cd} + d_{il}^{ab} d_{jk}^{cd} - d_{ij}^{ac} d_{kl}^{bd} + d_{ik}^{ac} d_{jl}^{bd} - d_{il}^{ac} d_{jk}^{bd} \\ & + d_{ij}^{ad} d_{kl}^{bc} - d_{ik}^{ad} d_{jl}^{bc} + d_{il}^{ad} d_{jk}^{bc} + d_{ij}^{cd} d_{kl}^{ab} - d_{ik}^{cd} d_{jl}^{ab} + d_{il}^{cd} d_{jk}^{ab} \\ & - d_{ij}^{bd} d_{kl}^{ac} + d_{ik}^{bd} d_{jl}^{ac} - d_{il}^{bd} d_{jk}^{ac} + d_{ij}^{bc} d_{kl}^{ad} - d_{ik}^{bc} d_{jl}^{ad} + d_{il}^{bc} d_{jk}^{ad} \end{aligned} \tag{11}$$

was given in Appendix III of Ref. [85] as (for an easier comparison we use the same spin orbital labelling as in Eq. (11))

$$\begin{bmatrix} a & b & c & d \\ i & j & k & l \end{bmatrix}_u = \sum_{r,s=1}^{3} (-1)^{r+s} \left(\langle V_r'' | t_2 | V_s' \rangle_A \langle \bar{V}_r'' | t_2 | \bar{V}_s' \rangle_A + \langle V_r'' | t_2 | \bar{V}_s' \rangle_A \langle \bar{V}_r'' | t_2 | V_s' \rangle_A \right) \tag{12}$$

where $|V'_1\rangle = |ij\rangle$, $|V'_2\rangle = |ik\rangle$, $|V'_3\rangle = |il\rangle$, with a bar designating a complementary bra or ket, e.g. $|\bar{V}'_1\rangle = |kl\rangle$, $|\bar{V}'_2\rangle = |jl\rangle$, $|\bar{V}'_3\rangle = |jk\rangle$, and similarly for the doubly primed quantities, i.e. $|V''_1\rangle = |ab\rangle$, $|V''_2\rangle = |ac\rangle$, $|V''_3\rangle = |ad\rangle$, etc. to the chagrin of many a reader. Clearly, both sets of amplitudes are related as

$$d_{ij}^{ab} = \langle ab|t_2|ij\rangle_A \tag{13}$$

Let me also comment on a purely formal distinction that strikes the uninitiated. We had chosen to let the 'time' run from right to left, while many other authors, particularly Bartlett's group, prefer it to run 'skywards'. This distinction is of little importance for CC diagrams that involve only one interaction vertex. However, the advantage of the right–left convention lies in the fact that the vertices appear in the same order as in the algebraic expression represented by the graph. Moreover, for high-order MBPT diagrams, the right–left or 'horizontal' convention requires less space on a printed page.

As the above narrative indicates, most of the ideas for the treatment of the many-electron problem were first developed by the nuclear and solid-state physicists. This is the case not only for perturbative methods, but also for variational ones, including the configuration interaction method, which nuclear physicists refer to as the shell model, or for the unitary group approach (see Ref. [90]; for additional references see Refs. [23, 78–80]). The same applies to the CC approach [70]. For this reason, quantum chemists, who were involved in the development of post-Hartree–Fock methods, paid a close attention to these works. However, with Čížek's 1966 paper the tables were turned around, at least as far as the CC method is concerned, since a similar development of the explicit CC equations, due to Lührmann and Kümmel [91] had to wait till 1972, without noticing that by that time quantum chemists were busily trying to apply these equations in actual computations.

It did not take long, however, for the leading figures in the nuclear CC problem to realize that this approach is "tailor-made" for the atomic and molecular correlation problem and a useful interchange of ideas and collaborative efforts followed. First, such "cross-fertilization" on a larger scale occurred during the Workshop on Coupled-Cluster Theory at the Interface of Atomic Physics and Quantum Chemistry, organized by Rod Bartlett at the Institute for Theoretical Atomic and Molecular Physics of the Harvard-Smithsonian Center for Astrophysics (the Proceedings were published in Volume 80 of Theoretica Chimica Acta). Nonetheless, the gulf still persists, as nicely characterized by Kümmel's observation at the just mentioned Workshop [69]:

> As much as the chemists must be praised for occasionally looking into nuclear physics journals, as much I have to regret that nuclear physicists (including myself) did not care to search in the other direction. This hasn't changed much, especially the particle physicists still believe that the higher the energy they are dealing with, the lesser they need to look down to the low energy people.

7.4 GROWING PAINS

...tu ne cede malis, sed contra audentior ito
[Publius Vergilius Maro (Virgil), Aeneid vi. 95][8]

Soon after my return to Prague, I joined Jiří in an attempt to test his CC theory in actual calculations. He had himself carried out a preliminary study of a very simplified model of the nitrogen and benzene molecules [73]. Since the computational facilities that were available at the time, particularly those accessible to us in Prague (even at the NRCC in Ottawa the only available machine during my first stay was an IBM 1620), were primitive by today's standards, we focussed on the semiempirical model Hamiltonians. We had considerable experience with the Pariser–Parr–Pople (PPP) Hamiltonian and the PPP model of linear polyenes, C_NH_N, $N = 4v + 2$, $v = 1, 2, \ldots$, having a non-degenerate ground state, seemed particularly appealing to us for the following reasons:

(i) The results obtained with the AMO method [41,92] were available [40] for comparison.
(ii) High symmetry (D_{Nh}) of the model, fully determining the Hartree–Fock orbitals—which are thus identical with the Brueckner (maximum overlap) orbitals—implying that CCSD reduces to CCD.
(iii) The $v = 1$ case corresponds to the benzene π-electron model, while the $v \rightarrow \infty$ limit gives a model of the one-dimensional metal.
(iv) Exact solution *via* full CI could be obtained for benzene and *via* Lieb and Wu equations [93] for the corresponding Hubbard Hamiltonian models.

For computational reasons, we first focused on the benzene and butadiene models [94], as well as on the Be atom using a very limited basis set.

In 1966 we were fortunate to obtain 40 hours on the CDC 3600 at the Centre de Méchanique Ondulatoire in Paris, thanks to the generosity of Carl Moser. We carefully prepared our codes in Prague and managed to get them running the very first day. But, alas, we ran into all sorts of convergence problems for large coupling constants (small resonance integrals). Moreover, due to the magnanimous hospitality of Professors Lefebvre and Chalvet, which included an obligatory visit to the 'caves' in the Sancerre region, I ended up at the emergency with excruciating pain, due to kidney stones, and had to return prematurely to Prague.

In the meantime, we realized that our problem might be related to what is nowadays referred to as the singlet instability. This peculiar behavior of the Hartree–Fock reference in the $\beta \rightarrow 0$ limit decidedly caught our interest. We found it very fascinating that, in addition to the standard symmetry-adapted solution, there would exist another ground-state closed-shell Hartree–Fock solution in that range of the coupling constant. The broken-spin unrestricted Hartree–Fock solutions (of the different orbitals for different spin or DODS type) were well known at the time, as well as the concept of the symmetry dilemma of Löwdin [36], but no space-symmetry breaking Hartree–Fock solutions were then known.

[8] "...do not yield to misfortunes, but advance all the more boldly against them" (Virgil, The Aeneid).

We thus embarked full time on the study of Hartree–Fock instabilities, introducing the concepts of the *singlet* and *triplet* (or *non-singlet*) stability [95], which soon became part of the quantum-chemical folklore. We then focused on the investigation of space-symmetry broken Hartree–Fock solutions implied by the singlet instabilities. For π-electron PPP models of planar hydrocarbons with conjugated double bonds these solutions could be classified into those displaying the diagonal and off-diagonal charge-density waves [96,97], the former ones being again singlet unstable. A more detailed classification was later developed by Fukutome and co-workers [98] (for general references see Refs. [99,100]). At the *ab initio* level, these effects were later explored for the oxygen doubly negative ion (O^{2-}), in which case broken space-symmetry Hartree–Fock solutions of both prolate and oblate type were again found [101].

For simple open-shell systems, we formulated analogous *doublet* stability conditions [102,103], and illustrated them on a series of polyenic radicals [104]. These type of solutions do not fit into Fukutome's classification, but have now been incorporated in its generalized version [100]. The doublet stability conditions were formulated just at the time of the 1968 Soviet invasion of Czechoslovakia [105], which greatly upset the everyday life of all citizens, not to mention the cultural and scientific activities in the entire country. For this reason, Jiří and myself gratefully welcomed the hospitality the University of Waterloo kindly provided, thanks to the efforts of Professor Sydney Davison, not knowing at the time that it would become our permanent abode. Thus, the bulk of our work on the stability problem was carried out at Waterloo.

In retrospect, our choice of cyclic polyenes, C_NH_N, for the study of correlation effects *via* CC methods was a most unfortunate one, since these systems involve one of the most difficult correlation problems. The main difficulty stems from the increasing quasi-degeneracy of the reference configuration with both the increasing coupling constant (or decreasing absolute value of the resonance integral β) and the size of the chain N (see, e.g. Ref. [106]). As a consequence, the role of higher-order connected cluster components, such as T_4, T_6, etc. becomes more prominent [107] and the standard CCSD or CCD methods completely break down for sufficiently large N, or even for small cycles when approaching the fully correlated limit (i.e. when the resonance integral $\beta \rightarrow 0$). On the other hand, the UHF solution yields the exact energy for the $\beta = 0$ limit. Let me note in passing that this problem was recently explored anew by the Warsaw group, who showed that the same breakdown is also encountered when using higher-order CC methods, such as CCSDT and CCSDTQ [108].

Only much later were we able to account for some of these difficulties [109–112] by developing methods that, under certain conditions, account for the T_4 clusters. In cases when the projected UHF method provides the exact pair clusters, it can be shown [109] that the T_4 contribution cancels the contribution from the non-linear T_2^2 terms representing the important exclusion-principle-violating (EPV) diagrams. Except for a numerical factor associated with the triplet-coupled pp–hh t_2-amplitudes, this method—referred to as ACPQ or CCDQ′—is identical to the ACCD approach that was developed independently by Dykstra's group [113–116] (see also Ref. [117]). This method is, in fact, simpler than the standard CCSD method, because it avoids computationally the most demanding terms. In spite of the fact that it invariably provides superior results to the standard CCSD, it is seldom used in actual applications.

Soon after our arrival in Waterloo, Jiří and myself were fortunate to make the personal acquaintance of Isaiah (Shi) Shavitt, whose work we knew well from the literature. We had learned much from him concerning the state of the art of *ab initio* calculations, especially those using the configuration interaction method. We were particularly interested in the role of higher than doubly excited configurations—and the related adequacy of truncation of the CC expansion at the pair-cluster level—when attempting to generate highly accurate energies. Thanks to the CC Ansatz, the four-body contributions are well described *via* the disconnected T_2^2 clusters. These are properly accounted for in both the CCD and CCSD methods. The connected T_4 clusters generally play a negligible role, except in highly quasi-degenerate cases, such as the metallic-like linear polyenic models, as can be verified by the cluster analysis of the relevant full CI wave functions [107]. However, the role of three-body clusters required a closer examination for at least two reasons: the role played by these triply excited configurations in the CI calculations and, on a more fundamental level, by the order of PT in which these clusters contribute for the first time to the energy.

Concerning the latter point, it is a simple exercise to see that, in contrast to the four-body clusters, where the disconnected T_2^2 clusters contribute for the first time in the fourth order and the connected T_4 clusters start contributing in the fifth order, the reverse situation occurs for the three-body terms: the disconnected T_1T_2 clusters contribute for the first time in the fifth order, while the connected T_3 clusters contribute already in the fourth order of PT. (The T_1^3 clusters will, of course, contribute in the eighth order and may be safely neglected, unless the T_1 clusters are prominent, as when using the localized orbitals.)

We were thus very pleased when Shi Shavitt became interested in this problem and offered us a helping hand. Since, at that time, the largest triply excited contribution to the CI wave function was found for the BH_3 molecule (when using Slater-type orbitals), we decided to focus on this system. Moreover, in order to examine the role of the various cluster components, Shi Shavitt kindly generated additional CI results (involving, say, only doubles and quadruples, for a comparison with the CCD energies, etc.). He kindly provided us with the required one- and two-electron integrals as a relatively small deck of IBM punched cards, since only a minimum basis set of Slater-type orbitals was employed, in order to be able to generate the full CI results for comparison.

On our side, we developed the required CC equations and codes, taking into account all the important terms involving one-, two- and three-body clusters. This so-called extended CPMET (E-CPMET) was not equivalent to the CCSDT method, since most cubic and quartic terms were neglected. This was certainly a *bona fide* approximation, because high-order terms play indeed a negligible role. These results [118], which represented the very first application of the CC method at the *ab initio* level, showed very clearly the capabilities of this approach, the E-CPMET energies agreeing with the full CI results to within a microhartree.

7.5 MATURATION

Vous m'écrivez, mon ange, des lettres de quatre page plus vite que je ne puis les lire[9]
(An epigraph from Alexander Pushkin's 'The Queen of Spades')

A "great leap forward" in the development and exploitation of the CC method came with the availability of first powerful work-stations, at the time best represented by the VAX machines (in fact, the very first VAX was installed in Pople's group). The first general-purpose codes for the CPMET method, today referred to by the acronym CCD, were developed almost simultaneously by the Pople and Bartlett groups. They were reported at the 1978 American Conference on Theoretical Chemistry in Boulder, Colorado and published back to back in the International Journal of Quantum Chemistry [119,120]. Another independently developed and truly efficient code, which exploited the matrix form of electron-pair operators and amplitudes in terms of the atomic orbital basis (*cf.* Ref. [121]), was soon produced by Chiles and Dykstra [89]. In fact, Chiles and Dykstra's code not only enabled efficient large basis set calculations, but also paved the way to a practical use of Brueckner orbitals that warrant the vanishing of one-body clusters. Eventually, these CCD codes were extended to the CCSD method, fully accounting for the one- and two-body cluster amplitudes [122]. Presently, the CCD, CCSD and CCSD(T) codes, often in both the spin-adapted or partially spin-adapted and spin-orbital forms, constitute a part of most *ab initio* program packages (*cf.* ACES II [123], GAUSSIAN [124], MOLCAS [125], MOLPRO [126], etc.; for a more complete list see Ref. [45]) and are being constantly updated and extended.

Undoubtedly, it has been Bartlett's group that has become synonymous with CC theory, as far as its development and promulgation is concerned. I believe it was the Czech émigré novelist Milan Kundera, who once claimed that writers are lucky if they possess—or, rather, are possessed by—one overriding obsession and their development as literary figures is to be judged by their probing and handling of their favorite pet subject. I believe that a similar statement could be made about other human endeavors, including science, and Rod Bartlett is an excellent example of someone who has steadily and admirably devoted his energies to the pursuit of CC theory in the broadest sense.

I vividly recall the Fifth School on Advanced Methods of Quantum Chemistry, organized by the Toruń group of the Nicholas Copernicus University at Bachotek in 1987, where I was presenting a series of lectures on the unitary group approach, which at the time was fully oriented towards large-scale CI calculations. Following my first lecture, Rodney approached me saying something like: "Why would you do this kind of stuff?". What he really meant, I believe, although he never used these words, was: "Real men don't do CI!". Anyway, I later returned to the 'CC fold' and, with the help of Bogumil Jeziorski and Xiangzhu Li, successfully used UGA even in the CC context [82–84,127] (see also later).

Let us now briefly mention some of the most important developments that originated in Bartlett's group, during the past two decades, and that are undoubtedly of lasting value.[10]

Since the late 1970s, significant attention has been devoted to the evaluation of analytical energy derivatives (gradients and hessians) with respect to atomic

[9] "To me, you write, my angel, four-page letters faster than I can read them."
[10] See Bartlett's contribution in this issue for a more complete account.

displacements, enabling an effective search of PESs for stationary points. By the early 1980s, these calculations were routine for most variational approaches (Hartree–Fock, CI, MC SCF, etc.), where the energy is given as an expectation value in terms of variationally optimized parameters (see, e.g. Ref. [128]). However, the corresponding calculations for the perturbative-type methods, such as the MBPT or CC, where the energy is expressed in terms of non-variational parameters, are much more challenging. The initial step in deriving the necessary expressions was taken by Jørgensen and Simons [129], who relied on response functions. Almost simultaneously, the required algebraic expressions for the CCSD gradients were developed in Bartlett's group [130]. Soon afterwards, they formulated and implemented the gradient calculation for the MP3 energies [131], which, however, involved a direct evaluation of the first-order changes in the cluster amplitudes.

Further development then avoided this direct evaluation and advanced a general strategy for the computation of both MBPT and CC gradients [132,133] by relying on the so-called 'relaxed' densities, a procedure that also enables the computation of other first-order properties, such as various multipole moments. In fact, the relevant codes based on this idea were first generated and exploited by Schaefer's group [134] (for numerous applications, see Refs. [18–26] in Ref. [135]).

Great progress was also made in the computation of higher-order derivatives (particularly hessians) that are needed for second-order properties (polarizabilities, spin–spin coupling constants, etc.). The required expressions were again first generated in Bartlett's group [136,137]. An alternative formulation, which essentially exploited the technique used in variational approaches by rewriting the CC energy in terms of a fully variational Lagrangian (with one Lagrange multiplier for each orbital rotation and cluster amplitude), was simultaneously published by Schaefer's group [135].

Bartlett and his collaborators also pushed the limits of the available computational tools by extending the truncation of the CC expansion to higher and higher orders, developing the CCSDT and CCSDTQ codes and their various approximate versions [138–140]. In this connection, they explored the perturbative account of triples *via* the CCSD[T] scheme [141], which is very close to the independently developed CCSD(T) method of Raghavachari [142], the latter also accounting for the singles–triples interaction terms. Although CCSDT, and especially CCSDTQ, methods are computationally very demanding and can presently be exploited only for relatively small model systems, their conceptual importance, as well as their role as benchmarks, cannot be underestimated. In due time, with ever-increasing computational power, they may eventually become as standard as the CCSD or CCSD(T) method is today. Very recently, even the pentuple [143] and higher [26,144] excitations have been considered. These brute-force-type extensions aim for higher and higher accuracy but, primarily, reflect the fact that the basic requirement of non-degeneracy of the SR CC reference is more and more violated, as one stretches genuine chemical bonds, so that higher and higher-order clusters are no longer negligible. Here the problems of the CC methodology approach those of the CI-type methods and call for a multireference-type formalism (see below).

The CCSD(T) approach is probably the most often exploited CC method in actual applications, and provides excellent results [145], as long as no quasi-degeneracy is present. This is usually the case for the closed-shell ground states near the equilibrium geometry. However, with the stretching of one or more genuine chemical bonds, the CCSD(T) method breaks down even more dramatically than does the standard CCSD approach. This deficiency has been largely avoided in the so-called renormalized or completely renormalized CCSD(T) method, R-CCSD(T) or CR-CCSD(T), respectively, developed by Piecuch and Kowalski [146,147]. This method will likely be soon embraced in lieu of the standard CCSD(T), at least when one is faced with quasi-degenerate situations. Indeed, a proper description of the entire PESs or curves continues to be one of the most challenging problems for CC theory, as we shall touch upon in the next section.

Another highly significant development, initiated by the Bartlett group, was the exploitation of the CC Ansatz in the equations-of-motion (EOM) approach, resulting in the EOM-CC method [148–150] that is, in fact, closely related to the CC linear-response theory, first exploited in this context by Monkhorst [151]. The same Ansatz for the wave function was already employed two decades earlier [152,153], as well as in the closely related SAC/SAC-CI method of Nakatsuji and collaborators [154,155], not to mention the relationship with Green function approaches [156–158] and the developments in nuclear physics [159]. The EOM-CC method enables not only the calculation of principal ionization potentials, but also of shake-up effects and, most importantly, of excitation energies that are dominated by single or double excitations, even though in the latter case there is a need to include connected three-body clusters in order to obtain a satisfactory accuracy. The EOM-CC method was further extended to its similarity transformed version STEOM-CC, as well as its perturbative analogue STEOM-PT exploiting the Fock space MR CC Ansatz for the second similarity transform [160,161]. This method enabled the handling of very large systems, such as the free base of porphin.

All these and other developments pioneered by Bartlett's group stimulated parallel activities elsewhere, often arousing beneficial competition. As already mentioned, the first CCD codes were developed and published at the same time. Likewise, CCSDT codes were developed at roughly the same time by Schaefer's group [162] and an analogue of the CCSDTQ method by Adamowicz's group [163] (for later developments along these lines, see Ref. [164]). As alluded to above, the evaluation of analytical derivatives also witnessed a parallel and fast development, which still continues, since the computation of certain properties, particularly NMR chemical shifts, requires additional attention to the gauge-origin problem [165].

7.6 QUO VADIS?

> *Scepticism alone is a cheap and barren affair. Scepticism in a man who has come nearer to the truth than anyone before, and yet clearly recognizes the narrow limits of his own mental construction, is great and fruitful, and does not reduce, but doubles the value of the discoveries.*
> (Erwin Schrödinger: Nature and the Greeks)

Although this very incomplete and biased account of the sources, emergence and rise of the CC theory focusses on its beginnings, let me point out some more recent developments and current trends, especially those that are most desirable, yet remain elusive. Here, more than before, personal bias will be the rule, if for no other reason than the fact that it is more and more difficult to follow all the literature, even in this very narrow subfield of quantum chemistry, not to mention developments exploiting the CC Ansatz in other fields of the physical sciences. A cursory idea of these activities may be found in various proceedings dedicated to this topic by quantum chemists and many-body theorists alike (in particular, see Refs. [166–170]). Concerning molecular applications, the reader is referred to recent reviews [45,171–178], and even monographs and textbooks (see, e.g. Refs. [179–181]). We should also be watching for the monograph by Shavitt and Bartlett [182] that will be appearing shortly.

While the CC method has been most widely exploited in atomic and molecular electronic structure calculations, as already pointed out, it has also received much attention in other branches of physics. In addition to the problem of nuclear matter, where it originated, and that of finite nuclei and pion-nucleon systems in general [183], it has been very extensively exploited in studies of quantum spin chain and lattice models and the related problems of quantum magnets, electron lattice models and lattice quantum field theories, namely, the lattice gauge theory [184]. It has been further applied in the exploration of charged impurities in polarizable media, in quantum fluid mechanics, quantum optics and solid-state optoelectronics (see, e.g. Ref. [185] for references), not to mention the investigation of numerous model systems, including anharmonic oscillators and various many-body models (Lipkin–Meshkov–Glick model, Heisenberg quantum spin Hamiltonian, Hubbard–Lieb model, polaron problem, etc.). Finally, the CC method was also employed in studies of one-component Coulomb plasmas and both continuous and relativistic quantum field theories, as well as in studies of critical phenomena in Φ^4 field theories. The standard and the so-called extended CC methods (ECCM) [186] were likewise generalized to the temperature-dependent case [187]. Needless to say, our cursory account completely ignores numerous applications in atomic physics, including the handling of relativistic effects (see, e.g. Refs. [188,189]) or parity violation in heavy atoms [190].

Of course, as in any field of endeavor, there have been a few 'blind paths' that were followed, and that have yet to bring the expected 'harvest'. These were particularly associated with various attempts to produce a variational version of CC theory. In most instances, it is the complexity of the resulting formalism and the implied computational difficulties that preclude a successful exploitation of some of these ideas [191]. This is also the case for the above-mentioned ECCM [186] that uses the exponential cluster Ansatz in both bra and ket states, each involving distinct amplitudes. Such an Ansatz would be desirable in studies of phase transitions and its truncated version is used in the response theoretical approach to properties.

Let us thus concentrate on one very important and most desirable development that has received much attention in the past, yet has produced very few actual applications and results in the meantime and, presently, seems to be largely abandoned. I am talking here about the multireference generalization of the standard single-reference CC theory, which is essential for a proper treatment of general open-shell systems or, generally, whenever the reference configuration is degenerate or quasi-degenerate. Needless to say, such situations always arise when we break genuine chemical bonds, as when computing the entire PESs or PECs, or when considering various radicals or excited states. The same problem also arises in the study of one-particle states in field theories.

The difficulty in constructing a genuine MR CC theory stems from the fact that the generalization of the SR CC Ansatz to the MR case is far from being unambiguous.

In principle, we have essentially two options: *either* to define a single valence-universal cluster operator that produces the desirable result when acting on any reference configuration spanning the model space, *or* to introduce distinct cluster operators, one for each model space configuration. Both options have been realized, the first one leading to the so-called *valence universal* (VU) MR CC method [30,192–194] and the second one yielding the *state universal* MR CC formalism [195]. The VU theory requires not only the normal product form of the exponential Anzatz, as first pointed out by Lindgren [192], but also the consideration of an entire family of model spaces that are associated with ionic species, obtained by sequentially removing or adding the valence electrons, until the closed-shell configuration involving only doubly occupied orbitals is reached. This latter aspect generated a number of heated discussions with various claims and counter-claims. The exact formulation of the valence universality conditions was given much later [196]. The necessity of considering the sequence of ionic species and the multiplicity of solutions of MR CC equations leads to a 'genealogy problem' [197], which is avoided in the SU CC type theories. The VU Ansatz has been primarily used in atomic calculations [198,199].

Both VU and SU MR CC methods employ the *effective Hamiltonian* formalism: the relevant cluster amplitudes are obtained by solving Bloch equations and the (in principle exact) energies result as eigenvalues of a non-Hermitian effective Hamiltonian that is defined on a finite-dimensional model space $\mathcal{M}_0$. An essential feature characterizing this formalism is the so-called *intermediate* or *Bloch normalization* of the projected target space wave functions $|\tilde{\Psi}_i\rangle$ with respect to the corresponding model space configurations $|\Phi_i\rangle$, namely $\langle\Phi_i|\tilde{\Psi}_j\rangle = \delta_{ij}$ (for details, see, e.g. Refs. [172,174]).

Moreover, once the cluster Ansatz is introduced (for an option of directly solving Bloch equations without invoking the cluster Ansatz, see Ref. [200]), it is essential that the so-called *complete model space* (CMS), spanned by configurations involving all possible occupancies of valence or active (spin) orbitals, be used, lest the desirable property of size-extensivity be violated. This requirement, however, leads not only to highly dimensional (and thus computationally demanding) model spaces, but, most importantly, to the occurrence of the so-called *intruder states*.

Similarly as the SR CC theory requires the reference configuration $|\Phi_0\rangle$ to be non-degenerate, the model space configurations $|\Phi_i\rangle \in \mathcal{M}_0$ of a MR theory should be well separated in energy from those spanning the orthogonal complement $\mathcal{M}_0^{\perp}$ of $\mathcal{M}_0$. However, as the target space $\mathcal{M}$ (of the same dimension as $\mathcal{M}_0$), that is spanned by the exact wave functions $|\Psi_i\rangle$ and is associated with a given model space $\mathcal{M}_0$, is far from being unique (since any $|\Psi_i\rangle$ that is not orthogonal to $\mathcal{M}_0$ can belong to $\mathcal{M}$), the energy of one or more excited configurations from $\mathcal{M}_0^{\perp}$ will often lie within the interval of the energies of the model space configurations $|\Phi_i\rangle$ or will be very close to it.

The intruder state(s) almost inevitably interfere when exploring the entire PESs involving the breaking of genuine chemical bonds leading to open-shell fragments. In such cases, the effective Hamiltonian formalism often breaks down, so that no physically meaningful solution to the Bloch equations can be found. In fact, the same failure may occur even at the equilibrium geometry. For example, when considering one of the simplest four-electron systems, represented by the LiH molecule, both the VU [201] and SU [202] MR CCSD approaches diverge when a complete model space is employed! For this reason, much attention has been given to the construction of CC theories that use truncated or incomplete model spaces (IMSs) of various kinds. In fact, most VU MR CC applications have employed such truncated spaces, often different ones for different geometries [201,203,204].

In the SU CC context, the IMS based formalism preserving the size-extensivity was first proposed by Meissner and Bartlett [205] in 1990 (see also Ref. [206]) and much hope was expressed at the time of the Cambridge workshop [207]:

> An important development presented at the meeting was the generalization of the Hilbert space MRCC to *incomplete* active spaces [205]. This makes the theory primed for a burst of computational activity. Although this helps to alleviate the severe problem of intruder states, the intruder state is still endemic to the method. Much activity in this area will be reported in the next few years.

Unfortunately, it was the cautionary statement concerning the endemicity of intruders that turned out to keep plaguing the actual applications. Indeed, very few applications have been accomplished and those few that were carried out used a CMS involving only two active orbitals. In fact, the genuine MR CC methods were mostly ignored during the past decade. The only exception was the formulation of the Brillouin–Wigner (BW) version of the MR CC theory [208–210], which avoids intruders on principle grounds *via* a suitable denominator shift, but, unfortunately, at the cost of a complete loss of size-extensivity. Even here, however, additional approximations were made (e.g. handling of the coupling constants and subsequent corrections for size-extensivity in the last iteration [211]), requiring separate computations for each state and thus reducing the approach to the SS-type one. Nonetheless, this approach produced many useful results [212] and its linkage with the standard RS-type version [213] offers much hope for future developments.

For these and other reasons, much attention was given to the so-called *state-selective* or *state-specific* (SS) MR CC approaches. These are basically of two types: (i) essentially SR CCSD methods that employ MR CC Ansatz to select a subset of important higher-than-pair clusters that are then incorporated either in a standard way [163,164], or implicitly [109–117], or *via* the so-called *externally corrected* (ec) approaches of either the amplitude [214–219] or energy [220,221] type, and (ii) those actually exploiting Bloch equations, but focusing on one state at a time [222]. The energy-correcting ec CC approaches [220,221] are in fact very closely related to the *renormalized* CCSD(T) method of Kowalski and Piecuch mentioned earlier [146,147].

The ec-type approaches exploit the fact that the electronic Hamiltonian involves at most two-body interactions, so that the energy is fully determined by one- and two-body amplitudes (of either the CI or CC type). For the same reason, those equations in the chain of CC equations that arise by projection onto singles and doubles involve at most connected three- and four-body clusters T_3 and T_4, respectively, so that the decoupling of the CCSD equations from the full CC chain results by setting $T_3 = T_4 = 0$. Thus, if we know the exact T_3 and T_4 clusters, we can recover the exact FCI or full CC (FCC) energy by accounting for them in the CCSD equations.

Now, let us recall the complementarity of variational and perturbative approaches, specifically of the CI and CC methods: while the former ones can simultaneously handle a multitude of states of an arbitrary spin multiplicity, accounting well for non-dynamic correlations in cases of quasi-degeneracy, they are not size-extensive, and are unable to properly describe dynamic correlation effects, unless excessively large dimensions can be afforded. On the other hand, CC approaches are size-extensive at any level of truncation and very efficiently account for dynamic correlations, yet encounter serious difficulties in the presence of significant non-dynamic correlation effects. In view of this complementarity, a conjoint treatment, if at all feasible, seems to be highly desirable. Moreover, in the limiting case, the FCI and FCC energies are identical, and the relationship between the CI and CC amplitudes is well defined (note that any MR CISD wave function can be easily rewritten as a SR CISDTQ… one relative to a chosen SR configuration).

We can thus conveniently exploit the CI-type wave functions as a source of approximate three- and four-body amplitudes. This is precisely the basis of the so-called reduced MR (RMR) CC method [216,218,219,221]. Modest-size MR CISD wave functions are nowadays computationally very affordable, and their cluster analysis provides us with a relatively small subset of the most important three- and four-body cluster amplitudes, which can be used to correct the standard CCSD equations. Moreover, such amplitudes implicitly account for higher than four-body amplitudes as well, as long as they are present in the MR CISD wave function. In this way, we were able to properly describe even the difficult triple-bond breaking in the nitrogen molecule [217]. Amplitude-type corrections are even more useful in the MR SU CCSD approach (see below). Very similar results are obtained with the energy-correcting CCSD, in which case we employ the MR CISD wave function in the asymmetric energy formula [220,221].

Finally, let me mention our recent contribution toward the MR CC methodology that is based on a completely general model space (GMS) [202,223]. In contrast to an IMS, which results by a truncation of the CMS, we define GMS as a model space that is spanned, at least in principle, by a set of arbitrarily chosen configurations.

Exploring the cluster analysis of a finite set of FCI wave functions based on the SU CC Ansatz [224], we realized that by introducing the so-called C-conditions ('C' implying either 'constraint' or 'connectivity', as will be seen shortly), we can achieve a unique representation of a chosen finite subset of the exact FCI wave functions, while preserving the intermediate normalization. (In fact, any set of MR CI wave functions can be so represented and thus reproduced *via* an MR CC formalism.) These C-conditions simply require that the *internal amplitudes* (i.e. those associated with the excitations within the chosen GMS) be set equal to the product of all lower-order cluster amplitudes, as implied by the relationship between the CI and CC amplitudes [223], rather than by setting them equal to zero, as was done in earlier IMS-based approaches [205,206] (see also Ref. [225]). Remarkably, these conditions also warrant that all disconnected contributions, in both the effective Hamiltonian and the coupling coefficients, cancel out, leaving only connected terms [202,223].

With the GMS-based SU CCSD method, we were able to carry out a series of test calculations for model systems that allow a comparison with full CI results, considering GMSs of as high a dimension as 14. These results are most promising. Moreover, we have formulated a generalization of the RMR CCSD method, resulting in the so-called (M, N)-CCSD approach [226] that employs an M-reference MR CISD wave functions as a source of higher-than-pair clusters in an N-reference MR SU CCSD (clearly, we require that $M \geqq N$). In this way, the effect of intruders can be taken care of *via* external corrections, which are even more essential at the MR level than in the SR theory, because, in contrast to the latter, where the energy is fully determined by the one- and two-body amplitudes, the effective Hamiltonian of the MR theory always involves higher-than-pair clusters. We also mention that the idea of C-conditions can be exploited in BW MR CC method (see Ref. [227] for a preliminary study).

Although the results obtained so far are very promising, the 'specter' of intruders will not be easy to overcome if we require that a single GMS be suitable for all the geometries considered. Yet, with a dose of 'healthy' scepticism, as expressed most fittingly in the motto to this section, we hope that the MR CC formalism will eventually deliver the expected 'harvest'.

In closing, let me take a fleeting glance back in time. There is no denying that during the past four decades tremendous advances have been made in our field of endeavor, and the CC method became the method of choice in many applications. Of course, these advancements could not have been achieved without the enormous progress of computing technology. Yet, in spite of their propitiousness—indeed necessity—they are not without pitfalls, as already alluded to above. They greatly influence not only our science, but our lives as well. The ever deepening commercialization of software is certainly worrisome, if hardly avoidable. I understand that some scientists, in fact entire institutions, are 'banned' from the use of competitor's commercially available software. But such ethical problems would take us too far astray from our topic.

Undeniably, during the past four decades, decisive progress has been made in our understanding of correlation effects and in the ways of how to account for them. The coupled-cluster approaches have certainly played their role here. Nonetheless, there is still enough work left for future generations of molecular quantum theorists. Hopefully, these will come not only from the brute-force exploitation of new and more powerful hardware, but also from theoretical and algorithmic developments. And, in our striving for greater and greater accuracy and reliability, we will hopefully not forget that the building and study of useful, simple models may often be more beneficial than the blind generation of ever more accurate numbers. We certainly look forward to these advances!

> ...
> *les sauriens du calcul se glissent pondéreux*
> *écrasant les tablogs les abaques les règles*
> *Leur mères les trieuses les pères binaires*
> *et l'oncle électronique avec son regard d'aigle*
> *admirent effarés ces athlètes modestes*
> *pulvérisant les records établis par les*
> *bipèdes qui pourtant savent compter parler*
> *soigner Soigner les sauriens du calcul et les*
> *bipèdes qui...*
> (Raymond Queneau: Petite Cosmogonie Portative, VIe chant)[11]

[11] "Sauriens of computation glide weightily/smashing log-tables abacuses rulers/Their mothers (punched card) sorting machines binary fathers/and electronic uncle with his eagle gaze/scarily admiring these modest athletes/pulverizing the records established by/bipeds who after all knew how to compute to speak/to care for To care for sauriens of computation and the/bipeds who after all knew etc." See also the Footnote 1 to the opening quotation from Queneau.

7.7 ACKNOWLEDGEMENTS

Continued support by the National Sciences and Engineering Research Council of Canada is gratefully acknowledged. I am also very much indebted to several of my colleagues for kindly reading the manuscript and their useful comments, and in particular to Professor Sydney Davison and to my daughter Barbara for helping me—in Hermann Weyl's words [228]—to avoid "the 'worst blunders' in a tongue that was not sung at my cradle".

7.8 REFERENCES

1 L. Pauling and E.B. Wilson, Introduction to quantum mechanics, McGraw-Hill, New York, 1935.
2 H. Hellmann, Einführung in die Quantenchemie, F. Deuticke, Leipzig, 1937. Note that the text by Pauling and Wilson [1] appeared 2 years earlier.
3 Die Quntenchemie ist eine junge Wissenschaft, die erst ein Jahrzehnt existiert. Dennoch ist das bei der genannten Begrenzung übrig bleibende Material immer noch so groß, daß im Stoff eine gewisse Beschränkung notwendig war, um den Umfang des Buches nicht zu sehr anwachsen zu lassen.
4 W. Heitler and F. London, Z. Phys., 44 (1927) 455.
5 O. Wichterle, General and inorganic chemistry, Academy of Sciences, Prague, 1950, in Czech.
6 O. Wichterle, Organic chemistry: Structure of organic molecules, Vol. 1, Reaction mechanisms, Vol. 2, Academy of Sciences, Prague, 1952, in Czech.
7 G.W. Wheland, The theory of resonance and its application to organic chemistry, Wiley, New York, 1944, Resonance theory in organic chemistry, Wiley, New York, 1955.
8 J.K. Syrkin and M.E. Djatkina, Chemical bond and molecular structure, Mir, Moscow, 1946, in Russian, English translation: Structure of molecules and the chemical bond, Interscience, New York, 1950.
9 See, e.g. I.V. Komarov, L.I. Ponomarev and S.Yu. Slavjanov, Spheroidal and Coulomb-spheroidal functions, Nauka, Moscow, 1976, in Russian, and references therein.
10 M. Born and J.R. Oppenheimer, Ann. Phys. (Leipzig), 84 (1927) 457.
11 J.H. van Vleck and A. Sherman, Rev. Mod. Phys., 7 (1935) 167.
12 R. McWeeny, Valence bond theory. A re-examination of concepts and methodology, in: Z.B. Maksić, W.J. Orville-Thomas (Eds.), Pauling's legacy. Modern modelling of the chemical bond, Elsevier, Amsterdam, 1999, pp. 365–401.
13 D.R. Hartree, Proc. Camb. Philos. Soc., 24 (1927) 89, 45 (1948) 230.
14 V. Fock, Z. Phys., 61 (1930) 126, 62 (1930) 795, 75 (1932) 622.
15 D.R. Hartree, The calculation of atomic structures, Wiley, New York, 1957.
16 M. Klobukowski and R. Carbó (Eds.), Self-consistent field: Theory and applications, Elsevier, Amsterdam, 1990.
17 D.R. Hartree and W. Hartree, Proc. R. Soc. (Lond.) A, 154 (1936) 588.
18 C. Froese Fischer, Douglas Rayner Hartree: His life in science and computing, World Scientific Publishing, New Jersey, 2003.
19 C.C.J. Roothaan, Rev. Mod. Phys., 23 (1951) 69.
20 A.C. Wahl, J. Chem. Phys., 41 (1964) 2600.
21 I. Shavitt, The method of configuration interaction, in: H.F. Schaefer, III (Ed.), Methods of electronic structure theory, Plenum Press, New York, 1977, pp. 189–275.
22 J. Paldus, J. Chem. Phys., 61 (1974) 5321.
23 J. Paldus, Many-electron correlation problem. A group theoretical approach, in: H. Eyring, D. Henderson (Eds.), Theoretical chemistry: Advances and perspectives, Vol. 2, Academic P, New York, 1976, pp. 131–290.
24 Chr. Møller and M.S. Plesset, Phys. Rev., 46 (1934) 618.
25 See, e.g. Tables 1 and 2 of Ref. [26]. Odd-order PT contributions often have an opposite sign (*cf.*, e.g. Table 3 of R.J. Bartlett, Annu. Rev. Phys. Chem., 32 (1981) 359) and for certain model Hamiltonians

even systematically vanish (*cf.*, e.g. M. Takahashi, P. Bracken, J. Čížek and J. Paldus, Int. J. Quantum Chem., 53 (1995) 457).
26 S. Hirata and R.J. Bartlett, Chem. Phys. Lett., 321 (2000) 216.
27 R.J. Bartlett and I. Shavitt, Chem. Phys. Lett., 50 (1977) 190, 57 (1978) 157 (E).
28 V.V. Tolmachev, Adv. Chem. Phys., 14 (1969) 421, 471.
29 B.H. Brandow, Rev. Mod. Phys., 39 (1967) 771, Adv. Quantum Chem., 10 (1977) 187.
30 I. Lindgren and J. Morrison, Atomic many-body theory, Springer, Berlin, 1982.
31 K.A. Brueckner, Phys. Rev., 100 (1955) 36.
32 J. Goldstone, Proc. R. Soc. (Lond.) A, 239 (1957) 267.
33 N.M. Hugenholtz, Physica (Utrecht), 23 (1957) 481.
34 J. Hubbard, Proc. R. Soc. (Lond.) A, 240 (1957) 539, 243 (1958) 336, 244 (1958) 199.
35 P.-O. Löwdin, Adv. Chem. Phys., 2 (1959) 207.
36 P.-O. Löwdin, Discussion of the Hartree–Fock approximation, in: P. Lykos, G.W. Pratt (Eds.), Rev. Mod. Phys., 35, 1963, p. 496.
37 P.-O. Löwdin, Adv. Chem. Phys., 14 (1969) 283, see also P.-O. Löwdin and I. Mayer, Adv. Quantum Chem., 24 (1992) 79.
38 P.-O. Löwdin, Phys. Rev., 97 (1955) 1474, 1490, 1509.
39 P.-O. Löwdin and O. Goscinski, Int. J. Quantum Chem. Symp., 5 (1971) 685, and references therein.
40 R. Pauncz, J. de Heer and P.-O. Löwdin, J. Chem. Phys., 36 (1962) 2247, 2257.
41 R. Pauncz, Alternant molecular orbital method, Saunders, Philadelphia, 1967, and references therein.
42 M. Gell-Mann and F. Low, Phys. Rev., 84 (1951) 350.
43 C. Bloch, Nucl. Phys., 7 (1958) 451.
44 J. Paldus, Perturbation theory, in: G.W.F. Drake (Ed.), Atomic, molecular and optical physics handbook, Section 2, Mathematical methods, Chapter 5, American Institute of Physics, New York, 1996, pp. 76–87.
45 J. Paldus and X. Li, Adv. Chem. Phys., 110 (1999) 1.
46 J. Paldus and J. Čížek, Adv. Quantum Chem., 9 (1975) 105.
47 J. Paldus, Diagrammatical methods for many-Fermion systems (lecture notes), University of Nijmegen, Nijmegen, The Netherlands, 1981.
48 J. Čížek, Adv. Chem. Phys., 14 (1969) 35.
49 J. Paldus and X. Li, Adv. Chem. Phys., 110 (1999) 1, see Appendix A.
50 A.P. Jucys, I.B. Levinson and V.V. Vanagas, Mathematical apparatus of the theory of angular momentum, Mintis, Vilnius, 1960, in Russian, English translations: Israel program for scientific translations, Jerusalem, 1962, and Gordon & Breach, New York, 1964.
51 E. El Baz and B. Castel, Graphical methods of spin algebras in atomic, nuclear and particle physics, Marcel Dekker, New York, 1972.
52 J. Paldus, B.G. Adams and J. Čížek, Int. J. Quantum Chem., 11 (1977) 813.
53 J. Paldus, J. Chem. Phys., 67 (1977) 303.
54 V.V. Tolmachev, Field theoretical form of the perturbation theory applied to atomic and molecular many-electron problems, University of Tartu, 1963, in Russian.
55 N.H. March, W.H. Young and S. Sampanthar, The many body problem in quantum mechanics, Cambridge University Press, Cambridge, 1967.
56 A.L. Fetter and J.D. Walecka, Quantum theory of many-particle systems, McGraw-Hill, New York, 1971.
57 J. Linderberg and Y. Öhrn, Propagators in quantum chemistry, Academic Press, New York, 1973.
58 E.K.U. Gross, E. Runge and O. Heinonen, Many-particle theory, Hilger, Bristol, 1991.
59 F.E. Harris, H.J. Monkhorst and D.L. Freeman, Algebraic and diagrammatic methods in many-Fermion theory, Oxford University Press, Oxford, 1992.
60 W. Macke, Z. Naturforsch., 5a (1950) 192.
61 M. Gell-Mann and K.A. Brueckner, Phys. Rev., 106 (1957) 364.
62 K.A. Brueckner and J.L. Gammel, Phys. Rev., 109 (1958) 1023.
63 H.P. Kelly, Adv. Chem. Phys., 14 (1969) 129, and references therein.
64 O. Sinanoğlu, J. Chem. Phys., 36 (1962) 706, Adv. Chem. Phys., 6 (1964) 315, 14 (1969) 237, and references therein.

65 O. Sinanoğlu and J. Čížek, Chem. Phys. Lett., 1 (1967) 337.
66 R.K. Nesbet, Adv. Chem. Phys., 14 (1969) 1, and references therein.
67 H. Primas, Separability in many-body systems, in: O. Sinanoğlu (Ed.), Modern quantum chemistry, part II, Academic, New York, 1965, pp. 45–74.
68 For original references, see M.J. Mayer and M.G. Mayer, Statistical mechanics, Wiley, New York, 1940, Chapter 13. For a quantum statistical version of the cluster expansion, see B. Khan and G.E. Uhlenbeck, Physica (Utrecht), 5 (1938) 399.
69 H. Kümmel, Theor. Chim. Acta, 80 (1991) 81.
70 F. Coester, Nucl. Phys., 7 (1958) 421; F. Coester and H. Kümmel, Nucl. Phys., 17 (1960) 477.
71 For a review and further references, see H.A. Bethe, Phys. Rev., 103 (1956) 1353.
72 R.J. Bartlett, Theor. Chem. Acc., 103 (2000) 273.
73 J. Čížek, J. Chem. Phys., 45 (1966) 4256, Also reprinted in: O. Sinanoğlu, K.A. Brueckner (Eds.), Three approaches to electron correlation in atoms, Yale University Press, New Haven, Connecticut, 1970, pp. 350–360.
74 R. McWeeny, Theoretical chemistry—a link between disciplines (inaugural lecture delivered 8 February 1967), University of Sheffield, Sheffield, 1967, pp. 8–9.
75 R.F. Bishop, T. Brandes, K.A. Gernoth, N.R. Walet and Y. Xian (Eds.), Recent progress in many-body theories (Proceedings of the 11th international conference), World Scientific Publishing, Singapore, 2002.
76 H.G. Kümmel, A biography of the coupled cluster method, in: R.F. Bishop, T. Brandes, K.A. Gernoth, N.R. Walet, Y. Xian (Eds.), Recent progress in many-body theories, Proceedings of the 11th international conference, World Scientific Publishing, Singapore, 2002, pp. 334–348.
77 M.A. Robb, Theor. Chem. Acc., 103 (2000) 317.
78 J. Paldus, Many-electron correlation problem and Lie algebras, Contemporary mathematics, Vol. 160, American Mathematical Society, Providence, RI, 1994, pp. 209–236.
79 F.A. Matsen and R. Pauncz, The unitary group in quantum chemistry, Elsevier, Amsterdam, 1986.
80 J. Paldus, Dynamical groups, in: G.W.F. Drake (Ed.), Atomic, molecular and optical physics handbook, Section 2, Mathematical methods, Chapter 4, American Institute of Physics, New York, 1996, pp. 65–75.
81 V.V. Tolmachev, The Fermi-gas theory, Moscow University Publishers, Moscow, 1973, in Russian.
82 B. Jeziorski, J. Paldus and P. Jankowski, Int. J. Quantum Chem., 56 (1995) 129.
83 X. Li and J. Paldus, J. Chem. Phys., 101 (1994) 8812.
84 P. Jankowski and B. Jeziorski, J. Chem. Phys., 111 (1999) 1857.
85 J. Čížek and J. Paldus, Int. J. Quantum Chem., 5 (1971) 359.
86 J. Paldus, J. Čížek and B. Jeziorski, J. Chem. Phys., 90 (1989) 4356, 93 (1990) 1485.
87 J.A. Pople, M. Head-Gordon and K. Raghavachari, J. Chem. Phys, 87 (1987) 5968, 90 (1989) 4635.
88 W. Kutzelnigg, Pair correlation theories, in: H.F. Schaefer, III (Ed.), Methods of electronic structure theory, Plenum, New York, 1977, pp. 129–188.
89 R.A. Chiles and C.E. Dykstra, J. Chem. Phys., 74 (1981) 4544.
90 M. Moshinsky, Group theory and the many-body problem, Gordon & Breach, New York, 1968.
91 K.H. Lührmann and H. Kümmel, Nucl. Phys. A, 194 (1972) 225.
92 R. Pauncz, The construction of spin eigenfunctions: An exercise book, Kluwer Academic, New York, 2000, and references therein.
93 E.H. Lieb and F.Y. Wu, Phys. Rev. Lett., 20 (1968) 1445.
94 J. Čížek, J. Paldus and L. Šroubková, Int. J. Quantum Chem., 3 (1969) 149.
95 J. Čížek and J. Paldus, J. Chem. Phys., 47 (1967) 3976.
96 J. Čížek and J. Paldus, J. Chem. Phys., 53 (1970) 821.
97 J. Paldus and J. Čížek, Phys. Rev. A, 2 (1970) 2268.
98 H. Fukutome, Int. J. Quantum Chem., 20 (1981) 955, and references therein.
99 J. Paldus, Hartree–Fock stability and symmetry breaking, in: M. Klobukowski, R. Carbó (Eds.), Self-consistent field: Theory and applications, Elsevier, Amsterdam, 1990, pp. 1–45.
100 J.L. Stuber and J. Paldus, Symmetry breaking in the independent particle model, in: E.J. Brändas, E.S. Kryachko (Eds.), Fundamental world of quantum chemistry, Vol. 1, Kluwer, Amsterdam, 2003, pp. 67–139.

101 J. Paldus and J. Čížek, Can. J. Chem., 63 (1985) 1803.
102 J. Paldus and J. Čížek, Chem. Phys. Lett., 3 (1969) 1.
103 J. Paldus and J. Čížek, J. Chem. Phys., 52 (1970) 2919.
104 J. Paldus and J. Čížek, J. Chem. Phys., 54 (1971) 2293.
105 J. Paldus and J. Čížek, Chem. Phys. Lett., 3 (1969) 1, see acknowledgment.
106 P. Piecuch and J. Paldus, Int. J. Quantum Chem., Symp., 25 (1991) 9.
107 J. Paldus and M.J. Boyle, Int. J. Quantum Chem., 22 (1982) 1281.
108 R. Podeszwa, S.A. Kucharski and Z. Stolarczyk, J. Chem. Phys., 116 (2002) 480.
109 J. Paldus, J. Čížek and M. Takahashi, Phys. Rev. A, 30 (1984) 2193.
110 J. Paldus, M. Takahashi and R.W.H. Cho, Phys. Rev. B, 30 (1984) 4267.
111 M. Takahashi and J. Paldus, Phys. Rev. B, 31 (1985) 5121.
112 P. Piecuch, R. Toboła and J. Paldus, Phys. Rev. A, 54 (1996) 1210.
113 R.A. Chiles and C.E. Dykstra, Chem. Phys. Lett., 80 (1981) 69.
114 S.M. Bachrach, R.A. Chiles and C.E. Dykstra, J. Chem. Phys., 75 (1981) 2270.
115 C.E. Dykstra, S.-Y. Liu, M.F. Daskalakis, J.P. Lucia and M. Takahashi, Chem. Phys. Lett., 137 (1987) 266.
116 C.E. Dykstra, Ab initio calculation of the structure and properties of molecules, Elsevier, Amsterdam, 1988.
117 C.E. Dykstra and E.R. Davidson, Int. J. Quantum Chem., 78 (2000) 226.
118 J. Paldus, J. Čížek and I. Shavitt, Phys. Rev. A, 5 (1972) 50.
119 J.A. Pople, R. Krishnan, H.B. Schlegel and J.S. Binkley, Int. J. Quantum Chem., 14 (1978) 545.
120 R.J. Bartlett and G.D. Purvis, III, Int. J. Quantum Chem., 14 (1978) 561.
121 W. Meyer, J. Chem. Phys., 64 (1976) 2901; C.E. Dykstra, H.F. Schaefer, III and W. Meyer, J. Chem. Phys., 65 (1976) 2740.
122 G.D. Purvis, III and R.J. Bartlett, J. Chem. Phys., 76 (1982) 1910.
123 ACES II, a CC and MBPT suite of codes by J.F. Stanton, J. Gauss, J.D. Watts, W.J. Lauderdale and R.J. Bartlett, Int. J. Quantum Chem. Symp., 26 (1992) 879.
124 GAUSSIAN 92, a system of codes written by M.J. Frisch, G.W. Trucks, M. Head-Gordon, P.M.V. Gill, M.W. Wong, J.B. Foresman, B.G. Johnson, H.B. Schlegel, M.A. Robb, E.S. Replogle, R. Gomperts, J.L. Andres, K. Raghavachari, J.S. Binkley, C. Gonzales, R.L. Martin, D.J. Fox, D.J. Defrees, J. Baker, J.J.P. Stewart and J.A. Pople, Gaussian, Inc., Pittsburgh, Pennsylvania.
125 MOLCAS-3, a system of programs by K. Andersson, M.R.A. Blomberg, M.P. Fülscher, V. Kellö, R. Lindh, P.-Å. Malmqvist, J. Noga, J. Olsen, B.O. Roos, A.J. Sadlej, P.E.M. Siegbahn, M. Urban and P.-O. Widmark, University of Lund, Lund, Sweden, 1994; MOLCAS-4, K. Andersson, M.R.A. Blomberg, M.P. Fülscher, G. Karlström, R. Lindh, P.-Å. Malmqvist, P. Neogrády, J. Olsen, B.O. Roos, A.J. Sadlej, M. Schütz, L. Seijo, L. Serrano-Andrés, P.E.M. Siegbahn and P.-O. Widmark, University of Lund, Lund, Sweden, 1997.
126 MOLPRO 94, a system of programs by H.-J. Werner and P.J. Knowles, with contributions by J. Almlöf, R.D. Amos, M. Deegan, S.T. Elbert, C. Hampel, W. Meyer, K.A. Peterson, R.M. Pitzer, E.-A. Reinsch, A.J. Stone and P.R. Taylor.
127 J. Paldus and B. Jeziorski, Theor. Chem. Acta, 73 (1988) 81.
128 P. Pulay, Analytical derivative techniques and the calculation of vibrational spectra, in: D.R. Yarkony (Ed.), Modern electronic structure theory, part II, World Scientific Publishing, Singapore, 1995, pp. 1191–1240.
129 P. Jørgensen and J. Simons, J. Chem. Phys., 79 (1983) 334.
130 L. Adamowicz, W.D. Laidig and R.J. Bartlett, Int. J. Quantum Chem. Symp., 18 (1984) 245.
131 G. Fitzgerald, R.J. Harrison, W.D. Laidig and R.J. Bartlett, J. Chem. Phys., 82 (1985) 4379.
132 R.J. Bartlett, Analytical evaluation of gradients in coupled-cluster and many-body perturbation theory, in: P. Jørgensen, J. Simons (Eds.), Geometric derivatives of energy surfaces and molecular properties, Reidel, Dordrecht, 1986, pp. 35–61.
133 G. Fitzgerald, R.J. Harrison and R.J. Bartlett, J. Chem. Phys., 85 (1986) 5143.
134 A.C. Scheiner, G.E. Scuseria, J.E. Rice, T.J. Lee and H.F. Schaefer, III, J. Chem. Phys., 87 (1987) 5361.

135 H. Koch, H.J.Aa. Jensen, P. Jørgensen, T. Helgaker, G.E. Scuseria and H.F. Schaefer, III, J. Chem. Phys., 92 (1989) 4924.

136 E.A. Salter, W. Trucks and R.J. Bartlett, J. Chem. Phys., 90 (1989) 1752.

137 E.A. Salter and R.J. Bartlett, J. Chem. Phys., 90 (1989) 1767.

138 J. Noga and R.J. Bartlett, J. Chem. Phys., 86 (1987) 7041, 89 (1988) 3401 (E).

139 J.D. Watts and R.J. Bartlett, J. Chem. Phys., 93 (1990) 6104.

140 S.A. Kucharski and R.J. Bartlett, J. Chem. Phys., 97 (1992) 4282.

141 M. Urban, J. Noga, S.J. Cole and R.J. Bartlett, J. Chem. Phys., 83 (1985) 4041.

142 K. Raghavachari, J. Chem. Phys., 82 (1985) 4607.

143 A.M. MusiaŁ, S.A. Kucharski and R.J. Bartlett, Chem. Phys. Lett., 320 (2000) 542.

144 M. Kállay and P. Surján, J. Chem. Phys., 113 (2000) 1359, P. Surján and M. Kállay, J. Mol. Struct. (THEOCHEM), 547 (2001) 145.

145 T.J. Lee and G.E. Scuseria, Achieving chemical accuracy with coupled-cluster theory, in: S.R. Langhoff (Ed.), Quantum mechanical electronic structure calculations with chemical accuracy, Kluwer, Dordrecht, The Netherlands, 1995, pp. 47–108.

146 P. Piecuch and K. Kowalski, In search of the relationship between multiple solutions characterizing coupled-cluster theories, in: J. Leszczynski (Ed.), Computational chemistry: Reviews of current trends, Vol. 5, World Scientific Publishing, Singapore, 2000, pp. 1–104; K. Kowalski and P. Piecuch, J. Chem. Phys., 113 (2000) 18.

147 K. Kowalski and P. Piecuch, J. Chem. Phys., 113 (2000) 5644, Chem. Phys. Lett., 344 (2001) 165; P. Piecuch, S.A. Kucharski and K. Kowalski, Chem. Phys. Lett., 344 (2001) 176; K. Kowalski and P. Piecuch, J. Mol. Struct. (THEOCHEM), 547 (2001) 191; P. Piecuch, I.S.O. Pimienta, P.-F. Fan and K. Kowalski, New alternatives for electronic structure calculations: Renormalized, extended, and generalized coupled-cluster theories, in: J. Maruani, R. Lefebvre, E. Brändas, (Eds.), Progress in theoretical chemistry and physics. Advanced topics in theoretical chemical physics, Vol. 12, Kluwer Dordrecht, The Netherlands, 2003, pp. 119–206; K. Kowalski and P. Piecuch, J. Chem. Phys., 120 (2004) 1715, and references therein.

148 H. Sekino and R.J. Bartlett, Int. J. Quantum Chem. Symp., 18 (1984) 255.

149 E.A. Salter, H. Sekino and R.J. Bartlett, J. Chem. Phys., 87 (1987) 502.

150 J.F. Stanton and R.J. Bartlett, J. Chem. Phys., 98 (1993) 7029.

151 H.J. Monkhorst, Int. J. Quantum Chem. Symp., 11 (1977) 421.

152 J. Paldus, J. Čížek, M. Saute and A. Laforgue, Phys. Rev. A, 17 (1978) 805.

153 M. Saute, J. Paldus and J. Čížek, Int. J. Quantum Chem., 15 (1979) 463.

154 H. Nakatsuji and K. Hirao, J. Chem. Phys., 68 (1978) 2053.

155 H. Nakatsuji, SAC-CI method: Theoretical aspects and some recent topics, in: J. Leszczynski (Ed.), Computational chemistry: Reviews of current trends, Vol. 2, World Scientific Publishing, Singapore, 1977, pp. 62–124, Chapter 2.

156 J. Paldus and J. Čížek, J. Chem. Phys., 60 (1974) 149.

157 J. Geertsen and J. Oddershede, J. Chem. Phys., 85 (1986) 2112.

158 J. Geertsen, S. Ericksen and J. Oddershede, Adv. Quantum Chem., 22 (1991) 167.

159 K. Emrich, Nucl. Phys. A, 351 (1981) 379.

160 M. Nooijen and R.J. Bartlett, J. Chem. Phys., 106 (1997) 6441, 6449.

161 M. Nooijen and R.J. Bartlett, J. Chem. Phys., 107 (1997) 6812.

162 G.E. Scuseria and H.F. Schaefer, III, Chem. Phys. Lett., 152 (1988) 382.

163 N. Oliphant and L. Adamowicz, J. Chem. Phys., 94 (1991) 1229.

164 P. Piecuch and L. Adamowicz, J. Chem. Phys., 100 (1994) 5792, 5857.

165 M. Kállay and J. Gauss, J. Chem. Phys., 120 (2004) 6841.

166 U. Kaldor (Ed.), Many-body methods in quantum chemistry, lecture notes in chemistry, Vol. 52, Springer, Berlin, 1989.

167 R.J. Bartlett (Ed.), Proceedings of the workshop on coupled cluster theory at the interface of atomic physics and quantum chemistry, Theor. Chim. Acta, Vol. 80, 1991, Nos. 2–6.

168 R.J. Bartlett (Ed.), Recent advances in coupled-cluster methods, Recent advances in computational chemistry, Vol. 3, World Scientific Publishing, Singapore, 1997.

169 T.L. Ainsworth, C.E. Campbell, B.E. Clements and E. Krotscheck (Eds.), Recent progress in many-body theories, Vol. 3, Plenum, New York, 1992.
170 R.F. Bishop, T. Brandes, K.A. Gernoth, N.R. Walet and Y. Xian (Eds.), Recent progress in many-body theories, Advances in quantum many-body theories, Vol. 6, World Scientific Publishing, Singapore, 2002.
171 R.J. Bartlett, J. Phys. Chem., 93 (1989) 1693.
172 J. Paldus, Coupled cluster theory, in: S. Wilson, G.H.F. Diercksen (Eds.), Methods in computational molecular physics, NATO ASI series, Series B: Physics, Vol. 293, Plenum Press, New York, 1992, pp. 99–194.
173 R.J. Bartlett and J.P. Stanton, Applications of post-Hartree–Fock methods: A tutorial, in: K.B. Lipkowitz, D.B. Boyd (Eds.), Reviews in computational chemistry, Vol. 5, VCH Publishers, New York, 1994, pp. 65–169.
174 J. Paldus, Algebraic approach to coupled cluster theory, in: G.L. Malli (Ed.), Relativistic and correlation effects in molecules and solids, NATO ASI series, Series B: Physics, vol. 318, Plenum Press, New York, 1994, pp. 207–282.
175 R.J. Bartlett, Coupled-cluster theory: An overview of recent developments, in: D.R. Yarkony (Ed.), Modern electronic structure theory, part II, World Scientific Publishing, Singapore, 1995, pp. 1047–1131.
176 T.D. Crawford and H.F. Schaefer, III, An introduction to coupled cluster theory for computational chemists, in: K.B. Lipkowitz, D.B. Boyd (Eds.), Reviews in computational chemistry, Vol. 14, Wiley, New York, 2000, pp. 33–136, Chapter 2.
177 J.F. Stanton and J. Gauss, Int. Rev. Phys. Chem., 19 (2000) 61.
178 J. Paldus, Coupled cluster methods, in: S. Wilson (Ed.), Handbook of molecular physics and quantum chemistry, Vol. 2, Wiley, Chichester, 2003, pp. 272–313, Part 3, Chapter 19.
179 S. Wilson, Electron correlation in molecules, Clarendon Press, Oxford, 1984.
180 R. McWeeny, Methods of molecular quantum mechanics, 2nd edn., Academic, London, 1989, Chapter 9.
181 J. Gauss, Coupled cluster theory, in: P. von R. Schleyer (Ed.), Encyclopedia of computational chemistry, Vol. 1, Wiley, New York, 1998, pp. 615–636.
182 I. Shavitt and R.J. Bartlett, Many-body methods in quantum chemistry, Cambridge University Press, Cambridge, 2005, in press.
183 G. Hasberg and H. Kümmel, Phys. Rev. C, 33 (1986) 1367.
184 R. McDonald and N.R. Walet, Coupled cluster calculations of the Schwinger model in Hamiltonian lattice gauge theory, in: R.F. Bishop, T. Brandes, K.A. Gernoth, N.R. Walet, Y. Xian (Eds.), Recent progress in many-body theories, Advances in quantum many-body theories, Vol. 6, World Scientific Publishing, Singapore, 2002, pp. 405–408.
185 R.F. Bishop, Theor. Chem. Acta, 80 (1991) 95.
186 J. Arponen, Ann. Phys. (NY), 151 (1983) 311.
187 D. Mukherjee, Chem. Phys. Lett., 192 (1992) 55; S.H. Mandal, R. Ghosh, G. Sanyal and D. Mukherjee, A finite-temperature generalization of the coupled-cluster method: A non-perturbative access to grand partition functions, in: R.F. Bishop, T. Brandes, K.A. Gernoth, N.R. Walet, Y. Xian (Eds.), Recent progress in many-body theories. Advances in quantum many-body theories, Vol. 6, World Scientific Publishing, Singapore, 2002, pp. 383–392.
188 I. Lindgren, Many-body problems in atomic physics, in: T.L. Ainsworth, C.E. Campbell, B.E. Clements, E. Krotscheck (Eds.), Recent progress in many-body theories, Vol. 3, Plenum, New York, 1992, pp. 245–276, and references therein.
189 U. Kaldor, Relativistic coupled cluster method and applications, in: R.J. Bartlett (Ed.), Recent advances in coupled-cluster methods, Recent advances in computational chemistry, Vol. 3, World Scientific Publishing, Singapore, 1997, pp. 125–153, and references 45–58 therein.
190 S.A. Blundell, A.C. Hartley, Z. Liu, A.-M. Mårtensson-Pendrill and J. Sapirstein, Theor. Chim. Acta, 80 (1991) 257.
191 B. Jeziorski and R. Moszynski, Int. J. Quantum Chem., 48 (1993) 161.
192 I. Lindgren, Int. J. Quantum Chem. Symp., 12 (1978) 33.
193 I. Lindgren and D. Mukherjee, Phys. Rep., 151 (1987) 93.

194 D. Mukherjee and S. Pal, Adv. Quantum Chem., 20 (1981) 292.
195 B. Jeziorski and H.J. Monkhorst, Phys. Rev. A, 24 (1981) 1668.
196 B. Jeziorski and J. Paldus, J. Chem. Phys., 90 (1989) 2714.
197 K. Jankowski, J. Paldus, I. Grabowski and K. Kowalski, J. Chem. Phys., 97 (1992) 7600, 101 (1994) 1759 (E), 101 (1994) 3085.
198 U. Kaldor, Theor. Chem. Acta., 80 (1991) 427, see also reference 5 of A. Landau, E. Eliav and U. Kaldor, Intermediate Hamiltonian Fock-space coupled cluster method and applications, in: R.F. Bishop, T. Brandes, K.A. Gernoth, N.R. Walet, Y. Xian (Eds.), Recent progress in many-body theories Advances in quantum many-body theories, Vol. 6, World Scientific Publishing, Singapore, 2002, pp. 355–364, for reviews of relativistic VU CC methods.
199 K. Jankowski and P. Malinowski, J. Phys. B, 27 (1994) 829, 1287; P. Malinowski and K. Jankowski, Phys. Rev. A, 51 (1995) 4583, and references therein.
200 H. Meißner and E.O. Steinborn, Int. J. Quantum Chem., 63 (1997) 257; H. Meißner and J. Paldus, J. Chem. Phys., 113 (2000) 2594, 2612, 2622; H. Meißner and J. Paldus, Coll. Czech. Chem. Commun., 66 (2001) 1164, and references therein.
201 S. Ben-Shlomo and U. Kaldor, J. Chem. Phys., 89 (1988) 956.
202 X. Li and J. Paldus, J. Chem. Phys., 119 (2003) 5346.
203 U. Kaldor, Atomic and molecular applications of the multireference coupled-cluster method, in: U. Kaldor (Ed.), Many-body methods in quantum chemistry, lecture notes in chemistry, Vol. 52, Springer, Berlin, 1989, pp. 199–213, J. Chem. Phys., 87 (1987) 467.
204 S. Pal, M. Rittby, R.J. Bartlett, D. Sinha and D. Mukherjee, J. Chem. Phys., 88 (1988) 4357.
205 L. Meissner and R.J. Bartlett, J. Chem. Phys., 92 (1990) 561.
206 L. Meissner, S.A. Kucharski and R.J. Bartlett, J. Chem. Phys., 91 (1989) 6187.
207 R.J. Bartlett, Theor. Chim. Acta., 80 (1991) 71.
208 I. Hubač, Size extensive Brillouin–Wigner coupled cluster theory, in: A. Tsipis, V.S. Popov, D.R. Herschbach, J.S. Avery (Eds.), New methods in quantum theory, NATO ASI series, Series 3: High technology, Vol. 8, Kluwer, Dordrecht, 1996, pp. 183–202.
209 J. Mášik and I. Hubač, Adv. Quantum Chem., 31 (1999) 75, and references therein.
210 J. Pittner, P. Nachtigall, P. Čársky, J. Mášik and I. Hubač, J. Chem. Phys., 110 (1999) 10 275.
211 I. Hubač and S. Wilson, J. Phys. B, 33 (2000) 365; I. Hubač, J. Pittner and P. Čársky, J. Chem. Phys., 112 (2000) 8779.
212 J. Pittner, O. Demel, P. Čársky and I. Hubač, Int. J. Mol. Sci., 2 (2002) 281; J. Pittner, P. Čársky and I. Hubač, Int. J. Quantum Chem., 90 (2002) 1031; J. Pittner, J. Šmydke, P. Čársky and I. Hubač, J. Mol. Struct. (THEOCHEM), 547 (2001) 239; J.C. Sancho-García, J. Pittner, P. Čársky and I. Hubač, J. Chem. Phys., 112 (2000) 8785; J. Pittner, P. Nachtigall, P. Čársky and I. Hubač, J. Phys. Chem. A, 105 (2001) 1354; I.S.K. Kerkines, J. Pittner, P. Čársky, A. Mavridis and I. Hubač, J. Chem. Phys., 117 (2002) 9733.
213 J. Pittner, J. Chem. Phys., 118 (2003) 10876.
214 J. Paldus and J. Planelles, Theor. Chim. Acta., 89 (1994) 13; J. Planelles, X. Li and J. Paldus, Theor. Chim. Acta, 89 (1994) 33, 59; P. Piecuch, R. Toboła and J. Paldus, Phys. Rev. A, 54 (1996) 1210; G. Peris, J. Planelles and J. Paldus, Int. J. Quantum Chem., 62 (1997) 137; X. Li, G. Peris, J. Planelles, F. Rajadell and J. Paldus, J. Chem. Phys., 107 (1997) 90; G. Peris, F. Rajadell, X. Li, J. Planelles and J. Paldus, Mol. Phys., 94 (1998) 235.
215 L.Z. Stolarczyk, Chem. Phys. Lett., 217 (1994) 1.
216 X. Li and J. Paldus, J. Chem. Phys., 107 (1997) 6257; 108 (1998) 637; 110 (1999) 2844; 118 (2003) 2470; Chem. Phys. Lett., 286 (1998) 145; Collect. Czech. Chem. Commun., 63 (1998) 1381; Mol. Phys., 98 (2000) 1185; Int. J. Quantum Chem., 80 (2000) 743.
217 X. Li and J. Paldus, J. Chem. Phys., 113 (2000) 9966.
218 J. Paldus and X. Li, Electron correlation in small molecules: Grafting CI onto CC, in: P. Surján (Ed.), Correlation and localization, Topics in current chemistry, Vol. 203, Springer, Berlin, 1999, pp. 1–20.
219 X. Li and J. Paldus, Simultaneous account of dynamic and nondynamic correlations based on complementarity of CI and CC approaches, in: M.R. Hoffmann, K.G. Dyall (Eds.), Low-lying potential-energy surfaces, ACS symposium series no. 828, ACS Books, Washington, 2002, pp. 10–30.

220 X. Li and J. Paldus, J. Chem. Phys., 115 (2001) 5759, 5774, 117 (2002) 1941, 118 (2003) 2470; J. Paldus and X. Li, Collect. Czech. Chem. Commun., 68 (2003) 554.
221 J. Paldus and X. Li, Coupled-cluster approach to correlation in small molecules. Energy vs. amplitude corrected methods, in: R.F. Bishop, T. Brandes, K.A. Gernoth, N.R. Walet, Y. Xian (Eds.), Recent progress in many-body theories, Advances in quantum many-body theories, Vol. 6, World Scientific Publishing, Singapore, 2002, pp. 393–404.
222 U.S. Mahapatra, B. Datta and J. Mukherjee, J. Chem. Phys., 110 (1999) 6171, and references therein.
223 X. Li and J. Paldus, J. Chem. Phys., 119 (2003) 5320, Int. J. Quantum Chem., 99 (2004) 914.
224 J. Paldus and X. Li, J. Chem. Phys., 118 (2003) 6769.
225 J. Paldus and X. Li, Collect. Czech. Chem. Commun., 69 (2004) 90.
226 X. Li and J. Paldus, J. Chem. Phys., 119 (2003) 5334.
227 J. Pittner, X. Li and J. Paldus, Mol. Phys., in press.
228 H. Weyl, Classical groups, Princeton University Press, Princeton, 1939.

Theory and Applications of Computational Chemistry: The First Forty Years
Edited by C. Dykstra et al.

CHAPTER 8

Controlling quantum phenomena with photonic reagents

Herschel Rabitz

Princeton University, Princeton, NJ, USA

Abstract

Efforts at controlling molecular dynamics and other quantum phenomena with lasers have a history going back to the early 1960s. This quest has followed a torturous evolution, but recent years have seen dramatic successes beginning to emerge. These laboratory advances utilize shaped ultra-fast laser pulses as a special class of photonic reagents having a fleeting existence, but with the capability of permanently altering molecules and materials in specific ways. Theoretical concepts and modeling are providing the basis for directing and analyzing the experiments. Although attempts at photonic reagent control of quantum dynamics is a subject with a 40-year history, the field may be viewed as just a few years young given only the recent emergence of successful experiments on physically and chemically interesting systems. It is anticipated that theory and modeling will continue to play leading roles in the further development of this field.

8.1 HOW CAN CONTROL OF QUANTUM DYNAMICS PHENOMENA BE ACHIEVED?

Light is well known to be capable of influencing chemical reactions, and this capability is the foundation of photochemistry [1]. The advent of lasers, starting in the early 1960s, was initially viewed in this context as a tool in the natural progression of the photochemical field. In particular, lasers were thought of as high-intensity monochromatic sources of radiation, which under favorable circumstances could resonantly excite a vibrational mode or chemical bond, subsequently influencing the molecule's reactivity [2]. Many efforts ensued in the 1960s and 1970s along these lines with various laser sources [3]. The ability of lasers to create molecular excitations was documented in

References pp. 162–163

many studies, including some involving chemical reactions [4]. A critical missing component of these efforts was controlled selectivity whereby, for example, one particular bond *versus* that of another is dissociated. A most challenging objective is the breaking of a strong bond over that of other weaker ones in the same molecule. Jumping ahead to the present time, it is now attractive to view a shaped laser pulse as a *photonic reagent* [5] as sketched out in Fig. 8.1, depicting the electric field of a laser pulse. The structure in the pulse implies that it is multi-spectral in contrast to the earlier approaches with continuous wave (cw) lasers. The detailed multi-spectral character of a shaped laser pulse is generally necessary to fully take over the dynamics of the molecule and steer the system from its initial state to the desired final state at time T. In this respect, the entire molecule can be thought of as an antenna with components simultaneously accepting different portions of the control pulse to ultimately create the desired evolution $|\psi(0)\rangle \rightarrow |\psi(T)\rangle$.

The discussion in the last paragraph jumps from the initial concepts of utilizing cw lasers for manipulating molecular scale chemical and physical events in the 1960s up to the current perspective employing tailored photonic reagents in 2004. However, a number of events occurred in between, which are significant for appreciating the present state of the field and perhaps where it might be heading. In the late 1970s and the early 1980s, considerable frustration was evident in the field, as extensive effort had already gone into attempting to redirect chemical and physical processes with lasers, but the goal of achieving good selectivity appeared difficult to reach [6]. It was recognized that the underling laser driven dynamics is quantum mechanical and therefore could involve quantum wave interferences [7], thereby suggesting that the desired selectivity could be attained by exploiting two constructively interfering quantum mechanical pathways for reaching the target. This proposal involved the use of radiation having two colors where each is associated with a particular pathway. Simulations and ultimately

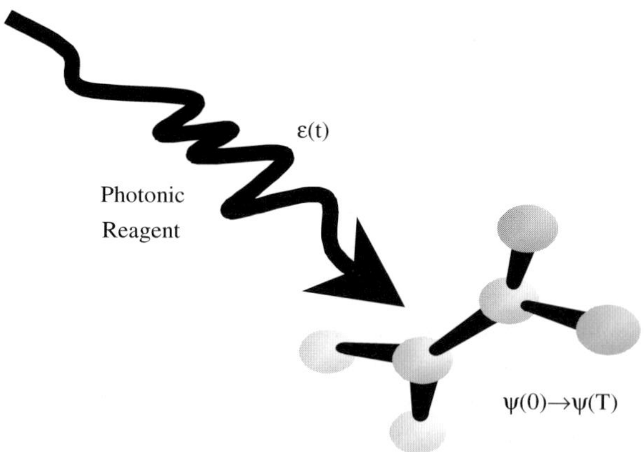

Fig. 8.1. A photonic reagent consisting of a shaped laser electric field $\varepsilon(t)$ interacts with a molecule. The photonic reagent envelops the molecule, which acts as an antenna to accept the various coherent spectral components of the shaped pulse. The ensuing coherent quantum dynamics steers the molecule towards a desired target.

experimental studies [8] verified the point that quantum wave interference could be exploited, but merely expanding the control from being monochromatic to having two or a few frequency components is too limited to generally deal with rich multi-spectral motion especially in polyatomic molecules and complex media. A different proposal was put forth involving a two step so-called pump–dump scheme whereby an initial ultra-fast pump pulse would create a molecular excitation possibly to an upper electronic state, and a second suitable time-delayed pulse would dump the excitation into the desired final target state [9]. This alternative scheme also has a realm of validity, and it has been verified as feasible under favorable conditions [10]. However, the mere timing of one pulse *versus* that of another is generally not sufficient to deal with the highly complex dynamics of polyatomic molecules.

Around the same time of the latter two developments, formal engineering design techniques were introduced within quantum mechanics in order to computationally identify an optimal control laser electric field $\varepsilon(t)$ for meeting the posed molecular dynamical objectives [11]. Considering design as a means of finding effective laser fields for this purpose is a natural perspective involving specification of the physical objective along with any other processes in competition with that objective. Virtually all realistic laser control applications involve competitive dynamical processes, minimally including the desire that laser control be performed efficiently or even more generally that the desired physical process occurs while simultaneously minimizing other accessible and undesirable outcomes from the dynamics. The notion of employing optimization is quite attractive in this context as one always aims to attain the best possible solution for the manipulated dynamics. A rigorous formulation of this task leads to a set of design equations for the control field $\varepsilon(t)$, which take on various forms depending on the nature of the physical objectives and dynamical processes to be avoided [11]. A simple form of these design equations is presented below,

$$i\hbar\frac{\partial}{\partial t}|\psi\rangle = [H_0 - \mu\varepsilon(t)]|\psi\rangle, \ |\psi(0)\rangle \tag{1}$$

$$i\hbar\frac{\partial}{\partial t}|\phi\rangle = [H_0 - \mu\varepsilon(t)]|\phi\rangle, \ |\phi(T)\rangle = \lambda O|\psi(T)\rangle \tag{2}$$

$$\varepsilon(t) = \mathrm{Im}\langle\phi(t)|\mu|\psi(t)\rangle \tag{3}$$

$$\lambda = \langle\psi(T)|O|\psi(T)\rangle - O_{\mathrm{target}} \tag{4}$$

Here H_0 is the free Hamiltonian of the system, under whose dynamics alone the desired physical goal is not met. The control field $\varepsilon(t)$ enters in this formulation coupled through the dipole operator μ. The evolving state of the system $|\psi(t)\rangle$ satisfies Eq. (1) starting from the initial condition $|\psi(0)\rangle$. Equation (1) cannot be standardly integrated forward from this initial condition as the desired control field $\varepsilon(t)$ is unknown upon initiation of the design process. Therefore, the remaining Eqs. (2–4) are there to close this process. In particular, $|\phi(t)\rangle$ satisfying Eq. (2) along with its final condition $|\phi(T)\rangle$ acts as a 'helper' function to guide the dynamics along the right path to the target O_{target}. The sought after control field $\varepsilon(t)$ is given by the matrix element in Eq. (3) and the constant λ characterizes the difference between the expectation value of the target operator O at time T and the actual desired target value O_{target}. The design equations (1–4) present a non-linear boundary value problem in time (i.e. Eq. (1) has an initial condition and Eq. (2) has a final condition with the non-linearity expressed in Eq. (3)), and they may be solved by appropriate iterative techniques [12]. Since the introduction of these quantum optimal control design procedures in the late 1980s, hundreds of simulated designs have been carried out on systems ranging from a few levels out through the excitation of rotational, vibrational, and electronic degrees of freedom and reactive processes. The first conclusion that may be drawn from these efforts is simply that successful controls may be designed on the computer. Furthermore, even the earliest design efforts [11] identified that successful controls would often be pulses with complex waveforms as indicated in Fig. 8.1. The pulse structures naturally arise as the laser control needs to cooperate with the delicate dynamical capabilities of the quantum system placed under control in order to effectively steer it to the target state. Numerous studies show this same complex pulse shape characteristic [13], and generally, it will only be a rare case where a successful control has one or a few frequency components.

Notwithstanding the ability to successfully design controls on the computer, the current experiments [14–25] generally do not operate by implementing the resultant designs. The reason for this lack of implementation is twofold [26]. First, we do not 'know' molecules well enough to design reliable laser controls. The notion of knowing molecules refers to quantatively specifying their Hamiltonian components H_0 and μ. The demand for high-quality Hamiltonian information is driven by the fact that typically the control processes rely on manipulating constructive and destructive quantum wave interferences, which can delicately depend on even slight errors in the Hamiltonian. Although, to zeroth order virtually all Hamiltonians for molecules and materials are known, very few of them could be categorized as quantitatively known for control purposes. Second, even with good knowledge of the system Hamiltonian, the design equations (1–4) need to be solved to high accuracy for meeting the same demands of manipulating delicate quantum wave interferences. Accurately solving Schrödinger's equation is quite difficult for complex systems and the design equations (1–4) are even more involved due to their coupled non-linear nature. These collective circumstances generally produce significant uncertainty in the control designs thereby preventing their direct successful implementation in the laboratory.

Design problems of the type above are common in the engineering disciplines, and the natural way to proceed is to close the loop and let the molecule subjected to control dictate the proper laser pulse shape to achieve its control [26]. This suggestion leads to a laboratory laser control apparatus architecture having the structure in Fig. 8.2. A key component of this machine architecture is the laser pulse shaper [27], which creates the photonic reagents for interaction with the sample. The duty cycle for proceeding from one laser pulse shape to another can be hundreds or even thousands or more per second thereby opening up an unprecedented capability for performing massive numbers of independent trial experiments in very short periods of laboratory time. The logic behind the laser control apparatus in Fig. 8.2 sidesteps the uncertainty in our knowledge of the Hamiltonian and the accuracy difficulties associated with solving Schrödinger's equation by using the actual quantum system under control to eliminate these issues. In particular,

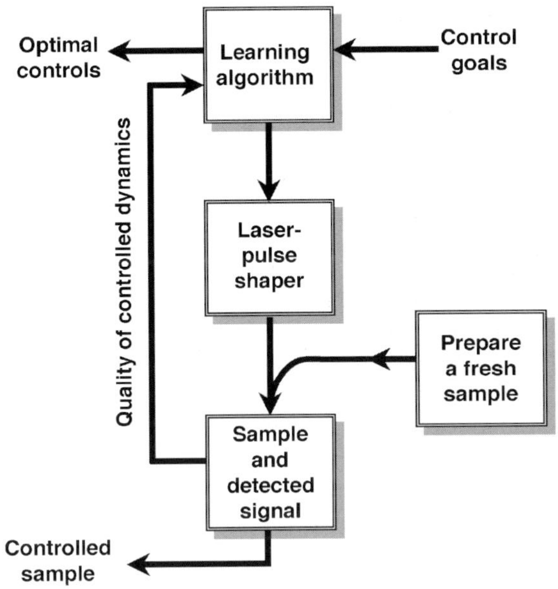

Fig. 8.2. A closed loop apparatus for manipulating quantum dynamics phenomena [26]. A learning algorithm guides the pulse shaper to optimize tailored laser pulses to act as photonic reagents. The tailored laser pulses induce quantum dynamic excursions in a sample. Under high-duty-cycle closed-loop operation, the process can home in on a particular pulse shape that steers the system as close as possible to the desired target. On each excursion of the loop, a new quantum system is prepared in the same initial state for controlled manipulation.

the actual quantum system certainly has full knowledge of its own Hamiltonian and secondly, solves its own Schrödinger equation to full precision and as rapidly as possible upon exposure to a trial laser control field! Each time a new solution of Schrödinger's equation is called for with a particular trial laser field, then the associated experiment is performed and the outcome recorded. The learning algorithm in the loop serves as a pattern recognition tool observing the outcomes from the prior experiments and suggesting new ones with the goal of homing in as rapidly as possible on the desired physical objective.

The apparatus sketched out in Fig. 8.2 now exists in many laboratories with a Ti:sapphire laser generally being the driving source functioning in the near fs regime [28]. The operation of the pulse shaping [27] aspect of the apparatus to create the photonic reagents can be readily understood as sketched in Fig. 8.3. The technology involved is still evolving, but it is in practical form for many applications. Since these experiments started [14] in 1997, many have been carried out [15–25] covering the categories listed in Table 8.1. Chemical transformations are amongst the items in this list, but it is evident that the objectives have expanded to include many other goals. Perhaps most important is the scope and range of the systems being brought under control, which covers atoms out through highly complex protein complexes. The system size itself is not proving to be a hindrance including the cases involving control in condensed phase environments.

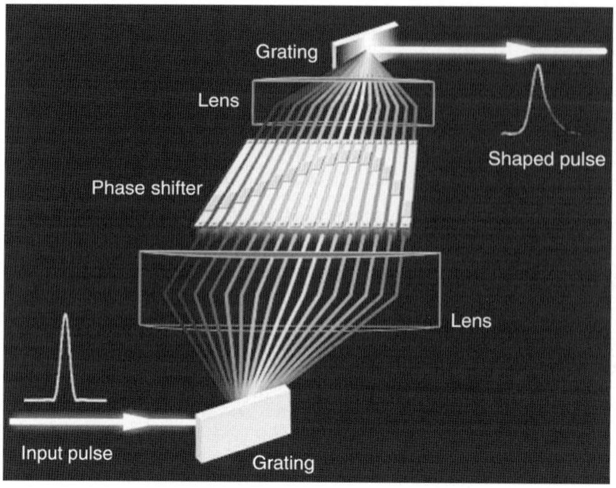

Fig. 8.3. Ultrashort optical pulses can be shaped by adjusting the phase and amplitude of each spectral component [27]. In the device, the input pulse is incident on a grating that disperses the different colors in different directions, as shown in the figure. The colors are collimated and focused by a lens or mirror. A second similar arrangement in reverse reconstitutes the pulse by redirecting the colors to another grating. At the mutual focal plane of the two lenses, the spectrum of the input pulse is completely resolved so that each spatial location corresponds to a single frequency (or a narrow band). By inserting at this plane a material that causes variations in the phase of each resolved frequency, one can construct a pulse of arbitrary shape, constrained only by the spatial resolution of the arrangement.

Two particular experiments hint at the novel capabilities of photonic reagents for creating molecular transformations. The photonic reagent driven reaction [21]

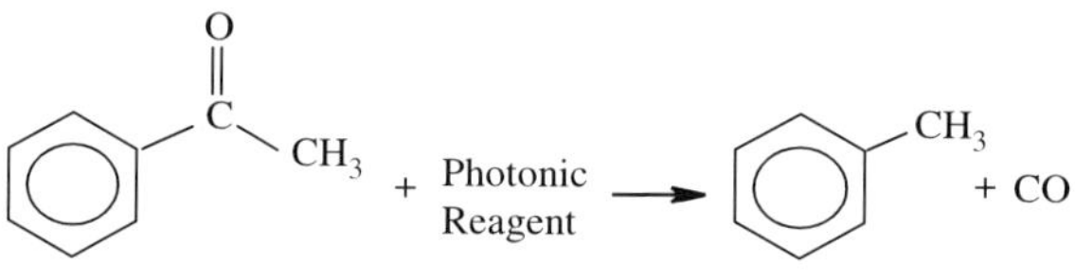

(5)

Table 8.1. Recent scope of adaptive quantum control experiments

(1) Atomic excitation tailoring
(2) Fluorescence spectrum manipulation
(3) Vibrational excitation tailoring in polymers
(4) Molecular fragmentation selectivity
(5) Molecular rearrangement selectivity
(6) Chemical (agent) discrimination
(7) High harmonic X-ray tailoring
(8) Ultrafast solid-state optical switching
(9) Distortion-free transmission of intense pulses in optical fibers
(10) Decoherence management
(11) High-resolution non-linear microscopy
(12) Photosynthetic bacteria electron transfer manipulation
(13) ?

resulted in the breaking of two chemical bonds and the simultaneous formation of a third bond to create the toluene product from the acetophenone. Furthermore, selective breakage of the methyl–carbonyl *versus* the phenyl–carbonyl bonds could also be achieved with distinctively different photonic reagents, all deduced from operating the experiments as in Fig. 8.2. The reaction in Eq. (5) is chemically novel and demonstrates that photonic reagents can have characteristics distinct from that of ordinary chemical reagents. Ordinary chemical reagents inherently attack substrates locally due to steric limitations. In contrast, photonic reagents envelop the substrate, considering that the wavelength of light is much larger than typical molecules. As explained earlier, the molecule acts as an antenna undergoing dynamical evolution directed by a multi-spectral photonic reagent inducing associated coherent quantum motion. The example in Eq. (5) is a novel reaction, which would be difficult to achieve by any single chemical reagent. Beyond novelty of action, a hint also exists that photonic reagents will have their own systematic behavior. An example suggesting this point [20] is shown in Eq. (6).

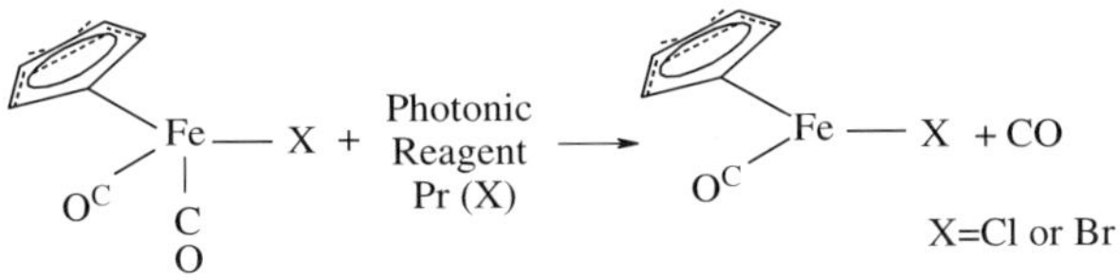

(6)

The photonic reagent Pr(Cl) optimized for removing the carbonyl group from the chlorinated compound was found to be effective for the brominated compound, but not optimal. The reverse situation was similarly found for Pr(Br) as the optimal photonic reagent for the brominated compound, which also worked to some degree in removing the carbonyl group from the chlorinated compound. This behavior is exactly what is typically found with ordinary chemical reagents where the systematics of their chemical transformations underlies practical chemistry. In summary, the two cases in Eqs. (5) and (6) indicate that photonic reagents (a) may have unique properties for making chemical and physical transformations, and (b) can do so in a systematic fashion.

Although, the theoretical concepts and computational simulations ensuing from the most general formulations of optimal control [29] (i.e. based on Eqs. (1–4) and its generalizations) continue to provide valuable insights into the control of quantum systems, these designs have proved to be wanting for laboratory implementations. Nonetheless, it was these same theoretical insights, which led to suggesting [26] the overall architecture in Fig. 8.2 and its application for successful control in the laboratory [14–25]. This situation speaks of the unusual roles that theory is playing in the successful control of quantum phenomena. Further roles for theory and simulation are emerging, and the remainder of this paper aims to highlight some of these continually growing developments. The greatest advances in this field are likely to come from drawing together the best capabilities of the laboratory control apparatus with those of theory and computation to guide and analyze the experiments.

8.2 WHY DOES QUANTUM CONTROL WITH PHOTONIC REAGENTS APPEAR TO BE SO EASY?

The ongoing 40-year effort at controlling quantum phenomena with lasers certainly indicates that the goal has been difficult to achieve. However, most of this period suffered under a lack of conceptual understanding as well as a lack of the appropriate laboratory tools to achieve the objectives. These latter circumstances began to change around 1990 and subsequently, rapid progress has been made culminating with a growing number of successful control experiments. Although the laser pulse designs coming from computer simulations are generally not being utilized in the laboratory, nonetheless, the simulations do describe legitimate model quantum systems on the computer. In the laboratory and on the computer, successful control outcomes have been reported for manipulating broad classes of phenomena including rotational, vibrational, and electronic degrees of freedom as well as reactive processes for systems ranging from atoms to highly complex molecules and materials [14–25]. Whether in the laboratory or on the computer, typically tens to hundreds of discrete control variables with respective gray scales often of at least 50 settings are searched over, potentially corresponding to $\sim 50^{100}$ accessible trial control fields! Yet, it is typically found that only hundreds to thousands of iterations (experiments) are needed to find successful controls for manipulating quantum phenomena. Performing thousands of experiments under normal conditions would require a quite prodigious effort, but the high-throughput nature of the apparatus sketched in Fig. 8.2 permits such experiments to be carried out in the order of minutes. Section 8.1 addressed *how* quantum control may be practically achieved, and the puzzling question [30] left by the collective observations above is *why* does the search for effective controls beat the so-called 'curse of dimensionality'? The curse of dimensionality refers to the potential exponential scaling (i.e. 50^{100} accessible experiments) of the search effort with respect to the number of variables involved. Under normal conditions, one would expect such searches to result in local trapping with a poor quality control outcome, or the search effort merely becoming lost in the vast domain of the search landscape. Global search procedures such as genetic algorithms [31] are effective, but they are also not a panacea for breaking the curse of dimensionality. If seeking the optimal control of quantum phenomena behaved as a normal high-dimensional search problem, a rational assessment would lead one to not even attempt an experiment involving a search over hundreds of laser pulse shaper phase and amplitude control variables. This reasonable assessment was reached by a number of researchers in the mid-1990s. Fortunately, other researchers proceeded ahead despite these forebodings, ultimately enjoying success in the laboratory.

An important challenge is to explain the evidently contradictory outcome that quantum control does work and it is easy to find good solutions! An analysis of this problem was recently undertaken [30] for various quantum mechanical objectives, with the generally most common case being

$$\max_{\varepsilon(t)} P_{\mathrm{i}\to\mathrm{f}}, \quad P_{\mathrm{i}\to\mathrm{f}} = |\langle \mathrm{f}|U|\mathrm{i}\rangle|^2 \tag{7}$$

where $P_{\mathrm{i}\to\mathrm{f}}$ is the probability of making a transition from state $|\mathrm{i}\rangle$ to state $|\mathrm{f}\rangle$ and U is the evolution operator describing the full dynamics of the system under control. Most of the computer simulations and experiments of controlled phenomena are concerned with the goal in Eq. (7). The basic assumption in the analysis of this problem is that the physical system is controllable [32], implying that in principle a field $\varepsilon(t)$ exists, which may arbitrarily steer about the quantum system. Although a rigorous establishment of controllability is difficult to come by in realistic applications, statistical arguments suggest that virtually all quantum systems are expected to be controllable either fully or to a high degree [33]. Thus, the assumption of controllability is a reasonable basic premise. The transition probability $P_{\mathrm{i}\to\mathrm{f}}[\varepsilon(t)]$ is a functional of the control $\varepsilon(t)$ through the generally highly complex mapping $\varepsilon(t) \longrightarrow U$. Little is presently understood about the latter mappings, especially under the circumstances of strong control implying that U has a highly non-linear dependence on the control field $\varepsilon(t)$. Nonetheless, we seek to understand the basic structure of control of landscape $P_{\mathrm{i}\to\mathrm{f}}[\varepsilon(t)]$ and the nature of its extrema satisfying

$$\frac{\delta P_{\mathrm{i}\to\mathrm{f}}}{\delta\varepsilon(t)} = 0 \tag{8}$$

The structure of the control landscape $P_{\mathrm{i}\to\mathrm{f}}[\varepsilon(t)]$ is the key to answering the question posed by the title of this section in this chapter. To proceed with the analysis it is convenient to introduce the transformation $U = \exp(\mathrm{i}A), A^{\dagger} = A$ where A is a Hermetian $N \times N$ matrix corresponding to the system having N states. Equation (8) may then be rewritten as

$$\frac{\delta P_{\mathrm{i}\to\mathrm{f}}}{\delta\varepsilon(t)} = \sum_{pq} \frac{\partial}{\partial A_{pq}} |\langle \mathrm{f}|\exp(\mathrm{i}A)|\mathrm{i}\rangle|^2 \frac{\delta A_{pq}}{\delta\varepsilon(t)} = 0. \tag{9}$$

The mapping $\varepsilon(t) \longrightarrow A$ is not expected to generally be any simpler than $\varepsilon(t) \longrightarrow U$. However, with the system being controllable each matrix element in A should be independently addressable by the control $\varepsilon(t)$ within hermeticity of the matrix. This statement implies that the functions $\delta A_{pq}/\delta\varepsilon(t)$ should be linearly independent over $0 \leq t \leq T$. Thus, the only way that Eq. (9) can generally be satisfied is from the requirement that the time-independent coefficients are zero,

$$\frac{\partial}{\partial A_{pq}} |\langle \mathrm{f}|\exp(\mathrm{i}A|\mathrm{i}\rangle|^2 = 0 \qquad \text{for all } p \text{ and } q. \tag{10}$$

The lack of explicit dependence upon the system Hamiltonian in Eq. (10) is consistent with the findings that successful control is being found across the board in both simple and highly complex systems, therefore suggesting that specific Hamiltonian details are not at the root of explaining *why* this is happening. A full analysis may be undertaken [30] of the non-linear algebraic equations in Eq. (10), leading to the rather surprising conclusion that Eq. (8), only has the solutions

$$P_{\mathrm{i}\to\mathrm{f}} = 0 \text{ or } 1. \tag{11}$$

The control landscape, $P_{\mathrm{i}\to\mathrm{f}}[\varepsilon(t)]$, only has extrema corresponding to either no control or perfect control! Furthermore, as the control problem $|\mathrm{i}\rangle \longrightarrow |\mathrm{f}\rangle$ only specifies traversal from the initial state $|\mathrm{i}\rangle$ to the final state $|\mathrm{f}\rangle$, generally an infinite number of optimal paths may be taken corresponding to the landscape being replete with an infinite number of perfect extrema $P_{i \to f} = 1$ with null valued extrema $P_{i \to f} = 0$ in between.

The control landscape analysis leading to the conclusion in Eq. (11), provides the basis to explain the many successful experiments and simulations of controlled quantum phenomena. Typically, an initial guess for the control field $\varepsilon(t)$ is made, and this trial will likely produce a very poor result for $P_{i \to f}$, landing at some interior low-value location in the landscape. However, virtually any reasonable algorithm should be able to sense the upward slope, towards the nearest prefect solution. Evidence shows that only a modest number of additional trial experiments or simulations are needed to rapidly reach the vicinity of $P_{i \to f} = 1$. Furthermore, additional detailed analysis of the Hessian at each of the extrema of $P_{i \to f} = 1$ satisfying Eq. (8) shows that there is an inherent degree of robustness to the control solutions. This observation is consistent with the laboratory successes, as they likely would not occur with even modest control errors, especially in the highly non-linear regime utilizing strong control fields.

Another important conclusion from this analysis concerns the significance of retaining as much flexibility as possible in the controls when seeking to manipulate quantum systems. This notion is perhaps intuitive, but its significance becomes more evident upon considering that despite the control landscape only having prefect extrema, typical quantum optimal control simulations produce less than perfect yields with common values like $P_{i \to f} \approx 0.9$. A closer examination of these calculations shows that they always contain either explicit or implicit constraints on the controls through the imposition of particular forms for the field, the fixing of the target time T, or the introduction of other costs to limit the field intensity. Inevitably, such limitations on the freedom for $\varepsilon(t)$ will lead to the search for the control taking a constrained path through the landscape producing convergence at a value of $P_{i \to f} < 1$ corresponding to less than the attainable perfect control yield. In extreme cases, the limitations on the freedom in the control can also give the impression that the landscape itself has false structures as the search traverses a torturous path up and down along the slopes of one or more of the landscape features. Maintaining the broadest freedom for the control is essential for achieving the best quality results, in lieu of the landscape structure findings.

8.3 WHAT IS OCCURRING DURING THE PROCESS OF CONTROLLING QUANTUM DYNAMICS PHENOMENA?

Sections 8.1 and 8.2, respectively, addressed *how* to control quantum phenomena in the laboratory and *why* it appears easy to achieve. However, left as a mystery is *what* happens during the control excursions or equivalently, what is the nature of the control mechanisms? It is evident that one may carry out successful control experiments using the procedure in Fig. 8.2 without either imposing or ultimately determining the underlying quantum control mechanism. Furthermore, this success may be achieved without any detailed knowledge of the specific system Hamiltonian structure [26]. These features of the overall procedure are very attractive for practical applications, but nonetheless the fundamental desire to understand the underlying mechanisms remains as a challenge. At the present time, little understanding exists about quantum control mechanisms, but the conceptual foundations for addressing this problem is coming into place [34].

The traditional route to achieving understanding about the control mechanism would be through the use quantum system simulations [25]. This approach to extracting the control mechanism builds upon the fact that the outcome from a successful experiment would consist of not only the observed objective yield, but also a determination of the optimal control field $\varepsilon(t)$ involved. Assuming (a) that full knowledge of the system Hamiltonian was available, the field determined in the laboratory could be folded in along with the further assumption (b) that the quantum system simulations could be reliably carried out describing the corresponding experiment. Lack of satisfaction of the latter two assumptions was exactly the critical reason leading to the suggestion [26] of sidestepping computational designs and performing the successful control experiments using the procedure in Fig. 8.2. Nonetheless, under favorable circumstances [25], for simple systems, these assumptions may be satisfied and the conditions for this to happen are likely to improve with better computational capabilities. Although much information about the mechanism could be learned in this fashion (when the simulations can be performed!), operating along these traditional lines is likely to still leave an incomplete understanding about quantum control mechanism in many cases. This point is evident from the many quantum optimal control simulations [11,13], which always have full access to the system dynamics including the wave function and yet they frequently have provided incomplete mechanistic insights into the control process. Thus, traditional simulations alone and even knowledge of the system wave function may not provide an easy route to understanding the mechanisms of controlling quantum phenomena.

The observations above leave fully open how to proceed in assessing quantum control mechanisms. A first issue to consider is the meaning of quantum coherences in mechanistic analyses. Normal chemical and physical dynamics processes (i.e. those without laser controls) are typically understood without the introduction of quantum coherence concepts. But, in the case of quantum control, coherence is likely to be a central enabling factor. A general perspective on quantum control mechanism analysis appears feasible to develop [34] by considering the quantum system under control as a functioning 'machine.' As in any area, the understanding of machine operation calls for the introduction of a disturbance and subsequent monitoring of the system response. With quantum control systems typically operating at the near fs time scale, a direct application of this procedure in the laboratory would call for even faster modulation than already contained in the control field to produce a discernable signal for mechanistic analysis in the experiment. This procedure would be technologically extremely difficult to execute. However, this difficulty can be overcome by taking advantage of the ability to perform massive numbers of experiments over short periods of laboratory time, using the high-duty cycle of the lasers and the pulse shapers. A pseudo-time-like variable $s \geq 0$ can be introduced, which in practice acts as an index labeling the sth experiment through a specific encoded modulation of the phase and amplitude control variables. For example, the nth control variable could be modulated in a Fourier fashion with the function $\cos(\nu_n s)$, where ν_n is an arbitrary encoding frequency having no relation to the quantum system frequencies. By choosing the collective set of frequencies $\{\nu_n\}$ to be incommensurate and simultaneously modulating some or all of the control variables over a sequence of experiments $s \rightarrow s + \Delta s \rightarrow s + 2\Delta s, \cdots$, the observable signal $O(s)$ may be recorded as a function of $s \geq 0$ discretized with resolution Δs. With this form of Hamiltonian encoding, upon inverse Fourier transform of the signal over $s \geq 0$, a spectrum will result, revealing various fundamental and combination bands associated with the frequencies $\{\nu_n\}$. In turn, the Fourier amplitudes from this analysis may be directly related to the interfering quantum amplitudes producing the desired physical objective.

At the present time, no laboratory experiments have been carried out using this proposed Hamiltonian encoding-dynamics decoding technique, although simulations indicate the capability of the procedure [34]. As an example, Fig. 8.4 presents a mechanistic analysis of the simple control process $|1\rangle \rightarrow |3\rangle$ in a four-level quantum system. Although an optimal field was easily found that successfully achieves this transition, as is typically the case, the mechanism of this process was not evident from examining the quantum amplitudes or the field itself. However, the encoding–decoding analysis sketched out above, demonstrated that the mechanism proceeded along the three dominant pathways shown in Fig. 8.4. The amplitudes associated with these pathways were found to constructively interfere reflecting the nature of their optimality. As the amplitude $\langle 3|U|1\rangle$ was nearly of unit magnitude, such that $P_{1\rightarrow 3} \approx 1.0$, it follows that the other transition amplitudes out of state $|1\rangle$ must essentially be zero. For example, $\langle 4|U|1\rangle \approx 0$, and the mechanistic analysis shows that the reason for this behavior does not lie in the fact that state $|4\rangle$ is not visited during the dynamics. Rather, a destructive interference is setup for the amplitudes arriving at this state as indicated in Fig. 8.4. This illustration demonstrates the fundamental principal that quantum control involves the search for an optimal field $\varepsilon(t)$ that can setup suitable constructive interferences in the desired state and destructive interferences in the others. This mechanistic phenomenon clearly points out the reason why simple intuition historically failed to identify successful controls: it is simply not possible to intuit and balance highly complex constructive and destructive quantum wave interferences in multi-state systems. Yet, the optimal control operations following the architecture in Fig. 8.2 are fully capable of discovering successful controls in diverse and even highly complex quantum systems.

Many questions and issues remain in the development of effective means to experimentally determine and understand quantum control mechanisms. Ultimately, the attainable knowledge about such mechanisms will be limited by the noise inherent in the laboratory. However, this limitation is already present in the underlying experiments attempting to achieve control over the dynamics, prior to extracting mechanistic insights. The optimal control experiments reach an objective by working as best as possible in the presence of laboratory noise, and nothing more can be expected of the mechanistic analysis than the determination of the key dynamical features, which survive in the presence of the noise. Any experiment achieving a reasonable level of control success should be amenable to a comparable level of mechanistic analysis using suitable encoding and decoding techniques. A further issue to consider is that both control and mechanistic analysis typically involve lasers that have spatial inhomogeneities implying that different subsets of the molecules of the sample may see fields of varying intensity. In addition, as the ultra-fast laser pulse passing through the medium will interact with the molecules effectively frozen and at various orientations, again, each molecule will

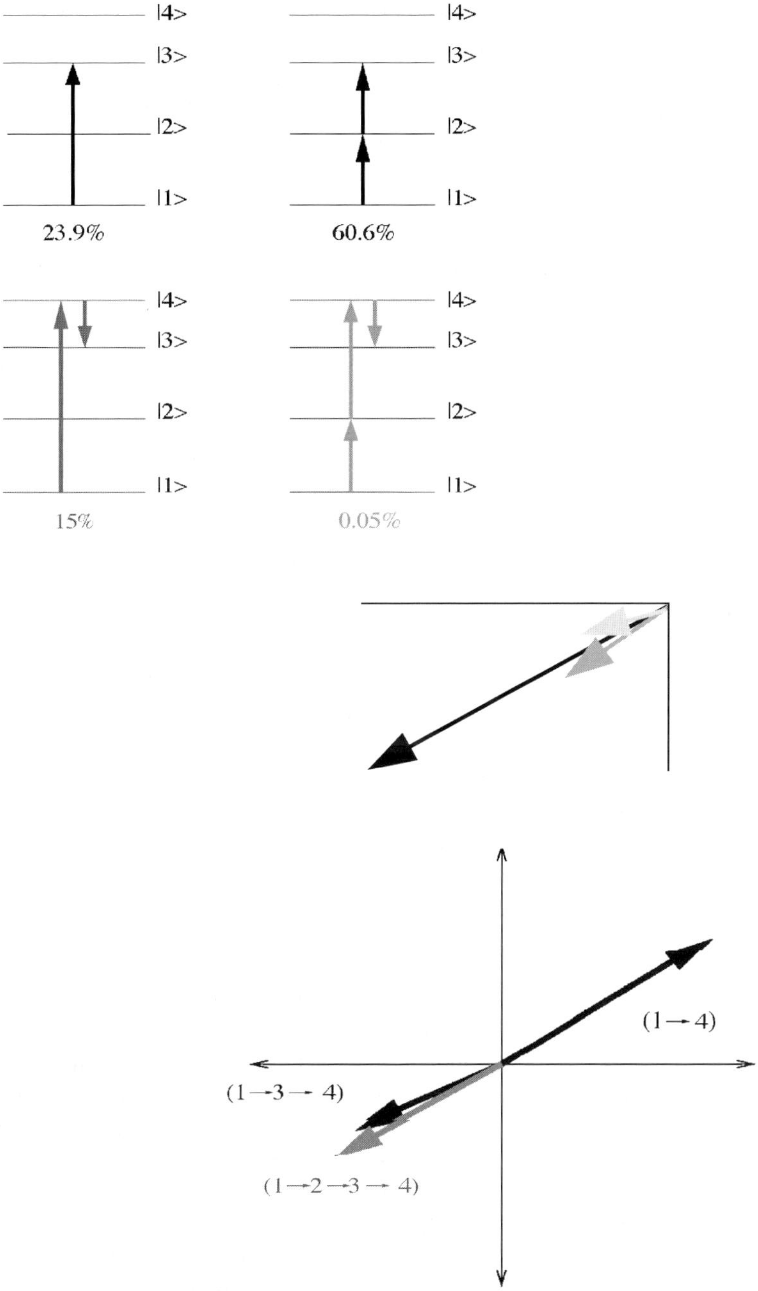

Fig. 8.4. A pathway mechanism analysis performed on a four-level quantum system where the goal is to make the transition $|1\rangle \rightarrow |3\rangle$. The analysis reveals three dominant pathways. In addition, it is seen that the complex amplitudes associated with these pathways align to constructively interfere in the state $|3\rangle$. In contrast a non-target state, such as $|4\rangle$, has destructively interfering amplitudes.

experience a distinctly different control field $\varepsilon(t)$. Theoretical analysis [35] reveals the rather surprising result that under these circumstances, at least one optimal field generally exists to fully control the entire ensemble of molecules, despite the fact that each member of the ensemble 'sees' the field in a different fashion. This outcome is very attractive from a control prospective, but this analysis also indicates that each member of the molecular ensemble may take a distinctly different pathway (i.e. have a different control mechanism) while still reaching the same final control objective. Mechanism analysis may ultimately have to be understood as an ensemble average process just as control itself is an ensemble average over molecules exposed to different local field environments.

8.4 CONCLUSION

Manipulating atomic and molecular scale phenomena is at the heart of many activities in chemistry and physics. The introduction of specific chemical reagents to produce suitable materials modifications constitutes the traditional means of achieving these manipulations. After some 40 years of effort, it now appears that photonic reagents have the capability of becoming a new tool in the arsenal for achieving such atomic and molecular scale manipulations [5]. Furthermore, experiments are beginning to reveal the unique nature of photonic reagents, suggesting that they have special capabilities beyond that of ordinary chemical reagents. Their uniqueness lies in the inherent feature of photonic reagents having the ability to envelop and totally speak to all dynamical aspects of a molecule. Furthermore, we may change from one photonic reagent to another at a blurring rate through the high-duty cycle of ultra-fast laser pulse shaping [27]. Although, the cost of shaped photonic reagents remains high, many applications and fundamental studies do not require the production of large amounts of materials. One could ultimately envision a mix and match situation, whereby the special features of photonic and ordinary reagents would be combined for fully flexible optimal manipulation of atomic and molecular scale processes. Proceeding along these lines may enable the creation of novel molecules and materials, as well as reveal deeper insights into quantum dynamics phenomena. Table 8.1 shows that applications in this area, although starting with chemistry, now go far beyond. It is reasonable to anticipate that the table will also expand in new directions in the future, as virtually any atomic or molecular scale process should be amenable to photonic reagent manipulation. The ultimate utility of photonic reagents remains open for assessment, but at this juncture, we may conclude that photonic reagent control is prime for development given that many of the basic principles and tools are now in hand.

8.5 REFERENCES

1 N. Turro, Modern molecular photochemistry, University Science Books, Sausalito, CA, 1991.
2 E. Teller, Lasers in chemistry, an address presented before 'The Robert A. Welch Foundation Conferences on Chemical Research. XI. Theoretical Chemistry', Houston, TX, November 20–22, 1972.
3 Physics Today Special Issue—Laser Chemistry, 6, 1980.

4 V.S. Letokhov, Phys. Today, 34 (1980).
5 H. Rabitz, Science, 299 (2003) 525.
6 A. Zewail, Phys. Today, 33 (1980) 11.
7 M. Shapiro and P. Brumer, Principles of the quantum control of molecular processes, *Wiley-Interscience*, Hoboken, NJ, 2003.
8 C. Chen, Y.-Y. Yin and D. Elliott, Phys. Rev. Lett., 64 (1990) 507.
9 D. Tannor and S. Rice, J. Chem. Phys., 83 (1985) 5013.
10 F. Baumert and G. Gerber, Isr. J. Chem., 34 (1994) 103.
11 A. Peirce, M. Dahleh and H. Rabitz, Phys. Rev. A, 37 (1988) 4950; S. Shi, A. Woody and H. Rabitz, J. Chem. Phys., 88 (1988) 6870; R. Kosloff, S. Rice, P. Gaspard, S. Tersigni and D. Tanner, Chem. Phys., 139 (1989) 201.
12 W. Zhu and H. Rabitz, Int. J. Quantum Chem., 93 (2003) 50.
13 R. Rice and M. Zhao, Optical control of molecular dynamics, Wiley, New York, 2000.
14 C. Bardeen, V. Yakovlev, K. Wilson, S. Carpenter, P. Weber and W. Warren, Chem. Phys. Lett., 280 (1997) 151.
15 A. Assion, T. Baumert, M. Bergt, T. Brixner, B. Kiefer, V. Seyfried, M. Strehle and G. Gerber, Science, 282 (1998) 919.
16 T. Brixner, N. Damrauer and G. Gerber, Adv. At. Mol. Opt. Phys., 46 (2001) 1.
17 R. Bartels, S. Backus, E. Zeek, L. Misoguti, G. Vdovin, I. Christov, M. Murnane and H. Kapteyn, Nature, 406 (2000) 164.
18 S. Vajda, A. Bartelt, E. Kaposta, T. Leisner, C. Lupulescu, S. Minemoto, P. Rosendo-Francisco and L. Woste, Chem. Phys., 267 (2001) 231.
19 T. Weinacht, J. Ahn and P. Bucksbaum, Nature, 397 (1999) 233.
20 M. Bergt, T. Brixner, C. Dietl, B. Kiefer and G. Gerber, Organomet. Chem., 661 (2002) 199.
21 R. Levis, G. Menkir and H. Rabitz, Science, 292 (2001) 709.
22 T. Weinacht, J. White and P. Bucksbaum, J. Phys. Chem. A, 103 (1996) 10166.
23 J. Kunde, B. Baumann, S. Arlt, F. Morier-Genoud, U. Siegner and U. Keller, Appl. Phys. Lett., 77 (2000) 924.
24 J. Herek, W. Wohlleben, R. Cogdell, D. Zeidler and M. Motzkus, Nature, 417 (2002) 533.
25 C. Daniel, J. Full, L. Gonzalez, C. Lupulescu, J. Manz, A. Merli, S. Vajda and L. Woste, Science, 299 (2003) 536.
26 R. Judson and H. Rabitz, Phys. Rev. Lett., 68 (1992) 1500.
27 A. Weiner, Prog. Quantum Electron., 19 (1995) 161; A. Weiner, Rev. Sci. Instrum., 71 (2000) 1929..
28 D. Strickland and G. Mourou, Opt. Commun., 56 (1985) 219.
29 E. Brown and H. Rabitz, J. Math. Chem., 31 (2002) 17.
30 H. Rabitz, M. Hsieh and C. Rosenthal, Science, 303 (2004) 998.
31 D. Goldberg, Genetic algorithms in search, optimization, and machine learning, Addison-Wesley, Reading, MA, 1989.
32 D. Luenberger, Introduction to dynamics systems, *Wiley*, New York, 1979.
33 R. Wu, H. Rabitz, G. Turinici and I. Sola, Phys. Rev. A, 70 (5) (2004).
34 A. Mitra and H. Rabitz, Phys. Rev. A, 67 (2003) 033407.
35 G. Turinici and H. Rabitz, Phys Rev. A, (2004) Submitted.

Theory and Applications of Computational Chemistry: The First Forty Years
Edited by C. Dykstra et al.

CHAPTER 9

First-principles calculations of anharmonic vibrational spectroscopy of large molecules

R.B. Gerber[1,2], G.M. Chaban[3], B. Brauer[1] and Y. Miller[1]

[1] *Department of Physical Chemistry and Fritz Haber Research Center, The Hebrew University of Jerusalem, Jerusalem 91904, Israel*
[2] *Department of Chemistry, University of California, Irvine, CA 92697, USA*
[3] *NASA Ames Research Center, Mail Stop T27B-1, Moffett Field, CA 94035, USA*

Abstract

This review describes methods for computing directly the anharmonic vibrational spectra of polyatomic molecules from potential surface points obtained from electronic structure theory. The focus is on the state of the art of the methodology, on the approximations and the algorithms involved and their limitations, and on the scaling of the computational effort with the number of vibrational modes. The performance of different electronic structure methods in obtaining accurate vibrational spectra is assessed by comparing the theoretical predictions with experiment for various test cases. Finally, some of the many open problems and challenges in this field are discussed.

9.1 INTRODUCTION

Vibrational spectroscopy is a tool of great importance for identifying molecular species, exploring their properties and learning about their potential energy surfaces. For these reasons, this discipline has been, since its birth, among the prime instruments of molecular science and a major research field in its own right. The theoretical interpretation and calculation of the vibrational spectra of polyatomic molecules was confined for many years to the harmonic approximation: within the harmonic model, combined with rigid-rotor treatment of molecular rotation, the vibrational Hamiltonian can be separated into a set of one-dimensional harmonic oscillators by using normal mode coordinates, thus yielding an analytic solution for the wave functions and energies.

The computational effort required thus involves the finding of the equilibrium configuration on the potential energy surface, then the calculation of the matrix of second derivatives of the potential with respect to the nuclear displacements. The normal modes, and the frequencies corresponding to the different normal modes are obtained by diagonalization of the Hessian. The normal modes are transformation vectors between Cartesian coordinates and normal mode coordinates. This elegant, conceptually simple approach has served for decades as the basis for the theoretical interpretation, within the harmonic approximation framework, of polyatomic vibrational spectroscopy. The principles of this approach, and elegant mathematical analysis and procedures are described by Wilson et al. [1] in their tour de force exposition of molecular vibrations. When the potential surface is available as an analytic function of the coordinates, the computational implementation of the harmonic normal-mode approach is very efficient. Normal modes have been determined and harmonic frequencies have been obtained for systems of the size of proteins, having thousands of vibrational modes. Calculations of the harmonic spectra from electronic structure methods are computationally more demanding, especially so is the evaluation of the matrix of second derivatives of the potential function with respect to the displacements of the atoms from equilibrium. However, algorithms for the calculation of the normal modes and harmonic frequencies have become available for many of the important electronic structure codes [2]. Thus, efficient algorithms for computing the harmonic frequencies are available for the *ab initio* methods such as Hartree–Fock (HF), second order Møller–Plesset Perturbation Theory (MP2), Quadratic Configuration Interaction including single and double substitutions (QCISD), and many others. Likewise, algorithms for obtaining the harmonic frequencies are available for several variants of Density Functional Theory (DFT), such as B-LYP, B-P86, B3-LYP and B3-PW91, and extensive applications and tests have been presented. Finally, algorithms for convenient evaluation of the harmonic frequencies were provided also for the semiempirical electronic structure methods AM1 and PM3 and these also have been extensively applied. These algorithms are conveniently implemented in important electronic structure code packages such as GAMESS [3], GAUSSIAN [4], MOLPRO [5], and others. It can thus be said that first-principles calculations of harmonic spectra have become essentially a matter of routine.

The situation is quite different for computation of anharmonic vibrational spectra, which despite the progress made is still to a large extent an open problem. The main difficulty is that realistic polyatomic anharmonic Hamiltonians are inherently nonseparable: there is no coordinate system in which the Hamiltonian becomes a sum of independent one-dimensional oscillators, as is the case for harmonic Hamiltonians when normal coordinates are used. The anharmonic vibrational problem is thus fundamentally a quantum-mechanical many-body problem. This problem admits no analytical solutions, and rigorous numerical approaches to the problem are a major challenge. Going beyond the harmonic approximation is, however, often crucial for the interpretation of experimental data. Very often anharmonic effects are so large that even assignment of the transitions becomes problematic when the harmonic approximation is employed. One pragmatic approach often used to bring harmonic *ab initio* calculations into closer accord with experiment is to modify the computed *ab initio* frequencies using calibration from a set of related compounds for which data on the anharmonic effects for the various transitions is available [6]. Several other non-first-principles prescriptions for modification of the harmonic frequencies have been proposed [7,8]. The Scaled Quantum Mechanical Method (SQM) due to Saebo and Pulay [7], for example, proved very successful in reproducing experimental frequencies in many cases [7–9]. While useful in a variety of applications, and the topic of much continuing activity [8,10], these methods are not considered in the present review. As pointed out by Pulay [9], these methods are essentially empirical. The introduction of the scaling factors to modify the computed harmonic frequencies may compensate not only for the anharmonic effects, but also for the inaccuracy of the *ab initio* method used at the harmonic level. Also, there seems to be no systematic way to improve the accuracy of the scaling predictions. The main limitation of these approaches is, however, from our perspective, the following: the experimental frequencies contain information on the anharmonic parts of the potential energy surface, and this anharmonicity is of great interest. To learn about the anharmonicity of molecular potentials, one must aim at computing the contribution of these anharmonicities to the observed spectrum. Thus it is essential for this purpose to aim at methods that compute the anharmonic spectrum systematically and on a first-principles basis from the given vibrational potential function. Such methods, and in particular those that are applicable directly to potential energy surfaces from electronic structure theory are the topic of this review.

The structure of this review is as follows. In Section 9.2, we briefly discuss methods for computing vibrational states of systems having several coupled vibrational degrees of freedom. This will also cover methods that were not yet adapted for direct use with *ab initio* potentials, since in our view, such extensions may be possible in the future, at least for some of the algorithms. The focus will be on methods that seem potentially applicable to large polyatomics, rather than those of great accuracy for small systems. Section 9.3 also deals with computational methods for anharmonic vibrational spectroscopy that are applicable to potential surfaces from electronic structure calculations. Our main focus will be on the Vibrational Self-Consistent Field (VSCF) approach in several variants and extensions. The performance of the available method in the present state of the art is discussed in Section 9.4. Future directions are outlined in Section 9.5.

9.2 ANHARMONIC VIBRATIONAL SPECTROSCOPY METHODS

Calculations utilizing the harmonic approximation are almost never of experimental accuracy, except of course, for model potentials fitted to the data. For any realistic potential not adjusted to the observed frequencies, the deviations from experiment are substantial, even for the fundamental transitions in the most rigid molecules. Nevertheless, the progress in anharmonic vibrational spectroscopy calculations was stimulated to a large extent by several advances in high-resolution experimental techniques that explored systems and transitions for which anharmonic effects are highly pronounced. This includes, for example, the vibrational spectroscopy of van der Waals and hydrogen-bonded clusters [11–13], vibrational overtone spectroscopy of OH, CH and similar groups in polyatomics [14,15], and spectroscopy that explores Intramolecular Vibrational Energy Redistribution (IVR), whether in the time or in the frequency domain [16].

A variety of methods for performing anharmonic vibrational spectroscopy computations were developed to address these and related systems. At the early stages, essentially all the methods were developed for potential surfaces available as explicit analytic functions of the coordinates.

The function assumed for the potential surface is generally different for each case. Indeed, application of a variety of methods involved attempts to fit parameters of the assumed analytic force field so as to closely reproduce the experimental spectroscopic data. Such a fitting strategy can result in some cases in superb agreement with experiment. However, in general the problem of fitting spectroscopic data for a polyatomic by an assumed potential, does not have a unique solution, especially when only a limited number of transitions are available from experiment. Therefore, a disadvantage of the fitting approach is that, in principle, the potential surface that is determined may differ greatly from the real one, despite the agreement with experiment. This is an important motivation for using potentials from first-principles calculations: the potentials used are likely to be realistic (though, of course approximate). On the other hand, agreement between spectra computed from *ab initio* potentials and experiment is likely to be limited, due to variations in the accuracy of electronic structure methods.

In briefly surveying several of the many methods proposed for anharmonic vibrational spectroscopy calculations, we mention also methods used so far only for analytic potential surfaces. For many of these methods, adaptations to algorithms may be feasible, and the methods seem promising in this respect.

9.2.1 Perturbation theory

Different forms of Perturbation Theory rank with the earliest methods applied to the calculation of anharmonic energy levels of coupled molecular vibrations [17,18], and these methods continue to be of great computational effectiveness in a range of applications. The simpler methods of this type are based on standard Rayleigh–Schrödinger Perturbation Theory, with the harmonic approximation as the zero-order Hamiltonian and with the entire anharmonic part of the potential as the coupling. This is conveniently pursued in the normal mode representation, with a polynomial expansion, usually up to quartic terms, for the perturbation. The method is suited for weak anharmonicity, and application is most commonly confined to first and second-order perturbation treatment. The method was applied to spectroscopic calculations using *ab initio* potentials, for example, see Refs. [19–23]. Recently, Barone [24,25] has introduced an efficient algorithm for the perturbation-theoretic calculation of anharmonic effects, interfaced with *ab initio* and DFT codes. His code has been included in the GAUSSIAN program package, and has been used to carry out anharmonic calculations for relatively large molecules such as the benzene analogs pyrrole and furan [25]. Important advantages of this type of perturbation theory are its simplicity and computational efficiency. An important issue in the method is the representation of the anharmonic part of the potential, which is the perturbation in the treatment. It is represented in most studies as a polynomial (up to quartic terms) in the normal mode displacements from equilibrium. For very anharmonic, floppy systems such a polynomial expansion may not be adequate or sufficient. Expansions to very high orders in the normal mode displacements may be required in demanding cases, and the computational generation of such expansions is difficult except for very simple analytic potential functions. The radius of convergence of the Rayleigh–Schrödinger perturbation expansion for realistic vibrational Hamiltonians is not known, so the approach may not be generally applicable. Also, in some cases, second-order perturbation theory may not suffice. When this is the case, the available algorithms do not seem applicable. High-order perturbation theory is expected to be computationally quite demanding. A very powerful form of perturbation theory that has a number of theoretical advantages over the Rayleigh–Schrödinger method is Canonical Van Vleck Perturbation Theory (CVPT), [26–28]. The origins of this variant of perturbation theory are not new, but it was elegantly developed by Sibert and McCoy [27,28]. The canonical transformation used in this approach provides insight into the connections between the energy level sequences and the underlying Hamiltonian, and is expected to provide improved convergence properties for the perturbation series. The method was used also with Morse coordinates, which possess a better limiting behavior. However, the algorithmic structure of CVPT is relatively complex, and so is the handling of the perturbation potentials in this approach. Many of the earlier applications address triatomic molecules, though more recently Sibert was able to treat larger systems [29,30], by making several approximations and simplifications specific to the systems. Except with such simplifying assumptions, CVPT has not been used so far for first-principles potentials.

9.2.2 The vibrational self-consistent field approach

This approach, and the variants derived from it, have their origin in work done in the late 1970s. Early contributions are from Bowman [31], Carney et al. [18], Cohen et al. [32] and Gerber and Ratner [33]. The approaches of Bowman [31] and of Carney et al. [18] were fully quantum mechanical, while Gerber and Ratner employed a semiclassical theory. As it turned out, the computational time saved by using semiclassical theory is negligible, although at the same time the additional error due to the semiclassical approximation is insignificant. Thus, it is preferable to use a fully quantum mechanical method in most applications, and only in special cases does the greater mathematical simplicity of semiclassical VSCF offer significant advantages. One example for which the semiclassical VSCF method is advantageous, is the direct inversion of vibrational spectroscopic data in order to obtain the multidimensional potential energy surface of a polyatomic system. Such an explicit mathematical inversion of the complete vibrational spectrum was demonstrated by Gerber, Roth and Ratner [34,35] in the framework of semiclassical VSCF. The necessary complete spectrum is, however, rarely available in practice.

The physical content of the VSCF approximation is simple: within this approximation, each vibrational mode is described as moving in the mean field of the other vibrational modes. The mean fields, and the wave functions of the different modes are determined self-consistently, so that the approximation is analogous to the Hartree method for many-electron systems. The total wave function within this simplest level of VSCF is thus separable into a product of single mode wave functions. When the coupling between different vibrational modes is strong enough, it can lead to breakdown of separability and result in important correlation effects between the modes. An important issue is which coordinates (or modes) one should employ for best mutual separability in the sense of VSCF. Near the bottom of the potential well, the choice of normal modes seems reasonable: for low energies, the vibrational amplitudes are expected to be small and the harmonic approximation should work reasonably well. Since within the harmonic approximation normal modes give rigorous separability, it seems natural to employ these modes also somewhat beyond the harmonic limit. For higher excitation energies, as for the ground state of extremely anharmonic systems (such as clusters of helium and hydrogen), this need not be the case, and the choice of good coordinates for VCSF is not obvious. This issue of optimal choice of coordinates for highly anharmonic systems was studied by Horn et al. [36], who for example, found that for treating a quantum cluster such as $XeHe_2$, VSCF in hyperspherical coordinates is far superior to VSCF in normal mode coordinates and gives excellent results: (approximate) separability works in hyperspherical modes in this system. Several VSCF calculations for specific systems were similarly carried out using non-normal coordinates, according to the apparent physically appropriate choice in these cases. Examples include: the HCN $\leftrightarrow$ HNC system, for states where isomerization takes place, for which Bačić et al. [37] showed that spheroid coordinates are appropriate for VSCF separability; H_2O and CO_2 molecules for which hyperspherical coordinates yield superior results [38]; the quantum cluster I_2He for which spheroid (ellipsoidal) modes were found suitable [36,39]; and excited vibrations of Ar_3, for which VSCF separability in hyperspherical modes works very well [40]. So far, the only widely applicable, computationally efficient, implementation of VSCF has been carried out in normal modes, since, as appealing as is the choice of physically motivated 'good coordinates' for VSCF separability, such coordinates could only be found for certain specific cases. Fortunately, most experimental spectroscopic data are available for fundamentals, or other rather relatively low excitations, where normal modes are usually a physically motivated good choice.

Since separability is always approximate, correcting VSCF by the introduction of correlation between different modes is desirable, and experience has shown that it is essential for satisfactory agreement with experiment even in cases where the potential employed is believed to be accurate [41–43]. The first approach used to improve VSCF by the introduction of correlation between modes is the Configuration Interaction (CI) method, analogous to the corresponding algorithm in electronic structure theory, for example see: Bowman et al. [44], Ratner et al. [45], and Thompson and Truhlar [46]. In this approach, primarily developed by the Bowman group [47–50], the vibrational wave function is written as a linear combination of separable terms. Each term of the CI wave function is thus a product of single-mode vibrational wave functions. The method was applied first to small systems such as H_2O and HCO, and then it became possible to tackle increasingly larger systems, such as CO adsorbed on a Cu(100) surface, with the latter represented by several vibrating atoms [50]. For small molecules, as for molecules of moderate size (number of atoms smaller than 10 or so) the method is very effective, and can be very accurate. Thus, Bowman and coworkers were able to incorporate vibration–rotation coupling (including Coriolis coupling) in the CI-VSCF treatment of vibrational state calculations. With the extensions provided, the method can be applied to the full Watson Hamiltonian [51] for the vibrational problem. The efficiency of the method depends greatly on the nature of the anharmonic potential that represents coupling between different vibrational modes. In favorable cases, the latter can be represented as a low-order polynomial in the normal-mode displacements. When this is not the case, the computational effort increases rapidly. The CI-VSCF is expected to scale as N^5 or worse with the number N of vibrational modes. The most favorable situation is obtained when only pairs of normal modes are coupled in the terms of the polynomial representation of the potential. The VSCF-CI method was implemented in MULTIMODE [47,52], a code for anharmonic vibrational spectra that has been used extensively. MULTIMODE has been successfully applied to relatively large molecules such as benzene [53]. Applications to much larger systems could be difficult in view of the unfavorable scalability trend mentioned above.

Another approach to the extension of VSCF by the inclusion of correlation effects was introduced by Jung and Gerber [54] and by Norris et al. [55]. Usually referred to as the Correlation-Corrected VSCF (CC-VSCF) method, this approach brings in the correlation effects, by applying second order perturbation theory to the anharmonic coupling interactions between the modes, while the VSCF Hamiltonian is treated as the zeroth-order approximation [54–56]. This approach to vibrational states is, of course, analogous to the MP2 method of electronic structure theory [57,58]. The CC-VSCF method was introduced with two alternative treatments to the anharmonic potential: either expansion in a power series of the normal modes (truncated in practice to a low order polynomial, such as cubic) or by assuming that the interaction can be represented by a sum of interactions between pairs of normal modes [54,56]. The latter treatment neglects couplings involving simultaneous triplets (or a greater number) of normal modes, but it eliminates the need for a truncated power series representation. Experience has shown that the pair representation is more accurate than a low-order polynomial for many systems, including many strongly anharmonic systems such as hydrogen-bonded clusters. The CC-VSCF method as presently formulated, ignores rotational–vibrational coupling effects, such as Coriolis coupling. Its great advantage, though, is a favorable scalability of the effort with the number of modes, N. Depending on whether the interaction potential is treated as a truncated polynomial, or as a sum of pairwise interactions between the different normal modes, the computational effort scales as N or as N^2. In practical terms, for simple analytic force fields, CC-VSCF is feasible for systems having up to hundreds or thousands of normal modes. Jung and Gerber, for example, used CC-VSCF to treat water clusters $(H_2O)_n$, up to $n = 8$ [54,56]. The assumption of pairwise interactions between different normal modes seemed to give much more satisfactory results in this case, than a low-order polynomial representation [56]. With some technical improvements of the VSCF algorithms, VSCF and CC-VSCF calculations were carried out for small peptides and peptide–water complexes [59], for sugars (monosacharides) [60] and even for a small protein BPTI [61,62]. The latter system, including a hydration shell of 196 water molecules, has approximately 3500 vibrational modes. The calculations reported were at the VSCF level, but unpublished partial CC-VSCF results are available. Another major advantage of CC-VSCF is that it proved convenient for interfacing with electronic structure codes. Thus the method became a basis for an *Ab initio* Vibrational

Spectroscopy algorithm, as will be outlined in the next Section. VSCF and CC-VSCF can fail for strongly anharmonic soft modes such as torsions. This is related to the fact that for large amplitude torsional motions, normal mode coordinates are not suitable, and when used may give rise to extremely large coupling due to their geometric nature. Other than such exceptions, the method seems in most cases of satisfactory accuracy.

In comparison with experimental frequencies, our estimates indicate that CC-VSCF method errors are usually smaller than the errors due to the inaccuracy of the potential. At the same time, it would be very useful to search for approaches of better accuracy, while preserving the computational merits of CC-VSCF, the comfortable moderate scaling of the effort with N and the convenient interfacing with electronic structure codes. One possibility is to pursue higher-order perturbation corrections. This was recently proposed by Senent and Dominguez-Gomez [63], however, since the authors present results for a reduced-dimensionality treatment, one cannot readily assess the computational effort involved for large systems. A very interesting new approach, related to VSCF but with prospects on improved accuracy, is found in the recent work of Christiansen [64,65]. This work pursues the development of a vibrational Coupled Cluster (CC) theory, in analogy to the CC methods of electronic structure theory [58]. Initial results for model systems are encouraging, and vibrational CC seems to out-perform vibrational CI in accuracy, for the same amount of effort [65].

9.2.3 Grid methods

Grid methods are among the most rigorous numerical approaches for solving the vibrational Schrödinger equation. These methods are, in principle, 'numerically exact', i.e. can be pursued systematically to any desired accuracy. Unfortunately, the demands on computer memory and CPU time are such that applications are feasible in the present state of the art for systems of only a few atoms. Available grid methods include those discussed by Bačić and Light [66], Hutson [67], Young and Peet [68], and Cohen and Saykally [69]. Perhaps the most widely used grid method is, at present, the Discrete Variable Representation (DVR) method. Some of the key papers are: Bačić and Light [70]; Light et al. [71]; Henderson et al. [72]; Bramley et al. [73]; Mandelshtam and Taylor [74]; and Wright and Hutson [75]. As powerful as the approach is for small molecules, extensions to larger systems are difficult. On the other hand, it may well be possible to combine it with *ab initio* codes for use in direct calculations of vibrational spectroscopy. A technique of this type could be very useful in studies of vib-rotational spectroscopy of small molecules.

9.2.4 Diffusion Quantum Monte Carlo

The Diffusion Quantum Monte Carlo (DQMC) algorithm and related methods such as the Vibrational Quantum Monte Carlo approach have the important property of scaling well with system size (number of degrees of freedom). At the same time the method can be pursued in principle to yield a 'numerically exact' energy. DQMC was introduced by Anderson [76] for electronic structure calculations. The method is both rigorous and simple for wave functions that have no nodes. It is thus very suitable for vibrational ground states. The method is robust even for the most delocalized, anharmonic systems such as ^{4}He clusters. Studies of vibrational ground states of molecular systems and clusters were performed by Suhm and Watts [77], Buch [78], Barnett and Whaley [79], Bačić et al. [80], Vegiri et al. [81] and several other authors. In general, the method is not applicable to excited vibrational states, since these have nodes. However, suggestions were made for extending the method to excited states, either by using approximations to determine the nodal structure of the excited wave functions, or by introducing systematic procedures to determine the nodes. With these extensions, application of DQMC to vibrational spectroscopy became possible [82–86]. The use of the method in molecular spectroscopy has so far been limited, probably due to the incompletely solved problem of the excited wave function nodes. To our knowledge, at present, there is no systematic algorithm for direct DQMC with *ab initio* potentials.

9.2.5 Semiclassical methods

These approaches are, of course, inherently approximate. However, the errors due to the semiclassical approximations are expected to be modest, and not a major concern. This expectation is supported by the fact that in one-dimensional systems the WKB approximation yields very accurate energies. Also, as previously mentioned, the results of semiclassical VSCF are extremely close to those of full quantum VSCF [33]. Moreover, calculations by semiclassical approximations for multidimensional systems, realistic ones or models, gave very accurate results. The main obstacle to more extensive applications of semiclassical methods for vibrational state calculations is, in our view, that computationally more convenient algorithms are desirable. As the field is extensive, we cannot survey the various approaches proposed for semiclassical treatment of polyatomic molecules. An interesting recent example is the calculations of Kaledin and Miller [87], who used the semiclassical Initial Value Representation [88,89] for the spectral density. The results show very good agreement with quantum calculations. Applications were presented for systems such as H_2CO, NH_3 and CH_4. Applicability of SC-IVR to *ab initio* potential energies appears to be still an open question.

9.3 *AB INITIO* VIBRATIONAL SPECTROSCOPY

9.3.1 Fitting *ab initio* potentials *versus* direct *ab initio* spectroscopy calculations

An obvious way for carrying out spectroscopic calculations for an *ab initio* potential is by fitting the latter to a suitable analytic function. With such an approach, any vibrational spectroscopy method can be used. Indeed, such calculations have been pursued since the first appearance of reliable *ab initio* potential surfaces, and this continues to be a very active and successful direction of research. There are, however, several problems that strongly limit the applicability of this approach. First, the requirements of the quality of the fit are rather stringent. High-quality fitting is essential for spectroscopy of good accuracy. Second, fitting is an extremely difficult numerical task, the complexity of which grows essentially exponentially with the dimensionality of the system. Thus, fitted *ab initio* potentials were produced for systems of a few atoms only, and not in many cases. We mention as example two recent studies of the spectroscopy of *cis* and *trans*-HONO [90,91] that combine good quality fitting with a rigorous method for the vibrational computations. The calculations are in very good agreement with the experimental frequencies, but it seems extremely hard to treat larger systems by a similar approach. We note that at present, it is hard to identify a generally accepted or agreed upon algorithm for the objective of fitting potential functions. The field of fitting methodology is rich, and useful algorithms and strategies are being introduced almost constantly [92]. A very nice example of a fitted *ab initio* potential is presented in the recent DQMC study of CH_5^+ [92b]. At the present, however, it seems that the main possibilities for *ab initio* vibrational spectroscopy calculations, for molecules having more than 4–5 atoms, are with direct methods, that is, algorithms that employ directly *ab initio* potential surface points in the computation of the spectroscopy.

9.3.2 *Ab initio* VSCF and CC-VSCF

These algorithms are at the core of the present review, as extensively used methods in the field. We briefly review the relevant equations. In addition, the methodology involved in the applications is discussed below.

9.3.2.1 VSCF equations

The description of the method follows closely that of Refs. [54,56,93]. We consider the system in a state of total angular momentum $J = 0$, and neglect all rotational coupling effects. Using an electronic structure code, the minimum energy configuration is obtained, and the normal-mode displacement coordinates from that minimum are determined [2]. The vibrational Schrödinger equation in mass weighted normal coordinates $Q_1, Q_2, \ldots, Q_N$ can be written:

$$\left[-\frac{1}{2}\sum_{j=1}^{N}\frac{\partial^2}{\partial Q_j^2} + V(Q_1,\ldots,Q_N)\right]\Psi_n(Q_1,\ldots,Q_N) = \mathrm{E}_n\Psi_n(Q_1,\ldots,Q_N) \qquad (1)$$

where V is the potential function of the system, n is the state number, and N is the number of vibrational modes. Although normal modes are chosen to describe the vibrations of the system, the harmonic approximation is not made, and the anharmonicity of the potential function $V(Q_1,\ldots,Q_N)$ is fully retained. The VSCF approximation is based on the ansatz:

$$\Psi_n(Q_1,\ldots,Q_N) = \prod_{j=1}^{N}\Psi_j^{(n)}(Q_j) \qquad (2)$$

where $\Psi_j^{(n)}(Q_j)$ are single-mode vibrational wave functions. The importance of the choice of coordinates in VSCF should be emphasized at this point. The accuracy of the separability approximation (2) obviously depends on the choice of the variables that are being factorized. As pointed out earlier, the choice of normal modes is a reasonable one for low-lying excitations, since the latter are not far from the bottom of the potential well where the harmonic approximation applies (and for which separability in normal modes is exact). Given the fact that very commonly normal modes are good choice for VSCF, the standard codes for *ab initio* VSCF were written for these coordinates. While some VSCF calculations were done for less common cases where other coordinates are advantageous, e.g. [36–40], standard codes for such systems are not available. The use of normal modes can lead to unsatisfactory results for, e.g. soft torsional modes that are better described by angular variables.

Using the variational principle for the separability ansatz (2) leads to the single-mode VSCF equations [41–43]:

$$\left[-\frac{1}{2}\frac{\partial^2}{\partial Q_j^2}+\bar{V}_j^{(n)}(Q_j)\right]\Psi_j^{(n)}(Q_j)=\varepsilon_j^{(n)}\Psi_j^{(n)}(Q_j) \tag{3}$$

where the effective potential $\bar{V}_j^{(n)}(Q_j)$ for the mode Q_j is given by:

$$\bar{V}_j^{n}(Q_j)=\left\langle\prod_{l\neq j}^{N}\Psi_l^{(n)}(Q_l)|V(Q_1,\ldots,Q_N)|\prod_{l\neq j}^{N}\Psi_l^{(n)}(Q_l)\right\rangle \tag{4}$$

Eqs. (3) and (4) for the single-mode wave functions, energies and effective potentials must be solved self-consistently. A variety of methods for solving the single-mode Schrödinger equations (3) can be used. The standard VSCF programs [54,93,94] employ a grid method for this purpose. Both ground and excited states of VSCF are computed in this way. The VSCF approximation for the total energy of the state n is given by:

$$E_n=\sum_{j=1}^{N}\varepsilon_j^{(n)}-(N-1)\left\langle\prod_{j=1}^{N}\psi_j^{(n)}(Q_j)|V(Q_1,\ldots,Q_N)|\prod_{j=1}^{N}\Psi_j^{(n)}(Q_j)\right\rangle \tag{5}$$

The main computational difficulty is solving Eqs. (3)–(5) for large systems is due to the multidimensional integrals involved, which must be evaluated numerically. This is accomplished by approximate (but sufficiently accurate) representations of the potential.

9.3.2.2 Representations of the potential

A very advantageous approach in many cases is to expand the potential function in powers of the normal modes. The expansion can be written:

$$V(Q_1,\ldots,Q_N)=\sum_{m1,\ldots,mN}V_{m1,\ldots,mN}(Q_1)^{m1}\ldots(Q_N)^{mN} \tag{6}$$

This obviously leads to a simplification in which only the evaluation of one-dimensional integrals is required in order to obtain the single-mode effective potentials. The numerical effort in evaluating the integrals is not large, except if high-order terms in the expansion must be retained. In many applications to molecular systems, a fourth-order polynomial provides a sufficient representation of the potential. The simple evaluation of the single-mode effective potentials is an important advantage, and many VSCF and CI-VSCF applications have used that approach [47–50,59–62,95]. When this is the case, the computational effort for the VSCF equations scales linearly with N. However, for highly anharmonic systems, the power series in normal modes may not converge at all, or converge very slowly. This was found to be the case for several weakly bound hydrogen-bonded and van der Waals clusters [54,56]. The use of the power series expansions is especially problematic in *ab initio* calculations, since the evaluation of high-order derivatives of the potential is computationally very costly. To deal with this difficulty, Jung and Gerber [54] have introduced an alternative approximate representation of the potential that proved very successful in *ab initio* VSCF and CC-VSCF [93,94]. In this approach the potential is written as a sum of terms that include single-mode potentials, and interactions between pairs of normal modes. Contributions to the potential from interactions of triplets and quartets (or higher order interactions) of normal modes are neglected. Thus, the potential is written as follows [54,56]:

$$V(Q_1, \ldots, Q_N) = \sum_{j=1}^{N} V_j^{\mathrm{diag}}(Q_j) + \sum_i \sum_{j>i} W_{ij}^{\mathrm{coup}}(Q_i, Q_j) \tag{7}$$

The potential at equilibrium is conveniently taken as zero. The 'diagonal' (single mode) potential functions $V_j^{\mathrm{diag}}(Q_j)$ are defined by:

$$V_j^{\mathrm{diag}}(Q_j) = V(0, \ldots, Q_j, \ldots, 0) \tag{8}$$

and the pairwise interactions are:

$$W_{ij}^{\mathrm{coup}}(Q_i, Q_j) = V(0, \ldots, Q_i, \ldots, Q_j, \ldots, 0) - V_i^{\mathrm{diag}}(Q_i) - V_j^{\mathrm{diag}}(Q_j) \tag{9}$$

Both the $V_j^{\mathrm{diag}}(Q_j)$ and the $W_{ij}^{\mathrm{coup}}(Q_i, Q_j)$ are readily obtained from *ab initio* codes by computing the potential function along Q_j while keeping the other modes at equilibrium (for $V_j^{\mathrm{diag}}(Q_j)$), and by computing the potential for different Q_i, Q_j while keeping all modes $l \neq i, j$ at equilibrium (for the coupling potential). The calculation of the effective VSCF potentials for the representation (7)–(9) requires evaluation of one-dimensional numerical integrals, and obtaining the VSCF energies necessitates also two-dimensional quadratures. The scaling of the computational effort with N becomes proportional to N^2, but this is still very satisfactory even for large systems. The representation (7) is an assumption that must be tested. Already for empirical potentials, it was found satisfactory in the case of water clusters $(H_2O)_n$ [54]. It has been used extensively in *ab initio* VSCF and CC-VSCF, for a wide range of molecules and clusters of various kinds, for example: H_2O, $Cl^-(H_2O)$ and $(H_2O)_2$ [93]; $(H_2O)_n$ clusters, $Cl^-(H_2O)_n$, $H^+(H_2O)_n$ and $H_2O\cdots CH_3OH$ [94]; glycine [96]; and *N*-methylacetamide [97]. So far, agreement of the computed spectra with experiment (direct tests or indirect evidence) support the pairwise interaction representation. Additional tests of this representation, especially by direct examination of the *ab initio* potential surfaces are nevertheless desirable. We note that it is not difficult to extend somewhat the representation (7), and include in the calculations a limited number of triplets or quartets of normal modes. In fact, so long as this is restricted to a small number of high order interaction terms, the scaling of the computational effort with N will not be affected, and the effort itself may only be moderately increased. Full inclusion of all higher order terms (e.g. quartets) in the potential will, however, render the method impractical. In the case of *ab initio* VSCF, inclusion of couplings of triplets of normal modes was tested for H_2O and $Cl^-(H_2O)$ [94], and the contributions were found to be non-negligible, but not significant enough to justify the much increased cost of *ab initio* computation of a large number of additional potential energy points.

9.3.2.3 CC-VSCF equations

The idea underlying this approach is that the difference between the (true) Hamiltonian of a system and the separable VSCF Hamiltonian must be small, since VSCF is found to be a good approximation. Thus, it should be reasonable to treat the difference as a perturbation, and use perturbation theory to a low order [54,93]. The Hamiltonian is written in the form:

$$H = H^{\mathrm{SCF},(n)} + \Delta V(Q_1, \ldots, Q_N) \tag{10}$$

where $H^{\mathrm{SCF},(n)}$ is the Hamiltonian in the VSCF approximation:

$$H^{\mathrm{SCF},(n)} = \sum_j \bar{H}_j^{(n)}(Q_j) \tag{11}$$

with

$$\bar{H}_j^{(n)}(Q_j) = -\frac{1}{2}\frac{\partial^2}{\partial Q_j^2} + \bar{V}_j^{(n)}(Q_j) \tag{12}$$

being the VSCF Hamiltonian for mode Q_j (in the state n). ΔV of Eq. (10) is given by:

$$\Delta V(Q_1, \ldots, Q_N) = V(Q_1, \ldots, Q_N) - \sum_{j=1}^{N} \bar{V}_j^{(n)}(Q_j) \tag{13}$$

The correlation effects are all included in ΔV, the difference between the correct Hamiltonian and the VSCF one. Assuming the term is sufficiently small, we apply second order perturbation theory and find:

$$E_n^{\mathrm{cc}} = E_n^{\mathrm{VSCF}} + \sum_{m \neq n} \frac{\left| \left\langle \prod_{j=1}^{N} \Psi_j^{(n)}(Q_j) | \Delta V | \prod_{j=1}^{N} \Psi_j^{(m)}(Q_j) \right\rangle \right|^2}{E_n^{(\mathrm{o})} - E_m^{(\mathrm{o})}} \tag{14}$$

where E_n^{cc} is the correlation-corrected energy of state n. As noted previously, this is analogous to the second order Møller–Plesset method in electronic structure theory [57, 58]. All the wave functions $\Psi_j^{(m)}$ and the energy $E_m^{(\mathrm{o})}$ in Eq. (14) are calculated from the VSCF Hamiltonian $H^{\mathrm{SCF},(n)}$, corresponding to state n. E_n^{VSCF} is the VSCF expectation energy, given by Eq. (5). $E_m^{(\mathrm{o})}$ is given by:

$$E_m^{(\mathrm{o})} = \sum_{j=1}^{N} \varepsilon_j^{(n),m} \tag{15}$$

where $\varepsilon_j^{(n),m}$ is the mth VSCF energy level of the j-mode, computed from the Hamiltonian $\bar{H}_j^{(n)}(Q_j)$.

These expressions were used by Chaban et al. [93,94] in the first *ab initio* VSCF calculations. Experience since then has shown that the correlation correction contributes significantly, in many cases, to improve agreement with experiment. In applications of the method since then, 8–16 grid points along each coordinate Q_j were used in computing the CC-VSCF energy levels, depending on the application. The computational effort involved is obviously dominated by the calculation of *ab initio* potential surface points for the required grid of points. Denoting by M_G the number of grid points along each normal mode, and by N the number of vibrational modes, then the number of *ab initio* potential points necessary for the anharmonic spectroscopy calculation is $NM_G + \frac{1}{2}N(N-1)M_G^2$.

9.3.2.4 Anharmonic infrared intensities

Ab initio anharmonic IR intensities are calculated using dipole moments estimated on grids by an *ab initio* method (at the same time when potential energies are computed).

For fundamental and overtone excitations:

$$I_i = \frac{8\pi^3 N_A}{3hc}\omega_i |\langle \Psi_i^{(0)}(Q_i)|\vec{u}(Q_i)|\Psi_i^{(m)}(Q_i)\rangle|^2 \tag{16}$$

where $\vec{u}$ is the dipole moment, ω_i is the CC-VSCF vibrational frequency for the normal mode i; $\Psi_i^{(0)}$ and $\Psi_i^{(m)}$ are the VSCF wave functions for the ground and the mth excited vibrational states.

For combination excitations of modes i and j:

$$I_{ij} = \frac{8\pi^3 N_A}{3hc}\omega_{ij} |\langle \Psi_i^{(0)}(Q_i)\Psi_j^{(0)}(Q_j)|\vec{u}(Q_i,Q_j)|\Psi_i^{(m)}(Q_i)\Psi_j^{(n)}(Q_j)\rangle|^2 \tag{17}$$

where m and n are excitation levels for modes i and j.

9.3.2.5 Electronic structure methods used with VSCF

The accuracy of the spectroscopic calculations obviously depends strongly on the electronic structure method used. High-level electronic structure methods are too costly, except for very small molecules. Fortunately the MP2 method gives satisfactory accuracy, while being computationally feasible to permit a fully coupled all-mode treatment of molecules such as glycine [96], *N*-methyl acetamide [97], and the glycine–H_2O complex [98]. Suitable basis sets for these MP2 calculations are typically of double-ζ or triple-ζ quality including polarization (DZP or TZP) [99,100]. For these basis sets, we estimate that the average inaccuracy of the potential makes a larger contribution to the errors of the computed frequencies than the inaccuracy of the vibrational method, but the magnitudes of the two contributions are not very different. The augmented correlation consistent triple-zeta (aug-cc-pVTZ) basis set [101] was also used, and in some cases (such as the benchmark H_2O molecule [93]) led to significantly improved results. For this basis set, we estimate that the potential-induced errors, in many cases, may be smaller than those due to the vibrational method. So far, methods of higher level than MP2 have rarely been used with CC-VSCF. One such case is $F^-(H_2O)$, a very challenging system with respect to both electronic structure theory and the vibrational dynamics [102]. For this system, the CCSD(T) method was shown to provide an improvement over MP2, but it is questionable whether the relatively modest improvement of the spectroscopic results justifies the substantially increased computational effort.

CC-VSCF was also used for computing anharmonic vibrational spectra from DFT potentials [103,104]. The most accurate spectroscopic results are obtained from the hybrid functionals B97 [105] and B3LYP [106], of the non-hybrid DFT functionals, HCTH [107] seems superior to BLYP [108]. The spectroscopic accuracy of B97 and B3LYP was about equal to that of MP2.

Not surprisingly, standard semiempirical electronic structure methods such as PM3 give poor spectroscopic results when used with CC-VSCF. However, Brauer et al. have recently succeeded with an improved PM3 potential [109]. This potential is improved by input from *ab initio* harmonic frequencies: the standard PM3 potential surface in normal coordinates $W^{\mathrm{PM3}}(Q_1, \ldots, Q_N)$ is modified by the scaling of each normal mode coordinate Q_j by a constant λ_j. The scaling coefficients are chosen so that the improved potential $W^{\mathrm{IMP}} = W^{\mathrm{PM3}}(\lambda_1 Q_1, \ldots, \lambda_N Q_N)$ reproduces at the harmonic level the harmonic frequencies from *ab initio* calculations. Application of the improved PM3 potentials in CC-VSCF calculations gave results in excellent accord with experiment for several amino acids [109]. The improved semiempirical potentials are extremely fast to compute, and can potentially be used for vibrational spectroscopy calculations of quite large systems (with hundreds of normal modes).

VSCF, CC-VSCF and related algorithms are incorporated in the *ab initio* code packages GAMESS [3] and MOLPRO [5]. In both cases, developers of the systems made important improvements to the codes (Dr. M.W. Schmidt for GAMESS, Dr. G. Rahut and Prof. H.-J. Werner for MOLPRO). There have been numerous applications of the VSCF codes through use of the GAMESS suite of programs.

9.3.2.6 Improvements and extensions of VSCF and CC-VSCF

Matsunaga et al. [110] introduced VSCF-DPT2, a method that includes the effects of degeneracies in the anharmonic vibrational spectra. The essential extension is to use Degenerate Perturbation Theory (as opposed to Non-degenerate Perturbation Theory) in introducing correlation effects. Also this method was interfaced with electronic structure codes, and is incorporated in GAMESS. There have been several applications of *ab initio* spectroscopy calculations with this method.

Wright and Gerber [111] have proposed another extension of VSCF that includes correlation effects, the Partly Separable VSCF (PS-VSCF) method. This method is based on an ansatz of the following type:

$$\Psi(Q_1, \ldots, Q_N) = \Phi(Q_1, \ldots, Q_M) \prod_{j=M+1}^{N} \Psi_j(Q_j) \tag{18}$$

The function $\Phi(Q_1, \ldots, Q_M)$ fully correlates the modes $Q_1, \ldots, Q_M$, while the $\Psi_j(Q_j)$ are VSCF-like single mode wave functions. PS-VSCF is very appropriate for dealing with strong correlation effects that are confined to a small subset of modes. The equations for $\Phi(Q_1, \ldots, Q_M)$ and $\Psi_j(Q_j)$ are solved by grid algorithms. Practically, only 2–3 modes can be treated as fully correlated in this approach. In the calculations with this method, also terms representing couplings of triplets of normal modes were included in the *ab initio* potential, and their inclusion was found to provide a small improvement.

An interesting improvement of the CC-VSCF algorithm was proposed by D.M. Benoit [112]. He suggests a method for estimating which pairwise couplings between the normal modes are negligibly small for the purpose of the spectroscopic calculation. In several test calculations, a large reduction of computational effort compared with the standard CC-VSCF was achieved. The speed-ups gained by this approach could be very useful in calculations of large molecules.

K. Yagi et al. [113] developed an algorithm for both CC-VSCF and vibrational CI calculations with quartic force fields (QFF) computed from *ab initio* potentials. In a more recent paper, K.Yagi et al. [114] implemented QFF representations for couplings of groups of normal modes, and applied these potentials in CC-VSCF. For the case of pairwise coupling between modes, as in standard CC-VSCF, great reduction of the computational effort is achieved, with only a modest loss of accuracy. The scheme of Yagi et al. [114] is formally applicable also to the treatment of couplings of triplets, quartets, etc. of normal modes. The approach referred to by the authors as '*n*-mode coupling representations of the quartic force field' (*n*MR-QFF) seems to open the way for major reduction of the computational effort in anharmonic spectroscopy calculations, especially for large molecules. This method has been recently implemented in GAMESS.

9.3.3 *Ab initio* anharmonic calculations using perturbation theory

Perturbation theory has been applied to anharmonic calculations of spectroscopy from *ab initio* potentials in a large number of studies [19–25,115–121]. In nearly all cases so far, second-order perturbation theory was employed. The representation of the anharmonic potential generally used in these studies is a polynomial in the normal modes, most often a quartic force field. A code implementing this vibrational method was recently incorporated by V. Barone in GAUSSIAN [24]. Calculations were carried out for relatively large molecules, such as pyrrole and furan [25], uracil and thiouracil [118], and azabenzenes [119]. We note that in addition to spectroscopy, the *ab initio* perturbation theoretic algorithms were also applied to the calculation of thermodynamic properties [120]. Finally, a very useful extension of the methodology is to the inclusion of Coriolis couplings [121]. These advances render the perturbation-theoretic method very competitive with the VSCF approaches.

9.4 APPLICATIONS AND PERFORMANCE

9.4.1 Performance for large molecules

Major progress was made in recent years in *ab initio* calculations of anharmonic vibrational spectroscopy. One of the important indicators is the good agreement with experiment found in calculations for relatively large molecules, having more than 10 atoms (24 vibrational modes). Treatment of such large systems at a good anharmonic level has become possible only in the last several years, as a consequence of methodological progress and increased computer power. Calculations of anharmonic vibrational spectroscopy, including all modes are feasible now for electronic structure methods such as MP2 or B3LYP, for molecules having up to 20 atoms. It seems that with the vibrational state methodologies available, errors of the computed frequencies are on the average mostly due to the electronic structure calculation (in the case of MP2/TZP) and to a somewhat lesser extent due to the vibrational algorithm. The difference between the contributions to the frequency errors is, however, probably not large on the average. In other words, the accuracy of the electronic structure and of the vibrational methodology is nearly balanced in typical calculations. Previous experience (based on comparison with available experimental data) shows that the average accuracy provided by the *ab initio* CC-VSCF method (in conjunction with MP2 or B3LYP potentials) is on the order of 30–50 cm^{-1}.

Table 9.1 shows an example of comparison between *ab initio* vibrational spectroscopy calculations (CC-VSCF/MP2-DZP) and experiment for the most stable conformer of glycine, the structure of which is shown in Fig. 9.1. The results are from Chaban et al. [96]. The agreement with experiment is good, much better than that provided by the

Table 9.1 Vibrational frequencies and IR intensities for the lowest energy conformer of glycine

Mode	Vibrational frequencies (cm^{-1})				Intensity (km/mol)	Description
	Harmonic	CC-VSCF	Exp.[a]	Exp.[b]		
1	3829	3598	3560	3585	109	OH stretch
2	3688	3382	3410		5	NH_2 asym stretch
3	3590	3343			3	NH_2 sym stretch
4	3213	2986			17	CH_2 asym stretch
5	3148	2959	2958		22	CH_2 sym stretch
6	1836	1805	1779		414	C=O stretch
7	1702	1669	1630		23	HNH bend
8	1495	1473	1429		13	HCH bend
9	1443	1410	1373		30	C–O(H) stretch
10	1410	1377			0	CCN bend
11	1327	1290			23	NCH_2 bend
12	1205	1185			1	CCN oop bend
13	1195	1167	1136		99	CN stretch
14	1155	1122	1101		232	CO_2 bend
15	975	970	907		184	CNH_2 umbrella
16	937	943	883		3	$NCCO_2$ tors
17	852	847	801		60	C–CO_2 stretch
18	665	613	619		143	CO_2 oop bend
19	636	633			16	NCCO(H) shear
20	516	514	500		23	OCOH tors
21	467	463	463		38	NCCOH shear
22	259	270			11	NCCO shear
23	240	352			54	HNHC tors
24	58	143			4	NCCO(H) tors

[a]Experimental data obtained in Ar matrix.
[b]Experimental O–H stretch obtained in He cluster.

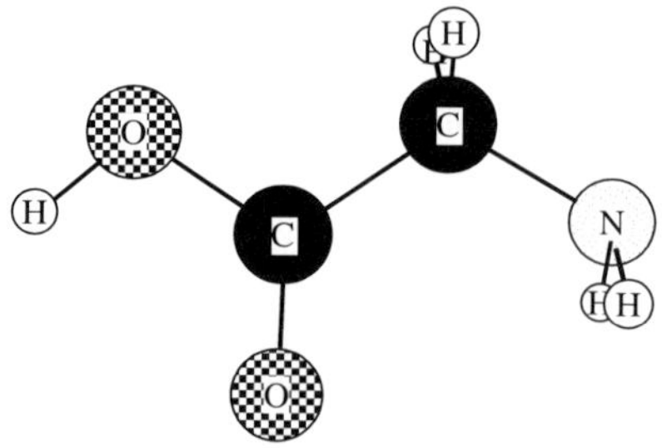

Fig. 9.1. Geometrical structure of the lowest energy conformer of glycine.

harmonic approximation. The most dramatic improvement is obviously for the OH and NH stretching modes, but there are significant improvements also for several other modes. As pointed out in Ref. [96], the *ab initio*-based results strongly outperform standard force fields, such as OPLS-AA [122], with respect to spectroscopic accuracy. This, of course, is a reflection of the fact that the *ab initio* force field is much more accurate. Table 9.2 is another comparison between CC-VSCF calculations (using the MP2/DZP *ab initio* potential) and matrix experiments, for a conformer of *N*-methylacetamide (NMA). The structure of the conformer is shown in Fig. 9.2, and the results are from Gregurick et al. [97]. Agreement with experiment for this molecule, an important model for peptides and for the peptide force field, is again much better for the CC-VSCF level than for the harmonic approximation. Agreement with experiment can be said to support the MP2 force field. Vibrational spectroscopy is thus emerging as a tool for testing the force fields of large polyatomic molecules.

9.4.2 Spectroscopy calculations as a test of *ab initio* and DFT force fields

In the spirit of the above observation, anharmonic vibrational spectroscopy calculations provide, by comparison with experiment, an evaluation of the quality of the potential surface used. It seems important to use spectroscopy as a way to compare the relative accuracies of different force fields. The results may, of course, differ from case to case. A crude general picture that seems to emerge from a limited set of small molecules (that includes H_2O, HCOOH, CH_3COOH) [104] indicates that MP2, B97 and B3LYP are very roughly comparable in the agreement with experimental spectroscopy, while HCTH and BLYP functionals do significantly less well (in this order). Much more can be learned about the quality of potential surfaces from different types of *ab initio* and DFT methods, but this will require systematic studies for various types of molecules.

9.4.3 Vibrational spectroscopy of hydrogen-bonded clusters

Vibrational spectroscopy is a powerful tool for probing the potential energy surfaces of weakly bound complexes. Both the intermolecular interactions of the species bound in

Table 9.2 Vibrational frequencies and IR intensities of *trans-N*-methylacetamide

Mode	Vibrational frequencies, (cm^{-1})			Intensity, (km/mol)	Mode description
	Harm	CC-VSCF	Exp.		
1	3751	3523	3498	44	NH stretch
2	3261	3014	3008	27	CH asym str (NCH_3)
3	3248	2993	3008	13	CH asym str (CCH_3)
4	3245	2985	2978	31	CH asym str (CCH_3)
5	3225	2979	2973	45	CH asym str (NCH_3)
6	3133	2939	2958	16	CH sym str (NCH_3)
7	3124	2940	2915	61	CH sym str (CCH_3)
8	1780	1751	1708	400	C=O str—amide I
9	1585	1547	1511	249	C–N str—amide II
10	1548	1566	1472	30	CH_2 bend (NCH_3)
11	1526	1541	1446	30	CH_2 bend (CCH_3)
12	1520	1557	1446	29	CH_2 bend (NCH_3)
13	1509	1515	1432	6	CH_2 bend (CCH_3)
14	1484	1468	1419	14	CH_3 umbr (NCH_3)
15	1438	1422	1370	21	CH_3 umbr (CCH_3)
16	1310	1283	1265	91	NCO bend—amide III
17	1208	1214	1181	2	CNC bend
18	1174	1184		6	NCH_3 rock
19	1133	1119	1089	7	CN str, NCH_3 rock
20	1074	1083	1037	9	CCH_3 rock
21	1020	1022	990	14	CCH_3 rock
22	890	891	857	3	
23	634	638	658	10	
24	624	637	626	4	
25	429	481	439	12	
26	363	439	391	90	
27	270	300	279	10	
28	153	244		28	
29	68	165		1	
30	52	266		3	

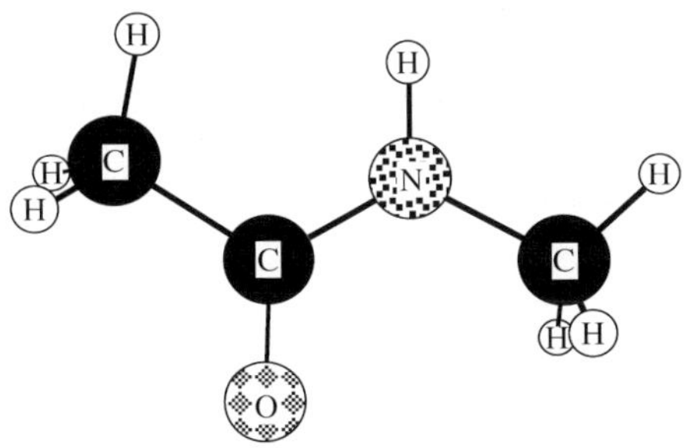

Fig. 9.2. Geometrical structure of *trans-N*-methylacetamide.

the complexes, and the intramolecular force fields, are reflected in different parts of the spectra. In applying *ab initio* vibrational spectroscopy calculations and comparing them to experiment, one is examining whether the electronic structure method is adequate for not only the intermolecular forces but also for the coupling between the intramolecular and intermolecular degrees of freedom, which is an issue of great importance. For these reasons, the study of hydrogen-bonded and other clusters by *ab initio* vibrational spectroscopy has been extensively pursued [54,93,94,98,102–104,123–126]. Table 9.3 shows results of CC-VSCF/MP2-TZP calculations for the complex CH_3OH-H_2O [94]. The structure of the complex is shown in Fig. 9.3. Good agreement is found between the experimental and theoretical fundamental frequencies (only transitions with $\omega > 1000\ cm^{-1}$ were measured). For the intensities, there is only a qualitative agreement between theory and experiment. Theoretical results are within a factor of 2–4 larger than the experimental intensities for the strong transitions. Theoretical intensities were obtained using RHF (rather than MP2) dipole moments, and this could be one of the reasons for the large difference with experiment. The origins of the discrepancy on intensities may be also partly due to experimental errors, since both theory and experiment are expected to be less accurate for intensities than for frequencies. For several very interesting hydrogen-bonded clusters, such as glycine–H_2O [98] and nitrile–water complexes [126], detailed experimental data for comparison are not available yet. The theoretical predictions may hopefully be useful in guiding future experiments. Both for glycine–H_2O and for the nitrile–H_2O complexes, some of the

Table 9.3 Vibrational frequencies and IR intensities for methanol–water complex

Mode	Vibrational frequencies (cm^{-1})			Intensity (kcal/mol)		Description
	Harmonic	CC-VSCF	Exp.	CC-VSCF	Exp.	
1	4005	3661	3714	78	44	H_2O asym str
2	3874	3645	3627	4	6	H_2O sym str
3	3812	3595	3536	347	100	meth OH str
4	3194	3007	2982	42	8	CH_3 str
5	3123	2904		71		CH_3 str
6	3061	2891	2835	68	12	CH_3 str
7	1619	1625	1601	77	23	H_2O bend
8	1555	1523	1475	2	1	CH_3 bend
9	1536	1516	1464	1	2	CH_3 bend
10	1531	1495	1448	12	1	CH_3 bend
11	1456	1420	1380	32	7	COH bend
12	1209	1189		1		OCH bend
13	1141	1128	1103	21	3	OCH bend
14	1094	1064	1048	94	30	CO str
15	721	827		167		
16	252	614		256		
17	209	432		8		
18	188	284		21		
19	99	270		15		
20	81	102		6		
21	75	192		23		

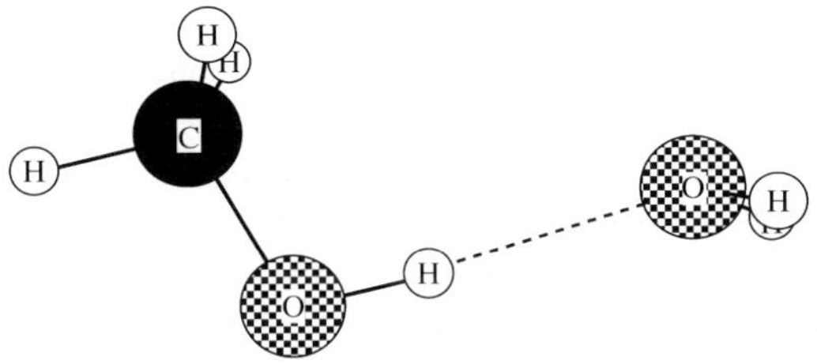

Fig. 9.3. Geometrical structure of methanol–water complex.

predicted features are surprising. For example, while complexation with water leads to very strong red shifts of the carboxyl group (O–H and C=O) stretching frequencies of glycine [98] and other organic acids, the C≡N stretches of the nitriles are predicted to shift to the blue upon complexation with water [126].

Small, very floppy hydrogen-bonded complexes reveal a failure of VSCF and CC-VSCF for soft torsional modes. The effect is quite pronounced for $(H_2O)_2$ [54,93,94]. This failure stems from the fact that the normal mode description (as used in the standard version of VSCF) is inadequate for the treatment of large-amplitude torsions. Angular coordinates are much more appropriate choices for such motions. When the standard normal modes are used to describe torsions, the anharmonic effects are extremely strong, resulting in breakdown of VSCF and CC-VSCF; this yields unphysical results for the torsional (angular) modes. The manifestation of such unphysical behavior is that the single-mode anharmonic correction in such cases is computed to be positive and very large, e.g. 20% or more, relative to the calculated harmonic frequency. Such breakdown of VSCF can also occur for molecular internal rotations or soft torsions. Results by Firsov et al. showed [127] that even extremely floppy systems, such as the van der Waals clusters SH...Ar, and SH...Kr, can be very successfully treated by CC-VSCF, when the rotational (bending) mode is described by an angular variable. Unfortunately, the standard VSCF codes do not have this option.

A comment is due regarding the performance of different electronic structure methods for cluster systems. For covalently bound molecules, the available spectroscopic evidence indicates, as was noted above, that the accuracy of the DFT B3LYP and B97 functionals roughly matches that of MP2. For hydrogen-bonded clusters, comparisons that can be made for systems such as $(H_2O)_n$ and $Cl^-(H_2O)_n$, (by using the results of Refs. [93,94,103, 104]), show that MP2 is in better accord with experiment than DFT. Such comparisons can be viewed as a spectroscopic test as to which potential surface is more accurate.

9.4.4 *Ab initio* spectroscopy and the identification of new molecular species

Ab initio vibrational spectroscopy can have unique advantages as a tool for the identification and characterization of new species. For existing molecules, available empirical force fields may be at hand, and can be used in interpreting the spectra. For new types of molecules, there may be no empirical force fields that one can rely on. A nice example is the novel rare gas molecules of the type of HRgY, where Rg is a noble gas atom, and Y is an electronegative group [128]. Synthesis of these molecules was pioneered by M. Räsänen and coworkers [129–131], who prepared them photochemically by photolysis of HY in rare-gas matrices. *Ab initio* spectroscopy, and specifically CC-VSCF calculations, played a key role in the development of this field [132–138]. The most important application of CC-VSCF in this area was its use in the identification and characterization of HArF [132], the first chemically bound compound of argon. CC-VSCF calculations were likewise essential in the identification of HKrF (also a new compound) and in understanding its properties [135]. CC-VSCF calculations contributed to the characterization of HKrCCH, the first (neutral) organic molecule containing krypton [138]. Currently, the spectroscopy of the new 'organo-rare gas' molecules HXeCCH and HXeCCXeH [128] is being studied with interesting implications to their properties. Most of the above calculations ignore matrix effects altogether, and these effects are indeed secondary in importance. They can, however, be important in identifying matrix sites, as was found in the cases of HKrF and HArF [135,137]. For these cases, vibrational spectroscopy calculations that incorporate a large number of matrix atoms were indeed carried out [135,137], using the approach of Bihary et al. [139]. However, the latter method is not yet feasible for *ab initio* potentials, so analytical force fields were used.

9.4.5 *Ab initio* spectroscopy and the elucidation of complex spectra

Ab initio vibrational spectroscopy calculations have been able to provide dramatic improvements in interpretation of spectra, or yield qualitatively new insights in a number of intriguing cases. We mention here a single example, the work of A.T. Kowal [140], in sorting out the vibrational spectra of hydroxylamine and its ^{15}N, ^{18}O, and deuterium isotopomers, by using CC-VSCF calculations.

9.4.6 Overtones and combination mode transitions

The calculation of overtone and combination transitions is a challenge to anharmonic vibrational spectroscopy algorithms. For such transitions, the anharmonic couplings are much stronger than for fundamental ones. For many high-overtone transitions, the total vibrational density of states at the overtone level is very high, therefore bringing up the concern that a dense manifold of resonances may play a role. Overtone transitions were successfully treated by useful *ad hoc* approaches [141–144], which involve model Hamiltonians. Such models probably describe realistically the mode being excited, but they cannot be used to predict other transitions in the same molecule, and the representation of coupling between the overtone-excited mode and the other modes is in any case empirically based. Useful as such models are, they cannot replace the need for algorithms using *ab initio* potentials. Table 9.4 presents results of recent CC-VSCF/MP2-TZP calculations for the OH overtone excitation frequencies of HNO_3 [145]. In all cases the results are, obviously, a large improvement over the harmonic approximation. Up to vibrational level $v = 4$ of the OH stretch, agreement between calculations and experiment is quite reasonable [14,15,146,147]. For $v = 5$ and beyond, the method clearly breaks down. The anharmonic coupling between the OH stretch and other modes is very strong

Table 9.4 OH-stretching overtone excitation frequencies for HNO_3

Vibrational state	Frequencies (cm^{-1})		
	Harmonic	CC-VSCF/MP2-TZP	Exp.
2	7568.59	6939.93	
3	11,352.88	10,221.54	10,173[a]
4	15,137.17	13,462.74	13,250[b], 13,245[a], 13,248[c]
5	18,921.46	16,949.52	16,160[d]
6	22,705.76	20,980.72	18,950[d]

[a]Ref. [146].
[b]Ref. [14].
[c]Ref. [15].
[d]Ref. [147].

for the high overtones, to the point of breakdown of the CC-VSCF method. Calculations were also carried out for HNO_4 (not shown here). The three lowest overtones were measured, and agreement between the CC-VSCF calculations and experiment is good for the three transitions. In summary, CC-VSCF works well for low overtones, but not for the high ones, probably due to the very strong anharmonic coupling. In Ref. [145], calculations were also carried out for several combination transitions of HNO_3 and the HNO_3–H_2O complex. The agreement with experimental frequencies is reasonable. An extensive CC-VSCF study was recently performed for overtone and combination transitions in the complexes of NH_3–HF and NH_3–DF [148]. For the couple of overtone transitions that were measured for these systems, agreement with experiment is reasonable (but somewhat less good than is common for fundamental transitions).

9.4.7 Open-shell systems

At the present state of the art, it seems that the available *ab initio* vibrational spectroscopy codes cannot yet be applied routinely to open-shell systems. In the case of CC-VSCF, the problem seems to be due to difficulties of convergence of the electronic structure codes employed, for at least some of the points used in the CC-VSCF calculations. A complete solution for this problem may well come from the development of improved and more versatile open-shell electronic structure codes. There are also, however, examples of successful applications. Table 9.5 lists the CC-VSCF/MP2-TZP calculations of Chaban

Table 9.5 Vibrational frequencies (cm^{-1}) and IR intensities (km/mol) for HOCO radical

Mode	Harmonic	CC-VSCF	Exp(gas)	Exp(Ne)	Exp(Ar)	Intensity
O–H stretch	3866	3654	3636	3628	3603	146
C=O stretch	1901	1877	1853	1848	1844	418
HCO bend	1254	1160		1210	1211	306
C–OH stretch	1100	1032		1050	1065	72
OCO bend	622	603				9
Out-of-plane torsion	560	454		508	515	84

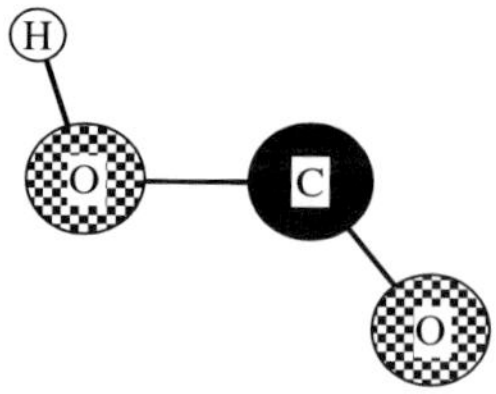

Fig. 9.4. Geometrical structure of HOCO radical.

[149] for HOCO radical (the structure of which is shown in Fig. 9.4). The agreement with gas phase and with matrix experiments is good, not less so than for typical closed-shell systems. *Ab initio* anharmonic vibrational spectroscopy of open-shell systems is still in its infancy, but these results are encouraging.

9.5 FUTURE DIRECTIONS

The field of *Ab initio* Anharmonic Vibrational Spectroscopy has seen rapid recent advancements. Already, several methods are at hand, and a variety of codes are available in several of the leading electronic structure suites of programs. It is tempting to speculate on the possible directions of progress in the next few years.

9.5.1 Larger systems

At present, the state of the art makes possible satisfactory anharmonic vibrational calculations for systems having more than 10 atoms, and probably with a major effort of up to 20 atoms. In the area of small biological molecules, the increased capability of the last several years is very substantial, yet far from what is needed. Several most recent methods seem to promise major speed-ups of the vibrational algorithms. One can draw much optimism also from the improvements in electronic structure methodology. For example, MP2 has been the electronic structure method in most applications to date. However, there have been important developments in fast, approximate MP2 algorithms, that have not yet been applied to anharmonic vibrational calculations. These include RI-MP2 and Local MP2 methods, see for example Refs [150–153]. With these electronic structure and vibrational methodology advances, it seems safe to expect rapid progress in the sizes of molecules amenable to spectroscopy calculations. For much larger molecules than hitherto treated, it may be necessary to apply non-first-principles methods as an intermediate step, until *ab initio* calculations become feasible. Improved semiempirical electronic structure methods can be a possible candidate for such an intermediate role. Another promising direction is ONIOM-like or QM/MM approaches [154–158], in which the chemically important part of the system is treated with more accurate methods (e.g. 'exact' MP2), while the rest of the molecular system is treated with more approximate *ab initio*, semiempirical, or empirical potentials. Development of new parallelization techniques within electronic structure packages GAMESS [3], MOLPRO [5], and NWCHEM [159] will also greatly improve the applicability of anharmonic vibrational spectroscopy to larger molecular systems.

9.5.2 Quest for increased accuracy

With regard to the electronic structure methodology, major obstacles must be surmounted before improvements can be made. Calculations with Coupled-Cluster methods, an obvious next step, are far more computationally costly than the presently used MP2, or B3LYP methods. In fact, there are extremely few direct *ab initio* calculations of anharmonic vibrational spectroscopy at higher than MP2 or DFT levels, even for small polyatomics. From the point of view of *ab initio* anharmonic spectroscopy, the leap from MP2 to the Coupled-Cluster method seems a bottleneck. One can draw encouragement from faster Coupled-Cluster implementations, so far employed with the perturbation theory anharmonic analysis [116,117].

The accuracy of the vibrational algorithms is probably sufficient at present in many cases, so long as MP2 is used with DZP or TZP basis sets. The use of improved basis sets, such as the correlation-consistent ones, may strongly motivate the formulation of higher-level vibrational algorithms. One can expect improvements of the present algorithms, but one can also expect the adaptation of other vibrational methods, so far used with analytically fitted *ab initio* potentials.

9.5.3 Time-domain spectroscopy

This review focused exclusively on frequency-domain spectroscopy. There is, however, major progress in time-domain spectroscopy. Methods such as 2D-IR spectroscopy offer exciting possibilities, especially for studies of large molecules, and of molecules in condensed phases [160]. Anharmonic interactions are of the essence in 2D-IR spectroscopy. The impressive experimental development of recent years [160], and pioneering theoretical studies, such as those conducted by Mukamel and coworkers [161], contribute to our impression that this may become a major direction for *ab initio* spectroscopic studies.

9.6 ACKNOWLEDGEMENTS

The work at the Hebrew University was supported under the auspices of the Saerree K. and Louis P. Fiedler Chair in Chemistry. Work at UC, Irvine was supported in part by NSF CRC grant CHE-0209719 and NSF Grant CHE-0101199.

We are very grateful to Dr. M.W. Schmidt and Professor M.S. Gordon for helpful suggestions and discussions, and for their useful input, especially in the context of the incorporation of VSCF into GAMESS. We thank also Dr. G. Rahut and Professor H-J.

Werner for their very helpful comments, partly in the context of the introduction of VSCF into MOLPRO.

We gratefully acknowledge helpful comments from Dr. T.L. Windus, Professor P. Hobza, and Dr. T. Lee. The review, and VSCF results presented here have benefited greatly from contributions and input from Dr. J.O. Jung, Professor S.M. Gregurick, Dr. N.J. Wright, Professor N. Matsunaga, Dr. Carrie Brindle, Dr. E. Brown, Dr. E. Fredji, A. Adesokan, L.Pele, and D. Shemesh.

9.7 REFERENCES

1 E.B. Wilson, Jr., J.C. Decius and P.C. Cross, Molecular vibrations, McGraw Hill, New York, 1955.
2 See, for example: (a) W.D. Gwinn, J. Chem. Phys., 55 (1971) 477. (b) J.A. Boatz and M.S. Gordon, J. Phys. Chem., 93 (1989) 1819. (c) C.E. Blum, C. Altana and A. Oskam, Mol. Phys., 34 (1977) 557. (d) G. Fogarasi and P. Pulay, Ann. Rev. Phys. Chem., 35 (1984) 191.
3 (a) M.W. Schmidt, K.K. Baldridge, J.A. Boatz, S.T. Elbert, M.S. Gordon, J.J. Jensen, S. Koseki, N. Matsunaga, K.A. Nguyen, S. Su, T.L. Windus, M. Dupuis and J.A. Montgomery, J. Comput. Chem., 14 (1993) 1347. (b) http://www.msg.ameslab.gov/GAMESS/GAMESS.html.
4 M.J. Frisch, G.W. Trucks, H.B. Schlegel, G.E. Scuseria, M.A. Robb, J.R. Cheeseman, J.A. Montgomery Jr., T. Vreven, K.N. Kudin, J.C. Burant, J.M. Millam, S.S. Iyengar, J. Tomasi, V. Barone, B. Mennucci, M. Cossi, G. Scalmani, N. Rega, G.A. Petersson, H. Nakatsuji, M. Hada, M. Ehara, K. Toyota, R. Fukuda, J. Hasegawa, M. Ishida, T. Nakajima, Y. Honda, O. Kitao, H. Nakai, M. Klene, X. Li, J.E. Knox, H.P. Hratchian, J.B. Cross, V. Bakken, C. Adamo, J. Jaramillo, R. Gomperts, R.E. Stratmann, O. Yazyev, A.J. Austin, R. Cammi, C. Pomelli, J.W. Ochterski, P.Y. Ayala, K. Morokuma, G.A. Voth, P. Salvador, J.J. Dannenberg, V.G. Zakrzewski, S. Dapprich, A.D. Daniels, M.C. Strain, O. Farkas, D.K. Malick, A.D. Rabuck, K. Raghavachari, J.B. Foresman, J.V. Ortiz, Q. Cui, A.G. Baboul, S. Clifford, J. Cioslowski, B.B. Stefanov, G. Liu, A. Liashenko, P. Piskorz, I. Komaromi, R.L. Martin, D.J. Fox, T. Keith, M.A. Al-Laham, C.Y. Peng, A. Nanayakkara, M. Challacombe, P.M.W. Gill, B. Johnson, W. Chen, M.W. Wong, C. Gonzalez, and J.A. Pople, Gaussian 03, revision, A.1, Gaussian Inc., Pittsburgh, PA, 2003.
5 H.-J. Werner, P.J. Knowles, M. Schutz, R. Lindh, P. Celani, T. Korona, G. Rauhut, F.R. Manby, R.D. Amos, A. Bernhardsson, A. Berning, D.L. Cooper, M.J.O. Deegan, A.J. Dobbyn, F. Eckert, C. Hampel, G. Hetzer, A.W. Lloyd, S. J. McNicholas, W. Meyer, M.E. Mura, A. Nicklas, P. Plamieri, R. Pitzer, U. Schumann, H. Stoll, A.J. Stone, R. Tarroni, and T. Thorsteinson, MOLPRO version 2002.6, a package of *ab initio* programs, Birmingham, UK, 2002.
6 A.P. Scott and L. Radom, J. Phys. Chem., 100 (1996) 16502.
7 S. Saebo and P. Pulay, J. Chem. Phys., 86 (1987) 914.
8 Y.N. Panchenko, J. Mol. Struct., 567 (2001) 217.
9 J.M. Coffin and P. Pulay, J. Phys. Chem., 95 (1991) 118.
10 M.D. Halls, J. Velkovski and H.B. Schlegel, Theor. Chem. Acc., 105 (2001) 413.
11 Z. Bačić and R.E. Miller, J. Phys. Chem., 100 (1996) 12945.
12 D.J. Nesbitt, Chem. Rev., 88 (1988) 843.
13 K. Liu, M.G. Brown, J.D. Cruzan and R.J. Saykally, Science, 271 (1996) 62.
14 (a) A. Sinha, R.L. Vander Wal and F.F. Crim, J. Chem. Phys., 92 (1990) 401. (b) A. Sinha, R.L. Vander Wal and F.F. Crim, J. Chem. Phys., 91 (1989) 2929.
15 P.R. Fleming, M. Li and T.R. Rizzo, J. Chem. Phys., 94 (1991) 2425.
16 D.J. Nesbitt and R.W. Field, J. Phys. Chem., 100 (1996) 12735.
17 S. Califano, Vibrational states, Wiley, New York, 1976.
18 G.D. Carney, L.L. Sprandel and C.W. Kern, Adv. Chem. Phys., 37 (1978) 305.
19 D.A. Clabo, Jr., W.D. Allen, R.B. Remington, Y. Yamaguchi and H.F. Schaefer, III, Chem. Phys., 123 (1988) 187.
20 W. Schneider and W. Thiel, Chem. Phys. Lett., 157 (1989) 367.

21 A. Willetts, N.C. Handy, W.H. Green, Jr. and D. Jayatilaka, J. Phys. Chem., 94 (1990) 5608.
22 S. Dressler and W. Thiel, Chem. Phys. Lett., 273 (1997) 91.
23 J. Neugebauer and B.A. Hess, J. Chem. Phys., 118 (2003) 7215.
24 V. Barone, J. Chem. Phys., 122 (2005) 014108.
25 V. Barone, Chem. Phys. Lett., 383 (2004) 528.
26 H.H. Nielsen, Rev. Mod. Phys., 23 (1951) 90.
27 A.B. McCoy, E.L. Sibert, III, in: J.M. Bowman and M.A. Ratner (Eds.), Advances on molecular vibrations and reaction dynamics, JAI press, Connecticut, USA, 1991, p. 255.
28 A.B. McCoy and E.L. Sibert, III, J. Chem. Phys., 92 (1990) 1893.
29 G.M. Florio, T.S. Zwier, E.M. Myshakin, K.D. Jordan and E.L. Sibert, III, J. Chem. Phys., 118 (2003) 1735.
30 S.G. Ramesh and E.L. Sibert, III J. Chem, Phys., 120 (2004) 11011.
31 J.M. Bowman, J. Chem. Phys., 68 (1978) 608.
32 M. Cohen, S. Greita and R.D. McEarchran, Chem. Phys. Lett., 60 (1979) 445.
33 R.B. Gerber and M.A. Ratner, Chem. Phys. Lett., 68 (1979) 195.
34 R.B. Gerber, R.M. Roth and R.A. Ratner, Mol. Phys., 44 (1981) 1335.
35 R.M. Roth, M.A. Ratner and R.B. Gerber, Phys. Rev. Lett., 52 (1984) 1288.
36 T.R. Horn, R.B. Gerber and M.A. Ratner, J. Chem. Phys., 91 (1989) 1813.
37 Z. Bačić, R.B. Gerber and M.A. Ratner, J. Phys. Chem., 90 (1986) 3606.
38 (a) R.M. Roth, R.B. Gerber and M.A. Ratner, J. Phys. Chem., 87 (1983) 2376. (b) L.L. Gibson, R.M. Roth, M.A. Ratner and R.B. Gerber, J. Chem. Phys., 85 (1986) 3425.
39 M.A. Ratner, R.B. Gerber, T.R. Horn and C.J. Williams, in: J.M. Bowman and M.A. Ratner (Eds.), Advances in molecular vibrations and collision dynamics, Vol. 1A, JAI Press, Connecticut, USA, 1991, p. 215.
40 T.R. Horn, R.B. Gerber, J.J. Valentini and M.A. Ratner, J. Chem. Phys., 94 (1991) 6728.
41 J.M. Bowman, Acc. Chem. Res., 19 (1986) 202.
42 M.A. Ratner and R.B. Gerber, J. Phys. Chem., 90 (1986) 20.
43 R.B. Gerber and M.A. Ratner, Adv. Chem. Phys., 70 (1988) 97.
44 J.M. Bowman, K. Christoffel and F. Tobin, J. Phys. Chem., 83 (1979) 905.
45 M.A. Ratner, V. Buch and R.B. Gerber, Chem. Phys., 53 (1980) 345.
46 T.C. Thompson and D.G. Truhlar, Chem. Phys. Lett., 75 (1980) 87.
47 H. Romanowski, J.M. Bowman and L.B. Harding, J. Chem. Phys., 82 (1985) 4155.
48 S. Carter, J.M. Bowman and L.B. Harding, Spectrochim. Acta A, 53 (1997) 1179.
49 S. Carter and J.M. Bowman, J. Chem. Phys., 108 (1998) 4397.
50 S. Carter, S.J. Culik and J.M. Bowman, J. Chem. Phys., 107 (1997) 10458.
51 J.K.G. Watson, Mol. Phys., 15 (1968) 479.
52 (a) S. Carter, J.M. Bowman and N. Handy, Theor. Chem. Acc., 100 (1998) 191. (b) D.A. Jelski, R.H. Haley and J.M. Bowman, J. Comput. Chem., 17 (1996) 1645. (c) H. Romanowski and J.M. Bowman, POLYMODE (QCPE 496), QCPE Bull., 5 (1985) 64.
53 M.-L. Senent, P. Palmieri, S. Carter and N.C. Handy, Chem. Phys. Lett., 354 (2002) 1.
54 J.O. Jung and R.B. Gerber, J. Chem. Phys., 105 (1996) 10332.
55 L.S. Norris, M.A. Ratner, A.E. Roitberg and R.B. Gerber, J. Chem. Phys., 105 (1996) 11261.
56 R.B. Gerber and J.O. Jung, in: P. Jensen, R.P. Bunker (Eds.), Computational molecular spectroscopy, Wiley, Sussex, UK, 2000, p. 365.
57 J.A. Pople, J.S. Binkley and R. Seeger, Int. J. Quantum Chem., S10 (1976) 1.
58 R. McWeeny, Methods of molecular quantum mechanics, 2nd edn., Academic Press, San Diego, CA, 1992.
59 S.K. Gregurick, E. Fredj, R. Elber and R.B. Gerber, J. Phys. Chem. B, 101 (1997) 8595.
60 S.K. Gregurick, J.H-Y. Liu, D.A. Brant and R.B. Gerber, J. Phys. Chem. B, 103 (1999) 3476.
61 A.E. Roitberg, R.B. Gerber, R. Elber and M.A. Ratner, Science, 268 (1995) 1319.
62 A.E. Roitberg, R.B. Gerber and M.A. Ratner, J. Phys. Chem. B, 101 (1997) 1700.
63 M.L. Senent and R. Dominguez-Gomez, Chem. Phys. Lett., 351 (2002) 251.
64 O. Christiansen, J. Chem. Phys., 120 (2004) 2140.

65 O. Christiansen, J. Chem. Phys., 120 (2004) 2149.
66 Z. Bačić and J.C. Light, Annu. Rev. Phys. Chem., 40 (1989) 469.
67 J.M. Hutson, Annu. Rev. Phys. Chem., 41 (1990) 123.
68 W. Yang and A.C. Peet, J. Chem. Phys., 92 (1990) 522.
69 R.C. Cohen and R.J. Saykally, Annu. Rev. Phys. Chem., 42 (1991) 369.
70 Z. Bačić and J.C. Light, J. Chem. Phys., 85 (1986) 4594.
71 J.C. Light, I.P. Hamilton and J.V. Lill, J. Chem. Phys., 82 (1985) 1400.
72 J.R. Henderson, J. Tennyson and B.T. Sutcliffe, J. Chem. Phys., 98 (1993) 7191.
73 M.J. Bramley, J.W. Tromp, T. Carrington, Jr. and G.C. Corey, J. Chem. Phys., 100 (1994) 6175.
74 V.A. Mandelshtam and H.S. Taylor, J.Chem. Soc. Faraday T, 93 (1997) 847.
75 N.J. Wright and J.M. Hutson, J. Chem. Phys., 110 (1999) 902.
76 J.B. Anderson, J. Chem. Phys., 63 (1975) 1499.
77 M.A. Suhm and R.O. Watts, Phys. Rep., 204 (1991) 293.
78 V. Buch, J. Chem. Phys., 97 (1992) 726.
79 R.N. Barnett and K.B. Whaley, J. Chem. Phys., 99 (1993) 9730.
80 Z. Bačić, M. Kennedy-Mandziuk, J.W. Moskowitz and K.E. Schmidt, J. Chem. Phys., 97 (1992) 6472.
81 A. Vegiri, M.H. Alexander, S. Gregurick, A.B. McCoy and R.B. Gerber, J. Chem. Phys., 101 (1994) 2577.
82 D.F. Coker and R.O. Watts, Mol. Phys., 58 (1986) 1113.
83 P. Sandler, V. Buch and J. Sadlej, J. Chem. Phys., 105 (1996) 10387.
84 J.K. Gregory and D.C. Clary, J. Chem. Phys., 103 (1995) 8924.
85 D. Blume, M. Lewerenz, P. Niyaz and K.B. Whaley, Phys. Rev. E, 55, Pt B (1997) 3664.
86 S. Broude, J.O. Jung and R.B. Gerber, Chem. Phys. Lett., 299 (1999) 437.
87 A.L. Kaledin and W.H. Miller, J. Chem. Phys., 119 (2003) 3078.
88 W.H. Miller, J. Phys. Chem. A, 105 (2001) 2942.
89 E. Jezek and N. Makri, J. Phys. Chem. A, 105 (2001) 2851.
90 D. Luckhaus, J. Chem. Phys., 118 (2003) 8797.
91 F. Richter, M. Hochlaf, P. Rosmus, F. Gatti and H.D. Meyer, J. Chem. Phys., 120 (2004) 1306.
92 (a) M.M. Law and J.M. Hutson, Comput. Phys. Commun., 102 (1997) 252. (b) A. Brown, A.B. McCoy, B.J. Braams, Z. Jin and J.M. Bowman, J. Chem. Phys., 121 (2004) 4105.
93 G.M. Chaban, J.O. Jung and R.B. Gerber, J. Chem. Phys., 111 (1999) 1823.
94 G.M. Chaban, J.O. Jung and R.B. Gerber, J. Phys. Chem. A, 104 (2000) 2772.
95 J.O. Jung and R.B. Gerber, J. Chem. Phys., 105 (1996) 10682.
96 G.M. Chaban, J.O. Jung and R.B. Gerber, J. Phys. Chem. A, 104 (2000) 10035.
97 S.K. Gregurick, G.M. Chaban and R.B. Gerber, J. Phys. Chem. A, 106 (2002) 8696.
98 G.M. Chaban and R.B. Gerber, J. Chem. Phys., 115 (2001) 1340.
99 T.H. Dunning, Jr., J.P. Hay, in: H.F. Schaefer, III (Ed.), Methods of electronic structure theory, Plenum Press, New York, 1977, pp. 1–27, Chapter 1.
100 T.H. Dunning, Jr., J. Chem. Phys., 55 (1971) 716.
101 T.H. Dunning, Jr., J. Chem. Phys., 90 (1989) 1007.
102 G.M. Chaban, S.S. Xantheas and R.B. Gerber, J. Phys. Chem. A, 107 (2003) 4952.
103 N.J. Wright and R.B. Gerber, J. Chem. Phys., 112 (2000) 2598.
104 N.J. Wright, R.B. Gerber and D.J. Tozer, Chem. Phys. Lett., 324 (2000) 206.
105 (a) H.L. Schmider and A.D. Becke, J. Chem. Phys., 108 (1998) 9624. (b) A.D. Becke, J. Chem. Phys., 107 (1997) 8554. (c) L.A. Curtiss, K. Raghavachari, P.C. Redfern and J.A. Pople, J. Chem. Phys., 106 (1997) 1063.
106 (a) A.D. Becke, J. Chem. Phys., 98 (1993) 5648. (b) P.J. Stephens, F.J. Devlin, C.F. Chabalowski and M.J. Frisch, J. Phys. Chem., 98 (1994) 11623.
107 F.A. Hamprecht, A.J. Cohen, D.J. Tozer and N.C. Handy, J. Chem. Phys., 109 (1998) 6264.
108 A.D. Becke, Phys. Rev. A, 38 (1988) 3098.
109 B. Brauer, G.M. Chaban and R.B. Gerber, Phys. Chem. Chem. Phys., 6 (2004) 2543.
110 N. Matsunaga, G.M. Chaban and R.B. Gerber, J. Chem. Phys., 117 (2002) 3541.
111 N.J. Wright and R.B. Gerber, J. Chem. Phys., 114 (2001) 8763.

112 D.M. Benoit, J. Chem. Phys., 120 (2004) 562.
113 K. Yagi, T. Taketsugu, K. Hirao and M.S. Gordon, J. Chem. Phys., 113 (2000) 1005.
114 K. Yagi, K. Hirao, T. Taketsugu, M.W. Schmidt and M.S. Gordon, J. Chem. Phys., 121 (2004) 1383.
115 O. Bludsky, J. Chocholousova, J. Vacek, F. Huisken and P. Hobza, J. Chem. Phys., 113 (2000) 4629.
116 J.M.L. Martin, T.J. Lee and P.R. Taylor, J. Chem. Phys., 108 (1998) 676.
117 T.J. Lee, J.M.L. Martin and P.R. Taylor, J. Chem. Phys., 102 (1995) 254.
118 V. Barone, G. Festa, A. Grandi, N. Rega and N. Sanna, Chem. Phys. Lett., 388 (2004) 279.
119 V. Barone, J. Phys. Chem. A, 108 (2004) 4146.
120 V. Barone, J. Chem. Phys., 120 (2004) 3059.
121 P. Carbonniere and V. Barone, Chem. Phys. Lett., 392 (2004) 365.
122 W.L. Jorgensen, D.S. Maxwell and J. Tirado-Rives, J. Am. Chem. Soc., 118 (1996) 11225.
123 G.M. Chaban, R.B. Gerber and K.C. Janda, J. Phys. Chem. A, 105 (2001) 8323.
124 G.M. Chaban and R.B. Gerber, Spectrochim. Acta A, 58 (2002) 887.
125 G.M. Chaban, W.M. Huo and T.J. Lee, J. Chem. Phys., 117 (2002) 2532.
126 G.M. Chaban, J. Phys. Chem. A, 108 (2004) 4551.
127 D.A. Firsov, A.A. Granovsky and A.V. Nemukhin, Chem. Phys., 244 (1999) 67.
128 R.B. Gerber, Annu. Rev. Phys. Chem., 55 (2004) 55.
129 M. Pettersson, J. Lundell and M. Räsänen, Eur. J. Inorg. Chem., (1999) 729.
130 M. Pettersson, L. Khriachtchev, J. Lundell, S. Jolkkonen and M. Räsänen, J. Phys. Chem. A, 104 (2000) 3579.
131 M. Pettersson, J. Lundell, L. Khriachtchev and M. Räsänen, J. Chem. Phys., 109 (1998) 618.
132 L. Khriachtchev, M. Pettersson, N. Runeberg, J. Lundell and M. Räsänen, Nature, 406 (2000) 874.
133 J. Lundell, G.M. Chaban and R.B. Gerber, J. Phys. Chem. A, 104 (2000) 7944.
134 J. Lundell, G.M. Chaban and R.B. Gerber, Chem. Phys. Lett., 331 (2000) 308.
135 M. Pettersson, L. Khriachtchev, A. Lignell, M. Räsänen, Z. Bihary and R.B. Gerber, J. Chem. Phys., 116 (2002) 2508.
136 J. Lundell, M. Pettersson, L. Khriachtchev, M. Räsänen, G.M. Chaban and R.B. Gerber, Chem. Phys. Lett., 322 (2000) 389.
137 Z. Bihary, G.M. Chaban and R.B. Gerber, J. Chem. Phys., 116 (2002) 5521.
138 L. Khriachtchev, H. Tanskanen, A. Cohen, R.B. Gerber, J. Lundell, M. Pettersson, H. Kiljunen and M. Räsänen, J. Am. Chem. Soc., 125 (2003) 6876.
139 Z. Bihary, R.B. Gerber and V.A. Apkarian, J. Chem. Phys., 115 (2001) 2695.
140 A.T. Kowal, Spectrochim. Acta A, 58 (2002) 1055.
141 B.R. Henry and H.G. Kjaergaard, Can. J. Chem., 80 (2002) 1635.
142 H.G. Kjaergaard, J. Phys. Chem. A., 106 (2002) 2979.
143 M.W.P. Petryk and B.R. Henry, Can. J. Chem., 79 (2001) 279.
144 L. Fono, D.J. Donaldson, R.J. Proos and B.R. Henry, Chem. Phys. Lett., 311 (1999) 131.
145 Y. Miller, G.M. Chaban and R.B. Gerber, Chem. Phys., 313 (2005) 213.
146 D.J. Donaldson, J.J. Orlando, S. Amann, G.S. Tyndall, R.J. Proos, B.R. Henry and V. Vaida, J. Phys. Chem. A., 102 (1998) 5171.
147 C.M. Roehl, S.A. Nizkorodov, H. Zhang, G.A. Blake and P.O. Wennberg, J. Phys. Chem. A., 106 (2002) 3766.
148 C.A. Brindle, G.M. Chaban, R.B. Gerber and K.C. Janda, Phys. Chem. Chem. Phys. 7(2005), 945.
149 G.M. Chaban, to be published.
150 F. Weigend, M. Haser, H. Patzelt and R. Ahlrichs, Chem. Phys. Lett., 294 (1998) 143.
151 D.E. Bernholdt and R.J. Harrison, J. Chem. Phys., 109 (1998) 1593.
152 H.J. Werner, F.R. Manby and P.J. Knowles, J. Chem. Phys., 118 (2003) 8149.
153 M.S. Lee, P.E. Maslen and M. Head-Gordon, J. Chem. Phys., 112 (2000) 3592.
154 M. Svensson, S. Humbel, R.D.J. Froese, T. Matsubara, S. Sieber and K. Morokuma, J. Phys. Chem., 100 (1996) 19357.
155 D.M. Philipp and R.A. Friesner, J. Comput. Chem., 20 (1999) 1468.
156 Q. Cui and M. Karplus, J. Chem. Phys., 112 (2000) 1133.
157 F. Maseras and K. Morokuma, J. Comput. Chem., 16 (1995) 1170.

158 J.R. Shoemaker, L.W. Burggraf and M.S. Gordon, J. Phys. Chem. A, 103 (1999) 3245.

159 E. Apra, T.L Windus, T.P. Straatsma, E.J. Bylaska, W. de Jong, S. Hirata, M. Valiev, M. Hackler, L. Pollack, K. Kowalski, R. Harrison, M. Dupuis, D.M.A. Smith, J. Nieplocha, V. Tipparaju, M. Krishnan, A.A. Auer, E. Brown, G. Cisneros, G. Fann, H. Fruchtl, J. Garza, K. Hirao, R. Kendall, J. Nichols, K. Tsemekhman, K. Wolinski, J. Anchell, D. Bernholdt, P. Borowski, T. Clark, D. Clerc, H. Dachsel, M. Deegan, K. Dyall, D. Elwood, E. Glendening, M. Gutowski, A. Hess, J. Jaffe, B. Johnson, J. Ju, R. Kobayashi, R. Kutteh, Z. Lin, R. Littlefield, X. Long, B. Meng, T. Nakajima, S. Niu, M. Rosing, G. Sandrone, M. Stave, H. Taylor, G. Thomas, J. van Lenthe, A. Wong, Z. Zhang NWChem, A computational chemistry package for parallel computers, version 4.7, Pacific Northwest National Laboratory, Richland, WA 99352-0999, USA, 2005.

160 (a) I.V. Rubtsov, J. Wang and R.M. Hochstrasser, J. Chem. Phys., 118 (2003) 7733. (b) I.V. Rubtsov, J.P. Wang and R.M. Hochstrasser, J. Phys. Chem. A., 107 (2003) 3384. (c) I.V. Rubtsov, J.P. Wang and R.M. Hochstrasser, Proc. Nat. Acad. Sci. USA, 100 (2003) 5601.

161 (a) J. Dreyer, A.M. Moran and S. Mukamel, J. Phys. Chem. B., 107 (2003) 5967. (b) A.M. Moran, S.M. Park and S. Mukamel, J. Chem. Phys., 118 (2003) 9971. (c) T. Hayashi and S. Mukamel, J. Phys. Chem. A., 107 (2003) 9113.

CHAPTER 10

Finding minima, transition states, and following reaction pathways on ab initio *potential energy surfaces*

Hrant P. Hratchian and H. Bernhard Schlegel

Department of Chemistry and Institute for Scientific Computing, Wayne State University, Detroit, MI 48202, USA

Abstract

Potential energy surfaces (PESs) form a central concept in the theoretical description of molecular structures, properties, and reactivities. In this chapter, recent advancements and commonly used techniques for exploring PESs are surveyed in the context of electronic structure methods. Specifically, minimization, transition state optimization, and reaction path following are discussed. In addition to reviewing current progress in these areas, this chapter offers a number of practical discussions regarding minimization, transition state optimization, and reaction path following, including suggestions for overcoming common pitfalls.

10.1 INTRODUCTION

Potential energy surfaces (PESs) play a central role in computational chemistry. The study of most chemical processes and properties by computational chemists begins with the optimization of one or more structures to find minima on PESs, which correspond to equilibrium geometries. To obtain reaction barriers and to calculate reaction rates using transition state theory (TST) [1,2], it is necessary to locate first-order saddle points on the PES, which correspond to transition states (TS). Often one needs to confirm that a TS lies on a pathway that actually connects the minima corresponding to reactants and products (i.e. a TS that is involved in the chemical process under investigation). This goal is typically accomplished by following the steepest descent reaction pathway downhill in each direction from the TS to the reactant minimum and to the product minimum. The reaction path can also be used in the computation of reaction rates using more

References pp. 243–249

sophisticated models such as variational transition state theory (VTST) and reaction path Hamiltonians (RPH) [3–7].

In this chapter, standard algorithms for these calculations—minimization, TS optimization, and reaction path following—are discussed. It should be noted that a number of reviews on these and related methods have appeared in the literature [8–24], including an excellent book on PESs and methods for exploring their landscapes by Wales [25]. Although an exhaustive review of the methods available to computational chemists would be of historical value, our presentation here, heeding to space limitations and a desire to maintain readability, focuses on algorithms commonly used in studies presented in the current literature. Nevertheless, approaches that are less often used in modern practice are included where they provide a map of methodology evolution and are pedagogically useful.

In the next section, essential background material is provided. In Section 10.3, methods for minimization are discussed followed by TS optimization methods in Section 10.4. Reaction path following is considered in Section 10.5. In Section 10.6, we conclude by summarizing the current state of this active area of research.

10.2 BACKGROUND

In this section, concepts that are common to multiple topics and those that form the foundation for the methods and algorithms presented here are discussed. We begin by developing the PES construct from the Born–Oppenheimer (BO) approximation. Next, we discuss the computation of analytic PES derivatives in the context of quantum chemical calculations. In the last part of this section, we consider the common coordinate systems used in optimization and reaction path following.

10.2.1 Potential energy surfaces

The PES arises naturally upon application of the BO approximation to the solution of the Schrödinger equation. We begin by considering the general Hamiltonian

$$H = T_{\mathrm{r}} + T_{\mathrm{R}} + V(r, R) \quad (1)$$

where T_{r} is the operator for the kinetic energy of electronic motion, T_{R} is the operator for the kinetic energy of nuclear motion, and $V(r, R)$ is the potential energy due to electrostatic interactions between all of the charged particles (electrons and nuclei). The BO approximation is applied by assuming that the three order of magnitude difference in the mass of nuclei and electrons renders the nuclei fixed in space on the time scale of electron motion. As a result, the nuclear kinetic energy term, T_{R}, in the molecular Hamiltonian vanishes and the electronic and nuclear degrees of freedom can be separated. This yields the time-independent Schrödinger equation for the electronic degrees of freedom

$$[T_{\mathrm{r}} + V(r, R)]\Phi(r; R) = E(R)\Phi(r; R) \quad (2)$$

In Eq. (2) $\Phi(r; R)$ is the electronic wavefunction, which depends parametrically on the nuclear positions, and the energy of the system, $E(R)$, is a function of the nuclear degrees of freedom. A plot of E *vs.* R gives the PES. For each non-degenerate electronic state, a different PES exists.

Fig. 10.1 shows a model PES. The potential energy of the system is given by the vertical axis and nuclear coordinates are given by the horizontal axes. A common analogy compares the topology of PESs to mountainous landscapes [26]. Molecular structures correspond to the positions of minima in the valleys. Reaction rates can be determined from the height and profile of the pathway connecting reactant and product valleys. Relative stabilities of isomers can be determined from the energies, or elevations, of the minima on the PES corresponding to each structure. From the shape of a valley, the vibrational spectrum of a molecule can be computed, and the response of the energy to electric and magnetic fields determines molecular properties such as dipole moment, polarizability, NMR shielding, etc. [8,27–29].

For simple systems, the PES can be fitted to experimental data. Molecular mechanics (MM) methods can also generate approximate PESs very quickly. However, the types of reactions that can be investigated using conventional MM methods are very limited. Thus, for more complex and reactive systems these options are not viable, and one must rely on PESs generated using quantum chemical calculations (i.e. semi-empirical, *ab initio*, density functional theory (DFT), etc.).

In this chapter, we will explore a number of methods designed to navigate through the mountain ranges of a PES to find local valleys (minimization), the highest point along a reaction path through a mountain pass connecting reactant and product valleys (TS optimization), and the path of least resistance down from the mountain pass to the valleys below (reaction path following). In general, we are concerned with reactive systems, i.e. chemistries involving bond making and breaking that cannot be treated using

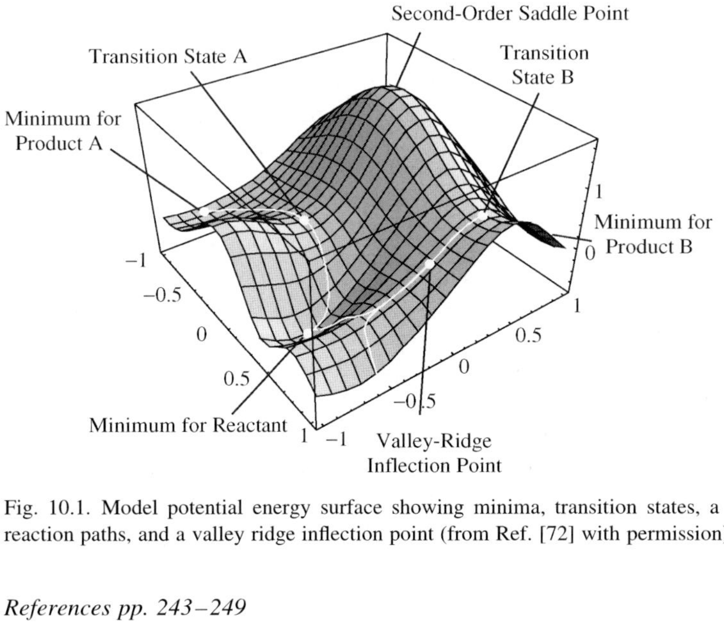

Fig. 10.1. Model potential energy surface showing minima, transition states, a second-order saddle point, reaction paths, and a valley ridge inflection point (from Ref. [72] with permission).

conventional MM methods. Therefore, the focus here is on PESs determined using quantum chemical calculations. Additionally, minimization algorithms designed specifically for use with hybrid quantum mechanical/molecular mechanics (QM/MM) methods are included.

10.2.2 Analytic PES derivatives

One point that is not always obvious after one's first encounter with the BO approximation is that the PES is not known *a priori*. As discussed above, one may fit a functional form to experimental and/or high-level quantum chemical calculation data, but in most applications of electronic structure methods this is not practical. All of the methodologies discussed in this chapter were designed with this in mind. When an algorithm is used to navigate a PES, the value of the energy at each new point is unknown and must be calculated.

Going back to the mountain range analogy [26], we can think of the starting structure supplied by the user for a minimization, TS optimization, or reaction path following calculation as a hiker who is dropped somewhere in the middle of the mountain range. The hiker's latitude and longitude correspond to the molecule's geometry and the elevation corresponds to the value of the PES. Depending on the type of calculation, we may want the hiker to head to the bottom of a valley (minimization), to the top of a mountain pass (TS optimization), or to follow the steepest pathway downhill from the top of a mountain pass to a valley floor (reaction path following). Since the global PES is not known, our hiker is essentially blind. Quantum chemical calculations provide the navigation tools. Energy calculations can be thought of as an altimeter, giving the current elevation. For the hiker to know the slope of the landscape, we also need to calculate the first derivatives of the energy with respect to the nuclear coordinates, or gradients. Note that the negative of the gradient on the PES is equal to the force. As a result, the terms 'gradient' and 'force' are often used interchangeably. If we compute the second derivatives of the energy with respect to the nuclear coordinates, or Hessian, the hiker now knows about the local curvature of the mountain range. The Hessian gives the force constants of the harmonic vibrations of the molecule, and is therefore often referred to as the 'force constant matrix.'

Although the hiker would certainly like as much help as possible, we must consider the relative cost of energy and derivative calculations and the usefulness of each computation. There is a definite trade-off involved between information and expense. To illustrate this point, let us consider minimization. The user's initial guess at the geometry of the structure places our hiker somewhere in the PES mountain range. Calculating the energy alone at this initial position does not provide the hiker with enough information to begin stepping toward the valley floor. Calculating the gradient, though, tells the hiker which direction is downhill. Now, the hiker can take a step towards the bottom of the valley. However, the optimal size of the step is not known. As a result, if too large a step is taken the hiker may overshoot the minimum and go to the other side of the valley. Or worse, the hiker may step through a mountain to a point in a different valley altogether! If the hiker is also given the force constants, the local quadratic character of the mountain range is known. Combined with the gradient, the hiker can take larger steps with greater confidence using the Hessian. Of course, in some applications the cost of the Hessian may be more than what we are willing to pay for the hiker's ability to take larger steps.

To discuss the form and cost of analytic gradient and Hessian evaluations, we consider the simple case of Hartree–Fock (HF) calculations. In nearly all chemical applications of HF theory, the molecular orbitals (MOs) are represented by a linear combination of atomic orbitals (LCAO). In the context of most electronic structure methods, the LCAO approximation employs a more convenient set of basis functions such as contracted Gaussians, rather than using actual atomic orbitals. Taken together, the collection of basis functions used to represent the atomic orbitals comprises a basis set.

Within the LCAO approximation, the HF energy for a spin restricted, closed shell system is given by

$$
\begin{aligned}
E &= \langle \Phi | H | \Phi \rangle \\
&= 2\sum_i (\phi_i | H_1 | \phi_i) + \sum_{ij} \left[2\left(\phi_i \phi_j \left| \frac{1}{r} \right| \phi_i \phi_j \right) - \left(\phi_i \phi_j \left| \frac{1}{r} \right| \phi_j \phi_i \right) \right] + V_{NN} \\
&= \sum_{\mu\nu} (\mu | H_1 | \nu) P_{\mu\nu} + \frac{1}{2} \sum_{\mu\nu\lambda\sigma} (\mu\nu|\lambda\sigma)(P_{\mu\nu} P_{\lambda\sigma} - \frac{1}{2} P_{\mu\sigma} P_{\lambda\nu}) + V_{NN}
\end{aligned} \tag{3}
$$

where Φ is the electronic wavefunction, H is the Hamiltonian, $\phi_i = \Sigma c_{\mu i} \chi_\mu$ are the MOs expanded as a linear combination of basis functions χ, H_1 is the one-electron Hamiltonian (kinetic energy and nuclear-electron attraction), $(\mu\nu|\lambda\sigma)$ are the electron–electron repulsion integrals, μ, ν, λ and σ denote the basis functions, and $\mathbf{P}$ is the density matrix given by

$$
\mathbf{P}_{\mu\nu} = 2 \sum_{i=1}^{\text{occupied}} c^*_{\mu i} c_{\nu i} \tag{4}
$$

where the summation is over occupied MOs and the factor of two comes from the fact that each occupied MO holds two electrons. Formally, calculation of the HF energy is $\mathcal{O}(N^4_{\text{basis}})$ computational work, where N_{basis} is the number of basis functions. Density functional methods are comparable in cost to HF and a number of developments have been made recently to achieve linear and near-linear scaling for large systems [30–38].

The first derivative of the energy with respect to a nuclear coordinate, q_i, (i.e. the gradient) is given by

$$
\frac{\partial E}{\partial q_i} = \left\langle \Phi \middle| \frac{\partial H}{\partial q_i} \middle| \Phi \right\rangle + 2\left\langle \frac{\partial \Phi}{\partial q_i} \middle| H \middle| \Phi \right\rangle \tag{5}
$$

The first term in Eq. (5) is the Hellman–Feynman term and the second term is the wavefunction derivative, or Pulay, term. The Hellman–Feynman portion of the gradient involves a basic computation of the expectation value of a one electron operator. The wavefunction derivative term, which arises because atom-centered basis functions are used, depends on the derivatives of the one- and two-electron integrals in Eq. (3).

Expanding Eq. (5) into a more convenient form for quantum chemical calculations gives

$$\frac{dE}{dq_i} = 2\sum_{\mu\nu} \frac{d(\mu|H_1|\nu)}{dq_i} P_{\mu\nu} + \sum_{\mu\nu\lambda\sigma} \frac{d(\mu\nu|\lambda\sigma)}{dq_i} (2P_{\mu\nu}P_{\lambda\sigma} - P_{\mu\lambda}P_{\nu\sigma}) - 2\sum_{\mu\nu} \frac{dS_{\mu\nu}}{dq_i} W_{\mu\nu} \quad (6)$$

where $S_{\mu\nu}$ are the overlap matrix elements and

$$\mathbf{W} = \mathbf{PFP} \quad (7)$$

where **F** is the Fock matrix. In terms of the basis functions, the elements of the Fock matrix are given by

$$F_{\mu\nu} = 2\sum_{\mu\nu} (\mu|H_1|\nu) P_{\mu\nu} + \sum_{\mu\nu\lambda\sigma} (\mu\nu|\lambda\sigma)(2P_{\mu\nu}P_{\lambda\sigma} - P_{\mu\sigma}P_{\lambda\nu}) \quad (8)$$

This form for the gradient was introduced by Pulay in 1969 [39], and it serves as the basis for the subsequent development of analytic energy derivatives for many different theoretical levels, including correlated methods [28]. The computational cost of the HF gradient calculation is formally $\mathcal{O}(N_{\text{basis}}^4)$. Linear scaling methods have also been developed [32,33,37]. The cost of an analytic gradient evaluation is roughly the same as for the energy calculation. Thus, analytic gradient calculations are relatively routine for *ab initio* PESs and do not generally represent a cost barrier for calculations.

Hessian calculations, on the other hand, are much more expensive and their use in PES exploration methods adds appreciable cost to the calculations. Therefore, as we will show in the later sections of this chapter, estimated and updated Hessians are often used where second derivatives are required by the equations directing movement on the PES. For some systems, the assumptions used to estimate the Hessian are not valid. In these cases, or for cases where very accurate force constants are necessary for vibrational energy calculations, computed (either numeric or analytic) Hessians may be necessary.

Prior to 1979, analytic calculation of second derivatives for *ab initio* methods was thought to be unreasonably expensive [40]. However, in that year Pople et al. [41] developed an efficient approach to solve the coupled perturbed HF (CPHF) equations making analytic Hessian calculation very practical. Indeed, analytic computation of the Hessian is generally cheaper than numeric evaluation of second derivatives and the calculation of *ab initio* force constants by analytic methods is typical for systems with a few hundred basis functions on commercially available computers. Analytic Hessians are routinely available in electronic structure programs for semi-empirical, HF, DFT, second-order Møller-Plesset (MP2), complete active space self-consistent field (CASSCF), configuration interaction singles (CIS), and for other levels of theory. Solution of the CPHF equations for force constants can be expanded to solve for third and higher order derivatives as well [42–47]. However, derivative calculations become increasingly expensive as the order is increased. Therefore, most PES exploration methods developed for use with moderate to large systems limit analytic derivative calculations to second order, and for large systems will often attempt to limit the number of analytic Hessian evaluations.

10.2.3 Coordinate systems

When defining the PES one can use a number of different coordinate systems to describe the geometry of the system being studied; the simplest of these being Cartesian coordinates. In Cartesian coordinate space, the position of each atom in the molecule (or cluster, etc.) is given by three spatial coordinates, x, y, and z. For a system having N_{atoms} atoms there will be a set of $3N_{\text{atoms}}$ Cartesian coordinates, $\{x_i\}$. A similar coordinate system, which naturally develops in the context of nuclear motion (i.e. molecular vibrations, reaction path following, molecular dynamics, etc.), is mass-weighted Cartesian coordinates, $\{\tilde{x}_i\}$, given by

$$\tilde{x}_i = x_i\sqrt{m_N} \tag{9}$$

where $\{m_N\}$ give the atomic masses. Note that the subscript i goes over all of the coordinates while the subscript N goes over the corresponding atomic centers.

An undesirable property of Cartesian coordinates (with or without mass-weighting) is that they generate PESs with strongly coupled coordinates. An alternative to the Cartesian coordinate systems is internal coordinates, sometimes referred to as Z-matrix coordinates. This set of coordinates defines the molecular structure in terms of bond lengths, angles, and dihedral angles. For a non-linear molecule, a unique structure is defined by $3N_{\text{atoms}} - 6$ internal coordinates; for a linear molecule $3N_{\text{atoms}} - 5$ internal coordinates are required. Because internal coordinates are based upon the connectivity of the molecule, they are very natural for chemical systems. Furthermore, nuclear motion on a PES defined in internal coordinates results in much weaker coupling between coordinates than when the PES is given in Cartesian coordinates.

It has been shown that adding some redundancy in the internal coordinates generates a more effective coordinate system, especially for cyclic molecules [48–52]. The molecule's geometry is described using all the chemically relevant bond lengths, angles, and dihedrals, often more than the minimal $3N_{\text{atoms}} - 5$ or $3N_{\text{atoms}} - 6$ internal coordinates needed to specify the structure. As a simple example, consider benzene. There are 12 atoms in benzene, which gives rise to 36 Cartesian coordinates and 30 internal coordinates. The number of redundant internal coordinates is 54—12 bonds, 18 angles, and 24 dihedrals. It should be noted that the number of redundant internal coordinates can be altered by using different degrees of redundancy, and a multitude of definitions have been proposed [49,50,52–55]. They all perform as well as or better than Cartesian and non-redundant internal coordinates.

Since the quantum chemical calculation of energy and derivatives is easiest in the Cartesian space, it is necessary to convert these values to, and from, internals. Although the transformation from Cartesian coordinates to internals (minimal or redundant) is straightforward for the positions, the transformation of Cartesian gradients and Hessians requires a generalized inverse of the transformation matrix [49] *viz.*

$$\Delta\mathbf{q} = \mathbf{B}\Delta\mathbf{x}, \quad \mathbf{g}_q = \mathbf{B}^{-1}\mathbf{g}_x, \quad \mathbf{H}_q = \mathbf{B}^{-t}\left(\mathbf{H}_x - \frac{\partial\mathbf{B}}{\partial\mathbf{x}}\mathbf{g}_q\right)\mathbf{B}^{-1}, \quad \mathbf{B} = \frac{\partial\mathbf{q}}{\partial\mathbf{x}} \tag{10}$$

In Eq. (10) $\mathbf{q}$ are the internal coordinates, $\mathbf{x}$ are the Cartesian coordinates, $\mathbf{g}$ is the gradient, $\mathbf{H}$ is the Hessian, and the Wilson B matrix is given by $\mathbf{B}$. Throughout this chapter, a superscript 't' denotes transpose. Finite displacements in redundant internal coordinates require that the back transformation of the positions to Cartesian coordinates be solved iteratively using Eq. (10) and

$$\mathbf{x}_1 = \mathbf{x}_0 + \mathbf{u}\mathbf{B}^{\mathrm{t}}\mathbf{G}^{-1}\Delta\mathbf{q} \tag{11}$$

where

$$\mathbf{G} = \mathbf{B}\mathbf{u}\mathbf{B}^{\mathrm{t}} \tag{12}$$

and $\mathbf{u}$ is an arbitrary matrix.

The calculation of the generalized inverses in Eqs. (10) and (11) scales as $\mathcal{O}(N_{\mathrm{atoms}}^3)$. Although the cost for this calculation can become significant for large molecules, it is often the case that this cost is far outweighed by the cost of energy and derivative calculations. Nevertheless, for large systems and studies where a low level of theory is used (resulting in relatively fast energy and derivative calculations) the computation of the generalized inverse can become a bottleneck. To make the redundant internal ↔ Cartesian coordinate transformations more tractable, techniques such as iterative solutions to linear equations, Cholesky decomposition, and sparse matrix methods have been developed and reported in the recent literature [56–63]. Using these methods, the redundant internal ↔ Cartesian coordinate transformations can be achieved with linear scaling. The costs of coordinate transformations are more than compensated for by the increased efficiency of the optimization algorithms that use them by decreasing the number of steps required by the algorithm, which in turn decreases the number of energy and derivative evaluations required to complete the job.

10.3 MINIMIZATION

At the start of nearly all chemical studies using electronic structure methods, geometry optimization is required. In this section, we explore some of the most utilized algorithms for minimization. As stated earlier, our focus here is on methods developed for use with quantum chemical calculations where simple functions of the energy and derivatives of the PES are not available but rather are calculated by electronic structure methods as needed. It should also be kept in mind that compared to the cost of the energy calculation a geometry optimization step in most cases is inexpensive. Topics related to global optimization [13–17] and methods catered toward specific advantages or disadvantages of fitted and empirical PESs [11,64–67] are beyond the scope of this chapter.

The problem of geometry optimization involves an unconstrained minimization on the PES. The numerical analysis literature abounds with methods for minimizing non-linear functions of many variables [68–71]. These methods can be placed in three general categories: methods using only the energy, gradient based methods, and methods employing second derivatives. Although energy-only algorithms are applicable across the widest range of levels of theory and problems, they tend to be the least efficient and require a large number of steps to converge. On the other hand, methods that use the Hessian are likely to converge in the least number of steps. However, as discussed in Section 10.2, Hessian evaluation can be quite costly and convergence in a small number of steps may not necessarily equate to the least expensive overall calculation. Gradient-based optimizers often give the best balance between the energy/derivative costs and rate of convergence. As a result, the most commonly used geometry minimization algorithms are gradient based.

Regardless of which method is used, all geometry minimizations are comprised of three basic elements. First, energy and derivatives are computed at the initial geometry. Second, the geometry is changed to take a step toward the minimum. Third, a test (or series of tests) is carried out to determine if the new position is close enough to the PES minimum. If it is, the minimization is complete. If not, the process is repeated to take another step toward the minimum.

Included in the methods discussed below are Newton-based methods (Section 10.3.1), the geometry optimization by direct inversion of the iterative subspace, or GDIIS, method (Section 10.3.2), QM/MM optimization techniques (Section 10.3.3), and algorithms designed to find surface intersections and points of closest approach (Section 10.3.4). We conclude the discussion of minimization methods in Section 10.3.5 with a discussion of practical considerations related to minimization, including suggestions for overcoming common problems.

10.3.1 Newton methods

It has been well established that Newton-based methods are the most efficient type for minimization problems [9,11,12,25,72]. The starting point for these algorithms is to approximate the PES by a Taylor series expansion about the current point, $\mathbf{x}_0$. Truncating the expansion at second order gives

$$E(\mathbf{x}) = E_0 + \mathbf{g}_0^t \Delta\mathbf{x} + \frac{1}{2}\Delta\mathbf{x}^t \mathbf{H}_0 \Delta\mathbf{x} \tag{13}$$

The gradient, $\mathbf{g}(\mathbf{x})$, for this Taylor series is

$$\mathbf{g}(\mathbf{x}) = \mathbf{g}_0 + \mathbf{H}_0 \Delta\mathbf{x} \tag{14}$$

In Eq. (13), E_0, $\mathbf{g}_0$, and $\mathbf{H}_0$ give the energy, gradient, and Hessian at the point $\mathbf{x}_0$ and

$$\Delta\mathbf{x} = \mathbf{x}_i - \mathbf{x}_0 \tag{15}$$

At the minimum, the gradient will be zero,

$$\mathbf{g}(\mathbf{x}) = \mathbf{g}_0 + \mathbf{H}_0 \Delta\mathbf{x} = 0 \tag{16}$$

Solving for $\Delta\mathbf{x}$ gives the step that leads to the minimum in the local quadratic region,

$$\Delta\mathbf{x} = -\mathbf{H}^{-1}\mathbf{g} \tag{17}$$

where the subscript '0' from (16) has been dropped for convenience. Eq. (17) is the Newton step, which is the basis of most minimization methods. Formally speaking, when analytic gradients and Hessians are both used to take steps according to Eq. (17) the algorithm is known as Newton–Raphson (NR). A general flowchart for Newton methods of optimization is shown in Fig. 10.2.

Since analytic Hessians can be rather expensive, especially for larger systems, it is advantageous to avoid computing second derivatives. Nevertheless, the NR algorithm is much more efficient than methods using only gradient or only energy information. The quasi-Newton (QN) approach satisfies both of these concerns. The QN step direction is still determined using Eq. (17), but instead of using an analytic Hessian at each step we begin with an approximate Hessian at the start of the calculation (i.e. an empirically estimated Hessian or a Hessian computed at a lower level of theory) and use Hessian updating at the subsequent steps in the optimization. Hessian updating approximates the Hessian using the change in position and gradient from the previous step. Commonly used updating schemes include Murtagh–Sargent (MS) or symmetric rank 1 (SR1), Powell symmetric Broyden (PSB), Davidson–Fletcher–Powell (DFP), and Broyden–Fletcher–Goldfarb–Shanno (BFGS) [68–71,73–78]. The BFGS update is generally

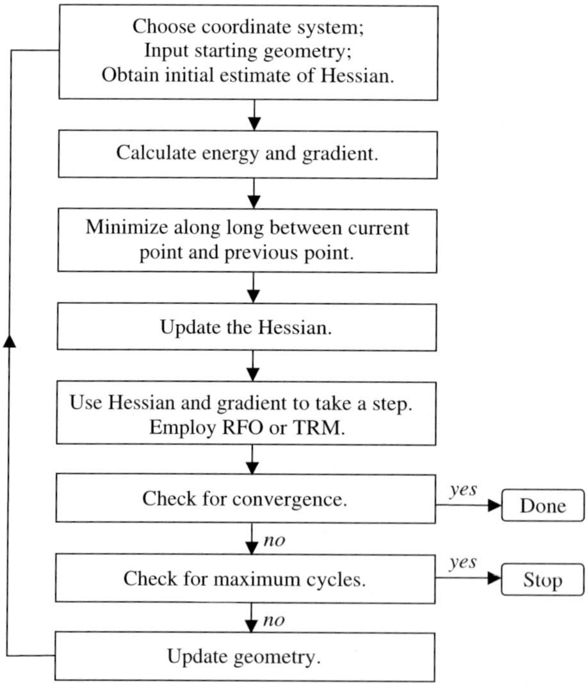

Fig. 10.2. Flowchart for quasi-Newton geometry optimization algorithms (from Ref. [72] with permission).

accepted as the best formula for minimizations and is given by

$$\mathbf{H}^{\mathrm{new}} = \mathbf{H}^{\mathrm{old}} + \frac{\Delta\mathbf{g}\Delta\mathbf{g}^{\mathrm{t}}}{\Delta\mathbf{x}^{\mathrm{t}}\Delta\mathbf{g}} - \frac{\mathbf{H}^{\mathrm{old}}\Delta\mathbf{x}\Delta\mathbf{x}^{\mathrm{t}}\mathbf{H}^{\mathrm{old}}}{\Delta\mathbf{x}^{\mathrm{t}}\mathbf{H}^{\mathrm{old}}\Delta\mathbf{x}} \tag{18}$$

Eq. (18) is symmetric and positive definite (i.e. the eigenvalues of the Hessian are all positive), and minimizes the norm of the change in the Hessian. Corresponding updating formulae also exist for the inverse of the Hessian [68–71,77], which allow the algorithm to avoid the inversion of **H** needed in Eq. (17).

A modification to Bofill's update method (which was originally designed for use in TS optimizations) [79] is also very useful for minimization. This update combines the BFGS and MS schemes, and is given by

$$\mathbf{H}^{\mathrm{new}} = \phi\Delta\mathbf{H}^{\mathrm{BFGS}} + (1-\phi)\Delta\mathbf{H}^{\mathrm{MS}} \tag{19}$$

where the MS update is

$$\Delta\mathbf{H}^{\mathrm{MS}} = \frac{(\Delta\mathbf{g} - \mathbf{H}^{\mathrm{old}}\Delta\mathbf{x})(\Delta\mathbf{g} - \mathbf{H}^{\mathrm{old}}\Delta\mathbf{x})^{\mathrm{t}}}{\Delta\mathbf{x}^{\mathrm{t}}(\Delta\mathbf{g} - \mathbf{H}^{\mathrm{old}}\Delta\mathbf{x})} \tag{20}$$

and the coefficient is computed according to

$$\phi = \frac{|\Delta\mathbf{x}^{\mathrm{t}}(\Delta\mathbf{g} - \mathbf{H}^{\mathrm{old}}\Delta\mathbf{x})|}{|\Delta\mathbf{x}||\Delta\mathbf{g} - \mathbf{H}^{\mathrm{old}}\Delta\mathbf{x}|} \tag{21}$$

Minimization of most small and moderately sized systems is handled very well by QN optimization. For more difficult cases, it is sometimes useful to calculate analytic Hessians at the beginning, every few steps, or even at every step, rather than using updated second derivatives. It may also be useful to compute key elements of the Hessian numerically, particularly those corresponding to coordinates changing rapidly in the optimization [80]. These approaches are discussed in more detail in Section 10.3.5.

For cases where the current structure is far from the minimum (where the magnitude of the gradient is large) or the PES is very flat (where the Hessian has one or more small eigenvalues), a Newton step may be very large and lead to a point where the model quadratic surface is no longer valid. Circumventing this problem is accomplished by limiting the size of each Newton step [68–71]. There are two closely related methods often used for this purpose: the trust radius method (TRM) and rational function optimization (RFO) [79,81–85].

TRM specifies a maximum step size, know as the trust radius, τ, and limits the size of each step in the optimization to this magnitude. Minimizing the energy in Eq. (13) subject to this constraint, $|\Delta\mathbf{x}| \le \tau$, gives

$$\Delta\mathbf{x} = -(\mathbf{H} - \lambda\mathbf{I})^{-1}\mathbf{g} \tag{22}$$

where **I** denotes the identity matrix. λ is less than the lowest eigenvalue of the Hessian, less than zero, and is adjusted in order to satisfy the constraint. This ensures that the step moves downhill on the PES. The TRM step and τ can also be written as

$$\Delta \mathbf{x} = -\sum_{i \geq 1} \frac{\mathbf{v}_i(\mathbf{v}_i^t \mathbf{g})}{b_i - \lambda} \tag{23}$$

$$\tau^2 = \sum_{i \geq 1} \frac{(\mathbf{v}_i^t \mathbf{g})^2}{(b_i - \lambda)^2} \tag{24}$$

where b_i and $\mathbf{v}_i$ are the Hessian eigenvalues and the corresponding eigenvectors. The value of the trust radius can be changed through the course of the optimization depending on how well the estimated energy difference (using Eq. (13)) after each step compares to the actual energy difference. When this difference is small, the trust radius is increased; when it is large, the trust radius is decreased.

RFO can also be used to control the size of the trust radius by minimizing the energy on a rational polynomial approximation to the PES

$$E = E_0 + \frac{\mathbf{g}_0^t \Delta \mathbf{x} + \frac{1}{2} \Delta \mathbf{x}^t \mathbf{H}_0 \Delta \mathbf{x}}{1 + \alpha \Delta \mathbf{x}^t \Delta \mathbf{x}} \tag{25}$$

The parameter α is adjusted to ensure that the step direction leads to a lower energy and that the trust radius is satisfied. This yields equations similar in form to Eqs. (22), (23), and (24). A principle advantage of TRM and RFO is that they step downhill even when the Hessian has one or more negative eigenvalues. Under the same conditions, a raw Newton step will move toward a saddle point. Nevertheless, the character of the structure resulting from a minimization using any algorithm should always be confirmed by calculating the second derivatives and checking that the Hessian has all positive eigenvalues (i.e. all real frequencies).

To ensure that the step lowers the energy and the magnitude of the gradient by a sufficient amount (e.g. the Wolfe condition), it is also important to include an approximate line search [70]. Often, satisfactory results can be achieved by fitting a cubic or constrained quartic polynomial to the energy and gradient at the beginning and end of the step [86]. If the minimum of the line search is within this interval, the energy and gradient can be obtained by interpolation and used in the next Newton step.

For large systems with several hundred atoms, the standard QN algorithms for determining the next geometry step can become a bottleneck. The QN step can also be a bottleneck when the energy evaluation is inexpensive, as is the case when low levels of theory such as molecular mechanics or semi-empirical MO methods are used. Storing the Hessian requires $\mathcal{O}(N^2)$ memory and solving the equations involves $\mathcal{O}(N^3)$ work. One alternative is to use conjugate gradient methods for minimization [70]. The storage and cpu requirements for these methods scale linearly with the system size, but their rate of convergence is significantly poorer than QN methods. Better convergence can be achieved with limited memory QN methods such as L-BFGS [87,88]. In this approach, the Hessian or its inverse is not stored or computed explicitly, but are constructed implicitly as needed. An initial diagonal Hessian, $\Delta \mathbf{x}$, and $\mathbf{g}$ from a fixed number of the most recent steps are stored. Thus, storage scales linearly with system size. The update of the inverse Hessian is formed implicitly and multiplied by the gradient to get the next step

$$\Delta\mathbf{x} = -\mathbf{H}^{-1}\mathbf{g}, \qquad \mathbf{H}_{k+1}^{-1} = \frac{(\mathbf{I} - \Delta\mathbf{x}_k \Delta\mathbf{g}_k^{\mathrm{t}})}{\Delta\mathbf{x}_k^{\mathrm{t}}\Delta\mathbf{g}_k}\mathbf{H}_k^{-1}\frac{(\mathbf{I} - \Delta\mathbf{g}_k \Delta\mathbf{x}_k^{\mathrm{t}})}{\Delta\mathbf{x}_k^{\mathrm{t}}\Delta\mathbf{g}_k} + \frac{\Delta\mathbf{x}_k \Delta\mathbf{x}_k^{\mathrm{t}}}{\Delta\mathbf{x}_k^{\mathrm{t}}\Delta\mathbf{g}_k}. \tag{26}$$

The computational work involves mostly dot products. If the maximum number of updates is fixed, the work scales linearly. The L-BFGS method has been used in a number of optimization problems in computational chemistry [62,64,89,90].

10.3.2 GDIIS

An alternative optimization method to NR and QN is geometry optimization by direct inversion of the iterative subspace, or GDIIS [91–94]. GDIIS is based on a linear interpolation/extrapolation approach and is very well suited for flat PESs (i.e. one or more eigenvalues of the Hessian are small), where NR can be less efficient. For other situations, the efficiency of GDIIS is roughly the same as NR when the initial structure is near the minimum. However, as discussed below, GDIIS can experience difficulties and a number of modifications to the initial GDIIS approach have been developed to overcome these impediments.

Using a linear combination of the structures from the previous n steps, $\mathbf{q}_1, \ldots, \mathbf{q}_n$, the guess for the next GDIIS structure is

$$\mathbf{q}^* = \sum_{i=1}^{n} c_i \mathbf{q}_i \tag{27}$$

where the coefficients c_i are defined by minimizing the estimated error in $\mathbf{q}^*$. This estimated error, or residuum vector, $\mathbf{r}$, is given by

$$\mathbf{r} = \sum_{i=1}^{n} c_i \mathbf{e}_i \tag{28}$$

where $\mathbf{e}_i$ is an error vector associated with $\mathbf{q}_i$. In practice the minimization of Eq. (28) is done with respect to $|\mathbf{r}|^2$, since $\mathbf{r}$ is a vector. This leads to a least-squares problem to solve for the coefficients, c_i. To estimate the error, two common definitions are used. The first is a NR step,

$$\mathbf{e}_i = -\mathbf{H}^{-1}\mathbf{g}_i \tag{29}$$

and the second is the gradient,

$$\mathbf{e}_i = \mathbf{g}_i \tag{30}$$

Clearly, if the point $\mathbf{q}_i$ is very good (e.g. very near the true PES minimum) then its gradient will be small as will the NR step from this point. Conversely, if $\mathbf{q}_i$ is far from the true PES minimum then $\mathbf{e}_i$ will be large regardless of whether Eq. (29) or (30) is used. Since an estimate of the error in $\mathbf{q}^*$ is available, the actual GDIIS step makes use of $\mathbf{r}$ in defining the next structure, $\mathbf{q}_{i+1}$.

$$\mathbf{q}_{i+1} = \sum_{i=1}^{n} c_i(\mathbf{q}_i - \mathbf{H}^{-1}\mathbf{g}_i) \tag{31}$$

The NR definition for $\mathbf{e}$, given in Eq. (29), has been used in Eq. (31). If optimization convergence is not met at the point $\mathbf{q}_{i+1}$, then it is added to the collection of structures and a new GDIIS step is taken.

As mentioned above, the efficiency of GDIIS is similar to, and at times greater than, NR. However, GDIIS comes with its share of problems. First, GDIIS tends to converge to the nearest stationary point on the PES, which may or may not be a minimum. For this reason, it is crucial to test the curvature (by computing the Hessian or harmonic frequencies) of an optimized structure that has been found using GDIIS. A second shortcoming of GDIIS is that it can fall into continuous oscillations if it steps near an inflection point (points where one or more Hessian eigenvalues are zero and the magnitude of the gradient is a minimum, but not zero) and the magnitude of the gradient is larger than the convergence criteria of the optimizer. An additional problem develops when many steps are taken. If a large number of points are used in the linear combinations shown in Eqs. (27) and (28) linear dependencies can appear and result in numerical instabilities in the least-squares solution for the expansion coefficients, c_i.

To overcome these deficiencies, Farkas and Schlegel [95] have developed a controlled GDIIS algorithm. Controlling numerical instabilities arising from linear dependencies in Eqs. (27) and (28) can be achieved by limiting the number of points used. Before taking the next GDIIS step, the linear combinations are built one term at a time beginning at the latest point and working back toward the first point. Before adding the next term, near linear dependency is tested for. If the addition of a point indicates a potential numerical instability, only the points used in the GDIIS expansion before the instability are employed.

Farkas and Schlegel [95] also suggested modifications to increase GDIIS's likelihood to converge on a minimum (or TS if desired; see below for further discussion), rather than a higher order saddle point, and to avoid oscillation problems near inflection points. The first modification employs a reference step, such as NR or any other standard minimization method. By comparing each GDIIS step to this reference, steps that head away from the stationary point of interest can be easily detected, and incorporation of RFO or TRM into the NR step (for computing $\mathbf{e}$) can be used to control the GDIIS step direction. Further stability in this regard can be attained by combining the GDIIS and NR (or other standard optimization method) steps. In this way, the actual step taken during the GDIIS optimization results from a weighted mixing of the standard optimization and GDIIS steps. A final modification involves Hessian updating. In general, the convergence of GDIIS is not dependent on the quality of the Hessian. For this reason, early implementations of GDIIS used a fixed Hessian for every step. However, it has been shown that updating the Hessian yields increased stability and efficiency for GDIIS. By implementing the modifications listed above, GDIIS minimization is an attractive optimization algorithm, especially for large systems and for flat PESs.

The GDIIS method and its modifications scale as $\mathcal{O}(N^3)$ with system size because of the need to invert the Hessian and to compute the reference step. If a diagonal Hessian is used and a fixed maximum number of structures are retained in the GDIIS equations, the method becomes linear in work and storage [95,96]. However, the convergence is somewhat slower than the full matrix GDIIS approach. A linear scaling GDIIS method with improved performance can be obtained by combining GDIIS with the L-BFGS approach [87,88] for updating and utilizing the inverse Hessian (see Eq. (26)).

10.3.3 QM/MM optimizations

In recent years much progress has been made to adapt electronic structure calculations to large systems, such as biochemical compounds. One of the more popular advances in this area is the QM/MM approach (QM—quantum mechanics; MM—molecular mechanics) [97–107]. Although there exists a wide variety of QM/MM methods, the basic principle behind them is the same. Namely, a large molecule or system is broken-up into two sections: one that will be treated at a high level of theory—the QM piece—and one that will be treated at a low level of theory—the MM piece. Typically the QM region of the system is defined as the sector where 'the chemistry' is occurring, and the rest of the system is then defined as the MM region. For instance, a QM/MM calculation on an enzyme generally places the active site in the QM region and the rest of the enzyme in the MM region. The ONIOM scheme is a particular type of QM/MM approach that has gained considerable attention because it is a generalized method that can break a system into many layers, which may consist of multiple QM levels, not just one QM region and one MM region [99,108–110].

Because the QM/MM energy, gradient, and Hessian are well defined, one can navigate the combined QM/MM PES. Optimization of the entire system on the QM/MM PES can require many optimization steps and be very costly. Most popular implementations of QM/MM optimization decrease the cost of minimization by employing microiteration schemes [111–114]. The idea here is to alternate between minimization of the QM and MM regions of the system. Since calculations in the MM region are cheap (in terms of computational cost), the typical use of microiterations fully optimizes the MM region after each QM step. Because the QM and MM regions are treated individually and are uncoupled during the optimizations, progress toward the QM/MM PES minimum can be problematic. As a result, traditional microiteration approaches can converge quite slowly or fail, especially when geometric constraints are imposed on the MM region.

One source of this problem is the choice and handling of the coordinate systems of each region during their independent optimizations. Since the MM energy and derivatives are very cheap, it may not be cost effective (in terms of computational time) to use internal coordinates due to the conversion to and from Cartesians (see Eqs. (10), (11), and (12)). However, within the QM region it is still useful to carry out the optimization in internal coordinates. To overcome the difficulties associated with QM/MM optimization using microiterations, Vreven et al. [112] chose a set of coordinates consisting of Cartesian coordinates for the MM region and a set of internal coordinates for the QM region. Furthermore, the Cartesian coordinates in the MM region are augmented to allow the QM region to undergo overall translation and rotation, which is imperative for cases where constraints in the MM region are present. They also found that this feature slightly improves efficiency when there are no constraints in the MM region. Lastly, their method takes special care to ensure the QM region remains in the same local minimum on the QM PES throughout the MM optimizations.

10.3.4 Finding surface intersections and points of closest approach

The search for conical intersections, avoided crossings and seams of intersection between two PESs are also tasks involving optimization [115–123]. If the two surfaces represent different spin states or have different spatial symmetry, they can cross. If they are the same symmetry and spin, they can interact and the crossing is avoided. Where the matrix element coupling the two surfaces is zero, they touch and give rise to a conical intersection, as illustrated in Fig. 10.3. To study the mechanisms of photochemical reactions, we often wish to find the lowest energy point on a seam or conical intersection. For a seam of intersection between two adiabatic surfaces E_1 and E_2, we require $E_1 = E_2$. Since a (non-linear) molecule has $3N_{\text{atoms}} - 6$ internal degrees of freedom, a seam of intersection has $3N_{\text{atoms}} - 7$ degrees of freedom because of the additional constraint. For a conical intersection, we also require the coupling matrix element, $\mathbf{H}_{12}$, to be zero. Hence conical intersections have $3N_{\text{atoms}} - 8$ degrees of freedom. For molecules larger than triatomic, finding the lowest point on a seam or conical intersection can be quite challenging because of the number of degrees of freedom in the constrained minimization.

One approach to finding the lowest point on a seam or conical intersection is to use Lagrangian multipliers [118–122]. The Lagrangian,

$$L = E_2 + \lambda_1(E_2 - E_1) + \lambda_2\mathbf{H}_{12} \tag{32}$$

is minimized with respect to λ_1, λ_2, and the geometric coordinates of the molecule so that the constraints $E_1 = E_2$ and $\mathbf{H}_{12} = 0$ are satisfied.

Alternatively, the constraints can be treated using projection methods [123]. Instead of minimizing the absolute value of the energy difference, $|E_2 - E_1|$, it is advantageous to use the square of the energy difference, $(E_2 - E_1)^2$, since this quantity is better suited for quasi-Newton optimization methods. For the remaining $3N_{\text{atoms}} - 7$ or $3N_{\text{atoms}} - 8$ degrees of freedom, the energy of the upper adiabatic surface is minimized. The gradient is given by

$$\mathbf{g} = \frac{\mathrm{d}(E_2 - E_1)^2}{\mathrm{d}\mathbf{x}} + \left(\mathbf{I} - \frac{\mathbf{v}_1\mathbf{v}_1^{\mathrm{t}}}{|\mathbf{v}_1|^2}\right)\left(\mathbf{I} - \frac{\mathbf{v}_2\mathbf{v}_2^{\mathrm{t}}}{|\mathbf{v}_2|^2}\right)\frac{\mathrm{d}E_2}{\mathrm{d}\mathbf{x}} \tag{33}$$

where

$$\mathbf{v}_1 = \frac{\mathrm{d}(E_2 - E_1)}{\mathrm{d}\mathbf{x}} \tag{34}$$

(a)

(b)

(c)

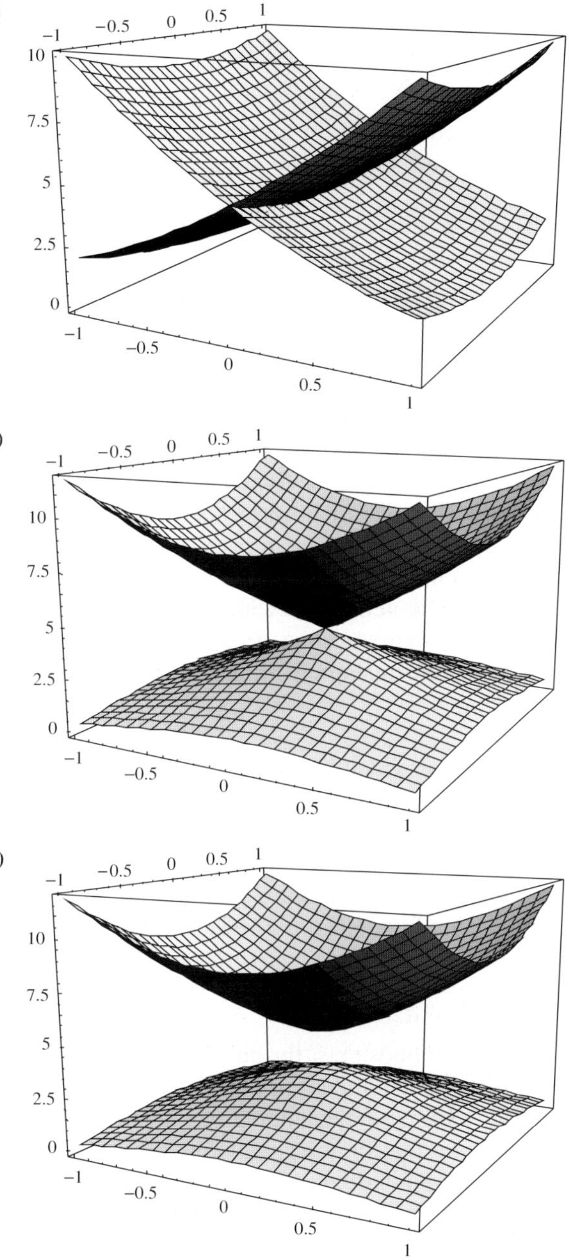

Fig. 10.3. Interactions between two model potential energy surfaces showing (a) a seam, (b) a conical intersection, and (c) a weakly avoided conical intersection (from Ref. [72] with permission).

and

$$\mathbf{v}_2 = \frac{d\mathbf{H}_{12}}{d\mathbf{x}} \tag{35}$$

The gradient given by Eq. (33) can be used directly in a conventional unconstrained quasi-Newton optimization algorithm to find the lowest points on seams, avoided crossings, and conical intersections.

10.3.5 Practical considerations

The principle factors affecting the success and efficiency of a minimization calculation are the starting structure provided by the user, the coordinate system, the algorithm to choose the direction and size of the step, the initial Hessian, and the quality of the updated Hessian at later steps. In this section, we discuss these, and other, practical aspects of minimization and offer suggestions for overcoming typical pitfalls. Illustrative calculations have been carried out with the Gaussian series of programs [124] to demonstrate some of these aspects, but the general considerations should also be relevant to other electronic structure codes. Although previous sections have included minimization techniques used with QM/MM calculations and algorithms for locating conical intersections and points of closest approach, the discussion here is focused primarily on standard minimization of structures using electronic structure methods.

10.3.5.1 Starting structure

Obviously the quality of the initial structure provided by the user will affect the success and efficiency of an optimization calculation. The closer the starting structure is to the PES minimum, the faster (in terms of the number of steps taken) the minimization will complete. On the other hand, a poor starting structure can lead to a lengthy calculation and even failure to converge. Preparing a starting structure for minimization is most readily accomplished with the aid of molecular modeling and visualization software. It is also common practice to use experimentally obtained structures (i.e. from crystal structures) when available. Additionally, starting structures can be generated by minimization at lower levels of theory.

While it is clear that a number of viable methods exist for building an initial structure for minimization, it is important to note that optimization calculations cannot yet be treated as black box operations. Instead, a basic understanding of the chemistry of the problem being studied is necessary to ensure that the user is able to properly diagnose problems and determine the appropriateness of an initial structure. For example, consider cyclohexane. From undergraduate organic chemistry classes we know that there are two conformations—chair and boat. Both conformations will have local minima on the PES and using a boat-like structure for the start of a minimization will result in a converged optimization at the boat conformation. If the goal of the optimization is to study the more stable chair conformation this result is not what we want. To get the chair structure, one must start with a chair-like structure. To further illustrate this point, Fig. 10.4 shows a one-dimensional slice of the PES of 1-chloro 2-fluoroethane along the Cl–C–C–F

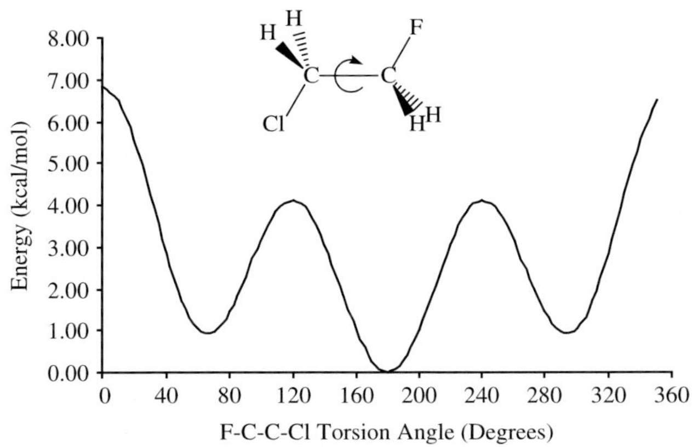

Fig. 10.4. A one-dimensional slice of the potential energy surface of 1-chloro 2-fluorethane along the Cl–C–C–F dihedral coordinate.

torsion angle. In this case there are three local minima on the PES, although the minima at 70 and 290° are mirror images. Again, the initial structure will determine which local minimum the optimizer converges to. Starting an optimization with a Cl–C–C–F dihedral of 30° will lead to the minimum at 70°, while starting at 160° will converge at the global minimum at 180°. One must employ chemical insight when generating the initial structure for minimization. For very complex systems, it may be necessary to generate multiple structures representing different potential conformations, which allows for the study of relative conformational energies, etc.

10.3.5.2 Coordinate system

Another aspect of geometry optimization affecting the success and efficiency of the calculation is the choice of coordinate system. Section 10.2.3 included definitions for different coordinate systems commonly used to describe PESs. Early work in geometry optimization was nearly always done in non-redundant internal coordinates. This coordinate system is easy to use when building acyclic structures and inherently removes rotational and translational degrees of freedom from the system, making it advantageous from an algorithmic standpoint. However, geometry optimization using non-redundant internal coordinates does require some thought since the efficiency, and sometimes the success or failure, of the optimization can depend on the choice of Z-matrix components. An overview of techniques for building effective Z-matrix inputs appeared in an earlier review, and interested readers are referred to that work for a more in depth discussion [80]. Non-redundant internals perform especially poorly in the optimization of polycyclic systems [48–53,63,125]. In these cases, it is better to use Cartesian coordinates than to employ a Z-matrix representation. It has also been shown that mixing non-redundant and Cartesian coordinates can also be useful at times [125].

The most efficient coordinate system, in general, for geometry optimization using electronic structure methods is redundant internal coordinates, especially for cyclic compounds [48–53,63,125]. Table 10.1 shows the number of steps necessary to minimize a set of nine structures using these four different systems. All of the optimizations were

Table 10.1 Comparison of the number of steps required to minimize geometries using various coordinate systems[a]

Molecule	Cartesian	Z-Matrix	Mixed	Redundant internals
2-Fluoro furan	7	7	7	6
Norbornane	5	7	5	5
Bicyclo[2.2.2]octane	19	11	14	7
Bicyclo[3.2.1]octane	6	6	7	5
Endo hydroxyl bicyclopentane	18	8	9	12
Exo hydroxyl bicyclopentane	20	10	11	11
ACTHCP	>80	65	72	27
1,4,5-Trihydroxy anthroquinone	11	10	17	8
Histamine H^+	>100	42	47	19

[a]For complete details see Ref. [51].

started from the same initial geometries [51]. It is clear from this data that for rigid compounds all four coordinate systems are essentially the same. As the flexibility of the molecules increases, the advantage of using redundant internal coordinates becomes increasingly apparent. Because of their obvious benefit, redundant internal coordinates are favored for nearly all geometry optimization calculations used with electronic structure methods.

There are two general problems that can arise in the course of an optimization calculation due to the use of redundant internal coordinates. The first problem is caused by a poorly chosen set of coordinates for the starting structure. Although most programs use an automated mechanism to determine the initial definitions for the redundant internal coordinates, these algorithms are not error proof. To the best of our knowledge, all of the available codes use tables of covalent radii or standard bond lengths, or their equivalent, to determine which atoms are bonded to each other. The user can also define additional internal coordinates. Once the connectivity of the molecule is established, angles and dihedrals can be readily defined. If fragments are detected (i.e. two or more parts of the system that are not connected by bonds), then one, or more, coordinates are added so that the fragments can interact and intermolecular distances can be accounted for. Nevertheless, the tables used to define the initial bonding can ignore important bonds, which can lead to considerably slower convergence. For this reason, it is essential to check the definitions of the redundant internals, which are usually printed out in an output file at the start of the optimization calculation. Our experience has shown that modifications to the redundant internal coordinate definitions are often necessary for transition and inner transition metal complexes.

Another problem that can occur during the course of a minimization in redundant internal coordinates is the internal forces being reported as infinity or undefined [126]. As discussed in Section 10.2.3, the energy derivatives are typically computed in Cartesian coordinates and later converted to internals using Eq. (10). If the redundant internal coordinate definitions become ill defined and/or include linear dependences, then the conversion of forces and Hessians to internal coordinates can become problematic. The easy fix to this problem begins by inspecting the latest structure in the optimization using visualization software to ensure that the structure is reasonable. If all is well, start the minimization again giving the last structure from the previous optimization as the initial structure for the current calculation. This allows for generation of new coordinate definitions. Again, these definitions should be checked to make sure that all of the critical definitions have been included. In extreme cases, it may be necessary to reduce the redundancy of the coordinate system.

10.3.5.3 Minimization algorithm

The third factor affecting the efficiency of an optimization calculation is the choice of the minimization algorithm. In the sections above we discussed the most common procedures used in practice—QN with RFO or TRM modifications, GDIIS, and controlled GDIIS. The relative efficiencies of these three approaches have appeared in the literature [95], and the results of that work are summarized in Table 10.2. As discussed in Section 10.3.2, regular GDIIS can fall into continuous oscillations near inflection points on the PES. This pathology is experienced in the optimization of five of the eight entries in Table 10.2. The controlled GDIIS method corrects these problems and converges to the appropriate minima. For the first three entries in Table 10.2, regular GDIIS optimization converges to saddle points instead of minima. Again, this problem is solved by using the controlled GDIIS algorithm. Minimization of the smaller systems is equally efficient using controlled GDIIS or QN (with step size and direction control). When larger and more flexible systems are studied, the controlled GDIIS method is often more efficient than QN. Controlled GDIIS is also the recommended method for optimizations on flat PESs and those employing tight convergence criteria.

10.3.5.4 Hessian quality

The final factors affecting optimization are the choice for the initial Hessian and the method used to form Hessians at later steps. As discussed in Section 10.3.1, QN methods avoid the costly computation of analytic Hessians by using Hessian updating. In that section, we also showed the mathematical form of some common updating schemes and pointed out that the BFGS update is considered the most appropriate choice for minimizations. What may not have been obvious from Section 10.3.1 is that the initial

Table 10.2 Comparison of the number of steps required to minimize geometries using QN with RFO, regular GDIIS, and modified GDIIS[a]

Molecule	QN with RFO	Regular GDIIS	Controlled GDIIS
Pterine	36	12[b]	36
Histamine H^+	24	85[b]	25
Hydrazobenzene	25	41[b]	25
ACTHCP	31	–[c]	32
Taxol	64	–[c]	67
For-$(Ala)_{10}$-NH_2	67	–[c]	59
For-$(Ala)_{20}$-NH_2	103	–[c]	93
Crambin	190	–[c]	150

[a]For complete details see Ref. [95].
[b]Attempted minimization yielded a transition state (i.e. first-order saddle point).
[c]Attempted minimization resulted in oscillations about an inflection point.

Hessian does not need to be very accurate. During the optimization the Hessian will be updated, and while the structure moves closer to the PES minimum the Hessian slowly approaches the correct second derivative matrix. In general, the closer the initial Hessian is to the correct Hessian, the faster the optimization will come to convergence.

Most QN optimizations use a Hessian generated from a valence force field or other empirical method. Alternatively, analytic second derivatives can be computed and used for the initial Hessian. This will often decrease the number of steps required to converge to the minimum. However, depending on the size of the system being studied the cost of the initial analytic Hessian calculation may yield a longer overall calculation time than an optimization starting with an estimated Hessian, even though this calculation might require a few more steps. For cases where the starting structure is far from the minimum or when the topology of the PES is not well approximated by a quadratic function (i.e. the Taylor series given in Eq. (13)), Hessian updating may be poor and convergence may be difficult. In these situations it may be necessary to calculate analytic Hessians at every step. In principle this should lead to fewer steps in the optimization, but the additional cost of analytic Hessians will severely increase the overall computational cost for moderate and large systems.

The data presented in Tables 10.3 and 10.4 illustrate some of these issues. Using a set of six compounds, ranging in size from 6 to 72 degrees of freedom, we have carried out optimizations using the identity matrix, an estimated empirical Hessian, and an analytic Hessian for the initial second derivative matrix. Additionally, we have included data for optimizations using analytic Hessians at every step, rather than updating the second derivatives. Table 10.3 shows the number of optimization steps required to complete an optimization using these different approaches to define the initial Hessian, and Table 10.4 contains the relative timing of each calculation. Clearly, using the identity matrix to estimate the initial Hessian is not efficient. For roughly the same cost per step, one can use an estimated Hessian based on empirical force fields [86,127] and achieve a significant improvement in the rate of convergence. However, in the case of pterine and caffeine the identity matrix incorrectly leads the minimization to a saddle point. Since the QN-RFO/TRM method for optimization shifts negative eigenvalues of the Hessian to positive

Table 10.3 Comparison of the number of steps required to minimize geometries (QN with RFO algorithm) using a unit matrix, empirically derived Hessian, and analytic Hessian for the initial Hessian followed by Hessian updating and using all analytic Hessians[a,b]

Molecule	Initial unit matrix with updating	Initial empirical Hessian with updating	Initial analytic Hessian with updating	All analytic Hessians
Ammonia	14	8	6	4
Pterine	22[c]	56	15	10
Caffeine	26[c]	13[c]	67	13
ACTHCP	>100	34	35	15
Histamine H^+	>100	23	30	11
Hydrazobenzene	90	30	29	19

[a]Starting structures taken from Ref. [51].
[b]All calculations carried out at the HF/STO-3G level of theory.
[c]Attempted minimization yielded a transition state (i.e. first-order saddle point).

Table 10.4 Relative timings to minimize calculations using a unit matrix, empirically derived Hessian, and analytic Hessian for the initial Hessian followed by Hessian updating and using all analytic Hessians[a]

Molecule	Initial unit matrix with updating	Initial empirical Hessian with updating	Initial analytic Hessian with updating	All analytic Hessians
Ammonia	2.2	1.3	1.1	1.0
Pterine	5.3[b]	13.8	4.3	8.8
Caffeine	8.5[b]	4.1[b]	12.1	23.0
ACTHCP	>34.5	8.0	8.5	10.7
Histamine H^+	>34.5	4.8	6.4	5.7
Hydrazobenzene	29.8	10.2	11.0	34.3

[a]Convergence data is given in Table 10.3.
[b]Attempted minimization yielded a transition state (i.e. first-order saddle point).

values, describing the curvature correctly for these structures allows the optimizer to properly converge to PES minima when analytic Hessians are used for the initial Hessian. Nevertheless, for most of the compounds studied, and for many structures in general, using an analytic Hessian at the start of the optimization provides very little, if any, improvement to an optimization carried out with an empirically derived Hessian. Moreover, calculating analytic Hessians only at the start of the minimization or at every step can severely increase the computational cost (see Table 10.4).

In general, it is advisable to use an empirical Hessian at the start of a minimization and to update the Hessian using any of the standard methods (see Section 10.3.1). After an optimization has completed, the nature of the stationary point must be confirmed by a calculation of the Hessian. This is not a significant extra burden, since the Hessian at the optimized geometry is also needed to calculate the zero point energy and thermal corrections to the enthalpy. If the Hessian calculation reveals that the optimization has converged to a saddle point (first order or higher), then using an analytic Hessian at the start of the minimization and Hessian updating at subsequent may be best. If this second optimization has also converged to a saddle point, then employing analytic Hessians at every step, or every few steps, may be necessary. What is clear from Table 10.4 is that the decrease in optimization steps using all analytic Hessians does not make up for the cost increase that such a calculation yields. Therefore, using all analytic Hessians in an optimization is best treated as a last resort approach.

10.3.5.5 Tips for difficult minimizations

In this, the last portion of Section 10.3.5, we provide suggestions for solving difficult minimization problems that have not yet been addressed. Specifically, this subsection focuses on problems generally resulting because of the topology of the PES and not because of numerical difficulties arising from the application of a particular method or algorithm.

In the previous section we suggested that analytic Hessians can be used to help converge to a minimum when minimizations using empirical and/or updated Hessians yield first- or higher-order saddle points. Often times, a more cost effective means to achieve the same goal is to slightly distort the offending coordinate and start another minimization using an empirical Hessian at the start with updating thereafter.

As an example, consider a minimization of 1-chloro 2-fluorethane starting with a Cl–C–C–F torsion angle of 125° that has converged to the saddle point at 120° in Fig. 10.4, even though our intent was to find the structure corresponding to the minimum at 180°. We now have a couple of options. One option is to start the calculation over from our initial starting geometry using an analytic Hessian for the initial second derivative matrix. We could also choose to use analytic Hessians at every step or every few steps in the optimization. Alternatively, we could use a molecular graphics package to inspect the imaginary frequency corresponding to the negative Hessian eigenvalue and distort the structure along that normal mode. In this particular case, visualization of this frequency will show motion along the Cl–C–C–F torsion angle distorting the molecule from an eclipsed geometry to a staggered one. Distorting the molecule by setting the Cl–C–C–F angle to roughly 150° will be enough to use an empirical Hessian with updating to converge to the intended minimum. This last option will be much more cost effective since it avoids analytic calculation of the Hessian during the optimization. Of course, a Hessian calculation will still be needed when the calculation has completed to ensure that the newly optimized structure is a minimum on the PES and to obtain the zero point energy.

Another method for removing imaginary frequencies is to utilize constrained optimizations. As the name implies, user defined coordinates are frozen and the remaining coordinates are allowed to minimize. If one or more coordinates have negative Hessian eigenvalues associated with them, it can be useful to freeze all of the other coordinates and allow the offending coordinates to relax. Once the constrained optimization has completed, a frequency calculation is in order to ensure that the imaginary frequency has been removed. If the imaginary frequency still exists, then the other suggestions given above may be helpful and applied together with constrained optimization. After the negative second derivative eigenvalues have been removed, a full optimization (i.e. without any constraints) should follow with a subsequent frequency calculation at the end.

In some very troublesome cases, where only one imaginary frequency remains, a method that can be used as an approach of last resort is to carry out a reaction path following calculation (details on reaction path following are given below in Section 10.5) for a few steps. Starting from a first-order saddle point, a reaction path following calculation will move downhill toward two minima, one of which should correspond to the structure of interest. It is not usually necessary to carry the calculation all the way to the endpoints. After the calculation has been completed the two final structures can be visualized and the appropriate one chosen. If enough steps have been taken this structure should now be near the quadratic region of the minimum and a QN minimization should be able to converge on the intended minimum. However, it may be necessary to calculate analytic second derivatives for the initial Hessian of this optimization.

10.4 TRANSITION STATE OPTIMIZATION

As discussed earlier, minima on the PES correspond to equilibrium geometries and chemical reactions can be described in terms of motion on the PES from one minimum corresponding to the reactant to a different minimum corresponding to the product. Along the way, the system may pass through other minima, which correspond to intermediates. The motion from one minimum to another can be approximated by the path of least resistance, or the minimum energy pathway (MEP). While moving along the MEP, the system will reach a point of highest energy—the transition state (TS). Turning back to the mountain range analogy from Section 10.2, the TS is at the top of the lowest mountain pass connecting two valleys. The TS will have one (and only one) direction of downward curvature, which points in the direction of the reactant minimum on one side and the product minimum on the other. In all other directions, the TS will have local upward curvature. A stationary point (a point where the first derivatives are zero) with this topology is known as a first-order saddle point. A nth-order saddle point has n directions of downward curvature. In terms of the Hessian, this means that the TS will have one (and only one) negative Hessian eigenvalue and all of the other Hessian eigenvalues will be positive. Since the Hessian is the system's force constant matrix and the vibrational frequencies are proportional to the square root of the eigenvalues of the mass-weighted Hessian, the TS will have one (and only one) imaginary frequency. The eigenvector corresponding to the imaginary frequency is known as the transition vector because it corresponds to molecular displacement along the reaction path through the TS.

A number of good reviews on TS optimization have appeared in recent years [9,11, 12,21,23–25,128]. In this section, we provide an overview of the three general classes of TS optimization methods—local schemes (Section 10.4.1), climbing, bracketing, and interpolation algorithms (Section 10.4.2), and path optimization approaches (Section 10.4.3). In Section 10.4.4 we discuss practical considerations related to TS optimization and offer suggestions for difficult cases.

10.4.1 Local methods

Many of the standard minimization algorithms presented in Section 10.3 can be modified to find TSs. Such methods are commonly referred to as local methods. Unfortunately, simple applications of QN methods are often unsuccessful in TS searches. The problem stems from the fact that they will only converge to the TS if the initial guess falls within, or very near, the quadratic region of the true TS, which is generally much smaller than for a minimum. This means the error tolerance in the starting structure is much less for TS optimization than for minimization. Therefore, it is necessary to ensure the Hessian has an appropriate eigenvector with a negative eigenvalue for a QN step to move closer to the desired TS. Despite these difficulties, good chemical intuition (sometimes along with a bit of luck) can provide adequate guesses for TSs to be found using local methods.

The adaptation of most minimization algorithms, such as QN and GDIIS, for TS optimization is rather straightforward [68–71,93–95]. Just as with minimization it is common to use Hessian updating. Unlike minimization, though, the BFGS updating scheme is unacceptable in TS optimization because it forms positive definite Hessians. When the Hessian becomes negative definite, as is the case near the TS, the BFGS formula becomes ill conditioned [129].

The Murtaugh–Sargent update, which was already discussed in Section 10.3.1, is one option [79,130,131]. Another common choice is the Powell-symmetric-Broyden (PSB) update, which was first recommended for TS optimization by Simons et al. [132]. A PSB updated Hessian is given by

$$\Delta \mathbf{H}^{\mathrm{PSB}} = \frac{(\Delta \mathbf{g} - \mathbf{H}^{\mathrm{old}}\Delta \mathbf{x})\Delta \mathbf{x}^{\mathrm{t}} + \Delta \mathbf{x}(\Delta \mathbf{g} - \mathbf{H}^{\mathrm{old}}\Delta \mathbf{x})^{\mathrm{t}}}{\Delta \mathbf{x}^{\mathrm{t}}\Delta \mathbf{x}} - \frac{\Delta \mathbf{x}^{\mathrm{t}}(\Delta \mathbf{g} - \mathbf{H}^{\mathrm{old}}\Delta \mathbf{x})\Delta \mathbf{x}\Delta \mathbf{x}^{\mathrm{t}}}{(\Delta \mathbf{x}^{\mathrm{t}}\Delta \mathbf{x})^2} \tag{36}$$

Bofill [79,89,129–131,133] has developed a hybrid updating scheme with better performance for TS optimization. The Bofill update mixes MS and PSB solutions giving

$$\Delta \mathbf{H}^{\mathrm{Bofill}} = \phi \Delta \mathbf{H}^{\mathrm{MS}} + (1 - \phi)\Delta \mathbf{H}^{\mathrm{PSB}} \tag{37}$$

where

$$\phi = \frac{(\Delta \mathbf{x}^{\mathrm{t}}(\Delta \mathbf{g} - \mathbf{H}^{\mathrm{old}}\Delta \mathbf{x}))^2}{\Delta \mathbf{x}^2(\Delta \mathbf{g} - \mathbf{H}^{\mathrm{old}}\Delta \mathbf{x})^2} \tag{38}$$

and $\Delta \mathbf{H}^{\mathrm{MS}}$ is given by Eq. (20).

10.4.2 Climbing, bracketing, and interpolation methods

Since local methods will, in general, only succeed in finding a TS if the initial geometry lies within the quadratic region of the first-order saddle point and the initial Hessian has an appropriate eigenvector with a negative eigenvalue, the fate of TS optimization by local methods rests in one's ability to apply chemical intuition to problems that, at an ever increasing rate, are anything but intuitive. To overcome this difficulty, a number of methods have been developed that automate the initial guess procedure for TS optimization. Using information provided by the user, these algorithms produce an initial guess at the TS by climbing uphill from one minimum, or by bracketing or interpolating a TS between the reactant and product minima. After the region of the TS is located by one of these methods, the structure can be improved using local TS optimization. Again, the literature abounds with climbing and walking, bracketing, and interpolation schemes [26, 84,132,134–152]. Here, we will consider coordinate driving, shallowest ascent and walking up valleys, linear and quadratic synchronous transit, synchronous transit-guided quasi-Newton, and ridge following.

For a limited number of reactions, the reaction pathway can be described by a scan of the PES along one (internal) coordinate. The most prevalent classes of reactions falling into this category are conformation and bond dissociation reactions, where a change in a torsion angle describes the reaction for the former and a bond stretch coordinate describes the latter. Climbing methods making use of this principle are often referred to as 'coordinate driving' algorithms. Beginning at a PES minimum (i.e. the reactant or product), the method traces a path along the coordinate of interest by incrementing its value from reactant to product minima. At each increment, the energy of the new structure is calculated and the highest energy structure on the pathway is taken as an estimate for the TS, which can be used in a local TS optimization calculation. Since other internal coordinates are affected by changes in the driven coordinate, most applications of this approach include a constrained optimization of the other $N-1$ coordinates at each increment. This provides a much better guess than a rigid scan. Although coordinate driving can yield a good estimate for the TS, it can be costly if small increments are necessary. Additionally, constrained optimization at each step will increase the computational cost since multiple energy and gradient calculations will be required at each point in the scan. If more than one coordinate are involved in the reaction, coordinate driving will not provide an adequate estimate of the TS. Further difficulties, such as discontinuities, can be encountered if the reaction path is strongly curved [153–155]. The reduced gradient following (RGF) method is an improved version of coordinate driving that is better able to handle curved reaction paths [156–161]. The RGF approach works by defining a path that connects stationary points on the PES according to the differential equation

$$\mathbf{r} - \frac{\mathbf{g}[\mathbf{x}(t)]}{|\mathbf{g}[\mathbf{x}(t)]|} = 0 \tag{39}$$

At each point in the RGF point, the gradient, $\mathbf{g}$, has a constant direction given by the unit vector $\mathbf{r}$. Typically, this direction is chosen in a similar manner to the coordinate driving algorithms. As we will see in Section 10.5, the second term in Eq. (39) comes from the steepest descent path definition used in reaction path following [162].

An alternative to coordinate driving is the ‘shallowest ascent’, ‘eigenvector following’ or ‘walking up valleys’ approach [26,84,132,134–137,143–145]. There are two general flavors of walking up valleys algorithms—one requires the structure of either the reactant or the product, while the other uses the structures of both minima. Starting at one minimum, one can walk uphill along the shallowest ascent direction by following the Hessian eigenvector corresponding to the lowest eigenvalue. The need for the Hessian at each step can make this method costly. As a result, most implementations employ Hessian updating. In order to assure that the shallowest ascent direction is followed, and to minimize along all other directions, methods similar to RFO and TRM have been implemented. In this way, each step is defined according to

$$\Delta\mathbf{x} = -(\mathbf{H} - \lambda\mathbf{I})^{-1}\mathbf{g} \tag{40}$$

The parameter λ is chosen such that $(\mathbf{H} - \lambda\mathbf{I})$ has only one negative eigenvalue and the step has an appropriate length. The use of Eq. (40) is also at the heart of successful local TS optimization methods and greatly expands the radius of convergence for TS optimization. Different values for λ can be used for the directions corresponding to the uphill climb and the downhill minimization [83,85,134]. However, following the shallowest ascent path may not necessarily lead to the correct TS. This issue is addressed by considering the model surface shown in Fig. 10.1. Although the shallowest ascent pathway from Product A correctly leads to the TS connected to the reactant, the shallowest ascent pathway from the reactant does not head toward the same TS. Instead, it leads to Product B.

Another interpolation method using structural information from the reactant and product minima is linear synchronous transit (LST) [139]. These two points on the PES can be used to form a rough approximation for the reaction pathway by interpolating a line between them. The maximum (in terms of energy) along this line serves as an upper bound to the TS. Although the Hessian at the LST maximum often has more than one negative eigenvalue, it is usually a satisfactory estimate for the TS and can be refined using a local optimization method.

An improvement to LST is incorporated in the quadratic synchronous transit (QST) method [139,148,149]. QST begins with the LST maximum and minimizes perpendicular to the LST path. Then, a new quadratic pathway is interpolated using this new approximate TS, the reactant, and the product. The maximum point on the QST pathway is located and serves as the QST estimate for the TS. For some reactions the LST and QST estimates can be quite similar. However, systems with curved reaction paths can show significant differences between the LST and QST estimates. As before, the estimated TS can be refined by a local method, using Eq. (40) to control the step size and direction. A variation of QST is synchronous transit-guided quasi-Newton (STQN) [138], which directly combines QN with LST or QST. STQN also uses the arc of a circle for the estimated path. The algorithm takes a limited number of initial steps to maximize the energy along the LST or QST path, and then heads toward the TS using the Hessian eigenvector that overlaps best with the LST/QST path. The Hessian eigenvalues are adjusted according to Eq. (40).

Fig. 10.5 shows the LST and QST pathways and their estimated TSs on a model PES. For some reactions, for instance Reactant → Product B, both pathways agree. For other reactions, for instance Reactant → Product A, the LST pathway can differ from the QST path. It is clear that for the latter case the QST estimate for the TS is much closer to the actual TS. For some reactions with strongly curved paths, QST and STQN interpolations can be enhanced by a user-supplied guess at the TS that differs from the automated search result [138].

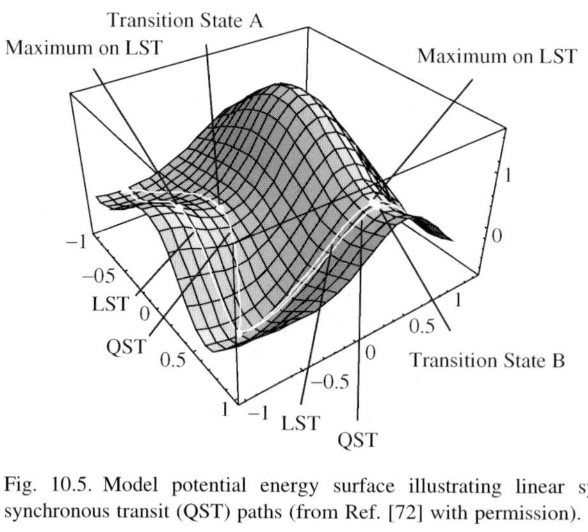

Fig. 10.5. Model potential energy surface illustrating linear synchronous transit (LST) and quadratic synchronous transit (QST) paths (from Ref. [72] with permission).

The ridge following method of Ionova and Carter [146,147] can be viewed as another modified version of LST. This approach begins with reactant and product structures and interpolates a line between them. In the same way as LST, an initial estimate for the TS is determined by finding the highest energy point on the linear path. Instead of feeding this point into a local method, the ridge following algorithm now picks two points on the LST pathway that lie on either side of the estimated TS. The distance between these points is kept small, and they are both allowed to move downhill toward the reaction path. Each step in the progression toward the reaction path and the TS is controlled by side step followed by a downhill step. Let $\mathbf{x}_i^a$ and $\mathbf{x}_i^b$ be the structures of the two points that straddle the TS ridge, where a maximum energy structure $\mathbf{x}_i^*$ exists on the line $(\mathbf{x}_i^a, \mathbf{x}_i^b)$. The side-step portion finds two new points, $\mathbf{x}_i^{a'}$ and $\mathbf{x}_i^{b'}$, which are on the line $(\mathbf{x}_i^a, \mathbf{x}_i^b)$ and are found according to

$$\mathbf{x}_i^{a'} = (\mathbf{x}_i^* + s\mathbf{p}_i), \qquad \mathbf{x}_i^{b'} = (\mathbf{x}_i^* - s\mathbf{p}_i) \tag{41}$$

where

$$\mathbf{p}_i = \frac{\mathbf{x}_i^{b'} - \mathbf{x}_i^{a'}}{|\mathbf{x}_i^{b'} - \mathbf{x}_i^{a'}|} \tag{42}$$

and s is a side-step step size. The downhill step is then given by

$$\mathbf{x}_{i+1}^a = \mathbf{x}_i^{a'} - u\mathbf{g}_i^{a'}, \qquad \mathbf{x}_{i+1}^b = \mathbf{x}_i^{b'} - u\mathbf{g}_i^{b'} \tag{43}$$

where u is the size of the downhill step, $\mathbf{g}_i^a$ and $\mathbf{g}_i^b$ are the gradients at $\mathbf{x}_i^{a'}$ and $\mathbf{x}_i^{b'}$, and $\mathbf{x}_{i+1}^a$ and $\mathbf{x}_{i+1}^b$ give the next set of structures along the ridge. Alternatively, the step can be taken along the gradient at $\mathbf{x}_i^*$. In either case, it can be shown [146] that unless the gradient at $\mathbf{x}_i^*$ is zero, values of s and u can be chosen such that there is a maximum of energy on the line connecting the next set of points, $\mathbf{x}_i^{a'}$ and $\mathbf{x}_i^{b'}$, and that this energy is lower than the maximum on the line connecting the previous set of points. One can think of this process as a constrained optimization following the ridge from a second- (or higher-) order saddle point to the first-order saddle point. When the component of the gradient perpendicular to the line between the two points is zero, or very near zero, they lie close to the reaction path and an intermediate point between them provides a very good estimate for the TS. A local TS optimization method can then be used to converge the ridge following TS to the stationary point.

A novel interpolation method proposed and refined by Jensen and others uses MM or valence bond PESs [140–142,150–152]. Earlier we mentioned that most MM methods do not properly describe reactive systems. This is due to the fact that atom types and molecular connectivity define the force fields generating the MM PES. Therefore, motion on the PES involving bond breaking or bond making is discontinuous. Jensen's approach makes use of this characteristic and treats the MM PESs of the reactant and product as separate surfaces that intersect and form a seam. One can then find the minimum on this seam and use the corresponding molecular configuration as the initial guess for the TS, which is optimized using a local TS optimization method on an electronic structure PES. The problem of finding this point on the intersecting seam is equivalent to the task of finding points of closest approach and minima on seams between different electronic state PESs (see Section 10.3.4) [115–123].

10.4.3 Path optimization methods

Clearly, the principle obstacle to interpolation methods is the choice of the approximate reaction pathway. A series of methods have been developed to find not just the TS, but the entire MEP. As a class of algorithms, these approaches have been called chain of states and path optimization methods [163]. Starting with the reactant and product structures as input, these methods search for the MEP by minimizing the energy of a number of points, known as images, which lie on an initial interpolated pathway. As a result, many energy and derivative calculations are required for each image making path optimization methods generally much more expensive than interpolation methods. However, path optimization provides a means for finding TSs while simultaneously elucidating the MEP (or at least a good approximation) without *a priori* knowledge of the TS or the curvature of the reaction path, making it an effective and robust alternative to the other approaches discussed above. Path methods are especially useful for difficult problems where the previously described methods fail to converge to a first-order saddle point or the TS of interest.

In this section, we begin by outlining the elastic band theory of Elber and Karplus [164], which forms the basis for most, if not all, of the development of chain-or-states methods. We then outline two popular methods that correct some of the failures of elastic band theory: the nudged elastic band (NEB) method [165] and the path optimization algorithm of Ayala and Schlegel [166]. Lastly, we consider approaches that develop the reaction pathway by growing separate paths from the reactant and product minima, which have been referred to as growing string methods [167]. Using any of these algorithms provides an estimate for the TS that is usually quite good (as opposed to the rough approximation to the TS provided by LST). Nevertheless, full TS optimization using a local approach is still advisable if one wishes to elucidate an accurate barrier height or characterize the TS geometry.

The elastic band method developed by Elber and Karplus [164] is based on the minimization of a line integral, which has the form

$$S(\mathbf{q}_\mathrm{i}, \mathbf{q}_\mathrm{f})_L = \frac{1}{L}\int_{\mathbf{q}_\mathrm{i}}^{\mathbf{q}_\mathrm{f}} [\mathbf{G}(\mathbf{q})\mathrm{d}\mathbf{l}(\mathbf{q})]_L \tag{44}$$

where S is the objective function (the function that will be optimized), $\mathbf{q}_\mathrm{i}$ and $\mathbf{q}_\mathrm{f}$ are the coordinates for the initial and final structures in the reaction (i.e. reactant and product configurations), $\mathbf{G}$ is a vector that is a function of the system coordinates given by

$$\mathbf{G}(\mathbf{q}_j) = E_j\mathbf{u}_j \tag{45}$$

and $\mathrm{d}\mathbf{l}(\mathbf{q})$ is an infinitesimal line segment on the path $\mathbf{L}$ of length L. In Eq. (45) $\mathbf{u}_j$ is a unit vector in the direction $\mathbf{q}_j - \mathbf{q}_{j-1}$. The path $\mathbf{L}$ begins as the initial guess at the reaction path, which we seek to modify and relax to the MEP. In order to find the path $\mathbf{L}$ that minimizes S, Eq. (44) must be discretized. This results in a series of M points, $\mathbf{q}_1$ through $\mathbf{q}_M$, being placed along $\mathbf{L}$ between the reactant and product structures, $\mathbf{q}_0$ and $\mathbf{q}_{M+1}$. Substituting Eq. (45) into (44) and discretizing yields

$$S(\mathbf{q}_0, \mathbf{q}_{M+1})_L = \frac{1}{L}\sum_{j=1}^{M} E_j \mathbf{u}_j \cdot \Delta\mathbf{l}_j \tag{46}$$

where

$$\Delta\mathbf{l}_j = l_j \mathbf{u}_j \tag{47}$$

Substitution of Eq. (47) into Eq. (46) gives

$$S(\mathbf{q}_0, \mathbf{q}_{M+1})_L = \frac{1}{L}\sum_{j=1}^{M} E_j l_j \tag{48}$$

In the original implementation, Elber and Karplus also found it necessary to add two penalty functions to Eq. (44) that avoid rotation by the images and also keep the images from collecting to one location on the path, i.e. in the minima wells and flat regions of the path [168]. Other pitfalls of the elastic band method include a failure to actually converge to a reaction pathway without a large number of images and for the path to develop kinks and to turn back on itself. Subsequent work by Elber and coworkers [164,168–171] and others [172–175] sought to correct these difficulties and resulted, most notably, in the development of the self-penalty walk [168–171] and locally updated planes algorithms [168,171]. However, most present-day implementations of the basic ideas put forth in elastic band theory are best encapsulated in discussions of the path optimization [166] and NEB approaches [165,176].

In the path optimization method [166] a series of constrained optimizations is used to find the transition state and points on the steepest descent reaction path. The path is relaxed by applying a QN-like optimization scheme on each image in turn, in a manner resembling the reaction path following algorithm developed by Gonzales and Schlegel [177,178] (see Section 10.5.2). Microiterations are used to simultaneously relax the images toward the reaction path and maintain even spacing. The highest energy point on the path is optimized to the TS using a modified version of the STQN optimization algorithm discussed in Section 10.4.2 [138]. Additionally, an initial guess for the TS can be provided by the user to better define the initial path.

The original elastic band approach places 'springs' between successive images to prevent them from falling down into the minima. Adding the springs, with a force constant k, gives the objective function, S, as

$$S(\mathbf{q}_0, \mathbf{q}_{M+1})_L = \sum_{j=1}^{M} E_j + \sum_{j=1}^{M} \frac{k}{2} l_j^2 \tag{49}$$

Many of the problems with elastic band theory are rooted in this definition and their effect on the relaxation of the path. The effective elastic band force acting on image j, $\mathbf{F}_j^{\mathrm{EB}}$, in this method is a modified force that is dependent upon k,

$$\mathbf{F}_j^{\mathrm{EB}} = \mathbf{F}_j + \mathbf{F}_j^{\mathrm{spring}} \tag{50}$$

where

$$\mathbf{F}_j^{\mathrm{spring}} = (k\mathbf{l}_{j+1} - k\mathbf{l}_j) \tag{51}$$

Recall that $\mathbf{F} = -\mathrm{d}E/\mathrm{d}\mathbf{x}$. The first term in Eq. (50) is the 'true' force and the second term is known as the 'spring' force. If k is too large the elastic band becomes stiff and the final path will not converge to the MEP. Instead, the elastic band path will cut corners. If k is too small, the images begin to slide down the path to the two minima wells.

A simple correction to these problems is achieved by projecting out the parallel (with respect to the path) component of the true force and the perpendicular (with respect to the path) component of the spring force before evaluating Eq. (50). In this way, the images are 'nudged' in the optimal direction, i.e. toward the MEP, while the springs carry out their intended responsibility. Hence, the method is known as NEB [165,179]. Now, the effective force is given by

$$\mathbf{F}_j^{\mathrm{EB}} = (\mathbf{I} - \hat{\tau}_{\|}\cdot\hat{\tau}_{\|}^{\mathrm{t}})\mathbf{F}_j + (\hat{\tau}_{\|}\cdot\hat{\tau}_{\|}^{\mathrm{t}})\mathbf{F}_j^{\mathrm{spring}} \tag{52}$$

where $\hat{\tau}_{\|}$ and $\hat{\tau}_{\perp}$ are unit vectors pointing parallel and perpendicular to the path. The definitions of $\hat{\tau}_{\|}$ and $\hat{\tau}_{\perp}$ can greatly affect the success of an NEB calculation and a number of different definitions have been contemplated and applied [165,179]. Regardless of the choice for the tangent vectors, NEB tends to maintain good spacing along the path. However, NEB is not without its problems. In order to get a smooth path a large number of images are often required, which in turn makes NEB a very costly method. Furthermore, the standard implementation of the method relaxes the path using the velocity Verlet algorithm, a scheme used in classical dynamics. The result is that many path relaxation iterations are often necessary before convergence is satisfied. Modifications to NEB have included dynamic adjustment of the end points to focus on regions of specific interest and choosing one image to climb uphill toward the TS [176, 180]. The L-BFGS algorithm (see Section 10.3.1) can be employed in the image relaxation steps to decrease the cpu and memory cost of NEB [181–183]. Another efficient extension of NEB, known as the replica path scheme, has been developed by Brooks and coworkers [184]. All of these developments dramatically cut down on the total cost of the calculation by leading to faster and more stable path convergence.

Another class of TS optimization algorithms is based on a 'burn the wick from both ends' principle. These methods begin at the reactant and product minima and grow two pathways that communicate and simultaneously head toward the TS. These approaches are possibly best described as hybrids of path optimization and bracketing philosophies. They have been included here in order to draw on the ideas presented above and also because the calculation yields an estimate for the reaction pathway that, in many cases, is a very good approximation to the MEP. In this regard, these methods are very similar to the path optimization schemes just discussed. The most basic application of this idea was originally encompassed by the saddle method [144], which begins by considering the coordinates of the reactant and product structures, $\mathbf{R}$ and $\mathbf{P}$, which are a distance δ apart on the PES. Let $\mathbf{R}'$ and $\mathbf{P}'$ denote the coordinates of the current step in the reactant and product valleys, respectively. One can simultaneously walk up these two valleys by letting the lower energy structure take a step that decreases δ by a small amount (i.e. 5–10%) followed by a minimization that constrains the distance between $\mathbf{R}'$ and $\mathbf{P}'$. This process is repeated enough times to converge to an estimate for the TS. Similar methods present in the earlier literature include the sphere optimization technique [185] and the locally updated planes method [168].

The most recent addition to this category of TS optimization methods is the growing string approach of Peters et al. [167]. The growing string algorithm is based on a modified version of the string path optimization scheme, which redistributes the images along the interpolated pathway after each minimization step [186]. In the simplest case, the redistribution step uniformly spaces the images along the path. This step can also distribute the images in an uneven manner. For instance, a higher density of images can be used near the TS or regions of the path with large curvature. Additionally, the number of images can be dynamically modified during the calculation. The growing string method makes use of all of these options by growing two strings, one from the reactant well and one from the product well. These individual strings are allowed to relax and the images along each string are redistributed after each step. When the force normal to the path is small at a frontier image (the image on each string closest to the TS), an additional image is added to the local string. After the two frontier images are close to each other, the two strings are merged and the TS optimization can be completed using a local method. This method has been shown to require fewer energy and gradient calculations than traditional path optimization schemes [167], making it an attractive alternative to the path optimization algorithms already discussed.

10.4.4 Practical considerations

In this section, we consider a few points related to the practical application of the methods described above for TS optimization. The factors affecting TS optimization are the same as for minimization: the starting structure provided by the user, coordinate system, algorithm choice, initial Hessian, and quality of the updated Hessian. Since the points made in Section 10.3.5 are also valid for TS optimization, we do not restate them here. Instead, we focus on specific issues unique to TS optimization and recommend that the reader first read Section 10.3.5. This section also contains suggestions for difficult TS problems.

10.4.4.1 Building an initial structure

For local methods, it can be difficult to build a guess at the TS structure. Using a guess at the TS is also useful for interpolating, bracketing, and path optimization methods. However, generating an initial structure of a TS is usually non-trivial. Unlike minima, there are no direct experimental observations of TS geometries. Instead, the best tools available to computational chemists for this purpose are chemical intuition and the theoretical literature. Over the past two decades, thousands of optimized TSs have been reported in the literature for a vast array of reaction classes. Here, we offer some general suggestions based on these data and our own experience.

Initial TS structures for unimolecular reactions, such as ring closures, hydrogen transfers, internal isomerizations, etc. are most easily built after the reactant and product structures have been minimized. Using these two structures, a TS guess can be made by choosing a molecular configuration lying between the two ground state structures that lines up along a reasonable reaction pathway. The requirement that the structure lie along a reasonable reaction pathway is crucial to the success of the TS optimization. For instance, consider the rearrangement reaction HCN → HNC. Both minima are linear, and choosing a structure that lies exactly between them (i.e. by averaging the Cartesian coordinates) produces a configuration that places the H atom in the middle of the C–N bond. Obviously, this is not a viable TS. A more reasonable reaction path has the H atom moving in a arc starting where the H atom is bonded to C and the H–C–N angle is 180°, passing through a structure where the H atom lies above the C–N bond and the three atoms form a triangle, and finishing with the H atom bonded to N and the C–N–H angle is 180°. A suitable guess for most bimolecular reactions can be generated by setting the lengths of the bonds being formed to 80–120% longer than equilibrium [52].

When the cost of a frequency calculation is reasonable, it can be useful to evaluate the Hessian for the TS guess structure. The purpose of this calculation is twofold: (1) to see if the initial guess has one, and only one, imaginary frequency; and (2) to see if this imaginary frequency corresponds to a reasonable displacement given the reaction being studied. If the guess structure does not have any imaginary frequencies, using a rigid scan along the perceived reaction coordinate and taking the highest energy structure in the scan can often yield an appropriate guess. For cases where multiple imaginary frequencies exist, there are two options. The first option, which is best when the largest magnitude imaginary frequency corresponds to the reaction coordinate and the other imaginary frequencies are much smaller in magnitude (i.e. one or more orders of magnitude difference), is to use this structure and carry out the TS optimization nevertheless. Often, the other imaginary frequencies will relax and the optimization algorithm will find the correct TS. The second option is to freeze the internal coordinate(s) corresponding to the reaction path and minimize the structure for a few steps ($\sim$10–20). This approach is best when the reaction coordinate has an imaginary frequency that is not the largest in magnitude or the additional imaginary frequencies are the same order of magnitude as the correct imaginary frequency. After a few steps of minimization the Hessian can be reevaluated. If the erroneous imaginary frequencies have become real or the conditions for the first option are met, the current structure can be used for the TS optimization. Otherwise, the process should be repeated.

10.4.4.2 Coordinate system

The coordinate system choice is also important in TS optimization. As with minimization, redundant internal coordinates have been shown to be the best choice for TS optimization [52]. Table 10.5 compares the number of optimization steps required for convergence using the three-structure STQN method with Z-matrix and redundant internal coordinates. Clearly, redundant internals work best. In Section 10.3.5.2, we advised that users check the redundant internal coordinate definitions to ensure all of

Table 10.5 Comparison of the number of steps required to optimize TS geometries using three point STQN with various coordinate systems

Reaction[a]	Z-Matrix internals	Redundant internals
$CH_4 + F \rightarrow CH_3 + HF$	6	5
$CH_3O \rightarrow CH_2OH$	9	9
$SiH_2 + H_2 \rightarrow SiH_4$	11	8
$C_2H_5F \rightarrow C_2H_4 + HF$	15	11
Diels–Alder reaction	23	14
Claisen reaction	15	15
Ene reaction	28	18

[a]For complete details see Ref. [51].

the coordinates are included in the definitions. The same issue arises in TS optimization. It is imperative that the redundant internal definitions include all of the coordinates relevant to the reactant and the product. Methods using these structures in their input can generate the union of the reactant and product internals for the TS coordinate definitions. However, local methods are not able to make use of that additional information and other coordinates will often need to be defined.

10.4.4.3 Algorithm choice

The algorithm chosen to carry out a TS optimization can dramatically affect the success of the optimization and the efficiency of the calculation. Unfortunately, a thorough comparison of all of the methods discussed above is not yet available in the literature and is beyond the scope of this review. Nonetheless, Table 10.6 is included to show a general comparison of the efficiency of local, interpolating, and path optimization methods. Specifically, we have used the QN local TS optimization, three point STQN, and path optimization approaches. The interpolating method has roughly the same cost or less cost than the local method. For the ene reaction, the local method is unable to optimize to the TS while the interpolating algorithm converges to the proper TS within 20 steps. The path optimization method also performs well. However, the path method requires many more energy and derivative evaluations since it must compute this information for each image each time a step is taken.

Table 10.6 Comparison of the number of gradient evaluations required to complete TS optimization using a QN with RFO, three point STQN methods, and the path optimization[a]

Reaction	QN with RFO	Three point STQN	Path optimization
$CH_3O \rightarrow CH_2OH$	12	9	51
$SiH_2 + H_2 \rightarrow SiH_4$	11	8	47
$C_2H_5F \rightarrow C_2H_4 + HF$	16	11	73
Diels–Alder reaction	56	14	41
Ene reaction	Fail	18	101

[a]For complete details see Refs. [138,166].

10.4.4.4 Hessian quality

The quality of the Hessian significantly affects the behavior of TS optimization. Generally speaking, the cost issues raised in Section 10.3.5.4 also hold for locating TSs. One notable difference for TS optimization is that the ability of the algorithm to converge to a first-order saddle point is much more sensitive to the initial Hessian quality. For this reason, using an analytic Hessian at the start of the optimization is very useful, provided that the cost of the Hessian is not too large. Analytic Hessian calculations can be prohibitively expensive for many systems. In these cases, an empirical Hessian can work well provided that it has a suitable negative eigenvalue and eigenvector. Another method often used to generate the Hessian is to use second derivatives computed at a lower, and cheaper, level of theory. For example, a HF/3-21G Hessian can be used at the start of a TS optimization at the HF/6-311G(d) level. An alternative approach is to analytically or numerically calculate the rows and columns of the Hessian that are important in the reaction coordinate [52,80,138]. The other elements can be determined using standard force field estimates [127].

10.4.4.5 Verifying TSs

After a TS optimization has completed, it is always necessary to verify the structure. Verification of a TS consists of two steps. First, the Hessian must be evaluated at the optimized structure and diagonalized to ensure that there is one, and only one, negative eigenvalue. The second step in verifying an optimized TS is to test if the saddle point lies on a path connecting the intended reactant and product minima. This task is readily accomplished by employing reaction path following (see Section 10.5) and/or by visualizing the displacement along the vibrational mode corresponding to the imaginary frequency. For instance, consider the rearrangement reaction of HCN $\rightarrow$ HNC. Visualization of the TSs imaginary frequency clearly shows movement of the H atom from C to N. Some reactions have curved reaction paths and visualization of the TSs imaginary frequency may not directly indicate that the TS is connected to the reactant and product. In these cases, reaction path following is required. If either verification test fails, the optimized TS is not a valid structure and the search for the proper structure must be restarted.

10.5 REACTION PATH FOLLOWING

After a TS has been located, it is necessary to confirm that it lies on a pathway connecting the requisite reactant and product. This can be done by following the path of steepest descent downhill from the TS to reactant and product PES minima. Following the reaction path can also show if any intermediates lie between the reactant and product. A reaction path determined in isoinertial coordinates (i.e. a coordinate system where all of the coordinates are scaled to have the same reduced mass) is known as a MEP [187–191]. Using the structure and vibrational frequencies at the TS, one can apply TST to determine rates of reaction. Knowing the MEP, especially near the TS, allows one to employ more sophisticated methods for determining reaction rates such as VTST and RPH methods [3–7]. Gradient extremals [137,192–196] also define paths across PESs, but because they do not necessarily connect stationary points as directly as MEPs we do not consider them further in this context. Also beyond the scope of this review are methods that determine the free energy along an MEP [114,197–205]. Interested readers should consult the literature.

The actual path mapped out by the MEP on the PES is dependent on coordinate system. However, changes in coordinate system do not alter the nature of the stationary points on the PES (i.e. minima, TSs, etc.). One coordinate system, mass-weighted Cartesian coordinates (see Section 10.2.3), is especially significant for reaction dynamics, and the MEP in this coordinate system is known as the intrinsic reaction coordinate (IRC) [162]. In this section, we use the terms MEP, IRC, steepest descent path, and reaction path synonymously.

The starting point for mathematically defining the MEP is a Taylor expansion of the PES.

$$E(\mathbf{x}) = E_0 + \mathbf{g}_0\Delta\mathbf{x} + \frac{1}{2}\Delta\mathbf{x}^{\mathrm{t}}\mathbf{H}_0\Delta\mathbf{x} + \cdots \tag{53}$$

In Eq. (53), $E(\mathbf{x})$ is the energy at point $\mathbf{x}$ and E_0, $\mathbf{g}_0$, and $\mathbf{H}_0$ are the energy, gradient, and Hessian at the point $\mathbf{x}_0$. It is convenient to think of the MEP as a one-dimensional slice through the PES. Defining the parameter s, which is dependent on $\boldsymbol{x}$, as the arc length along this one-dimensional slice gives rise to another Taylor series.

$$\mathbf{x}(s) = \mathbf{x}(0) + \boldsymbol{\nu}^0(0)s + \frac{1}{2}\boldsymbol{\nu}^1(0)s^2 + \frac{1}{6}\boldsymbol{\nu}^2(0)s^2 + \cdots \tag{54}$$

In Eq. (54), $\boldsymbol{\nu}^0$ and $\boldsymbol{\nu}^1$ are known as the tangent and curvature vectors. The tangent vector is given by

$$\boldsymbol{\nu}^0(s) = \frac{\mathrm{d}\mathbf{x}(s)}{\mathrm{d}s} = -\frac{\mathbf{g}(s)}{|\mathbf{g}(s)|} \tag{55}$$

Eq. (55) is the differential equation solved when following reaction paths. The curvature vector in Eq. (54) is given by

$$\boldsymbol{\nu}^1(s) = -\frac{\mathbf{H}\boldsymbol{\nu}^0 - (\boldsymbol{\nu}^{0\mathrm{t}}\mathbf{H}\boldsymbol{\nu}^0)\boldsymbol{\nu}^0}{|\mathbf{g}(\mathbf{x})|} \tag{56}$$

The magnitude of the curvature, κ, is equal to the inverse of the radius of curvature, R

$$\kappa = |\boldsymbol{\nu}^1| = \frac{1}{R} \tag{57}$$

Large curvature indicates the reaction path is undergoing a tight turn, and small curvature corresponds to a shallow turn.

At the TS, where the gradient is zero, Eqs. (55) and (56) become ill defined and the tangent is equal to the Hessian eigenvector corresponding to the negative eigenvalue—this eigenvector is known as the transition vector. The curvature at the TS is given by

$$\boldsymbol{\nu}^1(s) = -[\mathbf{H}\boldsymbol{\nu}^0 - (\boldsymbol{\nu}^{0\mathrm{t}}\mathbf{H}\boldsymbol{\nu}^0)\mathbf{I}]^{-1}[\mathbf{F}^1\boldsymbol{\nu}^0 - (\boldsymbol{\nu}^{0\mathrm{t}}\mathbf{F}^1\boldsymbol{\nu}^0)\boldsymbol{\nu}^0] \tag{58}$$

where

$$\mathbf{F}^1_{ij} = \sum_k \mathbf{F}_{ijk}\boldsymbol{\nu}^0_k \tag{59}$$

Since the MEP is defined by an ordinary differential equation (ODE), standard numerical integration techniques can be used. The basic idea behind any numerical integration algorithm is to rewrite $\mathbf{dx}$ and ds as $\Delta\mathbf{x}$ and Δs. In this way, the solution to Eq. (55) is given by a discrete set of points, $\{\mathbf{x}_i\}$, which are found by 'stepping' along the path with a step size of Δs. Although Eq. (55) appears to be rather benign, it can display stiff behavior [206] and be difficult to solve, especially in the regions where the gradient is very small. Therefore, solving Eq. (55) requires special care, and a number of specialized methods have been developed and reviewed in the literature [10,23,24,72, 128,177,178,207–217]. Generally speaking, numerical methods for integrating ODEs are classified as either explicit or implicit. Explicit methods use only information at the current point to define the position at the next point, while implicit methods use additional information from the next point, which typically means that these methods include some sort of iterative algorithm to converge the end point of each step.

Because the integration is numerical, different integration schemes yield different degrees of accuracy (i.e. the proximity of the points to the true MEP). The accuracy of a numerical integrator is defined by an order (i.e. first order, second order, etc.). The order of the integrator gives the highest order term in the Taylor expansion of the true solution. Hence, first-order methods give the correct first-order term in Eq. (54), second-order methods give correct first- and second-order terms in Eq. (54), and so on.

In the following sections, we describe some common methods for solving the reaction path equation and also discuss some interesting features and properties of MEPs. In Section 10.5.1, first-order methods are described. Included in the discussion are the explicit Euler integrator and its implicit and stabilized versions [212,213,216]. Second-order methods are considered in Section 10.5.2. There, we focus on the local quadratic approximation (LQA) method [214,215], the second-order Gonzalez–Schlegel algorithm [177,178], and the Hessian based predictor–corrector integrator [177,178,210,211]. Higher order methods are considered in Section 10.5.3 [209,214]. Path following methods based on classical dynamics, known as dynamic reaction path (DRP) methods [218–221], are discussed in Section 10.5.4. Lastly, in Section 10.5.5, we have included some practical considerations and common troubleshooting tips related to reaction path following calculations.

10.5.1 First-order methods

Perhaps the simplest integrator (both conceptually and in terms of coding) is the Euler method. This method is correct only to the first-order term in Eq. (54) and gives the next point in the integration, $\mathbf{x}_{i+1}$, as

$$\mathbf{x}_{i+1} = \mathbf{x}_i - \frac{\mathbf{g}(\mathbf{x}_i)}{|\mathbf{g}(\mathbf{x}_i)|}\Delta s = \mathbf{x}_i + \boldsymbol{\nu}^0_i\Delta s \tag{60}$$

In the limit of infinitesimally small step size, Euler integration yields the exact MEP. However, in practice Euler integration suffers from a number of pathologies. The steps taken in an Euler integration are linear, meaning that the Euler path will deviate from the true reaction path wherever the MEP is at all non-linear. Indeed, Euler integration is notorious for developing wild oscillations back and forth across the true solution for larger step sizes. This pathology is especially apparent in regions where the path curvature is large or where the slope along the path is small. As a result, reaction path following calculations using Euler integration must use very small steps. The problem, though, is that calculations using a small step size require a large number of energy and gradient evaluations (one energy and gradient evaluation is needed per step), which may be quite costly for moderate to large systems.

To combat this problem, Ishida, Morokuma, and Komornicki (IMK) developed the Euler stabilization (ES) algorithm [212]. The IMK algorithm is shown in Fig. 10.6a. Each ES step can be broken into two parts. First, an explicit Euler step, i.e. Eq. (60), is taken from $\mathbf{x}_i$ to $\mathbf{x}^*$, where the energy (E^*) and gradient ($\mathbf{g}^*$) are both evaluated. The second piece of ES is to stabilize the Euler step by minimizing the energy along the line bisecting the angle between $(\mathbf{x}_i - \mathbf{x}^*)$ and $\mathbf{g}^*$. The minimum on the bisector is chosen as $\mathbf{x}_{i+1}$, and the next ES step starts from there. In cases where the angle between $(\mathbf{x}_i - \mathbf{x}^*)$ and $\mathbf{g}^*$ is

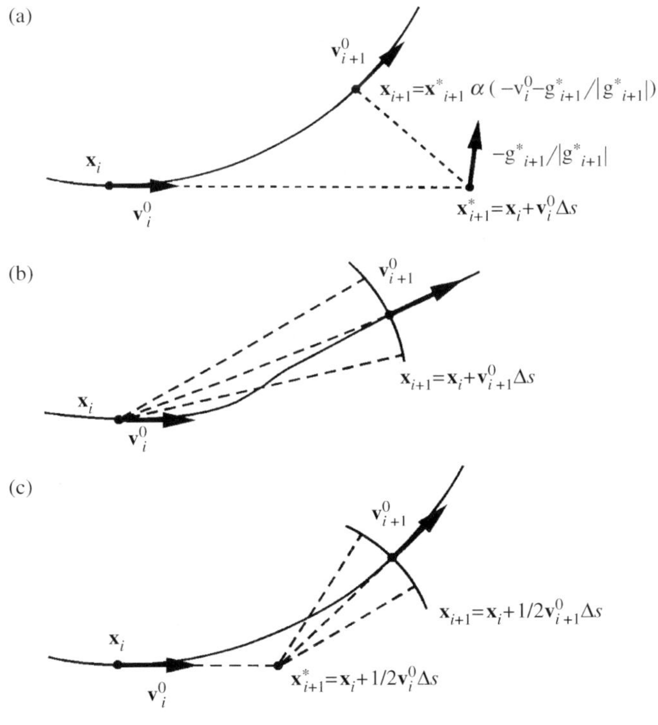

Fig. 10.6. Schematic representations of implicit reaction path following integrators: (a) Ishida, Morokuma, and Komornicki; (b) Müller–Brown; and (c) second-order Gonzalez and Schlegel (from Ref. [72] with permission).

nearly linear or when small step sizes are used, the stabilization step must be omitted. Although ES is much more stable than explicit Euler, difficult regions of the PES can still be problematic and cause oscillations in the ES pathway.

Another first-order method developed to overcome the problems of explicit Euler integration is the Müller–Brown (MB) method [213]. The MB method makes use of the implicit Euler integrator. Here, the step depends on the gradient at the endpoint, which is unknown. To find the endpoint of each step, a constrained optimization is required. The MB step begins with an explicit Euler step. Then, the energy is minimized according to the constraint that the distance between the starting and ending points remains constant

$$|\mathbf{x}_i - \mathbf{x}_{i+1}| = \Delta s \tag{61}$$

Fig. 10.6b shows, schematically, the MB method. The gradient at the endpoint after the constrained minimization will be parallel to the step direction, $(\mathbf{x}_i - \mathbf{x}_{i+1})$. Therefore, the MB step can be given by

$$\mathbf{x}_{i+1} = \mathbf{x}_i - \frac{\mathbf{g}_{i+1}}{|\mathbf{g}_{i+1}|}\Delta s = \mathbf{x}_i + \boldsymbol{\nu}_{i+1}^0\Delta s \tag{62}$$

10.5.2 Second-order methods

Second-order methods have also been proposed in the literature for integrating MEPs. Although a standard numerical integrator, such as the second-order Runge–Kutta method, could be used, it is generally accepted that a more effective approach is to directly expand the PES as a second-order Taylor series. Truncating the Taylor series in Eq. (53) at the second-order term and differentiating gives

$$\mathbf{g}(\mathbf{x}) = \mathbf{g}_0 + \mathbf{H}_0\Delta\mathbf{x} \tag{63}$$

Substituting Eq. (63) into Eq. (55) gives the LQA of Page and McIver [213–215], which is an explicit second-order integrator.

$$\frac{d\mathbf{x}(s)}{ds} = -\frac{\mathbf{g}_0 + \mathbf{H}_0\Delta\mathbf{x}}{|\mathbf{g}_0 + \mathbf{H}_0\Delta\mathbf{x}|} \tag{64}$$

In practice, Eq. (64) is integrated by parameterization and casting the problem in the Hessian eigenvector space. Sun and Ruedenberg [217] have modified the LQA algorithm by using each point, $\mathbf{x}_i$, as the midpoint in the integration range, rather than the endpoint.

Gonzalez and Schlegel developed an implicit second-order integrator for reaction path following (GS2) [177,178], which is shown in Fig. 10.6c. Each GS2 step consists of two components. First, an explicit Euler step of length $\frac{1}{2}\Delta s$ is taken from the current point, $\mathbf{x}_i$, to a pivot point, $\mathbf{x}^*$.

$$\mathbf{x}^* = \mathbf{x}_i + \frac{1}{2}\mathbf{v}_i^0\Delta s \tag{65}$$

The second piece of a GS2 step begins by taking an additional step of length $\frac{1}{2}\Delta s$ from the pivot point to the end point, $\mathbf{x}_{i+1}$. The energy, $E(\mathbf{x}_{i+1})$, is minimized using QN under the constraint that

$$|\mathbf{x}_{i+1} - \mathbf{x}^*| = \frac{1}{2}\Delta s \tag{66}$$

At the end of the optimization, the component of the gradient perpendicular to $|\mathbf{x}_{i+1} - \mathbf{x}^*|$ is zero. Just as with the MB implicit integrator, the GS2 step can be written in terms of the tangent at the initial point, $\mathbf{x}_i$, and at the final point, $\mathbf{x}_{i+1}$.

$$\mathbf{x}_{i+1} = \mathbf{x}_i - \frac{1}{2}\frac{\mathbf{g}_i}{|\mathbf{g}_i|}\Delta s - \frac{1}{2}\frac{\mathbf{g}_{i+1}}{|\mathbf{g}_{i+1}|}\Delta s = \mathbf{x}_i + \frac{1}{2}\boldsymbol{\nu}_i^0\Delta s + \frac{1}{2}\boldsymbol{\nu}_{i+1}^0\Delta s \tag{67}$$

Eq. (67) is similar to the implicit trapezoid approach for integrating stiff differential equations, except that the GS2 method utilizes optimization to obtain $\mathbf{x}_{i+1}$. It should be noted that no energy or derivative calculations are necessary at the pivot point. Furthermore, since

$$|\mathbf{x}^* - \mathbf{x}_i| = |\mathbf{x}_{i+1} - \mathbf{x}^*| = \frac{1}{2}\Delta s \tag{68}$$

the points $\mathbf{x}_i$, $\mathbf{x}^*$, and $\mathbf{x}_{i+1}$ form an isosceles triangle. By construction, two tangents to a circle form an isosceles triangle. Therefore, the GS2 algorithm will follow an arc of a circle exactly.

Recently, Hratchian and Schlegel (HS) introduced a second-order predictor–corrector reaction path following integrator [210,211]. A related algorithm has also been used for integrating *ab initio* classical trajectories [222,223]. Predictor–corrector integrators, as their name suggests, couple two different integration methods. The predictor integrator moves from the current point, $\mathbf{x}_i$, to a guess for the next point, $\mathbf{x}_{i+1}$. Using information (e.g. energy and/or derivatives) at the predicted $\mathbf{x}_{i+1}$, the corrector integrator re-integrates over the same interval and refines, or corrects, $\mathbf{x}_{i+1}$. The basic idea is illustrated in Fig. 10.7. The HS method uses LQA for the predictor step and a modified Bulirsch–Stoer integrator [206,224–227] for the corrector step. The corrector step increases the stability of LQA and allows for large steps without loosing accuracy. Bulirsch–Stoer integration requires several gradient evaluations per step, which would make direct use with electronic structure methods quite costly. To overcome this bottleneck, the HS integrator uses the positions, energies, gradients, and Hessians at $\mathbf{x}_i$ and the predicted point, $\mathbf{x}_{i+1}$, to construct a local analytic surface. The Bulirsch–Stoer corrector integration is carried out on this fitted surface. Relative to the cost of electronic structure energy and derivative

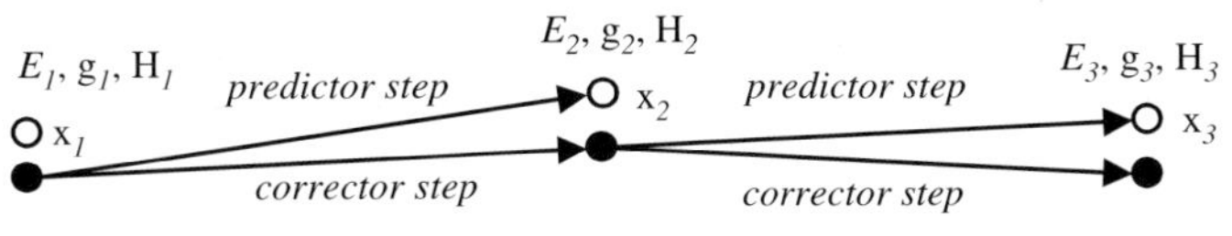

Fig. 10.7. Schematic representation of predictor–corrector integration, such as the integrator used in the Hratchian–Schlegel method (from Ref. [223] with permission).

calculations, energy and gradient evaluations on the fitted surface are free and the corrector step adds a negligible cost to the standard LQA calculation, which is more than compensated for by the larger step sizes that the HS algorithm allows. In the current implementation, HS fits the LQA data to a distance weighted interpolant surface for the corrector integration [228–231]. The energy on this surface, E_{DWI}, is given by

$$E_{\mathrm{DWI}} = w_i T_i + w_{i+1} T_{i+1} \tag{69}$$

where T_i and T_{i+1} are Taylor expansions about $\mathbf{x}_i$ and $\mathbf{x}_{i+1}$, respectively. These Taylor expansions,

$$\begin{aligned} T_i(\mathbf{x}) &= E_i + \mathbf{g}_i^{\mathrm{t}}(\mathbf{x} - \mathbf{x}_i) + \frac{1}{2}(\mathbf{x} - \mathbf{x}_i)^{\mathrm{t}}\mathbf{H}_i(\mathbf{x} - \mathbf{x}_i), \\ T_{i+1}(\mathbf{x}) &= E_{i+1} + \mathbf{g}_{i+1}^{\mathrm{t}}(\mathbf{x} - \mathbf{x}_{i+1}) + \frac{1}{2}(\mathbf{x} - \mathbf{x}_{i+1})^{\mathrm{t}}\mathbf{H}_i(\mathbf{x} - \mathbf{x}_{i+1}) \end{aligned} \tag{70}$$

are weighted by w_i and w_{i+1},

$$w_i = \frac{|(\mathbf{x} - \mathbf{x}_i)|^2}{|(\mathbf{x} - \mathbf{x}_i)|^2 + |(\mathbf{x} - \mathbf{x}_i)|^2}, \qquad w_{i+1} = \frac{|(\mathbf{x} - \mathbf{x}_{i+1})|^2}{|(\mathbf{x} - \mathbf{x}_{i+1})|^2 + |(\mathbf{x} - \mathbf{x}_{i+1})|^2} \tag{71}$$

The LQA and HS methods require second derivatives. Although the GS2 equations do not explicitly require the Hessian, second derivatives are used for the constrained optimization step. For the same reasons we discussed in Section 10.3.5.4, the calculation of second derivatives can greatly increase the cost of a reaction path following calculation and limit the usefulness of these methods for the study of moderate and large chemical systems. As before, Hessian updating can be employed, and previous studies have shown that Hessian updating is a viable option for reaction path following using these second-order integration schemes [177,211,232]. Also noteworthy is that Hessian updating methods designed for minimization cannot be used with reaction path following since the formulas for most of those updates become ill conditioned when the Hessian has one or more negative eigenvalues [129]. Recall that the same concern was encountered in our earlier discussion of TS optimization. Therefore, Hessian updating methods developed for TS optimization are also useful in reaction path following (i.e. MS, PSB, Bofill, etc.).

10.5.3 Higher order integrators

Higher order methods can also be envisioned. Here, we provide only a brief discussion of two sets of higher order integrators, since they are not often used in connection with electronic structure calculations. The first set of these methods, developed by Page et al. [214], consists of two explicit third-order integrators. One of these integrators finds each point on the reaction path by directly solving the first four terms of Eq. (54). The PES third derivatives are required for $\boldsymbol{\nu}^2$, which can either be computed analytically or numerically. A similar method provides a third-order analogue to LQA, the CLQA algorithm. After each LQA step, the Hessians at the initial and final points can be used to estimate $\boldsymbol{\nu}^2$ by finite difference by recognizing that $\boldsymbol{\nu}^2$ depends on the derivative of the force constants with respect to the arc length, s.

Gonzalez and Schlegel [209] have also developed a series of third- and fourth-order methods. All of their higher order approaches use implicit integrators and are extensions of the GS2 algorithm. One of these, a fourth-order method, uses the tangent and curvature vectors at the initial and final points of each step.

$$\mathbf{x}_{i+1} = \mathbf{x}_i + \frac{1}{2}\boldsymbol{\nu}_i^0 \Delta s + \frac{1}{2}\boldsymbol{\nu}_{i+1}^0 \Delta s + \frac{1}{12}\boldsymbol{\nu}_i^1 \Delta s^2 - \frac{1}{12}\boldsymbol{\nu}_{i+1}^1 \Delta s^2 \tag{72}$$

10.5.4 Dynamic reaction path

The solution for the MEP, as shown above, is time independent. Of course, molecular systems exist in a time dependent universe and they are constantly exercising motions on the PES different than the specific motion described by the MEP. An alternative picture for the reaction path is to allow the nuclei to move on the PES according to Newton's equations of motion. In terms of time, the coordinates of the system at time t_i, $\mathbf{x}_i$, can be given by the Taylor expansion

$$\mathbf{x}_i = \mathbf{x}_{i-1} + \mathbf{v}_{i-1}\Delta t + \frac{1}{2}\mathbf{a}_{i-1}\Delta t^2 + \cdots \tag{73}$$

where $\mathbf{v}$ and $\mathbf{a}$ are velocity and acceleration, respectively. The velocity is given by a similar Taylor series,

$$\mathbf{v}_i = \mathbf{v}_{i-1} + \mathbf{a}_{i-1}\Delta t + \cdots \tag{74}$$

and the acceleration is given by

$$\mathbf{F} = \mathbf{ma}$$

where $\mathbf{m}$ is a diagonal matrix of atomic masses and

$$\mathbf{F} = -\frac{\mathrm{d}V}{\mathrm{d}\mathbf{x}} = -\mathbf{g} \tag{75}$$

where V is the potential energy of the system, which is given by the value of PES. For the case where infinitesimal steps in t are taken and the kinetic energy is completely removed, or damped, from the system at every step, the path mapped out by classical dynamics (beginning at the TS) is identical to the IRC.

Starting at the TS and following a time-dependent path according to Newton's equation of motion yields the DRP of Stewart et al. [221]. Since a DRP without removing any kinetic energy is the same as a classical trajectory calculation, it can be used to study energy transfer processes during the course of a reaction, and understand how the energy in specific modes behaves relative to the reaction coordinate, or IRC [220]. By damping the kinetic energy at each step one can use the DRP as a means for finding qualitative MEPs. This is especially useful where the purpose for finding the MEP is to ensure that a located TS lies on a pathway connecting specific PES minima, and has been shown to be an efficient alternative to conventional integration of the MEP (i.e. using methods discussed in Sections 10.5.2 and 10.5.3). We have developed a method called damped velocity Verlet (DVV) that controls the path accuracy by employing a variable step size, Δt [219]. The DVV step size control is based on the third-order error scaling of velocity Verlet. Given the current step size of Δt_i and desired error in the path, Δ_0, the next step size, Δt_{i+1}, is set according to

$$\Delta t_{i+1} = \Delta t_i \left| \frac{\Delta_0}{\Delta_i} \right|^{1/3} \tag{76}$$

In Eq. (76) Δ_i is an estimated error in the path at step i. This estimated error is shown schematically in Fig. 10.8, and is obtained by propagating a double sized step from $\mathbf{x}_{i-2}$ to $\mathbf{x}_i'$ and comparing this to the point $\mathbf{x}_i$.

10.5.5 Practical considerations

We conclude Section 10.5 by discussing a few practical points related to reaction path following. This section begins with tips for choosing an appropriate reaction path following algorithm based on the application at hand. Projected frequencies and the relationship between path accuracy and errors in the path's tangent, curvature, and projected frequencies are considered in the second subsection. The third subsection is concerned with bifurcations, which are novel topological features of PESs and reaction paths. In the last subsection we suggest steps for difficult reaction path calculations.

10.5.5.1 Algorithm choice

We begin this section by briefly outlining some of the key points that need to be considered when choosing an algorithm for a reaction path following calculation. As mentioned earlier, reaction path following is typically employed to ensure that an optimized TS lies on a MEP connecting the correct reactant and product minima and/or to accurately determine reaction rates. These two applications of reaction path following calculations have different requirements on the quality and efficiency of the integration. In the former case, efficiency has primacy over strict accuracy, although it is essential that the MEP integration be trustworthy and able to qualitatively follow the true pathway. In the latter case, efficiency is desirable, but the accuracy of the path is paramount (see Section 10.5.5.2).

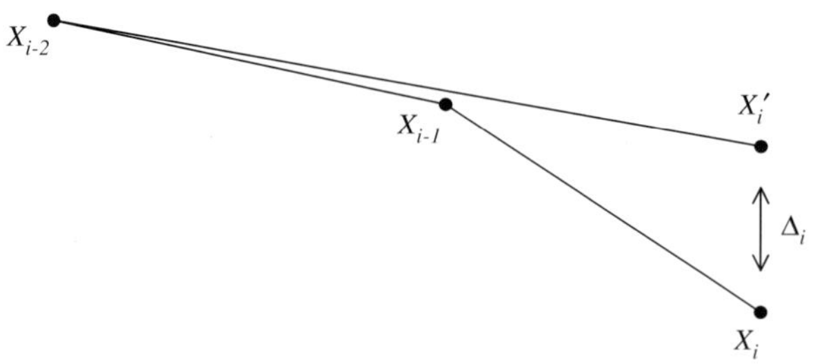

Fig. 10.8. Graphical depiction of damped velocity Verlet (DVV) path error estimation used for time variation (from Ref. [219] with permission).

First, we consider the ability of the various integrators discussed above to provide accurate pathways, such as those required for VTST and RPH calculations. DRP methods are not appropriate for these applications of reaction path following since they provide only an approximate path. A comparison of the accuracy of multiple integrators has been provided in a series of papers by Truhlar and coworkers [207,208,233]. Schlegel and colleagues have also provided comparisons of the accuracy afforded by a series of typical first- and higher-order integrators [209,234]. Fig. 10.9 shows results of reaction path following using a number of first- and second-order methods on the Müller–Brown surface. A step size of 0.2 has been used for all of the integrations shown. The solid line is the MEP computed by Euler integration using a very small step size (0.0001). The portion of the MEP considered in this example connects the TS at (−0.822, 0.624) and the minimum at (−0.558, 1.442). Because the reaction path is curved, this surface can be challenging for reaction path following integrators. It is clear from the figure that the first-order methods deviate most from the true MEP while the second-order methods perform very well. The accuracy of the first-order methods can be improved by decreasing the step size, but this option is undesirable since it means an increased number of energy and derivative evaluations. On the other hand, most second- and higher-order methods, such as LQA, CLQA, and HS, all require second derivatives that are often more expensive than multiple energy and gradient calculations, but this characteristic may not represent a bottleneck if the MEP is being determined for subsequent rate constant calculations since

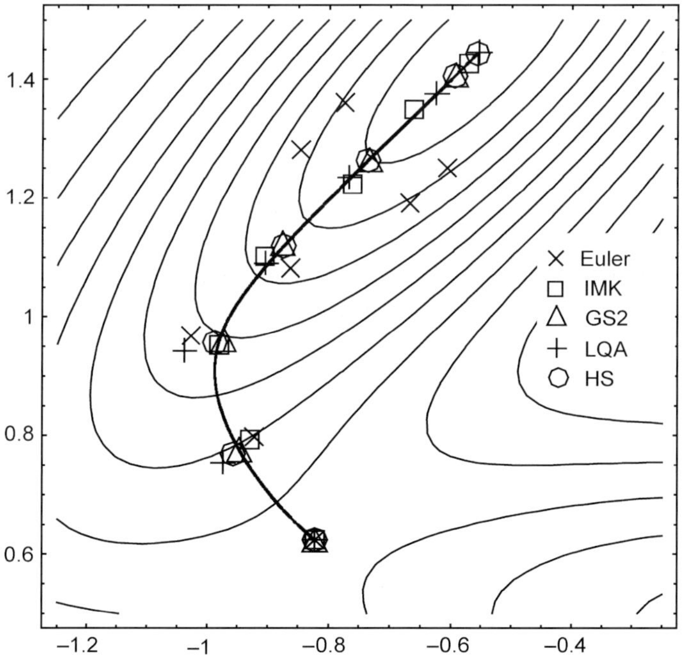

Fig. 10.9. Reaction path following on the Müller–Brown surface using Euler, Ishida, Morokuma, and Komornicki (IMK), local quadratic approximation (LQA), Hratchian–Schlegel (HS), and second-order Gonzalez–Schlegel (GS2) methods.

VTST and RPH calculations also require Hessians along the MEP. Formally, the GS2 method does not require second derivatives. However, in practice the constrained optimization at the end of each step uses Newton minimization and requires a Hessian matrix. Although updated Hessians can be used with GS2, it has been shown[234] that analytic Hessians are useful in providing very accurate pathways. One particularly advantageous feature of GS2 over LQA and CLQA is that larger steps can be used to give similar error in the path since GS2 is an implicit method. Additionally, the implicit nature of GS2 makes it much more stable and robust along flat PESs and in regions of the MEP where the gradient is small. This is especially important near the TS. Although this result makes GS2 an attractive method for accurate path following, the integration can be somewhat expensive due to the constrained optimizations, which typically require three or four cycles per step. The HS method is also able to accurately follow the MEP in difficult regions of the path, and only requires one evaluation of the energy and derivatives per step.

If reaction path following is used to confirm a TS lies on a pathway connecting appropriate reactant and product minima, the approaches used for kinetics studies are more expensive than the nature of the application demands. For qualitative path following, first-order methods are still not preferred since they require very small steps and consequently progress down the MEP relatively slowly. To make second-order methods more tractable for large systems Hessian updating can be employed [177,211, 232]. The use of updated Hessians only requires a slight decrease in the step size. As a result, methods such as LQA, CLQA, GS2, and HS progress down the MEP faster than the first-order approaches since they are able to take larger steps. To demonstrate the applicability of Hessian updating to reaction path following integrators requiring force constant matrices we have included Table 10.7, which shows the perpendicular distance between points on a path computed with the HS integrator using all analytic Hessians and using all updated Hessians. From this data, it is clear that Hessian updating can lead to qualitatively good pathways. DRP approaches are efficient methods for elucidating qualitative reaction pathways. Although these methods are able to take moderate step sizes, methods such as LQA, GS2, and HS are able to take considerably larger steps. The DVV method, which uses a dynamic time step, requires more integration steps than GS2,

Table 10.7 RMS errors in position (Å) for HS reaction path following Hessian updating[a,b]

Reaction	Step size (bohr)	RMS error
HNC → HCN	0.10	3.05×10^{-4}
	0.40	1.05×10^{-2}
$CH_3CH_2F \rightarrow CH_2CH_2 + HF$	0.10	4.73×10^{-3}
	0.40	8.46×10^{-2}
$ClCH_3 + Cl^- \rightarrow Cl^- + CH_3Cl$	0.10	3.04×10^{-3}
	0.40	3.46×10^{-2}
Diels–Alder	0.10	1.58×10^{-2}
	0.40	9.50×10^{-2}

[a] For complete details see Ref. [211].
[b] Bofill's updating scheme for transition state optimization has been employed.

Table 10.8 Comparison of the number of Fock matrix evaluations for damped velocity Verlet and second-order Gonzalez–Schlegel reaction path following[a]

Reaction	Fock evaluations for DVV[b]	Fock evaluations for GS2[c]
$CH_3 + HF \rightarrow CH_4 + F$	1276	716
$CH_2OH \rightarrow CH_3O$	1236	668
Diels–Alder reaction	2142	1352
$CH_3CH_2F \rightarrow CH_2CH_2 + HF$	1439	4281
Ene reaction	2134	7406
$[Ir(CO)_2I_3(CH_3)]^- \rightarrow [Ir(CO)I_3(COCH_3)]^-$	1078	9579

[a]For complete details see Ref. [219].
[b]DVV calculations were run with a damping factor of 0.04 au/fs.
[c]GS2 calculations were run with a step size of 0.1 amu$^{1/2}$ bohr.

but becomes more efficient because it is an explicit method and does not require constrained optimizations at each step. Table 10.8 shows the total number of Fock matrix evaluations, which is the bottleneck for large systems, necessary to follow a series of reaction paths using DVV and GS2 methods. The data clearly indicates that DVV is more efficient than GS2.

10.5.5.2 Projected frequencies and coupling matrix elements

As mentioned earlier, MEPs can be used to compute reaction rates using VTST or RPH methods. In order to use either of these approaches it is necessary to compute the vibrational frequencies lying perpendicular to the MEP [234]. The perpendicular vibrational frequencies are determined by projecting out motion along the tangent, $\boldsymbol{\nu}^0$, from the Hessian and then calculating the projected frequencies according to the normal procedures [235]. Mathematically, the projected Hessian, $\tilde{\mathbf{H}}$, is given by

$$\tilde{\mathbf{H}} = \mathbf{PHP} \tag{77}$$

where the projector, $\mathbf{P}$, is

$$\mathbf{P} = \mathbf{I} - \boldsymbol{\nu}^0 \boldsymbol{\nu}^{0t} \tag{78}$$

If the calculation is carried out in mass-weighted Cartesian coordinates then overall translation and rotation are also projected out of the Hessian.

Reaction rate calculations also require the coupling matrix terms $\mathbf{B}$. The vector $\mathbf{B}$ indicates the coupling between motion along the reaction path and the normal modes of vibration corresponding to the projected frequencies, i.e. the normal modes that are perpendicular to the MEP. The ith element of $\mathbf{B}$ is

$$\mathbf{B}_i = \boldsymbol{\nu}^{0t} \frac{d\mathbf{L}_i}{ds} = -\boldsymbol{\nu}^{1t} \mathbf{L}_i \tag{79}$$

where $\mathbf{L}_i$ is the ith eigenvector of $\tilde{\mathbf{H}}$. In order to compute reliable projected frequencies and coupling matrix elements, it is necessary to integrate the MEP accurately. This is especially important in the region very near the TS where the gradient is very small. Precise MEPs are also required for accurate projected frequencies and coupling matrix elements in regions where the valley has steep walls since small displacements from the MEP can cause large errors in the tangent and curvature vectors.

10.5.5.3 Bifurcation

Bifurcation is a novel topological feature that can be present on a PES [236], which occurs when one valley branches into two. More specifically, bifurcation can occur as a valley progresses down the MEP from the TS and splits into two different minima, or as a valley rises from a minimum and splits into two different MEPs heading toward two different TSs. Along the reaction path of a simple valley, all of the projected frequencies will be real. This corresponds to all of the eigenvalues of the projected Hessian being positive. If one or more of the projected frequencies are imaginary, this indicates that the PES is a maximum perpendicular to the reaction path and the path lies on a ridge. The development of the ridge is marked by a valley-ridge inflection (VRI) point. At the VRI point one projected frequency is zero. On one side of the VRI point this projected frequency is real and on the other side it is imaginary. Fig. 10.1 shows a model PES that has a VRI point on the MEP leading from Transition State B. At the VRI point the MEP integration is displaced slightly (along the Hessian eigenvector with a corresponding eigenvalue of zero) leading to the branching displayed in the figure. Bifurcation and methods for locating VRI points are active areas of research, and interested readers are referred to the current literature for more detailed discussion [156,158,236–251].

10.5.5.4 Tips for difficult reaction path calculations

In this subsection we offer suggestions for difficult reaction path following calculations. In conjunction with suggestions for correcting problems, we describe some useful diagnostics. Testing these diagnostics is especially important for calculations that terminate improperly (e.g. an SCF failure is encountered, etc.). Our principle focus here is on calculations where reaction path following is employed to ensure that an optimized TS lies on a pathway connecting the appropriate reactant and product minima. For situations where the primary concern is a very accurate path for determining reaction rates, these suggestions may also be useful.

The first test is to generate a plot of energy *vs.* reaction coordinate. This plot should be smooth and show a monotonic decrease in energy as the reaction path progresses from the TS to the reactant and product minima. A second diagnostic is to plot the r.m.s. force *vs.* reaction coordinate. This plot should show zero force at the TS and an increasing r.m.s. force for part of the progression to the minimum. Before reaching the minimum, the r.m.s. force will reach a maximum and begin to decrease until it goes to zero when the minimum is reached. Sharp spikes or sudden drops in either of these plots can indicate problems with the level of theory being used to study the chemistry under investigation. Mild undulations in either plot can indicate a problem with the actual integration of the MEP. In many cases, this problem can be corrected by taking smaller steps. If a first-order method is being used, an alternative solution is to switch to a second-order integrator.

For predictor–corrector and implicit integrators, the accuracy of the path can often be enhanced by tightening convergence criteria. Of the methods discussed in this chapter, this approach is applicable for MB, GS2, and HS. MB and GS2 both employ constrained optimizations to determine the tangent at the endpoint of each step. In the practical implementation of the constrained optimization the accuracy of the endpoint is based on convergence criteria, which can be tightened. In the case of HS, the Bulirsch–Stoer corrector step repeats the integration of the MEP with decreasing step size until an extrapolation to zero step size can be made with an estimated error in the energy below a threshold. Decreasing this acceptance threshold can lead to a more accurate integration by the HS algorithm.

DVV pathways can also display energy and force oscillations. This pathology can usually be corrected by modifying the various parameters involved in the calculation. Specifically, smaller ν_0 and Δ_0 values can lead to DVV paths more closely resembling the actual MEP. In fact, Andersen and Carter [252,253] have used values for these parameters that are much more stringent than those originally proposed. Note, though, that small values for either parameter will require more points along the MEP to progress to the minima, leading to more gradient evaluations and longer calculation times.

10.6 SUMMARY AND OUTLOOK

In this chapter, we have provided an overview of the current status of local minimization, transition state optimization, and reaction path following using electronic structure methods. This review also offered a number of suggestions for overcoming difficulties commonly encountered in geometry optimization and reaction path following. Clearly, great progress has been made in these areas in the past 40 years; however, the work is far from complete. Indeed, the development of new methods for exploring *ab initio* PES continues to be an active area of research in the computational quantum chemistry community, especially within the context of hybrid methods (i.e. QM/MM and QM/QM) and the study of very large systems.

In recent decades, electronic structure methods were used to study an ever-increasing diversity of chemistries ranging in size from a few atoms to thousands of nuclei. Doubtless, the future will bring faster computers and more efficient energy algorithms that will expand the limits of computational chemistry beyond the fringes of today's most optimistic estimates. As we have shown here, tools for exploring PES are invaluable and developments in this area will continue to respond to the demand for minimization algorithms featuring fast convergence, robust transition state optimization techniques, and reaction path following integrators coupling accuracy with efficiency.

10.7 REFERENCES

1 J.I. Steinfeld, J.S. Francisco and W.L. Hase, Chemical kinetics and dynamics, Prentice-Hall, Upper Saddle River, NJ, 1999.
2 S. Glasstone, K.J. Laidler and H. Eyring, The theory of rate processes; the kinetics of chemical reactions, viscosity, diffusion and electrochemical phenomena, McGraw-Hill, New York, 1941.
3 D.G. Truhlar and B.C. Garrett, Annu. Rev. Phys. Chem., 35 (1984) 159–189.
4 D.G. Truhlar, B.C. Garrett and S.J. Klippenstein, J. Phys. Chem., 100 (1996) 12771–12800.
5 B.C. Garrett and D.G. Truhlar, in Encyclopedia of computational chemistry, P.v.R. Schleyer, N.L. Allinger, T. Clark, J. Gasteiger, P.A. Kollman, H.F. Schaefer III and P.R. Schreiner, (Eds.), Wiley, Chichester, 1998, p. 3094–104.

6 W.H. Miller, N.C. Handy and J.E. Adams, J. Chem. Phys., 72 (1980) 99–112.
7 E. Kraka, in Encyclopedia of computational chemistry, P.v.R. Schleyer, N.L. Allinger, T. Clark, J. Gasteiger, P.A. Kollman, H.F. Schaefer III and P.R. Schreiner, (Eds.), Wiley, Chichester, 1998, p. 2437–63.
8 P. Pulay, in: D.R. Yarkony (Ed.), Modern electronic structure theory, World Scientific, Singapore, 1995, p. 1191.
9 H.B. Schlegel, in Encyclopedia of computational chemistry, P.v.R. Schleyer, N.L. Allinger, T. Clark, J. Gasteiger, P.A. Kollman, H.F. Schaefer III and P.R. Schreiner, (Eds.), Wiley, Chichester, 1998, pp. 1136–42.
10 H.B. Schlegel, in Encyclopedia of computational chemistry, P.v.R. Schleyer, N.L. Allinger, T. Clark, J. Gasteiger, P.A. Kollman, H.F. Schaefer III, and P.R. Schreiner, (Eds.), Wiley, Chichester, 1998, pp. 2432–37.
11 G. Henkelman, G. Jóhannesson and H. Jónsson, in: S.D. Schwartz (Ed.), Progress on theoretical chemistry and physics, Kluwer, Dordrecht, 2000.
12 F. Jensen, in Encyclopedia of computational chemistry, P.v.R. Schleyer, N.L. Allinger, T. Clark, J. Gasteiger, P.A. Kollman, H.F. Schaefer III, and P.R. Schreiner, (Eds.), Wiley, Chichester, 1998, pp. 3114–23.
13 C.A. Floudas and P.M. Pardalos, State of the art in global optimization: Computational methods and applications, Kluwer, Dordrecht, 1996.
14 C.A. Floudas and P.M. Pardalos, Optimization in computational chemistry and molecular biology: Local and global approaches, Kluwer, Dordrecht, 2000.
15 R. Horst and P.M. Pardalos, Handbook of global optimization, Kluwer, Dordrecht, 1995.
16 R. Horst, P.M. Pardalos and N.V. Thoai, Introduction to global optimization, Kluwer, Dordrecht, 2000.
17 A. Törn and A. Zhilinskas, Global optimization, Springer, Berlin, 1989.
18 M. Ben-Nun and T.J. Martinez, Adv. Chem. Phys., 121 (2002) 439–512.
19 G.A. Voth, J. Phys. Chem. A, 103 (1999) 9383.
20 R. Kosloff, Annu. Rev. Phys. Chem., 45 (1994) 145–178.
21 D. Heidrich, The reaction path in chemistry: Current approaches and perspectives, Kluwer, Dordrecht, 1995.
22 D. Heidrich, W. Kliesch and W. Quapp, Properties of chemically interesting potential energy surfaces, Springer, Berlin, 1991.
23 M.A. Collins, Adv. Chem. Phys., 93 (1996) 389–453.
24 M.L. McKee and M. Page, in: K.B. Lipkowitz, D.B. Boyd (Eds.), Reviews in computational chemistry, VCH, New York, NY, 1993, pp. 35–65.
25 D.J. Wales, Energy landscapes, Cambridge University Press, Cambridge, 2003.
26 J. Simons and J. Nichols, Int. J. Quantum Chem., Suppl. 24 (1990) 263–276.
27 C.E. Dykstra, Ab initio calculation of the structures and properties of molecules, Elsevier, Amsterdam, 1988.
28 P. Jørgensen, J. Simons and North Atlantic Treaty Organization, Scientific Affairs Division, Geometrical derivatives of energy surfaces and molecular properties, Reidel, Dordrecht, 1986.
29 F. Jensen, Introduction to computational chemistry, Wiley, Chichester, 1999.
30 X.S. Li, J.M. Millam, G.E. Scuseria, M.J. Frisch and H.B. Schlegel, J. Chem. Phys., 119 (2003) 7651–7658.
31 K.N. Kudin, G.E. Scuseria and E. Cances, J. Chem. Phys., 116 (2002) 8255–8261.
32 S. Goedecker, Rev. Mod. Phys., 71 (1999) 1085–1123.
33 G.E. Scuseria, J. Phys. Chem. A, 103 (1999) 4782–4790.
34 J.M. Millam and G.E. Scuseria, J. Chem. Phys., 106 (1997) 5569–5577.
35 W.Z. Liang and M. Head-Gordon, J. Chem. Phys., 120 (2004) 10379–10384.
36 Y. Shao, C. Saravanan, M. Head-Gordon and C.A. White, J. Chem. Phys., 118 (2003) 6144–6151.
37 D.R. Bowler, T. Miyazaki and M.J. Gillan, J. Phys.-Condens. Matter, 14 (2002) 2781–2798.
38 M.C. Strain, G.E. Scuseria and M.J. Frisch, Science, 271 (1996) 51–53.
39 P. Pulay, Mol. Phys., 17 (1969) 197–204.
40 K. Thompsen and P. Swanstron, Mol. Phys., 26 (1973) 735.

41 J.A. Pople, R. Krishnan, H.B. Schlegel and J.S. Binkley, Int. J. Quantum Chem. Quantum Chem. Symp., 13 (1979) 225–241.
42 J.D. Augspurger and C.E. Dykstra, J. Phys. Chem., 95 (1991) 9230–9238.
43 P.E. Maslen, D. Jayatilaka, S.M. Colwell, R.D. Amos and N.C. Handy, J. Chem. Phys., 95 (1991) 7409–7417.
44 S.M. Colwell, D. Jayatilaka, P.E. Maslen, R.D. Amos and N.C. Handy, Int. J. Quantum Chem., 40 (1991) 179–199.
45 J.F. Gaw, Y. Yamaguchi, H.F. Schaefer and N.C. Handy, J. Chem. Phys., 85 (1986) 5132–5142.
46 J.F. Gaw, Y. Yamaguchi and H.F. Schaefer, J. Chem. Phys., 81 (1984) 6395–6396.
47 P. Pulay, J. Chem. Phys., 78 (1983) 5043–5051.
48 P. Pulay, G. Fogarasi, F. Pang and J.E. Boggs, J. Am. Chem. Soc., 101 (1979) 2550.
49 P. Pulay and G. Fogarasi, J. Chem. Phys., 96 (1992) 2856–2860.
50 G. Fogarasi, X.F. Zhou, P.W. Taylor and P. Pulay, J. Am. Chem. Soc., 114 (1992) 8191–8201.
51 J. Baker, J. Comput. Chem., 14 (1993) 1085–1100.
52 C.Y. Peng, P.Y. Ayala, H.B. Schlegel and M.J. Frisch, J. Comput. Chem., 17 (1996) 49–56.
53 J. Baker, A. Kessi and B. Delley, J. Chem. Phys., 105 (1996) 192–212.
54 M. von Arnim and R. Ahlrichs, J. Chem. Phys., 111 (1999) 9183–9190.
55 K.N. Kudin, G.E. Scuseria and H.B. Schlegel, J. Chem. Phys., 114 (2001) 2919–2923.
56 K. Nemeth, O. Coulaud, G. Monard and J.G. Angyan, J. Chem. Phys., 113 (2000) 5598–5603.
57 Ö. Farkas and H.B. Schlegel, J. Chem. Phys., 109 (1998) 7100–7104.
58 Ö. Farkas and H.B. Schlegel, J. Chem. Phys., 111 (1999) 10806–10814.
59 B. Paizs, G. Fogarasi and P. Pulay, J. Chem. Phys., 109 (1998) 6571–6576.
60 B. Paizs, J. Baker, S. Suhai and P. Pulay, J. Chem. Phys., 113 (2000) 6566–6572.
61 K. Nemeth, O. Coulaud, G. Monard and J.G. Angyan, J. Chem. Phys., 114 (2001) 9747–9753.
62 S.R. Billeter, A.J. Turner and W. Thiel, Phys. Chem. Chem. Phys., 2 (2000) 2177–2186.
63 J. Baker, D. Kinghorn and P. Pulay, J. Chem. Phys., 110 (1999) 4986–4991.
64 X. Prat-Resina, M. Garcia-Viloca, G. Monard, A. Gonzalez-Lafont, J.M. Lluch, J.M. Bofill and J.M. Anglada, Theor. Chem. Acc., 107 (2002) 147–153.
65 T. Schlick and M. Overton, J. Comput. Chem., 8 (1987) 1025–1039.
66 T. Schlick, B.E. Hingerty, C.S. Peskin, M.L. Overton and S. Broyde, in: D.L. Beveridge, R. Lavery (Eds.), Theoretical biochemistry and molecular biophysics, Adenine Press, Schenectady, NY, 1991, pp. 39–58.
67 P. Derreumaux, G.H. Zhang, T. Schlick and B. Brooks, J. Comput. Chem., 15 (1994) 532–552.
68 R. Fletcher, Practical methods of optimization, Wiley, Chichester, 1987.
69 P.E. Gill, W. Murray and M.H. Wright, Practical optimization, Academic Press, New York, 1981.
70 J.E. Dennis and R.B. Schnabel, Numerical methods for unconstrained optimization and nonlinear equations, Prentice-Hall, Englewood Cliffs, NJ, 1983.
71 L.E. Scales, Introduction to non-linear optimization, Springer, New York, 1985.
72 H.B. Schlegel, in: D.R. Yarkony (Ed.), Modern electronic structure theory, World Scientific, Singapore, 1995, pp. 459–500.
73 D.F. Shanno, Math. Comput., 24 (1970) 647.
74 D. Goldfarb, Math. Comput., 24 (1970) 23.
75 R. Fletcher, Comput. J., 13 (1970) 317.
76 C.G. Broyden, J. Inst. Math. Appl., 6 (1970) 76.
77 M.J. Powell, Math. Program., 15 (1971) 1.
78 B.A. Murtagh and R.W.H. Sargent, Comput. J., 13 (1972) 185.
79 J.M. Bofill, J. Comput. Chem., 15 (1994) 1–11.
80 H.B. Schlegel, in: J. Bertrán, I.G. Csizmadia (Eds.), New theoretical concepts for understanding organic reactions, Kluwer, Dordrecht, 1989, pp. 33–53.
81 E. Besalu and J.M. Bofill, Theor. Chem. Acc., 100 (1998) 265–274.
82 J.M. Anglada and J.M. Bofill, Int. J. Quantum Chem., 62 (1997) 153–165.
83 P. Culot, G. Dive, V.H. Nguyen and J.M. Ghuysen, Theor. Chim. Acta, 82 (1992) 189–205.
84 A. Banerjee, N. Adams, J. Simons and R. Shepard, J. Phys. Chem., 89 (1985) 52–57.

85 T. Helgaker, Chem. Phys. Lett., 182 (1991) 503–510.
86 H.B. Schlegel, J. Comput. Chem., 3 (1982) 214–218.
87 J. Nocedal, Math. Comput., 35 (1980) 773–782.
88 D.C. Liu and J. Nocedal, Math. Program., 45 (1989) 503–528.
89 J.M. Anglada, E. Besalu, J.M. Bofill and J. Rubio, J. Math. Chem., 25 (1999) 85–92.
90 T.H. Fischer and J. Almlof, J. Phys. Chem., 96 (1992) 9768–9774.
91 P. Pulay, J. Comput. Chem., 3 (1982) 556–560.
92 P. Pulay, Chem. Phys. Lett., 73 (1980) 393–398.
93 P. Csaszar and P. Pulay, J. Mol. Struct., 114 (1984) 31–34.
94 Ö. Farkas, PhD (Csc) thesis, Eötvös Loránd University and Hungarian Academy of Sciences, Budapest, 1995.
95 O. Farkas and H.B. Schlegel, Phys. Chem. Chem. Phys., 4 (2002) 11–15.
96 O. Farkas and H.B. Schlegel, J. Mol. Struct. (THEOCHEM), 666 (2003) 31–39.
97 J. Gao, in Encyclopedia of computational chemistry, P.v.R. Schleyer, N.L. Allinger, T. Clark, J. Gasteiger, P.A. Kollman, H.F. Schaefer III, and P.R. Schreiner, (Eds.), Wiley, Chichester, 1998, pp. 1257–1263.
98 J. Gao, in: K.B. Lipkowitz, D.B. Boyd (Eds.), Reviews in computational chemistry, VCH, New York, 1996, pp. 119–185.
99 R.D.J. Froese and K. Morokuma, in Encyclopedia of computational chemistry, P.v.R. Schleyer, N.L. Allinger, T. Clark, J. Gasteiger, P.A. Kollman, H.F. Schaefer III, and P.R. Schreiner, (Eds.), Wiley, Chichester, 1998, pp. 1244–57.
100 K.M. Merz and R.V. Stanton, in Encyclopedia of computational chemistry, P.v.R. Schleyer, N.L. Allinger, T. Clark, J. Gasteiger, P.A. Kollman, H.F. Schaefer III and P.R. Schreiner, (Eds.), Wiley, Chichester, 1998, pp. 2330–2343.
101 J. Tomasi and C.S. Pomelli, in Encyclopedia of computational chemistry, P.v.R. Schleyer, N.L. Allinger, T. Clark, J. Gasteiger, P.A. Kollman, H.F. Schaefer III and P.R. Schreiner, (Eds.), Wiley, Chichester, 1998, pp. 2343–2350.
102 M.F. Ruiz-López and J.L. Rivail, in Encyclopedia of computational chemistry, P.v.R. Schleyer, N.L. Allinger, T. Clark, J. Gasteiger, P.A. Kollman, H.F. Schaefer III and P.R. Schreiner, (Eds.), Wiley, Chichester, 1998, pp. 437–48.
103 G. Monard and K.M. Merz, Acc. Chem. Res., 32 (1999) 904–911.
104 T.Z. Mordasini and W. Thiel, Chimia, 52 (1998) 288–291.
105 M.J. Field, P.A. Bash and M. Karplus, J. Comput. Chem., 11 (1990) 700–733.
106 U.C. Singh and P.A. Kollman, J. Comput. Chem., 7 (1986) 718–730.
107 A. Warshel and M. Levitt, J. Mol. Biol., 103 (1976) 227–249.
108 M. Svensson, S. Humbel, R.D.J. Froese, T. Matsubara, S. Sieber and K. Morokuma, J. Phys. Chem., 100 (1996) 19357–19363.
109 T. Vreven, B. Mennucci, C.O. da Silva, K. Morokuma and J. Tomasi, J. Chem. Phys., 115 (2001) 62–72.
110 S. Dapprich, I. Komaromi, K.S. Byun, K. Morokuma and M.J. Frisch, J. Mol. Struct. (THEOCHEM), 462 (1999) 1–21.
111 A.J. Turner, V. Moliner and I.H. Williams, Phys. Chem. Chem. Phys., 1 (1999) 1323–1331.
112 T. Vreven, K. Morokuma, O. Farkas, H.B. Schlegel and M.J. Frisch, J. Comp. Chem., 24 (2003) 760–769.
113 M. Sierka and J. Sauer, J. Chem. Phys., 112 (2000) 6983–6996.
114 Y.K. Zhang, H.Y. Liu and W.T. Yang, J. Chem. Phys., 112 (2000) 3483–3492.
115 D.R. Yarkony, Rev. Mod. Phys., 68 (1996) 985–1013.
116 D.R. Yarkony, Acc. Chem. Res., 31 (1998) 511–518.
117 D.R. Yarkony, J. Phys. Chem. A, 105 (2001) 6277–6293.
118 D.R. Yarkony, J. Chem. Phys., 92 (1990) 2457–2463.
119 D.R. Yarkony, J. Phys. Chem., 97 (1993) 4407–4412.
120 D.R. Yarkony, J. Phys. Chem. A, 108 (2004) 3200–3205.
121 M.R. Manaa and D.R. Yarkony, J. Chem. Phys., 99 (1993) 5251–5256.
122 J.M. Anglada and J.M. Bofill, J. Comput. Chem., 18 (1997) 992–1003.

123 M.J. Bearpark, M.A. Robb and H.B. Schlegel, Chem. Phys. Lett., 223 (1994) 269–274.
124 M.J. Frisch, G.W. Trucks, H.B. Schlegel, G.E. Scuseria, M.A. Robb, J.R. Cheeseman, J.A.M. Jr., T. Vreven, K.N. Kudin, J.C. Burant, J.M. Millam, S.S. Iyengar, J. Tomasi, V. Barone, B. Mennucci, M. Cossi, G. Scalmani, N. Rega, G.A. Petersson, H. Nakatsuji, M. Hada, M. Ehara, K. Toyota, R. Fukuda, J. Hasegawa, M. Ishida, T. Nakajima, Y. Honda, O. Kitao, H. Nakai, M. Klene, X. Li, J.E. Knox, H.P. Hratchian, J.B. Cross, C. Adamo, J. Jaramillo, R. Gomperts, R.E. Stratmann, O. Yazyev, A.J. Austin, R. Cammi, C. Pomelli, J.W. Ochterski, P.Y. Ayala, K. Morokuma, G.A. Voth, P. Salvador, J.J. Dannenberg, V.G. Zakrzewski, S. Dapprich, A.D. Daniels, M.C. Strain, O. Farkas, D.K. Malick, A.D. Rabuck, K. Raghavachari, J.B. Foresman, J.V. Ortiz, Q. Cui, A.G. Baboul, S. Clifford, J. Cioslowski, B.B. Stefanov, G. Liu, A. Liashenko, P. Piskorz, I. Komaromi, R.L. Martin, D.J. Fox, T. Keith, M.A. Al-Laham, C.Y. Peng, A. Nanayakkara, M. Challacombe, P.M.W. Gill, B. Johnson, W. Chen, M.W. Wong, C. Gonzalez and J.A. Pople, GAUSSIAN 03. Gaussian, Inc., Pittsburgh, PA, 2003.
125 H.B. Schlegel, Int. J. Quantum Chem. Quantum Chem. Symp., 24 (1992) 243–252.
126 When using programs written in FORTRAN these values are generally reported as infinity, 'inf', or not a number, 'nan', in the output.
127 H.B. Schlegel, Theor. Chim. Acta, 66 (1984) 333–340.
128 H.B. Schlegel, J. Comput. Chem., 24 (2003) 1514–1527.
129 J.M. Anglada and J.M. Bofill, J. Comput. Chem., 19 (1998) 349–362.
130 J.M. Bofill, Chem. Phys. Lett., 260 (1996) 359–364.
131 J.M. Bofill and M. Comajuan, J. Comput. Chem., 16 (1995) 1326–1338.
132 J. Simons, P. Jorgensen, H. Taylor and J. Ozment, J. Phys. Chem., 87 (1983) 2745–2753.
133 J.M. Bofill and J.M. Anglada, Theor. Chem. Acc., 105 (2001) 463–472.
134 J. Baker, J. Comput. Chem., 7 (1986) 385–395.
135 C.J. Cerjan and W.H. Miller, J. Chem. Phys., 75 (1981) 2800–2806.
136 J. Nichols, H. Taylor, P. Schmidt and J. Simons, J. Chem. Phys., 92 (1990) 340–346.
137 J. Pancir, Collect. Czech. Chem. Commun., 40 (1975) 1112–1118.
138 C.Y. Peng and H.B. Schlegel, Isr. J. Chem., 33 (1993) 449–454.
139 T. Halgren and W.N. Lipscomb, Chem. Phys. Lett., 49 (1977) 225–232.
140 F. Jensen, J. Chem. Phys., 119 (2003) 8804–8808.
141 F. Jensen and P.O. Norrby, Theor. Chem. Acc., 109 (2003) 1–7.
142 F. Jensen, J. Comput. Chem., 15 (1994) 1199–1216.
143 K. Mueller and L.D. Brown, Theor. Chim. Acta, 53 (1979) 75–93.
144 M.J.S. Dewar, E.F. Healy and J.J.P. Stewart, J. Chem. Soc. Faraday Trans. II, 80 (1984) 227–233.
145 C. Cardenas-Lailhacar and M.C. Zerner, Int. J. Quantum Chem., 55 (1995) 429–439.
146 I.V. Ionova and E.A. Carter, J. Chem. Phys., 98 (1993) 6377–6386.
147 I.V. Ionova and E.A. Carter, J. Chem. Phys., 103 (1995) 5437–5441.
148 S. Bell and J.S. Crighton, J. Chem. Phys., 80 (1984) 2464–2475.
149 A. Jensen, Theor. Chim. Acta, 63 (1983) 269–290.
150 J.M. Anglada, E. Besalu, J.M. Bofill and R. Crehuet, J. Comput. Chem., 20 (1999) 1112–1129.
151 F. Bernardi, J.J.W. McDouall and M.A. Robb, J. Comput. Chem., 8 (1987) 296–306.
152 J.J.W. McDouall, M.A. Robb and F. Bernardi, Chem. Phys. Lett., 129 (1986) 595–602.
153 I.H. Williams and G.M. Maggiora, J. Mol. Struct. (THEOCHEM), 6 (1982) 365–378.
154 M.J. Rothman and L.L. Lohr, Chem. Phys. Lett., 70 (1980) 405–409.
155 P. Scharfenberg, Chem. Phys. Lett., 79 (1981) 115–117.
156 W. Quapp, Chem. Phys. Lett., 253 (1996) 286–292.
157 W. Quapp, M. Hirsch and D. Heidrich, Theor. Chem. Acc., 105 (2000) 145–155.
158 W. Quapp, M. Hirsch, O. Imig and D. Heidrich, J. Comput. Chem., 19 (1998) 1087–1100.
159 W. Quapp, Comput. Math. Appl., 41 (2001) 407–414.
160 M. Hirsch and W. Quapp, J. Comput. Chem., 23 (2002) 887–894.
161 R. Crehuet, J.M. Bofill and J.M. Anglada, Theor. Chem. Acc., 107 (2002) 130–139.
162 K. Fukui, Acc. Chem. Res., 14 (1981) 363–368.
163 L.R. Pratt, J. Chem. Phys., 85 (1986) 5045–5048.
164 R. Elber and M. Karplus, Chem. Phys. Lett., 139 (1987) 375–380.

165 G. Henkelman and H. Jónsson, J. Chem. Phys., 113 (2000) 9978–9985.
166 P.Y. Ayala and H.B. Schlegel, J. Chem. Phys., 107 (1997) 375–384.
167 B. Peters, A. Heyden, A.T. Bell and A. Chakraborty, J. Chem. Phys., 120 (2004) 7877–7886.
168 C. Choi and R. Elber, J. Chem. Phys., 94 (1991) 751–760.
169 R. Czerminski and R. Elber, Int. J. Quantum Chem., (1990) 167–186.
170 R. Czerminski and R. Elber, J. Chem. Phys., 92 (1990) 5580–5601.
171 A. Ulitsky and R. Elber, J. Chem. Phys., 92 (1990) 1510–1511.
172 O.S. Smart, Chem. Phys. Lett., 222 (1994) 503–512.
173 E.M. Sevick, A.T. Bell and D.N. Theodorou, J. Chem. Phys., 98 (1993) 3196–3212.
174 R.E. Gillilan and K.R. Wilson, J. Chem. Phys., 97 (1992) 1757–1772.
175 T.L. Beck, J.D. Doll and D.L. Freeman, J. Chem. Phys., 90 (1989) 3181–3191.
176 G. Henkelman, B.P. Uberuaga and H. Jonsson, J. Chem. Phys., 113 (2000) 9901–9904.
177 C. Gonzalez and H.B. Schlegel, J. Chem. Phys., 90 (1989) 2154–2161.
178 C. Gonzalez and H.B. Schlegel, J. Phys. Chem., 94 (1990) 5523–5527.
179 H. Jónsson, G. Mills and W. Jacobsen, in: B.J. Berne, G. Cicotti, D.F. Coker (Eds.), Classical and quantum dynamics in condensed phase simulations, World Scientific, Singapore, 1998, p. 385.
180 P. Maragakis, S.A. Andreev, Y. Brumer, D.R. Reichman and E. Kaxiras, J. Chem. Phys., 117 (2002) 4651–4658.
181 D.J. Wales, Mol. Phys., 100 (2002) 3285–3305.
182 S.A. Trygubenko and D.J. Wales, J. Chem. Phys., 120 (2004) 2082–2094.
183 J.W. Chu, B.L. Trout and B.R. Brooks, J. Chem. Phys., 119 (2003) 12708–12717.
184 H.L. Woodcock, M. Hodoscek, P. Sherwood, Y.S. Lee, H.F. Schaefer and B.R. Brooks, Theor. Chem. Acc., 109 (2003) 140–148.
185 Y. Abashkin and N. Russo, J. Chem. Phys., 100 (1994) 4477–4483.
186 E. Weinan, W.Q. Ren and E. Vanden-Eijnden, Phys. Rev. B, 66 (2002).
187 B.C. Garrett and D.G. Truhlar, J. Phys. Chem., 83 (1979) 1079–1112.
188 A.D. Isaacson and D.G. Truhlar, J. Phys., 76 (1982) 1380–1381.
189 I. Shavitt, J. Chem. Phys., 49 (1968) 4048–4056.
190 D.G. Truhlar and A. Kuppermann, J. Am. Chem. Soc., 93 (1971) 1840–1851.
191 D.G. Truhlar and A. Kuppermann, J. Chem. Phys., 56 (1972) 2232–2252.
192 M.V. Basilevsky and A.G. Shamov, Chem. Phys., 60 (1981) 347–358.
193 D.K. Hoffman, R.S. Nord and K. Ruedenberg, Theor. Chim. Acta, 69 (1986) 265–279.
194 P. Jorgensen, H.J.A. Jensen and T. Helgaker, Theor. Chim. Acta, 73 (1988) 55–65.
195 H.B. Schlegel, Theor. Chim. Acta, 83 (1992) 15–20.
196 J.Q. Sun and K. Ruedenberg, J. Chem. Phys., 98 (1993) 9707–9714.
197 P.G. Bolhuis, C. Dellago and D. Chandler, Faraday Discuss., (1998) 421–436.
198 P.G. Bolhuis, C. Dellago, P.L. Geissler and D. Chandler, J. Phys.-Condens. Matter, 12 (2000) A147–A152.
199 B.R. Brooks, Y.S. Lee and T.E. Cheatham, Abstr. Pap. Am. Chem. Soc., 222 (2001) 63-COMP.
200 G.E. Crooks and D. Chandler, Phys. Rev. E, 6402 (2001) 026109–026112.
201 C. Dellago, P.G. Bolhuis and D. Chandler, J. Chem. Phys., 110 (1999) 6617–6625.
202 C. Dellago, P.G. Bolhuis, F.S. Csajka and D. Chandler, J. Chem. Phys., 108 (1998) 1964–1977.
203 A. Michalak and T. Ziegler, J. Phys. Chem. A, 105 (2001) 4333–4343.
204 L. Rosso, P. Minary, Z.W. Zhu and M.E. Tuckerman, J. Chem. Phys., 116 (2002) 4389–4402.
205 L. Rosso and M.E. Tuckerman, Mol. Simul., 28 (2002) 91–112.
206 C.W. Gear, Numerical initial value problems in ordinary differential equations, Prentice-Hall, Englewood Cliffs, NJ, 1971.
207 K.K. Baldridge, M.S. Gordon, R. Steckler and D.G. Truhlar, J. Phys. Chem., 93 (1989) 5107–5119.
208 B.C. Garrett, M.J. Redmon, R. Steckler, D.G. Truhlar, K.K. Baldridge, D. Bartol, M.W. Schidt and M.S. Gordon, J. Phys. Chem., 92 (1988) 1476–1488.
209 C. Gonzalez and H.B. Schlegel, J. Chem. Phys., 95 (1991) 5853–5860.
210 H.P. Hratchian and H.B. Schlegel, J. Chem. Phys., 120 (2004) 9918–9924.
211 H.P. Hratchian and H.B. Schlegel, J. Chem. Theory Comp. 1 (2004) 61–69.

212 K. Ishida, K. Morokuma and A. Komornicki, J. Chem. Phys., 66 (1977) 2153–2156.
213 K. Müller and L.D. Brown, Theor. Chim. Acta, 53 (1979) 75–93.
214 M. Page, C. Doubleday and J.W. McIver, J. Chem. Phys., 93 (1990) 5634–5642.
215 M. Page and J.M. McIver, J. Chem. Phys., 88 (1988) 922–935.
216 M.W. Schmidt, M.S. Gordon and M. Dupuis, J. Am. Chem. Soc., 107 (1985) 2585–2589.
217 J.Q. Sun and K. Ruedenberg, J. Chem. Phys., 99 (1993) 5269–5275.
218 M.S. Gordon, G. Chaban and T. Taketsugu, J. Phys. Chem., 100 (1996) 11512–11525.
219 H.P. Hratchian and H.B. Schlegel, J. Phys. Chem. A, 106 (2002) 165–169.
220 S.A. Maluendes and M.J. Dupuis, J. Chem. Phys., 93 (1990) 5902–5911.
221 J.J.P. Stewart, L.P. Davis and L.W. Burggraf, J. Comput. Chem., 8 (1987) 1117–1123.
222 V. Bakken, J.M. Millam and H.B. Schlegel, J. Chem. Phys., 111 (1999) 8773–8777.
223 J.M. Millam, V. Bakken, W. Chen, W.L. Hase and H.B. Schlegel, J. Chem. Phys., 111 (1999) 3800–3805.
224 R. Bulirsch and J. Stoer, Num. Math., 6 (1964) 413–427.
225 R. Bulirsch and J. Stoer, Num. Math., 8 (1966) 1–13.
226 R. Bulirsch and J. Stoer, Num. Math., 8 (1966) 93–104.
227 W.H. Press, Numerical recipes in FORTRAN 77: The art of scientific computing, Cambridge University Press, Cambridge, 1996.
228 R.P.A. Bettens and M.A. Collins, J. Chem. Phys., 111 (1999) 816–826.
229 M.A. Collins, Theor. Chem. Acc., 108 (2002) 313–324.
230 J. Ischtwan and M.A. Collins, J. Chem. Phys., 100 (1994) 8080–8088.
231 K.C. Thompson, M.J.T. Jordan and M.A. Collins, J. Chem. Phys., 108 (1998) 564–578.
232 F. Eckert and H.J. Werner, Theor. Chem. Acc., 100 (1998) 21–30.
233 V.S. Melissas, D.G. Truhlar and B.C. Garrett, J. Chem. Phys., 96 (1992) 5758–5772.
234 A.G. Baboul and H.B. Schlegel, J. Chem. Phys., 107 (1997) 9413–9417.
235 E.B. Wilson, J.C. Decius and P.C. Cross, The theory of infrared and Raman vibrational spectra, McGraw-Hill, New York, 1955.
236 P. Valtazanos and K. Ruedenberg, Theor. Chim. Acta, 69 (1986) 281–307.
237 J. Baker and P.M.W. Gill, J. Comput. Chem., 9 (1988) 465–475.
238 V. Bakken, D. Danovich, S. Shaik and H.B. Schlegel, J. Am. Chem. Soc., 123 (2001) 130–134.
239 M.V. Basilevsky, Theor. Chim. Acta, 72 (1987) 63–67.
240 Y. Kumeda and T. Taketsugu, J. Chem. Phys., 113 (2000) 477–484.
241 B. Lasorne, G. Dive, D. Lauvergnat and M. Desouter-Lecomte, J. Chem. Phys., 118 (2003) 5831–5840.
242 B. Peters, W.Z. Liang, A.T. Bell and A. Chakraborty, J. Chem. Phys., 118 (2003) 9533–9541.
243 W. Quapp, J. Theor. Comput. Chem., 2 (2003) 385–417.
244 W. Quapp, J. Comput. Chem., 25 (2004) 1277–1285.
245 W. Quapp, J. Mol. Struct., 695 (2004) 95–101.
246 W. Quapp, M. Hirsch and D. Heidrich, Theor. Chem. Acc., 100 (1998) 285–299.
247 M.N. Ramquet, G. Dive and D. Dehareng, J. Chem. Phys., 112 (2000) 4923–4934.
248 T. Taketsugu and T. Hirano, J. Mol. Struct. (THEOCHEM), 116 (1994) 169–176.
249 T. Taketsugu and Y. Kumeda, J. Chem. Phys., 114 (2001) 6973–6982.
250 T. Taketsugu, N. Tajima and K. Hirao, J. Chem. Phys., 105 (1996) 1933–1939.
251 T. Yanai, T. Taketsugu and K. Hirao, J. Chem. Phys., 107 (1997) 1137–1146.
252 A. Andersen and E.A. Carter, J. Phys. Chem. A, 107 (2003) 9463–9478.
253 A. Andersen and E.A. Carter, Isr. J. Chem., 42 (2002) 245–260.

Theory and Applications of Computational Chemistry: The First Forty Years
Edited by C. Dykstra et al.

CHAPTER 11

Progress in the quantum description of vibrational motion of polyatomic molecules

Joel M. Bowman[1], Stuart Carter[1,2] and Nicholas C. Handy[3]

[1] *Cherry L. Emerson Center for Scientific Computation and Department of Chemistry, Emory University, Atlanta, GA 30322, USA*
[2] *Department of Chemistry, University of Reading, Reading RG6 2AD, UK*
[3] *Department of Chemistry, University of Cambridge, Cambridge CB2 1EW, UK*

Abstract

We present a review of progress in the quantum description of the vibrational motion of polyatomic molecules, with a focus on developments made in the code 'MULTIMODE' (MM). Following a brief historical perspective of the use of normal modes in vibrational analysis we show how this approach has been extended to provide a computationally powerful approach to accurately describe the vibrations of fairly large molecules. The approach is based on an exact Hamiltonian in normal coordinates, a general *n*-mode representation of the full potential, and the implementation of Vibrational Self-Consistent Field/Configuration Interaction methods. Several case studies are presented based on what we refer to as the Single-Reference (SR) version of MM. Following that we describe a major generalization of MM which blends the key aspects of MM-SR with the Reaction Path Hamiltonian. This version is illustrated for the fluxional complex $(OH^-)H_2O$. We conclude with a critical examination of the current bottlenecks in the computational approach and prospects for future progress.

11.1 INTRODUCTION

Chemistry students learn the theory and technique of normal mode analysis of molecules in introductory courses in quantum chemistry or spectroscopy. Perhaps the simplest way to describe this approach is to start with a Taylor series expansion of the potential about a stationary point, i.e. where the first derivatives vanish. In this case the lowest order of the coordinate dependence of the potential appears in second order. Truncating the series at this order still leads to coupling (called bi-linear) in pairs of coordinates. Thus there is coupling even at the lowest order. The good news, however, is that this bi-linear coupling can be eliminated by a linear transformation of the original coordinates to a new set of mass-weighted coordinates, called *normal coordinates*. Doing this uncoupling transformation leads to a description of the motion as independent harmonic oscillators, which has a simple, analytical solution.

This basic approach, elaborated brilliantly in the classic book by Wilson et al. [1], was the only method used to perform vibrational analysis, both theoretically and experimentally during much of the 20th century. The approach is still used today in most of the popular electronic structure codes [2,3] and also has become a valuable tool in biomolecular chemistry [4].

11.2 BEYOND THE HARMONIC APPROXIMATION

Part of the reason for the longevity of the normal mode analysis is that it is quite difficult to go beyond it. However, there has been great progress in doing just that and we will review some of that progress in this chapter. We acknowledge that this is still a very active field and is very much a 'work in progress'. The most straightforward approach beyond the harmonic, normal mode approximation is to continue the expansion of the potential beyond second-order terms. Broadly speaking, there are two approaches that go beyond the harmonic approximation; one is based on Perturbation Theory (PT) and the other on Variational Theory (VT). Historically PT was developed first [5,6] and was widely used by experimentalists in the development of so-called *Spectroscopic Hamiltonians* [7]. PT has been elegantly and powerfully developed further recently [8–10] and is currently a very useful tool in computational chemistry [11–13].

We will focus mainly on Variational Theory in this chapter and use the Vibrational Self-Consistent Field (VSCF) theory as the starting point for this. This theory, however, also lends itself to a very workable form of PT, which we will briefly review as well. A much more detailed account of the PT approach is found in the chapter by Gerber and co-workers.

In our approach, we use the *Watson Hamiltonian* [14]. For the general case of non-linear molecules this Hamiltonian is given (in atomic units) by

$$H = \frac{1}{2}\sum_{\alpha\beta}(\hat{J}_\alpha - \hat{\pi}_\alpha)\mu_{\alpha\beta}(\hat{J}_\beta - \hat{\pi}_\beta) - \frac{1}{2}\sum_k \frac{\partial^2}{\partial Q_k^2} - \frac{1}{8}\sum_\alpha \mu_{\alpha\alpha} + V(\mathbf{Q}) \tag{1}$$

where $\hat{\mathbf{J}}$ is the total nuclear angular momentum, $\hat{\pi}$ is the so-called *vibrational angular momentum*, and μ is the inverse of the effective moment of inertia tensor and V is the potential. For simplicity consider the case of zero angular momentum and expanding the potential in a multinomial series beyond second order we get

$$H \equiv -\frac{1}{2}\left[\sum_k \frac{\partial^2}{\partial Q_k^2} + \omega_k^2 Q_k^2\right] + T_c + \sum_{ijk} F_{ijk}Q_iQ_jQ_k + \sum_{ijkl} G_{ijkl}Q_iQ_jQ_kQ_l + \cdots \tag{2}$$

where

$$T_{\mathrm{c}} \equiv \frac{1}{2}\sum_{\alpha\beta}\hat{\pi}_{\alpha}\mu_{\alpha\beta}\hat{\pi}_{\beta} - \frac{1}{8}\sum_{\alpha}\mu_{\alpha\alpha} \tag{3}$$

The Hamiltonian contains coupling among the normal modes and to obtain eigenvalues and eigenfunctions requires either an approach based on PT or VT.

In either the VT or PT approach it is convenient to write H as

$$H \equiv H^{(0)} + T_{\mathrm{c}} + \sum_{ijk} F_{ijk}Q_iQ_jQ_k + \sum_{ijk} G_{ijkl}Q_iQ_jQ_kQ_l + \cdots \tag{4}$$

where $H^{(0)}$ is just the first two (separable) terms in the right hand side of Eq. (2). Eigenfunctions of $H^{(0)}$ are simply direct products of harmonic-oscillator eigenfunctions and these can be used as a basis for a representation of the full Hamiltonian H. Matrix elements of H are analytical in this basis and this facilitates variational calculations as well as perturbation ones. Several codes making use of the above representation of the Hamiltonian have been developed. Examples based on VT are 'POLYMODE' [15] and 'ANHAR' [16].

Another approach to obtain approximate eigenvalues and eigenfunctions of H is the VSCF method [17–19]. In this approach a trial wavefunction is given by a simple Hartree product

$$\Psi^{\mathrm{VSCF}}_{n_1,n_2,\ldots,n_N}(Q_1, Q_2, \ldots Q_N) = \prod_{i=1}^{N}\phi^{(i)}_{n_i}(Q_i) \tag{5}$$

and the optimum modal wavefunctions $\phi^{(i)}_{n_i}(Q_i)$ are obtained from the solution of the VSCF equations for $J = 0$

$$\left[T_l + \langle\prod_{i\neq l}^{N}\phi^{(i)}_{n_i}|V + T_{\mathrm{c}}|\prod_{i\neq l}^{N}\phi^{(i)}_{n_i}\rangle - \epsilon^{(l)}_{n_l}\right]\phi^{(l)}_{n_l}(Q_l) = 0, \quad l = 1, N \tag{6}$$

where T_l denotes the kinetic energy operator $-(1/2)(\partial^2/\partial Q_l^2)$ and the integration is over the coordinates of $N - 1$ modals. These coupled equations are typically solved iteratively and for low-lying reference states convergence procedure is usually rapid and stable. If the modal wavefunctions are expanded in terms of a primitive harmonic-oscillator basis, the resulting procedure is analogous to the Hartree–Fock–Roothaan–Hall method in electronic structure calculations. In the 1978 formulation of the VSCF approach the coupled equations were solved on a grid [17]. Once convergence is achieved a VSCF Hamiltonian can be defined as

$$H^{\mathrm{VSCF}}_{n_1,n_2,\ldots n_N} \equiv \sum_{l=1}\left[T_l + \langle\prod_{i\neq l}^{N}\phi^{(i)}_{n_i}|V + T_{\mathrm{c}}|\prod_{i\neq l}^{N}\phi^{(i)}_{n_i}\rangle\right] \tag{7}$$

and the eigenfunctions of this Hamiltonian form an orthonormal set. The eigenfunction $\Psi^{\mathrm{VSCF}}_{n_1,n_2,\ldots,n_N}(Q_1, Q_2, \ldots Q_N)$ is the VSCF state and all other eigenfunctions are termed *virtual states*, which we discuss further below.

As noted the integration over the potential in the VSCF equations is $N - 1$ dimensional and so it is clear that if N is larger than 4 or so the integration becomes extremely computer intensive. Further it is usually necessary to go beyond the VSCF description to obtain accurate energies. Gerber and co-workers use second-order perturbation theory to correct VSCF energies [20,21]. This approach uses the virtual states defined above to correct the VSCF energies. The first-order correction vanishes since VSCF is correct to first order, and the second-order correction is

$$\Delta E_n^{(2)} = \sum_{n'} \frac{|\langle n'|\Delta V|n\rangle^2|}{(E_n^{\text{VSCF}} - E_{n'}^{\text{VS}})}$$

where E_n^{VSCF} the (zero-order) VSCF energy and $E_{n'}^{\text{VS}}$ is a corresponding virtual state energy and $\langle n'|$ and $|n\rangle$ are the corresponding bra and kets.

This theory is analogous to second-order Möller–Plesset theory of electronic structure theory. Gerber and co-workers term the resulting theory 'Correlation-corrected VSCF' or 'cc-VSCF'. Christiansen refers to the method as 'VMP' [22]. As noted already Gerber and co-workers use the 2-mode representation of the potential and so at most two-dimensional integrals have to be done in the approach. The use of second-order perturbation theory together with the 2-mode representation of the potential (see below for more details) renders this method very efficient and thus feasible to apply to large molecules [23]. Also it is important to note that the approach taken by Gerber and co-workers, although formulated in terms of normal coordinates, neglects the vibrational angular momentum and Watson terms and also is restricted to $J = 0$. A recent critical examination of convergence properties of vibrational MPn theory has been given by Christensen [22].

11.3 VIBRATIONAL CI THEORY

Another approach to go beyond the VSCF approximation is the Configuration Interaction (CI) variational approach. There are two schemes that can be used. In one, which we have denoted 'VSCF + CI', the many-mode wavefunction is expanded in a basis of VSCF states. These states are not orthogonal and this results in a generalized eigenvalue problem, which can be solved by standard methods. In the other scheme, denoted VCI, the many-mode wavefunction is expanded in terms of the virtual states of a given VSCF Hamiltonian $H_{n_1,n_2,\ldots n_N}^{\text{VSCF}}$. Usually the ground state VSCF Hamiltonian is used. This basis is orthonormal and results in a standard eigenvalue problem. Both approaches were proposed and applied to model problems, 20 years ago [24]. The method was incorporated into the code POLYMODE [15] and more recently into the code 'MULTIMODE' (MM) which we describe in detail here.

In the VSCF theory as well as the MP2 and CI methods, matrix elements of the Hamiltonian are required. For example, matrix elements of the potential occur in all methods. These elements are the most demanding ones in general. They require high-dimensional quadratures and for more than six degrees of freedom their straightforward evaluation can become prohibitively demanding.

One way to deal with this, which was exploited explicitly in POLYMODE, for example, is to expand the potential as in Eq. (2). In this case the matrix elements are simply products of one-dimensional integrals. This is certainly a viable approach; however, it does have some limitations, especially for larger molecules. First, it was recognized that this expansion converges slowly for variational approaches. That is, often terms to sixth order or higher are needed to obtain well-converged eigenvalues, even for low-lying ones. Second, evaluating the coefficients for the higher-order terms becomes problematic if one is using an *ab initio* approach because these coefficients are high-order derivatives of the potential evaluated at the reference stationary point. Third, the number of terms in the expansion is a rapidly increasing function of the number of degrees of freedom and so its utility becomes very limited for larger molecules.

11.3.1 The *n*-mode representation of the potential

If the multinomial expansion of the potential is abandoned, then in both the VT and PT approaches a major bottleneck is the evaluation of matrix elements. This severely limits the applicability of these methods to molecules beyond tetraatomics.

Recently, another representation of the potential has been proposed that extends the applicability of the VT and PT approaches to larger molecules. One way to think about this new representation is to sum the terms in the expansion given by Eq. (2) in a different way. Suppose we sum *all* terms involving two modes, three modes, four modes, etc. to infinite order. Then the potential for N modes would be given by the following *n-mode representation* [25]

$$V(Q_1,\cdots Q_N) = \sum_i V^{(i)}(Q_i) + \sum_{i\neq j} V^{(2)}(Q_i,Q_j) + \sum_{i\neq j\neq k} V^{(3)}(Q_i,Q_j,Q_k) + \cdots + \sum_{i\neq j\neq k,\ldots} V^{(n)}(Q_i,\ldots) \tag{8}$$

where $V_i^{(1)}(Q_i)$ terms are cuts through the hyperspace of normal coordinates with just one coordinate varying at a time, $V_{ij}^{(2)}(Q_i,Q_j)$ is an *intrinsic* 2-mode potential where only pairs of normal coordinates are non-zero, etc. Each summation in Eq. (8) is the overall combination of modes, thus there are $N!/[n!(N-n)!]$ terms in each sum. The above representation, limited to 2-mode coupling, was introduced by Jung and Gerber [20].

Obviously if n equals N there is no point in doing this expansion, which is exact in this case, and which offers no advantages over dealing directly with the full potential. Also note that these *intrinsic* potentials are not written as multinomial expansions in the coordinates and thus the matrix elements of the potential need to be done numerically. The advantage of using the above representation is that the dimensionality of the numerical quadratures is equal to n. However, the expectation is that accurate results can be obtained for n substantially less than N and thus matrix elements of the potential can be done much more efficiently. To summarize the state of progress with respect to the potential: the n-mode representation of the potential given by Eq. (8) has replaced the earlier expansion of the potential as a more general and robust approach for vibrational calculations. For example, in the case of a triatomic molecule, the 3-mode representation of the potential is exact.

Next consider how the methods used to treat vibrational motion exploit the n-mode representation of the potential. Clearly there can be a great simplification in these methods if n is small. We will examine the accuracy of this approach in the next section.

In the latest version of the code MULTIMODE [26] n can vary from 1 to 6. Having the option of a varying n-mode representation of the potential is very useful because it permits a systematic way to increase the accuracy of the calculation and also to monitor the convergence of results with respect to n. An important point to be made about the n-mode representation is that it does place limitations on the space of excitations in both MP2 theory and VT theory. To see this, consider the 2-mode representation of the potential and for simplicity for a triatomic molecule. General matrix elements of the potential are given by

$$\langle n'_1 n'_2 n'_3 | V | n_1 n_2 n_3 \rangle = \sum_{i,j,k=1}^{3} \langle n'_i | V^{(1)}(i) | n_i \rangle \delta_{n'_j n_j} \delta_{n'_k n_k} + \langle n'_1 n'_2 | V^{(2)}(1,2) | n_1 n_2 \rangle \delta_{n'_3 n_3}$$
$$+ \langle n'_1 n'_3 | V^{(2)}(1,3) | n_1 n_3 \rangle \delta_{n'_2 n_2} + \langle n'_2 n'_3 | V^{(2)}(2,3) | n_2 n_3 \rangle \delta_{n'_1 n_1} \quad (9)$$

Thus, if the primed numbers are all different from the unprimed ones this matrix element vanishes. So, the 2-mode representation limits excitations to pairwise ones. So, for example, a 'Fermi' resonance among three modes cannot be described with MP2 theory with a 2-mode representation of the potential. It is described in the lowest order by a 3-mode representation of the potential. Analogous limitations apply to any value of n less than N.

The VCI approach can quickly lead to very large Hamiltonian matrices, and perhaps it is worth noting that a 'benefit' of using a value of n much less than N in the VCI approach is that the Hamiltonian matrix is very sparse. This has not been exploited yet, and we return to this point in the last section. The major strategy to contain the size of the H-matrix in MM is based on a CI-selection scheme that depends on the type of mode excitation. The basic strategy is to restrict the sum of excitations involving two modes, three modes, etc. Details are given elsewhere [27] but the effect of the selection scheme is to have a more complete basis of 2-mode excitations than 3-mode excitations, etc. This strategy is consistent with the assumption that the importance of intrinsic n-mode terms in nMR representation of the potential decreases with n.

Before presenting results of some illustrative calculations using the n-mode representation, there are several important technical points regarding these n-mode potentials that are worth noting. In addition to greatly simplifying the calculation of potential matrix elements, this representation of the potential can result in great savings in doing *ab initio* calculations of the potential. One obvious saving occurs in doing 'direct-dynamics', i.e. calculating the potential directly at the configurations used in doing matrix element calculations. For example, consider a five-atom molecule, e.g. CH_4, where the potential depends on nine coordinates. Suppose that a modest quadrature grid of 10 points per degree of freedom is sufficient to do potential matrix elements. This would mean that 10^9 *ab initio* electronic energies would be required for this grid.

However, if a 3-mode representation of the potential were sufficiently accurate this would require $9!/(6!3!) \times 10^3 = 8.4 \times 10^4$ *ab initio* electronic energies. This is a dramatic reduction relative to 10^9, however, still a very large, barely feasible number of energies. For a 2-mode representation of the potential only 360 points are needed, which is a very feasible calculation. Even using a grid-based approach to obtain the VSCF and virtual eigenfunctions, as done by Gerber and co-workers, where the number of points would be perhaps 10 times larger the approach is still quite feasible.

Several variants of the 'direct-dynamics' approach have been coded in MM. Because a 3-mode and 4-mode representation of the potential is typically used in that code, the direct evaluation of *ab initio* energies on the 3 or 4-mode grids is not feasible, as the above example for CH_4 illustrates. Instead we use interpolation approaches on sparse grids. These methods have been described in detail elsewhere and so we only give a brief description here. In one approach local interpolation is used to obtain energies from relatively sparse grids where energies and gradients are known [28]. (The interpolation is termed 'reduced' Hermite interpolation.) In the second approach [29], least squares fitting is done for the n-mode (sparse) grids and then used to obtain approximations of the potential at quadrature points. (An added benefit of this approach is that the least-squares basis is a direct-product in the normal coordinates and so the matrix elements are products of 1d elements, as in the case of a multinomial expansion of the potential.)

As an aside, we note that the n-mode representation of the potential can be made in any set of coordinates and for use in the corresponding Hamiltonian. For triatomic and tetraatomic molecules it is possible to avoid this representation because the dimensionality of the internal space is 3 or 6 and exact treatments are possible in both normal [30,31] and various curvilinear coordinates which have been used in exact formulations and calculations for triatomic and tetraatomic molecules [32–34]. However, for larger molecules it could prove quite useful.

A summary of the points made above is shown graphically in Fig. 11.1. The figure emphasizes that accuracy depends on both the level of treatment and the value of n in the n-mode representation. Finally, because the above theory is based on a set of normal coordinates defined with respect to a single stationary point we refer hereafter to the version of MM based on this approach as MM-SR.

11.3.2 Results of selected calculations

In this section we present a small survey of calculations that illustrate the points discussed above. First, we examine the dependence of vibrational energies on the n-mode representation of the potential. The first example we consider is for H_2CN. This is a simple example of a 6-mode system, with a single minimum and localized states. This work was reported some time ago and made use of an *ab initio*-based force field [35]. Here we focus on both the accuracy of the VSCF method and the n-mode representation of the potential for fairly low-lying states. 'Exact' variational results were obtained for this molecule and reported as well. In Fig. 11.2 we show the error relative to these exact calculations of VSCF energies for 1–4 mode representations of the potential. As seen the 1MR results are very inaccurate, the 2MR are a very significant improvement and for

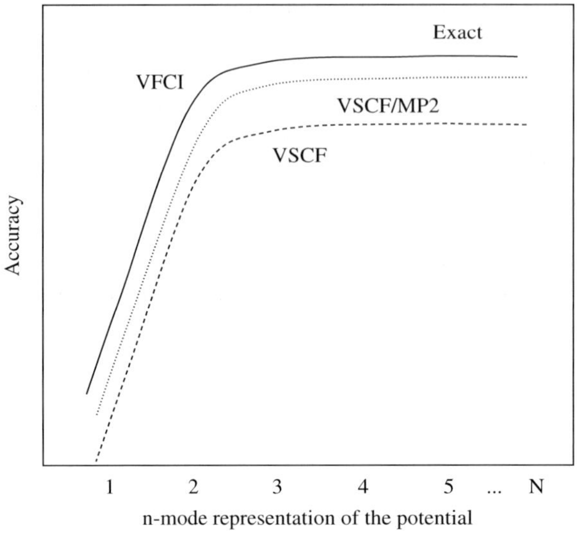

Fig. 11.1. Schematic of the accuracy of the n-mode representation of the potential and the method of treatment of the vibrational dynamics.

most states quite close to the 3 and 4MR energies, which are virtually identical. There are, however, two states for which the 2MR is substantially different from the 3 and 4MR results. Finally, note that the errors in the converged VSCF energies, relative to VCI ones, are in range of 10–40 cm^{-1}, which is typical for this type of system.

Consider next 9-mode CH_4. Several calculations have been reported for methane using MULTIMODE [36–37], including a very recent one for values of the total angular

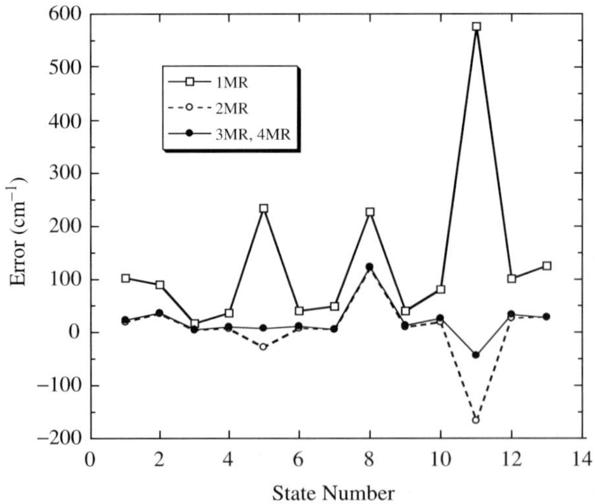

Fig. 11.2. Comparison of the errors in Vibrational SCF energies for H_2CN for the indicated n-mode representation of the potential. The 4MR results are essentially exact.

momentum up to 50 [38]. (These calculations made use of several different potentials. The ones in Refs. [36] and [37] were based on *ab initio* force fields in symmetry-adapted valence coordinates, whereas the one in Ref. [38] used a more global potential that is part of a reactive potential energy surface for $H + CH_4$.) This is clearly a challenging problem for 2, 3, or 4-mode representations of the potential. In Table 11.1 we show the convergence of low-lying states with respect to these levels of representation. We also include the harmonic normal-mode energies and results of an independent 'exact' calculation using the same potential energy. First, note that the 2MR of the potential generally produces energies that are quite close to the 3MR and 4MR ones. However, there are some exceptions (as seen in the previous example), where errors are of the order of 40–50 cm^{-1}. However, in all cases the 2MR results are a substantial improvement over the harmonic results. The 3MR results are quite close to the 4MR ones indicating good convergence of the eigenvalues with respect to the *n*-mode representation. Note that there is very good agreement with the independent calculations.

The final example in this series is H_3O^+. This is a more challenging system because like NH_3, it is a double-well molecule with a small barrier separating the two minima. The first calculations we did on this molecule [39] used a 'non-spectroscopically' accurate potential and so the results do not compare well with experiment. However, for the present purpose the results are adequate. In these calculations we also compared the 2MR, 3MR, and 4MR representations of the potential, but for two sets of calculations. In one set, one of the two equivalent minima was chosen as the reference, as is conventional. This reference, however, is not a good choice to describe the tunneling splittings, which are large. Thus, a second reference geometry at the (planar) saddle point was considered. Some results of these calculations are given in Table 11.2. First consider the zero-point energy (zpe) obtained using the minimum as the reference. The 2MR result is below the 4MR one by nearly 90 cm^{-1} whereas the 3MR result is only 5 cm^{-1} below the 4MR one. So, this indicates that the 2MR is quite inaccurate for the zpe. Also, there is no meaningful tunneling splitting of the zero-point level for any *n*MR for this choice of reference geometry. Looking now at the energies using the saddle point as the reference geometry, we do see reasonable tunneling splittings; however, we see even larger errors in the 2MR results. There are significant, although much smaller differences between the 4MR and 3MR results. Also, it is not clear that the 4MR results are converged, although

Table 11.1 Energies (cm^{-1}) of low-lying states of non-rotating CH_4

State	H.O.	2MR[a]	3MR[a]	4MR[a]	Lanczos[b]
ZPE	9835.0	9693.1	9707.4	9707.2	–
$\nu_4(F_2)$	1344.0	1311.7	1312.9	1313.3	1314.1
$\nu_2(E)$	1570.8	1531.1	1534.4	1534.5	1534.0
$\nu_1(A_1)$	3034.7	2925.7	2948.3	2949.4	2955.8
$2\nu_4$	2688.0	2626.1	2621.6	2623.9	2627.2
$\nu_2 + \nu_4$	2914.8	2881.6	2831.5	2836.4	2838.1
$\nu_3(F_2)$	3153.9	3004.3	3053.7	3053.1	3056.5
$2\nu_2$	3141.6	3067.3	3067.2	3067.3	3069.0

[a]Ref. [36].
[b]H.-G. Yu, J. Chem. Phys., 117 (200) 8190.

Table 11.2 Comparison of VCI zero-point energies and splitting of that energy (cm^{-1}) of H_3O^+ at the C_{3v} global minimum and at D_{3h} saddle point, using 2,3, and 4-mode representations (MR) of the potential

Energy (cm^{-1})	Minimum			Saddle point		
	4MR	3MR	2MR	4MR	3MR	2MR
Zero point energy	7118.9	7123.1	7030.3	7114.1	6967.0	7236.2
Splitting	0	0	0	15.2	72.7	143.0

No splitting exists for C_{3v} calculations, as discussed in the text.

the observation that the two 4MR zpe's are nearly equal suggests that these are near the correct answer. (That the zpe using the minimum reference is above the value using the saddle point reference is expected owing to the tunneling splitting, which would be expected to lower the single-well zpe estimate.) In fact, we showed in subsequent papers [40,41] that 4MR calculations are quite accurate compared with exact results using the specialized code 'RVIB4' [42]. In the latest set of calculations we also reported [26] the 5MR version of MULTIMODE and demonstrated improved accuracy compared to RVIB4 results. These calculations were done for angular momentum J equal to 0 and 1 and employed a much more accurate *ab initio* potential energy surface. (We note that Halonen and co-workers subsequently reported [43] an even more accurate potential energy surface for H_3O^+.)

MULTIMODE has been applied to a substantial body of molecules and complexes including H_2O and HO_2 where J greater than zero was considered and treated exactly and approximately [44], to CO–Cu(100) [25,45], NH_3^+ [46], NH_3 [47,48], benzene [49], Cl^-H_2O [50], $H_5O_2^+$ [51,52], thioformaldehyde [53], furan, pyrrole, and thiophene [54], and *cis* and *trans*-HOCO [55]. The code has also been used to obtain vibrational energies at transition states [56,57], which are usually done in the harmonic approximation. MM-SR has been adapted to obtain IR spectra and applications have been made to H_2O [58], furan, pyrrole, and thiophene [59].

11.3.3 The 'Reaction Path' version of MULTIMODE

The use of the saddle point as the reference geometry in H_3O^+ (and NH_3 [46,47]) has enabled MULTIMODE to provide an accurate treatment of a double minimum problem. However, as we have seen in the case of H_3O^+, a 4MR at least is required to obtain accurate results. For more challenging problems with more than one minimum, or where two minima are separated by a torsional mode, this approach cannot be expected to work as well.

Recently, Carter and Handy have addressed this very effectively by incorporating the approach taken in MULTIMODE into the 'Reaction Path Hamiltonian (RPH)' [60]. In this approach one special, large amplitude mode is singled out and the n-mode coupling idea is applied to the normal modes orthogonal to this mode. The kinetic energy operator is somewhat complex and is given elsewhere [60]. This version of MULTIMODE is denoted MULTIMODE-RPH or abbreviated as MM-RPH.

MM-RPH has been described and tested very successfully on H_2O_2 [61,62], using a high-quality, *ab initio*-based full-dimensional potential energy surface [63]. For this specific application a method very similar to that proposed by Hougen et al. [64] was adopted since we made use of an analytical potential in the six internal coordinates; for a given choice of torsional angle, it was possible to minimize the energy with respect to the remaining coordinates. The potential is represented as

$$V(\tau, Q_1, Q_2, \ldots) = V^{(0)}(\tau) + \sum_i V_i^{(1)}(\tau, Q_i) + \sum_{ij} V_{ij}^{(2)}(\tau, Q_i, Q_j) + \sum_{ijk} V_{ijk}^{(3)}(\tau, Q_i, Q_j, Q_k) + \sum_{ijkl} V_{ijkl}^{(4)}(\tau, Q_i, Q_j, Q_k, Q_l) + \ldots \quad (10)$$

The Cartesian coordinates of the resulting structure were then formed at the center-of-mass and were rotated to obey the Eckart conditions with the previous structure (the path was defined for torsional angles in half-degree intervals). Note that in general this approach does not generate the true reaction path as defined by Miller et al. [60] in that the path does not conform to that of steepest descent; it is expected to be very close to it, however, and their Hamiltonian is still valid. The choice of H_2O_2 was used for the initial test of this RPH version of MULTIMODE, where it was shown to produce results in excellent agreement with previous exact calculations [61,62].

The code has also very recently been applied to glyoxal [65], methanol and malonaldehyde [66]. These examples have all been studied by different variants of MM-RPH. For glyoxal, for which an analytical potential did not exist, the true reaction path was formed by deriving the energies and first derivatives from *ab initio* calculations, starting at the '*cis*' and '*trans*' transition states in turn, from which the path of steepest descent could be found. There are numerical difficulties associated with this approach, however, particularly in finding the true path close to the transition states where the first derivatives are almost zero. This leads to oscillations about the path in the initial searches, resulting in two equilibrium configurations which are not identical, depending on whether the starting point is '*cis*' or '*trans*'. In the initial calculations using this approach [65], the second-derivatives were calculated '*ab initio*' by finite differences to obtain the harmonic frequencies and normal coordinate vectors at each point along the path. The potential in the expression above is therefore truncated at the second term and is not expected to be very accurate. In subsequent calculations [67], partial anharmonicity was introduced by calculating third and semi-diagonal fourth derivatives at each path point.

Calculations on methanol [66] were carried out by *ab initio* techniques using a combination of the techniques adopted for H_2O_2 and glyoxal. The path was defined for one of the three equivalent torsional angles CHOH and the energy was minimized at each half-degree interval prior to the determination of the first and second derivatives. No anharmonicity was included in these calculations, but we now have a full-dimensional potential in all nine coordinates which will be used for more advanced studies. In the latter calculations on glyoxal and methanol, it was found possible to omit the Eckart rotations by modifying the moment-of-inertia tensor from that in the original (RPH). This then removed the ambiguities associated with rotation of the coordinate axes. However, we favor the use of the original steepest descent path coordinate where possible, because this removes coupling between the path and the vibrational modes in an optimum way.

Calculations on malonaldehyde [65] used a totally different path coordinate, where it was almost obvious to choose R1−R2, these distances being those from the migrating proton to the neighboring oxygen atoms. These calculations were in the full 21 dimensions, with all the fundamentals being split due to the tunneling motion (with one exception, namely the one vibration which is centered on the symmetric transition state). This is the first time that such a large study has been undertaken.

Very recently, MM-RPH has been applied to $H_3O_2^-$ (the mono-hydrated hydroxide ion sometimes written as $(OH^-)H_2O$ using a full-dimensional potential energy surface [67]. Recent fascinating, though somewhat puzzling, experiments by Johnson and co-workers [68] using the Ar-messenger technique stimulated us to apply the MM-RPH code to this molecule. The equilibrium structure is shown in Fig. 11.3. The bridging H atom is located roughly collinear with the OO axis with a minimum slightly closer to one O atom than the other. This implies, of course, that there is a double minimum in the O−H−O stretch coordinate. The barrier separating these two minima is roughly 100 cm^{-1} and was first reported by Samson and Klopper [69]. They argued (correctly) that the zero-point motion of the bridging H atom is delocalized over the two minima. The relaxed potential, from the full-dimensional one [67], along this coordinate is also shown in Fig. 11.3. Full-dimensional calculations of the zero-point state done with MM-RPH (see below for more details) as well as Diffusion Monte Carlo by McCoy [70] clearly show that the corresponding wavefunction is delocalized over the two minima. Thus the zero-point averaged structure has C_2-symmetry.

In addition to the delocalization in the OHO-stretch it was subsequently determined that, like H_2O_2, there is fairly facile torsional motion of the two monomer OH-groups [67]. For the application of MM-RPH, we chose a torsional path that passes through

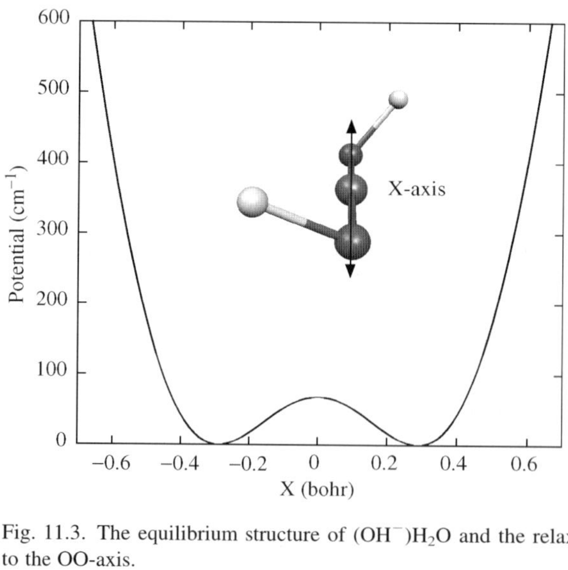

Fig. 11.3. The equilibrium structure of $(OH^-)H_2O$ and the relaxed potential for the H-atom motion parallel to the OO-axis.

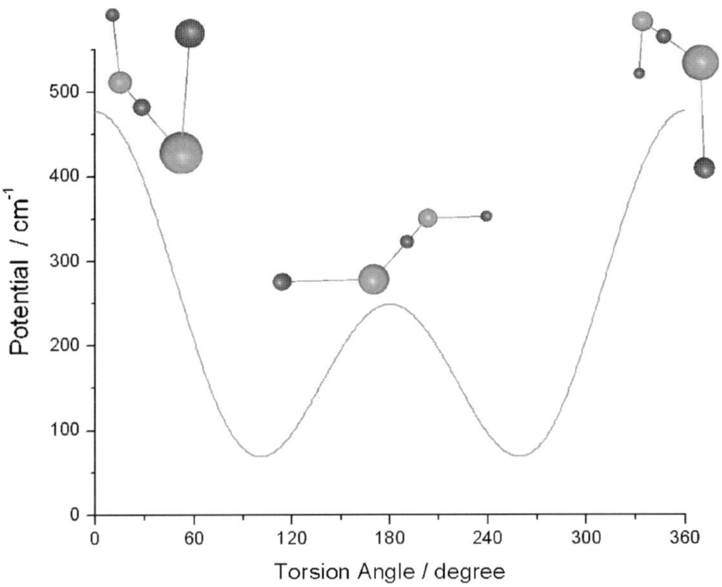

Fig. 11.4. The torsional potential of $(OH^-)H_2O$ along a C_2-symmetric path described in the text.

the bridging H atom transfer barrier (actually a second-order saddle point) and maintains C_2-symmetry. This places the path on the barrier shown in Fig. 11.3 where the zero-point wavefunction has its maximum value. The potential [67] along this path is shown in Fig. 11.4. As expected the fundamentals show significant splittings (of the order of 20 cm^{-1}) due to tunneling through the torsional barrier. The IR spectrum has also been obtained with MM-SR and this together with new MM-RPH and Diffusion Monte Carlo calculations have very recently been used to help assign a new band determined experimentally at around 1070 cm^{-1} [70, 71].

11.4 CURRENT BOTTLENECKS AND FUTURE PROGRESS

There are two major bottlenecks associated with MM calculations on large molecules. The first is that for nMR calculations the integration of the matrix elements is very time-consuming. This is particularly true for 'Reaction Path' calculations (even for $n = 4$) because the integration must always be carried out over the path coordinate, and the nMR refers only to the $3N - 7$ orthogonal normal coordinates. Hence 4MR for such calculations is equivalent to 5MR for standard calculations. Furthermore, for molecules such as methanol where the path coordinate is like a torsional angle, integration of the torsional basis expressed in $\sin(m\tau)$ and $\cos(m\tau)$ must be carried out by quadrature from 0 to 2π radians, and this typically involves between 90 and 120 integration points. Hence, until very recently, calculations above 3MR have been limited either to small molecules or low quanta of excitation in the CI basis. To overcome this problem [72] we have taken least squares fits of the intrinsic 4MR cuts of the potential (the last terms in Eq. (9) above) to polynomials in the coordinate $y = \tanh(\gamma Q)$ and then formed the one-dimensional integrals over the now separable terms in the 4MR intrinsic potentials. These fits must be performed at each torsional quadrature point required along the 'Reaction Path', and for all 4MR combinations of the $3N - 7$ orthogonal modes. However, as the order of nMR increases, the impact of the nMR cuts decreases, and the potentials required in the fits are weak and smoothly varying functions of Q which can be fit to high accuracy by low-degree polynomials. Use of such one-dimensional integrals effectively reduces the integration coupling from 4MR to 3MR, and this technique has recently been adopted in new calculations for $H_3O_2^-$.

The second bottleneck is also caused by the number of normal modes for large molecules, and this is the final size of the CI Hamiltonian matrix. In a way similar to the nMR coupling of the modes required for integration, we set up the CI matrix in terms of successive nMQ excitations (quanta) of the expansion set. For a molecule with M modes, one product function (corresponding to the zero point level) is present with no excited quanta in any of the M basis functions; there then follows M sets of 1MQ product functions in which only a single mode is excited; these are followed by $M!/[2!(M - 2)!]$ sets of 2MQ product functions in which any two modes are excited, and so on. Clearly for large molecules, the size of the matrix explodes as M and n increase. Depending on the nMR integration coupling, however, it is also clear that this matrix will also be sparse, unless the value of nMR is at least twice the maximum value of nMQ, in which case it will be full. Furthermore, many of the non-zero matrix elements, especially involving multiple excitations, will be small, which is the key to a new method we have used to reduce the size of the matrix [72]. For fundamentals (and possibly first-overtones and combination bands), the main contributors to the final wavefunctions will arise from the 1MQ and 2MQ basis, and we build this matrix to complete convergence, and diagonalize it, saving the eigenvectors. We set a second-order tolerance between all of these 2MQ eigenfunctions and any new 3MQ, 4MQ expansion function, etc. whereupon many of the additional nMQ functions are neglected. For a tolerance of 10^{-3} cm^{-1}, we have found that the $H_3O_2^-$ CI matrix is reduced from about 126,000 to 40,000, and this is then diagonalized by a Lanczos-like iteration procedure [27]. An optimum variant of this procedure is ongoing. There are operational hazards involved with this approach. Firstly, the path must be a smoothly varying function of all cartesian coordinates. When the coordinate axes are not rotated to conform to the Eckart conditions, it is not uncommon for the directions of axes (or even two different axes) to interchange. This leads to singularities along the path, which we have overcome by first normalizing the coordinates and then constructing the dot products between coordinates of adjacent points, for which the result should be almost unity. Any axis switch is then readily identified, which can be investigated and corrected. Secondly, the vectors of the $3N - 7$ normal coordinates orthogonal to the path must also vary smoothly around the path. We follow the progress of the vectors in a way similar to that adopted for the path points themselves, by defining the ordering of the vectors as that at the start point, and ensuring that this order is maintained at all subsequent points. We form the dot products of the adjacent (normalized) vectors, and if large deviations from unity arise, the most probable explanation is that the vector ordering has switched, whereupon we interchange vectors for modes 'm' and '$m + 1$' which usually corrects the discontinuity. Exceptions to this arise when two modes of the same symmetry appear to interchange (like an avoided intersection in electronic structure theory) when some trial-and-error method has to be introduced until the crossing point has been passed. A satisfactory solution to this important problem is still ongoing.

We are currently working on parallelizing both MM-SR and MM-RPH. Near linear speed-ups with the number of processors for the parts of the codes that evaluate the many n-mode terms on the n-mode representation of the potential, as well as matrix elements are being achieved. Further work on exploiting the sparse nature of the final H-matrix is also any area where progress can be made.

Finally, we expect to see many more 'direct' calculations using MULTIMODE and progress in making this more efficient. Very recently Rauhut [73] has reported an implementation of a code very similar to MULTIMODE interfaced to MOLPRO. The least squares fitting approach described is used and a 'screening' strategy is used to limit the number of 3-mode grids in the 3MR representation of the potential.

11.5 ACKNOWLEDGEMENTS

JMB thanks the Office of Naval Research (N00014-01-1-0235). SC and NCH thank the Office of Naval Research (N00014-01-1-0809). We also thank our many collaborators.

11.6 REFERENCES

1 E.B. Wilson Jr., J.C. Decius and P.C. Cross, Molecular vibrations, the theory of infrared and Raman vibrational spectroscopy, Dover, New York, 1980.

2 M.J. Frisch, G.W. Trucks, H.B. Schlegel, G.E. Scuseria, M.A. Robb, J.R. Cheeseman, J.A. Montgomery, Jr., T. Vreven, K.N. Kudin, J.C. Burant, J.M. Milliam, S.S. Iyengar, J. Tomasi, V. Barone, B. Mennucci, M. Cossi, G. Scalmani, N. Rega, G.A. Petersson, H. Nakatsuji, M. Hada, M. Ehara, K. Toyota, R. Fukuda, J. Hasegawa, M. Ishida, T. Nakajima, Y. Honda, O. Kitao, H. Nakai, M. Klene, X. Li, J.E. Knox, H.P. Hratchian, J.B. Cross, V. Bakken, C. Adamo, J. Jaramillo, R. Gomperts, R.E. Stratmann, O. Yazyev, A.J. Austin, R. Cammi, C. Pomelli, J.W. Ochterski, P.Y. Ayala, K. Morokuma, G.A. Voth, P. Salvador, J.J. Dannenberg, V.G. Zakrzewski, S. Dapprich, A.D. Daniels, M.C. Strain, O. Farkas, D.K. Malick, A.D. Rabuck, K. Raghavachari, J.B. Foreman, J.V. Ortiz, Q. Cui, A.G. Baboul, S. Clifford, J. Cioslowski, B.B. Stefanov, G. Liu, A. Liashenko, P. Piskorz, I. Komaromi, R.L. Martin, D.J. Fox, T. Keith, M.A. Al-Laham, C.Y. Peng, A. Nanayakkara, M. Challacombe, P.M.W. Gill, B. Johnson, W. Chen, M.W. Wong, C. Gonzalez and J.A. Pople, Gaussian, Inc., Pittsburgh, 2004.

3 MOLPRO is a package of *ab initio* programs written by H.-J. Werner, P.J. Knowles, M. Schütz, R. Lindh, P. Celani, T. Korona, G. Rauhut, F.R. Manby, R.D. Amos, A. Bernhardsson, A. Berning, D.L. Cooper, M.J.O. Deegan, A.J. Dobbyn, F. Eckert, C. Hampel, G. Hetzer, A.W. Lloyd, S.J. McNicholas, W. Meyer, M.E. Mura, A. NicklaB, P. Palmieri, R. Pitzer, U. Schumann, H. Stoll, A.J. Stone, R. Tarroni and T. Thorsteinsson. Birmingham, UK, 2002.

4 Q. Cui and I. Bahar (Eds.), Normal mode analysis: Theory and applications to biological and chemical systems, CRC Press, London, 2005.

5 H.H. Nielsen, Rev. Mod. Phys., 23 (1951) 90.

6 I. Mills, in: K.N. Rao, C.W. Mathews (Eds.), Molecular spectroscopy: Modern research, Academic Press, New York, 1972.

7 M.R. Aliev and D. Papousek, Molecular vibrational–rotational spectra, Academic Press, Prague, 1982.

8 A.B. McCoy and E.L. Sibert, in: R.E. Wyatt, J.Z.H. Zhang (Eds.), Dynamics of molecules and chemical reactions, Marcel Dekker, New York, 1996.
9 M.E. Kellman, Ann. Rev. Phys. Chem., 46 (1995) 395.
10 L.B. Harding and W.C. Ermler, J. Comput. Chem., 6 (1985) 13.
11 J.F. Gaw, A. Willetts, W.H. Green and N.C. Handy, in: J.M. Bowman, M.A. Ratner (Eds.), Advanced molecular vibrations and collision dynamics, Vol. IB, JAI, Greenwich, 1991.
12 ANHARM is a FORTRAN program written for VPT2 analyses, Y. Yamaguchi, H.F. Schaefer III, Center for Computational Chemistry, University of Georgia, Athens, GA 30602 (USA).
13 M.E. Varner, J. Vazquez and J.F. Stanton, private communication.
14 J.K.G. Watson, Mol. Phys., 15 (1968) 479.
15 H. Romanowski, J.M. Bowman and L. Harding, J. Chem. Phys., 82 (1985) 4155.
16 K.M. Dunn, J.E. Boggs and P. Pulay, J. Chem. Phys., 85 (1985) 5838.
17 J.M. Bowman, J. Chem. Phys., 68 (1978) 608.
18 For reviews, see J.M. Bowman, Acc. Chem. Res., 19 (1986) 202; M.A. Ratner and R.B. Gerber, J. Phys. Chem., 90 (1986) 20.
19 For a general code based on the VSCF method see D.A. Jelski, R.H. Haley and J.M. Bowman, J. Comp. Chem., 17 (1996) 1654.
20 J.O. Jung and R.B. Gerber, J. Chem. Phys., 105 (1996) 10332.
21 L.S. Norris, M.A. Ratner, A.E. Roitberg and R.B. Gerber, J. Chem. Phys., 105 (1996) 11261.
22 O. Christiansen, J. Chem. Phys., 120 (2004) 2149.
23 S.K. Gregurick, G.M. Chaban and R.B. Gerber, J. Phys. Chem. A, 106 (2002) 8696.
24 J.M. Bowman, K. Christoffel and F. Tobin, J. Phys. Chem., 83 (1979) 905.
25 S. Carter, S.J. Culik and J.M. Bowman, J. Chem. Phys., 107 (1997) 10458.
26 S. Carter, X. Huang, S. Carter and J.M. Bowman, J. Chem. Phys., 118 (2003) 5431.
27 S. Carter, J.M. Bowman and N. Handy, Theor. Chem. Acc., 100 (1998) 191.
28 S. Carter, J.M. Bowman and B.J. Braams, Chem. Phys. Lett., 342 (2001) 636.
29 S. Carter and N.C. Handy, Chem. Phys. Lett., 352 (2002) 1.
30 R.J. Whitehead and N.C. Handy, J. Mol. Spectrosc., 55 (1975) 356.
31 J.M. Bowman, J.S. Bittman and L.B. Harding, J. Chem. Phys., 85 (1986) 911.
32 T. Carrington, Jr., in: Paul von Rague Schleyer (Ed.), Encyclopedia of computational chemistry, Wiley, New York, 1998, Vol. 5, pp. 3157–3166.
33 J.C. Light and T. Carrington, Jr., Adv. Chem. Phys., 114 (2000) 263.
34 J. Tennyson, P. Barletta, M.A. Kostin, O.L. Polyansky and N.F. Zobov, Spectrochim. Acta A, 58 (2002) 663.
35 S. Carter, J.M. Bowman and L.B. Harding, Spectrochim. Acta A, 53 (1997) 1179.
36 S. Carter, H.M. Shnider and J.M. Bowman, J. Chem. Phys., 110 (1999) 8417.
37 S. Carter and J.M. Bowman, J. Phys. Chem., 104 (2000) 2443.
38 A. Chakraborty, D.G. Truhlar, J.M. Bowman and S. Carter, J. Chem. Phys., 121 (2004) 2071.
39 J.M. Bowman, X. Huang and S. Carter, Spectrochim. Acta Part A, 58 (2002) 839.
40 X. Huang, S. Carter and J.M. Bowman, J. Phys. Chem. B, 106 (2002) 8182.
41 X. Huang, S. Carter and J. Bowman, J. Chem. Phys., 118 (2003) 5431.
42 S.M. Colwell, S. Carter and N.C. Handy, Mol. Phys., 101 (2003) 523.
43 T. Rajamäki, A. Miani and L. Halonen, J. Chem. Phys., 118 (2003) 10929.
44 S. Carter and J.M. Bowman, J. Chem. Phys., 108 (1998) 4397.
45 F. Dzegilenko, J.M. Bowman and S. Carter, J. Chem. Phys., 109 (1998) 7506.
46 C. Leonard, S. Carter, N.C. Handy and P.J. Knowles, Mol. Phys., 99 (2001) 1335.
47 C. Leonard, N.C. Handy, S. Carter and J.M. Bowman, Spectrochim. Acta A, 58 (2002) 825.
48 C. Leonard, S. Carter and N.C. Handy, Phys. Chem. Chem. Phys., 4 (2002) 4087.
49 M.L. Senent, P. Palmieri, S. Carter and N.C. Handy, Chem. Phys. Lett., 354 (2002) 1.
50 S. Irle and J.M. Bowman, J. Chem. Phys., 113 (2000) 8401.
51 X. Huang, H.M. Cho, S. Carter, L. Ojamae, J.M. Bowman and S.J. Singer, J. Phys. Chem. A, 107 (2003) 7142.
52 J. Dai, Z. Bacic, X. Huang, S. Carter and J.M. Bowman, J. Chem. Phys., 119 (2003) 6571.

53 C. Leonard, G. Chambaud, P. Rosmus, S. Carter and N.C. Handy, Phys. Chem. Chem. Phys., 3 (2001) 508.
54 R. Burcl, S. Carter and N.C. Handy, Chem. Phys. Lett., 380 (2003) 237.
55 J.M. Bowman, K. Christoffel and G. Weinberg, J. Mol. Struct. (THEOCHEM), 461–462 (1999) 71.
56 J.M. Bowman and H.M. Shnider, J. Chem. Phys., 110 (1999) 4428.
57 K.M. Christoffel and J.M. Bowman, J. Phys. Chem. A, 103 (1999) 3020.
58 A.L. Kaledin, X. Huang and J.M. Bowman, Chem. Phys. Lett., 384 (2003) 80.
59 R. Burcl, S. Carter and N.C. Handy, Phys. Chem. Chem. Phys., 6 (2004) 340.
60 W.H. Miller, N.C. Handy and J.E. Adams, J. Chem. Phys., 72 (1980) 99.
61 S. Carter and N.C. Handy, J. Chem. Phys., 113 (2000) 987.
62 S. Carter and N.C. Handy, Spectrochim. Acta A, 60 (2004) 2107.
63 J. Koput, S. Carter and N.C. Handy, J. Phys. Chem. A, 102 (1998) 6325.
64 J.T. Hougen, P.R. Bunker and J.W.C. Johns, J. Mol. Spectrosc., 45 (1973) 136.
65 D.P. Tew, N.C. Handy and S. Carter, Mol. Phys., 99 (2001) 393.
66 D.P. Tew, N.C. Handy, S. Carter, S. Irle and J. Bowman, Mol. Phys., 101 (2003) 3513.
67 X. Huang, B.J. Braams, S. Carter and J.M. Bowman, J. Am. Chem. Soc., 126 (2004) 5042.
68 E.A. Price, W.H. Robertson, E.G. Diken, G.H. Weddle and M.A. Johnson, Chem. Phys. Lett., 366 (2002) 412.
69 C.C.M. Samson and W. Klopper, J. Mol. Struct. (THEOCHEM), 586 (2002) 201.
70 E.G. Diken, J.M. Headrick, J.R. Roscioli, J.C. Bopp, M.A. Johnson, A.B. McCoy, X. Huang, S. Carter and J.M. Bowman, J. Phys. Chem. A, 109 (2005) 571.
71 A.B. McCoy, X. Huang, S. Carter, J.M. Bowman and N.C. Handy, to be published.
72 N.C. Handy and S. Carter, Molec. Phys., 102 (2204) 2201.
73 G. Rauhut, J. Chem. Phys. 121 (2004) 9312.

Theory and Applications of Computational Chemistry: The First Forty Years
Edited by C. Dykstra et al.

CHAPTER 12

Toward accurate computations in photobiology

Adalgisa Sinicropi and Massimo Olivucci

Dipartimento di Chimica, Università di Siena, Siena, Italy

Abstract

In this chapter, we review the results of recent computational studies of the spectroscopy and molecular mechanism underlying the activity of very different photoactive proteins. Our aim is to show that, nowadays, high-level (i.e. accurate) *ab initio* quantum chemical methodologies may be used to unveil the mechanistic details of photochemical processes occurring in photobiological systems. Accordingly, we discuss some recent spectral simulations of the bovine visual pigment Rhodopsin (Rh) and of the green fluorescent protein (GFP) of the luminescent jellyfish *Aequorea victoria*. We will also discuss a recent attempt to map the photoisomerization path of the full Rh protein.

12.1 INTRODUCTION

There are two molecular events that may follow light energy absorption: energy wastage or energy exploitation (Scheme 12.1). The control of these events can be considered a basic requirement for the rational design of efficient photochemical reactions, artificial photosynthetic systems and for the design of novel materials, molecular devices and molecular level machines. In fact, technology often requires molecules where this energy is exploited to achieve specific chemical, conformational and electronic changes. In contrast, other applications, as those in the field of photoprotection or photostability, need molecules that eliminate the stored photon energy (i.e. be structurally unaffected by light absorption) efficiently *via* emission and/or internal conversion.

In this respect, during the last few years, computational methods have been successfully applied to explore photon energy wastage mechanisms (e.g. in fluorescent probes) [1–3] and the mechanism of fast internal conversion in the DNA basis [4]. Similarly, as an example of process where light is exploited to drive stereospecific

References pp. 288–289

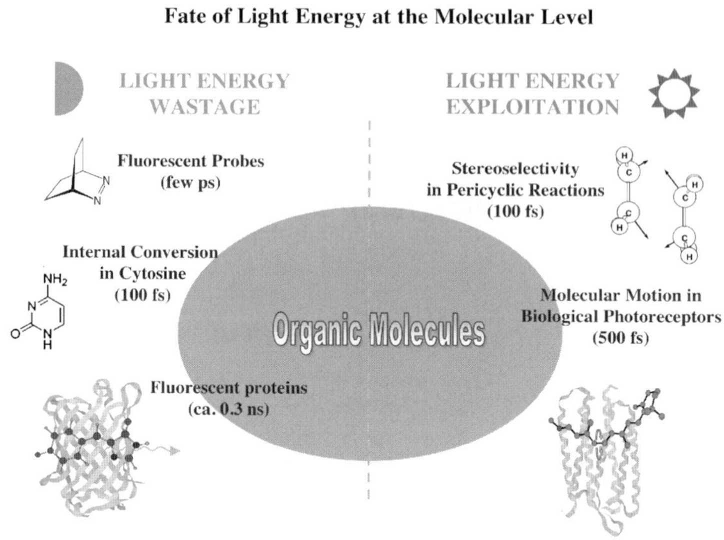

Scheme 12.1.

reactions, we can recall the photochemical pericyclic reactions [5]. The same types of processes can be found in photobiology. For instance, there are fluorescent proteins, such as the green fluorescent protein (GFP), where the energy of the photon is 'wasted' radiatively to produce fluorescence while there are other proteins, such as the visual pigment Rhodopsin (Rh), where the energy of the photon is exploited to produce a change in the protein conformation.

In the past, we have shown that a suitable approach to these mechanistic problems involves the computation of the photochemical reaction path. As illustrated in Fig. 12.1, after photoexcitation the reactant A starts to evolve on the excited state potential energy surface. As a consequence of such relaxation (that may even involve the overcoming of a transition state (TS) on the upper energy surface) the species reaches a point of conical intersection (CI) and decays. This process can be described by computing a minimum energy path (MEP) starting at the Franck–Condon (FC) point (i.e. at the ground state equilibrium structure) and ending at CI. The CI is a key mechanistic entity for the reaction as it provides a very efficient channel for the decay to ground state that, in the past, has been referred by the (organic) photochemists as the 'photochemical funnel' [6,7]. A photochemical funnel corresponds to a molecular structure that lives for only few femtoseconds (10^{-15} s). For this reason computer simulations based on modern quantum chemical methods appear to be the only practical source of direct information.

For a complete description of the reaction we need to compute also the ground state MEP describing photoproduct formation. In Fig. 12.1 we show that the entire photochemical reaction path is defined and computed in terms of a set of connected MEPs. In particular, the path starting at FC (structure A^*) on the potential energy surface of the spectroscopic excited state and ending at the photoproduct energy minimum B located on the ground state energy surface is constructed by joining two MEPs. A first

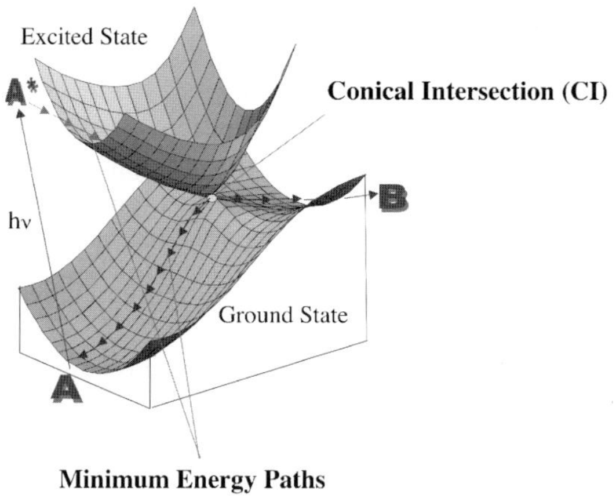

Fig. 12.1. Left side: Model intersecting the ground (S_0) and the first excited state (S_1) potential energy surfaces. The Franck–Condon point (A^*) is geometrically identical to the minimum on the ground state, but it is located on the excited state surface. The arrows indicate the direction of the minimum energy path connecting the FC point (A^*) to the conical intersection (CI) and then to A and to the photoproduct B (adapted from Ref. [8]).

MEP (grey arrows) connects the FC point to the conical intersection ($A^* \rightarrow$ CI). A second MEP (black arrows) connects the conical intersection to the photoproduct (CI $\rightarrow$ photoproduct B). A third MEP that starts at CI describes the reactant reconstitution process (CI $\rightarrow$ A) responsible for partial return to the ground state. As we will see below the mechanism of Fig. 12.1 appears to be general.

The nature of the CI has been a subject of research for at least three decades. Between 1966 and 1972, Zimmerman, Michl and Salem [9–14] were the first to propose, independently, that for a broad class of organic reactions the structure of the funnel could be determined by locating a 'cone shaped' crossing of the excited and ground state (potential) energy surfaces, known as conical intersection. More recently, Yarkony [15–17] and Ruendenberg [18] identified conical intersections geometries in small molecules.

At the end of the 1980s improved quantum chemical methods and faster computers became available which were suitable for computing excited state energy surfaces. In 1990 these methodologies were used to investigate the reaction path of the photochemical cycloaddition of ethylene [19,20]. It was shown that

(i) a conical intersection exists right at the bottom of the excited state energy surface of two interacting ethylene molecules.
(ii) the molecular structure of the conical intersection was intimately related to the observed production of cyclobutane, a hydrocarbon which is squared in shape.

These initial results suggested that conical intersections could indeed act as photochemical funnels.

Since there is no general theorem supporting the existence of low-lying conical intersections in organic molecules, the only way to prove the general validity of the hypothesis above was a painstaking systematic search for properties (i) and (ii) in different classes of organic molecules. For this reason, in 1992 Olivucci and Bernardi in Bologna and Robb in London started a long-term computational project. The excited energy surface of *ca.* 25 different organic chromophores was mapped to search for conical intersections [21]. The examination of this fairly large set of computed data allowed for the formulation of a few general results that lie at the basis of the 'chemistry' of conical intersections (as opposed to the 'usual' chemistry of transition structures):

1. Similar organic chromophores (e.g. conjugated hydrocarbons) have similar conical intersection structures.
2. All basic chemical events such as breaking, making and exchange of bonds between atoms can be mediated by conical intersections.

Computational tools and strategies have been developed to tackle the problem of determining the molecular structure and energy of mechanistically relevant conical intersections. These tools have the ability to

(a) trace the excited state MEP starting from the FC point (i.e. the ground state equilibrium structure of the system A^* in Fig. 12.1) or from an excited state reactant and ending at a conical intersection or excited state intermediate.
(b) locate the lowest energy conical intersection for a given, initial, molecular structure [17,22].
(c) determine the possible initial (steepest) relaxation directions starting from the located conical intersection and follow the associated MEPs connecting the intersection to the primary photoproduct energy minima [23,24].

12.2 *AB INITIO* QUANTUM CHEMICAL METHODS FOR EXCITED STATES

Among the possible types of available quantum chemical technologies (e.g. semiempirical, DFT and *ab initio*; see Scheme 12.2 for a rough classification in terms of computational cost—vertical—and accuracy—horizontal) one could use to compute

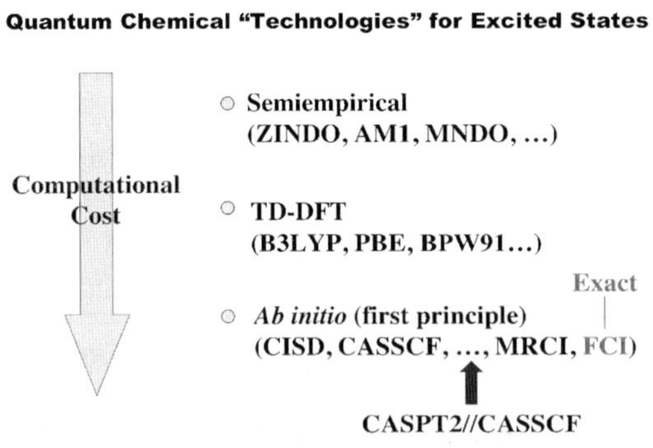

Scheme 12.2.

the potential energy surface of an excited state molecule, we adopt a 'mixed' *ab initio* CASPT2//CASSCF approach. This is considered a 'practical' compromise between computational cost and computational accuracy as well as a robust quantum chemical strategy when comparison with observed spectroscopic quantities and reaction barriers is needed. In fact, the CASPT2 (i.e. an implementation of second-order multireference perturbation theory) level ensures a correct balance of dynamic and non-dynamic electron correlation in the wavefunction. However, CASPT2 geometry optimizations are presently unpractical even for medium size organic molecules. Thus in CASPT2// CASSCF computations equilibrium geometries and reaction coordinates are computed at the CASSCF (Complete Active Space—Self Consistent Field) level of theory while the associated energy profile (determining the reaction energetics) is re-evaluated along a selected series of reaction coordinate points, performing single-point CASPT2 computations.

The potential accuracy of the CASPT2//CASSCF strategy has been evaluated taking as a benchmark the stability of a transient organic intermediate such as the diazomethyl radical [25] to fragmentation (i.e. N_2 loss). The CASSCF energy profile is consistent with a barrierless C–N bond breaking process. Very recently we set up for numerical derivative computations at the more accurate CASPT2 level [26] and found that the reaction coordinate clearly shows a 5 kcal mol^{-1} barrier. If we now compare the CASPT2 energy profile with the CASPT2//CASSCF profile in Fig. 12.2 we can clearly see that, at least for this process, the CASPT2//CASSCF strategy is substantially quantitative with respect to the CASPT2 level.

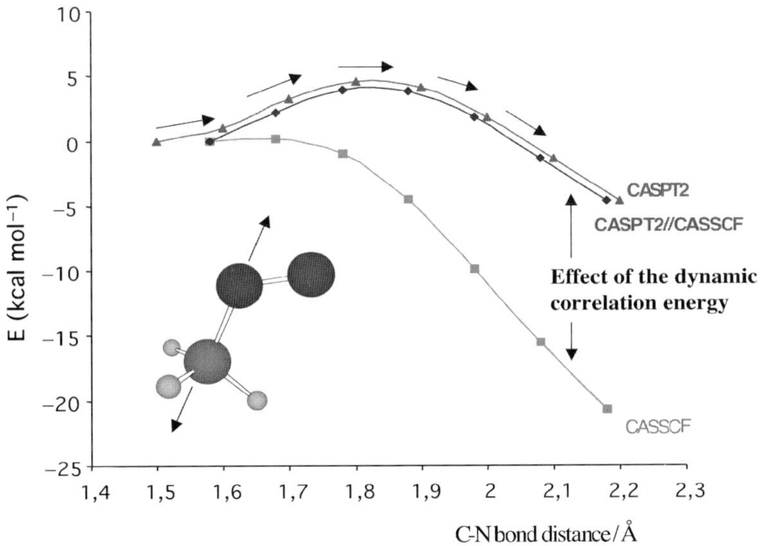

Fig. 12.2. Comparison of the CASSCF, CASPT2 (numerical gradients evaluation at the CASPT2 level) and CASPT2//CASSCF energy profiles for the denitrogenation of the methyl diazenyl diradical, which provides evidence of the correctness of the CASPT2//CASSCF strategy (redrawn with permission Ref. [25] © 2003 American Chemical Society).

References pp. 288–289

Notice that, in the presence of limited barriers (i.e. of short lifetimes of the reactive species) the CASPT2//CASSCF level appears to be required even if one is interested in a mechanistic (i.e. qualitative) rather than quantitative description of the reaction energetics. This is also demonstrated in the same figure where the CASSCF level of theory predicts an unbound radical species that will therefore denitrogenate spontaneously. In contrast the fact that the CASPT2//CASSCF level predicts a *ca.* 5 kcal mol^{-1} fragmentation barrier implies that the reactive species is bounded and that in certain experimental conditions may be detected.

The same type of conclusion is reached when one deals with the reactivity of excited state species. In Fig 12.3 we report the comparison between the CASPT2//CASSCF and CASSCF excited state reaction paths corresponding to two different isomerizations of the excited state GFP chromophore *in vacuo* [27]. It is clearly seen that, in contrast to CASSCF, the CASPT2//CASSCF level suggests a mechanism where there is no planar excited state intermediate (i.e. no planar energy minimum along the excited state energy surface). The effect of the dynamic correlation energy, only accounted for at the CASPT2//CASSCF level, is thus mechanistically relevant.

The use of TD-DFT methods for mapping the excited state potential energy surface represents a very attractive alternative to the *ab initio* methods due to the possibility to achieve similar accuracies at a much reduced computational cost. These methodologies, that, in principle, are valid only for excited states dominated by singly excited configurations (or when the linear response theory is valid) are currently under intensive testing. For the case of simple polyene protonated Schiff bases it has been possible to carry out a thorough test of the TD-DFT energetics within a TD-DFT//CASSCF scheme for the first excited state (a charge transfer state dominated by a single excitation) [28].

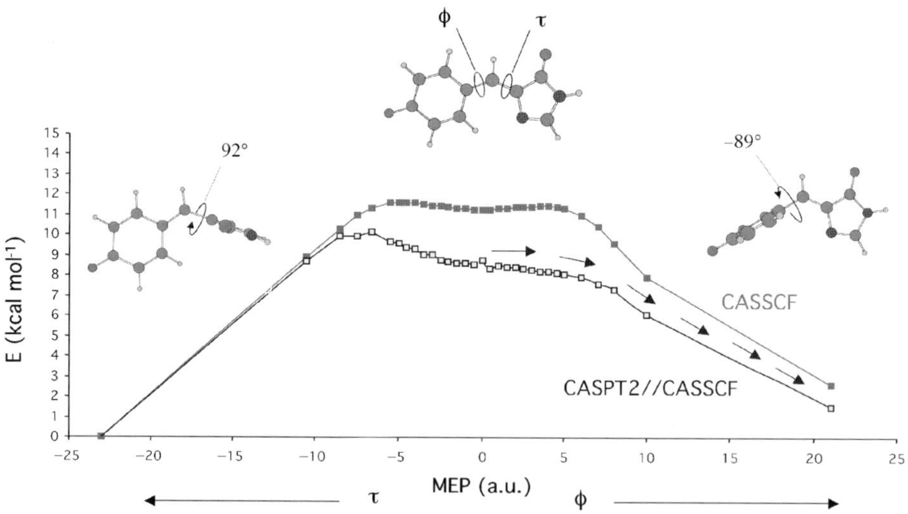

Fig. 12.3. CASSCF (full squares) and CASPT2 (open squares) energy profiles for the S_1 state along the computed τ and ϕ torsional reaction coordinates of the HBDI chromophore (redrawn with permission from Ref. [27] © 2004 American Chemical Society).

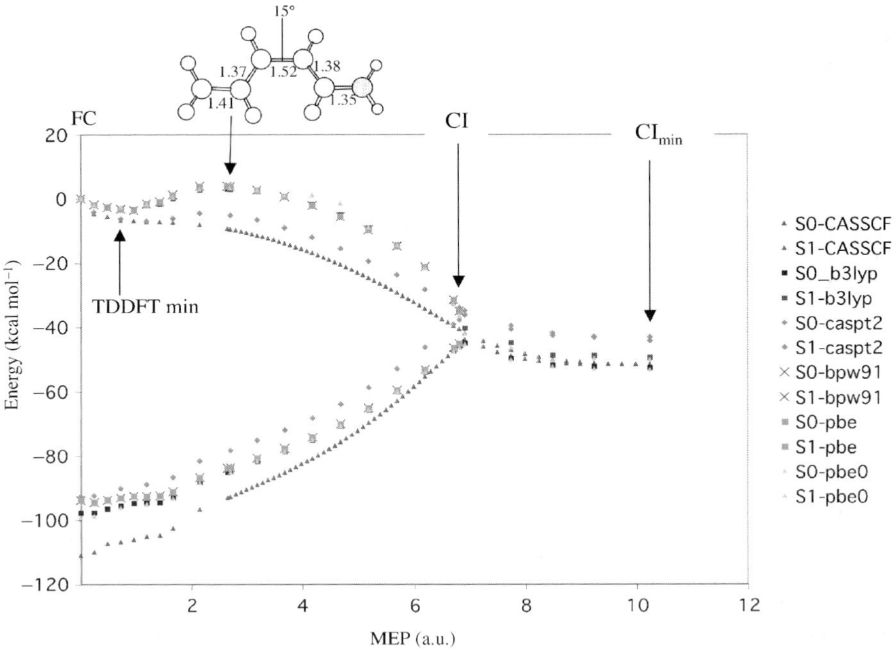

Fig. 12.4. Energy profiles along the S_1 photoisomerization coordinate of a PSB11 minimal model at CASPT2//CASSCF and TD-DFT//CASSCF levels of theory. Notice that both the reaction paths and the conical intersection family are qualitatively described at the TDDFT levels (redrawn with permission from Ref. [28] © 2004 American Chemical Society).

In other words, the TD-DFT energies are computed for each point along a reaction coordinate determined at the CASSCF level. With respect to the CASPT2//CASSCF scheme defined above, the practical advantage is that single point TD-DFT computations replace the more expensive CASPT2 computations. The result of this test is given in Fig. 12.4.

More work devoted to the testing of the TD-DFT methodologies can be found in different review articles [29,30]. A different approach that has been pursued with the target of computing both reaction paths and trajectories at a lower computational cost is based on *ad hoc* reparameterized multireference semiempirical methods. Indeed, recently these methods accounting for excited state calculations of specific chromophores have been reported [31–33].

A step forward along the route to the correct modelling of the spectroscopy and photochemical reactivity of photoreactive proteins is represented by the implementation of a Quantum Mechanics/Molecular Mechanics (QM/MM) computational strategy based on a suitable QM part coupled with a protein force field such as AMBER [34] (or CHARMM [35]). Very recently a CASPT2//CASSCF/AMBER method for rhodopsin has been implemented in our laboratory [36,37] within the QM/MM link-atom scheme [38]. Special care has been taken in the parametrization of the protonated Schiff base linkage region that describes the delicate border region between the MM (the protein)

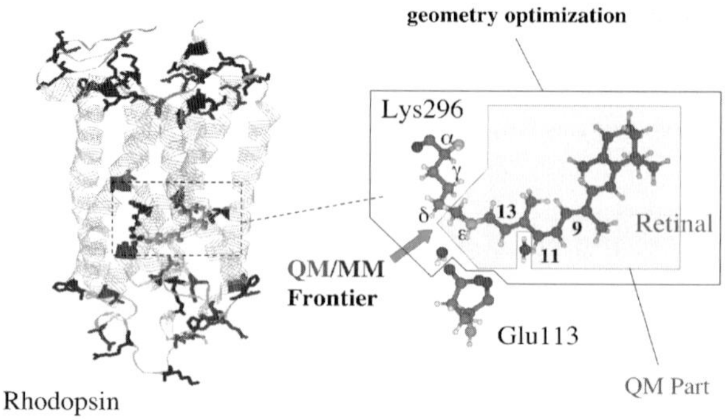

Scheme 12.3.

and QM (the chromophore) subunits [37] (see Scheme 12.3). While below we will review the results of this *ab initio* QM/MM methods other alternative QM/MM approaches, originally pioneered by Warshel [39], have been presented. In particular, Vreven and Morokuma [40] applied the ONIOM method to the photoisomerization of the Rh chromophore in the gas phase. Hayashi et al. [41,42] reported a CASSCF//HF/DZV/AMBER computation of the λ_{max} of the related pigment bacteriorhodopsin (bR). The same authors have reported a trajectory computation at the CASSCF/AMBER level for bR [42]. Warshel used a reparameterization of the QCFF/PI semiempirical method in their work on bR [33,43]. Martinez et al. implemented a semiempirical FOMO-CI to locate conical intersections and performing excited state dynamics in both condensed phase and protein environments [31,32]. Excited state dynamics calculations have also been reported by Robb and coworkers in a recent study of photoactive yellow protein (PYP) [44].

12.3 FATE OF LIGHT ENERGY IN PHOTOBIOLOGY

Below we revise the results of the application of the CASPT2//CASSCF strategy to the investigation of two very different spectroscopic problems. The first problem regards GFP, a protein that can waste light energy with great efficiency through fluorescence (left side of Fig. 12.5). The second protein, the visual pigment Rh, is instead designed to exploit light energy to drive a double bond photoisomerization reaction (right side of Fig. 12.5).

The GFP consists in a rigid β-sheet based structure and the prosthetic group responsible for the green fluorescence is the anionic form of a *p*-hydroxybenzilidene-imidazolone (HBDI) moiety located at the centre of the barrel-like protein backbone (left side of Fig. 12.5). The fluorophore is excited with UV light and fluoresces with an 80% quantum yield [45,46]. The structure of the visual pigment rhodopsin is totally different; it is more flexible and its secondary structure is mainly characterized by seven α-helix segments. The chromophore of the Rh is the protonated Schiff base of the 11-*cis* retinal (PSB11) and it is bounded to a lysine residue (Lys296) *via* a covalent linkage (right side

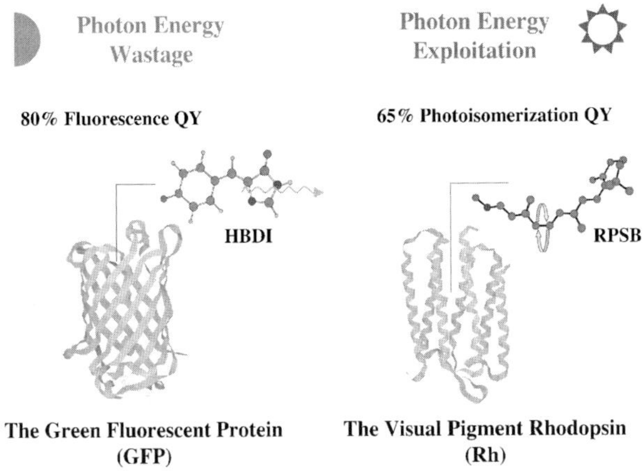

Fig. 12.5. Ball and stick representation of the green fluorescent protein chromophore (left side) as an example of molecule which can waste the photon energy and of the visual pigment rhodopsin chromophore (right side) which can efficiently convert the light energy into molecular motion (adapted from Ref. [8]).

of Fig. 12.5). PSB11 undergoes a 11-*cis* → *all-trans* Z/E photoisomerization, which triggers the Rh activity [47,48]. Notice that both the chromophores of GFP and Rh are ion pairs (see Fig. 12.6). For GFP, the HBDI anion is coupled to an Arg cation, while in the visual pigment the retinal PSB cation is interacting with a Glu113 carboxylate counterion.

The excited state lifetime of the GFP chromophore is very long in the protein (*ca.* 3 ns) but much shorter (less than 0.3 ps) in solution. The mechanistic hypothesis is that the decay is due to a Z/E isomerization. Thus, while in solution the fluorophore may undergo an ultrafast internal conversion, the protein should act by restraining the isomerization. In contrast in Rh the excited state lifetime is *ca.* 150 fs. However, if we look at the solution lifetime this is increased of one order of magnitude. Furthermore, one has 24% quantum yield in solution and 65% quantum yield in the protein. Thus, in this case the protein is 'catalyzing' the reaction. The absorption maxima (λ^a_{max}) of

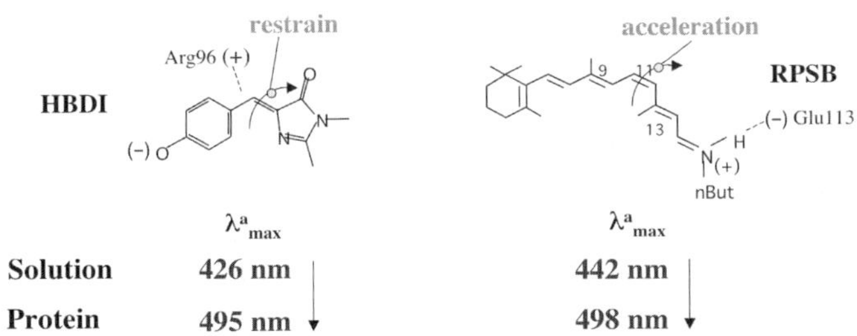

Fig. 12.6. Absorption maxima for the green fluorescent protein chromophore (left side) and the visual pigment rhodopsin chromophore (right side) in different environments (solution and protein). These features contribute to explain the different behaviour of the two proteins (adapted from Ref. [8]).

the two proteins are also environment dependent. In fact, while the absorption of HBDI in solution is 426 nm, GFP has a much red shifted (495 nm) absorption maximum. The same behaviour has been found in Rh: the absorption of retinal is 442 nm in solution and it becomes 498 nm in the protein. We begin our CASPT2//CASSCF investigation by trying to reproduce these spectral features. Why there is a red shifted absorption inside the protein?

12.3.1 GFP spectroscopy

The case of the GFP chromophore is important since its gas-phase spectra are available and one can make a direct comparison with the experiment at various level (gas phase, solution, protein matrix) [27,49]. Such comparison is schematically reported in Scheme 12.4. Inspection of these data reveals that the gas-phase absorption maximum is substantially closer to the protein absorption maximum than to the solution absorption maximum. This suggests the rather naïve idea that the GFP protein cavity offers an environment more similar to the gas phase than to the solution.

The simulation correctly reproduces the closeness of the gas phase and protein absorption λ^{a}_{max}. Geometry optimizations in the different environments indicate that the protein, gas phase and solution spectral features are related to the nature of the chromophore structure. Notice that, remarkably, as shown in Fig. 12.7, the equilibrium structures in the protein (Fig. 12.7a) and in the gas phase (Fig. 12.7c) display close absorption maxima in spite of the different geometrical structures. In contrast, the solution equilibrium structure (Fig. 12.7b) differs dramatically. Thus, even the geometrical analysis is consistent with the spectroscopy.

The CASSCF/AMBER method allows for geometry optimization on the excited state. Thus, we can also predict the emission maxima. The results, schematically illustrated in Scheme 12.5, indicate that the protein/gas-phase similarity also holds for the emission suggesting that the protein matrix mimics the gas phase even for the relaxed excited state chromophore.

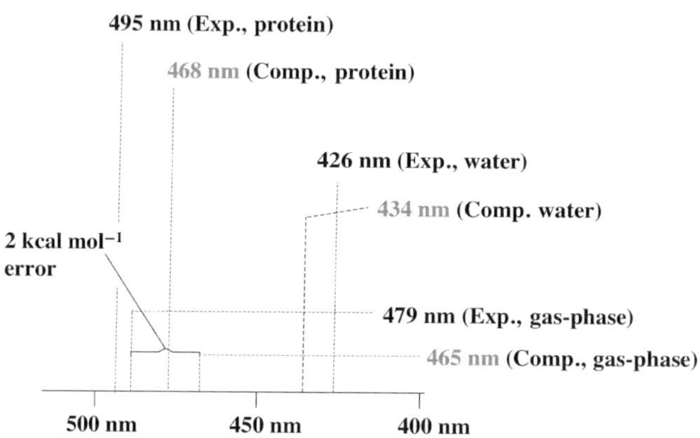

Scheme 12.4.

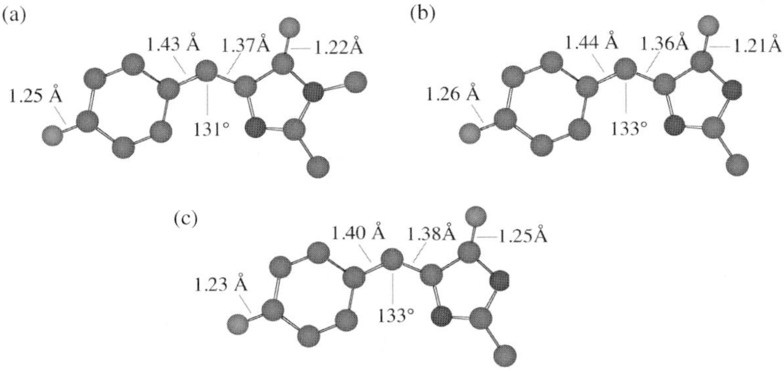

Fig. 12.7. Ground state CASSCF optimized structures for the GFP chromophore in three different environments: (a) protein, (b) water solution and (c) *in vacuo*. Geometrical parameters are in Å and degrees (adapted from Ref. [8]).

12.3.2 Rh spectroscopy

As already mentioned above, the visual pigment Rh [47,48] is a G-protein-coupled receptor containing an 11-*cis* retinal chromophore (PSB11) bounded to a lysine residue (Lys296) *via* a protonated Schiff base linkage (see dashed frame in the structure below). While the biological activity of Rh is triggered by the light-induced isomerization of PSB11, this reaction owes its efficiency (e.g. short time-scale and quantum yields) to the protein cavity.

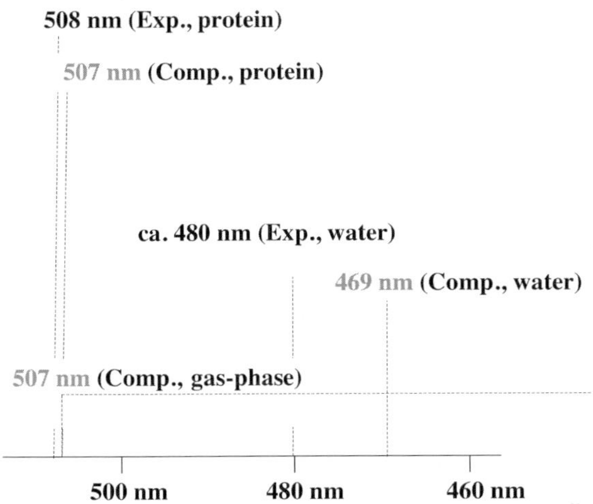

Scheme 12.5.

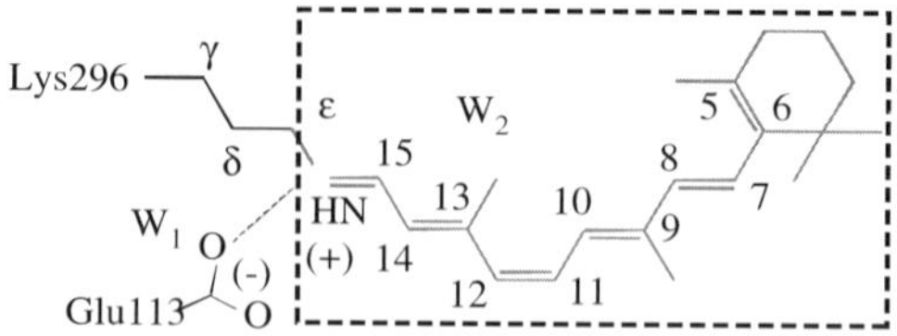

Accordingly, the investigation of the environment-dependent properties of PSB11 is a prerequisite for understanding Rh 'catalysis'. As for GFP the equilibrium geometry and absorption maximum (λ^{a}_{max}) are indicators of the environment effect.

We recently [36] provide evidence that our CASPT2//CASSCF/AMBER QM/MM strategy can be correctly applied to the investigation of the excited state of Rh with a computational error <5 kcal mol^{-1}. Thus, our simulation allows for a semiquantitative analysis of the factor determining the properties of the protein environment. Comparing the computed vertical excitation energy with the experimental values (see Scheme 12.6), we found that for Rh the absorption maximum is reproduced with only 3 kcal mol^{-1} (476 nm *vs.* 498 nm) while for the solution the error is only 1 kcal mol^{-1} (433 nm *vs.* 442 nm). These results confirm the quality of our approach and allow to reproduce the so-called 'opsin-shift' (the 445 nm λ_{max} observed for PSB11 in methanol is red shifted to 498 nm in Rh) with a 2 kcal mol^{-1} error.

As shown in Fig. 12.8, the S_0 Rh chromophore optimized structure (protein) is close to the crystallographic (X-ray) and NMR structures observed experimentally for bovine Rh. Comparing this geometry with the one obtained for PSB11 in methanol solution, using Cl^- as counterion, it is evident that the central segment of PSB11 is nearly planar and the β-ionone ring is 10° more twisted than in the protein.

The excitation energy of the chromophore–counterion (Glu113) system (RPSB + count., taken with its Rh geometry) is increased leading to a strongly blue shifted λ^{a}_{max}. The isolated RPSB displays instead an absorption maximum much closer to that of the protein. The results are illustrated in Scheme 12.7 (where we report the S_1–S_0 and S_2–S_0

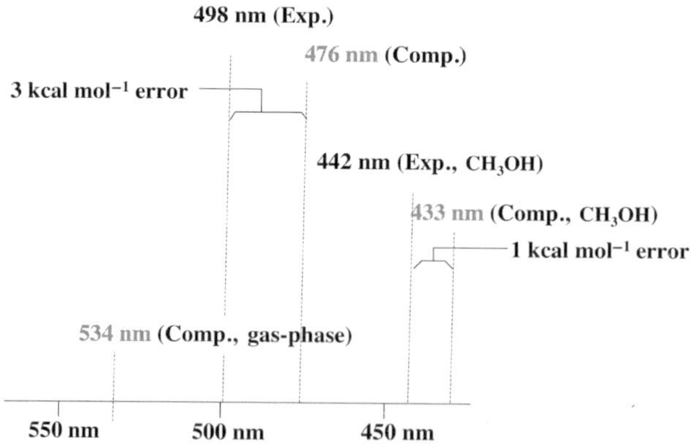

Scheme 12.6.

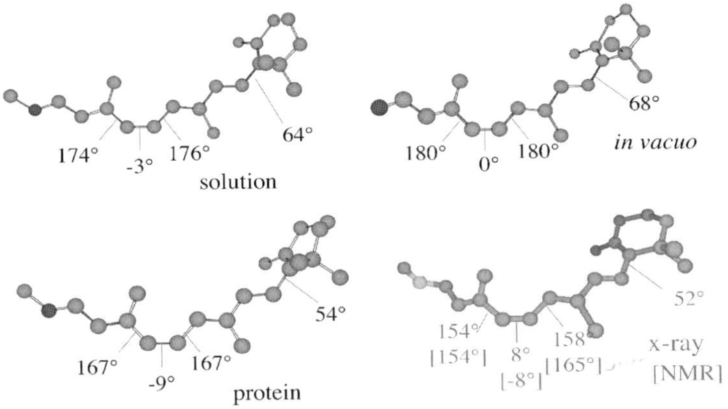

Fig. 12.8. Ground state CASSCF optimized structures for the Rh chromophore in solution, *in vacuo* and in the protein compared to the crystallographic and NMR structure (adapted from Ref. [36]).

energy gaps). From a more technical point of view, Scheme 12.7 suggests that the protein matrix offsets or counterbalances the electrostatic effect of the counterion. The excitation energy in solution is more similar to the chromophore–counterion couple than to the protein thus indicating that the offsetting effect of solvent environment is less marked. Furthermore, the same data reveal that the dynamic correlation energy provided by the CASPT2 correction is important both in the protein environment and in the gas phase (indeed in the specific case of Rh it is even larger in the protein).

Why is the counterion causing this effect? Quantum chemistry may help us to give an explanation. In fact (see Fig. 12.9) it has been shown that, upon photoexcitation, the positive charge of a simplified model of the Rh chromophore undergoes a 34% (with respect to the centre of the chain) displacement of the positive charge along the molecule

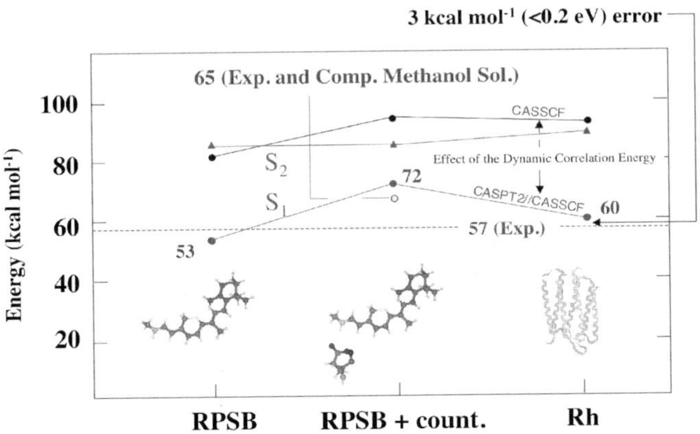

Scheme 12.7.

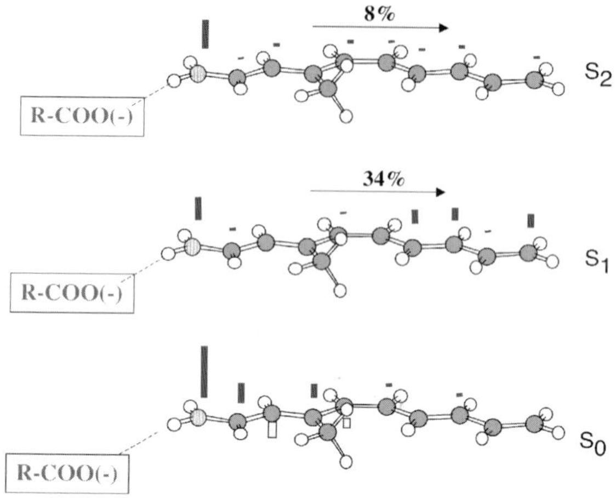

Fig. 12.9. Ground and excited state Mulliken charges along the backbone of Rh showing the displacement of the positive charge along the molecule and toward the C end (adapted from Ref. [8]).

backbone and toward the C end. When the counterion is close to the PSB, the Glu113 stabilizes preferentially the ground state, where the positive charge is located at the N-terminal, than the excited state. This effect yields a larger S_1-S_0 energy gap, leading to the observed blue shift. The specific charge distribution of the protein cavity counterbalances the effect of the counterion decreasing the S_1-S_0 energy gap. Notice also the effect of the counterion stabilization of the S_2 state. As reported in Fig. 12.9, the S_2 state does not show the same charge translocation phenomena seen in S_1. Thus, the counterion do not effectively destabilize S_2 with respect to S_0 and the S_2-S_1 energy gap decreases considerably in the ion pair of Scheme 12.7.

12.3.3 The photoisomerization path of Rh

We have recently demonstrated that the CASPT2//CASSCF level of theory can be used to get realistic information on the excited and ground state relaxation of PSB11 in the Rh cavity or, in other words, to compute the entire photochemical reaction path [50,51]. This information will enable us to understand the mechanism driving the excited state decay and photon energy storage in Rh. Time resolved spectroscopy studies have established that a first transient species (see Fig. 12.10) is formed after 50 fs. This fluorescent intermediate (FS-Rh) features a λ_{max} in the range of 600–700 nm. A second transient species is detected after *ca.* 200 fs (photoRh). The first ground state isolable intermediate is instead formed in 2 ps (bathoRh). In Fig. 12.10 we display the points recently computed along the photochemical reaction path of Rh. We have located the excited state structure corresponding to FS-Rh. Its energy is evaluated to be 1 kcal mol^{-1} of the experimental value. Then we located the structure of the lowest lying S_1/S_0 conical intersection (CI-Rh), which corresponds also to the absolute S_1 minimum. This structure

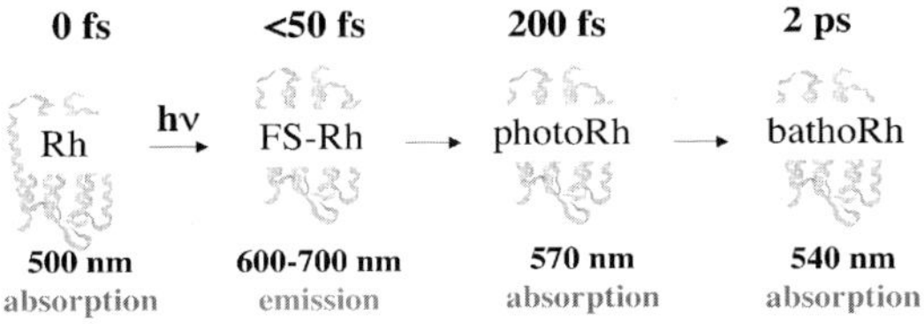

Fig. 12.10. Computed points along the photochemical reaction path of Rh (adapted from Ref. [8]).

displays a *ca.* 90° twisted central $C_{11}-C_{12}$ double bond. Starting at the conical intersection geometry and *via* standard geometry optimization it has been possible to locate the first stable ground state intermediate (bathoRh) with a 540 nm of absorption and featuring an *all-trans*-like chromophore structure (about 150° twisted $C_{11}-C_{12}$ double bond).

It has not yet been possible to unambiguously assign the photoRh transient that may well be a low energy excited state or an high energy ground state transient.

Concerning the energy storage problem (*ca.* 32 kcal mol^{-1} are efficiently stored in bathoRh), the results of our CASPT2//CASSCF/AMBER energy profile (Fig. 12.11) show that we are able to reproduce energy storage with an error of 5 kcal mol^{-1}.

What about the geometry evolution on the excited state? The change in the chromophore structure, occurring immediately after excitation, regards the bond lengths of the $-C_9{=}C_{10}-C_{11}{=}C_{12}-C_{13}{=}C_{14}-$ moiety. In fact, at FS-Rh there is a complete inversion of the single and double bond positions (Fig. 12.12). The CI-Rh displays a highly helical structure compared to Rh and FS-Rh and is mainly characterized by large structural change in the $-C_9{=}C_{10}-C_{11}{=}C_{12}-C_{13}{=}C_{14}-$ moiety. Thus, the motion driving the $S_1 \rightarrow S_0$ decay is mainly torsional with rotation of about 68° (22° → 90°) around the $C_{11}{=}C_{12}$ bond and 37 and 15° twisting around the $C_9{=}C_{10}$ and $C_{13}{=}C_{14}$ bonds, respectively. In contrast to the initial excited state relaxation, at conical

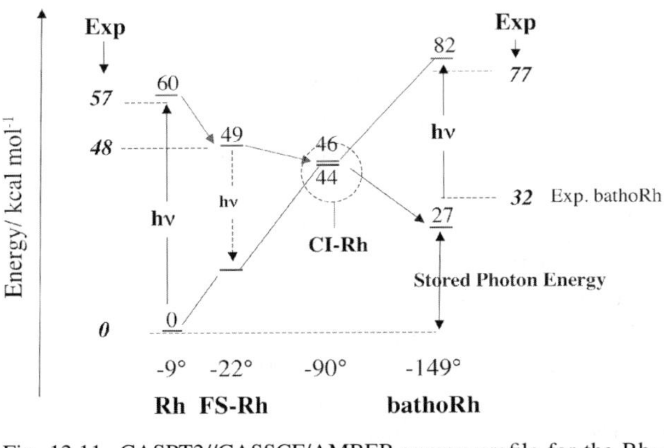

Fig. 12.11. CASPT2//CASSCF/AMBER energy profile for the Rh → bathoRh photochemical reaction compared to the experimental values (in bold) (redrawn with permission from Ref. [50] © 2004 by The National Academy of Sciences of the USA).

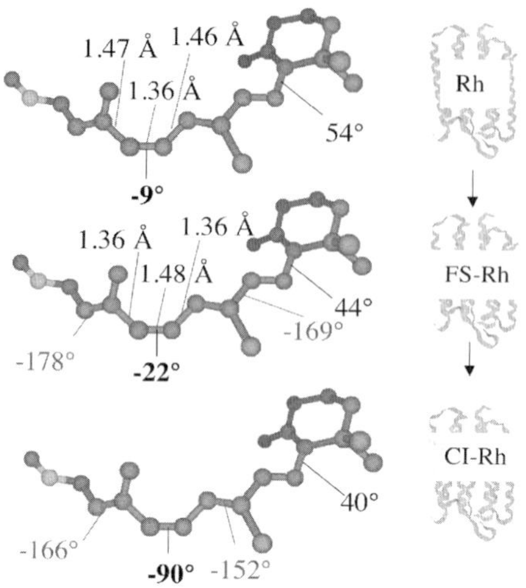

Fig. 12.12. Geometry evolution along the excited state reaction path connecting Rh to CI-Rh (adapted from Ref. [8]).

intersection all the other bonds remain substantially unchanged. Therefore, from a general point of view, during photoisomerization the structural changes do not occur exclusively at the central double bond but also involve the other two adjacent double bonds leading to a global change in the helicity of the chromophore.

12.3.4 Nature of the energy storage

By decomposing the computed 26 kcal mol^{-1} stored photon energy in terms of change in steric interaction between the chromophore and the protein cavity we have seen that this contributes for only *ca.* 7 kcal mol^{-1}. This is consistent with the fact that the photoisomerization requires little displacement/reorientation of the PSB11 backbone. The superposition of the optimized bathoRh and Rh ground state structures shows that the position of the PSB11 and β-ionone ring remains substantially unchanged while the structural change affects only the central fragment (Fig. 12.13).

Since the calculated change in the electrostatic interaction between the chromophore and the protein cavity reveals a contribution of other *ca.* +6 kcal mol^{-1}, the change in steric and electrostatic interaction must account for only half of the stored energy. So the photon energy has to be mainly stored in terms of the strain energy of the chromophore. In particular, it is due to the distorted π system of bathoRh (bond lengths and angles of the other moiety of the bathoRh are close to Rh) with respect to Rh.

In conclusion, the 32 kcal mol^{-1} photon energy stored in bathoRh are not fully related to a charge separation (a change in the electrostatic interaction between the chromophore and the protein cavity) but mostly 'reside' in the highly strained chromophore of bathoRh where three different double bonds are not completely reconstituted. Indeed the calculations reveal that three highly twisted double bonds in the first (isolable)

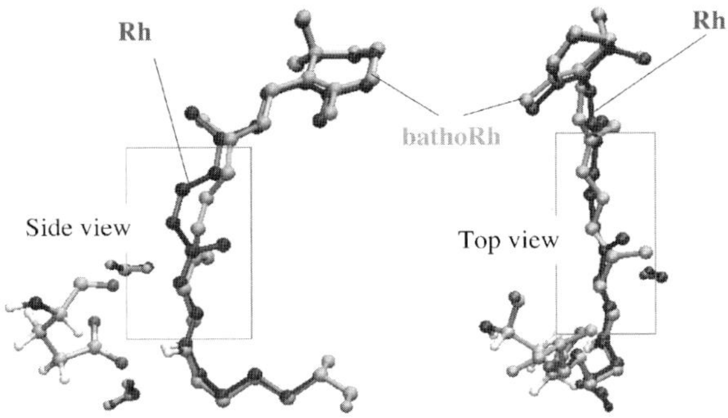

Fig. 12.13. Superposition of bathoRh and Rh structures. The frame indicates the part of a molecule that undergoes a structural change (redrawn with permission from Ref. [50] © 2004 by The National Academy of Sciences of the USA).

intermediate bathoRh are responsible for ~50% of the stored photon energy. The change in electrostatic and steric interaction of the chromophore with the protein cavity accounts each one for the 25% of the stored energy.

12.4 FROM PHOTOBIOLOGY TO BIOMIMETIC MOLECULAR SWITCHES

Can what we have learnt from photobiology be used to design novel and more efficient light-driven molecular devices? In other words, can we design biomimetic photochemical switches that exploit radiative energy to achieve a well-defined molecular structure deformation? The design, synthesis and characterization of this type of molecular devices appear to be of great importance in the field of modern technology such as the construction of light powered nanodevices and, in particular, in the field of the conformational control of biopolymers [52–54]. In this last field one can envision applications such as that of using light pulses to trigger protein folding/denaturation or to generate dynamical combinatorial libraries based on polypeptide backbones [55–57]. Prototypes for the preparation of fully efficient single-molecule switch may be found among molecules involved in photochemical reactions of biological photoreceptors, such as those occurring in Rh proteins. Indeed the photoisomerization of PSB11 can be seen as extremely efficient single-molecule molecular motors due to the fact that the photoisomerization is stereoselective, unidirectional, ultrafast and occurs with high quantum yield. Thus, molecules mimicking the PSB chromophore of Rh can be proposed as new types of biomimetic switches.

In the present context, we are currently studying peptidomimetic polycyclic structures derived from the thioredoxin reductase active site [56,57], which consist of a domain of peptidic nature containing the sequence for the recognition and of a domain of non-peptidic nature. In the present case, this last domain is formed by an unnatural amino acid of the size (i.e. isosteric) of a bi- or tri-peptide that can undergo *cis–trans* photoisomerization. Such isomerization induces conformational changes in

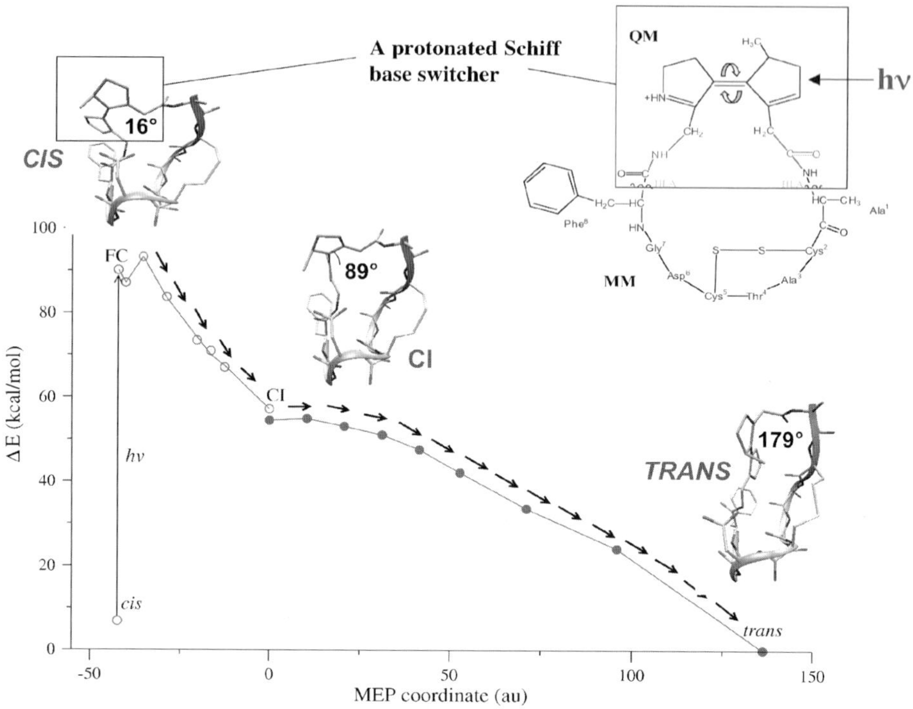

Fig. 12.14. Energy profiles along the S_1 and S_0 *cis* → *trans* photoisomerization coordinate of the chromophore in blue included in the oligopeptide chain (adapted from Ref. [58]).

the peptide domain covalently attached to the motor mainly due to changes in the skeletal strain tension (angular or torsional). In general, such a ligand will have at least two tunable states.

The structure in blue, that represents the non-peptidic domain currently investigated in our group, is displayed in Fig. 12.14 together with the computed excited state reaction path [58]. It is constituted by a protonated Schiff base in which the isomerization of the two terminal double bonds is blocked by the inclusion in two five-membered rings and that mimics the photoisomerization of Rh chromophore. Such chromophore is rigid because of the presence of the central double bond and of the rings. Obviously, a light triggered *cis–trans* isomerization in the switch will lead to remarkable conformational modifications in the peptide moiety. These conformational changes have been simulated using the same QM/MM method (CASPT2//CASSCF/AMBER) implemented in our laboratory [58]. In the diagram of Fig. 12.14 is shown the results of a photochemical reaction path calculations (*in vacuo*) for the whole system. The CI is 90° twisted as in the retinal chromophore. Along the excited state reaction path, the oligopeptide structure does not change, but following the ground state relaxation toward the *trans* form of the chromophore the oligopeptide structure is largely modified.

A series of systems chemically similar to the chromophore shown in Fig. 12.14 have already been synthesized in our laboratory [59].

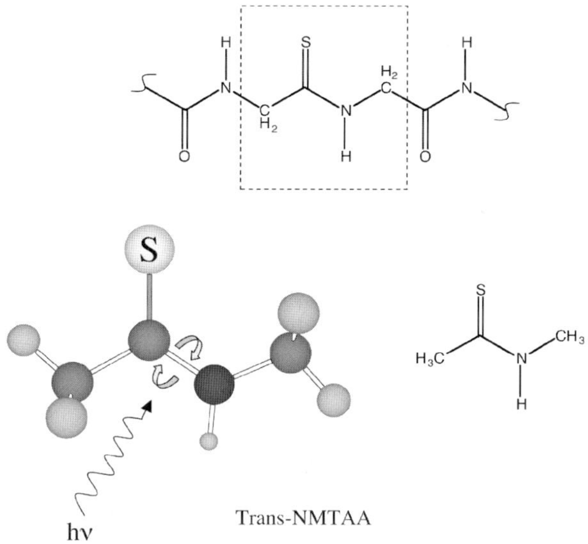

Fig. 12.15. Structure of *trans*-NMTAA. *Trans*-NMTAA can be inserted in an oligopeptide chain and photoisomerized to its *cis* form (adapted from Ref. [60]).

A different, minimal, non-natural aminoacid, that has been investigated in an attempt to control the conformation of an oligopeptide with light, is *N*-methylthioacetamide (NMTAA). As shown in Fig. 12.15, *trans*-NMTAA is isosteric to a dipeptide that contains a single thiocarbonyl group. The presence of such functional group in the amidic bond has two important consequences: (i) the λ^{a}_{max} of the thioamidic bond is distinct from the one of standard amidic bonds and (ii) light absorption at such wavelengths induces selective *cis* → *trans* isomerization of the thioamidic group. The mechanism of NMTAA has been studied in depth using the CASPT2//CASSCF approach. These studies have shown that both excited singlet and triplet states are involved in the reaction [60].

12.5 CONCLUSIONS

We hope we have provided some evidence that it is, nowadays, possible to investigate, at a semiquantitative level, photobiological problems using computational tools. In particular, the use of the *ab initio* CASPT2//CASSCF computational strategy hybridized with molecular mechanics force fields can be used to investigate both biological photoresponsive molecules and to design new biomaterials inducing conformational changes in oligopeptides. We believe that these two complementary aspects of modern research in computational chemistry will become more and more important in the near future.

12.6 ACKNOWLEDGEMENTS

Funds have been provided by the Università di Siena (Progetto di Ateneo 02/04) and HFSP (RG 0229/2000-M). We thank CINECA for granted calculation time.

12.7 REFERENCES

1 A. Sinicropi, U. Pischel, R. Basosi, W.M. Nau and M. Olivucci, Angew. Chem. Int. Ed., 39 (2000) 4582–4586.
2 A. Sinicropi, R. Pogni, R. Basosi, M.A. Robb, G. Gramlich, W.M. Nau and M. Olivucci, Angew. Chem. Int. Ed., 40 (2001) 4185–4189.
3 A. Sinicropi, W.M. Nau and M. Olivucci, Photochem. Photobiol. Sci., 1 (2002) 537–546.
4 N. Ismail, L. Blancafort, M. Olivucci, B. Kohler and M.A. Robb, J. Am. Chem. Soc., 124 (2002) 6818.
5 M. Garavelli, C.S. Page, P. Celani, M. Olivucci, W.E. Schmid, S.A. Trushin and W. Fuss, J. Phys. Chem. A, 105 (2001) 4458–4469.
6 F. Bernardi, M. Olivucci and M.A. Robb, Chem. Soc. Rev., 25 (1996) 321–328.
7 F. Bernardi, M. Olivucci, J. Michl and M.A. Robb, Spectrum, 9 (1996) 1–6.
8 A. Sinicropi, T. Andruniow, L.D. Vico, N. Ferré and M. Olivucci, Pure Appl. Chem., in press (2005).
9 H.E. Zimmerman, J. Am. Chem. Soc., 88 (1966) 1566.
10 H.E. Zimmerman, Science, 153 (1966) 837.
11 H.E. Zimmerman, J. Am. Chem. Soc., 88 (1966) 1564.
12 J. Michl, Mol. Photochem., 4 (1972) 243–255.
13 J. Michl, Top. Curr. Chem., 46 (1974) 1.
14 L. Salem, J. Am. Chem. Soc., 96 (1974) 3486–3501.
15 D.R. Yarkony, J. Chem. Phys., 92 (1990) 2457.
16 D.R. Yarkony, Rev. Mod. Phys., 68 (1996) 985.
17 D.R. Yarkony, J. Phys. Chem. A, 105 (2001) 6277.
18 G.J. Atchity, S.S. Xantheas and K.J. Ruendeberg, J. Chem. Phys., 95 (1991) 1862–1876.
19 F. Bernardi, S. De, M. Olivucci and M.A. Robb, J. Am. Chem. Soc., 112 (1990) 1737–1744.
20 F. Bernardi, M. Olivucci and M.A. Robb, Acc. Chem. Res., 23 (1990) 405–412.
21 See A. Migani and M. Olivucci, Conical intersections and organic reaction mechanisms, in: W. Domcke, D.R. Yarkony, H. Köppel (Eds.), Conical intersections: Electronic structure, dynamics and spectroscopy, *World Scientific*, Singapore, 2004, and references cited therein.
22 M.J. Bearpark, M.A. Robb and H.B. Schlegel, Chem. Phys. Lett., 223 (1994) 269.
23 P. Celani, M.A. Robb, M. Garavelli, F. Bernardi and M. Olivucci, Chem. Phys. Lett., 243 (1995) 1–8.
24 M. Garavelli, P. Celani, M. Fato, M.J. Bearpark, B.R. Smith, M. Olivucci and M.A. Robb, J. Phys. Chem. A, 101 (1997) 2023–2032.
25 A. Sinicropi, C.S. Page, W. Adam and M. Olivucci, J. Am. Chem. Soc., 125 (2003) 10945–10959.
26 C.S. Page and M. Olivucci, J. Comput. Chem., 24 (2003) 298–309.
27 M.E. Martin, F. Negri and M. Olivucci, J. Am. Chem. Soc., 126 (2004) 5452–5464.
28 S. Fantacci, A. Migani and M. Olivucci, J. Phys. Chem. A, 108 (2004) 1208–1213.
29 M.A.L. Marques and E.K.U. Gross, in: F. Fiolhas, M. Nogueira (Eds.), A primer in density functional theory LNP 620, 2003, pp. 144–84.
30 F. Furche and A. Reinhart, J. Chem. Phys., 117 (2002) 7433.
31 A. Toniolo, G. Granucci and T.J. Martinez, J. Phys. Chem. A, 107 (2003) 3822–3830.
32 A. Toniolo, S. Olsen, L. Manohar and T.J. Martinez, Faraday Discuss., 127 (2004) 149–163.
33 A. Warshel and Z.T. Chu, J. Phys. Chem. B, 105 (2001) 9857–9871.
34 W.D. Cornell, P. Cieplak, C.I. Bayly, I.R. Gould, K.M. Merz, Jr., D.M. Ferguson, D.C. Spellmeyer, T. Fox, J.W. Caldwell and P.A. Kollman, J. Am. Chem. Soc., 117 (1995) 5179–5197.
35 A.D. MacKerell, Jr., D. Bashford, M. Bellott, R.L. Dunbrack, Jr., J. Evanseck, M.J. Field, S. Fischer, J. Gao, H. Guo, S. Ha, D. Joseph, L. Kuchnir, K. Kuczera, F.T.K. Lau, C. Mattos, S. Michnick, T. Ngo,

D.T. Nguyen, B. Prodhom, W.E. Reiher, III, B. Roux, M. Schlenkrich, J. Smith, R. Stote, J. Straub and J. Watanabe, J. Phys. Chem. B, 102 (1998) 3586–3616.
36 N. Ferré and M. Olivucci, J. Am. Chem. Soc., 125 (2003) 6868–6869.
37 N. Ferré and M. Olivucci, J. Mol. Struct. (THEOCHEM), 632 (2003) 71–82.
38 U.C. Singh and P.A. Kollman, J. Comput. Chem., 7 (1986) 718.
39 A. Warshel and M. Levitt, J. Mol. Biol., 103 (1976) 227–249.
40 T. Vreven and K. Morokuma, Theor. Chem. Acc., 109 (2003) 125–132.
41 S. Hayashi, E. Tajkhorshid and K. Schulten, Biophys. J., 83 (2002) 1281–1297.
42 S. Hayashi, E. Tajkhorshid and K. Schulten, Biophys. J., 85 (2003) 1440–1449.
43 A. Warshel, Z.T. Chu and J.-K. Hwang, Chem. Phys. Lett., 158 (1991) 303–314.
44 G. Groenhof, M. Bouxin-Cademartory, B. Hess, S.P.d. Visser, H.J.C. Berendsen, M. Olivucci, A.E. Mark and M.A. Robb, J. Am. Chem. Soc., 126 (2004) 4228–4233.
45 R. Tsien, Annu. Rev. Biochem., 67 (1998) 510.
46 M. Zimmer, Chem. Rev., 102 (2002) 759–781.
47 R.A. Mathies and J. Lugtenburg, in: D.G. Stavenga, W.J. de Grip, E.N. Pugh (Eds.), Handbook of biological physics, Elsevier, Amsterdam, 2000.
48 H. Kandori, Y. Shichida and T. Yoshizawa, Biochemistry (Moscow), 66 (2001) 1197–1209.
49 A. Sinicropi, T. Andruniow, N. Ferré and M. Olivucci, submitted for publication (2005).
50 T. Andruniow, N. Ferré and M. Olivucci, PNAS USA, 101 (2004) 17908–17913.
51 A. Migani, A. Sinicropi, N. Ferré, A. Cembran, M. Garavelli and M. Olivucci, Faraday Discuss., 127 (2004) 179–191.
52 I. Wilner, Acc. Chem. Res., 30 (1997) 347–356.
53 H. Kessler, Angew. Chem. Int. Ed., 21 (1982) 512–523.
54 H.R. Haubner, D. Finsinger and H. Kessler, Angew. Chem. Int. Ed., 36 (1997) 1375–1389.
55 L. Ulysse, J. Cubillos and J. Chmielewski, J. Am. Chem. Soc., 117 (1995) 8466–8467.
56 R. Behrendt, C. Renner, M. Schenk, F. Wang, J. Wachtveitl, D. Oesterhelt and L. Moroder, Angew. Chem. Int. Ed., 38 (1999) 2771–2774.
57 C. Renner, R. Behrendt, S. Sporlein, J. Wachtveitl and L. Moroder, Biopolymers, 54 (2000) 489–500.
58 T. Andruniow, S. Fantacci, F.D. Angelis and M. Olivucci, submitted for publication (2005).
59 D.S. Ruiz, A. Migani, A. Pepi, E. Busi, R. Basosi, L. Latterini, F. Elisei, S. Fusi, F. Ponticelli, V. Zanirato and M. Olivucci, J. Am. Chem. Soc., 126 (2004) 9349–9359.
60 J. Helbing, H. Bregy, J. Bredenbeck, R. Pfister, P. Hamm, R. Huber, J. Wachtveit, L.D. Vico and M. Olivucci, J. Am. Chem. Soc., 126 (2004) 8823–8824.

Theory and Applications of Computational Chemistry: The First Forty Years
Edited by C. Dykstra et al.

CHAPTER 13

The nature of the chemical bond in the light of an energy decomposition analysis

Matthias Lein and Gernot Frenking

Fachbereich Chemie, Philipps-Universität Marburg, D-35043 Marburg, Hans-Meerwein-Strasse, Germany

Abstract

In this work we summarize the results of quantum chemical calculations where we investigate the nature of the chemical bond in main-group and transition metal compounds with an energy decomposition analysis (EDA). The EDA decomposes the instantaneous interaction energy A–B between two fragments A and B into three terms which can be interpreted in a chemically meaningful way. The three terms are the quasiclassical electrostatic interaction between the frozen charges of the fragments ΔE_{elstat}, the exchange (Pauli) repulsion between electrons possessing the same spin ΔE_{Pauli}, and the orbital interaction term ΔE_{orb}. The latter term can be divided into contributions of orbitals having different symmetry, which allows an estimate of the strength of σ, π, and δ bonding. The results show that the quasiclassical electrostatic interaction significantly contributes to the bonding interactions in all molecules. The trend of the bond strength is in most cases correctly predicted by the orbital term ΔE_{orb} but there are cases where the electrostatic attraction or the Pauli repulsion are more important for an understanding of the bonding interactions. The EDA is an unambiguously defined partitioning scheme, which considers all terms yielding a chemical bond. The EDA can be considered as a bridge between the classical heuristic bonding models of chemistry and the physical mechanism of chemical bond formation.

13.1 INTRODUCTION

Numerical quantum chemistry has been the most rapidly developing field of chemistry in the second half of the last century. Forty years ago, quantum chemistry was mainly practiced by theoretical physicists while very few chemists were actively engaged in the field. It was hardly considered by orthodox chemists to be part of chemical sciences because the foundations of quantum chemistry are deeply rooted in theoretical physics while chemistry in the past was nearly exclusively an experimental science [1]. After 20 years of ground-breaking method development, quantum chemistry made a first impact on a broad scale in organic chemistry in the 1970s when MO theoretical methods were used for explaining and predicting electrocyclic reactions which could not be understood with classical bonding models. The breakthrough came with the introduction of the frontier orbital model of Fukui [2] and the orbital symmetry model of Woodward and Hoffmann [3]. At the same time, the numerical accuracy of quantum chemical methods which was achieved due to method development and technical advances in computer hardware reached a level which led theoretical chemists to question experimental results. This was a daring undertaking in those days! The trust, which is nowadays placed on the results of high-level ab initio calculations stems from the time when the results of the decade long development of quantum chemical methods showed that theory and experiment can challenge each other. Chronicles of this development have been published by Henry F. Schaefer III who is one of the pioneers of Computational Chemistry [4].

At a conference on molecular quantum mechanics in Boulder, Colorado in 1960, Charles Coulson reminded the theoretical chemists that computational quantum chemistry did not pay enough attention to the heuristic chemical concepts and bonding models which were used by experimental chemists. He requested that "the role of quantum chemistry is to understand these concepts and show what are the essential features in chemical behavior". The statement "give us insight not numbers" is a quote from his famous after dinner speech of that conference. It points to the missing link between numerical quantum chemistry and experimental chemistry. While the former discipline impressed by ever higher accuracy in computing molecular geometries, energies and other properties, most theoretical chemists did not bother in the interpretation of the results in terms of chemical bonding models. A notable exception besides Coulson [5] was Linus Pauling [6], who underpinned the pre-quantum chemical bonding model of Gilbert Lewis [7] with quantum theoretical arguments which were not undisputed, however. Paulings dogmatic view in favor of valence bond theory and his refusal of recognizing molecular orbital theory as a valid theory of chemical bonding made it deceptively easy for experimental chemists to continue using their classical bonding models, but a true understanding of the nature of the chemical bond in terms of quantum chemical laws was not achieved.

An important contribution to the latter goal was made in the 1960s by Klaus Ruedenberg [8] who studied the contributions of potential and kinetic energy to the chemical bond. According to the quantum virial theorem, the total energy E of a molecule at equilibrium is related to the average values of the potential energy $\langle V \rangle$ and kinetic energy $\langle T \rangle$ as given by Eq. (1):

$$E = 1/2\langle V \rangle = -\langle T \rangle \qquad (1)$$

According to Eq. (1), the energy lowering caused by the formation of a chemical bond means lowering of the potential energy and increase of kinetic energy. Ruedenberg [8] showed that *the driving force for the chemical bond formation is actually a lowering of the kinetic energy and rising of the potential energy* [9]. This is because, during the bond formation, the valence electrons of the atoms are initially moving away from the nuclei toward the interatomic bonding region where the kinetic energy is lower than near the nucleus. The virial theorem at equilibrium [10] becomes fulfilled by the contraction of the atomic orbitals which is an intra-atomic process [11].

The ground breaking work in the 1970s by Fukui [2] and by Woodward and Hoffmann [3] pioneered the acceptance of MO theory by experimental chemists. Because of the big success in explaining chemical phenomena, most textbooks discuss now covalent bonds in terms of orbitals and orbital interactions. It is important to recognize that the model of a covalent bond consisting of a pair of electrons shared between two atoms was introduced into chemistry by Lewis already 50 years earlier and before modern quantum chemistry was developed [7]. The present understanding of chemical bonding considers orbital interactions as the driving force for covalent bonding while ionic bonds are discussed in terms of classical electrostatic attraction between opposite charges. It is not true, however, that classical electrostatic attraction is unimportant for covalent interactions. The latter impression which is held by many chemists stems from the analysis of the interatomic interactions in H_2 which goes back to the pioneering work of Heitler and London [12]. In this chapter, which is the work that gave birth to the field of quantum chemistry [13] the authors show that the attractive hydrogen–hydrogen interactions come mainly from "a characteristic quantum mechanical resonance phenomenon which is closely related to the resonance vibrations that were found by Heisenberg" [12]. But the bonding in dihydrogen is atypical for nonpolar bonds! Hirshfeld and Rzotkiewicz reported in 1974 that the net electrostatic attraction in the heavier first-row E_2 molecules provides more of the binding energy than those contributions that are the source of the strong binding in H_2 [14]. In 1986 Spackman and Maslen calculated the electrostatic energies of 148 polar and nonpolar diatomic compounds [15]. They showed that the quasiclassical electrostatic attraction is always very large *except for* H_2. The calculated values were in many cases even larger than the bond dissociation energies! For example, the quasiclassical electrostatic attraction between two spherical nitrogen atoms in the ^4S ground state at the equilibrium distance of N_2 was calculated as $\Delta E_{elstat} = -330.7$ kcal/mol [15] while the bond dissociation energy (BDE) of dinitrogen is $D_e = 228.5$ kcal/mol [16]. This is in striking contrast to H_2 which has a BDE of $D_e = 109.6$ kcal/mol [16] but the calculated value for ΔE_{elstat} is only -1.4 kcal/mol [15].

The work by Spackman and Maslen did not address the question about other energy contributions to the chemical bonding in diatomic molecules E_2. In particular, the question about the strength of the orbital interactions ΔE_{orb} which are frequently considered as the sole driving force of covalent bonding needs to be answered. Because the classical electrostatic attraction is already larger than the BDE, there must be significant Pauli repulsion ΔE_{Pauli} between electrons having the same spin which is the third contributor to the total interatomic interactions energy ΔE_{int} in E_2. The quantitative knowledge about the strength of ΔE_{elstat}, ΔE_{orb} and ΔE_{Pauli} would yield a complete picture of the nature of the chemical bond using three mathematically well-defined terms (see below) which can be meaningfully interpreted. Since ΔE_{elstat} can be identified with ionic bonding while ΔE_{orb} gives the conceptually different type of binding interactions in covalent bonds, the calculated size of the energy terms may be used as a bridge between exact quantum chemical calculations and the heuristic bonding models of chemistry.

In this chapter we summarize our systematic investigations of the nature of the chemical bond in transition metal (TM) molecules [17–37] and main group compounds [37–43] employing an energy decomposition analysis (EDA) which was originally developed by Morokuma [44] for ab initio methods and by Ziegler and Rauk [45] for DFT methods. The EDA method has been further developed by Baerends and coworkers [46]. Initially, the focus of our work was on the chemical bonding in donor–acceptor complexes which has been summarized in a review [47]. In more recent studies we analyzed the nonpolar bonds between main-group elements, which are classified as typical covalent bonds. The goal of this work is to give a rigorously defined quantum chemical analysis of the chemical bond that connects the established qualitative bonding models of chemistry with quantitative data which come from accurate calculations. The data are not exclusive in the sense that they give a definite answer about the nature of the chemical bond, because other partitioning schemes than the EDA may be suggested. The advantage of the EDA is that the energy terms can be interpreted in a plausible way which provides a quantitative ordering scheme for different chemical bonds.

We wish to comment on the interpretation and meaning of covalent bonding [48]. The definition of the latter according to Pauling is: "The covalent bond consists of a pair of electrons shared between two atoms, and occupying two stable orbitals, one of each atom" [49]. Paulings definition is based on the suggestion of Lewis who wrote: "The chemical bond is at all times and in all molecules merely a pair of electrons held jointly by two atoms" [50]. This type of bonding is distinguished from ionic bonding that is found, for example, in ionic solids. Covalent bonds thus arise from shared-electron interactions while ionic bonds come from closed-shell interactions between separated charged species. Purely, ionic bonds can be explained with classical electrostatic interactions. Nonpolar bonds like in N_2 are according to Paulings definition purely covalent, while polar bonds have ionic and covalent contributions [51]. As noted above, nonpolar bonds have significant contributions from classical electrostatic interactions besides the 'resonance phenomenon' which in modern terminology is called orbital interactions. In our previous work, we suggested that covalent bonding should be identified with orbital interactions (resonance interactions) because the difference between a shared-electron bond and an ionic bond lies in the quantum theoretical resonance term that is important for the former bond. Pauling identified ionic bonding with classical electrostatic bonding which according to the valence bond model [52] becomes important in polar molecules where the ionic resonance forms contribute to the bond energy [51]. It was later found [14,15] that classical electrostatic interactions are strong in nonpolar bonds, too. It is thus reasonable to identify covalent bonding with orbital interactions. This means, however, that the understanding of a covalent bond in our work is different from the definition given by Pauling. According to our definition, nonpolar bonds are only partially covalent and partially electrostatic. Ionic bonds are purely electrostatic in our and Paulings definition, but strong electrostatic interactions are also found in bonds which according to Pauling are purely covalent.

The work which is mentioned above and the results which are presented here show that the progress which was made during the first 40 years in theory and application of Computational Chemistry is not restricted to numerical quantum chemistry. Enormous advances have also been made in the interpretation and understanding of the chemical bond bridging the gap between the physical origin of the chemical bond [9] and chemical bonding models.

We wish to point out that other groups have also been very active in analyzing the chemical bond using the EDA. Most previous work was devoted to closed-shell donor–acceptor interactions. Early work particularly about metal–ligand interactions has been reported by Ziegler et al. [53]. More recent EDA studies which included studies about the interactions between open-shell fragments have been published by Bickelhaupt and coworkers [54].

13.2 ENERGY DECOMPOSITION ANALYSIS

The focus of the bonding analysis is the instantaneous interaction energy, ΔE_{int}, of the bond, which is the calculated energy difference between the molecule and the fragments in the frozen geometry of the compound. The interaction energy is divided into three main components:

$$\Delta E_{\mathrm{int}} = \Delta E_{\mathrm{elstat}} + \Delta E_{\mathrm{Pauli}} + \Delta E_{\mathrm{orb}} \tag{2}$$

$\Delta E_{\mathrm{elstat}}$ gives the electrostatic interaction energy between the fragments, which are calculated using the frozen electron density distribution of the fragments in the geometry of the molecules. Note that the choice of the electronic state of the fragments is not always trivial. It must be the proper reference state with regard to the electronic state of the molecule. The reference state may not be the electronic ground state of the fragment. For example, the electronic reference state of the CH fragment in acetylene, HC≡CH, is the $(^4\Sigma^-)$ quartet state but the electronic ground state is $^2\Pi$. The reader should, therefore, pay attention to the choice of the fragments in the EDA calculations. Details about the electronic states and the spatial arrangement of the fragments are given in the original publications. In some cases it was useful to analyze the bonding interactions twice using different fragments for the EDA calculation.

The second term in Eq. (2), $\Delta E_{\mathrm{Pauli}}$, refers to the repulsive interactions between the fragments, which are caused by the fact that two electrons with the same spin cannot occupy the same region in space. $\Delta E_{\mathrm{Pauli}}$ is calculated by enforcing the Kohn–Sham determinant on the superimposed fragments to obey the Pauli principle by antisymmetrization and renormalization. The stabilizing orbital interaction term, ΔE_{orb}, which is equivalent to the Heitler–London resonance phenomenon [12] is calculated in the final step of the energy partitioning analysis when the Kohn–Sham orbitals relax to their optimal form. This term can be further partitioned into contributions by the orbitals belonging to different irreducible representations of the point group of the interacting system. The interaction energy, ΔE_{int}, can be used to calculate the BDE, D_e, by adding ΔE_{prep}, which is the energy necessary to promote the fragments from their equilibrium geometry in the electronic ground state to the geometry and electronic reference state in the compounds (Eq. (3)). Further details of the energy partitioning analysis can be found in the literature [46].

$$-D_e = \Delta E = \Delta E_{\mathrm{prep}} + \Delta E_{\mathrm{int}} \tag{3}$$

It is important to recognize the physical meaning and the relevance of the energy terms given by the EDA, ΔE_{elstat}, ΔE_{Pauli} and ΔE_{orb}, in order to avoid misinterpretations of the calculated numbers. In particular, we want to point out that the calculation of ΔE_{elstat} uses the electron density distribution of the fragments without considering the polarization of the charge distribution that comes from the chemical interactions. The polarization is only considered in the final step of the EDA, which means that the stabilization due to polarization and relaxation is completely included in the ΔE_{orb} term. Thus, the calculated electrostatic interaction energy is not the same as the total potential energy change in the bond formation, because the final electron density differs from the superposition of the two fragment densities. While the ΔE_{elstat} term contains only quasiclassical electrostatic interactions which come from the frozen electron densities of the fragments, the ΔE_{orb} term contains electrostatic attraction coming from quantum interference, potential energy changes due to polarization and relaxation and kinetic energy contributions. The ΔE_{Pauli} term contains also potential energy contributions because electronic charge is removed from the overlap area of the fragments closer to the nuclei which is actually lowering the energy. The increase of the total energy due to the ΔE_{Pauli} term comes from the kinetic energy of the electrons which is much higher when they come closer to the nuclei. This is the reason that the ΔE_{Pauli} term is sometimes called kinetic repulsion [46].

The calculations which are reported have been carried out at the following levels of theory. The bond lengths and EDA data of the diatomic molecules which are discussed in Sections 3.1–3.3 have been calculated at the nonlocal DFT level of theory using the revised PBE exchange functional proposed by Hammer–Hansen–Norskov [55] in conjunction with the correlation functional of Perdew–Burke–Ernzerhof [56] (RPBE). The geometries of the other molecules have been optimized at the nonlocal DFT level using the exchange functional of Becke [57] in conjunction with the correlation functional of Perdew [58] (BP86). Uncontracted Slater-type orbitals (STOs) were employed as basis functions for the SCF calculations at the RPBE and BP86 levels [59]. The basis sets have triple-ζ quality augmented by either one (RPBE/TZP; BP86/TZP) or two (BP86/TZ2P) sets of polarization functions, i.e. p and d functions on hydrogen and d and f functions on the other atoms. The core electrons of the atoms were treated by the frozen core approximation [60]. An auxiliary set of s, p, d, f and g STOs was used to fit the molecular densities and to represent the Coulomb and exchange potentials accurately in each SCF cycle [61]. Scalar relativistic effects have been considered using the zero-order regular approximation (ZORA) [62]. All structures have been verified as minima on the potential energy surface by calculating the Hessian matrices. The EDA calculations were performed at the same level as the geometry optimizations. All calculations were carried out with the program package ADF [46].

13.3 BONDING IN MAIN-GROUP COMPOUNDS

We will first present the EDA results for the dihydrogen molecule which is the standard molecule in curricula and textbooks for discussing the nature of covalent bonding. Then we compare the results for H_2 with heavier diatomic molecules N_2 and isoelectronic CO and BF followed by a discussion of the dipnicogens N_2–Bi_2 and the dihalogens F_2–I_2. After this we will present the results for more complex molecules.

13.3.1 Diatomic molecules H_2, N_2, CO, BF [40]

We begin the discussion of the nature of the chemical bond with the archetypical example H_2. According to the definition of Pauling [6] H_2 should be a purely covalently bound species. The partitioning of the energy given in Table 13.1 shows, that this complies with our definition which identifies covalent bonding with orbital interactions.

The calculated BDE of $D_0 = 106.3$ kcal/mol is in good agreement with the experimental value of $D_0 = 109.5$ kcal/mol [16]. The interaction energy between the two hydrogen atoms in their respective doublet ground state yields $\Delta E_{int} = -112.9$ kcal/mol. Because there are only two electrons with opposite spin, there is no Pauli repulsion between the two fragments [63]. Quite interestingly the electrostatic term $\Delta E_{elstat} = 5.8$ kcal/mol at the equilibrium distance is repulsive. The quasiclassical electrostatic interactions become attractive only at longer H–H distances. Thus it is only the orbital term $\Delta E_{orb} = -118.7$ kcal/mol that leads to a chemical bond in H_2.

What about the bonding situation of heavier diatomic molecules? The EDA results for the isoelectronic species N_2, CO and BF are given for comparison in Table 13.1. The zero-point corrected interaction energy ΔE_{int} of N_2 is -232.2 kcal/mol which yields a BDE of $D_0 = 228.8$ kcal/mol. This is in excellent agreement with the experimental value of $D_0 = 225.0$ kcal/mol [16].

The individual terms that sum up to ΔE_{int} show, that the largest contribution to the interatomic interactions in N_2 comes from the Pauli repulsion ΔE_{Pauli}. The absolute value of the attractive orbital term $\Delta E_{orb} = -715.4$ kcal/mol is smaller than the Pauli term $\Delta E_{Pauli} = 791.7$ kcal/mol by 76.3 kcal/mol. This in turn means that the quasiclassical electrostatic attraction $\Delta E_{elstat} = -308.5$ kcal/mol is larger than the BDE $D_e = 232.2$ kcal/mol. This is in agreement with the work of Spackman and Maslen [15]. Thus, unlike in dihydrogen, the quasiclassical electrostatic attraction in dinitrogen

Table 13.1 Energy decomposition analysis of the H–H, N–N, C–O and B–F bonds at RPBE/TZP [40]

	H_2	N_2	CO	BF
ΔE_{int}	−112.9	−232.2	−258.4	−180.8
ΔE_{Pauli}	0.0	791.7	575.8	476.1
ΔE_{elstat}^{a}	+5.8	−308.5 (30.1%)	−240.0 (28.8%)	−210.5 (32.0%)
ΔE_{orb}^{a}	−118.7 (100%)	−715.4 (69.9%)	−594.2 (71.2%)	−446.4 (68.0%)
ΔE_{σ}^{b}	−118.7 (100%)	−470.0 (65.7%)	−301.7 (50.8%)	−396.4 (88.8%)
ΔE_{π}^{b}	0.0	−245.4 (34.3%)	−292.5 (49.2%)	−50.0 (11.2%)
R	0.745 (0.741)	1.105 (1.09768)	1.144 (1.128)	1.285 (1.262)
D_e	−112.9	−232.2	−258.4	−180.8
D_0	−106.3 (−103.3)	−228.8 (−225.0)	−255.4 (−255.7)	−178.9 (179.9)

Energy values in kcal/mol. Bond lengths R in Å. Experimental values are given in parentheses.
[a]The values in parentheses give the percentage contribution to the total attractive interactions $\Delta E_{elstat} + \Delta E_{orb}$.
[b]The values in parentheses give the percentage contribution to the orbital interactions ΔE_{orb}.

significantly contributes to the interatomic attraction. According to our definition of covalent bonding, N_2 is bound 30.1% electrostatic and 69.9% covalent.

The very large values for the quasiclassical attraction in N_2 and in other neutral molecules are sometimes met with scepticism. Referees called it 'unconceivable' or even 'unbelievable' because it seems difficult to imagine that the interaction of two overall neutral species yields strong attraction. The data have been criticized because the calculation of ΔE_{elstat} is obtained from a promolecule whose wave function is neither normalized nor antisymmetrized. A reply to the criticism is given in the following.

The first step of the EDA approach is an estimate of the hypothetical electrostatic interactions between the fragments A and B in terms of *classical* laws of electrostatic forces. The charge densities $\rho(\text{A})$ and $\rho(\text{B})$ are calculated from quantum chemical methods, but ΔE_{elstat} is determined by evaluating the integrals over the unperturbed charge distributions [64]. There are repulsive interactions between the nuclei of A and B which can be considered as point charges and between the electronic charges $\rho(\text{A})$ and $\rho(\text{B})$. There are attractive interactions between the nuclei and the electronic charges. At large distances $R(\text{A–B})$, the attractive and the repulsive terms cancel. At shorter distances when $\rho(\text{A})$ and $\rho(\text{B})$ start to overlap the two interpenetrating electronic charges have a repulsion which is smaller than the repulsion between two point charges at the center of the charges. This is the reason that at intermediate distances, when the overlap between $\rho(\text{A})$ and $\rho(\text{B})$ becomes significant the net interaction between the unmodified charges becomes attractive. Only at very short internuclear distances the nuclear repulsion term, which becomes singular at $R = 0$, dominates the other terms and ΔE_{elstat} becomes repulsive. It should be noted that the quasiclassical electrostatic interaction between polarizable fragments would become even more attractive at intermediate distances than for frozen fragments, because the polarization induces a more favorable charge distribution.

But is it reasonable to use the ΔE_{elstat} term of the EDA for estimating the quasiclassical electrostatic interaction in real molecules? We argue in favor of doing so and we will use dinitrogen as example. Fig. 13.1 shows the changes of the electron density $\Delta\rho(N_2)$ in the various steps of the EDA. Fig. 13.1a gives the change of the density distribution when the wave function becomes normalized and antisymmetrized, i.e. the difference between the density of the first step when ΔE_{elstat} is calculated and the second step when the ΔE_{Pauli} term is determined. It becomes obvious that charge density is removed (dashed lines) from the overlapping region between the nuclei to the backside regions. This is reasonable because, after the first step, electrons with the same spin overlapped in the same space. The charge movement shown in Fig. 13.1a should give a much lower charge attraction than it is given by the ΔE_{elstat} term. But the electron flow from the overlapping region to the nonoverlapping area is reversed in the third step of the EDA when the orbitals are allowed to relax (Fig. 13.1b). It becomes obvious that there is charge accumulation in the bonding region between the atoms which *enhances* the quasiclassical electrostatic attraction. Although Ruedenberg [8] has shown that the enhancement of the electrostatic attraction is not the driving force for the bond formation, the final result of the orbital relaxation is an increase of the electrostatic attraction. Fig. 13.1c shows the total electron flow which comes from steps 1–3, i.e. it is the sum of the changes which are displayed in Fig. 13.1a and b. It becomes obvious that there is an accumulation in

(a)

(b)

(c)

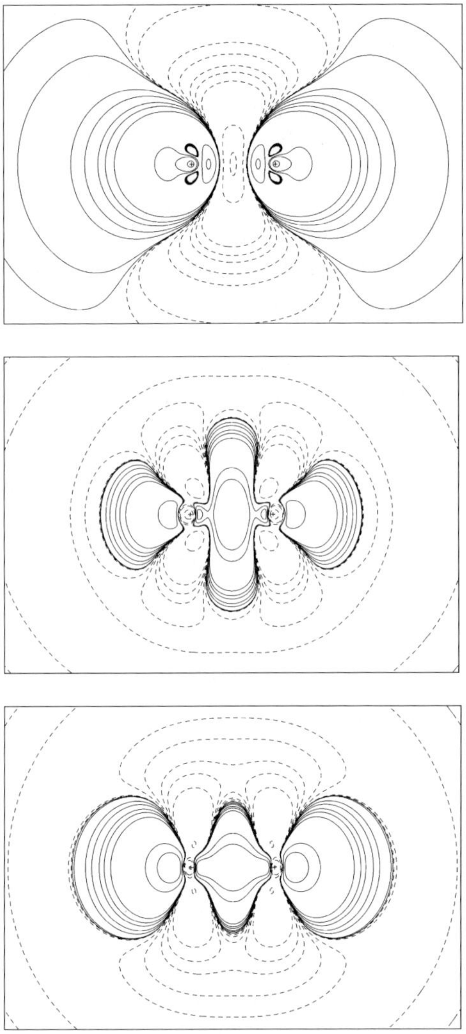

Fig. 13.1. Density difference plots of N_2 at various steps of the EDA calculated at BP86/TZ2P). Solid lines indicate charge accumulation while dashed lines give charge depletion. The contour line values [e/bohr^3] are $\pm$0.001, 0.002, 0.005, 0.01, 0.02, 0.05, 0.1, 0.2, 0.5. (a) Charge flow when the wavefunction of N_2 which is constructed from the superposition of two nitrogen atoms becomes normalized and antisymmetrized: $\Delta E_{elstat} \rightarrow \Delta E_{Pauli}$. (b) Charge flow when the wavefunction completely relaxes after normalization and antisymmetrization: $\Delta E_{Pauli} \rightarrow \Delta E_{orb}$. (c) Sum of the previous two steps: $\Delta E_{elstat} \rightarrow \Delta E_{orb}$.

the bonding region between the nuclei and at the σ lone pair region of the nitrogen atoms. The conclusion is that the values given by the ΔE_{elstat} term may rather underestimate the strength of the quasiclassical electrostatic attraction.

Table 13.1 shows that, in N_2, 65.7% of the orbital term ΔE_{orb} comes from orbitals with σ symmetry. The remaining 34.3% come from the contributions of the π orbitals. This is in nice agreement with the general concept of σ bonds being stronger than π bonds. The absolute values of the orbital interactions appear very large, however. It is important to recognize that the definition of orbital interactions given by the EDA formula for the ΔE_{orb} term is different from the orthodox understanding of orbital interactions which comes from orbital interaction diagrams. The latter model considers Pauli repulsion as part of the orbital interactions. This makes the interaction between two doubly occupied orbitals repulsive. The EDA separates the repulsive interactions between electrons having the same spin from the interactions between electrons having opposite spin which yields attractive ΔE_{orb} values even for the interaction between doubly occupied orbitals like in He_2. The overall repulsion then comes from the ΔE_{Pauli} values.

The separation of total orbital interactions into repulsion (ΔE_{Pauli}) and attraction (ΔE_{orb}) in the EDA must be considered when the results are compared with conclusions which are made from other work. For example, a previous examination of the strength of σ and π orbital interactions in N_2 came to the result that the σ bond between two nitrogen atoms is always much weaker than a σ bond [65]. This is at variance with the EDA results shown in Table 13.1 which suggest that σ bonding is much stronger than π bonding. However, the electrons in the σ orbitals encounter much stronger Pauli repulsion than the electrons in the π orbitals. Although the ΔE_{Pauli} term cannot be broken down into σ and π contributions it is possible to estimate the differences between them. N_2 has a much larger Pauli repulsion (791.7 kcal/mol) than $X(^1\Sigma_g^+)$ C_2 (252.3 kcal/mol) [66]. This is because the 2p(σ) AO of carbon is empty in the $2s^2 2p_x(\pi)^1 2p_y(\pi)^1 2p_z(\sigma)^0$ atomic reference state (^3P) which is used for the interaction in C_2 while the 2p(σ) AO of nitrogen in the ^4S reference state is singly occupied ($2s^2 2p_x(\pi)^1 2p_y(\pi)^1 2p_z(\sigma)^1$). The larger Pauli repulsion in N_2 comes mainly from the overlap of the $2p(\sigma)^1$ AO of nitrogen atom with the $2s^2$ AO of the other nitrogen. This means that the increase of the ΔE_{Pauli} term from C_2 to N_2 (539.4 kcal/mol) can be attributed to the σ orbitals which compensates the attractive value of $\Delta E_\sigma = -470.0$ kcal/mol for N_2.

The individual contributions to the interaction energy ΔE_{int} as a function of the N–N distance are shown in Fig. 13.2. On a large scale (Fig. 13.2a) it can be seen, that the Pauli term ΔE_{Pauli} almost cancels the orbital term ΔE_{orb}. The curve of the interaction energy ΔE_{int} follows the general trend of the electrostatic attraction ΔE_{elstat}, but the latter term has a much lower-lying energy minimum of ~ -450 kcal/mol at a rather short distance. If the quasiclassical electrostatic attraction between the two nitrogen atoms in their ^4S ground state would be the only contributing term, the nitrogen molecule would have a bond distance of just ~0.85 Å. Since the Pauli repulsion becomes dominant at short distances the equilibrium bond length becomes 1.1 Å. It is interesting to note that a quasiclassical calculation of the bond energy of N_2 yields a value which is close to the experimental data. According to the virial theorem (Eq. (1)) the kinetic energy at the equilibrium distance is equal to $-1/2$ of the potential energy (~ -450 kcal/mol) which would give a quasiclassical BDE of $D_e = \sim 225$ kcal/mol.

The results for the EDA analyses of CO and BF are also given in Table 13.1 [67]. CO is the strongest bound diatomic molecule. The experimental BDE $D_0 = 255.7$ kcal/mol [16] is nicely reproduced by our calculations which yield a value of $D_0 = 255.4$ kcal/mol. The interaction energy $\Delta E_{int} = -258.4$ kcal/mol is about 26 kcal/mol larger than in N_2. The individual terms that contribute to ΔE_{int} suggest, that 71.2% of the attractive

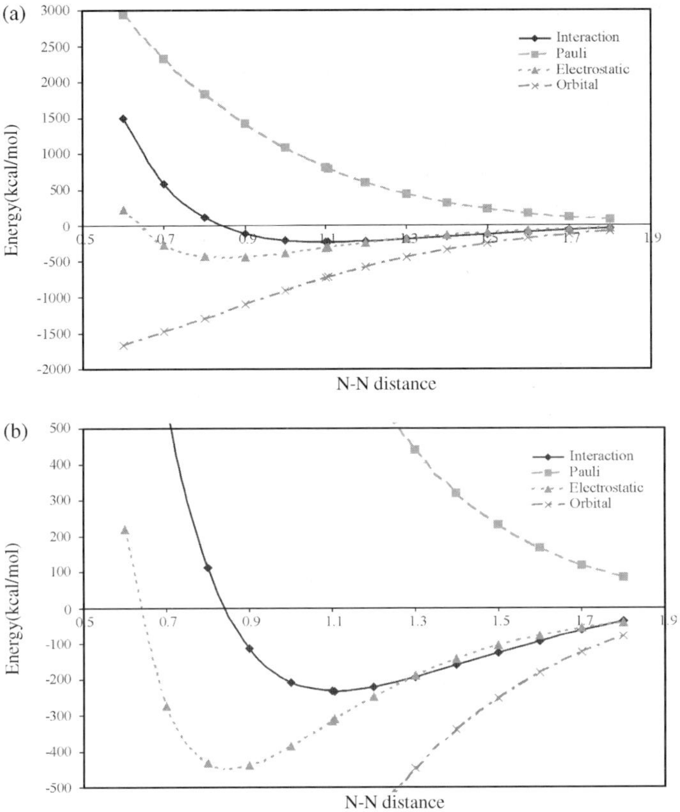

Fig. 13.2. Energy contributions to the bonding in N_2 as a function of N–N interatomic distance [Å] calculated at RPBE/TZP. (a) Small scale. (b) Large scale.

contributions come from the orbital term and only 28.8% come from the quasiclassical electrostatic attraction. Even though CO is heteroatomic, and therefore a polar molecule, the electrostatic attraction $\Delta E_{elstat} = -240.0$ kcal/mol is weaker than in the homopolar N_2. It is thus not only the relative contributions of ΔE_{elstat} and ΔE_{orb} that are smaller in CO than in N_2 but also their absolute values. *CO is stronger bonded than N_2 although the attractive terms in the former are weaker than in the latter*. The difference between the two molecules, that finally leads to the higher dissociation energy of CO, is the smaller Pauli repulsion in CO.

The molecule BF is isoelectronic with N_2 and CO but it has the lowest BDE of the three species. The calculated value $D_0 = 178.9$ kcal/mol is in excellent agreement with the experimental value of $D_0 = 179.9$ kcal/mol. The interaction energy $\Delta E_{int} = -180.8$ kcal/mol of BF is also smaller than for N_2 and CO. Is there an energy term which explains the weaker bond of BF? Inspection of the calculated data given in Table 13.1 shows that the values for all three main contributions ΔE_{elstat}, ΔE_{Pauli}, and ΔE_{orb} in BF are smaller than in N_2 and CO. An explanation for the weaker bonding in the former diatomic can be found when the σ and π contributions to the orbital term are considered. There is a significant difference between the strength of the π bonding in BF and the other species. The ΔE_{orb} term in BF is smaller than in CO although the σ contribution in the former molecule is actually larger. It is the dramatic decrease of the π bonding which leads to the comparatively weak B–F bond. Only 11.2% of the orbital term stems from the contributions of the π orbitals. If BF would have a π bond strength which is comparable to that in N_2 and CO it would have an even stronger bond than CO. The results in Table 13.1 clearly show that CO has a triple bond while BF should be written with a single bond. This is in agreement with the standard Lewis model of chemical bonding.

13.3.2 Dipnicogens N_2–Bi_2 [40]

The analysis of the bonding situation in the dipnicogens N_2–Bi_2 was carried out in order to investigate the difference between first octal-row elements and the heavier homologues in the light of the energy partitioning analysis. The EDA results of the diatomic group-15 molecules are given in Table 13.2.

The EDA results for N_2 have already been discussed in Section 13.3.1 where they were compared with the data for isoelectronic CO and BF. The calculated interaction energy of N_2 $\Delta E_{int} = -232.2$ kcal/mol is much larger than for the heavier analogues P_2–Bi_2. The bond dissociation energies and the interaction energies ΔE_{int} become smaller for the heavier elements. The results clearly show, that the electrostatic character of the E–E bonds increases from N_2 ($\Delta E_{elstat} = 30.1\%$) to Bi_2 ($\Delta E_{elstat} = 58.3\%$). This leads to the interesting result that, according to our definition based on the EDA results, the E–E bonds of Sb_2 and Bi_2 are more electrostatic than covalent although the molecules are nonpolar.

The relative contribution of the π orbitals to the ΔE_{orb} term increases from N_2 ($\Delta E_{\pi} = 34.3\%$) to P_2 ($\Delta E_{\pi} = 40.0\%$). Then it decreases slowly from As_2 ($\Delta E_{\pi} = 37.4\%$) and Sb_2 ($\Delta E_{\pi} = 35.6\%$) to Bi_2 ($\Delta E_{\pi} = 32.2\%$). Experimental experience shows that

Table 13.2 Energy decomposition analysis of the E–E bond for E = N–Bi at RPBE/TZP [40]

	N_2	P_2	As_2	Sb_2	Bi_2
ΔE_{int}	−232.2	−109.2	−80.6	−54.4	−48.4
ΔE_{Pauli}	791.7	299.3	247.9	182.3	168.1
ΔEE^{a}_{elstat}	−308.5 (30.1%)	−175.8 (43.0%)	−160.5 (48.9%)	−131.5 (55.6%)	−126.3 (58.3%)
ΔE^{a}_{orb}	−715.4 (69.9%)	−232.7 (57.0%)	−168.0 (51.1%)	−105.2 (44.4%)	−90.3 (41.7%)
ΔE^{b}_{σ}	−470.0 (65.7%)	−140.1 (60.0%)	−105.1 (62.6%)	−69.9 (66.4%)	−61.2 (67.8%)
ΔE^{b}_{π}	−245.4 (34.3%)	−92.6 (40.0%)	−62.9 (37.4%)	−35.3 (35.6%)	−29.1 (32.2%)
Overlap σ^{c}	1.58	1.46	1.35	1.26	1.15
Overlap π^{c}	0.74	0.61	0.55	0.48	0.47
R(E–E)	1.105 (1.0977)	1.935 (1.8931)	2.161 (2.103)	2.579 (2.48)	2.728 (2.660)
D_e	−232.2	−109.2	−80.6	−54.4	−48.4
D_0	−228.8 (−225.0)	−108.1 (−116.1)	−80.0 (−91.3)	−54.0 (71.3)	−48.1 (−47.0)

Energy values in kcal/mol. Bond lengths R in Å. Experimental values are given in parentheses.
[a]The values in parentheses give the percentage contribution to the total attractive interactions $\Delta E_{elstat} + \Delta E_{orb}$.
[b]The values in parentheses give the percentage contribution to the orbital interactions ΔE_{orb}.
[c]Reference [40] gives only the p_{σ}–p_{σ} values for the σ orbitals and only one component of the π orbitals.

molecules with multiple bonds between main group elements of the third period and higher are much less stable than comparable compounds of elements of the first octal row [68]. It is often assumed that this behavior is caused by weaker π interactions in multiple bonds between the heavier main-group elements. Table 13.2 shows that this assumption is not justified. The relative contribution of π bonding to the orbital term in N_2 is not higher than for the other elements of the pnicogen-group. On the contrary, only the value for Bi_2 is slightly lower than the value for N_2. The EDA results also do not agree with the statement that the σ bond in N_2 is very weak and that only the π bonds are actually bonding at the equilibrium distance [69]. The latter statement is often based on the size of the overlapping orbitals. Table 13.2 shows that the overlap of the σ orbitals which is the sum the p_σ and s AO_s in N_2 is rather large (1.58) much larger than the overlap of the p_π AO_s (0.61). The values for the overlaps of the σ and π orbitals become smaller when the atoms become heavier, but the ratio does not change very much. The main difference in the relative contributions to the E–E interactions between N_2 and the heavier homologues is the smaller percentage value of the ΔE_{orb} term in P_2–Bi_2.

13.3.3 Dihalogens F_2–I_2 [40]

The experimental bond dissociation energies of the dihalogens F_2–I_2 exhibit an unusual trend because the lightest member F_2 has a weaker bond than Cl_2 and Br_2. This is usually explained with the repulsion between the lone-pair electrons, which is assumed to become very large in difluorine because the F–F distance is rather short. On the other hand, the radius of fluorine is very small because the electron density of the highly electronegative atom is rather dense which means that the overlap of the p(π) orbitals in F_2 should be small. The EDA is an ideal method to investigate the validity of the standard textbook explanation because the Pauli repulsion is explicitly calculated. Table 13.3 gives the results for the dihalogens.

The calculated bond dissociation energies are in good agreement with experiment except for difluorine. The theoretical result $D_0 = 46.0$ kcal/mol is higher than the experimental value of $D_0 = 36.9$ kcal/mol. The trend however, that the BDE of F_2 is smaller than the bond dissociation energies of Cl_2 and Br_2 is preserved. Table 13.3 shows that the values of both ΔE_{Pauli} and ΔE_{orb} increase steadily from I_2 to F_2. We want to point out that the increase of the Pauli repulsion from Cl_2 to F_2 is not particularly steep. This becomes obvious from Fig. 13.3 which gives a graphical display of the trends of ΔE_{Pauli}, ΔE_{orb}, ΔE_{elstat} and the total interaction energy ΔE_{int}.

It becomes obvious that there is only a comparatively moderate increase of ΔE_{Pauli} and ΔE_{orb} from Cl_2 to F_2. The trend of the Pauli repulsion and orbital interaction terms thus behave normally. An abnormal trend is found, however, for the quasiclassical electrostatic interactions. The ΔE_{elstat} value steadily increases from I_2 to Cl_2 but then it decreases to F_2. Fig. 13.3 shows nicely that only the ΔE_{elstat} values show the same trend as the total interaction energies ΔE_{int}. If the quasiclassical electrostatic attraction would further increase from Cl_2 to F_2, the total interaction energy would also increase and thus, the BDE of difluorine would be larger than for dichlorine. The EDA results suggest that *F_2 has a weaker bond than Cl_2 because the quasiclassical Coulomb attraction is less attractive*

Table 13.3 Energy decomposition analysis of the E–E bond for E = F–I at RPBE/TZP [40]

	F_2	Cl_2	Br_2	I_2
ΔE_{int}	−47.4	−59.1	−47.8	−34.1
ΔE_{Pauli}	140.8	123.2	76.7	45.4
ΔE^{a}_{elstat}	−38.2 (20.3%)	−44.1 (24.2%)	−31.9 (25.6%)	−19.8 (24.9%)
ΔE^{a}_{orb}	−150.0 (79.7%)	−138.1 (75.8%)	−92.6 (74.4%)	−59.7 (75.1%)
ΔE^{b}_{σ}	−144.6 (96.4%)	−123.8 (89.6%)	−85.9 (92.8%)	−59.4 (99.5%)
ΔE^{b}_{π}	−5.4 (3.6%)	−14.3 (10.4%)	−6.7 (7.2%)	−0.3 (0.5%)
R(E–E)	1.424 (1.412)	2.037 (1.987)	2.381 (2.281)	2.860 (2.666)
D_e	−47.4	−59.1	−47.8	−34.1
D_0	−46.0 (−36.9)	−58.3 (−57.2)	−47.4 (−45.4)	−33.8 (−35.6)

Energy values in kcal/mol. Bond lengths R in Å. Experimental values are given in parentheses.
[a]The values in parentheses give the percentage contribution to the total attractive interactions $\Delta E_{elstat} + \Delta E_{orb}$.
[b]The values in parentheses give the percentage contribution to the orbital interactions ΔE_{orb}.

in the former molecule. This comes from the compact electron density distribution of fluorine atom which does poorly overlap with the nucleus of the other F atom.

A comparison of the EDA results for the dihalogens (Table 13.3) with the dipnicogens (Table 13.2) shows that the relative contributions of the orbital interactions in the former are much larger (74.4–79.7%) than in the latter (41.7–69.9%). According to our definition, bonding in the dihalogens F_2–I_2 is clearly more covalent than in the dipnicogens N_2–Bi_2. But although the contributions of the ΔE_{elstat} term in the dihalogens are rather

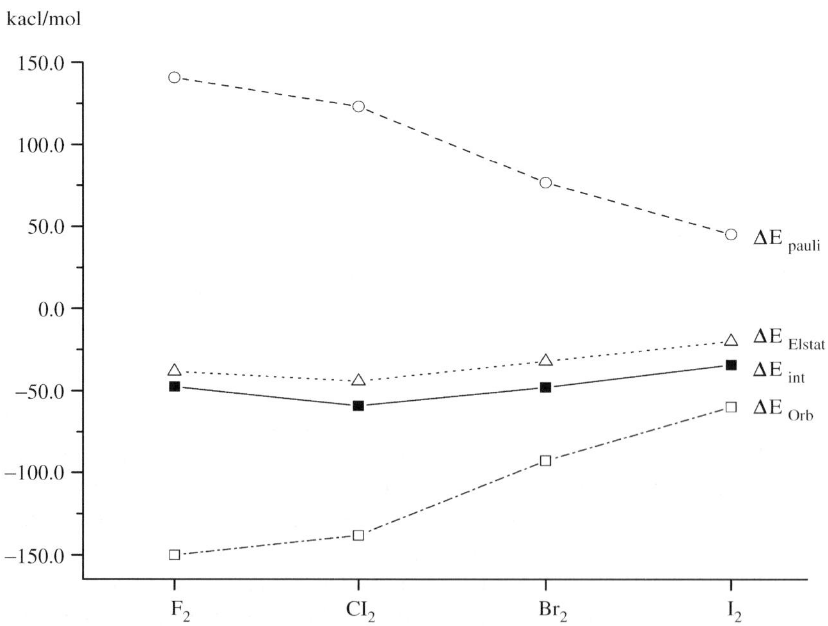

Fig. 13.3. Trends of the various terms of the energy decomposition analysis for the dihalogens F_2–I_2 calculated at RPBE/TZP.

small, they determine the trend of the BDE (Fig. 13.3). We want to comment on the small attractive contribution of the π orbitals to the binding interactions in F_2–I_2 (Table 13.3) which are surprising at first glance. The negative values for ΔE_{π} do not come from weak π bonding but from the relaxation of the frozen p(π) orbitals of the halogen atoms in the dihalogen molecules. The changes in the electronic structure of the fragments in the last step of the EDA procedure is not only the result of genuine orbital interactions. The interfragment interactions lead also to a rearrangement of those electrons which are not involved in chemical bonding.

13.3.4 Nonpolar single bonds of the first octal row H_nE–EH_n (E = Li–F; n = 0–3) [42]

In Sections 13.3.5–13.3.7 we discuss EDA results for polyatomic molecules which are representative for nonpolar bonds between main group elements. We first present the results of a 'first-row sweep' for the single bonds in hydrogen bonded molecules H_nE–EH_n. In Section 13.3.5 we discuss multiple bonding in HB=BH, H_2C=CH_2, HN=NH and HC≡CH. In Section 13.3.6 we give the results for the single bonds in the heavier group-14 homologues H_3E–EH_3 (E = C–Pb).

The EDA results for the H_nE–EH_n bonds (E = Li–F) are given in Table 13.4. The calculated geometries and bond dissociation energies are in good agreement with experimental data. As mentioned above, the only exception is difluorine where the calculated dissociation energy is too high.

Table 13.4 shows nicely, that the C–C bond of ethane has the largest interaction energy ($\Delta E_{int} = -114.8$ kcal/mol) of all molecules investigated in this section. The BDE ($D_e = 93.1$ kcal/mol), however, is significantly smaller because of the relatively large preparation energy ($\Delta E_{prep} = 21.7$ kcal/mol) of the two CH_3-fragments. This is partly due to the fact, that CH_3 has a planar ground state structure while it is pyramidal in ethane. Adding the zero-point energy correction yields $D_0 = 83.4$ kcal/mol which is in reasonable agreement with the experimentally determined value of 89.9 ± 0.5 kcal/mol.

The individual contributions to the interaction energy ΔE_{int} show, that the quasiclassical electrostatic term ΔE_{elstat} contributes 41.4% of the total attractive terms. The remaining 58.6% stem from the orbital term ΔE_{orb}. As shown before in the case of dinitrogen, the electrostatic attraction is even larger than the total interaction energy. Table 13.4 also shows, that the electrostatic character of the E–E bond increases monotonically from difluorine ($\Delta E_{elstat} = 20.7\%$) to HBe–BeH ($\Delta E_{elstat} = 58.4\%$) but then it decreases for Li–Li ($\Delta E_{elstat} = 37.3\%$). The irregular behavior of Li_2 can be explained with the rather diffuse 2s AO of lithium and with unusually strong induction forces [9]. We want to point out that the atomic partial charges must not be used to estimate the quasiclassical electrostatic interactions between the atoms E. The largest absolute values of q(E) are calculated for Be_2H_2 and N_2H_4. A naive consideration could lead to the suggestion that the quasiclassical E–E interaction in the two molecules should be repulsive. Table 13.4 shows that N_2H_4 has the strongest electrostatic attraction ($\Delta E_{elstat} = -178.3$ kcal/mol) while Be_2H_2 has the highest degree of electrostatic character (58.4%).

Table 13.4 Energy decomposition analysis of $H_nE{-}EH_n$ single bonds (E = Li–F) at BP86/TZ2P [42]

	Li_2	Be_2H_2	B_2H_4	C_2H_6	N_2H_4	O_2H_2	F_2
Symmetry	$D_{\infty h}$	$D_{\infty h}$	D_{2d}	D_{3d}	C_2	C_2	$D_{\infty h}$
ΔE_{int}	−20.6	−70.3	−114.5	−114.8	−76.8	−62.6	−52.8
ΔE_{Pauli}	1.9	41.1	161.7	201.6	407.9	384.6	145.8
ΔE_{elstat}^{a}	−8.4 (37.3%)	−65.1 (58.4%)	−140.5 (50.9%)	−130.9 (41.4%)	−178.3 (36.8%)	−145.8 (32.6%)	−41.1 (20.7%)
ΔE_{orb}^{a}	−14.1 (62.7%)	−46.3 (41.6%)	−135.7 (49.1%)	−185.5 (58.6%)	−306.4 (63.2%)	−301.4 (67.4%)	−157.5 (79.3%)
ΔE_{σ}^{b}		−46.3 (100%)	−119.7 (88.2%)	−175.5 (94.6%)			−151.3 (96.1%)
ΔE_{π}^{b}			−16.0 (11.8%)	−10.0 (5.4%)			−6.2 (3.9%)
ΔE_{prep}	0	1.6	6.1	21.7	4.2	3.2	0
D_e	20.6	68.7	108.4	93.1	72.6	59.4	52.8
D_0	20.1 (26.4 ± 1.0)	64.9 (71.7)	103.0 (104.0)	83.4 (89.9 ± 0.5)	62.8 (65.8)	53.6 (50.9 ± 1.0)	51.5 (37.9)
q(E)	0.0	0.196	0.081	−0.105	−0.214	−0.153	0.0
R(E–E)	2.737 (2.673)	2.101 (2.101)	1.623 (1.656)	1.532 (1.535)	1.443 (1.449)	1.473 (1.464)	1.421 (1.412)

Energy values in kcal/mol. Bond lengths R in Å. Experimental values are given in parentheses. Atomic partial charges q.
[a]The values in parentheses give the percentage contribution to the total attractive interactions $\Delta E_{elstat} + \Delta E_{orb}$.
[b]The values in parentheses give the percentage contribution to the orbital interactions ΔE_{orb}.

In $H_2N{-}NH_2$ the Pauli-repulsion term $\Delta E_{Pauli} = 407.9$ kcal/mol is very large. This strong destabilizing contribution comes mainly from the repulsion between the lone electron pairs of the nitrogen atoms in the equilibrium gauche conformation of hydrazine [42]. Obviously the attractive terms reach their largest values for hydrazine as well. This indicates that the comparatively low BDE of $H_2N{-}NH_2$ is due to the large destabilizing effect of the Pauli repulsion rather than small attractive contributions to the interaction energy ΔE_{int}. In fact the pure numerical values of both the interaction energy ΔE_{int} and the BDE D_e can be misleading when one tries to assess the nature of bonding. In the cases of $H_2N{-}NH_2$ and HBe–BeH the interaction energy as well as the BDE are comparable but the individual contributions to, e.g. the Pauli repulsion are 10 times higher in $H_2N{-}NH_2$.

It seems plausible that the weaker bond of F_2 compared to $H_2B{-}BH_2$ is caused by the larger ΔE_{Pauli} value due to the repulsion of the lone pairs in the former molecule which is missing in the latter. This is, however, not supported by the data given in Table 13.4. A comparison shows that in difluorine the strength of the orbital interactions is larger and the Pauli repulsion is weaker than in $H_2B{-}BH_2$. The weaker bonding is rather caused by the weak electrostatic attraction ($\Delta E_{elstat} = -41.1$ kcal/mol in F_2 as compared with $\Delta E_{elstat} = -140.5$ kcal/mol in $H_2B{-}BH_2$). In the discussion of the chemical bonding in the dihalogens it was already shown that the weak quasiclassical attraction in F_2 is the reason why difluorine has a lower BDE than Cl_2. The low value of ΔE_{elstat} can be rationalised by the compact 2p orbital of the most electronegative element fluorine. The same reasoning explains why the Pauli repulsion in F_2 is much weaker than in HO–OH and $H_2N{-}NH_2$ although difluorine has a much shorter bond. The absolute values for the orbital interactions in F_2 are also weaker than in HO–OH, $H_2N{-}NH_2$ and $H_3C{-}CH_3$ although the E–E bonds of the latter molecules are longer. Please note that the absolute values of the EDA calculation for F_2 given in Table 13.4 are slightly different from the data given in Table 13.3 because the latter values have been calculated at the RPBE/TZP level while the former data come from BP86/TZ2P calculations. However, the relative contributions of the energy components which are calculated at the two theoretical levels are very similar to each other.

Table 13.4 also contains the σ and π contributions to the E–E orbital interactions, for all cases where σ and π contributions are separable. As expected, the π contributions for B_2H_4, C_2H_6 and F_2 are rather small. In the case of B_2H_4 the only contribution with π symmetry is the hyperconjugative stabilization of the empty p(π) orbitals of boron. This interaction amounts to 11.8% of the total orbital stabilization. Each empty p(π) orbital of boron yields 4.0 kcal/mol adding to the total of 16.0 kcal/mol in $H_2B{-}BH_2$. It was already mentioned above that the stabilizing π contribution in F_2 comes from the relaxation of the p(π) AOs of fluorine rather than from genuine π bonding.

The most important conclusion of this section is that the quasiclassical electrostatic attraction significantly contributes to the chemical bonding of the single bonds between first octal-row elements $H_nE{-}EH_n$. The EDA results show that it is not sufficient to consider only orbital interactions for gaining an understanding of the trends of the bond strength. Besides the attractive energy terms ΔE_{elstat} and ΔE_{orb}, the Pauli repulsion ΔE_{Pauli} plays an important role for the formation of a chemical bond.

13.3.5 Nonpolar multiple bonds HB=BH, H_2C=CH_2, HN=NH and HC≡CH [42]

The EDA results of the single bonds in H_2B–BH_2, H_3C–CH_3 and H_2N–NH_2 shall now be compared with the multiple bonds in HB=BH, H_2C=CH_2, HN=NH and HC≡CH. The EDA data of the latter molecules are given in Table 13.5. To compare the strength of the double bonds in question, the boron compound was calculated at the excited ($^1\Sigma_g^+$) state. The ground state of HB=BH is ($^3\Sigma_g^-$) [70]. In addition to the double bonds in HB=BH, H_2C=CH_2 and HN=NH the triple bond of HC≡CH was also analyzed.

The contributions of the electrostatic attraction to the binding interactions in the multiple bonds are smaller than in the respective single bonds (see above) but they are still rather large. The ΔE_{elstat} term contributes between 34.6–40.1% to the attractive interactions in the respective double bonds. Thus, more than one third of the attractive forces stems from the quasiclassical electrostatic term. Even in the triply bound acetylene still 27.6% of the attractive force comes from the ΔE_{elstat} term.

The calculated bond length for HB=BH (1.526 Å) is much shorter than in H_2B–BH_2 (1.623 Å). Hence the interaction energy in the former compound ($\Delta E_{int} = -159.8$ kcal/mol) is significantly larger than in the latter species ($\Delta E_{int} = -114.5$ kcal/mol). The question that comes to mind is: How much of this increase can be attributed to the formation of the boron–boron π bond? Compared to the results of H_2B–BH_2 the orbital term ΔE_{orb} of HB=BH has a larger absolute ($\Delta E_{orb} = -135.7$ kcal/mol for H_2B–BH_2 compared to $\Delta E_{orb} = -165.5$ kcal/mol for HB=BH) and relative ($\Delta E_{orb} = 49.1\%$ for H_2B–BH_2 compared to $\Delta E_{orb} = 59.9\%$ for HB=BH) value. However, the σ contribution to ΔE_{orb} in HB=BH ($\Delta E_\sigma = -111.1$ kcal/mol) is actually *smaller* than in H_2B–BH_2

Table 13.5 Energy decomposition analysis of the HB=BH, H_2C=CH_2, *trans* HN=NH and HC≡CH multiple bonds at BP86/TZ2P [42]

	B_2H_2	C_2H_4	*trans* N_2H_2	C_2H_2
Symmetry	$D_{\infty h}$	D_{2h}	C_{2h}	$D_{\infty h}$
ΔE_{int}	−159.8	−191.1	−137.4	−280.0
ΔE_{Pauli}	116.3	281.9	599.4	255.4
ΔE_{elstat}^a	−110.6 (40.1%)	−181.6 (38.4%)	−254.6 (34.6%)	−147.5 (27.6%)
ΔE_{orb}^a	−165.5 (59.9%)	−291.4 (61.6%)	−482.2 (65.4%)	−387.9 (72.4%)
ΔE_σ^b	−111.1 (67.1%)	−212.2 (72.8%)	−392.6 (81.4%)	−215.5 (55.6%)
ΔE_π^b	−54.4 (32.9%)	−79.2 (27.2%)	−89.6 (18.6%)	−172.4 (44.4%)
ΔE_{prep}	4.4	12.9	5.6	32.8
D_e	155.4	178.2	131.8	247.2
D_0 [42]	149.8	168.3 (175.4 ± 2)	123.7 (123.8 ± 1)	238.7 (230.9 ± 2)
q(E)	0.031	−0.086	−0.106	−0.093
R(E–E)	1.526 (1.498)	1.333 (1.339)	1.249 (1.252)	1.206 (1.203)

Energy values in kcal/mol. Bond lengths *R* in Å. Experimental values are given in parentheses. Atomic partial charges *q*.
[a]The values in parentheses give the percentage contribution to the total attractive interactions $\Delta E_{elstat} + \Delta E_{orb}$.
[b]The values in parentheses give the percentage contribution to the orbital interactions ΔE_{orb}.

($\Delta E_\sigma = -119.7$ kcal/mol). So the formation of the π bond leads to a reduction of the σ contribution at the shorter B–B distance in HB=BH. The interaction energy ΔE_{int} is not lowered, because the π contribution overcompensates the loss of σ interaction. Table 13.5 also shows that ΔE_π contributes -54.4 kcal/mol (32.9%) to ΔE_{orb} while ΔE_σ contributes to -111.1 kcal/mol (67.1%). This leads to the conclusion that the π bonding in HB=BH is half as strong as the σ bonding while the electrostatic interactions contribute less than in $H_2B{-}BH_2$.

The interaction energy in ethylene is also higher than in ethane. In contrast to the boron systems discussed above, now all the contributions are larger than in the singly bound species. The covalent contribution of the ΔE_{orb} term is slightly larger in ethylene (61.6%) than in ethane (58.6%). The π bond strength in $H_2C{=}CH_2$ ($\Delta E_\pi = -79.2$ kcal/mol) contributes to 27.2% of the total orbital term ΔE_{orb}, while the σ contribution is much stronger $\Delta E_\sigma = -212.2$ kcal/mol. So the σ bond in $H_2C{=}CH_2$ is about three times as strong as the π bond. Unlike the boron systems, the σ contribution in $H_2C{=}CH_2$ is much larger than in $H_3C{-}CH_3$ ($\Delta E_\sigma = -175.5$ kcal/mol).

The EDA data in Table 13.5 show that the nitrogen–nitrogen interaction in *trans*-N_2H_2 ($\Delta E_{int} = -137.4$ kcal/mol) is also bigger than the N–N bond strength in hydrazine ($\Delta E_{int} = -76.8$ kcal/mol), but still clearly less than the interaction in ethylene. Although the absolute values of the contributions to ΔE_{int} are significantly higher than in ethylene, the strong Pauli repulsion leads to a smaller net attraction. It is important to distinguish between the *absolute* values of the π bonding interactions, which increase from HB=BH to $H_2C{=}CH_2$ and HN=NH, and the *relative* contributions to the ΔE_π term which are decreasing.

The much higher interaction energy of acetylene ($\Delta E_{int} = -280.0$ kcal/mol) compared with ethylene ($\Delta E_{int} = -191.1$ kcal/mol) comes from the increase in the strength of the orbital contribution in ΔE_{orb}. As can be seen in Table 13.5 the quasi-classical electrostatic interaction in HC≡CH ($\Delta E_{elstat} = -147.5$ kcal/mol) is weaker than in $H_2C{=}CH_2$ ($\Delta E_{elstat} = -181.6$ kcal/mol), whereas the orbital interaction of the triple bond ($\Delta E_{orb} = -387.9$ kcal/mol) is much higher than that of the double bond ($\Delta E_{orb} = -291.4$ kcal/mol). The increase in the orbital term ΔE_{orb} is almost exclusively due to the π bonding interactions in HC≡CH ($\Delta E_\pi = -172.4$ kcal/mol). This value is much higher than the corresponding value for $H_2C{=}CH_2$ ($\Delta E_\pi = -79.2$ kcal/mol). The reason for this is obvious. While in acetylene both components of the bonding p orbital are occupied, in ethylene only one of these orbitals is occupied. Hence the stabilizing effect is smaller. From a comparison of the two ΔE_π terms it is obvious that each individual component of the π bonds in acetylene has about the same strength as the single π bond in ethylene. The σ bond in HC≡CH is slightly stronger ($\Delta E_\sigma = -215.5$ kcal/mol) than in $H_2C{=}CH_2$ ($\Delta E_\sigma = -212.2$ kcal/mol).

If one finally compares the carbon–carbon triple bond in acetylene with the triple bond in dinitrogen (Table 13.2) the calculated terms indicate that the larger interaction energy of the triple bond in C_2H_2 comes from the weaker Pauli repulsion in the former bond. The attractive interactions in N_2 are almost twice as large as in C_2H_2 but the difference between the Pauli repulsion of the two molecules is even larger. This is another example which shows that it is not sufficient to consider attractive (orbital or quasiclassical) interactions for getting insight into the nature of chemical bonding.

13.3.6 Nonpolar group-14 bonds $H_3E–EH_3$ (E = C–Pb) [42]

The third part of the EDA investigation of nonpolar bonds between main group elements in polyatomic molecules focuses on the difference between atoms that come from different rows of the periodic system. The changes in the nature of the homopolar bonds in $H_3E–EH_3$ are given in Table 13.6. The results of the EDA analyses predict that the interaction energies ΔE_{int} decrease monotonically from the lighter elements ($\Delta E_{int} = -114.8$ kcal/mol for E=C) to the heaviest element ($\Delta E_{int} = -48.2$ kcal/mol for E=Pb). The dissociation energies D_0 follow the same trend because the preparation energies are rather small except for ethane.

The quasiclassical electrostatic contribution to the $H_3E–EH_3$ bond becomes more important compared to the ΔE_{orb} term when one goes from E=C to E=Sn. This trend is reversed in the last step. The lead–lead bond has a slightly larger contribution from the orbital term ΔE_{orb} than the tin–tin bond. This may be due to relativistic effects, which should be most important for such heavy elements as lead [71]. Apart from the latter result, the changes along the series in the $H_3E–EH_3$ bond are smooth. Only the absolute values of the Pauli repulsion ΔE_{Pauli} and the electrostatic interaction ΔE_{elstat} in $H_3Ge–GeH_3$ are larger than in the silicon compound. This is possibly caused by the, in comparison to silicon, contracted atomic radius of Ge, which is due to the filled 3d shell of the Ge atom. In summary, the EDA results suggest that the nature of the chemical bonding in $H_3E–EH_3$ (E=C–Pb) changes little. The weaker bonds can be explained with the fact that the valence electrons of the heavier atoms are further away from the nucleus and therefore, have less energy than the valence electrons of the lighter elements.

Table 13.6 Energy decomposition analysis of $H_3E–EH_3$ single bonds (E = C–Pb) at BP86/TZ2P [42]

	C_2H_6	Si_2H_6	Ge_2H_6	Sn_2H_6	Pb_2H_6
Symmetry	D_{3d}	D_{3d}	D_{3d}	D_{3d}	D_{3d}
ΔE_{int}	−114.8	−75.2	−69.7	−58.8	−48.2
ΔE_{Pauli}	201.6	101.2	114.4	98.3	84.0
ΔE^{a}_{elstat}	−130.9 (41.4%)	−83.8 (47.5%)	−95.2 (51.7%)	−86.6 (55.1%)	−68.3 (51.7%)
ΔE^{a}_{orb}	−185.5 (58.6%)	−92.6 (52.5%)	−88.9 (48.3%)	−70.5 (44.9%)	−63.9 (48.3%)
ΔE^{b}_{σ}	−175.5 (94.6%)	−87.8 (94.8%)	−84.3 (94.8)	−67.5 (95.7%)	−61.2 (95.8)
ΔE^{b}_{π}	−10.0 (5.4%)	−4.8 (5.2%)	−4.6 (5.2%)	−3.0 (4.3%)	−2.7 (4.2)
ΔE_{prep}	21.7	2.8	2.2	1.4	1.4
D_e	93.1	72.4	67.5	57.4(61.3)	46.8
D_0	83.4 (89.9 ± 0.5)	68.3 (74 ± 3)	63.8 (70.2)	54.4	43.6
q(E)	−0.105	0.208	0.184	0.263	0.250
R(E–E)	1.532 (1.535)	2.352 (2.331)	2.418 (2.403)	2.786 (2.758)	2.898 (2.897)

Energy values in kcal/mol. Bond lengths R in Å. Experimental values are given in parentheses. Atomic partial charges q.
[a]The values in parentheses give the percentage contribution to the total attractive interactions $\Delta E_{elstat} + \Delta E_{orb}$.
[b]The percentage values in parentheses give the percentage contribution to the orbital interactions ΔE_{orb}.

13.3.7 Donor–acceptor bonds $Y_3B{-}NX_3$ and $Y_3B{-}PX_3$ (X, Y = H, Me, Cl) [38,72]

Boron–nitrogen and boron–phosphorous compounds are classical textbook examples of donor–acceptor complexes. The chemical bonds of the Lewis-acid Lewis-base complexes are usually explained in terms of frontier orbital interactions and/or quasiclassical electrostatic attraction in the framework of the Hard and Soft Acid and Base (HSAB) model [73]. We were interested in seeing if the differences between the bond strengths of boron–nitrogen and boron–phosphorous complexes for different boranes, amines and phosphanes can be explained with the EDA method.

Table 13.7 shows the results of the EDA for borane–amine complexes $Y_3B{-}NX_3$ (X, Y = H, Me, Cl). Note that the preparation energies are rather large for all structures. It means that the bond dissociation energies may not be used for estimating the strength of the donor–acceptor interactions. For example, the D_e value of $H_3B{-}NMe_3$ is significantly larger (36.2 kcal/mol) than that of $Cl_3B{-}NMe_3$ (21.0 kcal/mol). This comes from the larger preparation energy of the latter complex (31.2 kcal/mol) than for the former (15.2 kcal/mol). It costs more energy to distort BCl_3 from its planar equilibrium structure to the pyramidal geometry in the complex than to distort BH_3. Unlike the D_e values, the interaction energies of $H_3B{-}NMe_3$ (−51.4 kcal/mol) and $Cl_3B{-}NMe_3$ (−52.2 kcal/mol) are similar. We will focus in our discussion of the donor–acceptor interactions on the analysis of the ΔE_{int} values rather than on the D_e data.

The Lewis base NH_3 has a significantly lower ΔE_{int} value with the Lewis acid BMe_3 (−27.7 kcal/mol) than with BH_3 (−44.6 kcal/mol) and BCl_3 (−41.3 kcal/mol). The same trend is found for the NMe_3 complexes where the ΔE_{int} value in $Me_3B{-}NMe_3$ (−31.3 kcal/mol) is much smaller than in $H_3B{-}NMe_3$ (−51.4 kcal/mol) and $Cl_3B{-}NMe_3$ (−52.2 kcal/mol). Inspection of the three energy terms (Table 13.7) shows that Me_3B complexes with NH_3 and NMe_3 are less covalent than the BH_3 and BCl_3 complexes. This is revealed by the lower percentage values of the ΔE_{orb} term in the former species. Since the orbital interactions come mainly from the $\sigma(a_1)$ orbitals it is reasonable to inspect the energy levels of the LUMO of BY_3. The LUMO of BMe_3 (−1.264 eV) is significantly higher in energy than the LUMO of BH_3 (−3.058 eV) and BCl_3 (−2.786 eV). The EDA results show that the weaker donor–acceptor interactions in $Me_3B{-}NH_3$ and $Me_3B{-}NMe_3$ can be explained with the frontier-orbital model [2]. The same reasoning explains why BCl_3 is a stronger Lewis acid than BF_3 [41].

The NCl_3 complexes with BY_3 are much more weakly bonded than the NH_3 and NMe_3 homologues (Table 13.7). The geometry optimizations of $Me_3B{-}NCl_3$ and $Cl_3B{-}NCl_3$ led to very weakly bonded van der Waals complexes which have long B–N distances. In order to present a meaningful comparison of the latter species with the bonding interactions in $H_3B{-}NCl_3$ and for explaining the weak bonding in $Me_3B{-}NCl_3$ and $Cl_3B{-}NCl_3$, the EDA analyses were carried out for structures which were optimized with an enforced B–N distance of 1.624 Å which is the optimized bond length $H_3B{-}NCl_3$. A comparison of the EDA data of $H_3B{-}NCl_3$ with the results for $H_3B{-}NH_3$ shows immediately that the weak quasiclassical electrostatic attraction in the formers species is responsible for the weak bonding. The values for ΔE_{orb} and ΔE_{Pauli} for the two compounds are very similar, but the ΔE_{elstat} term of $H_3B{-}NCl_3$ is much smaller.

Table 13.7 Energy decomposition analysis of the boron–amine complexes $X_3B{-}NY_3$ (X, Y = H, Me, Cl) at BP86/TZ2P [38,72]

	$H_3B{-}NH_3$	$Me_3B{-}NH_3$	$Cl_3B{-}NH_3$
ΔE_{int}	−44.6	−27.7	−41.3
ΔE_{Pauli}	108.8	128.5	192.8
ΔE^{a}_{elstat}	−77.3 (50.4%)	−83.8 (53.7%)	−120.5 (51.5%)
ΔE^{a}_{orb}	−76.0 (49.6%)	−72.4 (46.3%)	−113.6 (48.5%)
$\Delta E_{\sigma}(A_1)^{b}$	−68.4 (89.9%)	−65.1 (89.9%)	−102.2 (89.9%)
$\Delta E(A_2)$	0.0	−0.2	−0.2
$\Delta E_{\pi}(E_1)^{b}$	−7.7 (10.1%)	−7.1(9.8%)	−11.3 (9.9%)
ΔE_{prep}	12.7	15.1	21.5
$\Delta E(=-D_e)$	−31.9	−12.6	−19.8
R(B–N)	1.657	1.697	1.633
	$H_3B{-}NMe_3$	$Me_3B{-}NMe_3$	$Cl_3B{-}NMe_3$
ΔE_{int}	−51.4	−31.3	−52.2
ΔE_{Pauli}	119.8	115.1	187.4
ΔE^{a}_{elstat}	−89.4 (52.2%)	−79.8 (54.5%)	−123.7 (51.6%)
ΔE^{a}_{orb}	−81.8 (47.8%)	−66.6 (45.5)	−115.9 (48.4%)
$\Delta E_{\sigma}(A_1)^{b}$	−71.1 (87.9%)	−57.6 (86.5%)	−99.3 (85.7)
$\Delta E(A_2)$	0.0	−0.6	−0.8
$\Delta E_{\pi}(E_1)^{b}$	−10.7 (13.1%)	−8.5 (12.7%)	−15.8 (13.7%)
ΔE_{prep}	15.2	21.8	31.2
$\Delta E(=-D_e)$	−36.2	−9.5	−21.0
R(B–N)	1.651	1.777	1.677
	$H_3B{-}NCl_3$	$Me_3B{-}NCl_3{}^{c}$	$Cl_3B{-}NCl_3{}^{c}$
ΔE_{int}	−23.1	−10.3	−9.4
ΔE_{Pauli}	108.0	142.7	186.4
ΔE^{a}_{elstat}	−53.9 (41.1%)	−69.2 (45.2%)	−91.0 (46.4%)
ΔE^{a}_{orb}	−77.2 (58.9%)	−83.8 (54.8%)	−104.9 (53.6%)
$\Delta E_{\sigma}(A_1)^{b}$	−60.4 (78.2%)	−62.2 (74.2%)	−84.1 (80.2%)
$\Delta E(A_2)$	0.0	−0.3	−0.4
$\Delta E_{\pi}(E_1)^{b}$	−16.9 (21.8%)	−21.3 (25.4%)	−20.4 (19.4%)
ΔE_{prep}	9.2	17.1	21.0
$\Delta E(=-D_e)$	−13.9	6.8	11.6
R(B–N)	1.624	1.624[c]	1.624[c]

Energy values in kcal/mol. Bond lengths R in Å.

[a]The values in parentheses give the percentage contribution to the total attractive interactions $\Delta E_{elstat} + \Delta E_{orb}$.

[b]The values in parentheses give the percentage contribution to the orbital interactions ΔE_{orb}.

[c]Calculated with a fixed bond length R(B–N) = 1.624 Å.

The lone-pair σ donor orbital of NCl_3 is more compact and has a smaller radius than the donor orbitals of NH_3 and NMe_3. The donor–acceptor interactions in $Me_3B{-}NCl_3$ and $Cl_3B{-}NCl_3$ are further weakened by stronger Pauli repulsion, but the most important reason for the weak attraction is the weak quasiclassical attraction (Table 13.7).

Table 13.8 shows the EDA results of the boron–phosphorous bonds in complexes $Y_3B{-}PX_3$. Note that the preparation energies are rather high. The rather large ΔE_{prep}

Table 13.8 Energy decomposition analysis of the boron–phosphane complexes X_3B-PY_3 (X, Y = H, Me, Cl) at BP86/TZ2P [38,72]

	H_3B-PH_3	Me_3B-PH_3	Cl_3B-PH_3
ΔE_{int}	−38.6	−14.7	−25.1
ΔE_{Pauli}	113.3	112.1	166.4
ΔE^{a}_{elstat}	−58.0 (38.2%)	−59.4 (46.8%)	−89.8 (46.1%)
ΔE^{a}_{orb}	−93.9 (61.8%)	−67.4 (53.2%)	−104.7 (53.8%)
$\Delta E_{\sigma}(A_1)^{b}$	−79.3 (84.6%)	−56.9 (85.4%)	−92.0 (87.8%)
$\Delta E(A_2)$	0.0	−0.1	−0.1
$\Delta E_{\pi}(E_1)^{b}$	−14.7 (15.6%)	−10.4 (15.5%)	−12.6 (12.1%)
ΔE_{prep}	12.2	13.7	23.0
$\Delta E(= -D_e)$	−26.4	−1.0	−2.1
R(B–P)	1.939	2.070	2.026
	H_3B-PMe_3	Me_3B-PMe_3	Cl_3B-PMe_3
ΔE_{int}	−55.3	−32.1	−52.7
ΔE_{Pauli}	130.7	139.9	201.2
ΔE^{a}_{elstat}	−79.4 (42.7%)	−84.7 (49.2%)	−121.8 (48.0%)
ΔE^{a}_{orb}	−106.6 (57.3%)	−87.3 (50.8%)	−132.1 (52.0%)
$\Delta E_{\sigma}(A_1)^{b}$	−92.7 (87.0%)	−76.1 (87.2%)	−119.0 (90.1%)
$\Delta E(A_2)$	0.0	−0.2	−0.3
$\Delta E_{\pi}(E_1)^{b}$	−13.9 (13.0%)	−11.0 (12.6%)	−12.8 (9.7%)
ΔE_{prep}	15.5	20.1	31.4
$\Delta E(= -D_e)$	−39.8	−12.0	−21.3
R(B–P)	1.924	2.014	1.986
	H_3B-PCl_3	Me_3B-PCl_3	Cl_3B-PCl_3
ΔE_{int}	−30.8	−6.8	−8.4
ΔE_{Pauli}	112.4	94.0	109.7
ΔE^{a}_{elstat}	−48.7 (34.0%)	−44.0 (43.6%)	−50.2 (42.5%)
ΔE^{a}_{orb}	−94.4 (66.0%)	−56.8 (56.4%)	−68.0 (57.5%)
$\Delta E_{\sigma}(A_1)^{b}$	−73.9 (78.3%)	−43.4 (76.4%)	−56.2 (82.6%)
$\Delta E(A_2)$	0.0	−0.1	−0.1
$\Delta E_{\pi}(E_1)^{b}$	−20.5 (21.7%)	−13.4 (23.5%)	−11.7 (17.2%)
ΔE_{prep}	8.7	8.2	13.9
$\Delta E(= -D_e)$	−22.1	1.4	5.5
R(B–P)	1.909	2.110	2.166

Energy values in kcal/mol. Bond lengths R in Å.
[a]The values in parentheses give the percentage contribution to the total attractive interactions $\Delta E_{elstat} + \Delta E_{orb}$.
[b]The values in parentheses give the percentage contribution to the orbital interactions ΔE_{orb}.

values compensate the moderately strong interaction energies of Me_3B-PH_3 and Cl_3B-PH_3 and they wipe out the weakly attractive interactions of Me_3B-PCl_3 and Cl_3B-PCl_3.

The interaction energies of the PH_3 complexes with BY_3 are smaller compared to the ΔE_{int} values of the NH_3 analogues while the interaction energies of the PMe_3 and NMe_3 complexes are nearly the same (Tables 13.7 and 13.8). The common difference between the phosphane and amine complexes is that the former species have a higher covalent character than the latter. The rather weak interaction energy of Me_3B-PH_3 can again be explained with the energetically high-lying LUMO of BMe_3. It becomes obvious by comparing the EDA results for Me_3B-PH_3 with H_3B-PH_3 (Table 13.8). The absolute contributions of ΔE_{Pauli} and ΔE_{elstat} values are nearly the same but the ΔE_{orb} value of Me_3B-PH_3 (−67.4 kcal/mol) is much smaller than that of H_3B-PH_3 (−93.9 kcal/mol). It is difficult to identify a single factor which is responsible for the weaker bonding in H_3B-PH_3 and Cl_3B-PH_3 compared with the analogous PMe_3 complexes. Both attractive components, ΔE_{orb} and ΔE_{elstat} are weaker in the former species.

The still weaker bonding interactions in the PCl_3 complexes H_3B-PCl_3, Me_3B-PCl_3 and Cl_3B-PCl_3 can be explained with the weaker quasiclassical attraction which comes from the more compact lone-pair electron density at the phosphorous atom of PCl_3. We want to point out that H_3B-PCl_3 has the shortest B–P bond length of all Y_3-PX_3 complexes and yet, the bonding is not very strong. A comparison with the EDA data for H_3B-PH_3 shows that, in H_3B-PCl_3, the repulsive contribution of ΔE_{Pauli} is smaller and the attractive contribution of ΔE_{orb} is slightly larger but the ΔE_{elstat} value (−48.7 kcal/mol) is significantly smaller than that of H_3B-PH_3 (−58.0 kcal/mol). Note that the general trend of the strength of the phosphane and amine complexes given by the ΔE_{int} values, which is $PMe_3 > PH_3 > PCl_3$ and $NMe_3 > NH_3 > NCl_3$ correlates with a decrease of the *relative* contribution of quasiclassical attraction given by the percentage values of ΔE_{elstat} (Tables 13.7 and 13.8).

The EDA results for the phosphane complexes given in Table 13.8 thus give a plausible answer why PCl_3 is a good π acceptor while at the same time the BDE in donor–acceptor complexes is less than the BDE of the weaker π acceptors PH_3 and PMe_3. The EDA results do not support the suggestion that PCl_3 is a weak π acceptor [105].

13.3.8 Main group metallocenes ECp_2 (E = Be–Ba, Zn, Si–Pb) and ECp (E = Li–Cs, B–Tl) [39]

The fortuitous synthesis of the first metallocene compound ferrocene by Kealy and Pauson which was published in 1951 [74] was not only a landmark event in organometallic chemistry because it was the starting point for a new class of compounds. It was also a challenge for chemical bonding models. The initial suggestion of the structure having two $Fe-C_5H_5$ σ bonds [74a] was soon corrected in two independent publications by Fischer and Pfab [75] and by Wilkinson, Rosenblum, Whiting and Woodward [76] who showed that the molecule has a π bonded sandwich structure. A qualitative bonding model for ferrocene using symmetry adapted molecular orbitals, which is also valid for other metallocenes of the transition metals was introduced by Shustorovic and Dyatkina [77]. The orbital correlation model is now generally accepted and used as a standard in many textbooks. A quantitative investigation of the metal–ligand interactions in ferrocene and related compounds using the EDA will be given in Section 13.4.7.

Metallocenes have in the meantime also been synthesized with main group elements E which may be metals or other atoms. Recent reviews of main group metallocenes show that a large number of compounds with the formula ECp and ECp_2 particularly with elements E of the groups 1, 2, 13, 14 and 15 have been synthesized and structurally characterized by X-ray analysis [78]. Numerous quantum chemical calculations of main group metallocenes have also been published in recent years. The theoretical knowledge, which has been gained in these studies was recently reviewed by Kwon and McKee [79]. The bonding situation in the compounds has also been analyzed with qualitative bonding models. There is general agreement that, in main group metallocenes, π-type interactions are weaker than in transition metal metallocenes because the main group elements do not have d-type valence orbitals [78,79]. The bonding of the Cp ligand to the s-block elements (groups 1, 2) is considered to be mainly ionic, while complexes of the p-block elements should be predominantly covalent. Therefore, orbital correlation diagrams of main group metallocenes use only the s and p orbitals of E for a discussion of the covalent bonding in ECp and ECp_2 complexes [78,79].

It would be very helpful if the covalent and electrostatic contributions to the E–Cp bonding and the strength of the different orbital terms could be quantitatively given. A recent review by Kwon and McKee recognized this by saying "there is not enough data from high-level theoretical calculations to have a quantitative understanding of the factors involving covalent or ionic bonding between the Cp ligand and the main group element" [79]. We have therefore carried out a systematic investigation of the chemical bonding in main-group metallocenes with one and two Cp ligands ECp_2 (E = Be–Ba, Zn, Si–Pb) and ECp (E = Li–Cs, B–Tl) [39]. The most important results are summarized in this section.

Table 13.9 shows the EDA results for metallocenes of the alkali metals. A qualitative orbital correlation for the interaction between a singly charged metal atom E^+ with a singly charged cyclopentadiene anion Cp^- is shown in Fig. 13.4. The six E–Cp bonding electrons occupy the lowest lying a_1 and the degenerate e_1 orbitals.

In order to assess which of the two contributions to the covalent bond is stronger a series of energy decomposition analyses have been carried out. The results are shown in Table 13.9.

The bonding between E^+ and Cp^- is in all cases dominated by the quasiclassical electrostatic interactions. The ΔE_{elstat} term contributes between 80 and 90% of the total interaction energy ΔE_{int} for all alkali metals in this investigation. Although the trend is not perfectly smooth, as expected, the lowest amount of covalent interaction is found

Table 13.9 Energy decomposition analysis of group-1 metallocenes ECp (E = Li–Cs) using the fragments E^+ and Cp^- at BP86/TZP [39]

	Li	Na	K	Rb	Cs
ΔE_{int}	−164.1	−138.0	−120.8	−113.9	−105.9
ΔE_{Pauli}	22.6	21.5	27.5	24.1	22.9
ΔE_{elstat}^{a}	−148.7 (79.6%)	−138.0 (86.6%)	−128.2 (86.4%)	−121.0 (87.7%)	−114.7 (89.1%)
ΔE_{orb}^{a}	−38.1 (20.4%)	−21.4 (13.4%)	−20.1 (13.6%)	−17.0 (12.3%)	−14.1 (10.9%)
$\Delta E_{\sigma}(A_1)^{b}$	−9.6 (25.2%)	−6.4 (29.9%)	−5.2 (25.7%)	−4.8 (28.4%)	−4.2 (30.0%)
$\Delta E(A_2)$	0.0	0.0	0.0	0.0	0.0
$\Delta E_{\pi}(E_1)^{b}$	−24.0 (63.0%)	−12.3 (57.5%)	−13.0 (64.4%)	−10.5 (62.1%)	−8.5 (60.7%)
$\Delta E_{\delta}(E_2)^{b}$	−4.5 (11.8%)	−2.7 (12.6%)	−2.0 (9.9%)	−1.6 (9.5%)	−1.3 (9.3%)
q(E)	+0.90	+0.91	+0.91	+0.94	+0.93

The symmetry point group is C_{5v}. Energy values in kcal/mol. Atomic partial charges q.
[a]The values in parentheses give the percentage contribution to the total attractive interactions $\Delta E_{elstat} + \Delta E_{orb}$.
[b]The values in parentheses give the percentage contribution to the orbital interactions ΔE_{orb}.

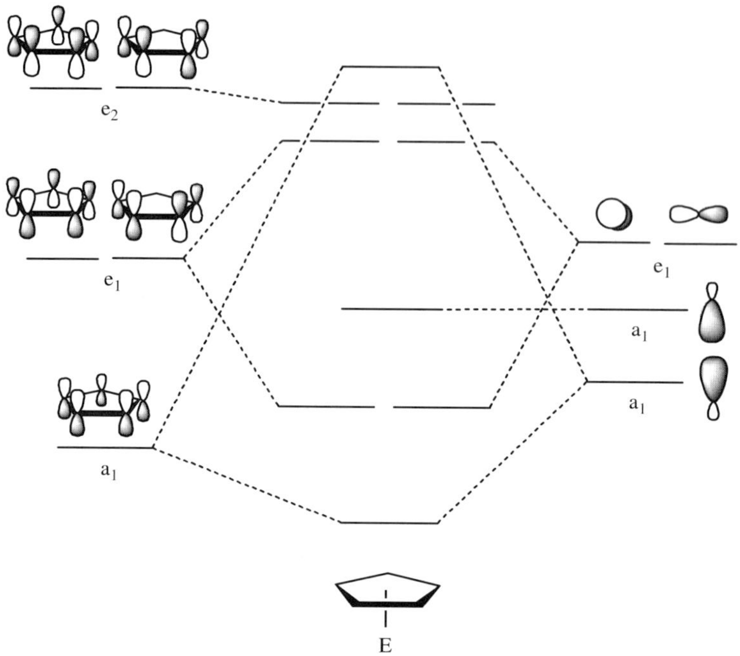

Fig. 13.4. MO correlation diagram for the interactions between a main group atomic ion E^+ and a Cp^- ligand in (C_{5v}) ECp.

for E = Cs (ΔE_{orb} = 10.9%) and the largest amount is found for E = Li (ΔE_{orb} = 20.4%). The largest contribution to the orbital term ΔE_{orb} comes always from the degenerate e_1 orbitals. Only between 25.2% (E = Li) and 30.0% (E = Cs) of the orbital term stem from the lowest lying orbital with a_1 symmetry.

Fig. 13.5 shows the orbital correlation diagram for sandwich complexes that includes the d orbitals of the main-group elements. Both, the parallel D_{5d} and the bent C_{2v} structures are shown. Usually the d orbitals are left out because they are considered unimportant for the covalent bonding. It will be shown below that this may not always be justified.

The group-2 metallocenes $E(Cp)_2$ with D_{5d} symmetry have 12 valence electrons which occupy the lowest lying $1a_{1g}$, $1a_{2u}$, $1e_{1u}$ and $1e_{1g}$ orbitals. Ligand to metal electron donation takes place *via* $(1a_{1g}) \rightarrow (s)$, $(1a_{2u}) \rightarrow (p_z)$ and $(e_{1u}) \rightarrow (p_{x,y})$, where z is the bonding axis. There is no valence orbital with e_{1g} symmetry on the central metal atom, and thus, the highest occupied e_{1g} orbital of the ligand should not significantly contribute to the covalent bonding. The only metal AOs with e_{1g} symmetry are the empty d_{xz} and d_{yz} orbitals. As will be shown below in Section 13.4 about transition metal compounds, the covalent bonding in transition metal metallocenes may have significant contributions from ligand to metal electron donation in the e_{1g} symmetry because the respective d orbitals are in the valence space for transition metals.

Table 13.10 shows the result of the energy partitioning analysis of the alkaline-earth metallocenes with D_{5d} symmetry. For all molecules the major contribution to

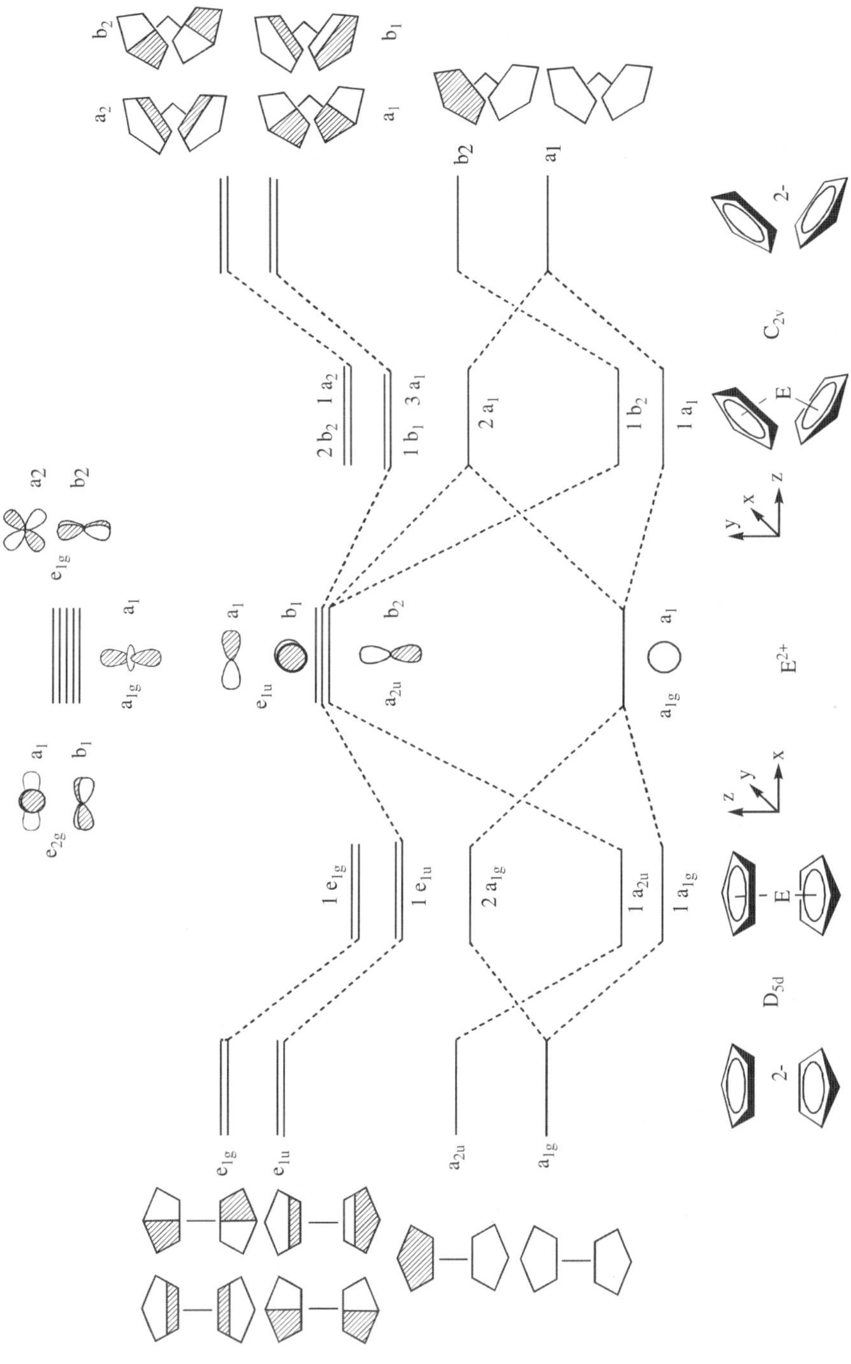

Fig. 13.5. MO correlation diagram for the interactions between a main group atomic ion E^{2+} and two Cp^- ligands in (D_{5d}) ECp_2.

Table 13.10 Energy decomposition analysis of group-2 metallocenes ECp_2 (E = Be–Ba) and $ZnCp_2$ using the fragments E^{2+} and Cp_2^{2-} at BP86/TZP [39]

	Be	Mg	Ca	Sr	Ba	Zn
ΔE_{int}	−780.9	−636.7	−549.0	−504.5	−461.0	−707.9
ΔE_{Pauli}	35.9	52.7	73.6	66.5	62.2	85.5
ΔE^{a}_{elstat}	−483.2 (59.2%)	−493.6 (71.6%)	−477.9 (76.8%)	−458.3 (80.3%)	−437.0 (83.5%)	−533.6 (67.3%)
ΔE^{a}_{orb}	−333.6 (40.8%)	−195.8 (28.4%)	−144.7 (23.2%)	−112.7 (19.7%)	−86.2 (16.5%)	−259.8 (32.7%)
$\Delta E_{\sigma}(A_{1g})^{b}$	−58.6 (17.6%)	−38.9 (19.9%)	−21.5 (14.9%)	−18.2 (16.1%)	−13.8 (16.0%)	−80.6 (31.0%)
$\Delta E(A_{2g})$	0.0	0.0	0.0	0.0	0.0	0.0
$\Delta E_{\pi}(E_{1g})^{b}$	−53.5 (16.0%)	−47.6 (24.3%)	−63.0 (43.5%)	−45.5 (40.4%)	−32.0 (37.1%)	−39.1 (15.0%)
$\Delta E_{\delta}(E_{2g})^{b}$	−15.8 (4.7%)	−12.5 (6.4%)	−9.1 (6.3%)	−7.1 (6.3%)	−5.5 (6.4%)	−15.1 (5.8%)
$\Delta E(A_{1u})$	0.0	0.0	0.0	0.0	0.0	0.0
$\Delta E_{\sigma}(A_{2u})^{b}$	−58.9 (17.6%)	−26.6 (13.6%)	−13.0 (9.0%)	−10.0 (8.9%)	−7.9 (9.2%)	−35.5 (13.7%)
$\Delta E_{\pi}(E_{1u})^{b}$	−132.8 (39.8%)	−59.0 (30.1%)	−29.6 (20.5%)	−24.7 (21.9%)	−21.2 (24.6%)	−78.4 (30.2%)
$\Delta E_{\delta}(E_{2u})^{b}$	−14.1 (4.2%)	−11.3 (5.8%)	−8.5 (5.9%)	−7.2 (6.4%)	−5.9 (6.8%)	−11.2 (4.3%)
q(E)	+1.68	+1.74	+1.72	+1.77	+1.77	+1.60

The symmetry point group is D_{5d}. Energy values in kcal/mol. Atomic partial charges q.

[a]The values in parentheses give the percentage contribution to the total attractive interactions $\Delta E_{elstat} + \Delta E_{orb}$.

[b]The values in parentheses give the percentage contribution to the orbital interactions ΔE_{orb}.

the attractive interaction comes from the quasiclassical electrostatic interaction ΔE_{elstat}. The covalent character of the bond ranges between 41% (E = Be) and just 17% (E = Ba) but is in general higher than for the alkaline complexes discussed above.

The individual contributions to the orbital term ΔE_{orb} show, that for beryllocene, the strongest interaction arises from the e_{1u} orbitals (39.8%) followed by the a_{1g} and the a_{2u} orbitals (each 17.6%). The smaller contributions in the e_{2g} and the e_{2u} symmetries stem from the relaxation of the corresponding ligand orbitals in the electrostatic field of the central metal atom. The contribution of the e_{1g} orbitals (16.0%) is, however, somewhat stronger than might have been expected. This contribution arises from the mixing of the ligand HOMO with the $d(e_{1g})$ polarization functions of the central metal atom which have a very large overlap with the former.

The change of the individual contributions to the orbital term ΔE_{orb} is very interesting. Table 13.10 indicates that the relative contribution of the e_{1g} orbitals increases for heavier central metal atoms. For the last three compounds $Ca(Cp)_2$, $Sr(Cp)_2$ and $Ba(Cp)_2$ this interaction is the largest contribution to ΔE_{orb}. This contribution, coming from the alkaline earth metal d orbitals, becomes even larger than the contributions from the a_{1g}, a_{2u}, and e_{1u} orbitals which come from covalent interactions of the ligand and the central metal atom. A similar observation has been made before [80] employing the Mulliken analysis. It is interesting to note, however, that although the *relative* contribution of the e_{1g} orbitals to the ΔE_{orb} term increases, the *absolute* value does not change very much. Furthermore, if one takes into account that, especially for the molecules with a large contribution of the e_{1g} orbitals, the overall contribution from the orbital term ΔE_{orb} is rather low, we think it is not justified to classify the d orbitals of the alkaline earth metals as true valence orbitals in the molecules. They rather play a very important role in polarizing the HOMO of the ligand and should therefore be included in an orbital correlation diagram.

To address the question whether zinc behaves like a main-group metal or like a transition metal $Zn(Cp)_2$ was included in the investigation. Table 13.10 shows that the bonding in zincocene is similar to the bonding in magnesocene. The electrostatic interaction (67%) is stronger than the covalent orbital interaction (33%). The contributions to the ΔE_{orb} term reveal, that there is only a small contribution from the e_{1g} orbitals (15%). Zinc has a filled d-shell and thus, there are no $d(e_{1g})$-acceptor orbitals in zincocene. Like in the lighter alkaline earth metals the main contribution to the ΔE_{orb} term comes from the orbitals with a_{1g} and e_{1u} symmetry.

How reliable are the results of an energy analysis of the D_{5d} forms of $Be(Cp)_2$, $Sr(Cp)_2$ and $Ba(Cp)_2$ when the equilibrium structures of these complexes have in fact a slipped sandwich or a bent geometry? Although the energy differences between the energy minima and the D_{5d} forms are small a dramatic change in the composition of the interaction energy could occur. The calculations showed that this is indeed not the case [39]. To assess whether the large contribution of the e_{1g} orbitals was special to the D_{5d} form of the complexes, $Sr(Cp)_2$ and $Ba(Cp)_2$ were calculated in bent structures with C_{2v} symmetry. The results are shown in Table 13.11.

The energy differences between the C_{2v} structures and the bent equilibrium forms of $SrCp_2$ (C_1) and $BaCp_2$ (C_s) are very small (<0.1 kcal/mol). In the C_{2v} structures it is possible to correlate the irreducible representations to their corresponding representations

Table 13.11 Energy decomposition analysis of group-2 metallocenes ECp_2 (E = Sr, Ba) using the fragments E^{2+} and Cp_2^{2-} at BP86/TZP [39]

	Sr	Ba
ΔE_{int}	−505.7	−461.3
ΔE_{Pauli}	79.8	88.1
ΔE_{elstat}^{a}	−466.9 (79.7%)	−452.8 (82.4%)
ΔE_{orb}^{a}	−118.6 (20.3%)	−96.6 (17.6%)
$\Delta E_{\sigma}(A_1)^{b}$	−35.9 (30.3%)	−30.8 (31.9%)
$\Delta E_{\delta}(A_2)^{b}$	−27.9 (23.5%)	−20.8 (21.5%)
$\Delta E_{\pi}(B_1)^{b}$	−16.9 (14.2%)	−15.3 (15.8%)
$\Delta E_{\pi}(B_2)^{b}$	−37.9 (32.0%)	−29.8 (30.8%)
$q(E)^{c}$	+1.77	+1.74

The symmetry point group is C_{2v}. Energy values in kcal/mol. Atomic partial charges q.
[a]The values in parentheses give the percentage contribution to the total attractive interactions $\Delta E_{elstat} + \Delta E_{orb}$.
[b]The values in parentheses give the percentage contribution to the orbital interactions ΔE_{orb}.
[c]The partial charges $q(E)$ have been calculated for the bent equilibrium structures.

in D_{5d}. As can be seen in Fig. 13.5, the p orbitals of the central metal atom have a_1, b_1 and b_2 symmetry, respectively. The ligand HOMO with e_{1g} symmetry splits into two orbitals with a_2 and b_2 symmetry. Since there is no corresponding metal orbital with a_2 symmetry, one can use the a_2 contribution as a probe for the interaction of the metal d orbitals in ΔE_{orb}.

A comparison of the data in Table 13.11 with those in Table 13.10 shows, that although the individual terms are slightly larger than for the C_{2v} structures, the relative weight of the individual terms does not change very much. The contributions to ΔE_{orb} are comparable to the D_{5d} forms. The a_2 contribution in particular is as important as the a_1, b_1 and b_2 contributions. So the d orbitals play an important role in the bent geometries as well. Note that the π and δ orbital contributions in the D_{5d} (Table 13.10) and C_{2v} (Table 13.11) forms can not directly be compared because different mirror planes are used to assign the symmetry of the orbitals.

The next molecules in this discussion are the group-13 half sandwich complexes. The results of the EDA are given in Table 13.12. In comparison to the group-1 half sandwich complexes the covalent contributions are larger. While in the boron complex 50% of the attractive interaction come from the ΔE_{orb} term, this relative contribution decreases steadily to a mere 28% for the thallium complex. But this value is still larger than the largest value of the group-1 complexes (ΔE_{orb} 20% for Li). The relative contributions of the a_1 orbitals are significantly higher in the group-13 complexes than in the group-1 half sandwiches. This can be rationalised by the energy decrease of the corresponding acceptor orbitals of the E^+ cations (E = B–Tl) compared to those of the E''^+ cations (E'' = Li–Cs).

Finally the bonding analysis for the group-14 sandwich complexes shall be discussed. The results for the EDA of the D_{5d} forms are given in Table 13.13. It can be seen that the covalent contributions of the ΔE_{orb} term are significantly larger than in the group-2 metallocenes, although the main attractive contribution to the interaction energy still

Table 13.12 Energy decomposition analysis of group-13 metallocenes ECp (E = B–Tl) using the fragments E^+ and Cp^- at BP86/TZP [39]

	B	Al	Ga	In	Tl
ΔE_{int}	−256.0	−188.4	−181.2	−164.6	−155.0
ΔE_{Pauli}	202.2	118.1	97.1	81.6	67.2
ΔE^{a}_{elstat}	−226.1 (49.3%)	−197.8 (64.6%)	−183.5 (65.9%)	−172.3 (70.0%)	−160.0 (72.0%)
ΔE^{a}_{orb}	−232.1 (50.7%)	−108.6 (35.4%)	−94.8 (34.1%)	−73.9 (30.0%)	−62.2 (28.0%)
$\Delta E_{\sigma}(A_1)^{b}$	−93.9 (40.4%)	−47.9 (44.1%)	−38.4 (40.5%)	−29.7 (40.2%)	−22.7 (36.5%)
$\Delta E(A_2)$	0.0	0.0	0.0	0.0	0.0
$\Delta E_{\pi}(E_1)^{b}$	−130.1 (56.1%)	−55.2 (50.8%)	−51.9 (54.8%)	−40.3 (54.5%)	−36.3 (58.4%)
$\Delta E_{\delta}(E_2)^{b}$	−8.10 (3.5%)	−5.6 (5.1%)	−4.5 (4.7%)	−3.9 (5.3%)	−3.2 (5.1%)
q(E)	+0.18	+0.61	+0.57	+0.63	+0.62

The symmetry point group is C_{5v}. Energy values in kcal/mol. Atomic partial charges q.
[a]The values in parentheses give the percentage contribution to the total attractive interactions $\Delta E_{elstat} + \Delta E_{orb}$.
[b]The values in parentheses give the percentage contribution to the orbital interactions ΔE_{orb}.

comes from the electrostatic term ΔE_{elstat}. It is interesting to note that the ratio of electrostatic *versus* covalent bonding is almost the same for the group-14 sandwich complexes $E(Cp)_2$ (E = Si–Pb) and the group-13 half sandwich complexes $E''(Cp)$ (E' = Al–Tl) for elements of the same period.

The largest contribution to the ΔE_{orb} term comes from the e_{1u} orbitals which yield 52–55% of the covalent interaction. The e_{1u} term resembles the donation from the occupied ligand orbitals into the vacant p(π)-AOs of E^{2+}. Overall the relative contributions from the different irreducible representations to the ΔE_{orb} term do not change very much from Si to Pb. In these complexes the e_{1g} orbitals have a rather small influence on the orbital interaction even for the heaviest element E = Pb. This leads to the conclusion that

Table 13.13 Energy decomposition analysis of group-14 metallocenes ECp_2 (E = Si, Ge, Sn, Pb) using the fragments E^{2+} and Cp_2^{2-} at BP86/TZP [39]

	Si	Ge	Sn	Pb
ΔE_{int}	−685.4	−661.4	−601.4	−577.2
ΔE_{Pauli}	120.4	108.7	101.6	91.3
ΔE^{a}_{elstat}	−506.7 (62.9%)	−501.3 (65.1%)	−490.0 (69.7%)	−480.2 (71.8%)
ΔE^{a}_{orb}	−299.1 (37.1%)	−268.7 (34.9%)	−213.0 (30.3%)	−188.3 (28.2%)
$\Delta E_{\sigma}(A_{1g})^{b}$	−16.6 (5.6%)	−13.2 (4.9%)	−12.2 (5.7%)	−10.3 (5.5%)
$\Delta E(A_{2g})$	0.0	0.0	0.0	0.0
$\Delta E_{\pi}(E_{1g})^{b}$	−48.9 (16.3%)	−37.7 (14.0%)	−28.9 (13.6%)	−24.9 (13.2%)
$\Delta E_{\delta}(E_{2g})^{b}$	−9.7 (3.2%)	−9.0 (3.3%)	−7.8 (3.6%)	−7.3 (3.9%)
$\Delta E(A_{1u})$	0.0	0.0	0.0	0.0
$\Delta E_{\sigma}(A_{2u})^{b}$	−58.7 (19.6%)	−54.2 (20.2%)	−41.5 (19.5%)	−36.3 (19.3%)
$\Delta E_{\pi}(E_{1u})^{b}$	−156.4 (52.3%)	−146.5 (54.5%)	−115.2 (54.1%)	−102.7 (54.5%)
$\Delta E_{\delta}(E_{2u})^{b}$	−8.8 (2.9%)	−8.2 (3.1%)	−7.5 (3.5%)	−6.8 (3.6%)
q(E)	+0.80	+0.86	+0.98	+0.99

The symmetry point group is D_{5d}. Energy values in kcal/mol. Atomic partial charges q.
[a]The values in parentheses give the percentage contribution to the total attractive interactions $\Delta E_{elstat} + \Delta E_{orb}$.
[b]The values in parentheses give the percentage contribution to the orbital interactions ΔE_{orb}.

Table 13.14 Energy decomposition analysis of group-14 metallocenes ECp_2 (E = Si, Ge, Sn, Pb) using the fragments E^{2+} and Cp_2^{2-} at BP86/TZP [39]

	Si	Ge	Sn	Pb
ΔE_{int}	−690.5	−663.1	−603.9	−577.8
ΔE_{Pauli}	144.8	118.9	109.8	93.8
ΔE^{a}_{elstat}	−511.6 (61.2%)	−504.1 (64.5%)	−492.5 (69.0%)	−481.0 (71.6%)
ΔE^{a}_{orb}	−323.7 (38.8%)	−278.0 (35.5%)	−221.1 (31.0%)	−190.6 (28.4%)
$\Delta E_{\sigma}(A_1)^{b}$	−117.4 (36.3%)	−97.9 (35.2%)	−79.6 (36.0%)	−66.8 (35.0%)
$\Delta E_{\delta}(A_2)^{b}$	−30.1 (9.3%)	−23.4 (8.4%)	−18.8 (8.5%)	−16.1 (8.4%)
$\Delta E_{\pi}(B_1)^{b}$	−85.4 (26.4%)	−78.9 (28.4%)	−62.6 (28.3%)	−55.4 (29.1%)
$\Delta E_{\pi}(B_2)^{b}$	−90.8 (28.1%)	−77.9 (28.0%)	−60.1 (27.2%)	−52.3 (27.4%)
$q(E)^{c}$	+0.88	+0.89	+1.01	+1.00

The symmetry point group is C_{2v}. Energy values in kcal/mol. Atomic partial charges q.
[a]The value in parentheses gives the percentage contribution to the total attractive interactions $\Delta E_{elstat} + \Delta E_{orb}$.
[b]The value in parentheses gives the percentage contribution to the total orbital interactions ΔE_{orb}.
[c]The partial charges q(E) have been calculated for the bent equilibrium structures.

the metal d orbitals are not very important for the description of the bond in group-14 sandwich complexes.

Table 13.14 finally gives the result for the bent structures of the group-14 metallocenes with C_{2v} symmetry. In contrast to the alkaline earth metallocenes, the bent structures exhibit a larger amount of covalent bonding than the D_{5d} forms although the difference is not very big. The degenerate e_{1u} orbitals split into a_1 and b_1. Table 13.14 shows that the sum of the orbital contributions $a_1 + b_1$ provides more than 60% of the ΔE_{orb} term. Thus the orbital correlation diagram shown in Fig. 13.5 provides a reasonable qualitative bonding model for the interactions of the central metal atom and the ligands in the group-14 metallocenes.

13.3.9 Bonding in SF_6 and XeF_6 and a comparison with WF_6 [43]

The closing chapter on main group compounds will address an old issue which connects the bonding in main-group elements with the bonding in transition metals. The pivotal questions concern the use of *d* functions for the bonding of heavier main group elements and the importance of p functions for TM compounds. The nature of the bonding in noble gas compounds, which is another longstanding topic will also be addressed. We compare the bonding in octahedral compounds EF_6 (E = S, Xe, W) between the central atom E and the fluorine atoms. We do not address the question about the equilibrium geometry of XeF_6 which has a fluctional structure where the C_{3v} distorted form is energetically nearly degenerate with the O_h structure [81]. We will first compare then discuss the chemical bonding in the main-group compounds. It has been shown before that a qualitative orbital picture leads to a good description of the chemical bonding in octahedral XeF_6 and SF_6 [82]. Fig. 13.6 shows such an MO diagram for an octahedral complex with a central atom E and six surrounding ligand-atoms F. For the bonding in XeF_6 and SF_6 it is usually assumed that only σ bonding takes place in these main-group

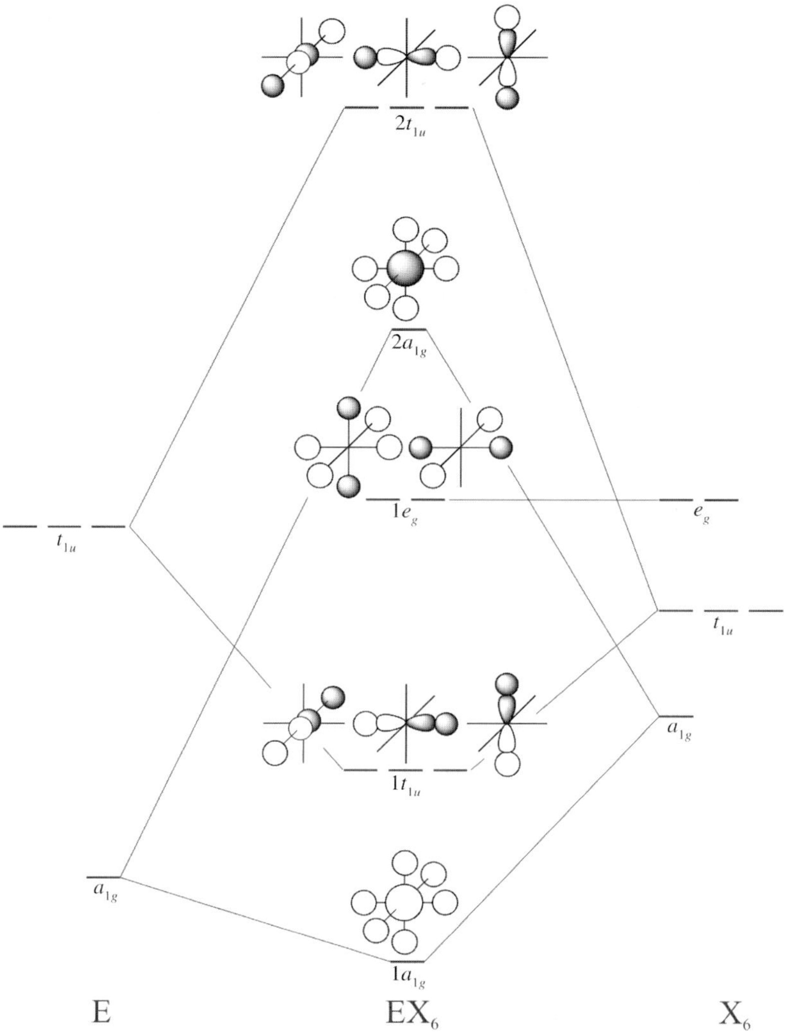

Fig. 13.6. MO correlation diagram for the interaction between a main group element E and a cage of six σ-bonded ligands X in octahedral EF_6.

compounds [82], therefore only orbitals of σ type are shown. The EDA makes it possible to see if the assumption is justified.

From the xenon 5s and 5p orbitals and one p orbital from each fluorine one can construct orbitals having a_{1g}, e_g and t_{1u} symmetry in an octahedral environment. The 14 valence electrons of XeF_6 occupy these orbitals according to the Aufbau-principle. In the a_{1g} symmetry the bonding combination $1a_{1g}$ is doubly occupied, as is the antibonding combination $2a_{1g}$ thus leaving no net stabilization in this irrep. The e_g orbitals of the fluorine ligand cage have no symmetry-counterpart on the central xenon atom, hence leaving four electrons occupying an essentially nonbonding $1e_g$ orbital. The only stabilizing contribution comes from the triply degenerate t_{1u} orbital. As SF_6 has only 12 valence electrons the antibonding combination of the a_{1g} orbitals, the HOMO in XeF_6, remains empty. This should lead to a bonding contribution from this irreducible representation.

Before looking into the bonding between the central atoms and their surrounding F_6 cages, one has to make sure that the assumption of an essentially σ bonding situation is justified. Therefore an EDA has been carried out in order to investigate the bonding interactions between the EF_5^+ fragment and F^-. The results are given in Table 13.15.

The EDA data given in Table 13.15 show that the binding interactions between EF_5^+ and F^- have a larger electrostatic character than covalent character. The contribution of ΔE_{elstat} to the attractive interactions is 60.0% for SF_6 and 65.1% for XeF_6. The bonding in WF_6 has an even higher electrostatic character (71.8%). The orbital term in the main-group hexafluorides is dominated by the σ term which comes from the a_1 orbitals (80.5% in SF_6 and 77.2% in XeF_6) while the π bonding contribution is much smaller. It is therefore justified to qualitatively analyze the orbital interactions in SF_6 and XeF_6 in terms of σ bonding only. This does not hold for WF_6! The σ and π bonding contributions in WF_6 have nearly the same strength.

The results of the energy partitioning analyses for the interaction of the central main group atoms with the surrounding hexafluoride cage are given in Table 13.16. The analysis was carried out between the charged species E^{6+} and $(F_6)^{6-}$ because on this way it is possible to compare the bonding using the same fragment configurations and spin states for the main-group compounds and WF_6.

Because of the choice of the highly charged fragments, the calculated values for the interaction energy are very large. The attractive interactions have a larger electrostatic character (57.1–68.0%) than the covalent bonding. EDA calculations using neutral fragments with different configurations gave a similar ratio of the ΔE_{orb} and ΔE_{elstat} terms [43]. The focus of this part of the analysis is on the different orbital terms, however. Since we compare the bonding in octahedral main-group compounds SF_6 and XeF_6 with the transition metal compound WF_6 we must include valence d orbitals in the analysis.

Table 13.15 EDA results for the binding interactions in EF_6 (E = S, Xe, W) using the fragments EF_5^+ and F^- at BP86/QZ4P [43]

	$SF_5^+ + F^-$	$XeF_5^+ + F^-$	$WF_5^+ + F^-$
ΔE_{int}	−283.0	−237.6	−277.4
ΔE_{Pauli}	343.0	235.5	214.4
ΔE^a_{elstat}	−375.6 (60.0%)	−307.8 (65.1%)	−353.1 (71.8%)
ΔE^a_{orb}	−250.2 (40.0%)	−165.2 (34.9%)	−138.7 (28.2%)
$\Delta E_\sigma(A_1)^b$	−201.5 (80.5%)	−127.6 (77.2%)	−70.4 (50.7%)
$\Delta E(A_2)^b$	−0.35 (0.1%)	−0.2 (0.1%)	−0.2 (0.2%)
$\Delta E(B_1)^b$	−1.1 (0.4%)	−0.5 (1.8%)	−1.0 (0.7%)
$\Delta E(B_2)^b$	−0.5 (0.2%)	−0.2 (0.1%)	−0.8 (0.6%)
$\Delta E_\pi(E)^b$	−46.8 (18.7%)	−36.8 (22.3%)	−66.3 (47.8%)

Energy values in kcal/mol.

[a]The value in parentheses gives the percentage contribution to the total attractive interactions $\Delta E_{elstat} + \Delta E_{orb}$.

[b]The value in parentheses gives the percentage contribution to the total orbital interactions ΔE_{orb}.

Table 13.16 Energy decomposition analysis of EF_6 using the fragments E^{6+} and F_6^{6-} (E = S, Xe, W) at BP86/QZ4P [43]

	$S^{6+} + F_6^{6-}$	$Xe^{6+} + F_6^{6-}$	$W^{6+} + F_6^{6-}$
ΔE_{int}	−8129.4	−6504.2	−6560.0
ΔE_{Pauli}	188.73	392.3	517.1
ΔE_{elstat}^{a}	−4746.9 (57.1%)	−4720.1 (68.4%)	−4815.3 (68.0%)
ΔE_{orb}^{a}	−3571.2 (42.9%)	−2176.4 (31.6%)	−2261.9 (32.0%)
$\Delta E(A_{1g})^{b}$ s(σ)	−672.3 (18.8%)	−54.6 (2.5%)	−131.2 (5.8%)
$\Delta E(A_{1u})^{b}$	0.0 (0.0%)	0.0 (0.0%)	−0.16 (<0.1%)
$\Delta E(A_{2g})$	0.0	0.0	0.0
$\Delta E(A_{2u})$	0.0	0.0	0.0
$\Delta E(E_{g})^{b}$ d(σ)	−622.8 (17.4%)	−316.6 (14.5%)	−738.4 (32.6%)
$\Delta E(E_{u})$	0.0	0.0	0.0
$\Delta E(T_{1g})^{b}$ p(ligand)	−84.5 (2.4%)	−70.9 (3.3%)	−77.9 (3.4%)
$\Delta E(T_{1u})^{b}$ p(σ and π)	−1760.3 (49.3%)	−1444.4 (66.4%)	−581.7 (25.7%)
$\Delta E(T_{2g})^{b}$ d(π)	−307.5 (8.6%)	−181.5 (8.3%)	−620.4 (27.4%)
$\Delta E(T_{2u})^{b}$ p(ligand)	−123.9 (3.5%)	−108.4 (5.0%)	−112.2 (5.0%)

Energy values in kcal/mol.
[a]The value in parentheses gives the percentage contribution to the total attractive interactions $\Delta E_{elstat} + \Delta E_{orb}$.
[b]The value in parentheses gives the percentage contribution to the total orbital interactions ΔE_{orb}.

An orbital correlation diagram for the splitting of the valence s, p and d orbitals of a transition metal in an octahedral ligand field is shown in Fig. 13.9 which is discussed in Section 13.4 which describes the bonding in TM hexacarbonyls. A schematic representation of all orbital terms which give a non zero contribution to the ΔE_{orb} term is presented there in Fig. 13.10.

The data in Table 13.16 indicate that the largest contribution to the orbital interaction in SF_6 and WF_6 comes from the t_{1u} term which comes from the p orbitals of the central atom. This is in agreement with the orbital correlation diagram shown in Fig. 13.6. However, Fig. 13.10 shows that the t_{1u} term comes not only from σ interactions but it may also come from π-orbital interactions. The latter interactions cannot be distinguished in the EDA when octahedral fragments are used. But the EDA results for the interactions between EF_5^+ and F^- have shown that σ bonding is much more important than π bonding. Thus, the largest orbital contributions in SF_6 and XeF_6 come indeed from the σ interactions of the valence p orbitals of the central atom. The a_{1g} contribution in SF_6 (−672.3 kcal/mol) is much larger than in XeF_6 (−54.6 kcal/mol) which is in agreement with the qualitative orbital interaction diagram (Fig. 13.6). Nevertheless, the relative contribution of the a_{1g} orbitals even in the former compound (18.8%) is much smaller than that of the t_{1u} orbitals. We want to point out that the contribution which comes from the d(σ) orbitals in SF_6 and XeF_6 (17.4 and 14.5%) is not negligible. The latter term becomes the largest orbital contribution in WF_6 (32.6%) followed by the t_{2g} term (27.4%) which comes from the d(π) orbitals (Table 13.16). This means that the d orbitals of tungsten yield 50.0% of the orbital interactions. Which valence orbitals contribute to the remaining 50% of ΔE_{orb}?

Table 13.16 shows the contribution of the t_{1u} term of WF_6 which comes from the p(σ and π) orbitals is 25.7% of ΔE_{orb}. This is much larger than the contribution of the tungsten s orbital which yields only 5.8% of the orbital interactions. The calculated energy values of the t_{1g} and t_{2u} terms makes it possible to estimate the strength of the relaxation of the ligand p orbitals (Fig. 13.10). The relative contributions are between 2.4 and 5.0%. It can therefore be estimated that >20% of the t_{1u} term which comes from the p(σ and π) orbitals are genuine orbital interactions. The EDA results of WF_6 indicate that the (n)p orbitals of tungsten are genuine valence orbitals.

13.4 BONDING IN TRANSITION METAL COMPOUNDS

The qualitative bonding models which are used for describing the bonding situation in transition metal compounds are very different from those which are employed for main group elements [83]. Covalent bonding of the latter is described in terms of sp hybridization and by obeying the octet rule. The valence orbitals of the transition metals are the $(n-1)$s and $(n-1)$d AOs which are supplemented by the (n)p orbitals. This leads to the 18 electron rule which is, however, less strict than the octet rule. It has recently been suggested that the bonds in TM compounds should be described using only the $(n-1)$s and $(n-1)$d AOs while the (n)p AOs should only be considered as polarization functions like the d AOs of main group elements but not as valence orbitals [84]. This view has been challenged, however, in more recent quantum chemical work which indicates that the (n)p AOs of the TMs are genuine valence orbitals [17,85].

The chemical bonds of TM compounds are frequently discussed in terms of donor–acceptor interactions between the metal TM and the ligand L using the synergistic bonding model of TM ← L σ donation and TM → L π-backdonation which was introduced by Dewar [86] and by Chatt and Duncanson (DCD) [87]. The synergistic bonding model was later extended to other metal–ligand bonds where the energetically high-lying occupied orbitals of the ligand, which have in most cases σ symmetry with regard to the whole complex, are considered as donor orbitals. The backdonation then occurs from occupied π orbitals of the metal to low-lying empty π^* orbitals of the ligand [88]. Since σ and π interactions can be quantitatively estimated using the EDA method it is possible to examine the validity of the DCD model. At the same time it is possible to estimate the strength of the quasiclassical interaction between the metal and the ligand which is neglected in the DCD model.

13.4.1 Carbonyl complexes $TM(CO)_6^q$ ($TM^q = Hf^{2-}, Ta^-, W, Re^+, Os^{2+}, Ir^{3+}$) [17]

It is reasonable to begin a discussion of bonding in transition metal compounds with carbonyl complexes because they can be used as a parent system for many TM molecules just like the hydrocarbons are used as parent compounds for organic molecules. Fig. 13.7 gives the standard textbook description of bonding between a transition metal and CO using the DCD model. The main bonding contributions should arise from the electron donation of the CO σ HOMO into the an empty d(σ) orbital of the metal and the backdonation from occupied d(π) orbitals of the metal into the unoccupied π^* orbital of CO.

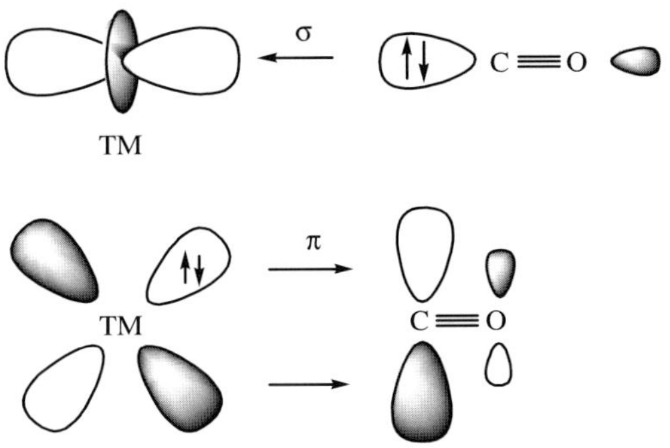

Fig. 13.7. Schematic representation of the synergistic TM ← CO σ-donation and TM → CO π-backdonation.

Numerous theoretical studies investigated the question which of the two contributions that are shown in Fig. 13.7 are more important for the bonding. Most investigations analyzed the charge distribution between the metal and the ligand [89] and only few papers analyzed the contributions of the different energy terms to the metal–CO interactions [90]. Nearly all studies focused on neutral compounds [91]. There is general agreement that, in most carbonyl complexes, the TM → CO π-backdonation is more important for the chemical bonding than the TM ← CO σ-donation. However, recent experimental research showed that carbonyl complexes which have been termed non-classical carbonyls can be prepared which are mainly bonded by σ-donation [92]. An ideal set of molecules for analyzing the individual contributions of both the σ-donation and the π-backdonation to the TM–CO bonding is the series of isoelectronic hexacarbonyls $TM(CO)_6^q$ ($TM^q = Hf^{2-}$, Ta^-, W, Re^+, Os^{2+}, Ir^{3+}) [17].

Table 13.17 gives the EDA results for the metal–ligand bonding in the hexacarbonyls $TM(CO)_6^q$ between the closed-shell fragments $TM(CO)_5^q$ and CO. The calculated data indicate that the $(CO)_5TM^q$ → CO π-backdonation which is given by the contribution of the orbitals having e symmetry decreases from $Hf(CO)_6^{2-}$ to $Ir(CO)_6^{3+}$. This is reasonable because the π-backdonation should become weaker when the electronic charge becomes less negative or more positive. However, the trend of the backdonation does not agree with the trend of the total interaction energy ΔE_{int} which exhibits a U-shaped form instead (Table 13.17). The ΔE_{int} values and the bond dissociation energies become smaller from $Hf(CO)_6^{2-}$ to $W(CO)_6$ but then they increase. The highly positively charged $Ir(CO)_6^{3+}$ even has the strongest $(CO)_5TM^q$–CO bond among the hexacarbonyls. Inspection of the various energy terms shows that the increase of the interaction energy in the cations comes from the large contribution of the $(CO)_5TM^q$ ← CO σ-donation which becomes very strong in $Ir(CO)_6^{3+}$. Note that the increase of ΔE_{int} from $W(CO)_6$ to $Hf(CO)_6^{2-}$ does not come from stronger attraction (Table 13.17). The electrostatic attraction and the orbital term decrease in the order, but the decrease is compensated by an even larger decrease of the Pauli repulsion ΔE_{Pauli}. Fig. 13.8 shows the trend of the energy terms which contribute to the total interaction energy. It becomes obvious that the ΔE_{orb} term correlates nicely with the ΔE_{int} values except for the trend from $Hf(CO)_6^{2-}$ to $W(CO)_6$. The EDA results indicate that the trend in the metal–CO bonding can be

Table 13.17 Energy decomposition analysis of $TM(CO)_5^q$ + CO at BP86/TZP [17]

	$Hf(CO)_6^{2-}$	$Ta(CO)_6^-$	$W(CO)_6$	$Re(CO)_6^+$	$Os(CO)_6^{2+}$	$Ir(CO)_6^{3+}$
ΔE_{int}	−56.6	−51.3	−49.6	−52.7	−61.9	−78.9
ΔE_{Pauli}	76.6	100.7	118.3	126.9	125.4	115.9
ΔE_{elstat}^{a}	−59.4 (44.6%)	−76.6 (50.4%)	−90.1 (53.6%)	−97.7 (54.4%)	−98.5 (52.6%)	−93.1 (47.8%)
ΔE_{orb}^{a}	−73.8 (55.4%)	−75.5 (49.6%)	−77.9 (46.4%)	−81.9 (45.6%)	−88.9 (47.4%)	−101.8 (52.2%)
$\Delta E_{\sigma}(A_1)^b$	−17.2 (23.3%)	−25.8 (34.3%)	−35.9 (46.1%)	−47.3 (57.8%)	−60.1 (67.6%)	−75.4 (74.2%)
$\Delta E(A_2)$	0.00	0.00	0.00	0.00	0.00	0.00
$\Delta E(B_1)$	0.05	0.02	−0.03	−0.07	−0.09	−0.10
$\Delta E(B_2)$	−0.05	−0.07	−0.07	−0.07	−0.06	−0.05
$\Delta E_{\pi}(E)^b$	−56.6 (76.7%)	−49.6 (65.8%)	−41.9 (53.8%)	−34.4 (42.1%)	−28.6 (32.2%)	−26.2 (25.8%)
ΔE_{prep}	5.8	3.1	3.7	4.4	5.0	5.2
$\Delta E(= -D_e)$	−50.8	−48.3	−46.0	−48.4	−56.9	−73.7
R(TM–C)	2.195	2.112	2.061	2.036	2.034	2.055

Energy values in kcal/mol. Bond lengths R in Å.

[a]The value in parentheses gives the percentage contribution to the total attractive interactions $\Delta E_{elstat} + \Delta E_{orb}$.

[b]The value in parentheses gives the percentage contribution to the total orbital interactions ΔE_{orb}.

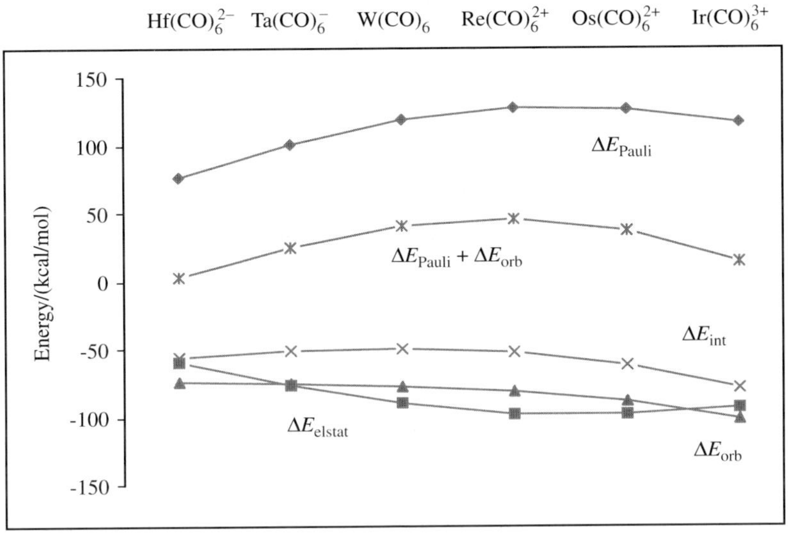

Fig. 13.8. Trends of the various terms of the energy decomposition analysis of $(CO)_5TM^q$–CO calculated at BP86/TZP.

discussed in terms of frontier–orbital interactions, while the electrostatic term does not agree with the changes of the bond strength. There are classes of compounds where the opposite correlation was found, such as the phosphane complexes which are presented and discussed below.

There is another surprising result which comes out of the energy analysis. Fig. 13.8 shows that, for the highest charged species $Hf(CO)_6^{2-}$ and $Ir(CO)_6^{3+}$, the covalent bonding is larger than the electrostatic bonding while for $W(CO)_6$, $Re(CO)_6^+$ and $Os(CO)_6^{2+}$ it holds that $\Delta E_{orb} < \Delta E_{elstat}$. This is a counterintuitive result because it means that the highest charged complexes have the smallest degree of electrostatic bonding. The explanation for this finding can be given when the energy levels of the interacting orbitals are considered. The highly charged pentacarbonyls have a very high lying HOMO ($Hf(CO)_5^{2-}$) and low lying LUMO ($Ir(CO)^{3+}$) which leads to very strong orbital interactions with the LUMO and HOMO of CO, respectively. Thus, the charge of the ionic hexacarbonyls has a larger effect on the covalent bonding through its raising or lowering the orbital energy levels than on the electrostatic interactions.

Even more insight into the metal–CO binding interactions can be obtained when the bonding in the octahedral complexes $TM(CO)_6^q$ is analyzed between the bare metal center TM^q and the ligand cage $(CO)_6$. The TM^q–$(CO)_6$ bonding is often discussed with the help of a qualitative orbital correlation diagram which is shown in Fig. 13.9 [81]. The σ donor orbitals of the six CO ligands split in an octahedral field into three sets of orbitals which have e_g, t_{1u} and a_{1g} symmetry. Fig. 13.9 shows that the metal acceptor orbitals are the s AO (a_{1g}), p AOs (t_{1u}) and the e_g set of d AOs. The remaining set of (t_{2g}) d AOs of a d^6 TM is occupied and serves as donor orbital for the TM $\rightarrow$ $(CO)_6$ π-backdonation. The advantage of the octahedral ligand field is that the contributions of the e_g, t_{1u} and a_{1g}

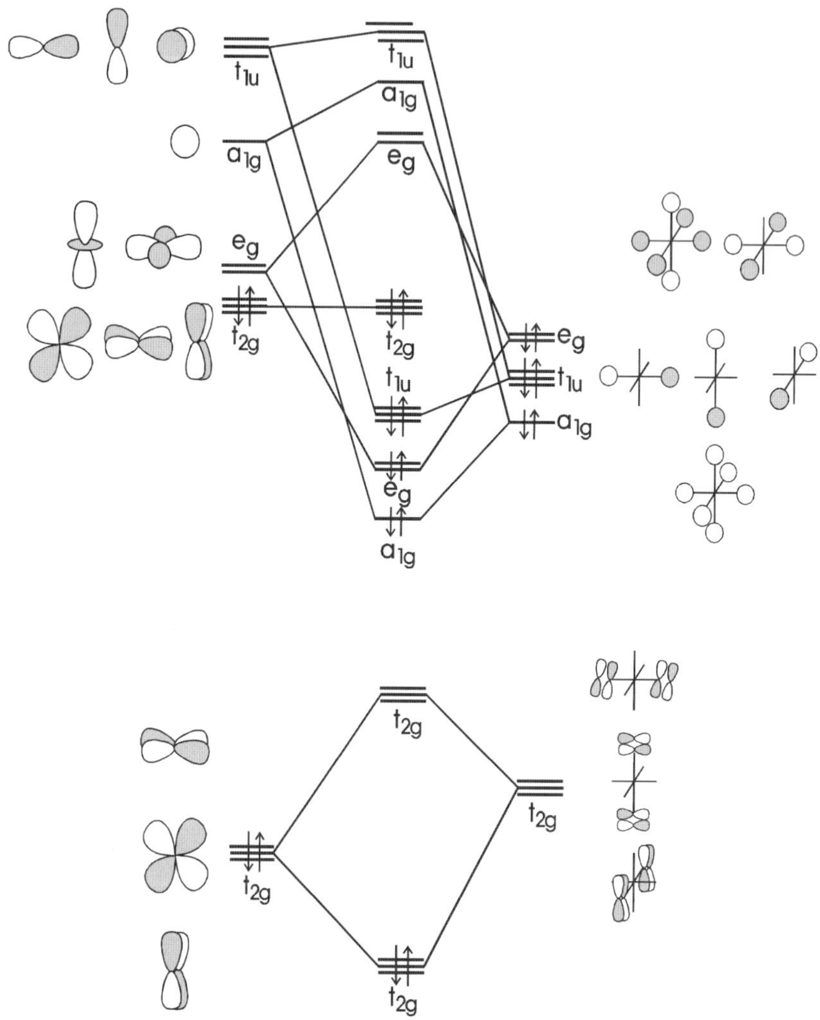

Fig. 13.9. Splitting of the orbital energy levels of an octahedral d^6 transition metal complex TML_6 where the ligand L has occupied donor orbitals with σ symmetry (top) and empty acceptor orbitals with π symmetry.

orbitals to the TM $\leftarrow$ $(CO)_6$ σ donation gives directly the relative importance of the metal valence orbitals to the metal ligand bonding.

Table 13.18 gives the results of the EDA of TM^q–$(CO)_6$. The most important results come from the breakdown of the orbital term into the contributions by the orbitals, which have different symmetry. The contribution of the t_{2g} orbitals which gives the TM $\rightarrow$ $(CO)_6$ π-backdonation is rather small in the trication $Ir(CO)_6^{3+}$ but it becomes stronger when the metal is less positively charged and it is the dominant orbital term in the neutral and negatively charged hexacarbonyls. The e_g orbitals make clearly the largest contribution to the TM $\leftarrow$ $(CO)_6$ σ donation. The e_g term is always much larger than the t_{1u}

Table 13.18 Energy decomposition analysis of $TM^q + (CO)_6$ at BP86/TZP [17]

	$Hf(CO)_6^{2-}$	$Ta(CO)_6^{-}$	$W(CO)_6$	$Re(CO)_6^{+}$	$Os(CO)_6^{2+}$	$Ir(CO)_6^{3+}$
ΔE_{int}	−543.9	−525.6	−473.9	−456.6	−544.4	−801.6
ΔE_{Pauli}	367.4	413.4	438.8	454.5	451.3	420.9
ΔE_{elstat}^{a}	−358.6 (39.4%)	−397.6 (42.3%)	−396.2 (43.4%)	−375.1 (41.2%)	−353.4 (35.5%)	−337.8 (27.6%)
ΔE_{orb}^{a}	−552.7 (60.6%)	−541.3 (57.7%)	−516.4 (56.6%)	−536.0 (58.8%)	−642.3 (64.5%)	−884.7 (72.4%)
$\Delta E_{\sigma}(A_{1g})^{b}$	−9.5 (1.7%)	−10.5 (1.8%)	−15.4 (2.8%)	−27.4 (4.6%)	−47.6 (6.4%)	−78.8 (7.5%)
$\Delta E(A_{2g})$	0.0	0.0	0.0	0.0	0.0	0.0
$\Delta E_{\sigma}(E_{g})^{b}$	−83.4 (14.6%)	−113.1 (20.3%)	−159.1 (29.3%)	−233.7 (39.6%)	−348.8 (46.9%)	−520.7 (49.5%)
$\Delta E(T_{1g})^{b}$ π(CO)	−1.3 (0.3%)	−1.0 (0.2%)	−2.9 (0.5%)	−8.9 (1.5%)	−19.4 (2.6%)	−33.9 (3.2%)
$\Delta E_{\pi}(T_{2g})^{b}$	−437.4 (76.6%)	−397.6 (71.2%)	−308.2 (56.8%)	−200.3 (34.0%)	−101.1 (13.6%)	−43.8 (4.2%)
$\Delta E(A_{1u})$	−0.03	−0.04	−0.03	0.0	−0.02	−0.02
$\Delta E(E_{u})$	0.0	−0.00	0.00	0.0	0.00	0.00
$\Delta E(T_{2u})^{b}$ π(CO)	−2.7 (4.8%)	−2.0 (0.4%)	−4.4 (0.8%)	−11.6 (2.0%)	−23.9 (3.2%)	−40.2 (3.8%)
$\Delta E_{\sigma,\pi}(T_{1u})^{b}$	−18.4 (3.2%)	−17.2 (3.1%)	−26.5 (4.9%)	−54.0 (9.2%)	−101.4 (13.6%)	−167.3 (15.9%)
$\Delta E_{\sigma}(T_{1u})^{c}$	−13.0	−12.1	−18.7	−38.5	−74.0	−125.7
$\Delta E_{\pi}(T_{1u})^{c}$	−5.4	−5.2	−7.9	−15.5	−27.4	−41.7

Energy values in kcal/mol.

[a] The value in parentheses gives the percentage contribution to the total attractive interactions $\Delta E_{elstat} + \Delta E_{orb}$.

[b] The value in parentheses gives the percentage contribution to the total orbital interactions ΔE_{orb}.

[c] The σ and π contributions of the T_{1u} orbitals have been estimated from the size of the overlaps.

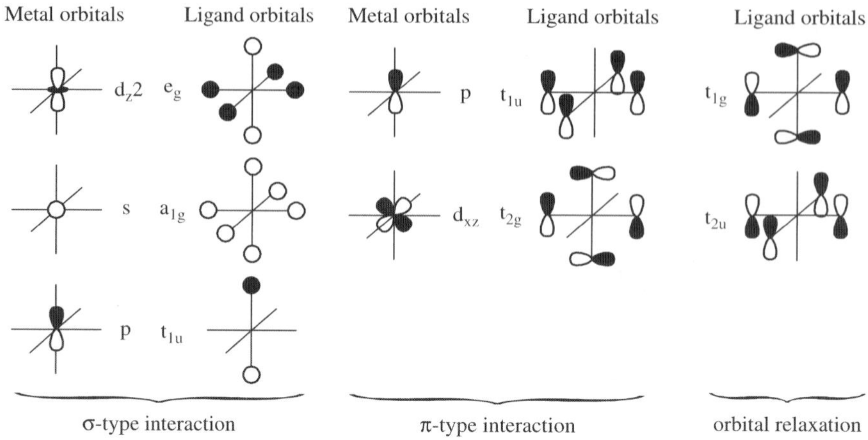

Fig. 13.10. Graphical representation of the orbital interaction terms of the energy decomposition analysis of TM^q–$(CO)_6$ given in Table 13.18.

and a_{1g} terms. Table 13.18 gives also energy contributions by t_{1g} and t_{2u} orbitals which are not shown in Fig. 13.9. Fig. 13.10 shows schematically all orbitals which contribute to ΔE_{orb}. The t_{1g} and t_{2u} orbitals are occupied ligand orbitals. The relaxation of the ligand orbitals in the final step of the energy partitioning analysis lowers the energy of the molecule. Thus, part of the stabilization energy, which comes from the orbital term is not related to the metal–ligand bonding. Fig. 13.10 shows also that the t_{1u} term does not only give the TM $\leftarrow$ $(CO)_6$ σ donation into the p(σ) AO but part of the t_{1u} stabilization energy comes from the TM $\leftarrow$ $(CO)_6$ π donation of the occupied π orbitals of CO into the empty p(π) AO of the metal. In order to estimate the σ and π contributions to the t_{1u} term, we used the size of the overlaps of the metal p orbitals with the σ and π orbitals of $(CO)_6$. Table 13.18 shows that the π contribution to the t_{1u} term is always much smaller than the σ contribution.

Concerning the question whether the metal p orbitals should be considered as valence orbitals or polarization function, the results in Table 13.18 show that the contribution of the t_{1u} term (metal p orbitals) is always larger than the a_{1g} term (metal s orbital). Thus, the p orbitals are as important as the s orbitals for the bonding in these compounds and thus, must be considered as valence functions.

13.4.2 Group-13 heteroleptic diyl complexes $(CO)_4$Fe–ER (E = B–Tl; R = Cp, Ph, Me) and homolytic diyl complexes Fe(EMe)$_5$ (E = B–Tl) and TM(EMe)$_4$ (TM = Ni, Pd, Pt; E = B–Tl) [18,21]

The group-13 diyl complexes are an example where quantum chemical studies determined the understanding of the metal–ligand bonding situation since the very first compound of the new class of molecules $(CO)_4$Fe–AlCp* was isolated and characterized by X-ray structure analysis [93]. The TM–ER bonding situation can be qualitatively

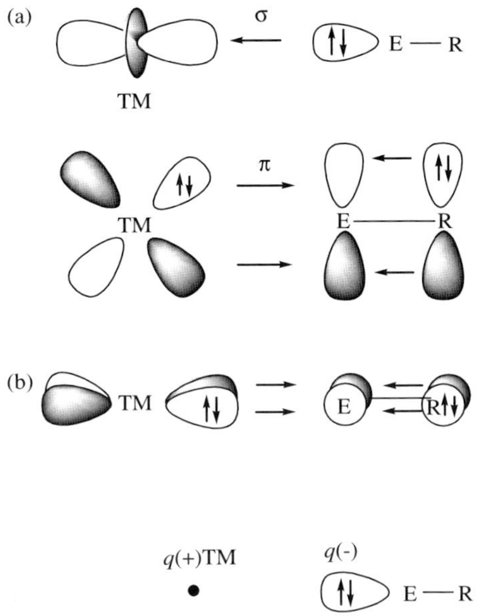

Fig. 13.11. (a) Schematic representation of the TM–ER orbital interactions when *R* has occupied p(π) orbitals. (b) Schematic representation of the dominant electrostatic interactions between the local electronic charge concentration at the donor atom E and the nucleus of the acceptor atom Fe. Note that the donor atom E has an overall positive partial charge and the TM atom an overall negative partial charge.

described with the DCD model in a similar way as the TM–CO bonding but there are significant differences. Fig. 13.11 shows a schematic representation of the TM–ER interactions in terms of TM ← ER σ-donation and TM → ER π-backdonation.

The difference between the TM–ER and TM–CO bonding (Fig. 13.7) lies in the π-backdonation which can be tuned in TM–ER by the π-donor ability of the group R which competes with the TM → ER π-backdonation. Early examples of stable group-13 diyl complexes had strong donor-groups R such as Cp* [93] and NR_2 [94] which indicated that the TM → ER π-backdonation in such compounds may be weak. The later synthesis of the stable complex $(CO)_4$Fe–GaAr* where Ar* is a bulky aryl substituent which is a poor π donor substituent let the authors suggest that there is strong Fe → Ga π-backdonation and that the compound would have a Fe≡Ga triple bond [95]. This view was soon challenged by other workers who analyzed the charge distribution in model compounds of $(CO)_4$Fe–GaAr* with less bulky aryl groups which indicated that there may only be a bond order of ~1 [96]. Questions like the strength of π-backdonation in TM complexes can be quantitatively addressed with the EDA which. Therefore, we undertook comprehensive studies of the bonding situation in group-13 diyl complexes using the EDA method [18,19,21,47].

As a first project we analyzed the metal–diyl interactions in the complexes $Fe(CO)_4$–ECp (E = B–Tl) which have the strong donor group Cp as substituent [18]. The results of the energy decomposition analyses are given in Table 13.19. Only the axial isomers are

Table 13.19 Energy decomposition analysis of the axial isomers of $Fe(CO)_4$–ECp and $Fe(CO)_5$ at BP86/TZP [18]

	BCp	AlCp	GaCp	InCp	TlCp	CO
ΔE_{int}	−90.3	−65.2	−31.7	−27.1	−33.1	−54.6
ΔE_{Pauli}	211.6	154.3	69.8	63.6	64.1	134.8
ΔE^{a}_{elstat}	−186.0 (61.6%)	−112.1 (51.1%)	−47.1 (46.6%)	−40.0 (44.1%)	−42.7 (44.0%)	−98.0 (51.7%)
ΔE^{a}_{orb}	−115.9 (38.4%)	−107.4 (48.9%)	−54.4 (53.4%)	−50.7 (55.9%)	−54.2 (56.0%)	−91.4 (48.3%)
ΔE^{b}_{σ}	−93.8 (80.9%)	−92.3 (85.9%)	−47.2 (86.8%)	−45.3 (89.3%)	−48.9 (89.4%)	−47.6 (52.1%)
ΔE^{b}_{π}	−22.1 (19.1%)	−15.1 (14.1%)	−7.2 (13.2%)	−5.4 (10.7%)	−5.8 (10.6%)	−43.8 (47.9%)
ΔE_{prep}	15.0	12.5	8.7	7.3	19.5	8.1
$\Delta E(= -D_e)$	−75.3	−52.7	−23.0	−19.8	−13.6	−46.5
q(Fe)	−0.56	−0.58	−0.51	−0.49	−0.45	–
q(E)	0.32	1.18	0.96	1.06	0.89	–
R(Fe–E)	1.968	2.253	2.395	2.548	2.578	–

Energy values in kcal/mol. Bond lengths R in Å. Atomic partial charges q.

[a]Values in parentheses give the percentage contribution to the total attractive interactions $\Delta E_{elstat} + \Delta E_{orb}$.

[b]Values in parentheses give the percentage contribution to the total orbital interactions ΔE_{orb}.

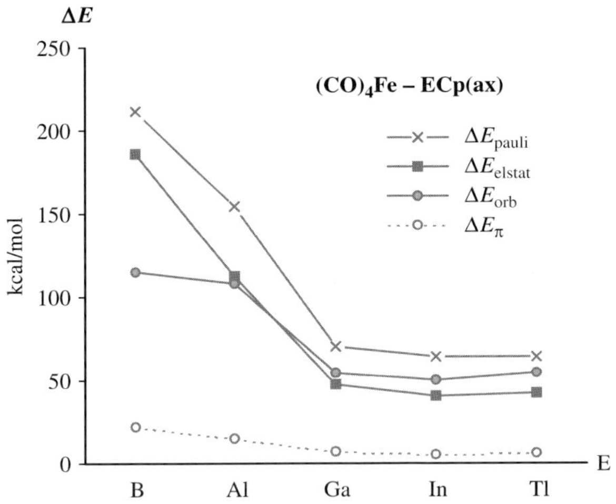

Fig. 13.12. Trends of the absolute values of the Pauli repulsion ΔE_{Pauli}, electrostatic interactions ΔE_{elstat}, total orbital interactions ΔE_{orb} and π-orbital interactions ΔE_{π} to the Fe–E bonding interactions in the axial isomers of $(CO)_4Fe$–ECp at BP86/TZP.

reported here. Fig. 13.12 shows the trend of the energy contributions ΔE_{Pauli}, ΔE_{elstat} and ΔE_{orb} for E = B–Tl. For comparison we give in Table 13.19 also the EDA result for $Fe(CO)_5$.

The most important results of the complexes $(CO)_4Fe$–ECp can be summarized as follows. The Fe–B bond between iron and the lightest group-13 element boron has a significantly higher electrostatic (61.6%) than covalent character (38.4%). The covalent contributions to the Fe–E bonds increase for the heavier group-13 elements E where it becomes as large as the electrostatic contribution. The covalent bonding comes mainly from the Fe ← ECp σ-donation. The contribution of the Fe → ECp π-backdonation is much smaller, i.e. <20% of the total ΔE_{orb} term.

Fig. 13.12 shows nicely that the values of ΔE_{Pauli}, ΔE_{elstat} and ΔE_{orb} for $(CO)_4Fe$–ECp run parallel from E = B–Tl except for the ΔE_{orb} value of the boron complex. There is a steep increase of ΔE_{Pauli} and ΔE_{elstat} from Al to B while the ΔE_{orb} value remains nearly the same. The larger values for ΔE_{Pauli} and ΔE_{elstat} for the boron complex can be explained with the much shorter Fe–B distance compared to Fe–Al which leads to stronger overlap repulsion between the occupied orbitals of the ligand and the metal fragment (ΔE_{Pauli}) and to stronger attraction between the Fe nucleus and the electron lone pair of the donor ligand (ΔE_{elstat}, see Fig. 13.11b) but why is there hardly any change in the ΔE_{orb} value? The explanation can be given when the nature of the dominant orbital interaction is analyzed. Table 13.19 shows that the dominant contribution of ΔE_{orb} comes from the Fe ← ECp σ donation. Fig. 13.11a displays the shape of the interacting orbitals. The σ donor orbital of E at first overlaps in a bonding fashion with the lobe of the d_{z^2} orbital of Fe, which has the same sign. At shorter distances, there is an overlap with the tubular-shaped lope of the d_{z^2} orbital which has an opposite sign, thus leading to antibonding orbital interactions. This cancels the increase of the bonding orbital interactions, which therefore remains nearly constant. The electrostatic interactions do not depend on the sign of the orbitals.

Table 13.19 gives also the atomic partial charges of the complexes which have been calculated with the NBO method. The charges suggest that (i) there is charge attraction between the positively charged atoms E and the negatively charged Fe; (ii) the charge attraction of the Fe–B bond should be significantly weaker compared to the Fe–E bonds of the heavier atoms E. The EDA reveals that and why both conclusions are not valid. The atomic partial charges are scalar properties, which do not reveal any information about the topography of the charge distribution of an atom in a molecule. The quasiclassical electrostatic attraction between Fe and E in $(CO)_4Fe$–ECp stems from *the interaction between the local concentration of negative charge of the overall positively charged atom E and the local depletion of negative charge of the overall negatively charged atom Fe*.

Table 13.19 gives also the results of the energy partitioning analysis of the Fe–CO(ax) bond of $Fe(CO)_5$. What is the difference between the nature of the Fe–ECp and Fe–CO bonds? The data show that the ratio between electrostatic to covalent bonding of the two ligands is quite similar. A significant difference, however, is found when the contributions by the Fe → L π-backdonation for L = CO and L = ECp are compared. The calculated values show that CO is a strong π-acceptor ligand while ECp is not.

In order to examine the statement that the Fe → ER π-backdonation becomes a significant part of the orbital interactions when the substituent R is a weak π donor such as an aryl group [95] we next carried out an energy partitioning analysis of $(CO)_4Fe$–EPh (Ph = phenyl) [18]. The results are given in Table 13.20. The trend of the energy contributions is displayed in Fig. 13.13.

The calculations show that the interaction energies ΔE_{int} and bond dissociation energies D_e of the $(CO)_4Fe$–EPh complexes are larger than those of the $(CO)_4Fe$–ECp molecules. A comparison of the calculated energy contributions of the axial Fe–EPh bonds (Table 13.20) with those of the Fe–ECp bonds (Table 13.19) reveals that the nature of the bonding with regard to the ratio of covalent and electrostatic bonding is very similar to each other. This becomes obvious when the trends of ΔE_{elstat} and ΔE_{orb} in the two sets of molecules are compared (Figures 11 and 12). For boron, the ΔE_{orb} value in $(CO)_4Fe$–EPh exhibits a similar anomaly as for the $(CO)_4Fe$–ECp compounds. The crucial information, however, concerns the degree of π bonding to the ΔE_{orb} term. Table 13.20 shows that the contribution of ΔE_{π} to the covalent bonding in the series $(CO)_4Fe$–EPh is indeed larger than in $(CO)_4Fe$–ECp. However, Fe–E π bonding remains much smaller than σ bonding in the former complexes. The largest contribution is found in the boron complex $(CO)_4Fe$–BPh where ΔE_{π} is 33.4% of the total covalent term. The value for ΔE_{π} in $(CO)_4Fe$–GaPh is only 17.2% of ΔE_{orb} (Table 13.20). It follows that the iron–gallium bond in $(CO)_4Fe$–GaAr* should not be considered as a triple bond. The stronger and shorter Fe–E bonds in the $(CO)_4Fe$–EPh complexes are caused by several factors of which enhanced π bonding is only a minor component.

Table 13.20 shows also the results of the energy analysis of the equatorial isomers of $(CO)_4Fe$–EPh. The calculated data show that the Fe–EPh bonding interactions in the axial and equatorial isomers are very similar. However, the latter isomers have C_{2v}

Table 13.20 Energy decomposition analysis of the axial and equatorial isomers of $Fe(CO)_4$–EPh at BP86/TZP [18]

	BPh		AlPh		GaPh		InPh		TlPh	
	ax	eq	ax	eq	ax	eq	ax	eq	ax	eq
ΔE_{int}	−110.3	−109.8	−73.2	−71.1	−61.0	−55.5	−48.8	−48.7	−49.4	−42.9
ΔE_{Pauli}	276.6	319.2	173.8	192.3	129.5	130.0	112.3	112.2	98.7	96.4
ΔE^{a}_{elstat}	−230.4	−258.8	−127.3	−147.6	−102.3	−107.5	−87.0	−91.7	−79.3	−81.3
	(59.6%)	(60.3%)	(51.5%)	(56.0%)	(53.7%)	(58.0%)	(54.0%)	(57.0%)	(53.5%)	(58.6%)
ΔE^{a}_{orb}	−156.5	−170.2	−119.7	−115.8	−88.2	−76.0	−74.1	−69.2	−68.8	−58.0
	(40.4%)	(39.7%)	(48.5%)	(44.0%)	(46.3%)	(42.0%)	(46.0%)	(43.0%)	(46.5%)	(41.4%)
ΔE^{b}_{σ}	−104.3	−110.3	−98.2	−91.6	−73.0	−61.7	−63.4	−57.7	−59.8	−48.6
	(66.6%)	(64.8%)	(82.0%)	(79.1%)	(82.8%)	(79.1%)	(85.6%)	(83.4%)	(86.9%)	(83.8%)
ΔE^{b}_{π}	−52.2	−59.9	−21.5	−24.2	−15.2	−16.3	−10.7	−11.5	−9.0	−9.4
	(33.4%)	(35.2%)	(18.0%)	(20.9%)	(17.2%)	(20.9%)	(14.4%)	(16.6%)	(13.1%)	(16.2%)
$\Delta E_{\pi}(b1)^{c}$	–	−39.3	–	−15.6	–	−11.4	–	−8.2	–	−6.7
$\Delta E_{\pi}(b2)^{c}$	–	−20.7	–	−8.6	–	−4.9	–	−3.3	–	−2.7
ΔE_{prep}	10.1	10.8	9.4	8.2	8.7	6.1	8.1	5.2	8.6	4.9
$\Delta E(= -D_e)$	−100.2	−99.0	−63.8	−62.9	−52.3	−49.4	−40.7	−43.5	−40.8	−38.0
R(TM–E)	1.803	1.800	2.217	2.206	2.296	2.304	2.478	2.488	2.478	2.544

Energy values in kcal/mol. Bond lengths R in Å.

[a]The value in parentheses gives the percentage contribution to the total attractive interactions $\Delta E_{elstat} + \Delta E_{orb}$.

[b]The value in parentheses gives the percentage contribution to the total orbital interactions ΔE_{orb}.

[c]π (b1)-orbital is in the Ph plane and π (b2)-orbital is perpendicular to the Ph plane.

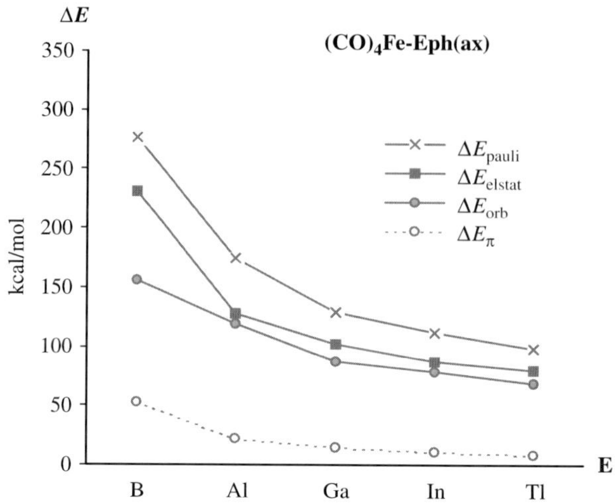

Fig. 13.13. Trends of the absolute values of the Pauli repulsion ΔE_{Pauli}, electrostatic interactions ΔE_{elstat}, total orbital interactions ΔE_{orb} and π-orbital interactions ΔE_{π} to the Fe–E bonding interactions in the axial isomers of $(CO)_4Fe$–EPh at BP86/TZP.

symmetry while the axial forms have only C_s symmetry. Thus, the calculated energy values of the orbitals having different symmetry can be used to distinguish between the in-plane (b_1) and out-of-plane (b_2) contributions to the Fe–EPh π bonding. The results show that the in-plane contributions are as expected larger than the out-of-plane values but the latter are not negligible. This is a nice example of how the EDA gives detailed information about the strength of different orbital interactions.

The EDA results show that the group-13 ligand ER is a weaker π acceptor than CO even when the substituent R is a poor π-donor group. What about the π-acceptor strength of group-13 diyl ligands ER when there is no other ligand which competes for the π electrons of the transition metal such as in homoleptic complexes $TM(ER)_n$? Examples of homoleptic complexes are only known for the group-12 elements Ni and Pt. Uhl and coworkers reported for the first time about the synthesis and X-ray structure analysis of $Ni(ER)_4$ and $Pt(ER)_4$ with E = In, Ga where R is a bulky silyl group [97]. The bonding situation in homoleptic model compounds was analyzed with the EDA method and the results were compared with the data for heteroleptic species [18]. In the investigation we first analyzed the compounds $(CO)_4Fe$–EMe and $Fe(EMe_3)_5$ (E = B–Tl) in order to compare the previous results of the iron complexes with the data which were obtained for the group-10 complexes $TM(EMe_3)_4$ (E = B–Tl, TM = Ni, Pd, Pt).

The EDA results for the complexes $(CO)_4Fe$–EMe (Table 13.21) are very similar to those of $(CO)_4Fe$–EPh (Table 13.20). This means that the methyl and phenyl groups behave as weak π-donor groups. A comparison of the bonding analysis of the homoleptic complexes $Fe(EMe)_5$ (Table 13.22) with the results of the heteroleptic species $(CO)_4Fe$–EMe (Table 13.21) shows that the BDEs of the former are higher than those of the latter. The EDA data suggest that the higher bond strength of

Table 13.21 Energy decomposition analysis of the axial and equatorial isomers of $Fe(CO)_4{-}ECH_3$ at BP86/TZP [18]

	BCH_3		$AlCH_3$		$GaCH_3$		$InCH_3$		$TlCH_3$	
	ax	eq	ax	eq	ax	eq	ax	eq	ax	eq
ΔE_{int}	−110.0	−108.8	−74.4	−72.9	−62.0	−56.7	−56.3	−50.8	−51.2	−51.6
ΔE_{Pauli}	274.2	322.4	178.9	201.6	133.4	138.7	119.1	120.6	104.8	103.6
ΔE^{a}_{elstat}	−228.2	−258.8	−131.5	−153.9	−106.1	−114.0	−93.7	−99.0	−83.2	−85.7
	(59.4%)	(60.0%)	(51.9%)	(56.1%)	(54.3%)	(58.3%)	(53.4%)	(57.8%)	(53.4%)	(55.2%)
ΔE^{a}_{orb}	−156.0	−172.4	−121.8	−120.6	−89.3	−81.4	−81.7	−72.4	−72.8	−69.5
	(40.6%)	(40.0%)	(48.1%)	(43.9%)	(45.7%)	(41.7%)	(46.6%)	(42.2%)	(46.6%)	(44.8%)
ΔE^{b}_{σ}	−105.5	−127.8	−101.0	−101.9	−75.0	−71.1	−71.4	−65.3	−64.3	−75.1
	(67.4%)	(74.1%)	(82.9%)	(84.5%)	(84.0%)	(87.3%)	(87.4%)	(90.2%)	(88.3%)	(93.1%)
ΔE^{b}_{π}	−50.5	−44.6	−20.8	−18.7	−14.3	−10.3	−10.3	−7.1	−8.5	5.6
	(32.4%)	(25.9%)	(17.1%)	(15.5%)	(16.0%)	(12.7%)	(12.6%)	(9.8%)	(11.7%)	(6.9%)
ΔE_{prep}	10.0	10.7	9.0	8.3	8.2	5.9	7.9	5.3	5.4	5.2
$\Delta E(= -D_e)$	−100.0	−98.1	−65.4	−64.6	−53.8	−50.8	−48.4	−45.5	−45.8	-46.4

Energy values in kcal/mol.

[a]Values in parentheses give the percentage contribution to the attractive interactions $\Delta E_{elstat} + \Delta E_{orb}$.

[b]Values in parentheses give the percentage contribution to the total orbital interactions ΔE_{orb}.

Table 13.22 Energy decomposition analysis of the equatorial Fe–E bonds of the complexes $Fe(ECH_3)_5$ at BP86/TZP [18]

	BCH_3	$AlCH_3$	$GaCH_3$	$InCH_3$	$TlCH_3$
ΔE_{int}	−119.2	−87.0	−67.0	−59.5	−54.1
ΔE_{Pauli}	247.8	140.2	120.8	113.9	113.0
ΔE^{a}_{elstat}	−228.4 (62.2%)	−135.4 (59.6%)	−115.2 (61.3%)	−107.8 (62.2%)	−103.8 (62.1%)
ΔE^{a}_{orb}	−138.6 (37.8%)	−91.8 (40.4%)	−72.6 (38.7%)	−65.6 (37.8%)	−63.3 (37.9%)
ΔE^{b}_{σ}	−74.6 (53.8%)	−55.0 (59.9%)	−45.5 (62.7%)	−41.7 (63.6%)	−42.9 (67.8%)
$\Delta E_{\pi}{}^{b}$	−64.0 (46.2%)	−36.8 (40.1%)	−27.1 (37.3%)	−23.9 (36.4%)	−20.4 (32.2%)
ΔE_{prep}	13.6	7.8	2.9	2.1	1.1
$\Delta E(= -D_e)$	−105.6	−79.2	−64.1	−57.4	−53.1
R(TM–E)	1.772	2.174	2.255	2.434	2.474

Bond lengths R in Å. Energy values in kcal/mol.
[a]The value in parentheses gives the percentage contribution to the total attractive interactions $\Delta E_{elstat} + \Delta E_{orb}$.
[b]The value in parentheses gives the percentage contribution to the total orbital interactions ΔE_{orb}.

the homoleptic species is caused by the intrinsic interactions energies ΔE_{int} which are significantly larger than those of the heteroleptic molecules. Further breakdown of the ΔE_{int} term into the attractive forces shows that, in $Fe(EMe)_5$, the electrostatic interactions contribute more to the chemical bonding than in $(CO)_4Fe$–EMe. Another difference between the homoleptic and heteroleptic species is the relative contribution of the π-backdonation. Table 13.22 shows that the Fe → ER π-backdonation in $Fe(EMe)_5$ is between 32–46% of the ΔE_{orb} term which means that it contributes significantly to the covalent bonding. The conclusion is that the group-13 diyl ligand ER is an intrinsically strong π acceptor but the π-acceptor strength is smaller than that of CO. In $(CO)_4Fe$–ER, the two forces compete and hence, Fe → ER π-backdonation is weak. In homoleptic complexes $Fe(EMe)_5$ there is no competition and therefore, Fe → ER π-backdonation becomes quite strong.

The EDA results of the group-10 homoleptic complexes $TM(EMe)_4$ (E = B–Tl, TM = Ni, Pd, Pt) are given in Table 13.23. The data indicate that the statement concerning significant Fe → ER π-backdonation holds also true for other transition metals. In $TM(EMe)_4$ (TM = Ni, Pd, Pt), the contribution of TM → EMe π-backdonation is between 33–49% of the covalent term ΔE_{orb}. However, the bonding interactions in the group-14 complexes $TM(EMe)_4$ is mainly electrostatic. The ΔE_{elstat} term contributes between 75–65% of the total attractive interactions. Table 13.23 gives also the results for the tetracarbonyls $TM(CO)_4$. The covalent bonding in the latter is larger than in the $TM(EMe)_4$ compounds but it remains smaller than the electrostatic bonding. Another difference between the two sets of compounds is the strength of the TM → L π-backdonation which is larger for L = CO than for L = EMe. The most important difference between the tetracarbonyls and the tetradiyl complexes is the total bond energy. Table 13.23 shows that the BDE of the TM–CO bond is always less than the BDE of the weakest TM–EMe bond which is TM–TlMe. This is the reason why $Ni(CO)_4$ but not $Pd(CO)_4$ and $Pt(CO)_4$ are stable complexes at room temperature while the tetradiyl species of the heavier group-14 elements can be isolated [21].

Table 13.23 Energy decomposition analysis of the TM–L bonds in $TM(EMe)_4$ and $TM(CO)_4$ (TM = Ni, Pd, Pt; E = B–Tl) at BP86/TZP [21]

	$Ni(BMe)_4$	$Ni(AlMe)_4$	$Ni(GaMe)_4$	$Ni(InMe)_4$	$Ni(TlMe)_4$	$Ni(CO)_4$
ΔE_{int}	−95.4	−66.2	−57.0	−49.8	−39.4	−39.5
ΔE_{Pauli}	236.2	140.1	130.9	113.5	94.3	123.0
ΔE_{elstat}	−215.5	−132.7	−123.7	−108.5	−89.2	−96.6
	(65.0%)	(64.4%)	(65.8%)	(67.5%)	(67.7%)	(59.4%)
ΔE_{orb}^{a}	−116.2	−73.5	−64.2	−54.7	−44.6	−65.9
	(35.0%)	(35.6%)	(34.2%)	(33.5%)	(33.3%)	(40.6%)
ΔE_{σ}	−59.9	−44.9	−38.9	−33.4	−26.9	−30.6
	(51.5%)	(61.0%)	(60.6%)	(61.0%)	(60.3%)	(46.4%)
ΔE_{π}^{b}	−56.4	−28.6	−25.3	−21.3	−17.7	−35.3
	(48.5%)	(39.0%)	(39.4%)	(39.0%)	(39.7%)	(53.6%)
ΔE_{prep}	3.3	1.3	3.6	3.1	4.1	10.6
$\Delta E(= -D_e)$	−92.1	−64.9	−53.4	−46.7	−35.3	−28.5
	$Pd(BMe)_4$	$Pd(AlMe)_4$	$Pd(GaMe)_4$	$Pd(InMe)_4$	$Pd(TlMe)_4$	$Pd(CO)_4$
ΔE_{int}	−80.5	−56.1	−46.7	−40.7	−30.4	−25.6
ΔE_{Pauli}	280.6	169.6	154.3	131.7	110.6	126.6
ΔE_{elstat}	−248.5	−162.6	−145.0	−127.6	−105.5	−94.7
	(68.8%	(72.0%)	(72.1%)	(74.0%)	(74.8%)	(62.2%)
ΔE_{orb}^{a}	−112.6	−63.1	−56.0	−44.8	−35.5	−57.5
	(31.2%)	(28.0%)	(27.9%)	(26.0%)	(25.2%)	(37.8%)
ΔE_{σ}	−60.2	−38.3	−34.0	−28.0	−22.6	−29.7
	(53.5%)	(60.7%)	(60.8%)	(62.5%)	(63.8%)	(51.6%)
ΔE_{π}^{b}	−52.4	−24.8	−22.0	−16.8	−12.8	−27.8
	(46.5%)	(39.3%)	(39.2%)	(37.5%)	(36.2%)	(48.4%)
ΔE_{prep}	6.6	2.4	4.7	3.9	4.7	12.8
$\Delta E(= -D_e)$	−73.90	−53.7	−42.0	−36.8	−25.7	−12
	$Pt(BMe)_4$	$Pt(AlMe)_4$	$Pt(GaMe)_4$	$Pt(InMe)_4$	$Pt(TlMe)_4$	$Pt(CO)_4$
ΔE_{int}	−96.9	−67.1	−57.7	−49.7	−38.1	−35.4
ΔE_{Pauli}	362.4	220.4	204.0	174.8	149.6	194.5
ΔE_{elstat}	−316.5	−206.9	−187.1	−164.5	−137.8	−140.3
	(68.9%)	(72.0%)	(71.5%)	(73.2%)	(73.4%)	(61.1%)
ΔE_{orb}^{a}	−142.9	−80.5	−74.6	−60.1	−49.9	−89.5
	(31.1%)	(28.0%)	(28.5%)	(26.8%)	(26.6%)	(38.9%)
ΔE_{σ}	−80.3	−51.9	−48.2	−39.8	−33.7	−49.2
	(56.2%)	(64.5%)	(64.7%)	(66.3%)	(67.5%)	(55.0%)
ΔE_{π}^{b}	−62.5	−28.6	−26.4	−20.2	−16.2	−40.3
	(43.8%)	(35.5%)	(35.3%)	(33.7%)	(32.5%)	(45.0%)
ΔE_{prep}	9.2	4.1	7.4	6.5	7.9	20.7
$\Delta E(= -D_e)$	−87.7	−63.0	−50.3	−43.2	−30.2	−14.7

Energy values in kcal/mol.

[a]The value in parentheses gives the percentage contribution to the total attractive interactions $\Delta E_{elstat} + \Delta E_{orb}$.

[b]The value in parentheses gives the percentage contribution to the total orbital interactions ΔE_{orb}.

13.4.3 Carbene and carbyne complexes and heavier homologues $(CO)_5W-CH_2$, $(CO)_5W-E(OH)_2$, Cl_4W-EH_2, $Cl(CO)_4W-EH$ and Cl_3W-EH (E = C, Si, Ge, Sn, Pb) [37]

Transition metal complexes with carbene (CR_2) or carbyne (CR) ligands are examples where two different orbital models are commonly used to describe the metal-ligand interactions. This comes from the observation that complexes with the general formula L_nTM-CR_2 and L_nTM-CR can be divided into two different categories which exhibit different chemical behavior. Fischer complexes usually contain the transition metal in a low oxidation state and the TM–EH bonding is well described in terms of donor–acceptor interactions. Schrock type carbenes and carbynes, on the other hand, usually contain a transition metal in a high oxidation state and are properly described in terms of covalent or electron-sharing interactions [98]. A more appropriate name for the latter compounds is therefore metal alkylidenes and metal alkylidynes rather than Schrock carbenes and carbynes. We will still use the latter names because they are quite common in the literature, and they emphasize the relationship with the Fischer complexes. The qualitative orbital models which are used to describe the metal–ligand interactions in Fischer and Schrock type carbene and carbyne complexes are shown in Fig. 13.14. The same models may be used for complexes containing ligands ER_2 and ER with the heavier analogues of carbon (E = Si, Ge, Sn, Pb). We analyzed the metal–ligand interactions in the series of model compounds $(CO)_5W-CH_2$, $(CO)_5W-E(OH)_2$, Cl_4W-EH_2, $Cl(CO)_4W-EH$ and Cl_3W-EH (E = C, Si, Ge, Sn, Pb) [37].

The results of the EDA for the Fischer-type carbene complexes $(CO)_5W-E(OH)_2$ are given in Table 13.24. Chemical experience shows that the substituent R must be a π-donor group in order that a Fischer-type complex becomes stable enough to be isolated. We included the model compound $(CO)_5W-CH_2$ in our study in order to analyze the influence of the π-donor group OH on the bonding situation.

The effect of the donor OH groups is indeed significant. The interaction energy for $(CO)_5W-C(OH)_2$ ($\Delta E_{int} = -56.0$ kcal/mol) is much smaller than for $(CO)_5W-CH_2$($\Delta E_{int} = -93.8$ kcal/mol). Not only the strength of the interaction is changed, the individual contributions to the ΔE_{int} term change as well. The covalent character of the former compound is higher ($\Delta E_{orb} = 67.3\%$) than in the latter ($\Delta E_{orb} = 61.2\%$) but at the same time the importance of the π contributions drops ($\Delta E_{\pi} = 33.9\%$ for $(CO)_5W-C(OH)_2$, $\Delta E_{\pi} = 53.7\%$ for $(CO)_5W-CH_2$). As expected, the major contribution to the π-orbital interactions comes from the b_1 orbitals which are the out-of-plane orbitals where the plane is defined by the ligand atoms. As the carbon homologues become heavier, the covalent character of the TM-E bond increases and the contribution of the σ interaction becomes more important.

Table 13.25 gives the EDA results of the Schrock-type carbene complexes Cl_4W-EH_2. The strength of the interaction is much higher than in the corresponding Fischer-type complexes, the absolute value decreases when the group-14 elements get heavier but remains rather high, even for the lead compound. Along the group-14 elements, the nature of the interaction changes very little. The interaction for all the complexes is about half electrostatic and half covalent. About two thirds of the orbital interaction stem from

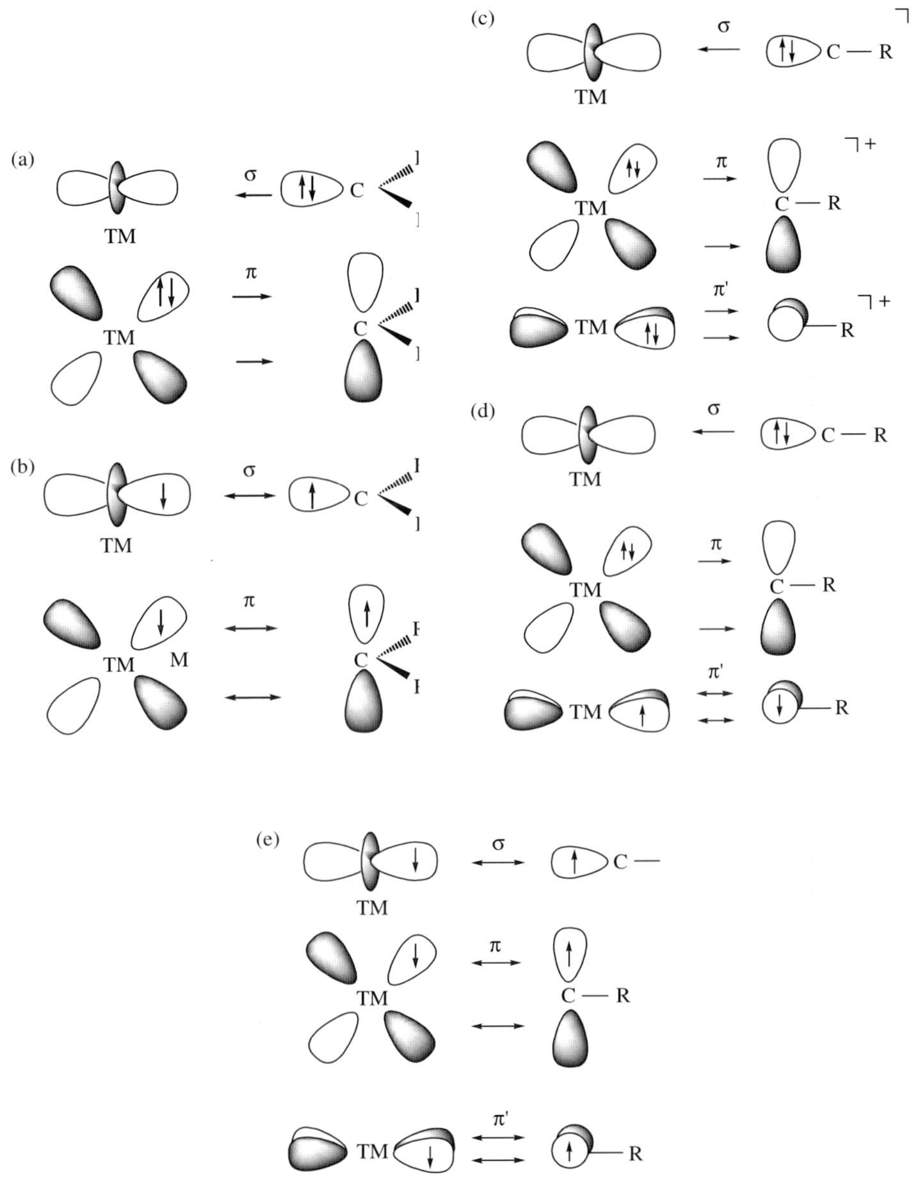

Fig. 13.14. Schematic representation of the dominant orbital interactions in (a) Fischer-type carbene complexes; (b) Schrock-type carbene complexes; (c) Fischer-type carbyne complexes using charged closed-shell fragments; (d) Fischer-type carbyne complexes using neutral open shell (doublet) fragments; (e) Schrock-type carbyne complexes using neutral open-shell (quartet) fragments.

Table 13.24 Energy decomposition analysis of the Fischer complexes $(CO)_5W{-}CH_2$ and $(CO)_5W{-}E(OH)_2$ using the metal fragments $(CO)_5W$ and the ligands CH_2 and $E(OH)_2$ in the singlet state (see Fig. 13.13a and text) at BP86/TZ2P [37]

	$(CO)_5W{-}CH_2$	$(CO)_5W{-}C(OH)_2$	$(CO)_5W{-}Si(OH)_2$	$(CO)_5W{-}Ge(OH)_2$	$(CO)_5W{-}Sn(OH)_2$	$(CO)_5W{-}Pb(OH)_2$
ΔE_{int}	−93.8	−56.0	−49.3	−30.8	−27.2	−15.0
ΔE_{Pauli}	188.1	129.2	120.8	74.7	63.7	35.0
ΔE^{a}_{elstat}	−172.4 (38.8%)	−124.7 (32.7%)	−103.9 (38.9%)	−56.7 (46.3%)	−48.6 (46.6%)	−20.9 (58.2%)
ΔE^{a}_{orb}	−109.5 (61.2%)	−60.5 (67.3%)	−66.2 (61.1%)	−48.8 (53.7%)	−42.4 (53.4%)	−29.1 (41.8%)
$\Delta E_{\sigma}(A_1)^{b}$	−50.6 (46.2%)	−39.6 (65.5%)	−44.7 (67.5%)	−33.8 (69.3%)	−31.5 (74.3%)	−22.1 (75.9%)
$\Delta E_{\delta}(A_2)^{b}$	−0.1 (0.2%)	−0.4 (0.7%)	−0.3 (0.5%)	−0.1 (0.2%)	−0.1 (0.2%)	−0.1 (0.3%)
$\Delta E_{\pi}(B_1)^{b}$	−53.1 (48.5%)	−14.2 (23.5%)	−12.7 (19.2%)	−9.8 (20.1%)	−6.9 (16.3%)	−4.6 (15.8%)
$\Delta E_{\pi}(B_2)^{b}$	−5.7 (5.2%)	−6.3 (10.4)	−8.5 (12.8%)	−5.1 (10.5%)	−3.8 (9.0%)	−2.4 (8.2%)
ΔE^{b}_{π}	−58.8 (53.7%)	−20.5 (33.9%)	−21.2 (32.0%)	−14.9 (30.5%)	−10.7 (25.3%)	−7.0 (23.9%)

Energy values in kcal/mol.

[a]The value in parentheses gives the percentage contribution to the total attractive interactions $\Delta E_{elstat} + \Delta E_{orb}$.

[b]The value in parentheses gives the percentage contribution to the total orbital interactions ΔE_{orb}.

Table 13.25 Energy decomposition analysis (kcal/mol) of the Schrock complexes Cl_4W-EH_2 using the fragments Cl_4W and EH_2 in the triplet state (see Fig. 13.13b and text) at BP86/TZ2P [37]

	Cl_4W-CH_2	Cl_4W-SiH_2	Cl_4W-GeH_2	Cl_4W-SnH_2	Cl_4W-PbH_2
ΔE_{int}	−132.5	−88.0	−83.3	−71.5	−65.3
ΔE_{Pauli}	260.7	156.3	150.2	129.1	120.1
ΔE^a_{elstat}	−190.0 (51.7%)	−120.4 (50.7%)	−114.8 (50.8%)	−100.8 (49.8%)	−86.9 (53.1%)
ΔE^a_{orb}	−203.2 (48.3%)	−123.9 (49.3%)	−118.6 (49.2%)	−99.9 (50.2%)	−98.5 (46.9%)
$\Delta E_\sigma(A_1)^b$	−129.7 (63.8%)	−80.7 (65.1%)	−79.1 (66.7%)	−67.4 (67.5%)	−64.2 (65.2%)
$\Delta E_\delta(A_2)^b$	−0.1 (<0.1%)	−0.1 (0.1%)	−0.1 (0.1%)	−0.1 (0.1%)	−0.1 (0.1%)
$\Delta E_\pi(B_1)^b$	−60.7 (29.9%)	−37.5 (30.3%)	−34.1 (28.8%)	−28.4 (28.4%)	−29.8 (30.3%)
$\Delta E_\pi(B_2)^b$	−12.7 (6.3%)	−5.5 (4.4%)	−5.4 (4.5%)	−4.0 (4.0%)	−4.4 (4.5%)
ΔE^b_π	−73.4 (36.1%)	−43.0 (34.7%)	−39.5 (33.3%)	−32.3 (32.5%)	−34.2 (34.8%)

Energy values in kcal/mol.
[a]Values in parentheses give the percentage of the attractive interactions $\Delta E_{elstat} + \Delta E_{orb}$.
[b]Values in parentheses give the percentage contribution to the total orbital interactions ΔE_{orb}.

the σ bonding coming from the a_1 orbitals. Another third comes from the π-bonding orbitals, mainly from the out-of-plane (b_1) and less from the in-plane (b_2) orbitals.

The orbital model for the bonding in Fischer-type carbyne complexes has the problem that the choice of the fragments is somewhat arbitrary. A description in terms of closed-shell fragments analogous to Fischer-type carbenes yields charged moieties as interacting species (Fig. 13.14c). Choosing neutral fragments one faces open-shell interactions in Fischer carbynes (Fig. 13.14d). The EDA method makes it possible to analyze Fischer carbynes in both ways. Table 13.26 gives the EDA results for Fischer-type carbyne complexes with charged fragments in a closed-shell state. Table 13.27 gives the EDA results with neutral fragments in their respective doublet states.

The interaction energies given in Table 13.26 are rather high. This is not surprising as the fragments carry opposite charges. In the case where E = C, the bond is much more electrostatic than covalent. As the group-14 element becomes heavier this effect levels out. For the heaviest elements, the chemical bond is about half electrostatic and half

Table 13.26 Energy decomposition analysis (kcal/mol) of the Fischer complexes $Cl(CO)_4W-EH$ using the charged fragments $Cl(CO)_4W^-$ and EH^+ in the singlet state (see Fig. 13.13c and text) at BP86/TZ2P [37]

	$Cl(CO)_4W-CH$	$Cl(CO)_4W-SiH$	$Cl(CO)_4W-GeH$	$Cl(CO)_4W-SnH$	$Cl(CO)_4W-PbH$
ΔE_{int}	−350.3	−217.5	−204.4	−181.5	−164.0
ΔE_{Pauli}	208.9	126.4	112.2	99.2	78.3
ΔE^a_{elstat}	−192.7 (65.5%)	−157.2 (54.3%)	−146.7 (53.7%)	−145.0 (48.4%)	−124.6 (48.6%)
ΔE^a_{orb}	−366.5 (34.5%)	−186.7 (45.7%)	−169.9 (46.3%)	−135.7 (51.6%)	−117.8 (51.4%)
$\Delta E_\sigma(A_1)^b$	−47.8 (13.0%)	−38.3 (20.5%)	−33.8 (19.9%)	−28.6 (21.1%)	−24.7 (21.0%)
$\Delta E(A_2)$	−0.4	−0.4	−0.4	−0.3	−0.3
$\Delta E(B_1)$	−0.7	−1.1	−1.1	−0.9	−0.9
$\Delta E(B_2)$	−1.8	−2.4	−2.1	−1.7	−1.6
$\Delta E_\pi(E)^b$	−315.7 (86.1%)	−144.5 (77.4%)	−132.5 (78.0)	−104.2 (76.8%)	−90.4 (76.7%)

Energy values in kcal/mol.
[a]Values in parentheses give the percentage contribution to the attractive interactions $\Delta E_{elstat} + \Delta E_{orb}$.
[b]Values in parentheses give the percentage contribution to the total orbital interactions ΔE_{orb}.

Table 13.27 Energy decomposition analysis (kcal/mol) of the Fischer complexes $Cl(CO)_4W{-}EH$ using the neutral fragments $Cl(CO)_4W$ and EH in the doublet state (see Fig. 13.13d and text) at BP86/TZ2P [37]

	$Cl(CO)_4W{-}CH$	$Cl(CO)_4W{-}SiH$	$Cl(CO)_4W{-}GeH$	$Cl(CO)_4W{-}SnH$	$Cl(CO)_4W{-}PbH$
ΔE_{int}	−161.3	−97.6	−88.1	−77.9	−70.5
ΔE_{Pauli}	294.8	178.31	154.6	134.4	103.5
ΔE^{a}_{elstat}	−226.9 (49.7%)	−133.0 (48.2%)	−114.0 (47.0%)	−104.6 (49.3%)	−76.5 (44.0%)
ΔE^{a}_{orb}	−229.2 (50.3%)	−143.0 (41.8%)	−128.7 (53.0%)	−107.7 (50.7%)	−97.4 (56.0%)
$\Delta E_{\sigma}(A_1)^{b}$	−71.2 (31.1%)	−51.2 (35.8%)	−46.9 (36.4%)	−41.6 (38.6%)	−35.4 (36.3%)
$\Delta E_{\delta}(A_2)^{b}$	−0.1 (<0.1%)	−0.4 (0.3%)	−0.3 (0.2%)	−0.3 (0.3%)	−0.2 (0.2%)
$\Delta E_{\pi}(B_1)^{b}$	−94.7 (41.3%)	−40.2 (28.1%)	−34.9 (27.1%)	−25.0 (23.2%)	−20.6 (21.1%)
$\Delta E_{\pi}(B_2)^{b}$	−63.2 (27.6%)	−51.2 (35.8%)	−46.6 (36.2%)	−40.8 (37.9%)	−41.3 (42.4%)
ΔE^{b}_{π}	−157.9 (68.9%)	−91.4 (63.9%)	−81.5 (63.3%)	−65.8 (61.2%)	−61.9 (63.4%)

Energy values in kcal/mol.
[a]The value in parentheses gives the percentage contribution to the total attractive interactions $\Delta E_{elstat} + \Delta E_{orb}$.
[b]The value in parentheses gives the percentage contribution to the total orbital interactions ΔE_{orb}.

covalent. The largest contributor to the orbital term ΔE_{orb} are the π orbitals (e symmetry). They contribute as much as 86.1% to the orbital term.

A comparison between the EDA results with different choices of interacting fragments given in Tables 13.26 and 13.27 shows that the interaction energies between the neutral fragments (Table 13.27) are much smaller. This effect is even more pronounced for the heavier elements. The nature of the interactions shifts only slightly, however. The electrostatic and covalent contributions have comparable strength for all compounds. The π interaction is, as before, the dominant contribution to the orbital term ΔE_{orb} but the relative contributions are slightly smaller than for the closed-shell interaction. Note that the EDA data make it possible to distinguish between the donor–acceptor interactions in π symmetry which are given by the b_1 term, and the electron sharing π interactions which are given by the b_2 term. It is interesting to note that the electron-sharing contribution is smaller than the electron donation when E = C while for the heavier atoms the electron-sharing contribution is the larger term.

The results of the energy decomposition analyses for the Schrock-type carbyne complexes are given in Table 13.28. The absolute values of the interaction energies are

Table 13.28 Energy decomposition analysis (kcal/mol) of the Schrock complexes $Cl_3W{-}EH$ using the fragments Cl_3W and EH in the quartet state (see Fig. 13.13e and text) at BP86/TZ2P [37]

	$Cl_3W{-}CH$	$Cl_3W{-}SiH$	$Cl_3W{-}GeH$	$Cl_3W{-}SnH$	$Cl_3W{-}PbH$
ΔE_{int}	−193.0	−128.0	−121.1	−103.6	−98.5
ΔE_{Pauli}	310.1	180.6	187.7	164.7	158.8
ΔE^{a}_{elstat}	−213.5 (57.6%)	−138.1 (55.3%)	−143.2 (53.7%)	−133.6 (50.2%)	−121.6 (52.7%)
ΔE^{a}_{orb}	−289.6 (42.4%)	−170.6 (44.7%)	−165.7 (46.4%)	−134.8 (49.8%)	−135.6 (47.3%)
$\Delta E_{\sigma}(A_1)^{b}$	−147.3 (50.9%)	−82.4 (48.3%)	−86.8 (52.4%)	−70.7 (52.5%)	−74.2 (54.7%)
$\Delta E(A_2)$	0.0	0.0	0.0	0.0	0.0
$\Delta E_{\pi}(E)^{b}$	−142.3 (49.1%)	−88.2 (51.7%)	−79.0 (47.6%)	−64.1 (47.5%)	−61.5 (45.3%)

Energy values in kcal/mol.
[a]The value in parentheses gives the percentage contribution to the total attractive interactions $\Delta E_{elstat} + \Delta E_{orb}$.
[b]The value in parentheses gives the percentage contribution to the total orbital interactions ΔE_{orb}.

very high for metal–ligand interactions between neutral fragments. The $Cl_3W{-}EH$ bonds are slightly more electrostatic than covalent but the differences to the Fischer-carbynes shown in Table 13.26 are very small. The σ and π interactions in the Schrock carbynes have almost the same strength while the latter clearly dominates the orbital interactions in the Fischer carbynes. This seems to be the major difference between the two classes of carbyne complexes.

13.4.4 Ethylene and acetylene complexes $(CO)_5TM{-}C_2H_x$ and $Cl_4TM{-}C_2H_x$ (TM = Cr, Mo, W), $(CO)_4TM{-}C_2H_x$ (TM = Fe, Ru, Os) and $TM^+{-}C_2H_x$ (TM = Cu, Ag, Au) [35]

The same dichotomy of bonding models for TM-ligand interactions which has proven to be very helpful for understanding the bonding situation in TM carbene and carbyne complexes in terms of donor-acceptor interactions or shared-electron bonding is also found in the theoretical literature about the structure and bonding of TM complexes with alkene and alkyne ligands. TM alkene complexes have actually been the first example of TM compounds for which Dewar suggested the model of synergistic TM $\leftarrow$ ligand σ-donation and TM $\rightarrow$ ligand π-backdonation [86]. Fig. 13.15a shows that the σ-donation arises from the occupied π MO of the alkene, which has σ symmetry in the complex, while the π-backdonation occurs through charge donation from the occupied d_{xz} MO of the metal to the vacant π^* orbital of the olefin. The alternative bonding model has two electron-sharing σ bonds between the metal and the carbon atoms which leads to a description of the molecule as a metallacyclopropane (Fig. 13.15b). We want to point out that the two bonding models should be considered as sketches of two extreme situations while the electronic situation of real molecules has components of both forms. The value of such dichotomic models lies in the fact that they establish an ordering scheme, which is very useful for describing the physical and chemical properties of molecules. A previous charge-decomposition analysis of alkene and alkyne complexes has shown that complexes exist which are intermediate between the two extreme cases [99b–d].

The chemical bonding in TM alkyne complexes can be discussed in a similar way as for the TM alkene complexes, i.e. the bonding may be considered either to arise from donor–acceptor interactions between the alkyne ligand and the TM or as a metallacyclic compound (Figure 14c and d). Both approaches have been shown to be helpful in terms of understanding the chemistry of transition metal ethylene and acetylene complexes [98]. The major difference between alkene and alkyne complexes is the fact that the alkyne ligand has a second set of occupied and vacant π orbitals orthogonal to the TMC_2 plane which may engage in TM–alkyne bonding. The latter TM $\leftarrow$ alkyne donation is denoted as $\pi_\perp$-interaction while the TM $\rightarrow$ alkyne backdonation into the out-of-plane $\pi^*_\perp$ orbital of the ligand has δ symmetry (Fig. 13.15c). Thus, alkynes may be 2e or 4e donors. Thus, three major topics arise in the bonding analysis of TM alkene and alkyne complexes: (a) Metallacyclic *versus* donor–acceptor bonding. (b) Participation of the $\pi_\perp$ orbitals of the alkyne in the binding interactions. (c) Difference between alkene and alkyne complexes. The topic has been addressed in an EDA study of the TM complexes $(CO)_5TM{-}C_2H_x$

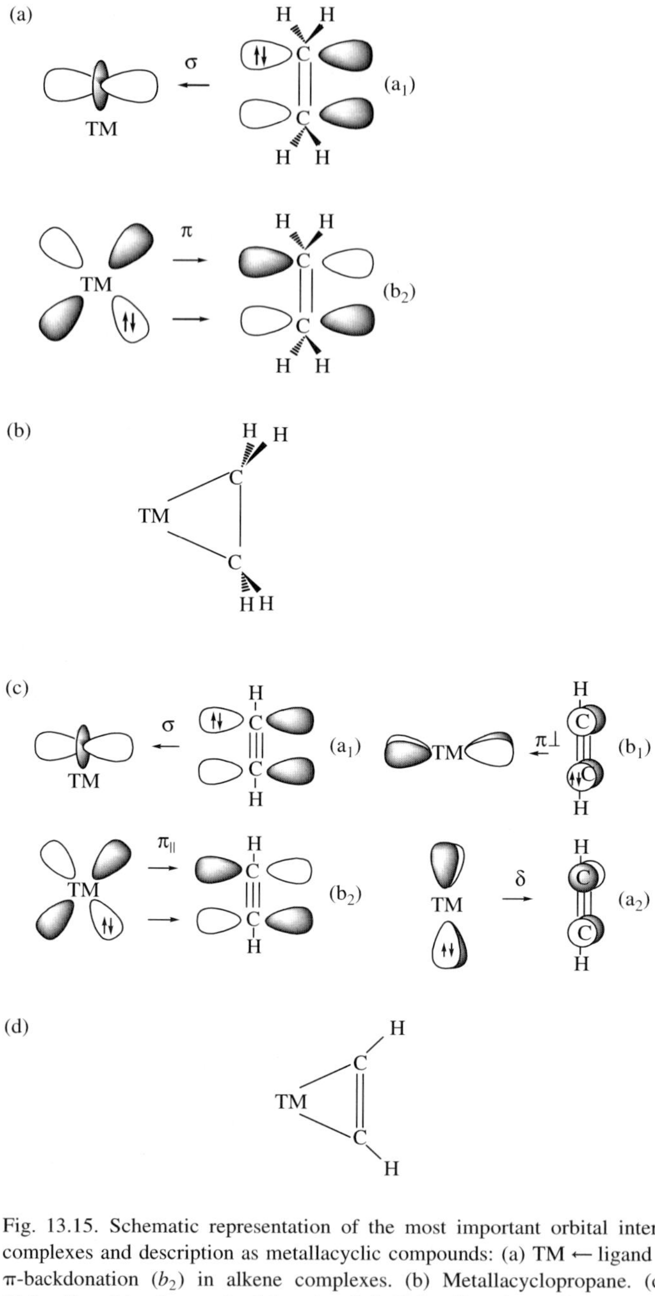

Fig. 13.15. Schematic representation of the most important orbital interactions in TM-alkene and alkyne complexes and description as metallacyclic compounds: (a) TM ← ligand σ-donation (a_1) and TM → ligand π-backdonation (b_2) in alkene complexes. (b) Metallacyclopropane. (c) TM ← ligand σ-donation (a_1), TM → ligand in-plane $\pi_{\|}$-backdonation (b_2), TM ← ligand out-of-plane $\pi_{\perp}$-donation (b_1) and TM → ligand δ-backdonation (a_2) in alkyne complexes. (d) Metallacycloropene. The symmetry assignments a_1, a_2, b_1, b_2 are given with respect to overall C_{2v} symmetry.

and $Cl_4TM-C_2H_x$ (TM = Cr, Mo, W), $(CO)_4TM-C_2H_x$ (TM = Fe, Ru, Os) and $TM^+-C_2H_x$ (TM = Cu, Ag, Au) [35].

The EDA results of the ethylene and acetylene complexes of the group-11 elements $TM^+-C_2H_x$ (TM = Cu, Ag, Au) are given in Table 13.29. The interaction energies and the bond-dissociation energies exhibit a U-shaped trend from Cu to Ag and Au. The slightly stronger bonding energies of the ethylene complexes are reflected in the interaction energies. They are therefore an intrinsic property of the metal–ligand interaction and not connected to the preparation energy. Yet the bond lengths of the acetylene complexes are smaller, hence there is no bond-length bond-strength correlation for this type of complexes.

As suggested by previous work [100] the electrostatic interaction ΔE_{elstat} is the major contributor to the attractive term of the interaction energy ΔE_{int} in the group-11 complexes $TM^+-C_2H_x$, but in contrast to the results of previous work, the influence of the covalent contribution ΔE_{orb} is not negligible. The ΔE_{orb} term contributes 41–44% of the attractive interactions. The principal orbital interactions shown in Fig. 13.15 can be seen in the individual contributions to the orbital term ΔE_{orb}. The different types of donation and backdonation are labeled next to the symbol of the irreducible representation in C_{2v} symmetry that contains them in Table 13.29. The largest contribution comes in all cases from the σ-donation in the a_1 orbitals. These orbitals contribute between 59.4 and 69.2% of the covalent orbital interaction ΔE_{orb} in ethylene complexes and between 55.7 and 68.4% in acetylene complexes. As expected the backdonation of the b_2 orbitals is the second most important term. It contributes between 20.6 and 30.5% for ethylene complexes and between 21.8 and 33.4% for acetylene complexes. The $(TM^+) \leftarrow C_2H_x$ $\pi_\perp$-donation (b_1) and the δ interaction (a_2) are, however, both small and appear to be negligible.

The EDA results of group-8 complexes $(CO)_4TM$-C_2H_x (TM = Fe, Ru, Os) are given in Table 13.30. The chemical bonding in these and similar complexes has been subjected to theoretical analysis before [101]. As the complexes discussed before the bond dissociation energies of the group-8 complexes exhibit a U-shaped trend from the first to the second and third TM row, but in contrast to the group-11 species, the BDEs of the first row TM-complexes are slightly higher than the BDEs of the third row TM-complexes. This is due to the larger values of the corresponding preparation energies and not to a change in the metal–ligand interaction. The metal–ligand interaction has a slightly higher electrostatic than covalent character. The orbital term contributes 43.7–47.5% of the total attractive interactions. Yet, the metal–ethylene bond has a lower covalent character than the metal–acetylene bond although the differences are minute. The ratio of quasiclassical electrostatic interaction and covalent orbital interaction is very similar to those observed in the cationic group-11 TM-complexes.

In the case of the group-8 TM-complexes the $(CO)_4TM \rightarrow C_2H_x$ $\pi_\parallel$-backdonation (b_2 symmetry) becomes the most important contribution to the ΔE_{orb} term. The contributions of the $(CO)_4TM \leftarrow C_2H_x$ $\pi_\perp$-donation (b_1) and the δ interaction (a_2) are both small. Because the preparation energies are higher for the acetylene ligands, the BDEs become larger for the ethylene ligands although their interaction energy ΔE_{int} is smaller.

The EDA results of the group-6 TM-complexes $(CO)_5TM-C_2H_x$ (TM = Cr, Mo, W) are shown in Table 13.31. In agreement with the results of the other complexes,

Table 13.29 Energy decomposition analysis of $TM{-}C_2H_x^+$ at BP86/TZP using closed-shell fragments $TM^+(d^{10})$ and C_2H_x [35]

	Cu^+		Ag^+		Au^+	
	C_2H_4	C_2H_2	C_2H_4	C_2H_2	C_2H_4	C_2H_2
ΔE_{int}	−63.5	−62.0	−40.8	−35.9	−79.1	−70.8
ΔE_{Pauli}	98.9	103.6	61.8	56.4	157.2	161.0
ΔE_{elstat} [a]	−90.6 (55.8%)	−94.6 (57.1%)	−59.8 (58.3%)	−54.8 (59.4%)	−134.7 (57.0%)	−134.1 (57.8%)
ΔE_{orb} [a]	−71.8 (44.2%)	−71.1 (42.9%)	−42.8 (41.7%)	−37.5 (40.6%)	−101.6 (43.0%)	−97.8 (42.2%)
$\Delta E(A_1)\sigma$ [b]	−42.6 (59.4%)	−39.5 (55.7%)	−29.6 (69.2%)	−25.6 (68.4%)	−68.0 (67.0%)	−63.1 (64.6%)
$\Delta E(A_2)\delta$ [b]	−2.2 (3.1%)	−0.9 (1.3%)	−1.6 (3.7%)	−0.3 (0.8%)	−2.6 (2.6%)	−1.3 (1.4%)
$\Delta E(B_1)\pi_\perp$ [b]	−5.1 (7.1%)	−6.9 (9.7%)	−2.7 (6.4%)	−3.4 (9.0%)	−5.5 (5.4%)	−6.9 (7.0%)
$\Delta E(B_2)\pi_\parallel$ [b]	−21.9 (30.5%)	−23.7 (33.4%)	−8.8 (20.6%)	−8.2 (21.8%)	−25.5 (25.0%)	−26.5 (27.1%)
E_{prep}	3.5	3.7	2.0	1.8	6.2	5.8
$\Delta E(= -D_e)$	−60.0	−58.3	−38.8	−34.1	−72.9	−64.9
$-D_0$	−58.5	−57.7	−37.5	−33.6	−71.2	−64.3
q(TM)	0.85	0.88	0.81	0.83	0.73	0.77
$q(C_2H_x)$	0.15	0.12	0.19	0.17	0.27	0.23
R(C–C)	1.390	1.238	1.374	1.225	1.414	1.250

Energy values in kcal/mol. Bond lengths R in Å. Atomic partial charges q.

[a] The value in parentheses gives the percentage contribution to the total attractive interactions $\Delta E_{elstat} + \Delta E_{orb}$.

[b] The value in parentheses gives the percentage contribution to the total orbital interactions ΔE_{orb}.

Table 13.30 Energy decomposition analysis of $(CO)_4TM{-}C_2H_x$ at BP86/TZP using closed-shell fragments $TM(CO)_4$ and C_2H_x [35]

	$Fe(CO)_4$		$Ru(CO)_4$		$Os(CO)_4$	
	C_2H_4	C_2H_2	C_2H_4	C_2H_2	C_2H_4	C_2H_2
ΔE_{int}	−48.3	−49.6	−41.7	−52.3	−62.2	−70.2
ΔE_{Pauli}	123.1	128.0	137.1	149.4	180.4	185.5
ΔE^{a}_{elstat}	−91.0 (53.1%)	−92.3 (51.9%)	−100.6 (56.3%)	−107.3 (53.2%)	−133.5 (55.0%)	−134.3 (52.5%)
ΔE^{a}_{orb}	−80.4 (46.9%)	−85.3 (48.1%)	−78.2 (43.7%)	−94.4 (46.8%)	−109.2 (45.0%)	−121.5 (47.5%)
$\Delta E(A_1)\sigma^{b}$	−35.2 (43.8%)	−34.1 (39.9%)	−31.3 (40.0%)	−32.6 (34.5%)	−46.3 (42.4%)	−45.5 (37.4%)
$\Delta E(A_2)\delta^{b}$	−0.9 (1.1%)	−1.1 (1.3%)	−0.8 (1.1%)	−1.0 (1.1%)	−1.1 (1.0%)	−1.3 (1.1%)
$\Delta E(B_1)\pi_{\perp}^{b}$	−2.9 (3.6%)	−3.5 (4.1%)	−2.7 (3.4%)	−3.4 (3.6%)	−3.6 (3.3%)	−4.4 (3.6%)
$\Delta E(B_2)\pi_{\parallel}^{b}$	−41.4 (51.5%)	−46.7 (54.7%)	−43.4 (55.5%)	−57.4 (60.8%)	−58.2 (53.3%)	−70.3 (57.9%)
E^{c}_{prep}	14.9 (5.3 + 9.6)	18.9 (6.0 + 12.9)	24.6 (14.3 + 10.3)	35.7 (18.2 + 17.5)	32.9 (17.9 + 15.0)	41.4 (21.3 + 20.1)
$\Delta E(=-D_e)$	−33.4	−30.8	−17.1	−16.6	−29.3	−28.8
$-D_0$	−30.6	−28.8	−15.3	−15.2	−26.9	−27.0
$q(TM)^{d}$	0.09	0.14	0.10	0.19	0.12	0.21
$q(C_2H_x)$	−0.09	−0.14	−0.10	−0.19	−0.12	−0.21
R(C–C)	1.411	1.258	1.415	1.269	1.438	1.278

Energy values in kcal/mol. Bond lengths R in Å. Atomic partial charges q.

[a]The value in parentheses gives the percentage contribution to the total attractive interactions $\Delta E_{elstat} + \Delta E_{orb}$.

[b]The value in parentheses gives the percentage contribution to the total orbital interactions ΔE_{orb}.

[c]The values in parentheses give the preparation energies of the metal fragment and ligand, respectively.

[d]Partial charge of $TM(CO)_4$.

Table 13.31 Energy decomposition analysis of $(CO)_5TM{-}C_2H_x$ at BP86/TZP using closed-shell fragments $TM(CO)_5$ and C_2H_x [35]

	$Cr(CO)_5$		$Mo(CO)_5$		$W(CO)_5$	
	C_2H_4	C_2H_2	C_2H_4	C_2H_2	C_2H_4	C_2H_2
ΔE_{int}	−29.9	−29.6	−25.6	−26.5	−35.4	−38.1
ΔE_{Pauli}	83.9	86.2	69.1	74.9	93.0	103.5
ΔE^{a}_{elstat}	−59.5 (52.3%)	−59.8 (51.7%)	−52.0 (55.0%)	−54.5 (53.8%)	−70.0 (54.5%)	−75.0 (52.9%)
ΔE^{a}_{orb}	−54.3 (47.7%)	−56.0 (48.3%)	−42.6 (45.0%)	−46.9 (46.3%)	−58.4 (45.5%)	−66.6 (47.1%)
$\Delta E(A_1)\sigma^{b}$	−29.0 (53.5%)	−27.1 (48.4%)	−20.4 (47.9%)	−19.8 (42.3%)	−28.8 (49.3%)	−28.4 (42.6%)
$\Delta E(A_2)\delta^{b}$	−0.6 (1.2%)	−1.3 (2.3%)	−0.6 (1.5%)	−1.1 (2.2%)	−0.7 (1.3%)	−1.4 (2.1%)
$\Delta E(B_1)\pi_{\perp}^{b}$	−2.2 (4.0%)	−2.5 (4.5%)	−2.1 (5.0%)	−3.0 (6.4%)	−2.7 (4.6%)	−4.1 (6.1%)
$\Delta E(B_2)\pi_{\parallel}^{b}$	−22.5 (41.4%)	−25.1 (44.9%)	−19.4 (45.6%)	−23.0 (49.0%)	−26.2 (44.9%)	−32.8 (49.2%)
E^{c}_{prep}	6.3 (0.9 + 5.4)	8.4 (1.0 + 7.4)	4.9 (1.3 + 3.6)	7.5 (1.4 + 6.1)	7.5 (1.9 + 5.6)	11.6 (2.2 + 9.4)
$\Delta E(= -D_e)$	−23.6	−21.2	−20.6	−19.0	−27.9	−26.6
$-D_0$	−20.9	−19.6	−18.8	−18.0	−25.9	−25.3
$q(TM)^{d}$	0.03	0.07	0.04	0.08	0.06	0.11
$q(C_2H_x)$	−0.03	−0.07	−0.04	−0.08	−0.06	−0.11
$R(C{-}C)$	1.383	1.237	1.375	1.235	1.388	1.245

Energy values in kcal/mol. Bond lengths R in Å. Atomic partial charges q.
[a]Values in parentheses give the percentage contribution to the attractive interactions $\Delta E_{elstat} + \Delta E_{orb}$.
[b]Values in parentheses give the percentage contribution to the total orbital interactions.
[c]The values in parentheses give the preparation energies of the metal fragment and ligand, respectively.
[d]Partial charge of $TM(CO)_5$.

the metal–ligand interaction is more electrostatic than covalent (ΔE_{elstat} = 51.7–55.0%) and the relative contribution of the covalent term ΔE_{orb} is stronger in the acetylene complexes than in the ethylene complexes. Again the out-of-plane $(CO)_5TM \leftarrow C_2H_x$ $\pi_\perp$-donation and the δ interaction are negligible. In these complexes the σ-donation and the π-backdonation are of comparable strength. Yet, the σ-donation is stronger in the ethylene complexes, while the π-backdonation is more important in the acetylene complexes. This effect is even more pronounced in the heavier complexes.

As before, the bond dissociation energies are slightly larger for the ethylene complexes than for the acetylene complexes. This is again caused by the effect of the preparation energy rather than the metal–ligand interaction. The deformation of the acetylene ligand always costs more energy than the geometric deformation of the ethylene ligand.

To compare previously discussed low-valent transition metal compounds with high-valent transition metal compounds, the former results can be compared with the energy decomposition analyses of the complexes $Cl_4TM–C_2H_x$ (TM = Cr, Mo, W). The EDA results are given in Table 13.32.

The high-valent compounds $Cl_4TM–C_2H_x$ show much larger interaction energies than the low-valent complexes discussed above. This is in agreement with the comparably small bond distances in these complexes. However, the bond dissociation energies are rather small. Note, that the predicted BDE of $Cl_4Mo–C_2H_4$ is positive. Thus, this molecule is only kinetically stable. This is caused by the large preparation energies which are in turn dominated by the large singlet to triplet excitation which is necessary

Table 13.32 Energy decomposition analysis of $Cl_4TM–C_2H_x$ at BP86/TZP using triplet fragments $TMCl_4$ and C_2H_x [35]

	$MoCl_4$		WCl_4	
	C_2H_4	C_2H_2	C_2H_4	C_2H_2
ΔE_{int}	−126.1	−153.1	−139.7	−169.4
ΔE_{Pauli}	209.1	268.7	257.9	313.9
ΔE^{a}_{elstat}	−132.6 (39.6%)	−165.9 (39.3%)	−172.8 (43.5%)	−206.8 (42.8%)
ΔE^{a}_{orb}	−202.6 (60.4%)	−256.0 (60.7%)	−224.7 (56.5%)	−276.5 (57.2%)
$\Delta E(A_1)\sigma^{b}$	−101.9 (50.3%)	−121.4 (47.4%)	−116.0 (51.6%)	−135.0 (48.8%)
$\Delta E(A_2)\delta^{b}$	−2.5 (1.2%)	−2.8 (1.1%)	−2.8 (1.2%)	−2.4 (0.9%)
$\Delta E(B_1)\pi_\perp^{b}$	−6.9 (3.4%)	−27.4 (10.7%)	−8.4 (3.7%)	−30.0 (10.9%)
$\Delta E(B_2)\pi_\parallel^{b}$	−91.3 (45.1%)	−104.4 (40.8%)	−97.5 (43.4%)	−109.1 (39.4%)
E^{c}_{prep}	133.9 (44.8 + 89.1)	141.2 (47.5 + 93.7)	126.8 (41.0 + 85.8)	131.5 (41.7 + 89.8)
$\Delta E(=-D_e)$	7.8	−11.9	−12.9	−37.9
$-D_0$	9.4	−9.8	−11.4	−35.6
$q(TM)^{d}$	0.03	0.07	0.04	0.08
$q(C_2H_x)$	−0.03	−0.07	−0.04	−0.08
R(C–C)	1.429	1.307	1.451	1.317

Energy values in kcal/mol. Bond lengths R in Å. Atomic partial charges q.
[a]Values in parentheses give the percentage contribution to the attractive interactions $\Delta E_{elstat} + \Delta E_{orb}$.
[b]Values in parentheses give the percentage contribution to the total orbital interactions.
[c]The values in parentheses give the preparation energies of the metal fragment and ligand, respectively.
[d]Partial charge of $TMCl_4$.

References pp. 367–372

to promote the fragment from its electronic ground state to the reference state used in the analysis.

In this case the contribution of the covalent interaction in the ΔE_{orb} term is more important than the electrostatic interaction. The ΔE_{elstat} term contributes only to 39–44% of the total attractive contributions. The individual contributions to the covalent orbital term ΔE_{orb} reveal, that although the σ-donation is dominant, the contribution of the backdonation in the b_2 symmetry is of almost comparable size. The largest difference between low- and high-valent transition metal complexes can be seen in the out-of-plane $\pi_{\perp}$-interaction in b_1 symmetry. In the ethylene complexes this contribution is small. It contributes to only up to 3.7% of the orbital term and is almost negligible, but in the acetylene complexes this contribution rises to 10.7–10.9% of the covalent contribution. Although comparably large, the amount of interaction is still insufficient to classify acetylene as a 4-electron donor in $Cl_4TM{-}C_2H_2$ complexes.

13.4.5 Phosphane complexes $(CO)_5TM{-}PX_3$ (TM = Cr, Mo, W; X = H, Me, F, Cl) [22]

The chemical bonding of transition metal phosphane complexes has been analyzed and reviewed many times in the chemical literature [102]. Usually the emphasis of the investigation is on the strength of the TM → PR_3 π-backdonation. This has led to a controversy because different methods which were used to estimate the π-acceptor strength of different phosphanes PR_3 came to conflicting conclusions particularly with regard to PCl_3. The interpretation of IR data of phosphane complexes by means of the Cotton-Kraihanzel force-field technique [103] led to the order of π acceptance $PF_3 > PCl_3 > P(OR)_3 > PR_3$ [104]. The interpretation of experimental NMR chemical shifts of $(CO)_nMo{-}(PR_3)_{6-n}$ ($n = 3{-}5$) and a re-examination of the various parameters used to evaluate σ and π contributions to the TM–P bond led to the suggestion that PCl_3 should be a weak π acceptor [105]. However, a subsequent theoretical study of [94] Mo and [32] P NMR chemical shifts indicated that PCl_3 is actually a very strong π acceptor which should be stronger than PF_3 and particularly PH_3 and PMe_3 [106]. The conflicting suggestions illustrate the dilemma which is often found when one tries to examine the nature of a chemical bond by correlating a particular property with specific components of the bonding interactions. A correlation is not the same as an explanation, which comes from the analysis of the electronic structure. The EDA is ideally suited to investigate the question about the correlation between the strength of the total interactions and the π-backdonation. Therefore, we carried out an EDA investigation of group-6 phosphane complexes $(CO)_5TM{-}PX_3$ (TM = Cr, Mo, W; X = H, Me, F, Cl) [23]. The results are given in Table 13.33.

The breakdown of the TM–P interaction energies into the contributions of ΔE_{elstat}, ΔE_{orb} and ΔE_{Pauli} shows that the repulsive term ΔE_{Pauli} has always the largest absolute values. For the TM–PH_3 and TM–PMe_3 bonds, the largest attractive contributions come from ΔE_{elstat}. The bonding in the PH_3 and PMe_3 complexes has between 56 and 65% electrostatic character. Thus, any discussion of the bonding in these molecules in terms of covalent bonding neglects the dominant part of the attractive interactions! The covalent

Table 13.33 Energy decomposition analysis of $(CO)_5TM{-}PX_3$ at BP86/TZP [22]

	$Cr(CO)_5PH_3$	$Mo(CO)_5PH_3$	$W(CO)_5PH_3$
ΔE_{int}	−33.7	−31.8	−36.4
ΔE_{Pauli}	81.5	70.7	83.7
ΔE^{a}_{elstat}	−64.9 (56.3%)	−59.4 (58.0%)	−71.5 (59.6%)
ΔE^{a}_{orb}	−50.3 (43.7%)	−43.1 (42.0%)	−48.5 (40.4%)
$\Delta E(A')$	−42.5	−35.9	−40.7
$\Delta E(A'')$	−7.8	−7.4	−7.9
ΔE^{b}_{σ}	−34.6 (68.8%)	−28.3 (65.7%)	−32.8 (67.6%)
ΔE^{b}_{π}	−15.9 (31.2%)	−14.8 (34.3%)	−15.7 (32.4%)
ΔE_{prep}	1.2	0.8	2.4
$\Delta E(=-D_e)$	−32.4	−31.0	−34.0
	$Cr(CO)_5PMe_3$	$Mo(CO)_5PMe_3$	$W(CO)_5PMe_3$
ΔE_{int}	−43.7	−40.9	−46.4
ΔE_{Pauli}	96.5	85.6	99.3
ΔE^{a}_{elstat}	−85.1 (60.7%)	−80.7 (63.8%)	−94.9 (65.1%)
ΔE^{a}_{orb}	−55.1 (39.3%)	−45.8 (36.2%)	−50.8 (34.9%)
$\Delta E(A')$	−48.0	−39.5	−44.0
$\Delta E(A'')$	−7.1	−6.3	−6.8
ΔE^{b}_{σ}	−41.0 (74.3%)	−33.1 (72.3%)	−37.3 (73.4%)
ΔE^{b}_{π}	−14.2 (25.7%)	−12.7 (27.7%)	−13.5 (26.6%)
ΔE_{prep}	2.5	3.0	2.6
$\Delta E(=-D_e)$	−41.2	−37.9	−43.8
	$Cr(CO)_5PF_3$	$Mo(CO)_5PF_3$	$W(CO)_5PF_3$
ΔE_{int}	−35.1	−33.5	−38.6
ΔE_{Pauli}	92.0	82.3	97.0
ΔE^{a}_{elstat}	−62.9 (49.5%)	−57.4 (49.6%)	−70.2 (51.8%)
ΔE^{a}_{orb}	−64.2 (50.5%)	−58.4 (50.4%)	−65.4 (48.2%)
$\Delta E(A')$	−48.5	−43.1	−48.9
$\Delta E(A'')$	−15.8	−15.3	−16.5
ΔE^{b}_{σ}	−32.7 (50.8%)	−27.8 (47.6%)	−32.4 (49.5%)
ΔE^{b}_{π}	−31.6 (49.2%)	−30.6 (52.4%)	−33.1 (50.5%)
ΔE_{prep}	1.3	2.7	3.1
$\Delta E(=-D_e)$	−33.8	−30.9	−35.4
	$Cr(CO)_5PCl_3$	$Mo(CO)_5PCl_3$	$W(CO)_5PCl_3$
ΔE_{int}	−27.8	−26.1	−31.1
ΔE_{Pauli}	78.4	70.1	83.3
ΔE^{a}_{elstat}	−49.6 (46.7%)	−45.3 (47.3%)	−55.8 (48.8%)
ΔE^{a}_{orb}	−56.6 (53.3%)	−51.0 (52.9%)	−58.5 (51.2%)
$\Delta E(A')$	−43.5	−38.6	−44.9
$\Delta E(A'')$	−13.1	−12.4	−13.6
ΔE^{b}_{σ}	−30.4 (53.7%)	−26.3 (51.5%)	−31.2 (53.3%)
ΔE^{b}_{π}	−26.2 (46.3%)	−24.7 (48.5%)	−27.3 (46.7%)
ΔE_{prep}	1.0	2.5	2.5
$\Delta E(=-D_e)$	−26.8	−23.7	−28.6

Energy values in kcal/mol.

[a]The value in parentheses gives the percentage contribution to the total attractive interactions $\Delta E_{elstat} + \Delta E_{orb}$.

[b]The value in parentheses gives the percentage contribution to the total orbital interactions ΔE_{orb}.

contributions become somewhat larger in the halophosphane complexes $(CO)_5TMPF_3$ and $(CO)_5TMPCl_3$ where the electrostatic forces and covalent forces have nearly the same strength (Table 13.33). This is one difference between the TM–PX_3 (X = F, Cl) bonds and the TM–PX_3 (X = H, Me) bonds. The second difference concerns the degree of π contributions to ΔE_{orb}. Table 13.33 shows that π bonding in the TM–PH_3 bonds (31–34%) and particularly in the TM–PMe_3 bonds (26–28%) is clearly weaker than σ bonding, while the σ and π contributions in the TM–PF_3 and TM–PCl_3 bonds have about equal strength. The latter result answers once and for all the question about the strength of the π contributions in PCl_3 complexes. The ligand PCl_3 is as strong a π acceptor than PF_3! However, the interaction energies ΔE_{int} and the bond dissociation energies D_e show clearly (Table 13.33) that PCl_3 is weaker bonded than PF_3.

Fig. 13.16 exhibits the trend of ΔE_{int} and the energy contributions ΔE_{elstat}, ΔE_{orb} and ΔE_{Pauli} to the W–PX_3 bonds. The π contribution to ΔE_{orb} is also given. The trends of the chromium and molybdenum complexes are nearly the same and thus, they are not shown here. The most important conclusion is that the trend of the electrostatic term ΔE_{elstat} shows a much better agreement with the total interaction energy ΔE_{int} than the orbital term ΔE_{orb}. In particular, the trend of the π-bonding values ΔE_{π} is very different from the curve of ΔE_{int}. The best correlation is actually found between ΔE_{int} and ΔE_{elstat}. Note that the stronger bonding of the PMe_3 ligand compared with the other PX_3 phosphane comes clearly from the larger ΔE_{elstat} values and not from ΔE_{orb} (Table 13.33). The weaker bonds of the PCl_3 ligands compared with the TM–PF_3 bonds is also mainly caused by the weaker electrostatic attraction. Only the relative values of ΔE_{int} and ΔE_{elstat} for PH_3 and PF_3 do not agree with each other. Nevertheless, the results of the energy analysis [22] suggest that *the trend of the metal-phosphane bond strength is determined by the electrostatic interactions and not by the orbital interactions*. Thus, focusing on the strength of π-backdonation in phosphane complexes can be misleading!

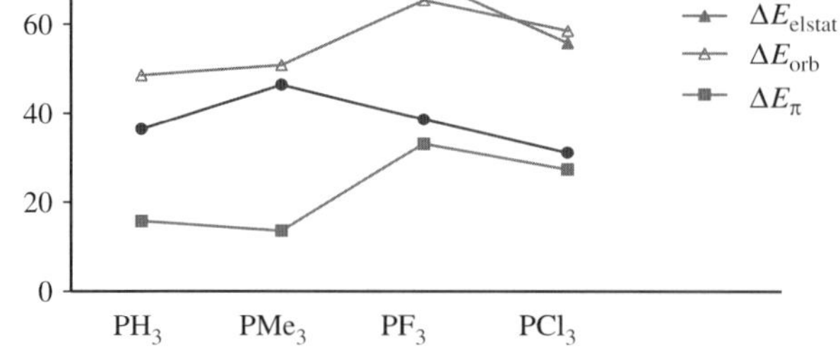

Fig. 13.16. Trends of the absolute values of the interaction energies ΔE_{int} and the energy contributions ΔE_{Pauli}, ΔE_{elstat}, total orbital interactions ΔE_{orb} and π-orbital interactions ΔE_{π} to the W–PX_3 bonds of $W(CO)_5PX_3$ at BP86/TZP.

13.4.6 Dihydrogen complexes $TM(CO)_5(\eta^2\text{-}H_2)$ (TM = Cr, Mo, W) and $W(CO)_3X_2(\eta^2\text{-}H_2)$ (X = CO, PH_3, PCl_3, PMe_3) [26]

The dichotomy of two bonding models for TM–ligand interactions which was successfully employed to describe TM complexes with carbene, carbyne, alkene and alkyne ligands is also helpful for describing the two classes of hydrid complexes which are known. One class are the metal hydrids $L_nTM(H)_m$ where one or more hydrogen atoms are directly bonded to the metal. They are called classical hydrides. Nonclassical hydrides have dihydrogen ligands H_2 in an η^2-bonded way $L_nTM(\eta^2\text{-}H_2)_m$. Since the first synthesis of such a complex by Kubas [107] in 1984 transition metal–dihydrogen complexes have been subject of many experimental and theoretical studies which have been reviewed recently [108]. The nature of the metal–dihydrogen interactions in nonclassical hydrides is still disputed. Therefore, we carried out an EDA investigation of the model complexes $TM(CO)_3X_2(\eta^2\text{-}H_2)$ (TM = Cr, Mo, W; X = CO, PH_3, PCl_3) [26]. We will first discuss the results of the parent systems $E(CO)_5\text{–}H_2$ (E = Cr, Mo, W). The results of the EDA are given in Table 13.34.

The EDA results of the parent compounds show, that all three components of ΔE_{int} vary parallel with the total interaction energy. Due to the larger $TM\cdots H_2$ distance in $Mo(CO)_5H_2$, both the repulsive Pauli interaction and the attractive electrostatic and orbital interactions are considerably weaker in this complex. The larger decrease in the attractive interactions as compared to that in the Pauli forces leads to ca. 3 kcal/mol destabilization with respect to $Cr(CO)_5H_2$ and $W(CO)_5H_2$. On the other hand, the ratio of the electrostatic and orbital interactions, and that of the components of ΔE_{orb}, is rather similar in all the three complexes. Thus, the nature of the $TM\text{–}H_2$ bonding in $TM(CO)_5H_2$ is slightly more electrostatic (53.1–54.4%) than covalent. We want to point out that the high electrostatic character and the trend of the electrostatic attraction is not obvious from the calculated partial charges. Table 13.34 shows that the NBO charges

Table 13.34 Results of the energy decomposition analysis of $(CO)_5TM\text{–}H_2$ at BP86/TZP [26]

	$Cr(CO)_5\text{–}H_2$	$Mo(CO)_5\text{–}H_2$	$W(CO)_5\text{–}H_2$
ΔE_{int}	−21.2	−18.3	−22.6
ΔE_{Pauli}	51.6	42.8	54.0
ΔE^{a}_{elstat}	−38.8 (53.2%)	−32.4 (53.1%)	−41.6 (54.4%)
ΔE^{a}_{orb}	−34.0 (46.8%)	−28.6 (46.9%)	−34.9 (45.6%)
$\Delta E(a_1)\sigma^{b}$	−21.8 (64.1%)	−18.1 (63.3%)	−22.3 (63.9%)
$\Delta E(A_2)\delta^{b}$	−0.1 (0.3%)	0.0	0.0
$\Delta E(B_1)\pi_{\perp}^{b}$	−1.3 (3.8%)	−1.2 (4.2%)	−1.3 (3.7%)
$\Delta E(B_2)\pi_{\parallel}^{b}$	−10.8 (31.8%)	−9.3 (32.5%)	−11.3 (32.4%)
ΔE_{prep}	2.5	2.3	3.2
$\Delta E(= -D_e)$	−18.7	−16.0	−19.4
q(TM)	−1.37	−1.03	−0.78
$q(H_2)$	0.22	0.18	0.18

Energy values in kcal/mol. Atomic partial charges q.
[a]The value in parentheses gives the percentage contribution to the total attractive interactions $\Delta E_{elstat} + \Delta E_{orb}$.
[b]The value in parentheses gives the percentage contribution to the total orbital interactions ΔE_{orb}.

of the metal and the H_2 ligand suggest that the electrostatic attraction in $Cr(CO)_5H_2$ should be much higher than in $W(CO)_5H_2$. The electrostatic interaction between the H_2 ligand and the $TM(CO)_5$ fragment comes mainly from the attraction between the electronic charge of the dihydrogen ligand, which serves as a Lewis base, and the positively charged nucleus of the metal. This becomes obvious by the topological analysis of the electron density distribution of the molecules, which has earlier been published by us [109]. There is an area of *local charge depletion* of the overall negatively charged metal atom and an area of *local charge concentration* of an overall positively charged ligand which leads to strong electrostatic attraction. The calculated partial charges give the wrong impression that the electrostatic attraction comes from the negatively charged metal atom and the positively charged ligand. The same problem was discussed in Section 13.4.2 about the group-13 diyl complexes.

On the basis of the energy components we can quantify the relative importance of donation and backdonation for the TM–H_2 interaction strength. Since the C_{2v} point group is preserved during the formation of the $TM(CO)_5H_2$ complexes, the energy contribution from the a_1 representation accounts for the donation from σ_{H_2} to the metal center, whereas the contribution from the b_1 representation can be attributed to the $d_\pi \rightarrow \sigma^*(H_2)$ backdonation. According to the data in Table 13.34, the σ-donation dominates over the π-backdonation in the $TM(CO)_5H_2$ complexes. The energy contribution of the former interaction is nearly twice of that of the π-backdonation. This phenomenon is reasonable in $TM(CO)_5L$ complexes because of the strong π-acceptor character of the CO groups which compete with the poor π-acceptor ligand H_2. A previous study on various $TML(CO)_4H_2$ complexes showed the importance of the trans ligand for the magnitude of the metal $\rightarrow \sigma^*(H_2)$ π-backdonation [110].

In order to assess the influence of ancillary ligands on the metal–dihydrogen bonding a series of EDA analyses of the model compounds $W(CO)_3X_2$–H_2 ($X = PH_3, PCl_3, PMe_3$) have been carried out. The dihydrogen ligand is *trans* to one CO ligand and *cis* to the phosphane ligands which have a *trans* position to each other [26]. The results of the energy analysis are shown in Table 13.35.

The EDA studies of $W(CO)_3(PR_3)_2(\eta^2\text{–}H_2)$ have been carried out with two different orientations of the H–H bond with respect to the P–W–P axis. In the global minimum structures of $W(CO)_3(PH_3)_2H_2$ and $W(CO)_3(PMe_3)_2H_2$ the H_2 bond vector is aligned along the P–W–P axis (||), whereas the alignment along the C–W–C axis ($\perp$) corresponds to a first-order saddle-point. On the other hand, the global minimum of the PCl_3 complex is the ($\perp$) conformer, whereas $W(CO)_3(PCl_3)_2H_2$ (||) is a slightly higher lying local minimum. The interaction energy in the (||) isomers is always considerably larger than in the ($\perp$) conformers (Table 13.35) but the preparation energy in the former species is also higher. This leads to a lower energy for the ($\perp$) form of $W(CO)_3(PCl_3)_2H_2$. Note that the (||) isomers have a significantly larger π-orbital contribution to the ΔE_{orb} term than the ($\perp$) conformers while the overall bonding character in terms of relative covalent/electrostatic bonding does not change very much (Table 13.35).

The data in Table 13.35 show that substitution of the strong π-acceptor ligand CO by the relatively poor π-acceptor phosphanes increases the W–H_2 interaction energy. In agreement with the very weak π-acceptor character of PH_3 (vide supra), the W–H_2 interaction becomes the strongest in $W(CO)_3(PH_3)_2H_2$. On the other hand,

Table 13.35 Results of the energy decomposition analysis of $(CO)_3X_2W{-}H_2$ at BP86/TZP

	X						
	CO	$PH_3(\parallel)$	$PH_3(\perp)$	$PCl_3(\parallel)$	$PCl_3(\perp)$	$PMe_3(\parallel)$	$PMe_3(\perp)$
ΔE_{int}	−22.6	−27.6	−21.2	−24.1	−20.8	−25.6	−20.5
ΔE_{Pauli}	54.0	66.0	54.4	61.6	57.5	65.3	54.9
ΔE_{elstat}^{a}	−41.6 (54.4%)	−50.8 (54.3%)	−41.8 (55.3%)	−45.8 (53.5%)	−42.3 (54.0%)	−49.8 (54.8%)	−42.2 (55.9%)
ΔE_{orb}^{a}	−34.9 (45.6%)	−42.7 (45.7%)	−33.8 (44.7%)	−39.9 (46.5%)	−36.0 (46.0%)	−41.0 (45.2%)	−33.3 (44.1%)
$\Delta E(A_1)\sigma^{b}$	−22.3 (63.9%)	−23.1 (54.1%)	−20.5 (60.6%)	−24.2 (60.6%)	−23.3 (64.7%)	−22.1 (53.9%)	−19.5 (58.6%)
$\Delta E(A_2)\delta^{b}$	0.0	−0.1 (0.2%)	−0.1 (0.3%)	−0.1 (0.3%)	−0.1 (0.3%)	−0.1 (0.2%)	−0.1 (0.3%)
$\Delta E(B_1)\pi_{\perp}^{b}$	−1.3 (3.7%)	−1.3 (3.0%)	−1.6 (4.7%)	−1.3 (3.3%)	−1.5 (4.2%)	−1.3 (3.2%)	−1.8 (5.4%)
$\Delta E(B_2)\pi_{\parallel}^{b}$	−11.3 (32.4%)	−18.2 (42.6%)	−11.6 34.3%)	−14.3 (35.8%)	−11.1 (30.8%)	−17.5 (42.7%)	−11.9 (35.7%)
ΔE_{prep}	3.2	8.3	3.5	8.1	3.9	5.9	3.4
$\Delta E(= -D_e)$	−19.4	−19.3	−17.7	−16.0	−16.9	−19.7	−17.1
q(W)	−0.78	−1.03	−1.01	−1.08	−1.08	−1.01	−1.01
$q(H_2)$	0.18	0.16	0.17	0.20	0.20	0.16	0.16

The two structures of $W(CO)_3X_2{-}H_2$ are characterized by the orientation of the H–H bond which is parallel (‖) or perpendicular (⊥) to the $P{-}W{-}P$ axis [26]; Energy values in kcal/mol. Atomic partial charges q.

[a] The value in parentheses gives the percentage contribution to the total attractive interactions $\Delta E_{elstat} + \Delta E_{orb}$.

[b] The value in parentheses gives the percentage contribution to the total orbital interactions ΔE_{orb}.

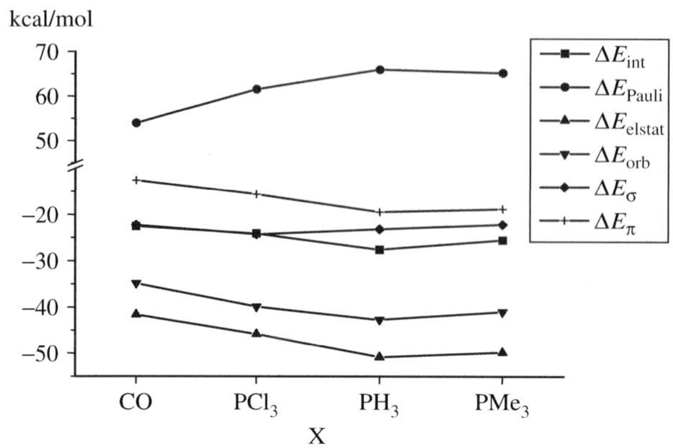

Fig. 13.17. Trend of the energy contributions to the metal-H_2 interaction energy ΔE_{int} in the complexes $W(CO)_3X_2(\eta^2\text{-}H_2)$ (X = CO, PH_3, PCl_3, PMe_3) at BP86/TZP.

the comparatively strong π-acceptor character of PCl_3 (see Section 13.4.5 about phosphane complexes) compared with the other phosphanes yields the weakest W−H_2 interaction in $W(CO)_3(PCl_3)_2H_2$. Inspection of the energy contributions to the total interaction energy shows that the relative strength of ΔE_{elstat} and ΔE_{orb} remains nearly the same in the complexes $W(CO)_3X_2H_2$ for X = CO and X = PR_3 (Table 13.35). The absolute values of the attractive interactions E_{elstat} and ΔE_{orb} are slightly higher for X = PR_3 than for X = CO, but also the values for E_{Pauli} are larger in the former complexes than in the latter. Fig. 13.17 shows nicely that the trend of the E_{elstat} and ΔE_{orb} values follow closely the trend of E_{int}. However, there is a noteworthy difference between the σ and π contributions to the two types of complexes. The increase of the ΔE_{orb} term in the PR_3 complexes is mainly due to the stronger π contributions (Table 13.35) while the $\Delta E\sigma$ values change very little. In fact, the change of the ΔE_π values follows closely the change of the total interaction energy ΔE_{int}. This is clearly shown in Fig. 13.17 where the trend of ΔE_π is graphically displayed. The calculations indicate that there is a correlation between the energy contribution of the W → $\sigma^*(H_2)$ π-backdonation and the total interaction energy.

13.4.7 Metallocene complexes $Fe(\eta^5\text{-}E_5)_2$ and $Ti(\eta^5\text{-}E_5)_2^{2-}$ (E = CH, N, P, As, Sb) and bis(benzene)chromium [20,23,29,31]

The landmark synthesis of ferrocene in 1951 [74] which was already mentioned in Section 13.3.8 and the equally important preparation of bis(benzene)chromium in 1955 [111] set the stage for the development of the chemistry of metallocene and arene complexes which are also known as sandwich compounds. The synthesis and application of metallocenes, which are important catalysts has been flourishing until today. It is thus appropriate to analyze the bonding situation in the two parent compounds in some more detail with the EDA method.

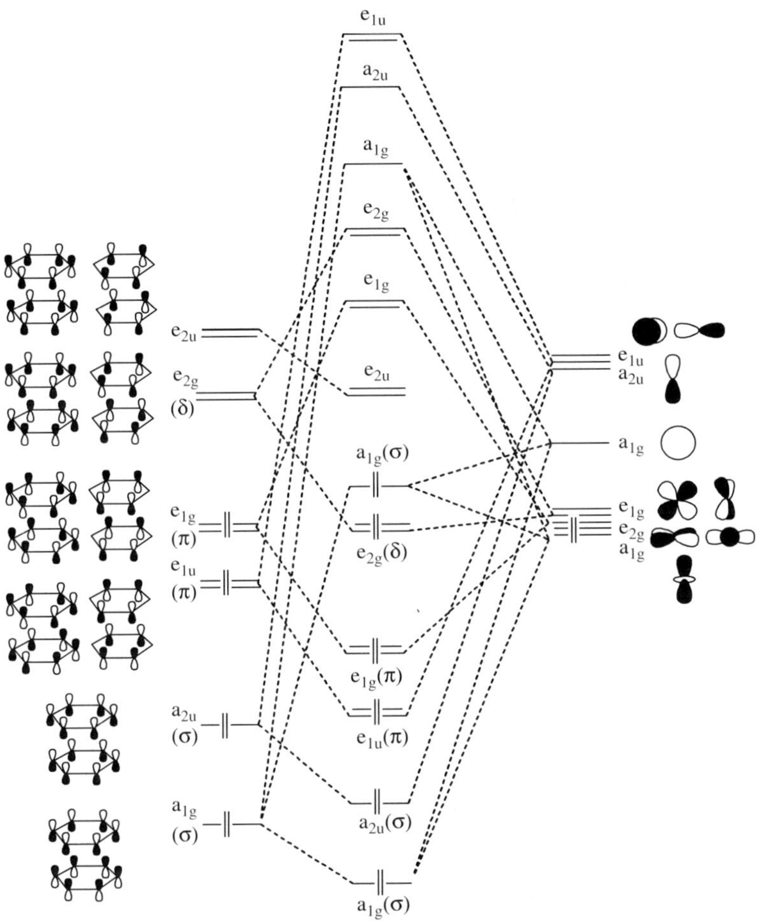

Fig. 13.18. MO correlation diagram between a d^6 transition metal with the electron configuration $(a_{1g})^2 \times (e_{2g})^4(e_{1g})^0$ and a cyclic 12 π-aromatic sandwich ligand. The shape of the ligand orbitals have been taken from the Bz_2 ligand. The orbitals of the Cp_2 ligands look very similar.

The metal–ligand interactions in metallocenes and metal arene complexes are usually discussed in terms of qualitative MO correlation diagrams. Fig. 13.18 shows a correlation diagram which gives the relevant orbitals of a d^6 transition metal with the reference electron configuration $(a_{1g})^2(e_{2g})^4(e_{1g})^0$ and a cyclic 12 π-aromatic sandwich ligand which can be used for ferrocene and bis(benzene)chromium. The symmetry assignments of the orbitals are related to the D_{6h} conformation of $Cr(Bz)_2$ and to the D_{5d} form of $Fe(Cp)_2$. Most textbooks discuss the orbital interactions in ferrocene using the D_{5d} form because it is more convenient for comparison with other transition metal complexes. We have also chosen to use the D_{5d} form of $Fe(Cp)_2$ because the molecular orbitals have the same symmetry as those of $Cr(Bz)_2$. Note that the a_{1g} and a_{2u} orbitals of the complexes have σ symmetry, the e_{1u} and e_{1g} orbitals have π symmetry, and

Table 13.36 Energy decomposition analysis of $Cr(Bz)_2$ and $Fe(Cp)_2$ at BP86/TZP [31]. The electron configurations of the metal fragments are given below. The electron configurations of the ligands are (a) neutral benzene dimer $(Bz)_2$, (b) Cp_2^{2-} and (c) triplet $Cp_2^{\beta\beta}$. Energy values in kcal/mol

(D_{6h}, D_{5d})	(a) $Cr(Bz)_2$ $Cr[d_{z^2}^2, d_{xy}, d_{x^2-y^2}]^6$	(b) $Fe(Cp)_2$ $Fe^{2+}[d_{z^2}^2, d_{xy}, d_{x^2-y^2}]^6$	(c) $Fe(Cp)_2$ $Fe^{2+}[d_{z^2}^2, d_{xy}, d_{x^2-y^2}]^6[d_{xz}, d_{yz}]^{\alpha\alpha}$
ΔE_{int}	−268.5	−893.9	−274.2
ΔE_{Pauli}	339.2	279.9	409.6
ΔE_{elstat}^{a}	−230.5 (37.9%)	−599.9 (51.1%)	−307.5 (45.0%)
ΔE_{orb}^{a}	−377.2 (62.1%)	−573.9 (48.9%)	−376.3 (55.0%)
$\Delta E_{\sigma}(A_{1g})^{b}$	−35.4 (9.4%)	−48.6 (8.5%)	−24.6 (6.4%)
$\Delta E(A_{2g})$	0.0	0.0	0.0
$\Delta E(B_{1g})^{b}$	−0.1 (<1%)	c	c
$\Delta E(B_{2g})^{b}$	−0.2 (<1%)	c	c
$\Delta E_{\pi}(E_{1g})^{b}$	−55.6 (14.7%)	−371.2 (64.7%)	−231.2 (61.4%)
$\Delta E_{\delta}(E_{2g})^{b}$	−277.0 (73.4%)	−47.8 (8.3%)	−111.9 (29.7%)
$\Delta E(A_{1u})$	0.0	0.0	0.0
$\Delta E_{\sigma}(A_{2u})^{b}$	−2.9 (<1%)	−28.3 (4.9%)	−2.9 (<1%)
$\Delta E(B_{1u})^{b}$	−0.2 (<1%)	c	c
$\Delta E(B_{2u})^{b}$	−0.1 (<1%)	c	c
$\Delta E_{\pi}(E_{1u})^{b}$	−3.8 (1.0%)	−61.5 (10.7%)	−4.8 (1.3%)
$\Delta E_{\delta}(E_{2u})^{b}$	−1.8 (<1%)	−16.6 (2.9%)	−1.0 (<1%)

[a] The value in parentheses gives the percentage contribution to the total attractive interactions $\Delta E_{elstat} + \Delta E_{orb}$.
[b] The value in parentheses gives the percentage contribution to the total orbital interactions ΔE_{orb}.
[c] There is no representation of this symmetry in the point group D_{5d}.

the e_{2g} and e_{2u} orbitals have δ symmetry. According to the qualitative bonding model shown in Fig. 13.18, the σ and π orbitals describe M ← L donation while the δ orbitals describe M → L backdonation.

The qualitative MO diagram shall now be discussed in the light of the results of the EDA of bis(benzene)chromium and ferrocene which are given in Table 13.36. The first column gives the results for $Cr(Bz)_2$ which were calculated using Cr atom with the valence electron configuration $(a_{1g})^2(e_{2g})^4(e_{1g})^0$ and $(Bz)_2$ as fragments [31]. The occupied valence orbitals of Cr are d_{z^2}, d_{xy} and $d_{x^2} - y^2$. The calculations predict that the interaction energy between the excited chromium atom and the $(Bz)_2$ ligand is $\Delta E_{int} = -268.5$ kcal/mol. The largest component of the interaction energy is the attractive orbital term $\Delta E_{orb} = -377.2$ kcal/mol. The electrostatic attraction $\Delta E_{elstat} = -230.5$ kcal/mol is clearly weaker than the covalent attraction. The energy partitioning analysis suggests that the metal–ligand bonding in bis(benzene)chromium is 62.1% covalent and 37.9% electrostatic. The largest contribution to the covalent interactions comes from the e_{2g} (δ) orbitals which give 73.4%. Thus, the energy partitioning analysis shows that bis(benzene)chromium is a δ bonded molecule.

The energy partitioning analysis of ferrocene has been carried out using two different fragmentation patterns. First we used the fragments which are isoelectronic with the fragments of $Cr(Bz)_2$, i.e. Fe^{2+} with the valence electron configuration $(a_{1g})^2(e_{2g})^4(e_{1g})^0$ and $(Cp^-)_2$. Table 13.36 shows that the interaction energy of the charged fragments is much larger ($\Delta E_{int} = -893.9$ kcal/mol) than for bis(benzene)chromium. The largest contribution comes from the electrostatic term ($\Delta E_{elstat} = -599.9$ kcal/mol) but the covalent attraction is only slightly weaker ($\Delta E_{orb} = -573.9$ kcal/mol). Thus, the calculations indicate that the strength of the covalent bonding in ferrocene (48.9%) is nearly as high as the electrostatic bonding (51.1%). The large value for ΔE_{orb} can be explained with the energetically low-lying acceptor orbitals of Fe^{2+} which strengthen the M ← L donation. The calculated values for the orbitals with different symmetry show that the contributions of the M ← L donor interactions (σ and π orbitals) are indeed stronger in $Fe(Cp)_2$ than in $Cr(Bz)_2$ while the M → L backdonation (δ orbitals) is weaker. The largest contribution to the covalent bonding in ferrocene comes from the e_{1g} (π) orbitals which give 64.7% of ΔE_{orb} (Table 13.36).

The analysis of the bonding situation in ferrocene may be biased towards M ← L donor interactions because of the choice of charged fragments Fe^{2+} and $(Cp^-)_2$. Therefore, we carried out a second energy partitioning analysis of $Fe(Cp)_2$ using neutral Fe and Cp_2 as interacting fragments. We solved the problem of occupying the degenerate e_{1g} orbitals with only two electrons by choosing the triplet states of Fe, which has a valence electron configuration $(a_{1g})^2(e_{2g})^4(e_{1g})^{\alpha\alpha}$ and Cp_2 which has a the valence occupation $(a_{1g})^2 \times (a_{2u})^2(e_{1u})^4(e_{1g})^{\beta\beta}$. Table 13.36 shows that the interaction energy ($\Delta E_{int} = -274.2$ kcal/mol) is much less than in case when the charged fragments are used. However, the *relative* contributions of ΔE_{elstat} and ΔE_{orb} are very similar in both analyses. The calculations show that the Fe–$(Cp)_2$ bonding between neutral Fe and Cp_2 is 45.0% electrostatic and 55.0% covalent. The breakdown of the latter term into orbitals having different symmetry indicates that the e_{1g} (π) orbitals contribute 61.4% of ΔE_{orb} which is not much less than the value of 64.7% which was calculated when charged fragments are used (Table 13.36). The contribution of the δ bonding orbitals are larger when neutral fragments are used (29.7%) than in case of charged fragments (8.3%) but the main conclusion remains that the covalent bonding in ferrocene comes mainly from π orbitals.

The recent interest in all-heteroatom analogues of ferrocene led us to investigate the isoelectronic nitrogen analogue of ferrocene, i.e. ironbispentazole $Fe(\eta^5\text{-}N_5)_2$ and its heavier homologues $Fe(\eta^5\text{-}E_5)_2$ with E = P, As, Sb [20,29]. The EDA was carried out for the interactions between $Fe(\eta^5\text{-}E_5)^+$ and E_5^- rather than between the bare metal and the ligands. This was done because in our study we also investigated the complexes $FeCp(\eta^5\text{-}E_5)$ where we were interested in the differences between the metal–ligand interactions in the homoleptic and heteroleptic complexes. The EDA was also carried out for ferrocene using the above fragments which makes it possible to compare the EDA results using different fragments.

According to the qualitative model shown in Fig. 13.19, the most important orbital interactions take place (i) between the occupied e_1 (π) orbital of the ligand and the empty e_1 metal fragment orbital mainly *via* the d_{xz} and d_{yz} orbitals of Fe and (ii) between the empty e_2 (π) ligand orbital and the occupied e_2 metal fragment orbital mainly *via* the d_{xy} and $d_{x^2-y^2}$ orbitals of Fe. The interactions between the occupied a_1 orbitals of the ligand (lowest lying π orbital) and metal fragment (mainly *via* the d_{z^2} Fe orbital) contributes to the bonding only by mixing in of the empty s and p_z orbitals of Fe. We will now discuss the results of the energy analysis in order to estimate quantitatively the strength of the orbital interactions with respect to each other and with respect to electrostatic attraction which is not considered in the orbital correlation model.

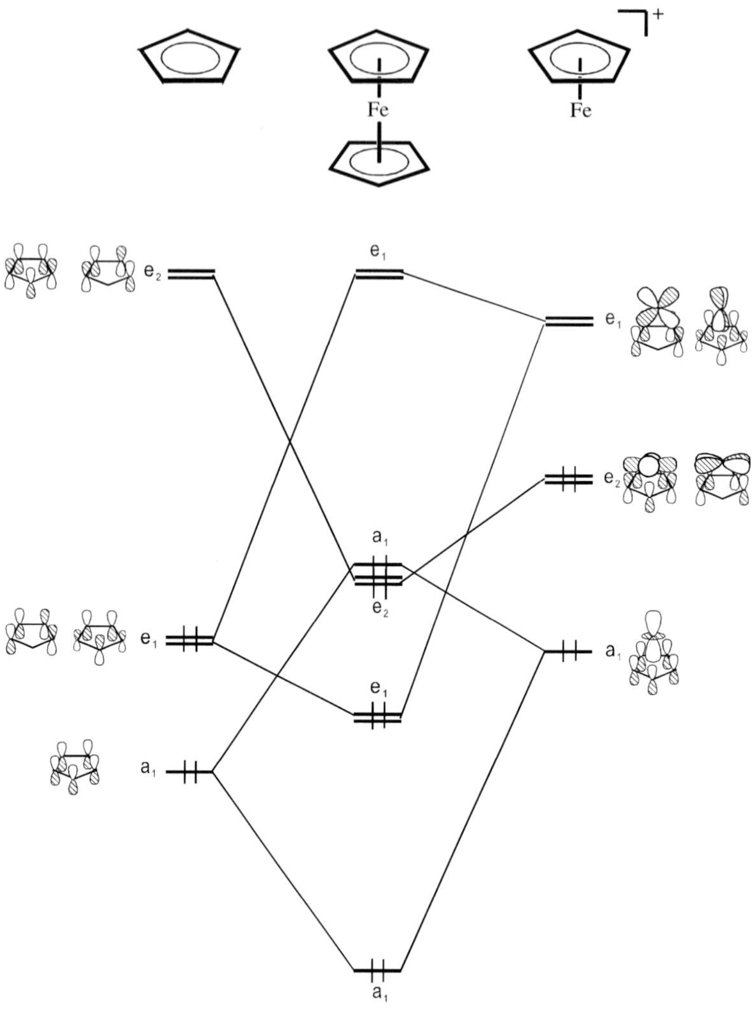

Fig. 13.19. MO correlation diagram showing the most important orbitals of the ligand *cyclo*-E_5^- and the fragment $Fe(\eta^5\text{-}E_5)^+$, which are relevant for the bonding in $Fe(\eta^5\text{-}E_5)_2$.

Table 13.37 shows the results of the energy partitioning analyses of the sandwich complexes of the form $Fe(E_5)_2$ with E = CH, N, P, As, Sb [23]. It is interesting to note, that the interaction energy ΔE_{int} is almost the same for the nitrogen and the phosphorous compounds. The larger BDE of the former stems from the contribution of the preparation energy ΔE_{prep}. The energy partitioning suggests that the bonding in all the complexes is half electrostatic and half covalent, although the electrostatic contribution ΔE_{elstat} is a little higher for ferrocene. The dominant contribution to the orbital term ΔE_{orb} is the π interaction in the e_1 symmetry. This term contributes between 65 and 69% to the total orbital interactions. The conclusion is that the nature of the bonding in the carbocyclic and heterocyclic $Fe(\eta^5\text{-}E_5)_2$ complexes does not vary significantly from E = CH to

Table 13.37 Energy decomposition analysis of $Fe(E_5)_2$ using the fragments $Fe(E_5)^+$ and E_5^- at BP86/TZ2P [23]

	E = CH	E = N	E = P	E = As	E = Sb
ΔE_{int}	−237.6	−198.0	−199.5	−183.8	−165.1
ΔE_{Pauli}	172.4	149.7	190.2	221.6	220.8
ΔE^a_{elstat}	−238.5	−184.3	−207.3	−223.1	−205.1
	(58.2%)	(53.0%)	(53.2%)	(55.0%)	(53.1%)
ΔE^a_{orb}	−171.5	−163.4	−182.5	−182.3	−180.9
	(41.8%)	(47.0%)	(46.8%)	(45.0%)	(46.9%)
$\Delta E_\sigma(A_1)^b$	−25.0	−22.4	−28.1	−29.1	−35.7
	(14.6%)	(13.7%)	(15.4%)	(16.0%)	(19.7%)
$\Delta E(A_2)$	0.0	0.0	0.0	0.0	0.0
$\Delta E_\pi(E_1)^b$	−109.3	−106.1	−120.7	−123.7	−125.6
	(63.8%)	(65.0%)	(66.1%)	(67.8%)	(69.4%)
$\Delta E_\delta(E_2)^b$	−37.1	−34.9	−33.7	−29.6	−19.6
	(21.6%)	(21.3%)	(18.5%)	(16.2%)	(10.8%)
ΔE_{prep}	2.8	13.1	28.0	23.3	46.3
$\Delta E(= -D_e)$	−234.8	−184.9	−171.5	−160.5	−118.8

Energy values in kcal/mol.
[a]The value in parentheses gives the percentage contribution to the total attractive interactions $\Delta E_{elstat} + \Delta E_{orb}$.
[b]The value in parentheses gives the percentage contribution to the total orbital interactions ΔE_{orb}.

Table 13.38 Energy decomposition analysis of $Ti(\eta^5\text{-}E_5)_2^{2-}$ at BP86/TZ2P using the fragments $Ti(\eta^5\text{-}E_5)^-$ and $(\eta^5\text{-}E_5)^-$ [29]

	Ti				
	E = CH	E = N	E = P	E = As	E = Sb
ΔE_{int}	−30.5	−27.5	−39.5	−33.3	−29.4
ΔE_{Pauli}	116.7	132.7	178.5	198.2	195.1
ΔE^a_{elstat}	−22.4	−3.7	−63.0	−84.5	−87.4
	(15.2%)	(2.3%)	(28.9%)	(36.5%)	(38.9%)
ΔE^a_{orb}	−124.8	−156.5	−154.9	−147.0	−137.2
	(84.8%)	(97.7%)	(71.1%)	(63.5%)	(61.1%)
$\Delta E_\sigma(A_1)^b$	−8.5	−12.7	−16.5	−18.9	−25.5
	(6.8%)	(8.1%)	(10.7%)	(12.9%)	(18.6%)
$\Delta E(A_2)$	0.0	0.0	0.0	0.0	0.0
$\Delta E_\pi(E_1)^b$	−30.6	−34.8	−48.8	−49.6	−52.0
	(24.5%)	(22.3%)	(31.5%)	(33.7%)	(37.9%)
$\Delta E_\delta(E_2)^b$	−85.7	−109.0	−89.6	−78.5	−59.6
	(68.7%)	(69.6%)	(57.8%)	(53.4%)	(43.5%)
ΔE_{prep}	83.0	44.4	27.4	27.5	30.5
$\Delta E(= -D_e)$	52.5	16.9	−12.1	−5.8	1.1

Energy values in kcal/mol.
[a]The value in parentheses gives the percentage contribution to the total attractive interactions $\Delta E_{elstat} + \Delta E_{orb}$.
[b]The value in parentheses gives the percentage contribution to the total orbital interactions ΔE_{orb}.

E = Sb. In the same paper [23] we reported also about the bonding situation in the mixed Cp/E_5 complexes FeCp(η^5-E_5). The EDA results showed that there are no major changes in the relative contributions of the energy components in the mixed complexes compared with the homoleptic molecules. Therefore we do not give the data here. The interested reader may read the original publication [23].

Shortly after the theoretical analysis of the bonding situation in Fe(η^5-E_5)$_2$ complexes was published a paper appeared which reported about the first synthesis of a homoleptic sandwich complex with a pentaphospholyl ligand, Ti(η^5-P_5)$_2^{2-}$ [112]. The authors reported also about a theoretical analysis of the metal–ligand bonding situation using a MO-correlation diagram. We extended our EDA work about the chemical bonds in metallocenes with heterocyclic ligands and investigated the compounds Ti(η^5-E_5)$_2^{2-}$ (E = CH, N, P, As, Sb). The calculations were done in the same fashion as for the neutral Fe(η^5-E_5)$_2$ species, which are 18-electron compounds while Ti(η^5-E_5)$_2^{2-}$ are formally 16-electron complexes. Thus, the highest-lying occupied a_1 orbital of Fe(η^5-E_5)$_2$ (Fig. 13.19) is empty in Ti(η^5-E_5)$_2^{2-}$. The EDA results using the fragments Ti(η^5-E_5)$^-$ and (η^5-E_5)$^-$ are shown in Table 13.38.

A comparison of the EDA results for Ti(η^5-E_5)$_2^{2-}$ (Table 13.38) with the data for Fe(η^5-E_5)$_2$ shows that the a_1 contribution to the orbital interactions is indeed smaller in the former compounds than in the latter. However, other differences are much more striking. The orbital interactions in the negatively charged titanium complexes Ti(η^5-E_5)$_2^{2-}$ come mainly from the E_2 orbitals which have δ symmetry while the largest contribution to ΔE_{orb} in Fe(η^5-E_5)$_2$ come from the E_1 orbitals which have π symmetries. The relative contributions of the δ orbitals in Ti(η^5-E_5)$_2^{2-}$ become smaller when E is a heavier element but the E_2 orbitals are always more important than the E_1 orbitals (Table 13.38). Thus, the complexes Ti(η^5-E_5)$_2^{2-}$ are δ-bonded compounds like Cr(Bz)$_2$. Note that, in Ti(η^5-E_5)$_2^{2-}$, the orbital interactions contribute between 61.1–97.7% to the total interaction energy which indicates a much higher degree of covalent bonding than in Fe(η^5-E_5)$_2$.

13.5 CONCLUSION

The results of the EDA which have been presented and discussed in this paper for a large variety of molecules and chemical bonds show that it is possible to address the question about the nature of the chemical bond in terms of familiar concepts which can be identified and quantitatively expressed using accurate quantum chemical methods. The three energy terms ΔE_{elstat}, ΔE_{Pauli} and ΔE_{orb} which are uniquely defined in the EDA method can be interpreted in a plausible way which connects heuristic bonding models with the physical mechanism of the chemical bonds. The EDA method can be used to investigate controversial topics such as the multiplicity of bonding interactions and the relative strength of quasiclassical electrostatic and covalent (orbital) interactions. The EDA results show that the repulsive forces, which come from the Pauli exclusion principle can be very important for understanding the strength of the interatomic interactions. It is also misleading to consider only orbital interactions as the crucial term for determining the trend of the bond strength. Another advantage of the EDA partitioning method is that the instantaneous interaction energy of a chemical bond is considered. The ΔE_{int} values can be very different from the bond dissociation energies because the preparation energy of the interacting fragments may be large.

There is still much room for progress in the field of energy decomposition of the chemical bond, both in application and method development. An important issue is the application of the EDA method for polar bonds. Another promising field, which is hardly touched is the bonding analysis of transition state structures, which could reveal important information about the reactivity of molecules. From a methodological point of view it is desirable that the EDA partitioning becomes developed for correlated ab initio wavefunctions. Another future avenue for further development is the choice of different interacting moieties. For example, the atomic basins as defined by the atom-in-molecule method of Bader [113] may be used as starting point of the analysis instead of using the undisturbed charge distributions of the fragments. It would also be useful for some molecules to study the interactions between more than two fragments at the same time. For example, the difference between the σ and π bonding interactions in benzene could then be studied with the EDA by taking three C_2H_2 fragments as interacting species. The energy decomposition of the interatomic interactions remains a promising tool in the large arsenal of quantum chemical methods for addressing the famous statement: "Give us insight not numbers" [114].

13.6 ACKNOWLEDGEMENTS

We are grateful to Andreas Krapp, Stefan Erhardt and Christoph Loschen for critically reading the manuscript. This work was supported by the Deutsche Forschungsgemeinschaft and by the Fonds der Chemischen Industrie. Excellent service by the Hochschulrechenzentrum of the Philipps-Universität Marburg is gratefully acknowledged. Additional computer time was provided by the HLRS Stuttgart and the HLR Darmstadt.

13.7 REFERENCES

1 Much information about the history of quantum chemistry is available through the highly entertaining website Early Ideas in the History of Quantum Chemistry by U. Anders: http://www.quantum-chemistry-history.com/index.html.
2 (a) K. Fukui, Acc. Chem. Res., 4 (1971) 57; (b) K. Fukui, Theory of orientation and stereoselection, Springer, Berlin, 1975.
3 R.B. Woodward and R. Hoffmann, The conservation of orbital symmetry, Verlag Chemie, Weinheim, 1970.
4 (a) H.F.S. Schaefer, Science, 231 (1986) 1100; (b) H.F.S. Schaefer, Chimia, 43 (1989) 1; (c) H.F.S. Schaefer, J. Mol. Struct. (THEOCHEM.), 398 (1997) 199.
5 C.A. Coulson, Valence, 2nd edn., Oxford University Press, London, 1963.
6 L. Pauling, The nature of the chemical bond, 3rd edn., Cornell University Press, Ithaca, NY, 1960.
7 G.N. Lewis, Valence and the structure of atoms and molecules, American Chemical Society Monograph Series, New York, 1923.
8 (a) K. Ruedenberg, Rev. Mod. Phys., 34 (1962) 326; (b) M. Feirberg, K. Ruedenberg and E. Mehler, Adv. Quart. Chem., 5 (1970) 27; (c) M. Feirberg and K. Ruedenberg, J. Chem. Phys., 54 (1971) 1495; (d) K. Ruedenberg, in: O. Chalvet, R. Daudel, S. Diner and J.P. Malrieu (Eds.), Localization and delocalization in quantum chemistry, Reidel, Dordrecht, 1975, p. 223.

9 A thorough discussion of the chemical bond is given by W. Kutzelnigg, The physical origin of the chemical bond, in: Z.B. Maksic (Ed.), The concept of the chemical bond, part 2. Springer, Berlin, 1990, p. 1.

10 The virial theorem for molecules at non-equilibrium distances has the expression: $1/2\langle V\rangle = -\langle T\rangle - R(\delta E/\delta R) = 0$ where R are the atomic coordinates. At equilibrium distances holds $R(\delta E/\delta R) = 0$ which leads to Eq. (1).

11 A nice didactical discussion about the paradoxical roles of kinetic and potential energy contributions to the chemical bond has been published by: F Rioux, Chem. Educ. 2 (1997) 1430. Another didactically skilful presentation is given by: J.K. Burdett, Chemical bonds: A dialog, Wiley, Chichester, 1997, p. 3f.

12 W. Heitler and F. London, Z. Physik, 44 (1927) 455.

13 G. Frenking, Theor. Chem. Acc., 103 (2000) 177.

14 F.L. Hirshfeld and S. Rzotkiewicz, Mol. Phys., 27 (1974) 1319.

15 M.A. Spackman and E.N. Maslen, J. Phys. Chem., 90 (1986) 2020.

16 K.P. Huber and G. Herzberg, Molecular spectra and molecular structure IV. Constants of diatomic molecules, Van Nostrand-Reinhold, New York, 1979.

17 A. Diefenbach, F.M. Bickelhaupt and G. Frenking, J. Am. Chem. Soc., 122 (2000) 6449.

18 J. Uddin and G. Frenking, J. Am. Chem. Soc., 123 (2001) 1683.

19 Y. Chen and G. Frenking, J. Chem. Soc., Dalton Trans., (2001) 434.

20 M. Lein, J. Frunzke, A. Timoshkin and G. Frenking, Chem. Eur. J., 7 (2001) 4155.

21 M. Dörr and G. Frenking, Z. Allg. Anorg. Chem., 628 (2002) 843.

22 G. Frenking, K. Wichmann, N. Fröhlich, J. Grobe, W. Golla, D. Le Van, B. Krebs and M. Läge, Organometallics, 21 (2002) 2921.

23 J. Frunzke, M. Lein and G. Frenking, Organometallics, 21 (2002) 3351.

24 M. Cases, G. Frenking, M. Duran and M. Solà, Organometallics, 21 (2002) 4182.

25 C. Esterhuysen and G. Frenking, Chem. Eur. J., 9 (2003) 3518.

26 D.S. Nemcsok, A. Kovács, V.M. Rayón and G. Frenking, Organometallics, 21 (2002) 5803.

27 K.K. Pandey, M. Lein and G. Frenking, J. Am. Chem. Soc., 125 (2003) 1660.

28 M. Lein, J. Frunzke and G. Frenking, Angew. Chem., 115 (2003) 1341; M. Lein, J. Frunzke and G. Frenking, Angew. Chem. Int. Ed., 42 (2003) 1303.

29 M. Lein, J. Frunzke and G. Frenking, Inorg. Chem., 42 (2003) 2504.

30 C. Massera and G. Frenking, Organometallics, 22 (2003) 2758.

31 V.M. Rayón and G. Frenking, Organometallics, 22 (2003) 3304.

32 C. Esterhuysen and G. Frenking, Chem. Eur. J., 9 (2003) 3518.

33 O. Dietz, V.M. Rayón and G. Frenking, Inorg. Chem., 42 (2003) 4977.

34 C. Loschen and G. Frenking, Inorg. Chem., 43 (2004) 778.

35 M.S. Nechaev, V.M. Rayón and G. Frenking, J. Phys. Chem. A, 108 (2004) 3134.

36 K.K. Pandey, M. Lein and G. Frenking, Organometallics, 23 (2004) 2944.

37 M. Lein, A. Szabó, A. Kovács and G. Frenking, Faraday Discuss., 124 (2003) 365.

38 C. Loschen, K. Voigt, J. Frunzke, A. Diefenbach, M. Diedenhofen and G. Frenking, Z. Allg. Anorg. Chem., 628 (2002) 1294.

39 V.M. Rayón and G. Frenking, Chem. Eur. J., 8 (2002) 4693.

40 C. Esterhuysen and G. Frenking, Theor. Chem. Acc., 111 (2004) 381.

41 F. Bessac and G. Frenking, Inorg. Chem., 42 (2003) 7990.

42 A. Kovács, C. Esterhuysen, G. Frenking, Chem. Eur. J., 11 (2005) 1813.

43 (a) M. Lein, G. Frenking, Austr. J. Chem., 57 (2004) 1191. (b) M Lein, G. Frenking, to be published.

44 K. Morokuma, J. Chem. Phys., 55 (1971) 1236.

45 T. Ziegler and A. Rauk, Theor. Chim. Acta, 46 (1977) 1.

46 (a) Bickelhaupt, F.M., Baerends, E.J., Boyd, D.B., Rev. Comput. Chem., Vol. 15, 2000, 1; (b) G.te Velde, F.M. Bickelhaupt, E.J. Baerends, S.J.A. van Gisbergen, C. Fonseca Guerra, J.G. Snijders, T. Ziegler, J. Comput. Chem., 22 (2001) 931.

47 G. Frenking, K. Wichmann, N. Fröhlich, C. Loschen, M. Lein, J. Frunzke and V.M. Rayón, Coord. Chem. Rev., 238–239 (2003) 55.

48 The term covalent bond was introduced into chemistry by I. Langmuir, J. Am. Chem. Soc, 41 (1919) 868.

49 L. Pauling, General chemistry, Dover Publications, New York, 1988, p. 152.

50 G.N. Lewis, Valence and the structure of atoms and molecules, American Chemical Society Monograph Series, New York, 1923, p. 78.

51 L. Pauling, The nature of the chemical bond, 3rd edn., Cornell University Press, Ithaca, NY, 1960, p. 64f.

52 (a) We wish to point out that the valence bond theory of chemical bonding has been revitalized in recent years mainly by new developments which have been made by Shaik, Hiberty and coworkers S. Shaik, P. Maitre, G. Sini and P.C. Hiberty, J. Am. Chem. Soc., 114 (1992) 7861; (b) D. Lauvergnat, P.C. Hiberty, D. Danovich and S. Shaik, J. Phys. Chem., 100 (1996) 5715; (c) A. Shurki, P.C. Hiberty and S. Shaik, J. Am. Chem. Soc., 121 (1999) 822.

53 Selected examples (a) T. Ziegler, V. Tschinke and A.D. Becke, J. Am. Chem. Soc., 109 (1987) 1351, (b) T. Ziegler, V. Tschinke and C. Ursenbach, J. Am. Chem. Soc., 109 (1987) 4825; (c) J. Li, G. Schreckenbach and T. Ziegler, J. Am. Chem. Soc., 117 (1995) 486.

54 (a) C. Fonseca Guerra, F.M. Bickelhaupt and E.J. Baerends, Chem. Phys. Chem., 5 (2004) 481; (b) C. Fonseca Guerra and F.M. Bickelhaupt, J. Chem. Phys., 119 (2003) 4262; (c) F.M. Bickelhaupt and E.J. Baerends, Angew. Chem., 115 (2003) 4315, Angew. Chem. Int. Ed. 42 (2003) 4183; (d) C. Fonseca Guerra, F.M. Bickelhaupt and E.J. Baerends, Cryst. Growth Des., 2 (2002) 239; (e) C. Fonseca Guerra and F.M. Bickelhaupt, Angew. Chem., 114 (2002) 2194, Angew. Chem. Int. Ed. 41 (2002) 2092; (f) F.M. Bickelhaupt, R.L. DeKock and E.J. Baerends, J. Am. Chem. Soc., 124 (2002) 1500; (g) A. Diefenbach and F.M. Bickelhaupt, J. Chem. Phys., 115 (2001) 4030; (h) C. Fonseca Guerra, F.M. Bickelhaupt, J.G. Snijders and E.J. Baerends, J. Am. Chem. Soc., 122 (2000) 4117; (i) C. Fonseca Guerra, F.M. Bickelhaupt, J.G. Snijders and E.J. Baerends, Chem. Eur. J., 5 (1999) 3581; (j) C. Fonseca Guerra and F.M. Bickelhaupt, Angew. Chem., 111 (1999) 3120, Angew. Chem. Int. Ed. 38 (1999) 2942; (k) A. Diefenbach and F.M. Bickelhaupt, Z. Anorg. Allg. Chem., 625 (1999) 892; (l) F.M. Bickelhaupt and F. Bickelhaupt, Chem. Eur. J., 5 (1999) 162; (m) F.M. Bickelhaupt, A. Diefenbach, S.V. de Visser, L.J. de Koning and N.M.M. Nibbering, J. Phys. Chem. A, 102 (1998) 9549; (n) A.W. Ehlers, E.J. Baerends, F.M. Bickelhaupt and U. Radius, Chem. Eur. J., 4 (1998) 210; (o) U. Radius, F.M. Bickelhaupt, A.W. Ehlers, N. Goldberg and R. Hoffmann, Inorg. Chem., 37 (1998) 1080; (p) F.M. Bickelhaupt, N.J.R. van Eikema Hommes, C. Fonseca Guerra and E.J. Baerends, Organometallics, 15 (1996) 2923; (q) F.M. Bickelhaupt, T. Ziegler and P.v.R Schleyer, Organometallics, 15 (1996) 1477; (r) F.M. Bickelhaupt, N.M.M. Nibbering, E.M. van Wezenbeek and E.J. Baerends, J. Phys. Chem., 96 (1992) 4864; (s) F.M. Bickelhaupt, E.J. Baerends and W. Raeneck, Inorg. Chem., 29 (1990) 350.

55 B. Hammer, L.B. Hansen and J.K. Norskov, Phys. Rev., B59 (1999) 7413.

56 J.P. Perdew, K. Burke and M. Ernzerhof, Phys. Rev. Lett., 77 (1996) 3865.

57 A.D. Becke, Phys. Rev. A, 38 (1988) 3098.

58 J.P. Perdew, Phys. Rev. B, 33 (1986) 8822.

59 J.G. Snijders, E.J. Baerends and P. Vernooijs, At. Nucl. Data Tables, (1982) 483.

60 E.J. Baerends, D.E. Ellis and P. Ros, Chem. Phys., 2 (1973) 41.

61 J. Krijn and E.J. Baerends, Fit Functions in the HFS-Method, Internal Report (in Dutch), Vrije Universiteit Amsterdam, The Netherlands, 1984.

62 (a) C. Chang, M. Pelissier and Ph. Durand, Phys. Scr., 34 (1986) 394; (b) J.-L. Heully, I. Lindgren, E. Lindroth, S. Lundquist and A.-M. Martensson-Pendrill, J. Phys. B, 19 (1986) 2799; (c) E. van Lenthe, E.J. Baerends and J.G. Snijders, J. Chem. Phys., 99 (1993) 4597; (d) E. van Lenthe, E.J. Baerends and J.G. Snijders, J. Chem. Phys., 105 (1996) 6505; (e) E. van Lenthe, R. van Leeuwen, E.J. Baerends and J.G. Snijders, Int. J. Quantum Chem., 57 (1996) 281.

63 The employed functional actually gives a non-zero value for the Pauli repulsion $\Delta E_{\mathrm{Pauli}} = -19.1$ kcal/mol which is an unphysical result. This energy has been added to the $\Delta E_{\mathrm{elstat}}$ term.

64 For a detailed discussion of the calculation of the electrostatic interaction term of the EDA approach see reference [46], page 14f.

65 (a) W. Kutzelnigg, Angew. Chem., 96 (1984) 262, Angew. Chem., Int. Ed. Engl., 23 (1984) 272; (b) A thorough discussion of the chemical bond is given by W. Kutzelnigg, The physical origin of the chemical bond, Z.B. Maksic,The concept of the chemical bond, part 2. Berlin Springer (1990) 35.

66 A. Krapp, G. Frenking, to be published.

67 The EDA of CO has been carried out in two different ways which are reported in reference 40. CO has a $^1\Sigma^+$ ground state which dissociates into (^{3}P) C and (^{3}P) O which are the atomic ground states. There are two different ways of combining the atoms with the electron configurations $1s^22s^22p^{\alpha}2p^{\alpha}2p^0$ (^{3}P) carbon and $1s^22s^22p^22p^{\beta}2p^{\beta}$ (^{3}P) oxygen yielding ($^1\Sigma^+$) CO. In the symmetry adapted approach the atomic configurations are the same as the point group of the CO molecule ($C_{\infty v}$) which leads to a donor–acceptor σ bond and two degenerate electron-sharing π bonds. The second approach has only C_{2v} symmetry for the atomic fragments which leads to one electron-sharing π bond, one donor–acceptor π bond and an electron-sharing σ bond. Both approached come to the same conclusion about the reason why CO has a stronger bond than N_2. Table 1 gives the EDA results for CO using the symmetry-adapted approach. The bonding analysis of BF was carried out using the ^{2}P ground states of B and F where the unpaired electrons yield the σ bond Table 13.1.

68 P.P. Power, Chem. Rev., 99 (1999) 3463.

69 A thorough discussion of the chemical bond is given byW. Kutzelnigg, The physical origin of the chemical bond, in: Z.B. Maksic (Ed.), The concept of the chemical bond. Part 2, Springer, Berlin, 1990, p. 35.

70 M. Perić, B. Ostojić and B. Engels, J. Mol. Spectrosc., 182 (1997) 295.

71 P. Pyykkö, Chem. Rev., 88 (1988) 563.

72 F. Bessac, doctoral thesis, Universities Marburg and Toulouse, 2004.

73 R.G. Pearson, Introduction to hard and soft acids and bases, Dowden, Hutchinson, Ross, Stroudsburg, PA, 1973.

74 (a) T.J. Kealy and P.L. Pauson, Nature, 168 (1951) 1039; (b)Another paper which reported about the synthesis of ferrocene was submitted earlier than the work of Kealy and Pauson, but it became published later S.A. Miller, J.A. Tebboth and J.F. Tremaine, J. Chem. Soc., (1952) 633.

75 E.O. Fischer and W. Pfab, Z. Naturforsch., 76 (1952) 377.

76 G. Wilkinson, M. Rosenblum, M.C. Whiting and R.B. Woodward, J. Am. Chem. Soc., 74 (1952) 2125.

77 E.M. Shustorovich and E.M. Dyatkina, Dok. Akad. Nauk. SSSR, 128 (1959) 1234.

78 (a) P. Jutzi and N. Burford, Chem. Rev., 99 (1999) 969; (b) P. Jutzi, N. Burford, A. Togni, R.L. Haltermann,Metallocenes, Vol. 1, Wiley–VCH, New York (1998) 3f.

79 O. Kwon, M.L. McKee, in: T.R. Cundari (Ed.), Computational organometallic chemistry, 397f, Marcel Dekker, New York, 2001, p. 397f.

80 A.J. Bridgeman, J. Chem. Soc. Dalton Trans., (1997) 2887.

81 M. Kaupp, Ch. Van Wüllen, R. Franke, F. Schmitz and W. Kutzelnigg, J. Am. Chem. Soc., 118 (1996) 11939.

82 T.A. Albright, J.K. Burdett and M.H. Whangbo, Orbital interactions in chemistry, Wiley, New York, 1985.

83 A review about the nature of the bonding in transition metal compounds has recently been presented by G. Frenking and N. Fröhlich, Chem. Rev., 100 (2000) 717.

84 (a) C.R. Landis, T.K. Firman, D.M. Root and T. Cleveland, J. Am. Chem. Soc., 120 (1998) 1842; (b) C.R. Landis, T. Cleveland and T.K. Firman, J. Am. Chem. Soc., 120 (1998) 2641; (c) T.K. Firman and C.R. Landis, J. Am. Chem. Soc., 120 (1998) 12650.

85 C.A. Bayse and M.B. Hall, J. Am. Chem. Soc., 121 (1999) 1348.

86 M.J.S. Dewar, Bull. Soc. Chim. Fr., 18 (1951) C79.

87 J. Chatt and L.A. Duncanson, J. Chem. Soc., (1953) 2929.

88 For a discussion of the DCD bonding model in the light of quantum chemical calculations see, G. Frenking, in: G.J. Leigh, N. Winterton (Eds.), Modern coordination chemistry: The legacy of Joseph Chatt, The Royal Society, London, 2002.

89 (a) C.W. Bauschlicher and P.S. Bagus, J. Chem. Phys., 81 (1984) 5889; (b) C.W. Bauschlicher, L.G.M. Petterson and P.E.M. Siegbahn, J. Chem. Phys., 87 (1987) 2129; (c) C.W. Bauschlicher, S.R. Langhoff and L.A. Barnes, Chem. Phys., 129 (1989) 431; (d) C.W. Bauschlicher, J. Chem. Phys., 84 (1986) 260; (e) C.W. Bauschlicher, P.S. Bagus, C.J. Nelin and B.O. Roos, J. Chem. Phys., 85 (1986) 354; (f) L.A. Barnes, M. Rosi and C.W. Bauschlicher, J. Chem. Phys., 94 (1991) 2031; (g) L.A. Barnes, M. Rosi and C.W. Bauschlicher, J. Chem. Phys., 93 (1990) 609; (h) L.A. Barnes and C.W. Bauschlicher, J. Chem. Phys., 91 (1989) 314; (i) M.R.A. Blomberg, U.B. Brandemark, P.E.M. Siegbahn, J. Wennerberg

and C.W. Bauschlicher, J. Am. Chem. Soc., 10 (1988) 6650; (j) M.B. Hall and R.F. Fenske, Inorg. Chem., 11 (1972) 1620; (k) P.E. Sherwood and M.B. Hall, Inorg. Chem., 18 (1979) 2325; (l) P.E. Sherwood and M.B. Hall, Inorg. Chem., 19 (1980) 1805; (m) R.L. Williamson and M.B. Hall, Int. J. Quantum Chem., 21S (1987) 503; (n) I.H. Hillier and V.R. Saunders, Mol. Phys., 23 (1971) 1025; (o) I.H. Hillier and V.R. Saunders, J. Chem. Soc., Chem. Commun., (1971) 642; (p) P.C. Ford and I.H. Hillier, J. Chem. Phys., 80 (1984) 5664; (q) P.C. Ford, I.H. Hillier, S.A. Pope and M.F. Guest, Chem. Phys. Lett., 102 (1983) 555; (r) G. Cooper, J.C. Green, M.P. Payne, B.R. Dobson and I.H. Hillier, Chem. Phys. Lett., 125 (1986) 97; (s) G. Cooper, J.C. Green, M.P. Payne, B.R. Dobson, I.H. Hillier, M. Vincent and M. Rosi, J. Chem. Soc., Chem. Commun., (1986) 438; (t) D. Moncrieff, P.C. Ford, I.H. Hillier and V.R. Saunders, J. Chem. Soc., Chem. Commun., (1983) 1108; (u) S. Smith, I.H. Hillier, W. von Niessen and M.C. Guest, Chem. Phys., 135 (1989) 357; (v) L.G. Vanquickenborne and J. Verhulst, J. Am. Chem. Soc., 109 (1987) 4825; (w) K. Pierfoot, J. Verhulst, P. Verbeke and L.G. Vanquickenborne, Inorg. Chem., 28 (1989) 3059; (x) S. Yamamoto and H. Kashiwagi, Chem. Phys. Lett., 205 (1993) 306; (y) M.R.A. Blomberg, P.E.M. Siegbahn, T.L. Lee, A.P. Rendell and J.E. Rice, J. Chem. Phys., 95 (1991) 5898.

90 (a) E.J. Baerends, A. Rozendaal, in: A. Veillard (Ed.), Quantum chemistry: The challenge of transition metals and coordination chemistry, D. Reidel Publishing Company, Dordrecht, 1986, p. 159; (b) E.R. Davidson, K.L. Kunze, F.B.C. Machado and S.J. Chakravorty, Acc. Chem. Res., 26, 1993, pp. 628; (c) K.L. Kunze and E.R. Davidson, J. Phys. Chem., 96, 1992, pp. 2129; (d) F.B.C. Machado and E.R. Davidson, J. Phys. Chem., 97, 1993, pp. 4397; (e) J. Li, G. Schreckenbach and T. Ziegler, J. Am. Chem. Soc., 117, 1995, pp. 486; (f) S.-C. Chung, S. Krüger, S.Ph. Ruzankin, G. Pacchioni and N. Rösch, Chem. Phys. Lett., 248, 1996, pp. 109.

91 A charge decomposition analysis of positively and negatively charged carbonyl complexes is presented by: R.K. Szilagy and G. Frenking, Organometallics, 16 (1997) 4807.

92 A.J. Lupinetti, S.H. Strauss, G. Frenking, Prog. Inorg. Chem. Vol. 49, p.1, K.D. Karlin (Ed), 2001.

93 J. Weiß, D. Stetzkamp, B. Nuber, R.A. Fischer, C. Boehme and G. Frenking, Angew. Chem., 109 (1997) 95, Angew. Chem., Int. Ed. Engl. 36 (1997) 70.

94 H. Braunschweig, C. Kollann and U. Englert, Angew. Chem., 110 (1998) 3355, Angew. Chem., Int. Ed. Engl. 37 (1998) 3179.

95 J. Su, X.-W. Li, R.C. Crittendon, C.F. Campana and G.H. Robinson, Organometallics, 16 (1997) 4511.

96 F.A. Cotton and X. Feng, Organometallics, 17 (1998) 128.

97 (a) W. Uhl, M. Pohlmann and R. Wartchow, Angew. Chem., 110 (1998) 1007, Angew. Chem., Int. Ed. Engl. 37 (1998) 961; (b) W. Uhl, M. Benter, S. Melle, W. Saak, G. Frenking and J. Uddin, Organometallics, 18 (1999) 3778; (c) W. Uhl and S. Melle, Z. Anorg. Allg. Chem., 626 (2000) 2043.

98 (a) K.H. Dötz, H. Fischer, P. Hofmann, F.R. Kreissl, U. Schubert and K. Weiss, Transition metal carbene complexes, Verlag Chemie, Weinheim, 1983; (b) R. Schrock, Acc. Chem. Res., 12, 1979, p. 98; (c) W.A. Nugent and J.M. Mayer, Metal-ligand multiple bonds, Wiley, New York, 1988; (d) H. Fischer, P. Hofmann, F.R. Kreissl, R.R. Schrock, U. Schubert and K. Weiss, Carbyne complexes, VCH, New York, 1988; (e) in: F.R. Kreissl (Ed.), Transition metal carbyne complexes, Kluwer Academic Publishers, Dordrecht, The Netherlands, 1993.

99 (a) A. Haaland, Angew. Chem. Int. Ed., 28 (1989) 992; (b) U. Pidun and G. Frenking, Organometallics, 14 (1995) 5325; (c) U. Pidun and G. Frenking, J. Organomet. Chem., 525 (1996) 269; (d) U. Pidun and G. Frenking, J. Chem. Soc., Dalton Trans., (1997) 1653; (e) N. Fröhlich and G. Frenking, Chem. Rev., 100 (2000) 717.

100 (a) R.H. Hertwig, W. Koch, D. Schröder, H. Schwarz, J. Hrusak and P. Schwerdtfeger, J. Phys. Chem., 100 (1996) 12253; (b) M. Böhme, T. Wagener and G. Frenking, J. Organomet. Chem., 520 (1996) 31; (c) M. Sodupe and C.W. Bauschlicher, J. Phys. Chem., 95 (1991) 8640; (d) J. Miralles-Sabater, M. Merchán, I. Nesbot-Gil and P. Viruela-Martn, J. Phys. Chem., 92 (1988) 4853.

101 (a) J. Li, G. Schreckenbach and T. Ziegler, Inorg. Chem., 34 (1995) 3245; (b) F.U. Axe and D.S. Marynick, J. Am. Chem. Soc., 106 (1984) 6230; (c) A.S. Decker and M. Klobukowski, J. Am. Chem. Soc., 120 (1998) 9342; (d) Y. Chen, M. Hartmann and G. Frenking, Z. Anorg. Allg. Chem., 627 (2001) 985; (e) B.R. Bender, J.R. Norton, M.M. Miller, O.P. Anderson and A.K. Rappé, Organometallics, 11 (1992) 3427; (f) D.L. Cedeno and E. Weitz, J. Am. Chem. Soc., 123 (2001) 12857.

102 (a) P. Fantucci, Comments Inorg. Chem., 13 (1992) 241; (b) P.B. Dias, M.E. Minas de Piedade and J.A. Martinho Simoes, Coord. Chem. Rev., 135/136 (1994) 738; (c) E.C. Alyea and S. Song, Comments Inorg. Chem., 18 (1996) 145; (d) E.C. Alyea and S. Song, Comments Inorg. Chem., 18 (1996) 189.

103 F.A. Cotton and C.S. Kraihanzel, J. Am. Chem. Soc., 84 (1962) 4432.

104 J.E. Huheey, E.A. Keiter and R.L. Keiter, Inorganic chemistry: Principles and structure of reactivity, 4th edn., Harper Collins College Publishers, New York, 1993, p. 431.

105 (a) E.C. Alyea and S. Song, Inorg. Chem., 34 (1995) 3864; (b) E.C. Alyea and S. Song, Comments Inorg. Chem., 18 (1996) 145; (c) E.C. Alyea and S. Song, Comments Inorg. Chem., 18 (1996) 189.

106 Y. Ruiz-Morales and T. Ziegler, J. Phys. Chem. A, 102 (1998) 3970.

107 G.J. Kubas, R.R. Ryan, B.I. Swanson, P.J. Vergamini and H.J. Wasserman, J. Am. Chem. Soc., 106 (1984) 451.

108 (a) D.M. Heinesky and W.J. Oldham, Jr., Chem. Rev., 93 (1993) 913; (b) G.J. Kubas, Metal dihydrogen and σ-bond complexes, Kluwer Academic/Plenum Publishers, New York, 2001; (c) J.K. Burdett, O. Eisenstein, S.A. Jackson, in: A. Dedieu (Ed.), Transition metal hydrides, VCH, New York, 1992, p. 149; (d) F. Maseras, A. Lledós, E. Clot and O. Eisenstein, Chem. Rev., 100 (2000) 601.

109 S. Dapprich and G. Frenking, Angew. Chem., 107 (1995) 383, Angew. Chem., Int. Ed. Engl. 34 (1995) 354.

110 S. Dapprich and G. Frenking, Organometallics, 15 (1996) 4547.

111 E.O. Fischer and W. Hafner, Z. Naturfosch., 10b (1955) 665.

112 E. Urnėžius, W.W. Brennessel, C.J. Cramer, J.E. Ellis and P.v.R. Schleyer, Science, 295 (2002) 832.

113 R.F.W. Bader, Atoms in molecules. A quantum theory, Oxford University Press, Oxford, 1990.

114 C.A. Coulson, After-dinner speech at the Conference on Molecular Quantum Mechanics in Boulder, Colorado, 1960.

Theory and Applications of Computational Chemistry: The First Forty Years
Edited by C. Dykstra et al.

CHAPTER 14

Superoperator many-body theory of molecular currents: non-equilibrium Green functions in real time

Upendra Harbola and Shaul Mukamel

Department of Chemistry, University of California, Irvine, CA 92697-2025, USA

Abstract

The electric conductance of a molecular junction is calculated by recasting the Keldysh formalism in Liouville space. Dyson equations for non-equilibrium many-body Green functions (NEGF) are derived directly in real (physical) time. The various NEGFs appear naturally in the theory as time-ordered products of superoperators, while the Keldysh forward/backward time loop is avoided.

14.1 INTRODUCTION

Recent advances in the fabrication and measurements of nanoscale devices have led to a considerable interest in non-equilibrium current-carrying states of single molecules. The tunneling of electrons between two metals separated by a thin oxide layer was first observed experimentally by Giaever [1] and later by others [2]. Vibrational resonances can be observed for molecules absorbed at the metal–oxide interface by analyzing the tunneling current as a function of the applied bias [3,4]. More recent development of scanning tunneling microscopy (STM) led to a direct, real space determination of surface structures. A metal tip is brought near the surface so that tunneling resistance is measurable. A contour map of the surface is obtained by recording the tunneling resistance as the tip scans the surface. The tunneling electrons interact and may exchange energy with the nuclear degrees of freedom of the absorbed molecule. This opens up inelastic channels for electron transmission from the tip to the surface, leading to inelastic electron tunneling (IET). IET may play an important role in manipulating molecules with STM [5,6]. Recently, IET was combined with STM for the chemical analysis of a single absorbed molecule with atomic spatial resolution [7,8]. The atomic-scale STM images have been useful in analyzing different binding configurations of the molecule chemisorbed on a semiconductor surface [9,10]. Recent advances in the theory of STM are reviewed in Ref. [11].

Electron tunneling was first analyzed by Bardeen [12] and Cohen et al. [13] using the perturbative transfer Hamiltonian (TH) approach and more recently by many other authors [14–16]. Although the TH gives, in many cases, a good description of the observed effects, it lacks a firm first principles theoretical basis and does not account properly for many-body effects [17]. An improved form of TH [18] that involved energy dependent transfer matrix elements was used to incorporate many-body effects. However, this model does not describe the electron–phonon interaction properly [19].

A many-body non-equilibrium Green functions (NEGF) formulation of electron tunneling was proposed by Caroli et al. [20]. The NEGF theory was originated by Schwinger [21] and Kaddanof and Baym [22], and developed further by Keldysh [23] and Craig [24]. This formalism involves the calculation of four basic Green functions, time ordered (G^T), anti-time ordered ($G^{\tilde{T}}$), lesser ($G^<$) and greater ($G^>$). Additional retarded (G^r) and advanced (G^a) Green functions are defined as specific combinations of these basic functions. At equilibrium it is sufficient to know only the retarded or advanced Green functions; all other Green functions simply follow from the fluctuation–dissipation theorem that connects the 'lesser' and 'greater' with the retarded Green function through the equilibrium Fermi distribution function ($f_0(E)$) [25]. However, for non-equilibrium measurements, where the distribution function is not known *a priori*, one needs to solve for the various NEGFs self-consistently.

Electronic transport in molecular wires and STM currents of single molecules have received considerable attention [26–31]. Electron transport through a single molecule [32–34] or a chain of several atoms [35] was studied. From a theoretical point of view, this is very similar to the electron tunneling in semiconductor junctions and various theories developed for STM [36,37] can directly be applied to molecular wires. The NEGF technique developed for tunneling currents has been used to analyze the electron conduction through a single molecule attached to electrodes [26,36,38–42]. The method has also been combined with density functional theory for the modeling of transport in molecular devices [43,44].

In this chapter, we develop a non-equilibrium superoperator Green function theory [23,25,45] (NESGFT) of molecular currents [46]. A notable advantage of working with superoperators in the higher dimensional Liouville space [44,47] is that we need to consider only time-ordered quantities in real (physical) time; all NEGFs show up naturally without introducing artificial time variables. Observables can be expressed in terms of various *Liouville space pathways* (LSP) [46]. The ordinary (causal) response function which represents the density response to an external field is one particular combination of these LSPs. Other combinations represent the spontaneous density fluctuations and the response of these fluctuations to the external field [47,48]. A simple time ordering operation of superoperators in real time is all it takes to derive the non-equilibrium theory, avoiding the Keldysh loop or Matsubara imaginary time. The NESGF theory provides new physical insights into the mechanism of the current. It can also be more naturally used to interpret time domain experiments involving external pulses.

In Section 14.2, we give a brief introduction to the superoperator formalism and recast the NEGFT in terms of the superoperator Green functions. Starting from the microscopic definitions for various NESGF, we construct equations of motion and obtain the Dyson matrix equation of Keldysh which couples the various NESGFs. In Section 14.3, we apply the NESGFT to the conduction through a molecular junction. In Section 14.4 we end with a discussion.

14.2 DYSON EQUATIONS FOR SUPEROPERATOR GREEN FUNCTIONS

We consider a system of externally-driven electrons and phonons described by the Hamiltonian [16,23,26,38],

$$H = H_0 + H_{\mathrm{ep}} + H_{\mathrm{ex}} \tag{1}$$

where H_0 represents the non-interacting electrons and phonons,

$$H_0 = \int \mathrm{d}\mathbf{r}\psi^\dagger(\mathbf{r})h_0(\mathbf{r})\psi(\mathbf{r}) + \int \mathrm{d}\mathbf{r}\phi^\dagger(\mathbf{r})\Omega_0(\mathbf{r})\phi(\mathbf{r}) \tag{2}$$

$h_0(\mathbf{r}) = (-\hbar^2/2m)\nabla^2$ is the kinetic energy and m is the electron mass. $\psi(\psi^\dagger)$ represent the annihilation (creation) operators which satisfy the Fermi anticommutation relations,

$$\begin{aligned} \psi(\mathbf{r})\psi^\dagger(\mathbf{r}') + \psi^\dagger(\mathbf{r}')\psi(\mathbf{r}) &= \delta(\mathbf{r} - \mathbf{r}') \\ \psi^\dagger(\mathbf{r})\psi^\dagger(\mathbf{r}') + \psi^\dagger(\mathbf{r}')\psi^\dagger(\mathbf{r}) &= 0 \\ \psi(\mathbf{r})\psi(\mathbf{r}') + \psi(\mathbf{r}')\psi(\mathbf{r}) &= 0 \end{aligned} \tag{3}$$

and $\phi(\phi^\dagger)$ are boson operators with the commutation relations,

$$\begin{aligned} \phi(\mathbf{r})\phi^\dagger(\mathbf{r}') - \phi^\dagger(\mathbf{r}')\phi(\mathbf{r}) &= \delta(\mathbf{r} - \mathbf{r}') \\ \phi^\dagger(\mathbf{r})\phi^\dagger(\mathbf{r}') - \phi^\dagger(\mathbf{r}')\phi^\dagger(\mathbf{r}) &= 0 \\ \phi(\mathbf{r})\phi(\mathbf{r}') - \phi(\mathbf{r}')\phi(\mathbf{r}) &= 0 \end{aligned} \tag{4}$$

The second term in Eq. (1) denotes the electron–phonon interaction,

$$H_{\mathrm{ep}} = \int \mathrm{d}\mathbf{r}\lambda(\mathbf{r})[\phi^\dagger(\mathbf{r}) + \phi(\mathbf{r})]\psi^\dagger(\mathbf{r})\psi(\mathbf{r}) \tag{5}$$

where $\lambda(\mathbf{r})$ is the coupling strength. Finally, H_{ex} represents the coupling to a time-dependent external potential $\xi(\mathbf{r}, t)$,

$$H_{\mathrm{ex}} = \int \mathrm{d}\mathbf{r}\xi(\mathbf{r}, t)\psi^\dagger(\mathbf{r})\psi(\mathbf{r}) \tag{6}$$

We next briefly survey some properties of Liouville space superoperators that will be useful in the following derivations [49]. The elements of the Hilbert space $N \times N$ density matrix, $\rho(t)$, are arranged as a Liouville space vector (bra or ket) of length N^2. Operators of $N^2 \times N^2$ dimension in this space are denoted as superoperators. With any Hilbert space operator A, we associate two superoperators A_{L} (left) and A_{R} (right) defined through their action on another operator X as

$$A_{\rm L}X \equiv AX \qquad \text{and} \qquad A_{\rm R}X \equiv XA \tag{7}$$

We further define symmetric and antisymmetric combinations of these superoperators

$$A_{+} = \frac{1}{2}(A_{\rm L} + A_{\rm R}) \qquad \text{and} \qquad A_{-} = (A_{\rm L} - A_{\rm R}) \tag{8}$$

The commutator and anticommutator operations in Hilbert space can thus be implemented with a single multiplication by a '−' and '+' superoperator, respectively. We further introduce the Liouville space–time ordering operator $\mathcal{T}$. This is a key ingredient for extending NEGFT to superoperators: when applied to a product of superoperators it reorders them so that time increases from right to left. We define $\langle A(t)\rangle \equiv \mathrm{Tr}\{A(t)\rho_{\rm eq}\}$ where $\rho_{\rm eq} = \rho(t=0)$ represents the equilibrium density matrix of the electron-phonon system. It is straightforward to see that for any two operators A and B we have

$$\langle \mathcal{T} A_{+}(t)B_{-}(t')\rangle = 0, \quad t' > t \tag{9}$$

$\langle \mathcal{T} A_{+}(t)B_{-}(t')\rangle$ is thus always a *retarded* function. This follows from the definitions (8): since a '−' superoperator corresponds to a commutator in Hilbert space, this implies that for $t < t'$, $\langle A_{+}(t)B_{-}(t')\rangle$ becomes a trace of a commutator and must vanish, i.e.

$$\begin{aligned}\langle \mathcal{T} A_{+}(t)B_{-}(t')\rangle &= \mathrm{Tr}\{B_{-}(t')A_{+}(t)\rho_{\rm eq}\} \quad t < t' \\ &= \frac{1}{2}\mathrm{Tr}\{[B(t'), A(t)\rho_{\rm eq} + \rho_{\rm eq}A(t)]\} = 0\end{aligned}$$

Similarly, it follows that the trace of two 'minus' operators always vanishes

$$\langle \mathcal{T} A_{-}(t)B_{-}(t')\rangle = 0 \qquad \text{for all } t \text{ and } t' \tag{10}$$

We shall make use of Eqs. (9) and (10) in discussing the retarded and advanced Green functions in Appendix 14D. Superoperator algebra was surveyed in Ref. [49].

In Liouville space the density matrix, $\rho(t)$ is a vector whose time dependence is given by

$$\rho(t) = \mathcal{G}(t, t_0)\rho(t_0) \tag{11}$$

with the Green function,

$$\mathcal{G}(t, t_0) = \mathcal{T}\exp\left\{-\frac{\mathrm{i}}{\hbar}\int_{t_0}^{t} \mathcal{H}_{-}(\tau)\mathrm{d}\tau\right\} \tag{12}$$

and $\mathcal{H}_{-}$ is the superoperator corresponding to the Hamiltonian (Eq. (1)). Note that unlike Hilbert space, where time dependence of the ket and the bra is governed by forward and backward time-evolution operators, respectively, in Liouville space one keeps track of both bra and ket simultaneously and the density matrix needs only to be propagated forward in time (Eq. (11)).

To introduce the interaction picture in Liouville space we partition $\mathcal{H}_{-} = \mathcal{H}_{0-} + \mathcal{H}'_{-}$ corresponding to the non-interacting ($\mathcal{H}_0$) and interacting ($\mathcal{H}' = \mathcal{H}_{\rm ep} + \mathcal{H}_{\rm ex}$)

Hamiltonians. With this partitioning, Eq. (12) can be written as

$$\mathcal{G}(t, t_0) = \mathcal{G}_0(t, t_0)\mathcal{G}_\mathrm{I}(t, t_0) \tag{13}$$

where $\mathcal{G}_0$ represents the time evolution with respect to H_0,

$$\mathcal{G}_0(t, t_0) = \theta(t - t_0) \exp\left\{-\frac{\mathrm{i}}{\hbar}\mathcal{H}_{0-}(t - t_0)\right\} \tag{14}$$

$\mathcal{G}_\mathrm{I}(t, t_0)$ is the time evolution operator in the interaction picture,

$$\mathcal{G}_\mathrm{I}(t, t_0) = \mathcal{T} \exp\left\{-\frac{\mathrm{i}}{\hbar}\int_{t_0}^{t} \tilde{\mathcal{H}}'_-(\tau)\mathrm{d}\tau\right\} \tag{15}$$

and $\tilde{\mathcal{H}}'_-$ is the interaction picture representation of $\mathcal{H}'_-$. We shall denote superoperators in the interaction picture by a ($\tilde{\ }$),

$$\tilde{A}_\alpha(t) \equiv \mathcal{G}_0^\dagger(t, t_0)A_\alpha(t_0)\mathcal{G}_0(t, t_0) \tag{16}$$

where $\alpha = +, -$ or L, R. Superoperators in the Heisenberg picture will be represented by a caret

$$\hat{A}_\alpha(t) \equiv \mathcal{G}^\dagger(t, t_0)A_\alpha(t_0)\mathcal{G}(t, t_0) \tag{17}$$

By adiabatic switching of the interaction $\mathcal{H}'_-$ starting at $t_0 = -\infty$ we have

$$\rho(t) = \rho_0 - \frac{\mathrm{i}}{\hbar}\int_{-\infty}^{t} \mathrm{d}\tau\mathcal{G}_0(t, \tau)\mathcal{H}'_-(\tau)\rho(\tau) \tag{18}$$

where $\rho_0 = \rho(-\infty)$ is the equilibrium density matrix of the non-interacting system

$$\rho_0 = \frac{\exp(-\beta H_0)}{\mathrm{Tr}\{\exp(-\beta H_0)\}} \tag{19}$$

An iterative solution of Eq. (18) yields

$$\rho(t) = \mathcal{G}_0(t, -\infty)\mathcal{G}_\mathrm{I}(t, -\infty)\rho_0 \tag{20}$$

which can also be obtained by applying the time evolution operator (13) to $\rho(t_0)$ and setting $t_0 = -\infty$. Using Eq. (20), the equilibrium density matrix of the interacting system can be generated from the non-interacting one by switching on the interactions adiabatically, starting at $t = -\infty$. The external potential is constant in time for $t < 0$ and is assumed to be time dependent only for $t > 0$ in the interaction picture. We then get

$$\tilde{\rho}_\mathrm{eq} = \mathcal{G}_\mathrm{I}(0, -\infty)\rho_0 \tag{21}$$

This *adiabatic connection formula* [49] is very useful for calculating expectation values using the interaction picture. In the corresponding Gellman-law expression in Hilbert space [50] there is an extra denominator that takes care of the phase of the wavefunction.

This is not necessary in Liouville space, since the density matrix does not acquire such a phase.

In the Heisenberg picture, the expectation value of an operator $\hat{A}_\alpha(t)$ is given by

$$\langle \hat{A}_\alpha(t) \rangle \equiv \mathrm{Tr}\{\hat{A}_\alpha(t)\rho_{\mathrm{eq}}\} \tag{22}$$

where $\rho_{\mathrm{eq}} = \rho(t=0)$. Using Eqs. (16), (17) and (20), this can be recast in the interaction picture as

$$\langle \hat{A}_\alpha(t) \rangle = \mathrm{Tr}\{\tilde{A}_\alpha(t)\mathcal{G}_{\mathrm{I}}(t,-\infty)\rho_0\} \equiv \langle \tilde{A}_\alpha(t)\mathcal{G}_{\mathrm{I}}(t,-\infty)\rangle_0 \tag{23}$$

Eq. (23) is a good starting point for developing a perturbation theory around the non-interacting system. Through Eqs. (22) and (23) we also define the expectation values $\langle \cdots \rangle$ and $\langle \cdots \rangle_0$. While the former represents the trace with respect to the interacting density matrix, the latter is defined with respect to the non-interacting density matrix. This will be used in the following.

Corresponding to the Hilbert space electron and phonon operators, $\hat{\psi}$, $\hat{\psi}^\dagger$, $\hat{\phi}$ and $\hat{\phi}^\dagger$ we define 'left' ($\alpha = \mathrm{L}$) and 'right' ($\alpha = \mathrm{R}$) superoperators, $\hat{\psi}_\alpha$, $\hat{\psi}^\dagger_\alpha$, $\hat{\phi}_\alpha$ and $\hat{\phi}^\dagger_\alpha$. The dynamics of a superoperator, $\hat{\psi}_\alpha$, is described by the generalized Liouville equation,

$$-\mathrm{i}\hbar\frac{\partial \hat{\psi}_\alpha(t)}{\partial t} = [\mathcal{H}_-(t), \hat{\psi}_\alpha(t)] = \mathcal{H}_-(t)\hat{\psi}_\alpha(t) - \hat{\psi}_\alpha(t)\mathcal{H}_-(t) \tag{24}$$

where $\mathcal{H}_-$ is the superoperator corresponding to the Hamiltonian given in Eq. (1). A similar equation can be written down for the phonon superoperators. In order to evaluate the commutator appearing in the RHS of Eq. (1), we need the commutation relations of superoperators [51]. The 'left' and the 'right' operators always commute. Thus, for $\alpha \neq \beta$ we have

$$\begin{aligned}
&[\psi_\alpha(\mathbf{r}), \psi_\beta(\mathbf{r}')] = [\psi^\dagger_\alpha(\mathbf{r}), \psi^\dagger_\beta(\mathbf{r}')] = [\psi^\dagger_\alpha(\mathbf{r}), \psi_\beta(\mathbf{r}')] = 0 \\
&[\phi_\alpha(\mathbf{r}), \phi_\beta(\mathbf{r}')] = [\phi^\dagger_\alpha(\mathbf{r}), \phi^\dagger_\beta(\mathbf{r}')] = [\phi^\dagger_\alpha(\mathbf{r}), \phi_\beta(\mathbf{r}')] = 0
\end{aligned} \tag{25}$$

For Fermi superoperators we have

$$\begin{aligned}
&\psi_\alpha(\mathbf{r})\psi_\alpha(\mathbf{r}') + \psi_\alpha(\mathbf{r}')\psi_\alpha(\mathbf{r}) = 0 \\
&\psi^\dagger_\alpha(\mathbf{r})\psi^\dagger_\alpha(\mathbf{r}') + \psi^\dagger_\alpha(\mathbf{r}')\psi^\dagger_\alpha(\mathbf{r}) = 0 \\
&\psi_\alpha(\mathbf{r})\psi^\dagger_\alpha(\mathbf{r}') + \psi_\alpha(\mathbf{r}')\psi^\dagger_\alpha(\mathbf{r}) = \delta(\mathbf{r}-\mathbf{r}')
\end{aligned} \tag{26}$$

Similarly for the boson operators

$$\begin{aligned}
&\phi^\dagger_\alpha(\mathbf{r})\phi^\dagger_\alpha(\mathbf{r}') - \phi^\dagger_\alpha(\mathbf{r}')\phi^\dagger_\alpha(\mathbf{r}) = 0 \\
&\phi_\alpha(\mathbf{r})\phi_\alpha(\mathbf{r}') - \phi_\alpha(\mathbf{r}')\phi_\alpha(\mathbf{r}) = 0 \\
&\phi_\alpha(\mathbf{r})\phi^\dagger_\alpha(\mathbf{r}') - \phi^\dagger_\alpha(\mathbf{r}')\phi_\alpha(\mathbf{r}) = \kappa_\alpha\delta(\mathbf{r}-\mathbf{r}')
\end{aligned} \tag{27}$$

Here $\kappa_\alpha = -1$ for $\alpha = \mathrm{R}$ and unity for $\alpha = \mathrm{L}$.

Using the commutation relations (25) and (26) and the identity

$$(XY...Z)_\alpha = X_\alpha Y_\alpha ... Z_\alpha, \quad \alpha = \mathrm{L}, \mathrm{R} \tag{28}$$

we can recast $\mathcal{H}_-$ in terms of the elementary field superoperators,

$$\mathcal{H}_- = \mathcal{H}_{0-} + \mathcal{H}_-^{\mathrm{ep}} + \mathcal{H}_-^{\mathrm{ex}} \tag{29}$$

with

$$\begin{aligned}
\mathcal{H}_{0-} &= \sum_{\alpha=\mathrm{L,R}} \kappa_\alpha \int \mathrm{d}\mathbf{r}(\psi_\alpha^\dagger(\mathbf{r})h_0(\mathbf{r})\psi_\alpha(\mathbf{r}) + \phi_\alpha^\dagger(\mathbf{r})\Omega_0(\mathbf{r})\phi_\alpha(\mathbf{r})) \\
\mathcal{H}_-^{e-p} &= \sum_{\alpha=\mathrm{L,R}} \kappa_\alpha \int \mathrm{d}\mathbf{r}\lambda(\mathbf{r})\Phi_\alpha(\mathbf{r})\psi_\alpha^\dagger(\mathbf{r})\psi_\alpha(\mathbf{r}) \\
\mathcal{H}_-^{\mathrm{ex}} &= \sum_{\alpha=\mathrm{L,R}} \kappa_\alpha \int \mathrm{d}\mathbf{r}\psi_\alpha^\dagger(\mathbf{r})\xi(\mathbf{r},t)\psi_\alpha(\mathbf{r})
\end{aligned} \tag{30}$$

where $\Phi_\alpha = \phi_\alpha + \phi_\alpha^\dagger$.

We next define electron and phonon superoperator Green functions

$$\begin{aligned}
G_{\alpha\beta}(\mathbf{r}t,\mathbf{r}'t') &= -\frac{\mathrm{i}}{\hbar}\langle \mathcal{T}\hat{\psi}_\alpha(\mathbf{r},t)\hat{\psi}_\beta^\dagger(\mathbf{r}',t')\rangle \\
D_{\alpha\beta}(\mathbf{r}t,\mathbf{r}'t') &= -\frac{\mathrm{i}}{\hbar}\langle \mathcal{T}\hat{\Phi}_\alpha(\mathbf{r},t)\hat{\Phi}_\beta^\dagger(\mathbf{r}',t')\rangle
\end{aligned} \tag{31}$$

As shown in Ref. [44] (see Appendix 14A), G_{LL}, G_{RR}, G_{LR} and G_{RL}, respectively, coincide with the standard Hilbert space–time ordered G^T, antitime ordered $G^{\tilde{T}}$, lesser $G^<$ and greater $G^>$ Green functions defined on a closed time loop.

Using the commutation relations (3), the Heisenberg equations of motion for superoperator $\hat{\psi}_\alpha(t)$ reads

$$\mathrm{i}\hbar\kappa_\alpha \frac{\partial \hat{\psi}_\alpha(\mathbf{r},t)}{\partial t} = h(\mathbf{r},t)\hat{\psi}_\alpha(\mathbf{r},t) + \lambda(\mathbf{r})\hat{\Phi}_\alpha(\mathbf{r},t)\hat{\psi}_\alpha(\mathbf{r},t) \tag{32}$$

where $h(\mathbf{r},t) \equiv h_0(\mathbf{r}) + \xi(\mathbf{r},t)$. By taking the time derivative of the electron Green function in Eq. (31) and using Eq. (32), we obtain the equation of motion for $G_{\alpha\beta}$,

$$\begin{aligned}
&\left(\mathrm{i}\hbar\frac{\partial}{\partial t} - \kappa_\alpha h(\mathbf{r},t)\right)G_{\alpha\beta}(\mathbf{r}t,\mathbf{r}',t') \\
&\quad = \delta_{\alpha\beta}\delta(\mathbf{x}-\mathbf{x}') - \frac{\mathrm{i}}{\hbar}\kappa_\alpha\lambda(\mathbf{r})\langle \mathcal{T}\hat{\Phi}_\alpha(\mathbf{r},t)\hat{\psi}_\alpha(\mathbf{r},t)\hat{\psi}_\beta^\dagger(\mathbf{r}',t')\rangle
\end{aligned} \tag{33}$$

In order to derive the equation of motion for the phonon Green function D_α, we add the following coupling term

$$\int \mathrm{d}\mathbf{r}J(\mathbf{r},t)\Phi(\mathbf{r}) \tag{34}$$

to the Hamiltonian. Here $J(\mathbf{r},t)$ is some artificial field that will be set to zero at the end of calculations. This new term does not effect the electron Green functions since Φ and ψ

commute. Thus, the superoperator $\mathcal{H}_-^{\mathrm{ex}}$ in Eq. (30) takes the form

$$\mathcal{H}_-^{\mathrm{ex}} = \sum_{\alpha=\mathrm{L,R}} \kappa_\alpha \int \mathbf{dr}[\psi_\alpha^\dagger(\mathbf{r})\xi(\mathbf{r},t)\psi_\alpha(\mathbf{r}) + J_\alpha(\mathbf{r},t)\Phi_\alpha(\mathbf{r})] \tag{35}$$

where fields J_L and J_R couple to the 'left' and 'right' boson operators, respectively. Using the boson commutation (Eq. (26)), the Heisenberg equation for the superoperator ϕ_α is obtained as

$$-\mathrm{i}\hbar\kappa_\alpha \frac{\partial\hat{\phi}_\alpha(\mathbf{r},t)}{\partial t} = \Omega_0(\mathbf{r})\hat{\phi}_\alpha(\mathbf{r},t) + \lambda(\mathbf{r})\hat{\psi}_\alpha^\dagger(\mathbf{r},t)\hat{\psi}_\alpha(\mathbf{r},t) + J_\alpha(\mathbf{r},t) \tag{36}$$

Using Eq. (36) we next obtain the equation of motion for operator Φ_α

$$-\frac{\hbar^2\kappa_\alpha}{\Omega_0(\mathbf{r})}\frac{\partial^2\hat{\Phi}_\alpha(\mathbf{r},t)}{\partial t^2} = \Omega_0(\mathbf{r})\hat{\Phi}_\alpha(\mathbf{r},t) + 2\lambda(\mathbf{r})\hat{\psi}_\alpha^\dagger(\mathbf{r},t)\hat{\psi}_\alpha(\mathbf{r},t) + 2J_\alpha(\mathbf{r},t) \tag{37}$$

Taking a trace with respect to the density matrix $\rho(t=0)$ we obtain

$$-\frac{1}{2}\left(\frac{\hbar^2\kappa_\alpha}{\Omega_0(\mathbf{r})}\frac{\partial^2}{\partial t^2} + \Omega_0(\mathbf{r})\right)\langle\hat{\Phi}_\alpha(\mathbf{r},t)\rangle = \lambda(\mathbf{r})\langle\hat{\psi}_\alpha^\dagger(\mathbf{r},t)\hat{\psi}_\alpha(\mathbf{r},t)\rangle + J_\alpha(\mathbf{r},t) \tag{38}$$

Using the interaction picture representation (Eq. (23)) with $\mathcal{H}_-^{\mathrm{ex}}$ given by Eq. (35), we can write

$$\langle\hat{\Phi}_\alpha(\mathbf{r},t)\rangle = \langle\tilde{\Phi}_\alpha(\mathbf{r},t)\mathcal{G}_\mathrm{I}(t,-\infty)\rangle_0 \tag{39}$$

By taking the functional derivative with respect to J_β, and setting $J_\mathrm{L} = J_\mathrm{R} = 0$, we obtain

$$\begin{aligned}\left|\frac{\delta}{\delta J_\beta(\mathbf{r}',t')}\langle\hat{\Phi}_\alpha(\mathbf{r},t)\rangle\right|_{J_\mathrm{L}=J_\mathrm{R}=0} &= -\frac{\mathrm{i}}{\hbar}\kappa_\beta\langle\mathcal{T}\tilde{\Phi}_\alpha(\mathbf{r},t)\tilde{\Phi}_\beta^\dagger(\mathbf{r}',t')\mathcal{G}_\mathrm{I}(t,-\infty)\rangle_0 \\ &= \kappa_\beta D_{\alpha\beta}(\mathbf{r}t,\mathbf{r}'t')\end{aligned} \tag{40}$$

Using Eqs. (38)–(40), the equation of motion for the phonon Green function is obtained as

$$\begin{aligned}&-\frac{1}{2}\left(\frac{\hbar^2\kappa_\beta}{\Omega_0(\mathbf{r})}\frac{\partial^2}{\partial t^2} + \Omega_0(\mathbf{r})\right)D_{\alpha\beta}(\mathbf{r}t,\mathbf{r}',t') \\ &= \delta_{\alpha\beta}\delta(\mathbf{x}-\mathbf{x}') + \frac{\mathrm{i}}{\hbar}\lambda(\mathbf{r})\kappa_\alpha\langle\mathcal{T}\hat{\psi}_\alpha^\dagger(\mathbf{r},t)\hat{\psi}_\alpha(\mathbf{r},t)\hat{\Phi}_\beta^\dagger(\mathbf{r}',t')\rangle\end{aligned} \tag{41}$$

We shall denote the space and time coordinates collectively by $\mathbf{x} = \mathbf{r}, t$; thus in Eqs. (33) and (41) $\delta(\mathbf{x}-\mathbf{x}') \equiv \delta(\mathbf{r}-\mathbf{r}')\delta(t-t')$.

Following Keldysh, we shall rearrange the superoperator Green functions in a 2×2 matrix $\bar{G}$,

$$\bar{G}(\mathbf{x},\mathbf{x}') = \begin{pmatrix} G_{\mathrm{LL}}(\mathbf{x},\mathbf{x}') & G_{\mathrm{LR}}(\mathbf{x},\mathbf{x}') \\ G_{\mathrm{RL}}(\mathbf{x},\mathbf{x}') & G_{\mathrm{RR}}(\mathbf{x},\mathbf{x}') \end{pmatrix} \tag{42}$$

and similarly for the phonon Green function matrix $\bar{D}$ with elements $D_{\alpha\beta}$. The corresponding Green functions of the non-interacting system described by the Hamiltonian (2) are denoted by $\bar{G}^0$ and $\bar{D}^0$, respectively. These are given by

$$G^0_{\alpha\beta}(\mathbf{r}t,\mathbf{r}'t') = \left(\mathrm{i}\hbar\frac{\partial}{\partial t} - \kappa_\alpha h(\mathbf{r},t)\right)^{-1}\delta_{\alpha\beta}\delta(\mathbf{x}-\mathbf{x}')$$
$$D^0_{\alpha\beta}(\mathbf{r}t,\mathbf{r}'t') = -\frac{1}{2}\left(\frac{\hbar^2\kappa_\beta}{\Omega_0(\mathbf{r})}\frac{\partial^2}{\partial t^2} + \Omega_0(\mathbf{r})\right)^{-1}\delta_{\alpha\beta}\delta(\mathbf{x}-\mathbf{x}') \tag{43}$$

Using our matrix notation, we can recast Eqs. (33) and (41) in the form of Dyson equations

$$\bar{G} = \bar{G}^0 + \bar{G}^0\bar{\Sigma}\bar{G},$$
$$\bar{D} = \bar{D}^0 + \bar{D}^0\bar{\Pi}\bar{D} \tag{44}$$

The effect of all interactions is now included in the electron ($\bar{\Sigma}$) and phonon ($\bar{\Pi}$) self-energies. Formal expressions for the self-energies are obtained by comparing Eqs. (33) and (41) with Eq. (44)

$$\Sigma_{\alpha\beta}(\mathbf{r}t,\mathbf{r}'t') = -\frac{\mathrm{i}}{\hbar}\kappa_\alpha\lambda(\mathbf{r})\sum_{\beta'}\int \mathrm{d}\tau\int \mathrm{d}\mathbf{r}_1\langle\mathcal{T}\hat{\Phi}^\dagger_\alpha(\mathbf{r},t)\hat{\psi}_\alpha(\mathbf{r},t)\hat{\psi}^\dagger_{\beta'}(\mathbf{r}_1,\tau)\rangle G^{-1}_{\beta'\beta}(\mathbf{r}_1\tau,\mathbf{r}'t')$$
$$\Pi_{\alpha\beta}(\mathbf{r}t,\mathbf{r}'t') = \frac{\mathrm{i}}{\hbar}\kappa_\alpha\lambda(\mathbf{r})\sum_{\beta'}\int \mathrm{d}\tau\int \mathrm{d}\mathbf{r}_1\langle\mathcal{T}\hat{\psi}^\dagger_\alpha(\mathbf{r},t)\hat{\psi}_\alpha(\mathbf{r},t)\hat{\Phi}^\dagger_{\beta'}(\mathbf{r}_1,\tau)\rangle D^{-1}_{\beta'\beta}(\mathbf{r}_1\tau,\mathbf{r}'t') \tag{45}$$

Eqs. (43)–(45) are exact and constitute the NESGFT.

In order to evaluate the self-energies perturbatively, we rewrite the Green functions, Eq. (31), in the interaction picture

$$G_{\alpha\beta}(\mathbf{r}t,\mathbf{r}'t') = -\frac{\mathrm{i}}{\hbar}\langle\mathcal{T}\tilde{\psi}_\alpha(\mathbf{r},t)\tilde{\psi}^\dagger_\beta(\mathbf{r}',t')\mathcal{G}_\mathrm{I}(t,-\infty)\rangle_0$$
$$D_{\alpha\beta}(\mathbf{r}t,\mathbf{r}'t') = -\frac{\mathrm{i}}{\hbar}\langle\mathcal{T}\tilde{\Phi}_\alpha(\mathbf{r},t)\tilde{\Phi}^\dagger_\beta(\mathbf{r}',t')\mathcal{G}_\mathrm{I}(t,-\infty)\rangle_0 \tag{46}$$

where $\mathcal{G}_\mathrm{I}(t,-\infty)$ is given by Eq. (15) with $t_0 = -\infty$. Using Eqs. (13), (16) and (20), the self-energies (45) can also be expressed in the interaction picture as

$$\Sigma_{\alpha\beta}(\mathbf{r}t,\mathbf{r}'t') = -\frac{\mathrm{i}}{\hbar}\kappa_\alpha\lambda(\mathbf{r})\sum_{\beta'}\int\int \mathrm{d}\tau \mathrm{d}\mathbf{r}_1\langle\mathcal{T}\tilde{\Phi}^\dagger_\alpha(\mathbf{r},t)\tilde{\psi}_\alpha(\mathbf{r},t)\tilde{\psi}^\dagger_{\beta'}(\mathbf{r}_1,\tau)\mathcal{G}_\mathrm{I}(t,-\infty)\rangle_0 G^{-1}_{\beta'\beta}(\mathbf{r}_1\tau,\mathbf{r}'t')$$

$$\Pi_{\alpha\beta}(\mathbf{r}t,\mathbf{r}'t') = \frac{\mathrm{i}}{\hbar}\kappa_\alpha\lambda(\mathbf{r})\sum_{\beta'}\int\int \mathrm{d}\tau \mathrm{d}\mathbf{r}_1\langle\mathcal{T}\tilde{\psi}^\dagger_\alpha(\mathbf{r},t)\tilde{\psi}_\alpha(\mathbf{r},t)\tilde{\Phi}^\dagger_{\beta'}(\mathbf{r}_1,\tau)\mathcal{G}_\mathrm{I}(t,-\infty)\rangle_0 D^{-1}_{\beta'\beta}(\mathbf{r}_1\tau,\mathbf{r}'t') \tag{47}$$

Eq. (47) together with Eq. (46) constitutes closed form equations for the self-energies where all the averages are given in the interaction picture, $\langle\cdots\rangle_0$, with respect to the non-interacting density matrix. By expanding $\mathcal{G}_\mathrm{I}$ (Eq. (15)) perturbatively in $\tilde{\mathcal{H}}'_-$ we can obtain perturbative expansion for the self-energies. Each term in the expansion can be calculated using Wick's theorem for superoperators [49] which is given in Appendix 14E. This results in a perturbative series in terms of the zeroth order Green functions.

14.3 THE CALCULATION OF MOLECULAR CURRENTS

We have applied NESGFT to study the charge conductivity of a molecular wire attached to two perfectly conducting leads. In the simplest approach the leads 'a' and 'b' are treated as two free electron reservoirs. Nuclear motions in the molecular region are described as harmonic phonons which interact with the surrounding electronic structure and the environment (secondary phonons) [26]. We first recast the general Hamiltonian, Eq. (1), in a single electron local basis and partition it as

$$H = H_\mathrm{f} + H_\mathrm{int} \tag{48}$$

where H_f represents the free, non-interacting electrons and phonons and with no coupling between molecule and leads

$$H_\mathrm{f} = \sum_{i,j} E_{i,j}\psi_i^\dagger\psi_j + \sum_{k\in \mathrm{a,b}} \epsilon_k\psi_k^\dagger\psi_k + \sum_l \Omega_l\phi_l^\dagger\phi_l + \sum_m \omega_m\phi_m^\dagger\phi_m \tag{49}$$

The indices (i,j) represent the electronic basis states corresponding to the molecule, k labels the electronic states in the leads (a and b), l denotes primary phonons which interact with the electrons and m denotes the secondary phonons which are coupled to the primary phonons and constitute a thermal bath. The applied external voltage V maintains a chemical potential difference, $\mu_\mathrm{a} - \mu_\mathrm{b} = eV$, between the two leads and also modifies the single electron energies. In addition it provides an extra term $\sum_i V_i\psi_i^\dagger\psi_i$ which is included in the zeroth order Hamiltonian, H_f, by modifying the single electron energies. The interaction Hamiltonian is given by

$$H_\mathrm{int} = \sum_{k\in \mathrm{a,b};i} (V_{ki}\psi_k^\dagger\psi_i + h.c.) + \sum_{l,i} \lambda_{li}\Phi_l\psi_i^\dagger\psi_i + \sum_{l,m} U_{lm}\Phi_l\Phi_m \tag{50}$$

The three terms represent the molecule/lead interaction, coupling of primary phonons with the molecule and the interaction of primary and secondary phonons, respectively.

The total current passing through the junction can be expressed in terms of the electron Green functions and the corresponding self-energies. At steady state it is given by (see Appendix 14B, Eq. (B22))

$$I_\mathrm{T} = \frac{2e}{\hbar}\sum_{ij'} \int \frac{d\omega}{2\pi}\left[\Sigma_\mathrm{LR}^{ij'}(\omega)G_\mathrm{RL}^{j'i}(\omega) - \Sigma_\mathrm{RL}^{ij'}(\omega)G_\mathrm{LR}^{j'i}(\omega)\right] \tag{51}$$

G_RL and G_LR are the electronic Green functions and Σ_LR and Σ_RL represent the corresponding self-energies that account for all interactions (Eq. (50)).

$\Sigma^{ij}_{\alpha\beta}$ has contributions from the electron–lead (σ) and electron–phonon (Ξ) interactions

$$\Sigma^{ij}_{\alpha\beta}(\omega) = \sigma^{ij}_{\alpha\beta}(\omega) + \Xi^{ij}_{\alpha\beta}(\omega) \tag{52}$$

These are given in Eqs. (C29) and (C38). The self-energy expressions (C38) and (C40) are calculated perturbatively to second order in the electron–phonon coupling in terms of the zeroth order Green functions (Eq. (55)). The simplest expression for current is obtained by substituting Eqs. (55), (C29) and (C38) in Eq. (51). This zeroth order result can be improved by using the renormalized Green functions obtained from the self-consistent solution of the Dyson equation (44).

In order to solve self-consistently for the electron Green functions that appear in the current formula, Eq. (51), the self-energy is calculated under the Born approximation by replacing the zeroth order Green functions, $G^0_{\alpha\beta}$ and $D^0_{\alpha\beta}$ with the corresponding renormalized Green functions, $G_{\alpha\beta}$ and $D_{\alpha\beta}$, as is commonly done in mode-coupling theories [52,53]. This approximation sums an infinite set of non-crossing diagrams [54,55] that appear in the perturbation expansion of the many-body Green function, $G_{\alpha\beta}$.

Since the electron self-energy (Eq. (C38)) also depends on the phonon Green function, the phonon self-energy, $\Pi^{ll'}_{\alpha\beta}$, is also required for a self-consistent solution of the electron Green functions. The phonon self-energy calculated in Appendix 14C is given by

$$\Pi^{ij}_{\alpha\beta}(\omega) = \gamma^{ij}_{\alpha\beta}(\omega) + \Lambda^{ij}_{\alpha\beta}(\omega) \tag{53}$$

where $\gamma^{ij}_{\alpha\beta}(\omega)$ (Eq. (C30)) and $\Lambda^{ij}_{\alpha\beta}(\omega)$ (Eq. (C40)) represent the contributions from the phonon–phonon and the electron–phonon interactions, respectively.

Computing the renormalized electron and phonon Green functions and the corresponding self-energies involves the self-consistent solution of the following coupled equations for the Green functions:

$$\begin{aligned}
G_{\mathrm{LR}}(\omega) &= G^0_{\mathrm{LL}}(\omega)\Sigma_{\mathrm{LL}}(\omega)G_{\mathrm{LR}}(\omega) + G^0_{\mathrm{LL}}(\omega)\Sigma_{\mathrm{LR}}(\omega)G_{\mathrm{RR}}(\omega)\\
G_{\mathrm{RL}}(\omega) &= G^0_{\mathrm{RR}}(\omega)\Sigma_{\mathrm{RL}}(\omega)G_{\mathrm{LL}}(\omega) + G^0_{\mathrm{RR}}(\omega)\Sigma_{\mathrm{RR}}(\omega)G_{\mathrm{RL}}(\omega)\\
G_{\mathrm{LL}}(\omega) &= G^0_{\mathrm{LL}}(\omega) + G^0_{\mathrm{LL}}(\omega)\Sigma_{\mathrm{LL}}(\omega)G_{\mathrm{LL}}(\omega) + G^0_{\mathrm{LL}}(\omega)\Sigma_{\mathrm{LR}}(\omega)G_{\mathrm{RL}}(\omega)\\
G_{\mathrm{RR}}(\omega) &= G^0_{\mathrm{RR}}(\omega) + G^0_{\mathrm{RR}}(\omega)\Sigma_{\mathrm{RL}}(\omega)G_{\mathrm{LR}}(\omega) + G^0_{\mathrm{RR}}(\omega)\Sigma_{\mathrm{RR}}(\omega)G_{\mathrm{RR}}(\omega)
\end{aligned} \tag{54}$$

Similarly, the equations for the phonon Green functions are obtained by replacing $G_{\alpha\beta}$ with $D_{\alpha\beta}$ and $\Sigma_{\alpha\beta}$ with $\Pi_{\alpha\beta}$. Here Green functions corresponding to the free Hamiltonian, $G^{0ij}_{\alpha\beta}$ and $D^{0ll'}_{\alpha\beta}$, are given by

$$G^{0ij}_{\alpha\beta}(\omega) = \frac{\delta_{\alpha\beta}}{\omega\delta_{ij} - \kappa_\alpha E_{ij} + i\eta}, \qquad D^{0ll'}_{\alpha\beta}(\omega) = \frac{2\Omega_l\delta_{\alpha\beta}\delta_{ll'}}{\kappa_\alpha\hbar^2\omega^2 - \Omega_l^2 + i\eta} \tag{55}$$

where we set $\hbar = 1$ and $\eta \rightarrow 0$. $E_{ij} = E_i - E_j$ is the energy difference between single electron i^{th} and j^{th} states. Ω_l denotes the molecular phonon eigenstates.

Once the Green functions, G_{LR}, G_{RL} and the corresponding self-energies Σ_{LR}, Σ_{RL}, are obtained from the self-consistent solution of Eq. (54) together with Eq. (52), Eq. (51) can be used to calculate the total current through the molecular junction.

14.4 DISCUSSION

In this chapter, we have developed the NESGFT and applied it to the computation of molecular current. The Liouville space–time ordering operator provides an elegant way for performing calculations in real time, thus avoiding the artificial backward and forward time evolution required in Hilbert space (Keldysh loop). Wick's theorem for superoperators is used to compute the self-energies perturbatively to the second order in phonon–electron coupling. Eqs. (54) have been derived earlier by many authors [19,43,55]. Recently, Galperin et al. [26] have used a fully self-consistent solution to study the influence of different interactions on molecular conductivity for a strong electron–phonon coupling. The main aim of the present work is to demonstrate that by doing calculations in Liouville space one can avoid the backward/forward time evolution (Keldysh loop) required in Hilbert space. This originates from the fact that in Liouville space both ket and bra evolve forward in time. Thus, one can couple the system with two independent, 'left' and 'right' fields. This property of Liouville space can be used to construct real (physical) time generating functionals for the non-perturbative calculation of the self-energies.

The present model [37–39] ignores electron–electron interactions. These may be treated using the GW technique [56–58] formulated in terms of the superoperators and extended to non-equilibrium situations. All non-equilibrium observables can be obtained from a single generating functional in terms of 'left' and 'right' operators. The retarded (advance) Green function that describes the forward (backward) motion of the system particle can also be calculated in terms of the basic Green functions, $G_{\alpha\beta}$ (see Appendix 14D).

The NESGFT can also be recast in terms of the + and − (rather than L/R) superoperators which are more directly related to observables. This is done in Appendix 14D where we focused on the primary quantities that are represented in terms of the 'left' and 'right' superoperators. All other quantities are obtained as the linear combination of these basic operators.

14.5 ACKNOWLEDGEMENTS

The support of Chemical Sciences, Geosciences and Biosciences Division, Office of Basic Energy Sciences, Office of Science, U.S. Department of Energy is gratefully acknowledged.

APPENDIX 14A: SUPEROPERATOR EXPRESSIONS FOR THE KELDYSH GREEN FUNCTIONS

The standard NEGFT formulated in terms of the four Hilbert space Green functions: time ordered (G^T), anti-time ordered ($G^{\tilde{T}}$), greater ($G^>$) and lesser ($G^<$) [23,25]. These are defined in the Heisenberg picture as

$$\begin{aligned}
G^T(\mathbf{x},\mathbf{x}') &\equiv -\frac{i}{\hbar}\langle T\hat{\psi}(\mathbf{x})\hat{\psi}^\dagger(\mathbf{x}')\rangle \\
&= -\frac{i}{\hbar}\theta(t-t')\langle\hat{\psi}(\mathbf{x})\hat{\psi}^\dagger(\mathbf{x}')\rangle + \theta(t'-t)\langle\hat{\psi}^\dagger(\mathbf{x}')\hat{\psi}(\mathbf{x})\rangle \\
G^{\tilde{T}}(\mathbf{x},\mathbf{x}') &\equiv -\frac{i}{\hbar}\langle \tilde{T}\hat{\psi}(\mathbf{x})\hat{\psi}^\dagger(\mathbf{x}')\rangle \\
&= -\frac{i}{\hbar}\theta(t'-t)\langle\hat{\psi}(\mathbf{x})\hat{\psi}^\dagger(\mathbf{x}')\rangle + \theta(t-t')\langle\hat{\psi}^\dagger(\mathbf{x}')\hat{\psi}(\mathbf{x})\rangle \\
G^>(\mathbf{x},\mathbf{x}') &\equiv -\frac{i}{\hbar}\langle\hat{\psi}(\mathbf{x})\hat{\psi}^\dagger(\mathbf{x}')\rangle \\
G^<(\mathbf{x},\mathbf{x}') &\equiv \frac{i}{\hbar}\langle\hat{\psi}^\dagger(\mathbf{x}')\hat{\psi}(\mathbf{x})\rangle
\end{aligned} \tag{A1}$$

These are known as T ($\tilde{T}$) is the Hilbert space–time (anti-time) ordering operator: when applied to a product of operators, it reorders them in ascending (descending) times from right to left.

The four Green functions that show up naturally in Liouville space are defined as

$$\begin{aligned}
G_{LL}(\mathbf{x},\mathbf{x}') &= -\frac{i}{\hbar}\langle\mathcal{T}\hat{\psi}_L(\mathbf{x})\hat{\psi}_L^\dagger(\mathbf{x}')\rangle \\
G_{RR}(\mathbf{x},\mathbf{x}') &= -\frac{i}{\hbar}\langle\mathcal{T}\hat{\psi}_R(\mathbf{x})\hat{\psi}_R^\dagger(\mathbf{x}')\rangle \\
G_{LR}(\mathbf{x},\mathbf{x}') &= -\frac{i}{\hbar}\langle\mathcal{T}\hat{\psi}_L(\mathbf{x})\hat{\psi}_R^\dagger(\mathbf{x}')\rangle \\
G_{RL}(\mathbf{x},\mathbf{x}') &= -\frac{i}{\hbar}\langle\mathcal{T}\hat{\psi}_R(\mathbf{x})\hat{\psi}_L^\dagger(\mathbf{x}')\rangle
\end{aligned} \tag{A2}$$

$\mathcal{T}$ is the Liouville space–time ordering operator that rearranges all superoperators in increasing order of time from right to left.

To establish the connection between Liouville space and Hilbert space Green functions we shall convert the superoperators back to ordinary operators [44]. For G_{LR}, and G_{RL}, we obtain

$$\begin{aligned}
G_{\mathrm{LR}}(\mathbf{x},\mathbf{x}') &\equiv -\frac{\mathrm{i}}{\hbar}\mathrm{Tr}\{\mathcal{T}\hat{\psi}_{\mathrm{L}}(\mathbf{x})\hat{\psi}^{\dagger}_{\mathrm{R}}(\mathbf{x}')\rho_{\mathrm{eq}}\}\\
&= -\frac{\mathrm{i}}{\hbar}\mathrm{Tr}\{\hat{\psi}(\mathbf{x})\rho_{\mathrm{eq}}\hat{\psi}^{\dagger}(\mathbf{x}')\}\\
&= -\frac{\mathrm{i}}{\hbar}\langle\hat{\psi}^{\dagger}(\mathbf{x}')\hat{\psi}(\mathbf{x})\rangle = G^{<}(\mathbf{x},\mathbf{x}')\\
G_{\mathrm{RL}}(\mathbf{x},\mathbf{x}') &\equiv -\frac{\mathrm{i}}{\hbar}\mathrm{Tr}\{\mathcal{T}\hat{\psi}_{\mathrm{R}}(\mathbf{x})\hat{\psi}^{\dagger}_{\mathrm{L}}(\mathbf{x}')\rho_{\mathrm{eq}}\}\\
&= -\frac{\mathrm{i}}{\hbar}\mathrm{Tr}\{\hat{\psi}^{\dagger}(\mathbf{x}')\rho_{\mathrm{eq}}\hat{\psi}(\mathbf{x})\}\\
&= -\frac{\mathrm{i}}{\hbar}\langle\hat{\psi}(\mathbf{x})\hat{\psi}^{\dagger}(\mathbf{x}')\rangle = G^{>}(\mathbf{x},\mathbf{x}')
\end{aligned} \tag{A3}$$

where ρ_{eq} is the fully interacting many body equilibrium density matrix.

For G_{LL} and G_{RR} we have two cases

(i) For $t > t'$, we get

$$\begin{aligned}
G_{\mathrm{LL}}(\mathbf{x},\mathbf{x}') &\equiv -\frac{\mathrm{i}}{\hbar}\mathrm{Tr}\{\mathcal{T}\hat{\psi}_{\mathrm{L}}(\mathbf{x})\hat{\psi}^{\dagger}_{\mathrm{L}}(\mathbf{x}')\rho_{\mathrm{eq}}\}\\
&= -\frac{\mathrm{i}}{\hbar}\mathrm{Tr}\{\hat{\psi}(\mathbf{x})\hat{\psi}^{\dagger}(\mathbf{x}')\rho_{\mathrm{eq}}\} = -\frac{\mathrm{i}}{\hbar}\langle\hat{\psi}(\mathbf{x})\hat{\psi}^{\dagger}(\mathbf{x}')\rangle\\
G_{\mathrm{RR}}(\mathbf{x},\mathbf{x}') &\equiv -\frac{\mathrm{i}}{\hbar}\mathrm{Tr}\{\mathcal{T}\hat{\psi}_{\mathrm{R}}(\mathbf{x})\hat{\psi}^{\dagger}_{\mathrm{R}}(\mathbf{x}')\rho_{\mathrm{eq}}\}\\
&= -\frac{\mathrm{i}}{\hbar}\mathrm{Tr}\{\rho_{\mathrm{eq}}\hat{\psi}^{\dagger}(\mathbf{x}')\hat{\psi}(\mathbf{x})\} = -\frac{\mathrm{i}}{\hbar}\langle\hat{\psi}^{\dagger}(\mathbf{x}')\hat{\psi}(\mathbf{x})\rangle
\end{aligned} \tag{A4}$$

(ii) For the reverse case, $t < t'$, we get

$$\begin{aligned}
G_{\mathrm{LL}}(\mathbf{x},\mathbf{x}') &\equiv -\frac{\mathrm{i}}{\hbar}\mathrm{Tr}\{\mathcal{T}\hat{\psi}_{\mathrm{L}}(\mathbf{x})\hat{\psi}^{\dagger}_{\mathrm{L}}(\mathbf{x}')\rho_{\mathrm{eq}}\}\\
&= -\frac{\mathrm{i}}{\hbar}\mathrm{Tr}\{\hat{\psi}^{\dagger}(\mathbf{x}')\hat{\psi}(\mathbf{x})\rho_{\mathrm{eq}}\} = -\frac{\mathrm{i}}{\hbar}\langle\hat{\psi}^{\dagger}(\mathbf{x}')\hat{\psi}(\mathbf{x})\rangle\\
G_{\mathrm{RR}}(\mathbf{x},\mathbf{x}') &\equiv -\frac{\mathrm{i}}{\hbar}\mathrm{Tr}\{\mathcal{T}\hat{\psi}_{\mathrm{R}}(\mathbf{x})\hat{\psi}^{\dagger}_{\mathrm{R}}(\mathbf{x}')\rho_{\mathrm{eq}}\}\\
&= -\frac{\mathrm{i}}{\hbar}\mathrm{Tr}\{\rho_{\mathrm{eq}}\hat{\psi}(\mathbf{x})\hat{\psi}^{\dagger}(\mathbf{x}')\} = -\frac{\mathrm{i}}{\hbar}\langle\hat{\psi}(\mathbf{x})\hat{\psi}^{\dagger}(\mathbf{x}')\rangle
\end{aligned} \tag{A5}$$

Combining Eqs. (A4) and (A5) we can write

$$\begin{aligned}
G_{LL}(\mathbf{x},\mathbf{x}') &\equiv -\frac{i}{\hbar}\mathrm{Tr}\{\mathcal{T}\hat{\psi}_L(\mathbf{x})\hat{\psi}_L^\dagger(\mathbf{x}')\rho_{eq}\} \\
&= -\frac{i}{\hbar}[\theta(t-t')\langle\hat{\psi}(\mathbf{x})\hat{\psi}^\dagger(\mathbf{x}')\rangle - \theta(t'-t)\langle\hat{\psi}^\dagger(\mathbf{x}')\hat{\psi}(\mathbf{x})\rangle] \\
&= G^T(\mathbf{x},\mathbf{x}') \\
G_{RR}(\mathbf{x},\mathbf{x}') &\equiv -\frac{i}{\hbar}\mathrm{Tr}\{\mathcal{T}\hat{\psi}_R(\mathbf{x})\hat{\psi}_R^\dagger(\mathbf{x}')\rho_{eq}\} \\
&= -\frac{i}{\hbar}[\theta(t-t')\langle\hat{\psi}^\dagger(\mathbf{x}')\hat{\psi}(\mathbf{x})\rangle - \theta(t'-t)\langle\hat{\psi}(\mathbf{x})\hat{\psi}^\dagger(\mathbf{x}')\rangle] \\
&= G^{\tilde{T}}(\mathbf{x},\mathbf{x}')
\end{aligned} \tag{A6}$$

Eqs. (A3) and (A6) establish the equivalence of Hilbert and Liouville space Green functions and they can be summarized as

$$\begin{aligned}
G_{LL}(\mathbf{x},\mathbf{x}') &= G^T(\mathbf{x},\mathbf{x}'), \quad G_{RR}(\mathbf{x},\mathbf{x}') = G^{\tilde{T}}(\mathbf{x},\mathbf{x}') \\
G_{LR}(\mathbf{x},\mathbf{x}') &= G^<(\mathbf{x},\mathbf{x}'), \quad G_{RL}(\mathbf{x},\mathbf{x}') = G^>(\mathbf{x},\mathbf{x}')
\end{aligned} \tag{A7}$$

APPENDIX 14 B: SUPEROPERATOR GREEN FUNCTION EXPRESSION FOR THE CURRENT

In this appendix, we present a formal microscopic derivation for the current flowing through a conductor. The conductor could be a molecule or a metal or any conducting material attached to two electrodes held at two different potentials.

In Hilbert space the charge–current density is given by

$$\mathbf{j}(\mathbf{r},t) = -\frac{ie\hbar}{2m}\langle[\hat{\psi}^\dagger(\mathbf{r},t)\nabla\hat{\psi}(\mathbf{r},t) - (\nabla\hat{\psi}^\dagger(\mathbf{r},t))\hat{\psi}(\mathbf{r},t)]\rangle \tag{B8}$$

where e and m are the electron charge and mass, respectively. Eq. (B8) can also be expressed in a slightly modified form as

$$\mathbf{j}(\mathbf{r},t) = \frac{ie\hbar}{2m}[\langle(\nabla-\nabla')\hat{\psi}^\dagger(\mathbf{r},t)\hat{\psi}(\mathbf{r}',t')\rangle]_{\mathbf{x}'=\mathbf{x}} \tag{B9}$$

where ∇' represents the derivative with respect to $\mathbf{r}'$.

Using relation (A7), the current density can be expressed in terms of the superoperator Green function as

$$\mathbf{j}(\mathbf{r},t) = -\frac{e\hbar^2}{2m}[(\nabla-\nabla')G_{LR}(\mathbf{r}t,\mathbf{r}'t')]_{\mathbf{x}'=\mathbf{x}} \tag{B10}$$

At steady state, the Green functions only depend on the time difference ($t - t'$) and the total current density ($\mathbf{J}_{\mathrm{T}}$) becomes time independent. Transforming to the frequency (energy) domain, the current density per unit energy is

$$\mathbf{j}(\mathbf{r}, E) = -\frac{e\hbar}{2m}[(\nabla - \nabla')G_{\mathrm{LR}}(\mathbf{rr}', E)]_{\mathbf{r}'=\mathbf{r}} \tag{B11}$$

and the total current density

$$\mathbf{J}_{\mathrm{T}}(\mathbf{r}) = \int \frac{dE}{2\pi}\mathbf{j}(\mathbf{r}, E) \tag{B12}$$

Eq. (B12) provides a recipe for calculating the current profile across the conductor once the Green function G_{LR} is known from the self-consistent solution of the Dyson equation. For computing the total current passing through the conductor, Eq. (B11) can be expressed in the form of Eq. (51). In order to get the total current per unit energy ($I_{\mathrm{T}}(E)$) passing between electrode and conductor we need to integrate the current density over the surface area of the conductor–electrode contact

$$I_{\mathrm{T}}(E) = \int_{s} \mathbf{j}(\mathbf{r}, E) \cdot \hat{\mathbf{n}} \mathrm{d}S = \int \nabla \cdot \mathbf{j}(\mathbf{r}, E) \mathrm{d}\mathbf{r} \tag{B13}$$

where $\hat{\mathbf{n}}$ is the unit vector normal to surface S. Substituting into Eq. (B13) from Eq. (B11), we get

$$I_{\mathrm{T}}(E) = -\frac{e\hbar}{2m}\mathrm{Tr}[(\nabla^2 - \nabla'2)G_{\mathrm{LR}}(\mathbf{rr}', E)] \tag{B14}$$

In general, a conductor–electrode system can be described by the Hamiltonian

$$H = H_0 + H_{\mathrm{int}} \tag{B15}$$

where H_0 represents the non-interacting part

$$H_0 = \int \mathrm{d}\mathbf{r}\psi^{\dagger}(\mathbf{r})h_0(\mathbf{r})\psi(\mathbf{r}) \tag{B16}$$

where $h_0(\mathbf{r}) = -\frac{\hbar^2}{2m}\nabla^2$ and all the interaction terms (conductor–electrode, electron–phonon) are included in H_{int}. The total current per unit energy, Eq. (B14), is

$$I_{\mathrm{T}}(E) = -\frac{e}{\hbar}\mathrm{Tr}[(h_0(\mathbf{r}) - h_0^{*}(\mathbf{r}'))G_{\mathrm{LR}}(\mathbf{rr}', E)] \tag{B17}$$

The Dyson equations for the retarded Green function (see Appendix 14D, Eq. (D48)) in frequency (energy) can be expressed in the matrix form as

$$h_0 G_{\mathrm{r}} = EG_{\mathrm{r}} - \mathbf{I} - \Sigma_{\mathrm{r}}G_{\mathrm{r}} \tag{B18}$$

where $\mathbf{I}$ is the identity matrix and Σ_{r} is the retarded self-energy, Eq. (D50). $E = \hbar\omega$ is a number. Henceforth, we write all the expressions in the matrix notation. Taking the complex conjugate of Eq. (B18), we obtain the Dyson equation for the advanced Green function,

$$G_\mathrm{a} h_0 = E G_\mathrm{a} - \mathbf{I} - G_\mathrm{a} \Sigma_\mathrm{a} \tag{B19}$$

with the corresponding advanced self-energy, Σ_a. From the matrix Dyson equation (D48), we also have the relation

$$G_\mathrm{LR} = G_\mathrm{r} \Sigma_\mathrm{LR} G_\mathrm{a} \tag{B20}$$

Using the relations (B18)–(B20), it is easy to see that

$$h_0 G_\mathrm{LR} - G_\mathrm{LR} h_0 = G_\mathrm{LR} \Sigma_\mathrm{a} + G_\mathrm{r} \Sigma_\mathrm{LR} - \Sigma_\mathrm{r} G_\mathrm{LR} - \Sigma_\mathrm{LR} G_\mathrm{a} \tag{B21}$$

Substituting this in Eq. (B17), the total current per unit energy becomes

$$I_\mathrm{T}(E) = \frac{2e}{\hbar} \mathrm{Tr}[\Sigma_\mathrm{LR}(E) G_\mathrm{RL}(E) - \Sigma_\mathrm{RL}(E) G_\mathrm{LR}(E)] \tag{B22}$$

where a factor of 2 is introduced to account for the spin degeneracy.

We have calculated the total current in real space. In practice, the Green functions and the self-energy matrices are calculated in an electronic basis (i,j). The total current through the conductor is obtained by integrating Eq. (B22) over energy resulting in Eq. (51).

APPENDIX 14C: SELF-ENERGIES FOR SUPEROPERATOR GREEN FUNCTIONS

The basic quantities required for describing the coupled molecule–lead system are the one particle electron and the phonon Green functions. Following the steps outlined in Section 14.2, the time development for various superoperators (Heisenberg equations) is (all primed indices should be summed over)

$$\begin{aligned}
i\hbar\kappa_\alpha \frac{\partial}{\partial t} \hat{\psi}_{i\alpha}(t) &= E_{ij'} \hat{\psi}_{j'\alpha}(t) + V_{k'i} \hat{\psi}_{k'\alpha}(t) + \lambda_{l'i} \hat{\phi}_{l'\alpha}(t) \hat{\psi}_{i\alpha}(t) \\
-i\hbar\kappa_\alpha \frac{\partial}{\partial t} \hat{\phi}_{l\alpha}(t) &= \lambda_{i'l} \hat{\psi}^\dagger_{i'\alpha}(t) \hat{\psi}_{i'\alpha}(t) + \Omega_l \hat{\phi}_{l\alpha}(t) + U_{lm'} \hat{\phi}_{m'\alpha}(t) \\
i\hbar\kappa_\alpha \frac{\partial}{\partial t} \hat{\psi}_{k\alpha}(t) &= \epsilon_k \hat{\psi}_{k\alpha}(t) + V_{ki'} \hat{\psi}_{i'}(t) \\
-i\hbar\kappa_\alpha \frac{\partial}{\partial t} \hat{\phi}_{m\alpha}(t) &= \omega_m \hat{\phi}_{m\alpha}(t) + U_{l'm} \hat{\phi}_{l'}(t)
\end{aligned} \tag{C23}$$

Using Eq. (C23) and following the procedure described in Section 14.2, it is straightforward to write the matrix Dyson equation (44) for the electron and phonon

Green functions defined as

$$\begin{aligned} G^{ij}_{\alpha\beta}(\mathbf{x},\mathbf{x}') &= -\frac{\mathrm{i}}{\hbar}\langle \mathcal{T}\psi_{i\alpha}(\mathbf{r},t)\psi^{\dagger}_{j\beta}(\mathbf{r}',t')\rangle \\ D^{ll'}_{\alpha\beta}(\mathbf{x},\mathbf{x}') &= -\frac{\mathrm{i}}{\hbar}\langle \mathcal{T}\Phi_{l\alpha}(\mathbf{r},t)\Phi^{\dagger}_{l'\beta}(\mathbf{r}',t')\rangle \end{aligned} \tag{C24}$$

with the corresponding self-energy matrix elements

$$\begin{aligned} \Sigma^{ij}_{\alpha\beta}(t,t') = -\frac{\mathrm{i}}{\hbar}\kappa_\alpha \sum_{\beta',j'} \int \mathrm{d}\tau \Bigg[\sum_{l'} \lambda_{l'i} \langle \mathcal{T}\hat{\Phi}_{l'\alpha}(t)\hat{\psi}_{i\alpha}(t)\hat{\psi}^{\dagger}_{j'\beta'}(\tau)\rangle + \sum_{k'} V_{k'i}\langle \mathcal{T}\hat{\psi}_{k'\alpha}(t)\hat{\psi}^{\dagger}_{j'\beta'}(\tau)\rangle \Bigg] \\ \times G^{j'j-1}_{\beta'\beta}(\tau,t') \equiv \Xi^{ij}_{\alpha\beta}(t,t') + \sigma^{ij}_{\alpha\beta}(t,t') \end{aligned} \tag{C25}$$

$$\begin{aligned} \Pi^{ll'}_{\alpha\beta}(t,t') = \frac{\mathrm{i}}{\hbar}\int \mathrm{d}\tau \sum_{\beta',l''} \Bigg[\sum_{m'} U_{lm'}\langle \mathcal{T}\hat{\Phi}_{m'\alpha}(t)\hat{\Phi}^{\dagger}_{l''\beta'}(\tau)\rangle + \sum_{i'} \lambda_{li'}\langle \mathcal{T}\hat{\psi}^{\dagger}_{i'\alpha}(t)\hat{\psi}_{i'\alpha}(t)\hat{\Phi}^{\dagger}_{l''\beta'}(\tau)\rangle \Bigg] \\ \times D^{l''l'-1}_{\beta'\beta}(\tau,t') \equiv \gamma^{ll'}_{\alpha\beta}(t,t') + \Lambda^{ll'}_{\alpha\beta}(t,t') \end{aligned}$$

The two terms in the electron self-energy represent the contributions from the phonon–electron (Ξ) and molecule–lead (σ) interactions. Similarly, the phonon self-energy has contributions from the electron–phonon (Λ) and the primary–secondary phonon (γ) couplings. The self-energy due to the molecule–lead coupling can be calculated exactly. To this end we need to obtain the quantity $\langle \mathcal{T}\tilde{\psi}_{k'\alpha}(t)\psi^{\dagger}_{j'\beta'}(\tau)\rangle$. By multiplying the third equation in Eq. (C23) by $\psi^{\dagger}_{j'\beta'}(\tau)$ from the left and from the right, taking trace and subtracting, we get (here primed indices are not summed over)

$$\begin{aligned} \left(\mathrm{i}\hbar\kappa_\alpha\frac{\partial}{\partial t} - \epsilon_{k'}\right)\langle \mathcal{T}\hat{\psi}_{k'\alpha}(t)\hat{\psi}^{\dagger}_{j'\beta'}(\tau)\rangle &= \sum_{i'} V_{k'i'}\langle \mathcal{T}\hat{\psi}_{i'\alpha}(t)\hat{\psi}^{\dagger}_{j'\beta'}(\tau)\rangle \\ \Rightarrow \langle \mathcal{T}\hat{\psi}_{k'\alpha}(t)\hat{\psi}^{\dagger}_{j'\beta'}(\tau)\rangle &= \mathrm{i}\hbar\sum_{i'} V_{k'i'}g_{k'}(t)G^{i'j'}_{\alpha\beta'}(t,\tau) \end{aligned} \tag{C26}$$

where $g_k(t) = (\mathrm{i}\hbar\kappa_\alpha\frac{\partial}{\partial t} - \epsilon_{k'})^{-1}$. Substituting expression (C26) in Eq. (C25) gives for the molecule–lead self-energy

$$\sigma^{ij}_{\alpha\beta}(t,t') = \kappa_\alpha\delta_{\alpha\beta}\sum_{k'\in \mathrm{a,b}} V_{k'i}V_{k'j}g_{k'}(t)\delta(t-t') \tag{C27}$$

Similarly, the contribution to the phonon self-energy from the interaction with secondary phonons can be calculated exactly

$$\gamma^{ll'}_{\alpha\beta} = -\kappa_\alpha\delta_{\alpha\beta}\sum_{m'} U_{lm'}U_{l'm'}g'_{m'}(t)\delta(t-t') \tag{C28}$$

where $g'_{m'}(t) = (\mathrm{i}\hbar\kappa_\alpha\frac{\partial}{\partial t} + \omega_{m'})^{-1}$. At steady state all Green functions and self-energies depend only on the time difference $(t_1 - t_2)$ and it is very convenient to express them in the frequency space. The self-energy contributions due to molecule–lead ($\sigma^{ij}_{\alpha\beta}$) and phonon–phonon ($\gamma^{l'}_{\alpha\beta}$) interactions, Eqs. (C27) and (C28), can be represented in frequency space as

$$\sigma^{ij}_{\alpha\beta}(\omega) = \kappa_\alpha \delta_{\alpha\beta} \sum_{k'} \frac{V_{k'i} V_{k'j}}{\kappa_\alpha \omega - \epsilon_{k'} + i\eta} \tag{C29}$$

$$\gamma^{ll'}_{\alpha\beta}(\omega) = -\kappa_\alpha \delta_{\alpha\beta} \sum_{m'} \frac{U_{lm'} U_{l'm'}}{\kappa_\alpha \omega + \omega_{m'} + i\eta} \tag{C30}$$

where $\eta \rightarrow 0$. However, in real calculations it is a common practice to calculate self-energies $\sigma_{\alpha\beta}$ and $\gamma_{\alpha\beta}$ in the wide band approximation, implying that the real parts of the self-energies can be ignored and the imaginary parts are considered as frequency independent. Eqs. (C29) and (C30) then reduce to simpler forms

$$\sigma^{ij}_{\alpha\beta} = \kappa_\alpha \delta_{\alpha\beta} \frac{\mathrm{i}}{2} \Gamma^{ij}, \quad \gamma^{ll'}_{\alpha\beta} = -\kappa_\alpha \delta_{\alpha\beta} \frac{\mathrm{i}}{2} \tilde{\Gamma}^{ll'} \tag{C31}$$

where $\Gamma^{ij} = 2\pi \sum_{k'} V_{k'i} V_{k'j}$ and $\tilde{\Gamma}^{ll'} = 2\pi \sum_{m'} U_{m'l} V_{m'l'}$.

The phonon contribution to the electronic self-energy is obtained perturbatively in the phonon–electron coupling. We recast the phonon contribution (first term on the RHS of Eq. (C25) for $\Sigma^{ij}_{\alpha\beta}$) in the interaction picture by writing

$$\langle \mathcal{T} \hat{\Phi}_{l\alpha}(t) \hat{\psi}_{i\alpha}(t) \hat{\psi}^\dagger_{j\beta}(t') \rangle = \langle \mathcal{T} \tilde{\Phi}_{l\alpha}(t) \tilde{\psi}_{i\alpha}(t) \tilde{\psi}^\dagger_{j\beta}(t') \mathcal{G}_{\mathrm{I}}(t, -\infty) \rangle_0 \tag{C32}$$

where

$$\begin{aligned} \mathcal{G}_{\mathrm{I}}(t, -\infty) = \exp\Bigg\{ -\frac{\mathrm{i}}{\hbar} \int \mathrm{d}\tau \sum_{i'\alpha'} \kappa_{\alpha'} \Bigg[\sum_{l'} \lambda_{l'i'} \tilde{\Phi}_{l'\alpha'}(\tau) \tilde{\psi}^\dagger_{i'\alpha'}(\tau) \tilde{\psi}_{i'\alpha'}(\tau) \\ + \sum_{k'} V_{k'i'} (\tilde{\psi}^\dagger_{k'\alpha'}(\tau) \tilde{\psi}_{i'\alpha'}(\tau) + \tilde{\psi}^\dagger_{i'\alpha'}(\tau) \tilde{\psi}_{k'\alpha'}(\tau)) \Bigg] \Bigg\} \end{aligned} \tag{C33}$$

Substituting Eq. (C33) in Eq. (C32), expanding the exponential to first order in λ_{li} and using Wick's theorem for superoperators [49] we obtain

$$\begin{aligned} \langle \mathcal{T} \hat{\Phi}_{l\alpha}(t) \hat{\psi}_{i\alpha}(t) \hat{\psi}^\dagger_{j\beta}(t') \rangle = -\hbar^2 \sum_{l'i'\alpha'} \kappa_{\alpha'} \lambda_{l'i'} \int \mathrm{d}\tau D^{0ll'}_{\alpha\alpha'}(t, \tau) \Big[G^{0ij}_{\alpha\beta}(t, t') G^{0i'i'}_{\alpha'\alpha'}(\tau, \tau^+) \\ + G^{0ii'}_{\alpha\alpha'}(t, \tau) G^{0i'j}_{\alpha'\beta}(\tau, t') \Big] \end{aligned} \tag{C34}$$

Here the superscript '0' represents the trace with respect to the non-interacting density matrix. The zeroth order Green functions are given in Eq. (55). The terms coming from the lead-molecule coupling (V_{ki}) vanish because they are odd in creation and annihilation operators. Substituting Eq. (C34) in Eq. (C25) gives for the phonon contribution to the self-energy

$$\Xi^{ij}_{\alpha\beta}(t,t') = i\hbar \sum_{l_1 l_2} \kappa_\alpha \lambda_{l_1 i}\Big[\kappa_\beta \lambda_{l_2 j} D^{0l_1l_2}_{\alpha\beta}(t,t') G^{0ij}_{\alpha\beta}(t,t') + \delta_{ij}\delta_{\alpha\beta}\delta(t-t')\sum_{i_1\alpha'} \lambda_{l_2 i_1}\kappa_{\alpha'}\int d\tau D^{0l_1l_2}_{\alpha\alpha'}(t,\tau) G^{0i_1i_1}_{\alpha'\alpha'}(\tau,\tau^+)\Big] \quad \text{(C35)}$$

In the derivation of Eq. (C35), we have used the identity

$$\int d\tau \sum_{\alpha' j'} G^{0ij'}_{\alpha\beta'}(t,\tau) G^{-1\,0j'j}_{\beta'\beta}(\tau,t') = \delta_{\alpha\beta}\delta_{ij}\delta(t-t') \quad \text{(C36)}$$

Similarly the contribution of the electron–phonon interaction to the phonon self-energy (second term in Eq. (C25) for $\Pi^{ij}_{\alpha\beta}$) can be obtained perturbatively. To second order in phonon–electron coupling, we obtain

$$\Lambda^{ll'}_{\alpha\beta}(t,t') = -i\hbar \sum_{ij} \kappa_\alpha \kappa_\beta \lambda_{li}\lambda_{l'j}[G^{0ji}_{\beta\alpha}(t',t) G^{0ij}_{\alpha\beta}(t,t') + G^{0ii}_{\alpha\alpha}(t,t^+) G^{0jj}_{\beta\beta}(t',t'^+)] \quad \text{(C37)}$$

The electronic self-energy depends on both the electron and phonon green functions while the phonon self-energy contains only the electron Green functions.

At steady state we transform to the frequency domain and obtain

$$\Xi^{ij}_{\alpha\beta}(\omega) = i\hbar \sum_{l_1 l_2} \kappa_\alpha \kappa_\beta \lambda_{l_1 i}\lambda_{l_2 j}\int \frac{d\omega'}{2\pi} D^{0l_1l_2}_{\alpha\beta}(\omega') G^{0ij}_{\alpha\beta}(\omega-\omega') + \delta_{ij}\delta_{\alpha\beta}\sum_{l_1,l_2,i_1,\alpha'} \kappa_{\alpha'}\lambda_{l_1 i}\lambda_{l_2 i_1}\rho^0_{i_1 i_1} D^{0l_1l_2}_{\alpha\alpha'}(\omega=0) \quad \text{(C38)}$$

where

$$\rho^0_{ii} \equiv i\hbar G^{ii}_{\alpha\alpha}(t=0) = i\int \frac{dE}{2\pi} G^{ii}_{\alpha\alpha}(E) \quad \text{(C39)}$$

The phonon self-energy becomes

$$\Lambda^{ll'}_{\alpha\beta}(\omega) = -i\hbar \sum_{ij} \kappa_\alpha \kappa_\beta \lambda_{li}\lambda_{l'j}\int \frac{d\omega'}{2\pi} G^{0ij}_{\alpha\beta}(\omega') G^{0ji}_{\beta\alpha}(\omega'-\omega) + \frac{i}{\hbar}\sum_{ij} \kappa_\alpha \epsilon_\beta \lambda_{li}\lambda_{l'j}\rho^0_{ii}\rho^0_{jj}\delta(\omega=0) \quad \text{(C40)}$$

APPENDIX 14D: DYSON EQUATIONS IN THE +/− REPRESENTATION

In this appendix, we define the retarded and advance Green's functions and the corresponding self-energies and relate them to the basic Green functions and self-energies obtained in Appendix 14C. The Liouville space retarded (G_r) and advance (G_a) Green functions are defined as

$$G_r^{ij}(t,t') \equiv -\frac{i}{\hbar}\langle \mathcal{T}\psi_{i+}(t)\psi_{j-}^{\dagger}(t')\rangle \tag{D41}$$

$$G_a^{ij}(t,t') \equiv -\frac{i}{\hbar}\langle \mathcal{T}\psi_{i-}(t')\psi_{j+}^{\dagger}(t)\rangle \tag{D42}$$

We further introduce the correlation function

$$G_c^{ij}(t,t') \equiv -\frac{2i}{\hbar}\langle \mathcal{T}\psi_{i+}(t')\psi_{j+}^{\dagger}(t)\rangle \tag{D43}$$

It follows from Eq. (10) that there are only three Green functions in the +/− representation. These are given by Eqs. (D41)–(D43). Using Eq. (8) these can be represented in terms of the basic Green functions (31) as

$$\begin{aligned}
G_r^{ij}(t,t') &= \frac{1}{2}[G_{LL}^{ij}(t,t') - G_{LR}^{ij}(t,t') + G_{RL}^{ij}(t,t') - G_{RR}^{ij}(t,t')] \\
&= G_{LL}^{ij}(t,t') - G_{LR}^{ij}(t,t') \\
G_a^{ij}(t,t') &= \frac{1}{2}[G_{LL}^{ij}(t,t') - G_{RR}^{ij}(t,t') - G_{RL}^{ij}(t,t') + G_{LR}^{ij}(t,t')] \\
&= -G_{RR}^{ij}(t,t') + G_{LR}^{ij}(t,t') = G_{LL}^{ij}(t,t') - G_{RL}^{ij}(t,t') \\
G_c^{ij}(t,t') &= \frac{1}{2}[G_{LL}^{ij}(t,t') + G_{RR}^{ij}(t,t') + G_{LR}^{ij}(t,t') + G_{RL}^{ij}(t,t')] \\
&= G_{LL}^{ij}(t,t') + G_{RR}^{ij}(t,t')
\end{aligned} \tag{D44}$$

where we have used the identity $G_{LL} + G_{RR} = G_{LR} + G_{RL}$ which can be verified using Eq. (10). A Dyson equation corresponding to G_r, G_a and G_c can be obtained from Eq. (44) using unitary transformation

$$G = S\bar{G}S^{-1} \tag{D45}$$

where G represents the matrix

$$G = \begin{pmatrix} 0 & G_a \\ G_r & G_c \end{pmatrix} \tag{D46}$$

and

$$S = \frac{1}{\sqrt{2}}\begin{pmatrix} 1 & -1 \\ 1 & 1 \end{pmatrix} \tag{D47}$$

The transformed Dyson equation (44) reads

$$G = G^0 + G^0 \tilde{\Sigma} G \tag{D48}$$

and the corresponding self-energy matrix reduces to

$$\tilde{\Sigma} = \begin{pmatrix} \Sigma_{\mathrm{c}} & \Sigma_{\mathrm{r}} \\ \Sigma_{\mathrm{a}} & 0 \end{pmatrix} \tag{D49}$$

with the matrix elements given by

$$\begin{aligned} \Sigma_{\mathrm{r}}^{ij}(t,t') &= \Sigma_{\mathrm{LL}}^{ij}(t,t') + \Sigma_{\mathrm{LR}}^{ij}(t,t') \\ \Sigma_{\mathrm{a}}^{ij}(t,t') &= \Sigma_{\mathrm{RR}}^{ij}(t,t') + \Sigma_{\mathrm{LR}}^{ij}(t,t') \\ \Sigma_{\mathrm{c}}^{ij}(t,t') &= \Sigma_{\mathrm{RR}}^{ij}(t,t') + \Sigma_{\mathrm{LL}}^{ij}(t,t') \end{aligned} \tag{D50}$$

Similar relations also hold for the phonon Green functions and self-energies.

Using Eqs. (D44) and (D50), the retarded self-energies for electron and phonon Green functions (retarded) coming from the electron–phonon coupling are obtained as

$$\begin{aligned} \Xi_{\mathrm{r}}^{ij}(\omega) &= \mathrm{i}\hbar \sum_{ll'} \lambda_{l'i}\lambda_{lj} \int \frac{d\omega'}{2\pi} [D_{\mathrm{r}}^{0ll'}(\omega') G_{\mathrm{r}}^{0ij}(\omega-\omega') + D_{\mathrm{r}}^{0ll'}(\omega') G_{\mathrm{LR}}^{0ij}(\omega-\omega') \\ &\quad + D_{\mathrm{LR}}^{0ll'}(\omega') G_{\mathrm{r}}^{0ij}(\omega-\omega')] \\ \Lambda_{\mathrm{r}}^{ll'}(\omega) &= -\mathrm{i}\hbar \sum ij \lambda_{li}\lambda_{l'j} \int \frac{d\omega'}{2\pi} \Big[G_{\mathrm{LR}}^{0ij}(\omega') G_{\mathrm{a}}^{0ij}(\omega-\omega') \\ &\quad + G_{\mathrm{r}}^{0ij}(\omega')(G_{\mathrm{RL}}^{0ij}(\omega-\omega') + G_{\mathrm{a}}^{0ij}(\omega-\omega')) \Big] \end{aligned} \tag{D51}$$

Similarly the retarded self-energies due to the lead and secondary phonons can be written in the wide band limit as

$$\sigma_{\mathrm{r}}^{ij} = \frac{\mathrm{i}}{2}\Gamma^{ij} \qquad \text{and} \qquad \gamma_{\mathrm{r}}^{ll'} = -\frac{\mathrm{i}}{2}\tilde{\Gamma}^{ll'} \tag{D52}$$

where Γ^{ij} includes contributions from both the leads a and b, i.e. $\Gamma^{ij} = \Gamma_{\mathrm{a}}^{ij} + \Gamma_{\mathrm{b}}^{ij}$.

APPENDIX 14E: WICK'S THEOREM FOR SUPEROPERATORS

Wick's theorem for superoperators was formulated in Ref. [49]. Using Eqs. (8) and (27), it can be shown that similar to the L and R superoperators, the commutator of '+' and '−' boson superoperators are also numbers. Thus, boson superoperators follow Gaussian statistics and Wick's theorem holds for both the L, R and '+', '−' representations. However, for Fermi superoperators life is more complicated. The anticommutator corresponding to only the 'left' or the 'right' Fermi superoperators are numbers but that for the 'left' and 'right' superoperators, in general, is not a number. Thus, the Fermi superoperators are not Gaussian. However, since the left and right superoperators always commute, the following Wick's theorem [49] can be applied to the time-ordered product of any number of 'left' and 'right' superoperators, e.g.

$$\langle \mathcal{T} A_{i_1\nu_1}(t_1)A_{i_2\nu_2}(t_2)\cdots A_{i_n\nu_n}(t_n)\rangle_0 = \sum_p \langle \mathcal{T} A_{i_a\nu_a}(t_a)A_{i_b\nu_b}(t_b)\rangle_0 \cdots \langle \mathcal{T} A_{i_p\nu_p}(t_p)A_{i_q\nu_q}(t_q)\rangle_0 \tag{E53}$$

Here $A_{i_n\nu_n}$, $\nu_n = \mathrm{L, R}$, represents either a boson or a fermion superoperator. $i_a\nu_a \ldots i_q\nu_q$ is a permutation of $i_1\nu_1\ldots i_n\nu_n$ and sum on p runs over all possible permutations, keeping the time ordering. In case of fermions, each term should be multiplied by $(-1)^P$, where P is the number of permutations of superoperators required to put them into a particular order. Only permutations among either 'left' or among 'right' superoperators count in P. The permutations among 'L' and 'R' operators leave the product unchanged.

14.6 REFERENCES

1 I. Giaever, Phys. Rev. Lett., 5 (1960) 147.
2 J. Nicol, S. Shapiro and P.H. Smith, Phys. Rev. Lett., 5 (1960) 461.
3 R.C. Jaklevik and J. Lambe, Phys. Rev. Lett., 17 (1966) 119.
4 W. Wang, T. Lee, I. Kretzschmar and M.A. Reed, Nano Lett., 4 (2004) 643; J. Chen and M.A. Reed, Chem. Phys., 281 (2002) 127.
5 R.E. Walkup, D.M. Newns and Ph. Avouris, Phys. Rev. B, 48 (1993) 1858.
6 S. Gao, M. Persson and B.I. Landqvist, Solid State Commun., 84 (1992) 271.
7 W. Ho, J. Chem. Phys., 117 (2002) 11033.
8 B.C. Stipe, M.A. Rezaei and W. Ho, Science, 280 (1998) 1732.
9 F. Wang, D.C. Sorescu and K.D. Jordan, J. Phys. Chem., 106 (2002) 1316.
10 P. Sautet, Chem. Rev., 97 (1997) 1097.
11 W.A. Hofer, A.S. Foster and A.L. Shluger, Rev. Mod. Phys., 75 (2003) 1287.
12 J. Bardeen, Phys. Rev. Lett., 6 (1961) 57.
13 M.H. Cohen, L.M. Felicov and J.C. Phillips, Phys. Rev. Lett., 8 (1962) 31.
14 J. Tersoff and D.R. Hamann, Phys. Rev. Lett., 50 (1983) 1998; Phys. Rev. B 31 (1985) 805.
15 N.D. Lang, Phys. Rev. Lett., 55 (1985) 230.
16 B.N.J. Persson and A. Baratoff, Phys. Rev. Lett., 59 (1987) 339; N. Lorent and M. Persson, Phys. Rev. Lett., 85 (2000) 2997; N. Lorent, M. Persson, L.J. Lauhon and W. Ho, Phys. Rev. Lett., 85 (2001) 2997.
17 J.A. Appelbaum and W.F. Brinkman, Phys. Rev., 183 (1969) 553.
18 J.A. Appelbaum and W.F. Brinkman, Phys. Rev., 186 (1969) 464.
19 C. Caroli, R. Combescot, P. Nozieres and D.S. James, J. Phys. C: Solid State Phys., 5 (1972) 21.
20 C. Caroli, R. Combescot, P. Nozieres and D.S. James, J. Phys. C: Solid State Phys., 4 (1971) 916.
21 J. Schwinger, J. Math. Phys., 2 (1961) 407.
22 L.P. Kaddanof and G. Baym, Quantum statistical mechanics, Benjamin, New York, 1992.
23 L.V. Keldysh, Sov. Phys. J., 20 (1965) 1018.
24 R.A. Craig, J. Math. Phys., 9 (1968) 407.
25 H. Haug and A.-P. Jauho, Quantum kinetics in transport and optics of semiconductors, Springer, Heidelberg, 1996.
26 M. Galperin, A. Nitzan, S. Sek and M. Majda, J. Electroanal. Chem., 550 (2003) 337; M. Galperin, M.A. Ratner and A. Nitzan, Nano Lett., 4 (2004) 1605.
27 P.S. Damle, A.W. Ghosh and S. Datta, Phys. Rev. B, 64 (2001) 201403.
28 M. Di Ventra, S.-G. Kim, S.T. Pantelides and N.D. Lang, Phys. Rev. Lett., 86 (2001) 288; Z. Yang and M. Di Ventra, Phys. Rev. B, 67 (2003) 161311.
29 W. Tian, S. Datta, S. Hong, R. Reifenberger, J.I. Henderson and C.P. Kubiak, J. Chem. Phys., 109 (1998) 2874.
30 R. Baer, T. Seideman, S. Llani and D. Neuhauser, J. Chem. Phys., 120 (2004) 3387.
31 H. Ness, H.A. Shevlin and A.J. Fisher, Phys. Rev. B, 63 (2001) 125422.

32 L.A. Bumm, J.J. Arnold, M.T. Cygan, T.D. Dunbar, T.P. Burgin, L. Jones, D.L. Allara, J.M. Taur and P.S. Weiss, Science, 271 (1996) 1705.
33 X.D. Cui, A. Primak, X. Zarate, J. Tomfohr, O.F. Sankey, A.L. Moore, T.A. Moore, D. Gust, G. Harris and S.M. Lindsay, Science, 294 (2001) 571.
34 J. Park, A.N. Pasupathy, J.I. Goldsmith, A.V. Soldatov, C. Chang, Y. Yaish, J.P. Sethna, H.D. Abruna, D.C. Ralph and P.L. McEuen, Thin Solid Films, 438 (2003) 457.
35 N. Agrait, C. Untiedt, G. Rubio-Bollinger and S. Vieira, Phys. Rev. Lett., 88 (2002) 216803.
36 S. Datta, Electronic transport in mesoscopic systems, Cambridge University Press, Cambridge, 1997.
37 A. Nitzan, Annu. Rev. Phys. Chem., 52 (2001) 681.
38 S. Tikhodeev, M. Natario, K. Makoshi, T. Mii and H. Ueba, Surf. Sci., 493 (2001) 63.
39 T. Mii, S. Tikhodeev and H. Ueba, Surf. Sci., 502 (2002) 26.
40 A. Di Carlo, Physica B, 314 (2002) 211.
41 P. Hanggi, M. Ratner and S. Yaliraki (Eds.) Chem. Phys. (special issue) 281 (2002).
42 T. Mii, S.G. Tikhodeev and H. Ueba, Phys. Rev. B, 68 (2003) 205406.
43 Y. Xue, S. Datta and M.A. Ratner, Chem. Phys., 281 (2002) 151.
44 O. Berman and S. Mukamel, Phys. Rev. B, 69 (2004) 155430.
45 K. Chau, Z. Su, B. Hao and L. Yu, Phys. Rep., 118 (1985) 1; B. Hao, Physica A, 109 (1981) 221.
46 S. Mukamel, Principles of nonlinear optical spectroscopy, Oxford University Press, New York, 1997.
47 A.E. Cohen and S. Mukamel, Phys. Rev. Lett., 91 (2003) 233202.
48 E. Wang and U. Heinz, Phys. Rev. D, 66 (2002) 025008.
49 S. Mukamel, Phys. Rev. E, 68 (2003) 021111.
50 J.W. Negele and H. Orland, Quantum many particle systems, Addison-Wesley, New York, 1988.
51 T.L.C. Jansen and S. Mukamel, J. Chem. Phys., 119 (2003) 7979.
52 J.-P. Bouchaud, L. Cagliandolo, J. Kurchan and M. Mezard, Physica A, 226 (1996) 243.
53 J. Schofield, R. Lim and I. Oppenheim, Physica A, 181 (1992) 89.
54 N.E. Bickers, Rev. Mod. Phys., 59 (1987) 845.
55 N.S. Wingreen and Y. Meir, Phys. Rev. B, 49 (1994) 11040.
56 L. Hedin, Phys. Rev., 139 (1965) 796; J. Phys.: Condens. Matter 11, (1999) 489.
57 G. Onida, L. Reining and A. Rubio, Rev. Mod. Phys., 74 (2002) 601.
58 M. Rohlfing and S.G. Louie, Phys. Rev. B, 62 (2000) 4927.

Theory and Applications of Computational Chemistry: The First Forty Years
Edited by C. Dykstra et al.

CHAPTER 15

Role of computational chemistry in the theory of unimolecular reaction rates

William L. Hase[1] and Reinhard Schinke[2]

[1] *Department of Chemistry and Biochemistry, Texas Tech University, Lubbock, TX 79409-1061, USA*
[2] *Max-Planck Institut für Strömungsforschung 37073 Göttingen, Germany*

Abstract

Initiated by the chemical dynamics simulations of Bunker in 1962, of the unimolecular decomposition of model triatomic molecules, computational chemistry has had an enormous impact on the development of unimolecular rate theory. These pioneering calculations showed that the molecular lifetime distribution $P(t)$ is the critical diagnostic for RRKM and non-RRKM dynamics. They also identified properties of molecules which lead to non-RRKM dynamics. Chemical dynamics simulations motivated the need to distinguish between *intrinsic* non-RRKM dynamics resulting from restricted intramolecular vibrational energy redistribution (IVR) and *apparent* non-RRKM dynamics arising from non-random excitation and a non-microcanonical ensemble of initial states. Extensive calculations and study of the classical dynamics of vibrationally excited molecules led to an understanding of the relationship between RRKM and non-RRKM dynamics and a molecule's intramolecular motion. RRKM theory requires irregular trajectories and ergodic dynamics on the time scale for unimolecular decomposition. A molecule, whose ensemble of trajectories includes a significant fraction with regular motion, will have non-ergodic and intrinsic non-RRKM dynamics. Small molecules, with a low density of states, decompose *via* isolated compound state resonances. If the dynamics of such a molecule are ergodic, the fluctuations in properties of the resonances (i.e. lifetime, wave function, etc.) are purely random. On the other hand, if the molecule's dynamics is totally or in part regular, the resonances will exhibit mode-specific characteristics and all or some will be assignable. Quantum dynamics calculations of the resonance states do not show 'steps' in the unimolecular rate constant with increase in E, as predicted by RRKM theory, and they have suggested there may not be a fundamental reason for expecting steps in $k(E)$. Direct dynamics simulations, which couple electronic structure theory to the methodology of classical trajectory calculations, have made

References pp. 420–423

it possible to investigate the unimolecular and intramolecular dynamics of a diverse group of molecules. An important application of direct dynamics is the study of post-transition state dynamics. The statistical model for post-transition state dynamics is one in which the transition state, multiple minima and product channels are connected *via* reaction paths, with RRKM theory used to determine the time the molecular system spends in each potential minimum and the probabilities for transitions between minima and forming products. The actual dynamics may be investigated by direct dynamics simulations.

15.1 INTRODUCTION

Different approaches have been followed to develop a theory of unimolecular reaction rates [1]. For example, one may make fundamental assumptions concerning the intramolecular dynamics of highly vibrationally excited molecules and then develop a theoretical model based on these assumptions. This was the approach used in the derivation of the Slater [2] and Rice–Ramsperger–Kassel–Marcus (RRKM) [3–5] theories. The former assumes the vibrational motion of an energized molecule is represented by a superposition of the molecule's uncoupled normal mode coordinates. In contrast, the latter is based on statistical mechanics and assumes that energy is randomized amongst the molecule's vibrational degrees of freedom. If it is further assumed that unimolecular decomposition is described by a reaction coordinate which attains a critical value representing the transition from reactant to products (i.e. the transition state), the latter yields the following well known RRKM/quasi-equilibrium theory (QET) expression for the unimolecular rate constant:

$$k(E) = \frac{N^{\neq}(E)}{h\rho(E)} \tag{1}$$

Here, $N^{\neq}(E)$ is the sum of vibrational states for the transition state and $\rho(E)$ is the density of states for the unimolecular reactant. (In the above equation and in the following discussion, total angular momentum J and its conservation is not explicitly included in representing the RRKM/QET rate constant. This is done to simplify the presentation.)

Experimental studies have had an enormous impact on the development of unimolecular rate theory. A set of classical thermal unimolecular dissociation reactions by Rabinovitch and co-workers [6–10], and chemical activation experiments by Rabinovitch and others [11–14], illustrated that the separability and symmetry of normal modes assumed by Slater theory is inconsistent with experiments. For many molecules and experimental conditions, RRKM theory is a substantially more accurate model for the unimolecular rate constant. Chemical activation experiments at high pressures [15,16] also provided information regarding the rate of vibrational energy flow within molecules. Experiments [17,18] for which molecules are vibrationally excited by overtone excitation of a local mode (e.g. C–H or O–H bond) gave results consistent with the chemical activation experiments and in overall good agreement with RRKM theory [19].

A much more detailed microscopic picture of a unimolecular reaction may be obtained from time-domain experiments in which the number of reactant and/or product molecules is followed in real time [20–26]. At this level of experimental measurement, differences with the RRKM model have been observed. The unimolecular reactions studied in this manner include $H_2O_2 \rightarrow 2OH$ [20], $NCNO \rightarrow CN + NO$ [21], $HOCO \rightarrow HO + CO$ [22,23], $NO_2 \rightarrow NO + O$ [24,25], and $CH_3CO \rightarrow CO + CH_3$ [26]. Pronounced non-RRKM kinetics were observed for the latter two reactions.

At the most fundamental level, unimolecular dissociation occurs *via* resonance states, which may be viewed as an extension of the bound states into the vibrational/rotational continuum of the energized molecule [27]. Computational chemistry has pioneered the study of these resonances and this is discussed in detail in Section 15.2.4. The unimolecular resonances for a molecule may be characterized experimentally, if the average line-width of the resonances is smaller than the average distances between neighboring resonances, i.e. the resonances are isolated. The line width for a resonance state gives its lifetime and, therefore, its unimolecular rate constant. Thus, resolving all of the resonance states, in a specific energy range for an energized molecule, gives a detailed, fundamental understanding of the molecule's state-specific unimolecular kinetics. This type of information has been obtained from experiments for D_2CO [28], HFCO [29], HCO [30], DCO [31], CH_3O [32], HOCl [33,34], HN_3 [35], and HONO [36] decomposition, and extensive fluctuations (i.e. orders of magnitude) were observed for the resonance state rate constants in a small energy range. In contrast to these fluctuations in rate constants at the quantum mechanical level, RRKM theory assumes the unimolecular rate constant $k(E)$ increases smoothly (or with small steps) [27] as E is increased. As discussed in Section 15.2.4, for molecules whose classical unimolecular dynamics agrees with RRKM theory, it is possible to find a relationship between the quantum mechanical state specific rates and the RRKM $k(E)$. Analyses of the computational quantum dynamics have been very helpful in this regard.

Initiated by the chemical dynamics simulations of Bunker [37,38] for the unimolecular decomposition of model triatomic molecules, computational chemistry has had an enormous impact on the development of unimolecular rate theory. Some of the calculations have been exploratory, in that potential energy functions have been used which do not represent a specific molecule or molecules, but instead describe general properties of a broad class of molecules. Such calculations have provided fundamental information concerning the unimolecular dissociation dynamics of molecules. The goal of other chemical dynamics simulations has been to accurately describe the unimolecular decomposition of specific molecules and make direct comparisons with experiment. The microscopic chemical dynamics obtained from these simulations is the detailed information required to formulate an accurate theory of unimolecular reaction rates. The role of computational chemistry in unimolecular kinetics was aptly described by Bunker [37] when he wrote "The usual approach to chemical kinetic theory has been to base one's decisions on the relevance of various features of molecular motion upon the outcome of laboratory experiments. There is, however, no reason (other than the arduous calculations involved) why the bridge between experimental and theoretical reality might not equally well start on the opposite side of the gap. In this paper... results are reported of the simulation of the motion of large numbers of 'triatomic' molecules by a very fast digital computer." In the following sections, descriptions are given of how computational chemistry has improved our understanding of the lifetime distribution, non-RRKM kinetics, the role of classical phase space structures, resonance states, and 'steps' in unimolecular rate theory. A discussion is also given of the important impact expected from direct dynamics simulations of unimolecular decomposition. The article concludes with a 'look' into the future.

15.2 ROLE OF COMPUTATIONAL CHEMISTRY

15.2.1 The lifetime distribution

The Hinshelwood–Lindemann model [5], in which molecules are activated and deactivated by collisions, is well accepted for describing the temperature T and pressure P dependence of thermal unimolecular reactions. The unimolecular rate constant, $k_{\text{uni}}(T, P)$, for this model is given by

$$k_{\text{uni}}(T,P) = \omega \int_{E_0}^{\infty} \frac{k(E)P(E)\mathrm{d}E}{k(E) + \omega} \tag{2}$$

where ω is the collision frequency proportional to pressure, $k(E)$ is the unimolecular rate constant for reactants with energy E, and $P(E)$ is the Boltzmann probability the reactants have energy E. Slater [2] and Bunker [37,38] showed that Eq. (2) requires the probability the reactant molecules have lifetime t, i.e. the lifetime distribution $P(t)$, is given by

$$P(t) = -\frac{1}{N(0)}\frac{\mathrm{d}N(t)}{\mathrm{d}t} = k(E)\mathrm{e}^{-k(E)t} \tag{3}$$

where $N(0)$ is the number of reactant molecules at $t = 0$. Eq. (3) assumes that the ensemble of reactant molecules has the same unimolecular rate constant $k(E)$ at all times as it dissociates, so

$$-\frac{\mathrm{d}N(t)}{\mathrm{d}t} = k(E)N(t) \tag{4}$$

Integration of this equation yields Eq. (3). For RRKM kinetics, both the intercept of $P(t)$ and its exponential decay are given by the RRKM $k(E)$.

The questions posed, at the time of the understanding of the relationship between Eqs. (2) and (3), are whether actual molecules behave according to Eq. (3) and if the rate constant $k(E)$ is given by the RRKM/QET model in Eq. (1). The latter was probed by experimental studies, in particular the classic work by Rabinovitch and co-workers [6–11]. However, experimental techniques were not sufficiently developed to test Eq. (3) and the validity of this expression was first tested in computer simulations by Bunker [37,38].

Bunker [37] surmised that the unimolecular rate constant $k(E)$ should correspond to that for a microcanonical ensemble of reactant molecules, for which every state in the energy window $E \rightarrow E + dE$ has equal probability of being populated. The classical mechanical representation of this ensemble is a uniform distribution over the reactant's phase space, a procedure still used today. Trajectories were then calculated for this ensemble of initial conditions to determine whether its lifetime distribution corresponded to Eq. (3).

The intercept of Eq. (3) is the unimolecular rate constant $k(E)$ for a microcanonical ensemble of reactant states. Bunker found that $k(E)$ is well represented by the RRKM expression in Eq. (1) if anharmonicity effects are included for $N^{\neq}(E)$ and $\rho(E)$ and if variational effects are included in identifying the transition state for reactions which do not have a barrier for the reverse association reaction [37–40]. Each of these two findings motivated extensive future studies [41–47].

Bunker's pioneering finding was that not all of his model triatomic molecules have lifetime distributions in accord with Eq. (3). Models with disparate force constants and/or masses, giving rise to a hierarchy of low and high vibrational frequencies, have a non-exponential $P(t)$, i.e.

$$P(t) = \sum_i c_i \, e^{-k_i t} \tag{5}$$

with an initial rate $d \ln P(t)/dt$ larger than that of RRKM theory and a long-time rate much smaller than the RRKM $k(E)$. (By definition the sum of the c_i in Eq. (5) equals $k(E)$, since the initial decay is for a microcanonical ensemble.) An initial rate constant larger than that of RRKM theory suggests the molecule is behaving smaller than its actual size. The possibility of such behavior was first proposed by Rice [48] of RRKM theory. A long-time rate, much smaller than the RRKM $k(E)$, suggests that the decomposition of the long-time trajectories is not controlled by the transition state for chemical reaction, but by a 'bottleneck' for the transfer of energy into the coordinate(s) leading to decomposition [38]. Bunker identified the phase space structure for his non-RRKM models as metrically decomposable [37]. His finding of non-RRKM dynamics was consistent with and followed the earlier simulation by Fermi et al. [49] of non-random energy transfer in a linear chain. Examples of RRKM [50] and non-RRKM [51] $P(t)$ distributions are given in Figs. 15.1 and 15.2.

Bunker's work was pioneering in that it documented, by computational chemistry simulations, that actual molecules may have non-RRKM unimolecular dynamics. The concept of intramolecular relaxation [i.e. intramolecular vibrational energy redistribution (IVR)] and its role in insuring RRKM dynamics was introduced. Oxtoby and Rice [52] later showed that Bunker's non-RRKM models do not have overlapping resonances, which are necessary for efficient IVR and RRKM dynamics. Preliminary discussions were also given by Bunker of the relationships between RRKM and non-RRKM dynamics and the structure of the phase space for a unimolecular reactant. The following statement by Bunker [37] is very insightful and a forerunner to later detailed analyses of the relationships between unimolecular dynamics and classical phase space structures, which are discussed in the next section, i.e. "In terms of the molecular phase space: a change in the character of the surface H, from harmonic to anharmonic, alters very little the number of representative points on trajectories which lead quickly (i.e. within a few vibrations periods) to dissociation. Points originally confined to non-dissociating trajectories require some little time to 'leak' into regions from which the critical surface

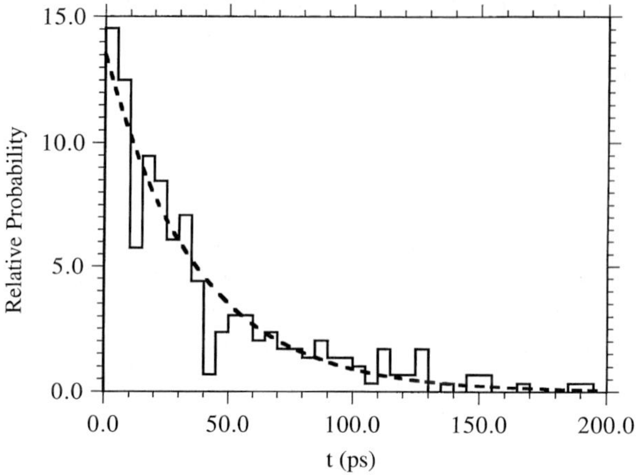

Fig. 15.1. Trajectory $Al_6 \rightarrow Al_5 + Al$ lifetime distribution, following microcanonical sampling. The histogram plot represents the number of Al_6 dissociations per time interval. The dashed line represents the random lifetime distribution of Eq. (3). The energy is 40 kcal/mol and the angular momentum is zero. Adapted from Ref. [50].

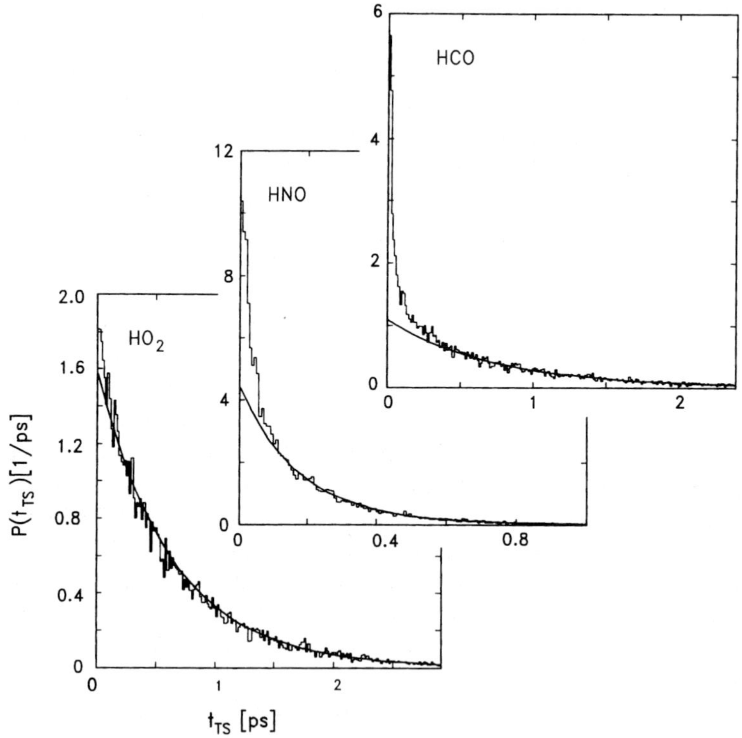

Fig. 15.2. Trajectory lifetime distributions for HCO, HNO, and HO_2 dissociation, following microcanonical sampling. Extensive intrinsic non-RRKM dynamics are present in HCO and HNO decomposition. The smooth curves are obtained by fitting the data to an exponential function, with the initial fast decomposition times being excluded. Adapted from Ref. [51].

is easily accessible. This interpretation, that anharmonic representative points stay near the corresponding harmonic trajectories for an appreciable time, may be verified by inspection of the details of trajectories."

15.2.2 Intrinsic and apparent non-RRKM behavior

Though RRKM theory assumes an excited molecule decomposes from a microcanonical ensemble of states, it is doubtful whether any experiment actually excites a molecule in this manner. For example, there is a non-microcanonical component in the ensemble of states prepared by thermal collisions. For excitation processes like chemical activation, i.e. $F + C_2H_5 \rightarrow C_2H_5F^*$ [53], overtone excitation, i.e. $HOO{-}H + h\nu \rightarrow HOOH^*$ [54], and internal conversion, i.e. $H_2CO\ (S_1) \rightarrow H_2CO\ (S_0)^*$ [55], the initially prepared ensemble of states is expected to be far from microcanonical. Computational chemistry simulations motivated the need to distinguish between non-RRKM dynamics resulting from restricted intramolecular relaxation, as discussed in the previous section, and non-RRKM dynamics arising from non-random excitation and a non-microcanonical ensemble of initial states [56–60].

Intrinsic RRKM behavior is defined by Eq. (3), where an initial microcanonical ensemble of states decomposes exponentially with the RRKM rate constant [56]. Such dynamics can be investigated by computational chemical dynamics simulations. Therefore, an intrinsic non-RRKM molecule is one for which the intercept in $P(t)$ is $k(E)$, as a result of the initial microcanonical ensemble, but whose decomposition probability *versus* time is not described by $k(E)$. For such a molecule there is a bottleneck (or bottlenecks) restricting energy flow into the dissociating coordinate. Intrinsic RRKM and non-RRKM dynamics are illustrated in Fig. 15.3(a), (b), and (e).

Apparent non-RRKM behavior arises from non-random initial excitation [56], which is present in all experiments, including collisional activation (Fig. 15.3(c)) [56]. As a result, the initial form of the lifetime distribution, including intercept will not be the exponential assumed by RRKM theory. Nevertheless, the RRKM assumption of rapid IVR requires the distribution of states to become microcanonical in a negligibly short time compared to the RRKM lifetime $1/k(E)$, resulting in a RRKM $P(t)$ following the initial non-RRKM short-time component. The extent of this component, and whether it enhances or suppresses reaction, depends on properties of the excitation process and where the energy is initially localized in the molecule. If the molecule is also intrinsically non-RRKM, its $P(t)$ will be a convolution of its apparent and intrinsic non-RRKM dynamics. Three possible $P(t)$ with apparent non-RRKM behavior are illustrated in Fig. 15.3(d). The pioneering chemical activation experiments of Rabinovitch and co-workers [15,16] are classic examples of short time apparent non-RRKM behavior and longer time RRKM dynamics.

Trajectory calculations have been used to study the intrinsic RRKM and apparent non-RRKM dynamics of ethyl radical dissociation, i.e. $C_2H_5 \rightarrow H + C_2H_4$ [61,62]. When C_2H_5 is excited randomly, with a microcanonical distribution of states, it dissociates with the exponential $P(t)$ of RRKM theory [61], i.e. it is an intrinsic RRKM molecule. However, apparent non-RRKM behavior is present in a trajectory simulation of C_2H_5

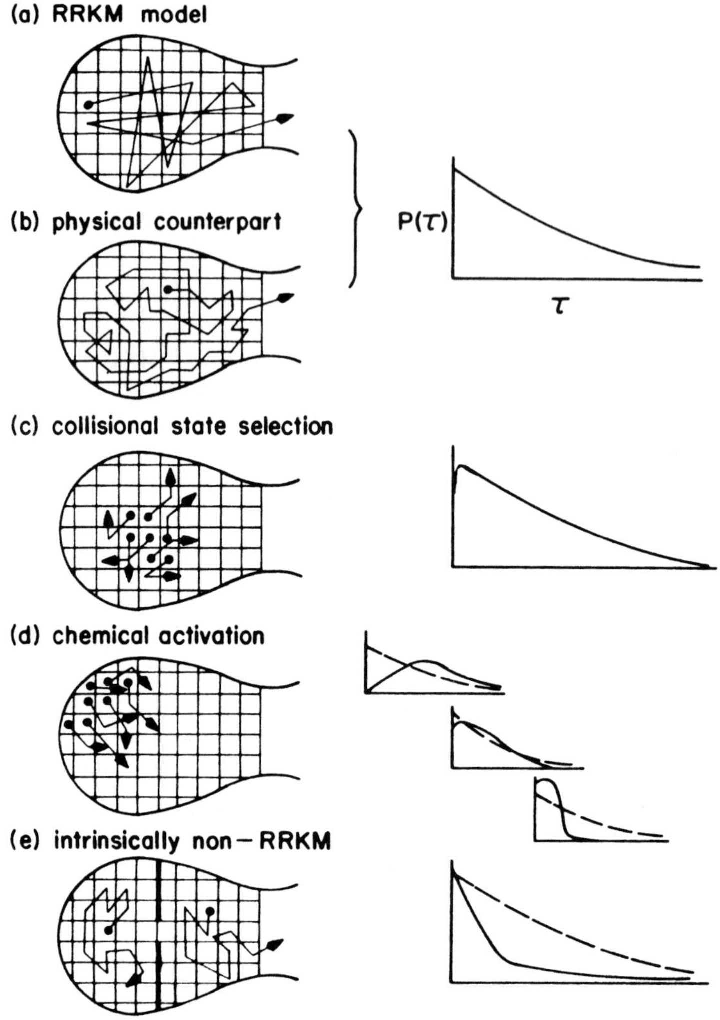

Fig. 15.3. Relation of state occupation, schematically shown at constant energy, to the lifetime distribution for RRKM theory and other situations. Dashed lines in the lifetime distributions indicate RRKM behavior. Adapted from Ref. [56].

non-randomly excited by $H + C_2H_4$ association [62]. For a $H + C_2H_4$ relative translational energy of 30 kcal/mol, association deposits both the 40 kcal/mol reaction exothermicity and the 30 kcal/mol relative translational energy non-randomly into C_2H_5, with preferential excitation of the formed C−H bond as compared to the other C−H bonds. As shown in Fig. 15.4, this leads to an initial dissociation rate that is faster than that of RRKM theory. This apparent non-RRKM component lasts for about 0.25 ps and then the rate constant (i.e. the slope of $\log N$ *versus* t) attains the RRKM value of $7 \times 10^{10}\ s^{-1}$.

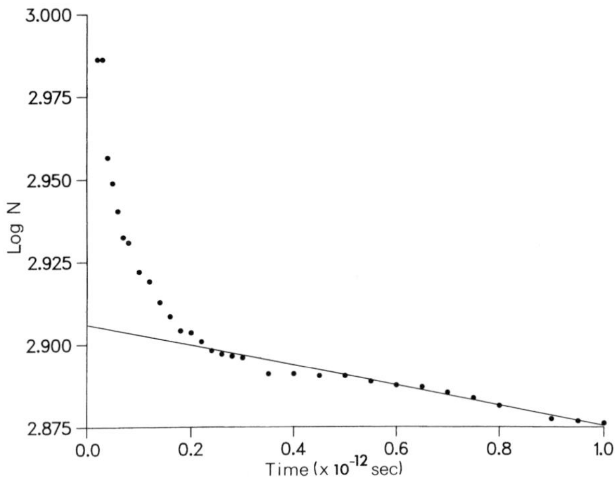

Fig. 15.4. Classical trajectory lifetime distribution for decomposition of C_2H_5, formed by $H + C_2H_4$ association. Adapted from Ref. [62].

15.2.3 Phase space structures

The above apparent non-RRKM and intrinsic RRKM and non-RRKM dynamics are reflections of a molecule's phase space structure. Extensive calculations and study of the classical dynamics of vibrationally excited molecules have shown that they may have different types of motions, e.g. regular and irregular [63]. A trajectory is regular if its motion may be represented by a separable Hamiltonian, for which each degree of freedom is uncoupled and moves independent of the other degrees of freedom. All trajectories are regular for the normal mode Hamiltonian, i.e.

$$H = \sum_i H_i = \sum_i (P_i^2 + \lambda_i Q_i^2)/2 \tag{6}$$

which is the model assumed by the Slater theory [2]. This Hamiltonian is only rigorously accurate for small displacements from equilibrium and as energy is increased anharmonic and vibration/rotation coupling terms become increasingly important and must be added to the Hamiltonian [47]. The simulations show that these terms, and the presence of 'internal' or 'anharmonic' resonances [64,65], begin destroying some of the regular motion as the energy is increased and the fraction of the trajectories in the phase space with regular motion, i.e. f_{reg}, becomes smaller. At some energy, the number of regular trajectories becomes negligibly small, leading to ergodic dynamics. The increase in chaotic dynamics, as the energy is increased, is illustrated in Fig. 15.5 for a two-dimensional model of HOCl; the OH bond distance is frozen in the calculations [66].

There are multiple ways to determine if a trajectory's motion is regular. For a large molecule a straightforward approach is to determine if it has well-defined vibration frequencies, a diagnostic of independent, uncoupled motion for each coordinate. This is

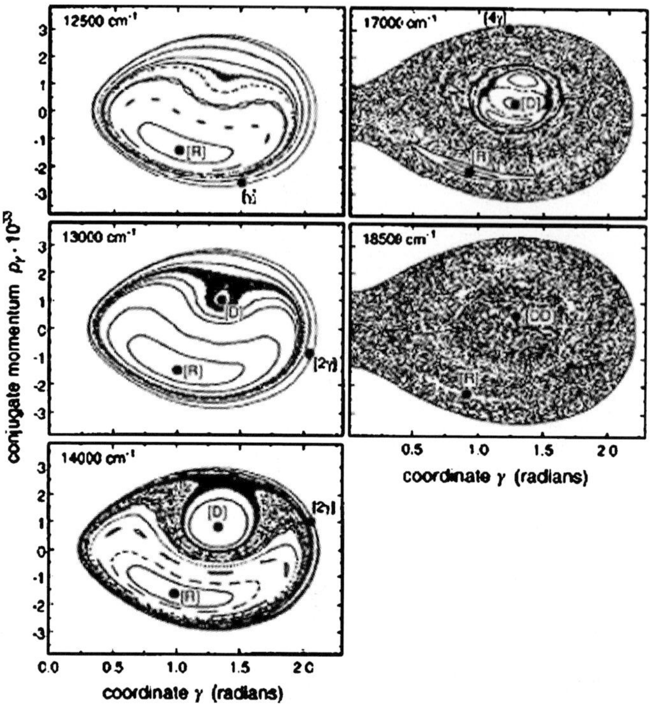

Fig. 15.5. Poincare' surface of section for a two-dimensional model of HOCl; the HO distance is frozen in the classical trajectory simulations. The symbol γ is the Jacobi angle and p_γ is the corresponding momentum. Different symbols in square brackets denote different types of regular trajectories. Adapted from Ref. [66].

done by taking the Fourier transform of the auto-correlation function for the molecule's coordinate displacements *versus* time [67]. In contrast, an irregular trajectory has a broad spectrum which cannot be characterized by specific vibration frequencies for individual coordinates [67]. An analysis of its motion shows chaotic dynamics.

If regular trajectories are important for a vibrationally excited molecule, its unimolecular dynamics is intrinsically non-RRKM. Regular and irregular trajectories are depicted in Fig. 15.6, where plots are given of the H–C *versus* C–C bond lengths, H–C bond length *versus* H–C–C angle, and C–C bond length *versus* H–C–C angle as a function of time for a Hamiltonian representing the model system H–C–C → H + C = C [68,69]. The trajectories are excited above the unimolecular threshold with the same total energy. The different motions for the trajectories are a reflection of where they are initiated in H–C–C's phase space. The regular trajectories are excited with a large fraction of the energy in the C–C bond, which leads to regular dynamics. On the other hand, energy is more equally distributed between the H–C and C–C stretch and H–C–C bend degrees of freedom for the irregular trajectories. The regular trajectories exhibit quasiperiodic motion and are trapped in the H–C–C phase space. They may never dissociate. Quantum numbers for each of the quasiperiodic trajectories' three degrees of freedom may be determined by quantizing their action,

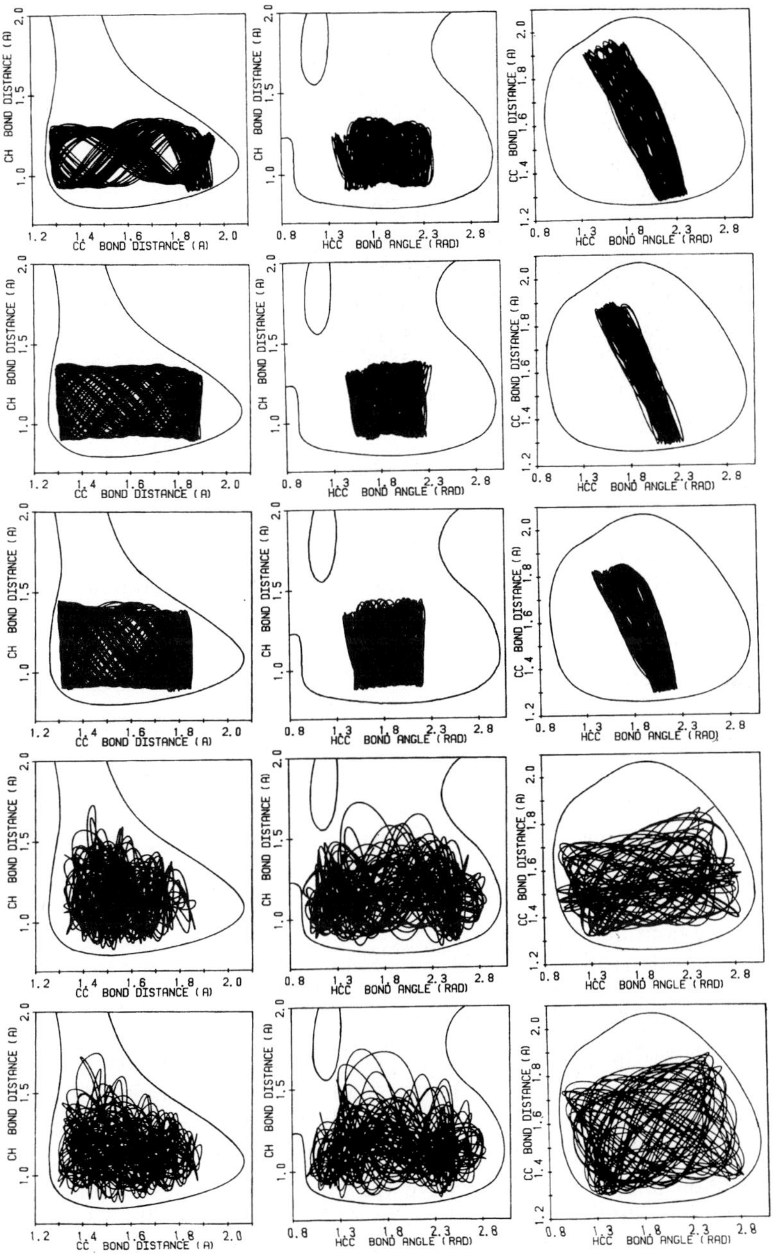

Fig. 15.6. Regular and irregular trajectories for the model H–C–C → H + C = C reaction. Adopted from Ref. [68].

as given by the Einstein–Brillouim–Keller (EBK) semiclassical quantization condition [70]

$$\oint p(q)\mathrm{d}q = (n + 1/2)h \tag{7}$$

The correspondence between classical and quantum mechanics tells us that this trajectory corresponds to a resonance state with a localized wave function in the H–C–C vibration continuum. The quantum mechanical resonance state (discussed in detail in Section 15.2.4) will have nearly the same energy and be assignable with the semiclassical quantum numbers. It is also expected to have a very long lifetime as a result of the classical quasiperiodic motion.

The irregular trajectories in Fig. 15.6 display the type of motion expected by RRKM theory. These trajectories moves chaotically throughout the coordinate space, not restricted to any particular type of motion. RRKM theory requires this type of irregular motion for all of the trajectories so that the intramolecular dynamics is ergodic [1]. In addition, for RRKM behavior the rate of intramolecular relaxation associated with the ergodicity must be sufficiently rapid so that a microcanonical ensemble is maintained as the molecule decomposes [1]. This assures the RRKM rate constant $k(E)$ for each time interval $t \rightarrow t + \mathrm{d}t$. If the ergodic intramolecular relaxation is slower than $1/k(E)$, the unimolecular dynamics will be intrinsically non-RRKM.

A simple phase space structure with intrinsic non-RRKM dynamics is one consisting of only regular and irregular trajectories, with the latter exhibiting chaotic intramolecular dynamics on a time scale shorter than that for unimolecular decomposition. Though this phase space structure has an exponential lifetime distribution $P(t)$, as assumed by RRKM theory for an initial microcanonical ensemble of states, the two $P(t)'$s are not the same. The intercept of $P(t)$ for the non-RRKM phase space is still $k(E)$ [37], but the rate constant in the exponential is larger reflecting the smaller volume of phase space in which the irregular trajectories move. The density of states for the irregular trajectories is $f_{\mathrm{irreg}}\rho(E)$, so that the rate constant becomes $k(E)/f_{\mathrm{irreg}}$.

Though the above analysis is instructive, computer simulations show that the actual intramolecular and unimolecular dynamics is significantly more complex for a molecule with a non-negligible fraction of regular trajectories. This is clearly illustrated by the non-exponential $P(t)$ for intrinsic non-RRKM molecules. The complexity of the molecule's dynamics arises from trajectories in the vicinity of regular trajectories and the non-ergodic structure of the molecule's multi-dimensional phase space. For a period of time, trajectories in the vicinity of regular trajectories retain some degree of regularity in their motion and, thus, do not have the chaotic dynamics assumed by RRKM theory [63]. As a result, their unimolecular lifetimes may be much longer than expected by RRKM theory. The chaotic regions of phase space, with irregular trajectories, are intermingled with the regions of phase space with regular and 'somewhat' regular trajectories and are connected *via* an Arnold web [64,65] of anharmonic resonances. This complex, non-ergodic phase space structure leads to a non-exponential $P(t)$. The general structure of the classical phase space is also reflected in the quantum mechanical decay rates discussed next.

15.2.4 Resonance states

In the quantum mechanical modeling of unimolecular reactions, resonances are the central quantities. Resonance states are the continuation of a molecule's bound states into its continuum above the dissociation threshold. Unlike the true bound states resonances have a finite width Γ and therefore a finite lifetime $\tau = \hbar/\Gamma$, where $\hbar$ is Planck's constant divided by 2π. The dissociation rate is then given by $k = \Gamma/\hbar$. However, these simple relations between width and dissociation rate are valid only when the resonances do not (significantly) overlap, i.e. when $\langle\Gamma\rangle/\Delta E = \langle\Gamma\rangle\rho(E)$ is markedly smaller than one. Here, $\langle\Gamma\rangle$ is the average width in a small energy interval, $\rho(E)$ is the density of states and ΔE is the average spacing between the resonances. For overlapping resonances the relationship between width and dissociation rate is more involved [71].

Resonances are prominent features in many fields of physics and chemistry such as nuclear [72] and atomic [73] physics, electron [74] and molecular [75] scattering, and photodissociation [76]. A detailed account of resonances in molecular systems, with particular emphasis on unimolecular reactions, has been given by Grebenshchikov et al. [27]. Experimentally, resonances in molecules can be studied in several ways; preparation by electronic excitation [28] (i.e. excitation to a bound state in an excited electronic state, which is mixed by non-adiabatic coupling with resonance states of the ground electronic state), overtone pumping [54] (i.e. direct excitation by several quanta of one bond, usually a bond which involves a hydrogen atom), stimulated emission pumping [77] (excitation of a long-lived excited electronic state and subsequent stimulated emission to a particular rovibrational state of the ground electronic state). These methods allow one to determine the widths of individual resonances states, even with rotational resolution [78]. Illuminating examples of molecules for which resonances have been studied are D_2CO [28], HFCO [29], HCO [30], DCO [31], CH_3O [32], HOCl [33,34], and HONO [36]. Results of resonance widths for the dissociation of HCO are depicted in Fig. 15.7. The pronounced state and mode dependence are significant and remarkable.

Because resonances are states in the continuum of a potential well, their numerical treatment is much more involved than the calculation of true bound states. For a given potential energy surface (PES) resonance energies E_r and widths Γ can be calculated by several methods [27]. In an indirect approach one calculates a quantity which includes the quantum mechanical wave function, e.g. the scattering cross section or an absorption-type spectrum, as a function of energy. These quantities show sharp features in the vicinity of a resonance, and provided the resonances do not significantly overlap, it is easy to extract E_r and Γ. This method is similar to a spectroscopic experiment. In a direct approach the positions and widths are obtained as complex eigenvalues of the Hamiltonian H with eigenfunctions satisfying special boundary conditions as the fragments separate. The two most common numerical approaches for obtaining the complex eigenvalues are complex scaling [79] and augmentation of the Hamiltonian with a complex absorbing potential [80–82], i.e. $\hat{H}_{\text{eff}} = \hat{H} - i\lambda W(R)$. In the complex-scaling method the dissociation coordinate R is rotated into the complex plane, which makes the kinetic and potential energy complex, but forces the eigenstates to become square integrable. The function of the absorbing potential is to damp the wave function in the different fragmentation channels, with the effect that the wave function becomes

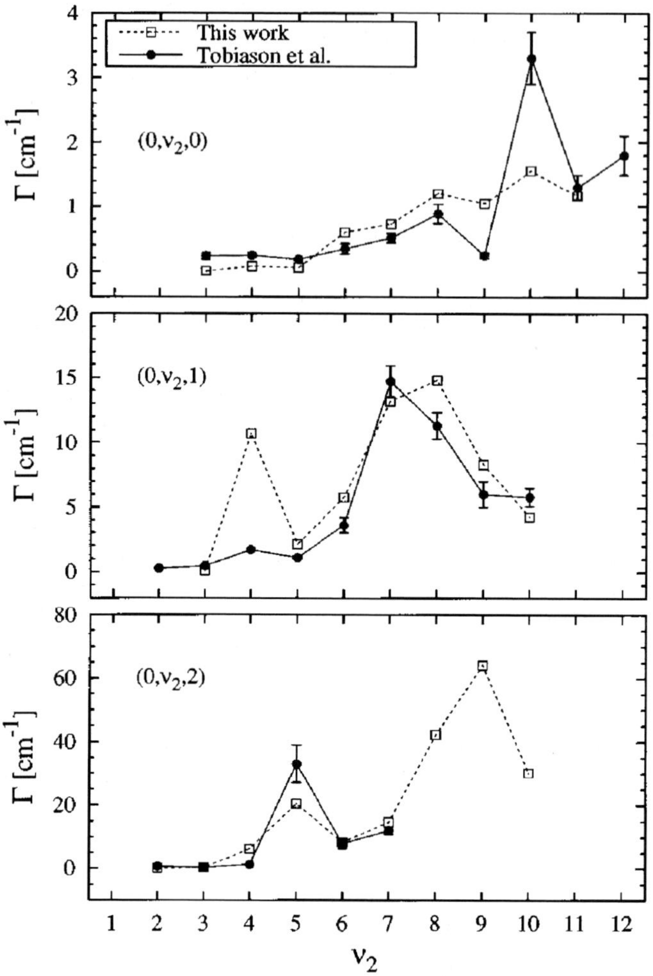

Fig. 15.7. Calculated (dashed lines and open squares) and measured (solid lines and filled circles) HCO resonance widths as a function of the CO stretching quantum number ν_2 and the bending quantum number ν_3 for three progressions. Adapted from Ref. [30].

square integrable. In both methods numerical procedures known from bound state calculations can be used with only small modifications. Examples of extensive resonance calculations on (accurate) ground state PESs include HCO [83], DCO [84], HOCl [85], H_2O [86], and NO_2 [25]. Although not trivial, resonance calculations for triatomic molecules are feasible.

Resonance energies and widths can be compared to experimental results. Since these are state-specific data and do not include any sort of averaging, in contrast to thermal rate constants, the accuracy of a PES can be assessed directly. Molecules for which such comparisons have been performed are HCO [83], DCO [84], and HOCl [85]. In Fig. 15.7 the calculated widths for HCO are compared to the experimental results. Although for a few states the agreement is not perfect, the general trends are nicely reproduced by the calculations. The maxima superimposed on a smoothly increasing 'background' are due to mixings with nearby 'dark' states [87]. These mixings are difficult to reproduce with accuracy and therefore deviations between experiment and theory are largest for these 'resonances' between resonance states. If an accurate PES is not available, valuable information can be deduced from fitting experimental data with an effective Hamiltonian model as done for DCO [88].

All spectroscopic experiments as well as all resonance calculations univocally revealed that the state-specific rate constant k_i is not a smooth function, which gradually increases with energy, but instead fluctuates over several orders of magnitude. A particularly intriguing example is shown in Fig. 15.8 [89]. Near threshold, the rates fluctuate over seven orders of magnitude and at higher energies k_i still varies over three orders. HOCl as well as HCO are examples of mode-specific decay. In such cases, the intramolecular coupling is comparatively weak with the result that many states, even resonance states, can be uniquely assigned, i.e. the corresponding wave functions have a clear nodal structure [83]. As a consequence, states which are very close in energy can have dissociation rates which differ by orders of magnitude. As energy becomes larger the average rate increases and the differences between states become generally smaller. For HOCl and HCO it is essentially the large mismatch between the HO and CO frequencies, respectively, and the other two frequencies which lead to the weak coupling. Also shown in Fig. 15.8 is the result of the statistical adiabatic channel model (SACM). Statistical theories cannot reproduce the large fluctuations in the rates. Especially the very small rates are not represented by the statistical theory.

If the intramolecular coupling is strong, i.e. if the molecule is more irregular, the differences between neighboring states are not so large and the fluctuations of the rates extend over a smaller range as compared to those for a regular molecule such as HOCl.

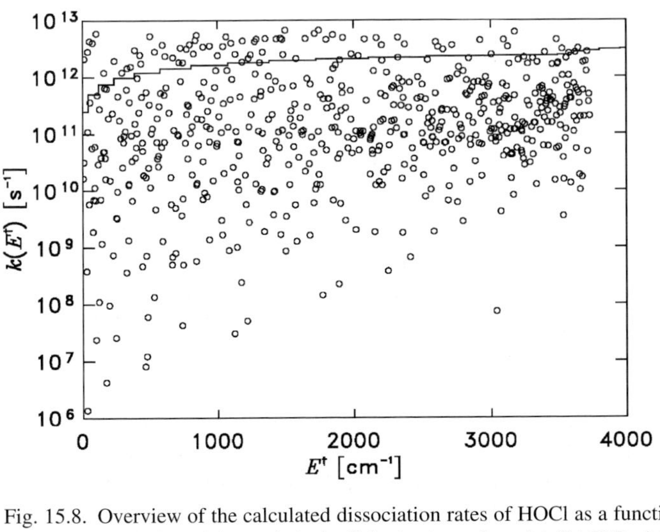

Fig. 15.8. Overview of the calculated dissociation rates of HOCl as a function of the excess energy. The solid line is the prediction of the statistical adiabatic channel model (SACM). Adapted from Ref. [89].

Nevertheless, even for a molecule like NO_2, which is known to be classically chaotic, the fluctuations near threshold are still more than two orders of magnitude [25]. The large fluctuations of the dissociation rates lead to non-exponential decay curves as has been demonstrated for NO_2 in a time-dependent pump-probe experiment [25]. The short-time decay is determined by the larger rates, whereas the long-time behavior is governed by the smallest rates. Statistical models reproduce the short-time behavior satisfactorily, but fail to correctly describe the decay at long times. The case of NO_2 can be classified as statistical state specific [47]. Other molecules which fall into this category are HO_2 [90] and D_2CO [8]. For molecules like NO_2 and D_2CO comparisons with experimental data on a state-specific level are difficult and one has to retreat to averaged quantities like the average dissociation rate or the distribution of rates. On the basis of random matrix theory, developed in nuclear physics, the rates follow a χ^2 distribution [91,92]. The measured rates for D_2CO as well as the calculated rates for NO_2 confirm this prediction.

The classical unimolecular dynamics is ergodic for molecules like NO_2 and D_2CO, whose resonance states are highly mixed and unassignable. As described above, their unimolecular dynamics is identified as statistical state specific. The classical dynamics for these molecules are intrinsically RRKM and a microcanonical ensemble of phase space points decays exponentially in accord with Eq. (3). The correspondence found between statistical state specific quantum dynamics and quantum RRKM theory is that the average of the N resonance rate constants $\langle k_i \rangle$ in an energy window $E \rightarrow E + \Delta E$ approximates the quantum RRKM rate constant $k(E)$ [27,90]. Because of the state specificity of the resonance rates, the decomposition of an ensemble of the N resonances is non-exponential, i.e.

$$P(t) = \sum_{i=1}^{N} k_i \, \mathrm{e}^{-k_i t}/N \tag{8}$$

and not the exponential, Eq. (3), of classical RRKM theory [47]. The χ^2 distribution of resonance rates, described above for statistical state specific decomposition [91,92], has a width specified by the parameter ν [47]. For unimolecular reaction, for energies in excess of the decomposition threshold, ν has been associated with the transition state sum of states of $N^{\neq}(E)$ [92]. The χ^2 distribution narrows as $N^{\neq}(E)$ increases. Inserting this distribution into Eq. (8) and taking the $\nu \rightarrow \infty$ limit, results in an exponential $P(t)$ with a rate constant given by $\langle k_i \rangle$ [47].

An interesting question is whether the large fluctuations in the quantum mechanical decay rates have an influence on the temperature and pressure dependent unimolecular rate constant $k_{\mathrm{uni}}(T, P)$ defined within the strong collision model, in Eq. (2). In the state-specific quantum mechanical approach the integral over the smooth temperature dependent rate $k(E)$ is replaced by a sum over the state-specific rates k_i. Applications have been done for HCO [93], HO_2 [94–96], and HOCl [97]. The effect of a broad distribution of widths is to decrease the observed pressure dependent rate constant as compared to the delta function-like distribution, assumed by statistical theories [98,99]. The reason is that broad distributions favor small decay rates and the overall dissociation slows down. This trend, pronounced in the fall-of region, was clearly seen in a study of thermal rate constants in the unimolecular dissociation of HOCl [97]. The extremely broad distribution of resonances in HOCl causes a decrease by a factor of two in the pressure-dependent rate, as compared to the RRKM predictions. The best chance to see the influence of the quantum mechanical fluctuations on unimolecular rate constants are studies performed close to the dissociation threshold, i.e. at low collision temperatures, because there the distribution of rates is the broadest. The state-specific ansatz for the unimolecular rate constant has been tested by Hippler et al. [100] by comparison with experimental rates for HCO measured over large ranges of T and P. They claimed: "An analysis of the pressure dependence of the rate constants shows that standard treatments like a Lindemann–Hinshelwood model or a master equation formalism are not able to describe the experimentally observed pressure dependence adequately. An isolated resonance model, however, which is based on calculated resonance lifetimes... allows for a description of the observed T- and P-dependence."

The main obstacle for applying the state-specific variant of Eq. (2) is the calculation of resonance widths for many rotational states (J, K). Even at room temperature many pairs (J, K) are populated and each rotational state has its own decay rate $k_i(J, K)$. Exact calculations for large angular momenta are prohibited because of 'exploding' computer times. Therefore, one has to rely on approximations such as the commonly used J-shifting approximation [101]. Efficient determination of reliable resonance widths for high rotational states, potentials with deep wells, and molecules like O_3 with heavy atoms is the bottleneck for calculating unimolecular reaction rates within the quantum mechanical approach.

15.2.5 Steps in unimolecular reaction rates

The RRKM/QET unimolecular rate constant $k(E)$ in Eq. (1) is derived from classical mechanics. The widely used quantum RRKM model is an *ad hoc* modification of the classical RRKM rate constant by simply replacing the transition state's sum of states $N^{\neq}(E)$ and the reactant molecule's density of states $\rho(E)$ by their quantum counterparts. Though this gives a 'working equation' it is not rigorously justified by quantum mechanics. An interesting feature of the quantum RRKM $k(E)$ expression is that $N^{\neq}(E)$ increases incrementally as E is increased. The minimum rate constant is at the threshold E_0 where $N^{\neq}(E_0) = 1$, i.e. $k(E_0) = 1/h\rho(E)$. A stepwise increase is then predicted for $k(E)$ as additional states become available at the TS with increase in E. The largest step is the first one for an increase in $N^{\neq}(E)$ from 1 to 2. As illustrated by the harmonic model for the TSs vibrational energy levels, the energy intervals, at which successive steps occur, are determined by the TSs vibrational frequencies, i.e.

$$E^{\neq}(n^{\neq}) = \sum_{i=1}^{3N-7} (n_i^{\neq} + 1/2)h\nu_i^{\neq} \qquad (9)$$

Thus, it would be possible to determine properties of PESs in TS regions by experimental measurements of such steps, if they exist and are indeed described by quantum RRKM theory.

The first report of the observation of steps in $k(E)$ was by Lovejoy et al. [102] for ketene dissociation on its triplet PES. This was followed by the study of Ionov et al. [103] for which steps were observed in $NO_2 \rightarrow NO + O$ dissociation. Leu et al. [104] also observed step-like structures in $k(E)$ for triplet acetaldehyde, CH_3CHO, dissociation and associated them with energy levels at the TS.

These were indeed exciting results, and the report by Lovejoy et al. lead to the pronouncement that the fundamental assumption of a TS for unimolecular dissociation with quantized energy levels had been proven [105,106]. However, the proposal that the above steps arise from TS quantum energy level structures has not withstood tests of computational chemistry. Triplet ketene dissociates by tunneling and RRKM modeling of the observed steps requires a broad barrier with an imaginary frequency of $\sim 100i\ cm^{-1}$ [107]. In contrast, high-level *ab initio* calculations give a much narrower barrier of $321i\ cm^{-1}$ [108], which gives rise to tunneling rates that would occlude any RRKM steps. A similar result was found for triplet acetaldehyde, for which RRKM modeling [104] gives a barrier frequency of $60i\ cm^{-1}$ in contrast to the *ab initio* prediction of frequency greater than $350i\ cm^{-1}$ [109]. Thus, RRKM analyses of the reported steps for both triplet ketene and acetaldehyde are inconsistent with TS properties of their PESs determined from *ab initio* calculations.

The reported steps for $NO_2 \rightarrow NO + O$ dissociation are inconsistent with a RRKM calculation based on an accurate PES for the reaction [110]. The observed steps for NO_2 dissociation are separated by $\sim 100\ cm^{-1}$ which, if resulting from a quantized TS, indicates a O–NO bending frequency at $\sim 100\ cm^{-1}$. However, adiabatic potential energy curves calculated by Grebenshchikov et al. for this dissociation show that the bending frequency at the TS (i.e. variational in this case) is much less than $100\ cm^{-1}$. This is expected given the looseness of the variational TS and the heavy masses. Thus, the reported steps for NO_2 dissociation are inconsistent with RRKM theory. The origin of the steps in the experimental studies may arise from fluctuations in the unimolecular decay of wave packets prepared in the experiments [25]. The wave packets are superpositions of the unimolecular resonance states and the fluctuations in the wave packet rates would originate from the extraordinary fluctuations in the resonance rates discussed in Section 15.2.4.

As pointed out above, RRKM theory has its roots in classical mechanics and the classical RRKM $k(E)$ is continuous without steps. Steps are introduced by the *ad hoc* introduction of the quantum $N^{\neq}(E)$ which is quantized. Computational chemistry dynamical calculations have provided the type of information needed to determine whether steps should be expected in the unimolecular rate constant $k(E)$. As pointed out by Grebenshchikov et al. [27] "In all quantum mechanical calculation which have been performed as well as in the few state-resolved experiments (D_2CO, CH_3O, HCO, and DCO) of unimolecular decomposition the resonance decay rates are found to fluctuate over several orders of magnitude. These fluctuations are especially pronounced and the distribution of rates is the widest near the dissociation threshold—exactly where the 'step' in the RRKM rate is expected to be the largest. The increase of the RRKM (i.e. average) rate by a factor of 2 is much smaller than the breadth of the distribution of rates—even if the narrowest resonances are not taken into account. Thus, the variation of the quantum mechanical rates in a small energy interval exceeds greatly the energy variation of $k(E)$ as predicted by RRKM theory at a 'step'. Moreover, an increase in E so that $N^{\neq}(E)$ increases by 1 has little effect on the quantum distribution of the resonance rates—even at threshold." Thus, the quantum RRKM prediction of steps in $k(E)$ is inconsistent with the findings from quantum dynamical calculations.

Grebenshchikov et al. [27] have also shown that "the existence of steps can be questioned by reviewing the main assumptions made in RRKM theory. First, the *ad hoc* replacement of the classical $N^{\neq}(E)$ by its quantum analog completely neglects the state-specific nature of the coupling of the resonance states to the continuum. Second, the energy levels at the TS are defined in the adiabatic approximation, in which the 'slow' dissociation mode is decoupled from the 'fast' vibrational modes. In a more accurate description non-adiabatic matrix elements, which couple the different adiabatic channels, have to be taken into account. They are usually not small at the TS where the potential 'perpendicular' to the reaction path and likewise the adiabatic vibrational wave functions change considerably as the reaction coordinate varies. Inclusion of non-adiabatic coupling would tend to smear out the 'steps'. Third, tunneling through the one-dimensional adiabatic potential curves also tends to round-off the step-like structures predicted by the RRKM expression as was convincingly demonstrated by model calculations for ketene by Gezelter and Miller [107]."

The conclusion one reaches is that quantum RRKM theory is an incomplete model for unimolecular decomposition. It does not describe fluctuations in state-specific resonance rates, which arise from the nature of the couplings between the resonance states and the continuum. It also predicts steps in $k(E)$, which appear to be inconsistent with the actual quantum dynamics as determined from computational chemistry. However, for molecules whose classical unimolecular dynamics is ergodic and intrinsically RRKM and/or whose resonance rates are statistical state specific (see Section 15.2.4), the quantum RRKM $k(E)$ gives an accurate average rate constant for an energy interval $E \rightarrow E + \Delta E$ [47].

15.2.6 Impact of direct dynamics simulations

Direct dynamics simulations, in which the methodology of classical trajectory simulations is coupled to electronic structure, have had and will continue to have an enormous impact on the use of computational chemistry to develop [111,112] the theory of unimolecular kinetics. In these simulations the derivatives of the potential, $\partial V/\partial q_i$, required for numerically integrating the classical trajectory, are obtained directly from electronic structure theory without the need for an analytic PES. Direct dynamics is particularly important for studying the unimolecular dynamics of molecules with many degrees of freedom, for which it is difficult to construct an accurate analytic PES.

Two methods, identified as Car–Parinello [113] and Born–Oppenheimer [114], have been advanced for performing direct dynamics simulations. For the former, the motions of the electrons are determined simultaneously as the nuclear classical equations of motion are integrated, to determine the change in the electronic wave function as the nuclei move. For the second method the electronic wave function is optimized during the numerical integration of the classical trajectory.

An important issue in direct dynamics simulations is the often large amount of computer time required to calculate the trajectories and the need to use a level of electronic structure theory which is computationally feasible and also gives a meaningful representation of the reaction's PES. Clearly, a higher level and more accurate theory may be used for a small molecule as compared to a larger one. A practical approach for studying the unimolecular decomposition of large molecules is to use a semiempirical electronic structure theory, but re-parameterized with specific reaction parameters (SRPs) [115,116] to fit properties of the reaction's PES obtained from experiment and/or high-level *ab initio* calculations. These properties often include structures, energies, and vibrational frequencies for the PES's stationary points and reaction path(s). The resulting model may also be supplemented with analytic potential energy functions to improve its accuracy [117].

Direct dynamics has made it possible to investigate the unimolecular decomposition of a broad group of molecules for different excitation processes, to compare with experiment and determine fundamental information concerning intramolecular and unimolecular dynamics. Summarized in Table 15.1 are the unimolecular direct dynamics simulations performed by the Hase research group [117–129]. Some degree of non-RRKM behavior is present in each of the reactions. It would not have been possible to determine this level of understanding of the unimolecular dynamics of these reactions without access to direct dynamics.

An important application of direct dynamics is the study of post-transition state intramolecular and unimolecular dynamics. To understand the issues at hand for these processes, consider starting at a high energy transition state and looking towards the reaction products and observing a rough landscape, with multiple potential energy minima, reaction pathways, low energy barriers, etc. connecting the transition state to multiple product channels. The statistical model for the exit-channel dynamics is one in which the transition state, multiple minima and product channels are connected *via*

Table 15.1 Direct dynamics simulations of unimolecular decomposition

Unimolecular reaction	Level of theory	Simulation result	Reference
Trimethylene decomposition	AM1-SRP	Apparent non-RRKM dynamics	[117–120]
Cl^{-}–CH_3Cl decomposition	MP2/6-31G*	$[Cl–CH_3–Cl]^{-}$ central barrier recrossing and intrinsic non-RRKM dynamics	[121]
$[HO–CH_3–F]^{-} \rightarrow HO^{-} + CH_3F$ decomposition	MP2/6-31+G*	Non-reaction path and non-statistical dynamics	[122]
$Cr^{+}(CO)_6$ and protonated peptide surface induced dissociation (SID)	AM1	Shattering fragmentation, apparent non-RRKM dynamics	[123–125]
Vinylcyclopropane → cyclopentene rearrangement	AM1-SRP	Apparent non-RRKM dynamics	[126,127]
$O(^3P) + C_2H_6 \rightarrow$ products	MSINDO, PM3-SRP	Shattering decomposition	[128,129]

reaction paths, with RRKM theory used to determine the time the molecular system spends in each potential minimum and the probabilities for transitions between minima and forming products. For the actual intramolecular and unimolecular dynamics to follow such a scenario, the potential energy that is released, as the system moves off the transition state, must be rapidly equilibrated between all the degrees of freedom by IVR. If instead, the potential energy is released to only a subset of the degrees of freedom and it remains there, extensive non-RRKM exit-channel dynamics is expected.

The question concerning exit-channel dynamics described above have been investigated for the $OH^- + CH_3F \rightarrow CH_3CH + F^-$ S_N2 reaction [122], whose reaction path potential energy curve is shown in Fig. 15.9. The geometry for the initial reaction path, from the central barrier toward products, has an approximate O–C–F collinear axis. At s of about 5 amu$^{1/2}$-Bohr the reaction path enters a flat region that is apparently a remnant of a backside potential energy minimum that is 'lost' because of the presence of the much deeper $CH_3OH–F^-$ hydrogen-bonded minimum. At s of approximately 12 amu$^{1/2}$-Bohr the reaction path leaves the flat region and starts its descent into the $CH_3OH–F^-$ minimum. RRKM theory assumes the molecular system becomes temporarily trapped in this minimum, with its vibrational energy randomized. Thus, the reaction path/RRKM mechanism is an indirect process.

MP2/6-31+G* direct dynamics was used to study the actual exit-channel dynamics for this reaction. Initial conditions for the trajectories were selected by sampling the $[HO–CH_3–F]^-$ central barrier's 300 K Boltzmann distribution. The simulations showed

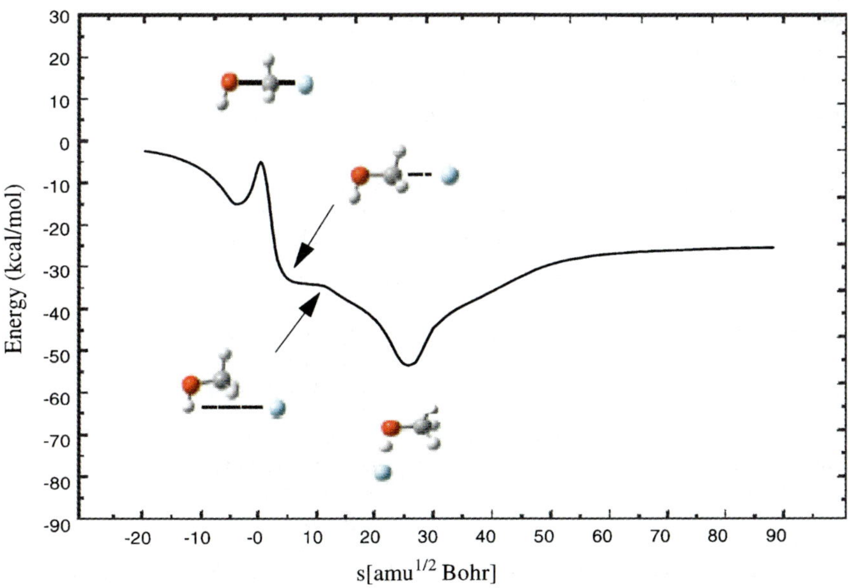

Fig. 15.9. Potential energy along the intrinsic reaction coordinate (IRC) for $OH^- + CH_3F \rightarrow CH_3OH + F^-$; s is the distance along the IRC. This figure shows the structures at the potential energy minima and at the saddle point barrier. Adapted from Ref. [121].

References pp. 420–423

that only a small fraction (roughly 10%) of the trajectories actually followed the above reaction path. The vast majority of them followed a direct dissociation pathway with a departure of the F^- ion approximately along the $O-C-F^-$ collinear axis. Fig. 15.10 depicts these two pathways for two of the trajectories. The origin of the propensity for this latter pathway is seen in Fig. 15.11, where the potential energy is plotted *versus* the $C-F^-$ distance and the $O-C-F^-$ angle. The release of potential energy to the asymmetric $O-C-F^-$ reaction coordinate's stretch motion, as the system moves off the central barrier, tends to propel F^- from CH_3OH, with the $O-C-F^-$ angle maintained at nearly 180°. The PES is rather flat for bending the $O-C-F^-$ angle; there is only a very weak force to pull the reactive system from the direct dissociation reaction path into the

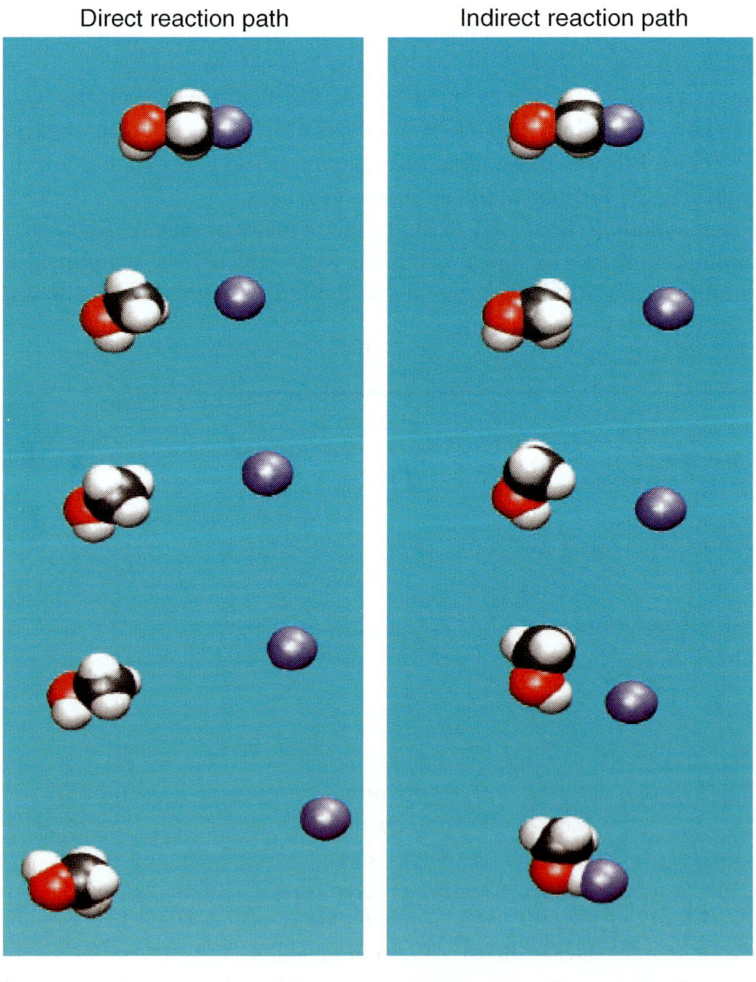

Fig. 15.10. The two pathways for motion from the $[HO-CH_3-F]^-$ central barrier to the $CH_3OH + F^-$ reaction products. Most of the trajectories follow the direct dissociation path. A small amount, roughly 10%, form the CH_3OH-F^- hydrogen-bonded intermediate and follow an indirect path. Adapted from Ref. [121].

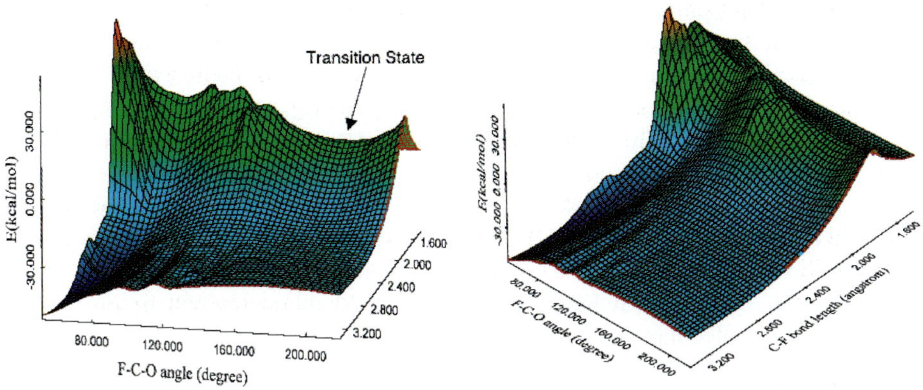

Fig. 15.11. Potential energy contour diagram for $[HO-CH_3-F]^-$ fragmentation as a function of the C–F distance and the O–C–F angle. The remaining coordinates are optimized at each point on the PES. Adapted from Ref. [121].

CH_3OH-F^- potential energy well, with an $O-C-F^-$ angle of 102.8°. As the system moves off the central barrier, it tends to move directly to products without forming an intermediate trapped in the CH_3OH-F^- potential energy well.

This simulation shows how the efficiency of IVR and formation of a statistical reaction intermediate are intimately linked to the hierarchy of time scales for intramolecular motions and structural transitions on the PES. Inefficient formation of the CH_3OH-F^- reaction immediate arises from rapid separation of the $CH_3OH + F^-$ products in comparison to the longer time scale for $O-C-F^-$ bending to form the intermediate. Such a hierarchy of time scales appears in previous work, in which a solvent cage's structural changes are too slow to allow a reactive system to access its deep potential energy minimum [130]. A hierarchy of time scales and inefficient access of deep potential energy minima could be important in enzyme catalysis, in which motions associated with the reaction center may be much faster than those associated with conformational changes and other multi-atom motions of the enzyme.

15.3 THE FUTURE

Computational chemistry will continue to make important contributions to unimolecular rate theory. The continual advance in computer technology and the development of computer programs with more sophisticated numerical algorithms and theoretical methods will allow the application of computational chemistry to larger scale and more complex unimolecular processes. Quantum dynamics provides a fundamental understanding of unimolecular rate theory and additional calculations of resonances in unimolecular decomposition are needed. These studies can address the role of resonances for larger molecules (e.g. the calculations to date have been for three atoms) and also the importance of vibrational/rotational coupling. Total angular momentum J is conserved, but is uncertain whether the projection of J onto a molecular axis, identified by the K quantum number, is also conserved [131]. Extensive vibrational/rotational coupling destroys K as a good quantum number. However it is possible that, though K may not be a good quantum number for long-time intramolecular dynamics, it may be conserved on the shorter time scale for unimolecular decomposition. Dynamics associated with the K quantum number may be investigated by classical as well as quantum dynamics.

Continual advances will be made in using direct dynamics to study unimolecular dynamics. By directly coupling classical dynamics with quantum mechanics allows one to probe the dynamics of a broad spectrum of unimolecular reactions. In future calculations, it is expected that RRKM and non-RRKM dynamics will be probed for many different unimolecular reactions to compare with and interpret experiments. The unimolecular dynamics of very large molecules may be studied by QM/MM methods [132,133], for which degrees of freedom directly participating in the reaction are represented by quantum mechanical electronic structure theory (QM) and analytic molecular mechanics (MM) functions are used for the remaining degrees of freedom. QM/MM direct dynamics may be used to study the intramolecular and unimolecular dynamics of large biomolecules such as peptides.

15.4 ACKNOWLEDGEMENTS

The authors' research in unimolecular dynamics is funded by the National Science Foundation (USA), Welch Foundation (USA), and Deutsche Forschungsgemeinschaft (Germany). Their understanding of unimolecular rate theory has benefited enormously from work with their students, post-doctorals, and collaborators.

15.5 REFERENCES

1 W.L. Hase, in: W.H. Miller (Ed.), Dynamics of molecular collisions, part B, Plenum, New York, 1976, p. 121.
2 N.B. Slater, Theory of unimolecular reactions, Cornell University Press, Ithaca, NY, 1959.
3 R.A. Marcus and O.K. Rice, J. Phys. Colloid Chem., 55 (1951) 894; R.A. Marcus, J. Chem. Phys., 20 (1952) 359.
4 H.M. Rosenstock, M.B. Wallenstein, A.L. Wahrhafting and H. Eyring, Proc. Natl Acad. Sci. USA, 38 (1952) 667.
5 In the ion-molecule community RRKM theory is widely known as quasiequilibrium theory (QET); see Ref. [6].
6 F.W. Schneider and B.S. Rabinovitch, J. Am. Chem. Soc., 84 (1962) 4215.
7 F.W. Schneider and B.S. Rabinovitch, J. Am. Chem. Soc., 85 (1963) 2365.
8 B.S. Rabinovitch, P.W. Gilderson and F.W. Schneider, J. Am. Chem. Soc., 87 (1965) 158.
9 K.M. Maloney and B.S. Rabinovitch, J. Phys. Chem., 73 (1969) 1652.
10 K.M. Maloney, S.P. Pavlou and B.S. Rabinovitch, J. Phys. Chem., 73 (1969) 2756.
11 D.W. Setser and B.S. Rabinovitch, Can. J. Chem., 40 (1962) 1425.
12 R.L. Johnson and D.W. Setser, J. Phys. Chem., 71 (1967) 4366.
13 J.W. Simons and G.W. Taylor, J. Phys. Chem., 73 (1969) 1274.
14 W.L. Hase and J.W. Simons, J. Chem. Phys., 52 (1970) 4004.
15 J.D. Rynbrandt and B.S. Rabinovitch, J. Phys. Chem., 75 (1971) 2164.

16 J.F. Meagher, K.J. Chao, J.R. Barker and B.S. Rabinovitch, J. Phys. Chem., 78 (1971) 2535.
17 J.M. Jasinski, J.K. Frisoli and C.B. Moore, J. Chem. Phys., 79 (1983) 1312.
18 F. Crim, Annu. Rev. Phys. Chem., 35 (1984) 657.
19 W.L. Hase, Chem. Phys. Lett., 116 (1985) 312.
20 N.F. Scherer and A.H. Zewail, J. Chem. Phys., 87 (1987) 97.
21 L.R. Khundkar, K.L. Knee and A.H. Zewail, J. Chem. Phys., 87 (1987) 77.
22 N.F. Scherer, C. Sipes, R.B. Bernstein and A.H. Zewail, J. Chem. Phys., 92 (1990) 5239.
23 S.I. Ionov, G.A. Brucker, C. Jaques, L. Valachovic and C. Wittig, J. Chem. Phys., 99 (1993) 6553.
24 S.I. Ionov, G.A. Brucker, C. Jaques, Y. Chen and C. Wittig, J. Chem. Phys., 99 (1993) 3420.
25 B. Kirmse, B. Abel, D. Schwarzer, S.Yu. Grebenshchikov and R. Schinke, J. Phys. Chem. A, 104 (2000) 10374.
26 S.K. Kim, J. Guo, J.S. Baskin and A.H. Zewail, J. Phys. Chem., 100 (1996) 9202.
27 S.Yu. Grebenshchikov, R. Schinke and W.L. Hase, Comprehensive chemical kinetics, in: N.J.B. Green (Ed.), Unimolecular kinetics part 1. The reaction step, Elsevier, Amsterdam, 2003, p. 105.
28 W.F. Polik, D.R. Guyer and C.B. Moore, J. Chem. Phys., 92 (1990) 3453.
29 Y.S. Choi and C.B. Moore, J. Chem. Phys., 97 (1992) 1010.
30 J.D. Tobiason, J.R. Dunlap and E.A. Rohlfing, J. Chem. Phys., 103 (1995) 1448.
31 C. Stöck, X. Li, H.-M. Keller, R. Schinke and F. Temps, J. Chem. Phys., 106 (1997) 5333.
32 S. Dertinger, A. Geers, J. Kappert, J. Wiebrecht and F. Temps, Faraday Discuss. Chem. Soc., 102 (1995) 31.
33 G. Dutton, R.J. Barnes and A. Sinha, J. Chem. Phys., 111 (1999) 4976.
34 A. Callegari, J. Rebstein, R. Jost and T.R. Rizzo, J. Chem. Phys., 111 (1999) 7359.
35 B.R. Foy, M.P. Casassa, J.C. Stephenson and D.S. King, J. Chem. Phys., 90 (1989) 7037.
36 F. Reiche, B. Abel, R.D. Beck and T.R. Rizzo, J. Chem. Phys., 116 (2002) 10267.
37 D.L. Bunker, J. Chem. Phys., 37 (1962) 393.
38 D.L. Bunker, J. Chem. Phys., 40 (1964) 1946.
39 D.L. Bunker and M. Pattengill, J. Chem. Phys., 48 (1968) 772.
40 D.L. Bunker, in: C. Schlier (Ed.), Proceedings of the International School of Physics "EnricoFermi" Course XLIV. Molecular beams and reaction kinetics, Academic, New York, 1970, p. 315.
41 L.B. Bhuiyan and W.L. Hase, J. Chem. Phys., 78 (1983) 5052.
42 W.L. Hase, J. Chem. Phys., 57 (1972) 730.
43 W.L. Hase, J. Chem. Phys., 64 (1976) 2442.
44 W.L. Hase, Acc. Chem. Res., 16 (1983) 258.
45 B.C. Garrett and D.G. Truhlar, J. Chem. Phys., 70 (1979) 1593.
46 D.G. Truhlar and B.C. Garrett, Acc. Chem. Res., 13 (1980) 440.
47 T. Baer and W.L. Hase, Unimolecular reaction dynamics. Theory and experiments, Oxford, New York, 1996.
48 O.K. Rice, Z. Phys. Chem. B, 7 (1930) 226, (quoted by L.S. Kassel, Kinetics of homogeneous reactions, Chemical Catalog, New York, 1932).
49 E. Fermi, J. Pasta and S. Ulam, Studies on non-linear problems I, LA-1940, Los Alamos Scientific Laboratory, Los Alamos, NM, 1955, November 2.
50 G.H. Peslherbe and W.L. Hase, J. Chem. Phys., 105 (1996) 7432.
51 M. Stumpf, A.J. Dobbyn, D.H. Mordaunt, H.-M. Keller, H. Fluethmann, R. Schinke, H.-J. Werner and K. Yamashita, Faraday Discuss., 102 (1995) 193.
52 D.W. Oxtoby and S.A. Rice, J. Chem. Phys., 65 (1976) 1676.
53 H.W. Chang, N.L. Craig and D.W. Setser, J. Phys. Chem., 76 (1972) 954.
54 F.F. Crim, Annu. Rev. Phys. Chem., 35 (1984) 659.
55 W.F. Polik, C.B. Moore and W.H. Miller, J. Chem. Phys., 89 (1988) 3584.
56 D.L. Bunker and W.L. Hase, J. Chem. Phys., 59 (1973) 4621.
57 W.L. Hase and D.F. Feng, J. Chem. Phys., 61 (1974) 4690.
58 C.S. Sloane and W.L. Hase, J. Chem. Phys., 66 (1977) 1523.
59 R.J. Wolf and W.L. Hase, J. Chem. Phys., 72 (1980) 316.
60 W.L. Hase, R.J. Duchovic, K.N. Swany and W.L. Hase, J. Chem. Phys., 80 (1984) 1462.

61 W.L. Hase and D.G. Buckowski, J. Comput. Chem., 3 (1982) 335.
62 W.L. Hase, D.G. Buckowski and K.N. Swamy, J. Phys. Chem., 87 (1983) 2754.
63 A.J. Lichtenberg and M.A. Lieberman, Regular and stochastic motion, Springer, New York, 1983.
64 G.S. Ezra, in: W.L. Hase (Ed.), Advances in classical trajectory methods, Vol. 1, JAI, London, 1992, p. 1.
65 S. Keshavamurthy and G.S. Ezra, J. Chem. Phys., 107 (1997) 156.
66 M. Joyeux, S.C. Farantos and R. Schinke, J. Phys. Chem. A, 106 (2002) 5407.
67 D.W. Noid, M.L. Koszykowski and R.A. Marcus, J. Chem. Phys., 67 (1977) 404.
68 R.J. Wolf and W.L. Hase, J. Chem. Phys., 73 (1980) 3779.
69 W.L. Hase, J. Phys. Chem., 86 (1982) 2873.
70 I.C. Percival, Adv. Chem. Phys., 36 (1977) 1.
71 Y. Alhassid and Y.V. Fyodorov, J. Phys. Chem. A, 102 (1998) 9577.
72 C. Mahaux and H.A. Weidenmüller, Shell-model approach to nuclear reactions, North-Holland, Amsterdam, 1969.
73 J.-M. Rost, Phys. Rep., 297 (1998) 271.
74 W. Domcke, Phys. Rep., 208 (1991) 97.
75 D.G. Truhlar (Ed.), Resonances in electron-molecule scattering, van der Waals complexes, and reactive chemical dynamics, American Chemical Society, Washington, DC, 1984.
76 R. Schinke, Photodissociation dynamics, Cambridge University Press, Cambridge, 1993.
77 H.-L. Dai and R.W. Field (Eds.), Molecular dynamics and spectroscopy by stimulated emission pumping, World Scientific, Singapore, 1995.
78 J. Wei, A. Tröllsch, C. Tesch and F. Temps, J. Chem. Phys., 120 (2004) 10530.
79 N. Moiseyev, Phys. Rep., 302 (1998) 211.
80 G. Jolicard and E.J. Austin, Chem. Phys., 103 (1986) 295.
81 I. Last, D. Neuhauser and M. Baer, J. Chem. Phys., 96 (1992) 2017.
82 U.V. Riss and H.-D. Meyer, J. Phys. B, 24 (1993) 4503.
83 H.-M. Keller, H. Flöthmann, A.J. Dobbyn, R. Schinke, H.-J. Werner and P. Rosmus, J. Chem. Phys., 105 (1996) 4893.
84 H.-M. Keller, T. Schröder, M. Stumpf, C. Stöck, F. Temps, R. Schinke, H.-J. Werner, C. Bauer and P. Rosmus, J. Chem. Phys., 106 (1997) 5359.
85 J. Weiss, J. Hauschildt, R. Schinke, O. Haan, S. Skokov, J.M. Bowman, V.A. Mandelshtam and K.A. Peterson, J. Chem. Phys., 115 (2001) 8880.
86 S.K. Gray and E.M. Goldfield, J. Phys. Chem. A, 105 (2001) 2634.
87 U. Brandt-Pollmann, J. Weiss and R. Schinke, J. Chem. Phys., 115 (2001) 8876.
88 R. Renth, F. Temps and A. Tröllsch, J. Chem. Phys., 118 (2003) 659.
89 J. Hauschildt, J. Weiss, C. Beck, S.Yu. Grebenshchikov, S. Düren, R. Schinke and J. Koput, Chem. Phys. Lett., 300 (1999) 569.
90 M. Stumpf, A.J. Dobbyn, H.-M. Keller, W.L. Hase and R. Schinke, J. Chem. Phys., 102 (1995) 5867; A.J. Dobbyn, M. Stumpf, H.-M. Keller and R. Schinke, J. Chem. Phys., 104 (1995) 8357.
91 R.D. Levine, Adv. Chem. Phys., 70, Part 1 (1988) 53.
92 W.F. Polik, D.R. Guyer, W.H. Miller and C.B. Moore, J. Chem. Phys., 92 (1990) 3471.
93 J. Qi and J.M. Bowman, J. Chem. Phys., 105 (1996) 9884.
94 V.A. Mandelshtam, H.S. Taylor and W.H. Miller, J. Chem. Phys., 105 (1996) 496.
95 T.C. German and W.H. Miller, J. Phys. Chem. A, 101 (1997) 6358.
96 K. Song and W.L. Hase, J. Phys. Chem. A, 102 (1998) 3648.
97 K. Song, L. Sun, W.L. Hase, S.Yu. Grebenshchikov and R. Schinke, J. Phys. Chem. A, 106 (2002) 8339.
98 W.H. Miller, J. Phys. Chem., 92 (1988) 4261.
99 D.-H. Lu and W.L. Hase, J. Phys. Chem., 93 (1989) 1681.
100 H. Hippler, N. Krasteva and F. Striebel, Phys. Chem. Chem. Phys., 6 (2004) 3383.
101 J.M. Bowman, J. Phys. Chem. A, 102 (1998) 3006.
102 E.R. Lovejoy, S.K. Kim and C.B. Moore, Science, 256 (1992) 1541.
103 S.I. Ionov, H.F. Davis, K. Mikhaylichenko, L. Lalachovic, R.A. Beaudet and C. Wittig, J. Chem. Phys., 109 (1994) 4809.
104 G.-H. Leu, C.-L. Huang, S.-H. Lee, Y.-C. Lee and I.-C. Chen, J. Chem. Phys., 109 (1998) 9340.

105 I.W.M. Smith, Nature, 358 (1992) 279.
106 R. Marcus, Science, 256 (1992) 1523.
107 J.D. Gezelter and W.H. Miller, J. Chem. Phys., 104 (1996) 3546.
108 R.A. King, W.D. Allen, B. Ma and H.F. Schaefer, III, Faraday Discuss., 110 (1998) 23.
109 D.A. King, W.D. Allen and H.F. Schaefer, III, J. Chem. Phys., 112 (2000) 5585.
110 S.Yu. Grebenshchikov, C. Beck, H. Flöthmann, R. Schinke and S. Kato, J. Chem. Phys., 111 (1999) 619.
111 L. Sun, W.L. Hase, in: K.B. Lipkowitz, R. Larter and T.R. Cundari (Eds.), Reviews in computational chemistry, Vol. 19, Wiley, Hoboken, NJ, 2003, p. 79.
112 W.L. Hase, K. Song and M.S. Gordon, Comput. Sci. Eng., 5 (2003) 36.
113 R. Car and M. Parrinello, Phys. Rev. Lett., 55 (1985) 2471.
114 I.S.Y. Wang and M. Karplus, J. Am. Chem. Soc., 95 (1973) 8160.
115 A. Gonzalez-Lafont, T.N. Truong and D.G. Truhlar, J. Phys. Chem., 95 (1991) 4618.
116 G.H. Peslherbe and W.L. Hase, J. Chem. Phys., 104 (1996) 7882.
117 C. Doubleday, K. Bolton, G.H. Peslherbe and W.L. Hase, J. Am. Chem. Soc., 118 (1996) 9992.
118 C. Doubleday, K. Bolton and W.L. Hase, J. Am. Chem. Soc., 119 (1997) 5251.
119 C. Doubleday, K. Bolton and W.L. Hase, J. Phys. Chem. A, 102 (1998) 3648.
120 K. Bolton, W.L. Hase and C. Doubleday, Ber. Bunsen-Ges. Phys. Chem., 101 (1997) 444.
121 L. Sun, W.L. Hase and K. Song, J. Am. Chem. Soc., 123 (2001) 5753.
122 L. Sun, K. Song and W.L. Hase, Science, 296 (2002) 875.
123 O. Meroueh, Y. Wang and W.L. Hase, J. Phys. Chem. A, 106 (2003) 9983.
124 Y. Wang, W.L. Hase and K. Song, J. Am. Soc. Mass Spectrom., 14 (2003) 1402.
125 O. Meroueh and W.L. Hase, Phys. Chem. Chem. Phys., 3 (2001) 2306.
126 C. Doubleday, J. Phys. Chem. A, 105 (2001) 6333.
127 C. Doubleday, G. Li and W.L. Hase, Phys. Chem. Chem. Phys., 4 (2002) 304.
128 T. Yan, C. Doubleday and W.L. Hase, J. Phys. Chem. A, 108 (2004) 9863.
129 D. Troya, R.Z. Pascual, D.J. Garton, T.K. Minton and G.C. Schatz, J. Phys. Chem. A, 107 (2003) 7161.
130 X. Hu and W.L. Hase, J. Phys. Chem., 96 (1992) 7535.
131 L. Zhu, W. Chen, W.L. Hase and E.W. Kaiser, J. Phys. Chem., 97 (1993) 311.
132 A. Warshel and M. Levitt, J. Mol. Biol., 103 (1976) 227.
133 U.C. Singh and P.A. Kollman, J. Comput. Chem., 7 (1986) 718.

Theory and Applications of Computational Chemistry: The First Forty Years
Edited by C. Dykstra et al.

CHAPTER 16

Molecular dynamics: an account of its evolution

Raymond Kapral[1] and Giovanni Ciccotti[2]

[1] *Chemical Physics Theory Group, Department of Chemistry, University of Toronto, Toronto, ON, Canada M5S 3H6*
[2] *INFM and Dipartimento di Fisica, Università 'La Sapienza', Piazzale Aldo Moro, 2, 00185 Roma, Italy*

Abstract

Molecular dynamics is a powerful theoretical method that is now routinely used to simulate the dynamics of complex physical and chemical systems. Its success stems from the development of a battery of algorithms and the availability of powerful computers. In this chapter we recount how molecular dynamics evolved from its very modest beginnings to its status as an indispensable tool for the investigation of the statistical mechanics of complex chemical and biological systems.

16.1 INTRODUCTION

Molecular dynamics and Monte Carlo computer simulation methods are now two of the most powerful tools that are used to explore the statistical mechanics of complex condensed phase and biochemical systems. A great deal of modern theoretical research is based on the results of such simulations or is devoted to the development of algorithms to extend the range of such simulation methods to larger systems and, in the case of molecular dynamics, to longer times. The list of applications is very broad: it ranges from studies of the dynamics of biomolecules and chemical reaction rates in the condensed phase to studies of solids under stress. Systems currently being investigated may contain up to 10^{10} particles or simulation times may be as long as μs. Molecular dynamics has been combined with *ab initio* electronic structure methods in order to be able to study systems with realistic interactions that include bond-making and bond-breaking processes. Molecular dynamics and Monte Carlo methods are being extended to treat quantum mechanical systems and large scale simulations are being carried out to study hydrodynamic flows and phase segregation processes.

This wide use and acceptance of computer simulation in statistical mechanics is relatively recent. In this chapter we adopt a narrow focus and present a description of how Molecular Dynamics (MD) simulation evolved to take such a prominent role in modern science.

Molecular dynamics had its modest beginnings in the late 1950s and early 1960s with a number of pioneering studies. The challenges were great because computers had limited power and capacity and suitable algorithms had to be developed. We document the evolution of molecular dynamics from a novel approach to solve selected problems in statistical mechanics to the pervasive powerful technique we have today. A parallel set of developments took place for the Monte Carlo method, which was introduced in a seminal paper by Metropolis et al. [1] in 1953, but we shall not review this work except where it is relevant to our description of molecular dynamics. Of course, the development of molecular dynamics simulation methods is not at an end; in fact, the challenges are as great as ever, since nature provides us with complexity that has been, thus far, beyond our grasp.

16.2 EARLY DAYS

The earliest research in a field builds on past work and this always makes it difficult to ascribe priority to important discoveries that lead to new directions or paradigms for future research. Certainly, there were many early investigations that used computer simulation of Newton's equations of motion to tackle important open problems. For instance, Hirschfelder et al. [2] studied the dynamics of the gas phase $H + H_2$ reaction on a model potential surface to determine the reaction rate. In later years this investigation spawned the field of gas phase molecular dynamics. The paper by Fermi et al. [3] on the simulation of the dynamics of a model one-dimensional solid was influential in the field of non-linear dynamics. Neither these papers nor the body of work they stimulated had an immediate important impact on statistical mechanics [4].

Alder and Wainwright [5,6] published the first paper of a molecular dynamics simulation of a condensed phase fluid system and this paper began a trend that did have a strong impact on statistical mechanics. These authors tackled one of the open questions of the day, whether a solid–fluid phase transition existed in a system of hard spheres. This problem could not be solved by existing analytical methods and Alder and Wainwright's simulation demonstrated that such analytically intractable problems could be studied and solved by direct MD simulation of the equations of motion of a many-body system. Of course, the simulation was modest by today's standards and was carried out on systems containing 32 and 108 hard spheres. This research set the stage for the development of MD as a basic tool in statistical mechanics.

Hard spheres are a highly idealized model for physical systems. The first MD simulations of a realistic fluid model were performed by Rahman in a now classic paper [7]. The system consisted of 864 Lennard-Jones particles with potential parameters fit to model liquid argon. The paper demonstrated that MD simulations can be used not only to address theoretical questions like the existence of a phase transition or the validity of the H-Theorem [8] in an idealized model, but to gain insight into and interpret experiments, in this case neutron scattering experiments on liquid argon. Because this work showed that MD could be used to model a real system, the paper has had a wide impact on the field and stimulated further work. It led to developments related to construction of potential models for other real systems and the use of MD to explore more complicated systems, in particular molecular fluids.

Molecular dynamics simulation investigations of Lennard-Jones fluids were continued by Verlet [9] and in the course of his studies he introduced several important techniques that have continued to be in use to the present day. The now-famous Verlet algorithm for integrating Newton's equations of motion was first presented in this paper. In this algorithm the positions of the particles are updated according to the following scheme: if $\mathbf{F}(\mathbf{r}_{ij})$ is the force on particle i due to particle j, the position of particle i at time $t+h$ can be found from the positions at two earlier times using,

$$\mathbf{r}_i(t+h) = -\mathbf{r}_i(t-h) + 2\mathbf{r}_i(t) + \sum_{j \neq i} \frac{1}{m}\mathbf{F}(\mathbf{r}_{ij}(t))h^2, \tag{1}$$

where h is the time step. This simple algorithm, precise to order h^4, preserves phase space volumes and is very stable. It forms the basis for many of the current MD codes in use today. The Lennard-Jones simulations of Rahman used a more elaborate predictor–corrector method to integrate the equations of motion. The predictor–corrector algorithm, although more precise, was much less reliable because of its non-symplectic character, a fact that was appreciated only relatively recently. Verlet also introduced the use of neighbor lists to evaluate the intermolecular forces, one of the most time consuming steps in MD codes. Neighbor lists are employed in many modern MD codes and can lead to savings of an order of magnitude in time.

The 1960s also saw the publication of Alder and Wainwright's important paper on the long time tail in the velocity correlation function [10]. This paper used MD simulation to show that hydrodynamics, constructed to describe behavior on macroscopic distance and time scales, can also explain phenomena related to the microscopic dynamics of particle motion. The $t^{-3/2}$ decay of the velocity correlation function at long times which they found has its origin in the coupling of the particle velocity to the collective hydrodynamic modes of the surrounding fluid. This paper stimulated a large literature in an attempt to provide a rigorous theoretical basis for this phenomenon. It also showed that hydrodynamic concepts could be pushed to scales that were previously thought to be inaccessible. In turn, this prompted modelling of experimental molecular relaxation data in terms of macroscopic concepts and also was a forerunner of developments in mode coupling theory that have had a large impact on the dynamics of critical phenomena and glasses.

These early MD studies were limited to hard sphere or Lennard-Jones interactions appropriate for simple fluids. In the early 1970s the first attempts to apply MD methods to molecular fluids were made. Harp and Berne [11] studied a model for carbon monoxide while Rahman and Stillinger [12] carried out MD simulations of liquid water. The MD simulations of water were especially important because they addressed questions related to the dynamics and structure of an ubiquitous solvent that would figure prominently in later work on biological and condensed phase systems. Furthermore, because of the nature of the long-range Coulomb forces in this system and the delicate nature of the hydrogen bonded structure of water, the construction of a suitable force field and its simulation presented a major challenge for the field at that time [13]. An earlier simulation of the structural properties of water was carried out by Barker and Watts [14] using the Monte Carlo method. For molecular systems like water one must deal with the internal rotational and vibrational degrees of freedom. Rahman and Stillinger considered a water model where the bonds were rigid so that vibrational degrees of freedom were frozen and rotational motion was described in terms of Euler angles. The long-range Coulomb forces were truncated. We now know that this procedure is not satisfactory and Ewald summation [15,16], although not free from artifacts and difficulties, is often currently used to treat such long-range forces. Other methods, such as the particle–particle/particle mesh method [17] and the fast multipole method [18] are computationally more efficient for large systems.

16.3 CLASSICAL PERIOD OF CLASSICAL MOLECULAR DYNAMICS

By the early 1970s MD computer simulation was evolving from its initial development phase and proving itself to be a powerful tool to explore the dynamics of dense fluids. The advances that occurred over the next 15–20 years helped to give MD the central role that it now plays in science. In this period MD methods for generalized ensembles were constructed, constraints that allowed one to simulate molecular systems using Cartesian coordinates were developed and methods for sampling rare events were introduced. In addition, simulations of more complicated condensed phase molecular systems, including the first MD simulations of proteins and chemical reactions in liquids, were performed.

The Monte Carlo method is easily carried out in any convenient ensemble since it simply requires the construction of a suitable Markov chain for the importance sampling. The simulations in the original paper by Metropolis et al. [1] were carried out in the canonical ensemble corresponding to a fixed number of molecules, volume and temperature, (N, V, T). By contrast, molecular dynamics is naturally carried out in the microcanonical ensemble, fixed (N, V, E), since the energy is conserved by Newton's equations of motion. This implies that the temperature of an MD simulation is not known a priori but is obtained as an output of the calculation. This feature makes it difficult to locate phase transitions and, perhaps, gave the first motivation to generalize MD to other ensembles.

In 1979 Andersen [19] constructed a MD scheme for systems with constant pressure by introducing the volume V as an extra coordinate. The volume changes couple to the system degrees of freedom since a change in the volume corresponds to a scaling of particle positions. The usual Newton's equations of motion were augmented with equations for the volume and an additional momentum conjugate to V. The volume variable was assigned a fictitious mass. The system energy is no longer conserved in this extended system; instead a quantity closely related to the total enthalpy is conserved. Thus, although the dynamics is fictitious, equilibrium properties may be determined from MD simulations on the extended system. Andersen also introduced a stochastic scheme for carrying out constant temperature and pressure simulations, (N, p, T) ensemble.

Parrinello and Rahman [20,21] extended Andersen's method to treat changes in the MD cell shape and in this way could study structural changes in solids and determine the crystal structure corresponding to the various pair potentials available at that time.

In well-known papers [22,23] Nosé showed how a deterministic canonical ensemble (N, V, T) MD simulation scheme could be constructed. Again the system phase space variables were augmented with an additional degree of freedom s. The physical system variables $(\mathbf{q}'_i, \mathbf{p}'_i t')$ were related to 'virtual'$_t$ variables $(\mathbf{q}_i, \mathbf{p}_i, t)$ by a non-canonical transformation $\mathbf{q}'_i = \mathbf{q}_i$, $\mathbf{p}'_i = \mathbf{p}_i/s$ and $t' = \int \mathrm{d}t/s$. The Hamiltonian of the extended system is

$$H = \sum_i \frac{\mathbf{p}_i^2}{2m_i s^2} + V(\mathbf{q}) = \frac{p_s^2}{2Q} + gk_\mathrm{B}T \ln s, \tag{2}$$

where P_s is the momentum associated with the variable s, Q is its mass and g is a free parameter in the model that is put equal to $3N + 1$ to reproduce the canonical ensemble. The equations of motion are constructed from this Hamiltonian and then reinterpreted in real variables. Hoover [24] showed that a different set of non-Newtonian equations of motion, that is free of the time scaling introduced by Nosé, can be used to carry out canonical ensemble MD simulations. Many subtle statistical mechanical issues arise in this approach which have been clarified later by Tuckerman et al. [25,26].

Most current applications of the MD method have as their aim an understanding of the dynamics of molecular systems with many components. Solvents of interest include water or other polar molecules and solutes can be very large proteins or other molecules. In such systems certain degrees of freedom, typically those associated with internal vibrational modes whose dynamics takes place on fast time scales, often do not affect the properties of interest but place heavy demands on computer time because of the small time step needed to accurately integrate the equations of motion. If the molecules are treated as rigid bodies, one may introduce a coordinate system consisting of the center of mass of the molecule and Euler angles describing its orientation. The singularities that appear in the evolution equations make it difficult to carry out simulations using such coordinates without the introduction of schemes to circumvent the singularities [27]. Another method, which is also applicable to flexible molecules, is to introduce constraints that remove certain degrees of freedom in the molecules. If the system is subjected to l holonomic constraints $\sigma_\alpha(\mathbf{r}) = 0$ $\alpha = 1, \ldots, l$, where the σ_α are bond constraints or any other set of constraints, the constrained equations of motion are

$$\dot{\mathbf{r}}_i = \frac{\mathbf{p}_i}{m_i}$$

$$\dot{\mathbf{p}}_i = \mathbf{F}_i - \lambda_\alpha \nabla_i \sigma_\alpha(\mathbf{r}) \tag{3}$$

where $\mathbf{F}_i = -\nabla_i V$ is the force on particle i due to the potential energy and the second term represents the constraint forces with Lagrange multipliers λ_α. These equations of motion can be integrated in Cartesian coordinates [28–30]. The use of constraints and the SHAKE algorithm for their implementation has proved very powerful for the simulation of complex molecular systems. The original iterative SHAKE algorithm could not be parallelized. Only recently, [31], by an elegant reformulation of the problem, was it possible to construct a parallel version of the scheme.

Another method for dealing with molecular systems, especially molecules where vibrational degrees of freedom are important and cannot be frozen by molecular constraints, is the use of multiple time step integrators [32,33]. In these integrators the classical Liouville operator is decomposed into parts pertaining to fast and slow degrees of freedom. A Trotter factorization of the propagator is made to derive reversible and measure preserving integration algorithms that treat the two types of degrees of freedom with different time steps. Such algorithms allow one to simulate systems with vibrational, or other rapidly varying, degrees of freedom efficiently.

In this period a thrust was made to develop MD techniques to simulate the rates of activated chemical reactions in the condensed phase. Activated rate processes typically take place on very long time scales because high-energy barriers must be traversed in passing between metastable states corresponding to chemical species. Direct MD simulation of the reaction rate is an impossible task in such situations. However, given that one knows the reaction coordinate $\xi(\boldsymbol{r})$, the statistical mechanical expression for the time-dependent rate of the reaction $A \rightleftharpoons B$,

$$k_f(t) = k_f^{\mathrm{TST}} \frac{\langle(\dot{\xi}\delta(\xi(\mathbf{r}) - \xi^{\ddagger}))\theta(\xi(\mathbf{r}(t)) - \xi^{\ddagger})\rangle}{\langle\dot{\xi}\delta(\xi(\mathbf{r}) - \xi^{\ddagger})\theta(\dot{\xi})\rangle} \tag{4}$$

where $k_f^{\mathrm{TST}} = \langle\dot{\xi}\delta(\xi(\mathbf{r}) - \xi^{\ddagger})\theta(\dot{\xi})\rangle/n_A^{\mathrm{eq}}$, shows that it can be computed from a knowledge of the transition state theory (TST) rate constant, k_f^{TST}, which involves the free energy along the reaction coordinate, and dynamical effects that may be determined from an average over trajectories starting from the free energy barrier top [34–36]. Keck [37] and Anderson [38] computed gas phase reaction rates using similar ideas involving trajectories initiated at the transition state. Bennett formulated the theory of activated defect migration in solids in these terms and carried out simulations of this rate process [35]. Early simulations of condensed phase reaction rates using these rare event sampling techniques include a study of butane isomerization in a model CCl_4 solvent [39], the $Cl + Cl_2$ reaction in a rare gas solvent [40], and ion association in a dipolar solvent [41].

Configurations at the barrier top may be obtained by Monte Carlo umbrella sampling [42], followed by MD trajectory segments to obtain dynamical corrections to the TST rate. One may also prepare ensembles of trajectories conditioned on starting from the barrier top using constrained MD trajectories instead of Monte Carlo sampling. The conditional average of the reaction coordinate in the expression for the rate constant may be expressed in terms of the ensemble of configurations generated from trajectories where $\xi(\mathbf{r})$ is fixed by a holonomic constraint on the equations (ξ-constrained ensemble). While the value $\xi^{\ddagger}$ one needs to sample is rare in the original ensemble, only configurations with $\xi = \xi^{\ddagger}$ are sampled in the ξ-constrained ensemble. The relation between conditional averages and averages in the constrained ensemble is

$$\frac{\langle A(\mathbf{r})\delta(\xi(\mathbf{r}) - \xi')\rangle}{\langle\delta(\xi(\mathbf{r}) - \xi')\rangle} = \frac{\langle |Z|^{-1/2} A(\mathbf{r})\rangle_{\xi'}}{\langle |Z|^{-1/2}\rangle_{\xi'}} \tag{5}$$

where $A(\mathbf{r})$ is any function of the configuration space [43,44]. The factor Z accounts for the bias introduced in the momentum distribution from constraining the reaction coordinate. This Blue Moon Ensemble method takes its name from the fact that it is designed to compute the rates of reactions that occur 'once in a blue moon' [43].

Computations of activated rate processes are now routine using rare event sampling techniques. However, often it is difficult to identify the reaction coordinate(s) in a complex (or even simple) condensed phase reaction and current research has focused on the development of methods to determine reaction paths. A number of methods [45], including an early application dealing with reaction paths in biochemical systems [46], have been suggested to determine reaction paths. One approach is to find the optimal TST estimate to the rate by finding the optimal hyperplane that determines the minimal TST rate [47–49]. Techniques have been introduced to determine the minimum energy path, such as that of Ulitsky and Elber [50], the Nudged Elastic Band (NEB) [51] and methods based on the minimization of the action [52]. Transition Path Sampling [53,54] does not assume a transition mechanism and determines dynamical trajectories that connect reactant and product metastable states. The finite temperature string method uses an adaptive blue moon sampling to determine the optimal reaction coordinate in terms of level sets that are the isoprobability surfaces for the system to reach one metastable state before the other [45,55,56]. Mile-Stoning is an algorithm to compute time scales of complex processes by following predetermined milestones along a reaction coordinate [57]. Methods to force the system to escape from free energy minima have been devised [58,59].

Proteins and many other biomolecules are often very large and have complex structures. Proteins are folded into characteristic shapes which give them specific properties; for example, they catalyze biochemical reactions. Structural fluctuations and solvent dynamics influence how proteins function and this fact has stimulated the development of molecular dynamics simulation methods for such biomolecules. The problems are challenging because of the size and complexity of proteins, the need to account for solvation effects and the disparity of time scales involved in the dynamics. While few relevant local motions occur on short molecular time scales, many important dynamical effects, such as protein folding or other activated conformational changes, may take place on very long macroscopic time scales. With the development of the molecular dynamics simulation techniques described above, in particular, stable integration algorithms, the construction of effective force fields, methods for dealing with long-range forces, introduction of constraints to remove inessential degrees of freedom, and rare event sampling methods, progress has been made also on these complex and important systems. Following the first protein simulations of bovine pancreatic trypsin inhibitor [60], the field of biomolecular simulation developed rapidly [61–64] and is one of the most currently active research areas where MD is providing insights into structure and function. Large molecular dynamics simulation programs with model force fields for biomolecule simulations have been developed [65–67] and are in wide use today for theoretical studies of biochemical problems.

Linear response theory provides autocorrelation function expressions for transport properties [68] and, starting with early investigations of hard sphere [69] and Lennard-Jones [70] systems, MD has been used, and continues to be used, to estimate transport coefficients using these Kubo formulas. However, very early it was recognized that MD could be employed to study non-equilibrium processes and transport beyond the linear response regime. In this more general perspective, linear response theory itself is considered to be an approximate theory, whose validity needs to be tested, rather than an exact algorithm from which the response of the system can be computed. More importantly, not only can molecular dynamics be used to simulate equilibrium phenomena but it also provides an approach to the study of non-equilibrium phenomena. Such simulations are particularly simple in the non-equilibrium stationary state (where fluxes are present but do not depend explicitly on time) and can also be carried out in more general situations, as was shown in the dynamic approach to non-equilibrium developed by Ciccotti and Jacucci [71].

The earliest efforts to develop non-equilibrium molecular dynamics (NEMD) methods used special boundary conditions and/or external fields to induce non-equilibrium behavior in the system. Important contributions to this development include those of Lees and Edwards [72], Gosling et al. [73], Hoover and Ashurst [74] and Ciccotti and Jacucci [71]. Later methods made adjustments to external forces to account for periodic boundary conditions and introduced suitable modifications of the Hamiltonian or the Newtonian equations of motion [75–78]. Considerable progress has been made since those early efforts, both with the original [79–83] and modified Hamiltonian approaches [84]. However, many subtle issues remain to be resolved. These issues concern the non-Hamiltonian nature of the models used in NEMD and the need to introduce a thermostat to obtain a stationary state. Recently Tuckerman et al. [25] have considered some statistical mechanical aspects of non-Hamiltonian dynamics and this work may provide a way to approach these problems. Although the field of NEMD has been extensively explored for simple atomic systems, its primary applications lie mainly in treating non-equilibrium phenomena in complex systems, such as transport in polymeric systems, colloidal suspensions, etc. We expect that there will be considerable activity and progress in these areas in the coming years [85].

16.4 QUANTUM MECHANICS AND MOLECULAR DYNAMICS

Thus far, the developments we described had as their aim the construction of algorithms for the study of some aspects of the physics of condensed phase classical systems. The calculation of the dynamical properties of condensed phase quantum systems remains one of the challenging problems in statistical mechanics. It is still impossible to solve the time-dependent Schrödinger equation for a large many-body system. Nevertheless, there has been some success in applying MD techniques to the study of quantum systems. Progress in this area has been made on diverse aspects of the problem. Molecular dynamics methods may be used to investigate equilibrium quantum statistical mechanics. Quantum electronic structure calculations have been married to MD to allow one to simulate the dynamics of complex systems using more realistic potential information. If only few degrees of freedom are treated quantum mechanically and the remainder is treated using classical mechanics, then algorithms for the solution of the dynamics may be constructed. We now briefly describe these advances.

Equilibrium properties can be determined from the partition function Z_Q and this quantity can, in turn, be computed using Feynman's path integral approach to quantum mechanics in imaginary time [86]. In this representation of quantum mechanics, quantum particles are mapped onto closed paths $\mathbf{r}(t)$ in imaginary time t, $0 \leq t \leq \beta\hbar$. The path integral expression for the canonical partition function of a quantum particle is given by the $P \rightarrow \infty$ limit of the quantum path discretized into P segments,

$$Z_Q = \lim_{P\rightarrow\infty} \left(\frac{mP}{2\pi\beta\hbar^2} \right)^{1/2} \int \prod_{i=1}^{P} \mathrm{d}r_i \, \mathrm{e}^{-\beta V_{\text{eff}}(r)} \tag{6}$$

where

$$V_{\text{eff}}(r) = \frac{Pm}{2(\beta\hbar)^2} \sum_{i=1}^{P} (r_{i+1} - r_i)^2 + \frac{1}{P} \sum_{i=1}^{P} V(r_i)$$

is an effective potential. In simulations a finite number P of segments or beads is chosen. Thus, considering the form that this potential takes, the single quantum particle with coordinate r is mapped to a closed or *ring polymer* with P beads with coordinates r_i[87]. The problem is then reduced to sampling from a classical canonical distribution involving V_{eff}. In molecular dynamics, this can be achieved by defining the Hamiltonian $H_{\text{eff}} = \sum_{i=1}^{P} (1/2) m_{\text{eff}} \dot{r}_i^2 + V_{\text{eff}}(r_i)$, where m_{eff} is an arbitrary mass assigned to the polymer beads.

The formulation is easily generalized to systems with many degrees of freedom, although the treatment of permutations needed to cope with the symmetrization postulate raises a difficult problem for MD (it is less severe, but still present, for MC and currently only bosonic systems can be treated satisfactorily [88]). The sign problem for fermionic systems is still essentially unsolved [89]. In the path integral formulation, classical particles are represented by single beads. It typically takes about 1000 polymer beads to represent an electron and about 30 beads for a proton. Methods for carrying out path integral simulations have been developed [90,91]. Monte Carlo or MD sampling of the configuration space described by V_{eff} is very difficult because the polymer spring stiffness increases with the number of beads. This problem has been solved by the introduction of the staging algorithm [92,93] where the size of the polymer chain is increased in a systematic fashion during the calculation. Early investigations concerned studies of quantum degrees of freedom in classical solvents and quantum bulk phases. In particular, electron solvation in molten potassium chloride [94] and in liquid ammonia [92] were investigated. Bulk quantum fluids were also studied; for example, bulk liquid water where only rotational degrees of freedom were treated quantum mechanically [95] and liquid ^{4}He [96]. The path integral method is now commonly used to investigate equilibrium properties of quantum fluids [88].

Almost all of the early MD studies used either model potential functions or effective potential functions constructed to reproduce structural data on condensed phase systems. Examples include the choice of Lennard-Jones parameters to reproduce argon fluids and the water potentials discussed earlier. A much more systematic approach is to combine a quantum chemical calculation of the potential energy function in the course ('on the fly') of an MD simulation. This is now termed *ab initio* molecular dynamics.

Car and Parrinello [97,98] proposed a scheme to combine density functional theory [99] with molecular dynamics in a paper that has stimulated a field of research and provided a means to explore a wide range of physical applications. In this scheme, the energy functional $E[\{\psi_i\}, \{R_I\}, \{\alpha_\nu\}]$ of the Kohn–Sham orbitals, ψ_i, nuclear positions, R_I, and external parameters such as volume or strain, α_ν, is minimized, subject to the ortho-normalization constraint on the orbitals, to determine the Born–Oppenheimer potential energy surface. The Lagrangian,

$$L = \sum_i \frac{\mu}{2} \int_\Omega \mathrm{d}\mathbf{r} |\dot{\psi}_i|^2 + \sum_I \frac{M_I}{2} \dot{R}_I^2 + \sum_\nu \frac{\mu_\nu}{2} \dot{\alpha}_\nu^2 - E[\{\psi_i\}, \{R_I\}, \{\alpha_\nu\}], \tag{7}$$

which depends on fictitious mass parameters for the ψ and α variables, is used to generate equations of motion that can be solved by classical MD to determine the potential energy surface. The original application dealt with the structure and dynamics of crystalline silicon [97] but now this method is widely used in MD simulations since it is able to determine the intermolecular potentials for complex materials and provide a realistic description of equilibrium structure and dynamical properties [98]. The technique is very powerful for the study of chemical properties, for example, association and dissociation processes, mixtures with reactive agents, etc. since it describes the bond-making and bond-breaking events at a quantum mechanical level. Classical MD with empirical potential functions is much more 'rigid' and it is difficult to account for such processes in a realistic and general fashion.

While the Feynman path integral method can be used to determine equilibrium properties of condensed phase systems, thus far, it has not led to a computationally feasible scheme for real time dynamics needed to compute transport properties. As a result a number of approaches have been used to study quantum dynamics, most often for a few quantum degrees of freedom interacting with a classical environment. The simplest such scheme is adiabatic dynamics where the Schrödinger equation for the quantum degrees of freedom is solved in the fixed field of the classical particles to determine the ground state energy. This energy is used as the potential energy to evolve the classical degrees of freedom by Newton's equations of motion. Molecular dynamics investigations of electron solvation [100] and proton transfer [101,102] in condensed phase systems have been carried out using this method. There is now a large literature employing this simulation method for studying reactive systems [103].

Various simulation schemes have been suggested to go beyond the adiabatic limit, such as mean field methods where the wave function obtained from the solution of the time-dependent Schrödinger equation is used to compute an effective force for the evolution of the classical equations of motion, and surface hopping schemes where evolution on single adiabatic surfaces is interspersed with quantum transitions that cause the system to hop from one surface to another. The most widely used surface hopping method is Tully's fewest switches algorithm [104]. In this algorithm, the probabilities for switching between adiabatic surfaces are determined from the off-diagonal elements of the density matrix which are, in turn, computed from the solution of the time-dependent

Schrödinger equation assuming that the classical degrees of freedom evolve by Newton's equations of motion on single adiabatic surfaces. This method has been used extensively to investigate non-adiabatic dynamics in condensed phase systems; for example, proton transfer reactions in molecular solvents have been studied using this MD scheme [105].

Variations on this surface hopping method that utilize Pechukas' [106] formulation of mixed quantum-classical dynamics have been proposed [107,108]. Surface hopping algorithms [109–111] for non-adiabatic dynamics based on the quantum-classical Liouville equation [109,111–113] have been formulated. In these schemes the dynamics is fully prescribed by the quantum-classical Liouville operator and no additional assumptions about the nature of the classical evolution or the quantum transition probabilities are made. Quantum dynamics of condensed phase systems has also been carried out using techniques that are not based on surface hopping algorithms, in particular, centroid path integral dynamics [114] and influence functional methods [115].

Currently, the above path integral, *ab initio* MD methods and quantum-classical surface hopping methods are being combined to produce powerful algorithms for studying physical problems. For example, proton transfer in ice has been studied by combining *ab initio* MD with path integrals [116] and *ab initio* MD 'spawning' methods [117] have been constructed to deal with dynamics on multiple energy surfaces encountered in photochemical reactions.

16.5 COARSE GRAINED AND MESOSCOPIC DYNAMICS

Classical MD simulations have been carried out on very large systems with up to 10^{10} particles using massively parallel computer resources. Such large-scale simulations demonstrated that brute force MD provides a way to study problems, such as hydrodynamic flows [118] and shock compression of solids [119], whose characteristic space and time scales are very long. Nevertheless, for most applications dealing with the dynamics of complex molecular systems, such as polymers, biomolecules and glasses, MD simulation times are too long to be able to extract the desired information. This is also true in most instances even for simple models of fluids if one is interested in pattern forming processes which also occur on long distance and time scales.

These facts have prompted the development of coarse grained and other mesoscopic MD methods that sacrifice details on short scales but are able to efficiently capture dynamics that occurs on longer scales. Such coarse grained methods have a long history dating at least to Langevin equation models where the effects of the solvent on particle motion is taken into account through frictional and random forces [120]. There have been many applications of Langevin dynamics to the study of complex molecular systems, too numerous to mention here. Langevin dynamics has also been used in conjunction with MD as a way to contract variables; for instance, in an early study of the dynamics of the active site of ribonuclease [121].

There is a rapidly expanding literature on techniques to construct coarse grained models for intermolecular potentials. A number of atoms are grouped together to form interaction sites that interact through effective potentials that are constructed to reproduce structural or dynamical data. A variety of different schemes have been used to construct such effective potential models. They have proved to be especially useful in investigations of lipid systems, polymers and colloids. Many of these applications and a discussion of the techniques used to determine the coarse grained models can be found in the articles in Refs. [122–124].

If one is interested in properties that vary on very long distance and time scales it is possible that a drastic simplification of the molecular dynamics will still provide a faithful representation of these properties. Hydrodynamic flows are a good example. As long as the dynamics preserves the basic conservation laws of mass, momentum and energy, on sufficiently long scales the system will be described by the Navier–Stokes equations. This observation is the basis for the construction of a variety of particle-based methods for simulating hydrodynamic flows and reaction-diffusion dynamics. (There are other phase space methods that are widely used to simulate hydrodynamic flows which are not particle-based, e.g. the lattice Boltzmann method [125], which fall outside the scope of this account of MD simulation.)

One of the earliest particle-based schemes is the Direct Simulation Monte Carlo (DSMC) method of Bird [126]. In DSMC simulations, particle positions and velocities are continuous variables. The system is divided into cells and pairs of particles in a cell are chosen for collision at times that are determined from a suitable distribution. This method has seen wide use, especially in the rarefied gas dynamics community where complex fluid flows can be simulated.

An even more drastic simplification of the dynamics is made in lattice-gas automaton models for fluid flow [127,128]. Here particles are placed on a suitable regular lattice so that particle positions are discrete variables. Particle velocities are also made discrete. Simple rules move particles from site to site and change discrete velocities in a manner that satisfies the basic conservation laws. Because the lattice geometry destroys isotropy, artifacts appear in the hydrodynamics equations that have limited the utility of this method. Lattice-gas automaton models have been extended to treat reaction-diffusion systems [129].

Another method that introduces a very simplified dynamics is the Multi-Particle Collision Model (or Stochastic Rotation Model) [130]. Like DSMC particle positions and velocities are continuous variables and the system is divided into cells for the purpose of carrying out collisions. Rotation operators, chosen at random from a set of rotation operators, are assigned to each cell. The velocity of each particle in the cell, relative to the center of mass velocity of the cell, is rotated with the cell rotation operator. After rotation the center of mass velocity is added back to yield the post-collision velocity. The dynamics consists of free streaming and multi-particle collisions. This mesoscopic dynamics conserves mass, momentum and energy. The dynamics may be combined with full MD for embedded solutes [131] to study a variety of problems such as polymer, colloid and reaction dynamics.

Dissipative Particle Dynamics (DPD) is a coarse graining method that groups several atoms into simulation sites whose dynamics is governed by conservative and frictional forces designed to reproduce thermodynamics and hydrodynamics [132,133]. Since the effective interactions are constructed to reproduce macroscopic properties soft repulsive forces can be used, thereby avoiding the small MD step sizes needed to integrate the system when full interactions are taken into account. In addition, random forces can be constructed to conserve momentum, consistent with dynamics leading to the Navier–Stokes equations. Smoothed Particle Hydrodynamics (SPH) [134] is a related technique that is used to discretize continuum equations by employing weight functions assigned to fictitious particles so that hydrodynamic simulations can be carried out in a particle based MD framework.

Particle-based methods have been linked to continuum models in order to exploit the efficiency of finite element schemes and yet account for boundary layers and other regions where continuum approaches may fail [135]. It is very likely that coarse grained methods will continue to be developed and used since many physical systems are too large or too complex to be treated by full MD.

16.6 CONCLUSION

Over the last several decades molecular dynamics evolved from a theoretical tool with very modest capability to the method of choice for exploring the dynamics of complex systems in chemistry, physics and biology. The results of molecular dynamics simulations now often play the role of 'thought experiments' in regimes where physical theory is lacking and cannot act as guide. The term 'computer experiment' reflects this usage. Molecular dynamics serves several functions. It is used to model and gain insight into real physical phenomena and to provide a microscopic dynamical picture, which forms the basis for the construction of a theoretical treatment or the interpretation of experiments. Molecular dynamics owes its success perhaps more to the construction of powerful algorithms rather than to the advances in the power of computers, although both are significant factors.

The success of molecular dynamics can be gauged by its very widespread use in science today. As a result, any account of the sort we have attempted here is bound to have left out many important and exciting developments in the field. Even by limiting the perspective to the impact of molecular dynamics on statistical mechanics, it is impossible to completely survey all contributions to this broad field. There are several texts [136,137] and proceedings [138,139] that provide access to the extensive literature in the field and give details of the methods used in MD as it is now practiced. While *ab initio* MD is now a well-developed method, there is no comprehensive text that summarizes the progress in this field. By contrast, quantum non-adiabatic molecular dynamics is a field that is still in flux and further advances will likely change the face of this topic in the coming years. Hopefully, our account pointed to some of the major developments that helped shape molecular dynamics into the powerful tool it is today and provided links to other work which was not explicitly described.

16.7 ACKNOWLEDGEMENTS

This work was supported in part by a grant from the Natural Sciences and Engineering Research Council of Canada. We thank Antonella Capogrossi and Susan Arbuckle for help with the references. We would also like to thank Eric Vanden-Eijnden for helpful comments.

16.8 REFERENCES

1 N. Metropolis, A. Rosenbluth, M.N. Rosenbluth, A.H. Teller and E. Teller, J. Chem. Phys., 21 (1953) 1087.
2 J. Hirschfelder, H. Eyring and B. Topley, J. Chem. Phys., 4 (1936) 170.
3 E. Fermi, J.G. Pasta and S.M. Ulam, Studies in non-linear problems, LASL Report, LA-1940, 1955; E. Amaldi, H.L. Anderson, E. Persico, F. Rasetti, C.S. Smith, A. Wattenberg and E. Segré, (Eds.), Collected works of Enrico Fermi, Vol. 2, University of Chicago Press, Chicago, 1965, 978–988.
4 The study of radiation damage by J.B. Gibson, A.N. Goland, M. Milgram and G.H. Vineyard, Phys. Rev., 120 (1960) 1229–1253, employed computer simulation methods and, while not focusing on statistical mechanical aspects of the problem, did introduce methods that were similar to those used in later simulations of Lennard-Jones fluids.
5 B.J. Alder, T.E. Wainwright, in: I. Prigogine (Ed.), Proceedings of the international symposium on statistical mechanical theory of transport processes, Brussels, 1956, Interscience, New York, 1958.
6 B.J. Alder and T.E. Wainwright, J. Chem. Phys., 27 (1957) 1280.
7 A. Rahman, Phys. Rev. A, 136 (1964) 405.
8 J. Orban and A. Bellemans, Phys. Lett. A, 24 (1967) 620.
9 L. Verlet, Phys. Rev., 159 (1967) 98.
10 B.J. Alder and T.E. Wainwright, Phys. Rev. A, 1 (1967) 18.
11 G.D. Harp and B.J. Berne, Phys. Rev. A, 2 (1970) 975.
12 A. Rahman and F.H. Stillinger, J. Chem. Phys., 55 (1971) 3336.
13 A number of effective water models are currently widely used [A. Wallqvist and R.D. Mountain, Rev. Comput. Chem., 13 (1999) 183]. These include the SPC/E [H.J.C. Berendsen, J.R. Gringera and T.P. Straatsma, J. Phys. Chem., 91 (1987) 6269] and TIP3P [W.L. Jorgensen, J. Chandrasekhar, J.D. Madura, R.W. Impey and M.L. Klein, J. Chem. Phys., 79 (1983) 926] models. The use of ab initio molecular dynamics methods has provided a more powerful and systematic route to determine potential functions for water [K. Laasonen, M. Sprik, M. Parrinello and R. Car, J. Chem. Phys., 99 (1993) 9080].
14 J.A. Barker and R.O. Watts, Chem. Phys. Lett., 3 (1969) 144.
15 S.W. De Leeuw, J.W. Perram and E.R. Smith, Proc. R. Soc. London A, 373 (1980) 27, 388 (1983) 177.
16 J.P. Hansen, in: G. Ciccotti and W.G. Hoover (Eds.), Molecular dynamics simulation of statistical–mechanical systems, North-Holland, Amsterdam, 1986.
17 J.W. Eastwood and R.W. Hockney, J. Comput. Phys., 16 (1974) 342.
18 L. Greengard and V. Rokhlin, J. Comput. Phys., 73 (1987) 325.
19 H.C. Andersen, J. Chem. Phys., 72 (1980) 2384.
20 M. Parrinello and A. Rahman, Phys. Rev. Lett., 45 (1980) 1196.
21 M. Parrinello and A. Rahman, J. Appl. Phys., 52 (1981) 7182.
22 S. Nosé, Mol. Phys., 52 (1984) 255.
23 S. Nosé, J. Chem. Phys., 81 (1984) 511.
24 W.G. Hoover, Phys. Rev. A, 31 (1985) 1695.
25 M.E. Tuckerman, C.J. Mundy and G.J. Martyna, Europhys. Lett., 44 (1999) 149.
26 M.E. Tuckerman, Y. Liu, G. Ciccotti and G.J. Martyna, J. Chem. Phys., 115 (2001) 1678.
27 D.J. Evans and S. Murad, Mol. Phys., 34 (1977) 327.
28 J.-P. Ryckaert, G. Ciccotti and H.J.C. Berendsen, J. Comput. Phys., 23 (1977) 327.
29 G. Ciccotti and J.-P. Ryckaert, Comput. Phys. Rep., 4 (1986) 345.
30 G. Ciccotti, M. Ferrario and J.-P. Ryckaert, Mol. Phys., 47 (1982) 1253.
31 Y. Weinbach and R. Elber, J. Comput. Phys., to be published.
32 M.E. Tuckerman, B.J. Berne and G.J. Martyna, J. Chem. Phys., 97 (1992) 1990.
33 G.J. Martyna, M.E. Tuckerman, D.J. Tobias and M.L. Klein, Mol. Phys., 87 (1996) 1117.
34 T. Yamamoto, J. Chem. Phys., 33 (1960) 281.

35 C.H. Bennett, in: R.E. Christofferson (Ed.), Algorithms for chemical computations, ACS symposium series no. 46, American Chemical Society, Washington, DC, 1977, p. 63.
36 D. Chandler, J. Chem. Phys., 68 (1978) 2959.
37 J.C. Keck, Faraday Soc. Discuss., 33 (1962) 173.
38 J.B. Anderson, J. Chem. Phys., 58 (1973) 4684.
39 R.O. Rosenberg, B.J. Berne and D. Chandler, Chem. Phys. Lett., 75 (1980) 162.
40 J.P. Bergsma, P.M. Edelsten, B.J. Gertner, K.R. Huber, J.R. Reimers, K.R. Wilson, S.M. Wu and J.T. Hynes, Chem. Phys. Lett., 123 (1986) 394.
41 G. Ciccotti, M. Ferrario, J.T. Hynes and R. Kapral, J. Chim. Phys., 85 (1988) 925.
42 G.M. Torrie and J.P. Valleau, J. Comput. Phys., 23 (1977) 187.
43 E. Carter, G. Ciccotti, J.T. Hynes and R. Kapral, Chem. Phys. Lett., 156 (1989) 472.
44 M. Sprik and G. Ciccotti, J. Chem. Phys., 109 (1998) 7737.
45 W. E and E. Vanden-Eijnden, in: S. Attinger, P. Koumoutsakos (Eds.), Lecture notes in computational science and engineering, Vol. 39, Springer, Berlin, 2004.
46 R. Elber and M. Karplus, Chem. Phys. Lett., 139 (1987) 375.
47 G.H. Jóhannesson and H. Jónsson, J. Chem. Phys., 115 (2001) 9644.
48 G. Mills, H. Jónsson and G.K. Schenter, Surf. Sci., 324 (1995) 305.
49 E. Vanden-Eijnden, Optimal transition state theory and the Bennett–Chandler two-step procedure, J. Chem. Phys., submitted.
50 A. Ulitsky and R. Elber, J. Chem. Phys., 92 (1990) 1510.
51 H. Jónsson, G. Mills, K.W. Jacobsen, in: B.J. Berne, G. Ciccotti and D.F. Coker (Eds.), Classical and quantum dynamics in condensed phase simulations, World Scientific, Singapore, 1998.
52 R. Elber, A. Ghosh and A. Cardenas, Acc. Chem. Res., 35 (2002) 396.
53 W. E, W. Ren and E. Vanden-Eijnden, Phys. Rev. B, 66 (2002) 05230.
54 R. Pratt, J. Chem. Phys., 85 (1986) 5045.
55 C. Dellago, P.G. Bolhuis and P.L. Geissler, Adv. Chem. Phys., 123 (2002) 1.
56 W. E, W. Ren and E. Vanden-Eijnden, Finite temperature string method for the study of rare events, J. Phys. Chem. B, submitted.
57 A.K. Faradjian and R. Elber, J. Chem. Phys., 120 (2004) 10880.
58 A. Laio and M. Parrinello, Proc. Natl. Acad. Sci. USA, 99 (2002) 12562.
59 M. Iannuzzi, A. Laio and M. Parrinello, Phys. Rev. Lett., 90 (2003) 238302.
60 J.A. McCammon, B.R. Gelin and M. Karplus, Nature, 267 (1977) 585.
61 W.F. van Gunsteren, H.J.C. Berendsen, J. Hermans, W.G.J. Hol and J.P.M. Postma, Proc. Natl. Acad. Sci., USA, 80 (1983) 4315.
62 W.F. van Gunsteren and H.J.C. Berendsen, Biochem. Soc. Trans., 10 (1982) 301.
63 M. Karplus and J.A. McCammon, Ann. Rev. Biochem., 52 (1983) 263.
64 J.A. McCammon, Rep. Prog. Phys., 47 (1984) 1.
65 B.R. Brooks, R.E. Bruccoleri, B.D. Olafson, D.J. States, S. Swaminathan and M. Karplus, J. Comput. Chem., 4 (1983) 187.
66 W.F. van Gunsteren and H.J.C. Berendsen, Angew. Chem. Int. Ed. Engl., 29 (1990) 992.
67 D.A. Pearlman, D.A. Case, J.W. Caldwell, W.R. Ross, T.E. Cheatham, III, S. DeBolt, D. Ferguson, G. Seibel and P. Kollman, Comput. Phys. Commun., 91 (1995) 1.
68 R. Kubo, J. Phys. Soc. (Jpn), 12 (1957) 570.
69 B.J. Alder, D.M. Gass and T.E. Wainwright, J. Chem. Phys., 53 (1970) 3813.
70 D. Levesque, L. Verlet and J. Kürkijarvi, Phys. Rev. A, 7 (1973) 1690.
71 G. Ciccotti and G. Jacucci, Phys. Rev. Lett., 35 (1975) 789.
72 A.W. Lees and S.F. Edwards, J. Phys. C, 5 (1972) 1921.
73 E.M. Gosling, I.R. McDonald and K. Singer, Mol. Phys., 26 (1973) 1475.
74 W.G. Hoover and W.T. Ashurst, Theor. Chem. Adv. Persp., 1 (1975) 1.
75 G. Ciccotti, G. Jacucci and I.R. McDonald, Phys. Rev. A, 13 (1976) 426.
76 G. Ciccotti, G. Jacucci and I.R. McDonald, J. Stat. Phys., 21 (1979) 1.
77 W.G. Hoover, D.J. Evans, R.B. Hickman, A.J.C. Ladd, W.T. Ashurst and B. Moran, Phys. Rev. A, 22 (1980) 1690.

78 D.J. Evans and P.G. Morriss, Comput. Phys. Rep., 1 (1984) 297.
79 A. Tenenbaum, G. Ciccotti and R. Gallico, Phys. Rev. A, 25 (1982) 2778.
80 C. Trozzi and G. Ciccotti, Phys. Rev. A, 29 (1984) 916.
81 M. Mareschal and E. Kestemont, Nature, 323 (1987) 427.
82 D.C. Rapaport, Phys. Rev. Lett., 60 (1988) 2480.
83 J. Koplik, J.R. Banavar and J.F. Willemsen, Phys. Rev. Lett., 60 (1988) 1282.
84 D.J. Evans and P.G. Morriss, Statistical mechanics of nonequilibrium liquids, Academic Press, New York, 1990.
85 H. Loewen and T.N. Likos (Eds.), Colloidal dispersion in external fields, J. Phys. Condens. Matter, 16, 2004, p. 3769.
86 R.P. Feynman, Statistical mechanics, Addison-Wesley, Reading, MA, 1972.
87 D. Chandler and P.G. Wolynes, J. Chem. Phys., 74 (1981) 4078.
88 D.M. Ceperley, Rev. Mod. Phys., 67 (1995) 279.
89 D.M. Ceperley, in: K. Binder, G. Ciccotti (Eds.), Monte Carlo and molecular dynamics of condensed matter systems, SIF, Bologna, 1996, p. 443.
90 M.F. Herman, E.J. Bruskin and B.J. Berne, J. Chem. Phys., 76 (1982) 5150.
91 B.J. Berne and D. Thirumalai, Ann. Rev. Phys. Chem., 37 (1986) 401.
92 M. Sprik, R. Impey and M.L. Klein, J. Chem. Phys., 83 (1985) 5802.
93 M.E. Tuckerman, A. Hughes, in: B.J. Berne, G. Ciccotti, D.F. Coker (Eds.), Classical and quantum dynamics in condensed phase simulations, World Scientific, Singapore, 1998.
94 M. Parrinello and A. Rahman, J. Chem. Phys., 80 (1984) 860.
95 R.A. Kuharski and P.J. Rossky, Chem. Phys. Lett., 103 (1984) 357.
96 D.M. Ceperley and E.L. Pollock, Phys. Rev. Lett., 56 (1986) 351.
97 R. Car and M. Parrinello, Phys. Rev. Lett., 55 (1985) 2471.
98 U. Röthlisberger, in: J. Leszczynski (Ed.), Computational chemistry: Reviews of current trends, World Scientific, Singapore, 2001, pp. 33–68.
99 W. Kohn and L.J. Sham, Phys. Rev., 140 (1965) A1133.
100 A. Selloni, P. Carnevali, R. Car and M. Parrinello, Phys. Rev. Lett., 59 (1989) 823.
101 D. Laria, G. Ciccotti, M. Ferrario and R. Kapral, J. Chem. Phys., 97 (1992) 378.
102 D. Borgis, G. Tarjus and H. Azzouz, J. Phys. Chem., 96 (1992) 3188.
103 B.J. Berne, G. Ciccotti and D.F. Coker (Eds.), Classical and quantum dynamics in condensed phase simulations, World Scientific, Singapore, 1998.
104 J.C. Tully, J. Chem. Phys., 93 (1990) 1061.
105 S. Hammes-Schiffer and J.C. Tully, J. Chem. Phys., 101 (1994) 4657.
106 P. Pechukas, Phys. Rev., 181 (1969) 166.
107 F. Webster, E.T. Wang, P.J. Rossky and R.A. Friesner, J. Chem. Phys., 100 (1994) 4835.
108 D.F. Coker and L. Xiao, J. Chem. Phys., 102 (1995) 496.
109 R. Kapral and G. Ciccotti, J. Chem. Phys., 110 (1999) 8919.
110 D. Mac Kernan, G. Ciccotti and R. Kapral, J. Phys. Condens. Matter, 14 (2002) 9069.
111 A. Donoso and C.C. Martens, J. Phys. Chem., 102 (1998) 4291.
112 I.V. Aleksandrov, Z. Naturforsch. A, 36 (1981) 902.
113 V.I. Gerasimenko, Rep. Ukranian Acad. Sci., 10 (1981) 65, Theor. Mat. Phys., 50 (1982).
114 G.A. Voth, Adv. Chem. Phys., 93 (1996) 135.
115 N. Makri, J. Phys. Chem. B, 103 (1999) 2823.
116 D. Marx, in: B.J. Berne, G. Ciccotti, D.F. Coker (Eds.), Classical and quantum dynamics in condensed phase simulations, World Scientific, Singapore, 1998, p. 361.
117 M. Ben-Nun, J. Quenneville and T.J. Martínez, J. Phys. Chem. A, 104 (2000) 5161.
118 A. Puhl, M. Mansour and M. Mareschal, Phys. Rev. A, 40 (1989) 1999.
119 T.C. Germann, B.L. Holian, P.S. Lomdahl and R. Ravelo, Phys. Rev. Lett., 84 (2000) 5351.
120 S. Chandrasekhar, Rev. Mod. Phys., 15 (1943) 1.
121 A.T. Brünger, C.L. Brooks, III and M. Karplus, Proc. Natl. Acad. Sci., USA, 82 (1985) 8458.
122 M.J. Kotelyanskii and D. Theodorou (Eds.), Simulation methods for polymers, Marcel Decker, New York, 2004.

123 M. Karttunen, I. Vattulainen and A. Lukkarinen (Eds.), Novel methods in soft matter simulations, Springer, Berlin, 2004.
124 P. Nielaba, M. Mareschal and G. Ciccotti (Eds.), Bridging time scales: Molecular simulations for the next decade, Springer, Berlin, 2002.
125 R. Benzi, S. Succi and M. Vergassola, Phys. Rep., 222 (1992) 145.
126 G.A. Bird, Molecular gas dynamics, Clarendon, Oxford, 1976.
127 U. Frisch, B. Hasslacher and Y. Pomeau, Phys. Rev. Lett., 56 (1986) 1505.
128 U. Frisch, D. d'Humières, B. Hasslacher, P. Lallemand, Y. Pomeau and J.P. Rivet, Complex Syst., 1 (1987) 649.
129 J.P. Boon, D. Dab, R. Kapral and A. Lawniczak, Phys. Rep., 273 (1996) 55.
130 A. Malevanets and R. Kapral, J. Chem. Phys., 110 (1999) 8605.
131 A. Malevanets and R. Kapral, J. Chem. Phys., 112 (2000) 7260.
132 P.J. Hoogerbrugge and J.M.V.A. Koelman, Europhys. Lett., 19 (1992) 155.
133 P. Español and P. Warren, Europhys. Lett., 30 (1995) 191.
134 J.J. Monaghan, Ann. Rev. Astrophys. Astron., 30 (1992) 543.
135 A.L. Garcia, J.B. Bell, W.Y. Crutchfield and B.J. Alder, J. Comput. Phys., 154 (1999) 134.
136 M.P. Allen and D.J. Tildesley, Computer simulation of liquids, Clarendon Press, Oxford, 1987.
137 D. Frenkel and B. Smit, Understanding molecular simulation, 2nd edn., Academic Press, San Diego, 2001.
138 G. Ciccotti and W.G. Hoover (Eds.), Molecular dynamics simulation of statistical mechanics, SIF/North-Holland, Amsterdam, 1986.
139 K. Binder and G. Ciccotti (Eds.), Monte Carlo and molecular dynamics of condensed matter systems, SIF, Bologna, 1996.

Theory and Applications of Computational Chemistry: The First Forty Years
Edited by C. Dykstra et al.

CHAPTER 17

Equations of motion methods for computing electron affinities and ionization potentials

Jack Simons

Chemistry Department and Henry Eyring Center for Theoretical Chemistry, University of Utah, Salt Lake City, UT 84112, USA

Abstract

The *ab initio* calculation of molecular electron affinities (EA) and ionization potentials (IP) is a difficult task because the energy of interest is a very small fraction of the total electronic energy of the parent neutral. For example, EAs typically lie in the 0.01–10 eV range, but the total electronic energy of even a small molecule is usually several orders of magnitude larger. Moreover, because the EA or IP is an intensive quantity but the total energy is an extensive quantity, the difficulty in evaluating EAs and IPs to within a fixed specified (e.g. ±0.1 eV) accuracy becomes more and more difficult as the size and number of electrons in the molecule grows. The situation becomes especially problematic when studying extended systems such as solids, polymers, or surfaces for which the EA or IP is an infinitesimal fraction of the total energy. Equations of motion (EOM) methods and other related approaches offer a route for calculating the intensive EAs and IPs directly as eigenvalues of a set of working equations. A history of the development of EOM theories as applied to EAs and IPs, their numerous practical implementations, and their relations to Greens function or propagator theories are given in this contribution. EOM methods based upon Møller–Plesset, multiconfiguration self-consistent field, and coupled-cluster reference wave functions are included in the discussion as is the application of EOM methods to metastable resonance states of anions.

17.1 INTRODUCTION

The vertical (i.e. at any given fixed molecular geometry) electron affinity (EA) of a molecule can be computed by (approximately) solving the Schrödinger equation for the energy $E(0,N)$ of the N-electron neutral molecule and the Schrödinger equation

References pp. 461–464

for the energy $E(K, N+1)$ of the Kth state of the $N+1$-electron anion and subtracting the two energies:

$$\mathrm{EA} = E(0, N) - E(K, N+1) \tag{1a}$$

The corresponding vertical ionization potential (IP) is given as

$$\mathrm{IP} = E(K, N-1) - E(0, N) \tag{1b}$$

Here, we use K to label the electronic state of the anion or cation that one wishes to study, and 0 to label the state of the neutral (usually but not necessarily the ground state) to which the electron is being attached or from which it is removed.

In using such an approach for obtaining the EA or IP, one is faced with a very difficult numerical challenge because $E(0, N)$, $E(K, N-1)$, and $E(K, N+1)$ tend to be extremely large (negative) numbers, whereas EA and IP nearly always lie in the range 0–20 eV. For example, the EA of the $^4S_{3/2}$ state of the carbon atom is [1] 1.262119 ± 0.000020 eV, whereas the total electronic energy of this state of C is − 1030.080 eV (relative to a C^{6+} nucleus and six electrons infinitely distant and not moving that defines the zero of energy). Since the EA is *ca.* 0.1% of the total energy of C, one needs to compute the C and C^- electronic energies to accuracies of 0.01% or better to calculate the EA to within 10%.

This observation shows only the 'tip of the iceberg', however, as the major problem relates to the fact that $E(0, N)$, $E(K, N-1)$, and $E(K, N+1)$ are extensive properties whereas EA and IP are intensive quantities. For example, the EA of C_2 in its X $^2\Sigma_g^+$ ground electronic state is [1] 3.269 ± 0.006 eV near the equilibrium bond length R_e but only 1.2621 eV at $R \to \infty$ (i.e. the same as the EA of a carbon atom). However, the total electronic energy of C_2 is − 2060.160 eV at $R \to \infty$ and lower by *ca.* 3.6 eV (the dissociation energy [2] of C_2) at R_e, so again EA is a very small fraction of the total energies. For buckyball C_{60}, the EA is [1] 2.666 ± 0.001 eV, but the total electronic energy is 60 times − 1030.080 eV minus the atomization energy (i.e. the energy change for $C_{60} \to 60$ C) of this compound. Clearly, the challenge of evaluating EA (or IP) to within even 50% becomes more and more difficult as the size (i.e. number of electrons) in the molecule grows, and it becomes impossible when the system of interest is an infinite solid, surface, or polymer. This same kind of difficulty (i.e. calculating an intensive quantity as the difference between to extensive energies) plagues the computation of EAs and IPs, bond energies, and electronic excitation energies.

The problems discussed in the preceding paragraph do not disappear if one uses a computer with higher numerical precision in its arithmetic (i.e. a longer word length) or algorithms that compute the one- and two-electron integrals needed for any quantum chemistry calculation to more significant figures. No mater how precise the integrals and how long the floating point word length (as long as they are finite), the evaluation of intensive properties such as IPs, EAs, and excitation energies as differences between pairs of extensive total electronic energies is doomed to fail.

Of course, much progress can be made in computing EAs and IPs as differences between anion and neutral or cation and neutral total energies [3] because of large systematic cancellation in energy errors [4,5]. For example, the pair correlation energies of the two 1s electron pairs in C_2 is quite large, but is very nearly the same as in C_2^-, so even a large percent error made in computing these contributions to the total energy may not greatly affect the EA computed by subtracting $E(K, N + 1)$ from $E(0, N)$. Some of the earliest high-quality *ab initio* calculations of EAs were in fact carried out by Professor Schaefer [6] during his PhD work with Frank Harris using wave function techniques and calculating separate neutral and anion energies. Nevertheless, in the late 1960s and early 1970s, workers were motivated to develop methods that would allow intensive energy differences such as EAs, IPs, and excitation energies (ΔEs) to be computed 'directly' rather than as differences in two very large numbers. This point of view is what led to the development of so-called equations of motion (EOM) methods as well as Greens function methods [7] and, more recently, response function approaches [8], within electronic structure theory. In all these theories, one performs a derivation in which the two total energies (i.e. ion and neutral or ground and excited states) are subtracted analytically (rather than numerically) thereby achieving an analytical expression for the desired intensive energy difference. It is by thus dealing with equations that involve only intensive energies that one can overcome the problems detailed earlier.

Among the earliest practitioners of EOM methods in the chemistry community were Prof. V. McKoy [9] and his group (including Thom Dunning) at Cal Tech. They imported many ideas and mathematical tools from the nuclear physics literature [10], where EOM theories had been used to study excited states of nuclei, and they focused their efforts on electronic excitation energies ΔE, not IPs or EAs. Because the present chapter is dedicated to how such methods are used to compute EAs and IPs, not much more will be said about the McKoy group's pioneering work on EOM theory for excitation energies, although its ultimate relationship to other excitation energy methods will be discussed briefly later.

In 1973, the author used the framework of EOM theory [11] as expressed by the McKoy group to develop a systematic (i.e. order-by-order in the Møller–Plesset perturbation theory sense) approach for directly computing molecular EAs and IPs as eigenvalues of the EOM working equations. It is this development and its subsequent improvement and extensions [12] by our group and others that we now wish to describe.

17.2 BASICS OF EOM THEORY AS APPLIED TO EAs AND IPs

17.2.1 The EA equations of motion

The fundamental working equations of any EOM theory are derived by writing the Schrödinger equations for the neutral and anion (or neutral or cation or ground and excited) states of interest and subtracting the two equations as a first step toward obtaining a single equation that will yield the EA or IP or ΔE. That is, the EOM theory produces the intensive energy difference directly as an eigenvalue of the working equation. As above, we use $|0, N\rangle$ to denote the 0th electronic state of the N-electron neutral and $|K, N + 1\rangle$ to denote the Kth state of the $N + 1$-electron anion and write the two Schrödinger equations as

$$H|0, N\rangle = E(0, N)|0, N\rangle \tag{2a}$$

$$H|K,N+1\rangle = E(K,N+1)|K,N+1\rangle \quad (2b)$$

Because $|0,N\rangle$ and $|K,N+1\rangle$ contain different numbers of electrons, it is convenient and most common in developing EOM theories of EAs to express the electronic Hamiltonian H in second-quantized form [13]:

$$H = \sum_{i,j} h(i,j)i^+j + 1/2 \sum_{i,j,k,l} \langle i,j|k,l\rangle i^+j^+lk \quad (3)$$

where $h(i,j)$ represents a matrix element of the one-electron operators (i.e. kinetic energy, electron-nuclear Coulomb attraction, etc.) within the orthonormal molecular spin-orbital basis $\{\phi_j\}$, $\langle i,j|k,l\rangle$ is a matrix element of the two-electron operators (i.e. electron–electron repulsion), and the set of Fermion creation operators $\{i^+\}$ create an electron in the $\{\phi_i\}$ spin-orbitals, whereas the $\{i\}$ operators destroy such an electron. Writing H in such a form allows us to use the same H in Eqs. (2a) and (2b) even though these two Schrödinger equations relate to N and $N+1$ electrons, respectively.

The next step in developing an EOM equation is to assume that the anion state $|K,N+1\rangle$ can be related to the neutral state $|0,N\rangle$ through an operator $Q^+(K)$

$$|K,N+1\rangle = Q^+(K)|0,N\rangle \quad (4)$$

that maps the neutral molecule wave function into the desired anion wave function. For the EA case at hand, the operator $Q^+(K)$ is usually written in terms of scalar coefficients $t(K,l)$ multiplied by operators $T^+(l)$, also expressed in second-quantization language, each of which involves adding an electron

$$Q^+(K) = \sum_l t(K,l)T^+(l) \quad (5)$$

It has been shown [14] that a complete set of such $T^+(l)$ operators consists of the union of sets of operators $\{p^+\}$ that add an electron to a spin-orbital ϕ_p, operators $\{p^+q^+a\}$ that add an electron to ϕ_p and excite another electron from ϕ_a to ϕ_q, operators $\{p^+q^+r^+ab\}$ that add an electron to ϕ_p, excite an electron from ϕ_a to ϕ_r and excite another electron from ϕ_b to ϕ_q as well as higher-level electron addition and excitation operators up to the highest-level operators that add an electron and induce N excitations. In labeling these operators, the indices a, b, c, d, etc. are used to denote spin-orbitals occupied in a so-called reference Slater determinant within $|0,N\rangle$ and p, q, r, s, etc. are used to denote unoccupied (i.e. virtual) spin-orbitals. The reference determinant, which is what defines the concept of occupied and unoccupied spin-orbitals, is usually chosen to be the determinant $|0\rangle$ within the neutral molecule wave function

$$|0,N\rangle = \sum_{J=0,M} C(0,J)|J\rangle \quad (6)$$

with the largest amplitude $C(0,0)$, but it has been shown [14] that $|0\rangle$ can actually be taken to be any determinant within $|0,N\rangle$ that possesses non-zero amplitude. Later we will deal with how one determines the $C(0,J)$ amplitudes in the wave function $|0,N\rangle$; for now, suffice it to say these amplitudes can, for example, be taken from Møller–Plesset (MP) perturbation theory, from multiconfiguration self-consistent field (MCSCF) theory, from configuration interaction (CI) theory or from coupled-cluster (CC) theory.

Using Eq. (4) in Eqs. (2a) and (2b) and subtracting Eq. (2a) from Eq. (2b) gives a single equation whose eigenvalue gives the desired EA:

$$(HQ^+(K) - Q^+(K)H)|0,N\rangle = (E(K,N+1) - E(0,N))Q^+(K)|0,N\rangle \tag{7a}$$

or, in terms of the commutator $[H, Q^+(K)]$

$$[H, Q^+(K)]|0,N\rangle = EQ^+(K)|0,N\rangle \tag{7b}$$

where the eigenvalue E is the negative of the EA. The key point is that one now has a single equation to be solved that produces the intensive EA as its eigenvalue. This equation appears to be of the conventional eigenvalue–eigenfunction form, but it is somewhat different because the operator that acts on the eigenfunction $Q^+(K)|0,N\rangle$ is not the Hamiltonian but a commutator involving the Hamiltonian. The fact that the commutator appears is what causes the eigenvalue to be an energy difference.

To progress further toward practical implementation, specific choices must be made for how one is going to approximate the neutral molecule wave function $|0,N\rangle$ and at what level one is going to truncate the expansion of the operator $Q^+(K)$ given in Eq. (5). It is also conventional to reduce Eq. (7) to a matrix eigenvalue equation by projecting this equation onto an appropriately chosen space of $N+1$-electron functions. Let us first deal with the latter issue.

Once the number of $T^+(l)$ operators used to construct $Q^+(K)$ has been chosen (we discuss this choice later), the total number $l^{\max}$ of $t(K,l)$ amplitudes has been determined. Multiplying Eq. (7) on the left by the adjoint $T(j)$ of any one of the T^+ operators, and then projecting the resultant equation against $\langle 0,N|$ gives one form of the working EOM-EA equations:

$$\sum_l \langle 0,N|T(j)[H,T^+(l)]|0,N\rangle t(K,l) = E\sum_l \langle 0,N|T(j)T^+(l)|0,N\rangle t(K,l) \tag{8}$$

To make use of this equation, the $\langle 0,N|T(j)[H,T^+(l)]|0,N\rangle$ and $\langle 0,N|T(j)T^+(l)|0,N\rangle$ matrices of dimension $l^{\max} \times l^{\max}$ must first be evaluated in terms of one- and two-electron integrals (appearing in H) and one-, two- and higher-body density matrices (depending upon the level at which the $\{T^+(l)\}$ operator expansion is truncated). Subsequently, the EA values (i.e. EAs for the various anion states, K, relative to the $|0,N\rangle$ state of the neutral) are computed as minus the eigenvalues E of Eq. (8).

17.2.2 The analogous equations of motion for ionization potentials

Before proceeding further, it is useful to explore how this same framework has been used to compute molecular ionization potentials (IPs). It is fairly straightforward to show that an equation analogous to Eq. (7) but reading

$$\langle 0,N|(HQ^+(K) - Q^+(K)H) = (E(0,N) - E(K,N-1))\langle 0,N|Q^+(K) \tag{9}$$

is valid if the operators $\{Q^+(K)\}$ are as given in Eq. (5) but with the $\{T^+(l)\}$ defined to include operators of the form $\{a^+, a^+b^+p, a^+b^+c^+qr, \text{etc.}\}$. Of course, in Eq. (9), the operators within $Q^+(K)$ act to the left on $\langle 0, N|$ to generate cationic states. As a result, neutral-cation energy differences appear in Eq. (9) and thus this offers a route for computing IPs. Multiplying this equation on the right by any one of the $T(j)$ operators and then projecting against $|0, N\rangle$ gives

$$\sum_l \langle 0, N|[H, T^+(l)]T(j)|0, N\rangle t(K, l) = E \sum_l \langle 0, N|T^+(l)T(j)|0, N\rangle t(K, l) \quad (10)$$

but now the eigenvalues E denote values of $(E(0, N) - E(K, N-1))$, which are the negatives of the IPs.

Thus far, we see that EOMs can be written that allow EAs or IPs to be computed. The fundamental constructs within these equations are as follows:

(i) for the EA case, matrix elements $\langle 0, N|T(j)[H, T^+(l)]|0, N\rangle$ involving the commutator of H with the $T^+(l)$ operators then multiplied on the left by a $T(j)$ operator, as well as an analogous overlap matrix element $\langle 0, N|T(j)T^+(l)|0, N\rangle$;

(ii) for the IP case, matrix elements $\langle 0, N|[H, T^+(l)]T(j)|0, N\rangle$ of the same commutator but with the $T(j)$ operator on the right, as well as the corresponding overlap matrix element $\langle 0, N|T^+(l)T(j)|0, N\rangle$;

(iii) the neutral molecule wave function $|0, N\rangle$ with respect to which the EA or IP is to be evaluated.

17.2.3 The rank of the operators

It is now useful to analyze the density matrix elements [15] that enter into these equations. Each of the $T^+(j)$ operators contains an odd number of creation or annihilation operators, and the Hamiltonian H contains two (i.e. i^+j) or four (i.e. i^+j^+lk) such operators. It can be seen that the commutator $[H, T^+(l)]$ does not contain four plus the number of creation or annihilation operators in $T^+(l)$, but two fewer operators. For example, the commutator $[i^+j^+lk, p^+q^+a]$ does not yield any terms with four creation and three annihilation operators but only terms with three creation and two annihilation operators. We say that the act of forming the commutator (which is what causes the higher order operators to cancel) gives rise to a reduction in the rank of the operators. As a result, both the operator products $T(j)[H, T^+(l)]$ and $[H, T^+(l)]T(j)$, which appear in the EA and IP-EOM, respectively, contain terms only involving both creation and annihilation operators equal to the number of creation operators in $T^+(l)$ plus one plus the number of creation operators in $T(j)$. For example, if $T^+(l) = p^+q^+a$ and $T(j) = b^+rs$, then $T(j)[H, T^+(l)]$ and $[H, T^+(l)]T(j)$ will contain terms with no more than four creation and four annihilation operators. This means that the density matrices needed from $\langle 0, N|T(j)[H, T^+(l)]|0, N\rangle$ and $\langle 0, N|[H, T^+(l)]T(j)|0, N\rangle$ will be, at most, fourth-order density matrices of the $\langle 0, N|\cdots|0, N\rangle$ density.

17.2.4 Equations of lower rank for both EAs and IPs

Indeed, in the early years of using EOM methods [16] to compute EAs and IPs, operator manifolds of the form $\{T^+(l)\} = \{p^+; p^+q^+a, p^+q^+r^+ba, \text{ etc.}\}$ or $\{T^+(l)\} = \{a^+, a^+b^+p, a^+b^+c^+qr, \text{ etc.}\}$ were employed with Møller–Plesset approximations to $|0, N\rangle$ (usually taken through first order) to form the kind of matrix elements appearing in Eqs. (8) and (10) and to then evaluate EAs and IPs from their eigenvalues E. However, it became more common to use a combination of the EA and IP-EOMs formed by adding Eqs. (8) and (10), while expanding the $\{T^+(l)\}$ operator manifold to include those needed to evaluate EAs $\{p^+; p^+q^+a, p^+q^+r^+ba, \text{ etc.}\}$ and those needed for the IPs $\{a^+, a^+b^+p, a^+b^+c^+qr, \text{ etc.}\}$, to simultaneously compute both such energy differences.

To understand why such a combination has proven beneficial, it suffices to examine the form and rank of the operators whose $\langle 0, N|\cdots|0, N\rangle$ matrix elements must be evaluated

$$\sum_l \langle 0, N|[H, T^+(l)]T(j) + T(j)[H, T^+(l)]|0, N\rangle t(K, l) = E \sum_l \langle 0, N|T^+(l)T(j) + T(j)T^+(l)|0, N\rangle t(K, l) \quad (11)$$

Recall that the $T^+(j)$ operators contain an odd number of creation or annihilation operators. Each of the products $[H, T^+(l)]T(j)$, $T(j)[H, T^+(l)]$, $T^+(l)T(j)$ and $T(j)T^+(l)$ thus contain an even number of such operators. However, because of the fundamental anti-commutation properties of these operators

$$i^+j + ji^+ = \delta_{i,j} \quad (12a)$$

$$ij + ji = 0 \quad (12b)$$

$$i^+j^+ + j^+i^+ = 0 \quad (12c)$$

it can easily be shown that the *operator combinations* $T^+(l)T(j) + T(j)T^+(l)$ and $[H, T^+(l)]T(j) + T(j)[H, T^+(l)]$ contain one fewer creation and one fewer annihilation operator than does either of the two terms in the sums. So, by combining the EA and IP-EOMs, one effects a rank reduction in the operators appearing in the equations although the dimensions of the matrices one needs to construct are doubled (because the $\{T^+(l)\}$ operator manifold was doubled when both EA and IP operators were included. The rank reduction is important because it means that the density matrices that need to be evaluated to compute the $\langle 0, N|\cdots|0, N\rangle$ matrix elements are of lower rank in Eq. (11) than in either Eq. (8) or Eq. (10). As we said, it has become more common to use the combined EA and IP equation (11) because lower-order density matrices are required.

17.2.5 Summary

Thus far, we have shown how one can obtain eigenvalue equations, in which the energy eigenvalues correspond to the intensive EAs (or IPs), by postulating that the anion (or cation) wave function can be related to the neutral molecule wave function through an operator. We have also shown how the EA and IP-EOM can be combined to generate a combined EOM from which both EAs and IPs can be obtained. The advantage to the latter approach is that the operators appearing in the resultant equations are of lower rank and thus lower-order density matrices must be evaluated to carry out the calculations. Let us now move on to address more specific embodiments of such EOM theories that result from different choices of the neutral molecule wave function and of the operator connecting the neutral and anion wave functions.

17.3 PRACTICAL IMPLEMENTATIONS OF EOM THEORIES FOR EAs AND IPs

The basic ideas underlying any EOM method for computing EAs or IPs appear above. However, as discussed earlier, in any specific embodiment of such a method, one must commit to

(i) a specific approximation to the neutral molecule wave function $|0, N\rangle$;
(ii) a specific choice of how large an operator manifold $\{T^+(l)\}$ to employ; and
(iii) how to solve the resultant EOM equations for the eigenvalues E that then produce the EAs or IPs.

In the following subsections, we describe the most commonly used choices for these three issues.

17.3.1 The Møller–Plesset based approximations

In the earliest implementation of EOM approaches to EAs, the author's group [11,16] chose to represent the $|0, N\rangle$ wave function in a Møller–Plesset expansion

$$|0, N\rangle = \psi^0 + \psi^1 + \psi^2 + \cdots \tag{13}$$

with the single-determinant unrestricted Hartree–Fock (HF) function being ψ^0 and the corresponding neutral molecule HF Hamiltonian being H^0. This choice was made because there existed substantial evidence that EAs and IPs computed at the Koopmans' theorem level would not meet the desired 0.1 eV accuracy. The evidence on atoms and small molecules also showed that EAs and IPs computed using standard second-order MP theory were much more accurate but not sufficient to approach the 0.1 eV standard. For this reason, the author's group set their sites on the next reasonable level, that of third-order MP theory.

The operator manifold $\{T^+(l)\}$ was taken to consist of $\{p^+; p^+q^+a\}$ and $\{a^+, a^+b^+p\}$. This choice of operator manifold was shown to be capable of producing EAs and IPs that were precise through third order [17] in the MP perturbation, which is why this choice was made.

The resultant variant of Eq. (11) was not solved by finding the eigenvalues of this matrix eigenvalue equation whose dimension is the sum of the dimensions of the $\{p^+; p^+q^+a\}$ and $\{a^+, a^+b^+p\}$ operator manifolds. Rather, that large matrix eigenvalue problem was partitioned [11] using a primary subspace defined by the $\{p^+, a^+\}$ operators and a secondary subspace defined by the $\{p^+q^+a, a^+b^+p\}$ operators. The partitioned eigenvalue problem

$$\sum_{j=a,p} H_{i,j}(E)X_j = EX_i \tag{14}$$

whose dimension was that of the $\{p^+, a^+\}$ operator space was used to find the eigenvalues E. Of course, the act of partitioning the higher-dimension matrix eigenvalue problem does not change the values of E that represent solutions to the equations. That is, the same E values that fulfill the original equations are also solutions to the partitioned equations. However, once one introduces approximations designed to evaluate elements of the partitioned $H_{i,j}(E)$ matrix to a chosen order in perturbation theory, this equivalence is lost. It is precisely by making such an order analysis (e.g. computing $H_{i,j}(E)$ through second or third order) that EOM theories capable of evaluating EAs or IPs to a given order were obtained.

When the elements of the partitioned matrices were evaluated through second order in the MP series, the following expression was obtained for the matrix elements $H_{i,j}$

$$\begin{aligned} H_{i,j}(E) = \varepsilon_i\delta_{i,j} &- \sum_{p,q,a} \langle i,a\|p,q\rangle\langle p,q\|j,a\rangle/(\varepsilon_p + \varepsilon_q - \varepsilon_a - E) \\ &+ \sum_{a,b,p} \langle i,p\|a,b\rangle\langle a,b\|j,p\rangle/(\varepsilon_a + \varepsilon_b - \varepsilon_p - E) \end{aligned} \tag{15}$$

Here, the ε_j denote the UHF spin-orbital energies of the neutral molecule and the $\langle i,j\|k,l\rangle$ denote differences in two-electron integrals ($\langle i,j\|k,l\rangle = \langle i,j|k,l\rangle - \langle i,j|l,k\rangle$).

The expression for $H_{i,j}(E)$ valid through third order in the MP series is more complicated and is derived in Refs. [11,17]. The partitioned matrix eigenvalue equation was written in those works as

$$\mathbf{H}(E)\mathbf{X} = E\mathbf{X}$$

where the elements of the $\mathbf{H}$ matrix were defined as follows:

$$\begin{aligned} H_{i,j}(E) = A_{i,j} &+ \sum_{\alpha<\beta,m} [B_{i,\alpha m\beta}B^*_{j,\alpha m\beta}]/[E^m_{\alpha\beta} + E] - \sum_{N+1<n,\alpha} [B_{i,n\alpha N+1}B^*_{j,n\alpha N+1}]/[E^{N+1n}_{\alpha} - E] \\ &- \sum_{N+1<m<n,\alpha} [B_{i,n\alpha m}B^*_{j,n\alpha m}]/[E^{mn}_{\alpha} - E] \end{aligned}$$

In turn, the elements of the A and B matrices are shown below:

$$B_{i,\alpha m\beta} = -\langle im|\alpha\beta\rangle - 1/2\sum_{p,q}\langle im|pq\rangle K^{(pq)}_{\alpha\beta} + \sum_{\gamma,p}[\langle i\gamma|p\alpha\rangle K^{(mp)}_{(\beta\gamma)} - \langle i\gamma|p\beta\rangle K^{(mp)}_{(\alpha\gamma)}]$$

$$B_{i,n\alpha m} = \langle i\alpha|mn\rangle + 1/2\sum_{\gamma,\delta}\langle i\alpha|\gamma\delta\rangle K^{mn}_{(\delta\gamma)} + \sum_{\gamma,p}[\langle ip|\gamma n\rangle K^{(mp)}_{(\alpha\gamma)} - \langle ip|\gamma m\rangle K^{(np)}_{(\alpha\gamma)}]$$

$$A_{i,j} = \delta_{i,j}\varepsilon_i + \sum_{k,l}\langle ik|jl\rangle F_{kl}$$

to which one adds the following E-independent terms

$$\delta A_{i,j} = \sum_{\delta,\beta,p,m,n} \langle jp||i\delta\rangle\langle\delta\beta||mn\rangle\langle mn||p\beta\rangle/[(\varepsilon_\delta - \varepsilon_p)(\varepsilon_\delta + \varepsilon_\beta - \varepsilon_m - \varepsilon_n)]$$

$$+ \sum_{\delta,\beta,p,m,n} \langle j\delta||ip\rangle\langle p\beta||mn\rangle\langle mn||\delta\beta\rangle/[(\varepsilon_\delta - \varepsilon_p)(\varepsilon_\delta + \varepsilon_\beta - \varepsilon_m - \varepsilon_n)]$$

$$+ \sum_{\delta,\alpha,\beta,p,n} \langle jp||i\delta\rangle\langle\delta n||\beta\alpha\rangle\langle\alpha\beta||pn\rangle[(\varepsilon_\delta - \varepsilon_p)(\varepsilon_\alpha + \varepsilon_\beta - \varepsilon_p - \varepsilon_n)]$$

$$+ \sum_{\delta,\alpha,\beta,p,n} \langle j\delta||ip\rangle\langle\delta n||\beta\alpha\rangle\langle\alpha\beta||pn\rangle[(\varepsilon_\delta - \varepsilon_p)(\varepsilon_\alpha + \varepsilon_\beta - \varepsilon_p - \varepsilon_n)]$$

The energy denominators appearing in the H_{ij} matrix elements are

$$E_\alpha^{mn} = \varepsilon_m + \varepsilon_n - \varepsilon_\alpha - \langle\alpha m|\alpha m\rangle - \langle\alpha n|\alpha n\rangle + \langle mn|mn\rangle$$

$$E_{\delta\gamma}^{p} = \varepsilon_p - \varepsilon_\delta - \varepsilon_\gamma - \langle\delta p|\delta p\rangle - \langle\gamma p|\gamma p\rangle + \langle\delta\gamma|\delta\gamma\rangle$$

Finally, the F quantities appearing above are given as

$$F_{kl} = \sum_{\alpha<\beta,p} [K_{\alpha\beta}^{pk}K_{\alpha\beta}^{pl} + K_{\alpha\beta}^{kp}K_{\alpha\beta}^{lp}] - \sum_{p<q,\alpha} [K_{\alpha l}^{pq}K_{\alpha k}^{pq} + K_{l\alpha}^{pq}K_{k\alpha}^{pq}]$$

where

$$K_{\alpha\beta}^{(pq)} = K_{\alpha\beta}^{pq} - K_{\alpha\beta}^{qp}$$

$$K_{(\alpha\beta)}^{pq} = K_{\alpha\beta}^{pq} - K_{\beta\alpha}^{pq}$$

$$K_{(\alpha\beta)}^{(pq)} = K_{\alpha\beta}^{pq} - K_{\alpha\beta}^{qp} - K_{\beta\alpha}^{pq} + K_{\beta\alpha}^{qp}$$

and the latter quantities are the MP expansion coefficients of the first-order wave function:

$$K_{\alpha\beta}^{mn} = \langle mn|\alpha\beta\rangle/(\varepsilon_\alpha + \varepsilon_\beta - \varepsilon_m - \varepsilon_n)$$

Although more complicated than their second-order counterparts, the basic structure of the above expressions for $H_{i,j}(E)$ are the same as those shown earlier.

These third-order equations have been used in many applications in which molecular EAs have been computed for a wide variety of species as illustrated in Ref. [16]. Clearly, all the quantities needed to form the second- or third-order EOM matrix elements $H_{j,k}$ are ultimately expressed in terms of the orbital energies $\{\varepsilon_k\}$ and two-electron integrals $\langle j,k|l,h\rangle$ evaluated in the basis of the neutral molecule's Hartree–Fock orbitals that form the starting point of the Møller–Plesset theory. However, as with most electronic structure theories, much effort has been devoted to recasting the working EOM equations in a manner that involves the atomic orbital (AO) two-electron integrals rather than the molecular orbital based integrals. Because such technical matters of direct AO-driven calculations are outside the scope of this work, we will not delve into them further.

17.3.2 Relationship to Greens functions/propagators

It turns out that in the early 1970s when the author was developing and implementing the EOM method for treating EAs and IPs, several groups had taken a different approach to the evaluation of atomic and molecular electronic energy differences using what were called Greens functions (GF) or propagators. Linderberg and Öhrn [18] pioneered the use of such methods in quantum chemistry, while Cederbaum and co-workers [19], Doll and Reinhardt [20], Taylor, Yaris, and co-workers [21] and Pickup and Goscinski [22] were among the first to apply the methods to EAs and IPs using an *ab initio* approach. Purvis and Öhrn [23] soon thereafter expanded the range of the theory to include open-shell systems. These workers as well as Jørgensen and Oddershede [24] and others [25] developed Møller–Plesset based GFs for evaluating electronic excitation energies but we will not discuss these developments further here because our emphasis is on IPs and EAs.

The GF EA and IP theories were derived from consideration of the following time-dependent matrix elements:

$$G_{j,k}(t) = (1/\mathrm{i}\hbar)\Theta(t)\langle 0,N|\exp(\mathrm{i}Ht/\hbar)j^{+}\exp(-\mathrm{i}Ht/\hbar)k|0,N\rangle + (1/\mathrm{i}\hbar)\Theta(-t)\langle 0,N|k\exp(\mathrm{i}Ht/\hbar)j^{+}\exp(-\mathrm{i}Ht/\hbar)|0,N\rangle \quad (16)$$

Here, $\Theta(t)$ is the Heaviside step function, which equals unity when t is positive and zero when t is negative, j^{+} and k are the same creation and annihilation operators discussed earlier and $|0,N\rangle$ is the neutral molecule reference wave function. Introducing complete sets of $N-1$ and $N+1$ electron Hamiltonian eigenfunctions into the first and second terms in Eq. (16), it is straightforward to see that one observes time dependences varying as $\exp(\mathrm{i}[E(0,N)-E(K,N-1)]t/\hbar)$ and $\exp(\mathrm{i}[E(K,N+1)-E(0,N)]t/\hbar)$, respectively.

Taking the time derivative of Eq. (16), one obtains expressions involving commutators of the form $[H,j^{+}]k$ and $k[H,j^{+}]$ just as one finds in Eq. (11). By analyzing the resulting time-derivative equations, workers in this field were able to obtain equations that such $G_{j,k}(t)$ matrix elements obey (n.b.: these were called the EOM for these quantities). The workers named above were able to express the resulting equations in terms of one- and two-electron integrals and corresponding density matrices much as the author had done within the EOM framework. In fact, it turned out that the final working equations of the so-called one-electron Greens function (GF) or electron propagator defined in Eq. (16), when Fourier transformed from the time to the energy domain, were exactly the same as the EOM equations given above, i.e. Eq. (15) and those reproduced from Refs. [11,17]. However, only the Cederbaum group achieved the full third-order expressions within the GF framework analogous to what we reproduced above.

It should be noted that although the GF equations are identical to the combined EA and IP equations (Eq. (15)), they are not equivalent to Eq. (8) or Eq. (10). In fact, only by making additional assumptions about the action of $Q^{+}(K)$ on $|0,N\rangle$ (i.e. that $Q(K)|0,N\rangle$ vanishes) is one able to fully justify adding Eqs. (8) and (10) to obtain Eq. (15) and thus to obtain an equivalence between the EOM and GF results.

Especially in recent years, much of the work aimed at calculating EAs and IPs using these direct-calculation EOM and GF methods has been performed within the notation of Greens functions and has been carried out by Vince Ortiz's group [26] as well as by the Cederbaum group. To further illustrate the impact that such advances have had within the quantum chemistry community, we note that the Ortiz group has implemented various (i.e. Møller–Plesset and other) variants of these theories within the highly successful Gaussian [27] suite of computer codes as a result of which many workers worldwide now employ EOM of GF-type methods to evaluate EAs and IPs.

17.3.3 The natural orbital or extended Koopmans' theorem approach

In the mid 1970s, Parr and co-workers [28] and, independently, Smith and co-workers [29] proposed to use an equation such as Eq. (10) for computing IPs and they referred to these methods as natural orbital or extended Koopmans' theorem theories. Subsequently, Andersen and the author [30] analyzed the working equations of this approach through second and third order in the MP series and noted the differences between them and the Greens function and equivalent EOM theories computed through these same orders. Of course, based on the discussion of Section 17.2.4, these differences relate to the ranks of the operators appearing in the working equations and are not surprising. More recently, Cioslowski et al. [31] have shown that these extended Koopmans' theorem approaches indeed offer a very efficient and reasonably accurate route to computing IPs or EAs, so it is likely that these methods will continue to develop. One of the more attractive aspects of the extended Koopmans approaches is that they have been shown [32] to be capable, at least in principle, of yielding the correct lowest ionization potential of a neutral molecule because they are capable of generating the proper asymptotic form for the density.

17.3.4 Multiconfiguration-based approximations

Following on the proof by Manne [14] that the operator spaces $\{T^+(l)\} = \{p^+; p^+q^+a, p^+q^+r^+ba, \text{etc.}\}$ and $\{T^+(l)\} = \{a^+, a^+b^+p, a^+b^+c^+qr, \text{ etc.}\}$ can be used (i.e. is capable of forming complete sets of ion states) even when no single determinant forms a dominant component of the neutral molecule wave function $|0,N\rangle$, the author's group extended the combined EA and IP-EOM theory to the case in which $|0,N\rangle$ is of an arbitrary MCSCF form [33] and the ionization operator manifold $\{T^+(l)\}$ included operators of the form $\{p^+; p^+q^+a\}$ and $\{a^+, a^+b^+p\}$. The resultant working equations were written as in Eq. (14), with the $H_{j,k}$ matrix elements given in Eq. (18) of Ref. [33], which we do not reproduce here because of their complexity. The primary additional difficulty involved in implementing these multiconfiguration-based equations is the fact that three-electron density matrices $\langle 0,N|i^+j^+k^+lhn|0,N\rangle$ taken with respect to the MCSCF wave function $|0,N\rangle$ are involved. These density matrices arise when the commutators $[H, p^+q^+a]$ and $[H, a^+b^+p]$ are evaluated.

To date, not much use has been made of the MCSCF based EOM theories as developed in the author's group. Instead, the framework of time-dependent response theory, which can treat essentially any kind of reference wave function $|0, N\rangle$ including the MCSCF variety, has superceded the EOM-based developments for such cases. It is important to keep in mind, however, that both the EOM and response function theories involve formulating and solving sets of equations whose solution (i.e. the unknown energy) is an intensive energy.

17.3.5 Coupled-cluster based EOM

The use of coupled-cluster (CC) wave functions within EOM theory for excitation energies, IPs and EAs has been developed [34,35] upon slightly different lines than outlined in Section 17.2. The CC wave function ansatz for $|0, N\rangle$ is written as usual in terms of an exponential operator acting on a single-determinant (e.g. unrestricted HF) 'reference function' $|0>$

$$|0, N\rangle = \exp(T)|0\rangle \tag{17}$$

The so-called cluster operator T is expressed in terms of spin-orbital excitation operators of the form $\{T_1\} = \{p^+a\}$, $\{T_2\} = \{p^+q^+ba\}$, $\{T_3\} = \{p^+q^+r^+cba\}$, etc. with T_k relating to the excitation of k electrons from occupied spin-orbitals (a, b, c, etc.) to virtual spin-orbitals (p, q, r, etc.). Prior to forming any EA EOM, the neutral molecule CC equations need to be solved for the amplitudes $\{t_n\}$ that multiply the $\{T_n\}$ operators to form the CC T operator. For completeness, let us briefly review how the conventional CC wave function evaluation is carried out.

We recall the CC equations are formed by manipulating the Schrödinger equation

$$H \exp(T)|0\rangle = E \exp(T)|0\rangle \tag{18}$$

to read

$$\exp(-T)\, H \exp(T)|0\rangle = E|0\rangle \tag{19}$$

and subsequently projecting this equation against the set of functions $\{\langle 0|T_n^+\}$. Because the T operator contains only creation operators for unoccupied spin-orbitals and annihilation operators for occupied spin-orbitals, it turns out that the commutator expansion

$$\exp(-T)\, H \exp(T) = H - [T, H] + 1/2[T, [T, H]] - 1/3![T, [T, [T, H]]] + 1/4![T, [T, [T, [T, H]]]] + \cdots \tag{20}$$

exactly truncates at the fourth-order term. So, the final working equations of CC theory can be written as

$$\begin{aligned}&\langle 0|T_n^+\{H - [T,H] + 1/2[T,[T,H]] - 1/3![T,[T,[T,H]]]\\&\quad + 1/4![T,[T,[T,[T,H]]]]\}|0\rangle = 0\end{aligned} \tag{21}$$

Once the CC amplitudes $\{t_n\}$ are determined by solving these quartic equations, the CC energy is computed as

$$\begin{aligned}&\langle 0|H - [T,H] + 1/2[T,[T,H]] - 1/3![T,[T,[T,H]]]\\&\quad + 1/4![T,[T,[T,[T,H]]]]|0\rangle = E\end{aligned} \tag{22}$$

The operator $Q^+(K)$ that maps the CC wave function $|0,N\rangle$ into an anion or cation state is expressed as in Eq. (5) with the $\{T^+(l)\}$ operators including, e.g. $\{T^+(l)\} = \{p^+; p^+q^+a, p^+q^+r^+ba, \text{ etc.}\}$ when EAs are to be computed and the adjoints of $\{a^+, a^+b^+p, a^+b^+c^+qr, \text{ etc.}\}$ when IPs are computed The basic EOM analogous to Eq. (7b) is then written as

$$[H, Q^+(K)]\exp(T)|0\rangle = EQ^+(K)\exp(T)|0\rangle \tag{23}$$

Multiplying on the left by $\exp(-T)$ and realizing that T and $Q^+(K)$ commute reduces this equation to

$$[H', Q^+(K)]|0\rangle = EQ^+(K)|0\rangle \tag{24}$$

where

$$H' = \exp(-T)\,H\,\exp(T) \tag{25}$$

which can be expanded as in Eq. (20) to involve at most quartic terms in the $\{t_n\}$ amplitudes. Then, multiplying on the left by $\langle 0|T(j)$ reduces the EOM equations to their final working form

$$\begin{aligned}&\sum_l \langle 0|T(j)[\{H - [T,H] + 1/2[T,[T,H]] - 1/3![T,[T,[T,H]]]\\&\quad + 1/4![T,[T,[T,[T,H]]]]\}, T^+(l)|0\rangle t(K,l)\\&\quad = E\sum_l \langle 0|T(j)T^+(l)|0\rangle t(K,l)\end{aligned} \tag{26}$$

This set of matrix eigenvalue equations are then solved to obtain E which gives the EA or the IP (depending on what operator set was used). Such so-called electron-attached and electron-removed EOM (EA-EOM and IP-EOM) approaches have proven highly successful [34,35] in computing EAs and IPs of a wide range of atoms and molecules primarily because the coupled-cluster treatment of electron correlation provides such a highly accurate treatment of the dynamical electron correlation. At present there is a great deal of activity within this framework of utilizing EOM theories for computing EAs, IPs, and ΔEs.

It should be noted that Eq. (26) does not contain an E-dependent matrix element on its left-hand side as do the MPn EOM equations (Eq. (14)) or the equivalent GF equations. This is because the EOM-CC equations are usually not subjected to the operator-space partitioning that the GF and MPn-based EOM theories commonly employ. It should also be noted that the operators appearing on the left-hand side of Eq. (26) produce non-Hermitian matrices. As a result, there arise non-orthogonal left- and right-eigenvectors in solving Eq. (26). As discussed in Refs. [34,35], it is important to compute both sets of eigenvectors if one wishes to compute, e.g. photo-electron intensities.

17.4 SOME SPECIAL CASES

17.4.1 Calculating EAs as IPs

In this discussion, we have focused on computing EAs and IPs by forming a neutral molecule wave function $|0, N\rangle$ and computing the EA or IP as an eigenvalue of an EOM matrix problem. Consider applying such an approach to evaluate the EA of the $X^2\Pi$ state of the NO molecule. Because the X-state wave function of NO is spatially degenerate (i.e. the π_x and π_y orbitals should be degenerate), one may encounter artifactual symmetry breaking when forming this neutral molecule wave function. That is, the π_x and π_y orbitals may not turn out to be degenerate; in fact, most commonly employed electronic structure codes are not able to guarantee this degeneracy as they should. It would then be unwise to use this symmetry-broken wave function to compute any property of this state of NO, including the EA. To overcome such difficulties, one could use the $X^3\Sigma^+$ state of NO^- as $|0, N\rangle$ and employ an EOM method to evaluate the IP of NO^- (actually the electron detachment energy of NO). The advantage to this approach is that the open-shell $^3\Sigma^+$ state of NO^- would not be susceptible to symmetry breaking because it is not spatially degenerate and has its π_x and π_y orbitals equivalently occupied. This example shows that it may sometimes be better to compute an EA of a molecule as the IP of the corresponding anion. Likewise, it may be better to compute an IP of a molecule as the EA of the molecule's cation in some cases.

Another example is provided by the EA of the ^2S state of the Na atom to generate the ^1S Na^- anion. Because the ^2S state is open shell, one would have to employ the unrestricted Hartree–Fock method to evaluate its orbitals and orbital energies to use in an EOM or GF EA calculation. However, one could, alternatively, compute the EA of Na by evaluating the IP of Na^-. The advantage would be that the Na^- is closed shell, so one could employ restricted Hartree–Fock methods to compute the requisite orbitals and orbital energies.

17.4.2 Metastable anion states

A different kind of problem arises when one attempts to compute the EA of a molecule whose anion is not electronically bound relative to the corresponding neutral. For example, the $X^1\Sigma_g^+$ state of the N_2 molecule does not bind an electron to form an electronically stable anion. Instead the $X^2\Pi_g$ state of N_2^-, formed by adding an electron to the π_g anti-bonding orbital of N_2 is a so-called resonance state that lies higher in energy than N_2 and can spontaneously eject its excess electron. One cannot simply employ conventional basis sets and *ab initio* electronic structure methods (including EOM or GF or response-function methods) to correctly determine the energies of such states.

The most common and powerful tool for studying such metastable states theoretically is the so-called stabilization method (SM). This method, pioneered by Professor Howard Taylor's group [36], involves embedding the system of interest (e.g. the N_2^{-1} anion) within a finite 'box' in order to convert the continuum of states corresponding, e.g. to $N_2 + e^-$, into discrete states that can be handled by conventional square-integrable basis functions using, e.g. the EOM method. By varying the size of the box, one can vary the energies of the discrete states that correspond to $N_2 + e^-$ (i.e. one varies the box size to vary the kinetic energy KE of the orbitals containing the excess electron). As the box size is varied, one eventually notices (e.g. by plotting the orbitals) that one of the $N_2 + e^-$ states obtained in the EOM process possesses a significant amount of valence character. That is, one such state has significant amplitude not only at large r but also in the region of the two nitrogen centers. It is this state that corresponds to the metastable resonance state, and it is the EOM eigenvalue E of this state that provides the stabilization estimate of the resonance state energy relative to that of the neutral N_2.

Let us continue using N_2^{-1} as an example for how one usually varies the box within which the anion is constrained. One uses a conventional atomic orbital basis set that likely includes s and p functions on each N atom, perhaps some polarization d functions and some conventional diffuse s and p orbitals on each N atom. These basis orbitals serve primarily to describe the motions of the electrons within the usual valence regions of space. To this basis, one appends an extra set of diffuse π-symmetry orbitals. These orbitals could be p_π (and maybe d_π) functions centered on each nitrogen atom, or they could be d_π orbitals centered at the midpoint of the N–N bond. Either choice can be used because one only needs a basis capable of describing the large-r $L = 2$ character of the metastable $^2\Pi_g$ state's wave function. One usually would not add just one such function; rather several such functions, each with an orbital exponent α_J that characterizes its radial extent, would be used. Let us assume, for example, that K such additional diffuse π functions have been used.

Next, using the conventional atomic orbital basis as well as the K extra π basis functions, one carries out an EOM calculation for the EA of the N_2 molecule. In this calculation, one tabulates the energies of many (say M) of the EOM-EA eigenvalues. One then scales the orbital exponents $\{\alpha_J\}$ of the K extra π basis orbitals by a factor $\eta: \alpha_J \rightarrow \eta\alpha_J$ and repeats the calculation of the energies of the M lowest EOM eigenvalues. This scaling causes the extra π basis orbitals to contract radially (if $\eta > 1$) or to expand radially (if $\eta < 1$). It is this basis orbital expansion and contraction that produces expansion and contraction of the 'box' discussed above. That is, one does not employ a box directly; instead, one varies the radial extent of the more diffuse basis orbitals to simulate the box variation.

If the conventional orbital basis is adequate, one finds that the extra π orbitals, whose exponents are being scaled, do not affect appreciably the energy of the neutral N_2 system. This can be probed by plotting the N_2 energy (computed as $\langle 0, N|H|0, N\rangle$) as a function of the scaling parameter η; if the energy varies little with η, the conventional basis is adequate.

In contrast to plots of the neutral N_2 energy *vs.* η, plots of the energies of the M N_2^{-1} anion states relative to the energy of N_2, obtained as EOM eigenvalues, show significant η-dependence as Fig. 17.1 illustrates.

What does such a stabilization plot tell us and what do the various branches of the plot mean? First, we notice that each of the plots of the energy of an anion state (relative to the neutral molecule's energy, which is independent of η) grows with increasing η. This η dependence arises from the η-scaling of the extra diffuse π basis orbitals. Because most of the amplitude of such basis orbitals lies outside the valence region, the kinetic energy is the dominant contributor to such states' relative energies. Because η enters into each orbital as $\exp(-\eta\alpha r^2)$, and because the kinetic energy operator involves the second derivative with respect to r, the kinetic energies of orbitals dominated by the diffuse π basis functions vary as η^2. It is this quadratic growth with η that appears as the basic trends in the energies *vs.* η plots in Fig. 17.1.

For small η, all the π diffuse basis functions have their amplitudes concentrated at large-r and have low kinetic energy. As η grows, these functions become more radially compact and their kinetic energies grow just as the particle-in-a-box energies grow as the box length decreases. For example, note the three lowest energies shown above in Fig. 17.1 increasing from near zero as η grows. As η further increases, one reaches a point at which the third and fourth anion-state energies in Fig. 17.1 undergo an avoided crossing. At higher η values, it is the second and third states and then the first and

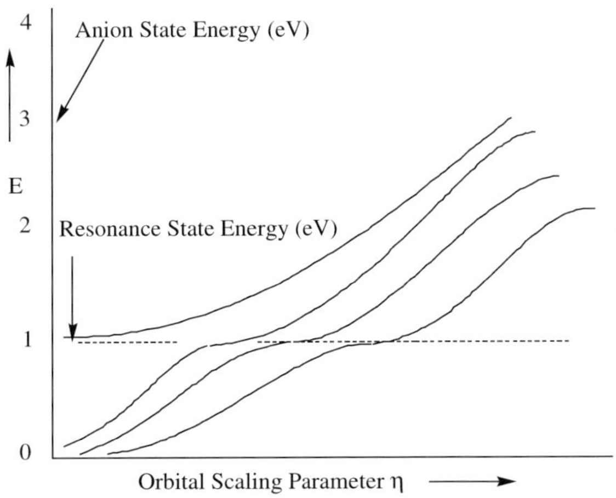

Fig. 17.1. Plots of the EOM-EA eigenvalues for several anion states *vs.* the orbital scaling parameter η. Note the avoided crossing of state energies near 1 eV.

second states whose energies undergo such avoided crossings. At such η values, if one examines the nature of the two anion wave functions (obtained as in Eq. (4)) whose energies avoid one another, one finds they contain substantial amounts of both valence and extra diffuse π function character. Just to the left of the avoided crossing, the lower-energy state (the third state in Fig. 17.1 for the smallest η at which an avoided crossing occurs) contains predominantly extra diffuse π orbital character, while the higher-energy state (the fourth state) contains largely valence π^* orbital character. To the right of the avoided crossing, the situation is reversed—the lower-energy state (the third state in Fig. 17.1 for small η) contains predominantly valence orbital character, while the higher-energy state (the fourth state) contains largely diffuse orbital character.

However, at the special values of η where the two states nearly cross, the kinetic energy of the diffuse state (as well as its radial size and de Broglie wavelength) are appropriate to connect properly with the valence state to form a single resonance state. By connect properly we mean that the two states have wave function amplitudes, phases and slopes that match. It is such boundary condition matching of valence-range and long-range character in the wave function that the stabilization method achieves. So, at such special η values, one can achieve a description of the resonance state that correctly describes this state both in the valence region and in the large-r region. Only by tuning the energy of the large-r states using the η scaling can one obtain this proper boundary condition matching.

Another observation helps to understand the content of such stabilization plots. One considers the density of states (i.e. how many states are there between energy E and $E + \mathrm{d}E$ for a fixed small value of $\mathrm{d}E$?) in a plot such as Fig. 17.1. Clearly, in the range of energies near the avoided crossings, there is an enhanced density of states, while the state density is lower at 'off resonance' energies. When viewed either from the point of view of state densities or avoided crossings, there is something special about the region of energies near such resonances. As noted above, it is the fact that the valence-range and continuum components of the wave function can be properly matched at such energies that is 'special'.

If one attempts to study metastable anion states without carrying out such a stabilization study, one is doomed to failure, even if one employs an extremely large and flexible set of diffuse basis functions. In such a calculation, one will certainly obtain a large number of anion 'states' with energies lying above that of the neutral, but one will not be able to select from these states the one that is the true resonance state because the true state will be buried in the myriad of 'states' representing the $N_2 + e^-$ continuum.

In summary, by carrying out a series of anion-state energy calculations for several states and plotting them *vs.* η, one obtains a stabilization graph. By examining this graph and looking for avoided crossings, one can identify the energies at which metastable resonances occur. It is absolutely critical to identify these resonance energies if one wishes to probe metastable anions. It is also possible [37] to use the shapes (i.e. the magnitude of the energy splitting between the two states and the slopes of the two avoiding curves) of the avoided crossings in a stabilization graph to compute the lifetimes of the metastable states. Basically, the larger the avoided-crossing energy splitting between the two states, the shorter is the lifetime of the resonance state.

17.5 SUMMARY

We have tried to illustrate how, by focusing on the intensive energies that one wishes to compute when studying EAs, IPs or electronic excitation energies, one can replace the solution of the Schrödinger equation by the solution of so-called EOM. It is the eigenvalues of these EOMs that produce the EAs and IPs directly. We have reviewed some of the history of the development of EOM theory, especially as it applied to EAs and IPs, and we have attempted to show its relationships to Greens functions and extended Koopmans' theorem approaches to these same intensive energies. We have shown that a wide variety of EOM theories can be developed depending on how one chooses to describe the neutral molecule's wave function (i.e. in MP, MCSCF or CC fashion). Finally, we discussed some of the pitfalls that one encounters when applying EOM theory to EAs of molecules whose anion states are not bound but are metastable resonance states. It is our hope and belief that EOM methods have proven useful computationally and for gaining insight and will continue to have a bright future.

Readers who wish to learn more about how molecular EAs (and to a lesser extent, IPs) have been studied theoretically are directed to this author's web site http://simons.hec.utah.edu as well as to a series [38] of his reviews and chapters. The species that this group have examined include dipole-bound anions, zwitterion ions, conventional valence anions, multiply charged anions as well as a wide variety of metastable anions.

17.6 ACKNOWLEDGEMENTS

This work was supported by NSF grants #9982420 and #0240387.

17.7 REFERENCES

1 J.C. Rienstra-Kiracofe, G.S. Tschumper, H.F. Schaefer, III, S. Nandi and G.B. Ellison, Acc. Chem. Res., 102 (2002) 231–282.

2 G. Herzberg, Molecular spectra and molecular structure I, Van Nostrand, New York, 1950, p. 512.

3 (a) G.I. Gutsev and A.I. Boldyrev, The theoretical investigation of the electron affinity of chemical compounds, Adv. Chem. Phys., 61 (1985) 169–221; (b) J. Baker, R.H. Nobes and L. Radom, The evaluation of molecular electron affinities, J. Comp. Chem., 7 (1986) 349–358; (c) J. Simons and K.D. Jordan, Ab initio electronic structure of anions, Chem. Rev., 87 (1987) 535–555; (d) J. Kalcher and A.F. Sax, Gas phase stabilities of small anions—theory and experiment in cooperation, Chem. Rev., 94 (1994) 2291–2318; (e) J. Kalcher, Gas-phase stabilities of small anions, J. Ann. Rep., Sec. C, R. Soc. Chem., 93 (1996) 147–186; (f) R.S. Berry, Small free negative ion, Chem. Rev., 69 (1969) 533–542, The most up-to-date overview is given in Ref. [1].

4 See Ref. [3e] and references contained therein. Some of the earlier successes are given in the following: (a) E. Clementi and A.D. McLean, Phys. Rev. A, 133 (1964) 419; E. Clementi, A.D. McLean, D.L. Raimondi and M. Yoshimine, Phys. Rev. A, 133 (1964) 1274; E. Clementi, Phys. Rev. A, 135 (1964) 980; (b) C.L. Pekeris, Phys. Rev., 112 (1958) 1649; (c) A.W. Weiss, Phys. Rev., 166 (1968) 70; (d) P.E. Cade, J. Chem. Phys., 47 (1967) 2390, Proc. R. Soc. A, 91 (1967) 842; (e) S. Taylor and F.E. Harris, J. Chem. Phys., 39 (1963) 1012; (f) V. Bondybey, P.K. Pearson and H.F. Schaefer, J. Chem. Phys., S7 (1972) 1123; (g) D. Feller and E.R. Davidson, The electron affinity of oxygen: A systematic configuration interaction approach, J. Chem. Phys., 90 (1989) 1024; D. Feller and E.R. Davidson, Ab initio multireference CI

determination of the electron affinity of carbon and oxygen; D. Feller and E.R. Davidson, J. Chem. Phys., 82 (1985) 1024, In addition to Ref. [1], good sources for EA and IP data include: (h) H. Hotop and W.C. Lineberger, Binding energies in atomic negative ions, J. Phys. Chem. Ref. Data, 4 (1975) 539–576, Binding energies in atomic negative ions: II, J. Phys. Chem. Ref. Data, 14 (1985) 731–750; (i) T. Andersen, H.K. Haugen and H. Hotop, Binding energies in atomic negative ions, J. Phys. Chem. Ref. Data, 28 (1999) 1511–1533; (j) B.K. Janousek and J.I. Brauman, Electron affinities, in: M.T. Bowers (Ed.), Gas phase ion chemistry, Vol. 2, Academic Press, New York, 1979, pp. 53–86; (k) T.M. Miller, Electron affinities, in: R.C. West, M.J. Astle and W.H. Beyer (Eds.), CRC handbook of chemistry and physics, 74th edn., CRC Press, Boca Raton, FL, 1993, pp. 10-180–10-191; (l) J.E. Bartmess, Negative ion energetics data, in: W.G. Ballard, P.S. Linstrom (Eds.), NIST chemistry web-book, NIST standard reference database number 69, National Institute of Standards and Technology, Gaithersburg, MD, 2000, February, http://webbook.nist.gov; (m) J.L. Franklin, J.G. Dillard and F.H. Field, Ionization potentials, appearance potentials, and heats of formation of gaseous positive ions, Natl Stand. Ref. Data Ser. Natl Bur. Stand. (US), 26 (1969).

5 Much of what is presently known about atomic and molecular EAs can be found in Refs. [3,4] as well as in the following: (a) G.I. Gutsev and A.I. Boldyrev, The theoretical investigation of the electron affinity of chemical compounds, Adv. Chem. Phys., 61 (1985) 169–221; (b) J. Baker, R.H. Nobes and L. Radom, The evaluation of molecular electron affinities, J. Comp. Chem., 7 (1986) 349–358; (c) J. Simons and K.D. Jordan, Ab initio electronic structure of anions, Chem. Rev., 87 (1987) 535–555; (d) J. Kalcher and A.F. Sax, Gas phase stabilities of small anions—theory and experiment in cooperation, Chem. Rev., 94 (1994) 2291–2318; J. Kalcher, Gas-phase stabilities of small anions, J. Ann. Rep., Sec. C, R. Soc. Chem., 93 (1996) 147–186; (e) R.S. Berry, Small free negative ion, Chem. Rev., 69 (1969) 533–542; (f) H. Massey, Negative ions, Cambridge University Press, Cambridge, England, 1976; (g) L.M. Branscomb, in: D.R. Bates (Ed.), Atomic and molecular processes, Academic Press, New York, 1962.

6 H.F. Schaefer, III and F.E. Harris, Phys. Rev., 170 (1968) 108–115; H.F. Schaefer, III, R.A. Klemm and F.E. Harris, J. Chem. Phys., 51 (1969) 4643–4650.

7 Some of the early chemistry developments of Greens functions include; (a) O. Goscinski and B. Lukman, Chem. Phys. Lett., 6 (1970) 573; (b) P. Jorgensen and J. Oddershede, J. Chem. Phys., 57 (1972) 277; (c) J. Linderberg, P. Jørgensen, J. Oddershede and M. Ratner, J. Chem. Phys., 56 (1972) 6213; (d) G. Csanak, H.S. Taylor and R. Yaris, Phys. Rev. A, 3 (1971) 6213; G. Csanak, H.S. Taylor and R. Yaris, Adv. At. Mol. Phys., 7 (1971) 288; (e) J. Linderberg and Y. Öhrn, J. Chem. Phys., 49 (1968) 716; Chem. Phys. Lett., 1 (1967) 295; Proc. R. Soc. A, 185 (1963) 445; Phys. Rev. A, 139 (1965) 1063; (f) J. Linderberg and M. Ratner, Chem. Phys. Lett., 6 (1970) 37; (g) J. Linderberg and E.W. Thulstrup, J. Chem. Phys., 49 (1968) 710; (h) W.P. Reinhardt and J.D. Doli, J. Chem. Phys., 50 (1969) 2767; (i) B. Schneider, H.S. Taylor and R. Yaris, Phys. Rev. A, I (1970) 855; (j) J.D. Doll and W.P. Reinhardt, J. Chem. Phys., 57 (1972) 1169; (k) P. Jørgensen and J. Simons, A complete treatment of the electron propagator through third order, J. Chem. Phys., 63 (1975) 5302; (l) J. Linderberg and Y. Öhrn, Propagators in quantum chemistry, Academic Press, London, 1973; (m) L.S. Cederbaum, G. Holneicher and S. Peyerimhoff, Chem. Phys. Lett., 11 (1971) 421; L.S. Cederbaum, G. Holneicher and W. von Niessen, Chem. Phys. Lett., 18 (1973) 503; L.S. Cederbaum, Theor. Chim. Acta, 31 (1973) 239; (n) B.S. Yarlagadda, Gy. Csanak, H.S. Taylor, B Schneider and R. Yaris, Phys. Rev., A7 (1973) 146; (o) B.T. Pickup and O. Goscinski, Mol. Phys., 36 (1973) 1013; (p) G.D. Purvis and Y. Öhrn, J. Chem. Phys., 60 (1974) 4063, 62 (1975) 2045; (q) For a more recent overview, see in: J.V. Ortiz and J. Leszczynski (Eds.), Computational chemistry: reviews of current trends, Vol. 2, World Scientific Publishing, Singapore, 1997, pp. 1–61.

8 A good overview of time-dependent response function theory, including linear and non-linear response functions is offered in J. Olsen and P. Jørgensen, Time-dependent response theory with applications to self-consistent field and multiconfigurational self-consistent field wave functions, in: D. Yarkony (Ed.), Modern electronic structure theory, World Scientific Publishing, Singapore, 1995, pp. 857–990.

9 T.H. Dunning and V. McKoy, J. Chem. Phys., 47 (1967) 1735, 48 (1968) 5263; T.I. Shibuya and V. McKoy, J. Chem. Phys., 53 (1970) 2208.

10 D.J. Rowe, Rev. Mod. Phys., 40 (1968) 153; D.J. Rowe, Nuclear collective motion—models and theory, Methuen and Co., London, 1970.

11 J. Simons and W.D. Smith, Theory of electron affinities of small molecules, J. Chem. Phys., 58 (1973) 4899. This work exploited developments made earlier in studies of excited electronic states and reduced density matrices: J. Simons, Direct calculation of first- and second-order density matrices. The higher RPA method, J. Chem. Phys., 55 (1971) 1218–1230, J. Simons, Energy-shift theory of low-lying excited electronic states of molecules, J. Chem. Phys., 57 (1972) 3787–3792.

12 J. Simons and P. Jørgensen, J. Chem. Phys., 64 (1976) 1413; T.-T. Chen, J. Simons and K.D. Jordan, Chem. Phys., 14 (1976) 145; J. Simons, J. Chem. Phys., 64 (1976) 4541; J. Simons, Int. J. Quantum Chem., XII (1977) 227; E. Dalgaard and J. Simons, J. Phys. B: At. Mol. Phys., 10 (1977) 2767; J. Simons, Ann. Rev. Phys. Chem., 28 (1977) 15; A. Banerjee, R. Shepard and J. Simons, Int. J. Quantum Chem.: Quantum Chem. Symp., 12 (1978) 389; R.A. Donnelly and J. Simons, J. Chem. Phys., 73 (1980) 2858.

13 See, for example, P. Jørgensen and J. Simons, Second quantization-based methods, Quantum chemistry, Academic Press, London, 1981; or T. Helgaker, P. Jørgensen and J. Olsen, Modern electronic structure theory, Wiley, New York, 2000.

14 R. Manne, Chem. Phys. Lett., 45 (1977) 470.

15 The first- and second-order density matrices, respectively, have elements given by $\langle 0,N|j^{+}k|0,N\rangle$ and $\langle 0,N|j^{+}k^{+}lh|0,N\rangle$.

16 J. Simons, T.-T. Chen and W.D. Smith, Chem. Phys. Lett., 26 (1974) 296; W.D. Smith, T.-T. Chen and J. Simons, Chem. Phys. Lett., 27 (1974) 499; K.M. Griffing and J. Simons, J. Chem. Phys., 62 (1975) 535; J. Kenney and J. Simons, J. Chem. Phys., 62 (1975) 592; K. Griffing, J. Kenney, J. Simons and K. Jordan, J. Chem. Phys., 63 (1975) 4073; K. Griffing and J. Simons, J. Chem. Phys., 64 (1976) 3610; K.D. Jordan, K.M. Griffing, J. Kenney, E.L. Andersen and J. Simons, J. Chem. Phys., 64 (1976) 4730; E. Andersen and J. Simons, J. Chem. Phys., 64 (1976) 4548; K.D. Jordan and J. Simons, J. Chem. Phys., 65 (1976) 1601; E. Andersen and J. Simons, J. Chem. Phys., 65 (1976) 5393; E. Andersen and J. Simons, J. Chem. Phys., 66 (1977) 2427; J. Simons, Ann. Rev. Phys. Chem., 28 (1977) 15.

17 P. Jørgensen and J. Simons, A complete treatment of the electron propagator through third order, J. Chem. Phys., 63 (1975) 5302.

18 J. Linderberg and Y. Öhrn, Propagators in quantum chemistry, Academic Press, London, 1973.

19 L.S. Cederbaum, G. Holneicher and S. Peyerimhoff, Chem. Phys. Lett., 11 (1971) 421; L.S. Cederbaum, G. Holneicher and W. von Niessen, Chem. Phys. Lett., 18 (1973) 503; L.S. Cederbaum, Theor. Chim. Acta, 31 (1973) 239.

20 J.D. Doll and W.P. Reinhardt, J. Chem. Phys., 67 (1972) 1169.

21 B.S. Yarlagadda, Gy. Csanak, H.S. Taylor, B. Schneider and R. Yaris, Phys. Rev., A7 (1973) 146.

22 B.T. Pickup and O. Goscinski, Mol. Phys., 36 (1973) 1013.

23 G.D. Purvis and Y. Öhrn, J. Chem. Phys., 60 (1974) 4063, 62 (1975) 2045.

24 P. Jorgensen and J. Oddershede, J. Chem. Phys., 57 (1972) 277.

25 J. Linderberg, P. Jorgensen, J. Oddershede and M. Ratner, J. Chem. Phys., 56 (1972) 6213; J. Linderberg and Y. Öhrn, J. Chem. Phys., 49 (1968) 716, Chem. Phys. Lett., 1 (1967) 295; Proc. R. Soc. A, 185 (1963) 445; Phys. Rev. A, 139 (1965) 1063; J. Linderberg and M. Ratner, Chem. Phys. Lett., 6 (1970) 37; J. Linderberg and E.W. Thulstrup, J. Chem. Phys., 49 (1968) 710.

26 See, for example, J.V. Ortiz and J. Leszczynski (Eds.), Computational chemistry: reviews of current trends, Vol. 2, World Scientific Publishing, Singapore, 1997, pp. 1–61.

27 M.J. Frisch, G.W. Trucks, H.B. Schlegel, G.E. Scuseria, M.A. Robb, J.R. Cheeseman, J.A. Montgomery, Jr., T. Vreven, K.N. Kudin, J.C. Burant, J.M. Millam, S.S. Iyengar, J. Tomasi, V. Barone, B. Mennucci, M. Cossi, G. Scalmani, N. Rega, G.A. Petersson, H. Nakatsuji, M. Hada, M. Ehara, K. Toyota, R. Fukuda, J. Hasegawa, M. Ishida, T. Nakajima, Y. Honda, O. Kitao, H. Nakai, M. Klene, X. Li, J.E. Knox, H.P. Hratchian, J.B. Cross, C. Adamo, J. Jaramillo, R. Gomperts, R.E. Stratmann, O. Yazyev, A.J. Austin, R. Cammi, C. Pomelli, J.W. Ochterski, P.Y. Ayala, K. Morokuma, G.A. Voth, P. Salvador, J.J. Dannenberg, V.G. Zakrzewski, S. Dapprich, A.D. Daniels, M.C. Strain, O. Farkas, D.K. Malick, A.D. Rabuck, K. Raghavachari, J.B. Foresman, J.V. Ortiz, Q. Cui, A.G. Baboul, S. Clifford, J. Cioslowski, B.B. Stefanov, G. Liu, A. Liashenko, P. Piskorz, I. Komaromi, R.L. Martin, D.J. Fox, T. Keith, M.A. Al-Laham, C.Y. Peng, A. Nanayakkara, M. Challacombe, P.M.W. Gill, B. Johnson, W. Chen, M.W. Wong, C. Gonzalez, J.A. Pople and Gaussian, Inc., Pittsburgh PA, 2003; J.V. Ortiz, Toward an exact one-electron picture of chemical bonding, Adv. Quantum Chem., 35 (1999) 33–52; J.V. Ortiz and

H.A. Kurtz (Eds.), Propagating insight: A tribute to the works of Yngve Öhrn, Advances in quantum chemistry, Vol. 35, Academic Press, New York, 1999.

28 M.M. Morrel, R.G. Parr and M. Levy, J. Chem. Phys., 62 (1975) 549.

29 O.W. Day, D.W. Smith and C. Garrod, Int. J. Quantum Chem., S8 (1974) 501; O.W. Day and D.W. Smith, J. Chem. Phys., 62 (1975) 113.

30 E. Andersen and J. Simons, J. Chem. Phys., 66 (1977) 1067.

31 J. Cioslowski, P. Piskorz and G. Liu, J. Chem. Phys., 107 (1997) 6804.

32 J. Katriel and E.R. Davidson, Asymptotic behavior of atomic and molecular wave functions, Proc. Natl. Acad. Sci., 77 (1980) 4403.

33 A. Banerjee, R. Shepard and J. Simons, One-particle Green's function with multiconfiguration reference states, Int. J. Quantum Chem.: Quantum Chem. Symp., 12 (1978) 389.

34 Among the first workers to develop the requisite EOM machinery for a CC reference wave function was the following: (a) H.J. Monkhorst, Int. J. Quantum Chem. Symp., 11 (1977) 421; (b) The framework for applying EOM-CC theory to EAs was developed in:M. Nooijen and R.J. Bartlett, Equation-of-motion coupled-cluster method for electron attachment, J. Chem. Phys., 102 (1995) 3629; The use of CC wave functions and EOM-type theories to compute excitation energies was advanced by (c) D. Mukhopadhyay, S. Mukhopadhyay, R. Chaudhuri and D. Mukherjee, Aspects of separability in the coupled cluster based direct method for energy differences, Theor. Chim. Acta, 80 (1991) 441, as well as; (d) J.F. Stanton and R.J. Bartlett, The equation of motion coupled-cluster method. A systematic biorthogonal approach to molecular excitation energies, transition probabilities, and excited state properties, J. Chem. Phys., 98 (1993) 7029.

35 A good overview is given in R.J. Bartlett and J.F. Stanton, in: K.B. Lipkowitz, D.B. Boyd (Eds.), Reviews in computational chemistry, Vol. 5, VCH, New York, 1994.

36 A.U. Hazi and H.S. Taylor, Phys. Rev. A, 1 (1970) 1109.

37 J. Simons, J. Chem. Phys., 75 (1981) 2465; R.F. Frey and J. Simons, J. Chem. Phys., 84 (1986) 4462.

38 (a) J. Simons, P. Skurski, in: J. Kalcher (Ed.), The roles of electrostatics in forming molecular anions and dianions, Recent Res. Dev. Phys. Chem., Theoretical prospect of negative ions, 2002a, pp. 117–138; (b) J. Simons, Theory and ion chemistry 2. Theory (energies and potential energy surfaces). Anions, Encyclopedia of mass spectrometry, Vol. 5, 2002, pp. 55–68; (c) M. Gutowski, P. Skurski, K.D. Jordan and J. Simons, Energies of dipole-bound anionic states, Int. J. Quantum Chem., 64 (1997) 183–191; (d) J. Simons, Electron propagator studies of molecular anions, Int. J. Quantum Chem. Symp., 16 (1982) 575–581; (e) J. Simons, Theoretical studies of negative molecular ions, Theoretical chemistry: Advances and perspectives, Vol. 3, Academic Press, New York, 1978; (f) J. Simons, Theoretical studies of negative molecular ions, Ann. Rev. Phys. Chem., 28 (1977) 15–45; (g) J. Simons and K.D. Jordan, Ab initio electronic structure of anions, Chem. Rev., 87 (1987) 535–555.

Theory and Applications of Computational Chemistry: The First Forty Years
Edited by C. Dykstra et al.

CHAPTER 18

Multireference coupled cluster method based on the Brillouin–Wigner perturbation theory

Petr Čársky[1], Jiří Pittner[1] and Ivan Hubač[2]

[1] *J. Heyrovský Institute of Physical Chemistry, Academy of Sciences of the Czech Republic, Dolejškova 3, 18223 Prague 8, Czech Republic*
[2] *Department of Chemical Physics, Faculty of Mathematics and Physics, Comenius University, 84215 Bratislava, Slovakia*

Abstract

This chapter deals with quantum chemical applications in which the electronic state of the system treated is degenerate or quasidegenerate. In chemistry this situation occurs in problems in which an accurate description of a large part of the potential energy surface is required to be treated on the same footing. Commonly used approaches such as the Møller–Plesset theory, configuration interaction, coupled cluster methods CCSD and CCSD(T), or DFT may fail in such cases. Instead, the multireference coupled cluster (MR CC) approach is recommended as the method of choice. A survey of MR CC methods of different kind is given, with a brief description of their main features, merits and drawbacks. Special problems of the MR CC methods, size extensivity and the intruder state problems, are discussed. A variant of the MR CC theory, based on the Brillouin–Wigner type resolvent, is presented. A few applications demonstrate its performance and utility in practical chemical applications.

18.1 INTRODUCTION

Recent progress in the development of quantum chemical computational methods [1] and quantum chemical software provided a user-friendly tool to chemists, helping them to explain many problems met in a chemical laboratory. The most frequent task is to determine the optimum structure of molecules, relative energies of reaction components, heats of formation, vibrational frequencies, energies of activation and energies of ionization. In all these cases, a standard theoretical approach is to perform first a Hartree–Fock (self-consistent field, SCF) calculation for obtaining a set of molecular orbitals (MOs). In the next step, the main defect of the Hartree–Fock approach is eliminated, which is the neglect of dynamic electron correlation. This is done by assuming the wave function for the ground state (or a particular excited state) as a linear combination of singly, doubly and even higher excited configurations. The respective expansion coefficients are fixed in one way or another. The most popular options are the variational approach in a form of the configuration interaction (CI), the perturbational approach by means of the Møller–Plesset (MP) theory and coupled cluster (CC) expansion. Alternatively, if computational feasibility is favored over the rigor of strictly *ab initio* treatments, density functional theory (DFT) can be applied. In DFT the two-electron part of the Hamiltonian is replaced by an adjusted coulomb and exchange-correlation functional and the MOs so obtained give electron density which takes account of electron correlation. All these approaches can be referred to as 'single-reference' approaches because the starting ('reference') wave function is taken in the form of a single Slater determinant or a symmetry adapted configuration state function constructed from several Slater determinants. Besides the availability of the user-friendly software, also the progress in the chemical education of undergraduates should be noted. Basic quantum chemistry and its applications to chemical reactivity and spectroscopy is contained in a curriculum which gives an opportunity to students to acquire the necessary basic knowledge and skill for performing routine quantum chemical calculations on their own. This favorable state in the field of theoretical chemistry may lead to a false impression that all important problems in quantum chemistry have already been solved and that the reliability and feasibility of quantum chemical calculations will increase in parallel with the anticipated continuing progress in the construction of computers in the near future. Unfortunately, there is a type of applications that does not conform to this optimistic outlook. It concerns problems in which an accurate description of a large part of the potential energy surface is required to be treated on the same footing. If the respective part of the surface is associated with a chemical bond breaking, the single-reference methods may fail, unless an enormously large CI or CC expansion is used, which prevents applications to all but the smallest molecular systems. We show an example of such a failure in Section 18.2. A more accessible solution to this problem is provided by a class of 'multireference' methods. In Section 18.3, we give a survey of such most commonly used methods that are based on the coupled cluster expansion and comment briefly on their merits and drawbacks. A version of the multireference coupled cluster method based on the Brillouin–Wigner perturbation theory (MR BWCC) is a method developed by the present authors. In Sections 18.4 and 18.5, we describe briefly the essence of the method and its properties and present the working equations for its computer implementation. A few selected MR BWCC applications are presented in Section 18.6 and in Section 18.7 we summarize our experience with the method in its present form and note on the prospects for its next development.

18.2 SINGLE-REFERENCE *VERSUS* MULTIREFERENCE METHODS

It is generally accepted that the family of coupled cluster methods, CCSD, CCSD(T) and CCSDT represents the most sophisticated class of single-reference methods that are commonly used for chemical applications. Unfortunately, even these sophisticated methods fail in some instances, as it has been demonstrated for potential curves of diatomic molecules on many occasions (see, for example, Ref. [2]). We present it in Fig. 18.1 for the F_2 molecule because of our MR BWCC study of that molecule [3]. The potential energy curves shown in Fig. 18.1 were obtained with the cc-pVTZ basis set. The CCSD curve overestimates the dissociation energy by a factor of about 2. CCSD(T) gives excellent result for geometries near to the equilibrium bond distance but a typical failure of the method is seen at intermediate and large interatomic distances. Only CCSDT and MR BWCCSD give correct potential curves for the whole range of interatomic distances. At the equilibrium bond length the MR BWCCSD energy is closer to the CCSD energy than to the CCSDT energy. This is caused by the fact that in the wave function at the equilibrium F–F bond length a single configuration dominates and hence MR BWCCSD represents only a slight improvement over CCSD. Inclusion of higher excitations by means of T_3 clusters in CCSDT is more effective than extension of the reference space to two configurations at the CCSD level. This is reflected in the obtained dissociation energies D_e of F_2: 1.40 eV with MR BWCCSD and 1.60 eV with CCSDT, to be compared with the experimental value of 1.66 eV [4]. Unfortunately, the cost of CCSDT calculations prevents their application to larger molecules. By using the ACES program [5] and the Xeon 4 2400 MHz/2 GB RAM, computer time (no use of symmetry was made) for a single-point calculation was 3 min for two-reference BWCCSD and 14.6 h for CCSDT.

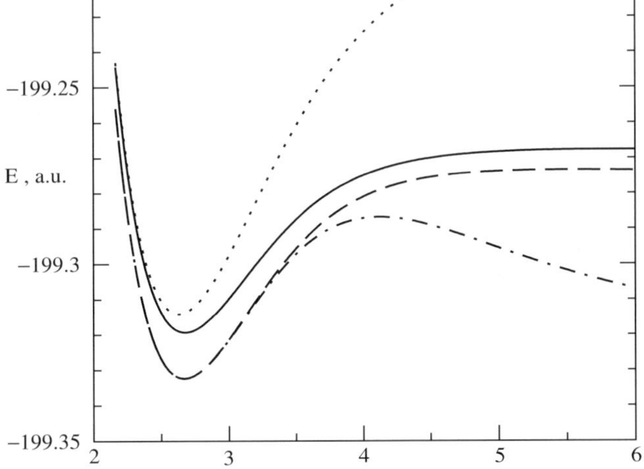

Fig. 18.1. Potential energy curves for the F_2 molecule calculated by CCSD (dotted line), CCSD(T) (staggered line), CCSDT (dashed line) and MR BWCCSD (solid line) methods. The cc-pVTZ basis set was used.

The origin of such a failure is well understood. The ground-state configuration of F_2 at the equilibrium bond length is $(1\pi_g)^4(3\sigma_g)^2$. As the interatomic F–F distance increases, the energy gap between the $(1\pi_g)^4(3\sigma_g)^2$ and $(1\pi_g)^4(3\sigma_u)^2$ configurations decreases to become zero at the dissociation limit. Evidently the $(1\pi_g)^4(3\sigma_g)^2$ configuration becomes a poor representation of the electronic state of F_2 at large interatomic distances. An obvious way how to treat this defect is to include highly excited configurations in the CI expansion or to include T_3 and higher clusters in the CC expansion. Since an approximate inclusion of triples by means of CCSD(T) is not applicable in this case, and a rigorous treatment of T_3 clusters by means of CCSDT is too costly, it is more profitable to apply a 'multireference' approach in which instead of the regular SCF procedure a multiconfiguration SCF (MC SCF) is used. In the case of F_2 it means to include both the $(1\pi_g)^4(3\sigma_g)^2$ and the $(1\pi_g)^4(3\sigma_u)^2$ configurations. This ensures that all parts of the potential curve are treated on the same footing. As Fig. 18.1 shows, extension of CCSD to two-configuration CCSD is successful in providing a realistic shape of the potential curve with the correct dissociation limit.

18.3 OVERVIEW OF MULTIREFERENCE CC METHODS

Over the years, many different CC-based approaches for treatment of systems with static correlation have been proposed:

1. The treatments based on single-reference CC expansion as CCSDtq [6], CASCC [7,8], and the recent state-specific method [9], which are based on a single (possibly formal) Fermi vacuum and include higher amplitudes with part of their indices confined to an active space. This has the advantage of retaining the formal simplicity of the single-reference CC ansatz, while being able to describe processes like dissociation, where one needs to consider higher than double or triple amplitudes. Since the active space is typically very small, the number of the higher amplitudes can be kept manageable, resulting in a reasonable disc storage and computational cost. However, so far we are not aware of a really efficient implementation of these techniques, which would allow us to compute chemically interesting molecules. The disadvantage of these approaches is that they do not treat all reference configurations on an equal footing and the results may be dependent, e.g. on the reference chosen as the Fermi vacuum.

2. The externally corrected CCSD and reduced multireference CCSD methods developed by Paldus and coworkers [10–13] make use of external data, typically from a CI calculation, to correct the CC calculation for multireference effects. They can be subdivided into two basic classes: energy corrected and amplitude corrected [14]. The amplitude-corrected methods analyze the CI wavefunction and take a selected set of T_3 and T_4 amplitudes from it and include them in the CC calculation, while using the standard CC energy formula. On the other hand, the energy-corrected methods use so-called asymmetric energy formula [14] to correct *a posteriori* the energy obtained from a standard CC calculation. Recently, the authors developed a general scheme that covers the state-universal multireference as well as amplitude-corrected CCSD methods [15]. A general disadvantage of any externally corrected CC method is that a potential

formulation of the analytic gradients for such a method becomes difficult, since the quantities which have been taken from the external source (like CI coefficients) depend on the molecular geometry and their derivatives would be needed.

3. The method of moments coupled clusters developed by Kowalski and Piecuch [16–18] belongs essentially also to the general scheme of energy-corrected methods described above, since it gives an *a posteriori* correction of a CC, EOMCC, or state-universal MR-CC energy based on an external wave function of CI or PT type. In principle, if the external wave function were full CI, this correction would recover the exact full CI energy, regardless on the level of approximation used as an input. In practice, the authors have shown a remarkable improvement of the CCSD(T) method, when applied to describe, e.g. bond breaking and dissociation of molecules like F_2 or N_2. As a disadvantage of this technique one can think of the lack of rigorous size-extensivity of the corrected energy. However, at least for some applications this should not be a serious obstacle, since on the basis of some test calculations the errors are expected to be small.

4. The spin-flip EOMCC method, suggested by Spilchenko and Krylov [19], describes a special class of systems exhibiting static correlation. It treats the state which has $M_S < S$ and has multireference character as an spin-changing excitation from the high spin $M_S = S$ component, which is well described by a single reference. This technique is not limited to CC method; it has been implemented also in the DFT framework. It is particularly suited for treatment of diradicals or triradicals; probably it is not able to describe whole sections of potential energy surfaces, where the nature of the wavefunction changes from single reference to multireference one.

5. We should also mention the conceptually very interesting general two-body cluster expansion [20–22], which—if a practicable numerical treatment can be found—would include the static correlation implicitly in a FCI-like manner. The assumption on which this approach is based is still a subject of discussion. Recently, some arguments against its validity [23] have been given. Even if a rigorous solution could not be found, the idea still might be a source of some useful approximations.

6. The Fock space multireference CC methods and the intermediate Hamiltonian techniques (see e.g. Refs. [24–29] and references therein), as well as closely related similarity transformed EOMCC [30–33] are methods particularly suited for calculation of excited/ionized states with a multireference character. Recently, a Brillouin–Wigner formulation of Fock space CC has also been derived [34].

As a disadvantage, we see particularly the very complex structure of the working equations resulting from these formalisms. This obstacle could be overcome by some automatized implementation tools like the 'Tensor Contraction Engine (TCE)' [35], however, at least some of the methods have also very big computational demands, since solutions for many sectors of the Fock space are required.

7. The Hilbert space multireference CC (see e.g. Refs. [36–40]), based on the Jeziorski–Monkhorst ansatz for the wave operator [36]. This ansatz can be either combined with the standard (Rayleigh–Schrödinger) Bloch equation, or with the Brillouin–Wigner Bloch equation (*cf.* Section 18.4), or with a linear combination of both [41]. Recently, an important progress has been achieved by Li and Paldus [15,42,43], who generalized the Jeziorski–Monkhorst formulation to arbitrary incomplete model spaces. This brings two advantages—computational savings due to smaller number of

amplitudes and decrease of the probability of occurrence of the intruder state problem. The Hilbert space MRCC methods are highly relevant to the subject of this review and will be mentioned in more detail in Section 18.4.

18.4 MULTIREFERENCE BRILLOUIN–WIGNER COUPLED CLUSTER METHOD

Our interest in Brillouin–Wigner perturbation theory was stimulated by our finding [44] that this theoretical tool proved very useful in the scattering theory. The fundamental equation, known in the scattering theory as Lippmann–Schwinger equation, expresses the scattering operator as

$$T = V + VB_0T \tag{1}$$

which can be rewritten in the Brillouin–Wigner perturbation series as

$$T = V + VB_0V + VB_0VB_0V + VB_0VB_0VB_0V + \cdots \tag{2}$$

This can be compared with the MP expansion,

$$E = \langle\Phi_0|V|\Phi_0\rangle + \langle\Phi_0|VQ_0V|\Phi_0\rangle + \langle\Phi_0|VQ_0VQ_0V|\Phi_0\rangle + \langle\Phi_0|VQ_0VQ_0VQ_0V|\Phi_0\rangle - \langle\Phi_0|VQ_0V|\Phi_0\rangle\langle\Phi_0|VQ_0VQ_0V|\Phi_0\rangle + \cdots \tag{3}$$

well known in the electronic structure theory and commonly used for the evaluation of MP2, MP3, MP4, and higher order ground state MP energies. In Eqs. (1)–(3) V stands for the perturbation

$$H = H_0 + V \tag{4}$$

and in the scattering theory it has the meaning of the interaction potential between the scattering particles. Eqs. (2) and (3) also differ by the form of the propagator B_0 or Q_0. As already noted, in scattering theory use is made of the Brillouin–Wigner perturbation theory and therefore the propagator is defined as

$$B_0 = \frac{1 - |\Phi_0\rangle\langle\Phi_0|}{\varepsilon - H_0} \tag{5}$$

where ε is the exact energy and H_0 is the unperturbed Hamiltonian with ground state Φ_0. MP theory is based on the Rayleigh–Schrödinger perturbation theory and the propagator is defined therefore as

$$Q_0 = \frac{1 - |\Phi_0\rangle\langle\Phi_0|}{E_0 - H_0} \tag{6}$$

where E_0 is the unperturbed ground state energy.

In contrast to the MP theory, interaction V in the Lippmann–Schwinger equation is strong and cannot be considered as a small perturbation. Hence, the scattering amplitude cannot be calculated by means of Eq. (2) order by order, as we are used to do so in the MP theory. Instead, Eq. (1) is used in the form

$$T = (1 - VB_0)^{-1}V \tag{7}$$

where contributions to all orders are summed up by means of the matrix inversion. We tried to exploit the merits of the Brillouin–Wigner for the evaluation of the correlation energy. On going from the Rayleigh–Schrödinger to the Brillouin–Wigner perturbation theory the 'renormalization' terms in Eq. (3) drop and we obtain

$$\begin{aligned} E =& \langle\Phi_0|V|\Phi_0\rangle + \langle\Phi_0|VB_0V|\Phi_0\rangle + \langle\Phi_0|VB_0VB_0V|\Phi_0\rangle \\ &+ \langle\Phi_0|VB_0VB_0VB_0V|\Phi_0\rangle + \cdots \end{aligned} \tag{8}$$

with B_0 defined by Eq. (5). Next we tried various approximations to Eq. (8) to obtain a practical method for calculations but all our attempts resulted [45] in one or another known version of the CI method [46–50]. We concentrated therefore on the development of the multireference coupled cluster method based on the Brillouin–Wigner perturbation theory. Derivation of working equations for the MR BWCC theory is beyond the scope of this chapter and therefore we refer the interested reader to our earlier papers [40,41,51–54]. Here we only present an outline of the method to explain its essence. As it is usual in the multireference coupled cluster theory we are using the concepts of the effective Hamiltonian and Bloch equation (see, for example, Ref. [55]).

Let us consider a complete model space spanned by M reference configurations, so that the model function for the ground state is expressed as

$$\Psi_0^{\mathrm{P}} = \sum_{\mu=0}^{M} C_\mu \Phi_\mu \tag{9}$$

Such a wave function was used for the two-reference MR BWCCSD calculation of F_2 in Section 18.2, where Φ_0 was the $(1\pi_g)^4(3\sigma_g)^2$ configuration and Φ_1 the $(1\pi_g)^4(3\sigma_u)^2$ configuration. Higher number of reference configurations has been employed in the study of IBr [56] and oxygen molecule [57]. The superscript P is used to indicate that Ψ_0^{P} is a 'projected' wave function. Its relation to the exact wave function Ψ_0 is provided by the (so far) unknown 'wave operator'

$$\Psi_0 = \Omega_0 \Psi_0^{\mathrm{P}} \tag{10}$$

Our task is then to find an effective Hamiltonian which would give us exact energy from the following equation

$$H^{\mathrm{eff}} \Psi_0^{\mathrm{P}} = \varepsilon_0 \Psi_0^{\mathrm{P}} \tag{11}$$

instead of solving rigorously the Schrödinger equation with the exact Hamiltonian and the exact wave function

$$H\Psi_0 = \varepsilon_0 \Psi_0 \tag{12}$$

As it is usual in the MR CC theory, we also used the wave operator in a form suggested by Jeziorski and Monkhorst [36]

$$\Omega_0 = \sum_{\mu=0}^{M} e^{T(\mu)} |\Phi_\mu\rangle\langle\Phi_\mu| \tag{13}$$

which is then substituted on the right side of the Brillouin–Wigner analogue of the Bloch equation [58]

$$\Omega_0 = 1 + B_0 V \Omega_0 \tag{14}$$

where V is the perturbation in the MP partitioning of the Hamiltonian (4) and B_0 is defined by a multireference generalization of Eq. (5) which reads as

$$B_0 = \sum_{q>M} \frac{|\Phi_q\rangle\langle\Phi_q|}{\varepsilon_0 - E_q} \tag{15}$$

Once we know the wave operator through Eqs. (13) and (14), we are ready to evaluate the matrix elements of the effective Hamiltonian in the basis of reference configurations because the effective Hamiltonian is given by the following relationship (see, for example, Ref. [55])

$$H^{\text{eff}} = PH\Omega_0 P \tag{16}$$

where P is the projection operator onto the model space

$$P = \sum_{\mu=1}^{M} |\Phi_\mu\rangle\langle\Phi_\mu| \tag{17}$$

We arrange the M^2 matrix elements $\langle\Phi_\mu|H^{\text{eff}}|\Phi_\nu\rangle$ in a square matrix, diagonalize it, and save the lowest root as the updated ground state energy to be used for a new resolvent (Eq. (15)). The updated resolvent is then used for M single-reference BWCCSD treatments for $\Phi_1, \ldots, \Phi_M$ for obtaining new H^{eff} matrix elements. Actually, the theory yields slightly modified single-reference-like equations, in particular additional Brillouin–Wigner specific terms are included to their right-hand sides. Construction of a new H^{eff} matrix completes a cycle of the iterative procedure for obtaining ε_0. Our approach may be referred to as MR BWCCSD because only T_1 and T_2 clusters are allowed to enter the exponential ansatz (13). The T_1 and T_2 amplitudes may be obtained by a small modification of the CCSD T_1 and T_2 equations [40].

18.5 INTRUDER STATES AND SIZE EXTENSIVITY

For avoiding the intruder state problem in BWCCSD we pay the price of losing size extensivity. Unfortunately, for the purpose of chemical application of MR BWCCSD, the size-extensivity error is not tolerable. For example, the MR BWCCSD value of the singlet–triplet gap in twisted ethylene exhibits an error of 3 kcal/mol with respect to full CI [40]. Even worse case is the dissociation of the F_2 molecule, where MR BWCCSD

Table 18.1 Size extensivity problem in different MR CC methods

Method	Resolvent	Active space	Disconnected diagrams	Size extensivity
MR BWCCSD uncorrected	BW	Complete	Yes	No
MR BWCCSD uncorrected	BW	Incomplete	Yes	No
MR BWCCSD last iteration corrected	BW	Complete	Yes	(No)[a]
MR BWCCSD last iteration corrected	BW	Incomplete	Yes	(No)[a,b]
MR BWCCSD iteratively corrected	BW	Complete	No	Yes
MR BWCCSD iteratively corrected	BW	Incomplete	No	Yes[b]
MR CCSD Kucharski–Bartlett [37]	RS	Complete	No	Yes
MR CCSD Kucharski–Bartlett [37]	RS	Incomplete	No	Yes[b]
MR CCSD Jeziorski–Monkhorst [36]	RS	Complete	No	Yes
MR CCSD Jeziorski–Monkhorst [36]	RS	Incomplete	No	Yes[b]
MR CCSD Mukherjee [38]	RS	Complete	No	Yes

[a] A small size-extensivity error can be expected.
[b] C-conditions for the incomplete model space [42,59] have to be employed.

overestimates the dissociation energy by 17 kcal/mol, but the error can be reduced to less than 0.5 kcal/mol by the size-extensivity correction (see below) [3].

Extent of the size-extensivity error depends on several factors: type of the resolvent, active space, derivation of working equations for amplitudes and the presence or absence of disconnected diagrams in the diagrammatic representation of working equations. In Table 18.1 we show some examples.

As in the single-reference CC approaches, the working equations for amplitudes in the MRCC approach are obtained by projection of the Schrödinger equation on the manifold of excited configurations (for a review see, e.g. Ref. [1,60,61]). This can be done in two different ways. Assume the CCSD approach and denote the manifold of singles and doubles as SD. Then the Schrödinger equation for the ground state CC wave function

$$He^T|0\rangle = Ee^T|0\rangle \tag{18}$$

can be projected on the manifold of singles and doubles as follows

$$\langle \mathrm{SD}|He^T|0\rangle = E\langle \mathrm{SD}|e^T|0\rangle \tag{19}$$

Alternatively, we can multiply by e^{-T} from the left and do the projection in the Baker–Campbell–Hausdorff manner [1]

$$\langle \mathrm{SD}|e^{-T}He^{T}|0\rangle = \langle \mathrm{SD}|e^{-T}Ee^{T}|0\rangle \tag{20}$$

For single-reference CC approaches the two equations, Eqs. (19) and (20), give the same set of diagrams and therefore the same set of equations for the amplitudes. It is profitable that only connected diagrams are obtained which in accordance with the linked cluster theorem guarantees size extensivity. With the MRCC approaches the situation is more complicated. First of all, for state-universal MRCC methods it is necessary to remove explicit occurrence of the energy E from the basic equation (analogy of Eq. (18)). One thus starts from the Bloch equation instead [55,60], which is, however, equivalent to the Schrödinger equation. The projection onto $\langle \mathrm{SD}|e^{-T}$ still gives only connected diagrams, whereas the plain $\langle \mathrm{SD}|$ projection gives both connected and disconnected diagrams and it therefore does not apparently ensure size extensivity. Hence the Baker–Campbell–Hausdorff approach may be viewed as the method of choice. There is a problem with this approach, however. The price one has to pay for removal of energy from the Bloch equation is the occurrence of a term quadratic in the wave operator, which leads to products of amplitudes as $T(\mu)T(\nu)$, where the indices μ and ν denote different reference configurations. Explicit formulas for evaluation of these coupling terms have been recently derived by Paldus and his coworkers [42,62], however, they are rather complicated, their implementation tedious, and evaluation computationally expensive. Their code is able to treat any number of reference configurations and incomplete model space, but is limited to rather small molecules. It seems therefore preferable to develop methods based on an MR analog of Eq. (19), since they lead to coupling terms, which are substantially easier to implement. Kucharski, Balková and Bartlett [37,63–65] proceeded along that way. They identified and dropped disconnected diagrams and achieved size extensivity. An implementation of this method for open-shell singlet two-reference case has been coded in the ACES program by Szalay [66]. General implementation for more than two reference configurations has been developed recently by Pittner [41], also within the ACES2 program. Moreover, based on a certain formula derived by Kowalski and Piecuch [18] in the context of the method of moments coupled clusters [16], it can be shown [67] that a method based on Eq. (19) is equivalent to the Jeziorski–Monkhorst approach based on Eq. (20) in the case of complete model space, thanks to an exact cancellation of the disconnected diagrams arising from He^{T} and the coupling terms.

In contrast, BWCCSD requires only one set of amplitudes for one reference configuration at a time and this feature makes the method easily amenable to tasks with more than two reference configurations and makes the extension of the method to connected triples (MR BWCCSDT) feasible.

Elimination of the size-extensivity error in the MRBWCC theory was suggested by Hubač and Wilson [68]. The Brillouin–Wigner perturbation theory has been known notoriously as a method not furnishing size extensivity. There was therefore every reason to believe that the size-extensivity error in MR BWCCSD originated from the use of the Brillouin–Wigner type resolvent (15) instead of the Rayleigh–Schrödinger type resolvent. Hubač and Wilson found the following identity for the two resolvents

$$\sum_{\vartheta>M} \frac{|\Phi_\vartheta\rangle\langle\Phi_\vartheta|}{\varepsilon_0 - E_\mathrm{q}} = \sum_{\vartheta>M} \frac{|\Phi_\vartheta\rangle\langle\Phi_\vartheta|}{E_0 - E_\mathrm{q}} + \sum_{\vartheta>M} \frac{(E_0 - \varepsilon_0)|\Phi_\vartheta\rangle\langle\Phi_\vartheta|}{(\varepsilon_0 - E_\mathrm{q})(E_0 - E_\mathrm{q})} \tag{21}$$

where M is the number of reference configurations. Since the BW resolvent on the l.h.s. is not size extensive, and the RS-type resolvent in the first term of the r.h.s. is size extensive, the last term on the r.h.s. cannot be size extensive. The idea was to find all terms corresponding to that last term and eliminate them in the equations for amplitudes. In practice we did it in several ways. In the simplest one [52] the MRCC procedure is iterated as described in Section 18.4 without paying any attention to the presence of size-inextensive terms. These are only eliminated in the last iteration. The amplitudes so obtained are free from disconnected diagrams. From them the H^{eff} matrix is constructed and its diagonalization gives the corrected energy. It should be realized that the size-extensivity error is not completely eliminated in this way. The amplitudes so obtained are not converged and they cannot be expected to give a rigorous result. A more sophisticated approach [41] is based on a continuous transition between Brillouin–Wigner and Rayleigh–Schrödinger perturbation theory. Computationally, a step-wise correction is performed in each CC iteration, finally leading to converged amplitude equations in the rigorously size-extensive Rayleigh–Schrödinger limit.

Absence of disconnected diagrams in the working equations alone is not a sufficient condition for size extensivity. In addition to that a stronger condition of a complete active space (CAS) is required. By a CAS we mean all configurations that can be formed by a given number of N electrons in a given number of n orbitals. For two electrons in two orbitals the CAS is formed by four configurations, unless the contribution of two singly excited configurations is vanishing on symmetry grounds, in which case the CAS is represented by the ground and doubly excited configuration. For example, for carbenes the important reference configurations are those that represent two electrons in the active space consisting of HOMO and LUMO orbitals ϕ_1 and ϕ_2 located on the carbenic center. Four spin-unrestricted reference configurations can be formed: $\Phi_1 = (\phi_1)^2(\phi_2)^0$, $\Phi_2 = (\phi_1)^0(\phi_2)^2$ and $\Phi_{3,4} = (\phi_1)^1(\phi_2)^1$, where the last two differ by the spin of the two electrons. However, except for systems of C_1 symmetry only two of these, Φ_1 and Φ_2, are required for the description of the singlet ground state.

There have been several attempts to avoid the CAS restriction and achieve size extensivity for an incomplete model space (see, for example, Refs. [69,70]). The methods based on abandoning the intermediate normalization condition resulted in an excessively complex formalism and as far as we know were actually never implemented. Only recently this problem has been attacked successfully by Li and Paldus, who introduced so called C-conditions for the amplitudes of internal excitations [15,42,43,71]. When these conditions are incorporated into the MRCC amplitude equations of the Jeziorski–Monkhorst method, the solutions with a general incomplete model space become also exactly size extensive. The C-conditions are, however, not limited to the original Jeziorski–Monkhorst formulation [36], but rather they apply to any Hilbert-space MRCC method, including BWCC, which can be corrected for size extensivity also in the incomplete model space [59].

As with the single-reference methods, size extensivity does not guarantee size consistency (a proper dissociation limit in the case of potential curves). The reference space must be carefully chosen to contain important configurations at any point of the potential curve as it was recognized long time ago [72]. But this is not a specific problem of MR BWCCSD.

18.6 PERFORMANCE OF THE MULTIREFERENCE BRILLOUIN–WIGNER CC METHOD AND APPLICATIONS

Computationally, MR BWCCSD may be viewed as a set of weakly coupled single-reference CC calculations. Equations for the determination of amplitudes depend only on a set of amplitudes (from a previous iteration) of a single reference configuration and coupling between reference configurations is only ensured by the energy, which is eigenvalue of the effective Hamiltonian matrix. This simplicity of the computational scheme is beneficial for the feasibility of calculations. The computer time for a MR BWCCSD run with M reference configurations is about M times higher than it is for a standard CCSD run. This permitted us to perform MR BWCCSD calculations on a desktop PC even for molecules that have been of interest to organic chemists and biochemists. The largest molecules we treated were tetramethyleneethane (TME) [73]

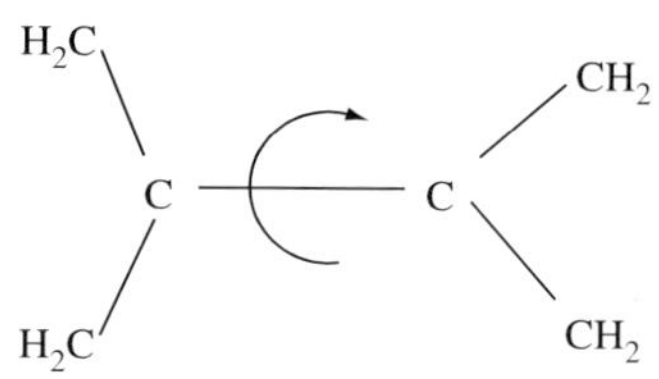

and benzyne [74]. The purpose of the calculations on TME was to bring some theoretical evidence on the ordering of the lowest singlet and triplet states. From the EPR spectra of TME measured in a matrix it was concluded (for references to experimental work, see Ref. [73]) that TME has D_{2d} structure at the triplet state. This was at variance with the gas-phase negative ion photoelectron spectra showing that the triplet state is about 2 kcal/mol above the singlet state. The MR BWCCSD/cc-pVDZ calculations support the latter assignment. At the optimum triplet geometry (dihedral angle of 49°) the calculated $S-T$ energy gap is 1.2 kcal/mol, and the triplet is predicted to lie above the singlet at any dihedral angle. However, a firm interpretation of the observed triplet state of the matrix-isolated TME would need more rigorous calculations, including triple excitations and using a larger basis set. In parallel, we also used the MR CCSD method by Kucharski and Bartlett [37]. The results obtained by the two methods were close in absolute value, showing that the deficiency of the MR BWCCSD method—its size inextensivity—may be eliminated by an *a posteriori* correction (correction done in the last CC iteration only). The calculations on benzyne [74] had a biochemical experimental background (see Ref. [74]). The natural products shown in Fig. 18.2 displayed high activity against a

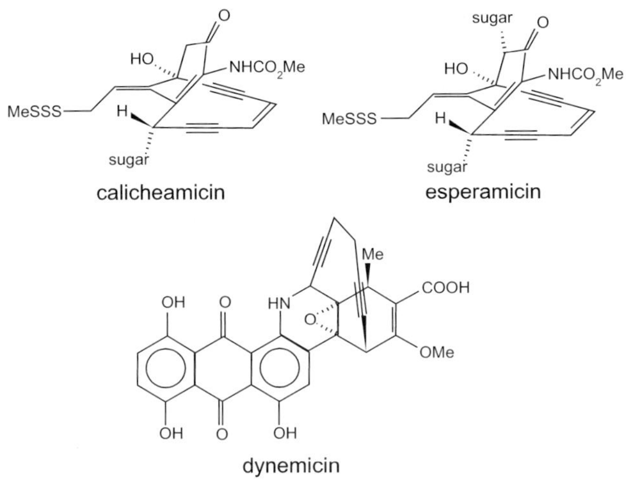

Fig. 18.2. Natural products with a high activity against a number of tumor cell lines.

number of tumor cell lines, both *in vivo* and *in vitro*, but unfortunately also showed unacceptable toxic effects in both animal and human trials. The active site is believed to be the hex-3-ene-1,5-diyne moiety, which yields 1,4-didehydrobenzene (benzyne) by the Bergman cyclization reaction (Fig. 18.3). Benzyne abstracts hydrogen from a saccharide phosphate backbone to form benzene, which denaturates the DNA and ultimately causes cell death. Benzyne is a diradical species and it is therefore difficult to calculate it by standard single-reference methods. The objective of MR BWCCSD calculations was to provide reliable data on the heat of reaction and enthalpy of activation for the Bergman reaction, which would be used as standards for less sophisticated (and less demanding) calculations, and to show the accuracy attainable by the present state-of-the-art techniques in designing new antitumor agents with an enediyne-like structure. The entries in Table 18.2 show that the MR BWCCSD method with the *a posteriori* correction for size extensivity improves greatly the CCSD results and it gives results in good agreement with the MR CI and experimental data. Good performance of CCSD(T) is probably fortuitous in this case.

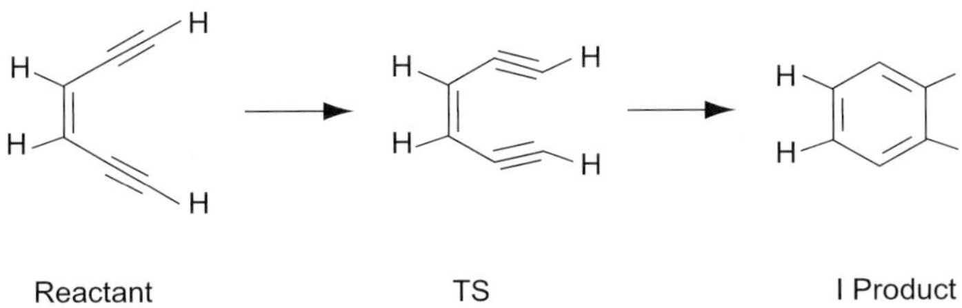

Fig. 18.3. Bergman reaction.

Table 18.2 Enthalpy of activation and heat of reaction of the Bergman reaction.[a]

Method	$\Delta H^{\dagger 298}$ (kcal/mol)	ΔH_0^{298} (kcal/mol)
CCSD	38.2	27.5
MR BWCCSD	32.7	12.9
MR CI	29.4	10.3
CCSD(T)	27.6	10.1
Experiment	28.2	8.5

[a]The cc-pVTZ basis set was used in all calculations; for details see Ref. [74].

We have also studied the automerization barrier in cyclobutadiene, where the transition structure has a diradical character [75] and the singlet–triplet gaps in alkylcarbenes [76].

Besides calculations of organic compounds, we have assessed the accuracy of the MR BWCCSD technique on benchmark systems, like the insertion of Be into hydrogen molecule [77], and compared its accuracy with other multireference methods. We have also employed the MR BWCCSD method for accurate treatment of diatomic molecules like F_2 [3] or IBr [56], which required four reference configurations to span the complete model space. We have also extended calculations by our method to low lying excited states and up to eight reference configurations, used for calculation of the oxygen molecule [57]. The isoelectronic, but heteronuclear NF molecule has been investigated along the same lines [78].

A series of carbide diatomics, CaC, ZnC, BeC, and MgC, were another challenge for the MR BWCCSD to treat systems of multireference nature caused by near-degeneracy effects. The task was to examine theoretically [79,80] the competing $^3\Sigma^-$ and $^5\Sigma^-$ states. In Table 18.3 we present results for CaC.

Single-reference methods CCSD and CCSD(T) give a wrong order of the two states, predicting the triplet to lie 10.0 and 0.4 kcal/mol above the quintet state. Experimentally, the ground state is not known but MR BWCCSD with the *a posteriori* size-extensivity correction and MR CI predicts consistently that the triplet is the ground state. The former gives $T_e = 0.9$ kcal/mol for the quintet and the later gives 3.0 kcal/mol.

Table 18.3 Spectroscopic constants of the CaC molecule

Method	D_e (kcal/mol)	r_e (Å)	ω_e (cm^{-1})	$\omega_e x_e$ (cm^{-1})
ROHF	1.2	2.838		
CCSD	35.5	2.368	464.0	3.67
CCSD(T)	46.1	2.383	464.2	2.39
MR BWCCSD	45.8	2.357	468.3	2.06
MR ACPF	48.0	2.365	465.1	2.65
MR CI	48.6	2.364	462.4	2.37
MR CI + Q	49.3	2.364	461.3	2.51

18.7 SUMMARY

We do not advocate MR BWCCSD as the method of choice for general use. Instead we wanted to show that in applications where a system with a quasidegenerate electronic state is to be calculated, it may be profitable to use a multireference approach as a more advantageous alternative to the single-reference treatment. This applies particularly to cases where the number of reference configurations needed is low. The computer time for MR BWCCSD increases about linearly with the number of reference configurations. Hence, if a regular single-reference CC calculation is feasible for the particular system, also the MR BWCCSD would be likely affordable. Of course, as with any other multireference method, some experimentation with the number of reference configurations would be necessary. In its present form and at the level of MR CCSD, the MR BWCC method seems to be slightly less accurate than large scale MR CI calculations with some sort of *a posteriori* correction [52,68] (such as the averaged coupled pair functional) for retaining size extensivity. However, MR BWCCSD has good prospects for its further development. Its inherent feature of being not size extensive is now well understood [41] which is a good starting point for the development of more accurate size-extensivity corrections. Also, development of the MR BWCCSDT computer program is in progress. First results obtained using an approximation of MR BWCCSDT for calculation of vibrational frequencies and anharmonicities of O_2 molecule [81] show that inclusion of T_3 clusters into the multireference CC expansion greatly improves the accuracy of the method. A profitable feature of MR BWCCSD is also the similarity of the MR BWCCSD working equations to those in CCSD, which enables to formulate analytically the MR BWCCSD energy gradient along the same lines as in CCSD. However, even in its present form the applications performed so far show that MR BWCCSD can be taken as a more economic alternative to MR CI for treatments of systems with a quasidegenerate electronic state.

18.8 ACKNOWLEDGEMENTS

This work has been supported by the COST D23 action (grant OC D23.001 of the Czech Ministry of Education). We acknowledge also the support by the Grant Agency of the Czech Republic (grant No. 203/04/0425) and the Grant Agency of the Academy of Sciences of the Czech Republic (Grants No. A4040401 and 1ET400400413). We also acknowledge the support by the Academy of Sciences of the Czech Republic (Project No. K4040110).

18.9 REFERENCES

1 P.v.R. Schleyer, N.L. Allinger, T. Clark, J. Gasteiger, P.A. Kollman, H.F. Schaefer, III and P.R. Scheiner (Eds.), The encyclopedia of computational chemistry, Wiley, Chichester, 1998.
2 X. Li and J. Paldus, J. Chem. Phys., 108 (1998) 637.
3 J. Pittner, J. Šmydke, P. Čársky and I. Hubač, J. Mol. Struct. (THEOCHEM), 547 (2001) 239.
4 K.P. Huber and G. Herzberg, Molecular spectra and molecular structure, Vol. 4, Van Nostrand Reinhold, New York, 1979.

5 ACES II is a program product of Quantum Theory Project, University of Florida, J.F. Stanton, J. Gauss, J.D. Watts, M. Nooijen, N. Oliphant, S.A. Perera, P.G. Szalay, W.J. Lauderdale, S.R. Gwaltney, S. Beck, A. Balková, D.E. Bernholdt, K.-K. Baeck, P. Rozyczko, H. Sekino, C. Huber and R.J. Bartlett. Integral packages included are VMOL (J. Almloef and P.R. Taylor); VPROPS (P. R. Taylor); and ABACUS (T. Helgaker, H.J.Aa. Jensen, P. Joergensen, J. Olsen and P.R. Taylor).
6 P. Piecuch, S.A. Kucharski and R.J. Bartlett, J. Chem. Phys., 110 (1999) 6103.
7 L. Adamowicz, J.-P. Malrieu and V.V. Ivanov, J. Chem. Phys., 112 (2000) 10075.
8 V.V. Ivanov and L. Adamowicz, J. Chem. Phys., 113 (2000) 8503.
9 M. Kállay, P.G. Szalay and R. Surján, J. Chem. Phys., 117 (2002) 980.
10 X. Li and J. Paldus, J. Chem. Phys., 107 (1997) 6257.
11 X. Li and J. Paldus, J. Mol. Struct. (THEOCHEM), 547 (2001) 69.
12 X. Li and J. Paldus, J. Chem. Phys., 115 (2001) 5759.
13 X. Li and J. Paldus, J. Chem. Phys., 115 (2001) 5774.
14 J. Paldus and X. Li, Collect. Czech. Chem. Commun., 68 (2003) 554.
15 X. Li and J. Paldus, J. Chem. Phys., 119 (2003) 5334.
16 K. Kowalski and P. Piecuch, J. Chem. Phys., 113 (2000) 18.
17 K. Kowalski and P. Piecuch, J. Chem. Phys., 115 (2001) 2966.
18 K. Kowalski and P. Piecuch, J. Mol. Struct. (THEOCHEM), 547 (2001) 191.
19 L.V. Spilchenko and A.I. Krylov, J. Chem. Phys., 117 (2002) 4694.
20 M. Nooijen, Phys. Rev. Lett., 84 (2000) 2108.
21 T.V. Voorhis and M. Head-Gordon, J. Chem. Phys., 115 (2001) 5033.
22 P. Piecuch, K. Kowalski, P.-D. Fan and K. Jedziniak, Phys. Rev. Lett., 90 (2003) 113001.
23 E.R. Davidson, Phys. Rev. Lett., 91 (2003) 123001.
24 J.-P. Malrieu, P. Durand and J.-P. Daudey, J. Phys. A, 18 (1985) 809.
25 L. Meissner, J. Chem. Phys., 108 (1998) 9227.
26 L. Meissner, P. Malinowski and A. Nowaczyk, Chem. Phys. Lett., 381 (2003) 441.
27 L. Meissner, P. Malinowski and J. Gryniakóv, J. Phys. B, 37 (2004) 1.
28 A. Landau, E. Eliav, Y. Ishikawa and U. Kaldor, J. Chem. Phys., 113 (2000) 9905.
29 N. Vaval, S. Pal and D. Mukherjee, Theor. Chem. Acc., 99 (1998) 100.
30 M. Nooijen, J. Chem. Phys., 104 (1996) 2638.
31 M. Nooijen and R.J. Bartlett, J. Chem. Phys., 106 (1997) 6441.
32 M. Nooijen and R.J. Bartlett, J. Chem. Phys., 107 (1997) 6812.
33 M. Nooijen and V. Lotrich, J. Chem. Phys., 113 (2000) 494.
34 N.D.K. Petraco, L. Horný, H.F. Schaefer, III and I. Hubač, J. Chem. Phys., 117 (2002) 9580.
35 P. Sadayappan, G. Baumgartner, D.E. Bernholdt, R.J. Harrison, S. Hirata, M. Nooijen, R.M. Pitzer and J. Ramanujam: The Tensor Contraction Engine, http://www.cse.ohio-state.edu:/~gb/TCE
36 B. Jeziorski and H.J. Monkhorst, Phys. Rev. A, 24 (1981) 1668.
37 S.A. Kucharski and R.J. Bartlett, J. Chem. Phys., 95 (1991) 8227.
38 U.S. Mahapatra, B. Datta and D. Mukherjee, J. Chem. Phys., 110 (1999) 6171.
39 I. Hubač, in: A. Tsipis, V.S. Popov, D.R. Herschbach and J.S. Avery (Eds.), New methods in quantum theory. NATO ASI series, Kluwer, Dordrecht, 1996, pp. 183–202.
40 J. Pittner, P. Nachtigall, P. Čársky, J. Mášik and I. Hubač, J. Chem. Phys., 110 (1999) 10275.
41 J. Pittner, J. Chem. Phys., 118 (2003) 10876.
42 X. Li and J. Paldus, J. Chem. Phys., 119 (2003) 5320.
43 X. Li and J. Paldus, J. Chem. Phys., 119 (2003) 5346.
44 P. Čársky, V. Hrouda, V. Sychrovský, I. Hubač, P. Babinec, P. Mach, J. Urban and J. Mášik, Collect. Czech. Chem. Commun., 60 (1995) 1419.
45 V. Sychrovský and P. Čársky, Mol. Phys., 88 (1996) 1137.
46 Z. Gershgorn and I. Shavitt, Int. J. Quantum Chem., 2 (1968) 751.
47 G.A. Segal and R.W. Wetmore, Chem. Phys. Lett., 32 (1975) 556.
48 L.E. Nitzsche and E.R. Davidson, J. Chem. Phys., 68 (1978) 3103.
49 T.H. Dunning, Jr., Chem. Phys., 42 (1979) 249.
50 J.J. Diamond, G.A. Segal and R.W. Wetmore, J. Phys. Chem., 88 (1984) 3532.

51 J. Mášik and I. Hubač, in: R. McWeeny, J. Maruani, Y.G. Smeyers, S. Wilson (Eds.), Quantum systems in chemistry and physics: Trends in methods and applications, Kluwer, Dordrecht, 1997, pp. 283–308.
52 I. Hubač, J. Pittner and P. Čársky, J. Chem. Phys., 112 (2000) 8779.
53 I. Hubač, J. Mášik, P. Mach, J. Urban and P. Babinec, in: J. Leszczynski (Ed.), Computational chemistry. Reviews of current trends, Vol. 3, World Scientific, Singapore, 1999, pp. 1–48.
54 P. Čársky, I. Hubač, P. Mach, J. Pittner and S. Wilson, in: J. Maruani, R. Lefebvre and E. Brandas (Eds.), Advanced topics in theoretical chemical physics, Progress in theoretical chemistry and physics, Vol. 12, Kluwer, Dordrecht, 2003, pp. 71–118.
55 I. Lindgren and J. Morrison, Atomic many-body theory, Springer, Berlin, 1982.
56 J. Pittner, O. Demel, P. Čársky and I. Hubač, Int. J. Mol. Sci., 2 (2002) 281.
57 J. Pittner, P. Čársky and I. Hubač, Int. J. Quantum Chem., 90 (2002) 1031.
58 J. Mášik, P. Mach and I. Hubač, J. Chem. Phys., 108 (1998) 5671.
59 J. Pittner, X. Li and J. Paldus, Mol. Phys., in press.
60 J. Paldus, in: S. Wilson, G.H.F. Diercksen (Eds.), Methods in computational molecular physics. NATO ASI series, Plenum, New York, 1992, pp. 99–194.
61 J. Paldus, in: G.L. Malli (Ed.), Relativistic and correlation effects in molecules and solids. NATO ASI series, Plenum, New York, 1994, pp. 207–282.
62 J. Paldus, X. Li and N.D. Petraco, J. Math. Chem. 35 (2004) 215.
63 A. Balková, S.A. Kucharski and R.J. Bartlett, Chem. Phys. Lett., 182 (1991) 511.
64 A. Balková, S.A. Kucharski, L. Meissner and R.J. Bartlett, J. Chem. Phys., 95 (1991) 4311.
65 A. Balková and R.J. Bartlett, J. Chem. Phys., 101 (1994) 8972.
66 P.G. Szalay and R.J. Bartlett, J. Chem. Phys., 101 (1994) 4936.
67 J. Pittner, K. Kowalski and P. Piecuch, to be published.
68 I. Hubač and S. Wilson, J. Phys. B, 33 (2000) 365.
69 L. Meissner, S.A. Kucharski and R.J. Bartlett, J. Chem. Phys., 91 (1989) 6187.
70 L. Meissner and R.J. Bartlett, J. Chem. Phys., 92 (1990) 561.
71 J. Paldus and X. Li, J. Chem. Phys., 118 (2003) 6769.
72 G.C. Lie and E. Clementi, J. Chem. Phys., 60 (1974) 1288.
73 J. Pittner, P. Nachtigall, P. Čársky and I. Hubač, J. Phys. Chem. A, 105 (2001) 1354.
74 O. Rey-Puiggros, J. Pittner, P. Čársky, P. Stampfuß and W. Wenzel, Collect. Czech. Chem. Commun., 68 (2003) 2309.
75 J.C. Sancho-García, J. Pittner, P. Čársky and I. Hubač, J. Chem. Phys., 112 (2000) 8785.
76 O. Demel, J. Pittner, P. Čársky and I. Hubač, J. Phys. Chem. A, 108 (2004) 3125.
77 J. Pittner, H. Valdes-González, R.J. Gdanitz and P. Čársky, Chem. Phys. Lett., 386 (2004) 211.
78 S. Kardahakis, J. Pittner, P. Čársky and A. Mavridis, Int. J. Quantum Chem., Int. J. Quantum Chem., in press.
79 I.S.K. Kerkines, J. Pittner, P. Čársky, A. Mavridis and I. Hubač, J. Chem. Phys., 117 (2002) 9733.
80 V.I. Teberekidis, J. Pittner, P. Čársky, A. Mavridis and I. Hubač, Int. J. Quantum Chem., in press.
81 J. Pittner and O. Demel, J. Chem. Phys., in press.

Theory and Applications of Computational Chemistry: The First Forty Years
Edited by C. Dykstra et al.

CHAPTER 19

Electronic structure: the momentum perspective

Ajit J. Thakkar

Department of Chemistry, University of New Brunswick, Fredericton, New Brunswick, Canada E3B 6E2

Abstract

This article provides an introduction to the momentum perspective of the electronic structure of atoms and molecules. After an explanation of the genesis of momentum–space wave functions, relationships among one-electron position and momentum densities, density matrices, and form factors are traced. General properties of the momentum density are highlighted and contrasted with properties of the number (or charge) density. An outline is given of the experimental measurement of momentum densities and their computation. Several illustrative computations of momentum–space properties are summarized.

19.1 INTRODUCTION

The Schrödinger equation is the foundation of the theory of the electronic structure of atoms and molecules. Within the Born–Oppenheimer approximation, it is usually formulated in position or r space as

$$\hat{H}\Psi_n(\vec{x}_1, \vec{x}_2, \ldots, \vec{x}_N) = E_n \Psi_n(\vec{x}_1, \vec{x}_2, \ldots, \vec{x}_N) \tag{1}$$

in which E_n is the electronic energy, the $\{\vec{x}_j = (\vec{r}_j, \sigma_j)\}_{j=1}^{N}$ are space–spin coordinates of the N electrons in the system, and $\Psi_n(\vec{x}_1, \vec{x}_2, \ldots, \vec{x}_N)$ is the wave function for the nth electronic state of the system defined by the Hamiltonian operator $\hat{H}$. The latter is obtained from its classical counterpart by replacing each Cartesian component of position q by a multiplicative operator $\hat{q}$ and each Cartesian component of linear momentum p_q by the corresponding operator $\hat{p}_q = -i(\partial/\partial q)$. Atomic units are used in the previous expression and throughout this chapter. The non-relativistic electronic Hamiltonian can be written as

$$\hat{H} = -\frac{1}{2}\sum_{j=1}^{N}\nabla_j^2 + V(\vec{r}_1, \vec{r}_2, \dots, \vec{r}_N) \tag{2}$$

in which the first term is a sum of the kinetic energy operators for each of the N electrons, and the second term is the potential in which the electrons move. Typically, the potential V can be decomposed into terms corresponding to the Coulombic interactions between electron–nucleus pairs, electron pairs and nuclear pairs. Techniques for the computation and interpretation of position–space wave functions $\Psi_n(\vec{x}_1, \vec{x}_2, \dots, \vec{x}_N)$ are the subject of many of the chapters in this book.

Unlike a trajectory in classical mechanics, the wave function depends only on the space–spin coordinates $\{\vec{x}_j\}_{j=1}^N$ of the electrons in the system, and not on their momenta $\{\vec{p}_j\}_{j=1}^N$. This lack of dependence of the wave function on the momenta reflects Heisenberg's uncertainty principle which in turn arises from the non-commutativity of conjugate pairs of position and momentum operators.

An equally valid but different approach is to work with momentum–space wave functions $\Phi_n(\vec{y}_1, \vec{y}_2, \dots, \vec{y}_N)$ which depend upon the momentum–spin coordinates $\{\vec{y}_j = (\vec{p}_j, \sigma_j)\}_{j=1}^N$ of the N electrons in the system but not on their positions. A momentum representation of the wave function does not yield any more or less information than the position representation of the wave function does. However, the momentum representation does provide a different perspective—one from the other end of Heisenberg's eyeglass.

In this chapter, an overview of the momentum perspective on the electronic structure of atoms and molecules is provided. The genesis of momentum–space wave functions is described in Section 19.2. Relationships among one-electron position and momentum densities, density matrices, and form factors are traced in Section 19.3. General properties of the momentum density are highlighted and contrasted with properties of the number density in Section 19.4. The experimental measurement of momentum densities is outlined in Section 19.5, and their computation in Section 19.6. Several illustrative computations of momentum–space properties are summarized in Section 19.7. Some concluding remarks are made in Section 19.8.

19.2 MOMENTUM–SPACE WAVE FUNCTIONS

There are three distinct ways by which the momentum–space wave function can be obtained: directly by solving either a differential or an integral equation in momentum or p space, or indirectly by transformation of the position–space wave function.

One direct method involves a differential Schrödinger-like equation, $\hat{H}\Phi_n = E_n\Phi_n$, which contains a Hamiltonian operator obtained from its classical counterpart by replacing each Cartesian component of linear momentum p_q by a multiplicative operator $\hat{p}_q$ and each Cartesian component of position q by the corresponding operator $\hat{q} = i(\partial/\partial p_q)$. This approach is not at all simple because the $\hat{q}$ operators transform Coulomb potentials, which involve r^{-1}, into fearsome operators. Nevertheless, Egil Hylleraas, a Norwegian physicist, did succeed in solving this equation for the hydrogen atom [1]. To my knowledge, nothing further has been done with this technique.

The other direct method is to Fourier transform the Schrödinger equation to obtain an integral equation for Φ that can be written, after suppression of spin variables, as follows:

$$\left(E_n - \frac{1}{2}\sum_{j=1}^{N}\vec{p}_j\cdot\vec{p}_j\right)\Phi_n(\vec{p}_1,\vec{p}_2,\ldots,\vec{p}_N)$$
$$= \int \mathrm{d}\vec{p}_1'\,\mathrm{d}\vec{p}_2'\ldots\mathrm{d}\vec{p}_N'\,W(\vec{p}_1-\vec{p}_1',\vec{p}_2-\vec{p}_2',\ldots,\vec{p}_N-\vec{p}_N')\Phi_n(\vec{p}_1',\vec{p}_2',\ldots,\vec{p}_N') \quad (3)$$

in which

$$W(\vec{p}_1,\vec{p}_2,\ldots,\vec{p}_N) = (2\pi)^{-3N}\int \mathrm{d}\vec{r}_1\,\mathrm{d}\vec{r}_2\ldots\mathrm{d}\vec{r}_N V(\vec{r}_1,\vec{r}_2,\ldots,\vec{r}_N)\exp\left[-i\sum_{j=1}^{N}\vec{r}_j\cdot\vec{p}_j\right]. \quad (4)$$

This equation was solved for the hydrogen atom by Vladimir Fock [2,3]. The solution in p space revealed the four-dimensional symmetry responsible for the degeneracy of states with the same n but different l quantum numbers in the hydrogen atom. This is a fine example where the momentum–space perspective led to fresh and deep insight. Fock's work spawned much further research on dynamical groups and spectrum-generating algebras.

Most work on the integral Eq. (3) has utilized an iterative method of Svartholm [4,5] for solving it. The integral equation was first applied to He and H_2^+ by Roy McWeeny and Charles Coulson [6,7] during McWeeny's doctoral work. At the 5th Canadian Symposium on Theoretical Chemistry, Ottawa, 1974, McWeeny delivered a tribute to Coulson who had passed away in January of that year. In that lecture, McWeeny described how he felt when he was first assigned the project by Coulson:

> I'd been given a thesis to read; by Svartholm, on 'The Binding Energies of the Lightest Atomic Nuclei'—which both flattered and frightened me. I could understand hardly a word of it; and my ignorance was so shameful (especially for a physicist in a chemistry department) that I couldn't ask for help—the only acceptable remedy was to follow the instructions: read the books by Pauling and Wilson, and by Dirac, and go to Maurice Pryce's lectures on quantum mechanics.

Students should take heart from the outstanding career in theoretical chemistry that McWeeny was able to build from this anxious start. Interesting work on exact solution of the integral equation for polyatomic one-electron systems was published by Shibuya and Wulfman [8] and Novosadov [9,10]. The integral equation approach has been developed and tested by several groups since then but the amount of effort devoted to this pales in comparison with the stupendous amount of work done on solving the position–space Schrödinger equation. A list of the relatively few references on approximate solution of the integral equation for Φ can be found in a recent review [11]. Beginning with the work of Armstrong [12], the integral Schrödinger equation has also been used to derive conditions to check and possibly improve the accuracy of approximate wave functions obtained by position–space methods; a bibliography can be found elsewhere [11].

The technology for solving the Schrödinger equation is so much more advanced in r-space than in p-space that it is most practical to obtain the momentum–space Φ from its position–space counterpart Ψ. The transformation theories of Dirac [13,14] and Jordan [15,16] provide the link between these representations:

$$\Phi = (2\pi)^{-3N/2} \int \Psi \exp\left[-i \sum_{k=1}^{N} \vec{p}_k \cdot \vec{r}_k \right] \mathrm{d}\vec{r}_1 \, \mathrm{d}\vec{r}_2 \ldots \mathrm{d}\vec{r}_N \tag{5}$$

and its inverse

$$\Psi = (2\pi)^{-3N/2} \int \Phi \exp\left[+i \sum_{k=1}^{N} \vec{p}_k \cdot \vec{r}_k \right] \mathrm{d}\vec{p}_1 \, \mathrm{d}\vec{p}_2 \ldots \mathrm{d}\vec{p}_N \tag{6}$$

In short, Ψ and Φ are related by a $3N$-dimensional, norm-preserving, Fourier transform.

The first use of the Fourier transformation technique for atoms or molecules was made by Boris Podolsky and Linus Pauling [17] for the hydrogen atom. Coulson [18] noted that if the r-space wave function is constructed from one-electron functions, then there is an isomorphism between Ψ and Φ. In particular, if the wave function Ψ can be written in terms of spin-orbitals $\{\psi_j\}$ as a single Slater determinant,

$$\Psi(\vec{x}_1, \vec{x}_2, \ldots, \vec{x}_N) = (1/\sqrt{N!}) \det|\psi_1(\vec{x}_1)\psi_2(\vec{x}_2)\cdots\psi_N(\vec{x}_N)| \tag{7}$$

then the corresponding p-space wave function is also a Slater determinant,

$$\Phi(\vec{y}_1, \vec{y}_2, \ldots, \vec{y}_N) = (1/\sqrt{N!}) \det|\phi_1(\vec{y}_1)\phi_2(\vec{y}_2)\cdots\phi_N(\vec{y}_N)| \tag{8}$$

in which the spin-*momentals*,[1] or p-space spin-orbitals, $\{\phi_j\}$ are obtained from the r-space spin-orbitals by the one-electron version of Eq. (5)

$$\phi_j(\vec{p}, \sigma) = (2\pi)^{-3/2} \int \psi_j(\vec{r}, \sigma) \mathrm{e}^{-i\vec{p}\cdot\vec{r}} \mathrm{d}\vec{r} \tag{9}$$

If the r-space wave function is a linear combination of Slater determinants constructed from a set of spin-orbitals $\{\psi_j\}$, then its p-space counterpart is the same linear combination of Slater determinants constructed from the spin-momentals $\{\phi_j\}$ obtained as Fourier transforms, Eq. (9), of the spin-orbitals. The overwhelming majority of contemporary r-space wave functions can be expressed as a linear combination of Slater determinants, and in these cases only three-dimensional Fourier transforms, Eq. (9), of the spin-orbitals are necessary to obtain the corresponding N-electron wave function in p space. Use of the full Eq. (5) becomes necessary only for wave functions, such as Hylleraas- or Jastrow-type wave functions, that are not built from a one-electron basis set. Examples of transformations to momentum space of such wave functions for He and H_2 can be found elsewhere [20–22].

[1] The term 'momental' for the p-space counterpart of an orbital is due to Robert S. Mulliken; in a private letter to me dated August 1, 1986, he confirmed that the term originated with him. The term was used in the 1970s in an unpublished manuscript by William (Bill) Henneker and Paul Cade, and then in print in a short paper of mine [19]. I have tried to keep this nomenclature alive since then in honor of Mulliken.

19.3 DENSITIES AND DENSITY MATRICES

If we are interested only in properties that can be expressed in terms of one-electron operators, then it is sufficient to work with the first-order reduced density matrix rather than the N-electron wave function [23–27].

We can write the r-space, first-order, density matrix as

$$\Gamma(\vec{x}|\vec{x}') = N \int \Psi^*(\vec{x}', \vec{x}_2, \ldots, \vec{x}_N)\Psi(\vec{x}, \vec{x}_2, \ldots, \vec{x}_N)\mathrm{d}\vec{x}_2 \ldots \mathrm{d}\vec{x}_N \quad (10)$$

and the p-space, first-order, density matrix as

$$\Pi(\vec{y}|\vec{y}') = N \int \Phi^*(\vec{y}', \vec{y}_2, \ldots, \vec{y}_N)\Phi(\vec{y}, \vec{y}_2, \ldots, \vec{y}_N)\mathrm{d}\vec{y}_2 \ldots \mathrm{d}\vec{y}_N \quad (11)$$

If electronic spin is not a focus of attention, then the spin-traced (spin-summed) versions of these density matrices can be used. The r-space, spin-traced, first-order, reduced density matrix is

$$\Gamma(\vec{r}|\vec{r}\,') = \int \Gamma(\vec{x}|\vec{x}')\delta(\sigma - \sigma')\mathrm{d}\sigma\, \mathrm{d}\sigma' \quad (12)$$

where δ is the Dirac delta function. Similarly, the p-space, spin-traced, first-order, reduced density matrix is

$$\Pi(\vec{p}|\vec{p}') = \int \Pi(\vec{y}|\vec{y}')\delta(\sigma - \sigma')\mathrm{d}\sigma\, \mathrm{d}\sigma' \quad (13)$$

Just as the N-electron wave functions in the two spaces are related by Fourier transformation, so are the density matrices in the two representations. Specifically, the first-order r- and p-space density matrices (whether spin-traced or not) are related by a six-dimensional Fourier transform [28,29]:

$$\Pi(\vec{p}|\vec{p}') = (2\pi)^{-3} \int \Gamma(\vec{r}|\vec{r}')\exp[-i(\vec{p}\cdot\vec{r} - \vec{p}'\cdot\vec{r}\,')]\mathrm{d}\vec{r}\, \mathrm{d}\vec{r}\,' \quad (14)$$

and its inverse

$$\Gamma(\vec{r}|\vec{r}\,') = (2\pi)^{-3} \int \Pi(\vec{p}|\vec{p}\,')\exp[+i(\vec{p}\,'\cdot\vec{r} - \vec{p}\cdot\vec{r}\,')]\mathrm{d}\vec{p}\, \mathrm{d}\vec{p}\,' \quad (15)$$

The physically meaningful quantities are the densities—the diagonal elements of the density matrices. We have the usual one-electron number or position density

$$\rho(\vec{r}) = \Gamma(\vec{r}|\vec{r}) = N \int |\Psi|^2 \delta(\vec{r}_1 - \vec{r})\mathrm{d}\vec{x}_1\, \mathrm{d}\vec{x}_2 \ldots \mathrm{d}\vec{x}_N \quad (16)$$

and the one-electron momentum density

$$\Pi(\vec{p}) = \Pi(\vec{p}|\vec{p}) = N \int |\Phi|^2 \delta(\vec{p}_1 - \vec{p})\mathrm{d}\vec{y}_1\, \mathrm{d}\vec{y}_2 \ldots \mathrm{d}\vec{y}_N \quad (17)$$

Letting $\vec{p}\,' = \vec{p}$ in Eq. (14) yields

$$\Pi(\vec{p}) = (2\pi)^{-3} \int \Gamma(\vec{r}|\vec{r}\,')\exp[+i\vec{p}\cdot(\vec{r}\,' - \vec{r})]\mathrm{d}\vec{r}\,\mathrm{d}\vec{r}\,' \tag{18}$$

and letting $\vec{r}\,' = \vec{r}$ in Eq. (15) gives

$$\rho(\vec{r}) = (2\pi)^{-3} \int \Pi(\vec{p}|\vec{p}\,')\exp[-i\vec{r}\cdot(\vec{p}\,' - \vec{p})]\mathrm{d}\vec{p}\,\mathrm{d}\vec{p}\,' \tag{19}$$

Eq. (18) was first pointed out and exploited for electron correlated wave functions by Robert Benesch and Vedene Smith [30,31] during Benesch's doctoral studies.

It is clear from Eqs. (18) and (19) that the number and momentum densities are not related by Fourier transformation. This is most readily understood for a one-electron system where the r-space density is just the squared magnitude of the orbital and the p-space density is the squared magnitude of the Fourier transform of the orbital. The densities are not Fourier transforms of one another because the operations of Fourier transformation and taking the absolute value squared do not commute. Moreover, there is no known direct and practical route from one density to the other even though the Hohenberg–Kohn theorem [32] guarantees that it must be possible to obtain the ground state $\Pi(\vec{p})$ from $\rho(\vec{r})$.

In fact, the Fourier transform of $\rho(\vec{r})$ is the form factor of X-ray crystallography:

$$F(\vec{\mu}) = \int \mathrm{e}^{i\vec{\mu}\cdot\vec{r}}\rho(\vec{r})\mathrm{d}\vec{r} \tag{20}$$

How is the form factor related to the p-space density matrix? Substitution of Eq. (19) into Eq. (20) and integration over $\vec{r}$ and $\vec{p}\,'$ yields [28,29]

$$F(\vec{\mu}) = \int \Pi(\vec{p}|\vec{p} + \vec{\mu})\mathrm{d}\vec{p} \tag{21}$$

What does Fourier transformation of the momentum density yield? This question has been considered [29] in some generality and detail. Here we merely summarize the outcome for the one-electron momentum density [28,29]. Consider the Fourier transform, or characteristic function in the terminology of probability theory, of $\Pi(\vec{p})$:

$$B(\vec{s}) = \int \mathrm{e}^{-i\vec{s}\cdot\vec{p}}\,\Pi(\vec{p})\mathrm{d}\vec{p} \tag{22}$$

Substitution of Eq. (18) in Eq. (22) followed by integration over $\vec{p}$ using the Fourier representation of the Dirac delta function

$$\delta(\vec{r}) = (2\pi)^{-3} \int \mathrm{e}^{i\vec{p}\cdot\vec{r}}\mathrm{d}\vec{p} \tag{23}$$

gives

$$B(\vec{s}) = \int \Gamma(\vec{r}|\vec{r}\,')\delta(\vec{r}\,' - \vec{r} - \vec{s})\mathrm{d}\vec{r}\,\mathrm{d}\vec{r}\,' \tag{24}$$

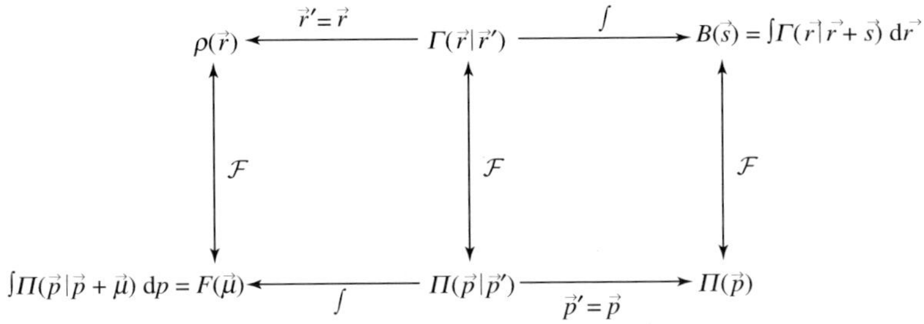

Fig. 19.1. Connections among $\vec{r}$- and $\vec{p}$-space densities, density matrices, and form factors. Two-headed arrows signify reversible Fourier transformations $\mathcal{F}$ whereas single-barbed arrows signify irreversible integrations or contractions.

and hence

$$B(\vec{s}) = \int \Gamma(\vec{r}|\vec{r}+\vec{s})\mathrm{d}\vec{r} \tag{25}$$

$B(\vec{s})$ is variously called the reciprocal form factor, the p-space form factor, and the internally folded density. $B(\vec{s})$ is the basis of a method for reconstructing momentum densities from experimental data [33,34], and it is useful for the r-space analysis of Compton profiles [35–37].

Fig. 19.1 provides a concise summary of these relationships. A more elaborate figure that adds the connections to the Wigner [38,39] and Moyal [40] mixed position–momentum representations of the first-order reduced density matrix can be found in an article that also works out all these functions in closed form for a simple harmonic model of the helium atom [41].

If the system is in a pure spin state, that is $\hat{S}^2\Psi = S(S+1)\Psi$ and $\hat{S}_z\Psi = M_s\Psi$, then the electron density can be decomposed into spin components as follows

$$\rho(\vec{r}) = \rho_\alpha(\vec{r}) + \rho_\beta(\vec{r}) \tag{26}$$

and it is then conventional to introduce the spin density defined by

$$\rho_s(\vec{r}) = \rho_\alpha(\vec{r}) - \rho_\beta(\vec{r}). \tag{27}$$

Similarly, the corresponding decomposition of the momentum density is

$$\Pi(\vec{p}) = \Pi_\alpha(\vec{p}) + \Pi_\beta(\vec{p}) \tag{28}$$

and the spin momentum density is

$$\Pi_s(\vec{p}) = \Pi_\alpha(\vec{p}) - \Pi_\beta(\vec{p}) \tag{29}$$

Sometimes a $(2M_s)^{-1}$ factor is included in the definitions of the spin densities, Eqs. (27) and (29).

19.4 PROPERTIES OF THE MOMENTUM DENSITY

Within the Born–Oppenheimer approximation, the nuclear configuration of a molecule may have geometrical symmetry described by the point group $\mathcal{G}_n$. In the absence of degeneracies, this guarantees that both the exact number density $\rho(\vec{r})$ and the exact momentum density $\Pi(\vec{p})$ transform as the totally symmetric irreducible representation of $\mathcal{G}_n$. If there are degeneracies, averages of the densities over the ensemble of degenerate states exhibit the same symmetry properties. Densities obtained from approximate wave functions also have this symmetry provided that the underlying basis set is closed under the symmetry operations of $\mathcal{G}_n$, and that there are no methodological symmetry-breaking artifacts.

The one-electron momentum density for bound states of atoms and molecules always has inversion symmetry as a consequence of time-reversal symmetry or the principle of microreversibility [42,43]:

$$\Pi(\vec{p}) = \Pi(-\vec{p}) \tag{30}$$

Most approximate wave functions lead to momentum densities that have inversion symmetry. Within the Born–Oppenheimer approximation, the total electronic system must be at rest, and the at-rest condition

$$\int \Pi(\vec{p})\vec{p}\,\mathrm{d}\vec{p} = 0 \tag{31}$$

follows from the inversion symmetry of the momentum density [44]. Another consequence of inversion symmetry is that $\vec{p} = \vec{0}$ is always a critical point of $\Pi(\vec{p})$ because the gradient of $\Pi(\vec{p})$ vanishes at $\vec{p} = \vec{0}$. By contrast, the number density always has critical points, in fact maxima, at the locations of the nuclei [45].

When the geometrical and inversion symmetries are considered together, it follows that the symmetry of the momentum density $\Pi(\vec{p})$ is $\mathcal{G}_n \otimes C_i$. If $\mathcal{G}_n$ does not contain the inversion operation $\hat{\imath}$, then the symmetry of the momentum density is higher than $\mathcal{G}_n$. For example, a heteronuclear diatomic molecule like HF has $C_{\infty v}$ symmetry as does its electron density $\rho(\vec{r})$, but its momentum density $\Pi(\vec{p})$ has $C_{\infty v} \otimes C_i = \mathcal{D}_{\infty h}$ symmetry, as seen graphically in the work of Thakkar et al. [46]. Further discussions of the symmetry of $\Pi(\vec{p})$ may be found elsewhere [47–50].

In the Born–Oppenheimer approximation, the nuclei are fixed and have zero momenta. So the momentum density $\Pi(\vec{p})$ is an intrinsically one-centered function whereas $\rho(\vec{r})$ is a multi-centered function. Thus one-center expansions in spherical harmonics work well for one-electron momentum densities [51–53]. The leading term of such an expansion is the spherically averaged momentum density $\Pi_0(p)$ defined by

$$\Pi_0(p) = (4\pi)^{-1} \int \Pi(\vec{p})\mathrm{d}\Omega_{\vec{p}} = (4\pi)^{-1} \int_0^{2\pi} \int_0^{\pi} \Pi(\vec{p})\sin\theta_p \mathrm{d}\theta_p \mathrm{d}\varphi_p \tag{32}$$

The small p behavior of $\Pi_0(p)$ is most easily treated with the MacLaurin expansion given by [19,54]:

$$\Pi_0(p) = \Pi_0(0) + \Pi_0''(0)p^2/2! + \mathcal{O}(p^4) \tag{33}$$

in which $\Pi_0(0) = \Pi(\vec{0}) = \Pi(0)$ is the electron momentum density at zero momentum, and $\Pi_0''(0)$ is its curvature at that point. Odd powers of p do not appear in the expansion because of the inversion symmetry of $\Pi(\vec{p})$.

The asymptotic expansion of $\Pi_0(p)$ for large p is given [19,55–57] by

$$\Pi_0(p) = b_8 p^{-8} + b_{10} p^{-10} + \mathcal{O}(p^{-12}) \tag{34}$$

in contrast with the exponential decay of the electron number density. Eq. (34) is best understood by recognizing that an orbital e^{-r} leads to a momental $c(1+p^2)^{-2}$. In atoms, the coefficient of the leading term in Eq. (34) can be expressed in terms of one- and two-electron r-space densities [56,57].

The moments of the electron momentum density are defined by

$$\langle p^k \rangle = \int p^k \Pi(\vec{p}) \mathrm{d}\vec{p} \tag{35}$$

Integration over the spherical polar angles leads to

$$\langle p^k \rangle = 4\pi \int_0^\infty p^{k+2} \Pi_0(p) \mathrm{d}p = \int_0^\infty p^k I(p) \mathrm{d}p \tag{36}$$

in which the radial momentum density $I(p)$ is defined by

$$I(p) = 4\pi p^2 \Pi_0(p). \tag{37}$$

An immediate consequence of the asymptotic behavior of $\Pi_0(p)$ shown in Eqs. (33) and (34) is that the moments $\langle p^k \rangle$ are finite only for $-3 < k < 5$. By contrast, the moments of the electron number density are finite for $-3 < k < \infty$. The $k = 0$ moment of $\Pi(\vec{p})$ is simply the number of electrons because of the normalization condition

$$\langle p^0 \rangle = \int \Pi(\vec{p}) \mathrm{d}\vec{p} = N \tag{38}$$

$\langle p^2 \rangle / 2$ is the electronic kinetic energy, and $-\alpha^2 \langle p^2 \rangle / 8$, where α is the fine structure constant, is the Breit–Pauli correction to the kinetic energy [58–60] due to the relativistic variation of mass with velocity.

19.5 EXPERIMENTAL DETERMINATION OF MOMENTUM DENSITIES

There are several experimental techniques for the measurement of electron momentum densities. Inelastic scattering [54,61–64] of high-energy electrons, X-rays or γ-rays by a molecule allows us to measure the intensity of the Compton scattering at wavelengths shifted, by a Doppler broadening-like mechanism, from the wavelength at which Compton scattering by a motionless electron would be predicted. Provided that the experiment can be analyzed within the impulse approximation, this intensity yields the probability of an electron having a certain momentum—the electron momentum density. The impulse approximation is valid whenever the energy of the incident projectile is much larger than the target electron's binding energy. In that case, the momentum transfer between the electron and incoming particle can be assumed to take place instantaneously, allowing the target electron to be treated as a free electron with the same momentum. Under these circumstances, we can imagine that the projectile catches the target electron between collisions with other particles in the system, and the main effect of the other electrons and nuclei in the molecule is to smear out the probability of finding the target electron with a given momentum $\vec{p}$.

Jesse DuMond [65,66] showed that, within the impulse approximation, the gas-phase Compton profile $J_0(q)$ is related to the isotropic momentum density by

$$J_0(q) = \frac{1}{2}\int_{|q|}^{\infty} p^{-1} I(p)\mathrm{d}p = 2\pi \int_{|q|}^{\infty} p\Pi_0(p)\mathrm{d}p \tag{39}$$

where q is the momentum transfer, and its inverse

$$\Pi_0(p) = \frac{-1}{2\pi p}\left(\frac{\mathrm{d}J_0(q)}{\mathrm{d}q}\right)_p \tag{40}$$

Comparison with Eq. (36) reveals that the peak height of the isotropic Compton profile is half of the $K = -1$ moment,

$$J_0(0) = (1/2)\langle p^{-1}\rangle \tag{41}$$

Other moments of momentum can also be obtained directly from the Compton profile without first going through the numerical differentiation of Eq. (40) which is prone to roundoff and truncation errors. Several groups [55,67–70] independently reported one or more of the sum rules

$$\langle p^k\rangle = 2(k+1)\int_0^{\infty} q^k J_0(q)\mathrm{d}q, \qquad 0 \le k \le 4 \tag{42}$$

The remaining integral moment $\langle p^{-2}\rangle$ can be obtained from a less widely known sum rule [71]

$$\langle p^{-2}\rangle = 2\int_0^{\infty} q^{-2}[J_0(0) - J_0(q)]\mathrm{d}q \tag{43}$$

The Compton profile and the stopping power are connected [72].

Typically the experimental measurements are most accurate at small q. Inserting the MacLaurin expansion, Eq. (33), of $\Pi_0(p)$ into Eq. (39) leads to the MacLaurin expansion [54,73] for the isotropic Compton profile:

$$J_0(q) = J_0(0) - [\pi\Pi_0(0)]q^2 - [\pi\Pi_0''(0)/4]q^4 + \mathcal{O}(q^6) \tag{44}$$

and hence fitting the measured profile at small momentum transfers q to a polynomial in q^2 yields values of $\Pi_0(0)$ and $\Pi_0''(0)$. These coefficients have been extracted from experimental Compton profiles for several atoms and molecules [54]. Insertion of the asymptotic expansion of the momentum density, Eq. (34), into Eq. (39) leads to the large-q expansion [55,73]

$$J_0(q) = (\pi b_8/3)q^{-6} + \mathcal{O}(q^{-8}) \tag{45}$$

No directional information is obtained from gas-phase experiments because the molecules are freely rotating. In the solid state, the rotational motion can be frozen, and the directional Compton profile can be written as:

$$J(q, \vec{q}/q) = \int_S \Pi(\vec{p})\mathrm{d}S \tag{46}$$

where S is the plane surface defined by $\vec{p}\cdot\vec{q} = q^2$. In other words, the Compton profile is given by the momentum density integrated over a plane perpendicular to the scattering vector. In the special case that the scattering vector is parallel to one of the coordinate axes, this expression looks much simpler. For example, if $\vec{q}$ is parallel to the z-axis, the directional Compton profile, expressed in Cartesian coordinates, is simply the marginal momentum density along the p_z-axis:

$$J(0, 0, p_z) = \int \Pi(p_x, p_y, p_z)\mathrm{d}p_x\ \mathrm{d}p_y \tag{47}$$

There are two main methods for the reconstruction of $\Pi(\vec{p})$ from the directional Compton profile. In the Fourier–Hankel method [33,51], a spherical harmonic expansion of the directional Compton profile is inverted term-by-term to obtain the corresponding expansion of $\Pi(\vec{p})$. In the Fourier reconstruction method [33,34], the reciprocal form factor $B(\vec{s})$ is constructed a ray at a time by Fourier transformation of the measured $J(\vec{q})$ along that same direction. Then the electron momentum density is obtained from $B(\vec{s})$ by using the inverse of Eq. (22). A vast number of directional Compton profiles have been measured for ionic and metallic solids, but none for free molecules. Nevertheless, several calculations of directional Compton profiles for molecules have been performed as another means of analyzing the momentum density.

In binary (e,2e) or electron momentum spectroscopy [54,74–77], an incoming electron collides with a molecule, removes an electron and leaves a cation behind. Within the plane-wave impulse approximation, the spherically averaged momentum density of the pertinent Dyson orbital can be extracted from the measured differential cross-section. A Dyson orbital ψ_k is defined by

$$\psi_k = S^{-1} \int \Psi^*_{k+}(\vec{x}_1, \ldots, \vec{x}_{N-1})\Psi(\vec{x}_1, \ldots, \vec{x}_{N-1}, \vec{x}_N)\mathrm{d}\vec{x}_1\ \mathrm{d}\vec{x}_2 \ldots \mathrm{d}\vec{x}_{N-1} \tag{48}$$

where S is a normalization constant, Ψ is the ground state wave function of the target molecule, and Ψ_{k+} is the wave function for some state of the resulting cation. If the neutral molecule is described within the Hartree–Fock model, then the Dyson orbital is a linear combination of the occupied Hartree–Fock orbitals. Further, if Koopmans' approximation is invoked to describe the cation by a Slater determinant constructed from the canonical Hartree–Fock orbitals of the parent molecule, then the Dyson orbital is just a canonical Hartree–Fock orbital of the target molecule. When all these approximations are valid, (e,2e) spectroscopy can be thought of as a technique for looking at or imaging orbitals in the laboratory.

Positron annihilation techniques [78] can be used to obtain information about the momentum density of the annihilating positron–electron pair. In solids, particularly metals, the distortion of the electron momentum density by the Coulomb interaction between the positron and electrons is relatively small, and this technique then gives us the electron momentum density.

19.6 *AB INITIO* COMPUTATIONS

Application of the Fourier transformation technique to the overwhelming majority of r-space wave functions that are constructed from a one-particle basis set $\{\chi_j(\vec{r}), j = 1, ..., n\}$ is straightforward. For such wave functions the r-space one-electron density matrix can be expressed in the underlying basis set as:

$$\Gamma(\vec{r}|\vec{r}\,') = \sum_{i=1}^{n} \sum_{j=1}^{n} P_{ij}\chi_j^*(\vec{r}\,')\chi_i(\vec{r}) \tag{49}$$

Substitution of Eq. (49) in Eq. (14) leads to an isomorphic expression for the p-space one-electron density matrix:

$$\Pi(\vec{p}|\vec{p}\,') = \sum_{i=1}^{n} \sum_{j=1}^{n} P_{ij}\omega_j^*(\vec{p}\,')\omega_i(\vec{p}) \tag{50}$$

where

$$\omega_j(\vec{p}) = (2\pi)^{-3/2} \int \chi_j(\vec{r})\exp(-i\vec{p}\cdot\vec{r})\mathrm{d}\vec{r} \tag{51}$$

The momentum density is then simply

$$\Pi(\vec{p}) = \sum_{i=1}^{n} \sum_{j=1}^{n} P_{ij}\omega_j^*(\vec{p})\omega_i(\vec{p}) \tag{52}$$

Many of the off-diagonal terms in Eq. (52) are complex-valued, even when the r-space basis functions and expansion coefficients are all real. However, the momentum density is always real because each off-diagonal ij term in Eq. (52) is the complex conjugate of the corresponding ji term.

The Fourier transforms of Eq. (51) can be performed in closed form for most commonly used basis sets. Moreover, formulas and techniques for the computation of the spherically averaged momentum density, isotropic and directional Compton profiles, and momentum moments have been worked out for both Gaussian- and Slater-type basis sets. Older work on the methods and formulas has been summarized in a review article by Kaijser and Smith [79]. A bibliography of more recent methodological work can be found in another review article [11]. Advantages and disadvantages of various types of basis sets, including many unconventional ones, have been analyzed from a momentum–space perspective [80–82]. Section 19.7 describes several illustrative computations chosen primarily from my own work for convenience.

19.7 ILLUSTRATIVE CALCULATIONS

The exact, non-relativistic, ground-state momental for the hydrogen atom and isoelectronic cations with nuclear charge Z, first worked out by Podolsky and Pauling [17], is important because it provides a hook for one's intuition. The r-space wave function is spherically symmetric and given by

$$\Psi(r) = Z^{3/2}\pi^{-1/2}\exp(-Zr) \tag{53}$$

and the corresponding p-space wave function is also spherically symmetric and given by

$$\Phi(p) = \frac{2\sqrt{2}}{\pi Z^{3/2}(1 + p^2/Z^2)^2} \tag{54}$$

The electron number density is exponentially decaying and given by

$$\rho(r) = Z^3\pi^{-1}e^{-2Zr} \tag{55}$$

whereas the electron momentum density has a p^{-8} decay and is

$$\Pi(p) = \frac{8}{\pi^2 Z^3(1 + p^2/Z^2)^4} \tag{56}$$

The isotropic Compton profile has a Lorentzian shape given by

$$J_0(q) = \frac{8}{3\pi Z(1 + q^2/Z^2)^3} \tag{57}$$

the momentum moments are given by

$$\begin{aligned} \langle p^{-2}\rangle &= 5/Z^2, & \langle p^{-1}\rangle &= 16/(3\pi Z), \\ \langle p\rangle &= 8Z/(3\pi), & \langle p^2\rangle &= Z^2, \\ \langle p^3\rangle &= 16Z^3/(3\pi), & \langle p^4\rangle &= 5Z^4 \end{aligned} \tag{58}$$

and the reciprocal form factor by

$$B(r) = e^{-Zr}(1 + Zr + Z^2r^2/3) \tag{59}$$

The reciprocal nature of r- and p-space densities is seen clearly from Eqs. (55) and (56)—as the nuclear charge is increased the momentum density becomes more and more diffuse whereas the electron number density becomes more and more compact.

For the hydrogen atom, the ground state wave function is the Dyson orbital. Remarkably detailed agreement has been found with non-coplanar symmetric (e,2e) experiments on atomic hydrogen over a wide range of total energies [83]. In this case, it is legitimate to say that an experimental determination of the p-space wave function for H has been carried out and agrees well with Eq. (54). This is a tremendously important experiment because it provides the first direct confirmation of the ground state wave function for the hydrogen atom obtained by solution of the Schrödinger equation.

Another manifestation of the reciprocity of densities in r- and p-space is provided by Fig. 19.2. It shows the radial electron number density $D(r) = 4\pi r^2\rho(r)$ and radial momentum density $I(p) = 4\pi p^2\Pi(p)$ for the ground state of the beryllium atom calculated within the Hartree–Fock model in which the Be ground state has a $1s^2 2s^2$ configuration. Both densities show a peak arising from the 1s core electrons and another from the 2s valence electrons. However, the origin of the peaks is reversed. The sharp,

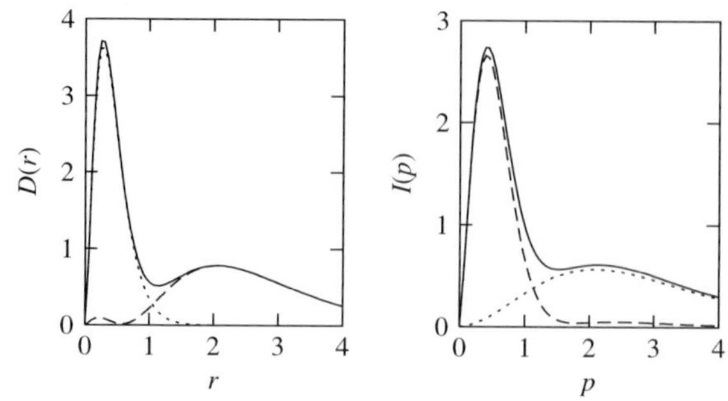

Fig. 19.2. The radial electron density $D(r) = 4\pi r^2 \rho(r)$ (left) and radial momentum density $I(p) = 4\pi p^2 \Pi(p)$ (right) for ground state Be. 1s contribution (dotted), 2s contribution (dashed), and total (solid). Adapted from Thakkar [11].

small-r peak in $D(r)$ is primarily due to the core electrons whereas the sharp, small-p maximum in $I(p)$ is primarily due to the valence shell electron pair. This is a general phenomenon and is often referred to as reciprocity of the number and momentum densities.

Numerical Hartree–Fock calculations, free from basis set artifacts, have been used to establish that the ground state momentum densities of all the atoms and their ions can be classified into three types [84,85]. Type I and III momentum densities are found almost exclusively in metal atoms: He, N, all atoms from groups 1–14 except Ge and Pd, and all the lanthanides and actinides. These momentum densities all have a global maximum at $p = 0$ and resemble the momentum density shown in Fig. 19.3 for the beryllium atom. The maximum at $p = 0$ comes mainly from the outermost s-subshell, 2s in this case. Type I and III densities differ in that the latter have a secondary maximum that is so small as to be invisible on a diagram such as Fig. 19.3. Type II densities are the norm for non-metallic atoms and are found in Ge, Pd and all atoms from groups 15–18 except He and N. Type II densities have a local minimum at $p = 0$ and a global maximum at $p_{\max} > 0$. A representative density of this type is shown in Fig. 19.3 for the neon atom.

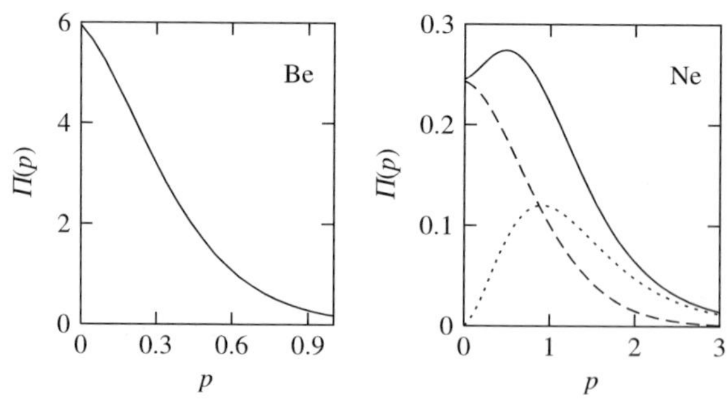

Fig. 19.3. Types of electron momentum densities $\Pi(\vec{p}) = \Pi_0(p)$ in atoms. Solid lines are used for the total density. Left: a Type I density for the beryllium atom; the contribution from the 2s orbital is indistinguishable from the total density. Right: a typical Type II density for the neon atom; the 2s and 2p contributions are shown as dashed and dotted lines, respectively.

The maximum arises because the unimodal contribution from the outermost p-subshell, 2p in this case, dominates the monotonic contribution from the outermost s-subshell. All Type I and III momentum densities have $\Pi_0''(0) < 0$ and all Type II densities have $\Pi_0''(0) > 0$.

The qualitative study of electronic structure through the electron momentum density $\Pi(\vec{p})$ relies heavily on graphical representations of the density and density differences. Coulson [18,86] made the pioneering effort in this direction. Early work was carried out by Bill Henneker and Paul Cade [87], and Irving Epstein and Bill Lipscomb [88–90]. Other early work by Henneker and Cade was unpublished but widely circulated in preprint form during the 1970s. A synthesis of the early work was made by Epstein and Tony Tanner [91] who abstracted some principles that they hoped would be generally applicable to chemical bonding. One of their abstractions pinned down an observation about the anisotropy of $\Pi(\vec{p})$ that had been made in several of the earlier studies. Epstein and Tanner called it the bond directional principle and they stated it as follows [91]:

> The momentum of an electron in a chemical bond is more likely to be directed perpendicular to than along the bond axis. Furthermore, in the chemical bond there is greater density at low momentum along the bond and greater density at high momentum perpendicular to the bond than was the case in the isolated atoms.

This principle served as an inspiration for many researchers interested in uncovering the links between the reorganization of electron momentum densities and chemical bonding. Unfortunately, it turned out to have many exceptions. Tanner [92] carefully reassessed it and formulated a new version as follows:

> In a chemical bond in a bound molecule at its equilibrium configuration, there are values $\vec{p}_m$ of momentum which are more probable, i.e. which correspond to local maxima of $\Pi(\vec{p})$. Those values are determined by both the geometric and electronic symmetries of the molecule. For momenta, $\vec{p} = \vec{p}_m + \delta\vec{p}$, near a maximum it is more likely that $\delta\vec{p}$ is perpendicular rather than parallel to the bond axis.

Tanner emphasized that although the revised principle has a few known exceptions, attempts to make it more precise would only serve to defeat its purpose which is to give a qualitative feel for $\Pi(\vec{p})$.

A parallel development to the study [45] of the topography of $\rho(\vec{r})$ is the consideration [49,93] of the topography of $\Pi(\vec{p})$. The topography is characterized in terms of the Hessian of $\Pi(\vec{p})$ at its critical points, that is the set of points $\{\vec{p}_c\}$ for which $\nabla\Pi(\vec{p}_c) = \vec{0}$. Recall that $\vec{p} = \vec{0}$, that is $p = 0$, is always a critical point because of the inversion symmetry of $\Pi(\vec{p})$. Topographic studies [49,94] of $\Pi(\vec{p})$ find that the $p = 0$ critical point is perhaps the most important one because it is a harbinger of the other critical points [95] in the sense that the nature of the other critical points can be predicted from the type of critical point found at $p = 0$. Connections between the zero-momentum critical point and the bond-directional principle [91,92] have been explored [95,96]. Unfortunately, simple rules are hard to find and all allowed types of topographies are found even in linear molecules [97]. The value of $\Pi(0)$ has been found to be a useful [98,99] measure for following a chemical reaction in p-space.

The high sensitivity of the topography of $\Pi(\vec{p})$ to computational details has been examined. A critical point is characterized by the rank r and signature s of the Hessian

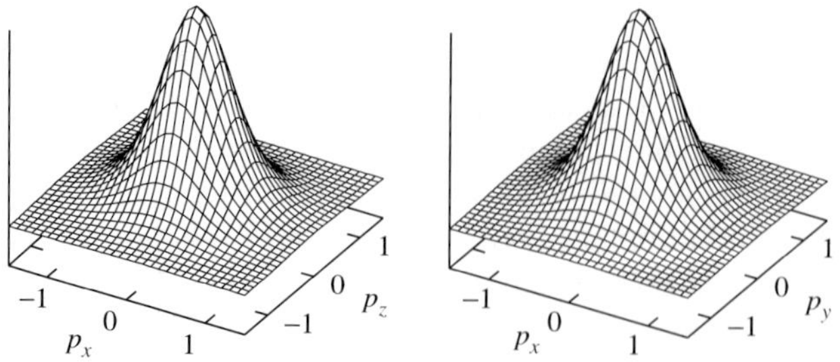

Fig. 19.4. Surface plots of the electron momentum density of H_2 illustrating a (3, −3) maximum at $\vec{p} = \vec{0}$. The left and right plots are in planes parallel and perpendicular to the molecular axis: $\Pi(p_x, 0, p_z)$ and $\Pi(p_x, p_y, 0)$, respectively.

matrix at $\vec{p}_c$. The rank is the number of non-zero eigenvalues and the signature is the number of positive eigenvalues minus the number of negative ones. Even the signature of a critical point can vary with the basis set and amount of electronic correlation included in a calculation [96,100–102]. Formaldehyde, H_2CO, is an egregious example in which the signature was found to take on three (±1, −3) of its four possible values (±1, ±3) depending on the basis set used and amount of electron correlation included in the computation [102]. Clearly, there is a need for careful development of basis sets for the computation of the zero-momentum Hessian of $\Pi(\vec{p})$. There are some numerical Hartree–Fock computations, free of basis set errors, for 78 diatomic molecules [103] that should be helpful for the calibration of such basis sets.

Provided that the Hessian is of full rank ($r = 3$), that is none of its eigenvalues vanish, there can be four types of critical points at $p = 0$ corresponding to the four possible values of the signature ($s = \pm 3, \pm 1$): maxima, minima, and ($r = 3$, $s = \pm 1$) saddle points. Graphical examples of each type were given earlier [97,11]. Fresh examples are given here. H_2 has a ($r = 3$, $s = -3$) global maximum at $p = 0$, and CO_2 has a ($r = 3$, $s = +3$) global minimum at $p = 0$ as seen in Figs. 19.4 and 19.5, respectively. Fig. 19.5 for CO_2 shows that in the vertical $p_x p_z$ plane of symmetry, the local minimum at $p = 0$ is surrounded by a ring of critical points including two local maxima located along the

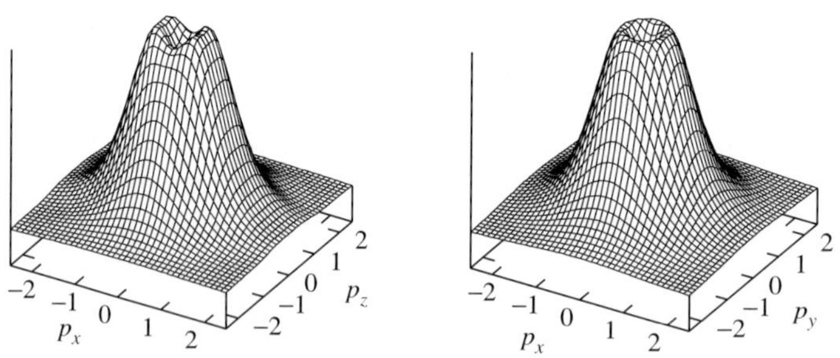

Fig. 19.5. Surface plots of the electron momentum density of CO_2 illustrating a (3, +3) minimum at $\vec{p} = \vec{0}$. The left and right plots are in planes parallel and perpendicular to the molecular axis: $\Pi(p_x, 0, p_z)$ and $\Pi(p_x, p_y, 0)$, respectively.

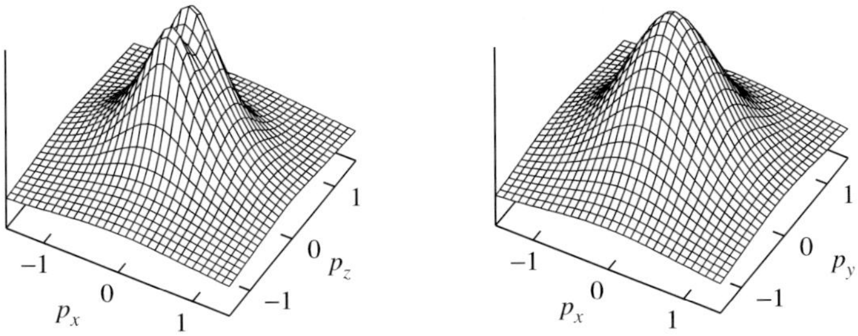

Fig. 19.6. Surface plots of the electron momentum density of MgO illustrating a (3, − 1) saddle point at $\vec{p} = \vec{0}$. The left and right plots are in planes parallel and perpendicular to the molecular axis: $\Pi(p_x, 0, p_z)$ and $\Pi(p_x, p_y, 0)$, respectively.

p_z axis. Critical points along the p_x axis must also appear in all other directions perpendicular to the p_z axis as seen in the plot for the horizontal p_xp_y plane of symmetry.

Typical electron momentum densities with (3, − 1) and (3, + 1) saddle points at zero momentum are found in MgO and acetylene (HCCH), respectively. A momentum density with a zero momentum (3, − 1) critical point is shown for MgO in Fig. 19.6. In the vertical plane of symmetry $\Pi(p_x, 0, p_z)$ has the structure of two hills separated by a ridge or col, and one sees two local (and global) maxima located symmetrically along the p_z axis. The plot in the horizontal symmetry plane has the structure of a hill.

The electron momentum density for acetylene is shown in Fig. 19.7. The plot in the p_xp_z plane has a complicated structure. The zero-momentum (3, + 1) critical point is a pass between two peaks located along the p_x axis, and is a barrier separating two troughs centered along the p_z axis. Critical points along the p_x axis must also appear in all other directions perpendicular to the p_z axis. The plot of $\Pi(\vec{p})$ in the horizontal p_xp_y plane resembles a volcano.

A good example of quantitative work is the calculation of the isotropic Compton profile for N_2. The deviations of five different gas-phase measurements from the latest calculations are shown in Fig. 19.8. The computations [104] used a multi-reference

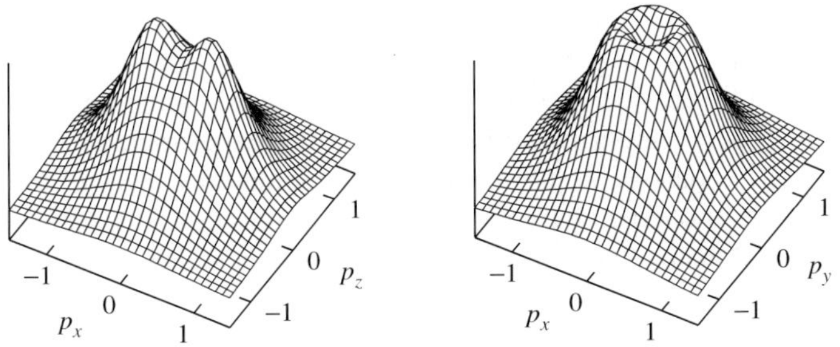

Fig. 19.7. Surface plots of the electron momentum density of HCCH illustrating a (3, + 1) saddle point at $\vec{p} = \vec{0}$. The left and right plots are in planes parallel and perpendicular to the molecular axis: $\Pi(p_x, 0, p_z)$ and $\Pi(p_x, p_y, 0)$, respectively.

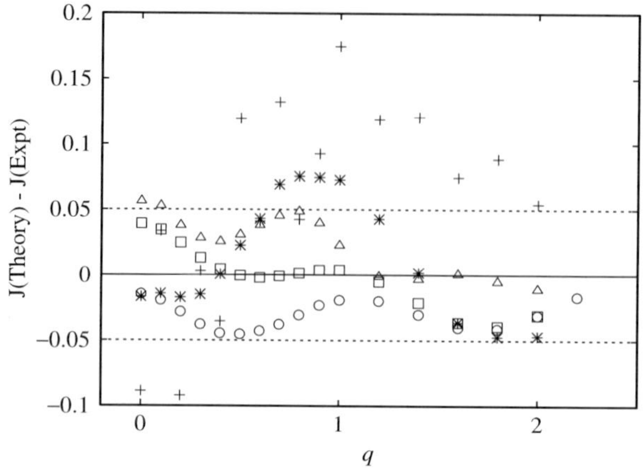

Fig. 19.8. Differences between the isotropic Compton profile computed from theory and various experiments. The theoretical data is from a multi-reference configuration interaction (MRCI) calculation in a (5s5p4d3f) basis set of Slater-type functions [104]. Experimental data: (+), 25 keV electron impact at 12° [109]; (∗), average of Ag $K\alpha$ and Mo $K\alpha$ X-ray scattering [108,105]; (□), 160 keV γ-ray scattering [105]; (○), 160 keV γ-ray scattering reanalyzed [105,106]; (△), 60 keV γ-ray scattering [107]. The dotted lines enclose the band of uncertainty in the experimental data.

configuration interaction (MRCI) wave function constructed from a (5s5p4d3f) basis set of Slater-type functions, referred to as T hereafter. The dotted lines in Fig. 19.8 indicate $\pm 0.01 J_0(0) = 0.05$ which is the estimated uncertainty in the experimental values. There is good agreement between theory and the three sets of γ-ray data [105–107]. However, the X-ray [108] and particularly the electron impact data [109] are not in totally satisfactory agreement with the calculations.

Table 19.1 shows a comparison between the experimental values and a representative subset of the theoretical values of $\Pi_0(0)$ and $\Pi_0''(0)$ for N_2. Comparison of the basis set-free numerical Hartree–Fock value [103] of $\Pi_0''(0)$ with the self-consistent field (SCF) value in a (5s5p4d3f) basis set of Slater-type functions shows that the latter is significantly in error. This is due to the lack of diffuse functions as shown by the significantly better value obtained with Dunning's aug-cc-pVTZ Gaussian basis set [110] which does contain diffuse functions. Comparison of the SCF and correlated values of $\Pi_0''(0)$ shows that electron correlation effects are about 50% of the SCF value or larger. All the correlated values of $\Pi_0(0)$ lie outside the error bars of the 160 keV γ-ray data but within the error bars of the other three experimental values. Further experimental work would be helpful. The correlated values of $\Pi_0''(0)$ lie within the wide error bars of the electron impact data but outside the error bars of the remaining data. This is less disturbing since numerical second derivatives extracted from experimental data cannot be expected to be very reliable.

There are many cases of good agreement between MRCI calculations of Dyson orbitals and those obtained from (e,2e) experiments. A perceptive survey of the insights into quantum chemistry gained from these computations has been given by Davidson [111].

Table 19.1 Coefficients in the small-p expansion, Eq. (33), of the momentum density for ground state N_2

Method/Basis	R	$\Pi_0(0)$	$\Pi_0''(0)$
SCF/T[a]	2.068	1.449	−3.76
SCF/aug-cc-pVTZ[b]	2.074	1.434	−3.22
Numerical HF[c]	2.068	1.436	−3.07
MRCI4/T[a]	2.068	1.528	−5.60
MRCI4/T+zero point vibration[a]	2.068	1.528	−5.55
SDQMP4/aug-cc-pVTZ[b]	2.074	1.534	−5.28
160 keV γ-rays[d]		1.39 ± 0.04	-2.9 ± 1.5
160 keV γ-rays[e]		1.40 ± 0.09	-2.8 ± 1.0
60 keV γ-rays[f]		1.40 ± 0.17	-2.7 ± 1.4
Ag $K\alpha$ and Mo $K\alpha$ X-rays[g]		1.51 ± 0.20	-3.1 ± 1.3
25 keV electron impact[h]		1.56 ± 0.31	-4.0 ± 2.0

[a]Ref. [104]. T is a (5s5p4d3f) basis set of Slater-type functions. SCF is self-consistent field, MRCI is multi-reference configuration interaction.
[b]Ref. [102]. The aug-cc-pVTZ basis set of Gaussian-type functions is from Dunning [110]. SDQMP4 is fourth-order Møller–Plesset perturbation theory excluding triples.
[c]Ref. [103].
[d]Ref. [105].
[e]Data of Ref. [105] reanalyzed in Ref. [106].
[f]Ref. [107].
[g]Data of Ref. [108] as reported in Ref. [105].
[h]Ref. [109].

Finally, we consider density functional theory (DFT) computations of p-space properties. A naive way of calculating p-space properties is to use the Kohn–Sham orbitals obtained from a DFT computation to form a one-electron, r-space density matrix $\Gamma(\vec{r}|\vec{r}')$, Fourier transform Γ according to Eq. (14), and proceed further. This approach is incorrect because the Kohn–Sham density matrix Γ is not the true one and, in fact, corresponds to a fictitious non-interacting system with the same $\rho(\vec{r})$ as the true system. On the other hand, Hamel and coworkers [112] have shown that if the exact Kohn–Sham exchange potential is used, then the spherically averaged momentum densities of the Kohn–Sham orbitals should be very close to those of the Hartree–Fock orbitals. Of course, in practical computations the exact Kohn–Sham exchange potential is not used since it is generally not known.

In practice, using currently available exchange and correlation potentials, this path leads to results [113] worse than those obtained with the Hartree–Fock method. This is illustrated for momentum moments in Table 19.2 which shows median absolute percent errors of $\langle p^k \rangle$ for 78 molecules relative to those computed by an approximate singles and doubles coupled-cluster method often called QCISD [114,115]. The molecules are mostly polyatomic, and contain H, C, N, O, and F atoms. The correlation-consistent cc-pVTZ basis set [110] was used for these computations. Table 19.2 shows the median errors for the Hartree–Fock method, for second-order Møller–Plesset perturbation theory (MP2), and for DFT calculations done with the B3LYP hybrid density functional [116,117] which is based upon the adiabatic connection, and uses a mixture of Hartree–Fock exchange and generalized gradient approximations (GGA) for exchange

Table 19.2 Mean absolute percent errors of $\langle p^k \rangle$ for 78 molecules relative to coupled cluster QCISD values in the cc-pVTZ basis set

Method	$\langle p^{-2} \rangle$	$\langle 1/p \rangle$	$\langle p \rangle$	$\langle p^2 \rangle$	$\langle p^3 \rangle$	$\langle p^4 \rangle$
HF	0.613	0.522	0.503	0.444	0.217	0.132
B3LYP	1.561	0.755	0.345	0.320	0.279	0.264
MP2	0.602	0.246	0.066	0.042	0.052	0.049
SDQMP4	0.059	0.020	0.007	0.006	0.006	0.006

Adapted from Ref. [113]

and correlation. Clearly the B3LYP results are worse than Hartree–Fock for $\langle p^k \rangle$, $k = -2, -1, 3, 4$, and worse than MP2 in all cases.

What is needed for a correct computation of momentum–space properties from DFT is an accurate functional for approximating the exact first-order reduced density matrix $\Gamma(\vec{r}|\vec{r}\,')$, or failing that, good functionals for each of the p-space properties of interest. Of course, a sufficiently good functional for $\langle p^2 \rangle$ would obviate the necessity of using Kohn–Sham orbitals and enable the formulation of an orbital-free DFT. Unfortunately, a kinetic energy functional sufficiently accurate for chemical purposes remains an elusive goal [118,119].

19.8 CONCLUDING REMARKS

The momentum perspective is not a transparent one and perhaps this is why it has attracted a minuscule amount of attention in comparison with that devoted to the position space perspective. Despite this, some useful insights into the electronic structure of molecules have been achieved by taking the electron momentum density viewpoint. Momentum space is an area of opportunity for young scientists. Applied mathematicians and numerical analysts should consider work on the integral Schrödinger equation in p-space. Quantum chemists should use recent developments in computer hardware, quantum chemical methods, software for generating wave functions, and visualization software to mount a sustained effort to understand momentum densities from a chemical perspective. A recent review article [11] that has more than 400 references is a good starting point for familiarizing oneself with the literature.

19.9 ACKNOWLEDGEMENTS

This work was supported by the Natural Sciences and Engineering Research Council of Canada.

19.10 REFERENCES

1 E.A. Hylleraas, Z. Phys., 74 (1932) 216.
2 V. Fock, Z. Phys., 98 (1935) 145.

3 M. Levy, Proc. R. Soc. London, 204 (1950) 145.
4 N. Svartholm, Ark. Mat. Astron. Fys., 35A (7) (1947) 1.
5 N. Svartholm, Ark. Mat. Astron. Fys., 35A (8) (1947) 1.
6 R. McWeeny and C.A. Coulson, Proc. Phys. Soc., London, 62A (1949) 509.
7 R. McWeeny, Proc. Phys. Soc., London, 62A (1949) 519.
8 T. Shibuya and C.E. Wulfman, Proc. R. Soc. London, Ser. A, 286 (1965) 376.
9 B.K. Novosadov, Opt. Spectrosc., 41 (1976) 490.
10 B.K. Novosadov, J. Mol. Struct., 52 (1979) 119.
11 A.J. Thakkar, Adv. Chem. Phys., 128 (2004) 303.
12 B.H. Armstrong, Bull. Am. Phys. Soc., 9 (1964) 401.
13 P.A.M. Dirac, Proc. R. Soc., A113 (1927) 621.
14 P.A.M. Dirac, The principles of quantum mechanics, 4th edn., Oxford University Press, London, 1958.
15 P. Jordan, Z. Phys., 40 (1927) 809.
16 P. Jordan, Z. Phys., 44 (1927) 1.
17 B. Podolsky and L. Pauling, Phys. Rev., 34 (1929) 109.
18 C.A. Coulson, Proc. Cambridge Philos. Soc., 37 (1941) 55.
19 A.J. Thakkar, J. Chem. Phys., 86 (1987) 5060.
20 R. Benesch, J. Phys. B, 9 (1976) 2587.
21 B. Jeziorski and K. Szalewicz, Phys. Rev. A, 19 (1979) 2360.
22 A.J. Thakkar, Chem. Phys. Lett., 381 (2003) 80.
23 K. Husimi, Proc. Phys. Math. Soc. Jpn, 22 (1940) 264.
24 P.O. Löwdin, Phys. Rev., 97 (1955) 1474.
25 R. McWeeny, Rev. Mod. Phys., 32 (1960) 335.
26 A.J. Coleman, Rev. Mod. Phys., 35 (1963) 668.
27 E.R. Davidson, Reduced density matrices in quantum chemistry, Academic Press, New York, 1976.
28 R. Benesch, S.R. Singh and V.H. Smith, Jr., Chem. Phys. Lett., 10 (1971) 151.
29 A.J. Thakkar, A.M. Simas and V.H. Smith, Jr., Chem. Phys., 63 (1981) 175.
30 R. Benesch and V.H. Smith, Jr., Int. J. Quantum Chem. Symp., 4 (1970) 131.
31 R. Benesch and V.H. Smith, Jr., Chem. Phys. Lett., 5 (1970) 601.
32 P. Hohenberg and W. Kohn, Phys. Rev., 136B (1964) 864.
33 P.E. Mijnarends, in: B.G. Williams (Ed.), Compton scattering: the investigation of electron momentum distributions, McGraw-Hill, New York, 1977, pp. 323–345.
34 F.M. Mueller, Phys. Rev. B, 15 (1977) 3039.
35 P. Pattison and B. Williams, Solid State Commun., 20 (1976) 585.
36 W. Schülke, Phys. Stat. Sol. (b), 82 (1977) 229.
37 P. Pattison, W. Weyrich and B. Williams, Solid State Commun., 21 (1977) 967.
38 E.P. Wigner, Phys. Rev., 40 (1932) 749.
39 M. Hillery, R.F. O'Connell, M.O. Scully and E.P. Wigner, Phys. Rep., 106 (1984) 121.
40 J.E. Moyal, Proc. Cambridge Philos. Soc., 45 (1949) 99.
41 A.J. Thakkar, A.C. Tanner and V.H. Smith, Jr., in: R.M. Erdahl, V.H. Smith, Jr. (Eds.), Density matrices and density functionals, Reidel, Dordrecht, Holland, 1987, pp. 327–337.
42 P.O. Löwdin, Adv. Quantum Chem., 3 (1967) 323.
43 A. Messiah, Quantum mechanics, North Holland, Amsterdam, 1970.
44 P. Kaijser, V.H. Smith, Jr., in: J.L. Calais, O. Goscinski, J. Linderberg, Y. Öhrn (Eds.), Methods and structure in quantum science, Plenum Press, New York, 1976, pp. 417–426.
45 R.F.W. Bader, Atoms in molecules: A quantum theory, Clarendon Press, Oxford, 1990.
46 A.J. Thakkar, A.M. Simas and V.H. Smith, Jr., J. Chem. Phys., 81 (1984) 2953.
47 D.C. Rawlings and E.R. Davidson, J. Phys. Chem., 89 (1985) 969.
48 M. Defranceschi and G. Berthier, J. Phys. (Paris), 51 (1990) 2791.
49 S.R. Gadre, A.C. Limaye and S.A. Kulkarni, J. Chem. Phys., 94 (1991) 8040.
50 J.L. Calais, M. Defranceschi, J.G. Fripiat and J. Delhalle, J. Phys. Condens. Matter, 4 (1992) 5675.
51 P.E. Mijnarends, Phys. Rev., 160 (1967) 512.
52 A. Seth and D.E. Ellis, Phys. Rev. A, 13 (1976) 1083.

53 E. Heuser-Hofmann and W. Weyrich, Z. Naturfors Sect. A, 40 (1985) 99.
54 R.A. Bonham, H.F. Wellenstein, in: B.G. Williams (Ed.), Compton scattering: the investigation of electron momentum distributions, McGraw-Hill, New York, 1977, pp. 234–272.
55 R. Benesch, V.H. Smith, Jr., in: W.C. Price, S.S. Chissick, T. Ravensdale (Eds.), Wave mechanics—the first fifty years, Butterworths, London, 1973, pp. 357–377.
56 J.C. Kimball, J. Phys. A, 8 (1975) 1513.
57 L. Mendelsohn, V.H Smith, Jr., in: B.G. Williams (Ed.), Compton scattering: the investigation of electron momentum distributions, McGraw-Hill, New York, 1977, pp. 103–138.
58 G. Breit, Phys. Rev., 34 (1929) 553.
59 H.A. Bethe and E.E. Salpeter, Quantum mechanics of one- and two-electron atoms, Springer, Berlin, 1957.
60 R. McWeeny and B.T. Sutcliffe, Methods of molecular quantum mechanics, Academic Press, London, 1969.
61 B.G. Williams (Ed.), Compton scattering: the investigation of electron momentum distributions, McGraw-Hill, New York, 1977.
62 M.J. Cooper, Rep. Prog. Phys., 48 (1985) 415.
63 M.J. Cooper, Radiat. Phys. Chem., 50 (1997) 63.
64 S. Manninen, J. Phys. Chem. Solids, 61 (2000) 335.
65 J.W.M. DuMond, Phys. Rev., 33 (1929) 643.
66 J.W.M. DuMond, Rev. Mod. Phys., 5 (1933) 1.
67 B.L. Hicks, Phys. Rev., 57 (1940) 665.
68 I.R. Epstein, Phys. Rev. A, 8 (1973) 160.
69 H.F. Wellenstein, R.A. Bonham and R.C. Ulsh, Phys. Rev. A, 8 (1973) 304.
70 N.H. March, J. Chem. Phys., 79 (1983) 3404.
71 A.J. Thakkar, A.M. Simas and V.H. Smith, Jr., Mol. Phys., 43 (1980) 1153.
72 V.H. Smith, Jr., Int. J. Quantum Chem. Symp., 23 (1989) 553.
73 A.J. Thakkar, A.L. Wonfor and W.A. Pedersen, J. Chem. Phys., 87 (1987) 1212.
74 I.E. McCarthy and E. Weigold, Phys. Rep., 27 (1976) 275.
75 M.A. Coplan, J.H. Moore and J.P. Doering, Rev. Mod. Phys., 66 (1994) 985.
76 E. Weigold and I.E. McCarthy, Electron momentum spectroscopy, Kluwer Academic Publishers, New York, 1999.
77 M.J. Brunger and W. Adcock, J. Chem. Soc. Perkin Trans., 2 (2002) 1.
78 S. Berko, in: B.G. Williams (Ed.), Compton scattering: the investigation of electron momentum distributions, McGraw-Hill, New York, 1977, pp. 273–322.
79 P. Kaijser and V.H. Smith, Jr., Adv. Quantum Chem., 10 (1977) 37.
80 A.M. Simas, A.J. Thakkar and V.H. Smith, Jr., Int. J. Quantum Chem., 21 (1982) 419.
81 A.M. Simas, A.J. Thakkar and V.H. Smith, Jr., Int. J. Quantum Chem., 24 (1983) 527.
82 P.E. Regier, J. Fisher, B.S. Sharma and A.J. Thakkar, Int. J. Quantum Chem., 28 (1985) 429.
83 B. Lohmann and E. Weigold, Phys. Lett. A, 86 (1981) 139.
84 T. Koga, H. Matsuyama, H. Inomata, E. Romera, J.S. Dehesa and A.J. Thakkar, J. Chem. Phys., 109 (1998) 1601.
85 T. Koga, H. Matsuyama, E. Romera, J.S. Dehesa and A.J. Thakkar, Phys. Rev. A, 59 (1999) 4805.
86 C.A. Coulson, Proc. Cambridge Philos. Soc., 37 (1941) 74.
87 W.H. Henneker and P.E. Cade, Chem. Phys. Lett., 2 (1968) 575.
88 I.R. Epstein, J. Chem. Phys., 52 (1970) 3838.
89 I.R. Epstein and W.N. Lipscomb, J. Chem. Phys., 53 (1970) 4418.
90 I.R. Epstein, J. Chem. Phys., 53 (1970) 4425.
91 I.R. Epstein and A.C. Tanner, in: B.G. Williams (Ed.), Compton scattering: the investigation of electron momentum distributions, McGraw-Hill, New York, 1977, pp. 209–233.
92 A.C. Tanner, Chem. Phys., 123 (1988) 241.
93 R.P. Sagar, A.C.T. Ku, V.H. Smith, Jr. and A.M. Simas, J. Chem. Phys., 90 (1989) 6520.
94 S.A. Kulkarni, S.R. Gadre and R.K. Pathak, Phys. Rev. A, 45 (1992) 4399.
95 S.A. Kulkarni and S.R. Gadre, Z. Naturforsch Sect. A., 48 (1993) 145.

96 J. Wang, B. Clark, H. Schmider and V.H. Smith, Jr., Can. J. Chem., 74 (1996) 1187.
97 A.J. Thakkar, Z. Phys. Chem., 215 (2001) 1265.
98 S.R. Gadre and S.A. Kulkarni, Proc. Ind. Acad. Sci., 105 (1993) 149.
99 S.A. Kulkarni and S.R. Gadre, J. Am. Chem. Soc., 115 (1993) 7434.
100 H. Schmider and M.H. Hô, J. Phys. Chem., 100 (1996) 17807.
101 S.A. Kulkarni and S.R. Gadre, Chem. Phys. Lett., 274 (1997) 255.
102 A.J. Thakkar and B.S. Sharma, J. Mol. Struct. (THEOCHEM), 527 (2000) 221.
103 A.K. Roy and A.J. Thakkar, Chem. Phys. Lett., 362 (2002) 428.
104 A.J. Thakkar and H. Tatewaki, Phys. Rev. A, 42 (1990) 1336.
105 P. Eisenberger and W.A. Reed, Phys. Rev. A, 5 (1972) 2085.
106 P. Eisenberger and W.A. Reed, Phys. Rev. B, 9 (1974) 3237.
107 T. Paakkari and M. Merisalo, Chem. Phys. Lett., 33 (1975) 432.
108 P. Eisenberger, Phys. Rev. A, 5 (1972) 628.
109 T.C. Wong, J.S. Lee, H.F. Wellenstein and R.A. Bonham, Phys. Rev. A, 12 (1975) 1846.
110 T.H. Dunning, Jr., J. Chem. Phys., 90 (1989) 1007.
111 E.R. Davidson, Can. J. Phys., 74 (1996) 757.
112 S. Hamel, P. Duffy, M.E. Casida and D.R. Salahub, J. Electron Spectrosc. Relat. Phenom., 123 (2002) 345.
113 J.R. Hart and A.J. Thakkar, Int. J. Quantum Chem., 102 (2005) 673.
114 J.A. Pople, M. Head-Gordon and K. Raghavachari, J. Chem. Phys., 87 (1987) 5968.
115 J. Paldus, B. Čížek and B. Jeziorski, J. Chem. Phys., 90 (1989) 4356.
116 A.D. Becke, J. Chem. Phys., 98 (1993) 5648.
117 P.J. Stephens, F. Devlin, C. Chabalowski and M. Frisch, J. Phys. Chem., 98 (1994) 11623.
118 A.J. Thakkar, Phys. Rev. A, 46 (1992) 6920.
119 G.K.-L. Chan and N.C. Handy, J. Chem. Phys., 112 (2000) 5639.

Theory and Applications of Computational Chemistry: The First Forty Years
Edited by C. Dykstra et al.

CHAPTER 20

Recent advances in ab initio, *density functional theory, and relativistic electronic structure theory*

Haruyuki Nakano[1,2], Takahito Nakajima[1,3], Takao Tsuneda[1,4] and Kimihiko Hirao[1]

[1] *Department of Applied Chemistry, School of Engineering, University of Tokyo, Tokyo 113-8656, Japan*
[2] *Department of Chemistry, Graduate School of Sciences, Kyushu University, Fukuoka 812-8581, Japan*
[3] *PRESTO, Japan Science and Technology Agency (JST), 4-1-8 Honcho Kawaguchi, Saitama 332-0012, Japan*
[4] *Department of Quantum Engineering and Systems Science, School of Engineering, University of Tokyo, Tokyo 113-8656, Japan*

Abstract

Recent advances in electronic structure theory achieved in our group have been reviewed. Emphasis is put on development of *ab initio* multireference-based perturbation theory, exchange and correlation functionals in density functional theory, and molecular theory including relativistic effects.

20.1 INTRODUCTION

Accurate theoretical/computational chemistry has evolved dramatically and has opened up a world of new possibilities. It can treat real systems with predictive accuracy. Computational chemistry is becoming an integral part of chemistry research. Our research group, the University of Tokyo (UT) group, was founded in 1993. Since then the UT group has continued to grow and now becomes one of the centers of theoretical/computational chemistry. We are interested in theory development and application calculations. We have continued our research in the following three directions: (i) development of new *ab initio* theory, particularly multireference-based perturbation theory; (ii) development of exchange and correlation functionals in density functional theory (DFT); and (iii) development of molecular theory including relativistic effects. We have enjoyed good progress in each of the above areas. We are very excited about our discoveries of new theory and algorithms, and new findings in chemistry. We would like to share this enthusiasm with readers. The present review is a summary of our research activities which have been achieved in the last 10 years at the University of Tokyo.

20.2 MULTIREFERENCE PERTURBATION THEORY AND VALENCE BOND DESCRIPTION OF ELECTRONIC STRUCTURES OF MOLECULES

The development of multireference methods represents important progress in electronic structure theory in the last decades. The multiconfiguration self-consistent field (MCSCF) method, and configuration interaction (CI), coupled cluster (CC), and perturbation methods based on the MCSCF functions play a central role in the studies of electronic structure of molecules and chemical reaction mechanisms, especially in those concerned with electronic excited states.

Among several types of the MCSCF method, the complete active space self-consistent field (CASSCF) method is commonly used at present. In fact, it has many attractive features: (1) applicable to excited state as well as the ground state in a single framework; (2) size-consistent; (3) well defined on the whole potential energy surface if an appropriate active space is selected. However, CASSCF takes into account only non-dynamic electron correlation and not dynamic correlation. The accuracy in the energy such as excitation energy and dissociation energy does not reach the chemical accuracy, that is, within several kcal/mol. A method is necessary which takes into account both the non-dynamic and dynamic correlations for quantitative description.

Our multireference Møller–Plesset (MRMP) perturbation method [1–4] and MC-QDPT quasi-degenerate perturbation theory (QDPT) with multiconfiguration self-consistent field reference functions (MC-QDPT) [5,6] are perturbation methods of such a type. Using these perturbation methods, we have clarified electronic structures of various systems and demonstrated that they are powerful tools for investigating excitation spectra and potential energy surfaces of chemical reactions [7–10]. In the present section, we review these multireference perturbation methods as well as a method for interpreting the electronic structure in terms of valence-bond resonance structure based on the CASSCF wavefunction.

20.2.1 Multireference perturbation theory

Many-body perturbation theory (MBPT) has been utilized as a convenient way of taking account of electron correlation beyond the Hartree–Fock (HF) approximation. In particular, its single-reference version is now fully established. Møller–Plesset perturbation method [11], up to the fourth order, is provided as a standard tool in most program systems such as GAUSSIAN or GAMESS. However, the application of the single-reference many-body perturbation theories (SR-MBPT) is limited only to the system where the Hartree–Fock approximation is a good starting point. It cannot describe open-shell molecules, dissociation to open-shell fragment, and transition state of chemical reaction.

Conventional QDPT [12–21] has been developed to be applied to open-shell systems and excited states. Although much effort has been made to develop the QDPT, it is not widely used among chemists. The major reason is that the QDPT often provides incorrect potential curves, since the perturbation series frequently diverges owing to the existence of intruder states [22].

In the 1980s and 1990s, multiconfigurational self-consistent field (MCSCF) reference perturbation theories [1–6,23–30] were proposed to overcome the defects of the single-reference PT and the QDPT, and now they are established as reliable methods that can be applied to wide variety of problems: potential energies surfaces of chemical reactions, excited spectra of molecules, etc. In fact, they have many advantages:

1. Generally applicable to a wide class of problems and a wide variety of molecules in a single framework;
2. Almost size-consistent;
3. Applicable to excited states and open-shells as well as the ground state;
4. Stable on the whole potential surface if reference CASSCF function is appropriately chosen;
5. Accurate enough to provide the chemical accuracy. Although MRMP and MC-QDPT at the lowest non-trivial order (the second order) does not yield very close total energy to the exact one, they are well balanced and the relative energies like dissociation energies, excitation energies, the activation energy are quite good.
6. Much more efficient and handy than MRCI and MRCC methods. The energy is computed as a sum of the product of molecular integrals and coupling constants between the target state and CSF divided by energy difference. The resource required does not depend strongly on the size of the active space and basis set. This presents a keen contrast to the case of MRCI and MRCC.

The MRMP method [1–4] and MC-QDPT [5,6] are perturbation methods in this category. In these methods, the CASSCF wavefunction(s) is(are) first determined, and the perturbation calculation is done with those wavefunctions used as reference (zeroth-order wavefunction) based on Rayleigh–Schrödinger PT in MRMP and van Vleck PT in MC-QDPT.

In the following sections, we show the formalism of the multireference perturbation theory and some applications to potential energy surfaces and electronic excited spectra.

20.2.1.1 Multireference Møller–Plesset perturbation method [1–4]

Our basic problem is to find approximations to some low-lying solutions of the exact Schrödinger equation

$$H\Psi = E\Psi \tag{1}$$

H is the Hamiltonian and it is decomposed into two parts, a zeroth-order Hamiltonian H_0 and a perturbation V

$$H = H_0 + V \tag{2}$$

We assume that a complete set of orthonormal eigenfunctions $\{\Psi_i^{(0)}\}$ and corresponding eigenvalues are available

$$H_0 \Psi_i^{(0)} = E_i^{(0)} \Psi_i^{(0)} \tag{3}$$

Then the state wavefunction Ψ_I is expanded in terms of basis functions $\Psi_k^{(0)}$ as

$$\Psi_I = \sum_k C_{Ik} \Psi_k^{(0)} \tag{4}$$

Some of the basis functions define an active space P and the remaining part of Hilbert space is called the orthogonal space $Q = 1 - P$. The active space is spanned by the basis functions that have a filled core and the remaining active electrons distributed over a set of active orbitals. The orthogonal complete space incorporates all other possible basis functions that are characterized by having at least one vacancy in a core orbital. The state wavefunction in an active space is written as

$$\Psi_I^{(0)} = \sum_k C_k \Phi_k \tag{5}$$

where the sum runs over active space basis functions $\{\Phi_i\}$ and C_k are the coefficients of only active space basis functions. It is convenient to use intermediate normalization, i.e.

$$\langle \Psi_I^{(0)} | \Psi_I^{(0)} \rangle = \langle \Psi_I^{(0)} | \Psi_I \rangle = 1 \tag{6}$$

We also assume that $\Psi_I^{(0)}$ is diagonal in P space

$$\langle \Psi_I^{(0)} | H | \Psi_J^{(0)} \rangle = \delta_{IJ}(E_I^{(0)} + E_I^{(1)}) \tag{7}$$

with

$$E_I^{(0)} = \langle \Psi_I^{(0)} | H_0 | \Psi_I^{(0)} \rangle \tag{8}$$

$$E_I^{(1)} = \langle \Psi_I^{(0)} | V | \Psi_I^{(0)} \rangle \tag{9}$$

The state-specific Rayleigh–Schrödinger perturbation theory based on the unperturbed eigenvalue equation

$$H_0 \Psi_I^{(0)} = E_I^{(0)} \Psi_I^{(0)} \tag{10}$$

leads to the first $E_I^{(k)}$ as

$$E_I^{(2)} = \langle \Psi_I^{(0)} | VRV | \Psi_I^{(0)} \rangle \tag{11}$$

$$E_I^{(3)} = \langle \Psi_I^{(0)} | VR(V - E_I^{(1)})RV | \Psi_I^{(0)} \rangle \tag{12}$$

$$E_I^{(4)} = \langle\Psi_I^{(0)}|VR(V - E_I^{(1)})R(V - E_I^{(1)})RV|\Psi_I^{(0)}\rangle - E_I^{(2)}[\langle\Psi_I^{(0)}|VR^2V|\Psi_I^{(0)}|$$
$$+ \langle\Psi_I^{(0)}|VRH_0SH_0RV|\Psi_I^{(0)}\rangle], \text{etc.} \tag{13}$$

The R and S are the resolvent operators

$$R = Q/(E_I^{(0)} - H_0) \tag{14}$$

$$S = P'/(E_I^{(0)} - H_0) \tag{15}$$

where $P' = P - |\Psi_I^{(0)}\rangle\langle\Psi_I^{(0)}|$.

The zeroth-order energies $E_I^{(0)}$ are given in terms of orbital energies as

$$E_I^{(0)} = \sum_{p\sigma} n_{p\sigma}(I)\varepsilon_{p\sigma} \tag{16}$$

where $\varepsilon_{p\sigma}$ and $n_{p\sigma}(I)$ are the energy of orbital p with spin σ and the occupation number of orbital p with spin σ in wavefunction Φ_I, respectively.

We have mainly used canonical Fock orbitals, which are defined so that the generalized Fock matrix

$$F_{pq} = h_{pq} + \sum_{rs} (D_{rs}^{\alpha} + D_{rs}^{\beta})[(pq|rs) - \tfrac{1}{2}(ps|rq)] \tag{17}$$

is partially diagonal in the core, active, and virtual orbital sub-blocks (D_{rs}^{α} and D_{rs}^{β} represent alpha and beta one-particle density matrices, respectively) and have adopted the diagonal elements of the generalized Fock matrix, F_{pp}, as orbital energies

$$\varepsilon_{p\alpha} = \varepsilon_{p\beta} = h_{pp} + \sum_{rs} (D_{rs}^{\alpha} + D_{rs}^{\beta})[(pp|rs) - \tfrac{1}{2}(ps|rp)] \tag{18}$$

However, in an open-shell system involving unpaired alpha electrons, the electron environment depends on whether the electron is in an alpha or a beta orbital. Thus, we have also proposed that we retain canonical Fock orbitals but adopt spin-dependent orbital energies defined by

$$\varepsilon_{p\alpha} = h_{pp} + \sum_{rs} \{D_{rs}^{\alpha}[(pp|rs) - (ps|rp)] + D_{rs}^{\beta}(pp|rs)\} \tag{19}$$

and

$$\varepsilon_{p\beta} = h_{pp} + \sum_{rs} \{D_{rs}^{\beta}[(pp|rs) - (ps|rp)] + D_{rs}^{\alpha}(pp|rs)\} \tag{20}$$

for alpha- and beta-spin orbitals, respectively [31].

When CASSCF wavefunction is used as the reference, the zeroth plus first-order energy $E_I^{(0)} + E_I^{(1)}$ is equal to the CASSCF energy. The lowest non-trivial order is, therefore, the second order. Let the reference function $|\Psi_\alpha^{(0)}\rangle$ be a CASSCF wavefunction

$$|\alpha\rangle = \sum_{\mathrm{A}} C_{\mathrm{A}}|\mathrm{A}\rangle \tag{21}$$

The energy up to the second order in MRMP is given by

$$E_\alpha^{(0-2)} = E_\alpha^{\text{CAS}} + \sum_I \frac{\langle\alpha|V|I\rangle\langle I|V|\alpha\rangle}{E_\alpha^{(0)} - E_I^{(0)}} \tag{22}$$

where $\{|I\rangle\}$ is the set of all singly and doubly excited configurations from the reference configurations in CAS. The first term of the RHS is the CAS-CI energy.

20.2.1.2 Multiconfigurational quasi-degenerate perturbation theory (MC-QDPT) [5,6] We have also proposed a multistate multireference perturbation theory, the QDPT with MCSCF reference functions (MC-QDPT). In this PT, state-averaged CASSCF is first performed to set reference functions, and then an effective Hamiltonian is constructed, which is finally diagonalized to obtain the energies of interest.

The van Vleck PT for the CASSCF reference wavefunctions gives the order-by-order expansion of the effective Hamiltonian

$$(H_{\text{eff}}^{(0-1)})_{\alpha\beta} = E_\beta^{\text{CAS}}\delta_{\alpha\beta} \tag{23}$$

$$(H_{\text{eff}}^{(2)})_{\alpha\beta} = \tfrac{1}{2}\langle\alpha|VRV|\beta\rangle + (\alpha \leftrightarrow \beta) \tag{24}$$

$$(H_{\text{eff}}^{(3)})_{\alpha\beta} = \tfrac{1}{2}\langle\alpha|VR(V - E_\beta^{(0)})RV|\beta\rangle + (\alpha \leftrightarrow \beta), \text{ etc.} \tag{25}$$

where R is the resolvent operator given by Eq. (14). Thus, the effective Hamiltonian to second order is given by

$$(H_{\text{eff}}^{(0-2)})_{\alpha\beta} = E_\beta^{\text{CAS}}\delta_{\alpha\beta} + \frac{1}{2}\sum_I\left\{\frac{\langle\alpha|V|I\rangle\langle I|V|\beta\rangle}{E_\beta^{(0)} - E_I^{(0)}} + \frac{\langle\beta|V|I\rangle\langle I|V|\alpha\rangle}{E_\alpha^{(0)} - E_I^{(0)}}\right\} \tag{26}$$

Substituting the second-quantized operator into V, we obtain an explicit formula using molecular integrals and orbital energies instead of matrix elements

$$\begin{aligned}(H_{\text{eff}}^{(0-2)})_{\alpha\beta} = {} & E_\alpha^{\text{CAS}}\delta_{\alpha\beta} - \sum_{pq,\text{B}}\langle\alpha|E_{pq}|\beta\rangle C_{\text{B}}(\beta)\sum_e \frac{u_{pe}u_{eq}}{\varepsilon_e - \varepsilon_q + \Delta E_{\text{B}\alpha}} \\ & - \sum_{pqrs,\text{B}}\langle\alpha|E_{pq,rs}|\text{B}\rangle C_{\text{B}}(\beta)\left[\sum_e \frac{u_{pe}g_{eqrs}}{\varepsilon_e - \varepsilon_q + \varepsilon_r - \varepsilon_s + \Delta E_{\text{B}\alpha}}\right. \\ & \left. + \sum_e \frac{g_{pers}u_{eq}}{\varepsilon_e - \varepsilon_q + \Delta E_{\text{B}\alpha}} + \frac{1}{2}\sum_{(a,b)} \frac{g_{parb}g_{aqbs}}{\varepsilon_a - \varepsilon_q + \varepsilon_b - \varepsilon_s + \Delta E_{\text{B}\alpha}}\right] \\ & - \sum_{pqrstu,\text{B}}\langle\alpha|E_{pq,rs,tu}|\text{B}\rangle C_{\text{B}}(\beta)\sum_e \frac{g_{pers}g_{eqtu}}{\varepsilon_e - \varepsilon_q + \varepsilon_t - \varepsilon_u + \Delta E_{\text{B}\alpha}} + (\alpha \leftrightarrow \beta)\end{aligned} \tag{27}$$

with

$$g_{pqrs} = (pq|rs) \tag{28}$$

$$u_{pq} = (h_{pq} - \delta_{pq}\varepsilon_p) - \sum_i^{\text{doc}}(2g_{pqii} - g_{piiq}) \tag{29}$$

and

$$\Delta E_{B\alpha} = E_B^{(0)} - E_\alpha^{(0)} \tag{30}$$

the difference between the energies of the zeroth-order state and configuration. The orbital labels $\{i\}$, $\{a\}$, and $\{e\}$ are for doubly occupied, active, and external orbitals, respectively, and $\{a', b'\}$ run over both active and external orbitals, and the suffix of the generator $\{p, q, r, s, t, u\}$ run over only active orbitals. The terms including doubly occupied orbitals are omitted in this equation. See Ref. [5] for the full formula.

This theory includes MRMP PT (the case that the set of reference functions reduces to a single function). Note that MRMP energy can be also calculated with the formula Eq. (27) by setting the number of the states to one.

20.2.1.3 Application of multireference perturbation theory to singlet–triplet splitting of CH_2 and CF_2 [31]

Many applications of the MRMP and MC-QDPT methods to chemical reactions and excitation spectra have been collected in our review articles [7–10]. Here we present the singlet–triplet splitting of CH_2 and CF_2: examples where the use of the spin-dependent orbital energies is crucial.

Various methods have been applied to the adiabatic singlet–triplet (ST) splitting of methylene with the same geometry, active space, and basis set given in Ref. [32]. We first calculated it with the same condition for comparison. The active space is a full-valence type CAS[6e,6o], and the basis set used is the double zeta plus polarization (DZP) basis. The results are listed in Table 20.1. The ST splitting by the original MRMP (MRMP with spin-averaged orbital energies, hereafter MRMP(SA)) is 15.9 kcal/mol; the deviation from the full CI value of 12.0 kcal/mol is 3.9 kcal/mol. This is improved by MRMP with spin-dependent orbital energies, MRMP(SD); the splitting is 12.6 kcal/mol, the discrepancy being only 0.6 kcal/mol.

We next carried out calculations with a larger basis set and active space to compare the calculated and experimental results. The splitting with Dunning's correlation consistent polarized valence triple zeta (cc-pVTZ) basis set is 10.1 kcal/mol, which is in good agreement with the experimental value of 9.4 kcal/mol [33]. In the calculations with CAS[6e,6o], even the reference CASSCF gives good results: 12.8 (DZP) and 10.5 (cc-pVTZ) kcal/mol. The deviation from the full CI and experimental values are only 0.8 and 1.1 kcal/mol, respectively. However, it is known that the energy splitting at the CASSCF level gets worse if the active space is enlarged to CAS[6e,12o]. We next calculated the splitting using CAS[6e,12o] as a further check. The result of CASSCF is 4.3 kcal/mol, which is rather poor compared with the CAS[6e,6o] value. On the contrary, the value 9.9 kcal/mol of MRMP(SD) is in much better agreement with experiment.

The geometry of CF_2 used in the calculations was determined with CASSCF [6e,6o]/cc-pVTZ. The active space and basis set for MRMP are CAS[12e,9o] and cc-pVTZ, respectively. The ST splitting energy calculated with MRMP(SD) is 52.6 kcal/mol. This is a fairly good estimate of the experimental value, 56.6 kcal/mol. [34] On the other hand, the value from MRMP(SA), 46.7 kcal/mol, is too small compared with experiment.

Table 20.1 Energy splitting between the $^1A_1$ and $^3B_1$ states in CH_2 and CF_2

Method	Energy (hartree)		ΔE (kcal/mol)
	$^1A_1$	$^3B_1$	
CH_2			
[6e,6o]CASSCF/DZP	−38.94532	−38.96578	12.8
[6e,6o]MRMP(SA)/DZP	−39.01106	−39.03636	15.9
[6e,6o]MRMP(SD)/DZP	−39.01106	−39.03115	12.6
Full CI/DZP	−39.02718	−39.04626	12.0
[6e,6o]CASSCF/cc-pVTZ	−38.95422	−38.97099	10.5
[6e,6o]MRMP(SA)/cc-pVTZ	−39.04461	−39.06778	14.5
[6e,6o]MROPT2/cc-pVTZ	−39.04461	−39.06138	10.5
[6e,6o]MRMP(SD)/cc-pVTZ	−39.04461	−39.06064	10.1
[6e,12o]CASSCF/cc-pVTZ	−39.02090	−39.02778	4.3
[6e,12o]MRMP(SA)/cc-pVTZ	−39.06931	−39.08723	11.2
[6e,12o]MROPT2/cc-pVTZ	−39.06931	−39.08513	9.9
[6e,12o]MRMP(SD)/cc-pVTZ	−39.06931	−39.08504	9.9
Exptl[a]			9.4
CF_2			
[12e,9o]CASSCF/cc-pVTZ	−236.85497	−236.76085	59.1
[12e,9o]MRMP(SA)/cc-pVTZ	−237.38184	−237.30738	46.7
[12e,9o]MROPT2/cc-pVTZ	−237.38184	−237.30391	48.9
[12e,9o]MRMP(SD)/cc-pVTZ	−237.38184	−237.29807	52.6
Exptl[b]			56.6

[a]Ref. [33].
[b]Ref. [34].

MROPT2 by Kozlowski and Davidson [29] also reproduces the splitting, 10.5 (CAS[6e,6o]) and 9.9 kcal/mol (CAS[6e,12o]) for CH_2. These numbers are very close to those of MRMP(SD). For CF_2, however, it gives a slightly smaller splitting of 48.9 kcal/mol.

20.2.1.4 Extension of reference wavefunctions—quasi-degenerate perturbation theory with quasi-complete active space self-consistent field reference functions (QCAS-QDPT) [35]

In the study of chemical reaction mechanisms, CAS-SCF method is a very useful approach and hence frequently used. However, CAS-SCF often generates far too many configurations, and the size of the active space outgrows the capacity of present technology. Perturbation methods using a selected reference configuration space but retaining the advantages of the CAS-based PTs are necessary.

We have proposed an MC-SCF method with a quasi-complete active space (QCAS), i.e. a QCAS-SCF method. In the MC-SCF method, we partition orbitals into core, active, and virtual, then construct the CI space by distributing active electrons among the active orbitals. Let us further divide the active electron and orbital sets into N sub-sets and fix the number of active electrons, m_i, and orbitals, n_i, in each sub-set

$$m_{\rm act} = \sum_i^N m_i, \qquad n_{\rm act} = \sum_i^N n_i \tag{31}$$

where $m_{\rm act}$ and $n_{\rm act}$ denote the number of active electrons and orbitals, respectively. We define the quasi-complete space as the product space of CAS spanned by the determinants or CSF as follows:

$$\mathrm{QCAS}(\{m_i\},\{n_i\}) = \mathrm{CAS}(m_1,n_1)\mathrm{CAS}(m_2,n_2)\cdots\mathrm{CAS}(m_N,n_N) \tag{32}$$

such that the number of electrons in each orbital group satisfies the restriction in Eq. (31).

Combining QCAS with MC-QDPT provides an effective tool for electronic structure theory. We present MC-QDPT using QCAS-SCF reference functions. (Hereafter, we call it QCAS-QDPT or QCAS-PT.)

Adopting (state-averaged) QCAS-SCF wave functions $\alpha(\beta)$ as reference functions (i), which define P space, we obtain the effective Hamiltonian to the second order

$$(H_{\rm eff}^{(0-2)})_{\alpha\beta} = E_\alpha^{\rm QCAS}\delta_{\alpha\beta} + \frac{1}{2}\left\{\sum_{I\notin \rm QCAS} \frac{\langle\alpha|V|I\rangle\langle I|V|\beta\rangle}{E_\beta^{(0)} - E_I^{(0)}} + (\alpha\leftrightarrow\beta)\right\} \tag{33}$$

which corresponds to Eq. (26) in the CAS-SCF reference case. Let us define a *corresponding complete active space* (CCAS) to a QCAS as the complete active space (CAS) that has the same active orbital set and electron but does not have the limitation (Eq. (31)). In other words, the corresponding CAS is the minimal CAS that includes the QCAS. Then the summation for I in Eq. (33) may be divided into the summations for determinants/CSFs outside the CAS and for the determinants/CSFs outside the QCAS but inside the corresponding CAS

$$\sum_{I\notin \rm QCAS} = \sum_{I\notin \rm CCAS} + \sum_{I\in {\rm CCAS}\wedge I\notin \rm QCAS} \tag{34}$$

and then the former second-order term in Eq. (33) may be written as

$$(H_{\rm eff}^{(2)})_{\alpha\beta} = \sum_{I\notin \rm CCAS} \frac{\langle\alpha|V|I\rangle\langle I|V|\beta\rangle}{E_\beta^{(0)} - E_I^{(0)}} + \sum_{I\in {\rm CCAS}\wedge I\notin \rm QCAS} \frac{\langle\alpha|V|I\rangle\langle I|V|\beta\rangle}{E_\beta^{(0)} - E_I^{(0)}} \tag{35}$$

The former term in Eq. (35) involves excitations from core orbitals and excitations to virtual orbitals in the intermediate states I (the *external* terms), while the latter term involves excitations where only active orbitals are involved (the *internal* terms).

The external term may be further written as

$$(H_{\rm ext}^{(2)})_{\alpha\beta} = \sum_{\rm A,B\in QCAS} C_{\rm A}(\alpha)C_{\rm B}(\beta)(H_{\rm ext}^{(2)})_{\rm AB} \tag{36}$$

with

$$(H_{\text{ext}}^{(2)})_{\text{AB}} = \sum_{I \notin \text{CCAS}} \frac{\langle \text{A}|H|I\rangle\langle I|H|\text{B}\rangle}{E_{\text{B}}^{(0)} - E_I^{(0)} + (E_\beta^{(0)} - E_{\text{B}}^{(0)})} \tag{37}$$

where $(H_{\text{ext}}^{(2)})_{\text{AB}}$ is the effective Hamiltonian in the determinant/CSFs basis in the conventional QDPT except for the energy shift, $E_\beta^{(0)} - E_{\text{B}}^{(0)}$, in the denominator. Since the second-order diagrams do not depend on the denominator, the second-order effective Hamiltonian equation (37) (hence, also Eq. (36)) is expressed by the same diagrams as the conventional QDPT. This situation is the same as MC-QDPT: the diagrams and the rule for translating them into mathematical expressions is described in detail in Ref. [35]. The internal terms are also expressed by diagrams. Since QCAS is a natural extension of CAS, computation of these diagrams can be done efficiently in a similar manner to CAS-QDPT.

20.2.1.5 Further extension of reference wavefunctions—quasi-degenerate perturbation theory with general-multiconfiguration space self-consistent field reference functions (GMC-QDPT) [36]

Adopting (state-averaged) general MC-SCF (or MC-CI) wavefunctions $\alpha(\beta)$ as reference functions $\Phi_{\text{A}}^{(0)}$ ($\Phi_{\text{B}}^{(0)}$), which define the P space, the effective Hamiltonian to the second order becomes

$$(H_{\text{eff}}^{(0-2)})_{\alpha\beta} = E_\alpha^{\text{MC}} \delta_{\alpha\beta} + \frac{1}{2}\left\{ \sum_{I \notin \text{GCS}} \frac{\langle \alpha|H|I\rangle\langle I|H|\beta\rangle}{E_\beta^{(0)} - E_I^{(0)}} + (\alpha \leftrightarrow \beta) \right\} \tag{38}$$

where I is now a determinant/CSF outside the general configuration space (GCS). The notation $(\alpha \leftrightarrow \beta)$ means interchange α with β from the first term in curly brackets. The complementary eigenfunctions of the MC-CI Hamiltonian and the determinants/CSFs generated by exciting electrons out of the determinants/CSFs in GCS are orthogonal to the reference functions and define the Q space. The functions in the space complementary to the P space in GCS, however, do not appear in Eq. (38) since the interaction between the complementary functions and the reference functions is zero.

The GMC-QDPT computation scheme is similar to that of QCAS-QDPT [35]. We define again the corresponding CAS (CCAS) as a CAS constructed from the same active electrons and orbitals, that is, the minimal CAS that includes the reference GCS. The summation over I in Eq. (38) may be divided into the summations over determinants/CSFs outside CCAS and over the determinants/CSFs outside the GCS but inside CCAS:

$$\sum_{I \notin \text{GCS}} = \sum_{I \notin \text{CCAS}} + \sum_{I \in \text{CCAS} \wedge I \notin \text{GCS}} \tag{39}$$

then the former second-order term in Eq. (38) may be written as

$$(H_{\text{eff}}^{(2)})_{\alpha\beta} = \sum_{I \notin \text{CCAS}} \frac{\langle \alpha|H|I\rangle\langle I|H|\beta\rangle}{E_\beta^{(0)} - E_I^{(0)}} + \sum_{I \in \text{CCAS} \wedge I \notin \text{GCS}} \frac{\langle \alpha|H|I\rangle\langle I|H|\beta\rangle}{E_\beta^{(0)} - E_I^{(0)}} \tag{40}$$

The first term in Eq. (40) represents external excitations, while the latter term represents internal excitations.

The external terms may be computed with diagrams in the same manner as the QCAS-QDPT case. On the other hand, for internal terms, the diagrammatic approach may not be applied. Instead, matrix operations for the Hamiltonian matrix are used

$$(H_{\mathrm{int}}^{(2)})_{\alpha\beta} = \mathbf{v}^{\mathrm{T}}(\alpha)\cdot\mathbf{w}(\beta) \tag{41}$$

with

$$\mathbf{v}_I(\alpha) = \sum_{\mathrm{A}\in\mathrm{GCS}} \langle I|H|\mathrm{A}\rangle C_{\mathrm{A}}(\alpha) \tag{42}$$

$$\mathbf{w}_I(\beta) = \sum_{\mathrm{B}\in\mathrm{GCS}} \langle I|H|\mathrm{B}\rangle C_{\mathrm{B}}(\beta)/(E_\beta^{(0)} - E_{\mathrm{B}}^{(0)}) \tag{43}$$

The intermediate determinants/CSFs I are constructed by exciting one or two electron(s) from the reference determinants/CSFs within the active orbital space. In general, the number of I is not large, and thus they may be managed in computer memory.

20.2.1.6 Application of QCAS- and GMC-QDPT

20.2.1.6.1 Transition state barrier height for the unimolecular dissociation reaction of formaldehyde $H_2CO \rightarrow H_2 + CO$ [35]. This reaction is Woodward–Hoffmann forbidden, and therefore, proceeds *via* the highly asymmetric transition structure. We examined in a previous paper [37] the barrier height using the MRMP method. In the present section, we show the QCAS-PT results and the comparison of them with the MRMP results.

The CAS we used for comparison is CAS(12,10), which is the full valence active space. We split the active orbitals into {CO(σ, σ^*)}, {CO(π, π^*)}, and {CH(σ, σ^*), CH(σ', σ'^*), O(lp, lp)}, where lp denotes a lone pair orbital, and then we distributed two, two, and eight electrons among the above groups, respectively, to construct QCAS[$(2,2)^2 \times (8,6)$]. The dimension of the CAS is 44,100, while that of the QCAS is 3600.

The results with cc-pVTZ and cc-pVQZ are shown in Table 20.2. First let us compare the results at the reference function (QCAS- and CAS-SCF) level. Although differences in the energy itself between QCAS-SCF and CAS-SCF are about 10 millihartree for both basis sets, the differences in the barrier height are 1.65 and 1.62 millihartree (1.0 and 1.0 kcal/mol) for cc-pVTZ and cc-pVQZ, respectively. The agreement of QCAS-SCF with CAS-SCF is very good.

Now let us compare the results at the multireference PT level. In total energy, there still remains a difference of about 8 millihartree between the results of QCAS-PT and MRMP. In relative energy, for the barrier height the QCAS-PT results are very close to those of CAS-PT in both basis sets. The barrier height of QCAS-PT is 83.7 kcal/mol in both basis sets, the differences from those of MRMP are only 0.1 and 0.3 kcal/mol for cc-pVTZ and cc-pVQZ, respectively. Moreover, the barrier height is also close to the experimental value, 84.6 kcal/mol [38]. The error of 0.9 kcal/mol is within twice the experimental uncertainty 0.8 kcal/mol.

Table 20.2 Transition state barrier height for the reaction $H_2CO \rightarrow H_2 + CO$

	Eq. (hartree)[a]	Tr. (hartree)[b]	ΔE (kcal/mol)	Error (kcal/mol)
cc-pVTZ				
CAS-SCF	−114.04696	−113.91381	83.6	−1.0
QCAS-SCF	−114.03786	−113.90306	84.6	0.0
MRMP	−114.30451	−114.17134	83.6	−1.0
QCAS-PT	−114.29674	−114.16338	83.7	−0.9
cc-pVQZ				
CAS-SCF	−114.05624	−113.92300	83.6	−1.0
QCAS-SCF	−114.04712	−113.91226	84.6	0.0
MRMP	−114.33057	−114.19763	83.4	−1.2
QCAS-PT	−114.32316	−114.18981	83.7	−0.9
Exptl. (classical)[c]			84.6 ± 0.8	

[a]Equilibrium structure.
[b]Transition state structure.
[c]Ref. [38]. Barrier height not including zero-point energy correction.

20.2.1.6.2 Valence excitation energies for formaldehyde [36]. The second example is the GMC-QDPT calculation of valence excitation energies for formaldehyde molecule. Calculations on formaldehyde were carried out at the ground state experimental geometry (i.e. r(CO) = 1.203 Å, r(CH) = 1.099 Å, and θ(HCH) = 116.5°). The basis set used was Dunning's cc-pVTZ.

Five reference spaces were constructed from eight electrons, 16 $[(a_1, a_2, b_1, b_2) = (7,1,3,5)]$, 18 $[= (7,1,4,6)]$, 20 $[= (8,1,5,6)]$, 22 $[= (8,2,5,7)]$, and 24 orbitals $[= (9,2,6,7)]$, by exciting one and two electrons from the following parent configurations:

$$
\begin{array}{llll}
{}^1\mathrm{A}_1 \text{ states}: & \ldots n^2(\mathrm{HF}); & \pi \rightarrow \pi^*; & n \rightarrow \sigma^* \\
{}^3\mathrm{A}_1 \text{ states}: & & \pi \rightarrow \pi^*; & n \rightarrow \sigma^* \\
{}^{1,3}\mathrm{A}_2 \text{ states}: & n \rightarrow \pi^*; & 1b_2(\sigma) \rightarrow \pi^* & \\
{}^{1,3}\mathrm{B}_1 \text{ states}: & 5a_1(\sigma) \rightarrow \pi^* & & \\
{}^{1,3}\mathrm{B}_2 \text{ states}: & n \rightarrow 6a_1(\sigma^*) & &
\end{array}
$$

All the calculations were done in each symmetry.

The results are summarized in Tables 20.3 and 20.4. The calculations with CAS-SCF and CAS-QDPT are far too large to be done. We, therefore, compare the results with available experimental results and some recent theoretical results, i.e. MR-CI results by Hachey et al. [39], the second-order complete active space perturbation theory (CASPT2) calculations by Merchán and Roos [40], and the equation of motion coupled cluster (EOM-CC) calculations by Gwaltney and Bartlett [41].

As can be computed from Table 20.3, the maximum differences in excitation energy for the largest three (two) numbers of active orbitals is 0.09 (0.05) eV. We can, therefore, consider that the excitation energies at the MC-SCF level are almost converged values for

Table 20.3 Valence excitation energies of H_2CO (eV)

State	Orbital picture	MC-SCF					GMC-QDPT					Exptl
		(8,16)	(8,18)	(8,20)	(8,22)	(8,24)	(8,16)	(8,18)	(8,20)	(8,22)	(8,24)	
$^1A_1$	$\pi \to \pi^*$; $n \to \sigma^*$	10.07	10.03	10.03	10.02	10.02	9.66	9.67	9.72	9.72	9.72	
	$n \to \sigma^*$; $\pi \to \pi^*$	11.01	11.01	11.00	11.03	11.01	10.65	10.63	10.65	10.63	10.64	10.70
$^1A_2$	$n \to \pi^*$	4.32	4.43	4.27	4.25	4.22	4.04	4.02	4.01	4.02	4.08	4.07
	$1b_2(\sigma) \to \pi^*$	10.96	11.16	10.97	10.98	10.94	10.30	10.34	10.33	10.37	10.43	
$^1B_1$	$5a_1(\sigma) \to \pi^*$	9.63	9.89	9.82	9.81	9.80	9.31	9.24	9.26	9.20	9.28	9.00
$^1B_2$	$n \to 6a_1(\sigma^*)$	7.73	8.16	8.22	8.32	8.31	8.31	8.23	8.41	8.45	8.45	
$^3A_1$	$\pi \to \pi^*$; $n \to \sigma^*$	6.18	6.28	6.19	6.13	6.11	6.13	6.13	6.18	6.17	6.18	6.00
	$n \to \sigma^*$;$\pi \to \pi^*$	9.66	9.64	9.70	9.74	9.75	9.60	9.61	9.62	9.62	9.62	
$^3A_2$	$n \to \pi^*$	3.84	3.95	3.78	3.75	3.71	3.63	3.58	3.58	3.61	3.63	3.50
	$1b_2(\sigma) \to \pi^*$	10.52	10.68	10.52	10.52	10.47	10.04	10.03	10.02	10.07	10.10	
$^3B_1$	$5a_1(\sigma) \to \pi^*$	8.78	9.02	8.92	8.91	8.90	8.45	8.41	8.39	8.27	8.50	8.50
$^3B_2$	$n \to 6a_1(\sigma^*)$	7.36	7.79	7.85	7.95	7.94	7.89	7.80	7.99	8.02	8.07	

Table 20.4 Valence excitation energies of H_2CO (eV)

State	Orbital picture	MC-SCF (8,24)	GMC-QDPT (8.24)	Exptl	MRCI[a]	CASPT2[b]	EOM-CC[c]	CCSD[d]	CIS-MP2[e]	SAC-CI[f]
$^1A_1$	$\pi \rightarrow \pi^*, n \rightarrow \sigma^*$	10.02	9.72		9.60	9.77	9.47	9.27	9.19	–
	$n \rightarrow \sigma^*, \pi \rightarrow \pi^*$	11.01	10.64	10.70						10.83
$^1A_2$	$n \rightarrow \pi^*$	4.22	4.08	4.07	4.05	3.91	3.98	3.95	4.58	4.16
	$1b_2(\sigma) \rightarrow \pi^*$	10.94	10.43				10.38		10.08	11.19
$^1B_1$	$5a_1(\sigma) \rightarrow \pi^*$	9.80	9.28	9.00	9.35	9.09	9.33	9.26	9.97	9.49
$^1B_2$	$n \rightarrow 6a_1(\sigma^*)$	8.31	8.45							
$^3A_1$	$\pi \rightarrow \pi^*; n \rightarrow \sigma^*$	6.11	6.18	6.00		5.99			6.72	6.10
	$n \rightarrow \sigma^*; \pi \rightarrow \pi^*$	9.75	9.62							
$^3A_2$	$n \rightarrow \pi^*$	3.71	3.63	3.50		3.48			4.15	3.70
	$1b_2(\sigma) \rightarrow \pi^*$	10.47	10.10						10.52	10.80
$^3B_1$	$5a_1(\sigma) \rightarrow \pi^*$	8.90	8.50	8.50					9.18	8.52
$^3B_2$	$n \rightarrow 6a_1(\sigma^*)$	7.94	8.07							

[a] Ref. [39].
[b] Ref. [40].
[c] Ref. [41].
[d] Ref. [42].
[e] Ref. [43].
[f] Ref. [44].

the change of the active orbital numbers. However, the agreement with the experimental values is not so good: the error is 0.32 eV on average and 0.80 eV at maximum.

At the GMC-QDPT level, the excitation energies are also almost converged (though the differences are a little larger than those at MC-SCF level are). Compared to the reference MC-SCF level, the results are somewhat improved. The error from the experimental value was reduced to 0.11 eV on average and 0.28 eV at maximum.

20.2.1.6.3 The most stable structure of SiC_3 [45]. Silicon-containing carbon clusters, Si_mC_n, have recently received much attention from various fields, such as astrophysics and nanoscience. In particular, SiC_n ($n = 1-4$) molecules have been well studied, both experimentally and theoretically. However, among the theoretical studies, the most stable structure of SiC_3 was in dispute. The most stable structure of the SiC_3 molecule was investigated using second-order perturbation theory with general multiconfiguration self-consistent field reference functions (GMC-PT).

The basis sets used were Dunning's cc-pVXZ (X = D, T, Q) and augmented cc-pVXZ (X = D, T, Q) basis sets. Using these basis sets, we first carried out Hartree–Fock (HF) calculations (unrestricted HF calculations for 1t and restricted HF calculations for 2s and 3s; see Fig. 20.1). The active spaces in the reference MC-SCF calculations were constructed from the HF configuration plus single and double excitation configurations among valence orbitals, that is, valence configuration interaction singles and doubles (CISD) space. The 16 electrons in the 16 orbitals were correlated in MC-SCF calculations; hereafter, we refer to these spaces as MC(16,16) following the CAS(n, m) notation.

Table 20.5 shows the relative stability of the 1t, 2s, and 3s isomers. The results indicate that the 2s isomer is most stable in all the basis sets and at both the CCSD(T) and

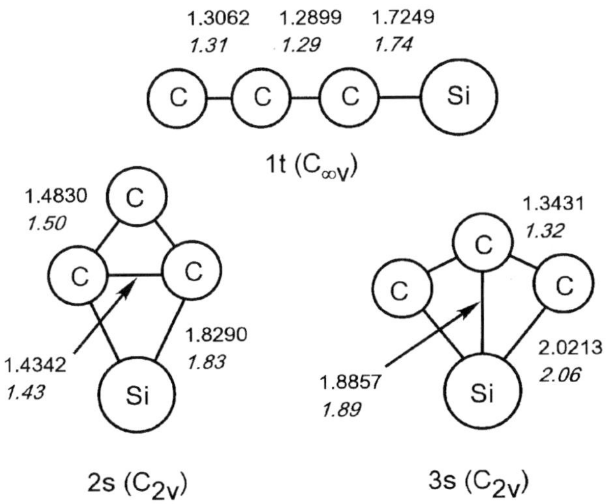

Fig. 20.1. Three isomers of SiC_3: 1t, 2s, and 3s. The numbers in roman and italic type represent the bond lengths in ångstrom optimized by CCSD(T)/cc-pCVQZ (Refs. [47,49]) and FORS-SCF/6-31G(d) (Ref. [46]) methods, respectively.

Table 20.5 Relative energy of the 1t and 3s isomers measured from the energy of the 2s isomer (in kcal/mol)

Structure	cc-pVXZ sets			aug-cc-pVXZ sets			CCSD(T)[a] (cc-pCVQZ)	MC-QDPT[b] (aug-cc-pVDZ)
	X = D	X = T	X = Q	X = D	X = T	X = Q		
CCSD(T) geometry								
1t (SD)	3.4	5.1	6.3	3.7	5.4	6.7 (7.2)[c]	7.5 (8.0)[c]	–
1t (SA)	−1.1	−0.4	0.7	−1.6	−0.4	0.9 (1.4)[c]		–
2s	0.0	0.0	0.0	0.0	0.0	0.0	0.0	–
3s	4.2	4.9	5.3	3.4	4.8	5.3 (5.1)[c]	6.2 (6.0)[c]	–
FORS-SCF geometry								
1t (SD)	4.6	7.3	8.8	5.5	7.9	9.9 (9.9)[d]	–	–
1t (SA)	0.1	1.8	3.0	0.3	2.1	3.9 (4.0)[d]	–	−4.4 (−4.3)[d]
2s	0.0	0.0	0.0	0.0	0.0	0.0	–	0.0
3s	7.3	8.5	9.0	6.6	7.7	9.0 (8.7)[d]	–	3.4 (3.1)[d]

[a]Refs. [47,49].
[b]Ref. [46].
[c]The numbers in the parentheses are zero point corrected energies. The zero point vibrational energies are taken from Ref. [47].
[d]The numbers in the parentheses are zero point corrected energies. The zero point vibrational energies are taken from Ref. [46].

CAS-SCF geometries except for some spin-averaged orbital energy numbers. The second most stable isomer is the 3s isomer, although the energy difference from the 1t isomer is quite small. The relative energy of the 3s isomer measured from the 2s isomer was 5.3 (9.0) kcal/mol in the calculations with the aug-cc-pVQZ basis and CCSD(T) (CAS-SCF) geometry, while the relative energy of the 1t isomer was 6.7 (9.9) kcal/mol. This order of the isomers is unchanged even though the zero-point vibrational energy (ZPVE) correction is included. The ZPVEs for the 1t, 2s, and 3s isomers are very close to one another: 7.41, 7.37, and 7.07 kcal/mol, respectively, at the CAS-SCF level [46]. (The differences of ZPVEs at the CCSD level are $E_{ZPV}(1t) - E_{ZPV}(2s) = 0.5$ kcal/mol and $E_{ZPV}(3s) - E_{ZPV}(2s) = -0.2$ kcal/mol at the CCSD level [47]. ZPVEs themselves were not reported in Ref. [47]). The zero-point corrected energies are also listed in Table 20.5.

Table 20.5 also tells us a clear trend of the basis set effect, with a larger basis set giving larger 2s–3s and 2s–1t energy separations. This implies that the most stable isomer will not change if we use larger basis sets than those used. To check the basis set effect on the order of the isomers, we calculated the complete basis set limits. The complete basis set extrapolation of the CCSD energy is given by the Gaussian dependence [48] on basis set sizes, i.e.

$$E(n) = E(\infty) + E_0 \exp[-(n-1)] + E_1 \exp[-(n-1)^2] \tag{44}$$

where n is the cardinal number of the basis set (2, 3, 4 for DZ, TZ, QZ, respectively), $E(\infty)$ is the complete basis set limit, and E_0 and E_1 are constants. By assuming the basis set dependence of GMC-PT is the same, apart from a multiplicative factor, we extrapolated the energy in the complete basis set. The relative energies at the complete basis set limit, computed through the above extrapolation, were 7.6 (11.1) and 5.6 (9.9) kcal/mol for the 1t and 3s isomer, respectively, in the CCSD(T) (FORS-SCF) geometry. The relative order of the 1t and 3s isomers was unchanged which supports our conclusion that the 2s isomer is most stable.

20.2.2 Valence bond description of complete active space self-consistent field function

A defect of the multireference-based methods is that the wavefunction is too complicated to extract chemical description from it. There are too many CI coefficients, cluster amplitudes, or terms corresponding to diagrams in those methods. The information of the chemical picture is hidden behind them and to extract it seems quite difficult.

Classical valence bond (VB) theory is very successful in providing a qualitative explanation for many aspects. Chemists are familiar with the localized molecular orbitals (LMO) and the classical VB resonance concepts. If modern accurate wave functions can be represented in terms of such well-known concepts, chemists' intuition and experiences will give a firm theoretical basis and the role of the computational chemistry will undoubtedly expand.

The CASVB functions [50,51] can be obtained by transforming the canonical CASSCF functions without loss of energy. First we transform the CASSCF delocalized MO to localized MO using the arbitrariness in the definition of the active orbitals. Then we perform a full CI again in the active space. The CASVB method provides an alternative tool for describing the correlated wave functions.

Similar approaches have been employed by various workers. Lam et al. [52] showed that wave functions in FORS can be expressed in terms of localized configuration-generating MOs which have essentially atomic character. Cundari et al. [53] extended the idea and used it successfully to study the high-valent transition metal complexes. They used the orthogonal spin functions generated by the Kotani–Yamanouchi branching diagrams. Also our method has some relation to the spin-coupled valence bond (SCVB) method of Cooper et al. [54,55], where the spin-coupled orbitals and the spin-coupling coefficients are optimized simultaneously. Goddard et al. [56,57] have proposed the generalized valence bond (GVB) method. GVB has the advantage of compactness, as the wave functions are generally assumed to be formally purely covalent. However, GVB does not offer the clear relationship between the wave function and the various Lewis structure. Hiberty et al. [58,59] have also developed a general VB method and discussed chemical reactivity and structure.

20.2.2.1 The CASVB method [50,51]

We have proposed two types of CASVB method. In one method the valence bond structures are constructed over *orthogonal* localized orbitals, and in the other the structures are written with *non-orthogonal* localized orbitals. These are henceforth referred as orthogonal CASVB and non-orthogonal CASVB, respectively.

The idea of CASVB is based on the fact that the densities of variational wave functions are invariant under the transformations which hold the variational space unchanged.

In the CASSCF case, a complete active space (CAS) is invariant under the linear transformation of active orbitals and also that of configuration state functions (CSFs).

One may re-define the active orbitals utilizing the invariance of the active orbital space. In the orthogonal CASVB method, the LMOs constructed by Boys' localization procedure are used; that is, active orbitals are transformed so as to have the minimum sum of r^2 expectation values. If the active orbitals are defined appropriately, the LMOs obtained nearly always turn out to be localized on a single atomic center with small localization tails on to neighboring atoms. In the non-orthogonal CASVB case, the atomic-like orbitals are constructed by Ruedenberg's projected localization procedure.

Let Ψ^{CASSCF} be a CASSCF wave function

$$\Psi^{\mathrm{CASSCF}} = \sum_i C_i \Phi_i^{\mathrm{CSF}}, \qquad \Phi_i^{\mathrm{CSF}} \equiv \Phi_i^{\mathrm{CSF}}(\{\varphi_i\}) \tag{45}$$

where Φ_i^{CSF} are the CSFs constructed by the orthogonal orbitals set $\{\varphi_i\}$ and C_i are the known CAS configuration interaction (CI) expansion coefficients. Similarly, one may define the CASVB function in terms of spin-paired functions as

$$\Psi^{\mathrm{CASVB}} = \sum_i \mathrm{A}_i \Phi_i^{\mathrm{VB}}, \qquad \Phi_i^{\mathrm{VB}} \equiv \Phi_i^{\mathrm{VB}}(\{\lambda_i\}) \tag{46}$$

where Φ_i^{VB} is are a spin-paired function constructed by LMOs. The spaces, spanned by $\{\Phi_i^{\mathrm{CSF}}\}$ and $\{\Phi_i^{\mathrm{VB}}\}$, are identical. Since Eqs. (45) and (46) are different expressions of the identical wave function, one may write

$$\sum_j \mathrm{A}_j \Phi_j^{\mathrm{VB}} = \sum_j C_j \Phi_j^{\mathrm{CSF}} \tag{47}$$

Multiplying Eqs. (45) and (46) by Φ_i^{CSF} and integrating the products, one has a linear equation

$$\sum_j \Omega_{ij} \mathrm{A}_j = C_i \qquad \text{with} \qquad \Omega_{ij} = \langle \Phi_i^{\mathrm{CSF}} | \Phi_j^{\mathrm{VB}} \rangle \tag{48}$$

whose dimension is equal to the dimension of CAS. Solving this linear equation, one obtains CASVB wave function Ψ^{CASVB}. In the orthogonal CASVB case, one can use the common set of (Boys') LMOs as $\{\varphi_i\}$ as well as $\{\lambda_i\}$ since the LMOs remain CASSCF MOs. In that case, the linear equation (48) is reduced to a set of linear equations for each *orbital* configuration, and the matrix Ω_{ij} for each linear equation becomes a triangular matrix depending only on spin configurations. The linear equation (48) can, therefore, be solved with ease, compared with the non-orthogonal CASVB case.

The occupation number (or weight) of a resonance structure is calculated with

$$n_i = A_i^* \sum_j S_{ij} A_j \tag{49}$$

where S_{ij} are overlaps between the structures i and j, defined by

$$S_{ij} = \langle \Phi_i^{\mathrm{VB}} | \Phi_j^{\mathrm{VB}} \rangle \tag{50}$$

and satisfies the normalization

$$\sum_i n_i = 1 \tag{51}$$

Note that the occupation number n_i could be negative because of the non-orthogonality of resonance structures.

20.2.2.2 Description of electronic structure of benzene

As an example, a CASVB description for benzene is given in Fig. 20.2. See Refs. [50,51] for the computational details. The CASVB affords a clear view of the wave functions for the various states. The excitation process is represented in VB theory in terms of the rearrangement of spin couplings and charge transfer. The former generates the covalent excited states and the latter gives rise to the ionic excited states, in which the covalent bond is broken and a new ionic bond is formed. Thus, the singly, doubly, etc. polar structures are generated from their respective *parent* ground state covalent (nonpolar), singly, etc., polar structures.

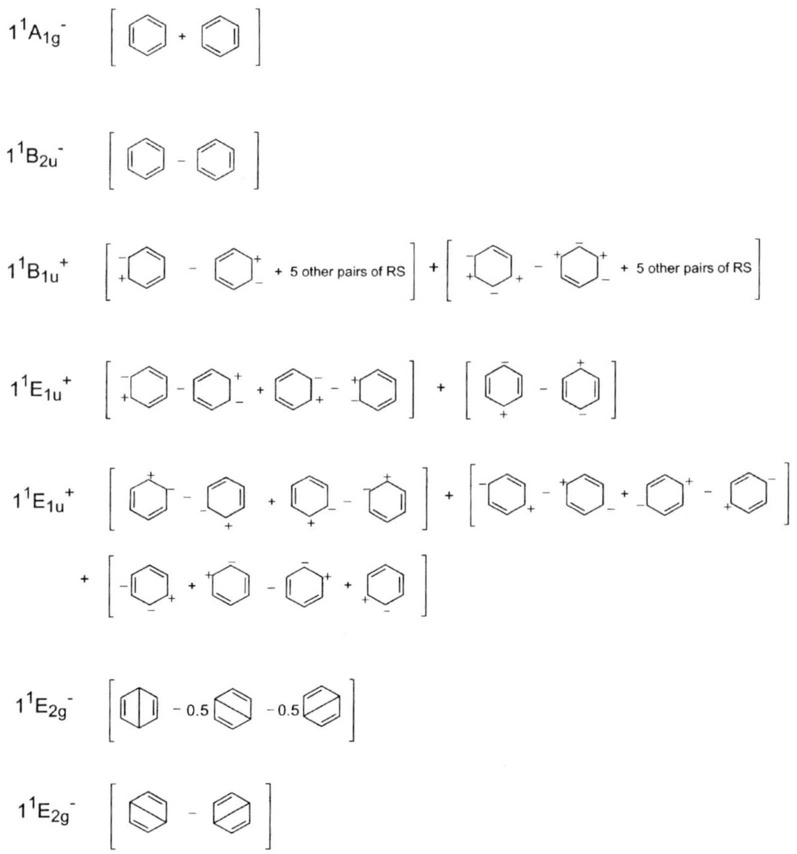

Fig. 20.2. CASVB description for the ground and $\pi \rightarrow \pi^*$ singlet excited states of benzene.

The ground state is represented by two covalent Kekulé structures as expected. The lowest excited $^1B_{2u}^-$ state is again described by a combination of the Kekulé structure. There are no significant contributions from the Dewar structures or the corresponding orthopolar structures. The linear combinations of the two equivalent Kekulé structures generate the plus and minus states. Their positive combination gives rise to the totally symmetric $^1A_{2g}^-$ ground state, while the negative combination yields the excited $^1B_{2u}^-$ state. The second and third $\pi \rightarrow \pi^*$ excited states are described by a number of ionic structures. There is no contribution from the covalent structures. The ionic character of these states can easily be found from a CASVB description. The highest valence excited states are the covalent $^1E_{2g}^-$ state. The state has a predominantly Dewar character with no contribution from the Kekulé structures. Thus, the Kekulé structures dominate the ground state and the singly excited $1B_{2u}^-$ state while the Dewar structures dominate the doubly excited degenerate $^1E_{2g}^-$ states. The states described by Dewar structures are described by doubly, triply, etc. excitations in an MO language.

20.2.2.3 Description of chemical reaction—hydrogen exchange reactions $H_2 + X \rightarrow H + HX$ (X = F, Cl, Br, and I) [60]

We examine a series of reactions including ionic bonds

$$H_2 + F \rightarrow H + HF \tag{R1}$$

$$H_2 + Cl \rightarrow H + HCl \tag{R2}$$

$$H_2 + Br \rightarrow H + HBr \tag{R3}$$

$$H_2 + I \rightarrow H + HI \tag{R4}$$

The reaction for fluorine (R1) is highly exothermic, while the reactions for chlorine (R2), bromine (R3), and iodine (R4) are endothermic. The heats of these reactions are 30.8, − 1.2, − 16.7, and − 32.7 kcal/mol for reactions (R1), (R2), (R3), and (R4), respectively. According to Hammond's postulate, reaction (R1) should have an early TS, and reactions (R2) and (R3) should have late TSs. What the electronic states are during these reactions, and how the CASVB method describes the electronic structure, are our interests in this section.

The active spaces were constructed by distributing three electrons in three orbitals consisting of $H_1(1s)$, $H_2(1s)$, and $X(2p\sigma)$, i.e. CAS(3,3). The dimension of the CAS is eight. According to this CAS, eight linearly independent VB structures

$$\varphi_{H_A}\varphi_{H_B}(\alpha\beta - \beta\alpha)\cdot\varphi_X\alpha, \qquad H_A - H_B\,\dot{X} \tag{I}$$

$$\varphi_{H_B}\varphi_{H_B}\alpha\beta\cdot\varphi_X\alpha, \qquad H_A^+\ {}^-H_B\,\dot{X} \tag{II}$$

$$\varphi_{H_A}\varphi_{H_A}\alpha\beta\cdot\varphi_X\alpha, \qquad H_A^-\ {}^+H_B\,\dot{X} \tag{III}$$

$$\varphi_{H_A}\alpha\cdot\varphi_{H_B}\varphi_X(\alpha\beta - \beta\alpha), \qquad \dot{H}_A\,H_B - X \tag{IV}$$

$$\varphi_{H_A}\alpha\cdot\varphi_{H_B}\varphi_{H_B}\alpha\beta, \qquad \dot{H}_A\,H_B^-\ {}^+X \tag{V}$$

$$\varphi_{H_A}\alpha\cdot\varphi_X\varphi_X\alpha\beta, \qquad \dot{H}_A\,H_B^+\ {}^-X \tag{VI}$$

$$\varphi_{H_B}\alpha\cdot\varphi_X\varphi_X\alpha\beta, \qquad H_A^+\ \dot{H}_B\ {}^-X \tag{VII}$$

and

$$\varphi_{H_A}\varphi_{H_A}\alpha\beta\cdot\varphi_{H_B}\alpha, \qquad H_A^-\ \dot{H}_B\ {}^+X \tag{VIII}$$

were used to construct CASVB functions, where the normalization constants and antisymmetrizers are omitted.

The contributions of the covalent H_AH_B bond, ionic H_AH_B bond, covalent H_BX bond, ionic H_BX bond, and ionic H_AX bond are defined by

$$n_{\text{Covalent } H_AH_B} = n_I, \qquad n_{\text{Ionic } H_AH_B} = n_{II} + n_{III} \tag{52}$$

$$n_{\text{Covalent } H_BX} = n_{IV}, \qquad n_{\text{Ionic } H_BX} = n_V + n_{VI} \tag{53}$$

and

$$n_{\text{Ionic } H_AX} = n_{VII} + n_{VIII} \tag{54}$$

Furthermore, the contributions of the total H_AH_B and H_BX bond structures are defined by the sums of the covalent structure (I)/(IV) and ionic structures (II)/(V) and (III)/(VI)

$$n_{H_AH_B} = n_{\text{Covalent } H_AH_B} + n_{\text{Ionic } H_AH_B}, \qquad n_{H_BX} = n_{\text{Covalent } H_BX} + n_{\text{Ionic } H_BX} \tag{55}$$

Let us first examine the electronic structure at the TS structure of the four reactions. Table 20.6 shows the VB structure at the TSs of $H_2 + X \rightarrow H + HX$. We can see that the covalent VB structures are dominant: the structures are well described by the superposition of the HH and HX covalent structure with small H^+H and H^+X^- ionic contributions. Using Eq. (55), these structures are further classified as the HH and HX bonds, as shown in Table 20.6. For X = F, the contribution of the HH bond (55.5%) is larger than that of the HX bond (39.4%). This relation is reversed for X = Cl, Br, and I. The contribution of the HH bond increases as the halogen atom becomes heavier (55.8 (Cl), 69.8 (Br), and 76.4% (I)). This means that *the TS of chemical bonds* (that is, the point where the occupation numbers of the two chemical bonds are equal) is placed in

Table 20.6 Occupation numbers of the VB structures at the TS

	H + H + F	H + H + Cl	H + H + Cl	H + H + Cl
H–H Ẋ (I)	0.485	0.328	0.217	0.172
H^+ $^-$H Ẋ (II)	0.053	0.059	0.043	0.034
H^- $^+$H Ẋ (III)	0.017	−0.022	−0.023	−0.019
Ḣ H–X (IV)	0.252	0.385	0.514	0.591
Ḣ H^- $^+$X (V)	−0.005	0.018	0.042	0.073
Ḣ H^+ $^-$X (VI)	0.147	0.155	0.142	0.100
H^-Ḣ $^+$X (VII)	0.006	0.013	0.011	0.010
H^+Ḣ $^-$X (VIII)	0.045	0.064	0.053	0.039
HH bond	0.555	0.365	0.237	0.187
HX bond	0.394	0.558	0.698	0.764
Others	0.051	0.077	0.064	0.049

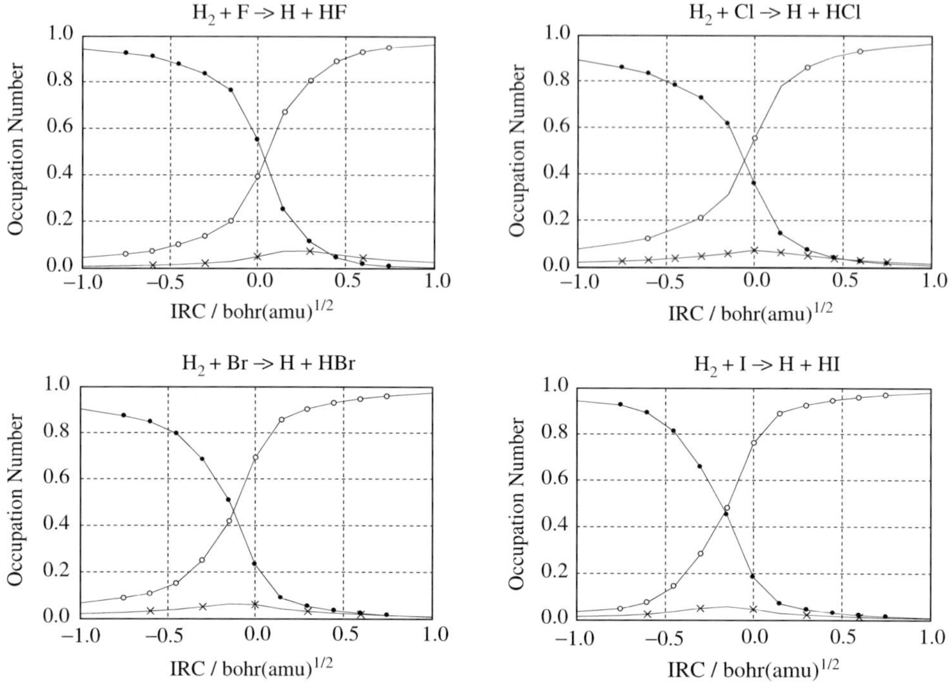

Fig. 20.3. Changes in the occupation numbers of the total HH bond (●), total HX bond (○), and the other (×) VB structures along IRC. The origin of the horizontal axis corresponds to the TS.

the reactant side in the X = F case and in the product side for the case of X = Cl, and it shifts more to the product side as the halogen atom becomes heavier.

We then examine the bond nature during the reactions. Fig. 20.3 shows the changes in the total occupation number of the HH and HX bond structures along the IRC. The occupation numbers of the HH and HX bond structures change rapidly and the curves cross near the TS. The crossing points are located at 0.07, −0.11, −0.25, and −0.33 bohr(a.m.u.)$^{1/2}$ for X = F, Cl, Br, and I, respectively, where a negative sign means the crossing point is located before the TS and a positive sign after the TS. We can see the trend that the crossing point shifts from the reactant side to the product side as the halogen atom gets heavier. The changes in the contents of the HH and HX bonds are plotted in Fig. 20.4. As expected in these reactions including ionic bonds, the contribution of ionic bond increases as that of the covalent bond increases. However, the crossing point of HH and HX covalent bonds are still close to that of the HH and HX bonds in Fig. 20.3. The covalent bonds are mainly responsible for determining the crossing points.

If we re-take *the TS of chemical bonds* as the origin, these facts well explain the shift of TS (from the early TS side to the late TS side) that Hammond's postulate predicts, indicating that the CASVB method is a powerful tool for describing the electronic structure and chemical bond during chemical reactions.

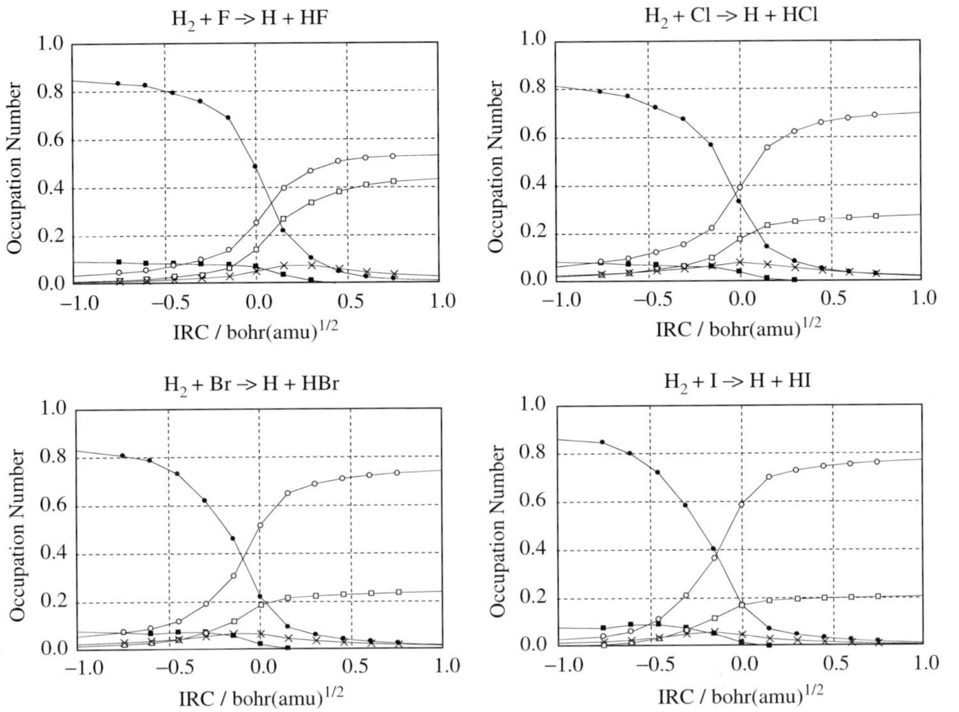

Fig. 20.4. Changes in the occupation numbers of the covalent HH bond (●), ionic HH bond (■), covalent HX bond (○), ionic HX bond (□), and the other (×) VB structures along the IRC.

20.3 LONG-RANGE AND OTHER CORRECTIONS FOR DENSITY FUNCTIONALS

20.3.1 Conventional correction schemes in density functional theory

In this century, the main concerns of theoretical chemistry obviously make the transition from accurate investigations of small molecules to the designs of complicated large molecular systems; e.g. proteins, nano-materials, environmental catalyses, and so forth. What is necessary for approaching these systems is an accurate theory of low-computational order. DFT [61–63] is expected to be a major candidate for such a theory at present, because this theory gives accurate chemical properties despite its low-computational order that may be reduced to order-N. In DFT, electronic states are usually determined by solving the nonlinear Kohn–Sham equation [61] with an exchange-correlation density functional. The most remarkable characteristic of DFT is the exchange-correlation energy part that is approximated by a one-electron potential functional. Hence, calculated DFT results depend on the form of this exchange-correlation functional.

In last two decades, various kinds of exchange functionals have been suggested especially for generalized gradient approximations (GGA) [64–67] beyond the local density approximation (LDA) functional [68]. Due to the requirement of order, these GGA exchange functionals are usually expressed as a functional of $x_\sigma = |\nabla\rho_\sigma|/\rho_\sigma^{4/3}$, where ρ_σ is the electron density of spin σ and $\nabla\rho_\sigma$ is the gradient of the density [69]. What should be noticed is that most GGA exchange functionals have unique behaviors only for large x_σ [69,70]. This is because small-x_σ behaviors of functionals are restricted by the physical condition for slowly-varying density [71], although there is no definite conditions for rapidly varying density [67,69,70]. Hence, GGA exchange functionals are usually characterized by the behaviors for large-x_σ (i.e. low-density-high-gradient) density. Conventional exchange-correlation functionals will be discussed by Prof. Scuseria in this book. We will give a summary account of correction schemes for exchange functionals in this section.

Since the latter half of 1990s, hybrid functionals have appeared in DFT calculations. In hybrid functionals, (pure) GGA functionals are combined with the Hartree–Fock (HF) exchange integral at a constant rate. This idea may have come from an observation that DFT calculations using pure GGA functionals often give opposite errors to those in HF calculations. In 1993, Becke suggested hybrid B3LYP functional [72]. Based on a concept of adiabatic connection, B3LYP exchange-correlation energy are expressed by a combination of Becke 1988 (B88) exchange [64] and Lee–Yang–Parr (LYP) correlation [73] GGA functionals, Slater (S) exchange [68] and Vosko–Wilk–Nusair (VWN) correlation [74] LDA functionals, and the HF exchange integral with three parameters:

$$E_{\mathrm{xc}}^{\mathrm{B3LYP}} = a_0 E_{\mathrm{x}}^{\mathrm{HF}} + (1 - a_0)E_{\mathrm{x}}^{\mathrm{S}} + a_{\mathrm{x}} E_{\mathrm{x}}^{\mathrm{B88}} + (1 - a_{\mathrm{c}})E_{\mathrm{c}}^{\mathrm{VWN}} + a_{\mathrm{c}} E_{\mathrm{c}}^{\mathrm{LYP}} \tag{56}$$

where $E_{\mathrm{x}}^{\mathrm{A}}$ and $E_{\mathrm{c}}^{\mathrm{B}}$ are exchange and correlation energies of A and B, and a_0, a_{x}, and a_{c} are 0.2, 0.72, and 0.81, respectively. Atomic units have been used ($\hbar = e^2 = m = 1$, energies are in hartree, and distances are in bohr). This adiabatic connection may have some incompatible parts; for example, parameters in B88 and LYP functionals were originally determined to reproduce exact exchange and correlation energies. Nevertheless, B3LYP becomes the most popular DFT functional in quantum chemistry, because it gives very accurate results for a wide variety of chemical properties. Becke 1997 (B97) [75] and Perdew–Burke–Ernzerhof 1996 (PBE0) [76] functionals are also hybrid functionals. Similar to B3LYP, these functionals combine GGA functionals with the HF exchange integral at a constant rate, and give accurate results for various chemical properties of molecules. However, inconsistencies in the adiabatic connection remains unsettled in these functionals.

Asymptotic corrections for exchange functionals have attracted attentions especially in time-dependent DFT (TDDFT) studies. In far regions from atomic nuclei, it is proved that exchange potential for σ-spin electrons, $v_{\mathrm{xc}}^{\sigma} = \delta E_{\mathrm{xc}}/\delta\rho_\sigma$ has the asymptotic relation [77]

$$\lim_{R\to\infty} v_{\mathrm{xc}}^{\sigma}(\mathbf{R}) = -\frac{1}{R} \tag{57}$$

where R = $|\mathbf{R}|$ and $\mathbf{R}$ is the distance vector from the nearest nucleus. On the ground of this relation, Van Leeuwen and Baerends suggested an exchange functional (LB) [78] by adapting the B88 exchange functional to the asymptotic behavior. Tozer and Handy suggested the asymptotic correction (AC) scheme that imposes, instead of Eq. (57) [79,80]

$$\lim_{R\to\infty} v_{\rm xc}^{\sigma}(\mathbf{R}) = -\frac{1}{R} + \varepsilon_{\sigma}^{\rm HOMO} + I_{\sigma} \tag{58}$$

where $\varepsilon_{\sigma}^{\rm HOMO}$ is the eigenvalue of the highest occupied σ-spin molecular orbital and I_{σ} is the ionization potential of the σ-spin electron. It has been reported that underestimations of Rydberg excitation energies in TDDFT calculations are modified by using LB and AC schemes.

Besides, self-interaction correction (SIC) is one of the most popular correction schemes. Perdew and Zunger suggested a scheme for the application of SIC to occupied orbitals where the self-interaction components of the Coulomb and exchange energies are simply subtracted from the total exchange-correlation energy [81]

$$E_{\rm xc}^{\rm SIC}[\rho_{\alpha},\rho_{\beta}] = E_{\rm xc}[\rho_{\alpha},\rho_{\beta}] - \sum_{i,\sigma}\left(\frac{1}{2}\int\frac{\rho_{i\sigma}(\mathbf{R})\rho_{i\sigma}(\mathbf{R}')}{|\mathbf{R}-\mathbf{R}'|}{\rm d}^3\mathbf{R}\,{\rm d}^3\mathbf{R}' + E_{\rm xc}[\rho_{i\sigma},0]\right) \tag{59}$$

and potential

$$v_{\rm xc}^{i\sigma,{\rm SIC}}(\mathbf{R}) = \frac{\delta E_{\rm xc}^{\rm SIC}}{\delta\rho_{i\sigma}(\mathbf{R})} = \frac{\delta E_{\rm xc}[\rho_{\alpha},\rho_{\beta}]}{\delta\rho_{i\sigma}(\mathbf{R})} - \int\frac{\rho_{i\sigma}(\mathbf{R}')}{|\mathbf{R}-\mathbf{R}'|}{\rm d}^3\mathbf{R}' - \frac{\delta E_{\rm xc}[\rho_{i\sigma},0]}{\delta\rho_{i\sigma}(\mathbf{R})} \tag{60}$$

where $\rho_{i\sigma}$ is the ith orbital component of ρ_{σ}. This SIC scheme has been frequently used in energy band calculations of solid states for improving underestimated band gap energies. However, this scheme essentially requires an orbital-localization process [82] or transformation of functionals to an orbital-dependent form [83] due to the degrees of freedom in unitary transformations of orbitals. Tsuneda, Kamiya, and Hirao suggested a regional self-interaction correction (RSIC) scheme as a simple SIC method requiring no additional processes [84]. On the ground that total kinetic energy density, $\tau_{\sigma}^{\rm total} = \sum_i^{\rm occ}|\nabla\psi_{i\sigma}|^2$, approaches the Weizsäcker kinetic energy density, $\tau_{\sigma}^{\rm W} = |\nabla\rho_{\sigma}|^2/(4\rho_{\sigma})$ for self-interacted electrons, an exchange functional is spatially replaced with a self-interaction energy density only for regions, where $\tau_{\sigma}^{\rm total}$ approaches $\tau_{\sigma}^{\rm W}$, in this scheme. As the self-interaction energy density, exact exchange self-interaction energy densities of 1s orbitals in hydrogen-like atoms, $\psi_{i\sigma}^{\rm 1s} = \sqrt{\alpha^3/\pi}\exp(-\alpha R)$, is employed such as

$$\varepsilon_{{\rm x}\sigma}^{\rm RSIC} = -\left(\frac{\rho_{\sigma}}{2R}\right)[1-(1+\alpha R)\exp(-2\alpha R)] \tag{61}$$

where ε_{σ} is defined by $E_{\rm x} \equiv \sum_{\sigma}\int\varepsilon_{{\rm x}\sigma}\,{\rm d}^3\mathbf{R}$ and $\alpha = \nabla\rho_{\sigma}/(2\rho_{\sigma})$. By applying the RSIC scheme to chemical reaction calculations, it was found that underestimated barrier energies of pure functionals were clearly improved for some reactions.

As mentioned above, various correction schemes have been developed up to the present. However, there is room for further improvement in conventional correction schemes. Conventional hybrid functionals give poor excitation energies in TDDFT calculations as mentioned later. Asymptotic and SICs have little (or worse) effect on reproducibilities of molecular chemical properties. Recently, it has been proved that a long-range correction for exchange functionals obviously brings solutions to various

DFT problems that have never been solved by other functionals or corrections. In later sections, we will briefly review the background of the long-range correction scheme and will reveal the applicabilities of this scheme.

20.3.2 Long-range correction schemes for exchange functionals

Pure DFT exchange-correlation functionals have been represented by using only local quantities at a reference point: e.g. electron density, gradient of density, and etc. (We are now describing 'local' quantity as a quantity determined at a reference point for clarity, although gradient of density is known as a 'nonlocal' quantity in common use.) It is, therefore, presumed that pure functionals overestimate local contributions and underestimate nonlocal contributions. The most significant nonlocal contribution neglected in pure functionals may be the long-range electron–electron exchange interaction, because it may be impossible to represent this interaction as a functional of a one-electron quantity.

In 1996, Savin suggested a long-range exchange correction scheme for LDA functional [85]. In this scheme, the two-electron operator, $1/r_{12}$, is separated into the short-range and long-range parts naturally by using the standard error function erf such that

$$\frac{1}{r_{12}} = \frac{1 - \mathrm{erf}(\mu r_{12})}{r_{12}} + \frac{\mathrm{erf}(\mu r_{12})}{r_{12}} \tag{62}$$

where $r_{12} = |\mathbf{r}_1 - \mathbf{r}_2|$ for coordinate vectors of electrons, $\mathbf{r}_1$ and $\mathbf{r}_2$, and μ is a parameter that determines the ratio of these parts. Based on Eq. (62), the long-range exchange interaction is described by the HF exchange integral

$$E_{\mathrm{x}}^{\mathrm{lr}} = -\frac{1}{2}\sum_{\sigma}\sum_{i}^{\mathrm{occ}}\sum_{j}^{\mathrm{occ}}\iint \psi_{i\sigma}^{*}(\mathbf{r}_1)\psi_{j\sigma}^{*}(\mathbf{r}_2)\frac{\mathrm{erf}(\mu r_{12})}{r_{12}}\psi_{j\sigma}(\mathbf{r}_1)\psi_{i\sigma}(\mathbf{r}_2)\mathrm{d}^3\mathbf{r}_1\,\mathrm{d}^3\mathbf{r}_2 \tag{63}$$

where $\psi_{i\sigma}$ is the ith σ-spin orthonormal molecular orbital. The LDA exchange functional is applied to the short-range exchange interaction such that

$$E_{\mathrm{x}}^{\mathrm{sr}} = -\frac{3}{2}\left(\frac{3}{4\pi}\right)^{1/3}\sum_{\sigma}\int \rho_{\sigma}^{4/3}\left\{1 - \frac{8}{3}a_{\sigma}\left[\sqrt{\pi}\,\mathrm{erf}\left(\frac{1}{2a_{\sigma}}\right) + \left(2a_{\sigma} - 4a_{\sigma}^{3}\right)\exp\left(-\frac{1}{4a_{\sigma}}\right) - 3a_{\sigma} + 4a_{\sigma}^{3}\right]\right\}\mathrm{d}^3\mathbf{R} \tag{64}$$

where $a_{\sigma} = \mu/(2k_{\sigma})$. The averaged relative momentum k_{σ} is written for LDA as the Fermi momentum, i.e. $k_{\mathrm{F}\sigma} = (6\pi^2\rho_{\sigma})^{1/3}$. Eq. (64) is derived by using the density matrix form corresponding to the LDA exchange functional

$$P_{1\sigma}^{\mathrm{LDA}}\left(\mathbf{R} + \frac{\mathbf{r}}{2}, \mathbf{R} - \frac{\mathbf{r}}{2}\right) = 3\frac{j_1(k_{\mathrm{F}\sigma}r)}{k_{\mathrm{F}\sigma}r}\rho_{\sigma}(\mathbf{R}) \tag{65}$$

where j_1 is the first-order spherical Bessel function.

However, Savin's scheme is inapplicable to conventional GGA exchange functionals, because GGA functionals usually have no corresponding density matrices unlike LDA. In 2001, Iikura et al. solved this problem by pushing gradient terms of GGA functionals into the momentum k_σ [86]. That is, the corresponding density matrix is determined for any GGA exchange functional by substituting $k_{F\sigma}$ in Eq. (65) with

$$k_\sigma^{GGA} = \left(\frac{9\pi}{K_\sigma^{GGA}} \right)^{1/2} \rho_\sigma^{1/3} \tag{66}$$

where K_σ^{GGA} is defined in an exchange functional used: $E_x^{GGA} = \int \rho_\sigma^{4/3} K_\sigma^{GGA} \, d^3\mathbf{R}$. Eq. (66) correctly reproduces the Fermi momentum $k_{F\sigma}$ for K_σ^{LDA}. By using k_σ^{GGA}, the short-range exchange energy in Eq. (64) is substituted by

$$E_x^{sr} = -\frac{3}{2}\left(\frac{3}{4\pi}\right)^{1/3} \sum_\sigma \int \rho_\sigma^{4/3} K_\sigma^{GGA} \left\{ 1 - \frac{8}{3} a_\sigma \left[\sqrt{\pi}\, \mathrm{erf}\left(\frac{1}{2a_\sigma}\right) + (2a_\sigma - 4a_\sigma^3)\exp\left(-\frac{1}{4a_\sigma}\right) - 3a_\sigma + 4a_\sigma^3 \right] \right\} d^3\mathbf{R} \tag{67}$$

It is easily confirmed that Eq. (67) reproduces the original GGA exchange functional for $\mu = 0$. Parameter μ is determined to optimize bond distances of homonuclear diatomic molecules up to the third period as $\mu = 0.33$. This scheme is called 'long-range correction (LC) scheme'. The applicabilities of the LC scheme will be discussed in the later section.

Besides the LC scheme, we should mention the screened Coulomb potential hybrid functional as an attempt to take account of the long-range exchange effect. Heyd et al. developed this functional by dividing the exchange terms of the hybrid PBE0 functional into short- and long-range parts and by omitting a part of long-range exchange term as [87]

$$E_{xc}^{\omega PBEh} = aE_x^{HF,sr}(\omega) + (1-a)E_x^{PBE,sr}(\omega) + E_x^{PBE,lr}(\omega) + E_c^{PBE} \tag{68}$$

where $a = 1/4$ is a mixing coefficient and ω is an adjustable parameter. The main characteristics of this functional are the inclusion of the short-range HF exchange integral and the exclusion of the long-range HF exchange integral. This functional gives more accurate chemical properties than those of B3LYP for G2 and G3 set of molecules [87,88]. It is, however, presumed that this functional may not solve DFT problems arising from the lack of long-range exchange effects due to the exclusion of the long-range HF exchange integral. Moreover, Leininger et al. extended the above Savin's scheme by using the long-range exchange integral for the multireference configuration interaction (MRCI) wavefunction [89]. Electron correlations in long-range interactions may hardly affect calculated properties of standard molecules. However, electron correlations may be important for the comparison of molecules that have much different spin-multiplicity or neardegeneracy.

20.3.3 Applicabilities of long-range correction scheme

In this section, the LC scheme are examined by illustrating its applicabilities to three DFT problems that have never been solved: (1) poor reproducibilities of van der Waals (vdW) bondings, (2) underestimations of Rydberg excitation energies, oscillator strengths, and charge transfer excitation energies in time-dependent density functional calculations, and (3) systematic overestimations of atomization energies of transition metal dimers.

20.3.3.1 Van der Waals calculations

One of the most critical DFT problems is the poor reproducibility of vdW bondings. Actually, conventional correlation functionals have incorporated no vdW interactions. Since vdW bondings, however, often determine structures of large-scale molecules, accurate calculations of vdW bondings are a pressing problem in DFT. Several DFT studies have been made on vdW calculations by using, e.g. a perturbation theory based on DFT [90]. The most effective and general way may be the use of a vdW functional. Up to the present, various types of vdW functionals have been suggested [91–93]. Some of these functionals reproduce accurate vdW C_6 coefficient comparable to the results of high-level *ab initio* methods [91]. However, these functionals give poor vdW bondings of, e.g. rare gas dimmers by simply combining with a conventional exchange-correlation functional in DFT calculations. It is presumed that this problem may be due to the lack of long-range interactions in exchange functionals, because vdW bondings are supposed to be in the balance between vdW attraction and long-range exchange repulsion interactions. On this ground, Kamiya et al. applied the LC scheme with a vdW functional to calculations of dissociation potentials of rare-gas dimmers [94]. Andersson–Langreth–Lundqvist (ALL) functional was used as the vdW functional [91]. This functional was developed to be correct for both separated electron gas regions and far-apart atoms. In this functional, a damping factor was used to diminish the vdW energy for regions at a short distances.

In Fig. 20.5, calculated dissociation potential energy curves of Ar_2 are shown for pure GGA functionals (BOP and PBEOP) [95] and LC functionals (LC-BOP and LC-PBEOP) with no vdW functionals. The 6-311++G(3df,3pd) basis functions was used [96–98]. The basis set superposition error was corrected by a counterpoise method [99]. As the figure shows, LC functionals give very close potential curves to each other, although pure GGA functionals provide obviously different curves. This may indicate that a long-range correction is necessary for exchange functionals to reproduce vdW bondings.

Next, Fig. 20.6 displays calculated dissociation potentials of Ar_2 by LC-BOP + ALL and conventional sophisticated functionals (mPWPW91 [100], mPW1PW91 [100], and B3LYP + vdW [101]). The second-order Møller–Plesset perturbation (MP2) and experimentally predicted (Expt) [16] potential curves are also shown for comparison. The figure clearly shows that LC-BOP + ALL functional gives accurate vdW potential curve in comparison with the results of MP2 and other conventional DFTs. It is, therefore, necessary for accurate DFT calculations of vdW bondings to use both long-range-corrected exchange and vdw-incorporated correlation functionals.

20.3.3.2 Time-dependent density functional calculations

Time-dependent density functional theory (TDDFT) becomes widely used as a simple method for rapid and accurate calculations of molecular excitation energies. It has, however, been reported that conventional TDDFT calculations underestimate Rydberg excitation energies, oscillator strengths, and charge-transfer excitation energies.

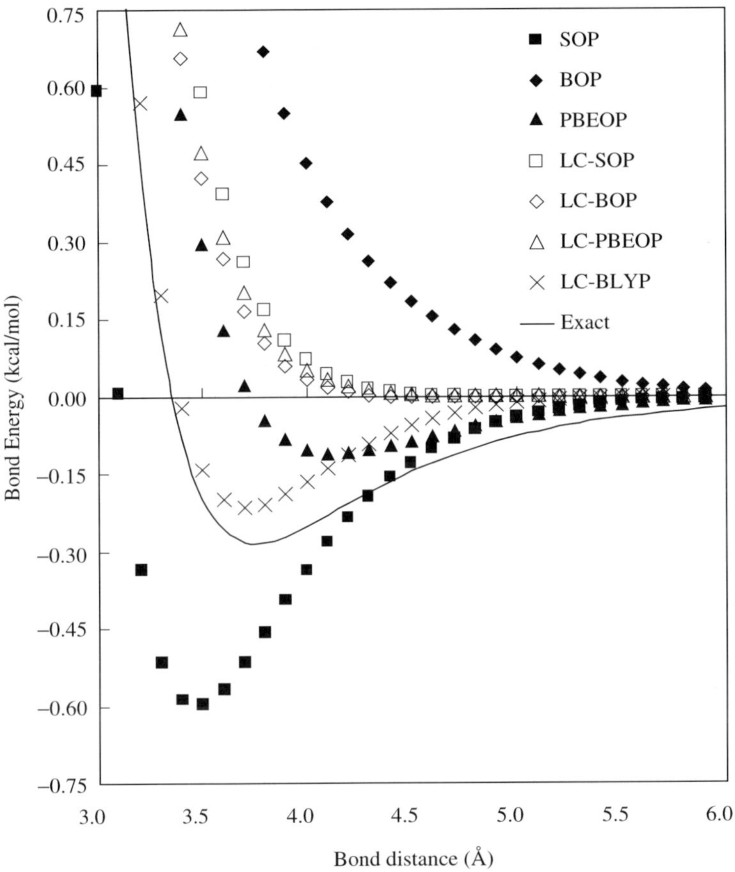

Fig. 20.5. Calculated bond energy potentials of argon dimer for long-range exchange corrected functionals (LC-SOP, LC-BOP, and LC-PBEOP). Pure functionals are also presented for comparison. Highly accurate potentials are also shown for comparison.

Tawada et al. supposed that this problem may also come from the lack of long-range exchange interaction, and applied the LC scheme to TDDFT calculations [102].

Table 20.7 summarizes mean absolute errors in calculated excitation energies of five typical molecules by TDDFT. The table also displays calculated results of asymptotically corrected AC [79] and LB [78] (AC-BOP and LBOP) and hybrid B3LYP [72] functionals, which are mentioned in the former section. The *ab initio* SAC-CI [103] results are also shown to confirm the accuracies. The 6-311G++(2d,2p) basis set was used in TDDFT calculations [104,105]. As the table indicates, the LC scheme clearly improves Rydberg excitation energies that are underestimated for pure BOP functional, at the same (or better) level as the AC scheme does. It should be noted that LC and AC schemes also provide improvements on valence excitation energies for all molecules. LC and AC results are comparable to SAC-CI results. The LB scheme clearly modifies

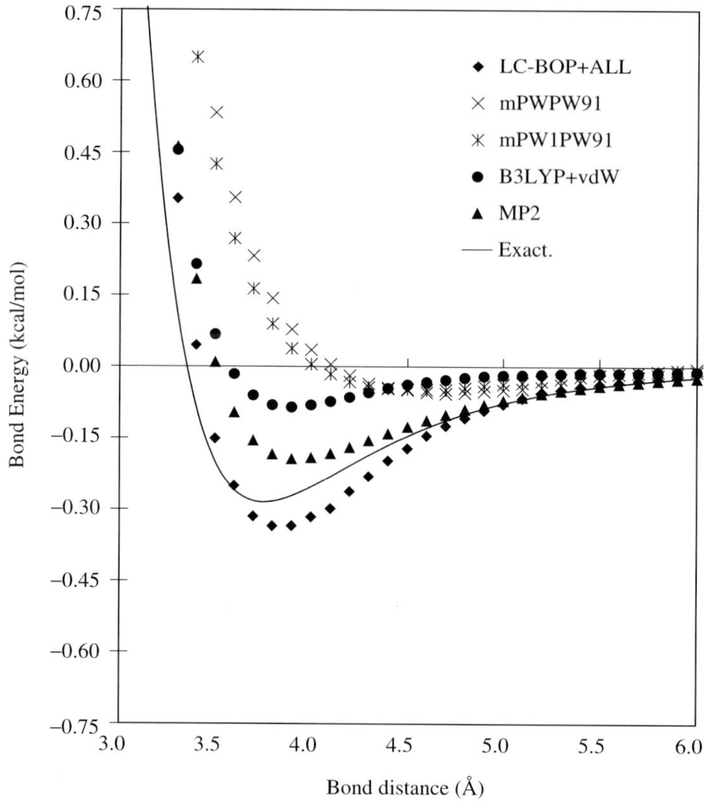

Fig. 20.6. Calculated bond energy potentials of argon dimer for LC-BOP+ALL functional. For comparison, calculated potentials of conventional sophisticated density functional schemes (mPWPW91, mPW1PW91, and B3LYP + vdW) and those of MP2 are also presented. Highly accurate potentials are also shown for comparison.

Rydberg excitation energies, and however brings underestimations of valence excitation energies. B3LYP results are obviously worse than LC and AC results for both valence and Rydberg excitation energies.

Next, calculated oscillator strengths of excited states by TDDFT are shown in Table 20.8. As is clearly shown in the table, LC scheme drastically improves oscillator strengths, which are underestimated for BOP as second to hundredth part of experimental values, to the same digit. Although AC-BOP, LBOP, and B3LYP also provide closer oscillator strengths to the experimental values than BOP do, the accuracies are unsatisfactory in comparison with LC-BOP ones. It is, therefore, concluded that the lack of long-range interactions in exchange functional may also cause the underestimations of oscillator strengths in TDDFT calculations.

Finally, calculated lowest charge transfer excitation energies of ethylene–tetrafluoroethylene dimer are shown in Fig. 20.7. Dreuw et al. recently suggested that poor charge transfer excitation energies of far-aparted molecules may be one of the main problems

Table 20.7 Mean absolute errors in calculated excitation energies of five typical molecules by TDDFT in eV

Molecule		LC-BOP	BOP	AC-BOP	LBOP	B3LYP	SAC-CI
N_2	Valence	0.36	0.40	0.27	1.48	0.54	0.33
	Rydberg	0.90	2.37	0.84	0.43	1.30	0.25
	Total	0.54	1.06	0.46	1.13	0.79	0.30
CO	Valence	0.19	0.28	0.17	1.02	0.36	0.26
	Rydberg	0.75	2.06	0.79	0.42	1.16	0.27
	Total	0.47	1.17	0.48	0.72	0.76	0.27
H_2CO	Valence	0.25	0.59	0.24	0.52	0.26	0.45
	Rydberg	0.47	1.66	0.59	0.07	0.84	0.13
	Total	0.40	1.30	0.47	0.22	0.64	0.24
C_2H_4	Valence	0.30	0.47	0.24	1.52	0.47	0.11
	Rydberg	0.18	1.41	0.58	0.69	0.92	0.17
	Total	0.20	1.28	0.53	0.80	0.85	0.16
C_6H_6	Valence	0.21	0.28	0.24	0.84	0.26	0.35
	Rydberg	0.24	1.01	0.88	0.35	0.56	0.15
	Total	0.23	0.74	0.64	0.53	0.44	0.22

of TDDFT [106]. They pointed out that intermolecular charge transfer excitation energies of far-aparted molecules should have the correct asymptotic behavior for long intermolecular distance. That is, for long molecular–molecular distances R and R_0 ($R > R_0$), charge transfer energy ω_{CT} should satisfy

$$\omega_{CT}(R) - \omega_{CT}(R_0) \geq -\frac{1}{R} + \frac{1}{R_0} \tag{69}$$

The figure shows that LC-BOP gives the correct asymptotic behavior as is different from AC-BOP and LBOP do. Although B3LYP recovers a part of this behavior, the degree is in proportion to the mixing rate of the HF exchange integral. Hence, this result may also indicate that problems in conventional TDDFT calculations come from the lack of long-range exchange interactions in exchange functionals rather than the poor far-nucleus asymptotic behavior of exchange functionals.

Table 20.8 Calculated oscillator strengths of excited states of typical molecules by TDDFT ($\times 10^{-2}$)

System	State	LC-BOP	BOP	AC-BOP	LBOP	B3LYP	SAC-CI	Exptl
N_2	$^1\Pi_u$	11.05	0.28	2.02	4.18	1.33	8.14	24.3
	$^1\Sigma_u^+$	24.06	0.69	6.07	3.60	3.84	15.67	27.9
CO	$^1\Pi$	19.76	8.66	6.68	5.97	11.24	9.63	17.6
H_2CO	$^1B_2$	2.19	1.68	1.02	3.02	2.71	1.88	4.13, 2.8, 3.8, 3.2
	$^1A_1$	6.94	2.11	2.62	1.80	3.64	4.26	6.05, 3.2, 3.8, 3.6
	$^1B_2$	6.50	1.75	2.42	2.23	2.32	2.95	2.81, 1.7, 1.9
C_2H_4	$^1B_{3u}$	12.85	3.49	4.77	5.08	6.75	8.20	4.00
	$^1B_{1u}$	73.85	12.85	24.41	32.38	34.67	40.65	29.00
C_6H_6	1^1E_{1u}	134.02	49.71	48.59	53.48	58.31	103.05	120, 90.0, 95.3

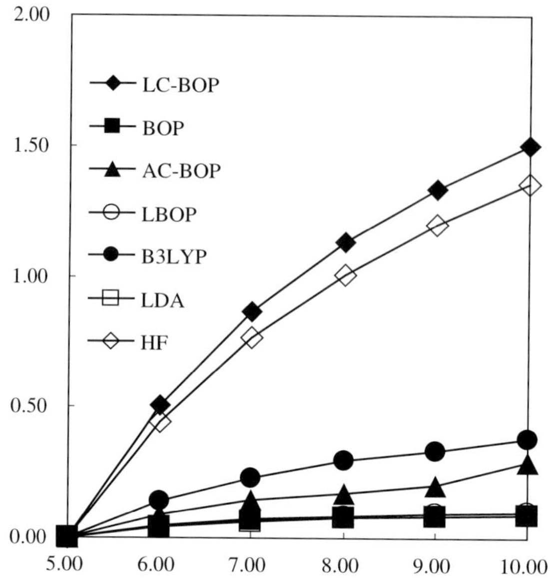

Fig. 20.7. The lowest charge transfer excitation energy of ethylene–tetrafluoroethylene dimer for the long intermolecular distance calculated by TDDFT employing various types of functionals. For all methods, the excitation energy at 5.0 Å is set to zero.

20.3.3.3 Transition metal dimer calculations

Yanagisawa et al. calculated the equilibrium geometries and the atomization energies for the first- to third-row transition metal dimers and concluded the following [107,108]:

1. Pure functionals tend to overstabilize electron configurations that contain orbitals in a high-angular momentum shell that is not fully occupied. This tendency is reduced from the first- to third-row transition metal dimers.
2. The overestimations of atomization energies of dimers are associated with the errors in outermost s–d interconfiguration transition energies of atoms. The latter errors may be due to the lack of long-range exchange interactions of outermost *s* and *d* orbitals that are fairly different in distributions. Hence, this lack may also cause the overestimations of the atomization energies.
3. B3LYP generally gives more accurate atomization energies than those of pure functionals, even if high-angular momentum orbitals are present in the configuration. However, B3LYP gives an erroneous energy gap between the configurations of fairly different spin-multiplicity probably due to the unbalance of the exchange and correlation functionals.

Based on this discussion, Tsuneda et al. applied the LC scheme to calculations of transition metal dimers [109].

Errors in calculated atomization energies of transition metal dimers are displayed in Fig. 20.8. The Wachters $+f$ basis set was used [110–112]. The figure shows that the LC scheme obviously improves the systematic overestimations calculated by pure

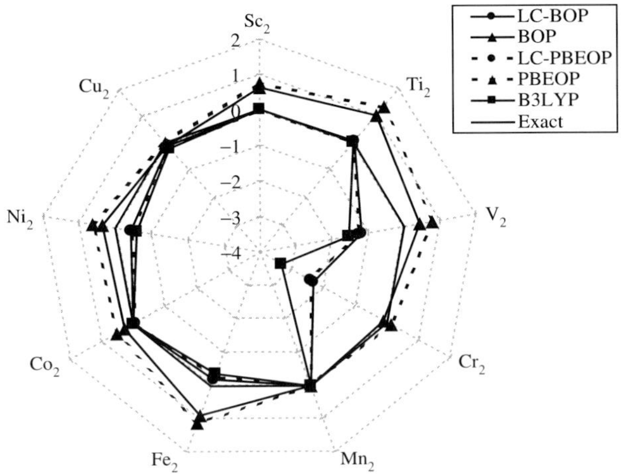

Fig. 20.8. Errors in calculated atomization energies of the first-row transition metal dimers for LC functionals (LC-BOP and LC-PBEOP), pure functionals (BOP and PBEOP), and B3LYP in eV. The line of no-error is also illustrated.

BOP functional. It was, therefore, proved that these overestimations maybe due to the lack of long-range exchange interactions in exchange functionals. It was also found that LC-BOP fairly underestimates the atomization energies of V_2 and Cr_2. As supposed from the similarity to B3LYP results, this underestimation may come from the errors in the HF exchange integral. That is, the HF exchange integral overstabilizes high-spin electronic states, because it only incorporates parallel-spin electron–electron interactions. It is, therefore, expected that this problem may be solved by taking well-balanced electron correlation into account for the long-range exchange part. The figure also shows that the LC scheme makes calculated atomization energies of different functionals much closer to each other. That is, the uniquenesses of functionals dissappeared after the long-range correction in this calculation. This may also support the conclusion that the lack of long-range exchange interactions is deeply committed to the overestimated atomization energies of dimers.

Table 20.9 Calculated barrier energies and bond distances of $H_2 + H \rightarrow H + H_2$ reaction

Molecule functional	Barrier height		Optimized geometry	
	Classical	ZPVC	R(H_2)	R(H_3)
LC-BOP	10.1	9.3	0.752	0.940
BOP	3.5	2.8	0.743	0.934
BLYP	2.9	2.2	0.745	0.930
B3LYP	4.3	3.5	0.742	0.934
Refs.	–	9.6	0.741	–

Barrier energies are in kcal/mol and optimized geometries are in Å. Reference values are the quantum Monte Carlo results.

Table 20.10 Calculated static isotropic polarizabilities by time-dependent Kohn–Sham theory in atomic unit

Molecule	LC-BOP	BOP	B3LYP	Exp.
Cl_2	30.87	31.69	31.16	30.35
CO_2	17.58	17.82	17.36	17.51
F_2	8.83	8.87	8.69	8.38
H_2O	10.03	10.49	9.95	9.64
H_2S	24.72	25.64	25.11	24.71
HCl	17.74	18.33	17.90	17.39
HF	5.99	6.17	5.83	5.60
N_2	11.99	12.07	11.88	11.74
SO_2	25.63	26.30	25.75	25.61

20.3.3.4 Other calculations

Besides the above-mentioned calculations, Tsuneda et al. are now applying the LC scheme to calculations of chemical reactions [113] and (hyper)polarizabilities [114]. We will exhibit some present works to show the wide applicabilities of this scheme.

First, calculated results of $H_2 + H \rightarrow H + H_2$ reaction are summarized in Table 20.9 [113]. The pV6Z basis set was used [115]. The table shows that the LC scheme remarkably improves underestimated reaction barrier energies of BOP. As far as we know, this result is certainly superior to results of other corrections that have ever reported. This may indicate that underestimations of reaction barriers in DFT calculations are also due to the lack of long-range interactions.

The LC scheme was also applied to overestimations of polarizabilities in DFT as shown in Table 20.10 [114]. The Sadlej valence triple zeta basis set was used [116,117]. The table shows that calculated polarizabilities of LC-BOP are obviously more accurate than those of BOP. Compared to B3LYP results, LC-BOP shows more improvements in many cases. Similar results were found in calculations of anisotropies and S_{-4} Cauchy moments of polarizabilities. Hence, we may say that these overestimations also come from the lack of long-range interactions in exchange functionals.

As mentioned above, the LC scheme was found to give better results for various chemical properties than the results of conventional corrections including hybrid functionals. For some properties, the LC scheme provided equivalent improvements in comparison with B3LYP. This may indicate that accurate B3LYP results may be due to the equivalency in the mixed HF exchange energy to the LC scheme, rather than the validity of the constant weight hybridization of the HF exchange. This argument may require further examination of the LC scheme.

20.4 RELATIVISTIC MOLECULAR THEORY

20.4.1 Introduction

The relativistic effect has been considered as an essential factor to figure out molecular structures, chemical activities, or various properties of heavy-element systems.

Recently many quantum chemists have dedicated a lot of efforts to the calculation and treatment of the electronic structures of polyatomic systems including heavy elements, which are involved in many interesting chemical and physical phenomena. They still present unique difficulties to the theoretical study. Until recently, the relativistic effect had ever been thought less important for chemical properties because the relativity appears primarily in the core electrons, which had been believed to be unlikely to affect chemically active valence regions dramatically. Recent studies, however, have revealed not only quantitatively but also qualitatively that the relativistic effect plays essential and comprehensive roles in total natures of molecular electronic structures for heavy-element systems. We are nowadays convinced that the relativistic effect is definitely important for the accurate theoretical treatment of heavy-element systems as well as the electron correlation effect.

To treat relativistic effects theoretically, the Dirac equation is usually solved rather than the non-relativistic Schrödinger equation. The one-electron Dirac Hamiltonian is written by

$$H_{\mathrm{D}} = c\boldsymbol{\alpha}\cdot\mathbf{p} + \beta c^2 + V_{\mathrm{ext}} \tag{70}$$

where the constant c is the speed of light, V_{ext} is the external potential, and $\mathbf{p}$ $(= -i\nabla)$ is the momentum operator. The 4×4 Dirac matrices $\boldsymbol{\alpha}$ and β in Eq. (70) are given by

$$\alpha_t \equiv \begin{pmatrix} 0_2 & \sigma_t \\ \sigma_t & 0_2 \end{pmatrix}, \qquad t = (x, y, z), \qquad \beta \equiv \begin{pmatrix} \mathbf{I}_2 & 0_2 \\ 0_2 & -\mathbf{I}_2 \end{pmatrix} \tag{71}$$

with the 2×2 Pauli spin matrices σ_t,

$$\sigma_x \equiv \begin{pmatrix} 0 & 1 \\ 1 & 0 \end{pmatrix}, \qquad \sigma_y \equiv \begin{pmatrix} 0 & -i \\ i & 0 \end{pmatrix}, \qquad \sigma_z \equiv \begin{pmatrix} 1 & 0 \\ 0 & -1 \end{pmatrix} \tag{72}$$

Since the Dirac equation is valid only for the one-electron system, the one-electron Dirac Hamiltonian has to be extended to the many-electron Hamiltonian in order to treat the chemically interesting many-electron systems. The straightforward way to construct the relativistic many-electron Hamiltonian is to augment the one-electron Dirac operator, Eq. (70) with the Coulomb or Breit (or its approximate Gaunt) operator as a two-electron term. This procedure yields the Dirac–Coulomb (DC) or Dirac–Coulomb–Breit (DCB) Hamiltonian derived from quantum electrodynamics (QED)

$$H = \sum_i H_{\mathrm{D}}(\mathbf{r}_i) + \sum_{i>j} g_{ij} \tag{73}$$

where

$$g_{ij}^{\mathrm{C}} = \frac{1}{r_{ij}} \tag{74}$$

and

$$g_{ij}^{\mathrm{CB}} = \frac{1}{r_{ij}} - \frac{1}{2}\left(\frac{(\alpha_i \cdot \alpha_j)}{r_{ij}} + \frac{(\alpha_i \cdot r_{ij})(\alpha_j \cdot r_{ij})}{r_{ij}^3} \right) \tag{75}$$

The DCB Hamiltonian is covariant to first order, and the presence of the Breit (or approximate Gaunt) interaction serves to increase the accuracy of calculated spectroscopic splittings and core binding energies.

Historically, approaches to treat the electronic structure relativistically have split into two camps: one is the four-component relativistic approach and another is the two-component one. Focusing on our recent studies, in this section, we will introduce these two types of relativistic approaches. The reader is referred to the detailed reviews for our recent relativistic works [118–120].

20.4.2 Four-component relativistic molecular theory

20.4.2.1 Dirac–Hartree–Fock and Dirac–Kohn–Sham methods

By an application of an independent-particle approximation with the DC or DCB Hamiltonian, the similar derivation of the non-relativistic Hartree–Fock (HF) method and Kohn–Sham (KS) DFT yields the four-component Dirac–Hartree–Fock (DHF) and Dirac–Kohn–Sham (DKS) methods with large- and small-component spinors.

The matrix DHF/DKS equation is generally written as

$$\mathbf{Fc} = \boldsymbol{\varepsilon}\mathbf{Sc} \tag{76}$$

where $\mathbf{c}$ is a matrix of molecular spinor coefficients, $\boldsymbol{\varepsilon}$ a spinor energy matrix, and $\mathbf{S}$ an overlap matrix

$$\mathbf{S}_{pq} = \begin{pmatrix} \mathbf{S}_{pq}^{\mathrm{LL}} & 0 \\ 0 & \mathbf{S}_{pq}^{\mathrm{SS}} \end{pmatrix} = \begin{pmatrix} \langle \chi_p^{\mathrm{L}} | \chi_q^{\mathrm{L}} \rangle & 0 \\ 0 & \langle \chi_p^{\mathrm{S}} | \chi_q^{\mathrm{S}} \rangle \end{pmatrix} \tag{77}$$

with two-component atomic spinors χ_p^{L} and χ_p^{S} for large (L) and small (S) components, respectively. Assuming the DC Hamiltonian, the Fock matrix $\mathbf{F}$ is given by

$$\begin{aligned}\mathbf{F}_{pq} &= \begin{pmatrix} \mathbf{F}_{pq}^{\mathrm{LL}} & \mathbf{F}_{pq}^{\mathrm{LS}} \\ \mathbf{F}_{pq}^{\mathrm{SL}} & \mathbf{F}_{pq}^{\mathrm{SS}} \end{pmatrix} \\ &= \begin{pmatrix} \mathbf{V}_{pq}^{\mathrm{LL}} + \mathbf{J}_{pq}^{\mathrm{LL}} - t_{\mathrm{ex}}\mathbf{K}_{pq}^{\mathrm{LL}} - t_{\mathrm{xc}}\mathbf{V}_{\mathrm{xc}pq}^{\mathrm{LL}} & c\boldsymbol{\Pi}_{pq}^{\mathrm{SL}} - t_{\mathrm{ex}}\mathbf{K}_{pq}^{\mathrm{SL}} \\ c\boldsymbol{\Pi}_{pq}^{\mathrm{LS}} - t_{\mathrm{ex}}\mathbf{K}_{pq}^{\mathrm{LS}} & \mathbf{V}_{pq}^{\mathrm{SS}} - 2c^2\mathbf{S}_{pq}^{\mathrm{SS}} + \mathbf{J}_{pq}^{\mathrm{SS}} - t_{\mathrm{ex}}\mathbf{K}_{pq}^{\mathrm{SS}} - t_{\mathrm{xc}}\mathbf{V}_{\mathrm{xc}pq}^{\mathrm{SS}} \end{pmatrix}\end{aligned} \tag{78}$$

Here, $\boldsymbol{\Pi}_{pq}^{\mathrm{X\bar{X}}}$, $\mathbf{V}_{pq}^{\mathrm{XX}}$, $\mathbf{V}_{\mathrm{xc}pq}^{\mathrm{XX}}$, $\mathbf{J}_{pq}^{\mathrm{XX}}$, and $\mathbf{K}_{pq}^{\mathrm{XY}}$ (X, Y = L or S, $\bar{\mathrm{L}}$ = S, and $\bar{\mathrm{S}}$ = L) are kinetic energy integral, electron–nuclear attraction integral, exchange-correlation potential,

Coulomb integral, and exchange integral matrices, respectively,

$$\mathbf{\Pi}_{pq}^{X\bar{X}} = \langle \chi_p^X | (\boldsymbol{\sigma} \cdot \mathbf{p}) | \chi_q^{\bar{X}} \rangle \tag{79}$$

$$\mathbf{V}_{pq}^{XX} = \langle \chi_p^X | V^{\text{nuc}} | \chi_q^X \rangle \tag{80}$$

$$\mathbf{V}_{\text{xc}pq}^{XX} = \left\langle \chi_p^X \left| \frac{\delta E_{\text{xc}}}{\delta \rho} \right| \chi_q^X \right\rangle \tag{81}$$

$$\mathbf{J}_{pq}^{XX} = \sum_{Y=L,S} \sum_{r,s} \mathbf{D}_{sr}^{YY} (\chi_p^X \chi_q^X | \chi_r^Y \chi_s^Y) \tag{82}$$

$$\mathbf{K}_{pq}^{XY} = \sum_{r,s} \mathbf{D}_{sr}^{XY} (\chi_p^X \chi_s^X | \chi_r^Y \chi_q^Y) \tag{83}$$

with the density matrix

$$\mathbf{D}_{sr}^{XY} = \sum_{i}^{N_{\text{occ}}} c_{si}^X c_{ri}^{Y^*} \tag{84}$$

The parameter t_{xc} is set to zero and one for the DHF and DKS approaches, respectively, and the constant t_{ex} is the parameter for the hybrid DFT approach, usually set to zero for the pure DFT approach.

The four-component DHF/DKS method is a theoretically straightforward relativistic approach. For heavy atoms four-component basis set expansion calculations are routine and attain spectroscopic accuracy together with extant correlation methods [121,122]. Recently the molecular DHF and DKS methods have become familiar and powerful relativistic approaches with the continuous development of efficient computational algorithms using the basis set expansion. Several four-component *ab initio* molecular orbital programs for polyatomics, e.g. MOLFDIR [123], DIRAC [124], BERTHA [125], and others [126–128], have been developed so far. Unfortunately, however, the treatment of more than one heavy atom within a molecule is not yet routine. The bottleneck in four-component calculations on heavy-element systems is evaluation of the two-electron electron repulsion integrals (ERIs). The number of relativistic integrals is greater than that of non-relativistic ones because the kinetic balance [129] between the large- and small-component primitive GTSs must be incorporated.

We have recently developed an efficient computational scheme for the four-component method that employs four-component contraction for molecular basis spinors and the new atomic spinor (AS) integral algorithm [130–132]. In the following sections we will briefly introduce our new relativistic scheme.

20.4.2.2 Generally contracted Gaussian-type spinors and kinetic balance

Accurate treatment of core spinors and of the valence spinors in the core region by a large basis set expansion is necessary, because most major relativistic effects, or the kinematic effects, come from the region near the nuclei. Because the core changes little with chemical environment, the extensive basis set contraction is possible. The difficulty in introducing contracted GTSs lies in the fact that the kinetic balance condition [129] between the large- and small-component primitive GTSs and spin–orbit splitting of spinors must be incorporated.

In our four-component molecular approach, thus, we use spin-coupled, kinetically balanced, generally contracted Gaussian-type spinors (GTSs) as basis functions. The basis expansion is

$$\begin{pmatrix} \psi_i^{2\mathrm{L}} \\ \psi_i^{2\mathrm{S}} \end{pmatrix} = \sum_{\mu}^{n} \begin{pmatrix} c_{\mu i}^{\mathrm{L}} \varphi_{\mu}^{2\mathrm{L}} \\ c_{\mu i}^{\mathrm{S}} \varphi_{\mu}^{2\mathrm{S}} \end{pmatrix} \tag{85}$$

where $\varphi_{\mu}^{2\mathrm{L}}$ and $\varphi_{\mu}^{2\mathrm{S}}$ are two-component basis spinors, and c_{μ}^{L} and c_{μ}^{S} are expansion coefficients. In Eq. (85), both scalar wavefunctions within a two-component basis spinor are multiplied by a common coefficient, thus, the dimensions of both the large and small components are n and the total number of variational parameters is $2n$. In the pioneering four-component program package, MOLFDIR, as well as in DIRAC, four-spinors are expanded in decoupled scalar spin-orbitals

$$\psi_i = \sum_{\mu}^{n^{\mathrm{L}}} c_{\mu i}^{\mathrm{L}\alpha} \varphi_{\mu}^{\mathrm{L}\alpha} \begin{pmatrix} 1 \\ 0 \\ 0 \\ 0 \end{pmatrix} + \sum_{\mu}^{n^{\mathrm{L}}} c_{\mu i}^{\mathrm{L}\beta} \varphi_{\mu}^{\mathrm{L}\beta} \begin{pmatrix} 0 \\ 1 \\ 0 \\ 0 \end{pmatrix} + \sum_{\mu}^{n^{\mathrm{S}}} c_{\mu i}^{\mathrm{S}\alpha} \varphi_{\mu}^{\mathrm{S}\alpha} \begin{pmatrix} 0 \\ 0 \\ 1 \\ 0 \end{pmatrix} + \sum_{\mu}^{n^{\mathrm{S}}} c_{\mu i}^{\mathrm{S}\beta} \varphi_{\mu}^{\mathrm{S}\beta} \begin{pmatrix} 0 \\ 0 \\ 0 \\ 1 \end{pmatrix} \tag{86}$$

There are $2n^{\mathrm{L}}$ large-component and $2n^{\mathrm{S}}$ small-component basis spinors. Imposing the kinetic balance implies that $2n^{\mathrm{S}} > n = 2n^{\mathrm{L}}$. Our scheme thus reduces the number of functions required for the small component.

The form of the large-component primitive set $\psi_k^{2\mathrm{L}}$ is chosen from large-component spinors obtained by analytical solution of the one-electron Dirac equation. The small-component set $\psi_k^{2\mathrm{S}}$ is derived so that it satisfies the accurate and rigorous kinetic balance condition *versus* $\psi_k^{2\mathrm{L}}$,

$$\psi_k^{2\mathrm{S}} = i(V - E - 2c^2)^{-1}(\boldsymbol{\sigma}\cdot\mathbf{p})\psi_k^{2\mathrm{L}} \tag{87}$$

rather than the condition

$$\psi_k^{2\mathrm{S}} = i(\boldsymbol{\sigma}\cdot\mathbf{p})\psi_k^{2\mathrm{L}} \tag{88}$$

20.4.2.3 Efficient evaluation of electron repulsion integrals

In construction of Coulomb and exchange integral matrices (Eqs. (82) and (83)), three types of ERIs, (LL|LL), (LL|SS) (or (SS|LL)), and (SS|SS), are required within the Coulomb approximation to the electron–electron interaction. Evaluation of ERIs

Table 20.11 CPU times (in hours) for computing four-component ERIs for Au_2, where the basis set used for Au is [19s14p10d5f]/(6s4p3d1f), which is commonly contracted between A = + and A = −

	LLLL + LLSS + SSSS	LLLL + LLSS	LLLL
Present[a]	1.37	0.77	0.21
DIRAC[b]	2.09	0.62	0.050
MOLFDIR[c]	76.35	21.16	1.63

[a]Number of basis spinors: 160 (for the large and small components).
[b]Number of basis spinors: 184 (for the large component) and 424 (for the small component).
[c]Number of basis spinors: 160 (for the large component) and 420 (for the small component).

includes a scaling with the forth power of the number of basis functions formally and is the most time-consuming step within the DHF/DKS calculation. To evaluate relativistic ERIs efficiently, we have recently developed a new integral evaluation method specialized for relativistic contracted GTSs [130,132]. The algorithm exploits the transfer relation of Head–Gordon and Pople (HGP) [133] and the accompanying coordinate expansion (ACE) formulas derived by Ishida [134] in the non-relativistic case. In this method, four-component ERIs (LL|LL), (LL|SS), and (SS|SS) reduce to several common two-center terms using the HGP transfer relation. The common integrals are evaluated rapidly using the ACE method.

We have performed comparative calculations of ERIs using MOLFDIR2000 and DIRAC version 3.2 in comparison with our REL4D program. MOLFDIR and DIRAC do not treat separately contracted REL4D-type basis sets. To make direct comparison possible, calculations with REL4D were done with the commonly contracted basis spinors employed in MOLFDIR and DIRAC, although the program is not optimized for such basis sets. Table 20.11 displays CPU times for computations on Au_2 with the [19s14p10d5f]/(6s4p3d1f) set. REL4D proved fastest for LLLL + LLSS + SSSS. LLLL + LLSS calculations with the present code were comparable to those of DIRAC. In the LLLL calculations, the present code worked about four times slower than DIRAC. Note that the numbers of spinors generated are different for each program: 160 for the large and small components in REL4D; 160 for the large component and 420 for the small component in MOLFDIR; 184 for the large component and 424 for the small component in DIRAC. The slightly larger basis size in DIRAC is caused by the fact that it uses, not spherical harmonic GTSs, but contracted Cartesian GTSs. This feature improves DIRAC's performance in some cases because the transformation from Cartesian to spherical harmonic is omitted. The reduced size of the small component basis renders our computational scheme efficient in storage, computation, and transformation of integrals, and in matrix manipulations.

20.4.2.4 Relativistic pseudospectral approach

Recently we have proposed more efficient relativistic molecular theory by an application of the pseudospectral (PS) approach [135]. In the PS approach [136,137], we use the mixed basis function between a grid representation in the physical space and spectral representation in the function space.

In the relativistic PS approach, the Coulomb matrix element (Eq. (82)) is given as

$$\mathbf{J}_{pq}^{\mathrm{XX}} = \sum_{\mathrm{Y=L,S}} \sum_{r,s} \mathbf{D}_{sr}^{\mathrm{YY}} (\chi_p^{\mathrm{X}} \chi_q^{\mathrm{X}} | \chi_r^{\mathrm{Y}} \chi_s^{\mathrm{Y}}) \cong \sum_g^M w_g \mathrm{A}_{pq}^{\mathrm{XX}}(g) \sum_{\mathrm{Y}}^{\mathrm{L,S}} \left(\sum_r^N \sum_s^N D_{rs}^{\mathrm{YY}} \chi_r^{\mathrm{Y}^*}(g) \chi_s^{\mathrm{Y}}(g) \right)$$

$$\equiv \sum_g^M w_g \mathrm{A}_{pq}^{\mathrm{XX}}(g) \rho(g) \tag{89}$$

with the three-center one-electron integral

$$\mathrm{A}_{pq}^{\mathrm{XY}}(g) = \int \chi_p^{\mathrm{X}^*}(1) \frac{1}{\left| r_1 - r_g \right|} \chi_q^{\mathrm{Y}}(1) \mathrm{d}r_1 \tag{90}$$

and $\rho(g)$ is the electronic density, which is calculated in terms of the density matrix and atomic spinors at a coordinate r_g. Likewise, the exchange contribution in the relativistic PS approach is given as

$$\mathbf{K}_{pq}^{\mathrm{XX}} = \sum_{r,s} \mathbf{D}_{sr}^{\mathrm{XX}} (\chi_p^{\mathrm{X}} \chi_s^{\mathrm{X}} | \chi_r^{\mathrm{X}} \chi_q^{\mathrm{X}}) \cong \sum_g^M w_g \chi_p^{\mathrm{X}^*}(g) \left[\sum_r^N \left(\sum_s^N D_{rs}^{\mathrm{XX}} \chi_s^{\mathrm{X}}(g) \right) \mathrm{A}_{rq}^{\mathrm{XX}}(g) \right] \tag{91}$$

and

$$\mathbf{K}_{pq}^{\mathrm{X\bar{X}}} = \sum_{r,s} \mathbf{D}_{sr}^{\mathrm{X\bar{X}}} (\chi_p^{\mathrm{X}} \chi_s^{\mathrm{X}} | \chi_r^{\bar{\mathrm{X}}} \chi_q^{\bar{\mathrm{X}}}) \cong \sum_g^M w_g \chi_p^{\mathrm{X}^*}(g) \left[\sum_r^N \left(\sum_s^N D_{rs}^{\mathrm{\bar{X}X}} \chi_s^{\mathrm{X}}(g) \right) \mathrm{A}_{rq}^{\mathrm{\bar{X}\bar{X}}}(g) \right] \tag{92}$$

for diagonal (LL and SS) and non-diagonal (LS and SL) parts in the DHF or DKS matrix, respectively. We note that no non-diagonal three-center one-electron integral is required in construction of both Coulomb and HF-exchange matrix elements within the DC approximation. Only diagonal $\mathrm{A}_{pq}^{\mathrm{LL}}(g)$ and $\mathrm{A}_{pq}^{\mathrm{SS}}(g)$ integrals are required. The high efficiency is hence achieved in the relativistic PS approach.

The features of the relativistic PS-DHF/DKS method are as follows:

(1) The computational scaling is reduced from $O(N^4)$ to $O(N^3)$ (N; the number of basis sets).
(2) Since the PS evaluation of HF-exchange matrix elements as well as Coulomb ones is efficient, post HF methods and hybrid-type DFT are applicable.
(3) It is possible to treat the large molecular systems that are compact and three-dimensional with high-quality basis sets in contrast to the fast multipole moment (FMM) method.
(4) The multigrid technique can powerfully save considerable CPU time in the direct SCF procedure.
(5) The PS program code is parallelized efficiently because of adoption of the numerical grid partition.

(6) It is possible to obtain the numerical result with arbitrary accuracy at adequate CPU time by careful choice of the number of grid points.

Table 20.12 shows the spectroscopic constants and total energies of the gold dimer calculated with the relativistic PS-DFT method using three types of grids. The details of computations are given in Ref. [135]. The results obtained by the conventional DFT method with the analytical ERIs and the experimental data [138,139] are also listed for comparison. The PS results for spectroscopic constants and total energies become closer to the analytical results as the level of grids is improved. The relativistic PS-DFT method with the ultrafine grid, the highest level of grid sets in this study, gives excellent good agreement with the analytical result. For the equilibrium bond length and the harmonic frequency for Au_2, the deviation of the ultrafine PS result from the analytical one is negligibly small. The discrepancies of the dissociation energy and the total energy between ultrafine PS and analytical results are $\Delta D_e = -0.4$ eV and $\Delta E = 0.0098$ a.u., respectively. It is also found that the fine grid yields satisfactory results; the difference of the equilibrium bond length, the harmonic frequency, the dissociation energy, and the total energy between fine PS and analytical results are $\Delta R_e = 0.005$ Å, $\Delta\omega_e = 1\ \text{cm}^{-1}$, $\Delta D_e = 0.3$ eV, and $\Delta E = 0.0376$ a.u., respectively.

The multigrid technique can be used in the SCF procedure of present PS calculations. This technique realizes the faster SCF calculation with the PS method. Average CPU times per one cycle for the direct SCF step in the DFT calculation including the ERI evaluation, the KS matrix construction, and the SCF diagonalization are also listed in Table 20.12. These times are taken from each single-point calculation at $R = 4.8$ a.u. In the present multigrid calculation, the coarse and medium grids are used in the first and second SCF stages, respectively. By adoption of the multigrid approach in this system, the PS methods with medium, fine, and ultrafine grids are 19, 13, and 9 times faster than the traditional analytical method, respectively.

Other PS applications to molecular systems also show that the relativistic PS-DHF/DKS approach is more efficient than the traditional approach without a loss of accuracy.

Table 20.12 Spectroscopic constants of the Au dimer calculated by conventional DFT and PS-DFT (B3LYP)

	Analytical	PS (medium)[a]	PS (fine)[b]	PS (ultra)[c]	Exptl
R_e (Å)	2.554	2.526	2.549	2.554	2.472
ω_e (cm^{-1})	168	191	169	168	191
D_e (eV)	1.98	2.11	2.01	1.94	2.36
Energy (a.u.)[d]	−0.7302	−0.8075	−0.7678	−0.7400	–
Time (s)[e]	17497	927	1352	1979	–

[a]Medium grid: 50 × 110 = 5500/atom.
[b]Fine grid: 75 × 194 = 14550/atom.
[c]Ultrafine grid: 96 × 302 = 28992/atom.
[d]Total DFT energy: − 38096 a.u.
[e]Average CPU time per one cycle for the direct SCF step in the DFT calculation including the ERI evaluation, the KS matrix construction, and the SCF diagonalization.

20.4.3 Two-component relativistic molecular theory

20.4.3.1 Approximate relativistic Hamiltonians

Despite recent implementations of an efficient algorithm for the four-component relativistic approach, the DC(B) equation with the four-component spinors composed of the large (upper) and small (lower) components still demands severe computational efforts to solve, and its applications to molecules are currently limited to small- to medium-sized systems. As an alternative approach, several two-component quasi-relativistic approximations have been proposed and applied to chemically interesting systems containing heavy elements, instead of explicitly solving the four-component relativistic equation.

An approximate relativistic Hamiltonian should include the following desirable features:

(1) It should be accurate enough to give a close result to the one-electron Dirac or many-electron Dirac–Coulomb(–Breit) Hamiltonian.
(2) It should be efficient and effective to apply to large molecular systems containing heavy elements.
(3) It should be well balanced so as to describe molecular systems containing a wide variety of atoms in the periodic table with the same quality.
(4) It should be variationally stable in order to avoid variational collapse in the sense that at least the non-relativistic limit is obtained correctly.
(5) It should be variational and not perturbative in order to evaluate various energy values and one-electron properties.

The Breit–Pauli (BP) approximation [140] is obtained truncating the Taylor expansion of the Foldy–Wouthuysen (FW) transformed Dirac Hamiltonian [141] up to the $(p/mc)^2$ term. The BP equation has the well-known mass–velocity, Darwin, and spin–orbit operators. Although the BP equation gives reasonable results in the first-order perturbation calculation, it cannot be used in the variational treatment.

One of the shortcomings of the BP approach is that the expansion in $(p/mc)^2$ is not justified in the case where the electronic momentum is too large, e.g. for a Coulomb-like potential. The zeroth-order regular approximation (ZORA) [142,143] can avoid this disadvantage by expanding in $E/(2mc^2 - V)$ up to the first order. The ZORA Hamiltonian is variationally stable. However, the Hamiltonian obtained by a higher order expansion has to be treated perturbatively, similarly to the BP Hamiltonian. Other quasi-relativistic methods have been proposed by Kutzelnigg [144,145] and Dyall [146].

We have developed two quasi-relativistic approaches. One is the RESC method [147–149], and the other is the higher order Douglas–Kroll (DK) method [150–152]. In the following sections we will introduce RESC and higher order DK methods briefly.

20.4.3.2 RESC method

The Dirac equation is equivalent to the Schrödinger–Pauli type equation composed of only the large component

$$\left[V + (\boldsymbol{\sigma}\cdot\mathbf{p})\frac{c^2}{2mc^2 - (V - E)}(\boldsymbol{\sigma}\cdot\mathbf{p})\right]\Psi^{\mathrm{L}} = e\Psi^{\mathrm{L}} \tag{93a}$$

with the normalization condition

$$\langle \Psi^{\mathrm{L}} | 1 + X^{+}X | \Psi^{\mathrm{L}} \rangle = 1 \tag{93b}$$

Here the X operator is defined by

$$X \equiv [2mc^2 - (V - E)]^{-1} c(\boldsymbol{\sigma}\cdot\mathbf{p}) \tag{94}$$

If Eq. (93a) could be solved with Eq. (93b), the solution to the Dirac equation can be obtained exactly. However, Eq. (93a) has the total and potential energies in the denominator, and an appropriate approximation is needed. In our strategy, $E - V$ in the denominator is replaced by the classical relativistic kinetic energy (relativistic substitutive correction)

$$T = (m^2c^4 + p^2c^2)^{1/2} - mc^2 \tag{95}$$

This simple approach is referred to as the relativistic scheme by eliminating small components (RESC). The derivation and the form of the RESC Hamiltonian are given in Ref. [147]. The RESC approach has several advantages. It is variationally stable. It can easily be incorporated in non-relativistic *ab initio* programs, and relativistic effects are considered on the same footing with electron correlation. RESC works well for a number of systems, and recent studies have shown it to give results similar to the Douglas–Kroll–Hess (DKH) method for chemical properties, although very large exponents in the basis set can lead to variational collapse in the current RESC approximation, which includes only the lowest truncation of the kinematic operator.

20.4.3.3 Douglas–Kroll method

The Douglas–Kroll (DK) approach [153] can decouple the large and small components of the Dirac spinors in the presence of an external potential by repeating several unitary transformations. The DK transformation is a variant of the FW transformation [141] and adopts the external potential V_{ext} as an expansion parameter instead of the speed of light, c, in the FW transformation. The DK transformation correct to second order in the external potential (DK2) has been extensively studied by Hess and co-workers [154], and has become one of the most familiar quasi-relativistic approaches. Recently, we have proposed the higher order DK method and applied the third-order DK (DK3) method to several systems containing heavy elements.

The first step in the DK transformation consists of a free-particle FW transformation to the Dirac Hamiltonian with the external potential V_{ext}

$$H_{\mathrm{D}} = \begin{pmatrix} V_{\mathrm{ext}} + c^2 & c\boldsymbol{\sigma}\cdot\mathbf{p} \\ c\boldsymbol{\sigma}\cdot\mathbf{p} & V_{\mathrm{ext}} - c^2 \end{pmatrix} \tag{96}$$

in momentum space. The resulting Hamiltonian yields the free-particle FW Hamiltonian and is also referred to as the first-order DK Hamiltonian. In successive DK transformations, in order to remove odd terms of arbitrary order in the external potential, the unitary operator defined by Douglas and Kroll [153],

$$U_n = (1 + W_n^2)^{1/2} + W_n \tag{97}$$

or the exponential-type unitary operator [150]

$$U_n = \exp(W_n) \tag{98}$$

is employed sequentially. Here W_n is an anti-Hermitian operator of nth order in V_{ext}. The resultant DK Hamiltonian is still a four-component formalism. Its two-component reduction is achieved by replacing β by the unit matrix and $\boldsymbol{\alpha}$ by the 2×2 Pauli spin matrix $\boldsymbol{\sigma}$. In order to correspond to the non-relativistic limit, the resulting two-component Hamiltonian is shifted by $-2c^2$.

The first-order, second-order, and third-order DK (DK1, DK2, and DK3) Hamiltonians in the two-component form are given as

$$H_{\text{DK1}} = E_p - c^2 + E_1 \tag{99}$$

$$H_{\text{DK2}} = H_{\text{DK1}} - \tfrac{1}{2}[W_1, [W_1, E_p]_+]_+ \tag{100}$$

$$H_{\text{DK3}} = H_{\text{DK2}} + \tfrac{1}{2}[W_1, [W_1, E_1]] \tag{101}$$

with

$$E_p = c[(\boldsymbol{\sigma}\cdot\mathbf{p})^2 + c^2]^{1/2} \tag{102}$$

$$E_1 = A(V_{\text{ext}} + RV_{\text{ext}}R)A \tag{103}$$

$$W_1 = A(Rv - vR)A \tag{104}$$

where the kinematical A and R operators and the v operator are defined by

$$A = \left(\frac{E_p + c^2}{2E_p}\right)^{1/2} \tag{105}$$

$$R = \frac{c\boldsymbol{\sigma}\cdot\mathbf{p}}{E_p + c^2} \tag{106}$$

$$v(p, p') = \frac{V_{\text{ext}}(p, p')}{E_p + E_{p'}} \tag{107}$$

and $[a, b]_+$ and $[a, b]$ denote the anti-commutator and the commutator, respectively.

20.4.3.4 Extended Douglas–Kroll transformations applied to the relativistic many-electron Hamiltonian

The DK approach satisfies all of the criteria in Section 20.4.3.1: the DK transformation avoids the high singularity in the FW transformation by adoption of the external potential as an expansion parameter, and thus the DK Hamiltonian is variationally stable. The DK Hamiltonian can be applied to the variational calculation in contrast to the Breit–Pauli Hamiltonian. Criterion (1) is also satisfied by the higher order DK method for the one-electron system. The DK3 Hamiltonian was shown to give excellent agreement with the one-electron Dirac Hamiltonian [150].

By an application of the DK transformation to the relativistic many-electron Hamiltonian, recently, we have shown that the many-electron DK Hamiltonian also gives satisfactory results for a wide variety of atoms and molecules compared with the DC(B) Hamiltonian [152]. To consider the higher order DK transformation to the two-electron interaction, the present approach adopts the effective one-electron potential in the DHF/DKS operator as an expansion parameter in the DK transformation.

The DHF/DKS operator, Eq. (78), can be written in the same form to the one-electron Dirac Hamiltonian, Eq. (96), by the following replacements:

$$V_{\mathrm{nuc}} + J^{\mathrm{LL}} + J^{\mathrm{SS}} - t_{\mathrm{ex}}K^{\mathrm{XX}} - t_{\mathrm{xc}}V_{\mathrm{xc}} \rightarrow V_{\mathrm{ext}} \tag{108}$$

$$\boldsymbol{\sigma}\cdot\mathbf{p} - \frac{t_{\mathrm{ex}}}{c}K^{\mathrm{XY}} \rightarrow \boldsymbol{\sigma}\cdot\mathbf{p} \tag{109}$$

By substituting these relations into Eqs. (99)–(107), we can straightforwardly obtain the DKn-Fock operators with the DK transformation to the DHF/DKS potential in the two-component DKn-HF/KS equation

$$F_{\mathrm{DK}n}\psi_i = \varepsilon_i\psi_i \tag{110}$$

where ψ_i is the (orthonormalized) two-component DKn spinor and ε_i is its spinor energy. The first-order DK (DK1) operator is given as

$$F_{\mathrm{DK1}} = E_0 - c^2 + E_1 \tag{111}$$

with

$$E_0 = c\left[\left(\boldsymbol{\sigma}\cdot\mathbf{p} - \frac{t_{\mathrm{ex}}}{c}K^{\mathrm{LS}}\right)\left(\boldsymbol{\sigma}\cdot\mathbf{p} - \frac{t_{\mathrm{ex}}}{c}K^{\mathrm{SL}}\right) + c^2\right]^{1/2} \tag{112}$$

$$\begin{aligned} E_1 &= A(V_{\mathrm{nuc}} + J^{\mathrm{LL}} + J^{\mathrm{SS}} - t_{\mathrm{ex}}K^{\mathrm{LL}} - t_{\mathrm{xc}}V_{\mathrm{xc}})A \\ &\quad + AR^{\mathrm{LS}}(V_{\mathrm{nuc}} + J^{\mathrm{LL}} + J^{\mathrm{SS}} - t_{\mathrm{ex}}K^{\mathrm{SS}} - t_{\mathrm{xc}}V_{\mathrm{xc}})R^{\mathrm{SL}}A \end{aligned} \tag{113}$$

where the A and R^{XY} operators are defined by

$$A = \left(\frac{E_0 + c^2}{2E_0}\right)^{1/2} \tag{114}$$

$$R^{\mathrm{XY}} = \frac{c}{E_0 + c^2}\left(\boldsymbol{\sigma}\cdot\mathbf{p} - \frac{t_{\mathrm{ex}}}{c}K^{\mathrm{XY}}\right) \tag{115}$$

In this approach, the density matrix is evaluated self-consistently with both the large and small component spinors, φ_i^{L} and φ_i^{S}, which can be reconstructed from the free-particle FW spinors ψ_i in the Schrödinger picture

$$\varphi_i^{\mathrm{L}} = A\psi_i \tag{116}$$

$$\varphi_i^{\mathrm{S}} = R^{\mathrm{SL}}A\psi_i \tag{117}$$

It is easy to verify that the DK1 operator, Eq. (111), is equivalent to the Fock operator derived from the no-pair or free-particle FW Hamiltonian. Likewise, the higher order DK operators are also derived straightforwardly by repeating the DK transformations, though their formulae are omitted only because of their lengthy forms.

As molecular applications of the extended DK approach, we have calculated the spectroscopic constants for At_2: equilibrium bond lengths (R_e), harmonic frequencies (ω_e), rotational constants (B_e), and dissociation energies (D_e). A strong spin–orbit effect is expected for these properties because the outer p orbital participates in their molecular bonds. Electron correlation effects were treated by the hybrid DFT approach with the B3LYP functional. Since several approximations to both the one-electron and two-electron parts of the DK Hamiltonian are available, we define that the DK$n1$ + DK$n2$ Hamiltonian ($n1$, $n2 = 1$–3) denotes the DK Hamiltonian with DK$n1$ and DK$n2$ transformations for the one-electron and two-electron parts, respectively. The DK$n1$ + DK1 Hamiltonian is equivalent to the no-pair DK$n1$ Hamiltonian. For the two-electron part the electron–electron Coulomb operator in the non-relativistic form can also be adopted. The DK$n1$ Hamiltonian with the non-relativistic Coulomb operator is denoted by the DK$n1$ + NR Hamiltonian.

Table 20.13 shows the results for At_2 obtained by approximate DK schemes in comparison with four-component DKS results. The DK results for the spectroscopic constants and the total energy in the equilibrium geometry (E_{tot}) become closer to the DKS results as the level of theory is improved. The highest level of theory, DK3–DK3, as well as DK3–DK2, gives fairly good agreement with the four-component result for At_2. The DK3–DK3 operator yields $R_e = 3.1102$ Å, $\omega_e = 102.3$ cm^{-1}, and $D_e = 0.546$ eV, the corresponding four-component DKS values being $R_e = 3.1121$ Å, $\omega_e = 102.0$ cm^{-1}, and $D_e = 0.542$ eV. The discrepancy between DK3–DK3 and DKS Hamiltonians is $\Delta R_e = 0.0019$ Å, $\Delta\omega_e = 0.3$ cm^{-1}, and $\Delta D_e = 0.004$ eV.

By comparison between the DK3–DK3 and DK3–NR results, it can be seen that two-electron relativistic effects are comparatively large, especially in the dimer; the bond length decreases by 0.06 Å, the frequency increases by 7 cm^{-1}, the rotational constant increases by 0.0006 cm^{-1}, and the dissociation energy increases by 0.12 eV. Neglect of the relativistic correction to the electron–electron interaction yields inferior results and gives relatively large deviations from the DK3–DK3 or DKS result. It is interesting that the importance of the two-electron DK correction for the bond length is shown, because it has been believed so far that the bond length is scarcely affected by the relativistic

Table 20.13 Bond lengths (R_e), harmonic frequencies (ω_e), rotational constants (B_e), dissociation energies (D_e), and total energies (E_{tot}) in the equilibrium geometry of At_2 with B3LYP

Hamiltonian	R_e (Å)	ω_e (cm^{-1})	B_e (cm^{-1})	D_e (eV)	E_{tot} (a.u.)
DKS	3.1121	102.0	0.0166	0.542	−45838.2314
DK3–DK3	3.1102	102.3	0.0166	0.546	−45841.9720
DK3–DK3	3.1080	102.3	0.0166	0.552	−45849.7971
(no mod. V_{xc})[a]					
DK3–DK2	3.1108	102.3	0.0166	0.545	−45842.2586
DK3–DK1	3.1074	102.6	0.0166	0.552	−45839.6417
DK3–NR	3.1697	95.5	0.0160	0.429	−45849.7240
DK2–DK2	3.1013	103.0	0.0167	0.561	−45773.7217

[a]Results without the relativistic modification to V_{xc}.

correction to the electron–electron interaction, while the harmonic frequency and the dissociation energy are often influenced.

The first-order DK correction to the electron–electron interaction is satisfactory also in molecular systems, as well as the atomic case. The deviation of DK3–DK1 from DK3–DK3 is $\Delta R_e = 0.0028$ Å, $\Delta\omega_e = 0.3\ \text{cm}^{-1}$, and $\Delta D_e = 0.006$ eV for At_2.

In the DFT approach with our general DK transformation, the exchange-correlation potential, V_{xc}, is corrected relativistically. The effect on the DK transformation to the exchange-correlation potential was estimated by comparison with the result without the relativistic modification to V_{xc} ((no mod. V_{xc}) in Table 20.13). Compared with the full DK3–DK3 approach, neglect of the relativistic DK correction to the exchange-correlation potential hardly affects the calculated spectroscopic values; its effect merely contributes 0.002 Å for R_e and 0.006 eV for D_e and does not affect ω_e and B_e for the At dimer. Thus, it is found that the relativistic correction to the electron–electron interaction contributes mainly to the Coulomb potential, not to the exchange-correlation potential.

In consequence, the several numerical results including the present results show that the third-order DK transformation to both one-electron and two-electron Hamiltonians gives excellent agreement with the four-component relativistic approach. The first-order DK correction to the two-electron interaction is shown to be satisfactory for both atomic and molecular systems.

20.5 SUMMARY

The recent advances in electronic structure theory achieved in our research group have been reviewed. New theory/algorithm has been implemented and incorporated into the program package, UTCHEM [155]. Software forms a basis for computational chemistry. It is not an easy task for an individual/group to develop a comprehensive new program package in quantum chemistry from scratch. Several years ago, we decided to accept this challenge. In view of the availability of such good programs as GAUSSIAN, GAMESS, MOLCAS, NWCHEM, etc., one may question the relevance of a new program package. We have three arguments for our project. (1) First, we believe that healthy competition is very important in science. (2) Second, we can have a good harvest by doing research using other programs, but it is an abortive flower. We could not make a true breakthrough if we were circumscribed by current software limitations. (3) Third, in spite of the excellent performance of other programs, there are important and powerful methods that others cannot yet handle. We have developed new methodologies in quantum chemistry, particularly the multireference-based perturbation theory for describing chemical reactions and excited states, relativistic molecular theory to treat heavy elements, parameter-free (less) and long-range corrected (LC) exchange and correlation functionals in DFT, highly efficient algorithms for calculating molecular integrals over generally contracted Gaussians, etc. UTCHEM is a research product of our work to develop new and better theoretical methods in quantum chemistry. Most of the codes have been developed recently by Hirao's group at the University of Tokyo. The basic philosophy behind UTCHEM is to develop methods that allow an accurate and efficient computational chemistry of electronic structure problems for molecular systems in both the ground and excited states. UTCHEM also contains codes for well-developed methods such as MPn, CI, CC, etc., which are standard in most quantum chemistry programs. We are aiming ultimately at better performance than other programs. UTCHEM has been released. If you want more information on UTCHEM, visit http://utchem.qcl.t.u-tokyo.ac.jp/.

20.6 REFERENCES

1 K. Hirao, Chem. Phys. Lett., 190 (1992) 374.
2 K. Hirao, Chem. Phys. Lett., 196 (1992) 397.
3 K. Hirao, Chem. Phys. Lett., 201 (1993) 59.
4 K. Hirao, Int. J. Quantum Chem. S, 26 (1992) 517.
5 H. Nakano, J. Chem. Phys., 99 (1993) 9873.
6 H. Nakano, Chem. Phys. Lett., 207 (1993) 372.
7 H. Nakano, M. Yamanishi and K. Hirao, Trends Chem. Phys., 6 (1997) 167.
8 K. Hirao, K. Nakayama, T. Nakajima and H. Nakano, in: J. Lesczynski (Ed.), Computational chemistry, Vol. 4, World Scientific, Singapore, 1999, pp. 227–270.
9 H. Nakano, T. Nakajima, T. Tsuneda and K. Hirao, J. Mol. Struct. (THEOCHEM), 573 (2001) 91.
10 T. Nakajima, T. Tsuneda, H. Nakano and K. Hirao, J. Theor. Comput. Chem., 1 (2002) 109.
11 C. Møller and M.S. Plesset, Phys. Rev., 46 (1934) 618.
12 B.H. Brandow, Rev. Mod. Phys., 39 (1967) 771.
13 I. Lindgren, Int. J. Quantum Chem. Symp., 12 (1978) 33.
14 B.H. Brandow, Int. J. Quantum Chem., 15 (1979) 207.
15 G. Hose and U. Kaldor, J. Phys. B, 12 (1981) 3827.
16 K.F. Freed and M.G. Sheppard, J. Chem. Phys., 75 (1981) 4507, see also p. 4525.
17 H. Primas, Helv. Phys. Acta, 34 (1961) 331.
18 H. Primas, Rev. Mod. Phys., 35 (1963) 710.
19 B. Kirtman, J. Chem. Phys., 49 (1968) 3890, 75 (1981) 798.
20 P.R. Certain and J.O. Hirschfelder, J. Chem. Phys., 52 (1970) 5977, 53 (1970) 2992.
21 J.O. Hirschfelder, Chem. Phys. Lett., 54 (1978) 1.
22 T.H. Schucan and H.A. Weidenmuller, Ann. Phys. N.Y., 76 (1973) 483.
23 K. Wolinski, H.L. Sellers and P. Pulay, Chem. Phys. Lett., 140 (1987) 225.
24 J.J.W. McDouall, K. Peasley and M.A. Robb, Chem. Phys. Lett., 148 (1988) 183.
25 K. Andersson, P. Malmqvist, B.O. Roos, A.J. Sadlej and K. Wolinski, J. Phys. Chem., 94 (1990) 5483.
26 K. Andersson, P. Malmqvist and B.O. Roos, J. Chem. Phys., 96 (1992) 1218.
27 R.B. Murphy and R.P. Messmer, Chem. Phys. Lett., 183 (1991) 443.
28 R.B. Murphy and R.P. Messmer, J. Chem. Phys., 97 (1992) 4170.
29 P.M. Kozlowski and E.R. Davidson, J. Chem. Phys., 100 (1994) 3672.
30 K.G. Dyall, J. Chem. Phys., 102 (1995) 4909.
31 Y. Kobayashi, H. Nakano and K. Hirao, Chem. Phys. Lett., 336 (2001) 529.
32 C.W. Bauschlicher and P.R. Taylor, J. Chem. Phys., 85 (1986) 6510.
33 P. Jensen and P.R. Bunker, J. Chem. Phys., 89 (1988) 1327.
34 S. Koda, Chem. Phys., 66 (1982) 383.
35 H. Nakano, J. Nakatani and K. Hirao, J. Chem. Phys., 114 (2001) 1133.
36 H. Nakano, R. Uchiyama and K. Hirao, J. Comput. Chem., 23 (2002) 1166.
37 H. Nakano, K. Nakayama, K. Hirao and M. Dupuis, J. Chem. Phys., 106 (1997) 4912.
38 W.F. Polik, D.R. Guyer and C.B. Moore, J. Chem. Phys., 92 (1990) 3453.
39 M.R.J. Hachey, P.J. Bruna and F. Grein, J. Phys. Chem., 99 (1995) 8050.
40 M. Merchán and B.O. Roos, Theor. Chim. Acta, 92 (1995) 227.
41 S.R. Gwaltney and R.J. Bartlett, Chem. Phys. Lett., 241 (1995) 26.
42 C.M. Hadad, J.B. Foresman and K.B. Wiberg, J. Phys. Chem., 97 (1993) 4293.
43 M. Head-Gordon, R.J. Rico, M. Oumi and T.J. Lee, Chem. Phys. Lett., 219 (1994) 21.
44 H. Nakatsuji, K. Ohta and K. Hirao, J. Chem. Phys., 75 (1981) 2952.

45 Y. Kurashige, H. Nakano and K. Hirao, J. Phys. Chem. A, 108 (2004) 3064.
46 J.M. Rintelman and M.S. Gordon, J. Chem. Phys., 115 (2001) 1795.
47 K.W. Sattelmeyer, H.F. Schaefer, III and J.F. Stanton, J. Chem. Phys., 116 (2002) 9151.
48 K.A. Peterson, D.E. Woon and T.H. Dunning, Jr., J. Chem. Phys., 100 (1994) 7410.
49 J.F. Stanton, J. Gauss and O. Christiansen, J. Chem. Phys., 114 (2001) 2993.
50 K. Hirao, H. Nakano, K. Nakayama and M. Dupuis, J. Chem. Phys., 105 (1996) 9227.
51 K. Hirao, H. Nakano and K. Nakayama, J. Chem. Phys., 107 (1997) 9966.
52 B. Lam, M.W. Schmidt and K. Ruedenberg, J. Phys. Chem., 89 (1985) 2221.
53 T.R. Cundari and M.S. Gordon, J. Am. Chem. Soc., 113 (1991), 114 (1992) 539.
54 D.L. Cooper, J. Gerratt and M. Raimond, Nature, 323 (1986) 699.
55 D.L. Cooper, J. Gerratt and M. Raimond, Adv. Chem. Phys., 27 (1987) 319.
56 W.J. Hunt, P.J. Hay and W.A. Goddard, J. Chem. Phys., 57 (1972) 738.
57 W.A. Goddard and L.B. Harding, Annu. Rev. Phys. Chem., 29 (1978) 363.
58 P.C. Hiberty, J.P. Flament and E. Noizet, Chem. Phys. Lett., 189 (1992) 259.
59 P. Maitre, J.M. Lefour, G. Ohanessian and P.C. Hiberty, J. Phys. Chem., 94 (1990) 4082.
60 H. Nakano, K. Sorakubo, K. Nakayama and K. Hirao, in: D.L. Cooper (Ed.), Valence bond theory, Theoretical and computational chemistry, Vol. 10, Elsevier, Amsterdam, 2001, pp. 55–77.
61 W. Kohn and L.J. Sham, Phys. Rev. A, 140 (1995) 1133.
62 R.G. Parr and W. Yang, Density-functional theory of atoms and molecules, Oxford University Press, New York, 1989.
63 R.M. Dreizler and E.K.U. Gross, Density-functional theory: An approach to the quantum many-body problem, Springer, Berlin, 1990.
64 A.D. Becke, Phys. Rev. A, 38 (1988) 3098.
65 J.P. Perdew and Y. Wang, in: P. Ziesche, H. Eschrig (Eds.), Electronic structure of solids '91, Akademie, Berlin, 1991.
66 J.P. Perdew, K. Burke and M. Ernzerhof, Phy. Rev. Lett., 77 (1996) 3865.
67 T. Tsuneda and K. Hirao, Phys. Rev. B, 62 (2000) 15527.
68 J.C. Slater, Phys. Rev., 81 (1951) 385.
69 T. Tsuneda, T. Suzumura and K. Hirao, J. Chem. Phys., 111 (1999) 5656.
70 T. Tsuneda, M. Kamiya, N. Morinaga and K. Hirao, J. Chem. Phys., 114 (2001) 6505.
71 L. Kleinman and S. Lee, Phys. Rev. B, 37 (1988) 4634.
72 A.D. Becke, J. Chem. Phys., 98 (1993) 5648.
73 C. Lee, W. Yang and R.G. Parr, Phys. Rev. B, 37 (1988) 785.
74 S.H. Vosko, L. Wilk and M. Nusair, Can. J. Phys., 58 (1980) 1200.
75 A.D. Becke, J. Chem. Phys., 107 (1997) 8554.
76 J.P. Perdew, K. Burke and M. Ernzerhof, J. Chem. Phys., 105 (1996) 9982.
77 M. Levy, J.P. Perdew and V. Sahni, Phys. Rev. A, 30 (1984) 2745.
78 R. van Leeuwen and E.J. Baerends, Phys. Rev. A, 49 (1994) 2421.
79 D.J. Tozer and N.C. Handy, J. Chem. Phys., 109 (1988) 10180.
80 D.J. Tozer, J. Chem. Phys., 112 (1999) 3507.
81 J.P. Perdew and A. Zunger, Phys. Rev. B, 23 (1981) 5048.
82 B.G. Johnson, C.A. Gonzales, P.M.W. Gill and J.A. Pople, Chem. Phys. Lett., 221 (1994) 100.
83 X. Tong and S. Chu, Phys. Rev. A, 55 (1997) 3406.
84 T. Tsuneda, M. Kamiya and K. Hirao, J. Comput. Chem., 24 (2003) 1592.
85 A. Savin, in: J.M. Seminario (Ed.), Recent developments and applications of modern density functional theory, Elsevier, Amsterdam, 1996, p. 327.
86 H. Iikura, T. Tsuneda, T. Yanai and K. Hirao, J. Chem. Phys., 115 (2001) 3540.
87 J. Heyd, G.E. Scuseria and M. Ernzerhof, J. Chem. Phys., 118 (2003) 8207.
88 J. Heyd and G.E. Scuseria, J. Chem. Phys., 120 (2004) 7274.
89 T. Leininger, H. Stoll, H.-J. Werner and A. Savin, Chem. Phys. Lett., 275 (1997) 151.
90 H.L. Williams and C.F. Chabalowski, J. Phys. Chem., 105 (2001) 646.
91 Y. Andersson, D.C. Langreth and B.I. Lundqvist, Phys. Rev. Lett., 76 (1996) 102.
92 K. Rapcewicz and N.W. Ashcroft, Phys. Rev. B, 44 (1991) 4032.

93 W. Kohn, Y. Meier and D.E. Makarov, Phys. Rev. Lett., 80 (1998) 4153.
94 M. Kamiya, T. Tsuneda and K. Hirao, J. Chem. Phys., 117 (2002) 6010.
95 T. Tsuneda, T. Suzumura and K. Hirao, J. Chem. Phys., 110 (1999) 10664.
96 R. Krishnan, J.S. Binkley, R. Seeger and J.A. Pople, J. Chem. Phys., 72 (1980) 650.
97 A.D. McLean and G.S. Chandler, J. Chem. Phys., 72 (1980) 5639.
98 M.J. Frisch, J.A. Pople and J.S. Binkley, J. Chem. Phys., 80 (1984) 3265.
99 S.F. Boys and F. Bernardi, Mol. Phys., 19 (1970) 553.
100 C. Adamo and V. Barone, J. Chem. Phys., 118 (1998) 664.
101 Q. Wu and W. Yang, J. Chem. Phys., 116 (2002) 515.
102 Y. Tawada, T. Tsuneda, S. Yanagisawa, T. Yanai and K. Hirao, J. Chem. Phys., 120 (2004) 8425.
103 H. Nakatsuji, Chem. Phys. Lett., 59 (1978) 362.
104 R. Krishnan, J.S. Binkley, R. Seeger and J.A. Pople, J. Chem. Phys., 72 (1980) 650.
105 A.D. McLean and G.S. Chandler, J. Chem. Phys., 72 (1980) 5639.
106 A. Dreuw, J.L. Weisman and M. Head-Gordon, J. Chem. Phys., 119 (2003) 2943.
107 S. Yanagisawa, T. Tsuneda and K. Hirao, J. Chem. Phys., 112 (2000) 545.
108 S. Yanagisawa, T. Tsuneda and K. Hirao, J. Comput. Chem., 22 (2001) 1995.
109 T. Tsuneda, S. Tokura and K. Hirao, in preparation.
110 A.J.H. Wachters, J. Chem. Phys., 52 (1970) 1033.
111 A.J.H. Wachters, IBM Tech. Rep., RJ584 (1969).
112 C.W. Bauschlicher, Jr., S.R. Langhoff and L.A. Barnes, J. Chem. Phys., 91 (1989) 2399.
113 T. Tsuneda, N. Kawakami and K. Hirao, in preparation.
114 M. Kamiya, H. Sekino, T. Tsuneda and K. Hirao, J. Chem. Phys., in press.
115 T.H. Dunning Jr., to be published.
116 A.J. Sadlej, Theor. Chim. Acta, 81 (1992) 339.
117 M.E. Casida, C. Jamorski, K.C. Casida and D.R. Salahub, J. Chem. Phys., 108 (1998) 4439.
118 T. Nakajima, T. Yanai and K. Hirao, J. Comput. Chem., 23 (2002) 847.
119 T. Yanai, T. Nakajima, Y. Ishikawa and K. Hirao, in: K. Hirao, Y. Ishikawa (Eds.), Recent advances in relativistic effects in chemistry, World Scientific, Singapore, 2004, pp. 221.
120 T. Nakajima, Bull. Kor. Chem. Soc., 24 (2003) 809.
121 Y. Ishikawa and K. Koc, Phys. Rev. A, 56 (1997) 1295; M.J. Vilkas, Y. Ishikawa and K. Koc, Phys. Rev. E, 58 (1998) 5096; M.J. Vilkas, Y. Ishikawa and K. Hirao, Chem. Phys. Lett., 321 (2000) 243.
122 E. Eliav, U. Kaldor and Y. Ishikawa, Phys. Rev. A, 53 (1996) 3050.
123 L. Visscher, O. Visser, H. Aerts, H. Merenga and W.C. Nieuwpoort, Comput. Phys. Commun., 81 (1994) 120.
124 T. Saue, K. Fægri, T. Helgaker and O. Gropen, Mol. Phys., 91 (1997) 937.
125 H.M. Quiney, H. Skaane and I.P. Grant, Adv. Quantum Chem., 32 (1999) 1; I.P. Grant and H.M. Quiney, Int. J. Quantum Chem., 80 (2000) 283.
126 A. Mohanty and E. Clementi, Int. J. Quantum. Chem., 39 (1991) 487.
127 L. Pisani and E. Clementi, J. Comput. Chem., 15 (1994) 466.
128 K.G. Dyall, P.R. Taylor, K. Fægrei and H. Partridge, J. Chem. Phys., 95 (1991) 2583.
129 Y. Ishikawa, R.C. Binning and K.M. Sando, Chem. Phys. Lett., 101 (1983) 111; Y. Ishikawa, R.C. Binning and K.M. Sando, Chem. Phys. Lett., 105 (1984) 189; Y. Ishikawa, R.C. Binning and K.M. Sando, Chem. Phys. Lett., 117 (1985) 444; R.E. Stanton and S. Havriliak, J. Chem. Phys., 81 (1984) 1910.
130 T. Yanai, T. Nakajima, Y. Ishikawa and K. Hirao, J. Chem. Phys., 114 (2001) 6525.
131 T. Yanai, H. Iikura, T. Nakajima, Y. Ishikawa and K. Hirao, J. Chem. Phys., 115 (2001) 8267.
132 T. Yanai, T. Nakajima, Y. Ishikawa and K. Hirao, J. Chem. Phys., 116 (2002) 10122.
133 M. Head-Gordon and J.A. Pople, J. Chem. Phys., 89 (1988) 5777.
134 K. Ishida, Int. J. Quantum Chem., 59 (1996) 209; K. Ishida, J. Chem. Phys., 109 (1998) 881; K. Ishida, J. Comput. Chem., 19 (1998) 923; K. Ishida, J. Chem. Phys., 111 (1999) 4913.
135 T. Nakajima and K. Hirao, J. Chem. Phys., 121 (2004) 3438.
136 R.A. Friesner, Chem. Phys. Lett., 116 (1985) 39.
137 R.A. Friesner, J. Chem. Phys., 85 (1986) 1462.

138 K.P. Huber and G. Herzberg, Molecular structure IV, constants of diatomic molecules, Van Nostrand, New York, 1979.
139 B. Simard and P.A. Hackett, J. Mol. Spectrosc., 412 (1990) 310.
140 H.A. Bethe and E.E. Salpeter, Quantum mechanics of one- and two-electron atoms, Springer, Berlin, 1957.
141 L.L. Foldy and S.A. Wouthuysen, Phys. Rev., 78 (1950) 29.
142 E. van Lenthe, E.J. Baerends and J.G. Snijders, J. Chem. Phys., 99 (1993) 4597.
143 Ch. Chang, M. Pelissier and Ph. Durand, Phys. Scr., 34 (1986) 394.
144 W. Kutzelnigg, Z. Phys. D, 11 (1989) 15.
145 W. Kutzelnigg, Z. Phys. D, 15 (1990) 27.
146 K.G. Dyall, J. Chem. Phys., 100 (1994) 2118.
147 T. Nakajima and K. Hirao, Chem. Phys. Lett., 302 (1999) 383.
148 T. Nakajima, T. Suzumura and K. Hirao, Chem. Phys. Lett., 304 (1999) 271.
149 D. Fedorov, T. Nakajima and K. Hirao, Chem. Phys. Lett., 335 (2001) 183.
150 T. Nakajima and K. Hirao, J. Chem. Phys., 113 (2000) 7786.
151 T. Nakajima and K. Hirao, Chem. Phys. Lett., 329 (2000) 511.
152 T. Nakajima and K. Hirao, J. Chem. Phys., 119 (2003) 4105.
153 M. Douglas and N.M. Kroll, Ann. Phys. N.Y., 82 (1974) 89.
154 B.A. Hess, Phys. Rev. A, 33 (1986) 3742; G. Jansen and B.A. Hess, Phys. Rev. A, 39 (1989) 6016.
155 T. Yanai, M. Kamiya, Y. Kawashima, T. Nakajima, H. Nakano, Y. Nakao, H. Sekino, T. Tsuneda, S. Yanagisawa and K. Hirao, UTCHEM.

Theory and Applications of Computational Chemistry: The First Forty Years
Edited by C. Dykstra et al.

CHAPTER 21

Semiempirical quantum-chemical methods in computational chemistry

Walter Thiel

Max-Planck-Institut für Kohlenforschung, Mülheim, Germany

Abstract

This review of semiempirical quantum-chemical methods outlines their development over the past 40 years. After a survey of the established methods such as MNDO, AM1, and PM3, recent methodological advances are described including the development of improved semiempirical models, new general-purpose and special-purpose parametrizations, and linear scaling approaches. Selected recent applications are presented covering examples from biochemistry, medicinal chemistry, and nanochemistry as well as direct reaction dynamics and electronically excited states. The concluding remarks address the current and future role of semiempirical methods in computational chemistry.

21.1 INTRODUCTION

Quantum mechanics provides the conceptual framework for understanding chemistry. The *ab initio* methods of nonrelativistic quantum mechanics aim at the solution of the time-independent Schrödinger equation, employing well-defined approximations that can be improved systematically on a convergent path to the exact solution. They do not use experimental data, except for the fundamental physical constants.

Standard dictionaries define 'semiempirical' as 'involving assumptions, approximations, or generalizations designed to simplify calculation or to yield a result in accord with observation' [1]. In this spirit, the semiempirical methods of quantum chemistry start out from the *ab initio* formalism and then introduce rather drastic assumptions to speed up the calculations, typically by neglecting many of the less important terms in the *ab initio* equations. In order to compensate for the errors caused by these approximations, empirical parameters are incorporated into the formalism and calibrated against reliable experimental or theoretical reference data. If the chosen semiempirical model retains the essential physics to describe the properties of interest, the parametrization may account for all other effects in an average sense, and it is then a matter of validation to establish the numerical accuracy of such methods.

Semiempirical approaches to quantum chemistry are thus characterized by the use of empirical parameters in a quantum mechanical framework. In this sense, many current methods contain semiempirical features. For example, some high-level *ab initio* treatments of thermochemistry employ empirical corrections for high-order correlation effects, and several advanced density functionals include a substantial number of empirical parameters that are fitted against experimental data. We shall not cover such approaches here, but follow the conventional classification by considering only semiempirical methods that are based on molecular orbital (MO) theory and make use of integral approximations and parameters already at the MO level.

In the following, Section 21.2 gives a brief historical overview, Section 21.3 surveys established semiempirical methods introduced before 1990, Section 21.4 addresses some more recent methodological developments, Section 21.5 presents selected recent applications, and Section 21.6 offers a summary and an outlook.

21.2 HISTORICAL OVERVIEW

One of the first semiempirical approaches in quantum chemistry was the π-electron method due to Hückel (1931) [2] which generates MOs essentially from the connectivity matrix of a molecule and provides valuable qualitative insights into the structure, stability, and spectroscopy of unsaturated molecules. This approach was extended to include all valence electrons by Hoffmann (1963) [3,4] and applied in many qualitative studies of inorganic and organometallic compounds. These early semiempirical methods have had a lasting impact on chemical thinking because they guided the development of qualitative MO theory which is commonly employed for rationalizing chemical phenomena in terms of orbitals interactions [5–7]. They are normally not used any longer as computational tools.

Hückel-type methods include only one-electron integrals and are therefore noniterative. Two-electron interactions are taken into account explicitly in semiempirical self-consistent-field (SCF) methods. Again, the first such approaches were restricted to π-electrons, most notably the Pariser–Parr–Pople (PPP) method (1953) [8,9] which describes the electronic spectra of unsaturated molecules quite reliably [10,11]. The generalization to valence electrons was given by Pople (1965) [12] who introduced a hierarchy of integral approximations that satisfy rotational invariance and other consistency criteria (CNDO complete neglect of differential overlap, INDO intermediate neglect of differential overlap, NDDO neglect of diatomic differential overlap). This landmark paper forms the basis of most current semiempirical SCF-MO methods.

The original parametrization of these all-valence-electron MO methods was designed to reproduce *ab initio* Hartree–Fock (HF) results obtained with a minimal basis set. This gives rise to approximate MO treatments which can at best reach the accuracy of the target *ab initio* HF methods. Prominent examples are the CNDO/2 [13] and INDO [14] methods.

A different parametrization strategy was pursued by Dewar. He realized that *ab initio* HF results with a minimal basis are not accurate enough for most chemical problems, and therefore advocated a calibration of semiempirical methods against experimental reference data which leads to semiempirical MO methods in the proper sense. Extensive developments and parametrizations of semiempirical SCF methods were performed in the Dewar group, both at the π-electron level (1965–1970) and the all-valence-electron level (1967–1990). This work aimed at a realistic description of ground-state potential surfaces, particularly for organic molecules. It culminated in an INDO-based method named MINDO/3 (1975) [15] and two NDDO-based methods labeled MNDO (1977) [16,17] and AM1 (1985) [18]. An independent later parametrization of the MNDO model gave rise to PM3 (1989) [19,20]. Formally, AM1 and PM3 differ from MNDO only in the choice of the empirical core repulsion function, and they can therefore be viewed as attempts to explore the limits of the MNDO electronic structure model through extensive parametrization.

In the time before 1990, there are two other noteworthy INDO-based developments. In the work by Jug (1973–1990), orthogonalization corrections to the one-electron integrals were included in an INDO scheme which was then parametrized against ground-state properties to yield the SINDO1 method [21,22]. A different direction was followed by Zerner (1973–1990) in the development of the INDO/S method [23,24] for calculating electronic spectra, particularly vertical excitation energies: INDO/S is a semiempirical CI method parametrized at the CIS level (configuration interaction with single excitations), like the older CNDO/S approach [25].

The established semiempirical methods developed until 1990 have been applied extensively in chemical research. There are thousands of publications with applications of semiempirical methods to solve chemical problems, as indicated by the Science Citation Index where the number of citations for the original papers on CNDO/2 [13], MINDO/3 [15], MNDO [16], AM1 [18], PM3 [19], and INDO/S [23] currently (June 2004) exceeds 2500, 2100, 5800, 9800, 4500, and 1200, respectively. This is further illustrated in Fig. 21.1 which shows the corresponding citations per year between 1966 and 2003. Assuming that the number of citations reflects the actual use of these methods, it is obvious that AM1 and PM3 are presently the most popular semiempirical tools in computational work.

During the past 40 years, the role of semiempirical calculations has changed. In the 1960s, the development of semiempirical SCF-MO methods was motivated to a large extent by the very limited capabilities of the available computer hardware: *ab initio* calculations could then be done only for small molecules at low levels of theory, and drastic simplifications were mandatory to handle somewhat larger systems at all. In the 1970s, computational chemistry seemed to split into two camps, and there were heated debates about the relative merits of *ab initio* and semiempirical calculations [26–29]: the *ab initio* side doubted the significance and accuracy of semiempirical SCF-MO results, whereas the semiempirical side maintained that, despite advances in methodology and computer hardware, *ab initio* calculations could not yet address most chemically interesting problems. During this time, several quantum-chemical computer programs became available that allowed nontheoreticians to carry out theoretical calculations by themselves and to judge the strengths and weaknesses of different approaches.

References pp. 577–580

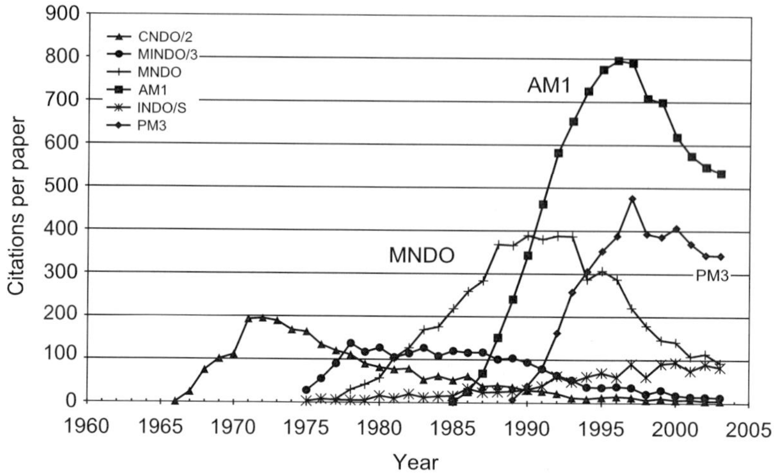

Fig. 21.1. Annual citations for original papers [13,15,16,18,19,23].

In the 1980s, computational chemistry thus gained widespread acceptance, and the *ab initio* and semiempirical branches moved towards peaceful coexistence: it was now generally recognized that *ab initio* calculations can give the right results for the right reason not only in principle, but often also in practice, and that semiempirical calculations can offer qualitatively correct results of useful accuracy for many larger and chemically interesting systems. Consequently, semiempirical SCF-MO methods (particularly MNDO and later AM1) became one of the workhorses of computational chemistry (see Fig. 21.1). In the 1990s, density functional theory (DFT) entered the mainstream of computational chemistry: gradient-corrected and hybrid functionals often provide fairly accurate results at relatively low computational costs, and corresponding DFT calculations have therefore replaced semiempirical SCF-MO calculations in many standard applications involving larger molecules (typically up to 100 atoms). However, semiempirical SCF-MO computations are still faster than DFT computations by more than two orders of magnitude, and therefore they often remain the method of choice in applications that involve really large molecules (e.g. in biochemistry) or a large number of molecules (e.g. in QSAR studies) or a large number of calculations (e.g. during direct dynamics and sampling). At the present stage, there seems to be no longer a sharp distinction between the *ab initio*, DFT, and semiempirical branches of computational chemistry, but rather a pragmatic coexistence: in practice, chemical problems are often solved by the combined use of these methods.

In the years since 1990, there have been some major new developments in semiempirical methodology which will be summarized in Section 21.4. To put these developments into perspective, the basic features of the established semiempirical methods (see above) will be outlined briefly in the following Section 21.3; for a more detailed account, the reader is referred to the available books [30–35] and review articles [36–45] on semiempirical MO theory.

21.3 ESTABLISHED METHODS

21.3.1 Basic concepts

A semiempirical *model* is defined by the underlying theoretical approach and the integral approximations that determine the types of interactions included. The established models employ a Hartree–Fock SCF-MO treatment for the valence electrons with a minimal basis set. The core electrons are normally only taken into account through a reduced nuclear charge (assuming complete shielding) although they are sometimes also represented by an effective core potential. Electron correlation is treated explicitly only if this is necessary for an appropriate zero-order description. Dynamic correlation effects are often included in an average sense by a suitable representation of the two-electron integrals and the overall parametrization.

In semiempirical approaches, the standard Hartree–Fock SCF-MO equations are simplified by *integral approximations* which are designed to neglect all three-center and four-center two-electron integrals. The CNDO, INDO, and NDDO schemes have been introduced for this purpose [12,31]. They are rotationally invariant generalizations of the zero-differential-overlap approximation from π-electron theory to valence-electron systems. The most refined of these schemes is NDDO which assumes

$$\mu^{\mathrm{A}}\nu^{\mathrm{B}}\mathrm{d}\tau = \delta_{\mathrm{AB}}\mu^{\mathrm{A}}\nu^{\mathrm{B}}\mathrm{d}\tau \tag{1}$$

during integral evaluation where the superscripts assign an AO (μ or ν) to an atom (A or B) and $\mathrm{d}\tau$ denotes the volume element. NDDO is equivalent to INDO for the one-center integrals, but keeps a much larger number of the two-center integrals and does not require any spherical averaging for maintaining rotational invariance. NDDO retains the higher multipoles of charge distributions in the two-center interactions (unlike CNDO and INDO which truncate after the monopole), and therefore accounts for anisotropies in these interactions. The CNDO, INDO, and NDDO approximations are applied to all integrals that involve Coulomb interactions, and to the overlap integrals that appear in the Hartree–Fock secular equations.

The *implementation* of a semiempirical model specifies the evaluation of all nonvanishing integrals and introduces the associate parameters. The integrals are either determined directly from experimental data or calculated exactly from the corresponding analytical formulas or represented by suitable parametric expressions. The first option is generally only feasible for the one-center integrals which may be derived from atomic spectroscopic data. The selection of appropriate parametric expressions is normally guided either by an analysis of the corresponding analytical integrals or by intuition.

The *parametrization* of a given implementation serves to determine optimum parameter values by calibrating against suitable reference data. The most widely used methods (see Section 21.2) adhere to the semiempirical philosophy and attempt to reproduce experiment. However, if reliable experimental reference data are not available, accurate theoretical data (e.g. from high-level *ab initio* calculations) are now generally considered acceptable as substitutes for experimental data. The quality of semiempirical results is strongly influenced by the effort put into the parametrization.

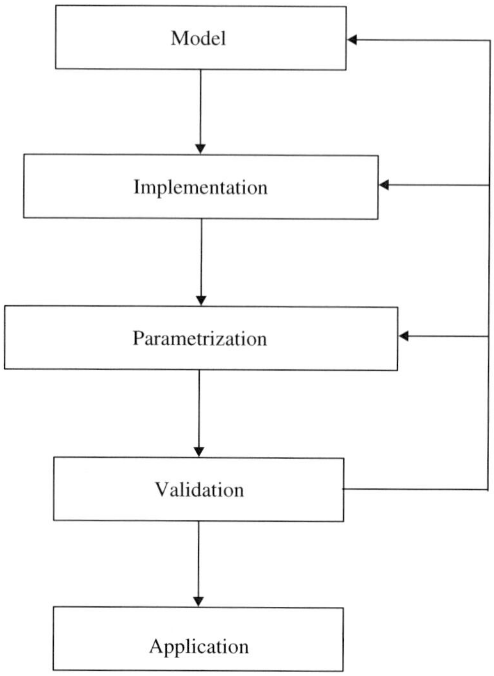

Fig. 21.2. Semiempirical method development.

Fig. 21.2 summarizes the various stages of semiempirical method development. The specification of the model and its implementation are followed by the parametrization and the subsequent validation for a broad range of systems not included in the parametrization. If all these steps are completed successfully, the method is ready for general use. If the results are not satisfactory at any stage, it is necessary to return to previous stages and look for improvements. Semiempirical method development is thus an iterative process which may be quite cumbersome: it is essential to explore the accuracy that can be reached by a given model and implementation, to determine the corresponding inherent limitations, and to move towards an improved model and/or implementation if necessary.

21.3.2 MNDO and related methods

The most popular semiempirical methods for studying ground-state potential surfaces are based on the MNDO model [16]. We shall therefore outline the MNDO formalism for closed-shell molecules as a point of reference for the following discussion.

The molecular orbitals ψ_i are expressed in the usual manner as linear combinations of atomic orbitals (AOs, ϕ_μ). The expansion coefficients $c_{\mu i}$ and the orbital energies ε_i are obtained from the solution of the secular equations.

$$\psi_i = \sum_{\mu} c_{\mu i} \phi_{\mu}, \tag{2}$$

$$0 = \sum_{\nu} (F_{\mu\nu} - \delta_{\mu\nu} \varepsilon_i) c_{\nu i}. \tag{3}$$

As a consequence of the NDDO integral approximation, the overlap integrals in the *ab initio* secular equations are replaced by Kronecker deltas, and the MNDO Fock matrix elements $F_{\mu\nu}$ contain only one-center and two-center terms. They are defined as

$$F_{\mu^A \nu^A} = H_{\mu^A \nu^A} + \sum_{\lambda^A} \sum_{\sigma^A} P_{\lambda^A \sigma^A} \left[(\mu^A \nu^A, \lambda^A \sigma^A) - \frac{1}{2} (\mu^A \lambda^A, \nu^A \sigma^A) \right]$$

$$+ \sum_{B} \sum_{\lambda^B} \sum_{\sigma^B} P_{\lambda^B \sigma^B} (\mu^A \nu^A, \lambda^B \sigma^B), \tag{4}$$

$$F_{\mu^A \nu^B} = H_{\mu^A \nu^B} - \frac{1}{2} \sum_{\lambda^A} \sum_{\sigma^B} P_{\lambda^A \sigma^B} (\mu^A \lambda^A, \nu^B \sigma^B), \tag{5}$$

where $H_{\mu\nu}$ and $P_{\lambda\sigma}$ are elements of the one-electron core Hamiltonian and the density matrix, respectively, and $(\mu\nu, \lambda\sigma)$ denotes a two-electron integral. The total energy E_{tot} of a molecule is the sum of its electronic energy E_{el} and the core-core repulsions $E_{\mathrm{AB}}^{\mathrm{core}}$ which are composed of an electrostatic term $E_{\mathrm{AB}}^{\mathrm{coul}}$ and an additional effective term $E_{\mathrm{AB}}^{\mathrm{eff}}$ (see below).

$$E_{\mathrm{el}} = \frac{1}{2} \sum_{\mu} \sum_{\nu} P_{\mu\nu} (H_{\mu\nu} + F_{\mu\nu}), \tag{6}$$

$$E_{\mathrm{tot}} = E_{\mathrm{el}} + \sum_{A} \sum_{< B} E_{\mathrm{AB}}^{\mathrm{core}}. \tag{7}$$

The MNDO model is defined by the equations given above. In its original implementation [16] the one-center terms are derived from atomic spectroscopic data, with the refinement that slight adjustments of the one-center one-electron energies are allowed in the parametrization. The one-center two-electron integrals provide the one-center limit ($R_{\mathrm{AB}} = 0$) of the two-center two-electron integrals while the asymptotic limit for $R_{\mathrm{AB}} \rightarrow \infty$ is determined by classical electrostatics. The semiempirical calculation of $(\mu^A \nu^A, \lambda^B \sigma^B)$ conforms to these limits and evaluates the integrals from semiempirical multipole–multipole interactions [46,47] damped according to the Klopman–Ohno formula. At small and intermediate distances, the semiempirical two-electron integrals are smaller than their analytical counterparts which is attributed to some average inclusion of electron correlation effects. Aiming for a reasonable balance between electrostatic attractions and repulsions within a molecule, the two-center core-electron attractions $(\mu^A \nu^A, \mathrm{B})$ and the core–core repulsions $E_{\mathrm{AB}}^{\mathrm{coul}}$ are expressed in terms of the two-electron integrals $(\mu^A \nu^A, s^B s^B)$ and $(s^A s^A, s^B s^B)$, respectively. The additional effective atom-pair potential $E_{\mathrm{AB}}^{\mathrm{eff}}$ (with an essentially exponential repulsion) attempts to account for Pauli exchange repulsions and also to compensate for errors introduced by other assumptions. Finally, following semiempirical tradition, the resonance integrals $\beta_{\mu\nu} = H_{\mu^A\nu^B}$ are taken to be proportional to the corresponding overlap integrals.

The parametrization of MNDO has focused on ground-state properties, mainly heats of formation and geometries, with the use of ionization potentials and dipole moments as additional reference data. The choice of heats of formation as reference data implies that the parametrization must account for zero-point vibrational energies and for thermal corrections between 0 and 298 K in an average sense. This is not satisfactory theoretically, but it has been shown empirically that the overall performance of MNDO is not affected much by this choice.

AM1 [18] and PM3 [19,20] are based on exactly the same model as MNDO and differ from MNDO only in one aspect of the implementation: the effective atom-pair term E_{AB}^{eff} in the core–core repulsion function is represented by a more flexible function with several additional adjustable parameters. The additional Gaussian terms in E_{AB}^{eff} are not derived theoretically, but justified empirically as providing more opportunities for fine tuning, especially for reducing overestimated nonbonded repulsions in MNDO. The parametrization in AM1 and PM3 follows the same philosophy as in MNDO. However, more effort has been spent on the parametrization of AM1 and PM3, and additional terms have been treated as adjustable parameters so that the number of optimized parameters per element has typically increased from 5 to 7 in MNDO to 18 in PM3.

21.4 SELECTED RECENT DEVELOPMENTS

In this section, we summarize advances in semiempirical MO theory that have been made in the years since 1990 in seven areas. In most of these cases, there are current activities aiming at further methodological improvements, and the discussion in this section should thus give an indication on where the field is moving. It should be stressed, however, that the selection of topics is subjective and not intended to be comprehensive.

21.4.1 Beyond the MNDO model: orthogonalization corrections

The *ab initio* SCF-MO secular equations include overlap and require a transformation from the chosen nonorthogonal to an orthogonal basis to arrive at the standard eigenvalue problem, **FC** = **CE**, where **F** and **C** denote the Fock and the eigenvector matrix in the orthogonal basis, respectively, and **E** is the diagonal matrix of orbital energies. The semiempirical integral approximations yield such secular equations without overlap directly (see Section 21.3) which suggests that the semiempirical Fock matrix implicitly refers to an orthogonalized basis and that the semiempirical integrals should thus be associated with theoretical integrals in an orthogonalized basis.

In the case of the two-electron integrals, this provides the traditional justification for the NDDO approximation [42]: Numerical studies have shown that the three- and four-center as well as certain two-center two-electron integrals are indeed small in a Löwdin-orthogonalized basis and may therefore be neglected. For the remaining two-center two-electron NDDO integrals, the orthogonalization transformation leads to a moderate reduction of these integrals at intermediate distances which is consistent with the Klopman–Ohno scaling of the semiempirical integrals.

The one-electron integrals are usually affected more strongly by the orthogonalization. The importance of orthogonalization corrections for the one-electron integrals is generally recognized. In an *ab initio* framework, the Pauli exchange repulsions and the asymmetric splitting of bonding and antibonding orbitals arise from these corrections. When using the standard semiempirical integral approximations, these and related effects are formally neglected, and many deficiencies of the established semiempirical SCF-MO methods may be attributed to this cause (see Ref. [42] for further details).

As mentioned (Section 21.3.2), the MNDO-type methods attempt to incorporate the effects of Pauli exchange repulsion in an empirical manner, through an effective atom-pair potential E_{AB}^{eff} that is added to the core–core repulsion. It would clearly be better to include the underlying orthogonalization corrections explicitly in the electronic calculation and to remove the effective atom-pair potential from the core–core repulsion. In a semiempirical context, the dominant one-electron orthogonalization corrections can be represented by parametric functions that reflect the second-order expansions of the Löwdin orthogonalization transformation in terms of overlap. These corrections can then be adjusted during the parametrization process.

Following previous INDO-based work [21,22], these basic ideas have been implemented at the NDDO level in three steps [48–52]. First, the Pauli exchange repulsions were introduced as valence-shell orthogonalization corrections only in the one-center part of the core Hamiltonian [48,49]. In the second step, they were also included in the two-center part of the core Hamiltonian [50,51], i.e. in the resonance integrals. In the third step, less important second-order correction terms were omitted [52].

The first approach (labeled OM1) contains only one-center and two-center terms, whereas the second and third approach (labeled OM2 and OM3) include three-center contributions in the corrections to the resonance integrals. These three-center contributions reflect the stereochemical environment of each electron pair bond and should thus be important for modeling conformational properties. OM1, OM2, and OM3 have been parametrized for the elements H, C, N, O, and F. They perform well for ground-state properties and offer consistent small improvements over the established MNDO-type methods in the usual statistical validations [49,50,52,53]. More significant qualitative advances are found in several other areas where the explicit inclusion of Pauli exchange repulsions is expected to be important, in particular for excited states (vertical excitation energies), conformational properties (rotational barriers, relative energies of conformers and isomers), and hydrogen bonds [49,50,52]. More work is needed to extend these approaches to other elements.

21.4.2 Implementation of d orbitals in MNDO-type methods

MNDO, AM1, and PM3 employ an sp basis without d orbitals in their original implementation. Therefore, they cannot be applied to most transition metal compounds, and difficulties are expected for hypervalent compounds of main-group elements where the importance of d orbitals for quantitative accuracy is well documented at the *ab initio* level. It is easier to include d orbitals in CNDO and INDO than in NDDO. Consequently, extensions to d orbitals have long been available for CNDO methods [31] and INDO-based methods [22,24], but only more recently for MNDO [47,54].

In the MNDO/d approach, the established MNDO formalism and parameters remain unchanged for hydrogen, helium, and the first-row elements. The inclusion of d orbitals for the heavier elements requires a generalized semiempirical treatment of the two-electron interactions. The two-center two-electron integrals for an spd basis are calculated by an extension [47] of the original point-charge model for an sp basis [46]. They are expanded in terms of semiempirical multipole–multipole interactions that are truncated at the quadrupole level and evaluated with the use of the Klopman–Ohno formula. All nonzero one-center two-electron integrals are retained to ensure rotational invariance. The implementation and parametrization of MNDO/d are analogous to MNDO, with only very minor variations in certain details [47,54]. MNDO/d parameters are available for the second-row elements, the halogens, and the zinc group elements. The corresponding results from MNDO/d are much improved over those from MNDO, as expected in particular for hypervalent compounds [54].

The two-electron integral scheme for an spd basis [47] can be implemented in combination with MNDO, AM1, or PM3. This has been done by several groups, and a number of corresponding parametrizations have become available, including PM3/tm for transition metals [55], AM1/d for Mo [56] and P [57] as well as AM1* for P, S, and Cl [58].

21.4.3 Modified general-purpose methods

In the MNDO model, the two-center core–core repulsions are composed of an electrostatic term E_{AB}^{coul} and an additional effective term E_{AB}^{eff} (see Section 21.3.2). The latter can easily be modified to allow for a more flexible parametrization. This basic idea has led to the development of AM1 and PM3, and corresponding parametrizations with improved optimization technology and more extended reference data sets are still being pursued in the quest for more accurate general-purpose MNDO-type methods that provide parameters for most elements in the periodic table; the latest product of these efforts is PM5 [59].

Following an analogous general strategy, another general-purpose implementation of the MNDO model has been introduced recently [60] which represents the effective core–core repulsion term E_{AB}^{eff} by Pairwise Distance Directed Gaussians (PDDG) and, moreover, differs from the established methods by treating the energy of each gaseous atom as an adjustable parameter (rather than computing it). PDDG/MNDO and PDDG/PM3 parameters have been published for H, C, N, and O [60] and for the halogens [61], and further work is in progress to extend the parametrization to other elements that are important in biochemistry. For the chosen large validation sets, PDDG/MNDO and PDDG/PM3 yield heats of formation with significantly smaller mean absolute errors than the parent methods [60,61], and PDDG/PM3 shows substantial improvement over density functional methods with large basis sets in this regard [60]. It will be interesting to see how these new general-purpose methods perform in other applications that were not considered in their development [62].

At the INDO level, a substantial modification of the established SINDO1 formalism [21,22] has led to a new treatment named MSINDO [63,64]. Both SINDO1 and MSINDO explicitly include orthogonalization corrections, but their implementation is different: MSINDO includes only the first-order terms, whereas SINDO1 also keeps the less important second-order terms. In addition, there are several other differences with regard to basis set (e.g. d orbitals included for Al–Cl in MSINDO) and integral evaluation [21,22,63,64]. As in MNDO-type methods, the parametrization of SINDO1 and MSINDO focuses on ground-state properties (employing binding energies rather than heats of formation as reference data). MSINDO has been parametrized for the elements H, C–F, Na–Cl, Sc–Zn, and Ga–Br [63–66]. The results generally appear to be of similar accuracy as those from MNDO-type methods.

The INDO/S method has recently been extended to the calculation of Rydberg states by adding a shell of diffuse orbitals at a single center in the molecule [67]. This required new integral approximations to take into account that a Rydberg orbital has a much greater radial extent than the valence basis functions and to allow mixing between the valence and Rydberg excited states [67]. Preliminary results for the vertical excitation energies of small unsaturated organic molecules seem encouraging.

To conclude this section, we mention an article [68] that discusses desirable features for 'next-generation' NDDO-based semiempirical methods. Apart from orthogonalization corrections and effective core potentials that are already included in some of the more recent developments (see above) it is proposed that an implicit dispersion term should be added to the Hamiltonian to capture intramolecular dispersion energies in large molecules. It is envisioned that dispersion interactions can be computed self-consistently from an additive polarizability model with some short-range scaling [68].

21.4.4 Special-purpose parametrizations

Up to this point, we have focused on general-purpose semiempirical methods which attempt to describe all classes of compounds and many properties simultaneously and equally well. This is an ambitious goal, and it is obvious that compromises cannot be avoided in such an endeavour. It is therefore an attractive idea to develop specialized semiempirical methods for certain classes of compounds or specific properties because such methods ought to be more accurate in their area of applicability than the general-purpose methods.

Such specialized semiempirical treatments exist (see Ref. [42] for a review on early work), many of which are based on the MNDO model in one of its standard implementations. For example, there are several early MNDO and AM1 variants with a special treatment of hydrogen bonds [42], which have recently been supplemented with an elaborate PM3-based parametrization for water–water interactions named PM3-PIF (pair interaction function) [69]. These special approaches exploit the flexibility offered in MNDO-type methods by the presence of the effective core–core repulsion term E_{AB}^{eff} which can be modified for fine tuning.

Other special-purpose approaches focus on chemical reactions and employ NDDO-based methods with specific reaction parameters (NDDO-SRP) [70]. This concept has been adopted by a number of groups for direct dynamics calculations. Typically, the parameters in a standard method such as AM1 are adjusted to optimize the potential surface for an individual reaction or a set of related reactions (allowing moderate parameter variations up to 10%). This is done by fitting against experimental data (reaction energies, barriers) and/or high-level *ab initio* data (relevant points on the potential surface of a suitable model system), often with the use of genetic algorithms [71]. The NDDO-SRP scheme serves as a robust and economic protocol for generating realistic potential surfaces in a cost-effective manner.

Special parametrizations are also available [42] for a number of properties including electrostatic potentials and effective atomic charges for use in biomolecular modeling. Another example concerns the semiempirical calculation of NMR chemical shifts at the GIAO-MNDO level [72]. In this approach, the NMR shielding tensor is evaluated in MNDO approximation using gauge-including atomic orbitals (GIAO) and analytic derivative theory. Calculations using the GIAO-MNDO formalism with standard MNDO parameters overestimate the variation of the paramagnetic contribution to the NMR chemical shifts which is caused by the systematic underestimation of excitation energies in MNDO. This failure can be rectified by a reparametrization [72] where some of the usual MNDO parameters are adjusted to increase the gap between occupied and unoccupied MOs and thus decrease the paramagnetic contribution to the NMR chemical shifts. This leads to significant improvements: the mean absolute errors of the computed shifts drop to less than 5% of the total chemical shift range of a given element, with values around 8, 29, and 43 ppm for C, N, and O, respectively [72]. The GIAO-MNDO approach also provides realistic nucleus-independent chemical shifts which are often used as a magnetic criterion for aromaticity [73]: the aromatic or antiaromatic character of a given system can normally be assigned correctly on the basis of the GIAO-MNDO results [74].

It is obvious from these examples that special-purpose parametrizations of established semiempirical models can be used in a pragmatic manner to enhance their accuracy for specific applications.

21.4.5 Computational aspects

Semiempirical SCF-MO methods are designed to be efficient. Due to the integral approximations commonly used (see Section 21.3) integral evaluation scales as $O(N^2)$ for N basis functions. For large molecules this becomes negligible compared with the $O(N^3)$ steps in the solution of the secular equations and the formation of the density matrix. Whenever possible, full diagonalizations of the Fock matrix are avoided by adopting a procedure [75] that involves transformation of the Fock matrix from the AO basis to the MO basis and subsequent noniterative annihilation of matrix elements in the occupied-virtual block. Therefore, most of the work in the $O(N^3)$ steps occurs in matrix multiplications. The small number of nonzero integrals implies that there are usually no input/output bottlenecks. The calculations can normally be performed completely in memory, with memory requirements that scale as $O(N^2)$.

Given these characteristics, it is evident that large-scale semiempirical SCF-MO calculations are ideally suited for vectorization and shared-memory parallelization: the dominant matrix multiplications can be performed very efficiently by BLAS library routines, and the remaining minor tasks of integral evaluation and Fock matrix construction can also be handled well on parallel vector processors with shared memory (see Ref. [43] for further details). The situation is less advantageous for massively parallel (MP) systems with distributed memory. In recent years, several groups have reported on the fine-grained parallelization of their semiempirical SCF-MO codes on MP hardware [76–79], but satisfactory overall speedups are normally obtained only for relatively small numbers of nodes (see Ref. [43] for further details).

Efficient explorations of potential surfaces require the derivatives of the energy with respect to the nuclear coordinates which can, in principle, be computed either analytically or numerically. Analytic derivatives generally contain contributions from integral derivatives and from density matrix derivatives (CPHF terms). At the semiempirical level, the evaluation of the former normally involves little computational effort since there are only few integral derivatives due to the neglect of most integrals, while the evaluation of the CPHF terms may be more demanding. Over the past decade, efficient analytic derivative codes have been developed for computing the gradient [80,81] and the harmonic force constants [82] in MNDO-type methods. The greatest practical benefits of these developments are found in the case of the gradient for semiempirical CI wavefunctions where a previous implementation [83] employed a procedure for solving the CPHF equations that scales as $O(N^4)$. This time-determining step can be reformulated [80] to scale as $O(N^3)$ when making use of the **Z**-vector method [84] which leads to dramatic speedups (by factors of about N). The analytic CI gradient code [81] is general and can therefore be used not only with the minimal CI treatments that are traditionally employed in semiempirical work, but also with elaborate multi-reference MR-CI approaches that have recently been implemented for semiempirical methods [85,86]. The evaluation of the analytic CI gradient is significantly faster than the underlying SCF and CI calculations, both for minimal CI and for large-scale MR-CI approaches [81,86]. These computational advances greatly facilitate semiempirical CI studies of electronically excited states of large molecules.

21.4.6 Linear scaling methods

In practice, conventional semiempirical SCF-MO calculations are easily done on current hardware for molecules containing up to about 1000 nonhydrogen atoms. For much larger molecules, it is advisable to employ alternative algorithms that attempt to achieve a linear scaling of the computational effort with system size, by exploiting the local character of most relevant interactions and the sparsity of the associate matrices. In semiempirical quantum chemistry, the primary objective of linear scaling methods is to avoid the bottlenecks related to diagonalization, i.e. to avoid the steps that scale as $O(N^3)$. Over the past decade, three different approaches have been developed for this purpose.

In the localized molecular orbital (LMO) approach [87], 2×2 rotations are applied to annihilate the interactions between occupied and virtual LMOs that are located within a certain cutoff radius, whereas all other interactions are considered to be negligible and therefore not treated. The resulting small numerical errors can be controlled by a renormalization of the LMOs and a suitable choice of the cutoff radius.

The divide-and-conquer methods [88–90] are based on a partitioning of the density matrix. The overall electronic structure calculation is decomposed into a series of relatively inexpensive, standard calculations for a set of smaller, overlapping subsystems. A global description of the full system is then obtained by combining the information from all subsystem density matrices.

The conjugate gradient density matrix search (CG-DMS) [91–93] avoids diagonalizations by a direct optimization of the density matrix. An acceptable density matrix must be normalized and idempotent, and it must commute with the Fock matrix after SCF convergence. In the CG-DMS treatment, a suitable functional of the density matrix is minimized with respect to the density matrix such that the resulting density matrix satisfies these criteria. By neglecting density and Fock matrix elements close to zero, sparse matrices are obtained, and linear scaling can be approached by applying sparse matrix techniques.

All these linear scaling methods introduce some approximations so that the results will show some deviations from the conventional results obtained by a full diagonalization (which can be controlled by the choice of suitable cutoffs). Furthermore, all require some overhead so that the conventional calculations remain faster for small molecules: the crossover point depends on a number of factors (e.g. the chosen cutoffs and the molecular shapes), but 200–300 atoms seems to be a typical range. Finally, it should be kept in mind that none of these algorithms can exhibit true linear scaling without addressing other parts of semiempirical computations that scale formally as $O(N^2)$, such as integral evaluation. However, in practice this is only a minor point since the $O(N^2)$ steps are generally very fast.

Applications of linear scaling semiempirical methods have focused on large biomolecules, and calculations have been reported for molecules up to about 20,000 atoms [93]. The merits of such calculations are most obvious for large systems with long-range charge transfer or long-range charge fluctuations since such effects can only be captured by quantum-chemical approaches that cover the complete system.

21.4.7 Hybrid methods

There are many processes in large systems where the electronically active part is quite localized, for example chemical reactions in a well-defined active site or electronic excitations in a chromophore. In such cases, hybrid methods may be appropriate where the electronically active region is described by quantum mechanics (QM) and the environment by molecular mechanics (MM). These QM/MM methods offer a versatile approach that can be tailored to the specific systems studied, by a suitable choice of the QM/MM partitioning and of the QM and MM methods employed.

A thorough review of QM/MM methods is beyond the scope of this article, and the reader is referred to the extensive literature for this purpose [94–104]. We shall only list some methodological issues that had to be resolved in the development of semiempirical QM/MM methods. These include the choice of a proper QM/MM coupling scheme [105] (mechanical, electronic, polarized, or self-consistent polarized embedding [106–108]), the treatment of the QM/MM boundary [109] (in terms of link atoms [95], localized orbitals from a local SCF scheme [110], generalized hybrid orbitals [111,112], or parametrized connection atoms [113]), the specific parametrization of QM/MM interaction terms (both for van-der-Waals [114] and electrostatic [115] contributions), and the treatment of long-range electrostatic interactions [116]. Based on these and other methodological developments, semiempirical QM/MM approaches have now become a popular tool for studying large (bio)molecules. Of course, *ab initio* and DFT methods can also serve as QM component in QM/MM calculations, and the choice between them and semiempirical methods depends on the required accuracy and the tolerable computational effort.

Semiempirical QM/MM methods are computationally much less demanding than pure QM methods, even compared with linear scaling semiempirical methods: factors of around 500 have been reported for the computation times in typical applications on enzymes [117]. In practice, semiempirical QM/MM methods are therefore used for extensive potential energy explorations and sampling in biomolecular simulations, while linear scaling semiempirical QM methods can be employed in single-point calculations to check the validity of the QM/MM approach (e.g. with regard to charge transfer and charge fluctuations). Both approaches are thus complementary since they may be used to address different questions. Their availability has made it possible to study complex systems with thousands of atoms such as enzymes at the semiempirical level.

21.5 SELECTED RECENT APPLICATIONS

In spite of the advent of density functional theory (see Section 21.2), the number of publications with semiempirical calculations remains high. In the Science Citation Index, one finds for each of the past ten years more than 1000 such papers under the topic 'semiempirical/MNDO/AM1/PM3', the actual numbers fluctuating between 1100 and 1500 (1994–2003); this should be regarded as a lower limit of the actual usage, for obvious reasons. Since DFT calculations have replaced semiempirical calculations in many studies on medium-sized molecules, the latter must have found new areas of application. In this section, we attempt to identify such areas from a survey of the recent literature. Given the diverse activities in this field, it is clear this cannot be a comprehensive overview and that the selection of topics will necessarily be subjective.

Large *biomolecules* with thousands of atoms are an attractive target for semiempirical work. Higher-level *ab initio* and DFT calculations are very costly for such systems even in the case of QM/MM approaches, especially when considering the need for an adequate sampling of conformational space, and the accuracy of semiempirical methods is normally best for organic compounds. Therefore, semiempirical QM/MM studies on large biomolecules have become very popular in recent years. In particular, many enzymatic reactions have been examined at this level (see Refs. [98,101] for corresponding reviews). Semiempirical QM/MM calculations are fast enough to compute potentials of mean force along the reaction coordinate (e.g. via umbrella sampling or free energy perturbation theory) so that free energy barriers of enzymatic reactions can be obtained. Comparing with the results from an analogous treatment of the reaction in aqueous solution and with gas-phase results allows an analysis of the origin of the rate enhancement in the enzyme. Examples of such semiempirical QM/MM work are the recent investigations on chorismate mutase [118,119] and on catechol O-methyl transferase [120]. Dynamical effects on enzyme kinetics have been incorporated on the basis of semiempirical QM/MM potential surfaces by a variety of techniques, including mixed quantum-classical molecular dynamics and variational transition state theory (see Ref. [101] for a review), and kinetic isotope effects have been evaluated for a number of enzymatic reactions [101,121,122]. Some of these QM/MM studies employ specific reaction parameters (mostly AM1-SRP or PM3-SRP) to improve the accuracy of the QM component [101,122,123].

Large biomolecules can also be treated by linear scaling semiempirical QM calculations, but at notably higher costs as compared to analogous QM/MM work [117]. Such calculations are thus done less often. Applications of the linear scaling divide-and-conquer approach to biological systems have been reviewed [124]. Topics that have been addressed by linear scaling semiempirical methods include charge-transfer interactions in macromolecular systems such as the protein/water interface [125], the electrostatic potential in the potassium channel [126], charge fluctuations in the nucleic acids DNA and RNA in solution [127], the regioselectivity and RNA-binding affinity of HIV-1 nucleocapsid protein [128], and the calculation of NMR chemical shifts in biomolecules using a divide-and-conquer implementation [129] of the GIAO-MNDO method [72]. In addition to such large-scale calculations on systems with thousands of atoms, there are of course conventional semiempirical calculations that address biochemical problems using smaller model compounds (e.g. to give just one example, the reaction mechanism of phosphoryl transfer in kinases [130,131]).

In *medicinal chemistry* and drug design, the focus is often not on the detailed understanding of one particular biomolecular system, but rather on the comparison and screening of a large number of complex molecules that may have certain pharmacological effects. Hence, quantitative structure-property and structure-activity relationships (QSPR and QSAR, respectively) are of interest that correlate properly chosen molecular descriptors with the desired properties and activities. Semiempirical MO methods are well suited for this purpose, and many different theoretical descriptors have been proposed and tested in semiempirical QSPR and QSAR work. Recent AM1-based applications include a QSPR study that uses 66 descriptors in an attempt to distinguish between drugs and nondrugs in a database of more than 2000 compounds [132], the introduction of new molecular descriptors based on local properties at the molecular surface [133], and a QSAR investigation aimed at the prediction of blood–brain barrier permeation [134].

Apart from biomolecules, there are many other *nanoparticles* that are of considerable experimental interest and are amenable to semiempirical calculations. Large fullerenes and carbon nanotubes are prime examples, and there have a large number of corresponding semiempirical studies which are too numerous to be listed here (see Ref. [42] for a review of early MNDO work). The performance of MNDO, AM1, and PM3 in fullerene chemistry has recently been validated by systematic comparisons with DFT results for a set of 153 fullerene species [135]. MSINDO has been employed to investigate the structure and the growth mechanism of siloxane-based nanotubes (up to 840 atoms) [136–138]. In molecular electronics, where large metal clusters have been proposed to serve as nanoelectrodes in complex devices, it has been shown that INDO/S provides an efficient computational scheme for studying a wide range of nanoparticle electronic properties [139]: due to the inclusion of exchange effects, INDO/S yields more realistic charge distributions and energetics than continuum electrostatics or classical atomistic electrostatics [139], and may even be superior to DFT in this area because of well-known DFT failures concerning dispersion, charge transfer, extended π conjugation, and bond cleavage [140].

Solids and surfaces can be modeled as large clusters, as embedded clusters, and as infinite periodic systems. Each of these strategies has been used in conjunction with semiempirical MO methods. The most straightforward semiempirical approach employs clusters of increasing size, taking advantage of the fact that semiempirical calculations can be extended easily to systems with more than 1000 atoms which facilitates convergence studies to the bulk limit. Semiempirical work in this field has been reviewed recently [45].

Given the availability of high-level *ab initio* and DFT methods, it may seem odd at first sight that there is one area where state-of-the-art applications involve semiempirical calculations on small molecules: this is *direct reaction dynamics* where energies and gradients are computed on the fly at each time step, which requires many thousands or even millions of such single-point calculations for proper sampling. The accuracy of the established semiempirical methods will normally not be sufficient for a dynamical study of a given reaction, and it is therefore common practice to derive specific reaction parameters (SRP) for this purpose [70,71]. Semiempirical direct dynamics studies have addressed, for example, the unimolecular dissociation of formaldehyde in the gas phase (AM1-SRP) [141], the nucleophilic substitution of methyl bromide by chloride (AM1-SRP) [142], the lifetime and stereomutation of cyclopropane (AM1-SRP) [143–145], the collision- and surface-induced dissociation of N-protonated glycine (AM1 + MM) [146], and the reactions of the triplet oxygen atom with small alkanes (PM3-SRP and MSINDO) [147–150].

The final topic in this section are *electronically excited states*. For vertical excitation energies, INDO/S has long been the method of choice, and it has been applied to many large systems including the bacteriochlorophyll *b* dimer of *Rhodopseudomonas viridis* (QM/MM study with 325 QM atoms) [106] and aggregates of bacteriochlorophylls that occur in light-harvesting complexes of photosynthetic bacteria (QM studies with up to 704 atoms for a model of the hexadecamer) [151]. For the computation of vertical excitation energies of smaller systems, time-dependent density functional theory (TDDFT) has become popular in recent years: it is often very successful, but currently available functionals are also known to fail for certain types of excitation, e.g. for those involving long-range charge transfer [152,153]. Such charge-transfer valence transitions do not pose any particular problems for semiempirical CI methods, and photoinduced charge separations have indeed been described reasonably well at this level, e.g. in recent AM1-CI studies on phenothiazine-pyrene dyads [154,155]. Moreover, substantial problems with TDDFT have been reported [153] in explorations of excited-state potential energy surfaces of unsaturated hydrocarbons, particularly polyenes, which do not appear in corresponding semiempirical CI calculations. Given such problems with TDDFT, semiempirical methods may be a good alternative for investigating the photochemistry even in medium-sized systems: NDDO-based approaches with orthogonalization corrections provide a balanced and reasonably accurate description of ground and excited states [50–52], efficient semiempirical MR-CI approaches have been developed [85,86] and applied in an OM2 study of butadiene [156], and analytic gradients [81,86] and algorithms for locating conical interactions [157] have been implemented at the semiempirical CI level. With all these ingredients available, one may expect large-scale applications of these methods to the electronically excited states and the photochemistry of complex systems.

21.6 SUMMARY AND OUTLOOK

Almost 40 years have passed since the introduction of all-valence-electron semiempirical SCF-MO methods in 1965 [12]. Subsequent developments and parametrizations have led to general-purpose methods, most notably MNDO, AM1, and PM3, that have served as some of the most popular tools in computational chemistry over the past decades (see Sections 21.2 and 21.3). Due to methodological advances and ever improving hardware performance, *ab initio* and DFT methods have replaced semiempirical methods in many applications on small and medium-sized molecules over the years. However, semiempirical calculations are so much faster than *ab initio* or DFT calculations that they have remained useful for applications involving large systems (biomolecules, nanoparticles), large numbers of compounds (QSPR, QSAR), and large numbers of calculations (direct dynamics, sampling), where higher-level calculations are still too costly. There are also some applications on medium-sized systems for which semiempirical approaches may offer a viable alternative to higher-level calculations, when the latter are still feasible but suffer from well-known problems (see Section 21.5).

Further developments in semiempirical theory should aim at improving the accuracy and applicability of semiempirical methods without compromising their computational efficiency. Such work is in progress (see Section 21.4). Theoretically guided improvements in the underlying semiempirical models would seem to offer the best strategy for developing new general-purpose treatments with better overall performance. Special parametrizations of established semiempirical models are a promising complementary approach for describing special classes of compounds, reactions, or properties as accurately as possible. Algorithmic and technical improvements extend the range of systems that can be studied at the semiempirical level. Finally, semiempirical methods are valuable components in hybrid approaches (QM/MM or QM/QM).

Experimental chemistry moves towards more complex systems, for example in biochemistry and nanochemistry. At any given point, there will be chemical problems that can only be tackled by lower-level electronic structure treatments within a practical time frame, and there will be others which are best solved by the synergetic use of several computational methods ranging from high-level electronic structure theory to classical force fields and molecular modeling. Being in the middle of this spectrum, the semiempirical methods of quantum chemistry are expected to remain useful, as in the first 40 years of computational chemistry.

21.7 ACKNOWLEDGEMENTS

The author wishes to thank his colleagues and coworkers whose names are listed in the references for their essential contributions to the work described.

21.8 REFERENCES

1 Merriam-Webster Online Dictionary, http://www.m-w.com/dictionary.htm
2 E. Hückel, Z. Phys., 70 (1931) 204–286.
3 R. Hoffmann, J. Chem. Phys., 39 (1963) 1397–1412.
4 R. Hoffmann, J. Chem. Phys., 40 (1964) 2745.
5 R. Hoffmann, Angew. Chem. Int. Ed., 21 (1982) 711–725.
6 K. Fukui, Angew. Chem. Int. Ed., 21 (1982) 801–809.
7 T.A. Albright, J.K. Burdett and M.-H. Whangbo, Orbital interactions in chemistry, Wiley, New York, 1985.
8 R. Pariser and R.G. Parr, J. Chem. Phys., 21 (1953) 466–471, 767–776.
9 J.A. Pople, Trans. Faraday Soc., 49 (1953) 1375–1385.
10 R.G. Parr, The quantum theory of molecular electronic structure, WA Benjamin, New York, 1963.
11 L. Salem, The molecular orbital theory of conjugated systems, WA Benjamin, New York, 1966.
12 J.A. Pople, D.P. Santry and G.A. Segal, J. Chem. Phys., 43 (1965) S129–S135.
13 J.A. Pople and G.A. Segal, J. Chem. Phys., 44 (1966) 3289–3296.
14 J.A. Pople, D.L. Beveridge and P.A. Dobosh, J. Chem. Phys., 47 (1967) 2026–2033.
15 R.C. Bingham, M.J.S. Dewar and D.H. Lo, J. Am. Chem. Soc., 97 (1975) 1285–1293.
16 M.J.S. Dewar and W. Thiel, J. Am. Chem. Soc., 99 (1977) 4899–4907.
17 M.J.S. Dewar and W. Thiel, J. Am. Chem. Soc., 99 (1977) 4907–4917.
18 M.J.S. Dewar, E. Zoebisch, E.F. Healy and J.J.P. Stewart, J. Am. Chem. Soc., 107 (1985) 3902–3909.
19 J.J.P. Stewart, J. Comput. Chem., 10 (1989) 209–220.
20 J.J.P. Stewart, J. Comput. Chem., 10 (1989) 221–264.
21 D.N. Nanda and K. Jug, Theor. Chim. Acta, 57 (1980) 95–106.
22 K. Jug, R. Iffert and J. Schulz, Int. J. Quantum Chem., 32 (1987) 265–277.
23 M.C. Zerner and J. Ridley, Theor. Chim. Acta, 32 (1973) 111–134.
24 A.D. Bacon and M.C. Zerner, Theor. Chim. Acta, 53 (1979) 21–54.
25 J. Del Bene and H.H. Jaffé, J. Chem. Phys., 48 (1968) 1807–1813.
26 J.A. Pople, J. Am. Chem. Soc., 97 (1975) 5306–5308.
27 W.J. Hehre, J. Am. Chem. Soc., 97 (1975) 5308–5310.
28 M.J.S. Dewar, J. Am. Chem. Soc., 97 (1975) 6591.
29 M.J.S. Dewar, J. Phys. Chem., 89 (1985) 2145–2150.
30 M.J.S. Dewar, The molecular orbital theory of organic chemistry, McGraw-Hill, New York, 1969.
31 J.A. Pople and D.L. Beveridge, Approximate molecular orbital theory, Academic Press, New York, 1970.
32 J.N. Murrell and A.J. Harget, Semiempirical self-consistent-field molecular orbital theory of molecules, Wiley, New York, 1972.
33 G.A. Segal (Ed.), Modern theoretical chemistry, Vols. 7–8, Plenum, New York, 1977.
34 M. Scholz, H.J. Köhler, Quantenchemie, Vol. 3, Hüthig, Heidelberg, 1981.

35 T. Clark, A handbook of computational chemistry, Wiley, New York, 1985.
36 M.J.S. Dewar, Science, 187 (1975) 1037–1044.
37 K. Jug, Theor. Chim. Acta, 54 (1980) 263–300.
38 W. Thiel, Tetrahedron, 44 (1988) 7393–7408.
39 J.J.P. Stewart, J. Comp.-Aided Mol. Design, 4 (1990) 1–103.
40 J.J.P. Stewart, in: K.B. Lipkowitz, D.B. Boyd (Eds.), Reviews in computational chemistry, Vol. 1, VCH Publishers, New York, 1990, pp. 45–81.
41 M.C. Zerner, in: K.B. Lipkowitz, D.B. Boyd (Eds.), Reviews in computational chemistry, Vol. 2, VCH Publishers, New York, 1991, pp. 313–365.
42 W. Thiel, Adv. Chem. Phys., 93 (1996) 703–757.
43 W. Thiel, in: J. Grotendorst (Ed.), Modern methods and algorithms of quantum chemistry, NIC Series, Vol. 3, John von Neumann Institute for Computing, Jülich, Germany, 2000, pp. 261–283.
44 W. Thiel, in: S. Wilson (Ed.), Handbook of molecular physics and quantum chemistry, Vol. 2, Wiley, Chicester, 2003, pp. 487–502.
45 T. Bredow and K. Jug, Theor. Chem. Acc., 113 (2005) 1–14.
46 M.J.S. Dewar and W. Thiel, Theor. Chim. Acta, 46 (1977) 89–104.
47 W. Thiel and A.A. Voityuk, Theor. Chim. Acta, 81 (1992) 391–404, 93 (1996) 315.
48 M. Kolb, Ph.D. Thesis, Universität Wuppertal, 1991.
49 M. Kolb and W. Thiel, J. Comput. Chem., 14 (1993) 775–789.
50 W. Weber, Ph.D. Thesis, Universität Zürich, 1996.
51 W. Weber and W. Thiel, Theor. Chem. Acc., 103 (2000) 495–506.
52 M. Scholten, Ph.D. Thesis, Universität Düsseldorf, 2003.
53 W. Thiel, in: K.K. Irikura, D.J. Frurip (Eds.), Computational thermochemistry, ACS Symposium Series 667, American Chemical Society, Washington, DC, 1998, pp. 142–161.
54 W. Thiel and A.A. Voityuk, J. Phys. Chem., 100 (1996) 616–626.
55 SPARTAN 4.0, Wavefunction Inc., 18401, Von Karman Avenue, Irvine, CA 92715, 1995.
56 A.A. Voityuk and N. Rösch, J. Phys. Chem. A, 104 (2000) 4089–4094.
57 X. Lopez and D.M. York, Theor. Chem. Acc., 109 (2003) 149–159.
58 P. Winget, A.H.C. Horn, C. Selcuki, B. Martin and T. Clark, J. Mol. Model., 9 (2003) 408–414.
59 J.J.P. Stewart, J. Mol. Model., 10 (2004) 6–12.
60 M.P. Repasky, J. Chandrasekhar and W.L. Jorgensen, J. Comput. Chem., 23 (2002) 1601–1622.
61 I. Tubert-Brohman, C.R.W. Guimaraes, M.P. Repasky and W.L. Jorgensen, J. Comput. Chem., 25 (2004) 138–150.
62 R. Casadenus, M. Moreno, A. Gonzalez-Lafont, J.M. Lluch and M.P. Repasky, J. Comput. Chem., 25 (2004) 99–105.
63 B. Ahlswede and K. Jug, J. Comput. Chem., 20 (1999) 563–571.
64 B. Ahlswede and K. Jug, J. Comput. Chem., 20 (1999) 572–578.
65 K. Jug, G. Geudtner and T. Homann, J. Comput. Chem., 21 (2000) 974–987.
66 T. Bredow, G. Geudtner and K. Jug, J. Comput. Chem., 22 (2001) 861–887.
67 J.D. Head, Int. J. Quantum Chem., 95 (2003) 580–592.
68 P. Winget, C. Selcuki, A.H.C. Horn, B. Martin and T. Clark, Theor. Chem. Acc., 110 (2003) 254–266.
69 M.I. Bernal-Uruchurtu, M.T.C. Martins-Costa, C. Millot and M.F. Ruiz-Lopez, J. Comput. Chem., 21 (2000) 572–581.
70 A. Gonzalez-Lafont, T.N. Truong and D.G. Truhlar, J. Phys. Chem., 95 (1991) 4618–4627.
71 I. Rossi and D.G. Truhlar, Chem. Phys. Lett., 233 (1995) 231–236.
72 S. Patchkovskii and W. Thiel, J. Comput. Chem., 20 (1999) 1220–1245.
73 P.v.R. Schleyer, C. Maerker, A. Dransfeld, H. Jiao and N.J.R. van Eikema Hommes, J. Am. Chem. Soc., 118 (1996) 6317–6318.
74 S. Patchovskii and W. Thiel, J. Mol. Model., 6 (2000) 67–75.
75 J.J.P. Stewart, P. Csaszar and P. Pulay, J. Comput. Chem., 3 (1982) 227–228.
76 W. Thiel and D.G. Green, in: E. Clementi, G. Corongiu (Eds.), Methods and techniques in computational chemistry, METECC-95, STEF, Cagliari, 1995, pp. 141–168.
77 H.A. Früchtl, R.H. Nobes and A. Bliznyuk, J. Mol. Struct. (THEOCHEM), 506 (2000) 87–97.

78 P.K. Berzigiyarov, V.A. Zayets, I.Y. Ginzburg, V.F. Razumov and E.F. Sheka, Int. J. Quantum Chem., 88 (2002) 409–462.
79 P.K. Berzigiyarov, V.A. Zayets, I.Y. Ginzburg, V.F. Razumov and E.F. Sheka, Int. J. Quantum Chem., 96 (2004) 73–79.
80 S. Patchkovskii and W. Thiel, Theor. Chim. Acta, 93 (1996) 87–99.
81 S. Patchkovskii and W. Thiel, Theor. Chem. Acc., 98 (1997) 1–4.
82 S. Patchkovskii and W. Thiel, J. Comput. Chem., 17 (1996) 1318–1327.
83 M.J.S. Dewar and D.A. Liotard, J. Mol. Struct. (THEOCHEM), 206 (1990) 123–133.
84 N.C. Handy and H.F. Schaefer, J. Chem. Phys., 81 (1984) 5031–5033.
85 P. Strodel and P. Tavan, J. Chem. Phys., 117 (2002) 4667–4676.
86 A. Koslowski, M.E. Beck and W. Thiel, J. Comput. Chem., 24 (2003) 714–726.
87 J.J.P. Stewart, Int. J. Quantum Chem., 58 (1996) 133–146.
88 W. Yang and T.-S. Lee, J. Chem. Phys., 103 (1995) 5674–5678.
89 T.S. Lee, D.M. York and W. Yang, J. Chem. Phys., 105 (1996) 2744–2750.
90 S.L. Dixon and K.M. Merz, J. Chem. Phys., 104 (1996) 6643–6649.
91 X.-P. Li, R.W. Nunes and D. Vanderbilt, Phys. Rev. B, 47 (1993) 10891–10894.
92 J.M. Millam and G.E. Scuseria, J. Chem. Phys., 106 (1997) 5569–5577.
93 A.D. Daniels, J.M. Millam and G.E. Scuseria, J. Chem. Phys., 107 (1997) 425–431.
94 A. Warshel and M. Levitt, J. Mol. Biol., 103 (1976) 227–249.
95 M.J. Field, P.A. Bash and M. Karplus, J. Comput. Chem., 11 (1990) 700–733.
96 J. Aqvist and A. Warshel, Chem. Rev., 93 (1993) 2523–2544.
97 J. Gao, in: K.B. Lipkowitz, D.B. Boyd (Eds.), Reviews in computational chemistry, Vol. 7, VCH Publishers, New York, 1995, pp. 119–185.
98 T.Z. Mordasini and W. Thiel, Chimia, 52 (1998) 288–291.
99 G. Monard and K.M. Merz, Acc. Chem. Res., 32 (1999) 904–911.
100 P. Sherwood, in: J. Grotendorst (Ed.), Modern methods and algorithms of quantum chemistry, NIC Series, Vol. 3, John von Neumann Institute for Computing, Jülich, Germany, 2000, pp. 285–327.
101 J. Gao and D.G. Truhlar, Ann. Rev. Phys. Chem., 53 (2002) 467–505.
102 M.J. Field, J. Comput. Chem., 23 (2002) 48–58.
103 G. Monard, X. Prat-Resina, A. Gonzalez-Lafont and J.M. Lluch, Int. J. Quantum Chem., 93 (2003) 229–244.
104 L. Ridder and A.J. Mulholland, Curr. Top. Med. Chem., 3 (2003) 1241–1256.
105 D. Bakowies and W. Thiel, J. Phys. Chem., 100 (1996) 10580–10594.
106 M.A. Thompson and G.K. Schenter, J. Phys. Chem., 99 (1995) 6374–6386.
107 M.A. Thompson, J. Phys. Chem., 100 (1996) 14492–14507.
108 J. Gao, J. Comput. Chem., 18 (1997) 1061–1071.
109 N. Reuter, A. Dejaegere, B. Maigret and M. Karplus, J. Phys. Chem. A, 104 (2000) 1720–1735.
110 V. Thery, D. Rinaldi, J.L. Rivail, B. Maigret and G. Ferenczy, J. Comput. Chem., 15 (1994) 269–282.
111 J. Gao, P. Amara, C. Alhambra and M.J. Field, J. Phys. Chem. A, 102 (1998) 4714–4721.
112 P. Amara, M.J. Field, C. Alhambra and J. Gao, Theor. Chem. Acc., 104 (2000) 336–343.
113 I. Antes and W. Thiel, J. Phys. Chem. A, 103 (1999) 9290–9295.
114 J.L. Gao and X.F. Xia, Science, 258 (1992) 631–635.
115 D. Bakowies and W. Thiel, J. Comput. Chem., 17 (1996) 87–108.
116 J. Gao and C. Alhambra, J. Chem. Phys., 107 (1997) 1212–1217.
117 S.J. Titmuss, P.L. Cummins, A.A. Bliznyuk, A.P. Rendell and J.E. Gready, Chem. Phys. Lett., 320 (2000) 169–176.
118 M.P. Repasky, C.R.W. Guimaraes, J. Chandrasekhar, J. Tirado-Rives and W.L. Jorgensen, J. Am. Chem. Soc., 125 (2003) 6663–6672.
119 C.R.W. Guimaraes, M.P. Repasky, J. Chandrasekhar, J. Tirado-Rives and W.L. Jorgensen, J. Am. Chem. Soc., 125 (2003) 6892–6899.
120 M. Roca, S. Marti, J. Andres, V. Moliner, I. Tunon, J. Bertran and I.H. Williams, J. Am. Chem. Soc., 125 (2003) 7726–7737.
121 S. Marti, V. Moliner, I. Tunon and I.H. Williams, Org. Biomol. Chem., 1 (2003) 483–487.

122 G. Tresadern, H. Wang, P.F. Faulder, N.A. Burton and I.H. Hillier, Mol. Phys., 101 (2003) 2775–2784.
123 P.A. Bash, L.L. Ho, A.D. MacKerrell, D. Levine and P. Hallstrom, Proc. Natl. Acad. Sci. USA, 93 (1996) 3698–3703.
124 A. van der Vaart, V. Gogonea, S.L. Dixon and K.M. Merz, J. Comput. Chem., 21 (2000) 1494–1504.
125 G. Nadig, L.C. van Zant, S.L. Dixon and K.M. Merz, J. Am. Chem. Soc., 120 (1998) 5593–5594.
126 A.A. Bliznyuk, A.P. Rendell, T.W. Allen and S.-H. Chung, J. Phys. Chem. B, 105 (2001) 12674–12679.
127 J. Khandogin and D.M. York, J. Phys. Chem. B, 106 (2002) 7693–7703.
128 J. Khandogin, K. Musier-Forsyth and D.M. York, J. Mol. Biol., 330 (2003) 993–1004.
129 B. Wang, E.N. Brothers, A. van der Vaart and K.M. Merz, J. Chem. Phys., 120 (2004) 11392–11400.
130 M.C. Hutter and V. Helms, Int. J. Quantum Chem., 95 (2003) 479–486.
131 M.C. Hutter and V. Helms, ChemBioChem, 3 (2002) 643–651.
132 M. Brustle, B. Beck, T. Schindler, W. King, T. Mitchell and T. Clark, J. Med. Chem., 45 (2002) 3345–3355.
133 B. Ehresmann, M.J. de Groot, A. Alex and T. Clark, J. Chem. Inf. Comp. Sci., 44 (2004) 658–668.
134 M.C. Hutter, J. Comput.-Aided Mol. Des., 17 (2003) 415–433.
135 Z. Chen and W. Thiel, Chem. Phys. Lett., 367 (2003) 15–25.
136 K. Jug and D. Wichmann, J. Comput. Chem., 21 (2000) 1549–1553.
137 K. Jug and I.P. Gloriozov, Phys. Chem. Chem. Phys., 4 (2002) 1062–1066.
138 K. Jug and I.P. Gloriozov, J. Phys. Chem. A, 106 (2002) 4736–4744.
139 J.R. Reimers and N.S. Hush, J. Phys. Chem. B, 105 (2001) 8979–8988.
140 J.R. Reimers, Z.L. Cai, A. Bilic and N.S. Hush, Ann. N.Y. Acad. Sci., 1006 (2003) 235–251.
141 G.H. Peslherbe and W.H. Hase, J. Chem. Phys., 104 (1996) 7882–7894.
142 G.H. Peslherbe, H. Wang and W.L. Hase, J. Am. Chem. Soc., 118 (1996) 2257–2266.
143 C. Doubleday, K. Bolton, G.H. Peslherbe and W.L. Hase, J. Am. Chem. Soc., 118 (1996) 9922–9931.
144 C. Doubleday, K. Bolton and W.L. Hase, J. Am. Chem. Soc., 119 (1997) 5251–5252.
145 C. Doubleday, K. Bolton and W.L. Hase, J. Phys. Chem. A, 102 (1998) 3648–3658.
146 S.O. Meroueh, Y. Wang and W.L. Hase, J. Phys. Chem. A, 106 (2002) 9983–9992.
147 T.-Y. Yan, C. Doubleday and W.H. Hase, J. Phys. Chem. A, 108 (2004) 9863–9875.
148 D.J. Garton, T.K. Minton, D. Troya, R. Pascual and G.C. Schatz, J. Phys. Chem. A, 107 (2003) 4583–4587.
149 D. Troya, R.Z. Pascual, D.J. Garton, T.K. Minton and G.C. Schatz, J. Phys. Chem. A, 107 (2003) 7161–7169.
150 D. Troya, R.Z. Pascual and G.C. Schatz, J. Phys. Chem. A, 107 (2003) 10497–10506.
151 M.G. Cory, M.C. Zerner, X.C. Xu and K. Schulten, J. Phys. Chem. B, 102 (1998) 7640–7650.
152 A. Dreuw and M. Head-Gordon, J. Am. Chem. Soc., 126 (2004) 4007–4016.
153 M. Wanko, M. Garavelli, F. Bernardi, T.A. Niehaus, T. Frauenheim and M. Elstner, J. Chem. Phys., 120 (2004) 1674–1692.
154 A. Stockmann, J. Kurzawa, N. Fritz, N. Acar, S. Schneider, J. Daub, R. Engl and T. Clark, J. Phys. Chem. A, 106 (2002) 7958–7970.
155 N. Acar, J. Kurzawa, N. Fritz, A. Stockmann, C. Roman, S. Schneider and T. Clark, J. Phys. Chem. A, 107 (2003) 9530–9541.
156 P. Strodel and P. Tavan, J. Chem. Phys., 117 (2002) 4677–4683.
157 A. Toniolo, G. Granucci and T.J. Martinez, J. Phys. Chem. A, 107 (2003) 3822–3830.

Theory and Applications of Computational Chemistry: The First Forty Years
Edited by C. Dykstra et al.

CHAPTER 22

Size-consistent state-specific multi-reference methods: a survey of some recent developments[☆]

Dola Pahari[1], Sudip Chattopadhyay[2], Sanghamitra Das[1], Debashis Mukherjee[1,3] and Uttam Sinha Mahapatra[4]

[1] *Department of Physical Chemistry, Indian Association for the Cultivation of Science, Calcutta 700032, India*
[2] *Department of Chemistry, Bengal Engineering and Science University, Shibpur, Howrah 711103, India*
[3] *Jawaharlal Nehru Centre for Advanced Scientific Research, Bangalore 560064, India*
[4] *Department of Physics, Darjeeling Government College, Darjeeling 734101, India*

Abstract

We present in this article a survey of the various approaches to state-specific multi-reference (SSMR) methods developed in our group, which use both complete and incomplete model spaces (IMSs). These SSMR methods satisfy the following desirable criteria: (a) they avoid instabilities due to intruder states in a natural manner; (b) they guarantee size-extensivity of the computed energy; (c) they ensure additivity of the energy for non-interacting subsystems when the model spaces are chosen in a separable manner; and (d) they are manifestly spin-free. For the general case involving open-shell functions in the model space, our spin-adaptation strategy is to choose them as some suitable configuration state functions (CSF), and use spin-free unitary generators to define the cluster operators. An important innovation in our spin-free formulation is the use of the entire portion $\phi_{0\mu}$ of the highest closed-shell component of a model function ϕ_μ as the vacuum to define all the excitations on ϕ_μ in normal order. This simplifies the expressions enormously. We discuss the allied issues of spin-adaptation in some detail. The size-extensive formulation in the IMS warrants abandoning the intermediate normalization convention for the wave operator, and we discuss the related issues in a concise manner. Although we do discuss the parent SSMR Coupled Cluster (SS-MRCC) formalism [J. Chem. Phys., 110 (1999) 6171] itself, albeit in a succinct fashion,

☆ It is our pleasure to dedicate this article to Henry Fritz Schaefer III on the happy occasion of his 60th birthday.

References pp. 631–633

our principal thrust here is to explore the efficacy and viability of the various physically motivated systematic approximations of graded complexity of the SS-MRCC theory. The wave operator in the SS-MRCC theory utilizes the Jeziorski–Monkhorst Ansatz, which amounts to the use of a separate exponential cluster operator $\exp(T^{\mu})$ for every ϕ_{μ}. The most computationally demanding component of the theory is the rather large number of cluster amplitudes involving inactive orbitals, distinct for each μ. We present an approximate variant, where the amplitudes of the inactive excitations involving the same set of inactive orbitals are assigned the same value independent of μ. This 'Anonymous Parentage for the Inactive excitations' (API) in SS-MRCC (API-SSMRCC) is expected to be very effective in drastically reducing the number of amplitudes without undue sacrifice in accuracy. Starting from the SS-MRCC theory, we will also develop the spin-free version of various Rayleigh–Schrödinger (RS) based perturbation theories in both complete model space and IMSs. Several reasonable choices of the one-particle unperturbed Hamiltonian H_0 in the Møller–Plesset (MP) partitioning will be introduced and discussed. The one-particle operators in H_0 in the MP partition are essentially various forms of diagonal elements of some Fock operator, defined for every model function. This corresponds to a multi-partitioning strategy, where unperturbed orbital energies depend on the model function. A special choice is, of course, the widely used state-averaged variant of the Fock operator, which has also been considered. In the Epstein–Nesbet (EN) partition, we additionally include all the diagonal direct and exchange ladders. Analogues of some of the MP partitionings for the spin-free EN partition are indicated. We also explore the performance of the perturbative variants of the API-SSMRCC, and establish their efficacy. We discuss the emergence of the CEPA-like approximants, starting from the SS-MRCC, where the space of excitations induced by the cluster operators is confined to the first order interactive virtual space, but the exclusion principle violating terms are included to various degrees. We consider two SS-MRCEPA schemes, one is basically the linear version of the SS-MRCC (SS-MRCEPA(0)), and the other is the orbital invariant version, termed as SS-MRCEPA(I). The performance of the various SS-MRPT and SS-MRCEPA-like methods is illustrated with example applications on the trapezoidal H_4 (H4 model), the PES of the BeH_2 complex and the dissociation curves of the ground states of LiH and BH, using various basis sets. Properties such as spectroscopic constants, dipole moment functions and polarizability functions are also studied. The electrical properties are calculated *via* the Finite-Field (FF) method. The results establish the SS-MRPT with the EN partition and the SS-MRCEPA methods as rather accurate viable tools for studying quasi-degeneracy of varying degrees in a size-extensive and spin-free manner, while bypassing intruders.

22.1 INTRODUCTION

It is now well established by numerous and extensive applications that the single reference (SR) based many-body methods, *viz.* many-body perturbation theory (PT) [1], coupled cluster (CC) theory [2], coupled electron-pair approximations (CEPA) [3], etc. provide rather accurate descriptions of the energy in and around the equilibrium geometry of the closed-shell states. In particular, the single reference coupled cluster (SRCC) method, with single and double (SD) excitations with the perturbative inclusion of triples (SR-CCSD(T)) [4] has turned out to be a very good compromise between computational demand and accuracy. The accuracy of the SR-based PT or CC theory decreases significantly in the presence of quasi-degeneracy, e.g. in the case of bond-breaking or avoided crossing in a potential energy surface (PES). The case of weakly avoided crossings are specially difficult to handle. Another special situation is the case of open-shell states which are ionic or radicaloid in nature requiring several determinants to have the proper spin symmetry. This warrants a multi-reference (MR) approach.

Although MR generalization is quite straightforward for the CI case, the MR-CISD [5] requires large reference space to minimize the size-extensivity error. Such difficulties can be avoided *via* the extension of SR-based CC, CEPA and PT to MR-based CC, CEPA and PT methods. The past two decades have witnessed that the MR generalization of non-variational methods is not a trivial aspect due to the various theoretical constraints and the attendant computational complications. Historically the oldest such formulations use the concept of model space (MS) and effective Hamiltonians (H_{eff}), generating simultaneously N eigenvalues from diagonalization of an effective Hamiltonian in an N-dimensional model space. The effective Hamiltonian approach separates the non-dynamical correlation effects due to near-degeneracy of the model functions from the dynamical and weaker correlation effects in an elegant manner.

We should mention here that the historically oldest MR many-body theories using H_{eff} have been the MR perturbation theories. Successful implementation of all the traditional multi-reference many-body perturbation theories (MR-MBPT), which were developed within the H_{eff} framework [6], was mainly confined to calculation of energy differences of spectroscopic interest, and not to the study of PES.

Prompted by the pre-eminent success of the SRCC method, its generalization to encompass open-shell and quasi-degenerate cases has been attempted by several authors. Again, the earlier MR coupled cluster (MRCC) formulations [7–10] also made use of effective Hamiltonian approach, though they were rather more varied in their scope of applications. There are two main classes of effective Hamiltonian-based MRCC methods: (i) Valence-Universal (VU), a Fock-space approach [7–9,11], which is very useful for the computation of spectroscopic energies; (ii) State-Universal (SU), a Hilbert-space approach [10], which has been used for the study of the state energies *per se* and also for computing PES over a wide range of nuclear geometries. In this approach, the model space is chosen in such a way so that it contains all strongly interacting configurations of interest, and separate cluster operators are introduced for each model space function. Size-extensivity of the computed energies is maintained by the use of Complete Active Space (CAS). Just as for the VU coupled cluster approach, a proper choice of the model space is not always straightforward for the SU coupled cluster approach as well. An additional complication of the SU coupled cluster formalims is the difficulty of spin-adaptation for non-singlet states. Nevertheless, since a major emphasis of the MR-based methods is to study PES, where the relative importance of the dynamical and non-dynamical effects may change quite strongly over changes in the nuclear geometry, the SU coupled cluster approach seems to be a more natural choice.

We should mention that there exists a straightforward, though tedious way, to incorporate the effects of non-dynamical correlations within the SR framework in CC method, *viz.* to identify the virtual orbitals which become quasi-degenerate with the occupied orbitals, and then to treat the strong effect of the quasi-degenerate functions with these virtual orbitals *via* the inclusion of the cluster operators higher than the two-body operators [12–15] involving those virtual orbitals. Such methods do not distinguish the dynamical and non-dynamical correlation effects contributions. The numerical implementation along this line is computationally quite demanding. Sometimes, these methods tend to produce artificial humps in the PES. Of course, the SR-CCSD(T) performs far worse in the presence of quasi-degeneracy [16]. We should also mention the reduced CC method where the higher-body cluster operators are computed by a CASSCF or allied methods, and are used in a SRCC theory to augment for quasi-degeneracy [17].

For the simple open-shell cases, straightforward CC generalizations starting from restricted open-shell Hartree–Fock functions (ROHF-based open-shell CC) have been attempted systematically [18], which, however, lead to spin-broken solutions. Suitable spin-restriction schemes have also been suggested [19], which, however, do not completely resolve the problem of having a spin-adapted theory in the case where genuine MR starting function has to be used, either for symmetry or for energetic reasons. The pioneering work of Jansen and Schaefer [20] as well as the related unitary group formulation of Li and Paldus [21] achieve an explicit spin-free formulation starting with a single configuration state function, but unlike in the case of closed shell, the CC series *does not terminate at the quartic power*, essentially due to the non-commuting nature of the spin-free generators containing both creation and destruction of electrons in the active orbitals. We will encounter this problem in our own spin-adaptation strategy as well.

Another class of method tries to generate MR states of interest *via* the action of an excitation operator on a simple base function, usually of the ground state. The linear response-based theories based on CC reference functions have been proposed quite some time ago to achieve this goal, starting from the HF or ROHF ground reference state [22–25]. An interesting variant to handle non-trivial open-shell states *via* a spin-flip operator [26] has revived the interest in the generalization of the CC-based linear response theory to open-shell states.

It now seems to be generally recognized that the simultaneous calculation of all the roots generated by an H_{eff} is seriously hampered when some virtual functions come very close in energy to some high-lying model functions. Owing to the coupling of all the eigenvalues *via* the H_{eff}, all the roots become poorly described as a consequence. This is the notorious intruder problem [27].

It seems that intruders can be avoided almost completely by working in an incomplete model space (IMS), for a fixed nuclear geometry, where the offending model functions mixing strongly with intruders can be left out of the model space. As shown by Mukherjee [28], size-extensive formulation can be obtained by abandoning the intermediate normalization of the wave operator in favor of a proper size-extensive normalization. Several applications of this formalism were made successfully [29], though it does not work that well for entire PES. PES computed *via* the IMS requires one to use different IMSs in the different regions of it.

To circumvent the difficulty of the intruders, Malrieu and co-workers [30] proposed, based on an earlier idea of Kirtman [31], partitioning the model space into a primary and a secondary (buffer) space to generate an effective operator, called the 'intermediate Hamiltonian' (IH), where the constraint is to demand that just M roots are genuine eigenvalues of H, with M being the dimension of the primary space. The rest of the roots are arbitrary. The various versions of the IH differ in the way the arbitrariness of the coupling of the secondary space to the virtual one is treated. The initial formulations of Malrieu and co-workers were matrix-oriented and were not size-extensive. More recently size-extensive formulations were given in the Fock-space context [32–35]. There have also been important developments by Hoffmann and co-workers [36].

An ideal MR method should be: (i) accurate, reliable and size-extensive; (ii) general (i.e. applicable to both ground and excited states of either closed-shell or open-shell systems); (iii) systematic or well ordered (i.e. enabling a systematic improvement over the SRCC theory, while reducing to it when the state of interest is strongly dominated by a single configuration); (iv) complete (i.e. recovering the full CI result when all clusters are taken into account); and (v) flexible (i.e. enabling the use of different reference spaces).

As an attractive route towards aspiring to reach most of the goals mentioned above, it seems that a state-specific (SS) method, starting with a multi-reference function—taking care of the non-dynamical correlation due to near-degeneracy, and targeting a specific state of interest *via* a state-specific wave operator bringing in the dynamical correlation by a cluster expansion inducing excitation to the virtual functions—would be a very good candidate. We want to call all such methods generically as state-specific multi-reference (SSMR) methods. The SSMR methods address the solution of specific states of interest one at a time and are thus free from the intruder state problem as long as the target state is well separated from the virtual one. Delineation of a class of SSMR formalisms, based on the CC approach and the systematic approximations thereof, along with some numerical applications, is the major concern of this article.

Although, as mentioned earlier, the SS method using an SR starting point with selective higher rank operators is being studied [12–15], in our opinion, an MR-based SS approach is more natural and more flexible to tackle the intruders, using rather low-body cluster operators. Among the many variants of SSMR methods proposed in the literature, the methods of Malrieu and co-workers [37], Mukherjee and co-workers [38, 39] and Hubač and co-workers [40] have been systematically studied both theoretically and in several applications. The developments of Mukherjee and co-workers [38,39] and Hubač and co-workers [40] are both based on the full Jeziorski–Monkhorst (JM) Ansatz [10] while Malrieu and co-workers [37] use the low order quasi-linearized truncation schemes of the JM wave operator. The method of Hubač and co-workers is structurally simpler than Mukherjee and co-workers, but the method is not rigorously size-extensive. In a later development, Hubač and co-workers [41] have proposed an *a posteriori* correction to eliminate inextensivity error, but this entails the danger of intruders. In all these SSMR many-body methods [37–40], the diagonalization of the effective operator in the model space generates the energies associated with altered coefficients for the model space functions, and hence the methods generate *relaxed* coefficients for the model space functions. An analysis indicating how a continuous transition from the state-universal theory of Jeziorski and Monkhorst [10] to the state-specific theories of Hubač and co-workers [40] and of Mukherjee and co-workers [38,39] can be accomplished has recently been done by Pittner [42].

The SSMR Coupled Cluster (SS-MRCC) method using CAS, proposed by Mukherjee and co-workers [38,39], is manifestly size-extensive, orbital invariant and hence size-consistent. The SS-MRCC [38,39] formalism is *not* just a trivial application of the state-universal MRCC (SU-MRCC) [10] multi-root formalism to one root of interest. It is altogether a new formalism, with the working equations quite different from that of the SU-MRCC theory. Since the SS-MRPT theory uses the same Ansatz as in SU-MRCC, it has more degrees of freedom than is needed to generate one root. This flexibility, in fact, is an advantage and is exploited to achieve the twin desirable goals: (1) avoiding intruders, and (2) generating an explicitly size-extensive formalism.

An alternative way to approach the problem is to start out with a fixed MR function, and develop a perturbation theory on it. This is thus an MRPT of the unrelaxed or contracted coefficients variety. The SSMR-based perturbation theories based on contracted description using CAS [43–48] have been widely used as efficient methods to treat quasi-degeneracy. There are usually two ways in which the virtual functions are handled: they can be contracted functions themselves or they can be simpler CSFs. Multi-state versions of the contracted variety has also been suggested [46]. An SS-based CC formulation of the *frozen* variety has been developed [38], where a wave operator is used to generate the exact state by its action on the entire MR function.

There are also hybrid solution strategies like the one suggested by Werner [49], where the double excitations are chosen as the composite functions as obtained by the first procedure, while the single excitations are taken as the singly excited determinants or CSFs with respect to the model CSFs. In this article, we shall also consider similar strategies, where all the cluster amplitudes for the excitation from the doubly occupied inactive orbitals to the unoccupied inactive orbitals are computed with respect to the entire MR starting function.

The various CAS-based perturbative methods can also differ by their choice of unperturbed Hamiltonian H_0, apart from the mode of representing the reference function relaxed or unrelaxed with respect to the coefficients. The original CASPT2 [43] was formulated with the generalized Fock operator as the unperturbed Hamiltonian. This and the related formalisms [43–48] may be viewed as generalizations of MP perturbation theory to a CAS reference function. To take a better account of the orbital occupancy in non-singlet cases, a portion of the two-body terms of H has also been added to H_0, with varied degrees of success [47]. Obviously, the most elaborate choice of such H_0 would be to include *the full two-body active part of interaction*, in addition to the standard generalized spin-average Fock operator [48]. Dyall [48] considered the entire active portion of H in H_0 in his development of CASPT2. Almost all the CASPT versions stop at the second order of energy (CASPT2) (see, however, Werner for a third order CASPT [49]). Most of the various versions of CASPT2 are not rigorously size-consistent. The extent of size-consistency error depends on the choice of the virtual functions (CSF or contracted excited functions), and also on the choice of the unperturbed Hamiltonian. In order to reduce the artifacts stemming from the lack of relaxation of the coefficients, they advocate the use of rather large CAS, which may be fraught with intruders (for a critique along this line, see Ref. [50]). Some recent formulations [50] of effective Hamiltonian perturbation theories have used a multi-reference MP approach with a relaxed description in the traditional effective Hamiltonian framework. Also Heully and Daudey [51] discussed a partitioning of the Hamiltonian where the size-consistency is imposed as an external constraint. Surján and co-workers [52] have investigated the MRPT using APSG (anti-symmetrized product of strongly orthogonal geminal) reference state. A recent review and analysis of various perturbation theories has been presented by Surján and Szabados [53]. The earlier CIPSI (Configuration Interaction with Perturbation Selection Iteratively) method [54] may be viewed as a forerunner of many of the more recent MR perturbation theories. This method can be viewed as a second order perturbation correction to CI energies *via* diagrammatic techniques using multi-configurational zeroth order wave functions.

The SS-MRCC method of Mukherjee and co-workers [38,39] allows transparent simplification *via* truncation of the working equations, leading to various perturbative methods (called by us as SS-MRPT) [55] of practical utility. Motivated by the success of the spin-free version of the SS-MRPT for model spaces spanned by only closed-shell systems, we have very recently formulated an explicitly spin-free SS-MRCC and, as its approximant, an SS-MRPT for general open-shell systems *via* cluster operators using spin-free unitary generators. We will discuss the salient features of the spin-adapted theory in this chapter, and refer to our forthcoming comprehensive papers [56,57] for the theoretical details. The important point in this development is the use of the generators with exchange spectator scattering involving singly occupied active orbitals in the open-shell model CSFs, which is necessary to span the full spin-space. We will present the essentials of the spin-adapted versions of both the SS-MRCC and the SS-MRPT in this article. Another useful approximant we will discuss later is where we ignore more than doubly excited CSFs from the virtual space, but retain the Exclusion Principle Violating (EPV) terms to various degrees in the formalism. This will lead to a suite of SS-MRCEPA-like approximants [58,59]. Applications of both the SS-MRPT and SS-MRCEPA will be shown. Viability of both the SS-MRPT [38,39,55] and the SS-MRCEPA [58,59] approximants have already been demonstrated in some preliminary applications, where the model functions were all closed-shell singlets.

Since the dimension of CAS grows very rapidly with the increase in the number of active orbitals, implementation of the SS-MRCC becomes prohibitively difficult. Since we use more cluster amplitudes than are necessary in describing the FCI function, this is an unwarranted situation. If we recall that the extra degrees of freedom were needed to ensure avoidance of intruders as well as guaranteeing size-extensivity, it is important to focus on these two aspects for any attempt to reduce the number of amplitudes. One good, physically motivated way of doing this is to assume that the cluster amplitudes of all the excitations involving inactive orbitals only are independent of the model functions ϕ_μ. Since inactive excitations are more numerous, such an assumption will lead to drastic reduction of the number of cluster amplitudes. This would amount to an 'Anonymous Parentage for the Inactive excitations' (API), and we will sketch the derivation of such an API-SSMRCC, and its perturbative realizations, API-SSMRPT in this article.

Another way to reduce the computational demand would be to work with an incomplete model (or active) space. Just like in the case of effective Hamiltonian-based theories, the choice of a complete model space (CMS) in a state-specific formalism is mainly dictated by the desire to achieve size-extensivity. An attractive alternative to a CMS-based formulation is to use some IMS which lead to proper fragmentation for some limited fragmentation channels. Very recently we have developed a size-extensive and size-consistent SS-MRCC approach based on an IMS and have also discussed simplification if the IMS is quasi-complete [60]. Methods such as perturbation theory or CEPA-like schemes follow naturally as suitable approximants. Just like in the case of VU- and SU-based IMS [28,61,62] method, intermediate normalization for the wave operator is abandoned in the formalism and suitable excitations, defined as open and quasi-open—which excite out of the IMS by their action on at least one of the model functions—are incorporated in the wave operator. The effective operator, leading to the energy on diagonalization, is a closed connected operator. This is the analogue of the hybrid strategy of Werner [49] for the situation using relaxed coefficients.

A second order MRPT for the restricted active space (RAS) and a general active space was proposed [63]. More recently another theory has been developed [64] using a QCAS (Quasi-Complete Active Space). These theories, unlike in Ref. [60], are not rigorously size-extensive.

Another approach for treating the quasi-degeneracy is adopted by the various MR-based CEPA methods, which have appeared parallely along with the MRCC and MRPT methods. The earlier developed state-specific MRCEPA methods [37,65–70] avoided the redundancy problem using non-redundant cluster operators to compute the dynamical correlation on the zeroth order MR wave function. The MR version of $(SC)^2$CI method, termed as MR-$(SC)^2$CI [37], can be viewed as the size-extensive dressing of the MR-CISD method just as the $(SC)^2$CI [71] is considered to be the size-extensive dressing of the SR-CISD method. Similar to the SR-case, they include all EPV terms in an exact manner.

Our group has recently suggested a suite of CEPA-like approximants [58,59] to their earlier formulated SS-MRCC [38,39] formalism based on a CAS, which show promise in their numerical performance *vis-à-vis* the parent theory. These theories, generically called SS-MRCEPA [58,59], were much simpler than the SS-MRCC, yet they captured most of the essential physics; in particular they bypassed intruders naturally in case the parent theory also had this property. They were all rigorously size-extensive. They also showed size-consistency with respect to fragments, but only with respect to localized orbitals on the fragments. The principal reason for this constraint is due to the lack of orbital invariance of the associated SS-MRCEPA equations within the subsets of doubly occupied, active and virtual orbitals defined by the choice of the CAS. This stemmed from the way EPV terms were treated, *viz.* by keeping only diagonal terms of a set of dressed operators in the SS-MRCEPA equations. However, we may mention that the simplest of the SS-MRCEPA proposed by us, SS-MRCEPA(0), does have the invariance properties, where the diagonal dressing is totally omitted. We have very recently reported a specific version of an SS-MRCEPA which displays the orbital invariance explicitly in addition to the SS-MRCEPA(0) method, while preserving size-extensivity and size-consistency rigorously [58,59]. The theory does not bring in empirical correction factors, nor does it require averaging of the pair-correlation terms. It instead relies on natural and automatic cancellation of the disconnected terms which are non-EPV in nature by certain counter terms intrinsically present in the parent SS-MRCC equations.

As a result, we may view the orbital invariant SS-MRCEPA, termed by us as SS-MRCEPA(I) (I, for invariant), as the *optimal approximation* to the parent SS-MRCC method, which includes all the EPV terms exactly and which utilizes only those counter terms of the equations which eliminate the lack of extensivity of the attendant non-EPV terms in an orbital invariant manner [59]. In this article, we will present a couple of invariant SS-MRCEPA methods, *viz.* SS-MRCEPA(0) and SS-MRCEPA(I), for general open-shell systems using spin-free unitary generator adapted cluster operators starting from explicitly spin-free full-blown parent SS-MRCC formalism. For a detailed discussion of the allied issues pertaining to all the SS-MRCEPA-like methods, we refer to our recent SS-MRCEPA papers [58,59] and an earlier exposé by Szalay [66].

This article is organized as follows: in Section 22.2.1 we will first describe the formulation of the spin-free SS-MRCC theory. We will next present the development of the API-SSMRCC method in Section 22.2.2. In Section 22.2.3 we will discuss the aspects of size-extensivity and consistency of the API-SSMRCC method. We will discuss the spin-free formulation of SS-MRPT and SS-MRCEPA methods starting from the SS-MRCC theory in Sections 22.3 and 22.4, respectively. The IMS version of the SS-MRCC method will be covered in Section 22.5. In Section 22.6, we will present the illustrative numerical applications, along with discussions. Finally, Section 22.7 will summarize our presentation.

22.2 THE SS-MRCC FORMALISM

22.2.1 General developments for the complete model space

We will start this section by introducing the basic issues related to the formulation of the SS-MRCC theory as originally formulated [38,39], and also briefly highlight the manipulations exploiting the redundancy of the cluster amplitudes which bypass the intruders and ensure size-extensivity.

In all of our earlier presentations [38,39], we did not address the theoretical aspects of spin-adaptation of the SS-MRCC equations, since we always worked previously with model functions which were closed-shell singlets. The spin-adaptation of the equations in such cases is straightforward and trivial, entirely analogous to the one for the case of SRCC. In this article, we will discuss the general case where the model functions are neither necessarily singlets nor single determinants. The model functions are, in the general formalism, CSFs rather than determinants, in a given spin-coupling scheme. The detailed nature of the coupling scheme is not important so long as each CSF is generated with respect to some base CSF by excitation *via* spin-free unitary generators.

In the SS-MRCC approach, the exact wave function of a given spin is expressed as:

$$\psi = \sum_{\mu} \exp(T^{\mu})\phi_{\mu}c_{\mu} \tag{1}$$

Each model space function $\{\phi_{\mu}\}$ is a CSF and together spans a CAS (equivalently called CMS).

Following the customary terminology, we will call 'inactive holes' the inactive occupied orbitals, doubly filled in every model CSF. The 'inactive particles' will refer to all the orbitals unoccupied in every CSF. Orbitals which are occupied in some (singly or doubly) but unoccupied in others are the 'active' orbitals. In our spin-free form, the labels are for orbitals only, and not for spin orbitals. From the mode of definition, no active orbital can be doubly occupied in every model CSF. We want to express the cluster operator T^{μ}, inducing excitations to the virtual functions, in terms of excitations of minimum excitation rank, and at the same time wish to represent them in a manifestly spin-free form. To accomplish this, we take as the vacuum—for excitations out of ϕ_{μ}—*the largest closed-shell portion* of it, $\phi_{0\mu}$. For each such vacuum, we redefine the holes and particles, respectively, as ones which are doubly occupied and unoccupied in $\phi_{0\mu}$. The holes are denoted by the labels $i_{\mu}, j_{\mu}, \ldots$, etc. and the particle orbitals are denoted as $a_{\mu}, b_{\mu}, \ldots$, etc. The particle orbitals are totally unoccupied in any ϕ_{μ}, or are necessarily active orbitals which are singly occupied in ϕ_{μ}. When we want to distinguish these singly occupied active orbitals, we shall denote them by u_{μ}, v_{μ}, ... etc. We note here that the holes in $\phi_{0\mu}$ include not only the doubly occupied inactive orbitals but also doubly occupied active orbitals of ϕ_{μ}. We will often use the terminology 'active' and 'valence' interchangeably. In most of our discussions, we will often discuss the various terms generated by the SS-MRCC theory referred to a particular vacuum $\phi_{0\mu}$, and we then drop the subscript μ from the orbital labels when it is clear which vacuum we are dealing with.

The general problem of spin-adaptation using multiple vacuua $\phi_{0\mu}$, depending upon the model function ϕ_{μ} the component of the wave operator $\exp(T^{\mu})$ acts upon, is a *non-trivial and rather involved exercise*. Here we will consider the simplest yet physically the most natural truncation scheme in the rank of cluster operators T^{μ}, where each such operator is truncated at the excitation rank two. For generating the working equations for the spin-adapted theory in this case, it is useful to classify the various types of excitation operators leading to various virtual CSFs as:

One-body:

(A) hole → particle (h → p): $t^{\mu a}_{i}\{E^{a}_{i}\}$
(B) hole → valence (h → v): $t^{\mu u}_{i}\{E^{u}_{i}\}$
(C) valence → particle (v → p): $t^{\mu a}_{u}\{E^{a}_{u}\}$

Two-body:

(D) 2h → 2p: $t^{\mu ab}_{ij}\{E^{ab}_{ij}\}$
(E) h,v → 2p: $t^{\mu ab}_{iu}\{E^{ab}_{iu}\}$
(F) 2h → p,v: $t^{\mu au}_{ij}\{E^{au}_{ij}\}$
(G) 2v → 2p: $t^{\mu ab}_{u\mathrm{v}}\{E^{ab}_{u\mathrm{v}}\}$
(H) h,v → p,v: $t^{\mu a\mathrm{v}}_{iu}\{E^{a\mathrm{v}}_{iu}\}$ and $t^{\mu \mathrm{v}a}_{iu}\{E^{\mathrm{v}a}_{iu}\}$
(I) 2v → p,v: $t^{\mu wa}_{u\mathrm{v}}\{E^{wa}_{u\mathrm{v}}\}$
(J) 2h → 2v: $t^{\mu u\mathrm{v}}_{ij}\{E^{u\mathrm{v}}_{ij}\}$
(K) h,v → 2v: $t^{\mu wx}_{iu}\{E^{wx}_{iu}\}$

In all the excitations above, the operators E in curly brackets denote the normal ordering with respect to $\phi_{0\mu}$, and the 'local' holes, valence and particles are all defined with respect to this vacuum. Let us emphasize at this stage now that the two-body operators of the type $\{E_{Yu}^{Xu}\}$ for arbitrary labels X, Y generate the same excitation as the one-body operator $\{E_Y^X\}$ on ϕ_μ, as can be easily verified:

$$\{E_{Yu}^{Xu}\}\phi_\mu = \{E_Y^X\}\{E_u^u\}\phi_\mu \tag{2}$$

$$= \{E_Y^X\}\phi_\mu \tag{3}$$

since $\{E_u^u\}$ is the number operator with occupancy 1 for ϕ_μ for singly occupied active orbitals. Hence we should include only one of them in our T^μ. We choose the strategy that T^μ never will have excitations of the type $\{E_{Yu}^{Xu}\}$ which we will henceforth term as excitations with direct spectators u. In contrast, excitations of the type $\{E_{Yu}^{uX}\}$, involving exchange type of spectator excitations are linearly independent of $\{E_Y^X\}$, as far as their respective actions on ϕ_μ are concerned, and we should keep both such operators in T^μ. Moreover, while considering excitations inducing $Y \rightarrow X$ from ϕ_μ, we will add together contributions from both the one-body and the two-body excitations $\{E_Y^X\}$ and $\{E_{Yu}^{Xu}\}$ for all u singly occupied in ϕ_μ. Although the Ansatz for the wave operator has a superficial resemblance to the JM Ansatz [10], unlike the JM Ansatz, it is in spin-free form, and also it has non-commuting operators involving exchange spectators. The detailed algebraic structure of our spin-free SS-MRCC formalism is thus going to be different from the more special case involving closed-shell model functions only.

For a function ψ with a given spin-multiplicity $(2S+1)$, we shall consider all the CSFs with $M_s = (2S+1)$, generated from some 'base' CSF ϕ_R with inactive occupied orbitals doubly filled, some active orbitals doubly filled, a set n_s of active orbitals with up-spin, a set n_s orbitals with down-spin, coupled to a singlet, and another set active orbitals, all with up-spin such that $M_s = (2S+1)$. With this generation scheme, every model CSF can be written as some spin-free excitation operator acting on the base function ϕ_R.

Just as in the case of our formulation with the closed-shell model functions, there is a redundancy in the cluster amplitudes of T^μ, in the sense that each virtual CSF χ_l can be generated from several model CSFs. We will impose suitable supplementary sufficiency conditions with the twin objectives of avoiding intruders and of maintaining size-extensivity.

To derive the spin-adapted SS-MRCC equations in the *SD* truncation scheme of the cluster operators, we rewrite the Schrödinger equation for ψ as follows:

$$H\sum_\mu \exp(T^\mu)|\phi_\mu\rangle c_\mu = \sum_\mu \exp(T^\mu)\bar{H}_\mu|\phi_\mu\rangle c_\mu = E\sum_\mu \exp(T^\mu)|\phi_\mu\rangle c_\mu \tag{4}$$

where $\bar{H}_\mu$ is the similarity-transformed Hamiltonian $\exp(-T^\mu)H\exp(T^\mu)$. Introducing the projectors P and Q, respectively, for the model and the virtual spaces, and noting the resolution of identity $I = (P+Q)$, we can write the above equation in long hand as:

$$\sum_\mu \exp(T^\mu)Q\bar{H}_\mu\phi_\mu + \sum_{\mu,\nu} \exp(T^\mu)|\phi_\nu\rangle\langle\phi_\nu|\bar{H}|\phi_\mu\rangle c_\mu = E\sum_\mu \exp(T^\mu)(P+Q)|\phi_\mu\rangle c_\mu \tag{5}$$

Using the same insight gleaned from our earlier SS-MRCC formulation in the spin orbital basis in terms of determinants, we interchange the dummy arguments in the second term on the left-hand side of the above equation, and get

$$\sum_{\mu} \exp(T^{\mu})Q\bar{H}_{\mu}\phi_{\mu} + \sum_{\mu,\nu} \exp T^{\nu}|\phi_{\mu}\rangle\langle\phi_{\mu}|\bar{H}_{\nu}|\phi_{\nu}\rangle c_{\nu} = E\sum_{\mu} \exp(T^{\mu})(P+Q)|\phi_{\mu}\rangle c_{\mu} \tag{6}$$

We now equate the Q projections on both sides of the above equation for each μ, and premultiplying with $\langle\chi_l|\exp(-T^{\mu})\ \forall\, l$, we get

$$\langle\chi_l|\bar{H}_{\mu}|\phi_{\mu}\rangle c_{\mu} + \sum_{\nu} \langle\chi_l|\exp(-T^{\mu})\exp(T^{\nu})|\phi_{\mu}\rangle\tilde{H}_{\mu\nu}c_{\nu} = 0 \ \ \forall\, l, \mu \tag{7}$$

where $\tilde{H}_{\mu\nu} = \langle\phi_{\mu}|\bar{H}_{\nu}|\phi_{\nu}\rangle$. The equations derived above are our principal working equations for the cluster amplitudes of T^{μ}, inducing single and double excitations out of each ϕ_{μ} to all possible χ_l. As we have emphasized earlier, for every single excitation of the type $\{E_Y^X\}$, with $Y \neq u$, we take both the one-body and the two-body excitations $\{E_Y^X\}$ and $\{E_{Yu}^{Xu}\}$ that are generated by $\{\bar{H}_{\mu}\}$ and $\exp(-T^{\mu})\exp(T^{\nu})$ in the above equation. The equation determining the model space coefficients $\{c_{\mu}\}$ and the target state energy is given by

$$\sum_{\nu} \tilde{H}_{\mu\nu}c_{\nu} = Ec_{\mu} \tag{8}$$

For computing all the matrix elements, we rewrite H in normal order with respect to the corresponding $\phi_{0\mu}$ as the vacuum.

We have

$$H = \langle\phi_{0\mu}|H|\phi_{0\mu}\rangle + \{F\} + \{V\} \tag{9}$$

where

$$\{F\} = \sum_{\mu,\nu} f_0^{XY}\{E_Y^X\} \tag{10}$$

with

$$f_0^{XY} = f_c^{XY} + \sum_{u_d} [2V_{Yu_d}^{Xu_d} - V_{Yu_d}^{u_dX}] \tag{11}$$

In the equation above, f_c is the core Fock operator, and sum over u_d runs over all the doubly occupied active orbitals of ϕ_{μ}. $\{V\}$ is the two-body portion of H in normal order with respect to $\phi_{0\mu}$. We do not explicitly indicate here and later which $\phi_{0\mu}$ has been used as the vacuum, since it would be clear from the functions the operators act upon.

The proof of the size-extensivity of the cluster amplitudes of T^{μ} follows essentially the same route as had been taken earlier by us in the spin orbital based formulation with determinants. We briefly recapitulate the steps here, both for the sake of completeness and for setting the scenario for the perturbative and the CEPA developments to follow.

If the cluster operator T^{μ} is connected, one can easily show that the dressed Hamiltonian $\bar{H}_{\mu}$ and the matrix elements $\tilde{H}_{\mu\nu}$ are also connected *via* multi-commutator expansion. Hence, the proof of the connectedness of the first term of Eq. (7) is quite straightforward. But this is non-trivial for the second term. The term $\langle\chi_l|\exp(-T^\mu)\exp(T^\nu)|\phi_\mu\rangle$ can be written as: $\langle\chi_l|\exp\{(T^\nu - T^\mu) + \frac{1}{2}[T^\nu, T^\mu] + \cdots\}|\phi_\mu\rangle$. All the terms involving commutators $[T^\mu, T^\nu]$ are non-vanishing when some active orbitals are occupied in ϕ_μ and unoccupied in ϕ_ν and *vice versa*, and thus must be labelled by orbitals distinguishing ϕ_μ and ϕ_ν. Since $\tilde{H}_{\mu\nu}$ is labelled by all the active orbitals distinguishing ϕ_μ and ϕ_ν, its products with the terms involving the multi-commutators have common active labels and are all connected. The connectivity of the term containing $(T^\mu - T^\nu)$ is rather subtle. Since in the SS-MRCC theory all the model space functions are treated on the same footing, all being generated from the base function ϕ_R in the same manner, the cluster operators T^ν and T^μ have the same functional form and consequently *the difference* $(T^\nu - T^\mu)$ also has common labels with the matrix $\tilde{H}_{\mu\nu}$. As a results of this, the second term of Eq. (7) is connected as a whole. This aspect is *very important* while formulating size-extensive approximate methods, such as SS-MRPT and SS-MRCEPA, from the parent SS-MRCC theory.

In this article, we will not present any results using the general SS-MRCC formalism involving generally open-shell CSFs, since we do not have these numbers yet. However, we will both derive and apply the size-extensive approximants like perturbation theories and CEPA-like schemes derived therefrom.

22.2.2 The use of anonymous parentage for inactive excitations in SS-MRCC method: API-SSMRCC theory

We now describe in a succinct manner a new theoretical innovation to reduce the number of cluster amplitudes in a physically sensible way. Since the number of the inactive excitations from the model CSFs are numerous, it seems, guided by perturbative arguments, to assume that the magnitudes of the inactive excitations are independent of the model CSFs ϕ_μ.

Again, we start with a CAS, but unlike the parent exact formalism, we now impose on the state $|\psi\rangle$, spanning the target space, the following Ansatz:

$$\psi = \exp(T) \sum_\mu \exp(T^\mu)|\phi_\mu\rangle c_\mu \tag{12}$$

In the Ansatz above, the full cluster operator is split into two parts T and T^μ. The cluster operator T, taken to be independent of μ, is restricted to purely inactive excitations of the type singles [T_1: (inactive h $\rightarrow$ inactive p)] and doubles [T_2: (inactive 2h $\rightarrow$ inactive 2p)] and the associated projector is denoted by the symbol Q_2. The excitations due to the other cluster operator T^μ involve at least inactive (3h $-$ 3p) type or at least one active orbital line. The corresponding projector is labeled by Q_1. In our formalism, the excitations due to the cluster operators T^μ and T are described by the symbol ex_1 and ex_2, respectively. The virtual space is spanned by the sets of ex_1 and ex_2 type of functions and hence $Q = Q_1 + Q_2$.

Since the cluster amplitudes for the inactive excitations are assumed to be independent of the parentage (ϕ_μ), we call this version of SS-MRCC theory as using anonymous parentage for inactive excitations (API-SSMRCC).

If we substitute the wave function ψ in the Schrödinger equation, then

$$H \exp(T) \sum_{\mu} \exp(T^{\mu})|\phi_{\mu}\rangle c_{\mu} = E \exp(T) \sum_{\mu} \exp(T^{\mu})|\phi_{\mu}\rangle c_{\mu} \tag{13}$$

The coefficients, c_{μ}, of the reference space and the required energy of the target ground state, E, can be determined by the following compactly written expression:

$$\sum_{\nu} \tilde{\bar{H}}_{\mu\nu} c_{\nu} = E c_{\mu} \tag{14}$$

where

$$\tilde{\bar{H}}_{\mu\nu} = \sum_{\nu} \langle \phi_{\mu} | \bar{\bar{H}}_{\nu} | \phi_{\nu} \rangle$$

$$\bar{\bar{H}}_{\nu} = \overline{\bar{H} \exp(T^{\nu})}$$

$$\bar{H} = \exp(-T) H \exp(T)$$

From Eq. (14), it is clear that to find T one needs to know the converged values of the cluster amplitudes corresponding to the operators T^{μ} as well as T and of the coefficients c_{μ}s.

To generate the working equations for both the cluster operators, T^{μ} and T, we now proceed in the following manner. Starting from Eq. (13), and using Eq. (15), we have

$$\begin{aligned} &\bar{H} \sum_{\mu} \exp(T^{\mu})|\phi_{\mu}\rangle c_{\mu} = E \sum_{\mu} \exp(T^{\mu})|\phi_{\mu}\rangle c_{\mu} \\ &\sum_{\mu} \exp(T^{\mu}) \bar{\bar{H}}_{\mu} |\phi_{\mu}\rangle c_{\mu} = E \sum_{\mu} \exp(T^{\mu})|\phi_{\mu}\rangle c_{\mu} \\ &\sum_{\mu} \exp(T^{\mu})(P+Q) \bar{\bar{H}}_{\mu} |\phi_{\mu}\rangle c_{\mu} = E \sum_{\mu} \exp(T^{\mu})|\phi_{\mu}\rangle c_{\mu} \end{aligned} \tag{15}$$

Using the expressions for Q and P, we get

$$\sum_{\mu} \exp(T^{\mu}) \bar{\bar{H}}_{\mu\ \mathrm{ex}} |\phi_{\mu}\rangle c_{\mu} + \sum_{\mu\nu} \exp(T^{\mu})|\phi_{\nu}\rangle\langle\phi_{\nu}| \bar{\bar{H}}_{\mu} |\phi_{\mu}\rangle c_{\mu} = E \sum_{\mu} \exp(T^{\mu})|\phi_{\mu}\rangle c_{\mu} \tag{16}$$

Inspired by the same manipulations as leading to the SS-MRCC equation [38,39] we interchange the labels μ and ν in the second term on the left-hand side of Eq. (16):

$$\begin{aligned} &\sum_{\mu} \exp(T^{\mu}) \bar{\bar{H}}_{\mu\ \mathrm{ex}} |\phi_{\mu}\rangle c_{\mu} + \sum_{\mu\nu} \exp(T^{\nu})|\phi_{\mu}\rangle\langle\phi_{\mu}| \bar{\bar{H}}_{\nu} |\phi_{\nu}\rangle c_{\nu} = E \sum_{\mu} \exp(T^{\mu})|\phi_{\mu}\rangle c_{\mu} \\ &\sum_{\mu} \exp(T^{\mu}) \bar{\bar{H}}_{\mu\ \mathrm{ex}} |\phi_{\mu}\rangle c_{\mu} + \sum_{\mu\nu} \exp(T^{\nu})|\phi_{\mu}\rangle \tilde{\bar{H}}_{\mu\nu} c_{\nu} = E \sum_{\mu} \exp(T^{\mu})|\phi_{\mu}\rangle c_{\mu} \end{aligned} \tag{17}$$

where ex stand for virtual excitations.

We have already mentioned that ex has two parts: (i) ex_1 due to the cluster operator T^μ; (ii) ex_2 for cluster operator T. Thus the above equation, Eq. (17) becomes:

$$\sum_\mu \exp(T^\mu)\bar{\bar{H}}_{\mu\ \mathrm{ex}_1}|\phi_\mu\rangle c_\mu + \sum_\mu \exp(T^\mu)\bar{\bar{H}}_{\mu\ \mathrm{ex}_2}|\phi_\mu\rangle c_\mu + \sum_{\mu\nu} \exp(T^\nu)|\phi_\mu\rangle \tilde{\bar{H}}_{\mu\nu} c_\nu$$
$$= E \sum_\mu \exp(T^\mu)|\phi_\mu\rangle c_\mu \tag{18}$$

Imposition of the sufficiency conditions as that in our SS-MRCC method [38,39] on the terms in Eq. (18) for each μ, yields the following expression:

$$\exp(T^\mu)\bar{\bar{H}}_{\mu\ \mathrm{ex}_1}|\phi_\mu\rangle c_\mu + \exp(T^\mu)\bar{\bar{H}}_{\mu\ \mathrm{ex}_2}|\phi_\mu\rangle c_\mu + \sum_\nu \exp(T^\nu)|\phi_\mu\rangle \tilde{\bar{H}}_{\mu\nu} c_\nu$$
$$= E \exp(T^\mu)|\phi_\mu\rangle c_\mu \tag{19}$$

Projection onto $Q_1 \exp(-T^\mu)$ from the left-hand side of the above equation gives:

$$Q_1 \bar{\bar{H}}_{\mu\ \mathrm{ex}_1}|\phi_\mu\rangle c_\mu + \sum_\nu Q_1\{\exp(-T^\mu)\exp(T^\nu)\}_{\mathrm{ex}_1}|\phi_\mu\rangle \tilde{\bar{H}}_{\mu\nu} c_\nu = 0 \tag{20}$$

Eq. (20) is our stipulated working equation for the computation of the cluster operator T^μ of our API-MRCCSD theory. The T-amplitudes appear in them, but they are not explicit, being buried within the dressed matrix elements. The equations above involve the coefficients c_μs explicitly, indicating that the cluster amplitudes depend on them, as it should be in a SS theory.

We now proceed to construct the equations for the T-amplitudes. Projecting Eq. (18) from the left-hand side onto $\langle\psi_0|Y_l$ we get:

$$\sum_\mu \langle\psi_0|Y_l \exp(T^\mu)\bar{\bar{H}}_{\mu\ \mathrm{ex}_1}|\phi_\mu\rangle c_\mu + \sum_\mu \langle\psi_0|Y_l \exp(T^\mu)\bar{\bar{H}}_{\mu\ \mathrm{ex}_2}|\phi_\mu\rangle c_\mu$$
$$+ \sum_{\mu\nu} \langle\psi_0|Y_l \exp(T^\nu)|\phi_\mu\rangle \tilde{\bar{H}}_{\mu\nu} c_\nu = E \sum_\mu \langle\psi_0|Y_l \exp(T^\mu)|\phi_\mu\rangle c_\mu \tag{21}$$

where Y_l envelopes (h − p) and $(2h - 2p)$ inactive de-excitations. Using the definition of ψ_0, we have:

$$\sum_{\mu\sigma} \langle\phi_\sigma|Y_l \exp(T^\mu)\bar{\bar{H}}_{\mu\ \mathrm{ex}_1}|\phi_\mu\rangle c_\mu c_\sigma + \sum_{\mu\sigma} \langle\phi_\sigma|Y_l \exp(T^\mu)\bar{\bar{H}}_{\mu\ \mathrm{ex}_2}|\phi_\mu\rangle c_\mu c_\sigma$$
$$+ \sum_{\mu\nu\sigma} \langle\phi_\sigma|Y_l \exp(T^\nu)|\phi_\mu\rangle \tilde{\bar{H}}_{\mu\nu} c_\nu c_\sigma = E \sum_{\mu\sigma} \langle\phi_\sigma|Y_l \exp(T^\mu)|\phi_\mu\rangle c_\mu c_\sigma \tag{22}$$

The matrix elements between the two functions ϕ_μ and ϕ_σ have been labeled by at least one active line, when the indices of model space determinants are not identical (i.e. for $\mu \neq \sigma$) and thus the corresponding ex is purely of ex_1 type. The remaining part (for $\mu = \sigma$) is of the type ex_2. Thus, we can write the above expression as follows:

$$\sum_{\mu\sigma}{}' \langle\phi_\sigma|Y_l \exp(T^\mu)\bar{\bar{H}}_{\mu\ \mathrm{ex}_1}|\phi_\mu\rangle c_\mu c_\sigma + \sum_{\mu\sigma}{}' \langle\phi_\sigma|Y_l \exp(T^\mu)\bar{\bar{H}}_{\mu\ \mathrm{ex}_2}|\phi_\mu\rangle c_\mu c_\sigma$$
$$+ \sum_{\mu\nu\sigma}{}' \langle\phi_\sigma|Y_l \exp(T^\nu)|\phi_\mu\rangle\tilde{\bar{H}}_{\mu\nu} c_\nu c_\sigma + \sum_{\mu} \langle\phi_\sigma|Y_l \exp(T^\mu)\bar{\bar{H}}_{\mu\ \mathrm{ex}_1}|\phi_\mu\rangle c_\mu^2$$
$$+ \sum_{\mu} \langle\phi_\sigma|Y_l \exp(T^\mu)\bar{\bar{H}}_{\mu\ \mathrm{ex}_2}|\phi_\mu\rangle c_\mu^2 + \sum_{\mu\nu} \langle\phi_\sigma|Y_l \exp(T^\nu)|\phi_\mu\rangle\tilde{\bar{H}}_{\mu\nu} c_\nu c_\mu$$
$$= E\sum_{\mu\sigma}{}' \langle\phi_\sigma|Y_l \exp(T^\mu)|\phi_\mu\rangle c_\mu c_\sigma + E\sum_{\mu} \langle\phi_\sigma|Y_l \exp(T^\mu)|\phi_\mu\rangle c_\mu^2 \qquad (23)$$

where the prime over the sum indicates that there is no common index between μ and σ.

It can be seen from Eq. (23), that the primed terms generate the same expression as that obtained by taking the Q_1 projection of Eq. (19). As we have already mentioned, the model space in our formalism is complete in nature. The projection manifolds, $Q_1 \exp(-T^\mu)$ and Q_1 span the same functional space, since they have the same number of linearly independent functions. Consequently, solving the cluster operator, T^μ determining equation, Eq. (20) is equivalent to solving the Q_1 projection of Eq. (19). It is rather convenient to prove and maintain the extensivity of the equation in the form of Eq. (20) than the other one. Projecting with $Q_1 \exp(-T^\mu)$ instead of Q_1 on the left-hand side of Eq. (19), we take the advantage of the Baker–Campbell commutator formula which automatically ensures connectivity of the resulting expansion.

Let us now consider the other terms ($\mu = \sigma$) in Eq. (23):

$$\sum_{\mu} \langle\phi_\mu|Y_l \exp(T^\mu)\bar{\bar{H}}_{\mu\ \mathrm{ex}_1}|\phi_\mu\rangle c_\mu^2 + \sum_{\mu} \langle\phi_\mu|Y_l \exp T^\mu\bar{\bar{H}}_{\mu\ \mathrm{ex}_2}|\phi_\mu\rangle c_\mu^2$$
$$+ \sum_{\mu\nu} \langle\phi_\mu|Y_l \exp(T^\nu)|\phi_\mu\rangle\tilde{\bar{H}}_{\mu\nu} c_\mu c_\nu = E\sum_{\mu} \langle\phi_\mu|Y_l \exp(T_\mu)|\phi_\mu\rangle c_\mu^2 \qquad (24)$$

Using Eq. (14) we get,

$$\sum_{\mu} \langle\phi_\mu|Y_l \exp(T^\mu)\bar{\bar{H}}_{\mu\ \mathrm{ex}_1}|\phi_\mu\rangle c_\mu^2 + \sum_{\mu} \langle\phi_\mu|Y_l \exp(T^\mu)\bar{\bar{H}}_{\mu\ \mathrm{ex}_2}|\phi_\mu\rangle c_\mu^2$$
$$+ \sum_{\mu\nu}{}' \langle\phi_\mu|Y_l(\exp(T^\nu) - \exp(T^\mu))|\phi_\mu\rangle\tilde{\bar{H}}_{\mu\nu} c_\mu c_\nu = 0 \qquad (25)$$

where the prime indicates $\mu \neq \nu$.

The cluster operators T^μ involve at least inactive (3h − 3p) excitations or excitations associated with active line(s)—purely ex_1 type and $\langle\phi_\mu|Y_l$ contains at most inactive (2h − 2p) excitations. Thus the contribution of the first and third terms of Eq. (25) will be zero. Using the same argument, it can be shown that the $\exp(T^\mu)$ will not appear in the second term.

Therefore, the above equation becomes:

$$\sum_{\mu} \langle\phi_\mu|Y_l\bar{\bar{H}}_{\mu\ \mathrm{ex}_2}|\phi_\mu\rangle c_\mu^2 = 0 \qquad (26)$$

Eq. (26) is our required working equation for the inactive cluster operators T of the type singles and doubles. This amplitude equation does not contain the renormalization terms, unlike the T^{μ} determining equations. This is due to the fact that, from the very mode of derivation, we have considered the T operator to be independent of reference determinants, ϕ_{μ}.

From the principal working equations (14), (20) and (26), it is clear that the set $\{c_{\mu}\}$, $\{T^{\mu}\}$ and T are coupled. Solving these coupled set of equations, we obtain cluster operators of both types and also converged coefficients from the diagonalization *via* Eq. (14). In our theory, the combining coefficients get iteratively updated to their values with the cluster amplitudes and hence generate the exact state ψ in relaxed form. But our formalism also provides an extra advantage, if one wishes that the combining coefficients be kept at their pre-assigned values during the iteration of the cluster amplitudes. A typical feature of our formalism in such a scheme is the use of the zeroth order coefficients $c_{\mu}^{(0)}$ to compute the cluster amplitudes and the dressed Hamiltonian, but allow the coefficients to relax while computing E, since this is obtained *via* diagonalization. Thus our method is completely flexible in the sense that we can use it to compute the energy either as an expectation value with respect to the unrelaxed (or frozen) function or by diagonalization in the relaxed form.

To show explicitly the avoidance of intruders of our API-SSMRCC theory, it is more convenient to rewrite the leading terms for the amplitudes of the cluster operators T^{μ} determining equation in the following *quasi-linearized* form:

$$\begin{aligned}
&\langle\chi_l|\bar{H}|\phi_{\mu}\rangle c_{\mu} + \Big[(\langle\chi_l|\bar{H}|\chi_m\rangle - \langle\phi_{\mu}|\bar{H}|\phi_{\mu}\rangle\delta_{lm})\langle\chi_m|T^{\mu}|\phi_{\mu}\rangle + \cdots\Big]c_{\mu} \\
&\quad + \sum_{\nu}\langle\chi_l|(T^{\nu} - T^{\mu} + \cdots)|\phi_{\mu}\rangle\tilde{\tilde{H}}_{\mu\nu}c_{\nu} = 0 \\
&\equiv \langle\chi_l|\bar{H}|\phi_{\mu}\rangle c_{\mu} + \left[\sum_{m}(\langle\chi_l|\bar{H}|\chi_m\rangle - \langle\phi_{\mu}|\bar{H}|\phi_{\mu}\rangle\delta_{lm})\langle\chi_m|T^{\mu}|\phi_{\mu}\rangle + \cdots\right]c_{\mu} \\
&\quad - E\langle\chi_l|T^{\mu}|\phi_{\mu}\rangle c_{\mu} + \sum_{\nu}\langle\chi_l|T^{\nu}|\phi_{\mu}\rangle\tilde{\tilde{H}}_{\mu\nu}c_{\nu} + \cdots = 0
\end{aligned} \tag{27}$$

where we use $Q_1 = \sum_l|\chi_l\rangle\langle\chi_l|$

The expression for the amplitudes of the cluster operators T^{μ} from the above equation can be written as follows:

$$\langle\chi_l|T^{\mu}|\phi_{\mu}\rangle = \frac{\langle\chi_l|\bar{H}|\phi_{\mu}\rangle + \sum_{\nu}\langle\chi_l|T^{\nu}|\phi_{\mu}\rangle\tilde{\tilde{H}}(c_{\nu}/c_{\mu}) + \sum_{m}(\langle\chi_l|\bar{H}|\chi_m\rangle + \cdots}{(E - \langle\chi_l|\bar{H}|\chi_l\rangle)} \tag{28}$$

The robustness of the energy denominators in the presence of the intruder is quite manifest in our formalism. The denominator is never small as long as the target state energy is well separated from the energies of the virtual functions. This situation is usually observed for the ground state.

For the inactive amplitudes, it becomes convenient to express them in terms of the MR starting function $\psi_0 = \sum_{\mu}\phi_{\mu}c_{\mu}$, and the contracted virtual functions $\langle\psi_0|Y_l$.

As an example, the amplitudes t_{ij}^{ab}, where the labels all refer to inactive orbitals (holes or particles), are given by

$$t_{ij}^{ab} = \frac{\langle\psi_0|Y_l\bar{H}|\psi_0\rangle}{[\langle\psi_0|\bar{H}|\psi_0\rangle - \langle\psi_0|Y_l\bar{H}Y_l^\dagger|\psi_0\rangle]} \tag{29}$$

Since the denominators involve the difference of state-energies for the functions ψ_0 and contracted virtual functions $\langle\psi_0|Y_l$, they are separated widely in energy, since the contracted virtual functions are excited by lifting inactive holes to inactive particles, which have a gap, in between which the active orbitals appear. Thus the denominators are again robust. The same is valid for the corresponding single excitations involving inactive orbitals. Extensivity of our newly developed API-SSMRCC theory will be discussed in the next section.

22.2.3 Proof of the connectivity of the API-SSMRCC formalism

We can prove the connectedness of the working equations by checking carefully the connectedness of each term present in the equations. If we assume that H, T^μ and T are connected in nature, we can proceed to prove the connectivity of our working equations using the following arguments.

To prove the connectivity of the working equation for the cluster operator T^μ, we rewrite Eq. (20) as follows:

$$Q_1\bar{\bar{H}}_{\mu\ \mathrm{ex}_1}|\phi_\mu\rangle + \sum_\nu Q_1\{\exp(-T^\mu)\exp(T^\nu)\}_{\mathrm{ex}_1}|\phi_\mu\rangle\tilde{\bar{H}}_{\mu\nu}(c_\nu/c_\mu) = 0 \tag{30}$$

Since ψ_0 is a CAS-type function, the first term and the dressed Hamiltonian $\tilde{\bar{H}}_{\mu\nu}$ of the second term of the left-hand side of Eq. (30) are manifestly connected involving H, T and T^μ *via* commutator expansion, and it is hence enough to show that the third term of the equation is connected. According to Baker–Campbell formula, the second term of the above equation can be written as follows:

$$\begin{aligned}
&\sum_\nu Q_1\{\exp(T^\nu)\exp(-T^\mu)\}_{\mathrm{ex}_1}|\phi_\mu\rangle\tilde{\bar{H}}_{\mu\nu}(c_\nu/c_\mu) \\
&= \sum_\nu Q_1\{\exp((T^\nu - T^\mu) + \frac{1}{2}[T^\nu, T^\mu] + \frac{1}{12}[[T^\nu, T^\mu], T^\mu] \\
&\quad - \frac{1}{12}[[T^\nu, T^\mu], T^\nu] + \cdots)\}_{\mathrm{ex}_1}|\phi_\mu\rangle\tilde{\bar{H}}_{\mu\nu}(c_\nu/c_\mu)
\end{aligned} \tag{31}$$

Just as for the parent SS-MRCC theory, special attention is needed towards proving the connectivity of the entire term $Y_{\mu\nu} = (T^\nu - T^\mu)\tilde{\bar{H}}_{\mu\nu}(c_\nu/c_\mu)$. The term $\tilde{\bar{H}}_{\mu\nu}(c_\nu/c_\mu)$ is labeled by all those orbitals that distinguish the reference CSFs ϕ_μ and ϕ_ν. If one considers the two terms $T^\nu\tilde{\bar{H}}_{\mu\nu}(c_\nu/c_\mu)$ and $T^\mu\tilde{\bar{H}}_{\mu\nu}(c_\nu/c_\mu)$ separately, it is not easy to prove their connectivity. To show the connectivity of $Y_{\mu\nu}$, one should consider both the terms on the same footing. Since the functional structure of T^ν and T^μ is identical, the difference would be labeled by those orbitals which distinguish the CSFs ϕ_μ and ϕ_ν.

All other components of T^ν and T^μ would cancel each other. Thus the terms $(T^\nu - T^\mu)$ and $\tilde{H}_{\mu\nu}(c_\nu/c_\mu)$ have common orbital labels and hence $Y_{\mu\nu}$ is connected in nature. The term $[T^\nu, T^\mu]$ is labeled by the group of orbitals distinguishing ϕ_μ and ϕ_ν. Thus the term $[T^\nu, T^\mu]\tilde{H}_{\mu\nu}(c_\nu/c_\mu)$ is connected in the sense of having common orbital indices between them. This is true for other higher commutators as well. Thus the terms in the working equation for the cluster operator T^μ are connected.

By construction, the matrix elements $\sum_\mu \langle\phi_\mu|Y_l\bar{\bar{H}}_{\mu\ \mathrm{ex}_2}|\phi_\mu\rangle$ in Eq. (26) are connected entities joining H with $\exp(T)$ and $\exp(T^\mu)$ and hence are extensive if cluster operators are connected. Therefore, the working equation for the inactive cluster amplitude T, Eq. (26), is also connected in nature. Hence the dressed Hamiltonian, $\tilde{\tilde{H}}_{\mu\nu}$ is also connected. Thus the energy obtained as an eigenvalue of Eq. (14) is also extensive, for the reference determinants span a CAS.

In this article, we will not present numerical applications of the full-blown API-MRCCSD theory, but only of its perturbative approximants.

22.3 EMERGENCE OF STATE-SPECIFIC MULTI-REFERENCE PERTURBATION THEORY SS-MRPT FROM SS-MRCC THEORY

A size-extensive perturbative (SS-MRPT) formalism starting from SS-MRCC theory can be generated by a partition of H into H_0 (a zeroth order part) and V (a perturbation), and an order-by-order expansion of T^μs. In the many-body problem, especially in evaluating the electron correlation energy, the most widely used partitionings are MP and EN. In EN partitioning all diagonal direct and exchange ladders are a part of the H_0, while the Fock operator plays the role of the H_0 in the MP scheme. Although such type of partitioning is straightforward for the SR-case, this is not so for the generalization within the framework of the MR-case.

For the development of SSMR-based perturbation theory starting from the full-blown SS-MRCC theory, we rewrite the equations determining the cluster amplitudes, Eq. (7), in the following quasi-linearized form:

$$\langle\chi_l|H|\phi_\mu\rangle c_\mu + \left[\left(\sum_m \langle\chi_l|H|\chi_m\rangle - \langle\phi_\mu|H|\phi_\mu\rangle\delta_{lm}\right)\langle\chi_m|T^\mu|\phi_\mu\rangle\right]c_\mu$$
$$-\left[\sum_\nu \langle\chi_l|T^\mu|\phi_\mu\rangle\tilde{H}_{\mu\nu}\right]c_\nu + \left[\sum_\nu \langle\chi_l|T^\nu|\phi_\mu\rangle\tilde{H}_{\mu\nu}\right]c_\nu = 0 \ \forall\, l, \mu \tag{32}$$

The first term of the above equation corresponds to the coupling of a virtual function to a model function, and is akin to the numerator in a simple perturbation theory. The second term is a commutator of T^μ and H, and with H_0 approximating H contributes an RS-like denominator of a traditional effective Hamiltonian-based theory. The third and the fourth terms together perform two inter-related but distinct functions: (a) to convert the usual RS-like denominators into one containing the actual state energies, to bypass intruders—as befitting a state-specific theory, and (b) to supply counter terms guaranteeing size-extensivity of the theory. The third term, in fact, supplies the term containing the state-energy, as shown below, while the fourth term, which couples the different model functions *via* the dressed Hamiltonian $\tilde{H}_{\mu\nu}$, containing T^ν, is, in conjunction with the third term, responsible to maintain size-extensivity. As we have mentioned earlier, the connectedness of the term $\langle\chi_l|(T^\nu - T^\mu)|\phi_\mu\rangle\tilde{H}_{\mu\nu}(c_\nu/c_\mu)$ has a direct bearing on the structure of the working equations in any size-extensive perturbative formalism.

Actually the form of working equation of the first order cluster amplitudes and hence the perturbative scheme depend on the treatment of the four above-mentioned terms. One can treat all the four terms consistently in the same partitioning scheme. This approach is rather inflexible, since this necessarily constrains us to use only a very specific partitioning strategy, *viz.* the one proposed by Dyall [48]. We follow here a somewhat hybrid strategy where we treat the first and the second terms strictly by perturbation theory, while we treat the third and fourth terms on the same footing, but not in the sense of strict perturbation. Since it is natural to have the unperturbed state-energy E_0 appear in the denominator in the RS version, we approximated $\tilde{H}_{\mu\nu}$ in these terms by $H_{\mu\nu}$ in $\sum_\nu\langle\chi_l|T^{\mu(1)}|\phi_\mu\rangle\tilde{H}_{\mu\nu}c_\nu$, since this leads to:

$$\sum_\nu\langle\chi_l|T^{\mu(1)}|\phi_\mu\rangle H_{\mu\nu}c_\nu = E_0\langle\chi_l|T^{\mu(1)}|\phi_\mu\rangle c_\mu \tag{33}$$

To treat the term containing $\langle\chi_l|T^{\nu(1)}|\phi_\mu\rangle$on the same footing, it should thus appear multiplied by $H_{\mu\nu}c_\nu$ in the RS version. Now we can write Eq. (32) as follows:

$$\begin{aligned}&[\langle\chi_l|H|\phi_\mu\rangle]c_\mu + \left[\left(\sum_m\langle\chi_l|H|\chi_m\rangle - \langle\phi_\mu|H|\phi_\mu\rangle\delta_{lm}\right)\langle\chi_m|T^\mu|\phi_\mu\rangle\right]c_\mu\\ &- [E_0\,\langle\chi_l|T^\mu|\phi_\mu\rangle]c_\mu + \left[\sum_\nu\langle\chi_l|T^\nu|\phi_\mu\rangle H_{\mu\nu}\right]c_\nu = 0\ \forall\, l,\mu\end{aligned} \tag{34}$$

As we mentioned above, the partitioning of H affects only the terms $\langle\chi_l|H_\mu|\phi_\nu\rangle$ and $\langle\chi_l|[H_\mu, T_\mu]|\phi_\nu\rangle$. In both MP and EN partitions, H_0 is a diagonal operator, and this lends a rather simple structure. Expanding the first two terms of Eq. (34) in orders of perturbation, and retaining only the terms of the first order, we have

$$\begin{aligned}&[\langle\chi_l|V_\mu|\phi_\mu\rangle]c_\mu + [\langle\chi_l|[H_0, T_\mu]|\phi_\nu\rangle]c_\mu - [E_0\langle\chi_l|T^\mu|\phi_\mu\rangle]c_\mu\\ &+\left[\sum_\nu\langle\chi_l|T^\nu|\phi_\mu\rangle H_{\mu\nu}\right]c_\nu = 0\ \forall\, l,\mu\end{aligned} \tag{35}$$

The actual expression of H_0 depends on the type of partitioning, MP or EN.

For actual applications, and to emphasize the organizational aspects of the theory, we rewrite the working equations, Eq. (34), in the following form:

$$t_\mu^{l(1)} = \frac{H_{l\mu} + \sum_\nu^{\nu\neq\mu}\langle\chi_l|T^{\nu(1)}|\phi_\mu\rangle H_{\mu\nu}(c_\nu/c_\mu)}{\left[(E_0 - H_{\mu\mu}) + (H^0_{\mu\mu} - H^0_{ll})\right]} \tag{36}$$

We note that the only coupling between the various Ts are *via* the sum over ν appearing in the numerator of Eq. (36) above. There is thus no coupling between the various excitation components in T^μs, and the coupling is present with only those T^νs which lead to the same excitation as by the product of excitation operators for the specific $t_\mu^{l(1)}$ under consideration. This leads to a very attractive computational scheme, where we consider each type of excitation involving a specific set of orbitals, and compute all the T amplitudes for various μ with the same set of orbitals using Eq. (36). The contributions of all these T amplitudes to the effective pseudo-operator $\tilde{H}$ are then computed, and a fresh set of excitations considered next. Thus, no T amplitudes need to be stored in this formulation. The computational resources required for our SS-MRPT are almost similar as that of the MRMP method [45]. It involves a small iterative step implied in Eq. (36) in contrast to MRMP method. However, this extra step leads to rigorous size-extensivity nature of our SS-MRPT.

After getting the first order cluster operator, the second order energy $E^{(2)}$, is obtained by diagonalizing $\tilde{H}_{\mu\nu}^{(2)}$:

$$\sum_\nu \tilde{H}_{\mu\nu}^{(2)} c_\nu^{(2)} = E^{(2)} c_\mu^{(2)} \tag{37}$$

with $\tilde{H}_{\mu\nu}^{(2)} = H_{\mu\nu} + \sum_l H_{\mu l} t_\nu^{l(1)}$

For the unrelaxed case we take expectation value: $E^{(2)} = \sum_{\mu,\nu} c_\mu^0 \tilde{H}_{\mu\nu}^{(2)} c_\nu^0$, where c_μ^0 stands for the frozen coefficients for the reference functions.

Eqs. (36) and (37) are our principal working equations for RS-based SS-MRPT. It is noteworthy that in the RS-based SS-MRPT formalism the zeroth order coefficients, c_μ^0s are used to evaluate the cluster operators in Eq. (36), but the coefficients are relaxed during the computation of $E^{(2)}$, since this is obtained by diagonalization *via* Eq. (37).

The robustness of the energy denominators in the presence of intruders is quite manifest in our SS-MRPT formalism. The denominator in Eq. (36) is never small as long as the unperturbed or the perturbed energy, E_0, is well separated from the energies of the virtual functions. Thus the SS-MRPT is intruder-free, and explicitly size-extensive and also size-consistent when we use orbitals localized on the separated fragments.

22.3.1 Choice of the zeroth order Hamiltonians

For the MP partitioning, we choose H_0 as a one-particle Fock-like operator. In a multi-partitioning scheme, we have tested quite a few physically motivated choices. They are described below. In the EN partition, there are additionally diagonal direct and exchange ladders. For the general situation involving open-shell model functions, we recall that the set of terms involving the same excitations on a given ϕ_μ have to be added together, and this leads to additional ladders in the EN partitioning.

The simplest choice of the Fock operator is the following:

$$f_\mu^{ij} = \sum_{ij} \left[f_{\text{core}}^{ij} + \sum_u \left(V_{iu}^{ju} - \frac{1}{2} V_{iu}^{uj} \right) D_{uu}^\mu \right] \{E_i^j\} \tag{38}$$

where u represents both a doubly occupied and a singly occupied active orbital in the ϕ_μ, and the D^μs are the densities labelled by the active orbitals. Since our H_0 is always diagonal for MP scheme, then, H$_0$ is: $H_0^\mu = \sum_i f_\mu^{ii} \{E_i^i\}$

In this article, we shall introduce another physically appealing Fock operator, which is defined with respect to $\phi_{0\mu}$, and it includes not only the Fock potential of the doubly occupied active orbitals included in $\phi_{0\mu}$, but also—*via* the blocks involving the direct exchange scattering of the singly occupied active orbitals—takes care of interaction of electrons in them. This is given by

$$\tilde{f}_{\mu}^{ij} = \sum_{ij} \left[f_{0\mu}^{ij} + \sum_{u_s} V_{iu_s}^{ju_s} \right] \{E_i^j\} \tag{39}$$

where the normal ordering is with respect to $\phi_{0\mu}$ and u_s represents a singly occupied active orbital in the ϕ_{μ}. $f_{0\mu}$ is the Fock operator for $\phi_{0\mu}$. With this choice for the Fock operator, the H_0 for MP partitioning is: $H_0^{\mu} = \sum_i \tilde{f}_{\mu}^{ii}\{E_i^i\}$.

In addition to the above type of Fock operators, f_{μ}^{ij} and $\tilde{f}_{\mu}^{ij}$, we also consider their generalized form $f^{ij} = \sum_{\mu} f_{\mu}^{ij} c_{\mu}^2$ and $\tilde{f}^{ij} = \sum_{\mu} \tilde{f}_{\mu}^{ij} c_{\mu}^2$, respectively. Then the corresponding H_0 is expressed as follows: $H_0^{\mu} = \sum_i f^{ii}\{E_i^i\}$ and $H_0^{\mu} = \sum_i \tilde{f}^{ii}\{E_i^i\}$.

For the EN partitioning the corresponding H_0 becomes:

$$H_0 = H_0^{\mu} + \sum [\text{all kinds of diagonal ladders (direct and exchange)}] \tag{40}$$

The additional ladders come in general from two sets of terms. In a given block with certain excitation structure, we add all diagonal 'direct and exchange ladders' on the diagrams in the block. In addition, for the CSFs with open-shells we also add the higher-body blocks with direct spectator scattering with singly occupied orbitals, and these blocks would necessarily include ladders involving the singly occupied active orbitals for a CSF (these active orbitals are the ones we called active for a vacuum $\phi_{0\mu}$).

For the case of closed-shell single reference functions, the above-mentioned Fock matrices become identical to the closed-shell Fock matrix, and our general SS-MRPT then reduces to ordinary second-order perturbation theory, introduced earlier by our group [55].

As we indicated in our general discussion of the API-SSMRCC theory, the leading terms for the inactive cluster amplitudes would have energy differences involving $[\langle\psi_0|H|\psi_0\rangle - \langle\psi_0|Y_l H Y_l^{\dagger}|\psi_0\rangle]$. In the SS-MRPT using the API approximation, depending on the partition, the corresponding expressions of H_0 will replace H in the corresponding expressions for the cluster amplitudes for the inactive excitations. Thus, the orbital energies for the generalized Fock operator (f) would appear if H_0 is defined as f in the multi-patitioning form for MP, and $\tilde{f}$ will appear in place of f if f is used in multi-partitioning in MP. In case the generalized Fock operator f itself is used as H_0, then its own orbital energies appear in the MP expression. In the EN partition, the situation is similar, except in so far as the additional ladders on top of the MP denominator have also to be considered.

22.4 EMERGENCE OF THE SS-MRCEPA(I) METHODS FROM SS-MRCC

The simplest CEPA-like approximation will follow from Eq. (7) if we keep only the linear terms in the cluster amplitudes. This CEPA has been called by us as the SS-MRCEPA(0).

Written in long hand, it amounts to:

$$\langle\chi_l|H|\phi_\mu\rangle c_\mu + \sum_m \langle\chi_l|(H - H_{\mu\mu})|\chi_m\rangle t_\mu^m c_\mu + \sum_\nu \langle\chi_l|(T^\nu - T^\mu)|\phi_\mu\rangle H_{\mu\nu} c_\nu = 0 \tag{41}$$

where the sum over m covers all the SD functions with respect to ϕ_μ, and $H_{\mu\mu} = \langle\phi_\mu|H|\phi_\mu\rangle$. If we start from a CAS-SCF energy E_0, with coefficients $\{c_\mu\}$ $(\sum_\mu H_{\mu\nu}c_\nu = E_0 c_\mu)$, then Eq. (41) can be simplified to:

$$\langle\chi_l|H|\phi_\mu\rangle c_\mu + \sum_m \langle\chi_l|(H - E_0)|\chi_m\rangle t_\mu^m c_\mu + \sum_{\nu\neq\mu} t_\mu^l(\nu) H_{\mu\nu} c_\nu = 0 \tag{42}$$

where $t_\mu^l(\nu)$ denotes the amplitude of the specific excitation present in T^ν, which excites to χ_l by its action on ϕ_μ. We have also used in the above equation the simplified notation t_μ^l for $t_\mu^l(\mu)$. The second term indicates that t_μ^l will be dominated (in the perturbative sense) by the ratio $\langle\chi_l|H|\phi_\mu\rangle/(E_0 - H_\mu)$ and will be intruder-free if the CAS energy E_0 is away from the virtual function energies $H_{ll} \equiv \langle\chi_l|H|\chi_l\rangle$. On the other hand, an equation containing only the first two terms for determining t_μ^l will have entailed disconnected non-EPV terms. Concretely speaking, the term $-E_0 t_\mu^l c_\mu$ can be written as:

$$E_0 t_\mu^l c_\mu = \sum_\nu t_\mu^l H_{\mu\nu} c_\nu \tag{43}$$

If this excitation $\mu \rightarrow l$ on ϕ_μ is also an allowed process on ϕ_ν by Pauli Exclusion Principle, then the corresponding excitation involves orbitals which are not among those distinguishing ϕ_μ and ϕ_ν. Such terms are then disconnected. However, for every such ϕ_ν where this is true, there is a *counter term in* $t_\mu^l(\nu)H_{\mu\nu}c_\nu$, which together leads to $(t_\mu^l(\nu) - t_\mu^l)H_{\mu\nu}c_\nu$ for all such ϕ_μs, and—as explained in Section 22.2.1—since the *difference* $(t_\mu^l(\nu) - t_\mu^l)$ is labelled by some or all orbitals distinguishing ϕ_μ and ϕ_ν, $(t_\mu^l(\nu) - t_\mu^l)H_{\mu\nu}c_\nu$ is a connected term. Hence, such an approximation, termed as SS-MRCEPA(0) by us, is the simplest among the CEPA-like approximants to the SS-MRCC theory which is extensive and also avoids intruders. However, the appearance of the CAS energy E_0, rather than the ground state energy E itself renders it rather approximate. As we already mentioned, the complete linearity of the SS-MRCEPA(0) equations in the cluster amplitudes lends the same invariance property to it as in the MR-CISD. This parallels the situation in the single reference CEPA(0), which also possesses the invariance. In our earlier papers [58,59], we suggested other schemes where E appears which, however, did not have the orbital invariance property as that of the SS-MRCEPA(0).

Recently we have proposed another variant where the state energy E appears in the approximation, which at the same time retains the orbital invariance [59]. In this formalism, there is a set of terms that are non-linear in the cluster amplitudes, but they have a special structure which again leads to the desired orbital-invariance. We approximate $\bar{H}_\mu$ in $\langle\chi_l|\bar{H}_\mu|\phi_\mu\rangle c_\mu$ by all terms which lead to single and double excitations out of ϕ_μ. We thus stop at the double commutator for T_1^μ for the single excitations, and at the single commutator for T_2^μ for the double excitations. These terms are just like the corresponding ones in an MR-CISD formalism, and would retain the orbital invariance. For the so-called coupling term $Y_{\mu\nu}\tilde{H}_{\mu\nu}c_{\nu}$, we approximate the first factor by all terms which can lead to χ_l from ϕ_{μ}, with each χ_l reached by single or double excitations from ϕ_{μ}. Thus, for single excitations, the first factor is $\langle\chi_l|(T_1^{\nu} - T_1^{\mu})|\phi_{\mu}\rangle$, while for the double excitations it is $\langle\chi_l|(\frac{1}{2}T_1^{\nu^2} + \frac{1}{2}T_1^{\mu^2} + T_2^{\nu} - T_2^{\mu} - T_1^{\mu}T_1^{\nu})|\phi_{\mu}\rangle$. This factor would then be form-invariant with respect to the rotations of the active orbitals, since the transformed functions in model and virtual spaces would still be connected by single and double excitations only. The second factor in the coupling term is $\tilde{H}_{\mu\nu}$, in which, in our previous formulations [59], we included diagonal parts of the dressed operator $\bar{H}_{\mu}$. The second factor was thus not invariant with respect to transformation of orbitals. In the formulation, we drop all such terms, and approximate $\tilde{H}_{\mu\nu}$ in the same way as the first term of Eq. (7). Thus, we approximate $\tilde{H}_{\mu\nu}$ as:

$$\tilde{H}_{\mu\nu} = \langle\phi_{\mu}|H + \left\{[H, T^{\nu}] + \frac{1}{2}[[H, T^{\nu}], T^{\nu}]\right\}_{\mathrm{SD}\nu}|\phi_{\nu}\rangle \tag{44}$$

with $\{\ \}_{\mathrm{SD}\nu}$ denoting single and double excitations out of ϕ_{ν}. The second factor of the coupling term is also thus invariant, since it has the structure as the set of model space projections of an MR-CISD equations. Collecting all the terms, the SS-MRCEPA(I) can be written as:

$$\underbrace{\langle\chi_l|H|\phi_{\mu}\rangle c_{\mu}}_{\mathrm{I}} + \underbrace{\langle\chi_l|\left\{[H, T^{\mu}] + \frac{1}{2}[[H, T^{\mu}], T^{\mu}]\right\}_{\mathrm{SD}\mu}|\phi_{\mu}\rangle c_{\mu}}_{\mathrm{II}}$$

$$+ \underbrace{\sum_{\nu}\langle\chi_l|\left\{(T^{\nu} - T^{\mu}) + \frac{1}{2}(T^{\nu} - T^{\mu})^2 + \frac{1}{2}[T^{\nu}, T^{\mu}]\right\}_{\mathrm{SD}\mu}|\phi_{\mu}\rangle\tilde{H}_{\mu\nu}c_{\nu}}_{\mathrm{III}} = 0 \tag{45}$$

with

$$\sum_{\nu}\tilde{H}_{\mu\nu}c_{\nu} = Ec_{\mu} \tag{46}$$

Clearly, all the components of III with only T^{μ} operator can be compactly written as:

$$X = \langle\chi_l|\left\{-T^{\mu} + \frac{1}{2}T^{\mu^2}\right\}_{\mathrm{SD}\mu}|\phi_{\mu}\rangle Ec_{\mu} \tag{47}$$

which has, apart from the EPV terms, all those non-EPV terms which are disconnected. The non-EPV terms are of the following types: (a) the set which are in the CISD space relative to the CAS, induced by products of powers of T^{μ} and T^{ν}, acting on ϕ_{μ} itself; (b) disconnected terms outside the CISD space by similar products of operators as in (a). The interesting aspect of our formalism is that the entire term of III in Eq. (45) above is a connected quantity, so that the inextensivity thus arising from the term X will all be eliminated by the counter terms containing at least one T^{ν}. We emphasize again that the terms I and II are manifestly invariant under the restricted orbital transformation discussed, while the entire term III is also similarly invariant since it involves a complete sum over all the functions ϕ_ν of the CAS. This approximation is thus *the minimal extension* of SS-MRCEPA(0) which uses the dressed Hamiltonian $\tilde{H}_{\mu\nu}$, rather than the bare $H_{\mu\nu}$ as in SS-MRCEPA(0), and uses a self-correcting set of terms in $Y_{\mu\nu}\tilde{H}_{\mu\nu}$ which leaves not only the EPV terms but also cancels the inextensivity in an orbital invariant manner arising out of the non-EPV terms. It should be noted that the non-EPV terms outside the CISD space, mentioned in (b) above, generated in the term III of Eq. (45) are not fully eliminated; rather, their size-inextensive component is eliminated. This is the minimal expansion into the space outside the MR-CISD space necessary to preserve orbital invariance. Eq. (45) define our SS-MRCEPA(I) approximant to the parent SS-MRCC theory. Since T^μ and T^ν are treated on the same footing while truncating $[\exp(-T^\mu)\exp(T^\mu)]$, size-extensivity is rigorously preserved in all our SS-MRCEPA methods.

We should mention here that, just as in the parent SS-MRCC method, the leading terms in the expression of any cluster amplitude t^l_μ exciting to χ_l from ϕ_μ will have in the denominator a difference of the state energy E itself and the energy of the virtual function χ_l. As long as the state energy does not come close to the virtual function energies, the theory is free from intruders. The analysis follows the same route as was taken to show avoidance of intruders in the parent SS-MRCC theory [38,39].

Since the SS-MRCEPPA(I) and SS-MRCEPA(0) methods treat all the model functions on the same footing, it is expected to provide a proper description of the PES of states with varying degrees of quasi-degeneracy among the model functions. They also would not be plagued by intruders. Moreover, they are completely flexible in the sense that we can either use them with frozen coefficients c^0_μ, or allow them to relax in a self-consistent manner. Most of the state-specific MRCEPA formalisms in use do not have the flexibility to relax the coefficients. In this sense our SS-MRCEPA methods share the desirable property of having relaxed coefficients as in the effective Hamiltonian-based multi-reference CEPA theories [72]. We have used both the frozen and the relaxed versions of the SS-MRCEPA(I) and SS-MRCEPA(0) in our numerical applications. They also have the additional advantages of the restricted orbital invariance, as discussed above.

Just as we have discussed—while considering the aspects of spin-adaptation for the general SS-MRCC case—where the spin-free excitations containing exchange spectator scatterings of the singly occupied active orbitals of a CSF ϕ_μ have to be included in the cluster operators of T^μ, and the blocks inducing the same scattering by their actions on ϕ_μ have to be added together, we will have to do exactly the same thing to generate the spin-free expressions for both the SS-MRCEPA(0) and the SS-MRCEPA(I). In fact, as we emphasized already, the SS-MRCEPA methods follow from some low order truncation of the spin-free versions of the SS-MRCC theory, so all the considerations discussed for the general SS-MRCC case remain operative and valid for the SS-MRCEPA case. In our SS-MRCEPA methods the number of unknowns are dominated by the number of inactive double excitations, the computation of the redundant cluster amplitudes is not the computationally dominant step. As a result of this, the computational cost of the SS-MRCEPA methods is similar to that required for an MR-CISD method.

We now present a size-extensive and size-consistent SS-MRCC approach based on an IMS.

22.5 THE SIZE-EXTENSIVE STATE-SPECIFIC MRCC FORMALISM USING AN IMS

A completely different route to reduce computational labor, yet retaining the rigor of formulation, is to have an SS-MRCC in an IMS.

Let us have a set of model functions $\{\phi_\mu\}$ spanning an IMS, with the projector P. The complementary functions $\{\overline{\phi}_\mu\}$ span the complement of the IMS, with the projector R. $(P+R)$ is the projector to the CMS, P_c. The rest of the functions in the Hilbert space are the virtual functions, having at least one inactive occupancy (hole or particle), characterized by the projector Q. $(P_c + Q) \equiv I$, the entire Hilbert space of a given N-electron problem in a finite basis.

We impose on the exact function ψ, the following Ansatz:

$$\psi = \sum_\mu \exp(T^\mu)\phi_\mu c_\mu \tag{48}$$

where the sum runs over all the model functions in IMS spanned by projector P. T^μ is the cluster excitation operator, which produces excitations out of ϕ_μ. As shown by Mukherjee [28,61], for deriving a size-extensive formalism using an IMS, one must define carefully the types of excitations which must be included in the wave operator. We recapitulate the essentials here, since they are directly relevant to the formalism we are going to describe.

We define an operator as 'closed', if its action on any model function $\phi_\mu \in P$ produces only internal excitations within the IMS. An operator is 'quasi-open', if there exists *at least one model function* which gets excited to the complementary model space R by its action. Obviously, both closed and quasi-open operators are all labeled by only active orbitals. An operator is 'open', if it involves at least one hole or particle excitation, leading to excitations to the Q-space by acting on any P-space function. It was shown by Mukherjee [28] that a size-extensive formulation within the effective Hamiltonians is possible for an IMS, if the cluster operators are chosen as all possible quasi-open and open excitations, and demand that the effective Hamiltonian is a closed operator. Mukhopadhyay et al. [61] developed an analogous Hilbert-space approach using the same idea. We note that the definition of the quasi-open and closed operators depends only on the IMS chosen by us, and not on any individual model function.

In our formalism, *we choose in every T^μ all open and quasi-open operators.* For an arbitrary IMS, a given quasi-open operator, acting on a given model function, may lead to excitation to some specific model function, but there would be at least one model function which, when acted upon by this quasi-open operator, would lead to excitations out of the IMS. A closed operator, by contrast, cannot lead to excitations out of the IMS by its action on any function in the IMS. Clearly, any pair of model functions ϕ_μ and ϕ_ν can be reached with respect to each other by either a quasi-open or a closed operator, but not both. This follows from the definition of these operators. For an arbitrary IMS, it is possible to remain within the IMS if a quasi-open operator acts on a specific model function. On another model function, it may lead to excitation out of the IMS. The QCMS (Quasi-Complete Model Space) is a special class of IMS, where we group orbitals into various subsets, labeled A, B, etc. and form a model space spanned by model functions where these subset of orbitals are filled by fixed number of electrons N_A, N_B, etc. in all possible manner. Closed operators then do not change number of electrons in any subset in the QCMS, and quasi-open operators move electrons from one or more subsets to others, so that occupancies of at least two subsets are different after the excitation. In this case, the action of a quasi-open operator on *any* function in the QCMS leads to excitation to R-space functions.

An important insight in the development of size-extensive formulations in a IMS was the realization that the 'intermediate normalization' convention for the wave operator, *viz.* $P\Omega P = P$, should be abandoned in favor of a more appropriate normalization [28,61]. For the IMS, in general, products of quasi-open operators may lead to internal excitations, or may even be closed, so that if we choose $\Omega = \sum_\mu \exp(T^\mu)|\phi_\mu\rangle\langle\phi_\mu|$, with $T^\mu = T^\mu_{op} + T^\mu_{q-op}$, then powers of T^μ_{q-op} coming from the exponential might lead from ϕ_μ to internal excitations to some other model function ϕ_ν, or it may contain closed operators. We would have to bear this in mind while developing our formalism, and would not force $P\Omega P = P$ in our developments.

We now introduce two effective operators. The first is a dressed Hamiltonian operator

$$\bar{H}_\mu = \exp(-T^\mu)H\exp(T^\mu) \tag{49}$$

The second one is an effective 'closed' operator W which is defined to produce as one of its eigenvalues, the desired energy E on diagonalization of its matrix representation in the IMS spanned by $\{\phi_\mu\}$. The coefficients $\{\phi_\mu\}$ come out as the components of its eigenvector. Thus W, by its very mode of definition, satisfies

$$\sum_\nu W_{\mu\nu} c_\nu = E c_\mu \tag{50}$$

Starting from the Schrödinger equation for ψ, we have:

$$\begin{aligned} H\psi = H\sum_\mu \exp(T^\mu)|\phi_\mu\rangle c_\mu &= \sum_\mu \exp(T^\mu)\bar{H}_\mu|\phi_\mu\rangle c_\mu \\ = E\sum_\mu \exp(T^\mu)|\phi_\mu\rangle c_\mu & \end{aligned} \tag{51}$$

Using Eq. (50), we have:

$$\sum_\mu \exp(T^\mu)\bar{H}_\mu|\phi_\mu\rangle c_\mu = \sum_{\mu\nu} \exp(T^\mu)|\phi_\mu\rangle W_{\mu\nu} c_\nu \tag{52}$$

We should note at this point that every virtual function $|\chi_l\rangle$ and also P or R space functions ϕ_ν or $\bar{\phi}_\nu$ may be reached from several ϕ_μs, so that there is a *redundancy* in the number of cluster operators. We will exploit it, following the same strategy as the one used for the CMS-based SS-MRCC theory [38,39], to satisfy the twin desirable goals: (a) to generate a theory free of intruders and (b) to generate rigorous size-extensivity.

We decompose $\bar{H}_\mu$ into its closed, open and quasi-open components. The open and the quasi-open components of Eq. (52) would be used in an appropriate manner to generate a set of sufficiency conditions to define the cluster determining equations. The closed components will be used to define $W_{\mu\nu}$.

We thus rewrite Eq. (52) as

$$\sum_{\mu} \exp(T^{\mu})[\bar{H}_{\mu_{op}} + \bar{H}_{\mu_{q-op}} + \bar{H}_{\mu_{cl}}]|\phi_{\mu}\rangle c_{\mu}$$
$$= \sum_{\mu} \exp(T^{\mu})[\bar{H}_{\mu_{op}} + \bar{H}_{\mu_{q-op}}]|\phi_{\mu}\rangle c_{\mu} + \sum_{\mu\nu} \exp(T^{\mu})|\phi_{\nu}\rangle\langle\phi_{\nu}|\bar{H}_{\mu_{cl}}|\phi_{\mu}\rangle c_{\mu}$$
$$= \sum_{\mu\nu} \exp(T^{\mu})|\phi_{\mu}\rangle W_{\mu\nu} c_{\nu} \tag{53}$$

Since $P_c + Q = I$, we may use a special resolution of identity introduced in Refs. [38,39]

$$\exp(T^{\mu})[P_c + Q]\exp(-T^{\mu}) = I \tag{54}$$

to be used for projection. Also, as shown in Refs. [38,39] to avoid intruders, it is imperative to interchange the labels μ and ν in the third term on the middle expression of Eq. (53). With this interchange, and with the use of Eq. (54) for projection, we get:

$$\sum_{\mu} \exp(T^{\mu})[P_c + Q][\bar{H}_{\mu_{op}} + \bar{H}_{\mu_{q-op}}]|\phi_{\mu}\rangle c_{\mu}$$
$$+ \sum_{\mu\nu} \exp(T^{\mu})[P_c + Q] \exp(-T^{\mu})\exp(T^{\nu})|\phi_{\mu}\rangle\langle\phi_{\mu}|\bar{H}_{\nu_{cl}}|\phi_{\nu}\rangle c_{\nu}$$
$$= \sum_{\mu\nu} \exp(T^{\mu})[P_c + Q]|\phi_{\mu}\rangle W_{\mu\nu} c_{\nu} \tag{55}$$

Equating the Q-projection on both sides of Eq. (55) for each ϕ_{μ} (note that this is a sufficiency condition), we have:

$$Q[\bar{H}_{\mu_{op}}|\phi_{\mu}\rangle c_{\mu} + \sum_{\nu} \{\exp(-T^{\mu}) \exp(T^{\nu})\}_{op}|\phi_{\mu}\rangle\tilde{H}_{\mu\nu} c_{\nu}] = 0 \ \forall \mu \tag{56}$$

where we used $Q|\phi_{\mu}\rangle = 0$ for the left side of the projection in Eq. (56). We also write $\tilde{H}_{\mu\nu}$ as $\langle\phi_{\mu}|\bar{H}_{\nu}|\phi_{\nu}\rangle$. Since every term $\exp(T^{\mu}|\chi_l\rangle$ for each l is linearly independent, we get:

$$\langle\chi_l|\bar{H}_{\mu_{op}}|\phi_{\mu}\rangle c_{\mu} + \sum_{\nu} \langle\chi_l|\{\exp(-T^{\mu}) \exp(T^{\nu})\}_{op}|\phi_{\mu}\rangle\tilde{H}_{\mu\nu} c_{\nu} = 0 \ \forall\, l, \mu \tag{57}$$

In Eq. (56) or Eq. (57), no quasi-open or closed operators can project ϕ_{μ} on to χ_l, since they have no inactive excitations in them, being labeled by active lines only. Hence, they do not appear in Eq. (56) or Eq. (57) for the Q-projection. Eq. (57) defines all the open components of T^{μ}, T^{μ}_{op}, for every μ. By an entirely similar reasoning, we project all the quasi-open components of both sides of Eq. (55) for each μ and equate them (as the sufficiency conditions for defining T^{μ}_{q-op}):

$$P_c[\bar{H}_{\mu_{q-op}}|\phi_{\mu}\rangle c_{\mu} + \sum_{\nu} \{\exp(-T^{\mu}) \exp(T^{\nu})\}_{q-op}|\phi_{\mu}\rangle\tilde{H}_{\mu\nu} c_{\nu}] = 0 \tag{58}$$

Of all the functions spanned by P_c, only a fixed subset of them will be reached by the action of the quasi-open operators on ϕ_{μ}. These functions will span the space of R, and a subset of P. The rest of the functions, all belonging to P, will be reached by the action of closed operators on ϕ_{μ}, since the action of a quasi-open and a closed operator connecting a given pair of functions in P_c is mutually exclusive.

Since $P_c|\phi_\mu\rangle \equiv |\phi_\mu\rangle$, and there are no quasi-open operators present to excite ϕ_μ on the right side of Eq. (55), the right side of the quasi-open projection of Eq. (55) is zero. We have:

$$\langle\overline{\phi}_\lambda|\bar{H}_{\mu_{q-op}}|\phi_\mu\rangle c_\mu + \sum_\nu \langle\overline{\phi}_\lambda|\{\exp(-T^\mu_{q-op})\exp(T^\nu_{q-op})\}_{q-op}|\phi_\mu\rangle\tilde{H}_{\mu\nu}c_\nu$$
$$= 0 \qquad \forall\lambda \in R, \mu \in P \tag{59}$$

$$\langle\phi_\lambda|\bar{H}_{\mu_{q-op}}|\phi_\mu\rangle c_\mu + \sum_\nu \langle\phi_\lambda|\{\exp(-T^\mu_{q-op})\exp(T^\nu_{q-op})\}_{q-op}|\phi_\mu\rangle\tilde{H}_{\mu\nu}c_\nu$$
$$= 0 \qquad \forall\lambda, \mu \in P \tag{60}$$

Again, in Eq. (60), not all $\langle\phi_\lambda|$ can be reached from $|\phi_\mu\rangle$ by a quasi-open operator, but only certain specific ones. These functions are those in the IMS which are unreachable by closed operators on ϕ_μ. Clearly, Eq. (60) is non-trivial only for those $\langle\phi_\lambda|$ reached by quasi-open operators on ϕ_μ; the rest are trivially zero. The non-trivial set of $\langle\phi_\lambda|$, for every μ, defines additional quasi-open operators of T^μ, besides those given by Eq. (59) which lead to excitations to the R-space.

Eqs. (57), (59) and (60) are the working equations for the cluster amplitudes. We should note that they are coupled also to the combining coefficients c_μs, which are obtained as the elements of the eigenvector from the diagonalization of W defined in the IMS, in Eq. (50). This is similar to what we had in the SS-MRCC theory for the CMS [38, 39]. Usually, one may get the coefficients from Eq. (50) in a macro-iteration, and get the cluster amplitudes in an inner, micro-iteration.

The proof of the connectivity of T^μ is somewhat involved, dependent on our proving the connectedness of certain products of pieces entering the definition of W and the cluster operators in a mutually consistent manner. We refer to our recent paper [60] for a concise proof, and to our more comprehensive forthcoming paper [73] for the details.

We now come to the expression of the closed operator W. By equating the closed projections of the entire Eq. (55), we have:

$$\sum_{\mu\nu} \exp(T^\mu_{q-op})P\{\exp(-T^\mu_{q-op})\exp(T^\nu_{q-op})\}_{cl}|\phi_\mu\rangle\tilde{H}_{\mu\nu}c_\nu$$
$$= \sum_{\mu\nu} \exp(T^\mu_{q-op})P|\phi_\mu\rangle W_{\mu\nu}c_\nu \tag{61}$$

To arrive at a size-extensive expression for W, and at the same time to ensure that our SS-MRCC theory using an IMS is intruder-free, it is necessary to write explicitly the projector P, and change the label of the dummy index μ on the right side of Eq. (61) appropriately. We thus have

$$\sum_{\mu\nu\lambda} \exp(T^\mu_{q-op})|\phi_\lambda\rangle\langle\phi_\lambda|\{\exp(-T^\mu_{q-op})\exp(T^\nu_{q-op})\}_{cl}|\phi_\mu\rangle\tilde{H}_{\mu\nu}c_\nu$$
$$= \sum_{\lambda\nu} \exp(T^\lambda_{q-op})|\phi_\lambda\rangle W_{\lambda\nu}c_\nu \tag{62}$$

To simplify further, we equate terms of both sides of Eq. (62) for each ν, pre-multiply by $\exp(-T^\nu)$, project with $\langle\phi_\sigma|$ and take closed projections of the products of exponentials thus generated. We then have,

$$\sum_{\mu\lambda} \langle\phi_\sigma|\{\exp(-T^{\nu}_{q-op})\exp(T^{\mu}_{q-op})\}_{cl}|\phi_\lambda\rangle\langle\phi_\lambda|\{\exp(-T^{\mu}_{q-op})\exp(T^{\nu}_{q-op})\}_{cl}|\phi_\mu\rangle\tilde{H}_{\mu\nu}$$

$$= \sum_{\lambda} \langle\phi_\sigma|\{\exp(-T^{\nu}_{q-op})\exp(T^{\lambda}_{q-op})\}_{cl}|\phi_\lambda\rangle W_{\lambda\nu} \tag{63}$$

which implicitly defines the matrix elements of $W_{\lambda\nu}$. By adducing the same reasoning as was advanced to show the connectivity of the pieces $\{\exp(-T^{\mu})\exp(T^{\nu})\}\tilde{H}_{\mu\nu}$, one may likewise show that W is a connected closed operator. More explicitly, up to the quartic power of cluster operators in the closed projection of the wave operator multiplying $\tilde{H}$, the matrix elements of W can be expressed as

$$\begin{aligned} W_{\sigma\nu} = \tilde{H}_{\sigma\nu} &+ \sum_{\mu} \langle\phi_\sigma|\{\exp(-T^{\mu}_{q-op})\exp(T^{\nu}_{q-op})\}^{[4]}_{cl}|\phi_\mu\rangle\tilde{H}_{\mu\nu} \\ &+ \sum_{\lambda\mu} \langle\phi_\sigma|\left\{\frac{1}{2}T^{\mu^2}_{q-op} - \frac{1}{2}T^{\lambda^2}_{q-op} - T^{\nu}_{q-op}(T^{\mu}_{q-op} - T^{\lambda}_{q-op})\right\}_{cl}|\phi_\lambda\rangle \\ &\times \langle\phi_\lambda|\left\{\frac{1}{2}T^{\mu^2}_{q-op} + \frac{1}{2}T^{\nu^2}_{q-op} - T^{\mu}_{q-op}T^{\nu}_{q-op}\right\}_{cl}|\phi_\mu\rangle\tilde{H}_{\mu\nu} \end{aligned} \tag{64}$$

which is a sufficiently accurate explicit expression of W for practical applications. The quantity $\{\cdots\}^{[4]}_{cl}$ denotes that up to quartic total powers of the cluster operators were taken. We note that there are no linear terms from the expression $\{\cdots\}_{cl}$ since, by construction, there are no closed projections of a quasi-open cluster operator.

We refer to our two papers [60,73] for more detailed discussions and for the demonstration of the avoidance of intruders. Here we just emphasize that, to avoid intruders, it would be necessary to choose the IMS appropriately. The special IMS such as QCMS, where one usually puts quasi-degenerate orbitals in one class, and the non-degenerate orbitals in a different class are very appropriate for separability. There are other desirable simplifications also if we work in the QCMS. For the QCMS, the quasi-open operators cannot lead to excitations into the QCMS itself. This automatically separates the contributions of the quasi-open and the closed operators by simply using the projectors $\exp(T^{\mu})R\exp(-T^{\mu})$ and $\exp(T^{\mu})P\exp(-T^{\mu})$, respectively. This is quite convenient for the practical applications. For the QCMS, we thus need to project Eq. (55) with $\exp(T^{\mu})|\overline{\phi}_\lambda\rangle\langle\overline{\phi}_\lambda|\exp(-T^{\mu})$ to get every quasi-open operator present in T^{μ} for each ϕ_μ, and project onto $\exp(T^{\mu})P\exp(-T^{\mu})$ to generate W. Eq. (60) are all trivially zero for a QCMS.

Although the approach described above is presented in its most general form, using a multiple coupled-cluster Ansatz for the SS-MRCC formalism, suitable approximants to it such as the state-specific multi-reference perturbation theory (SS-MRPT) or state-specific multi-reference CEPA (SS-MRCEPA) can be generated by straightforward approximations. Since the new closed component of the wave operator for IMS appear first at the quadratic power, it is evident that the expressions we have derived in this and the earlier papers for the CAS will remain valid if the quadratic powers of T^{μ} are ignored in the approximants to SS-MRCC for IMS. This implies that all the SS-MRPT at second order of energy as well as both the SS-MRCEPA(0) and SS-MRCEPA(I) will have the same expressions for IMS as for the CAS. This is an important simplification for these approximants.

The results, described in the following section, test the efficacies of the spin-free perturbative and CEPA approximants of the SS-MRCC method, along with the API-SSMRPT versions.

22.6 RESULTS AND DISCUSSION

We present now numerical tests of the performance of the various approximants to the SS-MRCC formalisms discussed in this article for some prototypical systems. We classify the applications in different groups, delineated in separate subsections, each discussing results of a given approximant. We have organized the sequence of the presentation of our results in such a way that more sophisticated approximants appear later. This will facilitate our discussions regarding the relative efficacies of the corresponding approximations, as more refined hierarchies are undertaken.

There are several different aspects of the numerical performance of the approximants to SS-MRCC, which we want to illustrate with our example applications:

(a) Since we as yet do not have the results for the spin-adapted full-blown SS-MRCC, we are unable at this stage to illustrate the extent to which the various approximants to it can capture electron correlation in the general situation relative to the SS-MRCC where open-shell model functions are present. However, in some example applications, the model functions are all of the closed-shell type, for which the spin-adaptation of the SS-MRCC is straightforward, and for these situations the SS-MRCC results are also discussed. However, comparison with the FCI benchmark exact results in a given basis will always be done to estimate the absolute accuracy of the numbers in a given theory.

(b) We first present the SS-MRPT results for PES with different partitionings, using both the relaxed and the frozen coefficients, and also under the physically motivated API approximation. Comparison with the full CI (FCI) results is used to test the efficacy of the approximations. Spectroscopic constants are also extracted from the PES, and are compared with the corresponding FCI results.

(c) It so turns out that the various SS-MRPT results do not differ significantly in the general trends for a given partitioning. We are talking of the trends here based not only on the sample results we are presenting in this article, but also of a whole lot of different sets of results obtained for different systems. The EN partitioning performs generally better in all the cases. It seems then justified to use the simplest of the formalisms, *viz.* the averaged Fock operator with the inclusion of diagonal ladders as H_0 in the EN partitioning as the best acceptable compromise between the computational demands and accuracy, and this we propose to adapt in our future extensive applications.

(d) We test the performance of the approximants for the spin-adapted SS-MRCC, and for this we show results for both the SS-MRCEPA(0) and SS-MRCEPA(I). Again comparison with the FCI is used as the yardstick of their efficacy. Spectroscopic constants are computed from the PES obtained from CEPA methods, and compared with the FCI counterparts.

(e) We have also computed the first and second order molecular properties, *viz.* the dipole moment and the polarizability functions as a function of geometry using just the relaxed version of SS-MRPT(MP) method with two partitions: the Fock operator in the sense of multi-partitioning as well as the averaged Fock operator and API-SSMRPT(MP) method.

In Section 22.3.1, we have discussed several variants of SS-MRPT, using Møller–Plesset (MP) as well as Epstein–Nesbet (EN) partitions, where the denominators for the EN partition contain additional terms from all the diagonal direct and exchange ladders. For the multi-partitioning strategy, we use the corresponding H_0 for each model function ϕ_μ, taking its highest closed-shell portion $\phi_{0\mu}$ as the vacuum. As we have emphasized in Section 22.3, we have two degrees of flexibility in our use of the SS-MRPT formalisms. We can, e.g. (1) use either the relaxed coefficient or the frozen coefficient description, and (2) either use the full-blown SS-MRPT, with μ-dependent first order cluster operators for all the excitations or we can use the API, which we termed as the API-SSMRPT. We have presented results from all these versions of SS-MRPT in our applications.

As we have mentioned earlier, we are also presenting the results of the size-extensive, size-consistent and invariant under restricted orbital rotations SS-MRCEPA methods developed from manifestly spin-free full-blown SS-MRCC method. We will display the values for both relaxed and frozen versions. In our preliminary applications of the SS-MRCEPA(I) method in this article as applied to model spaces with open-shell functions, we have not included the quadratic terms of the cluster operators. For the test case involving the closed-shell model functions only, we have included the quadratic operators. A comprehensive account considering the quadratic terms with the open-shell model functions also will be communicated in due course.

The modifications and the theoretical considerations needed for generating explicitly spin-free formalisms for open-shell model functions have been described in Section 22.2.1. We indicated that we need to add blocks of excitations which produce the same excitations by their action on an open-shell ϕ_μ to get the cluster-determining equations. We present results for four systems: the model H4, the PES for the perpendicular insertion of Be to H_2, and the PES of the bond-dissociation of LiH and BH molecules. Of these, the H4 problem and the PES of the BeH_2 complex use two active orbitals of different symmetries, with two active electrons. The two functions where these two active orbitals are, respectively, doubly occupied span a CAS due to symmetry reasons, and both the model functions are closed-shell type. Hence the special considerations of having pseudo-one body operators with spectator excitations involving the singly occupied active orbitals and adding contributions of blocks inducing the same excitations are not necessary for these two systems. The PES of LiH and BH molecules, on the other hand, have two active orbitals of σ symmetry, which is conceptually the minimal requirement for describing the breaking of the single bond, and the model space is three-dimensional, with an open-shell configuration, viz. $(core)\sigma^2$, $(core)\sigma^{*2}$ and $(core)\sigma\sigma^*$. The additional considerations for the open-shell cases described in Section 22.2.1 would be necessary for treating LiH and BH systems. The ground state PES of all the systems studied shows varying degrees of quasi-degeneracy and would be plagued by intruders with the change of geometries, and hence are appropriate to test the efficacy of the different SS-MRPT and SS-MRCEPA methods. The performance of the various variants is assessed by comparing our results with the FCI results using the same basis.

In all our applications we have employed the natural orbitals of the CASSCF calculations of appropriate state. The CASSCF and FCI calculations were performed using the GAMESS(US) electronic package. In all our calculations, we have employed the symbol SSMR to represent the results of frozen variant of various SSMR methods.

22.6.1 H4 model

The trapezoidal H_4 system, known as H4 model [39,40,74,75], consists of a square configuration of four H atoms, and the PES describes the energy curve as the two adjacent H atoms are moved out to trace a trapezoid with the deviation angle ϕ, which is the difference of the actual obtuse angle and $\pi/2$ [75]. The various geometries of the system are described by $\phi = \alpha\pi$. From the behaviour of the first few CSF energies as a function of the opening angle, it is found that H4 faces intruders at some points over the PES (see Fig. 22.1). The second model space function crosses a virtual function around $\alpha = 0.3$, and the standard effective Hamiltonian methods face intruder problem around this geometry. Our SS-MRPT and SS-MRCEPA methods are expected to be free of this shortcoming, and hence this system is a very good test case to study their potentiality. We have used the DZP basis [75] and CASSCF natural orbitals for the lowest root of the (2 × 2) CAS for computation of the ground state PES at each stage of the calculation within the framework of the C_{2v} symmetry.

We have plotted in Fig. 22.2, the difference energies (mH) for the SS-MRPT methods with respect to the FCI values of the PES of the ground state (1^1A_1) of the H4 model [39,40,74,75]. Two separate graphs are shown. Fig. 22.2a contains results

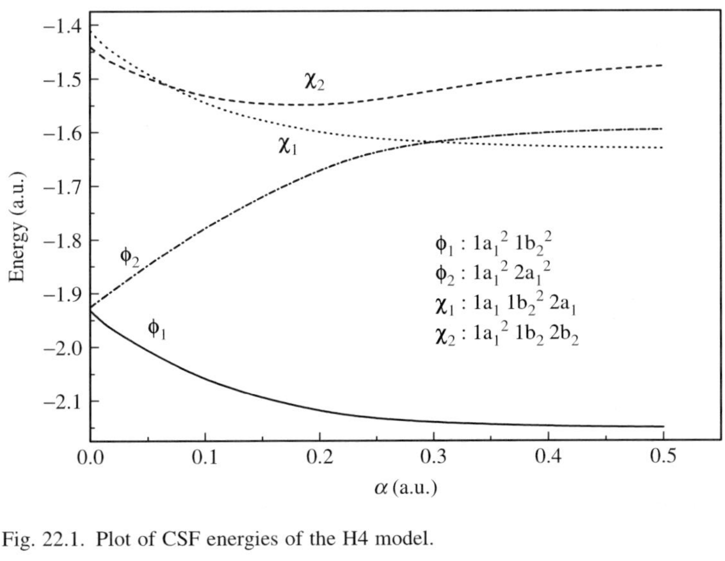

Fig. 22.1. Plot of CSF energies of the H4 model.

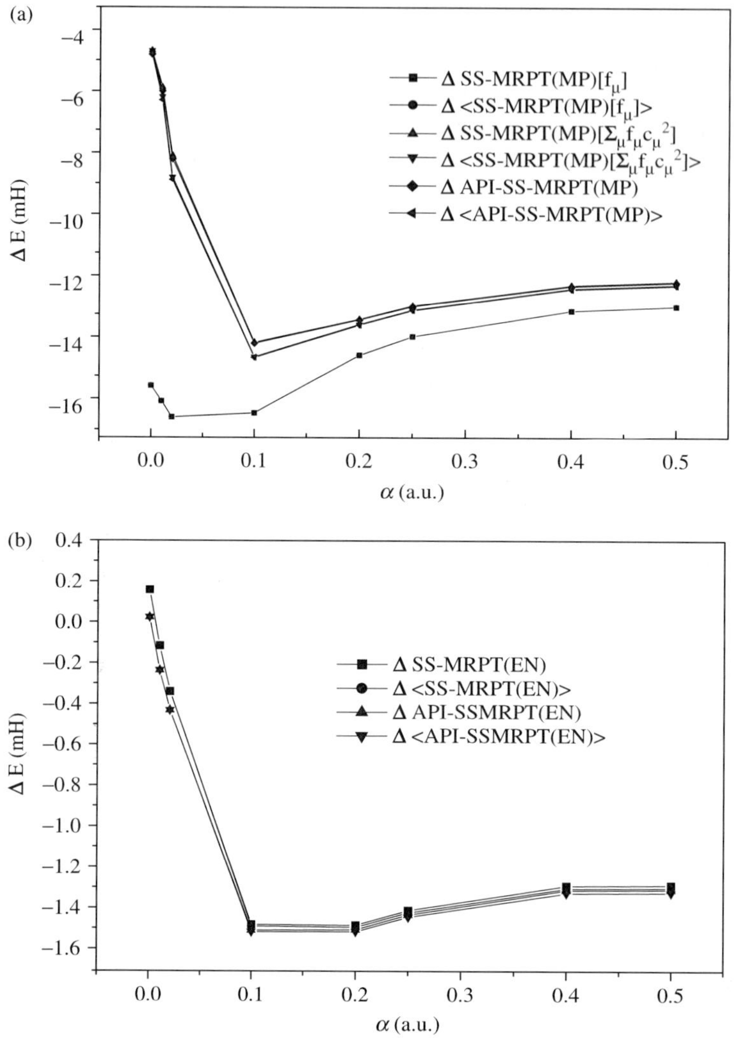

Fig. 22.2. Plot of energy difference [$\Delta E(\mathrm{mH}) = (E_{\mathrm{FCI}} - E_{\mathrm{method}})$] of the ground state of H4 model using DZP basis.

with the MP-like partition, while Fig. 22.2b displays the corresponding results with the EN partition. The various forms of the Fock operators discussed in Section 22.3.1 on perturbation theory are studied. Our particular interest here is to see whether the SS-MRPT can avoid intruders, and also to discern which partitions work better, and whether the most drastic perturbative approximant, *viz.* the API-SSMRPT can capture a major portion of the dynamical correlation energy despite the approximations made on the inactive excitations.

There are no significant deviations as compared to the FCI results in the general trend of the results for any of the choice of the Fock operators in either MP or EN partitions,

partitions used by us, as is evident from the plots shown in Fig. 22.2a and b. The results with the EN partition are significantly better. The figures also indicate clearly that our SS-MRPT methods are devoid of intruders. From the numbers plotted in Fig. 22.2a it is found that the SS-MRPT(MP) method with f_μ shows a minimum deviation of the order of 12.5 mH, a maximum of 16.3 mH, with the average deviation being 14.25 mH in both relaxed and frozen coefficients schemes. However, a switch over to the generalized Fock operator scheme reduces the minimum and maximum deviations to approximately 4 and 14 mH, respectively, with the average deviation of 10 mH. Use of the API-SSMRPT(MP) scheme shows almost identical deviation pattern as is observed for the scheme with generalized Fock operator. The performances of the SS-MRPT(EN) and API-SSMRPT(EN) methods are by far the best showing minimum and maximum deviations of the order of 0.1 and 1.5 mH, respectively, with an average of 0.9 mH. To highlight the numerical performance of the approximants of the parent SS-MRCC method, we mention that the average deviation from FCI for SS-MRCC is of the order of 0.3 mH. This analysis indicates that one can use the same generalized Fock operator ($\sum_\mu f_\mu c_\mu^2$) in our MRPT calculations for all the model functions, since this saves computation time as compared with the one using the multi-partitioning strategy. The performance of the API-SSMRPT for both the partitions is very close to the corresponding results from the full-blown SS-MRPT, although in the former we impose the physically motivated approximation of equal amplitudes for all the one and two-body inactive excitations from the MR reference function. Thus the scheme with the API approximation is computationally quite attractive.

The results with the SS-MRCEPA(0) and SS-MRCEPA(I) were computed by us earlier in another paper [59]. Here we summarize the deviation of the SS-MRCEPA results with respect to the FCI values reported earlier [59] to compare the relative performance of our SS-MRPT methods. The minimum and maximum deviations of SS-MRCEPA(I) method from FCI are approximately of the order of 0.45 and 0.7 mH, respectively, along with the average deviation being 0.6 mH whereas the corresponding values for the deviations in case of CEPA(0) are 0.1 and 0.5 mH, respectively. The average deviation for the CEPA(0) method is of the order of 0.4 mH. Hence we may conclude that the relative performance of our SS-MRPT(EN) is pretty close to the SS-MRCEPA(I) method for the H4 model system.

22.6.2 Insertion of Be into H_2: BeH_2 model

The ground state PES of the C_{2v} insertion reaction of Be into H_2 to form the BeH_2 complex over the reaction points defined in Ref. [76] is an important paradigm system designed to study theories for avoiding intruders [39,46,55,77]. In our calculations, we used the same basis as the one reported in Ref. [76]. The Be atom approaches the H_2 molecule from the direction perpendicular to the bond, and the bond is stretched to enhance the non-dynamical correlation effect as the distance of approach becomes shorter. This system poses serious challenge to any theory where both non-dynamical and dynamical correlations are important in varying degrees along the PES, the active orbitals cross along the distance of approach, thus necessitating theories which treat

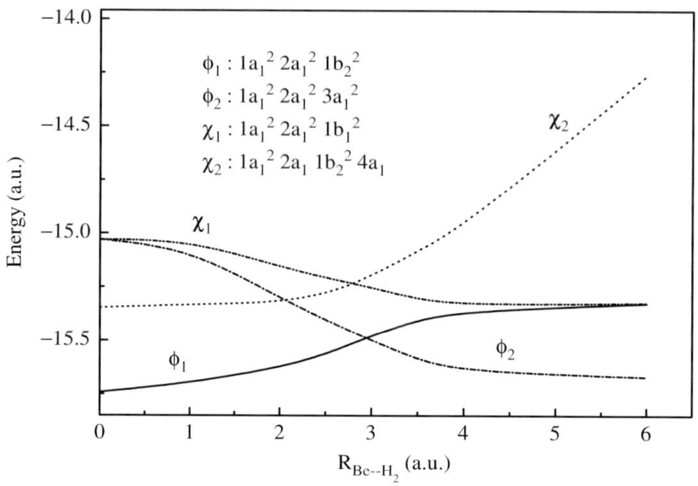

Fig. 22.3. Plot of CSF energies for insertion of Be into H_2.

all model functions on the same footing, and the two active orbitals face intruder orbitals at the two extreme geometries, *viz.* the large distance of approach, and short distance of approach. The qualitative nature of the behavior of the various low-lying model and virtual functions are displayed in Fig. 22.3, where the CSF energies are plotted against the Be–H_2 distance. State-specific formalism should handle this situation for the ground state PES.

It is observed from the coefficients of the FCI results that in the regions of the points: A–C and G–I, the dominant configurations are, respectively, $\phi_1 = 1a_1^2 2a_1^2 1b_2^2$ and $\phi_2 = 1a_1^2 2a_1^2 3a_1^2$. At the transition points, D–F, both configurations are equally important: *a quasi-degenerate reaction region*. Hence these two configurations constituting the CAS are sufficient for an accurate description of reference function of the ground state PES of the BeH_2 system. In our computation, we have used the orbitals of the CASSCF calculations corresponding to the lowest root of the (2 × 2) CAS over the entire reaction points.

Since the two active orbitals belong to different symmetries, the CAS is two-dimensional and both the model functions are closed. The SS-MRCC theory is trivially spin-adapted in this case, and the performance of the various SS-MRPT variants can be assessed with respect to both the FCI results and the SS-MRCC results.

We applied the SS-MRCC theory to the BeH_2 problem some years ago, and showed that it is very effective in generating the PES of the ground state [39], and we present these results along with the FCI and the corresponding results using the variants of the SS-MRPT to assess the efficacy of the various partitioning. As we emphasized above, from the point of view of intruders, the sensitive regions are the large and small distances of approach of the Be atom. Co-ordinates of the sample points are shown in Table 22.1. From the point of view of the relative shift of importance of the model functions, the region around $R = 2–3$ a.u. (i.e D, E, and F reaction points) is important, when the two active orbitals cross each other.

Table 22.1 Coordinates (a.u.) of points along the reaction path for the perpendicular insertion of Be into H_2

Reaction points	Coordinates for H_2(y,R)
A	2.54,0.0
B	2.08,1.0
C	1.62,2.0
D	1.39,2.5
E	1.275,2.75
F	1.16,3.0
G	0.93,3.5
H	0.7,4.0
I	0.7,6.0

Be is located at (0.0,0.0).

Around the region D–F, the unperturbed energy of the 'excited' orthogonal complement of the function with the same two model functions come close to that for the ground state, and the exact extent of splitting of the two states as well as the barrier height generated due to the weakly avoided crossing become very sensitive to the accuracy of the formalism used. The full-blown SS-MRCC theory performs very well in handling both the intruders and in predicting good barrier height [39]. Very recently Pittner et al. [77] have applied the MRBWCCSD method of Hubâc and co-workers [40] which is not rigorously size-extensive to study the perpendicular insertion reaction of Be into H_2 to form the BeH_2 complex. They observed that the performance of the MRBWCCSD method is promising, although the size-extensive correction of the method does not necessarily improve the results relative to the FCI values [77].

We have reported our SS-MRPT(MP) and SS-MRPT(EN) energies for the ground state of the BeH_2 system using various schemes as a deviation from the FCI values over the entire reaction path in Fig. 22.4a and b, respectively. The trend of the results indicate all the variants do rather well. As has been envisaged earlier with the H4 model, we observe that the performance of the SS-MRPT(EN) is better than the corresponding MP partitioning. On an average the deviation in the EN method is of the order of 2 mH, while that in the MP partition it is approximately 16 mH. The minimum and maximum deviations for EN partitioning are 0.4 and 7 mH, respectively, whereas for the MP scheme these values become 8 and 27 mH, respectively. From our earlier results with the SS-MRCC, we find that the corresponding average deviation for the full-blown SS-MRCC is of the order of 0.4 mH [59]. We note that, with both the partitions, the API approximation preserves the accuracy to an extent comparable to the parent full-blown MRPT methods with the same partitioning, except near the reaction point D. The deviation is slightly higher at the geometry F for the MP scheme. In all the SS-MRPT and API-SSMRPT methods, the relaxed values are comparatively better than the corresponding frozen values with respect to FCI. We again find that the trend of results in each partition is less sensitive to the actual Fock operator used. Considering the difficulty of both avoiding intruders and of providing a good potential barrier, the MRPT with the EN partition performs very well.

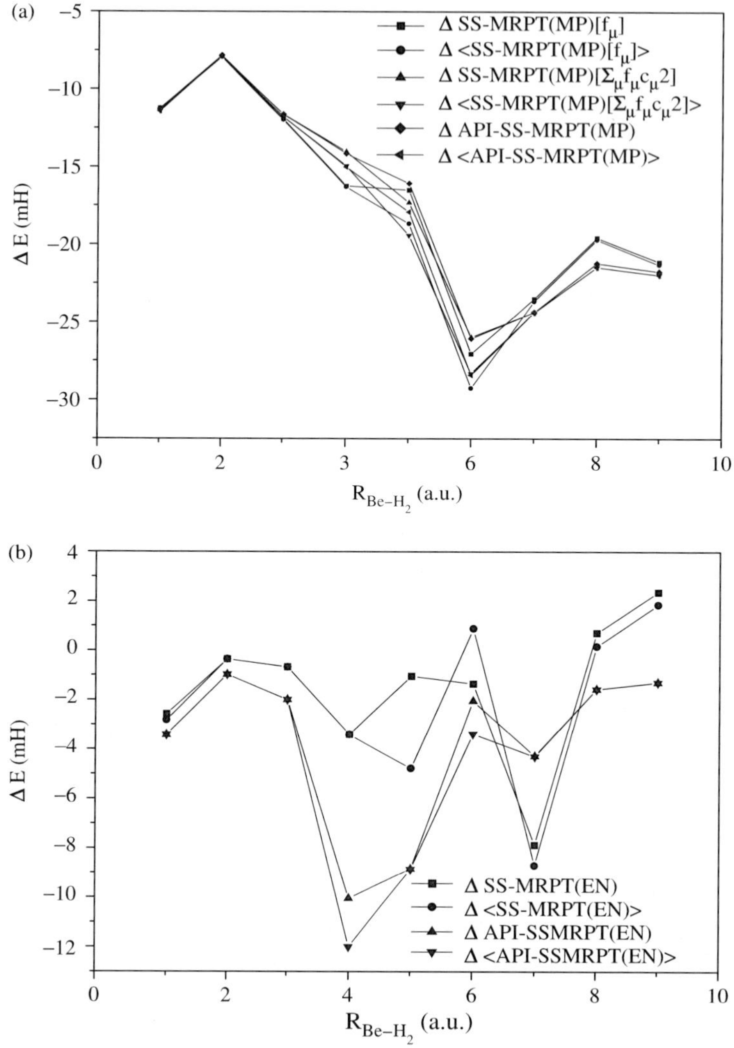

Fig. 22.4. Plot of energy difference [ΔE(mH) $= (E_{\mathrm{FCI}} - E_{\mathrm{method}})$] of the ground state of BeH_2 model using basis of Ref. [75].

For this system we have already published earlier results with the SS-MRCEPA methods. For the sake of completeness, we summarize the comparative performance of the various SS-MRPT methods *vis-à-vis* the SS-MRCEPA methods. Similar to the H4 model, for this system the performance of the SS-MRCEPA(0) with respect to the FCI values is slightly better than the CEPA(I) counterpart [59]. For example, for the SS-MRCEPA (0) method, the maximum and minimum deviations are of the order of 0.2 and 1.8 mH, respectively (except at reaction point F, the deviation is of the order of 4.6 mH) with the small average deviation of the order 1.5 mH while that for the SS-MRCCEPA (I) method, the deviations are approximately around 0.3 and 3.3 mH with the average deviation being 2.5 mH. Here we also observed that the performance of the SS-MRPT(EN) is very similar to the CEPA(I) counterpart of the SS-MRCC method.

22.6.3 LiH molecule

The ground state PES of LiH molecule [78,79] is again a paradigm system in the sense that (a) the effective Hamiltonian methods would fail to describe the equilibrium region very well since the doubly excited Σ states described by the excited roots of the H_{eff} would be close in energy with some virtual functions, and hence would be intruder-prone, and (b) a proper spin-adapted state-specific theory is called for since one of the CSFs is an open-shell singlet. The reference space (CASSCF) includes the following configurations: $\phi_1 = 1\sigma^2 2\sigma^2$, $\phi_2 = 1\sigma^2 3\sigma^2$ and $\phi_3 = 1\sigma^2 2\sigma 3\sigma$. The last CSF is an open-shell function. We have used the natural CASSCF orbitals of ground state from the three-dimensional CAS in our calculations. The basis used are the standard 6311G*, DZP, and cc-pVTZ [80].

We present the results of the spin-adapted versions of SS-MRPT first. The performance of the various MRPT methods with the two partitions is assessed *via* a comparison with the corresponding FCI values for all the basis sets. To judge the quality of the PES, we also evaluate the spectroscopic constants.

In Figs. 22.5–22.7 we present the ground state PES generated *via* various SS-MRPT methods for LiH using DZP, 6311G*, and cc-pVDZ basis, respectively. The results of the SS-MRCEPA methods are depicted in Figs. 22.8–22.10 for the DZP, 6311G*, and cc-pVDZ basis sets, respectively.

Since the model space for LiH contains an open-shell CSF, we use two different Fock operators, *viz.* f_μ and $\tilde{f}_\mu$, and their corresponding generalized versions ($f = \sum_\mu f_\mu c_\mu^2$ and $\tilde{f} = \sum_\mu f_\mu c_\mu^2$, respectively) in our partitioning strategy. As is evident from Figs. 22.5–22.7, with all the three basis sets, the SS-MRPT(EN) methods perform better in comparison to the corresponding MP scheme. In the DZP basis the average deviation for SS-MRPT(MP) method is approximately 1.2 mH, whereas the same for the corresponding EN strategy is close to 0.5 mH. In the 6311G* basis the deviation of SS-MRPT(EN) on an average from the FCI values is nearly 0.3 mH in contrast to 2.3 mH for the corresponding MP partition. A switch-over to a larger basis set shows that the average deviation for the SS-MRPT(EN) remains the same, whereas for the corresponding MP partition the deviation increases slightly. The performance of the API-SSMRPT is very similar to the parent SS-MRPT for the corresponding partitions as is evident from Figs. 22.5c, 22.6c and 22.7c where we plotted the results of API-SSMRPT along with results of SS-MRPT with $\tilde{f}$ for the sake of better insight into the relative performances of API-scheme. Again, as a general trend, we note that the performances with different Fock operators are pretty close to each other for SS-MRPT methods with MP and EN partitionings. The results for the relaxed and the frozen coefficient varieties are also almost identical.

We now discuss the performances of the two SS-MRCEPA methods. Unlike in the cases where the model functions are closed-shell, we have used only the linear terms

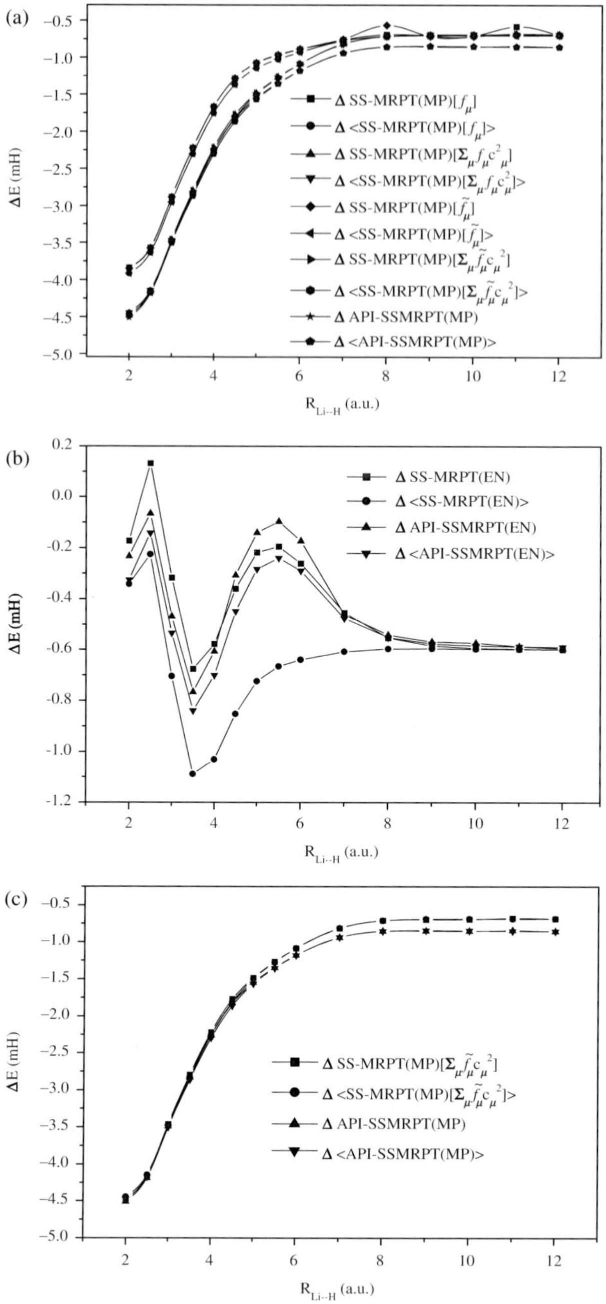

Fig. 22.5. Plot of energy difference [$\Delta E(\text{mH}) = (E_{\text{FCI}} - E_{\text{method}})$] of the ground state of LiH molecule using DZP basis.

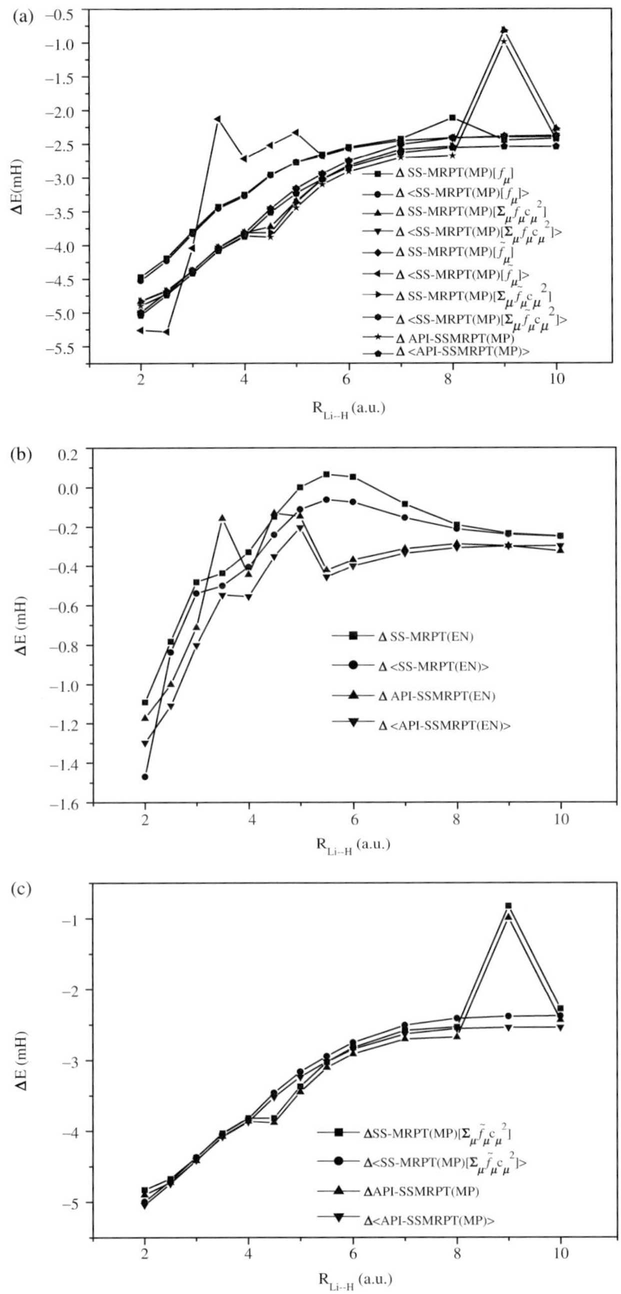

Fig. 22.6. Plot of energy difference [$\Delta E(\text{mH}) = (E_{FCI} - E_{method})$] of the ground state of LiH molecule using 6311G* basis.

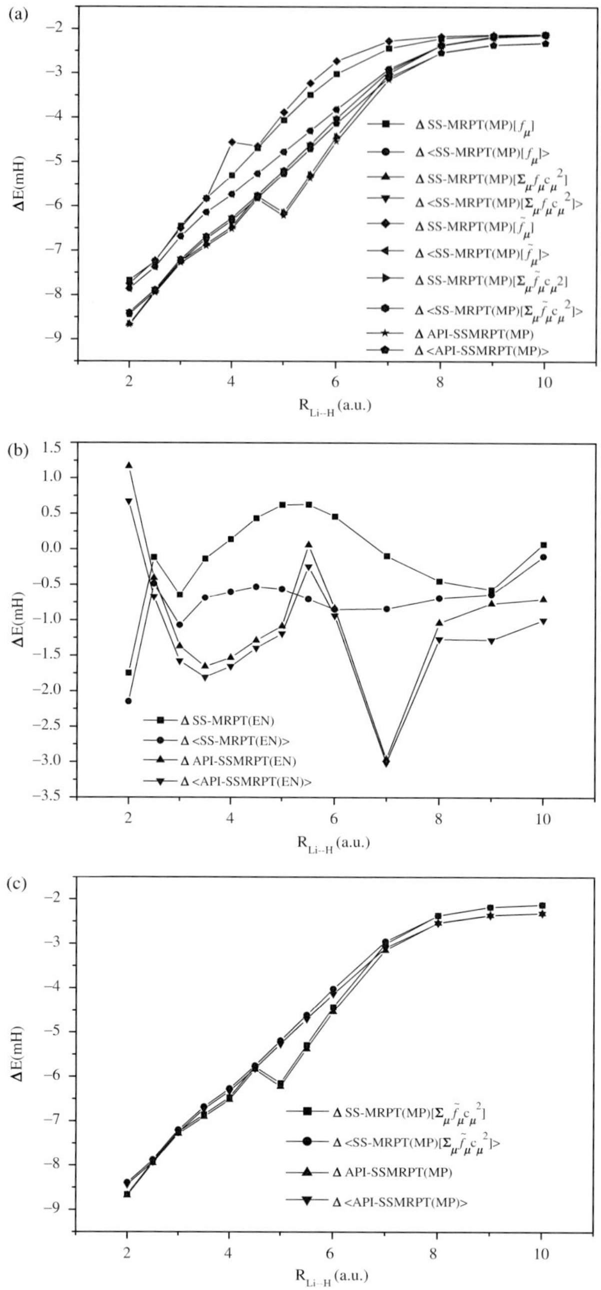

Fig. 22.7. Plot of energy difference [$\Delta E(\mathrm{mH}) = (E_{\mathrm{FCI}} - E_{\mathrm{method}})$] of the ground state of LiH molecule using cc-PVTZ basis.

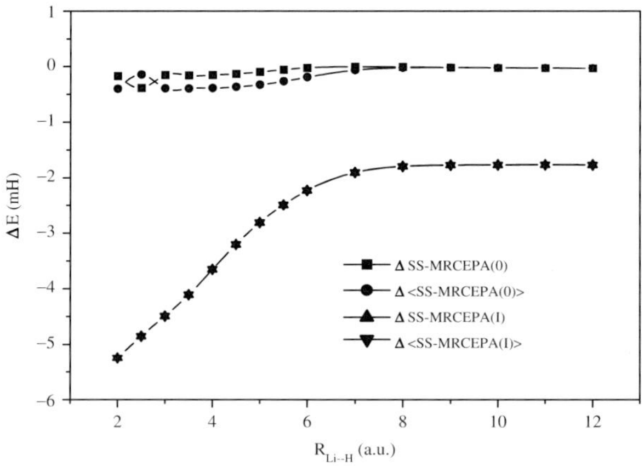

Fig. 22.8. Plot of energy difference [ΔE(mH) = ($E_{FCI} - E_{method}$)] of the ground state of LiH molecule using DZP basis.

in the direct and coupling terms while using SS-MRCEPA(I). This is still an invariant formalism, though it is more approximate, and its performance is not expected to be as good as that for the fully developed SS-MRCEPA(I) for the closed-shell situations.

From Figs. 22.8–22.10 it is clear that the performance of the SS-MRCEPA(0) method is generally better than the approximate version of the SS-MRCEPA(I) method. In the former the result of the relaxed scheme is better than the frozen one for all the basis sets. On the contrary, for the SS-MRCEPA(I) variant, the performance of the relaxed and frozen schemes are super-imposed on one another. The performance of SS-MRCEPA(0) is close to the SS-MRPT(EN) method over a wide range of geometry in all the three basis sets. The deviation from FCI for the SS-MRPT(MP) method is less than

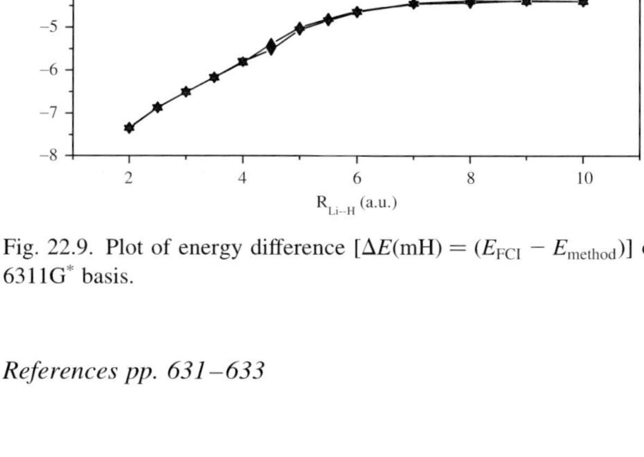

Fig. 22.9. Plot of energy difference [ΔE(mH) = ($E_{FCI} - E_{method}$)] of the ground state of LiH molecule using 6311G* basis.

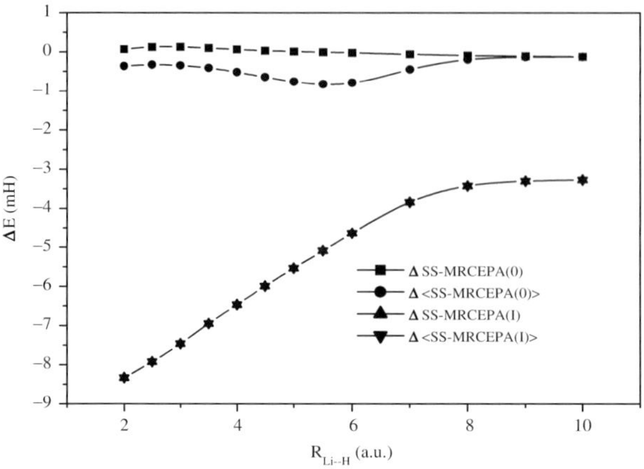

Fig. 22.10. Plot of energy difference [ΔE(mH) $= (E_{\mathrm{FCI}} - E_{\mathrm{method}})$] of the ground state of LiH molecule using cc-PVTZ basis.

the SS-MRCEPA(I) method for all the aforesaid basis. However, it should be borne in mind that the SS-MRCEPA(I) is invariant with respect to the restricted orbital transformation separately among the hole, active and the particle orbitals. We feel that the performance of the SS-MRCEPA(I) method, in particular the fully developed variant, needs to be studied in greater detail in future.

We have also calculated the equilibrium geometry r_{e}, harmonic vibrational frequency ω_{e}, anharmonicity constant $\omega_{\mathrm{e}} x_{\mathrm{e}}$, rotational constant B_{e}, rovibronic constant α_e, and centrifugal distortion constant $\bar{D}_{\mathrm{e}}$. r_{e} is expressed in Å and all other quantities are expressed in cm^{-1}.

Table 22.2 contains the spectroscopic constants for LiH in various basis sets with different SSMRPT methods using EN and MP scheme and SS-MRCEPA methods. The spectroscopic constants show that the performance of our all type of SS-MRPT methods as well as their API variants are quite similar and also pretty close to the FCI values. The values of the spectroscopic constants computed with both the SS-MRCEPA methods indicate that they also fare pretty well. Although we have not studied the performances of our SSMR-based CEPA and PT methods up to the basis set saturation limit, for the sake of completeness of comparison we have also tabulated the corresponding experimental values. Comparison with experimental data indicates that the values of the various spectroscopic constants improve with the increase in size of the basis set.

22.6.4 BH molecule

The pronounced multi-reference character of the BH molecule even around the equilibrium geometry is well known [79,81]. The computation of the ground state PES of BH thus should be a challenging test-case of an MR method, in particular of the relative shift of importance of the dynamical and the non-dynamical correlations.

Table 22.2 Spectroscopic constants of LiH molecule using various basis

Basis	Method	r_e	ω_e	$\omega_e x_e$	B_e	α_e	$D_e \times 10^{-3}$
DZP	SS-MRPT(MP)	1.619	1497.04	22.18	7.252	0.158	0.681
	SS-MRPT(MP) $[\sum_\mu f_\mu c_\mu^2]$	1.619	1470.90	25.51	7.252	0.188	0.705
	API-SSMRPT(MP)	1.619	1478.03	24.45	7.252	0.178	0.698
	SS-MRPT(EN)	1.621	1484.65	23.59	7.238	0.170	0.688
	API-SSMRPT(EN)	1.621	1468.56	25.79	7.238	0.190	0.703
	SS-MRCEPA(0)	1.621	1482.57	23.64	7.238	0.171	0.690
	SS-MRCEPA(I)	1.621	1498.96	21.29	7.238	0.150	0.675
	FCI	1.619	1505.29	21.06	7.252	0.147	0.673
$6311G^*$	SS-MRPT(MP)	1.630	1453.87	20.78	7.158	0.149	0.694
	SS-MRPT(MP) $[\sum_\mu f_\mu c_\mu^2]$	1.630	1442.33	21.53	7.149	0.156	0.702
	API-SSMRPT(MP)	1.631	1469.04	18.32	7.149	0.125	0.677
	SS-MRPT(EN)	1.672	1388.87	19.30	6.805	0.137	0.653
	API-SSMRPT(EN)	1.631	1474.69	17.68	7.149	0.119	0.672
	SS-MRCEPA(0)	1.621	1477.55	20.93	7.238	0.149	0.694
	SS-MRCEPA(I)	1.621	1467.08	22.16	7.238	0.161	0.705
	FCI	1.624	1406.25	28.97	7.215	0.223	0.760
cc-pVTZ	SS-MRPT(MP)	1.606	1430.10	29.22	7.372	0.226	0.783
	SS-MRPT(MP) $[\sum_\mu f_\mu c_\mu^2]$	1.606	1427.57	31.09	7.372	0.241	0.790
	API-SSMRPT(MP)	1.606	1441.44	28.39	7.372	0.218	0.771
	SS-MRPT(EN)	1.610	1453.84	24.23	7.334	0.182	0.746
	API-SSMRPT(EN)	1.582	1464.56	28.72	7.596	0.223	0.817
	SS-MRCEPA(0)	1.604	1459.49	24.39	7.385	0.183	0.756
	SS-MRCEPA(I)	1.605	1466.52	22.90	7.382	0.170	0.748
	FCI	1.601	1449.02	27.77	7.421	0.213	0.778
	Experiment	1.596	1405.65	23.20	7.513	0.213	0.861

r_e in Å and all other quantities in cm^{-1}. *Experiment*: K.P. Huber and G. Herzberg, Constants of diatomic of molecules, Van Norstrand, New York, 1979.

Just as in LiH, the three-dimensional reference space of BH contains one open-shell CSF, apart from the two closed-shell functions. The reference configurations are: $\phi_1 = [\text{core}]3\sigma^2$, $\phi_2 = [\text{core}]4\sigma^2$ and $\phi_3 = [\text{core}]3\sigma 4\sigma$. We use the orbitals obtained from CASSCF calculation corresponding to the lowest root of the (3×3) CAS reference function.

We have used the standard DZP basis set [79] for the calculations. We have calculated both PES and their relative performances have been compared with the standard FCI results in same basis.

Fig. 22.11a and b depicts the results for the different versions of the SS-MRPT(MP) and SS-MRPT(EN), respectively. A comparison of the figures clearly points out that the performance of the SS-MRPT(EN) is far better than the MP counterpart. While the average deviation from FCI for the former is close to 8 mH, for the latter the same is approximately 20 mH. Since we find that the performance of SS-MRPT(MP) with f_μ and $\tilde{f}_\mu$ as Fock operators are almost identical, we depict only the results corresponding to the former. For the same reason, Fig. 22.11a contains the results with the $\sum_\mu f_\mu c_\mu^2$ only. However, the performance of SS-MRPT(MP) using f_μ is marginally better than the one with $\sum_\mu f_\mu c_\mu^2$ for larger separations. The performance of the API schemes resembles closely the corresponding parent full-blown counterparts.

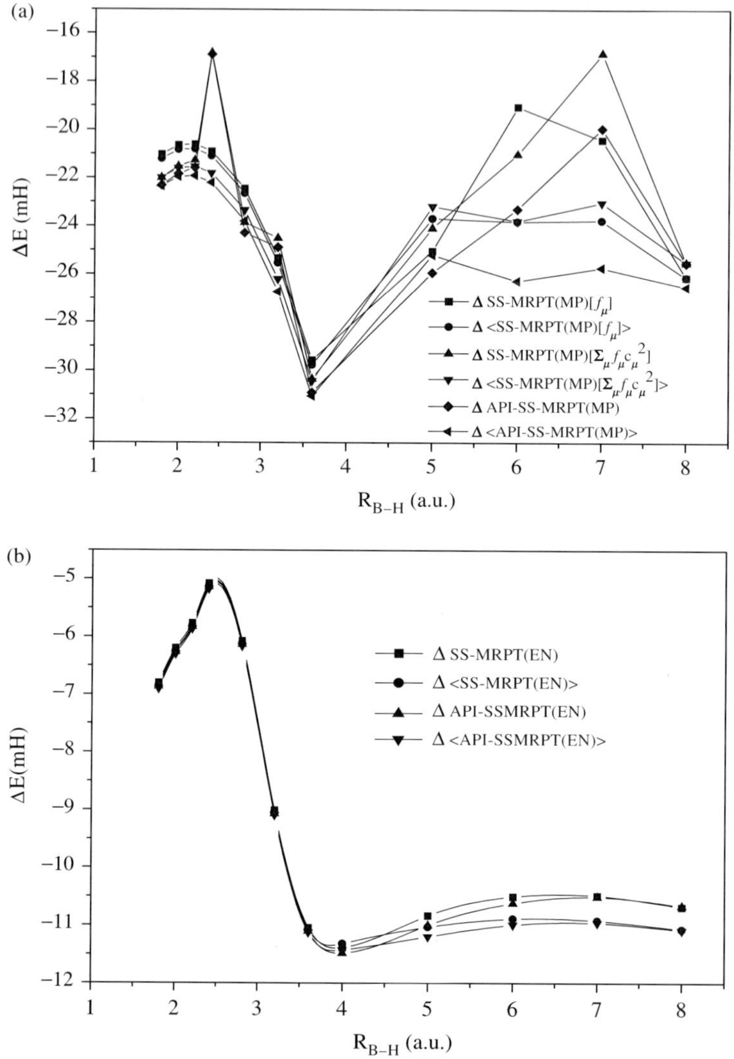

Fig. 22.11. Plot of energy difference [$\Delta E(\mathrm{mH}) = (E_{\mathrm{FCI}} - E_{\mathrm{method}})$] of the ground state of BH molecule using DZP basis.

Fig. 22.12 is the graphical representation of the results obtained using the different versions of SS-MRCEPA. At shorter distances the performance of SS-MRCEPA(0) and the approximate SS-MRCEPA(I) are very close to one another, whereas, for larger separations ($R \geq 7.0$ a.u.) the results of CEPA(I) are very close to the FCI as compared to CEPA(0). At intermediate internuclear distances, the SS-MRCEPA(0) performs better than the corresponding CEPA(I) approach. Akin to our observation for LiH, the approximate SS-MRCEPA(I) in the relaxed and frozen contexts show almost identical behaviour as is evident from Fig. 22.12. At shorter interatomic separations the frozen

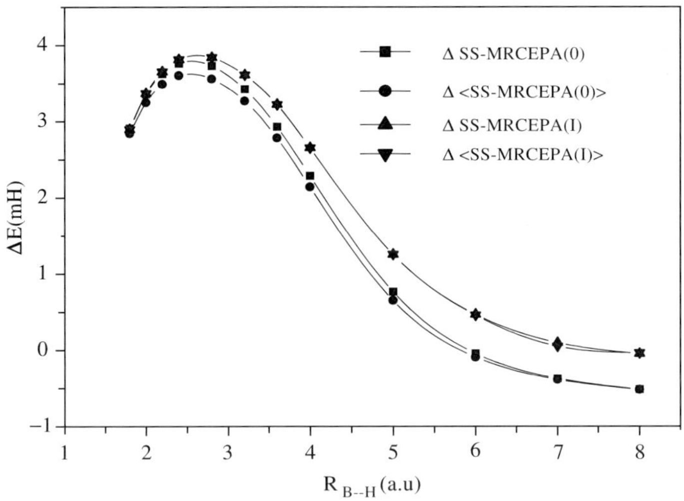

Fig. 22.12. Plot of energy difference [$\Delta E(\text{mH}) = (E_{\text{FCI}} - E_{\text{method}})$] of the ground state of BH molecule using DZP basis.

coefficients version of CEPA(0) is marginally better than the relaxed one, however, at larger distances they merge together.

With the example applications discussed here, it seems to us that it may not be fair at this stage to conclude definitively about the relative performance of SS-MRCEPA(0) and SS-MRCEPA(I) methods. At times CEPA(0) performs better than CEPA(I) and *vice versa* for the closed-shell case, as we found in more extensive applications, in the case of systems containing only closed-shell configurations [59]. More exhaustive calculations, in particular of the fully developed SS-MRCEPA(I), are needed to come to a definitive conclusion, which is on the way. For the SS-MRPT, we have more extensive applications, not all published yet, which definitely indicate the generally superior performance of the EN partitioning.

22.6.5 First and second order electrical property: LiH molecule

To test the quality of the perturbed wave function generated *via* our SS-MRPT(MP) method, we study the PES of LiH in the presence of an electric field, to extract the dipole moment and the polarizability functions [82]. Since the SS-MRPT(MP) is capable of yielding nearly uniform accuracy of the ground state energies for different nuclear geometries over the entire PES and recover a reasonably large portion of the pertinent correlation effects, the SS-MRPT(MP) method is also expected to provide reliable results for the property functions. The SS-MRPT(MP) method, with the Fock operator in the multi-partitioning sense as well as its generalized version, has been employed for the calculation of dipole moment and polarizability surface of the LiH molecule in its ground state. These calculations were carried out *via* the FF (Finite-Field) perturbations approach [83] using natural CASSCF orbitals for the lowest root of (3 × 3) CAS in the presence of the field. The external electric-field strength for each point has been chosen as equal to (±0.005) a.u. The importance of the appropriate choice of the basis set functions has also been carefully investigated.

It has been found that the relatively small three-dimensional active space is sufficient to generate good values of the dipole moment and polarizability surface of the parallel component over the wide range of geometries of the LiH molecule.

Fig. 22.13a and b represents the dipole and polarizability surfaces, respectively, using 6311G* basis. The shape of the dipole moment and polarizability functions for the parallel (α_{zz}) component of the polarizability tensor is qualitatively similar to the FCI

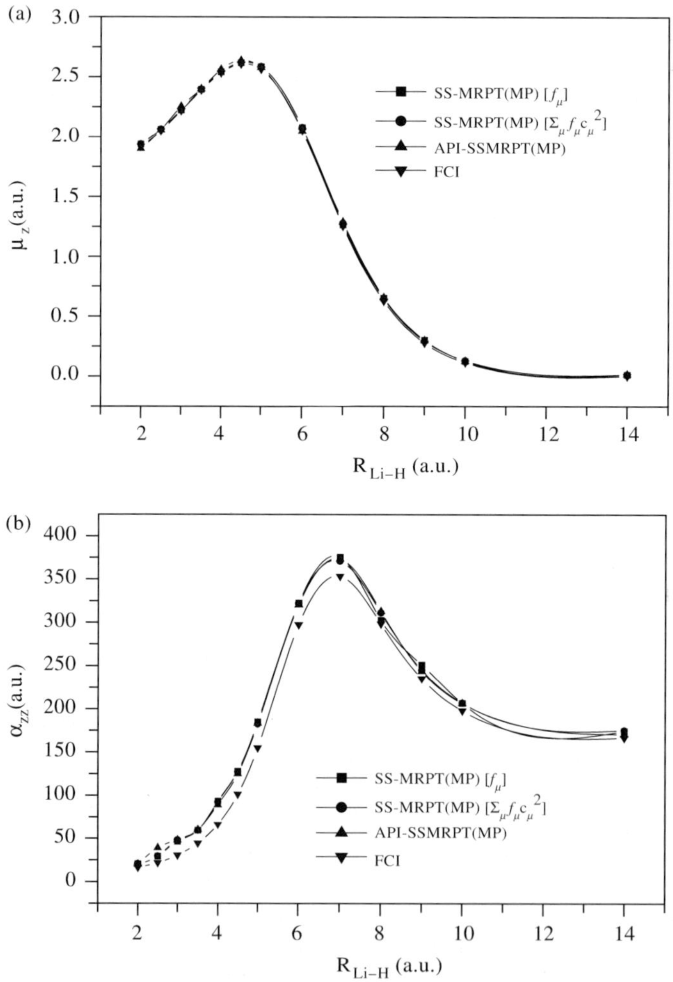

Fig. 22.13. Dipole moment function (μ_z) of the ground state of LiH molecule using 6311G* basis.

curves. It is evident that the deviation from the FCI values is quite small for the dipole moment in comparison to the polarizability values. The deviation for the polarizability function is also not substantial.

We have not reported the perpendicular component of the polarizability tensor, since the process of the computation of the corresponding perpendicular component clearly shows that other CSFs containing π orbitals should be included for convergent results, indicating that the three-dimensional CAS space is not a reasonable choice for computing the perpendicular component of the polarizability and this would require the reference space of larger dimension. LiH dissociates into the ground states of each atom, so they go to 2s of Li and 1s of H. The valence orbitals, σ and σ^*, have mixing of p_z orbitals of both Li and H, provided they are taken (they are a must for polarizability) along the bond axis, and no p_x or p_y because of symmetry. This is true for both the zero field case and when the field is along the z direction. So, when the bond is stretched, and finally broken, the p_x and p_y still do not mix with the active orbitals, and the orbitals get degenerate, leading essentially to an orthogonal combination of 2s of Li and 1s of H at large distance for the zero field case, and to sp hybrid (p in z directions) on Li and H in the case of the field in z directions. The situation is quite different when the field is in the x direction. In that case the active orbitals would have, apart from the 2s and 1s orbitals, the p_z orbitals, a mixture with the p_x orbitals because of charge polarization in the x direction. When the bond is stretched, the orbitals should lead to a combination of s and p_x orbitals in the dissociation limit. This requires the use of a larger model space with π orbitals as active orbitals, something we have not done in this calculation.

22.7 SUMMARY AND CONCLUSIONS

In this article we have presented an account of a class of SS-MR theories, which work in both CMS and IMS. The methods explicitly use spin-free formalisms, and are potentially capable of handling states which are non-singlets and consisting generally of spin-adapted CSFs. We have also discussed the emergence of the corresponding approximants like the spin-free perturbative and CEPA schemes, starting from the spin-adapted SS-MRCC. The use of the entire portion $\phi_{0\mu}$ of the highest closed-shell component of ϕ_μ as the vacuum to define all the excitations on ϕ_μ in normal order is rather powerful, and offers a simple yet convenient access not only to define the various excitation operators but also to simplify the resulting working equations in the spin-free formulation. Although we have presented in this article only the theory of the spin-free SS-MRCC method, we have illustrated the numerical efficacy of its various approximants by applying them in the systems where not only closed-shell but also open-shell functions are present in the model space. These systems possess quasi-degeneracy at some points on the PES and there are potential intruders at some other points. From the above discussion, we can say that the suite of SS-MRPT and SS-MRCEPA methods performs very well. Unlike most of the CAS-based perturbative state-specific methods, all of our SS-MRPT and CEPA counterparts are rigorously size-extensive and intruder-free in nature. Hence they are very effective to study the PES over the wide range of geometrical distortions.

In the perturbative versions, we have discussed and numerically applied two types of partitioning, *viz.* MP and EN in the context of perturbation theory. We have explored the various choices of H_0 in the MP-based SS-MRPT method. The performances of the SS-MRPT(MP) method with various types of Fock operators are very similar. We have observed that the SS-MRPT(EN) performs much better than the corresponding MP approach for the systems studied by us in this article.

In the case of CEPA methods, we have considered the orbital invariant versions. The simplest, the SS-MRCEPA(0), has no EPV terms, while in SS-MRCEPA(I), the full set of EPV terms are retained. The performance of SS-MRCEAP(0) methods in most of the cases is better than the SS-MRCEPA(I) version, though the latter has the advantage of preserving the desired restricted orbital invariance.

On the theoretical front, we have also presented a new approximate version of SS-MRCC theory, termed as API-SSMRCC method, using API where the inactive excitations are independent of μ, index of the configurations of the starting MR zeroth order function. There is a drastic reduction of the number of cluster amplitudes in this approximation. We have demonstrated the efficacies of the API-scheme by numerical implementation of the perturbative variant of the API-SSMRCC method using both type of partitionings (MP and EN). The performance of the API-SSMRPT method is quite good and it is evident from the first set of results that the API-based methods are computationally cost-effective due to the reduction of the cluster amplitudes without unduly sacrificing accuracy.

In addition to computation of the PES, we also observed that the SS-MRPT and SS-MRCEPA methods are also very effective for the calculations of various spectroscopic constants. The dipole moment and polarizability are also computed.

We have also presented a recently developed size-extensive and size-consistent SS-MRCC approach based on a general model space. For this, the intermediate normalization convention of the wave operator has to be abandoned in favor of some appropriate size-extensive normalization. Suitable operators, defined in Fock space—described as closed, open and quasi-open—have to be introduced to ensure that the effective operator furnishing the target energy on diagonalization is a closed operator.

Of course, the most stringent tests for the generality of our SS-MRCC formalism and its various approximants would be in situations where the orbitals change very rapidly as a function of minor geometrical distortions, as happens in weakly avoided crossings. We have shown the effectiveness of our formalisms in one such difficult test system. More extensive applications of the method are underway and will be reported in our forthcoming publications.

22.8 ACKNOWLEDGEMENTS

We thank Pradipta Ghosh for collaborating on the project leading to the results presented in this article and for his help and advice. DM thanks the CSIR [01(1624)//EMR-II, India] for the financial support. DP and SD acknowledge the CSIR for their Research Fellowships.

22.9 REFERENCES

1 R.J. Bartlett and W.D. Silver, Int. J. Quantum Chem., S9 (1975) 183.
2 J. Čížek, J. Chem. Phys., 45 (1966) 4256; Adv. Chem. Phys., 14 (1969) 35; J. Čížek, Adv. Chem. Phys., 9 (1975) 105; R.J. Bartlett, J. Phys. Chem., 93 (1989) 1697; R.J. Bartlett, D.R. Yarkony, Modern electronic structure theory, World Scientific, Singapore (1995).
3 W. Meyer, J. Chem. Phys., 58 (1973) 1017; Theor. Chim. Acta, 35 (1989) 277; R. Ahlrichs, H. Lischka, V. Staemmler and W. Kutzelnigg, J. Chem. Phys., 62 (1975) 1225; W. Kutzelnigg, H.F. Schaefer, III, Methods of electronic structure theory, Plenum Press, New York, 1977.
4 K. Raghavachari, J. Chem. Phys., 82 (1985) 4607; J. Noga, R.J. Bartlett and M. Urban, Chem. Phys. Lett., 134 (1987) 126; K. Raghavachari, G.W. Trucks, J.A. Pople and M. Head-Gordon, Chem. Phys. Lett., 157 (1989) 479.
5 R.J. Buenker and S.D. Peyerimhoff, Theor. Chim. Acta, 35 (1974) 33; P.E.M. Siegbahn, J. Chem. Phys., 72 (1980) 1674; R.J. Buenker, S. Krebs, K. Hirao,Recent advances in multi-reference methods, World Scientific, Singapore (1999) and references therein.
6 B. Brandow, Rev. Mod. Phys., 39 (1967) 771; I. Lindgren, J. Phys. B, 7 (1974) 2441.
7 D. Mukherjee, R.K. Moitra and A. Mukhopadhyay, Mol. Phys., 30 (1975) 1861; 33 (1977) 955; I. Lindgren and D. Mukherjee, Phys. Rep., 151 (1987) 93; D. Mukherjee and S. Pal, Adv. Quantum Chem., 20 (1989) 561.
8 I. Lindgren, Int. J. Quantum Chem., S12 (1978) 33.
9 W. Kutzelnigg, J. Chem. Phys., 77 (1982) 3081; 80 (1984) 822.
10 B. Jeziorski and H.J. Monkhorst, Phys. Rev. A, 24 (1981) 1668; B. Jeziorski and J. Paldus, J. Chem. Phys., 88 (1988) 5673.
11 M. Nooijen and R.J. Bartlett, J. Chem. Phys., 106 (1997) 6441.
12 M. Urban, J. Noga, S.J. Cole and R.J. Bartlett, J. Chem. Phys., 83 (1985) 4041; J. Noga, R.J. Bartlett and M. Urban, Chem. Phys. Lett., 134 (1987) 126; J.D. Watts and R.J. Bartlett, J. Chem. Phys., 93 (1990) 6104; S.A. Kucharski and R.J. Bartlett, J. Chem. Phys., 97 (1992) 4282; J.D. Watts, J. Leszczynski, Computational chemistry: reviews of current trends, Vol. 7, World Scientific, Singapore (2001) 89, and references therein.
13 G.E. Scuseria and H.F. Schaefer, III, Chem. Phys. Lett., 152 (1988) 382; H. Koch, O. Christiansen, P. Jørgensen, A. Sanchez de Merás and T. Helgaker, J. Chem. Phys., 106 (1997) 1808.
14 N. Oliphant and L. Adamowicz, J. Chem. Phys., 94 (1991) 1229; 96 (1992) 3739; V.V. Ivanov and L. Adamowicz, J. Chem. Phys., 112 (2000) 9258.
15 P. Piecuch, S.A. Kucharski and R.J. Bartlett, J. Chem. Phys., 110 (1999) 6103; P. Piecuch, K. Kowalski, I. Pimienta, S.A. Kucharski, M.R. Hoffmann, K.G. Dyall, Low-lying potential energy surfaces, ACS symposium series no. 828, Washington, DC (2002) and references therein.
16 K.B. Ghose, P. Piecuch and L. Adamowicz, J. Chem. Phys., 103 (1995) 9331.
17 X. Li and J. Paldus, J. Chem. Phys., 107 (1997) 6257.
18 G.E. Scuseria, Chem. Phys. Lett., 176 (1991) 27; T.J. Lee and D. Jayatilaka, Chem. Phys. Lett., 201 (1993) 1; D. Jayatilaka and T.J. Lee, J. Chem. Phys., 98 (1993) 9734.
19 P.G. Szalay and J. Gauss, J. Chem. Phys., 107 (1997) 9028.
20 C.L. Jansen and H.F. Schaefer, III, Theor. Chim. Acta, 79 (1991) 1.
21 X. Li and J. Paldus, Int. J. Quantum Chem., S27 (1993) 269, in: R.J. Bartlett (Ed.), Recent advances in coupled cluster methods, World Scientific, Singapore, 1997.
22 H.J. Monkhorst, Int. J. Quantum Chem., S11 (1977) 421.
23 D. Mukherjee and P.K. Mukherjee, Chem. Phys., 39 (1979) 325; B. Datta, P. Sen and D. Mukherjee, J. Phys. Chem., 99 (1995) 6441.
24 H. Nakatsuji and K. Hirao, J. Chem. Phys., 68 (1978) 2053; H. Nakatsuji, Acta Chim. Huang., 129 (1992) 719, and references therein.
25 H. Sekino and R.J. Bartlett, Int. J. Quantum Chem., S18 (1984) 255; H. Koch and P. Jørgensen, J. Chem. Phys., 93 (1990) 3333.
26 A.I. Krylov, Chem. Phys. Lett., 338 (2001) 375; A.I. Krylov and C.D. Sherill, J. Chem. Phys., 116 (2002) 3194; A.J.S. Sears, C.D. Sherill and A.I. Krylov, J. Chem. Phys., 118 (2003) 9084; S.V. Levchenko and A.I. Krylov, J. Chem. Phys., 120 (2004) 175.

27 T.H. Schucan and H.A. Weidenmüller, Ann. Phys., 73 (1972) 108; G. Hose and U. Kaldor, J. Phys. B, 12 (1979) 3827.
28 D. Mukherjee, Chem. Phys. Lett., 125 (1986) 207; D. Mukherjee, Int. J. Quantum Chem., S20 (1986) 409; D. Sinha, S.K. Mukhopadhyay and D. Mukherjee, Chem. Phys. Lett., 129 (1986) 369.
29 U. Kaldor, Theor. Chim. Acta, 80 (1991) 427, and references therein.
30 J.P. Malrieu, Ph. Durand and J.P. Daudey, J. Phys. A: Math. Gen., 18 (1985) 809.
31 B. Kirtman, J. Chem. Phys., 75 (1981) 798.
32 D. Mukhopadhyay, B. Datta and D. Mukherjee, Chem. Phys. Lett., 197 (1992) 236.
33 B. Datta, R. Chaudhuri and D. Mukherjee, J. Mol. Struct. (THEOCHEM), 361 (1996) 21.
34 S. Koch, Theor. Chim. Acta., 81 (1991) 169; L. Meissner, J. Chem. Phys., 108 (1998) 9227; M.R. Hoffmann and Y.G. Khait, Chem. Phys. Lett., 311 (1999) 372.
35 A. Landau, E. Eliav, Y. Ishikawa and U. Kaldor, J. Chem. Phys., 113 (2000) 9905.
36 M.R. Hoffmann, Chem. Phys. Lett., 210 (1993) 193; Y.G. Khait and M.R. Hoffmann, J. Chem. Phys., 108 (1998) 8317.
37 J.P. Malrieu, J.P. Daudey and R.J. Caballol, Chem. Phys., 101 (1994) 8908; J. Meller, J.P. Malrieu and J.L. Heully, Chem. Phys. Lett., 244 (1995) 440; J. Meller, J.P. Malrieu and R. Caballol, J. Chem. Phys., 104 (1996) 4068.
38 U.S. Mahapatra, B. Datta, B. Bandyopadhyay and D. Mukherjee, Adv. Quantum Chem., 30 (1998) 163.
39 U.S. Mahapatra, B. Datta and D. Mukherjee, Mol. Phys., 94 (1998) 157; J. Chem. Phys., 110 (1999) 6171; S. Chattopadhyay, U.S. Mahapatra and D. Mukherjee, J. Chem. Phys., 111 (1999) 3820.
40 J. Mášik and I. Hubač, in: R. McWeeny, J. Maruani, Y.G. Smeyers, and S. Wilson, (Eds.), Quantum systems in chemistry and physics: Trends in methods and applications, Kluwer, Dordrecht, 1997; J. Màšik, I. Hubač and P. Mach, J. Chem. Phys., 108, 1998, pp. 6571.
41 I. Hubač, J. Pittner and P. Čàrsky, J. Chem. Phys., 112 (2000) 8779.
42 J. Pittner, J. Chem. Phys., 118 (2003) 10876.
43 K. Andersson, P.Å. Malmqvist, B.O. Roos, A.J. Sadlej and K. Wolinski, J. Phys. Chem., 94 (1990) 5483; K. Andersson, P.Å. Malmqvist and B.O. Roos, J. Chem. Phys., 96 (1992) 1218.
44 C.W. Murray and E.R. Davidson, Chem. Phys. Lett., 187 (1991) 451; K. Wolinski and P. Pulay, J. Chem. Phys., 90 (1989) 3647.
45 K. Hirao, Chem. Phys. Lett., 190 (1992) 374, Int. J. Quantum Chem., S26 (1992) 517; R.B. Murphy and R.P. Messmer, J. Chem. Phys., 97 (1992) 4170; Y.-K. Choe, Y. Nakao and K. Hirao, J. Chem. Phys., 115 (2001) 621.
46 H. Nakano, J. Chem. Phys., 99 (1993) 7983.
47 P.M. Kozlowski and E.R. Davidson, Chem. Phys. Lett., 222 (1994) 615.
48 K.G. Dyall, J. Chem. Phys., 102 (1995) 4909.
49 H.J. Werner, Mol. Phys., 89 (1996) 645.
50 A. Zaitsevskii and J.P. Malrieu, Chem. Phys. Lett., 233 (1995) 597.
51 J.L. Heully and J.P. Daudey, J. Chem. Phys., 88 (1988) 1046.
52 E. Rosta and P.R. Surján, J. Chem. Phys., 116 (2002) 878.
53 P.R. Surján, Z. Rolik, Á. Szabados and D. Kőhalmi, Ann. Phys. (Leipzig), 13 (2004) 223; P.R. Surján, Á. Szabados, E.J. Brändas, E.S. Kryachko,Fundamental world of quantum chemistry: a tribute to the memory of Per-Olov Löwdin, Kluwer, Dordrecht, 2004, in press.
54 B. Huron, J.P. Malrieu and P. Rancurel, J. Chem. Phys., 58 (1973) 5745; R. Cimiraglia, J. Chem. Phys., 83 (1985) 1746.
55 U.S. Mahapatra, B. Datta and D. Mukherjee, J. Phys. Chem. A, 103 (1999) 1822; P. Ghosh, S. Chattopadhyay, D. Jana and D. Mukherjee, Int. J. Mol. Sci., 3 (2002) 733.
56 S. Das, P. Ghosh, S. Chattopadhyay, D. Mukherjee and U. S Mahapatra, in preparation.
57 P. Ghosh, D. Pahari, S. Chattopadhyay, D. Mukherjee and U. S Mahapatra, in preparation.
58 S. Chattopadhyay, U.S. Mahapatra, B. Datta and D. Mukherjee, Chem. Phys. Lett., 357 (2002) 426; S. Chattopadhyay, U.S. Mahapatra, P. Ghosh, D. Mukherjee, M.R. Hoffmann, K.G. Dyall, Low-lying potential energy surfaces, ACS symposium series no. 828, Washington, DC (2002) 109.
59 D. Pahari, S. Chattopadhyay, A. Dev and D. Mukherjee, Chem. Phys. Lett., 386 (2004) 307; S. Chattopadhyay, D. Pahari, D. Mukherjee and U.S. Mahapatra, J. Chem. Phys., 120 (2004) 5968.

60 D. Pahari, S. Chattopadhyay, S. Das and D. Mukherjee, Chem. Phys. Lett., 381 (2003) 223.

61 D. Mukhopadhyay and D. Mukherjee, Chem. Phys. Lett., 163 (1989) 171; D. Mukhopadhyay and D. Mukherjee, Chem. Phys. Lett., 177 (1991) 441.

62 L. Meissner and R.J. Bartlett, J. Chem. Phys., 92 (1990) 561.

63 P. Celani and H-J. Werner, J. Chem. Phys., 112 (2000) 5546.

64 H. Nakano, J. Nakatani and K. Hirao, J. Chem. Phys., 114 (2001) 1133.

65 R.J. Gdanitz and R. Ahlrich, Chem. Phys. Lett., 143 (1988) 413.

66 P.G. Szalay, in: R.J. Bartlett (Ed.), Recent advances in coupled-cluster methods, World Scientific, Singapore, 1997, and references therein.

67 P.G. Szalay and R.J. Bartlett, J. Chem. Phys., 103 (1995) 3600.

68 L. Füsti-Molnar and P.G. Szalay, J. Phys. Chem., 100 (1996) 6288.

69 P.J.A. Ruttink, J.H. van Lenthe, R. Zwaans and G.C. Groenenboom, J. Chem. Phys., 94 (1991) 7212.

70 R. Fink and V. Staemmler, Theor. Chim. Acta, 87 (1993) 129.

71 J.P. Daudey, J.L. Heully and J.P. Malrieu, J. Chem. Phys., 99 (1993) 1240; I. Nebot-Gil, J. Sanchez-Marin, J.P. Malrieu, J-L. Heully and D.J. Maynau, J. Chem. Phys., 103 (1995) 2576.

72 K. Tanaka, T. Sakai and H. Terashima, Theor. Chim. Acta, 76 (1989) 213; T. Sakai and K. Tanaka, Theor. Chim. Acta, 85 (1992) 451.

73 D. Pahari, S. Chattopadhyay, S. Das and D. Mukherjee, in preparation.

74 K. Jankowski and J. Paldus, Int. J. Quantum Chem., 18 (1980) 1243; G.D. Purvis and R.J. Bartlett, J. Chem. Phys., 75 (1981) 1284; S. Zarrabian and J. Paldus, Int. J. Quantum Chem., 38 (1990) 761; K. Jankowski, J. Paldus and J. Wasilewski, J. Chem. Phys., 95 (1991) 3549; X. Li and J. Paldus, J. Chem. Phys., 107 (1997) 6257.

75 J. Paldus, P.E.S. Wormer and M. Benard, Collect. Czech. Commun., 53 (1988) 1919.

76 G.D. Purvis and R.J. Bartlett, J. Chem. Phys., 76 (1982) 1910; G.D. Purvis, R. Sheppard, F.B. Brown and R.J. Bartlett, Int. J. Quantum Chem., 23 (1983) 835.

77 J. Pittner, H.V. Gonzalez, R.J. Gdanitz and P. Čársky, Chem. Phys. Lett., 386 (2004) 211.

78 B. Jönsson, B.O. Roos, P.R. Yalor and P.E.M. Siegbahn, J. Chem. Phys., 74 (1981) 4566; S. Liu, M.F. Daskalakis and C.E. Dykstra, J. Chem. Phys., 85 (1986) 5877; S. Ben-Shlomo and U. Kaldor, J. Chem. Phys., 89 (1988) 956; J. Noga, D. Tunega, W. Kolpper and W. Kutzelnigg, J. Chem. Phys., 1003 (1995) 309; X. Li and J. Paldus, J. Chem. Phys., 119 (2003) 5346.

79 G.E. Scuseria, T.P. Hamilton and H.F. Schaefer, III, J. Chem. Phys., 92 (1990) 568.

80 www.emsl.pnl.gov/forms/basisform.html.

81 I. Garcia-Cuesta, J. Sanchez-Marin Amor, A. Sanchez de Merás and N.B. Amor, J. Chem. Phys., 107 (1997) 6306; A. Krylov, C.D. Sherrill, E.F.C. Byrd and M. Head-Gordn, J. Chem. Phys., 109 (1998) 10669; B. Temelso, E.F. Valeev and C.D. Sherrill, J. Phys. Chem. A., 108 (2004) 3068.

82 B.O. Roos and A.J. Sadlej, J. Chem. Phys., 76 (1982) 4566; J. Kobus, D. Moncrieff and S. Wilson, J. Phys. B, 34 (2001) 5127, 37 (2004) 571; D. Tunega and J. Noga, Theor. Chem. Acc., 100 (1998) 78.

83 H.D. Cohen and C.C.J. Roothaan, J. Chem. Phys., 43 (1965) S34; J.A. Pople, J.W. McIver and N.S. Ostlund, J. Chem. Phys., 48 (1968) 2960.

Theory and Applications of Computational Chemistry: The First Forty Years
Edited by C. Dykstra et al.

CHAPTER 23

The valence bond diagram approach: a paradigm for chemical reactivity

Sason Shaik[1] and Philippe C. Hiberty[2]

[1] *Department of Organic Chemistry and Lise Meitner-Minerva Center for Computational Chemistry, Hebrew University, 91904 Jerusalem, Israel*
[2] *Laboratoire de Chimie Physique, Groupe de Chimie Théorique, Université de Paris-Sud, 91405 Orsay Cedex, France*

Abstract

Since the early 1970s, valence bond (VB) theory has started to enjoy some renaissance of methodologies and ideas. This chapter describes a new-generation VB paradigm, the so-called VB diagram model. With just two figures the VB diagram model describes both barrier formation in a single chemical step and chemical mechanisms. It thereby forms a basis for a unified system of thought about chemical reactivity. The emergence of new VB methodologies strengthens the model by providing a quantitative description of VB diagrams [J. Phys. Chem. 93 (1989) 5661; J. Chem. Soc. Perkin Trans 2 (1992) 1019; New J. Chem. 20 (1996) 1213; J. Am. Chem. Soc. 112 (1990) 1407; J. Chem. Soc. Chem. Commun. (1989) 772; Inorg. Chem. 29 (1990) 3047; J. Phys. Chem. 96 (1992) 4346; J. Phys. Chem. 94 (1990) 4089; J. Phys. Chem. A 105 (2001) 8226; J. Phys. Chem. A 106 (2002) 11361; J. Phys. Chem. A. 108 (2004) 6017; Chem. Eur. J. 9 (2003) 4540]. This chapter outlines a guide to the qualitative use of the model by discussing a variety of applications, to radical reactions, reactions of electrophiles and nucleophiles, cycloadditions, bond activation by transition metal catalysts, new reaction mechanisms, photochemical reactions, and so on.

23.1 INTRODUCTION

The age of computational quantum chemistry has brought with it a remarkable ability to produce accurate numbers and reliable chemical trends. This, however, has only emphasized the ever-growing need for models that can make sense of the numerical and experimental data, bridge between disciplines, and provide paradigms for the generation of new experimental problems. This model seeking, in chemical reactivity, has resulted in two traditional schools of thought, one rooted in 'classical physical organic chemistry' [1] and the other originated in 'MO theory' [2]. The two schools have crystallized their own unique concepts and defined matching 'observables'.

The classical discipline of chemical reactivity is based on rate–equilibrium relationships, and had its genesis in the first half of this century, in the works of Brønsted, Hammett, Evans, Polanyi, Bell, and the Ingold school. With early valence bond (VB) ideas about chemical bonding and a good deal of healthy empiricism there have emerged important concepts on transition state structure and energetics [1], such as; the Bell–Evans–Polanyi (BEP) principle, the reactivity–selectivity principle, the Leffler–Hammond postulate, and so on. Another type of rate–equilibrium relationships is the Marcus equation which has originated from electron transfer chemistry [3] and has been applied subsequently to a variety of chemical reactions, advancing the concept of an 'intrinsic barrier' which is the 'pure kinetic barrier' of the reaction in the absence of a thermodynamic driving force [1,3]. These concepts link experimental rate patterns with thermodynamic and 'kinetic' properties of the reagents, and, as such, provide a fruitful frame for organizing and predicting reactivity data.

The 'theoretical school' of chemical reactivity is based primarily on Molecular Orbital (MO) theory, [2] and had its genesis in the early work of Hückel followed by those of Fukui, Woodward and Hoffmann. By using perturbation theoretical ideas and appealing to the shape and nodal characters of the MOs, highly portable concepts have been formulated and triggered the development of corresponding experimental methodologies. More recently, density functional theory [4] has led to the derivation of a set of reactivity concepts such as absolute electronegativity and hardness–softness, by applying perturbation theoretic ideas to the connection between the interaction energy and the electron density. The concepts of the 'theoretical school' follow us from our early exposure to chemistry and provide an important link between experiment and quantum chemical theory, in its modern post World War II formulation.

Although the two schools seek the very same insights, understanding chemical reactivity, they still seem different like two pieces disjointed from a big jigsaw puzzle. It is not apparent, for example, where rate–equilibrium relationships or the Hammond–Leffler postulate fit in the theoretical MO and DFT concepts, nor can the principle of orbital symmetry, as such, be traced to anywhere in the underlying conceptual frame of the classical school. The advent of computational quantum chemistry has served to vindicate most if not all of the concepts. However, the ability of computational quantum chemistry techniques to generate accurate reactivity data has not narrowed the chasm, and in many instances has created more and more data, waiting to be conceptualized and patterned. A paradigm of chemical reactivity is still needed; a paradigm that creates bridges, forms a natural interface to computational quantum chemistry, and enables chemists to make predictions in an effective manner.

There are two fundamental questions that a model of chemical reactivity must answer as a prerequisite to becoming a paradigm: What is the origin of the barriers? And what are the factors that control reaction mechanisms? Once a quantum chemical mechanism of barrier formation is formulated, understanding of reactivity patterns is likely to follow, and conceptual means will become available for the generation and solution of new chemical problems. "What happens to Molecules as They React? A Valence Bond Approach to Reactivity" was a title of a 1981 paper [5] that utilized the connection between MO and VB wave functions to derive a general mechanism for barrier and transition state formation. This was a timely paradigm in some respects. Because, on the one hand, MO theory that has served as the main conceptual matrix could not offer a mechanism of barrier formation, with the exception of 'forbidden' reactions [6], and on the other hand, VB theory which was in principle capable of deriving such a general mechanism, was considered passé and its knowledge restricted to a handful of experts. However, the MO–VB relationship was essential to create bridges, and at the same time to generate a reactivity paradigm which enjoys the qualitative insights of both theories; these insights are the locality of the bond reorganization—best described by VB structures, and those of orbital symmetry and nodal features—best described by MO's. Hereafter, this is referred to as *the VB diagram approach*.

One advantage of representing reactions in terms of VB configurations is the unified insight that it brings to reactivity problems. The centerpiece of the VB diagram model is the VB correlation diagram that traces the energy of the VB configurations along the reaction coordinate. The subsequent configuration mixing reveals the cause of the barrier, the nature of the transition state, and the reasons for occurrence of intermediates. Furthermore, the diagram allows qualitative and semiquantitative predictions to be made about a variety of reactivity problems, ranging from barrier heights, stereo- and regio-selectivities, and mechanistic alternatives. Since its derivation, *via* the projection of MO-based wave functions along the reaction coordinate [5], the VB diagram model has been applied qualitatively [7–14] as well as quantitatively by direct computation of the VB diagram [15–24]; as such this is a qualitative model with an isomorphic quantitative analog. The ideas of curve crossing and avoided crossing were used in the early days of VB theory by London [25], Eyring, Polanyi [26], and Evans [27], who pioneered the implementation of VB computations as a means of generating potential energy surfaces

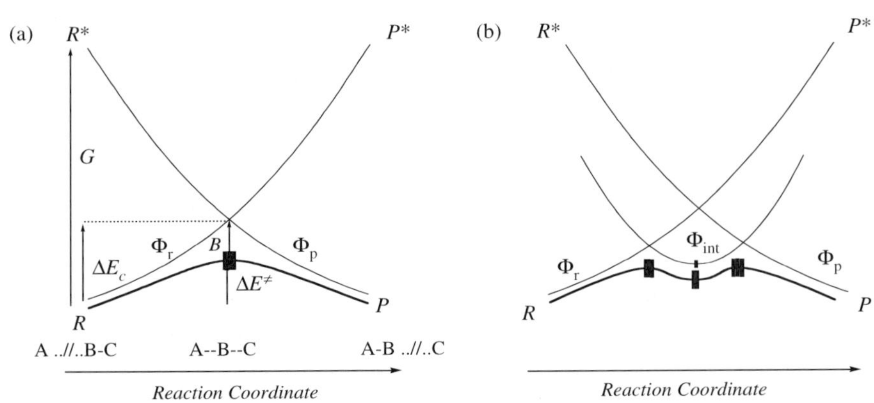

Fig. 23.1. VB diagrams for chemical reactivity: (a) VBSCD showing the mechanism of barrier formation by avoided crossing of two curves of reactant and product type states. (b) VBCMD showing the formation of a reaction intermediate by avoided crossing of a third curve. The final adiabatic states for the thermal reaction are drawn in bold curves (adapted from Ref. [52] with permission of Wiley, ©2004).

and locating transition states. In this respect, the VB diagrams (VBSCD and VBCMD, see Fig. 23.1) are developments of these ideas into a versatile system of thought that allows prediction of a variety of reactivity patterns from properties of the reactants and products.

This chapter reviews briefly some key elements of the VB diagram approach. More details and exhaustive treatments can be found in the various preceding reviews and monographs of these models [1,7–14,28–34].

23.2 VB DIAGRAMS FOR CHEMICAL REACTIVITY

The representation of the VB diagram focuses on the 'active bonds', those that are being broken or made during the reaction. For a chemist, the assignment of the 'active bonds' is straightforward, since in most chemical reactions the bond reorganization event is 'localized' in terms of changes in atomic connectivity. And what makes this focusing possible also at a theoretical level is the electron pair based formulation of the VB representation. In fact, an entire range of reactivity phenomena requires only the two generic diagrams that are schematized in Fig. 23.1. The first is a diagram of two interacting states, called a VB state correlation diagram (VBSCD), which describes the formation of a barrier in a single chemical step due to avoided crossing or resonance mixing of the VB states that describe reactants and products. The second is a three-curve diagram (or generally a many-curve diagram), called a VB configuration-mixing diagram (VBCMD), which describes a stepwise mechanism derived from the avoided crossing and VB mixing of three curves or more. This latter diagram enables one to predict and describe stepwise mechanisms, which involve the incursion of intermediates. In the case where the wave function, Φ_{int}, in Fig. 23.1b has different spin quantum number than the two other state curves, one then obtains a two-state reactivity (TSR) [35], where Φ_{int} mediates the transformation from R to P by a spin-forbidden process (characterized by very weak avoided crossing). In cases where the direct R-to-P transformation has a high barrier, and if the spin-inversion probability is not low, the crossover to Φ_{int} offers a low-energy path that catalyzes the transformation of Φ_{r} to Φ_{p}.

In what follows we describe VBSCDs and VBCMDs, and outline a brief account of their practical use.

23.3 VBSCD—THE ORIGINS OF BARRIERS IN CHEMICAL REACTIONS

VBSCDs apply to reactions that can be described as the interplay of two major VB structures, that of the reactants and that of the products (e.g. Fig. 23.1a). The diagram displays the ground state energy profile of the reacting species (bold curve), as well as the energy profile of each VB structure (thin curves) as a function of the reaction coordinate. Thus, starting from the reactant's geometry on the left-hand side of Fig. 23.1a, the VB structure that represents the reactant's electronic state, R, has the lowest energy and it merges with the ground state of the system. Then, as one deforms the molecular species towards the geometry of the products (on the right-hand side of Fig. 23.1a), the latter VB structure gradually rises in energy and finally reaches an excited state P^* that represents the VB structure of the reactants in the products' geometry. A similar diabatic curve can be traced from P, the VB structure of the products in its optimal geometry, to R^*, the same VB structure in the reactants' geometry. Consequently, the two curves cross somewhere in the middle of the diagram. The crossing is generally avoided, owing to the resonance energy B that results from the mixing of the two VB structures, and this mixing leads to the adiabatic ground state. As such, the barrier arises from avoided crossing between two diabatic curves, which represent the energy profiles of the VB structures of the reactants and products.

The above description is general and applies to all reactions that involve bond exchange, redox, or combination thereof. In fact, the excited states R^* and P^* are simply promoted states that have the same electron-pairing modes as the products and reactants, respectively. Thus, Fig. 23.1a is a state correlation diagram, and the state-correlation is defined only by the nature (bonding and atom connectivity properties) of R and P; *the correlation is independent of the trajectory of the reaction and will be valid in any reaction coordinate.*

The VBSCD is a portable tool for making predictions by assessing the magnitudes of barriers in terms of reactant properties that constitute fundamental parameters of the diagram in Fig. 23.1a. The first of these parameters is the vertical energy gap G that separates the ground state of the reactant, R, from the excited state R^*. This parameter, called *the promotion energy* of the reaction, can take different expressions, depending on which reaction is considered, but is always related to easily accessible energy quantities of the reactants. The promotion energy gauges the height of the crossing point, ΔE_c, of the diabatic curves in the diagram, relative to the energy of the reactants. Thus, ΔE_c can be expressed as a fraction f, smaller than unity, of the gap G (Eq. (1)).

$$\Delta E_c = fG, \qquad f < 1 \tag{1}$$

The f parameter is associated with the curvature of the diabatic curves, large upward curvatures is associated with large values of f, and *vice versa* for small upward curvature. For example, a curvature of two intersecting parabolas leads to $f = 0.25$, while that of two straight-line curves to $f = 0.5$, and so on [11,14]. The curvature depends on the descent of R^* and P^* toward the crossing point and on the relative pull of the ground states, R and P, so that f incorporates various repulsive and attractive interactions of the individual curves along the reaction coordinate. The remaining parameter, in Fig. 23.1a, is the resonance energy B arising from the mixing of the two VB structures in the geometry of the crossing point. The barrier $\Delta E^{\neq}$ can be given a rigorous expression as a function of the three physical quantities f, G and B as in Eq. (2):

$$\Delta E^{\neq} = fG - B \tag{2}$$

A similar expression can be written for the barrier of the reverse reaction as a function of the gap at the product side and the corresponding quantity f. One then distinguishes between the promotion gaps of the reactants and products, G_r and G_p, and the corresponding f factors, f_r and f_p, of the curves that emanate from reactants and products. A general expression for the barrier as a function of the two promotion gaps and the endo- or exo-thermicity of the reaction can be derived by making some simplifying algebraic approximations in the derivation of the height of the crossing point from two general curves [11,13,14]. One such simplified expression [24] is given in Eq. (3), which retains ΔE_{rp} effects up to first order.

$$\Delta E^{\neq} = f_0 G_0 + 0.5\Delta E_{\mathrm{rp}} - B, \qquad f_0 = 0.5(f_{\mathrm{r}} + f_{\mathrm{p}}), \qquad G_0 = 0.5(G_{\mathrm{r}} + G_{\mathrm{p}}) \quad (3)$$

Here the first term is 'an intrinsic factor' that is determined by the averaged f and G quantities, the second term accounts for the effect of the reaction thermodynamics, and the third term is the resonance energy of the transition state, due to the VB mixing (avoided crossing).

The barrier expressions, Eqs. (2) and (3), describe the interplay of three effects. The intrinsic factor f_0G_0 describes the energy cost due to unpairing of bonding electrons, so that new bonds can be made, the ΔE_{rp} factor accounts for the rate–equilibrium effect [13,14], while B involves information about the preferred stereochemistry of the reaction.

23.3.1 Bridges, causes, and causality: a VBSCD perspective

Fig. 23.2 articulates the model *via* Eq. (2), using four cartoons of intersecting curves, which outline the impact of the key factors on the barrier. Clearly, the VB diagram constitutes a unified and general structure–reactivity model that can in principle be applied to any reaction. Furthermore, Fig. 23.2 and Eq. (3) project the bridges between the VBSCD and other conceptual tools. Thus, as seen in the framed statements at the bottom of Fig. 23.2, the VBSCD incorporates rate–equilibrium effects, and thereby makes a connection to classical physical organic chemistry [1]. In addition,

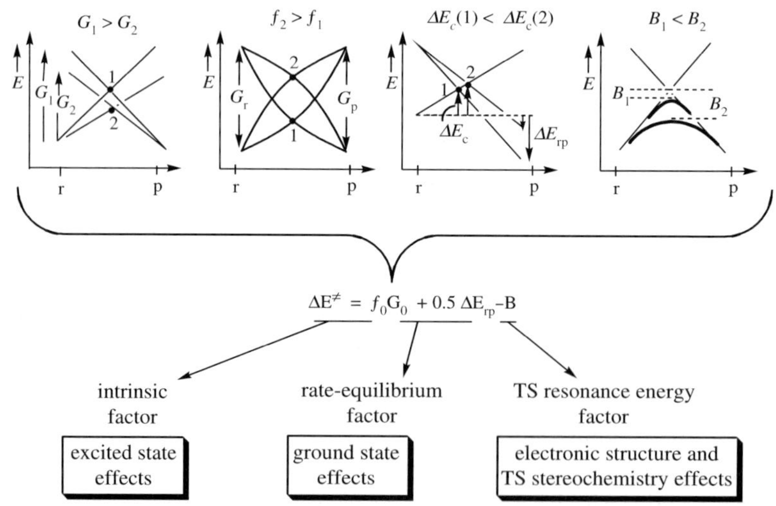

Fig. 23.2. Schematic illustrations of the VBSCD factors and their effects on the variation of the barrier height.

the VBSCD makes a connection to the Marcus equation [1,3] that analyzes the barrier in terms of an intrinsic barrier, $\Delta E_0^{\neq}$, and a thermodynamic attenuation factor, Eq. (4):

$$\Delta E^{\neq} = \Delta E_0^{\neq} + 0.5\Delta E_{rp} + (\Delta E_{rp})^2/16\Delta E_0^{\neq} \tag{4}$$

The intrinsic barrier in the Marcus equation plays an important role, but is essentially unknown and has either to be determined by averaging the barriers of pairs of identity reactions (when the reaction series possesses identity processes), or by assuming that it is a constant in a series. The VBSCD gives an explicit expression of the intrinsic barrier in Eq. (5):

$$\Delta E_0^{\neq} = f_0 G_0 - B \tag{5}$$

The first term, i.e. $f_0 G_0$, accounts for the deformation energy cost of undoing the electron pairing of the 'active' pairs in the reactants and newly re-pairing them as in the products. The second term is the resonance energy of the transition state, i.e. B.

Another bridge in the VBSCD is to orbital symmetry effects [6] and to frontier orbital theory [36]. The effect of orbital symmetry and frontier orbitals is implicit in the expression for B. It is the factor B that makes distinction between 'allowed' and 'forbidden' reactions, and determines the preferred trajectory of the reaction, thereby forming bridges between the VB diagram and MO-derived concepts of reactivity. This has been elaborated amply in previous reviews of the field [7,11,13,14,28], and a brief discussion, by way of an example, is given later.

As may be gleaned from Eq. (3) and Fig. 23.2, reactivity depends on a number of physical parameters, and is hence a multidimensional 'space'. One would have hoped that the key physical parameters are entirely unrelated, so that a correlation with one of the variables would have had a physical significance. Such a physical significance is attached in classical physical organic chemistry to the Brønsted parameters, α or β, which are derived from the correlation of the barrier with the reaction driving force, ΔE_{rp} [13]. Within the BEP principle, and the Hammond postulate, these Brønsted parameters signify the 'location of the transition state' along the reaction coordinate between reactants and products [13,37,38]. However, chemistry does not make life so easy, and usually in any reaction series the parameters change simultaneously and often have significant interdependence [13,37]. Thus, a correlation of the rate constant with a given quantity does not necessary establish causality (a cause and effect relationship).

Fig. 23.3 shows schematically the two different cases. Fig. 23.3a shows an idealized situation where the only change in a reaction series is caused by stabilizing the product (see $\Delta(\Delta E_{rp})$); *the promotion gap, G, and hence also the intrinsic barrier, are constants in the series*. In such a case, the changes in the height of the crossing point along the series are controlled *only* by the change in the reaction driving force, and the Brønsted parameter can be attached with a meaning of a progress parameter, giving the 'location' of the transition state between reactants and products. In such an idealized series, the barriers will decrease and the transition states will become progressively 'earlier' as the reaction becomes more exothermic. But in such an ideal series the Brønsted parameter will also change and converge to zero at the limit of the series [1,13,14]. In such a rare

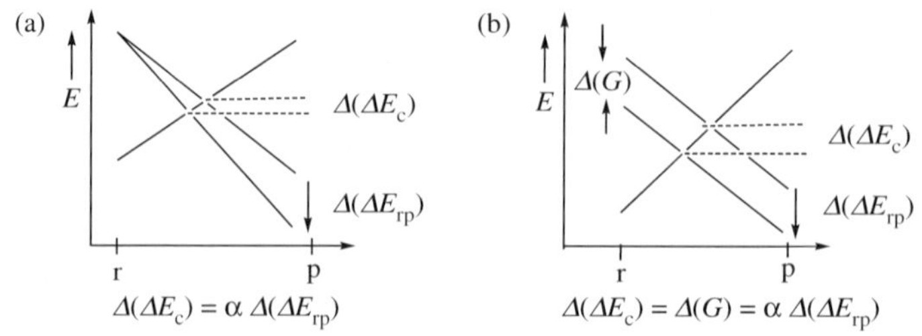

Fig. 23.3. Schematic illustrations of the effect of changing the relative energy of reactants and products, $\Delta(\Delta E_{rp})$, in a reaction series: (a) Only ΔE_{rp} changes in the series, G is a constant in the series. (b) ΔE_{rp} changes as a result of a change of G. Only in part (a) the Brønsted parameter (α) is meaningful as a 'progress parameter'. In any event, the cause of the barrier is G.

case, one can establish causality, namely, the change of ΔE_{rp} is the cause of the change in the barrier and in the structure of the transition states. However, the more common situation is shown in Fig. 23.3b, where it is shown that the change of ΔE_{rp}, ($\Delta(\Delta E_{rp})$), is caused by the change of the promotion gap ($\Delta(G)$). In such a case, the Brønsted parameter does not indicate the progress of the transition state. The correlation of the barriers with the ΔE_{rp} quantity is simply a correlation with the promotion gap, G, which is the true cause of the barrier. Thus, it must be recognized that in any event, *the meaningful causality in structure–reactivity correlations is between the barrier and its originating quantity, the promotion energy, G.*

Having clarified the nature of possible structure–reactivity correlations, we can turn to discuss the application of the VBSCD to chemical reactivity. These can in turn be quantitative, by means of quantum mechanical calculations, or qualitative whenever one has knowledge about variations of the quantities in a series of reactions.

23.3.2 Comments on quantitative applications of VBSCDs

A quantitative application of the diagram requires calculations of both the quantities ΔE_c and B or of the f, G, and B trio. The following comments may be useful in this sense.

The height of the crossing point incorporates effects of bond deformations (bond stretching, angular changes, etc.) in the reactants and non-bonded repulsions between them at the geometry corresponding to the crossing point of the lowest energy, on the seam of crossing between the two state curves (Fig. 23.1a). This, in turn, can be computed by means of *ab initio* calculations, e.g. straightforwardly by use of a VB method [15,22–24,39]. Among the currently available modern *ab initio* VB codes [40], the breathing orbital VB (BOVB) and VB configuration interaction (VBCI) methods have been specifically devised to provide reliable energetics for diabatic as well as adiabatic quantities, by including dynamic correlation effects. The BOVB [41] method expresses the wave function as linear combination of a minimal set of VB structures, precisely as done during the VBSCD construction. The wave function is determined in the usual variational procedure by optimizing the structural coefficients and the orbitals of the structures. However, during the orbital optimization, each of the VB structures is allowed to possess its specific set of orbitals, different from one VB structure to the other. This improvement relative to classical VB methods allows some dynamic electron correlation to be taken into account, a necessary condition for an accurate energetic performance. The VBCI method [42] is similar in spirit to BOVB. It starts from a minimal set of VB structures, built with a common set of orbitals (obtained by VBSCF [40b]), and improves the wave function by subsequent configuration interaction, truncated at any desired level, involving excitations from the fundamental VB structures to virtual orbitals. To preserve the unambiguous interpretability of the final wave function in terms of VB structures, the virtual orbitals, generated by use of a projector, are kept strictly localized on the same atom as the corresponding occupied orbital of the reference fundamental VB structure. Like BOVB, VBCI also accounts for dynamic electron correlation. Both BOVB and VBCI method have been shown to provide bonding energies and reaction barriers with an accuracy comparable to the best MO–CI method, like CCSD(T) [40–43].

Alternatively, the height of the crossing point can be calculated with any MO-based method, by determining the energy of the reactant wave function at zero iteration (see Appendix 23A), by constrained optimization of block-localized wavefunctions [44], or by an energy decomposition scheme of the Morokuma-analysis type [45]. Lastly, the height of the crossing point can be computed by means of molecular mechanical methods [46], or related empirical VB calculations [47,48].

Except for VB theory that calculates B explicitly, in all other methods this quantity is obtained as the difference between the energy of transition state and the computed height of the crossing point. In a few cases it is possible to use analytical formulas to derive expressions for the parameters f and B [7,11,23,49].

As we shall see later, it is straightforward to compute the gap factor, G, for any kind of process, by simply defining the dependence of the nature of the promoted states R^* and P^* on the reaction type. Thus, in principle, the VBSCD is computable at any desired accuracy, and a recent application to the identity exchange reaction $H + H_2$ demonstrates that one can obtain barriers with experimental quality using VB theory itself [50].

23.3.3 Comments on and some qualitative applications of VBSCDs

The purpose of this section is to teach an effective way for using the diagrams in a qualitative manner. The simplest way starts with the G parameter, *which is the origin of the barrier*, since it serves as a promotion energy needed to undo the electron pairing of the bonds of the reactants and pair the electrons in the mode required by the products. In certain families of related reactions the curvatures of the diabatic curves (the parameter f) and the avoided crossing resonance energy (the parameter B) can be assumed as quasi-constant quantities, while in other reaction series f and B vary in the same manner as G. In such cases, the parameter G is the crucial quantity that governs the changes of reaction barriers in the series: the larger the gap G, the larger the barrier. Let us proceed with a few applications of this type.

23.3.3.1 Radical exchange reactions

Fig. 23.4 describes the VB correlation diagram for a reaction which involves cleavage of a bond A–Y by a radical $X^{\cdot}$ (X, A, Y = any atom or molecular fragment):

$$X^{\cdot} + A{-}Y \rightarrow X{-}A + Y^{\cdot} \quad (6)$$

As long as A–Y and X–Y are polar-covalent bonds, we can focus on the covalent parts of the wave function, denoted also as the Heitler–London wave function, called so after the names of the originators of this wave function [51]. Designating the active orbitals of the fragments X, A, and Y, respectively, as x, a and y, the wave functions of R is given in Eq. (7a), without the normalization constant. This equation describes $X^{\cdot}$ and the A–Y bond. The wave function for R^* is given in Eq. (7b):

$$\psi(R) = |xa\bar{y}| - |x\bar{a}y| \quad (7a)$$

$$\psi(R^*) = |x\bar{a}y| - |\bar{x}ay| \quad (7b)$$

Since R^* is just the VB image of the product in the geometry of the reactants, this excited state displays a covalent bond coupling between the infinitely separated fragments X and A, and an uncoupled fragment $Y^{\cdot}$ in the vicinity of A. It is seen, from Eq. (7b), that the electrons in a and y, which are not coupled to a singlet, have a dominant triplet relationship. In one determinant, the right-hand side one in Eq. (7b), they are pure triplet and in the left-hand side one, they are neither singlet nor triplet, but an average of both spin-pairing patterns. Using the rules of qualitative VB theory (neglecting overlap in the normalization constant) [28,52], the energy of R^* relative to the separated X, A, Y fragments is repulsive by the quantity $-\beta_{ay}S_{ay}$, where β is the resonance integral and S the overlap between the active orbitals a and y. On the other hand, the energy of R is just the bonding energy of the A–Y fragment, i.e. $2\beta_{ay}S_{ay}$. Thus the energy gap G for any radical exchange reaction of the type in Eq. (6) is $-3\beta_{ay}S_{ay}$, which is just three quarters

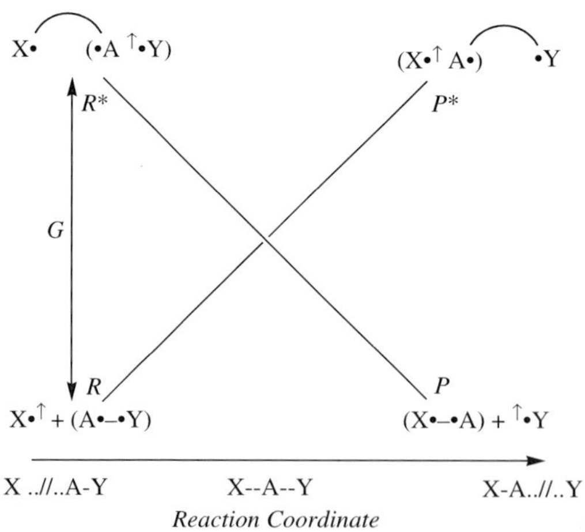

Fig. 23.4. The state correlation in the VBSCD that describes a radical exchange reaction. Avoided crossing as in Fig. 23.1a will generate the final adiabatic profile. The lines connecting dots signify that the two electrons (dots) are singlet-paired (adapted from Ref. [52] with permission of Wiley, ©2004).

of the singlet–triplet gap ΔE_{ST} of the A–Y bond [28,52], namely,

$$G \approx 0.75\Delta E_{ST}(A-Y) \tag{8}$$

The state R^* in Eq. (7b) keeps strictly the wavefunction of the product P, and is hence a quasi-spectroscopic state. Since this R^* has a finite overlap with R, the pair of states, R and R^*, can be mutually orthogonalized by, for example, a Graham–Schmidt procedure. Doing so, the excited state R^* becomes a pure spectroscopic state in which the A–Y moiety is in a triplet state and is coupled to $X^{\cdot}$ to yield a doublet state. In such an event, one could simply use, instead of Eq. (8), the gap G_{spect} in Eq. (9) that is simply the singlet–triplet energy gap of the A–Y bond:

$$G_{spect} = \Delta E_{ST}(A-Y) \tag{9}$$

Each formulation of the state R^* has its own advantages [45,53,54]: for computational purposes, the formulation in terms of the quasi spectroscopic state (Eqs. (7b) and (8)) is exceedingly more convenient, because one can use the variational procedure to calculate all the diabatic curve straightforwardly [22–24,50]. On the other hand, using a spectroscopic state (Eq. (9)) appears more elegant in terms of a 'physical' state-to-state correlation. What is essential for the moment is that both formulations use a gap that is either the singlet-to-triplet excitation of the bond that is broken during the reaction, or the same quantity scaled by approximately a constant 0.75.

As an application, let us consider a typical class of radical exchange reactions, the hydrogen abstractions from alkanes. Eq. (10) describes the identity process of hydrogen abstraction by an alkyl radical:

$$R^{\cdot} + H{-}R \rightarrow R{-}H + R^{\cdot} \tag{10}$$

Identity reactions proceed without a thermodynamic driving force, and project, therefore, the role of promotion energy as the origin of the barrier, what we referred to above as the 'intrinsic factor' (see Fig. 23.2).

The barriers for a series of radicals have been computed by Yamataka and Nagase [55a], and were found to increase as the R–H bond energy D increases; the barrier is the largest for $R{=}CH_3$ and the smallest for $R{=}C(CH_3)_3$. This relationship seems puzzling, since at first glance it suggests that only the bond that is broken matters despite the fact that the transition state has symmetric R···H···R structure, such that the bond that is made is identical to the one that is broken. The correlation of the barrier has been interpreted by Pross et al. [55b] using the VBSCD model. The promotion gap G that originates the barrier (Eqs. (2), (8), and (9)) involves the singlet–triplet excitation of the R–H bond. Now, according to qualitative VB theory [22,28,49,52], this singlet–triplet gap is proportional to the bonding energy of the R–H bond, i.e. $\Delta E_{ST} \approx 2D$. Therefore, the correlation of the barrier with the bond strength is equivalent to a correlation with the singlet–triplet promotion energy (Eq. (9)), a correlation that reflects the electronic reorganization that is required during the reaction. In fact, the barriers calculated by Pross et al. [55] were shown [23] to fit the barrier equation (Eq. (2)), as follows in Eq. (11):

$$\Delta E^{\neq} = 0.3481G - 50\ \text{kcal/mol}, \qquad G = 2D_{RH} \tag{11}$$

which indicates that this is a 'reaction family' with constant $f = 0.3481$ and $B =$ 50 kcal/mol.

Recently, *ab initio* VB computations demonstrated that the ΔE_{ST} quantity [23,43] is the factor that organizes the trends for the barriers for the hydrogen exchange identity reaction, $R^{\cdot} + R'H \rightarrow RH + {}^{\cdot}R'$ $(R = R')$, when R varies down the column of the Periodic Table, i.e. $R = CH_3$, SiH_3, GeH_3, and PbH_3. Thus, in this series, the barriers decrease down the column since the ΔE_{ST} quantity decreases. In this series too, the corresponding bond energies correlate usually with the ΔE_{ST} quantity, and therefore, the barriers correlate with the bond energies. Nevertheless, it must be kept in mind that the correlation with the bond energy disguises a physically meaningful correlation with the ΔE_{ST} quantity, the real cause of the barrier.

Reaction series where reactivity obeys the promotion energy gap abound [7,11], and a few more examples are discussed below.

23.3.3.2 Electrocyclic and transition metal catalyzed bond activation reactions

Once the meaning of the promoted excited state, R^*, is assimilated, one can construct the VBSCD without doing any calculations, and then proceed to make predictions based on the variation of *G*. Some examples of electrocyclic reactions (which involve breaking and making of more than one bond) are shown in Figs. 23.5 and 23.6. In Fig. 23.5a we show a Woodward–Hoffmann 'forbidden reaction', while in Fig. 23.5b an 'allowed' Diels–Alder reaction. In Fig. 23.6a and b we show bond activation processes: a carbene (and its heavier analogs) insertion reaction in a C–H bond and an oxidative addition of d^{10}-PtL_2 or d^{10}-PdL_2 complexes to a C–X bond; the latter is a key step in catalytic cycle of, e.g. the Heck reaction and Suzuki coupling [56]. In all the four examples, the promoted state involves undoing the pairing of the electron pairs of the ground state into triplet pairs, and coupling the electrons to covalent bonds across the new linkages that will become the bonds in the product. This description is invariant to the details of the trajectory (reaction coordinate) or reaction mechanism (concerted or stepwise), since the state correlation is defined solely by the nature of reactants and products.

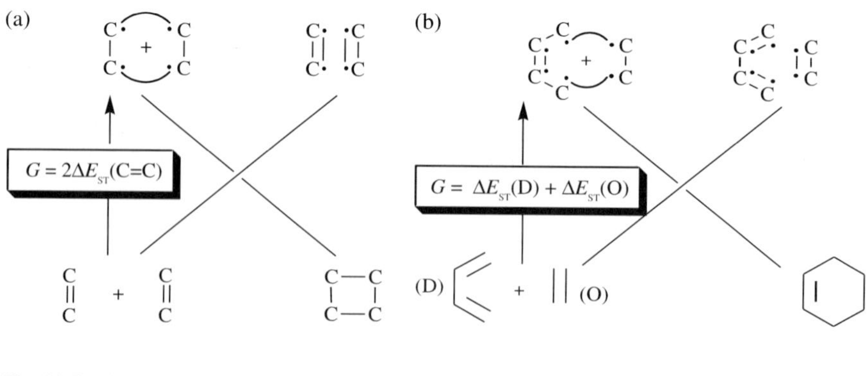

Fig. 23.5. The state correlations in VBSCDs that describe the 'forbidden' 2 + 2 cycloaddition (a), and the 'allowed' Diels–Alder reaction (b). Avoided crossing as in Fig. 23.1a will generate the final adiabatic profiles.

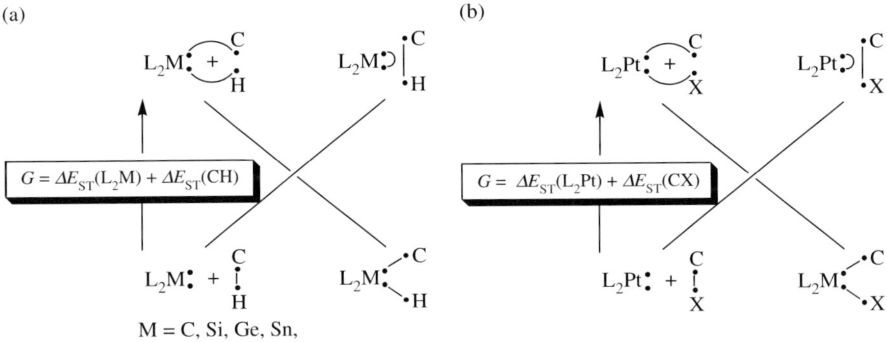

Fig. 23.6. The state correlations in VBSCDs that describe the insertion of a carbene-type reagent into a C–H bond (a), and the oxidative addition of a platinum complex into a C–X bond (b). Avoided crossing as in Fig. 23.1a will generate the final adiabatic profiles.

Consider first a series of Woodward–Hoffmann forbidden 2 + 2 dimerizations, as in Fig. 23.5a, where the promotion gap is proportional to the sum of the $\Delta E_{ST}(\pi\pi^*)$ quantities of the two π-bonds. If reactivity changes are controlled primarily by changes in the promotion energy gap, *G*, it is expected that the dimerization barrier will decrease as the singlet–triplet excitation energies get lower. Indeed, the barrier decreases from 42.2 kcal/mol for the dimerization of ethylene where $\sum\Delta E_{ST}(\pi\pi^*)$ is large (*ca.* 200 kcal/mol) down to less than 10 kcal/mol for the dimerization of disilene for which $\sum\Delta E_{ST}(\pi\pi^*)$ is small (*ca.* 80 kcal/mol) [11]. A similar trend occurs for Woodward–Hoffmann allowed reactions (4 + 2 or 2 + 2 + 2), where *G* is given by $\sum\Delta E_{ST}(\pi\pi^*)$. Here, there is a high barrier of 62 kcal/mol for the trimerization of acetylene, where $\sum\Delta E_{ST}(\pi\pi^*)$ is very large (*ca.* 297 kcal/mol). The barrier gets lowered to 22 kcal/mol for the Diels–Alder reaction where $\sum\Delta E_{ST}(\pi\pi^*)$ *is comparatively much smaller* (*ca.* 173 kcal/mol). Thus, with differences of 120 kcal/mol in *G*, an *f* factor of 0.3 (as quantified recently for radical reaction [22–24]) lowers the barrier by 40 kcal/mol.

Consider now the bond activation reactions in Fig. 23.6a and b; in both cases, the insertion into the bond, C–H or C–X, requires singlet–triplet unpairing on the two reactants. These reactions have been treated extensively by Su et al. [57–59] using the VBSCD model. In all cases, an excellent correlation was obtained between the computed barriers of the reaction and the ΔE_{ST} quantity, including the relative reactivity of carbene and its heavier analogs, and of PtL_2 *vs.* PdL_2 [57–59]. A similar treatment led to the same reactivity patterns for C–F bond activation reactions by *trans*-$Rh(PR_3)_2X$ and *trans*-$Ir(PR_3)_2X$ d^8-complexes [58].

23.3.3.3 Reactions between nucleophiles and electrophiles

Fig. 23.7 illustrates the VB correlation diagram for a reaction between a nucleophile and an electrophile, Eq. (12):

$$X:^- + A{-}Y \rightarrow X{-}A + Y:^- \qquad (12)$$

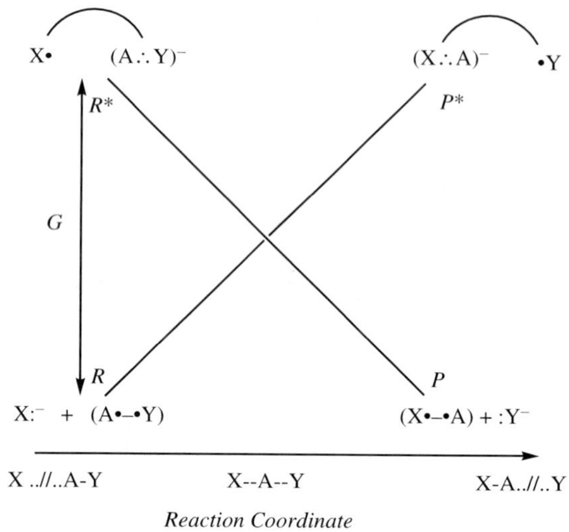

Fig. 23.7. The state correlation in the VBSCD that describes a nucleophilic substitution reaction (S_N2). Avoided crossing as in Fig. 23.1a will generate the final adiabatic profile (adapted from Ref. [52] with permission of Wiley, ©2004).

The equation represents a typical S_N2 reaction where the nucleophile, $X:^-$, shifts an electron to the A–Y electrophile, forms a new X–A bond, while the leaving group Y departs with the negative charge.

Let us now examine the nature of the R^* excited state for this process. In this state, the fragments A and Y are spatially close together (exactly as in the ground state R) and separated from X by a long distance. The X fragment, which is neutral in the product P, must remain so in R^*, and therefore, carries a single active electron. The electron that is lost from $X:^-$ is located on the A–Y moiety, so that the R^* state is the result of a charge transfer from the nucleophile to the electrophile, and is depicted as $X^{\cdot}/(A\therefore Y)^-$, where the $(A\therefore Y)^-$ species shares 3-electrons in the active orbitals of A and Y. It follows that the promotion from R to R^* is made of two events: an electron detachment from the nucleophile, e.g. $X:^-$, and an electron attachment to the electrophile, e.g. A–Y. The promotion energy G is, therefore, the difference between the vertical ionization potential of the nucleophile, $I^*_{X:}$, and the vertical electron affinity of the electrophile, A^*_{A-Y}, given by Eq. (13)

$$G = I^*_{X:} - A^*_{A-Y} \tag{13}$$

where the asterisk denotes a vertical quantity with respect to molecular as well as solvent configurations [13,14,31,60,61]. Precisely symmetric arguments exist for the reverse reaction, where the promotion gap is given by $G = I^*_{Y:} - A^*_{A-X}$. Thus, the mechanism of a nucleophilic substitution may be viewed as an electron transfer from the nucleophile to the electrophile, and a coupling of the supplementary electron of the electrophile to the remaining electron of the nucleophile.

A monograph and many reviews were dedicated to discussion of S_N2 reactivity based on the VBSCD model [1,11,13,14,31]. One key feature that emerged from these treatments was the insight into variations of f. Thus, whether or not the two odd electrons in the R^* state are easily accessible to couple to a bond determines the size of the factor f; the easier the bond coupling along the reaction coordinate, in Fig. 23.7, the smaller the f and vice versa (see also Fig. 23.2b). For example, in delocalized nucleophiles (e.g. acetate, phenyl thiolate, etc.), the active electron is not 100% located on the atom that is going to be eventually linked to the fragment A in the reaction in Eq. (12). So the diabatic curve will slowly descend from R^* toward P and one may expect a large f factor. On the other hand, localized nucleophiles will correspond to smaller f factors. Of course the same distinction can be made between localized and delocalized electrophiles, leading to the same prediction regarding the magnitude of f.

In general, all reactions between closed-shell electrophiles and nucleophiles are describable by the same diagram type [11] with R^* and P^* states, *which are vertical charge transfer states that involve an electron transfer from the nucleophile to the electrophile*, while coupling the single electron on the oxidized nucleophile to that on the reduced electrophile to form a bond-pair. One of the many examples is the nucleophilic assisted cleavage of an ester where the rate-determining step [62,63] is the formation of a tetrahedral intermediate, as depicted in Fig. 23.8.

The promotion energy for the rate-determining step is then, the difference between the vertical ionization potential of the nucleophile and the electron attachment energy of the carbonyl group. The latter quantity is a constant for a given ester, and therefore, the correlation of barriers with the promotion energy becomes a correlation with the vertical ionization energy of the nucleophiles. Fig. 23.9 shows the correlation for the nucleophilic cleavage of a specific ester based on the VBSCD analysis of Buncel et al. [64]. It is seen that the free energies of activation correlate with the vertical ionization energies of the nucleophile in the reaction solvent. Furthermore, localized and delocalized nucleophiles generate correlation lines of different slopes. The two correlation lines obtained for the experimental data in Fig. 23.9 are readily understood based on Eq. (14)

$$\Delta G^{\neq} = f(I^*_{X:} - A^*_{C=O}) - B \tag{14}$$

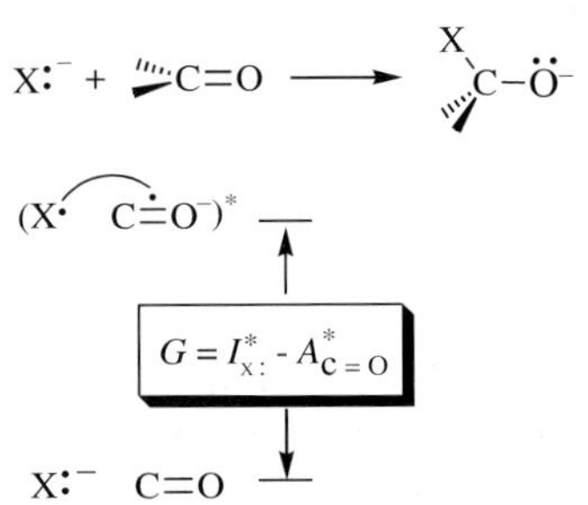

Fig. 23.8. The ground and vertical charge transfer states, of the reactants, in the VBSCD that describes a nucleophilic attack on a carbonyl group (adapted from Ref. [11] with permission of Wiley-VCH STM-Copyright and Licenses).

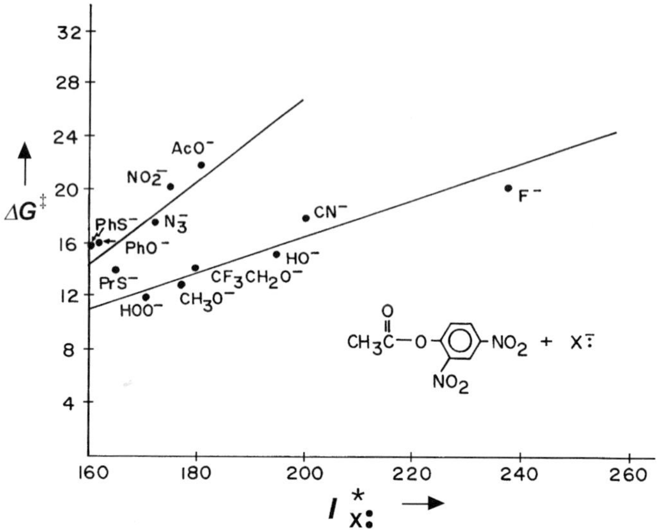

Fig. 23.9. A plot of the free energy barriers for nucleophilic cleavage of an ester *vs.* the vertical ionization potential of the nucleophile (adapted from Ref. [11] with permission of Wiley-VCH STM-Copyright and Licenses).

as corresponding to different *f* values, where the localized nucleophiles possess the smaller *f* value and hence a smaller slope in comparison with the delocalized nucleophiles. The dependence of the *f* factor on delocalization properties of the promoted states in the diagram has been demonstrated also for radical addition reactions [45,54].

23.3.4 Making stereochemical predictions with the VBSCD model

Making stereochemical predictions is easy using VB wave function based on fragment orbitals (FOs) rather than atomic orbitals, hence FO–VB configurations [5,11,28]. To illustrate the manner by which this can be practiced, let us take a simple example with well-known stereochemistry, the nucleophilic substitution reaction, Eq. (12). The corresponding R^* state is depicted in Fig. 23.10 in its FO–VB formulation, where the nucleophile appears here in its one-electron reduced form $X^{\cdot}$, with a single electron in ϕ_X, while the substrate has an extra electron in its σ^*_{AY} orbital. The two single electrons are coupled into a $\phi_X-\sigma^*_{AY}$ bond-pair, as indicated by the line connecting these FOs.

The R^* state correlates to product, X–A/:Y^- ($A = L_3C$), since it contains a $\phi_X-\sigma^*_{AY}$ bond-pair that becomes the X–A bond, and at the same time the occupancy of the σ^*_{CY} orbital causes the cleavage of the C–Y bond to release the :Y^- anion. Furthermore, the R^* state contains information about the stereochemical pathway. Since the bond-pair involves a $\phi_X-\sigma^*_{AY}$ overlap, due to the nodal properties of the σ^*_{AY} orbital the bond-pair will be optimized when the $X^{\cdot}$ is coupled to the substrate in a collinear X–A–Y fashion. Thus, the steepest descent of the R^* state, and the lowest crossing point will occur along a backside trajectory of the nucleophile toward the substrate.

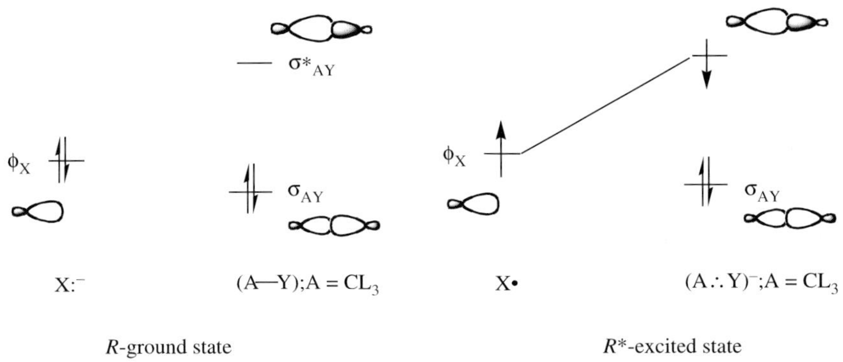

Fig. 23.10. The ground (R) and charge transfer (R^*) states in the VBSCD of S_N2 reactions using the FO–VB representation. The line connecting the orbitals ϕ_X and σ^*_{AY} in the R^* state represents a bond-pair that will become the new X–A bond in the products (adapted from Ref. [52] with permission of Wiley, ©2004).

If we assume that the charge transfer state remains the leading configuration of R^* near the crossing point, then the matrix element between R and R^* will dominate the size of the resonance energy *B*, and will enable making predictions about *B*. In fact, as argued before, generally, the other configurations simply optimize the mixing of the two leading configurations [5,28]. Since these two VB configurations differ by the occupancy of one spin orbital (ϕ_X in R is replaced by σ^*_{AY} in R^*) then following the qualitative rules for matrix elements [11,28], the resonance energy of the transition state (TS) will be proportional to the overlap of these orbitals, i.e.

$$B \propto \langle \phi_X | \sigma^*_{CY} \rangle \tag{15}$$

It follows, therefore, that in a backside trajectory, we obtain both the lowest crossing point as well as the largest TS resonance energy. Computationally the backside barrier is smaller by *ca.* 10–20 kcal/mol compared with a front-side attack [65]. Eq. (15) constitutes *an orbital selection rule for an S_N2 reaction.* Working out this somewhat trivial prediction is nevertheless necessary since it constitutes a prototypical example for deriving orbital selection rules in other reactions, using FO–VB configurations [11]. Thus, the intrinsic bonding features of R^* provide information about the reaction trajectory, while the $\langle R|R^* \rangle$ overlap provides information about the geometric dependence of the resonance energy, *B*, of the TS.

Using this approach, it is possible to *derive orbital selection rules* for cases that are ambiguous in qualitative MO theory [11]. For example, for radical cleavage of σ-bonds, using the R^* with a triplet $\sigma^1\sigma^{*1}$ configuration on the substrate leads to the prediction that the course of the reaction and the resonance interaction in the transition state will be determined by the product of overlaps between the orbital of the attacking radical, ϕ_R, and the σ and σ^* orbitals of the substrate, namely, $\langle \phi_R | \sigma \rangle \langle \phi_R | \sigma^* \rangle$. This product is optimized once again in a backside attack, and therefore, one can predict that radical cleavage of σ-bonds will proceed with inversion of configuration. All known experimental data [66–71] conform to this prediction. Another area where successful predictions have been made involves nucleophilic attacks on radical cations. Here using the corresponding R and R^* states [72], it was predicted that stereoselectivity and regioselectivity of nucleophilic attack should be controlled by the LUMO of the radical cation. Both regioselectivity and stereospecificity predictions were verified by experiment [73,74] and computational means [65].

Finally, using VB–FO representations of R and R^* for an electrocyclic reaction (Fig. 23.5), the B factor for an electrocylic reaction is simply proportional to the product of the HOMO–LUMO overlaps for the two reactants [11]. As such, the following relationship holds for the 'allowed' *vs.* the 'forbidden' pathways:

$$B\,(\text{'allowed'}) > B\,(\text{'forbidden'}) \tag{16}$$

Thus, as we stated above by reference to Fig. 23.2, the *resonance energy of the transition state holds the stereochemical information of the reaction*. For a more in depth discussion the reader may consult the most recent review of the VBSCD model [11].

23.4 VALENCE BOND CONFIGURATION MIXING DIAGRAMS

The VBCMD is an alternative and a complementary diagram to the VBSCD [7,8–11,13, 14], typified by more than two curves as shown above in Fig. 23.1b. A few examples are discussed below, while a more in-depth description can be found in a recent review [11].

23.4.1 General features of the VBCMD

Fig. 23.1b, above, shows the generic VBCMD that features two fundamental curves, labeled as Φ_r and Φ_p, and an intermediate curve denoted by Φ_{int}. In those situations where the intermediate curve lies higher than the crossing point of the fundamental curves, the VB mixing will be prone to generate a single transition state that has a mixed character of the fundamental and intermediate VB structures [13]. However, the diagram in Fig. 23.1b describes a situation where the intermediate curve, being significantly more stable than the crossing point of the fundamentals will generate, though not always [16] an intermediate state in a stepwise mechanism. This intermediate structure provides a low-energy pathway that mediates the transformation of R to P ($\Phi_r \rightarrow \Phi_p$). There are two types of intermediate curves: (i) the intermediate curve is an ionic structure, and (ii) the intermediate curve is a third state that differs from the R–R^* and P–P^* state curves.

23.4.2 VBCMD with ionic intermediate curves

Any two-state VBSCD can be transformed into a VBCMD where the Heitler–London (HL) and ionic VB structures are plotted explicitly as independent curves, instead of being combined into state curves [11]. As a rule, ionic structures, which are the secondary VB configurations of polar-covalent bonds, lie above the covalent Heitler–London (HL) structures at the reactant and product geometries, and generally they cross the two HL structures above their own crossing point. In many cases, the ionic curve is low enough in energy in the hypercoordinated geometry near the transition state, so that solvation can further stabilize the ionic situation and cause it to cross the HL-curves significantly below their own crossing point; in such an event a stepwise mechanism mediated by an ionic intermediate will transpire (e.g. S_N1 mechanism in the reactants of Eq. (12), etc.). The following examples serve to illustrate the impact of ionic VB structures on the reactivity of covalent bonds.

23.4.2.1 Proton transfer processes

The small size of H^+ enables very tight ion-pair geometries with large electrostatic energies. Consequently, the triple ionic structure $X:^-H^+:X^-$ in a proton transfer process will usually possess a deep energy minimum [75]. In some instances it may become the dominant VB structure of the reaction and give rise to 'low-barrier hydrogen bonds', a topic that has been associated with controversies in enzyme catalysis [76]. An analysis of the case of the $(FHF)^-$ anion, which is a stable symmetric hydrogen bond, can illustrate the importance of the ionic-covalent crossing in this and analogous cases. It should be emphasized that most other hydrogen bonded dihalide anions are nearly symmetric and feature double well minima separated by a tiny barrier for the proton transfer [11,75,77].

Fig. 23.11a depicts the HL and ionic structures for a proton transfer process between bases, $X:^-$ which have moderate or low stability as anions (e.g. carbanions with significant pK_a). In such a case, the ionic structure lies above the HL state (the mixture of the two HL structures), and the avoided crossing leads generally to a single transition state separating the hydrogen bonded clusters. Nevertheless, the ionic structure is seen to have a deep minimum near the crossing point of the HL curves, and as such the transition state will be expected to possess a significant triple ion character. As the anion $X:^-$ gets increasingly more stable, so will the ionic structure descend more and more in energy and may dominate the region near the transition state. This is seen in Fig. 23.11b that depicts the computed [11,75,78] VB configurations for the F^- exchange along the reaction coordinate, at the BOVB/6-31G* level. At the diagram onset, the ionic structure lies

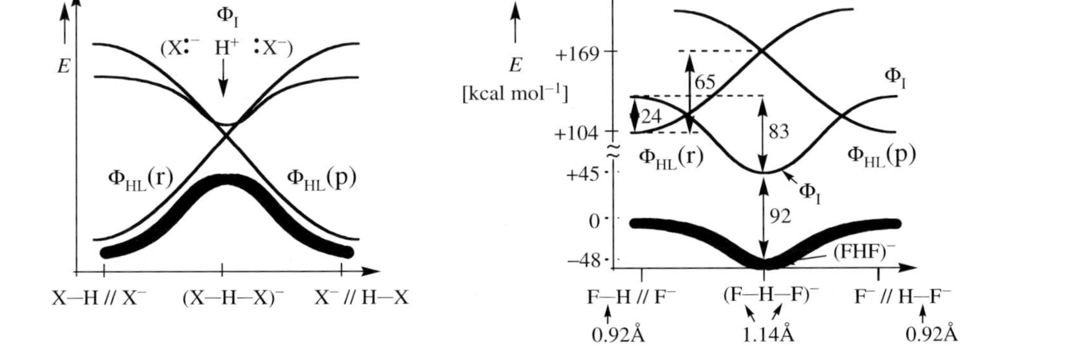

Fig. 23.11. VBCMDs, like Fig. 23.1b, for proton transfer between X^- bases: (a) The X^- base is not very stable (strong base). (b) The X^- base is F^- (adapted from Ref. [11] with permission of Wiley-VCH STM-Copyright and Licenses).

above the HL structure by a moderate energy gap of only *ca.* 24 kcal/mol. However, at the symmetric F–H–F geometry, the lowest VB curve is the ionic structure that undergoes 83 kcal/mol of stabilization relative to its onset at the reactant geometry. The origins of this remarkable stability of the ionic structure is, as already noted, the small size of H^+ that leads to short F^-H^+ distances at the cluster geometry, of $(FHF)^-$, and thereby to very large electrostatic stabilization. This electrostatic stability along with its low initial energy makes the ionic structure the dominant configuration at the cluster geometry.

The short H–F distance is associated with yet another outcome and this is the inception of very large resonance energy due to the mixing of the resonating HL-state (which by itself is a resonating mixture of the two HL structures) with the ionic structure. This ionic-covalent resonance energy is seen to be *ca.* 90 kcal/mol, which contributes a significant fraction to the bonding in $(FHF)^-$. Thus, the symmetric hydrogen bonded species is neither fully ionic nor fully covalent, it is virtually a resonating mixture of the two structures [11].

The question whether or not the symmetric $(FHF)^-$ species will be a minimum on the adiabatic (bold) curve, is a question of balance between the difference in electrostatic stabilization and ionic-covalent resonance energies at the cluster geometry relative to the reactant and product geometries. It is seen, from Fig. 23.11b, that the ionic-covalent resonance energy is largest at the reactant/product geometries. It follows, therefore, that the crucial factor why the symmetric $(FHF)^-$ species is so stable, is the electrostatic stabilization that lowers the ionic structure well below the onset of the HL structures. It is this difference that causes the final adiabatic state-profile (in bold) to retain the shape of the ionic curve, and to exhibit a minimum. The relatively small size of the F^- anion is also important for the electrostatic stabilization, and we may expect that, as the anion increases in size (e.g. I^-) or becomes delocalized (e.g. aspartate, etc.), the intrinsic stabilization of the ionic structure at the cluster geometry will decrease, and the symmetric geometry may cease to be a minimum of the energy profile [11,75].

Finally, the impact of the ionic structure is fleshed out by comparison of $(FHF)^-$ with the corresponding radical species, $(FHF)^{\cdot}$. Thus, with one electron less, in the $(FHF)^{\cdot}$ species, the triple ionic structure is replaced by the $F^-H^{+\cdot}F$ structures which loses at least half of the electrostatic stabilization, and therefore, rises above the HL curves. This loss has a tremendous impact on the reaction profile, and the $>$40 kcal/mol energy well of $(FHF)^-$ becomes an $(FHF)^{\cdot}$ transition state *ca.* 18 kcal/mol [21,79] above the reactants. For the same reason, it is expected, therefore, that $(XHX)^{\cdot}$ species will generally be transition states for the hydrogen abstraction process with a barrier significantly larger than the corresponding proton transfer process *via* the $(XHX)^-$ species. Experimental data show that this is indeed the case [80].

23.4.2.2 Nucleophilic substitution on silicon—stable hypercoordinated species

Another demonstration of the role of ionic structures is the nucleophilic substitution on Si, which proceeds *via* pentacoordinated intermediates [81,82], in contrast to the situation in carbon where the pentacoordinated species is a transition state. Recently, Lauvergnat et al. [83], Shurki et al. [84], Sini et al. [85], and Shaik et al. [86] have performed BOVB/6-31G* (and a few other basis sets) calculations for a C–X and Si–X bonds (X = H, F, Cl) and made an interesting observation that *the minimum of the ionic curve* *for Si^+X^- is significantly shorter than the corresponding minimum for C^+X^-*. By contrast, the minima of the HL curves for Si–X were found to be longer than the corresponding C–X minima. Since X is common for the two bonds, these differences mean that while the covalent radius of Si is longer than that of C, the opposite is true for the ionic radii; thus the ionic radius of CH_3^+ was determined as 0.64 Å, compared with 0.35 Å for SiH_3^+ [83]. It was concluded that the origins of these effective ionic sizes is the charge distribution of the corresponding ions. In CH_3^+ and generally in CL_3^+ (L, a ligand) the charge is distributed over the ligands, while the central carbon possesses a relatively small positive charge [84]. Consequently, the minimum distance of approach of an anion X^- toward CL_3^+ will be relatively long and the electrostatic energy will be small. In contrast, in SiL_3^+, the charge is localized on Si [84], and consequently the minimum distance of approach of an X^- anion will be relatively short and the electrostatic stabilization large. Indeed, the depth of the ionic curve $H_3Si^+X^-$ was found to exceed the depth of $H_3C^+X^-$, by >50 kcal/mol [83,84]. In conclusion, therefore, the silicenium ion L_3Si^+ is expected to behave more like the small proton, whereas the corresponding carbenium ion CL_3^+ will be bulky.

Based on these differences, it is possible to represent the VBCMDs for typical nucleophilic substitution reactions on Si *vs.* C as shown in Fig. 23.12a and b. In Fig. 23.12a, the ionic curve is very stable in the pentacoordinated geometry due to the electrostatic energy of the triple ion structure, much like the case of the $(FHF)^-$ species discussed before. Consequently, the pentacoordinated silicon species will generally be a stable entity, which depending on the energy cost, may or may not mediate the exchange of the anions. By the same analogy, to the $(FHF)^-$ species, the pentacoordinated silicon species will be neither ionic nor covalent, but rather a resonating mixture of the two structures with bonding augmented by significant ionic-covalent resonance.

Fig. 23.12b shows the typical situation for the carbon analog where the ionic structure is slightly higher in energy, relative to the HL-state at the crossing point. The consequent VB mixing leads to a single step reaction with a pentacoordinated transition state.

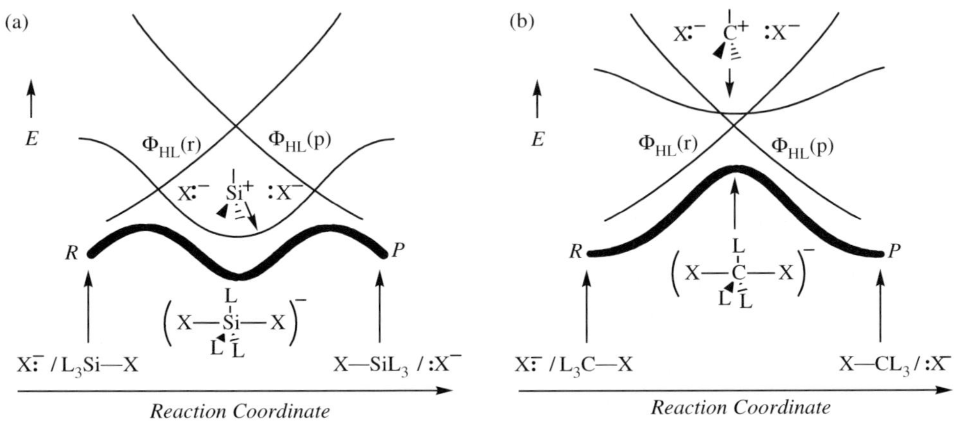

Fig. 23.12. VBCMDs, like in Fig. 23.1b, for nucleophilic displacements on (a) silicon and (b) carbon (adapted from Ref. [11] with permission of Wiley-VCH STM-Copyright and Licenses).

In a solvent, we might expect that in some cases the ionic structure will cross slightly below the HL-state and give rise to a transient triple-ion intermediate species (or will proceed by the related S_N1 mechanisms). A more stable pentacoordinated carbon may be envisioned if the reaction is conducted in the presence of a charge template that will stabilize the ionic ($X^- L_3C^+X^-$) structure well below the crossing point of the HL structures. However, even in such an event the pentacoordinated species of carbon will be significantly different than the corresponding silicon species. Thus, the larger size of the CL_3^+ species will lead to smaller ionic-covalent resonance energy compared with the silicon analog [83]. Consequently, should a pentacoordinated intermediate of carbon become stable, it will generally be highly ionic. A study of ion-pair S_N2 reactions by Harder et al. [87] shows that the presence of a single Li^+ cation or of an XLi_2^+ moiety makes the transition states virtually ionic (but still not as a stable species). One wonders if a stable pentacoordinated carbon can be stabilized in a charged template.

23.4.3 VBCMD with intermediates nascent from 'foreign states'

Every reaction system possesses, in addition to the promoted excited states, R^* and P^*, which are localized in the active bonds, numerous 'foreign' excited states which involve electronic excitations in orbitals and bonds that do not belong to the active bonds [11,13]. Some of these 'foreign' states are high in energy, but some which are of low energy can become accessible along the reaction coordinate. As already stated, mixing of foreign states provides means by which complex molecules find low-energy pathways for otherwise difficult transformations. To elucidate this mechanistic feature of the 'foreign' states, we have chosen two novel mechanisms in which the 'foreign' state plays a cardinal role. Others can be found in a recent review [11].

23.4.3.1 The $S_{RN}2$ and $S_{RN}2^c$ mechanisms

Foreign excited states are capable of creating novel reaction mechanisms. A case in point is the recent proposition of new nucleophilic substitution mechanisms, shown in Fig. 23.13. These mechanisms were termed $S_{RN}2$ and $S_{RN}2^c$, by Zipse as part of

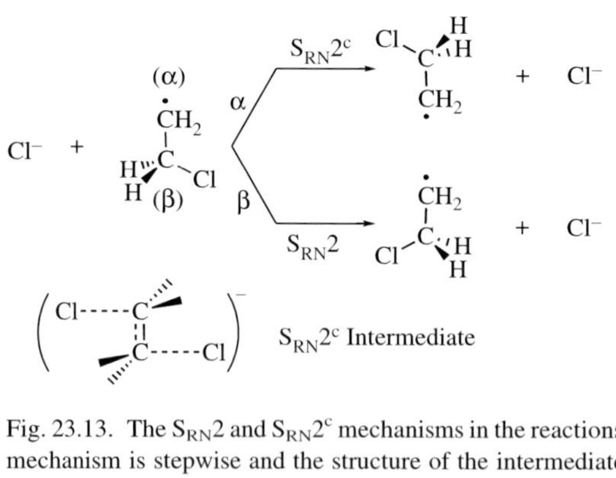

Fig. 23.13. The $S_{RN}2$ and $S_{RN}2^c$ mechanisms in the reactions between Cl^- and β-chloroethyl radical. The $S_{RN}2^c$ mechanism is stepwise and the structure of the intermediate is shown at the bottom of the figure.

the strategy he conceived of catalyzing rates of classical reactions by the presence of a radical centre adjacent to the reaction centers [88–90]. As illustrated in Fig. 23.13, for the identity reactions of β-chloro ethyl radical, these mechanisms involve a Cl^- exchange *via* attack on either the β or α positions of the radical. It was found by Zipse [88] that the $S_{RN}2$ process occurs in a single step reaction with a transition state which is very similar to the corresponding S_N2 transition state for the Cl^-/ethyl chloride reaction. At the same time, the $S_{RN}2$ barrier was shown to be lower by *ca.* 11 kcal/mol in comparison with the S_N2 barrier. Even more intriguing are the findings by Zipse [89], triggered by application of the VBCMD model, which revealed that the $S_{RN}2^c$ mechanism proceeds in fact in a stepwise manner *via* a C_{2h} intermediate which is *ca.* 3 kcal/mol lower than the transition state of the $S_{RN}2$ mechanism. Thus, the adjacent radical center on the one hand, lowers considerably the barriers for the Cl^- exchange reaction, and on the other, leads to a novel intermediate species.

Fig. 23.14 shows the VBCMDs for the two mechanisms. In both diagrams there exist two fundamental curves identical to those of the classical S_N2 reaction (Fig. 23.7), and a low-lying intermediate curve in which the C_2H_4 moiety is π-bonded. According to the calculations of Zipse [89], the intermediate curve is low lying relative to the fundamental curves. It is the mixing of this intermediate structure into the fundamental curves that accounts for the much lower energetic of the $S_{RN}2$, and $S_{RN}2^c$ mechanisms in comparison with S_N2.

The difference between the $S_{RN}2$, and $S_{RN}2^c$ mechanisms is rooted in the relationship between the intermediate structure and the fundamental curves. The placement of the

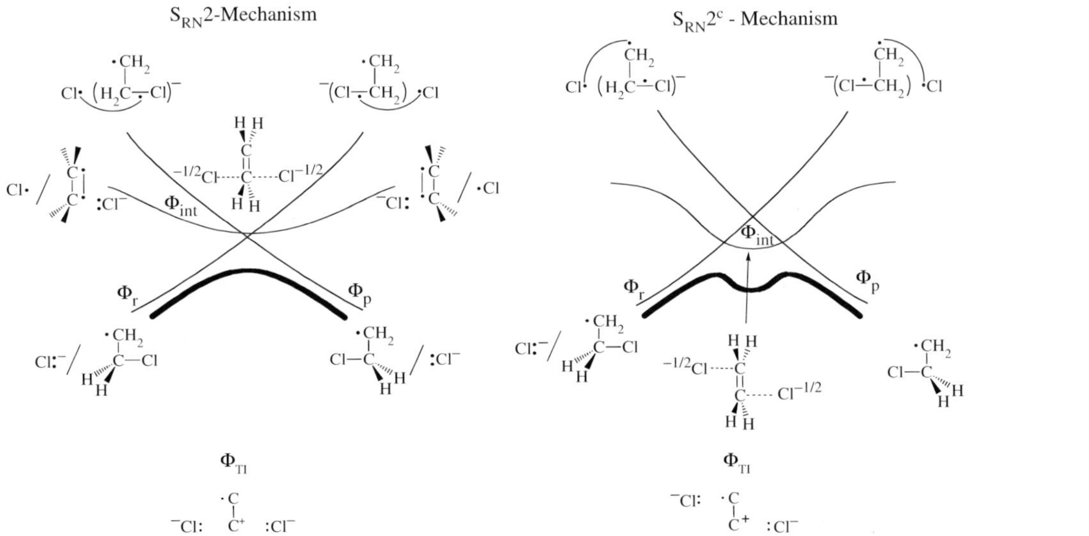

Fig. 23.14. VBCMDs, like in Fig. 23.1b, for the $S_{RN}2$ and $S_{RN}2^c$ mechanisms. Below the VBCMDs we show the respective triple-ionic configuration (Φ_{TI}) that affect the height of the crossing point of the reactant and product curves (adapted from Ref. [11] with permission of Wiley-VCH STM-Copyright and Licenses). The three electron bonds in the charge transfer states are indicated here by a combination of a line and a dot.

intermediate curve, in Fig. 23.14, is based on spin and charge density analysis of Zipse [88,89]. Thus, the crossing point of the fundamental curves is lower for the $S_{RN}2$ mechanism, because the triple ion structure (Φ_{TI}), which mixes into the two principal curves (drawn below the diagram), in Fig. 23.14, can optimize better the electrostatic interactions in comparison with the corresponding Φ_{TI} structure of the $S_{RN}2^c$ mechanism in Fig. 23.14. This effect is enhanced by the somewhat preferred intrinsic stability of the intermediate structure in the $S_{RN}2^c$ case (smaller repulsion between negative charges). Consequently, the crossing point of the fundamental curves for the $S_{RN}2$ mechanism is lower than the intermediate state as shown in the left-hand diagram in Fig. 23.14, while the opposite is true for the $S_{RN}2^c$ mechanism, as shown in right-hand diagram. Indeed, based on the computed [89] C–C distance in the C_2H_4 moiety, the charge development on this moiety, and on the spin density development on the chlorine moieties, it is apparent that the intermediate state character is more dominant in the $S_{RN}2^c$ mechanism. Thus, $S_{RN}2^c$ is *a stepwise mechanism mediated by a low-energy state due to strong electronic coupling with the accessory radical centre*. It appears then, that a radical centre adjacent to the reaction centre is a novel strategy to generate low energy pathways *via* intermediate states [88,90]. Such a process may transpire in damage mechanisms of DNA bases [91].

23.5 ADDITIONAL APPLICATIONS OF VB DIAGRAMS

23.5.1 VBSCD: A general model for electronic delocalization

TheVBSCD serves as a model for elucidating trends of electronic delocalization in isoelectronic series. Consider, for example, the following bond exchange process, between monovalent atoms. The process passes through an $X_3^{\cdot}$ species in which three electrons are delocalized over three centers. This species may be a transition state or a stable cluster.

$$X^{\cdot} + X{-}X \rightarrow [X_3^{\cdot}] \rightarrow X{-}X + {}^{\cdot}X \qquad (17)$$

We can imagine a variety of such species, e.g. X = H, F, Cl, Li, Na, Cu, etc. The question is: when do we expect the $X_3^{\cdot}$ species to be a transition state for the exchange process, and when will it be a stable cluster, an intermediate en route to exchange? The answer to this question comes from the VBSCD model, that describes all these process in a single diagram where G is given by Eq. (8), i.e. $G \approx 0.75\Delta E_{ST}(X{-}X)$. Thus, as shown in Fig. 23.15, a very large triplet promotion energy for X = H (250 kcal/mol) results in an $H_3^{\cdot}$ transition state, while the small promotion energy for X = Li (32 kcal/mol) results in a stable $Li_3^{\cdot}$ cluster. The VB computations of Maître et al. [22] in Fig. 23.15 show that, as the promotion gap drops drastically, the avoided crossing state changes its status from a transition state for $H_3^{\cdot}$ to a stable cluster for $Li_3^{\cdot}$. In fact, this transformation in the nature of the $X_3^{\cdot}$ species was predicted by a semiquantitative application of the VBSCD barrier expression [23], using experimental ΔE_{ST} and bond energy data.

The spectacular relationship between the nature of the $X_3^{\cdot}$ species and the promotion energy shows that the VBSCD is in fact a general and successful model for

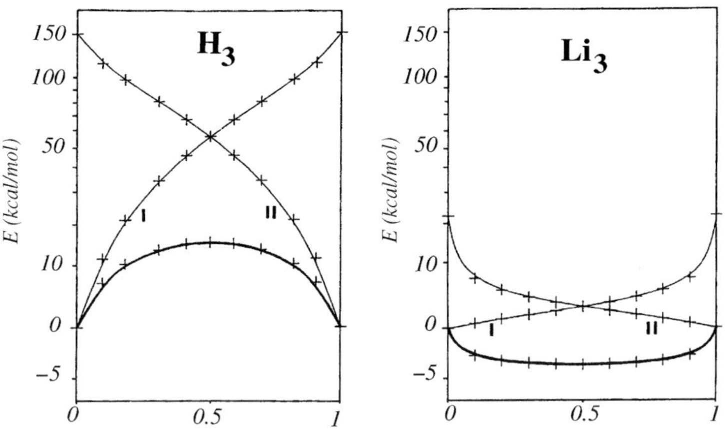

Fig. 23.15. *Ab initio* computations of VBSCDs for the exchange processes, $X^{\cdot} + X{-}X \rightarrow X{-}X + {}^{\cdot}X$, for X = H and Li (adapted with permission from Ref. [22], ©1990, American Chemical Society). The reaction coordinate is $0.5(n_1 - n_2 + 1)$, where n_1 and n_2 are the X–X bond orders in X–X–X. The computations were based on a VBCI type method with single excitations, using the 6-31G** basis set for H_3 and 6-31G for Li_3.

the pseudo-Jahn—Teller effect (PJTE). Note that a qualitative application of PJTE would predict all the $X_3^{\cdot}$ species to be transition state structures that relax to the distorted $X^{\cdot}\cdots X{-}X$ and $X{-}X\cdots{}^{\cdot}X$ entities. By contrast, the VBSCD makes a distinction between strong binders that form transition states and weak binders that form stable bond-symmetric intermediate clusters. Thus, the VBSCD model is in tune with the general observation that as one moves in the Periodic Table from strong binders to weak ones (e.g. metallic) the stable molecular entity changes from a discrete molecule to an extended delocalized structure.

The variable nature of the $X_3^{\cdot}$ species in the isoelectronic series forms a general model for electronic delocalization, enabling one to classify the species either as distortive or as stable. Using the isoelectronic analogy, one might naturally ask about the isoelectronic π-species in allyl radical; does it behave by itself like $H_3^{\cdot}$ or like $Li_3^{\cdot}$? Moreover, the same transformation displayed for $X_3^{\cdot}$ in Fig. 23.15 can be shown for X_3^+, X_3^-, X_4 and X_6 species [29]. Likewise one might wonder about the status of the corresponding isoelectronic π-components in allyl cation, anion, cyclobutadiene or benzene. These questions were answered in detail elsewhere and the reader is advised to consult a recent review [29].

23.5.2 VBSCD: The twin-state concept and its link to photochemical reactivity

The pioneering work of van Der Lugt and Oosterhoff [92] and Michl [93] highlighted the importance of avoided crossing regions as decay channels in photochemistry. Köppel and coworkers [94,95] showed that conical intersections are the most efficient decay channels, from excited states to ground states. Indeed, Robb and his coworkers [96,97] have amply demonstrated this role of conical intersections in organic photochemistry using *ab initio* calculations that are now routinely performed with the GAUSSIAN package.

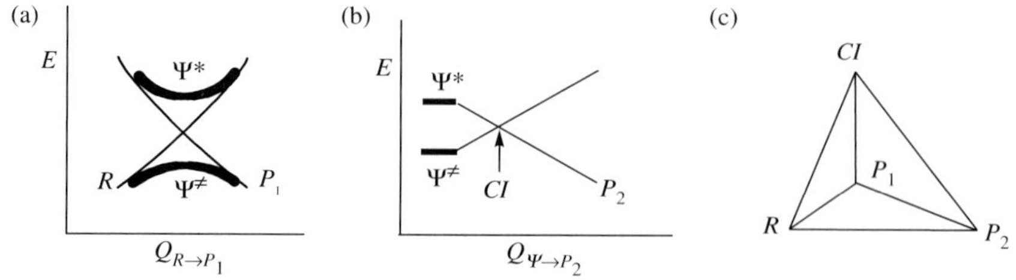

Fig. 23.16. (a)VBSCD showing the twin-states ($\Psi^{\neq}$ and Ψ^{*}) formed by avoided crossing along the reaction coordinate ($Q_{R \to P_1}$) of the thermal reaction leading to product P_1. (b) Crossover of the twin-states along a coordinate that stabilizes the twin-excited state and generates to generate a conical intersection (CI) that leads to product P_2. (c) The conical intersection will be anchored in three minima (or more): reactants (R), P_1 and P_2 (adapted from Ref. [52] with permission of Wiley, ©2004).

Bernardi et al. [96] have further shown that VB notions can be useful to rationalize the location of conical intersections.

As was subsequently argued by Shaik and Reddy [30], the VBSCD is a portable model for discussing the relation between thermal and photochemical reactions and between the avoided crossing region and a conical intersection. Thus, the avoided crossing region of the VBSCD leads to the twin-states $\Psi^{\neq}$ and Ψ^{*} (Fig. 23.16); the first corresponds to the resonant state of the VB configurations and the second to the antibonding, i.e. antiresonant state [11]. Since the extent of this VB mixing depends on geometry, there should exist in principle, a specific distortion mode that converts the avoided crossing region into a conical intersection where the twin-states $\Psi^{\neq}$ and Ψ^{*} become degenerate, and thereby enable the excited reaction complex to decay into the ground state surface. As such, the conical intersection will usually be anchored at three structures [97]; two of them are the reactant (R) and product (P_1) of the thermal reaction, and the third is the product (P_2) generated by the distortion mode that causes the degeneracy of the twin-states $\Psi^{\neq}$ and Ψ^{*}. The new product would, therefore, be characteristic of the distortion mode that is required to convert the avoided crossing region into a conical intersection. In those cases where most of the excited species roll down eventually to the Ψ^{*} funnel, the major photoproduct would be P_2. If, however, there exist other excited state funnels near the twin-excited state, Ψ^{*}, other products will be formed, which are characteristic of these other excited states and can be predicted in a similar manner provided one knows the identity of these excited states. Our following description is restricted to the analysis of the twin-excited states.

An instructive, albeit trivial, example is the hydrogen exchange reaction, $H_a - H_b + H_c \rightarrow H_a + H_b - H_c$ where the thermal transition state has a collinear geometry, $H_a - H_b - H_c$. In the linear structure, the ground state $\Psi^{\neq}$ is the resonating combination of R and P, and it constitutes the transition state for the thermal reaction, while the twin-excited state Ψ^{*} is the corresponding antiresonant combination that forms the photochemical funnel:

$$\Psi^{\neq} = R - P,$$
$$R = |a\bar{b}c| - |\bar{a}bc|, \qquad P = |ab\bar{c}| - |a\bar{b}c| \tag{18}$$

$$\Psi^* = \mathrm{R} + \mathrm{P} = |\mathrm{ab\bar{c}}| - |\mathrm{\bar{a}bc}| \qquad (19)$$

where the orbitals a, b and c belong to H_a, H_b and H_c, respectively.

It is clear from Eq. (19) that Ψ^* involves a bonding interaction [28,52] between H_a and H_c and will be lowered by the bending mode that brings H_a and H_c together. Furthermore, the expression for the avoided crossing interaction B (Eq. (20)), based on qualitative VB theory [28], shows that this quantity will shrink to zero in an equilateral triangular

$$B = \langle \mathrm{R}|\mathrm{H}^{\mathrm{eff}}|\mathrm{P}\rangle = -2\beta_{\mathrm{ab}}S_{\mathrm{ab}} - 2\beta_{\mathrm{bc}}S_{\mathrm{bc}} + 4\beta_{\mathrm{ac}}S_{\mathrm{ac}} \qquad (20)$$

structure where the H_a-H_c, H_a-H_b and H_b-H_c interactions are identical. As such, the equilateral triangle will define a conical intersection with a doubly degenerate state, in the crossing point of the VBSCD. This D_{3h} structure will relax to the isosceles triangle with short H_a-H_c distance that will give rise to a 'new' product $H_b + H_a - H_c$. The photocyclization of allyl radical to cyclopropyl radical is precisely analogous. The ground state of allyl is the resonating combination of the two Kekulé structures, while the twin-excited state, Ψ^*, is their antiresonant combination with the long bond between the allylic terminals [28]. As such, rotation of the two allylic terminals will lower Ψ^*, raise the ground state, and establish a conical intersection that will channel the photoexcited complex to the cyclopropyl radical, and *vice versa*. This structural dichotomy of the resonant and antiresonnt states in the VBSCD accounts for the thermal–photochemical dichotomy as stated in the Woodward–Hoffmann rules [98], and as amply observed.

The photostimulation of S_N2 systems such as $X^- + A - Y$ (A = alkyl) is a nice example for the utility of the VBSCD. This photoreaction was analyzed before, using the VBSCD model, for predicting the location of conical intersections [30]. Here, the transition state for the thermal reaction is the collinear $[X-A-Y]^-$ structure, which is the $\Psi^{\neq}(A')$ resonating combination of the two Lewis structures, while the twin-excited state, $\Psi^*(A'')$, is their antiresonating combination; the symmetry labels refer to C_s symmetry. This latter excited state is readily written as an A'' symmetry-adapted combination of Lewis structures, Eq. (21):

$$\Psi^* = (|\mathrm{x\bar{x}a\bar{y}}| - |\mathrm{x\bar{x}\bar{a}y}|) - (|\mathrm{y\bar{y}x\bar{a}}| - |\mathrm{y\bar{y}\bar{x}a}|) \qquad (21)$$

where the orbitals x, a and y belong to the fragments X, A and Y, respectively.

Rearranging Eq. (21) to Eq. (22) reveals that the two terms of the wave function involve a stabilizing three-electron bonding interaction between X and Y, of the type $(X{:}^{-}\dot{}Y \leftrightarrow X\dot{}{:}Y^{-})$:

$$\Psi^* = (|\mathrm{\bar{x}x\bar{y}a}| + |\mathrm{\bar{x}y\bar{y}a}|) + (|\mathrm{x\bar{x}y\bar{a}}| + |\mathrm{x\bar{y}y\bar{a}}|) \qquad (22)$$

As such, the bending mode that brings the X and Y fragments together destabilizes the $[X-A-Y]^-$ structure and stabilizes the twin-excited state, until they establish a conical intersection that correlates down to $X\therefore Y^-$ and $R^{\cdot}$, as shown in Fig. 23.17. This analysis is supported by experimental observation that the irradiation of the I^-/CH_3I cluster at the charge transfer band leads to I_2^- and $CH_3^{\cdot}$, while for or I^-/CH_3Br such excitation generates IBr^- and $CH_3^{\cdot}$ [99].

The presence of excited state minima above the thermal transition state is well known [5,92,93,96,100,101]. The VBSCD model merely characterizes this phenomenon using

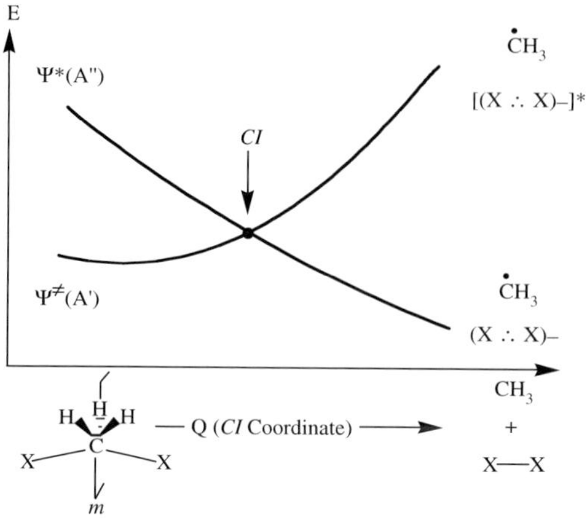

Fig. 23.17. Generation of a conical intersection (CI) by crossing of the twin-states along the bending distortion mode, in S_N2 systems. Symmetry labels refer to the mirror plane m. The products of these processes are the $CH_3^{\cdot}$ radical and the $X_2^{\cdot-}$ radical anion (adapted from Ref. [52] with permission of Wiley, ©2004).

a lucid mechanism of avoided crossing of VB structures; the model is portable and enables one to make systematic predictions. Other important applications of the twin-states concern the possibility of probing the twin-excited state that lies directly above the transition state of a thermal reaction. Thus, *any chemical reaction will possess a transition state,* $\Psi^{\neq}$ *and a twin-excited state,* Ψ^{*} [5,101]. For some cases, the twin-excited state should be stable and hence observable; its geometry is approximately coincident with the thermal transition state and its electronic state symmetry is identical to the symmetry of the reaction coordinate of the ground state process [102], namely

$$\Gamma(\Psi^{*}) = \Gamma(Q_{RC}) \tag{23}$$

In addition, the twin-excited state should possess a real and greatly increased frequency of the reaction coordinate mode. Thus, since the twin-pair has coincident geometry, a spectroscopic characterization of Ψ^{*} will provide complementary information on the transition state $\Psi^{\neq}$ and will enable resolution of the transition state structure.

As a proof of principle, we show in Fig. 23.18 the twin-states characterized by Zilberg et al. [102] for the semibullvalene rearrangement. It is seen that the geometries of the two states are similar, and the twin-excited state possesses B_2 symmetry, identical to the symmetry of the reaction coordinate for the thermal process. Furthermore, the transition state mode, b_2, which is imaginary for $\Psi^{\neq}(A_1)$ becomes real for $\Psi^{*}(B_2)$ [102]. While offering a proof of principle, these calculations also match some intriguing experimental findings of Quast et al. [103]. Thus, Quast [103,104] has designed semibullvalene and barbaralene derivatives in which the barrier for the rearrangement could be lowered drastically, to a point where it almost vanishes. He and his coworkers [103,104] found that these molecules exhibit thermochromism without having a chromophore; they are colorless at low temperatures but highly colored at 380 K. According to Quast, the thermochromism is a result of the low-energy transition from the transition state ($\Psi^{\neq}$)

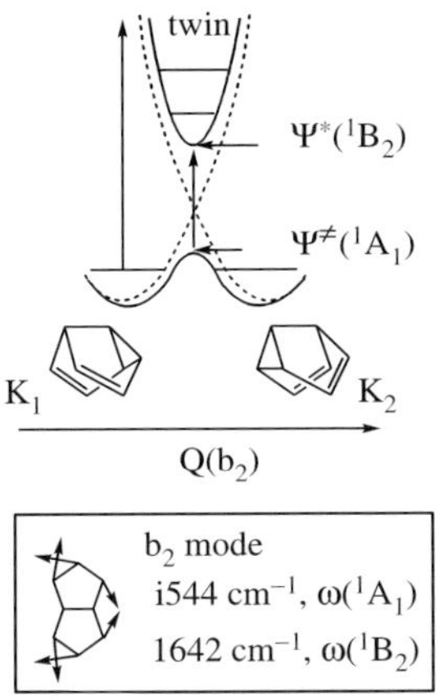

Fig. 23.18. The twin-states along the b_2 reaction coordinate for the semibullvalene rearrangement. When the thermal barrier is not much higher than the zero point levels of the two isomers, the transition state ($\Psi^{\neq}$) region becomes available thermally. Absorption in the transition state region is in the visible, leading to thermochromism at elevated temperature (adapted with permission from Ref. [29], Copyright 2001, American Chemical Society).

to the twin excited state (Ψ^*), Fig. 23.18. Thus, since the thermal barrier is low, at elevated temperatures the transition state becomes thermally populated. Since the $\Psi^{\neq}-\Psi^*$ gap is small, one observes color due to absorption within the visible region. However, at lower temperatures, the molecules reside at the bottom of the reactant/product wells, where the gap between the ground and excited state is large and hence, the absorption is in the UV region and the color is lost. To quote Quast, "thermochromic … semibullvalene allows the observation of transition states even with one's naked eye " [104]. Of course, identifying appropriate systems where the twin-excited state is observable is required for the eventual 'observation' of the transition states of thermal reactions.

Coherent control [105] is a new powerful method that makes use of the availability of the twin-excited state to control the course of chemical reactions by laser excitation. Thus, laser excitation from $\Psi^{\neq}$ to Ψ^* (Fig. 23.16a), using two different and complementary photons, causes the decay of Ψ^* to occur in a controlled manner either to the reactant or products. In the case where the reactants and products are two enantiomers, the twin-excited state is achiral, and the coherent control approach leads to chiral resolution [105].

In summary, the twin-excited state plays an important role in photochemistry as well as in thermal chemistry.

23.6 PROSPECTIVE

Valence bond (VB) theory is a new-old theory. Until the mid-1950s, it has dominated the mental map of chemists. Then it fell into disrepute, the mental map was recharted, and the wisdom of VB theory has been written over by other theories [52,106–109].

In this sense VB theory is old, though in the field of molecular dynamics it never really ceased to be applied [110,111]. By the early 1970s and onwards, VB theory has started to enjoy a renaissance, [52,112] in terms of new methodologies [40] and ideas. There is a new beginning in many respects, and in this sense VB is a new or renewed theory. This chapter describes a new-generation VB paradigm [5], so-called the VB diagram model that is summarized by Fig. 23.1a and b, and can form a basis for a wide-ranging system of thought about chemical reactivity. Applications of the model are discussed in this chapter, but many more exist in previous and recent reviews of the topic [7,8–14,28,31]. One of the essential features of the model is that it combines lucidity and relative simplicity with quantum mechanical rigor. Indeed, the emergence of new VB methodologies [40] strengthens the model by providing a quantitative description of VB diagrams, as may be deduced from some recent applications [15–24,39,50].

The model lends itself to articulation of new ideas and new applications. Some potential applications are mentioned above, e.g. the concept of 'twin-states' [5,11,102] and its applications to photochemistry, etc., the concept of catalysis by spin crossover [35] and its application to bond activation processes, the application to problems of electron delocalization [29], and so on. Other articulations of the diagram serve to solve chemical puzzles, such as the recent applications [53] to dissociation of alkoxyradicals, to the invention of a new mechanism, $S_{RN}2^{c}$ and its wide range of applications [88–90,113], to transition metal catalyzed reactions [57–59], to the concept of entangled mechanisms [11], and so on. The future acceptance of the VB diagram model as a general reactivity paradigm depends on such articulations.

23.7 ACKNOWLEDGEMENTS

The research at the Hebrew University is supported in part by a grant (to SS) from the Israeli Science Foundation. The two authors are thankful to all their coworkers during their years of collaboration (1985–present). SS is especially thankful to A. Pross for the intense collaboration in the period 1981–1985.

APPENDIX 23A: COMPUTING MONO-DETERMINANT VB WAVE FUNCTIONS WITH STANDARD *AB INITIO* PROGRAMS

This technique utilizes a possibility that is offered by most *ab initio* standard programs to compute the energy of the guess function even if it is made of non-orthogonal orbitals. The technique orthogonalizes the orbitals without changing the Slater determinant, then computes the expectation energy by use of Slater's rules. In the course of the subsequent optimization of the Hartree–Fock orbitals, this expectation value of the energy appears as the energy at iteration zero. If the guess determinant is made of localized bonding orbitals that typify a given VB structure, then the expectation energy of this wavefunction at iteration zero defines the energy of this VB structure. Practically, the localized bonding orbitals that are used to construct the guess determinant can be determined by the most convenient means. For example, a Kekulé structure of benzene will display a set of three two-centered π-bonding MOs that can arise from the Hartree–Fock calculation of an ethylene molecule [114,115]. In a VBSCD calculation, the energy of the crossing point will be the energy of a guess function made of the orbitals of the reactants, but in the geometry of the transition state, without further orbital optimization. The zero-iteration technique has also been used to estimate the energy of spin-alternated determinants (quasi-classical state) [53,116–118].

23.8 REFERENCES

1 For a summary of concepts of physical organic chemistry, see: S.S. Shaik, H.B. Schlegel and S. Wolfe, Theoretical aspects of physical organic chemistry, The S_N2 mechanism, Wiley, New York, 1992, pp. 1–41.
2 T.A. Albright, J.K. Burdett and M.H. Whangbo, Orbital interactions in chemistry, Wiley, New York, 1985.
3 L. Eberson, Electron transfer reactions in organic chemistry, Springer, Heidelberg, 1987.
4 R.G. Pearson, in: Z.B. Maksic (Ed.), Theoretical concepts of chemical bonding, Springer, Heidelberg, 1990, pp. 45–76, Part 2.
5 S.S. Shaik, J. Am. Chem. Soc., 103 (1981) 3692.
6 R.B. Woodward and R. Hoffmann, Angew. Chem., 81 (1969) 797; Angew. Chem. Int. Ed. Engl., 8 (1969) 781.
7 S. Shaik and P.C. Hiberty, Adv. Quant. Chem., 26 (1995) 99.
8 A. Pross and S.S. Shaik, Acc. Chem. Res., 16 (1983) 363.
9 A. Pross, Theoretical and physical principles of organic reactivity, Wiley, New York, 1995, pp. 83–124; see also pp. 235–290.
10 S. Shaik and P.C. Hiberty, in: Z.B. Maksic (Ed.), Theoretical models of chemical bonding, Vol. 4, Springer, Berlin, 1991, pp. 269–322.
11 S. Shaik and A. Shurki, Angew. Chem. Int. Ed., 38 (1999) 586.
12 A. Pross, Acc. Chem. Res., 18 (1985) 212.
13 S.S. Shaik, Prog. Phys. Org. Chem., 15 (1985) 197.
14 An exhaustive treatment of S_N2 reactivity is given in Ref. [1], Chapters 3–6.
15 G. Sini, S.S. Shaik, J.-M. Lefour, G. Ohanessian and P.C. Hiberty, J. Phys. Chem., 93 (1989) 5661.
16 G. Sini, S. Shaik and P.C. Hiberty, J. Chem. Soc. Perkin Trans., 2 (1992) 1019.
17 J. Mestres and P.C. Hiberty, New J. Chem., 20 (1996) 1213.
18 G. Sini, G. Ohanessian, P.C. Hiberty and S.S. Shaik, J. Am. Chem. Soc., 112 (1990) 1407.
19 G. Sini and P.C. Hiberty, J. Chem. Soc. Chem. Commun., (1989) 772.
20 P. Maître, F. Volatron, P.C. Hiberty and S.S. Shaik, Inorg. Chem., 29 (1990) 3047.
21 P.R. Benneyworth, G.G. Balint-Kurti, M.J. Davis and I.H. Williams, J. Phys. Chem., 96 (1992) 4346.
22 P. Maître, P.C. Hiberty, G. Ohanessian and S.S. Shaik, J. Phys. Chem., 94 (1990) 4089.
23 S. Shaik, W. Wu, K. Dong, L. Song and P.C. Hiberty, J. Phys. Chem. A, 105 (2001) 8226.
24 L. Song, W. Wu, K. Dong, P.C. Hiberty and S. Shaik, J. Phys. Chem. A, 106 (2002) 11361.
25 F. London, Z. Elektrochem., 35 (1929) 552.
26 H. Eyring and M. Polanyi, Z. Phys. Chem., B12 (1931) 279.
27 M.G. Evans, Trans. Faraday Soc., 35 (1939) 824.
28 S. Shaik, in: J. Bertran, I.G. Csizmadia (Eds.), New theoretical concepts for understanding organic reactions, NATO ASI Series, C267, Kluwer, Dordrecht, 1989, pp. 165–217.
29 S. Shaik, A. Shurki, D. Danovich and P.C. Hiberty, Chem. Rev., 101 (2001) 1501.
30 S. Shaik and A.C. Reddy, J. Chem. Soc. Faraday Trans., 90 (1994) 1631.
31 S. Shaik, Acta Chem. Scand., 44 (1990) 205.
32 A. Pross, Adv. Phys. Org. Chem., 21 (1985) 99.
33 S. Shaik, J. Mol. Liq., 61 (1994) 49.
34 S. Shaik, in: P.v.R. Schleyer, N.L. Allinger, T. Clark, J. Gasteiger, P.A. Kollman, H.F. Schaefer, III (Eds.), Encyclopedia of computational chemistry, Vol. 5, Wiley, Chichester, 1998, pp. 3143–3156.

35 (a) S. Shaik, D. Danovich, A. Fiedler, D. Schröder and H. Schwarz, Helvetica Chem. Acta, 78 (1995) 1393; (b) D. Schröder, S. Shaik and H. Schwarz, Acc. Chem. Res., 33 (2002) 139.
36 K. Fukui, T. Yonezawa and H. Shingu, J. Chem. Phys., 20 (1952) 722.
37 S.S. Shaik, Nouv. J. Chim., 7 (1983) 201.
38 A. Pross and S.S. Shaik, Nouv. J. Chim., 13 (1989) 427.
39 L. Song, W. Wu, Q. Zhang and S. Shaik, J. Phys. Chem. A., 108 (2004) 6017.
40 (a) The BOVB and VBCI methods are incorporated in the XMVB package: L. Song, W. Wu, Y. Mo, Q. Zhang, XMVB, an ab initio nonorthogonal valence bond program; Xiamen University: Xiamen, China, 1999; (b) The VBSCF and BOVB methods are incorporated in the TURTLE program: J. Verbeek, J.H. Langerberg, C.P. Byrman, F. Dijkstra, J.H. van Lenthe, TURTLE: An ab initio VB/VBSCF program (199–2000); (c) A general purpose VB program is incorporated in VB2000: J. Li, R. McWeeny, VB2000: An ab initio valence bond program based on product function method and the algebrant algorithm., 2000; (d) The CRUNCH package incorporates multiconfigurational (MCVB) methods: G.A. Gallup, The _RUNCH Suite of Atomic and Molecular Structure Programs, 2001. gallup@phy-ggallup.unl.edu; (e) the Spin-Coupled method, incorporated in the MOLPRO package, is described in D.L.Cooper, J.P.Gerratt, M.Raimondi, Adv. Chem. Phys.69, (1987) 319.
41 (a) P.C. Hiberty, in: D.L. Cooper (Ed.), Valence bond theory, Elsevier, Amsterdam, 2002, pp. 187–226; (b) P.C. Hiberty and S. Shaik, Theor. Chem. Acc., 108, 2002, pp. 255.
42 L. Song, W. Wu, Q. Zhang and S. Shaik, J. Comput. Chem., 25 (2004) 472.
43 P. Su, L. Song, W. Wu, P.C. Hiberty and S. Shaik, J. Am. Chem. Soc., 126 (2004) 13539.
44 Y. Mo and S.D. Peyerimhoff, J. Chem. Phys., 109 (1998) 1687.
45 S.S. Shaik and E. Canadell, J. Am. Chem. Soc., 112 (1990) 1446.
46 (a) F. Jensen and P.-O. Norrby, Theor. Chim. Acc., 109 (2003) 1; (b) Y. Kim, J.C. Corchado, J. Villa, J. Xing and D.G. Truhlar, J. Chem. Phys., 112 (2000) 2718; (c) F. Jensen, Introduction to computational chemistry, Wiley, New York, 1999, pp. 48–49.
47 (a) A. Warshel and R.M. Weiss, J. Am. Chem. Soc., 102 (1980) 6218; (b) A. Warshel and S.T. Russell, Q. Rev. Biophys., 17 (1984) 282; (c) A. Warshel, J. Phys. Chem., 83 (1979) 1640.
48 H.J. Kim and J.T. Hynes, J. Am. Chem. Soc., 114 (1992) 10508.
49 S.S. Shaik, E. Duzy and A. Bartuv, J. Phys. Chem., 94 (1990) 6574.
50 L. Song, W. Wu, P.C. Hiberty and S. Shaik, Chem. Eur. J., 9 (2003) 4540.
51 W. Heitler and F. London, Zeits. für Physik., 44 (1927) 455.
52 S. Shaik and P.C. Hiberty, in: K.B. Lipkowitz, D.B. Boyd (Eds.), Reviews in computational chemistry, Vol. 20, 2004, pp. 1–100.
53 R. Méreau, M.T. Rayez, J.C. Rayez and P.C. Hiberty, Phys. Chem. Chem. Phys., 3 (2001) 3650.
54 D.J. Henry, M.L. Coote, R. Gomez-Balderas and L. Radom, J. Am. Chem. Soc., 126 (2004) 1732.
55 (a) H. Yamataka and S. Nagase, J. Org. Chem., 53 (1988) 3232; (b) A. Pross, H. Yamataka and S. Nagase, J. Phys. Org. Chem., 4 (1991) 135.
56 C. Amatore and A. Jutand, Acc. Chem. Res., 33 (2000) 314.
57 M. der Su and S.-Y. Chu, Inorg. Chem., 37 (1998) 3400.
58 M. der Su and S.-Y. Chu, J. Am. Chem. Soc., 119 (1997) 10178.
59 M. der Su, Inorg. Chem., 34 (1995) 3829.
60 S. Shaik, J. Am. Chem. Soc., 106 (1984) 1227.
61 S. Shaik, J. Org. Chem., 52 (1987) 1563.
62 I.M. Kovach, J.P. Elrod and R.L. Schowen, J. Am. Chem. Soc., 102 (1980) 7530.
63 D.G. Oakenfull, T. Riley and V. Gold, J. Chem. Soc. Chem. Commun., (1966) 385.
64 E. Buncel, S.S. Shaik, I.-H. Um and S. Wolfe, J. Am. Chem. Soc., 110 (1988) 1275.
65 S. Shaik, A.C. Reddy, A. Ioffe, J.P. Dinnocenzo, D. Danovich and J.K. Cho, J. Am. Chem. Soc., 117 (1995) 3205.
66 J.H. Incremona and C.J. Upton, J. Am. Chem. Soc., 94 (1972) 301.
67 C.J. Upton and J.H. Incremona, J. Org. Chem., 41 (1976) 523.
68 B.B. Jarvis, J. Org. Chem., 35 (1970) 924.
69 G.G. Maynes and D.E. Appliquist, J. Am. Chem. Soc., 95 (1973) 856.
70 K.J. Shea and P.S. Skell, J. Am. Chem. Soc., 95 (1973) 6728.

71 M.L. Poutsma, J. Am. Chem. Soc., 87 (1965) 4293.
72 S.S. Shaik and J.P. Dinnocenzo, J. Org. Chem., 55 (1990) 3434.
73 J.P. Dinnocenzo, W.P. Todd, T.R. Simpson and I.R. Gould, J. Am. Chem. Soc., 112 (1990) 2462.
74 L. Eberson, R. Gonzalez-Luque, M. Merchan, F. Radner, B.O. Roos and S. Shaik, J. Chem. Soc. Perkin Trans., 2 (1997) 463.
75 W. Wu, S. Shaik and W.H. Saunders, Jr., J. Phys. Chem. A., 106 (2002) 11616.
76 (a) A. Warshel, A. Papazyan and P.A. Kollman, Science, 269 (1995) 102; (b) W.W. Cleland and M.M. Kreevoy, Science, 269 (1995) 104; P.A. Frey, Science, 269 (1995) 104.
77 B.S. Ault, Acc. Chem. Res., 15 (1992) 103.
78 G. Sini, PhD Thesis, Université de Paris-Sud, Orsay, France, 1991.
79 S. Shaik, S.P. de Visser, W. Wu, L. Song and P.C. Hiberty, J. Phys. Chem. A, 106 (2002) 5043.
80 D.M. Neumark, Acc. Chem. Res., 26 (1993) 33.
81 Y. Apeloig, in: S. Patai, Z. Rappoport (Eds.), The chemistry of organic silicon compounds, Wiley, Chichester, 1989, Chapter 2.
82 (a) R.J.P. Corriu and M. Henner, J. Organometal. Chem., 74 (1974) 1; (b) A.E. Reed and P.v.R. Schleyer, J. Am. Chem. Soc., 112 (1990) 1434.
83 D. Lauvergnat, P.C. Hiberty, D. Danovich and S. Shaik, J. Phys. Chem., 100 (1996) 5715.
84 A. Shurki, P.C. Hiberty and S. Shaik, J. Am. Chem. Soc., 121 (1999) 822.
85 G. Sini, P. Maitre, P.C. Hiberty and S.S. Shaik, J. Mol. Struct. (Theochem), 229 (1991) 163.
86 S. Shaik, P. Maitre, G. Sini and P.C. Hiberty, J. Am. Chem. Soc., 114 (1992) 7861.
87 S. Harder, A. Streitwieser, J.T. Petty and P.v.R. Schleyer, J. Am. Chem. Soc., 117 (1995) 3253.
88 (a) H. Zipse, Angew. Chem., 106 (1994) 2019, Angew. Chem. Int. Ed. Engl., 33 (1994) 1985; (b) H. Zipse, J. Am. Chem. Soc., 116 (1994) 10773.
89 H. Zipse, J. Chem. Soc. Perkin Trans., 2 (1997) 2691.
90 H. Zipse, J. Am. Chem. Soc., 119 (1997) 1087, 119 (1997) 2889; J. Chem. Soc. Perkin Trans., 2 (1996) 1797.
91 B. Geise, X. Beyrich-Graf, J. Burger, C. Kesselheim, M. Senn and T. Schäfer, Angew. Chem. Int. Ed. Engl., 32 (1993) 1742, Angew. Chem., 105 (1993) 1850.
92 A.Th. Van der Lugt and L.J. Oosterhoff, J. Am. Chem. Soc., 91 (1969) 6042.
93 J. Michl, Top. Curr. Chem., 46 (1974) 1.
94 U. Manthe and H. Köppel, J. Chem. Phys., 93 (1990) 1658.
95 H. Köppel, W. Domcke and L.S. Cederbaum, Adv. Chem. Phys., 57 (1984) 59.
96 F. Bernardi, M. Olivucci and M. Robb, Isr. J. Chem., 33 (1993) 265.
97 M.A. Robb, M. Garavelli, M. Olivucci and F. Bernardi, in: K.B. Lipkowitz, D.B. Boyd (Eds.), Reviews in computational chemistry, Vol. 15, 2000, pp. 87–146.
98 R.B. Woodward and R. Hoffmann, The conservation of orbital symmetry, Verlag Chemie, Weinheim, 1971.
99 D.M. Cyr, G.A. Bishea, M.G. Scranton and M.A. Johnson, J. Chem. Phys., 97 (1992) 5911.
100 W. Gerhartz, R.D. Poshusta and J. Michl, J. Am. Chem. Soc., 99 (1977) 4263.
101 I.B. Bersuker, Nouv. J. Chim., 4 (1980) 139.
102 S. Zilberg, Y. Haas, D. Danovich and S. Shaik, Angew. Chem. Int. Ed., 37 (1998) 1394.
103 H. Quast, K. Knoll, E.-M. Peters, K. Peters and H.G. von Schnering, Chem. Ber., 126 (1993) 1047.
104 H. Quast and M. Seefelder, Angew. Chem. Int. Ed., 38 (1999) 1064.
105 M. Shapiro and P. Brumer, Adv. At. Mol. Opt. Phys., 42 (2000) 287.
106 S.G. Brush, Stud. Hist. Phil. Sci., 30 (1999) 21.
107 S.G. Brush, Stud. Hist. Phil. Sci., 30 (1999) 263.
108 R. Hoffmann, S. Shaik and P.C. Hiberty, Acc. Chem. Res., 36 (2003) 750.
109 S. Shaik and P.C. Hiberty, Helv. Chem. Acta, 86 (2003) 1063.
110 D.G. Truhlar and R.E. Wyatt, Adv. Chem. Phys., 36 (1977) 141.
111 J.N. Murrell, S. Carter, S.C. Farantos, P. Huxley and A.J.C. Varandas, Molecular potential energy functions, Wiley, New York, 1984.
112 See, for example in: D.L. Cooper (Ed.), Valence bond theory, Elsevier, Amsterdam, 2002.
113 H. Zipse, Acc. Chem. Res., 32 (1999) 571–578.

114 S.S. Shaik, P.C. Hiberty, J.-M. Lefour and G. Ohanessian, J. Am. Chem. Soc., 109 (1987) 363.
115 H. Kollmar, J. Am. Chem. Soc., 101 (1979) 4832.
116 P.C. Hiberty, D. Danovich, A. Shurki and S. Shaik, J. Am. Chem. Soc., 117 (1995) 7760.
117 J.M. Galbraith, E. Blank, S. Shaik and P.C. Hiberty, Chem. Eur. J., 6 (2000) 2425.
118 W. Wu, D. Danovich, A. Shurki and S. Shaik, J. Phys. Chem. A., 104 (2000) 8744.

Theory and Applications of Computational Chemistry: The First Forty Years
Edited by C. Dykstra et al.

CHAPTER 24

Progress in the development of exchange-correlation functionals

Gustavo E. Scuseria and Viktor N. Staroverov

Department of Chemistry, Rice University, Houston, Texas 77005, USA

Abstract

This review provides a comprehensive account of the recent progress in constructing practical exchange-correlation approximations of Kohn–Sham density functional theory. The emphasis is on the general techniques of density functional design that have been particularly successful in quantum chemistry. Nearly all density functionals embraced nowadays by computational chemists are discussed. Persistent misconceptions about several widely used functionals are clarified.

24.1 INTRODUCTION

The rise of density functional theory (DFT) to the prominence and popularity it enjoys today was hardly anticipated by computational chemists 40 or even 30 years ago. As recently as 1983, when DFT was but a footnote in quantum chemistry textbooks, Robert Parr was writing a review "to alert the physical chemistry community to the promise and the charm of the density functional theory of electronic structure of atoms and molecules" [1]. Two decades later, DFT is a household tool for computing everything from atoms to biopolymers. How did this extraordinary reversal of fortunes come about?

Electronic structure methods that use the electron density as the basic variable trace their origin to the Thomas–Fermi [2], Thomas–Fermi–Dirac [3], and related models [4–7] developed in the early years of quantum mechanics. Many similarities with the present day DFT can be also found in Gáspár's exchange potential [8] and Slater's Xα-methods [9–11]. By the 1960s, these precursors of DFT were fully developed and used extensively for the calculations of atoms and solids, but their impact on molecular quantum chemistry remained insignificant. The Thomas–Fermi and Thomas–Fermi–Dirac models proved to be of little use in chemistry because they can never yield a lower total energy for a molecule than for separated atoms (the 'no bonding theorem') [12–14]. The experience of quantum chemists with other DFT prototypes was not so discouraging, but the accuracy of those approximations could not match the accuracy of the increasingly sophisticated wave function techniques.

Certain skepticism toward the budding DFT existed also on theoretical grounds. It seemed improbable that a quantitative theory based solely on the electron density could be exact for anything other than model systems like the free-electron gas, simply because the density appears to contain not enough information about electron-electron interactions. The two-electron reduced density matrix was deemed the simplest mathematical object necessary to describe a many-electron system with Coulombic interactions. It was not until 1964 that Hohenberg and Kohn put these doubts to rest by proving their now famous theorem [15]. The Hohenberg–Kohn paper contains two fundamental results: (i) the ground-state electron density uniquely determines the Hamiltonian and, therefore, the ground-state electronic wave function (or a family of *degenerate* ground-state wave functions [16]) and all properties of the system; (ii) the true density functional for the electronic energy assumes its minimum for the correct ground-state density. These propositions effectively reduce the problem of solving the many-body Schrödinger equation to the problem of minimizing a density functional. This idea can be put to work in various ways, of which the approach of Kohn and Sham [17] has been embraced more often than others.

The challenge of DFT consists in the determination or, rather, approximation of the *unknown* energy density functional. Compared to quantum chemists, solid-state physicists have it easy. The simple formulas derived in the theory of a uniform electron gas (UEG) work quite well for typical crystals. Not so in chemistry, where the UEG is not a good approximation for rapidly varying, shell-structured, electron densities. In order to achieve useful accuracy for molecules, even small ones, much more sophisticated approximations are required. The absence of such approximations until the mid-1980s was the single most important reason why DFT conquered chemistry many years after it took a prominent place in solid-state physics [18]. Once the first successful density functional approximations for molecules were developed, interest in DFT surged, prompting the discovery of new fundamental results, stimulating the development of scores of density functionals, and generating countless applications. To appreciate this amazing progress, the reader only needs to compare Parr's 1983 review [1] with its follow-up published 12 years later [19]. For more details, we recommend many excellent introductory [20–24] and advanced [25–31] expositions of DFT.

To an uninitiated user of quantum chemistry programs, mathematical expressions of density functionals may appear esoteric. The analytic form of many functionals is indeed complicated and nonintuitive, but it often conceals beautifully simple ideas. Lack of familiarity with these ideas and a 'black-box' attitude toward the alphabet soup of density functional approximations are in part responsible for the wide-spread sentiment that DFT is effectively an empirical method with no prescription for systematic convergence to the right answer. We hope to convince the reader here that this view is unfair and that the development of density functionals can be, in its own way, a rigorous procedure. We will do so by systematizing and explaining the principal ideas behind modern density functional approximations. Because the most important developments in DFT relevant to chemistry occurred in the last quarter of a century, it is this later period that is primarily covered here.

Although this review focuses on general principles of density functional design, many particular approximations will be discussed in detail. It is not our purpose, however, to present an exhaustive survey of density functionals or to characterize their performance. Such a task would be nearly impossible at a time when the number of published functionals continues to grow at an accelerating pace. Nonetheless, all 'popular' exchange-correlation functionals for Kohn–Sham DFT (i.e. functionals available in program packages like GAUSSIAN 03 [32] and enjoying wide use) will be covered. Nor shall we touch upon indirectly related subjects such as orbital-free kinetic-energy DFT [33], quantal DFT [34], local-scaling transformation version of DFT [35], and others. Finally, we realize that this review is biased toward our own work and the work of our collaborators and hope that the reader will forgive us for any unintended omissions.

24.2 KOHN–SHAM DENSITY FUNCTIONAL THEORY

DFT aspires to predict *exactly* properties of many-electron systems without recourse to the wave function, using only the information contained (explicitly or implicitly) in the ground-state electron density. This section reviews the basic DFT formalism and introduces fundamental relations that will recur throughout this work.

24.2.1 Motivation for density functional theory

Consider the problem of solving the nonrelativistic, stationary nucleus, Schrödinger equation

$$\hat{H}\Psi = E\Psi \tag{1}$$

involving the N-electron Hamiltonian operator (in atomic units)

$$\hat{H} = -\frac{1}{2}\sum_{i=1}^{N}\nabla_i^2 + \sum_{i=1}^{N} v(\mathbf{r}_i) + \sum_{i<j}^{N}\frac{1}{|\mathbf{r}_i - \mathbf{r}_j|} = \hat{T} + \hat{V} + \hat{V}_{\mathrm{ee}} \tag{2}$$

where $v(\mathbf{r}_i)$ is a multiplicative external potential in which the electrons move. For atoms, molecules, and solids, $v(\mathbf{r}_i)$ is simply the Coulombic potential of the nuclei with charges Z_{A} at positions $\mathbf{R}_{\mathrm{A}}$,

$$v(\mathbf{r}_i) = -\sum_{A}^{\text{nuclei}}\frac{Z_{\mathrm{A}}}{|\mathbf{r}_i - \mathbf{R}_{\mathrm{A}}|} \tag{3}$$

although DFT is not restricted to potentials of this form. Multiply Eq. (1) from the left by Ψ^* and integrate each term over the spatial ($\mathbf{r}_i$) and spin (σ_i) coordinates not acted upon by the operators $\hat{T}$, $\hat{V}$, and $\hat{V}_{ee}$. Assuming that Ψ is normalized, the result is

$$E = -\frac{1}{2}\int [\nabla^2_{\mathbf{r}}\rho(\mathbf{r},\mathbf{r}')]_{\mathbf{r}'=\mathbf{r}}\, d\mathbf{r} + \int v(\mathbf{r})\rho(\mathbf{r})d\mathbf{r} + \int\int \frac{P_2(\mathbf{r}_1,\mathbf{r}_2)}{|\mathbf{r}_1-\mathbf{r}_2|}d\mathbf{r}_1\, d\mathbf{r}_2 \tag{4}$$

where we have introduced the one-electron reduced density matrix $\rho(\mathbf{r},\mathbf{r}')$ and the *pair density*

$$P_2(\mathbf{r}_1,\mathbf{r}_2) = \frac{N(N-1)}{2}\sum_{\sigma_1\ldots\sigma_N}\int\cdots\int|\Psi(\mathbf{r}_1\sigma_1,\mathbf{r}_2\sigma_2,\mathbf{r}_3\sigma_3,\ldots,\mathbf{r}_N\sigma_N)|^2 d\mathbf{r}_3\ldots d\mathbf{r}_N \tag{5}$$

The density, reduced density matrix, and the pair density are related by

$$\rho(\mathbf{r}_1) = \rho(\mathbf{r}_1,\mathbf{r}_1) = \frac{2}{N-1}\int P_2(\mathbf{r}_1,\mathbf{r}_2)d\mathbf{r}_2 \tag{6}$$

The three terms in Eq. (4) represent the kinetic energy of the electrons

$$T = -\frac{1}{2}\int [\nabla^2_{\mathbf{r}}\rho(\mathbf{r},\mathbf{r}')]_{\mathbf{r}'=\mathbf{r}}d\mathbf{r} \tag{7}$$

the electron-nuclear attraction

$$V = \int v(\mathbf{r})\rho(\mathbf{r})d\mathbf{r} \tag{8}$$

and electron–electron interaction

$$V_{ee} = \int\int \frac{P_2(\mathbf{r}_1,\mathbf{r}_2)}{r_{12}}d\mathbf{r}_1\, d\mathbf{r}_2 \tag{9}$$

The last term includes the classical Coulomb repulsion and quantum-mechanical exchange-correlation effects. The separation of the classical and quantum-mechanical parts can be made explicit by writing the pair density as

$$P_2(\mathbf{r}_1,\mathbf{r}_2) = \frac{1}{2}\rho(\mathbf{r}_1)[\rho(\mathbf{r}_2) + h_{xc}(\mathbf{r}_1,\mathbf{r}_2)] \tag{10}$$

which effectively defines $h_{xc}(\mathbf{r}_1,\mathbf{r}_2)$, the *exchange-correlation* (xc) *hole* of an electron at $\mathbf{r}_1$. Using Eq. (10) we can rewrite Eq. (9) as

$$V_{ee} = J + E^{(c)}_{xc} \tag{11}$$

where J is the classical Coulomb repulsion energy

$$J = \frac{1}{2}\int\int \frac{\rho(\mathbf{r}_1)\rho(\mathbf{r}_2)}{r_{12}}d\mathbf{r}_1\, d\mathbf{r}_2 \tag{12}$$

and $E^{(c)}_{xc}$ is the conventional (in wave function-based methods) exchange-correlation energy

$$E^{(c)}_{xc} = \frac{1}{2}\int\int \frac{\rho(\mathbf{r}_1)h_{xc}(\mathbf{r}_1,\mathbf{r}_2)}{r_{12}}d\mathbf{r}_1\, d\mathbf{r}_2 \tag{13}$$

Now Eq. (4) becomes

$$E = T + V + J + E_{xc}^{(c)} \tag{14}$$

Observe that V and J are explicit functionals of $\rho(\mathbf{r})$, but T and $E_{xc}^{(c)}$ are not. One might suppose that T and $E_{xc}^{(c)}$ cannot be determined from $\rho(\mathbf{r})$ in principle, since they appear to require the knowledge of the density matrix $\rho(\mathbf{r}, \mathbf{r}')$ and the pair function $P_2(\mathbf{r}_1, \mathbf{r}_2)$, respectively. Intuition is wrong here, because the Hohenberg–Kohn theorem [15] asserts that the ground-state energy of a real many-electron system in a static external potential $v(\mathbf{r})$ is a *unique* functional of the density

$$E[\rho] = \int v(\mathbf{r})\rho(\mathbf{r})\mathrm{d}\mathbf{r} + F[\rho] \tag{15}$$

where F $[\rho]$ absorbs the kinetic energy and electron–electron interaction terms

$$F[\rho] = T[\rho] + J[\rho] + E_{xc}^{(c)}[\rho] \tag{16}$$

The Hohenberg–Kohn theorem assures only that the functional $F[\rho]$ exists, but the actual form of $F[\rho]$ is unknown (except for the term $J[\rho]$) and must be approximated. Once the number of electrons N is fixed, Hamiltonian operators for any two systems differ only by the external potential $v(\mathbf{r})$. The functional $F[\rho]$ is therefore *universal*.

The partitioning of $F[\rho]$ into the three components T, J, and $E_{xc}^{(c)}$ in Eq. (16) is by no means unique. Different partitioning schemes give rise to different variants of DFT. In fact, the particular partitioning of Eq. (16), although very natural, is *not* the one that is normally used. The most popular variant of DFT is the Kohn–Sham formulation, to which we now turn our attention.

24.2.2 Kohn–Sham scheme

The idea of the Kohn–Sham method is best understood as follows. Consider a generalized Hamiltonian of Eq. (2) in which the term $\hat{V}_{ee}$ is scaled by an electron–electron coupling constant λ. We are interested in values of λ between 0 and 1. Each value of λ corresponds to a distinct universal functional of the density. In Levy's constraint search formulation [36] of the Hohenberg–Kohn principle, this is explicitly stated as

$$F_\lambda[\rho] = \langle \Psi_\rho^{\min,\lambda} | \hat{T} + \lambda \hat{V}_{ee} | \Psi_\rho^{\min,\lambda} \rangle \tag{17}$$

where $\Psi_\rho^{\min,\lambda}$ is the N-electron wave function that minimizes the expectation value of $\hat{T} + \lambda\hat{V}_{ee}$ and simultaneously yields the density $\rho(\mathbf{r})$. For real systems, $\lambda = 1$, so that $F_1[\rho] = F[\rho]$ is the universal functional of interest. This is the problem we wish to solve but cannot.

The value $\lambda = 0$ corresponds to a system of *noninteracting* electrons moving in the external potential $v(\mathbf{r})$. The noninteracting Schrödinger equation is, of course, trivially solvable. The solution is $\Phi_0 = \Psi_\rho^{\min,0}$, a single Slater determinant of one-electron wave functions (orbitals) ϕ_i obtained from the single-particle equations

$$\left[-\frac{1}{2}\nabla^2 + v(\mathbf{r})\right]\phi_i(\mathbf{r}) = \epsilon_i\phi_i(\mathbf{r}) \tag{18}$$

The universal density functional for this noninteracting system is therefore

$$F_0[\rho] = T_s[\rho] = -\frac{1}{2}\sum_{i=1}^{N}\langle\phi_i|\nabla^2|\phi_i\rangle \tag{19}$$

where the density is given by

$$\rho(\mathbf{r}) = \sum_{i=1}^{N}|\phi_i(\mathbf{r})|^2 \tag{20}$$

Note that $T_s[\rho]$ is written in terms of orbitals, and so is an implicit functional of $\rho(\mathbf{r})$.

What Kohn and Sham did was to assume that for any real (interacting) system with the ground-state density $\rho(\mathbf{r})$ there always exists a noninteracting system with the same ground-state $\rho(\mathbf{r})$. Then one can rewrite Eq. (16) as

$$F[\rho] = T_s[\rho] + J[\rho] + E_{xc}[\rho] \tag{21}$$

where $T_s\,[\rho]$ is the kinetic energy of the noninteracting system computed exactly by Eq. (19), and $E_{xc}[\rho]$ is the Kohn–Sham exchange-correlation energy, formally defined as

$$E_{xc}[\rho] = T[\rho] - T_s[\rho] + E_{xc}^{(c)}[\rho] \tag{22}$$

Application of the variational principle, $\delta E/\delta\rho(\mathbf{r}) = 0$, to the Kohn–Sham functional

$$E[\rho] = \int \rho(\mathbf{r})v(\mathbf{r})d\mathbf{r} + T_s[\rho] + J[\rho] + E_{xc}[\rho] \tag{23}$$

subject to the orthonormality constraints $\langle\phi_i|\phi_j\rangle = \delta_{ij}$, yields N Hartree-type one-electron equations

$$\left[-\frac{1}{2}\nabla^2 + v(\mathbf{r}) + \int\frac{\rho(\mathbf{r}')}{|\mathbf{r}-\mathbf{r}'|}d\mathbf{r}' + v_{xc}(\mathbf{r})\right]\phi_i(\mathbf{r}) = \epsilon_i\phi_i(\mathbf{r}) \tag{24}$$

where $\phi_i(\mathbf{r})$ are Kohn–Sham orbitals, ϵ_i are Kohn–Sham orbital energies, and $v_{xc}(\mathbf{r})$ is the exchange-correlation potential

$$v_{xc}(\mathbf{r}) = \frac{\delta E_{xc}[\rho]}{\delta\rho(\mathbf{r})} \tag{25}$$

that is, the functional derivative of $E_{xc}\,[\rho]$ with respect to the density. The orbitals $\phi_i(\mathbf{r})$ form a Slater determinant $\Phi_\rho^{\min}$, called the Kohn–Sham wave function.

Eqs. (20), (24), and (25), known as the Kohn–Sham equations, are formally exact and contain only one unknown term, $E_{xc}[\rho]$. It is E_{xc} that is approximated in Kohn–Sham DFT, not the conventional exchange-correlation energy $E_{xc}^{(c)}$. Exact treatment of the kinetic energy as an orbital-dependent functional is crucial to the practicality of this scheme because $T[\rho]$ and $T_s[\rho]$ are notoriously difficult to approximate as explicit functionals of the density [33,37]. One can readily show [20] that $T_s \leq T$, so that by Eq. (22) $E_{xc} \geq E_{xc}^{(c)}$, where the equality holds only for one-electron systems. For a historical account of the developments leading to the announcement of the Hohenberg–Kohn theorem and the formulation of the Kohn–Sham equations, we refer the reader to a memoir of Hohenberg et al. [38] and Kohn's Nobel lecture [39]. For a modern perspective on DFT fundamentals, we recommend a recent review by Savin et al. [40].

24.3 EXCHANGE AND CORRELATION DENSITY FUNCTIONALS

The problem of finding accurate approximations to $E_{xc}[\rho]$ is the biggest challenge of Kohn–Sham DFT. The better we understand the exact functional, the better approximations we can design. The following section presents a closer look at the exchange-correlation functional.

24.3.1 Exchange-correlation energy

The formal definition of the exchange-correlation energy by Eq. (22) is not very helpful for designing approximate density functionals. Fortunately, there exist more constructive exact formulas for $E_{xc}[\rho]$. Observe that by the Hellmann–Feynman theorem [41]

$$\frac{\partial F_\lambda[\rho]}{\partial \lambda} = \langle \Psi_\rho^{\min,\lambda} | \hat{V}_{ee} | \Psi_\rho^{\min,\lambda} \rangle \tag{26}$$

where $F_\lambda[\rho]$ is given by Eq. (17). Let us integrate Eq. (26) over λ from 0 to 1 keeping the density $\rho(\mathbf{r})$ fixed at all λ. This procedure is called 'adiabatic integration' and its result is

$$\int_0^1 \frac{\partial F_\lambda[\rho]}{\partial \lambda} \mathrm{d}\lambda = F_1[\rho] - F_0[\rho] = E_{xc}[\rho] + J[\rho] \tag{27}$$

where we have used Eqs. (19) and (21) for $F_1[\rho] \equiv F[\rho]$ and $F_0[\rho]$, respectively. Combining Eqs. (26) and (27) we obtain the *adiabatic connection formula* [42–45][1]

$$E_{xc}[\rho] = \frac{1}{2}\int_0^1 \langle \Psi_\rho^{\min,\lambda} | \hat{V}_{ee} | \Psi_\rho^{\min,\lambda} \rangle \mathrm{d}\lambda - J[\rho] = \int_0^1 E_{xc}^\lambda[\rho] \mathrm{d}\lambda \tag{28}$$

Generalizing the definition of the pair function and exchange-correlation hole in Eq. (10) to intermediate coupling strengths, we now introduce $P_2^\lambda(\mathbf{r}_1, \mathbf{r}_2)$ and $h_{xc}^\lambda(\mathbf{r}_1, \mathbf{r}_2)$, and use Eqs. (9)–(11) to rewrite Eq. (28) as

$$E_{xc}[\rho] = \frac{1}{2}\int_0^1 \mathrm{d}\lambda \int\int \frac{\rho(\mathbf{r}_1) h_{xc}^\lambda(\mathbf{r}_1, \mathbf{r}_2)}{r_{12}} \mathrm{d}\mathbf{r}_1 \, \mathrm{d}\mathbf{r}_2 \tag{29}$$

In this equation, $\rho(\mathbf{r}_1)$ does not have the subscript λ because the electron density is fixed. Since the integration over λ is decoupled from the integration over space

[1] The density was held fixed for *all* λ in Refs. [43–45]; it was held fixed only for $\lambda = 0$ and 1 in Ref. [42].

coordinates, we can define

$$\bar{h}_{\mathrm{xc}}(\mathbf{r}_1, \mathbf{r}_2) = \int_0^1 h_{\mathrm{xc}}^{\lambda}(\mathbf{r}_1, \mathbf{r}_2)\, \mathrm{d}\lambda \tag{30}$$

and rewrite Eq. (29) as

$$E_{\mathrm{xc}}[\rho] = \frac{1}{2} \int\int \rho(\mathbf{r}_1) \frac{\bar{h}_{\mathrm{xc}}(\mathbf{r}_1, \mathbf{r}_2)}{r_{12}}\, \mathrm{d}\mathbf{r}_1\, \mathrm{d}\mathbf{r}_2 \tag{31}$$

Eq. (31) can be regarded as a Kohn–Sham DFT analog of Eq. (13). Let us now make a change of variables $\mathbf{r} = \mathbf{r}_1$, $\mathbf{u} = \mathbf{r}_2 - \mathbf{r}_1$ and integrate over the angular coordinates of $\mathbf{u}$. This gives the spherically-averaged exchange-correlation hole

$$\bar{h}_{\mathrm{xc}}(\mathbf{r}, u) = \frac{1}{4\pi} \int_0^{2\pi} \mathrm{d}\phi_u \int_0^{\pi} \bar{h}_{\mathrm{xc}}(\mathbf{r}, \mathbf{r} + \mathbf{u}) \sin\theta_u\, \mathrm{d}\theta_u \tag{32}$$

Using this quantity, Eq. (31) can be written as

$$E_{\mathrm{xc}}[\rho] = \frac{1}{2} \int \rho(\mathbf{r})\, \mathrm{d}\mathbf{r} \int_0^{\infty} 4\pi u^2 \frac{\bar{h}_{\mathrm{xc}}(\mathbf{r}, u)}{u}\, \mathrm{d}u \tag{33}$$

This equation serves as the starting point for many density functionals approximations.

In practical Kohn–Sham DFT, the exchange-correlation functional E_{xc} is usually separated into the exchange and correlation parts,

$$E_{\mathrm{xc}}[\rho] = E_{\mathrm{x}}[\rho] + E_{\mathrm{c}}[\rho] \tag{34}$$

The exchange energy is *defined* by

$$E_{\mathrm{x}}[\rho] = \langle \Phi_\rho^{\mathrm{min}} | \hat{V}_{\mathrm{ee}} | \Phi_\rho^{\mathrm{min}} \rangle - J[\rho] \tag{35}$$

where Φ_ρ^{min} is the Kohn–Sham determinant, while the correlation energy is taken formally as the difference

$$E_{\mathrm{c}}[\rho] = E_{\mathrm{xc}}[\rho] - E_{\mathrm{x}}[\rho] = \langle \Psi_\rho^{\mathrm{min}} | \hat{V}_{\mathrm{ee}} | \Psi_\rho^{\mathrm{min}} \rangle - \langle \Phi_\rho^{\mathrm{min}} | \hat{V}_{\mathrm{ee}} | \Phi_\rho^{\mathrm{min}} \rangle \tag{36}$$

where Ψ_ρ^{min} is the exact interacting wave function. The last term in Eq. (36) is nothing but the integral of Eq. (9) with the pair density P_2 derived from Φ_ρ^{min}. For a single-determinant wave function like Φ_ρ^{min}, the pair density is [46]

$$P_2(\mathbf{r}_1, \mathbf{r}_2) = \frac{1}{2}\rho(\mathbf{r}_1)\rho(\mathbf{r}_2) - \frac{1}{2}[\rho_\alpha(\mathbf{r}_1, \mathbf{r}_2)\rho_\alpha(\mathbf{r}_2, \mathbf{r}_1) + \rho_\beta(\mathbf{r}_1, \mathbf{r}_2)\rho_\beta(\mathbf{r}_2, \mathbf{r}_1)] \tag{37}$$

For spin-compensated systems, substitution of Eq. (37) into Eqs. (9) and (35) yields the following *exact* expression for the exchange energy

$$E_{\mathrm{x}}^{\mathrm{exact}} = -\frac{1}{4} \int\int \frac{|\rho(\mathbf{r}_1, \mathbf{r}_2)|^2}{r_{12}}\, \mathrm{d}\mathbf{r}_1\, \mathrm{d}\mathbf{r}_2 \tag{38}$$

where $\rho(\mathbf{r}_1, \mathbf{r}_2)$ is the Kohn–Sham one-electron density matrix

$$\rho(\mathbf{r}_1, \mathbf{r}_2) = \sum_{i=1}^{N} \phi_i(\mathbf{r}_1)\phi_i^*(\mathbf{r}_2) \tag{39}$$

Eq. (38) looks exactly like the expression for the exchange energy in the Hartree–Fock (HF) theory, E_X^{HF}. It should be emphasized, however, that $\phi_i(\mathbf{r})$ in Eq. (39) are the Kohn–Sham, not Hartree–Fock, orbitals. The two sets of orbitals are different, because they arise from solving different equations. Thus, in general $E_X^{exact} \neq E_X^{HF}$. This distinction is important and should always be kept in mind, especially because the terms 'exact exchange' and 'Hartree–Fock exchange' are often loosely used as synonyms.

The true exchange functional is therefore known exactly. Then why do we need approximate exchange functionals? The problem with Eq. (38) is that it is not an explicit functional of the density, so the corresponding exchange potential cannot be readily evaluated as the functional derivative of E_X^{exact} with respect to $\rho(\mathbf{r})$. For this purpose, an indirect procedure, called the optimized effective potential (OEP) method [47–55] can be used. Another reason for avoiding the exact exchange functional is that in practice it is very difficult to achieve useful accuracy by combining exact exchange with approximate correlation functionals.

24.3.2 Ingredients of density functional approximations

In principle, the original Kohn–Sham formalism applies to both spin-compensated ($\rho_\alpha = \rho_\beta$) and spin-polarized ($\rho_\alpha \neq \rho_\beta$) systems. In the case of spin-polarized systems, however, the electronic energy is extremely difficult to approximate as a functional of the total density alone. In practice, one always prefers to describe spin-polarized systems using spin-density functional theory. Spin-DFT is a generalization of the original Hohenberg–Kohn principle and Kohn–Sham method to systems in the presence of a non-zero external magnetic field. It was first discussed in the original Kohn–Sham paper [17] and elaborated by von Barth and Hedin [56] and by Pant and Rajagopal [57]. In spin-DFT, the basic variables are the spin-up and spin-down electron densities $\rho_\alpha(\mathbf{r})$ and $\rho_\beta(\mathbf{r})$, and the exchange-correlation energy is a functional of both. In the absence of a magnetic field, spin-DFT gives the same results for spin-polarized systems as the spin-independent DFT, but still operates with functionals of the type $E_{xc}[\rho_{\alpha,}\rho_\beta]$. This offers a significant practical advantage because approximate spin-density functionals $E_{xc}[\rho_{\alpha,}\rho_\beta]$ usually provide a much better description of spin-polarized systems than functionals $E_{xc}[\rho]$.

Let us emphasize again that in spin-DFT the exchange-correlation energy E_{xc} is a functional of $\rho_\alpha(\boldsymbol{r})$ and $\rho_\beta(\boldsymbol{r})$ alone, and, in principle, these are the only ingredients needed. This appealing picture is gravely complicated by the fact that the dependence of E_{xc} on $\rho_\alpha(\mathbf{r})$ and $\rho_\beta(\mathbf{r})$ is highly nonlocal, meaning that: (i) small variations of the densities may cause large variations of the exchange-correlation potential v_{xc}; (ii) v_{xc} at a given point $\mathbf{r}$ may be very sensitive to changes of the density not only in the vicinity of $\mathbf{r}$, but also at very distant points $\mathbf{r}'$. Of course, for practical reasons we want approximate functionals that are explicit and local (or semilocal). To compensate for their locality, such functionals must use some other (local or semilocal) ingredients in addition to the local $\rho_\sigma(\mathbf{r})$. These can be derivatives of $\rho_\sigma(\mathbf{r})$ (semilocal ingredients) and quantities that are *indirectly* (nonlocally) determined by the densities. Examples of ingredients of the latter type include Kohn–Sham orbitals and variables constructed from them, provided they are gauge-independent and invariant with respect to unitary transformation of the orbitals.

The most important examples of orbital-dependent ingredients are the Kohn–Sham reduced density matrix of Eq. (39) and the noninteracting kinetic energy density

$$\tau_\sigma(\mathbf{r}) = \frac{1}{2}\sum_{k=1}^{\text{occ.}} |\nabla \phi_{k\sigma}(\mathbf{r})|^2 \tag{40}$$

The name of the latter derives from the fact that $\tau_\sigma(\mathbf{r})$ integrates to the Kohn–Sham kinetic energy T_s of Eq. (19) for σ-spin electrons, as can be readily verified via integration by parts. The definition of $\tau_\sigma(\mathbf{r})$ by Eq. (40) is gauge-invariant only in the absence of an external magnetic field [58]. Furthermore, it is not unique, because any function that integrates to zero, such as $\nabla^2\rho_\sigma(\mathbf{r})$, can be added to $\tau_\sigma(\mathbf{r})$ without changing the value of T_s. The noninteracting kinetic energy density is not a far-fetched ingredient as may seem at first sight. It naturally arises in the Taylor series expansion of the Kohn–Sham density matrix (see Section 24.8.1). Chemical content of the kinetic energy density has been interpreted by Schmider and Becke [59,60]. The possibility of constructing functionals of even more complicated variables like $\tau_{2\sigma} = \frac{1}{4}\sum_{k=1}^{\text{occ.}} |\nabla^2 \phi_{k\sigma}(\mathbf{r})|^2$, which appear in higher-order density matrix expansions, has also been considered [61].

Another ingredient whose importance is being increasingly recognized [62–67] is the paramagnetic current density, defined in atomic units as

$$\mathbf{j}_\sigma(\mathbf{r}) = -\frac{i}{2}\sum_{k=1}^{\text{occ.}} [\phi^*_{k\sigma}(\mathbf{r})\nabla\phi_{k\sigma}(\mathbf{r}) - \phi_{k\sigma}(\mathbf{r})\nabla\phi^*_{k\sigma}(\mathbf{r})] \tag{41}$$

This quantity arises in current-dependent DFT, an extension of the Hohenberg–Kohn theory to strong magnetic fields [68–71]. The paramagnetic current density is nonvanishing only for degenerate states that are described with complex wave functions (Kohn–Sham determinants in DFT), such as the three configurations of the B atom, where the unpaired electron can occupy either the $2p_1$, $2p_0$, or $2p_{-1}$ orbital. Such states have the same energy but different density distributions. All standard functionals in existence predicts a large artificial separation between the $M_L = 0$ and $M_L = \pm 1$ levels. This unphysical splitting can be drastically reduced by including the current density $\mathbf{j}_\sigma(\mathbf{r})$ into approximate functionals [65–67].

Ingredients other than the density do more than just provide additional degrees of freedom for designing density functional approximations. They are necessary if we want an approximation to satisfy exact constraints that are impossible to impose using spin-densities and their derivatives alone. The type of ingredients of $E_{xc}[\rho_\alpha, \rho_\beta]$ forms the basis for Perdew's 'Jacob's ladder' classification [72] of density functionals. The lower three rungs of this ladder are, in the ascending order: (i) the local spin-density approximation (LSDA), which employs only ρ_σ; (ii) the generalized gradient approximation (GGA), whose ingredients are ρ_σ and $\nabla\rho_\sigma$; (iii) the meta-GGA, which makes use of τ_σ (or $\nabla^2\rho_\sigma$) in addition to the GGA ingredients. The ascent of the ladder consists in embedding increasingly complex ingredients and exact properties in $E_{xc}[\rho_\alpha,\rho_\beta]$ and results, in practice, in better functionals [73].

24.3.3 Analytic properties of exchange-correlation functionals

Although the exact functional $E_{xc}[\rho_\alpha,\rho_\beta]$ remains an enigma, many of its analytic properties, ranging from obvious to subtle, are known [74]. For any possible electron densities, the exchange energy is strictly negative, while the correlation energy is nonpositive:

$$E_x < 0, \qquad E_c \leq 0 \tag{42}$$

Lieb and Oxford [75] showed that the exchange-correlation energy of electrons in Coulombic systems is bounded also from below

$$E_x[\rho_\alpha,\rho_\beta] \geq E_{xc}[\rho_\alpha,\rho_\beta] \geq C_{LO}\int \rho^{4/3}(\mathbf{r})\,d\mathbf{r} \tag{43}$$

where $-1.44 \geq C_{LO} \geq -1.68$ [76].

For one-electron densities $\rho_1(\mathbf{r})$, E_x must cancel the spurious Coulomb self-repulsion energy:

$$E_x[\rho_1,0] + J[\rho_1] = 0 \tag{44}$$

while E_c must vanish altogether:

$$E_c[\rho_1,0] = 0 \tag{45}$$

Eqs. (44) and (45) were stressed in the self-interaction correction scheme of Perdew and Zunger [77].

For uniform electron densities, $E_{xc}[\rho_\alpha,\rho_\beta]$ should reduce to the formulas for the exchange-correlation energy of a uniform free-electron gas (the LSDA):

$$E_{xc}[\rho_\alpha,\rho_\beta] = E_{xc}^{LSDA}[\rho_\alpha,\rho_\beta], \qquad \text{if} \qquad \rho_\sigma(\mathbf{r}) = \text{const} \tag{46}$$

A large number of known exact properties of density functionals involve coordinate scaling transformations of the density. Most of such relations have been derived by Levy and coworkers [78–84]. The *uniform* scaling of the density is defined by

$$\rho_\gamma(\mathbf{r}) = \gamma^3\rho(\gamma\mathbf{r}) \tag{47}$$

The effect of this transformation is to contract or thin the density while preserving its normalization. Coordinate scaling constraints for exchange and correlation functionals are reviewed by Levy [85]. The most important among these constraints are the following:

$$E_x[\rho_\gamma] = \gamma E_x[\rho] \tag{48}$$

$$\lim_{\gamma \to \infty} E_c[\rho_\gamma] > -\infty \tag{49}$$

The exchange functionals appropriate to spin-compensated and to spin-polarized systems are related to each other by the spin-scaling relation [86]

$$E_x[\rho_\alpha, \rho_\beta] = \frac{1}{2}(E_x[2\rho_\alpha] + E_x[2\rho_\beta]) \tag{50}$$

where $E_x[\rho] \equiv E_x[\rho/2, \rho/2]$. The spin-scaling relation effectively defines the resolution of the exchange energy into two parallel-spin contributions, $E_x^{\sigma\sigma} = \frac{1}{2}E_x[2\rho_\sigma]$, where $\sigma = \alpha, \beta$. Note that E_x has no opposite-spin components. Eq. (50) is respected by all exchange energy approximations in existence. In the literature, exchange functionals are often stated in the form $E_x[\rho]$ only, while the form $E_x[\rho_\alpha, \rho_\beta]$ (the one actually used) is implied through Eq. (50). For the correlation energy, there is no simple spin-scaling relation similar to Eq. (50).

24.4 STRATEGIES FOR DESIGNING DENSITY FUNCTIONALS

DFT would be the ultimate practical theory if one could derive the exact exchange-correlation functional in a closed form, as were the Thomas–Fermi kinetic energy and Dirac exchange energy functionals for the UEG. This appears to be a hopeless task. DFT would still be almost perfect if, by analogy with wave function methods, we had a mechanical prescription for the systematic improvement of approximations that guarantees their convergence to the right answer. In principle, Görling–Levy perturbation theory [87–89] and '*ab initio* DFT' [90–92] offer such prescriptions, but these approaches abandon the idea of a universal density functional and, for all practical purposes, are hardly different from wave function techniques. In practice, one has to devise density functional approximations relying on many little bits of information about the true functional and compensating for lack of mechanical recipes with insight and ingenuity. As the success of DFT attests, this is a workable plan. We distinguish the following six approaches to designing density functional approximations.

1. *Local density approximations (LDA).* This group includes functionals derived in the analytic theory of a UEG and applied, directly or with empirical modifications, to nonuniform densities. All LDA functionals have the form

$$E_{xc}^{LDA}[\rho] = \int e_{xc}(\rho) d\mathbf{r} \tag{51}$$

where the exchange-correlation energy density $e_{xc}(\rho)$ is a function of $\rho(\mathbf{r})$ only.

2. *Density-gradient expansion (DGE).* These are formal analogs of the three-dimensional Taylor expansion of the exchange-correlation energy in derivatives of the density:

$$E_{xc}^{DGE}[\rho] = \int [e_{xc}^{(0)}(\rho) + e_{xc}^{(1)}(\rho)\nabla\rho + e_{xc}^{(2)}(\rho)|\nabla\rho|^2 + \ldots] d\mathbf{r} \tag{52}$$

Derivation of the coefficients $e_{xc}^{(k)}(\rho)$ is very involved mathematically, and the performance of such nonempirical functionals is reasonable only for slowly varying densities.

3. *Constraint satisfaction.* A more successful approach consists in designing functionals of the form

$$E_{xc}[\rho] = \int e_{xc}(\rho, \nabla\rho, \nabla^2\rho, \tau, \ldots)\mathrm{d}\mathbf{r} \tag{53}$$

where the integrand is constructed to satisfy chosen exact constraints. The constraints in question may concern the asymptotic behavior of e_{xc} and v_{xc}, upper and lower bounds on the energy, density scaling transformations, and other properties of the true functional. A number of approximations of this type are fully nonempirical.

4. *Modeling the exchange-correlation hole.* Functionals of this type are based on Eqs. (31) or (38) and form one of the largest and most diverse groups. For example, the exchange hole may be approximated by a truncated Taylor series expansion in $\rho, \nabla\rho, \nabla^2\rho$, and τ. The correlation hole may be derived from an approximate correlated wave function or modeled after the correlation hole of an analytically solvable problem. The general analytic form of these functionals is the same as in Eq. (53). Functionals based on an exchange-correlation hole model may be fully nonempirical or contain fitted parameters.

5. *Empirical fits.* Functionals of this type are designed by fitting reasonably chosen analytic forms of $E_{xc}[\rho]$ to experimental values of thermochemical and/or other properties of atoms and molecules. The analytic form may be borrowed directly from functionals of any other group or simply postulated without a rigorous derivation. Some fitted ('optimized') functionals are linear combinations of pre-existing functionals. The general form of such approximations is

$$E_{xc}[\rho] = \sum_k C_k \int e_{xc}^{(k)}(\rho, \nabla\rho, \nabla^2\rho, \tau, \ldots; a_k, b_k, \ldots)\, \mathrm{d}\mathbf{r} \tag{54}$$

where $C_k, a_k, b_k, \ldots$ are adjustable parameters.

6. *Mixing exact and approximate exchange.* These functionals, termed hybrids, have the form

$$E_{xc}^{\text{hybrid}}[\rho] = \int [a e_x^{\text{exact}}(\mathbf{r}) + b e_x^{\text{DFT}}(\mathbf{r}) + e_c^{\text{DFT}}(\mathbf{r})]\, \mathrm{d}\mathbf{r} \tag{55}$$

where the mixing coefficients a and b may be constants or depend on $\mathbf{r}$. In the latter case, functionals of this form are called 'local hybrids' [93].

The majority of density functional approximations in current use fall into one of these categories. The scheme presented above, however, is neither perfect nor complete. For example, the LDA for exchange may be viewed as a functional derived from the exact exchange hole $h_x^{\text{LDA}}(\mathbf{r}, u)$ of a UEG [it is given by Eq. (57)] and placed in group 4. Functionals of any group may have adjustable parameters, not just optimized functionals. Exact constraints are often explicitly imposed on density functionals based on an exchange-correlation hole model, not only on functionals of group 3. The above scheme does not include several interesting but less common methods such as the weighted-density approximation [94–97], the phase-space approach [98,99], Laurent series expansions in terms of homogeneous functionals [100,101] or moments of the density [102], and many others. Nevertheless, we believe that the classification of density functionals by the method of their derivation is instructive. In what follows, we will use it as a map to navigate the realm of approximate density functionals.

24.5 LOCAL DENSITY APPROXIMATIONS

In a narrow sense, the LDA consists in applying the exact results of the theory of a UEG to real nonuniform densities. More generally, an LDA is any approximation of the form

$$E_{\mathrm{xc}}^{\mathrm{LDA}}[\rho] = \int \rho(\mathbf{r})\epsilon_{\mathrm{xc}}(\rho)\mathrm{d}\mathbf{r} \tag{56}$$

where $\epsilon_{\mathrm{xc}}(\rho) = \epsilon_{\mathrm{x}}(\rho) + \epsilon_{\mathrm{c}}(\rho)$ is the exchange-correlation energy per particle of the electron gas, which is a function of the density only.

24.5.1 Local density approximation for exchange

The exchange energy of a UEG can be evaluated analytically by the method of Bloch [103] or Dirac [3]. Details of both approaches are discussed by Gombás [4], Bethe [104], Slater [105], and Parr and Yang [20]. The outline of the derivation is as follows. The Kohn–Sham orbitals for a UEG are plane waves, $\phi_{\mathbf{k}}(\mathbf{r}) = V^{-1/2}e^{\mathrm{i}\mathbf{k}\cdot\mathbf{r}}$, where V is the volume of the box. Given the orbitals, one calculates the Kohn–Sham density matrix of Eq. (39) by replacing the sum over occupied orbitals $\mathbf{k}$ with an integral over a sphere of radius $k_{\mathrm{F}} = (3\pi^2\rho)^{1/3}$. Then a transformation to relative coordinates $\mathbf{r} = \mathbf{r}_1$, $\mathbf{u} = \mathbf{r}_2 - \mathbf{r}_1$ and angle-averaging yield the LDA exchange hole

$$h_{\mathrm{x}}^{\mathrm{LDA}}(\mathbf{r}, u) = -\frac{9}{2}\rho(\mathbf{r})\left[\frac{\sin(k_F u) - k_F u\cos(k_F u)}{(k_F u)^3}\right]^2 \tag{57}$$

which determines the exchange energy by Eq. (33). The final result is

$$E_{\mathrm{x}}^{\mathrm{LDA}} = -C_{\mathrm{x}}\int \rho^{4/3}(\mathbf{r})\mathrm{d}\mathbf{r}, \qquad \text{where} \qquad C_{\mathrm{x}} = \frac{3}{4}\left(\frac{3}{\pi}\right)^{1/3} \tag{58}$$

Equivalently, the LDA exchange energy per particle is

$$\epsilon_{\mathrm{x}}^{\mathrm{LDA}}(\rho) = -C_{\mathrm{x}}\rho^{1/3} = -\frac{3}{4}\left(\frac{3}{2\pi}\right)^{2/3}\frac{1}{r_{\mathrm{s}}} \tag{59}$$

where $r_{\mathrm{s}} = (3/4\pi\rho)^{1/3}$ is the radius of a sphere that contains the charge of one electron. Eq. (59) was first obtained by Wigner and Seitz [106].

The LDA exchange formula (58) is exact for a UEG but underestimates the exchange energy of inhomogeneous systems. Generally, LDA is more accurate than the Hartree–Fock method but falls far short of chemical accuracy (1 kcal/mol).

The extension of Eq. (58) to spin-polarized systems is called the LSDA. According to Eq. (50), the LSDA exchange energy is

$$E_{\mathrm{X}}^{\mathrm{LSDA}}[\rho_\alpha, \rho_\beta] = -2^{1/3} C_{\mathrm{X}} \int \left(\rho_\alpha^{4/3} + \rho_\beta^{4/3}\right) \mathrm{d}\mathbf{r} \tag{60}$$

Although Eq. (60) is all one needs for practical purposes, let us cast it in a different form (for reasons that will become clear when we discuss the LSDA for correlation). Introducing the relative spin-polarization

$$\zeta = \frac{\rho_\alpha - \rho_\beta}{\rho_\alpha + \rho_\beta} \tag{61}$$

and using $\rho_\alpha = \frac{1}{2}(1+\zeta)\rho$ and $\rho_\beta = \frac{1}{2}(1-\zeta)\rho$, we rewrite Eq. (60) as

$$E_{\mathrm{X}}^{\mathrm{LSDA}}[\rho_\alpha, \rho_\beta] = \int \rho \epsilon_{\mathrm{X}}(\rho, \zeta)\, \mathrm{d}\mathbf{r} \tag{62}$$

where

$$\epsilon_{\mathrm{X}}(\rho, \zeta) = -\frac{1}{2} C_{\mathrm{X}} \rho^{1/3} [(1+\zeta)^{4/3} + (1-\zeta)^{4/3}] \tag{63}$$

For a spin-compensated ('paramagnetic', $\zeta = 0$) electron gas,

$$\epsilon_{\mathrm{X}} = \epsilon_{\mathrm{X}}^{P} = -C_{\mathrm{X}} \rho^{1/3} \tag{64}$$

and for fully polarized ('ferromagnetic', $\zeta = \pm 1$)

$$\epsilon_{\mathrm{X}} = \epsilon_{\mathrm{X}}^{F} = -2^{1/3} C_{\mathrm{X}} \rho^{1/3} \tag{65}$$

For intermediate spin-polarizations $0 < \zeta < 1$, one can write $\epsilon_{\mathrm{X}}(\rho, \zeta)$ as an exact interpolation between the para- and ferromagnetic cases,

$$\epsilon_{\mathrm{X}}(\rho, \zeta) = \epsilon_{\mathrm{X}}^{P}(\rho) + [\epsilon_{\mathrm{X}}^{F}(\rho) - \epsilon_{\mathrm{X}}^{P}(\rho)]\, f(\zeta) \tag{66}$$

where the interpolating function is readily shown to be

$$f(\zeta) = \frac{1}{2} \frac{[(1+\zeta)^{4/3} + (1-\zeta)^{4/3} - 2]}{(2^{1/3} - 1)} \tag{67}$$

Ernzerhof and Scuseria [107] suggested approximating E_{X} using the LDA functional in which the actual $\rho(\mathbf{r})$ is replaced by a fictitious density $\tilde{\rho}(\mathbf{r})$ defined by $\tau(\mathbf{r}) = C_F \tilde{\rho}^{5/3}(\mathbf{r})$, where $\tau(\mathbf{r})$ is the kinetic energy density of the actual nonuniform system. This leads to the local τ-approximation (LTA)

$$E_{\mathrm{X}}^{\mathrm{LTA}}[\tau] = \frac{C_{\mathrm{X}}}{C_F^{4/5}} \int \tau^{4/5}(\mathbf{r})\, \mathrm{d}\mathbf{r} \tag{68}$$

where $C_F = \frac{10}{3}(3\pi^2)^{2/3}$. Eq. (68) can be viewed as an alternative to the conventional LDA. In numerical tests, the LTA was found to be 'complementary' to the LDA. It predicts more accurately exchange energy contributions to the atomization energies in those cases where LDA is in large error, and *vice versa*. For a UEG, LDA is equivalent to LTA.

24.5.2 Local density approximation for correlation

Correlation is a much more difficult problem than exchange, so exact analytic forms of $\epsilon_c^{\mathrm{LDA}}(\rho)$ are known only for two limiting cases. The first is the high-density (weak correlation) limit of a spin-compensated UEG

$$\epsilon_c^P(r_s) = A_{\mathrm{GB}} \ln r_s + B + r_s(C \ln r_s + D), \qquad r_s \ll 1 \tag{69}$$

The constants A_{GB} and B were evaluated by Gell-Mann and Brueckner [108], C and D by Carr and Maradudin [109]. Specifically, in hartree units (E_h),

$$A_{\mathrm{GB}} = \frac{1 - \ln 2}{\pi^2} \approx 0.031091 \tag{70}$$

The second case is the low-density (strong correlation) limit obtained by Nozières and Pines [110] and Carr [111]

$$\epsilon_c^P(r_s) = \frac{1}{2}\left(\frac{U_0}{r_s} + \frac{U_1}{r_s^{3/2}} + \frac{U_2}{r_s^2} + \ldots\right), \qquad r_s \gg 1 \tag{71}$$

where U_k are again known constants. Similar formulas exist for $\epsilon_c^F(r_s)$.

The exact numerical values of $\epsilon_c^P(r_s)$ and $\epsilon_c^F(r_s)$ are known, with small statistical uncertainties, for several intermediate values of r_s from Monte Carlo simulations of the UEG carried out by Ceperley and Alder [112]. Based on these results, several interpolation formulas for $\epsilon_c^P(r_s)$ and $\epsilon_c^F(r_s)$ have been devised to connect the high- and low-density limits [Eqs. (69) and (71)] and simultaneously reproduce the Ceperley–Alder data for intermediate r_s. Three such parametrizations are widely used in quantum-chemical codes.

Perdew and Zunger [77] (PZ81) suggested the following parametrization of the Ceperley–Alder data for the spin-compensated and spin-polarized cases

$$\epsilon_c^{\mathrm{PZ81,i}}(r_s) = \begin{cases} \dfrac{\gamma}{1 + \beta_1 r_s^{1/2} + \beta_2 r_s}, & \text{if } r_s \geq 1, \\ A \ln r_s + B + C r_s \ln r_s + D r_s, & \text{if } r_s < 1, \end{cases} \tag{72}$$

where $\gamma, \beta_1, \beta_2, A, B, C$, and D are parameters, different for $i = P$ and $i = F$. In particular, $A^P = 2A^F = A_{\mathrm{GB}}$. The PZ81 parametrization has several shortcomings, such as an artificial discontinuity of second and higher derivatives at $r_s = 1$.

Vosko *et al.* [113] (VWN) proposed a more accurate but less transparent representation

$$\epsilon_c^{\mathrm{VWN,i}}(x) = A\left\{\ln \frac{x^2}{X(x)} + \frac{2b}{Q} \tan^{-1} \frac{Q}{2x + b}\right.$$

$$-\frac{bx_0}{X(x_0)}\left[\ln\frac{(x-x_0)^2}{X(x)}+\frac{2(2x_0+b)}{Q}\tan^{-1}\frac{Q}{2x+b}\right]\Bigg\} \quad (73)$$

where $x = r_s^{1/2}$, $X(x) = x^2 + bx + c$, $Q = (4c - b^2)^{1/2}$, and A, b, c, and x_0 are parameters. For an explanation of this form, see Ref. [113].

The best available analytic representation of the Ceperley–Alder data was devised by Perdew and Wang [114] (PW92)

$$\epsilon_c^{\mathrm{PW92,i}}(r_s) = -2A(1+\alpha_1 r_s)\ln\left[1+\frac{1}{2A(\beta_1 r_s^{1/2}+\beta_2 r_s+\beta_3 r_s^{3/2}+\beta_4 r_s^2)}\right] \quad (74)$$

where A, p, α_1, β_1, β_2, β_3, and β_4 are parameters assuming different values for each $\epsilon_c^{\mathrm{i}}(r_s)$. Unlike the VWN form, form (74) properly allows for a nonzero coefficient C in Eq. (69) and avoids other deficiencies of the PZ81 and VWN parametrizations.

Even with accurate representations of $\epsilon_c^P(r_s)$ and $\epsilon_c^F(r_s)$ at our disposal, we need a general formula applicable to spin-polarized systems. Without loss of generality we can assume that, in analogy with Eq. (62),

$$E_c^{\mathrm{LSDA}}[\rho_\alpha, \rho_\beta] = \int \rho\epsilon_c(r_s, \zeta)\mathrm{d}\mathbf{r} \quad (75)$$

where the function $\epsilon_c(r_s, \zeta)$ is to be determined. Unfortunately, unlike for exchange, there is no simple exact formula relating $\epsilon_c(r_s, \zeta)$ to $\epsilon_c^P(r_s)$, $\epsilon_c^F(r_s)$, and ζ. Von Barth and Hedin [56] proposed using the same interpolation formula for $\epsilon_c(r_s, \zeta)$ as for $\epsilon_x(\rho, \zeta)$, that is,

$$\epsilon_c^{\mathrm{BH}}(r_s, \zeta) = \epsilon_c^P(r_s) + [\epsilon_c^F(r_s) - \epsilon_c^P(r_s)]f(\zeta) \quad (76)$$

where $f(\zeta)$ is given by Eq. (67). In practice, Eq. (76) is not very accurate [45]. Vosko et al. [113] examined several alternatives to the Barth–Hedin interpolation formula and recommended the following expression

$$\epsilon_c^{\mathrm{VWN}}(r_s, \zeta) = \epsilon_c^P(r_s) + \alpha_c(r_s)\left[\frac{f(\zeta)}{f''(0)}\right](1-\zeta^4) + [\epsilon_c^F(r_s) - \epsilon_c^P(r_s)]f(\zeta)\zeta^4 \quad (77)$$

where $\alpha_c(r_s)$ is a new function called spin stiffness. The spin stiffness is formally defined as $\alpha_c(r_s) = [\partial^2\epsilon_c(r_s, \zeta)/\partial\zeta^2]_{\zeta=0}$ and fitted to the same analytic form as $\epsilon_c^P(r_s)$ and $\epsilon_c^F(r_s)$.

The PW92 parametrization of the LDA correlation energy of Eq. (74) combined with VWN interpolation formula (77) is the most accurate representation of the LSDA correlation energy functional currently available.[2] It is used as part of the PW91 correlation functional implemented in the GAUSSIAN program [32].[3] The PZ81 parametrization of $\epsilon_c^{\mathrm{i}}(r_s)$ in combination with the Barth–Hedin interpolation formula

[2] In the GAUSSIAN program [32], the keyword LSDA requests the Barth–Hedin interpolation formula in which $\epsilon_c^P(r_s)$ and $\epsilon_c^F(r_s)$ have the analytic form of Eq. (73) but employ parameters that were fitted to the correlation energy of a UEG calculated in the random-phase approximation (RPA).

[3] Currently, there is no stand-alone keyword for the PW92 representation of LSDA in the GAUSSIAN program. In GAUSSIAN 03, one can perform LSDA-PW92 calculations by specifying the combination of keywords SPW91 IOp(3/78 = 0000010000) in the input section.

(76) is used in the GAUSSIAN implementation of the Perdew 1986 correlation functional (P86) and as a standalone LDA correlation functional (keyword PL). The VWN parametrization of $\epsilon_c^i(r_s)$ of Eq. (73) is meant to be used with the VWN interpolation formula (77).

24.6 DENSITY-GRADIENT EXPANSION

Although the LSDA is exact for a UEG and quite accurate for solids, it is less than satisfactory for atoms and molecules. The natural step beyond the LDA is a formal expansion of E_{xc} in gradients of the density. This idea was suggested already by Hohenberg and Kohn [15]. In general, the DGE of the exchange-correlation energy has the form of Eq. (52), but the requirement that E_{xc} be invariant under rotations about $\mathbf{r}$ makes the coefficients of $\nabla\rho$ and other nonscalar terms vanish, yielding [15]

$$E_{xc}[\rho] = \int \rho\epsilon_{xc}^{LDA}(\rho)d\mathbf{r} + \int |\nabla\rho|^2 \epsilon_{xc}^{(2)}(\rho)d\mathbf{r} + \ldots \tag{78}$$

where $\epsilon_{xc}^{(2)}(\rho)$ are coefficients of appropriate dimensionality. The $\nabla^2\rho$ term in Eq. (78) has been eliminated *via* integration by parts.

It is customary to discuss gradient expansions for *exchange* in terms of dimensionless reduced density gradients

$$x = \frac{|\nabla\rho|}{\rho^{4/3}}, \qquad s = \frac{x}{2(3\pi^2)^{1/3}} \tag{79}$$

The form of the gradient expansion for E_x is fixed by dimensional analysis:

$$E_x[\rho] = E_x^{LDA}[\rho] - \beta_x \int \rho^{4/3}x^2 d\mathbf{r} + \ldots \tag{80}$$

where β_x is the second-order DGE coefficient. Equivalently, Eq. (80) can be written as

$$\begin{aligned} E_x[\rho] &= \int \rho\epsilon_x^{LDA}(\rho)\left[1 + \frac{\beta_x}{C_x}x^2 + \ldots\right]d\mathbf{r} \\ &= \int \rho\epsilon_x^{LDA}(\rho)\left[1 + \mu_x s^2 + \ldots\right]d\mathbf{r} \end{aligned} \tag{81}$$

The coefficients of s^2 and x^2 are related by

$$\mu_x = \frac{[2(3\pi^2)^{1/3}]^2}{C_x}\beta_x \tag{82}$$

Derivation of gradient expansion coefficients is very involved even for the second-order terms [115–118]. According to Antoniewicz and Kleinman [115], the exact second-order coefficient in Eq. (81) is

$$\mu_x = \mu_{AK} = \frac{10}{81} \tag{83}$$

which is different from Sham's result [116]

$$\mu_X = \mu_S = \frac{7}{81}, \qquad \beta_S = 0.00166721... \tag{84}$$

Coefficients of the fourth-order terms in the DGE for the exchange energy for slowly-varying densities have been obtained only recently [119,120].

The gradient expansion for the *correlation* energy begins as

$$E_c[\rho] = E_c^{LDA}[\rho] + \int C_c(r_s)\rho^{4/3}x^2 d\mathbf{r} + \cdots$$

$$= \int \rho\Big[\epsilon_c^{LDA}(\rho) + \beta_c(\rho)t^2 + \ldots\Big]d\mathbf{r} \tag{85}$$

where

$$t = \frac{1}{4(3/\pi)^{1/6}}\frac{|\nabla\rho|}{\rho^{7/6}} \tag{86}$$

The value of the coefficient β_c for the high-density limit was obtained analytically by Ma and Brueckner [121] (in a.u.)

$$\lim_{\rho\to\infty} \beta_c = \beta_{MB} = 16(3/\pi)^{1/3}C_c(0) \approx 1.97563/3\pi^2 = 0.0667244... \tag{87}$$

An analytic representation of the function $C_c(r_s)$ has been given by Rasolt and Geldart [122]. Observe that Eqs. (81) and (85) correctly reduce to LDA for uniform densities.

The second-order DGE improving upon the LTA of Eq. (68) has been derived by Ernzerhof and Scuseria [123]

$$E_X^{\tau DGE}[\tau] = E_X^{LTA}[\tau] + \beta_{ES}\int \tau^{4/5}\left(\frac{|\nabla\tau|}{\tau^{6/5}}\right)^2 d\mathbf{r} \tag{88}$$

where is β_{ES} is a nonempirical coefficient. Because β_{ES} turns out to be positive, Eq. (88) can give unphysical positive exchange energies.

The second-order DGE results only in modest improvements over LDA for exchange energies. Herman et al. [124] found empirically that the optimal value of the second-order gradient coefficient β_x in the second-order DGE approximation is about 2.5 times greater than the exact value. For correlation, the second-order DGE overestimates the correction needed to reproduce the exact correlation energy by a factor of 5 and so predicts positive correlation energies [121]. This implies that density variations in atoms and molecules are too rapid to be approximated by Eqs. (81) and (85). Because of such problems, truncated DGEs are not used as practical density functionals. Instead, they are regarded as the exact forms to which approximate exchange-correlation functionals should reduce in the limit of slowly varying densities.

24.7 CONSTRAINT SATISFACTION

Improvement of the second-order density gradient expansion became the focus of many research efforts in the 1970s and 1980s. Some developers of density functionals [125–132] saw the root of problems in the wrong analytic behavior of the exchange-correlation potential of the DGE functional and concentrated on designing semiempirical gradient corrections with desired analytic properties. Others [133–135] explained failures of the second-order DGE (divergence of the exchange-correlation potential, overcorrection of the correlation energy, etc.) by the fact that the exchange-correlation hole corresponding to the second-order DGE exhibits undamped oscillation at large u values and thereby severely violates the important normalization constraints and sign properties of the correct hole. Accordingly, functionals based on explicitly normalized and properly signed exchange-correlation holes have been proposed. It is not hard to see that the two approaches are essentially different routes to the same goal: satisfaction of important exact constraints. As we shall see in this section, the process of identifying important analytic properties of the exact exchange-correlation functional and imposing them on approximate constructions is the most reliable strategy of density functional design.

24.7.1 Corrections on the asymptotic behavior

The second-order density gradient expansion for exchange performs well only in the limit of small reduced density gradients x. The assumption that x is small may be justified for an infinite electron gas but not for finite systems, where x diverges in far-out regions. This can be demonstrated by assuming a spherically symmetric exponential density $\rho(r) = e^{-ar}$,

$$\lim_{r \to \infty} x = \lim_{r \to \infty} \frac{|\partial\rho/\partial r|}{\rho^{4/3}} = \lim_{r \to \infty} a e^{ar/3} = \infty \tag{89}$$

The second-order gradient contribution to the energy density, proportional to $\rho^{4/3}x^2$ [see Eq. (80)], remains finite but the leading term of the exchange potential $v_{\mathrm{x}}(\mathbf{r})$ [see Eq. (25)] scales like $\rho^{1/3}x^2$, and so is unbounded at large r. Various modifications of the second-order DGE have been suggested to eliminate the asymptotic divergence of $v_{\mathrm{x}}(\mathbf{r})$. Most of these functionals fall into the category of GGAs [136]

$$E_{\mathrm{X}}^{\mathrm{GGA}}[\rho] = \int \rho\epsilon_{\mathrm{X}}^{\mathrm{LDA}}(\rho)F_{\mathrm{x}}(x)\,\mathrm{d}\mathbf{r} = \int \rho\epsilon_{\mathrm{X}}^{\mathrm{GGA}}(\rho, x)\mathrm{d}\mathbf{r} \tag{90}$$

in which $F_{\mathrm{x}}(x)$ [or $F_{\mathrm{x}}(s)$] approximates the exact enhancement factor of Eq. (81).

One of the first successful functionals of this type was proposed in 1986 by Becke [125]. Its enhancement factor is modeled after that of Eq. (81) and has a damped second-order gradient term

$$F_X^{B86}(s) = 1 + \frac{\beta c_2 (c_1 s)^2}{1 + \gamma (c_1 s)^2} \tag{91}$$

where $c_1 = 2(6\pi^2)^{1/3}$ and $c_2 = (2^{1/3}C_x)^{-1}$, while $\beta = 0.0036$ and $\gamma = 0.004$ are empirical parameters.[4] Shortly thereafter, Becke observed [126] that the large-gradient ($s \rightarrow \infty$) limit of Eq. (91) is $\sim s^0$, whereas the correct limit should be $\sim s^{2/5}$. To ensure the correct behavior *both* in the low- and high-gradient limits, he proposed the following modification

$$F_X^{mB86}(s) = 1 + \frac{\beta c_2 (c_1 s)^2}{[1 + \gamma (c_1 s)^2]^{4/5}} \tag{92}$$

where $\beta = 0.00375$ and $\gamma = 0.007$ are again optimized parameters.

With an empirical value of β, neither B86 nor mB86 recovers the correct second-order gradient expansion of Eq. (81). DePristo and Kress [127] explicitly imposed this constraint by using the formula

$$F_X^{DK87}(s) = 1 + \mu_S s^2 \frac{1 + \gamma_1 (c_1 s)^m}{1 + \gamma_2 (c_1 s)^2} \tag{93}$$

where $\mu_S = 7/81$ is Sham's exact second-order coefficient, but γ_1, γ_2, and $m \leq 1$ are still adjustable parameters. Note that the large-gradient limit of Eq. (93) is $\sim s^m$, which is correct only if $m = 2/5$.

It is also known that in a finite many-electron system, the true exchange potential $v_x(\mathbf{r})$ and exchange energy density $e_x(\mathbf{r})$ have the following asymptotic behavior [45,137]:

$$v_x(\mathbf{r})|_{r \rightarrow \infty} = -\frac{1}{r} + C \tag{94}$$

where C is a constant that vanishes everywhere except in nodal surfaces of the highest occupied orbital [138], and

$$e_x(\mathbf{r})|_{r \rightarrow \infty} = -\frac{\rho(r)}{2r} \tag{95}$$

Although the B86, mB86, and DK87 exchange potentials no longer diverge and, like energy densities, go to zero in the $r \rightarrow \infty$ limit, their asymptotic decay is not of the form prescribed by Eqs. (94) and (95). Becke [128] argued that the asymptotic limit of $e_x(\mathbf{r})$ is an important constraint and proposed a functional which satisfies Eq. (95) for exponentially decaying densities

$$F_X^{B88}(s) = 1 + \frac{\beta c_2 (c_1 s)^2}{1 + 6\beta (c_1 s) \sinh^{-1}(c_1 s)} \tag{96}$$

[4] The factors $c_1 = 2(6\pi^2)^{1/3}$ and $c_2 = (2^{1/3}C_x)^{-1}$ in Eqs. (91)–(93), (96), (99), (101), (110), and (180) arise in the transition from the $E_x[\rho_\alpha, \rho_\beta]$ form of the original definitions, where F_x is written in terms of $x_\sigma = |\nabla\rho_\sigma|/\rho_\sigma^{4/3}$, to the standard form $E_x[\rho]$, where F_x is written in terms of s only. The transformation is based on the spin-scaling relation of Eq. (50). Note that $c_1 s = 2^{1/3}x$.

where $\beta = 0.0042$ is an empirical parameter determined by a least squares fit to exact Hartree–Fock exchange energies of six noble gas atoms (He through Rn). It is the function $\sinh^{-1} y = \ln(y + \sqrt{1 + y^2})$ that does the trick of ensuring the correct behavior of $e_x^{B88}(\mathbf{r})$ at large r. To see this, use the expansion [139]

$$\sinh^{-1} y = \ln(2y) + \frac{1}{2 \cdot 2y^2} - \frac{1 \cdot 3}{2 \cdot 4 \cdot 4y^4} + \cdots, \quad |y| > 1 \tag{97}$$

and the large-r form of $c_1 s = 2^{1/3} x$, where x is given by Eq. (89). The B88 functional proved very successful and remains one of the most popular approximations for the exchange energy.

In the early 1990s, Lacks and Gordon (LG) [140] observed that the LDA, B86, mB86, DK87, and B88 exchange functionals all give a poor description of van der Waals interactions between rare gas atoms, especially at large internuclear distances. Even at the energy minima, exchange contributions to the binding energy are in errors exceeding 100%. In an effort to refine the description of low-density, high-gradient regions responsible for long-range interactions, they considered the form

$$F_x^{LG}(s) = \frac{\left[1 + (a_d + \mu_{AK})s^2/b + a_4 s^4 + a_6 s^6 + a_8 s^8 + a_{10} s^{10} + a_{12} s^{12}\right]^b}{1 + a_d s^2} \tag{98}$$

where $a_4, a_6, a_8, a_{10}, a_{12}$, and b are empirical parameters obtained by least squares fitting to the total exchange energies of the He, Ne, and C^+ atoms and the exchange contributions to the binding energies of He_2 and Ne_2. The value of a_d was fixed at 10^{-8}. In more extensive tests, Adamo and Barone [141] found that the combination of the LG exchange and LYP correlation is not a substantial improvement over BLYP.

In 1997, Filatov and Thiel [142] observed that the correct asymptotic limit of the exchange energy per density is also obtained with functionals of the general form

$$F_x^{FT97}(s) = 1 + \frac{\beta c_2 (c_1 s)^2}{\left[1 + (6\beta/m)^m (c_1 s)^m \sinh^{-1}(c_1 s)^m\right]^{1/m}} \tag{99}$$

Using $m = 2$ and allowing the adjustable parameter β to assume different values for spin-up and spin-down densities, they managed to reduce the mean absolute error (MAE) in the exchange energy by about one third relative to B88.

Van Leeuwen and Baerends [129] achieved the correct asymptotic behavior of $v_{xc}(\mathbf{r})$ by modeling the Kohn–Sham exchange-correlation potential directly rather than by approximating the energy density. Their first model $v_{xc}(\mathbf{r})$ had an analytic form inspired by the B88 functional [129]. More recent shape-corrected potentials of Baerends and coworkers [143–147] involve explicit dependence on Kohn–Sham orbitals and satisfy even more exact constraints, such as invariance with respect to shifting the external potential by a constant and correct short-range behavior. The advantage of modeling $v_{xc}(\mathbf{r}) = v_x(\mathbf{r}) + v_c(\mathbf{r})$ is that potentials, unlike energy densities, are uniquely determined by $\rho(\mathbf{r})$. Given an exchange potential $v_x(\mathbf{r})$, the exchange energy can be evaluated by the Levy–Perdew formula [78]. More recently, Tozer and Handy [131,132] proposed an explicit asymptotic correction for conventional exchange-correlation potentials and obtained encouraging results for sensitive properties. The correct asymptotic behavior of the exchange energy density and potential appears to be much less important for obtaining accurate atomic exchange energies.

What about satisfying constraints (94) and (95) simultaneously? Engel and coworkers [148] proved that no GGA of the form of Eq. (90) can reproduce the asymptotic limits of Eqs. (94) and (95) at the same time. Jemmer and Knowles [130] and Filatov and Thiel [149] attempted to satisfy Eqs. (94) and (95) simultaneously by going *beyond* the conventional gradient approximation and introducing the dependence on the Laplacian of the density:

$$E_x[\rho] = \int \rho\epsilon_x^{\mathrm{LDA}}(\rho)F_x(\rho, |\nabla\rho|, \nabla^2\rho)\mathbf{dr} \tag{100}$$

Such functionals suffer, however, from numerical instabilities with respect to small density changes, which makes it practically impossible to obtain variational solutions of the Kohn–Sham equations [130]. Neumann and Handy [150] investigated the possibility of including terms of up to fourth order in ∇ and also arrived at decidedly disappointing conclusions.

Gill [151] disputed the claim that the asymptotic divergence of the exchange potential is of practical significance and argued in favor of a minimalistic empirical functional with a readjusted fractional power of the gradient correction

$$F_x^{\mathrm{G96}}(s) = 1 + \beta c_2(c_1 s)^{3/2} \tag{101}$$

where $\beta = 1/137$ is a parameter whose value was chosen to reproduce the Hartree–Fock exchange energy of the Ar atom (one also cannot help suspecting a playful reference to the fine structure constant). The exchange potential corresponding to Eq. (101) again diverges at large x. Contrary to what one might expect, G96 shows an accuracy on par with B88 in test calculations of thermochemical molecular properties, a better performance than of some much more sophisticated functionals.

GGAs based on the local-τ approximation of Eq. (68) have been constructed by Ernzerhof et al. [152]. These workers focused on improving the original LTA of Eq. (68) and second-order τ-DGE of Eq. (88) by imposing several important constraints on the enhancement factor: (a) the correct homogeneous density limit; (b) negativity of E_x; (c) correct asymptotic behavior of the exchange energy density [Eq. (95)]. The resulting τ-GGA shows a performance comparable to that of conventional GGA functionals.

Intuitively designed damped gradient corrections have been also used to improve the LDA for correlation. The first attempt of this kind was made by Ma and Brueckner [121] in their paper on the exact second-order expansion of $E_c[\rho]$, where they also propose the functional

$$E_c^{\mathrm{MB68}}[\rho] = \int \rho\epsilon_c^{\mathrm{LDA}}(\rho)\left[1 - \beta_{\mathrm{MB}}\frac{t^2}{\nu\rho\epsilon_c^{\mathrm{LDA}}(\rho)}\right]^{-\nu}\mathbf{dr} \tag{102}$$

The constant $\nu = 0.32$ was fitted to the empirical correlation energies of several atoms.

24.7.2 Normalization of the exchange-correlation hole

Similar to the partitioning of E_{xc} into exchange and correlation contributions, the exchange-correlation hole at a coupling strength λ can be partitioned as

$$h_{\mathrm{xc}}^{\lambda}(\mathbf{r}_1, \mathbf{r}_2) = h_{\mathrm{x}}(\mathbf{r}_1, \mathbf{r}_2) + h_{\mathrm{c}}^{\lambda}(\mathbf{r}_1, \mathbf{r}_2) \tag{103}$$

where the exchange hole is defined as the noninteracting limit

$$h_{\mathrm{x}}(\mathbf{r}_1, \mathbf{r}_2) \equiv h_{\mathrm{xc}}^{\lambda=0}(\mathbf{r}_1, \mathbf{r}_2) \tag{104}$$

The basic properties of the exact exchange and correlation holes are:

$$h_{\mathrm{x}}(\mathbf{r}_1, \mathbf{r}_2) \leq 0, \qquad \int h_{\mathrm{x}}(\mathbf{r}_1, \mathbf{r}_2)\mathrm{d}\mathbf{r}_2 = -1 \tag{105}$$

$$\int h_{\mathrm{c}}^{\lambda}(\mathbf{r}_1, \mathbf{r}_2)\mathrm{d}\mathbf{r}_2 = 0 \tag{106}$$

Langreth and Perdew [133] and Perdew [135] explained the failure of the second-order DGE by a gross violation of Eqs. (105) and (106) by the DGE exchange and correlation holes. Such violations have been traced to spurious large-u behavior of the spherically averaged $h_{\mathrm{xc}}(\mathbf{r}, u)$. Perdew showed [135] that by cutting off large-u parts of $h_{\mathrm{xc}}(\mathbf{r}, u)$ and positive parts of $h_{\mathrm{x}}(\mathbf{r}, u)$ one can obtain successful density functional approximations. The essence of the cut off procedure for exchange in real space is as follows [153–155].

Start with the analytic second-order density gradient expansion for the coupling constant- and angle-averaged exchange hole, $h_{\mathrm{x}}^{\mathrm{DGE}}(\mathbf{r}, u)$. Convert the diverging DGE hole into the normalized GGA hole by applying a sharp cutoff:

$$h_{\mathrm{x}}^{\mathrm{GGA}}(\mathbf{r}, u) = h_{\mathrm{x}}^{\mathrm{DGE}}(\mathbf{r}, u)\theta(h_{\mathrm{x}}^{\mathrm{DGE}})\theta(u_0(\mathbf{r}) - u) \tag{107}$$

where

$$\theta(y) = \begin{cases} 1, & \text{if } y \geq 0, \\ 0, & \text{if } y < 0. \end{cases} \tag{108}$$

The first step function in Eq. (107) enforces the negativity constraint. The second function involves a cutoff radius $u_0(\mathbf{r})$ chosen to satisfy the hole normalization constraint of Eq. (105). The hole of Eq. (108) is substituted into Eq. (33) to give the numerical enhancement factor $F_{\mathrm{x}}(x)$, which is then fitted to an analytic form.[5] The real-space cut-off procedure determines only the general features of the functional but not its analytic representation. The latter may be chosen in different ways depending on which exact constraints one decides to impose. The first practical exchange functional of this type was

[5] Strictly speaking, the cutoff procedure uses the angle- and *system*-averaged GGA exchange hole [156], i.e., the hole around an electron averaged over the position of that electron. We omit here technical details for the sake of clarity.

derived in 1986 by Perdew and Wang [136]

$$F_{\mathrm{X}}^{\mathrm{PW86}}(s) = \left(1 + \frac{\mu_{\mathrm{S}} s^2}{m} + b s^4 + c s^6\right)^m \tag{109}$$

in which $m = 1/15$, $\mu_{\mathrm{S}} = 7/81$, $b = 14$, and $c = 0.2$ are nonempirical parameters.

In 1991, Perdew and Wang [76,155,157] used the same approach to derive a nonempirical exchange functional whose analytic representation was inspired by the B88 form:

$$F_{\mathrm{X}}^{\mathrm{PW91}}(s) = 1 + \frac{\beta c_2 (c_1 s)^2 - [\beta c_2 (c_1 s)^2 - \mu_{\mathrm{AK}}] e^{-100 s^2} - 0.004 s^4}{1 + 6\beta (c_1 s) \sinh^{-1}(c_1 s) + 0.004 s^4} \tag{110}$$

where $c_1 = 2(6\pi^2)^{1/3}$, $c_2 = (2^{1/3} C_{\mathrm{x}})^{-1}$, $\beta = 0.0042$ (the B88 coefficient), and $\mu_{\mathrm{AK}} = 10/81$. The particular form of Eq. (110) was chosen to satisfy a large number of exact constraints. Although PW91 is modeled after the B88 functional, it does not yield the correct asymptotic behavior of the exchange energy density, a property abandoned in favor of more desirable constraints. The function $\sinh^{-1}(c_1\, s)$ in Eq.(110) is only a B88 relic.

The exact DGE for the correlation hole is not known except in the high-density limit, so away from this limit approximate gradient-corrected models have to be used instead. Perdew and coworkers [153,158] constructed a real-space model for the spherically-averaged GGA correlation hole starting from an accurate analytic representation $A_{\mathrm{c}}(r_{\mathrm{s}}, \zeta, u)$ of the LSDA correlation hole, adding a gradient correction $B_{\mathrm{c}}(r_{\mathrm{s}}, \zeta, u)$, and truncating the sum to satisfy the normalization constraint of Eq. (106)

$$\bar{h}_{\mathrm{c}}^{\mathrm{GGA}}(r_{\mathrm{s}}, \zeta, t, u) = \phi^3 (\phi k_{\mathrm{s}})^2 [A_{\mathrm{c}}(r_{\mathrm{s}}, \zeta, u) + t^2 B_{\mathrm{c}}(r_{\mathrm{s}}, \zeta, u)] \theta(u_0 - u) \tag{111}$$

where $k_{\mathrm{s}} = (4k_{\mathrm{F}}/\pi)^{1/2}$, t is the reduced gradient of Eq. (86) generalized to spin-polarized systems

$$t = \frac{|\nabla\rho|}{2\phi k_{\mathrm{s}} \rho} \tag{112}$$

u_0 is the cutoff radius, and

$$\phi(\zeta) = \frac{(1+\zeta)^{2/3} + (1-\zeta)^{2/3}}{2} \tag{113}$$

The hole of Eq. (111) is substituted into Eq. (33) and the integral is evaluated numerically. The result is written as

$$E_{\mathrm{c}}^{\mathrm{GGA}}[\rho_\alpha, \rho_\beta] = \int \rho \epsilon_{\mathrm{c}}^{\mathrm{LSDA}}(r_{\mathrm{s}}, \zeta)\, \mathrm{d}\mathbf{r} + \int \rho H(r_{\mathrm{s}}, \zeta, t)\, \mathrm{d}\mathbf{r} \tag{114}$$

where $H(r_{\mathrm{s}}, \zeta, t)$ is a numerically defined gradient correction. The last step in this derivation is to find a suitable analytic representation for $H(r_{\mathrm{s}}, \zeta, t)$. Two different solutions to this problem have been proposed, leading to the PW91 [76,155,157] and

PBE [159] correlation functionals. The PW91 gradient correction has the form

$$H^{\mathrm{PW91}}(r_s, \zeta, t) = H_0^{\mathrm{PW91}}(r_s, \zeta, t) + H_1(r_s, \zeta, t) \tag{115}$$

The H_1 term in Eq. (115) is negligible, unless $s \ll 1$ [155]. In PBE,

$$H^{\mathrm{PBE}}(r_s, \zeta, t) = H_0^{\mathrm{PBE}}(r_s, \zeta, t) \tag{116}$$

The analytic representations of H_0^{PW91} and H_0^{PBE} have the same form given by

$$H_0(r_s, \zeta, t) = \gamma\phi^3(\zeta)\ln\left[1 + \frac{\beta_{\mathrm{MB}}}{\gamma}t^2\left(\frac{1 + At^2}{1 + At^2 + A^2t^4}\right)\right] \tag{117}$$

where

$$A = \frac{\beta_{\mathrm{MB}}}{\gamma}\frac{1}{e^{-\epsilon_c^{\mathrm{LSDA}}(r_s,\zeta)/\gamma\phi^3} - 1} \tag{118}$$

The only difference between H_0^{PW91} and H_0^{PBE} is in the values of γ used in Eqs. (117) and (118). In PW91,

$$\gamma_{\mathrm{PW91}} = \frac{\beta_{\mathrm{MB}}^2}{2\alpha} \approx 0.024734 \tag{119}$$

where $\alpha = 0.09$ is a constant chosen to approximate the t^4 dependence of the numerical $\epsilon_c^{\mathrm{GGA}}$. In PBE,

$$\gamma_{\mathrm{PBE}} = A_{\mathrm{GB}} \approx 0.031091 \tag{120}$$

which is the exact second-order gradient coefficient of the LDA correlation energy expansion [see Eq. (70)].

The analytic form of the gradient correction $H(r_s, \zeta, t)$ was motivated by the following three conditions [159]: (1) H must reduce to the second-order term in the DGE expansion for correlation in the slowly-varying limit, $H \rightarrow \beta_{\mathrm{MB}}\phi^3t^2$; (2) if the density changes infinitely fast ($t \rightarrow \infty$), the correlation energy must vanish, that is, $H \rightarrow -\epsilon_c^{\mathrm{LSDA}}$; (3) under uniform scaling to the high-density limit, H must cancel the logarithmic singularity of $\epsilon_c^{\mathrm{LDA}}$ to satisfy the uniform scaling constraint of Eq. (49).

The H_0^{PBE} recovers the first constraint by itself, but H_0^{PW91} does not. In order to recover the correct second-order gradient expansion for the correlation energy in the $s \rightarrow 0$ limit, another term is used:

$$H_1(r_s, \zeta, t) = \nu\left[C_c(r_s) - C_c(0) + \frac{3}{7}\beta_S\right]\phi^3(\zeta)t^2e^{-100\phi^4(\zeta)(k_st/k_F)^2} \tag{121}$$

where $\nu = 16(3/\pi)^{1/3}$ is the conversion factor from the units of ρt^2 to $\rho^{4/3}x^2$.

The numerical GGA based on the model hole of Eq. (111) and the PBE analytic representation of this functional both satisfy Levy's uniform scaling constraint for correlation, but the form $E_c^{\mathrm{PW91}}[\rho_\alpha, \rho_\beta]$ does not (the culprit here is the small H_1 term). In molecular calculations, however, these subtle differences never show up, so the PBE and PW91 correlation functionals are equivalent for most practical purposes.

24.7.3 Systematic constraint satisfaction

Simply imposing correct asymptotic limits on the second-order density-gradient expansion (Section 24.7.1) proved to be a very effective strategy for designing exchange functionals. It is reasonable to assume that the more exact constraints an approximate density functional satisfies, the more accurate and universal it will be. This idea is behind many existing density functional approximations.

Langreth and Mehl [134] used the sharp cut-off procedure in momentum space to eliminate spurious contributions to E_c *and* an empirical exponential function to damp the gradient contribution to the energy. The Langreth–Mehl (LM) functional [134,160,161] has now mostly a historical significance. A few years later, Perdew [162] improved the LM functional by imposing two additional requirements that it recover the correct second-order DGE in the slowly varying density limit and reduce in the uniform density limit to LDA, not to the random-phase approximation (RPA), as the LM functional does. Perdew's 1986 correlation functional is

$$E_c^{\mathrm{P86}}\left[\rho_\alpha,\rho_\beta\right]=E_c^{\mathrm{LSDA}}\left[\rho_\alpha,\rho_\beta\right]+\int\frac{e^{-\Phi(\rho,\nabla\rho)}}{\mathrm{d}(\zeta)}C_c(r_s)\frac{|\nabla\rho|^2}{\rho^{4/3}}\,\mathrm{d}\mathbf{r} \tag{122}$$

where the function

$$\mathrm{d}(\zeta)=2^{1/3}\left[\left(\frac{1+\zeta}{2}\right)^{5/3}+\left(\frac{1-\zeta}{2}\right)^{5/3}\right]^{1/2} \tag{123}$$

interpolates between the spin-compensated and spin-polarized forms [160], while

$$\Phi(\rho,\nabla\rho)=(9\pi)^{1/6}\tilde{f}\frac{C_c(0)}{C_c(r_s)}\frac{|\nabla\rho|}{\rho^{7/6}} \tag{124}$$

with the parameter $\tilde{f}=0.11$ chosen to fit the exact correlation energy of the neon atom. $C_c(r_s)$ is the gradient coefficient of Eq. (85) in the parametrization by Rasolt and Geldart [122]. In the original paper, $E_c^{\mathrm{LSDA}}[\rho_\alpha,\rho_\beta]$ was taken in the PZ81 parametrization of the Ceperley–Alder data, and this is how P86 is implemented in the GAUSSIAN program. The P86 correlation functional correctly reduces to LDA for uniform densities, but does not scale to a constant under the uniform scaling transformation of the density.

Wilson and Levy [163] were the first to explicitly impose the high-density scaling constraint of Eq. (49) on an approximate correlation functional

$$E_c^{\mathrm{WL}}[\rho_\alpha,\rho_\beta]=\int\rho\frac{(a+bx)(1-\zeta^2)^{1/2}}{c+d(x_\alpha+x_\beta)+r_s}\,\mathrm{d}\mathbf{r} \tag{125}$$

where $x=|\nabla\rho|/\rho^{4/3}$, $x_\sigma=|\nabla\rho_\sigma|/\rho_\sigma^{4/3}$, and a, b, c, and d are empirical parameters. Some prior correlation functionals, such as B88c and LYP (see Section 24.8 below), also respect Eq. (49) for all densities, although they predate the formal proof of it. More recent nonempirical functionals described in the rest of this section are purposely constructed to satisfy Eq. (49) among many other exact constraints. Satisfaction of density scaling constraints for the exchange and correlation energies may not be important for 'usual' systems, but it becomes crucial for a proper description of high-Z atomic ions [164].

In 1996, Perdew, Burke, and Ernzerhof (PBE) [159] constructed a simple exchange functional by imposing several energetically significant exact constraints. The enhancement factor of the PBE exchange functional is similar to that of B86 [Eq. (91)],

$$F_{\mathrm{X}}^{\mathrm{PBE}}(s) = 1 + \frac{\mu s^2}{1 + \mu s^2/\kappa} \tag{126}$$

Here $\kappa = 0.804$ is a nonempirical parameter chosen to satisfy the Lieb–Oxford bound in its local form: $F_{\mathrm{x}}(s) \leq 1.804$ for any s. The value of $\mu = \beta_{\mathrm{MB}}(\pi^2/3) \approx 0.21951$ is determined from the condition that the second-order gradient term for exchange cancel that for correlation (i.e. $\rho\beta_{\mathrm{MB}}t^2 = \mu C_{\mathrm{x}}\rho^{4/3}s^2$). This choice, rather than $\mu = \mu_{\mathrm{AK}} = \frac{10}{81}$, assures that in the $s \rightarrow 0$ limit the PBE exchange-correlation reduces to LDA for nearly uniform densities faster than PW91 does. Although Eq. (126) is not based on any particular hole model, it is numerically very similar to the PW91 exchange functional of Eq. (110) and, in fact, produces nearly the same results. An empirical adjustment of the parameter κ from 0.804 to 1.245 in the PBE exchange functional has been proposed [165], but Perdew and coworkers [166] defended the original nonempirical value. Hammer et al. [167] attempted to reconcile both points of view by suggesting a minor revision of the analytic form of Eq. (126).

The PW91 and PBE exchange-correlation approximations exhaust the number of exact constraints that can be practically imposed on GGAs, that is, functionals whose ingredients are $\rho_\sigma(\mathbf{r})$ and $\nabla\rho_\sigma(\mathbf{r})$. Bringing into play the kinetic energy density $\tau_\sigma(\mathbf{r})$ of Eq. (40) opens new opportunities (the meta-GGA level). One particular use of $\tau_\sigma(\mathbf{r})$ is based on the following property:

$$\tau_\sigma^W(\mathbf{r}) \leq \tau_\sigma(\mathbf{r}) \tag{127}$$

where

$$\tau_\sigma^W = \frac{1}{8}\frac{|\nabla\rho_\sigma|^2}{\rho_\sigma} \tag{128}$$

is the von Weizsäcker kinetic energy density.[6] In Eq. (127), the strict equality holds only if $\rho_\sigma(\mathbf{r})$ is represented by a single real Kohn–Sham orbital. Therefore, the quantity

$$\eta_\sigma^{\mathrm{SCC}} = 1 - \frac{\tau_\sigma^W}{\tau_\sigma} \tag{129}$$

vanishes for any one-electron system and is strictly positive in systems with more than

[6] Proof. Let us introduce the symbolic vectors $\mathbf{f} \equiv (\phi_1, \phi_2, \ldots, \phi_N)$ and $\mathbf{g} \equiv (\nabla\phi_1, \nabla\phi_2, \ldots, \nabla\phi_N)$, where the spin-index σ has been suppressed for brevity. By definition

$$\rho = |\mathbf{f}|^2, \qquad \tau = \frac{1}{2}|\mathbf{g}|^2, \qquad \tau_W = \frac{1}{8}\frac{|\nabla\rho|^2}{\rho} = \frac{1}{8}\frac{|\mathbf{g}^*\cdot\mathbf{f} + \mathbf{f}^*\cdot\mathbf{g}|^2}{|\mathbf{f}|^2} \leq \frac{1}{2}\frac{|\mathbf{f}^*\cdot\mathbf{g}|^2}{|\mathbf{f}|^2}$$

where the strict equality holds only if all ϕ_i are real. The result $\tau_W \leq \tau$ follows from the Schwarz inequality $|\mathbf{f}^*\cdot\mathbf{g}|^2 \leq |\mathbf{f}|^2|\mathbf{g}|^2$. Thus, if $N = 1$ and ϕ_1 is real, then $|\mathbf{f}^*\cdot\mathbf{g}|^2 = |\mathbf{f}|^2|\mathbf{g}|^2$, and so $\tau_W = \tau$.

one σ-spin electron [168,169]. In practical terms, this means that $\eta_\sigma^{\mathrm{SCC}}$ can be used as a multiplicative self-correlation correction (SCC) ensuring that E_c vanishes for any one-electron density [Eq. (45)]. Satisfaction of this constraint is not possible at the GGA level. The functionals discussed in the rest of this section all make use of this trick.

Becke's 1995 correlation functional (B95) [168] was constructed to satisfy the following set of conditions: (a) the correct uniform density limit; (b) separation of the correlation energy into parallel-spin and opposite-spin components; (c) zero correlation energy for one-electron systems; (d) good fit to the atomic correlation energies. These requirements are met by the following analytic form

$$E_\mathrm{c}^{\mathrm{B95}} = E_\mathrm{c}^{\alpha\alpha} + E_\mathrm{c}^{\beta\beta} + E_\mathrm{c}^{\alpha\beta} \tag{130}$$

where

$$E_\mathrm{c}^{\sigma\sigma} = \int \frac{\tau_\sigma - \tau_\sigma^W}{\tau_\sigma^{\mathrm{unif}}} \frac{e_{c\sigma\sigma}^{\mathrm{UEG}}(\rho_\sigma)}{(1 + c_{\sigma\sigma}x_\sigma^2)^2} \mathbf{dr} \tag{131}$$

$$E_\mathrm{c}^{\alpha\beta} = \int \frac{e_{c\alpha\beta}^{\mathrm{UEG}}(\rho_\alpha, \rho_\beta)}{1 + c_{\alpha\beta}(x_\alpha^2 + x_\beta^2)} \mathbf{dr} \tag{132}$$

in which $\tau_\sigma^{\mathrm{unif}} = \frac{3}{10}(6\pi^2)^{2/3}\rho_\sigma^{5/3}$, and $c_{\sigma\sigma} = 0.038$ and $c_{\alpha\beta} = 0.0031$ are fitted parameters. The same-spin correlation energy density of the UEG is simply

$$e_{c\sigma\sigma}^{\mathrm{UEG}}(\rho_\sigma) = e_\mathrm{c}^{\mathrm{LSDA}}(\rho_\sigma, 0) \tag{133}$$

The opposite-spin energy density is freed from self-correlation by the method of Stoll et al. [170,171]:

$$e_{c\alpha\beta}^{\mathrm{UEG}}(\rho_\alpha, \rho_\beta) = e_\mathrm{c}^{\mathrm{LSDA}}(\rho_\alpha, \rho_\beta) - e_\mathrm{c}^{\mathrm{LSDA}}(\rho_\alpha, 0) - e_\mathrm{c}^{\mathrm{LSDA}}(\rho_\beta, 0) \tag{134}$$

where $e_\mathrm{c}^{\mathrm{LSDA}}(\rho_\alpha, \rho_\beta) = \rho\epsilon_\mathrm{c}^{\mathrm{LSDA}}(\rho_\alpha, \rho_\beta)$ is the LSDA correlation energy density.

Krieger, Chen, Iafrate, and Savin (KCIS) [172] revisited the LDA for correlation and attributed the LDA overestimation of correlation energies for nonuniform densities to the fact that, unlike an electron gas, finite many-electron systems have a nonzero energy gap between the Fermi level and the continuum (the gap is equal to the ionization potential [173]). To improve the LDA description of systems with a finite orbital gap, they made use of the formula of Rey and Savin [174] for the correlation energy per particle of the electron gas with an energy gap G,

$$\tilde{\epsilon}_\mathrm{c}^i(\rho, \nabla\rho) = \frac{\epsilon_\mathrm{c}^i(\rho, \nabla\rho) + c_1^i(\rho)G}{1 + c_2^i(\rho)G + c_3^i(\rho)G^2} \tag{135}$$

where $i = P, F, c_k^i(\rho)$, are density-dependent coefficients, and $\epsilon_\mathrm{c}^i(\rho, \nabla\rho)$ is defined in Ref. [175]. They also assumed that the local value of the gap G is determined by the density and its gradient, $G = |\nabla\rho|^2/8\rho^2 = \tau_W/\rho$, and used the Barth–Hedin-type interpolation formula for $\tilde{\epsilon}_\mathrm{c}$, the correlation energy per particle of a spin-polarized electron gas *with a gap*. The final expression for the KCIS correlation functional was designed to satisfy the same three conditions as those respected by the PBE correlation functional, plus the requirement that the correlation energy vanish for one-electron densities [Eq. (45)]

$$E_{\mathrm{c}}^{\mathrm{KCIS}}[\rho_\alpha, \rho_\beta] = \int \left[\rho\tilde{\epsilon}_{\mathrm{c}}(\rho_\alpha, \rho_\beta, \nabla\rho_\alpha, \nabla\rho_\beta) - \sum_\sigma \frac{\tau_\sigma^W}{\tau_\sigma} \rho_\sigma \tilde{\epsilon}_{\mathrm{c}}(\rho_\sigma, 0, \nabla\rho_\sigma, 0) \right] d\mathbf{r} \qquad (136)$$

Perdew, Kurth, Zupan, and Blaha [176] (PKZB) constructed a τ-dependent functional by preserving and extending the list of constraints satisfied by the PBE exchange-correlation GGA. The kinetic energy density appears both in the exchange and correlation components of the PKZB functional. In the exchange part, τ is an argument of the enhancement factor

$$F_{\mathrm{x}}^{\mathrm{PKZB}} = 1 + \frac{\chi}{1 + \chi/\kappa} \qquad (137)$$

where $\chi = \chi(\rho, \nabla\rho, \tau)$ is a new variable replacing the variable μs^2 of PBE. The particular form of χ (not shown here) was chosen to reproduce the fourth-order gradient expansion of F_{x} for slowly varying densities. The analytic representation of χ contains only one empirical parameter optimized by minimizing the MAE in the atomization energies of a 20-molecule training set [176].

The PKZB correlation functional combine the ingredients $\epsilon_{\mathrm{c}}^{\mathrm{PBE}}$ and τ_σ,

$$\begin{aligned} E_{\mathrm{c}}^{\mathrm{PKZB}}[\rho_\alpha, \rho_\beta] = \int \Bigg\{ & \rho\epsilon_{\mathrm{c}}^{\mathrm{PBE}}(\rho_\alpha, \rho_\beta, \nabla\rho_\alpha, \nabla\rho_\beta) \left[1 + C\left(\frac{\sum_\sigma \tau_\sigma^W}{\sum_\sigma \tau_\sigma} \right)^2 \right] \\ & -(1 + C) \sum_\sigma \left(\frac{\tau_\sigma^W}{\tau_\sigma} \right)^2 \rho_\sigma \epsilon_{\mathrm{c}}^{\mathrm{PBE}}(\rho_\sigma, 0, \nabla\rho_\sigma, 0) \Bigg\} d\mathbf{r} \end{aligned} \qquad (138)$$

where C is a nonempirical constant chosen to reproduce PBE surface correlation energies of a UEG. Note that PKZB is self-correlation free for one-electron densities.

PKZB is much more accurate than PBE for atomization energies [177] but inferior for equilibrium bond lengths, vibrational frequencies [177], dissociation energies and, especially, geometries of hydrogen-bonded complexes [178]. These concerns prompted Tao, Perdew, Staroverov, and Scuseria (TPSS) [179] to search for additional constraints to be imposed at the meta-GGA level in order to improve upon PBE and PKZB for *all* properties. The principal problem with PKZB turned out to be with the exchange component. In PBE and PKZB, the enhancement factors are such that the exchange potentials $v_{\mathrm{x}\sigma}(\mathbf{r})$ diverge at the nuclei (see Fig. 1 of Ref. [180]). In the case of PBE, the divergence of $v_{\mathrm{x}\sigma}(\mathbf{r})$ is harmless, but in the case of PKZB it is manifested in overstretched bond lengths. This problem was addressed in the TPSS approximation by requiring that $v_{\mathrm{x}\sigma}(\mathbf{r})$ be finite at the nucleus whenever $\tau_\sigma^W = \tau_\sigma$, a condition that covers compact iso-orbital densities (i.e. one- and spin-compensated two-electron densities represented by real Kohn–Sham orbitals). In effect, the new constraint eliminates the divergence for nearly all realistic many-electron systems, because $\rho_\sigma(\mathbf{r})$ near the nucleus is dominated by the 1s orbital, so that $\tau_\sigma \longrightarrow \tau_\sigma^W$. This proved to be the last and key constraint that made it possible to obtain accurate atomization energies *and* bond lengths from a nonempirical meta-GGA [180].

The TPSS exchange enhancement factor has the same form as in PKZB [Eq. (137)], but the function $\chi(\rho, \nabla\rho, \tau)$ in TPSS is much more complicated (which is necessary to satisfy additional constraints). Aside from producing a finite $v_{x\sigma}(\mathbf{r})$ at the nucleus, the TPSS exchange functional reproduces the fourth-order density gradient expansion [120] for slowly varying densities and yields the correct exchange energy for ground-state one-electron hydrogenic densities (for the H atom, $E_x^{\text{exact}} = -\frac{5}{16}E_h$). Improvements in the correlation part of TPSS are of a technical character [179]:

$$E_c^{\text{TPSS}}[\rho_\alpha, \rho_\beta] = \int \rho\epsilon_c^{\text{revPKZB}}\left[1 + d\epsilon_c^{\text{revPKZB}}\left(\frac{\tau_W}{\tau}\right)^3\right]\mathbf{dr} \tag{139}$$

Here d is a nonempirical constant and $\epsilon_c^{\text{revPKZB}}$ is the revised PKZB correlation energy per particle

$$\begin{aligned}\epsilon_c^{\text{revPKZB}} &= \epsilon_c^{\text{PBE}}(\rho_\alpha, \rho_\beta, \nabla\rho_\alpha, \nabla\rho_\beta)\left[1 + C(\zeta,\xi)\left(\frac{\tau_W}{\tau}\right)^2\right] - [1 + C(\zeta,\xi)]\left(\frac{\tau_W}{\tau}\right)^2 \\ &\times \sum_\sigma \frac{\rho_\sigma}{\rho}\max[\epsilon_c^{\text{PBE}}(\rho_\sigma, 0, \nabla\rho_\sigma, 0), \epsilon_c^{\text{PBE}}(\rho_\alpha, \rho_\beta, \nabla\rho_\alpha, \nabla\rho_\beta)]\end{aligned} \tag{140}$$

In TPSS, C is no longer a constant but a function of the spin polarization ζ and a variable $\xi = |\nabla\zeta|/2(3\pi^2\rho)^{1/3}$. $C(\zeta, \xi)$ is designed to make $E_{xc}^{\text{TPSS}}[\rho_\alpha, \rho_\beta]$ properly independent of ζ in the low-density limit. The max() function in Eq. (140) ensures that $\epsilon_c^{\text{revPKZB}}$ is strictly negative everywhere, a property that is weakly violated by PKZB. Extensive tests on molecules, hydrogen-bonded complexes [180], and solids [181] indicate that, property by property, TPSS overcomes all the shortcomings of PKZB and closely follows or exceeds in accuracy nearly all other density functionals, including B3LYP.

24.8 MODELING THE EXCHANGE-CORRELATION HOLE

The sharp cutoff procedure introduces spurious kinks into the GGA exchange hole of Eq. (107), which complicate the derivation of the P86 and PW91 exchange functionals. In this section, we will consider functionals that are based on smooth analytic hole models normalized from the outset.

24.8.1 Exchange functionals based on a model hole

Becke and Roussel [182] constructed a model exchange hole starting with the second-order Taylor expansion of the exact spherically averaged σ-spin hole [183]

$$h_x^{\sigma\sigma}(\mathbf{r}, u) = -\rho_\sigma(\mathbf{r}) - \frac{1}{6}[\nabla^2\rho_\sigma(\mathbf{r}) - 4\tau_\sigma(\mathbf{r}) + 4\tau_\sigma^W(\mathbf{r})]u^2 + \cdots \tag{141}$$

and fitting it to a generalized analytic form of the exact hydrogenic exchange hole. The exact normalized hydrogenic 1s density is $\rho_H(\mathbf{r}) = (\alpha^3/8\pi)e^{-\alpha|\mathbf{r}|}$, where $\alpha = 2Z$.

Hence the spherically averaged exchange hole for a hydrogenic atom is

$$h_{\mathrm{x}}^{\mathrm{H}}(\mathbf{r},u) = -\frac{1}{4\pi}\int_0^{2\pi}\mathrm{d}\phi_u\int_0^{\pi}\rho_{\mathrm{H}}(\mathbf{r}+\mathbf{u})\sin\theta_u\mathrm{d}\theta_u$$

$$= -\frac{1}{4\pi}\int_0^{2\pi}\mathrm{d}\phi_u\int_0^{\pi}\frac{\alpha^3}{8\pi}e^{-\alpha\sqrt{r^2+u^2-2ru\cos\theta_u}}\sin\theta_u\mathrm{d}\theta_u$$

$$= -\frac{\alpha}{16\pi ru}\left[(\alpha|r-u|+1)e^{-\alpha|r-u|} - (\alpha|r+u|+1)e^{-\alpha|r+u|}\right] \quad (142)$$

The analytic form of Eq. (142) is sufficiently flexible to parametrize the exact second-order expansion of the exchange hole for *any* many-electron system, if it is generalized as

$$h_{\mathrm{x}}^{\mathrm{BR}}(a,b;u) = -\frac{a}{16\pi bu}\left[(a|b-u|+1)e^{-a|b-u|} - (a|b+u|+1)e^{-a|b+u|}\right] \quad (143)$$

where a and b are positive scalar functions of ρ_σ, $|\nabla\rho_\sigma|$, $\nabla^2\rho_\sigma$, and τ_σ, without any physical significance. The underlying hydrogenic model still ensures that the generalized hole of Eq. (143) is nonpositive and normalized to -1. For a given reference point $\mathbf{r}$, the values of $a(\mathbf{r})$ and $b(\mathbf{r})$ are obtained by expanding Eq. (143) in a Taylor series to second order in u and comparing its zeroth- and second-order coefficients with those of Eq. (141). Then the substitution of Eq. (143) into Eq. (33) yields

$$E_{\mathrm{x}}^{\mathrm{BR}}[\rho_\alpha,\rho_\beta] = -\frac{1}{2}\sum_{\sigma=\alpha,\beta}\int\frac{\rho_\sigma(\mathbf{r})}{b(\mathbf{r})}\left[1-e^{-ab}\left(1+\frac{1}{2}ab\right)\right]\mathrm{d}\mathbf{r} \quad (144)$$

In contrast to the gradient expansion of the exchange hole of Perdew [135], the BR functional does not reduce to LSDA in the uniform density limit. To recover this limit *approximately*, Becke and Roussel multiplied the term $(\tau_\sigma - \tau_\sigma^W)$ in Eq. (141) by an adjustment factor of 0.8. At the same time, the BR exchange energy density has the correct $-\rho(r)/2r$ asymptotic behavior in the $r\to\infty$ limit.

Around the same time, Becke [184] proposed a real-space normalized model for the correlation hole. He started from the observation that, for a given coupling strength λ, the spin-polarized components of the spherically averaged pair densities have well-defined short-range behavior near the reference point [185]:

$$P_2^{\sigma\sigma,\lambda}(\mathbf{r},u) = A^{\sigma\sigma}(\mathbf{r})\left(1+\frac{\lambda}{2}u\right)u^2 + \cdots \quad (145)$$

$$P_2^{\alpha\beta,\lambda}(\mathbf{r},u) = A^{\alpha\beta}(\mathbf{r})(1+\lambda u) + \cdots \quad (146)$$

where $A^{\sigma\sigma}(\mathbf{r})$ and $A^{\alpha\beta}(\mathbf{r})$ are certain functions. The pair densities trivially determine the exchange-correlation hole by Eq. (10). By using Eqs. (10), (103), the Taylor series expansion for the exchange hole [Eq. (141)], and going through a number of algebraic manipulations, Becke obtained explicit formulas for $h_{\mathrm{c}}^{\sigma\sigma,\lambda}(\mathbf{r},u)$ and $h_{\mathrm{c}}^{\alpha\beta,\lambda}(\mathbf{r},u)$. Substitution of these expressions into the adiabatic connection formula of Eq. (29)

yields the correlation energy functional

$$E_c^{B88c} = E_c^{\alpha\alpha} + E_c^{\beta\beta} + E_c^{\alpha\beta} \tag{147}$$

$$E_c^{\sigma\sigma} = C_{\sigma\sigma} \int \rho_\sigma (\tau_\sigma - \tau_\sigma^W) z_{\sigma\sigma}^4 \left[1 - \frac{2}{z_{\sigma\sigma}} \ln\left(1 + \frac{z_{\sigma\sigma}}{2}\right)\right] \mathrm{d}\mathbf{r} \tag{148}$$

$$E_c^{\alpha\beta} = C_{\alpha\beta} \int \rho_\alpha \rho_\beta z_{\alpha\beta}^2 \left[1 - \frac{1}{z_{\alpha\beta}} \ln\left(1 + z_{\alpha\beta}\right)\right] \mathrm{d}\mathbf{r} \tag{149}$$

where $z_{\sigma\sigma}$ and $z_{\alpha\beta}$ are the so-called correlation lengths, defined in Ref. [184], while $C_{\sigma\sigma}$ and $C_{\alpha\beta}$ are adjustable constants.

Performance of the BR and B88c functionals has been studied by Becke [186], as well as by Neumann and Handy [187]. A comprehensive assessment of these functionals, completed recently by Izmaylov et al. [188], indicates that the accuracy of BR is comparable to that of the B88 exchange, while the B88c approximation is less accurate than LYP, PW91, and similar correlation functionals.

Approximate functionals derived from a model exchange-correlation hole are relatively few in number. The majority of functionals are not based on any explicit model hole. The holes corresponding to such functionals are not even known and, if needed, have to be 'reverse-engineered' from the functional itself. Ernzerhof and Perdew [189] did this to obtain a smooth analytic representation for the angle- and system-averaged PBE GGA exchange hole. When substituted into Eq. (33), their model gives the hole-based PBE functional [189]. For practical purposes, the hole-based PBE exchange is numerically equivalent to the energy-based functional of Eq. (126). Explicit exchange-correlation holes are of considerable interest on their own. The work on the TPSS hole is currently in progress [190].

24.8.2 Functionals based on a correlated wave function

The best-known example of approximations of this type is the correlation functional of Lee, Yang, and Parr (LYP) [191]. LYP is nothing but a DFT adaptation of the orbital-dependent correlation energy formula of Colle and Salvetti (CS) [192]. The starting point of the derivation is a correlated wave function of the form

$$\Psi(\mathbf{x}_1, \mathbf{x}_2, \ldots, \mathbf{x}_N) = \Psi^{HF}(\mathbf{x}_1, \mathbf{x}_2, \ldots, \mathbf{x}_N) \prod_{i<j} [1 - f(\mathbf{r}_i, \mathbf{r}_j)] \tag{150}$$

where Ψ^{HF} is the Hartree–Fock wave function (a single determinant) and $f(\mathbf{r}_i, \mathbf{r}_j)$ is a model two-electron correlation function. Arguing that $f(\mathbf{r}_i, \mathbf{r}_j)$ is small for typical positions of electrons, Colle and Salvetti obtained an approximate formula for the corresponding electron pair density

$$P_2^{CS}(\mathbf{r}_1, \mathbf{r}_2) = P_2^{HF}(\mathbf{r}_1, \mathbf{r}_2)[1 - f(\mathbf{r}_1, \mathbf{r}_2)]^2 \tag{151}$$

where $P_2^{HF}(\mathbf{r}_1, \mathbf{r}_2)$ is the spin-free Hartree–Fock electron pair density of Eq. (37).

The conventional quantum-mechanical correlation energy is then given by

$$E_c^{CS} = \int\int \frac{P_2^{CS}(\mathbf{r}_1, \mathbf{r}_2) - P_2^{HF}(\mathbf{r}_1, \mathbf{r}_2)}{r_{12}} d\mathbf{r}_1 d\mathbf{r}_2$$
$$= \int\int P_2^{HF}(\mathbf{r}_1, \mathbf{r}_2) \frac{f^2(\mathbf{r}_1, \mathbf{r}_2) - 2f(\mathbf{r}_1, \mathbf{r}_2)}{r_{12}} d\mathbf{r}_1 d\mathbf{r}_2 \quad (152)$$

By making a series of approximations, Colle and Salvetti then obtained the following expression in terms of interparticle coordinates $\mathbf{R} = (\mathbf{r}_1 + \mathbf{r}_2)/2$ and $\mathbf{u} = \mathbf{r}_1 - \mathbf{r}_2$

$$E_c^{CS} = -a \int \rho(\mathbf{R}) \frac{1 + b\rho^{-8/3}(\mathbf{R})[\nabla_{\mathbf{u}}^2 P_2^{HF}(\mathbf{R}, \mathbf{u})]_{u=0} \exp[-c\rho^{-1/3}(\mathbf{R})]}{1 + d\rho^{-1/3}(\mathbf{R})} d\mathbf{R} \quad (153)$$

where the values of the parameters $a = 0.04918$, $b = 0.132$, $c = 0.2533$, and $d = 0.349$ were determined by fitting to certain exact data relevant to the He atom.

Lee, Yang, and Parr [191] carried out this analysis further by rewriting Eq. (153) in an equivalent form

$$E_c^{CS}[\rho] = -a \int \rho(\mathbf{r}) \frac{1 + b\rho^{-5/3}(\mathbf{r})[\tau_{HF}(\mathbf{r}) - 2\tilde{\tau}_W(\mathbf{r})] \exp[-c\rho^{-1/3}(\mathbf{r})]}{1 + d\rho^{-1/3}(\mathbf{r})} d\mathbf{r} \quad (154)$$

where

$$\tilde{\tau}_W = \frac{1}{8}\frac{|\nabla\rho|^2}{\rho} - \frac{1}{8}\nabla^2\rho \quad (155)$$

and τ_{HF} is the Hartree–Fock kinetic energy density [note also the addition of the Laplacian term which was not included in the definition of τ_W in Eq. (127)]. Eq. (154) is still not a conventional density functional, because it involves dependence on the orbitals through the τ_{HF} term. Lee, Yang, and Parr simplified this expression by replacing τ_{HF} with its second-order gradient expansion

$$\tau_{HF} \approx \tau_{unif} + \frac{1}{9}\tilde{\tau}_W + \frac{1}{18}\nabla^2\rho \quad (156)$$

where $\tau_{unif} = \frac{3}{10}(3\pi^2)^{2/3}\rho^{5/3}$ is the Thomas–Fermi kinetic energy density of a UEG. The final expression,

$$E_c^{LYP}[\rho] = -a \int \frac{\rho + b\rho^{-2/3}\left(C_F\rho^{5/3} - 2\tilde{\tau}_W + \frac{1}{9}\tilde{\tau}_W + \frac{1}{18}\nabla^2\rho\right)\exp(-c\rho^{-1/3})}{1 + d\rho^{-1/3}} d\mathbf{r} \quad (157)$$

where $C_F = \frac{3}{10}(3\pi^2)^{2/3}$, proved to be an excellent approximation to Eq. (153).

The LYP functional as given by Eq. (157) is a meta-GGA, because it depends on the Laplacian of the density. Miehlich et al. [193] pointed out that $\nabla^2\rho(\mathbf{r})$ can be eliminated from the LYP formula by partial integration. Their transformed version of LYP, reproduced here in the spin-polarized form, is given by [194]

$$E_c^{\mathrm{LYP}}[\rho_\alpha, \rho_\beta] = -a \int \Bigg[\frac{4}{1 + d\rho^{-1/3}} \frac{\rho_\alpha \rho_\beta}{\rho} + 2^{11/3} C_F b\omega \rho_\alpha \rho_\beta (\rho_\alpha^{8/3} + \rho_\beta^{8/3}) \\ + L_{\alpha\alpha} |\nabla \rho_\alpha|^2 + L_{\alpha\beta} \nabla \rho_\alpha \cdot \nabla \rho_\beta + L_{\beta\beta} |\nabla \rho_\beta|^2 \Bigg] \mathbf{dr} \tag{158}$$

where

$$L_{\alpha\alpha} = b\omega \left\{ \frac{1}{9} \rho_\alpha \rho_\beta \left[1 - 3\delta - (\delta - 11) \frac{\rho_\alpha}{\rho} \right] - \rho_\beta^2 \right\} \tag{159}$$

$$L_{\alpha\beta} = b\omega \left[\frac{1}{9} \rho_\alpha \rho_\beta (47 - 7\delta) - \frac{4}{3} \rho^2 \right] \tag{160}$$

$L_{\beta\beta}$ is obtained by interchanging subscripts α and β in Eq. (159),

$$\omega = \frac{e^{-c\rho^{-1/3}}}{1 + d\rho^{-1/3}} \rho^{-11/3}, \qquad \delta = c\rho^{-1/3} + \frac{d\rho^{-1/3}}{1 + d\rho^{-1/3}} \tag{161}$$

and a, b, c, d are the same parameters as in Eq. (153).

Although the LYP functional does not reduce to LSDA for uniform densities (only about 25% of the true correlation energy of a UEG is recovered [195]), it is an accurate approximation for atomic correlation energies. The good performance of LYP, however, appears to be a fluke. In a critical analysis, Singh et al. [196] demonstrated that the Colle–Salvetti correlation energy formula is seriously flawed. For example, the wave function of Eq. (150) is not normalized, and the Kohn–Sham correlation potential corresponding to the Colle–Salvetti formula is grossly inaccurate [197]. Their analysis was continued by Imamura et al. [198].

The idea of generating density functionals from correlated wave functions continues to attract attention [199]. Imamura and Scuseria [200] recently derived a correlation functional starting from a Colle–Salvetti type correlated wave function and using the transcorrelated method of Boys and Handy [201,202]. Colle–Salvetti-type correlation functionals that treat parallel-spin and opposite-spin contributions to the correlation energy separately have been also developed by Tsuneda and Hirao [203], Tsuneda et al. [204].

24.8.3 Functionals based on a model pair correlation function

The exchange-correlation hole $h_{\mathrm{xc}}^\lambda(\mathbf{r}_1, \mathbf{r}_2)$ is closely associated with the *pair distribution function* $g^\lambda(\boldsymbol{r}_1, \boldsymbol{r}_2)$ defined by

$$P_2^\lambda(\mathbf{r}_1, \mathbf{r}_2) = \frac{1}{2} \rho(\mathbf{r}_1) \rho(\mathbf{r}_2) g^\lambda(\mathbf{r}_1, \mathbf{r}_2) \tag{162}$$

where P_2^λ is the pair density at coupling strength λ. Comparison with Eq. (10) shows that

$$h_{xc}^\lambda(\mathbf{r}_1, \mathbf{r}_2) = \rho(\mathbf{r}_2)[g^\lambda(\mathbf{r}_1, \mathbf{r}_2) - 1] \tag{163}$$

Note that $g^\lambda(\mathbf{r}_1, \mathbf{r}_2)$ is symmetric in its variables, unlike $h_{xc}^\lambda(\mathbf{r}_1, \mathbf{r}_2)$. At $\lambda = 0$, $g^\lambda = g_x + g_c^\lambda$ reduces to the exchange-only pair-distribution function g_x. This implies that $h_x(\mathbf{r}_1, \mathbf{r}_2) = \rho(\mathbf{r}_2)[g_x(\mathbf{r}_1, \mathbf{r}_2) - 1]$ and $h_c^\lambda(\mathbf{r}_1, \mathbf{r}_2) = \rho(\mathbf{r}_2) g_c^\lambda(\mathbf{r}_1, \mathbf{r}_2)$. It is easy to establish that the pair distribution function determines the exchange-correlation energy by a formula similar to Eq. (31), and that

$$E_c = \frac{1}{2}\int_0^1 d\lambda \int\int \rho(\mathbf{r}_1)\rho(\mathbf{r}_2)\frac{g_c^\lambda(\mathbf{r}_1, \mathbf{r}_2)}{r_{12}} d\mathbf{r}_1\, d\mathbf{r}_2 \tag{164}$$

Although $g_c^\lambda(\mathbf{r}_1, \mathbf{r}_2)$ is trivially related to $h_c^\lambda(\mathbf{r}_1, \mathbf{r}_2)$, it is instructive to treat density functionals based on $g_c^\lambda(\mathbf{r}_1, \mathbf{r}_2)$ as a separate subgroup. The function $g_c^\lambda(\mathbf{r}_1, \mathbf{r}_2)$ is either derived (approximately) from a correlated wave function (the Colle–Salvetti formula can be viewed in this light) or postulated as a 'model'.

Proynov, Salahub, and coworkers have developed several correlation functionals [205–209] starting from a GAUSSIAN model of the spherically-averaged pair distribution function for opposite-spin electrons

$$g_{c,\alpha\beta}^\lambda(\mathbf{R}, u) = -e^{-z_{\alpha\beta}^2 u^2}[F_1(\mathbf{R}, \lambda) + F_2(\mathbf{R}, \lambda)] \tag{165}$$

where $F_1(\mathbf{R}, \lambda)$ and $F_2(\mathbf{R}, \lambda)$ are certain parametrized functions and $z_{\alpha\beta}$ is the correlation length. Substitution of $g_{c,\alpha\beta}^\lambda$ into Eq. (164), followed by analytic integration over λ, yields

$$E_c^{\alpha\beta} = \frac{1}{2}\int d\mathbf{R}\int_0^\infty \langle \rho_\alpha(\mathbf{R} + \mathbf{u}/2)\rho_\beta(\mathbf{R} - \mathbf{u}/2)\rangle \bar{g}_{c,\alpha\beta}(\mathbf{R}, u) u\, du, \tag{166}$$

where $\langle\cdots\rangle$ indicates averaging over the spherical components of $\mathbf{u}$. In the LAP1 and LAP2 models [206], the opposite-spin contribution to the correlation energy ($E_c^{\alpha\beta}$) is obtained by Eq. (166), the $E_c^{\sigma\sigma}$ terms are neglected, and the total correlation energy is approximated as $E_c = 2E_c^{\alpha\beta}$. The LAP3 model [208] takes into account the parallel spin contributions and approximates them by the formula $\epsilon_c^{\sigma\sigma} = (1 - 1/N_\sigma)C_p\epsilon_c^{\sigma\sigma'}$, where N_σ is the number of spin-σ electrons and C_p is an empirical factor. The total correlation energy is then given as the sum of these components. The spherical average in the LAP1–LAP3 models is approximated by the first term of the Taylor series expansion about $\mathbf{R}$:

$$\langle \rho_\alpha(\mathbf{R} + \mathbf{u}/2)\rho_\beta(\mathbf{R} - \mathbf{u}/2)\rangle \approx 4\pi\rho_\alpha(\mathbf{R})\rho_\beta(\mathbf{R}) \tag{167}$$

but in a recent revision [209], called the τ1-model, the right-hand side of Eq. (167) is augmented by a correction term that depends on ρ, $|\nabla\rho|$, and $\nabla^2\rho$.

24.8.4 Functionals based on a density matrix expansion

The exact exchange energy of a closed-shell system can be obtained from the density matrix $\rho(\mathbf{r}_1, \mathbf{r}_2)$ by Eq. (38). Replacing $\rho(\mathbf{r}_1, \mathbf{r}_2)$ in this formula by a truncated expansion in ρ and its gradients is a natural path to an exchange functional. This idea was pioneered by Negele and Vautherin [210] in the framework of nuclear matter theory and developed into practical exchange functionals for quantum chemistry by Scuseria and coworkers [211–214]. Originally, Negele and Vautherin expanded $\rho(\mathbf{r}_1, \mathbf{r}_2)$ in transformed coordinates $\mathbf{R} = (\mathbf{r}_1 + \mathbf{r}_2)/2$ and $\mathbf{u} = \mathbf{r}_1 - \mathbf{r}_2$. Koehl et al. [211] generalized this transformation to

$$\mathbf{R} = a\mathbf{r}_1 + (1 - a)\mathbf{r}_2, \qquad \mathbf{u} = \mathbf{r}_1 - \mathbf{r}_2 \tag{168}$$

where $0 \le a \le \frac{1}{2}$. After performing the Taylor expansion of $\rho(\mathbf{R}, \mathbf{u})$ around $\mathbf{u} = 0$, averaging the series over the angular coordinates of $\mathbf{u}$, expanding the result in Bessel functions $j_n(u)$ and Legendre polynomials $P_n(\mathbf{R})$, and truncating the series after the second-order terms, one obtains [211]

$$\rho(\mathbf{R},u) = \frac{3j_1(ku)}{ku}\rho(\mathbf{R}) + \frac{35j_3(ku)}{2k^3u}\left[\left(a^2 - a + \frac{1}{2}\right)\nabla^2\rho(\mathbf{R}) - 2\tau(\mathbf{R}) + \frac{3}{5}k^2\rho(\mathbf{R})\right] \tag{169}$$

where k is the relative momentum of two electrons. Substitution of this DME into Eq. (38) gives

$$E_{\mathrm{x}}^{\mathrm{DME}}[\rho] = -\pi\int\left\{\frac{9}{4k^2}\rho^2 + \frac{35}{12k^4}\rho\left[\left(a^2 - a + \frac{1}{2}\right)\nabla^2\rho - 2\tau + \frac{3}{5}k^2\rho\right]\right\}\mathbf{dr} \tag{170}$$

Note that the choice $k = k_{\mathrm{F}}$, regardless of the value of a, yields the LDA (Dirac) exchange formula (58) as the first term of $E_{\mathrm{x}}^{\mathrm{DME}}[\rho]$. The remaining terms containing $\nabla\rho$, $\nabla^2\rho$, and τ, naturally arise as corrections to the LDA.

Surprisingly, the nonempirical functional of Eq. (170) with $a = 0$ (the implicit choice of Negele and Vautherin) proved to be a worse approximation than LDA, with an MAE of 3.66 E_h for a test set of 32 molecules, compared to 1.79 E_h for LDA. Koehl, Odom, and Scuseria (KOS) suggested treating a as an adjustable parameter and found that at its optimal value ($a = 0.0000638$) the error of Eq. (170) drops to 0.087 E_h, which is only about twice as large as the error of B88 (MAE = 0.041 E_h). By setting $a = 0$ and optimizing the coefficients of the ρ^2 and ρ terms they found

$$E_{\mathrm{x}}^{\mathrm{KOS}}[\rho] = \int\left[\frac{A}{k_{\mathrm{F}}^2}\rho^2 + \frac{B}{k_{\mathrm{F}}^4}\rho\left(\frac{1}{2}\nabla^2\rho - 2\tau + \frac{3}{5}k_{\mathrm{F}}^2\rho\right)\right]\mathbf{dr} \tag{171}$$

where $A = -7.31275$ and $B = -5.43182$. The MAE of the KOS functional for the same 32-molecule test set is only 0.026 E_h.

Recently, Maximoff and Scuseria [214] worked out the full fourth-order density matrix expansion (DME4) and proposed a corresponding exchange functional called DME4*x*. In test calculations, DME4*x* produced some of the most accurate atomic and molecular exchange energies. Developing exchange functionals by incorporating higher orders of the DME may look like a safe bet, but in practice this approach is wrought with difficulties. High-order DMEs involve unusual ingredients like $\nabla^n\rho(\mathbf{r})$ and $\nabla^n\tau(\mathbf{r})$ with $n > 1$, whose presence causes an extreme numerical sensitivity of exchange-correlation matrix elements and makes self-consistent calculations difficult to converge.

24.9 EMPIRICAL FITS

A pragmatic approach to developing density functional approximations is to expand the post-LDA correction in a set of suitably chosen parametrized functions and optimize the parameters by training the functional to reproduce certain calibration data as accurately as possible. Of course, fitted parameters appear in many of the previously discussed functionals, but there they were used only to 'clean up' the construction. In this survey, we call 'empirical' or 'optimized' only those functionals whose design is avowedly empirical.

Optimization of empirical parameters is often accompanied by relaxing exact constraints. As a result, many optimized functionals violate some of the basic analytic properties, such as the uniform density limit of Eq. (46). For instance, Slater's $X\alpha$ method [9–11], which is nowadays regarded as an empirical variant of Dirac's exchange, clearly does so.

Systematic procedures for optimizing GGA and meta-GGA exchange-correlation functionals have been developed by Becke [215,216], Adamson et al. [217], Handy and coworkers [218–221]. In Becke's method, the total exchange-correlation energy is written as

$$E_{\mathrm{xc}} = \sum_{\sigma} E_{\mathrm{x}}^{\sigma} + E_{\mathrm{c}}^{\alpha\beta} + \sum_{\sigma} E_{\mathrm{c}}^{\sigma\sigma} \tag{172}$$

where the individual energy components have the form

$$E_{\mathrm{x}}^{\sigma} = \int e_{\mathrm{x}\sigma}^{\mathrm{UEG}}(\rho_{\sigma}) F_{\mathrm{x}}^{\sigma}\, d\mathbf{r} \tag{173}$$

$$E_{\mathrm{c}}^{\alpha\beta} = \int e_{\mathrm{c}\alpha\beta}^{\mathrm{UEG}}(\rho_{\alpha}, \rho_{\beta}) F_{\mathrm{c}}^{\alpha\beta}\, d\mathbf{r} \tag{174}$$

$$E_{\mathrm{c}}^{\sigma\sigma} = \int e_{\mathrm{c}\sigma\sigma}^{\mathrm{UEG}}(\rho_{\sigma}) \eta_{\sigma}^{\mathrm{SCC}} F_{\mathrm{c}}^{\sigma\sigma}\, d\mathbf{r} \tag{175}$$

Here, $e_{\mathrm{x}\sigma}^{\mathrm{UEG}}(\rho_{\sigma}) = -2^{1/3} C_{\mathrm{x}} \rho_{\sigma}^{4/3}$, $e_{\mathrm{c}}^{\mathrm{UEG}}$ are the correlation energy densities of a UEG given by Eqs. (133) and (134); F are enhancement factors (functions of ρ_{σ},$\nabla\rho_{\sigma}$ and, possibly, $\nabla^2\rho_{\sigma}$ and τ_{σ}), and $\eta_{\sigma}^{\mathrm{SCC}}$ is the SCC of Eq. (129). Each F is represented by a polynomial

$$F = \sum_{m=0}^{n} c_m w^m(\rho_{\sigma}, \nabla\rho_{\sigma}, \ldots) \tag{176}$$

where w is a suitably chosen function derived from approximate expressions for gradient corrections [215]; c_m are empirical coefficients determined by least-squares fitting to training sets of accurate experimental and/or theoretical data. Eqs. (172)–(176) underlie several empirical functionals, such as Becke's 1997 (B97) exchange-correlation [215], Schmider–Becke's 1998 hybrid GGA (SB98h) [222] and hybrid meta-GGA (τ-SB98h) [223], the GGA of Hamprecht, Cohen, Tozer, and Handy (HCTH) [219], its various reparametrizations (HCTH/120, HCTH/147 [224], and HCTH/407 [225], and τ-HCTH [226].

In 1997, Van Voorhis and Scuseria [212] argued that, since electron densities in chemistry are not uniform, one may generalize the DME of Eq. (169) by replacing the fixed $k = k_F$ by a variable $k^2 \rightarrow k_F^2 w(x, z)$, where $w(x, z)$ is a function of $x = |\nabla\rho|/\rho^{4/3}$ and $z = 2(\tau/\rho^{5/3} - C_F)$. Accordingly, they proposed two exchange functionals of the form

$$F_\mathrm{x}^{\mathrm{VS97}}(x, z) = a + \frac{b_1 x^2 + b_2 z}{w(x, z)} \tag{177}$$

where $w(x, z) = 1 + \alpha_1 x^2 + \alpha_2 z$, and $a, b_1, b_2, \alpha_1, \alpha_2$ are empirical parameters. The resulting approximations were found to be in a better agreement with the Hartree–Fock values than most of the other exchange functionals available at the time.

In 1998, Van Voorhis and Scuseria extended this approach [213] to include more terms

$$F_\mathrm{x}^{\mathrm{VS98}}(x, z) = \frac{a}{w(x, z)} + \frac{b_1 x^2 + b_2 z}{w^2(x, z)} + \frac{c_1 x^4 + c_2 x^2 z + c_3 z^2}{w^3(x, z)} \tag{178}$$

where a, b_i, and c_i are empirical parameters. The form of the function $w(x, z) = 1 + \alpha(x^2 + z)$ was chosen to satisfy certain nonuniform scaling relations. Notice that neither VS97 nor VS98 reduces to the LSDA for uniform densities.

For the VS98 correlation, Van Voorhis and Scuseria [213] assumed the general form of Eq. (174) and (175) with the Perdew–Wang parametrization [114] of $\epsilon_\mathrm{c}^{\mathrm{LSDA}}(\rho_\alpha, \rho_\beta)$. Hoping for cancellation of errors, they chose the factors $F_\mathrm{c}^{\alpha\beta}(x, z)$ and $F_\mathrm{c}^{\sigma\sigma}(x, z)$ to have the same analytic form as the exchange enhancement factor of Eq. (178). The VS98 exchange-correlation functional (also referred to as VSXC, after the keyword in GAUSSIAN) has 21 empirical parameters (the exchange component, the parallel-, and opposite-spin correlation components have 7 parameters each), and is the most accurate functional in existence for atomization energies, surpassing even the hybrid functionals (see Table 24.2). Also, because the parameters of VS98 were optimized for exchange and correlation together, neither component is accurate separately. Nevertheless, the analytic form of Eq. (178) is very flexible, so that a suitable reparametrization of the stand-alone VS98 exchange functional can produce excellent approximations to the exact exchange energies [213].

Another approach is to optimize linear combinations of several existing functionals,

$$E_\mathrm{xc} = \sum_k c_k E_\mathrm{xc}^{(k)} \tag{179}$$

possibly with a focus on a particular property. Internal parameters of individual functionals $E_\mathrm{xc}^{(k)}$ may also be subject to reoptimization. This type of approximations is represented by EDF1 ('empirical density functional 1') of Adamson et al. [217], which was optimized to yield accurate thermochemistry, and EDF2 of Lin, George, and Gill, optimized to give accurate vibrational frequencies [227].

In search for a functional that would accurately reproduce the Hartree–Fock energies of first and second-row atoms, Handy and Cohen [228] refined Becke's 1986 exchange approximation [Eq. (91)] by optimizing the coefficient of the LSDA term and revising the gradient correction. The enhancement factor of their functional, called

OPTX, is given by

$$F_X^{OPTX}(s) = a_1 + a_2 c_2 \left[\frac{\gamma(c_1 s)^2}{1 + \gamma(c_1 s)^2} \right]^2 \tag{180}$$

where s is the reduced gradient of Eq. (79), $a_1 = 1.05151$, $a_2 = 1.43169$, $\gamma = 0.006$ are empirical parameters, and $c_1 = 2(6\pi^2)^{1/3}$ and $c_2 = (2^{1/3} C_X)^{-1}$. Tests of the OPTX functional paired with LYP indicate [229] that the OLYP model is perceptibly more accurate than BLYP, although not as accurate [180] as meta-GGAs or hybrid functionals (see also Table 24.2). Several reparametrizations of standard functionals like the B97 GGA and the VS98 meta-GGA have been discussed in Ref. [230].

Xu and Goddard (XG) [231] proposed an exchange functional of the form

$$F_X^{XG}(s) = 1 + a_1[F_X^{B88}(s) - 1] + a_2[F_X^{PW91}(s) - 1] \tag{181}$$

where the mixing parameters $a_1 = 0.722$ and $a_2 = 0.347$ were determined by least squares fitting to the total energy of 10 atoms and atomization energies of 38 molecules. In combination with the LYP correlation, the XG exchange functional is reported to show superior performance for transition metals, hydrogen- and van der Waals complexes.

Purely empirical functionals often work well, because satisfaction of many secondary constraints (the uniform, high- and low-density limits, density scaling transformations, and so on) is not essential for a good performance for usual properties ('molecules do not behave like the UEG'). The price for avoiding the hard work of constraint satisfaction is that optimized functionals may fail badly in situations for which they were not 'trained'.

24.10 MIXING EXACT AND APPROXIMATE EXCHANGE

Functionals that combine GGAs or meta-GGAs with exact exchange of Eq. (38) are called hybrids. At present, hybrid functionals outnumber any other group of exchange-correlation approximations. 'Hybridization' has been embraced widely because it greatly improves performance of 'pure DFT' functionals and is easy to implement.

24.10.1 Global hybrids

The idea of mixing density functional approximations with exact (Hartree–Fock-like) exchange rests on theoretical considerations involving the adiabatic connection formula [Eq. (28)]. Becke [232] reasoned that, since $E_{XC}^{\lambda=0} = E_X^{exact}$ and $E_{XC}^{\lambda=1} \approx E_{XC}^{LSDA}$, the integral over λ in Eq. (28) can be approximated by the mean value theorem:

$$E_{XC} \approx \frac{1}{2}\left(E_X^{exact} + E_{XC}^{LSDA}\right) \tag{182}$$

Eq. (182) has been termed the 'half-and-half' theory [232]. The extent and validity of this approximation were analyzed in detail by Levy et al. [233], as well as Proynov et al. [207].

Soon thereafter, Becke proposed [234] a more empirical yet much more accurate three-parameter hybrid model

$$E_{xc} = a_0 E_x^{exact} + (1 - a_0) E_x^{LSDA} + a_x \Delta E_x^{GGA} + E_c^{LSDA} + a_c \Delta E_c^{GGA} \quad (183)$$

where ΔE_x^{GGA} and ΔE_c^{GGA} are gradient correction parts of the GGA exchange and correlation functionals, and a_0, a_x, and a_c are adjustable parameters. The original three-parameter hybrid functional of Becke [234] has the form

$$E_{xc}^{B3PW91} = a_0 E_x^{exact} + (1 - a_0) E_x^{LSDA} + a_x \Delta E_x^{B88} + E_c^{LSDA(PW92)} + a_c \Delta E_c^{PW91} \quad (184)$$

where ΔE_x^{B88} is the beyond-LSDA part of the B88 exchange, $E_c^{LSDA(PW92)}$ is the Perdew–Wang parametrization [114] of the LSDA correlation, ΔE_c^{PW91} is the gradient correction of the PW91 correlation [the second term on the right-hand side of Eq. (114)], and a_0, a_x, and a_c are empirical constants. The optimal values of these constants were determined by a fit to a series of atomization energies, ionization potentials, and proton affinities to give $a_0 = 0.20$, $a_x = 0.72$, and $a_c = 0.81$ [234].

Using Eq. (184) as a template, the developers of the GAUSSIAN program [32] introduced two other well-known three-parameter hybrid functionals, B3LYP and B3P86. Unfortunately, these two functionals are scantily documented in the literature. The original paper by Becke [234] contains no mention of either B3LYP or B3P86. B3LYP debuted very inconspicuously in a paper titled '*Ab initio* calculation of vibrational absorption and circular dichroism spectra using density functional force fields' [235], while B3P86 is defined solely by its implementation in the GAUSSIAN program. To complicate the matter, the B3LYP functional is coded in GAUSSIAN not exactly as intended in Ref. [235]. We would like to use this opportunity to clarify several issues concerning these functionals.

The B3LYP hybrid, as implemented in the GAUSSIAN program, is given by

$$E_{xc}^{B3LYP} = a_0 E_x^{exact} + (1 - a_0) E_x^{LSDA} + a_x \Delta E_x^{B88} + (1 - a_c) E_c^{VWN-RPA} + a_c E_c^{LYP} \quad (185)$$

where a_0, a_x, and a_c have the same values as in B3PW91. Because the LYP correlation functional is *not* of the 'LSDA + gradient correction' form, it cannot be naturally separated into local and nonlocal parts. This is why the coefficient a_c in Eq. (185) multiplies the entire LYP correlation energy, not the gradient correction for correlation, as in B3PW91. To avoid double counting, the amount of local correlation is reduced by a_c. It is this VWN term that became the source of confusion. When defining B3LYP, Stephens et al. [235] had in mind: (a) Eq. (73) with the parameters fitted to the exact Ceperley–Alder data; (b) the VWN interpolation scheme of Eq. (77). Instead, B3LYP was implemented: (a) with the parameters of Eq. (73) fitted to reproduce the RPA values of the correlation energy of the UEG; (b) using the obsolete Barth–Hedin interpolation formula of Eq. (76). The Barth–Hedin interpolation is 'form I' of the VWN paper, not 'form III', as is stated in the GAUSSIAN manual and in the literature [236]. The correct VWN parametrization of the LSDA correlation energy was added to the GAUSSIAN program later along with the keyword VWN5 (the parametrization was referred to as 'form V' in the VWN paper). The parameters of the intended VWN and the implemented VWN-RPA forms are listed in Table 24.1. All other terms in Eq. (185) are unambiguous. Hertwig and Koch [236] compared both variants of B3LYP and found that, luckily, the 'wrong' B3LYP is actually more accurate!

The B3P86 hybrid also uses VWN-RPA as the local correlation component

$$E_{\mathrm{xc}}^{\mathrm{B3P86}} = a_0 E_{\mathrm{x}}^{\mathrm{exact}} + (1 - a_0) E_{\mathrm{x}}^{\mathrm{LSDA}} + a_{\mathrm{x}} \Delta E_{\mathrm{x}}^{\mathrm{B88}} + E_{\mathrm{c}}^{\mathrm{VWN-RPA}} + a_{\mathrm{c}} \Delta E_{\mathrm{c}}^{\mathrm{P86}} \quad (186)$$

where $\Delta E_{\mathrm{c}}^{\mathrm{P86}}$ is the second term on the right-hand side of Eq. (122), and the coefficients a_0, a_{x}, and a_{c} have the same values as in B3PW91. The original B3PW91 functional was designed to be exact for a uniform density. In contrast, neither B3LYP nor B3P86 has this property, because LYP and VWN-RPA energies are not exact for a UEG.

In the wake of the success of B3PW91, Becke also introduced [168] a one-parameter simplification of Eq.(183),

$$E_{\mathrm{xc}} = a_0 E_{\mathrm{x}}^{\mathrm{exact}} + (1 - a_0) E_{\mathrm{x}}^{\mathrm{DFT}} + E_{\mathrm{c}}^{\mathrm{DFT}} \quad (187)$$

which proved to be almost as accurate as three-parameter hybrids. Perdew et al. [237] provided a theoretical justification for one-parameter GGA hybrids and gave a theoretical estimate of the mixing coefficient suitable for calculations of atomization energies of typical molecules: $a_0 = 1/4$. A rigorous formal treatment of hybrid schemes was also given by Görling and Levy [238].

Adamo and Barone [239] empirically readjusted two parameters of the PW91 exchange based on an analysis of the low-density, large-gradient, regions important in van der Waals systems, and combined their modified functional (*m*PW91) with the PW91 correlation to form a one-parameter hybrid, named *m*PW1PW91. This functional is given by Eq. (187) with $E_{\mathrm{x}}^{\mathrm{DFT}} = E_{\mathrm{x}}^{m\mathrm{PW91}}$, $E_{\mathrm{c}}^{\mathrm{DFT}} = E_{\mathrm{c}}^{\mathrm{PW91}}$, and $a_0 = 0.25$.

A one-parameter hybrid version of the PBE exchange-correlation functional, called here PBEh (other names of the same functional include PBE0 and PBE1PBE) was introduced by Ernzerhof and Scuseria [240] and extensively studied by Adamo and Barone [241]. It has the form of Eq. (187) with DFT = PBE and $a_0 = 0.25$.

The hybrid TPSS functional (TPSSh) [180] is also given by Eq. (187) with DFT = TPSS and the mixing parameter $a_0 = 0.10$, which was determined by minimizing the MAE in the enthalpies of formation of 223 G3/99 molecules. TPSSh satisfies the same

Table 24.1 Parameters (in E_h) of the VWN and VWN–RPA correlation functionals [113] of Eq. (73). The value of A_{GB} is given by Eq. (70)

Parameter	VWN			VWN–RPA	
	$\epsilon_{\mathrm{c}}^{P}(r_{\mathrm{s}})$	$\epsilon_{\mathrm{c}}^{F}(r_{\mathrm{s}})$	$-\alpha_{\mathrm{c}}(r_{\mathrm{s}})$	$\epsilon_{\mathrm{c}}^{P}(r_{\mathrm{s}})$	$\epsilon_{\mathrm{c}}^{F}(r_{\mathrm{s}})$
A	A_{GB}	$A_{\mathrm{GB}}/2$	$1/6\pi^2$	A_{GB}	$A_{\mathrm{GB}}/2$
x_0	−0.10498	−0.32500	−0.00475840	−0.409286	−0.743294
b	3.72744	7.06042	1.13107	13.0720	20.1231
c	12.9352	18.0578	13.0045	42.7198	101.578

exact constraints as the original TPSS. The fact that the value of a_0 in TPSSh is smaller than for a typical GGA hybrid suggests that meta-GGAs are better approximations to the exact exchange functional than GGAs.

Owing to ease of construction and agreeable performance, empirical hybrid functionals of the 'mix-and-optimize' type have been proliferating in the literature over the past few years. These are functionals of the form

$$E_{xc} = E_x + E_c + a_x E_x^{exact} \tag{188}$$

where E_x and E_c, are optimized simultaneously with a_x, an empirical parameter. Examples of such hybrids include B97-1 [219], revised in Ref. [230], B97-2 [221], the Schmider–Becke hybrid GGA (SB98h) [222], τ-SB98h [223], the τ-HCTH hybrid [226], and the Boese–Martin functional for kinetics (BMK) [242].

Cohen and Handy [243] presented a hybrid functional similar to B3LYP that is based on the exchange functional OPTX of Eq. (180). This functional, named O3LYP, is given by

$$E_{xc}^{O3LYP} = a_0 E_x^{exact} + b_0 E_x^{LSDA} + a_x \Delta E_x^{OPTX} + (1 - a_c) E_c^{VWN} + a_c E_c^{LYP} \tag{189}$$

where $a_0 = 0.1161$, $b_0 = 0.9262$, $a_x = 0.8133$, and $a_c = 0.81$. Numerical studies show that O3LYP is overall more accurate than B3LYP for atoms and molecules, although not by much [229,244–246]. Note also that the fractions of approximate and exact exchange in Eqs. (188) and (189) no longer add up to 1. This means that the exchange components of these functionals do not reduce to E_x^{LSDA} for uniform densities.

Very recently, Xu and Goddard [231] introduced a B3LYP-style hybrid version of their 'extended' GGA functional of Eq. (181). Termed X3LYP, this functional uses reoptimized parameters in the ΔE_x^{XG} part [$a_1 = 0.675$, $a_2 = 0.235$ in Eq. (181)] and is given by

$$E_{xc}^{X3LYP} = a_0 E_x^{exact} + (1 - a_0) E_x^{LSDA} + a_x \Delta E_x^{XG} + (1 - a_c) E_c^{VWN-RPA} + a_c E_c^{LYP} \tag{190}$$

where $a_0 = 0.218$, $a_x = 0.709$, and $a_c = 0.871$ are also reoptimized. The performance of X3LYP appears to be comparable or better than that of other hybrid functionals [231].

A one-parameter hybrid functional combining the B88 exchange and B95 correlation has been proposed by Zhao et al. [247]. In another recent work, Zhao et al. [248] initiated the development of 'doubly hybrid' functionals—empirical linear combinations of the Hartree–Fock, approximate DFT, and *ab initio* (e.g. MP2) energy terms.

24.10.2 Local hybrids

Although conventional hybrid functionals offer the best accuracy, they suffer from the same problems that plague GGAs and meta-GGAs, especially the self-interaction error (SIE). The SIE arises from inexact cancellation of spurious Coulomb self-interaction energy by approximate exchange. The exact exchange functional is, of course, self-interaction free. Inclusion of a fraction of exact exchange in global hybrid functionals alleviates, but does not solve completely, this problem. Adding a self-interaction correction [77] to the existing functionals to satisfy the constraint of Eq. (44) is an interesting method [249], but it is not straightforward to implement and apply.

A different approach to this problem is offered by local hybrid (Lh) [93] or hyper-GGA [72] scheme, in which the amount of approximate and exact exchange is determined by the local mixing function

$$E_{\mathrm{xc}}^{\mathrm{Lh}} = \int \{a(\mathbf{r}) e_{\mathrm{x}}^{\mathrm{exact}}(\mathbf{r}) + [1 - a(\mathbf{r})] e_{\mathrm{x}}^{\mathrm{DFT}}(\mathbf{r}) + e_{\mathrm{c}}^{\mathrm{DFT}}(\mathbf{r})\} \mathrm{d}\mathbf{r} \tag{191}$$

where $e_{\mathrm{x}}^{\mathrm{exact}}(\mathbf{r})$ is the exact exchange energy density. The mixing function $a(\mathbf{r})$ must assume values between 0 and 1 and reduce to 1 for any one-electron density. This ensures that the SIE is eliminated in one-electron-like regions which are particularly sensitive to the SIE. The local hybrid scheme can also be viewed as a generalization of global hybrid functionals.

The simplest function that fulfills these requirements is due to Becke [169]

$$a(\mathbf{r}) = \frac{\tau_W(\mathbf{r})}{\tau(\mathbf{r})} \tag{192}$$

Its use in the above local hybrid scheme was first reported by Jaramillo et al. [93]. This particular choice of $a(\mathbf{r})$ improves equilibrium geometries and binding energies of two-center, odd-electron symmetric cations such as H_2^+, He_2^+, $(HF)_2^+$ (the notorious problem discussed, for example, in Refs. [250] and [251]), and hydrogen abstraction barriers, but gives disappointing results for atomization energies [93]. The search for better mixing functions will certainly continue in the future [72]. It is clear that a local hybrid, if designed well, cannot be worse than the parent global hybrid.

24.10.3 Screened hybrids

Screened Coulomb interaction techniques find many interesting applications in modern quantum chemistry. For example, Gill and coworkers [252–256] describe an algorithm aimed at speeding up the computation of the Coulomb energy based on splitting the Coulomb operator $1/u$ (where $u \equiv r_{12}$) into short-range (SR) and long-range (LR) components:

$$\frac{1}{u} = \underbrace{\frac{1 - \mathrm{erf}(\omega u)}{u}}_{\mathrm{SR}} + \underbrace{\frac{\mathrm{erf}(\omega u)}{u}}_{\mathrm{LR}} \tag{193}$$

where $\mathrm{erf}(\omega u)$ is the error function and ω is an adjustable parameter. Their idea is to approximate the slowly decaying $1/u$ operator with the fast-decaying SR term plus corrections for the LR part. This reduces dramatically the required number of two-electron integrals. Savin and coworkers [257–260] use Coulomb screening to handle atomic and molecular near-degeneracy effects by combining multideterminantal wave functions with density functionals.

In the context of DFT, the partitioning of $1/u$ is equivalent to representing E_{xc} as

$$E_{xc} = E_{xc}^{SR}(\omega) + E_{xc}^{LR}(\omega) \tag{194}$$

The short-range component of exact exchange is obtained from electron repulsion integrals calculated with the screened Coulomb potential:

$$\langle \mu\nu|\lambda\sigma\rangle^{SR} = \int d\mathbf{r}_1 \int \chi_\mu^*(\mathbf{r}_1)\chi_\nu^*(\mathbf{r}_2)\frac{1-\mathrm{erf}(\omega r_{12})}{r_{12}}\chi_\lambda(\mathbf{r}_1)\chi_\sigma(\mathbf{r}_2)d\mathbf{r}_2 \tag{195}$$

To evaluate the long-range part of the DFT exchange energy, one inserts the screened Coulomb potential into Eq. (33) to obtain

$$E_x^{DFT,LR}(\omega) = 2\pi \int d\mathbf{r} \int_0^\infty \mathrm{erf}(\omega u) u h_x(\mathbf{r}, u) du \tag{196}$$

For example, for the LDA exchange this gives [258,261]

$$E_x^{LDA,LR}(\omega) = C_x \int \rho^{4/3} \frac{2\xi}{3}\left[2\sqrt{\pi}\,\mathrm{erf}\left(\frac{1}{\xi}\right) - 3\xi + \xi^3 + (2\xi - \xi^3)e^{-1/\xi^2}\right] d\mathbf{r} \tag{197}$$

where $\xi = \omega/k_F$.

Recently, Heyd et al. [262] used the screening of $1/u$ to adapt global hybrid functionals to calculations of periodic systems with small band gaps. Such an adaptation is necessary because direct evaluation of exact exchange in periodic systems with metallic character is prohibitively expensive, rendering conventional hybrids utterly impractical for many solids. Heyd et al. start with rewriting the one-parameter global hybrid functional of Eq. (187) as

$$E_{xc}^{\omega DFTh} = a\left[E_x^{exact,SR}(\omega) + E_x^{exact,LR}(\omega)\right] + (1-a)E_x^{DFT} + E_c^{DFT} \tag{198}$$

choose DFT = PBE, and then replace $E_x^{exact,LR}(\omega)$, the slowest-decaying term, by $E_x^{PBE,LR}(\omega)$:

$$E_{xc}^{\omega PBEh} = a\left[E_x^{exact,SR}(\omega) + E_x^{PBE,LR}(\omega)\right] + (1-a)E_x^{PBE} + E_c^{PBE} \tag{199}$$

Since the explicit PBE exchange hole is known [189], $E_x^{PBE,LR}$ can be obtained analytically (for details, see Ref. [262]). For $a = 1/4$ and $\omega = 0$, the ωPBEh functional is equivalent to PBEh, while in the limit $\omega \to \infty$ it reduces to the PBE GGA. Extensive tests [262–264] show that ωPBEh is an excellent approximation to the original PBEh functional and that the computational cost of ωPBEh for solids is much closer to that of PBE than PBEh. A revised version of ωPBEh, called HSE [263], gives spectacular results in calculations of band gaps of semiconductors [264].

Other researchers [265–268] have used Coulomb screening to describe long-range interactions between atoms and molecules within hybrid DFT. In these methods, the LR component of exact exchange is kept, yielding the correct tails of the exchange potential and energy density, while the SR part is approximated by GGA or meta-GGA functionals. The fraction of exact exchange *increases* with r_{12}, which is precisely the opposite of the trend in Eq.(199). The resulting LR-corrected approximations, called 'Coulomb-attenuated hybrids' [267], show improved performance for polarizabilities of long chains, Rydberg excitations, charge transfer, and van der Waals interactions.

Eq. (196) represents the proper way of screening exchange but, unfortunately, it requires knowing the explicit expression for the exchange hole. For most exchange functionals, $h_x(\mathbf{r}, u)$ is not known. In such cases, the screened DFT exchange energy can be approximated as

$$E_x^{\mathrm{DFT,LR}}(\omega) \approx \int e_x^{\mathrm{LDA,LR}}(\rho, \omega) F_x^{\mathrm{DFT}}(\rho, \nabla\rho, \ldots) d\mathbf{r} \tag{200}$$

where $e_x^{\mathrm{LDA,LR}}$ is the exchange energy density in Eq. (197) and F_x is a GGA or meta-GGA enhancement factor. Although Coulomb screening of a density functional by Eq. (200) is only an approximation to the exact procedure, it is easy to implement for almost any exchange functional and has been commonly used in practice [265–268].

24.11 IMPLEMENTATION AND PERFORMANCE

The self-consistent field procedure in Kohn–Sham DFT is very similar to that of the conventional Hartree–Fock method [269]. The main difference is that the functional $E_{xc}[\rho]$ and matrix elements of $v_{xc}(\mathbf{r})$ have to be evaluated in Kohn–Sham DFT numerically, whereas the Hartree–Fock method is entirely analytic. Efficient formulas for computing matrix elements of $v_{xc}(\mathbf{r})$ in finite basis sets have been developed [270, 271], along with accurate numerical integration grids [272–277] and techniques for real-space grid integration [278,279].

For an exchange-correlation functional of the form

$$E_{xc}[\rho] = \int e_{xc}(\rho, |\nabla\rho|, \nabla^2\rho) d\mathbf{r} \tag{201}$$

the exchange-correlation potential is given by the rules of calculus of variations [280]

$$v_{xc} \equiv \frac{\delta E_{xc}}{\delta\rho} = \frac{\partial e_{xc}}{\partial\rho} - \nabla\cdot\left(\frac{\partial e_{xc}}{\partial\nabla\rho}\right) + \nabla^2\left(\frac{\partial e_{xc}}{\partial\nabla^2\rho}\right) \tag{202}$$

If e_{xc} also depends on ϕ_i or τ, then explicit differentiation with respect to ρ is no longer possible and the OEP method [47–55] should be used. In practice, the determination of the OEP is often avoided by departing slightly from the true Kohn–Sham scheme and minimizing the energy functional with respect to Kohn–Sham orbitals [271], that is, by assuming

$$v_{xc}\phi_i = \frac{\delta E_{xc}}{\delta\phi_i^*} \tag{203}$$

Table 24.2 Mean absolute deviations (MAD) from experiment for standard enthalpies of formation ($\Delta_f H^\circ_{298}$), ionization potentials (IP), electron affinities (EA), proton affinities (PA), equilibrium bond lengths (r_e), and harmonic vibrational frequencies (ω_e) computed with approximate functionals using the 6-311++G(3df,3pd) basis set. The fully nonempirical functionals in this table are HF, LSDA, PW91, PBE, and TPSS

Method	Property/(units of MAD)/Test set (size)					
	$\Delta_f H^\circ_{298}$ (kcal/mol) G3 (223)	IP (eV) G3/IP (86)	EA (eV) G3/EA (58)	PA (kcal/mol) G3/PA (8)	r_e (Å) T-96R (96)	ω_e (cm^{-1}) T-82F (82)
HF	211.5	1.03	1.10	3.1	0.025[a]	136.2[a]
LSDA[b]	121.4	0.23	0.24	5.9	0.013[c]	48.9[d]
Generalized gradient approximations						
BLYP	9.5	0.29	0.12	1.6	0.022	55.2
BPW91	9.0	0.24	0.11	1.4	0.017	41.4
BP86	26.3	0.21	0.19	1.3	0.017	45.5
PW91	23.6	0.22	0.14	1.6	0.014	39.8
PBE	22.2	0.24	0.12	1.6	0.016	42.0
HCTH/407	7.2	0.23	0.19	1.9	0.014	39.9
OLYP	5.9	0.29	0.15	1.7	0.018	40.2
Meta-generalized gradient approximations						
VS98	3.5	0.23	0.13	1.6	0.013	33.9
BKCIS	7.2	0.22	0.20	1.5	0.019[e]	45.1[e]
PKZB	7.0	0.31	0.15	1.8	0.027	51.7
TPSS	5.8	0.24	0.14	1.8	0.014	30.4
BRxB88c[f]	14.6	0.22	0.14	2.7	0.026	48.7
Hybrid functionals						
B3PW91	3.9	0.19	0.14	1.1	0.009	36.2
B3LYP	4.9	0.18	0.12	1.2	0.010	33.5
B3P86	26.1	0.55	0.59	1.0	0.008	37.0
*m*PW1PW91	4.1	0.19	0.16	1.1	0.010	42.9
PBEh	6.7	0.20	0.17	1.1	0.010	43.6
HSE	6.0	0.22	0.12	1.0	0.009	43.9
O3LYP	3.7	0.22	0.13	1.5	0.012	33.7
B97-1	4.9	0.19	0.11	1.0	0.012	32.5
B97-2	4.7	0.18	0.14	2.2	0.012	40.9
SB98h[g]	3.9	0.18	0.11	0.9	0.012	33.8
TPSSh	3.9	0.23	0.16	1.8	0.010	26.7

[a] Excludes Be_2 (no binding).
[b] Using the Perdew–Wang representation of the LSDA correlation energy [114].
[c] Excludes F_2^+ and SF (no convergence).
[d] Excludes F_2^+ (no convergence).
[e] Excludes C_2 (no convergence).
[f] From Ref. [188].
[g] Fit 2c in Table III of Ref. [222]. The GAUSSIAN keyword is B98.

For functionals having the form of Eq. (201), both approaches are equivalent because by the chain rule

$$\begin{aligned}\frac{\delta E_{\mathrm{xc}}}{\delta\phi_i^*} &= \int \frac{\delta E_{\mathrm{xc}}}{\delta\rho(\mathbf{r}')}\frac{\delta\rho(\mathbf{r}')}{\delta\phi_i^*(\mathbf{r})}\mathrm{d}\mathbf{r}' = \int \frac{\delta E_{\mathrm{xc}}}{\delta\rho(\mathbf{r}')}\frac{\partial\rho}{\partial\phi_i^*}\delta(\mathbf{r}-\mathbf{r}')\mathrm{d}\mathbf{r}' = \frac{\delta E_{\mathrm{xc}}}{\delta\rho}\frac{\partial\rho}{\partial\phi_i^*} \\ &= \frac{\delta E_{\mathrm{xc}}}{\delta\rho}\phi_i\end{aligned} \tag{204}$$

In the general case of orbital-dependent functionals, minimization with respect to orbitals is only an approximation to the true Kohn–Sham scheme [281–285] (see also Ref. [58] concerning the gauge invariance problem with conventional τ-dependent functionals).

Table 24.2 compares the performance of most density functionals discussed in this review for selected atomic and molecular properties. The list of molecules comprising the G3 thermochemical test set is given in Refs. [286] and [287]. The G3/IP, G3/EA, and G3/PA test sets are described in Ref. [288]. The T-96R and T-82F sets are defined in Ref. [180]. All other computational details can be found in Ref. [180]. The results in Table 24.2 show that LSDA is already a great improvement over the Hartree–Fock method. GGA functionals further increase the accuracy for molecular binding energies, electron and proton affinities. Meta-GGAs and hybrid functionals generally provide the most accurate predictions for all properties. Note that the recent nonempirical functionals are now vying with the best fitted functionals for the top places in the performance ranking. Systematic numerical studies of density functionals for other properties regularly appear in the literature [181,194,246,288–293].

24.12 CONCLUSION

Back in the 1960s, hopes for future progress in electronic structure theory were associated with correlated wave function techniques and the tantalizing possibility of variational calculations based on the two-electron reduced density matrix [294]. DFT was not on the quantum chemistry agenda at that time. The progress of wave function techniques has been remarkable, as documented elsewhere in this volume. In contrast, the density matrix approach has not yet materialized into a competitive computational method, despite many persistent efforts [295]. Meanwhile, approximate DFT has become the most widely used method of quantum chemistry, offering an unprecedented accuracy-to-cost ratio.

A few personal reminiscences of the time when DFT was entering mainstream quantum chemistry are appropriate here. In March of 1992, while the first author of this review was actively working on coupled cluster theory, John Pople visited Rice University to deliver the Franklin Memorial Lecture. The subject of his talk was the impressive performance of the BLYP functional for thermochemistry, especially as judged by its low computational cost relative to the G2 theory. Pople was on his way to the Sanibel Symposium and gave the author a preprint of Ref. [296], which later became Pople's first publication on DFT (see also Ref. [297]). Pople credited the lecture by Axel Becke at the 7th International Congress of Quantum Chemistry in Menton in the Summer of 1991 as a turning point in his views on DFT. With the help of Peter Gill by email correspondence, the author promptly added a numerical quadrature code to his Hartree–Fock program and was comparing coupled-cluster results to DFT in a matter of weeks [298].

Impressive as the performance of modern density functionals is, none of the practical approximations is still suitable for general use. LSDA, GGAs, meta-GGAs, and global hybrids all systematically underperform or simply fail for several broad classes of problems such as: left-right correlation, van der Waals interactions, negative ions, orbital spectra and band structure, open-shell singlet diradicals, etc. Even for thermochemistry of small molecules, none of the currently available density functionals has yet achieved an accuracy better than a few kcal/mol. In principle, there is no doubt that DFT can describe all such systems, provided that accurate Kohn–Sham exchange-correlation potentials and functionals are used. This has been demonstrated convincingly by generating the exact $v_{xc}(\mathbf{r})$ *via ab initio* methods and solving the corresponding Kohn–Sham equations [52,91,299–304]. Given the formidable complexity of the true exchange-correlation functional and the magnitude of the reward, development of better approximations is certain to remain a 'promising and charming' subject for years to come.

24.13 ACKNOWLEDGEMENTS

We are very grateful to John Perdew, Jianmin Tao, Andreas Savin, Sergey Maximoff, Oleg Vydrov, and Artur Izmaylov for comments and suggestions about our manuscript. This work was supported by the National Science Foundation, the Welch Foundation, and the Department of Energy.

24.14 REFERENCES

1 R.G. Parr, Annu. Rev. Phys. Chem., 34 (1983) 631.
2 E. Fermi, Z. Phys., 48 (1928) 73, reprinted in English in Ref. [6].
3 P.A.M. Dirac, Proc. Camb. Phil. Soc., 26 (1930) 376.
4 P. Gombás, Die statistische Theorie des Atoms und ihre Anwendungen, Springer, Wien, 1949.
5 N.H. March, Adv. Phys., 6 (1957) 1.
6 N.H. March, Self-consistent fields in atoms: Hartree and Thomas–Fermi atoms, Pergamon, Oxford, 1975.
7 E.H. Lieb, Rev. Mod. Phys., 53 (1981) 603; E.H. Lieb, Rev. Mod. Phys., 54 (1982) 311(E).
8 R. Gáspár, Acta Phys. Hung., 3 (1954) 263.
9 J.C. Slater, Phys. Rev., 81 (1951) 385.
10 J.C. Slater, Adv. Quantum Chem., 6 (1972) 1.
11 J.C. Slater, Int. J. Quantum Chem. Symp., 9 (1975) 7.
12 E. Teller, Rev. Mod. Phys., 34 (1962) 627.
13 N.L. Balàzs, Phys. Rev., 156 (1967) 42.
14 I. Catto and P.L. Lions, Commun. Part. Diff. Eq., 17 (1992) 1051.
15 P. Hohenberg and W. Kohn, Phys. Rev., 136 (1964) B864.
16 W. Kohn, in: F. Bassani, F. Fumi, M.P. Tosi (Eds.), Highlights of condensed-matter theory, North-Holland, Amsterdam, 1985.
17 W. Kohn and L.J. Sham, Phys. Rev., 140 (1965) A1133.
18 W. Kohn, A.D. Becke and R.G. Parr, J. Phys. Chem., 100 (1996) 12974.

19 R.G. Parr and W. Yang, Annu. Rev. Phys. Chem., 46 (1995) 701.
20 R.G. Parr and W. Yang, Density-functional theory of atoms and molecules, Oxford University Press, New York, 1989.
21 P.M.W. Gill, in: P. von R. Schleyer (Ed.), Encyclopedia of computational chemistry, Wiley, New York, 1998, p. 678.
22 W. Koch and M.C. Holthausen, A chemist's guide to density functional theory, Wiley-VCH, Weinheim, 2000.
23 J.P. Perdew and S. Kurth, in: C. Fiolhais, F. Nogueira, M. Marques (Eds.), A primer in density functional theory, Springer, Berlin, 2003.
24 J. Kohanoff and N.I. Gidopoulos, in: S. Wilson (Ed.), Handbook of molecular physics and quantum chemistry, Molecular electronic structure, Vol. 2, Wiley, Chichester, 2003.
25 J.P. Dahl and J. Avery (Eds.), Local density approximations in quantum chemistry and solid state physics, Plenum, New York, 1984.
26 R.M. Dreizler and J. da Providência (Eds.), Density functional methods in physics, NATO ASI Series, Vol. B123, Plenum, New York, 1985.
27 R.M. Dreizler and E.K.U. Gross, Density functional theory, Springer, Berlin, 1990.
28 E.S. Kryachko and E.V. Ludeña, Energy density functional theory of many-electron systems, Kluwer, Dordrecht, 1990.
29 E.K.U. Gross and R.M. Dreizler (Eds.), Density functional theory, NATO ASI Series, Vol. B337, Plenum, New York, 1995.
30 J.M. Seminario and P. Politzer (Eds.), Modern density functional theory: A tool for chemistry, Elsevier, Amsterdam, 1995.
31 J.M. Seminario (Ed.), Recent developments and applications of modern density functional theory, Elsevier, Amsterdam, 1996.
32 M.J. Frisch, G.W. Trucks, H.B. Schlegel, G.E. Scuseria, M.A. Robb, J.R. Cheeseman, J.A. Montgomery, Jr., T. Vreven, K.N. Kudin, J.C. Burant, J.M. Millam, S.S. Iyengar, J. Tomasi, V. Barone, B. Mennucci, M. Cossi, G. Scalmani, N. Rega, G.A. Petersson, H. Nakatsuji, M. Hada, M. Ehara, K. Toyota, R. Fukuda, J. Hasegawa, M. Ishida, T. Nakajima, Y. Honda, O. Kitao, H. Nakai, M. Klene, X. Li, J.E. Knox, H.P. Hratchian, J.B. Cross, C. Adamo, J. Jaramillo, R. Gomperts, R.E. Stratmann, O. Yazyev, A.J. Austin, R. Cammi, C. Pomelli, J.W. Ochterski, P.Y. Ayala, K. Morokuma, G.A. Voth, P. Salvador, J.J. Dannenberg, V.G. Zakrzewski, S. Dapprich, A.D. Daniels, M.C. Strain, O. Farkas, D.K. Malick, A.D. Rabuck, K. Raghavachari, J.B. Foresman, J.V. Ortiz, Q. Cui, A.G. Baboul, S. Clifford, J. Cioslowski, B.B. Stefanov, G. Liu, A. Liashenko, P. Piskorz, I. Komaromi, R.L. Martin, D.J. Fox, T. Keith, M.A. Al-Laham, C.Y. Peng, A. Nanayakkara, M. Challacombe, P.M.W. Gill, B. Johnson, W. Chen, M.W. Wong, C. Gonzalez and J.A. Pople, GAUSSIAN 03, Revision C.2, Gaussian, Inc., Wallingford, CT, 2004.
33 Y.A. Wang and E.A. Carter, in: S.D. Schwartz (Ed.), Theoretical methods in condensed phase chemistry, Kluwer, Dordrecht, 2000.
34 V. Sahni, Quantal density functional theory, Springer, Berlin, 2004.
35 E.V. Ludeña, E.S. Kryachko, T. Koga, R. López-Boada, J. Hinze, J. Maldonado and E. Valderrama, in: J.M. Seminario, P. Politzer (Eds.), Modern density functional theory: A tool for chemistry, Theoretical and computational chemistry, Vol. 2, Elsevier, Amsterdam, 1995, p. 75.
36 M. Levy, Proc. Natl. Acad. Sci. USA, 76 (1979) 6062.
37 S.S. Iyengar, M. Ernzerhof, S.N. Maximoff and G.E. Scuseria, Phys. Rev. A, 63 (2001) 052508.
38 P.C. Hohenberg, W. Kohn and L.J. Sham, Adv. Quantum Chem., 21 (1990) 7.
39 W. Kohn, Rev. Mod. Phys., 71 (1999) 1253.
40 A. Savin, F. Colonna and R. Pollet, Int. J. Quantum Chem., 93 (2003) 166.
41 S.T. Epstein, in: B.M. Deb (Ed.), The force concept in chemistry, Van Nostrand, New York, 1981.
42 J. Harris and R.O. Jones, J. Phys. F, 4 (1974) 1170.
43 D.C. Langreth and J.P. Perdew, Solid State Commun., 17 (1975) 1425.
44 D.C. Langreth and J.P. Perdew, Phys. Rev. B, 15 (1977) 2884.
45 O. Gunnarsson and B.I. Lundqvist, Phys. Rev. B, 13 (1976) 4274.
46 R. McWeeny and B.T. Sutcliffe, Methods of molecular quantum mechanics, Academic Press, London, 1969.

47 R.T. Sharp and G.K. Horton, Phys. Rev., 90 (1953) 317.
48 J.D. Talman and W.F. Shadwick, Phys. Rev. A, 14 (1976) 36.
49 T. Grabo, T. Kreibich, S. Kurth and E.K.U. Gross, in: V.I. Anisimov (Ed.), Strong coulomb correlations in electronic structure calculations: Beyond the local density approximation, Gordon and Breach, Amsterdam, 2000.
50 S. Ivanov, S. Hirata and R.J. Bartlett, Phys. Rev. Lett., 83 (1999) 5455.
51 S. Ivanov, S. Hirata and R.J. Bartlett, J. Chem. Phys., 116 (2002) 1269.
52 E. Engel and R.M. Dreizler, J. Comput. Chem., 20 (1999) 31.
53 O.V. Gritsenko and E.J. Baerends, Phys. Rev. A, 64 (2001) 042506.
54 W. Yang and Q. Wu, Phys. Rev. Lett., 89 (2002) 143002.
55 S. Kümmel and J.P. Perdew, Phys. Rev. B, 68 (2003) 035103.
56 U. von Barth and L. Hedin, J. Phys. C: Solid State Phys., 5 (1972) 1629.
57 M.M. Pant and A.K. Rajagopal, Solid State Commun., 10 (1972) 1157.
58 S.N. Maximoff and G.E. Scuseria, Chem. Phys. Lett., 390 (2004) 408.
59 H.L. Schmider and A.D. Becke, J. Mol. Struct. (THEOCHEM), 527 (2000) 51.
60 H.L. Schmider and A.D. Becke, J. Chem. Phys., 116 (2002) 3184.
61 S.N. Maximoff, M. Ernzerhof and G.E. Scuseria, J. Chem. Phys., 117 (2002) 3074.
62 S.M. Colwell and N.C. Handy, Chem. Phys. Lett., 217 (1994) 271.
63 A.M. Lee, S.M. Colwell and N.C. Handy, Chem. Phys. Lett., 229 (1994) 225.
64 A.M. Lee, N.C. Handy and S.M. Colwell, J. Chem. Phys., 103 (1995) 10095.
65 A.D. Becke, Can. J. Chem., 74 (1996) 995.
66 A.D. Becke, J. Chem. Phys., 117 (2002) 6935.
67 S.N. Maximoff, M. Ernzerhof and G.E. Scuseria, J. Chem. Phys., 120 (2004) 2105.
68 G. Vignale and M. Rasolt, Phys. Rev. Lett., 59 (1987) 2360; G. Vignale and M. Rasolt, Phys. Rev. Lett., 62 (1989) 115(E).
69 G. Vignale, M. Rasolt and D.J.W. Geldart, Phys. Rev. B, 37 (1988) 2502.
70 G. Vignale and M. Rasolt, Phys. Rev. B, 37 (1988) 10685.
71 G. Vignale, M. Rasolt and D.J.W. Geldart, Adv. Quantum Chem., 21 (1990) 235.
72 J.P. Perdew and K. Schmidt, in: V. Van Doren, C. Van Alsenoy, P. Geerlings (Eds.), Density functional theory and its application to materials, AIP, Melville, NY, 2001.
73 J.P. Perdew, A. Ruzsinszky, J. Tao, V.N. Staroverov, G.E. Scuseria and G.I. Csonka, J. Chem. Phys. (to appear).
74 J.P. Perdew and M. Levy, Int. J. Quantum Chem. Symp., 49 (1994) 539.
75 E.H. Lieb and S. Oxford, Int. J. Quantum Chem., 19 (1981) 427.
76 J.P. Perdew, in: P. Ziesche, H. Eschrig (Eds.), Electronic structure of solids '91, Akademie, Berlin, 1991.
77 J.P. Perdew and A. Zunger, Phys. Rev. B, 23 (1981) 5048.
78 M. Levy and J.P. Perdew, Phys. Rev. A, 32 (1985) 2010.
79 M. Levy, Int. J. Quantum Chem. Symp., 23 (1989) 617.
80 H. Ou-Yang and M. Levy, Phys. Rev. A, 42 (1990) 155.
81 M. Levy and H. Ou-Yang, Phys. Rev. A, 42 (1990) 651.
82 M. Levy, Phys. Rev. A, 43 (1991) 4637.
83 A. Görling and M. Levy, Phys. Rev. A, 45 (1992) 1509.
84 M. Levy and J.P. Perdew, Phys. Rev. B, 48 (1993) 11638.
85 M. Levy, in: E.K.U. Gross, R.M. Dreizler (Eds.), Density functional theory, Plenum, New York, 1995.
86 G.L. Oliver and J.P. Perdew, Phys. Rev. A, 20 (1979) 397.
87 A. Görling and M. Levy, Phys. Rev. A, 47 (1993) 13105.
88 A. Görling and M. Levy, Phys. Rev. A, 50 (1994) 196.
89 A. Görling and M. Levy, Int. J. Quantum Chem. Symp., 29 (1995) 93.
90 S. Ivanov and R.J. Bartlett, J. Chem. Phys., 114 (2001) 1952.
91 I. Grabowski, S. Hirata, S. Ivanov and R.J. Bartlett, J. Chem. Phys., 116 (2002) 4415.
92 S. Ivanov, S. Hirata, I. Grabowski and R.J. Bartlett, J. Chem. Phys., 118 (2003) 461.
93 J. Jaramillo, G.E. Scuseria and M. Ernzerhof, J. Chem. Phys., 118 (2003) 1068.
94 J.A. Alonso and L.A. Girifalco, Phys. Rev. B, 17 (1978) 3735.

95 O. Gunnarsson, M. Jonson and B.I. Lundqvist, Phys. Rev. B, 20 (1979) 3136.
96 M.D. Glossman, A. Rubio, L.C. Balbás and J.A. Alonso, Int. J. Quantum Chem. Symp., 26 (1992) 347.
97 O.V. Gritsenko, N.A. Cordero, A. Rubio, L.C. Balbás and J.A. Alonso, Phys. Rev. A, 48 (1993) 3197.
98 S.K. Ghosh, M. Berkowitz and R.G. Parr, Proc. Natl Acad. Sci. USA, 81 (1984) 8028.
99 S.K. Ghosh and R.G. Parr, Phys. Rev. A, 34 (1986) 785.
100 S. Liu and R.G. Parr, Phys. Rev. A, 55 (1996) 1972.
101 Y.A. Wang, S. Liu and R.G. Parr, Chem. Phys. Lett., 267 (1997) 14.
102 A. Nagy, S. Liu and R.G. Parr, Phys. Rev. A, 59 (1999) 3349.
103 F. Bloch, Z. Phys., 57 (1929) 545.
104 A. Sommerfeld and H. Bethe, in: H. Geiger, K. Scheel (Eds.), Handbuch der Physik, 2nd edn., Springer, Berlin, 1933, p. 485.
105 J.C. Slater, Quantum theory of atomic structure, Vol. 2, McGraw-Hill, New York, 1960.
106 E. Wigner and F. Seitz, Phys. Rev., 46 (1934) 509.
107 M. Ernzerhof and G.E. Scuseria, J. Chem. Phys., 111 (1999) 911.
108 M. Gell-Mann and K.A. Brueckner, Phys. Rev., 106 (1957) 364.
109 W.J. Carr, Jr. and A.A. Maradudin, Phys. Rev., 133 (1964) A371.
110 P. Nozières and D. Pines, Phys. Rev., 111 (1958) 442.
111 W.J. Carr, Jr., Phys. Rev., 122 (1961) 1437.
112 D.M. Ceperley and B.J. Alder, Phys. Rev. Lett., 45 (1980) 566.
113 S.H. Vosko, L. Wilk and M. Nusair, Can. J. Phys., 58 (1980) 1200.
114 J.P. Perdew and Y. Wang, Phys. Rev. B, 45 (1992) 13244.
115 P.R. Antoniewicz and L. Kleinman, Phys. Rev. B, 31 (1985) 6779.
116 L.J. Sham, in: P.M. Marcus, J.F. Janak, A.R. Williams (Eds.), Computational methods in band theory, Plenum, New York, 1971, p. 458.
117 L. Kleinman and S. Lee, Phys. Rev. B, 37 (1988) 4634.
118 J.A. Chevary and S.H. Vosko, Phys. Rev. B, 42 (1990) 5320.
119 E. Engel and S.H. Vosko, Phys. Rev. B, 50 (1994) 10498.
120 P.S. Svendsen and U. von Barth, Phys. Rev. B, 54 (1996) 17402.
121 S.-K. Ma and K.A. Brueckner, Phys. Rev., 165 (1968) 18.
122 M. Rasolt and D.J.W. Geldart, Phys. Rev. B, 34 (1986) 1325.
123 M. Ernzerhof and G.E. Scuseria, J. Chem. Phys., 112 (2000) 5270.
124 F. Herman, J.P. Van Dyke and I.B. Ortenburger, Phys. Rev. Lett., 22 (1969) 807.
125 A.D. Becke, J. Chem. Phys., 84 (1986) 4524.
126 A.D. Becke, J. Chem. Phys., 85 (1986) 7184.
127 A.E. DePristo and J.D. Kress, J. Chem. Phys., 86 (1987) 1425.
128 A.D. Becke, Phys. Rev. A, 38 (1988) 3098.
129 R. van Leeuwen and E.J. Baerends, Phys. Rev. A, 49 (1994) 2421.
130 P. Jemmer and P.J. Knowles, Phys. Rev. A, 51 (1995) 3571.
131 D.J. Tozer and N.C. Handy, J. Chem. Phys., 109 (1998) 10180.
132 D.J. Tozer and N.C. Handy, Mol. Phys., 101 (2003) 2669.
133 D.C. Langreth and J.P. Perdew, Phys. Rev. B, 21 (1980) 5469.
134 D.C. Langreth and M.J. Mehl, Phys. Rev. B, 28 (1983) 1809.
135 J.P. Perdew, Phys. Rev. Lett., 55 (1985) 1665; J.P. Perdew, Phys. Rev. Lett., 55 (1985) 2370(E).
136 J.P. Perdew and Y. Wang, Phys. Rev. B, 33 (1986) 8800.
137 N.H. March, Phys. Rev. A, 36 (1987) 5077.
138 F. Della Sala and A. Görling, Phys. Rev. Lett., 89 (2002) 033003.
139 M. Abramowitz and I.A. Stegun, Handbook of mathematical functions with formulas, graphs, and mathematical tables, US Government, Washington, DC, 1972.
140 D.J. Lacks and R.G. Gordon, Phys. Rev. A, 47 (1993) 4681.
141 C. Adamo and V. Barone, J. Comput. Chem., 19 (1998) 418.
142 M. Filatov and W. Thiel, Mol. Phys., 91 (1997) 847.
143 O.V. Gritsenko, R. van Leeuwen, E. van Lenthe and E.J. Baerends, Phys. Rev. A, 51 (1995) 1944.
144 O.V. Gritsenko, R. van Leeuwen and E.J. Baerends, Int. J. Quantum Chem., 61 (1997) 231.

145 O.V. Gritsenko, P.R.T. Schipper and E.J. Baerends, Chem. Phys. Lett., 302 (1999) 199.
146 P.R.T. Schipper, O.V. Gritsenko, S.J.A. van Gisbergen and E.J. Baerends, J. Chem. Phys., 112 (2000) 1344.
147 O.V. Gritsenko, P.R.T. Schipper and E.J. Baerends, Int. J. Quantum Chem., 76 (2000) 407.
148 E. Engel, J.A. Chevary, L.D. Macdonald and S.H. Vosko, Z. Phys. D, 23 (1992) 7.
149 M. Filatov and W. Thiel, Phys. Rev. A, 57 (1998) 189.
150 R. Neumann and N.C. Handy, Chem. Phys. Lett., 266 (1997) 16.
151 P.M.W. Gill, Mol. Phys., 89 (1996) 433.
152 M. Ernzerhof, S.N. Maximoff and G.E. Scuseria, J. Chem. Phys., 116 (2002) 3980.
153 J.P. Perdew, Physica B, 172 (1991) 1.
154 K. Burke and J.P. Perdew, Int. J. Quantum Chem., 56 (1995) 199.
155 K. Burke and J.P. Perdew, Y. Wang, in: J.F. Dobson, G. Vignale, M.P. Das (Eds.), Electronic density functional theory: Recent progress and new directions, Plenum, New York, 1998.
156 J.P. Perdew, K. Burke and Y. Wang, Phys. Rev. B, 54 (1996) 16533; J.P. Perdew, K. Burke and Y. Wang, Phys. Rev. B, 57 (1998) 14999(E).
157 J.P. Perdew, J.A. Chevary, S.H. Vosko, K.A. Jackson, M.R. Pederson, D.J. Singh and C. Fiolhais, Phys. Rev. B, 46 (1992) 6671; J.P. Perdew, J.A. Chevary, S.H. Vosko, K.A. Jackson, M.R. Pederson, D.J. Singh and C. Fiolhais, Phys. Rev. B, 48 (1993) 4978(E).
158 Y. Wang and J.P. Perdew, Phys. Rev. B, 43 (1991) 8911.
159 J.P. Perdew, K. Burke and M. Ernzerhof, Phys. Rev. Lett., 77 (1996) 3865; J.P. Perdew, K. Burke and M. Ernzerhof, Phys. Rev. Lett., 78 (1997) 1396(E).
160 C.D. Hu and D.C. Langreth, Phys. Scr., 32 (1985) 391.
161 C.D. Hu and D.C. Langreth, Phys. Rev. B, 33 (1986) 943.
162 J.P. Perdew, Phys. Rev. B, 33 (1986) 8822; J.P. Perdew, Phys. Rev. B, 34 (1986) 7406(E).
163 L.C. Wilson and M. Levy, Phys. Rev. B, 41 (1990) 12930.
164 V.N. Staroverov, G.E. Scuseria, J.P. Perdew, J. Tao and E.R. Davidson, Phys. Rev. A, 70 (2004) 012502.
165 Y. Zhang and W. Yang, Phys. Rev. Lett., 80 (1998) 890.
166 J.P. Perdew, K. Burke and M. Ernzerhof, Phys. Rev. Lett., 80 (1998) 891.
167 B. Hammer, L.B. Hansen and J.K. Nørskov, Phys. Rev. B, 59 (1999) 7413.
168 A.D. Becke, J. Chem. Phys., 104 (1996) 1040.
169 A.D. Becke, J. Chem. Phys., 109 (1998) 2092.
170 H. Stoll, C.M.E. Pavlidou and H. Preuß, Theor. Chim. Acta, 49 (1978) 143.
171 H. Stoll, E. Golka and H. Preuß, Theor. Chim. Acta, 55 (1980) 29.
172 J.B. Krieger, J. Chen, G.J. Iafrate and A. Savin, in: A. Gonis, N. Kioussis, M. Ciftan (Eds.), Electron correlations and materials properties, Kluwer/Plenum, New York, 1999.
173 J.P. Perdew, R.G. Parr, M. Levy and J.L. Balduz, Jr., Phys. Rev. Lett., 49 (1982) 1691.
174 J. Rey and A. Savin, Int. J. Quantum Chem., 69 (1998) 581.
175 J. Toulouse, A. Savin and C. Adamo, J. Chem. Phys., 117 (2002) 10465.
176 J.P. Perdew, S. Kurth, A. Zupan and P. Blaha, Phys. Rev. Lett., 82 (1999) 2544.
177 C. Adamo, M. Ernzerhof and G.E. Scuseria, J. Chem. Phys., 112 (2000) 2643.
178 A. Rabuck and G.E. Scuseria, Theor. Chem. Acc., 104 (2000) 439.
179 J. Tao, J.P. Perdew, V.N. Staroverov and G.E. Scuseria, Phys. Rev. Lett., 91 (2003) 146401.
180 V.N. Staroverov, G.E. Scuseria, J. Tao and J.P. Perdew, J. Chem. Phys., 119 (2003) 12129; ibid. 121 (2004) 11507 (E).
181 V.N. Staroverov, G.E. Scuseria, J. Tao and J.P. Perdew, Phys. Rev. B, 69 (2004) 075102.
182 A.D. Becke and M.R. Roussel, Phys. Rev. A, 39 (1989) 3761.
183 A.D. Becke, Int. J. Quantum Chem., 23 (1983) 1915.
184 A.D. Becke, J. Chem. Phys., 88 (1988) 1053.
185 A.K. Rajagopal, J.C. Kimball and M. Banerjee, Phys. Rev. B, 18 (1978) 2339.
186 A.D. Becke, Int. J. Quantum Chem. Symp., 28 (1994) 625.
187 R. Neumann and N.C. Handy, Chem. Phys. Lett., 246 (1995) 381.
188 A.F. Izmaylov, S.N. Maximoff and G.E. Scuseria, in preparation.
189 M. Ernzerhof and J.P. Perdew, J. Chem. Phys., 109 (1998) 3313.

190 L. Constantin and J.P. Perdew, work in progress.
191 C. Lee, W. Yang and R.G. Parr, Phys. Rev. B, 37 (1988) 785.
192 R. Colle and O. Salvetti, Theor. Chim. Acta, 37 (1975) 329.
193 B. Miehlich, A. Savin, H. Stoll and H. Preuss, Chem. Phys. Lett., 157 (1989) 200.
194 B.G. Johnson, P.M.W. Gill and J.A. Pople, J. Chem. Phys., 98 (1993) 5612.
195 J. Tao, P. Gori-Giorgi, J.P. Perdew and R. McWeeny, Phys. Rev. A, 63 (2001) 032513.
196 R. Singh, L. Massa and V. Sahni, Phys. Rev. A, 60 (1999) 4135.
197 C. Filippi, X. Gonze and C.J. Umrigar, in: J.M. Seminario (Ed.), Recent developments and applications of modern density functional theory, Theoretical and computational chemistry, Vol. 4, Elsevier, Amsterdam, 1996, p. 295.
198 Y. Imamura, G.E. Scuseria and R.M. Martin, J. Chem. Phys., 116 (2002) 6458.
199 R. Colle and O. Salvetti, J. Chem. Phys., 79 (1983) 1404.
200 Y. Imamura and G.E. Scuseria, J. Chem. Phys., 118 (2003) 2464.
201 S.F. Boys and N.C. Handy, Proc. Roy. Soc. A, 310 (1969) 43.
202 A.C. Hurley, Electron correlation in small molecules, Academic Press, London, 1976.
203 T. Tsuneda and K. Hirao, Chem. Phys. Lett., 268 (1997) 510.
204 T. Tsuneda, T. Suzumura and K. Hirao, J. Chem. Phys., 110 (1999) 10664.
205 E.I. Proynov and D.R. Salahub, Phys. Rev. B, 49 (1994) 7874; E.I. Proynov and D.R. Salahub, Phys. Rev. B, 57 (1998) 12616(E).
206 E.I. Proynov, A. Vela and D.R. Salahub, Chem. Phys. Lett., 230 (1994) 419; E.I. Proynov, A. Vela and D.R. Salahub, Chem. Phys. Lett., 234 (1995) 462(E).
207 E.I. Proynov, E. Ruiz, A. Vela and D.R. Salahub, Int. J. Quantum Chem. Symp., 29 (1995) 61.
208 E.I. Proynov, S. Sirois and D.R. Salahub, Int. J. Quantum Chem., 64 (1997) 427.
209 E. Proynov, H. Chermette and D.R. Salahub, J. Chem. Phys., 113 (2000) 10013.
210 J.W. Negele and D. Vautherin, Phys. Rev. C, 5 (1972) 1472.
211 R.M. Koehl, G.K. Odom and G.E. Scuseria, Mol. Phys., 87 (1996) 835.
212 T. Van Voorhis and G.E. Scuseria, Mol. Phys., 92 (1997) 601.
213 T. Van Voorhis and G.E. Scuseria, J. Chem. Phys., 109 (1998) 400.
214 S.N. Maximoff and G.E. Scuseria, J. Chem. Phys., 114 (2001) 10591.
215 A.D. Becke, J. Chem. Phys., 107 (1997) 8554.
216 A.D. Becke, J. Comput. Chem., 20 (1999) 63.
217 R.D. Adamson, P.M.W. Gill and J.A. Pople, Chem. Phys. Lett., 284 (1998) 6.
218 D.J. Tozer and N.C. Handy, J. Chem. Phys., 108 (1998) 2545.
219 F.A. Hamprecht, A.J. Cohen, D.J. Tozer and N.C. Handy, J. Chem. Phys., 109 (1998) 6264.
220 G.K.-L. Chan and N.C. Handy, J. Chem. Phys., 112 (2000) 5639.
221 P.J. Wilson, T.J. Bradley and D.J. Tozer, J. Chem. Phys., 115 (2001) 9233.
222 H.L. Schmider and A.D. Becke, J. Chem. Phys., 108 (1998) 9624.
223 H.L. Schmider and A.D. Becke, J. Chem. Phys., 109 (1998) 8188.
224 A.D. Boese, N.L. Doltsinis, N.C. Handy and M. Sprik, J. Chem. Phys., 112 (2000) 1670.
225 A.D. Boese and N.C. Handy, J. Chem. Phys., 114 (2001) 5497.
226 A.D. Boese and N.C. Handy, J. Chem. Phys., 116 (2002) 9559.
227 C.Y. Lin, M.W. George and P.M.W. Gill, Aust. J. Chem., 57 (2004) 365.
228 N.C. Handy and A.J. Cohen, Mol. Phys., 99 (2001) 403.
229 W.-M. Hoe, A.J. Cohen and N.C. Handy, Chem. Phys. Lett., 341 (2001) 319.
230 A.J. Cohen and N.C. Handy, Chem. Phys. Lett., 316 (2000) 160.
231 X. Xu and W.A. Goddard, Proc. Natl Acad. Sci. USA, 101 (2004) 2673.
232 A.D. Becke, J. Chem. Phys., 98 (1993) 1372.
233 M. Levy, N.H. March and N.C. Handy, J. Chem. Phys., 104 (1996) 1989.
234 A.D. Becke, J. Chem. Phys., 98 (1993) 5648.
235 P.J. Stephens, F.J. Devlin, C.F. Chabalowski and M.J. Frisch, J. Phys. Chem., 98 (1994) 11623.
236 R.H. Hertwig and W. Koch, Chem. Phys. Lett., 268 (1997) 345.
237 J.P. Perdew, M. Ernzerhof and K. Burke, J. Chem. Phys., 105 (1996) 9982.
238 A. Görling and M. Levy, J. Chem. Phys., 106 (1997) 2675.

239 C. Adamo and V. Barone, J. Chem. Phys., 108 (1998) 664.
240 M. Ernzerhof and G.E. Scuseria, J. Chem. Phys., 110 (1999) 5029.
241 C. Adamo and V. Barone, J. Chem. Phys., 110 (1999) 6158.
242 A.D. Boese and J.M.L. Martin, J. Chem. Phys., 121 (2004) 3405.
243 A.J. Cohen and N.C. Handy, Mol. Phys., 99 (2001) 607.
244 J. Baker and P. Pulay, J. Chem. Phys., 117 (2002) 1441.
245 J. Baker and P. Pulay, J. Comput. Chem., 24 (2003) 1184.
246 Y. Zhao, J.Z. Pu, B.J. Lynch and D.G. Truhlar, Phys. Chem. Chem. Phys., 6 (2004) 673.
247 Y. Zhao, B.J. Lynch and D.G. Truhlar, J. Phys. Chem. A, 108 (2004) 2715.
248 Y. Zhao, B.J. Lynch and D.G. Truhlar, J. Phys. Chem. A, 108 (2004) 4786.
249 O.A. Vydrov and G.E. Scuseria, J. Chem. Phys., 121 (2004) 8187.
250 B. Braïda, P.C. Hiberty and A. Savin, J. Phys. Chem. A, 102 (1998) 7872.
251 M. Grüning, O.V. Gritsenko, S.J.A. van Gisbergen and E.J. Baerends, J. Phys. Chem. A, 105 (2001) 9211.
252 J.P. Dombroski, S.W. Taylor and P.M.W. Gill, J. Phys. Chem., 100 (1996) 6272.
253 R.D. Adamson, J.P. Dombroski and P.M.W. Gill, Chem. Phys. Lett., 254 (1996) 329.
254 P.M.W. Gill and R.D. Adamson, Chem. Phys. Lett., 261 (1996) 105.
255 P.M.W. Gill, Chem. Phys. Lett., 270 (1997) 193.
256 R.D. Adamson and P.M.W. Gill, J. Mol. Struct. (THEOCHEM), 398 (1997) 45.
257 A. Savin and H.-J. Flad, Int. J. Quantum Chem., 56 (1995) 327.
258 A. Savin, in: J.M. Seminario (Ed.), Recent developments and applications of modern density functional theory, Theoretical and computational chemistry, Vol. 4, Elsevier, Amsterdam, 1996, p. 327.
259 T. Leininger, H. Stoll, H.-J. Werner and A. Savin, Chem. Phys. Lett., 275 (1997) 151.
260 R. Pollet, A. Savin, T. Leininger and H. Stoll, J. Chem. Phys., 116 (2002) 1250.
261 P.M.W. Gill, R.D. Adamson and J.A. Pople, Mol. Phys., 88 (1996) 1005.
262 J. Heyd, G.E. Scuseria and M. Ernzerhof, J. Chem. Phys., 118 (2003) 8207.
263 J. Heyd and G.E. Scuseria, J. Chem. Phys., 120 (2004) 7274.
264 J. Heyd and G.E. Scuseria, J. Chem. Phys., 121 (2004) 1187.
265 H. Iikura, T. Tsuneda, T. Yanai and K. Hirao, J. Chem. Phys., 115 (2001) 3540.
266 M. Kamiya, T. Tsuneda and K. Hirao, J. Chem. Phys., 117 (2002) 6010.
267 T. Yanai, D.P. Tew and N.C. Handy, Chem. Phys. Lett., 393 (2004) 51.
268 Y. Tawada, T. Tsuneda, S. Yanagisawa, T. Yanai and K. Hirao, J. Chem. Phys., 120 (2004) 8425.
269 A. Szabo and N.S. Ostlund, Modern quantum chemistry, Macmillan, New York, 1982.
270 J.A. Pople, P.M.W. Gill and B.G. Johnson, Chem. Phys. Lett., 199 (1992) 557.
271 R. Neumann, R.H. Nobes and N.C. Handy, Mol. Phys., 87 (1996) 1.
272 C.W. Murray, N.C. Handy and G.J. Laming, Mol. Phys., 78 (1993) 997.
273 P.M.W. Gill, B.G. Johnson and J.A. Pople, Chem. Phys. Lett., 209 (1993) 506.
274 V.I. Lebedev, Sib. Mat. Zh., 18 (1977) 132.
275 V.I. Lebedev and A.L. Skorokhodov, Dokl. Akad. Nauk, 324 (1992) 519.
276 V.I. Lebedev, Dokl. Akad. Nauk, 338 (1994) 454.
277 V.I. Lebedev and D.N. Laikov, Dokl. Akad. Nauk, 366 (1999) 741.
278 A.D. Becke, J. Chem. Phys., 88 (1988) 2547.
279 R.E. Stratmann, G.E. Scuseria and M.J. Frisch, Chem. Phys. Lett., 257 (1996) 213.
280 I.M. Gelfand and S.V. Fomin, Calculus of variations, Prentice-Hall, Englewood Cliffs, NJ, 1963.
281 A.V. Arbuznikov, M. Kaupp, V.G. Malkin, R. Reviakine and O.L. Malkina, Phys. Chem. Chem. Phys., 4 (2002) 5467.
282 A.V. Arbuznikov and M. Kaupp, Chem. Phys. Lett., 381 (2003) 495.
283 A.V. Arbuznikov and M. Kaupp, Chem. Phys. Lett., 386 (2004) 8.
284 A.M. Teale and D.J. Tozer, Chem. Phys. Lett., 383 (2004) 109.
285 W. Hieringer, F. Della Sala and A. Görling, Chem. Phys. Lett., 383 (2004) 115.
286 L.A. Curtiss, K. Raghavachari, P.C. Redfern and J.A. Pople, J. Chem. Phys., 106 (1997) 1063.
287 L.A. Curtiss, K. Raghavachari, P.C. Redfern and J.A. Pople, J. Chem. Phys., 112 (2000) 7374.
288 L.A. Curtiss, P.C. Redfern, K. Raghavachari and J.A. Pople, J. Chem. Phys., 109 (1998) 42.

289 J. Andzelm and E. Wimmer, J. Chem. Phys., 96 (1992) 1280.
290 J. Baker, M. Muir and J. Andzelm, J. Chem. Phys., 102 (1995) 2063.
291 A.C. Scheiner, J. Baker and J.W. Andzelm, J. Comput. Chem., 18 (1997) 775.
292 S. Kurth, J.P. Perdew and P. Blaha, Int. J. Quantum Chem., 75 (1999) 889.
293 M. Grüning, O.V. Gritsenko and E.J. Baerends, J. Phys. Chem. A, 108 (2004) 4459.
294 C.A. Coulson, Rev. Mod. Phys., 32 (1960) 170.
295 A.J. Coleman and V.I. Yukalov, Reduced density matrices: Coulson's challenge, Springer, Berlin, 2000.
296 P.M.W. Gill, B.G. Johnson, J.A. Pople and M.J. Frisch, Int. J. Quantum Chem. Symp., 26 (1992) 319.
297 P.M.W. Gill, B.G. Johnson, J.A. Pople and M.J. Frisch, Chem. Phys. Lett., 197 (1992) 499.
298 G.E. Scuseria, J. Chem. Phys., 97 (1992) 7528.
299 S. Hirata, S. Ivanov, I. Grabowski and R.J. Bartlett, J. Chem. Phys., 116 (2002) 6468.
300 R.J. Bartlett, I. Grabowski, S. Hirata and S. Ivanov, J. Chem. Phys., 122 (2005) 034104.
301 E. Engel, A. Höck and R.M. Dreizler, Phys. Rev. A, 61 (2001) 032502.
302 O.V. Gritsenko and E.J. Baerends, Theor. Chem. Acc., 96 (1997) 44.
303 E.J. Baerends and O.V. Gritsenko, J. Phys. Chem. A, 101 (1997) 5383.
304 P.R.T. Schipper, O.V. Gritsenko and E.J. Baerends, Phys. Rev. A, 57 (1998) 1729.

Theory and Applications of Computational Chemistry: The First Forty Years
Edited by C. Dykstra et al.

CHAPTER 25

Multiconfigurational quantum chemistry

Björn O. Roos

Department of Theoretical Chemistry, Chemical Center, P.O. Box 124, S-221 00 Lund, Sweden

Abstract

The chapter gives a brief overview of the multiconfigurational approach in quantum chemistry. The method has been developed for studies of problems where a single configuration will not give a qualitatively correct description of the electronic structure. Examples of such cases are discussed and it is illustrated how a valid wave function may be constructed. The concept of inactive and active orbitals is introduced and the complete active space (CAS) wave function is defined. The choice of the active space is a crucial part of a CASSCF calculation. Without attempting to be complete, we illustrate how this choice can be made for some typical electron structure problems. The problem of calculating the dynamic correlation energy is discussed with special emphasis on second-order perturbation theory (CASPT2). How relativistic effects may be introduced is briefly discussed. Finally, we work through three-electron structure problems: ozone, the allyl radical, and the PbF molecule.

25.1 INTRODUCTION

The molecular orbital (MO) is the fundamental quantity in contemporary quantum chemistry. Almost all of our understanding of "what the electrons are doing in molecules" is based on the molecular orbital concept. Also, most of the computational methods used today start by a calculation of the molecular orbitals of the system using some quantum chemical method.

The molecular orbital describes the "motion" of one electron in the electric field generated by the nuclei and some average distribution of the other electron. It is in the simplest model occupied by zero, one, or two electrons. In the case of two electrons occupying the same orbital, the Pauli principle demands that they have opposite spin.

Such an approach leads to a total wave function for the system, which is an anti-symmetrized product of molecular spin orbitals (spin orbital = molecular orbital × a spin function). The Hartree–Fock (HF) method is obtained by applying the variation principle to the corresponding energy functional.

The concept of electrons moving in orbitals in the mean-field generated by the other electrons was introduced first for atoms by Hartree in 1928 [1]. A similar model was actually proposed for molecules already in 1927 by Hund [2]. The approach was immediately used by Mulliken and others for the interpretation of electronic spectra of small (diatomic) molecules [3].

The success of the Hartree–Fock method in describing the electronic structure of most closed-shell molecules has made it natural to analyze the wave function in terms of the molecular orbitals. The concept is simple and has a close relation to experiment through Koopmans theorem.

The two fundamental building blocks of Hartree–Fock theory are the molecular orbital and its occupation number. In closed-shell systems each occupied molecular orbital carries two electrons, with opposite spin. The occupied orbitals themselves are only defined as an occupied one-electron subspace of the full space spanned by the eigenfunctions of the Fock operator. Transformations between them leave the total HF wave function invariant. Normally the orbitals are obtained in a delocalized form as the solutions to the HF equations. This formulation is the most relevant one in studies of spectroscopic properties of the molecule, that is, excitation and ionization. The invariance property, however, makes a transformation to localized orbitals possible. Such localized orbitals can be valuable for an analysis of the chemical bonds in the system.

The concept of the molecular orbital and their occupation is, however, not restricted to the HF model. It has much wider relevance and is applicable also for more accurate wave functions. For each wave function we can form the first-order reduced density matrix. This matrix is Hermitian and can be diagonalized. The basis for this diagonal form of the density matrix are the *Natural Orbitals* first introduced in quantum chemistry by Per-Olof Löwdin [4].

Molecular orbital theory starts with the definition of a set of molecular spin orbitals (SOs):

$$\phi_i(\mathbf{r}, s) = \varphi_i(\mathbf{r})\theta_i(s) \tag{1}$$

where $\varphi_i(\mathbf{r})$ is a molecular orbital and $\theta_i(s)$ is a spin function. Usually, the molecular orbitals are expanded in a basis set, χ_p, $p = 1, m$:

$$\varphi_i = \sum_p C_{ip}\chi_p \tag{2}$$

The dimension of this basis set, m, sets the dimension of the quantum chemical problem. Normally, the basis functions are centered on the atoms of the molecular system (the Linear Combination of Atomic Orbitals (LCAO) approximation). Thus, the size is approximately proportional to the number of atoms in the system.

Having established the basis of one-electron functions, we construct a basis set for our N-electron wave functions as the set of Slater determinants that we can construct from the SOs:

$$\Phi_K = \hat{A}\{\phi_{K1}(x_1), \phi_{K2}(x_2), \ldots, \phi_{KN}(x_N)\} \tag{3}$$

where $x = \mathbf{r}, s$ and $\hat{A}$ is an anti-symmetrizer. The number of such determinants is

$$K = \binom{2m}{N} \tag{4}$$

The total wave function for the system can be expanded in these determinants:

$$\Psi = \sum_K C_K \Phi_K \tag{5}$$

Applying the variation principle we obtain the standard secular equation for the determination of the expansion coefficients and the energy:

$$\sum_L (H_{KL} - E\delta_{KL})C_L = 0 \tag{6}$$

This way to solve the Schrödinger equation has obtained the name *Full CI-FCI.* It becomes an exact solution in the limit of a complete basis set. The Full CI wave function is the "best" solution to the Schrödinger equation that can be obtained with a given basis set. This is true both for the ground state and for the excited states following the variation principle and McDonald's theorem [5]. The above formulation was made in terms of Slater determinants. It is, however, also possible to construct spin-adapted functions, which are linear combinations of determinants with a given orbital occupation. We shall call such functions as Configuration State Functions (CSF).

The Full CI method can, of course, not be used in practical applications. The size of the secular problem is simply too large. It defines, however, a calibration point to which all molecular orbital derived method should converge. There exists today a number of FCI calculations on small molecules with limited basis sets, which can be used to test different approximations to the FCI wave function.

In this chapter we shall describe how we can obtain wave functions that are of the FCI type but in a limited orbital space defined as the active orbitals of the system. But before developing such a model we shall describe how the molecular orbital concept can be extended to any type of wave functions.

25.2 THE DENSITY MATRIX AND THE NATURAL ORBITALS

For any wave function we can define the first-order reduced density matrix [4] as

$$\rho_1(x, x') = N \int \Psi^*(x', x_2, \ldots, x_N)\Psi(x, x_2, \ldots, x_N)\mathrm{d}x_2 \cdots \mathrm{d}x_N \tag{7}$$

The diagonal elements of this matrix give the probability density for the electrons in a point $\mathbf{r}$ with the spin s. A matrix representation of the density matrix can be obtained by an expansion in the basis of SOs used to construct the total wave function (orbitals which are not occupied in the wave function will not appear in the expansion):

$$\rho_1(x,x') = \sum_{i,j} \rho_{ij}\phi_i(x)\phi_j^*(x') \tag{8}$$

The density matrix is Hermitian and we can transform it to a diagonal form by a unitary transformation of the SOs:

$$\rho_1(x,x') = \sum_i \eta_i\sigma_i(x)\sigma_i^*(x') \tag{9}$$

The orbitals $\sigma_i^*(x')$ are called the Natural Spin Orbitals (NSOs), and η_i are their *occupation numbers*. They fulfill the condition

$$0 \leq \eta_i \leq 1 \tag{10}$$

and measure the probability to find an electron in the corresponding NSO.

A spin-free formulation is usually preferred where the spin has been integrated out of the above equations. If the Hamiltonian does not contain any spin-dependent terms, we can always write the density matrix as

$$\rho_1(x,x') = \rho_{\alpha\alpha}(\mathbf{r},\mathbf{r}')\alpha\alpha'* + \rho_{\beta\beta}(\mathbf{r},\mathbf{r}')\beta\beta'* \tag{11}$$

The *charge density matrix* is obtained by integrating over the spin-variable

$$\rho_q(\mathbf{r},\mathbf{r}') = \sum_{i,j} q_{ij}\varphi_i(\mathbf{r})\varphi_j^*(\mathbf{r}') \tag{12}$$

This matrix is often referred to as the density matrix or the 1-matrix. The diagonal element gives the probability distribution of the electronic charge in the system. The spin distribution is expressed in terms of the *spin density matrix*:

$$\rho_s(\mathbf{r},\mathbf{r}') = \rho_{\alpha\alpha}(\mathbf{r},\mathbf{r}') - \rho_{\beta\beta}(\mathbf{r},\mathbf{r}') \tag{13}$$

which measures the spin-density distribution in the system.

We can diagonalize the 1-matrix by a unitary transformation of the molecular orbitals and obtain in such a representation:

$$\rho_q(\mathbf{r},\mathbf{r}') = \sum_i \eta_i\lambda_i(\mathbf{r})\lambda_i^*(\mathbf{r}') \tag{14}$$

where $\lambda_i(\mathbf{r})$ are the Natural Orbitals (NOs), and η_i are the *occupation numbers*. They fulfill the condition

$$0 \leq \eta_i \leq 2 \tag{15}$$

and is a measure of the probability of finding an electron in the corresponding NO independently of the spin. Such natural orbitals can be defined for any wave function. They will have the occupation number 2 or 0 for a closed shell Hartree–Fock wave function, but in the general case all occupation numbers will be smaller than 2. For wave functions that are well described by HF theory there will be one set of orbitals that have occupation numbers close to 2, while the others have small occupation numbers. The first set of NOs are then characterized as strongly occupied while the others are labeled as weakly occupied. Open shell systems will have one or more NOs with occupation numbers close to 1 (or fractions of the degeneracy number for degenerate open shells). As we shall see below more complicated situation often occur where some occupation numbers are not close to 2, 1, or 0. These are the cases where multiconfigurational wave function are needed for a qualitatively correct description of the electronic structure.

The natural orbitals and their occupation numbers are stable with respect to the approximations we make in our construction of the total wave function. Once we have established a qualitatively correct description of the electronic structure, further improvements will not lead to substantial modifications of the NOs with the highest occupation numbers.

As an example we present in Fig. 25.1 the eight first NOs of the water molecule (excluding the trivial oxygen 1s orbital). As can be seen in the figure, the first set of four orbitals has occupation numbers in the range 1.97–1.99. This is typical for a system that is well described by a single determinant. The other four orbitals describe the most important effects of electron correlations. One notices that each of these orbitals is closely linked to one of the strongly occupied orbitals such that the sum of the two occupation numbers is close to 2. It is a reflection of the fact that electron correlation in a HF stable system can to a good approximation be described as a sum of pair correlations. Thus, for example, the $4a_1$ orbital describes correlation of the electron pair in $3a_1$, which is mainly built from an oxygen 2p orbital with some contribution from the hydrogen 1s orbitals. This type of electron correlation leads to a separation of the electron pair, such that if one electron is close to the oxygen atom, the other prefers to be further away (and closer to the hydrogen atom). The correlation is, therefore, often called *horizontal* (along a

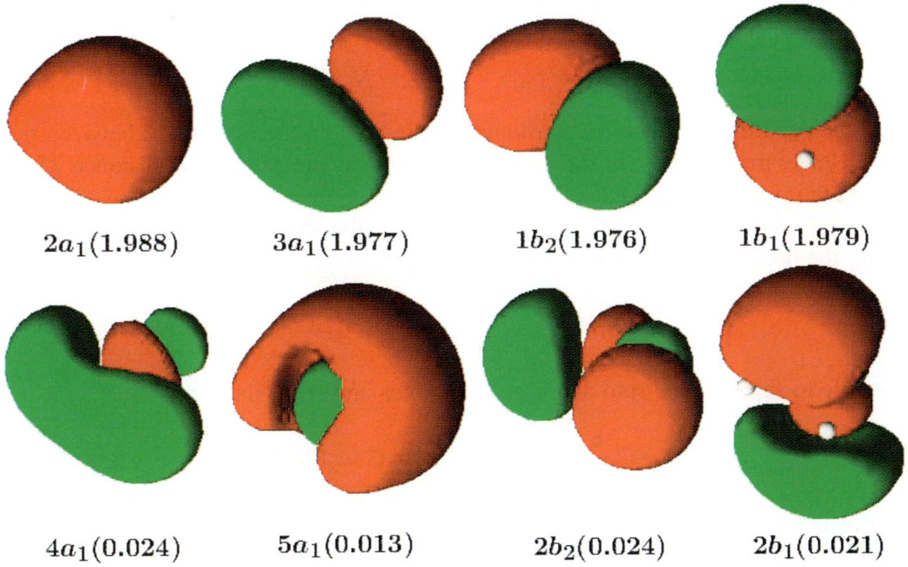

Fig. 25.1. The eight first active orbitals in the water molecule obtained from a CASSCF calculation with these orbitals active and an ANO basis set of the size: O/4s2p2d, H/3s2p.

chemical bond). The effect is obtained with an orbital that is antibonding, that is, it has a node in the bonding region. The separation of the electron pair in a chemical bond is more clearly seen in the $1b_2$, $2b_2$ pair. The orbital $2b_1$ describes *radial* correlation in the 2pπ lone-pair of the oxygen atom. The correlating orbital has the shape of a 3p atomic orbital but it should not be confused with the 3p Rydberg orbital, which is much more diffuse. Similarly, the $5a_1$ orbital describes radial correlation in the $2a_1$ orbital, which is mainly built from the oxygen 2s atomic orbital.

This type of analysis of the natural orbitals in terms of electron correlation can be performed for any wave function that is accurate enough to include the appropriate orbitals. In this chapter we shall describe molecular systems and chemical processes where the NOs are not trivially partitioned into strongly and weakly occupied and we shall show that in such cases the HF wave function is not a good starting point. We need more than one configuration for a qualitatively correct description of the electronic structure. As a first example we shall consider the simplest of all chemical bonds: the hydrogen molecule.

25.3 THE HYDROGEN MOLECULE

We shall use the hydrogen molecule to illustrate how the electronic structure changes during the formation of a chemical bond.

$$\mathrm{H_A} \underset{R}{\text{———}} \mathrm{H_B}$$

To illustrate the qualitative features we do not need an extended atomic orbital basis set. It suffices to use the minimal set constructed from the two 1s orbitals: $(1s_A, 1s_B)$. The molecular orbitals are then given by symmetry

$$\sigma_g = N_g(1s_A + 1s_B), \qquad \sigma_u = N_u(1s_A - 1s_B) \tag{16}$$

where N_g and N_u are normalization constants that depend on the overlap, S, between the two AOs. Notice that the form of the molecular orbitals is independent of distance. Only the normalization constants change.

The closed shell HF electronic configuration for the ground state is obtained by occupying the bonding σ_g orbital with two electrons:

$$\Phi_1 = (\sigma_g)^2 = \frac{1}{\sqrt{2}}|\sigma_g\alpha, \sigma_g\beta| \tag{17}$$

It is possible to separate the spin and space variables and write the function as:

$$\Phi_1 = \frac{1}{\sqrt{2}}|\sigma_g\alpha, \;\; \sigma_g\beta| = \sigma_g(1)\sigma_g(2)\Theta_{2,0} \tag{18}$$

where $\Theta_{2,0}$ is a singlet spin function for two electrons. We can expand the space part of the wave function in the AOs and obtain four terms:

$$\Phi_1 = N_g^2\{1s_A(1)1s_A(2) + 1s_B(1)1s_B(2) + s_A(1)1s_B(2) + s_B(1)1s_A(2)\} \tag{19}$$

The last two terms give the famous Heitler–London (HL) wave function for the hydrogen molecule that was suggested as early as 1927, only one year after the formulation of the Schrödinger equation [6]. The above wave function contains also two ionic terms that correspond to a situation where both electrons reside on one of the atoms. We know that the HF wave function gives a good representation of the electronic structure at the equilibrium geometry. Thus, the HL wave function cannot be accurate. The structure of the HF solution is on the other hand independent of the internuclear distance between the hydrogen atoms. Thus, we have for large R:

$$\Phi \propto \Phi(\mathrm{H}+\mathrm{H}) + \Phi(\mathrm{H}^{+}+\mathrm{H}^{-}) \tag{20}$$

which is clearly wrong and corresponds to an energy 6.7 eV above the true energy. The HL wave function is correct at large values of R. This is not a satisfactory situation if we want to describe the full potential. The closed shell HF method will simply not dissociate the chemical bond properly. It is clear that this will be true for all covalent bonds.

C. Coulson and I. Fischer Suggested in 1949 [7] a method that solves the problem. Instead of using the same orbitals, they introduced a parameter, λ, that splits the SOs generated from the σ_g orbital into two different orbitals (the CF orbitals):

$$\phi_1 = N(1s_A + \lambda 1s_B)\alpha, \qquad \phi_2 = N(\lambda 1s_A + 1s_B)\beta \tag{21}$$

This wave function thus uses different orbitals for different spin (DODS). The total wave function is a determinant with one electron in each of the two SOs. We can express the SOs in terms of the symmetry adapted orbitals σ_g and σ_u. The corresponding wave function will be a linear combination of four determinants. After some algebra we obtain:

$$\Phi_{CF} = c_1^2(\sigma_g)^2 - c_1c_2\sqrt{2}(\sigma_g\sigma_u)_T - c_2^2(\sigma_u)^2 \tag{22}$$

where: $c_1 = N(1+\lambda)/2N_g$ and $c_2 = N(1-\lambda)/2N_u$. The second term in this wave function is a linear combination of two determinants corresponding to triplet coupling of the two orbitals with $M_s = 0$. This term also has the wrong symmetry (ungerade instead of gerade). We shall return to this problem later, but for the moment we accept this wave function and compute the energy as a function of the parameter λ. We introduce $a_1 = c_1^2$, $a_2 = c_2^2$, $a_1 + a_2 = 1$. Then:

$$E(\lambda) = a_1^2E_1 + 2a_1a_2(E_T - K_{gu}) + a_2^2E_2 \tag{23}$$

where $E_1 = E(\sigma_g^2)$, $E_2 = E(\sigma_u^2)$, $E_T = E((\sigma_g\sigma_u)_T)$; and $K_{gu} = (\sigma_g\sigma_u|\sigma_g\sigma_u) = (J_{AA} - J_{AB})/2(1-S^2)$, where J_{AA} and J_{AB} are the one- and two-center Coulomb integrals.

The stationary points (the CF orbitals) are obtained by solving the equation:

$$\mathrm{d}E/\mathrm{d}\lambda = 0 \tag{24}$$

The stability of the solution is obtained from the second derivative. We obtain after some algebra

Solution 1, if $(E_T - E_1) > K_{gu}$:

$\lambda = 1(a_1 = 1)$	$d^2E/d\lambda^2 > 0$	$\Phi_{CF} = (\sigma_g)^2$
$\lambda = -1(a_1 = 0)$	$d^2E/d\lambda^2 < 0$	$\Phi_{CF} = (\sigma_u)^2$

Solution 2, if $(E_T - E_1) \leq K_{gu}$:

$\lambda = 1(a_1 = 1)$	$d^2E/d\lambda^2 < 0$	Unstable
$\lambda = -1(a_1 = 0)$	$d^2E/d\lambda^2 < 0$	Unstable
$-1 < \lambda < 1(a_1 < 1)$	$d^2E/d\lambda^2 \geq 0$	Stable

Thus, we obtain the closed shell HF wave function as a stable solution when the excitation energy to the triplet state is larger than the exchange integral K_{gu}. This is the situation around equilibrium geometry. The excitation energy will, however, decrease when the distance increases and finally go to zero at infinite separation. This is not true for the exchange integral that remains finite and positive for all distances. Thus, at some distance we pass over to Solution 2 where the only stable solution is for a λ value less than one in magnitude.

For infinite distance we obtain the stable solution $\lambda = 0$, which corresponds to the wave function:

$$\Phi_{UHF} = \frac{1}{\sqrt{2}}|1s_A\alpha, 1s_B\beta| \tag{25}$$

which we recognize as one term in the HL wave function. It has the correct energy but is not symmetric.

What have we learned from the above exercise? First of all that in order to obtain a correct potential curve for the dissociation of a covalent chemical bond with a single determinant wave function, we have to sacrifice both the spin and space symmetry. Secondly, we have seen that the closed shell HF wave function is stable around the equilibrium geometry. Even if we allow different orbitals for different spin, the wave function will converge to a solution where the electrons arrange themselves in closed shells. This is quite general and occurs for most molecules around their equilibrium geometry.

The Coulson–Fischer wave function for H_2 can be considered as the start of the Unrestricted Hartree–Fock (UHF) approach in quantum chemistry, which is the most general single determinant method. We shall not proceed further along this line, but instead ask ourselves if there is a way to correct the situation such that we obtain a wave function that dissociates correctly while preserving the spin and space symmetry of the wave function. The CF wave function gives actually a hint. What happens if we simply skip the trouble-some triplet term in Eq. (22). This gives rise to a wave function that is a linear combination of two closed shell determinants:

$$\Phi_{MC} = C_1(\sigma_g)^2 + C_2(\sigma_u)^2 \tag{26}$$

where we have introduced two new coefficients C_1 and C_2 ($C_1^2 + C_2^2 = 1$). Let us take a closer look at the second term. If we expand in AOs, we obtain the same expression as in

Eq. (19) but with a negative sign for the covalent terms. At infinite separation we want only these terms to survive. We can achieve this if we simply put $C_2 = -C_1$ in Eq. (26). On the other hand we know that $\Phi_{MC}$is dominated by the HF configuration for geometries close to equilibrium. Thus, by introducing the expansion coefficients as new variational parameters we can describe the wave function for all distances.

Φ_{MC} is a qualitatively correct wave function for all internuclear distances. It introduces *horizontal* electron correlation by occupying the antibonding orbital σ_u. This orbital separates the electrons in the pair such the probability to find one on each atom increases. For large internuclear distances the probability will go to one. We can thus conclude that

The Quantum Chemical Description of a Chemical Bond Involves both the Bonding and the Antibonding Orbital!

The occupation of the two orbitals is seen in Fig. 25.2. The occupation of the bonding orbital is close to two around equilibrium. It decreases continuously towards one when the distance increases. The occupation of the antibonding orbital is the mirror image, going from a value close to zero to one. This picture will of course be slightly modified when we include more NOs in the wave function but the general behavior will persist.

We have thus been able to construct a wave function that describes the qualitative behavior of the electronic structure for all internuclear distances. The price we have paid is to leave the single configurational description and construct the wave function as a linear combination of several configurations (determinants) with expansion coefficients to be determined by the variational principle together with the molecular orbital coefficients. This is the multiconfigurational approach in quantum chemistry. Before we end this section let us take a look at a more complex chemical bond, that in the Cr_2 molecule.

The Cr atom has the ground state $(3d)^5\,(4s)^1$, ^7S, with six unpaired electrons. Thus, we can in principle form a sextuple bond between two chromium atoms. This is also what happens, even if the bond is weak with a bond energy of only 1.53 eV. In order to

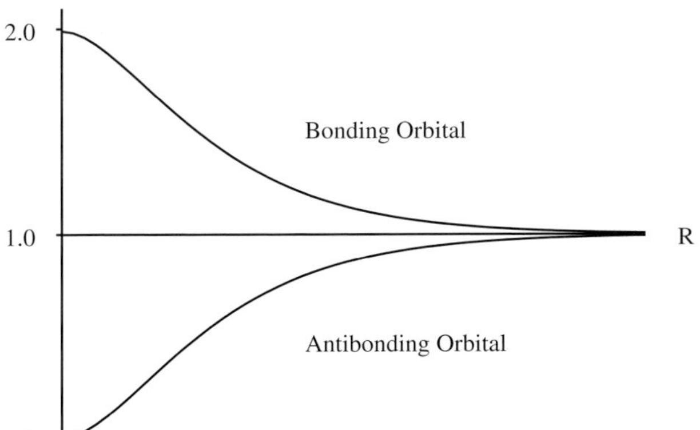

Fig. 25.2. The occupation numbers for the σ_g and σ_u orbitals in the H_2 molecule as a function of the internuclear distance.

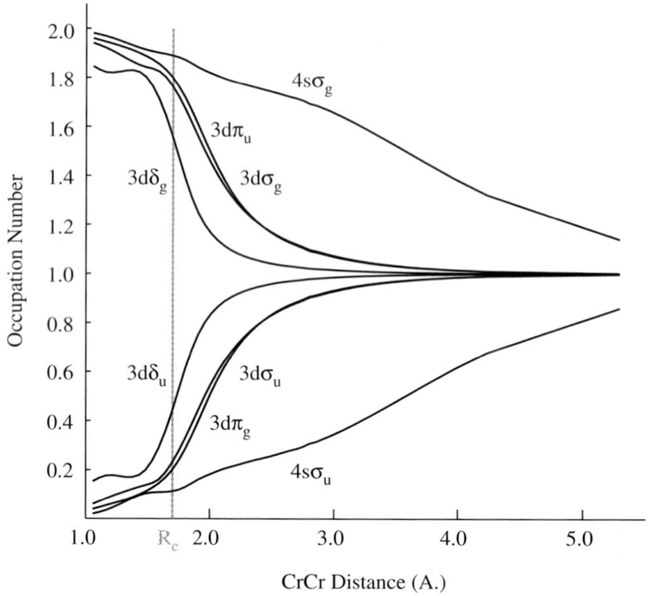

Fig. 25.3. Variation of the NO occupation numbers for the Cr_2 molecule as a function of the internuclear distance. The vertical line shows the equilibrium bond distance.

describe the formation of six bonds we need a complex wave function comprising many hundred CSFs. In Fig. 25.3 we show how the occupation of the NOs varies with the bond distance for such a wave function. We can see the formation of a bond between the 4s orbitals starting at rather large distances and slowly progressing when the distance decreases. The formation of the bonds between the 3d orbitals starts at a considerably shorter distance and are not fully developed even at equilibrium geometry. The occupation number of the $3d\delta_g$ orbital is only about 1.5 electrons at this geometry. Notice the nice pairing between bonding and antibonding orbitals also in this complex case. The large difference in distance between 4s and 3d bonding results in a potential with an inner sharp minimum and a broad plateau at larger distance [8]. The Cr_2 molecule is an example of a system where a single configurational wave function is not valid even at equilibrium geometry.

25.4 DEGENERACY AND NEAR DEGENERACY

Suppose that two electronic configurations Φ_1 and Φ_2 have same energy and that the Hamiltonian matrix element them is non-zero and positive. The lowest energy solution is then

$$\Psi = \frac{1}{\sqrt{2}}(\Phi_1 - \Phi_2) \tag{27}$$

This is exactly the situation we have in the H_2 molecule at large distance. If there is an energy difference but it is small, we obtain the solution

$$\Psi = C_1\Phi_1 + C_2\Phi_2 \tag{28}$$

with both coefficients different from zero. This is the situation at intermediate distances and also the situation for Cr_2 at equilibrium but in that case more than two configurations are involved.

Can we think of more situations of this kind? There are many. Consider, for example, a chemical reaction:

$$A + B \rightarrow C + D \tag{29}$$

Suppose that there is an energy barrier for this reaction at some geometry (the transition state). At this point one or more of the bonds will be weakened (longer bond distance). We are on a section of the energy surface where antibonding orbitals play a large role in the wave function and the wave function will be multiconfigurational. A chemical reaction can either proceed with retention or change of the electronic configuration. In the first case we say that the reaction is allowed and we expect a low barrier, in the second case the reaction is said to be forbidden and we normally expect a high barrier. In this case will the MOs of the reactants be qualitatively different from those of the product. It we call the configuration for the reactants $\Phi_{\rm I}$ and that for the product $\Phi_{\rm II}$ we shall have the situation illustrated in Fig. 25.4. The energy, $E_{\rm I}$ corresponding will be low for the reactants but high for the product and vice versa for $E_{\rm II}$. At some point the two energy surface cross and a barrier is generated. The two energies become equal and we obtain the same situation as in Eq. (27) with total mixing of the two configurations.

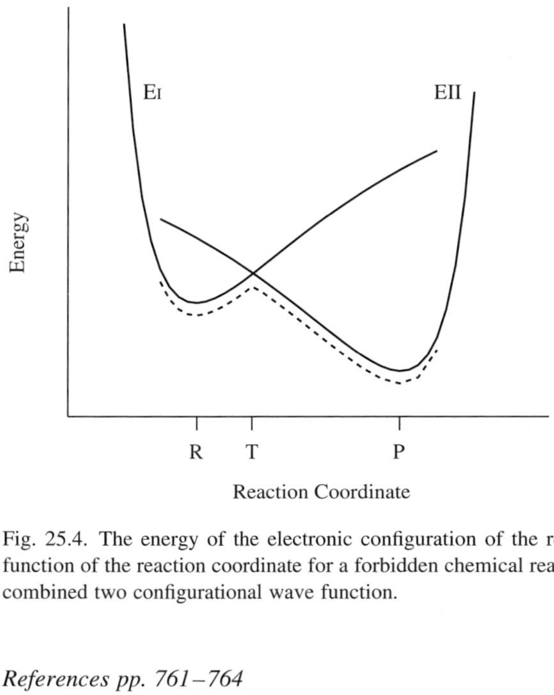

Fig. 25.4. The energy of the electronic configuration of the reactant ($E_{\rm I}$) and that of the product ($E_{\rm II}$) as a function of the reaction coordinate for a forbidden chemical reaction. The dotted line shows the energy for the combined two configurational wave function.

Degeneracy or near degeneracy is common for excited states. The ground state is often well separated from the excited states of a molecule, at least close to the equilibrium geometry, while different excited states can be close in energy. When we study photochemical processes and thus compute energy surfaces at geometries away from equilibrium we often find crossing energy surfaces with change of dominating electronic configurations. Such studies must be performed using multiconfigurational wave functions.

25.4.1 Static and dynamic electron correlation

The correlation energy is usually defined as the difference between the exact eigenvalue of the Hamiltonian and the Hartree–Fock energy:

$$E_{\text{corr}} = E_{\text{exact}} - E_{\text{HF}} \tag{30}$$

This definition depends on the possibility to define a unique single configurational energy. As we have seen above, that might often not be possible if we do not want to fall back on the unsatisfactory UHF approach. We can as an alternative use the statistical definition of electron correlation. The electronic motion is uncorrelated if the probability to find electron one in a small volume element around x_1 and electron two around x_2 is the product of the one-electron probability distributions:

$$\rho_2(x_1;x_2) = \rho_1(x_1,x_1)\rho_1(x_2,x_2) \tag{31}$$

where $\rho_2(x_1;x_2)$ is the diagonal element of the second order reduced density matrix. We can express the second order matrix in terms of the first order matrix for a single determinant wave function (for a more detailed analysis of reduced density matrices, see, for example, Ref. [4]):

$$\rho_2(x_1;x_2) = \rho_1(x_1,x_1)\rho_1(x_2,x_2) - \rho_1(x_1,x_2)\rho_1(x_2,x_1) \tag{32}$$

We note that this relation is not the same as Eq. (31). The electronic motion is thus correlated also for the HF wave function. A deeper analysis shows that for electrons with opposite spin will the second term (the exchange term) disappear when we integrate over the spin variable. This is not the case for electrons with the same spin (compare the exchange term in the HF energy expression). The electrons thus create a hole around themselves were other electrons with the same spin are forbidden. We call this *Fermi* correlation. Such a hole is not created for electrons with the opposite spin. There is a finite probability to find them in the same point in space but the magnitude of the exact wave function has a minimum for $r_{12} = 0$ (r_{12} being the distance between a pair of electrons). The derivative of the exact wave function is discontinuous at this point. We call this behavior a *cusp*. The wave function has the form

$$\Psi(r_{12}) \propto \exp(r_{12}/2) \tag{33}$$

in the neighborhood of the singularity. The feature is called a *Coulomb* hole (cf. Fig. 25.5). The description of this hole is difficult in molecular orbital theory and makes the correlation energy converge slowly with an increased basis set.

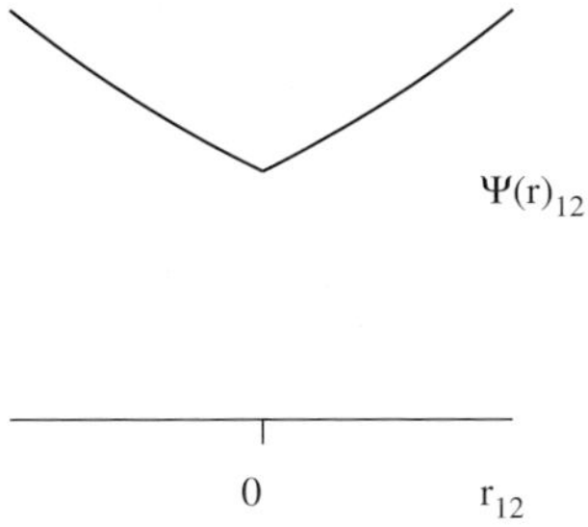

Fig. 25.5. The Coulomb cusp, showing the functional form of the exact wave function, $\Psi(r_{12})$, in an area where the distance between two electrons is small.

The above definition of electron correlation makes a generalization to the multiconfigurational case possible. To explain this, we return to the hydrogen molecule. We can easily compute the second order density from the two-configurational wave function (26); it is just the square of the wave functions. We obtain after spin integration:

$$\rho_2(1;2) = C_1^2\sigma_g^2(1)\sigma_g^2(2) + C_2^2\sigma_u^2(1)\sigma_u^2(2) + 2C_1C_2\sigma_g(1)\sigma_g(2)\sigma_u(1)\sigma_u(2) \quad (34)$$

Now, let us assume that both electrons are located close to atom A where the amplitude of $1s_B$ is small. The probability for this is P_{AA}:

$$P_{AA} = 1s_A(1)^2 1s_A(2)^2 (C_1 + C_2)^2 \quad (35)$$

This probability is large when C_1 is much larger than C_2 (note that C_2 has the opposite sign of C_1). This happens close to equilibrium where the HF wave function gives a good description of the electronic structure. For a large internuclear distance will both coefficients have the same magnitude and P_{AA} goes ton zero.

The probability to find electron two on atom B can in the same way be written as:

$$P_{AB} = 1s_A(1)^2 1s_B(2)^2 (C_1 + C_2)^2 \quad (36)$$

We notice that this probability will increase when C_2 increases in magnitude. The second configuration thus has the effect of separating the electron pair over a large distance. The effect is caused by the negative sign between the two atomic orbitals on the σ_u orbital (the node). This is an example of *Static Correlation*.

A similar effect can be found in atoms where the *n*s shell is filled but the *n*p shell is empty or only partially filled. The symmetry then allows the multiconfigurational wave function:

$$\Psi = C_1(\mathrm{ns})^2(\mathrm{np})^x + C_2(\mathrm{np})^{2+x} \quad (37)$$

where we have moved two electrons form the *n*s to the *n*p shell in the second term. The different nodal structure to the *n*p orbital compared to the *n*s orbital will separate the electron pair such that the second electron has an increased probability of being found on the other side of the atom compared to the first electron. This effect can be large. In the Be atom it accounts for 95% of the correlation energy of the 2s electron pair.

Static correlation is thus used to describe correlation effects which leads to a separation of an electron pair in space. It originates in the incapability of the HF wave function to describe such situations (without breaking space and spin symmetry). The remaining correlation effects are called *Dynamic Correlation* and describes the situation in the Coulomb hole when two electron collide. The separation of electron correlation into static and dynamic is not very precise but is of great practical value. Static correlation effects may be treated using multiconfigurational wave functions, while other methods are more suited to treat dynamic correlation, such as: configuration interaction, coupled cluster, perturbation theory, explicit r_{12} methods, and even density functional theory. The combination of multiconfigurational wave functions and such methods to treat dynamic correlation is an active area of research.

We could of course have given many more examples where it is necessary to use a multiconfigurational approach to the wave function. A few more will be given at the end of the chapter, but we now turn to the question of how to construct and evaluate multiconfigurational wave functions.

25.5 MULTICONFIGURATIONAL WAVE FUNCTIONS

25.5.1 A brief historical expose

The first multiconfigurational (MC) SCF calculation was probably performed by Hartree and co-workers in 1939 [9]. They realized (as discussed above) that for the ^1S state of the oxygen atom there where two possible configurations, s^2p^4 and p^6, and constructed the two configurational wave function

$$\Psi = C_1\Phi(s^2p^4) + C_2\Phi(p^6) \tag{38}$$

The atomic orbitals were determined (numerically) together with the two co-efficients. Similar MCSCF calculations on atoms and negative ions were simultaneously performed in Kaunas, Lithuania, by A. Jucys [10]. The possibility was actually suggested already in 1934 in the book by J. Frenkel [11]. Further progress was only made with the advent of the computer. A.C. Wahl and G. Das developed the *Optimized Valence Configuration (OVC) Approach*, which was applied to diatomic and some triatomic molecules [12,13].

An important methodological step forward was the formulation of the *Extended Brillouin's (BLB) theorem* by B. Levy in 1968 [14,15]. This theorem states that for any CI wave function, which is stationary with respect to orbital rotations, we have:

$$\langle\Psi|\hat{H}(\hat{E}_{pq} - \hat{E}_{qp})\Psi\rangle = 0 \tag{39}$$

where $\hat{E}_{pq}$ is a spin averaged excitation operator:

$$\hat{E}_{pq} = a^\dagger_{p\alpha}a_{q\alpha} + a^\dagger_{p\beta}a_{q\beta} \tag{40}$$

The theorem is an extension to the multiconfigurational regime of the Brillouin theorem, which gives the corresponding condition for an optimized HF wave function.

A forerunner to the BLB theorem can actually be found already in Löwdin's 1955 article [4,16].

The early MCSCF calculations were tedious and often difficult to converge. The methods used were based on an extension of the HF theory formulated for open shells by Roothaan in 1960 [17]. An important paradigm change came with the *Super-CI* method, which was directly based on the BLB theorem [18]. One of the first modern formulations of the MCSCF optimization problem was given by J. Hinze in 1973 [19]. He also introduced what may be called an approximate second order (Newton–Raphson) procedure based on the partitioning: $U = 1 + T$, where U is the unitary transformation matrix for the orbitals and T is an anti-Hermitian matrix. This was later to become $U = \exp(T)$. The full exponential formulation of the orbital and CI optimization problem was given by Dalgaard and Jørgensen in 1978 [20]. Variations in orbitals and CI coefficients were described through unitary rotations expressed as the exponent of anti-hermitian matrices. They formulated a full second order optimization procedure (Newton–Raphson, NR), which has since then become the standard. Other methods (e.g. The Super-CI method) can be considered as approximations to the NR approach.

25.5.2 The MCSCF wave function

One of the problems that the early applications of the MCSCF method faced was the construction of the wave function. It was necessary to keep it short in order to make the calculations feasible. Thus, one had to decide beforehand which where the most important CSFs to include in the CI expansion. Even if this is quite simple in a molecule like H_2 it quickly becomes ambiguous for larger systems. However, the development of more efficient techniques to solve large CI problems made another approach possible. Instead of having to choose individual CSFs, one could choose only the orbitals that were involved and then make a full CI expansion in this (small) orbital space. In 1976 Ruedenberg introduced the *orbital reaction space* in which a complete CI expansion was used (in principle). All orbitals were optimized—the *Fully Optimized Reaction Space—FORS* [21].

An important prerequisite for such an approach was the possibility to solve large CI expansions. A first step was taken with the introduction of the *Direct CI* method in 1972 [22]. This method solved the problem of performing large scale SDCI calculations with a closed shell reference wave function. It was not useful for MCSCF where a more general approach is needed that allows an arbitrary number of open shells and all possible spin-couplings. The generalization of the direct CI method to such cases was made by J. Paldus and I. Shavitt through the *Graphical Unitary Group Approach (GUGA)*.Two Papers by Shavitt explained how to compute CI coupling coefficients using GUGA [23,24]. Shavitt's approach was directly applicable to full CI calculations. It formed the basis for the development of the *Complete Active Space (CAS) SCF* method, which has become the standard for performing MCSCF calculations [25,26].

25.5.3 The complete active space SCF method

The CASSCF method is best understood if we go back to our examples. We learned that to compute the potential curve for the H_2 molecule we needed two orbitals: the bonding σ_g and the antibonding σ_u. We constructed an MC wave function as a linear combination of the two configurations $(\sigma_g)^2$ and $(\sigma_u)^2$. They are the only configurations that we can construct with the two orbitals and two electrons, which have the correct spin and space symmetry. For the Cr_2 molecule we had twelve orbitals and twelve electrons and it becomes difficult to find the electronic configurations that contribute to the total wave function for the ${}^1\Sigma_g^+$ ground state. So, we do not try. Instead, we tell our CASSCF program to find all possible CSFs that have the correct spin and space symmetry. The wave function is expanded in all of them, corresponding to a full CI in the chosen orbital space.

The generalization is obvious. We partition our orbital space into three sub-spaces: *inactive*, *active*, and *external*. The inactive orbitals are doubly occupied in all CSFs (like the 1s–3p orbitals in Cr_2). The choice of the active orbitals must be based on our knowledge about the system and the chemical process we are studying. This is not a trivial task and a CASSCF program can never be treated as a black box. Once we have chosen the active orbitals, we distribute the *active electrons* among the active orbitals in all ways possible for a given total spin and space symmetry. This generates a set of CSFs in which we expand the wave function:

$$\Psi = \sum_{\nu} C_{\nu} \Phi_{\nu}, \tag{41}$$

where Φ_ν are the CSFs. The expansion corresponds to a full CI in the active orbital space. It can in modern applications contain several million terms. Most softwares use CSFs in the original wave function but transform to determinants for the solution of the secular problem because this leads to more efficient code.

CASSCF wave function may be constructed for virtually any type of electronic structure, closed or open shell, ground or excited state, neutral or ionic, etc. The only limitation is the size of the active space. The wave function is invariant to orbital transformations within each subspace, which simplifies optimization and makes it possible to construct the same wave function from the natural orbitals, or to use localized orbitals.

In electronic spectroscopy one is often interested in obtaining wave functions and energies for a set of electronic states of the same symmetry. It might be difficult or even impossible to make a separate calculation for each state. A compromise, which is often used with success, is to use the same orbitals for all states of that symmetry. The optimization is then carried out for the average energy (state average CASSCF).

It is not meaningful to attempt to describe here how the optimization of the wave function parameters is carried out. We refer instead to the existing literature [27,28].

25.5.4 Choosing the active space

To choose the correct active space for a specific application is not trivial and many times one has to make several experiments. It is difficult to set up any general rules because each chemical problem poses its own problems. There is also a tight connection to the choice of AO basis, which must be extensive enough to be able to describe the occupied molecular orbitals properly. The size of the active space is limited. The maximum size in most softwares is around 15 for the case where the number of orbitals and electrons are equal. This is the most severe limitation of the CASSCF method and makes it sometimes difficult or even impossible to carry out a specific study.

During the 25 years that the CASSCF method has been used in practical application we have learned much about the choice of active orbitals. Without attempting to be complete, we give some examples below.

25.5.4.1 Atoms and atomic ions

This is the simplest case. It is usually sufficient to have the valence orbitals active, perhaps with added Rydberg type orbitals for studies of excited states. One can normally leave the *n*s orbital inactive for main group atoms with more than three *n*p electrons. First row transition metals are, however, more demanding. It has been shown that in order to be able to accurately describe the relative correlation effects in atomic states, which differ in the number of 3d electrons, one needs to use two sets of d-orbitals, 3d and 3d′ where the second set describes the strong radial correlation effects in the 3d shell [29]. Adding the 4s and 4p orbital one is faced with an active space of 14 orbitals. The importance of the second 3d orbital decreases for second and, in particular, for third row transition metals.

Lanthanide atoms and ions need the active space 4f, 5s, and 5p, in all 11 active orbitals (for some elements also 5d has to be added). It is possible that a double shell effect obtains also here for atoms with many 4f electrons. Actinides are the most complicated atoms. The orbitals 5f, 6d, 7s, and 7p have similar energies and are occupied in low lying electronic states. One should, therefore, ideally use 16 active orbitals. Our experience in this part of the periodic table is, however, yet rather limited.

25.5.4.2 Small molecules

Diatomic and triatomic systems can often be treated with an active space comprising the valence orbitals. A recent study of all main group dimers showed that accurate results were possible with *n*s and *n*p active (8 orbitals) [30]. We shall, however, give an example below, the PbF molecule, which illustrates that this rule may sometimes be an oversimplification. Transition metal compounds are more difficult. One set of *n*d-orbitals is enough for the low lying electronic states, but it is necessary to add the $(n + 1)$s orbital. Thus, the minimum active space is 12. It was shown in a recent study of the Cr dimer that one also had to add orbitals of 4p character to be able to describe the ground state potential accurately [8]. The situation becomes even more complicated in studies of many excited states. A recent study of the electronic spectrum of CrH used 16 active orbitals [31]. One the other hand, only seven active orbitals were needed in a study of the electronic spectrum of the PbO molecule [32]. Most difficult are molecules containing actinide atoms. A recent study of the electronic spectrum of UO_2 illustrates the problem. It was here not possible to saturate the active space. This would have needed 18 active orbitals with 12 electrons, a calculation which is beyond the capacity of the software available today. Instead, the spectrum was studied using a sequence of active spaces, where the choice was guided by the known nature of the most important excited states [33].

Energy surfaces for chemical reactions involving three main group atoms can be performed with the 12 valence orbitals active. This will cover all possible reaction channels. Tetra-atomic molecules would need 12–16 orbitals depending on whether the *ns* orbitals have to be included or not. The problem is of course simplified if one or more of the atoms is hydrogen. Additional active orbitals may be needed for excited states surfaces where Rydberg states may become important.

25.5.4.3 Electronic spectroscopy for organic molecules

The calculation of excitation energies and transition moments for unsaturated organic molecules has been one of the more successful applications of multiconfigurational quantum chemistry since the first application to the benzene molecule in 1992 [34]. Many hundred molecules have been studied. The CASSCF method allows optimization of excited state energy surfaces and this has been used to compute vibrationally resolved electronic spectra [35,36]. The method is used by several research groups for studies of photochemical reactions, including the localization of conical intersections, etc.

The choice of the active space is non-trivial in most such calculation, in particular if one is interested in many excited states at energies where Rydberg states become important. Nine active orbitals are needed to describe the 3s, 3p, and 3d Rydberg states. In addition a number of valence orbitals of σ- and π-type are needed. Thus, the active space can easily become too large. When the molecule is planar one can use different sets of Rydberg orbitals for electronic states of Σ and Π type, respectively. It is safest of try to keep all the valence π-orbitals in the active space. For large aromatic systems this will not be possible and one then has to experiment to see if it is possible to converge the results using an increasing number of active orbitals. The number needed depends on how many excited state one is interested in. Often it is only the low-lying states, and the demands on the active space is then smaller. In systems with hetero atoms one also needs to add lone-pair orbitals to the active space. It is clear that it might be difficult or even impossible to saturate the active space for large aromatic systems.

Discussions of the choice of active orbitals for organic molecules can be found in the literature [37–39]. One more word about Rydberg states and AO basis sets. Do not represent Rydberg orbitals by adding diffuse functions to all non-hydrogen atoms in the molecule. This creates a lot of diffuse functions, which causes convergence problems and ill defined Rydberg orbitals. Add instead diffuse functions to the charge center of the positive ion. A recipe how one can produce will defined Rydberg orbitals this way can be found in Ref. [37]. We give a detailed example below.

25.5.4.4 Transition metal compounds

Multiconfigurational methods have been particularly successful in studies of transition metal complexes, both for ground state and excited states. This is an area where the alternative methods are few. Many open shells with varying values of the spin together with many close lying electronic states (not only singly excited) makes it difficult to use simple methods like Density Functional Theory, or other methods which assume a Hartree–Fock like ground states.

To choose the active space for these compounds is not easy and one sometimes has to realize that the calculation one was hoping to do cannot be performed with the software available today. However, there are some rules that can be useful. One is the "10-orbital/10-electron" rule. As pointed out above one needs two d-type orbitals to represent a doubly occupied d shell. In a transition metal complex the situation becomes more complex, but the principle is the same. Here, the second d-type orbital it needed only for doubly occupied orbitals that do not interact directly with ligand orbitals. If it does, it is replaced by the corresponding ligand orbital and the addition of one more orbital does not affect the result. Instead the active orbital describing the interaction with the ligand will acquire a certain amount of d'-character, how much depends on the covalency of the metal ligand bond. A general rule is to include in the active space all orbitals, which have some d-character. A more detailed discussion of the 10/10 active space can be found in Ref. [39] where a number of specific examples are given. This active space may have to be supplemented with extra ligand orbitals in studies of excited states of charge transfer character.

It is helpful to use the formal oxidation number of the ion when deciding about the active space for a TM complex. Consider as an example the permanganate ion MnO_4^-. The formal oxidation state is Mn^{7+} with 3d electrons. Thus the active space is to be chosen as the five 3d-orbitals and five corresponding ligand orbitals. It has, however, been shown by K. Pierloot that for atoms in a high formal oxidation state, charge transfer to the metal ion will occur also from orbitals that are not directly interacting with the metal d-orbitals. For MnO_4^- one therefore needs to include all the oxygen 2p-orbitals. Thus, in the end we have 17 active orbitals with 24 electrons. More details about this and similar difficult but interesting cases can be found in Ref.[40]. The same author has treated a number of transition metal compounds. A discussion of static correlation effects can be found in Ref. [41] and a review of applications in electronic spectroscopy in Ref. [42].

25.5.4.5 Lanthanide and actinide chemistry

The experience in this area is still limited. Lanthanide in are characterized by an open 4f shell. Thus, the seven 4f-orbitals need to be active even if they remain localized and do not participate in the chemical bonding. In addition, one needs to include some ligand orbitals in the active space. The specific choice depends on the case and no general rules can be given. It is clear, however, that the need to include the 4f-orbitals puts more severe constraints on the number of ligand orbitals that may be used. An extension of the CASSCF method to allow for singly occupied inactive orbitals would be quite helpful for such applications.

The situation is similar in actinide compounds but here the 5f-orbitals play a more active role and contribute to the chemical bond. Most applications have so far been to systems with a high oxidation state of the actinide ion. A typical case is the uranyl ion, UO_2^{2+} where the uranium ion has a formal charge of +6. As a result, three strongly covalent bonds are formed to each of the oxygen atoms. The resulting active space consists of 12 electrons in 12 orbitals [43]. This active space can also be used when the uranyl ion forms complexes with other ligands, such as carbonate [44]. Additional active orbitals are needed for the neutral UO_2 molecule [45]. Wahlgren and co-workers have studied electron transfer reactions for uranyl (V)-uranyl(VI) complexes in solution [46]. K. Pierloot has studied the electronic spectrum of the uranyl ion and the complex with chlorine, $UO_2Cl_4^{2-}$ with excellent agreement with experiment. The recipe for choosing the active space was based on the considerations made above [40].

We shall stop here. Many more examples can be found in the literature. It should be clear to the reader that the choice of an active space is a non-trivial exercise. At the same time it is exciting because it teaches us something about the electronic structure and the important correlation effects in the system we are studying. It is not a black box, but this has the virtue that we understand more about our system than we would do form a simple DFT calculation. Many of the examples given above could anyway not have been handled by this approximate and semi-empirical method.

25.6 DYNAMIC CORRELATION AND THE CASPT2 METHOD

So far, we have only discussed the CASSCF method, but all the examples given above also included dynamic correlation effects. This is as necessary in the multiconfigurational approach as it would be if we started from the HF approximation. The CASSCF calculation will only include the static correlation effects and we need to complement it with a treatment of the dynamic correlation. How do we do that? In a single configurational approach, the obvious choices are preferably coupled cluster (CC) methods, or if the system is too large, second order perturbation theory (MP2), which is already quite accurate. A practical multiconfigurational CC theory does not exist yet. A method that has been used with great success since the 1980s is *Multi-Reference CI (MRCI)* where the most important of the CSFs of the CAS wave function are used as reference configuration in a CI expansion that includes all CSFs that can be generated by single and double replacements of the orbitals in the reference CSFs [47]. The method is still used thanks to recent technological developments [48]. It becomes, however, time consuming for systems with many electrons and has also the disadvantage of lacking size-extensivity, even if this can be approximately corrected for.

Another way to treat the dynamic correlation effects is so use perturbation theory. Such an approach has the virtue of being size-extensive and ought to be computationally more efficient than the MRCI approach. Møller–Plesset second order perturbation theory (MP2) has been used for a long time to treat electron correlation for ground states, where the reference function is a single determinant. It is known to give accurate results for structural and other properties of closed shell molecules. Could such an approach work also for a multiconfigurational reference function like CASSCF? The idea was suggested soon after the introduction of the CASSCF method [49] but technical difficulties delayed a full implementation until the late 1980s [50,51].Today it is the most widely used method to compute dynamic correlation effects for multiconfigurational (CASSCF) wave functions.

The principle is simple. We compute the second order energy with a CASSCF wave function as the zeroth order approximation. Having said that, one easily realizes that there are some problem to be solved that do not occur in MP2. We need to define a zeroth order Hamiltonian with the CASSCF function as an eigenfunction. It should preferably be a one-electron Hamiltonian in order to avoid a too complicated formalism. We need to define an interacting space of configurations. This is straightforward. They are given as:

$$\hat{E}_{pq}\hat{E}_{rs}|\mathrm{CASSCF}\rangle \tag{42}$$

This is an internally contracted configuration space, doubly excited with respect to the CAS reference function $|0\rangle = |\mathrm{CASSCF}|$. One or two of the four indices p, q, r, s must be outside the active space. Functions (42) are linear combinations of CSFs and span the entire configuration space that interacts with the reference function. Labeling the compound index $pqrs$ as μ or ν, we can write the first order equation as:

$$\sum_{\mu} [H^{(0)}_{\mu\nu} - E_0 S_{\mu\nu}] C_\nu = -V_{0\mu} \tag{43}$$

Here, $H^{(0)}_{\mu\nu}$ are matrix elements of a zeroth order Hamiltonian, which is chosen as a one-electron operator in the spirit of MP2. $S_{\mu\nu}$ is an overlap matrix: the excited functions are not in general orthogonal to each other. Finally, $V_{0\mu}$ represents the interaction between the excited function and the CAS reference function. The difference between Eq. (43) and ordinary MP2 is the more complicated structure of the matrix elements of the zeroth order Hamiltonian. In MP2 it is a simple sum of orbital energies. Here it is a complex expression involving matrix elements of a generalized Fock operator, $\hat{F}$, combined with up to fourth order density matrices of the CAS wave function. We do not give further details here but refer to the original papers. The zeroth order Hamiltonian is written as a sum of projections of $\hat{F}$ onto the reference function $|0\rangle$

$$\hat{H}_0 = \hat{P}_0\hat{F}\hat{P}_0 + \hat{P}_{\mathrm{SD}}\hat{F}\hat{P}_{\mathrm{SD}} + \hat{P}_{\mathrm{X}}\hat{F}\hat{P}_{\mathrm{X}} \tag{44}$$

where $\hat{P}_0$ projects onto the reference function, $\hat{P}_{\mathrm{SD}}$ onto the *interacting configurations space* (42), and $\hat{P}_{\mathrm{X}}$ onto the remaining configuration space that does not interact with $|0\rangle$. $\hat{F}$ has been chosen as the *generalized Fock operator*:

$$\hat{F} = \sum_{p,q} f_{pq}\hat{E}_{pq} \tag{45}$$

with

$$f_{pq} = h_{pq} + \sum_{r,s} D_{rs}\left[(pq|rs) - \frac{1}{2}(pq|qs)\right] \tag{46}$$

It has the property that $f_{\mathrm{pp}} = -IP_{\mathrm{p}}$ when the orbital p is doubly occupied and $f_{\mathrm{pp}} = -EA_{\mathrm{p}}$ when the orbital is empty. The value will be somewhere between these two extremes for active orbitals. Thus, we have for orbitals with occupation number one: $f_{\mathrm{pp}} = -\frac{1}{2}(IP_{\mathrm{p}} + EA_{\mathrm{p}})$. This formulation is somewhat unbalanced and will favor systems with open shells, leading, for example, to somewhat low binding energies [52]. The problem is that one would like to separate the energy connected with excitation out from an orbital from that of excitation into the orbital. This cannot be done within a one-electron formulation of the zeroth order Hamiltonian. K. Dyall has suggested to use a two-electron operator for the active part [53], but this leads to a too complicated formalism and also breaks important orbital invariance properties (the result is, for example, different if a doubly occupied orbital is labeled inactive or active). Andersson has suggested to use a shift that stabilizes the active orbitals [54]. This method works well in cases where excitations from the inactive to the active orbital space are unimportant. A more general shift technique has recently been suggested by Ghigo and Roos [55]. This should not be confused with the level shift technique used to remove intruder states.

Perturbation theory like MP2 of CASPT2 should only be used when the perturbation is small. Orbitals that give rise to large coefficients for the states Eq. (42) should ideally be included in the active space. Such large coefficients are the result of small energy differences between the CAS reference state and one or more of the excited functions. We call these functions *intruder states*. In cases where the interaction term, $V_{0\mu}$, is also small one can remove the intruder using a level shift technique that does not affect the contributions for the other states. We shall not show in detail how this is done, but refer instead to the original papers [56,57]. This method to solve the intruder state problem has been used successfully in a large number of applications.

The CASPT2 method adds a correlation function to a fixed CAS function. The orbitals need not be optimized for this specific state. The CASPT2 space includes also singly excited functions and can thus correct for small errors in the orbitals. The method is often used with reference functions that have been state average optimized. The fact that the reference function is fixed can, however, sometimes be a problem, in particular when two or more electronic states of the same symmetry are close in energy. Such situations are quite common for excited states. One can then expect the dynamic correlation to affect also the reference function. We can deal with this problem by extending the method to include all the electronic states that are close in energy. This is the *Multi-State CASPT2* method [58]:

Assume a number of CASSCF wave functions, Ψ_i, $I = 1, N$, obtained in a state average calculation. The corresponding (single state) CASPT2 functions are: ξ_i, $i = 1, N$. The functions $\Psi_i + \chi_i$ are used as basis functions in a "variational" calculation where all terms higher than second order are neglected. The corresponding effective Hamiltonian has the elements:

$$(H_{\text{eff}})_{ij} = \delta_{ij} E_i + \langle \Psi_i | \hat{H} | \chi_j \rangle \tag{47}$$

E_i is the CASSCF energy for state i. This Hamiltonian is not symmetric. In practice a symmetrized matrix is used. This may cause problems if the non-Hermiticity is large and it is then advisable to extend the active space. A detailed analysis of this and related problems in photochemical applications has recently been given by M. Merchán and L. Serrano-Andrés [59].

The multi-state method becomes particularly important when electronic states of different character are close in energy. One can then expect that the effects of dynamic correlation is different and large errors in relative energies can occur at the CASSCF level of theory, which may result in an erratic mixing of the different wave function types in the reference functions. A typical example is valence-Rydberg mixing the unsaturated organic molecules. Another is mixed d–d and charge transfer excitations in transition metal complexes. The multi-state CASPT2 method has been successful in solving problems in a number of applications.

25.7 THE RELATIVISTIC REGIME

In the discussion of active spaces we mentioned lanthanides and actinides. These are heavy atoms and we cannot expect a non-relativistic method to work well for such cases. The multiconfigurational approach has to be extended to the relativistic regime in order to be able to treat heavy atom systems. Can this be done without a complete rethinking? Relativistic quantum chemistry should start from the Dirac equation. Much work has also been done in recent years to develop a 4-component quantum chemistry. It can today be made quite accurate, for example, when combined with the coupled-cluster approach for electron correlation. The problem is, as in the non-relativistic case, that one needs to start from a single determinant formulation. It is, however, possible to transform the 4-component Dirac operator to a 2-component form where one simultaneously analyzes the magnitude of the different terms and keeps only the most important. Today, the most used such transformation leads to the Douglas–Kroll–Hess Hamiltonian [60,61]. The DKH Hamiltonian can be divided into a scalar part and a spin-orbit coupling part. The scalar part includes the mass–velocity term and also modifies the potential close to the nucleus so that the relativistic weak singularity of the orbital is removed. The effect on energies is similar to that of the Darwin term, but the resulting operator is variationally stable. This part of the relativistic corrections can easily be included in a non-relativistic treatment. Usually, only contributions to the one-electron Hamiltonian are included. For lighter atoms, the scalar relativistic effects will be dominating and calculations on, say, first row transition metal compounds, can safely be performed by adding only this term to the one-electron Hamiltonian used in non-relativistic quantum chemical methods.

The DKH Hamiltonian has recently been implemented into the CASSCF/CASPT2 version of the multiconfigurational approach [62]. A two-step procedure is used to account for the relativistic effects

- The scalar part of the DKH Hamiltonian is added to the one-electron integrals. All scalar relativistic effects are then automatically included at all levels of theory. Such calculations have to be performed using basis sets that are adapted to such a Hamiltonian because of the contraction of the core orbitals. A new set of ANO-type basis sets have, therefore, been constructed, called ANO-RCC. They are available for all atoms of the periodic table [30,63].
- The treatment of the spin-orbit part of the DKH Hamiltonian is based on the assumption that the strongest effects of SOC arise from the interaction of *electronic states* that are close in energy. For these states we perform independent CASSCF/CASPT2 calculations. The resulting CASSCF wave functions are then used as basis functions for the calculation of the spin-orbit coupling. The diagonal elements of the spin-orbit Hamiltonian can be modified to account for dynamic correlation effects on the energy by, for example, replacing the CASSCF energies with CASPT2 or MRCI energies.

The second part of the procedure outlined above depends on the possibility to compute matrix elements between complex CASSCF wave functions, which is not trivial because the orbitals of two different CASSCF wave functions are usually not orthogonal.

A method to deal with this problem was solved by P.-Å. Malmqvist 20 years ago [64,65]. The method has become known as the CASSCF State Interaction (CASSI) method and is effective also for long CAS-CI expansions. It was recently extended to deal also with the integrals of the spin–orbit Hamiltonian [66]. The whole approach has been implemented in the latest version of the MOLCAS quantum chemistry software [67].

A number of studies have been performed on atoms and small molecules, which show that the approach is capable of describing relativistic effects in molecules containing most atoms of the periodic system with good accuracy, maybe with one exception: The method does not take into account that the shape of the spin-orbitals depend on the *j* quantum number. For example, the orbitals $p_{1/2}$ and $p_{3/2}$ in main group have different radial shape. This is not accounted for in the present approach. It is particularly important for the fifth row elements Tl–At. Here the method give larger errors than for any other atom in the periodic system [30]. Still, the method is capable of giving rather accurate results for compounds including, at least Tl and Pb atoms [36,62]. We shall give an example later in the text.

The present method to study heavy element compounds in new and our experience is so far limited to atoms and some small molecules. It has the virtue that we can now use the machinery of CASSCF/CASPT2 for the entire periodic system. The method has been tested for all alkaline, alkaline earth, main group, transition metal atoms and in addition for some of the lanthanides and actinides. The results are promising. One example: It has recently been possible to assign the electronic spectrum of the UO_2 molecule (more than 150 electronic levels were computed) [33]. A drawback is that for the heaviest elements one has to include a large number of electronic states in order to fully account for the effects of spin–orbit coupling.

25.8 THREE EXAMPLES

We shall finish by analyzing the electronic structure of three small molecules that provide good examples of multiconfigurational effects on the electronic structure. The first is the ozone molecule. It is a prototype for molecules with unfilled valencies. Most molecules made up of main group elements have all their valencies filled in the ground electronic state. The electronic structure is characterized by core orbitals, bonding orbitals, and lone-pairs, all doubly occupied. Such molecules are well described by a closed shell single determinant wave function. There is, however, a class of molecules that do no fulfill these simple structural rules. All valencies cannot be saturated in a natural way. Instead, there are competing structural possibilities. The ozone molecule is such a case.

Another simple system will be used to illustrate how multiconfigurational methods are used to describe radicals and at the same time to compute the electronic spectrum of an organic molecule. It is the allyl radical.

Finally, we give one example of a heavy element compound, where we need to include relativistic effects. The molecule is PbF and we shall compute the potential for the ground state and a number of excited states.

25.8.1 The ozone molecule

Ozone is a triangular molecule with a bond distance of 1.27 Å, and a bond angle of 117°. This could be compared with the bond distance in the double bonded O_2 molecule, 1.21 Å, or to the single bond in HOOH, 1.46 Å. We conclude that partial double bonds are formed between the end and central oxygen atoms. How can we explain this? One would expect the ozone molecule to be a bi-radical because the end atoms can only use one of their free valencies to form a bond with the central atom. But ozone is not a bi-radical, even if it is quite reactive. The radical structure is quite high in energy and there are ionic structures that may compete. We show the three most important valence structures in Fig. 25.6.

We note that the ionic structures result in some double bond character. It is clear that we cannot easily use this information to build a wave function and we cannot a priori expect it to be well represented by a single determinant. But we can easily see which orbitals will be involved. It is the three π-orbitals with four electrons. We can formally write them as (with the central atom labeled A):

$$\pi_1(1b_1) = c_{11}\pi_A + C_{12}(\pi_B + \pi_C), \quad \pi_2(1a_2) = c_{22}(\pi_B - \pi_C),$$
$$\pi_3(2b_1) = c_{31}\pi_A + c_{32}(\pi_B + \pi_C)$$

The first orbital is bonding between all three atoms, the second is nonbonding, and the third is antibonding. A single determinant can be formed by occupying the two first orbitals. This will clearly yield a molecule with strong ionic character. A biradical character can be introduced by adding a configuration where orbital three is doubly occupied instead of two:

$$\Psi = C_1(\pi_1)^2(\pi_2)^2 + C_2(\pi_1)^2(\pi_3)^2 \tag{49}$$

We notice the similarity between this wave function and the one we used to explain the dissociation of a chemical bond. The biradical character was a result of the mixing of two configurations, one with a bonding and one with an antibonding orbital doubly occupied. The orbital π_3 is bonding between the end atoms, while π_2 is antibonding. The situation here is complicated by the orbital π_1, which is also bonding but the principle is the same.

In order to find out how large the effect is, we need to perform a calculation. We shall make it a little bit more extensive just for the fun of it. We choose as active orbitals all nine orbitals generated from the oxygen 2p orbitals with 12 active electrons. The calculations are run in C_{2v} symmetry where the orbital labels are a_1 and b_2 for the σ-orbitals. The active space is then $3a_1$, $3b_2$, $2b_1$, and $1a_2$ orbital, which we write as (3321). We use a basis set of the Atomic Natural Orbital (ANO) type with the contraction

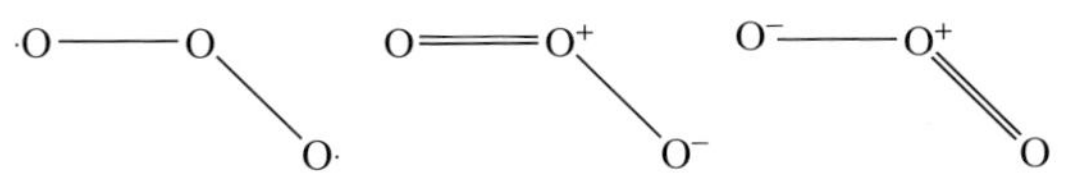

Fig. 25.6. The most important bonding structures for the ozone molecule.

Table 25.1 Natural orbital occupation numbers for the ozone molecule with two different active spaces. The third column shows the occupation numbers for the triangular form with the 3321 active space

Orbital	(3321)	(4421)	Tri. Form
$5a_1$	1.991	1.982	1.901
$6a_1$	1.956	1.954	1.931
$7a_1$	–	0.015	–
$8a_1$	0.063	0.062	0.077
$3b_2$	1.992	1.980	1.901
$4b_2$	1.947	1.945	0.123
$5b_2$	–	0.011	–
$6b_2$	0.052	0.051	0.077
$1b_1$	1.963	1.963	1.999
$2b_1$	0.258	0.230	1.995
$1a_2$	1.780	1.807	1.995

4s3p2d1f and run a geometry optimization at the CASPT2 level of theory. It is a trivial calculation. Each energy calculations takes less than a minute on a Pentium 4 laptop.

The resulting bond distance is 1.281(1.272) Å and the bond angle 117.0 (116.8) degrees (experimental values within parenthesis). The somewhat long bond distance is mainly due to remaining basis set effects. The CASPT2 method usually gives bond distances that are accurate to better than 0.01 Å.

The NO occupation numbers are shown in Table 25.1. Note that the third π-orbital, $2b_1$ has an occupation of 0.26 electrons. The coefficient of the second term in (49) is -0.298. Clearly, this second configuration will increase the biradical nature of the electronic structure. We can see that also on the Mulliken charge, which increases form -0.11 to -0.09 for each of the end atoms.

Table 25.1 also shows that there is some multiconfigurational nature in the σ part of the electronic structure with occupation numbers as large as 0.06 for two of the antibonding orbitals. The two orbitals $5a_1$ and $3b_2$, which describe the in-plane lone-pairs of the end oxygens have occupation numbers close to 2.0. The reason is that there are no correlating orbitals within the valence shell. But we can add two extra valence orbitals to the active space, which is then extended to (4421). The result of such a calculation is shown in the second column of Table 25.1. The new orbitals, which are of 3p character, obtain occupation numbers between 0.01 and 0.02. All the active orbitals are shown in Fig. 25.7. The first line shows the strongly occupied σ-orbitals, the second the corresponding correlating orbitals. The π-orbitals are shown on the third line.

We conclude from this little exercise that the electronic structure of ozone is clearly multiconfigurational. It will be difficult to treat it using methods based on a HF reference function (like CC or DFT).

Let us make one more little exercise. The triangular form of ozone has all valencies filled but for steric reasons we expect the energy to be high. The electronic structure is different with six π-electrons instead of four. We optimize the structure of this form (CASPT2) using the smaller valence orbital generated active space. The computed bond distance is 1.451 Å, close to the 1.46 Å in HOOH. The NO occupation numbers

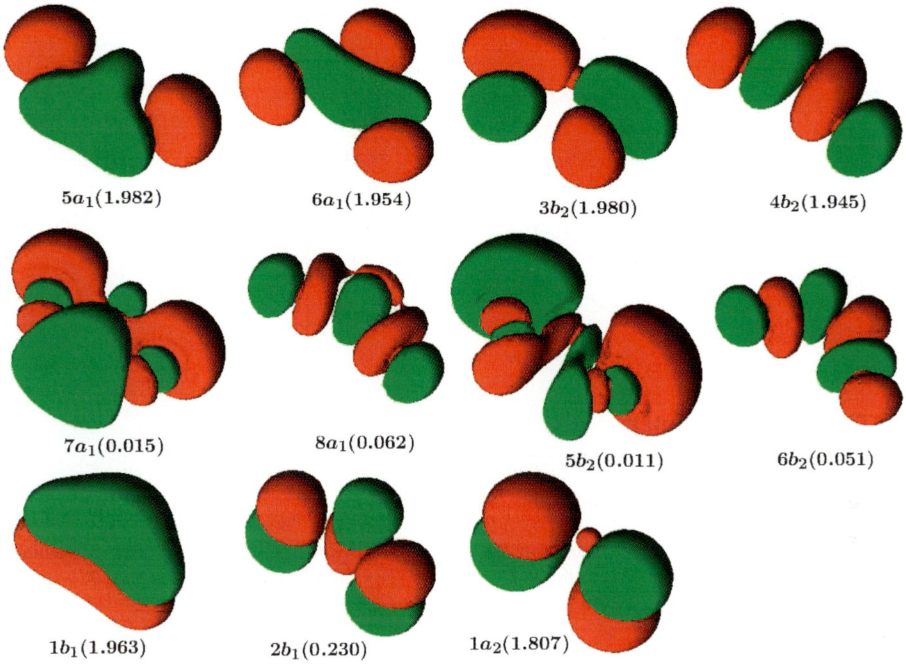

Fig. 25.7. The most important natural orbitals in the ozone molecule. The iso-surface shown is ± 0.04.

are shown in the third column of Table 25.1. We notice the large occupation of the antibonding orbitals, altogether 0.23 electrons. Clearly, this form of ozone is also highly multiconfigurational. The triangular form is found to be 35.7 kcal/mol higher in energy than the open form. Earlier studies have given values around 30 kcal/mol (see, for example, Ref. [68]).

Next, we make the following exercise. We interpolate along a straight line between the closed and the open structure and use MS-CASPT2 to compute the energy of two roots, one with four and one with six π-electrons. This is not the minimum energy path between the two stationary points, but it is close enough to illustrate the origin of the energy barrier and the nature of the transition state.

We show the energy surfaces in Fig. 25.8. The energy gap at the transition state is small (about 4 kcal/mol). The wave function will change character abruptly at the transition state, from a system four π electrons to one with six. Only a multiconfigurational treatment can capture this change. A single determinant treatment (e.g. DFT or CC) would give two crossing potentials.

25.8.2 The allyl radical

As a second example we choose the allyl radical, $(CH_2)(CH)(CH_2)$. Its electronic structure is similar to that of ozone, but with only one electron in the $1a_2$ orbital.

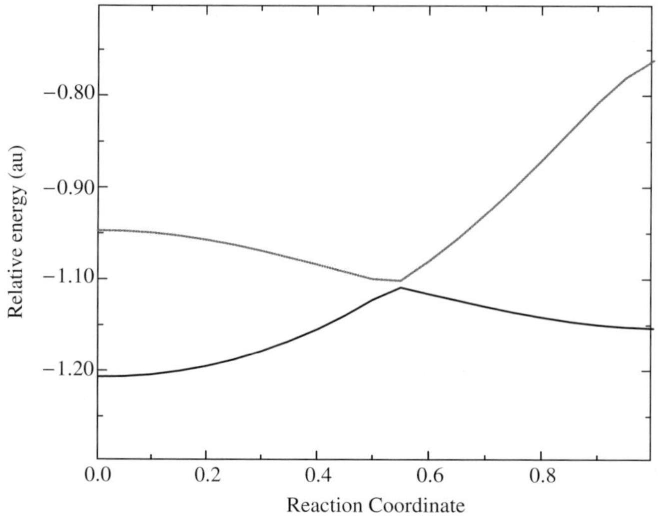

Fig. 25.8. The energy of two states of ozone (one with four and one with six π electrons) along a reactions coordinate leading from the open ($x = 0.0$) to the closed form ($x = 1.0$). The reaction coordinate is a simple interpolation between the geometries of the two forms, keeping C_{2v} symmetry.

The molecule is planar with C_{2v} symmetry, so the electronic ground state is $^2A_2$. The dominant electronic configuration is $(1b_1)^2$ $(1a_2)$. It is not obvious that the electronic structure should have a multiconfigurational nature, but a calculation using as active the three π-orbitals with three active electrons shows that the leading configuration has a weight of only 92%, to be compared with 91% in ozone (see below).

We shall use the allyl radical to illustrate how one can treat planar unsaturated organic molecules using multiconfigurational methods. Some properties of the ground state will be studied and, in addition, the electronic spectrum. The system has actually been used as a example in a course given at the Department of Theoretical Chemistry in Lund called *Quantum Chemistry at Work* and a number of students have performed the calculations. We shall use their results. Some of them were recently published [69].

The basis set used was again of the ANO type with the contraction C(4s3p2d)/H(3s2p). We shall discuss the addition of Rydberg like functions later. As active space we use the three π-orbitals with three active electrons. There is no meaning to add σ-orbitals to the active space. The σ skeleton in hydrocarbons is nicely described as a closed shell and we do not expect any appreciable occupation of the antibonding orbitals.

25.8.2.1 The ground state

First we optimize the ground state assuming a planar molecule with C_{2v} symmetry. The optimization is performed at the CASPT2 level using numerical gradients, which is quite feasible for an eight atom system.

The results are shown in Table 25.2. We see that there is agreement between theory and experiment. This is typical for unsaturated hydrocarbons with the π-orbitals active.

Table 25.2 Equilibrium geometries for the ground and first excited state of the allyl radical C_1 is the central carbon atom (bond distances in Å)

State	$\angle C_2C_1C_3$	C_1-C_2	C_1-H_1	C_2-H_2	C_2-H_2
1^2A_2	124.8°	1.384	1.081	1.075	1.078
Expt.[a]	124.0°	1.386	1.087	1.082	1.085
1^2B_1	123.1°	1.455	1.077	1.074	1.076

[a]Ref. [82].

We are going to compute the electronic spectrum and it is advised to do this using a geometry that has been optimized using a good theoretical method instead of an experimental geometry. In this case the two geometries agree. If they would not, it is likely that the theoretical result is more accurate.

Another property which is interesting for a radical is the spin-density. We can compute the spin populations on the carbon atoms using a Mulliken analysis. The CASSCF wave function is used. The spin-density will come entirely from the π system because we have used a closed shell for the σ electrons.

Electron Paramagnetic Resonance (EPR) can be used to measure the spin-densities in radicals. It is then assumed that the hyperfine coupling constants for the hydrogen atoms are proportional to the spin-density of the adjacent carbon atom [70]. Measurements on the allyl radical [71] give with such an analysis ratio of -0.282 between the spin-densities of the central and the end carbon atom. The CASSCF value is 0.311. One would suspect that methods that include spin polarization of the σ skeleton would give better values. The UHF value is, however, -0.717. What is the reason for this large difference? Let us take a closer look at the CAASCF wave function. It contains three terms:

$$\Psi = 0.96(1b_1)^2(1a_2) - 0.14(2b_1)^2(1a_2) - 0.24(1b_1 2b_1)_T(1a_2) \tag{50}$$

It is the last term in this wave function, which give rise to the spin polarization. The two b_1 orbitals are coupled to a triplet, which is then in turn coupled with the a_2 orbital to form a total doublet. We can write this configuration in terms of Slater determinants as:

$$(1b_1 2b_1)_T(1a_2) = \frac{1}{\sqrt{3}}(|1b_1\alpha, 2b_1\beta, 1a_2\alpha| + |1b_1\beta, 2b_1\alpha, 1a_2\alpha| + |1b_1\alpha, 2b_1\alpha, 1a_2\beta|) \tag{51}$$

Such a wave function cannot be described by an UHF determinant, which will assign only and α spin to the $1a_2$ orbital. As a result, the spin density on the central atom will be grossly overestimated. DFT uses the same Slater determinant as UHF and one would not expect much improvement. The UB3LYP value of the ratio is, however, -0.394, which is considerably smaller than the UHF value, but still larger than the experimental value. Neither UHF nor UDFT can fully account for the multiconfigurational nature of the wave function.

25.8.2.2 The electronic spectrum

Let us next try to compute the excited states of the radical at the ground state equilibrium geometry (the vertical excitation energies). What would we expect to find in such a calculation? We have three π-orbitals, one doubly and one singly occupied, and one empty. This generates three singly excited doublet states: $1b_1 \rightarrow 1a_2(1^2B_1)$, $1a_2 \rightarrow 2b_1(2^2B_1)$, and $1b_1 \rightarrow 2b_1(2^2A_2)$. The third state is expected to be quite high in energy, but the other two should appear at lower energies. Simple π-electron theory would actually predict the two configurations to be degenerate and two states form as a plus and minus combination, one with low energy and intensity and the other with higher energy and large intensity. We shall see that this is also what comes out of a CASSCF/MS-CASPT2 calculation.

With so few valence excited states we expect an electronic spectrum dominated by Rydberg like transitions. In order to include such states in our calculation, we need to extend the AO basis set with diffuse functions. To add more basis functions with lower exponents to each of the atoms is not a recommended procedure. It easily introduces linear dependencies in the basis set and does not give a good representation of the Rydberg orbitals. Instead, we do the following. First we do a calculation on the positive ion. We localize the center of charge. Here, we add a number of primitive Gaussian functions of s-, p-, and d-type. The method is described in detail in Ref. [37], which also gives the exponents of the new functions. They are based on the universal Gaussian basis sets devised by Kaufmann et al. [72]. With these primitives added, we perform a new calculation on the ground state of the positive ion. To avoid mixing of Rydberg and valence orbitals, we use a small valence basis set. The lower virtual orbitals can now be selected as the Rydberg basis functions after projecting out the valence basis set and spherically averaging the angular momentum components (codes are available, for example, in the MOLCAS software to perform these tasks automatically).

Here, we select one s-, one set of three p-, and one set of five d-type Rydberg functions and add them to our valence basis set. They will be added to the active space together with the three valence π-orbitals. Only Rydberg states originating from excitation for the $1a_2$ orbital will be included in the calculation. The radical has C_{2v} symmetry, so the calculation can be split into four, one for each of the symmetries $^2A_1$, $^2B_1$, $^2B_2$, and $^2A_2$. It is only needed to have the appropriate Rydberg orbitals active in each case. For example, the states of $^2A_1$ symmetry will correspond to the excitations $1a_2 \rightarrow 3s$, $3p_z$, $3d_{z^2}$, $3d_{x^2} - y^2$ and only five Rydberg orbitals need to be active. A state average CASSCF calculation is now performed for the appropriate number of roots, followed by an MS-CASPT2 calculation in the space of the CASSCF wave functions.

The excitation energies are obtained as the MS-CASPT2 energy difference between the excited state and the ground state computed with the same active space. The transition moments have been computed from the CASSCF wave functions. This is usually a reasonably accurate procedure. If the MS-CASPT2 treatment shows appreciable mixing between different CASSCF wave functions, we use instead these perturbation mixed functions (PM-CAS) to compute the transition properties. As we shall see, such a procedure becomes necessary for the $^2B_1$ states. All calculations were performed using the MOLCAS quantum chemistry software [67].

The results are shown in Table 25.3. The two $^2B_1$ valence states are found 3.32 and 6.90 eV above the ground state. Only the first states have been localized experimentally. This state was studied in more detail. The wave function is the minus combination of the two configurations shown above. The low intensity is due to cancellation of the transition moments to the two configurations. The geometry was optimized (at the CASPT2 level) assuming C_{2v} symmetry. The resulting parameters are given in Table 25.2. The main difference from the ground state is the longer CC bond, due to the excitation of one electron from either a nonbonding to an antibonding orbital or from a bonding to a nonbonding orbital. The calculated "adiabatic" excitation energy is 3.07 eV, a stabilization of 0.25 eV compared to the vertical energy. The experimental band maximum at 3.07 is expected somewhere between the adiabatic and vertical energy.

The second state is found at 6.90 eV at the MS-CASPT2 level, but at 6.59 eV at the CASPT2 level. There is thus a large effect of the multi-state treatment. This is not atypical for valence states that appear in the Rydberg region of the spectrum. Such a situation often lead to CASSCF waver functions that are mixtures of valence and Rydberg states. Two Rydberg states have $^2B_1$ symmetry. They are also affected. The MS-CASPT2 energies are in better agreement with experiment.

Energies for valence excited states are usually lowered when dynamic electron correlation effects are included. This is particularly the case when the ionic character of the electronic structure increases in the excited state. Ionic structures give rise to large dynamic polarization effects, which reflects the response of the *sigma* electrons to the ionicity of the π system. The opposite is true for the Rydberg states, where the effect of dynamic correlation decrease in the excited states. The excitation energies are thus larger at the CASPT2 level compared to CASSCF. The reason is obvious. A Rydberg states is characterized as a diffuse, singly occupied, orbital, which only weakly interacts with the rest of the molecule. The correlation energy resembles that of the positive ion.

With this, we leave the allyl radical. The electronic spectra of many hundred organic molecules have today been treated with the method we have exemplified above, including

Table 25.3 Vertical excitation energies for the allyl radical (oscillator strengths within parenthesis)[a]

State	CASSCF	CASPT2	MS-CASPT2	Expt.
Valence excited states				
1^2B_1	3.70	3.33	3.32(0.0006)	3.07(0.0013)
4^2B_1	7.57	6.59	6.90(0.118)	–
Rydberg excited states				
1^2A_1 ($1a_2 \rightarrow 3s$)	4.47	5.11	5.11(f)	4.97
2^2A_1 ($1a_2 \rightarrow 3p_z$)	4.94	5.65	5.65(f)	–
1^2B_2 ($1a_2 \rightarrow 3p_x$)	5.10	5.65	5.65(0.013)	–
2^2B_1 ($1a_2 \rightarrow 3p_y$)	5.37	5.83	5.73(0.11)	(5.4–5.9)(0.23)
3^2B_1 ($1a_2 \rightarrow 3d_{yz}$)	6.04	6.55	6.36(0.08)	(6.2–6.3)(0.05)
3^2A_1 ($1a_2 \rightarrow 3d_{x^2-y^2}$)	5.76	6.51	6.51(f)	–
4^2A_1 ($1a_2 \rightarrow 3d_{z^2}$)	5.82	6.61	6.61(f)	–
2^2B_2 ($1a_2 \rightarrow 3d_{xz}$)	5.77	6.56	6.56(0.0000)	–
2^2A_2 ($1a_2 \rightarrow 3d_{xy}$)	5.75	6.60	6.61(0.0000)	–

[a]Results from Ref. [69] where more details can be found.

neutral molecules, cation and anions, radicals and biradicals [37–39,73]. The method has also been used to treat transition metal complexes and heavy atom systems, where relativistic effects become important [74,62].

25.8.3 The PbF molecule

As an example of such a study we pick a small molecule: PbF. Suppose we want to compute the potentials for the lower electronic states of this molecule, with the relativistic CASSCF/CASPT2 approach, how do we proceed? Well, it should not be difficult. Relativistic basis sets (ANO-RCC) are available for Pb and F [30] and we choose a reasonably extended set: Pb:25s22p16d12f4g/9s8p6d4f3g and F:14s9p4d3f2g/5s4p3d2f1g. It is of quadruple zeta accuracy.

Now, the active space. Well, that appears to be simple. For main group elements one needs the *n*s and *n*p orbitals. Actually, the 2s of fluorine can be left in the inactive space. It does not contribute to the chemical bond. So, we have seven active orbitals with $4 + 5 = 9$ electrons.

Which electronic states do we include in the calculation? Here starts the more difficult part. It has been shown [30] that in order to account for the spin-orbit effects in the Pb atom, one has to include all electronic states of the $6s^2 6p^2$ configuration, that is, ^3P, ^1D and, ^1S. The result of such a calculation is shown in Table 25.4. The accuracy is reasonable but all computed numbers are on the low side. Thus we have not been fully successful in computing the stabilization of the ground state due to spin–orbit coupling. This will affect the computed bond energy of the molecule because we can expect the SOC to be much smaller here, resulting in a too small value of D_0. The quenching of the SOC in the molecule is due to three effects: the change of spin and orbital angular momentum, the usually larger separation of the excited electronic states from the ground state and the delocalization of the spin onto other (less heavy) atoms. Table 25.4 also shows the atomic energies obtained in the molecular calculation at a large internuclear separation (50 au). We note that they are the same as the calculations on the free atom. It should also be added that the computed separation between $^2P_{1/2}$ and $^2P_{3/2}$ in the fluorine atom is 0.052 eV to be compared with the experimental value 0.051 eV.

What will be the ground electronic state of the molecule. Fluorine has one open shell electron and we can, therefore, expect a single bond to be formed between the singly

Table 25.4 The energies of the electronic states of the Pb atom corresponding to the electronic configuration $6s^2 6p^2$ (in eV)[a]

State	Calc. 1	Calc. 2	Expt.[b]
^3P $(J = 0)$	0.00	0.00	0.00
^3P $(J = 1)$	0.83	0.85	0.97
^3P $(J = 2)$	1.20	1.19	1.32
^1D $(J = 2)$	2.37	2.35	2.66
^1S $(J = 0)$	3.53	3.42	3.65

[a]Calc. 1 has been performed on the free atom, Calc. 2 on PbF at large separation of the two atoms.
[b]Experimental results from Ref. [83].

occupied 2pσ orbital of F and a 6s, 6pσ hybrid of Pb. This leaves the other electron of the $6p^2$ configuration in a π-orbital and the ground state is $^2\Pi$ which will be split into two Kramer pairs with $\Omega = 1/2$ and $\Omega = 3/2$ by the SOC.

So, in order to account for the effects of SOC at the dissociation limit we need to include a large number of electronic states. It is left as an exercise to the reader to figure out that if we combine the above electronic states of Pb with the ^2P state of the F atom we end up with the electronic states: $1 \times {}^4\Delta$, $2 \times {}^4\Pi$, $1 \times {}^4\Sigma^+$, $2 \times {}^4\Sigma^-$, $1 \times {}^2\Phi$, $3 \times {}^2\Delta$, $6 \times {}^2\Pi$, $4 \times {}^2\Sigma^+$, $3 \times {}^2\Sigma^-$. The calculations are best run in C_2 symmetry because it allows us to average over the degenerate components of states with $M_z \neq 0$. We end up with the following set of calculations: 5 quartets and 13 doublets in symmetry 1, 4 quartets and 14 doublets in symmetry 2. So, for each internuclear distance we need to make four calculations where we state average over the given number of states. All the states are then mixed in the calculation of the SOC. There is one danger with this type of calculation. One has to check that the same type of electronic state is obtained for all distances. If this is not the case one has to change the number of states. Otherwise, the result may be discontinuous potential curves. Actually, this happens here for the doublets in symmetry 2, but at distances shorter than equilibrium, so it does not affect the analysis of the spectroscopic constants.

The calculations are straightforward once we have set up the input, which consists of an integral calculation, four CASSCF/CASPT2 calculations and a final CASSI calculation for the SOC. It takes about 2 h for each internuclear distance on the Pentium 4 laptop I am writing this chapter on. We write a shell script that loops over 16 chosen distances, most densely spaced around the equilibrium geometry, and let the computer work a couple of days. The program MOLCAS-6.0 is used [67].

Having generated all the 23 potential curves we analyze the most important of them, by solving the ro-vibrational Schrödinger equation numerically using the program VIBROT in MOLCAS. Let us first take a look at the results obtained without SOC. We find all six quartet states to be repulsive, as expected because no bond is formed. The 16 doublet states are shown in Fig. 25.9. We see two bound states. The lowest is the expected $^2\Pi$ state and the next is a $^2\Sigma^+$ state obtained by moving the odd electron to the nonbonding 6s, 6pσ hybrid. The higher states are either repulsive or only weakly bonding. They are also densely spaced, which makes an assignment of the electronic spectrum in this region almost impossible.

From these electronic states we generate 45 pairs of levels when we include SOC. It is impossible to discuss all of them. The seven lowest potentials are shown in Fig. 25.10. We shall discuss the lower states in a while but let us first take a look at the experimental data [75]. Huber and Herzberg lists eight levels up to 6.0 eV. The three uppermost are uncertain and some bands are diffuse. Nevertheless, we have listed the T_e values in Table 25.5 together with the excitation energies computed at an internuclear distance of 3.80 au (2.01 Å), which is close to the equilibrium distance of the ground state, 2.06 Å. The assignments are from the calculation. They agree with experiment for the three lowest levels. Only one more experimental level has been assigned. We are comparing vertical energies with T_e values so only a semi-quantitative comparison can be made. Nevertheless we see a general agreement between computed and experimental energies with one exception: The fourth state is missing in the calculation. This is according to Huber and Herzberg a bound $^2\Sigma^+_{1/2}$ level with a short bond distance, 1.98 Å, and a

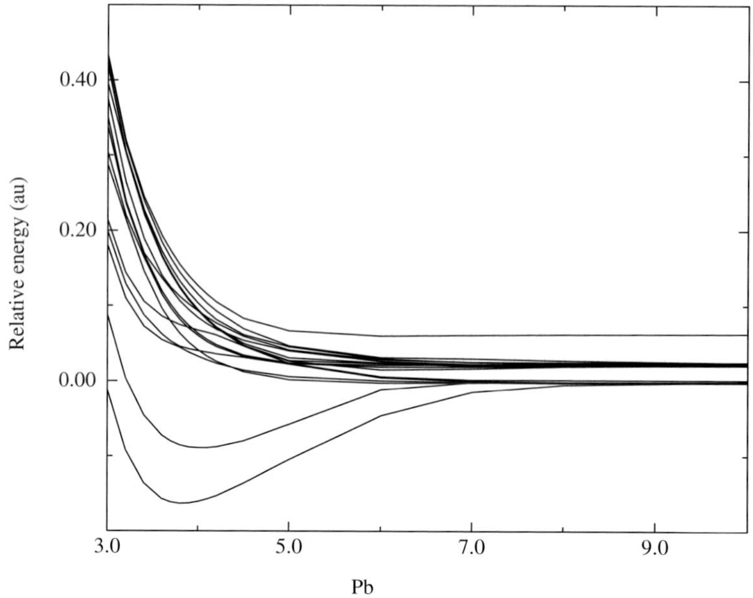

Fig. 25.9. Potential curves for 16 doublet states of PbF dissociating to Pb(^{3}P, ^{1}D, ^{1}S) and F(^{2}P). Results without spin–orbit coupling. The lowest curve is $^2\Pi$ and the next is $^2\Sigma^+$.

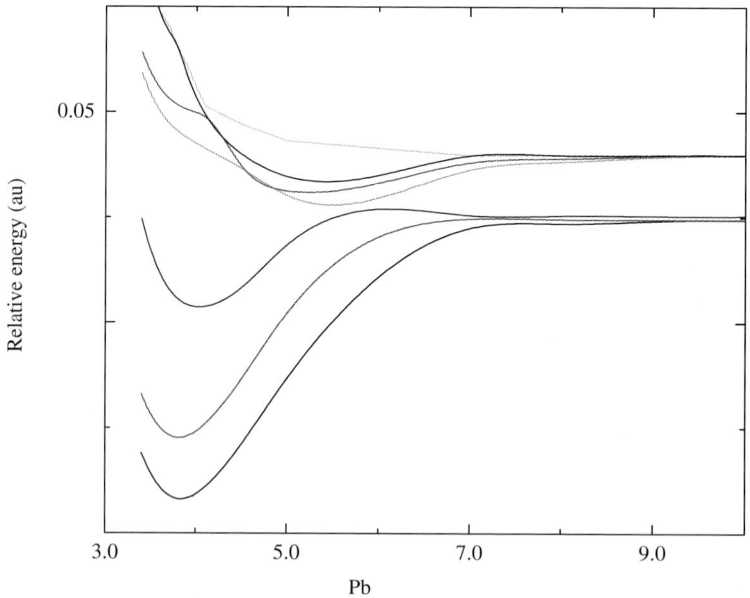

Fig. 25.10. Potential curves for the seven lowest levels of PbF including spin–orbit coupling. The three lowest levels are in order $^2\Pi_{1/2}$, $^2\Pi_{3/2}$ and $^2\Sigma^+_{1/2}$.

Table 25.5 Calculated energies of the electronic levels of the PbF molecule computed at an internuclear distance of 3.80 au (2.01 Å) compared to experimental T_e values[a] (in eV)

Level	Calc.	Expt.
$^2\Pi_{1/2}$	0.00	0.00
$^2\Pi_{3/2}$	0.79	1.02
$^2\Sigma^+_{1/2}$	2.59	2.80
$^2\Sigma^+_{1/2}$	–	4.42
$^4\Sigma^-_{1/2}$	4.74	4.72
$^4\Sigma^+_{3/2}$	5.09	5.43
$^2\Delta_{3/2}$	5.81	5.63
$^4\Pi_{1/2}$	5.81	5.93

[a]Experimental results from Ref. [75].

large frequency, 606 cm^{-1}. There is no trace if this state in the calculation as can be seen in Figs. 25.9 and 25.10. We have included all states that dissociate to $6s^26p^2$ Pb plus F, so this state must dissociate to a higher dissociation limit. The next level for the separated atoms is $Pb^+(^2P_{1/2}) + F^-(^1S_0)$, which lies 4.01 eV above the ground level. To include this level, we need to add one more doublet state to our calculation. This is probably not enough. We may also need to extend the active space because the state is clearly not included among the states covered with the present active space. A mystery! We shall not try to resolve it in this little exercise, but leave the problem to the interested reader. It might be added that a recent large spin–orbit free MRCI calculation finds a $^2\Sigma^+$ state with a minimum at a short bond distance with a T_e of about 5 eV. The minimum does, however, disappear when spin–orbit coupling is added [76].

Let us now take a closer look at the ground state. Inspection of the wave function shows that it is dominated by one configuration corresponding to the valence shell occupation $(6s)^2(\sigma)^2(2p\pi)^4(6p\pi)^1$. This is what we expected with a single σ bond formed between Pb and F. The computed spectroscopic constants are presented in Table 25.6. We notice a slightly too short bond distance and a vibrational frequency that is a little bit too large. The perfect agreement for the bond energy is obviously fortuitous. We know that there are errors in the computed SOC energies of the order of 0.1 eV, which should give a too low bond energy, so there is clearly some cancellation of errors. But the overall agreement between theory and experiment is acceptable, considering the rather trivial calculation we have performed. The same is true for the other two low lying levels, $^2\Pi_{3/2}$

Table 25.6 Calculated and experimental (within parentheses) spectroscopic constants for the three lowest levels of the PbF molecule[a]

	$^2\Pi_{1/2}$	$^2\Pi_{3/2}$	$^2\Sigma^+_{1/2}$
r_e (Å)	2.023(2.058)	2.019(2.034)	2.133(2.160)
D_0 (eV)	3.61(3.64)	–	–
w_e cm^{-1}	540(502)	550(528)	412(395)
T_e (eV)	–	0.79(1.02)	2.48(2.80)

[a]Experimental results from Ref. [75].

and $^2\Sigma^+_{1/2}$. Also here is the bond distance too short and the frequencies too high. Computed T_e values are about 0.2 eV too small, an error which is within the limits for the present method.

With this we leave PbF. The calculations presented have not been exhaustive. We even missed one electronic state. A more accurate study would need a larger active space and more electronic states to account also for the $Pb^+ + F^-$ dissociation limit. Calculations on small molecules are often not as trivial as they seem.

25.9 CONCLUSIONS

We have shown in this chapter how one can construct wave function for electron structure problems where a single electronic configuration will not give a valid description. The basic concept is the partitioning of the occupied orbital space into an inactive and an active part. The choice of the active subspace is a matter of experience combined with some insight into the electronic structure of the problem under study. A number of examples were given ranging form the atoms of the periodic table, small molecules, electron spectroscopy of organic molecules, to transition metal compounds and finally also lanthanide and actinide chemistry. It has been illustrated that the approach can be used for compounds containing almost all atoms of the periodic system (the possible exception being the heaviest main group elements Bi–At). The relativistic formulation of the method, which becomes necessary for heavy atoms, is based on a partitioning of the relativistic effects into a scalar part and spin-orbit coupling. Higher order relativistic effects can be neglected in most chemical applications.

It is clear that the presentation has been biased towards the work of the author himself. Much of the important development made by other research groups may have been neglected. Maybe the best way to acknowledge the modern work in multiconfigurational quantum chemistry is to mention the most important codes: The MOLPRO code contains a very effective CASSCF code. Dynamic correlation can be treated using CASPT2 and even CASPT3, or the internally contracted MRCI method [77]. A similar code is DALTON that reflects much of the important early development by the Aarhus group in the early 1980s. DALTON has specialized in various electric and magnetic properties [78]. Another code that should be mentioned is COLUMBUS, which includes the probably most efficient MRCI code available today [48].

What about the future? The multiconfigurational approach will most likely remain an important tool in quantum chemistry. For many quantum chemical problems there exists no real alternative, in particular not for excited states and photochemistry. The method is, however, not without problems. The fact that it is not a black box, but is rather demanding on the users chemical and quantum chemical knowledge, is an obstacle for many potential users. Attempts have been made to make the procedure more automatic but they have not been very successful. Some codes solve this problem by an oversimplification of the input, removing the necessary flexibility. It is rarely possible to perform a sensible CASSCF calculation by only specifying the number of active orbitals and electrons. To suggest so is simply cheating the user. Thus, there is still a need for improvement in simplifying the use of the CASSCF method.

The number of active orbitals that can be used in a CASSCF calculation is limited. Some problems needs larger active spaces than can be used today. It is then possible to use a less general way to construct the wave function, for example, the restricted active space (RAS) SCF method [79]. Such methods can use larger active spaces, but there exist no corresponding RASPT2 program yet. Such a code would be of great value.

The treatment of dynamic correlation effects is still a problem. The CASPT2 method has proven to be a very useful tool in many applications, but it has its limitations both in accuracy and in applicability. It is not possible to extend the method to higher orders without great difficulty and it is even not clear that such an extension would be meaningful because we do not know if the perturbation expansion will converge. The MRCI method can be made very accurate for small molecules but has severe limitations, and the lack of size-extensivity is a serious obstacle. Multireference cluster expansions are studied but this approach has not yet been shown to be practical for large-scale applications. The coupled cluster method can be applied to some simpler multiconfigurational wave functions using the "spin–flip" method where the reference function is chosen as a closed shell and the state of interest is treated as an excited state [80].

A tempting possibility is to attempt to combine the CASSCF (or RASSCF) method with density functional theory. One would then include all long-range (static) correlation effects into a reference CASSCF wave function, while describing dynamic correlation effects using a correlation potential. It has been shown that such a method is theoretically sound and that it is possible to define a universal functional for the dynamic part of the correlation energy. Several groups are currently attempting to construct such a functional (see, for example, Ref. [81]).

The examples given in the previous section were all for small molecules. This should not be misunderstood. The approach is not more restricted in size than other quantum chemical methods. The limit is not the size of the basis set but rather the size of the active space. It is possible to develop linear scaling procedures that work well with the CASSCF/CASPT2 method. It is expected that it will be possible to use multiconfigurational methods to study molecular systems comprising several hundred atoms in the near future.

25.10 ACKNOWLEDGEMENTS

This work has been supported by a grant from the Swedish Science Research Council, VR and the Swedish Foundation for Strategic Research (SSF). I am grateful to Professor Per-Åke Malmqvist for valuable comments on the manuscript.

25.11 REFERENCES

1 D.R. Hartree, Proc. Cambridge Philos. Soc., 24 (1928) 89, see also 111,426.
2 F. Hund, Z. Phys., 40 (1927) 742.
3 R.S. Mulliken, Phys. Rev., 32 (1928) 186.
4 P.-O. Löwdin, Phys. Rev., 97 (1955) 1474.
5 J.K.L. McDonald, Proteins, 43 (1933) 830.

6 W. Heitler and F London, Z. Phys., 44 (1927) 455.
7 C.A. Coulson and I. Fischer, Philos. Mag., 40 (1949) 386.
8 B.O. Roos, Collect. Czech. Chem. Commun., 68 (2003) 265.
9 D.R. Hartree, W. Hartree and B. Swirles, Phil. Trans., A238 (1939) 229.
10 A. Jucys, Proc. R. Soc. (London), Ser. A, 173 (1939) 59.
11 J. Frenkel, Wave mechanics, advanced general theory, Clarendon Press, Oxford, 1934.
12 A.C. Wahl and G. Das, J. Chem. Phys., 44 (1966) 87.
13 A.C. Wahl and G. Das, in: H.F. Schaefer, III (Ed.), Methods of electronic structure theory, Plenum Press, New York, 1977, p. 51.
14 B. Levy and G. Berthier, Int. J. Quantum Chem., 2 (1968) 307.
15 B. Levy and G. Berthier, Int. J. Quantum Chem., 3 (1969) 247.
16 B.O. Roos, Theor. Chem. Acc., 103 (2000) 228–230.
17 C.C.J. Roothaan, Rev. Mod. Phys., 32 (1960) 179.
18 F. Grein and T.C. Chang, Chem. Phys. Lett., 12 (1971) 44.
19 J. Hinze, J. Chem. Phys., 59 (1973) 6424.
20 E. Dalgaard and P. Jørgensen, J. Chem. Phys., 69 (1978) 3833.
21 K. Ruedenberg and K.R. Sundberg, in: J.-L. Calais (Ed.), Quantum science; methods and structure, Plenum Press, New York, 1976.
22 B.O. Roos, Chem. Phys. Lett., 15 (1972) 153.
23 I. Shavitt, Int. J. Quantum Chem., S11 (1977) 131.
24 I. Shavitt, Int. J. Quantum Chem., S12 (1978) 5.
25 B.O. Roos, P.R. Taylor and P.E.M. Siegbahn, Chem. Phys., 48 (1980) 157.
26 B.O. Roos, The complete active space self-consistent field method and is applications in electronic structure calculations, in: K.P. Lawley (Ed.), Advances in Chemical Physics; Ab Initio Methods in Quantum Chemistry–II, 399, Wiley, Chichester, 1987, Chapter 69.
27 T. Helgaker, P. Jørgensen, J. Olsen, Molecular electronic structure theory, Wiley, Chichester, 2000.
28 B.O. Roos, Multiconfigurational (MC) self-consistent (SCF) theory, in: B.O. Roos, P.-O. Widmark (Eds.), European summer school in quantum chemistry, book II, Lund University, Lund, Sweden, 2000.
29 K. Andersson and B.O. Roos, Chem. Phys. Lett., 191 (1992) 507.
30 B.O. Roos, R. Lindh, P.-Å. Malmqvist, V. Veryazov and P.-O. Widmark, J. Phys. Chem. A, 108 (2004) 2851.
31 G. Ghigo, B.O. Roos and P.-Å. Malmqvist, Chem. Phys. Lett., 396 (2004) 142.
32 B.O. Roos and P.-Å. Malmqvist, Adv. Quantum Chem., 47 (2004) 41.
33 L. Gargliardi, M.C. Heaven, J. Wisborg Krogh and B.O. Roos, J. Am. Chem. Soc., 127 (2005) 86–91.
34 B.O. Roos, K. Andersson and M.P. Fülscher, Chem. Phys. Lett., 192 (1992) 5–13.
35 A. Bernhardsson, N. Forsberg, P.-Å. Malmqvist, B.O. Roos and L. Serrano-Andrés, J. Chem. Phys., 112 (2000) 2798–2809.
36 L. Gagliardi, G. Orlandi, V. Molina, P.-Å. Malmqvist and B.O. Roos, J. Phys. Chem. A, 106 (2002) 7355.
37 B.O. Roos, M.P. Fülscher, P.-Å. Malmqvist, M. Merchán and L. Serrano-Andrés, Theoretical studies of electronic spectra of organic molecules, in: S.R. Langhoff (Ed.), Quantum mechanical electronic structure calculations with chemical accuracy, Understanding Chemical Reactions, Kluwer, Dordrecht, 1995, pp. 357–438.
38 B.O. Roos, M.P. Fülscher, P.-Å. Malmqvist, M. Merchán and L. Serrano-Andrés, Understanding Chem. React., 13 (1995) 357.
39 B.O. Roos, K. Andersson, M.P. Fülscher, P.-Å. Malmqvist, L. Serrano-Andrés, K. Peirloot and M. Merchán, in: I. Prigogine, S.A. Rice (Eds.), Multiconfigurational perturbation theory: Applications in electronic spectroscopy, Advances in chemical physics: New methods in computational quantum mechanics, Vol. XCIII, Wiley, New York, 1996, pp. 219–331.
40 K. Pierloot, Mol. Phys., 101 (2004) 2083.
41 K. Pierloot, in: T. Cundari (Ed.), Nondynamic correlation effects in transition metal coordination compounds, Computational organometallic chemistry, Marcel Dekker, New York, 2001, p. 123.
42 K. Pierloot, Computation of electronic spectroscopy of transition metal complexes, in: J. Michl, M. Olivucci (Eds.), Computational photochemistry, Elsevier, Amsterdam, 2004.

43 L. Gagliardi and B.O. Roos, Chem. Phys. Lett., 331 (2000) 229–234.
44 L. Garliardi, I. Grenthe and B.O. Hoos, Inorg. Chem., 40 (2001) 2976–2978.
45 L. Gagliardi, B.O. Roos, P.-Å. Malmqvist and J.M. Dyke, J. Phys. Chem. A, 105 (2001) 10602–10606.
46 T. Privalov, P. Macak, B. Schimmelpfennig, E. Fromager, I. Grentheand and U. Wahlgren, J. Am. Chem. Soc., 126 (2004) 9801.
47 P.E.M. Siegbahn, J. Chem. Phys., 72 (1980) 1647.
48 H. Lischka, R. Shepard, I. Shavitt, R.M. Pitzer, M. Dallos, Th. Mller, P.G. Szaley, F.B. Brown, R. Ahlrichs, H.J. Bhm, A. Chang, D.C. Comueau, R. Gdanitz, H. Dachsel, C. Ehrhardt, M. Ernzerhof, P. Hchtl, S. Irle, G. Kedziora, T. Kovar, V. Parasuk, M.J.M. Pepper, P. Scharf, H. Schiffer, M. Schindler, M. Schler, M. Seth, E.A. Stahlberg, J.-G. Zhao, S. Yabushita and Z. Zhang, COLUMBUS, an ab initio electronic structure program, release 5.9, (2004).
49 B.O. Roos, P. Linse, P.E.M. Siegbahn and M.R.A. Blomberg, Chem. Phys., 66 (1982) 197.
50 K. Andersson, P.-Å. Malmqvist, B.O. Roos, A.J. Sadlej and K. Wolinski, J. Phys. Chem., 94 (1990) 5483–5488.
51 K. Andersson, P.-Å. Malmqvist and B.O. Ross, J. Chem. Phys., 96 (1992) 1218–1226.
52 K. Andersson and B.O. Ross, Int. J. Quantum Chem., 45 (1993) 591–607.
53 K.G. Dyall, J. Chem. Phys., 102 (1995) 4909.
54 K. Andersson, Theor. Chim. Acta, 91 (1995) 31.
55 G. Ghigo, B.O. Ross, P.C. Stancil and P.F. Weck, J. Chem. Phys., 121 (2004) 8194–8200.
56 B.O. Ross and K. Andersson, Chem. Phys. Lett., 245 (1995) 215–223.
57 B.O. Ross, K. Andersson, M.P. Fülscher, L. Serrano-Andrés, K. Pierloot, M. Merchán and V. Molina, J. Mol. Struct. (THEOCHEM), 388 (1996) 257–276.
58 J. Finley, P.-Å. Malmqvist, B.O. Roos and L. Serrano-Andrés, Chem. Phys. Lett., 288 (1998) 299–306.
59 M. Merchán and L. Serrano-Andrés, Ab initio methods for excited states, in: J. Michl, M. Olivucci (Eds.), Computational photochemistry, Elsevier, Amsterdam, 2004.
60 N. Douglas and N.M. Kroll, Ann. Phys., 82 (1974) 89.
61 B.A. Hess, Phys. Rev. A, 33 (1986) 3742.
62 B.O. Roos and P.-Å. Malmqvist, Phys. Chem. Chem. Phys., 6 (2004) 2919–2927.
63 B.O. Ross, V. Veryazov and P.-O. Widmark, Theor. Chim. Acta, 111 (2004) 345.
64 P.-Å. Malmqvist, Int. J. Quantum Chem., 30 (1986) 479.
65 P.Å. Malmqvist and B.O. Ross, Chem. Phys. Lett., 155 (1989) 189–194.
66 P.-Å. Malmqvist, B.O. Roos and B. Schimmelpfinnig, Chem. Phys. Lett., 357 (2002) 230.
67 G. Karlström, R. Lindh, P.-Å. Malmqvist, B.O. Ross, U. Ryde, V. Veryazov, P.-O. Widmark, M. Cossi, B. Schimmelpfennig, P. Neogrady and L. Seijo, Comput. Mater. Sci., 28 (2003) 222.
68 T. Lee, Chem. Phys. Lett., 169 (1990) 529.
69 F. Aquilante, K.P. Jensen and B.O. Ross, Chem. Phys. Lett., 380 (2003) 689–698.
70 H.M. McConnell, J. Chem. Phys., 24 (1956) 632.
71 T. Bally, D.A. Hrovat and W.T. Borden, Chem. Phys. Phys. Chem., 2 (2000) 3363.
72 K. Kaufmann, W. Baumeister and M. Jungen, J. Phys. B: At. Mol. Opt. Phys., 22 (1989) 2223.
73 B.O. Ross, Acc. Chem. Res., 32 (1999) 137–144.
74 B.O. Ross, P.-Å. Malmqvist and L. Gagliardi, Heavy element quantum chemistry—the multiconfigurational approach, in: E. Brändas, E. Kryachko (Eds.), Fundamental world of quantum chemistry, Kluwer, Dordrecht, 2003, p. 425.
75 K.P. Huber and G. Herzberg, Molecular spectra and molecular structure 4: Constants of diatomic molecules, Van Nostrand Reinhold Company, Princeton, 1979.
76 K.K. Das, I.D. Petsalakis, H.-P. Liebermann, A.B. Alekseyev and R.J. buenker, J. Chem. Phys., 116 (2002) 608.
77 H.-J. Werner, P.J. Knowles, R. Lindh and M. Schütz, et al. Molpro, version 2002.6, a package of ab initio programs, 2003. see http://www.molpro.net.
78 T. Helgaker, H.J.Aa. Jensen, P. Jrgensen, J. Olsen, K. Ruud, H. gren, A.A. Auer, K.L. Bak, V. Bakken, O. Christiansen, S Coriani, P. Dahle, E.K. Dalskov, T. Enevoldsen, B. Fernandez, C. Httig, K. Hald, A. Halkiere, H. Heiberg, H. Hettema, D. Jonsson, S. Kirpekar, R. Kobayashi, H. Koch, K.V. Mikkelsen, P. Norman, M.J. Packer, T.B. Pedersen, T.A. Ruden, A. Sanchez, T. Saue, S.P.A. Sauer,

B. Schimmelpfennig, K.O. Sylvester-Hvid, P.R. Taylor and O. Vahtras. Dalton, a molecular electronic structure program, Release 1.2, 2001.

79 J. Olsen, B.O. Ross, P. Jørgensen and H.J.Aa. Jensen, J. Chem. Phys., 89 (1988) 2185–2192.

80 A.I. Krylov, Chem. Phys. Lett., 338 (2001) 375.

81 S. Gusarov, P.-Å. Malmqvist, R. Lindh and B.O. Ross, Theor. Chim. Acta, (2004) 112.

82 E. Hiraota, C. Yamada and M. Okunishi, J. Chem. Phys., 97 (1992) 2963.

83 J.E. Sansonetti, W.C. Martin and S.L. Young, Handbook of basic atomic spectroscopic data (version 1.00), National Institute of Standards and Technology, Gaithersburg, MD, 2003, Available: http://physics.nist.gov/Handbook.

Theory and Applications of Computational Chemistry: The First Forty Years
Edited by C. Dykstra et al.

CHAPTER 26

Concepts of perturbation, orbital interaction, orbital mixing and orbital occupation

Myung-Hwan Whangbo

Department of Chemistry, North Carolina State University, Raleigh, NC 27695-8204, USA

Abstract

In understanding structure–property relationships in a variety of molecules and solids, the concepts of perturbation, orbital interaction, orbital mixing and orbital occupation are widely employed. Theoretical bases of these qualitative concepts are briefly surveyed.

26.1 INTRODUCTION

An important role of an electronic structure theory is to provide quantitative predictions. In this role theoretical predictions require developments of efficient programs for theoretical computations. During the past few decades the scientific community has witnessed the advent of powerful computers and efficient programs for electronic structure calculations and dynamic simulations. These theoretical tools are affordable at modest costs, and the menu-driven operation of these programs is highly user friendly. As a consequence, even neophytes can now perform theory at many different levels by running commercially available programs although their command of theory is limited to the available options of the programs.

Another important role of an electronic structure theory is to provide a conceptual framework in which to think and organize [1,2]. In this role theoretical predictions need not be quantitative but should provide a bias toward correct thinking about further experimental and theoretical studies. When combined with the ideas of symmetry and overlap, the concepts of perturbation, orbital interaction, orbital mixing and orbital occupation have been indispensable not only in understanding structure–property relationships in various chemical compounds but also in interpreting results of electronic structure calculations. These qualitative concepts work at all levels of electronic structure descriptions from one-electron theory neglecting self-consistent-field (SCF) adjustments of orbitals to theories including electron correlation and to those including relativistic effects. In this chapter, we briefly survey the theoretical bases of these concepts.

26.2 ORBITAL INTERACTION ON THE BASIS OF EFFECTIVE ONE-ELECTRON HAMILTONIAN

26.2.1 Exact relationship between two sets of molecular orbitals [3,4]

Consider a molecular system with a set of atomic orbitals $\{\chi_1, \chi_2, \chi_3, \ldots, \chi_M\}$. The molecular orbitals (MOs) of this system, written as

$$\phi_i = \sum_{\mu=1}^{M} C_{\mu i}\chi_\mu \quad (i = 1, 2, \ldots, M) \tag{1}$$

are eigenfunctions of the effective one-electron Hamiltonian H^{eff},

$$H^{\text{eff}}\phi_i = e_i\phi_i \tag{2}$$

where the eigenvalue e_i is the MO energy. For simplicity, we assume that the MOs ϕ_i and the MO energies e_i do not depend on the number of electrons in the system, as in the case of extended Hückel tight binding calculations [5]. (See Section 26.3 for cases when this constraint is relaxed.) By defining the overlap integrals $S_{\mu\nu}$ and the interaction energy integrals $H_{\mu\nu}$ in terms of the atomic orbitals,

$$S_{\mu\nu}\langle\chi_\mu|\chi_\nu\rangle, \qquad H_{\mu\nu} = \langle\chi_\mu|H^{\text{eff}}|\chi_\nu\rangle \tag{3}$$

the matrix representation of Eq. (2) is given by

$$\mathbf{HC} = \mathbf{SCe} \tag{4}$$

where

$$\mathbf{H} = \begin{pmatrix} H_{11} & H_{12} & \cdots & H_{1M} \\ H_{21} & H_{22} & \cdots & H_{2M} \\ \cdots & \cdots & \cdots & \cdots \\ H_{M1} & H_{M2} & \cdots & H_{MM} \end{pmatrix}, \quad \mathbf{S} = \begin{pmatrix} S_{11} & S_{12} & \cdots & S_{1M} \\ S_{21} & S_{22} & \cdots & S_{2M} \\ \cdots & \cdots & \cdots & \cdots \\ S_{M1} & S_{M2} & \cdots & S_{MM} \end{pmatrix}$$
$$\mathbf{C} = \begin{pmatrix} C_{11} & C_{12} & \cdots & C_{1M} \\ C_{21} & C_{22} & \cdots & C_{2M} \\ \cdots & \cdots & \cdots & \cdots \\ C_{M1} & C_{M2} & \cdots & C_{MM} \end{pmatrix}, \quad \mathbf{e} = \begin{pmatrix} e_1 & 0 & \cdots & 0 \\ 0 & e_2 & \cdots & 0 \\ \cdots & \cdots & \cdots & \cdots \\ 0 & 0 & \cdots & e_M \end{pmatrix} \tag{5}$$

Suppose that another system closely related to the one discussed above is also described by the same set of atomic orbitals $\{\chi_1, \chi_2, \chi_3, \ldots, \chi_M\}$. The MOs of this system may be written as

$$\phi_i^0 = \sum_{\mu=1}^{M} C_{\mu i}^0 \chi_\mu \quad (i = 1, 2, \ldots, M) \tag{6}$$

which satisfy the eigenvalue equation

$$H^{\text{eff}} \phi_i^0 = e_i^0 \phi_i^0 \tag{7}$$

Given the matrices $\mathbf{H}^0$, $\mathbf{S}^0$, $\mathbf{C}^0$ and $\mathbf{e}^0$ of this system, the matrix representation of Eq. (7) becomes

$$\mathbf{H}^0\mathbf{C}^0 = \mathbf{S}^0\mathbf{C}^0\mathbf{e}^0 \tag{8}$$

For convenience of our discussion, the system described by Eq. (4) may be referred to as the perturbed system, and that described by Eq. (8) as the unperturbed system. The MOs of the perturbed system can be expressed in terms of those of the unperturbed system, i.e.

$$\phi_i = \sum_{j=1}^{M} T_{ji} \phi_j^0 \quad (j = 1, 2, \ldots, M) \tag{9}$$

where the transformation matrix $\mathbf{T}$ is expressed as [3]

$$\mathbf{T} = (\mathbf{C}^0)^+ \mathbf{S}^0 \mathbf{C} \tag{10}$$

To relate the MO energies of the perturbed system to those of the unperturbed system, we define the matrices

$$\delta\mathbf{H} = \mathbf{H} - \mathbf{H}^0, \quad \delta\mathbf{S} = \mathbf{S} - \mathbf{S}^0 \tag{11}$$

and evaluate their matrix representations in terms of the MOs of the unperturbed system,

$$\tilde{\mathbf{H}} = (\mathbf{C}^0)^+ \delta\mathbf{H}\mathbf{C}, \quad \tilde{\mathbf{S}} = (\mathbf{C}^0)^+ \delta\mathbf{S}\mathbf{C} \tag{12}$$

Then, Eq. (4) is converted to [3]

$$(\mathbf{e}^0 + \tilde{\mathbf{H}})\mathbf{T} = (1 + \tilde{\mathbf{S}})\mathbf{T}\mathbf{e} \tag{13}$$

Eqs. (9) and (13) provide the exact relationships between the perturbed and unperturbed systems, although they are hardly informative. Useful concepts are obtained on the basis of approximate relationships between the two systems.

26.2.2 Perturbation analysis and orbital interaction [3,6]

The concept of orbital interaction arises naturally when approximate relationships between the MOs of the perturbed and unperturbed systems are probed in terms of perturbation theory. It should be noted that the matrix representation of the perturbation

Geometry perturbation

Electronegativity perturbation

Intermolecular perturbation

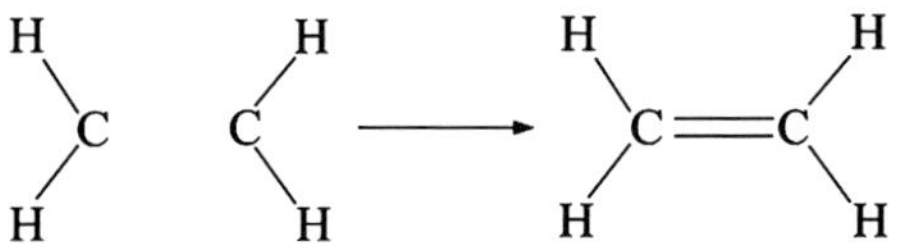

Fig. 26.1. Three classes of perturbation leading to orbital interactions.

is given by Eq. (11) in terms of the atomic orbitals, and by Eq. (12) in terms of the MOs of the unperturbed system. In general, there are three classes of perturbation (Fig. 26.1) [6]. In the 'geometry perturbation' of a molecule, one compares lower symmetry and higher symmetry structures, and considers that the lower symmetry structure is perturbed, and the higher symmetry structure is unperturbed. In the 'electronegativity perturbation', one compares two molecules that differ in one atom, and considers that the molecule with a more electronegative atom is perturbed, and that with the less electronegative atom is unperturbed. In the 'intermolecular perturbation', one deals with the formation of a composite molecule from two molecular fragments (or two molecules), and considers that the composite molecule is perturbed, and the two non-interacting molecular fragments as a whole are unperturbed.

26.2.2.1 Non-degenerate perturbation

In relating the MOs ϕ_i and the MO energies e_i of the perturbed system to the MOs ϕ_i^0 and the MO energies e_i^0 of the unperturbed system, it is convenient to employ the column vectors $\mathbf{C}_i$ of the MOs ϕ_i and the column vectors $\mathbf{C}_i^0$ of the MOs ϕ_i^0,

$$\mathbf{C}_i = \begin{pmatrix} C_{1i} \\ C_{2i} \\ \cdots \\ C_{Mi} \end{pmatrix}, \qquad \mathbf{C}_i^0 = \begin{pmatrix} C_{1i}^0 \\ C_{2i}^0 \\ \cdots \\ C_{Mi}^0 \end{pmatrix} \tag{14}$$

and define the overlap integrals $\tilde{S}_{ij}$ and the interaction integrals $\tilde{H}_{ij}$ in terms of the MOs of the unperturbed systems,

$$\tilde{S}_{ij} = (\mathbf{C}_i^0)^+ \delta\mathbf{S}\mathbf{C}_j^0, \quad \tilde{H}_{ij} = (\mathbf{C}_i^0)^+ \delta\mathbf{H}\mathbf{C}_j^0 \tag{15}$$

If there is no degeneracy in the MO energies e_i^0 of the unperturbed system, the non-degenerate perturbation theory leads to the results,

$$e_i = e_i^0 + \tilde{H}_{ii} + \sum_{j(\neq i)} \frac{(\tilde{H}_{ij} - e_i^0 \tilde{S}_{ij})^2}{e_i^0 - e_j^0} \tag{16}$$

$$\phi_i = \left[1 - \sum_{j(\neq i)} \tilde{S}_{ij} t_{ji} - \frac{1}{2} \sum_{j(\neq i)} (t_{ji})^2 \right] \phi_i^0 + \sum_{j(\neq i)} t_{ji} \phi_j^0 \tag{17}$$

where the mixing coefficient t_{ji} is given by

$$t_{ji} = \frac{\tilde{H}_{ji} - e_i^0 \tilde{S}_{ji}}{e_i^0 - e_j^0} \tag{18}$$

26.2.2.2 Degenerate perturbation

If there is degeneracy in the MO energies of the unperturbed system, use of the degenerate perturbation theory is necessary to obtain the first-order corrected orbitals and orbital energies that arise from the set of degenerate orbitals. Suppose that some eigenvalues e_i^0 are degenerate, e.g. $e_1^0 = e_2^0 = \cdots = e_n^0$. In such a case, the orbitals constructed as linear combinations of the degenerate MOs ϕ_i^0 $(i = 1, 2, \ldots, n)$,

$$\phi_i' = \sum_{\nu=1}^{n} C_{\nu i}' \phi_\nu^0 \ (i = 1, 2, \ldots, n) \tag{19}$$

are used to solve the eigenvalue equation,

$$H^{\text{eff}} \phi_i' = e_i' \phi_i' \ (i = 1, 2, \ldots, n) \tag{20}$$

This leads to the matrix representation,

$$\tilde{\mathbf{H}}'\tilde{\mathbf{C}}' = \tilde{\mathbf{S}}'\tilde{\mathbf{C}}'\mathbf{e}' \tag{21}$$

where

$$\tilde{\mathbf{H}}' = \begin{pmatrix} \tilde{H}_{11} & \tilde{H}_{12} & \cdots & \tilde{H}_{1n} \\ \tilde{H}_{21} & \tilde{H}_{22} & \cdots & \tilde{H}_{2n} \\ \cdots & \cdots & \cdots & \cdots \\ \tilde{H}_{n1} & \tilde{H}_{n2} & \cdots & \tilde{H}_{nn} \end{pmatrix}, \quad \tilde{\mathbf{S}}' = \begin{pmatrix} \tilde{S}_{11} & \tilde{S}_{12} & \cdots & \tilde{S}_{1n} \\ \tilde{S}_{21} & \tilde{S}_{22} & \cdots & \tilde{S}_{2n} \\ \cdots & \cdots & \cdots & \cdots \\ \tilde{S}_{n1} & \tilde{S}_{n2} & \cdots & \tilde{S}_{nn} \end{pmatrix}$$

$$\mathbf{C}' = \begin{pmatrix} C'_{11} & C'_{12} & \cdots & C'_{1n} \\ C'_{21} & C'_{22} & \cdots & C'_{2n} \\ \cdots & \cdots & \cdots & \cdots \\ C'_{n1} & C'_{n2} & \cdots & C'_{nn} \end{pmatrix}, \quad \mathbf{e}' = \begin{pmatrix} e'_1 & 0 & \cdots & 0 \\ 0 & e'_2 & \cdots & 0 \\ \cdots & \cdots & \cdots & \cdots \\ 0 & 0 & \cdots & e'_n \end{pmatrix} \tag{22}$$

The orbitals ϕ'_i and the orbital energies e'_i are first-order corrections. These orbitals can interact with other unperturbed orbitals ϕ_j^0 ($j \neq 1, 2, \ldots, n$) to give rise to second-order energy corrections (i.e. those similar to the last term of Eq. (16)).

As an example, consider that $e_a^0 = e_b^0$. Then, under the degenerate perturbation, the orbitals ϕ_a^0 and ϕ_b^0 lead to the following new energy levels

$$e'_+ = \frac{e_a^0 + \tilde{H}_{ab}}{1 + \tilde{S}_{ab}}, \quad e'_- = \frac{e_a^0 - \tilde{H}_{ab}}{1 - \tilde{S}_{ab}} \tag{23}$$

and the corresponding orbitals

$$\phi'_+ = \frac{1}{\sqrt{2 + 2\tilde{S}_{ab}}}(\phi_a^0 + \phi_b^0) \quad \phi'_+ = \frac{1}{\sqrt{2 - 2\tilde{S}_{ab}}}(\phi_a^0 - \phi_b^0) \tag{24}$$

The e'_+ level is lower than the e'_- level if the overlap $\tilde{S}_{ab}$ is positive, and the opposite is true if $\tilde{S}_{ab}$ is negative.

26.2.3 Normal *versus* counterintuitive orbital interaction [7]

Consider that two orbitals ϕ_i^0 and ϕ_j^0 with different energies ($e_i^0 < e_j^0 < 0$) interact to give rise to two orbitals ϕ_i and ϕ_j as depicted in Fig. 26.2. Then, the lower lying orbital ϕ_i (i.e. the bonding level) and its energy e_i are written as

$$e_i = e_i^0 + \tilde{H}_{ii} + \frac{(\tilde{H}_{ij} - e_i^0 \tilde{S}_{ij})^2}{e_i^0 - e_j^0} \tag{25}$$

$$\phi_i = [1 - \tilde{S}_{ij} t_{ji} - (t_{ji})^2/2]\phi_i^0 + t_{ji}\phi_j^0 \tag{26}$$

Since the terms $\tilde{S}_{ji}$ and t_{ji} are small in magnitude compared with 1, the coefficient $[1 - \tilde{S}_{ij} t_{ji} - (t_{ji})^2/2]$ is positive. Suppose that the phases of the two orbitals ϕ_i^0 and ϕ_j^0 are chosen such that their overlap $\tilde{S}_{ij}$ is positive, and hence the associated interaction energy

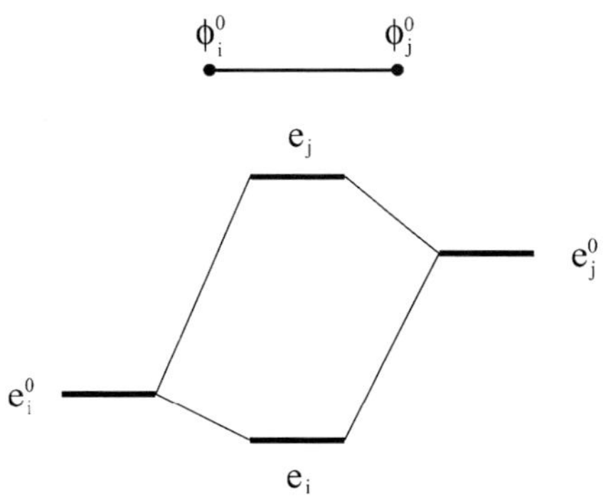

Fig. 26.2. Interaction of orbitals ϕ_i^0 and ϕ_j^0 with different energies ($e_i^0 < e_j^0 < 0$) leading to orbitals ϕ_i and ϕ_j.

$\tilde{H}_{ij}$ is negative. Then, ϕ_j^0 combines in-phase with ϕ_i^0 if the mixing coefficient t_{ji} is positive, but out of phase if t_{ji} is negative. Since $(e_i^0 - e_j^0) < 0$, Eq. (18) shows that

$$\begin{aligned} t_{ji} > 0, \quad \text{if } (\tilde{H}_{ji} - e_i^0 \tilde{S}_{ji}) < 0 \\ t_{ji} < 0, \quad \text{if } (\tilde{H}_{ji} - e_i^0 \tilde{S}_{ji}) > 0 \end{aligned} \tag{27}$$

Note that the third term on the right-hand side of Eq. (25) is negative (i.e. energy lowering) regardless of the sign of $(\tilde{H}_{ji} - e_i^0 \tilde{S}_{ji})$, i.e. regardless of whether ϕ_j^0 combines in phase or out of phase with ϕ_i^0. The case of $t_{ji} > 0$ describes the normal orbital interaction, and that of $t_{ji} < 0$ the counterintuitive orbital interaction.

To understand the difference between the normal and counterintuitive orbital interactions, it is necessary to consider an alternative expression of the MO energy e_i, i.e. $e_i = \langle \phi_i | H^{\text{eff}} | \phi_i \rangle$ which, after combining with Eq. (26), leads to the expression

$$e_i = [1 - 2\tilde{S}_{ij}t_{ji} - (t_{ji})^2]e_i^0 + \tilde{H}_{ii} + 2t_{ji}\tilde{H}_{ji} + (t_{ji})^2 e_j^0 \tag{28}$$

where the terms higher than the second order in $\tilde{S}_{ij}$, t_{ji} and $\tilde{H}_{ii}$ are neglected. It should be noted that Eq. (28) is equivalent to Eq. (25), although they look different. The terms of Eq. (28) whose signs depend on that of t_{ji} are $-2\tilde{S}_{ij}t_{ji}e_i^0$ and $2t_{ji}\tilde{H}_{ji}$. Their signs are as follows:

$$\begin{aligned} \text{For } t_{ji} > 0: \quad -2\tilde{S}_{ij}t_{ji}e_i > 0, \ 2t_{ji}\tilde{H}_{ji} < 0 \\ \text{For } t_{ji} < 0: \quad -2\tilde{S}_{ij}t_{ji}e_i < 0, \ 2t_{ji}\tilde{H}_{ji} > 0 \end{aligned} \tag{29}$$

That is, the in-phase orbital mixing in the normal bonding interaction accumulates electron density between ϕ_i^0 and ϕ_j^0 leading to energy lowering by the term $2t_{ji}\tilde{H}_{ji}$, and depletes electron density from the region of ϕ_i^0 leading to energy raising by the term $-2\tilde{S}_{ij}t_{ji}e_i^0$. The out-of-phase orbital mixing in the counterintuitive bonding interaction depletes electron density between ϕ_i^0 and ϕ_j^0 leading to energy raising by the term $2t_{ji}\tilde{H}_{ji}$, and accumulates electron density in the region of ϕ_i^0 leading to energy lowering by the term $-2\tilde{S}_{ij}t_{ji}e_i^0$.

The normal bonding interaction occurs when the energy difference between the two interacting orbitals is not large ($e_i^0 < e_j^0$). In such a case the extent of orbital contraction is similar in ϕ_i^0 and ϕ_j^0, so that it is energetically favorable to accumulate more electron density in the bonding region between ϕ_i^0 and ϕ_j^0 because the orbitals overlap efficiently. The counterintuitive bonding interaction occurs when the energy difference between the two interacting orbitals is very large ($e_i^0 \ll e_j^0$). In such a case the orbital ϕ_i^0 is contracted and the orbital ϕ_j^0 is diffuse so that it is energetically more preferable to accumulate more electron density in the contracted region of ϕ_i^0 (i.e. the lower potential region) because the contracted and diffuse orbitals do not overlap efficiently [7,8].

26.3 EFFECT OF ELECTRON–ELECTRON REPULSION

Electron–electron repulsion can have a profound effect on the electronic structure of a system. For a closed-shell system described by one Slater determinant, in which the up-spin and down-spin electrons of a given MO are restricted to have an identical spatial function, the effective one-electron Hamiltonian H^{eff} employed in Section 26.2 is given by the Fock operator [3]. When one Slater determinant is used to describe the electronic structure of an open-shell system, the up-spin and down-spin electrons are allowed to have different spatial functions. For a certain open-shell system (e.g. diradical), a proper description of its electronic structures even on a qualitative level requires the use of a configuration interaction (CI) wave function [6], i.e. a linear combination of Slater determinants. In this section, we probe how electron–electron repulsion affects the concepts of orbital interaction, orbital mixing and orbital occupation by considering a dimer that is made up of two identical sites with one electron and one orbital per site (Fig. 26.3).

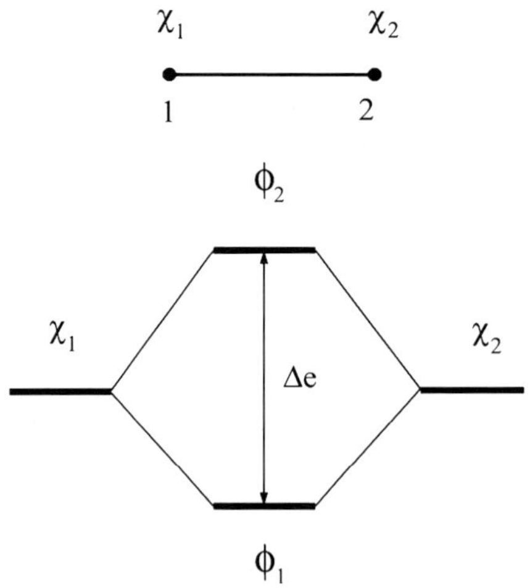

Fig. 26.3. Interaction between the orbitals χ_1 and χ_2 of a dimer with two equivalent sites leading to orbitals ϕ_1 and ϕ_2.

26.3.1 Configuration interaction

The interaction between the two site orbitals χ_1 and χ_2 of a dimer leads to two molecular levels ϕ_1 and ϕ_2 with energies e_1 and e_2, respectively (Fig. 26.3). The MOs ϕ_1 and ϕ_2 represent bonding and antibonding levels, respectively. Three electron configurations of interest for this dimer are the triplet configuration Φ_T and the singlet configurations Φ_1 and Φ_2 (Fig. 26.4).

When chemical bonding interaction between the spin sites is weak, the orbital energy difference Δe between the levels ϕ_1 and ϕ_2 is small, so that the two electrons of the dimer are considered as localized. In this case, the two singlet configurations Φ_1 and Φ_2 become close in energy, so the CI between Φ_1 and Φ_2 becomes strong and the ground singlet state Φ_S is described by the linear combination [6,9]

$$\Phi_S = d_1\Phi_1 - d_2\Phi_2 \tag{30}$$

where the mixing coefficients d_1 and d_2 have the same sign and are similar in magnitude (e.g. $d_1 = d_2 = 1\sqrt{2}$, if $\Delta e = 0$). Using this CI wave function for the singlet state, the electronic energy difference between the triplet and singlet electronic states, $\Delta E_e = E_T - E_S$, can be calculated under the assumptions that the orbitals ϕ_1 and ϕ_2 are determined from SCF calculations for the triplet state, and that these orbitals are linear combinations of two orthogonal atomic orbitals χ'_1 and χ'_2 (instead of non-orthogonal orbitals χ_1 and χ_2) located at the spin sites 1 and 2, respectively [9].

$$\phi_1 = \frac{1}{\sqrt{2}}(\chi'_1 + \chi'_2), \quad \phi_2 = \frac{1}{\sqrt{2}}(\chi'_1 + \chi'_2), \quad \langle \chi'_1 | \chi'_2 \rangle = 0 \tag{31}$$

Then the electronic energy difference ΔE_e between the singlet and triplet state is expressed as

$$\Delta E_e = E_S - E_T \approx 2K_{12} - \frac{(\Delta e)^2}{U_{11} - J_{12}} \tag{32}$$

where K_{12} is the exchange repulsion integral, and U_{11} and J_{12} are the Coulomb

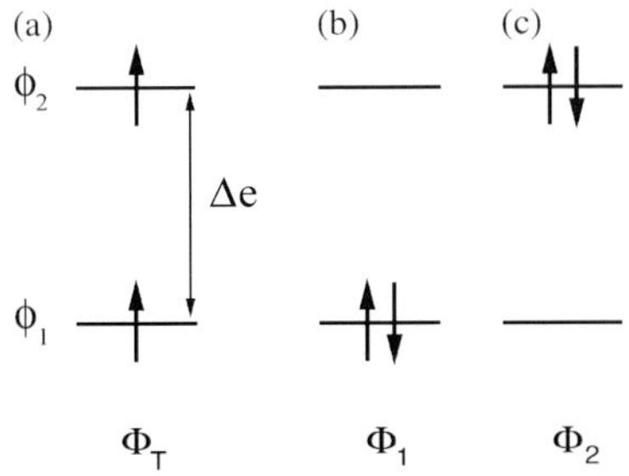

Fig. 26.4. Three electron configurations of a dimer with two equivalent sites and two electrons.

repulsion integrals, associated with the site orbitals χ'_1 and χ'_2

$$K_{12} = \langle 12|21\rangle, \quad U_{11} = \langle 11|11\rangle, \quad J_{12} = \langle 12|12\rangle \tag{33}$$

When there occurs a strong chemical bonding between the two sites of a dimer, the orbital energy difference $\Delta e = e_2 - e_1$ becomes large, which increases the energy difference $\Delta E = E_2 - E_1 > 0$ between the configurations Φ_2 and Φ_1, where E_2 and E_1 are the energies of the configurations and Φ_2 and Φ_1, respectively. Thus, using the non-degenerate perturbation theory, the ground singlet state Φ_S can be written as [6]

$$\Phi_S \approx \Phi_1 - \frac{K_{12}}{\Delta E}\Phi_2 \tag{34}$$

Thus, when Δe is large, the ground singlet state Φ_S is well approximated by Φ_1. In this case of large Δe, the singlet state Φ_S is more stable than the triplet state Φ_T, and the two electrons are regarded as delocalized [10].

26.3.2 States with different orbital occupancy

It is important to consider the effect of electron–electron repulsion on electronic configurations from the viewpoint of one-electron orbital theory. In the case of large Δe, the ground singlet state Φ_S is well approximated by the configuration Φ_1, in which the orbital ϕ_1 is doubly occupied (Fig. 26.4b). This occupancy is energetically favorable in terms of filling the lower lying bonding level, but unfavorable because the double occupancy gives rise to on-site repulsion $U_{11} = \langle 11|11\rangle$. The triplet configuration Φ_T, in which the orbitals ϕ_1 and ϕ_2 are each singly occupied (Fig. 26.4a), avoids the on-site repulsion at the expense of occupying the bonding and antibonding levels equally. The singlet state Φ_1 is more stable than the triplet state Φ_T if $2\Delta e > U_{11}$, and the reverse is true otherwise [10]

$$\begin{aligned} E_S - E_T < 0, &\quad \text{if } 2\Delta e > U_{11} \\ E_S - E_T > 0, &\quad \text{if } 2\Delta e < U_{11} \end{aligned} \tag{35}$$

The singlet configuration Φ_1 can be used to refer to the electron-delocalized state. In the case of small Δe, the ground singlet state Φ_S is represented by the CI wave function, Eq. (30). Although the configurations Φ_1 and Φ_2 each consist of doubly occupied orbitals, their weights in Φ_S are nearly the same when $\Delta e \approx 0$ so that the occupancy of the orbitals ϕ_1 and ϕ_2 orbital is essentially close to unity. In essence, this situation is similar to that of the triplet state Φ_T in which the orbitals ϕ_1 and ϕ_2 are each singly occupied. In this sense, the electron configuration Φ_T can be used to refer to the electron-localized state, regardless of whether the ground state is triplet or singlet [10].

It is important to recognize the solid-state counterpart of the above observations. Consider a one-dimensional (1D) chain with one electron and one orbital per site (Fig. 26.5a). If electron–electron repulsion is neglected, the levels of the bottom half of the band are each doubly filled, thereby leading to a metallic state (Fig. 26.5b). Non-spin-polarized electronic band structure calculations predict that a system with a half-filled

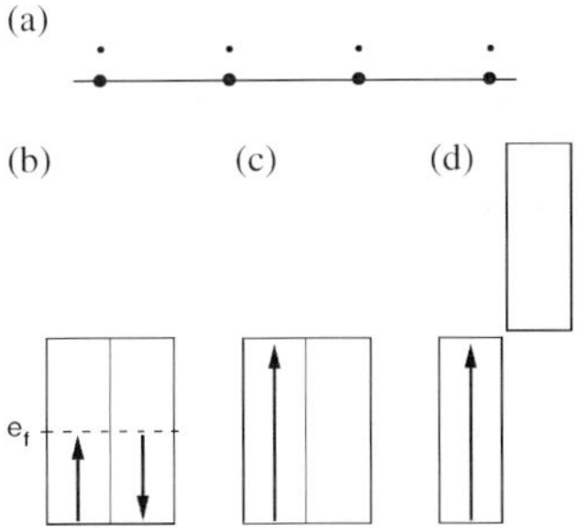

Fig. 26.5. Metallic and magnetic insulating states of a 1D chain with one electron and one orbital per site: (a) schematic representation of a 1D chain, (b) metallic state, (c) magnetic insulating state in non-spin-polarized representation, (d) magnetic insulating state in spin-polarized representation.

band is always metallic, which is obviously incorrect. By analogy with the dimer example discussed above, a magnetic insulating state can be represented by a band filling in which all the band levels are each singly filled with up-spin electrons (Fig. 26.5c) [10]. From the viewpoint of spin-polarized representation, the latter is equivalent to filling all the up-spin band levels singly while leaving all the down-spin band levels unoccupied. In spin-polarized electronic band structure calculations for this formally ferromagnetic insulating state, the up-spin and down-spin bands become split in energy (Fig. 26.5d) [11]. The metallic and magnetic insulating states are similar in that they possess a partially filled band, but they differ in the way the band levels are occupied. Given the bandwidth of the 1D chain as W, the metallic state is more stable than the magnetic insulating state if $W > U_{11}$, while the reverse is true if $W < U_{11}$ [10,12]

$$\begin{aligned} E_{\text{metal}} - E_{\text{insulator}} &< 0, \quad \text{if } W > U_{11} \\ E_{\text{metal}} - E_{\text{insulator}} &> 0, \quad \text{if } W < U_{11} \end{aligned} \tag{36}$$

The bandwidth W is equal to $2\Delta e$. Therefore, the delocalized and localized-electron states of a dimer are similar in nature to the metallic and magnetic insulating states of a solid, respectively [10].

26.3.3 Mapping between electronic and spin Hamiltonians [11]

The energy states of the dimer (Fig. 26.3) can be described by a phenomenological spin Hamiltonian, e.g. the Heisenberg Hamiltonian

$$\hat{H}_{\text{spin}} = -J\hat{S}_1 \cdot \hat{S}_2 \tag{37}$$

where $\hat{S}_1$ and $\hat{S}_2$ are the spin angular momentum operators at the sites 1 and 2, respectively, and J is the spin exchange parameter. The spin states to consider for this dimer are the singlet state $|S\rangle$ and the triplet states $|T\rangle$ shown below

$$|S\rangle = (|\uparrow\downarrow\rangle - |\downarrow\uparrow\rangle)/\sqrt{2}, \quad |T\rangle = |\uparrow\uparrow\rangle,\ |\downarrow\downarrow\rangle,\ (|\uparrow\downarrow\rangle + |\downarrow\uparrow\rangle)/\sqrt{2} \tag{38}$$

These spin states are eigenfunctions of the spin Hamiltonian $\hat{H}_{\text{spin}}$ such that

$$E_{\text{S}}^{\text{spin}} = \langle S|\hat{H}_{\text{spin}}|S\rangle = 3J/4, \quad E_{\text{T}}^{\text{spin}} = \langle T|\hat{H}_{\text{spin}}|T\rangle = -J/4 \tag{39}$$

Therefore, the energy difference between the singlet and triplet states is given by

$$\Delta E_{\text{spin}} = E_{\text{S}}^{\text{spin}} - E_{\text{T}}^{\text{spin}} = J \tag{40}$$

Thus, by equating ΔE_{spin} to the electronic energy difference ΔE_{e} calculated in Eq. (32), we obtain

$$J = \Delta E_{\text{e}} \approx 2K_{12} - \frac{(\Delta e)^2}{U_{11} - J_{12}} \tag{41}$$

The above mapping analysis reveals that the spin exchange parameter J consists of two terms of opposite signs, so it is convenient to write J as

$$J = J_{\text{F}} + J_{\text{AF}} \tag{42}$$

where the ferromagnetic term $J_{\text{F}}(>0)$ and the antiferromagnetic term $J_{\text{AF}}(<0)$ are

$$J_{\text{F}} = 2K_{12} \tag{43}$$

$$J_{\text{AF}} = -\frac{(\Delta e)^2}{U_{11} - J_{12}} \tag{44}$$

Eqs. (42)–(44) provide a basis for discussing spin exchange interactions in terms of chemical concepts such as symmetry and overlap [9,11,13]. In general, the exchange integral K_{12} is very small so that the spin exchange cannot be ferromagnetic (i.e. $J > 0$) unless the antiferromagnetic term J_{AF} either vanishes or is very small in magnitude by symmetry. For various spin exchange paths of a magnetic solid and for a series of closely related magnetic solids, the $(U_{11} - J_{12})$ term of J_{AF} is nearly constant and can be considered as the effective on-site repulsion U_{eff}. Thus, Eq. (44) is rewritten as

$$J_{\text{AF}} \approx -\frac{(\Delta e)^2}{U_{\text{eff}}} \tag{45}$$

and the variation of the J_{AF} values is mainly governed by that of the $(\Delta e)^2$ values. In the tight-binding approximation, the orbital energy difference Δe is related to the hopping integral β between the spin sites as

$$\Delta e = 2\beta = 2\langle \chi_1|H^{\text{eff}}|\chi_2\rangle \tag{46}$$

so that

$$J_{\text{AF}} \approx -\frac{4\beta^2}{U_{\text{eff}}} \tag{47}$$

If the orbitals χ_1 and χ_2 at the spin sites 1 and 2 are not orthogonal to each other, the hopping integral is proportional to the overlap integral S_{12} between them, i.e. $\beta \propto -S_{12}$, so that

$J_{AF} \propto -(\Delta e)^2 \propto -(S_{12})^2$. For an antiferromagnetic exchange interaction (i.e. $J < 0$), the J_{AF} term dominates over the J_F term because the latter is very small. Therefore, trends in antiferromagnetic exchange interactions can be discussed on the basis of the trends in the corresponding J_{AF} values [11]. The latter can be estimated by calculating the associated Δe or β values. Note that J_{AF} is a quantity resulting from electron correlation (i.e. CI), while β is a one-electron integral with no electron correlation. Nevertheless, it is interesting that the trends in J_{AF} can be reproduced by those in β.

An alternative mapping between the electronic and spin Hamiltonians can be carried out by employing broken-symmetry (BS) states [14–16]. The triplet states $|\uparrow\uparrow\rangle$ and $|\downarrow\downarrow\rangle$ of the dimer are referred to as the high spin states $|\mathrm{HS}\rangle$, and the Neél states $|\uparrow\downarrow\rangle$ and $|\downarrow\uparrow\rangle$ as the broken-symmetry states $|\mathrm{BS}\rangle$

$$|\mathrm{HS}\rangle = |\uparrow\uparrow\rangle, |\downarrow\downarrow\rangle, \qquad |\mathrm{BS}\rangle = |\uparrow\downarrow\rangle, |\uparrow\uparrow\rangle \tag{48}$$

where the $|\mathrm{BS}\rangle$ states are not an eigenfunction of the spin Hamiltonian $\hat{H}_{\mathrm{spin}}$. The expectation values of these states are given by

$$E_{\mathrm{BS}}^{\mathrm{spin}} = \langle \mathrm{BS}|\hat{H}_{\mathrm{spin}}|\mathrm{BS}\rangle = J/4, \qquad E_{\mathrm{HS}}^{\mathrm{spin}} = \langle \mathrm{HS}|\hat{H}_{\mathrm{spin}}|\mathrm{HS}\rangle = -J/4 \tag{49}$$

so that

$$\Delta E'_{\mathrm{spin}} = E_{\mathrm{BS}}^{\mathrm{spin}} - E_{\mathrm{HS}}^{\mathrm{spin}} = J/2 \tag{50}$$

Thus, when the corresponding electronic energy difference $\Delta E'_{\mathrm{e}}$ between the $|\mathrm{BS}\rangle$ and $|\mathrm{HS}\rangle$ states is evaluated by appropriate electronic structure calculations, the value of J is determined by

$$\Delta E'_{\mathrm{e}} = J/2 \tag{51}$$

Typically, the $\Delta E'_{\mathrm{e}}$ values are determined on the basis of density functional theory (DFT) calculations.

26.3.4 Spin polarization

This phenomenon is best illustrated by considering a π radical system (Fig. 26.6a), in which the C–H bond lies on the nodal plane of the carbon p_π orbital that carries some unpaired spin density. As depicted in Fig. 26.6b, the local electronic structure around the carbon center can be described in terms of the σ_{CH} and σ^*_{CH} orbitals (ϕ_1 and ϕ_3, respectively) of the C–H bond and the p_π orbital (ϕ_2) of the carbon atom. To a first approximation, the up- and down-spin electrons in σ_{CH} are regarded to have an identical spatial function, which leads to no unpaired spin density on the H atom. From the viewpoint of perturbation theory, this corresponds to the 'unperturbed' electronic structure. However, this picture is not complete, because EPR experiments show the presence of a small amount of unpaired spin density on the H atom.

The up-spin electrons of the orbitals ϕ_1 and ϕ_2 stabilize the system by $-K_{12}$, where K_{12} is the exchange integral between ϕ_1 and ϕ_2. The down-spin electron of ϕ_1 and the up-spin electron of ϕ_2 lead to the Coulomb repulsion J_{12}. Thus, the energy

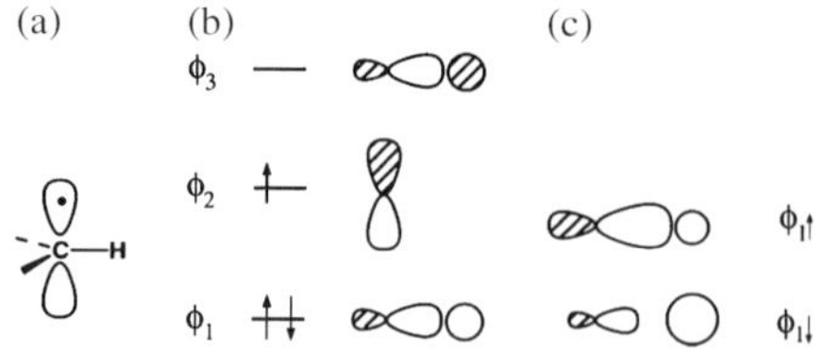

Fig. 26.6. (a) A local environment of an organic π radical system. (b) The C–H sigma bonding orbital ϕ_1, the p_π orbital ϕ_2, and the sigma antibonding orbital ϕ_3 in the absence of spin polarization. (c) The $\phi_{1\uparrow}$ and $\phi_{1\downarrow}$ orbitals in the presence of spin polarization.

of the radical system can be lowered by enhancing K_{12} and reducing J_{12}. The exchange integral K_{12} is the self-repulsion associated with the overlap density $\phi_1\phi_2$. The ϕ_1 orbital is an in-phase combination of C sp^2 and H 1s orbitals, while the ϕ_3 orbital is an out-of-phase combination between them (Fig. 26.6b). At the carbon center the p_π orbital is closer to the C sp^2 hybrid than to the H 1s orbital so that the overlap density $\phi_1\phi_2$ (and hence the associated exchange integral K_{12}) is increased by increasing the weight of C sp^2 in the up-spin orbital $\phi_{1\uparrow}$. From the viewpoint of perturbation theory, it is a perturbation to relax the constraint that $\phi_{1\uparrow}$ and $\phi_{1\downarrow}$ have an identical spatial function. The spatial function of low-lying orbital ϕ_1 can be modified by mixing it with a small amount of the high-lying orbital ϕ_3 (i.e. the orbitals ϕ_1 and ϕ_3 interact under non-degenerate perturbation). To increase the weight of C sp^2 in the up-spin orbital $\phi_{1\uparrow}$, ϕ_3 should be combined with ϕ_1 as

$$\phi_{1\uparrow} = \frac{\phi_1 + \lambda\phi_3}{\sqrt{1+\lambda^2}} \tag{52}$$

where λ is a small positive mixing coefficient ($0 < \lambda \ll 1$). The Coulomb repulsion J_{12} is the repulsion between the charge density distributions $\phi_1\phi_1$ and $\phi_2\phi_2$. J_{12} can be reduced by decreasing the weight of C sp^2 in the down-spin orbital $\phi_{1\downarrow}$, which is achieved by mixing a small amount of ϕ_3 into ϕ_1 as

$$\phi_{1\downarrow} = \frac{\phi_1 - \kappa\phi_3}{\sqrt{1+\kappa^2}} \tag{53}$$

where κ is a small positive mixing coefficient ($0 < \kappa \ll 1$). As a consequence, the spatial functions of $\phi_{1\uparrow}$ and $\phi_{1\downarrow}$ are modified such (Fig. 26.6c) that the hydrogen atom has more down-spin density than up-spin density, resulting in a small amount of unpaired down-spin density on it. Namely, the up-spin electron in the carbon p_π orbital polarizes the distribution of the up- and down-spin electrons in the C–H bond to have a small net down-spin density on the H atom.

26.3.5 Non-equivalent orbital interactions in an open-shell system [17]

Electronic structures of crystalline solids are mostly calculated on the basis of DFT. In this approach an open-shell system is described by spin polarized electronic band structures, in which the up-spin and down-spin bands are allowed to have different orbital compositions. The difference between the up-spin and down-spin bands can be quite substantial in the d-block bands of solids containing high-spin transition metal cations, as discussed [17] for $CaFeO_3$ in which the high spin Fe^{4+} (d^4) cations in octahedral sites have the $(t_{2g})^3(e_g)^1$ configuration. In this section, we discuss why the up-spin and down-spin levels of an open-shell system have non-equivalent orbital interactions by considering $CaFeO_3$ as an example.

In the occupied up-spin e_g bands of $CaFeO_3$, the Fe 3d and the O 2p orbitals contribute almost equally, which implies that the Fe 3d and the O 2p levels are nearly the same for the up-spin bands. In the unoccupied down-spin e_g bands of $CaFeO_3$, the Fe 3d orbital contribution is greater than the O 2p orbital contribution, which implies that the Fe 3d level lies considerably above the O 2p level for the down-spin bands. In other words, the Fe–O bonds are more covalent in the up-spin bands than in the down-spin bands [17].

To account for these observations, it is necessary to examine how orbital interactions are affected by electron–electron repulsion and in what way the effect is different for up-spins and down-spins. Consider that the energies of Fe 3d and O 2p levels in the absence of electron–electron repulsion are given as depicted in Fig. 26.7a ($e_d^0 < e_p^0$), which shows four singly filled Fe 3d levels and a doubly filled 2p orbital of a ligand O atom. When electron–electron repulsion is taken into consideration [18–20], the up-spin and down-spin levels of the Fe 3d and O 2p orbitals are expressed as (Fig. 26.7b)

$$\begin{aligned} e_{p\uparrow} &= e_p^0 + U_{pp} + 4J_{pd} - 4K_{pd} \\ e_{d\uparrow} &= e_d^0 + 3J_{dd} - 3K_{dd} + 2J_{pd} - K_{pd} \\ e_{p\downarrow} &= e_p^0 + U_{pp} + 4J_{pd} \\ e_{d\downarrow} &= e_d^0 + 4J_{dd} + 2J_{pd} - K_{pd} \end{aligned} \tag{54}$$

where U_{pp} is the on-site Coulomb repulsion for the O 2p orbital, J_{dd} is the Coulomb repulsion between two different Fe 3d orbitals, and J_{dp} is the Coulomb repulsion between a Fe 3d and the O 2p orbital. Therefore, for the up-spin and down-spin electrons, the energy gaps between the Fe 3d and the O 2p levels are written as

$$\begin{aligned} e_{d\uparrow} - e_{p\uparrow} &= (e_d^0 - e_p^0) + (-3K_{dd} + 3K_{pd}) + (3J_{dd} - U_{pp} - 2J_{pd}) \\ e_{d\downarrow} - e_{p\downarrow} &= (e_d^0 - e_p^0) + (J_{dd} - K_{pd}) + (3J_{dd} - U_{pp} - 2J_{pd}) \end{aligned} \tag{55}$$

Consequently, the energy gap between the Fe 3d and the O 2p levels for down-spin electrons is larger than that for the up-spin electrons by

$$(e_{d\downarrow} - e_{p\downarrow}) - (e_{d\uparrow} - e_{p\uparrow}) = J_{dd} + 3K_{dd} - 4K_{pd} \approx U_{dd} + K_{dd} - 4K_{pd} \tag{56}$$

where U_{dd} is the on-site Coulomb repulsion for two electrons in a metal 3d level and is approximately related to J_{dd} and K_{dd} as $J_{dd} \approx U_{dd} - 2K_{dd}$ [20]. In general, the magnitudes of the various Coulomb and exchange integrals are expected to decrease as $U_{dd} > U_{pp} > J_{dd} > J_{dp} > K_{dd} > K_{dp}$. It is clear from Eq. (56) that, due to electron–electron repulsion, the energy gap between the Fe 3d and the O 2p levels

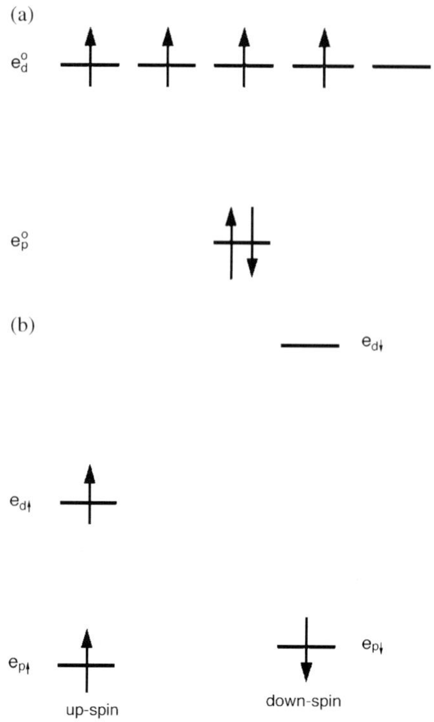

Fig. 26.7. Effect of electron–electron repulsion on the energy gap associated with the orbital interaction of a doubly filled O 2p orbital with partially filled 3d orbitals of Fe^{4+} (d^4) cation. (a) The energy levels of the Fe 3d and O 2p orbitals in the absence of electron–electron repulsion, and (b) those in the presence of electron–electron repulsion.

becomes larger for the down-spin orbital interactions than for the up-spin orbital interactions. This explains why, for $CaFeO_3$, the covalency of the Fe–O bond is greater in the up-spin than in the down-spin bands.

26.3.6 Orbital ordering in magnetic solids

Electron–electron repulsion has an important effect on the arrangement of singly filled orbitals between adjacent spin sites (i.e. orbital ordering) [11,21–23]. Suppose that the sites 1 and 2 of a dimer (Fig. 26.3) are represented by the non-orthogonal orbitals χ_1 and χ_2, respectively. If the two sites are equivalent, the energies of the orbitals χ_1 and χ_2 are the same (i.e. e_1^0). The interaction between the two orbitals gives rise to the bonding level e_1 and the antibonding level e_2 such that the stabilization of the bonding level is smaller in magnitude than the destabilization of the antibonding level, i.e. $(e_2 - e_1^0) > (e_1^0 - e_1)$ (Fig. 26.8a) [6]. As discussed in the previous section, the delocalized state (Fig. 26.8b) is more stable than the localized state (Fig. 26.8c) if $U_{11} < 2\Delta e$ whereas the reverse is true

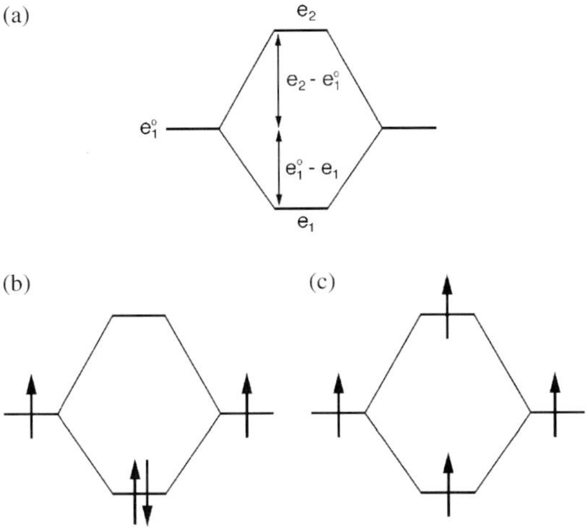

Fig. 26.8. Orbital occupations of the electron-delocalized and electron-localized states of a dimer consisting of one electron and one orbital per site, where the two sites are equivalent: (a) orbital interaction between the two sites leading to the bonding and antibonding levels of the dimer, (b) orbital occupation of the electron-delocalized state, (c) orbital occupation of the electron-localized state.

if $U_{11} > 2\Delta e$ [10]. In the delocalized state, the two-electron two-orbital interaction stabilizes the system by $\Delta E = (2e_1 - 2e_1^0) \propto -S_{12}$, which is enhanced by increasing the overlap [6]. For the localized state, the two-electron two-orbital interaction destabilizes the system by $\Delta E = (e_2 + e_1) - 2e_1^0 \propto (S_{12})^2$, which is reduced by decreasing the overlap [6]. As an example, consider the arrangements of two singly filled p orbitals. When $U_{11} < 2\Delta e$, the stabilization is maximum in the parallel arrangement (Fig. 26.9a) but vanishes in the orthogonal arrangement (Fig. 26.9b). When $U_{11} > 2\Delta e$, however, the destabilization is maximum in the parallel arrangement (Fig. 26.9a) and vanishes in the orthogonal arrangement (Fig. 26.9b). The orbital ordering phenomenon is observed in magnetic solids of transition metal elements when their magnetic ions have partially filled degenerate d-block levels. This phenomenon manifests the fact that in electron-localized states (i.e. $U_{11} > 2\Delta e$) a net destabilization results from a two-electron two-orbital interaction, and this destabilization is reduced by decreasing the overlap between the two orbitals.

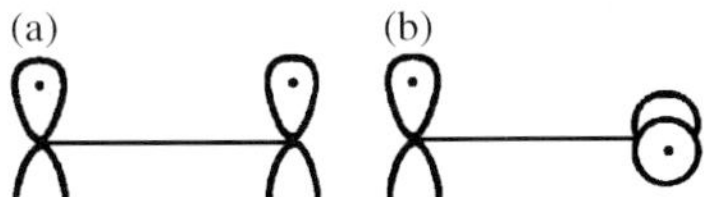

Fig. 26.9. Arrangements of two singly filled p orbitals of a dimer: (a) parallel arrangement, (b) perpendicular arrangement.

26.4 SPIN–ORBIT COUPLING AND ORBITAL MIXING

In relativistic theory a wave function has four components, two large components and two small components. For simplicity, we consider only the two large components of an atomic p orbital. Due to spin–orbit coupling brought about by the relativistic effect, the p orbitals of an atom are given by the atomic $p_{1/2}$ and $p_{3/2}$ orbitals [24–26]

$$P_{1/2}(m_s = 1/2): \ \phi_a \propto \begin{bmatrix} -\sqrt{1/3}p_z \\ -\sqrt{1/3}(p_x + ip_y) \end{bmatrix} \tag{57}$$

$$P_{3/2}(m_s = 1/2): \ \phi_b \propto \begin{bmatrix} \sqrt{2/3}p_z \\ -\sqrt{1/6}(p_x + ip_y) \end{bmatrix} \tag{58}$$

$$P_{3/2}(m_s = 3/2): \ \phi_c \propto \begin{bmatrix} \sqrt{1/2}(p_x + ip_y) \\ 0 \end{bmatrix} \tag{59}$$

where the p_x, p_y and p_z are the p orbitals in the absence of spin–orbit coupling. Suppose a homonuclear diatomic molecule with the z-axis taken along the internuclear axis. For simplicity, we consider relativistic wave functions of a dimer resulting only from the $p_{1/2}$ and $p_{3/2}$ orbitals. By identifying the two sites of a dimer with the indices 1 and 2, the 'equal-weight' linear combinations of the relativistic atomic orbitals ϕ_a, ϕ_b and ϕ_c can be written as

$$\phi_a(1) + \phi_a(2) \propto \begin{bmatrix} -\sqrt{1/3}\sigma_z^* \\ -\sqrt{1/3}(\pi_x + i\pi_y) \end{bmatrix} \tag{60a}$$

$$\phi_a(1) - \phi_a(2) \propto \begin{bmatrix} -\sqrt{1/3}\sigma_z \\ -\sqrt{1/3}(\pi_x^* + i\pi_y^*) \end{bmatrix} \tag{60b}$$

$$\phi_b(1) + \phi_b(2) \propto \begin{bmatrix} \sqrt{2/3}\sigma_z^* \\ -\sqrt{1/6}(\pi_x + i\pi_y) \end{bmatrix} \tag{61a}$$

$$\phi_b(1) - \phi_b(2) \propto \begin{bmatrix} \sqrt{2/3}\sigma_z \\ -\sqrt{1/6}(\pi_x^* + i\pi_y^*) \end{bmatrix} \tag{61b}$$

$$\phi_c(1) + \phi_c(2) \propto \begin{bmatrix} \sqrt{1/2}(\pi_x + i\pi_y) \\ 0 \end{bmatrix} \tag{62a}$$

$$\phi_c(1) - \phi_c(2) \propto \begin{bmatrix} \sqrt{1/2}(\pi_x^* + i\pi_y^*) \\ 0 \end{bmatrix} \tag{62b}$$

where σ and σ^* represent the sigma-bonding and sigma-antibonding orbitals of the dimer, respectively, in the absence of spin–orbit coupling. Likewise, π and π^* represent the pi-bonding and pi-antibonding orbitals of the dimer, respectively, in the absence of spin–orbit coupling. The relativistic orbitals of a dimer such as Bi_2^{2-} calculated by relativistic DFT calculations are composed of the orbitals listed in Eqs. (60)–(62) [27].

Eqs. (60) and (61) reveal that some relativistic dimer orbitals have the character of both σ and π^* orbitals, and others have the character of both σ^* and π orbitals. Namely, the spin–orbit coupling brought about by the relativistic effect causes orbital mixing between the sigma and pi orbitals of a non-relativistic Hamiltonian. It is of interest to see how this orbital mixing comes about. The orbital angular momentum of a diatomic molecule along the internuclear axis is designated by λ. Namely, $\lambda = 0$ for σ and σ^* orbitals, and $\lambda = 1$ for π and π^* orbitals. The coupling of the spin angular momentum and the angular momentum λ leads to a net angular momentum denoted by $\omega = |\lambda \pm 1/2|$ for a spin-1/2 electron. Therefore, under spin–orbit coupling, the σ and σ^* ($\lambda = 0$) levels become $\omega = 1/2$ levels, whereas the π and π^* ($\lambda = 1$) levels split into $\omega = 1/2$ and $\omega = 3/2$ levels. The levels of an identical ω value can interact and hence mix together, if their symmetry (gerade or ungerade) is the same. Thus, the gerade $\omega = 1/2$ levels combine σ and π^* orbitals (Eqs. (60b) and (61b)), while the ungerade $\omega = 1/2$ levels combine σ^* and π orbitals (Eqs. (60a) and (61a)). The σ and σ^* orbitals with $\omega = 1/2$ do not mix due to their difference in symmetry. For the same reason, the π and π^* orbitals with $\omega = 3/2$ do not mix (Eqs. (62a) and (62b)).

26.5 CONCLUDING REMARKS

As briefly surveyed in this chapter, the concepts of perturbation, orbital interaction, orbital mixing and orbital occupation work at all levels of electronic structure descriptions. These qualitative concepts provide a conceptual framework in which to rationalize experimental/theoretical observations and to generate qualitative predictions that can be tested by further experimental/theoretical studies.

26.6 ACKNOWLEDGEMENTS

This work was supported by the Office of Basic Energy Sciences, Division of Materials Sciences, US Department of Energy, under Grant DE-FG05-86ER45259.

26.7 REFERENCES

1 R. Hoffmann, Chem. Eng. News, 52 (1974) 32, No. 30, July 29.
2 M.-H. Whangbo, Theor. Chem. Acc., 103 (2000) 252.

3 M.-H. Whangbo, H.B. Schlegel and S. Wolfe, J. Am. Chem. Soc., 99 (1977) 1296.
4 M.-H. Whangbo, Computational theoretical organic chemistry, Reidel, Dordrecht, The Netherlands, 1981, p. 233.
5 R. Hoffmann, J. Chem. Phys., 39 (1963) 1397.
6 T.A. Albright, J.K. Burdett and M.-H. Whangbo, Orbital interactions in chemistry, Wiley, New York, 1985.
7 M.-H. Whangbo and R. Hoffmann, J. Chem. Phys., 68 (1978) 5498.
8 J.H. Ammeter, H.-B. Bürgi, J. Thibeault and R. Hoffmann, J. Am. Chem. Soc., 100 (1978) 3686.
9 P.J. Hay, J. Thibeault and R. Hoffmann, J. Am. Chem. Soc., 97 (1975) 4884.
10 M.-H. Whangbo, J. Chem. Phys., 70 (1979) 4963.
11 M.-H. Whangbo, H.-J. Koo and D. Dai, J. Solid State Chem., 176 (2003) 417.
12 N.F. Mott, Metal–insulator transitions, 2nd edn., Taylor & Francis, New York, 1990.
13 O. Kahn, Molecular magnetism, VCH, New York, 1993.
14 L. Noodleman, J. Chem. Phys., 74 (1981) 5737.
15 D. Dai and M.-H. Whangbo, J. Chem. Phys., 114 (2001) 2887.
16 D. Dai and M.-H. Whangbo, J. Chem. Phys., 118 (2003) 29.
17 M.-H. Whangbo, H.-J. Koo, A. Villesuzanne and M. Pouchard, Inorg. Chem., 41 (2002) 1920.
18 B.H. Brandow, Adv. Phys., 26 (1977) 651.
19 M.-H. Whangbo, Inorg. Chem., 19 (1980) 1728.
20 M. Pouchard, A. Villesuzanne and J.-P. Doumerc, J. Solid State Chem., 162 (2001) 282.
21 K.I. Kugel and D.I. Khomskii, Sov. Phys. Usp., 25 (1982) 231.
22 K. Terakura, J. Lee, J. Yu, I.V. Solovyev and H. Sawada, Mater. Sci. Eng. B, 63 (1999) 11.
23 M.-H. Whangbo and H.-J. Koo, Solid State Sci., 4 (2002) 335.
24 R.E. Powell, J. Chem. Ed., 45 (1969) 558.
25 K.S. Pitzer, Acc. Chem. Res., 12 (1979) 271.
26 P. Pyykkö and J.-P. Desclaux, Acc. Chem. Res., 12 (1979) 276.
27 D. Dai, M.-H. Whangbo, A. Ugrinov, S.C. Sevov, F. Wang, L. Li, A. Villesuzanne, A.B. Alekseyev, H.-P. Liebermann and R.J. Buenker, J. Phys. Chem. A, 109 (2005) 1675.

Theory and Applications of Computational Chemistry: The First Forty Years
Edited by C. Dykstra et al.

CHAPTER 27

G2, G3 and associated quantum chemical models for accurate theoretical thermochemistry

Krishnan Raghavachari[1] and Larry A. Curtiss[2]

[1] *Department of Chemistry, Indiana University, Bloomington, IN 47405, USA*
[2] *Materials Science and Chemistry Divisions, Argonne National Laboratory, Argonne, IL 60439, USA*

Abstract

We present a brief review of G2 and G3 theories which are composite techniques for the accurate prediction of experimental thermochemical data for molecules. We discuss the components of G2 and G3 theories as well as approximate versions such as G2(MP2), G3(MP2) and G3(MP3). Additional methods such as extended G3 theory (G3X) as well as scaled G3 theory (G3S) are also discussed. The methods are assessed on the comprehensive G2/97 and G3/99 test sets of experimental energies (heats of formation, ionization energies, electron affinities and proton affinities) that we have assembled. The most accurate method, G3X, has a mean absolute deviation of 0.95 kcal/mol from experiment for the 376 energies in the G3/99 test set. Some illustrative applications of the methods to resolve experimental data for other systems are also discussed.

27.1 INTRODUCTION

The last two decades have witnessed enormous progress in the development and application of *ab initio* electronic structure techniques. The energies of small molecules and ions can now be routinely evaluated to an accuracy of better than 1 kcal/mol ('chemical accuracy') and the computation of the energies of larger molecules to a similar accuracy is increasingly becoming feasible [1–4]. In addition to the enormous impact from the ever-increasing hardware performance of high-speed computers, two important areas of theoretical development have contributed to this progress. The first is the development and implementation of accurate electron correlation methods such as augmented coupled cluster theory [5,6] and the associated computational models [7]. The second is the understanding of the effects of large basis sets and their convergence behavior with respect to the inclusion of higher angular momentum functions [8]. Together they have led to several successful approaches for the accurate evaluation of thermochemical quantities from first principles [9–66].

Even though the demands placed on the theoretical methods are well understood, the practical realization of chemical accuracy for more than very small molecules is still a formidable task. This is due to the high scaling of the cost of accurate calculations with molecular size. For example, it is known that we can achieve high accuracy with coupled cluster calculations such as CCSDTQ (coupled cluster with single, double, triple, and quadruple excitations) [13]. However, such a calculation is iterative and scales as the 10th power of molecular size ($\sim N^{10}$) and is not practical, particularly in conjunction with large basis sets. Somewhat simpler CCSDT calculations [7,12] are still iterative and scale as $\sim N^8$ and are also impractical for most systems. The practical success in the development of thermochemical models came with the development of perturbative (non-iterative) evaluation of the effects of triple excitations [5,6] (N^7 scaling) subsequent to an iterative evaluation of the effects of single and double excitations (CCSD or quadratic configuration interaction, QCISD). The resulting models, denoted as CCSD(T) [5] and QCISD(T) [6], appear to provide the best compromise between accuracy and computational feasibility. In particular, the CCSD(T) method has been successfully used with progressively larger basis sets containing high angular momentum functions [9–19]. The most commonly used basis sets in this approach belong to the correlation-consistent family developed by Dunning et al. [8]. After explicit calculations are carried out to the extent possible (typically up to quintuple zeta basis sets), the results are then extrapolated to the complete basis set (CBS) limit [14,15] and corrected for some smaller effects such as core-valence and relativistic effects. Such *ab initio* methods without using any parameters have been used successfully for molecules up to the size of benzene [18]. Examples of this approach include the W1 and W2 models of Martin and co-workers [16,17]. In addition, higher order effects of electron correlation have been considered carefully in the recently proposed W3 model [19], which targets kJ/mol accuracy.

An alternative successful strategy that makes it possible to approach chemical accuracy for significantly larger molecules involves a composite multilevel approach typically combined with a small number of empirical parameters. The Gaussian-n series [20–45] that we have developed exploits this idea to predict thermochemical data for molecules containing first- and second-row elements. Our objective in these studies was to develop a general predictive procedure, applicable to any molecular system in an unambiguous manner, that can reproduce known experimental data to a prescribed accuracy of ± 2 kcal/mol (somewhat less ambitious than 'chemical accuracy'), and can be applied with similar accuracy to other species where the experimental data are unknown or uncertain. In our approach a suite of methods with different levels of accuracy is used with practical basis sets to attempt to approach the exact result. For example, high-level correlation calculations (e.g. QCISD(T), CCSD(T)) with moderate-sized basis sets can be combined with results from lower level calculations (e.g. MP4, MP2 or even HF) using larger basis sets to approximate the results of more expensive calculations. In addition, molecule-independent empirical parameters are used in these methods to estimate the remaining deficiencies in the calculations. This avoids the use of very large basis sets containing high angular momentum functions and such an approach using 'higher level corrections (HLCs)' (additive parameters that depend on the number of paired and unpaired electrons in the system) has been quite successful and the latest version, Extended Gaussian-3 (G3X) theory [42], achieves an overall accuracy of better than 1 kcal/mol for the G3/99 test set [37]. Petersson et al. [49–56] have developed a related series of methods, referred to as CBS methods, for the evaluation of accurate energies of molecular systems. The central idea in the CBS methods is an extrapolation procedure to determine the projected second-order (MP2) energy in the limit of a complete basis set. Several empirical corrections, similar in spirit to the HLC used in the Gaussian-n series, are added to the resulting energies in the CBS methods to remove systematic errors in the calculations. Another approach for calculating thermochemical data that has been proposed is a scaling of the calculated correlation energy using multiplicative parameters [57–66] determined by fitting to experimental data. This work has been pioneered by Truhlar and co-workers who have derived a series of methods using this approach. Finally, hybrid density functionals [67] are being used increasingly to predict the thermochemistry of molecules with reasonable accuracy [1–4].

In this chapter, we review the elements of G2 and G3 theories and related techniques to computational thermochemistry. An important part of the development of such quantum chemical methods is their critical assessment on test sets of accurate experimental data. Section 27.2 provides a brief description of the comprehensive G3/99 test set [37] of experimental data that we have collected. Sections 27.3 and 27.4 discuss the components of G2 and G3 theories as well as the approximate versions such as G2(MP2), G3(MP2) and G3(MP3) and their performance for the G3/99 test set. Section 27.5 discusses the G3X method [42] that corrects for some of the deficiencies of G3 theory for larger molecules. The G3S method [41] that includes multiplicative scale factors is discussed in Section 27.6 along with other related variants. Extensions of G3 theory to the third-row elements is discussed in Section 27.7. Finally, a summary is presented in Section 27.9.

27.2 THERMOCHEMICAL TEST SETS

A key to the development and assessment of new composite quantum chemical methods for predicting accurate thermochemistry is the availability of test sets containing reliable experimental data. An ideal test set should comprise accurate experimental values of heats of formation for a wide variety of molecules in many different bonding environments. It should contain experimental values for molecules with light as well as heavy elements, main group elements as well as transition metals. However, the number of molecules having highly accurate heats of formation, within chemical accuracy (± 1 kcal/mol), is small and most of the available data are dominated by molecules containing the first- and second-row elements. In this review, we restrict most of our discussion to test sets of molecules of this type.

In our own work, we have assembled three such test sets of good, credible experimental data to perform such assessments [23,35–37]. The original small molecule test set (now referred to as G2-1 set) includes the atomization energies for 55 neutral molecules containing 1–3 heavy atoms (systems such as H_2O, C_2H_4, CO_2 and SO_2) [23]. While small, it has nevertheless proved to be very useful in deriving parameters needed for the development of new thermochemical models. In fact, the parameters in the popular B3LYP density functional [67] were derived by fitting to the molecules in this test set. The test set was augmented in 1997 to 148 molecules by including medium-sized molecules containing 3–6 heavy atoms (systems such as C_3H_6, C_4H_4O, C_6H_6, etc.) [35]. We also added other thermodynamic quantities such as ionization energies, electron affinities and proton affinities to this subset (now referred to as G2-2) [36]. The two subsets, G2-1 and G2-2, are together referred to as G2/97 and contain 301 test energies [35,36]. An additional 75 molecules (referred to as G3-3 subset) were added in 1999. The G3-3 subset [37] contains 75 enthalpies of formation for molecules that are, on average, larger (containing 3–10 heavy atoms). The largest molecule in the G3-3 subset contains 10 non-hydrogen atoms (e.g. naphthalene or azulene). It also contains some larger hypervalent molecules such as PF_5 or SF_6 that provide a challenge for many theoretical models. The cumulative current test set, referred to as G3/99 [37], contains 376 energies (222 enthalpies of formation, 88 ionization energies, 58 electron affinities and 8 proton affinities) that are known experimentally [68–71] to an accuracy of better than ± 1 kcal/mol.

The 222 enthalpies of formation included in the G3/99 test set correspond to a wide variety of molecules in many bonding environments. They are conveniently classified into subgroups of molecules in order to identify trends within the different groups. They include 47 non-hydrogen-containing molecules, 38 hydrocarbons, 91 substituted hydrocarbons, 15 inorganic hydrides and 31 open-shell radicals. Overall, the test set is dominated by organic molecules, while the number of molecules containing metallic elements is relatively modest. However, they provide a comprehensive assessment of new theoretical models in a wide variety of bonding environments.

Additional test sets have been derived by other authors. For example, Cioslowski et al. [72] have compiled a set of 600 enthalpies of formation for a broad range of molecules up to the size of C_{60}. This compilation includes many molecules containing Li, Be, B, Na, Mg and Al where the experimental uncertainties are larger. In addition, it includes many large molecules such as polycondensed benzenoid hydrocarbons and relatively unstable molecules such as carbenes and molecules with dative bonds where the inherent uncertainties are larger. Though significantly more diverse, the presence of significant experimental uncertainties makes it difficult to evaluate the performance of theoretical methods in such cases. An alternative procedure may be to use weighting factors in cases where the paucity of accurate experimental data makes assessment difficult.

Another area where significant work has been done is in the prediction of activation barriers of chemical reactions. Truhlar and co-workers have assembled a set of 44 barrier heights [65] for assessment of the performance of theoretical methods for 'thermo-chemical kinetics'. Finally, a limited set of thermochemical information has been collected for molecules containing third-row main group elements and for transition metal systems [27–29,73,74].

The collection of such a large test set of experimental data is indeed a formidable challenge. All the experimental values that are included in the G3/99 test set have a quoted uncertainty of less than 1 kcal/mol [35–37]. However, a critical evaluation of the experimental uncertainties for all these species is a daunting task. Cioslowski et al. [72] have discussed the possible sources of errors in such thermochemical compilations. It is possible (and perhaps likely) that some of the values included in our G3/99 test set may turn out to be incorrect. For example, the G2/97 test set originally contained 302 energies, but the enthalpy of formation of COF_2 was deleted because a new experimental upper limit [71] cast doubt on the original value used. While there are a few other cases where the experimental data may not be as accurate as cited, in our analysis we have chosen not to throw out experimental data unless there is new *experimental* evidence that warrants it. In our opinion, it is important not to use any theoretical input in the collection of experimental test sets in order to avoid circular arguments and to have an unbiased assessment process [44]. If the accuracy of theory improves and becomes demonstrably better than that of experiment, theoretical values may be included in the future to assemble test sets of molecules for critical assessment.

27.3 G2 THEORY

The Gaussian-n theories are composite techniques in which a sequence of well-defined *ab initio* molecular orbital calculations [75] is performed to arrive at a total energy of a given molecular species. G2 theory [23] was the second-generation model and was developed to correct significant non-additivity problems of an earlier model (G1 theory) [21,22]. G2 theory and several approximate versions have been extensively applied to a variety of problems. The detailed steps involved in G2 theory are described in the following paragraphs with some editorial comments. The calculations that are needed to derive the G2 energy are also shown pictorially in Fig. 27.1(a).

(1) An initial equilibrium structure is obtained at the Hartree–Fock (HF) level with the 6-31G(d) basis [75]. Spin-restricted (RHF) theory is used for singlet states and spin-unrestricted Hartree–Fock theory (UHF) for others. The HF/6-31G(d) equilibrium structure is used to calculate harmonic frequencies, which are then scaled by a factor of 0.8929 to take account of known deficiencies at this level [76]. These frequencies are used to evaluate the zero-point energy (E(ZPE)) and thermal effects.

The accuracy of the zero-point energies used in G2 theory has been the subject of some discussion. While the scale factor used (0.8929) is appropriate for the reproduction of experimental fundamental frequencies, the optimum scale factor for obtaining accurate zero-point energies from HF/6-31G(d) harmonic frequencies was determined by Scott and Radom [77] to be larger (0.9135). This can cause significant errors in the computed zero-point energies. However, the resulting errors are systematic and it turns out that the HLC used in G2 theory (*vide infra*) makes up for any such systematic deficiencies. Nevertheless, it is desirable to avoid such errors in computational models since they can make an analysis of the residual deficiencies difficult or impossible. In one of our more recent thermochemical models, G3X [42], we have used B3LYP/6-31G(2df,p) harmonic frequencies with a scale factor of 0.9854 that is indeed appropriate for the prediction of zero-point energies.

(a)

G2	MP2	MP4	QCISD (T)
6-311G (d,p)	X	X	X
6-311 + G (d,p)	X	X	
6-311G (2df,p)	X	X	
6-311 + G (3df,2p)	X		"QCISD (T)/ 6-311+G (3df,2p)"

HLC: based on electron pairs and fit to G2

(b)

G3	MP2	MP4	QCISD (T)
6-31G (d)	X	X	X
6-31 + G (d)	X	X	
6-31G (2df,p)	X	X	
G3Large	X (Full)		"QCISD (T, FULL)/ G3Large"

Spin-orbit for atoms

HLC: based on electron pairs and unpaired electrons and fit to G2/97 test

(c)

G3X	HF	MP2	MP4	QCISD (T)
6-31G (d)	X	X	X	X
6-31 + G (d)	X	X	X	
6-31G (2df,p)	X	X	X	
G3Large	X	X (Full)		
G3XLarge	X			"QCISD (T, FULL)/ G3XLarge"

Spin-orbit for atoms

HLC: based on electron pairs and unpaired electrons and fit to G3/99 test

Fig. 27.1. Pictorial representation of the terms that are included in G2, G3 and G3X theories.

(2) The equilibrium geometry is refined at the MP2(full)/6-31G(d) level, using all electrons for the calculation of correlation energies. This is the final equilibrium geometry in the theory and is used for all single-point calculations at higher levels of theory in step 3. These subsequent calculations include only valence electrons in the treatment of electron correlation.

Two points should be noted here. First, the use of all electrons in the MP2 optimization is due to historical reasons in the development of analytical gradient techniques. The geometry change due to the inclusion of core electrons in the optimization is likely to be very small. Secondly, the geometry used for the correlation energy calculations (MP2(full)/6-31G(d)) is different from the geometry used for zero-point energies (HF/6-31G(d)). While it does not seem to add to the errors significantly, it may be more satisfactory to use the same model for both. Indeed, in the G3X model, the same level of theory (B3LYP/6-31G(2df,p)) is used for both.

(3) A series of single-point energy calculations is carried out at higher levels of theory. The first higher level calculation is triples-augmented quadratic configuration interaction with the 6-311G(d,p) basis set, i.e. QCISD(T)/6-311G(d,p).

Several points should be mentioned in this context. First, while we use QCISD(T) in our basic definition of G2 and G3 theories, analogous methods have been defined where the CCSD(T) method replaces QCISD(T). Both variations seem to yield very similar mean absolute deviations in most cases. However, in our most recent work on transition metal systems [78], it appears that CCSD(T) has a clear advantage over QCISD(T) and will thus become the preferred method. For the first- and second-row molecules, however, there is no clear preference. The key point to note is that the accuracy and predictive capability of these methods comes from the inherent accuracy of QCISD(T) or CCSD(T). Finally, this is one of the N^7 steps in the calculation and is likely to be rate-limiting if carried out with very large basis sets. Indeed, it is the bottleneck in CCSD(T) calculations with large correlation-consistent basis sets. In G2 theory, QCISD(T) calculations are carried out with a polarized valence triple-zeta basis set. This is a very modest basis set and this makes it possible to carry out G2 calculations on molecules of the size of naphthalene on small workstations. In our later work on G3 theory, we use even smaller 6-31G(d) calculations that makes these methods applicable for even larger molecules.

This energy is then modified by a series of corrections from additional calculations using larger basis sets. The key assumption in G2 theory is that such contributions can be derived using cheaper models such as perturbation theory (MP4 or MP2). This will be an excellent approximation if the behavior of these methods with respect to basis set increase is parallel to that of QCISD(T) (additivity approximation). Two such corrections are included in G2 theory at the MP4 level.

(a) A correction for diffuse functions, *E*(plus):

$$E(\text{plus}) = \text{MP4/6-311} + \text{G(d, p)} - \text{MP4/6-311G(d, p)} \tag{1}$$

(b) A correction for higher polarization functions on non-hydrogen atoms and p-functions on hydrogens, $E(\text{2df, p})$:

$$E(\text{2df, p}) = \text{MP4/6-311G(2df, p)} - \text{MP4/6-311G(d, p)} \tag{2}$$

The evaluation of the two independent corrections is based on the assumption that contributions to larger basis set effects can be evaluated independently. However, careful analysis shows that there is some linear dependence and this is corrected by the next term at the MP2 level.

(c) A correction for larger basis set effects and for the non-additivity caused by the assumption of separate basis set extensions for diffuse functions and higher polarization functions, $E(\Delta)$:

$$E(\Delta) = \text{MP2/6-311} + \text{G(3df, 2p)} - \text{MP2/6-311G(2df, p)} \\ - \text{MP2/6-311} + \text{G(d, p)} + \text{MP2/6-311G(d, p)} \quad (3)$$

Thus 6-311 + G(3df,2p) is the largest basis set used in G2 theory.

(4) A 'higher level correction' (E(HLC)) is added to take into account the remaining deficiencies in the energy calculations: the HLC is $-An_\beta - B(n_\alpha - n_\beta)$ where n_β and n_α are the number of β and α valence electrons, respectively, with $n_\alpha \geq n_\beta$. The number of valence electron pairs corresponds to n_β. Thus, A is the correction for pairs of valence electrons in the system and B is the correction for unpaired electrons in the system. For G2 theory, $A = 4.81$ mhartrees, $B = 0.19$ mhartrees. The parameter B was derived to give the exact energy for H atom and the parameter A was derived to give zero mean deviation for the 55 molecules in the G2-1 test set.

The use of empirical correction factors in such *ab initio* models has been discussed extensively in the literature. However, the parameters in G2 theory are 'molecule independent' in the sense that they do not depend on the constituent atoms in the molecule. This is in contrast to the standard semiempirical methods (such as AM1) that have parameters depending on the type of atom and its neighbors. In G2 theory, however, they only depend on the number of electrons in the molecule. Thus the error in the energies is assumed to be the same for any electron pair implying some kind of universal behavior of electron pairs. The slow convergence of the electron correlation energy with basis set size is now understood to result from the inherent difficulty in describing the cusp in the interelectronic wavefunction. This is a fundamental property of electron pairs and may perhaps be captured by universal correction factors as used in G2 theory. Nevertheless, the presence of such HLC parameters sometimes leads to the labeling of G2 and G3 theories as 'semiempirical', though with only two parameters to describe all of the chemistry we prefer to call our methods 'slightly empirical'.

(5) Finally, the total energy at 0 K ('G2 energy') is obtained by adding all the individual energy corrections in an additive manner

$$E_0[\text{G2}] = \text{QCISD(T)/6-311G(d, p)} + E(\text{plus}) + E(\text{2df, p}) + E(\Delta) \\ + E(\text{HLC}) + E(\text{ZPE}) \quad (4)$$

The final total energy (excluding the HLC and ZPE terms) is effectively at the QCISD(T)/6-311 + G(3df,2p) level if the different additivity approximations work well. The validity of such approximations has been previously investigated for G2 theory on the G2-1 subset of G2/97 by performing complete QCISD(T)/6-311 + G(3df,2p) calculations directly. The additivity approximation works well for most of the energies in the test set [25].

The MP4/6-311G(2df,p) calculation is the most time-consuming step (N^7) and represents the bottleneck limiting the applicability of G2 theory. A simpler version of G2 theory, denoted as G2(MP2), was developed to avoid the MP4 step and to extend the applications of G2 theory for larger molecules. In this case, the effect of all large basis set extensions is evaluated in a single step at the MP2 level as

$$E(\Delta\text{MP2}) = \text{MP2/6-311+G(3df, 2p)} - \text{MP2/6-311G(d, p)} \quad (5)$$

$$E_0[\text{G2(MP2)}] = \text{QCISD(T)/6-311G(d, p)} + E(\Delta\text{MP2}) + E(\text{HLC}) + E(\text{ZPE}) \quad (6)$$

The only N^7 step in a G2(MP2) calculation is QCISD(T)/6-311G(d,p) which now becomes the bottleneck step.

27.3.1 Assessment of G2 theory

Before performing an assessment of computational models such as G2 theory, well-defined procedures have to be developed to derive thermodynamic quantities from theory. The enthalpies of formation for most molecules in the G2/97 and G3/99 test sets have been measured at 298 K. In order to compare with experiment, the heats of formation for molecules have to be calculated using a general procedure that is applicable for any molecular system. In our own work, we follow a systematic procedure described in detail previously [35]. Briefly, thermal corrections (298 K) are first evaluated using the calculated (scaled) vibrational frequencies and standard statistical mechanical methods. The calculated total energies of the given molecule and its constituent atoms are used to evaluate its atomization energy. This value is then used along with the thermal corrections and the known experimental enthalpies of formation for the atomic species to calculate the enthalpy of formation for the molecule (298 K). The electron affinities are calculated as the difference in total energies at 0 K of the anion and the corresponding neutral. Likewise, the ionization potentials are calculated as the difference in total energies at 0 K of the cation and the corresponding neutral. Similarly, the proton affinities are evaluated using the 0 K energies of the molecule and its protonated analogue.

G2 theory and its approximate version G2(MP2) have been assessed carefully for the 301 energies in the G2/97 test set. The results are summarized in Table 27.1. The overall mean absolute deviation is 1.50 kcal/mol for G2 and 1.91 kcal/mol for G2(MP2). While they satisfy the target accuracy of 2 kcal/mol, the errors clearly increase significantly for the larger molecules. For example, the G2 heats of formation for the molecules in the G2-1 subset have a mean absolute deviation of 1.21 kcal/mol while the larger molecules in the G2-2 subset have a deviation of 1.77 kcal/mol (overall 1.56 kcal/mol). The largest errors at the G2 and G2(MP2) levels are seen for C_2F_4 (8.2 and 10.1 kcal/mol, respectively), though the accuracy of its experimental value has been questioned. The breakdown of the deviations for the different types of molecules is also given in Table 27.1. It is clearly seen that the largest deviations occur for non-hydrogen species. Again they get worse for the larger molecules. The non-hydrogen species in the smaller G2-1 subset have a mean absolute deviation of 1.73 kcal/mol but the corresponding species in the G2-2 subset have a mean absolute deviation of 3.01 kcal/mol (overall 2.48 kcal/mol). In particular, significant errors are seen in the heats of formation for compounds containing multiple halogens (3–8 kcal/mol) and for unsaturated ring compounds (3–4 kcal/mol) [35]. A careful analysis revealed the importance of larger

Table 27.1 Summary of mean absolute deviations (kcal/mol) for G2 theories

	G2	G2(MP2)
G2/1 test set		
Enthalpies of formation (55)	1.21	1.34
Non-hydrogens (14)	1.73	1.85
Hydrocarbons (5)	0.88	0.86
Substituted hydrocarbons (5)	0.95	1.32
Inorganic hydrides (13)	0.99	1.29
Radicals (18)	1.18	1.13
Complete G2/97		
Enthalpies of formation (147)	1.56	2.03
Non-hydrogens (34)	2.48	3.31
Hydrocarbons (22)	1.29	1.83
Substituted hydrocarbons (47)	1.48	1.89
Inorganic hydrides (15)	0.95	1.20
Radicals (29)	1.16	1.36
Ionization energies (88)	1.45	1.75
Electron affinities (58)	1.41	1.94
Proton affinities (8)	1.08	1.25
All (301)	1.50	1.91

basis set effects as well as spin–orbit interactions, particularly for halogenated species. This led to the development of G3 theory described in the next section.

Several other variants of G2 theory have been proposed. A higher level correlation treatment was examined using coupled cluster (CCSD(T)) or Brueckner doubles (BD(T)) energies instead of QCISD(T). Referred to as G2(CC) and G2(BD), their performance was found to be very similar to G2 theory. Other variants include the use of QCISD geometries instead of the MP2 geometry, and the use of MP2 zero-point energies. Finally, a completely different approach using isodesmic bond separation reactions has been used by Raghavachari et al. [30] to evaluate the heats of formation for a test set of 40 molecules containing H, C, O, and N. In contrast to standard G2 theory where atomization energies are used along with the experimental heats of formation of the atoms, bond separation reaction energies are used in this approach along with the experimental heats of formation of the reference molecules to determine the unknown heats of formation. The mean absolute deviation for the 40 molecules test set with the isodesmic approach (0.5 kcal/mol) is an improvement of a factor of three over standard G2 theory (1.5 kcal/mol). This shows that a significant part of the residual error in G2 theory is systematic and that the intrinsic errors for similar bonds in different molecules are transferable. Though such an approach is not generally applicable for all molecules, it may be useful to build accurate heats of formation for some large molecules.

27.4 G3 THEORY

G3 theory was designed to correct some of the deficiencies of G2 theory for systems such as halogen-containing molecules, unsaturated hydrocarbons, etc. It also contains important physical effects such as core-valence correlation and spin–orbit contributions that were not included in G2 theory. G3 theory is computationally less demanding than G2 theory though it is significantly more accurate. The detailed steps involved in G3 theory are given below. The calculations that are needed to derive the G3 energy are also shown pictorially in Fig. 27.1(b).

(1) Same as in G2 theory.

(2) Same as in G2 theory.

(3) As in G2 theory, a series of single-point energy calculations is carried out at higher levels of theory. However, the basis sets used are different and somewhat smaller. The first higher level calculation is QCISD(T) with the 6-31G(d) basis set, i.e. QCISD(T)/6-31G(d). Note that this is a smaller double-zeta basis set compared to the triple-zeta set used in G2 theory. In addition, p-type polarization functions on hydrogen atoms are not included. This energy is then modified by a series of corrections from additional calculations.

(a) A correction for diffuse functions, *E*(plus):

$$E(\text{plus}) = \text{MP4/6-31+G(d)} - \text{MP4/6-31G(d)} \tag{7}$$

(b) A correction for higher polarization functions on non-hydrogen atoms and p-functions on hydrogens, *E*(2df, p):

$$E(\text{2df, p}) = \text{MP4/6-31G(2df, p)} - \text{MP4/6-31G(d)} \tag{8}$$

(c) A correction for larger basis set effects and for the non-additivity caused by the assumption of separate basis set extensions for diffuse functions and higher polarization functions, *E*(G3Large):

$$\begin{aligned} E(\Delta\text{G3L}) = {} & \text{MP2(full)/G3Large} - \text{MP2/6-31G(2df, p)} \\ & - \text{MP2/6-31+G(d)} + \text{MP2/6-31G(d)} \end{aligned} \tag{9}$$

While steps (7) and (8) are very similar to the corresponding steps in G2 theory, the last step is somewhat different. The basis set used, denoted as G3Large [32], is significantly larger than the corresponding basis set used in G2 theory and includes some core polarization functions as well as multiple sets of valence polarization functions. In addition, it should be noted that the MP2 calculation in step (9) is carried out at the MP2(full) level. This is done to take account of core-related correlation contributions to total energies.

(4) An additional important term not present in G2 theory is now included in G3 theory. Spin–orbit correction, *E*(SO), is included for *atomic species only*. The spin–orbit correction is taken from experiment [79], where available, and accurate theoretical calculations [80] in other cases. These are particularly important for halide-containing systems [32]. Molecular spin–orbit corrections are not included in G3 theory.

(5) A 'higher level correction' (*E*(HLC)) is added to take into account the remaining deficiencies in the energy calculations. The HLC used in G3 theory is somewhat different: it is $-An_\beta - B(n_\alpha - n_\beta)$ for molecules and $-Cn_\beta - D(n_\alpha - n_\beta)$ for atoms (including atomic ions). The n_β and n_α are the number of β and α valence electrons, respectively, with $n_\alpha \geq n_\beta$. The number of valence electron pairs corresponds to n_β.

Thus, A is the correction for pairs of valence electrons in molecules, B the correction for unpaired electrons in molecules, C the correction for pairs of valence electrons in atoms, and D the correction for unpaired electrons in atoms. The use of different corrections for atoms and molecules can be justified, in part, by noting that effects of basis functions with higher angular momentum are likely to be of more importance in molecules than in atoms. The A, B, C, D values are chosen to give the smallest average absolute deviation from experiment for the G2/97 test set. For G3 theory, $A = 6.386$ mhartrees, $B = 2.977$ mhartrees, $C = 6.219$ mhartrees, $D = 1.185$ mhartrees.

The use of different HLC parameters for atoms and molecules, though it improves the overall performance relative to G2 theory, causes some problems. For example, a single bond formation (or breaking) between two atoms (e.g. two F atoms) is treated differently than that between two molecules (e.g. two methyl radicals). Bond formation between atoms with many unpaired spins may not be described correctly. While this is not a major source of problem normally, preliminary work on transition metal systems [78] indicates that the HLC contribution to the binding energy involving atoms such as Cr may be unreasonably large. There can also be a significant discontinuity in the HLC correction that may cause difficulties for weakly bound systems.

(6) Finally, the total energy at 0 K ('G3 energy') is obtained by adding all the individual energy corrections in an additive manner

$$
\begin{aligned}
E_0[\mathrm{G3}] = {} & \mathrm{QCISD(T)/6\text{-}31G(d)} + E(\mathrm{plus}) + E(\mathrm{2df,p}) + E(\Delta\mathrm{G3L}) \\
& + E(\mathrm{SO}) + E(\mathrm{HLC}) + E(\mathrm{ZPE})
\end{aligned} \tag{10}
$$

The final total energy is effectively at the QCISD(T,full)/G3Large level if the different additivity approximations work well.

The correlation methods in G3 theory are still computationally demanding and it is of interest to find modifications to reduce the computational requirements. Two approximate versions of G3 theory have been proposed to make the methods more applicable. The first is G3(MP3) [33] that eliminates the expensive MP4/6-31G(2df,p) calculation by evaluating the larger basis set effects at the MP3 level. It also eliminates the MP4/6-31 + G(d) calculation

$$
\begin{aligned}
E_0(\mathrm{G3(MP3)}) = {} & \mathrm{QCISD(T)/6\text{-}31G(d)} + [\mathrm{MP3/6\text{-}31G(2df,p)} - \mathrm{MP3/6\text{-}31G(d)}] \\
& + [\mathrm{MP2(full)/G3L} - \mathrm{MP2/6\text{-}31G(2df,p)}] + E(\mathrm{SO}) \\
& + E(\mathrm{HLC}) + E(\mathrm{ZPE})
\end{aligned} \tag{11}
$$

The second is G3(MP2) theory [34] that evaluates the larger basis set effects at the MP2 level, similar to the successful G2(MP2) theory

$$
\begin{aligned}
E_0(\mathrm{G3(MP2)}) = {} & \mathrm{QCISD(T)/6\text{-}31G(d)} + [\mathrm{MP2/G3MP2L} - \mathrm{MP2/6\text{-}31G(d)}] \\
& + E(\mathrm{SO}) + E(\mathrm{HLC}) + E(\mathrm{ZPE})
\end{aligned} \tag{12}
$$

In G3(MP2) theory, the MP2(full)/G2Large calculation in G3 is replaced with a frozen core calculation with the G3MP2Large basis set [34] that does not contain the core polarization functions in the G2Large basis set.

27.4.1 Assessment of G3 theory

The performance of G3, G3(MP3) and G3(MP2) theories for the energies in the G2/97 and G3/99 test sets are summarized in Table 27.2. For the 301 energies in the G2/97 test set, the mean absolute deviation for G3 theory (1.01 kcal/mol) is substantially improved from G2 theory (1.50 kcal/mol). As mentioned earlier, G3 theory was designed to correct for some of the deficiencies in G2 theory. The histograms in Fig. 27.2 show the range of deviations of G2 and G3 theories from experiment for the G2/97 test set. Nearly 88% of the G3 deviations fall within the range of -2.0 to $+2.0$ kcal/mol. This is substantially better than G2 theory for which about 74% of the deviations fall in this range. In addition to improving the accuracy, the use of the 6-31G(d) basis set based calculations in G3 theory substantially decreases the computer time as well as the disk space requirements relative to G2 theory (which uses the larger 6-311G(d,p) basis set based calculations). For example, the computer time for a G3 calculation on benzene is nearly a factor two faster than a G2 calculation.

Overall the mean absolute deviations increase slightly for the G3/99 test set compared to that of the G2/97 test set. The mean absolute deviation of G3 theory increases from 1.01 to 1.07 kcal, G3(MP3) theory increases from 1.21 to 1.27 kcal/mol and G3(MP2) theory remains at 1.31 kcal/mol. While most types of molecules behave similarly in both test sets, larger deviations are seen for some of the non-hydrogen species in the expanded test set. In fact, the mean absolute deviation of 3.24 kcal/mol for the 13 non-hydrogen species in the G3-3 subset is nearly twice that of the 35 non-hydrogens in the G2/97 set

Table 27.2 Summary of mean absolute deviations (kcal/mol) for G3 theories

	G3	G3(MP3)	G3(MP2)
G2/97 test set			
Enthalpies of formation (147)	0.92	1.19	1.17
Non-hydrogens (34)	1.68	2.09	2.06
Hydrocarbons (22)	0.68	0.86	0.70
Substituted hydrocarbons (47)	0.56	0.78	0.74
Inorganic hydrides (15)	0.87	1.18	1.03
Radicals (29)	0.84	1.05	1.23
All (301)[a]	1.01	1.21	1.31
Complete G3/99			
Enthalpies of formation (222)	1.05	1.29	1.22
Non-hydrogens (47)	2.11	2.74	2.45
Hydrocarbons (38)	0.69	0.77	0.71
Substituted hydrocarbons (91)	0.75	0.86	0.83
Inorganic hydrides (15)	0.87	1.18	1.03
Radicals (31)	0.87	1.06	1.21
Ionization energies (88)	1.14	1.24	1.46
Electron affinities (58)	0.98	1.24	1.46
Proton affinities (8)	1.34	1.25	1.02
All (376)[a]	1.07	1.27	1.31

[a]The mean absolute deviation for the ionization energies, electron affinities and proton affinities in the G2/97 test set are the same as in G3/99.

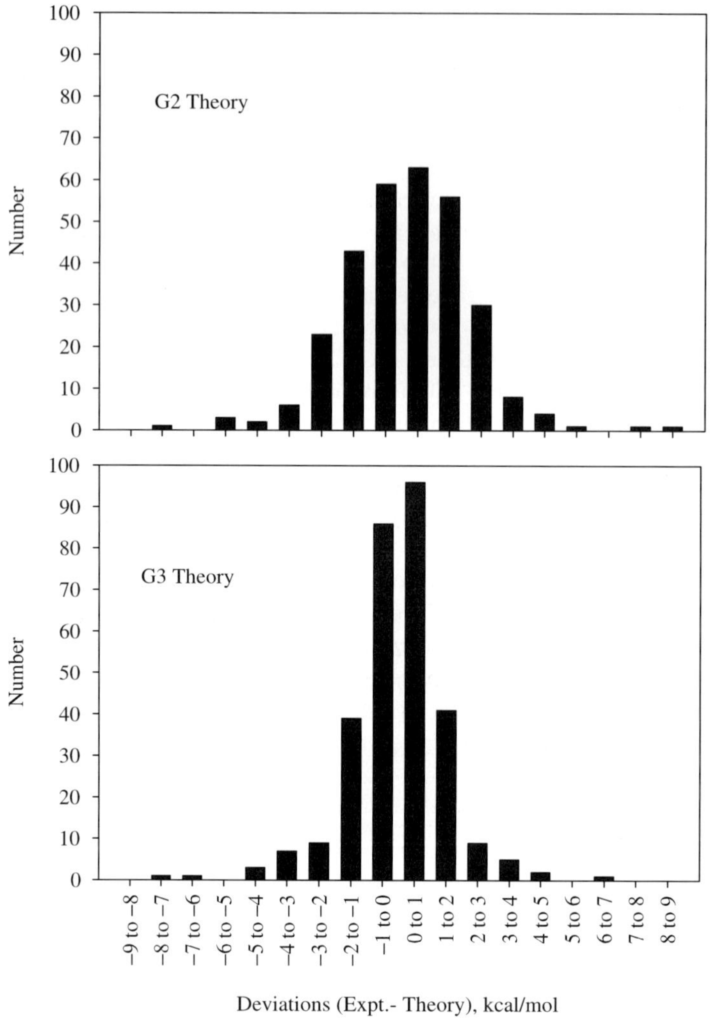

Fig. 27.2. Histograms of G2 and G3 deviations for the G2/97 test set. Each vertical bar represents deviations in a 1 kcal/mol range.

(1.68 kcal/mol). These errors were traced to the deficiencies in the MP2 geometries as well as basis set deficiencies and eventually led to the development of G3X theory discussed in the next section.

Other variations of G3 theory have been proposed that use alternate geometries, zero-point energies or higher order correlation methods. As described earlier, G2 and G3 theories use MP2(full)/6-31G(d) geometries and scaled HF/6-31G(d) frequencies and zero-point energies. A method using B3LYP/6-31G(d) geometries and scaled B3LYP/6-31G(d) zero-point energies (0.96) was considered to make it more uniform. Denoted as

G3//B3LYP [27], its performance was very similar to that of G3 theory though it may be useful in cases where the MP2 theory is deficient for geometries. Another variation involves the use of CCSD(T)/6-31G(d) instead of QCISD(T)/6-31G(d) to evaluate the contribution of higher order correlation effects. The resulting G3(CCSD) method [28] has an accuracy very similar to that of G3 theory and may be useful in cases where the QCISD(T) method is not available or deficient.

Apart from these minor variations, two major variations (G3X and G3S) [29,30] have been proposed to address some of the main deficiencies of G3 theory. These are discussed in detail in Sections 27.5 and 27.6.

27.5 G3X THEORY

As noted earlier, the G3 mean absolute deviation from experiment (1.07 kcal/mol) found for the G3/99 test set is slightly larger than the corresponding value of 1.01 kcal/mol found for the smaller G2/97 test set (Fig. 27.3). However, the larger non-hydrogen systems in the G3/99 test set have deviations (3.24 kcal/mol) almost twice as large as those in the smaller G2/97 test set (1.68 kcal/mol). Especially large deviations (G3 theory) occur for SF_6 (−6.2 kcal/mol), PF_5 (−7.1 kcal/mol), SO_3 (−5.1 kcal/mol), P_4 (−4.2 kcal/mol) and Cl_2O_2S (−4.4 kcal/mol). These are hypervalent molecules except P_4 which is unusually strained. Overall, the G3 deviations for nearly all the non-hydrogen species are negative indicating underbinding. Part of the source of error for these non-hydrogen species was identified as resulting from the deficiency of the MP2(full)/6-31G(d) geometries used for the single-point energies. Use of experimental geometries in a small subset of non-hydrogens reduced the deviations in those molecules, but they still remained around 3–4 kcal/mol. The remainder of the error appeared to arise from basis set deficiencies in such molecules. A family of G3 methods, referred to as G3X (*G3 eXtended*), has been developed [42] to remedy these problems and improve the accuracy of the results for larger molecules.

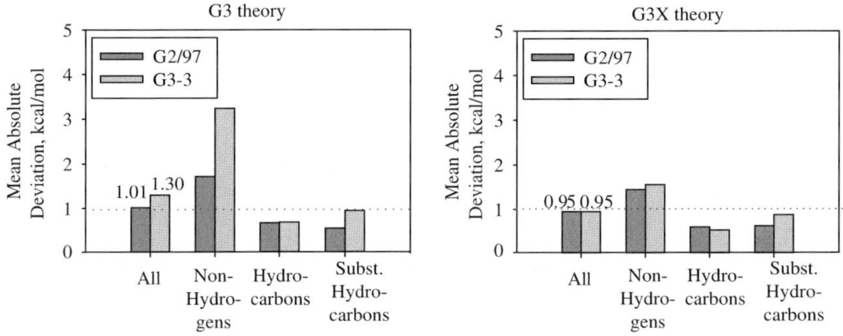

Fig. 27.3. Mean absolute deviation from experiment for G3 and G3X theories on the G3/99 test set broken down into the G2/97 and G3-3 subsets.

Three modifications have been included in the G3X method to correct these deficiencies:

(1) B3LYP/6-31G(2df,p) geometries are used in place of the MP2(full)/6-31G(d) geometries. These new geometries were found to be significantly improved relative to the original geometries. For example, the mean absolute deviation from experiment in the bond lengths for a subset of seven representative molecules (PF_3, PF_5, P_4, PCl_5, SO_2, SO_3 and SF_6) decreased from 0.027 Å (MP2(full)/6-31G(d)) to 0.011 Å (B3LYP/6-31G(2df,p)).

(2) B3LYP/6-31G(2df,p) zero-point energies (scaled by 0.9854) are used in place of the HF/6-31G(d) zero-point energies (scaled by 0.8929). The scale factor (0.9854) was derived from fitting the set of zero-point energies compiled by Scott and Radom [77]. This is in contrast to the HF/6-31G(d) scale factor (used in G3 theory) that was more appropriate for fundamental vibrational frequencies. The procedure used in G3X theory is clearly more appropriate. Moreover, the use of the same level of theory for both the geometry and the zero-point energy in G3X is more logical than the use of two different levels of theory (MP2(full)/6-31G(d) and HF/6-31G(d), respectively) in the original G3 theory. Finally, the small correction suggests that B3LYP/6-31G(2df,p) is accurate for zero-point energies as well as geometries.

(3) A set of *g* valence polarization functions [42] is added to the G3Large basis set for second-row atoms at the Hartree–Fock level. The addition of these extra basis functions was clearly motivated by the deficiencies noted earlier for several hypervalent molecules involving second-row elements. Significant improvement in the calculated atomization energies was found for some representative molecules, particularly those involving a mixture of first- and second-row elements. For example, the addition of a single *g* function to the second-row atoms (Al–Cl) increases the binding in SiF_4, PF_5 and SF_6 by 3.6, 5.1 and 5.5 kcal/mol, respectively, at the Hartree–Fock level. Addition of more polarization functions (2 *g*, 2 *gh*) on the second-row atoms results in substantially smaller changes in the atomization energies [42]. Thus, a single set of *g* polarization functions (seven pure functions) is added to the second-row G3Large basis set at the HF level in G3X theory. The *g* exponents for Al–Cl are taken from Dunning's correlation-consistent cc-pvqz basis set [8]. No *g* function was used on Na or Mg. This new basis set is referred to as G3XLarge. It should be noted that similar basis set deficiencies may occur at correlated levels also. However, correcting such deficiencies at the correlated level explicitly is more difficult due to their slow convergence, though the HLC parameters may offer partial remedy if the resulting errors are systematic. Thus, the effect of *g* polarization functions at the HF level is evaluated as

$$E(\Delta\mathrm{HF}) = \mathrm{HF/G3XL} - \mathrm{HF/G3L} \tag{13}$$

where G3L = G3Large basis set, G3XL = G3XLarge basis set. The total G3X energy incorporating all three features is given by the equation

$$\begin{aligned} E_0(\mathrm{G3X}) = {} & \mathrm{QCISD(T)/6\text{-}31G(d)} + E(\mathrm{plus}) + E(\mathrm{2df,p}) + E(\Delta\mathrm{G3L}) \\ & + E(\Delta\mathrm{HF}) + E(\mathrm{SO}) + E(\mathrm{HLC}) + E(\mathrm{ZPE}) \end{aligned} \tag{14}$$

Eq. (14) is the same as for G3 theory except for the addition of the $E(\Delta HF)$ term. This term extends the HF/G3L energy, which is part of the MP2(full)/G3L energy, to the G3XL basis set. The calculations that are needed to derive the G3X energy are also shown pictorially in Fig. 27.1(c).

The HLC parameters for G3X theory were obtained by fitting to the full G3/99 test set. Fitting of the HLC parameters to the smaller G2/97 test set gives nearly the same values for the four parameters indicating that there is little sensitivity to the increase in the data set size. The G3X method takes about 10–15% more time than G3 due to the B3LYP/6-31G(2df,p) frequency calculation.

The three new features of G3X theory can also be easily included in the G3(MP3) and G3(MP2) methods. The resulting methods are referred to as G3X(MP3) and G3X(MP2), respectively. The G3X(MP3) energy is given by

$$
\begin{aligned}
E_0(\mathrm{G3X(MP3)}) = {} & \mathrm{QCISD(T)/6\text{-}31G(d)} + [\mathrm{MP3/6\text{-}31G(2df,p)} - \mathrm{MP3/6\text{-}31G(d)}] \\
& + [\mathrm{MP2(full)/G3L} - \mathrm{MP2/6\text{-}31G(2df,p)}] + E(\Delta\mathrm{HF}) + E(\mathrm{SO}) \\
& + E(\mathrm{HLC}) + E(\mathrm{ZPE})
\end{aligned} \tag{15}
$$

The G3X(MP2) energy is given by

$$
\begin{aligned}
E_0(\mathrm{G3(MP2)}) = {} & \mathrm{QCISD(T)/6\text{-}31G(d)} + [\mathrm{MP2/G3MP2L} - \mathrm{MP2/6\text{-}31G(d)}] \\
& + E(\Delta\mathrm{HF}) + E(\mathrm{SO}) + E(\mathrm{HLC}) + E(\mathrm{ZPE})
\end{aligned} \tag{16}
$$

Eqs. (15) and (16) are the same as for G3(MP3) and G3(MP2) theories, except for the addition of the $E(\Delta HF)$ term. The HLC parameters were again obtained by fitting to the G3/99 test set.

27.5.1 Assessment of G3X theory

G3X theory gives significantly better agreement with experiment for the G3/99 test set of 376 energies (Table 27.3). Overall, the mean absolute deviation from experiment decreases from 1.07 kcal/mol (G3) to 0.95 kcal/mol (G3X). Thus the mean absolute deviation in G3X not only beats our target accuracy of 2 kcal/mol but also meets the more rigorous definition of chemical accuracy (1 kcal/mol). Even more impressively, the mean absolute deviation for the 222 enthalpies of formation decreases from 1.05 kcal/mol (G3) to 0.88 kcal/mol (G3X). This is clearly due to the improvement for non-hydrogen systems where the mean absolute deviation decreases from 2.11 to 1.49 kcal/mol.

Unlike G3 theory, there is very little degradation of G3X theory on going from the smaller G2/97 to the larger G3/99 test set. The overall mean absolute deviation from experiment is nearly the same for the larger G3/99 test set (0.95 kcal/mol) as for the smaller G2/97 test set (0.96 kcal/mol). Similarly, the mean absolute deviation from experiment for enthalpies is nearly the same for the larger G3/99 test set (0.88 kcal/mol) as for the smaller G2/97 test set (0.86 kcal/mol). This is somewhat surprising since we do expect larger errors in larger molecules. Perhaps it is due to the presence of a relatively large number of hydrocarbons and substituted hydrocarbons which tend to have smaller deviations overall.

Table 27.3 Summary of mean absolute deviations (kcal/mol) for G3X theories

	G3X	G3X(MP3)	G3X(MP2)
Enthalpies of formation (222)	0.88	1.07	1.05
Non-hydrogens (47)	1.49	2.05	1.75
Hydrocarbons (38)	0.56	0.68	0.76
Substituted hydrocarbons (91)	0.75	0.76	0.78
Inorganic hydrides (15)	0.81	1.12	1.01
Radicals (31)	0.76	0.96	1.17
Ionization energies (88)	1.07	1.16	1.36
Electron affinities (58)	0.98	1.29	1.51
Proton affinities (8)	1.21	1.09	0.79
All (376)	0.95	1.13	1.19

However, it is important to note that this result is not dependent on the data set that is used to obtain the HLC parameters since both sets give essentially the same values.

Some examples of the changes on going from G3 to G3X theory for selected non-hydrogen systems can be seen in Table 27.4. Significant improvements are seen in most molecules though errors in the range of 2 or 3 kcal/mol still remain in some cases. The reliability of the G3X method can be epitomized by the fact that only one molecule (C_2F_4) has an error of more than 4 kcal/mol in the entire test set and even that experimental value has been recently called into question.

Summaries of G3X(MP3) and G3X(MP2) mean absolute deviations from experiment for the G3/99 test set are also given in Table 27.3. The overall mean absolute deviations for G3X(MP3) and G3X(MP2) for the 376 energies are 1.13 and 1.19 kcal/mol, respectively. These are improvements over the corresponding values of G3(MP3) and G3(MP2) for the same set (1.27 and 1.31 kcal/mol, respectively). For enthalpies of formation the mean absolute deviations decrease from 1.29 to 1.07 kcal/mol (G3X(MP3)) and from 1.22 to 1.05 kcal/mol (G3X(MP2)). As in the case of G3X theory, much of the improvement in enthalpies is due to non-hydrogen species. The G3X(MP3) and G3X(MP2) methods save considerable computational time without significant loss

Table 27.4 Deviations from experiment for the calculated enthalpies of formation of some non-hydrogen systems (experiment—theory, kcal/mol)

Molecule	G3	G3X
SO_2	−3.81	−0.73
SO_3	−5.17	−1.54
PF_3	−4.84	−1.85
PF_5	−7.05	−1.80
SF_6	−6.22	−0.47
SiF_4	−1.12	2.27
$SiCl_4$	0.02	−0.63
P_4	−4.15	−2.18
PCl_3	−3.19	−3.30
PCl_5	2.40	1.74
$POCl_3$	−3.07	−2.32
Cl_2O_2S	−4.37	−2.55

in accuracy. The ratio of the cpu times for G3X:G3X(MP3):G3X(MP2) is approximately 5:2:1 for a molecule such as benzene.

27.6 G3S THEORY

G3 theory and the other variants discussed thus far include a HLC term to correct for the remaining deficiencies in the computed energies. The HLC term in G3 theory consists of four molecule-independent additive parameters that depend only on the number of paired and unpaired electrons in the system. While basis set deficiencies are the biggest source of errors in the computed energies, the HLC parameters can also correct for any other systematic errors (such as those from zero-point energies or from relativistic effects). Such an approach will work if such deficiencies are systematic and scale as the number of electrons. While it may make it difficult to identify the source of systematic errors, this approach is indeed successful as indicated by the overall mean absolute deviation of <1 kcal/mol for the G3X method with the large G3/99 test set.

G3 theory was developed to derive the ground state energies of molecules at their equilibrium geometries. However, a major topic of importance in chemistry is the investigation of chemical reactions and the study of potential energy surfaces. Thus it is desirable to extend these methods to investigate structures other than minima on the potential energy surface. The computation of the component energies in such methods can be easily performed at any other geometry. However, one of the deficiencies of G3 theory is that the HLC parameters do not depend on the geometry and thus do not vary on the potential energy surface. This may cause deficiencies for regions near transition states that contain partially broken bonds. Somewhat surprisingly, a careful study of activation barriers for chemical reactions [63] has yielded accurate results for G3 theory though further analysis of the results may be interesting. Even more importantly, G3 theory cannot be used to study potential energy surfaces for reactions where the reactants and products have a different number of electron pairs.

An alternative approach for calculating accurate thermochemical data is to scale the calculated correlation energy using multiplicative parameters determined by fitting to the experimental data. The central idea in all such methods is that in a balanced treatment, roughly the same percentage of the correlation energy components is obtained for every system. Pioneering methods using such an approach include the scaling all correlation (SAC) method of Gordon and Truhlar [57], the parameterized correlation (PCI-X) method of Siegbahn et al. [58] and the multi-coefficient correlation methods (MCCM) of Truhlar and co-workers [59–66]. Such methods can be used to yield continuous potential energy surfaces even for reactions where the reactants and products contain different numbers of electron pairs. It is also possible to derive analytical gradient techniques for such methods that may make it easier to investigate potential energy surfaces.

A family of methods, referred to as G3S (*G3 S*caled), has been developed [41] where the additive HLC is replaced by a multiplicative scaling of the different component energies that comprise the G3 energy. The scale factors have been obtained by fitting to the G2/97 test set of energies. The use of such a large fitting test set enables a reliable assessment of the use of such a scaling approach to computational thermochemistry.

References pp. 810–812

As discussed earlier, the G3 energy is written in terms of corrections (basis set extensions and correlation energy contributions) to the QCISD(T)/6-31G(d) energy. Alternatively, the G3 energy can be specified in terms of HF and different correlation energy components. Denoting E2, E3 and E4 as the second, third, and fourth-order contributions from perturbation theory, and ΔQCI as the contributions beyond fourth order in a QCISD(T) calculation, the G3 energy expression can be expressed as

$$\begin{aligned} E_0[\mathrm{G3}] = {} & \mathrm{HF/d} + [\mathrm{HF/G3L} - \mathrm{HF/d}] + E2/\mathrm{d} + E3/\mathrm{d} + E4/\mathrm{d} \\ & + \Delta\mathrm{QCI/d} + [E2(\mathrm{full})/\mathrm{G3L} - E2/\mathrm{d}] + [E3/\mathrm{plus} - E3/\mathrm{d}] \\ & + [E3/\mathrm{2df,p} - E3/\mathrm{d}] + [E4/\mathrm{plus} - E4/\mathrm{d}] + [E4/\mathrm{2df,p} - E4/\mathrm{d}] \\ & + E(\mathrm{SO}) + E(\mathrm{HLC}) + E(\mathrm{ZPE}) \end{aligned} \quad (17)$$

where we use the short-hand notation: d = 6-31G(d); plus = 6-31 + G(d); 2df,p = 6-31G(2df,p); G3L = G3Large basis set; E(SO) = spin–orbit correction for atoms only; E(HLC) = higher level correction and E(ZPE) = zero-point energy correction. In deriving the scaled methods, the HLC term is set to zero and parameters are introduced that scale the different terms in this energy expression. A systematic study has been performed to investigate the performance of different scaled methods as the number of parameters is increased. In each case, the parameters have been optimized to give the smallest root mean square deviation from experiment for the energies in the G2/97 test set.

The simplest scaled scheme can be obtained by using a single parameter to scale all the correlation energy terms in Eq. (17). Such a single-parameter scaling of G3 theory is similar to the SAC method of Gordon and Truhlar [57] and the PCI-X method of Siegbahn et al. [58]. Such a method gives a mean absolute deviation of 1.43 kcal/mol for the energies in the G2/97 test set (compared to 1.01 kcal/mol for G3 theory). On the other extreme, the addition of scale factors to all the 11 terms in Eq. (17) yields a method with a mean absolute deviation of only 0.97 kcal/mol. However, most of this improvement is obtained using only six parameters (mean absolute deviation of 0.99 kcal/mol). Such a method is referred to as G3S theory.

The G3S energy expression is given in Eq. (18)

$$\begin{aligned} E_0[\mathrm{G3S}] = {} & \mathrm{HF/d} + S_{E234}[E2/\mathrm{d} + E3/\mathrm{d} + E4/\mathrm{d}] + S_{\mathrm{QCI}}[\Delta\mathrm{QCI/d}] \\ & + S_{\mathrm{HF'}}[\mathrm{HF/G3L} - \mathrm{HF/d}] + S_{E2'}[E2(\mathrm{full})/\mathrm{G3L} - E2/\mathrm{d}] \\ & + S_{E3'}\{[E3/\mathrm{plus} - E3/\mathrm{d}] + [E3/\mathrm{2df,p} - E3/\mathrm{d}]\} \\ & + S_{E4'}\{[E4/\mathrm{plus} - E4/\mathrm{d}] + [E4/\mathrm{2df,p} - E4/\mathrm{d}]\} \\ & + E(\mathrm{SO}) + E(\mathrm{ZPE}) \end{aligned} \quad (18)$$

The common scale factor S_{E234} is used for the second, third and fourth-order perturbation corrections at the 6-31G(d) level, S_{QCI} scales the QCI correction beyond MP4/6-31G(d) and $S_{\mathrm{HF'}}$, $S_{E2'}$, $S_{E3'}$, $S_{E4'}$ are the scale factors for the larger basis set extension terms at those levels. Optimization of all six parameters in Eq. (18) gives a mean absolute deviation of 0.99 kcal/mol, which is slightly better than the standard G3 theory with the HLC correction (mean absolute deviation of 1.01 kcal/mol). The optimized values for the parameters in the six-parameter fit are all of reasonable magnitude and range from 0.95 to 1.38. The largest scale factor occurs for the basis set extensions at the third order of perturbation theory. Only one scale factor is less than unity—the scale factor for the basis set extensions at fourth-order perturbation theory (0.95). Overall, an accurate version of G3 theory with scaling of energies can clearly be obtained as also shown by Truhlar and co-workers [61].

In a similar manner, the approximate G3(MP3) method can be modified to use multiplicative scale factors. The resulting G3S(MP3) energy expression is

$$\begin{aligned} E_0[\mathrm{G3S(MP3)}] = {} & \mathrm{HF/d} + S_{E234}[E2/\mathrm{d} + E3/\mathrm{d} + E4/\mathrm{d}] + S_{\mathrm{QCI}}[\Delta\mathrm{QCI/d}] \\ & + S_{\mathrm{HF'}}[\mathrm{HF/G3L} - \mathrm{HF/d}] + S_{\mathrm{E2'}}[E2(\mathrm{full})/\mathrm{G3L} - E2/\mathrm{d}] \\ & + S_{\mathrm{E3'}}[E3/\mathrm{2df,p} - E3/\mathrm{d}] + E(\mathrm{SO}) + E(\mathrm{ZPE}) \end{aligned} \tag{19}$$

Eq. (19) contains five parameters and yields a mean absolute deviation of 1.16 kcal/mol for the energies in the G2/97 test set (compared to the corresponding G3(MP3) deviation of 1.22 kcal/mol).

Finally, the G3(MP2) method can also be modified using multiplicative scale factors. The resulting G3S(MP2) energy expression is

$$\begin{aligned} E_0[\mathrm{G3S(MP2)}] = {} & S_{\mathrm{HF}}[\mathrm{HF/d}] + S_{E2}[E2/\mathrm{d}] + S_{E34}[E3/\mathrm{d} + E4/\mathrm{d}] \\ & + S_{\mathrm{QCI}}[\Delta\mathrm{QCI/d}] + S_{\mathrm{HF'}}[\mathrm{HF/G3MP2L} - \mathrm{HF/d}] \\ & + S_{\mathrm{E2'}}[E2(\mathrm{FC})/\mathrm{G3MP2L} - E2/\mathrm{d}] + E(\mathrm{SO}) + E(\mathrm{ZPE}) \end{aligned} \tag{20}$$

However, the nature of the scaling parameters in this case is somewhat different. In particular, the addition of scale factors to the HF/d and *E*2/d terms was found to be important to yield good results. The resulting six-parameter fit yields a mean absolute deviation of 1.35 kcal/mol for the G2/97 test set, only slightly larger than the 1.30 kcal/mol for G3(MP2).

A summary of the mean absolute deviations of the G3S, G3S(MP3) and G3S(MP2) theories is given in Table 27.5 for the entire G3/99 test set. It is of interest to see if the behavior with the larger test set is similar to that obtained with the fitting set (G2/97). Overall, the mean absolute deviations increase slightly for the G3/99 test set compared

Table 27.5 Summary of mean absolute deviations (kcal/mol) for G3S theories

	G3S	G3S(MP3)	G3S(MP2)
Complete G3/99			
Enthalpies of formation (222)	1.12	1.19	1.29
Non-hydrogens (47)	2.09	2.49	2.37
Hydrocarbons (38)	0.79	0.98	0.86
Substituted hydrocarbons (91)	0.92	0.75	0.99
Inorganic hydrides (15)	0.63	0.79	1.06
Radicals (31)	0.86	0.96	1.17
Ionization energies (88)	1.09	1.27	1.54
Electron affinities (58)	0.90	1.24	1.56
Proton affinities (8)	1.17	1.10	0.74
All (376)	1.08	1.21	1.38

to the G2/97 test set for all three methods. The mean absolute deviation of G3S theory increases from 0.99 to 1.08 kcal/mol, G3S(MP3) increases from 1.15 to 1.21 kcal/mol and G3S(MP2) increases from 1.36 to 1.38 kcal/mol on going from the G2/97 to the G3/99 test set.

The increase in the mean absolute deviation for all three methods is primarily due to large deviations in the calculated enthalpies of formation of some of the non-hydrogen species in the expanded G3/99 test set. This is similar to the results for the G3 methods based on the HLC per electron pair. The G3S mean absolute deviation of 3.37 kcal/mol for the 13 non-hydrogen species in the G3-3 subset is more than twice that of the 35 non-hydrogens in the G2/97 set (1.60 kcal/mol). Similar increases in the mean absolute deviations occur for the G3S(MP3) and G3S(MP2) theories.

The mean absolute deviation of G3S for the G3-3 subset of larger molecules is 1.43 kcal/mol compared 1.30 kcal/mol for G3. The larger increase for G3S suggests that the scaling approach based on six parameters may not work as well on molecules outside the parameterization test set compared to an approach based on the four-parameter HLC. For example, P_4, a molecule with unusual bonding, has a deviation of 10 kcal/mol with G3S theory. This is perhaps due to the nature of the convergence of the individual correlation energy components being somewhat different in such cases. Further analysis of the source of such discrepancies is clearly warranted.

The three new features of G3X theory can also be included in the G3S method. The resulting method is referred to as G3SX and the energy is given by

$$\begin{aligned} E_0[\mathrm{G3SX}] = {} & \mathrm{HF/d} + S_{E234}[E2/\mathrm{d} + E3/\mathrm{d} + E4/\mathrm{d}] + S_{\mathrm{QCI}}[\Delta\mathrm{QCI/d}] \\ & + S_{\mathrm{HF'}}[\mathrm{HF/G3XL} - \mathrm{HF/d}] + S_{E2'}[E2(\mathrm{full})/\mathrm{G3L} - E2/\mathrm{d}] \\ & + S_{E3'}\{[E3/\mathrm{plus} - E3/\mathrm{d}] + [E3/\mathrm{2df,p} - E3/\mathrm{d}]\} \\ & + S_{E4'}\{[E4/\mathrm{plus} - E4/\mathrm{d}] + [E4/\mathrm{2df,p} - E4/\mathrm{d}]\} \\ & + E(\mathrm{SO}) + E(\mathrm{ZPE}) \end{aligned} \tag{21}$$

Eq. (21) is identical to that of the corresponding G3S method, except for the use of the G3XLarge (G3XL) basis set in the Hartree–Fock term instead of the G3Large basis. Also, the single-point energies are calculated at the B3LYP/6-31G(2df,p) geometry and the zero-point energy E(ZPE) is obtained from scaled B3LYP/6-31G(2df,p) frequencies (scaled by 0.9854). The scaling parameters were obtained by fitting to the G3/99 test set. As in the case of G3S, G3SX has six parameters, one for the Hartree–Fock energy extension and five for the correlation terms. Note that ideally the parameters should be close to one and should indicate the level of underestimation of that component; however, the scale factor for the $E4$ term is 0.66. It is likely that it results from the perturbation series being oscillatory in some cases and monotonically convergent in other cases. This may cause problems in some cases (see below). In a similar manner, the methods based on reduced perturbation orders, G3SX(MP3) and G3SX(MP2), are derived by adding the three new features to the G3S(MP3) and G3S(MP2) methods, respectively (not shown).

A summary of G3SX mean absolute deviations from experiment for the G3/99 test set is given in Table 27.6. The mean absolute deviation for G3SX for the 376 energies is 0.95 kcal/mol. This is a substantial improvement over G3S theory, which has a mean

Table 27.6 Mean absolute deviations (kcal/mol) from experiment for the G3SX methods

	G3SX	G3SX(MP3)	G3SX(MP2)
Enthalpies of formation (222)	0.88	0.90	1.26
Non-hydrogens (47)	1.60	1.70	2.23
Hydrocarbons (38)	0.64	0.66	0.67
Substituted hydrocarbons (91)	0.72	0.65	1.12
Inorganic hydrides (15)	0.61	0.65	0.98
Radicals (31)	0.67	0.88	1.06
Ionization energies (88)	1.05	1.16	1.38
Electron affinities (58)	1.02	1.32	1.65
Proton affinities (8)	1.23	1.29	0.70
All (376)	0.95	1.04	1.34

absolute deviation of 1.08 for the same set of energies. The mean absolute deviation for enthalpies of formation decreases substantially from 1.12 to 0.88 kcal/mol. The improvement is due to the non-hydrogen species (2.09 to 1.60 kcal/mol) as well as the substituted hydrocarbons (0.92 to 0.72 kcal/mol), hydrocarbons (0.79 to 0.64 kcal/mol) and radicals (0.86 to 0.67 kcal/mol). The overall behavior G3SX is close to that of G3X. However, consideration of the specific deviations for several non-hydrogens indicates that G3SX does not do as well as G3X for these systems. The scaling approach is especially poor for P_4, which has an error of 8.8 kcal/mol. This is probably due to the small scaling factor for the MP4 term. Otherwise, the overall accuracy of G3SX theory, as assessed on the G3/99 test set, is very similar in terms of the mean absolute deviations to that attained by G3X theory, suggesting that both additive and multiplicative types of parameterizations work equally well. All the G3SX methods have the advantage that they can be used to study potential energy surfaces.

27.7 G3 THEORY FOR THIRD-ROW ELEMENTS

As proposed originally, G2 and G3 theories are applicable only to molecules containing atoms of the first (Li–F) and second (Na–Cl) rows of the periodic chart. They have now been extended [27–29] to molecules containing the third-row non-transition elements K, Ca, Ga–Kr. Basis sets compatible to those used in G3 theory for molecules containing first- and second-row atoms have also been derived. The G3 mean absolute deviation from experiment for a set of 47 test energies containing these elements is 0.94 kcal/mol. This is a substantial improvement over G2 theory for the third row, which has a mean absolute deviation of 1.43 kcal/mol for the same set [27,28]. G3(MP2) theory for third-row molecules has a mean absolute deviation from experiment of 1.30 kcal/mol, and is significantly more accurate than G2(MP2). The G3 method based on third-order perturbation theory, G3(MP3), has an average absolute deviation of 1.24 kcal/mol. In addition, these methods have been assessed on a set of molecules containing K and Ca for which the experimental data are not accurate enough for them to be included in the test set [29]. Results for this set indicate that G3 theory performs significantly better than G2 for molecules containing Ca.

We have also performed preliminary investigations about extending G3 theory to molecules containing the first-row transition metal elements. We have found some significant deficiencies that suggest that important modifications to our standard approach may be needed to reach an accurate model. First, the MP2(full)/6-31G(d) geometries are deficient for many species and density functional geometries (such as B3LYP geometries as in the G3X method) offer significant improvement. Secondly, CCSD(T) performs significantly better than QCISD(T) even for relatively simple diatomic hydrides and is clearly the method of choice, unlike the first- and second-row molecules where their behavior was essentially identical. Thirdly, multiple unrestricted Hartree–Fock (UHF) solutions are obtained in many cases indicating that care should be exercised in arriving at stable solutions. The stable UHF solution in several such cases has significant spin contamination that makes the additive approximations using perturbation theory to be somewhat deficient. Fourthly, the use of different HLC corrections for atoms and molecules causes unphysically large contribution to the binding energies involving atoms with many unpaired electrons. The use of simpler HLC parameters (as in G2 theory) or the use of scaled approaches such as G3S may be more appropriate in such cases. Overall, despite the observations made above, our explorations show promise in deriving an accurate model for transition metal systems [78].

27.8 APPLICATIONS

The Gn methods have been used in numerous quantum chemical studies. In a review of G2 theory [3] published in 1995 we reported a summary of many of the applications of G2 theory up to that time. Since then there have been many additional calculations reported using G2 theory as well as the G3 methods. In this section we present several examples of studies where Gn methods have been used to help resolve experimental thermochemical data.

The first example involves the adiabatic ionization potential of methoxy radical and C–H bond dissociation energy of methanol. In 1991 Ruscic and Berkowitz [81] reported a value of 10.726 eV from photoionization studies that was in sharp disagreement with a previous photoelectron value of 7.37 eV [82]. The value of Ruscic and Berkowitz was confirmed in a study reported by Curtiss et al. [83] who found the G2 ionization potential to be 10.78 eV. This same study [83] also provided theoretical thermodynamic data for the energies of CH_2OH, CH_3O and related compounds (CH_3OH, CH_2OH^+ and H_2CO). The G2 value for the C–H bond dissociation energy of CH_3OH of 96.2 kcal/mol was in disagreement with the experimental value of 92.6 kcal/mol from kinetic measurements [84]. A subsequent theoretical study [85] concurred with the G2 value. The theoretical results motivated a photoionization study by Ruscic and Berkowitz [86] who derived a new value of 94.6 $\pm$ 0.1 kcal/mol for the C–H bond dissociation energy of methanol, which is within the error limits ($\pm$2 kcal/mol) of G2 theory.

Another example is a recent study of proton affinities of a furan, methylphenols and related anisoles by van Beelen et al. [87]. The proton affinities were determined by ion cyclotron resonance mass spectroscopy and theoretically by G3(MP2) theory. The G3(MP2) proton affinity provided support for revision of the proton affinity of furan to 194.1 kJ/mol from the recommended literature value of 191.9 kJ/mol. The experimental study also determined proton affinities for the methylphenols and methylanisoles. The G3(MP2) calculations were used to confirm the order of basicity of the phenols and the anisoles.

Law et al. [88] used Gaussian-3 (G3) and Gaussian-3X (G3X) theories to calculate thermochemical data for chlorine fluorides ClF_n, $n = 1-7$, as well as for their singly charged cations and anions. The quantities calculated included enthalpies of formation and bond dissociation energies of all species, as well as ionization energies and electron affinities of the neutrals. The authors compared the well-established experimental data of ClF and ClF_3 with the G3 and G3X results. They found that the G3X method yielded more accurate enthalpy values. On the basis of these findings, they used the G3X results to appraise the widely scattered experimental for various species such as the enthalpy of formation of ClF. The G3X results were also used for predictions for some missing experimental thermochemical data of chlorine fluorides and their ions.

The final example involves the dissociative photoionization of CH_2Br_2 investigated with photoionization mass spectroscopy by Chiang et al. [89]. An adiabatic ionization energy of 10.25 eV determined for CH_2Br_2 agreed with predictions of 10.26 and 10.25 eV with G2 and G3 methods, respectively. The Gn results, along with this new photoionization study, helped to establish that the ionization energy of CBr in the literature was inaccurate. The Gn results also confirmed their new experimental value for the ionization energy of CHBr.

27.9 SUMMARY AND CONCLUDING REMARKS

G2 and G3 theories are general predictive procedures for thermochemical calculations of molecules containing first- and second-row atoms. They have been extended to molecules containing third-row non-transition elements also. G3 theory is a significant improvement over G2 theory while also being computationally more efficient. Overall, G3 theory has a mean absolute deviation of 1.07 kcal/mol from experiment for the comprehensive collection of 376 energies in the G3/99 test set. This is a slight increase compared to the corresponding deviation of 1.01 kcal/mol for the 301 energies in the smaller G2/97 test set. G3 theory does about as well for the larger hydrocarbons and substituted hydrocarbons in the expanded test set as does for those in the G2/97 test. However, it does poorly for some of the new and larger non-hydrogen systems in the G3/99 test set such as SF_6 and PF_5, which have errors of 6–7 kcal/mol. G3X theory corrects for most of the deficiencies of G3 theory for larger molecules. It includes better geometries as well as *g* polarization functions on second-row atoms to correct for the deficiencies of G3 theory for hypervalent molecules. G3X theory gives significantly better agreement with experiment for the G3/99 test set of 376 energies. Overall the mean absolute deviation from experiment decreases from 1.07 kcal/mol (G3) to 0.95 kcal/mol (G3X). The largest improvement occurs for non-hydrogens for which the mean absolute deviation from experiment decreases from 2.11 to 1.49 kcal/mol. G3X has a mean absolute deviation of 0.88 kcal/mol for the 222 enthalpies of formation in the G3/99 test set. G3S theory based on multiplicative scaling of the energy terms instead of the additive HLC has a mean absolute deviation of 1.08 for the G3/99 test set, an increase from 0.99 for the G2/97 test set. As in the case of G3X theory, the increase is largely due to the new non-hydrogen species in the test set. However, systems such as the highly strained P_4 molecule perform poorly with the scaled methods. G3S and the related G3SX methods have the advantage that they can be used to study potential energy surfaces.

While significant progress has been made in the prediction of accurate thermochemical properties of molecules, much remains to be done in the future. The principal challenge is the extension of the thermochemical models to heavier elements. The paucity of accurate experimental data for the heavier elements makes the calibration and assessment of theoretical models particularly difficult. The first-row transition metals pose a significant challenge and many modifications of the standard procedures appear necessary to arrive at a successful and accurate model. The development of model chemistries with better than kcal/mol accuracy is another important challenge. The W1–W3 models [17–19] that attempt to reach kJ/mol accuracy are significant milestones in this area, though they are applicable to only small molecules. Finally, the development of methods and computational procedures that are applicable to significantly larger molecules is critical to extend the usefulness of these methods to emerging areas such as the study of novel materials and biological systems.

27.10 ACKNOWLEDGEMENT

The Gaussian-n quantum chemical methods reviewed in this chapter have been part of a long-term collaboration with John Pople who passed away this year. We gratefully acknowledge his seminal contributions to the developments covered in this review. We also acknowledge the contributions of many other people who have been co-authors on the papers referenced herein. One of us (LAC) acknowledges support from the US Department of Energy, BES-Division of Materials Science under Contract No. W-31-109-ENG-38.

27.11 REFERENCES

1 J. Cioslowski (Ed.), Quantum-mechanical prediction of thermochemical data, understanding chemical reactivity, Vol. 22, Kluwer Academic Publishers, Dordrecht, 2001.
2 K.K. Irikura and D.J. Frurip (Eds.), Computational thermochemistry, ACS Symposium Series 677, ACS, Washington, DC, 1998.
3 S.R. Langhoff (Ed.), Quantum mechanical electronic structure calculations with chemical accuracy, Kluwer Academic Publishers, Dordrecht, 1995.
4 D.R. Yarkony (Ed.), Modern electronic structure theory, parts I and II, World Scientific Publishing, Singapore, 1995.
5 K. Raghavachari, G.W. Trucks, J.A. Pople and M. Head-Gordon, Chem. Phys. Lett., 157 (1989) 479.
6 J.A. Pople, M. Head-Gordon and K. Raghavachari, J. Chem. Phys., 87 (1987) 5968.
7 For a review of coupled cluster theory see R.J. Bartlett, in: D.R. Yarkony (Ed.), Modern electronic structure theory, parts I and II, World Scientific Publishing, Singapore, 1995, pp. 1047–1131.
8 For a review of correlation consistent basis sets see T.H. Dunning, Jr., K.A. Peterson, D.E. Woon, in: P.v.R. Schleyer (Ed.), Encyclopedia of computational chemistry, Wiley, New York, 1998.

9 T.H. Dunning, J. Phys. Chem. A, 104 (2000) 9062.
10 T. Helgaker, W. Klopper, A. Halkier, K.L. Bak, P. Jorgensen and J. Olsen, in: J. Cioslowski (Ed.), Quantum-mechanical prediction of thermochemical data, understanding chemical reactivity, Vol. 22, Kluwer Academic Publishers, Dordrecht, 2001, pp. 1–30.
11 D. Feller and K.A. Peterson, J. Chem. Phys., 108 (1998) 154; 110 (1999) 8384.
12 D. Feller and D.A. Dixon, J. Chem. Phys., 115 (2001) 3484; J. Phys. Chem. A, 104 (2000) 3048.
13 T.A. Ruden, T. Helgaker, P. Jorgensen and J. Olsen, Chem. Phys. Lett., 371 (2003) 62.
14 J.M.L. Martin, Chem. Phys. Lett., 259 (1996) 669.
15 A. Halkier, T. Helgaker, P. Jorgensen, W. Klopper, H. Koch, J. Olsen and A.K. Wilson, Chem. Phys. Lett., 286 (1998) 243.
16 For a recent review of W1 and W2 methods see J.M.L. Martin and S. Parthiban, in: J. Cioslowski (Ed.), Quantum-mechanical prediction of thermochemical data, understanding chemical reactivity, Vol. 22, Kluwer Academic Publishers, Dordrecht, 2001, pp. 31–66.
17 J.M.L. Martin and G. de Oliveira, J. Chem. Phys., 111 (1999) 1843.
18 S. Parthiban and J.M.L. Martin, J. Chem. Phys., 114 (2001) 6014; 115 (2001) 2051.
19 A. Boese, M. Oren, O. Atasoylu, J.M.L. Martin, M. Kállay and J. Gauss, J. Chem. Phys., 120 (2004) 4129.
20 For recent reviews of G3 theory see K. Raghavachari, L.A. Curtiss, in: J. Cioslowski (Ed.), Quantum-mechanical prediction of thermochemical data, understanding chemical reactivity, Vol. 22, Kluwer Academic Publishers, Dordrecht, 2001, pp. 67–98; L.A. Curtiss and K. Raghavachari, Theor. Chem. Acc., 108, 2002, pp. 61.
21 J.A. Pople, M. Head-Gordon, D.J. Fox, K. Raghavachari and L.A. Curtiss, J. Chem. Phys., 90 (1989) 5622.
22 L.A. Curtiss, C. Jones, G.W. Trucks, K. Raghavachari and J.A. Pople, J. Chem. Phys., 93 (1990) 2537.
23 L.A. Curtiss, K. Raghavachari, G.W. Trucks and J.A. Pople, J. Chem. Phys., 94 (1991) 7221.
24 L.A. Curtiss, K. Raghavachari and J.A. Pople, J. Chem. Phys., 98 (1993) 1293.
25 L.A. Curtiss, J.E. Carpenter, K. Raghavachari and J.A. Pople, J. Chem. Phys., 96 (1992) 9030.
26 L.A. Curtiss, K. Raghavachari and J.A. Pople, J. Chem. Phys., 103 (1995) 4192.
27 L.A. Curtiss, M.P. McGrath, J.-P. Blaudeau, N.E. Davis and R. Binning, J. Chem. Phys., 103 (1995) 6104.
28 J.-P. Blaudeau, M.P. McGrath, L.A. Curtiss and L. Radom, J. Chem. Phys., 107 (1997) 5016–5021.
29 P.C. Redfern,, J.-P. Blaudeau and L.A. Curtiss, J. Phys. Chem., 101 (1997) 8701.
30 K. Raghavachari, B.B. Stefanov and L.A. Curtiss, J. Chem. Phys., 106 (1997) 6764.
31 K. Raghavachari, B.B. Stefanov and L.A. Curtiss, Mol. Phys., 91 (1997) 555.
32 L.A. Curtiss, K. Raghavachari, P.C. Redfern, V. Rassolov and J.A. Pople, J. Chem. Phys., 109 (1998) 7764.
33 L.A. Curtiss, P.C. Redfern, K. Raghavachari and J.A. Pople, Chem. Phys. Lett., 313 (1999) 600.
34 L.A. Curtiss, P.C. Redfern, K. Raghavachari, V. Rassolov and J.A. Pople, J. Chem. Phys., 110 (1999) 4703.
35 L.A. Curtiss, K. Raghavachari, P.C. Redfern and J.A. Pople, J. Chem. Phys., 106 (1997) 1063.
36 L.A. Curtiss, P.C. Redfern, K. Raghavachari and J.A. Pople, J. Chem. Phys., 109 (1998) 42.
37 L.A. Curtiss, K. Raghavachari, P.C. Redfern and J.A. Pople, J. Chem. Phys., 112 (2000) 7374.
38 A.G. Baboul, L.A. Curtiss, P.C. Redfern and K. Raghavachari, J. Chem. Phys., 110 (1999) 7650.
39 P.C. Redfern, P. Zapol, L.A. Curtiss and K. Raghavachari, J. Phys. Chem. A, 104 (2000) 5850.
40 L.A. Curtiss, K. Raghavachari, P.C. Redfern, A.G. Baboul and J.A. Pople, Chem. Phys. Lett., 314 (1999) 101.
41 L.A. Curtiss, K. Raghavachari, P.C. Redfern and J.A. Pople, J. Chem. Phys., 112 (2000) 1125.
42 L.A. Curtiss, P.C. Redfern, K. Raghavachari and J.A. Pople, J. Chem. Phys., 114 (2001) 108.
43 L.A. Curtiss, P.C. Redfern, V. Rassolov, G. Kedziora and J.A. Pople, J. Chem. Phys., 114 (2001) 9287.
44 L.A. Curtiss, K. Raghavachari, P.C. Redfern, G.S. Kedziora and J.A. Pople, J. Phys. Chem., 105 (2001) 227.
45 L.A. Curtiss, P.C. Redfern, K. Raghavachari and J.A. Pople, Chem. Phys. Lett., 359 (2002) 390.
46 For a review on the thermochemistry of radicals see D.J. Henry, L. Radom, in: J. Cioslowski (Ed.), Quantum-mechanical prediction of thermochemical data, understanding chemical reactivity, Vol. 22, Kluwer Academic Publishers, Dordrecht, 2001, pp. 161–198.
47 D.J. Henry, M.B. Sullivan and L. Radom, J. Chem. Phys., 118 (2003) 4849.

48 T.I. Solling, D.M. Smith, L. Radom, M.A. Freitag and M.S. Gordon, J. Chem. Phys., 115 (2001) 8758.
49 For a recent review of CBS methods see G.A. Petersson, in: J. Cioslowski (Ed.), Quantum-mechanical prediction of thermochemical data, understanding chemical reactivity, Vol. 22, Kluwer Academic Publishers, Dordrecht, 2001, pp. 99–130.
50 J.A. Montgomery, Jr., J.W. Ochterski and G.A. Petersson, J. Chem. Phys., 101 (1994) 5900.
51 J.W. Ochterski, G.A. Petersson and J.A. Montgomery, Jr., J. Chem. Phys., 104 (1996) 2598.
52 G.A. Petersson, D.K. Malick, W.G. Wilson, J.W. Ochterski, J.A. Montgomery, Jr. and M.J. Frisch, J. Chem. Phys., 109 (1998) 10570.
53 J.A. Montgomery, Jr., M.J. Frisch, J.W. Ochterski and G.A. Petersson, J. Chem. Phys., 110 (1999) 2822.
54 J.A. Montgomery, Jr., M.J. Frisch, J.W. Ochterski and G.A. Petersson, J. Chem. Phys., 112 (2000) 6532.
55 L.A. Curtiss, K. Raghavachari, P.C. Redfern and J.A. Pople, J. Chem. Phys., 108 (1998) 692.
56 J.A. Montgomery, Jr., M.J. Frisch, J.W. Ochterski, G.A. Petersson, K. Raghavachari and V. Zakrzewski, J. Chem. Phys., 109 (1998) 6505.
57 M.S. Gordon and D.G. Truhlar, J. Am. Chem. Soc., 108 (1986) 5412.
58 P.E.M. Siegbahn, R.A.M. Blomberg and M. Svensson, Chem. Phys. Lett., 223 (1994) 35.
59 P.L. Fast, M.L. Sanchez and D.G. Truhlar, J. Chem. Phys., 111 (1999) 2921.
60 P.L. Fast, M.L. Sanchez, J.C. Corchado and D.G. Truhlar, J. Chem. Phys., 110 (1999) 11679.
61 P.L. Fast, M.L. Sanchez and D.G. Truhlar, Chem. Phys. Lett., 306 (1999) 407.
62 C.M. Tratz, P.L. Fast and D.G. Truhlar, Phys. Chem. Commun., 2 (1999) 14.
63 P.L. Fast and D.G. Truhlar, J. Phys. Chem. A, 104 (2000) 6111.
64 B.J. Lynch and D.G. Truhlar, J. Phys. Chem. A, 106 (2002) 842.
65 B.J. Lynch and D.G. Truhlar, J. Phys. Chem. A, 107 (2003) 3898.
66 P.L. Fast, J.C. Corchado, M.L. Sanchez and D.G. Truhlar, J. Phys. Chem. A, 103 (1999) 5129.
67 A.D. Becke, J. Chem. Phys., 98 (1993) 5648.
68 J.B. Pedley, R.D. Naylor and S.P. Kirby, Thermochemical data of organic compounds, 2nd edn., Chapman and Hall, New York, 1986.
69 M.W. Chase, Jr., C.A. Davies, J.R. Downey, Jr., D.J. Frurip, R.A. McDonald and A.N. Syverud, J. Phys. Chem. Ref. Data, 14, Suppl. 1 (1985) JANAF thermochemical tables third edition.
70 S.G. Lias, J.E. Bartmess, J.F. Liebman, J.L. Holmes, R.D. Levin and W.G. Mallard, J. Phys. Chem. Ref. Data, 17, Suppl. 1 (1988) Gas-phase ion and neutral thermochemistry.
71 R.L. Asher, E.H. Appelman and B. Ruscic, J. Chem. Phys., 105 (1996) 9781.
72 J. Cioslowski, M. Schimeczek, G. Liu and V. Stoyanov, J. Chem. Phys., 113 (2000) 9377.
73 J.F. Harrison, Chem. Rev., 100 (2000) 679.
74 N. Frohlich and G. Frenking, in: J. Cioslowski (Ed.), Quantum-mechanical prediction of thermochemical data, understanding chemical reactivity, Vol. 22, Kluwer Academic Publishers, Dordrecht, 2001, pp. 199–234.
75 W.J. Hehre, L. Radom, J.A. Pople and P.v.R. Schleyer, Ab initio molecular orbital theory, Wiley, New York, 1987.
76 J.A. Pople, H.B. Schlegel, R. Krishnan, D.J. Defrees, J.S. Binkley, M.J. Frisch, R.A. Whiteside, R.F. Hout and W.J. Hehre, Int. J. Quantum Chem. Symp., 15 (1981) 269.
77 A.P. Scott and L. Radom, J. Phys. Chem., 100 (1996) 16502.
78 L.A. Curtiss, P.C. Redfern, K. Raghavachari and J.A. Pople, to be published.
79 C. Moore, Natl Bur. Stand. (US), Circ 467 (1952).
80 J.-P. Blaudeau and L.A. Curtiss, Int. J. Quantum Chem., 60 (1997) 943.
81 B. Ruscic and J. Berkowitz, J. Chem. Phys., 95 (1991) 4033.
82 J.M. Dyke, J. Chem. Soc. Faraday Trans. 2, 83 (1987) 69.
83 L.A. Curtiss, L.D. Kock and J.A. Pople, J. Chem. Phys., 95 (1991) 40440.
84 D.M. Golden and S.W. Benson, Chem. Rev., 69 (1969) 125.
85 C.W. Bauschlicher, S.R. Langhoff and S.P. Walch, J. Chem. Phys., 96 (1992) 450.
86 B. Ruscic and J. Berkowitz, J. Phys. Chem., 97 (1993) 11451.
87 E. Van Beelen, T.A. Koblenz, S. Ingemann and S. Hammerum, J. Phys. Chem. A, 108 (2004) 2787.
88 C.N. Law, S.H. Chien, W.K. Li and Y.S. Cheung, J. Phys. Chem. A, 106 (2002) 11271.
89 S.Y. Chiang, Y.S. Fang, K. Sankaran and Y.P. Lee, J. Chem. Phys., 120 (2004) 3270.

Theory and Applications of Computational Chemistry: The First Forty Years
Edited by C. Dykstra et al.

CHAPTER 28

Factors that affect conductance at the molecular level

Charles W. Bauschlicher Jr. and Alessandra Ricca

Mail Stop 230-3, Center for Nanotechnology, NASA Ames Research Center, Moffett Field, CA 94035, USA

Abstract

The current flow through molecules is discussed. We report on some recent calculations of current–voltage ($I-V$) curves for molecules between two metal surfaces. The change in the density of states of carbon nanotubes when molecules are allowed to interact with their sidewalls is also discussed. We speculate on the future of this area.

28.1 INTRODUCTION

Many people would point to Feynman's 1959 talk "There is room at the bottom" as the beginning of nanotechnology; however, even in 1990 most people thought of nanotechnology as science fiction, or maybe even fantasy. However, this view has changed rapidly as more and more experiments have shown that is it possible to manipulate and/or measure the properties of one or a few molecules. In fact, now most scientists expect nanotechnology to make significant changes in the world in the next 20 years.

While there have been impressive experimental accomplishments, working at the nanoscale is not easy. Many of the experimental techniques cannot be applied to very small systems and therefore there can be uncertainty in the exact structure that has been made or what a specific measurement corresponds to. Computational modeling can give a basic understanding of the chemistry or physics of these nanoscale systems. While these nanoscale systems might be very small by engineering standards, they may still be very large on a molecular scale, and therefore computational modeling of these systems can stretch the current computational techniques to the limit. In this chapter we report on two areas of computational research in nanotechnology that are related to current flowing through molecules. In the first, we consider our recent work on using molecules as components in electrical circuits, for the application of computing, and in the second we consider how gases adsorbed on the side walls of carbon nanotubes (CNTs) affect their electrical properties, for application as gas sensors. After describing our recent work in these two areas, we speculate on the future directions of the application of computational chemistry to nanotechnology.

28.2 MOLECULAR ELECTRONICS

Silicon-based technologies have brought impressive improvements in computer technologies. The computers that most of us have on our desks, which we use for manuscript processing and email, are much more powerful than the state-of-the-art computers of 20 years ago, and cost *vs* performance of workstations that do many of our largest calculations makes access to computer power much less of a bottleneck than at any time in the past. While we can expect silicon-based technologies to continue to improve, in the long term it seems that either the feature size, variations in doping concentration, or other problems will eventually be a road block for silicon-based technologies. Nanoelectronics could provide a way to overcome the limitations of silicon-based technologies. There has been a tremendous effort to design devices and wires. Understanding molecular conduction for a wire between two metal contacts is an important step toward building a molecular device.

Experimentally it is possible to measure the current–voltage ($I-V$) curves for one (or a few) molecule(s) between two metal surfaces using several different techniques [1–5]. Stimulated by such experiments, computationalists have developed techniques [6–19] to compute such $I-V$ curves from first principles. The approach that we use is the self-consistent, non-equilibrium, Green's function approach described by Xue, Datta, and Ratner [15–19]. In this approach, the molecule of interest, X, is placed between two metal clusters, M1 and M2, forming what we refer to as an extended molecule M1–X–M2. The extended molecule is treated using density functional theory (DFT). The extended molecule is connected to the two bulk electrodes using the Green's function approach. In our work, the coupling between the bulk contacts and the extended molecule is determined using a semi-empirical tight-binding approach [20], but we note that other approaches to couple the molecule and contacts exist [8].

We first consider the example of benzene-1,4-dithiol between two Au(111) surfaces—see the top half of Fig. 28.1. The bottom part of the figure shows the extended molecule that is used in our calculations. Since these calculations are quite new, some calibration calculations [21,22] were performed and are illustrated in Fig. 28.2. Not unexpectedly, the $I-V$ curves show some dependence on the basis set used. The valence double zeta (VDZ) is qualitatively correct. Adding polarization functions (VDZP) increases the current at all biases. Adding diffuse functions to the S (VDZ + P) increases the current still further. Replacing the VDZ + P basis by a valence triple zeta set (VTZ + P) actually decreases the current slightly, but the VTZ + P and VDZ + P sets are in reasonable agreement. Increasing the polarization set, to yield the VTZ + (2df,2p) set, makes a small increase in the current. The general increase in current with basis set is consistent with

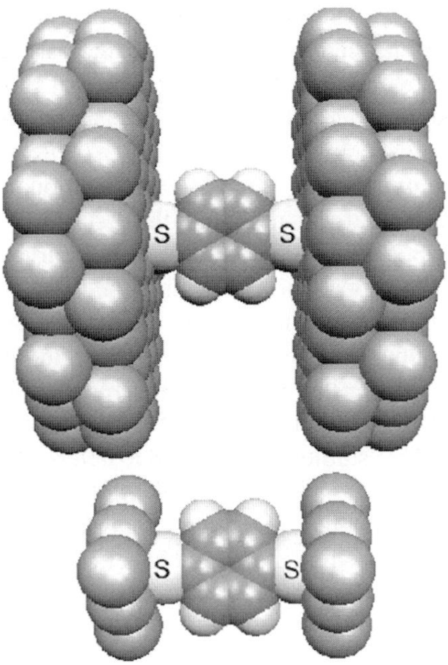

Fig. 28.1. Benzene-1,4-dithiol between two Au(111) surfaces. The bottom part of the figure shows the extended molecule using the Green's function calculation of the $I-V$ curves.

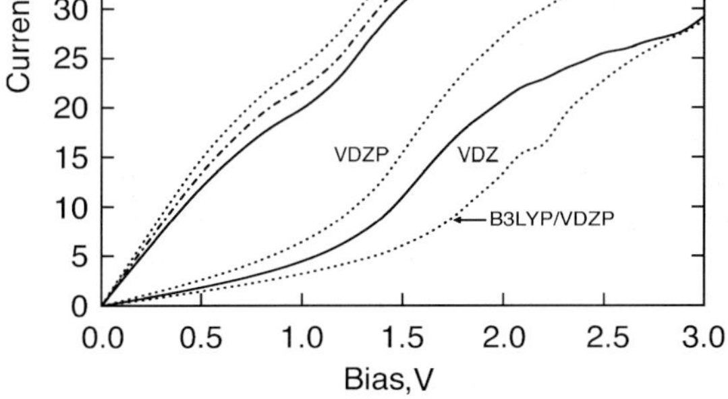

Fig. 28.2. A comparison of the $I-V$ curves obtained using different basis sets and different functionals. The BPW91 functional is used unless otherwise noted.

the view that the bottleneck in the current flow is the Au–S interaction; thus as the basis set is increased, the Au–S interaction is stronger and hence the current is increased. Clearly the three biggest basis sets are in reasonable agreement. This suggests that this property does not have any unusual basis set dependences that make the calculation prohibitively expensive.

Most of the calculations reported in this manuscript use the BPW91 functional; however, in one set of calculations, the B3LYP functional was used, and the results [21] obtained using this functional are shown in Fig. 28.2. The BPW91 and B3LYP curves have a similar shape, but the B3LYP results are smaller by about 25%. While the BPW91 is a pure functional, the B3LYP is a hybrid functional containing some Hartree–Fock exchange. Considering that the functionals are rather different, this small difference suggests that the results may not depend too strongly on the choice of the functional. Since BPW91 tends to overbind molecules to metal atoms and B3LYP tends to underbind them, the true answer probably lies between them. More studies are required to determine which functional is better for this class of problem.

Our calibration study [21] suggests that the computed results are not strongly dependent on the basis set or functional. We have also used 21 atom Au clusters on each end of the benzene-1,4-dithiol instead of only 6, and while this makes some changes in the $I-V$ curves, the variation is smaller than the difference between the BPW91 and B3LYP functionals [23]. Datta and co-workers [8], using the B3PW91 approach in conjunction with a VZP basis set, but with a different approach for the treatment of the metal atoms in the extended molecule and their interaction with the contacts, find an $I-V$ curve that is between our B3LYP and BPW91 curves. From the results so far, it does not appear that the results are too sensitive to the exact treatment of the metal surface.

Despite the small dependence on the computed $I-V$ curves on the level of theory used, the computed curves are about two orders of magnitude larger than experiment [24]. Using a perturbative approach to the calculation of $I-V$ curves, we investigated [25] the sensitivity of the $I-V$ curve to the assumption about the shape of the metal surface, namely, the break junctions used in experiment [24] are not smooth like the (111) surface used in most calculations. Several changes in the surface were tried, but the one change that made a significant reduction in the computed current was bonding the S atom to a small FCC cluster on top of the (111) surface—see Fig. 28.3. That is, the small FCC cluster creates a bottleneck that reduces the current; however an order of magnitude reduction in the computed current still leaves the computed $I-V$ curves an order of magnitude larger than experiment.

While great progress has been made in computing the $I-V$ curves for benzene-1,4-dithiol between Au surfaces, the computed results are still not in complete agreement with experiment, and the origin of the difference is not known. It is possible that despite the calibration calculations the theory is not as refined as believed. It is also possible that we have not yet modeled exactly what is measured in experiment or that experiment is not as accurate as believed. Clearly additional work on this system, probably both theory and experiment, will be required to completely understand the origin of these differences, but the work to date suggests that rigorous theoretical methods can compute $I-V$ curves that are in at least qualitative agreement with experiment.

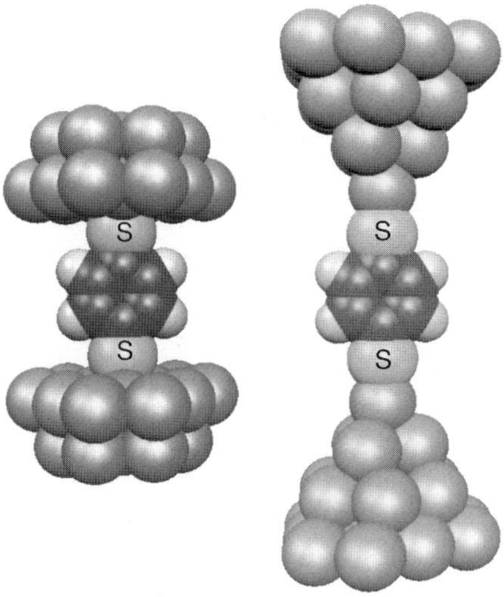

Fig. 28.3. Two models for benzene-1,4-dithiol on a metal surface. The cluster on the left is used to study benzene-1,4-dithiol between two Au(111) surfaces, while the cluster on the right is used to model benzene-1, 4-dithiol bonded to two small clusters, which are sitting on top of the Au(111) surface.

One of the more interesting experimental discoveries has been the negative differential resistance (NDR) found for a conjugated phenylene ethynylene trimer with an NO_2 side group, see molecule IV in Fig. 28.4. Reed et al. [26] have shown, that since this molecule can have a high or low current flow, it could be used to store data. While these experiments have demonstrated that such a system can be used to store a bit of data, the current state of the molecule has a retention time measured in minutes, and therefore, the data stored will be lost after some relatively short time. NDR has been demonstrated for other similar molecules, and the data retention time appears to depend on the specific side groups and their location [27]. It has been suggested [27] that rotation of the benzene rings about the C_2 units can lead to a change in state, and hence a loss of the data. Simple calculations have shown [28] that the rotational barriers of the anions are consistent with the observed retention times.

Given the potential use of this class of molecules as molecular memories, we have studied a series of related molecules. If rotation is in fact the mechanism by which data is lost, it should be possible to dramatically increase the retention time by using molecules that do not have internal rotational degrees of freedom with low barriers. It is also of interest to better understand the NDR mechanism, leading to, perhaps, a better choice of side groups. Molecules I, II, and III shown in Fig. 28.4, and some derivatives have been investigated [22,23] to gain insight into the design of molecules producing NDR.

2,7-Dithiolpyrene, molecule I in Fig. 28.4, cannot rotate, while 4,4′-dithiol-diphenylacetylene (molecule II) has a low barrier for rotation about the central C_2 unit, like molecule IV. An inspection of the HOMO, LUMO, and other nearby orbitals

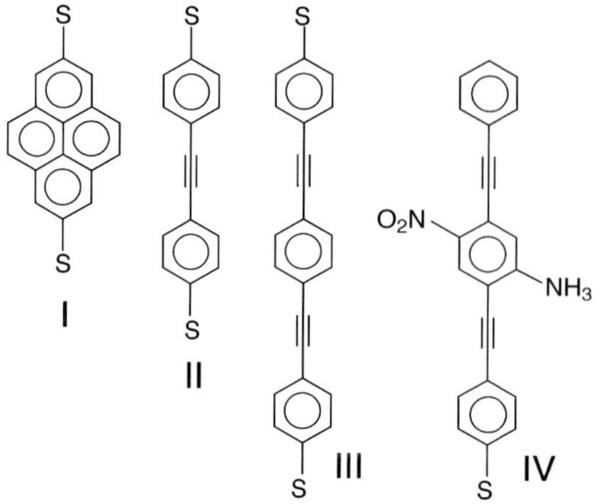

Fig. 28.4. Conducting molecules: (I) 2,7-dithiolpyrene; (II) 4,4′-dithiol-diphenylacetylene; (III) 1-thiol-4-ethynylphenyl-4′-ethynylphenyl-1′-benzenethiolate; and (IV) 2′-amino-4-ethynylphenyl-4′-ethynylphenyl-5′-nitro-1′-benzenethiolate.

for 4,4′-dithiol-diphenylacetylene, shows that the HOMO has a sizeable density on the S atoms and on the carbon atoms. That is, there is an orbital that connects the two Au contacts. There are no similar orbitals for 2,7-dithiolpyrene, suggesting that the current-carrying ability of 2,7-dithiolpyrene will be smaller than that of 4,4′-dithiol-diphenylacetylene. As shown in Fig. 28.5, the computed $I-V$ curves support this view, but the difference is only about a factor of two for biases less than 2.5 eV.

While the planar 4,4′-dithiol-diphenylacetylene can carry more current than the 2,7-dithiolpyrene it must be remembered that rotation about the C_2 central unit will decrease

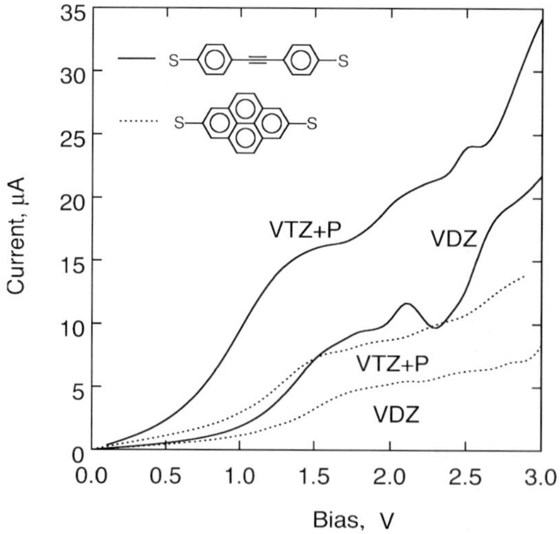

Fig. 28.5. The $I-V$ curves computed for 2,7-dithiolpyrene and 4,4′-dithiol-diphenylacetylene.

the current for 4,4′-dithiol-diphenylacetylene. In Fig. 28.6 we show the $I-V$ curves for 4,4′-dithiol-diphenylacetylene as a function of the angle between the two benzene rings. As expected, the current decreases with the angle, until when the rings are perpendicular, the current is essentially zero. Thus, there is a potential trade-off between using molecules like 4,4′-dithiol-diphenylacetylene and 2,7-dithiolpyrene, where the higher current-carrying ability must be balanced against reduction in current with rotation about the C_2, which could mean a balance between current *vs* data retention time.

One question that arises is how does the NDR work. It has been suggested [29] that the NDR is related to the formation of the anion of the bridging molecule. If this is true, then perhaps other side groups, that stabilize anions, might work better than NO_2. With this in mind, we computed [23] the $I-V$ curves of the benzene-1,4-dithiol and the perchloro derivative and 1-thiol-4-ethynylphenyl-4′-ethynylphenyl-1′-benzenethiolate (molecule III in Fig. 28.4) and its perchloro derivative. For the benzene-1,4-dithiol species, its $I-V$ curve is very similar to that of the perchloro analog. However, for 1-thiol-4-ethynylphenyl-4′-ethynylphenyl-1′-benzenethiolate the $I-V$ curve for the perchloro derivative is less than half of the parent. For the short benzene-1,4-dithiol molecule most of the current bottleneck comes from the Au–S junction, so changes in the energy levels of the C_6 ring due to substitution are minimal. However, for the larger, three-ring species, changes in the rings have a larger effect on the $I-V$ curves. This suggests that both the length and nature of the side group could be important in designing molecules producing NDR. While this is still a very new area of research, we strongly suspect that computational approaches will help elucidate the principles that govern the NDR, and will, therefore, be very helpful in determining the best class of molecules for molecular memories.

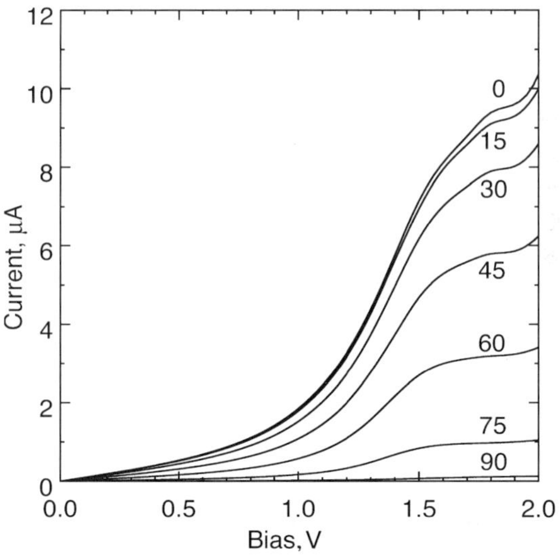

Fig. 28.6. $I-V$ curves for 4,4′-dithiol-diphenylacetylene as a function of the angle between the benzene rings.

28.3 CARBON NANOTUBES AS MOLECULAR SENSORS

Chemical sensors have a wide range of applications in health and environmental sciences. From a NASA prospective, chemical sensors are important in monitoring the health and welfare of humans in space and in exploration, e.g. in determining the makeup of the atmosphere of a distant planet. Ideally, sensors should be accurate, light weight, energy efficient, and have a fast recovery time. It was therefore of great interest when Kong et al. [30] found that CNTs might be very sensitive chemical sensors. They observed changes in the current of a single-walled carbon nanotube (SWCNT) when it was exposed to NO_2 and NH_3. This has led to numerous experiments and calculations, but still a full picture of how the CNTs function as a chemical sensor has not emerged.

Before considering NO_2 or NH_3, we consider O_2 for which more definitive experiments and calculations have been performed. The extreme oxygen sensitivity of SWCNT bundles and thin films was first reported by Collins et al. [31]. They showed that exposure to air or to oxygen greatly modifies the SWCNT's electrical resistance and thermoelectric power. To explain this behavior Jhi et al. [32] performed DFT calculations using the local density approximation (LDA). They showed that O_2 binds to SWCNTs with an adsorption energy of about 0.25 eV and dopes semiconducting tubes with hole carriers. Same conclusions were obtained by Zhao et el. [33] who also used the LDA approach. One major drawback of the LDA approach is that it tends to overbind. When the generalized-gradient approximation (GGA) is used instead of LDA O_2 is only bound by 0.87 kcal/mol with almost no charge transfer [34,35]. The DFT approach does not account for long-range dispersion interactions and is clearly not suited to describe weakly bound systems, and correlated methods such as MP2 are required to treat such systems. We have studied [36] the interaction of O_2 with a metallic (9,0) tube using the MP2 approach within the ONIOM scheme [37–39] and have shown that MP2 is indeed more suited to describe weakly bound systems. O_2 weakly physisorbs both inside and outside with a binding energy of 2 kcal/mol and with almost no charge transfer.

Recent experimental work has begun to question the hypothesis that O_2 dopes SWCNTs. Avouris and co-workers [40,41] have shown that for SWCNT field-effect transistors (FETs), the main effect of oxygen adsorption is not to dope the bulk of the tube but to modify the barrier at the metal–semiconductor contacts. Using photoemission spectroscopy Goldoni et al. [42] have shown that the presence of Na contaminants is responsible for the high sensitivity of SWCNT bundles to oxygen. Clearly a synergy between theory and experiment is required to shed light on the sensing mechanism of SWCNTs.

We have also studied [43] a semiconducting (10,0) SWCNT interacting with O_2 using the ONIOM method of Morokuma and co-workers [37–39] which allows the division of a molecular system into two or three layers and the treatment of each layer using a different theoretical method. We use the MP2 method as the high-level treatment and the Universal Force Field (UFF) [44] as the low-level treatment. We use two different models, denoted as models A and B, which are shown in Fig. 28.7. In model A the double bond connecting carbon 1 to carbon 2 and O_2 are treated as the high-level system and we use the 6-311++G(3df,3pd) basis set to describe it. The rest of the tube is treated as the low-level system using the UFF force field. Model A is used to study the adsorption

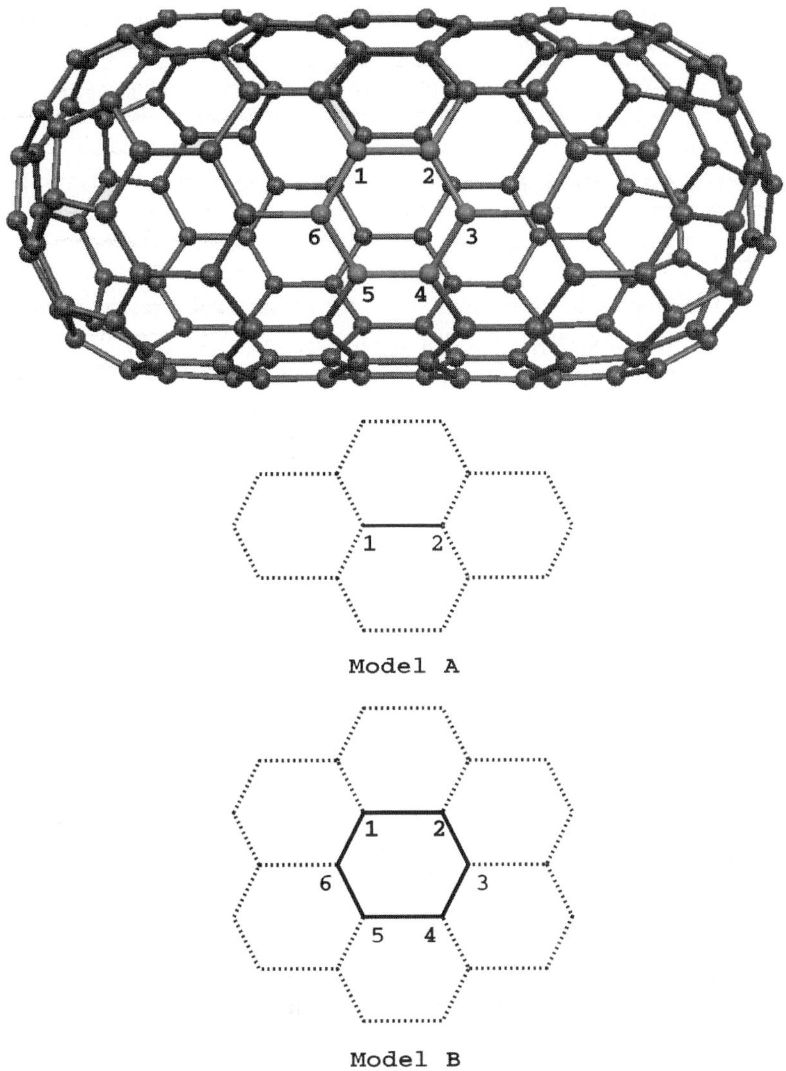

Fig. 28.7. Models used in the ONIOM calculations. For each model, the numbered carbons denote the high-level atoms and carbon atoms directly connected to them are treated as the hydrogen link atoms in the ONIOM calculations. Oxygen atoms are not shown in the figure and are considered as high-level atoms. The bonds shown in bold represent the high-level system and the remaining bonds drawn with dotted lines are just shown to better locate the high-level system in the nanotube.

of O_2 on a double bond. In model B we use a six-membered ring (numbered from 1 to 6 in Fig. 28.7) and O_2 as the high-level system and the 6-311 + G(d,p) to describe it. The rest of the tube is treated as the low-level system using UFF. Model B is used to model the adsorption of O_2 above a six-membered ring. Transition state geometry optimizations are performed by computing analytic second derivatives at every point using the 6-31 + G(d,p) basis set.

Table 28.1 Structures (Å) and binding energies (kcal/mol) of O_2 triplet physisorbed inside and outside a (10,0) nanotube obtained using the ONIOM method

Configuration	Basis set	R(C–O)[a]	R(O–O)	Binding energy[b]
Inside paral. model B	6-311 + G(d,p)	3.592	1.225	2.48
Outside paral. model B	6-311 + G(d,p)	3.422	1.225	2.44
Outside paral. model A	6-311++G(3df,3pd)	3.449	1.218	0.81

[a]Average C–O distance.
[b]Computed without including the zero-point energy.

The physisorption of O_2 triplet both outside and inside a SWCNT was studied and the results are summarized in Table 28.1. The most favorable site outside the tube is above a six-membered ring as shown in Fig. 28.8(b). The oxygen molecule is at 3.13 Å above the center of the hexagon. The oxygen atoms of O_2 are neutral which shows that no charge transfer occurs from the nanotube to O_2. The binding energy of 2.44 kcal/mol is weak.

(a)
(b)

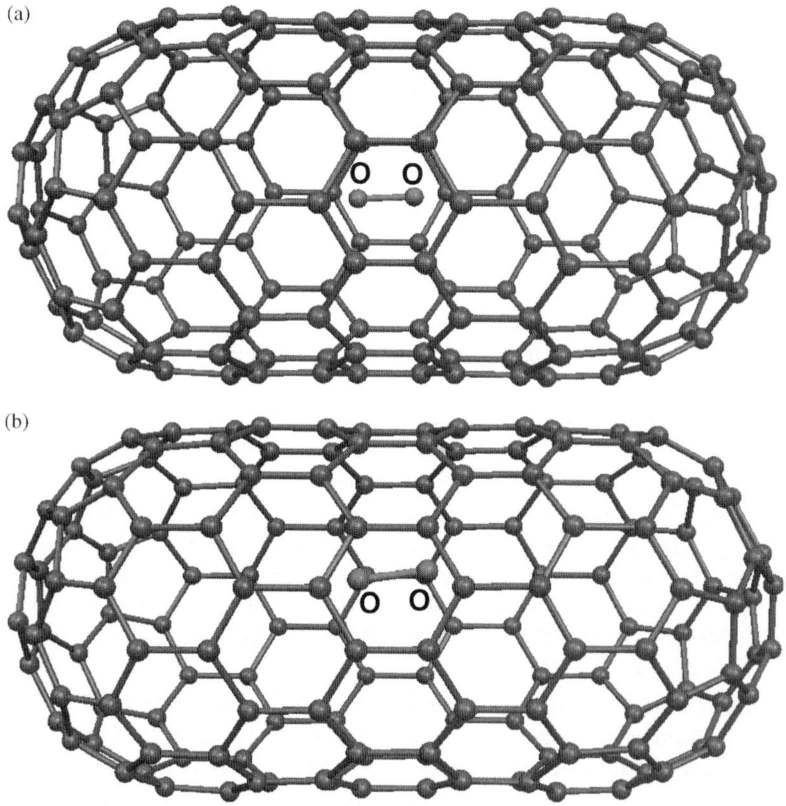

Fig. 28.8. Most stable configurations for the physisorption of O_2 on a (10,0) nanotube: (a) O_2 is inside the tube; (b) O_2 is outside the tube.

Table 28.2 Structure (Å) and binding energy (kcal/mol) of O_2 chemisorbed outside a double bond of a (10,0) nanotube obtained using the ONIOM method

Configuration	Basis set	Spin[a]	R(C–O)[b]	R(O–O)	Binding energy[c]
Paral. model A	6-311++G(3df,3pd)	T	1.407	1.942	– 16.90
Paral. model A	6-311++G(3df,3pd)	S	1.451	1.496	17.50

[a]Indicates the multiplicity of the system.
[b]Average C–O distance.
[c]Computed with respect to tube (singlet) + O_2 (triplet). The zero-point energy is not included.

Inside the tube the most favorable binding site is also above a six-membered ring with the O_2 molecule almost aligned along the nanotube axis at 3.54 Å above the center of the hexagon (see Fig. 28.8(a)). The binding energy of 2.48 kcal/mol for O_2 inside the tube is very similar to that for O_2 outside the tube. No charge transfer occurs.

We also study the chemisorption of triplet and singlet O_2 outside a SWCNT and we report the results in Table 28.2. The chemisorption of singlet O_2 on a double bond is exothermic by 17.50 kcal/mol relative to the ground-state tube and $^3\Sigma_g^-$ O_2 and leads to the formation of a four-membered ring (see Fig. 28.9(b)). The reaction barrier computed with respect to the nanotube and singlet O_2 is 41 kcal/mol. The chemisorption of O_2 triplet is unfavorable by 16.90 kcal/mol (see Fig. 28.9(a)). Overall the results show that the physisorption of O_2 triplet is weak with no charge transfer occurring. The chemisorption of O_2 singlet is energetically favorable but the reaction barrier is high and the reaction is unlikely to occur under normal conditions. The chemisorption of O_2 triplet is energetically unfavorable.

We have computed [43] the electronic density of states of a bare (10,0) tube and of a (10,0) tube interacting with O_2 at the DFT level using plane-waves and periodic bound conditions. The plots are shown in Fig. 28.10. The upper plot corresponds to the bare tube. The band gap is approximately 2 eV, which is somewhat overestimated at this level of theory [45]. In the presence of O_2 a new peak appears at about 0 eV which corresponds to an unoccupied β orbital of O_2. The density of states of the tube is not affected by the presence of O_2 and no new conducting channel is present in the band gap. We can conclude that the presence of O_2 is not going to increase the conductance of the nanotube.

As noted above, the addition of NH_3 was reported to have a significant effect on the current carried by a single SWCNT. Experiments showed [46] that most of the effect was associated with NH_3 interacting with the CNT and not the NH_3 interacting with the contacts. A computational study [47] was performed on NH_3 interacting with graphite and with the side walls of a (9,0) CNT. The calculations were performed at the MP2 level and basis set superposition error (BSSE) was accounted for. The MP2 results were scaled based on CCSD(T) calculations for a small model system. The results are summarized in Table 28.3. The BSSE-corrected binding energy increases with basis set improvement. For a C_6H_6 model of graphite, the best value is 2.02 kcal/mol, taken from the aug-cc-pVTZ basis set. Expanding the C_6H_6 model of graphite to $C_{24}H_{12}$ (coronene) increases the binding energy; compare the results obtained using the 6-31G(2d,p) basis set, 1.13 *vs* 1.88 kcal/mol. Adding the effect of expanding the graphite model to the C_6H_6 aug-cc-pVTZ result yields an estimated binding energy of 2.77 kcal/mol.

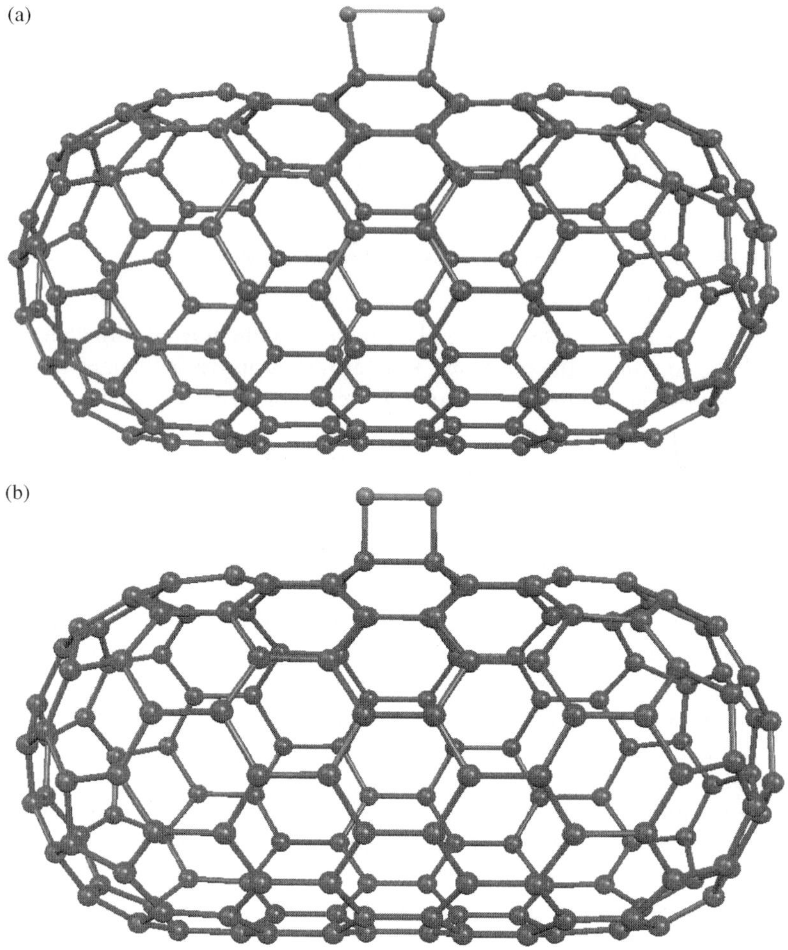

Fig. 28.9. Optimized configurations for the chemisorption of O_2 on a (10,0) nanotube: (a) triplet conformation with O_2 adsorbed on a double bond located in the center of the tube; (b) singlet conformation with O_2 adsorbed on a double bond located in the center of the tube.

A sizeable fraction of the binding is electrostatic in origin and it is known [48] that the quadrupole moments of small hydrocarbons are about twice that of graphite. Using the computed NH_3 dipole and quadrupole moments and the computed quadrupole moments for benzene and coronene yields electrostatic contributions of 0.33 and 1.08 kcal/mol for the benzene and coronene systems, respectively. Applying a correction for an overestimation of the electrostatic bonding yields binding energies of 1.86 and 2.23 kcal/mol for the benzene and coronene models of graphite, respectively. On the basis of these calculations and on the uncertainty in the calculations and corrections, an estimate of 2.0 ± 2.0 kcal/mol was made for the NH_3 binding energy to graphite. It is very difficult to measure [49] the binding energy since the NH_3–NH_3 interaction is of

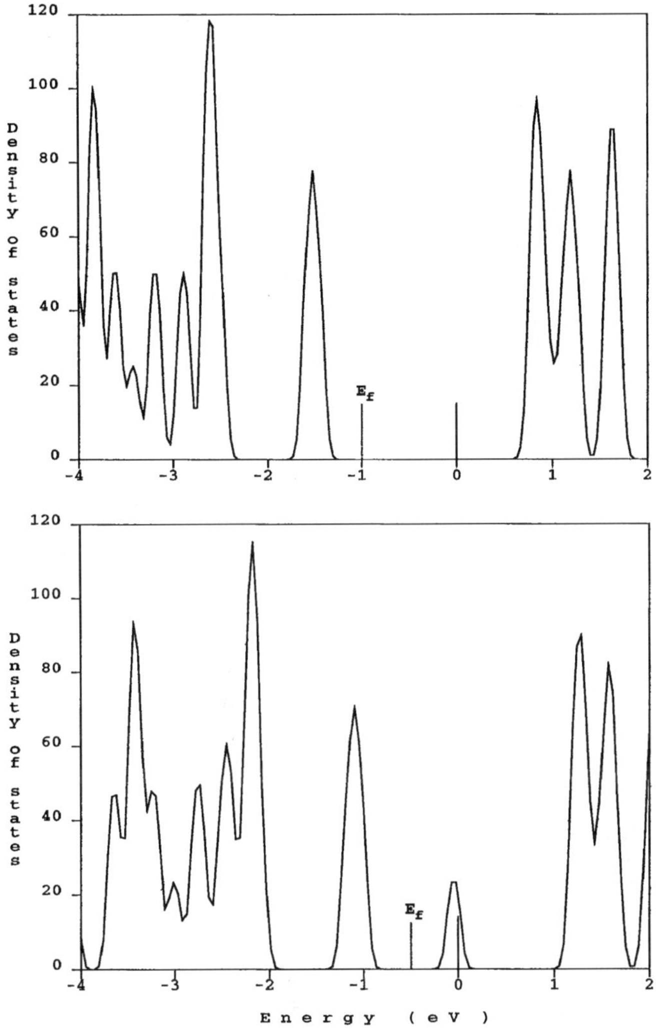

Fig. 28.10. Electronic density of states computed using 4k points and a 0.05 eV Gaussian broadening. The upper plot is for a bare (10,0) tube and the lower plot is for a (10,0) interacting with O_2.

the same size as the NH_3–graphite interaction, but the estimate based on the calculations is reasonably consistent with previous work.

Using a curved coronene to model a (9,0) CNT, the NH_3 binding energy is computed to be about 0.6 kcal/mol smaller than that of graphite—see Table 28.3. However, we note that curving the coronene induces a dipole moment and a CNT does not have a dipole moment. Thus the computed reduction in the binding energy is too large. This problem could be eliminated by treating a larger model for (9,0) that does not have a dipole moment, e.g. a ring of carbons or a small capped tube, see Fig. 28.11. Unfortunately

Table 28.3 NH_3 binding energy, in kcal/mol, computed using the MP2 level of theory

	D_e	D_e-BSSE	Scaled[a]
Planar C_6H_6 graphite model			
6-31G*	2.74	0.91	0.76
6-31 + G*	3.02	1.08	0.90
6-31G(2d,p)	3.01	1.35	1.13
6-311G(2df,2p)	3.66	1.91	1.59
6-311 + G(2df,2p)	2.89	2.11	1.76
aug-cc-pVTZ	3.17	2.42	2.02
Planar $C_{24}H_{12}$ graphite model			
6-31G*	3.26	1.51	1.26
6-31G(2d,p)	4.08	2.25	1.88
Curved $C_{24}H_{12}$ model perpendicular			
6-31G*	3.60	0.79	0.66
6-31G(2d,p)	4.44	1.45	1.21

[a]Scaled by 0.83, which is computed using the ratio of the BSSE-corrected MP2 binding energy (2.06 kcal/mol) to the CCSD(T) (1.72 kcal/mol) value for the aug-cc-pVDZ basis set.

it is not easy to perform MP2 calculations for species this large, and hence one is essentially restricted to DFT calculations. In Table 28.4 we report the results of B3LYP calculations on the ring and tube, as well as on the three models used in the MP2 calculations described above. We first note that the B3LYP results are similar to the MP2, but there is no quantitative agreement. For example, the NH_3 binding energy is larger for

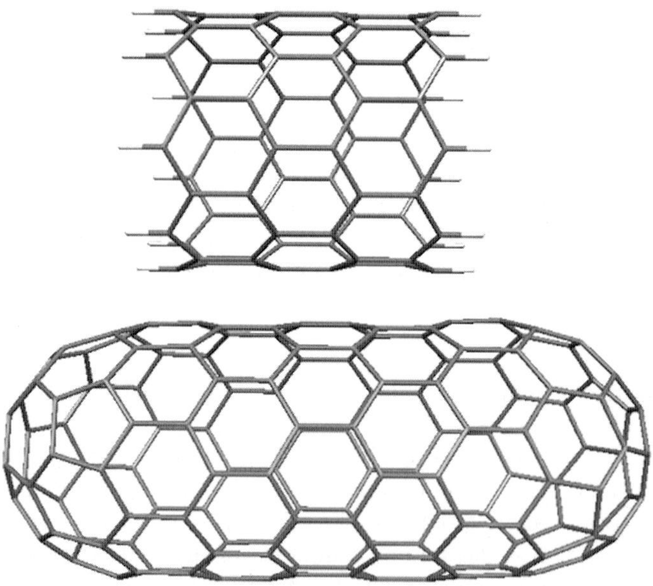

Fig. 28.11. Models for a (9,0) carbon nanotube.

Table 28.4 NH_3 binding energy, in kcal/mol, computed using the B3LYP/6-31G* level of theory

Planar C_6H_6	1.37
Planar $C_{24}H_{12}$	1.25
Curved $C_{24}H_{12}$	1.43
Ring	3.54
Full tube (C_{150}) model	0.51

curved coronene than for planar coronene at the B3LYP level, whereas the opposite is true for the MP2 level. It is perhaps surprising that the ring is so much more bound than the other systems. An analysis of the ring systems shows that the terminal H atoms result in a very large quadrupole moment for the ring and hence in a too large binding energy. The full tube removes both the unphysical dipole moment of the curved coronene and the very large quadrupole moment of the ring and the binding energy is now very small. However, the B3LYP does not account for all the bonding as evidenced by the planar species, where the B3LYP is up to 1 kcal/mol smaller than the MP2. Thus, while the B3LYP can treat a very large system removing some of the non-physical effects observed in smaller cluster models, it does not describe the bonding with sufficient accuracy to yield an accurate binding energy.

The sensor experiments measure the change in current when NH_3 is added. It has been suggested [30] that charge transfer between the NH_3 and CNT affects the CNT density of states to produce the observed change in the current. Our calculations do not support this view; the maximum charge transfer observed in our calculations is 0.008 electrons, which is consistent with the weak bonding. We should note, however, that the NH_3 dipole and quadrupole moment can induce some polarization of the charge on the SWCNT. We therefore consider the change in the band structure with the addition of NH_3. A (10,0) tube is used for this purpose since this is a semiconductor and one expects a larger change in a semiconductor tube than a conducting tube. In Fig. 28.12 we present the computed electronic density of states of a bare (10,0) tube and a (10,0) tube with NH_3 bound to its side wall. An inspection of the two plots shows only small peak at approximately -2.1 eV, which is associated with the NH_3 molecule. We can conclude that the presence of NH_3 alone does not modify the density of states of the nanotube and should not lead to a change in the conductance of the nanotube.

28.4 CONCLUSIONS AND OUTLOOK

We have given examples of our recent calculations on two problems in nanotechnology. The first is in the calculation of $I-V$ curves of molecules between metal surfaces. This is an area where chemistry and physics meet, and the problem is being solved by merging typical quantum chemical calculations with Green's function approaches. We believe that such methods will become more common in the future as engineering-like applications reach the molecular size. However, as this application shows, there are still

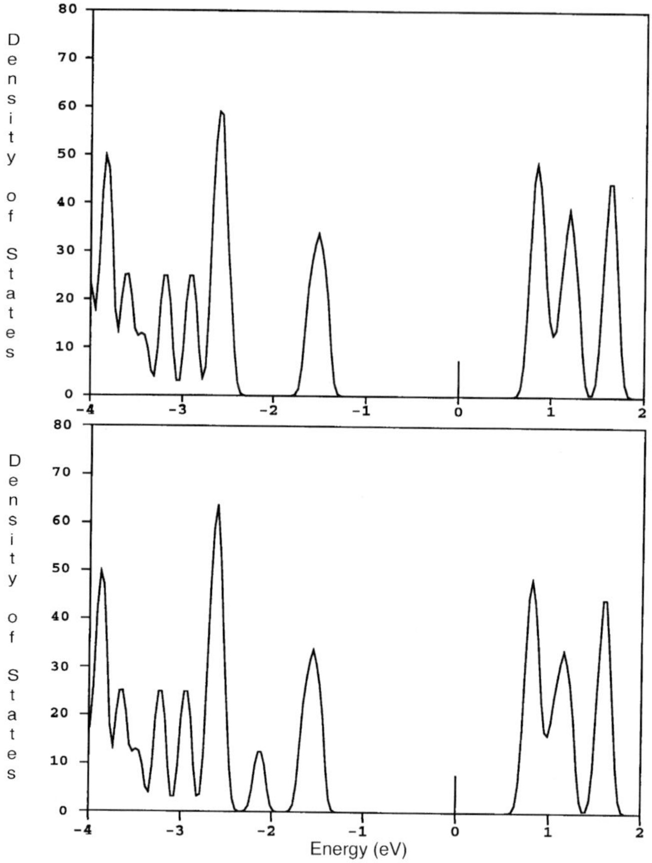

Fig. 28.12. Electronic density of states computed using 4k points and a 0.05 eV Gaussian broadening. The upper plot is for a bare (10,0) tube and the lower plot is for a (10,0) tube interacting with NH_3.

many remaining questions: for example, how best to interface the description of the bulk contacts with the molecular system or how to include forces in the calculations so the geometry can be optimized and frequency computed under bias conditions with a current flowing. To put this simply, so far only the evaluation of energies and densities has been implemented in the calculation of current flowing through molecules between two metal surfaces, but computational chemists have developed many powerful tools, such as the calculation of analytic first and second derivatives that need to be applied to this problem. The close collaboration of physicists and chemists will probably be required to bring this area of nanotechnology to its full potential.

The CNT sensor work shows that current calculations can be used to help answer some of the existing experimental problems; however, as the NH_3 calculations demonstrated, the size of the problems requires some compromises in the level of theory used. Clearly the continued improvement in computers will lead to the ability to treat larger systems at the MP2 level. However, it seems likely that bigger enhancements will come from improved ONIOM methods [37–39], where different levels of theory can be more accurately interfaced together to produce high accuracy for large systems.

28.5 ACKNOWLEDGEMENTS

C.W.B is a civil servant in the Space Technology Division. A.R. is employed by UC Santa Cruz.

28.6 REFERENCES

1 W. Wang, T. Lee and M.A. Reed, Phys. Rev. B, 68 (2003) 035416.
2 L.A. Bumm, J.J. Arnold, T.D. Dunbar, D.L. Allara and P.S. Weiss, J. Phys. Chem. B, 103 (1999) 8122.
3 D.J. Wold, R. Haag, M.A. Rampi and C.D. Frisbie, J. Phys. Chem. B, 106 (2002) 2813.
4 X.D. Cui, X. Zarate, J. Tomfohr, O.F. Sankey, A. Primak, A.L. Moore, T.A. Moore, D. Gust, G. Harris and S.M. Lindsay, Nanotechnology, 13 (2002) 5.
5 R. Holmlin, R. Haag, M.L. Chabinyc, R.F. Ismagilov, A.E. Cohen, A. Terfort, M.A. Rampi and G.M. Whitesides, J. Am. Chem. Soc., 123 (2001) 5075; M.A. Rampi and G.M. Whitesides, Chem. Phys., 281 (2002) 373..
6 P.A. Derosa and J.M. Seminario, J. Phys. Chem. B, 105 (2001) 471.
7 E.G. Emberly and G. Kirczenow, Phys. Rev. B, 64 (2001) 235412.
8 P. Damle, A.W. Ghosh and S. Datta, Chem. Phys., 281 (2002) 171.
9 M. Di Ventra, S.T. Pantelides and N.D. Lang, Phys. Rev. Lett., 84 (2000) 979.
10 L.E. Hall, J.R. Reimers, N.S. Hush and K. Silverbrook, J. Chem. Phys., 112 (2000) 1510.
11 J. Taylor, H. Guo and J. Wang, Phys. Rev. B, 63 (2001) 245407.
12 J. Taylor, M. Brandbygre and K. Stokbro, Phys. Rev. B, 68 (2003) 121101.
13 H. Ness, S.A. Shevlin and A.J. Fisher, Phys. Rev. B, 63 (2001) 125422.
14 M. Magoga and C. Joachim, Phys. Rev. B, 56 (1997) 4722.
15 Y. Xue, S. Datta and M.A. Ratner, J. Chem. Phys., 115 (2001) 4292.
16 Y. Xue, S. Datta and M.A. Ratner, Chem. Phys., 281 (2002) 151.
17 Y. Xue and M.A. Ratner, Phys. Rev. B, 68 (2003) 115406.
18 Y. Xue and M.A. Ratner, Phys. Rev. B, 68 (2003) 115407.
19 Y. Xue, PhD thesis, School of Electrical and Computer Engineering, Purdue University, 2000.
20 D.A. Papaconstantopoulos, Handbook of the band structure of elemental solids, Plenum Press, New York, 1986.
21 C.W. Bauschlicher, J.W. Lawson, A. Ricca, Y. Xue and M.A. Ratner, Chem. Phys. Lett., 388 (2004) 427.
22 C.W. Bauschlicher, A. Ricca, Y. Xue and M.A. Ratner, Chem. Phys. Lett., 390 (2004) 246.
23 C.W. Bauschlicher, A. Ricca and J.W. Lawson, unpublished.
24 M.A. Reed, C. Zhou, C.J. Muller, T.P. Burgin and J.M. Tour, Science, 278 (1997) 252.
25 C.W. Bauschlicher, A. Ricca, N. Mingo and J. Lawson, Chem. Phys. Lett., 372 (2003) 723.
26 M.A. Reed, J. Chen, A.M. Rawlett, D.W. Price and J.M. Tour, Appl. Phys. Lett., 78 (2001) 3735.
27 C. Li, D. Zhang, X. Liu, S. Han, T. Tang, C. Zhou, W. Fan, J. Koehne, J. Han, M. Meyyappan, A.M. Rawlett, D.W. Price and J.M. Tour, Appl. Phys. Lett., 82 (2003) 645.
28 C.W. Bauschlicher and A. Ricca, Chem. Phys. Lett., 375 (2003) 459.
29 J.M. Seminario, A.G. Zacarias and P.A. Derosa, J. Chem. Phys. A, 105 (2001) 791.
30 J. Kong, N.R. Franklin, C. Zhou, M.G. Chapline, S. Peng, K. Cho and H. Dai, Science, 287 (2000) 623.
31 P.G. Collins, K. Bradley, M. Ishigami and A. Zettl, Science, 287 (2000) 1801.
32 S.-H. Jhi, S.G. Louie and M.L. Cohen, Phys. Rev. Lett., 85 (2000) 1710.
33 J. Zhao, A. Buldum, J. Han and J.P. Lu, Nanotechnology, 13 (2002) 195.
34 D.C. Sorescu, K.D. Jordan and Ph. Avouris, J. Phys. Chem. B, 105 (2001) 11227.
35 P. Giannozzi, R. Car and G. Scoles, J. Chem. Phys., 118 (2003) 1003.

36 A. Ricca and J.A. Drocco, Chem. Phys. Lett., 362 (2002) 217.
37 M. Svensson, S. Humbel, R.D.J. Froese, T. Matsubara, S. Sieber and K. Morokuma, J. Phys. Chem., 100 (1996) 19357.
38 S. Humbel, S. Sieber and K. Morokuma, J. Chem. Phys., 105 (1996) 1959.
39 F. Maseras and K. Morokuma, J. Comp. Chem., 16 (1995) 1170.
40 V. Derycke, R. Martel, J. Appenzeller and Ph. Avouris, Appl. Phys. Lett., 80 (2002) 2773.
41 S. Heinze, J. Tersoff, R. Martel, V. Derycke, J. Appenzeller and Ph. Avouris, Phys. Rev. Lett., 89 (2002) 106801.
42 A. Goldoni, R. Larciprete, L. Petaccia and S. Lizzit, J. Am. Chem. Soc., 125 (2003) 11329.
43 A. Ricca, C.W. Bauschlicher and A. Maiti, Phys. Rev. B, 68 (2003) 035433.
44 A.K. Rappe, C.J. Casewit, K.S. Colwell, W.A. Goddard and W.M. Skiff, J. Am. Chem. Soc., 114 (1992) 10024.
45 V. Barone, J. Heyd and G.E. Scuseria, Chem. Phys. Lett., 389 (2004) 289.
46 K. Bradley, J.-C.P. Gabriel, A. Star and G. Grüner, Appl. Phys. Lett., 83 (2003) 3821.
47 C.W. Bauschlicher and A. Ricca, Phys. Rev. B, 70 (2004) 115409.
48 D.B. Whitehouse and A.D. Buckingham, J. Chem. Soc. Faraday Trans., 89 (1993) 1909.
49 P. Rowntree, G. Scoles and J. Xu, J. Chem. Phys., 92 (1990) 3853.

Theory and Applications of Computational Chemistry: The First Forty Years
Edited by C. Dykstra et al.

CHAPTER 29

The CH···O hydrogen bond: a historical account

Steve Scheiner

Department of Chemistry and Biochemistry, Utah State University, Logan, UT 84322-0300, USA

Abstract

Conventional hydrogen bonds usually contain a pair of electronegative atoms, bridged by a H atom. The possibility that a C atom might be involved as a proton donor has a long history of inquiry, going back to the 1930s, and was reinforced by crystal structural information in the 1950s. However, the issue had not been resolved until very recently, when advances in experimental and theoretical methods have permitted detailed and systematic study of putative CH···O H-bonds. The issue had become particularly intriguing, as infrared spectra had indicated that the C–H stretching frequency of at least some such CH···O interactions shifts to the blue, instead of to the red which is the normal shift of almost every H-bond that had ever been investigated. The causes of this unexpected phenomenon are explored, as are the various other aspects of the interaction, so that the CH···O interaction can be properly characterized.

29.1 INTRODUCTION

The story of the hydrogen bond goes back to around 1920 [1], or perhaps even earlier [2], as the concept was in its infancy. As computational chemical techniques were being developed, it was frequently H-bonded systems to which they were applied as both a test of the methods, and as a means of probing the underlying nature of this interaction. Indeed, the very productive interplay between experiment and computational chemistry has long been a centerpiece of our developing understanding of the hydrogen bond.

This chapter focuses not on the standard H-bond, but rather on its weaker cousin, wherein CH acts as proton donor, instead of the usual OH or NH. Because of its weaker nature, the CH···O H-bond offers a particularly stringent test of computational methods, and their ability to make real contributions to the solution of a pressing chemical problem. Of perhaps greater importance, this sort of H-bond was largely unexpected, and therefore unrecognized, by the greater chemical community for many years. It hence serves as a unique forum to showcase the ability of quantum chemistry to address a current issue, as it is developing, and to in fact contribute to that development.

The chapter first presents a history of the problem, how the CH···O H-bond went from a nonissue to one of recognized importance. This history also discusses the contributions made by computational chemistry along the way, at each stage. We then turn our attention to a unique and surprising property of the CH···O bond (or at least certain of them) that led some to initially deny its characterization as a H-bond at all, and others to go so far as to dub it an 'anti-H-bond'. This property is then analyzed and placed into proper perspective through the power of modern quantum chemistry. The chapter closes with a perspective on the paths that future work on this problem is apt to take.

29.2 EARLY THINKING

From its very earliest days, quantum chemistry played an instrumental role in the development of the theory of hydrogen bonding. Perusal of a number of monographs might provide the interested reader with a fascinating chronological history of theoretical ideas, and how they related to contemporary experimental data at each point in time [3–14]. The early definition of a H-bond paired a proton donor group, typically OH or NH, with an acceptor that contained a nonbonded electron pair. And indeed, it was the OH···O, OH···N, and NH···O sorts of H-bonds that dominated most thinking about H-bonds. Nonetheless, the weaker sorts of H-bonds were not completely ignored: a smaller number of studies considered the H-bonding abilities of F, Cl, S, and so on.

There was little thought that the C–H group could be a satisfactory substitute for O–H or N–H as the proton donor, as the similar electronegativity of the C and H atoms is not conducive to charge buildup on the proton. Nevertheless, the concept of a CH···O H-bond is far from a new one, but goes back many years. In 1937, Glasstone [15] rationalized certain observed deviations from ideal behavior in the liquid state *via* a C–H···O H-bond between the CH of $CHCl_3$ and the oxygen of ether. Some support was soon gained for this sort of H-bond in the form of surprisingly high dissociation constants of *o*-toluic and *n*-butyric acids [16]. The issue lay largely fallow for some time until NMR measurements in 1955 [17] confirmed chloroform's ability to act as a proton donor.

The development of experimental means of obtaining molecular structures in crystals opened a new window into this idea. Rather short R(C···O) distances, consistent with a CH···O H-bond, were noted in dimethyl oxalate [18] and uracil [19] in the mid-1950s. The 1960s witnessed a larger set of molecular structures, again supportive of the short C···O contact, and in a geometrical disposition consistent with a H-bond [20–23], including a number of biological systems [24,25].

At about this same time, Brown and Smith's [26] measurements of liquid–vapour equilibrium indicated the presence of a C–H···O bond between the methyl groups of nitromethane or acetonitrile and the O of acetone. Also in the 1960s, there were a number of opportunities to investigate this question *via* spectroscopic inquiry. Allerhand and

Schleyer's [27] comprehensive study of infrared spectroscopic data supported the idea of a CH···O H-bond, and further refined the minimum acidity requirement for the CH donor. Much of the evidence concerning C–H···O H-bonds that had accumulated prior to 1974, based on a range of properties including vapor pressure, azeotrope formation, second virial coefficients, solubility, freezing-point diagrams, enthalpies of mixing, dipole moments, viscosity, refractive indices, electronic spectroscopy, vibrational spectra, and NMR, was summarized in a treatise that appeared in that year [28].

29.2.1 1970s: the beginning of quantum chemical study

At this point in time, in the early 1970s, quantum calculations were still making wide use of semiempirical methods. *Ab initio* studies were coming into their own, but computational limitations restricted them for the most part to small basis sets, and without account for electron correlation. It is now understood that semiempirical calculations are generally unreliable for treatment of weak interactions such as H-bonds [29,30]. The situation is exacerbated for weak H-bonds, such as CH···O, and any such semiempirical data would be of little use. Hartree–Fock minimal basis set calculations are of some utility for standard H-bonds, at least as far as gross features are concerned. However, they could not be relied upon for any of the finer details, and would be suspect when considering H-bonds that are weaker than normal. Another point concerns the use of electron correlation. This effect is important for quantitative accuracy of standard H-bonds, but one could hope to achieve some measure of H-bond strength, and likely compute relative strengths of comparable H-bonds, even without it. However, it is expected that dispersion, and other correlation-related forces play a disproportionately larger role in the weaker molecular interactions, such as CH···O bonds. Density functional theory (DFT) was not a real option at that time. These methods developed rapidly in the 1990s, and were applied to H-bonds, albeit with mixed success. The failure of these methods to include dispersion argued against their ability to provide a trustworthy treatment of weak H-bonds. Taking into account these various factors, most conceded that any sort of reliable, quantitatively accurate treatment of weak H-bonds would almost certainly require an *ab initio* treatment, with a reasonable account of electron correlation, and using a fairly extended basis set, including diffuse functions.

It is nonetheless interesting to consider some of the earlier computations that bucked the odds and piloted theoretical investigations of CH···O interactions. A pioneering set of computations by Morokuma in 1971 [29] considered the CH···O interaction between formaldehyde and water using an *ab initio* approach. It what was likely a remarkable cancellation of errors, in that he applied a minimal Slater basis set with neither correlation nor counterpoise corrections, nor was there much geometry optimization, he nonetheless achieved a quite reasonable estimate of 0.6 kcal/mol for the interaction energy. That same year witnessed the appearance of an attempt to apply a semiempirical approach to H-bonds in general, and CH···O in particular [31]. The authors' consideration of CH_4 as a potential proton donor was plagued by the uncertainties of their method, and they were hence unable to draw any conclusions.

Kollman and coworkers picked up the *ab initio* baton shortly thereafter, still at the SCF level, and with small basis sets. A couple of 1972 papers [32,33] considered HCN as a proton donor. This strong acid is of course expected to form rather strong H-bonds, but the calculations further suggested it is weaker than HF, i.e. HCN···HF is more stable than is HF···HCN. The computations further supported the inability of semiempirical methods like CNDO/2 to treat these systems reliably. Kollman's group raised the bar a bit in 1975, applying a slightly larger 4-31G basis set [34]. Their pairing of the very weak donor CH_4 with NH_3 as a base presented what appeared to be a CH···N H-bond, bound by more than 1 kcal/mol; the much stronger acid CF_3H was bound by 7.6 kcal/mol. These results are now known to be an overestimate of the bonding by a factor of about 2, reflecting first the exaggeration of electrostatic effects by 4-31G, and also the failure to remove basis set superposition error. Nonetheless, there were theoretical confirmations of the potential binding ability of these supposedly weak interactions.

Despite earlier contraindications concerning CNDO/2, Bonchev and Cremaschi [35] nonetheless carried out calculations of CH···X interactions involving a number of different donors and acceptors. Perhaps surprisingly, their work did indeed suggest the presence of a H-bond, even if weak in a number of cases. Impressively, these primitive computations supported the idea that hybridization of the C atom is important, and that sp-hybridized carbon forms the strongest H-bonds, then sp^2, followed by sp^3. The latter idea was buttressed several years later by *ab initio* calculations by Vishveshwara [36], with the minimal STO-3G basis set. This author suggested that substituents play a role as well, albeit less so than hybridization. This work also predicted that HCN would be an even stronger proton donor than is HOH, attributing this feature in large measure to an alignment of dipole moments of the two molecules.

During this decade of the 1970s, experimental work had been progressing as well. Structural analyses of crystals had suggested the presence of CH···O interactions in acetic acid chains [37], as well as implicated the involvement of crown ethers in such bonds [38,39]. Technical developments had facilitated structural information on single molecules in the gas phase, which had in turn led to finding intramolecular CH···O contacts in diethers [40]. The evidence for CH···O H-bonds in biomolecules continued to mount as well, particularly with respect to nucleic acid structures [41–45]. Advances in IR and NMR spectral determinations lent further confirmation to the idea of a CH···O H-bond [46–48].

29.2.2 1980s: more accurate calculations

The release of the Gaussian-80 *ab initio* package around 1980 was a revolution of sorts, permitting much more definitive identifications of minima in potential energy surfaces, as well as facilitating the incorporation of electron correlation. Many of the ensuing computations in the 1980s dealt with alkynes like acetylene, and other triply bonded species such as HCN [49–53]. As an example, a 1983 report [54] assessed the ability of HC≡CH to donate a proton using larger, polarized basis sets, and including correlation. HF and H_2O did not readily form a CH···X interaction, whereas NH_3 was apparently just strong enough of a base so as to form a CH···N bond of surprisingly high strength,

3.6 kcal/mol; however, this result is likely an overestimate as no counterpoise corrections were added. The ability of a hydrocarbon like acetylene to act as a proton donor found experimental support from IR measurements in matrices [55–58] and gas phase structures [59,60], just as the stronger acid HCN forms such bonds [61,62].

More interesting in some ways were carbon donors that were nominally weaker acids. Although restricted to a rather small 4-31G basis set, and without correlation, Hobza and Sandorfy [63] found that the substitution of even one H atom of methane by the more electronegative Cl might be sufficient to facilitate a CH···O H-bond; this bond becomes even stronger with higher degrees of substitution. Their contention was consistent with earlier spectroscopic measurements involving $CHCl_3$ [64] and by low-T inert gas matrix findings [65]. Additional support followed quickly *via* a gas-phase rotational spectrum of $CF_3H\cdots NH_3$ [66], crystal structural work [67], and another spectroscopic indication of a $CCl_3H\cdots O$ H-bond [68].

CH activation is not limited to Cl only but applies to other electronegative groups like NO_2 [69–71]. This idea of 'activation' of the CH group by neighboring groups found contemporary support from crystal structures [72–74], including a comprehensive survey of 113 such systems [75]. Meanwhile, NMR chemical shift measurements were detecting a rapidly growing number of CH···X interactions in a range of different systems [76–78].

On the biological front, the evidence of CH···O interactions in large biomolecules accelerated in the 1980s. A survey of 32 neutron diffraction structures of amino acid crystals [79] yielded convincing evidence for these interactions, most notably involving the $C^{\alpha}H$ group of amino acids. A genuine structural role was attributed to these interactions in determining molecular packing and conformation [80], as for example in the Met-Phe dipeptide [81] where they lead to parallel (*vs.* antiparallel) β-sheet structure. Nucleic acids, too, were observed to make use of these interactions [82,83].

29.2.3 1990s: proliferation and diversification

Theoretical investigations of these weak H-bonds greatly intensified in the ensuing years. Indeed, one can visualize the growing interest in these systems *via* Fig. 29.1 which represents the number of papers published each year that addressed some aspect of CH···X interactions making use of theoretical methods. (This list is not the result of an exhaustive search but reflects the author's own compilation of papers of which he is aware.) One can see a background level of perhaps 3–10 papers annually through the early 1990s, but this number increased to the 17–24 range at the end of that decade. It is interesting to observe a rapid acceleration in 2001 and 2002, up to 39 papers in the latter year.

There was extended consideration of unsubstituted methane as a proton donor [84–92], but all such interactions were found to be quite weak regardless of the proton acceptor, unless of course the acceptor is anionic [93,94]. Further confirmation was established for the notion that the replacement of at least one H atom of CH_4 by a halogen was sufficient to yield an interaction strong enough to be characterized as a H-bond [95,96], and that the interaction strengthened with a growing number of such replacements [97–99]. It was reiterated that the CH might be activated by some other neighboring

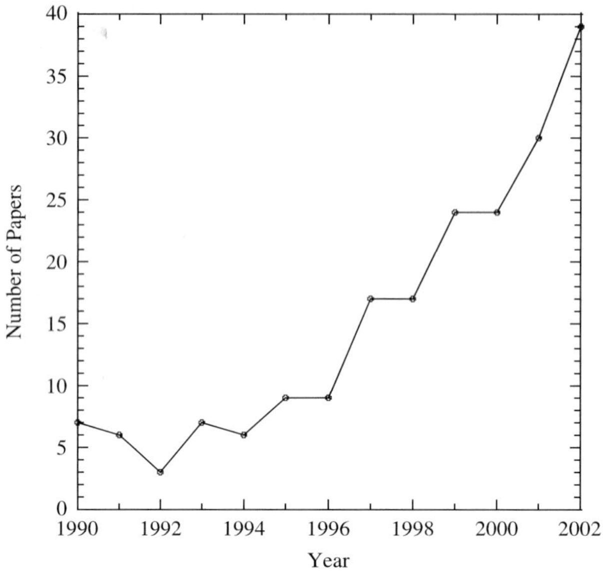

Fig. 29.1. Numbers of peer-reviewed papers published in each year, that apply quantum chemical methods to an inquiry into the issue of CH···X H-bonding.

electronegative group [100–104], although interpretation of the data was commonly clouded by the fact that the CH···O component of interest was secondary to a stronger interaction. This sort of relationship attained a quantitative linear correlation between acidity, interaction energy, and H-bond length [105,106]. Other studies reaffirmed the growing strength of the interaction as the hybridization of the C atom was modified [107], and there were indications that the aryl CH group might act as a proton donor [108,109].

The growing number of crystal determinations facilitated more definitive and quantitative structural information about these weak interactions. Consideration of a large group of alkenes and alkynes affirmed a strong tendency for CH···O linearity [110], and yielded average H-bond lengths of $R(C\cdots O) = 3.46$ Å for alkynes and 3.64 Å for alkenes [111,112], ideas that were further supported by a series of matrix studies and spectroscopic measurements [113–116], as well as the later determination of additional structures [110]. The latter work and others [117] noted that shorter contacts are associated with the more acidic donors. Other workers concluded that the nature of the proton acceptor played less of a role in terms of H-bond geometry than does the donor [118], and offered support to the suggestion that aryl CH groups can participate in H-bonds as well [119–126], as can CH groups that appear in other rings such as cyclopropane and cyclopropene [127], cyclohexane [128], cubyl groups [129], or those in carboranes [130]. Mascal's [131] statistical analysis of a large number of crystals provided 965 examples of the CH···N H-bond and led him to conclude that it is a common occurrence. A 1997 paper summarized some of the findings to that point [132], and concluded that CH···O are "important as secondary interactions and in many instances even play dominant roles in determining crystal packing and molecular conformation, in molecular recognition processes, in the stabilization of inclusion complexes and in the stability and possibly even in the activity of biological macromolecules". Steiner went on to suggest that "a large volume of quantum chemical calculations is required to obtain a broad picture of interaction energies under various conditions".

Meanwhile, the evidence for the presence of these weak H-bonds in biologically important molecules became progressively more overwhelming. Analysis of the structures of 590 oxohydrocarbons [133] identified numerous O···H distances less than 2.7 Å, some as short as 2.3 Å. A survey based on neutron diffraction of 32 zwitterionic amino acids indicated numerous CH···O=C interactions, with r(H···O) between 2.35 and 2.66 Å [8]. Sixteen high-precision neutron diffraction structures of α-amino acids [134] successfully correlated r(CH) with the H-bond length, and concluded that $C^{\alpha}H$ participates in short CH···O interactions. $C^{\alpha}H$···O H-bonds were identified as well in L-prolyl-L-leucine monohydrate [135] and in the collagen triple helix [136]. These interactions appear to be widespread in β-sheets as well [137]. Another study of protein structures solved at the time [138] provided strong evidence that a large percentage of short CH···O=C contacts are cohesive, and that many of these were connected with the $C^{\alpha}H$ of peptides, and went so far as to postulate that "CH···O may be ubiquitous in macromolecular structures". Nucleic acids were also associated with many such interactions [139–147] that also played a role in the binding of drugs [148,149]. An analysis of 26 carbohydrate crystals [150] counted 22% of H atoms that are engaged in CH···O contacts with r(H···O) $<$ 2.5 Å. Another survey found that 93% of CH groups are within 3 Å of a nearby O atom, implying the presence of a stabilizing interaction [145].

29.3 A SURPRISING OBSERVATION

Over the years a set of features had come to be accepted as common to H-bonds, a fingerprint of sorts. Among these standard aspects of a X–H···Y H-bond was the red shift of the X–H stretching frequency, denoting a weakening of this covalent bond within the proton donor molecule that arises upon formation of the H-bond. In addition to the shift toward lower frequencies, this stretching band was typically accompanied by an intensification and broadening. Nonetheless, even some of the earliest indications of the ability of the CH group to act as a proton donor also yielded a surprising finding that the C–H stretching frequency was shifting in the opposite direction, to the blue. Some of the first such observations [151–154] dealt with aldehydes, but there were a number of others as well [64,68]. On the other hand, there was little consistency to this trend, as a number of other CH···O H-bonds seemed to shift in the more traditional, red, direction [27,47,48,65,155]. There were some attempts to rationalize this contrary behavior. Forbes, for example, proposed that steric crowding might be the cause [156] as did Sammes and Harlow [46]. There were on the other hand other observations of a blue shift that could not be explained on this basis [78]. Nonetheless, the occasional blue shift had been relegated to categorization as an odd anomaly that appeared from time to time, but did not warrant a detailed analysis.

It was in the mid-1990s that the continued recurrence of this blue shift became too pressing to ignore any longer. Adcock and Zhang [128], referring to this contrary behavior as the 'Pinchas Effect' in deference to the first reference to it [151], mused as to whether their own observation of a blue shift might be caused by large bends in the CH···O alignment, or whether some sort of electric field effect might be operative. In addition, theoretical calculations had begun to demonstrate that the blue shift that occurs in the C–H stretching frequency [157–159] is connected with a contraction in the equilibrium bond length, again contrary to the stretch that had been found typical of conventional H-bonds [91,160–162].

A contemporary paper [163] initiated the process of analyzing for the source of the contraction. While they did not provide an answer, Giribet et al. began a process of elimination, offering their belief that the shift could not be attributed to the electric field of the acceptor molecule. Barnes and coworkers [164] combined their IR measurements of haloform blue-shifting with semiempirical calculations and proposed the cause might be a repulsion between the acceptor atom and the halogens, which would enlarge the relevant angle, and thereby rehybridize the C atom. The ensuing greater s-character in the C–H bond was thought to strengthen it, causing it to contract and shift its stretch to higher frequency. Indeed, these early themes, involving electric fields and rehybridization, recur later in more modern analyses (*vide infra*).

The issue of these H-bonds that shifted to the blue gained increasing import around 1998 with the publication of a series of papers by Hobza and coworkers. It was pointed out first [165] that HCN shifts to the red (and r_{CH} elongates) when donating a proton to the π cloud of the benzene molecule, whereas both CH_4 and $CHCl_3$ shift to the blue and shorten their CH bond. This group christened the latter set as 'anti-H-bonds' due to their supposed opposite behavior to the red-shifting classical H-bonds. Focusing on the $CHCl_3$ interactions, the blue shift/bond contraction was at first attributed to dispersion phenomena [166]. Continued work in this direction led to expansion of the so-called anti-H-bond concept to nonaromatic systems, as for example the pairing of CHF_3 with ethylene oxide [167]. The authors at this point reversed their earlier claim that the blue shift was dispersion related, as they observed very similar results even when correlation was ignored. They instead proposed that the bond compression might be due to electrostatic interactions, since the molecular dipole moment of CHF_3 increased when the CH bond was shortened.

A contemporary calculation [168] verified the importance of electrostatics in this complex but argued that this interaction was indeed a true H-bond based on the topology of the electron density. This conclusion confirmed Novoa et al. [169] who claimed that even the weak interaction between CH_4 and formaldehyde constitutes a H-bond, based again on the properties of the electron density and its bond critical points. Similar treatments further buttressed this conclusion for other donor–acceptor combinations [170,171]. Finally, a combination of *ab initio* calculations and statistical analysis of crystal structures [172] suggested that while the CH···O interaction polarizes its partner proton acceptor molecule less than would a traditional OH donor, one should nonetheless categorize both as H-bonds.

29.3.1 Status at the end of the 20th century

Thus, it was that after some 60 years from the earliest suggestions of a possible CH···O H-bond, much had been learned, but there remained some pressing questions. Quite a substantial amount of structural data had accumulated from crystal refinements that were suggestive of H-bond geometries. While much of the earlier structural work had been concerned with small molecules, similar sorts of geometrical information was rapidly accumulating for much larger, biological macromolecular systems as well. However, this structural data, despite its volume, was largely mute as to the question of whether the CH donor group was truly attracted toward the acceptor. There was always the possibility that the two groups were commonly forced into proximity to one another by the myriad of other structural restraints of the system, particularly in the larger, more complex biological systems. Of course, this idea was certainly becoming increasingly more remote, given the very large number of such interactions, coupled with the findings of bimolecular interactions in the gas phase and in inert matrices.

From as far back as the early 1970s, quantum chemistry had begun to contribute to the question. Although not quantitatively reliable, these early calculations provided more and progressively stronger indications that the CH donor and O acceptor groups do in fact attract one another. As time wore on, and calculations became more rigorous, correlations were developed between the strength of the interaction, the length of the H-bond, and the acidity of the donor, based on the number of electronegative groups surrounding it. The calculations also confirmed structural indications that the hybridization of the donor C atom is an important factor in the strength of the H-bond.

Other, and perhaps equally compelling, evidence for the presence of these weak H-bonds had come from spectroscopic data, most notably vibrational and NMR. Indeed, it was the former sort of spectral information that had led to the surprising observation of the shift of the C–H stretching vibration in a direction opposite to that of O–H and N–H donors, which had become a virtual trademark of the H-bond. Although dismissed in its earliest manifestations, the increasing frequency of observations of such blue shifts could no longer be ignored. This idea gained further import because empirical correlations had been developed between the degree of red shift on one hand, and H-bond length and energy on the other [173]; how might blue shifts affect this validity of such correlations?

Ab initio calculations had made some progress on addressing the issue, first by connecting these blue shifts with a small contraction of the pertinent CH bond length. But the underlying question as to *why* these interactions undergo a blue shift remained a mystery. There had been a number of proposals, including mutual repulsion, rehybridization, dispersion, and electrostatic phenomena, but these represented unproven hypotheses. Perhaps more fundamentally, there remained the question as to whether the CH···O interaction is in fact a genuine H-bond, or represents some intrinsically different and distinct sort of interaction.

29.4 THE 21ST CENTURY

It was at about this time that our own research group became involved in this issue. We hoped to first address the question as to whether a CH···O interaction constituted a true H-bond. In order to accomplish this task, we planned to make use of the quantum chemical methods that had become progressively more available to researchers in the years leading up to the end of the 20th century. Calculations had established a ‘fingerprint’ of sorts for a H-bond, one that went beyond the sorts of information available from experiment. In other words, while experimental methods of structure determination had led to a picture of preferred geometry for a AH···B H-bond, one in which the θ(AH···B) angle tends toward 180°, and the R(A···B) distance is perhaps 2.9 Å, calculations were able to go beyond this single point, and determine how quickly the stability of the complex is eroded when the geometry deviates from this optimum. Spectroscopic data had shown how formation of a H-bond might alter the characteristics of one or two bands in the IR spectrum, but calculations had been able to establish the effect on a larger group of vibrational bands, and had been able to pinpoint precisely what atomic displacements each correspond to.

In addition to simply amplifying on experimental observations, quantum chemistry had been able to consider aspects of a H-bond which were simply unavailable to experiment. For example, the determination of the actual energetics of H-bonding had been notoriously difficult to pin down in an *in vacuo* setting. Quantum calculations were well suited to evaluate the energetics of this process. From a deeper standpoint, means had been devised to dissect the binding energy into various components, e.g. electrostatic attraction and steric repulsion. No less important, quantum calculations had been shown able to examine the perturbations imposed on the electron cloud of each subunit as the H-bond is formed.

29.4.1 H-bond: to be or not to be?

In an attempt to determine which aspects of the fingerprint of a conventional H-bond are satisfied by a CH···O interaction, our group devised a set of complexes that varied in a systematic manner from extremely weakly bound to others that are comparable to a traditional H-bond [174]. As proton donors, we considered the set of fluorinated methanes. CH_4 itself is of course a very weak acid, and its interaction with a base is expected to be only marginally binding. But as each H of methane is changed to F, one by one, all the way up to CF_3H, the acidity increases as does the molecule’s likely ability to contribute a bridging H to a H-bond. The proton acceptors all contained an oxygen atom, but this O was considered in both a hydroxyl environment, as in HOH and CH_3OH, and as a carbonyl in H_2CO. Altogether, the study encompassed the 12 different complexes involving each of four different proton donor molecules with three different acceptors. All results were compared to data computed for the OH···O H-bond in the water dimer, as the paradigm of classical H-bonding.

Before embarking on a detailed comparison, it was first necessary to establish the level of theory necessary to achieve reliable results. Table 29.1 illustrates the binding energies computed for each of the three different proton acceptor molecules with CH_4 as proton donor. The results indicate first that the energetics are fairly insensitive to basis set. Whether SCF or MP2, the binding energies change very little from one row to the next. Correlation does appear to be important as the values in the lower section of data are considerably higher than those in the SCF section. On the other hand, the precise means of incorporating correlation is much less important, as MP2, MP4, CCSD(T), and QCISD

Table 29.1 Binding energies ($-\Delta E$, in kcal/mol) computed for various acceptor molecules, all with CH_4 as proton donor; data corrected by counterpoise procedure

Method	Basis set	OH_2	CH_3OH	H_2CO
SCF	6-31 + G*	0.21	0.20	0.15
	6-31 + G**	0.21	0.22	0.16
	6-31++G**	0.21	0.22	0.16
	6-311 + G**	0.20	0.22	0.17
	aug-cc-pVDZ	0.16	0.16	0.14
MP2	6-31 + G*	0.24	0.37	0.35
	6-31 + G**	0.29	0.43	0.37
	6-31++G**	0.30	0.43	0.37
	6-311 + G**	0.35	0.50	0.38
	aug-cc-pVDZ	0.43	0.63	0.46
CCSD(T)	6-31 + G**	0.29	–	0.35
MP4	6-31 + G**	0.29	0.43	0.37
QCISD	6-31 + G**	0.23	0.34	0.31
B3LYP	6-31 + G**	0.25	0.29	0.13
	6-311 + G**	0.26	0.30	0.14

all yield similar results. It might be noted finally that the hydroxyl O acceptors are not quite as strong as the carbonyl O in H_2CO, although the differences are not dramatic.

These same trends are in evidence when a common acceptor molecule, H_2O, is paired with a series of progressively stronger proton donors, as may be noted in Table 29.2. Again, basis set sensitivity is low, both at the SCF and correlated levels. And once more the precise means of including electron correlation is unimportant. It is noted parenthetically that DFT computation of the binding energetics provides surprisingly accurate results, in both Tables 29.1 and 29.2. The ability of fluorosubstitution to enhance

Table 29.2 Binding energies ($-\Delta E$, in kcal/mol) for various donor molecules, all with OH_2 as acceptor; data corrected by counterpoise procedure

Method	Basis set	H_3CH	FH_2CH	F_2HCH	HOH
SCF	6-31 + G*	0.21	1.40	2.23	4.17
	6-31 + G**	0.21	1.14	2.28	4.21
	6-31++G**	0.21	1.15	2.27	4.20
	6-311 + G**	0.20	1.11	2.23	4.10
	aug-cc-pVDZ	0.16	0.89	1.85	3.53
MP2	6-31 + G*	0.24	1.33	2.57	4.61
	6-31 + G**	0.29	1.34	2.53	4.51
	6-31++G**	0.30	1.34	2.53	4.56
	6-311 + G**	0.35	1.28	2.34	4.30
	aug-cc-pVDZ	0.43	1.23	2.24	4.11
CCSD(T)	6-31 + G**	0.29	1.32	2.50	4.36
MP4	6-31 + G**	0.29	1.31	2.47	4.34
QCISD	6-31 + G**	0.23	1.26	2.45	4.25
B3LYP	6-31 + G**	0.25	1.32	2.45	4.80
	6-311 + G**	0.26	1.28	2.48	4.77

the proton-donating ability of the donor molecule is obvious in Table 29.2. Each replacement of a H atom by F increases the binding energy by roughly 1 kcal/mol. This quantity is rapidly approaching the value for the water dimer, in the final column of Table 29.2, which is nearly reached when F_3CH is bound to water.

Geometry optimizations of each complex revealed a simple correlation between the strength of each H-bond and its length. That is, each fluorosubstitution leads not only to an approximate 1 kcal/mol strengthening, but also to a contraction of R(C···O). For example, at the MP2/6-311 + G** level, the equilibrium values of R(C···O) for H_3CH, FH_2CH, and F_2HCH are 3.649, 3.511, and 3.379 Å, respectively, amounting to an approximate contraction of 0.13 Å for each F atom.

There are a number of other facets of a H-bond that are amenable to computation as well. For example, it is typically observed that formation of a H-bond leads to the transfer of a certain amount of electron density from the proton acceptor molecule to the donor. The first row of Table 29.3 illustrates the progressive increase of this quantity, from 4 me for $H_3CH\cdots OH_2$, up to 11 me for $F_3CH\cdots OH_2$, rapidly approaching the value of 13 me for the water dimer. (Charges were computed by a natural bond orbital analysis [175,176], uncorrected for basis set superposition error.) Associated with this transfer of electron density is an enhancement of the dipole moment of the complex. The second row of Table 29.3 lists the increase in dipole moment (along the CH···O axis), as compared to the moment that would result from simple vector addition of the two isolated monomers. Reading across from left to right, one can again see the steady progression of increasing values as each H is replaced by F, approaching the value in $HOH\cdots OH_2$.

Another typical feature of conventional H-bonds is the loss of electron density from the bridging H nucleus, making this hydrogen more positively charged. The third row of Table 29.3 reports the magnitude of this charge loss, varying between 23 and 32 me for the four CH···O H-bonds. This property is somewhat larger than in the conventional H-bond of the water dimer, where the charge loss on the bridging hydrogen is 19 me. The proton donor and acceptor atoms (C and O, respectively) both undergo increases in electron density upon H-bond formation, resulting in greater negative charge. The next two rows of Table 29.3 document these enhanced charges, and once more obey the same

Table 29.3 Electronic properties[a] of different donor molecules when engaged in H-bond with OH_2 as an acceptor

	H_3CH	FH_2CH	F_2HCH	F_3CH	HOH
Charge transfer (me)	4	6	8	11	13
$\Delta\mu_z$[b] (D)	0.217	0.277	0.336	0.405	0.409
Δq_H (me)	23	26	29	32	19
Δq_C (me)	−7	−9	−13	−20	−24
Δq_O (me)	−2	−5	−9	−12	−18
$\Delta\sigma_H$[c] (ppm)	−1.12	−1.15	−1.18	–	−2.60

[a]Natural population analysis [175,176].

[b]Dipole moment of complex, along XH···O axis, relative to vector sum of two unperturbed monomers, in same orientation as in complex.

[c]Isotropic NMR shift, at MP2/GIAO level.

Table 29.4 Morokuma decomposition contributions to binding energies of different donor molecules when engaged in H-bond with OH_2 as an acceptor

	H_3CH	FH_2CH	F_2HCH	F_3CH	HOH
ES	−0.42	−1.96	−3.83	−7.06	−7.58
EX	0.38	1.17	2.06	4.14	4.24
POL	−0.13	−0.24	−0.36	−0.69	−0.71
CT	−0.11	−0.32	−0.53	−0.97	−0.93
CORR	−0.08	−0.20	−0.25	−0.25	−0.30

trends of increasing magnitude with each progressive F-substitution, approaching the water dimer case. In summary, the electron density shifts in the CH···O systems all show progressive increases that parallel the acidity of the donor, and approach what is observed in the traditional H-bond of the water dimer.

The final row of Table 29.3 lists the change in the isotropic NMR chemical shift calculated for the bridging proton as a result of H-bond formation [177]. The values for all CH···O H-bonds are negative, consistent with the same sign for the OH···O bond, another indicator of the similarity between these different interactions.

It is commonly accepted that the H-bond is composed largely of electrostatic interactions, with smaller contributions from factors that involve redistribution of electrons within the subunits, such as polarization. One way of assessing this assertion is *via* an energy decomposition scheme which is possible *via* quantum chemistry. There are a number of means of performing such a decomposition; one of the oldest and most reliable schemes was derived by Morokuma and coworkers [178,179]. The electrostatic (ES) and exchange repulsion (EX) result from the interactions of the 'frozen' charge clouds of the two subunits, prior to any redistribution that results from a mutual interaction. Polarization (POL) and charge transfer (CT) energies are a consequence of the subsequent electronic redistribution, which allows density to move from one section of the complex to another. The latter terms are computed at the SCF level; correlation contributions can be gathered into a CORR term which includes dispersion as its largest segment.

The various contributions are collected in Table 29.4 where it may be noted that the negative signs of ES, POL, CT, and CORR all represent attractive interactions. It is the exchange energy, with its positive sign, which keeps the two molecules from collapsing together any further. Note that these sign patterns are common to all systems in Table 29.4, whether CH···O or OH···O. As observed repeatedly above, one sees again the same progressive magnification of each term as the H atoms of methane are changed one by one to F. And again the strengthening of the CH···O bond leads ultimately to values quantitatively similar to those observed in the conventional OH···O bond of the water dimer.

29.4.2 Underlying reasons for blue shift

Up to this point, all properties of the CH···O bonds have been similar to those computed for OH···O, even if smaller in magnitude for these weaker bonds. We now turn to a set of properties which violate this principle. The first row of Table 29.5 illustrates that the CH bonds of the proton donors contract when engaged in a H-bond with the water molecule. This behavior contrasts with the elongation of the OH bond of a water donor. Associated with these changes are the shifts to higher (lower) frequency of the stretching vibrations of the corresponding CH (OH) bonds. Moreover, as indicated in the last row of Table 29.5, the intensities of the former CH bonds are reduced ($I/I_0 < 1$) while that of OH is magnified. We are thus confronted with a set of properties, all related to the XH bond of the donor molecule, where the behavior of the CH···O bond seems contrary to that of the conventional H-bond. It was indeed this 'antisocial' behavior that spurred some to classify this system as an 'anti-H-bond', so it behooves us to study it in some detail, *particularly* as all the other features of this sort of system fall so nicely into line with traditional H-bonds.

Our group's first attempt [174] to examine the underlying reasons for this contrary behavior made use once more of the Morokuma energy partitioning protocol. The CH···O system of $F_2HCH\cdots OH_2$ was compared directly with $HOH\cdots OH_2$. For each system, the effect of stretching the bridging H a small distance from its equilibrium position, along the XH···O line was computed. The results are presented in Table 29.6 where some strong similarities are readily apparent. For *either* system, the electrostatic terms are negative, indicating that this component favors the stretching of the H atom away from the donor atom, whether C or O. Likewise, the polarization, charge transfer and correlation energies, all tend to push the H away from the donor atom in *both* systems, albeit not as strongly as does ES. Working in the opposite direction, the exchange energy disfavors this stretching motion in CH···O and OH···O, as indicated by the positive signs for EX in Table 29.6, as does the MIX term (which 'mixes' together all contributions not neatly categorized as ES, EX, POL, or CT in the Morokuma scheme).

So the behavior of the CH···O bond is identical to that of the OH···O interaction in terms of the tendencies of each and every component. Why then does the CH bond contract whereas OH elongates upon formation of these bonds? The answer rests in the quantitative aspects of each component. For example, the ES term is three times greater for the OH···O system, while the exchange term which works in opposition is less than twice as large. In summary, when all terms are added together, the total force (last row of Table 29.6) is negative for $HOH\cdots OH_2$ but positive for $F_2HCH\cdots OH_2$. It is for this reason that the former system tends to stretch its OH bond whereas the CH bond in the latter contracts. In other words, the fact that the CH bond contracts while the OH bond elongates does not represent any fundamental difference between the two types of interactions, but reflects merely the far less profound result that the forces pushing toward

Table 29.5 Changes in properties of CH(OH) bond of proton donor molecules when engaged in H-bond with OH_2 as an acceptor

	H_3CH	FH_2CH	F_2HCH	F_3CH	HOH
Δr_{XH} (mÅ)	−0.5	−1.7	−2.7	−2.3	+5.2
$\Delta \nu_{XH}$ (cm^{-1})	10	22	26	42	−31
I/I_0	0.07	0.09	0.16	0.70	1.89

Table 29.6 Changes in Morokuma decomposition terms (kcal/mol) when bridging proton is stretched 0.01 Å away from equilibrium position toward proton acceptor

	$F_2HCH\cdots OH_2$	$HOH\cdots OH_2$
ES	−0.07	−0.20
EX	+0.13	+0.22
POL	−0.01	−0.03
CT	−0.03	−0.07
MIX	+0.01	+0.02
CORR	−0.02	−0.02
Total	+0.01	−0.08

contraction in one case are slightly larger than the elongation forces, while the opposite is true in the other case.

Our work thus found that the CH···O H-bond fits the classical fingerprint of conventional H-bonds in a large number of ways. The energetics of the interaction, and their trends as the proton donor is rendered more acidic, are quite classical, as are the behavior of the intermolecular separation, and the shifts of electron density that accompany formation of the bond, not only on a molecular level, but with respect to individual atoms. The NMR chemical shifts of the bridging H are similar for CH···O and OH···O, as are the magnitudes of the individual components to the total interaction energy. These relationships were confirmed by a contemporary paper [170] that extended the list of similarities between CH···O and conventional H-bonds to computed electron densities at key locations in the complexes. Where the two sorts of interactions differ is in the question of whether the covalent XH bond contracts or elongates when the H-bond occurs, with accompanying changes in stretching frequency and intensity. However, the calculations indicated this distinction does not reflect any profound differences, but results merely from certain quantitative differences that result in tipping the already delicate balance from one direction to the other.

The similarity between CH···O and OH···O H-bonds is further reinforced when one recognizes that it is only a subset of CH···O bonds that shift to the blue. It had long been recognized that sp-hybridized alkynic CH bonds, as in HCN or HCCH, shift to the red, as do conventional H-bonds [55–57,114]. Most of the CH···O systems that were noted to shift to the blue tended to contain sp^3-hybridized carbon. These patterns were reproduced by calculations [107,171,180–183]. Our own contribution was to first verify this distinction between sp and sp^3 systems, and then to study relations between the magnitude of the shift (either red or blue) and other properties such as H-bond energetics, intermolecular separation, and NMR spectral shifts [184]. We also considered the intermediate sp^2 situation, and found that such CH···O bonds undergo a very small change in either the C–H bond length, or its stretching frequency. What changes are observed can be in either direction, but again, are quite small. In conclusion, one can make the following predictions about the CH···O bonds: alkynes will shift to the red, alkanes to the blue, and alkenes will undergo very little shift. This conclusion has been verified by other calculations [185–191].

29.4.3 Other approaches—similar findings

Our group's analysis based upon energy decomposition was of course not the only means of assessing the underlying reason for the differing direction of shift of the XH vibrational frequency. Concurrent with a study which found grounds to assert that the direction of shift has at least part of its origin in electrostatic effects [192], another set of computations [193] arrived at a related conclusion. More precisely, Masunov et al. were able to reproduce contractions or elongations of CH bonds that occur in certain molecules when they act as proton donors in a complex, simply by placing each molecule in an external electric field, even in the absence of the acceptor molecule. This work was admittedly limited to only three CH donors but was nonetheless intriguing in that it was suggestive that the direction of shift was an intrinsic property of the donor molecule, in and of itself, independent of the precise nature of the acceptor molecule. A second important conclusion had to do with a limit on CH bond contraction. It was predicted that the contraction of the CH bond in an alkane like methane was limited; this trend was predicted to reverse itself beyond a certain field strength, such that the CH bond would begin to elongate.

The attribution of the stretching or shortening of the CH bond to the intrinsic properties of the donor molecule itself confirmed our own earlier contention [174] that, opposite to the behavior of most proton donor molecules where the dipole moment is enhanced when the OH bond is elongated, the dipole of alkane donors tends to increase when the CH bond is shortened. This trend is consistent with the observed CH bond contraction in that the resulting increased dipole moment can then interact more strongly with the field generated by the acceptor molecule. This same idea was further developed by Hermansson [194] who associated the direction of spectroscopic shift with the sign of $d\mu/dr_{XH}$. This author went further and discussed the exchange overlap effects. The analysis supported the earlier contention by Masunov et al. [193] that blue-shifting molecules will in fact shift to the red in the presence of a sufficiently strong electric field, which might be generated by an anionic proton acceptor.

Pejov and Hermansson [195] later delved somewhat more deeply and attributed the blue shifts of the CF_3H proton donor, and its negative value of $d\mu/dr_{CH}$, to the exchange repulsion segment of the total interaction energy, claiming that the other terms in the Morokuma decomposition, i.e. electrostatic and polarization effects, tend toward red shifts. This result is in consonance with our earlier decomposition analysis which indicated that it is only the exchange energy (along with MIX) which pulls the bridging H atom toward the donor atom, with the other components tending toward a longer X–H bond, and hence toward a red shift (see Table 29.6).

Very similar conclusions were drawn by Qian and Krimm [196,197] who deduced that the direction of shift of the CH vibration in formic acid dimers was due entirely to the response of the donor molecule to an external electric field *via* its dipole derivative $d\mu/dE$ (they considered both the permanent and induced dipole derivatives). They were also able to derive the behavior of the band intensities from these same first principles. These authors showed that these conclusions were not limited to highly idealized uniform electric fields: extension to nonuniform electric fields, including also exchange repulsion, led to similar conclusions. The ability of an external electric field to substitute for the full proton acceptor molecule was further buttressed by a combination of experimental and theoretical work by Delanoye et al. [198,199] who successfully reproduced the direction of shift in a number of fluorochloromethanes. These authors went so far as to reproduce trends in intensity *via* the electric field. To be fair, however, not all calculations concur on the importance of the electric field. A study of formaldehyde dimers [200] concluded that this field plays an unimportant role in CH bond contraction.

Li et al. [201] later arrived at a nearly identical conclusion as did we earlier, that both blue and red-shifting H-bonds are governed by the same interactions, and that their different directions of shift lie only in the proportions of each. Their work provided a caution to others who might wish to study such interactions *via* semiempirical approaches, since AM1 and PM3 did not predict blue shifts at all; a similar caveat applies to minimal basis sets like STO-3G. This set of computations expanded the range of interactions that are predicted to shift to the blue beyond just CH···X, incorporating certain cases of NH···O, SiH···O, NH···F, PH···N, and PH···O. In terms of identifying the 'culprit', their calculations confirmed our own findings that electron correlation and its related phenomena, such as dispersion, are not responsible for the direction of shift, nor is it a result of charge transfer phenomena or HOMO–LUMO interactions. They ruled out a direct correlation between the magnitude of the frequency shift and the amount of charge transferred from one molecule to the other. The electron density redistributions in the full complexes are essentially indistinguishable from those resulting from placement of a simple dipole in the vicinity of the donor molecule, although the authors did caution that the direction of shift is not solely the province of electrostatic interactions. Their calculations supported the idea that the direction of frequency shift is largely determined by the nature of the proton donor, not the acceptor molecule, confirming other work.

Adopting an alternative perspective on the situation, Alabugin et al. [202,203] analyzed the complexes in terms of two competing effects. On one hand, the transfer of density from a lone pair of the proton acceptor molecule to the σ^* antibonding orbital of the CH bond ($n_Y > \sigma^*_{CH}$), characterized as a hyperconjugation effect, weakens the CH bond, acting to lengthen it. Factors which rehybridize the C orbital responsible for CH bonding *via* an increase in its s-character tend to strengthen and shorten this bond. Like most of the other means of addressing the problem, this analysis too yielded the similar conclusion that the direction of shift is a result of a delicate balance between competing effects, and these authors likewise concluded that "there is no fundamental difference between classic and improper H-bonding" [202].

29.4.4 Another interpretation

In contrast to the view enunciated above, that red and blue-shifting systems are fundamentally rather similar, an alternative interpretation had been evolving, particularly amongst a group of researchers in Prague. A series of papers [204–209] opined that there is a physically important difference as follows. In either sort of H-bond, there is a certain amount of electron density that is transferred from the proton acceptor molecule to the donor. In conventional H-bonds of the OH···O variety, the bulk of this density appears in the σ^* antibonding orbital of the O–H bond, accounting for its weakening and lengthening. In the so-called 'improper blue-shifting' H-bonds that might be typified by CF_3H as proton donor, this density bypasses the C–H region of the donor, and makes its way to the lone pairs of the peripheral F atoms. This increased lone pair density supposedly causes a stretch of the C–F bonds, which in turn forces the C–H bond to contract, and thereby shifts its frequency to the blue.

These ideas were, however, based largely upon partitioning the total electron density into localized orbitals, *viz.* natural bond orbitals. The changes in population used to justify their conclusions were fairly small, and potentially questionable, particularly since different means of localizing the total electron density leave different sort of 'tails', which have implications for numerical data. Indeed, a set of calculations by Kryachko and Zeegers-Huyskens [210] noted a situation where a CH bond was shortened, but at the same time the σ^*_{CH} population *rose*, an apparent contradiction to the principles laid out by the Prague group. More importantly, these same authors later stressed that the σ^*_{CH} population increases for *all* H-bonds, blue-shifting as well as red [211], so cannot be used as a discriminator. This contradiction was reinforced by Kolandaivel and Nirmala [212] who showed that contraction of the CH bond can take place despite the buildup of additional charge in the σ^*_{CH} orbital. This same work, involving both red and blue-shifting H-bonds, observed that the presumed buildup of charge on the peripheral sections of the proton donor molecule is not a general phenomenon at all, but occurs only in certain cases. Another study of natural bond orbitals [213] identified an important complicating factor, in that density can be dumped into the σ^*_{CH} orbital not only from the other molecule, but can arise also from internal shifts wholly within the proton donor molecule. These two sources of density can compete with one another, rendering this sort of interpretation oversimplistic.

In an effort to clarify this situation, our group performed a set of calculations designed to answer some simple questions [214]. While consideration of small changes in certain sorts of localized molecular orbitals might lend qualified support for one interpretation or another, it is far less ambiguous to consider the total electron density, with no partitioning into particular orbitals. Are shifts of the total density distinct and characteristic for red and blue-shifting H-bonds? Fig. 29.2a illustrates the shifts of electron density undergone by the water dimer as a result of the two molecules coming together to form the H-bonded complex. Blue regions designate buildups of density, and depletions are signified in red. Some principal features of this diagram are the loss of density around the bridging H atom and the gain in the area to the left of the proton-accepting O atom. Importantly, the area around the covalent O–H bond, to the left of the bridging hydrogen, is blue, consistent with a buildup of density.

Turning next to a CH···O interaction, Fig. 29.2b illustrates the density shifts in the HC≡CH···OH_2 complex. Like the traditional OH···O H-bond in the water dimer, the CH···O bond involving this sp-hybridized alkyne also shifts to the red. It is hence not surprising that the patterns in Fig. 29.2b look very similar to those in Fig. 29.2a: both manifest the characteristic red region around the bridging H, the X–H covalent bonding region is distinctly blue, as is the area to the left of the proton-accepting O atom. All of these same patterns are common as well to the $H_3CH\cdots OH_2$ system depicted in Fig. 29.2c. It is important to stress, however, that this complex, with its sp^3-hybridized C,

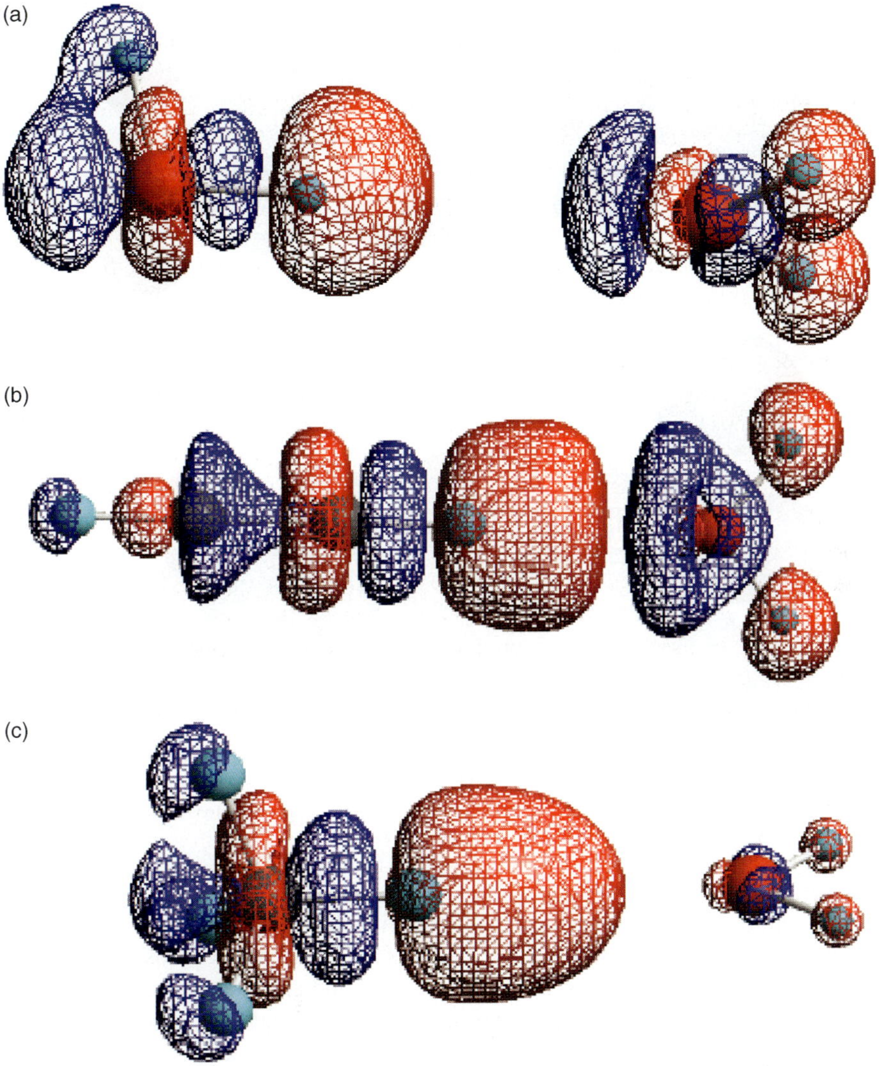

Fig. 29.2. Shifts of electron density occurring in (a) water dimer, (b) HC≡CH···OH_2, and (c) H_3CH···OH_2, as a result of formation of the complex. Blue regions denote gain and red regions represent loss. Contour illustrated corresponds to change by 0.0005 au.

shifts to the *blue*, not to the red. The similarities in the density shifting patterns in the three sections of Fig. 29.2 therefore belie the notion that red and blue-shifting H-bonds are characterized by different types of density shifts.

It had been argued by the Prague group that the density shift differences are related to the presence of electronegative F atoms on the proton-donating molecule. The requisite replacements of the peripheral H atoms by F, leading to the FC≡CH···OH_2 and

(a)

(b)

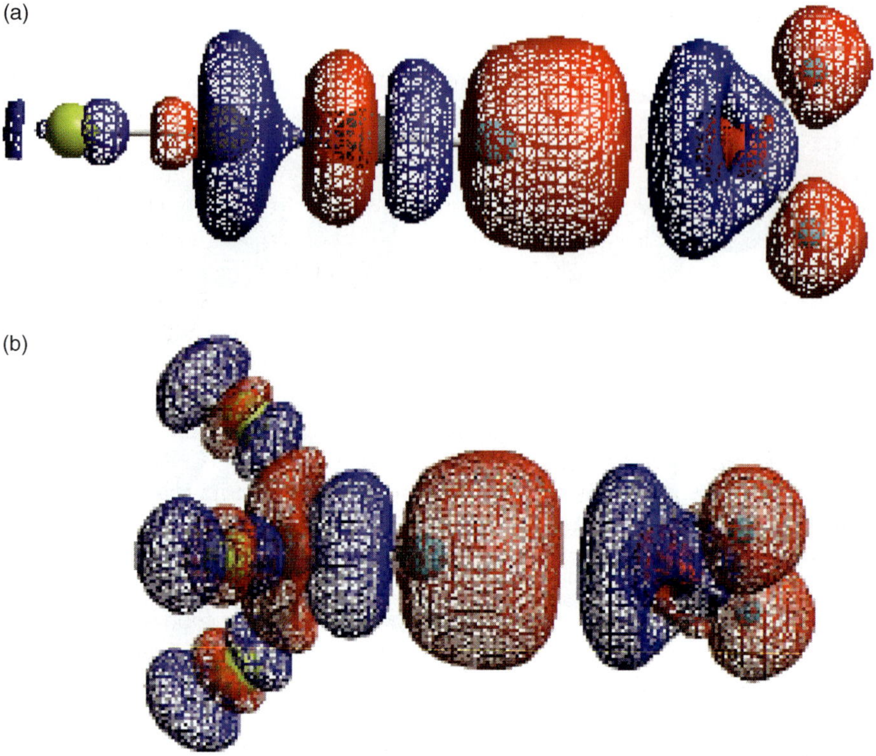

Fig. 29.3. Shifts of electron density occurring in (a) FC≡CH···OH_2, and (b) F_3CH···OH_2, as a result of formation of the complex. Blue regions denote gain and red regions represent loss. Contour illustrated corresponds to change by 0.0005 au.

F_3CH···OH_2 complexes, are represented by the density shifting patterns in Fig. 29.3a,b, respectively. Despite these substitutions, the patterns are little changed: loss is observed around the bridging proton, and gain in both the C–H bond and the region to the left of the O atom. The diagrams also undercut the proposal that a great deal of density builds up on the F atoms. There is little to distinguish the buildup on these atoms in Fig. 29.3 as compared to the original H atoms in Fig. 29.2. Nor is there any evidence that what charge buildup there is on these F atoms occurs on their lone pairs.

Another aspect of the argument for a fundamentally different blue-shifting bond contended that the peripheral C–F bonds are stretched by the formation of the H-bond. While a consideration of geometry changes confirmed such stretches in blue-shifting H-bonds, these same stretches were observed in red-shifting H-bonds as well, and of similar magnitudes [214], again arguing against any fundamental difference. In addition, whereas the elongations of the C–F bonds had been claimed as the root cause of the C–H bond contraction, our calculations demonstrated the converse as well. That is, a contraction of the C–H bond length results in elongation of the C–F bond lengths. One hence cannot argue that one is the cause and the other the effect.

29.5 FUTURE PERSPECTIVES

Why is all of this important? Given the fact that these CH···O interactions tend to be rather weak, why should we be concerned with them at all? In addition to provoking our curiosity as scientists about their fundamental nature, and about expanding the limits on what we call H-bonds, these interactions may be of intense importance in biology.

The 21st century has thoroughly erased any real doubts that CH···O interactions are a common component of biological systems. As some recent examples, a crystal survey of 125 helix–helix interfaces in 11 membrane proteins found that CH···O contacts favor parallel right-handed packing of helices [215], supported by later conclusions that CH···O bonds help to register α-helices and β-strands [216]. These interactions also play a role in β-turns and sheets in proteins, and novel chain reversal motifs [217–223]. Aravinda et al. [224] go so far as to claim these bonds are not only an important part of the Schellman motif (an unusual chain reversal at the C-terminus), but that they can also determine the direction of chain folding. This group believes there is "firm evidence in support of the emerging view that CH···O interactions may contribute significantly in energetic terms in determining folded structures". Jiang and Lai [225] estimated that these bonds contribute ~17% of the total interaction energy at protein–protein interfaces, as high as 50% in certain cases and that they are indispensable to forming a bifurcated H-bond motif between β-strands. These authors maintained that empirical force fields must be adjusted, as minimization without account of CH···O bonds leads to error in C=O distances. In addition to structural contributions, NMR evidence points to CH···O interactions as a part of enzymatic activity [226–228].

These bonds are thought to perhaps provide a driving force for ligand selectivity [229] as well. Researchers have found what they believe to be compelling evidence for the existence of CH···O bonds in protein–ligand interactions [229]. Based on crystal data for 200 liganded kinase structures, Pierce et al. [230], suggest that the impact of these CH···O bonds could be 'tremendous' to the field of ligand design; Klaholz and Moras [229] believe that "CH···O H-bonds may turn out to be the driving force for the introduction of ligand selectivity". A crystal study of photosystem I [231] found numerous CH···O interactions, indeed more than conventional H-bonds, leading Loll et al. to conjecture that the former play a prominent role in association and orientation, and that they add stability and increase the rigidity of the system.

Quantum chemistry has begun to delve more deeply into biological aspects of these weak H-bonds in recent years. Calculations have gone beyond the simplest models and considered the CH···O bonds that might occur in more realistic biological systems. Computations of the CH···O bonds of nucleic acid bases [192,232–237] led to estimates that were surprisingly high at the time. Model interpeptide CH···O interaction energies fell into the 2–4 kcal/mol range [238–240]. Our own work has suggested that, in particular, the CH···O interaction involving the $C^{\alpha}H$ of amino acids amounts to some 2.5 kcal/mol in the gas phase, roughly 40% of a conventional H-bond [241]. The H-bonds that involve the CH donors of aromatic residues run the range of 1–1.3 kcal/mol for Phe and Tyr, to 2–2.5 kcal/mol for His and Trp [242]. While each such H-bond is clearly weaker than its traditional OH···O or NH···O counterparts, there are a great number of CH···O H-bonds which can be present in biological systems such as proteins and nucleic acids. In that sense, these individually weaker interactions can collectively play a major role in our understanding of macromolecular structure. Moreover, the calculations are providing experimentalists with signposts that would signal the presence of a CH···O H-bond in a system under study, e.g. NMR or IR shifts.

There are likely two (not mutually exclusive) paths that future quantum chemical calculations will take to probe this issue more deeply. On one hand, continuing developments in computer software and hardware will permit quantitatively meaningful calculations of larger and increasingly more realistic models of biological systems, e.g. larger 'pieces' of the proteins. A second approach will be the transfer of these systems from an *in vacuo* environment, to more realistic surroundings, which might include solvent and/or other segments of the protein. The incorporation of dynamics and statistical mechanics into the mix would add another very interesting layer of information. In any case, it seems certain that the next few years will witness a more thorough and painstaking study of these weak H-bonds, and permit an incorporation of their effects into our current understanding of biomolecular structure and function.

29.6 REFERENCES

1 W.M. Latimer and W.H. Rodebush, J. Am. Chem. Soc., 42 (1920) 1419.
2 T.S. Moore and T.F. Winmill, J. Chem. Soc., 101 (1912) 1635.
3 G.C. Pimentel and A.L. McClellan, The hydrogen bond, Freeman, San Francisco, 1960.
4 M.D. Joesten and L.J. Schaad, Hydrogen bonding, Marcel Dekker, New York, 1974.
5 P. Schuster, G. Zundel and C. Sandorfy (Eds.), The hydrogen bond. Recent developments in theory and experiments, North-Holland, Amsterdam, 1976.
6 H. Umeyama and K. Morokuma, J. Am. Chem. Soc., 99 (1977) 1316.
7 S. Scheiner, Molecular orbital treatment of hydrogen bonded systems, in: E. Wyn-Jones, J. Gormally (Eds.), Aggregation processes in solution, Elsevier, Amsterdam, 1983, p. 462.
8 G.A. Jeffrey and W. Saenger, Hydrogen bonding in biological structures, Springer, Berlin, 1991.
9 S. Scheiner, Ab initio studies of hydrogen bonding, in: Z.B. Maksic (Ed.), Theoretical models of chemical bonding, Vol. 4, Springer, Berlin, 1991, p. 171.
10 S. Scheiner, Annu. Rev. Phys. Chem., 45 (1994) 23.
11 D. Hadzi (Ed.), Theoretical treatments of hydrogen bonding, Wiley, Chichester, 1997, p. 297.
12 G.A. Jeffrey, An introduction to hydrogen bonding, Oxford University Press, New York, 1997.
13 S. Scheiner, Hydrogen bonding. A theoretical perspective, Oxford University Press, New York, 1997.
14 G.R. Desiraju and T. Steiner, The weak hydrogen bond in structural chemistry and biology, Oxford, New York, 1999.
15 S. Glasstone, Trans. Faraday Soc., 33 (1937) 200.
16 J.F.J. Dippy, Chem. Rev., 25 (1939) 151.
17 C.M. Huggins, G.C. Pimentel and J.N. Shoolery, J. Chem. Phys., 23 (1955) 1244.
18 M.W. Dougill and G.A. Jeffrey, Acta Crystallogr., 6 (1953) 831.
19 G.S. Parry, Acta Crystallogr., 7 (1954) 313.
20 D.J. Sutor, Nature, 195 (1962) 68.
21 D.J. Sutor, J. Chem. Soc., (1963) 1105.
22 G. Ferguson and J. Tyrrell, Chem. Commun., (1965) 195.
23 A.L. Bednowitz and B. Post, Acta Crystallogr., 21 (1966) 566.
24 S. Krimm, Science, 158 (1967) 530.
25 G.N. Ramachandran and R. Chandrasekharan, Biopolymers, 6 (1968) 1649.
26 I. Brown and F. Smith, Aust. J. Chem., 13 (1960) 30.
27 A. Allerhand and P.V.R. Schleyer, J. Am. Chem. Soc., 85 (1963) 1715.

28 R.D. Green, Hydrogen bonding by C–H groups, Wiley, New York, 1974.
29 K. Morokuma, J. Chem. Phys., 55 (1971) 1236.
30 J.J. Dannenberg, J. Mol. Struct. (THEOCHEM), 401 (1997) 279.
31 J.G.C.M. van Duijneveldt-van de Rijdt and F.B. van Duijneveldt, J. Am. Chem. Soc., 93 (1971) 5644.
32 A. Johansson, P. Kollman and S. Rothenberg, Chem. Phys. Lett., 16 (1972) 123.
33 A. Johansson, P. Kollman and S. Rothenberg, Theor. Chim. Acta, 26 (1972) 97.
34 P. Kollman, J. McKelvey, A. Johansson and S. Rothenberg, J. Am. Chem. Soc., 97 (1975) 955.
35 D. Bonchev and P. Cremaschi, Theor. Chim. Acta, 35 (1974) 69.
36 S. Vishveshwara, Chem. Phys. Lett., 59 (1978) 26.
37 P.-G. Jönsson, Acta Crystallogr., B27 (1971) 893.
38 I. Goldberg, Acta Crystallogr., B31 (1975) 754.
39 R. Kaufmann, A. Knöchel, J. Kopf, J. Oehler and G. Rudolph, Chem. Ber., 110 (1977) 2249.
40 E.E. Astrup and A.M. Aomar, Acta Chem. Scand., A29 (1975) 794.
41 J.L. Sussman, N.C. Seeman, S.-H. Kim and H.M. Berman, J. Mol. Biol., 66 (1972) 403.
42 J. Rubin, T. Brennan and M. Sundaralingam, Biochemistry, 11 (1972) 3112.
43 W. Saenger, Angew. Chem. Int. Ed., 12 (1973) 591.
44 Å. Kvick, T.F. Koetzle and R. Thomas, J. Chem. Phys., 61 (1974) 2711.
45 F. Takusagawa, T.F. Koetzle, T. Srikrishnan and R. Parthasarathy, Acta Crystallogr., B35 (1979) 1388.
46 M.P. Sammes and R.L. Harlow, J. Chem. Soc., Perkin Trans. II, (1976) 1130.
47 F.M. Slasinski, J.M. Tustin, F.J. Sweeney, A.M. Armstrong, Q.A. Ahmed and J.P. Lorand, J. Org. Chem., 41 (1976) 2693.
48 M.A. Hussein and D.J. Millen, J. Chem. Soc., Faraday Trans. 2, 72 (1976) 693.
49 R.J. Boyd and S.C. Choi, Chem. Phys. Lett., 120 (1985) 80.
50 K. Somasundram, R.D. Amos and N.C. Handy, Theor. Chim. Acta, 69 (1986) 491.
51 A. Hinchliffe, J. Mol. Struct. (THEOCHEM), 136 (1986) 193.
52 I.L. Alberts, T.W. Rowlands and N.C. Handy, J. Chem. Phys., 88 (1988) 3811.
53 J.S. Craw, W.B. De Almeida and A. Hinchliffe, J. Mol. Struct. (THEOCHEM), 201 (1989) 69.
54 M.J. Frisch, J.A. Pople and J.E. Del Bene, J. Chem. Phys., 78 (1983) 4063.
55 A. Engdahl and B. Nelander, Chem. Phys. Lett., 100 (1983) 129.
56 A.M. DeLaat and B.S. Ault, J. Am. Chem. Soc., 109 (1987) 4232.
57 M.-L.H. Jeng, A.M. DeLaat and B.S. Ault, J. Phys. Chem., 93 (1989) 3997.
58 M.-L.H. Jeng and B.S. Ault, J. Phys. Chem., 93 (1989) 5426.
59 G.T. Fraser, K.R. Leopold and W. Klemperer, J. Chem. Phys., 80 (1984) 1423.
60 N.W. Howard and A.C. Legon, J. Chem. Phys., 85 (1986) 6898.
61 A.J. Fillery-Travis, A.C. Legon and L.C. Willoughby, Proc. Roy. Soc. Lond. A, 396 (1984) 405.
62 E.J. Goodwin and A.C. Legon, J. Chem. Phys., 87 (1987) 2426.
63 P. Hobza and C. Sandorfy, Can. J. Chem., 62 (1984) 606.
64 G. Trudeau, J.-M. Dumas, P. Dupuis, M. Guerin and C. Sandorfy, Top. Curr. Chem., 93 (1980) 91.
65 S.L. Paulson and A.J. Barnes, J. Mol. Struct., 80 (1982) 151.
66 G.T. Fraser, F.J. Lovas, R.D. Suenram, D.D. Nelson, Jr. and W. Klemperer, J. Chem. Phys., 84 (1986) 5983.
67 G.R. Desiraju, J. Chem. Soc., Chem. Commun., (1989) 179.
68 M. Budesinsky, P. Fiedler and Z. Arnold, Synthesis, (1989) 858.
69 S.J. Cole, K. Szalewicz, G.D. Purvis and R.J. Bartlett, J. Chem. Phys., 84 (1986) 6833.
70 P. Seiler, G.R. Weisman, E.D. Glendening, F. Weinhold, V.B. Johnson and J.D. Dunitz, Angew. Chem. Int. Ed., 26 (1987) 1175.
71 R.A. Kumpf and J.R. Damewood, J. Chem. Soc., Chem. Commun., (1988) 621.
72 J.A. Bandy and M.R. Truter, Acta Crystallogr., B37 (1981) 1568.
73 I. Goldberg and E.M. Kosower, J. Phys. Chem., 86 (1982) 332.
74 A. Elbasyouny, H.J. Brügge, K. von Deuten, M. Dickel, A. Knöchel, K.U. Koch, J. Kopf, D. Melzer and G. Rudolph, J. Am. Chem. Soc., 105 (1983) 6568.
75 R. Taylor and O. Kennard, J. Am. Chem. Soc., 104 (1982) 5063.
76 C. Li and M.P. Sammes, J. Chem. Soc., Perkin Trans. I, (1983) 1303.

77 C. Li and M.P. Sammes, J. Chem. Soc., Perkin Trans. I, (1983) 2193.
78 H. Satonaka, K. Abe and M. Hirota, Bull. Chem. Soc. Jpn., 61 (1988) 2031.
79 G.A. Jeffrey and H. Maluszynska, Int. J. Biol. Macromol., 4 (1982) 173.
80 Z. Berkovitch-Yellin and L. Leiserowitz, Acta Crystallogr., B40 (1984) 159.
81 R. Parthasarathy, S.M. Fridey and T. Srikrishnan, Int. J. Pept. Protein Res., 33 (1989) 308.
82 W. Saenger, Principles of nucleic acid structure, Springer, New York, 1984.
83 J.M. Benevides and G.J. Thomas, Biochemistry, 27 (1988) 3868.
84 D.E. Woon, P. Zeng and D.R. Beck, J. Chem. Phys., 93 (1990) 7808.
85 A.C. Legon, B.P. Roberts and A.L. Wallwork, Chem. Phys. Lett., 173 (1990) 107.
86 J.J. Novoa, M.-H. Whangbo and J.M. Williams, Mol. Crystallogr. Liq. Crystallogr., 181 (1990) 25.
87 J.J. Novoa, B. Tarron, M.-H. Whangbo and J.M. Williams, J. Chem. Phys., 95 (1991) 5179.
88 K.B. Wiberg, R.F. Waldron, G. Schulte and M. Saunders, J. Am. Chem. Soc., 113 (1991) 971.
89 P.G. Sennikov, R.I. Sharibdjanov and K. Khoudoinazarov, J. Mol. Struct., 270 (1992) 87.
90 M.M. Szczesniak, G. Chalasinski, S.M. Cybulski and P. Cieplak, J. Chem. Phys., 98 (1993) 3078.
91 C. Vizioli, M.C.R. de Azúa, C.G. Giribet, R.H. Contreras, L. Turi, J.J. Dannenberg, I.D. Rae, J.A. Weigold, M. Malagoli, R. Zanasi and P. Lazzeretti, J. Phys. Chem., 98 (1994) 8858.
92 J.J. Novoa, M. Planas and M.C. Rovira, Chem. Phys. Lett., 251 (1996) 33.
93 J.J. Novoa, M.-H. Whangbo and J.M. Williams, Chem. Phys. Lett., 180 (1991) 241.
94 J.J. Novoa, M.-H. Whangbo and J.M. Williams, Chem. Phys. Lett., 177 (1991) 483.
95 C.H. Reynolds, J. Am. Chem. Soc., 112 (1990) 7903.
96 J.A. Erickson and J.I. McLoughlin, J. Org. Chem., 60 (1995) 1626.
97 I. Alkorta and S. Maluendes, J. Phys. Chem., 99 (1995) 6457.
98 J.J. Novoa and F. Mota, Chem. Phys. Lett., 266 (1997) 23.
99 G.L. Sosa, N. Peruchena, R.H. Contreras and E.A. Castro, J. Mol. Struct. (THEOCHEM), 401 (1997) 77.
100 L. Turi and J.J. Dannenberg, J. Phys. Chem., 97 (1993) 12197.
101 S. Tsuzuki, T. Uchimaru, K. Tanabe and T. Hirano, J. Phys. Chem., 97 (1993) 1346.
102 R.L. Jaffe, J. Phys. Chem., 97 (1993) 12745.
103 T. Neuheuser, B.A. Hess, C. Reutel and E. Weber, J. Phys. Chem., 98 (1994) 6459.
104 L. Turi and J.J. Dannenberg, J. Phys. Chem., 99 (1995) 639.
105 A. Masunov and J.J. Dannenberg, J. Mol. Struct. (THEOCHEM), 371 (1996) 17.
106 R.L. Ornstein and Y. Zheng, J. Biomol. Struct. Dyn., 14 (1997) 657.
107 M.C. Rovira, J.J. Novoa, M.-H. Whangbo and J.M. Williams, Chem. Phys., 200 (1995) 319.
108 D. Feller and M.W. Feyereisen, J. Comput. Chem., 14 (1993) 1027.
109 K. Kim and R.A. Friesner, J. Am. Chem. Soc., 119 (1997) 12952.
110 T. Steiner and G.R. Desiraju, Chem. Commun., (1998) 891.
111 G.R. Desiraju, J. Chem. Soc., Chem. Commun., (1990) 454.
112 G.R. Desiraju, Acc. Chem. Res., 24 (1991) 290.
113 M.-L.H. Jeng and B.S. Ault, J. Phys. Chem., 94 (1990) 13232.
114 M.-L.H. Jeng and B.S. Ault, J. Phys. Chem., 94 (1990) 4851.
115 A.V. Afonin, M.V. Sigalov, S.E. Korostova, I.A. Aliev, A.V. Vashchenko and B.A. Trofimov, Magn. Reson. Chem., 28 (1990) 580.
116 P.A. Block, M.D. Marshall, L.G. Pedersen and R.E. Miller, J. Chem. Phys., 96 (1992) 7321.
117 V.R. Pedireddi and G.R. Desiraju, J. Chem. Soc., Chem. Commun., (1992) 988.
118 T. Steiner, J. Chem. Soc., Chem. Commun., (1994) 2341.
119 D. Braga, P.J. Dyson, F. Grepioni, B.F.G. Johnson and M.J. Calhorda, Inorg. Chem., 33 (1994) 3218.
120 D. Braga, F. Grepioni, H. Wadepohl, S. Gebert, M.J. Calhorda and L.F. Veiros, Organometallics, 14 (1995) 5350.
121 D. Braga, A.L. Costa, F. Grepioni, L. Scaccianoce and E. Tagliavini, Organometallics, 15 (1996) 1084.
122 F.D. Lewis, J.-S. Yang and C.L. Stern, J. Am. Chem. Soc., 118 (1996) 12029.
123 C.V.K. Sharma and M.J. Zaworotko, Chem. Commun., (1996) 2655.
124 T. Steiner, E.B. Starikov and M. Tamm, J. Chem. Soc., Perkin Trans. II, (1996) 67.
125 C.E. Marjo, M.L. Scudder, D.C. Craig and R. Bishop, J. Chem. Soc., Perkin Trans. II, (1997) 2099.
126 F.C. Pigge, F. Ghasedi and N.P. Rath, Tetrahedron Lett., 40 (1999) 8045.

127 F.H. Allen, J.P.M. Lommerse, V.J. Hoy, J.A.K. Howard and G.R. Desiraju, Acta Crystallogr., B52 (1996) 734.
128 J.L. Adcock and H. Zhang, J. Org. Chem., 60 (1995) 1999.
129 S.S. Kuduva, D.C. Craig, A. Nangia and G.R. Desiraju, J. Am. Chem. Soc., 121 (1999) 1936.
130 G. Harakas, T. Vu, C.B. Knobler and M.F. Hawthorne, J. Am. Chem. Soc., 120 (1998) 6405.
131 M. Mascal, Chem. Commun., (1998) 303.
132 T. Steiner, Chem. Commun., (1997) 727.
133 A. Gavezzotti, J. Phys. Chem., 95 (1991) 8948.
134 T. Steiner, J. Chem. Soc., Perkin Trans. II, (1995) 1315.
135 K. Panneerselvam and K.K. Chacko, Int. J. Pept. Protein Res., 35 (1990) 460.
136 J. Bella and H.M. Berman, J. Mol. Biol., 264 (1996) 734.
137 G.F. Fabiola, S. Krishnaswamy, V. Nagarajan and V. Pattabhi, Acta Crystallogr., D53 (1997) 316.
138 Z.S. Derewenda, L. Lee and U. Derewenda, J. Mol. Biol., 252 (1995) 248.
139 G.A. Leonard, K. McAuley-Hecht, T. Brown and W.N. Hunter, Acta Crystallogr., D51 (1995) 136.
140 I. Berger, M. Egli and A. Rich, Proc. Natl Acad. Sci. USA, 93 (1996) 12116.
141 M.C. Wahl, S.T. Rao and M. Sundaralingam, Nat. Struct. Biol., 3 (1996) 24.
142 S. Metzger and B. Lippert, J. Am. Chem. Soc., 118 (1996) 12467.
143 J.H. Cate, A.R. Gooding, E. Podell, K. Zhou, B.L. Golden, C.E. Kundrot, T.R. Cech and J.A. Doudna, Science, 273 (1996) 1678.
144 P. Fan, A.K. Suri, R. Fiala, D. Live and D.J. Patel, J. Mol. Biol., 258 (1996) 480.
145 M.C. Wahl and M. Sundaralingam, Trends Biochem. Sci., 22 (1997) 97.
146 R. Biswas and M. Sundaralingam, J. Mol. Biol., 270 (1997) 511.
147 C.C. Corell, B. Freeborn, P.B. Moore and T.A. Steitz, Cell, 91 (1997) 705.
148 J.P. Glusker, Acta Crystallogr., D51 (1995) 418.
149 P.M. Takahara, C.A. Frederick and S.J. Lippard, J. Am. Chem. Soc., 118 (1996) 12309.
150 T. Steiner and W. Saenger, J. Am. Chem. Soc., 114 (1992) 10146.
151 S. Pinchas, Anal. Chem., 27 (1955) 2.
152 S. Pinchas, Anal. Chem., 29 (1957) 334.
153 S. Pinchas, J. Phys. Chem., 67 (1963) 1862.
154 W.G. Schneider and H.J. Bernstein, Trans. Faraday Soc., 52 (1956) 13.
155 M.-L.H. Jeng and B.S. Ault, J. Mol. Struct., 246 (1991) 33.
156 W.F. Forbes, Can. J. Chem., 40 (1962) 1891.
157 T.A. Ford and L. Glasser, J. Mol. Struct. (THEOCHEM), 398–399 (1997) 381.
158 J.M. Hermida-Ramon and M.A. Rios, Chem. Phys. Lett., 290 (1998) 431.
159 M. Masella and J.-P. Flament, J. Chem. Phys., 110 (1999) 7245.
160 H. Yoshida, T. Harada, T. Murase, K. Ohno and H. Matsuura, J. Phys. Chem. A, 101 (1997) 1731.
161 K. Ohno, A. Tonegawa, H. Yoshida and H. Matsuura, J. Mol. Struct., 435 (1997) 219.
162 N. Karger, A.M. Amorim da Costa and J.A. Ribeiro-Claro, J. Phys. Chem. A, 103 (1999) 8672.
163 C.G. Giribet, C.V. Vizioli, C. Ruiz de Azua, R.H. Contreras, J.J. Dannenberg and A. Masunov, J. Chem. Soc., Faraday Trans., 92 (1996) 3029.
164 I.E. Boldeskul, I.F. Tsymbal, E.V. Ryltsev, Z. Latajka and A.J. Barnes, J. Mol. Struct., 436 (1997) 167.
165 P. Hobza, V. Spirko, H.L. Selzle and E.W. Schlag, J. Phys. Chem. A, 102 (1998) 2501.
166 P. Hobza, V. Spirko, Z. Havlas, K. Buchhold, B. Reimann, H.-D. Barth and B. Brutschy, Chem. Phys. Lett., 299 (1999) 180.
167 P. Hobza and Z. Havlas, Chem. Phys. Lett., 303 (1999) 447.
168 E. Cubero, M. Orozco and F.J. Luque, Chem. Phys. Lett., 310 (1999) 445.
169 J.J. Novoa, P. Lafuente and F. Mota, Chem. Phys. Lett., 290 (1998) 519.
170 I. Alkorta and J. Elguero, J. Phys. Chem. A, 103 (1999) 272.
171 E. Cubero, M. Orozco, P. Hobza and F.J. Luque, J. Phys. Chem. A, 103 (1999) 6394.
172 J. van de Bovenkamp, J.M. Matxain, F.B. van Duijneveldt and T. Steiner, J. Phys. Chem. A, 103 (1999) 2784.
173 M. Rozenberg, A. Loewenschuss and Y. Marcus, Phys. Chem. Chem. Phys., 2 (2000) 2699.
174 Y. Gu, T. Kar and S. Scheiner, J. Am. Chem. Soc., 121 (1999) 9411.

175 A.E. Reed, L.A. Curtiss and F. Weinhold, J. Chem. Phys., 83 (1985) 735.
176 A.E. Reed, L.A. Curtiss and F. Weinhold, Chem. Rev., 88 (1988) 899.
177 Y. Gu, T. Kar and S. Scheiner, J. Mol. Struct. (THEOCHEM), 500 (2000) 441.
178 K. Kitaura and K. Morokuma, Int. J. Quantum Chem., 10 (1976) 325.
179 K. Morokuma and K. Kitaura, Energy decomposition analysis of molecular interactions, in: P. Politzer, D.G. Truhlar (Eds.), Chemical applications of atomic and molecular electrostatic potentials, Plenum, New York, 1981, p. 215.
180 B. Schulz and P. Botschwina, Mol. Phys., 89 (1996) 1553.
181 A. Heikkilä, M. Pettersson, J. Lundell, L. Khriachtchev and M. Räsänen, J. Phys. Chem. A, 103 (1999) 2945.
182 M. Hartmann and L. Radom, J. Phys. Chem. A, 104 (2000) 968.
183 S.D. Wetmore, R. Schofield, D.M. Smith and L. Radom, J. Phys. Chem. A, 105 (2001) 8718.
184 S. Scheiner, S.J. Grabowski and T. Kar, J. Phys. Chem. A, 105 (2001) 10607.
185 Y. Wang and P.B. Balbuena, J. Phys. Chem. A, 105 (2001) 9972.
186 M. Hartmann, S.D. Wetmore and L. Radom, J. Phys. Chem. A, 105 (2001) 4470.
187 T.A. Ford and L. Glasser, Int. J. Quantum Chem., 84 (2001) 226.
188 A. Vila, A.M. Graña and R.A. Mosquera, Chem. Phys., 281 (2002) 11.
189 T. Harada, H. Yoshida, K. Ohno and H. Matsuura, Chem. Phys. Lett., 362 (2002) 453.
190 P. Raveendran and S.L. Wallen, J. Am. Chem. Soc., 124 (2002) 12590.
191 J.L. Alonso, S. Antolínez, S. Blanco, A. Lesarri, J.C. López and W. Caminati, J. Am. Chem. Soc., 126 (2004) 3244.
192 M. Brandl, M. Meyer and J. Sühnel, J. Biomol. Struct. Dyn., 18 (2001) 545.
193 A. Masunov, J.J. Dannenberg and R.H. Contreras, J. Phys. Chem. A, 105 (2001) 4737.
194 K. Hermansson, J. Phys. Chem. A, 106 (2002) 4695.
195 L. Pejov and K. Hermansson, J. Chem. Phys., 119 (2003) 313.
196 W. Qian and S. Krimm, J. Phys. Chem. A, 106 (2002) 6628.
197 W. Qian and S. Krimm, J. Phys. Chem. A, 106 (2002) 11663.
198 S.N. Delanoye, W.A. Herrebout and B.J. van der Veken, J. Am. Chem. Soc., 124 (2002) 7490.
199 S.N. Delanoye, W.A. Herrebout and B.J. van der Veken, J. Am. Chem. Soc., 124 (2002) 11854.
200 A. Kovács, A. Szabó, D. Nemcsok and I. Hargittai, J. Phys. Chem. A, 106 (2002) 5671.
201 X. Li, L. Liu and H.B. Schlegel, J. Am. Chem. Soc., 124 (2002) 9639.
202 I.V. Alabugin, M. Manoharan, S. Peabody and F. Weinhold, J. Am. Chem. Soc., 125 (2003) 5973.
203 I.V. Alabugin, M. Manoharan and F.A. Weinhold, J. Phys. Chem. A, 108 (2004) 4720.
204 B. Reimann, K. Buchhold, S. Vaupel, B. Brutschy, Z. Havlas, V. Spirko and P. Hobza, J. Phys. Chem. A, 105 (2001) 5560.
205 P. Hobza and Z. Havlas, Chem. Rev., 100 (2000) 4253.
206 B. van der Veken, W.A. Herrebout, R. Szostak, D.N. Shchepkin, Z. Havlas and P. Hobza, J. Am. Chem. Soc., 123 (2001) 12290.
207 P. Hobza, Phys. Chem. Chem. Phys., 3 (2001) 2555.
208 P. Hobza, Int. J. Quantum Chem., 90 (2002) 1071.
209 P. Hobza and Z. Havlas, Theor. Chem. Acc., 108 (2002) 325.
210 E.S. Kryachko and T. Zeegers-Huyskens, J. Phys. Chem. A, 105 (2001) 7118.
211 E.S. Kryachko and T. Zeegers-Huyskens, J. Phys. Chem. A, 106 (2002) 6832.
212 P. Kolandaivel and V. Nirmala, J. Mol. Struct., 694 (2004) 33.
213 G.L. Sosa, N.M. Peruchena, R.H. Contreras and E.A. Castro, J. Mol. Struct. (THEOCHEM), 577 (2002) 219.
214 S. Scheiner and T. Kar, J. Phys. Chem. A, 106 (2002) 1784.
215 A. Senes, I. Ubarretxena-Belandia and D.M. Engelman, Proc. Natl Acad. Sci. USA, 98 (2001) 9056.
216 S.K. Singh, M.M. Babu and P. Balaram, Proteins Struct. Funct. Genet., 51 (2003) 167.
217 A.K. Thakur and R. Kishore, Biopolymers, 53 (2000) 447.
218 G.G. Fabiola, V. Bobde, L. Damodharan, V. Pattabhi and S. Durani, J. Biomol. Struct. Dyn., 18 (2001) 579.
219 P.W. Baures, A.M. Beatty, M. Dhanasekaran, B.A. Helfrich, W. Perez-Segarra and J. Desper, J. Am. Chem. Soc., 124 (2002) 11315.

220 E.Y. Cheung, E.E. McCabe, K.D.M. Harris, R.L. Johnston, E. Tedesco, K.M.P. Raja and P. Balaram, Angew. Chem. Int. Ed., 41 (2002) 494.
221 M.M. Babu, S.K. Singh and P. Balaram, J. Mol. Biol., 322 (2002) 871.
222 K.M. Lee, H.-C. Chang, J.-C. Jiang, J.C.C. Chen, H.-E. Kao, S.H. Lin and I.J.B. Lin, J. Am. Chem. Soc., 125 (2003) 12358.
223 F. Cordier, M. Barfield and S. Grzesiek, J. Am. Chem. Soc., 125 (2003) 15750.
224 S. Aravinda, N. Shamala, A. Bandyopadhyay and P. Balaram, J. Am. Chem. Soc., 125 (2003) 15065.
225 L. Jiang and L. Lai, J. Biol. Chem., 277 (2002) 37732.
226 E.L. Ash, J.L. Sudmeier, R.M. Day, M. Vincent, E.V. Torchilin, K.C. Haddad, E.M. Bradshaw, D.G. Sanford and W.W. Bachovchin, Proc. Natl Acad. Sci. USA, 97 (2000) 10371.
227 W.W. Bachovchin, Magn. Reson. Chem., 39 (2001) S199.
228 C.A. Olson, Z. Shi and N.R. Kallenbach, J. Am. Chem. Soc., 123 (2001) 6451.
229 B.P. Klaholz and D. Moras, Structure, 10 (2002) 1197.
230 A.C. Pierce, K.L. Sandretto and G.W. Bemis, Proteins Struct. Funct. Genet., 49 (2002) 567.
231 B. Loll, G. Raszewski, W. Saenger and J. Biesiadka, J. Mol. Biol., 328 (2003) 737.
232 M. Kratochvil, O. Engkvist, J. Sponer, P. Jungwirth and P. Hobza, J. Phys. Chem. A, 102 (1998) 6921.
233 P. Hobza and J. Sponer, Chem. Rev., 99 (1999) 3247.
234 D. Barsky, E.T. Kool and M.E. Colvin, J. Biomol. Struct. Dyn., 16 (1999) 1119.
235 P. Hobza, J. Sponer, E. Cubero, M. Orozco and F.J. Luque, J. Phys. Chem. B, 104 (2000) 6286.
236 N.G. Fidanza, F.D. Suvire, G.L. Sosa, R.M. Lobayan, R.D. Enriz and N.M. Peruchena, J. Mol. Struct. (THEOCHEM), 543 (2001) 185.
237 G. Louit, A. Hocquet and M. Ghomi, Phys. Chem. Chem. Phys., 4 (2002) 3843.
238 R. Vargas, J. Garza, D.A. Dixon and B.P. Hay, J. Am. Chem. Soc., 122 (2000) 4750.
239 R. Vargas, J. Garza, R.A. Friesner, H. Stern, B.P. Hay and D.A. Dixon, J. Phys. Chem. A, 105 (2001) 4963.
240 R. Vargas, J. Garza, B.P. Hay and D.A. Dixon, J. Phys. Chem. A, 106 (2002) 3213.
241 S. Scheiner, T. Kar and Y. Gu, J. Biol. Chem., 276 (2001) 9832.
242 S. Scheiner, T. Kar and J. Pattanayak, J. Am. Chem. Soc., 124 (2002) 13257.

Theory and Applications of Computational Chemistry: The First Forty Years
Edited by C. Dykstra et al.

CHAPTER 30

Ab initio *and DFT calculations on the Cope rearrangement, a reaction with a chameleonic transition state*

Weston Thatcher Borden

Department of Chemistry, University of Washington, PO Box 351700, Seattle, WA 98195-1700, USA

Abstract

The contribution of two diradical extremes to the transition structure (TS) for the Cope rearrangement results in the energy of this structure being calculated to be rather insensitive to changes in the interallylic bond lengths in it. The variability of the TS with different *ab initio* methods made calculating its geometry challenging, and inclusion of dynamic electron correlation proved essential for obtaining an unbiased potential energy surface. The variability of the geometry of the TS with radical stabilizing substituents was shown by B3LYP calculations to be responsible for the cooperative and competitive effects of phenyl substituents on the rates of Cope rearrangements, which have been found experimentally. A simple mathematical model, which embodies the chameleonic nature of the TS and the strong effect of substituents on its geometry, has been developed and shown to predict quite well the phenyl substituent effects that have been revealed by these experiments.

30.1 INTRODUCTION

Many experiments and calculations have been performed in order to elucidate the nature of the transition structure (TS) for the Cope rearrangement [1]. This chapter is devoted to a review of these studies. Emphasis is placed on how *ab initio* and density functional calculations have contributed to the current understanding of the mechanism for this reaction and the effects of substituents on it.

Three possible chair-like TSs for the degenerate chair Cope rearrangement of 1,5-hexadiene are depicted in Fig. 30.1. If C–C bond breaking were to lag far behind

C–C bond making, the TS would resemble cyclohexane-1,4-diyl (**A**). On the other hand, if C–C bond breaking were to precede C–C bond making, the TS would resemble two allyl radicals (**C**). Finally, if C–C bond breaking and C–C bond making were to occur synchronously, the six cyclically delocalized electrons in structure **B** would render the TS 'aromatic' [2]. The Cope rearrangement is, indeed, a thermally allowed, pericyclic reaction [3], but the degree of bond making and bond breaking in TS for this reaction has been hotly debated [1].

A correlation diagram shows that the filled MOs in two weakly interacting allyl radicals and in cyclohexane-1,4-diyl are correlated with each other [4]. Therefore, these two structures may both be regarded as resonance contributors to the TS for the Cope rearrangement. From this fact it is possible to make two qualitative predictions.

The first is that the chair TS for the Cope rearrangement of 1,5-hexadiene should resemble more closely the lower energy of the two diradical extremes, **A** and **C**. The second prediction is that to the extent structures **A** and **C** contribute to the Cope TS, radical stabilizing substituents should be capable of lowering the energy of the TS, relative to the reactants, and thus accelerating the Cope rearrangement. Substituents at C2 and C5 of 1,5-hexadiene should stabilize structure **A** in Fig. 30.1, and substituents placed at C1, C3, C4, and C6 should stabilize structure **C**.

The second prediction raises a question. Are the relative contributions of **A** and **C** to the Cope TS fixed, or do their relative contributions vary with the number of radical stabilizing substituents and the carbons to which they are attached? Professor William von Eggers Doering, who has arguably contributed more than any other experimentalist to the understanding of the Cope rearrangement, has formulated this question succinctly in terms of two adjectives that he invented. Is the Cope TS 'centauric' or 'chameleonic' [5]?

If, like the contributions of man and horse to the mythical centaur, the relative contributions of structures **A** and **C** to the Cope TS are fixed, then Doering would describe the TS as 'centauric'. However, if, in the same way that a chameleon changes

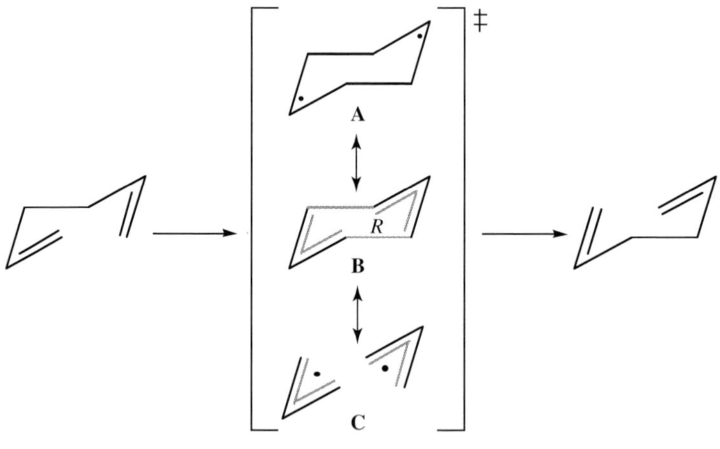

Fig. 30.1. Schematic depiction of the bonding in the transition structure (TS) for the chair Cope rearrangement, showing the diradical resonance contributors (**A** and **C**) and the aromatic representation (**B**). *R* is the interallylic distance.

color with its environment, the contributions of structures **A** and **C** to the Cope TS change with the number and placement of radical stabilizing substituents, then Doering would call the TS 'chameleonic'.

30.2 INFORMATION FROM EXPERIMENTS ABOUT THE COPE TS

By combining the heat of formation of cyclohexane with the then-current value of a secondary C–H bond dissociation enthalpy (BDE), in 1971 Doering obtained an estimated heat of formation for cyclohexane-1,4-diyl (**A**) [6]. The heat of formation of this fictional diradical, in which the two radical centers interact neither through bonds nor through space, was estimated by Doering to be far lower than that of two allyl radicals (**C**). Indeed, the 1971 value for the estimated heat of formation of diradical **A** was only slightly higher than that of the TS for the Cope rearrangement (Fig. 30.2) that equilibrates 1,5-hexadiene-1,1-d_2 (**1**) with 1,5-hexadiene-3,3-d_2 (**4**) [6].

Subsequent upward revision of the accepted value for the BDE of a secondary C–H bond increased the estimated heat of formation of diradical **A** [7], and the currently accepted value for the BDE of a secondary C–H bond [8] places **A** about 11 kcal/mol above the Cope TS. The heat of formation of two allyl radicals [9] puts **C** 26 kcal/mol higher than the Cope TS.

Experimental confirmation of the prediction that the Cope TS resembles **A** much more than **C** came from an unlikely source—M.J.S. Dewar, a chemist much better known for his contributions to the development of methodology for semiempirical electronic structure calculations, than for his experimental research. Nevertheless, the first experimental evidence that the Cope TS really does resemble structure **A** more than structure **C** was published by Dewar and Wade [10].

The Cope rearrangement of 2-phenyl-1,5-hexadiene-1,1-d_2 (**2**) to 2-phenyl-1,5-hexadiene-3,3-d_2 (**5**) has no thermodynamic driving force; whereas the Cope rearrangement of 3-phenyl-1,5-hexadiene (**3**) to 1-phenyl-1,5-hexadiene (**6**) does. However, the kinetic studies by Dewar and Wade [10] found the former reaction to be faster than the latter by a factor of four at 190 °C. The larger effect of a C2 than a C3 radical-stabilizing phenyl substituent on the rate of the Cope rearrangement indicates

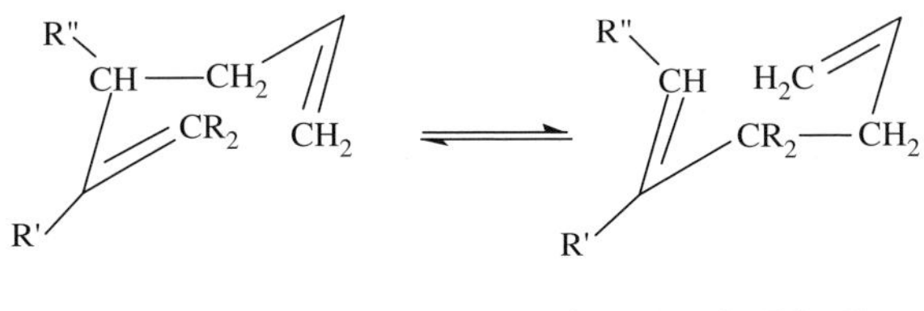

1, R = D, R' = R" = H

2, R = D, R' = Ph, R" = H

3, R = R' = H, R" = Ph

4, R = D, R' = R" = H

5, R = D, R' = Ph, R" = H

6, R = R' = H, R" = Ph

Fig. 30.2. Two Cope rearrangements (**1** → **4** and **2** → **5**) that are, except for the deuterium labels, degenerate and one Cope rearrangement (**3** → **6**) that is non-degenerate.

that, as expected, the lower energy diradical (**A**) contributes to the Cope TS more than the higher energy diradical (**C**) does.

The question of whether the Cope TS is centauric or chameleonic was first addressed by the now-classic experiments, performed by Gajewski and Conrad [11]. They found that the size of the secondary H/D kinetic isotope effects for bond making and bond breaking in the Cope rearrangement depends on the carbons to which radical stabilizing substituents are attached. This finding is consistent with the balance between bond making in structure **A** of Fig. 30.1 and bond breaking in structure **C** being substituent dependent, and this variability means that the TS is chameleonic.

30.3 MINDO/3, AM1, AND CASSCF CALCULATIONS ON THE COPE TS

On the basis of MINDO/3 calculations, Dewar argued for a non-concerted mechanism for the parent Cope rearrangement, involving formation of cyclohexane-1,4-diyl (**A**) as an intermediate [12]. Several years later, Dewar published another paper in which he boldly claimed that multi-bond reactions, such as the Cope rearrangement, cannot, *in general*, involve synchronous bond making and bond breaking [13].

Testing Dewar's claims, by performing unbiased *ab initio* calculations on the Cope rearrangement, proved to be challenging. For example, an RHF calculation on the Cope rearrangement is not unbiased, because, with a wave function that consists of a single configuration, only the 'aromatic' TS (**B**) can be adequately described. To be unbiased, a calculation must also be capable of describing both of the diradical extremes, cyclohexane-1,4-diyl (**A**) and two weakly interacting allyl radicals (**C**).

Although both of these diradical extremes have the same dominant electronic configuration as the 'aromatic' TS [4], a proper wave function for each of these diradical extremes requires a different 'second' configuration. Therefore, at least a three-configuration wave function is demanded for an unbiased study of the Cope rearrangement. Even better is a (6/6)CASSCF wave function in which three virtual orbitals are used to correlate the six electrons in the three bonding orbitals (two pi and one sigma) that are 'active' in this reaction.

It now seems hard to believe that, 20 years ago, (6/6)CASSCF/3-21G calculations were beyond the capabilities of the computers available to most computational chemists. Nevertheless, in the first *ab initio* study of the Cope rearrangement, published in 1984, (6/6)CASSCF/3-21G calculations could not actually be performed; they could only be simulated [14]. Smaller MCSCF calculations were employed to obtain a partially optimized set of orbitals. Then those orbitals were used to carry out full CI calculations for the 52 configurations that comprise the complete (6/6) active space at the C_{2h} stationary point for the chair Cope rearrangement of 1,5-hexadiene.

The simulated (6/6)CASSCF calculations were performed at several partially optimized C_{2h} geometries. The calculations found that the Cope TS has an interallylic separation of $R = 2.062$ Å, which is close to that of $R = 2.023$ Å for the fully optimized RHF/3-21G TS. Thus, the simulated (6/6)CASSCF/3-21G calculations found that bond making and bond breaking occur synchronously in the Cope rearrangement *via* an 'aromatic' TS (**B**).

The obvious shortcomings of these simulated CASSCF calculations were not lost on Michael Dewar, who in the 1980s had taken upon himself the role of *bête noire* of computational chemists who did *ab initio* electronic structure calculations. Dewar criticized the simulated CASSCF calculations in a paper that described his own AM1 study of the Cope rearrangement [15]. These AM1 calculations found a much shallower well for cyclohexane-1,4-diyl (**C**) than Dewar's MINDO/3 calculations had, and Dewar even conceded that this intermediate might even turn out to be the TS for a concerted Cope rearrangement. However, the interallylic distance in this C_{2h} species was only $R = 1.646$ Å, indicating that the Cope rearrangement was highly asynchronous, with bond formation preceding bond breaking.

Subsequent AM1 calculations by Dewar actually did find a synchronous TS, which he had initially missed; but the energy of this TS was higher than that of the C_{2h} species with $R = 1.646$ Å [16]. The AM1 potential energy surface (PES) for the chair Cope rearrangement supported Dewar's early hypothesis that there are two discrete types of reaction pathways by which the Cope rearrangement can occur [10a]. The lower energy pathway passes though a TS that leads to formation of a cyclohexane-1,4-diyl intermediate (**A**), and the higher involves a TS in which bond making and bond breaking occur synchronously (**B**).

In response to Dewar's criticisms [15] of the simulated (6/6)CASSCF calculations [14], in 1988 Morokuma published the results of genuine (6/6)CASSCF/3-21G calculations on the chair and boat Cope rearrangements [17]. The calculations predicted $R = 2.086$Å for the chair TS, and the computed activation enthalpy was only about 4 kcal/mol higher than Doering's experimental value of $\Delta H^{\ddagger} = 33.5$ kcal/mol [6]. Based on these results, a review article, published that year proclaimed, "*Ab initio* calculations and experimental evidence both indicate that multi-bond reactions not only can be synchronous but often are..." [1b].

This apparent victory of *ab initio* over semiempirical calculations was short lived. It was snatched from the *ab initio* computational chemists and given back to Michael Dewar by none other than Ernest Davidson, who had hithertofore been regarded by the *ab initio* community as one of its most stalwart defenders against the onslaughts of the semiempirical barbarians.

In 1991, Dupuis et al. [18] revisited the Cope rearrangement with (6/6)CASSCF/6-31G* calculations. Their (6/6)CASSCF/6-31G* PES showed two C_{2h} stationary points—one the TS for a concerted Cope rearrangement, with $R = 2.189$ Å, and the other an intermediate in a highly asynchronous reaction, with $R = 1.641$ Å. The latter C_{2h} stationary point was calculated to be 1.9 kcal/mol lower in energy than the former. Thus, the results of Davidson's then 'state-of-the-art' *ab initio* calculations actually validated Dewar's AM1 results [15,16]!

30.4 INCLUSION OF DYNAMIC ELECTRON CORRELATION

However, there still remained some hope that *ab initio* calculations on the Cope rearrangement might provide a different PES than AM1. The (6,6)CASSCF/6-31G* value of $\Delta H^{\ddagger}$ was higher than the experimental value [6] by >12 kcal/mol [18]. Therefore, it was possible that the (6,6)CASSCF/6-31G* PES was also grossly incorrect.

In order to go beyond the (6/6)CASSCF level, while maintaining the ability of a (6/6)CASSCF wave function to describe the two possible diradical extremes for the Cope TS, CASPT2 calculations proved ideal. This method, which had been developed by Roos and co-workers [19], applies multi-reference, second-order perturbation theory to CASSCF wave functions; and in 1993 CASPT2 had become available in the MOLCAS suite of *ab initio* calculations from the Roos group [20].

Therefore, the Cope rearrangement was revisited again, but this time with inclusion of electron correlation at the CASPT2 level of theory [21]. The (6/6)CASPT2/6-31G* value of $\Delta H^{\ddagger}$ was within 3 kcal/mol of Doering's experimental value [6], and use of larger basis sets reduced this difference to *ca.* 1 kcal/mol. In addition, with all the basis sets tested, only one C_{2h} stationary point was found, and it proved to be the CASPT2 TS for the chair Cope rearrangement.

Because the Cope PES is very flat along a C_{2h} cut [18,21], the size of R in the TS is highly dependent on the type of calculation performed. Thus, increasing the size of the basis set from 6-31G* to 6-311G(2d,2p) increases R by 0.14 Å, from $R = 1.745$ Å to $R = 1.885$ Å. Subsequent B3LYP calculations [22,23] have given values of R that are 0.08 Å larger than the CASPT2/6-311G(2d,2p) value.

Both the CASPT2 and B3LYP interallylic bond lengths in the TS indicate that bond making and bond breaking occur synchronously in the Cope rearrangement of unsubstituted 1,5-hexadiene. In addition, Jiao and Schleyer's [23] NICS calculations have found that the B3LYP TS has a large, induced, diamagnetic ring current. Hence, by this magnetic criterion the Cope TS really is aromatic.

Subsequently, Kozlowski et al. [24] also revisited the Cope rearrangement with inclusion of dynamic correlation between the 'active' and 'inactive' electrons. However, they used Davidson's own version of multi-reference, second-order perturbation theory [25], which allows the coefficients of the configurations in the CASSCF wave function to be recalculated after inclusion of dynamic electron correlation. Kozlowski et al. found that the addition of dynamic correlation to the (6/6)CASSCF wave function for the Cope TS causes the weight of the RHF configuration to increase at the expense of the pair configurations that are necessary to describe the two diradical extremes in Fig. 30.1. Thus, without the inclusion of dynamic electron correlation in the wave function, (6/6) CASSCF overestimates the diradical character of the C_{2h} wave function [24].

As pointed out in a 1996 review of the importance of dynamic electron correlation [1e], in general, addition of correlation between the 'active' and 'inactive' electrons stabilizes the ionic terms in the CASSCF wave functions for the 'active' electrons. Not only in the Cope TS, but also in the TSs for other Woodward–Hoffmann-allowed, pericyclic reactions [3], the ionic terms in the RHF configuration benefit from this type of stabilization. Consequently, with the inclusion of dynamic electron correlation, the weight of the RHF configuration in the wave function for the TS increases and the energy of the TS is lowered, relative to that of the reactants and products. It is for this reason that calculations which include dynamic electron correlation usually give much lower activation enthalpies for allowed pericyclic reactions than either RHF or CASSCF calculations do [1e,21].

Subsequently, Stavoverov and Davidson [26] computed the number, n, of 'effectively unpaired electrons' in multi-configurational wave functions for the chair Cope rearrangement at different C_{2h} geometries. In a single configuration, RHF wave function there are no effectively unpaired electrons (i.e. $n = 0$); but in the multi-configurational wave functions for the chair Cope TS Stavoverov and Davidson found that n is large at both very small and very large values of the interallylic distance (R). However, at the B3LYP/6-31G* Cope TS geometry ($R = 1.965$ Å), Stavoverov and Davidson found that n is small, because at this geometry the RHF configuration dominates the wave function.

The results of this study again call attention to the chameleonic nature of the C_{2h} wave function for the Cope rearrangement. At small values of R the CASSCF wave function is essentially that for cyclohexane-1,4-diyl (structure **A** in Fig. 30.1); whereas, at large values of R the CASSCF wave function approaches that for two allyl radicals (structure **C**). The variable contributions of structures **A** and **C** to the wave function as a function of R is what makes the change in the energy of the (6/6)CASSCF wave function so small (*ca.* 2 kcal/mol) over a range of R values that span >0.5 Å [18,21].

30.5 SUBSTITUENT EFFECTS ON THE COPE REARRANGEMENT

Not only do the relative contributions of structures **A** and **C** to the C_{2h} wave function for the Cope rearrangement change with the interallylic distance, R, their relative contributions also change with the number and placement of radical stabilizing substituents, such as cyano and phenyl. Since the relative contributions of structures **A** and **C** and the optimal value of R in the Cope TS are functionally related, the value of R in the TS is also highly substituent dependent.

The variations of the secondary H/D kinetic isotope effects with substituents, found by Gajewski and Conrad [11], provide indirect experimental evidence for the connection between the number and placement of radical stabilizing substituents and the TS geometry. Direct computational evidence comes from the calculated dependence of R in the Cope TS on the number and placement of cyano [27] and phenyl [28] groups. This dependence is illustrated for phenyl substituents by the calculated values of R in Table 30.1 [28].

The experimental activation enthalpies in Table 30.1 show that the effects of phenyl substituents on $\Delta H^{\ddagger}$ for Cope rearrangement can be either cooperative or competitive. For example, a single phenyl substituent at C2 of 1,5-hexadiene lowers $\Delta H^{\ddagger}$ for Cope rearrangement by 4.2 kcal/mol, but a second phenyl group, attached to C5, provides an additional 8.0 kcal/mol lowering of the barrier height. Clearly, there is a strong cooperative effect between phenyl substituents at C2 and C5, both of which stabilize structure **A** in Fig. 30.1.

The experimental values of $\Delta H^{\ddagger}$ in Table 30.1 show that phenyl groups at C1, C3, C4, and C6, which stabilize structure **C**, also give rise to a strongly cooperative substituent effect. When a pair of phenyl groups is attached to C1 and C3 of 1,5-hexadiene, the barrier to the Cope rearrangement is decreased by 3.0 kcal/mol from that for the unsubstituted molecule. If the phenyl substituent effects on the Cope rearrangement were additive, augmentation of a pair of phenyl substituents at C1 and C3 by another pair at C4

Table 30.1 Calculated and experimental activation enthalpies (kcal/mol) for the Cope rearrangements of phenyl substituted 1,5-hexadienes and the calculated interallylic distance, R (Å), in each TS and intermediate

Substituents	Method[a]	$\Delta H^{\ddagger}$ (calc.)	$\Delta H^{\ddagger}$ (exp.)	R[b]
None	B3LYP	33.2	33.5 ± 0.5[c]	1.965
2-Phenyl	B3LYP	30.4		1.837, 1.821
2-Phenyl	UB3LYP	30.3		1.777, 1.700
2-Phenyl	UB3LYP	29.4[d]	29.3 ± 1.6[e]	1.599
1,3-Diphenyl	B3LYP	30.2	30.5 ± 0.2[f]	2.218
2,5-Diphenyl	B3LYP	25.1		1.794, 1.680
2,5-Diphenyl	UB3LYP	24.8		1.839, 1.667
2,5-Diphenyl	UB3LYP	21.3[d]	21.3 ± 0.3[e,g]	1.576
1,3,5-Triphenyl	B3LYP	29.2	27.8 ± 0.2[f]	2.113, 2.106
1,3,4,6-Tetraphenyl	B3LYP	19.1	21.3 ± 0.1[h]	2.649

[a]The 6-31G* basis set was used for all of the calculations in this table.
[b]Length of the forming bond and the breaking bond, respectively. Just one bond length is given for those species having a plane of symmetry.
[c]Ref. [6].
[d]Enthalpy, relative to the reactant, of the diradicaloid intermediate.
[e]Ref. [10].
[f]Ref. [5].
[g]Ref. [29].
[h]Ref. [30].

and C6 would be anticipated to lower $\Delta H^{\ddagger}$ by another 3.0 kcal/mol. However, the measured effect of the additional pair of phenyl groups is three times larger, lowering $\Delta H^{\ddagger}$ for Cope rearrangement by an additional 9.2 kcal/mol. As in the case of phenyl substituents at C2 and C5, the effect of phenyl substituents at C1, C3, C4, and C6 is obviously cooperative.

In contrast to the cooperative substituent effects described above, placement of phenyl groups at C1, C3, and C5 of 1,5-hexadiene results in a competitive substituent effect. By stabilizing structure **A**, a single phenyl group at C2/C5 lowers $\Delta H^{\ddagger}$ by 4.2 kcal/mol; and, by stabilizing structure **C**, phenyl groups at C1 and C3 lower $\Delta H^{\ddagger}$ by 3.0 kcal/mol. However, the simultaneous presence of phenyl groups at C1, C3, and C5 lowers $\Delta H^{\ddagger}$ by 5.7 kcal/mol, 1.5 kcal/mol *less* than the 7.2 kcal/mol expected on the basis of substituent effect additivity. Clearly, positioning the phenyl substituents, so that they simultaneously stabilize both structures **A** and **C**, results in a small negative deviation from substituent effect additivity.

Table 30.1 shows that B3LYP/6-31G* calculations give activation enthalpies for the Cope rearrangements of phenyl-substituted 1,5-pentadienes that are in very good agreement with experiment. This agreement gives credibility to the predicted variation with substitution of the interallylic distance, R, in the Cope TS. The dependence of R on the number and placement of phenyl substituents is responsible for the cooperative and competitive phenyl substituent effects that have been both observed in and calculated for the Cope rearrangement [28].

Consider, for example, the effect of the pair of phenyl substituents in the TS for Cope rearrangement of 1,3-diphenyl-1,5-hexadiene. The calculated interallylic distance in this

TS is $R = 2.218$ Å, which is 0.253 Å longer than the value calculated for the TS for Cope rearrangement of unsubstituted 1,5-hexadiene ($R_0 = 1.965$ Å). The larger value of R in the 1,3-diphenyl-1,5-hexadiene Cope TS increases the contribution of structure **C** in Fig. 30.1 to the TS wave function, thus allowing the phenyl substituents at C1 and C3 to provide more stabilization for this wave function at $R = 2.218$ Å than at $R_0 = 1.965$ Å.

On addition of a second pair of phenyl substituents at C4 and C6, the interallylic distance further lengthens so that the contribution of structure **C** to the TS wave function is further enhanced. The ability of the second pair of phenyl substituents to increase the optimal value of R from $R = 2.218$ Å to $R = 2.649$ Å allows each of the phenyl substituents in 1,3,4,6-tetraphenyl-1,5-hexadiene to provide more stabilization for the Cope TS than the pair of phenyl substituents in 1,3-diphenyl-1,5-hexadiene. This cooperative effect of the four phenyl groups in 1,3,4,6-tetraphenyl-1,5-hexadiene lowers $\Delta H^{\ddagger}$ for Cope rearrangement by, not twice, but by four times as much as the pair of phenyl substituents in 1,3-diphenyl-1,5-hexadiene.

The optimal value of R in the TS for Cope rearrangement of a substituted 1,5-hexadiene can be viewed as a compromise between maximizing the stabilizing effect of the substituent(s) ($\Delta E_{\mathrm{subst}}$) on structure **A** or structure **C** in the TS and minimizing the energetic cost (ΔE_{dist}) of distorting the interallylic bond lengths from the value of $R_0 = 1.965$ Å for the Cope rearrangement of unsubstituted 1,5-hexadiene [30]. ΔE_{dist} can be easily calculated, since it is just the increase in the energy of the TS for the Cope rearrangement of unsubstituted 1,5-hexadiene on going from $R_0 = 1.965$ Å to the value of R in the TS for the Cope rearrangement of a substituted 1,5-hexadiene.

$\Delta E_{\mathrm{subst}}$ cannot be computed directly. However, it can be obtained by calculating ΔE_{dist} and the net stabilization energy ($\Delta E^{R}_{\mathrm{stab}}$) that results from allowing the symmetrical chair geometry of the substituted diene to distort from $R_0 = 1.965$ Å to the optimal value of R in the TS for its Cope rearrangement. Since $\Delta E^{R}_{\mathrm{stab}} = \Delta E_{\mathrm{dist}} + \Delta E_{\mathrm{subst}}$, calculating $\Delta E^{R}_{\mathrm{stab}}$ and ΔE_{dist} allows $\Delta E_{\mathrm{subst}}$ to be obtained from the relationship, $\Delta E_{\mathrm{subst}} = \Delta E^{R}_{\mathrm{stab}} - \Delta E_{\mathrm{dist}}$.

The calculated values of $\Delta E^{R}_{\mathrm{stab}}$, ΔE_{dist}, and $\Delta E_{\mathrm{subst}}$ are given in Table 30.2 for the Cope rearrangements of 2-phenyl, 1,3- and 2,5-diphenyl-, 1,3,5-triphenyl-, and 1,3,4,6-tetraphenyl-1,5-hexadiene. Also given is $\Delta E^{0}_{\mathrm{stab}}$, the stabilization energy provided by the phenyl substituents at $R_0 = 1.965$ Å. $\Delta E^{0}_{\mathrm{stab}}$ is calculated as the difference between the energies of the symmetrical chair geometry of a substituted 1,5-hexadiene at $R_0 = 1.965$ Å and at the equilibrium geometry of the diene, minus the difference between the energies of unsubstituted 1,5-hexadiene at the TS for its Cope rearrangement and at its equilibrium geometry.

When $\Delta E^{0}_{\mathrm{stab}}$ is added to $\Delta E^{R}_{\mathrm{stab}}$, their sum gives, $\Delta E_{\mathrm{total}}$—the total amount by which the substituents make the energy difference between the TS and the reactant in the Cope rearrangement of a substituted 1,5-hexadiene smaller than the size of this energy difference in the Cope rearrangement of an unsubstituted 1,5-hexadiene. Because the values of $\Delta E_{\mathrm{total}}$ in Table 30.2 are uncorrected for differences in zero-point energies or integrated heat capacities, the values of $\Delta E_{\mathrm{total}}$ in Table 30.2 are similar to but not exactly the same as the differences between the $\Delta H^{\ddagger}$ values in Table 30.1.

A number of observations can be made from inspection of Table 30.2. First, at R_0, the interallylic bond length in the TS for the parent Cope rearrangement, a phenyl group at C2 provides 1.7 kcal/mol of stabilization, which is slightly more than twice that provided

Table 30.2 Dissection of the effects of phenyl substituents on lowering the energy of the TS for the chair Cope rearrangement of 1,5-hexadiene. Energies (kcal/mol) were obtained from B3LYP/6-31G* calculations at different interallylic distances (R) in the manner described in the text

Substituents	R (Å)	ΔE_{dist}	ΔE_{subst}	ΔE^{R}_{stab}	ΔE^{0}_{stab}	ΔE_{total}
2-Phenyl[a]	1.599	3.3	−6.2	−2.9	−1.7	−4.6
2-Phenyl[b]	2.210	0.7	1.4	2.1	−1.7	0.4
2,5-Diphenyl[a]	1.576	3.7	−12.9	−9.2	−3.7	−12.9
1,3-Diphenyl	2.218	2.0	−4.0	−2.0	−0.7	−2.7
1,3-Diphenyl[b]	2.110	0.7	−2.4	−1.7	−0.7	−2.4
1,3,5-Triphenyl	2.110	0.7	−1.1	−0.4	−2.8	−3.2
1,3,4,6-Tetraphenyl	2.649	12.0	−23.7	−11.7	−1.4	−13.1
1,3,4,6-Tetraphenyl[c]	2.218	2.0	−9.3	−7.3	−1.4	−8.7

[a]Values are from UB3LYP calculations.

[b]Calculations performed at the same interallylic bond distance, R, as in the TS for Cope rearrangement of 1,3,5-diphenyl-1,5-pentadiene.

[c]Calculations performed at the same interallylic bond distance, R, as in the TS for Cope rearrangement of 1,3-diphenyl-1,5-pentadiene.

by a *pair* of phenyl groups at C1 and C3. The larger stabilization provided by a single C2 phenyl group supports the conjecture, based on the relative enthalpies of the two diradical extremes in Fig. 30.1, that the TS for the Cope rearrangement of unsubstituted 1,5-hexadiene resembles cyclohexane-1,4-diyl (**A**) more than two allyl radicals (**C**).

The values of ΔE^{0}_{stab} for 2,5-diphenyl, 1,3,5-triphenyl, and 1,3,4,6-tetraphenyl substituents are each close to being the sum of the appropriate ΔE^{0}_{stab} values for 2-phenyl and 1,3-diphenyl substituents. It is noteworthy that the ΔE^{0}_{stab} values for 2-phenyl- and 1,3-diphenyl-1,5-hexadiene are, respectively, only about one-third to one-quarter of ΔE_{total}, the total lowering of the Cope TS energy, relative to the energy of the reactants.

The fact that substituents provide a comparatively small amount of stabilization at $R_0 = 1.965$ Å was first found for cyano rather than for phenyl substituents [27]. This finding is consistent with the results of Stavoverov and Davidson. They used local spin analysis to show that not only does the TS for the parent Cope rearrangement have little diradical character [26] but also that, at the geometry of this TS, radical stabilizing substituents do little to enhance the diradical character of the wave function [31].

Another observation, which can be made from the results in Table 30.2, is that at the optimal values of R for the intermediate in the Cope rearrangement of 2-phenyl-1,5-hexadiene and for the TSs in the Cope rearrangements of 1,3-diphenyl-, and 1,3,4,6-tetraphenyl-1,5-hexadiene, $\Delta E_{subst} \approx -2\Delta E_{dist}$. Therefore, since $\Delta E^{R}_{stab} = \Delta E_{dist} + \Delta E_{subst}$, it is also true that $\Delta E^{R}_{stab} \approx \Delta E_{subst}/2 \approx -\Delta E_{dist}$. This finding—that at the optimal value of R the magnitude of ΔE_{subst} is approximately twice the size of ΔE_{dist}—can be shown to be a consequence of the fact that the leading term in ΔE_{dist} is quadratic in $R - R_0$; whereas, the leading term in ΔE_{subst} must be linear in $R - R_0$ [32].

When a second phenyl group is added to C5 of 2-phenyl-1,5-hexadiene, the increase in the magnitude of ΔE_{subst} in the TS is about a factor of two; but when a second pair of phenyls is added to C4 and C6 of 1,3-diphenyl-1,5-hexadiene, the increase is about a factor of six. The reason for this difference is that the intermediates in the Cope rearrangements of 2-phenyl- and 2,5-diphenyl-1,5-hexadiene have UB3LYP values of R

that differ by only 0.023 Å. In contrast, when a second pair of phenyl groups is added at C4 and C6 of 1,3-diphenyl-1,5-hexadiene, the value of R in the TS increases by 0.431 Å. This large increase in R, accompanying addition of a second pair of phenyl substituents, results in a very large increase in the magnitude of $\Delta E_{\rm subst}$.

In the 1-phenylcyclohexane-1,4-diyl diradical intermediate, formed in the Cope rearrangement of 2-phenyl-1,5-hexadiene, the UB3LYP/6-31G* value of $R = 1.599$ Å approaches that of a standard C–C single bond; so addition of a second phenyl substituent at C4 of the diradical (C5 of the diene) has only a small effect on further decreasing R. Because addition of a second phenyl group to C4 of 1-phenyl-cyclohexane-1,4-diyl leaves the value of R in this intermediate largely unchanged, the effect of the second phenyl group is approximately just to double the value of $\Delta E_{\rm subst}$.

The fact that R is nearly the same in 1,4-diphenylcyclohexane-1,4-diyl as in the diradical that lacks the second phenyl substituent has another consequence—the approximate doubling of $\Delta E_{\rm subst}$ upon adding the second phenyl group occurs without the significant increase in $\Delta E_{\rm dist}$; so $\Delta E^R_{\rm stab}$ increases in magnitude by approximately $\Delta E_{\rm subst}$. In contrast, for addition of the first phenyl substituent, $\Delta E_{\rm dist} \approx -\Delta E_{\rm subst}/2$; so $\Delta E^R_{\rm stab} = \Delta E_{\rm dist} + \Delta E_{\rm subst} \approx \Delta E_{\rm subst}/2$. Therefore, addition of a second phenyl group to C4 of the diradical increases the magnitude of $\Delta E^R_{\rm stab}$ by approximately twice as much as addition of the first phenyl group does. Put colloquially, the second phenyl group gets for free the $\Delta E_{\rm subst} \approx -6.2$ kcal/mol for which the first phenyl group has to pay $\Delta E_{\rm dist} = 3.3$ kcal/mol.

When a second pair of phenyl groups is added to C4 and C6 of 1,3-diphenyl-1,5-hexadiene at the value of $R = 2.218$ Å in the TS for Cope rearrangement of the disubstituted diene, Table 30.1 shows that $\Delta E_{\rm subst}$ increases in magnitude by 5.3 kcal/mol. This increase in magnitude is a little more than the factor of 2 that is expected for doubling the number of phenyl substituents at a fixed value of R. It seems likely that the 1.3 kcal/mol of extra stabilization comes from a reduction in the steric repulsion between the pairs of phenyl groups at C1 and C6 and C3 and C4 upon increasing R from $R_0 = 1.965$ Å in the TS for the parent Cope rearrangement to $R = 2.218$ Å in the TS for Cope rearrangement of 1,3-diphenyl-1,5-hexadiene.

Upon increasing the interallylic distance from $R = 2.218$ Å to $R = 2.649$ Å, the optimal value of R in the TS for Cope rearrangement of 1,3,4,6-tetraphenyl-1,5-hexadiene, the magnitude of $\Delta E_{\rm subst}$ increases by 14.4 kcal/mol. However, on stretching the interallylic bonds by this additional 0.431 Å, $\Delta E_{\rm dist}$ increases by 10.0 kcal/mol. Thus, the net change in the magnitude of $\Delta E^R_{\rm stab}$ is only 4.4 kcal/mol. Therefore, of the 9.7 kcal/mol increase in the magnitude of $\Delta E^R_{\rm stab}$ that results from adding a pair of phenyl groups to C4 and C6 of the TS for Cope rearrangement of 1,3-diphenyl-1,5-hexadiene, 55% comes from the two additional phenyl groups increasing the magnitude of $\Delta E_{\rm subst}$ at $R = 2.218$ Å, and 45% comes from the increase in the size of the difference between $\Delta E_{\rm subst}$ and $\Delta E_{\rm dist}$ when R increases from 2.218 to 2.649 Å.

As the discussion above demonstrates, the cooperative substituent effects, which have been both observed in and computed for the Cope rearrangements of 2,5-diphenyl- and 1,3,4,6-tetraphenyl-1,5-hexadiene, can be understood on the basis of the dependence of the size of $\Delta E^R_{\rm stab}$ on R. The competitive substituent effect, both observed in and computed for the Cope rearrangement of 1,3,5-triphenyl-1,5-hexadiene, can similarly be understood.

Table 30.2 shows that at $R = 1.599$ Å the C2 phenyl group in 2-phenyl-1,5-hexadiene provides a net stabilization of $\Delta E^{R}_{\text{stab}} = -2.9$ kcal/mol, and at $R = 2.218$ Å the C1 and C3 phenyl groups in 1,3-diphenyl-1,5-pentadiene provide a net stabilization of $\Delta E^{R}_{\text{stab}} = -2.0$ kcal/mol. However, the TS for Cope rearrangement of 1,3,5-triphenyl-1,5-pentadiene occurs with interallylic bond lengths of about $R = 2.110$ Å (Table 30.1). At this TS geometry, neither the phenyl groups at C1 and C3 nor the phenyl group C5 provides as much stabilization as these phenyl groups furnish in the TSs for the Cope rearrangements of, respectively, 1,3-diphenyl-1,5-pentadiene at $R = 2.218$ Å and 2-phenyl-1,5-hexadiene at $R = 1.599$ Å.

Table 30.2 reveals that at an average value of $R = 2.110$ Å the C2 phenyl group in 2-phenyl-1,5-hexadiene provides a net TS stabilization energy that is 5.0 kcal/mol smaller in magnitude than at $R = 1.599$ Å. At $R = 2.110$ Å the C1 and C3 phenyl groups in 1,3-diphenyl-1,5-pentadiene provide a net TS stabilization energy that is 0.3 kcal/mol smaller in magnitude than at $R = 2.218$ Å. Thus, based on strict additivity, one might expect the three phenyl groups in 1,3,5-triphenyl-1,5-pentadiene would provide 5.3 kcal/mol less net TS stabilization than the total amount they furnish in the Cope rearrangements of 1,3-diphenyl-1,5-pentadiene and 2-phenyl-1,5-hexadiene. In fact, Table 30.2 shows that $\Delta E^{R}_{\text{stab}}$ for Cope rearrangement 1,3,5-triphenyl-1,5-pentadiene actually is 4.5 kcal/mol smaller in magnitude than the sum of the $\Delta E^{R}_{\text{stab}}$ values for the Cope rearrangements of 2-phenyl-1,5-pentadiene and 1,3-diphenyl-1,5-pentadiene.

Therefore, the competitive substituent effect, both predicted and found for the Cope rearrangement of 1,3,5-triphenyl-1,5-hexadiene, is a consequence of the fact the value of R in the TS is a compromise between the value of $R = 1.599$ Å, at which the phenyl group at C5 can provide optimal stabilization for the contribution of structure **A**, and the value of $R = 2.218$ Å, at which the phenyl groups at C1 and C3 provide optimal stabilization for the contribution of structure **C**. At the compromise value of $R = 2.110$ Å the three phenyl groups in 1,3,5-triphenyl-1,5-hexadiene are calculated to provide 4.5 kcal/mol less TS stabilization than the total amount they furnish in the Cope rearrangements of 2-phenyl-1,5-hexadiene at $R = 1.599$ Å and in 1,3-diphenyl-1,5-hexadiene at $R = 2.218$ Å.

30.6 SUMMARY

The ability of diradicals **A** and **C** in Fig. 30.1 to contribute to the electronic structure of the C_{2h} species in the Cope rearrangement makes it possible to lengthen or shorten the interallylic bonds in the chair TS without significantly affecting the energy. The pathological flatness of the PES with respect to changes in this coordinate is largely responsible for the large variations in the optimal value of R in the TS that are found with changes in computational methodology and/or basis set. Put another way, the chameleonic nature of the Cope TS results in its TS geometry responding to the biases built into different types of calculations by changing the value of R in the TS.

Given the insensitivity of the energy of the C_{2h} geometry to changes in R, the values of $R = 1.885$ Å, obtained by (6/6)CASPT2/6-311G (2d,2p) calculations [21], and $R = 1.965$ Å, found by B3LYP/6-31G* calculations [22,23] are in reasonable agreement. Both are on the short side of the range of C–C bond lengths commonly computed for the TSs in concerted pericyclic reactions [1b–d]. The reason for the slightly shorter than usual bond lengths in the Cope TS is almost certainly that cyclohexane-1,4-diyl (**A**) [6,7,27] is considerably lower in enthalpy than two allyl radicals (**C**) [9]. Therefore, the electronic structure of the TS resembles **A** more than **C**.

This conjecture is supported by the calculated values of ΔE^0_{stab} in Table 30.2. Phenyl groups at C2 and C5 of 1,5-hexadiene, which stabilize structure **A**; provide twice as much stabilization per phenyl at $R_0 = 1.965$ Å as each *pair* of phenyl substituents at C1, C3, C4, and C6, which stabilize structure **C**.

However, really large phenyl stabilization energies come, not at $R_0 = 1.965$ Å, where the TS is best described by structure **B** in Fig. 30.1, but at values of R at which the TS has a larger contribution from either structure **A** or **C**. Small values of R enhance the contribution of diradical **A**, and large values of R enhance the contribution of diradical **C**. Thus, as shown in Tables 30.1 and 30.2, phenyl groups at C2 and C5 of 1,5-hexadiene favor a small value of R in the Cope TS, and phenyl substituents at C1, C3, C4, and C6 favor a large value of R in the TS.

The ability of radical stabilizing substituents to have a large effect on R in the TS geometry is what makes the Cope TS chameleonic; and the chameleonic nature of the TS is responsible for the cooperative and competitive phenyl substituent effects that have been both calculated [27,28] and observed [5,29,30]. Multiple phenyl substituents, attached where they either all stabilize structure **A** or all stabilize structure **C**, distort the TS geometry toward one of these two diradical extremes, thus enhancing the ability of all the substituents to stabilize the TS. In contrast, when one set of substituents stabilizes one of these resonance structures and another substituent stabilizes the other, the result is a compromise TS geometry, with a value of R that is optimal for neither set of substituents.

The calculations of the geometry of the TS for the Cope rearrangement of 1,5-hexadiene, at levels of *ab initio* theory that were unbiased and which included dynamic electron correlation [21,24], established that bond making and bond breaking occur synchronously in the TS. The ability of DFT calculations, particularly B3LYP/6-31G*, to compute not only activation enthalpies [22,23] but also kinetic isotope effects [22] that are in very good agreement with those that have been measured [6,11], gave further credibility to a value of $R \approx 1.9$–2.0 Å for the interallylic distance in the TS for the Cope rearrangement of unsubstituted 1,5-hexadiene. In fact, even without the (6/6)CASPT2 calculations [21], the B3LYP calculations might have settled the controversy about the value of R in and the nature of the TS for the parent Cope rearrangement.

The combination of relatively high accuracy and low cost of B3LYP/6-31G* calculations made possible the computational investigation of the cooperative and competitive substituent effects in the Cope rearrangements of phenyl-substituted 1,5-hexadienes. The B3LYP calculations established that the geometry of the TS in the Cope rearrangement depends on the number and placement of radical-stabilizing substituents, i.e. that the Cope TS is chameleonic [27,28].

References pp. 872–873

30.7 ACKNOWLEDGEMENTS

I thank Professors Ernest Davidson, Keiji Morokuma, and Ken Houk for the very enjoyable collaborations that provided most of the contents of this chapter, and I am very grateful to Dr. David Hrovat for his many calculations on Cope rearrangements during the 20 years that we have been working together. I am also indebted to Professor William von Eggers Doering for first piquing my interest in the Cope rearrangement in 1970 and for keeping me interested for the ensuing 35 years by discussing his experimental studies with me. Finally, I thank the National Science Foundation for supporting my research during this period.

30.8 REFERENCES

1 Reviews:(a) J.J. Gajewski, Hydrocarbon thermal isomerizations, Academic Press, New York, 1981, pp. 166–176; (b) W.T. Borden, R.J. Loncharich and K.N. Houk, Annu. Rev. Phys. Chem., 39 (1988) 213; (c) K.N. Houk, Y. Li and J.D. Evanseck, Angew. Chem. Int. Ed., 31 (1992) 682; (d) K.N. Houk, J. Gonzalez and Y. Li, Acc. Chem. Res., 28 (1995) 81; (e) W.T. Borden and E.R. Davidson, Acc. Chem. Res., 29 (1996) 67; (f) O. Wiest, D.C. Montiel and K.N. Houk, J. Phys. Chem. A, 101 (1997) 8378; (g) W.T. Borden, Mol. Phys., 100 (2001) 337.
2 (a) H.E. Zimmerman, J. Am. Chem. Soc., 88 (1966) 1564; (b) M.J.S. Dewar, Tetrahedron Suppl., 8 (1966) 75.
3 R.B. Woodward and R. Hoffmann, Angew. Chem. Int. Ed., 8 (1969) 781.
4 W.T. Borden, Modern molecular orbital theory for organic chemists, Prentice-Hall, Englewood Cliffs, NJ, 1975, pp. 129–132.
5 W. von E. Doering and Y. Wang, J. Am. Chem. Soc., 121 (1999) 10112.
6 W. von E. Doering, V.G. Toscano and G.H. Beasley, Tetrahedron, 27 (1971) 5299.
7 W. von E. Doering, Proc. Natl Acad. Sci., USA, 78 (1981) 5279.
8 J.J. Russell, J.A. Seetula and D. Gutman, J. Am. Chem. Soc., 110 (1988) 3092.
9 W.R. Roth, F. Bauer, A. Beitat, T. Ebbrecht and M. Wüstefeld, Chem. Ber., 124 (1991) 1453.
10 (a) M.J.S. Dewar and L.E. Wade, J. Am. Chem. Soc., 95 (1973) 290; (b) M.J.S. Dewar and L.E. Wade, J. Am. Chem. Soc., 99 (1977) 4417.
11 J.J. Gajewski and N.D. Conrad, J. Am. Chem. Soc., 101 (1979) 6693.
12 M.J.S. Dewar, G.P. Ford, M.L. McKee, H.S. Rzepa and L.E. Wade, J. Am. Chem. Soc., 99 (1977) 5069.
13 M.J.S. Dewar, J. Am. Chem. Soc., 106 (1984) 209.
14 Y. Osamura, S. Kato, K. Morokuma, D. Feller, E.R. Davidson and W.T. Borden, J. Am. Chem. Soc., 106 (1984) 3362.
15 M.J.S. Dewar and C. Jie, J. Am. Chem. Soc., 109 (1987) 5893.
16 M.J.S. Dewar and C. Jie, J. Chem. Soc., Chem. Commun., (1987) 1451.
17 K. Morokuma, W.T. Borden and D.A. Hrovat, J. Am. Chem. Soc., 110 (1988) 4474.
18 M. Dupuis, C. Murray and E.R. Davidson, J. Am. Chem. Soc., 113 (1991) 9756.
19 K. Andersson, P.-A. Malmqvist and B. Roos, J. Chem. Phys., 96 (1992) 1218.
20 K. Andersson, M.R.A. Blomberg, M.P. Fülscher, V. Kellö, R. Lindh, P.-A. Malmqvist, J. Noga, J. Olsen, B.O. Roos, A. Sadlej, P.E.M. Siegbahn, M. Urban and P.-O. Widmark, MOLCAS, University of Lund, Sweden, 1991, Version 2.
21 D.A. Hrovat, K. Morokuma and W.T. Borden, J. Am. Chem. Soc., 116 (1994) 1072.
22 O. Wiest, K.A. Black and K.N. Houk, J. Am. Chem. Soc., 116 (1994) 10336.
23 H. Jiao and P. von. R. Schleyer, Angew. Chem. Int. Ed., 34 (1995) 334.
24 P.M. Kozlowski, M. Dupuis and E.R. Davidson, J. Am. Chem. Soc., 117 (1995) 774.
25 P.M. Kozlowski and E.R. Davidson, J. Chem. Phys., 100 (1992) 3672.
26 V.K. Stavoverov and E.R. Davidson, J. Am. Chem. Soc., 122 (2000) 186.

27 D.A. Hrovat, B.R. Beno, H. Lange, H.-Y. Yoo, K.N. Houk and W.T. Borden, J. Am. Chem. Soc., 121 (1999) 10529.

28 D.A. Hrovat, J. Chen, K.N. Houk and W.T. Borden, J. Am. Chem. Soc., 122 (2000) 7456.

29 W.R. Roth, H.-W. Lennartz, W. von E. Doering, L. Birladeanu, C.A. Guyton and T. Kitagawa, J. Am. Chem. Soc., 112 (1990) 1722.

30 W. von E. Doering, L. Birladeanu, K. Sarma, G. Blaschke, U. Scheidemantel, R. Boese, J. Benet-Bucholz, F.-G. Klärner, J.-S. Gehrke, B.-U. Zinny, R. Sustmann and H.-G. Korth, J. Am. Chem. Soc., 122 (2000) 193.

31 V.K. Stavoverov and E.R. Davidson, J. Am. Chem. Soc., 122 (2000) 7377.

32 D.A. Hrovat and W.T. Borden, J. Chem. Theory Comput., 1 (2005) 87.

Theory and Applications of Computational Chemistry: The First Forty Years
Edited by C. Dykstra et al.

CHAPTER 31

High-temperature quantum chemical molecular dynamics simulations of carbon nanostructure self-assembly processes

Stephan Irle[1], Guishan Zheng[2], Marcus Elstner[3] and Keiji Morokuma[1,2]

[1] *Cherry L. Emerson Center for Scientific Computation, Emory University, Atlanta, GA 30322, USA*
[2] *Department of Chemistry, Emory University, Atlanta, GA 30322, USA*
[3] *Fachbereich Physik, Universität Paderborn, 33095 Paderborn, Germany*

Abstract

Very little is known either experimentally or theoretically concerning the mechanism of formation of fullerenes and carbon nanotubes (CNTs). We present an analysis of our recent fully quantum chemical high-temperature molecular dynamics simulations for (a) self-assembly capping processes of open-ended single-walled CNT models of different diameter, chirality, and lengths, and (b) self-assembly formation of fullerene molecules from ensembles of C_2 *without imposing a designed reaction pathway*. Density functional tight binding is used to compute the quantum chemical potential energy surfaces in direct trajectory calculations, and its accuracy is estimated in benchmark calculations. Capping of open-ended CNTs is observed to be a rapid process at temperatures of 2000 and 3000 K involving long-lived "wobbling C_2" and longer chains, typically within 14 ps simulation time. The self-assembly formation mechanism of fullerenes from ensembles of randomly oriented C_2 molecules was discovered by periodically adding batches of more C_2 molecules to the simulations, modeling an open environment. Three distinct steps of fullerene formation can be identified: *nucleation* of polycyclic structures by entangled polyyne chains, *growth* by ring condensation of attached chains, and *cage closure*. In this "size-up" roadmap, giant fullerenes C_n with $n > 120$ appear to be the dominant species, and a subsequent "size-down" roadmap leading to smaller fullerene cages by C_2 elimination is suggested based on prolonged heating of these large carbon cages. The combined two-stage "size-up/size-down" mechanism explains readily the abundance of buckminsterfullerene C_{60} in experiment as well as the distribution of larger fullerenes obtained by typical combustion processes.

31.1 INTRODUCTION

Today carbon nanostructures are the most intensively researched species in the context of molecular engineering on the nanometer scale, with the widest range of promising applications in materials science, molecular electronics, and molecular machinery. The elementary building blocks of carbon-based nanotechnology are all members of the third allotrope form of carbon, namely fullerenes and single- (SWNT) and multi-walled (MWNT) carbon nanotubes (CNTs). From these 0- (fullerene) and 1D (CNT) units it is hoped that 2D and 3D structures can be readily engineered with atomic precision to create any kind of desirable material or objects on the nanoscale, either in a "top-down" or "bottom-up" approach or a combination thereof, thereby extending the range of traditional chemistry to true molecular manufacturing. It is generally believed that self-assembly processes [1] are the most likely successful candidates for the bottom-up approach; the more traditional top-down approach of positional assembly is certainly more cumbersome for large-scale manufacturing processes [2]. However, at present it remains unclear what the nature of these self-assembly mechanisms should be, and how one could facilitate them in practical applications.

Remarkably, almost 20 years after the discovery of C_{60} buckminsterfullerene, it is still a much debated topic as to how fullerene molecules with their remarkable structural order are formed under the extremely chaotic high-energy conditions of carbon arc, laser ablation, or fuel rich, low-pressure aromatic hydrocarbon/oxygen combustion processes. Due to the large number of different carbon clusters present under the conditions of all of these synthetic methods, and the difficulty in tracking down single molecules in front of this background, spectroscopic methods have been of little value to follow formation mechanism pathways experimentally. Yet, it is a vital prerequisite to understand these reaction mechanism pathways in more detail to control self-assembly processes required for carbon nanostructure bottom-up-based molecular engineering. Several hypothetical models of fullerene formation have been described in the literature, some of which seemingly are more plausible than others. The most basic and prominent ones for fullerene formation mechanisms are, in chronological order, the "nautilus model" [3,4], the "party line" mechanism [5], the "pentagon road" [6–8], the "fullerene road" [9], the "ring-stacking" mechanism [10], and the "ring fusion spiral zipper" mechanism [11–13]. All of them, however, are more or less guesswork and share the assumption of an underlying principal of structural order, in a sense that such order is achieved either by combination of well-defined molecular "building blocks" like macrocyclic carbon rings or by collapse of a highly organized precursor structure.

Theoretical approaches, developed over the past 40 years of quantum chemistry, have recently become very helpful tools for developing an atomic-level understanding of the processes involved in high-temperature carbon chemistry. Interestingly, a combination of two theoretical approaches developed at opposite ends of this time-scale has proven to be extremely fruitful for such studies, namely the relatively new quantum chemical molecular dynamics (QM/MD) approach [14], using improved versions of early-day Extended Hückel electronic structure method [15–17] for the calculation of potential energy surfaces (PESs) on the fly in direct trajectories calculations. We have successfully used QM/MD simulations based on the density functional tight binding (DFTB) method [18–20] in our previous work to simulate the capping of open-ended CNTs [21,22], as well as the formation processes leading to fullerene molecules from randomly oriented ensembles of C_2 molecules [23,24]. Key to understanding high-temperature small-fragment carbon chemistry is the inclusion of non-equilibrium conditions in QM/MD simulations. Ergodic creation of order in dynamical, non-equilibrium systems appears to be inherent to many natural phenomena [25], and it seems quite logical to take into account the open environment in the study of fullerene and CNT formation processes in QM/MD simulations. Theoretical MD and quantum chemical studies following equilibrium pathways and investigating relative stabilities of fullerene precursor structures have been carried out before, as described in Section 31.2, but no definitive conclusion on fullerene or CNT formation mechanisms has been reached from these studies. Theoretical studies described in the literature by other groups suffer frequently from the choice of poor to very poor potential energy functions, and none of those studies considered the constantly changing, dynamic environment of high-temperature systems far from chemical equilibrium.

31.2 PREVIOUS THEORETICAL INVESTIGATIONS TOWARD FULLERENE FORMATION MECHANISMS

In the light of the experimental difficulties to identify pathways for fullerene formation processes, numerous quantum chemical electronic structure calculations on purely hypothetical intermediates of these proposed fullerene formation mechanisms have been performed in the past [26–29]. Moreover, Mishra et al. attempted to locate transition state structures connecting intermediate structures and describing entire pathways for the formation of C_{28} fullerene through their "circumscribing ring-stacking" mechanism at the semi-empirical AM1 [30] level of theory, starting from small monocyclic carbon rings such as C_9 and C_{13} [31]. We performed similar studies and found in accordance with Mishra et al. that due to the requirement of forcing the growing carbon cluster out of planarity, a significant energy barrier has to be overcome (see Fig. 31.1), before "3D-aromaticity" lowers the energy and makes the entire process exothermic. However, it is more than questionable whether an orderly growth process along a single or a few reaction pathways with well-defined transition states and intermediate structures could take place under the high-temperature non-equilibrium conditions characteristic for fullerene formation. The large amount of kinetic energy available under the experimental conditions allows carbon clusters to climb upward on hills on the PES rather than to cling to a specific minimum energy reaction pathway.

To address this problem, Monte Carlo simulations [32] and MD studies [33–37] have been carried out before, all of them using reactive empirical bond-order (REBO) Tersoff-type [38,39] interatomic carbon–carbon potentials developed originally for studying the vapor deposition of diamond [40,41]. Unlike traditional molecular mechanics force fields, the REBO potential allows for the formation and dissociation of covalent chemical bonds by determination of next neighbors and on-the-fly switching

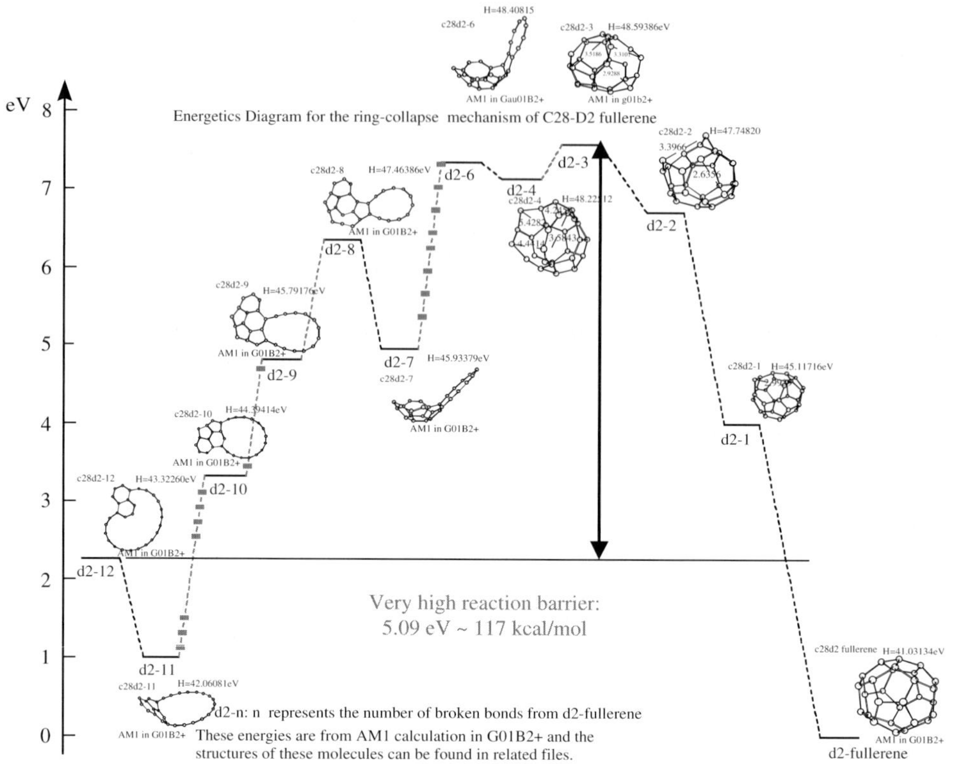

Fig. 31.1. AM1 reaction path for an assumed formation pathway of C_{28} fullerene following the approach of Mishra et al. [31].

bond functions. However, it has been criticized that the REBO potential is only successful in describing *intramolecular* interactions in carbon and hydrocarbon materials and that it lacks a mechanism for treating *intermolecular* interactions [42], which is a severe flaw for the description of a hot mixture of small carbon fragments. Even more seriously, REBO-type molecular force fields do not take into account π-conjugational effects, since REBO was designed to only describe the formation and breaking of carbon–carbon σ-bonds. However, aromaticity and π-conjugational stabilization are important features of carbon nanochemistry, and therefore we felt that straightforward QM/MD simulation (without imposing physically unrealistic spherical potentials such as used, e.g. in Ref. [43]) is a more appropriate computational tool for the theoretical investigation of carbon nanostructure formation mechanisms. So far, none of the published tight binding MD calculations on carbon nanosystems [44] except for ours have addressed the formation mechanism issue of fullerenes in a physically reasonable manner, including the very recent study of dynamic pathway models for the formation of C_{60} by Lee et al. [45]. Moreover, the potential energy functions employed in these studies are inferior to semi-empirical quantum chemical methods like AM1 and PM3 [46] and even far more inferior to the DFTB method employed by us in our CNT capping [21,22] and fullerene formation studies [23,24].

31.3 COMPUTATIONAL METHODOLOGY

The DFTB method is the central method employed in our direct trajectory QM/MD calculations. All DFTB calculations are carried out with the program package developed by Frauenheim, Seifert, and Elstner [18–20]. DFTB is an approximate density functional theory method based on the tight binding approach, and utilizes an optimized minimal LCAO Slater-type all-valence basis set in combination with a two-center approximation for Hamiltonian matrix elements. DFTB yields excellent agreement with far more costly B3LYP/6-31G(d) energetics in the comparison of relative energetics and geometries

Table 31.1 Relative energies of C_{28} isomers in eV and corresponding linear regression coefficient R^2 for correlation between B3LYP/6-31G(d), and DFTB, B3LYP/6-31G, AM1, and PM3 methods. All energetics are obtained using individually optimized molecular structures at respective levels

	B3LYP/6-31G(d)	DFTB	B3LYP/6-31G	AM1	PM3
buckyD2	0.00	0.00	0.00	0.00	0.00
Ring	3.32	8.10	0.78	−7.69	−2.15
c24-6	3.17	3.56	1.99	0.43	1.77
2 + 2r14	5.08	9.66	2.90	−3.34	0.91
2 + 2r16	6.01	10.25	3.87	−3.37	0.90
c20-6o	5.41	5.52	4.34	3.42	4.23
c20-6m	5.57	5.62	4.48	3.43	4.24
2 + 4	7.97	10.28	6.00	0.10	3.60
Central7	5.86	6.07	4.84	2.86	4.08
8 + 8	7.43	9.43	5.31	−3.24	0.79
4 + 4	9.91	14.27	8.52	1.52	4.97
R^2		0.7566	0.9238	0.0322	0.3640

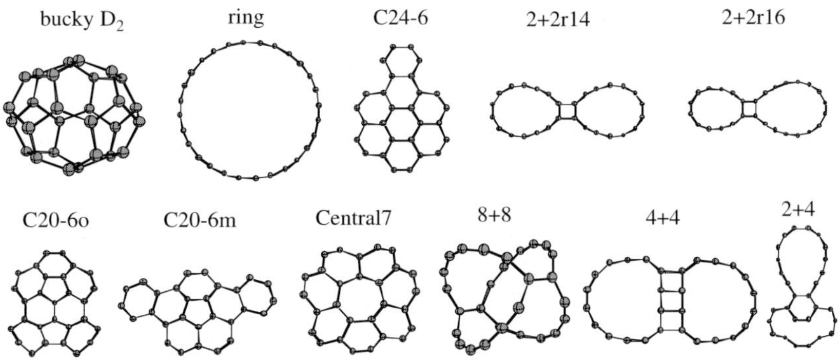

Structures of C_{28} used in the benchmark calculations present in Table 31.1.

of isomers of C_{28} [21], delivering at the same time superior accuracy over similarly inexpensive AM1 and PM3 semi-empirical methods [47]. In fact, single point B3LYP/ 6-31G(d) energies at DFTB optimized geometries showed a linear correlation coefficient R^2 of close to 1.0, while R^2 at, respectively, optimized geometries is still impressive with 0.7566 [21]. These benchmarks indicate that DFTB is highly suitable for quantum chemical MD studies of carbon clusters, as was also found before by other groups [48–53]. For details on these benchmark calculations, see Table 31.1. In particular, AM1 and PM3 predict the C_{28} fullerene structure **buckyD2** to be higher in energy than the C_{28} macrocycle, which is not acceptable for the study of fullerene formation mechanisms as carried out, e.g. by Mishra et al. [31].

Direct DFTB MD trajectories were run by calculating energy and the analytical energy gradient on the fly with a velocity Verlet integrator, using 1.209 fs (equals 50 a.u.) as the time step Δt. This time-step, larger than what is often used, is adopted since the system contains no light hydrogen atoms. We have shown that the error introduced by our choice of Δt amounts only to a few kcal/mol, which is a negligible amount at temperatures of 1000 K and higher [21]. Temperature is kept constant by scaling of atomic velocities in two ways. For one, scaling is regularly performed after 12 fs, and additional random scaling is performed with a probability of 10%, thereby the overall probability of scaling velocities is 20% for the entire length of the simulations.[1] Initial velocities are assigned randomly. All calculations were carried out with periodic boundary conditions to maintain constant densities.

31.4 SELF-ASSEMBLY CAPPING PROCESS OF OPEN-ENDED CARBON NANOTUBES

The studies described here are presented in detail in Refs. [21,22], and we will only focus on their most important aspects for self-assembly processes in the context of CNT capping. We chose to run QM/MD trajectories starting with open-ended CNTs of different chirality and lengths at various temperatures. In total, we ran more than 70 trajectories, most of them for at least 12 ps. The low-temperature 1000 K regime was found to be non-reactive and the high-temperature regime above 4000 K led to immediate fragmentation. Therefore we mainly used target temperatures between 2000 and 4000 K. Three different types of (n,m) nanotubes were chosen with about the same diameter d: armchair (5,5), $d = 6.88$ Å, chiral (7,3), $d = 7.15$ Å, and zigzag (9,0), $d = 7.06$ Å. In addition a (10,5) nanotube with a much larger diameter of $d = 10.5$ Å was studied. Three different tube lengths: 7.5, 10, and 20 Å were adopted for each species. An important finding is that the diameter of the (10,5) nanotube appears to be too large to allow its openings to be closed within 12 ps, regardless of the tube length. Table 1 of Ref. [22] summarizes or results for 2000, 3000, and 4000 K for (5,5), (7,3), and (9,0) open nanotubes. Fig. 31.2 displays a typical DFTB QM/MD trajectory of a (5,5) 7.5 Å

[1] We have also tested scaling of temperature at a lower rate (5%) and found that a scaling probability of 20% does not artificially speed up the dynamics simulation.

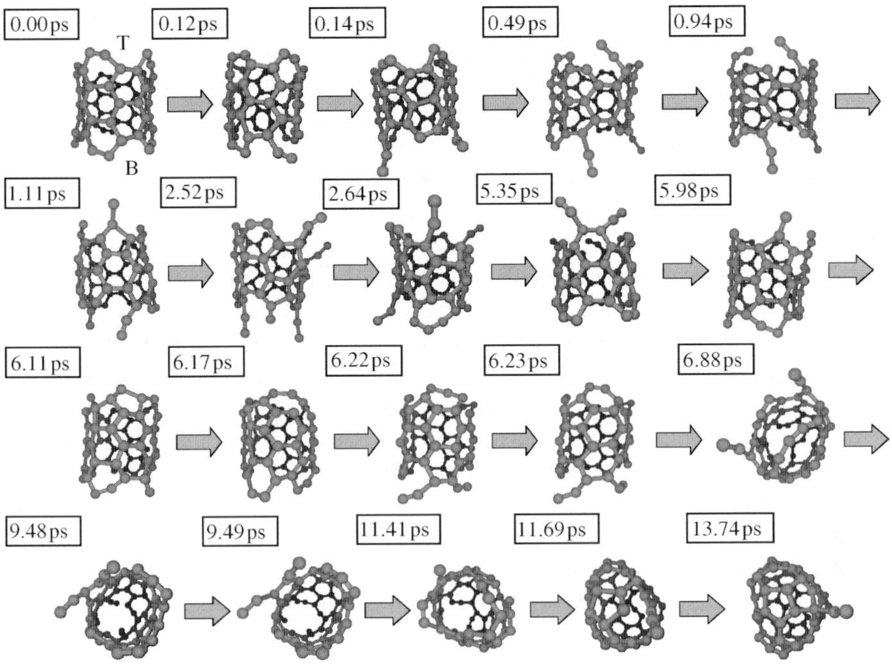

Fig. 31.2. DFTB QM/MD simulation of a (5,5) open-ended SWNT of 7.5 Å length at 3000 K. Wobbling C_2 units are the predominant structural feature of this type of system, not at all appearing in comparable REBO type simulations (see Fig. 31.3).

open-ended armchair tube at 3000 K, forming a fullerene molecule by capping of both ends within 14 ps.

Since only one or a few trajectories were run for each case, we make no claims on the statistical validity of the findings. However, we were able to extract important dynamic events common to many of these trajectories. The most striking finding of the present study is that many trajectories at 3000 and 4000 K closed both ends and formed fullerene structures within 14 ps. The key dynamical feature is an interplay between wobbling C_2 fragments, which are formed at the ends of the tube and occasionally catch neighboring hexagons to form pentagons, and creation, migration and isomerization of pentagons. This finding is in noticeable agreement with the observations of Car and co-workers in their computationally much more expensive CPMD studies on similar systems [54]. Trajectories run at 4000 K, compared to 3000 K, show higher C_2 formation activities and tend to close faster, accompanied with loss of C_1 to C_3 fragments. Trajectories at 2000 K also formed wobbling C_2 units at lower rate but did not close at either end; longer simulation times are required in this temperature regime for capping of the open-ended tubes. Chiral tubes like (7,3) show a greater tendency to develop even longer polyyne chain defects at the open-ends, in particular at those defects where the open ends resemble a stair case, but especially at higher temperatures, wobbling C_4 units are also developed in the case of the (5,5) armchair tubes [21]. At about 3000 K, armchair tubes typically develop only wobbling C_2 units from the breaking of cisoid rim-bonds, but very rapidly. (9,9) zigzag chains are somewhat different, as wobbling C_2 units only develop after Stone–Wales rearrangements have introduced heptagon/pentagon pairs at the openings, heptagons being more prone to develop such carbon chain fragments due to their greater flexibility. Once they have been created, these long-lived wobbling carbon chains occasionally make contact with defect structures (other wobbling C_2 units or 10-membered macrocycles created from adjacent five- and seven-membered rings) on the opposite site of the opening, leading eventually to rapid tube closure within typically 14 ps. We noted that once a structural defect like "wobbling C_2 unit" creation occurs, π-delocalization in that region of the opening is decreased, and more bonds are likely to break in its vicinity, leading to a high concentration of wobbling C_2 units in that area. At final stages of tube closure, [2 + 4] cycloaddition "zipper" type reactions occur when the opening size has been reduced and only consists of 12-membered rings.

These studies already indicate that linear, sp-hybridized carbon chains are favorable species under high-temperature conditions, due to their high entropy factors and their remarkable C–C bond strengths, while maintaining enough flexibility to move around considerably, especially at high temperatures. They are essential species in the self-assembly capping mechanism of nanotubes, forming easily new bonds with other chains or defects, and subsequently lead to restructured graphene sheets, which include

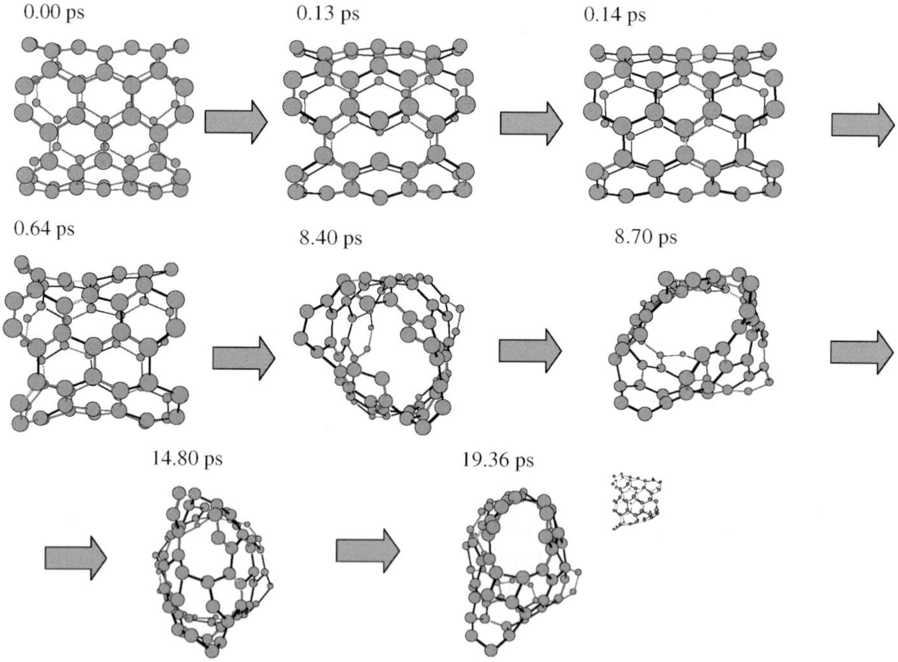

Fig. 31.3. REBO MD of a (5,5) open-ended SWNT of 7.5 Å length at 3000 K. Structural transformations occur mainly in the center of the tube, almost never at the edges. No wobbling C_2 units are observed at all.

pentagons and hexagons in approximately 2:1 ratio, leading to the formation of curved graphite spheres.

Remarkably, none of these reactions could be observed when similar trajectories were run using the REBO potential (see Fig. 31.3); instead, unreasonable transformation in the center parts of the tubes were observed, which prolonged the tube closing processes substantially into the nanosecond regime. Obviously, quantum mechanical hybridization and delocalization effects play a synergetic effect in carbon chemistry that cannot be included without full electronic structure calculations.

31.5 SELF-ASSEMBLY OF FULLERENE MOLECULES FROM ENSEMBLES OF RANDOMLY ORIENTED C_2 MOLECULES

Under experimental conditions of fullerene formation, a constant flow of carbon material is certainly present during the microsecond long process, forcing initially formed carbon clusters constantly out of thermodynamic equilibrium. We chose to model such an open environment [23,24] by adding more C_2 units to a trajectory once an equilibrium state is reached where no apparent structural reorganizations take place. For a schematic diagram of the exact schedule for C_2 additions in our previous simulations (Fig. 31.4). We ran about 130 such trajectories for up to 160 ps. Snapshots for two of these trajectories are shown in Figs. 31.5 and 31.6, where giant fullerenes emerge after approximately 140 ps. At initial stages, long linear sp-hybridized carbon chains entangle and form initial clusters of condensed "small" rings (which we call nucleus), which can attain lifetimes long enough at around 2000 K temperature to form bonds with neighboring carbon atoms of attached chains to form larger condensed ring systems by the ring-collapse

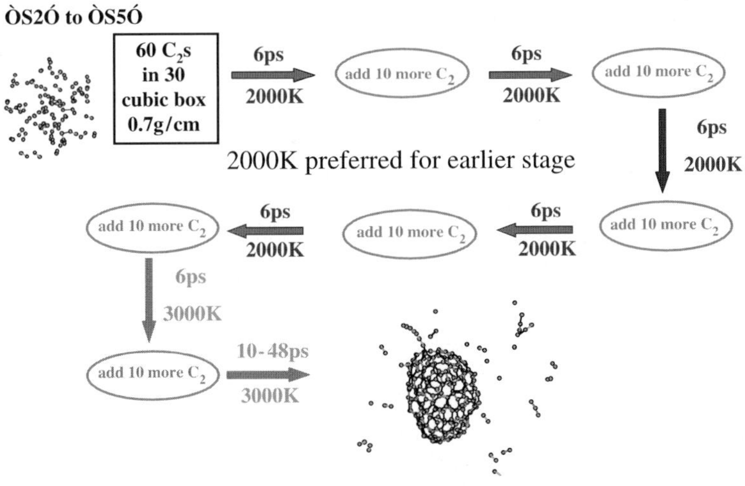

Fig. 31.4. Schematic diagram for QM/MD trajectories starting from ensembles of randomly oriented C_2 molecules leading to fullerene formation in a self-assembly mechanism under periodic addition of 10 C_2's along the trajectory. Temperature is increased at the final stage to accelerate cage closure.

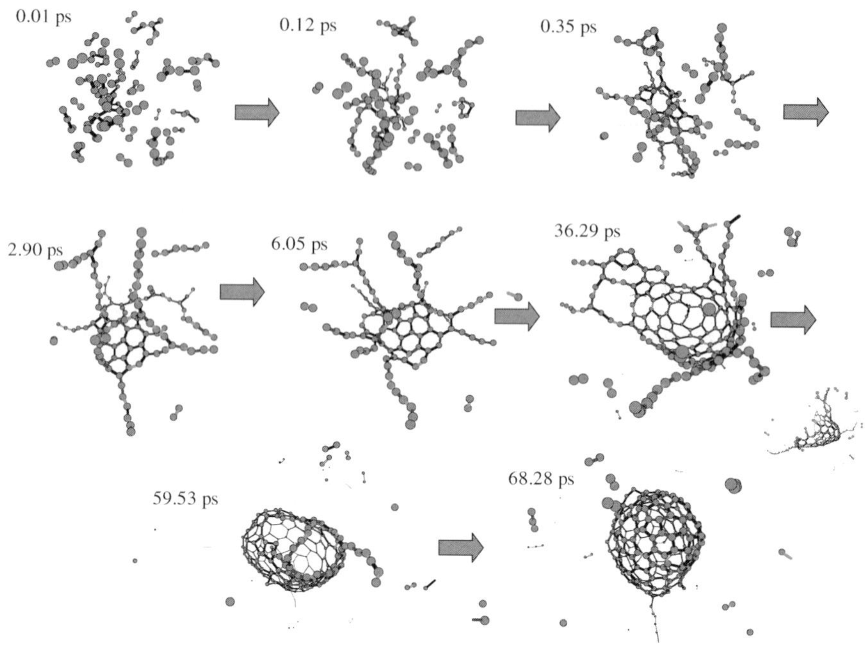

Fig. 31.5. S2 trajectory from Ref. [23], starting from 60 C_2 units with 0.09 g/cm^3 initial carbon density, leading to a C_{184} giant fullerene.

mechanism. We call this initial event "nucleation", which is the first step in fullerene formation. We found that the possibility for ring destruction is greatly reduced once two to three rings are forming a condensed ring system. At temperatures of 2000 K and higher, the energetic difference between pentagons and hexagons does not matter very much, and we found that both species are readily created with a ratio of about 1:2, respectively, similar as in the case of self-assembled nanotube caps. While the growing condensed ring system gains energy non-linearly by growing π-delocalization in the ring condensation step (the second step in fullerene formation), the embedded pentagons force the growing slab to adapt a uniform curvature. At the same time, attached polyyne chains are growing by catching additional C_n molecules from the carbon mixture of the environment, and retain thereby their flexibility to bend so much that ring condensation between slab border atoms and the nearest chain neighbor atoms can continue to proceed. The result is typically a basket-shaped carbon cluster with several long linear carbon chains attached to its opening, and these chains are able to reach over to the other ends, forcing the system to eventually close to a fullerene cage. This is the third and last step of fullerene formation. The energy profile associated with this mechanism is constantly downhill, because catching high-energetic small carbon fragment molecules and forming bonds to attach them at the attached carbon chains releases constantly energy. The entire process appears to be a fascinating interplay between growing stabilization of an expanding π-conjugated, polyaromatic system and the dynamics of the polyyne chains attached to it.

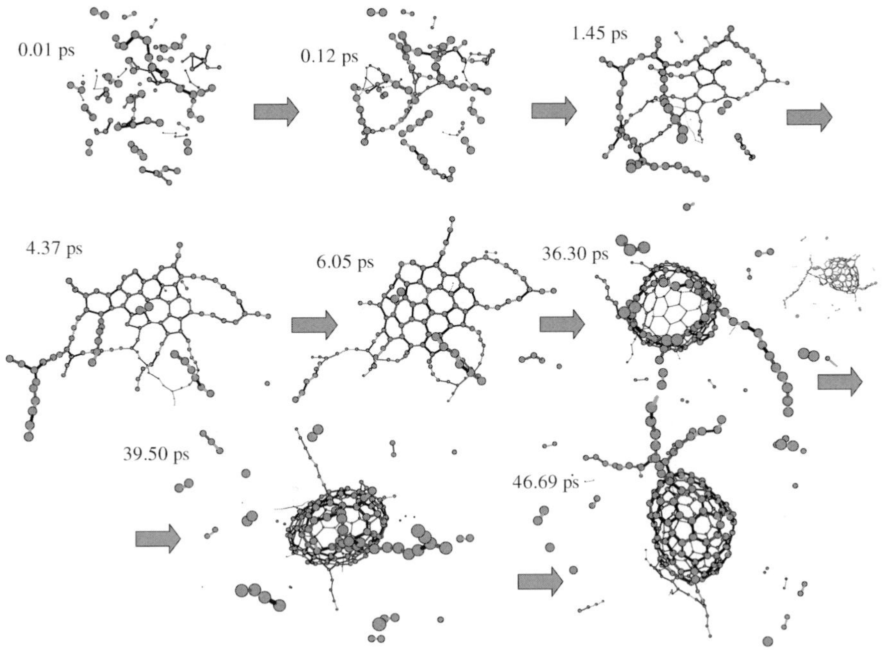

Fig. 31.6. S3 trajectory from Ref. [23], starting from 60 C_2 units with 0.09 g/cm^3 initial carbon density, leading to a C_{146} giant fullerene.

We found that the gradual addition of C_2 is essential for the formation of a significantly curved carbon cluster, and that entanglement of carbon chains at initial stages is crucial for two reasons: (a) formation of a nucleus of polycyclic structures, and (b) formation of a 3D-scaffold along which ring-condensation growth can occur subsequently. We found that if one starts simulations with the final number of C_2 molecules of our successful trajectories from the beginning without adding more C_2 molecules subsequently, only very few linear chains grow very long, as opposed to many shorter chains created by the subsequent addition of C_2 molecules. The ultra-long carbon chains created by the "all-at-once" approach are likely to lead eventually to large graphene sheets with lesser curvature than when starting from more but shorter polyyne chains, due to the lack of initial entanglement.

We have shown that our QM/MD isokinetic trajectories, although not representing statistical ensembles, are valid exploratory investigations into the dynamics of carbon clusters at high temperature, capable of detecting key features in the formation mechanisms of carbon nanostructures. However, in Ref. [23], our QM/MD simulations led only to the formation of so-called "giant fullerenes", namely C_{146}, C_{184}, and C_{208} by self-assembly from ensembles of C_2 molecules. We, therefore, call this stage of fullerene formation the size-up roadmap. Starting at higher carbon densities reduced the size of resulting fullerenes [24] down to C_{72}, but somewhat surprisingly, neither C_{60} nor C_{70}, the most important and abundant representative of fullerenes, were found in our simulations.

Moreover, the product distribution among higher fullerenes C_{72}–C_{100} is in thermodynamic equilibrium [55], but so far we have been unable to obtain such a product distribution. Of course, one may not expect from running over a hundred trajectories to obtain the correct product distribution, but at least we should have been able to see indicators for the preference of small fullerenes over large ones. Yet, our trajectories so far never led to the self-assembly of C_{60} or C_{70} directly, indicating that perhaps giant fullerenes are formed first following the size-up roadmap, followed by "fall off" of attached polyyne chains and subsequent shrinking caused by expelling C_2 units, which is an energetically feasible process [56] and has been observed and described experimentally by Smalley and co-workers [57]. We call this second stage of fullerene formation size-down roadmap. We have preliminary theoretical results on the rate at which C_2 units are expelled in what we termed "pop-out" processes from giant fullerenes under prolonged heating at 3000 K, and our estimate is currently 0.5 C_2 per picosecond (see Fig. 31.7) [24]. C_2 loss is expected to slow down when the size of the cage becomes smaller, until more rigid cages like C_{60} are reached, which are kinetically extraordinarily stable. These smaller fullerene molecules become then "trapped" and their quantity increases, enriching the fraction of smaller fullerenes as the main reaction product.

Additionally, we expect that heptagon, fused pentagon, and four-membered ring defects will eventually give way to the regular structures of isolated pentagon rule (IPR) fullerenes. The pathways from eight-membered ring defect fullerenes down to IPR fullerenes has already been described in the literature using a generalized Stone–Wales (GSW) approach [58], and Maruyama and Yamaguchi found formation of perfectly I_h-symmetric C_{60} in their simulated annealing simulations using the REBO semiclassical

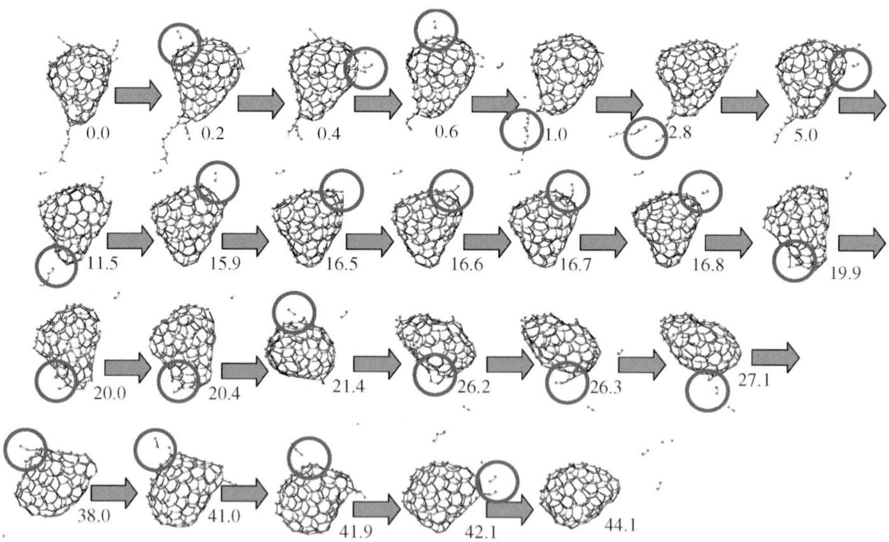

Fig. 31.7. Snapshots of trajectory S3 from Ref. [24]. Roman number below each snapshot is the time t in picosecond. Red circles indicate locations of "fall-off" or "pop-out" processes.

molecular mechanics method [37]. The combined two-stage mechanism consisting of size-up, followed by size-down, explains the abundance of buckminsterfullerene C_{60} in experiment as well as the distribution of larger fullerenes obtained by typical combustion processes. We are currently continuing nanosecond long QM/MD simulations along these lines.

31.6 CONCLUSIONS

In performing QM/MD high-temperature simulations for the capping process of open-ended CNTs, we learned about the importance of polyyne chains for structural rearrangements of the carbon network, and their amazing abilities to introduce curvature into graphene sheets by making use of their abundance at high temperatures, high C–C bond strengths, long lifetimes, and flexibility. Their structural and dynamic importance already suggests that at initial stages of fullerene formation, the party line mechanism is most likely to dominate. The concept of "self-assembly" in fullerene formation emerges somewhat surprisingly but naturally as a consequence of our non-equilibrium dynamics, and we found that fullerenes might in fact represent "frozen" dissipative dynamic carbon structures commonly found in non-linear dynamic systems [25], trapped by rapid cooling thanks to their kinetic stability. A size-up roadmap has been described by which fullerene molecules are formed from random ensembles of small carbon fragments in QM/MD simulations when periodically batches of C_2 units are added. Three steps are clearly identified which are required for fullerene formation: (1) nucleation; (2) ring condensation growth; and (3) cage closure. Depending on the density, different sizes and formation times are observed, but so far all trajectories lead to fullerene cages larger than C_{60}. In a size-down roadmap, these giant fullerenes are exposed to prolonged heating at 3000 K, and "pop-out" events are observed, consistent with experimental observations. The combined two-stage size-up/size-down mechanism provides a natural explanation for the dominant appearance of kinetically stable C_{60} as final fate of giant fullerenes, and the existence of larger fullerenes in experiment.

31.7 ACKNOWLEDGEMENTS

This work was partially supported by grants from the Mitsubishi Chemical Corporation and from the Petroleum Research Fund, the American Chemical Society. Acknowledgment is made to the Cherry L. Emerson Center of Emory University for the use of its resources. We also thank the Pacific Northwest Laboratory's EMSL for valuable computer time through its Grand Challenge project, GC3564.

31.8 REFERENCES

1 G. Nicolis and I. Prigogine, Self-organization in nonequilibrium systems: From dissipative structures to order through fluctuations, Wiley, New York, 1977.
2 R.C. Merkle, Nanotechnology, 11 (2000) 89–99.

3 H.W. Kroto and K. McKay, Nature, 331 (1988) 328–331.
4 H.W. Kroto, Science, 242 (1988) 1139–1145.
5 H.W. Kroto, Carbon, 30 (1992) 1139–1141.
6 R.F. Curl and R.E. Smalley, Sci. Am., 265 (1991) 54–63.
7 R.E. Haufler, Y. Chai, L.P.F. Chibante, J. Conceicao, C. Jin, L.-S. Wang, S. Maruyama and R.E. Smalley, Mater. Res. Soc. Symp. Proc., 206 (1991) 627–637.
8 R.E. Smalley, Acc. Chem. Res., 25 (1992) 98–105.
9 J.R. Heath, ACS Symp. Ser., 481 (1991) 1–23.
10 T. Wakabayashi, H. Shiromaru, K. Kikuchi and Y. Achiba, Chem. Phys. Lett., 201 (1993) 470–474.
11 G. van Helden, N.G. Gotts and M.T. Bowers, Nature, 363 (1993) 60–63.
12 J. Hunter, J. Fye and M.F. Jarrold, Science, 260 (1993) 784–786.
13 R.J. Lagow, J.J. Kampa, H.-C. Wei, S.L. Battle, J.W. Genge, D.A. Laude, C.J. Harper, R. Bau, R.C. Stevens, J.F. Haw and E. Munson, Science, 267 (1995) 362–367.
14 A. Gonzalez-Lafont, T.N. Truong and D.G. Truhlar, J. Phys. Chem., 94 (1991) 5544–5558.
15 R.S. Mulliken, J. Chim. Phys., 46 (1949) 497.
16 R.S. Mulliken, J. Chim. Phys., 46 (1949) 675.
17 M. Wolfsberg and L. Helmholtz, J. Chim. Phys., 20 (1952) 837.
18 D. Porezag, T. Frauenheim, T. Köhler, G. Seifert and R. Kaschner, Phys. Rev. B, 51 (1995) 12947–12957.
19 G. Seifert, D. Porezag and T. Frauenheim, Int. J. Quantum Chem., 58 (1996) 185–192.
20 M. Elstner, D. Porezag, G. Jungnickel, J. Elsner, M. Haugk, T. Frauenheim, S. Suhai and G. Seifert, Phys. Rev. B, 58 (1998) 7260–7268.
21 G. Zheng, S. Irle, M. Elstner and K. Morokuma, J. Phys. Chem. A, 108 (2004) 3182–3194.
22 S. Irle, G. Zheng, M. Elstner and K. Morokuma, Nano Lett., 3 (2003) 465–470.
23 S. Irle, G. Zheng, M. Elstner and K. Morokuma, Nano Lett., 3 (2003) 1657–1664.
24 G. Zheng, S. Irle, M. Elstner and K. Morokuma, J. Chim. Phys., 122 (2005) 014708.
25 I. Prigogine and I. Stengers, Order out of Chaos: Man's new dialogue with nature, Bantam Books, Toronto, 1984.
26 D.L. Strout and G.E. Scuseria, J. Phys. Chem., 100 (1996) 6492–6498.
27 K.R. Bates and G.E. Scuseria, J. Phys. Chem. A, 101 (1997) 3038–3042.
28 R.O. Jones, J. Chem. Phys., 110 (1999) 5189–5200.
29 S. Portmann, J.M. Galbraith, H.F. Schaefer, G.E. Scuseria and H.P. Lüthi, Chem. Phys. Lett., 301 (1999) 98–104.
30 M.J.S. Dewar, E.G. Zoebisch, E.F. Healy and J.J.P. Stewart, J. Am. Chem. Soc., 107 (1985) 3902–3909.
31 R.K. Mishra, Y.-T. Lin and S.-L. Lee, J. Chem. Phys., 112 (2000) 6355–6364.
32 V.A. Schweigert, A.L. Alexandrov, Y.N. Morokov and V.M. Bedanov, Chem. Phys. Lett., 235 (1995) 221–229.
33 J.R. Chelikowsky, Phys. Rev. B, 45 (1992) 12062–12070.
34 Y. Yamaguchi and S. Maruyama, Trans. JSME, 63-611B (1997) 2298–2404.
35 S. Maruyama and Y. Yamaguchi, Trans. JSME, 63-611B (1997) 2405–2412.
36 Y. Yamaguchi and S. Maruyama, Chem. Phys. Lett., 286 (1998) 336–342.
37 S. Maruyama and Y. Yamaguchi, Chem. Phys. Lett., 286 (1998) 343–349.
38 J. Tersoff, Phys. Rev. B, 37 (1988) 6991.
39 J. Tersoff, Phys. Rev. B, 39 (1989) 5566.
40 D.W. Brenner, Phys. Rev. B, 42 (1990) 9458–9471.
41 D.W. Brenner, Phys. Rev. B, 46 (1992) 1948.
42 S.J. Stuart, A.B. Tutein and J.A. Harrison, J. Chem. Phys., 112 (2000) 6472–6486.
43 C.Z. Wang, C.H. Xu, C.T. Chan and K.M. Ho, J. Phys. Chem., 96 (1992) 3563.
44 C.Z. Wang and K.M. Ho, J. Comput. Theor. Nanosci., 1 (2004) 3–17.
45 I.-H. Lee and H. Kim, J. Chem. Phys., 120 (2004) 4672–4676.
46 J.J.P. Stewart, J. Comput. Chem., 10 (1989) 209–220.
47 G. Zheng, S. Irle, M. Elstner, K. Morokuma, unpublished.
48 T. Heine, Die Berechnung von Struktur, Energetik, und kernmagnetischen Abschirmungen von Fullerenen und ihren Derivaten, PhD, Technische Universität Dresden, Dresden (2000).

49 P.W. Fowler, T. Heine, K.M. Rogers, J.P.B. Sandall, G. Seifert and F. Zerbetto, Chem. Phys. Lett., 300 (1999) 369–378.
50 P.W. Fowler, T. Heine, D. Mitchell, G. Orlandi, R. Schmidt, G. Seifert and F. Zerbetto, J. Chem. Soc. Farad. Trans., 92 (1996) 2203–2210.
51 P.W. Fowler, T. Heine, D. Mitchell, G. Orlandi, R. Schmidt, G. Seifert and F. Zerbetto, J. Phys. Chem., 100 (1996) 6984–6991.
52 M.C. Domene, P.W. Fowler, D. Mitchell, G. Seifert and F. Zerbetto, J. Phys. Chem. A, 101 (1997) 8339–8344.
53 G. Jungnickel, D. Porezag, T. Köhler, T. Frauenheim and M.R. Pedersen, From icosahedral clusters to polymeric systems—a theoretical approach, in: H. Kuzmany (Ed.), Fullerenes and fullerene nanostructures, World Scientific, Singapore, 1996, pp. 305–318.
54 J.-C. Charlier, A. De Vita, X. Blase and R. Car, Science, 275 (1997) 646–649.
55 Z. Slanina, X. Zhao, P. Deota and E. Osawa, in: K.M. Kadish, R.S. Ruoff (Eds.), Fullerenes—chemistry, physics, and technology, Wiley-Inerscience, New York, 2000, pp. 283–330.
56 A.D. Boese and G.E. Scuseria, Chem. Phys. Lett., 294 (1998) 233–236.
57 S.C. O'Brien, J.R. Heath, R.F. Curl and R.E. Smalley, J. Chem. Phys., 88 (1988) 220–230.
58 E. Osawa, H. Ueno, M. Yoshida, Z. Slanina, X. Zhao, M. Nishiyama and H. Saito, J. Chem. Soc. Perkin Trans., 2, 1998 (1998) 943–950.

Theory and Applications of Computational Chemistry: The First Forty Years
Edited by C. Dykstra et al.

CHAPTER 32

Computational chemistry of isomeric fullerenes and endofullerenes

Zdeněk Slanina[1,2] and Shigeru Nagase[1]

[1] *Department of Theoretical Molecular Science, Institute for Molecular Science, Myodaiji, Okazaki 444-8585, Aichi, Japan*
[2] *Institute of Chemistry, Academia Sinica, 128 Yen-Chiu-Yuan Rd., Sec. 2, Nankang, Taipei 11529, Taiwan, ROC*

Abstract

Fullerenes and metallofullerenes were for the first time observed in the gas phase about 20 years ago and then prepared in crystalline form about 15 years ago. An enormous amount of observed and computed data has been obtained during the period. The chapter surveys various computational aspects of fullerene science including rich isomerism and the enthalpy–entropy interplay both of which represent essential features of fullerene and metallofullerene formations.

32.1 INTRODUCTION

Fullerenes—or cage compounds built exclusively from carbon atoms—and their metal-containing derivatives, metallofullerenes, were first observed in the gas phase by Kroto et al. [1,2] less than 20 years ago and prepared in crystalline form by Krätschmer et al. [3] less than 15 years ago. Still, an enormous amount of observed and computed data has been obtained during this time (see, e.g. recent surveys on fullerenes [4–8] and endohedral metallofullerenes [9,10]). In addition to spheroidal fullerene cages, fullerene science also deals with other objects like elongated cylindrical bodies known as nanotubes, prepared by Iijima [11] soon after mastering the fullerene synthesis, nanocones [12] or peapods [13]. All the species exhibit a substantial application potential, especially for molecular electronics [14].

Fullerenes and metallofullerenes have attracted a considerable interest from experiment [15] as well as computations and theory [16,17]. Their experimental characterization is mostly based [18–20] on ^{13}C NMR spectroscopy and more recently

References pp. 908–917

also on ^{3}He NMR spectroscopy. Fullerenes are commonly defined as polyhedral cages containing only carbon atoms arranged into five- and six-membered rings. *Quasi*-fullerenes [21] are cages containing other types of cycles, too, like (topological) squares and heptagons.

The very early history [22–26] of carbon clusters starts with a mass-spectrometric observation of clusters up to C_{15} by Hahn and his co-workers [27] 60 years ago. In the 1950s and 1960s, experiments [28–30] could expand up to C_{33}. At this stage, initial simple computations had also been performed on small carbon clusters by Pitzer and Clementi [31,32], Hoffmann [33] and others [34,35]. Moreover, various qualitative estimations for larger carbon cages including C_{60} were presented [36–40].

At the time of the C_{60} discovery [1], computations could support fullerene [16,17] and metallofullerene [41] research with increasingly reliable data. It is of historic interest that the four IR bands computationally predicted [42–47] for C_{60} were used [48] in the analysis of the carbon-arc C_{60} synthesis. Fullerene research [49,50] has indeed been characterized by a close theory–experiment cooperation and application of large-scale computations [51,52]. Although numerous surveys are available [53–80], the sole computations of fullerenes and metallofullerenes are reviewed relatively rarely [16,17,41,81–90].

Structural elucidations for the known higher fullerenes C_n [19,91–93] have been based on the so-called isolated pentagon rule (IPR), which claims [94,95] that especially stable fullerenes should have all pentagons surrounded just by hexagons. Connectivity patterns for such cages can be generated by various topological schemes [96–99] and in fact the cages can be quite numerous. Beyond $n = 76$ isomerism of the IPR structures should generally play a role in observations. Several such mixtures of empty fullerene isomers have indeed been computed and an agreement with experiments found: C_{76} [100–110], C_{78} [111–114], C_{80} [115–119], C_{82} [120–125], C_{84} [126–132], C_{86} [91,133–136], C_{88} [91,134–136], C_{90} [91,137,138], C_{92} [139], C_{94} [140] and C_{96} [141], while there is still only a computational prediction [142] for C_{98}. A similar isomeric interplay has also been described for smaller fullerene systems like [143–146] C_{32} or C_{36}. Various metallofullerenes can coexist [41] in several isomeric forms as well, e.g. $Ca@C_{72}$ [147–150], $Mg@C_{72}$ [151,152], $Ca@C_{74}$ [147,149,153,154], $Ca@C_{82}$ [155–161], $La@C_{82}$ [162–166], $Tm@C_{82}$ [167], $Ti_2@C_{80}$ [168], $Sc_2@C_{84}$ [169], $Ti_2@C_{84}$ [170] or $Sc_2@C_{76}$ [171]. Still other interesting isomeric sets can be created [172–176] by derivatives of fullerenes or nanotubes.

Although the inter-isomeric separation energies are important, they alone cannot predict the relative stabilities of the isomers. Owing to very high temperatures, entropy contributions can even over-compensate the enthalpy terms. Hence, the enthalpy–entropy interplay represents an essential feature of fullerenes and metallofullerenes.

32.2 RELATIVE STABILITIES OF ISOMERS

The present quantum-chemical calculations of fullerenes deal with the optimized geometries [177–181] obtained at semiempirical, *ab initio* Hartree–Fock Self-Consistent Field (HF SCF) or density functional theory (DFT) level while *ab initio* correlated treatments are very rare. The most common semiempirical methods used in fullerene research are [182–185] MNDO, AM1, PM3 and SAM1 implemented in the MOPAC and AMPAC programs [186–188]. Various *ab initio* procedures are mostly applied using GAUSSIAN [189,190] and SPARTAN [191] program packages. Stability of the SCF wave function [192,193] can be an issue for some fullerenic structures. The geometry optimizations are frequently followed by harmonic vibrational analysis to check the nature of stationary points found and also to construct vibrational partition functions. Vibrational frequency scaling would be relevant for the spectral simulations though it is not important for the high-temperature partition functions [194].

The geometry optimizations will end with a set of m isomeric structures. Their relative concentrations can be expressed as their mole fractions, w_i, using the isomeric partition functions q_i. In terms of q_i and the ground-state energy changes $\Delta H^{\text{o}}_{0,i}$ the mole fractions are given [195–197]:

$$w_i = \frac{q_i \exp[-\Delta H^{\text{o}}_{0,i}/(RT)]}{\sum_{j=1}^{m} q_j \exp[-\Delta H^{\text{o}}_{0,j}/(RT)]} \tag{1}$$

where R stands for the gas constant and T for the absolute temperature. Eq. (1) is an exact formula that can be directly derived [198] from the standard Gibbs energies of the isomers, supposing the conditions of the inter-isomeric thermodynamic equilibrium. Although the partition functions are to be constructed within the rigid-rotor and harmonic-oscillator (RRHO) approximation, Eq. (1) itself is essentially exact. The partition functions q_i reflect the rotational, vibrational, electronic, symmetry and chirality [199] features of the individual isomers.

If the partition functions are neglected and the vibrational zero-point energy is extracted from $\Delta H^{\text{o}}_{0,i}$, the relative potential energies $\Delta E_{\text{r},i}$ remain, giving the simple Boltzmann factors:

$$w'_i = \frac{\exp[-\Delta E^{\text{o}}_{\text{r},i}/(RT)]}{\sum_{j=1}^{m} \exp[-\Delta E^{\text{o}}_{\text{r},j}/(RT)]} \tag{2}$$

entirely based only on the potential energy terms with no reference to entropy contributions. However, the simple Boltzmann factors can never cross with a temperature change.

32.3 ENERGETICS AND THERMODYNAMICS OF CARBON CLUSTERS

The conventional fullerenes are understood [21,97] as cages built from three-coordinated (sp^2) carbon atoms arranged into two types of rings, pentagons and hexagons. As any other polyhedra, fullerenes have to obey Euler's polyhedral theorem [200,201] (more precisely, the theorem for convex polyhedra):

$$V + F = E + 2 \tag{3}$$

where V denotes the number of vertexes (atoms), F the number of faces (rings) and E the number of edges (bonds), e.g. for C_{60}: $V = 60$, $F = 32$, $E = 90$. As only pentagons and hexagons are allowed, their numbers, n_5 and n_6, give the total count of faces:

$$F = n_5 + n_6 \tag{4}$$

As all atoms are three-coordinated, it must hold for the number of edges:

$$E = \frac{3V}{2} \tag{5}$$

as each bond is accounted twice. One can also count the edges through rings:

$$E = \frac{5n_5 + 6n_6}{2} \tag{6}$$

Combining Eqs. (5) and (6) yields

$$V = \frac{5n_5 + 6n_6}{3} \tag{7}$$

Eq. (5) can be placed in Euler's theorem:

$$V + F = \frac{3V}{2} + 2 \tag{8}$$

and combination with Eqs. (7) and (4) gives

$$n_5 + n_6 = \frac{5n_5 + 6n_6}{6} + 2 \tag{9}$$

which can immediately be reduced to:

$$n_5 = 12 \tag{10}$$

In other words, in any conventional fullerene C_n the number of five-membered rings must be equal to 12. The number of six-membered rings is variable; however, Eq. (7) gives a relationship to the number of carbon atoms:

$$n_6 = \frac{n - 20}{2} \tag{11}$$

According to Eq. (11), the smallest possible fullerene has the stoichiometry C_{20} (12 pentagons, no hexagons). Among all possible fullerenes, a special position is given to the IPR cages [94,95] (that have all the pentagons surrounded just by hexagons) as the arrangements should be especially stable. In other words, there are no pentagon–pentagon junctions in the IPR fullerenes. As there are always 12 pentagons, then the smallest cage that could be of the IPR type should have 12 × 5 carbon atoms. Indeed, 60 carbon atoms can create an IPR cage (which geometrically is a truncated icosahedron), and the structure is the well-known buckminsterfullerene C_{60}.

Already the MINDO/2 computations [34,202] of small carbon clusters C_n pointed out a simple, smooth dependency of the relative heats of formation $\Delta H^{\circ}_{f,298}/n$ on the number of carbons n. Later on, the curve was extended into the fullerene domain [51,86,203–206] and it was still a smoothly decreasing dependency. Xu and Scuseria [206] could find such behavior even with carbon clusters up to C_{8640}.

There is a simple way to rationalize the finding. Let us limit our reasoning to the IPR fullerenes. Then, we deal with two types of bonds, frequently called as the 5/6 (between pentagons and hexagons) and 6/6 bonds (shared by two hexagons). Let us suppose moreover that those two types of bonds can be represented [207] by some uniform dissociation energies, $H_{5/6}$ and $H_{6/6}$. In a general IPR fullerene C_n we have always sixty 5/6 bonds, while the number of the 6/6 bonds is variable, $(3n/2) - 60$. Now, we can readily write for the atomization heat of the considered C_n:

$$\Delta H_{\mathrm{at}} = 60H_{5/6} + \left(\frac{3n}{2} - 60\right)H_{6/6} \tag{12}$$

The atomization and formation heats for carbon aggregates are linked by the heat of vaporization of carbon, $\Delta H^{\mathrm{o}}_{\mathrm{vap}}$:

$$\frac{\Delta H^{\mathrm{o}}_{\mathrm{f}}}{n} = -\frac{\Delta H_{\mathrm{at}}}{n} + \Delta H^{\mathrm{o}}_{\mathrm{vap}} \tag{13}$$

or simply in terms of the bond energies:

$$\frac{\Delta H^{\mathrm{o}}_{\mathrm{f}}}{n} = -\frac{3}{2}H_{6/6} + \Delta H^{\mathrm{o}}_{\mathrm{vap}} + \frac{60}{n}(H_{6/6} - H_{5/6}) \tag{14}$$

This functional dependency can formally be expressed by:

$$\frac{\Delta H^{\mathrm{o}}_{\mathrm{f}}}{n} = A + \frac{B}{n} \tag{15}$$

where B is a positive constant and A a relatively small number. Hence, Eq. (15) is indeed a smoothly decreasing curve. From such a curve, one cannot see a particular stability of C_{60} or C_{70}. Just the opposite—it might suggest a straightforward rule that the stability would increase with the dimension which is, however, not observed.

In its most general form, the fullerene synthesis could be treated as a complex kinetic scheme described by a huge number of kinetic differential equations. The equilibrium composition comes as the limiting case for infinite time. If we treat the problem from a thermodynamic point of view, we should realize that the conventional standard pressure of 1 atm is considerably different from the actual fullerene synthesis conditions. We should expect lower cluster pressures in the carbon-arc synthesis. The actual entropy and Gibbs free energy change with pressure as can be demonstrated [208–212] on the C_{60} and C_{70} cases based on computed or observed [213] data. For example, the equilibrium constant $K_{60/70}$ for an interconversion between the two clusters, expressed in partial pressures p, offers a deeper insight into the problem [208–212]:

$$K_{60/70} = \frac{p_{70}^{6/7}}{p_{60}} = \frac{(1 - x_{60})^{6/7}}{x_{60}}P^{-1/7} \tag{16}$$

where P stands for the total pressure of the two clusters and x_{60} is the mole fraction of buckminsterfullerene. At higher pressures, C_{70} is more populated than C_{60}, but at the conditions of a saturated carbon vapor (simulated simply as a mixture of seven cluster at present) the stability order is reversed in favor of C_{60} so that an agreement with experiment is obtained. Although the isomeric stability problem is much better understood [198,214–216] at present than the relative stabilities of non-isomeric carbon clusters, there are some interesting results available for the non-isomeric situation also. The numerical simulations [208–212] pointed out a temperature increase of the clustering degree under the saturation conditions. This somewhat surprising result can in fact be easily rationalized [217]. While the equilibrium constants for cluster formation decrease with temperature, the saturated pressure increases. It is just the competition between these two terms which decides the final temperature behavior. The finding actually has a more general validity as it was also reported for other vapors [218–221].

32.4 SMALL CARBON CLUSTERS

The relative stabilities of small carbon clusters can be treated in a similar way as those of fullerenes and thus they represent a useful model system. They have been studied by both theoretical [222–240] and experimental [241–264] techniques. Raghavachari and Binkley [223] concluded that linear isomers exist for both even- and odd-numbered species (the even clusters being in the triplet electronic state) while planar cyclic isomers exist for the even species. Ion chromatography observations by von Helden et al. [261, 263,264] showed that starting from about $n = 7$ cyclic rings always exist in addition to the linear forms. Consequently, C_7 was computed [236,265] at the MP2/6-31G* level and found to be about 2 kcal/mol above the linear form. A non-planar cyclic structure was found for C_8 [237] and C_9 [266] at the same level so that an inversion is possible [237, 267]. The non-planar C_8 cycle was, however, not found in the coupled-cluster computations by Martin and Taylor [240]. The MP2/6-31G* calculations are also available [268–270] for C_{10}, C_{11}, C_{12} and C_{13}. Overall [271], the computations agree with the findings of ion chromatography [261,263,264]. C_{11} should be the first species for which the cyclic structure becomes dominant. Switch from cyclic to polyhedral species is expected [272] around $n = 45$.

The linear and cyclic (rhombic, bicyclic) C_4 isomers were frequently computed. The cyclic *vs.* linear difference varied considerably; however, newer estimates [273–277] suggest only a small separation between the linear and cyclic forms. Then, the entropy contributions can lead to one dominant species at higher temperatures [245,278–280], namely the linear isomer in agreement with ion chromatography [261].

Role of entropy is also important for C_6. Raghavachari et al. [222] found the cyclic ground state for C_6. Still, the linear isomer is computed [214,240,281] to form more than 95% of the equilibrium mixture at high temperatures in agreement with ion chromatography [261] (*cf.* Refs. [282,283]).

32.5 GENERATION OF CAGES

Exhaustive topological enumerations and generations of fullerene cages represent a crucial step, systematically studied by Fowler et al. [96,97,284–302] and other groups as well [95,98,99,303–331]. Several concepts are useful in the treatments, e.g. Goldberg polyhedra [286], leapfrog transformation [296], topological duals [306], Stone–Wales transformation [304]. Stone and Wales [304] introduced processes for ring rearrangements in fullerene cages. The topological concept can further be generalized [332] by inclusion of other types of rings. The kinetic processes can physically be realized [333–335] by means of catalysis or autocatalysis. The Stone–Wales transformation can in any case be applied as a formal topological structure generator [132,143,296].

A versatile enumeration technique for fullerene cages was developed by Fowler and Manolopoulos [96,97,308]. They represent fullerenes by a ring spiral as fullerenes can be peeled like an orange—each face, after the first, borders its immediate predecessor, so a single continuous spiral is created. Some exceptional cases are, however, possible [302] like a tetrahedral fullerene with 380 atoms.

In the spiral algorithm, the bonding topology is reconstructed from the sequence of rings in the spiral. Thus, all possible C_n fullerene graphs can be generated by considering all the ways in which 12 pentagons and $((n/2) - 10)$ hexagons can be combined into a spiral:

$$S_n = \frac{\left(\frac{n}{2} + 2\right)!}{12!\left(\frac{n}{2} - 10\right)!} \qquad (17)$$

A uniqueness test is based on the eigenvalues of the adjacency matrix.

It was found [308–313] that, e.g. the number of C_{60} isomers is 1812. Odd-numbered cages are usually not considered in the enumerations though they are known [314,336, 337]. Among all the possible isomers, the IPR structures are particularly important species. The second smallest IPR fullerene after C_{60} is C_{70}. It is still a unique structure like those for $n = 72$ and 74. For any higher carbon atom content, we already deal with IPR isomers, e.g. [96,97], for $n = 76$, 78, 80, 82, 84, 86, 88 and 90 there are 2, 5, 7, 9, 24, 19, 35 and 46 IPR structures, respectively.

Other enumeration algorithms were developed [98,99,132,309,310,320]. There are also interesting applications of Pólya's enumeration theorem [198,338]. The theorem was, e.g., applied [321,324,330,331] to enumerations of isomers for various substituted fullerenes.

32.6 SMALLER FULLERENES

Although fullerene science is focused at $n = 60$ and above, smaller fullerenes have also been studied. von Helden et al. [261,272] observed gas-phase fullerene-like structures around C_{30} (5% for C_{32}, 1% for C_{30}). Even the smallest fullerene C_{20} is now available [339–342] (for its computations, see, e.g. Refs. [216,343–346]). C_{36} fullerene was also isolated [347], computed [144,145,348,349] and linked to narrow nanotubes [350–353].

An interesting illustrative example is supplied with computations [143] of C_{32}. There are just six conventional C_{32} cages built from pentagons and hexagons [348]. As no IPR structures are possible for C_{32}, a pentagon/pentagon junction will always be present. Their number can be reduced by introduction of four-membered and/or seven-membered rings. Euler's network closure requirement reads then:

$$2n_4 + n_5 - n_7 = 12 \tag{18}$$

where n_i is the number of rings with i vertices. An exhaustive generation [143] of all C_{32} cages built from four-, five-, six- and/or seven-membered rings with a limitation, $2n_4 + n_7 \leq 4$, gives 199 cages. All the 199 structures were computed [143] and a D_{4d} structure with two four-membered rings (and no heptagons) was found as the ground state. Among five structures lowest in energy, only two are conventional fullerenes, the other cages contain at least one square. However, the entropy contributions reverse [143] the stability order. The structure lowest in energy is the most populated species till a temperature of about 2500 K. Beyond that point it is surpassed by a conventional fullerene. At still higher temperatures, another *quasi*-fullerene becomes a leading isomer. An interesting enthalpy–entropy interplay was also computed [144–146] for C_{36}.

32.7 HIGHER FULLERENES

Studies of isomerism of the IPR cages from C_{76} till C_{98} are available [100–142]. In fact, C_{72} is the last member among higher fullerenes [19,92] C_{60}–C_{96} that has not been isolated in condensed phase. C_{74} has also been counted as a 'missing fullerene', however, some access to C_{74} is already possible [354,355]. A low solubility [356] in conventional solvents may be one reason for the difficulties. Hence, as C_{72} could only be recorded in gas phase [357,358], its structure is not known. Nevertheless, its computations can still serve as an interesting illustrative example. C_{72} can also act as a host cage for some metallofullerenes [41,147,148] like $Ca@C_{72}$ or $La_2@C_{72}$.

There is just one IPR-satisfying structure for C_{72}, namely with D_{6d} symmetry [97]. It was, however, pointed out in conjunction with the $Ca@C_{72}$ computations [148] that a non-IPR (i.e. IPR-violating) structure with one pentagon–pentagon junction is by a few kcal/mol lower in energy than the IPR cage. Moreover, it was demonstrated [359] on the Si_{60} case that the IPR/non-IPR stability order can be reversed by the entropy factor. Hence, the C_{72} system was investigated at semiempirical [360] and DFT levels [361]. At the DFT level, the geometry optimizations were carried out [361] using the B3LYP/3-21G treatment. In the optimized B3LYP/3-21G geometries the harmonic vibrational analysis was carried out and also the B3LYP/6-31G* separation energies were evaluated. The electronic excitation energies were evaluated by means of the ZINDO method [362,363].

The following structures were considered [361]: the IPR cage (a), two non-IPR cages [148] with one pentagon–pentagon junction (b) and (c), a structure [148] with one heptagon (d), a cage [149] with two heptagons (e) and two structures [356] each with two pentagon–pentagon junctions (f) and (g). Fig. 32.1 presents the B3LYP/3-21G optimized structures of the seven C_{72} isomers. The (c) structure of C_{2v} symmetry with just one pentagon/pentagon fusion represents the lowest energy isomer, being followed

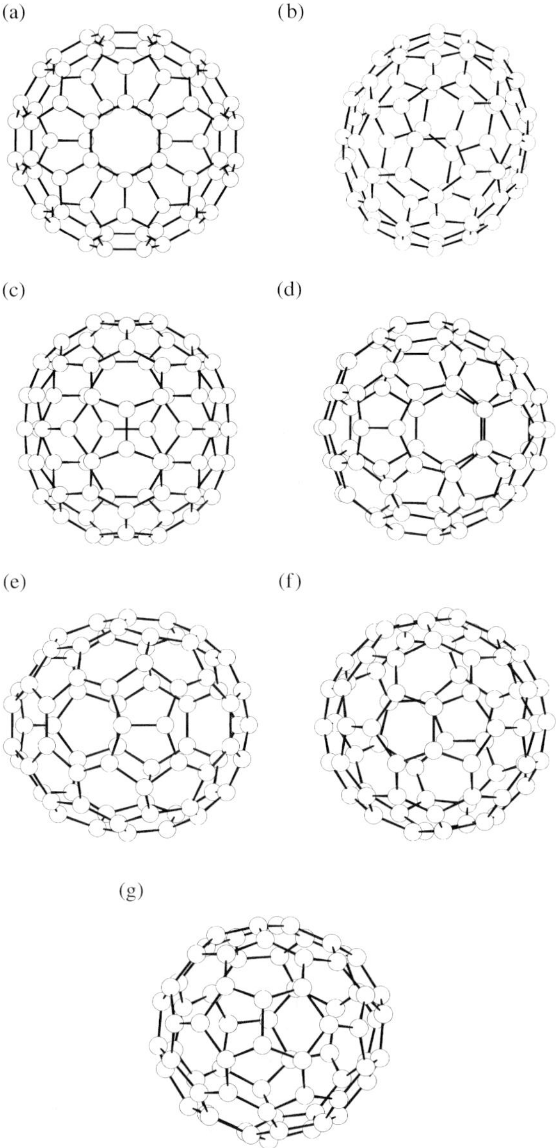

Fig. 32.1. B3LYP/3-21G optimized structures of C_{72} isomers: (a) IPR, (b) 5/5 pair, (c) 5/5 pair, (d) seven-membered ring, (e) two seven-membered rings, (f) two 5/5 pairs, (g) two 5/5 pairs. (Reproduced from Ref. [361] with permission of Elsevier BV.)

by the IPR structure (a). The other structure with just one pentagon–pentagon junction, (b) isomer of C_2 symmetry, comes as the third lowest species. The B3LYP/6-31G*// B3LYP/3-21G relative energies are 0.0, 11.3, 18.4, 31.1, 34.9, 35.5 and 43.7 kcal/mol for the (c), (a), (b), (e), (g), (d) and (f) structures, respectively.

Fig. 32.2 presents the DFT computed temperature development of the relative concentrations of the seven C_{72} isomers in a high-temperature region. The lowest energy structure (c) is the most populated species at any temperature. On the other hand, the IPR structure (a) is always negligible. Its elimination cannot be ascribed only to the potential energy. In fact, several structures higher in potential energy than the (a) isomer are more populated at high temperatures. This is in particular true for the (b) structure with one pentagon–pentagon junction and the (e) structure with two heptagons (the (e) cage

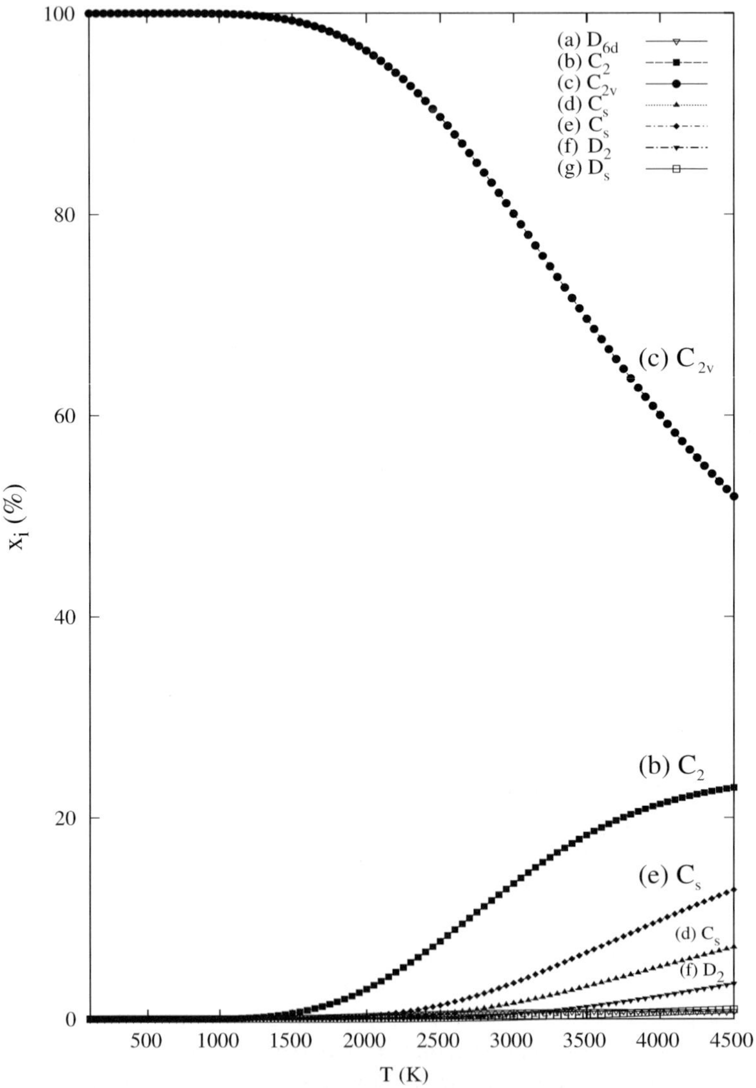

Fig. 32.2. Relative concentrations of the C_{72} isomers based on the B3LYP/6-31G* energetics and the B3LYP/3-21G and ZINDO entropy. (Reproduced from Ref. [361] with permission of Elsevier BV.)

is located in the B3LYP/6-31G* scale about 20 kcal/mol above the (a) isomer). Clearly enough, the IPR structure is suppressed not only by its energy but also by its unfavorable entropy term (especially by the vibrational entropy).

The C_{74} case is, however, different. Recently, Shinohara and co-workers [355] recorded electronic spectrum of C_{74} anion and suggested that the cage could have D_{3h} symmetry (i.e. the only one available IPR structure [97]). This interesting finding prompted DFT calculations [364] that treated a set of six isomers, five of them being non-IPR species. The computational evaluation, however, shows that the IPR structure prevails at any relevant temperature in agreement with the experimental finding [355].

32.8 ENDOHEDRAL METALLOFULLERENES

The combined quantum-chemical and statistical-mechanical stability computations have been applied to a few metallofullerenes also, especially to Ca@C_{72}, Ca@C_{74}, Ca@C_{82} or La@C_{82}. Ca@C_{72} was isolated [147] though its observed structure is not yet available. It follows from its very first computations [148,149] that there are four isomers especially low in potential energy (though C_{72} has only one [97] IPR structure). The endohedral Ca@C_{72} species created by putting Ca inside the sole IPR cage has been labeled [148] by (a). The other three Ca@C_{72} isomers considered in Ref. [148] are related to two non-IPR C_{72} cages (b) and (c), and to a C_{72} structure with one heptagon (d)—see Fig. 32.3.

The extended computations [150] started from the four optimized structures [148] derived using *ab initio* HF treatment with a combined basis set: 3-21G basis for C atoms and a dz basis set [365] with the effective core potential (ECP) on Ca (for the sake of simplicity, the treatment is coded by HF/3-21G ~ dz). The structures [148] were reoptimized at the B3LYP/3-21G ~ dz level. The B3LYP/6-31G ~ dz//B3LYP/3-21G ~ dz relative energies for the (b), (c), (d) and (a) structures are 0.0, 0.8, 18.9 and 35.3 kcal/mol, respectively. In the optimized B3LYP/3-21G ~ dz geometries the harmonic vibrational analysis was carried out with the analytical force-constant matrix. In the same geometries single-point energy calculations were also performed at the B3LYP/6-31G* level. The electronic excitation energies were evaluated by means of time-dependent DFT response theory [366] at the B3LYP/3-21G ~ dz level.

Fig. 32.4 presents [150] the temperature development of the relative concentrations of the four Ca@C_{72} isomers in a high-temperature region. At very low temperatures (not shown in Fig. 32.4) the structure lowest in the $\Delta H^{o}_{0,i}$ scale must be prevailing. However, already at a temperature of 226 K (though such a low temperature is of no importance for the fullerene synthesis) the relative concentrations of the (c) and (b) structures are interchanged and beyond the point the (b) structure is always somewhat more populated. Even more interesting is the behavior of the IPR-satisfying (a) structure. As the structure is the highest in the potential energy, it must be the least populated species at low temperatures. However, later on the entropy contributions (low symmetry, some lower vibrational frequencies and some lower electronic excitation energies) elevate the (a) isomer into the status of a minor isomer that could also be observed. On the other hand, the (d) isomer has the least chances to be detected. Interestingly enough,

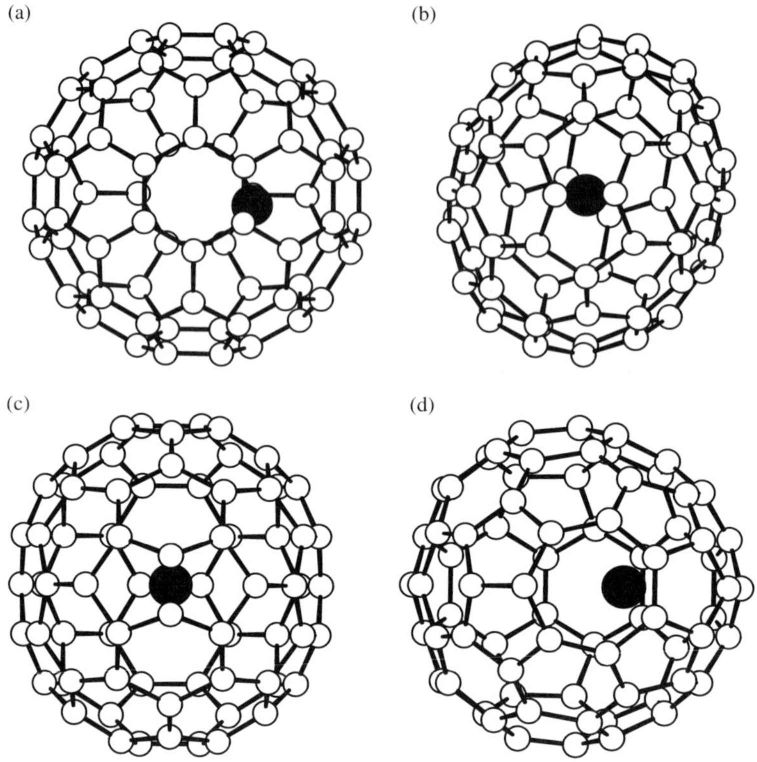

Fig. 32.3. B3LYP/3-21G ~ dz optimized structures of $Ca@C_{72}$ isomers: (a) IPR, (b) 5/5 pair, (c) 5/5 pair, (d) seven-membered ring. (Reproduced from Ref. [150] with permission of Elsevier BV.)

the concentration order at high temperatures for $Ca@C_{72}$ is quite similar to that previously computed [151] for $Mg@C_{72}$.

In contrast to $Ca@C_{72}$, $Ca@C_{74}$ was not only isolated [147] but its structure was also determined [153]. According to the ^{13}C NMR spectra recorded by Achiba and co-workers [153], $Ca@C_{74}$ exhibits D_{3h} symmetry of its cage. There is only one IPR structure possible [97] for the empty C_{74} and the sole C_{74} IPR cage has D_{3h} symmetry. A set of altogether six $Ca@C_{74}$ isomers was subjected to the stability computations [154]. It turns out, in agreement with the experiment, that the encapsulate with the IPR cage has not only favorable enthalpy but also entropy term and thus all the remaining isomers can act as minor species at best.

The third illustrative system, $Ca@C_{82}$, exhibits the richest isomerism among the Ca endohedrals [147,155,157,366–369]. Shinohara and co-workers [155] isolated four isomers of $Ca@C_{82}$ and labeled the isomers (I), (II), (III) and (IV). Dennis and Shinohara concluded [158,370] from the ^{13}C NMR spectra of $Ca@C_{82}$ (III) its symmetry as C_2. The ultraviolet photoelectron spectra measured by Hino et al. [159] support the finding; a similarity with $Tm@C_{82}$ (II) was also noted [167]. Very recently, Achiba and co-workers

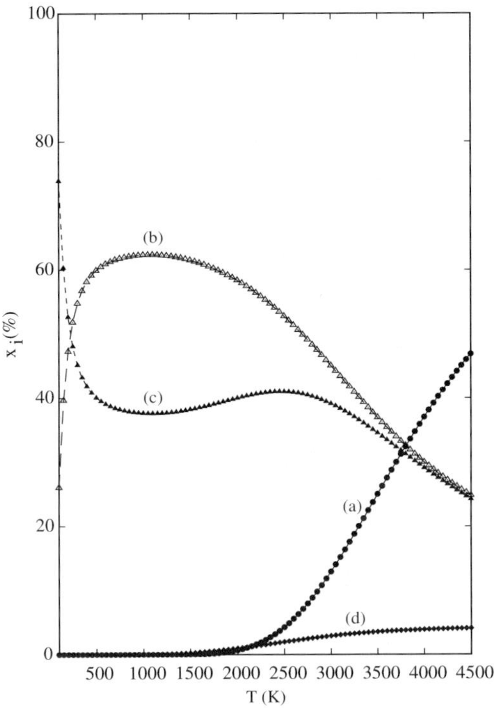

Fig. 32.4. Relative concentrations of the Ca@C_{72} isomers based on the B3LYP/6-31G* energetics and the B3LYP/3-21G ~ dz entropy. (Reproduced from Ref. [150] with permission of Elsevier BV.)

[160] measured the ^{13}C NMR spectra of all the four isomers and assigned the symmetry for the isomers (I), (II), (III) and (IV) as C_s, C_{3v}, C_2 and C_{2v}, respectively.

The Ca@C_{82} structure–energetics relationships were also computed [156] and a qualitative agreement with the experiment found [41]. The computations were performed at the HF and DFT levels and in both cases the C_{2v} structure was the lowest isomer in the potential energy. There were still three other low energy species—C_s, C_2, and C_{3v}. The combined stability computations are also available [161] for the full set of nine isomers [156] of Ca@C_{82}.

The nine C_{82} IPR cages [97] produce nine Ca@C_{82} endohedrals. Fig. 32.5 presents their structures optimized [161] at the B3LYP/3-21G ~ dz level. It has turned out that in five cases the original HF structures after the DFT reoptimizations within the same symmetry lead to saddle points with imaginary vibrational frequencies, not to the required local energy minima. When the five saddle points are relaxed and reoptimized, the following local minima are obtained: C_{3v} (b) → C_s, C_{2v} → C_s, C_2 (a) → C_1, C_2 (b) → C_1, C_s (b) → C_1.

Fig. 32.6 presents the temperature development [161] of the relative concentrations of the nine Ca@C_{82} isomers in a wide temperature region. The enthalpy part of the Gibbs energy is taken from the B3LYP/6-31G*//B3LYP/3-21G ~ dz calculations, the entropy

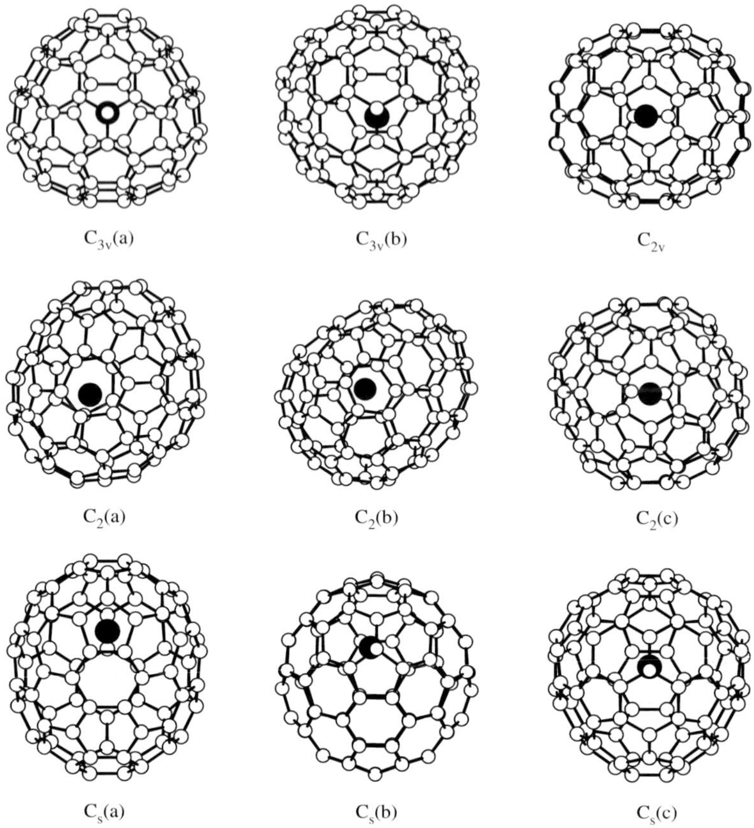

Fig. 32.5. B3LYP/3-21G ~ dz optimized structures of Ca@C_{82} isomers. (Reproduced from Ref. [161] with permission of American Institute of Physics.)

part is evaluated at the B3LYP/3-21G ~ dz level. At very low temperatures the structure lowest in the $\Delta H^{o}_{0,i}$ scale must be prevailing. However, at a temperature of 1700 K the relative concentrations of the $C_{2v} \rightarrow C_s$, and C_s, (c) structures are interchanged and beyond the point the C_s (c) structure is always somewhat more populated. The C_s (c) isomer and also C_2 (c) exhibit a temperature maximum. Then, there are still two other structures with significant populations at high temperatures: C_{3v} (b) $\rightarrow C_s$ and C_s (b) $\rightarrow C_1$. Although the former species is a bit more populated, their concentrations are rather close. In contrast to the five species with significant populations at least in some temperature regions, the remaining four isomers are computed to be negligible at any temperature. The C_{3v} (a) structure is remarkable as it is lower in potential energy than C_{3v} (b) $\rightarrow C_s$, however, it is actually suppressed by entropy. The observed yields [160] of the isomers were nearly equal except for the considerably less-produced C_{3v} species (though the HPLC chromatograms [155] could indicate somewhat larger production differences). Fig. 32.6 is in reasonable agreement with the qualitative population

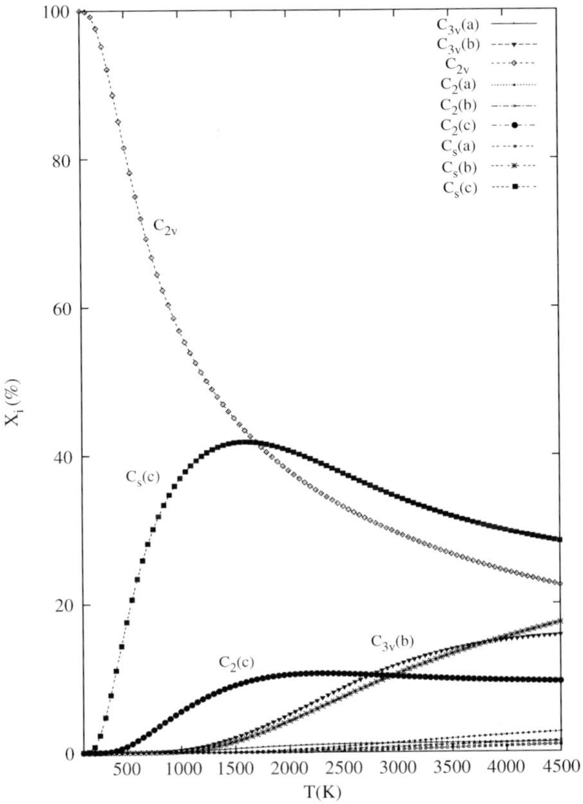

Fig. 32.6. Relative concentrations of the $Ca@C_{82}$ isomers based on the B3LYP/6-31G* energetics and the B3LYP/3-21G ~ dz entropy. (Reproduced from Ref. [161] with permission of American Institute of Physics.)

information in a relatively wide temperature interval. However, the fifth isomer also pointed out by the computations, C_s (b) → C_1, has not been observed. The fifth species could be hidden in a chromatographic fraction as a minor component.

The last illustrative example deals with $La@C_{82}$, i.e. an electronic open-shell system. The $La@C_{82}$ metallofullerene is one of the very first endohedrals that was macroscopically produced [371]. Recently, structures of two of its isomers were clarified [163,164] using ^{13}C NMR spectra of their monoanions generated electrochemically. The major isomer [163] was thus assigned C_{2v} symmetry and the minor species [164] C_s. The C_{2v} structure was moreover confirmed by an X-ray powder diffraction study [372]. The findings stand in contrast to $Ca@C_{82}$ with four known isomers. Computations at *ab initio* HF and DFT levels pointed out [121,162,373] just three IPR cages with a sufficiently low energy after La atom encapsulation: C_{2v}, C_{3v} (b) and C_s (c). The fourth lowest La endohedral species, C_2 (a), is actually already rather high in energy to be significant in experiment.

A partial agreement with experiment can be reached [374] for temperatures roughly from 1000 to 1300 K. However, the agreement can be further improved by

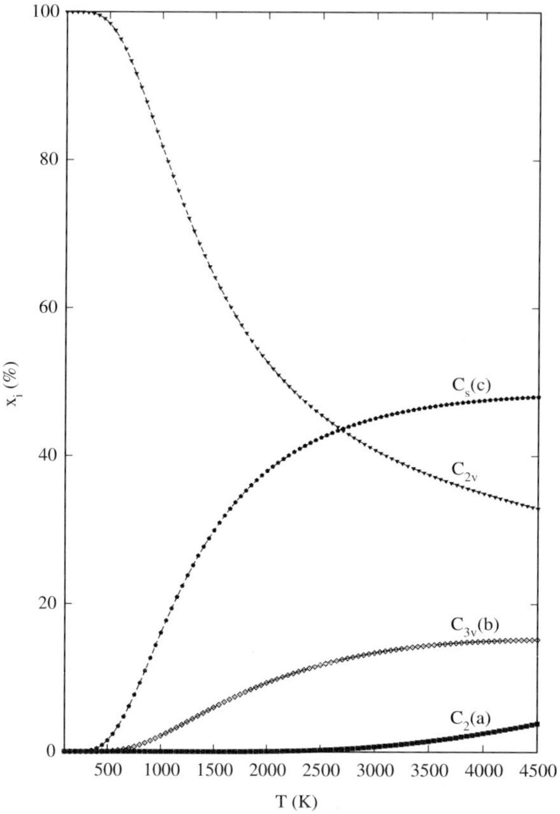

Fig. 32.7. Relative concentrations [364] of the La@C_{82} isomers derived within the FEM approach.

an alternative approach to description of the encapsulate motions. One can expect that if the encapsulate is relatively free to move within the cage then, at sufficiently high temperatures, its behavior in different cages will bring about the same contribution to the partition functions. The contributions would then cancel out in Eq. (1). This simplification can be called free, fluctuating or floating encapsulate model (FEM). In the FEM model, in addition to the removal of the three lowest vibrational frequencies, the symmetries of the cages should be treated as the highest possible, considering the averaging effect of the large amplitude motions of the encapsulate. Fig. 32.7 shows the results for the La@C_{82} isomeric system evaluated within the FEM approach. A good agreement with the observed facts [163,164] is reached in the treatment.

There is still another aspect to be considered. The fullerene and metallofullerene production is not always close to the inter-isomeric equilibrium. This factor may be pertinent to the La@C_{82} case. Lian et al. [375] reported a Ni-catalyzed production of La@C_{82} with a considerably variable isomeric ratio, indicating a possible kinetic control.

32.9 CONCLUDING REMARKS

This survey shows the ongoing productive interaction between theory and experiment in fullerene research that allows for a deeper insight into the observed facts. For fullerenes and metallofullerenes the mutual interaction is indeed essential, given the complexity of the systems. The considerable thermal effects on the relative isomeric populations revealed by the quantum-chemical and statistical-mechanical computations result from a complex interplay between rotational, vibrational, electronic, potential energy terms, chirality factors, etc. Such effects would never be seen if only energetics is considered and entropy neglected. The treatment is, however, built upon the presumption of the inter-isomeric thermodynamic equilibrium. We do not know yet to which degree this presumption is satisfied in reality though there a few indications that the equilibrium is not necessarily always established. We can only acknowledge that the thermodynamic equilibrium treatment has already produced a reasonable computation–observation agreement for the isomeric systems from C_{76} till C_{96}, with no serious failure, and also for a few first metallofullerenes. This relatively large tested set supports the belief in a still wider applicability of the equilibrium treatment. There are also other aspects of the combined treatment that should further be studied—especially anharmonicity of vibrations, motions of encapsulates in cages and reliability of inter-isomeric energetics. The non-IPR structures have not been sufficiently explored yet. Heterofullerenes, like cages based on boron nitride, should also attract more attention. Once the thermodynamic treatment is completely exposed, the interest will gradually shift towards complex kinetic schemes and to fullerene sets with variable stoichiometry. The steadily growing family of fullerene and metallofullerene cages will further attract interest of more applied branches of nanoscience and nanotechnology [376], including quantum computing [377] and superconductivity [378]. Further developments are ahead for the cages, the first public exposure [379] of which (namely in Konstanz just 15 years ago) W. Krätschmer[1] commented with a modest: "Fullerene molecules must belong to a very frequently overlooked molecular species".

32.10 ACKNOWLEDGEMENTS

The reported research has been supported by a Grant-in-aid for NAREGI Nanoscience Project, Scientific Research on Priority Area (A), and Scientific Research (B) from the Ministry of Education, Culture, Sports, Science and Technology of Japan, and also by the Japan Society for the Promotion of Science. The authors also wish to thank the following organizations for kindly permitting the reprinting of copyrighted material: American Institute of Physics; Elsevier Scientific Publishing Company.

[1] Interestingly enough, the first association [23,24,380] which came to W. Krätschmer's mind, when asked by R. Smalley to come to the ISSPIC-5 conference in Konstanz [26], was the fact that, in 1415, the Czech reformer and president of Charles University of Prague, Jan Hus, was also invited to a conference in Konstanz, and then burnt there at the stake as a heretic for the revolutionary theses he presented. Times are changing—575 years later, the Krätschmer–Huffman cross-disciplinary breakthrough was in Konstanz loudly applauded [26,381,382].

32.11 REFERENCES

1 H.W. Kroto, J.R. Heath, S.C. O'Brien, R.F. Curl and E. Smalley, Nature, 318 (1985) 162.

2 J.R. Heath, S.C. O'Brien, Q. Zhang, Y. Liu, R.F. Curl, H.W. Kroto, F.K. Tittel and E. Smalley, J. Am. Chem. Soc., 107 (1985) 7779.

3 W. Krätschmer, L.D. Lamb, K. Fostiropoulos and D.R. Huffman, Nature, 347 (1990) 354.

4 S. Yoshimura and R.P.H. Chang (Eds.), Supercarbon: Synthesis, properties and applications, Springer, Berlin, 1998.

5 A. Hirsch (Ed.), Fullerenes and related structures, Springer, Berlin, 1999.

6 M.S. Dresselhaus, G. Dresselhaus and Ph. Avouris (Eds.), Carbon nanotubes: Synthesis, structure, properties and applications, Springer, Berlin, 2001.

7 G. Benedek, P. Milani and V.G. Ralchenko (Eds.), Nanostructured carbon for advanced applications, Kluwer Academic Publishers, Dordrecht, 2001.

8 M. Buhl and A. Hirsch, Chem. Rev., 101 (2001) 1153.

9 S. Nagase, K. Kobayashi and T. Akasaka, Bull. Chem. Soc. Jpn, 69 (1996) 2131.

10 S. Nagase, K. Kobayashi, T. Akasaka, T. Wakahara, in: K.M. Kadish, R.S. Ruoff (Eds.), Fullerenes: Chemistry, physics, and technology, Wiley, New York, 2000, p. 395.

11 S. Iijima, Nature, 354 (1991) 56.

12 M.H. Ge and K. Sattler, Chem. Phys. Lett., 220 (1994) 192.

13 B.W. Smith, M. Monthioux and D.E. Luzzi, Nature, 396 (1998) 323.

14 S.J. Tans, A.R.M. Verschueren and C. Dekker, Nature, 393 (1998) 49.

15 M.S. Dresselhaus, G. Dresselhaus and P.C. Eklund, Science of fullerenes and carbon nanotubes, Academic Press, San Diego, CA, 1996.

16 J. Cioslowski, Electronic structure calculations on fullerenes and their derivatives, Oxford University Press, Oxford, UK, 1995.

17 G.E. Scuseria, in: D.R. Yarkony (Ed.), Modern electronic structure theory, part I, World Scientific Publishing, Singapore, 1995, p. 279.

18 K. Jinno and Y. Saito, Adv. Chromatogr., 36 (1996) 65.

19 Y. Achiba, K. Kikuchi, Y. Aihara, T. Wakabayashi, Y. Miyake, M. Kainosho, in: W. Andreoni (Ed.), The chemical physics of fullerenes 10 (and 5) years later, Kluwer Academic Publishers, Dordrecht, 1996, p. 139.

20 M. Saunders, R.J. Cross, H.A. Jiménez-Vázquez, R. Shimshi and A. Khong, Science, 271 (1996) 1693.

21 R. Taylor, in: R. Taylor (Ed.), The chemistry of fullerenes, World Scientific Publishing, Singapore, 1995, p. 1.

22 K. Heinzinger and Z. Slanina, MPG-Spiegel No. 2, (1993) 11.

23 J. Baggott, Perfect symmetry, The accidental discovery of buckminsterfullerene, Oxford University Press, Oxford, UK, 1994.

24 H. Aldersey-Williams, The most beautiful molecule, An adventure in chemistry, Aurum Press, London, 1995.

25 Z. Slanina, Chem. Intell., 4, 2 (1998) 52.

26 Z. Slanina, Int. J. Hist. Eth. Nat. Sci. Technol. Med. NTM, 9 (2001) 41.

27 J. Mattauch, H. Ewald, O. Hahn and F. Strassmann, Z. Phys., 20 (1943) 598.

28 R.E. Honig, J. Chem. Phys., 22 (1954) 126.

29 J. Drowart, R.P. Burns, G. DeMaria and M.G. Inghram, J. Chem. Phys., 31 (1959) 1131.

30 H. Hintenberger, J. Franzen and K.D. Schuy, Z. Naturforsch., 18a (1963) 1236.

31 K.S. Pitzer and E. Clementi, J. Am. Chem. Soc., 81 (1959) 4477.

32 S.J. Strickler and K.S. Pitzer, in: B. Pullman, P.-O. Löwdin (Eds.), Molecular orbitals in chemistry, physics, and biology, Academic Press, New York, NY, 1964.

33 R. Hoffmann, Tetrahedron, 22 (1966) 521.

34 Z. Slanina, Radiochem. Radioanal. Lett., 22 (1975) 291.

35 Z. Slanina and R. Zahradník, J. Phys. Chem., 81 (1977) 2252.

36 H.P. Schultz, J. Org. Chem., 30 (1965) 1361.

37 D.E.H. Jones, New Sci., 32 (1966) 245.

38 E. Ōsawa, Kagaku, 25 (1970) 854, Chem. Abstr., 74 (1971) 75698v.

39 D.A. Bochvar and E.G. Gal'pern, Dokl. Akad. Nauk SSSR, 209 (1973) 610.
40 R.A. Davidson, Theor. Chim. Acta, 58 (1981) 193.
41 K. Kobayashi and S. Nagase, in: T. Akasaka, S. Nagase (Eds.), Endofullerenes—a new family of carbon clusters, Kluwer Academic Publishers, Dordrecht, 2002, p. 99.
42 Z.C. Wu, D.A. Jelski and F. George, Chem. Phys. Lett., 137 (1987) 291.
43 V. Elser and R.C. Haddon, Nature, 325 (1987) 792.
44 R.E. Stanton and M.D. Newton, J. Phys. Chem., 92 (1988) 2141.
45 D.E. Weeks and W.G. Harter, Chem. Phys. Lett., 144 (1988) 366.
46 D.E. Weeks and W.G. Harter, J. Chem. Phys., 90 (1989) 4744.
47 Z. Slanina, J.M. Rudziński, M. Togasi and E. Ōsawa, J. Mol. Struct. (THEOCHEM), 61 (1989) 169.
48 D.R. Huffman and W. Krätschmer, Mater. Res. Soc. Proc., 206 (1991) 601.
49 T. Braun, Angew. Chem. Int. Ed., 31 (1992) 588.
50 T. Braun, A. Schubert, H. Maczelka and L. Vasvári, Fullerene research 1985–1993, World Scientific Publishing, Singapore, 1995.
51 D. Bakowies and W. Thiel, J. Am. Chem. Soc., 113 (1991) 3704.
52 M. Häser, J. Almlöf and G.E. Scuseria, Chem. Phys. Lett., 181 (1991) 497.
53 W. Weltner, Jr. and R.J. van Zee, Chem. Rev., 89 (1989) 1713.
54 R.F. Curl and R.E. Smalley, Science, 242 (1988) 1017.
55 H.W. Kroto, Science, 242 (1988) 1139.
56 R.C. Haddon, Acc. Chem. Res., 21 (1988) 243.
57 D.A. Jelski and T.F. George, J. Chem. Educ., 65 (1988) 879.
58 D.J. Klein and T.G. Schmalz, in: I. Hargittai (Ed.), Quasicrystals, networks, and molecules of fivefold symmetry, VCH Publishers, New York, NY, 1990.
59 E. Ōsawa, Kagaku, 45 (1990) 552.
60 D.R. Huffman, Phys. Today, 44 (1991) 22.
61 H.W. Kroto, A.W. Allaf and S.P. Balm, Chem. Rev., 91 (1991) 1213.
62 H.W. Kroto, Angew. Chem. Int. Ed., 31 (1992) 111.
63 J.R. Heath, in: G.S. Hammond, V.J. Kuck (Eds.), ACS symposium series no. 481—fullerenes. Synthesis, properties, and chemistry of large carbon clusters, ACS, Washington, DC, 1992.
64 J.E. Fischer, P.A. Heiney, D.E. Luzzi and D.E. Cox, in: G.S. Hammond, V.J. Kuck (Eds.), ACS symposium series no. 481—fullerenes. Synthesis, properties, and chemistry of large carbon clusters, ACS, Washington, DC, 1992.
65 Z. Slanina, Chem. Listy, 86 (1992) 327.
66 R.E. Smalley, Acc. Chem. Res., 25 (1992) 98.
67 J.P. Hare and H.W. Kroto, Acc. Chem. Res., 25 (1992) 106.
68 J.E. Fischer, P.A. Heiney and A.B. Smith, III, Acc. Chem. Res., 25 (1992) 112.
69 F. Diederich and R.L. Whetten, Acc. Chem. Res., 25 (1992) 119.
70 R.C. Haddon, Acc. Chem. Res., 25 (1992) 127.
71 P.J. Fagan, J.C. Calabrese and B. Malone, Acc. Chem. Res., 25 (1992) 134.
72 J.H. Weaver, Acc. Chem. Res., 25 (1992) 143.
73 J.M. Hawkins, Acc. Chem. Res., 25 (1992) 150.
74 F. Wudl, Acc. Chem. Res., 25 (1992) 157.
75 S.W. McElvany, M.M. Ross and J.H. Callahan, Acc. Chem. Res., 25 (1992) 162.
76 R.D. Johnson, D.S. Bethune and C.S. Yannoni, Acc. Chem. Res., 25 (1992) 169.
77 D. Koruga, S. Hameroff, J. Withers, R. Loutfy and M. Sundareshan, Fullerene C_{60}—history, physics, nanobiology, nanotechnology, Elsevier, Amsterdam, 1993.
78 M.L. Cohen and V.H. Crespi, in: W.E. Billups, M.A. Ciufolini (Eds.), Buckminsterfullerenes, VCH Publishers, New York, NY, 1993.
79 S.C. Erwin, in: W.E. Billups, M.A. Ciufolini (Eds.), Buckminsterfullerenes, VCH Publishers, New York, NY, 1993.
80 M.S. Dresselhaus, G. Dresselhaus and P.C. Eklund, J. Mater. Res., 8 (1993) 2054.
81 T.G. Schmalz and D.J. Klein, in: W.E. Billups, M.A. Ciufolini (Eds.), Buckminsterfullerenes, VCH Publishers, New York, NY, 1993.

82 G.E. Scuseria, in: W.E. Billups, M.A. Ciufolini (Eds.), Buckminsterfullerenes, VCH Publishers, New York, NY, 1993.

83 C.T. White, J.W. Mintmire, R.C. Mowrey, D.W. Brenner, D.H. Robertson, J.A. Harrison and B.I. Dunlap, in: W.E. Billups, M.A. Ciufolini (Eds.), Buckminsterfullerenes, VCH Publishers, New York, NY, 1993.

84 R.C. Haddon and K. Raghavachari, in: W.E. Billups, M.A. Ciufolini (Eds.), Buckminsterfullerenes, VCH Publishers, New York, NY, 1993.

85 J. Cioslowski, Rev. Comput. Chem., 4 (1993) 1.

86 C.Z. Wang, B.L. Zhang, K.M. Ho and X.Q. Wang, Int. J. Mod. Phys. B, 7 (1993) 4305.

87 J. González, F.F. Guinea and M.A.H. Vozmediano, Int. J. Mod. Phys. B, 7 (1993) 4331.

88 W. Andreoni, in: H. Kuzmany, J. Fink, M. Mehring, S. Roth (Eds.), Electronic properties of fullerenes, Springer, Berlin, 1993.

89 Z. Slanina, S.-L. Lee and C.-H. Yu, Rev. Comput. Chem., 8 (1996) 1.

90 J. Cioslowski, N. Rao and D. Moncrieff, J. Am. Chem. Soc., 122 (2000) 8265.

91 Y. Achiba, K. Kikuchi, Y. Aihara, T. Wakabayashi, Y. Miyake and M. Kainosho, in: P. Bernier, D.S. Bethune, L.Y. Chiang, T.W. Ebbesen, R.M Metzger, J.W. Mintmire (Eds.), Science and technology of fullerene materials, Materials Research Society, Pittsburgh, PA, 1995.

92 Y. Achiba, Kagaku, 52, 5 (1997) 15.

93 R. Mitsumoto, H. Oji, Y. Yamamoto, K. Asato, Y. Ouchi, H. Shinohara, K. Seki, K. Umishita, S. Hino, S. Nagase, K. Kikuchi and Y. Achiba, J. Phys. IV, 7 (1997) C2-525.

94 H.W. Kroto, Nature, 329 (1987) 529.

95 T.G. Schmalz, W.A. Seitz, D.J. Klein and G.E. Hite, J. Am. Chem. Soc., 110 (1988) 1113.

96 D.E. Manolopoulos and P.W. Fowler, J. Chem. Phys., 96 (1992) 7603.

97 P.W. Fowler and D.E. Manolopoulos, An atlas of fullerenes, Clarendon Press, Oxford, UK, 1995.

98 M. Yoshida and E. Ōsawa, Bull. Chem. Soc. Jpn, 68 (1995) 2073.

99 M. Yoshida and E. Ōsawa, Bull. Chem. Soc. Jpn, 68 (1995) 2083.

100 Z. Slanina, M.-L. Sun, S.-L. Lee and L. Adamowicz, in: K.M. Kadish, R.S. Ruoff (Eds.), Recent advances in the chemistry and physics of fullerenes and related materials, Vol. 2, The Electrochemical Society, Pennington, NJ, 1995, p. 1138.

101 R. Ettl, I. Chao, F.N. Diederich and R.L. Whetten, Nature, 353 (1991) 149.

102 D.E. Manolopoulos, J. Chem. Soc. Faraday Trans., 87 (1991) 2861.

103 H.P. Cheng and R.L. Whetten, Chem. Phys. Lett., 197 (1992) 44.

104 Q. Li, F. Wudl, C. Thilgen, R.L. Whetten and F. Diederich, J. Am. Chem. Soc., 114 (1992) 3994.

105 S. Hino, K. Matsumoto, S. Hasegawa, H. Inokuchi, T. Morikawa, T. Takahashi, K. Seki, K. Kikuchi, S. Suzuki, I. Ikemoto and Y. Achiba, Chem. Phys. Lett., 197 (1992) 38.

106 G. Orlandi, F. Zerbetto, P.W. Fowler and D.E. Manolopoulos, Chem. Phys. Lett., 208 (1993) 441.

107 J.M. Hawkins and A. Meyer, Science, 260 (1993) 1918.

108 S.J. Austin, P.W. Fowler, G. Orlandi, D.E. Manolopoulos and F. Zerbetto, Chem. Phys. Lett., 226 (1994) 219.

109 R.H. Michel, H. Schreiber, R. Gierden, F. Hennrich, J. Rockenberger, R.D. Beck, M.M. Kappes, C. Lehner, P. Adelmann and J.F. Armbruster, Ber. Bunsen-ges. Phys. Chem., 98 (1994) 975.

110 R.H. Michel, M.M. Kappes, P. Adelmann and G. Roth, Angew. Chem. Int. Ed., 33 (1994) 1651.

111 F. Diederich, R.L. Whetten, C. Thilgen, R. Ettl, I. Chao and M.M. Alvarez, Science, 254 (1991) 1768.

112 K. Kikuchi, N. Nakahara, T. Wakabayashi, S. Suzuki, H. Shiromaru, Y. Miyake, K. Saito, I. Ikemoto, M. Kainosho and Y. Achiba, Nature, 357 (1992) 142.

113 D. Bakowies, A. Geleßus and W. Thiel, Chem. Phys. Lett., 197 (1992) 324.

114 Z. Slanina, J.-P. François, D. Bakowies and W. Thiel, J. Mol. Struct. (THEOCHEM), 279 (1993) 213.

115 S.J. Woo, E. Kim and Y.H. Lee, Phys. Rev. B, 47 (1993) 6721.

116 K. Nakao, N. Kurita and M. Fujita, Phys. Rev. B, 49 (1994) 11415.

117 M.-L. Sun, Z. Slanina, S.-L. Lee, F. Uhlík and L. Adamowicz, Chem. Phys. Lett., 246 (1995) 66.

118 F.H. Hennrich, R.H. Michel, A. Fischer, S. Richard-Schneider, S. Gilb, M.M. Kappes, D. Fuchs, M. Bürk, K. Kobayashi and S. Nagase, Angew. Chem. Int. Ed., 35 (1996) 1732.

119 C.-R. Wang, T. Sugai, T. Kai, T. Tomiyama and H. Shinohara, J. Chem. Soc. Chem. Commun., (2000) 557.

120 K. Kikuchi, N. Nakahara, T. Wakabayashi, M. Honda, H. Matsumiya, T. Moriwaki, S. Suzuki, H. Shiromaru, K. Saito, K. Yamauchi, I. Ikemoto and Y. Achiba, Chem. Phys. Lett., 188 (1992) 177.
121 S. Nagase, K. Kobayashi, T. Kato and Y. Achiba, Chem. Phys. Lett., 201 (1993) 475.
122 G. Orlandi, F. Zerbetto and P.W. Fowler, J. Phys. Chem., 97 (1993) 13575.
123 S. Nagase and K. Kobayashi, Chem. Phys. Lett., 214 (1993) 57.
124 Z. Slanina, S.-L. Lee, K. Kobayashi and S. Nagase, J. Mol. Struct. (THEOCHEM), 339 (1995) 89.
125 M. Saunders, H.A. Jiménez-Vázquez, R.J. Cross, W.E. Billups, C. Gesenberg, A. Gonzalez, W. Luo, R.C. Haddon, F. Diederich and A. Herrmann, J. Am. Chem. Soc., 117 (1995) 9305.
126 F. Diederich, R. Ettl, Y. Rubin, R.L. Whetten, R. Beck, M. Alvarez, S. Anz, D. Sensharma, F. Wudl, K.C. Khemani and A. Koch, Science, 252 (1991) 548.
127 R.D. Beck, P.S. John, M.M. Alvarez, F. Diederich and R.L. Whetten, J. Phys. Chem., 95 (1991) 8402.
128 K. Kikuchi, N. Nakahara, M. Honda, S. Suzuki, K. Saito, H. Shiromaru, K. Yamauchi, I. Ikemoto, T. Kuramochi, S. Hino and Y. Achiba, Chem. Lett., (1991) 1607.
129 K. Raghavachari and C.M. Rohlfing, J. Phys. Chem., 95 (1991) 5768.
130 Z. Slanina, J.-P. François, M. Kolb, D. Bakowies and W. Thiel, Fullerene Sci. Technol., 1 (1993) 221.
131 T.J.S. Dennis, T. Kai, T. Tomiyama and H. Shinohara, Chem. Commun., (1998) 619.
132 E. Ōsawa, H. Ueno, M. Yoshida, Z. Slanina, X. Zhao, M. Nishiyama and H. Saito, J. Chem. Soc. Perkin Trans., 2 (1998) 943.
133 Z. Slanina, S.-L. Lee, M. Yoshida and E. Ōsawa, Chem. Phys., 209 (1996) 13.
134 Z. Slanina, S.-L. Lee, M. Yoshida and E. Ōsawa, in: H. Kuzmany, J. Fink, M. Mehring, S. Roth (Eds.), Physics and chemistry of fullerenes and their derivatives, World Scientific Publishing, Singapore, 1996, p. 389.
135 Z. Slanina, S.-L. Lee, M. Yoshida and E. Ōsawa, in: K.M. Kadish, R.S. Ruoff (Eds.), Recent advances in the chemistry and physics of fullerenes and related materials, Vol. 3, The Electrochemical Society, Pennington, NJ, 1996, p. 967.
136 Z. Slanina, S.-L. Lee and L. Adamowicz, Int. J. Quantum Chem., 63 (1997) 529.
137 Z. Slanina, X. Zhao, S.-L. Lee and E. Ōsawa, in: K.M. Kadish, R.S. Ruoff (Eds.), Recent advances in the chemistry and physics of fullerenes and related materials, Vol. 4, The Electrochemical Society, Pennington, NJ, 1997, p. 680.
138 Z. Slanina, X. Zhao, S.-L. Lee and E. Ōsawa, Chem. Phys., 219 (1997) 193.
139 Z. Slanina, X. Zhao, P. Deota and E. Ōsawa, J. Mol. Model., 6 (2000) 312.
140 X. Zhao, Z. Slanina, H. Goto and E. Ōsawa, J. Chem. Phys., 118 (2003) 10534.
141 X. Zhao and Z. Slanina, in: P. Kamat, D. Guldi, K. Kadish (Eds.), Fullerenes, The exciting world of nanocages and nanotubes, Vol. 12, The Electrochemical Society, Pennington, NJ, 2002, p. 679.
142 X. Zhao and Z. Slanina, J. Mol. Struct. (THEOCHEM), 636 (2003) 195.
143 X. Zhao, Z. Slanina, M. Ozawa, E. Ōsawa, P. Deota and K. Tanabe, Fullerene Sci. Technol., 8 (2000) 595.
144 Z. Slanina, X. Zhao and E. Ōsawa, Chem. Phys. Lett., 290 (1998) 311.
145 Z. Slanina, F. Uhlík, X. Zhao and E. Ōsawa, J. Chem. Phys., 113 (2000) 4933.
146 S.A. Varganov, P.V. Avramov, S.G. Ovchinnikov and M.S. Gordon, Chem. Phys. Lett., 362 (2002) 380.
147 T.S.M. Wan, H.W. Zhang, T. Nakane, Z.D. Xu, M. Inakuma, H. Shinohara, K. Kobayashi and S. Nagase, J. Am. Chem. Soc., 120 (1998) 6806.
148 K. Kobayashi, S. Nagase, M. Yoshida and E. Ōsawa, J. Am. Chem. Soc., 119 (1997) 12693.
149 S. Nagase, K. Kobayashi and T. Akasaka, J. Mol. Struct. (THEOCHEM), 462 (1999) 97.
150 Z. Slanina, K. Kobayashi and S. Nagase, Chem. Phys. Lett., 372 (2003) 810.
151 Z. Slanina, X. Zhao, X. Grabuleda, M. Ozawa, F. Uhlík, P.M. Ivanov, K. Kobayashi and S. Nagase, J. Mol. Graph. Mod., 19 (2001) 252.
152 Z. Slanina, F. Uhlík, L. Adamowicz, K. Kobayashi and S. Nagase, Int. J. Quantum Chem., 96 (2004).
153 T. Kodama, R. Fujii, Y. Miyake, S. Suzuki, H. Nishikawa, I. Ikemoto, K. Kikuchi and Y. Achiba, Fullerenes, in: D.M. Guldi, P.V. Kamat, F. D'Souza (Eds.), Fullerenes and nanotubes: The building blocks of next generation nanodevices, Vol. 13, The Electrochemical Society, Pennington, NJ, 2003, p. 548.
154 Z. Slanina, K. Kobayashi and S. Nagase, Chem. Phys., 301 (2004) 153.

155 Z.D. Xu, T. Nakane and H. Shinohara, J. Am. Chem. Soc., 118 (1996) 11309.
156 K. Kobayashi and S. Nagase, Chem. Phys. Lett., 274 (1997) 226.
157 T. Kimura, T. Sugai and H. Shinohara, Int. J. Mass Spectrom., 188 (1999) 225.
158 T.J.S. Dennis and H. Shinohara, Appl. Phys. A, 66 (1998) 243.
159 S. Hino, K. Umishita, K. Iwasaki, M. Aoki, K. Kobayashi, S. Nagase, T.J.S. Dennis, T. Nakane and H. Shinohara, Chem. Phys. Lett., 337 (2001) 65.
160 T. Kodama, R. Fujii, Y. Miyake, K. Sakaguchi, H. Nishikawa, I. Ikemoto, K. Kikuchi and Y. Achiba, Chem. Phys. Lett., 377 (2003) 197.
161 Z. Slanina, K. Kobayashi and S. Nagase, J. Chem. Phys., 120 (2004) 3397.
162 K. Kobayashi and S. Nagase, Chem. Phys. Lett., 282 (1998) 325.
163 T. Akasaka, T. Wakahara, S. Nagase, K. Kobayashi, M. Waelchli, K. Yamamoto, M. Kondo, S. Shirakura, S. Okubo, Y. Maeda, T. Kato, M. Kako, Y. Nakadaira, R. Nagahata, X. Gao, E. van Caemelbecke and K.M. Kadish, J. Am. Chem. Soc., 122 (2000) 9316.
164 T. Akasaka, T. Wakahara, S. Nagase, K. Kobayashi, M. Waelchli, K. Yamamoto, M. Kondo, S. Shirakura, Y. Maeda, T. Kato, M. Kako, Y. Nakadaira, X. Gao, E. van Caemelbecke and K.M. Kadish, J. Phys. Chem. B, 105 (2001) 2971.
165 T. Wakahara, S. Okubo, M. Kondo, Y. Maeda, T. Akasaka, M. Waelchli, M. Kako, K. Kobayashi, S. Nagase, T. Kato, K. Yamamoto, X. Gao, E. van Caemelbecke and K.M. Kadish, Chem. Phys. Lett., 360 (2002) 235.
166 K. Kobayashi and S. Nagase, Mol. Phys., 101 (2003) 249.
167 T. Kodama, N. Ozawa, Y. Miyake, K. Sakaguchi, H. Nishikawa, I. Ikemoto, K. Kikuchi and Y. Achiba, J. Am. Chem. Soc., 124 (2002) 1452.
168 B. Cao, M. Hasegawa, K. Okada, T. Tomiyama, T. Okazaki, K. Suenaga and H. Shinohara, J. Am. Chem. Soc., 123 (2001) 9679.
169 S. Nagase and K. Kobayashi, Chem. Phys. Lett., 276 (1997) 55.
170 B. Cao, K. Suenaga, T. Okazaki and H. Shinohara, J. Phys. Chem. B, 106 (2002) 9295.
171 C.R. Wang, P. Georgi, L. Dunsch, T. Kai, T. Tomiyama and H. Shinohara, Curr. Appl. Phys., 2 (2002) 141.
172 Y.K. Kwon, D. Tomanek and S. Iijima, Phys. Rev. Lett., 82 (1999) 1470.
173 Z. Slanina, K. Kobayashi and S. Nagase, Chem. Phys. Lett., 382 (2003) 211.
174 Z. Slanina and F. Uhlík, Chem. Phys. Lett., 374 (2003) 100.
175 Z. Slanina, F. Uhlík and L. Adamowicz, Fullerene Nanotubes Carbon Nanostruct., 11 (2003) 219.
176 E. Dietel, A. Hirsch, B. Pietzak, M. Waiblinger, K. Lips, A. Weidinger, A. Gruss and K.P. Dinse, J. Am. Chem. Soc., 121 (1999) 2432.
177 T. Clark, A handbook of computational chemistry, A practical guide to chemical structure and energy calculations, Wiley, New York, NY, 1985.
178 D.B. Boyd, Rev. Comput. Chem., 1 (1990) 321.
179 J.J.P. Stewart, Rev. Comput. Chem., 1 (1990) 45.
180 M.C. Zerner, Rev. Comput. Chem., 2 (1991) 313.
181 W.J. Hehre, L. Radom, P.R. von Schleyer and J.A. Pople, Ab initio molecular orbital theory, Wiley, New York, NY, 1986.
182 M.J.S. Dewar and W. Thiel, J. Am. Chem. Soc., 99 (1997) 4899.
183 M.J.S. Dewar, E.G. Zoebisch, E.F. Healy and J.J.P. Stewart, J. Am. Chem. Soc., 107 (1985) 3902.
184 J.J.P. Stewart, J. Comput. Chem., 10 (1989) 209.
185 M.J.S. Dewar, C. Jie and J. Yu, Tetrahedron, 49 (1993) 5003.
186 J.J.P. Stewart, MOPAC 5.0, QCPE 455, Indiana University, IN, 1990.
187 J.J.P. Stewart, MOPAC 2002, Fujitsu Ltd, Tokyo, Japan, 1999.
188 AMPAC 6.0, Semichem, Shavnee, KS, 1997.
189 M.J. Frisch, G.W. Trucks, H.B. Schlegel, G.E. Scuseria, M.A. Robb, J.R. Cheeseman, V.G. Zakrzewski, J.A. Montgomery, Jr., R.E. Stratmann, J.C. Burant, S. Dapprich, J.M. Millam, A.D. Daniels, K.N. Kudin, M.C. Strain, O. Farkas, J. Tomasi, V. Barone, M. Cossi, R. Cammi, B. Mennucci, C. Pomelli, C. Adamo, S. Clifford, J. Ochterski, G.A. Petersson, P.Y. Ayala, Q. Cui, K. Morokuma, D.K. Malick, A.D. Rabuck, K. Raghavachari, J.B. Foresman, J. Cioslowski, J.V. Ortiz, B.B. Stefanov, G. Liu, A. Liashenko,

P. Piskorz, I. Komaromi, R. Gomperts, R.L. Martin, D.J. Fox, T. Keith, M.A. Al-Laham, C.Y. Peng, A. Nanayakkara, C. Gonzalez, M. Challacombe, P.M.W. Gill, B. Johnson, W. Chen, M.W. Wong, J.L. Andres, C. Gonzalez, M. Head-Gordon, E.S. Replogle and J.A. Pople, Gaussian 98, Revision A.11.1, Gaussian, Inc., Pittsburgh, PA, 1998.
190 M.J. Frisch, G.W. Trucks, H.B. Schlegel, G.E. Scuseria, M.A. Robb, J.R. Cheeseman, J.A. Montgomery, Jr., T. Vreven, K.N. Kudin, J.C. Burant, J.M. Millam, S.S. Iyengar, J. Tomasi, V. Barone, B. Mennucci, M. Cossi, G. Scalmani, N. Rega, G.A. Petersson, H. Nakatsuji, M. Hada, M. Ehara, K. Toyota, R. Fukuda, J. Hasegawa, M. Ishida, T. Nakajima, Y. Honda, O. Kitao, H. Nakai, M. Klene, X. Li, J.E. Knox, H.P. Hratchian, J.B. Cross, C. Adamo, J. Jaramillo, R. Gomperts, R.E. Stratmann, O. Yazyev, A.J. Austin, R. Cammi, C. Pomelli, J.W. Ochterski, P.Y. Ayala, K. Morokuma, G.A. Voth, P. Salvador, J.J. Dannenberg, V.G. Zakrzewski, S. Dapprich, A.D. Daniels, M.C. Strain, O. Farkas, D.K. Malick, A.D. Rabuck, K. Raghavachari, J.B. Foresman, J.V. Ortiz, Q. Cui, A.G. Baboul, S. Clifford, J. Cioslowski, B.B. Stefanov, G. Liu, A. Liashenko, P. Piskorz, I. Komaromi, R.L. Martin, D.J. Fox, T. Keith, M.A. Al-Laham, C.Y. Peng, A. Nanayakkara, M. Challacombe, P.M.W. Gill, B. Johnson, W. Chen, M.W. Wong, C. Gonzalez and J.A. Pople, Gaussian 03, Revision B.02, Gaussian, Inc., Pittsburgh, PA, 2003.
191 W.J. Hehre, L.D. Burke and A.J. Schusterman, Spartan, Wavefunction, Inc., Irvine, CA, 1993.
192 H.B. Schlegel and J.J.W. McDouall, in: C. Ögretir, I.G. Csizmadia (Eds.), Computational advances in organic chemistry, Kluwer Academic Publishers, Dordrecht, 1991.
193 F. Jensen, Introduction to computational chemistry, Wiley, Chichester, UK, 1999.
194 Z. Slanina, F. Uhlík and M.C. Zerner, Rev. Roum. Chim., 36 (1991) 965.
195 Z. Slanina, Int. Rev. Phys. Chem., 6 (1987) 251.
196 Z. Slanina, Comput. Chem., 13 (1989) 305.
197 Z. Slanina, Theor. Chim. Acta, 83 (1992) 257.
198 Z. Slanina, Contemporary theory of chemical isomerism, Kluwer, Dordrecht, 1986.
199 Z. Slanina, F. Uhlík, J.-P. François and E. Ōsawa, Croat. Chem. Acta, 73 (2000) 1047.
200 H.S.M. Coxeter, Regular polytopes, Macmillan, New York, NY, 1963.
201 H.S.M. Coxeter, Introduction to geometry, Wiley, New York, NY, 1969.
202 Z. Slanina, Chem. Intell., 4, 2 (1998) 52.
203 R.F. Curl, Philos. Trans. R. Soc. Lond. A, 343 (1993) 19.
204 G.E. Scuseria, Unpublished results, 1993.
205 M.-L. Sun, Z. Slanina and S.-L. Lee, Chem. Phys. Lett., 233 (1995) 279.
206 C.H. Xu and G.E. Scuseria, Chem. Phys. Lett., 262 (1996) 219.
207 Z. Slanina and E. Ōsawa, Fullerene Sci. Technol., 5 (1997) 167.
208 J.M. Rudziński, Z. Slanina, M. Togasi, E. Ōsawa and T. Iizuka, Thermochim. Acta, 125 (1988) 155.
209 Z. Slanina, J.M. Rudziński and E. Ōsawa, Carbon, 25 (1987) 747.
210 Z. Slanina, J.M. Rudziński and E. Ōsawa, Collect. Czech. Chem. Commun., 52 (1987) 2381.
211 Z. Slanina, J.M. Rudziński, M. Togasi and E. Ōsawa, Thermochim. Acta, 140 (1989) 87.
212 Z. Slanina, X. Zhao, N. Kurita, H. Gotoh, F. Uhlík, J.M. Rudziński, K.H. Lee and L. Adamowicz, J. Mol. Graph. Mod., 19 (2001) 216.
213 T. Kiyobayashi and M. Sakiyama, Fullerene Sci. Technol., 1 (1993) 269.
214 Z. Slanina, Chem. Phys. Lett., 145 (1987) 512.
215 Z. Slanina, L. Adamowicz, D. Bakowies and W. Thiel, Thermochim. Acta, 202 (1992) 249.
216 Z. Slanina and L. Adamowicz, Thermochim. Acta, 205 (1992) 299.
217 Z. Slanina, Z. Phys. Chem., 217 (2003) 1119.
218 Z. Slanina, Chem. Phys., 150 (1991) 321.
219 P.S. Dardi and J.S. Dahler, J. Chem. Phys., 93 (1990) 3562.
220 Z. Slanina, Thermochim. Acta, 207 (1992) 9.
221 Z. Slanina, Thermochim. Acta, 209 (1992) 1.
222 K. Raghavachari, R.A. Whiteside and J.A. Pople, J. Chem. Phys., 85 (1986) 6623.
223 K. Raghavachari and J.S. Binkley, J. Chem. Phys., 87 (1987) 2191.
224 J.M.L. Martin, J.P. François and R. Gijbels, J. Chem. Phys., 90 (1989) 3403.
225 V. Parasuk and J. Almlöf, J. Chem. Phys., 91 (1989) 1137.

226 C. Liang and H.F. Schaefer, III, J. Chem. Phys., 93 (1990) 8844.
227 J.M.L. Martin, J.P. François and R. Gijbels, J. Chem. Phys., 93 (1990) 8850.
228 J. Kurtz and L. Adamowicz, Astrophys. J., 370 (1991) 784.
229 J.M.L. Martin, J.P. François and R. Gijbels, J. Chem. Phys., 94 (1991) 3753.
230 J.M.L. Martin, J.P. François and R. Gijbels, J. Comput. Chem., 12 (1991) 52.
231 D.W. Ewing, Z. Phys. D, 19 (1991) 419.
232 Z. Slanina, J.M. Rudziński and E. Ōsawa, Z. Phys. D, 19 (1991) 431.
233 J.M.L. Martin, J.P. François, R. Gijbels and J. Almlöf, Chem. Phys. Lett., 187 (1991) 367.
234 J.D. Watts and R.J. Bartlett, Chem. Phys. Lett., 190 (1992) 19.
235 V. Parasuk and J. Almlöf, Theor. Chem. Acta, 83 (1992) 227.
236 Z. Slanina, J. Kurtz and L. Adamowicz, Chem. Phys. Lett., 196 (1992) 208.
237 Z. Slanina, J. Kurtz and L. Adamowicz, Mol. Phys., 76 (1992) 387.
238 L.A. Curtiss and K. Raghavachari, in: S.R. Langhoff (Ed.), Quantum mechanical electronic structure calculations with chemical accuracy, Kluwer, Dordrecht, 1995.
239 K. Raghavachari and L.A. Curtiss, in: S.R. Langhoff (Ed.), Quantum mechanical electronic structure calculations with chemical accuracy, Kluwer, Dordrecht, 1995.
240 J.M.L. Martin and P.R. Taylor, J. Phys. Chem., 100 (1996) 6047.
241 W. Krätschmer, N. Sorg and D.R. Huffman, Surf. Sci., 156 (1985) 814.
242 S.W. McElvany, W.R. Creasy and A. O'Keefe, J. Chem. Phys., 85 (1986) 632.
243 S.W. McElvany, B.I. Dunlap and A. O'Keefe, J. Chem. Phys., 86 (1987) 715.
244 W. Krätschmer and K. Nachtigall, in: A. Léger, L. d'Hendecourt, N. Boccara (Eds.), Polycyclic aromatic hydrocarbons and astrophysics, D. Reidel, Dordrecht, 1987.
245 R.J. van Zee, R.F. Ferrante, K.J. Zerinque, W. Weltner, Jr. and D.W. Ewing, J. Chem. Phys., 88 (1988) 3465.
246 S.W. McElvany, J. Chem. Phys., 89 (1988) 2063.
247 J.R. Heath, A.L. Cooksy, H.W. Gruebele, C.A. Schmuttenmaer and R.J. Saykally, Science, 244 (1989) 564.
248 M. Vala, T.M. Chandrasekhar, J. Szczepanski, R. van Zee and W. Weltner, Jr., J. Chem. Phys., 90 (1989) 595.
249 M. Vala, T.M. Chandrasekhar, J. Szczepanski and R. Pellow, High Temp. Sci., 27 (1989) 19.
250 L.N. Shen and W.R.M. Graham, J. Chem. Phys., 91 (1989) 5115.
251 H.M. Cheung and W.R.M. Graham, J. Chem. Phys., 91 (1989) 6664.
252 J. Kurtz and D.R. Huffman, J. Chem. Phys., 92 (1990) 30.
253 J.R. Heath, R.A. Sheeks, A.L. Cooksy and R.J. Saykally, Science, 249 (1990) 895.
254 J. Szczepanski and M. Vala, J. Phys. Chem., 95 (1991) 2792.
255 L.N. Shen, P.A. Withey and W.R.M. Graham, J. Chem. Phys., 94 (1991) 2395.
256 J.R. Heath, A. van Orden, E. Kuo and R.J. Saykally, Chem. Phys. Lett., 182 (1991) 17.
257 J.R. Heath and R.J. Saykally, J. Chem. Phys., 94 (1991) 1724.
258 J.R. Heath and R.J. Saykally, J. Chem. Phys., 94 (1991) 3271.
259 P.A. Withey, L.N. Shen and W.R.M. Graham, J. Chem. Phys., 95 (1991) 820.
260 Q. Jiang and W.R.M. Graham, J. Chem. Phys., 95 (1991) 3129.
261 G. von Helden, M.-T. Hsu, P.R. Kemper and T. Bowers, J. Chem. Phys., 95 (1991) 3835.
262 D. Zajfman, H. Feldman, O. Heber, D. Kella, D. Majer, Z. Vager and R. Naaman, Science, 258 (1992) 1129.
263 G. von Helden, P.R. Kemper, N.G. Gotts and T. Bowers, Science, 259 (1993) 1300.
264 N.G. Gotts, G. von Helden and M.T. Bowers, Int. J. Mass Spectrom. Ion Process., 149/150 (1995) 217.
265 Z. Slanina, F. Uhlík and L. Adamowicz, J. Radioanal. Nucl. Chem., 219 (1997) 69.
266 Z. Slanina, S.-L. Lee, J.-P. François, J. Kurtz, L. Adamowicz and M. Smigel, Mol. Phys., 81 (1994) 1489.
267 Z. Slanina, S.-L. Lee, J.-P. François, J. Kurtz and L. Adamowicz, Chem. Phys. Lett., 223 (1994) 397.
268 Z. Slanina, S.-L. Lee, M. Smigel, J. Kurtz and L. Adamowicz, in: P. Bernier, D.S. Bethune, L.Y. Chiang, T.W. Ebbesen, R.M. Metzger, J.W. Mintmire (Eds.), Science and technology of fullerene materials, Materials Research Society, Pittsburgh, PA, 1995.
269 Z. Slanina, X. Zhao, E. Ōsawa and L. Adamowicz, Fullerene Sci. Technol., 8 (2000) 369.

270 T.F. Giesen, A. van Orden, H.J. Hwang, R.S. Fellers, R.A. Provençal and R.J. Saykally, Science, 265 (1994) 756.
271 D.A. Plattner and K.N. Houk, J. Am. Chem. Soc., 117 (1995) 4405.
272 G. von Helden, N.G. Gotts and M.T. Bowers, Nature, 363 (1993) 60.
273 Z. Slanina, Chem. Phys. Lett., 173 (1990) 164.
274 M.W. Chase, Jr., C.A. Davies, J.R. Downery, Jr., D.J. Frurip, R.A. McDonald and A.N. Syverud, J. Phys. Chem. Ref. Data 14, Supplement 1, JANAF Thermochemical Tables, 3rd edn., Vols. 1 and 2 (1985).
275 M.A. Nygren and L.G.M. Pettersson, Chem. Phys. Lett., 191 (1992) 473.
276 B.G. Johnson, Modern density functional theory: A tool for chemistry, Elsevier, Amsterdam, 1995.
277 J.D. Watts, J. Gauss, J.F. Stanton and R.J. Bartlett, J. Chem. Phys., 97 (1992) 8372.
278 Z. Slanina, Chem. Phys. Lett., 161 (1989) 265.
279 Z. Slanina, Chem. Phys. Lett., 166 (1990) 317.
280 D. Michalska, H. Chojnacki, B.A. Hess, Jr. and L.J. Schaad, Chem. Phys. Lett., 141 (1987) 376.
281 Z. Slanina, Thermochim. Acta, 127 (1988) 237.
282 A. van Orden and R.J. Saykally, Chem. Rev., 98 (1998) 2313.
283 Z. Slanina, Skept. Inq., 24, 4 (2000) 18.
284 P.W. Fowler and J. Woolrich, Chem. Phys. Lett., 127 (1986) 78.
285 P.W. Fowler, Chem. Phys. Lett., 131 (1986) 444.
286 P.W. Fowler and J.I. Steer, J. Chem. Soc. Chem. Commun., (1987) 1403.
287 P.W. Fowler, J.E. Cremona and J.I. Steer, Theor. Chim. Acta, 73 (1988) 1.
288 A. Ceulemans and P.W. Fowler, Phys. Rev. A, 39 (1989) 481.
289 A. Ceulemans and P.W. Fowler, J. Chem. Phys., 93 (1990) 1221.
290 P.W. Fowler, J. Chem. Soc. Faraday Trans., 86 (1990) 2073.
291 P. Fowler, Nature, 350 (1991) 20.
292 P.W. Fowler, J. Chem. Soc. Faraday Trans., 87 (1991) 1945.
293 P.W. Fowler, R.C. Batten and D.E. Manolopoulos, J. Chem. Soc. Faraday Trans., 87 (1991) 3103.
294 A. Ceulemans and P.W. Fowler, Nature, 353 (1991) 52.
295 P.W. Fowler and D.E. Manolopoulos, Nature, 355 (1992) 428.
296 P.W. Fowler, D.E. Manolopoulos and R.P. Ryan, J. Chem. Soc. Chem. Commun., (1992) 408.
297 P.W. Fowler, J. Chem. Soc. Perkin Trans., II (1992) 145.
298 P.W. Fowler, D.E. Manolopoulos and R.P. Ryan, Carbon, 30 (1992) 1235.
299 D.E. Manolopoulos, D.R. Woodall and P.W. Fowler, J. Chem. Soc. Faraday Trans., 88 (1992) 2427.
300 P.W. Fowler and V. Morvan, J. Chem. Soc. Faraday Trans., 88 (1992) 2631.
301 P.W. Fowler, D.E. Manolopoulos, D.B. Redmond and R.P. Ryan, Chem. Phys. Lett., 202 (1993) 371.
302 D.E. Manolopoulos and P.W. Fowler, Chem. Phys. Lett., 204 (1993) 1.
303 S.-L. Lee, Theor. Chim. Acta, 81 (1992) 185.
304 A.J. Stone and D.J. Wales, Chem. Phys. Lett., 128 (1986) 501.
305 C. Coulombeau and A. Rassat, J. Chim. Phys., 88 (1991) 173.
306 W.N. Lipscomb and L. Massa, Inorg. Chem., 31 (1992) 2297.
307 A.C. Tang, Q.S. Li, C.W. Liu and J. Li, Chem. Phys. Lett., 201 (1993) 465.
308 D.E. Manolopoulos, J.C. May and S.E. Down, Chem. Phys. Lett., 181 (1991) 105.
309 X. Liu, D.J. Klein, T.G. Schmalz and W.A. Seitz, J. Comput. Chem., 12 (1991) 1252.
310 X. Liu, D.J. Klein, W.A. Seitz and T.G. Schmalz, J. Comput. Chem., 12 (1991) 1265.
311 X. Liu, T.G. Schmalz and D.J. Klein, Chem. Phys. Lett., 188 (1992) 550.
312 D.E. Manolopoulos, Chem. Phys. Lett., 192 (1992) 330.
313 X. Liu, T.G. Schmalz and D.J. Klein, Chem. Phys. Lett., 192 (1992) 331.
314 M.-L. Sun, Z. Slanina and S.-L. Lee, Fullerene Sci. Technol., 3 (1995) 627.
315 P.W. Fowler, C.M. Quinn and D.B. Redmond, J. Chem. Phys., 95 (1991) 7678.
316 S. Wei, Z. Shi and A.W. Castleman, Jr., J. Chem. Phys., 94 (1991) 3268.
317 D.J. Klein, W.A. Seitz and T.G. Schmalz, J. Phys. Chem., 97 (1993) 1231.
318 D.J. Klein, Chem. Phys. Lett., 217 (1994) 261.
319 A.T. Balaban, D.J. Klein and C.A. Folden, Chem. Phys. Lett., 217 (1994) 266.
320 J.R. Dias, Chem. Phys. Lett., 209 (1993) 439.

References pp. 908–917

321 K. Balasubramanian, Chem. Phys. Lett., 182 (1991) 257.
322 K. Balasubramanian, Chem. Phys. Lett., 183 (1991) 292.
323 K. Balasubramanian, J. Chem. Inf. Comput. Sci., 32 (1992) 47.
324 K. Balasubramanian, Chem. Phys. Lett., 198 (1992) 577.
325 K. Balasubramanian, Chem. Phys. Lett., 197 (1992) 55.
326 K. Balasubramanian, J. Mol. Spectrosc., 157 (1992) 254.
327 K. Balasubramanian, Chem. Phys. Lett., 201 (1993) 306.
328 K. Balasubramanian, Chem. Phys. Lett., 206 (1993) 210.
329 K. Balasubramanian, Chem. Phys. Lett., 202 (1993) 399.
330 K. Balasubramanian, J. Phys. Chem., 97 (1993) 6990.
331 K. Balasubramanian, J. Phys. Chem., 97 (1993) 4647.
332 D. Babić, S. Bassoli, M. Casartelli, F. Cataldo, A. Graovac, O. Ori and B. York, J. Mol. Simulat., 14 (1995) 395.
333 B.R. Eggen, M.I. Heggie, G. Jungnickel, C.D. Latham, R. Jones and P.R. Briddon, Science, 272 (1996) 87.
334 B.R. Eggen, M.I. Heggie, G. Jungnickel, C.D. Latham, R. Jones and P.R. Briddon, Fullerene Sci. Technol., 5 (1997) 727.
335 Z. Slanina, X. Zhao, F. Uhlík, M. Ozawa and E. Ōsawa, J. Organometall. Chem., 599 (2000) 57.
336 S.W. McElvany, J.H. Callahan, M.M. Ross, L.D. Lamb and D.R. Huffman, Science, 260 (1993) 1632.
337 J.-P. Deng, D.-D. Ju, G.-R. Her, C.-Y. Mou, C.-J. Chen, Y.-Y. Lin and C.-C. Han, J. Phys. Chem., 97 (1993) 11575.
338 S. Fujita, Symmetry and combinatorial enumeration in chemistry, Springer, Berlin, 1991.
339 F. Wahl, J. Wörth and H. Prinzbach, Angew. Chem. Int. Ed. Engl., 32 (1993) 1722.
340 H. Prinzbach, A. Weller, P. Landenberger, F. Wahl, J. Worth, L.T. Scott, M. Gelmont, D. Olevano and B. von Issendorff, Nature, 407 (2000) 60.
341 Z. Wang, X. Ke, Z. Zhu, F. Zhu, M. Ruan, H. Chen, R. Huang and L. Zheng, Phys. Lett. A, 280 (2001) 351.
342 Z. Iqbal, Y. Zhang, H. Grebel, S. Vijayalakshmi, A. Lahamer, G. Benedek, M. Bernasconi, J. Cariboni, I. Spagnolatti, R. Sharma, F.J. Owens, M.E. Kozlov, K.V. Rao and M. Muhammed, Eur. Phys. J. B, 31 (2003) 509.
343 Z. Slanina and L. Adamowicz, Fullerene Sci. Technol., 1 (1993) 1.
344 Z. Slanina and L. Adamowicz, J. Mol. Struct. (THEOCHEM), 281 (1993) 33.
345 J. Lu, S. Re, Y. Choe, S. Nagase, Y.S. Zhou, R.S. Han, L.M. Peng, X.W. Zhang and X.G. Zhao, Phys. Rev. B, 67 (2003) 125415.
346 Z.X. Zhang, Z.Y. Pan, Y.X. Wang, Z.J. Li and Q. Wei, Mod. Phys. Lett., 17 (2003) 877.
347 C. Piskoti, J. Yarger and A. Zettl, Nature, 393 (1998) 771.
348 R.L. Murry, J.R. Colt and G.E. Scuseria, J. Phys. Chem., 97 (1993) 4954.
349 S.G. Louie, Nature, 384 (1996) 612.
350 L.F. Sun, S.S. Xie, W. Liu, W.Y. Zhou, Z.Q. Liu, D.S. Tang, G. Wang and L.X. Qian, Nature, 403 (2000) 384.
351 L.C. Qin, X.L. Zhao, K. Hirahara, Y. Miyamoto, Y. Ando and S. Iijima, Nature, 408 (2000) 50.
352 N. Wang, Z.K. Tang, G.D. Li and J.S. Chen, Nature, 408 (2000) 50.
353 Z. Slanina, F. Uhlík and L. Adamowicz, J. Mol. Graph. Mod., 21 (2003) 517.
354 M.D. Diener and J.M. Alford, Nature, 393 (1998) 668.
355 H. Moribe, T. Inoue, H. Kato, A. Taninaka, Y. Ito, T. Okazaki, T. Sugai, R. Bolskar, J.M. Alford and H. Shinohara, The 25th Fullerene-Nanotubes Symposium, Awaji, Japan, 2003, paper 1P-1.
356 H. Kato, A. Taninaka, T. Sugai and H. Shinohara, J. Am. Chem. Soc., 125 (2003) 7782.
357 O.V. Boltalina, E.V. Dashkova and L.N. Sidorov, Chem. Phys. Lett., 256 (1996) 253.
358 O.V. Boltalina, I.N. Ioffe, L.N. Sidorov, G. Seifert and K. Vietze, J. Am. Chem. Soc., 122 (2000) 9745.
359 Z. Slanina, S.-L. Lee, K. Kobayashi and S. Nagase, J. Mol. Struct. (THEOCHEM), 312 (1994) 175.
360 Z. Slanina, X. Zhao, F. Uhlík and E. Ōsawa, in: H. Kuzmany, J. Fink, M. Mehring, S. Roth (Eds.), Electronic properties of novel materials—science and technology of molecular nanostructures, AIP, Melville, NY, 1999, p. 179.

361 Z. Slanina, K. Ishimura, K. Kobayashi and S. Nagase, Chem. Phys. Lett., 384 (2004) 114.
362 D.R. Kanis, M.A. Ratner, T.J. Marks and M.C. Zerner, Chem. Mater., 3 (1991) 19.
363 R.D. Bendale and M.C. Zerner, J. Phys. Chem., 99 (1995) 13830.
364 Z. Slanina, F. Uhlík, X. Zhao and S. Nagase, to be published.
365 P.J. Hay and W.R. Wadt, J. Chem. Phys., 82 (1985) 299.
366 M.E. Casida, C. Jamorski, K.C. Casida and D.R. Salahub, J. Chem. Phys., 108 (1998) 4439.
367 L.S. Wang, J.M. Alford, Y. Chai, M. Diener, J. Zhang, S.M. McClure, T. Guo, G.E. Scuseria and R.E. Smalley, Chem. Phys. Lett., 207 (1993) 354.
368 Y. Kubozono, T. Ohta, T. Hayashibara, H. Maeda, H. Ishida, S. Kashino, K. Oshima, H. Yamazaki, S. Ukita and T. Sogabe, Chem. Lett., (1995) 457.
369 F.G. Hopwood, K.J. Fisher, P. Greenhill, G.D. Willett and R. Zhang, J. Phys. Chem. B, 101 (1997) 10704.
370 H. Shinohara, in: K.M. Kadish, R.S. Ruoff (Eds.), Fullerenes: Recent advances in the chemistry and physics of fullerenes and related materials, Vol. 4, The Electrochemical Society, Pennington, NJ, 1997, p. 467.
371 Y. Chai, T. Guo, C. Jin, R.E. Haufler, L.P.F. Chibante, J. Fure, L. Wang, J.M. Alford and R.E. Smalley, J. Phys. Chem., 95 (1991) 7564.
372 E. Nishibori, M. Takata, M. Sakata, H. Tanaka, M. Hasegawa and H. Shinohara, Chem. Phys. Lett., 330 (2000) 497.
373 S. Nagase and K. Kobayashi, Chem. Phys. Lett., 228 (1994) 106.
374 Z. Slanina, K. Kobayashi and S. Nagase, Chem. Phys. Lett., 388 (2004) 74.
375 Y.F. Lian, S.F. Yang and S.H. Yang, J. Phys. Chem. B, 106 (2002) 3112.
376 D.M. Guldi and N. Martin (Eds.), Fullerenes: From synthesis to optoelectronic properties, Kluwer Academic Publishers, Dordrecht, 2002.
377 A. Ardavan, M. Austwick, S.C. Benjamin, G.A.D. Briggs, T.J.S. Dennis, A. Ferguson, D.G. Hasko, M. Kanai, A.N. Khlobystov, B.W. Lovett, G.W. Morley, R.A. Oliver, D.G. Pettifor, K. Porfyrakis, J.H. Reina, J.H. Rice, J.D. Smith, R.A. Taylor, D.A. Williams, C. Adelmann, H. Mariette and R.J. Hamers, Philos. Trans. R. Soc. Lond. A, 361 (2003) 1473.
378 C.C. Chancey and M.C.M. O'Brien, The Jahn–Teller effect in C_{60} and other icosahedral complexes, Princeton University Press, Princeton, 1997.
379 W. Krätschmer, Z. Phys. D, 19 (1991) 405.
380 Z. Slanina, Jpn. Fullerene Quart., 5, 4 (1997) 9.
381 W. Krätschmer and D.R. Huffman, Carbon, 30 (1992) 1143.
382 R.F. Curl, Carbon, 30 (1992) 1149.

Theory and Applications of Computational Chemistry: The First Forty Years
Edited by C. Dykstra et al.

CHAPTER 33

On the importance of many-body forces in clusters and condensed phase

Krzysztof Szalewicz[1], Robert Bukowski[1] and Bogumil Jeziorski[2]

[1] *Department of Physics and Astronomy, University of Delaware, Newark, DE 19716, USA*
[2] *Department of Chemistry, University of Warsaw, Pasteura 1, 02-093 Warsaw, Poland*

Abstract

The interaction energies of clusters of molecules can be decomposed into pair contributions and pairwise-nonadditive contributions. The emphasis of this chapter is on the latter components. Both the historical and current investigations are reviewed. The physical mechanisms responsible for the existence of the many-body forces are described using symmetry-adapted perturbation theory of intermolecular interactions. The role of nonadditive effects in several specific trimers, including some open-shell trimers, is discussed. These effects are also discussed for the condensed phases of argon and water.

33.1 INTRODUCTION

Aggregates of molecules (or closed-shell atoms)—which can range from clusters to molecular condensed phase—are bound by forces of electromagnetic origin that are much weaker than the forces acting within molecules. The former are often called van der Waals forces or interactions. All the properties of systems bound by van der Waals forces can, at least in principle, be predicted by solving the Schrödinger equation for the motion of nuclei on the potential energy surface of a given system. Such surfaces can be obtained from solutions of the clamped-nuclei electronic Schrödinger equation. This model does involve some approximations, as certain physical effects such as nonadiabatic couplings of electronic and nuclear motions or relativistic contributions are neglected. However, for systems containing more than a couple of atoms, these effects would typically contribute much below numerical uncertainties of the model resulting from the use of finite basis sets and approximate methods of solving both the electronic and nuclear Schrödinger equations. Several properties of condensed phase can be described reasonably well by using the classical rather than quantum equations for nuclear motion, a significant further simplification.

Obviously, accuracy of predictions for clusters and condensed phase depends critically on the accuracy of the potential energy surface used. Although solutions of quantum or classical equations for the motion of nuclei are far from trivial, it turns out that uncertainties resulting from various approximations included in these solutions are usually smaller than those originating from inaccuracies of the potentials. In fact, up to recently, properties of only a very few condensed-phase systems have been accurately predicted based on first-principle potentials, i.e. potentials obtained from solutions of the clamped-nuclei electronic Schrödinger equation. Therefore, a common practice in physics, chemistry, and biophysics/biochemistry has been to use empirical potentials. These potentials are usually fairly simple functions of interatomic distances with parameters adjusted to reproduce experimental results. Although a lot of effort has been invested in fitting such potentials, there is no assurance that any of them resembles the true potential. In fact, we now know that in some cases deviations can be large. Some exceptions to this rule are the potentials fitted to spectroscopic data.

Electronic structure calculations have recently become capable to provide interaction potentials for medium-size dimers that can be used to predict properties of such dimers with accuracy approaching, and in a few favorable cases even surpassing, experimental accuracies [1–4]. The dimer (pair) potentials are the basic building blocks of potential surfaces for larger clusters represented in the form of many-body expansions. Such expansions decompose each surface into intramonomer contributions, i.e. the potentials within single molecules (monomers), the pair potentials, and the so-called nonadditive potentials. Since derivatives of potential energy surfaces define forces, one may alternatively use the term 'force fields' equivalently with 'potentials'.

This chapter will be devoted to analysis of many-body potentials in clusters and condensed phase with emphasis on elucidating the physical origins of interactions. It appears that the only reasonable physical interpretation of many-body interactions comes from perturbation theory that treats separated monomers as the zeroth-order approximation [2,4]. In such theory, it is possible to split the interaction energy into several contributions that can be related to distinct physical mechanisms, such as the electrostatic, induction (sometimes called polarization), dispersion, and exchange interactions. To describe the latter interactions, one has to include in this theory some symmetry projectors, leading to the name symmetry-adapted perturbation theory (SAPT). For pair interactions, the shapes of potential energy surfaces are resulting from a balancing act between induction and dispersion forces which are always attractive, exchange forces which are always repulsive, and the electrostatic forces which can be of either sign but for polar molecules provide large attractive contributions in the regions of potential minima. The nonadditive forces, i.e. those beyond pair interactions (precise definitions will be given below) are much less uniform in their behavior. In particular, the pair potentials for each fixed mutual orientation of monomers have in most regions of configuration space the well-known shape, starting from small negative values at large intermonomer separations R, slowly increasing the depth as R decreases until a minimum is reached, and then rapidly becoming positive and very large upon a further decrease of R. This behavior does not apply at all to nonadditive forces which at a given point can be of either the same, or opposite sign as the sum of pair interactions. Also the physical components of nonadditive forces are, understandably, different from pair interactions. In particular, the electrostatic interactions are completely pair additive, i.e. do not give any nonadditive effects. Similarly, there is no nonadditive contribution from the dispersion energy in the second order of perturbation theory. Such a contribution appears only in the third order and is usually almost negligible (a notable exception are rare-gas clusters). The induction and exchange effects do give nonadditive contributions appearing in the same orders as in the case of additive interactions (i.e. the second and the first order, respectively), but both contributions can be of either sign. The nonadditive exchange forces are important for all systems. For polar systems these forces determine, together with the induction ones, the total nonadditive effect rather completely. For less polar systems, the nonadditive dispersion interactions may have to be taken into account.

The empirical intermolecular force fields are in most cases built of terms that are in a close correspondence with the interaction energy components described above. One may say that such force fields are simplest possible implementations of the SAPT approach. The functional forms used are based on SAPT analysis of the asymptotic behavior of the components. The electrostatic interactions are usually approximated by interactions of fractional charges located on atoms in each monomer. In simplest cases, the induction effects are not included explicitly but some more sophisticated force fields use the classical polarization model. The dispersion forces are accounted for by linear combinations of $1/R_{ab}^6$ terms where R_{ab} are interatomic distances and the exchange forces by either exponential or $1/R_{ab}^{12}$ terms.

33.2 DEFINITIONS

Consider the potential energy surface for a system consisting of N molecules. Such a surface will be denoted by $E_{\text{tot}}(\mathbf{Q}_1, \ldots, \mathbf{Q}_N)$, where $\mathbf{Q}_i = (\mathbf{R}_i, \boldsymbol{\omega}_i, \boldsymbol{\xi}_i)$ stands for the set of all coordinates needed to specify the spatial position $\mathbf{R}_i$, orientation $\boldsymbol{\omega}_i$, and the internal geometry $\boldsymbol{\xi}_i$ of the ith monomer. $E_{\text{tot}}(\mathbf{Q}_1, \ldots, \mathbf{Q}_N)$ is defined as an eigenvalue of the clamped-nuclei electronic Hamiltonian for a given configuration of nuclei. In most applications, the total potential is conventionally separated into the sum of internal energies $E_i(\boldsymbol{\xi}_i)$ of the monomers and the interaction energy $E_{\text{int}}(\mathbf{Q}_1, \ldots, \mathbf{Q}_N)$:

$$E_{\text{tot}}(\mathbf{Q}_1, \ldots, \mathbf{Q}_N) = E_{\text{int}}(\mathbf{Q}_1, \ldots, \mathbf{Q}_N) + \sum_i E_i(\boldsymbol{\xi}_i) \tag{1}$$

The physical insight into the nature of an N-body interaction potential is facilitated by expressing it in the form of the many-body expansion, i.e. as a sum of terms involving interactions of 2,3,... monomers at a time

$$E_{\text{int}} = E_{\text{int}}[2, N] + E_{\text{int}}[3, N] + \cdots + E_{\text{int}}[N, N] \tag{2}$$

where $E_{\text{int}}[K,N]$ is the K-body contribution to the N-mer energy. The contributions $E_{\text{int}}[K,N]$ can be written as the following sums

$$E_{\text{int}}[2,N] = \sum_{i<j} E_{\text{int}}(\mathbf{Q}_i,\mathbf{Q}_j)[2,2] \tag{3}$$

$$E_{\text{int}}[3,N] = \sum_{i<j<k} E_{\text{int}}(\mathbf{Q}_i,\mathbf{Q}_j,\mathbf{Q}_k)[3,3] \tag{4}$$

etc. The two-body or *pairwise-additive* interaction energies $E_{\text{int}}[2,2]$ are simply the dimer interaction energies, i.e. are defined by Eq. (1) for $N=2$. The sum of all pair interactions, $E_{\text{int}}[2,N]$, is the (pairwise) additive component of the interaction energy of an N-mer. The higher body terms in Eq. (2) are called the *(pairwise-) nonadditive* contributions to the N-mer interaction energy and are defined recursively. The sum of all these contributions ($K>2$) constitutes the (pairwise-) nonadditive component of the interaction energy of an N-mer.

The three-body contribution to a trimer interaction energy, $E_{\text{int}}[3,3]$, is the difference between the total interaction energy of a given trimer and the sum of all pair interaction energies within this trimer. This definition allows determination of the three-body contributions $E_{\text{int}}(\mathbf{Q}_i,\mathbf{Q}_j,\mathbf{Q}_k)[3,3]$ for each of the trimers within an N-mer. For the general case, each K-body contribution $E_{\text{int}}(\mathbf{Q}_{i_1},\mathbf{Q}_{i_2},\ldots,\mathbf{Q}_{i_K})[K,K]$ is formally defined as the difference between the interaction energy of a given K-member cluster (with monomers placed at fixed positions and orientations $\mathbf{Q}_{i_1},\mathbf{Q}_{i_2},\ldots,\mathbf{Q}_{i_K}$) and this part of the K-mer interaction energy which can be built from all 2,3,..., $(K-1)$-body contributions.

In fact, the K-body contribution to the N-mer energy can be expressed directly in terms of the total energies of all subclusters containing up to K monomers [5]

$$E_{\text{int}}[K,N] = \sum_{I=1}^{K} (-1)^{K-I} \binom{N-I}{K-I} S_{\text{tot}}[I,N] \tag{5}$$

where $S_{\text{tot}}[I,N]$ is the sum of the total energies of all I-monomer subclusters of the whole N-mer, e.g.

$$S_{\text{tot}}[3,N] = \sum_{i<j<k} E_{\text{tot}}(\mathbf{Q}_i,\mathbf{Q}_j,\mathbf{Q}_k) \tag{6}$$

when $I=3$. A somewhat similar equation expressing $E_{\text{int}}[K,N]$ in terms of $S_{\text{tot}}[K,N]$ and all I-body components $E_{\text{int}}[I,N]$ for $I<K$

$$E_{\text{int}}[K,N] = S_{\text{tot}}[K,N] - \sum_{I=1}^{K-1} \binom{N-I}{K-I} E_{\text{int}}[I,N] \tag{7}$$

where $E_{\text{int}}[1,N] \equiv S_{\text{tot}}[1,N]$, has been given by Kaplan et al. [6,7]. The recipe of Eq. (5) or (7) can be applied to compute $E_{\text{int}}[K,N]$ within the supermolecular [1] approach, however, the number of calculations that need to be performed increases rapidly with N and is equal to 7 already for trimers (assuming that counterpoise corrected [8] approach is used). Another problem with using Eq. (5) or (7) is a substantial loss of significant figures resulting from the performed subtractions. In contrast, in the perturbative approach discussed in Section 33.4, this contribution can be obtained directly, without the knowledge of interaction energies of clusters containing less than K monomers. The perturbation theory approach shows that the $E_{\text{int}}[K, K]$ contributions with different K have different physical interpretation (i.e. appear due to physically different mechanisms of intermolecular interactions).

Obviously, the many-body expansion of the interaction energy can be defined only when the quantum states of all subsystems can be unambiguously specified. For strongly interacting systems, such as metals or chemically bound molecules, this condition is not fulfilled and the suitability of the many-body expansion can be questioned. In most applications of Eq. (2), the number of molecules N will be fixed and therefore the index N in $[K, N]$ will be omitted from now on.

The complete interaction potential $E_{\text{int}}(\mathbf{Q}_1, \ldots, \mathbf{Q}_N)$ depends on internal geometries of all molecular monomers [9]. However, in most physical applications, the dependence of $E_{\text{int}}(\mathbf{Q}_1, \ldots, \mathbf{Q}_N)$ on the intramonomer coordinates $\xi_1, \ldots, \xi_N$ can be neglected and the resulting intermolecular potentials are referred to as rigid-monomer potentials. Such potentials depend on a much smaller number of coordinates and therefore are significantly easier to calculate and apply. For example, for the water trimer, the complete potential is 21-dimensional, whereas the rigid-monomer potential is 'only' 12-dimensional. Further discussions in this chapter will be limited to rigid-monomer potentials. For issues connected with calculations of flexible-monomer potentials, see Refs. [9–11]. The main use of flexible-monomer potentials could be to predict spectral shifts of intramolecular vibrations due to intermolecular interactions. Such shifts are very often studied experimentally. Theoretical predictions of the shifts have been mainly based on calculations of harmonic constants, which is a reasonable approximation in this case since the intramolecular vibrations are highly harmonic.

33.3 HISTORICAL PERSPECTIVE

One may say that investigations of nonadditive interactions started in 1943 with the famous papers by Axilrod and Teller [12] and by Muto [13]. The authors of these two papers independently derived the leading asymptotic term (the so-called 'triple dipole' or ATM term) of the three-body dispersion energy for atoms. This term was found to make an important contribution to the cohesion energies of rare-gas crystals. It is interesting to remark that, as we know now, the success of the ATM term in reproducing the measured values of these energies is due to quite a complete cancellation of the other nonadditive components which are of the same order of magnitude as the ATM term (see Refs. [14, 15] and Section 33.9.2). Other than the dispersion types of nonadditive forces have achieved much less attention. It appears that the exchange nonadditivity was investigated for the first time by Rosen [16] in 1953 and by Shostak [17] in 1955 for the helium trimer. The activities intensified in the early 1960s and in particular the work of Jansen [18] on rare-gas trimers became well known. The number of papers on nonadditive forces increased gradually in the 1970s and 1980s. The first review devoted exclusively to the subject of nonadditive interactions was probably that by Meath and Aziz [19] in 1984.

Later reviews were written by Meath and Koulis [20], Elrod and Saykally [21], and by Szczesniak and Chalasinski [22]. Some of the monographs on the intermolecular forces contain chapters on many-body interactions [23–25]. Spectroscopic measurements for trimers and larger cluster started in the late 1980s and provided very precise data on such systems [21,26–28], from which information about three-body effects could potentially be extracted. Theoretical calculations enabling such extractions have been possible for simplest of the experimentally investigated systems such as Rg_2–HX (where Rg and X denote rare-gas and halogen atoms, respectively) [29,30] and for systems as large as the water trimer (with some degrees of freedom frozen) [31]. In the late 1990s, computational power became sufficient to obtain some complete three-body potentials [32,33]. At the same time, the first empirical trimer potentials fitted to spectroscopic data were developed by Ernesti and Hutson [29,30]. Finally, recently it was possible to develop a three-body potential for the water [34]. Some of this work will be discussed in more detail later on.

Until very recently, the subject of nonadditive forces was almost synonymous with the field of rare gases (see Ref. [35] for a recent review of this field). The reason was not that these forces are particularly large or important for rare gases compared to other systems, but the availability of rather accurate empirical two-body potentials derived by fitting mainly gas-phase properties of the gases. Then, one could find out the values of many-body effects by comparing measured values for condensed phases of rare gases (since these effects are appreciable only in the condensed phase) with the theoretical predictions based on two-body potentials. The apparent reasonably good representation of the three-body nonadditivity by the triple-dipole [12,13] approximation to the dispersion energy gave generally sensible predictions for the values inferred from experiments. However, in the 1980s, the field went through a period of intensive disputes when some researchers realized that the first-order exchange effects cancel most of the contributions to the cohesion energies produced by the ATM term [19,36]. Indeed, one has to include several more complicated nonadditive components to 'again' bring theory to agreement with experiment [14]. The importance of many-body effects for condensed phase of rare gases depends strongly on the property investigated. A concise discussion of these issues has been given in the review by Huber et al. [37]. It has been found [38] that while the structure, transport properties, heat capacities, compressibilities, and sound velocities are determined essentially by the two-body potentials, the nonadditive forces significantly influence the internal energies [14] and pressures [39]. Some of the effects of many-body forces can lead even to qualitative differences in observed properties. For example, pair potentials predict that heavier rare gases crystallize in hcp lattice whereas the inclusion of three-body effects favors the fcc lattice [14], in agreement with measurements. More recently it has been found that these forces are also very essential for predictions of vapor–liquid equilibria parameters [15,40–42].

For systems more complex than rare-gas atoms, little has been known about many-body effects until very recently. This led Ernesti and Hutson [29] to state that *for systems containing molecules, there are no reliable models of nonadditive forces*. In fact, for these systems, the prerequisite of knowing a sufficiently accurate pair potential has been difficult to achieve. The simplest nonatomic systems of interest from the point of view of condensed-phase many-body effects are diatomic molecules. The four-dimensional rigid-monomer potentials for dimers consisting of diatomics can now be computed in a reliable way. However, only a few accurate potentials for homogeneous dimers of this type are known. Among them are the potentials for $(HF)_2$ obtained by Klopper et al. [43] (this potential, in fact, accounts for the flexibility of monomers and is therefore six-dimensional), for $(H_2)_2$) by Diep and Johnson [44], and for $(CO)_2$ by Vissers et al. [45]. The largest group of dimers for which accurate potentials are known are atom–diatom complexes. With the rigid-monomer approximation, these potentials are two-dimensional. However, such potentials are not yet helpful for studying the nonadditive effects in the bulk mixtures since accurate diatom–diatom potentials would also be needed. On the other hand, trimers of the form Rg_2-diatom, for which atom–diatom pair potentials are needed, can be studied in molecular beams and the investigation of such clusters has become one of the more fruitful subjects in the field of nonadditive forces in the last decade.

There is only one dimer involving two polyatomic monomers for which six-dimensional rigid-monomer potentials accurate enough to enable studies of nonadditive effects are available, the water dimer. Water has been the most often investigated system due to its general importance and the fact that the nonadditive contributions to the interaction energy are exceptionally large for this system, as large as about 15% of pair contributions [46]. Investigations of the interactions between two water molecules date back to 1970s when rotational spectroscopy measurements determined the structure of this system. This structure [47] agrees very well with results of modern *ab initio* calculations [48]. Since late 1980s, numerous measurements of the infrared spectra of the water dimer have been published. These spectra, in particular in the far-infrared region (see Refs. [49–51] for recent examples), provided a very detailed characterization of the system. The spectroscopic data were then used to fit an empirical water dimer potential [52,53]. One should mention that fits to spectroscopic data for systems as large as the water dimer have become possible only recently, since only in the late 1990s the six-dimensional calculations of quantum dynamics were done for the first time [31, 54–56]. Water dimer has been also one of the systems most investigated by *ab initio* calculations [48,57–63].

The leading nonadditive term in the many-body expansion of a potential is the three-body interaction. Similarly like dimers, trimers (and larger clusters) can be selectively studied by molecular beam spectroscopy. A number of such trimers have been the subjects of investigations. Among them are the Rg_2-diatom trimers mentioned above, with the most extensive data available for Ar_2–HF [64]. Both empirical [29,30] and *ab initio* [33] nonadditive potentials have been obtained for this system. A large number of spectral data are available also for the water trimer [65,66]. An accurate three-body potential for water has recently been developed [34].

It has been long recognized that pairwise-nonadditive forces are critical for describing bulk materials. As one can already see from the discussion above, determinations of trimer nonadditive forces are at the current edge of research capabilities. Thus, for condensed phase (and for clusters larger than trimers), first-principle predictions can be reliable only if four- and higher body interactions are relatively unimportant. The three-body effects are in most cases from about 10 to 100 times smaller in magnitude than the two body effects, as it will be discussed below. Thus, if this rate of convergence of the many-body expansion continues, the four-body and higher effects should give contributions negligible at the levels of accuracy that one is currently interested in. However, the convergence beyond $K = 3$ may not be that fast as the step from $K = 2$ to $K = 3$ suggests. This subject will be discussed in more detail later on.

Despite the question of the importance of the higher than three-body effects, first-principle simulations of bulk have been pursued by many research groups. The best-known work is probably that of Clementi, Corongiu, and collaborators [67] on water. Recently, the number of such simulations has been increasing rapidly, see for example work by Huber and collaborators [37,68,69]. Several first-principle simulations have recently been performed based on SAPT potentials [15,46,70,71].

Another type of first-principle simulations are those performed using the Car–Parrinello molecular dynamics (CPMD) method [72]. In CPMD the total interaction potential is computed for the whole cluster 'on the fly', without using many-body expansions. This can be done using only the simplest electronic structure methods such as the density functional theory (DFT). The interaction energy is computed for each step in molecular dynamics simulations, although in practice this is done completely only in the first step, later the electron density is propagated. Some CPMD 'on-the-fly' simulations for water have been rather successful in predicting properties of the liquid [73]. However, recent work shows [74] that a part of this success originates from choosing a variant of the DFT method which happens to work well for water. Other variants of DFT may predict the properties of water poorly [74] and it is not possible to decide *a priori* which of the variants to choose. In fact, for the ammonia dimer, none of the known functionals seems to be able to give results that would agree with experiment and the authors of Ref. [75] had to optimize an ammonia-specific functional.

Although there have been a fairly large number of first-principle simulations of condensed phase published to date, this number is completely dwarfed by simulations based on empirical potentials. The popular empirical pair potentials for water [76–78] have been used in many (thousands) research projects. The empirical potentials are usually fitted in simulations for liquids to reproduce measured properties of this phase. Thus, these potentials mimic the nonadditive effects by distortions of two-body potentials [79]. Since there is no rigorous way that such a transformation can be performed, one may say that the successful applications of empirical potentials happen 'for wrong reasons'. In fact, such potentials typically fail beyond the range of thermophysical parameter space that was used in the fitting process. The effective pair potentials are also not able to predict some measured properties even within the range, for example, the second virial coefficients. It is in fact not possible to simultaneously fit such potentials to the virial coefficients and to typical bulk properties that depend on the pair-nonadditive effects. One solution of this problem is to use polarizable empirical potentials. Such potentials have to a lesser extent the effective character due to the explicit modeling of the nonadditive induction energy. Thus, the pairwise-additive component is less biased by efforts to mimic nonadditive forces. The polarizable empirical potentials can be made to reproduce dimer properties. Thus, virial coefficients can be used in the fits. Even if this is not the case, as for example for the polarizable point-charge (PPC) potential of Svishchev et al. [80], the computed virial coefficients are much better than obtained with the empirical pair potentials, but still significantly less accurate than those predicted by SAPT *ab initio* potentials [48,63,81].

33.4 PERTURBATION THEORY OF INTERMOLECULAR INTERACTIONS

As mentioned before, interactions between molecules are much weaker than the chemical bonds inside molecules. Thus, van der Waals interactions can naturally be described in terms of a perturbation theory treating isolated monomers as the zeroth-order approximation. Each monomer is perturbed due to the electric fields of the partners. The simplest type of perturbative approach that can be applied is the standard Rayleigh–Schrödinger perturbation theory, which in the context of intermolecular interactions is often called the polarization theory. The major problem with this method is that the resulting wave functions of the N-mer do not fulfill the Pauli's exclusion principle, i.e. are not antisymmetric upon exchange of electrons between different monomers. This problem can be overcome in the simplest way by applying explicit antisymmetrization operators to the wave functions of polarization theory. This leads to the SAPT method. For pair interactions, SAPT is now a mature and routinely used method. Reviews of SAPT can be found in Refs. [2,4,82,83] and a general purpose computer program can be downloaded from the web [84]. More recently, SAPT of three-body interactions has been developed [85–88]. The computer programs realizing this theory are also available on the SAPT web site [84], although these are not yet as robust and mature as the two-body programs. One should mention here that Tachikawa and Iguchi [89] developed some nonadditive SAPT corrections including the exchange terms, initially only in the S^2 approximation (i.e. including only the terms proportional to the square of the overlap integrals) which is not satisfactory for trimers, but later work by Tachikawa et al. [90] accounted also for the S^3 terms.

The perturbation theory describing nonadditive effects is a relatively straightforward extension of the SAPT approach for pair interactions except that each additional monomer increases the complexity of the resulting expressions. In the SAPT expansion of the interaction energy for an N-mer, the total Hamiltonian of the N-mer is partitioned as

$$H = F + V + W \tag{8}$$

where $F = F_A + F_B + F_C + \cdots$ is the sum of the Fock operators for all monomers, $V = V_{AB} + V_{BC} + V_{AC} + \cdots$ is the sum of all binary intermolecular interaction operators V_{XY}, and $W = W_A + W_B + W_C + \cdots$ is the sum of the intramolecular correlation operators (Møller–Plesset fluctuation potentials) for all monomers. The latter operators are defined as $W_X = H_X - F_X$, where H_X is the complete Hamiltonian for monomer X. This partitioning of the total Hamiltonian allows the interaction energy to be written as a double perturbation expansion in V and W. The latter expansion is analogous to that used in the standard many-body perturbation theory and coupled-cluster methods [91,92]. The K-body interaction energy defined before can now be written as

$$E_{\text{int}}[K] = \sum_{i=1;\, j=0} (E^{(ij)}_{\text{pol}}[K] + E^{(ij)}_{\text{exch}}[K]) \tag{9}$$

where subscripts 'pol(exch)' denote the polarization (exchange) energy components. The former components result from the application of the polarization perturbation theory, whereas the latter from the action of the antisymmetrization operators. The exchange energy components are formally defined as the differences between the SAPT and polarization energy contribution, but are derived and computed as separate terms. The superscripts i and j in Eq. (9) refer to the order with respect to the intermolecular interaction V and the intramolecular correlation operator W, respectively. We will also consider the interaction energy components which fully include intramonomer correlation effects

$$E^{(i)}_{\mathrm{pol(exch)}}[K] = \sum_{j=0}^{\infty} E^{(ij)}_{\mathrm{pol(exch)}}[K] \tag{10}$$

Very recently, a new and very efficient version of SAPT has been developed using the DFT description of monomers [93,94]. This method, which will be denoted by SAPT(DFT), has not yet been applied to three-body interactions, but such an application will be straightforward. The use of SAPT(DFT) not only results in a very significant reduction of the needed computational resources, but often gives more accurate interaction energies than does SAPT. There are two reasons for the latter. First, SAPT(DFT) converges much faster in basis set size than does SAPT. This feature together with the computational efficiency enables calculations of results nearly saturated in the basis set size. Second, the truncation of the SAPT expansion in W leads to the neglect of a part intramonomer correlation effects. In contrast, SAPT(DFT) is formally of infinite order in W.

33.5 OVERVIEW OF PAIR CONTRIBUTIONS

Before we proceed to the discussion of nonadditive effects, a short overview of pair interactions is needed to provide a proper reference. Numerical experience has shown that for most purposes it is sufficient to restrict the expansion (9) for $K = 2$ to the components up to the second order in V. Such a truncated expansion can be written in terms of the well-known electrostatic, induction, dispersion, and exchange contributions:

$$E_{\mathrm{int}}[2] = E^{(1)}_{\mathrm{elst}}[2] + E^{(1)}_{\mathrm{exch}}[2] + E^{(2)}_{\mathrm{ind}}[2] + E^{(2)}_{\mathrm{disp}}[2] + E^{(2)}_{\mathrm{exch-ind}}[2] + E^{(2)}_{\mathrm{exch-disp}}[2] + \cdots \tag{11}$$

These components provide a clear physical picture of the two-body interactions. As an example let us analyze the components for the water dimer. The physical interpretation of the energy for this dimer has been the subject of great interest as it advances our understanding of hydrogen bond [57,95–97]. Table 33.1 shows various contributions on a cut through the potential energy surface passing near the global minimum. The subscripts 'resp' appearing in some terms listed in Table 33.1 indicate that this contribution was computed with orbital relaxation effects [98,99]. The correction $E^{(1)}_{\mathrm{exch}}$(CCSD) was computed with the monomer wave functions correlated at the coupled-cluster level with single and double excitations [100]. The term $\delta E^{\mathrm{HF}}_{\mathrm{int,resp}}$ collects contributions beyond the second order in V to the supermolecular Hartree–Fock

Table 33.1 Components of the SAPT interaction energy of the water dimer for four intermolecular separations R, computed in the [5s3p3d2f1g1h/3s3p1d1f] basis taken from Ref. [103], containing the 3s2p1d1f set of midbond functions. The angular coordinates are close to those of the global minimum configuration. All energies in kcal/mol

Component	$R = 2.3$ Å	$R = 2.5$ Å	$R = 3.0$ Å	$R = 3.6$ Å
$E^{(10)}_{\text{elst}}[2]$	−31.422	−19.779	−7.217	−3.058
$E^{(12)}_{\text{elst,resp}}[2]$	−0.262	−0.158	0.097	0.145
$E^{(13)}_{\text{elst,resp}}[2]$	0.390	0.336	0.116	0.012
$E^{(1)}_{\text{elst}}[2]$	−31.294	−19.601	−7.005	−2.901
$E^{(20)}_{\text{ind,resp}}[2]$	−30.059	−13.615	−2.180	−0.346
${}^{\text{t}}E^{(22)}_{\text{ind}}[2]$	−3.751	−1.849	−0.322	−0.043
$E^{(2)}_{\text{ind}}[2]$	−33.810	−15.464	−2.502	−0.389
$E^{(20)}_{\text{disp}}[2]$	−11.201	−6.910	−2.142	−0.587
$E^{(21)}_{\text{disp}}[2]$	0.910	0.377	0.025	−0.003
$E^{(22)}_{\text{disp}}[2]$	−1.810	−1.165	−0.412	−0.129
$E^{(2)}_{\text{disp}}[2]$	−12.100	−7.698	−2.529	−0.719
$E^{(10)}_{\text{exch}}[2]$	62.605	31.081	5.160	0.573
$E^{(1)}_{\text{exch}}(\text{CCSD})[2]-E^{(10)}_{\text{exch}}[2]$	4.858	3.278	0.917	0.153
$E^{(1)}_{\text{exch}}[2]$	67.463	34.359	6.077	0.726
$E^{(20)}_{\text{exch-ind,resp}}[2]$	19.300	8.631	1.088	0.091
${}^{\text{t}}E^{(22)}_{\text{exch-ind}}[2]$	2.408	1.172	0.161	0.011
$E^{(20)}_{\text{exch-disp}}[2]$	2.806	1.635	0.358	0.050
$\delta^{\text{HF}}_{\text{int,resp}}[2]$	−7.269	−4.074	−0.696	−0.080
$E^{(2)}_{\text{exch}}[2] + \delta^{\text{HF}}_{\text{int,resp}}[2]$	17.245	7.363	0.910	0.072
$E_{\text{int}}[2]$	7.506	−1.040	−5.049	−3.210

interaction energy $E^{\text{HF}}_{\text{int}}$ [101,102]:

$$\delta E^{\text{HF}}_{\text{int,resp}}[2] = E^{\text{HF}}_{\text{int}}[2] - E^{(10)}_{\text{elst}}[2] - E^{(10)}_{\text{exch}}[2] - E^{(20)}_{\text{ind,resp}}[2] - E^{(20)}_{\text{exch-ind,resp}}[2] \quad (12)$$

In the asymptotic region, i.e. for large intermonomer separations R, the interaction energy is well described by the polarization terms alone. Moreover, at such distances one can make an additional approximation and represent the operator V in terms of its multipole expansion containing terms inversely proportional to powers of R. As R increases, the interaction energy will eventually be well represented by just the term with the lowest power of $1/R$. For polar dimers such as water, the lowest power equal to three is coming from the electrostatic interactions and the water dimer potential is completely dominated at very large R by the classical interaction of the permanent dipole moments of water molecules. Already at $R = 3.6$ Å, the electrostatic component constitutes 90% of the interaction energy (cf. Table 33.1). As R approaches the minimum configuration, the induction and dispersion energies enter the picture. For water, these terms happen to be of nearly equal magnitude. At the minimum, their sum is only slightly smaller than the electrostatic contribution. As R decreases further, the exponentially increasing exchange components begin to quench the attractive interactions originating from the polarization components. At the minimum distance, the exchange component is about 40% larger in magnitude than the total interaction energy and therefore it cancels more than a half of the attractive effect. For shorter separations, due to the faster growth of the exchange energy, the value of the potential starts to increase with the decrease of R, i.e. the potential enters the repulsive region. At $R = 2.5$ Å, the interaction energy is still negative but the exchange contribution is about as large as the sum of the attractive components. For still smaller separations the exchange repulsion prevails and the interaction energy becomes positive, increasing very fast with the decrease of R. Thus, this region is often called the potential 'wall'.

Another possible decomposition of the interaction energy is into the Hartree–Fock and correlation effects. This decomposition can be considered for practical reasons, as calculations at the Hartree–Fock level are much less expensive than the calculations of the terms including electron correlation. For the water dimer near the minimum (column $R = 3.0$ Å in Table 33.1), the Hartree–Fock interaction energy of −3.85 kcal/mol provides a reasonable, but far from accurate approximation to the interaction energy. The total electron correlation contribution near the minimum is −1.20 kcal/mol or 24% of interaction energy. The largest correlation effect comes from the dispersion energy equal to −2.53 kcal/mol. This attractive component is partly quenched by the intramolecular correlation correction to the first-order exchange energy equal to 0.92 kcal/mol. The intramonomer correlation correction to the electrostatic energy is also positive but much smaller, as it amounts to only 0.21 kcal/mol. Another correlation component, the exchange–dispersion energy, contributes 0.36 kcal/mol. The overall correlation correction to the induction and exchange–induction energy is −0.16 kcal/mol. These results are in very good agreement with the ones of Ref. [103], computed in the same basis set but for slightly different geometry than the one used in Table 33.1.

33.6 PERTURBATION THEORY OF NONADDITIVE FORCES

Theoretical analysis and numerical calculations show that a given correction appearing in Eq. (9) in the case of pair interactions ($K = 2$) and nonadditive three-body interactions ($K = 3$) may be of completely different importance for the total interaction energy. First of all, some contributions that are usually large for dimers are exactly equal to zero in the expansion of the nonadditive energy. In particular, the electrostatic interactions and the dispersion interactions in the second order are in this category. Furthermore, for a reasonable description of the nonadditive interactions, one has to include terms up to the third order in V:

$$E_{\rm int}[3] = E^{(1)}_{\rm exch}[3] + E^{(2)}_{\rm ind}[3] + E^{(2)}_{\rm exch-ind}[3] + E^{(2)}_{\rm exch-disp}[3] + E^{(3)}_{\rm disp}[3] + E^{(3)}_{\rm ind-disp}[3] + E^{(3)}_{\rm ind}[3] + \cdots \quad (13)$$

For polar systems, the Hartree–Fock three-body contribution, $E^{\rm HF}_{\rm int}[3]$, may give a reasonable approximation of the total nonadditive effect. This quantity can be decomposed as

$$E^{\rm HF}_{\rm int}[3] = E^{(10)}_{\rm exch}[3] + E^{(20)}_{\rm ind}[3] + E^{(20)}_{\rm exch-ind}[3] + \delta E^{\rm HF}_{\rm int}[3] \quad (14)$$

where $\delta E^{\rm HF}_{\rm int}[3]$ contains mostly $E^{(30)}_{\rm ind}[3]$ and its exchange counterpart.

As discussed in Section 33.3, the first nonvanishing nonadditive dispersion component, the dispersion energy of the third order in V, has probably been by far the most studied nonadditive contribution. This component provides a very significant contribution to the nonadditive energy of nonpolar and weakly polar species. It has been extensively investigated for rare gases, in particular when approximated by the leading term in its large R asymptotic expansion, i.e. the ATM term [12,13]. In its complete (nonexpanded) form, including the effect of the charge overlap, the third-order dispersion nonadditivity is given by the following formula [85]

$$E^{(3)}_{\rm disp}[3] = 2\langle \Phi^{(1)}_{\rm disp,AB}\Phi^{\rm C}_0 | V_{\rm BC} \Phi^{(1)}_{\rm disp,AC}\Phi^{\rm B}_0\rangle + \cdots \quad (15)$$

In this equation, $\Phi^{(1)}_{\rm disp,XY}$ denotes the two-body dispersion function [2] of dimer XY and $\Phi^{\rm X}_0$ is the exact unperturbed wave function of monomer X. Notice that we use here the notation introduced in Eq. (10), i.e. we consider interactions of exact (fully correlated) monomers. The dots in Eq. (15) and in the remaining equations in this section denote two further terms obtained by cyclic permutations of A, B, and C.

The other very often considered nonadditive component is the induction energy. This component in its asymptotic form is the basis of the 'polarizable' empirical potentials described in Section 33.3. For strongly polar systems, the second- and third-order nonadditive induction terms can indeed be expected to provide the largest nonadditive contribution except for small intermonomer separations [46] and to constitute the major part of the Hartree–Fock nonadditive contribution. The second-order terms have a very simple physical interpretation: a multipole on system A induces multipole moments on B and C which interact with the permanent multipoles on C and B, respectively (see a more extensive discussion below). The second-order induction nonadditivity can be written as [85,86]

$$E^{(2)}_{\rm ind}[3] = 2\langle \Phi^{\rm B}_0 | \Omega^{\rm B}_{\rm A} | \Phi^{(1)}_{\rm ind,B\leftarrow C}\rangle + \cdots \quad (16)$$

The operator $\Omega^{\rm Y}_{\rm X}$ represents here the electrostatic potential generated by the unperturbed monomer X and depends on electron coordinates of monomer Y

$$\Omega^{\rm Y}_{\rm X} = \sum_{i\in {\rm Y}} \omega_{\rm X}(\mathbf{r}_i) \quad (17)$$

were

$$\omega_X(\mathbf{r}) = \int \frac{1}{|\mathbf{r} - \mathbf{r}'|} \rho_X(\mathbf{r}')d\mathbf{r}' \tag{18}$$

is determined in terms of ρ_X, the unperturbed charge distribution (including the nuclear charges) of monomer X. The two-body induction function of, for example, monomer A in the field of monomer B, $\Phi^{(1)}_{\mathrm{ind,A}\leftarrow\mathrm{B}}$, is defined by equation

$$(H_A - E_A)\Phi^{(1)}_{\mathrm{ind,A}\leftarrow\mathrm{B}} = (\langle \Phi_0^A | \Omega_B^A | \Phi_0^A \rangle - \Omega_B^A)\Phi_0^A \tag{19}$$

where E_A is the ground-state energy of the exact Hamiltonian of monomer A.

The nonadditive exchange components have been investigated much less often than the dispersion and induction nonadditive energies. This is simply due to the fact that the exchange components are much more difficult to compute than the dispersion and induction ones, particularly if the latter are considered in the multipole approximation. From the importance point of view, the exchange nonadditive term should never be neglected, except for large intermonomer separations. This is not surprising in view of the fact that the exchange nonadditivity is the only component that gives a contribution already in the first order in V. It often is the largest nonadditive term near the minimum configuration. Fortunately, the importance of the exchange nonadditivity for bulk properties is somewhat smaller than might be expected from these statements, since a significant fraction of the total nonadditive effect originates from trimers with larger than equilibrium separations, for which the exchange nonadditivity is small due to the exponential decay [15,46].

Each exchange correction, both pair additive and nonadditive, can be expanded into powers of the overlap integrals S. The S^2 component of the first-order exchange nonadditivity can be written in the following form [85]

$$E^{(1)}_{\mathrm{exch}}[3](S^2) = -\langle \Phi_0 | (V_{BC} - \langle V_{BC} \rangle)(\mathcal{P}_{AB} + \mathcal{P}_{AC})\Phi_0 \rangle + \cdots \tag{20}$$

where $\mathcal{P}_{XY}$ is the sum of all transpositions of electrons between monomers X and Y and $\Phi_0 = \Phi_0^A \Phi_0^B \Phi_0^C$. The expression for $E^{(1)}_{\mathrm{exch}}[3](S^3)$ can be obtained from Eq. (20) by replacing the operator $\mathcal{P}_{AB} + \mathcal{P}_{AC}$ by $-\mathcal{P}_{ABC}$ where $\mathcal{P}_{ABC}$ is the sum of cyclic permutations exchanging three electrons among monomers A, B and C. The exchange nonadditive contribution in the second order can be split into the exchange–induction and exchange–dispersion energies. The definitions and the explicit working expressions (in terms of one- and two-electron integrals) for these contributions—obtained with the neglect of intramonomer electron correlation—have been given in literature [85,86].

In practical calculations, one has to expand the nonadditive terms discussed above in powers of intramolecular correlation operator W. Such an expansion was developed in Ref. [85]. The currently programmed nonadditive corrections are $E^{(10)}_{\mathrm{exch}}[3]$, $E^{(20)}_{\mathrm{ind}}[3]$, $E^{(20)}_{\mathrm{exch-ind}}[3]$, $E^{(30)}_{\mathrm{ind}}[3]$, $E^{(30)}_{\mathrm{disp}}[3]$, $E^{(31)}_{\mathrm{disp}}[3]$, $E^{(40)}_{\mathrm{disp}}[3]$, $E^{(30)}_{\mathrm{ind-disp}}[3]$, and $E^{(20)}_{\mathrm{exch-disp}}[3]$ [85]. More recently the effects of intramonomer correlation on the first-order exchange nonadditivity were taken into account by including the corrections $E^{(11)}_{\mathrm{exch}}[3]$ and $E^{(12)}_{\mathrm{exch}}[3]$ [87,88]. Examples of applications of these theoretical developments will be discussed in Sections 33.7 and 33.11.2.

33.7 COMPARISON OF NONADDITIVE EFFECTS FOR SELECTED SYSTEMS

It has been already alluded in several places that the importance of various nonadditive components varies dramatically from system to system. In this section, we will analyze these trends on a few examples. The particularly important case of water will be additionally discussed in Section 33.11.2. The knowledge of relative importance of nonadditive components of interaction energy is needed to decide which terms from the expansion (2) have to be included for a given system. For example, the dispersion nonadditivity is completely dwarfed by the induction effects in the case of water, whereas it is the leading term for rare gases. Some predictions for the polarization components can be made from the knowledge of the multipole moments and polarizabilities of monomers. However, qualitative predictions seem near to impossible for the exchange terms. Therefore, in practice, the only way for predicting trends must be based on *ab initio* calculations. For some systems an empirical approach—greatly helped by whatever theoretical information can be gathered and included in potential fits to spectroscopic data—can also provide useful information about the size of nonadditive effects [29,30].

We will now examine the major nonadditive components of interaction energy for four different near-equilibrium trimers. These components are displayed in Fig. 33.1. The first observation that can be made is that the ratio of the three-body component to the sum of two-body interactions changes dramatically from system to system: from 0.4% for He_3 to 16% for the water trimer. The total nonadditive contribution can also be of either sign, even for similar systems like rare gas trimers, *cf.* He_3 *vs.* Ar_3. The increased role of nonadditive terms is clearly correlated with the polarity of a system: the more polar a system is, the larger are the relative contributions of nonadditive terms with respect to two-body terms.

The relative importance of the various components of nonadditive interaction energies appearing in Eq. (13) also vary dramatically between different trimers. Notice that since only the largest components are shown for a given trimer (plus all non-negligible first and second-order components), the set of components displayed in Fig. 33.1 is not always the same. For rare gas trimers, the nonadditive contribution is dominated by the first-order exchange and third-order dispersion energies, although for Ar_3 the exchange–dispersion contribution is quite large. This contribution is critical for predicting the correct crystal structure of argon [14]. For Ar_2–HF, all these components are still important but a very significant role is played also by the induction and exchange–induction nonadditive contributions. Although these components cancel nearly exactly in the second order, the slightly smaller third-order components (not shown in Fig. 33.1) add up and together make an important contribution. The situation changes completely in the water trimer. The dispersion contributions are dwarfed by the very significant induction effects. Even the third-order induction nonadditivity, a major part of $\delta E_{\mathrm{int}}^{\mathrm{HF}}[3]$, is very important. Fig. 33.1 shows also that the first-order exchange nonadditivity is important for all systems.

Table 33.2 presents numerical values of the three-body interaction components for the water trimer. The decomposition is presented for various stationary points on the trimer

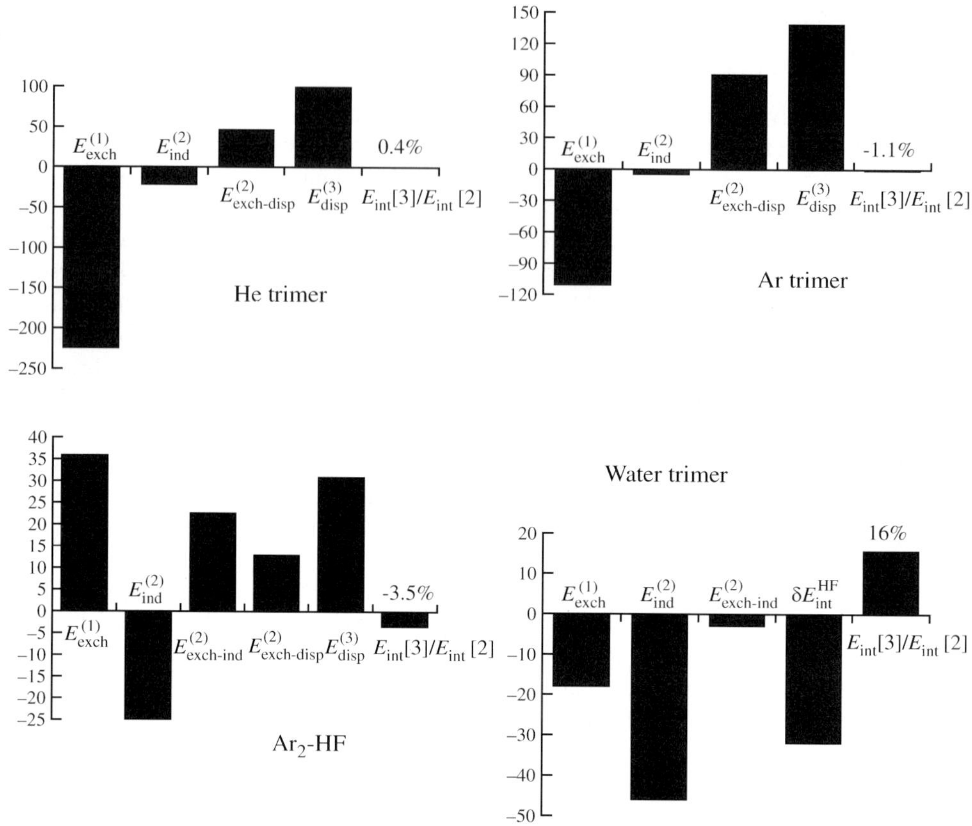

Fig. 33.1. Comparison of the relative importance of various three-body interactions. The bars show the percentage contribution of a given three-body component to the *magnitude* of the total three-body interaction energy except for the last bar which shows percentage ratio of the three-body to two-body contribution. The data are from Ref. [87] for He_3, Ref. [32] for Ar_3, Ref. [33] for Ar_2–HF, and Ref. [34] for the water trimer.

potential surface. The configurations [34] are as follows: uud: global minimum (a cyclic hydrogen-bonded structure), bif: the bifurcated structure, ada and dad: two quasi-linear structures where 'a' stands for acceptor and 'd' for donor. The terms displayed in the upper part of Table 33.2, corresponding to the Hartree–Fock level of theory, show several interesting characteristics. As discussed above, the first-order nonadditive exchange energy can be decomposed into terms resulting from different types of electron exchanges: the contributions from single exchanges of a pair of electrons, given by Eq. (20), as well as from the cyclic permutations involving three electrons at a time, which can be obtained from Eq. (20) as described in Section 33.6. The resulting contributions differ by the powers of S and can be quadratic (single exchanges), cubic (three-electron cycles), quartic (double exchanges), etc. in S. All these terms are listed in Table 33.2. One can see that for the water trimer at cyclic configurations the quadratic and cubic terms

Table 33.2 SAPT decomposition of *ab initio* three-body interaction energies (in kcal/mol) for various stationary points on the fitted water trimer potential surface of Ref. [34]. All calculations in trimer-centered [5s3p2d1f/3s2p] basic set of Ref. [63] (without bond functions)

	uud	bif	ada	dad
Two-body (SAPT-5s)	−13.197	−11.595	−9.101	−8.909
$E^{(10)}_{exch}[3](S^2)$	−0.047	−0.006	0.034	−0.078
$E^{(10)}_{exch}[3](S^3)$	−0.317	−0.225	0.005	0.006
$E^{(10)}_{exch}[3](S^4)$	−0.007	0.000	−0.011	−0.010
$E^{(10)}_{exch}[3]$	−0.371	−0.231	0.027	−0.082
$E^{(20)}_{ind}[3]$	−0.960	−0.752	0.186	0.143
$E^{(20)}_{exch-ind}[3]$	−0.063	−0.027	0.011	−0.064
$E^{(30)}_{ind}[3]^a$	−0.312	−0.219	0.055	0.099
$\delta E^{HF}_{int}[3]$	−0.676	−0.444	0.197	0.189
SCF three-body ($E^{HF}_{int}[3]$)	−2.070	−1.454	0.421	0.186
Total (SAPT-5s + $E^{HF}_{int}[3]$)	−15.267	−13.048	−8.681	−8.723
$E^{(20)}_{exch-disp}[3]$	0.093	0.074	0.007	0.008
$E^{(30)}_{ind-disp}[3]$	0.015	0.002	−0.022	0.029
$E^{(30)}_{disp}[3]$	0.052	0.046	−0.003	−0.002
$E^{(31)}_{disp}[3]$	0.005	0.005	−0.001	−0.001
$E^{(40)}_{disp}[3]$	−0.057	−0.046	−0.005	−0.007
Total correlation (SAPT)	0.108	0.081	−0.024	0.027
$E^{CCSD(T)}_{int}[3]-E^{HF}_{int}[3]$	0.057	0.066	−0.010	0.040

[a]This correction is included in $\delta E^{HF}_{int}[3]$.

dominate. Moreover, for all the configurations where the exchange nonadditive terms are important, like in the equilibrium uud trimer, the S^3 terms give by far the largest contribution. For large trimers, the S^3 contribution increases in importance due to its slower decay (see Section 33.8). In the equilibrium configuration, the overall first-order nonadditive exchange energy contributes as much as 18% to the total nonadditive effect computed at the Hartree–Fock level.

Although the first-order exchange energy is quite large, Table 33.2 shows that for all cyclic configurations the negative second-order induction energy provides the largest contribution. The third-order induction energy, included in $\delta E^{HF}_{int}[3]$, is of comparable size. One should point out that for the noncyclic trimers ada and dad, the situation is quite different. First, the induction contribution is positive and the effects contained in $\delta E^{HF}_{int}[3]$ are even larger than the second-order one. The nonadditivity of exchange interactions is almost negligible for these trimers.

The lower part of Table 33.2 shows the SAPT correlation contributions to the three-body energy of the water trimer. These contributions include the electron correlation effects at the level approximately comparable to the supermolecular many-body perturbation theory at the fourth order (MBPT4). Among these contributions, $E^{(20)}_{exch-disp}[3]$ dominates for the cyclic and bifurcated configurations, providing a repulsive effect larger in magnitude than $E^{(20)}_{exch-ind}[3]$ and the first-order S^2 exchange corrections. About a factor of two smaller are the dispersion components $E^{(30)}_{disp}[3]$ and $E^{(40)}_{disp}[3]$, which are of opposite signs and cancel one another almost exactly. For the remaining two structures, the correction $E^{(30)}_{ind-disp}[3]$ is the largest among the correlation effects. As shown by the analysis performed in Ref. [33] for the Ar_2–HF trimer, some correlation corrections, especially $E^{(30)}_{ind-disp}[3]$ and $E^{(40)}_{disp}[3]$, are strongly quenched by their exchange counterparts, not yet implemented in the SAPT code. Therefore, the sum of all the correlation corrections from Table 33.2 is generally somewhat different from the total correlation effect computed in the same basis at the coupled cluster level of theory with single, double, and noniterative triple excitations [CCSD(T)] using the supermolecular approach. It should be noted that the correlation effect typically constitutes only a few percent of the total three-body nonadditive interaction and reaches the largest relative value of 18% only for the dad configuration, where this interaction is anyway small. In any case, the correlation component of the three-body nonadditivity does not exceed 0.5% of the total trimer energy (including the two-body component).

The results of Table 33.2 suggest that, as expected, the bulk of nonadditive interactions in the water trimer is reproduced quite well already at the SCF level of theory. This observation is further quantified in Fig. 33.2, where the three-body energies computed at the SCF and CCSD(T) levels are compared for ten trimers with maximum O–O separations less than 5 Å extracted from a snapshot of a Monte Carlo (MC) simulation of liquid water at ambient conditions. Typically, trimers for which the nonadditive

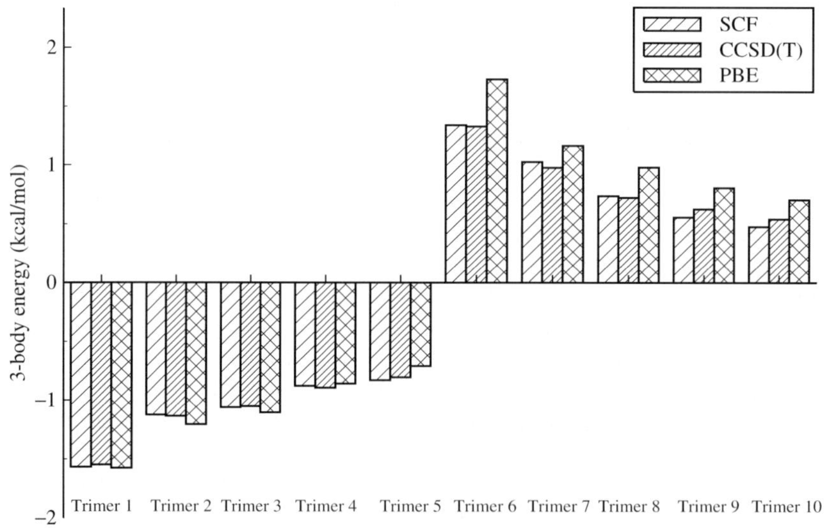

Fig. 33.2. Comparison of the SCF, CCSD(T), and PBE three-body interaction energies for trimers extracted from an MC simulation of liquid water.

interaction is attractive are characterized by a quasi-cyclic arrangement of hydrogen bonds, i.e. one of the molecules acts as both a donor and an acceptor. Other types of trimers, with one molecule acting as a double donor or double acceptor, as well as trimers with deficit of hydrogen bonds, tend to experience repulsive nonadditive interactions. One can see that for all trimers considered, the SCF nonadditivities are very close to the ones computed at the CCSD(T) level. Fig. 33.2 also shows the three-body energies obtained using the DFT with the PBE exchange-correlation potential, as employed in a recent Car–Parrinello study of liquid water [104]. While in the case of trimers with attractive three-body interactions the agreement between the PBE and CCSD(T) results is quite good, the former method seems to significantly overestimate the repulsive nonadditive effects. While our CCSD(T) results are not completely saturated with respect to the basis set, it is rather unlikely that an increase in basis quality would produce CCSD(T) energies systematically higher than the SCF ones. This suggests that the three-body energies computed at the SCF level are more accurate than the PBE results. Since the distortion of the PBE three-body potential observed in Fig. 33.2 favors tetrahedral arrangement of molecules, in a simulation it may lead to an overstructured liquid. Such an effect has indeed been observed in Car–Parrinello simulations of water at ambient conditions [105].

As mentioned in Section 33.2, the many-body expansion cannot be expected to work for metals. One reason is that most atoms forming metals have open-shell ground states of symmetry other than *S*, therefore it is difficult to determine quantum states of the subsystems needed in the definition of the expansion, *cf.* Section 33.10. The second reason is that the complete delocalization of the conduction electrons results in the electronic structure of a metal that is very far from that of monomers. The first problem does not occur for alkaline-earth metals or for high-spin alkali-metal clusters, and the many-body expansion can be defined for such clusters. However, this expansion appears to be very slowly convergent [106–108]. For some specific information about the spin-polarized sodium trimer, see Section 33.10.2.

33.8 PHYSICAL INTERPRETATION OF NONADDITIVE COMPONENTS

As already stated, perturbation theory approach allows the individual components of a *K*-body contribution to the interaction energy to be given a physical interpretation. Such a physical interpretation is important for several reasons. First, it allows one to better understand the mechanisms of nonadditive forces. Second, it relates components to monomer properties. Third, it makes possible to propose analytic forms of fitting functions which are the most appropriate for a given type of interactions.

Fig. 33.3 illustrates the physical interpretation of the three-body components important for the water trimer. Let us start from the first-order nonadditive exchange energy. The first term in the expansion of this energy into powers of *S* is the quadratic term, resulting from single exchanges of a pair of electrons and therefore denoted in the literature as the 'SE' term [3]. Such exchanges can involve only two water molecules at a time, and lead to deformations of the wave function of each dimer within the trimer. This effect can be described in the first approximation as leading to

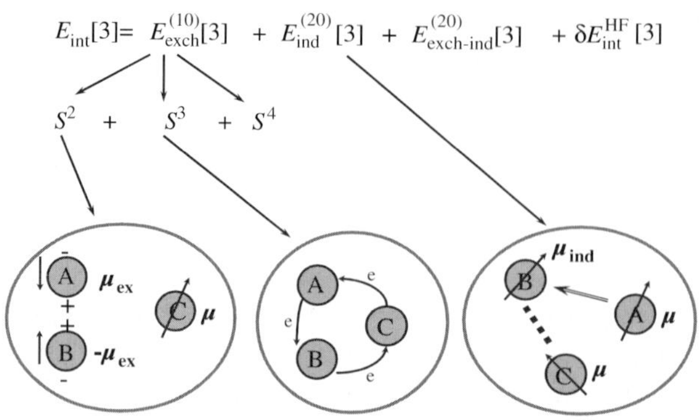

Fig. 33.3. Physical interpretation of major components of three-body interaction energy for water.

the creation of a quadrupole and higher multipole moments on a given dimer, as shown in Fig. 33.3. These moments interact with multipole moments of a third monomer, giving a nonadditive contribution. This contribution has an interesting property that it decays exponentially in one of the intermonomer separations and as an inverse power of the two other separations (due to the interaction of the exchange multipoles with the permanent multipoles), a rather atypical feature for an exchange component.

The next term in the expansion of $E^{(1)}_{exch}[3]$, proportional to S^3, results from a pure exchange interaction: a triple of electrons is cyclically permuted among the monomers A, B, and C, see Fig. 33.3. This process is realized if, e.g. two electrons are exchanged between monomers A and B and then the electron originally from monomer A is exchanged with an electron from monomer C. In the terminology used by Chalasinski, Szczesniak, and co-workers [3,109,110], the S^3 term is referred to as the 'triple exchange' (TE) term.

The most important component of the water trimer nonadditive energy is the induction interaction of the second order in V. Its simple mechanism is shown in Fig. 33.3: a permanent multipole moment on monomer A induces multipole moments on monomer B which in turn interact with the permanent multipole moments of monomer C. Higher orders involve interactions between induced moments. The nonadditive induction energy is in general the most important nonadditive component for hydrogen-bonded systems. As already mentioned, it is the only term used—and only in the asymptotic approximation, i.e. neglecting charge-overlap effects—in the polarizable empirical potentials.

The interpretation of the third-order nonadditive induction contribution is somewhat more involved and will be discussed in the following subsection.

The dispersion energy is a pure quantum effect, more difficult to interpret than those terms for which one can apply classical concepts. In general terms, the dispersion energy is due to the correlation of the electronic motions in different monomers. Using the time-dependent point of view, one can loosely say that instantaneous multipole moments in monomer X—created by fluctuations of electronic density—induce multipole moments in monomer Y, which interact with the original moments of monomer X. The moments in X and Y are correlated by the instantaneous induction process. Always favorable alignment of the instantaneous and induced moments leads to a lowering of energy for second-order pair interactions. The instantaneous induction is in the first approximation a linear response process and it is closely related to the dynamic polarizabilities of monomers. In fact, both the dispersion energy and the dynamic polarizabilities can be expressed *via* the frequency-dependent density susceptibility functions which describe the linear response of the density of an atom or molecule to a time-dependent external perturbation [111]. The leading nonadditive component of the dispersion interaction, a quantity of the third order in the intermolecular potential, is identified as consisting of terms containing correlations between electrons of all three monomers [85]. This effect can be visualized as resulting from interactions of the dispersion-deformed wave functions of the subsystem AB with those of subsystem AC *via* the potential V_{BC} (*cf.* Eq. (C17) in Ref. [85]).

33.8.1 Third-order induction energy

The third-order nonadditive induction energy is given by Eqs. (C14) and (C18) in Ref. [85]. These equations lead to a simple physical interpretation of the particular components of this energy which, however, seems to disagree with the interpretation proposed by Piecuch [112,113]. We shall show below that both interpretations are possible, i.e. the same terms may be viewed as resulting from several different mechanisms. This correspondence is in general related to two ways of expressing the third-order energy corrections *via* the second-order perturbed functions or exclusively *via* the first-order perturbed functions.

To see the equivalence, one has to associate asymptotic expansions with the expressions given by Eqs. (C14) and (C18) in Ref. [85]. To this end, let us write the asymptotic expansion of the interaction potential in the form [4,25]

$$V_{\mathrm{AB}} = \sum_{l_1,l_2} \sum_{k_1,k_2} \hat{Q}^{\mathrm{A}}_{l_1k_1} \hat{Q}^{\mathrm{B}}_{l_2k_2} T^{\mathrm{AB}}_{l_1k_1,l_2k_2} \qquad (21)$$

where $\hat{Q}^{\mathrm{X}}_{lk}$ is the spherical multipole moment operator of monomer X in monomer-fixed coordinate system and T denotes the interaction function dependent on the vector $\mathbf{R}$ describing the relative position of the monomers and on two sets of Euler angles describing the orientation of monomers. Precise definitions of both quantities can be found in the monograph by Stone [25]. In what follows, the sums of this type will also be written with all the subscripts suppressed. Eq. (19) for the first-order induction function can be written in the following alternative form

$$\Psi^{(1)}_{\mathrm{ind,A}\leftarrow\mathrm{B}} = -R^{\mathrm{A}}_0 \Omega^{\mathrm{A}}_{\mathrm{B}} \Phi^{\mathrm{A}}_0 \qquad (22)$$

where $R_0^{\rm A}$ is the reduced resolvent of $H_{\rm A} - E_{\rm A}$. The expansion of $\Omega_{\rm B}^{\rm A}$ in the asymptotic series gives

$$\Psi^{(1)}_{\rm ind,A\leftarrow B} \approx -R_0^{\rm A}\hat{Q}^{\rm A}_{l_1k_1}\Phi_0^{\rm A}Q^{\rm B}_{l_2k_2}T^{\rm AB}_{l_1k_1,l_2k_2} \tag{23}$$

where $Q^{\rm X}_{lk} = \langle\Phi_0^{\rm X}|\hat{Q}^{\rm X}_{lk}|\Phi_0^{\rm X}\rangle$ denote the permanent multipole moments of monomer X and the summation over repeated indices is implied.

To analyze the nonadditive induction energy in more detail, one has to introduce a notation where orders in each of the operators constituting V: $V_{\rm AB}$, $V_{\rm BC}$, and $V_{\rm AC}$, are explicitly shown, rather than the overall order in V as it was done in Eq. (9). Thus, each correction can now be written as $E^{(i_1,i_2,i_3;j)}$ with the consecutive i_k referring to the operators $V_{\rm AB}$, $V_{\rm BC}$, and $V_{\rm AC}$, respectively. The overall order in V is $i = i_1 + i_2 + i_3$. With these definitions, let us analyze the expression for $E^{(111;0)}_{\rm ind}$ given by Eq. (C18) in Ref. [85]. Since we neglect here the expansion in W, we will drop the last superscript in the spirit of the definition of Eq. (10). All the terms in the equation for $E^{(111)}_{\rm ind}$ are equivalent and can be represented by

$$I_1 = \langle\Phi_0^{\rm A}\Psi^{(1)}_{\rm ind,B\leftarrow C}|V_{\rm AB}|\Psi^{(1)}_{\rm ind,A\leftarrow C}\Phi_0^{\rm B}\rangle \tag{24}$$

Expansion of the operators and of the induction functions gives

$$\begin{aligned} I_1 \approx\ & \langle\Phi_0^{\rm A}|\hat{Q}^{\rm A}_{l_1'k_1'}R_0^{\rm A}\hat{Q}^{\rm A}_{l_1k_1}\Phi_0^{\rm A}\rangle Q^{\rm C}_{l_3k_3}T^{\rm AC}_{l_1k_1,l_3k_3} \\ & \times\langle\Phi_0^{\rm B}|\hat{Q}^{\rm B}_{l_2'k_2'}R_0^{\rm B}\hat{Q}^{\rm B}_{l_2k_2}\Phi_0^{\rm B}\rangle^* Q^{\rm C}_{l_3'k_3'}T^{\rm BC}_{l_2k_2,l_3'k_3'}T^{\rm AB}_{l_1'k_1',l_2'k_2'} \end{aligned} \tag{25}$$

Since $\langle\Phi_0^{\rm A}|\hat{Q}^{\rm A}_{l_1'k_1'}R_0^{\rm A}\hat{Q}^{\rm A}_{l_1k_1}\Phi_0^{\rm A}\rangle = \alpha^{\rm A}_{l_1'k_1',l_1k_1}$ is the polarizability of A, we can interpret the first line of this equation as the multipole moments induced on A by the field due to the permanent multipole moments of C. With an analogous interpretation of the second line, we see that I_1 represents mutual interaction of the multipole moments on A and on B induced by the electric field of C. This mechanism is illustrated by diagram (f) in Fig. 33.4.

Eq. (25) can, however, be also interpreted in an alternative way. Notice that the polarizabilities are pure monomer properties. Let us interpret the first line of Eq. (25) as above, but then multiply the induced moments on A by $T^{\rm AB}_{l_1'k_1',l_2'k_2'}$. This product represents the field on B due to the presence of the induced moments on A. This field coupled with the polarizability $\alpha^{\rm B}_{l_2'k_2',l_2k_2}$ creates a set of induced multipole moments on B. These moments now interact with the permanent moments on C. This interpretation is the same as the one given by Piecuch [112] for the same component (${}^{\rm II}E^{(3){\rm ind}}_{Q,\alpha\alpha Q}$ in Piecuch's notation). This alternative interpretation of I_1 is shown as diagram (g) in Fig. 33.4 (except that the roles of A and B are interchanged). Notice that this interpretation appears more naturally if the third-order induction energy is expressed *via* the second-order induction function (*cf.* Eqs. (115)–(117) in Ref. [85]).

Consider now $E^{(210)}_{\rm ind}$, Eq. (C14) in Ref. [85]. The first term is (the fourth integral is also of the same form)

$$J_1 \approx \langle\Psi^{(1)}_{\rm ind,A\leftarrow B}\Phi_0^{\rm B}|V_{\rm AB}|\Phi_0^{\rm A}\Psi^{(1)}_{\rm ind,B\leftarrow C}\rangle = (\alpha^{\rm A*}Q^{\rm B}T^{\rm AB})(\alpha^{\rm B}Q^{\rm C}T^{\rm BC})T^{\rm AB} \tag{26}$$

in a short-hand notation. The terms in parentheses are multipole moments on A and B induced by fields due to permanent moments of B and C, respectively. These induced

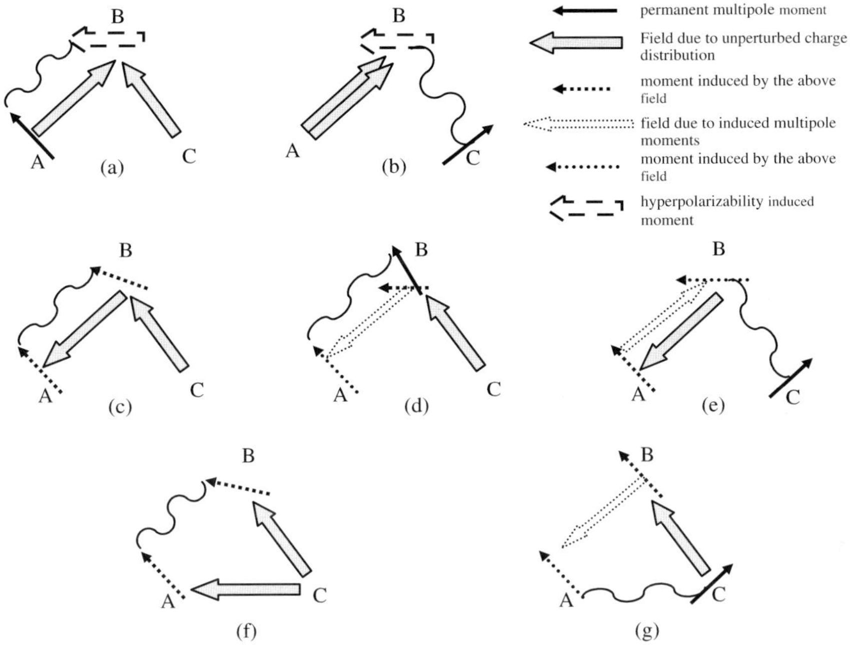

Fig. 33.4. Physical interpretation of third-order nonadditive induction energy.

moments interact with each other, as it is illustrated in diagram (c) of Fig. 33.4. This term can also be interpreted in two other ways. The induced moments on B can be coupled with the polarizability of A to induce a moment on A. These moments then interact with permanent moments on B, as shown in Fig. 33.4(d). One more possible interpretation is that of Fig. 33.4(e). The two alternative interpretations were proposed by Piecuch [112] who denoted those terms by ${}^{\mathrm{I}}E^{(3)\mathrm{ind}}_{Q,\alpha\alpha Q}$.

The second term in Eq. (C14) of Ref. [85] is

$$J_2 = \langle \Phi_0^{\mathrm{A}} \Psi^{(1)}_{\mathrm{ind,B\leftarrow A}} | V_{\mathrm{AB}} | \Phi_0^{\mathrm{A}} \Psi^{(1)}_{\mathrm{ind,B\leftarrow C}} \rangle$$

$$\approx Q^{\mathrm{A}}_{l_1'k_1'} T^{\mathrm{AB}}_{l_1'k_1',l_2'k_2'} Q^{\mathrm{A}}_{l_1k_1} T^{\mathrm{AB}}_{l_1k_1,l_2k_2} \langle \Phi_0^{\mathrm{B}} | \hat{Q}^{\mathrm{B}}_{l_2k_2} R_0^{\mathrm{B}} \hat{Q}^{\mathrm{B}}_{l_2'k_2'} R_0^{\mathrm{B}} \hat{Q}^{\mathrm{B}}_{l_2''k_2''} | \Phi_0^{\mathrm{B}} \rangle Q^{\mathrm{C}}_{l_3k_3} T^{\mathrm{BC}}_{l_2''k_2'',l_3k_3} \quad (27)$$

The integral displayed above is the hyperpolarizability $\beta_{l_2k_2,l_2'k_2',l_2''k_2''}$. After multiplying by the terms representing fields due to the permanent moments on A and C, it gives the hyperpolarizability induced multipole moments on B. These moments then interact with the permanent moments of A as shown in Fig. 33.4(a). Alternatively, if the hyperpolarizability is multiplied by the square of the field on A, the resulting induced moments interact with permanent moments on C (see Fig. 33.4(b)). The latter interpretation is more natural for the third integral in Eq. (C14) of Ref. [85]. Piecuch interpreted these terms in the same way and denoted them as $E^{(3)\mathrm{ind}}_{Q,\beta QQ}$. The interpretation presented in this subsection can also be carried out at the level of the self-consistent field

(Hartree–Fock) theory provided that all multipole moments and polarizabilities are replaced by their (coupled) Hartree–Fock counterparts [114].

33.9 CASE STUDIES OF NONADDITIVE EFFECTS IN CLUSTERS

In this section, a few systems for which some knowledge about nonadditive effects is available will be discussed. The water clusters will be studied separately in subsequent sections. The number of systems considered here is actually a large fraction of all systems for which a sufficiently reliable information about nonadditive effects is available. The most studied systems are Rg_3 and Rg_2–HX, where X is a halogen atom. As mentioned, the latter systems are not necessarily relevant from the point of view of modeling the condensed phase, but are well suited for spectroscopic investigations providing stringent limits on the theoretical results.

33.9.1 Helium trimer

The only bound state of the helium dimer is known to be very diffuse, the average R is size of about 50 Å [115,116]. The ground state of the trimer is much more compact (about 6.5 Å [117]), however, the so-called Efimov excited states [118] are over 100 Å in size [119]. The latter states are very sensitive to the depth of the two-body potential and changes of this depth of the order of 1% may decide about their number [119]. Although the nonadditive energy in helium trimer is of comparable magnitude, numerous investigations of Efimov states (see Refs. [119,120]) usually neglected the nonadditive potential.

The unique properties of helium at low temperatures result in a continuous interest in this system. The three-body nonadditive interactions are small for helium (*cf.* Fig. 33.1), but become important for condensed helium at high pressures. Simulations of such conditions performed in Ref. [121] used the Bruch–McGee three-body nonadditive potential [122]. Later several other nonadditive potentials became available for He_3 [123–125]. Recently, Moroni et al. [126] (see also Ref. [127]) applied the SAPT two-body potential [115,116] and the three-body potential of Cohen and Murrell [125] to predict properties of solid and liquid helium. Excellent agreement with experiment in most regions has been obtained in the latter work. In fact, the only discrepancy with experiment has been attributed by Moroni et al. [126] to an improper choice of the reference energy in the experimental equation of state. However, as shown in Ref. [87], the nonadditive potential used by these authors is a poor approximation to the true potential. The use of inappropriate three-body potentials may also be responsible for the deterioration of the agreement with experiment upon inclusion of three-body effects observed by Chang and Boninsegni [128].

The many-body SAPT approach was used in Ref. [87] to compute the nonadditive interactions in the helium trimer. The computed points were fitted by a global potential function. The SAPT potential agrees well with accurate literature calculations at selected points on the surface [129,130] but disagrees significantly with a previous helium trimer potential [125].

33.9.2 Argon trimer and condensed phase

Argon properties have probably been investigated more than those of any other rare gas. This is partly due to the fact that argon is the largest rare-gas atom for which accurate calculations are still possible and at the same time the nonadditive effects are larger for argon than for lighter atoms (*cf.* Fig. 33.1). At the present, the helium and argon three-body potentials are known better than those for other trimers. The most accurate argon trimer potential to date was developed in Ref. [32] using the three-body SAPT approach. The nonadditive part of the interaction energy was computed at more than 30 configurations. An analytic potential was then fitted to the computed values. This was the first *ab initio* potential for the argon trimer (in fact, for any trimer) that would include nonadditive components (all very important for argon) such as the exchange–dispersion energy, third-order dispersion energy with overlap, and several others. Therefore, this potential allowed an insight into physical origins of the three-body forces in trimers of heavier rare gas atoms. As already mentioned, this issue has been the subject of heated literature discussions in the past (for reviews, see Refs. [19–21]). In particular, the work of Ref. [32] confirms that the first-order exchange nonadditive energy [131–134] is large and cancels a significant part of the ATM contribution to the dispersion energy, as it was pointed out by Meath and Aziz [19] in 1984. This cancellation takes place for most configurations near the minimum of the trimer, although eventually for large trimers the ATM term will be larger in magnitude due to its slower decay. However, several other terms, not considered before, are similar in importance to these two terms (in particular, at the minimum, $E_{\rm int}^{\rm HF}[3]$ and $E_{\rm exch-disp}^{(20)}[3]$ are of the same order of magnitude as $E_{\rm disp}^{(3)}[3]$). Ref. [32] has shown that due to a startling and quite complete cancellation among several of these terms, the total nonadditive interaction energy is fairly well approximated by the third-order dispersion energy at the minimum and for larger separations. This finding rationalizes the success of the ATM term in predicting most of the properties of the condensed phases of argon, although it points out to its somewhat fortuitous character. Of course, the fact that asymptotically the ATM term becomes dominant and that large trimers do make a significant contribution to the cohesion energy explains the part of the success which is not fortuitous. At the intermolecular separations in Ar_3 smaller than the minimum, the cancellations do not take place and the nonadditivity of the Hartree–Fock interaction energy, $E_{\rm int}^{\rm HF}[3]$, dominates the total nonadditive potential. This component is in turn dominated by the first-order exchange nonadditive energy $E_{\rm exch}^{(10)}[3]$.

The nonadditive SAPT components have been compared in Ref. [32] order by order with results from supermolecular MBPT4 and CCSD(T) calculations. The agreement between individual MBPT*n* nonadditive contributions and the sum of the corresponding SAPT terms varied depending on the order of perturbation theory and on the configuration of the trimer. However, the total nonadditive interaction energies computed by the two methods agreed very well. Typically, the differences between the two approaches amounted to only a few percent (except for obvious and unavoidable larger differences near the points where the value of the total nonadditive energy crosses zero). It has been estimated in Ref. [32] that the total error of the nonadditive potential is of the order of 10%.

Argon trimer has an interesting property that some spectral transitions that would have been forbidden if only two-body interactions were present are allowed due to three-body forces, as pointed out by Suhm and Nesbitt [26]. Investigations performed by Cooper et al. [135], who calculated the bound rovibrational state energies of the argon trimer, suggest that information about nonadditive three-body effects should be extractable from the ground state rotational constants or from Raman spectra of the argon trimer if these quantities were measured. However, this has not been done so far. The rovibrational states of Ar_3 are also difficult to compute, despite the small dimensionality of the problem [136].

The availability of the new three-body nonadditive potential made it possible to investigate the structure of the argon crystal at a much increased level of accuracy compared to that possible earlier and resolve an outstanding problem, often called the rare-gas crystal structure 'paradox'. The cohesion energy of the argon crystal computed [14] using the nonadditive potential from Ref. [32] and the two-body potential of Aziz [137] was in very good agreement, within 0.3%, with experiment [138]. The estimated 10% error in the trimer nonadditive potential translates into 0.7% error in the cohesion energy. The inaccuracies in the two-body potential and the neglected four- and higher body effects are expected to give uncertainties of a similar size. Calculations were performed both for the fcc and hcp lattices, allowing a resolution of the paradox mentioned above, consisting in an inability of theory to agree with experiment on the lattice structure of argon. All reasonable two-body potentials applied over the years predicted the hcp structure to be more stable. The addition of the ATM term did make the energy difference between the hcp and fcc structures smaller, but the former lattice remained lower in energy. The study of Ref. [14] has shown that one has to include the nonadditive exchange components of the first and second order to make the fcc crystal more stable. Also, the complete dispersion energy (not utilizing the multipole expansion) was used, rather than the ATM term alone. In this way, the correct crystal structure has been predicted, the goal which eluded theory for 50 years. The role of the three-body interactions in the crystal was shown to be significantly amplified compared to the argon trimer: the addition of the three-body effects changes the interaction energy in the trimer by only 1%, whereas the analogous change of the crystal cohesion energy is as large as 7%. This amplification results from an interesting balance of two competing effects. On one hand, there should be an increase of the relative role of three-body forces due to the fact that there are many more triples than pairs in a crystal. On the other hand, for larger trimers, the relative importance of nonadditive forces gets smaller since these forces decay faster than the two-body terms. Thus, one may expect the amplification factor to be strongly system dependent.

The knowledge of the three-body potential allows calculations of the third virial coefficient for a substance, an experimentally accessible property that depends *only* on two- and three-body potentials. The classical third virial coefficient of argon has been computed in Ref. [139] for temperatures between 113.15 and 1223.15 K using the *ab initio* three-body potential of Ref. [32]. The results agree with the virial coefficients extracted directly from experimental data to within about 5% for temperatures between 220 and 420 K, theory almost touching the experimental error bars. For lower temperatures the discrepancies become larger, but results from different experiments are also significantly apart. The calculated values agreed very well with those of Dymond and Alder [140] determined by fitting experimental data with theoretical constraints. However, the disagreement for low T with the data of Gilgen et al. [141] was somewhat worrisome. Later it turned out that the experimental values in Ref. [141] have been incorrect (too large for low T) due to the evaluation procedure applied to extract them from the ρPT data. The new, corrected coefficients [142] are in much better agreement with the theoretical values extending the range of agreement down to 170 K. The equation of state obtained using the computed second and third virial coefficients reproduced experimental PV data over a wide range of pressure.

Liquid argon is a benchmark system in thermal physics. The ability of theory to predict thermophysical properties of this system from first principles may be important for calibration of experimental techniques. MC simulations of liquid, gas, and liquid–vapor equilibria of argon have been performed in Ref. [15]. The complete *ab initio* three-body nonadditive potential of Ref. [32]—including the short-range contributions—has been used for the first time in such type of simulations. The two-body potential used was the empirical potential of Aziz [137], so that the simulations were not completely first principle. However, Aziz's potential could now be substituted by an *ab initio* potential from Ref. [143], so close to the former potential that the results of the simulation should not change in any appreciable way. The simulations of Ref. [15] have achieved very good overall agreement with experiment for phase equilibrium parameters and pV data of pure phases. Importance of various three-body interaction components has been assessed and their relationship to the liquid structure has been investigated. Although the short-range contributions to the nonadditive potential are quite large, cancellations occurring between these and other components at the interatomic distances typical for the liquid make the total three-body nonadditive effect very similar to that given by the ATM potential. Performance of the three-term virial equation of state [139] was assessed based on the results of simulations.

Some specific properties where the nonadditive effects could possibly play a much larger role than usual are defects [144] and surface phenomena in rare gas crystals. In the case of defects, Losee and Simmons [145] pointed out already in 1967 that vacancies may provide important information about many-body contributions since the number of pairs, triples, etc. of atoms differ between the perfect lattice and the lattice containing vacancies. However, it is difficult to obtain precise information about these effects since estimates of the entropic terms are uncertain. Guarini et al. [146] have studied liquid krypton by neutron scattering. These authors have found that in order to explain experimental findings in some regions of scattering parameters, apparently related to surface scattering, one had to assume that the many-body interactions contribute as much as 40% to the scattering. This estimate, if confirmed, would be an extreme case of the importance of the many-body effects in the condensed phase of rare gases. As mentioned above, for the argon crystal the nonadditive effects are responsible for about 7% of the cohesion energy. For the krypton crystal, the contribution of nonadditive effects can be estimated by subtracting from the experimental cohesion energy of 11,192 ± 17 J/mol

(Ref. [138]) the theoretical energy computed using a pair potential only, equal to 12,186 J/mol (Ref. [147]). The obtained value of 8.8%, although slightly larger than for argon, is very much smaller than the value inferred in Ref. [146].

33.9.3 Ar–Ar–HF trimer

The Ar_2HF trimer has been the subject of a number of experimental investigations [64] starting from 1987. Recently, the spectra of this trimer have been measured in superfluid helium nanodroplets [148]. Ernesti and Hutson [29,30] constructed a model nonadditive potential for Ar_2HF using asymptotic information as well as results of Hartree–Fock calculations. This potential was applied, together with the best available two-body empirical potentials, to compute rovibrational levels and rotational constants of the complex. By calculating spectra using only the pair potentials, Ernesti and Hutson showed that the differences between the computed and the experimental spectra are too large to be attributed to inaccuracies of the additive potentials and therefore must be due to three-body nonadditive effects. The authors then examined the effects of various nonadditive contributions and concluded that the S^2 exchange term, modeled by the exchange quadrupole [149], was the dominant nonadditive component. Notice that Ernesti and Hutson did not fit their nonadditive model to the spectra. They have constructed the asymptotic terms from monomer properties and fitted the exchange terms to the Hartee–Fock calculations. Although for the majority of data known from experiments the inclusion of the nonadditive potential improved agreement with experiment, for some quantities the improvement was small.

The dynamical calculations performed by Ernesti and Hutson were restricted to the rotational quantum number $J = 0$. Recently, Kozin et al. [150] developed a method enabling predictions for higher J levels. These authors used the potential from Ref. [30]. The rotational constants could now be computed from differences between the $J = 1$ and $J = 0$ levels, leading to an improved agreement with experiment.

The work of Ernesti and Hutson was followed by *ab initio* calculations using the nonadditive SAPT method [33,151]. These calculations, performed at a large number of configurations, have shown that the nonadditive effects in this system are dominated for most geometries by the Hartree–Fock contribution. All the first- and second-order components, as well as the leading higher order one are depicted in Fig. 33.1 for the equilibrium configuration of the trimer. For this configuration, $E_{\text{int}}^{\text{HF}}[3]$ constitutes 86% of the total nonadditive energy. The third-order nonadditive dispersion effect is for most configurations a factor of three or so smaller in magnitude. The other types of interactions play a still smaller role. However, to obtain a quantitatively correct potential, it is essential to include the exchange–dispersion and the (quenched) induction–dispersion nonadditive effects. The Hartree–Fock nonadditive contribution was shown to be very well approximated by the sum of the first-order nonadditive exchange term and the third-order induction and exchange–induction terms. The second-order nonadditive induction component vanishes exponentially for Ar_2HF and is nearly canceled by its exchange counterpart (see Fig. 33.1). In contrast, the third-order induction and exchange–induction terms (not shown in Fig. 33.1) are both positive and together make an important contribution. The third-order induction term consists mainly of the component representing the electrostatic interaction of the dipoles and higher multipoles induced on the argon atoms by the electrostatic field of the HF molecule (*cf.* Fig. 33.4(f)). Both the exchange and induction components are highly anisotropic. The overall contribution of the nonadditive component to the total interaction energy of the trimer is largest at the equilibrium configuration (HF on the C_2 axis of Ar_2 with H pointing towards argons) where this contribution constitutes 3.5% of the additive part. For this geometry and for most of other configurations that were studied, the nonadditive energy is positive, i.e. it destabilizes the complex.

The total nonadditive energies from the work of Ernesti and Hutson [29,30] and Lotrich et al. [33] agree reasonably well. This agreement is due to the dominant role of the Hartree–Fock nonadditive energy and to the fact that the next largest contribution, from the dispersion energy, is well approximated by the asymptotic formula used by Ernesti and Hutson. However, the individual components of the Hartree–Fock nonadditive energy obtained by Ernesti and Hutson are significantly different from those predicted by SAPT. In particular, Ernesti and Hutson completely neglected the important exchange terms proportional to S^3.

The three-body potential developed by Ernesti and Hutson [30] for Ar_2HF was successfully used by Hutson et al. [152] in calculations for larger Ar_nHF clusters. In another simulation of such clusters, the three-body dispersion nonadditivity was shown to directly affect the vibrational red-shifts for large n [153].

33.9.4 $(H_2O)_2$ HCl trimer

The weakly bound $(H_2O)_2$ HCl trimer, whose properties were predicted first by electronic structure calculations [154,155], was observed by Kisiel et al. [156] using high-resolution rotational spectroscopy. Subsequently, Milet et al. [157,158] employed SAPT to compute a number of points on the trimer potential energy surface and to interpret the nature of three-body interactions in this cluster. They found that the three-body energy amounts to about 13–20% of the pairwise contribution. The nonadditive induction energy plays a dominant role, but it is substantially quenched by the first- and higher-order exchange nonadditive terms. The convergence of SAPT for this system was tested by a comparison with the results of supermolecular CCSD(T) calculations. It was found that the deviation of the SAPT interaction energies from the CCSD(T) values was only of the order of 3%. The structural predictions (i.e. the minimum geometry) obtained from SAPT calculations agreed well with the microwave measurements [156]. To aid interpretation of future experiments on $(H_2O)_2HCl$, Struniewicz et al. [159] developed a two-dimensional SAPT potential energy surface describing flipping motions in the trimer and predicted the trimer's vibration-rotation-tunneling (VRT) levels and the line strengths of the far-infrared transitions. In a similar investigation based on a CCSD(T) surface, Wormer et al. [160] identified the effect of three-body forces on the transition frequencies and tunneling splittings and found it to be important. It should be noted that, although the four-body effects in the $(H_2O)_3HCl$ cluster were found to be small [161], the many-body expansion for larger clusters of this type has a limited utility because of the protolytic dissociation of HCl solvated by more than three water molecules [155,162].

33.10 THREE-BODY EFFECTS IN OPEN-SHELL CLUSTERS

Nonadditive effects in open-shell clusters have been investigated only recently and relatively little information is available on their importance and physical origin. From the theoretical point of view, open-shell systems are more difficult to study since the conventional, size-consistent computational tools of the theory of intermolecular forces, like the Møller–Plesset perturbation theory, coupled cluster theory, or SAPT, are less suitable or less developed for applications to open-shell systems than to closed-shell ones. Moreover, there are many types of qualitatively different open-shell states, exhibiting different behavior and requiring different theoretical treatments.

33.10.1 $Ar_2NO^-(^3\Sigma^-)$ trimer

Open-shell clusters of a relatively simple electronic structure arise when an open-shell molecule without orbital degeneracy is solvated by closed-shell atoms. An example of a system of this class is the $Ar_2NO^-(^3\Sigma^-)$ trimer studied recently by Jakowski et al. [163]. The knowledge of the two- and three-body components of the interaction energy in this system is relevant for an understanding of the photoelectron studies of the solvation of NO^- by argon atoms [164]. Using the unrestricted Møller–Plesset perturbation theory through the fourth order, Jakowski et al. [163] have found that the influence of the three-body nonadditivity on the structure of the Ar_2NO^- trimer is negligible. The nonadditive effect destabilizes, however, the complex reducing its dissociation energy (predicted to be equal to 1260 cm^{-1}) by about 2%. The three-body interactions play larger role in the tetramer Ar_3NO^-, where they influence the energetic ordering of two lowest energy structures and contribute about 7% to the binding energy determined in Ref. [163] by the rigid body diffusion Monte Carlo (RBDMC) method [165,166]. It is expected [163] that for still larger clusters the three-body potential will play an even more important (destabilizing) role, but it was not included thus far in the RBDMC simulations for such clusters [163].

33.10.2 High-spin sodium trimer

The nonadditive effects are much more important if more than one monomer is in an open-shell state. For example, it has been found both experimentally [107] and theoretically [108,167] that the three-body force plays a critical role in stabilizing the sodium trimer in the spin-polarized $^4A_2'$ state. Similar, although gradually less and less pronounced stabilization effect has been predicted [108] for heavier high-spin alkali-metal trimers, whereas for the lithium trimer the three-body effect has been found to be really dramatic—the complete potential is at least four times more attractive than the pairwise additive one [108]. The very strong three-body effect in the lithium trimer, amounting to about 15 kcal/mol, should probably be interpreted as being of a chemical rather than of the van der Waals nature, and is not expected that it can be well understood in terms of the SAPT interaction energy components discussed in Sections 33.6 and 33.8.

A detailed physical analysis of the three-body interaction in the sodium trimer was recently carried out by Jakowski et al. [168]. This analysis shows that in the van der Waals well region the three-body effect is dominated by the first-order exchange term $E^{(1)}_{\rm exch}[3]$, i.e. it originates from the Pauli exclusion principle imposed on the unperturbed monomers. This means that a serious net reduction of the exchange repulsion is caused by the very presence of the third monomer, without any contribution from the polarization or exchange–polarization effect. At the minimum distance of 4.8 Å, about 94% of the total nonadditivity is due to this effect. Only at larger distances, the repulsive ATM (triple-dipole) nonadditivity, correctly accounted for at the third order of the Møller–Plesset theory [1], dominates the total three-body effect. The three-body induction effects are of secondary importance for all distances. Among the components of $E^{(1)}_{\rm exch}[3]$ discussed in Section 33.8, the TE S^3 terms dominate. However, the S^2 SE terms are always significant and cannot be neglected. The angular dependence of the S^2 and S^3 terms is very different and the specific balance of these terms determines the sign of $E^{(1)}_{\rm exch}[3]$ for different configuration of the trimer. It may be noted here that also for the simplest high-spin trimer—the quartet state of H_3—the first-order exchange energy is attractive for the equilateral configurations and dominates the total three-body effect [168,169] (except for the large distances), the S^3 contribution being larger than the S^2 terms. The overall importance of the nonadditive component is, however, much smaller for H_3 than for high-spin metallic clusters.

33.10.3 Ar_2O^- ionic trimer—the case of orbital degeneracy

When the solvated chromophore exhibits an orbital degeneracy, the quantum states of the two-body subsystems of the trimer are not clearly defined and a fundamental difficulty arises when one tries to perform the many-body expansion of the interaction energy. This problem appears, for example, in the case the $Ar_2OH(X^2\Pi)$ trimer [170], high-spin Na_2B trimer (with the boron atom in the ^2P state) [168] or the Ar_2O^- trimer (with the oxygen ion in the ^2P state). The latter system, treated in considerable detail in Ref. [171], is important for the modeling of the structure and energetics of Ar_nO^- clusters studied by Bowen and co-workers [172,173] using the photoelectron spectroscopy. The linear structure of the trimer, postulated by Arnold et al. [172], was confirmed by Roszak et al. [174] in *ab initio* calculations, but no two- and three-body potentials were extracted in that work.

The trimer Ar_2O^- includes as a subunit an open-shell dimer ArO^-, which in isolation can be in either a Σ or a doubly degenerate Π state. These states differ by the orientation (specified by the angle θ) of the singly occupied oxygen 2p orbital, which is parallel to the dimer axis ($\theta = 0$) for the Σ state and perpendicular ($\theta = \pi/2$) for the Π state. For a trimer of an arbitrary geometry, the p orbital is usually oriented at an angle different than zero or $\pi/2$. One can easily see that for the isosceles triangle configuration (the C_{2v} point group) characterized by the angle γ between the vectors pointing from the O^- ion to the Ar atoms, the angle θ equals $\gamma/2$ for the A_1 state, $\pi/2 - \gamma/2$ for the B_2 state, and $\pi/2$ for the B_1 state. In the first two cases, the 2p orbital is in the trimer plane and in the third case it is perpendicular to this plane. Thus, only for the B_1 state, the two-body part of the trimer energy can be constructed from the potential $V_\Pi(R)$ of the Π state of ArO^-. Since $V_\Pi(R)$ and the potential $V_\Sigma(R)$ for the Σ state of ArO^- are very different, it matters which of these pair potentials is used as the two-body part of trimer's A_1 and B_2 potential energy surfaces. One could think of using the arithmetic average $V_0(R)$ of the $V_\Pi(R)$ and $V_\Sigma(R)$ potentials but, as shown by Jakowski et al. [168,171], a much smaller three-body effect is obtained if a suitable θ-dependent two-body potential is used. Such a potential, dependent on the orientation of the singly occupied 2p orbital, can be obtained by the simplest sensible trigonometric interpolation between the Π and Σ potentials:

$$V(R,\theta) = V_\Pi(R) + [V_\Sigma(R) - V_\Pi(R)]\cos^2\theta \tag{28}$$

For the isosceles triangle, the pairwise contribution to the potential originating from a single $O^- \cdots Ar$ interaction becomes then [168,175]

$$V_{A_1}(R,\gamma) = V_\Sigma(R)\cos^2\frac{\gamma}{2} + V_\Pi(R)\sin^2\frac{\gamma}{2} \tag{29}$$

$$V_{B_2}(R,\gamma) = V_\Sigma(R)\sin^2\frac{\gamma}{2} + V_\Pi(R)\cos^2\frac{\gamma}{2} \tag{30}$$

The total pairwise contribution is twice as large as in the two equations above, so the total interaction potentials for the A_1, B_1, and B_2 states can be written in the form:

$$E_{\text{int}}^{A_1}(R,\gamma) = 2V_0(R) + V_{\text{Ar-Ar}}(R\sin\frac{\gamma}{2}) + \Delta V(R)\cos\gamma + E_{\text{3B}}^{A_1}(R,\gamma) \tag{31}$$

$$E_{\text{int}}^{B_1}(R,\gamma) = 2V_0(R) + V_{\text{Ar-Ar}}(R\sin\frac{\gamma}{2}) - \Delta V(R) + E_{\text{3B}}^{B_1}(R,\gamma) \tag{32}$$

$$E_{\text{int}}^{B_2}(R,\gamma) = 2V_0(R) + V_{\text{Ar-Ar}}(R\sin\frac{\gamma}{2}) - \Delta V(R)\cos\gamma + E_{\text{3B}}^{B_2}(R,\gamma) \tag{33}$$

where $V_{\text{Ar-Ar}}(r)$ is the pair potential for argon, $\Delta V(R)$ is the difference potential $\Delta V(R) = V_\Sigma(R) - V_\Pi(R)$, while $E_{\text{3B}}^{A_1}(R,\gamma)$, $E_{\text{3B}}^{B_1}(R,\gamma)$, and $E_{\text{3B}}^{B_2}(R,\gamma)$ are the so-called 'genuine three-body contributions', defined essentially through Eqs. (31)–(33). The terms $\Delta V(R)\cos\gamma$, $-\Delta V(R)$, and $-\Delta V(R)\cos\gamma$, expressed through the conventional Σ and Π pair potentials of ArO^-, but dependent, in general, on the position of the third atom *via* the angle γ, are referred to as the 'orientational nonadditivity' [168,171]. It should be noted that mixing of different diatomic states to define an effective two-body potential has been used earlier by Aquilanti et al. [176], by Danilychev and Apkarian [177], and by Naumkin and Knowles [175]. Jakowski et al. [171] have found that the orientational nonadditivity is usually much larger than the genuine one and that it can be viewed as responsible for the collinear structure of the Ar_2O^- trimer. The small genuine nonadditivity is of similar nature as in the analogous closed-shell complexes like, e.g. Ar_2Cl^-. It is repulsive at the intermediate and long range and favors the collinear structure of the complex. Jakowski et al. [171] have also found that the differences between the genuine three-body terms in the A_1, B_1, and B_2 potentials are of secondary importance and can be neglected in cluster simulations [178]. It should be noted that in their simulations, Jakowski et al. [178] included also the spin–orbit coupling in the two-body potential, which turned out to be important only for small clusters. The three-body contribution to the cluster energy resulting from the inclusion of the spin–orbit interaction has apparently not been considered in the literature so far.

The genuine nonadditivity can be interpreted using the open-shell SAPT approach of Cybulski [179]. It has been found to be dominated by the induction component at large distances, with the exchange contribution increasing its significance at a shorter range [171]. It should be mentioned that for larger clusters the genuine nonadditivity can be more important than for trimers and can reach as much 30% of the total interaction energy in the cluster [178].

For arbitrary triangles (the C_s point group), the angle θ is not uniquely defined, but can be obtained by an analysis of the wave function in a way similar to that used to construct the diabatic states from the adiabatic ones [180,181]. It should also be noted that for arbitrary trimer configurations, the resulting two A' states are coupled nonadiabatically. This coupling can be viewed as a pure three-body effect.

33.11 WATER CLUSTERS AND CONDENSED PHASE

33.11.1 Two-body potentials for water

As already mentioned, the prerequisite for investigations of nonadditive effects is the knowledge of an accurate pair potential for a given system. Therefore, in this section, we will briefly discuss the water dimer potentials before starting an extensive discussion of the water trimer potentials in Section 33.11.2 and simulations of liquid water in Section 33.11.3.

The number of potentials developed for the water dimer is probably larger than for any other system. However, most of these potentials are 'effective' pair potentials fitted in simulations of liquid water or ice such that the results of these simulations match bulk measurements for the investigated systems. Thus, these potentials are of no help in investigations of nonadditive effects. There was a number of *ab initio* potentials for water published, the best known are those by Clementi and coworkers [67,182]. More recently the ASP potentials of Millot et al. [183] have been very popular. However, only in the last few years it has become possible to develop interaction potentials accurate enough for investigations of nonadditive effects.

Several accurate potentials for the water dimer have been developed by the present authors and collaborators using the regular SAPT approach [48,63] and its DFT-based version [81]. The most extensively tested potential, dubbed SAPT-5s for the five symmetry-distinct sites used in the fit, was based on about 2500 *ab initio* grid points—a much larger number of points than in any previous work on this dimer. The calculations employed a medium-size, interaction-optimized spdf-symmetry basis set with bond functions. By comparisons to the most elaborate single-point calculations [184], the SAPT interaction energies have been estimated to be accurate to about 0.3 kcal/mol or 5% with respect to basis set and theory level extensions. Recent SAPT calculations

[81] in very large basis sets generally confirmed this estimate, but have shown that the deviations are not uniform, ranging from less than 0.1 to about 0.5 kcal/mol in different regions of potential energy surface. Nevertheless, the stationary points on the surface and the barriers to proton tunneling are recovered by SAPT-5s quite accurately, as found earlier by van Duijneveldt-van de Rijdt et al. [185]. The SAPT-5s potential allowed to resolve some controversies related to the equilibrium structure of the water dimer. For example, the average value of $1/R^2_{\mathrm{O-O}}$—obtained from the two lowest energy rovibrational water dimer wave functions computed using the SAPT-5s potential—has been combined with the value of $\langle 1/R^2_{\mathrm{O-O}} \rangle$ inferred from experiments [47] to obtain a new empirical estimate of the equilibrium O–O separation equal to 5.50 ± 0.01 bohr [48], significantly shorter than the previously accepted value, but in a very good agreement with high-level *ab initio* calculations of Klopper et al. [184]. The zero-point energy computed from SAPT-5s was subtracted from the depth of the potential from Ref. [184] to give a new prediction of the dissociation energy D_0 equal to $1165 \pm 54\ \mathrm{cm}^{-1}$ [48], close to but significantly more accurate than the best empirical value [186]. The SAPT-5s potential of Ref. [48] and the earlier SAPT-pp potential from Ref. [63] were used to predict the second virial coefficient for water and reproduced the experimental data with discrepancies of the same order of magnitude as estimated errors of measurements. The agreement with experiment was further improved by the recent SAPT(DFT) potential for the water dimer [81].

The most stringent tests of the accuracy of the SAPT potentials [48,63] came from comparisons of the spectra calculated using these potentials with the experimental ones. The majority of measured transitions [49,50,187,188] among the vibrational ground-state manifold were predicted to within about $0.01\ \mathrm{cm}^{-1}$ [31,56] and the agreement with experiment for vibrational excitations was also very good [189]. The latter agreement allowed reassignment of some spectral lines. One of the few exceptions, where SAPT-5s predictions were somewhat less accurate, was the sum of acceptor splittings (a single experimental number). This sum was used together with the SAPT computed interactions energies to tune the SAPT-5s potential. The tuned potential, denoted by SAPT-5st, reproduced this value to $0.06\ \mathrm{cm}^{-1}$ and at the same time did not lose the high accuracy of SAPT-5s for other transitions.

The spectral results computed from the SAPT-5s potential could be compared to those produced by the VRT(ASP-W) potential of Fellers et al. [52], which was fitted to the dimer spectra. In general, SAPT-5s, SAPT-5st, and VRT(ASP-W) agree with experiment equally well, except for the sum of acceptor splittings in the case of SAPT-5s. However, when the VRT(ASP-W) potential was used together with a three-body nonadditive SAPT potential to calculate the spectra of the water trimer (see below), the predictions from VRT(ASP-W) were significantly worse than those given by the SAPT-5s plus a three-body potential. Also, a recent paper by Keutsch et al. [51] has found that SAPT-5st reproduces the complete set of the vibrational ground-state properties of the water dimer very well, whereas the refined versions of the empirical potential, VRT(ASP-W)III [53] and VRT(MCY-5f) [190], show larger discrepancies. Thus, among published potentials SAPT-5s/st provides the best current characterization of the water dimer interaction.

33.11.2 Three-body potentials for water

Attempts to represent the three-body interactions for water in terms of an analytic function fitted to *ab initio* results date back to the work of Clementi and Corongiu [191] and Niesar et al. [67]. These authors used about 200 three-body energies computed at the Hartree–Fock level and fitted them to parametrize a simple polarization model in which induced dipoles were generated on each molecule by the electrostatic field of other molecules. Thus, the induction effects were distorted in order to describe the exchange effects. The three-body potentials obtained in this way and their many-body polarization extensions have been used in simulations of liquid water. We know now that the two-body potentials used in that work were insufficiently accurate for a meaningful evaluation of the role of three-body effects.

The more recent ASP potentials of Millot et al. [183] and their counterparts fitted to the experimental dimer spectra, VRT(ASP-W) [52] and VRT(ASP-W)III [53], have been utilized in diffusion Monte Carlo (DMC) simulation of water clusters of different sizes [192,193]. The three- and higher body effects were described by a polarization model only, similarly as in empirical polarizable potentials. While polarization models are quite efficient in describing the nonadditive induction in the asymptotic regime, they fail to properly model the short-range nonadditivities, which are definitely non-negligible in smaller trimers.

Free from this deficiency is the three-body potential obtained recently by Mas et al. [34], based on *ab initio* calculations using the three-body version of SAPT developed in Ref. [85]. The *ab initio* three-body energies computed at 7533 trimer geometries were fitted to a 12-dimensional analytic form motivated by the physical interpretation of various SAPT corrections. The nonadditive exchange components proportional to the second power of the overlap integrals S are represented using an exchange-quadrupole model and a flexible expansion in products of Legendre polynomials and exponential factors is used to model the short-range terms of the S^3 type. The long-range nonadditive induction effects are accounted for up to the third order by a damped one-center polarization model. The physical content and flexibility of such a functional form allowed for a very small root mean square error, only 0.07 kcal/mol. With the substantial effort devoted to proper modeling of the short-range nonadditive effects, the three-body SAPT potential of Ref. [34] is significantly more sophisticated than any other published potential of this type. The three-body fit, combined with the SAPT-5s pair potential, yielded characteristic structures and energies of water trimer in very good agreement with other accurate calculations [194–196]. The nonadditive effects were found to affect mostly intermolecular distances in the minimum energy structures, while the influence of these effects on the orientations of monomers was smaller. However, the barriers on the trimer potential surface turned out to be quite sensitive to the presence of nonadditive effects. For example, the barrier uud–upd for flipping of a nonbonded hydrogen is 50% higher on the nonadditive surface than the corresponding barrier on the pairwise-additive surface (u, d, and p refer to up, down, and planar free-hydrogen positions, respectively).

The three-body potential computed earlier on a grid of 568 geometries, at the same level as the data used in the fit discussed above, was utilized in Ref. [31] in a three-dimensional calculation of the vibrational spectrum of the water trimer. Here again,

References pp. 958–962

the SAPT-5s representation of the pair interaction was adopted. In this calculation, the centers of mass of the three molecules together with the bonding hydrogen atoms were forming a symmetric rigid planar frame, while the free hydrogens were allowed to undergo flipping motion. The resulting spectrum of the trimer agreed with experiment very well. Recently, the combination of SAPT-5s and the three-body potential of Ref. [34] has been used in DMC simulations of water clusters [193]. For clusters larger than the trimer, the four- and higher many-body induction effects were taken into account by means of the polarization model inherent in the VRT(ASP-W)III [53] potential. Good agreement with experiment was obtained for the ground-state structures and rotational constants for all clusters through the pentamer.

33.11.3 Simulations of liquid water

MC simulations of liquid water, incorporating the SAPT-5s pair potential and the three-body potential described above, were carried out in Ref. [46] with 512 molecules in the canonical ensemble at ambient temperature and density. These simulations will be referred to as the SAPT-5s + 3B simulations. In addition, the four-body and higher nonadditive terms were included *via* a one-center polarization model (SAPT-5s + NB). The effect of the nonadditive forces on the hydrogen-bonded structure of liquid water is reflected in the shape of radial distribution functions, shown in Fig. 33.5. If the interaction potential was represented by the two-body SAPT-5s term only, these functions were very far from the measured ones. In particular, the first peak of the O–H function was strongly suppressed, which corresponds to one hydrogen bond per molecule less than observed experimentally. The first peak of the O–O function was shifted far to the right with respect to its experimental position, and integrating this function over the first coordination shell shows that this shell contained as many as 10 molecules. When three-body effects were added to the SAPT-5s pair potential, the structure became significantly closer to experiment. In particular, it was found that (in contrast to an isolated trimer) the three-body effects result in a reorientation of water molecules leading to significantly increased number of hydrogen bonds, 3.34 compared to the experimental estimate of 3.58. The H–H and O–H pair distribution functions were found to be in very good agreement with the experimental functions of Soper [197]. Also the O–O function agreed well in the region of first peak. However, the agreement in the regions of the first minimum and the second maximum was still rather poor, indicating that the simulated liquid structure lacked the signatures of the experimentally observed tetrahedral coordination. The addition of higher nonadditive terms in SAPT-5s + NB simulation brought virtually no change to the radial distribution functions, suggesting that the structural properties may be already saturated at the 2 + 3-body level.

The recent development of the SAPT(DFT) pair potential [81] and preliminary simulations for the liquid with this potential allowed to shed some new light on the issues discussed above. Comparison of SAPT(DFT) with SAPT-5s showed that the accuracy of the latter potential was somewhat uneven in different regions of configuration space [81]. In particular, the depth of SAPT-5s for the hydrogen-bonded configurations was underestimated compared to the depth in other regions. As it turned out, such details of the pair potential have a direct effect on how much the nonadditive forces affect the calculated liquid structure. The radial distribution functions obtained with the SAPT(DFT) potential combined with different representations of nonadditivities are presented in Fig. 33.6. Comparing these functions to those of Fig. 33.5, one can see that SAPT(DFT) and SAPT-5s give very similar structures when used without nonadditive terms. Addition of the three-body potential [curves denoted by SAPT(DFT) + 3B] results in the O–H and H–H functions with peak and minima positions agreeing with experiment better that those from the analogous SAPT-5s simulation, but no significant differences in shapes of the curves can be observed. These functions seem to be well saturated at the three-body level, as the addition of higher nonadditive terms [curves denoted by SAPT(DFT) + NB] does not bring much change. The situation is more interesting in the case of the O–O function. While the SAPT(DFT) + 3B simulations still lead to an insufficiently structured liquid, the inclusion of nonadditivities beyond three-body appears to induce a qualitative change. Although the depth of the first minimum and the height of the second maximum on the SAPT(DFT) + NB curve are still somewhat too small compared to the experiment, both these features appear in right locations, and integration of the O–O function over the first coordination shell gives the value of 4.4, indicating a local tetrahedral structure. A recent work by Allesch et al. [198] has shown that the lifting of the monomer rigidity approximation leads to an increase of the

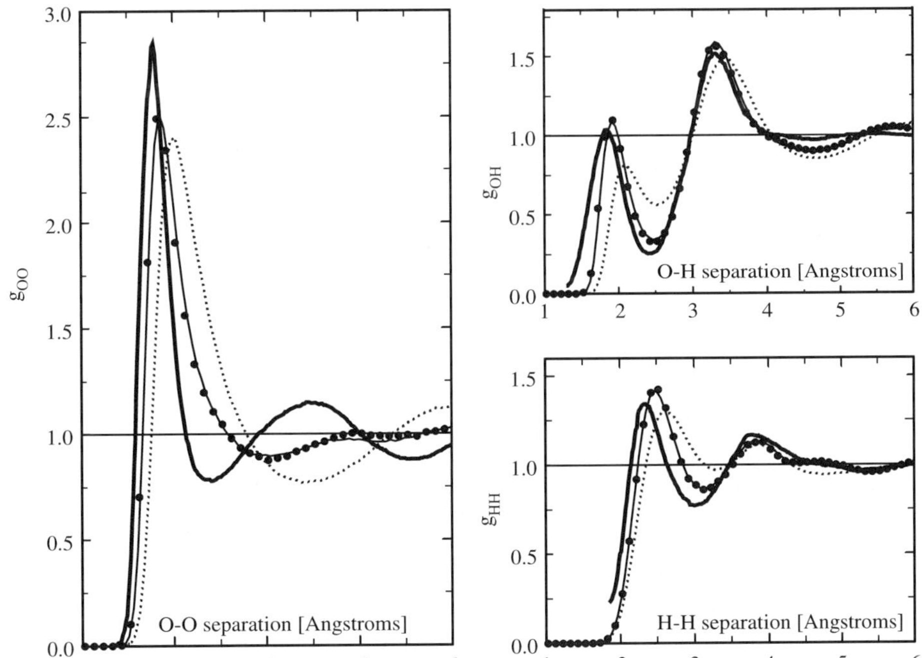

Fig. 33.5. Comparison of radial distribution functions predicted by SAPT-5s (thin dotted line), SAPT-5s + 3B (thin solid line), and SAPT-5s + NB (circles) with the experimental curve of Soper [197] (thick solid line).

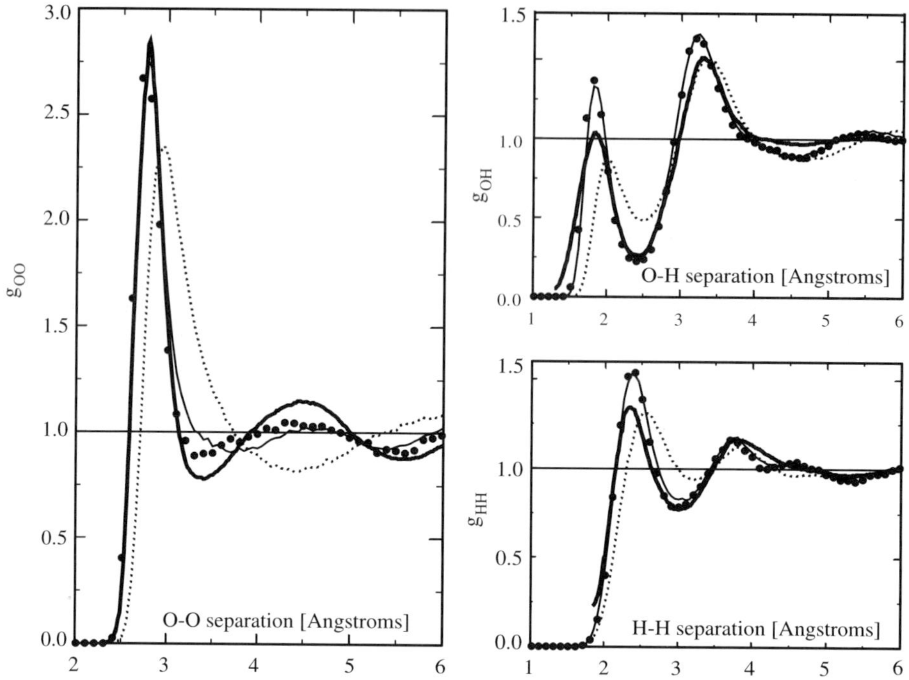

Fig. 33.6. Comparison of radial distribution functions predicted by SAPT(DFT) (thin dotted line), SAPT(DFT) + 3B (thin solid line), and SAPT(DFT) + NB (circles) with the experimental curve of Soper [197] (thick solid line).

amplitude of the O–O function. If this increase is added to the SAPT(DFT) + NB values, the agreement with experiment becomes near prefect.

One of the problems investigated in Ref. [46] was the performance of the simple one-center polarization model, without any short-range nonadditive terms, in representing the many-body interactions. In this model, the full three-body SAPT potential was replaced by the classical three-body induction interactions in the form of the polarization model applied to trimers and limited to the third order (or two iterations). Alternatively, the total nonadditive polarization energy of the whole system could be obtained by iteratively solving the equations for induced molecular dipoles and subtracting the analogous two-body induction energies (as the two-body induction is already included in the pair potential used). In simulations with the SAPT-5s pair potential, both variants of the polarization model led to liquid structure and energetics very similar to those obtained in the SAPT-5s + 3B simulation, i.e. at the three-body level. These conclusions do not carry over to the case of SAPT(DFT) representing the two-body interactions. While the first variant leads, somewhat surprisingly, to an agreement with the SAPT(DFT) + NB structure, the fully iterated variant generates too much structure in the O–O function.

The MC simulations of Ref. [46] as well as the recent ones, performed with the SAPT(DFT) pair potential [81], show a profound effect of nonadditive interactions on the energetics of liquid water. If these effects are neglected, the average potential energy of the liquid in ambient conditions is equal to − 8.8 and − 9.1 kcal/mol for SAPT-5s and SAPT(DFT), respectively, which can be compared to the experimental value of − 9.92 kcal/mol. As expected, the SAPT(DFT) potential—generally deeper than SAPT-5s—provides more attraction in the liquid. Augmenting SAPT-5s and SAPT(DFT) pair potentials with many-body interactions (in the form of the three-body potential plus the rest of many-body effects from polarization model) lowers the average energies to − 10.76 and − 11.9 kcal/mol, respectively. While comparing these values to experiment one should bear in mind that those are results of classical MC simulations and the agreement would improve upon adding the appropriate quantum correction. According to various literature estimates [199,200], the value of such correction may range from 0.7 to 1.5 kcal/mol.

The simulations of Ref. [46] indicated that three-body effects contribute − 1.54 kcal/mol or 14.5% to the internal energy of water, whereas four- and higher body effects contribute − 0.15 kcal/mol (1.4%). These contributions change to − 2.3 kcal/mol (19%) and − 0.2 kcal/mol (2%), respectively, if SAPT(DFT) pair potential is used instead of SAPT-5s. Although the contribution of the effects beyond three body (calculated as an ensemble average) is still quite small, the presence of such effects causes significant rearrangements in the two- and three-body energies compared to the SAPT(DFT) + 3B simulation. The three-body contribution becomes about 0.2 kcal/mol more attractive and the two-body contribution becomes about 0.1 kcal/mol more repulsive, an effect which was not observed in simulations of Ref. [46]. Clearly, the distribution of energy over various terms of the many-body expansion is quite sensitive to details of the pair potential, similarly as in the case of liquid structure.

Although the contribution of higher than three-body interactions was obtained from a relatively simple polarization model, the magnitude of this contribution is consistent with the results of *ab initio* calculations for small water clusters. Tetramers and pentamers have been investigated in Refs. [201,202], where it was found that the four-body effects contribute between − 1.2 and 3.6% of the total interaction energy. The five-body contribution in pentamers was practically negligible, between − 0.11 and 0.25%. Since nonadditive effects in water saturate faster than pair interactions with the increase of the basis set, the percentage contributions quoted above would become even smaller if a larger than double-zeta basis sets were used in the calculations. The study of water hexamers by Pedulla et al. [203] using basis sets of up to quadruple-zeta quality revealed that also for these clusters the sum of the many-body effects beyond three-body does not exceed 4%.

Numerical experience accumulated so far indicates that the structure and energetics of liquid water are determined by a complex balance between various components of the interaction potential with nonadditive interactions playing a major role. This balance is certainly affected by thermodynamic conditions. Therefore, one cannot expect the empirical 'effective' pair potentials, in which the nonadditive effects are included by distorting the pair interactions, to provide a universal model of water capable of correct predictions outside of the range of conditions which determined the potential parameters.

13.12 ACKNOWLEDGEMENTS

This work was partly supported by the NSF grant CHE-0239611 and by a grant from the Foundation for Polish Science.

13.13 REFERENCES

1 G. Chalasinski and M.M. Szczesniak, Chem. Rev., 94 (1994) 1723.
2 B. Jeziorski, R. Moszynski and K. Szalewicz, Chem. Rev., 94 (1994) 1887.
3 G. Chalasinski and M.M. Szczesniak, Chem. Rev., 100 (2000) 4227.
4 B. Jeziorski and K. Szalewicz, in: S. Wilson (Ed.), Handbook of molecular physics and quantum chemistry, Vol. 3, Wiley, Chichester, 2003, p. 232.
5 J. Jakowski, PhD Thesis, Warsaw, 2001.
6 I.G. Kaplan, R. Santamaria and O. Novaro, Mol. Phys., 84 (1995) 105.
7 I.G. Kaplan, J. Hernandez-Cobos, I. Ortega-Blake and O. Novaro, Phys. Rev. A, 53 (1996) 2493.
8 S.F. Boys and F. Bernardi, Mol. Phys., 17 (1970) 553.
9 K. Szalewicz and B. Jeziorski, J. Chem. Phys., 109 (1998) 9518.
10 G. Murdachaew and K. Szalewicz, Faraday Discuss., 118 (2001) 121.
11 G. Murdachaew, K. Szalewicz and R. Bukowski, Phys. Rev. Lett., 88 (2002) 123202.
12 B.M. Axilrod and E. Teller, J. Chem. Phys., 11 (1943) 299.
13 Y. Muto, J. Phys. Math. Soc. Jpn., 17 (1943) 629, (in Japanese).
14 V.F. Lotrich and K. Szalewicz, Phys. Rev. Lett., 79 (1997) 1301.
15 R. Bukowski and K. Szalewicz, J. Chem. Phys., 114 (2001) 9518.
16 P. Rosen, J. Chem. Phys., 21 (1953) 1007.
17 A. Shostak, J. Chem. Phys., 23 (1955) 1808.
18 L. Jansen, Adv. Quantum Chem., 2 (1965) 119.
19 W.J. Meath and R.A. Aziz, Mol. Phys., 52 (1984) 225.
20 W.J. Meath and M. Koulis, J. Mol. Struct. (THEOCHEM), 226 (1991) 1.
21 M.J. Elrod and R.J. Saykally, Chem. Rev., 94 (1994) 1975.
22 M.M. Szczesniak and G. Chalasinski, in: S. Scheiner (Ed.), Molecular interactions—from van der Waals to strongly bound complexes, Wiley, New York, 1996, p. 45.
23 H. Margenau and N.R. Kestner, Theory of intermolecular forces, Pergamon Press, Oxford, 1969.
24 I.G. Kaplan, Intermolecular interactions, Elsevier, Amsterdam, 1987.
25 A.J. Stone, The theory of intermolecular forces, Clarendon Press, Oxford, 1996.
26 M.A. Suhm and D.J. Nesbitt, Chem. Soc. Rev., 24 (1995) 45.
27 K.R. Leopold, G.T. Fraser, S.E. Novick and W. Klemperer, Chem. Rev., 94 (1994) 1807.
28 F.N. Keutsch, J.D. Cruzan and R.J. Saykally, Chem. Rev., 103 (2003) 2533.
29 A. Ernesti and J.M. Hutson, Phys. Rev. A, 51 (1995) 239.
30 A. Ernesti and J.M. Hutson, J. Chem. Phys., 106 (1997) 6288.
31 G.C. Groenenboom, E.M. Mas, R. Bukowski, K. Szalewicz, P.E.S. Wormer and A. van der Avoird, Phys. Rev. Lett., 84 (2000) 4072.
32 V.F. Lotrich and K. Szalewicz, J. Chem. Phys., 106 (1997) 9688.
33 V.F. Lotrich, P. Jankowski and K. Szalewicz, J. Chem. Phys., 108 (1998) 4725.
34 E.M. Mas, R. Bukowski and K. Szalewicz, J. Chem. Phys., 118 (2003) 4386.
35 R.A. Aziz, A.R. Janzen and R.O. Simmons, in: G. Trigg (Ed.), Encyclopedia of applied physics, Vol. 16, VCH, New York, 1996, pp. 71–96.
36 Faraday Discuss. Chem. Soc. 80 (1985) 59–65, 67, 68.
37 H. Huber, A.J. Dyson and B. Kirchner, Chem. Soc. Rev., 28 (1999) 121.
38 E. Ermakova, J. Solca, H. Huber and M. Welker, J. Chem. Phys., 102 (1994) 4942.
39 J.A. Barker, R.A. Fisher and R.O. Watts, Mol. Phys., 21 (1971) 657.
40 G. Marcelli and R.J. Sadus, J. Chem. Phys., 111 (1999) 1533.

41 K. Leonhard and U.K. Deiters, Mol. Phys., 98 (2000) 16037.
42 P.S. Vogt, R. Liapine, B. Kirchner, A.J. Dyson, H. Huber, G. Marcelli and R.J. Sadus, Phys. Chem. Chem. Phys., 3 (2001) 1297.
43 W. Klopper, M. Quack and M.A. Suhm, Chem. Phys. Lett., 261 (1996) 35.
44 P. Diep and J.K. Johnson, J. Chem. Phys., 112 (2000) 4465; P. Diep and J.K. Johnson, J. Chem. Phys., 113 (2000) 3480E.
45 G.W.M. Vissers, A. Hesselmann, G. Jansen, P.E.S. Wormer and A. van der Avoird, J. Chem. Phys., 122 (2005) 054306.
46 E.M. Mas, R. Bukowski and K. Szalewicz, J. Chem. Phys., 118 (2003) 4404.
47 J.A. Odutola and T.R. Dyke, J. Chem. Phys., 72 (1980) 5062.
48 E.M. Mas, R. Bukowski, K. Szalewicz, G.C. Groenenboom, P.E.S. Wormer and A. van der Avoird, J. Chem. Phys., 113 (2000) 6687.
49 L.B. Braly, J.D. Cruzan, K. Liu, R.S. Fellers and R.J. Saykally, J. Chem. Phys., 112 (2000) 10293.
50 L.B. Braly, K. Liu, M.G. Brown, F.N. Keutsch, R.S. Fellers and R.J. Saykally, J. Chem. Phys., 112 (2000) 10314.
51 F.N. Keutsch, N. Goldman, H.A. Harker, C. Leforestier and R.J. Saykally, Mol. Phys., 101 (2003) 3477.
52 R.S. Fellers, C. Leforestier, L.B. Braly, M.G. Brown and R.J. Saykally, Science, 284 (1999) 945.
53 N. Goldman, R.S. Fellers, M.G. Brown, L.B. Braly, C.J. Keoshian, C. Leforestier and R.J. Saykally, J. Chem. Phys., 116 (2002) 101487.
54 C. Leforestier, L.B. Braly, K. Liu, M.J. Elrod and R.J. Saykally, J. Chem. Phys., 106 (1997) 8527.
55 H. Chen, S. Liu and J.C. Light, J. Chem. Phys., 110 (1999) 168.
56 G.C. Groenenboom, P.E.S. Wormer, A. van der Avoird, E.M. Mas, R. Bukowski and K. Szalewicz, J. Chem. Phys., 113 (2000) 6702.
57 S. Scheiner, in: Z.B. Maksic (Ed.), Theoretical models of chemical bonding, Springer, Berlin, 1991, p. 171; S. Scheiner, Annu. Rev. Phys. Chem., 45, 1994, p. 23.
58 S. Scheiner, Hydrogen bonding. A theoretical perspective, Oxford University Press, Oxford, 1997.
59 J.G.C.M. van Duijneveldt-van de Rijdt and F.B. van Duijneveldt, in: D. Hadzi (Ed.), Theoretical treatments of hydrogen bonding, Wiley, New York, 1997, p. 13.
60 C.J. Burnharm, J. Li, S.S. Xantheas and M. Leslie, J. Chem. Phys., 110 (1999) 4566.
61 A. Famulari, R. Specchio, M. Sironi and M. Raimondi, J. Chem. Phys., 108 (1998) 3296.
62 Y.-P. Liu, K. Kim, B.J. Berne, R.A. Friesner and S.W. Rick, J. Chem. Phys., 108 (1998) 4739.
63 E.M. Mas, K. Szalewicz, R. Bukowski and B. Jeziorski, J. Chem. Phys., 107 (1997) 4207.
64 H.S. Gutowsky, T.D. Klots, C. Chuang, C.A. Schmuttenmaer and T. Emilsson, J. Chem. Phys., 86 (1987) 569; A. McIlroy, R. Lascola, C.M. Lovejoy and D.J. Nesbitt, J. Chem. Phys., 95 (1991) 2636; J.T. Farrell, Jr. and D.J. Nesbitt, J. Chem. Phys., 105 (1996) 9421; C.-C. Chuang, S.N. Tsang, J.G. Hanson, W. Klemperer and H.-C. Chuang, J. Chem. Phys., 107 (1997) 7041.
65 M.R. Viant, M.G. Brown, J.D. Cruzan, R.J. Saykally, M. Geleijns and A. van der Avoird, J. Chem. Phys., 110 (1999) 4369.
66 M.G. Brown, M.R. Viant, R.P. McLaughlin, C.J. Keoshian, E. Michael, J.D. Cruzan, R.J. Saykally and A. van der Avoird, J. Chem. Phys., 111 (1999) 7789.
67 U. Niesar, G. Corongiu, M.-J. Huang, M. Dupuis and E. Clementi, Int. J. Quantum Chem. Symp., 23 (1989) 421; G. Corongiu and E. Clementi, J. Chem. Phys., 97 (1992) 2030, 8818(E).
68 R. Eggenberger, H. Huber and M. Welker, Chem. Phys., 187 (1994) 317; E. Ermakova, J. Solca, H. Huber and D. Marx, Chem. Phys. Lett., 246 (1995) 204.
69 G. Steinebrunner, A.J. Dyson, B. Kirchner and H. Huber, J. Chem. Phys., 109 (1998) 3153.
70 A.K. Sum, S.I. Sandler, R. Bukowski and K. Szalewicz, J. Chem. Phys., 116 (2002) 7627.
71 A.K. Sum, S.I. Sandler, R. Bukowski and K. Szalewicz, J. Chem. Phys., 116 (2002) 7637.
72 R. Car and M. Parrinello, Phys. Rev. Lett., 55 (1985) 2471.
73 K. Laasonen, M. Spirk, M. Parrinello and R. Car, J. Chem. Phys., 99 (1993) 9080; M. Spirk, J. Hutter and M. Parrinello, J. Chem. Phys., 105 (1996) 1142; P.L. Silvestrelli and M. Parrinello, Phys. Rev. Lett., 82 (1999) 3308.
74 J. VandeVondele, F. Mohamed, M. Krack, J. Hutter, M. Spirk and M. Parrinello, 122 (2005) 014515.
75 A.D. Boese, A. Chandra, J.M.L. Martin and D. Marx, J. Chem. Phys., 119 (2003) 5965.

76 W.L. Jorgensen, J. Am. Chem. Soc., 103 (1981) 335.
77 W.L. Jorgensen, J. Chandrasekhar, J.D. Madura, R.W. Impey and M.L. Klein, J. Chem. Phys., 79 (1983) 926.
78 H.J.C. Berendsen, J.R. Grigera and T.P. Straatsma, J. Phys. Chem., 91 (1987) 6269.
79 B.J. Mhin, J. Kim, S. Lee, J.Y. Lee and K.S. Kim, J. Chem. Phys., 100 (1994) 4484.
80 I.M. Svishchev, P.G. Kusalik, J. Wang and R.J. Boyd, J. Chem. Phys., 105 (1996) 4742.
81 R. Bukowski et al., to be published.
82 K. Szalewicz and B. Jeziorski, in: S. Scheiner (Ed.), Molecular interactions—from van der Waals to strongly bound complexes, Wiley, New York, 1997, p. 3.
83 B. Jeziorski and K. Szalewicz, in: P. von Ragué Schleyer, P.R. Schreiner, N.L. Allinger, T. Clark, J. Gasteiger, P. Kollman and H.F. Schaefer III (Eds.), Encyclopedia of computational chemistry, Vol. 2, Wiley, New York, 1998, p. 1376.
84 R. Bukowski, W. Cencek, P. Jankowski, B. Jeziorski, M. Jeziorska, S.A. Kucharski, A.J. Misquitta, R. Moszynski, K. Patkowski, S. Rybak, K. Szalewicz, H.L. Williams and P.E.S. Wormer, SAPT2002: An ab initio program for many-body symmetry-adapted perturbation theory calculations of intermolecular interaction energies, University of Delaware and University of Warsaw: http://www.physics.udel.edu/~szalewic/SAPT/SAPT.html
85 V.F. Lotrich and K. Szalewicz, J. Chem. Phys., 106 (1997) 9668.
86 R. Moszyński, P.E.S. Wormer, B. Jeziorski and A. van der Avoird, J. Chem. Phys., 103 (1955) 8058, E: 107 (1997) 672.
87 V.F. Lotrich and K. Szalewicz, J. Chem. Phys., 112 (2000) 112.
88 P.E.S. Wormer, R. Moszyński and A. van der Avoird, J. Chem. Phys., 112 (2000) 3159.
89 M. Tachikawa and K. Iguchi, J. Chem. Phys., 101 (1994) 3062.
90 M. Tachikawa, K. Suzuki and K. Iguchi, Struct. Chem., 6 (1995) 287.
91 J. Paldus and J. Cizek, Adv. Quantum Chem., 9 (1975) 105.
92 R.J. Bartlett, Annu. Rev. Phys. Chem., 32 (1981) 359.
93 A.J. Misquitta and K. Szalewicz, Chem. Phys. Lett., 357 (2002) 301; A. Hesselmann and G. Jansen, Chem. Phys. Lett., 357 (2002) 464.
94 A.J. Misquitta, B. Jeziorski and K. Szalewicz, Phys. Rev. Lett., 91 (2003) 033201; A. Hesselmann and G. Jansen, Chem. Phys. Lett., 367 (2003) 778.
95 B. Jeziorski and M.C. van Hemert, Mol. Phys., 31 (1976) 7137.
96 K. Szalewicz, S.J. Cole, W. Kolos and R.J. Bartlett, J. Chem. Phys., 89 (1988) 3662.
97 K. Szalewicz, in: R.A. Meyers, R. Bajcsy, A.J. Bard, W.-K. Chen, J.T. Christian, R. Coleman, et al. (Eds.), Encyclopedia of physical science and technology, 3rd edn., Vol. 7, Academic Press, San Diego, CA, 2002, p. 505.
98 R. Moszynski, B. Jeziorski, A. Ratkiewicz and S. Rybak, J. Chem. Phys., 99 (1993) 8856.
99 R. Moszynski, S.M. Cybulski and G. Chalasinski, J. Chem. Phys., 100 (1994) 4998.
100 R. Moszynski, B. Jeziorski and K. Szalewicz, J. Chem. Phys., 100 (1994) 1312.
101 M. Jeziorska, B. Jeziorski and J. Cizek, Int. J. Quantum Chem., 32 (1987) 149.
102 R. Moszynski, T.G.A. Heijmen and B. Jeziorski, Mol. Phys., 88 (1996) 741.
103 E.M. Mas and K. Szalewicz, J. Chem. Phys., 104 (1996) 7606.
104 M. Sharma, Y. Wu and R. Car, Int. J. Quantum Chem., 95 (2003) 821.
105 J.C. Grossman, E. Schwegler, E.W. Draeger, F. Gygi and G. Galli, J. Chem. Phys., 120 (2004) 300.
106 I.G. Kaplan, S. Roszak and J. Leszczynski, J. Chem. Phys., 113 (2000) 6245.
107 J. Higgins, T. Hollebeek, J. Reho, T.S. Ho, K.K. Lehmann, H. Rabitz, G. Scoles and M. Gutowski, J. Chem. Phys., 112 (2000) 5751.
108 P. Soldan, M.T. Cvitas and J.M. Hutson, Phys. Rev. A, 67 (2003) 054702.
109 G. Chalasinski, J. Rak, M.M. Szczesniak and M. Cybulski, J. Chem. Phys., 106 (1997) 3301.
110 J. Jakowski, G. Chalasinski, M.M. Szczesniak and M. Cybulski, Chem. Phys., 239 (1998) 573.
111 H.C. Longuet-Higgins, Discuss. Faraday Soc., 40 (1965) 7; R. McWeeny, Croat. Chem. Acta, 57 (1984) 865.
112 P. Piecuch, Mol. Phys., 59 (1986) 1067.
113 P. Piecuch, Int. J. Quantum Chem., 47 (1993) 261.

114 S.M. Cybulski, Chem. Phys. Lett., 238 (1995) 261.
115 H.L. Williams, T. Korona, R. Bukowski, B. Jeziorski and K. Szalewicz, Chem. Phys. Lett., 262 (1996) 431.
116 T. Korona, H.L. Williams, R. Bukowski, B. Jeziorski and K. Szalewicz, J. Chem. Phys., 106 (1997) 5109.
117 M.V. Rama Krishna and K.B. Whaley, J. Chem. Phys., 93 (1990) 6738.
118 V. Efimov, Commun. Nucl. Part. Phys., 19 (1990) 271.
119 B.D. Esry, C.D. Lin and C.H. Greene, Phys. Rev. A, 54 (1996) 394.
120 K.B. Whaley, in: J.M. Bowman (Ed.), Advances in molecular vibrations and collision dynamics, Vol. 3, JAI Press, Stamford, CT, 1998, p. 397.
121 M. Boninsegni, C. Pierleoni and D.M. Ceperley, Phys. Rev. Lett., 72 (1994) 1854.
122 L.W. Bruch and I.J. McGee, J. Chem. Phys., 59 (1873) 409.
123 C.A. Parish and C.E. Dykstra, J. Chem. Phys., 98 (1993) 437.
124 C.A. Parish and C.E. Dykstra, J. Chem. Phys., 101 (1994) 7618.
125 M.J. Cohen and J.N. Murrell, Chem. Phys. Lett., 260 (1996) 371.
126 S. Moroni, F. Pederiva, S. Fantoni and M. Boninsegni, Phys. Rev. Lett., 84 (2000) 2650.
127 F. Pederiva, S. Moroni, S. Fantoni and M. Boninsegni, Physica B, 284 (2000) 365.
128 S.-Y. Chang and M. Boninsegni, J. Chem. Phys., 115 (2001) 2629.
129 G. Chalasinski, M.M. Szczesniak and R. Kendall, J. Chem. Phys., 101 (1994) 8860.
130 I. Roeggen and J. Almlof, J. Mol. Struct. (THEOCHEM), 388 (1996) 331.
131 B. Jeziorski, M. Bulski and L. Piela, Int. J. Quantum Chem., 10 (1976) 281.
132 M. Bulski, Mol. Phys., 29 (1975) 1171.
133 M. Bulski and G. Chalasinski, Theor. Chim. Acta, 56 (1980) 199.
134 M. Bulski and G. Chalasinski, Chem. Phys. Lett., 89 (1982) 450.
135 A.R. Cooper, S. Jain and J.M. Hutson, J. Chem. Phys., 98 (1993) 2160.
136 N.J. Wright and J.M. Hutson, J. Chem. Phys., 110 (1999) 902.
137 R.A. Aziz, J. Chem. Phys., 99 (1993) 4518.
138 C. Tessier, A. Terlain and Y. Larher, Physica A, 113 (1982) 286.
139 E.M. Mas, V.F. Lotrich and K. Szalewich, J. Chem. Phys., 110 (1999) 6694.
140 J.H. Dymond and B.J. Alder, J. Chem. Phys., 54 (1971) 3472.
141 R. Gilgen, R. Kleinrahm and W. Wagner, J. Chem. Thermodyn., 26 (1994) 383.
142 C. Tegeler, R. Span and W. Wagner, J. Phys. Chem. Ref. Data, 28 (1999) 779.
143 K. Patkowski, G. Murdachaew, C.-M. Fou and K. Szalewicz, Mol. Phys., in press.
144 R.O. Simmons, Mater. Chem. Phys., 50 (1997) 124.
145 D.L. Losee and R.O. Simmons, Phys. Rev. Lett., 18 (1967) 451.
146 E. Guarini, R. Magli, M. Tau, F. Barochi, G. Casanova and L. Reatto, Phys. Rev. E, 63 (2001) 052201.
147 A.K. Dham, A.R. Alnatt, W.J. Meath and R.A. Aziz, Mol. Phys., 67 (1989) 1291.
148 K. Nauta and R.E. Miller, J. Chem. Phys., 115 (2001) 10138.
149 L. Jansen, Phys. Rev., 125 (1962) 1798.
150 I.N. Kozin, M.M. Law, J.M. Hutson and J. Tennyson, J. Chem. Phys., 118 (2003) 4896.
151 R. Moszynski, P.E.S. Wormer, T.G.A. Heijmen and A. van der Avoird, J. Chem. Phys., 108 (1998) 579.
152 J.M. Hutson, S. Liu, J.W. Moskowitz and Z. Bacic, J. Chem. Phys., 111 (1999) 8378.
153 C.E. Dykstra, J. Chem. Phys., 108 (1998) 6619.
154 M.J. Packer and D.C. Clary, J. Phys. Chem., 99 (1995) 14323.
155 S. Re, Y. Osamura, Y. Suzuki and H.F. Scheafer, III, J. Chem. Phys., 109 (1998) 973.
156 Z. Kisiel, E. Bialkowska-Jaworska, L. Pszczolkowski, A. Milet, C. Struniewicz, R. Moszynski and J. Sadlej, J. Chem. Phys., 112 (2000) 5767.
157 A. Milet, C. Struniewicz, R. Moszynski and P.E.S. Wormer, Theor. Chem. Acc., 104 (2000) 195.
158 A. Milet, C. Struniewicz, R. Moszynski, J. Sadlej, Z. Kisiel, E. Bialkowska-Jaworska and L. Pszczolkowski, Chem. Phys., 271 (2001) 267.
159 C. Struniewicz, T. Korona, R. Moszynski and A. Milet, Chem. Phys. Lett., 343 (2001) 588.
160 P.E.S. Wormer, G.C. Groenenboom and A. van der Avoird, J. Chem. Phys., 115 (2001) 3604.
161 C. Struniewicz, A. Milet, J. Sadlej and R. Moszynski, Int. J. Quantum Chem., 90 (2002) 1151.
162 A. Milet, C. Struniewicz, R. Moszynski and P.E.S. Wormer, J. Chem. Phys., 115 (2001) 349.

163 J. Jakowski, J. Klos, G. Chalasinski, M.W. Severson, M.M. Szczesniak and S.M. Cybulski, J. Chem. Phys., 112 (2000) 10895.
164 A.W. Castelman and K.H. Bowen, J. Phys. Chem., 100 (1996) 12911.
165 M.A. Suhm and R.O. Watts, Phys. Rep., 204 (1991) 293.
166 V. Buch, J. Chem. Phys., 97 (1992) 726.
167 M. Gutowski, J. Chem. Phys., 110 (1999) 4695.
168 J. Jakowski, G. Chalasinski, M.M. Szczesniak and S.M. Cybulski, Collect. Czech. Chem. Commun., 68 (2003) 587.
169 T. Korona, R. Moszynski and B. Jeziorski, J. Chem. Phys., 105 (1996) 8178.
170 M. Xu, Z. Bacic and J.M. Hutson, J. Chem. Phys., 117 (2002) 4777.
171 J. Jakowski, G. Chalasinski, S.M. Cybulski and M.M. Szczesniak, J. Chem. Phys., 118 (2003) 2731.
172 S.T. Arnold, J.H. Hendricks and K.H. Bowen, J. Chem. Phys., 102 (1995) 39.
173 H. de Clercq, J.H. Hendricks and K.H. Bowen, J. Chem. Phys., 117 (2002) 2619.
174 S. Roszak, R. Gora and J. Leszczynski, Chem. Phys. Lett., 313 (1999) 198.
175 F.Y. Naumkin and P.J. Knowles, J. Chem. Phys., 103 (1995) 3392.
176 V. Aquilanti, G. Liuti, F. Pirani and F. Vecchiocattivi, J. Chem. Soc., Faraday Trans., 85 (1989) 955.
177 A.V. Danilychev and V.A. Apkarian, J. Chem. Phys., 100 (1994) 5556.
178 J. Jakowski, G. Chalasinski, J. Gallegos, M.W. Severson and M.M. Szczesniak, J. Chem. Phys., 118 (2003) 2748.
179 M. Cybulski, TRURL-94 Package, Miami University, OH, 1994.
180 M.H. Alexander, J. Chem. Phys., 99 (1993) 6014.
181 K. Stark and H.-J. Werner, J. Chem. Phys., 104 (1996) 6515.
182 O. Matsuoka, E. Clementi and M. Yoshimine, J. Chem. Phys., 64 (1976) 1351.
183 C. Millot, J.C. Soetens, M.T.C.M. Costa, M.P. Hodges and A.J. Stone, J. Phys. Chem., 102 (1998) 754.
184 W. Klopper, J.G.C.M. van Duijneveldt-van de Rijdt and F.B. van Duijneveldt, Phys. Chem. Chem. Phys., 2 (2000) 2227.
185 J.G.C.M. van Duijneveldt-van de Rijdt, W.T.M. Mooij and F.B. van Duijneveldt, Phys. Chem. Chem. Phys., 5 (2003) 1169.
186 L.A. Curtiss, D.J. Frurip and M. Blander, J. Chem. Phys., 71 (1979) 2703.
187 E. Zwart, J.J. ter Meulen, W.L. Meerts and L.H. Coudert, J. Mol. Spectrosc., 147 (1991) 27.
188 G.T. Fraser, Int. Rev. Phys. Chem., 10 (1991) 189.
189 M.J. Smit, G.C. Groenenboom, P.E.S. Wormer, A. van der Avoird, R. Bukowski and K. Szalewicz, J. Phys. Chem. A, 105 (2001) 6212.
190 C. Leforestier, F. Gatti, R.S. Fellers and R.J. Saykally, J. Chem. Phys., 117 (2002) 8710.
191 E. Clementi and G. Corongiu, Int. J. Quantum Chem. Symp., 10 (1983) 31.
192 J.K. Gregory and D.C. Clary, J. Phys. Chem., 100 (1996) 18014.
193 N. Goldman and R.J. Saykally, J. Chem. Phys., 120 (2004) 4777.
194 T. Bürgi, S. Graf, S. Leutwyler and W. Klopper, J. Chem. Phys., 103 (1995) 1077.
195 J.G.C.M. van Duijneveldt-van de Rijdt and F.B. van Duijneveldt, Chem. Phys., 175 (1993) 271; J.G.C.M. van Duijneveldt-van de Rijdt and F.B. van Duijneveldt, Chem. Phys. Lett., 273 (1995) 560.
196 I.M.B. Nielsen, E.T. Seidl and C.L. Janssen, J. Chem. Phys., 110 (1999) 9435.
197 A.K. Soper, Chem. Phys., 258 (2000) 121.
198 M. Allesch, E. Schwegler, F. Gygi and G. Galli, J. Chem. Phys., 120 (2004) 5192.
199 G.S. Del Buono, P.J. Rossky and J. Schnitker, J. Chem. Phys., 95 (1991) 3728.
200 H.A. Stern and B.J. Berne, J. Chem. Phys., 115 (2001) 7622.
201 M.P. Hodges, A.J. Stone and S.S. Xantheas, J. Phys. Chem. A, 101 (1997) 9163.
202 A. Milet, R. Moszyński, P.E.S. Wormer and A. van der Avoird, J. Phys. Chem. A, 103 (1999) 6811.
203 J.M. Pedulla, K. Kim and K.D. Jordan, Chem. Phys. Lett., 291 (1998) 78.

Theory and Applications of Computational Chemistry: The First Forty Years
Edited by C. Dykstra et al.

CHAPTER 34

Clusters to functional molecules, nanomaterials, and molecular devices: theoretical exploration

Kwang S. Kim, P. Tarakeshwar and Han Myoung Lee

National Creative Research Initiative Center for Superfunctional Materials, Department of Chemistry, Division of Molecular and Life Sciences, Pohang University of Science and Technology, San-31, Hyojadong, Namgu, Pohang 790-784, South Korea

Abstract

Since theoretical investigations of gas-phase clusters have enabled the evaluation of assembling structures with detailed insight of intermolecular interactions and the understanding of their intrinsic molecular properties, this information has helped predict novel nanomaterials. We have focused our attention on designing functional molecular systems, nanomaterials, and nanodevices toward the experimental realization. In this review, we show how insights obtained from theoretical investigations of various cluster systems (ranging from simple water clusters to π-system-containing complexes) have enabled us to predict structures and properties of novel functional molecular systems like endo/exohedral fullerenes, nanotori, nonlinear optical materials, ionophores/receptors, polypeptides, enzymes, organic nanotubes, nanowires, electronic and nanomechanical molecular devices. Interestingly, most of these theoretically predicted systems have been experimentally realized.

34.1 INTRODUCTION

The demand for novel nanomaterials and nanodevices cuts across almost every sector of world high-tech industries. The immense potential of nanodevices in the fields of communication, information storage, materials, and biological sciences has heightened the quest for novel functional nanomaterials. Conventional methods of designing novel nanomaterials involve tedious experimentation with poor success rates. On the other hand, one of the most attractive and challenging features of contemporary research in science is the ability to rationally design and fabricate novel functional nanomaterials. Essential to the success of this strategy is to have a detailed knowledge of the relationship between structure and properties of the designed nanomaterials. As one of the important missions of computational research in chemistry, computational chemistry has helped design and invent novel molecular systems possessing desired specific properties toward realization [1–4]. It should be emphasized that the success of design strategies based on computational techniques is highly interlinked to the dramatic advances in evaluating and analysing interaction forces and molecular properties in computational chemistry in the last four decades.

The essence of computational design of new materials is how atoms and molecules are assembled to dictate interesting properties. The goal of this work can also be seen in light of John Maddox's 1988 Nature editorial (Nature, 335 (1988) 201) wherein he indicated that "One of the continuing scandals in the physical sciences is that it remains in general impossible to predict the structure of even the simplest crystalline solids from a knowledge of their chemical composition". In computational terms, this involves the evaluation of properties of systems having thousands of atoms. Despite the spectacular advances in computational methods, the sizes of systems which can be handled using conventional computational methods is still of the order of a few thousands of atoms. When the properties of the system under consideration are dominated by quantum effects, the system size shrinks to a few hundred atoms. In this context, atomic and molecular clusters are useful model systems, because they bridge the gap between atom and bulk-like behaviour. However, what is much more important is that small-size clusters show quantum effects which are essential for novel nanodevices.

Clusters are self-assembled structures comprised of a number of monomers under the given condition. Apart from aiding the development of novel materials, clusters are very useful for understanding the intrinsic and fundamental nature of molecular recognition and self-assembling phenomena [5–8]. This is amply illustrated in a number of publications on a wide variety of atomic and molecular clusters, ranging from H-bonded clusters [9–12], π-system-containing clusters [8,13–16], and metal clusters [17–20]. These investigations not only provide pertinent information useful for nanomaterial design but also highlight some of the important similarities and differences in their physical characteristics. These characteristics include structures, magnitudes of both attractive and repulsive interaction energies, vibrational frequencies, and charge redistributions. Additionally one also obtains an insight into the contributions of cooperative and competitive forces, both of which govern self-assembly and molecular recognition [21].

Given the ability of nanochemistry to provide functional materials of practical utility in the near future, the elaborative quantum-chemical calculations of clusters are of significant value because interesting individual molecules or small clusters can be developed as viable functional materials and devices. The prediction of properties of nanomaterials, from a nanoscopic determination of the properties of individual atoms, molecules, or clusters, is based on molecular interactions (with ion/atom/molecule/electron/photon) and the resulting molecular structures (clustering/aggregation by self-assembly and self-synthesis process). In this context, we have carried out detailed theoretical investigations of the properties of a large number of cluster systems, ranging from simple water clusters to large π-systems. As an effective design strategy requires a thorough understanding of various interaction forces and mechanisms, we have theoretically characterized novel interaction forces. We have employed various theoretical methods ranging from traditional *ab initio* theory, density functional theory (DFT), tight-binding theory, Monte Carlo to molecular dynamics (MD) simulations. In the course of these investigations, we have been successful in elucidating the properties of a diverse range of novel molecular systems. These include molecular clusters (water clusters, solvated cations/anions/electrons/chemicals/biomolecules, and inorganic/metal clusters), endo/exohedral fullerenes/nanotori, nonlinear optical materials, ionophores/receptors, polypeptides/membranes, enzymes, organic nanotubes, metal nanowires, photo/electro-nanodevices, and nanomechanical molecular devices.

The success of our design strategy is validated by our experimental characterization of novel ionophores, organic nanotubes, and molecular flippers as well as other experimental demonstration by other groups. For example, we have designed receptors with high affinity and selectivity for specific cations (acetylcholine, NH_4^+) or anions (F^-, Cl^-, $H_2PO_4^-$, CH_3COO^-, ATP, GDP) which are biologically important. The designed quantum nanostructures have been synthesized, the predicted photo-electronic devices would be useful for computer memory with nonlinear optical switch phenomena, and the designed nanomechanical molecular devices and molecular vehicles/tweezers would be a stepping stone for the design of practical systems for drug delivery and nanosurgery.

In the course of this review, we elaborate on the theoretical accomplishments which eventually led to the *de novo* design of these novel functional molecular systems. We illustrate how computational strategies based on our understanding of intermolecular interactions and cluster formation aid the design of novel nanomaterials and molecular devices. Although there are many studies on computational material science, most of them are different from our aims because their views have centred on simple prediction of cluster properties (though interesting results), understanding of known nanomaterials, and analysing of new experimental results; otherwise, the predictions are often too hypothetical beyond experimental realization at the present status. In this regard, instead of reviewing other groups' work, the present mini-review will centre on our aim at designing novel functional molecules, nanomaterials, and nanodevices, toward experimental realization. It should be emphasized here that the highlight of our work is that these designed nanomaterials have been experimentally characterized, and are found to be very novel because of the utilization of novel interactions and subtle difference between competing interactions, with their structures and properties very close to the theoretical predictions.

The review is organized as follows. After this introduction, the theoretical background with a short description for various existing computational methods employed for the design of functional nanomaterials is given in Section 34.2. The salient features of our work on aqueous, metallic, and weakly bound clusters are described in Section 34.3. We highlight our work on the design and experimental realization of novel ionophores, receptors, sensors, and enzymes in Section 34.4. Our work on nanomaterials and molecular devices are highlighted in Sections 34.5 and 34.6. Finally, concluding remarks are given in Section 34.7.

34.2 THEORETICAL BACKGROUND

Before we proceed any further, it is useful to briefly highlight various theoretical methodologies employed to obtain the results discussed in this review. The large variation in the nature and sizes of the systems imply that a wide variety of theoretical strategies have to be employed to obtain the results. In general, computational calculations of large systems can be carried out using semi-empirical or empirical methods. This is the norm in simulations of large biological systems. However, in systems dominated by quantum effects, it is imperative that one has to employ quantum-chemical *ab initio* methods, to obtain a reasonable and realistic description of the systems.

Quantum-chemical *ab initio* calculations of large systems can generally be carried out using a judicious combination of supermolecular (SM) and perturbational methods [22,23]. While the former method is often employed to obtain total interaction energies, perturbational methods yield the magnitudes of the various repulsive and attractive energy terms constituting the total interaction energy. As a result, perturbational methods provide important information on the factors responsible for cluster or complex formation. A major difference between the SM and perturbational methods is that in the former, the interaction energy is evaluated as the difference of the energy of the complex and the energy of the isolated monomers, while the interaction energy is obtained as a sum of the individual electrostatic, exchange, dispersion, and induction energies, in the latter.

Given the subtle effects of structure on various calculated properties, it is desirable that inferences are made on the geometry-optimized structures of the clusters or complexes under investigation. This also facilitates the evaluation of the vibrational frequencies, because one can readily compare the calculated numbers with the experimentally observed frequencies or shifts. As a consequence, one can readily test the efficacy of the theoretical method employed in the calculations.

SM calculations are broadly based on either the (i) Hartree–Fock method; (ii) Post-Hartree–Fock methods like the Møller–Plesset level of theory (MP), configuration interaction (CI), complete active space self-consistent field (CASSCF), coupled cluster singles and doubles (CCSD); or (iii) methods based on DFT [24–27]. Since the inclusion of electron correlation is vital to obtain an accurate description of nearly all the calculated properties, it is desirable that SM calculations are carried out at either the second-order Møller–Plesset (MP2) or the coupled cluster with single, double, and perturbative triple substitutions (CCSD(T)) levels using basis sets composed of both diffuse and polarization functions.

However, with an increase in the size of the system, it is impossible to carry out the calculations at the levels of theory employed in investigations of smaller systems. In such cases, DFT methods have been widely employed in a large number of studies because of the relatively smaller computational resources needed to describe very large systems. Though DFT calculations are effective in description of systems dominated by hydrogen bonding interactions, care should be taken in using them in systems dominated by weak intermolecular complexes. This is because of the inability of most of the current density functionals to describe dispersion energies.

Beyond a certain system size, even DFT methods using conventional basis sets become computationally very intensive. In such situations, one has to take recourse to the use of solid-state physics methods like the pseudopotential plane wave or tight-binding methods [28,29]. As the systems become larger, Monte Carlo (MC) simulations and molecular dynamics simulations based on effective pair potentials (including two-body to multi-body interactions) are carried out.

One of the most important approaches in theoretical design strategy is to utilize novel intermolecular interactions. These interactions can be found from the theoretical investigation of various atomic and molecular clusters. We investigate the electrostatic interactions, induction, dispersion, and exchange repulsions. We particularly utilize the subtle difference between competing interactions among various types of interactions such as hydrogen bonding (normal, weak, strong, short strong H-bonds (SSHBs), or charged or ionic H-bonds, π–H-bonds, etc.), ionic interactions, interactions involving π-systems (π–π interactions, π–H interactions, cation–π interactions, anion–π interactions), metallic interactions, and interactions involving quantum species (electron binding/releasing or charge transfer related to molecule–electron/hole interactions, photo-excitation related to molecule–photon interactions) [1,2,21]. The competing interactions and the cooperative interactions for the given systems are related to nanorecognition, and the quest of nanorecognition is essential for nanomaterial design [21].

34.3 CLUSTERS

Investigations of clusters have been a pervading theme of research being carried out in computational chemistry for the last several years. In early years, most of the computational effort was expended in the evaluation of the interaction energies of a small number of structures of these clusters. However, the vast number of structures feasible for even small-sized clusters implies that one has to carry out an elaborate conformational search to obtain reliable values, which could be compared to the available experimental data.

This has been one of the persistent themes of research being carried out in our group over the last several years. Though several other research groups in the world have been working on similar themes, we have been one of the few groups which have carried out a systematic evaluation of structures for several classes of clusters, *viz*., H-bonded clusters, π-system-containing clusters, and metal clusters. This enabled us to obtain detailed information on the similarities and differences of several classes of clusters. Additionally, we have also probed the nature of intermolecular interactions prevailing in these clusters by an evaluation of the magnitudes of various attractive and repulsive components of the interaction energies. This, together with information on the magnitudes of the many-body interaction energies, was useful in obtaining reasonable inferences on the factors dictating cluster formation.

34.3.1 Aqueous clusters

Aqueous clusters have been widely investigated because of their ability to provide a microscopic understanding of the properties of water and their ability to yield molecular level details of solvation. In particular, a greater understanding of the nonadditive interactions can be obtained because one can incrementally increase the number of water molecules solvating a solute. As a result, molecular level details can be distinguished from dielectric effects. This not only helps understand the properties of biologically important ions and molecules in their natural environments, but is also very useful in the *de novo* design of ionophores and receptors.

In one of the earliest works carried out by the leading author, a large number of initial structures were randomly generated for small-sized water clusters, and their energies were evaluated using a Monte Carlo annealing technique [30]. Interestingly this investigation, which also pioneered the use of two-body, three-body, and four-body interaction energies, was successful in identifying the lowest-energy structures of noncyclic water hexamer to octamer, which were experimentally confirmed about a decade after the prediction [31–34]. It is useful to mention in this context that the lowest-lying energy structures of the water hexamer was also confirmed later from a detailed theoretical investigation of about 800 topologically different conformers (which can be categorized into 39 distinctive types of structures) [35–38]. This investigation was greatly facilitated with the use of an in-house graphics program (POSMOL) [39,40]. An excellent agreement of the calculated –OH stretch vibrational spectra, tunnelling splitting, and rotational constants of these predicted structures with the experimental data indicates that these structures are indeed being observed in experiments. Our extensive calculations show that with an increase in the size of the water cluster, there is a gradual structural transition from linear one-dimensional (1D; dimer) to two-dimensional (2D; trimer to the pentamer), to three-dimensional (3D; hexamer and above) structures (Fig. 34.1).

One of our important contributions in the field of aqueous clusters has been to provide quantitative estimates of the effects of anharmonicity on zero point vibrational energies (ZPVE) and thermodynamic energies at finite temperatures [41]. For the water dimer, MP2/[13s8p4d2f/8s4p2d] calculations yielded values of enthalpy ($\Delta H = -3.19$ kcal/mol), free energy ($\Delta G = 3.39$ kcal/mol), and entropy ($\Delta S = -17.7$ cal/mol/K) at 373 K, which are well within the experimental bounds ($\Delta H = -3.59 \pm 0.5$ kcal/mol,

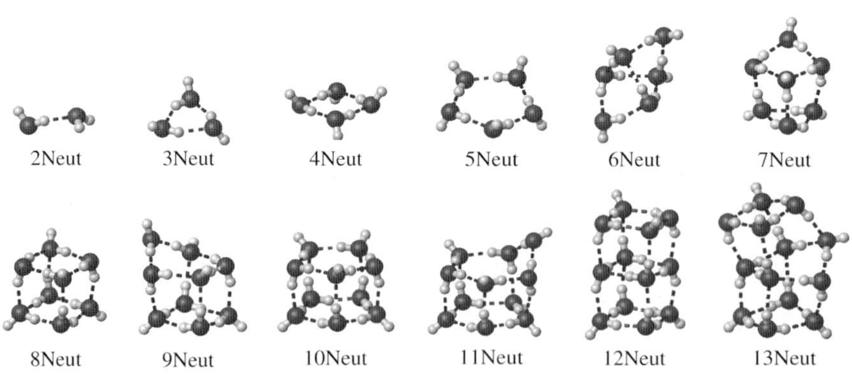

Fig. 34.1. Calculated minimum energy structures of the neutral water cluster structures. More details on these structures can be obtained from Refs. [43,48].

$\Delta G = 3.34$ kcal/mol, $\Delta S = -18.59 \pm 1.3$ cal/mol/K) [42]. However, it is difficult to carry out similar calculations on large water clusters.

Our calculations indicate that the lowest-energy structures of the water tetramer, pentamer, and octamer have compact shapes and are energetically more stable when compared to the water dimer, hexamer, and heptamer [43]. The structure of the water hexamer is very interesting because of the coexistence of five isoenergetic conformations. This is because the hexamer heralds the transition from 2D to 3D structures. Calculations carried out using very large basis sets indicate that the cage conformer has the lowest energy followed by the book, prism, cyclic, and bag conformers [36]. It has been shown experimentally that the cage conformer is observed at near 0 K [31,32]. The nearly isoenergetic conformer, book structure, was recently observed [44]. In addition, the slightly higher energy conformer of cyclic structure was also observed in Ar matrix [45]. In terms of many-body interactions, the total interaction energy of the water hexamer is made up of 75–90% contributions from two-body interactions. While three-body interactions contribute the remaining 10–25% energy, the contribution of higher order contributions is very small [37]. It should be noted that planar structures have large three-body contributions, which should be taken into account before generating effective two-body water potentials.

A fact which emerges from the structures of the neutral water clusters is that the individual water monomers are linked by different kinds of H-bonds: single proton acceptor (‘a’), double acceptor (‘aa’), single donor (‘d’), single donor–single acceptor (‘da’), single donor–double acceptor (‘daa’), double donor (‘dd’), double donor–single acceptor (‘dda’), and double donor–double acceptor (‘ddaa’). The vibrational frequency shifts are found to be strongly dependent on the number of donors, and much less dependent on the number of acceptors (Fig. 34.2). In the low-lying energy conformers of these small water clusters ($n = 1-11$), ‘ddaa’ water molecules, which are characteristic of bulk water, are not observed [43,46,47]. However, in the water dodecamer, four of the water molecules are of the ‘ddaa’ type [48].

It is useful to mention that the major distinction between small water clusters and bulk water is the presence of free or dangling hydrogens in the former. Thus, the spectra of these small water clusters are quite different from those of bulk water. However, as the size of the water cluster increases, one can expect that the spectral characteristics of larger water clusters would be quite similar to that of bulk water. Interestingly the H-bond types appear as signatures in the O–H vibrational spectra, and hence can provide detailed information on the structure of water clusters. It can be seen from Figs. 34.1 and 34.2 that the fingerprints of the hydrogen bonding networks of the water cluster structures can be observed in the calculated –OH stretch vibrational spectra. These fingerprints which are based on the frequency shifts can distinguish proton-donating and proton-accepting water molecules.

Generally, single H-donor water molecules (‘da’ or ‘daa’) have strong H-bond interactions in neutral water clusters. The red shifts of O–H stretch frequencies monotonically increase up to the hexamer ring structure because of the increased H-bond strength by the decreased bond angle strain. However, this H-bond strength is saturated at the hexamer, and so the heptamer and octamer ring conformers have similar red shifts. The cyclic pentamer cluster exhibits a red shift of ~ 540 cm^{-1}, with respect to ‘da’ water,

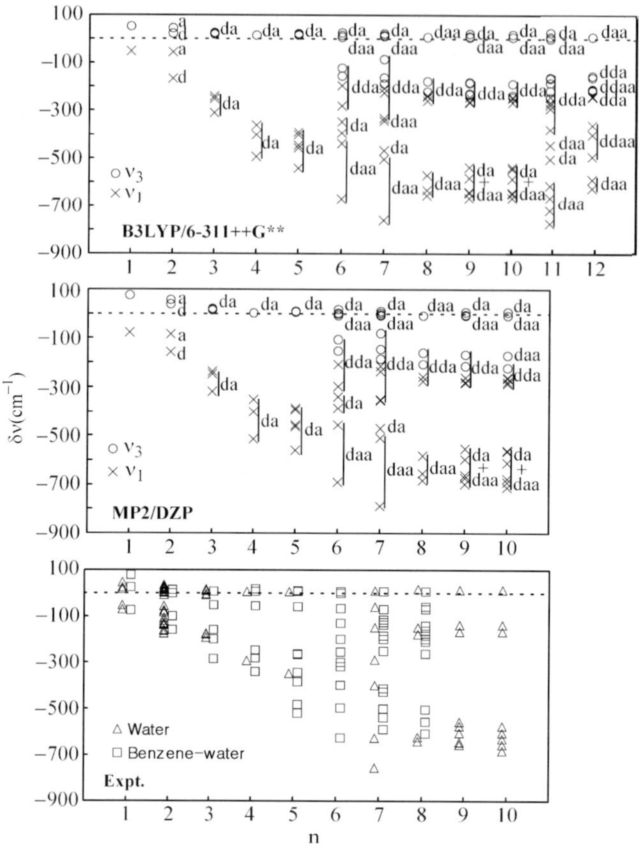

Fig. 34.2. Frequency shifts of the different types of O–H stretching modes predicted at both the B3LYP and MP2 levels for the lowest-energy conformers of the water clusters ($n = 1-12$). (Reproduced with permission from American Institute of Physics [43,48].)

while the hexamer-Cage and heptamer-Prism conformers exhibit red shifts of ~ 670 and 760 cm^{-1}, with respect to the 'daa'-type water. The large shifts are due to the strong polarization effect by the relayed H-bonds.

In addition to the red shifts observed in case of the –OH stretching modes, the HOH bending modes of the neutral water clusters display characteristic blue shifts with respect to that of the pure water monomer. Thus a transition from 'da' to 'dda' type leads to an increase in the blue shifts, with the 'dda' waters in the hexamer and heptamer cages and cubic D_{2d} conformers exhibiting the largest shifts (~ 120 cm^{-1}) for the bending modes.

The order of red shifts of OH stretching frequencies with respect to the average value of ν_3 and ν_1 of the water molecule in the water dimer to dodecamer is 'da'(ν_3) $<$ 'daa'(ν_3) $<$ 'dda'(ν_3) $<$ 'ddaa'(ν_3) $<$ 'ddaa'(ν_1) $\approx<$ 'dda'(ν_1) $<$ 'da'(ν_1) $<$ 'daa'(ν_1). The IR intensities of double proton donor-type waters ('dda' and 'ddaa') in asymmetric OH stretching modes (ν_3) are strong, while the intensities of single donor-type waters ('da' and 'daa') are strong in symmetric OH stretching modes (ν_1). The order of red shifts of bending frequencies with respect to that of monomer is 'ddaa' > 'dda' > 'daa' ≈ 'da'. In the cases of undecamer and dodecamer, the ranges of $-\Delta\nu_3$ and $-\Delta\nu_1$ of 'ddaa' type are 209–299 and 245–496 cm^{-1}, respectively, and that of $\Delta\nu_2$ of 'ddaa' type is 93–135 cm^{-1}. The values of $-\Delta\nu_3$ and $-\Delta\nu_1$ of 'dda' type are 66–242 and 181–436 cm^{-1}, respectively, and that of $\Delta\nu_2$ of 'dda' type is 46–138 cm^{-1} [43,48].

With this background, it is useful to examine how the structures and properties of these neutral water clusters are modulated in the presence of cations, anions, electrons, and π-systems. Most theoretical investigations seem to indicate that cationic water clusters are better described than the corresponding anionic water clusters. This contrasting behaviour is due to the widely different nature of interactions prevailing in these systems. The interaction between water molecules and cations is dominated by ion–dipole interactions. Consequently the hydrogens of the water molecule are oriented away from the ion, and so the water–water H-bonding around the cation is rather suppressed unless the cation (such as Cs^+) has large ionic radius. In addition, the involvement of the oxygen lone pairs in the interaction with the cation also tends to suppress the formation of inter-water H-bonding. As a result, most of the cationic water clusters, if the ionic size is small, tend to be somewhat symmetrically hydrated (Fig. 34.3) [12,49–56]. However, as the size of the cation increases (like in Rb^+ and Cs^+), inter-water H-bonding becomes very important.

The interaction of anions with water clusters is much more complicated because the electron clouds of the anion are very diffuse. This can be noted from the ionic radii of the halide anions (F^- = 1.36 Å; Cl^- = 1.81 Å; Br^- = 1.95 Å; I^- = 2.16 Å) as compared to those of the alkali cations (Li^+ = 0.60 Å; Na^+ = 0.95 Å; K^+ = 1.33 Å; Cs^+ = 1.69 Å).

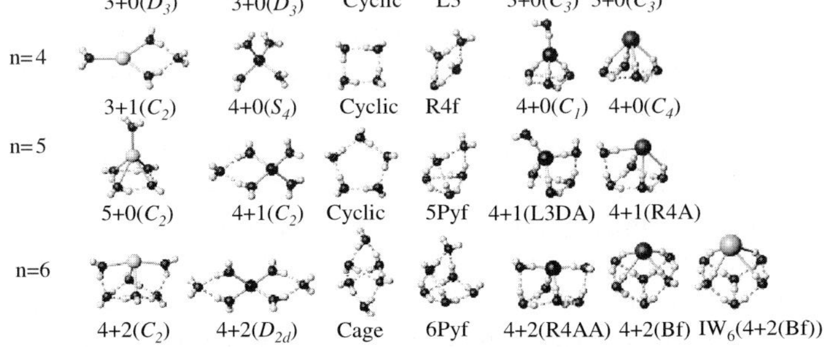

Fig. 34.3. Calculated lowest-energy structures of $M\cdot(H_2O)_{1-6}$ ($M = K^+, Na^+, e^-, F^-, Cl^-, Br^-, I^-$) at 0 K.

Since anions interact with hydrogens of water and each water molecule has two hydrogens, there can be strong repulsions between hydrogens, as the hydrogens point toward the anion. Therefore, anion–water dipole interaction is suppressed, and the inter-water H-bonding interactions around the anion become important. More importantly, the excess electron in an anion needs a large empty space around the anion in which the excess electron can be stabilized by reducing its kinetic energy due to the uncertainty principle. Therefore, while internally bound compact spherical structures of hydrated anions are not stable, a surface-bound nonspherical structure with a large vacant space on the opposite site of the dipole direction tends to be much more stable, as can be seen from the anion water cluster structures in Fig. 34.3 [57–69].

Thus, cation water clusters favour internal structures in contrast to the surface structures favoured by anionic water clusters. This critical difference in the structural preferences of hydrated cation and anion clusters provides important cues for the design of cation- and anion-specific ionophores and receptors. Indeed, we note that most cation receptors have spherical structures, while almost all anion receptors do not have compact spherical structures but have a vacant space around the anion binding site without full coordination (which might be exceptional for the F^- ion with strong electronegativity for which the excess electron is strongly bound to F^- due to its small ion radius). However, as the temperature increases, the hydration structure tends to be more spherical due to entropy effects.

In light of some recent experimental work, it is useful to discuss one of the salient observations of our calculations on e^--water clusters. Apart from the issue of the internal *vs.* surface solvation, one of the enduring enigmas of small e^--water clusters is the geometry of the water molecules in the vicinity of the excess electron. We were the first to propose that the excess electron binds in the vicinity of a water molecule which accepts two hydrogen bonds from adjacent molecules but does not itself donate any hydrogen bonds to the hydrogen bonding network [67,69]. Interestingly, this prediction of ours was confirmed by some elegant experiments carried out very recently [70].

In the above discussion, we only elaborated on the structures and properties of these water clusters, when only a single cation or anion is present. However, from both the chemical and biological points of view, it is interesting to examine the modulation of the structures of these water clusters, when both cations and anions are simultaneously present in the system. One of the classic cases which can be discussed in this context is the dissociation of acids [71,72] and bases [73,74]. In the case of hydrogen halides, we have shown that in the presence of four water molecules, the dissociated states of hydrogen chloride, bromide, and iodide are more stable than the corresponding undissociated forms, while an undissociated hydrogen fluoride is more stable than the dissociated form even when it is hydrated by six water molecules (Fig. 34.4) [71].

The structures of aqueous clusters formed by the interaction of a cation, an electron, or an anion, with various sizes of water clusters bear no structural resemblance to the parent neutral water clusters because of the dominance of electrostatic interactions. However, things are very different in the case of interactions involving π-systems with varying sizes of water clusters. This is because of the fact that the interactions existing between

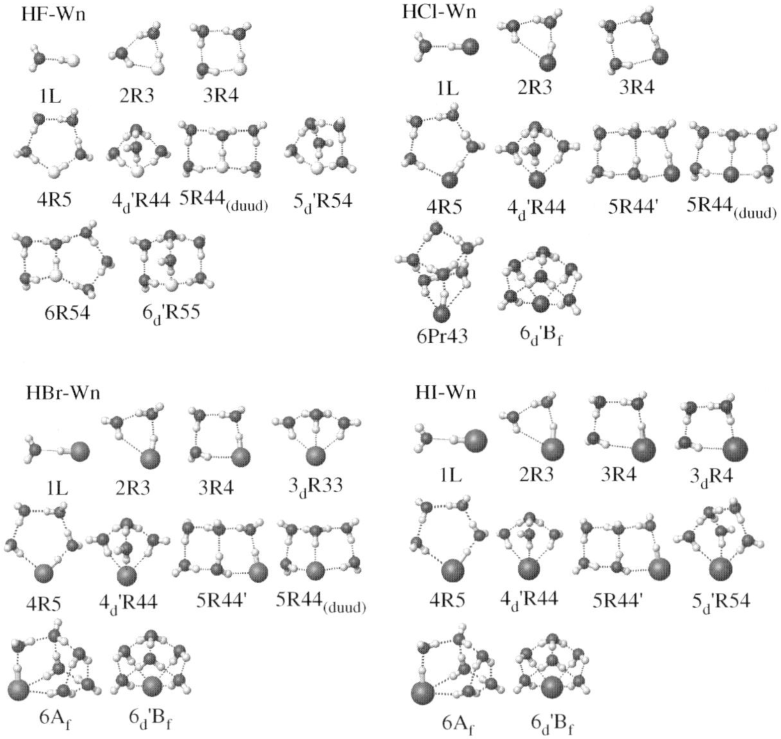

Fig. 34.4. Lowest-energy structures of hydrated hydrogen halides. (Reproduced with permission from American Institute of Physics [71].)

the water molecules in these clusters are much stronger than the π or σ type of interactions which exist between the water clusters and the π-systems. As a result, there are little changes in the gross structural features of the water clusters in the neutral state and in their complexed states with the π-systems. This feature can particularly be noted in the work of Dykstra and Zwier [75] on benzene–water clusters and our work on the structure and vibrational spectra of the π-$(H_2O)_{1-6}$ and π-$(CH_3OH)_1$ (π = benzene, toluene, fluorobenzene, chlorobenzene, *p*-difluorobenzene, anisole) systems [76–83], wherein the experimental infrared spectra were reproduced by theoretical calculations. Similar observations were made in the case of benzene–$(H_2O)_6$ and benzene–$(H_2O)_8$, wherein a cage water hexamer and a cubic water octamer interact with the π-system [33,84].

Given the scope of this review, we desist from a detailed discussion of the energetics and vibrational spectra of the solvated aqueous clusters. However, the study of solvation phenomena can be extended to the hydration phenomena of biologically important chemical systems using Monte Carlo and molecular dynamics simulations [85–94]. More information on aqueous clusters can be obtained from our extensive work on these systems cited above.

34.3.2 Metallic clusters

Unlike aqueous clusters with predominating electrostatic interaction, the interactions in metallic clusters can be classified as either continuous transitions or metallic [95,96]. In the former interaction, there is a continuous shift from one type of bonding to another type as function of a chemical or physical variable. These types of bondings include covalent, ionic, metallic, and van der Waals interactions. Thus, the accuracy of the results on metallic clusters to a large extent depends on the ability of the theoretical method to describe several types of interactions. In the case of the clusters of higher transition metals, the theoretical method should also be able to describe relativistic effects accurately.

Noble metal clusters are of interest because they serve as prototypical models of nanoelectrodes, nanowires, etc. Until very recently, most studies of noble metal clusters were focussed on the pure metallic species. However, in light of recent observations which indicate that these mixed clusters exhibit unique electronic, magnetic, optical, and mechanical properties, it is useful to compare their properties with the corresponding pure metallic clusters. One of the offshoots of such a comparison is that it allows one to obtain a microscopic understanding of alloying.

Theoretical studies of varying sizes of neutral and anionic gold (Au) and silver (Ag) clusters indicate that they exhibit an even–odd oscillation in their stability and electronic properties (Fig. 34.5) [17]. Thus, clusters which have an even number of atoms tend to be more stable in the neutral state, while those having an odd number of atoms tend to be more stable in the anionic state. Since the 6s orbital energy of Au is almost as low as 5d orbitals, the strong s–d hybridization in Au favours 1D and 2D structures in the case of the gold clusters.

This structural preference, due to the relativistic effects, explains the ductility of small gold clusters. In sharp contrast, silver clusters exhibit a strong preference to exist as 3D structures with spherical coordination because the valence orbitals are predominantly of the s-type. This preference can be noted for clusters larger than the hexamer. A similar argument can also be employed to explain the lower coordination number of the Au atom in the gold clusters as compared to the Ag atom in the silver clusters. This preference in coordination reflects itself in the location of the Au and Ag atoms in the corresponding binary clusters of gold and silver, with the Au atoms being located on the boundary, while Ag atoms are generally located on the inner side.

In the anionic systems of both pure and mixed clusters, there is marked tendency to adopt lower dimensional conformations as compared to the corresponding neutral clusters. However, it should be noted that in the mixed clusters, the conformational preferences is strongly correlated to the percentage of Au and Ag atoms in the cluster. Thus, small Au clusters tend to exist as 1D or 2D structures, which is in contrast to the 3D structures for the Ag clusters. Given the higher energy of the Ag 5s as compared to the Au 6s orbital, charge transfer from the Au to Ag atom is noted in the mixed clusters. This together with the predilection for the Au atom to be located on the boundary indicates that the core of the mixed clusters is positively charged, and the surface, negatively charged. The marked ease for the formation of the mixed gold–silver clusters and, as

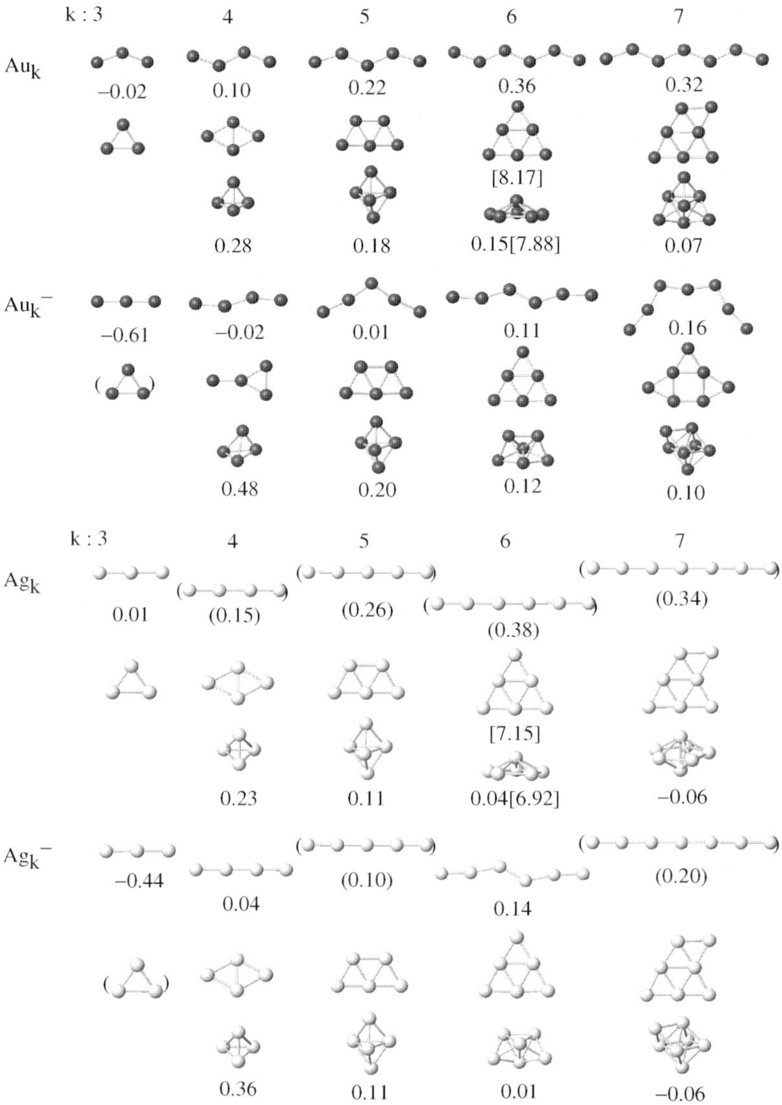

Fig. 34.5. Lowest-energy structures of neutral and anionic $(M)_n$ (M = Au, Ag) clusters. (Reproduced with permission from American Chemical Society [17].)

a consequence, their alloys is due to the significant electrostatic stabilization accruing from the charge transfer from the Au to the Ag atom.

In order to obtain more insight into the role of dimensionality, these theoretical investigations of small clusters were extended to nanowires, thin films, and the bulk systems [97]. The 1D and 2D structures, which are obtained in isolated and small gold clusters, are no longer preferred in the bulk. Thus, in low-dimensional structures of gold, there is a strong sharpening of the d bands, which vastly enhances the corresponding density of states and raises the band edges. Given the technological need for 1D and 2D structures, we note that one can obtain lower dimensional structures by modulating the charge transfer in gold alloy nanowires by injecting s electrons into the gold wire without distorting its band structure [98]. The s band of the alloying metal should therefore possess energy levels similar to the 6s orbital of gold. Of all the s band metals, beryllium, magnesium, zinc, cadmium, and mercury have their valence energy levels closest to that of the 6s orbital of gold, and hence can form alloys with gold. Indeed, density functional calculations carried out on free-standing infinite monoatomic gold wires alloyed with both magnesium and zinc indicate that in sharp contrast to pure gold nanowires, the 1D and 2D structures of both gold–magnesium and gold–zinc alloys exhibit distinct minima in the plots of the cohesive energy, indicating that both forms are energetically accessible. We note that both zinc 4s and magnesium 3s bands display p character near the Fermi energy. This sp hybridization, which is absent in pure gold nanowires, favours a linear structure in the case of these alloyed nanowires.

Our group has also investigated the 1D, 2D, and 3D systems of most transition metals [99]. Only Fe, Co, and Ni show the ferromagnetism in the bulk, while many atoms in 2D show ferromagnetism. In particular, in 1D all the transition metals except for Ir are predicted to be ferromagnetic. Indeed, our results are in agreement with the recent experimental observation that 1D systems of Co, Pd, and Pt are ferromagnetic [100].

34.3.3 Weakly bound clusters

Unlike both aqueous and metallic clusters, clusters containing π-systems are relatively weak. Indeed the weakness of these interactions was responsible for doubts about their existence and relevance in chemical and biological systems for a very long time. Extensive work in our laboratory over the last several years, however, indicates that the magnitude of this interaction depends both on the nature of the π-system and that of the countermolecule with which it is interacting. Thus, when the countermolecule is a metal cation, electrostatic and induction energies dominate the interaction energy [101–103]. Dispersion interactions dominate when the countermolecule is either a rare gas atom or a nonpolar molecule (gas dimers, hydrocarbons) [104]. A combination of electrostatic and dispersion interactions govern the interaction energy, when the countermolecule is polar (water or methanol clusters, Lewis acids) [76,77,82,105–107]. Unlike most other clusters, the magnitude of repulsive interactions plays a critical role in governing the observed geometry of the cluster [104]. The presence of electron-donating substituents on the π-system leads to complexes with enhanced interaction energies, when compared to benzene. This enhancement is due to an increase in both dispersion and electrostatic interactions. On the other hand, the geometries of complexes formed with aromatic rings having electron-withdrawing substituents are extremely dependent on the characteristics of the interaction energy. Thus, when the interaction energy is dominated by dispersion interactions, a π type of complex is formed. When electrostatic energies dominate the interaction energy, a σ type of complex involving the electron-withdrawing substituent is formed. When the countermolecule is a water or methanol cluster, the size of the cluster also influences the nature of the interaction with the π-system, resulting in a crucial influence on the nature of the resulting structure (σ or π).

In the context of nanomaterial design, cation–π interactions [101–103,108,109] are important interactions involving π-systems. Since the interactions involving cations and π-systems are mediated through the negatively charged π-electron cloud, early investigations were of the opinion that purely electrostatic considerations could explain the observed trends of their interaction energies. However, it was soon realized that a purely electrostatic model could not explain the exceptionally high interaction energies observed in the case of the ethene systems. Thus, the most recent experimental estimates of the enthalpies of binding (ΔH_0) of Na^+ to ethene and benzene are -10.7 ± 1.0 and -21.5 ± 1.0 kcal/mol, respectively. Even after accounting for geometrical considerations, the general impression would be that the enthalpy of binding of Na^+ to benzene is three times that to ethene, because benzene has three times the number of π-electrons. The issue is more complicated when one compares the interaction energy of the benzene-Na^+ and pyrrole-Na^+ complexes. Though both of them have equal number of π-electrons, the cation–π interaction in the latter is stronger than in the former. It therefore becomes apparent that one has to take into account the differential electron density distributions of these π-systems to explain the trends in interactions. In one of the early works on this subject, it was shown that dispersion energies play an important role in the interaction of organic cations with benzene [101].

A detailed theoretical investigation of the cation–π interaction involving both metallic and organic cations indicated that both electrostatic and induction energies are major contributors to cation–π interactions involving alkali metal cations [103]. The interaction between the highest occupied molecular orbital (HOMO) of the π-system and the empty s orbital of the metal cation is responsible for the large induction energy. Consequently, systems exhibiting an enhanced overlap of the constituting molecular orbitals (MO) would exhibit larger induction and hence larger interaction energies. A smaller cation can approach much closer to the π-system because of the reduced exchange-repulsion. However, in the case of the interaction of these π-systems with transition metal cations like Ag^+, a closer approach of the small-sized Ag^+ cation (1.26 Å) leads to a large increase in the repulsive energies [17]. The higher repulsive energies are to a large extent compensated by the higher charge density and the presence of d electrons, which leads to large interaction energies. One also notes that the nature of the π-system has only a small effect on the magnitude of interaction energy.

The magnitudes of the electrostatic and induction energies in the case of the organic cation complexes of these π-systems are much smaller than those observed in case of the π–alkali metal cation complexes, and hence the contribution of dispersion energies becomes vital. The inclusion of electron correlation in calculations involving π-systems and organic cations is important, because the inductive energies described by the π–σ^* interaction (where π is the HOMO of the π-system and σ^* is the LUMO of the organic cation) are magnified [101]. This magnification of the induction energies upon inclusion of electron correlation is due to the depletion in the electron density from the centres of the bonds and a concurrent increase in the shells around the atomic nuclei. Consequently the organic cation can have a closer approach to the π-system and as a result both induction and dispersion energies are magnified.

Since we have already talked about cation–water interactions, it is useful to compare their characteristics with the cation–π interaction. The distinguishing factor between interactions of cations with water and π-systems is the magnitude of the electrostatic energy, with it being dominant in case of the former (Fig. 34.6). This marked difference can be attributed to the difference in the nature of the donor, with the donor lone pair of the water oxygen atom being more localized than the diffuse electron cloud of the π-system. This leads to a larger electrostatic contribution and a much smaller induction contribution in the case of the water complexes. The repulsive energies are higher in the case of water complexes because the cation encounters another atom in contrast to a bond in ethene or the centre of the aromatic ring (no atoms or bonds).

In the context of nanomaterial design, these findings are significant because a greater electrostatic contribution implies that the magnitudes of the interaction energies are more susceptible to the dielectric of the environment. It was also shown in the course of our studies that the nearly similar interaction energies of benzene with the ammonium and potassium cations result from a balance of dispersion and induction energies because the electrostatic and exchange energies are nearly similar and hence cancel out. However, the ammonium cation complexes exhibit a larger contribution of dispersion energies

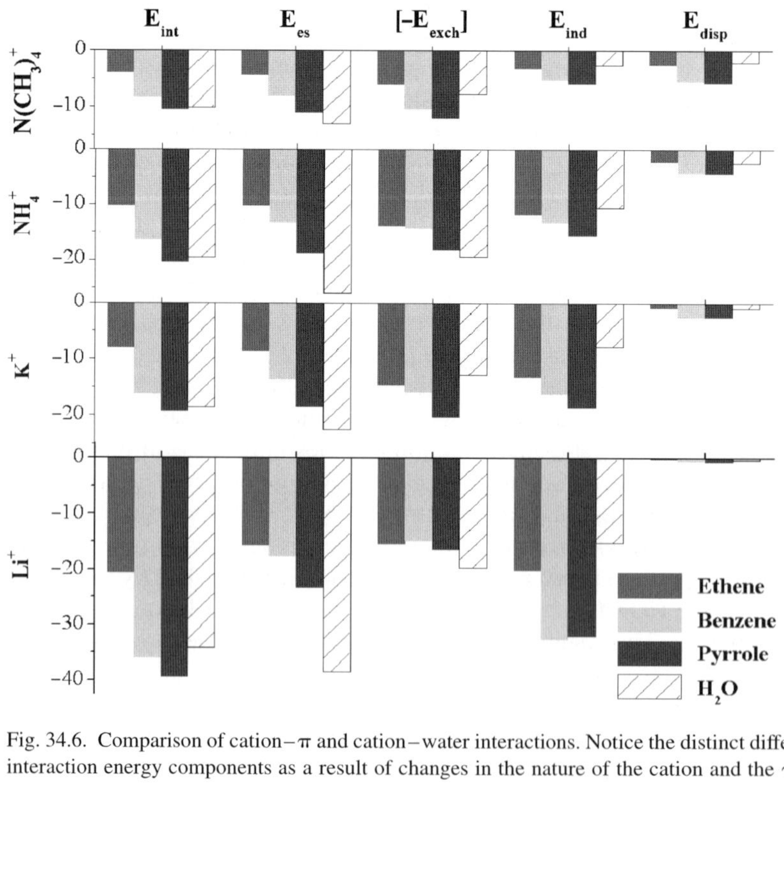

Fig. 34.6. Comparison of cation–π and cation–water interactions. Notice the distinct differences in the various interaction energy components as a result of changes in the nature of the cation and the π-system.

[101–103]. In a subsequent section, we show how suitable receptors specific for the ammonium cation could be designed by enhancing the contribution of the dispersion energies.

In our recent work, we have also investigated the interactions of anions with electron-deficient π-systems. Except for the greater relevance of dispersion energies, the anion–π interactions are in many respects similar to the cation–π interactions [110].

Apart from a few experimental reports detailing the interaction of Lewis acids with π-systems in the context of donor–acceptor complexes and electrophilic aromatic substitution reactions, little was known about either the magnitudes or the exact nature of their interaction. However, in the course of our extensive investigations, it was found that the halides of aluminium, boron, and hydrogen form fairly strong noncovalent complexes with both benzene and ethene [16,105–107]. The magnitude of the interaction depends on the nature of the Lewis acid, with the benzene complexes of aluminium halides exhibiting interaction energies (8 ~ 15 kcal/mol), which are significantly larger than the corresponding complexes of the boron halides (3 ~ 6 kcal/mol). As far as our knowledge goes, the interaction energies exhibited by the aluminium halide complexes of benzene are the highest for a neutral (uncharged species) binding to it. Recent works by Olah and Seddon highlight the importance of our proposed π–Lewis acid interactions in understanding the nature of catalysts and ionic liquids [111,112].

Complexes exhibiting the π–H interaction are of interest because they are the prototypes of a typical hydrogen bond in aromatic compounds [8,113,114]. Apart from their biological and chemical relevance, an investigation of these interactions offers the best opportunity to highlight the distinct differences between olefinic and aromatic π–H interactions [16]. Our studies of the π–H interaction were done by carrying our theoretical investigations of the complexes of the first-row hydrides with ethene and benzene [16]. Most of these differences stem from the widely differing electron density profiles of ethene and benzene. Thus, in the benzene complexes, the hydride aims at the midpoint of the ring (where there are no nuclei or bonds), but in ethene, at the midpoint of the double bond. As we progress from CH_4 to HF, the increase in the repulsive exchange energies is more pronounced in the ethene than in the benzene complexes. This leads to a smaller variation in the intermolecular distances in the benzene complexes. Consequently, the magnitude of the dispersion energies is nearly independent of the nature of the hydride, because the magnitude of the dispersion energies is nearly proportional to the number of electrons participating in the interaction and the intermolecular distance. Similar to what is observed in the cation–π complexes, one also notes an increase in the magnitude of the induction energies of the hydride complexes of both ethene and benzene upon inclusion of electron correlation.

The π–π interactions are one of the most intriguing noncovalent interactions, in the sense that the negatively charged and diffuse electron clouds of the π-systems exhibit an attractive interaction [113–119]. This interaction is predominated by dispersion interactions, when the π-systems possess nearly similar electron densities. However, when one of the systems is electron-rich (benzene) and the other electron-deficient (hexafluorobenzene), the resulting complexes are bound by induction interactions with the negative charge being transferred from benzene to hexafluorobenzene [120,121]. The acetylene dimer is one of the simplest systems exhibiting a π–π interaction [122,123].

Though a parallel displaced conformer, which involves a stacking of the two acetylene π-systems exhibiting a $\pi-\pi$ interaction, the most stable conformer is a T-shaped structure, which involves the interaction of hydrogen of acetylene with the π-system of the other [122]. In the case of the ethene dimer, the most stable structure has a D_{2d} symmetry, which involves the formation of a quadruple hydrogen bond [124]. The interaction of two benzene rings (benzene dimer) has been widely investigated both experimentally and theoretically. The experimental estimates of the interaction energy is of the order of ~2 kcal/mol, which indicates that the attraction is appreciable and significantly influences the interaction of phenyl rings in solution or other environments, in addition to other factors such as solvophobic effects [116–119,125]. The interaction is predominated by dispersion interactions and as in the case of the acetylene dimer, the benzene dimer can manifest itself in any one of the structural forms: stacked, T-shaped, or parallel-displaced. The available evidence seems to indicate that the T-shaped and parallel-displaced conformations are most stable [116–119]. However, it should be noted that the isolated benzene dimer is highly fluxional and can coexist in any of the three structural forms. In a subsequent section, we take advantage of the fact that the interconversion between different conformers or the orientations of two interacting π-systems can be exquisitely controlled [113]. In order to have maximum control on this interconversion, we use the electrochemically and photochemically active π-systems (quinone and hydroquinone) [126,127].

While we have investigated several other interactions involving π-systems, we believe that they have little relevance in our subsequent discussion on nanomaterial design. Mention should, however, be made of the interaction of paramagnetic atoms with π-systems, which is discussed in our account of $A@C_{60}$ (A: N, P, As, O, or S) [128].

34.4 IONOPHORES, RECEPTORS, AND CHEMICAL SENSORS

A lot of interest has been evinced in the design and synthesis of receptors capable of binding anionic or cationic guests because of their potential applications in environmental and biological processes. Unlike nanoclusters and nanowires, the theoretical challenge in these systems is to describe the affinity and selectivity of an organic system towards a charged or neutral species. The situation is complicated because the affinity and selectivity have to be displayed in the presence of a number of competing factors like counterions, solvents, etc. We begin the discussion of our work on ionophore/receptor design with one of the seemingly intractable problems of contemporary biochemistry: the selective recognition of the ammonium cation (NH_4^+). Much of the problem is due to the nearly equivalent sizes of NH_4^+ and the potassium cation (K^+). The first step in the receptor design was that high selectivity for NH_4^+ could be achieved with cation–π interactions, if the receptors have an optimal space to capture NH_4^+ and exhibit strong interactions towards NH_4^+. However, the ionic radius of K^+ is nearly similar to that of NH_4^+, so spatial differentiation is not useful. On the other hand, the difference in coordination numbers can be utilized. K^+ favours a coordination number of six, while NH_4^+ favours only four. Furthermore, one also has to take into account the directional H-bonds involving NH_4^+ cations, to describe the higher selectivity for

NH_4^+ over K^+. Our initial calculations indicated that a benzene-based tripodal system with imidazoline arms (Fig. 34.7a) possessed vacant sites for the interaction with only one solvent molecule, while the K^+ ion has three vacant sites for three solvent molecules. In order to maximize the affinity and selectivity of these receptors for NH_4^+, it becomes important to maximize the π-electron density of the receptor. Indeed, receptors with enhanced π-electron density by trimethylated phenyl ring with the strong proton-withdrawing subunits exhibit much higher affinities and selectivities [129].

Given this background, an extended concept has been applied to the receptor design for a biologically important molecule, acetylcholine [130]. The receptor should have higher affinity and selectivity for acetylcholine over NH_4^+. This requires enhanced dispersion interactions and diminished ionic interactions, which is met by replacing the imidazole arms of the NH_4^+ receptors with pyrrole (Fig. 34.7b). These theoretical inferences were confirmed by experiments.

Interactions involving anions are very different from those of cations. Since anions are more polarizable and hence more susceptible to polar solvents than cations, it becomes important to take into account solvent effects. Based on molecular dynamics simulations combined with *ab initio* calculations, highly selective anion ionophores have been designed [131,132]. Enhanced dipole moments (Fig. 34.7c) were employed by attaching a strategically placed electron-withdrawing group. This approach would also aid the design of novel functional molecular systems and biologically important chemosensors. Utilizing the $CH^+ \cdot X^-$ H-bonds, fluorescent photoinduced electron transfer chemosensors for the recognition of $H_2PO_4^-$ have also been designed and synthesized (Fig. 34.7d) [133, 134]. In addition, we have also been successful in designing cyclopeptides as amphi-ionophores [135,136]. Recently, we have extended this concept to design receptors specific for the biologically important phosphates (ATP and GTP) (Fig. 34.8) [137].

It would be appealing to explore the possibility of π-based materials being used as ionophores [138–140]. Belt-shaped carbocyclic conjugated systems (annulenes,

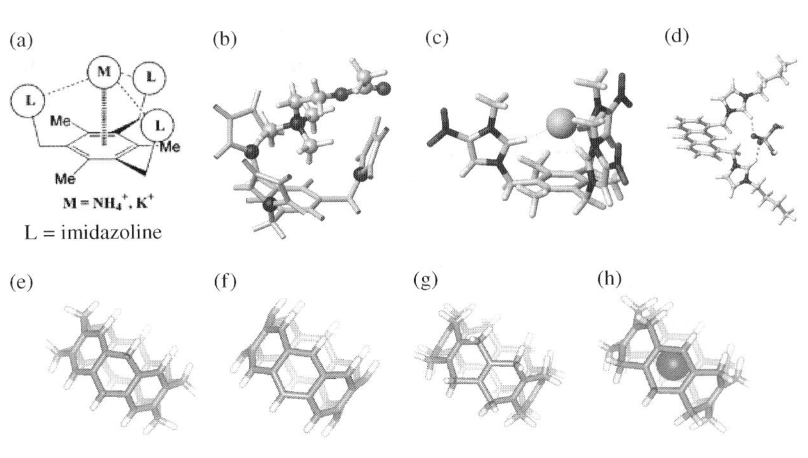

Fig. 34.7. Receptors for NH_4^+ (a), acetylcholine (b), Cl^- (c), $H_2PO_4^-$ (d), and structures of collarenes (e), cyclacenes (f), beltenes (g), and Rb^+-complexed [8]beltene (h).

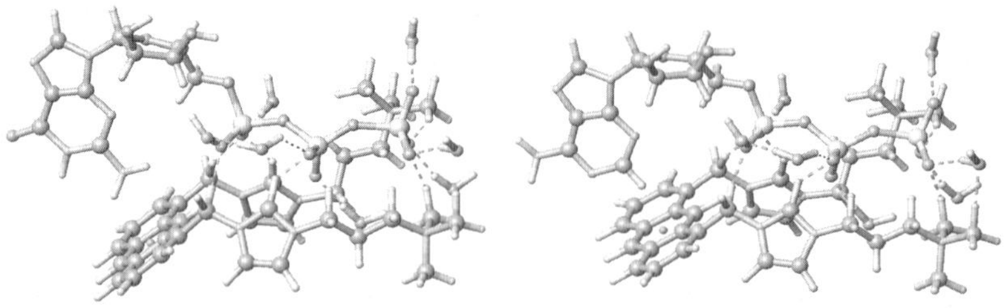

Fig. 34.8. Computed geometries of 1-GTP and 1-ATP complexes. (Reproduced with permission from American Chemical Society [137].)

beltenes, cyclacenes, and collarenes) are closely related to other carbon-based systems containing curved surfaces. These include fullerenes and carbon nanotubes. The ion binding characteristics of these carbon materials have been unravelled through *ab initio* calculations, Monte Carlo simulations, and molecular dynamics simulations of collarenes (benzene rings linked by methylene linkages), cyclacenes (comprised of only benzene rings), and beltenes (ethene groups linked by methylene linkages) and of their complexes with various cations (alkali, alkaline earth metal, and organic cations) in both gas and aqueous phases (Fig. 34.7e). Additionally, suitable substituents could also enhance their binding affinities and selectivities. In particular, the designed molecules could be modified to be soluble in polar solvents by adding hydrophilic groups on the edges of the molecules.

Till very recently, the role of intermolecular interactions in catalytic mechanisms of enzymes has never been discussed. In the course of our work on ionophores and receptors, we realized that the catalytic mechanism of some enzymatic reactions, which involve charged H-bonds, could also be explained using theoretical methods. Towards this end, we have investigated the catalytic mechanism of one of the most proficient enzymes, Ketosteroid Isomerase (KSI). Our calculations reveal that in addition to preorganization, SSHB, charge buffering/dissipation plays a very important role in the mechanism of the enzyme (Fig. 34.9) [141–144]. Charge buffering/dissipation is crucial

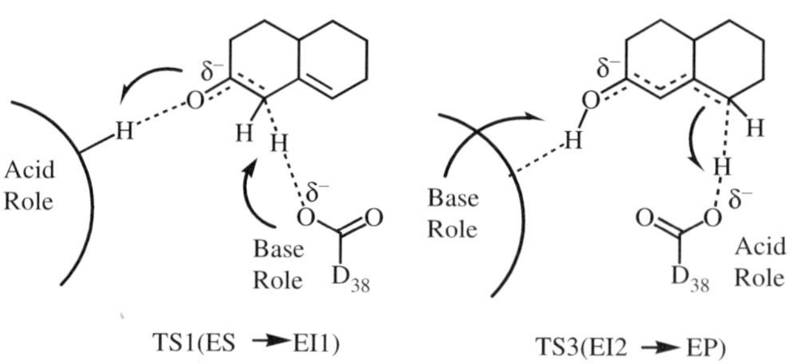

Fig. 34.9. Catalytic role in stabilizing the transition state *via* an amphoteric role through the short strong H-bond along with preorganized structure of and charge buffering by (or dissipation to) catalytic residues. (Reproduced with permission from American Chemical Society [141].)

because it lowers the transition barrier, by dissipating a portion of the excess electronic charge build-up around the transition state to the catalytic residues. This is mostly mediated by the SSHB [145,146].

34.5 NANOMATERIALS

Since the discovery of fullerenes and carbon nanotubes, a lot of effort has gone into the discovery of other interesting allotropes of carbon with unusual structural characteristics and novel physical properties. Towards this end, the geometries, electronic structures, and energetics of small carbon nanotori were investigated employing both tight-binding and semi-empirical quantum-chemical methods [147].

It should be mentioned here that the very large size of these carbon-based materials precludes the use of high-level quantum methods. One has to therefore take recourse to the use of semi-empirical or tight-binding methods. As can be seen from Fig. 34.10, the structures and electronic properties of smallest nanotori exhibit interesting metal, semiconductor, and insulator characteristics depending on nanotube building blocks.

An interesting offshoot in the context of carbon-based nanomaterials is the role of external perturbations in modulating their physical and chemical characteristics. These perturbations can include cations or neutral atoms. In this context, we examined the magnetic properties of exohedral fullerenes of alkali metal fullerides (A_xC_{60}, A = Na, K, Rb, Cs) [148–150] and the spin properties of endohedral fullerenes ($A@C_{60}$, A = N, P, As, O, S) [128,151] (Fig. 34.11). The most interesting aspect of the experimental

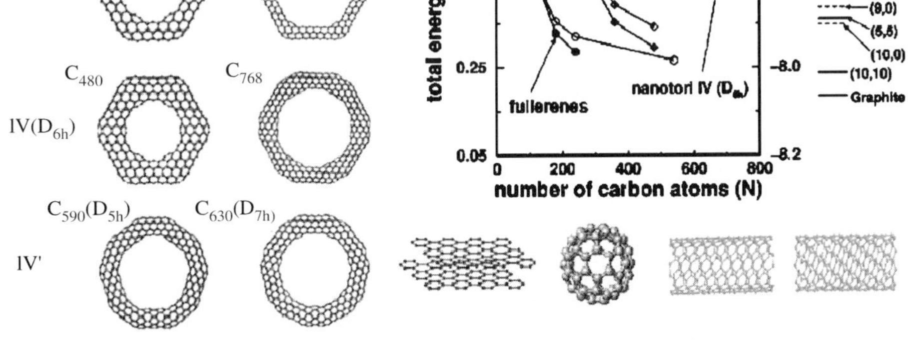

Fig. 34.10. Optimized structures of various types of carbon nanotori and their stability compared with graphene, fullerenes, and nanotubes. (Reproduced with permission from American Physical Society [147].)

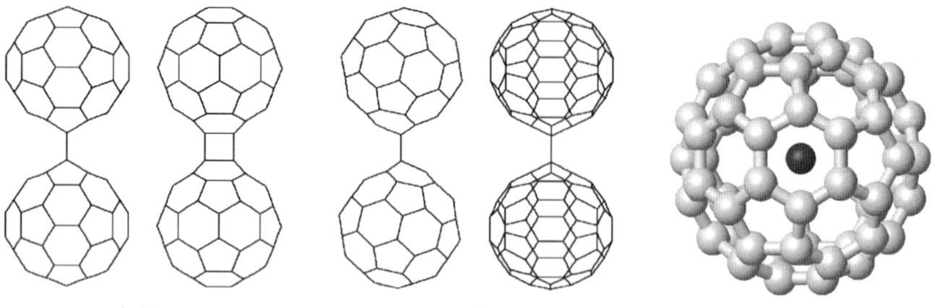

d_D_{2h} s_C_{2h}

Fig. 34.11. Ground triplet (d_D_{2h}) and excited singlet (s_C_{2h}) states of $(C_{60})_2^{2-}$ with two views (for the alkali cation doped exohedral fullerenes), and spin-containing endohedral fullerenes [128,148].

investigation of endohedral fullerenes was the fact that the encapsulated nitrogen, which possesses three unpaired electrons and is paramagnetic in nature, is totally inert within C_{60}. Our calculations, however, indicated that the interaction observed in the case of N@C_{60} is predominantly dispersive in nature. These systems are interesting because endohedral fullerenes containing paramagnetic atoms could be utilitized to design the quantum computers [128].

There are several advantages of using hydrogen bonds to design nanomaterials [126, 127,152–156], and in particular nanotubes, because these nanotubes have potential applications as artificial biological channels, drug delivery systems, nanochemical reactors, etc. One of the spectacular aspects of a recent report on the self-assembly of an organic nanotube from nontubular units of calix-4-hydroquinone (CHQ) was that the theoretical design preceded the actual experimental synthesis and investigation of the X-ray structure. Apart from highlighting the robustness of the theoretical approach, this study also provided several insights into the mechanism of self-assembly of CHQ nanotubes.

In the absence of water, for each CHQ monomer, the number of dangling H atoms is 4, while in the presence of water, these dangling H atoms of CHQs form chains HQ–(water–HQ–HQ–)$_n$water. Although the strength of 1D short H-bonding interaction ($\sim$10 kcal/mol) is similar or slightly stronger than the strength of the $\pi-\pi$ stacking interaction, the assembling along the 1D short H-bonds relay should be much more favourable because the number of H-bonds is three times the number of $\pi-\pi$ stacks. Indeed, in experiments with water, CHQs are assembled to form long tubular structures with four infinitely long SSHB arrays.

The CHQ tubes assemble to form long tubular structures in the presence of water, which in turn form bundles with intertubular $\pi-\pi$ stacking interactions (Fig. 34.11), resulting in crystals with well-ordered 2D arrays of pores. The structures of these pores were utilized for the synthesis of encapsulated nanowires [154], which would be described in the next section. A needle-like nanotube bundle exhibits the infinitely long 1D H-bonding network between hydroxyl groups of CHQs and water molecules and well-ordered intertubular $\pi-\pi$ stacking pairs (Fig. 34.12) [126,127]. The geometries of the calculated $\pi-\pi$ stacks are very close to the $\pi-\pi$ stacks in the X-ray structure [127].

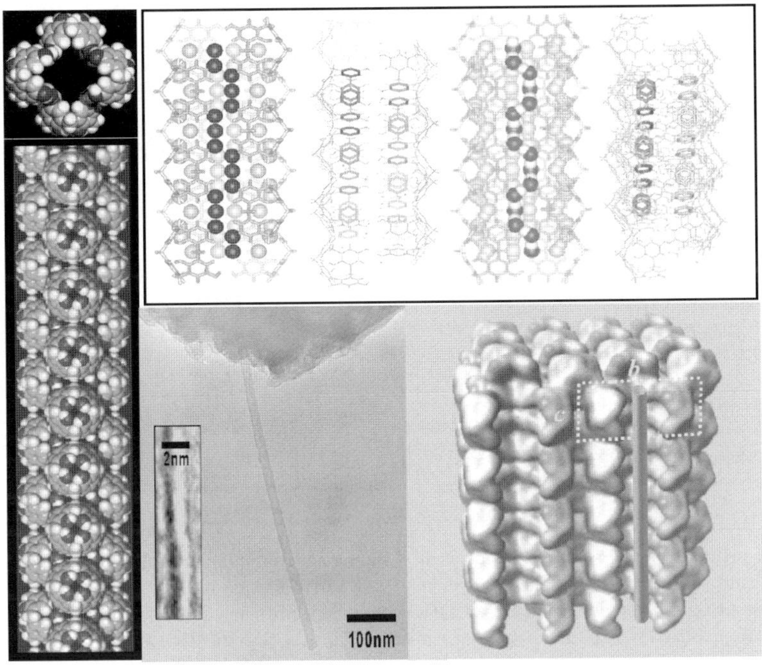

Fig. 34.12. Calix[4]hydroquinone nanotubes: structure, longitudinal one-dimensional H-bond relay *vs.* intertubular $\pi-\pi$ stacking, the HREM image of a single nanotube, and the water-accessible surface of the tubes. Each tube has four pillar frames of short H-bonds, and the pore size is 8×8 Å^2. The unit cell is drawn by the dashed lines. (Reproduced with permission from American Chemical Society [126,127].)

As was mentioned earlier, CHQ nanotube arrays can be utilized in promising templates for nanosynthesis. Redox reaction of the nanotube in the presence of silver nitrate leads to the formation of silver nanowire arrays in the pores (pore size of 8×8 Å^2) of the CHQ nanotube. The wires exist as uniformly oriented 3D arrays of ultrahigh density. The driving force for the formation of these nanowires is the free energy gain due to the reduction–oxidation process [154,155]. It was experimentally observed that the resulting nanowire is comprised of four dumbbells, each of which contains two silver atoms, superimposed on one another, and crisscrossed in their length.

The theoretical characterization of the reduced form of the CHQ nanotube was carried out using plane-wave pseudopotential methods [156]. Our calculations indicated that upon reduction with silver nitrate, the CHQ nanotubes get transformed to the corresponding calix[4]quinone-hydroquinone (CQHQ) nanotubes, whose band gaps of 0.3 eV indicate that they are semiconducting in nature. The gross structural feature of CQHQ nanotubes is similar to that of CHQ nanotubes, with well-ordered H-bond arrays and intertubular $\pi-\pi$ stacking pairs. In the CQHQ nanotubes, there are only two infinitely long 1D H-bond arrays per nanotube because two hydroxyl groups are transformed to the corresponding reduced forms. Simultaneously, silver cations get transformed to metallic silver. Upon reduction of the CHQ nanotubes, the silver atoms

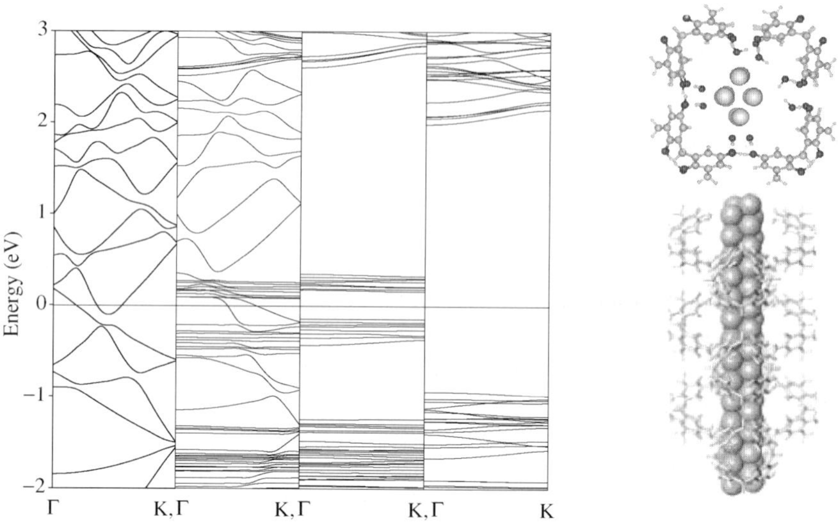

Fig. 34.13. Top and side views of a silver nanowire inside a calix[4]quinone-hydroquinone (CQHQ) nanotubes (right figures), and the band structures of an isolated silver nanowire (left—first on the left figure), a silver nanowire encapsulated in a CQHQ nanotube (second), a CQHQ nanotube (third), and a calix[4]hydroquinone (CHQ) nanotube (last). (Reproduced with permission from American Physical Society [156].)

are located within the reduced CQHQ nanotube (Fig. 34.13). In the case of 2/2 nanowire, the predicted cohesive energy is 2.4 eV, which is 0.9 eV smaller than the bulk value. The encapsulation of a silver nanowire within the CQHQ nanotube leads to several additional states in the band gap region, which are similar to that of an isolated silver nanowire. In this case two s channels cross the Fermi energy level, which indicates the existence of quantum conductance.

We had talked about harnessing the interaction of photons, electrons, protons, or charged species with molecular systems in the design of new nanodevices [157–160]. While photochemical and electrochemical switching devices have been extensively investigated, most of the photochemical switching devices are limited to the linear regime. However, it would be of interest to use nonlinear optical properties for an efficient memory device. In investigations of the interaction of photons with matter, it is essential that the employed theoretical method should be capable of characterizing both open-shells and excited states. The theoretical investigations of 1,2-bis-(3-thienyl)-ethene derivatives indicate that after photoswitching, the resulting π-conjugated closed forms exhibit highly nonlinear optical properties [161,162]. The substitution of suitable donors and acceptors on certain strategic positions of these ethene derivatives, however, makes the closed form optically nonlinearly active and the resulting molecular system behaves as an efficient nonlinear optical switch. It should be noted that the above discussion on nonlinear optical devices involves the breaking and formation of bonds. It is interesting to explore the possibility of devices, which rely entirely on conformational or enantiomeric changes.

Peptides are well-known biological systems, whose conformational characteristics are well understood. To date, the conventional wisdom was that a polypeptide can exist only as right-handed helix. However, calculations reveal the feasibility of a left-handed helix [163]. Most of these calculations on these peptide systems were carried out using molecular dynamics simulations based on empirical potentials. The calculations indicate that the diameter of the left-handed helix is larger than the conventional right-handed helix. The left-handed helix is stabilized when the terminal residues are charged, because the dipole moments of carbonyl groups for the former are aligned opposite to those of the latter. Thus, a molecular dynamics simulation of a poly-alanine peptide capped with neutral amino and methyl groups (CH_3–$(Ala)_n$–NH_2) under neutral terminal charge conditions yielded a right-handed α-helix pattern, in about ~1.8 ns. However, when the terminals are charged (NH_3^+–$(Ala)_{30}$–COO^-), a left-handed λ-helix is formed in about ~4.0 ns. During the formation of the left-handed λ-helix, helix nucleation first occurs at the terminal sites (in particular, near the N-terminus in the case of the left-handed λ-helix) and it promotes the propagation of the helix pattern along the segment. It should be noted that in both the molecular dynamics simulations, only the terminal composition is different. Therefore, the handedness of the final conformation is related to the terminal charge conditions. The propagation of the helix pattern along the segment clearly shows that sequential local interactions determine the nascent folding patterns of the protein. The initial folding in the left-handed λ-helix arises from the electrostatic interactions of the positively charged NH_3^+ group with the adjacent carbonyl dipole moiety, followed by the dipole–dipole interactions between two adjacent carbonyl moieties. These results were further confirmed with more accurate calculations using a density functional approach. The preceding discussion implies that a transition between left-handed and right-handed helix motifs can be triggered by the presence of charged species near the end of the helix terminals. Such a possibility holds immense promise in the development of novel chiral switches and in the control of the protein folding/unfolding.

34.6 MOLECULAR DEVICES

Up to now, our discussion was only centred on static systems. However, the quest for nanodevices implies that one has to induce motion in a system using external or internal means. The external means could include changes in pH, radiation, etc. We discuss one such device (a molecular flipper), which has been designed, synthesized, and characterized [164]. The flipping/flapping motion, in the case of designed device, is due to the changes of edge-to-face and face-to-face aromatic interactions [165]. It is interesting to note that this conformational change can be electrochemically controlled by reduction/oxidation of the quinone moiety in the molecular system.

The strategy for the design of nanodevices is to harness the subtle changes in the π-electron densities of a quinone moiety as a result of changes in the electronic environment. Quinones are particularly suited for this endeavour because their electronic characteristics can be electrochemically or photochemically controlled. Based on a theoretical investigation of the conformational characteristics of *p*-benzoquinone–benzene

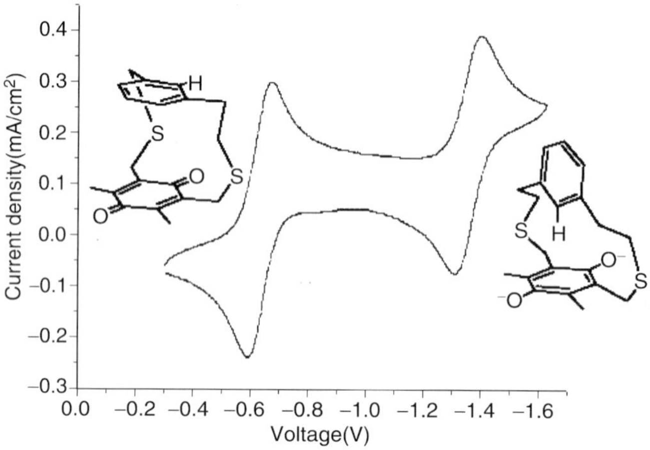

Fig. 34.14. Cyclic voltammogram of MHQC (left)/MQC (right) (1 mM) in acetonitrile with tetrabutylammonium dihydrogen phosphate (0.1 M) at 25 °C (scan rate 100 mV/s). (Reproduced with permission from American Chemical Society [164].)

complexes, we found that the energy difference between the stacked and edge-to-face conformations of cyclophane molecules is substantial.

Thus, if one could subtly control the conformational characteristics of 2,11-dithio[4,4]metametaquinocyclophane (MQC) (stacked conformer is 7 kcal/mol more stable than the edge-to-faced conformer) and 2,11-dithio[4,4]metametahydroquinocyclophane (MHQC) (edge-to-face is 9 kcal/mol more stable than the edge-to-face conformer) by electrochemical and/or photochemical means, we can have a very interesting model of a potential molecular device.

The cyclic voltammograms of MQC exhibits two clear reversible redox reactions (Fig. 34.14). In aprotic media, quinones exhibit two reduction peaks separated by 0.7 V, which corresponds to the formation of a radical anion species and a dianion species of quinones, respectively. This is in agreement with the reduction characteristics of MQC. Two well-separated reduced states of MQC are formed in the aprotic solvent of acetonitrile upon reduction. Therefore, the electronic states of MQC and MHQC can be easily transformed into each other by simple electrochemical control of the redox reaction, which results in large conformational flapping motions due to a preference for the stable conformation caused by the change in the electronic state of the quinone moiety.

Thus, a cyclophane system composed of quinone and benzene rings exhibits a flapping motion involving squeezing and thrusting motions in the presence of solvent molecules by electrochemical redox process. This case illustrates a promising pathway of harnessing the differences in the relative magnitudes of different kinds of intermolecular interactions to design a nanomechanical device. The large flapping/flipping motion from the edge-to-face and stacked conformations and *vice versa* is a first step toward a propelling molecular vessel or a molecular flipper that can be electrochemically or photochemically controlled. It could be applied to design molecular hinges, molecular switches, and eventually to design mobile nanomechanical devices for drug delivery and nanosurgery.

34.7 CONCLUDING REMARKS

In the course of this review, we have presented cogent reasons for carrying out high-level theoretical investigations of clusters. Apart from providing a detailed knowledge of intermolecular interaction forces including novel molecular interactions, such studies have proven to be very useful in the *de novo* design of functional nanomaterials and nanodevices which can be realized experimentally. Additionally, one can also understand the controlled assembly phenomena of novel functional molecular systems. Therefore, in order to describe subtle differences in competing interaction forces and cooperative interaction forces for nanomaterial design, the accurate description of interaction forces including new type of interactions and the systematic comparison of these forces would emerge as one of the utmost important subjects in the recent computational quantum chemistry. In the context of nanomaterial design, we have illustrated it using several experimentally realized examples of our work on nanowires/nanotubes, ionophores/receptors/sensors, electron/proton/molecular tweezers, molecular vehicles, etc. that theoretical design provides important cues in the search for novel molecules and materials. We expect that some of the ideas, which emerge from this work, could be useful in the search for molecular nanoelectronic/mechanical devices, quantum computing devices, biomolecular sensors, and nanosurgery. Though the idea of using theoretical methods as a tool for nanomaterial design is still in its infancy, the advance of fast computers and extremely powerful programs would dramatically increase its pace in the near future. Finally, we believe that the present work on microscopic interactions would also help obtain an enhanced understanding of important chemical and biological processes in the macroscopic world.

34.8 ACKNOWLEDGEMENTS

Much of the work reported in this review has been supported by the Korean Ministry of Science and Technology under the Creative Research Initiatives Program and the Korean Ministry of Education under the Brain Korea (BK21) program.

34.9 REFERENCES

1 K.S. Kim, P. Tarakeshwar and H.M. Lee, in: J.A. Schwarz, C. Contescu, K. Putyera (Eds.), Dekker encyclopedia of nanoscience, nanotechnology, Vol. 3, Marcel Dekker, New York, 2004, pp. 2423–2433.
2 P. Tarakeshwar, D. Kim, H.M. Lee, S.B. Suh and K.S. Kim, in: J. Leszczynski (Ed.), Theoretical, computational chemistry (computational materials science), Vol. 15, Elsevier, Amsterdam, 2004, pp. 119–170.
3 M.S. Dresselhaus, G. Dresselhaus and P.C. Eklund, Science of fullerenes, carbon nanotubes, Academic Press, San Diego, 1996.

4 K.S. Kim, Bull. Korean Chem. Soc., 24 (2003) 757.
5 H. Kuhn and H.-D. Försterling, Principles of physical chemistry: Understanding molecules, molecular assemblies, supramolecular machines, Wiley, New York, 1999.
6 A.W. Castleman, Jr. and K.H. Bowen, Jr., J. Phys. Chem., 100 (1996) 12911.
7 H. Haberland (Ed.), Clusters of atoms, molecules, Springer, Berlin, 1994.
8 K.S. Kim, P. Tarakeshwar and J.Y. Lee, Chem. Rev., 100 (2000) 4145.
9 U. Buck and F. Huisken, Chem. Rev., 100 (2000) 3863.
10 P.E.S. Wormer and A. van der Avoird, Chem. Rev., 100 (2000) 4109.
11 S. Scheiner, Hydrogen bonding: A theoretical perspective, University Press, Oxford, 1997.
12 P. Tarakeshwar, H.M. Lee, K.S. Kim, in: K.D. Sen (Ed.), Reviews in modern quantum chemistry—a celebration of the contributions of R.G. Parr, World Scientific Publishing, Singapore, 2002, pp. 1642–1683.
13 B. Brutschy, Chem. Rev., 100 (2000) 3891.
14 P. Hobza and Z. Havlas, Chem. Rev., 100 (2000) 4253.
15 P. Tarakeshwar and K.S. Kim, J. Mol. Struct., 615 (2002) 227.
16 P. Tarakeshwar, H.S. Choi and K.S. Kim, J. Am. Chem. Soc., 123 (2001) 3323.
17 H.M. Lee, M. Ge, B.R. Sahu, P. Tarakeshwar and K.S. Kim, J. Phys. Chem. B, 107 (2003) 9994.
18 J. Yoon, K.S. Kim and K.K. Baeck, J. Chem. Phys., 112 (2000) 9335.
19 P. Pyykkö and N. Runeberg, Angew. Chem. Int. Ed., 41 (2002) 2174.
20 X. Li, B. Kiran, J. Li, H.-J. Zhai and L.-S. Wang, Angew. Chem. Int. Ed., 41 (2002) 4786.
21 P. Tarakeshwar and K.S. Kim, in: H.S. Nalwa (Ed.), Encyclopedia of nanoscience, nanotechnology, Vol. 7, American Science Publishers, California, 2004, pp. 367–404.
22 B. Jeziorski, R. Moszynski and K. Szalewicz, Chem. Rev., 94 (1994) 1887.
23 K. Szalewicz and B. Jeziorski, in: S. Scheiner (Ed.), Molecular interactions—from van der Waals to strongly bound complexes, Wiley, New York, 1997, p. 3.
24 H.F. Schaefer, III (Ed.), Methods of electronic structure theory, Plenum Press, New York, 1977.
25 P.v.R. Schleyer, N.L. Allinger, T. Clark, J. Gasteiger, P.A. Kollman, H.F. Schaefer, III and P.R. Schreiner, Encyclopedia of computational chemistry, Wiley, Chichester, 1998.
26 G.H.F. Diercksen and S. Wilson (Eds.), Methods in computational molecular physics, D. Reidel, Dordrecht, 1983.
27 R.G. Parr and W. Yang, Density functional theory of atoms, molecules, Oxford University Press, Oxford, 1989.
28 G. Kresse and J. Furthmüller, Comput. Mater. Sci., 6 (1996) 15.
29 W.A. Harrison, Solid State Commun., 124 (2002) 443.
30 K.S. Kim, M. Dupuis, G.C. Lie and E. Clementi, Chem. Phys. Lett., 131 (1986) 451.
31 K. Liu, M.G. Brown, C. Carter, R.J. Saykally, J.K. Gregory and D.C. Clary, Nature, 381 (1996) 501.
32 R.N. Pribble and T.S. Zwier, Science, 265 (1994) 75.
33 C.J. Gruenloh, J.R. Carney, C.A. Arrington, T.S. Zwier, S.Y. Fredericks and K.D. Jordan, Science, 276 (1997) 1678.
34 U. Buck, I. Ettischer, M. Melzer, V. Buch and J. Sadlej, Phys. Rev. Lett., 80 (1998) 2578.
35 S. Lee, J. Kim, S.J. Lee and K.S. Kim, Phys. Rev. Lett., 79 (1997) 2038.
36 J. Kim and K.S. Kim, J. Chem. Phys., 109 (1998) 5886.
37 B.J. Mhin, J. Kim, S. Lee, J.Y. Lee and K.S. Kim, J. Chem. Phys., 100 (1994) 4484.
38 B.J. Mhin, H.S. Kim, H.S. Kim, J.W. Yoon and K.S. Kim, Chem. Phys. Lett., 176 (1991) 41.
39 S.J. Lee, H.Y. Chung and K.S. Kim, Bull. Korean Chem. Soc., 25 (2004) 1061.
40 S.J. Lee and K.S. Kim, Computing Code POSMOL (Reg. No. 2000-01-12-4239), Postech Licensing Center, Pohang, Korea 2000 (anonymous ftp address; ftp://csm50.postech.ac.kr/posmol).
41 K.S. Kim, B.J. Mhin, U.-S. Choi and K. Lee, J. Chem. Phys., 97 (1992) 6649.
42 L.A. Curtiss, D.J. Frurip and M. Blander, J. Chem. Phys., 71 (1979) 2703.
43 H.M. Lee, S.B. Suh, J.Y. Lee, P. Tarakeshwar and K.S. Kim, J. Chem. Phys., 112 (2000) 9759, 114 (2001) 3343.
44 C. Steinbach, P. Andersson, M. Melzer, J.K. Kazimirski, U. Buck and V. Buch, Phys. Chem. Chem. Phys., 6 (2004) 3320.

45 R.E. Miller and K. Nauta, Science, 283 (1999) 1895.
46 J. Kim, B.J. Mhin, S.J. Lee and K.S. Kim, Chem. Phys. Lett., 219 (1994) 243.
47 J. Kim, D. Majumdar, H.M. Lee and K.S. Kim, J. Chem. Phys., 110 (1999) 9128.
48 H.M. Lee, S.B. Suh and K.S. Kim, J. Chem. Phys., 114 (2001) 10749, 115 (2001) 7331.
49 H.M. Lee, P. Tarakeshwar, J.W. Park, M.R. Kolaski, Y.J. Yoon, H.-B. Yi, W.Y. Kim and K.S. Kim, J. Phys. Chem., 108 (2004) 2949.
50 J. Kim, S. Lee, S.J. Cho, B.J. Mhin and K.S. Kim, J. Chem. Phys., 102 (1995) 839.
51 S. Lee, J. Kim, J.K. Park and K.S. Kim, J. Phys. Chem., 100 (1996) 14329.
52 H.M. Lee, J. Kim, S. Lee, B.J. Mhin and K.S. Kim, J. Chem. Phys., 111 (1999) 3995.
53 B.J. Mhin, S. Lee, S.J. Cho, K. Lee and K.S. Kim, Chem. Phys. Lett., 197 (1992) 77.
54 B.J. Mhin, J. Kim and K.S. Kim, Chem. Phys. Lett., 216 (1993) 305.
55 J. Park, M. Kolaski, H.M. Lee and K.S. Kim, J. Chem. Phys., 121 (2004) 3108.
56 E.C. Lee, H.M. Lee, P. Tarakeshwar and K.S. Kim, J. Chem. Phys., 119 (2003) 7725.
57 J. Baik, J. Kim, D. Majumdar and K.S. Kim, J. Chem. Phys., 110 (1999) 9116.
58 D. Majumdar, J. Kim and K.S. Kim, J. Chem. Phys., 112 (2000) 101.
59 J. Kim, H.M. Lee, S.B. Suh, D. Majumdar and K.S. Kim, J. Chem. Phys., 113 (2000) 5259.
60 H.M. Lee and K.S. Kim, J. Chem. Phys., 114 (2001) 4461.
61 H.M. Lee, D. Kim and K.S. Kim, J. Chem. Phys., 116 (2002) 5509.
62 H.M. Lee and K.S. Kim, Mol. Phys., 100 (2002) 875.
63 J. Kim, J.M. Park, K.S. Oh, J.Y. Lee, S. Lee and K.S. Kim, J. Chem. Phys., 106 (1997) 10207.
64 J. Kim, J.Y. Lee, K.S. Oh, J.M. Park, S. Lee and K.S. Kim, Phys. Rev. A, 59 (1999) R930.
65 S.B. Suh, H.M. Lee, J. Kim, J.Y. Lee and K.S. Kim, J. Chem. Phys., 113 (2000) 5273.
66 H.M. Lee and K.S. Kim, J. Chem. Phys., 117 (2002) 706.
67 H.M. Lee, S.B. Suh and K.S. Kim, J. Chem. Phys., 118 (2003) 9981.
68 H.M. Lee, S. Lee and K.S. Kim, J. Chem. Phys., 119 (2003) 187.
69 H.M. Lee, S.B. Suh and K.S. Kim, J. Chem. Phys., 119 (2003) 7685.
70 N.I. Hammer, J.-W. Shin, J.M. Headrick, E.G. Diken, J.R. Roscioli, G.H. Weddle and M.A. Johnson, Science, 306 (2004) 675.
71 S. Odde, B.J. Mhin, S. Lee, H.M. Lee and K.S. Kim, J. Chem. Phys., 120 (2004) 9524.
72 S. Re, Y. Osamura, Y. Suzuki and H.F. Schaefer, III, J. Chem. Phys., 109 (1998) 973.
73 S. Odde, C. Park, H.M. Lee, K.S. Kim and B.J. Mhin, J. Chem. Phys., 121 (2004) 204.
74 S. Odde, H.M. Lee, M. Kolaski, B.J. Mhin and K.S. Kim, J. Chem. Phys., 121 (2004) 4665.
75 J.D. Augspurger, C.E. Dykstra and T.S. Zwier, J. Phys. Chem., 96 (1992) 7252.
76 K.S. Kim, J.Y. Lee, H.S. Choi, J. Kim and J.H. Jang, Chem. Phys. Lett., 265 (1997) 497.
77 P. Tarakeshwar, H.S. Choi, S.J. Lee, J.Y. Lee, K.S. Kim, T.-K. Ha, J.H. Jang, J.G. Lee and H. Lee, J. Chem. Phys., 111 (1999) 5838.
78 P. Tarakeshwar, K.S. Kim and B. Brutschy, J. Chem. Phys., 110 (1999) 8501.
79 P. Tarakeshwar, K.S. Kim and B. Brutschy, J. Chem. Phys., 112 (2000) 1769.
80 P. Tarakeshwar, K.S. Kim and B. Brutschy, J. Chem. Phys., 114 (2001) 1295.
81 P. Tarakeshwar, K.S. Kim, S. Djafari, K. Buchhold, B. Reimann, H.-D. Barth and B. Brutschy, J. Chem. Phys., 114 (2001) 4016.
82 C. Riehn, B. Reimann, K. Buchhold, H.-D. Barth, S. Vaupel, B. Brutschy, P. Tarakeshwar and K.S. Kim, J. Chem. Phys., 115 (2001) 10045.
83 B. Reimann, K. Buchhold, H.-D. Barth, B. Brutschy, P. Tarakeshwar and K.S. Kim, J. Chem. Phys., 117 (2002) 8805.
84 J.Y. Lee, J. Kim, H.M. Lee, P. Tarakeshwar and K.S. Kim, J. Chem. Phys., 113 (2000) 6160.
85 K.S. Kim and E. Clementi, J. Am. Chem. Soc., 107 (1985) 227.
86 K.S. Kim and E. Clementi, J. Phys. Chem., 89 (1985) 3655.
87 K.S. Kim and E. Clementi, J. Am. Chem. Soc., 107 (1985) 5504.
88 K.S. Kim, H.L. Nguyen, P.K. Swaminathan and E. Clementi, J. Phys. Chem., 89 (1985) 2870.
89 K.S. Kim, J. Comput. Chem., 6 (1985) 256.
90 K.S. Kim, D.P. Vercauteren, M. Welti, S. Chin and E. Clementi, Biophys. J., 47 (1985) 327.
91 B. Roux and R. Mackinnon, Science, 285 (1999) 100.

References pp. 989–993

92 J. Åqvist and V. Luzhkov, Nature, 404 (2000) 881.
93 C. Cui, S.J. Cho and K.S. Kim, J. Phys. Chem. A, 102 (1998) 1119.
94 C. Cui and K.S. Kim, J. Phys. Chem. A, 103 (1999) 2751.
95 P. Pyykkö, Chem. Rev., 97 (1997) 597.
96 M. Laing, Educ. Chem., (1993) 160.
97 T. Nautiyal, S.J. Youn and K.S. Kim, Phys. Rev. B, 68 (2003) 033407.
98 W.-T. Geng and K.S. Kim, Phys. Rev. B, 67 (2003) 233403.
99 T. Nautiyal, T.H. Rho and K.S. Kim, Phys. Rev. B, 69 (2004) 193404.
100 V. Rodrigues, J. Bettini, P.C. Silva and D. Ugarte, Phys. Rev. Lett., 91 (2003) 096801.
101 K.S. Kim, J.Y. Lee, S.J. Lee, T.-K. Ha and D.H. Kim, J. Am. Chem. Soc., 116 (1994) 7399.
102 J.Y. Lee, S.J. Lee, H.S. Choi, S.J. Cho, K.S. Kim and T.-K. Ha, Chem. Phys. Lett., 232 (1995) 67.
103 D. Kim, S. Hu, P. Tarakeshwar, K.S. Kim and J.M. Lisy, J. Phys. Chem. A, 107 (2003) 1128.
104 P. Tarakeshwar, K.S. Kim, E. Kraka and D. Cremer, J. Chem. Phys., 115 (2001) 6018.
105 P. Tarakeshwar, J.Y. Lee and K.S. Kim, J. Phys. Chem. A, 102 (1998) 2253.
106 P. Tarakeshwar, S.J. Lee, J.Y. Lee and K.S. Kim, J. Phys. Chem. B, 103 (1999) 184.
107 P. Tarakeshwar and K.S. Kim, J. Phys. Chem. A, 103 (1999) 9116.
108 R.A. Kumpf and D.A. Dougherty, Science, 261 (1993) 1708.
109 J.C. Ma and D.A. Dougherty, Chem. Rev., 97 (1997) 1303.
110 D. Kim, P. Tarakeshwar and K.S. Kim, J. Phys. Chem. A, 108 (2004) 1250.
111 G.A. Olah, B. Torok, J.P. Joschek, I. Bucsi, P.M. Esteves, G. Rasul and G.K. Suryaprakash, J. Am. Chem. Soc., 124 (2002) 11379.
112 M.J. Earle and K.R. Seddon, Pure Appl. Chem., 72 (2000) 1391.
113 B.H. Hong, J.Y. Lee, S.J. Cho, S. Yun and K.S. Kim, J. Org. Chem., 64 (1999) 5661.
114 T. Ren, Y. Jin, K.S. Kim and D.H. Kim, J. Biomol. Struct. Dyn., 15 (1997) 401.
115 S.H. Gellman, Chem. Rev., 97 (1997) 1231.
116 S. Tsuzuki, T. Uchimaru and K. Tanabe, Chem. Phys. Lett., 287 (1998) 202.
117 M.O. Sinnokrot, E.F. Valeev and C.D. Sherrill, J. Am. Chem. Soc., 124 (2002) 10887.
118 H. Krause, B. Ernstberger and H.J. Neusser, Chem. Phys. Lett., 184 (1991) 411.
119 P. Hobza, H.L. Selzle and E.W. Schlag, J. Am. Chem. Phys., 116 (1994) 3500.
120 M.I. Cabaço, Y. Danten, M. Besnard, Y. Guissani and B. Guillot, J. Phys. Chem. B, 102 (1998) 10712.
121 A.P. West, Jr., S. Mecozzi and D.A. Dougherty, J. Phys. Org. Chem., 10 (1997) 347.
122 A. Karpfen, J. Phys. Chem. A, 103 (1999) 11431.
123 B.H. Cardelino, C.E. Moore, D.O. Frazier, D.G. Musaev and K. Morokuma, Int. J. Quantum Chem., 66 (1998) 189.
124 M.C. Chan, P.A. Block and R.E. Miller, J. Chem. Phys., 102 (1995) 3993.
125 J.R. Grover, E.A. Walters and E.T. Hui, J. Phys. Chem., 91 (1987) 3233.
126 B.H. Hong, J.Y. Lee, C.-W. Lee, J.C. Kim and K.S. Kim, J. Am. Chem. Soc., 123 (2001) 10748.
127 K.S. Kim, S.B. Suh, J.C. Kim, B.H. Hong, E.C. Lee, S. Yun, P. Tarakeshwar, J.Y. Lee, Y. Kim, H. Ihm, H.G. Kim, J.W. Lee, J.K. Kim, H.M. Lee, D. Kim, C. Cui, S.J. Youn, H.Y. Chung, H.S. Choi, C.-W. Lee, S.J. Cho, S. Jeong and J.-H. Cho, J. Am. Chem. Soc., 124 (2002) 14268.
128 J.M. Park, P. Tarakeshwar, K.S. Kim and T. Clark, J. Chem. Phys., 116 (2002) 10684.
129 K.S. Oh, C.-W. Lee, H.S. Choi, S.J. Lee and K.S. Kim, Org. Lett., 2 (2000) 2679.
130 S. Yun, Y.-O. Kim, D. Kim, H.G. Kim, H. Ihm, J.K. Kim, C.-W. Lee, W.J. Lee, J. Yoon, K.S. Oh, J. Yoon, S.-M. Park and K.S. Kim, Org. Lett., 5 (2003) 471.
131 H. Ihm, S. Yun, H.G. Kim, J.K. Kim and K.S. Kim, Org. Lett., 4 (2002) 2897.
132 S. Yun, H. Ihm, H.G. Kim, C.-W. Lee, B. Indrajit, K.S. Oh, Y.J. Gong, J.W. Lee, J. Yoon, H.C. Lee and K.S. Kim, J. Org. Chem., 68 (2003) 2467.
133 S.K. Kim, N.J. Singh, S.J. Kim, H.G. Kim, J.K. Kim, J.W. Lee, K.S. Kim and J. Yoon, Org. Lett., 5 (2003) 2083.
134 J. Yoon, S.K. Kim, N.J. Singh, J.W. Lee, Y.J. Yang, K. Chellappan and K.S. Kim, J. Org. Chem., 69 (2004) 581.
135 S.B. Suh, C. Cui, H.S. Son, J.S. U, Y. Won and K.S. Kim, J. Phys. Chem., 106 (2002) 2061.
136 K.S. Kim, C. Cui and S.J. Cho, J. Phys. Chem. B, 102 (1998) 461.

137 J.Y. Kwon, N.J. Singh, H.N. Kim, S.K. Kim, K.S. Kim and J. Yoon, J. Am. Chem. Soc., 126 (2004) 8892.
138 H.S. Choi, S.B. Suh, S.J. Cho and K.S. Kim, Proc. Natl Acad. Sci. USA, 95 (1998) 12094.
139 H.S. Choi, D. Kim, P. Tarakeshwar, S.B. Suh and K.S. Kim, J. Org. Chem., 67 (2002) 1848.
140 H.S. Choi and K.S. Kim, Angew. Chem. Int. Ed., 38 (1999) 2256.
141 K.S. Kim, D. Kim, J.Y. Lee, P. Tarakeshwar and K.S. Oh, Biochemistry, 41 (2002) 5300.
142 K.S. Kim, K.S. Oh and J.Y. Lee, Proc. Natl Acad. Sci. USA, 97 (2000) 6373.
143 K.S. Oh, S.-S. Cha, D.-H. Kim, H.-S. Cho, N.-C. Ha, G. Choi, J.Y. Lee, P. Tarakeshwar, H.S. Son, K.Y. Choi, B.-H. Oh and K.S. Kim, Biochemistry, 39 (2000) 13891.
144 H.-S. Cho, N.-C. Ha, G. Choi, H.-J. Kim, D. Lee, K.S. Oh, K.S. Kim, W. Lee, K.Y. Choi and B.-H. Oh, J. Biol. Chem., 274 (1999) 32863.
145 W.W. Cleland and M.M. Krevoy, Science, 264 (1994) 1887.
146 P.A. Frey, Science, 269 (1995) 104.
147 D.-H. Oh, J.M. Park and K.S. Kim, Phys. Rev. B., 62 (2000) 1600.
148 K.S. Kim, J.M. Park, J. Kim, S.B. Suh, P. Tarakeshwar, K.H. Lee and S.S. Park, Phys. Rev. Lett., 84 (2000) 2425.
149 G. Oszlányi, G. Bortel, G. Faigel, M. Tegze, L. Gránásy, S. Pekker, P.W. Stephens, G. Bendele, R. Dinnebier, G. Mihály, A. Jánossy, O. Chauvet and L. Forró, Phys. Rev. B, 51 (1995) 12228.
150 F. Bommeli, L. Degiorgi, P. Wachter, Ö. Legeza, A. Jánossy, G. Oszlanyi, O. Chauvet and L. Forro, Phys. Rev. B, 51 (1995) 14794.
151 T.A. Murphy, Th. Pawlik, A. Weidinger, M. Höhne, R. Alcala and J. Spaeth, Phys. Rev. Lett., 77 (1996) 1075.
152 S.B. Suh, J.C. Kim, Y.C. Choi, S. Yun and K.S. Kim, J. Am. Chem. Soc., 126 (2004) 2186.
153 K.S. Kim, Curr. Appl. Phys., 2 (2002) 65.
154 B.H. Hong, S.C. Bae, C.-W. Lee, S. Jeong and K.S. Kim, Science, 294 (2001) 348.
155 B.H. Hong, C.-W. Lee and K.S. Kim, European Patent No. EP 1264919, 2002.
156 S.B. Suh, B.H. Hong, P. Tarakeshwar, S.J. Youn, S. Jeong and K.S. Kim, Phys. Rev. B, 67 (2003) 241402(R).
157 J.Y. Lee, B.J. Mhin, S. Mukamel and K.S. Kim, J. Chem. Phys., 119 (2003) 7519.
158 H.M. Lee, J. Kim, C.-J. Kim and K.S. Kim, J. Chem. Phys., 116 (2002) 6549.
159 H.-B. Yi, X.-H. Duan, J.Y. Lee, H.M. Lee, X.-Y. Li and K.S. Kim, J. Chem. Phys., 119 (2003) 8854.
160 J.Y. Lee, K.S. Kim and B.J. Mhin, J. Chem. Phys., 115 (2001) 9484.
161 D. Majumdar, H.M. Lee, J. Kim and K.S. Kim, J. Chem. Phys., 111 (1999) 5866.
162 S. Shim, T. Joo, S.C. Bae, K.S. Kim and E. Kim, J. Phys. Chem. A, 107 (2003) 8106.
163 H.S. Son, B.H. Hong, C.-W. Lee, S. Yun and K.S. Kim, J. Am. Chem. Soc., 123 (2001) 514.
164 H.G. Kim, C.-W. Lee, S. Yun, B.H. Hong, Y.-O. Kim, D. Kim, H. Ihm, J.W. Lee, E.C. Lee, P. Tarakeshwar, S.-M. Park and K.S. Kim, Org. Lett., 4 (2002) 3971.
165 T.K. Manojkumar, H.S. Choi, B.H. Hong, P. Tarakeshwar and K.S. Kim, J. Chem. Phys., 121 (2004) 841.

Theory and Applications of Computational Chemistry: The First Forty Years
Edited by C. Dykstra et al.

CHAPTER 35

Monte Carlo simulations of the finite temperature properties of $(H_2O)_6$

R.A. Christie and K.D. Jordan

Department of Chemistry and Center for Materials and Molecular Simulations, University of Pittsburgh, Pittsburgh, PA 15215, USA

Abstract

The thermodynamic behavior of the $(H_2O)_6$ cluster has been the subject of several theoretical studies employing model potentials. This article reviews the earlier work and presents new results on the finite temperature properties of the $(H_2O)_6$ cluster calculated using a 2 + 3-body MP2-level intermolecular potential in a canonical Monte Carlo simulation at $T = 220$ K. Book and ring configurations are found to dominate the sampled configurations. The finite histogram method of Ferrenberg and Swendsen [Phys. Rev. Lett., 61 (1988) 2635] is used to calculate the energy and energy fluctuations over a range of temperatures. The resulting heat capacity curve displays a broad peak near $T = 135$ K.

35.1 INTRODUCTION

Water clusters have been studied extensively both experimentally and theoretically over the past two decades. Among the issues that have been addressed are the geometrical structures of the low-energy isomers [1–19], the role of cooperative effects in the bonding [20–26], the sensitivity of the vibrational and rotational spectra on the H-bonding arrangements [27–36], the dynamics of isomer interconversion [37–39], and the finite temperature behavior of the clusters [39–47].

The water hexamer is one of the most thoroughly studied water clusters. Nevertheless, while it is well established that the smaller $(H_2O)_{3-5}$ clusters have cyclic global minimum structures, and that $(H_2O)_7$ and larger clusters have three-dimensional fused-ring global minima [13,31,33,34,48,49], the situation regarding $(H_2O)_6$ has taken much longer to sort out.

References pp. 1006–1009

During the late 1980s and early 1990s several groups investigated theoretically the low-lying isomers of $(H_2O)_6$ as characterized by model potentials. In particular, Kim et al. [2] used the MCY potential [50] both by itself as well as augmented with 3- and 4-body interactions [51,52], Belford and Campbell [53] used the Campbell–Mezei water model [54], Schröder [55] employed the QPEN/B2 [56] potential, Dykstra and coworkers [57,58] carried out calculations with the MCY, MMC [59,60] and TIP4p [61] models, and Vegiri and Farantos [44] investigated $(H_2O)_6$ using the CKL potential [62]. One of the most thorough studies of the energy landscape of the water hexamer was carried out by Tsai and Jordan [12] using an eigenmode-following algorithm with the TIP4p model potential. The energy disconnectivity diagram of $(H_2O)_6$ was first plotted by Wales et al. [63] for the rigid TIP4P and ASP-W4 [6,25] potentials, and by Burnham and Xantheas [64] for the flexible TTM2-F potential [65]. The topology of the H-bonding network of the low-lying isomers of $(H_2O)_6$ was studied by Tissandier et al. [66].

The energy ordering of the low-energy isomers of $(H_2O)_6$ varies from one model potential to another, clearly indicating the limitations of the water models available at the time these studies were carried out. Depending on the model potential employed the cage (**A**, Fig. 35.1), prism (**C**, Fig. 35.1), open-book (**D**, Fig. 35.1) and ring isomers (**B**, Fig. 35.1) have all been predicted to be the global minimum. Pedulla et al. [26] have compared the relative energies of the various isomers calculated with several model potentials and MP2-level electronic structure calculations.

Kim et al. [2] were apparently the first to apply *ab initio* methods to $(H_2O)_6$, carrying out SCF optimizations of the hexamer as early as 1988. However, as it became possible to carry out MP2 calculations with suitable flexible basis sets on water clusters, it became clear that inclusion of electron correlation effects is crucial for describing the bonding in these species [5,11,12,20,23,67,68]. The first MP2 level calculations on $(H_2O)_6$ were by Tsai and Jordan [12]. Subsequently, Kim and coworkers [69,70], Xantheas et al. [71],

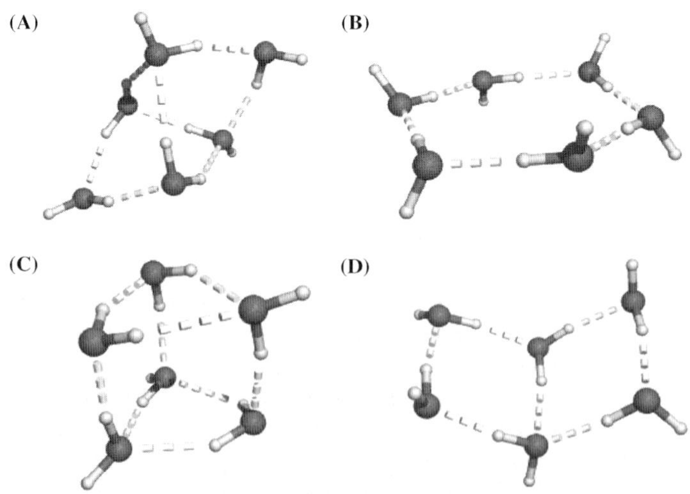

Fig. 35.1. Four low-lying isomer types of $(H_2O)_6$; the cage (**A**), ring (**B**), prism (**C**) and book (**D**) isomers.

Pedulla et al. [26], Kryachko [72], and Losada and Leutwyler [73] have characterized the low-energy isomers of $(H_2O)_6$ at the MP2 level and employing large, flexible basis sets. There have also been several other *ab initio* studies of the hexamer [74,75]. In addition, the role of vibrational zero-point energy has been investigated in both the harmonic [12,13,73] and anharmonic approximations [22,76–78]. These calculations reveal that the cage, book and ring isomers are of comparable energy, with the most stable prism-like isomer being a few tenths of a kcal/mol less stable. Over the past few years the cage [76,79], book [80,81] and ring [82,83] structures have all been observed experimentally.

The thermodynamic properties of $(H_2O)_6$ have also long been of interest, with Monte Carlo and molecular dynamics simulations having been carried out using several different model potentials [44,45,47,84]. Although some of these studies concluded that $(H_2O)_6$ undergoes a melting transition (e.g. Ref. [44]), others (e.g. Ref. [47]) concluded that this system does not display a clear-cut melting transition. Obviously the sensitivity of the relative energies of the various isomeric forms of $(H_2O)_6$ on the model potential employed implies that none of the finite temperature simulations can be considered as definitive. While the simulations on $(H_2O)_6$ are unclear as to whether this species undergoes a melting transition, the consensus is that the slightly larger $(H_2O)_8$ cluster does undergo a relatively sharp transformation between 'solid-like' and 'liquid-like' phases [39–47]. Although more sophisticated water models have been introduced recently [64,85], these have yet to be applied to finite temperature simulations of water clusters.

To date, theoretical studies of the thermodynamic behavior of water clusters have been limited to model potentials [41–45,47,84], although there have been studies that examined the temperature dependence of selected isomers [67,73] and others that have used DFT-based MD simulations to optimize structures and to calculate vibrational spectra [7]. In the absence of either experimental or *ab initio* data on the thermodynamic properties of small water clusters, it is difficult to assess the reliability of the simulations carried out with various model potentials. Although simulations of the thermodynamic properties of small water clusters are feasible with DFT methods, it has been found that DFT calculations with commonly employed functionals such as Becke3LYP [86,87] incorrectly order various isomers of small water clusters [88]. This appears to be due to the inability of current density functionals to describe long-range dispersion interactions [89,90]. For this reason, it is preferable that finite temperature simulations of water clusters be carried out using an appropriate wavefunction-based electronic structure method.

The MP2 method has been found to accurately describe the energetics of water clusters, providing that sufficiently flexible atomic basis sets are employed [5,11,20, 67,68]. However, while MP2 calculations are feasible for clusters containing up to 30 or so water molecules, the steep ($O(N^5)$) computational scaling of conventional MP2 calculations with system size precludes their use in carrying out Monte Carlo or molecular dynamics simulations of water clusters containing six or more monomers.

It is known from studies using model potentials that even for a cluster as small as $(H_2O)_6$, Monte Carlo simulations at temperatures between 50 and 200 K (the range typically considered to examine the issue of cluster melting) need to be carried out for on the order of 10^6 moves to achieve convergence. Such simulations at the MP2 level would take on the order of 1 year of CPU time (although this could be reduced to a few weeks or even days of wall clock time with an efficient parallel MP2 code and enough CPUs). For larger clusters, even longer simulation runs would be required and it would also be necessary to employ an algorithm such as parallel tempering [91] to achieve convergence, particularly at low temperatures. The parallel tempering algorithm requires the use of a grid of temperatures, further increasing the costs of the calculations. For these reasons, simulations of the thermodynamic properties of water clusters the size of the hexamer or larger using traditional MP2 methods would be a computationally daunting task even on parallel computers. A significant reduction in the CPU time could be achieved by use of localized second-order Møller–Plesset perturbation theory (LMP2 [92–95]) or resolvent of the identity second-order Møller–Plesset perturbation theory (RIMP2 [96,97]) methods, but there would still be the bottleneck introduced by the Hartree–Fock (HF) portion of the calculations. Although there are linear scaling HF codes [98], they have not been combined with fast MP2 methods such as LMP2 or RIMP2.

These considerations have led us to consider an alternative approach based on a n-body decomposition procedure for carrying out Monte Carlo simulations on small water clusters with MP2-level energies. The n-body decomposition procedure is described in the following section. This is followed by application to a Monte Carlo simulation of the study of the $(H_2O)_6$ cluster. As noted above, model potential simulations have predicted the $(H_2O)_6$ cluster to exist in a wide range of hydrogen bonding topologies, making this a good system for testing the n-body decomposition procedure.

35.2 METHODOLOGY

In the n-body decomposition procedure, the total binding energy is written as

$$E = E^{(1)} + E^{(2)} + E^{(3)} + \cdots + E^{(n)} \tag{1}$$

where $E^{(i)}$ denotes the 'ith' body contribution to the interaction energy and the various energy terms depend on the appropriate nuclear coordinates. $E^{(1)}$ is simply the relaxation energy, which reflects the change in the energies of the monomers due to the geometrical distortions that occur when they are incorporated in the cluster. $E^{(2)}$ is the 2-body interaction energy which can be obtained by performing calculations on all dimer pairs and $E^{(3)}$ is the 3-body energy obtained by performing calculations on all possible trimers in the cluster, etc.

A significant body of work has shown that, for water clusters, the many-body effects are dominated by the 3-body terms, i.e. that

$$E \approx E^{(1)} + E^{(2)} + E^{(3)} \tag{2}$$

provides a good approximation to the total binding energy of a water cluster [21,23]. This suggests a strategy of basing an MP2-level Monte Carlo simulation procedure in Eq. (2). Truncated n-body approaches have been used in water model potentials (most notably, the NCC [99,100] and SAPT [101–103] potentials), however we are unaware of this approach having been employed in conjunction with *ab initio* Monte Carlo simulations. For simplicity, in the remainder of this work we will assume that the monomers are kept rigid (i.e. $E^{(1)} = 0$ in Eq. (2)).

For a cluster containing N molecules, the evaluation of the $E^{(2)}$ contributions requires $N(N-1)/2$ separate dimer calculations and $E^{(3)}$ requires $N(N-1)(N-2)/6$ separate trimer calculations. In the absence of other computational savings, the n-body decomposition procedure truncated at the $E^{(3)}$ term, would not offer an advantage over supermolecule LMP2 calculations with a code such as Jaguar [104] where both the HF and LMP2 steps are carried out using pseudospectral algorithms.

Thus far, we have focused on the cost of performing a single total energy calculation. In Monte Carlo simulations in which each step in the simulation involves a displacement (translation or rotation) of a single molecule, one needs only to compute the contributions to $E^{(2)}$ and $E^{(3)}$ involving the monomer that was moved. In this case, the number of terms that need to be evaluated is of the order $(N-1)$ and $(N-1)(N-2)/2$, for $E^{(2)}$ and $E^{(3)}$, respectively. As a result, the truncated 2 + 3-body approach provides a significant computational savings compared to that required with the more conventional supermolecule approach for carrying out MP2-level Monte Carlo simulations of cluster systems.

There are several additional opportunities for computational savings in the 2 + 3-body Monte Carlo simulation procedure. These include: (i) use of the LMP2 or RIMP2 methods for calculating the correlated energies, (ii) use of different levels of theory for the 2- and 3-body interaction energies, and (iii) use of different approaches for treating short-range and long-range interactions. In the present study, the first two of these cost-saving features were adopted. Namely, the 2-body energies were calculated using Saebø and Pulay's [92–95] implementation of the LMP2 method, and the Hartree–Fock method was used to calculate the 3-body interactions. Compared with the standard MP2 approach, LMP2 has the advantages of lower computational cost and reduced basis set superposition error (BSSE [105]) in the correlation energy. The adoption of the HF method for calculating the 3-body interaction energies was motivated by the observation that electron correlations play only a small role in describing such interactions in H-bonded clusters [16,26,103]. No attempt was made in this study to exploit the use of a less computationally demanding method for the long-range 2- or 3-body energies. The various 2- and 3-body interactions were farmed out to multiple CPUs on a Beowulf cluster. Care was taken in dividing up the various interaction terms to ensure that the CPUs were efficiently used.

The simulations were carried out with the 6-31+G(d) basis set [106–108] as well as with a more flexible mixed cc-pVDZ [109]/aug-cc-pVDZ [110] (for H/O atoms, respectively) basis set, hereafter referred to as apVDZ. The electronic structure calculations were carried out using the MOLPRO program package (version 2002.3) [111].

The Monte Carlo simulations were carried out in the canonical ensemble and at temperatures of $T = 100$ and 220 K with the 6-31 + G(d) basis set and at $T = 220$ K for the larger apVDZ basis set. A constraining sphere with a radius of $r = 5.5$ Å was employed to prevent evaporative events during the simulation. Each simulation was carried out for 1.1×10^5 Monte Carlo moves, with the first 1×10^4 moves being for equilibration and the last 10^5 moves being employed in the averaging. Based upon earlier work with model potentials, it is known that the $(H_2O)_6$ system is relatively easy to equilibriate, especially at a temperature as high as $T = 220$ K. Nonetheless, it would be desirable to carry out longer simulations (e.g. for 10^6 Monte Carlo moves) to ensure better convergence. This was not done because of the limitations of the Beowulf cluster available for the calculations (1 GHz Pentium III CPUs), and because the primary purpose of this study is to demonstrate the feasibility of 2 + 3-body MP2-level simulations on $(H_2O)_n$ clusters rather than to do definitive calculations.

The average interaction energies $\langle E\rangle$, $\langle E^{(2)}\rangle$ and $\langle E^{(3)}\rangle$ and their fluctuations (in the form of dimensionless heat capacities) were calculated. The constant volume, configurational heat capacity was calculated from the standard expression

$$\bar{C}_V = C_V/k_B = \frac{\sigma^2(E)}{(k_B T)^2} = \frac{\langle E^2\rangle - \langle E\rangle^2}{(k_B T)^2} \tag{3}$$

In a similar manner, the 2- and 3-body components to the total heat capacity, were calculated as

$$\bar{C}_V^{(2)} = C_V^{(2)}/k_B = \frac{\langle E^{(2)}(E^{(2)} + E^{(3)})\rangle - \langle E^{(2)}\rangle\langle E^{(2)} + E^{(3)}\rangle}{(k_B T)^2} \tag{4}$$

and

$$\bar{C}_V^{(3)} = C_V^{(3)}/k_B = \frac{\langle E^{(3)}(E^{(2)} + E^{(3)})\rangle - \langle E^{(3)}\rangle\langle E^{(2)} + E^{(3)}\rangle}{(k_B T)^2} \tag{5}$$

where these equations were obtained by differentiating $\langle E^{(2)}\rangle$ and $\langle E^{(3)}\rangle$ with respect to T. To provide insight into the interdependence of $E^{(2)}$ and $E^{(3)}$, the covariance

$$c^{(2),(3)} = \frac{\langle \delta E^{(2)} \delta E^{(3)}\rangle}{\sigma(E^{(2)})\sigma(E^{(3)})} \tag{6}$$

where $\delta E^{(2)} = E^{(2)} - \langle E^{(2)}\rangle$, $\delta E^{(3)} = E^{(3)} - \langle E^{(3)}\rangle$, $\sigma(E^{(2)}) = \sqrt{\langle (E^{(2)})^2\rangle - \langle E^{(2)}\rangle^2}$, and $\sigma(E^{(3)}) = \sqrt{\langle (E^{(3)})^2\rangle - \langle E^{(3)}\rangle^2}$, were calculated. In a similar fashion, the covariance between the HF 2-body and LMP2 2-body energy components was calculated from

$$c^{\mathrm{HF,LMP2}} = \frac{\langle \delta E_{\mathrm{HF}}^{(2)} \delta E^{(2)}\rangle}{\sigma(E_{\mathrm{HF}}^{(2)})\sigma(E^{(2)})} \tag{7}$$

The single histogram method of Ferrenberg and Swendsen [112] was used to estimate the average energies over a range of temperatures. In this approach, configurations saved from a simulation at a temperature T were used to calculate the energy at other temperatures T' using

$$\langle E(\beta')\rangle = \frac{\sum_v E_v(\beta)\exp\{-(\beta' - \beta)E_v\}}{\sum_v \exp\{-(\beta' - \beta)E_v\}} \tag{8}$$

where the sums are over all configurations sampled in the simulation at T, and $\beta = 1/kT$ and $\beta' = 1/kT'$.

35.3 RESULTS

35.3.1 Energetics of $(H_2O)_6$; basis set and thermal effects

Table 35.1 presents the energies from the 2 + 3-body LMP2 simulation of $(H_2O)_6$ carried out at 220 K. The average binding energy from the simulation is −31.39 kcal/mol. Of this, the average 2-body binding energy is −26.98 kcal/mol, −22.25 kcal/mol of which is recovered at the HF level. The average 3-body energy is −4.41 kcal/mol, which accounts for about 14.0% of the total average binding energy and which is close to the percentage contribution of 3-body interactions to the binding in bulk liquid water at 298 K, where the 3-body contribution has been estimated to be 14.5% of the net cohesive energy [113].

The 3-body terms are even more important for the energy fluctuations than for the total interaction energy, accounting for over 32% of the total energy fluctuations in the $T = 220$ K MP2/apVDZ simulation. $|c^{(2),(3)}| = 0.08$, which indicates that there is little statistical correlation between the 2- and 3-body energies of $(H_2O)_6$ at $T = 220$ K. $\sigma^2(E^{(2)})$ is nearly the same for the Hartree–Fock and LMP2 energies. Also, $|c^{\text{LMP2,HF}}| = 0.89$, which implies that (not surprisingly) there is a strong correlation between the HF and LMP2 2-body energies.

35.3.2 Error analysis of the truncated n-body approximation for E

In order to evaluate the accuracy of the truncated 2 + 3-body approach employed in this study, supermolecule LMP2 calculations with the mixed cc-pVDZ/aug-cc-pVDZ basis set were carried out on 200 structures selected from the apVDZ simulation trajectory. A comparison of the energies from the two methods is presented in Fig. 35.2, from which it is seen that the errors in the total energies due to the use of the 2 + 3-body approach are relatively small. The average absolute error is calculated to be 0.22 kcal/mol, which is less than 0.7% of the total interaction energy. The largest error in the 2 + 3-body energy

Table 35.1 Energy and energy fluctuations of $(H_2O)_6$ from mixed cc-pVDZ/aug-cc-pVDZ LMP2 Monte Carlo simulation at $T = 220$ K

Property	HF	LMP2
$\langle E^{(2)}\rangle$ (kcal/mol)	−22.25	−26.98
$\langle E^{(3)}\rangle$ (kcal/mol)	−4.41	−4.41[a]
$\langle E\rangle$ (kcal/mol)	−26.66	−31.39
$\sigma^2(E^{(2)})$ (kcal/mol)^2	2.86	2.70
$\sigma^2(E^{(2)})$ (kcal/mol)^2	1.28	–
$\sigma^2(E)$ (kcal/mol)^2	3.52	3.98

[a]The 3-body contributions were calculated only at the HF level of theory.

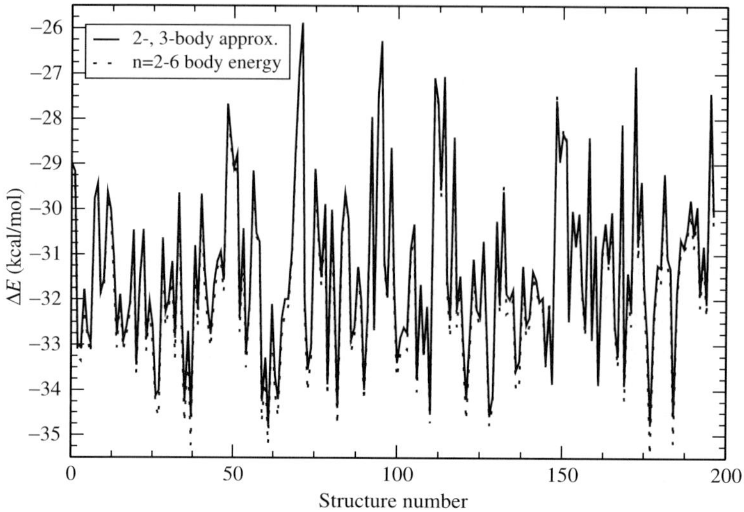

Fig. 35.2. Error analysis of the 2 + 3-body approximation to E. The binding energies of 200 structures selected from the LMP2/apVDZ Monte Carlo simulation trajectory at $T = 220$ K were calculated using supermolecule LMP2 calculations (which also recover the $n = 4-6$ body interactions).

for an individual structure is 0.64 kcal/mol. As indicated by Fig. 35.2, in the vast majority (97%) of structures, the truncated n-body approach slightly underestimates, in magnitude, the full 6-body binding energy.

35.3.3 Inherent structures

To gain further insight into the nature of the $(H_2O)_6$ cluster at $T = 220$ K, every 100th configuration from the simulation using the apVDZ basis set was saved and subsequently optimized to its inherent structure using the flexible water model potential of Ren and Ponder [85]. Ring-based isomers (**B** from Fig. 35.1, 65.5%) are found to be the dominant inherent structure. Book-based isomers (**D** from Fig. 35.1, 24.7%), prism-based (**C** from Fig. 35.1, 4.5%), and cage-based (**A** from Fig. 35.1, 3.3%) isomers also have significant population. The remaining inherent structures (<2.0%) consist of isomers with one or more rings (but distinct from the isomers **A**–**D**). This distribution of inherent structures is similar to that found from Monte Carlo simulations [47] at $T = 220$ K and using the TIP4p water model.

35.3.4 Radial distribution function

Fig. 35.3 reports the oxygen–oxygen partial radial distribution function, $g_{OO}(r)$, determined from the three 2 + 3-body MP2 simulations of $(H_2O)_6$. The maximum in the first peak in $g_{OO}(r)$ obtained from the $T = 220$ K simulation with the apVDZ basis set

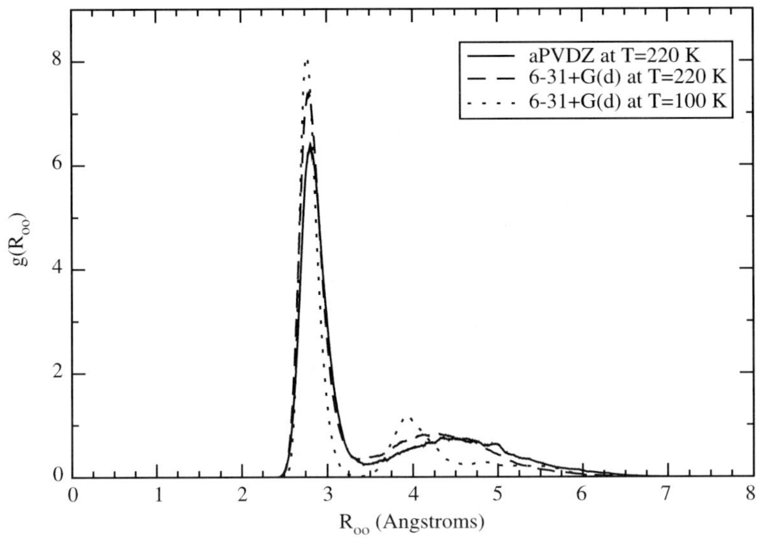

Fig. 35.3. The oxygen–oxygen partial radial distribution functions, $g_{OO}(r)$, from Monte Carlo simulations using the apVDZ (at $T = 220$ K) and 6-31 + G(d) (at $T = 100$ and 220 K) basis sets.

occurs at $r = 2.82$ Å, which is close to that (2.8 Å) observed experimentally for liquid water [114] under ambient conditions.

The value of $g_{OO}(r)$ for the first peak in Fig. 35.3, 6.4, is appreciably higher than that ($\approx$2.8) determined experimentally [114] or calculated for bulk water [113]. This difference is a consequence of the normalization of $g_{OO}(r)$ for an ideal gas of the same density. The $g_{OO}(r)$ curve for $(H_2O)_6$ also displays a second peak centered near 4.40 Å in the $T = 220$ K simulations and near 3.95 Å in the $T = 100$ K simulations.

35.3.5 Temperature dependence of the energy and heat capacity of $(H_2O)_6$

Plots of the temperature dependence of the total binding energy and of the 2- and 3-body contributions obtained from the histogram procedure are presented in Fig. 35.4. The total binding energy varies from -37.69 kcal/mol at $T = 25$ K to -29.41 kcal/mol at $T = 270$ K, with there being an inflection point near 130 K. Both $E^{(2)}$ and $E^{(3)}$ decrease in magnitude with increasing temperature, with the result being that their sum decreases more rapidly than does $E^{(2)}$ alone. The shape of the $E^{(2)}$ curve calculated at the MP2 level is nearly identical to that calculated at the Hartree–Fock level, with the former being lower by about 5 kcal/mol. This is surprising given the contribution of dispersion interactions to the MP2 energies. The 3-body contribution ranges from 19.8% of the total interaction energy at $T = 25$ K, to 12.9% at $T = 270$ K.

The calculated (dimensionless) heat capacity curve is presented in Fig. 35.5. The total heat capacity, $\bar{C}_V$ increases from 0.22 at $T = 25$ K to a maximum of 24.35 near $T = 135$ K, and then decreases slowly between $T = 135$ and 270 K. In addition, $\bar{C}_V$ also has

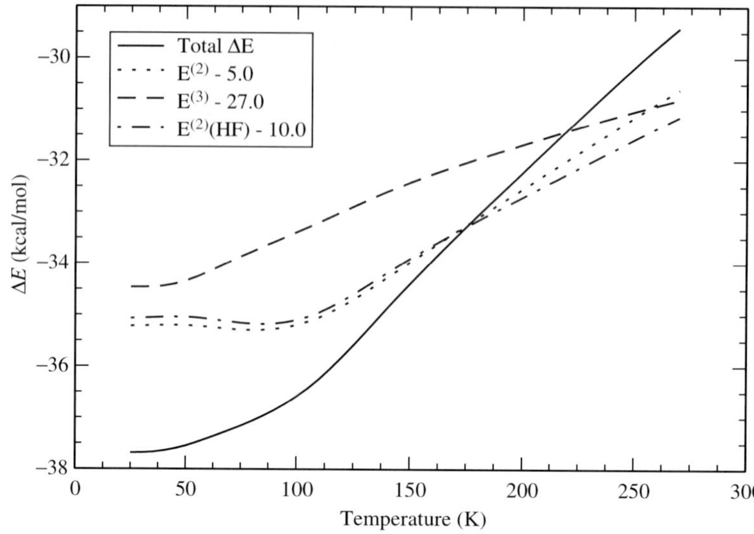

Fig. 35.4. Finite temperature behavior of the binding energy components of the LMP2/apVDZ Monte Carlo simulation at $T = 220$ K. To facilitate comparison with the total energy, the 2-body, 3-body and HF 2-body energy components were shifted by constant values of 5.0, 27.0 and 10.0 kcal/mol, respectively.

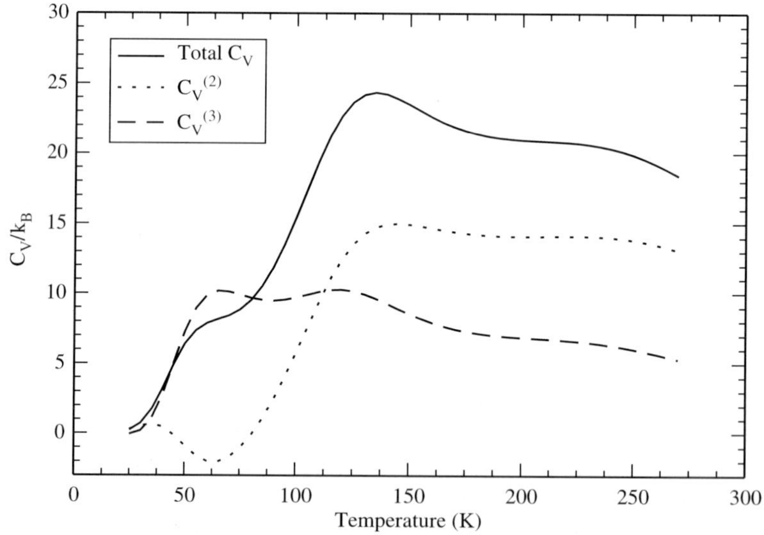

Fig. 35.5. Finite temperature behavior of the components of the dimensionless, configurational constant volume heat capacity (from Eqs. (3)–(5) from the LMP2/apVDZ simulation at $T = 220$ K).

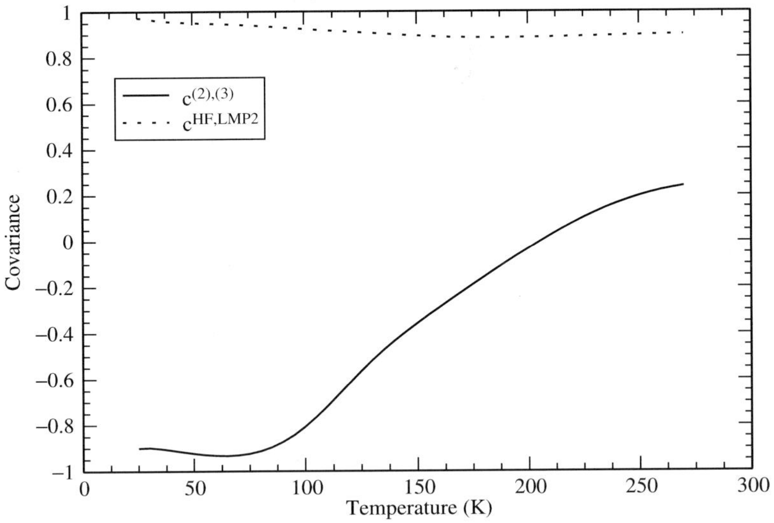

Fig. 35.6. Covariance function $c^{(2),(3)}$ for the 2- and 3-body energies, and $c^{\text{LMP2,HF}}$ for the HF and LMP2 2-body energies.

a shoulder near $T = 65$ K. The temperature dependence of $\bar{C}_V$ for $(H_2O)_6$, calculated by using the histogram procedure and the 2 + 3-body LMP2 simulations, is close to that obtained from J-walk Monte Carlo simulations [45] of $(H_2O)_6$ using the TIP4p [61] and NCC [99,100] model potentials.

The 2-body component to the heat capacity, $\bar{C}_V^{(2)}$, has a minimum at $T = 65$ K, close to the temperature of the shoulder in the $\bar{C}_V$ curve. The shoulder, in fact, arises from the 3-body contribution to $\bar{C}_V$ which is calculated to have maxima near $T = 60$ and 120 K. For temperatures between 30 and 110 K, the 3-body contribution to the energy fluctuations actually exceeds the 2-body contributions.

The temperature dependencies of $c^{\text{HF,LMP2}}$ and $c^{(2),(3)}$ deduced from the histogram analysis are reported in Fig. 35.6. $c^{\text{HF,LMP2}}$ shows little variation in temperature, ranging from 0.88 to 0.98. On the other hand, $c^{(2),(3)}$ is relatively flat between $T = 25$ and 70 K and grows rapidly from about -0.90 to 0.22, as T increases from 70 to 270 K.

35.4 CONCLUSIONS

The water hexamer has been studied extensively experimentally and theoretically. Although there have been several calculations on the dependence of various properties of $(H_2O)_6$ as a function of temperature, the results have proven to be quite sensitive to the model potential employed. In this study, a new approach was adopted in which the finite temperature properties of the hexamer were studied using an *ab initio* potential.

A 2 + 3-body MP2-level procedure was used to characterize the $(H_2O)_6$ cluster at $T = 220$ K. These results were used in a histogram procedure to predict the potential energy and heat capacity of the cluster over a range of temperatures. At $T = 220$ K, the 3-body contributions are found to account for 14% of the total interaction energy, whereas the 3-body contribution to the total fluctuations is 32%. Comparison of the 2 + 3-body energies with full supermolecule LMP2 energies for a subset of sampled structures reveals that the average discrepancy between the two sets of energies is only 0.22 kcal/mol.

At $T = 220$ K, book and ring structures are predicted to account for over 90% of the configurations sampled in the simulation. This preponderance of ring and book isomers at $T = 220$ K is consistent with the results of previous model potential simulations [47]. The 2-body component of the interaction energy varies by over 8 kcal/mol from $T = 25$ to 270 K, while the 3-body component varies by about 3 kcal/mol over the same temperature range.

The percentage contribution of the 3-body component to the total potential energy varies from 20% at $T = 25$ K to 13% at $T = 270$ K. There is a weak maximum in the $\bar{C}_V$ *vs.* T curve at $T = 135$ K. The functional behavior of $\bar{C}_V$ is of a similar form to that obtained from previous simulations employing the NCC and TIP4P water potentials [45]. Based on earlier model potential studies [47], this maximum is due to the transformation from cage and prism structures to the ring and open-book structures.

The 2 + 3-body LMP2/apVDZ simulations reported in this study required about 3 weeks on eight 1.0 GHz CPUs of a Beowulf computer cluster. By use of current, high-end CPUs and by running over 16–32 CPUs, 2 + 3-body MP2-level simulations for as many as 10^6 Metropolis moves would be feasible for clusters as large as $(H_2O)_{10}$. To extend the simulations to appreciably larger clusters requires the introduction of additional approximations, e.g. using a cut-off for calculation of 3-body interactions. The simplest approach for reducing the computational cost of the calculation of the 3-body interactions is to skip the evaluation of structures in which the distance of one of the monomers from the other two in a trimer exceeds a threshold distance. It is also possible to implement a scheme in which the 3-body interactions are divided into three classes: (1) the short-range interactions, treated *via ab initio* electronic structure theory, (2) intermediate range interactions, treated using a classical polarization model, and (3) long-range interactions, which are neglected. With these strategies the 2 + 3-body LMP2 procedure could be used to carry out Monte Carlo simulations on clusters as large as $(H_2O)_{20}$, at least if the temperature is high enough that quasi-ergodicity is not a serious problem. Extension to still larger clusters would be possible by adoption of approximate procedures for estimating the 2-body interactions between distant monomers.

35.5 ACKNOWLEDGEMENTS

This research was supported by a grant (CHE-0078528) from the National Science Foundation.

35.6 REFERENCES

1 J. Del Bene and J.A. Pople, J. Chem. Phys., 52 (1970) 4858.
2 K.S. Kim, M. Dupuis, G.C. Lie and E. Clementi, Chem. Phys. Lett., 131 (1986) 451.

3 E. Honegger and S. Leutwyler, J. Chem. Phys., 88 (1988) 2582.
4 B.J. Smith, D.J. Swanton, J.A. Pople, H.F. Schaefer, III and L. Radom, J. Chem. Phys., 92 (1990) 1240.
5 D. Feller, J. Chem. Phys., 96 (1992) 6104.
6 C. Millot and A.J. Stone, J. Mol. Phys., 77 (1992) 439.
7 K. Laasonen, M. Parinello, R. Car, C. Lee and D. Vanderbilt, Chem. Phys. Lett., 207 (1993) 208.
8 S.S. Xantheas and T.H. Dunning, Jr., J. Chem. Phys., 99 (1993) 8774.
9 S.J. Chakravorty and E.R. Davidson, J. Phys. Chem., 97 (1993) 6373.
10 D.J. Wales, J. Am. Chem. Soc., 115 (1993) 11180.
11 S.S. Xantheas and T.H. Dunning, Jr., J. Chem. Phys., 98 (1993) 8037.
12 C.J. Tsai and K.D. Jordan, Chem. Phys. Lett., 213 (1993) 181.
13 K. Kim, K.D. Jordan and T.S. Zwier, J. Am. Chem. Soc., 116 (1994) 11568.
14 J.E. Fowler and H.F. Schaefer, III, J. Am. Chem. Soc., 117 (1995) 446.
15 M.W. Feyereisen, D. Feller and D.A. Dixon, J. Phys. Chem., 100 (1996) 2993.
16 S.S. Xantheas, J. Chem. Phys., 104 (1996) 8821.
17 D.J. Wales, Science, 271 (1996) 925.
18 M.P. Hodges, A.J. Stone and S.S. Xantheas, J. Phys. Chem. A, 101 (1997) 9163.
19 I.M.B. Nielsen, E.T. Siedi and C.L. Janssen, J. Chem. Phys., 110 (1999) 9435.
20 O. Mó, M. Yánez and J. Elguero, J. Chem. Phys., 97 (1992) 6628.
21 S.S. Xantheas, J. Chem. Phys., 100 (1994) 7523.
22 J.K. Gregory and D.C. Clary, J. Phys. Chem., 100 (1996) 18014.
23 J.M. Pedulla, F.D. Vila and K.D. Jordan, J. Chem. Phys., 105 (1996) 11091.
24 M. Masella and J.P. Flament, J. Chem. Phys., 107 (1997) 9105.
25 C. Millot, J.C. Soetens, M.T.M.C. Costa, M.P. Hodges and A.J. Stone, J. Phys. Chem. A, 102 (1998) 754.
26 J.M. Pedulla, K. Kim and K.D. Jordan, Chem. Phys. Lett., 291 (1998) 78.
27 T.R. Dyke and J.S. Muenter, J. Chem. Phys., 57 (1972) 5011.
28 T.R. Dyke and J.S. Muenter, J. Chem. Phys., 60 (1974) 2929.
29 T.R. Dyke, K.M. Mack and J.S. Muenter, J. Chem. Phys., 66 (1977) 498.
30 J.A. Odutola and T.R. Dyke, J. Chem. Phys., 72 (1980) 5062.
31 N. Pugliano and R.J. Saykally, Science, 257 (1992) 1937.
32 N. Pugliano, J.D. Cruzan, J.G. Loeser and R.J. Saykally, J. Chem. Phys., 98 (1993) 6600.
33 R.N. Pribble and T.S. Zwier, Science, 265 (1994) 75.
34 J.D. Cruzan, L.B. Braly, K. Liu, M.G. Brown, J.G. Loeser and R.J. Saykally, Science, 271 (1996) 59.
35 J.D. Cruzan, L.B. Braly, K. Liu, M.G. Brown, J.G. Loeser and R.J. Saykally, J. Phys. Chem. A, 101 (1997) 9022.
36 M.R. Viant, J.D. Cruzan, D.D. Lucas, M.G. Brown, K. Liu and R.J. Saykally, J. Phys. Chem. A, 101 (1997) 9032.
37 T. Walsh and D. Wales, J. Chem. Soc. Faraday Trans., 92 (1996) 2505.
38 D.J. Wales, Dynamics and rearrangements of water clusters, in: J. Bowman, Z. Bacic (Eds.), Advances in molecular vibrations and collision dynamics, JAI Press, Stamford, CT, 1998, p. 365.
39 D. Laria, J. Rodriguez, C. Dellago and D. Chandler, J. Phys. Chem. A, 105 (2001) 2646.
40 C.W. David, J. Chem. Phys., 73 (1980) 5395.
41 C.J. Tsai and K.D. Jordan, J. Chem. Phys., 95 (1991) 3850.
42 C.J. Tsai and K.D. Jordan, J. Chem. Phys., 99 (1993) 6957.
43 D.J. Wales and I. Ohmine, J. Chem. Phys., 98 (1993) 7245.
44 S. Vegiri and S.C. Farantos, J. Chem. Phys., 98 (1993) 4059.
45 J.M. Pedulla and K.D. Jordan, Chem. Phys., 239 (1998) 593.
46 P. Nigra, M. Carignano and S. Kais, J. Chem. Phys., 115 (2001) 2621.
47 A. Tharrington and K.D. Jordan, J. Phys. Chem. A, 107 (2003) 7380.
48 K. Liu, J.D. Cruzan and R.J. Saykally, Science, 271 (1996) 929.
49 R.J. Saykally and G.A. Blake, Science, 259 (1993) 1937.
50 O. Matsuoka, E. Clementi and M. Yoshimine, J. Chem. Phys., 64 (1976) 1351.
51 E. Clementi and G. Corongiu, Int. J. Quantum Chem., S10 (1983) 31.
52 J. Detrich, G. Corongiu and E. Clementi, Chem. Phys. Lett., 112 (1984) 426.

53 D.E. Belford and E.S. Campbell, J. Chem. Phys., 86 (1987) 7013.
54 E. Campbell and M. Mezei, J. Chem. Phys., 67 (1977) 2338.
55 K.-P. Schröder, Chem. Phys., 123 (1988) 91.
56 J. Sauer, Ph.D. thesis, Academy of Sciences of the GDR, Berlin, 1985.
57 C.E. Dykstra, J. Chem. Phys., 91 (1989) 6472.
58 K.A. Franken, M. Jalaie and C.E. Dykstra, Chem. Phys. Lett., 198 (1992) 59.
59 C.E. Dykstra, J. Am. Chem. Soc., 111 (1989) 6168.
60 C.E. Dykstra, J. Comput. Chem., 9 (1988) 476.
61 W.L. Jorgensen, J. Chandrasekhar, J.D. Madura, R.W. Impey and M.L. Klein, J. Chem. Phys., 79 (1983) 926.
62 P. Cieplak, P. Kollman and T. Lybrand, J. Chem. Phys., 92 (1990) 6755.
63 D.J. Wales, J.P.K. Doye, M.A. Miller, P.N. Mortensen and T.R. Walsh, Adv. Chem. Phys., 115 (2000) 1.
64 C.J. Burnham and S.S. Xantheas, J. Chem. Phys., 116 (2002) 1500.
65 C.J. Burnham and S.S. Xantheas, J. Chem. Phys., 116 (2002) 5115.
66 M.D. Tissandier, S.J. Singer and J.V. Coe, J. Phys. Chem. A, 104 (2000) 752.
67 K.S. Kim, B.J. Mhin, U.-S. Choi and K. Lee, J. Chem. Phys., 97 (1992) 6649.
68 C.J. Tsai and K.D. Jordan, unpublished results.
69 B.J. Mhin, J. Kim, S. Lee, J.Y. Lee and K.S. Kim, J. Chem. Phys., 100 (1994) 4484.
70 J. Kim and K.S. Kim, J. Chem. Phys., 117 (1998) 5886.
71 S.S. Xantheas, C.J. Burnham and R.J. Harrison, J. Chem. Phys., 116 (2002) 1493.
72 E.S. Kryachko, Chem. Phys. Lett., 314 (1999) 353.
73 M. Losada and S. Leutwyler, J. Chem. Phys., 117 (2002) 2003.
74 P.N. Krishnan, J.O. Jensen and L.A. Burke, Chem. Phys. Lett., 217 (1994) 311.
75 C. Lee, H. Chen and G. Fitzgerald, J. Chem. Phys., 101 (1994) 4472.
76 K. Liu, M.G. Brown, C. Carter, R.J. Saykally, J.K. Gregory and D.C. Clary, Nature, 381 (1996) 501.
77 M.W. Severson and V. Buch, J. Chem. Phys., 111 (1999) 10866.
78 K. Diri, E.M. Myshakin, and K.D. Jordan, J. Phys. Chem. A, in press.
79 J.K. Gregory, D.C. Clary, K. Liu, M.G. Brown and R.J. Saykally, Science, 275 (1997) 814.
80 E.G. Diken, W.H. Robertson and M.A. Johnson, J. Phys. Chem. A, 108 (2004) 64.
81 C. Steinbach, P. Andersson, M. Melzer, J.K. Kazimirski, U. Buck and V. Buch, Phys. Chem. Chem. Phys., 6 (2004) 3320.
82 K. Nauta and R.B. Miller, Science, 287 (2000) 293.
83 C.J. Burnham, S.S. Xantheas, M.A. Miller, B.E. Applegate and R.E. Miller, J. Chem. Phys., 117 (2002) 1109.
84 J. Rodriguez, D. Laria, E.J. Marceca and D.A. Estrin, J. Chem. Phys., 110 (1999) 9039.
85 P. Ren and J.W. Ponder, J. Phys. Chem. B, 107 (2003) 5933.
86 A.D. Becke, J. Chem. Phys., 98 (1993) 5648.
87 C. Lee, W. Yang and R. Parr, Phys. Rev. B, 37 (1993) 785.
88 K.D. Jordan, unpublished results.
89 S. Kristyán and P. Pulay, Chem. Phys. Lett., 229 (1994) 175.
90 J.M. Pérez-Jordá and A.D. Becke, Chem. Phys. Lett., 233 (1995) 134.
91 J.P. Neirotti, F. Calvo, D.L. Freeman and J.D. Doll, J. Chem. Phys., 112 (2000) 10340.
92 P. Pulay, Chem. Phys. Lett., 100 (1983) 151.
93 P. Pulay and S. Saebø, Theor. Chim. Acta, 69 (1986) 357.
94 S. Saebø and P. Pulay, J. Chem. Phys., 86 (1987) 914.
95 S. Saebø and P. Pulay, Annu. Rev. Phys. Chem., 44 (1993) 213.
96 F. Weigend and M. Häser, Theor. Chem. Acc., 97 (1997) 331.
97 F. Weigend, M. Häser, H. Patzelt and R. Ahlrichs, Chem. Phys. Lett., 294 (1998) 143.
98 M. Challacombe and E. Schwegler, MondoSCF a suite of programs for linear scaling SCF theory, 1999.
99 U. Niesar, C. Corongiu, M.-J. Huang, M. Dupuis and E. Clementi, Int. J. Quantum Chem. Symp., 23 (1989) 421.
100 U. Niesar, G. Corongiu, E. Clementi, G.R. Kneller and D.K. Bhattacharya, J. Phys. Chem., 94 (1990) 7949.

101 E.M. Mas, K. Szalewicz, R. Bukowski and B. Jeziorski, J. Chem. Phys., 107 (1997) 4207.
102 E.M. Mas, R. Bukowski, K. Szalewicz, G. Groenenboom, P.E.S. Wormer and A. van der Avoird, J. Chem. Phys., 113 (2000) 6687.
103 E.M. Mas, R. Bukowski and K. Szalewicz, J. Chem. Phys., 118 (2003) 4386.
104 Jaguar 4.2, Schrödinger, Inc., Portland, OR, 1991–2000.
105 S.F. Boys and F. Bernardi, Mol. Phys., 19 (1970) 553.
106 W.J. Hehre, R. Ditchfield and J.A. Pople, J. Chem. Phys., 56 (1972) 2257.
107 T. Clark, J. Chandrasekhar, G.W. Spitznagel and P.v.R. Schleyer, J. Comput. Chem., 4 (1983) 294.
108 M.J. Frisch, J.A. Pople and J.S. Binkley, J. Chem. Phys., 80 (1984) 3265.
109 T.H. Dunning, Jr., J. Chem. Phys., 90 (1989) 1007.
110 R.A. Kendall, T.H. Dunning, Jr. and R.J. Harrison, J. Chem. Phys., 96 (1992) 6796.
111 R.D. Amos, A. Bernhardsson, A. Berning, P. Celani, D.L. Cooper, M.J.O. Deegan, A.J. Dobbyn, F. Eckert, C. Hampel, G. Hetzer, P.J. Knowles, T. Korona, R. Lindh, A. Lloyd, S.J. McNicholas, F.R. Manby, W. Meyer, M.E. Mura, A. Nicklass, P. Palmieri, R. Pitzer, G. Rauhut, M. Schtz, U. Schumann, H. Stoll, A.J. Stone, R. Tarroni, T. Thorsteinsson and H.-J. Werner, MOLPRO, a package of ab initio programs designed by H.-J. Werner and P.J. Knowles, version 2002.3, 2002.
112 A.M. Ferrenberg and R.H. Swendsen, Phys. Rev. Lett., 61 (1988) 2635.
113 E.M. Mas, R. Bukowski and K. Szalewicz, J. Chem. Phys., 118 (2003) 4404.
114 G. Hura, J.M. Sorenson, R.M. Glaeser and T. Head-Gordon, J. Chem. Phys., 113 (2000) 9140.

Theory and Applications of Computational Chemistry: The First Forty Years
Edited by C. Dykstra et al.

CHAPTER 36

Computational quantum chemistry on polymer chains: aspects of the last half century

Jean-Marie André

FUNDP, Laboratoire de Chimie Théorique Appliquée, 61, rue de Bruxelles, 5000 Namur, Belgium

Abstract

Some aspects of computational quantum chemistry applied to the analysis of the electronic structure of polymers are reviewed in connection with the timely trends observed in their electrical and optical properties. The paper is organized as follows: after an introduction (Section 36.1), the basic theory of the quantum chemical methodologies as applied to periodic chains is summarized (Section 36.2). Several fields of applications are then presented: photoelectron spectra (Section 36.3), conducting and semi-conducting conjugated polymers (Section 36.4), linear and non-linear optical properties (Section 36.5) and the role of charge transfer in organic chains (Section 36.6). Possible developments for the near future are also sketched.

> Cultivate science: there is no safer path for man than that which wise men have always chosen
> Ludwig van Beethoven [1]
>
> When we consider what others have done better than ourselves, we come to hate our own lives
> Vincent Van Gogh [2]
>
> Does not maturity consist, perhaps, in allowing others to exist?
> Liv Ullmann [3]

36.1 INTRODUCTION

Since the first pioneering theoretical works on LCAO techniques in polymer quantum chemistry were written by Ladik and André in the 1960s [4–9] this field has rapidly developed. Periodic systems are computationally easier to deal with than aperiodic ones since the translational symmetry can be fully exploited in order to reduce to manageable dimensions the formidable task of computing electronic states of an extended system.

Ab initio programs for polymers are available and are currently applied by several groups such as Erlangen [10], Vienna [11], Budapest [12], Torino [13], Kingston [14,15] and Namur [16,17]. The actual versions of these programs are available on the sites mentioned in the references:

GAUSSIAN03 [18]
ADF [19]
CRYSTAL [20]
PLH [21] with Sun's MP2 part [22]
POLYMER [23]
DJPOL [24]

However, due to different computational strategies, these programs do not rigorously produce the same numbers, even for polymers studied using the same basis sets and within exactly the same geometrical conformations. This is a result of different cut-off procedures for the integrals and originates mainly from an internal structure, which is, in all cases, a logical extension of molecular programs. It is interesting to note that all polymer packages use standard 'molecular' strategies taken from the IBMOL, KGNMOL, GAUSSIAN or other series.

Furthermore, it is important to realize that sophisticated *ab initio* methods, which are already time consuming for medium- and large-sized molecules, become very onerous when applied to polymers of chemical or biological interest. In this sense, double-zeta or polarization LCAO techniques have only routinely been used in latter years since the development of efficient *ab initio* programs for polymers which require much shorter computing times than was previously the case. Traditionally, this has been achieved by implementing into a general system the new fast techniques for evaluating integrals on a Gaussian basis, through the explicit use of helical symmetry and thanks to efficient methods of computing long-range electrostatic effects. This has resulted in a general methodology that is more 'polymer-minded' than 'molecule-minded'.

36.2 ELECTRONIC STRUCTURE OF POLYMERS: METHODOLOGY (1965–TILL DATE)

Pioneering quantum mechanical calculations [25] were made on significant polymers such as polyethylene or polyacetylene chains in the late 1960s with the development of the first *ab initio* programs such as POLYMOL [26].

The standard theory in molecules and polymers is based on the Hartree–Fock theory. In this model, a single electron moves in the field of the nuclei and in the mean Coulomb and exchange field of all the other electrons. A set of molecular orbitals (MOs) is obtained to describe the occupied and unoccupied one-electron wave functions. In molecular quantum chemistry, the molecular orbitals are drawn as single levels, which are at most doubly occupied by a pair of electrons of opposite spin. On the other hand, the solid-state physicist takes advantage of the translation symmetry of the lattice and uses the concept of Brillouin zones introduced by Bloch in 1928 [27]. In this theory, the so-called Bloch functions (molecular orbitals for an infinite 1D chain) are eigenfunctions of a translation operator. Bloch's theorem is a direct consequence of the periodicity of the electron density:

$$\rho(\mathbf{r} + ja) = \rho(\mathbf{r})$$

where a is the length of the polymer unit cell in direct space. Bloch's theorem states the phase relation of the orbitals at periodically related points:

$$\phi_{\mathrm{n}}(\mathbf{r} + ja) = \mathrm{e}^{ikja}\phi_n(\mathbf{r})$$

Given that the argument of an exponential is a pure number, a is a length and j is a pure number (the counter index of a given unit cell), k must have the dimensions of an inverse length. Thus, the orbitals and their associated energies are functions of that k:

$$\varepsilon_n = \varepsilon_n(k)$$

and can be plotted with respect to k. The representation of the corresponding dispersion curves as a function of k is called an energy band (Fig. 36.1).

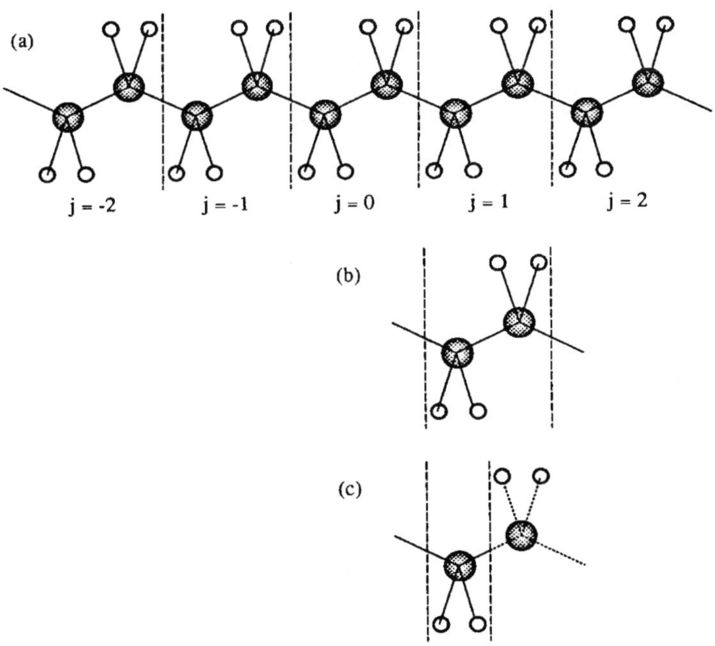

Fig. 36.1. Representation of a polyethylene polymer chain (a) polyethylene polymer chain (b) polyethylene CH_2–CH_2 unit cell (c) polyethylene CH_2 half unit cell.

Standard theorems of solid-state physics demonstrate that those energy bands are periodic in the 1D-reciprocal space:

$$\varepsilon_n(k) = \varepsilon_n(k + lg)$$

where g is the reciprocal translation unit length ($= 2\pi/a$). Due to the periodicity of the energy bands in k-space the full k-energy dependence is reduced to a single unit cell of the reciprocal space. The search is simplified if we use a symmetrized part of the reciprocal space with respect to $k = 0$, the so-called first Brillouin zone ranging from $-\pi/a$ to $+\pi/a$. Due to further symmetries, the reciprocal space must be only explored from $k = 0$ to $+\pi/a$ (half the first Brillouin zone) (Fig. 36.2).

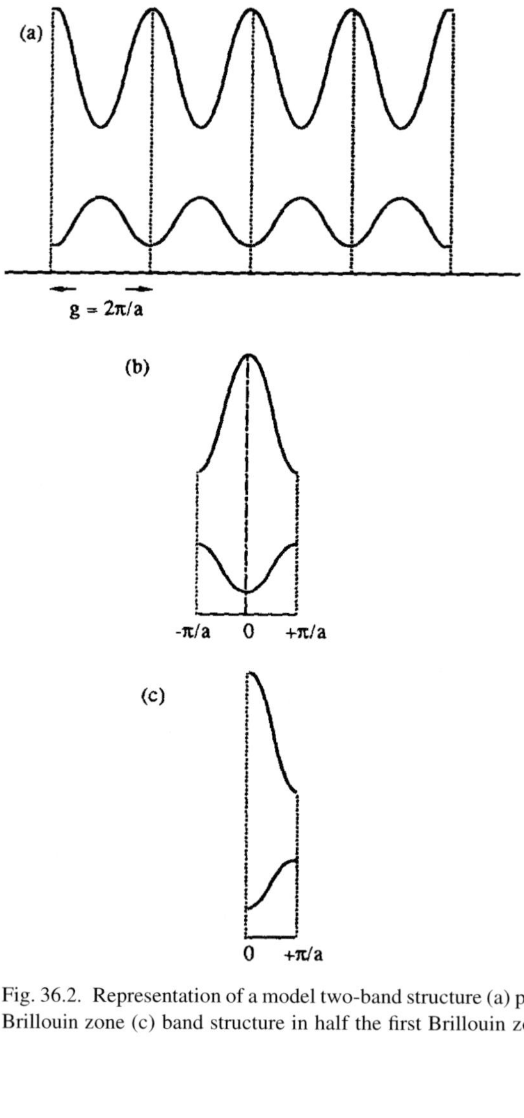

Fig. 36.2. Representation of a model two-band structure (a) periodic band structure (b) band structure in the first Brillouin zone (c) band structure in half the first Brillouin zone.

From a conceptual point of view, it appears that polymer quantum chemistry is an ideal field for cooperation between condensed matter physicists and molecular quantum chemists. There exists a common interpretation in the discussions concerning orbital energies, orbital symmetry, and gross charges by chemists and solid-state physicists. These physicists use terms less familiar to the chemist, such as first Brillouin zone, dependence of 'wave function' with respect to wave vector k (the one-electron wave function is called an 'orbital' by the chemist), Fermi surfaces, Fermi contours, and density of states (DOS).

However, it is important to note that polymer quantum chemistry is not a 1D, solid-state physical science. In strictly 1D physics, the systems are periodic in 1D and have 1D wave functions. In polymer quantum chemistry, the systems and their wave functions are 3D but periodic only in 1D. Usual theorems of 1D physics are consequently no longer valid [28]. A typical example is that of extrema of the energy bands, which should only occur at the center and the edges of the Brillouin zone in a strictly 1D system. For polymers, even in simple cases like the linear zigzag polyethylene chain, some extrema of the energy bands are encountered at arbitrary positions in the first Brillouin zone that are not points of high symmetry (Fig. 36.3).

Turning to actual applications, the numerical procedure combines the equations of the methods of molecular quantum chemistry and of solid-state physics. In molecular quantum chemistry, a molecular orbital is expanded in terms of basis functions. Secular systems of equations and determinants are solved. Their eigenvalues are the orbital energies. Charges and bond orders (projections of the density matrices onto the limited basis used) are calculated from the LCAO coefficients. In polymer quantum chemistry, we take into account the lattice periodicity; the orbitals, the systems of equations and the determinants are no longer real but have imaginary components. This introduces lattice sums that require to be evaluated using appropriate procedures. The key problem is to obtain the matrix elements over the basis functions. Fortunately, these matrix elements roughly exponentially decrease with the distance between the orbital centers and force the natural convergence properties of the lattice sums. This is true for overlap-like and kinetic electrostatic integrals, and for combinations of nuclei- and electron–electron electrostatic integrals. The behavior of exchange integrals is more complex.

The methods are classified as self-consistent and non-self-consistent. In the self-consistent methods, the effect of electron–electron interactions is explicitly taken into account. In *ab initio* methods, as soon as the geometry and the basis functions are defined all necessary integrals are explicitly evaluated. It has been proven that long-range effects become highly significant as soon as the unit cell contains permanent dipoles. Multipole expansion must, therefore, be used in order to obtain a satisfactory balance of the

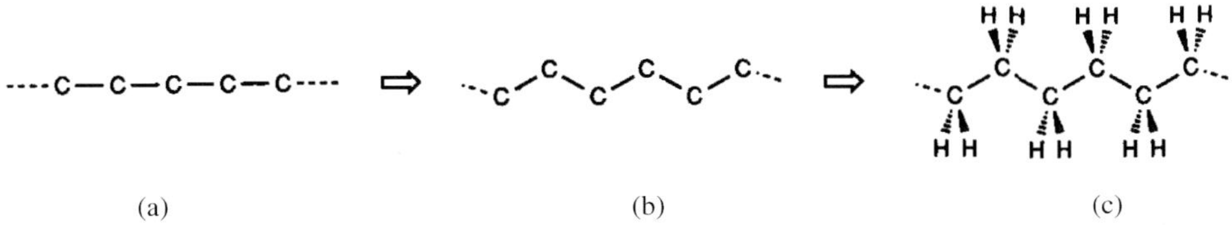

Fig. 36.3. 3D spatial configuration of the 1D-periodic polyethylene chain.

electrostatic interactions. Less sophisticated *ab initio* methods are available and produce results of good quality in much less computer time. The use of semi-empirical techniques like Extended Hückel or CNDO has sometimes been advocated. In these cases, reliable results are obtained with little attention to the parameterization procedure. Initial sketches of the conjugated bands are easily obtained from simple Hückel calculations. Those techniques that endeavor to simulate the *ab initio* results are also of particular interest. For polymeric band structures, good results have also been obtained with simple model potential techniques [29] (the so-called Valence Effective Hamiltonian, VEH, technique which has been popularized by a series of calculations by Brédas in his Mons, Tucson and Atlanta groups). In this procedure, the Hartree–Fock operator is approximated by a sum of atomic potentials that are, in general, anisotropic projectors determined on model molecules.

The practice of polymer computations is rather standard. The Fock and overlap matrices are computed at a given level of approximation (semi-empirical or *ab initio* [30]). The calculation is made in direct space. The k-dependent matrices are diagonalized and, if necessary, a self-consistent procedure is used. The output consists of the energy bands and of the form of the molecular orbitals of the polymer. An interactive graphical communication can be initiated for plotting the standard electronic properties, such as band structures, bandwidths and DOS. It can also be used for ordering the energy bands, simulating electron spectra and electron densities, calculating electron indexes as charges and bond orders, and determining conformations or other properties. Automatic programs taking into account the effects of long-range interactions are fully implemented and described in the literature [31]. The methodology of band structure calculations is described in detail in a few summer school proceedings [32] and monographs [33].

36.3 BAND STRUCTURE CALCULATIONS AND PHOTOELECTRON SPECTRA

Since the beginning of the 1970s, quantum mechanical methods have increasingly been applied to polymers. One of several reasons for this is the improvement of experimental techniques for investigating the electronic properties of polymers, such as photoelectron spectroscopy (PS) and, especially, X-ray induced photoelectron spectroscopy (XPS), sometimes known as ESCA. The development of the latter has been mainly due to the efforts of Kai Siegbahn (Nobel prize in 1981) in Uppsala. In polymer chemistry, two pioneering papers were published in 1972. They refer to the experimental ESCA analysis of the core levels of fluoropolyethylenes [34] and of the valence band of polyethylene [35]. Theoretical interpretations rapidly followed to support the analysis of the ESCA core levels [36]. The theoretical analysis of ESCA valence spectra has been less direct. The ESCA spectrum and a semi-empirical band structure of polyethylene devised by Wood et al. are sketched in Fig. 36.4. It shows that band structure is not directly measurable and that transformations have to be applied such that calculated data is in a form readily comparable to experiments.

In the author's opinion, band structure plots do not offer the best representation of valence band properties, particularly in those systems where a large number of bands lie

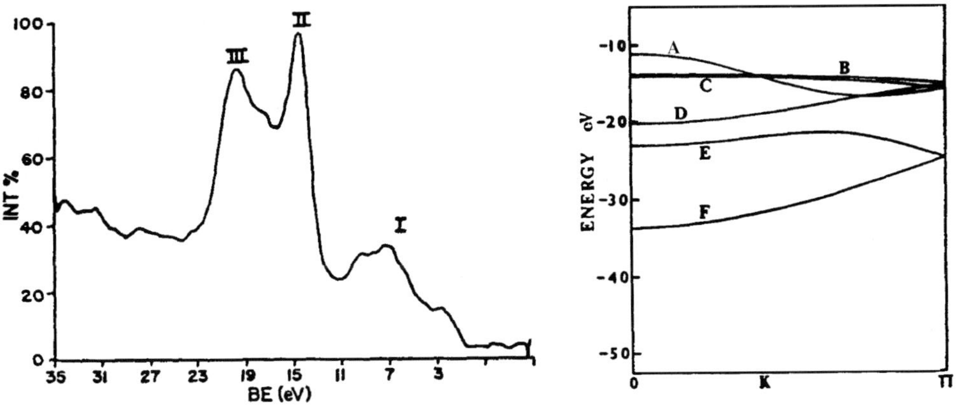

Fig. 36.4. Sketch of the theoretical band structure and experimental density of states of polyethylene. (After M.H. Wood, M. Barber, I.H. Hillier and J.M. Thomas, J. Chem. Phys., 56 (1972) 1788, reproduced with permission).

in a narrow energetic region, as is the case for polymers. For this reason we have implemented and systematized a three-step procedure [37] that leads to a theoretical simulation of an XPS valence spectrum of a regular polymer as illustrated in Fig. 36.5. In its presently simplified form, this procedure (a) corresponds to the calculation of the DOS histograms, (b) takes into account cross-section effects, and (c) uses convolution for simulating experimental resolution. In Fig. 36.5, this procedure has been applied to the test case of the FSGO valence bands of an all-trans polyethylene chain [38]. The positions of both theoretical and experimental peaks tally to a surprising degree and both fine structures are directly comparable. From the bond order analysis of the theoretical calculations, the four highest bands can be labeled as contributing to the C–H bonds while the two lowest bands correspond to the C–C bonds. Fig. 36.5 is also of fundamental interest since it states the existence of an experimental basis for the concept of electron bands in polymers.

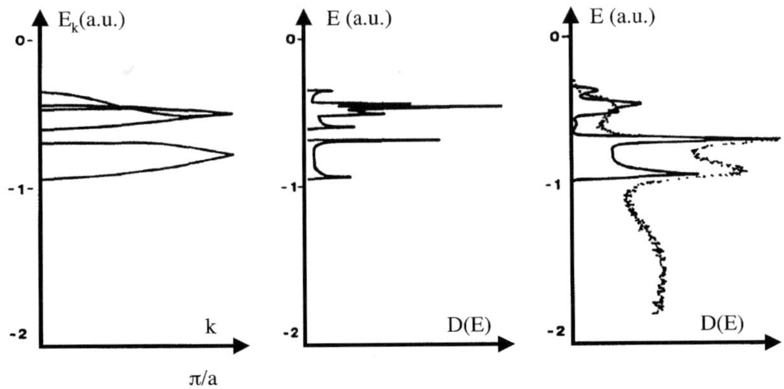

Fig. 36.5. FSGO theoretical simulation of the valence photoelectron spectrum of polyethylene.

This was expertly demonstrated by Angle Resolved UPS (ARUPS) experiments [39] at the beginning of the 1990s, as illustrated in Fig. 36.6. A natural continuation of this work was to study the effects of structural perturbations such as conformational changes on the valence bands of stereoisomers. To illustrate this approach, a computer experiment has been conducted in the case of polyethylene [40]. A vibrational analysis [41] of crystalline samples suggests the possibility of four conformations of polyethylene: trans (T), gauche (G), trans–gauche (TG) and trans–gauche–trans–gauche (TGTG′). The theoretical calculations reveal major differences in the shape of the densities of states of those four conformers of polyethylene, as illustrated in Fig. 36.7.

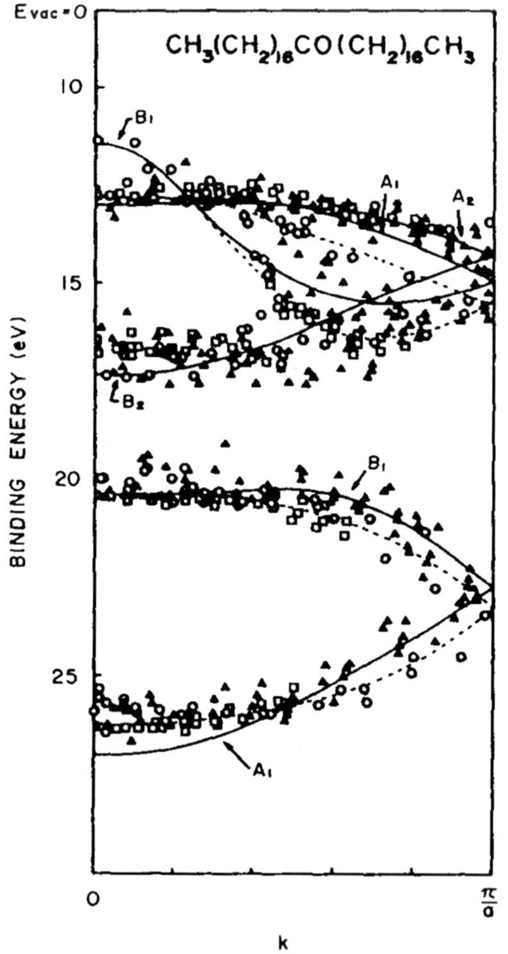

Fig. 36.6. Experimental ARUPS band structure of polyethylene.
(After N. Ueno, K. Seki, H. Fujimoto, T. Kuramochi, K. Sigita and H. Inokuchi, Phys. Rev. B, 41 (1990) 1176, reproduced with permission).

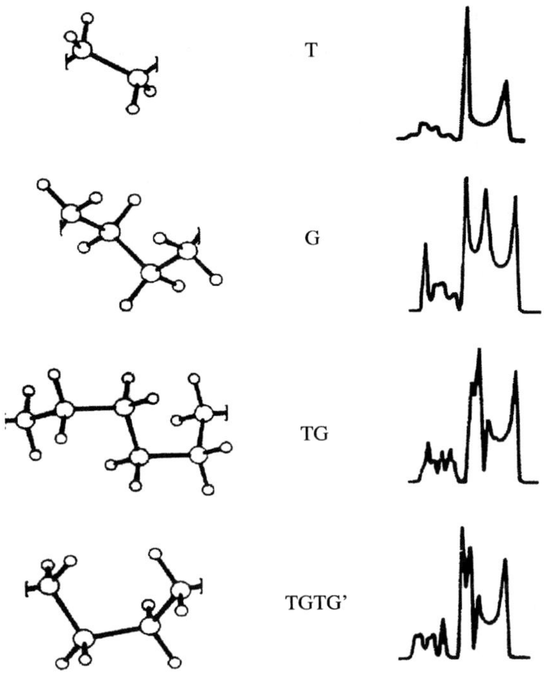

Fig. 36.7. Four conformations of polyethylene (from top to bottom: T, G, TG and TGTG′) and their simulated VEH photoelectron spectra.

The conformational changes do affect the C–C bands. Their bottom energy values are almost constant while a modification takes place at the top of the bands giving rise to important changes in the bandwidths. Furthermore, an energy gap appears in the C–C band for which there are successive trans–gauche conformations (TG and TGTG′). The T and G forms are easily distinguished by their fine structure. Unfortunately, the corresponding samples are not available and this computer experiment only supports the presumption that conformational effects could be experimentally observable using photoelectron spectroscopy measurements of valence bands. This is an incentive to investigate a real case and prove the existence of observable conformational effects on the valence electronic DOS. A tentative interpretation of the experimental ESCA spectrum of isotactic polypropylene has been reported [42], assuming a fully extended zigzag chain conformation in the calculation. No satisfactory agreement as to the peak structure was obtained. The problem has been reinvestigated by explicitly considering the DOS of polypropylene in its actual isoclined $2 \times 3/1$ helical form. As a consequence, the correct structure appears in the theoretical spectrum as illustrated in Fig. 36.8. The conformational effects have also been studied in the case of syndiotactic polypropylenes [43], which exist in a zigzag and in a two-order helical form.

These few examples demonstrate the mutual enrichment that has been gained from a close interrelation between theory and experiment in the field of photoelectron spectroscopy.

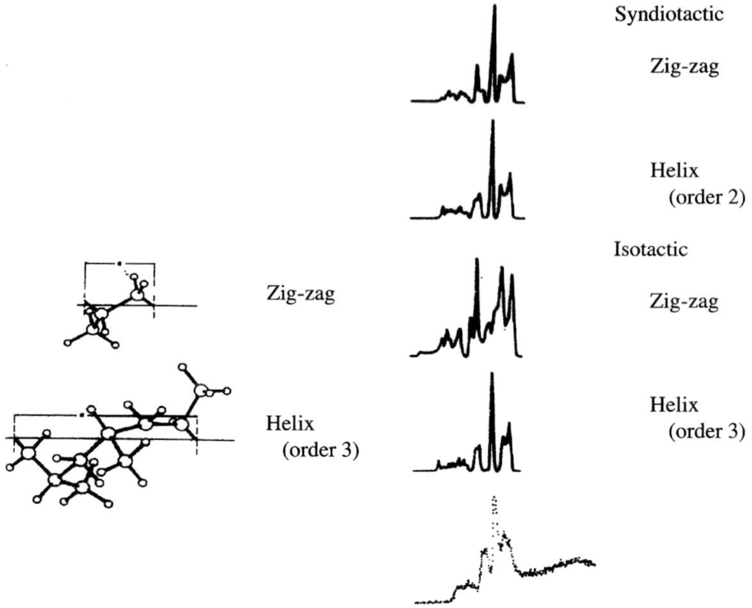

Fig. 36.8. Theoretical (VEH) and experimental (when available) spectra of isotactic zigzag and helical polypropylene.

36.4 BAND STRUCTURE CALCULATIONS AND (SEMI)CONDUCTING PROPERTIES (1978–TILL DATE)

The discovery in the late 1970s that doped organic polymers display high electrical conductivity generated a substantial renewed research interest in the electronic structure of conjugated chains among physicists and chemists alike [44]. On the one hand, the design of new materials combining the processability, lightweight, and durability of plastics with the electrical conductivity of metals is a driving force in the development of conducting polymers. On the other hand, doped organic polymers constitute a new and fascinating area of condensed-matter physics in which non-linear phenomena play an important role. A number of polymers with doped derivatives possessing conductivities greater than 1 S cm^{-1} have been reported. They include conjugated systems such as polyacetylene, polyparaphenylene, polythiophene and polypyrrole. The doping process can be performed chemically or electrochemically and involves exposure of the polymer to electron donors (such as alkali metals) or acceptors (such as I_2 or AsF_5). Although doped polymers display phenomena in some ways similar to conventional doped inorganic semiconductors, their basic physics is very different. One fundamental difference is that these polymers are organic materials. Therefore, it is expected that charge-transfer processes (or electron excitation) will result in significant local modifications (relaxations) of both geometries and electronic structures (strong electron–phonon coupling). In such systems, the electrical conductivity can increase upon doping by several orders of magnitude to reach a metallic-like regime. When MacDiarmid, Heeger and Shirakawa prepared the first polyacetylene doped with iodine in the second half of the 1970s [45], they obtained a conductivity of around 200 S cm^{-1}. Naarman and his group at BASF have synthesized a doped polyacetylene with a conductivity of the order of, or larger than, that of copper at room temperature [46] (6.10^5 S cm^{-1}).

The standard idea about electrical conductivity is that a metal should have a zero energy gap. If a gap exists, depending on its size, a semi-conducting or insulating state is observed. In the case of organic polymers, the existence or non-existence of energy gaps was related very early on to the concept of bond alternation by the pioneering work done by Kuhn [47] in the late 1940s.

Kuhn has shown that in the series of polymethine dyes, the bond lengths between all the carbon atoms are equal due to a resonance balance between equivalent extreme forms:

$$>N-(CH=CH)_n-CH=N^+< \leftrightarrow >N^+ = CH-(CH=CH)_n-N<$$

All carbon–carbon bonds in the skeleton have 50% double bond character. This fact was later confirmed by X-ray diffraction studies. A simple free-electron model calculation shows that there is no energy gap between the valence and conduction bands and that the limit of the first UV–visible transition for an infinite chain is zero. Thus a simple free-electron model correctly reproduces the first UV transition with a metallic extrapolation for the infinite system. Conversely, in the polyene series, $CH_2=CH-(CH=CH)_n-CH=CH_2$, he had to disturb the constant potential using a sinusoidal potential in order to cover the experimental trends. The role of the sinusoidal potential is to take into account the structural bond alternation between bond lengths of single- and double-bond character. When applied to the infinite system, in this type of disturbed free-electron model or Hückel-type theory, a non-zero energy gap is obtained (about 1.90 eV in Kuhn's calculation), as illustrated in Fig. 36.9.

From that work, it became clear that the metallic character of the infinite chain heavily depended on geometrical parameters; an alternating chain being semi-conductive,

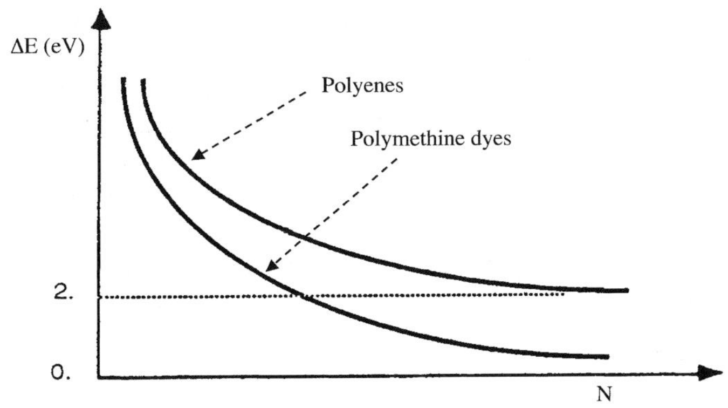

Fig. 36.9. Sketch of the UV transition energies in polymethine dyes and in polyenes.

a regular one being possibly metallic. An energy gap opens up in the band structure if alternating bond interactions are present. This forbidden gap is also related to Peierls distortion: a 1D metallic chain is unstable and tends to distort in order to become semiconducting or insulating and decreases its total energy (this behavior is similar to a Jahn–Teller distortion). In a first approximation, the presence of high conductivity could, therefore, mean some equalization of the bonds: upon doping, the charge transfer process would perturb the electronic distribution and make the chains more regular. This idea that modification of orbital occupancy strongly perturbs the electronic density and the geometrical parameters is not new. It was, for example, already formulated by Pullman and Daudel [48] some time around 1946 in the completely different field of chemical reactivity. In their elegant paper, these authors show by a simple summarized Hückel calculation that a local excitation of a polyene molecule (butadiene in their calculation), leads to a rearrangement of the electronic density and constrains the molecule in a geometry ideally suited for cyclization into cyclobutene. The equalization of bonds in π-conjugated chains under the effect of charge transfer or of electronic excitation HOMO → LUMO is easily explained by the topology of molecular orbitals since the LUMO has nodes at intermediate sites with respect to those of the HOMO. Thus, simple bonds exhibit a larger double character while double bonds become more simple.

The fact that doping does indeed largely perturb the geometry of 1D chains has been confirmed by several computer experiments on isolated chains of equidistant hydrogen atoms [49] and on a more realistic model of polyacetylene [50]. In both cases, the doping is shown to have the same effect, i.e. the amount of bond alternation is calculated to be drastically reduced: bond lengths of 1.41 and 1.43 Å *versus* 1.33 and 1.48 Å for the double and single bonds, respectively (already in minimal basis sets calculations).

It is essential to note that, in these calculations, the energy gap is never fully removed and that other mechanisms must be involved in order to explain the high electrical conductivity. Pople and Walmsley [51] introduced implicitly but unconsciously the idea of solitons in polyacetylene in a classical and important paper in 1962. The neutral soliton is a radical misfit which exists in the middle of a long polyene chain containing an odd number of conjugated carbons, and which consists of several successive bonds of similar lengths near which the unpaired electron is localized, as illustrated in Fig. 36.10.

It is clear from Fig. 36.10 that the neutral radical defect localized at the center of the molecule separates the chain into two alternating regions of the same energy per unit cell, i.e. (a) on the left, a phase (phase A) where the double bonds are 'up' oriented when looking from left to right and (b) on the right, a phase (phase B) where the double bonds are 'down' oriented. The middle, 'regular' defect should have a less stable energy since

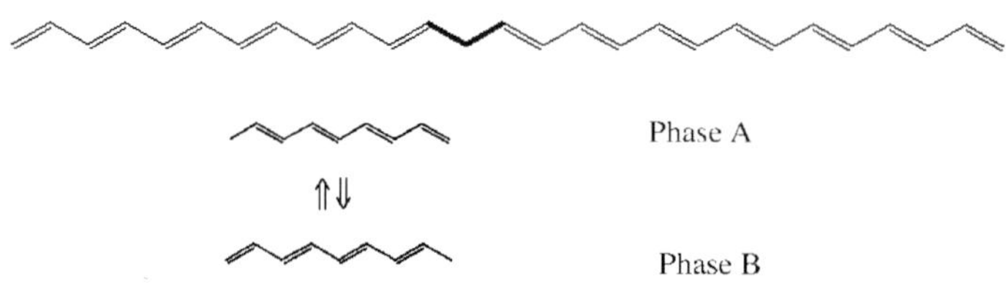

Fig. 36.10. Sketch of a neutral soliton defect.

it corresponds to a phase in which the bonds are more equal. Thus, such a defect delimits polyacetylene regions that present different phases, and due to the strong electron–phonon coupling produces a geometrical distortion of the regular lattice. In this way Δr describes the bond length difference between right- and left-oriented bonds. Thus, an important result is that the motion in going from left (phase A) to right (phase B) is controlled by a double-well potential as illustrated in Fig. 36.11.

From simple mathematical analysis, the double-well potential is represented by a fourth-order polynomial with opposite signs between the quadratic and the fourth terms as, in a dimensionless form

$$V = a(\Delta r)^4 + b(\Delta r)^2 + c \propto (\Delta r)^4 - (\Delta r)^2$$

and that corresponds to a cubic dimensionless force for the motion

$$F = -\frac{\partial V}{\partial(\Delta r)} \propto (\Delta r)^3 - (\Delta r)$$

That form will be important for the formulation of the Su, Schrieffer and Heeger (SSH) theory as we will show later. Note that the neutral radical defect has a non-zero spin that is experimentally detected by ESR measurements. Charged solitons are defects that may be generated by the reduction or the oxidation of the chain, i.e. the monoanions or monocations created, respectively, have no spin as illustrated in the following formulae and in Fig. 36.12 for an oxidative doping such as

$$[\mathrm{CH}]_n + \frac{3x}{2}\mathrm{I}_2 \rightarrow [\mathrm{CH}]_n^{x+} + x\mathrm{I}_3^-$$

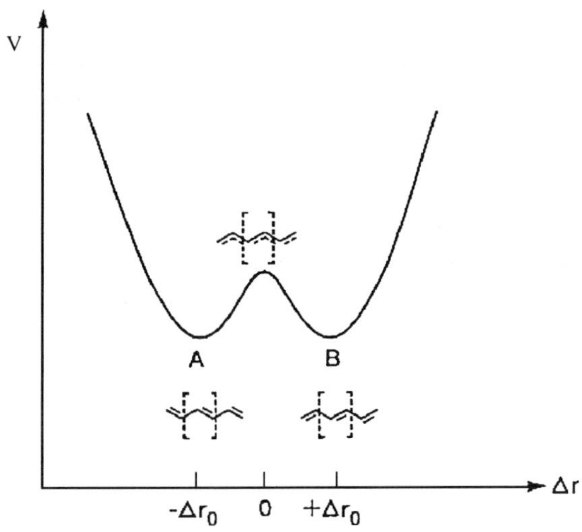

Fig. 36.11. Double-well potential controlling the motion in a polyacetylene chain and sketch of the two phases A and B.

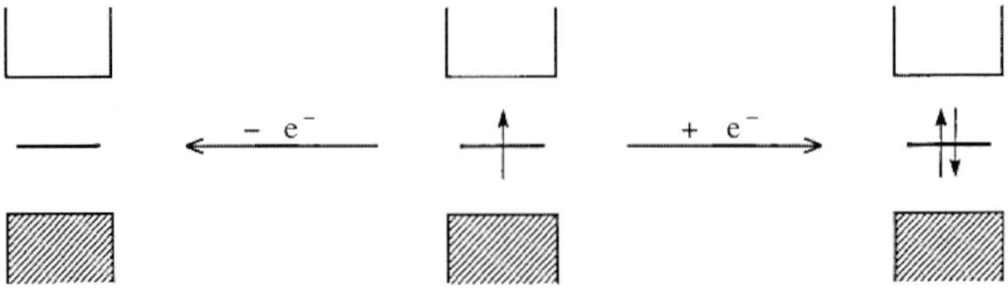

Fig. 36.12. Charged solitons without spin produced from oxidative or reductive doping.

or a reductive doping such as

$$[CH]_n + xNa \rightarrow [CH]_n^{x-} + xNa^+$$

As far back as in 1962, Pople and Walmsley suggested that such a defect could be mobile and thus, if charged, could be responsible for a high electrical conductivity without spin. This idea has been largely developed in the SSH theory [52] which has established the connection between the topological electronic defect and the related moving geometrical distortion. The technique they used corresponds to a Hückel or tight binding method with the lattice motions taken into account. In practice, they solved the dimensionless equation of motion submitted to the force mentioned above and corresponding to the double-well potential

$$\frac{\partial^2 u}{\partial t^2} - \frac{\partial^2 u}{\partial x^2} = u^3 - u$$

The soliton misfit is no longer localized on a single carbon site but delocalized on a given domain whose length is estimated by comparison with experimental data as shown in Fig. 36.13.

This equation is the so-called ϕ^4-equation, the first analysis of which was provided in the first half of the 19th century by Scott-Russel [53] together with pictorial comments: "I was observing the motion of a boat which was rapidly drawn along a narrow channel by a pair of horses, when the boat suddenly stopped—not so the mass of water in the channel which it had put in motion; it accumulated round the prow of the vessel in a state of violent agitation, then suddenly leaving it behind, rolled forward with great velocity, assuming the form of a large solitary elevation, a rounded, smooth and well-defined heap of water, which continued its course along the channel apparently without change of form or diminution of speed. I followed it on horseback, and overtook it still rolling on at a rate of some eight or nine miles an hour preserving its original figure some thirty feet long and a foot to a foot and half in height. Its height gradually diminished, and after a chase of one or two miles I lost it in the windings of the channel. Such, in the month of August 1834 was my first chance interview with that singular and beautiful phenomenon...".

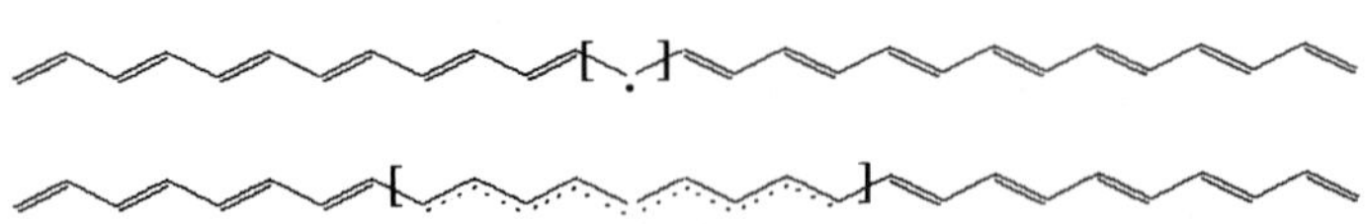

Fig. 36.13. Top: Pople and Walmsley misfit, bottom: Su, Schrieffer and Heeger soliton defect.

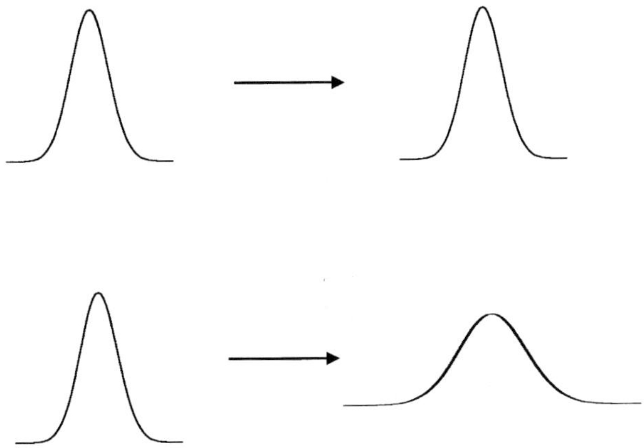

Fig. 36.14. Top: solitonic waves keep their size and shape, bottom: normal waves spread and disperse.

The solution is solitonic waves, i.e. waves that do not spread or disperse like normal waves but instead retain their shape and size as shown in Fig. 36.14.

It is striking to note that bond-length equalization occurs in the middle of a soliton. The soliton can, therefore, be viewed as a kind of metallic island on an otherwise semi-conducting alternating chain in agreement with Kuhn's original ideas, presented above.

An important part of André and Brédas's work during the 1980s was to analyze and extend the soliton, polaron and bipolaron models to a number of cases [54]: polypyrrole, polyparaphenylene, polyanilines, polythiophenes, etc. More details on the theory of conducting polymers can be found in specialized monographs [55].

As an introduction to Section 36.5, let us note that solitons, polarons and bipolarons are excitations of major importance which are inherently nonlinear and thus will favor nonlinear responses in the presence of external fields. Also, it must be realized that the quasi one-dimensionality of the polymer chains allows them to easily undergo structural distortions that result in a significant lowering of the first electronic excitation. Thus, in a perturbation scheme, the linear and nonlinear polarizabilities must be significantly enhanced, since they are inversely proportional to the energy of the electronic excitation. Since 1983, these facts have drawn our attention to the field of polarization of polymeric materials, which is described in detail in Section 36.5.

36.5 BAND STRUCTURE CALCULATIONS AND NON-LINEAR OPTICAL PROPERTIES

Based on arguments such as those developed in Section 36.4, the organic solid state has recently gained much interest in the field of nonlinear optics. The advantages of organics over inorganics are the occurrence of much greater effects due to higher optical damage thresholds; electronic effects inducing quasi-instantaneous responses; and ultra-fast signal processing. The excellent mechanical and molding properties of organic polymers combined with the virtually unlimited potential of organic synthesis have generated many studies and the development of a new physics.

The polarization $\vec{P}$ in an external field $\vec{E}$ is related to the polarizability and the refraction index by

$$\vec{P} = N\left(\alpha + \frac{\mu^2}{3k_\mathrm{B}T}\right)\vec{E}$$

$$\alpha(\omega) = \left(\frac{3M\varepsilon_0}{\rho N}\right)\left(\frac{n^2(\omega) - 1}{n^2(\omega) + 2}\right)$$

The expansion series of the macroscopic polarization $\vec{P}$ and of the microscopically equivalent dipole moment is expressed by the (macroscopic) susceptibilities and the (microscopic) polarizability and hyperpolarizabilities:

$$\vec{P} = \vec{P}_0 + \chi^{(1)}\cdot\vec{E} + \chi^{(2)} : \vec{E}\vec{E} + \chi^{(3)}\vdots\vec{E}\vec{E}\vec{E} + \cdots$$

$$\vec{\mu} = \vec{\mu}_0 + \overleftrightarrow{\alpha}\cdot\vec{E} + \tfrac{1}{2}\overleftrightarrow{\vec{\beta}}: \vec{E}\vec{E} + \tfrac{1}{6}\overleftrightarrow{\vec{\gamma}}\vdots\vec{E}\vec{E}\vec{E} + \cdots$$

Depending on whether the external field is static or not, one obtains a variety of phenomena, such as Degenerate Four Wave Mixing (DFWM $= -\omega; \omega, -\omega, \omega$), Electric Field Induced Second Harmonic Generation (EFISH or ESHG $= -2\omega; \omega, \omega, 0$), the Electro-Optical Kerr Effect (EOKE $= -\omega; \omega, 0, 0$) or Third Harmonic Generation (THG $= 3\omega; \omega, \omega, \omega$), among others. For example, the effect of a second harmonic generation (SHG) is easily understood using the following simplified proof:

$$\begin{aligned} P &= P_0 + \chi^{(1)}E_0 \sin(\omega t) + \chi^{(2)}[E_0 \sin(\omega t)]^2 + \cdots \\ &= \chi^{(1)}E_0 \sin(\omega t) + \chi^{(2)}E_0^2 \sin^2(\omega t) + \cdots \\ &= \chi^{(1)}E_0 \sin(\omega t) + \frac{1}{2}\chi^{(2)}E_0^2[1 - \cos(2\omega t)] + \cdots \end{aligned}$$

where one recognizes a direct component $(1/2\chi^{(2)}E_0^2)$, the fundamental component $(\chi^{(1)}E_0 \sin(\omega t))$ and the SHG component $(1/2\chi^{(2)}E_0^2 \cos(2\omega t))$. This type of procedure can be extended to the cubic, quadratic and larger terms.

In quantum mechanical terms, the interaction Hamiltonian associated with a perturbing external field is approximated by the multipole expansion

$$H_1 = -(\vec{\mu}\cdot\vec{E} + \vec{m}\times\vec{H} + \overleftrightarrow{Q}\cdot\vec{\nabla}\vec{E} + \cdots)$$

from which, in the approximations traditionally used, only the electric dipolar contributions are retained. In particular, an external static electric field modifies the total energy of the system as follows:

$$E(\vec{E}) = E_{\vec{E}=0} - \vec{\mu}\cdot\vec{E} \Leftrightarrow -\vec{\mu} = \left(\frac{\partial E(\vec{E})}{\partial\vec{E}}\right)$$

As a consequence, the linear polarizability can be viewed as the linear response of the dipole moment to the electric field or as the opposite of the second field derivatives of the energy. Similarly, the first hyperpolarizability is the second derivative of the dipole moment or minus the third derivative of the energy with respect to the electric field.

The significance of polyenes (polyacetylenes) in NLO experiments is historically based on FE calculations [56] (which, due to the intrinsic delocalizability of free electrons, produce artificially strong superlinear dependence in N, with the number of double bonds in conjugated systems of the type: $\alpha \approx N^3$ and $\gamma \approx N^3$) and on pioneering Hückel calculations made by Davies [57], Risser [58], Hameka [59], and Flytzanis [60].

In the 1980s and in the early 1990s, the majority of calculations were made based on a static electrical field. All of those approaches have now been generalized to create dynamic processes [61]. A detailed review of many of the technical aspects can be found in the review paper [62].

There are different ways to address the properties, and in particular the (hyper)polarizabilities, of stereoregular conjugated polymers. Indeed, due to electron delocalization, the longitudinal component of α and γ of polyacetylene chains grows supralinearly with the chain length until a linear regime is attained. The response per unit cell increases for smaller chains and then saturates towards an asymptotic 'polymeric' value. These trends are represented in Fig. 36.15. The polymer asymptotic limit should grow when the system exhibits increased geometrical regularity (reduced bond alternation, i.e. metallic situation). In most cases, procedures are needed which directly compute the asymptotic values. This problem is not a trivial one. When a static external electric field is applied along the periodicity axis of the polymer, the potential becomes non-periodic. Bloch's theorem is no longer applicable and the single-electron wave functions cannot be represented in the form of crystalline orbitals. For example, in the simple case of the free electron in a 1D box with an external electric field, the solutions of the Schrödinger equation are given as combinations of the first- and second-species Airy functions and do not show any periodic character [63]. Consequently, the transformation of dipole integrals from molecular limited systems to infinite periodic polymers is not a trivial matter and is detailed, for example, in Ref. [62]. Such procedures have been developed in our laboratory since 1990. They are also reviewed in Ref. [62].

An example of a result that can be obtained is given below. It concerns the comparative calculation of the static polarizability per unit cell of polyethylene and polysilane chains

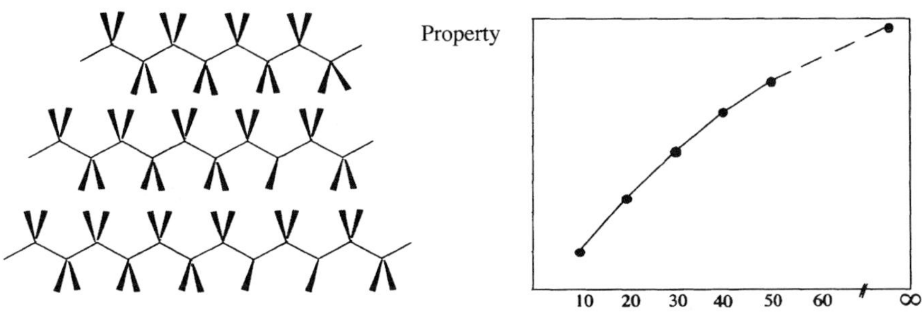

Fig. 36.15. Left: models of polyethylene chains of increasing size, right: sketch of the evolution of an electronic property with chain size.

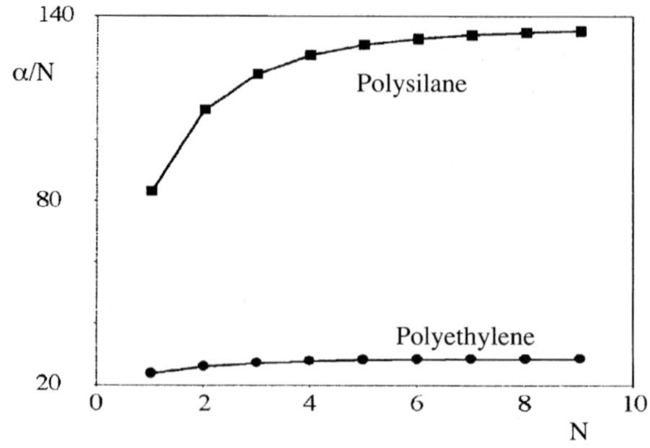

Fig. 36.16. Comparative polarizabilities (in a.u., RHF/3-21G calculation) per unit cell of polyethylene $H(-CH_2-CH_2-)_nH$ and of polysilane $H(-SiH_2-SiH_2-)_nH$.

illustrated in Fig. 36.16. As expected, the more highly polarizable polysilane chain leads to a slower asymptotic convergence.

One of the undeniable advantages of polymer calculation is that one can identify the contribution of each electronic band. As an illustration, the results for polyacetylene are given in Fig. 36.17.

The total electronic polarizability per unit cell (obtained by a RPA/3-21G approximation) is found to be 217.1 a.u. In the theoretical estimation, the contribution of the last occupied π-band is 181.1 a.u., or in other words 83% of the total electronic polarizability. The experimental determination is 203 a.u. [64]. Note that the theoretical calculation does not include correlation and vibrational contributions.

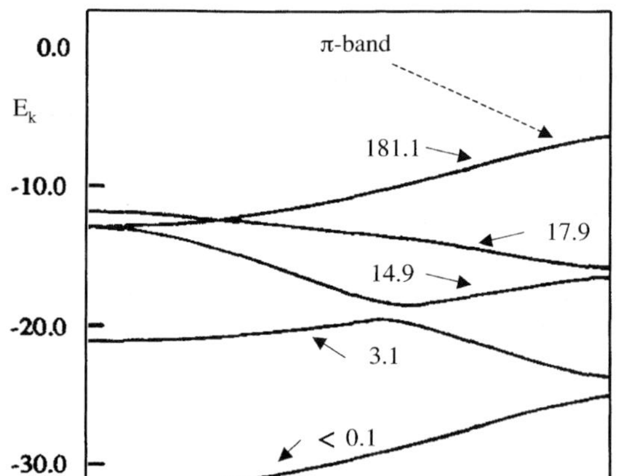

Fig. 36.17. Electronic band polarizability per unit cell of polyacetylene $H(-CH{=}CH-)_nH$–energies in eV, polarizability contributions in a.u.

Today interest in nonlinear polarizabilities is of prime importance. These systems require delicate engineering to create a perfect balance between the various factors involved. In order to maximize a property linked to the first hyperpolarizability β, and more precisely β related to the size of the compound, electron delocalization and asymmetry have to be efficiently combined. Indeed, on the one hand it is known that the (hyper)polarizabilities of organic systems are enhanced by the presence of mobile π-electrons, a problem already addressed in the search for organic conducting polymers. Polyacetylene skeletons are thus excellent candidates. On the other hand, the first hyperpolarizability β is strictly zero for centrosymmetric molecules, as it corresponds to an odd term in the dipole moment Taylor expansion. A second factor is thus an intrinsic asymmetry as discovered in the pioneering work of Zyss et al. [65] on para-nitroaniline (pNA) and the related compounds, 2-methyl-4-nitroaniline (MNA) and 3-methyl-4-nitropyridine-1-oxide (POM). The combination of the two has led to the suggestion of push–pull polyenes [66]. In these systems, electron-acceptor/donor groups are added at the extremities of a symmetric delocalizable segment. Thus the end groups provide the asymmetry to the mobile electrons of a conjugated bridge. It is clear that as the size of the system increases, the centrosymmetry becomes important and annihilates the dissymmetric end effects as illustrated in Fig. 36.18. Thus, the most efficient asymmetry/delocalization combination is usually difficult to achieve due to the antagonism of the two factors: the delocalization increases as the asymmetry decreases and *vice versa*. Once a strong push–pull pair has been selected, the only way to increase β/N is to use longer conjugated segments. Nevertheless, for a large N, the end groups no longer interact with each other as outlined in Fig. 36.18. As a result, the first hyperpolarizability β becomes constant and β/N decreases sharply. Consequently, the β/N *versus* N curve presents a maximum.

An alternative way of obtaining high β/N has been recently investigated by Jacquemin et al. [67]. He has focused on AB systems $[-(A{=}B-)_n]$ in which each unit contains π-electrons and is asymmetric. Different AB compounds have been studied. An old

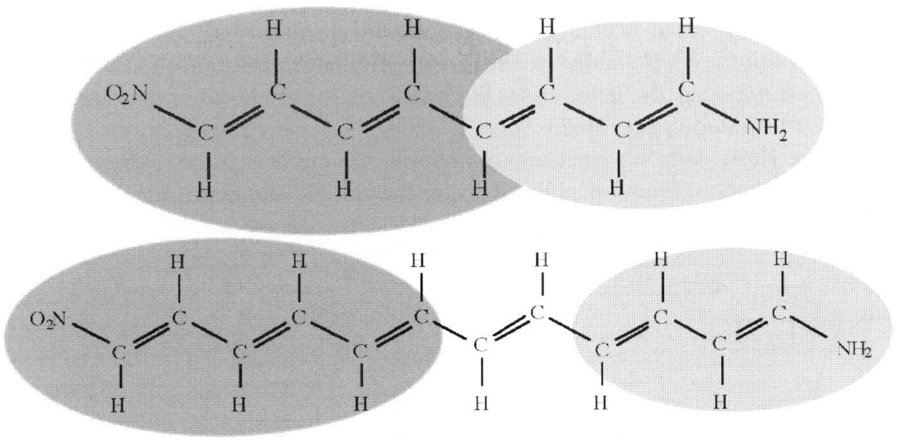

Fig. 36.18. Scheme of interactions of end groups in push–pull polyenes of increasing size.

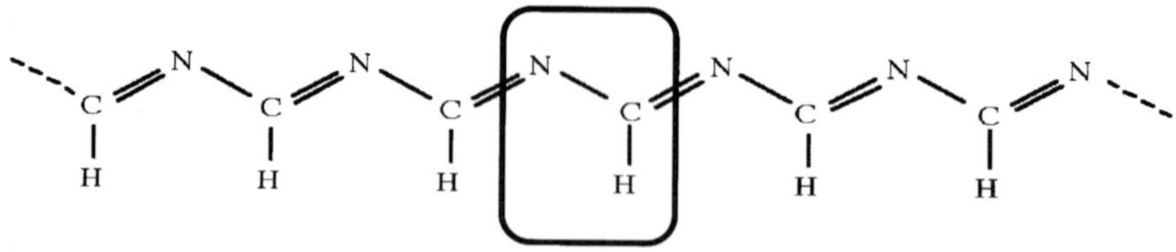

Fig. 36.19. Polymethine chain.

model of this type already studied in the 1960s for its potential characteristics of semi-conductivity [68] is polymethineimine (PMI), represented in Fig. 36.19. As shown, an intrinsic asymmetry is characteristic of the unit cell.

For smaller sizes of the polymer chain, the asymmetry of the two different end groups (CH_2 on one end of the chain and NH on the other end) adds to the intrinsic asymmetry of the –CH=N unit cell. However, if the two bonds CH=N and N–CH would be strictly regular, the intrinsic asymmetric character of the chain would disappear as the size of the chain increases, as indicated in Fig. 36.20.

In a model based on such conditions, such as a Hückel calculation made with equivalent bond lengths, the β curve would exhibit a maximum and tend asymptotically to zero as obtained in Fig. 36.21.

In reality, the CH=N and N–CH bonds are not equivalent and the calculated behavior is reproduced in Fig. 36.22.

Other interesting systems such as polysilacetylenes [69], polyphosphinoboranes, and polyphosphazene-based materials [70] are currently being investigated.

The importance of the effect of vibrations on (hyper)polarizabilities has also become increasingly apparent since the mid-1990s. It is clear that vibrational effects are at crossroads between electronic and geometric effects and are also a manifestation of the electron–phonon interaction, which has already been shown to be of prime importance in estimating conducting and semi-conducting properties.

Indeed, in addition to the electronic contribution to α, β, and γ, a vibrational counterpart (α^{v}, β^{v}, and γ^{v}) also exists which involves nuclear relaxation (α^{r}, β^{r}, and γ^{r}) and curvature contributions (α^{c}, β^{c}, and γ^{c}).. The former originates from field-induced nuclear relaxation effects, whereas the latter arises from the field-dependence of the vibrational energy [71] as illustrated in Fig. 36.23.

Indeed, in the Born–Oppenheimer approximation, the effects can be interpreted at a first-order level of approximation referred to as the double harmonic approximation.

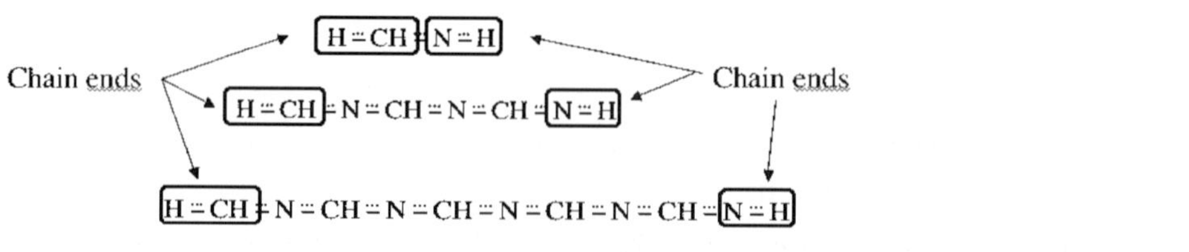

Fig. 36.20. Asymmetric behavior of end groups and tendency toward centrosymmetry in large PMI chains.

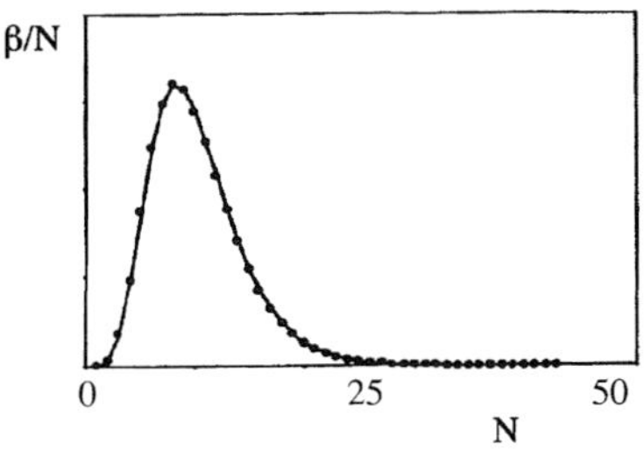

Fig. 36.21. Hückel β/N hyperpolarizability (in arbitrary units) of a PMI chain with equivalent bonds.

A simple rule for the harmonic oscillator is that a perturbation with a linear potential in the stretching coordinates will not change the oscillation frequency or the energy spacing between the states. A linear potential term leaves the force constant unchanged but it will, however, change the equilibrium point and the energy of the potential minimum. In practice, vibrational effects are not trivial due to the close connections between nuclei and electrons. It is obvious that an external electrical field has interaction with charged particles and thus induces reorganizations of the density (positions) of these particles leading to both electronic polarizability and vibrational polarizability.

Nowadays, the relative importance of the vibrational response in linear and non-linear processes is clear but does depend on the optical frequencies concerned. The results obtained for polyacetylene chains in the static limit are sketched in Fig. 36.24.

The vibrational contribution to the polarizability of polyacetylene amounts to 10% of its electronic counterpart and mainly originates from low-frequency transverse acoustic modes (TAMs) [73]. It tends towards zero in the optical regime. The nature of the frequency dispersion, therefore, constitutes a major difference between electronic and vibrational (hyper)polarizability contributions. The response time is also different: the electronic processes are faster and their response time is of the order of 1 fs, whereas vibrational phenomena are slower and have response times ranging between 10 and 100 fs as a function of the associated vibrational normal mode frequency.

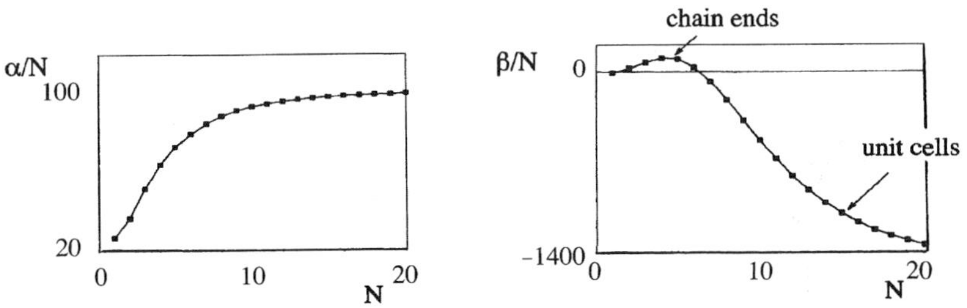

Fig. 36.22. Diagram of the ab initio α/N polarizability and of the β/N hyperpolarizability of a PMI real chain (data in a.u.)

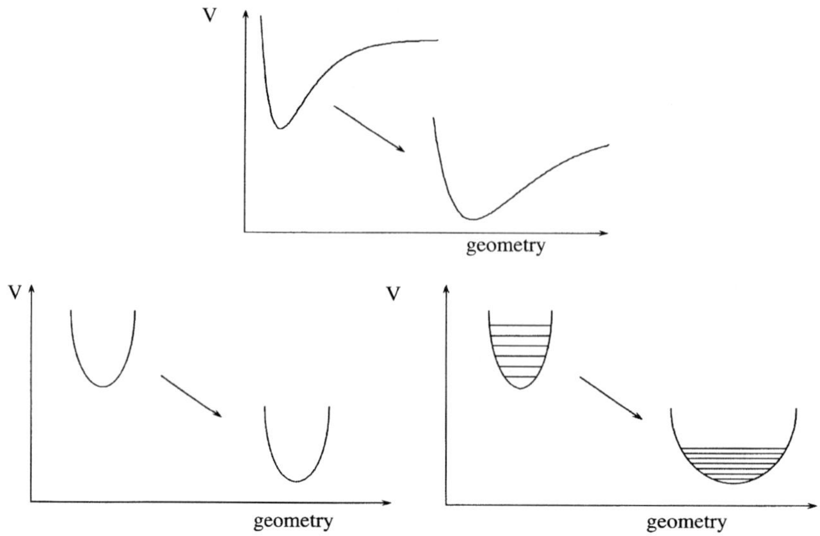

Fig. 36.23. Illustration of nuclear curvature and relaxation vibrational effects.

Recent investigations summarized in Fig. 36.25 have also demonstrated that the importance of the ratio varies markedly with both chemical nature and optical process [74].

The vibrational contribution to the second hyperpolarizability of polyacetylene is mainly supported by the Raman intensity-related term. Its importance varies according to the optical process: 0.5% for EFISHG, 45% for EOKE, 87% for DFWM and 129% for the static second hyperpolarizability.

This new field of vibrational linear and non-linear optics has also recently and rather unexpectedly generated a renewed and significant interest in calculations and interpretations of IR, Raman and Hyper-Raman experiments [75].

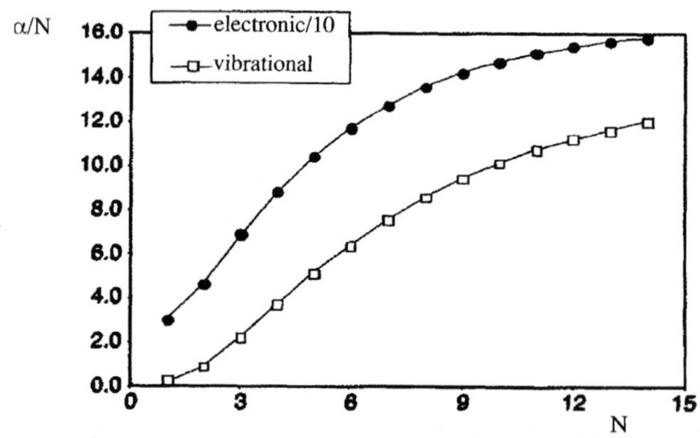

Fig. 36.24. Relative importance of electronic and vibrational effects in polyacetylene chains [72]—data in a.u.

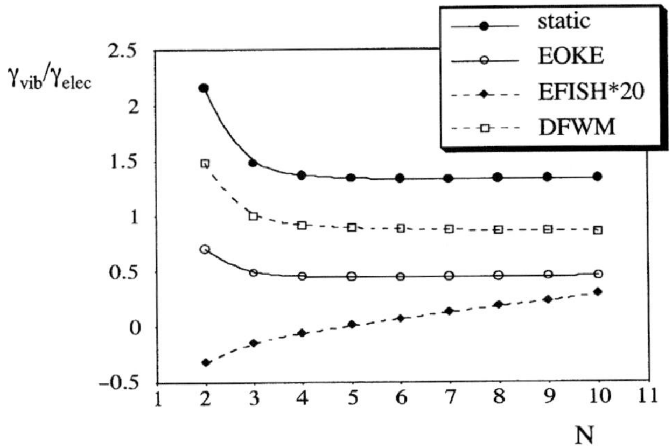

Fig. 36.25. Relative importance of electronic and vibrational contributions for various third-order non-linear optical effects.

36.6 BAND STRUCTURE CALCULATIONS AND ELECTRON TRANSFER MARCUS THEORY

On the one hand, there is a clear need for this section in the light of the advanced information released by the Nobel committee for the 2000 Nobel Prize in Chemistry on conductive polymers, which is extensively discussed in Section 36.4. Indeed, this information contains a full section entitled 'molecular electron transfer theory' [76].

On the other hand, it is striking to note that transition state theory has been applied extensively in quantum chemistry ever since the computational facilities needed for this were made available, while Marcus Electron Transfer (ET) theory has only been used systematically since the last decade.

We believe that there are two basic reasons for this situation. Firstly, the Marcus theory has only been experimentally validated since the beginning of the 1980s. Secondly, the advent of organic material science and, in particular, the development of Organic Electroluminescent Diodes (OLED's) has given significant impetus to the development of quantum chemical calculations related to electron and energy transfer in and between molecules.

Excellent reviews on the theory of ET are available in the literature. Among them, the Nobel lecture of Marcus [77], the 1996 reviews of Barbara et al. [78] and of Balzani et al. [79], both the 1999 Jortner and Bixon issues of Advances in Chemical Physics [80], and the 2000 monograph of May and Kühn [81] are worthy of note.

In this section, we summarize the Marcus theory in its classical formulation. We would like to draw the reader's attention to aspects that are not generally presented in the literature and that, in our opinion, are valuable to note. We show by a few examples how Marcus' ideas can be successfully applied to the elucidation of practical problems in conducting, semi-conducting, photo- and electro-luminescent systems.

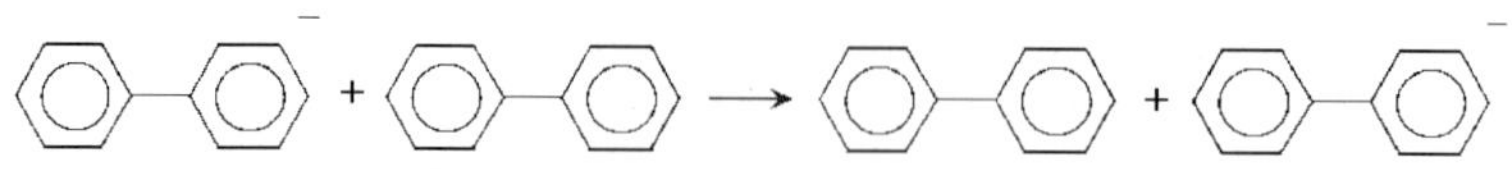

Fig. 36.26. Self-exchange charge transfer in biphenyl oligomers.

ET processes are common in chemistry. In redox processes, for example, the oxidation of a molecule means the removal of electrons while reduction means the addition of electrons. In other terms, an electron jumps from the donor (D), which is oxidized, to the acceptor (A), which is reduced

$$D + A \rightarrow D^{+} + A^{-}$$

In a self-exchange reaction, an electron leaps between two identical molecules

$$A^{-} + A \rightarrow A + A^{-}$$

In the case of conjugated conducting polymers, which were previously mentioned, the ET theory applies since the conductivity is not a simple band type conductivity but, depending on the doping level, merely an intrachain and/or interchain motion of charged soliton-like or polaron-like defects. We can sketch this motion as indicated in Fig. 36.26.

This shows nothing more than an electron transfer in the sense advocated by Marcus. In the case of polyparaphenylene, the next example illustrates the Nobel information already mentioned (Fig. 36.27).

Let us note that the first reaction corresponds to a self-exchange reaction where the reagent and the products have the same stability, whereas in the second example, the products are better stabilized by delocalization effects.

The history of ET processes goes back to Arrhenius, 1903 Nobel Prize winner in Chemistry, who established that salts in aqueous solutions exist as positive and negative ions and not as neutral molecules. For example, trivalent Co ions oxidize bivalent Cr ions and form bivalent Co ions and trivalent Cr ions:

$$Co^{3+} + Cr^{2+} \rightarrow Co^{2+} + Cr^{3+}$$

A few years later, Werner, 1913 Nobel Prize winner in Chemistry, postulated that metal ions in solution are surrounded by a fixed number of neighboring negative ions or neutral molecules, arranged in a certain way, e.g. at the corners of an octahedron if there are six of them:

$$[Co^{3+}(NH_3)_5Cl^{-}]^{2+} + [Cr^{2+}(H_2O)_6]^{2+} \rightarrow [Co^{2+}(NH_3)_5(H_2O)]^{2+} + [Cr^{3+}Cl^{-}(H_2O)_5]^{2+}$$

The first systematic studies of ET reactions between the oxidation states of Pb^{212}(II) and Pb^{212}(IV) were performed in 1920 by (von) Hevesy and Zechmeister. They used

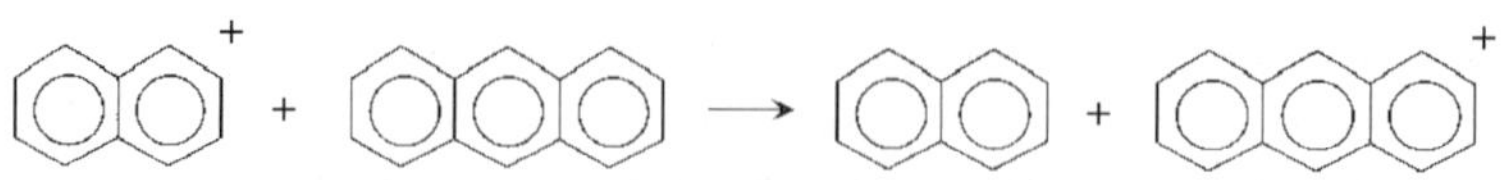

Fig. 36.27. ET exchange in polyparaphenylene units.

a new method of chemical research using natural radioactive elements as tracers. The post-war advent of the isotopic radioactive tracer technique was the driving force behind many experimental studies. The study of the self-exchange reaction

$$Fe^{3+*} + Fe^{2+} \rightarrow Fe^{2+*} + Fe^{3+}$$

provoked initial surprise since it turned out to be a very slow reaction occurring over a period of hours. This was an unexpected result since no chemical bond is broken. The electron is a very light particle compared to the nucleus, and only one electron changes place. Thus the changes are insignificant and should not give rise to any large energy barrier. Note here that the advantage of studying self-exchange reactions such as the previously mentioned one is that the driving force of the reaction (i.e. the standard free energy of reaction ΔG^0) is equal to zero. Thus, one of the factors, which usually influences the rate of a chemical reaction in a major way, namely, the relative thermodynamic stability of the reactants and products, is eliminated. Note, however, that such cases are of no interest for chemiluminescence which requires energy differences between the partners involved.

A further surprise came from the study of ET 'cross-section' reactions like

$$Fe^{2+} + Ce^{4+} \rightarrow Fe^{3+} + Ce^{3+}$$

where the reactants and the products do not have the same stability. Very different rates of reaction were observed. The differences spanned 15 orders of magnitude, from very slow reactions like

$$Co(NH_3)_6^{2+} + Co(NH_3)_6^{3+} (10^{-6}\ M^{-1}\ s^{-1})$$

to very fast ones

$$MeL_3^{2+} + MeL_3^{3+}\ (10^6 - 10^7\ M^{-1}\ s^{-1})$$

where the ligand L stands for bipyridine or 1–10 phenanthroline and the metal Me = Fe, Ru, Os. Libby was the first to propose a theoretical model for the interpretation of the large variety of rate constants observed. In the simplest self-exchange ET processes—the transfer of an electron between two molecules—no chemical bonds are broken. However, since the nuclei do not have time to move during the rapid electron jump the new species are formed in the solvent molecules, which is the wrong kind of environment. It was Marcus who cleverly remarked that were such a reaction scheme to occur in the dark, the energy would not be conserved. Since the ions would be formed in an incorrect high-energy environment, the only way such a non-energy-conserving event could happen would be by the absorption of light provoking a 'vertical' transition, and this could not take place in the dark. In light of this, Marcus proposed that, by their thermal fluctuations, the solvent molecules change their positions in the immediate vicinity of the ions, thus increasing the energy in the molecular system. The electron can only jump between two states that have the same energy, and this condition is only satisfied by increasing the energy for both molecules. A summary of these opposing viewpoints is provided in Fig. 36.28.

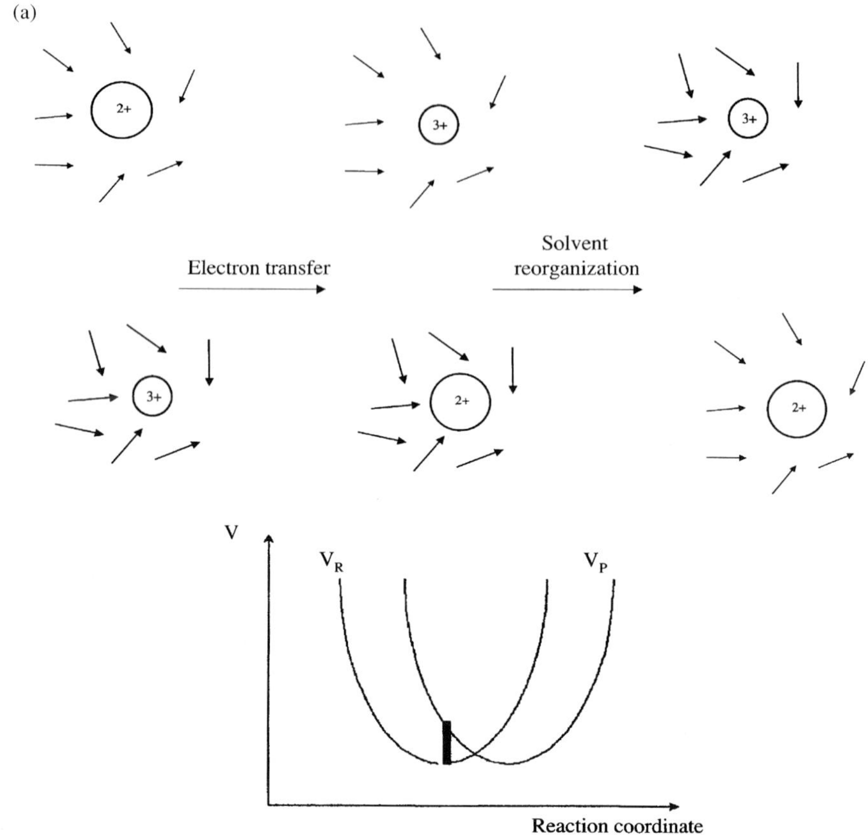

Fig. 36.28. (a) Libby's hypothesis: the electron transfer is anterior to the solvent reorganization, the energy required is λ. (b) Marcus's hypothesis: the solvent reorganization is anterior to the electron transfer, the energy required is $\lambda/4$.

Thermal fluctuations from the original equilibrium ensemble of configurations are thus needed prior to the electron transfer. They are followed by a relaxation to the equilibrium ensemble for the products, after electron transfer. Marcus's theory states that the potential energy V_R of the entire system—reactants plus solvent—is a function of the many hundreds of relevant coordinates of the system. The latter include, among others, the position and orientation of the individual solvent molecules and the vibrational coordinates of the reactants. After the electron transfer, the reacting molecules have the ionic charges appropriate to the reaction products, and so the relevant potential energy function V_P is that for the products plus solvent. Ignoring entropy changes, the free energies become energies or potential energies.

These two potential energy surfaces will intersect if the electronic coupling that leads to electron transfer is neglected. This intersection then constitutes, in Marcus's approximation, the transition state of the reaction. Due to the effect of the previously neglected electronic coupling and to the coupling of electronic and nuclear motion near

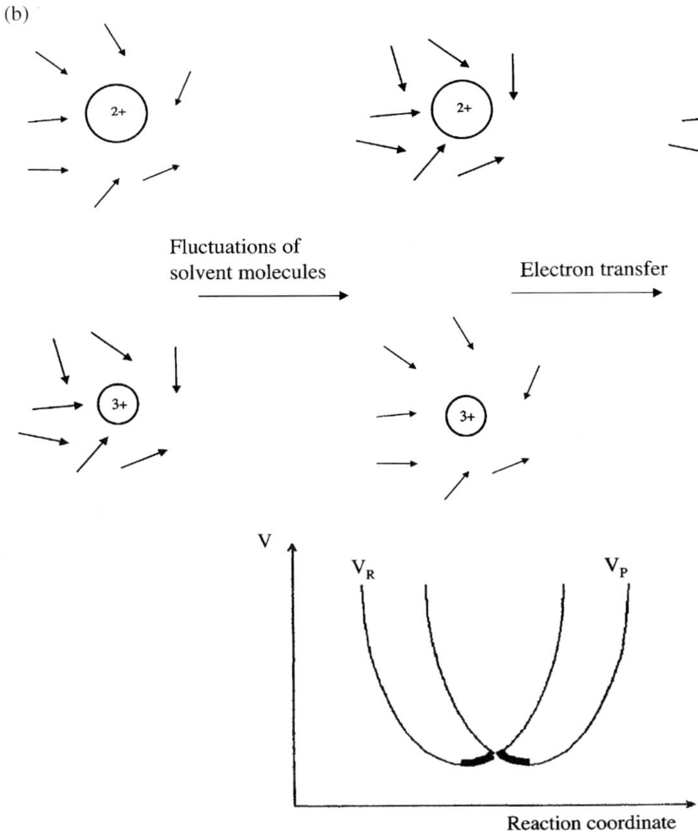

Fig. 36.28. (*continued*)

the intersection surface, an electron transfer can occur at the crossing of the two curves in Fig. 36.29. This view has the enormous advantage that it satisfies two basic principles of quantum mechanics: firstly, the Franck Condon Principle is satisfied, since the transfer between the two curves occurs at fixed positions and momenta of atoms; secondly, the principle of conservation of energy is also fulfilled.

Note also that, in Marcus's original idea, the parabolas represent free energy plots during the ET reaction. It is, therefore, correct to assume that they represent energy plots if working at 0 K. In this sense, they are legitimate as single-coordinate plots with respect to the degree of progression of a chemical reaction. They should not be confused with potential energy plots that are at most a profile of the complicated V_R and V_P in N-dimensional space.

With these assumptions, the original Marcus theory gives a simple mathematical formula for calculating the energy change along the reaction profile, as well as the size of the energy barrier. It also introduces the important concept of reorganization energy λ. The origin of the latter concept lies in the fact that, although no bonds are broken during the reaction, there are still small changes in structure when electrons are added or removed.

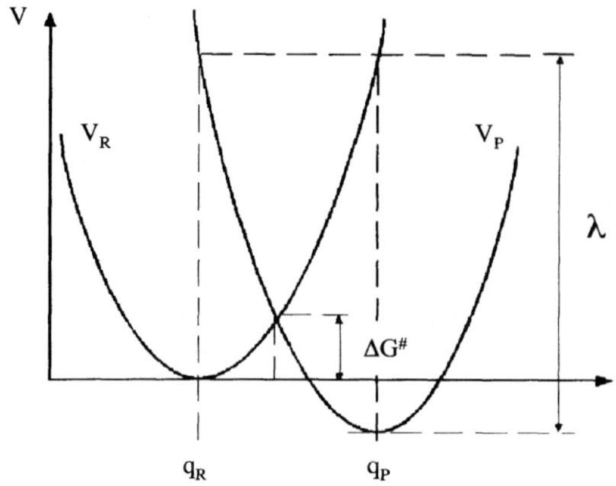

Fig. 36.29. General plot of (free) energies of the reactant and product states with respect to the degree of advancement of the ET reaction.

The lengths of the chemical bonds are altered and the molecules of the solvent are reoriented. Such structural changes require reorganization energy. The meaning of λ can also be interpreted as follows: initially the electron is with the donor and the vibrational coordinate has the value $q = q_R$. Then a sudden change of the electronic state occurs. In order to reorganize the vibrational coordinate (nuclear configuration) to the new equilibrium value $q = q_P$, the energy λ has to be removed from the system. Therefore, this energy is usually called reorganization energy. If the ET reaction proceeds in a solvent, the change of the electronic charge density in the DA complex is accompanied by a rearrangement of the solvent polarization field. Thus, the name polarization energy is also common for λ. Note that the definition of λ

$$\lambda = \tfrac{1}{2} f(q_R - q_P)^2$$

has the important conceptual consequence that λ increases in parallel to the separation distance $(q_P - q_R)$ between the reactants and the products and with the force constant f. Other cases that we have studied, such as soliton or polaron transport models, concern the transport of composite defects consisting of an electronic charge and of its associated lattice polarization, and are thus related to electron–phonon coupling. We refer to our systematic studies of bipolaron defects in conjugated polymers to provide examples of quantum mechanical studies of these reorganization effects [82].

The free energy barrier is obtained as

$$\Delta G^{\#} = G_R(q_c) - G_R(q_R) = V_R(q_c) - V_R(q_R) = \frac{1}{2} f(q_c - q_R)^2 = \frac{1}{4\lambda}(\lambda + \Delta G^0)^2$$

The classical Marcus theory is thus of paramount importance since it defines the free energy barrier in terms of two quantities only: the free energy of reaction ΔG^0 and the reorganization energy λ. Consequently, when obeying Marcus's law, ET processes are activated by temperature according to a Boltzmann-dependent activation process:

$$k \propto \exp\left(-\frac{E_a}{RT}\right)$$

The plot of the mobilities with respect to the inverse of temperature produces a U-like curve separating the domain of band regimes (low temperature) from that of hopping between molecules controlled by Marcus's activation energy between molecules (high temperature). The mobility decreases with temperature in the band regime domain while it is activated in the hopping regime. This dependence was recently publicized by the measurements of electron and hole mobilities in polyacenes [83].

Quantum chemistry is also of paramount importance to estimation of the reorganization energy. In the case of polyacenes, an estimation of the relaxation energy has been obtained by computing the energy difference of the cation states of the oligoacenes between the neutral and cation geometries [84]. It was found that the theoretical and experimental estimates agree well and are remarkably small (theoretically from 0.069 eV for anthracene to 0.049 eV for pentacene, experimentally from 0.071 ± 0.002 eV for anthracene to 0.061 ± 0.002 eV for pentacene). This observation is clearly important in the rationalization of the exceptionally high mobilities observed in organic systems. According to the previous remark, it is also understood that the reorganization energies will be greater for smaller molecules (e.g. anthracene) than for larger ones (e.g. pentacene), where the electronic structures containing more electrons are less perturbed by the transfer of a single charge.

The final important aspect of Marcus's theory that we would like to mention is the curious square dependence of the activation energy with respect to the free energy of reaction:

$$\Delta G^{\#} = \frac{1}{4\lambda}(\lambda + \Delta G^0)^2$$

A first statement of fact is that this formula *a posteriori* confirms Marcus's view of ET reactions as opposed to Libby's understanding. Indeed (see Fig. 36.28), in self-exchange reactions ($\Delta G^0 = 0$), Libby's mechanism would require an activation energy of λ, while in Marcus's scheme, the leap of the electron *via* the crossing point of the parabolas only requires the energy $\lambda/4$.

Another important point is that the actual value of the activation energy is determined by the balance between the value of the reorganization energy and that of the free energy of reaction. If $|\Delta G^0| < \lambda$, $\Delta G^{\#}$ decreases and k increases with a growing driving force. This is an expected trend in chemical reactions, observed in Bronsted plots of acid or base catalyzed reactions or in the Tafel plots of electrochemical reactions. If $|\Delta G^0| = \lambda$, $\Delta G^{\#}$ is equal to zero; the liberated energy is equal to the reorganization energy and the theory predicts a very fast barrier-free reaction, i.e. an activation-free case observed experimentally when the rate becomes independent of temperature.

The most surprising fact applies to largely exothermic reactions when $|\Delta G^0| > \lambda$. The square in Marcus's expression of the activation energy completely neutralizes the large driving force of the reaction. The activation energy increases with the exothermicity of the reaction. The reaction is slower the greater the amount of energy liberated,

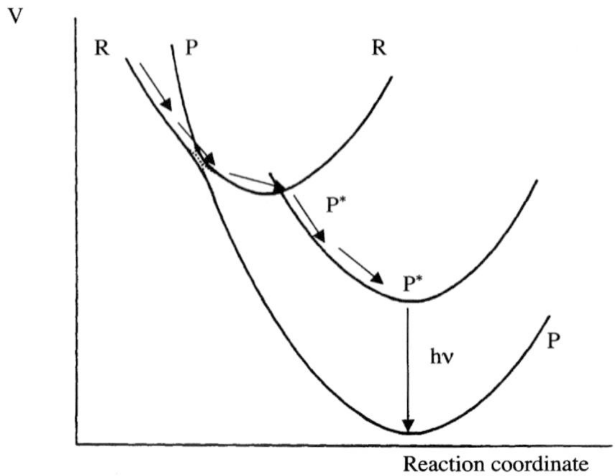

Fig. 36.30. Crossing of ground state reactants and products in Marcus's inverted region allowing transition from the ground state of the reactants to the excited state of products and photoluminescence to the ground state of the products.

a behavior as unexpected as were a ball to roll more slowly down a slope the steeper it was. Chemists have long found this remarkable counterintuitive result difficult to accept. That is perhaps the main reason why Marcus's theory has not been broadly exploited in quantum chemical calculations. It was only experimentally confirmed in the 1980s [85]. Due to its anomalous character, this domain has been called the 'inverted' region.

An indirect consequence of the considerable interest in Marcus's theory is that, as brilliantly suggested by Marcus himself in 1965, the inverted region is responsible for most chemiluminescent effects. Indeed, as shown in Fig. 36.30, when the V_R and V_P curves intersect with a high activation barrier $\Delta G^{\#}$ because of the inverted region effect, there may be an electron transfer to a more easily accessible V_{P^*} curve. In this case, one of the products is electronically excited and intersects the V_R curve in the normal region with a low activation barrier.

Organic Light-Emitting Diodes (OLED) and Polymer Light-Emitting Diodes (PLED) based on the mechanism foreseen by Marcus have been realized in practice. PLEDs are successively formed from an Al, Mg or Ca cathode, an organic conductive polymer and an ITO (Indium, Tin, Oxide) anode, as schematically illustrated in Fig. 36.31.

The electrons are generated at the cathode, transmitted by an electron transport layer (Alq3, tris-8-hydroxyquinoline) Al and recombine with holes generated at the transparent ITO anode by emitting light. The organic polymers are derivatives of polyparaphenylene vinylene or polyfluorene containing organic fluorescent material emitting in the three main colors ((RGB synthesis): Red (e.g. nile dye), Green (e.g. coumarin), Blue (e.g. perylene dye)). Information on the progress of OLEDs and PLEDs is regularly found in general literature such as Chemical and Engineering news [86] or on specialized sites [87].

In an interview for the 2000 Nobel Prize, Alan MacDiarmid concluded: "Research on conductive polymers has also fueled the rapid *development of molecular electronics.*

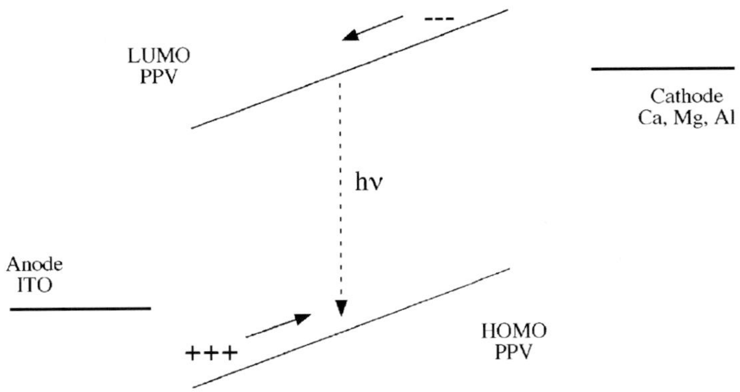

Fig. 36.31. General overview of the energy levels of an OLED.

In the future scientists may be able to produce transistors and other electronic components consisting of individual molecules, dramatically increasing the speed and reducing the size of computers: a computer corresponding to the laptops we now carry around suddenly fits inside a wristwatch".

To conclude this section, it is important to note that, in its classical form, Marcus's theory only implies two parameters: the activation energy to the free energy of the reaction and the reorganization energy. It does not explicitly depend on the importance of the energetic coupling between the initial (reactant) and final (product) state. This effect is explored in the semi-classical Marcus theory [88] and will not be detailed here.

36.7 CONCLUSIONS

This paper presents research work carried out during the last 50 years. It deals with the theoretical understanding of the structure of macromolecules employing the methods of quantum chemical physics, a field which is today known as 'quantum chemistry of polymers'. It summarizes the formulation of a theoretical approach as well as the applications relevant to contemporary technology: work (a) on photoelectron spectra in the early 1970s, (b) on highly conducting polymers, from the latte 1970s, (c) on systems of interest in nonlinear optics since 1983, with a strong emphasis to original methods for large polymer chains since 1990 including since 1995 the too rarely studied vibrational effects, (d) on charge and energy transfers in conjugated polymers since 2001.

The main conclusion is that the success of this field is due to a close interconnection between analytical and computational approaches. The paper has clearly demonstrated that we need to take into account electron–phonon interactions. In other terms, there exists a timely need to include translational, rotational, vibrational contributions and non-adiabatic approaches in our models. Then the road will be open to complete non-equilibrium interpretations of the electronic properties of macromolecules [89].

36.8 ACKNOWLEDGEMENTS

The author would like to thank many of his collaborators and colleagues of the last 40 years, in particular: Dr Marie-Claude André, Dr Joseph G. Fripiat, Prof. Joseph Delhalle, Prof. Jean-Luc Brédas, Dr Benoît Champagne, Dr Eric Perpète, Dr Denis Jacquemin and Mrs Joanne O'Donnell for proofreading the full paper. He also acknowledges the support provided by the Interuniversity Attraction Poles Program on 'Supramolecular Chemistry and Supramolecular Catalysis (IUAP No. P5-03)'—Belgian State Federal Office for Scientific, Technical and Cultural Affairs.

36.9 REFERENCES

1 L. van Bethoven, carnets intimes, Editions Buchet Chastel, Paris (1991): Cultive la science: il n'est pas de sentier plus sûr pour l'homme Que celui dont toujours mes sages ont fait le choix.
2 P. Leprohon, Vincent Van Gogh, Editions du Valhermeil, p. 60 (2001): Lorsqu'on regarde ce que d'autres ont fait mieux que soi, on en vient à haïr sa propre vie.
3 L. Ullmann, Devenir, Editions Stock (1977) p. 263 (Original Title: Changing): La maturité ne consiste-t-elle pas, peut-être, à laisser les autres exister?
4 J. Ladik, Acta Phys. Hung., 18 (1965) 173.
5 J. Ladik, Acta Phys. Hung., 18 (1965) 185.
6 J.M. André, L. Gouverneur and G. Leroy, Int. J. Quantum Chem., 1 (1967) 427.
7 J.M. André, L. Gouverneur and G. Leroy, Int. J. Quantum Chem., 1 (1967) 451.
8 G. Del Re, J. Ladik and G. Biczo, Phys. Rev., 155 (1967) 997.
9 J.M. André, J. Chem. Phys., 50 (1969) 1536.
10 S. Suhai and J. Ladik, Solid State Commun., 22 (1977) 227.
11 A. Karpfen, Int. J. Quantum Chem., 19 (1981) 1297.
12 M. Kertesz, Acta Phys. Acad. Sci. Hung., 41 (1976) 127.
13 C. Pisani, R. Dovesi and C. Roetti, Hartree–Fock treatment of crystalline systems, Springer, Berlin, 1988.
14 P. Otto, E. Clementi and J. Ladik, IBM Technical Report, POK-13 (1982).
15 J.M. André, D.P. Vercauteren, V.P. Bodart, J.L. Brédas, J. Delhalle and J.G. Fripiat, Documentation for an ab initio polymer program (PLH), IBM IS&TG, POK-28 (1983).
16 J.M. André, V.P. Bodart, J.L. Brédas, J. Delhalle and J.G. Fripiat, in: J. Ladik J.M. André (Eds.), Quantum theory of polymers: Solid state aspects, NATO-ASI Series C123, Reidel, Dordrecht, 1984, pp. 1–21.
17 J.M. André, J.L. Brédas, J. Delhalle, D.J. Vanderveken, D.P. Vercauteren and J.G. Fripiat, LCAO ab initio band structure calculations for polymers, in: E. Clementi (Ed.), MOTECC-90, modern techniques in computational chemistry, ESCOM, Leiden, 1990, pp. 745–783, Chapter 15; J.G. Fripiat, D.J. Vanderveken, D.P. Vercauteren and J.M. André, PLH-90 from MOTECC-90, in: E. Clementi (Ed.), MOTECC-90, modern techniques in computational chemistry, input/output documentation, ESCOM, Leiden, 1990, pp. 619–635; J.M. André, J.L. Brédas, J. Delhalle, D.J. Vanderveken, D.P. Vercauteren and J.G. Fripiat, LCAO ab initio band structure calculations for polymers, in: E. Clementi (Ed.), MOTECC-91, modern techniques in computational chemistry, ESCOM, Leiden, 1991, pp. 793–828, Chapter 19; J.G. Fripiat, D.J. Vanderveken, D.P. Vercauteren and J.M. André, PLH-91 from MOTECC-91, in: E. Clementi (Ed.), MOTECC-91, modern techniques in computational chemistry, input/output documentation, ESCOM, Leiden, 1991, pp. 195–207; J.M. André, D.H. Mosley, B. Champagne, J. Delhalle, J.G. Fripiat, J.L. Brédas, D.J. Vanderveken and D.P. Vercauteren, LCAO ab initio band structure calculations for polymers, in: E. Clementi (Ed.), Methods and techniques in computational chemistry: METECC-94, Volume B: Medium size systems, STEF, Cagliari, 1993, pp. 423–480, Chapter 10.
18 http://www.gaussian.com/
19 http://www.scm.com/
20 http://www.cse.clrc.ac.uk/cmg/CRYSTAL/

21 J.G. Fripiat, B. Champagne, F. Castet, T.D. Poulsen, D.H. Mosley, J.L. Brédas, V. Bodart, J. Delhalle, D.P. Vercauteren and J.M. André, PLH-2000, Facultés Universitaires N.D. de la Paix, B-5000 Namur, Belgium.
22 J.-Q. Sun and R.J. Bartlett, J. Chem. Phys., 104 (1996) 8553.
23 http://emslbios.pnl.gov/bios/biosketch.nsf/All+Public+Documents/Hirata,+So?Opendocument
24 D. Jacquemin, DJPol, a Crystalline Orbital (CO) code for evaluating the properties of stereoregular polymers, FUNDP, Namur, (1999–2003).
25 J.M. André and G. Leroy, Chem. Phys. Lett., 5 (1970) 71; J.M. André and G. Leroy, Int. J. Quantum Chem., 5 (1971) 557
26 J.M. André, Comput. Phys. Commun., 1 (1970) 391.
27 F. Bloch, Z. Physik., 52 (1928) 555.
28 W.L. McCubbin and F.A. Teemull, Phys. Rev. A, 6 (1972) 2478.
29 J.M. André, L.A. Burke, J. Delhalle, G. Nicolas and Ph. Durand, Int. J. Quantum Chem., S13 (1979) 283; J.M. André, J.G. Fripiat and J.L. Brédas, VEH-91 from MOTECC-91, E. Clementi, (Ed.), MOTECC-91, modern techniques in computational chemistry, input/output documentation, Part 3, ESCOM, Leiden (1991), pp. 185–193.
30 J.M. André, in: J. Bicerano (Ed.), Computational modeling of polymers, Marcel Dekker, New York, 1992, Chapter 10, pp. 483–520.
31 J.M. André, B. Champagne, J. Delhalle, J.G. Fripiat and D.H. Mosley, Lecture notes in computer science, in: W. Gentzch and U. Harms (Eds.), High-Performance Computing and Networking, Vol. 796, Springer, Berlin, 1994, pp. 183–188; G. Paddeu, D.H. Mosley, J.M. André and J.G. Fripiat, Vercauteren, in: R. Scateni (Ed.), Scientific visualization 95, World Scientific, Singapore, 1995, pp. 79–91.
32 J.M. André, J. Ladik (Eds.), Electronic structure of polymers and molecular crystals, Plenum Press, New York, 1975; J.M. André J. Delhalle J. Ladik (Eds.), Quantum theory of polymers, Reidel, Dordrecht, 1978; J.M. André, J.L. Brédes (Eds.), Recent advances in the quantum theory of polymers, Springer, Berlin, 1980; J. Ladik, J.M. André (Eds.), Quantum chemistry of polymers: Solid state aspects, Reidel, Dordrecht, 1984.
33 J. Ladik, Quantum theory of polymers as solids, Plenum Press, New York, 1988; C. Pisani, R. Dovesi and C. Roetti, Hartree–Fock treatment of crystalline systems, Springer, Berlin, 1988; J.M. André, J. Delhalle and J.L. Brédas, Quantum chemistry aided design of organic polymers for molecular electronics, World Scientific, New York, 1991.
34 C.R. Ginnard and W.M. Riggs, Anal. Chem., 44 (1972) 1310.
35 M.H. Wood, M. Barber, I.H. Hillier and J.M. Thomas, J. Chem. Phys., 56 (1972) 1788.
36 J.M. André and J. Delhalle, Chem. Phys. Lett., 17 (1972) 145.
37 J. Delhalle, D. Thelen and J.M. André, Comput. Chem., 3 (1979) 1.
38 J.L. Brédas, J.M. André and J. Delhalle, Chem. Phys., 45 (1980) 109.
39 K. Seki, N. Ueno, U.O. Karlsson, R. Engelhardt and E.E. Koch, Chem. Phys., 105 (1986) 247; N. Ueno, K. Seki, H. Fujimoto, T. Kuramochi, K. Sigita and H. Inokuchi, Phys. Rev. B, 41 (1990) 1176.
40 J.M. André, J. Delhalle and J.J. Pireaux, in: D.W. Dwight, T.J. Fabish, H.R. Thomas (Eds.), Photon, electron, and ion probes of polymer structure and properties, ACS Symposium Series, Vol. 162, 1981, p. 151.
41 R.G. Snyder, J. Chem. Phys., 47 (1967) 1316.
42 J.J. Pireaux, J. Riga, R. Caudano, J.J. Verbist, J. Delhalle, S. Delhalle, J.M. André and Y. Gobillon, Phys. Scripta, 16 (1977) 329.
43 J.M. André, D.P. Vercauteren, V.P. Bodart and J.G. Fripiat, J. Comput. Chem., 5 (1984) 535.
44 T.A. Skotheim (Ed.), Handbook of conducting polymers, Marcel Dekker, New York, 1986.
45 C.K. Chiang, C.R. Fincher, Y.W. Park, A.J. Heeger, H. Shirakawa, E.J. Louis, S.C. Gau and A.G. MacDiarmid, Phys. Rev. Lett., 39 (1977) 1089.
46 H. Naarman and N. Theophilou, Synth. Met., 22 (1987) 1.
47 H. Kuhn, J. Chem. Phys., 17 (1949) 1198.
48 A. Pullman and R. Daudel, C.R. Acad. Sci., 222 (1946) 288.
49 J. Delhalle, J.L. Brédas and J.M. André, Chem. Phys. Lett., 78 (1981) 93.
50 J.M. André, in: R. Carbo (Ed.), Current aspects of quantum chemistry, Elsevier, Amsterdam, 1982, pp. 273–291; J.L. Brédas, B. Thémans, J.M. André, R.R. Chance, D.S. Boudreaux and R. Silbey, J. Phys., 44, C3 1983, pp. 273.

51 J.A. Pople and S.H. Walmsley, Mol. Phys., 5 (1962) 15.

52 W.P. Su, J.R. Schrieffer and A.J. Heeger, Phys. Rev. B, 22 (1980) 2099.

53 J. Scott-Russel, Reports on waves, Proc. R. Soc. Edinburgh, (1844) 319.

54 J.L. Brédas, B. Thémans and J.M. André, Phys. Rev. B, 26 (1982) 6000; J.L. Brédas, B. Thémans and J.M. André, J. Chem. Phys., 78 (1983) 6137; J.L. Brédas, B. Thémans and J.M. André, Phys. Rev. B, 27 (1983) 7827; J.M. André, D.P. Vercauteren, G.B. Street and J.L. Brédas, J. Chem. Phys., 80 (1984) 5643; J.L. Brédas, B. Thémans, J.G. Fripiat, J.M. André and R.R. Chance, Phys. Rev. B, 29 (1984) 6761; J.L. Brédas, B. Thémans, J.M. André, R.R. Chance and R. Silbey, Synth. Met., 9 (1984) 265; B. Thémans, J.M. André and J.L. Brédas, Solid State Commun., 50 (1984) 1047; J.L. Brédas, B.G. Street, B. Thémans and J.M. André, J. Chem. Phys., 83 (1985) 1323; B. Thémans, J.M. André and J.L. Brédas, in: H. Kuzmany, M. Mehring and S. Roth, (Eds.), "Electronic properties of Polymers and Related Compounds", Springer series in solid state sciences, Vol. 63 (1985) 107; D.S. Boudreaux, R.R. Chance, J.F. Wolf, L.W. Shacklette, J.L. Brédas, B. Thémans, J.M. André and R. Silbey, J. Chem. Phys., 85 (1986) 4584; J. Riga, Ph. Snauwaert, A. De Pryck, R. Lazzaroni, J.P. Boutique, J.J. Verbist, J.L. Brédas, J.M. André and C. Taliani, Synth. Met., 21 (1987) 223; R. Lazzaroni, A. De Pryck, Ch. Debaisieux, J. Riga, J. Verbist, J.L. Brédas, J. Delhalle and J.M. André, Synth. Met., 21 (1987) 189; B. Thémans, J.M. André and J.L. Brédas, Synth. Met., 21 (1987) 149.

55 J.M. André, J. Delhalle and J.L. Brédas, Quantum chemistry aided design of organic polymers for molecular electronics, World Scientific, New York, 1991; W.R. Salaneck, S. Stafström and J.L. Brédas, Conjugated polymer surfaces and interfaces, Cambridge University Press, Cambridge, 1996.

56 J.P. Hermann, D. Ricard and J. Ducuing, Appl. Phys. Lett., 23 (1973) 178; J.P. Hermann and J. Ducuing, J. Appl. Phys., 45 (1974) 5100; K.C. Rustagi and J. Ducuing, Opt. Commun., 10 (1974) 258.

57 P.L. Davies, Trans. Faraday Soc., 47 (1952) 789..

58 S. Risser, S. Klemm, D.W. Allender and M.A. Lee, Mol. Cryst. Liq. Cryst., 150b (1987) 631.

59 H.F. Hameka, J. Chem. Phys., 67 (1977) 2935.

60 C. Flytzanis, in: D.S. Chemla, J. Zyss (Eds.), Nonlinear optical properties of organic molecules and crystals, Vol. 2, Academic Press, New York, 1987, p. 121.

61 B.J. Orr and J.F. Ward, Mol. Phys., 20 (1971) 513; D.M. Bishop, J. Chem. Phys., 100 (1994) 6535.

62 J.M. André and B. Champagne, in: J.L. Brédas (Ed.), Nonlinear optical hyperpolarizabilities and electronic properties of oligomers and polymers: the polyene story, from dyes to conducting and NLO systems, Conjugated oligomers, polymers, and dendrimers: From polyacetylene to DNA, Bibliothèque Scientifique Francqui, De Boeck Université, 1999, pp. 349–394.

63 J. Zak, Phys. Rev., 168 (1968) 686; J.E. Churchill and F.E. Holmström, Am. J. Phys., 50 (1982) 848; E. Churchill and F.E. Holmström, Physica B, 123 (1983) 1.

64 J. Fink and G. Leising, Phys. Rev. B, 34 (1986) 5320; G. Leysing, Phys. Rev. B, 38 (1988) 10313.

65 J. Zyss, D.S. Chemla and J.F. Nicoud, J. Chem. Phys., 74 (1981) 4800.

66 K. Clays, E. Hendrickx, M. Triest, T. Verbiest, A. Persoons, C. Dehu and J.L. Brédas, Science, 262 (1993) 1419; F. Meyers, S.R. Marder, B.M. Pierce and J.L. Brédas, J. Am. Chem. Soc., 116 (1994) 10703.

67 D. Jacquemin, B. Champagne and J.M. André, Synthetic metals, 80 (1996) 205; D. Jacquemin, B. Champagne and J.M. André, J. Chem. Phys., 108 (1998) 1023; D. Jacquemin, B. Champagne and J.M. André, J. Mol. Struct. (THEOCHEM), 425 (1998) 69; D. Jacquemin, B. Champagne and J.M. André, Chem. Phys. Lett., 284 (1998) 24; D. Jacquemin, B. Champagne and J.M. André, Synth. Met., 101 (1999) 490; D. Jacquemin, D. Beljonne, B. Champagne, V. Geskin, J.L. Brédas and J.M. André, J. Chem. Phys., 115 (2001) 6766.

68 J.M. André, L. Gouverneur and G. Leroy, Bull. Soc. Chim. Belges, 76 (1967) 661.

69 D. Jacquemin, E.A. Perpete and J.M. André, J. Chem. Phys., 120 (2004) 10317.

70 D. Jacquemin, O. Quinet, B. Champagne and J.M. André, J. Chem. Phys., 120 (2004) 9401.

71 C.E. Dykstra, J. Chem. Educ., 65 (1988) 198; D.M. Bishop, Rev. Mod. Phys., 62 (1990) 343.

72 B. Champagne, E.A. Perpète and J.M. André, J. Chem. Phys., 101 (1994) 10796.

73 B. Champagne, E.A. Perpète and J.M. André, J. Chem. Phys., 100 (1994) 10796; B. Champagne, H. Vanderhoeven, E.A. Perpète and J.M. André, Chem. Phys. Lett., 254 (1996) 301; B. Kirtman, B. Champagne and J.M. André, J. Chem. Phys., 104 (1996) 4125; B. Champagne, E.A. Perpète, J.M. André and B. Kirtman, Synthetic Metals, 85 (1997) 1047.

74 B. Kirtman and B. Champagne, Int. Rev. Phys. Chem., 16 (1997) 389.

75 D. Jacquemin, J.M. André and B. Champagne, J. Chem. Phys., 118 (2003) 3956.

76 see http://www.nobel.se/chemistry/laureates/2000/chemadv.pdf, p. 11.

77 R.A. Marcus, Rev. Mod. Phys., 65 (1993) 599.

78 P.F. Barbara, T.J. Meyer and M.A. Ratner, J. Phys. Chem., 100 (1996) 13148.

79 V. Balzani, A. Juris, M. Venturi, S. Campagna and S. Serroni, Chem. Rev., 96 (1996) 759.

80 J. Jortner and M. Bixon, Electron transfer from isolated molecules to biomolecules, Part 1, Advances in chemical physics, Vol. 106, 1999, Part 2., Vol. 107, 1999.

81 V. May and O. Kühn, Charge and energy transfer dynamics in molecular systems, Wiley-VCH, London, 2000.

82 J.L. Brédas, B. Thémans and J.M. André, Phys. Rev. B, 26 (1982) 6000; J.L. Brédas, B. Thémans and J.M. André, Phys. Rev. B, 27 (1983) 7827; J.L. Brédas, B. Thémans, J.G. Fripiat, J.M. André and R.R. Chance, Phys. Rev. B, 29 (1984) 6761.

83 J.H. Schön, C. Kloc and B. Batlogg, Phys. Rev. Lett., 86 (2001) 3843, However the following retraction has been published in Phys. Rev. Lett., 89 (28) 28 (December 31, 2002): Universal Crossover from Band to Hopping Conduction in Mol. Org. Semicond. [Phys. Rev. Lett. 86 (2001) 3843], Jan Hendrik Schön, Christian Kloc, and Bertram Batlogg (Received 18 December 2002; published 27 December 2002) DOI: 10.1103/PhysRevLett.89.289902 PACS numbers: 72.80.Le, 71.38−k, 72.20.Fr, 99.10+g. On 25 September 2002, Bell Laboratories, Lucent Technologies, announced the findings of an independent committee it formed to investigate the validity of certain research reported from 1998 to 2002 by teams of Bell Labs and other scientists. In its report [1], the committee concludes that 'based on the preponderance of the evidence, Hendrik Schön committed scientific misconduct as defined by the falsification or fabrication of data, such that the research is not accurately represented in the research record.' The above-mentioned manuscript was not reviewed by the committee and therefore did not appear in their report. Nevertheless, all of the authors of the Letter with the exception of J.H. Schön have agreed to a complete retraction of the paper. [1] http://dx.doi.org/10.1103/APS.Reports.Lucent

84 H. Gruhn, D.A. da Silva Filho, T.G. Bill, M. Malagoli, V. Coropceanu, A. Kahn and J.L. Brédas, J. Am. Chem. Soc., 124 (2002) 7918.

85 G.L. Gloss and J.R. Miller, Science, 240 (1988) 440.

86 see, for example in 2003 and 2004: C&N, p. 10, July 7, 2003; C&EN, p. 13, December 15, 2003; C&EN, p. 9, February 2, 2004; C&EN, p. 12, March 15, 2004; C&EN, p. 9, May 3, 2003.

87 http://www.universaldisplay.com; http://www.uniax.com; http://www.kodak.com/go/oled; http://www.cdtltd.co.uk

88 see, for example: V. Coropceanu, J.M. André, M. Malagoli and J.L. Brédas, Theor Chem. Acc., 110 (2003) 59.

89 see, for example, J.M. André, in: J. Leszczynski (Ed.), Chaos and chemistry: simple models to understand chaos in chemistry, Theoretical and computational chemistry, Vol. 15, Elsevier, Amsterdam, 2004; J.M. André, in: E. Brandäs, E. Kryachko (Eds.), Silico chemistry, past, present and future, Fundamental world of quantum chemistry, A tribute to the memory of P.O. Löwdin, Vol. 3, Kluwer, Dordrecht, 2004.

Theory and Applications of Computational Chemistry: The First Forty Years
Edited by C. Dykstra et al.

CHAPTER 37

Forty years of ab initio *calculations on intermolecular forces*

Paul E.S. Wormer and Ad van der Avoird

Institute of Theoretical Chemistry, IMM, University of Nijmegen, Toernooiveld 1, 6525 ED Nijmegen, The Netherlands

Abstract

This review sketches the development of methods for the computation of intermolecular forces; emphasis is placed on dispersion forces. The last 40 years, which saw the birth, growth, and maturation of *ab initio* methods, are reviewed.

37.1 INTRODUCTION

Intermolecular forces, sometimes called non-covalent interactions, are caused by Coulomb interactions between the electrons and nuclei of the molecules. Several contributions may be distinguished: electrostatic, induction, dispersion, exchange that originate from different mechanisms by which the Coulomb interactions can lead to either repulsive or attractive forces between the molecules. This review deals with the *ab initio* calculation of complete intermolecular potential surfaces, or force fields, but we focus on dispersion forces since it turned out that this (relatively weak, but important) contribution took longest to understand and still is the most problematic in computations. Dispersion forces are the only attractive forces that play a role in the interaction between closed-shell (^1S) atoms. We will see how the understanding of these forces developed, from complete puzzlement about their origin, to a situation in which accurate quantitative predictions are possible.

The subfield of quantum chemistry concerned with the computation of intermolecular forces has always depended very much on computer technology, not unlike most of the other subfields. Because of this strong influence, we will divide the following history along the lines of hardware development. The first attempt of an *ab initio* calculation on the interaction between two closed-shell atoms was made in 1961. Rather than let

References pp. 1072–1077

the story begin there, we will first review briefly the precomputer era of the theory of intermolecular forces. Then the infancy of computers and computational quantum chemistry will be reviewed, followed by the era dominated by mainframes. We will end with the present democratic times in which every household has at its disposal the power of a 1980 supercomputer and ordinary research groups possess farms of powerful computers.

37.2 PREHISTORY: BEFORE COMPUTERS

When on the 10th of July 1908 Kamerling Onnes and his coworkers saw the meniscus of liquid helium in their apparatus [1], it was proved to them that two helium atoms attract each other—so that a liquid can be formed—but also that they repel each other—so that the liquid does not implode. Of course, this is what they had known all along, because their work was guided by the van der Waals law of corresponding states, which gave them a fair idea of the temperature and pressure at which the liquefaction of helium would take place. In deriving his law van der Waals (1873) had to assume the existence of attractive and repulsive forces.

Until the advent of quantum mechanics it was an enigma why two S-state atoms would repel or attract each other. Shortly after the introduction of the Schrödinger equation in 1926, Wang [2] solved this equation perturbatively for two ground-state hydrogen atoms at large interatomic distance R. Approximating the electronic interaction by a Taylor series in $1/R$ he found an attractive potential with a leading term $-C_6/R^6$. A few years later Eisenschitz and London (E&L) [3] systematized this work by introducing a perturbation formalism in which the Pauli principle is consistently included. They showed that the intermolecular antisymmetrization of the electronic wave function (electron exchange), which is required by the Pauli principle, can give rise to repulsion. This is why the intermolecular repulsion is often referred to as exchange (or Pauli) repulsion.

Considering distances long enough that intermolecular differential overlap and exchange can be neglected (the so-called long-range regime), Wang and E&L showed that the same dipole matrix elements that give rise to transitions in the monomer spectrum appear in the equations for the interaction. E&L pointed out that these transition dipole moments are closely related to the oscillator strengths arising in the classical theory of the dispersion of light (associated with the names of Drude and Lorentz) and in the quantum mechanical dispersion theory of Kramers and Heisenberg. Oscillator strengths, being simply proportional to squares of transition moments, are known experimentally, enabling E&L to give reasonable estimates of C_6. In 1930 London [4] published another paper in which he coined the name 'dispersion effect' for the attraction between S state atoms, which is why it is common today to refer to these attractive long-range forces as 'London' or 'dispersion forces'.

Apropos of nomenclature: the forces between closed-shell molecules (exchange repulsion, electrostatics, induction, and dispersion) are nowadays usually referred to as van der Waals forces. A stable cluster consisting of closed-shell molecules bound by these forces is called a van der Waals molecule. This terminology was introduced in the early 1970s (see Refs. [5–9]).

After the pioneering quantum mechanical work not much new ground was broken until computers and software had matured enough to try fresh attacks. In the meantime the study of intermolecular forces was mainly pursued by thermodynamicists who fitted model potentials, often of the Lennard–Jones form [10]: $4\epsilon[(\sigma/R)^{12} - (\sigma/R)^6]$, to quantities like second virial coefficients, viscosity and diffusion coefficients, etc. Much of this work is described in the authoritative monograph of Hirschfelder et al. [11] who, incidentally, also gave a good account of the relationship of Drude's classical work to that of London.

37.3 ANTIQUITY: THE SIXTIES

Around 1960 the computer began to enter quantum chemistry. This was the beginning of a very optimistic era; expectations of the new tool were tremendous. All over the western world quantum chemists were appointed in the belief that many of the problems in chemistry could be solved by computation within a decade or so. However, computers and *ab initio* methods were not received with this great enthusiasm by everyone. Coulson [12] one of the outstanding quantum chemists of his day, stated in his after-dinner speech of the June 1959 Conference on Molecular Quantum Mechanics in Boulder, Colorado: "It is in no small measure due to the success of these [Coulson here refers to *ab initio*] programs that quantum chemistry is in its present predicament."

37.3.1 Supermolecular methods

At the same 1959 Boulder conference Ransil, working in the Chicago Laboratory of Molecular Structure and Spectra, one of the leading quantum chemistry groups of the time, announced a research program [13] on the computation of properties of diatomic molecules. With the benefit of hindsight one can say that his program was overambitious and far too optimistic, because he intended to use self-consistent field (SCF) methods with atomic orbital (AO) minimum basis sets, albeit of Slater type. The fourth paper [14] of this research program was devoted to He_2. Here Ransil considered the dispersion-bound dimer as a molecule amenable to ordinary molecular computational methods. Nowadays this method is referred to as a 'supermolecule' approach. Ransil writes in his abstract that "remarkable good agreement with the available experimental data is obtained for distances greater than 1.5 Å". We now know that his van der Waals minimum was spurious and solely due to the so-called basis set superposition error (BSSE). This BSSE is the lowering of the energy of monomer A, caused by the distance dependent improvement of the basis by the approaching AOs on B, and *vice versa*: the basis of B is improved by basis functions on A. How much the difficulties of *ab initio* calculations on intermolecular forces were still underestimated is witnessed by another paper on He_2, also originating from the Chicago group. Phillipson [15] attributing the deviation of the energy for $R < 1.5$ Å to correlation effects, introduces configuration interaction (CI) including 10–64 configurations, but still uses a minimum basis set and does not correct for the BSSE.

Six years later Kestner [16] published a paper, containing SCF-MO results on He_2, in which he stresses the importance of the choice of AO basis sets, even for systems as small as He_2. Using large basis sets he finds completely repulsive curves by the SCF method. According to Kestner in 1968: “it is generally believed, but nobody has proved, that this should be the case for two closed-shell atoms”. This statement exhibits the great advance made in understanding *ab initio* results in the early 1960s. Since Kestner used the Chicago computer codes and thanks Chicago staff members (Roothaan, Ransil, Cade, and Wahl) for guidance and support, it is clear that the Chicago work on programming *ab initio* codes for diatomics was instrumental in gaining this understanding, in contrast to Coulson’s doubts.

A correction of the BSSE appearing in supermolecular calculations was proposed by Boys and Bernardi [17] in 1970. A similar correction was already applied somewhat earlier by Jansen and Ros [18]. At present, 35 years later, the ‘counterpoise’ procedure of Boys and Bernardi is still regularly applied, although—especially for smaller systems—we now can afford AO basis sets that are so large that the SCF counterpoise correction is essentially zero. In correlated supermolecular methods the counterpoise correction is usually still needed. In essence, Boys and Bernardi proposed to perform all calculations (energy of monomer A and B and energy of dimer A–B) in the same dimer basis set by the same computational method. Although the sum of the monomer energies, which serves as zero point, becomes distance dependent, vast experience has shown [19–22] that this procedure yields the most reliable (basis set independent) results.

As stated above, it was already known in 1968 (and confirmed by calculation) that the SCF method applied to He_2 gives a purely repulsive interaction. Recall that by Löwdin’s definition [23] the SCF energy serves as the zero of electron correlation, or in other words, the SCF method does not give any correlation. Sinanoğlu [24] was the first to observe that interatomic sp pair correlation yields London R^{-6} dispersion. By the converse of this finding it seems plausible that without interatomic correlation dispersion effects are not accounted for. Since these effects contribute so significantly to the attraction of S-state atoms, one may conjecture that for non-polar systems there is no binding without inclusion of interatomic correlation. And indeed, we will show this below. As a matter of fact, Pauli repulsion is now usually taken for granted and attention is focused usually on the explanation of observed minima in intermolecular potentials.

The fact that interatomic pair correlation gives dispersion was semi-quantitatively confirmed in a calculation [25] on He_2. This Kestner–Sinanoğlu work on He_2 gave a well depth of $4.32\ \mathrm{K} = 3.00\ \mathrm{cm}^{-1}$, which is about 2.5 times lower than the presently accepted value. The discrepancy is due to an inadequate AO basis.

Sinanoğlu’s method was later improved [26] by adaptation of the pair functions to the spin-operator S^2. The He_2 potential was recomputed [27] by this method with the use of a much larger AO basis. This paper, and a simultaneous paper published in the same issue of Phys. Rev. Lett. by Bertoncini and Wahl [28] describing MCSCF calculations, are the first that report within one consistent supermolecule formalism a complete van der Waals curve that shows a physically meaningful well. In Ref. [27] the depth of this well is $D_e = 12.0\ \mathrm{K}$ at the equilibrium distance $R_e = 2.96\ \text{Å}$ and in Ref. [28] $D_e = 11.4\ \mathrm{K}$

at $R_e = 2.99$ Å. The presently accepted values [29,30] are $D_e = 11.008 \pm 0.008$ K and $R_e = 5.6$ bohr $= 2.963$ Å. The choice of configurations in the MCSCF calculation was inspired by the London long-range theory.

So, while the 1960s started with the belief that SCF could give a complete potential curve for closed-shell atoms, at the end of the decade it was known that the inclusion of interatomic correlation is essential for obtaining the dispersion attraction. The new decade saw the light with the two papers just mentioned [29,28] proving this quantitatively.

37.3.2 Perturbation methods

Independent of the work on coding *ab initio* programs, other workers in the 1960s carried further the torch of London. In the first place, methods were improved to compute better long-range C_6 coefficients and the corresponding higher coefficients C_8, C_{10}, etc. in the expansion of the interaction energy $-\sum_{n=6}^{\infty} C_n R^{-n}$.

On the other hand, people took a closer look at the symmetrized perturbation theory of Eisenschitz and London, which in principle can give a full potential energy surface (PES), not just the long range of it.

A variety of techniques has been employed for the estimation of dispersion coefficients. Good reviews are those by Dalgarno and Davison [31] and Dalgarno [32]. The semi-empirical methods based on oscillator strengths f_s were refined by using sum rules for Cauchy moments. A Cauchy moment $S(k)$ is defined by the following sum over monomer states ψ_s with energies E_s

$$S(k) = \sum_{s>0} f_s (E_s - E_0)^k$$

where the oscillator strength f_s is given as a squared matrix element of the dipole operator $\boldsymbol{\mu}$

$$f_s = \frac{2}{3}(E_s - E_0) \sum_{i=x,y,z} \langle \psi_0 | \mu_i | \psi_s \rangle \langle \psi_s | \mu_i | \psi_0 \rangle$$

The even Cauchy moments arise in the expansion of a frequency-dependent polarizability $\alpha(\omega)$

$$\alpha(\omega) = \sum_{s>0} \frac{f_s}{(E_s - E_0)^2 - \omega^2} = \sum_{k=0}^{\infty} S(-2k-2)\omega^{2k} \tag{1}$$

The coefficient C_6 in the long-range interaction energy between two molecules A and B is given in terms of the oscillator strengths f_s^A and f_s^B by

$$C_6 = \frac{3}{2} \sum_{ss'} \frac{f_s^A f_{s'}^B}{(E_s^A - E_0^A)(E_{s'}^B - E_0^B)(E_s^A - E_0^A + E_{s'}^B - E_0^B)}$$

One can factorize the denominator of this expression by invoking the identity ($\Delta_X > 0$ is an excitation energy on X):

$$\frac{1}{\Delta_A \Delta_B (\Delta_A + \Delta_B)} = \frac{2}{\pi} \int_0^\infty \left(\frac{1}{\Delta_A^2 + \omega^2} \right) \left(\frac{1}{\Delta_B^2 + \omega^2} \right) d\omega \tag{2}$$

Casimir and Polder [33] have shown that this identity, which can be proved easily by contour integration, can be used to express the long-range coefficient C_6 in terms of frequency-dependent monomer polarizabilities

$$C_6 = \frac{3}{\pi} \int_0^\infty \alpha^A(i\omega) \alpha^B(i\omega) d\omega$$

Many values of f_s are known empirically; their reliability can be checked by the sum rules: $S(-2) = \alpha(0)$, $S(0) = N$ (number of electrons) and $S(-1) = \frac{2}{3}\langle \psi_0 | \boldsymbol{\mu} \cdot \boldsymbol{\mu} | \psi_0 \rangle$.

Effectively summing the power series in Eq. (1) by means of different Padé approximants makes it possible to give upper and lower bounds on the C_6 values. Much work in the 1960s was done on calculating such bounds (see Ref. [34] for more on this).

The 1960s also saw the first *ab initio* calculations of $\alpha(\omega)$ by the time-dependent uncoupled Hartree–Fock (TDUHF) method [35] and by the time-dependent coupled Hartree–Fock (TDCHF) method [36].

Intermolecular electron exchange does not play a role in the long range, since all integrals that would arise by intermolecular antisymmetrization vanish by virtue of vanishing intermolecular differential overlap. However, for shorter distances where this overlap may not be neglected, the electrons on the monomers can no longer be distinguished and the wave functions must be antisymmetric under permutations of all electrons. As we saw earlier, Eisenschitz and London considered this problem as early as 1930 and it was revived in the late 1960s by Murrell et al. [37], Hirschfelder and Silbey [38], Hirschfelder [39,40], van der Avoird [41–44], Murrell and Shaw [45], and Musher and Amos [46].

From the Pauli principle follows that the projected function $\mathcal{A}_{AB}\Phi_0$, rather than Φ_0, should be considered as the correct zeroth-order wave function in the perturbation theory of intermolecular interactions. Here $\mathcal{A}_{AB}$ is the usual intermolecular antisymmetrization operator and $\Phi_0 = \Phi_0^A \Phi_0^B$ is (the lowest) eigenfunction of $H^{(0)} \equiv H^A + H^B$, the sum of monomer Hamiltonians. We assume here that Φ_0 is antisymmetric under monomer permutations, i.e. $\mathcal{A}_X \Phi_0 = \Phi_0$ for X = A, B. Unfortunately, since intermolecular permutations do not commute with $H^{(0)}$, $[\mathcal{A}_{AB}, H^{(0)}] \neq 0$, it follows that $\mathcal{A}_{AB}\Phi_0$ is not an eigenfunction of $H^{(0)}$. This has the consequence that conventional Rayleigh–Schrödinger (RS) perturbation theory is not applicable for those intermolecular distances where the effect of $\mathcal{A}_{AB}$ is non-negligible. The RS perturbation treatment must be adapted to permutation symmetry. The workers just quoted proposed procedures to achieve this symmetrization. From a practical point of view their theories can be divided into two categories [47]: first project with $\mathcal{A}_{AB}$ then perturb ('strong symmetry forcing'), or perturb first and project later ('weak symmetry forcing'). In strong symmetry forcing the symmetry operators enter the perturbation equations, significantly complicating their solution. In weak symmetry forcing the perturbed wave functions are obtained by minor modifications of RS perturbation theory: the operator $\mathcal{A}_{AB}$ only enters overlap terms and perturbation energies.

Although symmetry-adapted perturbation theories were well studied in the second half of the 1960s, numerical applications were scarce and restricted to H_2^+ and H_2. See Ref. [48] for a review of the exchange perturbation theories up to the beginning of the 1970s.

37.4 THE MIDDLE AGES: ERA OF MAINFRAMES

Around 1970 every self-respecting university possessed a mainframe computer (in the majority of cases an IBM 360, sometimes a CDC 6600). This was usually placed in a stronghold well defended by brave knights (the computer center staff). A scientist who wanted access to the machine had to master a strange and difficult tongue (job control language) and to cross swords with computer personnel to conquer CPU cycles, RAM, tapes, and disk space. This medieval state of affairs lasted until workstations arrived at the end of the 1980s.

The development of *ab initio* methods, such as speeding up the computation of Gaussian integrals, improving convergence of SCF procedures, and theory and programming of correlation methods was vigorously pursued on mainframes. Electron correlation can be included by CI or by coupled cluster (CC) methods. Especially the work on the development of CC methods proved later to be significant for the study of intermolecular forces, because the CC method, in contrast to the CI method, is size extensive. Size extensivity in the thermodynamic sense of this word (energy linear in amount of substance) implies that in the limit of zero density the energy of a system of molecules converges to the sum of energies of the individual molecules. Much work on the coupled cluster doubles (CCD) method (a supermolecule correlation method) was performed in the 1960s and 1970s by Paldus and Čížek and their coworkers [49–51] and from the late 1970s onward by Bartlett and coworkers [52,53], who added single excitations to the method. Also Pople [54] recognized the importance of the CC method at a rather early stage.[1]

However, in the 1970s the supermolecule correlation methods—and the computers on which they ran—were not yet powerful enough to have much significance in the field of intermolecular forces, except for small systems like He_2. It was already known that dimer SCF gives a fair description of electrostatic interactions (dipole–dipole, dipole–quadrupole, etc.), of induction forces (dipole-induced-dipole, etc.), and also of Pauli repulsion, but not of dispersion. So, for systems where dispersion was expected to be dominant, other paths than SCF supermolecule computations had to be followed. A well-known procedure was separate computation of SCF and perturbative dispersion (without exchange effects) and to add the two. Dispersion is known to be affected by exchange and so for shorter distance the dispersion has to be damped [55–57]. When the multipole-expanded form of the dispersion is used, this damping must also correct for the divergent character of the multipole series. Instead of computing converged dimer SCF energies, one often stopped the SCF procedure after the first cycle. This makes sense when the start orbital set is the direct sum of the sets of

[1] See Chapter 6 in this book for more on the development of the CC method.

occupied MOs of the monomers. In that case the first SCF cycle gives the expectation value $N\langle \mathcal{A}_{AB}\Phi_0^A\Phi_0^B|H^A + H^B + V^{AB}|\mathcal{A}_{AB}\Phi_0^A\Phi_0^B\rangle$, where N is the corresponding normalization constant and the intermolecular interaction operator $V^{AB} = H - H^A - H^B$ contains the Coulomb interactions between the electrons and nuclei of different molecules. This expectation value accounts for exchange repulsion and electrostatic interaction, albeit *without any intramolecular corrrelation*. Induction effects are obtained by cycling the dimer SCF, but because this cycling introduces BSSE and the counterpoise correction was deemed fairly expensive, as it requires three calculations for each geometry of the dimer, the neglect of induction was either accepted, or induction was added later in the multipole expanded form.

37.4.1 Unexpanded dispersion

Dispersion energy can be computed without invoking the multipole expansion. This was done in the 1970s by, among others, Kochanski [58,59], Jeziorski and van Hemert [60], and by van Duijneveldt and coworkers [61]. It is natural to assume that the supermolecular second-order Møller–Plesset (MP) [62] energy E_{MP2}^{AB} also accounts for dispersion energy. And, indeed, in their study on the connection between MP energies and the perturbation theory of intermolecular forces Chałasiński and Szczęśniak [63] showed that for large R the supermolecule second-order (MP2) energy becomes equal to a sum of two monomer MP2 energies plus an uncoupled HF dispersion term. They showed further that for two monomers possessing permanent multipoles the long-range MP2 energy also contains a correlation contribution to the electrostatic energy. In Appendix 37.A we provide a short alternative derivation of the asymptotic (large R) limit of E_{MP2}^{AB} for the special case of two S-state atoms. Fig. 37.1 gives a diagrammatic representation of the connection between dispersion and MP2 energy. The diagrams in

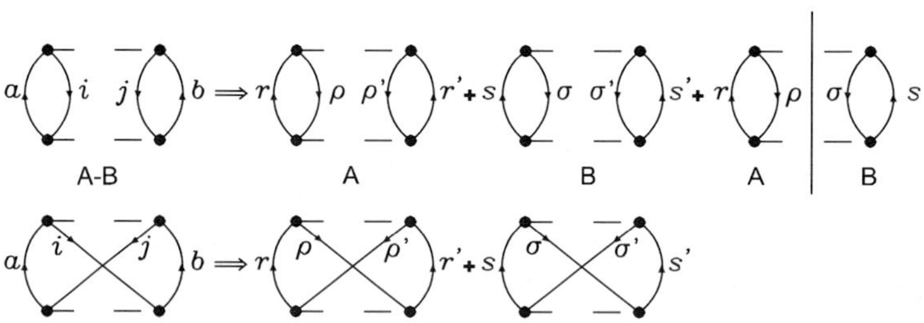

Fig. 37.1. Goldstone diagrams depicting the large-R behavior of E_{MP2}^{AB}. See text for definition of orbital labels. Particle orbitals run upward, hole orbitals downward. Each dashed line represents a two-electron integral. Closed lines are summed over. The first row shows Coulomb and the second row exchange interactions. Diagrams in the first row have $h = 2$ (two hole lines) $l = 2$ (two loops); diagrams in the second row have $h = 2$ and $l = 1$. The overall factor is $(-1)^{l+h}2^l w$ where 2^l is from spin integration. All diagrams, except the one with the bar in the middle, have weight $w = 1/2$ (because of a vertical symmetry plane). An imaginary horizontal line in each diagram gives the energy denominator. The diagrams in the first three columns give the MP2 energy of A–B, A, and B, respectively. The fourth column gives the London dispersion energy. In the large R limit integrals containing differential overlap between A and B vanish.

this figure are similar to those of Ref. [64], which was the first work to apply many-body diagrammatic techniques to symmetry-adapted perturbation theory (SAPT). Note that the dispersion energy and the Coulombic part of the supermolecule energy $E_{\mathrm{MP2}}^{\mathrm{AB}}$ are represented by diagrams that are topologically the same, so that diagrammatically their relationship seems obvious. It is tacitly assumed, however, that for large R the dimer orbital energies are equal to monomer orbital energies. This fact, which holds for S-state systems, is proved explicitly in Appendix 37A. The appendix gives the following equation for the unexpanded dispersion energy,

$$E_{\mathrm{disp}} = 4 \sum_{\rho,r,\sigma,s} \frac{|\langle \rho(1)\sigma(2)|r_{12}^{-1}|r(1)s(2)\rangle|^2}{\epsilon_\rho + \epsilon_\sigma - \epsilon_r - \epsilon_s} \tag{3}$$

where ρ is an occupied ('hole') and r a virtual ('particle') spatial orbital on A. The definition of σ and s on B is analogous. The denominator contains the corresponding orbital energy differences. This equation can also be extracted from the fourth diagram in the first row of Fig. 37.1 by application of the diagrammatic rules.

In the mid-1970s valence bond studies [65,66] on He_2 and $(C_2H_4)_2$ were performed. This work was based on valence bond structures (configurations) that account for most of the important dispersion effects. The VB structures were constructed from pure monomer MOs, which are orthogonal on each monomer but have intermolecular overlap. The exact, unexpanded, Hamiltonian was used and a secular problem was solved. Because of the non-orthogonality of the orbitals, the main drawback of the VB method is that only a relatively small subset of the electrons can participate in the binding, while the majority of electrons reside in closed shells. The all-electron VB work [65] on He_2 brought to light very clearly the considerable size of BSSEs, not only in VB results, but also in full CI results. Until that time BSSE was mainly discussed at the SCF level.

Two *ab initio* methods, which were well known and much discussed in the 1970s and 1980s, were the pair natural orbital CI (PNO-CI) method and the coupled electron pair approximation (CEPA) method. They were proposed by Meyer [67] in 1973 and 2 years later improved by Ahlrichs et al. [68]. In 1983, Burton and Senff [69] applied the method of Ahlrichs et al. to an analysis of the anisotropy of $(H_2)_2$ interaction near the minimum in the van der Waals interaction energy.

In Eq. (3) we find orbital energy differences in the denominator. This is due to the fact that we took the Fock operator as the zeroth-order operator, as did Møller and Plesset [62] in 1934. An alternative zeroth-order operator is due to Epstein [70] saved from oblivion by Nesbet [71]. This operator can be written as [72]

$$H^{(0)} = \sum_I |I\rangle\langle I|H|I\rangle\langle I|$$

where $|I\rangle$ is an excited Slater determinant consisting of localized HF orbitals. A simplification, compatible with Eq. (3), is obtained by restricting the summation to determinants that are singly excited on each of the monomers. Working out the energy denominators $\langle 0|H|0\rangle - \langle I|H|I\rangle$ we find orbital energy differences, as in Eq. (3), but shifted by a few additional Coulomb and exchange integrals.

The pros and cons of Møller–Plesset *versus* Epstein–Nesbet (EN) partitioning were of some interest all through the 1970s. Especially the French school [58,59,73–75] strongly preferred EN over MP partitioning, although later French work [72] criticizes the use of EN partitioning with delocalized orbitals. See Kelly [76] for a proof that EN partitioning gives an (infinite) number of diagonal ladder diagrams in addition the diagrams accounted for by the MP partitioning. Today EN partitioning is rarely applied, mainly for pragmatic reasons, because most standard *ab initio* packages only have the MP option.

37.4.2 Multipole-expanded dispersion

Before 1970 the multipole expansion (by which we mean the expansion in powers of $1/R$) of the interaction operator V^{AB} was usually truncated after the R^{-3} dipole–dipole term, so that the only dispersion interaction term was $-C_6R^{-6}$. Around 1970 it became clear that this approximation was not sufficient and that more terms were needed. However, the straightforward application of the Taylor expansion, and its natural formulation in terms of Cartesian tensors [77], soon becomes cumbersome. Nineteenth century potential theory [78,79] came to the rescue. In this theory the multipole series is rephrased in terms of associated Legendre functions, which enables a closed form of it. Multipole operators are defined as

$$Q_m^{l_\mathrm{X}} = \sum_{\xi\in\mathrm{X}} Z_\xi S_m^{l_\mathrm{X}}(\vec{r}_{X\xi}) - \sum_{i=1}^{N_\mathrm{X}} S_m^{l_\mathrm{X}}(\vec{r}_{Xi}) \qquad \text{with} \qquad \mathrm{X} = \mathrm{A}, \mathrm{B}$$

where $\vec{r}_{X\xi}$ and $\vec{r}_{Xi}$ are the coordinates of the nuclei ξ, with charges Z_ξ, and the electrons i of molecule X with respect to a frame with its origin on the nuclear center of mass of X. The function $S_m^l(\vec{r}) \equiv r^l C_m^l(\hat{r})$ is a regular solid harmonic; $C_m^l(\hat{r})$ is a Racah normalized spherical harmonic.

The intermolecular interaction operator is

$$V^{\mathrm{AB}} = -\sum_{i=1}^{N_\mathrm{A}} \sum_{\beta\in\mathrm{B}} \frac{Z_\beta}{r_{i\beta}} - \sum_{j=1}^{N_\mathrm{B}} \sum_{\alpha\in\mathrm{A}} \frac{Z_\alpha}{r_{j\alpha}} + \sum_{i=1}^{N_\mathrm{A}} \sum_{j=1}^{N_\mathrm{B}} \frac{1}{r_{ij}} + \sum_{\substack{\alpha\in\mathrm{A}\\ \beta\in\mathrm{B}}} \frac{Z_\alpha Z_\beta}{r_{\alpha\beta}} \tag{4}$$

Introducing the notation $\vec{r}_{\mathrm{PQ}}$ for a vector pointing from P to Q we substitute into Eq. (4)

$$\vec{r}_{i\beta} = -\vec{r}_{\mathrm{A}i} + \vec{R}_{\mathrm{AB}} + \vec{r}_{\mathrm{B}\beta}$$

$$\vec{r}_{j\alpha} = -\vec{r}_{\mathrm{B}j} - \vec{R}_{\mathrm{AB}} + \vec{r}_{\mathrm{A}\alpha}$$

$$\vec{r}_{ij} = -\vec{r}_{\mathrm{A}i} + \vec{R}_{\mathrm{AB}} + \vec{r}_{\mathrm{B}j}$$

$$\vec{r}_{\alpha\beta} = -\vec{r}_{\mathrm{A}\alpha} + \vec{R}_{\mathrm{AB}} + \vec{r}_{\mathrm{B}\beta}$$

where A and B are the nuclear centers of mass of the respective monomers. Upon using the expansions given in Eqs. (A4) and (A5) of Appendix 37A, we get the multipole expansion

$$V^{\mathrm{AB}} = \sum_{l_{\mathrm{A}} l_{\mathrm{B}} m} (-1)^{l_{\mathrm{B}}+m} \binom{2l_{\mathrm{A}} + 2l_{\mathrm{B}}}{2l_{\mathrm{B}}}^{1/2} Y_{-m}^{l_{\mathrm{A}}+l_{\mathrm{B}}}(\vec{R})[Q^{l_{\mathrm{A}}} \otimes Q^{l_{\mathrm{B}}}]_m^{l_{\mathrm{A}}+l_{\mathrm{B}}}$$

where $\vec{R} = \vec{R}_{\mathrm{AB}}$ and we recall from the appendix that $Y_{-m}^{l_{\mathrm{A}}+l_{\mathrm{B}}}(\vec{R})$ is proportional to $R^{-(l_{\mathrm{A}}+l_{\mathrm{B}}+1)}$.

When we substitute this expansion into the RS second-order expression, we get a numerator that contains two Clebsch–Gordan coupled products of transition matrix elements on A and B. They are of the type

$$[\langle \Phi_0^{\mathrm{A}} | Q^{l_{\mathrm{A}}} | \Phi_n^{\mathrm{A}} \rangle \otimes \langle \Phi_0^{\mathrm{B}} | Q^{l_{\mathrm{B}}} | \Phi_{n'}^{\mathrm{B}} \rangle]_m^{l_{\mathrm{A}}+l_{\mathrm{B}}}$$

This coupling is not convenient. As will become clear below it is better to first couple the transition moments on each center. The dispersion energy becomes a sum of which the summand can be expressed with the use of Eq. (2) as a Casimir–Polder integral

$$\int_0^\infty \alpha_{M_{\mathrm{A}}}^{(l_{\mathrm{A}} l'_{\mathrm{A}})L_{\mathrm{A}}}(i\omega) \alpha_{M_{\mathrm{B}}}^{(l_{\mathrm{B}} l'_{\mathrm{B}})L_{\mathrm{B}}}(i\omega) \mathrm{d}\omega$$

over a product of irreducible frequency-dependent polarizabilities. The latter are given by

$$\alpha_{M_{\mathrm{X}}}^{(l_{\mathrm{X}} l'_{\mathrm{X}})L_{\mathrm{X}}}(i\omega) \equiv \sum_{n>0} \frac{2(E_n^{\mathrm{X}} - E_0^{\mathrm{X}})}{(E_n^{\mathrm{X}} - E_0^{\mathrm{X}})^2 + \omega^2} \left[\langle \Phi_0^{\mathrm{X}} | Q^{l_{\mathrm{X}}} | \Phi_n^{\mathrm{X}} \rangle \otimes \langle \Phi_n^{\mathrm{X}} | Q^{l'_{\mathrm{X}}} | \Phi_0^{\mathrm{A}} \rangle \right]_{M_{\mathrm{X}}}^{L_{\mathrm{X}}}$$

The recoupling of the transition moments requires a 9*j*-symbol (see Refs. [80–82]). Later the same recoupling was performed in Ref. [83]. These references give the expression without introduction of the Casimir–Polder integral. See Refs. [84–88] for the recoupled expression containing the Casimir–Polder integral. We have now a closed form of all the terms in the multipole expansion of the dispersion energy. This energy can be computed once the irreducible monomer polarizabilities $\alpha^{L_{\mathrm{A}}}(i\omega)$ and $\alpha^{L_{\mathrm{B}}}(i\omega)$ are known. Unfortunately, this series does not converge; in fact it is divergent and 'therefore we may be able to do something with it' [O. Heaviside (1899), as quoted in Ref. [89]]. However, the series is asymptotic [90–92] in the sense of Poincaré.

A very similar equation holds for the multipole-expanded induction energy. The difference is that the polarizabilities are static, so that there is no Casimir–Polder integral, and that one of the irreducible polarizabilities is replaced by a Clebsch–Gordan coupled product of permanent multipole moments.

The problem of computing dispersion energies is reduced to the computation of polarizabilities for a sufficient number of frequencies, so that the Casimir–Polder integral can be obtained by numerical quadrature [93]. An alternative to this quadrature is the substitution of the product of the polarizabilities by a sum over Hartree–Fock orbitals [94,95] or a sum over effective (pseudo) states of the monomers [85,96]. The pseudo states can be obtained from time-dependent coupled Hartree–Fock calculations [85, 96–98], or from CI calculations [99]. The CI calculations of Ref. [99] were at the single and double excitation level. They gave very good results for the frequency-dependent polarizabilities of He and H_2—where SDCI is equivalent to full CI—and very poor results for N_2, O_2 and the neon atom. The failure of the SDCI method for the response properties of more than two-electron systems was shown to be caused by unlinked clusters [100]. Addition of triply excited states removes the most important unlinked clusters and was shown for Ne_2 to improve the results considerably [100].

Doran [101] was the first to apply Goldstone diagrammatic techniques to the computation of frequency-dependent polarizabilities and dispersion coefficients. He applied his method to Ne_2 and heavier noble gases, but owing to an inadequate basis, got results of fairly poor quality. Later Wormer and coworkers [87,93,102] derived and programmed all polarizability diagrams through second order of intramolecular correlation, so that dispersion (by definition second order in V^{AB}) is completely correlated to second order on each monomer. Their programs are in practice hardly limited by the rank of the multipoles: up to $l = 63$ can be computed.

37.4.3 Applications

At the beginning of the 1980s quantum chemical methods and computer hardware had developed to a stage that the computation of properties depending on PESs of systems larger than two atoms could be contemplated. Examples are thermodynamic properties, such as virial coefficients [11,103] and moments of collision-induced infrared spectral densities [104,105]. The computation of spectroscopic properties of van der Waals molecules came into reach [106–111] and also of molecular crystals [112].

Intramonomer vibrations have in general a much higher energy than intermolecular vibrations, i.e. the intramolecular motions are much 'faster' than the intermolecular motions, so that an adiabatic separation of the two motions is reasonable. In practice this means that we can consider the monomers to be frozen in their vibrationally averaged geometry and that it is a good approximation to consider the interaction energy as a function (referred to as PES) of the relative coordinates of the rigid monomers. Examples of intermolecular coordinates are the well-known Jacobi coordinates R, θ for an atom–diatom system, while for a system consisting of two rigid diatoms R (the distance between the respective mass centers), θ_A and θ_B (the colatitude angles of the diatomics) and ϕ (the dihedral angle) are very common.

An early computation of a full (i.e. depending on all intermolecular coordinates) PES of two diatomics is the work by Berns and van der Avoird [113] on $(N_2)_2$. Their approach is in essence the one sketched above: one-cycle SCF to account for first-order exchange and electrostatics (including charge overlap effects) plus multipole-expanded dispersion. The dispersion coefficients were taken from Ref. [114].

At that time this was a formidable calculation. It was performed on an IBM 370/158, which was not a supercomputer, but nevertheless a respectable mainframe. A basis of 144 Gaussian type orbitals (GTOs) was used; for restart purposes integrals were stored on tape, requiring two tapes of 170 Mb for one point on the PES. About 150 points were computed, so that 300 tapes were needed. A tape reel had a diameter of 10.5 in. and, including its case, was about 1 in. wide, so that a rack of about 7.5 m long and 30×30 cm wide had to be used to store the 50 Gb of information. One point on the surface took 2.5–3.5 h CPU time and since the whole university—from sociology to solid state physics—used the same mainframe for time sharing during the day, at most one point per night could be done. The computational part of the project, therefore, took about half a year.

Two fits of the PES were made: one in terms of products of spherical harmonics (coupled to a rotational invariant) and one as an atom–atom potential

$$\Delta E^{\mathrm{AB}} = \sum_{a \in \mathrm{A}} \sum_{b \in \mathrm{B}} \left[\frac{q_a q_b}{r_{ab}} - \frac{C_{ab}}{r_{ab}^6} + A_{ab} e^{-B_{ab} r_{ab}} \right]$$

This potential was subsequently used in self-consistent phonon lattice dynamics calculations [115] for α and γ nitrogen crystals. And although the potential—and its fit—were crude by present day standards, lattice constants, cohesion energy and frequencies of translational phonon modes agreed well with experimental values. The frequencies of the librational modes were less well reproduced, but this turned out to be a shortcoming of the self-consistent phonon method. When, later [116,117], a method was developed to deal properly with the large amplitude librational motions, also the librational frequencies agreed well with experiment.

Ten years later van der Pol et al. [118] published similar calculations of the CO–CO interaction potential, also performed on a mainframe (NAS 9160). The GTO basis was of dimension 148; 315 points on the PES were computed. One point took 30 min CPU time so that there was no need to save integrals. The dispersion, computed in the multipole expansion at the MP2 level of intramonomer correlation [87], was damped by the Tang–Toennies [57] damp function. Notice, parenthetically, that the decrease in computer time from the calculations on $(N_2)_2$ to $(CO)_2$, a decade apart, was certainly not revolutionary. Judging by these calculations the speed of mainframes improved less than an order of magnitude; Moores law [119] (doubling of speed every 18 months) predicts two orders of magnitude. The $(CO)_2$ potential of van der Pol was applied [120] to the computation of properties of solid CO and gave good agreement with experimental values. A later application [121] to the rotation–vibration spectrum of the dimer showed, however, that the potential was not of spectroscopic accuracy.

37.5 MODERN TIMES: REVOLUTION AND DEMOCRACY

Around 1990 the advent of workstations initiated a revolution in scientific computing. Until that time batch processing was the norm for longer running jobs. The user estimated an upper limit for the CPU time that his/her computation would take, submitted the job, prayed that it contained no trivial errors causing an immediate crash, and then settled down to wait until there was room on the central computer to run the computation. When the workstations arrived, which had the computational speed of mainframes and were cheap enough that research groups could afford one or more, the mode of operation was revolutionized. In the first place, jobs went into execution immediately, so that the user had the chance to weed out trivial errors instantaneously. In the second place, there was no longer a need to chop the calculations into chunks of a few hours CPU time.

An example of a calculation, performed on an IBM RS/6000-320 workstation, is the study of the collisions of argon and NH_3 by van der Sanden et al. [122] with the use of an *ab initio* calculated Ar–NH_3 potential [123]. The program Hibridon [124] was used to compute the elastic and rotationally inelastic scattering cross sections and the probability that the collisions with Ar invert the ammonia umbrella. A single (one collisional energy) coupled channel calculation on *para* NH_3 colliding with argon took 241 CPU hours and was finished in about 2 weeks. On a mainframe this would have been a matter of months.

The increase of computer power made it possible not only to employ the most refined *ab initio* methods in the computation of potentials, but also to solve the nuclear motion problem sufficiently often to tune the *ab initio* potentials to the experimental results. This was the *Leitmotif* of the past decade: compute the best possible PES, fit it, solve the appropriate nuclear motion Schrödinger equation for the corresponding van der Waals complex, and compare with experiment. The remaining discrepancies between theory and experiment may be removed by scaling one or more of the parameters in the analytical fit of the potential surface. This procedure has helped in disentangling complicated spectra, for instance the ν_3 (asymmetric stretch) spectrum of CH_4 in interaction with the argon atom [125,126] and the ν_4 (asymmetric bend) spectrum of the same system [127,128]. At the same time, this provided an assessment of the quality of the *ab initio* Ar–CH_4 potential. An *ab initio* calculated water pair potential [129] was tested and improved [130,131] by the calculation of vibrational–rotational-tunneling spectra of the water dimer and comparison with experimental high-resolution spectra [132,133]. Again, the calculated energy levels and transition intensities [134] could be used to assign the bands in the measured spectrum to specific intermolecular vibrations.

For the computation of the interaction between two closed-shell monomers there are at present two excellent computational methods, both implemented in black box programs. The first is based on SAPT [135] and the second is the supermolecule CCSD method [136,137] with triply excited terms added in a non-iterative fashion.

37.5.1 The SAPT method

The SAPT method was mainly developed by workers in the Warsaw quantum chemistry group. Jeziorski and his former supervisor Kolos [47,138], believing in the prospects of SAPT, continued and extended the work of Refs. [37,40–46,48]; later Szalewicz [139] joined forces in this development. These workers came to the conclusion that symmetrized Rayleigh–Schrödinger theory (weak symmetry forcing—see above) was the most viable of the different variants of SAPT.

We saw earlier that a very simple form of the dispersion energy is obtained from frequency-dependent polarizabilities at the so-called uncoupled Hartree–Fock level. The sum over states appearing in second order RS perturbation theory is simply a sum over (occupied and virtual) orbitals. A first improvement of this simple model is obtained by including apparent correlation [140], i.e. by using frequency-dependent polarizabilities obtained from the TDCHF method [36,141]. This method was initially proposed in the context of the multipole expansion, but could be generalized [142–146] to charge density susceptibility functions (or polarization propagators), which avoids the use of the multipole expansion. It is possible to graft intramonomer correlation corrections onto TDCHF theory, but this is not the road taken by the Warsaw group. Instead they worked top down, from an exact formulation to (approximate) equations in terms of one- and two-electron integrals that are coded in the SAPT program.

We now present the basic philosophy of symmetrized RSPT as implemented in the SAPT program [135], see for more details Refs. [88,147]. Referring to Eq. (4) for the definition of V^{AB}, we rewrite the Schrödinger equation $(H^{(0)} + V^{AB})\Psi_{pol} = E\Psi_{pol}$. Projection with the eigenfunction $\Psi^{(0)}_{pol}$ of $H^{(0)}$ with eigenvalue $E^{(0)}$ gives the following exact expression for the interaction energy

$$E - E^{(0)} \equiv E_{pol} = \frac{\langle \Psi^{(0)}_{pol} | V^{AB} | \Psi_{pol} \rangle}{\langle \Psi^{(0)}_{pol} | \Psi_{pol} \rangle}$$

The subscript pol (polarization [40]) indicates that no intermolecular antisymmetry has been introduced, or, in other words, that Ψ_{pol} is expanded in products of monomer wave functions. See Ref. [88] about the convergence characteristics of this expansion. The convergence to a state satisfying the Pauli principle is greatly improved by introducing the intermolecular antisymmetrizer $\mathcal{A}_{AB}$. Hence we define, in the spirit of weak symmetry forcing, the energy expression

$$E_{SRS} \equiv \frac{\langle \Psi^{(0)}_{pol} | V^{AB} | \mathcal{A}_{AB} \Psi_{pol} \rangle}{\langle \Psi^{(0)}_{pol} | \mathcal{A}_{AB} \Psi_{pol} \rangle} \tag{5}$$

If we introduce the intramonomer correlation W^A, cf. Eq. (A1), multiplied by the perturbation parameter μ, the Schrödinger equation for monomer A becomes

$$H^A(\mu)\Phi^A(\mu) = (F^A + \mu W^A)\Phi^A(\mu) = E^A(\mu)\Phi^A(\mu)$$

Clearly, $\Phi^A(0)$ is the Hartree–Fock function of monomer A with energy $E^A(0)$ (a sum of orbital energies) and $\Phi^A(\mu)$ can be developed in a power series in μ. Analogously we assume for monomer B a power expansion in ν. Multiplying V^{AB} with the perturbation parameter λ, we may expand the eigenfunction of $H = F^A + F^B + \mu W^A + \nu W^B + \lambda V^{AB}$

$$\Psi_{pol}(\lambda, \mu, \nu) = \sum_{i,j,k=0}^{\infty} \lambda^i \mu^j \nu^k \Psi^{ijk}_{pol}$$

Here the subscript pol indicates that $\Psi_{pol}(\lambda, \mu, \nu)$ is obtained from PT equations that do not contain intermolecular exchange. Observing that $\Psi_{pol} \equiv \Psi_{pol}(1,1,1)$, and $\Psi^{(0)}_{pol} \equiv \Psi_{pol}(0,1,1)$, we may analytically continue Eq. (5) by substituting $\Psi_{pol}(1,1,1) \to \Psi_{pol}(\lambda, \mu, \nu)$ and $\Psi_{pol}(0,1,1) \to \Psi_{pol}(0, \mu, \nu)$ into this equation. The resulting expression of E_{SRS} is a function of λ, μ and ν. After expanding also the denominator in powers of λ, μ and ν, followed by collecting the powers of λ, μ and ν arising from numerator and denominator, E_{SRS} gets the form

$$E_{SRS}(\lambda, \mu, \nu) = \sum_{i,j,k=0}^{\infty} \lambda^i \mu^j \nu^k E^{ijk}_{SRS} \tag{6}$$

Obviously, the exact antisymmetrized interaction is equal to $E_{SRS}(1, 1, 1)$. This is the basis of SRS theory—a weak-symmetry-forcing variant of SAPT.

Exchange effects have a clear operational definition in SAPT, because E_{pol} can be expanded in the same RSPT manner, leading to terms E_{pol}^{ijk}. The difference $E_{SRS}^{ijk} - E_{pol}^{ijk}$ is the exchange contribution to the (i, j, k) term. The terms linear in λ are electrostatic terms and those quadratic in λ can be divided in induction and dispersion terms (including their exchange corrections).

As we stated above, SAPT is formulated in a 'top down' manner. Eq. (6) then forms the top; going down to workable equations, one is forced to introduce a multitude of approximations. In practice, i is restricted to the values 1 and 2: interactions of first and second order in V^{AB}. Different truncation levels for $j + k$ are applied, depending on the importance of the term (and the degree of complexity of the formula). Working out the equations to the level of one- and two-electron integrals is a far from trivial job. This has been done in a long series of papers that use techniques from coupled cluster theory and many-body PT; see Refs. [147,148] for references to this work and a concise summary of the formulas resulting from it.

Some of the earliest potentials computed by the SRS variant of SAPT were for Ar–H_2 [149] and for He–HF [150,151]. An application of the latter potential in a calculation of differential scattering cross sections [152] and comparison with experiment shows that this potential is very accurate, also in the repulsive region. Some other SAPT results are for Ar–HF [153], Ne–HCN [154], CO_2 dimer [155], and for the water dimer [129,156]. The accuracy of the water pair potential was tested [130,131] by a calculation of the various tunneling splittings caused by hydrogen bond rearrangement processes in the water dimer and comparison with high resolution spectroscopic data [132,133]. Other complexes studied are He–CO [157,158], and Ne–CO [159]. The pair potentials of He–CO and Ne–CO were applied in calculations of the rotationally resolved infrared spectra of these complexes measured in Refs. [160,161]. They were employed [162–165] in theoretical and experimental studies of the state-to-state rotationally inelastic He–CO and Ne–CO collision cross sections and rate constants. It was reaffirmed that both potentials are accurate, especially the one for He–CO.

Small organic molecules in interaction with noble gases were studied in Refs. [166] (He–C_2H_2), [167] (Ne–C_2H_2), [125] (Ar–CH_4), and [168] (He–CO_2). For He–C_2H_2, Ne–C_2H_2, and Ar–CH_4 the SAPT potentials were applied [126,166,167] in *ab initio* calculations of the infrared spectra of these complexes. A typical feature of all these potentials for weakly interacting systems is that their shape is determined by a subtle balance between the geometry dependence of the repulsive short-range interactions and that of the long-range forces, which mostly are attractive. All these results demonstrate that the pair potentials from *ab initio* SAPT calculations are accurate. Another, more global, comparison with experiment, confirming this finding, was made by computations of the (mixed) second virial coefficients of most of these dimers over a wide range of temperatures [169].

An extension of SAPT that includes also third-order interactions [170–174] permits the explicit calculation and analysis of three-body interactions. For details about this development and a survey of its applications we refer to Chapter 33 in this book on many-body interactions written by Szalewicz, Bukowski, and Jeziorski.

37.5.2 The coupled cluster method

Above we referred to the development of the CC method by Čížek and Paldus [49–51]. The CC method may be viewed as a consistent summation to infinite order of certain type of linked correlation (MBPT, MP) diagrams. Thus, there is a clear relationship between many-body perturbation theory [based on the MP operator of Eq. (A1) in Appendix 37A] and coupled cluster theory. Both are supermolecule methods that give size-extensive energies.

Around 1980 MP calculations at second-order of perturbation (MP2) came within computational reach, while around 1990 third- (MP3) and fourth-order (MP4) calculations became feasible. For some time MP4 calculations were widely applied to weakly bound complexes, but soon it was discovered that a full MP4 computation (including terms that include sums over triply excited states) is hardly cheaper than a CCSD(T) computation. Since the latter is in general more reliable, MP4 lately lost much ground to the CCSD(T) method.

We can be very brief about the CC method since the chapters by Paldus and others in this book give in-depth treatments of it. As is well known, the exact N-electron wave function Ψ is written as

$$|\Psi\rangle = e^{T_1+T_2+\cdots+T_N}|\Phi_0\rangle \tag{7}$$

where Φ_0 is a closed-shell Hartree–Fock reference function. In most applications $T_3, T_4, \ldots$ are neglected and only $T_1 = t^i_a E^a_i$ and $T_2 = t^{ij}_{ab} E^{ab}_{ij}$ (summation convention is used here) are included. Here E^a_i and E^{ab}_{ij} are orbital replacement operators, where orbital i and j are occupied in Φ_0 and a, b refer to virtual orbitals. The cluster amplitudes t^i_a and t^{ij}_{ab} are obtained from the solution of equations that are quadratic in the t^{ij}_{ab} and fourth order in t^i_a. For the closed-shell (spin singlet) case the projection of the CC equation on doubly excited states (the CCD method) yields coupled equations of dimension $K(K+1)/2$ where K is the product of the number n_{occ} of occupied and the number n_{vir} of virtual orbitals.

Naively one could expect that the solution of these equations scales as $O(K^4)$. For if one linearizes the equations according to the Newton–Raphson method, a set of $O(K^2)$ linear equations must be repeatedly solved, which takes $O(K^4)$ operations per solution. Fortunately, the scaling is not that bad. In the first place the sums in the equations do not run over all four orbital labels of t^{ab}_{ij} simultaneously, but at most over two. In the second place a quasi-Newton method, in which the linear equations are approximated by a partially diagonal form, usually converges well. See for more details about the computational aspects of the CCSD method the recent book by Helgaker et al. [175]. This book also shows that the exponential ansatz, Eq. (7), leads in the long range to a factorization of the wave function and a corresponding decomposition of the dimer energy into a sum of monomer energies.

In total [176], the solution of the CCSD equation scales as $n^2_{\text{occ}} n^4_{\text{vir}} N_{\text{it}}$, where N_{it} is the number of quasi-Newton iterations needed. Once the amplitudes t^a_i and t^{ab}_{ij} have been solved, they can be used to compute additional perturbation terms that include triply excited states [177] [a non-iterative $O(n^7)$ process], which are not accounted for in the CCSD method; their inclusion is indicated by (T) in CCSD(T).

We will end this section by mentioning a dozen or so illustrative examples of modern supermolecule calculations on dispersion-bound complexes. Of course, it is hopeless to strive for completeness, almost daily new calculations are published, and hence the following list of references is far from exhaustive.

As stated above, around 1990 many workers used the MP4 method, see, for instance, Ref. [178] for the potential of CH_4–H_2O, Ref. [179] for MP4 applications to CO_2–Ar, and Ref. [180] for argon in interaction with Cl_2 and ClF. Later the MP4 and CCSD(T) methods were compared, in calculations on Ar–H_2 and Ar–HCl [181], on N_2–HF [182], and for CO–CO [183].

A few examples of recent CCSD(T) computations on intermolecular potentials are by Cybulski and coworkers, who computed potentials of the noble gas dimers He_2, Ne_2, Ar_2, He–Ne, He–Ar, and Ne–Ar (Ref. [184]) and Ne–Kr, Ar–Kr, and Kr_2 (Ref. [185]). Further they considered HCN in interaction with He, Ne, Ar, Kr [186], and Ar–CO [187]. Computational and experimental studies of intermolecular states and forces in the benzene–He complex were reported in Ref. [188]. A thorough CCSD(T) study on benzene–Ar is by Koch et al. [189] and on Ne–HCl by Fernández and coworkers [190].

37.5.3 Latest developments

Lately two completely different topics in the field of intermolecular forces have drawn attention and are now actively being studied. In the first place there is the possible application of density functional theory (DFT) to van der Waals molecules. The second topic concerns van der Waals molecules of which the electronic state of one or more of the monomers is spatially degenerate.

DFT in the standard Kohn–Sham (KS) formulation has its limitations in application to dispersion forces. Standard local and gradient-corrected functionals are not appropriate for the description of dispersion, which is inherently a non-local correlation effect [there is no such thing as a dispersion potential $V_{\text{disp}}(\vec{r})$]. Despite the search for functionals capable of describing London forces (cf. Ref. [191] and references therein), there still is no generally applied solution in the framework of KS-DFT. However, *via* a detour DFT can play an important role. Earlier we discussed an approach for obtaining non-expanded dispersion by the Casimir–Polder integration of a product of two polarization propagators. This approach can easily and seamlessly be interwoven with DFT [192, 193], because DFT is known to give accurate response properties, provided functionals with correct asymptotics are used [194–197]. Complete intermolecular interaction potentials can be obtained from the so-called DFT-SAPT method that substitutes KS orbitals and exchange-correlation kernels into the SAPT expressions for the interaction energies with $j = k = 0$, cf. Eqs. (5) and (6). The evaluation of these expressions is computationally much cheaper than the inclusion of monomer correlation effects by calculation of the SAPT terms with $j + k > 0$.

In Ref. [193] various non-hybrid and hybrid exchange-correlation potentials and suitable adiabatic local density approximations for the exchange-correlation kernel were compared for the dimers He_2, Ne_2, Ar_2, NeAr, NeHF, ArHF, $(H_2)_2$, $(HF)_2$, and $(H_2O)_2$. This comparison showed that the effects of intramonomer electron correlation on the dispersion energy are most accurately reproduced with an asymptotically corrected [197] version of the exchange-correlation potential of Perdew et al. [198]. In Ref. [199] the importance of asymptotically correct exchange-correlation potentials in DFT-SAPT was emphasized particularly. In Ref. [192] dispersion energies of He, Ne, and H_2O dimers were obtained by the DFT-SAPT approach to within 3% or better.

Earlier we also discussed the uncoupled HF approach to dispersion, where the sum over states is performed at the orbital level. Of course, this approach can also be applied with KS orbitals. However, Heßelmann and Jansen [193] found that the uncoupled sum-over-states approximation yields unacceptable errors. These are mainly due to neglect of the Coulomb and exchange-correlation kernels and are not substantially improved through an asymptotic correction of the exchange-correlation potential.

The DFT-SAPT approach has been very recently applied [200] to the notoriously difficult case of the CO dimer. Earlier computations [183] of the $(CO)_2$ PES by means of MP4 and CCSD(T) methods encountered some unexpected complications. It was shown that high-order correlation effects are important and that both CCSD(T) and CCSDT formally do not have a correct asymptotic (large R) behavior. Later [201,202] it was pointed out that on top of this problem also very large basis sets are needed for an accurate description of the CO–CO PES. Notwithstanding this problem, a full four-dimensional PES (rigid monomers) was computed in Ref. [203] by the CCSD(T) method as a springboard for further refining. The potential was fitted in terms of analytic functions, and the fitted potential was used to compute the lowest rovibrational states of the dimer. It gave semi-quantitative agreement with the experimental infrared and millimeter wave spectra of McKellar, Winnewisser and coworkers [204–210]. Application of a fit of the recent DFT-SAPT potential [200] gave rovibrational results that differed somewhat from the CCSD(T) data, and were also in semi-quantitative agreement with the measured spectra. It was decided to combine the two potentials, CCSD(T) and DFT-SAPT, and it was shown that a weighted average of the DFT-SAPT (30%) and the CCSD(T) potential (70%) gives results that are in very good agreement with experimental data, for both $(^{12}CO)_2$ and $(^{13}CO)_2$.

The second topic of recent interest—dimers that dissociate into a degenerate open-shell monomer and a non-degenerate closed-shell monomer or into two open-shell monomers—has two intrinsic difficulties that are both due to spatial degeneracy. The dimer is an open-shell system in such cases and it has multiple PESs that become degenerate for large intermolecular separations and in many cases also for other geometries. In the first place, it is fair to say that at present there are no generally applicable size-extensive electronic structure methods for open-shell, spatially degenerate, systems. From a theoretical point of view, the complete active space multi-configuration SCF (CASSCF) method [211] is probably the most satisfactory, as it handles electron spin correctly and is size extensive. However, the active spaces that can be handled in practice are too small to give a reliable account of dynamic correlation effects like dispersion. The CASSCF method has been extended to CAS perturbation theory (CASPT) in order to include dynamical correlation effects [212–217]. The CASPT approach is almost size extensive when the CASSCF reference function is dominated by a single determinant. However, for reference wave functions in which several determinants have large weights, as is the case for spatially degenerate open-shells, size extensivity is broken [175].

An alternative electron-correlation method is the multi-reference configuration interaction (MRCI) method. This method is plagued by unlinked diagrams, the presence of which break the size extensivity of the MRCI energy. Often MRCI results are corrected by a simple formula introduced 30 years ago by Langhoff and Davidson [218], who derived it by inspection of a CI wave function consisting of all double excitations (DCI) from a single Slater determinant. One can look upon this 'Davidson correction' as an approximate formula for the unlinked diagram that enters the DCI energy.

Paldus, elsewhere in this book, discusses that there is as yet no generally applicable, open-shell, size-extensive, coupled cluster method, and the same holds for open-shell SAPT methods. Therefore, for the computation of potentials of open-shell van der Waals molecules one has the choice between CASSCF followed by a Davidson-corrected MRCI calculation of the interaction energy, or the single reference, high spin, method RCCSD(T). When the ground state of the open-shell monomer is indeed a high spin state, then RCCSD(T) is the method of choice. With regard to the latter method we recall that a major difficulty in open-shell systems is the adaptation of the wave function to the total spin operator S^2; for the CCSD method a partial spin adaptation was published by Knowles et al. [219,220] who refer to their method as 'partially spin restricted'. When non-iterative triple corrections [221] are included, the spin restricted CCSD(T) method, RCCSD(T), is obtained.

Even when free monomers are in degenerate states, the RCCSD(T) method is often employed, because for most points on the PES the symmetry is lowered to Abelian symmetry, so that degeneracies are lifted and RCCSD(T) is formally applicable. But it can be applied only to the lowest state of a given symmetry, while one needs to know also the potential surfaces of the higher dimer states that become asymptotically degenerate with the ground state. Moreover, it is clear that the method fails for points on the PES that have symmetry higher than Abelian and states that belong to more-dimensional representations of the non-Abelian point group.

The second problem that often occurs in open-shell van der Waals molecules is the breakdown of the Born–Oppenheimer (BO) approximation. As is well known, the BO approximation can be trusted when the PESs are well separated in energy. However, when certain points on the PES are degenerate this condition is not fulfilled, not in the degenerate points themselves, but also not in nearby points. This breakdown of the BO approximation can be shown as follows. Let us write **R** for the collection of nuclear coordinates and **r** for the electron coordinates. Indicating electronic and nuclear interactions by subscripts e and n, respectively, the Schrödinger equation takes the form

$$(T_{\mathrm{n}} + T_{\mathrm{e}} + V_{\mathrm{nn}} + V_{\mathrm{ne}} + V_{\mathrm{ee}})\Psi(\mathbf{R}, \mathbf{r}) = E\Psi(\mathbf{R}, \mathbf{r})$$

where the kinetic energy terms T_{n} and T_{e} have the usual form. In particular $T_{\mathrm{n}} = \sum_{\alpha} P_{\mathrm{n}}^{\alpha} P_{\mathrm{n}}^{\alpha}/(2M_{\alpha})$ with the nuclear momentum $P_{\mathrm{n}}^{\alpha} = -i\partial/\partial R_{\alpha}$. The wave function is expanded in eigenfunctions $\chi_k(\mathbf{r};\mathbf{R})$ of $H_e \equiv T_e + V_{ee} + V_{ne}$

$$\Psi(\mathbf{R},\mathbf{r}) = \sum_k \chi_k(\mathbf{r};\mathbf{R})\phi_k(\mathbf{R})$$

with $\langle\chi_{k'}(\mathbf{r};\mathbf{R})|\chi_k(\mathbf{r};\mathbf{R})\rangle_{(\mathbf{r})} = \delta_{k'k}$ and where the subscript $(\mathbf{r})$ indicates that the integration is over electronic coordinates only. By definition

$$\langle\chi_{k'}(\mathbf{r};\mathbf{R})|H_e|\chi_k(\mathbf{r};\mathbf{R})\rangle_{(\mathbf{r})} = (\mathbb{H}_e(\mathbf{R}))_{k'k} = \delta_{k'k}E_k(\mathbf{R})$$

and we assume that $\chi_k(\mathbf{r};\mathbf{R})$ is real (invariant under time reversal).

After multiplication by $\chi_{k'}(\mathbf{r};\mathbf{R})$ and integration over $\mathbf{r}$ the Schrödinger equation is turned into a set of coupled equations depending on nuclear coordinates only

$$[\mathbb{H}_n(\mathbf{R}) + \mathbb{H}_e(\mathbf{R})]\boldsymbol{\phi}(\mathbf{R}) = E\boldsymbol{\phi}(\mathbf{R})$$

where the column vector $\boldsymbol{\phi}(\mathbf{R})$ has elements $\phi_k(\mathbf{R})$. The matrix $\mathbb{H}_e(\mathbf{R})$ is diagonal and

$$(\mathbb{H}_n(\mathbf{R}))_{k'k} = \langle\chi_{k'}(\mathbf{r};\mathbf{R})|T_n|\chi_k(\mathbf{r};\mathbf{R})\rangle_{(\mathbf{r})} + \delta_{k'k}V_{nn}$$

Suppressing the coordinates in the notation, we can write the matrix elements of T_n as

$$\langle\chi_{k'}|T_n|\chi_k\rangle_{(\mathbf{r})} = \delta_{k'k}T_n + \sum_\alpha \frac{1}{M_\alpha}\langle\chi_{k'}|(P_n^\alpha\chi_k)\rangle_{(\mathbf{r})}P_n^\alpha + \langle\chi_{k'}|(T_n\chi_k)\rangle_{(\mathbf{r})} \tag{8}$$

The diagonal ($k' = k$) matrix elements $\langle\chi_k|(P_n^\alpha\chi_k)\rangle_{(\mathbf{r})}$ of the operator P_n^α vanish, because this operator is Hermitian and odd with respect to time reversal. The off-diagonal matrix elements satisfy

$$\langle\chi_{k'}|(P_n^\alpha\chi_k)\rangle_{(\mathbf{r})} = \frac{\langle\chi_{k'}|[P_n^\alpha, H_e]|\chi_k\rangle_{(\mathbf{r})}}{E_k(\mathbf{R}) - E_{k'}(\mathbf{R})}$$

We see that whenever two surfaces come close, $E_k(\mathbf{R}) \approx E_{k'}(\mathbf{R})$, the nuclear momentum coupling term is no longer negligible. Conversely, if all surfaces are well separated, all off-diagonal terms can be neglected and hence the whole matrix of P_n^α is effectively zero. The third term on the right-hand side of Eq. (8) can be written as the matrix of P_n^α squared and, accordingly, is then negligible also. Only the first (diagonal) kinetic energy term in Eq. (8) survives and a diagonal, uncoupled, set of nuclear motion equations results. These are the normal second-step of the BO approximation equations.

Let us, for the sake of argument, assume now that only two surfaces 1 and 2 approach each other and that all other surfaces are well separated; the argument is easily generalized to more surfaces. We then have to solve a set of two coupled nuclear Schrödinger equations with non-negligible coupling element $\langle\chi_1(\mathbf{r};\mathbf{R})|T_n|\chi_2(\mathbf{r};\mathbf{R})\rangle_{(\mathbf{r})}$. Define two new orthonormal states by a rotation of χ_1 and χ_2 (for clarity reasons we suppress the coordinates)

$$(\varphi_1, \varphi_2) = (\chi_1, \chi_2)\mathbb{R}(\gamma) \tag{9}$$

where $\mathbb{R}(\gamma(\mathbf{R}))$ is a 2×2 rotation matrix and $\gamma(\mathbf{R})$ is the 'diabatic angle'. Transformation of the matrix of nuclear momentum $\langle \chi_{k'} | (P_{\rm n}^{\alpha} \chi_k) \rangle_{(\mathbf{r})}$ for $k', k = 1, 2$ gives

$$\langle \varphi_k | (P_{\rm n}^{\alpha} \varphi_k) \rangle_{(\mathbf{r})} = 0 \qquad \text{for} \qquad k = 1, 2,$$

i.e. the diagonal matrix elements remain zero, and

$$\langle \varphi_2 | (P_{\rm n}^{\alpha} \varphi_1) \rangle_{(\mathbf{r})} = (P_{\rm n}^{\alpha} \gamma(\mathbf{R})) - \langle \chi_2 | (P_{\rm n}^{\alpha} \chi_1) \rangle_{(\mathbf{r})}$$

We search for a $\gamma(\mathbf{R})$, such that to a good approximation

$$(P_{\rm n}^{\alpha} \gamma(\mathbf{R})) - \langle \chi_2 | (P_{\rm n}^{\alpha} \chi_1) \rangle_{(\mathbf{r})} \approx 0 \tag{10}$$

i.e. φ_1 and φ_2 diagonalize the 2×2 matrix of the nuclear momentum. By the definition of Smith [222] φ_1 and φ_2 are *diabatic states*. Smith was the first to define this concept. (Earlier the term 'diabatic' was used somewhat loosely by Lichten [223]). The nuclear motion problem takes the following 'generalized Born–Oppenheimer' form

$$\begin{aligned} &\begin{pmatrix} T_{\rm n} + V_{\rm nn} + \dfrac{E_1(\mathbf{R}) + E_2(\mathbf{R})}{2} & 0 \\ 0 & T_{\rm n} + V_{\rm nn} + \dfrac{E_1(\mathbf{R}) + E_2(\mathbf{R})}{2} \end{pmatrix} \boldsymbol{\phi}(\mathbf{R}) \\ &+ \frac{E_2(\mathbf{R}) - E_1(\mathbf{R})}{2} \begin{pmatrix} \cos 2\gamma & \sin 2\gamma \\ \sin 2\gamma & -\cos 2\gamma \end{pmatrix} \boldsymbol{\phi}(\mathbf{R}) \\ &= E \boldsymbol{\phi}(\mathbf{R}) \end{aligned} \tag{11}$$

The surfaces $E_1(\mathbf{R})$ and $E_2(\mathbf{R})$ are BO energies obtained from electronic structure calculations and $T_{\rm n}$ is the first term of Eq. (8). The (transformed) third term in this equation is neglected. The determination of $\gamma(\mathbf{R})$ is the remaining problem before a solution of Eq. (11) can be attempted.

Several methods for the determination of $\gamma(\mathbf{R})$ have been proposed [224,225]. One is the direct computation of the non-adiabatic coupling matrix element $\langle \chi_1 | (P_{\rm n}^{\alpha} \chi_2) \rangle_{(\mathbf{r})}$ by finite difference techniques, which gives the derivative of γ (cf. Eq. (10)). Another is by supposing that the diabatic states φ_1 and φ_2 are states of the free monomers and by using Eq. (9) backwards. This is obviously only possible when the adiabatic states χ_k and $\chi_{k'}$ are (almost) pure linear combinations of the two monomer states. This approximation can be made at the orbital level or at the N-electron level (or at both levels simultaneously). Also mixing matrix elements of molecular properties over adiabatic states may be used.

We will end this section by mentioning some recent representative calculations on van der Waals molecules consisting of a closed- and open-shell monomer. The simplest closed-shell monomer is of course the ground state helium atom. Its interaction with $NO(X^2\Pi)$ [226], $CO(a^3\Pi)$ [227,228], $CaH(^2\Sigma^+)$ [229], and $NH(X^3\Sigma^-)$ [230] was studied recently. In the case of the $He-CO(^3\Pi)$ complex the potential was applied in computing the spectrum of the bound complex [227] and in photodissociation processes [228]. The $He-CaH(^2\Sigma^+)$ interaction was employed in the study of collisions at cold and ultracold temperatures [231], and the $He-NH(X^3\Sigma^-)$ potential was used in calculations on low temperature collisions in the presence of a magnetic field [232]. Further work is on $Cl(^2P)$–HCl [233] and its bound states [234]. Finally, we refer to the work on the diabatic intermolecular potential and bound states of the H_2–$F(^2P)$ complex [235].

APPENDIX 37A: RELATIONSHIP BETWEEN DISPERSION AND $E_{\mathrm{MP2}}^{\mathrm{AB}}$

Often dispersion energy is described as the interaction between mutually induced dipoles, one on each atom. One can see this as a 'correlation' between two dipoles. It is not obvious how this 'correlation' is related to Löwdin's 'beyond-Hartree–Fock-correlation' [23]. In this appendix it is shown how the latter correlation and dispersion are interrelated. Earlier this connection was shown [63] in a somewhat different manner.

The MP2 energy, the simplest correlation correction, is obtained from RS perturbation theory with the perturbation

$$W \equiv H - F - \langle \Phi_0 | H - F | \Phi_0 \rangle \tag{A1}$$

where the Slater determinant Φ_0 is the lowest eigenfunction of the Fock operator $F = \sum_{k=1}^{N} f(k)$ with eigenvalue $2\sum_{i=1}^{N/2} \epsilon_i$. The Fock operator serves as the unperturbed (zeroth-order) operator. Since the first order MP energy $\langle \Phi_0 | W | \Phi_0 \rangle$ is obviously zero, the lowest order MP energy appears in second order. We write the MP2 energy formula for a supermolecule A–B with closed-shell monomers A and B. After application of the Slater–Condon rules for the simplification of N-electron matrix elements and integrating out spin, it becomes

$$E_{\mathrm{MP2}}^{\mathrm{AB}} = \sum_{i,j,a,b} \langle \phi_i(1)\phi_j(2) | r_{12}^{-1} | \phi_a(1)\phi_b(2) \rangle \times \frac{2\langle \phi_a(1)\phi_b(2) | r_{12}^{-1} | \phi_i(1)\phi_j(2) \rangle - \langle \phi_a(1)\phi_b(2) | r_{12}^{-1} | \phi_j(1)\phi_i(2) \rangle}{\epsilon_i + \epsilon_j - \epsilon_a - \epsilon_b} \tag{A2}$$

where ϕ_i and ϕ_j are occupied and ϕ_a and ϕ_b are virtual orbitals of the *dimer* A–B. We consider the limit of this expression for R large enough that the differential overlap between wave functions of A and B can be neglected. We recall that we can localize SCF orbitals and write $|\rho\rangle$ and $|r\rangle$ for the occupied and virtual spatial orbitals localized on A and $|\sigma\rangle$ and $|s\rangle$ for the occupied and virtual orbital localized on B. These orbitals are expressed in the dimer basis. The Fock operator is invariant under unitary localization of the $\{\phi_i\}$, i.e.

$$\sum_{i=1}^{(N_{\mathrm{A}}+N_{\mathrm{B}})/2} \left\langle \phi_i(2) \left| \frac{2-P_{12}}{r_{12}} \right| \phi_i(2) \right\rangle = \sum_{\rho=1}^{N_{\mathrm{A}}/2} \left\langle \rho(2) \left| \frac{2-P_{12}}{r_{12}} \right| \rho(2) \right\rangle + \sum_{\sigma=1}^{N_{\mathrm{B}}/2} \left\langle \sigma(2) \left| \frac{2-P_{12}}{r_{12}} \right| \sigma(2) \right\rangle$$

In general, the Fock operator is no longer diagonal when the orbitals $\{\phi_i\}$ are localized, but we will show below that we can still use its eigenvalues, i.e. the dimer orbital energies ϵ_i, which under specific conditions applicable here, become equal to the orbital energies of monomers A and B.

Let us consider two S-state atoms and the action of the dimer Fock operator on, for instance, $|\rho(1)\rangle$

$$f^{\mathrm{AB}}(1)|\rho(1)\rangle = \left[-\frac{1}{2}\nabla^2 - \frac{Z_{\mathrm{A}}}{r_{\mathrm{A}1}} + \sum_{\rho}\left\langle \rho(2)\left|\frac{2-P_{12}}{r_{12}}\right|\rho(2)\right\rangle\right]|\rho(1)\rangle$$

$$+\left[-\frac{Z_{\mathrm{B}}}{r_{\mathrm{B}1}} + \sum_{\sigma}\left\langle \sigma(2)\left|\frac{2-P_{12}}{r_{12}}\right|\sigma(2)\right\rangle\right]|\rho(1)\rangle \tag{A3}$$

Because of zero differential overlap the P_{12} contribution can be dropped in the second term of Eq. (A3). The terms that remain in the second expression between large square brackets cancel each other. This is because the electronic charge distribution $Q(\vec{r}_{\mathrm{B}2}) \equiv 2\sum_{\sigma}|\sigma(2)|^2$ is spherically symmetric and screens completely the nucleus of B.

We will prove this intuitive statement and to that end we need the following two expansions, dating back to the 19th century [78,79] (see for a modern version, e.g. Appendix VI of Ref. [236]). Together they give the multipole expansion of $1/r_{12}$ (for $R > r$)

$$\frac{1}{|\vec{R}-\vec{r}|} = \sum_{l=0}^{\infty}\sum_{m=-l}^{l}(-1)^m Y^{l}_{-m}(\vec{R})S^{l}_{m}(\vec{r}) \tag{A4}$$

$$S^{l}_{m}(\vec{r}_1-\vec{r}_2) = \sum_{L=0}^{l}(-1)^L\binom{2l}{2L}^{1/2}[S^{l-L}(\vec{r}_1)\otimes S^{L}(\vec{r}_2)]^{l}_{m} \tag{A5}$$

Here $Y^{l}_{m}(\vec{R}) \equiv R^{l-1}C^{l}_{m}(\hat{R})$ is an irregular solid harmonic function and $S^{l}_{m}(\vec{r}) \equiv r^{l}C^{l}_{m}(\hat{r})$ is a regular solid harmonic function. The function $C^{l}_{m}(\hat{r})$ is a spherical harmonic function normalized to $4\pi/(2l+1)$ (Racah normalization). The expression between square brackets in Eq. (A5) is a Clebsch–Gordan coupled product. We write $\vec{r}_{12} = -\vec{r}_{\mathrm{A}1} + \vec{R}_{\mathrm{AB}} + \vec{r}_{\mathrm{B}2}$, and find, assuming that $|\vec{R}_{\mathrm{AB}}| > |\vec{r}_{\mathrm{B}2} - \vec{r}_{\mathrm{A}1}|$

$$-\frac{Z_{\mathrm{B}}}{r_{\mathrm{B}1}} + 2\sum_{\sigma}\langle\sigma(2)|\frac{1}{r_{12}}|\sigma(2)\rangle$$

$$= -\frac{Z_{\mathrm{B}}}{r_{\mathrm{B}1}} + 2\sum_{L,l,m}(-1)^{L+m}\binom{2l}{2L}^{1/2} Y^{l}_{-m}(\vec{R}_{\mathrm{AB}})\sum_{\sigma}[S^{l-L}(\vec{r}_{\mathrm{A}1})\otimes\langle\sigma|S^{L}(\vec{r}_{\mathrm{B}2})|\sigma\rangle]^{l}_{m} \tag{A6}$$

The expression $\langle S^{L}_{M}\rangle \equiv 2\sum_{\sigma}\langle\sigma|S^{L}_{M}(\vec{r}_{\mathrm{B}2})|\sigma\rangle$ is the Hartree–Fock expectation value of the (L, M) multipole moment of the S-state atom B. When the charge distribution $Q(\vec{r}_{\mathrm{B}2})$ is spherical symmetric around B $\langle S^{L}_{M}\rangle = N_{\mathrm{B}}\delta_{L0}\delta_{M0}$. Eq. (A6) becomes under this condition

$$-\frac{Z_{\mathrm{B}}}{r_{\mathrm{B}1}} + N_{\mathrm{B}}\sum_{l,m}(-1)^m Y^{l}_{-m}(\vec{R}_{\mathrm{AB}})S^{l}_{m}(\vec{r}_{\mathrm{A}1}) = -\frac{Z_{\mathrm{B}}}{r_{\mathrm{B}1}} + \frac{N_{\mathrm{B}}}{r_{\mathrm{B}1}}$$

The simplification of this result follows from Eq. (A4). Since for neutral atoms $N_{\mathrm{B}} = Z_{\mathrm{B}}$ the second term of Eq. (A3) indeed vanishes. It follows that the dimer Fock operator, when it acts on orbital $|\rho\rangle$ localized on monomer A, is equivalent to the atomic

Fock operator of A

$$f^{\mathrm{AB}}|\rho\rangle = f^{\mathrm{A}}|\rho\rangle \tag{A7}$$

Under the conditions of our derivation, i.e. S-state atoms A and B with vanishing differential overlap, we can show that the localized orbitals $|\rho\rangle$ and $|\sigma\rangle$ are identical (apart from mixing possibly degenerate orbitals) to the orbitals obtained by solving the *monomer* Hartree–Fock equations (in the dimer basis)

$$f^{\mathrm{A}}|\rho\rangle = \epsilon_\rho|\rho\rangle \qquad \text{and} \qquad f^{\mathrm{B}}|\sigma\rangle = \epsilon_\sigma|\sigma\rangle$$

These Fock equations yield solutions for A and B with corresponding charge distributions that are spherically symmetric around A and B, respectively [i.e. the solutions span irreps of $SO(3)$]. Hence the spherical symmetry of S-state atom A is not disturbed by the presence of S-state atom B and *vice versa*, so that Eq. (A7) holds. Expand the solution of A in dimer MOs $|k\rangle$

$$|\rho\rangle = \sum_k |k\rangle U_{k\rho} \qquad \text{with } f^{\mathrm{AB}}|k\rangle = \epsilon_k|k\rangle$$

Then

$$\epsilon_\rho|\rho\rangle = \epsilon_\rho \sum_k |k\rangle U_{k\rho} = f^{\mathrm{AB}}|\rho\rangle = \sum_k \epsilon_k|k\rangle U_{k\rho}$$

so that

$$\epsilon_\rho U_{k\rho} = \epsilon_k U_{k\rho}$$

If $\epsilon_\rho \neq \epsilon_k$ it follows that $U_{k\rho} = 0$, so that, in general, $|\rho\rangle$ is a linear combination of degenerate dimer orbitals $|k\rangle$ with orbital energy $\epsilon_k = \epsilon_\rho$. If there is no degeneracy, then $|\rho\rangle$ is identical to $|k\rangle$.

The same argument may be applied to the other localized dimer orbitals $|r\rangle$, $|\sigma\rangle$ and $|s\rangle$. In other words, we can solve the monomer HF equations in the dimer basis and get the same orbital energies as from the solution of the dimer HF equations.

When we now replace the sums over the canonical orbitals by sums over localized orbitals and the dimer orbital energies by monomer orbital energies in Eq. (A2), we obtain

$$\lim_{R\to\infty} E_{\mathrm{MP2}}^{\mathrm{AB}} = E_{\mathrm{MP2}}^{\mathrm{A}} + E_{\mathrm{MP2}}^{\mathrm{B}} + 4 \sum_{\rho,r,\sigma,s} \frac{|\langle \rho(1)\sigma(2)|r_{12}^{-1}|r(1)s(2)\rangle|^2}{\epsilon_\rho + \epsilon_\sigma - \epsilon_r - \epsilon_s}$$

See Fig. 37.1 for a diagrammatic representation of this limit. The third term on the right-hand side is the non-expanded 'Hartree–Fock' expression [59] for dispersion. Incidentally, this equation shows that the MP2 method is size extensive. That is, when the distance R_{AB} between A and B is so large that the interaction term vanishes, the dimer MP2 energy becomes the sum of the monomer MP2 energies. Although this statement sounds obvious, it is not. The singles and doubles configuration interaction method forms a counter example.

The equivalence between the interaction energy from dimer MP2 calculations and the simple expression for the dispersion interaction does not hold when the interacting systems are molecules or non-S-state atoms. The second term of Eq. (A3) does not vanish in that case, because of non-vanishing multipole moments contributing to the expansion in Eq. (A6). Even for large distances *R*, where all differential overlap between A and B vanishes and the dimer orbitals can be localized, these orbitals are not equal to the unperturbed monomer orbitals. This is due to the polarization of each monomer, induced by the multipole moments of the other monomer. This gives long-range electrostatic and induction interactions, which thus are accounted for by the supermolecule HF method. Conversely, for spherically symmetric systems the HF method does not give any interaction at distances where differential overlap is negligible.

37.6 REFERENCES

1 H. Kamerlingh Onnes, Proc. Sec. Sci., KNAW, XI (1908) 168.
2 S.C. Wang, Phys. Z., 28 (1927) 663.
3 R. Eisenschitz and F. London, Z. Phys., 60 (1930) 491.
4 F. London, Z. Phys., 60 (1930) 245.
5 T.R. Dyke, G.R. Tomasevich, W. Klemperer and W.E. Falconer, J. Chem. Phys., 57 (1972) 2277.
6 C.A. Long and G.E. Ewing, J. Chem. Phys., 58 (1973) 4824.
7 S.E. Novick, P.B. Davies, T.R. Dyke and W. Klemperer, J. Am. Chem. Soc., 57 (1973) 2277.
8 G. Henderson and G.E. Ewing, Mol. Phys., 27 (1974) 903.
9 G.E. Ewing, Acc. Chem. Res., 8 (1975) 185.
10 J.E. Jones, Proc. Roy. Soc. (London), A106 (1924) 463.
11 J.O. Hirschfelder, C.F. Curtiss and R.B. Bird, Molecular theory of gases and liquids, Wiley, New York, 1954.
12 C.A. Coulson, Rev. Mod. Phys., 32 (1960) 170.
13 B.J. Ransil, Rev. Mod. Phys., 32 (1960) 239.
14 B.J. Ransil, J. Chem. Phys., 34 (1961) 2109.
15 P.E. Phillipson, Phys. Rev., 125 (1962) 1981.
16 N.R. Kestner, J. Chem. Phys., 48 (1968) 252.
17 S.F. Boys and F. Bernardi, Mol. Phys., 19 (1970) 553.
18 H.B. Jansen and P. Ros, Chem. Phys. Lett., 3 (1969) 140.
19 M. Bulski and G. Chałasiński, Theor. Chim. Acta, 44 (1977) 399.
20 M. Gutowski, F.B. van Duijneveldt, G. Chałasiński and L. Piela, Chem. Phys. Lett., 129 (1986) 325.
21 M. Gutowski and G. Chałasiński, J. Chem. Phys., 98 (1993) 5540.
22 F.B. van Duijneveldt, J.G.C.M. van Duijneveldt-van der Rijdt and J.H. van Lenthe, Chem. Rev., 94 (1994) 1873.
23 P.-O. Löwdin, Adv. Chem. Phys., 2 (1959) 207.
24 O. Sinanoğlu, Adv. Chem. Phys., 6 (1964) 315.
25 N.R. Kestner and O. Sinanoğlu, J. Chem. Phys., 45 (1966) 194.
26 J.W. Viers, F.E. Harris and H.F. Schaefer, Phys. Rev. A, 1 (1970) 24.
27 H.F. Schaefer, D.R. McLaughlin, F.E. Harris and B.J. Alder, Phys. Rev. Lett., 25 (1970) 988.
28 P. Bertoncini and A.C. Wahl, Phys. Rev. Lett., 25 (1970) 991.
29 M. Jeziorska, R. Bukowski, W. Cencek, M. Jaszunski, B. Jeziorski and K. Szalewicz, Collect. Czech. Chem. Commun., 68 (2003) 463.
30 W. Cencek, M. Jeziorska, R. Bukowski, M. Jaszunski, B. Jeziorski and K. Szalewicz, J. Phys. Chem. A, 108 (2004) 3211.
31 A. Dalgarno and W.D. Davison, Adv. At. Mol. Phys., 2 (1966) 1.
32 A. Dalgarno, Adv. Chem. Phys., 12 (1967) 143.

33 H.B.G. Casimir and D. Polder, Phys. Rev., 73 (1948) 360.
34 P.W. Langhoff and M. Karplus, Application of Padé approximants to dispersion force and optical polarizability computations, in: G.A. Baker, J.L. Gammel (Eds.), The Padé approximant in theoretical physics, Academic Press, New York, 1970, pp. 41–97.
35 M. Karplus and H.J. Kolker, J. Chem. Phys., 41 (1966) 3955.
36 A. Dalgarno and G.A. Victor, Proc. Roy. Soc. (London), A291 (1966) 291.
37 J.M. Murrell, M. Randić and D.R. Williams, Proc. Roy. Soc. (London), A284 (1965) 566.
38 J.O. Hirschfelder and R. Silbey, J. Chem. Phys., 45 (1966) 2188.
39 J.O. Hirschfelder, Chem. Phys. Lett., 1 (1967) 325.
40 J.O. Hirschfelder, Chem. Phys. Lett., 1 (1967) 363.
41 A. van der Avoird, Chem. Phys. Lett., 1 (1967) 24.
42 A. van der Avoird, Chem. Phys. Lett., 1 (1967) 411.
43 A. van der Avoird, J. Chem. Phys., 47 (1967) 3649.
44 A. van der Avoird, Chem. Phys. Lett., 1 (1967) 429.
45 J.M. Murrell and G.J. Shaw, J. Chem. Phys., 46 (1967) 1768.
46 J.I. Musher and A.T. Amos, Phys. Rev., 164 (1967) 31.
47 B. Jeziorski and W. Kołos, Int. J. Quantum Chem., S12 (1977) 91.
48 D.M. Chipman, J.D. Bowman and J.O. Hirschfelder, J. Chem. Phys., 59 (1973) 2830.
49 J. Čížek, J. Chem. Phys., 45 (1966) 4256.
50 J. Čížek and J. Paldus, Int. J. Quantum Chem., 5 (1971) 359.
51 J. Paldus, J. Čížek and I. Shavitt, Phys. Rev., 164 (1972) 50.
52 R.J. Bartlett and G.D. Purvis, Int. J. Quantum Chem., 14 (1978) 561.
53 G.D. Purvis, III and R.J. Bartlett, J. Chem. Phys., 76 (1982) 1910.
54 J.A. Pople, R. Krishnan, H.B. Schlegel and J.S. Binkley, Int. J. Quantum Chem., 14 (1978) 545.
55 R. Ahlrichs, R. Penco and G. Scoles, Chem. Phys., 19 (1977) 119.
56 C. Douketis, G. Scoles, S. Marchetti, M. Zen and A.J. Thakkar, J. Chem. Phys., 76 (1982) 3057.
57 K.T. Tang and J.P. Toennies, J. Chem. Phys., 80 (1984) 3726.
58 E. Kochanski, Chem. Phys. Lett., 10 (1971) 543.
59 E. Kochanski, J. Chem. Phys., 58 (1973) 5823.
60 B. Jeziorski and M. van Hemert, Mol. Phys., 31 (1976) 713.
61 P.H. Smit, J.L. Derissen and F.B. van Duijneveldt, Mol. Phys., 37 (1979) 501.
62 C. Møller and M.S. Plesset, Phys. Rev., 46 (1934) 618.
63 G. Chałasiński and M.M. Szczęśniak, Mol. Phys., 63 (1988) 205.
64 M.V. Basilevsky and M.M. Berenfeld, Int. J. Quantum Chem., 6 (1972) 23.
65 P.E.S. Wormer, T. van Berkel and A. van der Avoird, Mol. Phys., 29 (1975) 1181.
66 P.E.S. Wormer and A. van der Avoird, J. Chem. Phys., 62 (1975) 3326.
67 W. Meyer, J. Chem. Phys., 58 (1973) 1017.
68 R. Ahlrichs, H. Lischka, V. Staemmler and W. Kutzelnigg, J. Chem. Phys., 62 (1975) 1225.
69 P.G. Burton and U.E. Senff, J. Chem. Phys., 79 (1983) 526.
70 P.S. Epstein, Phys. Rev., 28 (1926) 695.
71 R.K. Nesbet, Proc. Roy. Soc. (London), A230 (1955) 322.
72 J.P. Malrieu and F. Spiegelmann, Theor. Chim. Acta, 52 (1979) 55.
73 P. Claverie, S. Diner and J.P. Malrieu, Int. J. Quantum Chem., 1 (1967) 751.
74 A. Masson, B. Levy and J.P. Malrieu, Theor. Chim. Acta, 18 (1970) 197.
75 J.P. Daudey, P. Claverie and J.P. Malrieu, Int. J. Quantum Chem., 8 (1974) 1.
76 H.P. Kelly, Adv. Chem. Phys., 14 (1969) 129.
77 A.D. Buckingham, Adv. Chem. Phys., 12 (1967) 107.
78 F.E. Neumann, Vorlesungen über die theorie des potentials und der kugelfunktionen, B.G. Teubner, Leipzig, 1887.
79 E. Heine, Handbuch der kugelfunctionen, 2nd edn., G. Reimer, Berlin, 1878.
80 P.E.S. Wormer, Intermolecular forces and the group theory of many-body systems, PhD thesis, University of Nijmegen, 1975.
81 P.E.S. Wormer, F. Mulder and A. van der Avoird, Int. J. Quantum Chem., 11 (1977) 959.

82 A. van der Avoird, P.E.S. Wormer, F. Mulder and R.M. Berns, Top. Curr. Chem., 93 (1980) 1.
83 R.P. Leavitt, J. Chem. Phys., 72 (1980) 3472.
84 A. Koide, J. Phys. B, 11 (1978) 633.
85 F. Visser, P.E.S. Wormer and P. Stam, J. Chem. Phys., 79 (1983) 4973.
86 A.J. Stone and R.J.A. Tough, Chem. Phys. Lett., 110 (1984) 123.
87 W. Rijks and P.E.S. Wormer, J. Chem. Phys., 90 (1989) 6507.
88 B. Jeziorski, R. Moszynski and K. Szalewicz, Chem. Rev., 94 (1994) 1887.
89 M. Kline, Mathematical thought from ancient to modern times, Oxford University Press, Oxford, 1972.
90 R.H. Young, Int. J. Quantum Chem., 9 (1975) 47.
91 R. Ahlrichs, Theor. Chim. Acta, 41 (1976) 7.
92 J.D. Morgan and B. Simon, Int. J. Quantum Chem., 17 (1980) 1143.
93 W. Rijks and P.E.S. Wormer, J. Chem. Phys., 88 (1988) 5704.
94 F. Mulder, M.C. van Hemert, P.E.S. Wormer and A. van der Avoird, Theor. Chim. Acta, 46 (1977) 39.
95 F. Mulder, A. van der Avoird and P.E.S. Wormer, Mol. Phys., 37 (1979) 159.
96 F. Visser, P.E.S. Wormer and P. Stam, J. Chem. Phys., 81 (1984) 3755.
97 G.P. Arrighini, F. Biondi and C. Guidotti, Chem. Phys., 2 (1973) 85.
98 F. Visser and P.E.S. Wormer, Chem. Phys., 92 (1985) 129.
99 F. Visser, P.E.S. Wormer and W.P.J.H. Jacobs, J. Chem. Phys., 82 (1985) 3753.
100 P.E.S. Wormer and W. Rijks, Phys. Rev. A, 33 (1986) 2928.
101 M.B. Doran, J. Phys. B, 7 (1974) 558.
102 P.E.S. Wormer and H. Hettema, J. Chem. Phys., 97 (1992) 5592.
103 R.T. Pack, J. Chem. Phys., 78 (1983) 7217.
104 J.D. Poll and J. van Kranendonk, Can. J. Phys., 39 (1961) 189.
105 P.E.S. Wormer and G. van Dijk, J. Chem. Phys., 70 (1979) 5695.
106 J. Tennyson and B.T. Sutcliffe, J. Chem. Phys., 77 (1982) 4061.
107 J. Tennyson and A. van der Avoird, J. Chem. Phys., 77 (1982) 5664.
108 J. Tennyson and B.T. Sutcliffe, Mol. Phys., 46 (1982) 97.
109 J. Tennyson and A. van der Avoird, J. Chem. Phys., 76 (1982) 5710.
110 J. Tennyson and J. Mettes, Chem. Phys., 76 (1983) 195.
111 A. van der Avoird, J. Chem. Phys., 79 (1983) 1170.
112 A. van der Avoird, Faraday Discuss. Chem. Soc., 73 (1982) 33.
113 R.M. Berns and A. van der Avoird, J. Chem. Phys., 72 (1980) 6107.
114 F. Mulder, G. van Dijk and A. van der Avoird, Mol. Phys., 39 (1980) 407.
115 T. Luty, A. van der Avoird and R.M. Berns, J. Chem. Phys., 73 (1980) 5305.
116 A.P.J. Jansen, W.J. Briels and A. van der Avoird, J. Chem. Phys., 81 (1984) 3648.
117 W.J. Briels, A.P.J. Jansen and A. van der Avoird, J. Chem. Phys., 81 (1984) 4118.
118 A. van der Pol, A. van der Avoird and P.E.S. Wormer, J. Chem. Phys., 92 (1990) 7498.
119 G.E. Moore, Electronics 8.
120 W.B.J.M. Janssen, J. Michiels and A. van der Avoird, J. Chem. Phys., 94 (1991) 8402.
121 P.R. Bunker, P. Jensen, S.C. Althorpe and D.C. Clary, J. Mol. Spectrosc., 157 (1993) 208.
122 G.C.M. van der Sanden, P.E.S. Wormer, A. van der Avoird, J. Schleipen and J.J. ter Meulen, J. Chem. Phys., 97 (1992) 6460.
123 M. Bulski, P.E.S. Wormer and A. van der Avoird, J. Chem. Phys., 94 (1991) 491.
124 HIBRIDON is a package of programs for the time-independent quantum treatment of inelastic collisions and photodissociation written by M.H. Alexander, D. Manolopoulos, H.-J. Werner, and B. Follmeg, with contributions by P. Vohralik, G. Corey, B. Johnson, T. Orlikowski, P. Valiron.
125 T.G.A. Heijmen, P.E.S. Wormer, A. van der Avoird, R.E. Miller and R. Moszynski, J. Chem. Phys., 110 (1999) 5639.
126 R.E. Miller, T.G.A. Heijmen, P.E.S. Wormer, A. van der Avoird and R. Moszynski, J. Chem. Phys., 110 (1999) 5651.
127 M. Wangler, D.A. Roth, V.M. Krivtsun, I. Pak, G. Winnewisser, M. Geleijns, P.E.S. Wormer and A. van der Avoird, Spectrochim. Acta A, 58 (2002) 2499.

128 M. Wangler, D.A. Roth, I. Pak, G. Winnewisser, P.E.S. Wormer and A. van der Avoird, J. Mol. Spectrosc., 222 (2003) 109.

129 E.M. Mas, R. Bukowski, K. Szalewicz, G.C. Groenenboom, P.E.S. Wormer and A. van der Avoird, J. Chem. Phys., 113 (2000) 6687.

130 G.C. Groenenboom, E.M. Mas, R. Bukowski, K. Szalewicz, P.E.S. Wormer and A. van der Avoird, Phys. Rev. Lett., 84 (2000) 4072.

131 G.C. Groenenboom, P.E.S. Wormer, A. van der Avoird, E.M. Mas, R. Bukowski and K. Szalewicz, J. Chem. Phys., 113 (2000) 6702.

132 L.B. Braly, J.D. Cruzan, K. Liu, R.S. Fellers and R.J. Saykally, J. Chem. Phys., 112 (2000) 10293.

133 L.B. Braly, K. Liu, M.G. Brown, F.N. Keutsch, R.S. Fellers and R.J. Saykally, J. Chem. Phys., 112 (2000) 10314.

134 M.J. Smit, G.C. Groenenboom, P.E.S. Wormer, A. van der Avoird, R. Bukowski and K. Szalewicz, J. Phys. Chem. A, 105 (2001) 6212.

135 R. Bukowski, P. Jankowski, B. Jeziorski, M. Jeziorska, S.A. Kucharski, R. Moszynski, S. Rybak, K. Szalewicz, H. L. Williams, and P.E.S. Wormer, SAPT96: An *ab initio* Program for Many-Body Symmetry-Adapted Perturbation Theory Calculations of Intermolecular Interaction Energies, University of Delaware and University of Warsaw, 1996.

136 Gaussian 98, Revision A.5, is written by M.J. Frisch, G.W. Trucks, H.B. Schlegel, G.E. Scuseria, M. A. Robb, J.R. Cheeseman, V.G. Zakrzewski, J.A. Montgomery, Jr., R.E. Stratmann, J.C. Burant, S. Dapprich, J.M. Millam, A.D. Daniels, K.N. Kudin, M.C. Strain, O. Farkas, J. Tomasi, V. Barone, M. Cossi, R. Cammi, B. Mennucci, C. Pomelli, C. Adamo, S. Clifford, J. Ochterski, G.A. Petersson, P.Y. Ayala, Q. Cui, K. Morokuma, D.K. Malick, A.D. Rabuck, K. Raghavachari, J.B. Foresman, J. Cioslowski, J.V. Ortiz, B.B. Stefanov, G. Liu, A. Liashenko, P. Piskorz, I. Komaromi, R. Gomperts, R.L. Martin, D.J. Fox, T. Keith, M.A. Al-Laham, C.Y. Peng, A. Nanayakkara, C. Gonzalez, M. Challacombe, P.M.W. Gill, B. Johnson, W. Chen, M.W. Wong, J.L. Andres, C. Gonzalez, M. Head-Gordon, E.S. Replogle, and J.A. Pople. Gaussian, Inc., Pittsburgh PA, 1998.

137 MOLPRO is a package of *ab initio* programs written by H.-J. Werner and P.J. Knowles, with contributions from J. Almlöf, R.D. Amos, A. Berning, D.L. Cooper, M.J.O. Deegan, A.J. Dobbyn, F. Eckert, S.T. Elbert, C. Hampel, R. Lindh, A.W. Lloyd, W. Meyer, A. Nickla, K. Peterson, R. Pitzer, A.J. Stone, P.R. Taylor, M.E. Mura, P. Pulay, M. Schütz, H. Stoll and T. Thorsteinsson.

138 B. Jeziorski and W. Kolos, Molecular interactions, Vol. 3, Wiley, New York, 1982, pp. 1–46.

139 K. Szalewicz and B. Jeziorski, Mol. Phys., 38 (1979) 191.

140 A.J. Sadlej, J. Chem. Phys., 75 (1981) 320.

141 T.C. Caves and M. Karplus, J. Chem. Phys., 50 (1969) 3649.

142 H.C. Longuet-Higgins, Faraday Discuss. Chem. Soc., 40 (1965) 7.

143 E. Zaremba and W. Kohn, Phys. Rev. B, 13 (1976) 2270.

144 Y. Dmitriev and G. Peinel, Int. J. Quantum Chem., 19 (1981) 763.

145 R. McWeeny, Croat. Chem. Acta, 57 (1984) 865.

146 M. Jaszunski and R. McWeeny, Mol. Phys., 57 (1985) 1257.

147 B. Jeziorski and K. Szalewicz, Intermolecular interactions by perturbation theory, in: P. von Ragué Schleyer, N.L. Allinger, T. Clark, J. Gasteiger, P.A. Kollman, H.F. Schaefer, P.R. Schreiner (Eds.), Encyclopedia of computational chemistry, Vol. 2, Wiley, New York, 1998, pp. 1376–1398.

148 R. Moszynski, P.E.S. Wormer and A. van der Avoird, Symmetry adapted perturbation theory applied to the computation of intermolecular forces, in: P.R. Bunker, P. Jensen (Eds.), Computational molecular spectroscopy, Wiley, New York, 2000, pp. 69–108.

149 H.L. Williams, K. Szalewicz, B. Jeziorski, R. Moszynski and S. Rybak, J. Chem. Phys., 98 (1993) 1279.

150 R. Moszynski, B. Jeziorski, S. Rybak, K. Szalewicz and H.L. Williams, J. Chem. Phys., 100 (1994) 5080.

151 R. Moszynski, P.E.S. Wormer, B. Jeziorski and A. van der Avoird, J. Chem. Phys., 101 (1994) 2811.

152 R. Moszynski, F. de Weerd, G.C. Groenenboom and A. van der Avoird, Chem. Phys. Lett., 263 (1996) 107.

153 V.F. Lotrich, H.L. Williams, K. Szalewicz, B. Jeziorski, R. Moszynski, P.E.S. Wormer and A. van der Avoird, J. Chem. Phys., 103 (1995) 6076.

154 G. Murdachaew, A.J. Misquitta, R. Bukowski and K. Szalewicz, J. Chem. Phys., 114 (2001) 764.

155 R. Bukowski, J. Sadlej, B. Jeziorski, P. Jankowski, K. Szalewicz, S.A. Kucharski, H.L. Williams and B.M. Rice, J. Chem. Phys., 110 (1999) 3785.
156 E.M. Mas, K. Szalewicz, R. Bukowski and B. Jeziorski, J. Chem. Phys., 107 (1997) 4207.
157 R. Moszynski, T. Korona, P.E.S. Wormer and A. van der Avoird, J. Chem. Phys., 103 (1995) 321.
158 T.G.A. Heijmen, R. Moszynski, P.E.S. Wormer and A. van der Avoird, J. Chem. Phys., 107 (1997) 9921.
159 R. Moszynski, T. Korona, P.E.S. Wormer and A. van der Avoird, J. Phys. Chem. A, 101 (1997) 4690.
160 C.E. Chuaqui, R.J.L. Roy and A.R.W. McKellar, J. Chem. Phys., 101 (1994) 39.
161 M.C. Chan and A.R.W. McKellar, J. Chem. Phys., 105 (1996) 7910.
162 S. Antonova, A. Lin, A.P. Tsakotellis and G.C. McBane, J. Chem. Phys., 110 (1999) 2384.
163 S. Antonova, A. Lin, A.P. Tsakotellis and G.C. McBane, J. Chem. Phys., 110 (1999) 11742.
164 T.C. Smith, D.A. Hostutler, G.D. Hager, M.C. Heaven and G.C. McBane, J. Chem. Phys., 120 (2004) 2285.
165 D.A. Hostutler, T.C. Smith, G.D. Hager, G.C. McBane and M.C. Heaven, J. Chem. Phys., 120 (2004) 7483.
166 R. Moszynski, P.E.S. Wormer and A. van der Avoird, J. Chem. Phys., 102 (1995) 8385.
167 R.J. Bemish, L. Oudejans, R.E. Miller, R. Moszynski, T.G.A. Heijmen, T. Korona, P.E.S. Wormer and A. van der Avoird, J. Chem. Phys., 109 (1998) 8968.
168 T. Korona, R. Moszynski, F. Thibault, J.-M. Launay, B. Bussery-Honvault, J. Boissoles and P.E.S. Wormer, J. Chem. Phys., 115 (2001) 3074.
169 R. Moszynski, T. Korona, T.G.A. Heijmen, P.E.S. Wormer, A. van der Avoird and B. Schramm, Pol. J. Chem., 72 (1998) 1479.
170 R. Moszynski, P.E.S. Wormer, B. Jeziorski and A. van der Avoird, J. Chem. Phys., 103 (1995) 8058.
171 R. Moszynski, P.E.S. Wormer, B. Jeziorski and A. van der Avoird, J. Chem. Phys., 107 (1997) E672.
172 V.F. Lotrich and K. Szalewicz, J. Chem. Phys., 106 (1997) 9668.
173 V.F. Lotrich and K. Szalewicz, J. Chem. Phys., 112 (2000) 112.
174 P.E.S. Wormer, R. Moszynski and A. van der Avoird, J. Chem. Phys., 112 (2000) 3159.
175 T. Helgaker, P. Jørgensen and J. Olsen, Molecular electronic structure theory, Wiley, Chichester, 2000.
176 R.J. Bartlett and J.F. Stanton, in: K.B. Lipkowitz, D.B. Boyd (Eds.), Applications of post-Hartree–Fock methods: A tutorial, Reviews in computational chemistry, Vol. V, VCH Publishers, New York, 1994, pp. 65–169.
177 K. Raghavachari, G.W. Trucks, J.A. Pople and M. Head-Gordon, Chem. Phys. Lett., 157 (1989) 479.
178 M.M. Szczęśniak, G. Chałasiński, S.M. Cybulski and P. Cieplak, J. Chem. Phys., 98 (1993) 3078.
179 P.J. Marshall, M.M. Szczęśniak, J. Sadlej, G. Chałasiński, M.A. ter Horst and C.J. Jameson, J. Chem. Phys., 104 (1996) 6569.
180 F.M. Tao and W. Klemperer, J. Chem. Phys., 97 (1992) 440.
181 D.E. Woon, K.A. Petersen and T.H. Dunning, J. Chem. Phys., 109 (1998) 2233.
182 D.E. Woon, T.H. Dunning and K.A. Peterson, J. Chem. Phys., 104 (1996) 5883.
183 M. Rode, J. Sadlej, R. Moszynski, P.E.S. Wormer and A. van der Avoird, Chem. Phys. Lett., 314 (1999) 326.
184 S.M. Cybulski and R.R. Toczyłowski, J. Chem. Phys., 111 (1999) 10520.
185 T.P. Haley and S.M. Cybulski, J. Chem. Phys., 119 (2003) 5487.
186 R.R. Toczyłowski, F. Doloresco and S.M. Cybulski, J. Chem. Phys., 114 (2001) 851.
187 R.R. Toczyłowski and S.M. Cybulski, J. Chem. Phys., 112 (2000) 4604.
188 S. Lee, J.S. Chung, P.M. Felker, J. López Cacheiro, B. Fernández, T.B. Pedersen and H. Koch, J. Chem. Phys., 119 (2003) 12956.
189 H. Koch, B. Fernández and J. Makarewicz, J. Chem. Phys., 111 (1999) 198.
190 J.L. Cagide Fajín, J. López Cacheiro and B. Fernández, J. Chem. Phys., 121 (2004) 4599.
191 E. Engel, A. Höck and R.M. Dreizler, Phys. Rev. A, 61 (2000) 32502.
192 A.J. Misquitta, B. Jeziorski and K. Szalewicz, Phys. Rev. Lett., 91 (2003) 33201.
193 A. Heßelmann and G. Jansen, Chem. Phys. Lett., 367 (2003) 778.
194 V.P. Osinga, S.J.A. van Gisbergen, J.G. Snijders and E.J. Baerends, J. Chem. Phys., 106 (1997) 5091.
195 S.J.A. van Gisbergen, V.P. Osinga, O.V. Gritsenko, R. van Leeuwen, J.G. Snijders and E.J. Baerends, J. Chem. Phys., 105 (1996) 3142.

196 S.J.A. van Gisbergen, J.G. Snijders and E.J. Baerends, J. Chem. Phys., 103 (1995) 9347.
197 M. Grüning, O.V. Gritsenko, S.J.A. van Gisbergen and E.J. Baerends, J. Chem. Phys., 114 (2001) 652.
198 J.P. Perdew, K. Burke and M. Ernzerhof, Phys. Rev. Lett., 77 (1996) 3865.
199 A.J. Misquitta and K. Szalewicz, Chem. Phys. Lett., 357 (2002) 301.
200 G.W.M. Vissers, A. Heßelmann, G. Jansen, P.E.S. Wormer and A. van der Avoird, J. Chem. Phys., 122 (2004) 054306.
201 T.B. Pedersen, B. Fernández and H. Koch, Chem. Phys. Lett., 334 (2001) 419.
202 M. Rode, J. Sadlej, R. Moszynski, P.E.S. Wormer and A. van der Avoird, Chem. Phys. Lett., 334 (2001) 424.
203 G.W.M. Vissers, P.E.S. Wormer and A. van der Avoird, Phys. Chem. Chem. Phys., 5 (2003) 4767.
204 D.A. Roth, M. Hepp, I. Pak and G. Winnewisser, Chem. Phys. Lett., 298 (1998) 381.
205 D.A. Roth, L.A. Surin, B.S. Dumesh, G. Winnewisser and I. Pak, J. Chem. Phys., 113 (2000) 3034.
206 M.D. Brookes and A.R.W. McKellar, J. Chem. Phys., 111 (1999) 7321.
207 K.A. Walker, C. Xia and A.R.W. McKellar, J. Chem. Phys., 113 (2000) 6618.
208 A.R.W. McKellar, J. Chem. Phys., 115 (2001) 3571.
209 K.A. Walker and A.R.W. McKellar, J. Mol. Spectrosc., 208 (2001) 209.
210 J. Tang, A.R.W. McKellar, L.A. Surin, D.N. Fourzikov, B.S. Dumesh and G. Winnewisser, J. Mol. Spectrosc., 214 (2002) 87.
211 B.O. Roos, P.R. Taylor and P.E.M. Siegbahn, Chem. Phys., 48 (1980) 157.
212 B.O. Roos, P. Linse, P.E.M. Siegbahn and M.R.A. Blomberg, Chem. Phys., 66 (1982) 197.
213 K. Wolinski, H.L. Sellers and P. Pulay, Chem. Phys. Lett., 140 (1987) 225.
214 K. Wolinski and P. Pulay, J. Chem. Phys., 90 (1989) 3647.
215 K. Andersson, P.-A. Malmqvist, B.O. Roos, A.J. Sadlej and K. Wolinski, J. Phys. Chem., 94 (1990) 5483.
216 K. Andersson, P.-A. Malmqvist and B.O. Roos, J. Chem. Phys., 96 (1992) 1218.
217 H.-J. Werner, Mol. Phys., 89 (1996) 645.
218 S.R. Langhoff and E.R. Davidson, Int. J. Quantum Chem., 8 (1974) 61.
219 P.J. Knowles, C. Hampel and H.-J. Werner, J. Chem. Phys., 99, 7 (1993) 5219.
220 P.J. Knowles, C. Hampel and H.-J. Werner, J. Chem. Phys., 112 (2000) E3106.
221 J.D. Watts, J. Gauss and R.J. Bartlett, J. Chem. Phys., 98 (1993) 8718.
222 F.T. Smith, Phys. Rev., 179 (1969) 111.
223 W. Lichten, Phys. Rev., 131 (1963) 229.
224 H.-J. Werner, B. Follmeg and M.H. Alexander, J. Chem. Phys., 89 (1988) 3139.
225 A.J. Dobbyn and P.J. Knowles, Mol. Phys., 91 (1997) 1107.
226 J. Kłos, G. Chałasiński, M.T. Berry, R. Bukowski and S.M. Cybulski, J. Chem. Phys., 112 (2000) 2195.
227 W.B. Zeimen, G.C. Groenenboom and A. van der Avoird, J. Chem. Phys., 119 (2003) 131.
228 W.B. Zeimen, G.C. Groenenboom and A. van der Avoird, J. Chem. Phys., 119 (2003) 141.
229 G.C. Groenenboom and N. Balakrishnan, J. Chem. Phys., 118 (2003) 7380.
230 H. Cybulski, R.V. Krems, H.R. Sadeghpour, A. Dalgarno, J. Kłos, G.C. Groenenboom, A. van der Avoird, D. Zgid and G. Chałasiński, J. Chem. Phys., 122 (2005) 094307.
231 N. Balakrishnan, G.C. Groenenboom, R.V. Krems and A. Dalgarno, J. Chem. Phys., 118 (2003) 7386.
232 R.V. Krems, H.R. Sadeghpour, A. Dalgarno, D. Zgid, J. Kłos and G. Chałasiński, Phys. Rev. A, 68 (2003) 51401.
233 J. Kłos, G. Chałasiński, M.M. Szczęśniak and H.-J. Werner, J. Chem. Phys., 115 (2001) 3085.
234 W.B. Zeimen, J.A. Kłos, G.C. Groenenboom and A. van der Avoird, J. Phys. Chem. A, 107 (2003) 5110.
235 W.B. Zeimen, J.A. Kłos, G.C. Groenenboom and A. van der Avoird, J. Chem. Phys., 118 (2003) 7340.
236 D.M. Brink and G.R. Satchler, Angular momentum, 3rd edn., Clarendon, Oxford, 1993.

Theory and Applications of Computational Chemistry: The First Forty Years
Edited by C. Dykstra et al.

CHAPTER 38

Applied density functional theory and the deMon codes 1964–2004

D.R. Salahub[1], A. Goursot[2], J. Weber[3], A.M. Köster[4] and A. Vela[4]

[1] *University of Calgary, 2500 University Drive NW, Calgary, Alberta, Canada, T2N 1N4*
[2] *UMR 5618 CNRS, Ecole Nationale Supérieure de Chimie, 8 rue de l'Ecole Normale, 34296 Montpellier, Cédex 5, France*
[3] *Département de Chimie Physique, Université de Genève, Sciences II, 30 quai Ernest Ansermet, 1211 Genève 4, Switzerland*
[4] *Departamento de Química, Cinvestav, Avenida Instituto Politécnico Nacional 2508, A.P. 14-740 México DF 07000, México*

Abstract

Advances in density functional theory and its applications over the past four decades are reviewed from the perspective of developers of the methodology and codes embodied in the deMon software.

38.1 INTRODUCTION. FROM THE 1920s TO THE 1960s

One can consider that applied DFT goes back to the early 1930s when Dirac [1] and Wigner and Seitz [2,3] treated the problem of the exchange potential in the Thomas–Fermi atom [4–7], arriving at the expression for local exchange, proportional to the 1/3 power of the density. A few years later, Slater [8] proposed a model of spherically symmetric atomic potentials embedded in a region of constant potential expanded in plane waves, which was later called the augmented plane wave (APW) method for the calculation of energy bands in solids.

After the long gap generated by World War II, a new route to applications was opened by Slater's 1951 paper [9], which introduced the idea of approximating the complicated non-local Hartree–Fock exchange operator by an average local potential. Slater's derivation, which represents a generalization and extension of Wigner and Seitz, defined the properties of the exchange charge density associated with an electron and used the free-electron-gas model to approximate the exchange potential in terms of the local electron density.

The magnitude of Slater's exchange term in the Hamiltonian was questioned three years later by Gaspar [10] who, through a different derivation, obtained the same $\rho^{1/3}$ form, but with a factor of 2/3. This new value was confirmed later by Kohn and Sham. The factor under debate, then named α, was used later on as an atom-dependent parameter [11] in the so-called Xα methodologies. The first among these was derived from the original muffin-tin approximation of Slater [8] and generalized by Johnson [12, 13] to treat molecular clusters. The multiple scattering-Xα (MS-Xα) method, also known as the Xα-Scattered Wave (Xα-SW) method is thus based on the local Hartree–Fock–Slater (HFS) approximation, with eigensolutions obtained using the multiple-scattering approximation. This was an important step as it represented one of the first attempts to adapt a method of theoretical solid-state physics to the study of molecular systems.

From a more technical point of view the casting of the Hartree–Fock equations into eigenvalue equations, the so-called Roothan–Hall equations [14,15], was of fundamental importance for the future development of quantum chemical codes. The well-defined structure of the mathematical problem allowed the use of early computational facilities. Moreover, quantum chemistry codes could benefit directly from the fast growing experience in the implementation of linear algebra methods. Boys introduced Gaussian type orbital (GTO) functions for the calculation of molecular integrals [16]. With these functions a general strategy for the calculation of molecular integrals could be developed. In terms of GTO functions, all integrals for the Hartree–Fock energy calculation can either be solved analytically or reduced to the incomplete gamma function. Thus, with the introduction of GTOs the task of solving complicated multi-center molecular integrals was reformulated into the problem of solving huge numbers of one-center Gaussian integrals. This represented a considerable simplification and opened the door to more systematic approaches. Soon it was realized that recurrence relations can be of great help to treat the huge number of integrals efficiently. However, many years were still to come before the first general integral algorithms for GTOs appeared on the scene. Over the years the linear combination of Gaussian type orbitals (LCGTO) approximation became a standard for *ab initio* Hartree–Fock methods.

On the other hand, modern density functional theory (DFT) started with the famous 1964 paper by Hohenberg and Kohn [17], followed by the method of implementation by Kohn and Sham (KS) [18]. With these contributions, a new conceptual way of approaching the many-body problem was opened. On the basis of a formally rigorous theory, the electron density of any system was recognized as containing all the necessary information to describe its ground state. In addition, the idea of mapping the exact density of an interacting system to that of a non-interacting model, which is more easily solved, provided an alternative to the conventional wave function approach. At this moment, the so-called statistical methods like Thomas–Fermi, Thomas–Fermi–Dirac, etc. and the KS method merged to open the new avenue of DFT. Thus, 1964 marks the introduction of one of the most important concepts in KS theory: the exchange-correlation energy functional, an unknown and universal quantity which contains all the information about an electronic system. It is also worth mentioning that immediately after the presentation of the HK theorems, several extensions appeared in the literature, like the finite temperature extension of Mermin [19], that opened the door to the DFT description of inhomogeneous fluids in classical statistical mechanics [20]. In relation to the basis sets mentioned in the previous paragraph, for first-principles DFT methods, the form of the exchange-correlation potential resulted in integrals that could not be solved analytically. Therefore, a much larger variety of basis functions was introduced in DFT methods than in Hartree–Fock methods.

Thus, the 1960s was a very fruitful decade, with new concepts developed through the HFS and KS approaches. Even if the first KS calculations, within the local density approximation (LDA), appeared in this decade [21], it is only in the 1970s that the systematic development of local correlation functionals as well as applications to a growing number and variety of systems appeared.

38.2 THE 1970s

Even though both Hohenberg–Kohn and Kohn–Sham papers have been subsequently recognized as extremely important for Chemistry, that recognition came late in the community of theoretical chemists. Meanwhile, the MS-Xα method received much more attention. For example, in 1970, Johnson and Smith addressed polyatomic molecules such as perchlorate and sulphate ions for the first time [13]. A landmark application of MS-Xα was the first investigation by Johnson and Smith of the electronic structure of a coordination compound, namely the permanganate ion [22]. The interest in the MS-Xα method for calculating the electronic structure of transition metal complexes increased rapidly and realistic results were soon obtained [23–25].

This was the starting point for a large range of MS-Xα applications, including valence band and ESCA photoemission spectra of molecules [26–28], photoelectron spectra [29–32], chemisorption and catalysis [24,33,34], geometrical and electronic structures of metal clusters [35–37], metal dimers [38–40], inorganic species [41–46], and the inclusion of relativistic corrections [47,48] for heavy elements. Further ambitious investigations were devoted to the study of biosystems such as the ferrodoxin active site [49] or porphine systems [50]. Finally, in the late 1970s Karplus and Case developed a general formalism and performed the first one-electron property calculations using the MS-Xα method [51]. Their contributions also constituted a basis for numerous further publications of molecular properties such as hyperfine tensors [52].

Despite its great success in describing one-electron properties of molecules and solids, the MS-Xα method was unreliable in the description of geometries, mainly due to the muffin-tin approximation. The main contribution to further progress was the introduction of LCAO functions, based on GTOs [53,54] or Slater type orbitals (STOs) [55]. Since then, methodological developments have been possible, allowing the computation of properties for large systems both Xα (HFS) and, later, DFT-based methods (for a review of the use of GTOs with Xα or DFT models, see Ref. [56]). Charge density fitting, introduced first by Baerends et al. [55] in the discrete variational method (DVM) [57], allows one to reduce the N^4 problem to N^3, without losing accuracy. Sambe and Felton [54] proposed to fit the exchange potential also using an entire set of auxiliary functions, in addition to the charge density fitting. The contribution made by Dunlap et al. [58] in improving the fitting procedure yielded more accurate total energies using finite basis sets. In this approach, which has been used in all versions of the DFT program deMon, which has been developed in our groups, the self-interaction error of the Coulomb energy from the density and an auxiliary function density is variationally minimized. The variational nature of the fit ensures that only second order errors enter the energy expression. Moreover, the approximate energy expression remains variational. Therefore, analytic derivatives can be accurately calculated within this approach. Originally introduced within the Xα method, this approximation was taken over by many LCGTO–DFT implementations. Many years later, the variational fitting was recast into the so-called resolution of the identity by Almlöf and co-workers [59]. Based on this work the variational fitting has today entered the Hartree–Fock-based methods, too. It is still a very active research field. These technical aspects originally developed in LCAO-Xα methods have been of great benefit for the later development of DFT codes.

It is interesting to note that, for many years, Xα was presented as an independent self-contained method, and it is only in 1977 that the review of Connolly [26] proposed that it should be viewed as an approximation to the 'exact' DFT.

In this decade, the formal development of DFT was also an active area of research. One relevant work was the spin DFT extension of von Barth and Hedin, who also presented a local functional [60]. An important contribution to chemistry in this decade came from the Gordon and Kim approach to treat molecular interactions [61]. It was surprising, as well as encouraging, that this non-variational theory where the kinetic and exchange-correlation energy contributions were described by local density functionals provided results that were very acceptable and computationally affordable. It was almost immediately recognized that the HK universal energy functional was not free of formal difficulties. The most significant was the v-representability problem that was solved by Levy with his constrained-search approach [62]. As it was mentioned before, the roots of DFT *à la* HK are in condensed matter physics; thus, it is not surprising that many of the important contributions in the early 1970s came from this branch of Physics, such as the presentation by Gunnarsson and Lundqvist of the adiabatic connection method for obtaining the exchange-correlation energy functional of the KS method [63]. Parallel to these developments in Solid-State Physics, the problem of evaluating the correlation contribution to the energy was a very active field of research in Theoretical Chemistry. Among the many works in this topic, the modelling of the second order radial distribution by Colle and Salvetti deserves a special mention [64]. This work inspired several researchers to propose new exchange-correlation energy functionals. The wave-vector analysis of the exchange-correlation hole paved the way to the development of several functionals [65]. Several local correlation functionals were proposed in the framework of the KS equations for applications to solids [66,67]. Initiated by a calculation on H_2 in 1976 [68], there has been a series of impressive Kohn–Sham–type density functional calculations on molecules performed by Gunnarsson, Harris and Jones [69–75]. These results showed that the KS-LDA method was able to describe molecular bonding reasonably well, in contrast to the well-known non-bonding effects in the Thomas–Fermi theory. The following decade brought improved descriptions of the exchange-correlation energy functional [76,77] and, consequently, better results.

It is also pertinent to look a little bit outside the main stream of DFT development and to review other developments in this period that were important for the development of deMon. At the beginning of the 1970s non-linear optimization methods were already thoroughly investigated in applied mathematics. The broader availability of microcomputers inspired interest in numerical studies of such methods. As a result, iterative optimization algorithms with considerably improved numerical stability were developed. For future versions of deMon, and most other quantum chemistry programs, the development of numerically stable quasi-Newton methods [78] was very important. The most popular second derivative update, named after its authors BFGS [79–83], was published in 1970. Most modern geometry optimizers in quantum chemical programs rely on the BFGS update in order to avoid the explicit calculation of second derivatives for the geometry optimization. At the same time, and most likely motivated by similar reasons, Lebedev developed a two-dimensional Gauss-type quadrature scheme (Gauss–Markov) for the unit sphere [84–88]. These grids exactly integrate real spherical harmonics up to a maximum degree that is used to characterize them. Today, Lebedev grids, as they are named now, are used in most DFT programs for the numerical integration of the exchange-correlation energy and potential.

Also in the same time period, one sees the development of the first systematic molecular integral algorithms for LCGTO approximations [89,90]. For the future deMon development the introduction of Hermite–Gaussian functions as basis functions [91,92] is certainly important, too. These early works inspired us much later to use atom-centered Hermite–Gaussian functions for the expansion of the auxiliary density. This results in short and, therefore, very efficient integral recurrence relations for the three-center Coulomb integrals [93]. In the new version of deMon, Hermite–Gaussian functions are also used for the expansion of the Cartesian Gaussian orbital basis. This part of our integral algorithm is closely related to the original formulation of McMurchie and Davidson [90]. However, a much earlier transformation to Cartesian Gaussian functions is used. A good overview of the early systematic integral algorithms for Gaussian type functions is provided by the 1983 review by Saunders [94].

38.3 THE 1980s

At the beginning of this decade, the Xα methods, i.e. without explicit correlation, were still used in Chemistry. MS-Xα was used for the study of large systems, in particular for magnetic properties [95–101], and LCAO-Xα (also called HFS) for electronic structure investigations of small metallic systems [102–106]. A big improvement was reached when the correlation energies of the spin-paired and spin-polarized uniform electron gas, calculated accurately by Ceperley and Alder [76], were incorporated in the LDA correlation functionals.

By the early 1980s, the first reviews on DFT appeared [107–110], underlining its great potential for applications to atoms and molecules. Also the first publications about analytic gradients for the Kohn–Sham method appeared in the literature [111,112]. This clearly moved the Kohn–Sham method into the computational chemistry arena. Still it took almost a decade more before stable geometry optimization with analytic gradients [113] was available within the framework of the Kohn–Sham method.

During this decade, electronic structure calculations of metal dimers [114–117], clusters [118–123] and organometallics [124–128] led perhaps to the largest success and progress beyond the traditional *ab initio* methods, over a broad range of molecular studies. For the first time, the delicate balance between exchange and correlation effects was proved to be responsible for previous failures in the quantitative treatments of metal–metal bonds [114]. From this period, incorporation of explicit correlation in the KS equations became more and more the standard. Despite the overestimation of binding energies, LCAO-LSD calculations were able to provide good geometries, even for metal–metal bonds [115], reasonable ionization potentials [129] and UV spectra [130–132].

Hence, LCAO-LSD studies followed naturally the pioneering LCAO-Xα results for a very large variety of molecular applications, including sufficiently large metallic systems and taking into account, when necessary, relativistic and/or spin orbit corrections [133, 134]. Incorporation of model core potentials in LCGTO–LSD [135–137] allowed larger metallic systems to be handled (including eventually relativistic effects in an approximate, but very inexpensive, way) and thus further work was possible on chemisorption models on metallic surfaces [138–141].

The 1980s also witnessed a continuous development of new exchange and correlation functionals, incorporating the effects of the non-locality of the density. Indeed, quite soon, it was recognized that the local exchange energy term was introducing a substantial error in the total energy. The so-called generalized gradient approximation (GGA) model of Perdew and Wang [142] initiated a long series of gradient corrected functionals, which led to much more accurate energetic properties. The introduction in this later work of the enhancement function that measures the deviations from the electron gas behavior deserves special mention. Plots of these enhancement functions with respect to the exchange dimensionless scale length, allowed a classification of the exchange energy functionals into three categories. First, those that did not fulfill the Lieb–Oxford bound, like Becke [143], second, those that approached this bound asymptotically, like the family later introduced by Perdew et al. [144], and finally, those that either go to zero or to a value smaller than the Lieb–Oxford bound for large density gradients. To this latter class belongs one of the most commonly used exchange functional in deMon, Perdew86 [142]. The implementation in deMon of this energy functionals uses a cut off for large gradients that makes this functional different from the original one [145].

At the end of the 1970s and the beginning of the 1980s attempts to go beyond the LDA were made through the gradient expansion approximation (GEA). The divergence of the functional derivative of the exchange-correlation energy functional was a serious problem. It was necessary to wait until the GGA made its appearance to have an exchange-correlation functional that was really incorporating the inhomogeneities of the electron density. The real-space analysis of the exchange-correlation hole [146] provided working functionals that satisfy many known conditions. Along this line it is important to note that the functional developed by Perdew et al. [147], that is commonly known as PW91, is the GGA functional that satisfies the most conditions. A very important step in our understanding of the exchange-correlation contribution to the energy was made with the series of works published mainly by the New Orleans school, led by Levy and Perdew. In a series of very important papers, these authors presented rigorous scaling and virial conditions that the exact exchange and correlation energy functionals have to satisfy [148–151]. The discovery of these conditions was used in the development of new functionals.

The view at the end of the 1980s was very optimistic. The formation energies reported by Becke with his functional and his basis-set-free program, NuMol, showed that the GGAs were in reality capable of providing results with chemical accuracy. It is commonly said that it was these results that turned the heads of many theoretical chemists to DFT. In the mean time, the North Carolina team, led by Robert Parr made many important contributions to the application of DFT to chemical problems. He focused his attention on using DFT to justify and explain the origin of many chemical concepts, opening what is now sometimes misleadingly called conceptual DFT. Parr's group also contributed to the practical use of DFT in chemistry with the development of the correlation functional known as LYP that used an analysis of the radial distribution function together with the local thermodynamics approach to model the pair distribution function. To the surprise of some people, and the delight of others, the combination of the GGA from Becke for the exchange and LYP for the correlation, that produces the very well known BLYP method, provided atomization energies that were in very good agreement with the experimental values. The first implementations of the GGA were not self-consistent. The SCF was done within the local approximation and then, perturbatively, the gradient corrections were incorporated to the total energy. In the early 1990s full self-consistency was implemented in several codes, deMon amongst them [145]. The results obtained with the GGAs were certainly encouraging. The achievement of chemical accuracy together with the low computational cost compared with conventional wave function methods that produce the same thermodynamic accuracy was a strong motivation to establish the limits of applicability of these methodologies. Actually, quantitative comparisons will appear in the next decades, after Becke's proposals of the 'half-and-half'[152] and the B3LYP hybrid schemes [153].

Finally, in 1985, the technique introduced by Car and Parrinello [154] to minimize simultaneously the electronic and nuclear coordinates, together with the use of a plane wave basis, has been at the origin of the explosion of applications in material sciences in the next decades.

Important developments for the LCGTO–DFT methods using auxiliary functions for the fitting of the Coulomb potential during the 1980s were the introduction of (Cartesian) auxiliary function sets with shared exponents [155,136]. With this technique, the computational effort for the calculation of the three-center Coulomb integrals was considerably reduced [156]. In the now increasing number of calculations with LCGTO–DFT type methods using auxiliary functions, self-consistent field (SCF) convergence problems frequently appeared [157]. It took some time before it was realized that the variational fitting of the Coulomb potential not only reduces the scaling of the integral calculation but also influences the SCF procedure.

The 1980s also saw an impressive revival of integral algorithm development for the LCGTO approximation. It was initialized by the work of Obara and Saika in 1986 [158]. In fact, the first deMon and DGAUSS versions had explicit integral routines for different shell combinations based on the Obara–Saika algorithm. Even though these routines were quite fast, they were later substituted by recurrence relations in order to allow higher angular momentum basis and auxiliary functions. The early deMon versions were restricted to d type orbitals and auxiliary functions only.

Another important technical contribution to the development of LCGTO–DFT programs was the introduction of the so-called fuzzy Voronoi polyhedra for the calculation of atomic weights in the numerical integration by Becke [159]. Due to its complicated form, the exchange-correlation potential integrals cannot be solved analytically. Therefore, numerical integration schemes are necessary. This was a large technical drawback for early LCGTO–DFT implementations. The numerical integration scheme of Becke solved this problem in a way that was easy to implement. Based on this scheme, very efficient numerical integrators were developed within a decade. Today, the numerical integration of the exchange-correlation potential is a minor problem in LCGTO–DFT codes. Nevertheless, the accuracy of this integration is still a matter of discussion [160,161].

An important technical development for the geometry optimization and transition state search was the investigation of the Levenberg–Marquard algorithm [162,163] by Banerjee et al. [164]. In this work the authors proposed theoretically well-founded step size selections for quasi-Newton methods, a problem particularly important for the transition state search. The algorithm was recast by Baker [165] substituting the quadratic Taylor series expansion by a Pade expansion. The resulting algorithm has entered the literature as (partitioned) rational function optimization (P)-RFO. It is the basic algorithm for the step selection in the geometry optimization and transition state search in the current version of deMon. It has proven very stable, even for large scale optimizations involving hundreds of atoms.

The intense efforts of the DFT community during the 1980s, both in numerical and methodological improvements allowed the development of more sophisticated programs, which, in some cases, became available in the early 1990s.

38.4 THE 1990s

The first DFT codes appearing in 1990 were DMOL [166,167] and deMon [168,169], followed by DGAUSS [170–172] in 1991, ADF [173,174] and Gaussian [175] in 1992. The deMon code was originally published as a new LCGTO–MCP code, based on the same solution of the KS equations as in previous work [54,56], incorporating MCPs, [135–137], as well as a new algorithm for geometry optimization. This algorithm, called hybrid, adopted the Car–Parrinello approach of minimizing simultaneously the orbitals and the nuclear coordinates, but adding a few SCF iterations at each geometry to keep the system close to the Born–Oppenheimer surface. It turned out that this approach was not competitive to the traditional quasi-Newton optimizer that was also included in deMon. As already mentioned, the first version of deMon had explicit integral routines for the three-center Coulomb integrals. The geometry optimization worked already with analytic gradients introduced by Fournier et al. [113]. Local and gradient corrected functionals were available. Shortly after its appearance, the original deMon code was substantially modified for commercialization by BIOSYM Technologies. The beta release of this version appeared in 1993. It inspired the deMon-KS1 [176] series of programs developed in Montreal until 1997. In combination with DFT-optimized basis sets and auxiliary functions [177] this deMon version was used for a large number of applications on small systems with up to 20 atoms. It also served as a basis for the implementation of property calculations within the Kohn–Sham method, including the calculation of NMR shielding tensors and spin–spin coupling constants [178–184], EPR parameters [185–187], nuclear quadrupole coupling constants [188], simulation of photoelectron [189–192], IR, and Raman spectra [193–195], and the calculation of molecular polarizabilities [196,197]. For visualization, interfaces to MOLDEN [198], MOLEKEL [199], and VU [200] were implemented. The topological analysis of molecular fields, primarily the molecular electrostatic potential and the density, was also realized with this deMon version [201].

At the same time, the original deMon version was further developed in Montpellier and in Stockholm. Larger independent developments included the implementation of the orbital symmetry analysis and the calculation of X-ray spectra [202,203]. Both of these developments were first independent from the Montreal version. In 1997, they merged to the deMon-KS3 [176,202–204] series of programs. The improved implementation in deMon-KS3 allowed calculations of systems with roughly 50 atoms. This version allowed, already in the early 1990s, the study of a broad range of properties. Applications of DFT in various domains of chemistry were thus performed, with particular interests in NMR and EPR properties of bioorganic molecules [205,206], materials [183], transition metal compounds [207–210], transition metal clusters, naked or interacting with small molecules [138,139,141,211–227], Mössbauer [228], and ZEKE [229] spectroscopies, optical properties with TD–DFT [192,230–232], and reactivity [233,234].

In the exchange-correlation functional development arena in this decade it is worth mentioning the detailed analysis of the relation between the exchange functional and the kinetic energy functional that was presented by Roy Gordon [235,236].

Since the 1970s it was very well known that the LSDA or GEA energy functionals did not contain the van der Waals contribution. The underbinding of the local and gradient corrected energy functionals when applied to the description of rare gas interactions was well known. But, their behavior in other weak bonding situations was unknown. Thus, establishing the limits of applicability was a very important issue. In the first half of the 1990s deMon was used to test these functionals in the description of hydrogen bonding [237] and charge transfer complexes [238,239]. In the first case, two intermolecular and two intramolecular hydrogen-bonded systems were tested, including the basis set superposition error using the counterpoise method. LDA was found to be seriously deficient, and the non-local corrections provided an encouraging improvement. For the charge transfer complexes the set of electron donor–acceptor systems, formed from ethylene or ammonia interacting with a halogen molecule ($C_2H_4 \ldots X_2$, $NH_3 \ldots X_2$; $X = F$, Cl, Br, and I) were tested. Similar to the hydrogen-bonded systems, it was found that the LDA provides a strong overestimation of the intermolecular interaction. The GGA moved the results in the right direction but not nearly enough; large errors remain. Hybrid functionals were tested, and it was found that the parameters related to the intermolecular interaction for the so-called half-and-half potential are in very good agreement with those obtained through second-order Moller–Plesset calculations and with available experimental data. Interestingly, the widely used and well-known three-parameter B3LYP functional does not perform well; the hybrid methods are not a panacea.

These contributions showed that DFT has to be used with caution when the system under study has important weak inter or intra molecular bonds. During this decade it became clear that it was necessary to go beyond the GGA. In deMon this was done with the LAP [240–244] and τ functionals [245] in which the correlation functional was derived from an explicit integration over the coupling constant in the adiabatic connection approach.

Independent of the deMon development, the AllChem project [246] was started in Hannover in 1995. The main aims of this project were the development of recursive integral routines for the three-center Coulomb integrals [156,93], a stable and efficient numerical integrator for the exchange-correlation potential [247] and an improvement in the SCF convergence behavior of LCGTO–DFT programs using the variational approximation of the Coulomb potential. The recursive integral algorithm developed for AllChem was based on the PRISM algorithm of Pople and co-workers [248,249] and took into account the special structure of the auxiliary function sets used in deMon. For the calculation of the incomplete gamma function an algorithm that ensures close to machine precision (16 decimals) was implemented [250]. The main focus in the development of AllChem was numerical stability in order to avoid noise in the SCF procedure. This was believed to be one of the main reasons for the SCF convergence problems of previous deMon versions. With the same aim an adaptive numerical integrator for the exchange-correlation potential was developed. It was motivated by the work of Perez–Jorda et al. [251]. However, the radial integration was excluded from the adaptive grid construction. This was the key to success in obtaining an efficient and reliable adaptive numerical integrator. Very recently, the adaptive radial integration has again attracted attention [252,253]. The abscissas and weights of the Lebedev grids were recalculated using quadruple precision [250]. The stability of this integrator was extensively tested in numerically sensitive hyperpolarizability calculations [254,255]. In this calculation, it was shown that the adaptive grid automatically adapts to basis sets. For the generated number of grid points this effect can be large. The adaptive grid has also been used for the geometry optimization and frequency analysis. For these calculations, weight derivatives [256] are implemented in deMon. However, our experience has shown that these derivatives are only important if very small adaptive grids are generated. This is different from recent studies with fixed grids [161]. Because very small adaptive grids are not reliable, weight derivatives are not used in the standard setting for the numerical integration in deMon. Nevertheless, they can be activated by a grid keyword option.

Despite these efforts to improve the numerical stability of the LCGTO–DFT implementation, only little improvement was seen in the SCF convergence behavior. A major breakthrough for the solution of this problem was the recasting of the SCF procedure into a MinMax problem if the variational fitting of the Coulomb potential is used [257]. Based on this approach efficient convergence accelerators were developed and implemented, first in AllChem and later in new deMon versions. To improve the SCF convergence, a GTO-based DFT tight-binding approach [258] has been implemented in deMon for the generation of start densities. The advantage of this start density generator is that it adapts to the basis set and does not involve any parameterization. Therefore, it can be used for all elements of the periodic table. Very recently, it has been extended for the use of effective core potentials.

38.5 THE 2000s

In March 2000, the first deMon Developers meeting was held in Ottawa. At this meeting the deMon and AllChem developers agreed to merge their codes in order to keep a Tower of Babel from propagating. As a result, the new code deMon2K couples the deMon functionality with the stable and efficient integral and SCF part from AllChem. Besides the merging of the two codes the new deMon version possesses improved integral recurrence relations [93], also for effective core potential and model core potential integrals, as well as an improved numerical integration scheme [259]. The main difference to the adaptive integrator of AllChem is the grid generating function. In AllChem the diagonal elements of the overlap matrix were used as the grid generating function. Because the convergence of the numerical integration is directly related to the absolute value of the corresponding quantity this approach ensured the convergence of the full overlap matrix and, therefore, the convergence of the numerical integration of the electron density. From the experiences with this approach we learned that the reliable numerical integration of the electron density is not sufficient for the reliable numerical integration of the exchange-correlation potential matrix. Therefore, in deMon2K the diagonal elements of the exchange-correlation potential matrix are used as the grid generating function. Because a density is needed for the calculation of these elements, the adaptive grid is generated twice in deMon2K. First, it is generated with the start density and after the convergence of this SCF the grid is rebuilt using the converged SCF density. With this grid the energy is then converged again. This usually takes only three or four SCF cycles. Because the grid generation is very fast this step does not slow down the overall performance of the program.

Besides the implementation of a new grid generating function, the adaptive grid in deMon2K also works with a new cell function that can be screened for large molecules. It has been noticed in the literature [260] that the calculation of the cell function according to Becke [159] scales cubically with respect to the number of atoms. Already for systems with a few hundred atoms this step can become dominant in the calculation. By introducing a cell function that can be screened, this problem is avoided. This also opens the door for a direct grid generation step that would avoid the use of a grid tape. Work in this direction is currently under way in our laboratory.

Due to the introduction of Hermite–Gaussian auxiliary functions, asymptotic expansion for the three-center Coulomb integrals could be derived [93]. With this expansion the calculation of the three-center Coulomb integrals scales nearly linearly. By using the approximated density for the calculation of the exchange-correlation energy and potential [261] a very fast construction of the Kohn–Sham matrix is obtained. Therefore, for systems with more than 3000 basis functions, the linear algebra part of the program becomes dominant. In combination with the MinMax SCF procedure, molecules with several thousand basis functions can be routinely calculated. For the optimization of these systems delocalized internal coordinates [262–264] have been implemented in the new deMon2K version. The optimization of systems with several hundred atoms is now feasible. In fact, the previous time bottleneck has now been shifted to a memory bottleneck. In Fig. 38.1 the requested RAM size for some benchmark systems is depicted. As this figure shows, a system with 5000–6000 basis functions can still be calculated

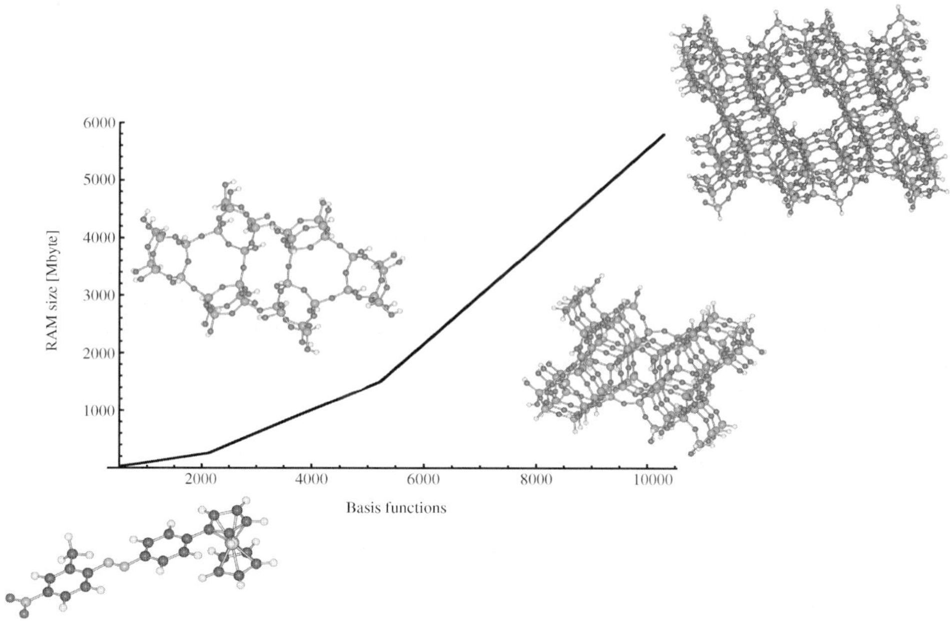

Fig. 38.1. Core memory allocation of deMon for SCF calculations with 1000 up to more than 11,000 basis functions.

on a 32-bit architecture because the RAM request is less than 2 GB. For larger systems, like the depicted double unit cell of the ZSM-5 zeolite in the upper right corner, a 64-bit architecture is required. This system possesses around 11,000 basis functions and more than 700 atoms.

The improved performance of the new deMon2K version is also very useful for Born–Oppenheimer molecular dynamics (BOMD) simulations. This technique was already introduced into deMon-KS1 in the 1990s [265–268]. However, due to the computational limitations of this early deMon version only small systems could be treated. With the new deMon version BOMD simulations may be performed for systems with 50–100 atoms over several picoseconds [269]. With the incorporation of a QM/MM embedding scheme [270,271] the first step towards multi-scale modeling has been performed in deMon2K.

The improved numerical stability of the new deMon2K version also opened the possibility for accurate harmonic Franck–Condon factor calculations. Based on the combination of such calculations with experimental data from pulsed-field ionization zero-electron-kinetic energy (PFI-ZEKE) photoelectron spectroscopy, the ground state structure of V_3 could be determined [272]. Very recently, this work has been extended to the simulation of vibrationally resolved negative ion photoelectron spectra [273]. In both works the use of newly developed basis sets for gradient corrected functionals was the key to success for the ground state structure determination. These basis sets have now been developed for all 3d transition metal elements. With the simulation of vibrationally resolved photoelectron spectra of small transition metal clusters reliable structure and electronic state predictions become possible. This work nicely demonstrates how closely applied DFT and experiment can come together.

38.6 RÉSUMÉ

In this review we tried to follow the development of applied DFT over the last 40 years from the perspective of those involved in developing the deMon suite of LCGTO–DFT programs and their underlying methodologies. Modern DFT programs are based on the fundamental work of Hohenberg and Kohn that provides a solid theoretical foundation. However, many technical developments originally proposed within the Xα methodology have found their way into these programs, too. With the introduction of LCGTO DFT programs, a bridge between the Hartree–Fock and DFT worlds was built. Both worlds have profited considerably from each other. This process is still under way.

The application of DFT methods in chemistry was for a long time, exotic. The Xα method occupied this space for some years. The situation changed rapidly with the introduction of LCAO DFT methods. The possibility of structure optimization brought these methods into mainstream computational chemistry. The late 1980s and early 1990s saw the consolidation of DFT methods in Chemistry. Today, they serve as standard tools for most computational chemists.

The development of deMon also demonstrates how different areas impact each other. The wide availability of more and more powerful microcomputers also triggers continuous software development. This is also true for scientific software development. The deMon Developers come from diverse backgrounds and from all parts of the world. They have come together in a loose cooperative consortium that, in our view, represents a situation where the whole is very much greater than the sum of its parts. While the present authors are responsible for errors or omissions in this small review, the credit for the advances is shared with a much wider group cited in the references here and elsewhere. We predict a bright and rich future not only for this group, but for applied DFT in general.

38.7 ACKNOWLEDGEMENTS

We would like to acknowledge the following persons for their contributions to the development of deMon. They are, in alphabetical order and with their home countries in parentheses, Yuri Abashkin (Russia), Jan Andzelm (Poland), Alexei Arbuznikov (Russia), Patrizia Calaminici (Italy), Mark Casida (USA), Henry Chermette (France), Steeve Chrétien (Canada), Clemence Corminboeuf (Switzerland), Claude Daul (France), Helio Duarte (Brazil), Marcin Dulak (Poland), Roberto Flores-Moreni (México), Elisa Fadda (Italy), René Fournier (Canada), Gerald Geudtner (Germany), Nathalie Godbout (Canada), Jingang Guan (China), Sébastien Hamel (Canada), Thomas Heine (Germany), Klaus Hermann (Germany), Christine Jamorski (France), Florian Janetzko (Germany), Martin Kaupp (Germany), Mariusz Klobukowski (Poland), Hisayoshi Kobayashi (Japan), Matthias Krack (Germany), Martin Leboeuf (Canada), Vladimir Malkin (Russia), Olga Malkina (Russia), Gabriel Merino (México), Tzonka Mineva (Bulgaria), Piotr Mlynarski

(Poland), Seongho Moon (Korea), Benoît Ozell (Canada), Serguei Patchkovskii (Russia), Luca Pedocchi (Italy), Lars G. M. Pettersson (Sweden), Emil Proynov (Bulgaria), José Ulises Reveles (México), Eliseo Ruiz (Spain), Nino Russo (Italy), Emilia Sicilia (Italy), Fiona Sim (Scotland), Suzanne Sirois (Canada), Alain St-Amant (Canada), Luciano Triguero (Cuba), Knut Vietze (Germany), Dongqing Wei (China), Tomasz Wesolowski (Poland), Bernd Zimmermann (Germany).

38.8 REFERENCES

1 P.A.M. Dirac, Proc. Cambridge Phil. Soc., 26 (1930) 376.
2 E. Wigner and F. Seitz, Phys. Rev., 43 (1933) 804.
3 E. Wigner and F. Seitz, Phys. Rev., 46 (1934) 509.
4 E. Fermi, Rend. Accad. Lincei, 6 (1927) 602.
5 L.H. Thomas, Proc. Cambr. Phil. Soc., 23 (1927) 542.
6 E. Fermi, Z. Physik., 48 (1928) 73.
7 E. Fermi, Rend. Accad. Lincei, 7 (1928) 342.
8 J.C. Slater, Phys. Rev., 51 (1937) 846.
9 J.C. Slater, Phys. Rev., 81 (1951) 385.
10 R. Gaspar, Acta Phys. Acad. Scientiarum Hungaricae, 3 (1954) 263.
11 K. Schwarz, Phys. Rev. B, 5 (1972) 2466.
12 K.H. Johnson, J. Chem. Phys., 45 (1966) 3085.
13 K.H. Johnson and F.C. Smith, Phys. Rev. Lett., 24 (1970) 139.
14 G.G. Hall, Proc. Roy. Soc. Ser. A, 205 (1951) 541.
15 C.C.J. Roothaan, Rev. Mod. Phys., 23 (1951) 69.
16 S.F. Boys, Proc. Roy. Soc. Ser. A, 200 (1950) 542.
17 P. Hohenberg and W. Kohn, Phys. Rev., 136 (1964) B864.
18 W. Kohn and L.J. Sham, Phys. Rev., 140 (1965) A1133.
19 N.D. Mermin, Phys. Rev., 137 (1965) 1441.
20 J.S. Rowlinson and B. Widom, Molecular theory of capillarity, Oxford University Press, New York, 1982.
21 B.Y. Tong and L.J. Sham, Phys. Rev., 144 (1966) 1.
22 K.H. Johnson and F.C. Smith, Chem. Phys. Lett., 10 (1971) 219.
23 K.H. Johnson and F.C. Smith, Phys. Rev. B, 5 (1972) 831.
24 K.H. Johnson and R.P. Messmer, J. Vac. Sci. Technol., 11 (1974) 236.
25 R.P. Messmer, C.W. Tucker and K.H. Johnson, Chem. Phys. Lett., 36 (1975) 423.
26 J.W.D. Connolly, Int. J. Quantum Chem., (1972) 201.
27 J.W.D. Connolly, in: G.A. Segal (Ed.), Semiempirical methods of electronic structure; Part A, Plenum, New York, 1977, p. 105.
28 R.P. Messmer, D.R. Salahub and J.W. Davenport, Chem. Phys. Lett., 57 (1978) 29.
29 J. Weber, H. Berthou and C.K. Jorgensen, Chem. Phys. Lett., 45 (1977) 1.
30 D.R. Salahub, J. Chem. Soc., Chem. Commun., (1978) 385.
31 E.L. Anderson, T.P. Fehlner, A.E. Foti and D.R. Salahub, J. Am. Chem. Soc., 102 (1980) 7422.
32 R.P. Messmer, S.H. Lamson and D.R. Salahub, Solid State Commun., 36 (1980) 265.
33 R.P. Messmer and D.R. Salahub, Chem. Phys. Lett., 49 (1977) 59.
34 D.R. Salahub, M. Roche and R.P. Messmer, Phys. Rev. B, 18 (1978) 6495.
35 F.A. Cotton and G.G. Stanley, Inorg. Chem., 16 (1977) 2668.
36 K.H. Johnson, C.Y. Yang, D. Vvedensky, R.P. Messmer, D.R. Salahub, in: W.C. Johnson, J.M. Blakely (Eds.), Interfacial segregation, American Society for Metals, Metals Park, Ohio, 1979, p. 25.
37 D.R. Salahub, L.C. Niem and M. Roche, Surf. Sci., 100 (1980) 199.
38 F.A. Cotton, G.G. Stanley, B.J. Kalbacher, J.C. Green, E. Seddon and M.H. Chisholm, Proc. Natl. Acad. Sci. USA, 74 (1977) 3109.
39 J.G. Norman and D.J. Gmur, J. Am. Chem. Soc., 99 (1977) 1446.

40 D.R. Salahub and R.P. Messmer, Phys. Rev. B, 16 (1977) 2526.
41 J. Weber, Chem. Phys. Lett., 40 (1976) 275.
42 J. Weber, Chem. Phys. Lett., 45 (1977) 261.
43 J. Weber and M. Geoffroy, Theor. Chim. Acta, 43 (1977) 299.
44 A. Goursot, E. Penigault, J. Weber and J.G. Fripiat, New J. Chem., 2 (1978) 469.
45 J. Weber, M. Geoffroy, A. Goursot and E. Penigault, J. Am. Chem. Soc., 100 (1978) 3995.
46 A. Goursot, E. Penigault and J. Weber, New J. Chem., 3 (1979) 675.
47 D.D. Koelling, D.E. Ellis and R.J. Bartlett, J. Chem. Phys., 65 (1976) 3331.
48 J.G. Norman, H.J. Kolari, H.B. Gray and W.C. Trogler, Inorg. Chem., 16 (1977) 987.
49 C.Y. Yang, K.H. Johnson, R.H. Holm and J.G. Norman, J. Am. Chem. Soc., 97 (1975) 6596.
50 H. Sambe and R.H. Felton, Chem. Phys. Lett., 61 (1979) 69.
51 D.A. Case and M. Karplus, Chem. Phys. Lett., 39 (1976) 33.
52 D.A. Case and M. Karplus, J. Am. Chem. Soc., 99 (1977) 6182.
53 H. Sambe, Int. J. Quantum Chem., (1975) 95.
54 H. Sambe and R.H. Felton, J. Chem. Phys., 62 (1975) 1122.
55 E.J. Baerends, D.E. Ellis and P. Ros, Chem. Phys., 2 (1973) 41.
56 B.I. Dunlap and N. Rösch, Adv. Quantum Chem., 21 (1990) 317.
57 D.E. Ellis and G.S. Painter, Computational methods in band theory, Plenum, New York, 1971, pp. 271.
58 B.I. Dunlap, J.W.D. Connolly and J.R. Sabin, J. Chem. Phys., 71 (1979) 3396.
59 O. Vahtras, J. Almlöf and M.W. Feyereisen, Chem. Phys. Lett., 213 (1993) 514.
60 U.V. Barth and L. Hedin, J. Phys. C, 5 (1972) 1629.
61 R.G. Gordon and Y.S. Kim, J. Chem. Phys., 56 (1972) 3122.
62 M. Levy, Proc. Natl. Acad. Sci. USA, 76 (1979) 6062.
63 O. Gunnarsson and B.I. Lundqvist, Phys. Rev. B, 13 (1976) 4274.
64 R. Colle and O. Salvetti, Theor. Chim. Acta, 37 (1975) 329.
65 D.C. Langreth and J.P. Perdew, Phys. Rev. B, 15 (1977) 2884.
66 L. Hedin and B.I. Lundqvist, J. Phys. C, 4 (1971) 2064.
67 L. Hedin, B.I. Lundqvist and S. Lundqvist, Solid State Commun., 9 (1971) 537.
68 O. Gunnarsson and B.I. Lundqvist, Phys. Rev. B, 15 (1977) 6006.
69 O. Gunnarsson, J. Harris and R.O. Jones, J. Chem. Phys., 67 (1977) 3970.
70 J. Harris and R.O. Jones, Phys. Rev. A, 18 (1978) 2159.
71 J. Harris and R.O. Jones, J. Chem. Phys., 68 (1978) 3316.
72 J. Harris and R.O. Jones, J. Chem. Phys., 68 (1978) 1190.
73 J. Harris and R.O. Jones, Phys. Rev. A, 19 (1979) 1813.
74 J. Harris and R.O. Jones, J. Chem. Phys., 70 (1979) 830.
75 O. Gunnarsson and R.O. Jones, Phys. Scr., 21 (1980) 394.
76 D.M. Ceperley and B.J. Alder, Phys. Rev. Lett., 45 (1980) 566.
77 R.O. Jones, J. Chem. Phys., 76 (1982) 3098.
78 R. Fletcher, Practical methods of optimization, Wiley, New York, 1988.
79 C.G. Broyden, J. Inst. Maths. Applns., 6 (1970) 76.
80 C.G. Broyden, J. Inst. Maths. Applns., 6 (1970) 222.
81 R. Fletcher, Comput. J., 13 (1970) 317.
82 D. Goldfarb, Math. Comput., 24 (1970) 23.
83 D.F. Shanno, Math. Comput., 24 (1970) 647.
84 V.I. Lebedev, Zh. Vychisl. Mat. Mat. Fiz., 15 (1975) 48.
85 V.I. Lebedev, Zh. Vychisl. Mat. Mat. Fiz., 16 (1976) 293.
86 V.I. Lebedev, Sibirsk. Mat. Zh., 18 (1977) 132.
87 V.I. Lebedev and A.L. Skorokhodov, Russ. Acad. Sci. Dokl. Math., 45 (1992) 587.
88 V.I. Lebedev, Russ. Acad. Sci. Dokl. Math., 50 (1995) 283.
89 M. Dupuis, J. Rys and H.F. King, J. Chem. Phys., 65 (1976) 111.
90 L.E. McMurchie and E.R. Davidson, J. Comput. Phys., 26 (1978) 218.
91 T. Zivkovic and Z.B. Maksic, J. Chem. Phys., 49 (1968) 3083.
92 A. Golebiew and J. Mrozek, Int. J. Quantum Chem., 7 (1973) 623.

93 A.M. Köster, J. Chem. Phys., 118 (2003) 9943.
94 V.R. Saunders, in: G.H.F. Diercksen, S. Wilson (Eds.), Methods in computational molecular physics, Reidel, Dordrecht, 1983.
95 C. Daul, C.W. Schlapfer, A. Goursot, E. Penigault and J. Weber, Chem. Phys. Lett., 78 (1981) 304.
96 C.A. Daul and J. Weber, Chem. Phys. Lett., 77 (1981) 593.
97 D.R. Salahub and R.P. Messmer, Surf. Sci., 106 (1981) 415.
98 C.Y. Yang, K.H. Johnson, D.R. Salahub, J. Kaspar and R.P. Messmer, Phys. Rev. B, 24 (1981) 5673.
99 J. Weber, A. Goursot, E. Penigault, J.H. Ammeter and J. Bachmann, J. Am. Chem. Soc., 104 (1982) 1491.
100 J. Kaspar and D.R. Salahub, J. Phys.-Metal Phys., 13 (1983) 311.
101 D.R. Salahub and F. Raatz, Int. J. Quantum Chem., (1984) 173.
102 C. Famiglietti and E.J. Baerends, Chem. Phys., 62 (1981) 407.
103 D. Post and E.J. Baerends, Chem. Phys. Lett., 86 (1982) 176.
104 D.R. Salahub, S.H. Lamson and R.P. Messmer, Chem. Phys. Lett., 85 (1982) 430.
105 A. Pellegatti, B.N. McMaster and D.R. Salahub, Chem. Phys., 75 (1983) 83.
106 M. Morin, A.E. Foti and D.R. Salahub, Can. J. Chem., 63 (1985) 1982.
107 A.K. Rajagopal, Adv. Quantum Chem., 41 (1980) 59.
108 W. Kohn and P. Vashishta, Theory of the inhomogenous electron gas, Plenum, New York, 1983, p. 79.
109 S. Lundqvist and N.H. March, Theory of the inhomogenous electron gas, Plenum, New York, 1983, p. 410.
110 R.G. Parr, Annu. Rev. Phys. Chem., 34 (1983) 631.
111 C. Satoko, Chem. Phys. Lett., 83 (1981) 111.
112 P. Bendt and A. Zunger, Phys. Rev. Lett., 50 (1983) 1684.
113 R. Fournier, J. Andzelm and D.R. Salahub, J. Chem. Phys., 90 (1989) 6371.
114 N.A. Baykara, B.N. McMaster and D.R. Salahub, Mol. Phys., 52 (1984) 891.
115 D.R. Salahub and N.A. Baykara, Surf. Sci., 156 (1985) 605.
116 D.R. Salahub, Adv. Chem. Phys., 69 (1987) 447.
117 T. Ziegler, V. Tschinke and A. Becke, Polyhedron, 6 (1987) 685.
118 B. Delley, D.E. Ellis, A.J. Freeman, E.J. Baerends and D. Post, Phys. Rev. B, 27 (1983) 2132.
119 D. Post and E.J. Baerends, J. Chem. Phys., 78 (1983) 5663.
120 J. Andzelm and D.R. Salahub, Int. J. Quantum Chem., 29 (1986) 1091.
121 R. Fournier and D.R. Salahub, Int. J. Quantum Chem., 29 (1986) 1077.
122 D.R. Salahub, in: J. Davenas, P. Rabette (Eds.), Contribution of clusters physics to materials science and technology, NATO ASI series, Nijhoff, Amsterdam, 1986, p. 143.
123 A. Selmani, J. Andzelm and D.R. Salahub, Int. J. Quantum Chem., 29 (1986) 829.
124 T. Ziegler, J. Am. Chem. Soc., 105 (1983) 7543.
125 T. Ziegler, J. Am. Chem. Soc., 106 (1984) 5901.
126 T. Ziegler, Organometallics, 4 (1985) 675.
127 T. Ziegler, V. Tschinke and C. Ursenbach, J. Am. Chem. Soc., 109 (1987) 4825.
128 A. Peluso, D.R. Salahub and A. Goursot, Inorg. Chem., 29 (1990) 1544.
129 J. Andzelm, N. Russo and D.R. Salahub, Chem. Phys. Lett., 142 (1987) 169.
130 E. Radzio and D.R. Salahub, Int. J. Quantum Chem., 29 (1986) 241.
131 A. Selmani and D.R. Salahub, J. Chem. Phys., 89 (1988) 1529.
132 T. Ziegler, J.K. Nagle, J.G. Snijders and E.J. Baerends, J. Am. Chem. Soc., 111 (1989) 5631.
133 M. Morin, D.R. Salahub, S. Nour, C. Mehadji and H. Chermette, Chem. Phys. Lett., 159 (1989) 472.
134 T. Ziegler, V. Tschinke, E.J. Baerends, J.G. Snijders and W. Ravenek, J. Phys. Chem., 93 (1989) 3050.
135 J. Andzelm, E. Radzio and D.R. Salahub, J. Chem. Phys., 83 (1985) 4573.
136 J. Andzelm, N. Russo and D.R. Salahub, J. Chem. Phys., 87 (1987) 6562.
137 N. Russo, J. Andzelm and D.R. Salahub, Chem. Phys., 114 (1987) 331.
138 J. Andzelm, A. Rochefort, N. Russo and D.R. Salahub, Surf. Sci., 235 (1990) L319.
139 R. Fournier, J. Andzelm, A. Goursot, N. Russo and D.R. Salahub, J. Chem. Phys., 93 (1990) 2919.
140 I. Papai, D.R. Salahub and C. Mijoule, Surf. Sci., 236 (1990) 241.
141 A. Rochefort, J. Andzelm, N. Russo and D.R. Salahub, J. Am. Chem. Soc., 112 (1990) 8239.
142 J.P. Perdew and W. Yue, Phys. Rev. B, 33 (1986) 8800.

143 A.D. Becke, Phys. Rev. A, 38 (1988) 3098.
144 J.P. Perdew, K. Burke and M. Ernzerhof, Phys. Rev. Lett., 77 (1996) 3865.
145 P. Mlynarski and D.R. Salahub, Phys. Rev. B, 43 (1991) 1399.
146 J.P. Perdew, Phys. Rev. Lett., 55 (1985) 1665.
147 J.P. Perdew, J.A. Chevary, S.H. Vosko, K.A. Jackson, M.R. Pederson, D.J. Singh and C. Fiolhais, Phys. Rev. B, 46 (1992) 6671.
148 M. Levy and J.P. Perdew, Int. J. Quantum Chem., (1985) 743.
149 M. Levy and J.P. Perdew, Phys. Rev. A, 32 (1985) 2010.
150 M. Levy, Int. J. Quantum Chem., (1989) 617.
151 M. Levy and O.Y. Hui, Phys. Rev. A, 42 (1990) 651.
152 A.D. Becke, J. Chem. Phys., 98 (1993) 1372.
153 A.D. Becke, J. Chem. Phys., 98 (1993) 5648.
154 R. Car and M. Parrinello, Phys. Rev. Lett., 60 (1988) 204.
155 J. Andzelm, E. Radzio and D.R. Salahub, J. Comput. Chem., 6 (1985) 520.
156 A.M. Köster, J. Chem. Phys., 104 (1996) 4114.
157 B.I. Dunlap, Phys. Rev. A, 25 (1982) 2847.
158 S. Obara and A. Saika, J. Chem. Phys., 84 (1986) 3963.
159 A.D. Becke, J. Chem. Phys., 88 (1987) 2547.
160 J.M.L. Martin, C.W. Bauschlicher and A. Ricca, Comput. Phys. Commun., 133 (2001) 189.
161 M. Malagoli and J. Baker, J. Chem. Phys., 119 (2003) 12763.
162 K. Levenberg, Quart. Appl. Math., 2 (1944) 164.
163 D.W. Marquardt, J. Soc. Ind. App. Math., 11 (1963) 431.
164 A. Banerjee, N. Adams, J. Simons and R. Shepard, J. Phys. Chem., 89 (1985) 52.
165 J. Baker, J. Comput. Chem., 7 (1986) 385.
166 B. Delley, J. Chem. Phys., 92 (1990) 508.
167 B. Delley, DMOL, Biosym Technologies, San Diego, CA, 1990.
168 A. St-Amant and D.R. Salahub, Chem. Phys. Lett., 169 (1990) 387.
169 A. St-Amant, Université de Montréal, Montreal, (1991).
170 J. Andzelm, in: J.K. Labanowski, J. Andzelm (Eds.), Density functional methods in chemistry, Springer, New York, 1991.
171 J. Andzelm, DGAUSS, Cray Research, Eagan, Minnesota, 1991.
172 J. Andzelm and E. Wimmer, J. Chem. Phys., 96 (1992) 1280.
173 P.M. Boerrigter, G.T. Velde and E.J. Baerends, Int. J. Quantum Chem., 33 (1988) 87.
174 G.T. Velde, F.M. Bickelhaupt, E.J. Baerends, C.F. Guerra, S.J.A. Van Gisbergen, J.G. Snijders and T. Ziegler, J. Comput. Chem., 22 (2001) 931.
175 M.J.T. Frisch, G.W. Trucks, M. Head-Gordon, P.M.W. Gill, M.W. Wong, J.B. Foresman, B.G. Johnson, H.B. Schlegel, M.A. Robb, E.S. Repiogle, R. Gomperts, J.L. Andres, K. Raghavachari, J.S. Binkley, C. Gonzalez, R.L. Martin, D.J. Fox, D.J. Defrees, J. Baker, J.J.P. Stewart and J.A. Pople, Gaussian 92, Gaussian, Inc., Pittsburgh, PA, 1992.
176 M.E. Casida, C. Daul, A. Goursot, A.M. Köster, L.G.M. Pettersson, E. Proynov, A. St-Amant, D.R. Salahub, H. Duarte, N. Godbout, J. Guan, C. Jamorski, M. Leboeuf, V. Malkin, O. Malkina, F. Sim, A. Vela, deMon-KS Version 3.4, 4, deMon Software, Montreal, 1996.
177 N. Godbout, D.R. Salahub, J. Andzelm and E. Wimmer, Can. J. Chem., 70 (1992) 560.
178 V.G. Malkin, O.L. Malkina and D.R. Salahub, Chem. Phys. Lett., 204 (1993) 80.
179 V.G. Malkin, O.L. Malkina and D.R. Salahub, Chem. Phys. Lett., 204 (1993) 87.
180 V.G. Malkin, O.L. Malkina, M.E. Casida and D.R. Salahub, J. Am. Chem. Soc., 116 (1994) 5898.
181 V.G. Malkin, O.L. Malkina and D.R. Salahub, Chem. Phys. Lett., 221 (1994) 91.
182 G. Cuevas, E. Juaristi and A. Vela, J. Mol. Struct. (THEOCHEM), 418 (1997) 231.
183 G. Valerio, A. Goursot, R. Vetrivel, O. Malkina, V. Malkin and D.R. Salahub, J. Am. Chem. Soc., 120 (1998) 11426.
184 G. Cuevas, E. Juaristi and A. Vela, J. Phys. Chem. A, 103 (1999) 932.
185 L.A. Eriksson, V.G. Malkin, O.I. Malkina and D.R. Salahub, J. Chem. Phys., 99 (1993) 9756.
186 L.A. Eriksson, V.G. Malkin, O.L. Malkina and D.R. Salahub, Int. J. Quantum Chem., 52 (1994) 879.

187 L.A. Eriksson, O.L. Malkina, V.G. Malkin and D.R. Salahub, J. Chem. Phys., 100 (1994) 5066.
188 A.M. Köster, P. Calaminici and N. Russo, Phys. Rev. A, 53 (1996) 3865.
189 V. Russier, D.R. Salahub and C. Mijoule, Phys. Rev. B, 42 (1990) 5046.
190 C. Daul, H.U. Gudel and J. Weber, J. Chem. Phys., 98 (1993) 4023.
191 M.E. Casida, in: D.P. Chong (Ed.), Recent advances in density functional methods, World Scientific, Singapore, 1995.
192 C. Jamorski, M.E. Casida and D.R. Salahub, J. Chem. Phys., 104 (1996) 5134.
193 I. Papai, A. St-Amant and D.R. Salahub, Surf. Sci., 240 (1990) L604.
194 I. Papai, A. St-Amant, J. Ushio and D.R. Salahub, Int. J. Quantum Chem., (1990) 29.
195 A. Stirling, J. Chem. Phys., 104 (1996) 1254.
196 F. Sim, D.R. Salahub and S. Chin, Int. J. Quantum Chem., 43 (1992) 463.
197 J.G. Guan, M.E. Casida, A.M. Köster and D.R. Salahub, Phys. Rev. B, 52 (1995) 2184.
198 G. Schaftenaar and J.H. Noordik, J. Computer-Aided Mol. Design, 14 (2000) 123.
199 S. Portmann and H.P. Luthi, Chimia, 54 (2000) 766.
200 B. Ozell, R. Camarero, A. Garon and F. Guibault, Finite Elem. Des., 19 (1995) 295.
201 A.M. Köster, M. Leboeuf, D.R. Salahub, in: J.S. Murray, K. Sen (Eds.), Theoretical and computational chemistry 3, Elsevier, Amsterdam, 1996.
202 L. Triguero and L.G.M. Pettersson, Surf. Sci., 398 (1998) 70.
203 L. Triguero, L.G.M. Pettersson and H. Agren, J. Phys. Chem. A, 102 (1998) 10599.
204 M.E. Casida, C. Daul, A. Goursot, A.M. Köster, L.G.M. Pettersson, E. Proynov, A. St-Amant, D.R. Salahub, H. Duarte, N. Godbout, J. Guan, K. Hermann, C. Jamovski, M. Leboeuf, V. Malkin, O. Malkina, M. Nyberg, L. Pedocchi, F. Sim, L. Triguero and A. Vela, deMon Software, deMon Software, Montreal, 2001.
205 V.G. Malkin, O.L. Malkina and D.R. Salahub, J. Am. Chem. Soc., 117 (1995) 3294.
206 T.B. Woolf, V.G. Malkin, O.L. Malkina, D.R. Salahub and B. Roux, Chem. Phys. Lett., 239 (1995) 186.
207 K. Bellafrouh, C. Daul, H.U. Gudel, F. Gilardoni and J. Weber, Theor. Chim. Acta, 91 (1995) 215.
208 M. Kaupp, V.G. Malkin, O.L. Malkina and D.R. Salahub, J. Am. Chem. Soc., 117 (1995) 8492.
209 M. Kaupp, V.G. Malkin, O.L. Malkina and D.R. Salahub, Chem. Phys. Lett., 235 (1995) 382.
210 M. Kaupp, V.G. Malkin, O.L. Malkina and D.R. Salahub, Chem.-Eur. J., 2 (1996) 24.
211 B. Coq, A. Goursot, T. Tazi, F. Figueras and D.R. Salahub, J. Am. Chem. Soc., 113 (1991) 1485.
212 P. Mlynarski and D.R. Salahub, J. Chem. Phys., 95 (1991) 6050.
213 A. Goursot, I. Papai and D.R. Salahub, J. Am. Chem. Soc., 114 (1992) 7452.
214 M. Castro and D.R. Salahub, Phys. Rev. B, 47 (1993) 10955.
215 L. Goodwin and D.R. Salahub, Phys. Rev. A, 47 (1993) R774.
216 A. Goursot, I. Papai and D.R. Salahub, Studies Surf, Sci. Catal., 75 (1993) 1547.
217 M. Castro and D.R. Salahub, Phys. Rev. B, 49 (1994) 11842.
218 M. Castro, D.R. Salahub and R. Fournier, J. Chem. Phys., 100 (1994) 8233.
219 A. Martinez, A. Vela, D.R. Salahub, P. Calaminici and N. Russo, J. Chem. Phys., 101 (1994) 10677.
220 P. Calaminici, A.M. Köster, N. Russo and D.R. Salahub, J. Chem. Phys., 105 (1996) 9546.
221 C. Blanchet, H.A. Duarte and D.R. Salahub, J. Chem. Phys., 106 (1997) 8778.
222 M. Castro, C. Jamorski and D.R. Salahub, Chem. Phys. Lett., 271 (1997) 133.
223 H.A. Duarte and D.R. Salahub, J. Phys. Chem. B, 101 (1997) 7464.
224 C. Jamorski, A. Martinez, M. Castro and D.R. Salahub, Phys. Rev. B, 55 (1997) 10905.
225 A. Martinez, A.M. Köster and D.R. Salahub, J. Phys. Chem. A, 101 (1997) 1532.
226 D.S. Yang, M.Z. Zgierski, A. Berces, P.A. Hackett, A. Martinez and D.R. Salahub, Chem. Phys. Lett., 277 (1997) 71.
227 A. Martinez, C. Jamorski, G. Medina and D.R. Salahub, J. Phys. Chem. A, 102 (1998) 4643.
228 L.A. Eriksson, O.L. Malkina, V.G. Malkin and D.R. Salahub, Int. J. Quantum Chem., 63 (1997) 575.
229 D.S. Yang, M.Z. Zgierski, A. Berces, P.A. Hackett, P.N. Roy, A. Martinez, T. Carrington, D.R. Salahub, R. Fournier, T. Pang and C.F. Chen, J. Chem. Phys., 105 (1996) 10663.
230 M.E. Casida, K.C. Casida and D.R. Salahub, Int. J. Quantum Chem., 70 (1998) 933.
231 M.E. Casida, C. Jamorski, K.C. Casida and D.R. Salahub, J. Chem. Phys., 108 (1998) 4439.
232 M.E. Casida and D.R. Salahub, J. Chem. Phys., 113 (2000) 8918.

233 A. Rochefort, P.H. McBreen and D.R. Salahub, Surf. Sci., 347 (1996) 11.
234 H. Guo and D.R. Salahub, Angew. Chem. Int. Ed., 37 (1998) 2985.
235 D.J. Lacks and R.G. Gordon, Phys. Rev. A, 47 (1993) 4681.
236 D.J. Lacks and R.G. Gordon, J. Chem. Phys., 100 (1994) 4446.
237 F. Sim, A. St-Amant, I. Papai and D.R. Salahub, J. Am. Chem. Soc., 114 (1992) 4391.
238 E. Ruiz, D.R. Salahub and A. Vela, J. Am. Chem. Soc., 117 (1995) 1141.
239 E. Ruiz, D.R. Salahub and A. Vela, J. Phys. Chem., 100 (1996) 12265.
240 E.I. Proynov, A. Vela and D.R. Salahub, Chem. Phys. Lett., 230 (1994) 419.
241 E.I. Proynov, E. Ruiz, A. Vela and D.R. Salahub, Int. J. Quantum Chem., (1995) 61.
242 E.I. Proynov, S. Sirois and D.R. Salahub, Int. J. Quantum Chem., 64 (1997) 427.
243 S. Sirois, E.I. Proynov, D.T. Nguyen and D.R. Salahub, J. Chem. Phys., 107 (1997) 6770.
244 H.A. Duarte, E. Proynov and D.R. Salahub, J. Chem. Phys., 109 (1998) 26.
245 E. Proynov, H. Chermette and D.R. Salahub, J. Chem. Phys., 113 (2000) 10013.
246 A.M. Köster, M. Krack, M. Leboeuf and B. Zimmermann, AllChem, Universität Hannover, 1998.
247 M. Krack and A.M. Köster, J. Chem. Phys., 108 (1998) 3226.
248 M. Head-Gordon and J.A. Pople, J. Chem. Phys., 89 (1988) 5777.
249 B.G. Johnson, P.M.W. Gill, J.A. Pople and D.J. Fox, Chem. Phys. Lett., 206 (1993) 239.
250 A.M. Köster: Habilitation thesis, Universität Hannover, Hannover, 1998.
251 J.M. Perez-Jorda, A. Becke and E. San-Fabian, J. Chem. Phys., 100 (1994) 6520.
252 R. Lindh, P.A. Malmqvist and L. Gagliardi, Theor. Chem. Acc., 106 (2001) 178.
253 P.M.W. Gill and S.H. Chien, J. Comput. Chem., 24 (2003) 732.
254 P. Calaminici, K. Jug and A.M. Köster, J. Chem. Phys., 109 (1998) 7756.
255 P. Calaminici, A.M. Köster, A. Vela and K. Jug, J. Chem. Phys., 113 (2000) 2199.
256 J. Baker, J. Andzelm, A. Scheiner and B. Delley, J. Chem. Phys., 101 (1994) 8894.
257 A.M. Köster, P. Calaminici, Z. Gomez, J.U. Reveles, in: G. Parr, K. Sen (Eds.), Reviews of modern quantum chemistry, a celebration of the contribution of Robert, World Scientific, Singapore, 2002.
258 G. Seifert, D. Proezag and T. Frauenheim, Int. J. Quantum Chem., 58 (1996) 185.
259 A.M. Köster, R. Flores-Moreno and J.U. Reveles, J. Chem. Phys., 121 (2004) 681.
260 R.E. Stratmann, G.E. Scuseria and M.J. Frisch, Chem. Phys. Lett., 257 (1996) 213.
261 A.M. Köster, J.U. Reveles and J.M. del Campo, J. Chem. Phys., 121 (2004) 3417.
262 J. Baker, A. Kessi and B. Delley, J. Chem. Phys., 105 (1996) 192.
263 Ö. Farkas and H.B. Schlegel, J. Chem. Phys., 111 (1999) 10806.
264 J.U. Reveles and A.M. Köster, J. Comput. Chem., 25 (2004) 1109.
265 D.Q. Wei and D.R. Salahub, J. Chem. Phys., 101 (1994) 7633.
266 D.Q. Wei and D.R. Salahub, Chem. Phys. Lett., 224 (1994) 291.
267 D.Q. Wei and D.R. Salahub, J. Chem. Phys., 106 (1997) 6086.
268 D.Q. Wei, E.I. Proynov, A. Milet and D.R. Salahub, J. Phys. Chem. A, 104 (2000) 2384.
269 S. Krishnamurty, T. Heine and A. Goursot, J. Phys. Chem. B, 104 (2003) 5728.
270 S. Moon, S. Patchkovskii and D.R. Salahub, J. Mol. Struct. (THEOCHEM), 632 (2003) 287.
271 N. Gresh, S.A. Kafafi, J.F. Truchon and D.R. Salahub, J. Comput. Chem., 25 (2004) 823.
272 P. Calaminici, A.M. Köster, T. Carrington, P.N. Roy, N. Russo and D.R. Salahub, J. Chem. Phys., 114 (2001) 4036.
273 P. Calaminici, A.M. Köster and D.R. Salahub, J. Chem. Phys., 118 (2003) 4913.

Theory and Applications of Computational Chemistry: The First Forty Years
Edited by C. Dykstra et al.

CHAPTER 39

SAC–CI method applied to molecular spectroscopy

M. Ehara[1], J. Hasegawa[1] and H. Nakatsuji[1,2]

[1] *Department of Synthetic Chemistry and Biological Chemistry, Graduate School of Engineering, Kyoto University, Kyoto-Daigaku-Katsura, Nishikyo-ku, Kyoto 615-8510, Japan*
[2] *Fukui Institute for Fundamental Chemistry, Kyoto University, 34-4 Takano-Nishihiraki-cho, Sakyo-ku, Kyoto 606-8103, Japan*

Abstract

This chapter gives a brief review of the SAC–CI (Symmetry-Adapted Cluster–Configuration Interaction) method applied to molecular spectroscopy. The SAC–CI method was proposed in 1978 as an accurate electronic-structure theory for the ground, excited, ionized, electron-attached, and high-spin states of atoms and molecules. Since then, it has been successfully applied to various chemistries including more than 150 molecules and established to be a useful method for studying chemistry and physics involving various electronic states. The topics covered in this review are electronic excitation spectra, ionization spectra, collision-induced absorption processes, photochemical reactions, inner-shell ionizations, equilibrium geometries of excited states, and ESR hyperfine splitting constants of radicals. The applied systems are organic and inorganic compounds, van der Waals complexes, transition metal complexes, phthalocyanines, and the bacterial photosynthetic reaction center. These results show the reliability and applicability of the SAC–CI method for studying molecular spectroscopy.

39.1 INTRODUCTION

Electronic excitation and ionization spectra have been widely used in various fields of molecular sciences and technologies. These spectroscopies provide useful information about the electronic structures of molecules. The electronic spectrum is sometimes called as molecular finger print and used for identification of molecules. The electronic-structure theory is necessary to understand and identify the electronic states involved in the spectra. Without the help of the reliable theories, it is sometimes very difficult to analyze and understand complex spectra of molecules.

Theoretical studies of the molecular excited states were extensively advanced by Buenker and Peyerimhoff [1,2] from the end of 1960s. Many pioneering works were done by their multireference-configuration interaction (MR-CI) method. A different approach for describing excited and ionized states was proposed by one of the present authors in 1978 as SAC–CI (Symmetry Adapted Cluster–Configuration Interaction) method [3–5] based on the SAC description of the ground state [6,7]. Fig. 39.1 shows an overview of the current SAC–CI program. The SAC–CI method is based on the cluster expansion approach and is an accurate electronic-structure theory for ground, excited, ionized and electron-attached states of molecules. This method has been developed mainly by the Nakatsuji laboratory, both in theories and algorithms, and has been successfully applied to various chemical phenomena involving more than 150 molecules. The energy gradient method has also implemented and therefore, we can calculate the energy gradient (force) acting on nuclei for all of these electronic states. We can study molecular structures and chemical reaction dynamics involving these different electronic states.

The SAC–CI method has been applied to a wide variety of chemistry; electronic excitation spectroscopy of valence, Rydberg and inner-core states, ionization (photoelectron) spectroscopy in outer-valence, inner-valence and core regions, molecular spectroscopies including multi-electron processes, molecular structure and spectroscopic constants in the excited states, electronic structures of adsorbates, catalytic reactions and surface sciences, photochemical reaction dynamics, biological photochemistry and electron transfer, atmospheric chemistry, molecular interactions, and ESR hyperfine splitting constants [8,9]. The size of the applied system ranges from small atoms and molecules to biologically important systems composed of more than hundred atoms. Through these applications, the SAC–CI method has been established to be a reliable method for studying the electronic structure of atoms and molecules in their ground, excited, ionized and electron-attached states.

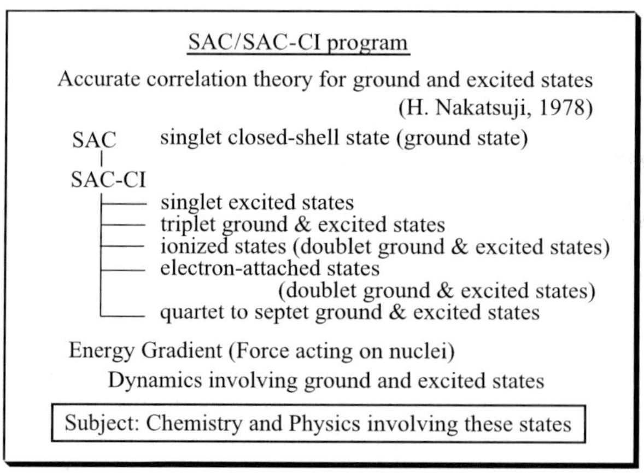

Fig. 39.1. Overview of the current SAC–CI program system.

Historically, the SAC–CI method was published in 1978 as a theory for calculating the wave functions of excited states, ionized states and electron-attached states. Since then, the SAC–CI method has been applied to the calculations of the excited and ionized states of various molecules and has been established as a reliable useful theory. In the calculations of properties with the coupled-cluster (CC) method, Monkhorst [10] gave a linear response formulation of the excitation energy from a consideration of the poles of the dynamic polarizability, a property of the ground state, but the wave functions of the excited states were not explicitly formulated. Mukherjee [11] gave a more extensive formulation of the coupled cluster linear response theory (CC-LRT). Hirao derived the CC-LRT equation starting from the SAC–CI theory [12]. The applications of the CC-LRT theories were not done before Koch, Jorgensen and others reformulated it and applied to some molecular systems [13,14]. The equation of motion coupled cluster (EOM-CC) method published around 1989 [15,16] was essentially equivalent to the SAC–CI method: the basic theories were identical and the differences were only in the way of approximations in actual calculations. Thus, the SAC–CI theory gives the identical numerical results with the CC-LRT and EOM-CC theories when the same approximations are done [17].

Recently, the time-dependent density functional theory (TD-DFT) [18,19] has been developed based on the DFT [20–22]. Although the TDDFT has been successfully applied to many systems, it is also generally recognized that there are serious problems in describing van der Waals interactions [23,24], Rydberg states[25,26] and charge-transfer excitations[27]. These problems originate from the semi-empirical nature of the functionals and there is no way to improve the results when the method turns out to be inappropriate. In this sense, DFT/TDDFT is not an exact theory. On the other hand, the SAC/SAC–CI theoretical framework is exact, and there is a unique way to make the results converge to the full-CI limit. Therefore, the underlying physics of the approximations, such as the truncation of the excitation operators at some level, is very clear: if it does not work, one can simply improve the results by including higher-order excitation operators, as we do for example in the SAC–CI general-*R* formalism.

Recently, our research group has collaborated with Gaussian Inc. for incorporating our SAC–CI program system into the Gaussian program package. In Gaussian98 [28], the *ab initio* methods for the excited states were limited to the CI singles and CASSCF, although there were many reliable methods for describing the ground-state electronic structures. In the spring of 2003, the SAC–CI program system was incorporated into the Gaussian03 program package [29] and distributed worldwide. With this program, we can calculate the electronic structures and energy gradients of the ground and excited states of singlet to septet spin multiplicities of both one-electron and multi-electron excitation natures.

In this review, we explain the SAC–CI applications to molecular spectroscopy with some examples. In Section 2, we briefly explain the theoretical and computational aspects of the SAC–CI method. Then, we show some SAC–CI applications to molecular spectroscopy; the excitation and ionization spectra of π-conjugated organic molecules (Section 3), collision-induced absorption spectra of van der Waals complex (Section 4), excitation spectra and NMR chemical shifts of transition metal complexes (Section 5), photofragmentation reaction of $Ni(CO)_4$ (Section 6), absorption spectrum of free-base phthalocyanine (FBPc) and bacterial photosynthetic reaction center

(Section 7), inner-shell ionization spectra (Section 8), geometries of the molecular excited states (Section 9), and ESR hyperfine splitting constants (Section 10).

39.2 SAC–CI METHOD

In this section, we briefly explain the SAC–CI theory and its computational algorithms. For detailed descriptions, we refer to the original papers [3–7] and the earlier review articles [8,9].

The SAC theory [6,7] is based on the cluster expansion approach. The excitation operators included are spin-symmetry and space-symmetry adapted. Therefore, the SAC method calculates a pure spin state and is free from the spin-contamination problem that may appear in the CC method [30,31]. The SAC method is size-consistent, which means the method can be applicable to large systems. When we apply the variational principle to the SAC wave function, we obtain the variational SAC equation, which is just the generalized-Brillouin theorem. This theorem implies the following important possibility. Based on the SAC wave function, we can define a set of basis functions for describing the excited states; this basis function satisfies orthogonality and Hamiltonian orthogonality to the ground state SAC wave function. Therefore, the excited state wave function can be expanded by a linear combination of these basis functions, which is the SAC–CI theory [3–5]. It is evident that the SAC–CI wave function satisfies the orthogonality and Hamiltonian-orthogonality to the ground state. Because of this theoretical foundation, the SAC–CI method performs quite well for both ground and excited states, which is confirmed by comparing the SAC–CI results with the full-CI ones [32] and also by applying to numerous examples [8,9]. Another important property of the SAC–CI method is that we can study various electronic states in the same accuracy. This property is very difficult to realize with the independent theories for the ground and excited states.

The SAC and SAC–CI theories are the exact theories and the excitation operators R can be included up to the exact limit. For the practical calculations, there are two standards with respect to the R-operators. For calculating one-electron excitation and ionization processes, it is sufficient to include singles and doubles linked excitation operators in the SAC and SAC–CI wave functions (SAC–CI SD-R methods). On the other hand, for describing many-electron processes like shake-up ionizations, we include general excitation operators in the SAC–CI linked operators, which is the general-R method [33–36]. This approach has been successfully applied to the valence ionization spectra with satellites, molecular structure of multi-electron processes, and the excited states of open-shell systems [17]. Further, the R-operators of the SAC–CI wave function can be defined for arbitrary spin-multiplicity. In the original papers [3–5], the SAC–CI method was performed for singlet, triplet, cation doublet, and anion doublet states. Later, the high-spin states up to septet spin multiplicity were implemented [37].

In many aspects of molecular chemistry and physics, the energy gradient of the potential energy surface (the force acting on the nucleus) is a very useful quantity. We formulated the analytical energy gradients of the SAC–CI method, not only for the SD-R method [38,39] but also for the general-R method [40,41] and not only for ground, excited, ionized and electron-attached states of singlet to triplet multiplicities but also for high-spin multiplicities of up to septet spin-multiplicities [42]. With these developments, the SAC–CI method can be applicable to the calculations of molecular geometries and reaction dynamics in the excited states of various spin multiplicity and the high-energy states.

The computational program code of the SAC–CI method was first completed in 1978 [4,5] and published in 1985 (SAC85) [43]. Our targets of interest have been chemically or physically interesting molecules that were often large for the SAC–CI method if no approximation was adopted in the computations. Actually, in the SAC–CI program, the perturbation-selection algorithm for the double-excitation operators was adopted to apply the method to larger systems [44]. The linked doubles operators can be effectively selected in the SAC and SAC–CI wave functions by evaluating the second-order perturbation energy to the reference states. This method reduces the number of doubles without losing much accuracy. Owing to these advantages, the SAC–CI method has been successfully applied to larger systems such as porphyrin systems. In the Gaussian03 program, we prepared three levels of energy thresholds: LevelOne, LevelTwo, and LevelThree. LevelThree (default) uses (1×10^{-6} au, 1×10^{-7} au) for (ground, excited) states. LevelTwo and LevelOne are defined as (5×10^{-6} au, 5×10^{-7} au) and (1×10^{-5} au, 1×10^{-6} au), respectively. LevelThree calculation is most accurate of the three and used in the default condition. Calculations with the lower levels are more approximate, but computationally easier to apply the SAC–CI method to larger systems. We generally observed that the relative energies among the excited states were rather insensitive among these three threshold sets.

When we calculate potential energy surface, the perturbation selection causes a problem: the perturbation selection at different coordinate space may lead to a small discontinuity of the potential and its gradient. To perform geometry optimizations with this perturbation selection of operators, we developed the GSUM method [45] and the Minimum Orbital-Deformation (MOD) method [46,47]. The GSUM method takes a group sum of the selected operators within the manifold of the geometries under investigation. First, we select several representative points in the nuclear configuration space which covers the potential energy surface under consideration. At each geometry, the ordinary perturbation selection is performed to determine the linked and unlinked operators [44]. Next, the Group SUM (GSUM) of the selected operators is taken for all the geometries and used in the SAC/SAC–CI calculations to calculate the potential energy and the related properties within the energy surface first designed. This method is useful to study the potential energy curves and the related properties of many electronic states in a wide range of configuration space [45].

The perturbation selection breaks the invariance of the SAC/SAC–CI energies on the unitary transformation of orbitals among the occupied and/or unoccupied manifolds. This is the source of the discontinuity in the potential energy surface, if the external perturbation induces a large (sudden) mixing of orbitals within the occupied and unoccupied orbitals. Suppose a deformation of benzene from the D_{6h} geometry to less symmetric one, the degenerate set of orbitals in occupied or unoccupied manifold makes a sudden (discontinuous) deformation of MOs (see Ref. [44] for some details]. The MOD method [46] can solve this problem that may occur in the optimization process with respect to the external parameters. This method minimizes the deformation of the MOs induced by the perturbation at the neighborhood of the geometry under consideration. The perturbation dependence of the unitary transformation among MOs is attributed to $\mathbf{T}^x$, the anti-symmetric part of the CPHF coefficients [46]. Therefore, it is required that the matrix elements between any real occupied [unoccupied] MOs $T_{ij}^x[T_{ab}^x]$ vanish.

$$T_{ij}^x = U_{ij}^x - U_{ji}^x = 0 \tag{1}$$

The matrix U^x is the so-called CPHF coefficients at the given geometry x and satisfies

$$U^x = C^T S \frac{\mathrm{d}C}{\mathrm{d}x}, \text{ or } \frac{\mathrm{d}C}{\mathrm{d}x} = CU^x \tag{2}$$

The matrices C and S denote the MO coefficient and overlap, respectively. Next, we introduce 'displaced' overlap matrix defined by

$$M^{x_0} = C^{x_0 T} S^{x_0} C \tag{3}$$

where x_0 denotes the reference geometry. During the geometry optimization, we assume that the coordinate x is restricted within the neighborhood of the reference coordinate and the Taylor series of the MO coefficient C converges well to the first-order in the nuclear displacement, $x - x_0$. This would lead to an approximate equation

$$M^{x_0} = 1 + (x - x_0)U^x \tag{4}$$

Inserting Eq.(4) to Eq.(1) yields

$$0 = T^x = U^x - (U^x)^T = (x - x_0)^{-1}(M^{x_0} - (M^{x_0})^T) \tag{5}$$

Hence, a condition

$$(M^{x_0} - (M^{x_0})^T) = 0 \tag{6}$$

diminishes the unphysical orbital rotations. The MO ψ_i satisfying Eq. (6) is obtained by the orthogonal transformations of the canonical MOs $\{\tilde{\psi}_i\}$ as

$$\psi_j = \sum_i \tilde{\psi}_i W_{ij}^{x_0} \tag{7}$$

where the transformation W^{x_0} leading to a MOD is given by

$$W^{x_0} = \{(\tilde{M}^{x_0})^T \tilde{M}^{x_0}\}^{-(1/2)}(\tilde{M}^{x_0})^T \tag{8}$$

and

$$\tilde{M}^{x_0} = (C^{x_0})^T S^{x_0} \tilde{C} \tag{9}$$

where $\tilde{C}$ stands for the orbital coefficient matrix for the canonical MOs. The displaced overlap matrix $\tilde{M}^{x_0}$ is asymmetric and the transformation (8) rotates the canonical orbitals $\tilde{\psi}_i$ so as to make one-to-one correspondence to the reference MOs.

The expression of the energy gradient can be written in a general form as

$$\frac{\partial E_{\mathrm{corr}}}{\partial \alpha} = \sum_{pq} \gamma_{pq} F_{pq}^{\alpha} + \sum_{pq,rs} \Gamma_{pq,rs} (pq|rs)^{\alpha} - \sum_{pq} X_{pq} S_{pq}^{\alpha} + \sum_{p>q} (X_{pq} - X_{qp}) T_{pq}^{\alpha} \quad (10)$$

where γ_{pq} and $\Gamma_{pq,rs}$ are one and two electron reduced density matrices, respectively. The F_{pq}^{α}, $(pq|rs)^{\alpha}$, and S_{pq}^{α} are the skeleton Fock, two-electron, and overlap derivative integrals, respectively. The indices p, q, r, s denote the atomic orbitals. Eq. (10) explicitly includes the anti-symmetric part of the CPHF coefficient T_{pq}^{α}. This entity can be obtained from the coupled-perturbed (CP)-MOD equations [47].

$$\frac{\partial}{\partial \alpha} (M^{x_0} - (M^{x_0})^T) = 0 \quad (11)$$

which is the first-order derivative of Eq. (6).

Combining the perturbation selection, MOD and GSUM methods, the SAC/SAC–CI geometry optimization becomes efficient and applicable to moderately large molecules. These methods have been incorporated in the SAC–CI code connected with the Gaussian suit of programs [29]. The functions of the current SAC–CI program system are overviewed in Fig. 39.1. We also refer to the SAC–CI guide for the detailed explanation of the theory, program, performance, and applications [48].

Before closing this section, the extensions of the SAC/SAC–CI method to the quasi-degenerate and multi-reference cases are described. The first such proposal was made in 1985 as the MR–SAC method, in which the reference function was extended from the Hartee–Fock single-determinant to a general CI type wave function [34]. Further, based on the consideration on the nature of electron correlations in quasi-degenerate situations, a generalization of the exponential operator was made; i.e. $\mathscr{EXP}$ and EXP, etc. beside the standard exp operator, and the general exponentially generated wave functions were proposed [35]. The EGCI (Exponentially Generated CI)/Excited (Ex)-EGCI method was applied to the potential surface of some small molecules in the ground and excited states and compared with the full-CI results [36]. This method was also applied to the high-spin multiplicities [49]. The exponential-type operator used in the EGCI method is size-consistent, and efficiently generates the higher-order excitation operators. Therefore, this operator was also used in generating higher-order excitation operators in the SAC–CI general-*R* method [33]. The MEG (Mixed Exponentially Generated) and EX (Excited)-MEG methods were proposed as the generalization of the SAC–CI method to the multi-reference formalism [50]. The method uses the EGCI wave function for the reference functions. The MEG/EX-MEG method was examined for the potential energy curves of the ground and excited states of several small molecules including multiple bond and was shown to be accurate for the quasi-degenerate system by comparing with the full-CI results [50]. Recently, the MEG/EX-MEG method was applied to the excited and ionized states of ozone, in which the calculated spectra well reproduced the experimental observations.

39.3 EXCITED AND IONIZED STATES OF π-CONJUGATED ORGANIC COMPOUNDS

The SAC–CI method can be applicable to various kinds of molecular spectroscopy; excitation, ionization and electron-attached states in high accuracy. First, we overview the excitation [51–53] and ionization spectra [51,54] of five-membered ring compounds, furan and thiophene. For these spectra, pioneering theoretical spectroscopy was performed by the SAC–CI method [51] and has been followed by many theoretical works [52,53,55–58]. Comparisons of our old and new calculations show the importance of using the sufficiently flexible basis sets with valence and Rydberg functions as well as the reliable theory for the accurate descriptions of these spectra. We also explain the spectroscopy of *p*-benzoquinone (*p*-BQ), its anion radical [59], and aniline [60].

39.3.1 Excitation and ionization spectra of furan and thiophene

The electronic spectra of the five-membered ring compounds have been intensively studied by the experimental and theoretical works. These molecules are fundamental units in many important biological systems. Furthermore, their excitation spectra are benchmark examples for theoretical studies of molecular excited states [51,55–58]. For furan and thiophene, various types of excitation spectra were measured: the vacuum ultraviolet (VUV) spectrum, electron energy-loss (EEL) spectrum and magnetic circular dichroism (MCD) spectrum. The SAC–CI method offered consistent interpretations of these electronic spectra [51–53].

For furan, the underlying interesting complexities have been demonstrated in calculating the excited states; (i) strong valence-Rydberg mixing is essential, (ii) electron correlations are crucial and (iii) σ electron correlation occurs for some excitations. Though various high-level electron correlation theories were applied to the excitation spectra of furan, there were inconsistencies in the assignments. The SAC–CI calculation with the extensive basis sets clarified the details of these spectra [52]. Theoretical spectrum [52] together with the VUV spectrum [61] are shown in Fig. 39.2. The first band system occurs in the region of 5.5–7.0 eV. The 1^1B_2 state was computed at 6.40 eV for this peak. This state is characterized as the π-π^* excitation mixed with the Rydberg transitions, which is also seen in the second moment of this state $< r^2 > = 93\ au$ in comparison with the ground state $< r^2 >= 65\ au$. The largest intensity in the VUV spectrum occurs from 7.2 to 8.4 eV. The SAC–CI method calculated the 6^1A_1 state at 8.34 eV with the large oscillator strength of 0.48. This state also has the valence-Rydberg mixing character. For the band III at 8.7 eV, 9^1B_2 state was calculated at 9.08 eV with large second moment 130 au, which clearly shows that this state has significant diffuse character. In the excited states furan, two series ($1a_2$ and $2b_1$) of low-lying Rydberg states and the valence π–π^* excited states strongly influence to each other.

For thiophene, very interesting changes occur in both ground and excited states by a replacement of oxygen with sulfur. The SAC–CI [53] and VUV [62] spectra of thiophene are compared in Fig. 39.2. The first VUV band, historically denoted as bands A and B, is

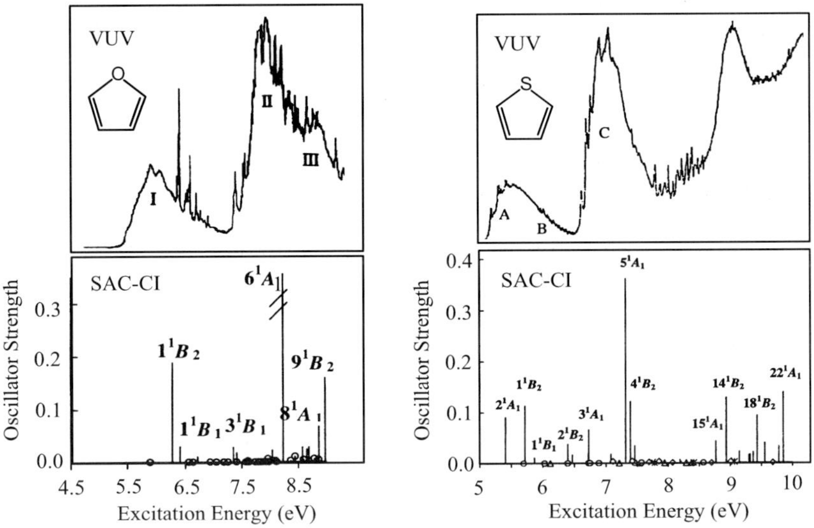

Fig. 39.2. VUV [61,62] and SAC–CI excitation spectra [52,53] of furan and thiophene.

located in the energy region of 5–6.5 eV. The MCD spectrum shows two peaks at 5.27 and 5.64 eV. Correspondingly, the 2^1A_1 and 1^1B_2 states are calculated at 5.41 and 5.72 eV. The second moments of these states, 85 and 86 au, in comparison with that of the ground state, 82 au, show that these excited states are $\pi-\pi^*$ valence excitations. The second VUV band (band C) exists in the region of 6.5–7.7 eV and the band maximum is at around 7.05 eV. The SAC–CI calculation gives the 5^1A_1 and 4^1B_2 states at 7.32 eV and 7.40 eV, respectively, for this band. The fine structure in 7.7–8.8 eV was suggested by the calculations to represent the Rydberg excitations. Most of them were identified as two Rydberg series converging to the ionized states $1a_2^{-1}$ (8.87 eV) and $3b_1^{-1}$ (9.52 eV).

Accurate investigation of the valence ionization spectra is important subject to elucidate the electronic structure of molecules. Ionization spectra of five-membered aromatic compounds have also been intensively studied. The high-resolution synchrotron radiation photoelectron spectra (SRPES) of furan and thiophene were measured and analyzed with asymmetry parameter up to about 40 eV [63,64]. The electron momentum spectroscopy (EMS) was also applied to furan up to 30–40 eV [65]. The ionization spectra of these molecules were also studied by several theoretical methods. However, there were some controversial assignments even for the outer-valence region, in particular for the peak position of $1b_1(\pi_1)$ state and the inner-valence spectra have not been theoretically reproduced.

The SAC–CI general-*R* method is useful for investigating the satellite spectrum of molecules; it describes the multi-electron process very accurately. The SAC–CI spectrum [54] of furan is shown with the SRPES spectrum [64] in Fig. 39.3. The SAC–CI method reproduces the spectrum very accurately in both peak positions and intensities. In the outer-valence region, some congested main peaks exist and the theory gives accurate assignments for them. In particular, the π_1 state was calculated to split into two peaks at

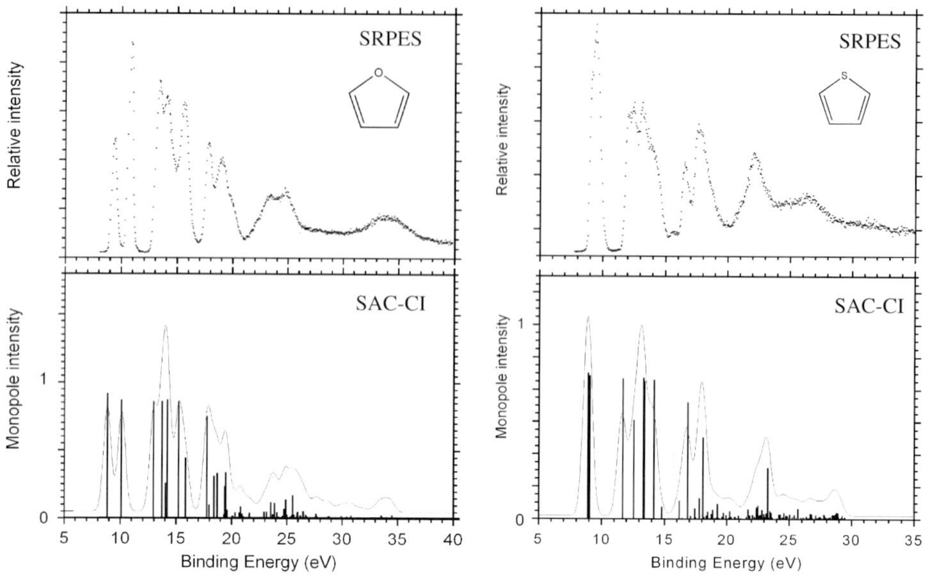

Fig. 39.3. Photoelectron [64] and SAC–CI ionization spectra [54] of furan and thiophene.

14.03 and 15.84 eV by the interaction with the two-electron process, $1a_2(\pi)^{-2}3b_1(\pi^*)$. This agrees with the experimental observation [65], in which two peaks were suggested at 13.6 and 15.6 eV. The effect of electron correlations of furan is remarkable in higher outer-valence region of 18–20 eV and many split peaks appear. In the inner-valence region, the breakdown of the orbital picture is prominent and the intensities due to the $3b_2$ and $5a_1$ orbitals are distributed to many shake-up states.

The valence ionization spectrum of thiophene was also studied up to about 30 eV [54]. One of the important features should be a satellite peak observed at 15.66 eV by the He I PES and Penning ionization electron spectroscopy [66]. Accordingly, shake-up states of b_1 symmetry were calculated in the outer-valence region with the considerable intensity at 14.78 and 16.19 eV: these states were characterized as $1a_2(\pi)^{-2}4b_1(\pi^*)$ and $3b_1(\pi)^{-2}4b_1(\pi^*)$. In the higher outer-valence region of 17–19 eV, three main peaks were calculated and some shake-up states with small intensities were obtained around it, which shows the electron correlations are not so significant in comparison with furan. The characteristic spectrum in the inner-valence region is also well reproduced in the SAC–CI calculation. A prominent peak was observed at 22.1 eV by the SRPES [63] and the strong peak characterized as the ionization from the $4b_2$ orbital was also obtained by the SAC–CI calculation. This feature is different from the satellite peaks of furan and pyrrole in this region.

39.3.2 *p*-Benzoquinone and its anion radical

Quinones work as the electron-acceptor in chemical and biological systems. Ubiquinone is seen in the photosynthetic reaction center in the photosynthetic bacteria and the

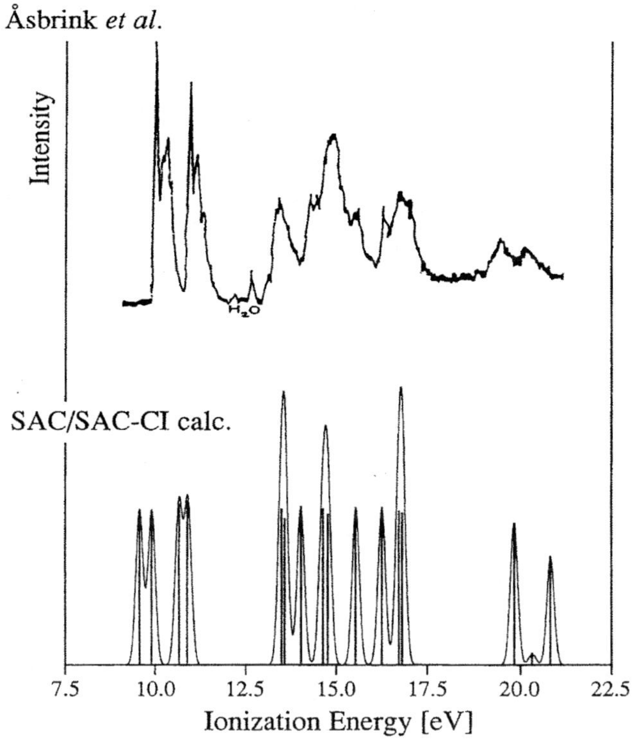

Fig. 39.4. SAC–CI ionization spectrum [59] of *p*-benzoquinone compared with the experimental spectrum observed by Åsbrink et al. [68].

photosystem II in the green plants [67]. Among the quinones, *p*-benzoquinone (*p*-BQ) is well known as a key compound, and there are many experimental and theoretical studies focused on the electronic structure of this molecule. However, for the ionization spectrum, it was difficult to make a definitive assignment, especially in the high energy region, due to the lack of the fine agreement between the theoretical and experimental results [68–70]. In addition, the anion radical of *p*-BQ has several optically forbidden excited states below the lowest allowed state [71]. Since these states lie in a narrow energy range, a highly accurate calculation is necessary to give a reliable assignment.

The SAC–CI calculations were performed for the excited and ionized states of *p*-BQ in the neutral and anion states [59]. In Fig. 39.4, the SAC–CI ionization spectrum of *p*-BQ is compared with the experiment. The calculated ionization spectrum shows good agreement with the experimental spectrum and the reliable assignments were given for the ionization peaks up to *ca.* 20 eV. For the lowest-four peaks, our result supported the previous assignments given by MR–CISD [69] and EOM–CCSD [72] studies: n_g^-, n_u^+, π_u^+, and π_g^+ in increasing order of energy. For the higher ionized states, the SAC–CI study gave new assignments [59], which were different from those of the previous calculations.

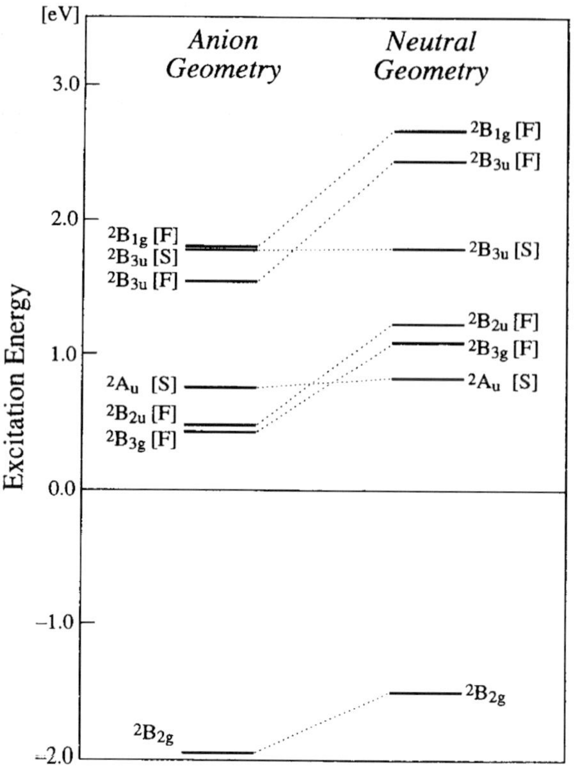

Fig. 39.5. Geometry dependence of the excited states of the *p*-benzoquinone anion radical calculated by the SAC–CI method [59]. The energy origin is the singlet ground state at the neutral geometry. [S] and [F] represent shape and Feshbach resonance states, respectively.

The excited states of the *p*-BQ anion radical were also studied [59]. In Fig. 39.5, the energy levels of the ground and excited states are shown. We compared the energy levels for the two geometries: equilibrium structures of anion and neutral species. The ground state of the anion radical is $^2B_{2\ g}$ state. The adiabatic electron affinity was calculated to be 1.96 eV in comparison with the experimental values, 1.86 [71], 1.89 [73], and 1.99 eV [74]. The lowest allowed excitation of *p*-BQ anion radical was assigned to 1^2A_u ($\pi_g^{-*}{}_{\mathrm{SOMO}} - \pi_u^{-*}$ shape resonance) state. It seemed that the previous assignment of the lowest allowed transition to the $^2B_{3u}$ state [75] was due to the exchange of the ^2A_u and $^2B_{3u}$ states by the hydrogen bonds with water solvent, which was supported by the SAC–CI calculations for *p*-benzosemiquinone (*p*-BQH) radical [59]. There are dipole forbidden excited states below the first allowed state [71]. Two forbidden n-π^*_{SOMO} transitions, $^2B_{3g}$ and $^2B_{2u}$ states, were calculated below the lowest peak (^2A_u) in the anion geometry. In the neutral geometry, however, these transitions were calculated to be less stable than the ^2A_u state. There are two states observed above the ^2A_u state. They are assigned to $^2B_{3u}$ states ($\pi_u^{+} - \pi_g^{-*}{}_{\mathrm{SOMO}}$ and $\pi_g^{-*}{}_{\mathrm{SOMO}} - \pi_u^{+*}$): the order of these states depends on the geometries, i.e. $\pi_g^{-*}{}_{SOMO} - \pi_u^{+*}$ (Feshbach resonance) was lower in the anion geometry, whereas $\pi_u^{+} - \pi_g^{-*}{}_{SOMO}$ (shape resonance) was lower in the neutral geometry. This ordering was shown to be explained by the effect of geometry on the orbitals.

39.3.3 Aniline: Effect of the amino-group conformation to the excitation spectrum

Aniline is one of the important monosubstituted benzenes. The electronic interaction between the NH_2- and benzene ring causes the new bands in the absorption spectrum. Some of the new bands include charge-transfer (CT) character between the benzene ring and the substituent [60]. There was controversy in the literature on the character of the low-lying excited states [76–80]. The significant problem was that the previous calculations used a planar conformation, although the amino group is pyramidal in the equilibrium geometry [60]. Actually, the conformation of the amino group plays a key role in the character of the excited states. These CT states can be related to the twisted intramolecular CT (TICT) states which are extensively studied in *N*, *N*-dimethylaminobenzonitrile (DMABN). The SAC–CI calculations were performed for aniline having typical three conformations of the amino group, in order to investigate how the conformation of the amino group affects the character of the excited states [60]. The SAC–CI ionization potentials were also compared with the previous experiments and the OVGF (Outer Valence Green's function) [81] calculations.

In Fig. 39.6, the SAC–CI excitation spectra of two different conformations are compared with the experimental spectrum. In the spectrum of equilibrium form, five peaks (I–V) were assigned to $1^1A''$ ($\sim {}^1B_2$), $3^1A'$ ($\sim {}^1A_1$), $4^1A''$ ($\sim {}^1B_2$), $6^1A'$ ($\sim {}^1A_1$), and $7^1A''$ ($\sim {}^1B_2$) states [60], respectively, with the average error of 0.14 eV. These states are characterized as $\pi-\pi^*$ excited states. The $1^1A''$ state (Peak I) was characterized as the charge-resonance (CR) state, and the $3^1A'$ (Peak II) and $6^1A'$ (Peak IV) states were the local excitation (LE) within the benzene ring. Based on the Mulliken charges and the dipole moments, the $4^1A''$ (Peak III) and $7^1A''$ (Peak V) states have back CT (BCT) character from the ring to amino group. With the planar geometry which was adopted in the previous studies, the resultant spectrum was similar to that of the pyramidal equilibrium structure. However, the geometry greatly affects the amount of charge-transfer. For example, the Peak II was assigned to a mixed state of the CT + LE characters. This implies that the assumption of planar conformation may cause incorrect assignments, in particular for the CT character. Change in the CT character was explained in terms of mixing of π and σ natures in the pyramidal structure. Such $\sigma-\pi$ interactions are, however, forbidden in the planar structure.

The CT character of the excited states is significantly affected by the rotation of the amino group. Fig. 39.6(c) shows the excitation spectrum in the twisted form. The amino group was rotated by 90°. Both spectral shape and nature of the excited states were largely altered by twisting. The $2^1A''$ and $4^1A''$ states have explicit CT character (NH_2– to C_6H_5–), while $5^1A'$–$8^1A'$ states BCT character. This indicates that aniline can have TICT states as in DMABN.

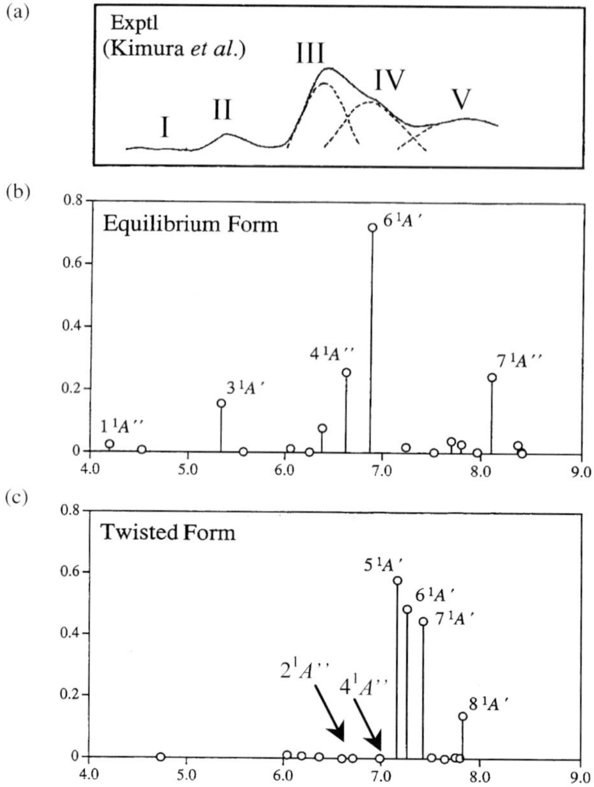

Fig. 39.6. Excitation spectra of aniline; (a) Experimental spectrum in the gas phase, (b) SAC–CI spectrum in the pyramidal structure (equilibrium structure), and (c) SAC–CI spectrum in the twisted form [60].

39.4 COLLISION-INDUCED ABSORPTION SPECTRA OF CsXe SYSTEM

The dipole forbidden transitions, S–S and S–D transitions of alkali atoms, are induced by the collisions with the rare gas atoms [82]. This phenomenon is called collision-induced absorption. The position and the profile of the collision-induced absorption sensitively reflect the interatomic potentials of the ground and excited states of the alkali-noble gas system and the (induced) transition moments between them. The SAC–CI method elucidated the detailed mechanism of the collision-induced absorption spectra of CsRg (Rg = Ne, Ar, Kr, Xe) system [83,84].

For calculating the reduced absorption coefficients directly observed in the spectra, the quasi-static approximation [85] was used. From this approximation, the collision-induced absorption spectra are determined by the following three quantities, (1) excitation energies, (2) Boltzmann distribution of the ground state and (3) induced transition moments as the functions of the internuclear distance.

The SAC–CI potential curves of the 6S, 6P, 5D and 7S states of CsXe are shown in Fig. 39.7. The $6p\Sigma$ and $5d\Sigma$ states are the excitations from the non-bonding MO

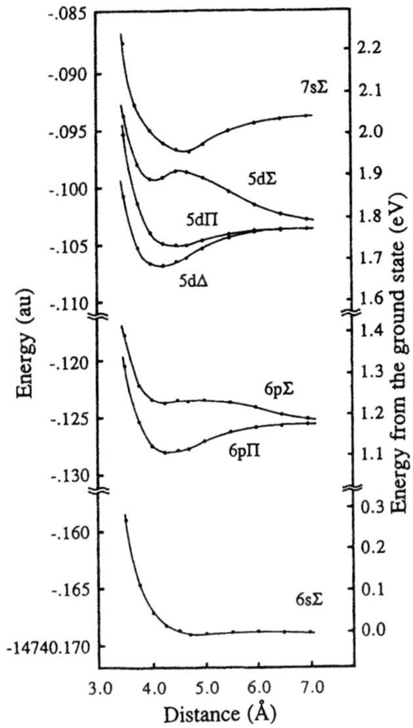

Fig. 39.7. Potential energy curves of the 6s, 6p, 5d and 7s states of the CsXe system without spin-orbit coupling [84].

to the weakly anti-bonding MOs, so that the potential curves become repulsive. The 5dΣ state has a characteristic hump at 4.6 Å, which is caused by the avoided crossing with the 7sΣ state, and the 6pΣ state has a shoulder at 5.0 Å. The 6pΠ, 5dΠ, and 5dΔ states have shallow minima in their potential curves. The ion core of the Cs atom polarizes the electron density of the rare gas atom, which is responsible for the attractive force.

The induced transition dipole moments by the SAC–CI method are shown in Fig. 39.8 as a function of the internuclear distance. These induced moments are mainly due to the intra-atomic orbital mixing of the p-component of the Cs atom, caused by a reduction of the spatial symmetry of the system. The 6sΣ–5dΣ transition moment is induced at larger internuclear distances than that of 6sΣ–7sΣ. The magnitude of the transition moment is reversed between the 5dΣ and 7sΣ states at the avoided crossing point.

The calculated collision-induced absorption spectra associated with the 5D line are shown in Fig. 39.9(a) with the experimental one. The experimental peak in the blue side of this transition was attributed to the 6sΣ → 5dΣ transition of the collision complex. In the theoretical spectrum, R denotes, in the classical picture, the internuclear distance of the collision complex at which the absorption occurs. The steep cusp exists in both experimental and theoretical spectra and is due to the extremum in the excitation energy

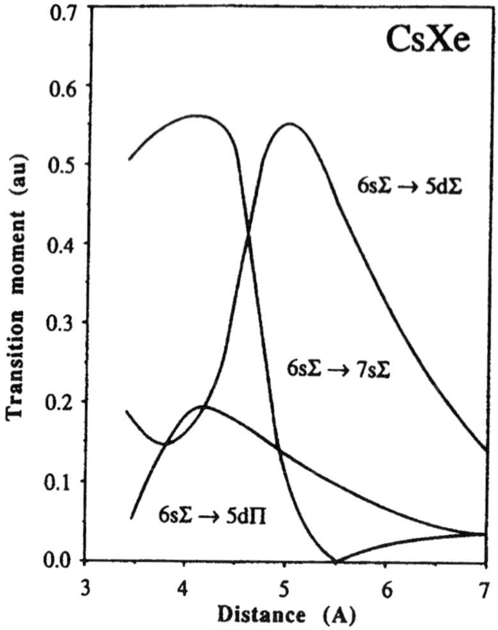

Fig. 39.8. Transition dipole moments for the $6s\Sigma \rightarrow 5d\Sigma$, Σ, $6s\Sigma \rightarrow 5d\Pi$, and $6s\Sigma \rightarrow 7s\Sigma$ excitations of the CsXe system [84].

dependence on R. From the R-values in the theoretical spectrum and the potential curves, we see that this extremum is due to the hump of the $5d\Sigma$ potential lying between 4.5 and 5.0 Å. Namely, the cusp in the spectrum is assigned as being due to the hump of the $5d\Sigma$ potential. Both experimental and theoretical spectra have shoulders on the right-hand side of the cusp. From the theoretical spectra, we see that this shoulder is due to the absorptions in the region of $R = 5-6$ Å. The absorption band observed in the red side of the 6S–5D transition of Cs is due to the $6s\Sigma \rightarrow 5d\Pi$ transition of the collision complex. This transition was calculated to have smaller intensity than the $6s\Sigma \rightarrow 5d\Sigma$ transition, in agreement with the experiment.

Fig. 39.9(b) shows the collision-induced absorption spectra owing to the $6s\Sigma \rightarrow 7s\Sigma$ transition of the collision complex. The 6S–7S atomic line was calculated at 607 nm in comparison with 540 nm of the experimental spectrum. The induced absorption peak separated by about 30 nm in the experimental spectrum was reproduced in the theoretical one separated by 23 nm. This peak is due to the potential minimum of the $7s\Sigma$ state. The vibrational structure of the experimental spectrum indicates an existence of the attractive well in the excited state, which certainly exists in the potential curve shown in Fig. 39.7.

It was shown that the spectral peaks, shoulders and shapes reflect the detailed characters of the CsXe collision complex. Having such an understanding of the absorption spectra, we can increase the insight about the nature of the collision dynamics and the formed complexes.

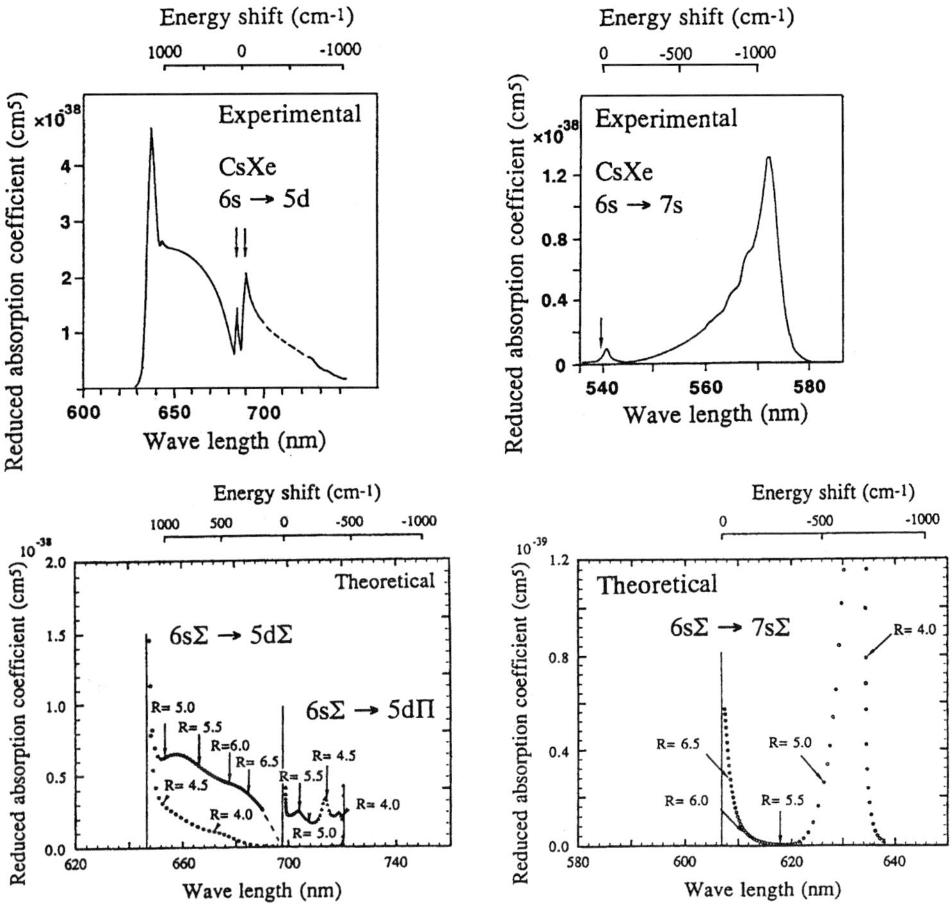

(a) 5d line (b) 7s line

Fig. 39.9. Reduced absorption coefficients for the (a) 5d and (b) 7s states of CsXe. The experimental spectrum is due to Moe et al. [82], and the SAC–CI theoretical spectrum is from ours [84].

39.5 TRANSITION METAL COMPLEXES

39.5.1 CrO_2Cl_2

Chromyl chloride (CrO_2Cl_2) has been intensively studied because of its strong and specific ability to oxidize various organic compounds [86]. Its absorption spectrum has also been the subject of various experimental investigations [87]. Fig. 39.10 shows the comparison between the experimental [87] and SAC–CI spectra [88] of chromyl chloride CrO_2Cl_2. The right-hand side is the spectrum in the lower-energy region and the left-hand side is for higher energy region. The overall feature of the experimental spectrum is well

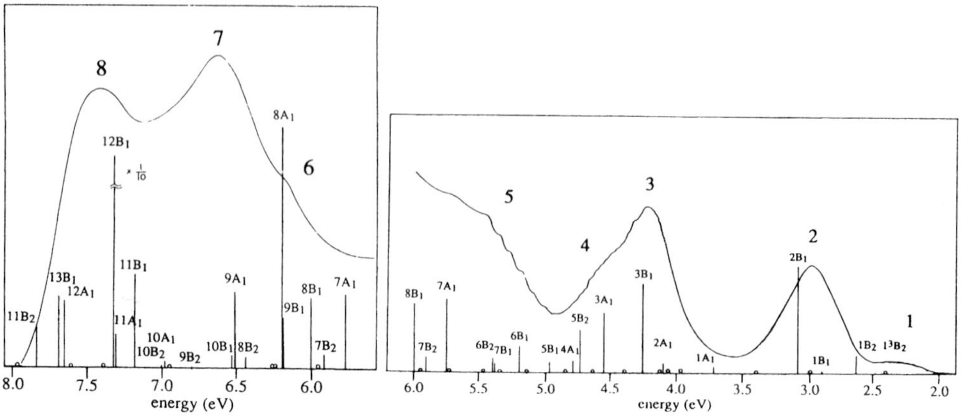

Fig. 39.10. SAC–CI theoretical excitation spectrum of CrO_2Cl_2 [88] compared with the experimental absorption spectrum [87].

reproduced by the SAC–CI spectrum. Since this molecule is C_{2v}, there are many dipole allowed peaks in comparison with those of T_d or O_h symmetric molecules. Partially, for this reason the observed peaks are composed of many transitions.

The excitation spectrum is classified into three energy regions with respect to the excitation character. Lower energy excited states below ~5 eV are characterized as transitions from the non-bonding ligand MOs to the anti-bonding Cr-ligand MOs, while the higher energy excited states above ~7 eV are characterized as the transitions from the bonding Cr + ligand MOs to the anti-bonding Cr − ligand MOs. In the medium energy region of ~6 eV, a strong mixing between these two types of transitions occurs and the spectrum becomes rather complicated.

For this molecule, few theoretical studies [87,89] have been applied to the relatively low excited states. Fig. 39.11 shows a comparison of the theoretical results for the excitation spectrum of CrO_2Cl_2 with the experimental spectrum. Among the theoretical spectra, the SAC–CI result agrees well with the experimental spectrum and the quantitative assignments are possible in the whole energy region.

39.5.2 Tetraoxo complexes: CrO_4^{2-}, MoO_4^{2-}, MnO_4^{2-}, TcO_4^{2-}, RuO_4^{2-} and OsO_4^{2-}

Four coordinated oxo-metal complexes are important as oxidizing agent. Their visible and ultraviolet spectra have been reported in 1960s [90]. Some complexes show interesting photochemistry as observed for MnO_4^{2-} [91]. Systematic SAC–CI studies on the electronic spectra of the tetraoxo metal complexes, CrO_4^{2-} [92], MoO_4^{2-} [93], MnO_4^{2-} [94], TcO_4^{2-} [95], RuO_4^{2-} [96] and OsO_4^{2-} [96] were reported.

Fig. 39.12 summarizes the SAC–CI spectra for the lowest three excited states of six tetraoxo complexes. The SAC–CI calculations were the first applications that successfully reproduced the experimental absorption spectra of the transition metal complexes. The singly excited CI (SECI) [97] and Xα [98] calculations often gave even qualitatively

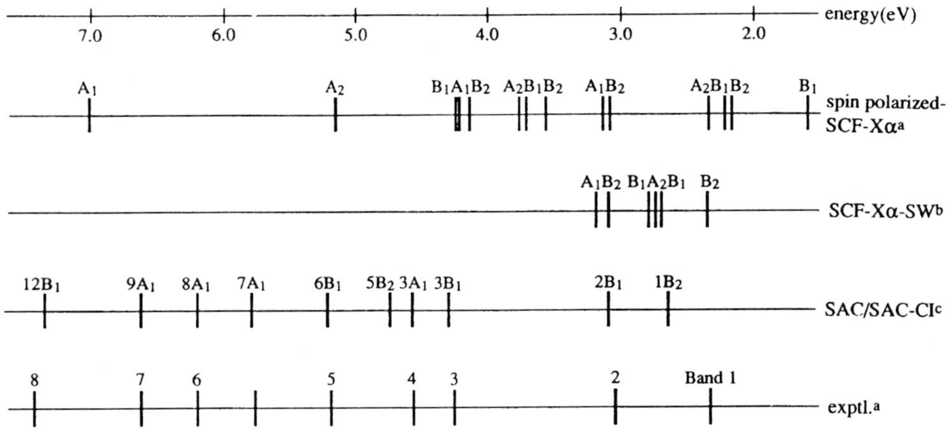

Fig. 39.11. Summary of the *ab initio* calculations of the excitation energies of CrO_2Cl_2; (a) Jasinski et al. [87] (b) Miller et al. [89], and (c) SAC–CI [88].

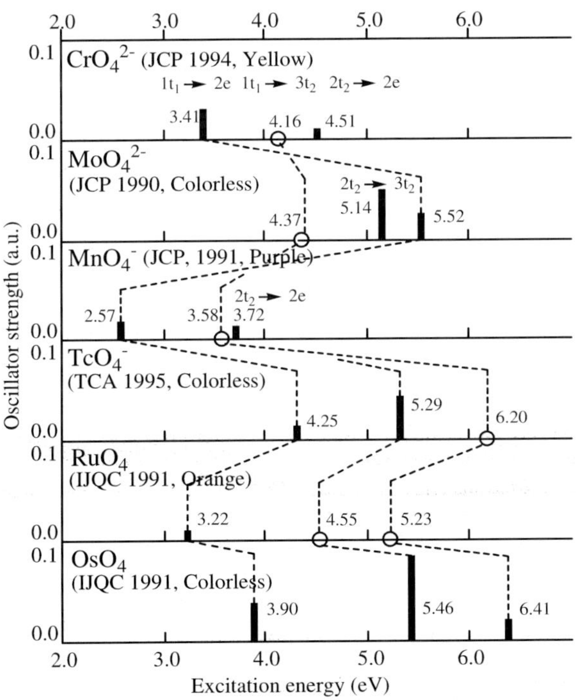

Fig. 39.12. The lowest three excited states of the tetraoxo transition metal complexes. The dotted lines show the correspondence of the excited states among the complexes.

wrong assignments. The averaged deviation of the SAC–CI excitation energies from the experiment was 0.24 eV. These results were rather accurate considering that these calculations were performed in the early 1990s. The largest error was found in the case of RuO_4^{2-} and OsO_4^{2-}, in which the effective core potential (ECP) was used.

The color of the metal complexes reflects the excitation energy of the first optically allowed 1^1T_2 state. Only MnO_4^- has the dipole allowed transition in the visible region (1.6–2.9 eV). Except for MoO_4^{2-}, the lowest allowed 1^1T_2 state was assigned to the excitation from $1t_2$ orbital to 2e orbital. The $1t_2$ and 2e orbitals are characterized as the lone-pair orbital of oxygen, O(p), and the anti-bonding interaction between the metal d- and oxygen p-orbitals, M(d)–O(p), respectively. On the other hand, in MoO_4^{2-}, the main configuration of this state is $1t_2 \rightarrow 3t_2$, in which the $3t_2$ orbital originates from Mo s- and p-orbitals. In Fig. 39.12, we can see that the excitation energy of the $1T_2$ state characterized as $1t_2 \rightarrow 2e$ decreases as the central metal atom is replaced to the upper-right-hand side in the periodic table. This tendency is explained by the structure of the complexes; the bond distance between the metal and the oxygen ligands [92].

39.5.3 Excited states and ^{95}Mo NMR chemical shift of $MoO_{4-n}S_n^{2-}$ ($n = 0$–4) and $MoSe_4^{2-}$

A systematic measurement of the electronic spectra was reported for the molybdenum complexes, $MoO_{4-n}S_n^{2-}$ ($n = 0$–4) and $MoSe_4^{2-}$ [99]. The change in the ligand environment causes a variety of the spectral feature, due to the difference in the nature of metal (M)–ligand (L) bonding. For the six molybdenum complexes studied [93], the SAC–CI calculations gave new reliable assignments. All of the observed peaks except for the peak at 5.99 eV of MoS_4^{2-} are due to the electron transfer excitations from the L to M or from L to the $(M-L)^*$ (anti-bonding). The electronic spectrum of MoO_4^{2-} is much different from those of MoS_4^{2-} and $MoSe_4^{2-}$, since the M–L bonds of MoO_4^{2-} are stronger and more ionic than those of MoS_4^{2-} and $MoSe_4^{2-}$. As the number of the soft ligands (S and Se) increases, the excitation energies of $L \rightarrow (M-L)^*$ transitions decreases, since the M–L bonds become weaker. In the MoO_3S^{2-}, $MoO_2S_2^{2-}$ and $MoOS_3^{2-}$, the observed bands consist of many dipole allowed peaks because of the splitting of the excited states due to the symmetry lowering from T_d to C_{3v} and C_{2v}.

In the study of the electronic spectra of Mo compounds, we concentrate on the dipole allowed transitions. However, the magnetically allowed states, which are not necessarily to be optically allowed, play very important role in the chemical shift [100]. In the case of the ^{95}Mo NMR chemical shifts of $MoO_{4-n}S_n^{2-}$ ($n = 0$–4) and $MoSe_4^{2-}$ [101], the paramagnetic term σ^{para} was dominant and is expressed in the perturbation form as,

$$\sigma_{Mtu}^{para} = -\frac{e^2}{4m^2c^2}\sum_{n\neq 0}\frac{1}{(E_n - E_0)}\left[\left\langle 0\left|\sum_j r_{Mj}^{-3}L_{Mjt}\right|n\right\rangle\left\langle n\left|\sum_j L_{ju}\right|0\right\rangle + \left\langle 0\left|\sum_j L_{ju}\right|n\right\rangle\left\langle n\left|\sum_j r_{Mj}^{-3}L_{Mjt}\right|0\right\rangle\right] \quad (12)$$

where $|0\rangle$ and $|n\rangle$ denote the ground and the excited states, respectively, L_{jt} is the angular momentum operator of the j-th electron in t direction, and the subscript M denotes the nucleus under consideration. The summation is taken over all the excited states. As seen in Eq.(12), only the magnetically allowed transition to the excited state $|n\rangle$ contributes to the chemical shift. The SAC–CI excitation energy was compared with the NMR chemical shift of $MoO_{4-n}S_n^{2-}$ $(n = 0–4)$ and $MoSe_4^{2-}$ [100]. In Fig. 39.13, the relationship between the inverse of the lowest magnetically allowed d–d* excitation energies and the chemical shifts was shown. The experimental values [102] were used for the chemical shifts. The 4dσ → 4dπ* transitions belong to the magnetically allowed transition. There is a nice linear relationship between the chemical shift δ and the inverse of the ΔE. This result indicates that the origin of the Mo chemical shift is attributed to the variations in the excitation energy of the magnetically allowed d–d* transition, namely,

$$\delta = \mathrm{A}(1/\Delta E_{\mathrm{ref}} - 1/\Delta E) = \alpha + \beta/\Delta E \tag{13}$$

Although a linear relationship with the lowest *optically allowed* transition was sometimes discussed in the literature [103], the optically allowed states are not necessarily magnetically allowed. We have shown that the relationship between the inverse of the lowest *optically allowed* transition energy and the chemical shift showed worse linearity [100].

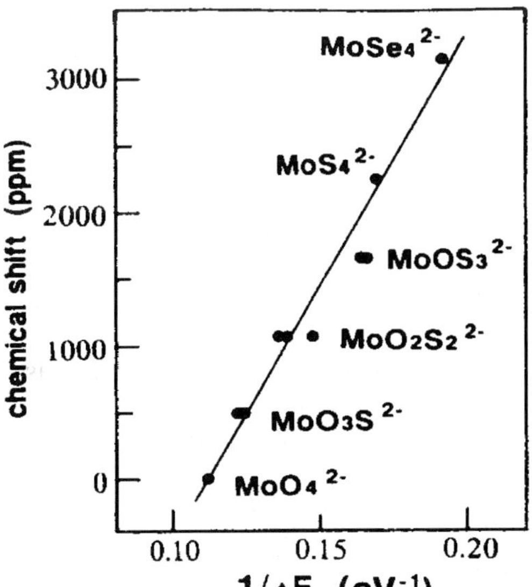

Fig. 39.13. Relationship between the inverse of the lowest magnetically allowed d–d* excitation energies ΔE and the chemical shifts [100]. The chemical shifts are the experimental values reported by Gheller et al. [102] and the excitation energies are obtained by the SAC–CI method [93].

39.6 PHOTOCHEMISTRY OF TRANSITION METAL COMPLEX, $Ni(CO)_4$

Transition metal carbonyl is important in laser chemistry as sources of metal atoms or as precursors for chemical vapor deposition (CVD) [104]. $Ni(CO)_4$ shows a typical photofragmentation reaction initiated by the XeCl laser (308 nm, 4.03 eV), and the knowledge on the mechanism is valuable for the design and control of the laser-induced CVD. The SAC–CI method was applied to the excitation spectrum and the potential energy curves relevant to the photofragmentation reaction [105].

In Fig. 39.14, the SAC–CI theoretical spectrum is compared to the experimental one. The calculated excitation energies and the oscillator strengths showed a good agreement with the experimental absorption spectrum. On the other hand, the SECI calculation could not reproduce the experimental spectrum particularly in the energy region higher than 5.2 eV (~240 nm) as shown in Fig. 39.14. All the observed peaks were assigned to the excitations to the dipole allowed T_2 states. The strong third peak was assigned to the $3T_2$ and $4T_2$ states. The main configurations for the $1T_2$ and $2T_2$ states are the excitations, $9t_2 \rightarrow 12t_2$ and $9t_2 \rightarrow 4e$, respectively, and both are assigned to Ni 3d $\rightarrow$ [Ni–CO] π^*. Invisible low-lying states 1E and $1T_1$ states exist below the $1T_2$ state and both are

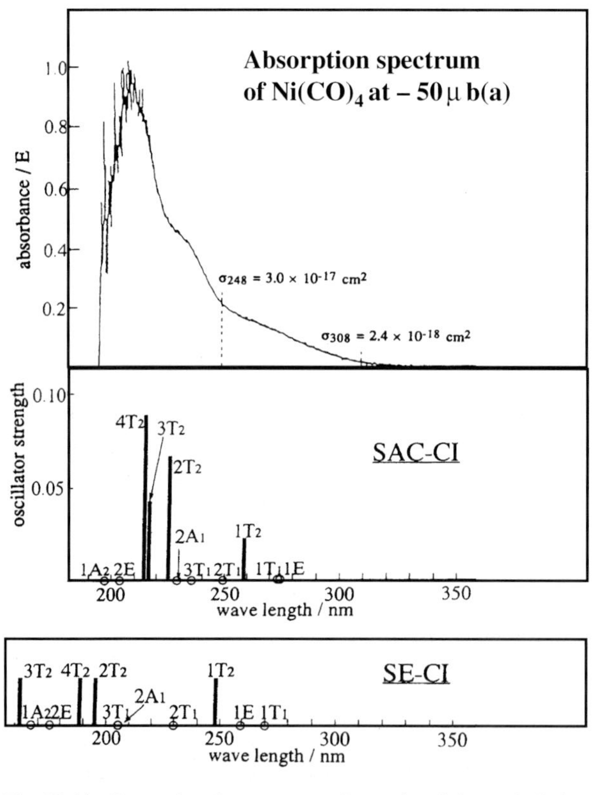

Fig. 39.14. Comparison between experimental and theoretical absorption spectra of $Ni(CO)_4$ [105].

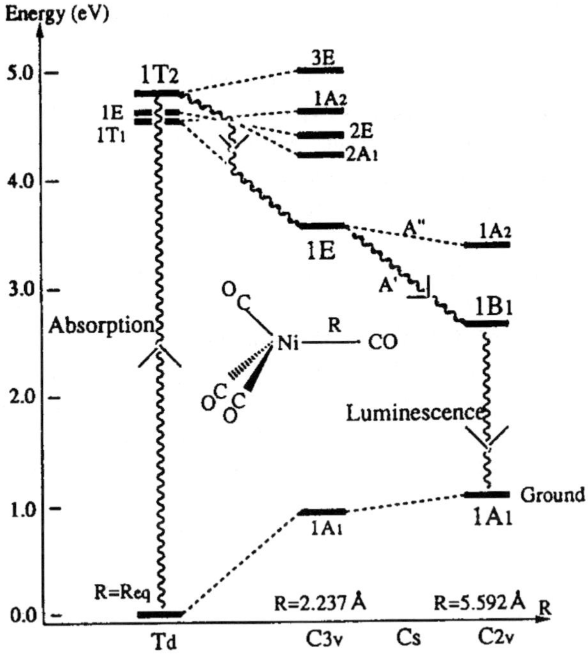

Fig. 39.15. Overall energy diagram and the pathway for the photofragmentation reaction of $Ni(CO)_4$ [105].

characterized as $9t_2 \rightarrow 12t_2$. These invisible states play important roles in the photofragmentation reaction.

The mechanism of the photofragmentation reaction and the luminescence is illustrated in Fig. 39.15. First, $Ni(CO)_4$ is excited to the allowed $1T_2$ state by the 308 nm laser. The geometry of the complex is distorted along the repulsive potential of the $2A_1$ state. The excited complex in the $2A_1$ state makes an internal conversion to the lowest excited state 1E, as the Kasha's rule implies. In this 1E state, the $[Ni(CO)_3-CO]^*$ is further distorted, partially by the Jahn–Teller effect, and fragmented into the $1B_1$ excited state of $Ni(CO)_3$ and the ground state of CO. The excited $Ni(CO)_3$ has the bonding potential curve for the Ni–CO bond, so that further CO dissociation does not occur in this process. Then, the complex is de-excited to the ground state by the radiative decay. The calculated emission energy 1.54 eV agree well with the experimental luminescence energy, 1.7–2.0 eV.

39.7 PORPHYRINS AND RELATED COMPOUNDS

In this section, we briefly review the SAC–CI applications to the excited states of porphyrin compounds: FBPc [106] and bacterial photosynthetic reaction center of *Rhodopseudomonas viridis* [107]. Phthalocyanine is a famous chromophore that has a large absorbance in the visible region of the spectrum. By systematically comparing with the excited states of FB porphin (FBP) and FB tetrazaporphin (FBTAP), the electronic mechanism how phthalocyanine shows such a strong absorption was explained [106]. The SAC–CI method was also applied to the excited states and electron transfers of a bacterial photosynthetic reaction center of *Rhodopseudomonas viridis* [107–109]. The absorption spectrum of the reaction center was successfully assigned [107,109], and the enigmatic unidirectionality of the electron transfer mechanism was analyzed by using the SAC–CI wave functions [108]. The TD–DFT method is often applied to the excited states of these large molecular systems because of its low computational cost. However, it should be noticed that the TD–DFT method is not as reliable as the SAC–CI method. The TD–DFT assignment for the excited states of some porphyrins is sometimes problematic; for example, the oscillator strength of B band of free-base porphin (FBP) was calculated to be very small, while that of N band was extraordinary large [110]. This strange behavior depends on the functional used. There is no simple way to achieve a convergent limit. For the reliable theoretical spectroscopy, it is important to calculate both energy and oscillator strength accurately.

39.7.1 Excited states of free-base phthalocyanine

FBPc, the well-known pigment, has much larger Q-band absorption than FBP. FBPc has tetraza-substitution at the four meso-positions and the butadiene units are attached to the four pyrrole rings. Therefore, free-base tetrazaporphin (FBTAP) is the structure intermediate between FBPc and FBP. In Fig. 39.16, the experimental and SAC–CI spectra of FBPc [106] are compared. The SAC–CI calculations successfully reproduced the experimental spectrum, which realized the detailed assignment and interpretation of the electronic mechanism for the strong absorption in FBPc.

As seen in Fig. 39.17, the orbital energy levels of FBTAP and FBPc change from those of FBP especially in the HOMO and next-HOMO levels. Since 2p level of nitrogen atom is relatively low, next-HOMO levels are stabilized in FBTAP and FBPc. Therefore, near degeneracy in the main configuration of Q-bands is somehow relaxed, which results in incomplete cancellation of transition moment, as schematically shown in Fig. 39.17. In FBPc, orbital-specific interaction with the butadiene unit raises HOMO level. Consequently, the weight of the two main configurations becomes more unbalanced. Therefore, FBPc shows the strongest absorption of the three compounds in the visible region of the spectrum.

Generally, in the porphyrin compounds, such breakdown of the quasi-degeneracy of the two main configurations causes an increase in the intensities of the Q bands due to incomplete cancellation of the two contributions to the transition dipole moment. This can be a general strategy for designing pigment. Tetraza substitution in meso position and tetrabenzo substitution in the pyrrole rings changed the HOMO and next-HOMO levels. On the other hand, relaxing the degeneracy of LUMO and next-LUMO levels also can be the subject of molecular design. Newly designed pigment using such a strategy also has a possibility for the new useful pigment.

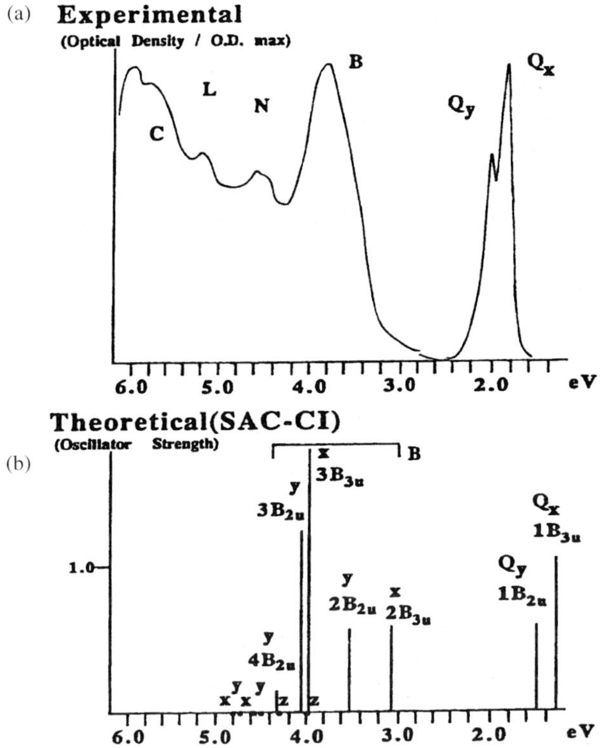

Fig. 39.16. Excitation spectra of free-base phthalocyanine; (a) Experimental spectrum in a gas phase [146] and (b) SAC–CI theoretical spectrum [106].

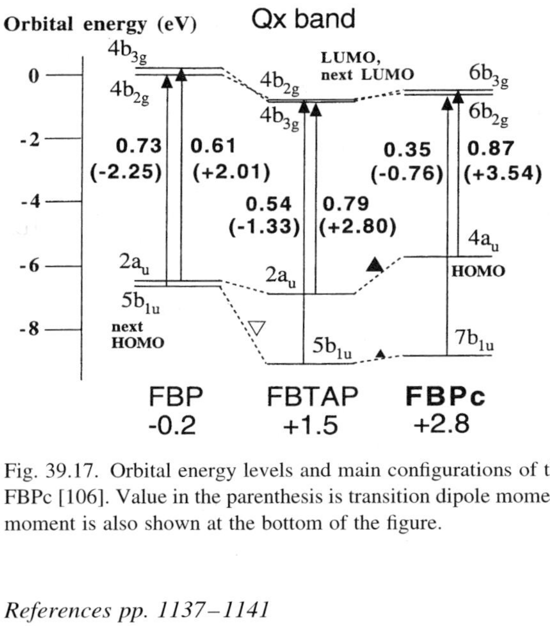

Fig. 39.17. Orbital energy levels and main configurations of the 1^1B_{3u} states (Q_x band) of FBP, FBTAP, and FBPc [106]. Value in the parenthesis is transition dipole moment of the configuration. The net transition dipole moment is also shown at the bottom of the figure.

39.7.2 Bacterial photosynthetic reaction center

Developments in structural biology have realized the detailed analysis of biological systems in molecular level. Increasing amount of the interesting biological-molecular structures is available in the electronic database as represented by the Protein Data Bank. Therefore, in addition to the various experimental studies, theoretical investigation can play an important role for interpreting the bio-chemical/physical phenomena. Actually, increasing numbers of theoreticians have contributed to elucidate the mechanism of bio-chemical reactions.

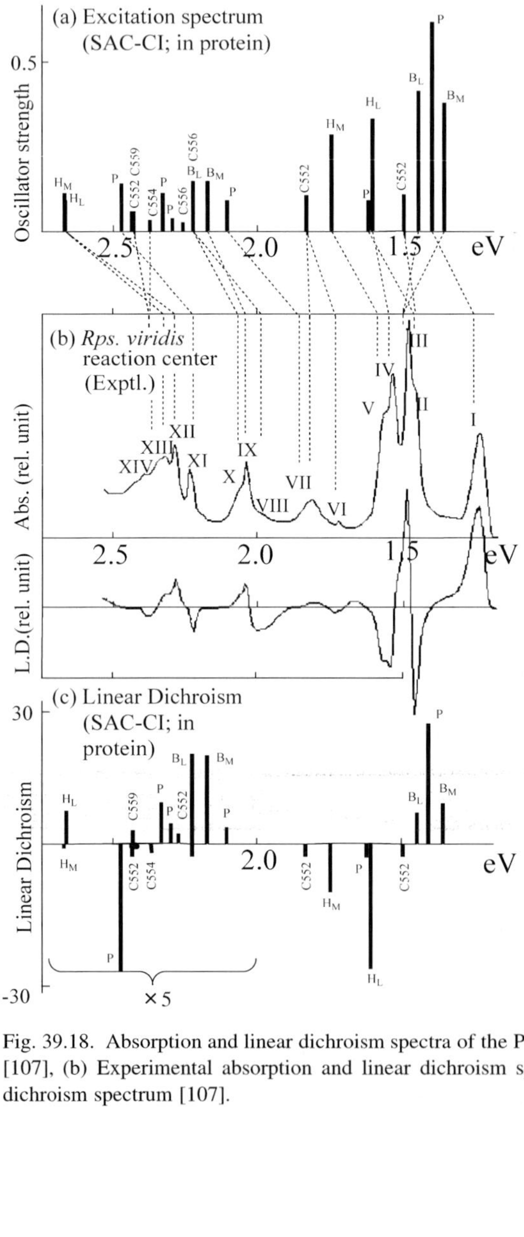

Fig. 39.18. Absorption and linear dichroism spectra of the PSRC; (a) SAC–CI theoretical excitation spectrum [107], (b) Experimental absorption and linear dichroism spectra [111], and (c) SAC–CI theoretical linear dichroism spectrum [107].

There are totally 11 chromophores in the bacterial photosynthetic reaction center (PSRC) of *Rhodopseudomonas (Rps.) virids*. Since the excitation process of the reaction center is the primary event of the photo-induced electron transfer in the reaction center, the detailed analysis of the absorption spectrum is one of the key steps for the understanding of photochemistry of the system. The chromophores included in the PSRC are bacteriochlorophyll *b* dimer (special pair, P), bacteriochlorophyll *b* in L- and M-branches (B_L and B_M), bacteriopheophytin *b* in L- and M-branches (H_L and H_M), menaquinone (MQ), ubiquinone (UQ) and four different hemes, c-552, c-554, c-556, and c-559 in c-type cytochrome subunit.

The chromophores were calculated within the environment of protein and waters which were dealt with by the point-charge electrostatic model [107]. The calculation was performed for each chromophore and the total theoretical spectrum was obtained by summing up those of chromophores. In Fig. 39.18, the calculated excitation and linear-dichroism spectra of the reaction center are compared with the experimental ones. The experimental absorption spectrum [111] has very complex structure having 14 peaks within the energy range of 1.5 eV. The SAC–CI calculations gave about 25 states in this energy region. We have assigned successfully all the peaks in the experimental spectrum in the energy range from 1.2 to 2.5 eV. The average discrepancy between the calculated and experimental excitation energies was 0.13 eV. The assignment was performed by comparing the SAC–CI theoretical spectrum with the experimental one in excitation energy, oscillator strength, linear dichroism data (angle of transition moment), and other experimental information available. Almost all the peaks were red shifted due to the effect of proteins.

Using the SAC–CI wave functions of the chromophores, the mechanism and the unidirectionality of the electron transfer in the PSRC of *Rps. Viridis* was analyzed by calculating the electronic factor in the rate constant: the enigmatic unidirectionality was clearly explained from the electronic origin [108].

39.8 INNER-SHELL IONIZATION SPECTROSCOPY

Core-electron ionization spectra contain the information not only about inner-core electrons but also about valence electrons and chemical bonds. Extensive experimental studies have measured the core-electron binding energies (CEBE) of numerous molecules [112,113] and the recent development of X-ray photoelectron spectroscopy (XPS) has enabled the detailed analysis of the satellites accompanied by the inner-shell ionization.

The SAC–CI method is useful to study both the main and satellite peaks in the core-ionization spectra as well as valence-ionization spectra. In this section, we explain the SAC–CI applications to the CEBE [114], inner-shell satellite spectrum [114,115] and vibrational spectrum of inner-shell ionization [116].

39.8.1 Core-electron binding energy

The SAC–CI method was systematically applied to the CEBE of C, N, O and F atoms of 22 molecules [114]. For calculating the CEBE, the *R*-operators up to triples are

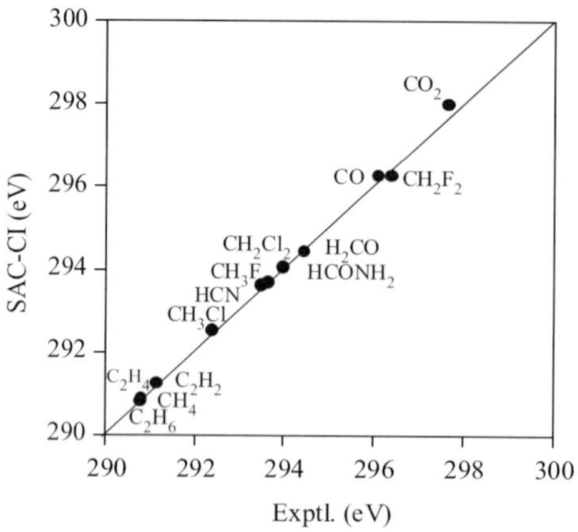

Fig. 39.19. C1s CEBEs: SAC–CI [114] *vs.* experiments [113].

necessary for describing orbital relaxations as well as electron correlations. In Fig. 39.19, we show a comparison between the theoretical and experimental values for the C1s ionizations of 13 molecules. The SAC–CI method reproduces the CEBE of these molecules quite accurately. In general, the CEBE shifts to higher energy, when the adjacent atom is electron withdrawing. This feature is correctly reproduced and the average absolute deviation from the experiment is 0.09 eV for these molecules.

39.8.2 Inner-shell satellite spectrum

Shake-up satellites appearing in the inner-shell photoelectron spectra are challenging spectroscopic subject. Theoretically, a proper description of the satellite spectra is possible only with accurate theoretical methods, since the spectra reflect complex electron-correlation and orbital-reorganization effects; shake-up satellites are described by the multi-electron process. Experimentally, weak intensities of the inner-shell photoelectron satellites make high-resolution XPS observation difficult. Therefore, intensive cooperative investigations are necessary for elucidating fine details of the satellite spectra. In this section, we show the cooperative study by the SAC–CI and XPS on the inner-shell satellite spectrum of formaldehyde [115]. Formaldehyde has both valence and Rydberg excitations accompanying the inner-shell ionization, and therefore its satellite spectrum becomes complex.

Fig. 39.20 shows the observed XPS and the SAC–CI general-*R* spectra for the C1s satellites of H_2CO. The general-*R* method finely reproduces the details of the experimental spectra, namely, both peak positions and relative intensities, and enables the quantitative assignments for the seven satellite bands 1–7. The shake-up transitions are mainly attributed to the valence excitations such as $\pi-\pi^*$, $n-\sigma^*$, and $\sigma-\sigma^*$ accompanying the inner-shell ionization. The Rydberg excitations are minor and they contribute

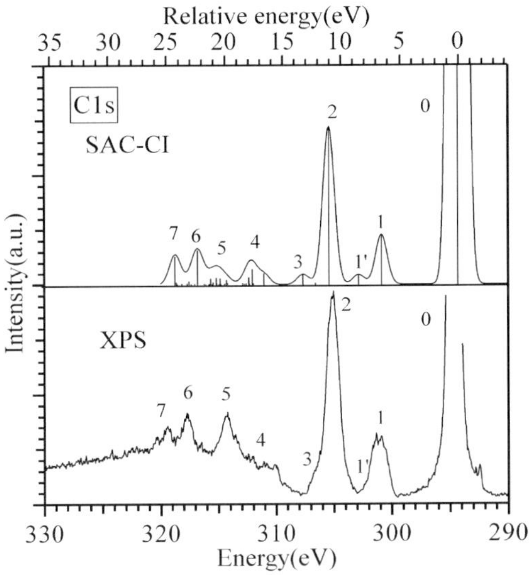

Fig. 39.20. C1s photoelectron satellite spectrum of formaldehyde [115].

to the spectrum only in the high-energy region. Three-electron processes such as $1s^{-1}n^{-2} - \pi^{*2}$ and $1s^{-1}\pi^{-2} - \pi^{*2}$ were also predicted in the low-energy region.

39.8.3 Vibrational spectrum of inner-shell ionization

The geometrical relaxation caused by the inner-shell ionization is not small, though the ionized electron does not significantly contribute to the chemical bonding.

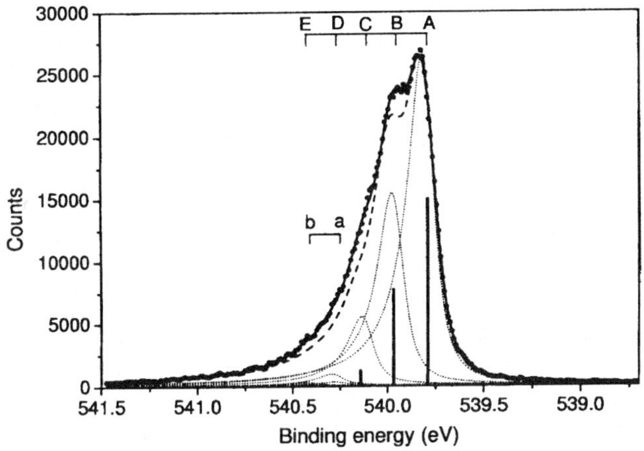

Fig. 39.21. Photoelectron vibrational spectrum of O1s for H_2O. Solid line is the XPS spectrum and dashed line is the SAC–CI spectrum [116]. Theoretical Franck–Condon factors are given by the vertical lines.

This geometrical relaxation appears in the vibrational spectrum of the inner-shell ionization. The vibrational spectrum of the O1s ionized state of H_2O was studied by the SAC–CI method [116]. The three-dimensional potential energy surfaces of the ground and O1s core-ionized states were calculated and the vibrational spectrum was simulated from them. In Fig. 39.21, the photoelectron spectrum of O1s of water excited at 590 eV is compared with the theoretical spectrum. The vibrational structure can be attributed mostly to the bending mode of the core-ionized molecule. Actually, the geometry relaxation was calculated to be remarkably large for the HOH angle θ, $\Delta\theta = 13.5°$, but, insignificant for r, $\Delta r = 0.007$ Å. In this way, the SAC–CI method can calculate the accurate potential energy surface, and hence can perform the precise analysis for the vibrational spectra of the ionized or excited states.

39.9 GEOMETRIES OF MOLECULAR EXCITED STATES

The analytical energy gradients of the SAC–CI method are useful to investigate the geometries, vibrations, chemical reactions, energy relaxation processes and dynamics of molecules in their ground and excited states. In particular, it is generally rather difficult to obtain this information for the excited states by the experimental techniques alone. The SAC–CI method has been applied to the geometry and spectroscopic constants of many molecular systems. In this section, we present the results for malonaldehyde using the localized molecular orbitals (LMOs) [47] and applications to multi-electron processes by the general-*R* method [40].

39.9.1 Malonaldehyde

The SAC–CI method formally scales as $O(N^6)$, where N is the number of basis functions. This scaling property means that the computational requirement rapidly increases with the size of the system. Therefore, the SAC–CI program adopted the perturbation selection method. With this method, the scaling property can be relaxed by the suitable selections of the excitation operator. In this case, LMOs are a rational choice for the reference orbitals. However, the selection may cause the discontinuity in the potential energy surface. This problem can be solved by the MOD method [46,47], as introduced in the Section 2.

Table 39.1 summarizes the optimized structures and adiabatic excitation energies of the ground and excited states of malonaldehyde. The structure of malonaldehyde with the specifications of atoms is shown in Fig. 39.22. The Pipek–Mezey localization [117] was used for obtaining the reference orbitals. The perturbation selection effectively reduces the dimension of the excitation operators without losing the accuracy of the result. The number of the excitation operators is reduced to 5% in comparison with the conventional CCSD method. However, the average deviations of the structure parameters remain only within 0.009 Å in the bond-length and 0.5° in the angles. The calculated geometries and adiabatic excitation energy agree well with the experimental values [118,119], showing that the SAC–CI method with the MOD approach works effectively. In the local approach,

Table 39.1 Spectroscopic constants of ground and excited states of malonaldehyde [47]

Structure parameters		Ground state			Excited state	
		SAC[a]	CCSD	Exptl.[c]	SAC–CI [a]	Exptl.[d]
	λ^b	10^{-7}	No selection		10^{-8}	
Dimensions		60,811	1,186,569		46,225	
(%)		5	100		5	
C(2)=O(1) (Å)		1.233	1.240	1.234	1.344	
C(3)–C(2) (Å)		1.471	1.455	1.454	1.434	
C(4)=C(3) (Å)		1.363	1.365	1.348	1.369	
O(5)–C(4) (Å)		1.339	1.337	1.320	1.367	
H(6)–O(5) (Å)		0.971	0.982	0.969	0.958	
Average deviation (Å)		0.009 Å	0			
O(1)=C(2)–C(3) (°)		123.6	123.4	123.0	122.6	
C(2)–C(3)=C(4) (°)		120.8	120.4	119.4	127.4	
C(3)=C(4)–O(5) (°)		125.7	125.2	124.5	127.9	
C(4)–O(5)–H(6) (°)		107.4	106.7	106.3	111.0	
Average deviation (°)		0.5°	0			
Adiabatic excitation energy (eV)					3.39	3.504

[a]Localized MOs were used for the reference orbitals. The perturbation selection was carried out for selecting the double-excitation operators.
[b]Energy threshold for the perturbation selection.
[c]Ref. [118].
[d]Ref. [119].

singularity can appear in the derivatives with respect to external parameter, since the localizations are not guaranteed to be smooth functions of the parameters and hence lead to the discontinuity of the potential energy surface. The MOD method would be one of the solutions and is expected to be applicable to other post Hartree–Fock methodologies.

39.9.2 Multi-electron processes; C_2 and CO^+

The SAC–CI analytical energy gradient was extended to the general-*R* method [33,40] and the high-spin multiplicities [37,42]. Formulations and implementations were recently

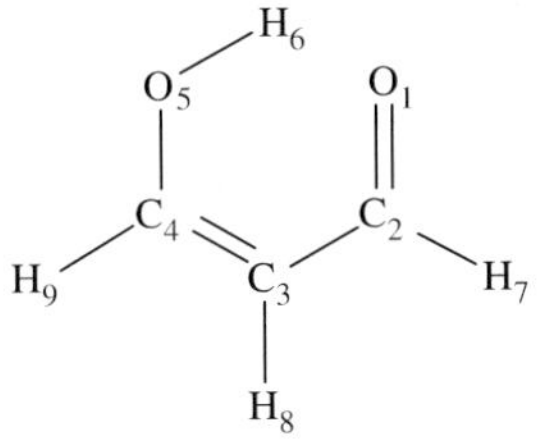

Fig. 39.22. Structure of malonaldehyde with the specification of atoms [47].

summarized with some numerical applications [41]. The general-*R* method is quantitatively applicable to the various electronic states described by the multi-electron processes. It is also useful for investigating the excited states of open-shell systems, since they are often described by the two-electron process from the closed-shell SAC state. First, we show the applications to the ground and excited states of C_2 and CO^+.

The spectroscopic constants in the various excited states of C_2 have been determined by experiment [120] and some of them are described by the two-electron processes. Table 39.2 summarizes r_e and T_e for nine singlet and triplet excited states of C_2,

Table 39.2 Excitation levels, equilibrium internuclear distances (r_e), adiabatic excitation energies (T_e) and main configurations for the singlet and triplet states of C_2

State	Method	Excitation level	r_e (Å)	T_e (eV)	Main configurations[a] (|C| > 0.3)
$X^1\Sigma_g^+$	General-*R*	0	1.268	–	
	CCSD(T)[b]		1.245	–	
	MRCI[c]		1.248	–	
	Exptl.[d]		*1.243*	–	
$a^3\Pi_u$	General-*R*	1	1.307	0.094	$0.93(1\pi_u^{-1}3\sigma_g)$
	CCSD(T)[b]		1.311	0.104	
	MRCI[c]		1.318	0.081	
	Exptl.[d]		*1.312*	*0.089*	
$b^3\Sigma_g^-$	General-*R*	2	1.366	0.842	$0.95(1\pi_u^{-2}3\sigma_g^2)$
	CCSD(T)[b]		1.370	0.809	
	MRCI[c]		1.374	0.911	
	Exptl.[d]		*1.369*	*0.798*	
$A^1\Pi_u$	General-*R*	1	1.317	0.993	$0.97(1\pi_u^{-1}3\sigma_g)$
	Exptl.[d]		*1.318*	*1.040*	
$c^3\Sigma_u^+$	General-*R*	1	1.200	1.037	$0.92(2\sigma_u^{-1}3\sigma_g)$
	Exptl.[d]		*1.23*	*1.650*	
$d^3\Pi_g$	General-*R*	2	1.262	2.362	$0.89(1\pi_u^{-1}2\sigma_u^{-1}3\sigma_g^2)$
	Exptl.[d]		*1.266*	*2.482*	
$C^1\Pi_g$	General-*R*	2	1.250	4.181	$0.90(1\pi_u^{-1}2\sigma_u^{-1}3\sigma_g^2)$
	Exptl.[d]		*1.255*	*4.248*	
$e^3\Pi_g$	General-*R*	2	1.543	4.997	$0.83(1\pi_u^{-2}1\pi_g3\sigma_g)$ $- 0.42(1\pi_u^{-1}2\sigma_u^{-1}3\sigma_g^2)$ $+ 0.39(1\pi_u^{-2}1\pi_g3\sigma_g)$
	Exptl.[d]		*1.535*	*5.058*	
$D^1\Sigma_u^+$	General-*R*	1	1.243	5.342	$0.84(2\sigma_u^{-1}3\sigma_g)$
	Exptl.[d]		*1.238*	*5.361*	
$g^3\Delta_g$	General-*R*	2	1.355	9.130	$0.64(1\pi_u^{-1}2\sigma_u^{-1}1\pi_g3\sigma_g)$ $- 0.63(1\pi_u^{-1}2\sigma_u^{-1}1\pi_g3\sigma_g)$ $+ 0.31(1\pi_u^{-2}1\pi_g^2)$
	Exptl.[d]		*1.358*	*9.073*	
RMS[e]			0.009	0.058	

[a]Hartree–Fock configuration for the ground state is $(\text{core})^4(2\sigma_g)^2(2\sigma_u)^2(1\pi_u)^4$.
[b]Ref. [122].
[c]Ref. [121].
[d]Ref. [120].
[e]Root mean square (RMS) is the difference between experimental and theoretical values.

calculated by the general-*R* method in comparison with the experimental [120] and other theoretical values [121,122]. Since the ground state of C_2 has quasi-degenerate character, the general-*R* method was used for the calculation of the ground state. The $C^1\Pi_g$, $b^3\Sigma_g^-$, $d^3\Pi_g$, $e^3\Pi_g$ and $g^3\Delta_g$ states are described by the two-electron excitations from the ground state, while other states are described by the one-electron excitations. The agreements of the general-*R* results with the experimental values are satisfactory regardless of the excitation level of each state: the discrepancies of the general-*R* results from the experimental values were 0.009 Å and 0.058 eV for r_e and T_e, respectively. The $a^3\Pi_u$ and $b^3\Sigma_g^-$ states were also studied by the multi-reference CI [121] and CCSD(T) [122] methods in agreement with experiment.

Table 39.3 summarizes r_e and T_e for the doublet states of CO^+ by the SD-*R* and general-*R* methods in comparison with the experiments [120,123]. For $D^2\Pi$ and $3^2\Sigma^+$ states, there are no experimental values of r_e. The $X^2\Sigma^+$, $A^2\Pi$ and $B^2\Sigma^+$ states of CO^+ are characterized by the one-electron process, while the $C^2\Delta$, $D^2\Pi$ and $3^2\Sigma^+$ states are represented by the two-electron processes, which correspond to the satellite states observed in the valence ionization spectrum of CO. For the $C^2\Delta$ and $3^2\Sigma^+$ states, the general-*R* method gives good results for both r_e and T_e in comparison with experiment:

Table 39.3 Excitation levels, equilibrium internuclear distances (r_e), adiabatic excitation energies (T_e) and main configurations for the doublet states of CO^+

State	Method	Excitation level	r_e (Å)	T_e (eV)	Main configurations[a] (|C| > 0.3)
$X^2\Sigma^+$	SD-*R*	1	1.115	–	$0.97(5\sigma^{-1})$
	General-*R*	1	1.115	–	$0.95(5\sigma^{-1})$
	Exptl.[b]		*1.115*	–	
$A^2\Pi$	SD-*R*	1	1.249	2.382	$0.96(1\pi^{-1})$
	General-*R*	1	1.237	2.557	$0.93(1\pi^{-1})$
	Exptl.[b]		*1.244*	*2.571*	
$B^2\Sigma^+$	SD-*R*	1	1.155	5.594	$0.93(4\sigma^{-1})$
	General-*R*	1	1.162	5.646	$0.86(4\sigma^{-1})$
	Exptl.[b]		*1.169*	*5.688*	
$C^2\Delta$	SD-*R*	2	1.300	11.22	$0.71(1\pi^{-1}2\pi5\sigma^{-1}) - 0.71(1\pi^{-1}2\pi5\sigma^{-1})$
	General-*R*	2	1.333	8.004	$0.71(1\pi^{-1}2\pi5\sigma^{-1}) - 0.71(1\pi^{-1}2\pi5\sigma^{-1})$
	Exptl.[b]		*1.340*	*7.812*	
$D^2\Pi$	SD-*R*	2	1.248	12.28	$0.84(5\sigma^{-2}2\pi) - 0.35(4\sigma^{-1}2\pi1\pi^{-1})$
	General-*R*	2	1.249	8.885	$0.82(5\sigma^{-2}2\pi) - 0.37(4\sigma^{-1}2\pi1\pi^{-1})$ $- 0.30(5\sigma^{-2}3\pi)$
	Exptl.[b]		–	*8.087*	
	Exptl.[c]		–	*8.407*	
$3^2\Sigma^+$	SD-*R*	2	1.270	11.88	$0.24(4\sigma^{-1}) - 0.70(5\sigma^{-1}2\pi1\pi^{-1})$ $- 0.70(5\sigma^{-1}2\pi1\pi^{-1})$
	General-*R*	2	1.270	9.140	$0.41(4\sigma^{-1}) - 0.65(5\sigma^{-1}2\pi1\pi^{-1})$ $- 0.65(5\sigma^{-1}2\pi1\pi^{-1})$
	Exptl.[b]		–	*9.074*	
	Exptl.[c]		–	*9.011*	

[a]Hartree–Fock configuration for the ground state is $(\text{core})^4 (3\sigma)^2(4\sigma)^2 (1\pi)^4 (5\sigma)^2$.
[b]Ref. [120].
[c]Ref. [123].

the deviations from the experimental values are within 0.007 Å and 0.19 eV for r_e and T_e, respectively.

39.9.3 Acetylene and CNC

The general-*R* method was also applied to the ground and excited states of polyatomic molecules; acetylene and CNC. Some of the excited states of these molecules are described by the two-electron process. The DZP basis sets were used for acetylene and the DZP basis sets augmented with s- and p-type Rydberg functions ($\zeta_s = 0.023$, $\zeta_p = 0.021$ for C and $\zeta_s = 0.028$, $\zeta_p = 0.025$ for *N*) were used for CNC: these Rydberg functions are important for T_e of CNC. The general-*R* method including *R*-operators up to triples was performed.

The ground and lowest singlet excited (A^1A_u) states have been extensively studied both experimentally and theoretically. Recent theoretical calculation has investigated the trans-bent structure in the A^1A_u state using EOM–CCSD method with DZP and VDZ2P basis sets and well reproduced the experimental structure. The EOM–CCSD method is essentially identical to the SAC–CI SD-*R* method. In 1992, the trans-bent structure in the $C'A_g$ state was spectroscopically measured by Lundberg et al. [124]. The A^1A_u state is described by the one-electron process, while the $C'A_g$ state is described two-electron process. Therefore, we are interested in the spectroscopic constants of these excited states especially for the $C'A_g$ state. The results are shown in Table 39.4. The SAC method well reproduced the experimental geometries [125]. For the A^1A_u state [126], both SD-*R* and general-*R* methods gave the results of the same quality. On the other hand, for the $C'A_g$ state, the general-*R* method drastically improves the results of the SD-*R* method, especially for T_e. The optimized geometries are in excellent agreement with experimental values; the deviations are within 0.02 Å and 0.03°. The A^1A_u and $C'A_g$ states are described as $(1a_u^{-1}4a_g)$ and $(1a_u^{-2}4a_g^2)$, respectively, in which $1a_u$ orbital corresponds to valence $1\pi_u$ orbital for the linear

Table 39.4 Spectroscopic constants of ground and excited states of acetylene [41]

State	Method	Excitation level	R_{CH} (Å)	R_{CC} (Å)	θ_{CCH} (degree)	T_e (eV)
$X\ ^1A_g$	SAC	–	1.068	1.219	180	–
(linear)	Exptl.[a]		1.063	1.203	180	–
$A\ ^1A_u$	SD-*R*	1	1.098	1.377	122.6	5.485
(trans)	General-*R*	1	1.097	1.385	121.7	5.329
	Exptl.[b]		1.097	1.375	122.5	5.232
$C'\ ^1A_g$	SD-*R*	2	1.105	1.634	103.8	10.098
(trans)	General-*R*	2	1.111	1.643	103.0	7.844
	Exptl.[c]		1.14	1.65	103	7.723

[a] Ref. [125].
[b] Ref. [126].
[c] Ref. [124].

Table 39.5 Excitation levels, equilibrium internuclear distances (R_{CN}), adiabatic excitation energies (T_e) and main configurations for the doublet ground and excited states of CNC

State	Method	Excitation level	R_{CN} (Å)	T_e (eV)	Main configurations[a] (|C| > 0.3)
$X^2\Pi_g$	SD-*R*	1	1.256	–	$0.97(1\pi_u)$
	General-*R*	1	1.251	–	$0.92(1\pi_u)$
	CASPT2[b]		1.253	–	
	Exptl.[c]		*1.245*	–	
$A^2\Delta_u$	SD-*R*	2	1.257	7.197	$0.65(3\sigma_u^{-1}1\pi_g^2) + 0.65(3\sigma_u^{-1}1\pi_g^2)$
	General-*R*	2	1.255	3.675	$0.65(3\sigma_u^{-1}1\pi_g^2) + 0.65(3\sigma_u^{-1}1\pi_g^2)$
	CASPT2[b]		1.257	3.425	
	Exptl.[c]		*1.249*	*3.761*	
$B^2\Sigma_u^+$	SD-*R*	2	1.262	7.694	$0.65(3\sigma_u^{-1}1\pi_g^2) + 0.65(3\sigma_u^{-1}1\pi_g^2)$
	General-*R*	2	1.260	4.298	$0.65(3\sigma_u^{-1}1\pi_g^2) + 0.65(3\sigma_u^{-1}1\pi_g^2)$
	CASPT2[b]		1.259	4.001	
	Exptl.[c]		*1.259*	*4.315*	

[a]The electronic configuration and the excitation level relative to the closed shell CNC^+, namely, $(\text{core})^6(3\sigma_g)^2(2\sigma_u)^2(1\pi_u)^4(4\sigma_g)^2(3\sigma_u)^2$.

[b]Molecular geometries are optimized by CASSCF/6s4p2d2f (Ref. [128]).

[c]Ref. [127]

structure. Thus, trans-bent structures become stable in these excited states by the single or double excitation from the $1\pi_u$ orbital.

The open-shell molecule CNC was also studied by the general-*R* method; the excited states are described by the two-electron processes from the closed-shell SAC state. In Table 39.5, the SAC–CI SD-*R* and general-*R* results of r_e and T_e are compared with the experiment [127] and with CASPT2 results [128]. The ground state is described by the one-electron process, while the other two excited states, $A^2\Delta_u$ and $B^2\Sigma_u^+$, are represented by the two-electron process. For r_e, both SD-*R* and general-*R* methods gave excellent results for the ground and excited states: the deviations from the experiment were within 0.009 Å. For T_e, the general-*R* method drastically improves the results of the SD-*R* method. The effect of including triples in the *R*-operators is as large as 3.4–3.5 eV. The adiabatic excitation energies calculated by the general-*R* method agree well with the experiment; T_e were slightly underestimated by 0.086 and 0.017 eV for $A^2\Delta_u$ and $B^2\Sigma_u^+$, respectively. The CASPT2 method underestimated these values by 0.34 and 0.31 eV, though the basis sets were different [128].

39.10 HYPERFINE SPLITTING CONSTANTS

Fermi contact hyperfine splitting constants (HFSCs) give the important information of the molecular electronic structure of radicals. They are proportional to the spin densities at the nuclei of an open-shell molecule and are observed by the electron-spin resonance (ESR) and microwave spectroscopy. Since this is a very local property, *ab initio* calculations of the HFSCs are more difficult than the other electronic properties such as dipole moments, polarizabilities, etc. It has been clarified that the following factors are important for the adequate descriptions of the HFSCs: (i) spin-polarization correction [129–132], (ii) electron-correlation correction [133–140], (iii) cusp condition at the nuclei [141].

The SAC–CI method is very effective for describing both (i) spin-polarization and (ii) electron correlation corrections [136,140]. For calculating the HFSCs, we have to be careful about perturbation selection [44], since the energy and spin density are very different properties. In the SAC–CI method, we can avoid the perturbation selection, since the sizes of the matrices to be diagonalized are small without selection, in contrast to ordinary CI methods. The perturbation selection method taking into account of the HFSCs is also possible [142].

For the cusp condition, we proposed the method [141] using the Slater-type orbitals (STOs) by the STO–GTO expansion method [143,144]. Following Kato [145], the exact

Table 39.6 Hyperfine splitting constants (HFSCs) calculated with STO basis (in Gauss) [141].

Molecule	Nucleus	SAC–CI	Exptl.[a]
H_2	H	327.4	333.7
H_2O^+	O	−26.2	(−)29.7
	H	−24.1	(−)26.1
CH_3	C	25.6	28.7
	H	−29.6	−24.7
CH_3CH_2	Cβ	−12.1	−13.6
	Hβ	24.5	26.9
	Cα	24	29.5
	Hα	−25.9	−24.5
CH_3NH	C	−12.3	
	Hβ	29.6	34
	N	7.8	13
	Hα	−21.2	(−)22
CH_3O	C	−13.7	−15.6
	H	39.4	43.7
	O	−18.3	
$CH_3OCH_3^+$	C	−8.5	
	H	39.8	43
	O	−23.9	
H_2CO^+	C	−29.8	−38.8
	H	119.4	132.7
	O	−19.9	
CH_2CH	Cβ	−1.1	−8.6
	H_{trans}	51.4	68.5
	H_{cis}	34.1	34.2
	Cα	102.2	107.6
	Hα	12.1	13.3
HCO	C	119	131
	H	119	127
	O	−16.3	

[a]The sign (−) is based on the theoretical result.

density ρ should satisfy the following equation at the position $\{R_A\}$ of the nucleus A:

$$\lim_{r \to R_A} \left\{ \frac{d}{dr} \ln \rho(r) \right\} = -2Z_A \tag{14}$$

where Z_A is the nuclear charge. The wave function calculated with Gaussian-type orbitals (GTOs) does not satisfy this condition. The method of using the STOs as a basis set is therefore desirable, since they have cusps at the position of the nuclei. By using the STO–GTO expansion method[143,144], we can do such STO calculations using conventional GTO programs in enough accuracy for the integrals necessary for the SCF and SAC–CI calculations. The HFSCs and the cusp values are calculated with the original STO basis functions [141].

The results of this method applied to the HFSCs of some radicals [141] are shown in Table 39.6. Fig. 39.23 shows the correlation between the theoretical results and the experimental values. Generally speaking, the agreement between theory and experiment is fairly good, considering that the HFSC's is the relatively difficult properties to be calculated. The averaged discrepancies are 3.6, 8.2, 5.2 and 2.3G for proton, carbon, nitrogen and oxygen, respectively. From Fig. 39.23, we see that the theoretical values tend to be smaller in absolute value than the experimental values.

The convergence of the calculated HFSCs with respect to the order of *R*-operators was also examined [142]. In Fig. 39.24, the HFSCs calculated by the SAC–CI SD-*R* and general-*R* methods are compared with the full-CI values. The calculations are due to the double-zeta basis sets because of the limitation of the full-CI calculations. The SAC–CI values almost converge to the full-CI values when the *R*-operators are included up to

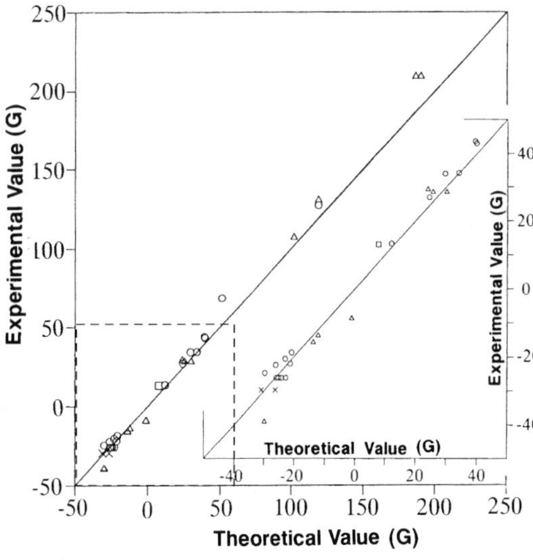

Fig. 39.23. Comparison between experimental and theoretical values [141] of the HFSCs for hydrogen (○), carbon (Δ), nitrogen (□), and oxygen (×) nuclei of the doublet radicals. The region of small HFSCs is enlarged in the figure.

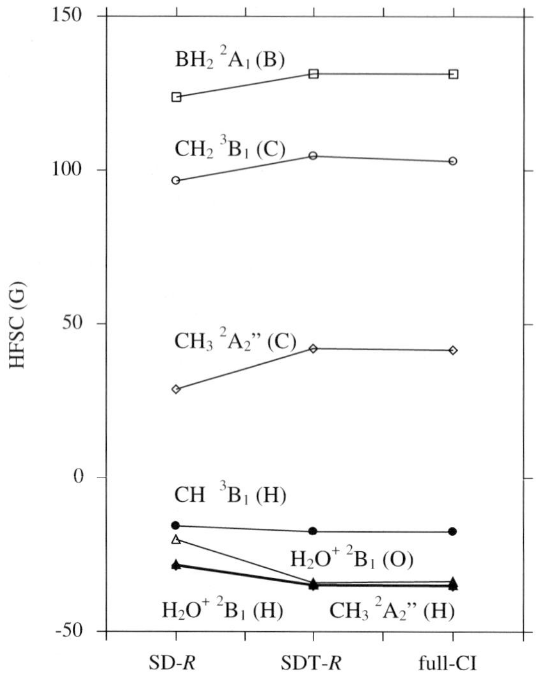

Fig. 39.24. HFSC calculated by the SAC–CI SD-*R*, SDT-*R*, and full-CI method [142].

triples. The SD-*R* method is also accurate for these values, since the electronic states of these radicals are described by one-electron processes.

39.11 SUMMARY

The SAC–CI method was proposed in 1978 as an accurate electronic-structure theory for the ground, excited, ionized and electron-attached states of atoms and molecules. The method has been successfully applied to various photochemistry involving more than 150 molecules and established to be a useful method for studying chemistry and physics involving various electronic states. In this article, we gave a brief overview of our SAC–CI applications to the molecular spectroscopy.

The theoretical and computational backgrounds were briefly explained. The SAC–CI program system, now released through Gaussian03, can calculate the ground, excited, ionized and electron-attached states of the singlet to septet spin multiplicities. The program can include the higher order excitation operator up to 6-ple excitations. Therefore, Gaussian users can study the various kinds of electronic states with the SAC–CI program system.

The SAC–CI method calculates the excitation and ionization spectra which are comparable in accuracy to the experimental spectra. The applications reviewed in this article are the excitation and/or ionization spectra of furan, thiophene, *p*-benzoquinone, aniline, phthalocyanine and bacteriochlorophylls in the photosynthetic reaction center. The SAC–CI study on the collision-induced absorption spectra clarifies the mechanism of the transient absorption which is optically forbidden in the isolated atoms. The SAC–CI method was also shown to be applicable to the inner-shell spectroscopies.

Various kinds of the theoretical spectroscopies for the transition metal complexes were also reviewed. For the excitation spectrum of CrO_2Cl_2, the SAC–CI method simulated accurate spectrum. For tetraoxo metal complexes, the systematic studies explained the spectral differences when the central metal was substituted. In the analysis of the NMR chemical shift, not only the optically allowed states but also the magnetically allowed states are important. In the molybdenum complexes, the inverse of the d–d* excitation energy is proportional to the experimental ^{95}MO chemical shift. The photofragmentation reaction of $Ni(CO)_4$ was also studied and the reaction mechanism was clarified.

In the SAC–CI program, the analytical energy gradient is available for all the electronic states calculated by the program. Therefore, the program can calculate the force acting on the nuclei and the equilibrium geometry of all kinds of electronic states. We reviewed the results for the ground and excited states of malonaldehyde, acetylene and some small molecules. This first example showed that the combined use of LMOs and the perturbation selection effectively reduces the computational cost without losing the accuracy. In acetylene, the geometry of the low-lying two-electron excited states was computed by the SAC–CI general-*R* method. The calculated structure parameters and adiabatic excitation energy agree well with the experimental values, showing that the SAC–CI method is useful for the studies of the molecular structures and reaction dynamics in the excited states.

The SAC–CI method is also useful for studying the ESR hyperfine splitting constants which characterize the electronic structure of radicals. It was shown that the STO–GTO expansion method is useful for calculating the HFSCs of radicals with the conventional GTO program: the HFSCs are calculated with the STO basis set which satisfies the cusp condition at nuclei.

39.12 ACKNOWLEDGEMENTS

These studies were supported by the Grant for Creative Scientific Research from the Ministry of Education, Science, Culture and Sports of Japan.

39.13 REFERENCES

1 R.J. Buenker and S.D. Peyerimhoff, Theor. Chem. Acc., 12 (1968) 183.
2 R.J. Buenker and S.D. Peyerimhoff, J. Chem. Phys., 53 (1970) 1368.
3 H. Nakatsuji, Chem. Phys. Lett., 59 (1978) 362.
4 H. Nakatsuji, Chem. Phys. Lett., 67 (1979) 329.
5 H. Nakatsuji, Chem. Phys. Lett., 67 (1979) 334.
6 H. Nakatsuji and K. Hirao, Chem. Phys. Lett., 47 (1977) 569.
7 H. Nakatsuji and K. Hirao, J. Chem. Phys., 68 (1978) 2053.

8 H. Nakatsuji, Acta Chim. Hung., Models Chem., 129 (1992) 719.

9 H. Nakatsuji, in: J. Leszczynski (Ed.), Computational chemistry—reviews of current trends, World Scientific, Singapore, 1997.

10 H. Monkhorst, Int. J. Quantum Chem. Symp., 11 (1977) 421.

11 D. Mukherjee and P.K. Mukherjee, Chem. Phys., 39 (1979) 325.

12 K. Hirao, J. Chem. Phys., 79 (1983) 5000.

13 H. Koch and P. Jørgensen, J. Chem. Phys., 93 (1990) 3333.

14 H. Koch, H.J.A. Jensen, T. Helgaker and P. Jørgensen, J. Chem. Phys., 93 (1990) 3345.

15 J. Geertsen, M. Rittby and R.J. Bartlett, Chem. Phys. Lett., 164 (1989) 57.

16 J.F. Stanton and R.J. Bartlett, J. Chem. Phys., 98 (1993) 7029.

17 M. Ehara, M. Ishida, K. Toyota and H. Nakatsuji, in: K.D. Sen (Ed.), Reviews in modern quantum chemistry, World Scientific, Singapore, 2002, pp. 293–319.

18 M.E. Casida, in: D.P. Chong (Ed.), Recent advances in density functional methods: Part I, World Scientific, Singapore, 1995.

19 R. Bauernschmitt and R. Ahlrichs, Chem. Phys. Lett., 256 (1996) 454.

20 P. Hohenberg and W. Kohn, Phys. Rev., 136 (1964) B864.

21 W. Kohn and L.J. Sham, Phys. Rev., (1965) A1133.

22 R.G. Parr and W. Yang, Density-functional theory of atoms and molecules, Oxford University Press, Oxford, UK, 1989.

23 W. Kohn, A.D. Becke and R.G. Parr, J. Phys. Chem., 100 (1996) 12974.

24 Y. Zhang, W. Pan and W. Yang, J. Chem. Phys., 107 (1997) 7921.

25 M.E. Casida, C. Jamorski, K.C. Casida and D.R. Salahub, J. Chem. Phys., 108 (1998) 4439.

26 M.E. Casida and D.R. Salahub, J. Chem. Phys., 113 (2000) 8918.

27 A. Dreuw, J.L. Weisman and M. Head-Gordon, J. Chem. Phys., 119 (2003) 2943.

28 M.J. Frisch, G.W. Trucks, H.B. Schlegel, G.E. Scuseria, M.A. Robb, J.R. Cheeseman, V.G. Zakrzewski, J.J.A. Montgomery, R.E. Stratmann, J.C. Burant, S. Dapprich, J.M. Millam, A.D. Daniels, K.N. Kudin, M.C. Strain, O. Farkas, J. Tomasi, V. Barone, M. Cossi, R. Cammi, B. Mennucci, C. Pomelli, C. Adamo, S. Clifford, J. Ochterski, G.A. Petersson, P.Y. Ayala, Q. Cui, K. Morokuma, N. Rega, P. Salvador, J.J. Dannenberg, D.K. Malick, A.D. Rabuck, K. Raghavachari, J.B. Foresman, J. Cioslowski, J.V. Ortiz, A.G. Baboul, B.B. Stefanov, G. Liu, A. Liashenko, P. Piskorz, I. Komaromi, R. Gomperts, R.L. Martin, D.J. Fox, T. Keith, M.A. Al-Laham, C.Y. Peng, A. Nanayakkara, M. Challacombe, P.M.W. Gill, B. Johnson, W. Chen, M.W. Wong, J.L. Andres, C. Gonzalez, M. Head-Gordon, E.S. Replogle and J.A. Pople, Gaussian 98 (Revision A.11.3), Gaussian, Inc., Pittsburgh, PA 2002.

29 M.J. Frisch, G.W. Trucks, H.B. Schlegel, G.E. Scuseria, M.A. Robb, J.R. Cheeseman, J.J.A. Montgomery, T. Vreven, K.N. Kudin, J.C. Burant, J.M. Millam, S.S. Iyengar, J. Tomasi, V. Barone, B. Mennucci, M. Cossi, G. Scalmani, N. Rega, G.A. Petersson, H. Nakatsuji, M. Hada, M. Ehara, K. Toyota, R. Fukuda, J. Hasegawa, M. Ishida, T. Nakajima, Y. Honda, O. Kitao, H. Nakai, H. Nakai, M. Klene, X. Li, J.E. Knox, H.P. Hratchian, J.B. Cross, C. Adamo, J. Jaramillo, R. Gomperts, R.E. Stratmann, O. Yazyev, R. Cammi, C. Pomelli, J. Ochterski, P.Y. Ayala, K. Morokuma, W.L. Hase, G. Voth, P. Salvador, J.J. Dannenberg, V.G. Zakrzewski, S. Dapprich, A.D. Daniels, M.C. Strain, O. Farkas, D.K. Malick, A.D. Rabuck, K. Raghavachari, J.B. Foresman, J.V. Ortiz, Q. Cui, A.G. Baboul, S. Clifford, J. Cioslowski, B.B. Stefanov, G. Liu, A. Liashenko, P. Piskorz, I. Komaromi, R.L. Martin, D.J. Fox, T. Keith, M.A. Al-Laham, C.Y. Peng, A. Nanayakkara, M. Challacombe, P.M.W. Gill, B. Johnson, W. Chen, M.W. Wong, C. Gonzalez and J.A. Pople, Gaussian development version (Revision A.03), Gaussian, Inc., Pittsburgh, PA, 2003.

30 J. Cizek, J. Chem. Phys., 45 (1966) 4256.

31 J. Cizek, Adv. Chem. Phys., 14 (1969) 35.

32 H. Nakatsuji, K. Hirao and Y. Mizukami, Chem. Phys. Lett., 179 (1991) 555.

33 H. Nakatsuji, Chem. Phys. Lett., 177 (1991) 331.

34 H. Nakatsuji, J. Chem. Phys., 83 (1985) 713.

35 H. Nakatsuji, J. Chem. Phys., 83 (1985) 5743.

36 H. Nakatsuji, J. Chem. Phys., 94 (1991) 6716.

37 H. Nakatsuji and M. Ehara, J. Chem. Phys., 98 (1993) 7179.

38 T. Nakajima and H. Nakatsuji, Chem. Phys. Lett., 280 (1997) 79.
39 T. Nakajima and H. Nakatsuji, Chem. Phys., 242 (1999) 177.
40 M. Ishida, K. Toyoda, M. Ehara and H. Nakatsuji, Chem. Phys. Lett., 347 (2001) 493.
41 M. Ishida, K. Toyoda, M. Ehara, M.J. Frisch and H. Nakatsuji, J. Chem. Phys., 120 (2004) 2593.
42 M. Ishida, K. Toyoda, M. Ehara and H. Nakatsuji, Chem. Phys. Lett., 350 (2001) 351.
43 H. Nakatsuji, Program Library SAC85 (No. 1396), Computer center of the institute for molecular science, Okazaki, Japan, 1986.
44 H. Nakatsuji, Chem. Phys., 75 (1983) 425.
45 H. Nakatsuji, J. Ushio and T. Yonezawa, Can. J. Chem., 63 (1985) 1857.
46 K. Toyota, M. Ehara and H. Nakatsuji, Chem. Phys. Lett., 356 (2002) 1.
47 K. Toyota, M. Ishida, M. Ehara, M.J. Frisch and H. Nakatsuji, Chem. Phys. Lett., 367 (2003) 730.
48 The SAC–CI guide is available at http://www.sbchem.kyoto-u.ac.jp/nakatsuji-lab/
49 H. Nakatsuji and M. Ehara, J. Chem. Phys., 99 (1993) 1952.
50 H. Nakatsuji, J. Chem. Phys., 95 (1991) 4296.
51 H. Nakatsuji, O. Kitao and T. Yonezawa, J. Chem. Phys., 83 (1985) 723.
52 J. Wan, J. Meller, M. Hada, M. Ehara and H. Nakatsuji, J. Chem. Phys., 113 (2000) 7853.
53 J. Wan, M. Hada, M. Ehara and H. Nakatsuji, J. Chem. Phys., 114 (2001) 842.
54 M. Ehara, Y. Ohtsuka, H. Nakatsuji, M. Takahashi, Y. Udagawa, (submitted for publication).
55 L. Serrano-Andres, M. Merchan, I. Nebot-Gil, B.O. Roos and M. Fülscher, J. Am. Chem. Soc., 115 (1993) 6184.
56 H. Nakano, T. Tsuneda, T. Hashimoto and K. Hirao, J. Chem. Phys., 104 (1996) 2312.
57 A.B. Trofimov and J. Schirmer, Chem. Phys., 214 (1997) 153.
58 O. Christiansen, J. Gauss, J. Stanton and P. Jørgensen, J. Chem. Phys., 111 (1999) 525.
59 Y. Honda, M. Hada, M. Ehara and H. Nakatsuji, J. Phys. Chem. A, 106 (2002) 3838.
60 Y. Honda, M. Hada, M. Ehara and H. Nakatsuji, J. Chem. Phys., 117 (2002) 2045.
61 M.H. Palmer, I.C. Walker, C.C. Ballard and M.F. Guest, Chem. Phys., 192 (1995) 111.
62 M.H. Palmer, I.C. Walker and M.F. Guest, Chem. Phys., 241 (1999) 275.
63 A.D.O. Bawagan, B.J. Olsson, K.H. Tan, J.M. Chen and B.X. Yang, Chem. Phys., 164 (1992) 283.
64 D.M.P. Holland, L. Karlsson and W.v. Niessen, J. Electron Spectrosc. Relat. Phenom., 113 (2001) 221.
65 M. Takahashi, K. Otsuka and Y. Udagawa, Chem. Phys., 227 (1995) 375.
66 N. Kishimoto, H. Yamakado and K. Ohno, J. Phys. Chem., 100 (1996) 8204.
67 D. Voet and J. Voet, Biochemistry, 2nd edn., Wiley, New York, 1995.
68 L. Åsbrink, G. Bieri, C. Fridh, E. Lindholm and D.P. Chong, Chem. Phys., 43 (1979) 189.
69 T.-K. Ha, Mol. Phys., 49 (1983) 1471.
70 P.E. Stevenson, J. Phys. Chem., 64 (1976) 3791.
71 J. Schiedt and R. Weinkauf, J. Chem. Phys., 110 (1999) 304.
72 J.F. Stanton and J. Gauss, J. Chem. Phys., 101 (1994) 8938.
73 C.D. Cooper, W.T. Naff and R.N. Compton, J. Chem. Phys., 63 (1975) 2752.
74 P.B. Comita and J.I. Brauman, J. Am. Chem. Soc., 109 (1987) 7591.
75 G.N.R. Tripathi, Q. Sun and R.H. Schuler, Chem. Phys. Lett., 156 (1989) 51.
76 K. Kimura and S. Nagakura, Mol. Phys., 9 (1965) 117.
77 J.N. Murrell, Proc. Phys. Soc., London, Sect. A, 68 (1955) 969.
78 H. Baba, Bull. Chem. Soc. Jpn, 34 (1961) 76.
79 I. Fisher-Hjalmars, Ark. Fys., 21 (1962) 123.
80 S. Suzuki and T. Fujii, Bull. Chem. Soc. Jpn, 48 (1975) 835.
81 N. Kishimoto, M. Furuhashi and K. Ohno, J. Electron Spectrosc. Relat. Phenom., 113 (2000) 35.
82 G. Moe, A.C. Tam and W. Happer, Phys. Rev. A, 14 (1976) 349.
83 H. Nakatsuji and M. Ehara, Chem. Phys. Lett., 172 (1990) 261.
84 M. Ehara and H. Nakatsuji, J. Chem. Phys., 102 (1995) 6822.
85 S.Y. Chen and M. Takeo, Rev. Mod. Phys., 29 (1957) 20.
86 A.K. Rappe and W.A. Goddard, III, J. Am. Chem. Soc., 102 (1980) 5114.
87 P. Jasinski, S.L. Holt, J.H. Wood and L.B. Asprey, J. Chem. Phys., 63 (1975) 757.
88 K. Yasuda and H. Nakatsuji, J. Chem. Phys., 99 (1993) 1945.

89 R.M. Miller, D.S. Tinti and D.A. Case, Inorg. Chem., 28 (1989) 2738.
90 A. Carrington and M.C. Symmons, Chem. Rev., 63 (1963) 443.
91 D.G. Lee, C.R. Moylan, T. Hayashi and J.I. Brauman, J. Am. Chem. Soc., 109 (1987) 3003.
92 S. Jitsuhiro, H. Nakai, M. Hada and H. Nakatsuji, J. Chem. Phys., 101 (1994) 1029.
93 H. Nakatsuji and S. Saito, J. Chem. Phys., 93 (1990) 1865.
94 H. Nakai, Y. Ohmori and H. Nakatsuji, J. Chem. Phys., 95 (1991) 8287.
95 J. Hasegawa, K. Toyota, M. Hada, H. Nakai and H. Nakatsuji, Theoret. Chim. Acta, 92 (1995) 351.
96 H. Nakatsuji and S. Saito, Int. J. Quantum Chem., 39 (1991) 93.
97 I.H. Hillier and V.R. Saunders, Chem. Phys. Lett., 9 (1971) 219.
98 V.A. Gubanov, J. Weber and J.W.D. Connoly, J. Chem. Phys., 63 (1975) 1455.
99 A. Müller, E. Diemann, A.C. Ranade, P.J. Aymonino and Z. Naturforsch, 24b, (1969) 1247.
100 H. Nakatsuji, M. Sugimoto and S. Saito, Inorg. Chem., 29 (1990) 3095.
101 H. Nakatsuji and M. Sugimoto, Inorg. Chem., 29 (1990) 1221.
102 S.F. Gheller, T.W. Hambley, J.R. Rodgers, R.T.C. Brownlee, M.J. O'Connor, M.R. Snow and A.G. Wedd, Inorg. Chem., 23 (1984) 2519.
103 R.G. Kidd and R.J. Goodfellow, in: R.K. Harris, B.E. Mann (Eds.), NMR and the Periodic Table, Academic, London, 1978.
104 S. Boughaba and G. Auvert, Appl. Surf. Sci., 69 (1993) 79.
105 M. Hada, Y. Imai, M. Hidaka and H. Nakatsuji, J. Chem. Phys., 103 (1995) 6993.
106 K. Toyota, J. Hasegawa and H. Nakatsuji, J. Phys. Chem. A, 101 (1997) 446.
107 J. Hasegawa, K. Ohkawa and H. Nakatsuji, J. Phys. Chem. B, 102 (1998) 10410.
108 J. Hasegawa and H. Nakatsuji, J. Phys. Chem. B, 102 (1998) 10420.
109 K. Ohkawa, M. Hada and H. Nakatsuji, J. Porphyrins Phthalocyanines, 5 (2001) 256.
110 K.A. Nguyen, P.N. Day and R. Pachter, J. Phys. Chem. A, 104 (2000) 4748.
111 J. Breton, Biochim. Biophys. Acta, 810 (1985) 235.
112 K. Siegbahn, C. Nordling, G. Johansson, J. Hedman, P.F. Heden, K. Hamrin, U. Gelius, T. Bergmark, L.O. Werme, R. Manne and Y. Baer, ESCA applied to free molecules, North-Holland, Amsterdam, 1969.
113 A.A. Bakke, A.W. Chen and W.L. Jolly, J. Electron Spectrosc. Relat. Phenom., 20 (1980) 333.
114 K. Kuramoto, M. Ehara and H. Nakatsuji, J. Chem. Phys., J, Chem. Phys. 122 (2005) 014304.
115 K. Kuramoto, M. Ehara, H. Nakatsuji, M. Kitajima, H. Tanaka, A.D. Fanis, Y. Tanemori and K. Ueda, J. Electron Spectrosc. Relat. Phenom., 142 (2005) 253.
116 R. Sankari, M. Ehara, H. Nakatsuji, Y. Senba, K. Hosokawa, H. Yoshida, A.D. Fanis, Y. Tamenori, S. Aksela and K. Ueda, Chem. Phys. Lett., 380 (2003) 647.
117 J. Pipek and P.G. Mezey, J. Chem. Phys., 90 (1989) 4916.
118 S.L. Baughcum, R.W. Duerst, W.F. Rowe, Z. Smith and E.B. Wilson, J. Am. Chem. Soc., 103 (1981) 6296.
119 C.J. Seliskar and R.E. Hoffmann, J. Mol. Spectrosc., 88 (1981) 30.
120 K.P. Huber and G. Herzberg, Molecular spectra and molecular structure, IV, Constants of diatomic molecules, Van Nostrand, New York, 1979.
121 J.C.W. Baushlicher and S.R. Langhoff, J. Chem. Phys., 87 (1987) 2919.
122 J.D. Watts and R.J. Bartlett, J. Chem. Phys., 96 (1992) 6073.
123 R.C. Shiell, M. Evabs, S. Stimson, C.-W. Hsu, C.Y. Ng and J.W. Hepburn, Chem. Phys. Lett., 315 (1999) 390.
124 J.-K. Lundberg, Y. Chen, J.-P. Pique and R.W. Field, J. Mol. Spectrosc., 156 (1992) 104.
125 R.J. Berry and M.D. Harmony, Struct. Chem., 1 (1990) 49.
126 T.R. Huet, M. Godefroid and M. Herman, J. Mol. Spectrosc., 144 (1990) 32.
127 G. Herzberg, Electronic spectra and electronic structure of polyatomic molecules, Van Nostrand, New York, 1966.
128 R. Pd and P. Chandra, J. Chem. Phys., 114 (2001) 1589.
129 H. Nakatsuji, H. Kato and T. Yonezawa, J. Chem. Phys., 51 (1969) 3175.
130 W. Meyer, J. Chem. Phys., 51 (1969) 5149.
131 H. Nakatsuji, J. Chem. Phys., 59 (1973) 2586.
132 H. Nakatsuji and K. Hirao, Chem. Phys. Lett., 47 (1977) 569.

133 S.Y. Chang, E.R. Davidson and G. Vincow, J. Chem. Phys., 52 (1970) 1740.
134 S.Y. Chang, E.R. Davidson and G. Vincow, J. Chem. Phys., 52 (1970) 5596.
135 K. Ishida and H. Nakatsuji, Chem. Phys. Lett., 19 (1973) 268.
136 H. Nakatsuji, K. Ohta and T. Yonezawa, J. Phys. Chem., 87 (1983) 3068.
137 D. Feller and E.R. Davidson, J. Chem. Phys., 80 (1984) 1006.
138 D. Feller and E.R. Davidson, Theor. Chim. Acta, 68 (1985) 57.
139 H. Sekino and R.J. Bartlett, J. Chem. Phys., 82 (1985) 4225.
140 T. Momose, H. Nakatsuji and T. Shida, J. Chem. Phys., 89 (1988) 4185.
141 H. Nakatsuji and M. Izawa, J. Chem. Phys., 91 (1989) 6205.
142 H. Nakatsuji, M. Ehara and T. Momose, J. Chem. Phys., 100 (1994) 5821.
143 K. O-ohata, H. Taketa and S. Huzinaga, J. Phys. Soc. Jpn, 21 (1966) 2306.
144 H. Taketa, S. Huzinaga and K. O-ohata, J. Phys. Soc. Jpn, 21 (1966) 3313.
145 T. Kato, Commun. Pure Appl. Math., 10 (1957) 151.
146 L. Edwards and M. Gouterman, J. Mol. Spectrosc., 33 (1970) 292.

Theory and Applications of Computational Chemistry: The First Forty Years
Edited by C. Dykstra et al.

CHAPTER 40

Forty years of Fenske–Hall molecular orbital theory

Charles Edwin Webster and Michael B. Hall

Department of Chemistry, Texas A&M University, College Station, TX 77843, USA

Abstract

In this contribution, we revisit the so-called 'Fenske–Hall' molecular orbital (MO) method, an approximate self-consistent-field (SCF) *ab initio* method that contains no empirical parameters, which began almost 40 years ago in the research group of Richard F. Fenske. We demonstrate for transition metal complexes that the non-empirical Fenske–Hall (FH) approach provides qualitative results that are quite similar to the more rigorous treatment given by density functional theory (DFT) and are quite different from Hartree–Fock–Roothaan (HFR) calculations which have no electron correlation. For example, the highest occupied molecular orbital of ferrocene is metal based for both DFT and FH while it is ligand (cyclopentadienyl) based for HFR. In the doublet ($S = 1/2$) cluster, $Cp_2Ni_2(\mu\text{-}S)_2(MnCO)_3$, the unpaired electron is delocalized over the complex in agreement with the DFT and FH results, but localized on Mn in the HFR calculation. A brief description of the theory of FH calculations is used to rationalize the origin of its similarity to DFT.

40.1 INTRODUCTION

This special issue celebrating 40 years of computational chemistry seems an appropriate venue to revisit the so-called 'Fenske–Hall' molecular orbital (MO) method. This approximate molecular orbital theory had its origins almost exactly 40 years ago when Richard F. Fenske joined the Department of Chemistry at the University of Wisconsin, Madison. Thus, the method might be more properly called the 'Fenske' method or the Fenske, Radtke, Caulton, DeKock, Hall method as the early development [1] involved many students in addition to one of the authors above (MBH). The name 'Fenske–Hall' was generated not by the first papers on the development of the method, but by the last paper on its theoretical development [2]. The method is still in use today and has

References pp. 1163–1165

remained essentially unchanged for the past 30 years. However, important extensions have been made in adding elements containing f electrons [3] and extending the method to solid-state band-structure calculations [4].

The method is an approximate self-consistent-field (SCF) *ab initio* method, as it contains no empirical parameters. All of the SCF matrix elements depend entirely on the geometry and basis set, which must be orthonormal atomic orbitals. Originally, the impetus for its development was to mimic Hartree–Fock–Roothaan [5] (HFR) calculations especially for large transition metal complexes where full HFR calculations were still impossible (40 years ago). However, as we will show here, the method may be better described as an approximate Kohn–Sham (KS) density functional theory (DFT) [6]. Early hints that Fenske–Hall (FH) calculations had some advantage over full HFR calculations came from comparisons of the FH molecular orbital energies with the experimental ionization energies from gas-phase ultraviolet photoelectron spectroscopy [2,7,8], where the order of MOs paralleled the order of states from the PES better for FH calculations than for HFR calculations. In other words, Koopmans' theorem [9] seemed to work better for Fenske–Hall than for HFR calculations.

40.2 ILLUSTRATIVE EXAMPLE

Before describing the Fenske–Hall method itself, we will examine an example of the behavior described above. Our example will be ferrocene $(\eta^5\text{-}C_5H_5)_2Fe$, whose qualitative molecular orbital diagram is shown in Scheme 40.1. The principal bonding

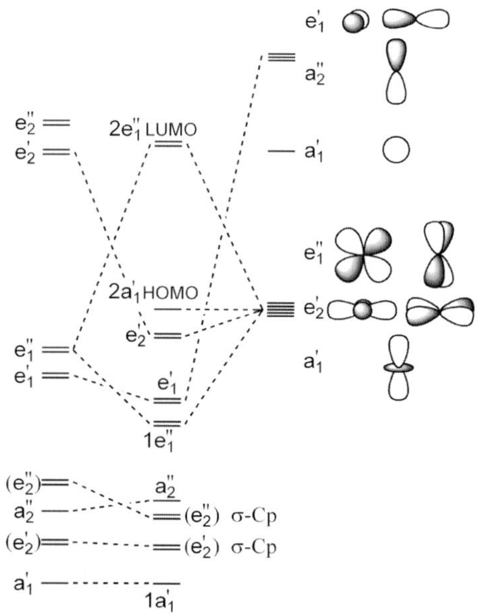

Scheme 40.1.

interactions occur between the π orbitals of the C_5H_5 (Cp) ring and the 3d, 4s, and 4p orbital of the Fe. The formal oxidation states are usually assigned such that the Cp ring is the 6π-electron anion, Cp^-, and the iron is the 6d-electron dication, Fe^{2+}. The π orbitals of a single Cp^- transform in D_{5h} point group as the a_2'' (the low-lying symmetric linear combination), the e_1'' (the doubly degenerate, highest occupied molecular orbital, HOMO) and the e_2'' (the doubly degenerate, lowest unoccupied molecular orbital, LUMO). Two of these rings then 'sandwich' the Fe between them in forming the D_{5h} ferrocene structure. Thus, the π orbitals of two Cp^- rings form symmetric and antisymmetric combinations (with respect to the molecular plane through the Fe atom). The two lowest lying π combinations, a_1' and a_2'', are stabilized by a $3d_{z^2} + 4s$ combination and the $4p_z$ orbital of Fe, respectively. The Cp^- HOMOs form the e_1'' and the e_1', which are strongly and weakly stabilized by the $3d_{xz}(3d_{yz})$ and $4p_x(4p_y)$, respectively. Finally, the Fe $3d_{xy}(3d_{x^2-y^2})$ pair are stabilized by the Cp LUMO combination of e_2' symmetry. These interactions leave the $3d_{z^2} - 4s$ combination as the non-bonding HOMO of ferrocene.

Fenske–Hall calculations and most qualitative descriptions correspond to this MO description [10]. Table 40.1 and Scheme 40.2 compare the experimental ionization energies for ferrocene [11] with theoretical determinations [12] based on Koopmans' theorem for DFT(B3LYP/6-31G), DFT(BVP86/6-31G), FH, and HFR/6-31G. In terms of the absolute ionization energies (IE), the DFT calculations are much closer to the experimental values and have the correct order for the first two ion states (lowest IEs) [13]. Both B3LYP and BVP86 produce the same orbital ordering and very similar orbital plots (not shown). Except for these two close lying, mainly metal orbitals, the Fenske–Hall results show the same order for the orbitals as the DFT results. The absolute energies from the Fenske–Hall calculations are closer to the HFR results.

However, the HFR results have very different order to the MOs than either DFT or FH, which are rather similar in order. Because of the very non-local nature of the Hartree–Fock exchange and compact 3d orbitals, which produces large exchange integrals for the 3d electrons, the Fe orbitals lie lower in energy than the more diffuse Cp^- MOs. Thus, the incorporation of exact exchange (Fermi correlation) and the neglect of true electron correlation (Coulomb correlation) in the HFR calculations result in more ionic bonding and less mixing between the Fe and Cp orbitals. These differences can be seen in the orbital plots shown in Fig. 40.1, especially in the plots of

Table 40.1 Vertical ionization energy (eV) [11] and orbital energies (eV) from various methods

Region [11]	PES (eV) [11]		B3LYP	BVP86	FH	HFR
		LUMO	0.01	−1.33	−1.29	4.19
A′	6.86	HOMO	−5.23	−3.99	−9.00	−9.31
	7.21		−6.18	−4.26	−10.28	−9.33
A″	8.77		−6.76	−6.10	−13.32	−11.63
	9.28		−7.28	−6.63	−14.37	−13.53
B	12.2		−9.78	−8.84	−18.37	−14.06
	13.3		−10.25	−9.13	−18.52	−14.20

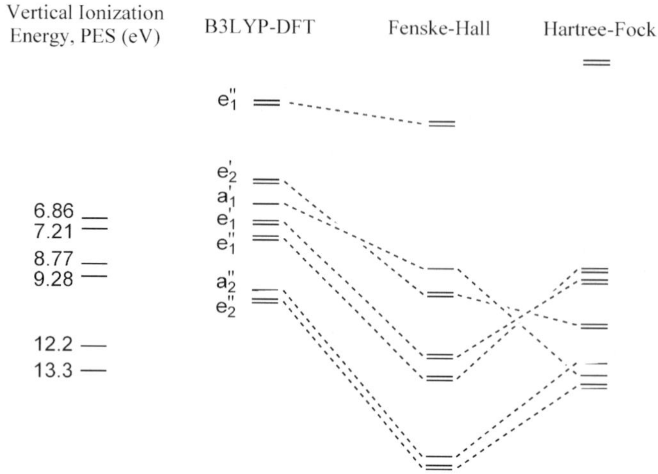

Scheme 40.2.

the e_2', which shows this metal combination to be stabilized by in-phase mixing with the LUMOs of the Cp's in the DFT and FH MOs, but to be destabilized by out-of-phase mixing with low-lying σ Cp orbitals in the HFR MO.

At the time of its development no thought was given to the idea that FH might be an approximation to DFT, as HFR theory seemed to be the 'holy grail'. However, now it is apparent that the FH method produces MOs and orbital energy orderings that seem to be more like those from DFT than those from HFR. Some of this observed similarity is built into the FH method as we will illustrate below.

40.3 THEORY

Here, we will not provide a detailed derivation of the Fenske–Hall theory beginning from the HFR equations as the reader may refer to the original reference [2]. Instead, we will provide a conceptual derivation that will allow us more freedom to comment on its relationship with DFT.

The FH method assumes that the Fock/Kohn Sham (KS) operator can be written as

$$F = -\frac{1}{2}\nabla^2 + \sum_{A} V_A \tag{1}$$

i.e. the electron's kinetic energy plus a potential due to the nuclei and all the other electrons on each atom in the system. Thus, the FH method must be able to partition the electron density and assign it to individual atomic centers. Traditionally, FH did this through the Mulliken populations analysis [14], but one could use some other procedure.

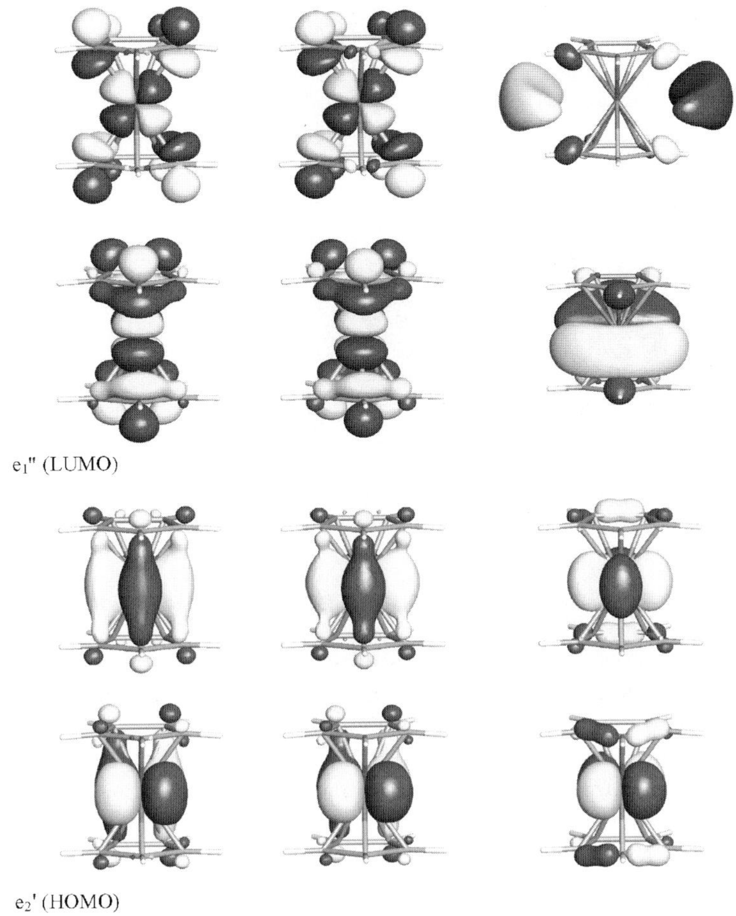

Fig. 40.1. A comparison of the DFT-B3LYP/6-31G (left), FH (center), and HFR/6-31G (right) e_1'' (LUMO), e_2' (HOMO), a_1', e_1', e_1'', a_2'' calculated orbitals of ferrocene. The orbital plots from DFT-BVP86/6-31G (not shown) are virtually identical to the DFT-B3LYP/6-31G orbital plots.

For a diagonal matrix element, orbital a on center A

$$F_{aa} = \langle a|F|a\rangle = \langle a| -\frac{1}{2}\nabla^2 + V_A|a\rangle + \sum_{B\neq A}\langle a|V_B|a\rangle \tag{2}$$

The first term is the atomic orbital (AO) energy for a. This AO energy, ε_a, represents the kinetic energy, nuclear attraction, coulomb repulsion, and exchange term for the electron in AO a on center A, and

$$\varepsilon_a = \langle a| -\frac{1}{2}\nabla^2 + V_A|a\rangle = \langle a| -\frac{1}{2}\nabla^2 - \frac{Z_A}{r_A} + V_{co} + V_{ex}|a\rangle \tag{3}$$

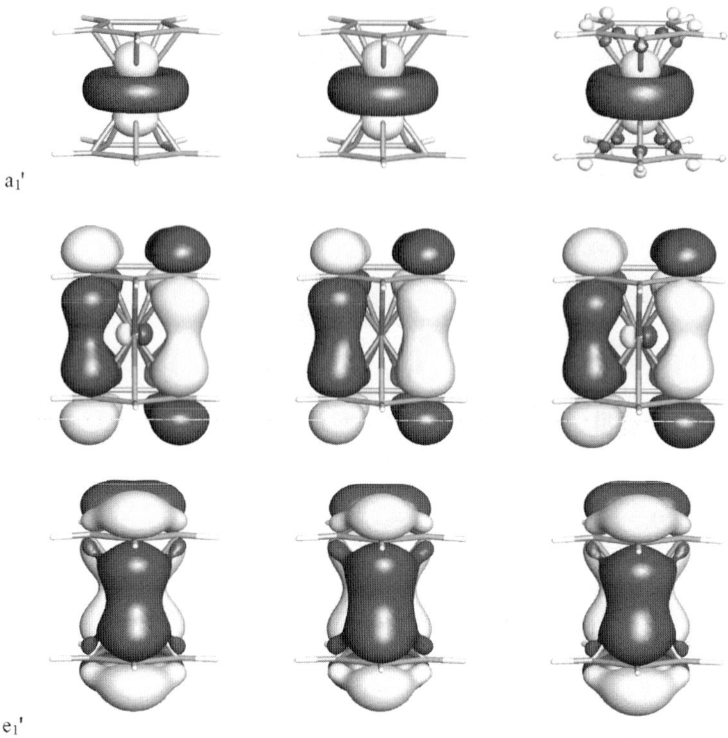

Fig. 40.1. (*continued*).

The kinetic energy and nuclear attraction are calculated exactly for the AO a, while the coulomb term is

$$\langle a|V_{\mathrm{co}}|a\rangle = \sum_{a'(\mathrm{A})} P_{a'} g(a', a) \tag{4}$$

where a' runs over other AOs (or subshell) on center A, $P_{a'}$ is the Mulliken population of the AO (or subshell), and $g(a', a)$ is the state averaged electron repulsion integral for each atomic subshell [15]. Since many metal–ligand bonds are relatively weak compared to the strong covalent bonds in first-row main group molecules, it was believed important that these AO energies were correct for a single free atom. Thus, the exchange term instead of being weighted by a Mulliken population, as it is in typical zero differential overlap (ZDO) procedures, was taken to represent the complete removal of one-electron (removal of all self-exchange)

$$\langle a|V_{\mathrm{ex}}|a\rangle = -g(a, a) \tag{5}$$

Clearly this exchange term is quite different from the HFR situation, in which the exchange integrals contain multicenter terms and are weighted by the density matrix

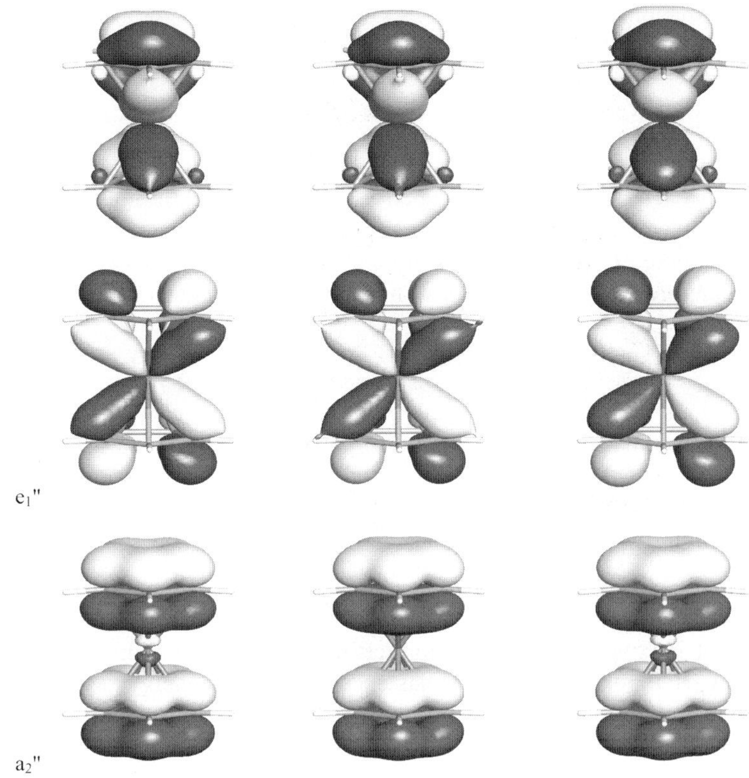

Fig. 40.1. (*continued*).

terms. Furthermore, because a full (unweighted by the density) exchange integral is subtracted, the FH term and HFR diagonal term will have similar coulomb and exchange *only* when the Mulliken population of AO a (P_a) is 2.0. In other words, FH exchange has local character, like DFT, and in most cases somewhat more exchange than HFR. The exact exchange in HFR theory leads to molecular orbitals that are less covalently mixed between the metal and ligands because more ionic bonds have larger exchange, especially for first-transition row metals [16]. Both FH and DFT have more mixing in the metal–ligand bonding MOs, as shown above for ferrocene.

The two-center term in these diagonal matrix elements contains the nuclear attraction and electron repulsion for all the other atoms in the systems. In FH these were both set to the point-charge classical limit. Thus,

$$\langle a|V_{\mathrm{B}}|a\rangle = \frac{-q_{\mathrm{B}}}{R_{\mathrm{AB}}} \tag{6}$$

where q_{B} is the Mulliken charge on atom B. This approximation is similar to that used in other approximate MO schemes, where the $1/R_{\mathrm{AB}}$ term is often replaced with a reduced integral, γ_{AB}, especially in many ZDO schemes.

For the two-center off-diagonal matrix elements, a kinetic energy operator is added to and subtracted from Eq. (1) so that

$$F_{ab} = \langle a| - \frac{1}{2}\nabla^2 + V_A|b\rangle + \langle a| - \frac{1}{2}\nabla^2 + V_B|b\rangle - \langle a| - \frac{1}{2}\nabla^2|b\rangle + \sum_{C \neq A,B} \langle a|V_C|b\rangle \quad (7)$$

Then, it is assumed that each AO is an eigenfunction (ε) of the kinetic energy and potential term for that center [17] and the Mulliken and point charge approximation are applied to the three-center potential terms. In the Fenske–Hall calculation, the two-center kinetic energy is calculated for the given basis set and subtracted as shown in Eq. (8)

$$F_{ab} = S_{ab}\varepsilon_a + S_{ab}\varepsilon_b - \langle a| - \frac{1}{2}\nabla^2|b\rangle + \frac{1}{2}S_{ab} \sum_{C \neq A,B} \left(\frac{-q_B}{R_{AC}} + \frac{-q_C}{R_{BC}} \right) \quad (8)$$

Because the Mulliken approximation is less accurate for the kinetic energy, i.e.

$$\langle a| - \frac{1}{2}\nabla^2|b\rangle \neq \left(S_{ab}\left[\left\langle a \middle| -\frac{1}{2}\nabla^2 \middle| a \right\rangle + \left\langle b \middle| -\frac{1}{2}\nabla^2 \middle| b \right\rangle \right] \right) \Big/ 2 \quad (9)$$

the eigenvalue approximation used above convolutes the kinetic and potential energy in an odd way, again reminiscent of DFT. One-center off-diagonal terms are set to zero.

Although FH does not have any explicit or implicit operator corresponding to the correlation functional of DFT, for molecules FH has excess exchange energy, which to some degree mimics the missing correlation functional. This excess exchange energy arises from the fact that FH subtracts a full one-center coulomb integral to account for removal of self-exchange, while for HFR theory the magnitude of self-exchange in a delocalized MO is less because it is weighted by the density and contains multi-center exchange integrals. The result of this excess exchange can be seen in the orbital energies of the π system of the Cp ring, which are stabilized more than those in the HFR calculation. However, FH does not have the variational freedom to contract the AOs in response to this excess exchange. Furthermore, because FH treats the exchange somewhat like a simple exchange correlation functional, such as X_α [18], the virtual orbitals in FH see the potential due to $\sim(n-1)$ electrons rather than n electrons as in HFR theory. Thus, like DFT calculations, the HOMO–LUMO gap for a given compound is smaller in FH than that in HFR calculations.

40.4 TRANSITION METAL CLUSTERS

As the system becomes larger one might expect to see larger deviations between Fenske–Hall and DFT results. However, as we will illustrate here the results are surprisingly similar. Fenske–Hall almost always has the same HOMO and the same LUMO as DFT. In the following two subsections, we will describe the bonding description derived from Fenske–Hall calculations and then for one complex compare the FH results with those produced by both HFR and DFT (B3LYP) calculations and for the other complex compare to results of DFT (B3LYP) calculations.

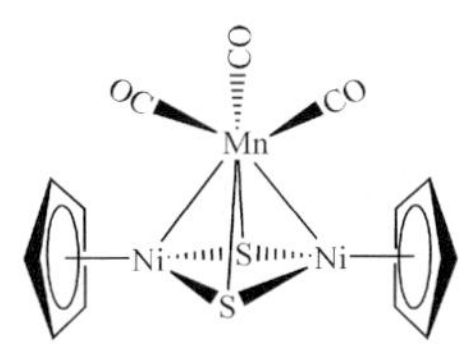

The electronic structure of $Cp_2Ni_2(\mu\text{-}S)_2(MnCO)_3$ ($S = 1/2$ ground state) has been described by Fenske–Hall calculations [19]. Here, we will describe those results first and then compare them with DFT and HFR calculations. The MO diagram shown in Scheme 40.3 was obtained by using a suitable combination of the requisite atomic and fragment orbitals described below. The left-hand side of Scheme 40.3 shows the molecular orbital structure of the two principal transition metal fragments, $Mn(CO)_3$ and CpNi. As expected for pseudo-octahedral fragments, the MO splittings show three low-lying occupied orbitals from the pseudo-octahedral 't_{2g}' set and two higher lying orbitals from the 'e_g' set, which contain three electrons for CpNi and one electron for $Mn(CO)_3$. The next lowest-lying orbital for each fragment has a_1 symmetry. The right hand of Scheme 40.3 shows the combination of two CpNi fragments with two bridging sulfurs to form the $Cp_2Ni_2(\mu\text{-}S)_2$ fragment, followed by the attachment of $Mn(CO)_3$. The key frontier orbitals of the $Cp_2Ni_2(\mu\text{-}S)_2$ fragment are pictured in Scheme 40.4; these representations

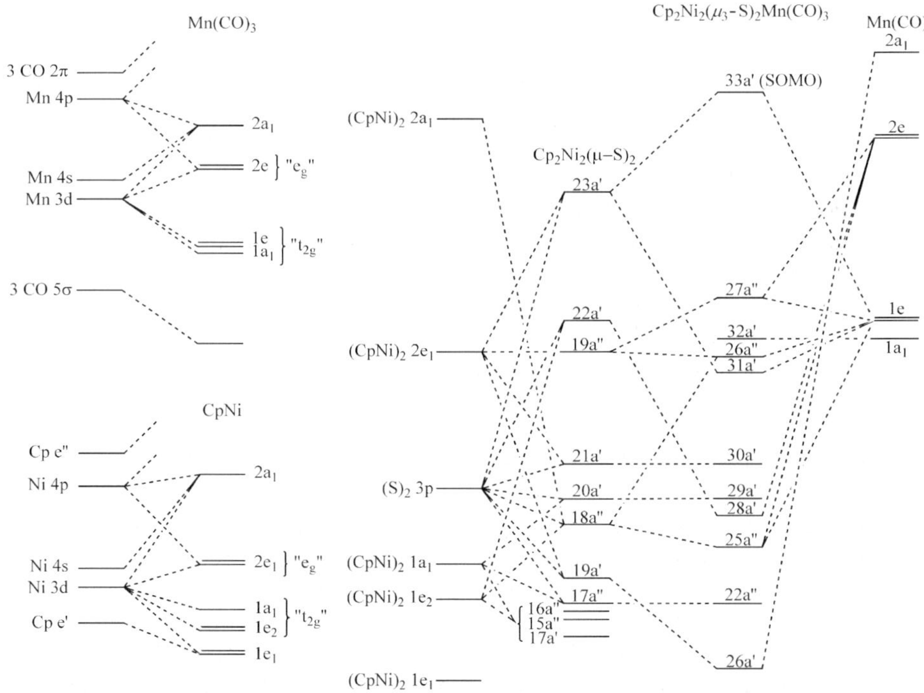

Scheme 40.3. The MO diagram for $Cp_2Ni_2(\mu\text{-}S)_2(MnCO)_3$ for its experimentally determined geometry.

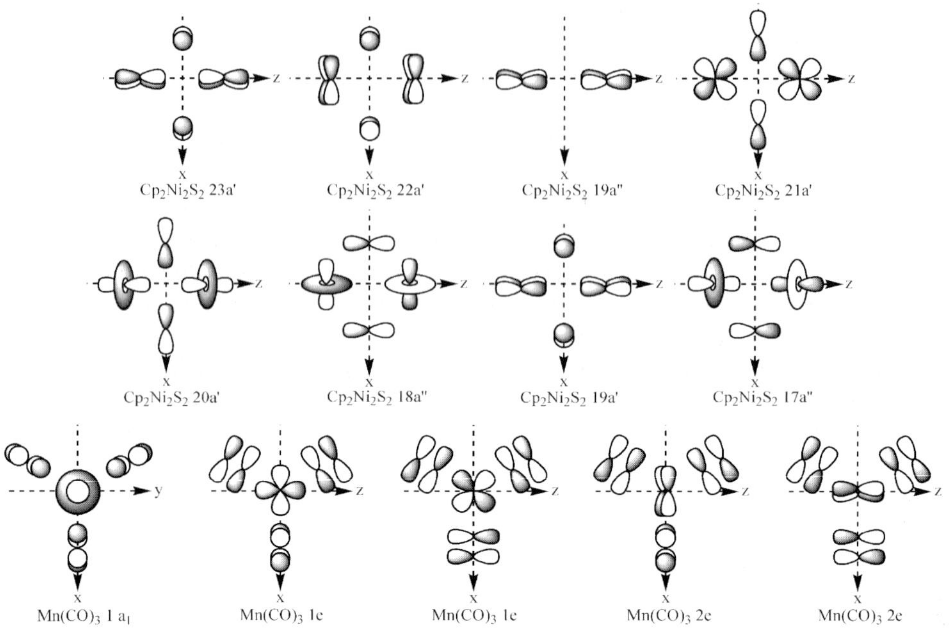

Scheme 40.4. Diagrams of frontier orbitals of $Cp_2Ni_2(\mu\text{-}S)_2$ and $Mn(CO)_3$.

are schematics of the principal orbital character as derived from 3D visualizations of the MOs. In forming the $Cp_2Ni_2(\mu\text{-}S)_2$ fragment, the principal bonding interactions occur between the CpNi $2a_1$ and $2e_1$ orbitals and S 3p orbitals. The 't_{2g}' orbitals ($1e_2$ and $1a_1$) and CpNi bonding orbitals ($1e_1$) of the CpNi fragment remain occupied and essentially non-bonding in $Cp_2Ni_2(\mu_3\text{-}S)_2$ and the cluster, but they complicate the MO diagram because they occupy the same energy region as the cluster bonding pairs. The 17a′, 15a″, and 16a″ $Cp_2Ni_2S_2$ fragment orbitals correspond to the higher lying occupied combinations of these 't_{2g}' like orbitals and are essentially non-bonding with respect to the cluster. The four Ni–S bonding orbitals in the Ni_2S_2 'plane' are 17a″ [composed of $(CpNi)_2$ $2e_1$, $1a_1$ and S 3p], 18a″ [$(CpNi)_2$ $2a_1$, $1e_2$ and S 3p], 20a′ [$(CpNi)_2$ $2a_1$, $1a_1$, and S 3p], and 21a′ [$(CpNi)_2$ $2e_1$ and S 3p]. The remaining $Cp_2Ni_2S_2$ fragment orbitals, 19a′ [composed of $(CpNi)_2$ $2e_1$ and S 3p], 22a′ [$(CpNi)_2$ $1e_2$ and S 3p], 19a″ [$(CpNi)_2$ $2e_1$], and 23a′ [$(CpNi)_2$ $2e_1$ and S 3p], are mainly perpendicular to the Ni_2S_2 'plane'.

As shown in Scheme 40.3, combining the frontier orbitals of $Cp_2Ni_2(\mu\text{-}S)_2$ with the frontier orbitals of $Mn(CO)_3$ (shown in Scheme 40.4) generates 11 high-lying orbitals for $Cp_2Ni_2(\mu\text{-}S)_2(MnCO)_3$; for clarity, the non-bonding 't_{2g}' orbitals of the $Cp_2Ni_2S_2$ in this energy region are not shown. Seven of these final orbitals (26a′, 22a″, 25a″, 28a′, 29a′, 30a′, and 31a′) are used for cluster bonding. Three of the seven cluster bonding orbitals, 22a″ (17a″ $Cp_2Ni_2S_2$ fragment orbital), 29a′ (20a′ $Cp_2Ni_2S_2$ fragment orbital), and 30a′ (21a′ $Cp_2Ni_2S_2$ fragment orbital), primarily contribute to cluster bonding by providing interactions between the two CpNi and two S fragments. The four remaining of these seven cluster bonding orbitals, 26a′ [composed of 19a′ $Cp_2Ni_2S_2$ and $2a_1$ $Mn(CO)_3$

fragment orbitals], 25a″ [18a″ $Cp_2Ni_2S_2$ and 2e $Mn(CO)_3$], 28a′ [22a′ $Cp_2Ni_2S_2$ and 2e $Mn(CO)_3$], and 31a′ [23a′ $Cp_2Ni_2S_2$ and 1e $Mn(CO)_3$], contribute to cluster bonding between the $Cp_2Ni_2(\mu\text{-}S)_2$ and $Mn(CO)_3$ fragments. The principal Mn–Ni bonding arises from molecular orbital 25a″ and 31a′ shown in Fig. 40.2. It is somewhat unexpected that one of the 't_{2g}' pairs of the $Mn(CO)_3$ fragment enters into cluster bonding, as it does in MO 31a′, by donating electron density to the empty $Cp_2Ni_2S_2$ 23a′ orbital because standard skeletal electron pair counting usually does not count 't_{2g}' electron pairs from the $Mn(CO)_3$ fragment as contributors [19]. Three of the four remaining orbitals (26a″, 32a′, and 27a″) are essentially non-bonding with respect to the $Cp_2Ni_2(\mu\text{-}S)_2$ and $Mn(CO)_3$ fragment interactions. There is a small contribution to cluster bonding from donation of electrons from the 19a″ to one of the 'e_g' orbital of the $Mn(CO)_3$ fragment in forming the cluster MO 27a″. The remaining orbital, 33a′, is the singly occupied MO (shown in Fig. 40.3) and has antibonding character with respect to the Mn–Ni interactions, but bonding character between the Mn and the two S. The percent Mn character derived for the singly occupied

(a)

(b)

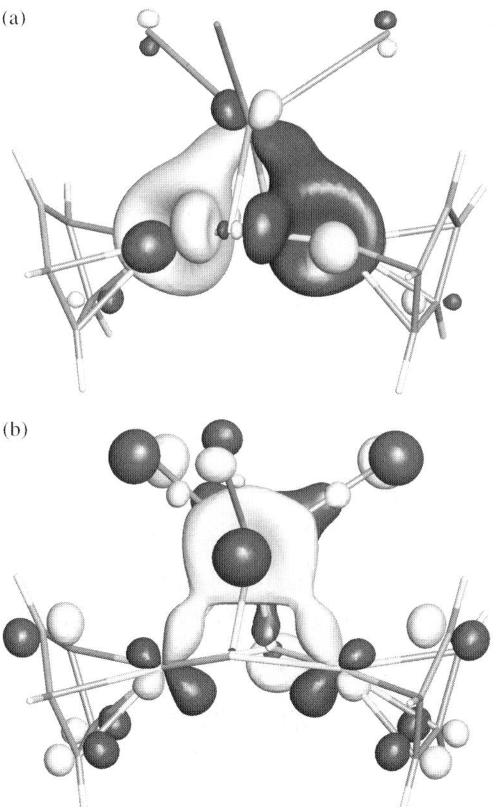

Fig. 40.2. The two MOs containing the principal Mn–Ni bonding of $Cp_2Ni_2(\mu\text{-}S)_2(MnCO)_3$ from FHMO calculations: (a) is 25a″; (b) is 31a′.

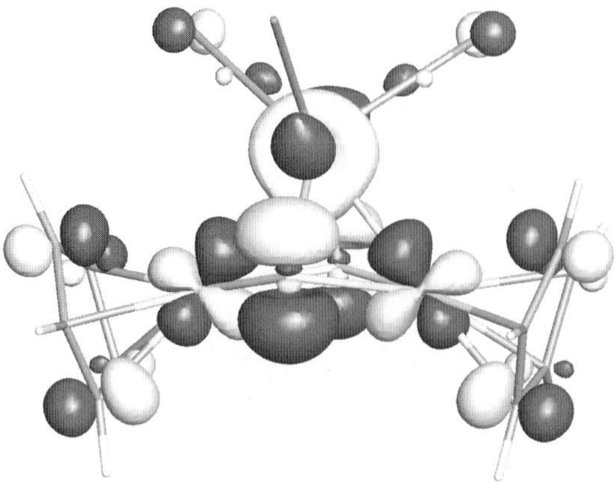

Fig. 40.3. The FH singly-occupied HOMO 33a′ of $Cp_2Ni_2(\mu\text{-}S)_2(MnCO)_3$.

HOMO is approximately 20% for the FH calculation, which is in reasonable agreement with the value derived from EPR measurements (approximately 13%) [20].

One can see from the comparison of the B3LYP, FH, and HFR frontier regions (Figs. 40.4 and 40.5) that the calculated B3LYP and FH orbitals are quite similar, but both are very unlike the HFR orbitals. The low symmetry of this complex (C_s) allows for a great deal of mixing for these closely spaced orbitals (spanning only approximately 2 kcal mol^{-1}) and accounts for the minor differences between the B3LYP and FH frontier

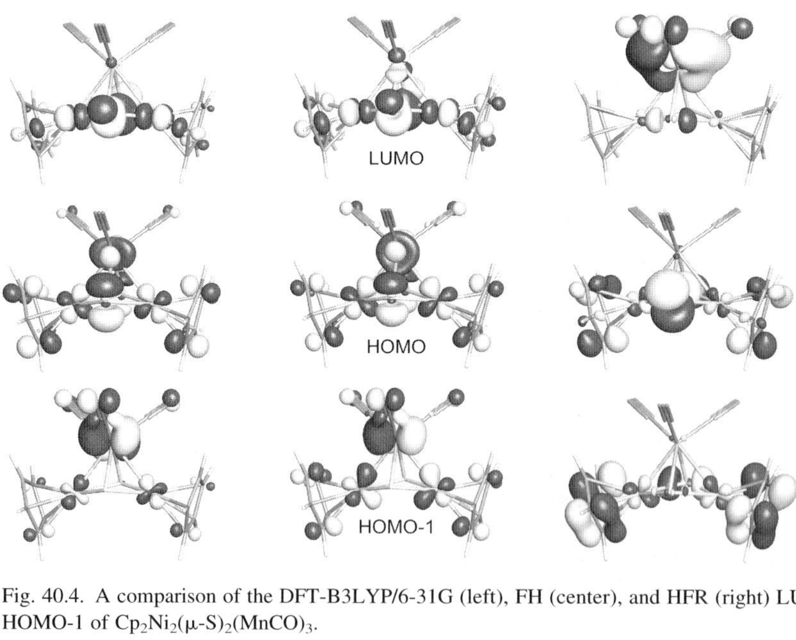

Fig. 40.4. A comparison of the DFT-B3LYP/6-31G (left), FH (center), and HFR (right) LUMO, HOMO, and HOMO-1 of $Cp_2Ni_2(\mu\text{-}S)_2(MnCO)_3$.

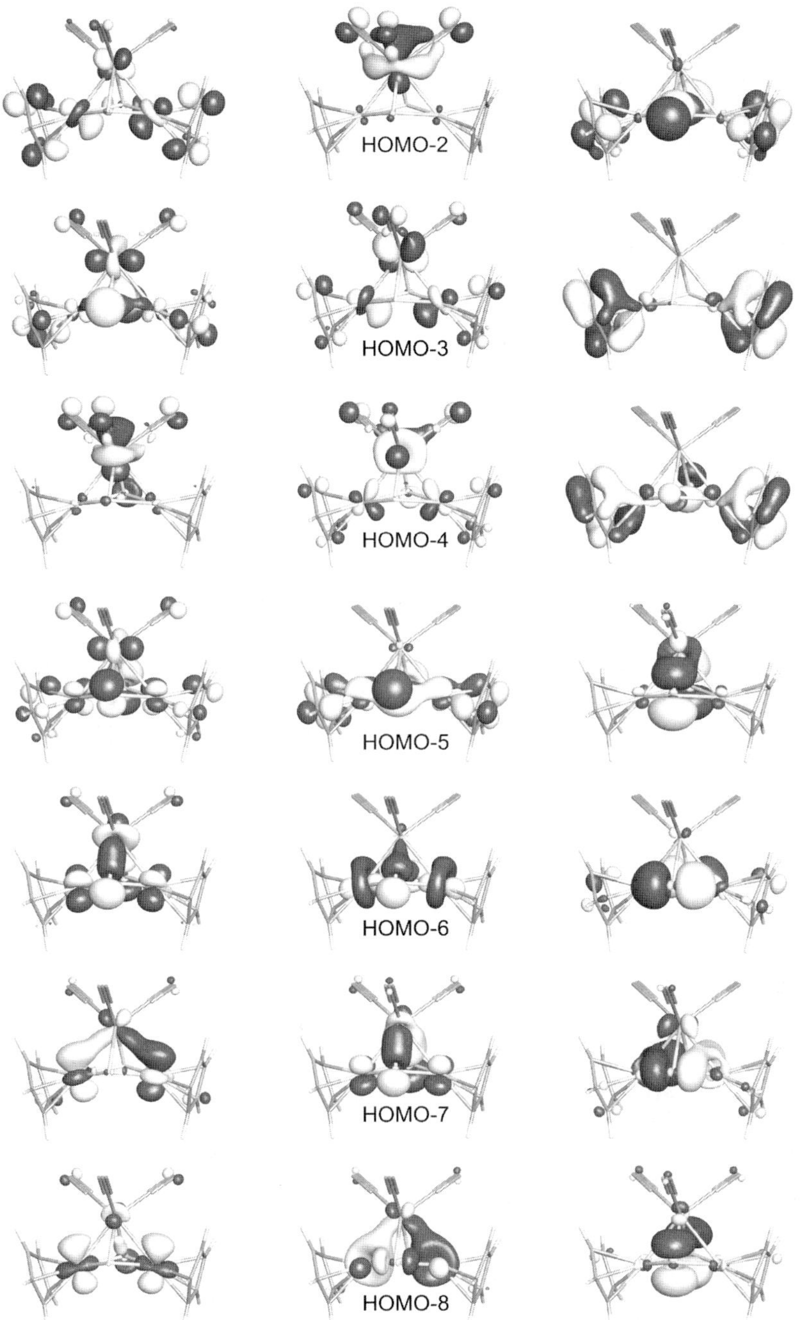

Fig. 40.5. A comparison of the DFT-B3LYP/6-31G (left), FH (center), and HFR (right) HOMO-2 to HOMO-8 of $Cp_2Ni_2(\mu\text{-}S)_2(MnCO)_3$.

ordering. The B3LYP and FH LUMO, HOMO, and HOMO-1 are quite similar and are in the same ordering, but the ordering of the deeper orbitals differ slightly. However, there are still many closely corresponding orbitals: B3LYP HOMO-2 and FH HOMO-3; B3LYP HOMO-4 and FH HOMO-2; B3LYP HOMO-5 and FH HOMO-5; and B3LYP HOMO-6 and FH HOMO-7. The remaining differences can be accounted for by the mixing mentioned above.

The B3LYP and FH singly occupied HOMOs are delocalized across the complex and have very similar percentages of Mn character, and the B3LYP and FH LUMO are similar except for slightly less Mn character in the B3LYP LUMO. However, HFR calculations provide frontier orbitals that are very heavily Cp and S based with very little metal character. The Mulliken atomic spin density for B3LYP gives the unpaired electron on Mn at about 7% (FH 20%, exp. 13%, see above); but the Mulliken atomic spin density for the HFR calculation provides a very different picture: the excess spin is very localized on the Mn, approximately 100%. The HFR orbital (shown in Fig. 40.6) that contains this Mn character is not in the frontier region, but 25 orbitals below the HOMO, and is lower in energy than the HOMO (by approximately 7.3 eV) [21]. The ground state HFR wavefunction (a doublet, one unpaired electron) also has a large amount of spin contamination, while the ground state B3LYP wavefunction (again, a doublet) has very little spin contamination. For the HFR wavefunction, the value of S^2 ($S^2 = n/2((n/2)+1)$) where n = number of unpaired electrons, $S^2 = 0.75$ for the pure doublet ($S = 1/2$) state is 2.75 and 5.04 before annihilation of the quartet and after annihilation of the quartet, respectively, while for the B3LYP wavefunction it is 0.809 and 0.752 before annihilation of the quartet and after annihilation of the quartet, respectively. Therefore, higher spins (such as the sextet) are strongly contaminating the HFR ground state, but not the B3LYP wavefunction (FH produces a restricted result without any spin contamination).

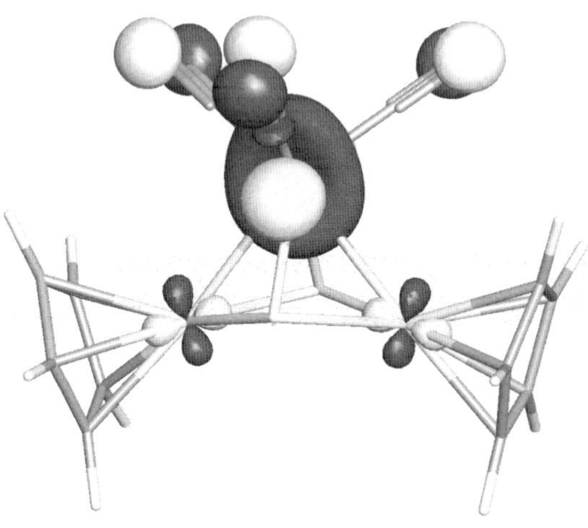

Fig. 40.6. The HFR/6-31G orbital of $Cp_2Ni_2(\mu\text{-}S)_2(MnCO)_3$ that contains the unpaired electron which from HFR is mostly localized on the Mn.

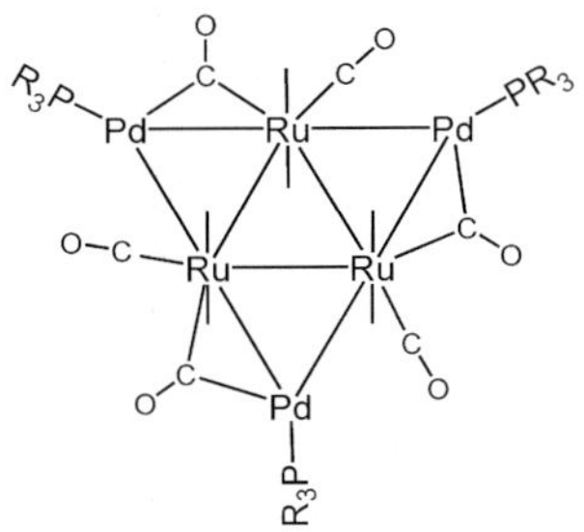

Here, we will describe the FH results of the $Ru_3(CO)_{12}[Pd(PR_3)]$ cluster [22] first and then compare those with DFT calculations. One can easily envision how this cluster can be assembled from three $Ru(CO)_4$ units and three $Pd(PR_3)$ units, but one wonders how this is accomplished in an orbital sense and why the dimer of trimers is distorted to a C_3-like structure rather than having a higher symmetry D_3-like structure. The bridging CO was assigned to Ru because its σ-donor orbital was directed more toward Ru than Pd. The key low-lying orbitals are shown diagrammatically in Scheme 40.5. The $Ru(CO)_4$ appears most like a trigonal bipyramid with a missing ligand. As this neutral fragment would have a d^8 configuration, the HOMO is the d_{xz}, the d orbital that would have been stabilized by the missing CO in the fully ligated $Ru(CO)_5$ molecule (see Scheme 40.5). The LUMO is the 'sp^{2}' hybrid that would have been destabilized by accepting the fifth COs lone pair (see Scheme 40.5). The last orbital of importance on the $Ru(CO)_4$ fragment is the low-lying π^* orbital on the distorted CO (LUMO + 1). This distortion (a Ru–C–O angle of ~140° rather than 180°) arises from the clockwise (or counter clockwise) twisting of each $Ru(CO)_4$ unit to place one CO closer to each Pd. The distortion lowers the energy of this LUMO + 1 orbital because it is now less effective in back bonding to the Ru. We will return to the origin of this twisting and subsequent distortion later after describing the bonding in the structure as found. The key orbitals of the $Pd(PR_3)$ fragment

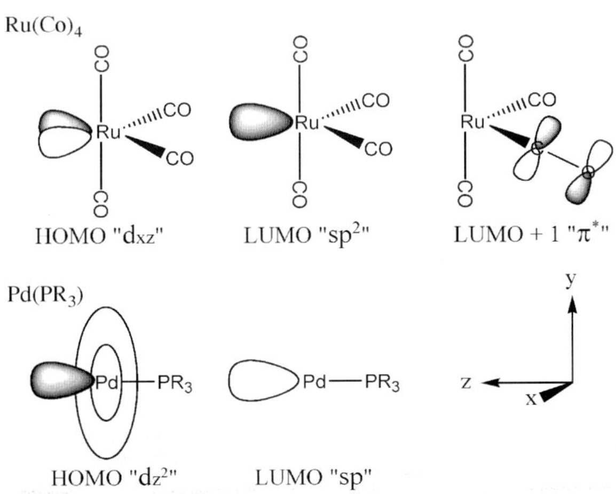

Scheme 40.5.

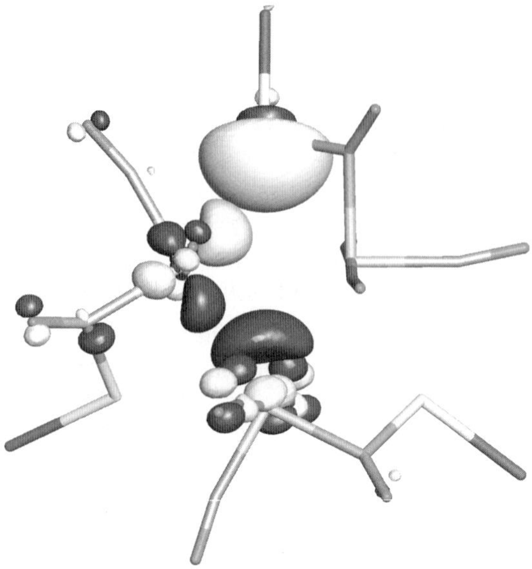

Fig. 40.7. Fragment molecular orbitals that produce the three-center/two-electron bonds (one of three by symmetry), which constitute the principal metal-to-metal bonding interactions of $Ru_3(CO)_{12}[Pd(PR_3)]$.

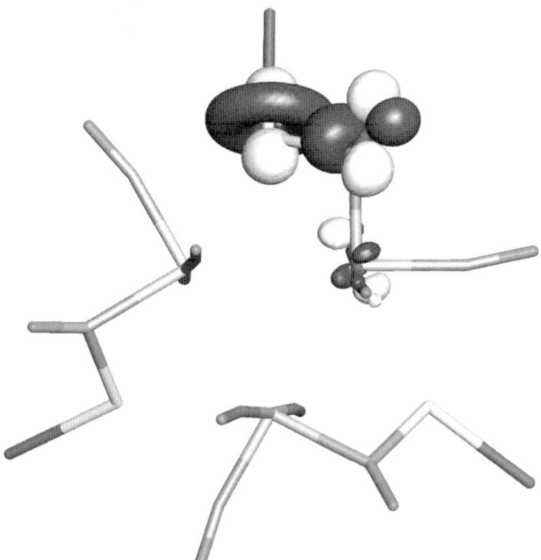

Fig. 40.8. $Pd(PR_3)$ fragment HOMO that donates electron density to the LUMO + 1 of the $Ru(CO)_4$ fragment, primarily π^* orbital of the semibridging CO.

are somewhat simpler (see Scheme 40.5). The HOMO is the z^2, the d orbital pointing away from the only ligand in this d^{10} fragment, while the LUMO is the 'sp' hybrid pointing in the same direction.

When three $Ru(CO)_4$ and three $Pd(PR_3)$ fragments assemble into the cluster, the principal metal–metal bonding arises from the HOMO of one Ru donating electron density into both the LUMO of the Pd opposite the bridging CO and the LUMO of the Ru on the other side. Three of these three-center, two-electron bonds constitute the principal metal-to-metal bonding molecular orbitals, one of these interactions is shown in Fig. 40.7, which shows the actual fragment MOs from fragment Fenske–Hall calculations (these are the same fragment orbitals shown diagrammatically in Scheme 40.5, here, they are shown in their correct relationship to each other). An important secondary interaction is shown in Fig. 40.8, where the HOMO of the $Pd(PR_3)$ fragment (in particular the 'donut'

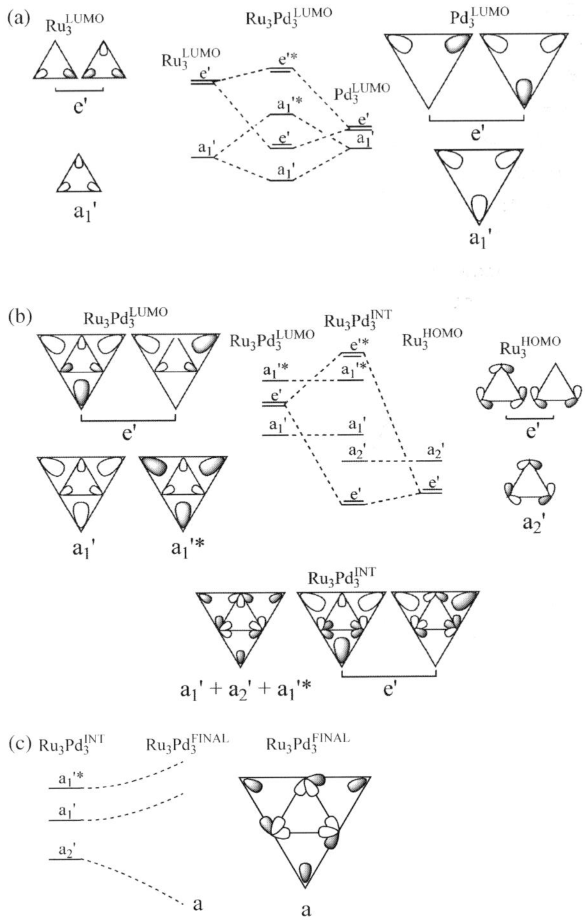

Scheme 40.6.

of the d_{z^2}) donates electron density to the LUMO + 1 of the $Ru(CO)_4$, the π^* orbital of the semibridging CO.

Now, one may wonder why the $Ru(CO)_4$ distorts such that the secondary interaction described above is unsymmetrical. In a more symmetrical structure the COs on both sides of the Ru could accept electron density from the Pd HOMO. So the distortion most likely has another origin, i.e. the bent Ru–C–O structure and the subsequent 'one-side' Pd to CO(Ru) back bonding is a response to the distortion not the driving force.

The driving force for the distortion arises from a second-order Jahn–Teller (JT) effect [23], because in a high-symmetry point group, such as D_{3h} or C_{3v}, the irreducible representations of symmetry-adapted linear combinations of the $Ru(CO)_4$ HOMOs have an incomplete correspondence to irreducible representations of the symmetry adapted LUMOs of the neighboring fragments. Thus, the $Ru(CO)_4$ twists to reduce the symmetry and maximize its interaction with its neighbors. A complete analysis of the second-order JT effect is illustrated in Scheme 40.6. For the purpose of this analysis, the molecular electronic structure will be represented simply by two triangular metal clusters one with three Ru fragments and one larger triangle, rotated 60°, with three Pd fragments. For the main metal-to-metal bonding there are three donor orbitals on the Ru (Ru^{HOMO}) and six acceptor orbitals, three on Ru (Ru^{LUMO}) and three on Pd (Pd^{LUMO}). In the highest possible local symmetry, D_{3h}, the Ru_3^{LUMO} and Pd_3^{LUMO} orbitals transform as a_1' and e', as shown in Scheme 40.6a. These two sets of LUMOs interact with each other (as shown by the orbital interaction diagram in Scheme 40.6a) to form in-phase, lower lying combinations and out-of-phase higher lying combinations.

Thus, one now has a set of strong acceptor orbitals involving all six metal atoms that transform as a_1', $a_1'^*$, and e' (the $Ru_3Pd_3^{LUMO}$ orbitals shown in Scheme 40.6a). The primary

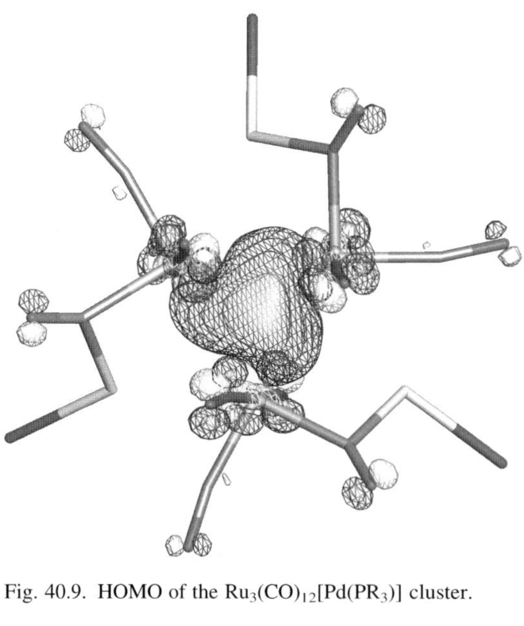

Fig. 40.9. HOMO of the $Ru_3(CO)_{12}[Pd(PR_3)]$ cluster.

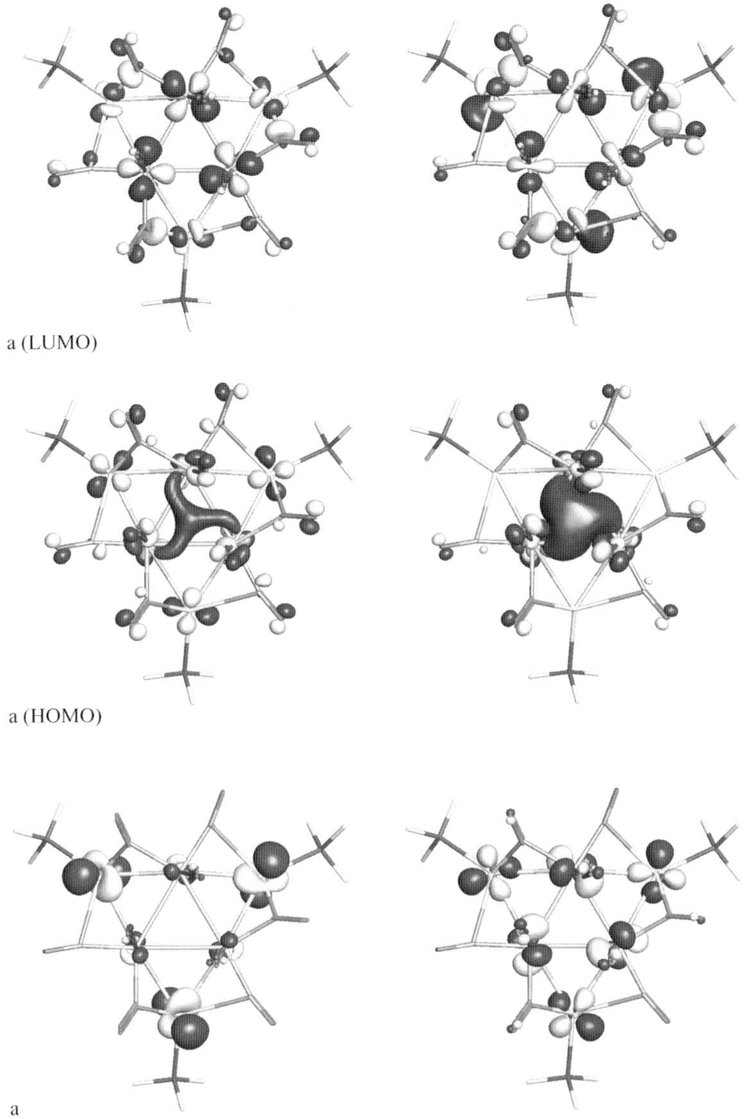

Fig. 40.10. A comparison of the DFT-B3LYP/BS2 (left) and FH (right) LUMO to HOMO-4 of $Ru_3(CO)_{12}[Pd(PR_3)]_3$.

metal-to-metal bonding occurs when these $Ru_3Pd_3^{LUMO}$ orbitals accept electron density from the Ru_3^{HOMO} orbitals, the symmetry-adapted linear combinations of the latter orbitals are shown interacting with the $Ru_3Pd_3^{LUMO}$ set in Scheme 40.6b. However, since the Ru_3^{HOMO} set transforms as a_2' and e', not a_1' and e', there is a symmetry mismatch and in D_{3h} or C_{3v} (a_2, a_1, and e) point groups. Thus, all three pairs of electrons from the Ru_3^{HOMO}

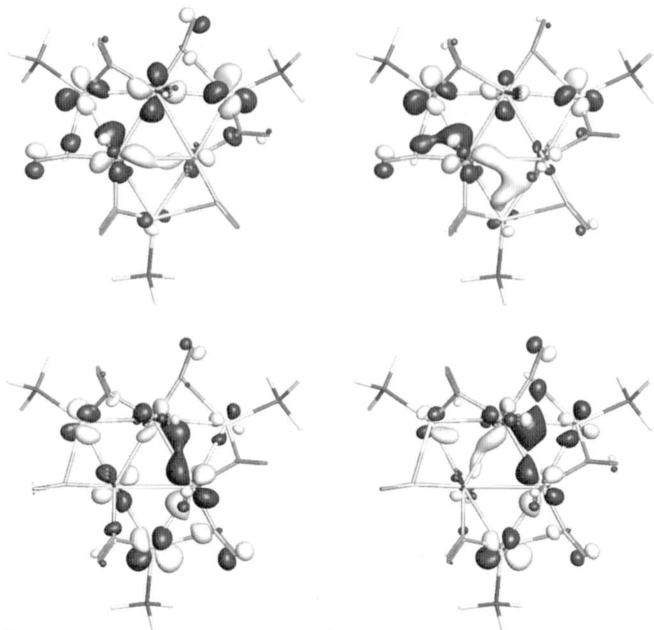

e

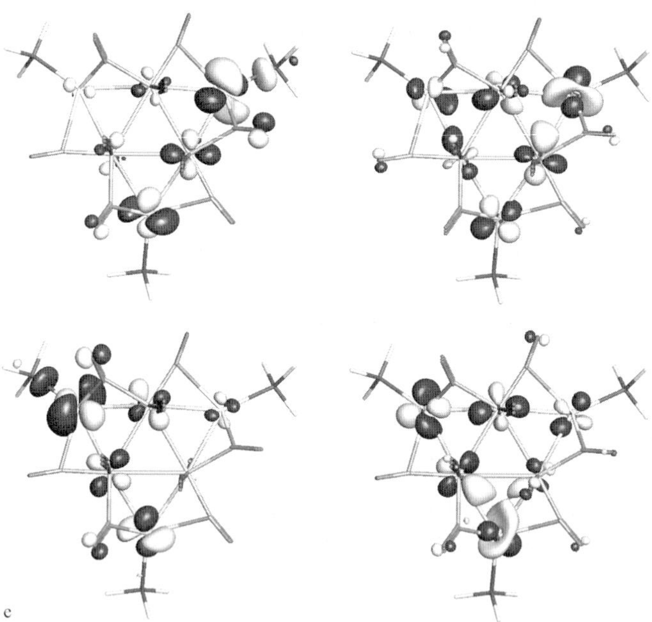

e

Fig. 40.10. (*continued*).

set cannot be used for bonding in a high symmetry situation. This dilemma is illustrated in the orbital interaction diagram in Scheme 40.6b, which shows a suitable bonding $Ru_3Pd_3^{INT}$ interaction for the e′ arising from the e′ of the Ru_3^{HOMO} donating to the e′ of the $Ru_3Pd_3^{LUMO}$. However, in this high symmetry both the occupied Ru_3^{HOMO} a_2' and the unoccupied $Ru_3Pd_3^{LUMO}$ a_1' and $a_1'^*$ must remain non-bonding as the energy diagram shows and as is illustrated in the linear combination $a_1' + a_2' + a_1'^*$, where one can see that the a_2' combination has zero overlap by symmetry with a_1' and $a_1'^*$. The dilemma can be resolved by twisting the $Ru(CO)_4$ so that the symmetry drops to C_3 point group and the a_1' and a_2' both become a symmetry in $Ru_3Pd_3^{FINAL}$ and their mutual interaction will stabilize the occupied combination as shown in Scheme 40.6c; the HOMO for $Ru_3Pd_3^{FINAL}$ is shown in Fig. 40.9.

One can see from the comparison of the frontier B3LYP and FH orbitals (Fig. 40.10) that the calculated orbitals are quite similar. Comparing the two sets of orbitals at the same contour levels, the LUMOs are similar, but there is more Pd character in the FH orbital. For the HOMO, the B3LYP orbital has less $(Ru)_3$ overlap than the FH orbital, and the FH orbital has virtually no Pd character. The two sets of e orbitals have similarities, but the Pd character is less consistent between the two calculations (because these two orbitals are mixing slightly with each other). The non-degenerate HOMO-3 is again quite similar for the two calculations. We also know from DFT geometry optimizations of this complex that the observed reduction in symmetry is from intramolecular forces (presumably those described in Scheme 40.6c) and not from crystal packing. The higher symmetry C_{3v} species is slightly higher in energy (~ 1 kcal mol^{-1}) than the C_3 geometry, and the approximate D_{3h} species is even higher in energy because optimal overlap is reduced since all of the equatorial atoms must be in the same plane.

40.5 CONCLUSIONS

We have shown through several examples that the non-empirical Fenske–Hall approach provides qualitative results that are quite similar to the more rigorous treatment given by DFT for transition metal complexes, both for single transition metal complexes as well as for larger cluster complexes.

40.6 ACKNOWLEDGEMENTS

The National Science Foundation (Grant Nos. CHE 98-00184 and DMS 02-16275) and the Welch Foundation (Grant No. A-648) supported this work.

40.7 REFERENCES

1 (a) R.F. Fenske, K.G. Caulton, D.D. Radtke and C.C. Sweeny, Inorg. Chem., 5 (1966) 951; (b) R.F. Fenske and D.D. Radtke, Inorg. Chem., 7 (1968) 479; (c) K.G. Caulton and R.F. Fenske, Inorg. Chem., 7 (1968) 1273; (d) R.L. DeKock, PhD Dissertation, University of Wisconsin, 1970.

2 M.B. Hall and R.F. Fenske, Inorg. Chem., 11 (1972) 768.

3 (a) B.E. Bursten, unpublished work, 1980; (b) T.A. Barckholtz, B.E. Bursten, G.P. Niccolai and C.P. Case, J. Organometallic Chem. 478 (1994) 153.

4 A. Tan and S. Harris, Inorg. Chem., 37 (1998) 2205.

5 C.C.J. Roothaan, Rev. Mod. Phys., 23 (1951) 69.

6 W. Koch and M.C. Holthausen, A chemist's guide to density functional theory, 2nd edn., Wiley, Weinheim, 2001; R.G. Parr and W. Yang, Density functional theory of atoms and molecules, Oxford University Press, New York, 1989.

7 R.F. Fenske and R.L. DeKock, Inorg. Chem., 9 (1970) 1053.

8 (a) D.L. Lichtenberger, M.L. Hoppe, L. Subramanian, E.M. Kober, R.P. Hughes, J.L. Hubbard and D.S. Tucker, Organometallics, 12 (1993) 2025; (b) D.L. Lichtenberger, S.K. Renshaw, A. Wong and C.D. Tagge, Organometallics, 12 (1993) 3522; (c) D.L. Lichtenberger, M.A. Lynn and M.H. Chisholm, J. Am. Chem. Soc., 121 (1999) 12167.

9 T. Koopmans, Physica, 1 (1933) 104.

10 T.A. Albright, J.K. Burdett and M.-H. Whangbo, Orbital interactions in chemistry, Wiley, New York, 1985, pp. 392–394.

11 C. Cauletti, J.C. Green, M.R. Kelly, P. Powell, J.V. Tilborg, J. Robbins and J. Smart, J. Electron. Spectrosc. Relat. Phenom., 19 (1980) 327. Experimentalists have defined three regions in the PES spectrum, with designations of A′, A″, and B.

12 The single-point HFR/6-31G, FH, BVP86/6-31G, and B3LYP/6-31G calculations used the B3LYP/6-31G(d) optimized geometry for ferrocene. The FH and B3LYP calculations used experiment geometries for $Cp_2Ni_2(\mu\text{-}S)_2(MnCO)_3$[19] (with all-electron 6-31G basis sets for all atoms) and $[Ru(CO)_4]_3[Pd(PR_3)]_3$ (a basis set denoted BS2 with all-electron 6-31G basis sets for C, O, and H; LANL2DZ(d) [24a] for phosphorus; and Couty and Hall modified LANL2DZ [24b] for Ru and Pd) [22]. All HFR and DFT single-point calculations used standard double zeta Pople-style basis sets. The DFT calculations were performed with the B3LYP [26] and BVP86 [27] functionals as implemented in Gaussian 03 [28]. Fenske-Hall calculations were performed utilizing a graphical user interface (JIMP) developed to build inputs and view outputs from stand-alone Fenske-Hall (version 5.2) and MOPLOT2 binary executables. Contracted double-zeta basis sets were used for the Mn, Fe, and Ni 4d; Ru and Pd 5d; S and P 3p; and C and O 2p atomic orbitals. Orbital plots from Fenske-Hall calculations were created with MOPLOT2; orbital plots from B3LYP and Hartree-Fock calculations were created with Gaussian; and all orbital plots were visualized with JIMP. MOPLOT2: for orbital and density plots from linear combinations of Slater or Gaussian type orbitals, version 2.0, June 1993; Dennis L. Lichtenberger, Department of Chemistry, University of Arizona, Tucson, AZ 85721; JIMP Development Version 0.1 (built for Windows PC and Redhat Linux); J Manson, C.E. Webster, and M.B. Hall, Department of Chemistry, Texas A&M University, College Station, TX 77842 http://www.chem.tamu.edu/jimp. FH binaries are available from http://www.chem.tamu.edu/jimp

13 (a) O.V. Gritsenko and E.J. Baerends, J. Chem. Phys., 117 (2004) 9154; (b) O.V. Gritsenko and E.J. Baerends, J. Chem. Phys., 120 (2004) 8364.

14 R.S. Mulliken, J. Chem. Phys., 23 (1955) 1841.

15 J.C. Slater, Quantum theory of atomic structure, Vol. I, McGraw-Hill, New York, 1960, pp. 322–331.

16 R.F. Fenske and J.R. Jensen, J. Chem. Phys., 71 (1979) 3374.

17 M. Goeppert-Mayer and A.L. Sklar, J. Chem. Phys., 6 (1938) 645.

18 W. Kohn and K.J. Sham, Phys. Rev., 140 (1965) A1133.

19 M.B. Hall, Metal-metal bonds and clusters in chemistry and catalysis, Fackler, J.P., Jr., Ed., Plenum Press New York (1990) pp. 265–273.

20 R.D. Adams, S. Miao, M.D. Smith, H. Farach, C.E. Webster, J. Manson and M.B. Hall, Inorg. Chem., 43 (2004) 2515.

21 There are actually two nearly degenerate alpha HFR orbitals that contain this Mn character, the HOMO-25 and HOMO-26; they are both a symmetry and mix with two different sets of Ni d combinations. We have rotated these two nearly degenerate orbitals (they differ by only 0.24 kcal mol^{-1}) by 40° to produce two new orbitals, one with mainly $Mn(CO)_3$ character (shown in Fig. 40.6) and the other with mainly Ni character (not shown).

22 R.D. Adams, B. Captain, W. Fu, M.B. Hall, J. Manson, M.D. Smith and C.E. Webster, J. Am. Chem. Soc., 126 (2004) 5253.

23 I.B. Bersuker, The Jahn–Teller effect and vibronic interactions in modern chemistry, Plenum Press, New York, 1984; T.A. Albright, J.K. Burdett and M.W. Whangbo, Orbital interactions in chemistry, Wiley, New York, 1985, pp. 95–100; R.G. Pearson, Symmetry rules for chemical reactions: orbital topology and elementary processes, Wiley, New York, 1976, pp. 75–82.

24 (a) A. Höllwarth, M. Böhme, S. Dapprich, A.W. Ehlers, A. Gobbi, V. Jonas, K.F. Köhler, R. Stegmann, A. Veldkamp and G. Frenking, Chem. Phys. Lett., 208 (1993) 237, 224 (1994) 603; (b) M. Couty and M.B. Hall, J. Comput. Chem., 17 (1996) 1359; P.J. Hay and W.R. Wadt, J. Chem. Phys., 82 (1985) 270–283; W.R. Wadt and P.J. Hay, J. Chem. Phys., 82 (1985) 284–298.

25 V.A. Rassolov, J.A. Pople, M.A. Ratner and T.L. Windus, J. Chem. Phys., 109 (1998) 1223; R. Ditchfield, W.J. Hehre and J.A. Pople, J. Chem. Phys., 54 (1971) 724; P.C. Hariharan and J.A. Pople, Theor. Chim. Acta, 28 (1973) 213; M.M. Francl, W.J. Pietro, W.J. Hehre, J.S. Binkley, M.S. Gordon, D.J. DeFrees and J.A. Pople, J. Chem. Phys., 77 (1982) 3654.

26 Becke three-parameter exchange functional (B3) and the Lee–Yang–Parr correlation functional (LYP), A.D. Becke, J. Chem. Phys., 98 (1993) 5648; C. Lee, W. Yang and R.G. Parr, Phys. Rev. B, 37 (1988) 785. Also, see R.H. Hertwig and W. Koch, Chem. Phys. Lett., 268 (1997) 345, concerning the various implementations of B3LYP.

27 Becke exchange functional (B) and the Perdew correlation functional (P86). In Gaussian, when "P86" correlation functional is specified, this keyword combines non-local Perdew correlation with Perdew–Zunger81 local correlation. When "VP86" correlation functional is specified, Gaussian uses VWN5 local correlation instead of Perdew–Zunger81, A.D. Becke, Phys. Rev. A, 38 (1988) 3098; J.P. Perdew, Phys. Rev. B, 33 (1986) 8822, 34 (1986) 7406; J.P. Perdew and A. Zunger, Phys. Rev. B, 23 (1981) 5048.

28 M.J. Frisch, G.W. Trucks, H.B. Schlegel, G.E. Scuseria, M.A. Robb, J.R. Cheeseman, J.A. Montgomery Jr., T. Vreven, K.N. Kudin, J.C. Burant, J.M. Millam, S.S. Iyengar, J. Tomasi, V. Barone, B. Mennucci, M. Cossi, G. Scalmani, N. Rega, G.A. Petersson, H. Nakatsuji, M. Hada, M. Ehara, K. Toyota, R. Fukuda, J. Hasegawa, M. Ishida, T. Nakajima, Y. Honda, O. Kitao, H. Nakai, M. Klene, X. Li, J.E. Knox, H.P. Hratchian, J.B. Cross, C. Adamo, J. Jaramillo, R. Gomperts, R.E. Stratmann, O. Yazyev, A.J. Austin, R. Cammi, C. Pomelli, J.W. Ochterski, P.Y. Ayala, K. Morokuma, G.A. Voth, P. Salvador, J.J. Dannenberg, V.G. Zakrzewski, S. Dapprich, A.D. Daniels, M.C. Strain, O. Farkas, D.K. Malick, A.D. Rabuck, K. Raghavachari, J.B. Foresman, J.V. Ortiz, Q. Cui, A.G. Baboul, S. Clifford, J. Cioslowski, B.B. Stefanov, G. Liu, A. Liashenko, P. Piskorz, I. Komaromi, R.L. Martin, D.J. Fox, T. Keith, M.A. Al-Laham, C.Y. Peng, A. Nanayakkara, M. Challacombe, P.M.W. Gill, B. Johnson, W. Chen, M.W. Wong, C. Gonzalez and J.A. Pople, Gaussian 03, Revision B.05, Gaussian, Inc., Pittsburgh, PA, 2003.

Theory and Applications of Computational Chemistry: The First Forty Years
Edited by C. Dykstra et al.

CHAPTER 41

Advances in electronic structure theory: GAMESS a decade later

Mark S. Gordon and Michael W. Schmidt

Department of Chemistry and Ames Laboratory, Iowa State University, Ames, IA 50011, USA

Abstract

Recent developments in advanced quantum chemistry and quantum chemistry interfaced with model potentials are discussed, with the primary focus on new implementations in the GAMESS electronic structure suite of programs. Applications to solvent effects and surface science are discussed.

41.1 INTRODUCTION

The past decade has seen an extraordinary growth in novel new electronic structure methods and creative implementations of these methods. Concurrently, there have been important advances in 'middleware', software that enables the implementation of efficient electronic structure algorithms. Combined with continuing improvements in computer and interconnect hardware, these advances have extended both the accuracy of computations and the sizes of molecular systems to which such methods may be applied.

The great majority of the new computational chemistry algorithms have found their way into one or more broadly distributed electronic structure packages. Since this work focuses on new features of the GAMESS (General Atomic and Molecular Electronic Structure System [1]) suite of codes, it is important at the outset to recognize the many other packages that offer both similar and complementary features. These include ACES [2], CADPAC [3], DALTON [4], GAMESS-UK [5], HYPERCHEM [6], JAGUAR [7], MOLCAS [8], MOLPRO [9], NWCHEM [10], PQS [11], PSI3 [12], Q-CHEM [13], SPARTAN [14], TURBOMOLE [15], and UT-CHEM [16]. Some [1,3,4,10] of these packages are distributed at no cost to all, or to academic users, while others are commercial packages, but all are generally available to users without restriction or constraint. Indeed, the developers of these codes frequently collaborate to share features, a practice that clearly benefits all of their users.

The remainder of this chapter is organized as follows. In Section 41.2, recently developed and implemented methods in quantum mechanics (QM) are discussed. New scalable methods for correlated wavefunctions are presented in Section 41.3, and several approaches for interfacing quantum mechanics with molecular mechanics (MM) are considered in Section 41.4. Conclusions and projections into the future are provided in Section 41.5.

41.2 QM METHODS

41.2.1 Variational methods

The simplest variational methods are Hartree–Fock (HF) and density functional theory (DFT), both of which are available in GAMESS. In principle, DFT is an exact, *ab initio* method, but in practice, since one does not know the exact density, it has become a very successful semi-empirical method, with several parameters that are fitted either to experiment or to simple models like the non-interacting homogeneous electron gas. DFT has been implemented in GAMESS using both the traditional grid-based approach and a grid-free method, due originally to Almlof [17a] and extended by Glaesemann and Gordon [17b], that employs the resolution of the identity (RI) to facilitate formulating DFT in algebraic form. Several of the most commonly used functionals are available in GAMESS, including B3LYP and BLYP.

In the 1950s, Löwdin [18] showed that a wavefunction that includes all possible excitations from the reference wavefunction (usually the electronic ground state) is the exact wavefunction for the given atomic basis. Therefore, this level of theory, commonly called full configuration interaction (full CI), is the benchmark against which all advanced QM methods that include electron correlation may be measured. Indeed any level of CI, perturbation theory, or coupled cluster theory can be extracted from a full CI wavefunction and compared with the exact result. It is therefore very useful to develop and implement a full CI method that can be applied to as large an array of atomic and molecular species as possible. Such a full CI code based on a determinant, rather than a configuration, expansion has been developed by Ivanic and Ruedenberg [19] and implemented into GAMESS.

A special case of full CI is the complete active space self-consistent field (CASSCF) or fully optimized reaction space (FORS) approach in which one defines an active space of orbitals and corresponding electrons that are appropriate for a chemical process of interest [20]. The FORS wavefunction is then obtained as a linear combination of all possible electronic excitations (configurations) from the occupied to the unoccupied (virtual) orbitals in the active space, so a FORS wavefunction is a full CI within the specified active space. Since a full CI provides the exact wavefunction for a given atomic basis, there is no need to re-optimize the component molecular orbitals. On the other hand, a FORS wavefunction generally corresponds to an incomplete CI, in the sense that only a subset of configuration (or determinant) space is included. Therefore, one also optimizes the molecular orbital coefficients to self-consistency. The calculation of a full CI wavefunction is extremely computationally demanding, scaling exponentially with the number of atomic basis functions. Its use is therefore currently limited to either very small molecules with extended basis sets or molecules of modest size with modest basis sets. Therefore, even though new algorithms and computer hardware are expanding the range of applicability of full CI wavefunction, the FORS/CASSCF multi-configurational self-consistent-field (MCSCF) method is very powerful for a variety of applications. The method is especially important when one encounters near degeneracies. There are a great many phenomena in chemistry and related fields in which near degeneracies occur. Examples include:

- crossings or near-crossings of potential energy surfaces, in which non-adiabatic interactions become important;
- diradicals, such as those that occur during bond-breaking processes and near transition states in chemical reactions;
- atomic and molecular species that have multiple close-lying electronic states, such as unsaturated transition metal compounds and high-energy metastable compounds.

Because of the associated resource (e.g. memory and disk) requirements for MCSCF calculations, the size of the active space is effectively limited to approximately 16 active electrons in 16 active orbitals. This seems somewhat limiting at first thought, since such an (16,16) active space is only slightly larger than a full valence active space for a molecule like ethane. However, in many chemical processes that involve large molecules only a small number of electron pairs are changing in any significant manner. So, MCSCF methods can frequently be applied to rather large problems, as long as the active space remains tractable. In recent years, FORS calculations have been applied to such complex species as 7-azaindole (7AI) [21], 7-AI interacting with water in ground and excited electronic states [22], the 7-AI dimer [23], large clusters that simulate the Si (100) surface, including up to five dimers ($Si_{33}H_{28}$) [24], and the oxidation and etching reaction mechanisms of these silicon clusters [25]. Such large applications are made possible, in part, by a suite of MCSCF algorithms in GAMESS that render the iterative process more effective and efficient. These methods range from very simple first-order methods that have small resource requirements and require more iterations to highly resource-demanding full Newton–Raphson methods that are rapidly convergent but considerably more resource-demanding. A compromise method employs an approximate second-order iterative (SOSCF [26]) procedure that is the default convergence option in GAMESS.

Even with efficient algorithms, MCSCF calculations eventually run out of steam as the size of the required active space increases. In order to expand the sizes of active spaces that can be included in a molecular calculation, Ivanic has developed and implemented into GAMESS the occupation restricted multiple active spaces (ORMAS) method [27]. In the spirit of methods like restricted active space SCF [28] (RASSCF) and quasi CAS (QCAS) [29], ORMAS divides the desired active space into multiple active subspaces that are chemically separable, thereby expanding the effective size of the orbitals and electrons in an active space. The advantage of the ORMAS approach is the complete flexibility in the number of active spaces and the manner in which they are defined. Fig. 41.1 illustrates the efficiency of the ORMAS method on an important biological compound, oxo(Mn)Salen [27]. It is clear from this figure that subdividing

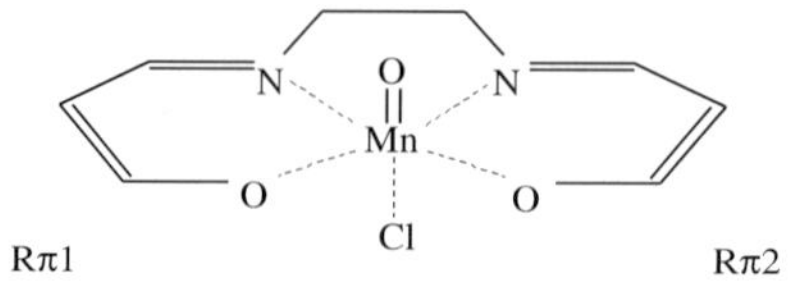

CASSCF Active space = 12 electrons in 11 orbitals:
(Rπ1, Rπ1*, Rπ2, Rπ2*, 3xMn-O, 3xMn-O*, d[Mn])
Predicts spontaneous dissociation of O when going from S=0 (d^2) to S=1 ($d^1\pi^{*1}$) transition.

ORMAS Same active space as CASSCF, divided into 6 groups. Also predicts spontaneous dissociation when going from S=0 to S=1.

	No. of Dets ($S = 0$)	($S = 1$)	Mn(Salen) ($S = 2$)
CAS-SCF	63,504	152,460	784
ORMAS	2,424	8,836	168
Differences in energy, *E*(ORMAS) – *E*(CASSCF)			
	($S = 0$)	($S = 1$)	($S = 2$)
ΔE (millihartree)	1.5	1.6	3.5
Dissociation energies (kcal/mol): oxoMn(salen) → Mn(salen) + O			
	($S = 0$)	($S = 1$)	
CAS-SCF	−33.4	−31.5	
ORMAS	−31.4	−29.6	

Fig. 41.1. Test of ORMAS method on oxoMn(Salen) complex.

the full CASSCF (12,11) active space six subspaces results in errors of the order of only 1–3 mh in the total energy and 2 kcal/mol in the dissociation energy for different electronic spin states. This is extraordinary, given the fact that ORMAS requires two orders of magnitude fewer determinants than the full CASSCF!

Since, as noted above, a full CI provides the exact wavefunction for a given atomic basis, it is useful to expand the size of molecules for which a full CI wavefunction is accessible. Because of the large resource demands for full CI calculations, it is therefore important to devise ways in which one can approach the accuracy of a full CI wavefunction with a significantly reduced effort. The Ruedenberg group has devised two approaches that achieve this goal, both of which have been implemented in GAMESS. In the first, Ruedenberg, Ivanic and Bytautas have used the full CI code and a systematic analysis of single, double, triple,... excitations to develop a general method for eliminating the 'deadwood' from the full CI wavefunction [30]. Making use of localized MCSCF orbitals (LMOs), they have shown for several test cases that roughly 90% of the configurations in a full CI list can be eliminated while retaining millihartree accuracy. Fig. 41.2 illustrates the effectiveness of this approach by comparing the model energies *vs.* the known CCSD(T) correlation energies for 38 small to moderate size molecules, with a mean absolute deviation of less than 3 mh. The error in *relative* energies, for example for chemical reactions, is likely to be much less.

The use of localized orbitals leads to the recognition that interactions that are further from each other than vicinal (three bonds) can be safely ignored in a correlation

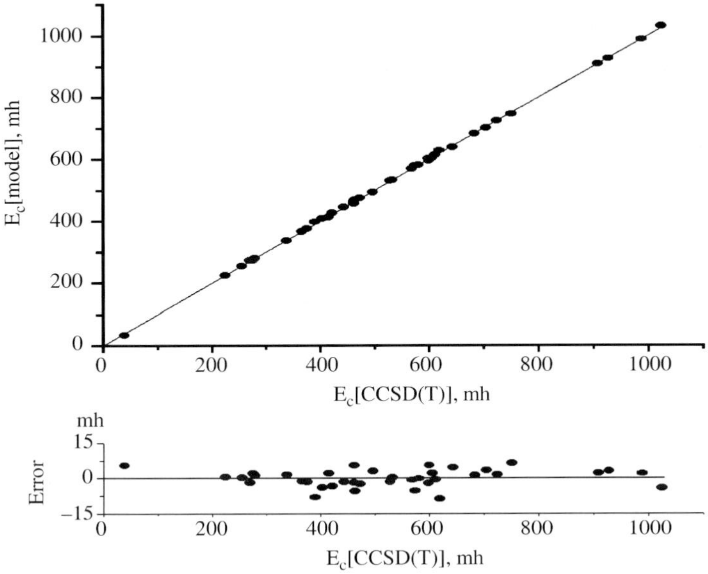

Fig. 41.2. Model *vs.* CCSD(T) correlation energies for 38 molecules from H_2 to C_6H_6 (Mean absolute deviation = 2.85 mh).

calculation. The extension of a general CI method such as that developed by the Ruedenberg group to a general MCSCF method in which the molecular orbital coefficients, as well as the CI coefficients, is non-trivial, but it has been accomplished and implemented into GAMESS in collaboration with the Ruedenberg group [31]. It is important to recognize that there are clear advantages and some disadvantages to the general MCSCF approach. The obvious advantage is the dramatic reduction in computation time that one attains by eliminating most of the configurations. What one gives up in this approach is the built-in size-consistency that is guaranteed by the complete active space approach. Until the method has been extensively tested, it is not clear how serious a matter this is. Indeed, it is possible that eliminating essentially non-contributing configurations has only a small effect on size-consistency. Similarly, it is not clear how the MCSCF convergence will be affected when a complete active space is not used, but one suspects that the use of less than a FORS active space will lead to deterioration of the MCSCF convergence.

An alternative approach to achieving full CI at a reduced computational cost has been developed by Bytautas and Ruedenberg [32]. This method is built upon a careful analysis of single, double, triple, ... excitations, and the observation of nearly linear relationships between lower and higher excitations. These linear relationships allow one to use extrapolation techniques to very accurately predict the energies due to higher excitations without performing the actual calculations. The impact of this method is that one can predict the full CI energy for a given atomic basis to submillihartree accuracy. Combined with an extrapolation to the complete basis set limit, this method is capable of predicting bond energies to within chemical (1 kcal/mol) accuracy. The method has already been applied to the first row homonuclear diatomics [32].

In general, an MCSCF calculation is based on a compact active space and provides the correct zeroth-order wavefunction for a chemical process, such as bond dissociation or a more complex chemical reaction. In this sense, the MCSCF wavefunction may be thought of as a replacement for Hartree–Fock (HF) when a single electron configuration cannot provide a qualitatively correct description of a process of interest. That is, an MCSCF calculation accounts only for 'non-dynamic' correlation. To obtain accurate energies, and sometimes accurate molecular structures, one must add dynamic correlation, just as one must add dynamic correlation to a HF wavefunction when accurate structures and energies are desired. Several methods are available in GAMESS for recovering dynamic correlation on top of MCSCF wavefunctions. These include multi-reference (MR) configuration interaction (MRCI) and multi-reference second-order perturbation theory (MRPT2). The MRCI wavefunctions can include either single or single and double excitations out of the active space, referred to as first and second-order CI, respectively. These methods may be applied to both FORS and ORMAS wavefunctions. The version of MRPT2 in GAMESS is that developed by Hirao, Nakano and co-workers. When only one electronic state is included in the calculation, the method is referred to as MRMP2 [33]. For multiple state calculations, the appropriate method is second-order quasi-degenerate perturbation theory (MCQDPT2 [34]). A multi-reference perturbation theory that is built upon ORMAS wavefunctions is in progress. Whereas both FORS and full CI wavefunctions are size consistent and size extensive, none of these multi-reference methods for recovering dynamic correlation have this property. Interestingly, it appears that both the choice of the unperturbed Hamiltonian H_0 and the manner in which the excitations are included (configuration state functions (CSF) *vs.* internally contracted configurations (ICC)) impacts how close to size consistency a particular MRPT method is. The ICC approach appears to provide a much more nearly size consistent result than does the choice of CSF [35]. The MRMP2 and MCQDPT2 methods in GAMESS are based on CSFs.

41.2.2 Many-body methods

All of the methods discussed above are based on a multi-reference (MR) approach to obtaining wavefunctions and properties. Such MR approaches are often necessary, because many chemical problems involve species with considerable configurational mixing due the existence of near degeneracies (diradical character). However, the amount of diradical character in a chemical system can span a very broad range, from essentially zero (e.g. HOMO occupancy ~ 2, LUMO occupancy ~ 0) to fully diradical (HOMO occupancy ~ 1, LUMO occupancy ~ 1). As one approaches fully diradical character, all single reference methods break down, but they do not break down at the same rate as one approaches this limit. In particular, there is considerable evidence that coupled cluster (CC) methods, particularly those like CCSD(T) that incorporate a triples correction, can overcome the deficiency of a single reference wavefunction for problems with non-trivial diradical character. This has been demonstrated, for example, by examining the N_2 dissociation curves for MP2 and CCSD(T) *vs.* various MR methods [36]. The breakdown in the CCSD(T) calculation appears much later in the dissociation process than does the MP2 breakdown. Recent developments by Piecuch et al. [37] are particularly exciting, since they extend this breakdown even further out in the dissociation curve. Termed re-normalized and completely re-normalized methods (e.g. R-CCSD(T) and CR-CCSD(T)), these methods are designed to account for an increasing amount of diradical character. Although they do eventually break down at large distances for multiple bonds, they are clearly more robust for intermediate cases. The full suite of closed shell CC, R-CC and CR-CC methods are now available in GAMESS. Since the coupled cluster methods are not fully variational, one must use the relaxed density [38] that requires the gradient of the wavefuction (non-Hellman–Feynman term) to evaluate molecular properties such as the dipole moment and electrostatic potential. The formulation of the relaxed density is very similar to that of the energy gradient, so the evaluation of properties is considerably more complex for non-variational methods than it is for variational methods. The Piecuch group has completed the derivation and coding of the relaxed density and associated properties for the coupled cluster and renormalized coupled cluster methods [39]. These new features will be in a new GAMESS release shortly.

Even though single reference second-order perturbation theory (MP2) cannot describe bond breaking very well, this level of theory is still the most efficient *ab initio* approach that includes electron correlation [40]. For closed shells, the restricted MP2 energy is the standard choice, while for open shells there are two possible choices, unrestricted and restricted MP2. Unrestricted methods (UMP2) have the advantage that the orbitals and orbital energies are well defined, but these methods are generally not spin correct. It is common to encounter modest 'spin contamination', in which the spin expectation value differs from the correct value by 0.05–0.1. However, during bond breaking processes, this deviation can become very large, so that the identity of the spin state is effectively lost. The orbitals and orbital energies in restricted open shell Hartree–Fock are not uniquely defined. Consequently, there are several ways in which restricted open shell MP2 can be formulated. The most popular of these methods, referred to as RMP2 was independently developed by Bartlett and co-workers [41] and Pople, Handy and co-workers [42]. This method is spin correct only through second order in the energy. A more complex and more rigorously spin correct method, called *Z*-averaged perturbation theory (ZAPT2), was developed by Lee and Jayatilaka [43]. Both methods are available in GAMESS, while the default is ZAPT2.

41.2.3 Excited states, non-adiabatic and relativistic methods

The simplest approach to the calculation of excited electronic states is to perform a singles configuration interaction (CIS) calculation. While the accuracy of CIS is limited, especially if an excited state of interest has significant contributions from doubly excited configurations, such as charge transfer states, it is a computationally efficient method that can frequently provide at least useful qualitative trends correctly. The new CIS code in GAMESS (developed by Webb [44]) provides energies and analytic gradients, so one can predict excited state geometries, determine transition states, and follow reaction paths at this level of theory with approximately the computational cost of Hartree–Fock. At the other end of the computational complexity spectrum, one can use the full CI option to obtain energies and wavefunctions for both ground and excited states. While this level of theory provides the exact wavefunction for the atomic basis used, it is generally limited to atoms and small molecules. For this reason, the Bytautas–Ruedenberg method for extrapolating to the full CI solution is very exciting, since the computational complexity is reduced by several orders of magnitude. As noted in Section 41.2.1, this method is capable of predicting ground state dissociation energies of diatomic molecules to within chemical accuracy. The ability of this method to predict full potential energy curves and surfaces is currently under investigation, and the ability of this method to treat excited states with equivalent accuracy is promising.

There are several choices for excited state calculations that are intermediate in accuracy and computational complexity between singles CI and full CI. These include the MCSCF, multi-reference CI (called first order or second order CI in GAMESS) and second order multi-reference perturbation theory methods discussed above. Excited states may be analyzed for each of these methods simply by requesting multiple roots. The most recent excited state method to be implemented in GAMESS is the suite of equations-of-motion (EOM) coupled cluster methods that have been developed by Piecuch and co-workers [37a,45]. EOM-CCSD and EOM-CCSD(T) and their completely renormalized analogs (using the method of moments to obtain triples corrections) are available. The EOM methods are usually initiated by performing a CI singles (CIS) calculation to obtain the starting wavefunction. This can be somewhat limiting if the state(s) of interest have significant contributions from two-electron excitations. Using a full singles + doubles CI (CISD) would be prohibitively expensive in most cases. The EOM-CC implementation in GAMESS permits an alternative approach called CISd, in which a subset of double excitations is added to the CIS wavefunction. The choice of which double excitations to include is flexible, but the most sensible approach is to treat the choice of doubles in a manner that is similar to choosing an MCSCF active space. That is, one would choose that subset of orbitals that are most likely to be involved in the excited state(s) of interest.

While the electronic ground state is usually, although not always, well separated from excited electronic states, there are frequently multiple excited states within a small energy gap. When electronic states come close to, or cross, each other, the adiabatic assumption can break down. Non-adiabatic effects can be manifested in either spin-orbit coupling (a fundamentally relativistic effect) or derivative (sometimes called vibronic) coupling. The former phenomenon arises from the interactions of orbital angular momentum and the magnetic moment due to spin, while the latter derives from a breakdown in the Born–Oppenheimer approximation according to which one ignores the changes in the electronic wavefunction that are induced by nuclear motion. Several methods for dealing with spin-orbit coupling have been implemented in GAMESS, while derivative coupling codes are planned for the near future.

The array of methods in GAMESS for treating spin-orbit coupling effects has recently been the subject of two reviews [46,47]. These methods include the full Breit–Pauli spin-orbit operator and approximations to it, primarily developed by Koseki and Fedorov. All of the methods require a multi-reference wavefunction as a starting point. This can be MCSCF, first or second order CI, or MRPT2. The simplest method is a semi-empirical Breit–Pauli implementation, in which the most time-consuming two-electron part of the spin-orbit operator is neglected. This approximation is made viable by replacing the nuclear charge Z in the one-electron operator is replaced by an effective nuclear charge Z_{eff} that is fitted to atomic spectra. This method has been parametrized for virtually the entire periodic table, for both all electron and effective core potential basis sets, and then applied to a number of interesting problems [48]. Of course, the accuracy of this semi-empirical approach is inherently limited. For greater accuracy, one can use the full Breit–Pauli approach in which both one- and two-electron operators are included. An interesting and effective compromise is the partial two-electron (P2E) method [49], in which only the core-valence two-electron term is included. This method is accurate because the core-valence term is the most important one, and it is computationally efficient because this term can be reformulated to resemble an effective one-electron interaction.

Scalar relativistic effects (e.g. mass–velocity and Darwin-type effects) can be incorporated into a calculation in two ways. One of these is simply to employ effective core potentials (ECPs), since the core potentials are obtained from calculations that include scalar relativistic terms [50]. This may not be adequate for the heavier elements. Scalar relativity can be variationally treated by the Douglas–Kroll (DK) [51] method, in which the full four-component relativistic *ansatz* is reduced to a single component equation. In GAMESS, the DK method is available through third order and may be used with any available type of wavefunction.

41.2.4 Properties related to nuclear energy derivatives

Analytic energy derivatives are as important as the energies themselves. One needs first derivatives for geometry optimizations, reaction path following, dynamics simulations, and (if analytic second derivatives are not available) second derivatives *via* finite differencing. Second derivatives are necessary for the computation of vibrational frequencies and, subsequently, thermodynamic properties *via* the appropriate partition functions.

Analytic first derivatives (gradients) are available in GAMESS for open and closed shell Hartree–Fock, open and closed shell DFT, generalized valence bond (GVB), MCSCF and ORMAS, CIS, and MP2 for closed and (unrestricted) open shells. Analytic second derivatives (Hessians) are available for open and closed shell HF, GVB and MCSCF wavefunctions. One can also calculate Hessians using numerical finite differencing of analytic gradients using any method for which analytic gradients are available. Fully numerical first and second derivative codes have very recently been implemented, so one can optimize molecular geometries with any level of theory. Fully numerical derivatives are inherently less efficient than analytic derivatives, however, the numerical derivatives are more amenable to very efficient coarse-grained parallel computing.

The most common method for determining vibrational frequencies is the normal mode analysis, based on the harmonic force constant matrix of energy second derivatives (Hessians). Of course, vibrations are not truly harmonic, and the anharmonicity generally increases as the frequency of the vibration (steepness of the potential) decreases. That is, the more anharmonic a motion is, the less applicable is the traditional approach to determining vibrational frequencies. In such cases, one can use the vibrational self-consistent field (VSCF) method to obtain anharmonic frequencies [52–54]. This method uses energies and gradients along a given normal mode direction to obtain the anharmonic cubic and quartic terms. In analogy with electronic structure theory, the VSCF method can be augmented by a CI or perturbation theory, in order to obtain improved vibrational frequencies. The VSCF and related methods in GAMESS have been developed by Chaban, Matsunaga, and Taketsugu.

Energy derivatives are essential for the computation of dynamics properties. There are several dynamics-related methods available in GAMESS. The intrinsic reaction coordinate (IRC) or minimum energy path (MEP) follows the infinitely damped path from a first-order saddle point (transition state) to the minima connected to that transition state. In addition to providing an analysis of the process by which a chemical reaction occurs (e.g. evolution of geometric structure and wavefunction), the IRC is a common starting point for the study of dynamics. Example are variational transition state theory (VTST [55]) and the modified Shepard interpolation method developed by Collins and co-workers [56]. Indeed, interfaces are maintained between GAMESS and the VTST codes (GAMESS-PLUS) [57] and between GAMESS and the Collins Grow code (GAMESS–Grow) [58]. The GAMESS–Grow interface, in particular, benefits greatly from the availability of analytic energy derivatives. Since Grow builds global potential energy surfaces and GAMESS has multi-state capability *via* its multi-reference methods, this interface can build multiple potential energy surfaces, a precursor for the dynamical analysis of such phenomena as conical intersections.

In addition to the IRC functionality, GAMESS also has a direct dynamics capability, the dynamic reaction path, DRC [59]. The DRC allows one to perform dynamics 'on-the-fly', by performing classical trajectories at any level of theory for which analytic gradients are available. One can, for example, put an amount of energy equal to n quanta into any vibrational mode(s), in order to model mode specific chemistry.

The gradient extremal is another trajectory, available in GAMESS, defined as that curve for which the nuclear gradient vector is an eigenvector of the nuclear Hessian [60]. Gradient extremals may lead from minima to any of the following: dissociation products, minima, transition states, or higher order saddle points (stationary points); or united atom collisions. For small numbers of atoms, tracing all gradient extremals may lead to a global understanding of all relevant parts of the potential energy surface [61].

41.2.5 Other properties

Although the generation of wavefunctions and corresponding potential energy surfaces is the primary use for quantum chemistry, other molecular properties are also of interest. In addition to the set of electrostatic properties and localized orbitals described earlier, GAMESS now contains a number of other useful properties or analysis procedures.

Analytic computation of IR intensities [62] and numerical computation of Raman intensities [63] for the harmonic vibrational spectra have been programmed. The analytic computation [64] of closed shell static or frequency-dependent polarizabilities is possible, and numerical computation [65] of the static polarizabilities of most wavefunctions can be performed. A program to explore solvent effects on NMR shifts has been developed [66].

Analysis of molecular wavefunctions is less directly tied to experiment but is of considerable interest to chemists looking for trends or explanations. Atomic charges are of considerable interest to force field developers, and a procedure for fitting them to the computed electrostatic potential and total molecular dipole and quadrupole is available [67]. The localized orbital programs have been extended to include a detailed orbital analysis of energy contributions, called the Localized Charge Distribution, available for closed shell SCF and MP2 wavefunctions [68]. As part of the effective fragment potential (EFP) solvent project (see below), the Stone distributed multipole analysis (DMA) [69] and the Kitaura–Morokuma or Stevens–Fink dimer energy analysis [70] have been added, and generalized to any number of weakly interacting monomers.

41.3 SCALABLE ELECTRONIC STRUCTURE THEORY

One approach to grow the size of a chemical system that can be realistically treated by the most sophisticated electronic structure methods is to devise new methods that are inherently more efficient, as discussed in the previous section. Another, complementary approach is to devise algorithms in such a manner that the calculations are scalable; that is, the computationally most demanding tasks may be distributed among whatever processors are available. Often referred to as parallel programming, this approach is *relatively* straightforward for low-level methods like Hartree–Fock and DFT energies and gradients using a replicated data (RD) algorithm, in which the necessary data is either replicated on each available processor (e.g. density and Fock matrices) or recomputed on the fly each iteration (e.g. 'direct' calculation of the two-electron integrals). However, the design of scalable algorithms becomes increasingly complicated for the more sophisticated, correlated methods. The disadvantage of the RD approach is that although a calculation proceeds more rapidly than it would on a single processor, the feasible size of a chemical system is limited by the amount of memory and disk on the smallest node. Therefore, the RD approach is sensible when only two-dimensional matrices are involved, but becomes much less viable for correlated methods for which the four-dimensional electron repulsion integrals must be manipulated (i.e. transformed between the AO and MO basis).

A major advance in the manner in which QM (especially correlated QM) calculations may be performed on parallel computers was provided by the development at PNNL of the global array (GA) tools [71], a one-sided message passing library that facilitates the distribution of large sets of data across all available nodes. The development of the distributed data interface (DDI) [72] in GAMESS, initially led by Fletcher and more recently by Olson, benefited considerably from the prior development of GA technology. DDI performs best when it can take advantage of the SHMEM library, especially on Cray or SGI systems, but it has also been very successful on IBM computers running under LAPI and clusters of UNIX and Linux computers. The point-to-point messages required for the implementation of DDI on such hardware are carried by TCP/IP socket messages or, sometimes, an MPI-1 library. In a related development, DDI now makes optimal use of the processors in an SMP box by avoiding message passing between CPUs in a given node [73], by using System V memory regions accessible by all processes within a multiprocessor node.

The initial implementation of DDI was for closed shell MP2 energies and gradients [74]. This has been extremely successful. As long as the size of the system of interest is increased as the number of CPUs is increased, the method scales almost linearly up through 512 T3E processors [75]. For species with unpaired electrons, the implementation of restricted open-shell energies is equally efficient, and the UMP2 energy and gradient code [76] scales as well as the closed shell analog. Restricted open-shell gradients using the ZAPT *ansatz* have been derived [43d], and the coding of both sequential and parallel codes is in progress. DDI has also been used to develop a distributed parallel MR second-order perturbation method in collaboration with the Koseki group [77]. It appears that the parallel MRMP2 method currently scales well up to about 32 processors. Of course, the most demanding many-body method in GAMESS is the set of coupled cluster methods. The coupled cluster codes in GAMESS are currently sequential, but the development of parallel coupled cluster methods is in progress.

Since MCSCF is an important starting point for so many chemical problems, it is very important to develop parallel MCSCF methods as well. The initial attempt at this was a RD approach which scaled well only to ~4–8 processors [78]. Very recently, a DD parallel MCSCF algorithm has been developed using the full Newton–Raphson convergence algorithm [79]. This DD MCSCF method addresses the integral transformation and orbital rotation steps, but not the CI coefficient optimization, which is discussed below. Initial tests suggest that this algorithm will scale well up to ~32–64 processors, a major advance over the RD algorithm. Particularly exciting is the very recent implementation of a parallel MCSCF analytic Hessian code [80]. Analytic derivatives are generally preferable to numerical finite differencing schemes, and this is especially so for MCSCF wavefunctions. The reason for this is that numerical schemes usually decrease the molecular symmetry upon some subset of the coordinate displacements. Even for high-symmetry species, some displacements reduce the symmetry all the way down to C_1. So, for example, if one is interested in an excited state that belongs to an irreducible representation that is different from that of the ground state (say, $^1B_2$ *vs.* $^1A_1$ in C_{2v} symmetry), both states will become ^1A in C_1 symmetry. Because of this, one will frequently encounter root flipping during a numerical Hessian calculation, such that the $^1B_2$ state flips to the $^1A_1$ state when the symmetry is reduced to C_1. This is avoided if the Hessian calculation is done analytically. The implementation of an analytic Hessian code is also an important step in the development of a derivative (i.e. vibronic) coupling code that is in progress. Derivative coupling plays an important role in the chemistry and physics of excited states, since the crossing of potential energy surfaces and the associated conical intersections are common phenomena in photochemistry, photophysics and photobiology. Since excited state calculations can be very computationally intensive, especially for molecules that are important in biochemistry, the availability of an MCSCF Hessian code that scales well with the number of processors is also important. The scalability of the MCSCF analytic Hessian code in GAMESS, developed by Dudley, is illustrated in Fig. 41.3 for H_2C_4O, using a 10 electrons in nine orbitals active space.

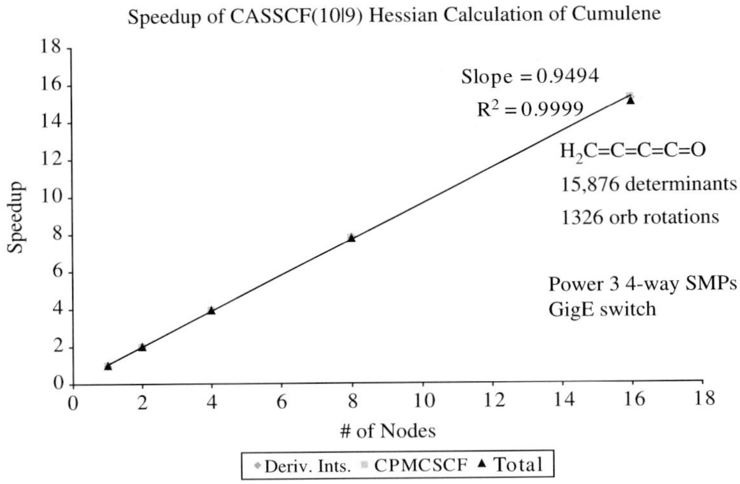

Fig. 41.3. Parallel speedup of analytic MCSCF Hessian.

As noted in Section 41.1, the ultimate wavefunction for a given basis is the full CI wavefunction, so it is important to extend the sizes of chemical species that can be realistically approached using full CI. Equally important is the recognition that a full CI within a specified set of orbitals and corresponding electrons is just a FORS/CASSCF wavefunction. So, the development of a scalable Full CI method serves a dual purpose. Both RD and DD full CI codes have been developed and implemented into GAMESS by Gan [81]. The algorithm uses a CI driven approach, in which communication is controlled by a string-driven method. The success of the DD/FCI method is especially encouraging, as is illustrated in Fig. 41.4. This figure demonstrates a test on a cluster of 64-bit IBM Power 3II dual processor computers running AIX. The illustrated problems are CH_3OH (14,14) and H_2O_2 (14,15), where the numbers in parentheses signify the number of electrons and orbitals, respectively. These problems include ~11,800,000 and 40,400,000 determinants, respectively, and the scalability through 32 processors is excellent. Similar performance is observed on Linux clusters up through the 16 processors that were available for testing.

One can think of the parallel methods discussed above as fine-grained parallelism, in that each subtask in a single energy or energy + gradient evaluation is individually distributed among available processors. There are also problems for which a very coarse-grained approach is appropriate. Examples are the computation of numerical derivatives (e.g. gradients and Hessians) for which each displaced geometry is separate from the others, and all displacements may be identified at the beginning of the calculation. Other examples are multiple Monte Carlo simulations and multiple classical trajectories, since the associated energy and gradient evaluations are independent of one another. A development underway in GAMESS is the GDDI (generalized DDI) method which makes use of the concept of groups and subgroups (in a computational science sense) to make use of both fine-grained and coarse-grained parallelism [82]. For example, if one

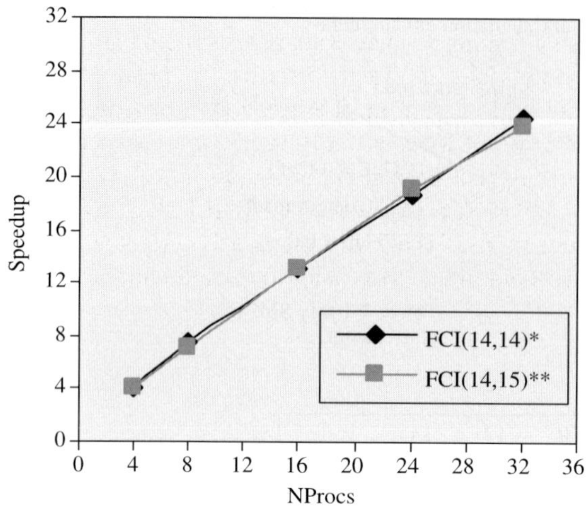

Fig. 41.4. Parallel speedup of full CI.

wishes to perform a CCSD(T) fully numerical Hessian, one can distribute the large number of CCSD(T) energy evaluations among all available nodes. At the same time, if each node is a multi-processor (e.g. SMP) computer, each CCSD(T) energy calculation can itself, in principle be distributed among the processors on its node.

Of course parallel algorithms are only one way in which one can improve the scalability of a particular method. Two other approaches that have recently been implemented into GAMESS are linear scaling methods for HF and DFT [83] and the fragment molecular orbital (FMO) method [84]. Linear scaling methods are based on the premise that the further away atoms are from each other, the more their mutual interactions become similar to classical electrostatic interactions that can be accounted for by multipolar expansions. The key is to be able to devise a procedure in which the linear scaling interaction is turned on when it is appropriate. Among the leaders in this field have been Head-Gordon [85], Scuseria [86], and others [87]. The algorithm in GAMESS, developed by Choi, follows the lead of Wilhite and Head-Gordon [85], but implements an optimization scheme that determines the optimal parameters in a self-consistent manner, depending on the desired accuracy. The latter is an input parameter, so that one can design the accuracy to fit the application. The linear scaling method in GAMESS is also scalable in the parallel sense, with nearly linear scaling with the number of processors up to at least 16 nodes [83b].

The basic idea of the FMO method is to take advantage of the observation that exchange interactions and self-consistency in molecular calculations are generally local phenomena. This is even more true for extended systems, such as clusters, crystals and polymers. This locality permits one to treat interactions between remote parts of a system by purely Coulombic operators, ignoring exchange interactions. One can therefore perform the usual quantum calculations on separate units (fragments) in the near field regions, while fragment–fragment interactions are reduced to Coulombic interactions. In practice, one divides the system of interest into fragments and performs *n*-mer calculations of these fragments in the Coulomb field of the other fragments. There are no empirical parameters, and the only approximation is the fragmentation itself. So, in this sense, there are common features with the linear scaling method discussed in the previous paragraph. The fragmentation scheme introduces little error as long as the fragments are chosen in a physically sensible manner. The most important impact on accuracy is the size of the fragments. The FMO method has been enabled by Fedorov for SCF [84b], DFT [84c], and MP2 [84d] methods at present. Parallelization of the FMO method is implemented by assigning each monomer, dimer, or trimer computation to a subgroup of the full processor set [84a].

41.4 QM/MM METHODS

Even with the most clever and efficient methods and scalable algorithms, as the size of the system of interest grows, sooner or later the available compute power is not up to the task if one uses fully QM methods, especially correlated ones. Two important areas of research that fall into this category are solvent effects (more generally liquid behavior) and surface science. An effective alternative to fully QM methods is the combination of QM with molecular mechanics (MM) methodology. MM is a term that generally suggests that one is using classical techniques with no wavefunction; such methods vary broadly in sophistication. Two types of MM methods that are very different in their level of sophistication are discussed here.

41.4.1 Discrete solvent approaches

The approach taken in GAMESS to study solvation is a multi-layer one in which the innermost layer consists of the solute plus some number of solvent molecules that one feels must be treated with explicit QM. Examples of the latter are water molecules that act as conduits in H-transfer reactions. The second layer consists of a sophisticated potential, the EFP that is derived from rigorous QM [88]. The outermost layer is represented by a reliable continuum method to represent the bulk liquid. In its original EFP1/HF version, this method described solvent molecules (i.e. water) by three terms that are added to the QM (i.e. HF) Hamiltonian. The first term represents the Coulomb interactions by a distributed multipole analysis (DMA) expanded through octopoles. The entire Coulomb term is multiplied by a distance-dependent cutoff to account for overlapping charge distributions. The second, induction, term accounts for the polarization of charge densities in a self-consistent manner using localized molecular orbitals (LMOs). The third term is fitted to the remainder of the HF water dimer potential, determined from 192 points on the dimer surface, after subtracting the Coulomb and induction contributions. This remainder represents exchange repulsion and (implicitly) charge transfer. This EFP1/HF method has been very successful for problems that are well described by the HF method, but it is limited in two respects. First, HF includes no electron correlation which invades all of the terms mentioned above and introduces entirely new interactions, most notably dispersion. Second, the process of fitting to obtain the exchange repulsion/ charge transfer term is not something one wants to do for every solvent or liquid of interest.

The first of these limitations has been partially addressed by reformulating the EFP1 approach with DFT, using the popular B3LYP functional [89]. Denoted EFP1/DFT, this method, developed by Adamovic, includes some correlation, although not long-range dispersion, and therefore produces much better binding energies, for example, in water clusters. So, this approach only partially accounts for the correlation problem. Likewise, an MP2 EFP1 analog has been developed by Song in a similar manner, except that the dispersion contribution is fitted in a separate term using the usual C_n/R^n expansion [90]. So far, only the fragment–fragment interaction has been coded, while the fragment–QM interaction is in progress. So far, the EFP1/MP2 method looks very promising: a molecular dynamics simulation using this method maps almost exactly onto the experimental g(OO), g(OH) and g(HH) spectra, whereas the EFP1/HF method does very poorly and EFP1/DFT not nearly as well as EFP1/MP2 [91]. A parallel code has also been developed by Netzloff for all of the EFP1 methods. The scaling is very good as long as a sufficient number of fragments is included in the calculation.

All of the EFP1 methods involve fitting the repulsive and dispersion terms, so it is difficult to extend the method to other species, such as solvents that may be of interest to organic or inorganic chemists. In this sense, it is desirable to derive the exchange repulsion, dispersion, and charge transfer from 'first principles' instead of employing fitting procedures. This has been accomplished for the exchange repulsion by expanding this interaction as a power series in the intermolecular overlap. This is not a new idea, but combining this approach with highly transferable LMOs to calculate these integrals and the related intermolecular kinetic energy integrals has been very successful for a wide variety of solvents [88]. The exchange repulsion calculated by this method maps the exact HF intermolecular exchange typically to within 0.5 kcal/ mol. More recently, the dispersion interaction has been derived by Adamovic, again from first principles, based on the frequency-dependent polarizabilities in the imaginary frequency range obtained from the time-dependent Hartree–Fock method [92]. In keeping with the other terms in the interaction energy, the dispersion is obtained in a distributed manner, by summing over individual interactions between LMOs located on the two interacting fragments to obtain the final C_6 terms. Combined with an estimate of the C_8 term, this provides an accurate evaluation of the dispersion energy. An important feature of this new EFP2 implementation is that one can generate an effective fragment for any species, simply by performing a single 'MAKEFP' run to generate all of the pieces for a given fragment. Indeed, performing calculations on mixed species (e.g. methanol–water) is just as simple, because no parameter fits are required. At this time, EFP2 is primarily an EFP only method, since not all of the QM–EFP interactions have been developed, but these terms are in progress. In addition, the EFP2 method has been extended by Jensen and co-workers [93] to the treatment of intramolecular covalent interactions, in addition to intermolecular interactions. This is accomplished by interfacing a 'buffer region' constructed from frozen localized orbitals that connects the QM part of the molecule with the EFP part. This allows one to study very large molecules (e.g. proteins and polymers) by treating the 'action region' with quantum mechanics and the remainder of the system with EFP. This is similar to traditional QM/MM methods, except that the EFP provides a much more sophisticated classical potential that captures most of the physics.

Although the cost of an EFP calculation is several orders of magnitude smaller than that of a corresponding QM (e.g. HF, DFT, MP2) calculation, the cost can rise considerably if one incorporates a large number of solvent molecules. This cost reflects not only the inherent cost of a single energy + gradient calculation, but also the fact that the number of arrangements of solvent or liquid molecules expands rapidly with the number of molecules. To put this in context, a molecular dynamics run to predict bulk properties of some species should include at least 64 molecules, and 256–512 molecules are considered to be more realistic. As another example, the process of finding a global minimum for some reacting system that consists of a QM 'solute' and some number of solvent molecules requires a Monte Carlo or similar calculation that frequently requires tens of thousands of energy evaluations, plus geometry optimizations at a subset of these points. Both Monte Carlo/simulated annealing [94] and molecular dynamics [95] codes have been implemented in GAMESS, combined with the EFP methods. To make such calculations more feasible for several hundred fragments, each term in the EFP method has been made scalable by Netzloff [96]. As for any other application, the scalability relies on the size of the problem: the larger the system, the more effective in the parallel code. There are two scalability considerations with the EFP code. One is the inherent scaling of the EFP code itself. The other is the composite scalability of a QM + EFP calculation. For small numbers of water molecules, the pure EFP calculation does not scale very well, since the time requirements are very small. However, the scalability of up to 16 processors improves dramatically as the number of EFPs increases [96]. When a QM molecule is present, the inherent QM scalability [1] plus the greater CPU demands of the QM part cause the scalability to improve for even relatively small numbers of solvent molecules.

41.4.2 Surface chemistry

For surface chemistry, a more traditional QM/MM approach, SIMOMM [97] (surface integrated molecular orbital molecular mechanics), has been developed and implemented in GAMESS. SIMOMM is an embedded cluster approach in which the QM part of the system is embedded into a much larger MM cluster to represent the bulk. Any level of QM theory in GAMESS can be used for the QM part, while the TINKER [98] code is used for the MM part. The interface between the QM and MM parts is represented by link atoms that appear in the QM part (typically as hydrogens, although this is not a requirement) and in the MM part as the actual surface atoms of interest. Gradients for both the QM and MM methods are generally available, so full geometry optimizations are both feasible and recommended. The method has been most extensively applied to problems that involve the Si(100) surface, including addition of organic molecules to the surface [99], etching [25], and diffusion of metal atoms along the surface [100]. More recently, it has been applied to the growth of diamond [101] and silicon carbide [102] surfaces.

41.4.3 Continuum solvent methods

As noted above, solvent effects may be modeled in GAMESS as discrete solvent molecules using the EFP model. Continuum solution models are also useful, in that these remove the solvent structural problem completely (no need to sample many solvent minima structures). The most sophisticated continuum solvent model in GAMESS is the Polarizable Continuum Model (PCM) of Tomasi and Persico, which models the molecule as a space-filling surface surrounded by continuum [103]. PCM permits the computation of the effect of solvation on the energy [104], nuclear gradients [105], and even frequency-dependent polarizibilities [106]. The present PCM implementation has been greatly extended by the Jensen group, to use the conductor-like PCM model [107], and to treat all SCF wavefunctions from RHF to MCSCF and their DFT counterparts. The numerical solution of the Poisson equation to obtain apparent surface charges has been made possible for large systems by using an iterative linear equation solver [108]. Numerically stable nuclear gradients have been obtained by using an 'area scaling' tessellation [109] that deals with surface elements that are too close to each other by rescaling their sizes, permitting the study of solute geometry optimization. The PCM computations just described are enabled for parallel computation [108], and they have been interfaced with the EFP [110].

Two other approaches are used less often. The original SCRF spherical cavity model [111] is less preferable than the space filling cavity used in the PCM code. The COSMO model has been implemented in GAMESS to describe the electrostatic interactions between the solute and the continuum, for RHF and its corresponding MP2 [112].

41.5 SUMMARY AND PROGNOSIS

The focus of this chapter has been on the new developments in electronic structure theory during the past decade. These developments include new methods in quantum mechanics, including approaches for extrapolating to the full CI and complete basis set limits, novel methods for CASSCF calculations, new coupled cluster techniques, methods for evaluating non-adiabatic and relativistic interactions, new approaches for distributed parallel computing, and QM/MM methods for describing solvent effects and surface science. It is useful to note in this regard that GAMESS is a general purpose suite of electronic structure and QM/MM methods (including open- and closed-shell Hartree–Fock which has been essentially ignored here) that can be run on virtually any computer, cluster, massively parallel system, or for that matter a desktop Mac or PC. Indeed, GAMESS is used at many universities as an educational tool, making use of its graphical back end MacMolPlt [113]. GAMESS and MacMolPlt can be downloaded at no cost from www.msg.ameslab.gov, with only a simple license required. A great many people from essentially every region of the world have contributed to the development of GAMESS. This is due in no small measure to the philosophy that has stood us in good stead for more than two decades: science is to be shared openly and freely, and the development of computer codes should be driven by the science that the community wishes to accomplish. This approach to disseminating GAMESS will continue into the foreseeable future.

It is anticipated that several new features will be implemented into GAMESS in the near future. These include:

- Restricted open shell second-order perturbation theory (ZAPT2) analytic gradients
- Properties and gradients for the closed shell coupled cluster methods that are already in GAMESS
- Parallel coupled cluster methods
- Derivative coupling at the MCSCF level of theory
- Determinant based multi-reference perturbation theory
- Model core potentials with the correct nodal behavior
- Interfaces between MP2, CIS, MRMP2, CCSD(T), and EOM-CCSD(T) codes with EFP, in order to facilitate the study of solvent effects on correlated systems and in excited electronic states
- Molecular dynamics capability for both EFP1 and EFP2
- An interface between EFP and MM codes

As always, additional code developments will be driven by the interests of our many colleagues who contribute to GAMESS and by the chemistry that we and our many users wish to study.

41.6 ACKNOWLEDGEMENTS

The work described in this chapter was supported by the Air Force Office of Scientific Research *via* a software development (CHSSI) grant, by the Department of Energy through a SciDAC grant, and by IBM *via* multiple SUR grants. The authors are most grateful to our many collaborators who have contributed their efforts to GAMESS. We mention in particular, Professors Klaus Ruedenberg, Kim Baldridge, Jan Jensen, Shiro Koseki, and Piotr Piecuch, and Drs. Dmitri Fedorov, Graham Fletcher, Joe Ivanic, and Simon Webb, and of course all the present and past graduate students who have contributed to GAMESS.

41.7 REFERENCES

1 M.W. Schmidt, K.K. Baldridge, J.A. Boatz, S.T. Elbert, M.S. Gordon, J.H. Jensen, S. Koseki, N. Matsunaga, K.A. Nguyen, S. Su, T.L. Windus, M. Dupuis and J.A. Montgomery, J. Comput. Chem., 14 (1993) 1347.

2 ACES II is a program product of the Quantum Theory Project, University of Florida. Authors: J.F. Stanton, J. Gauss, J.D. Watts, M. Nooijen, N. Oliphant, S.A. Perera, P.G. Szalay, W.J. Lauderdale, S.R. Gwaltney, S. Beck, A. Balkova, D.E. Bernholdt, K.-K. Baeck, P. Rozyczko, H. Sekino, C. Huber and R.J. Bartlett.

3 The Cambridge Analytic Derivatives Package Issue 6, Cambridge, 1995. A suite of quantum chemistry programs developed by R.D. Amos with contributions from I.L. Alberts, J.S. Andrews, S.M. Colwell,

N.C. Handy, D. Jayatilaka, P.J. Knowles, R. Kobayashi, K.E. Laidig, G. Laming, A.M. Lee, P.E. Maslen, C.W. Murray, J.E. Rice, E.D. Simandiras, A.J. Stone, M.-D. Su and D.J. Tozer.

4 Dalton, a molecular electronic structure program, Release 1.2 (2001), written by T. Helgaker, H.J.Aa. Jensen, P. Jørgensen, J. Olsen, K. Ruud, H. Agren, A.A. Auer, K.L. Bak, V. Bakken, O. Christiansen, S. Coriani, P. Dahle, E.K. Dalskov, T. Enevoldsen, B. Fernandez, C. Hättig, K. Hald, A. Halkier, H. Heiberg, H. Hettema, D. Jonsson, S. Kirpekar, R. Kobayashi, H. Koch, K.V. Mikkelsen, P. Norman, M.J. Packer, T.B. Pedersen, T.A. Ruden, A. Sanchez, T. Saue, S.P.A. Sauer, B. Schimmelpfennig, K.O. Sylvester-Hvid, P.R. Taylor and O. Vahtras.

5 M.F. Guest, J.M.H. Thomas, P. Sherwood, I.J. Bush, H.J.J. van Dam, J.H. van Lenthe, R.W.A. Havenith and J. Kendrick, Mol. Phys, in press.

6 HyperChem Molecular Modeling Software, Release X.Y, a product of Hypercube, Inc., 1115 NW 4th St., Gainesville, FL.

7 http://www.scienceserve.com/Software/Schroedinger/Jaguar.htm.

8 MOLCAS. G. Karlstrom, R. Lindh, P.-E. Malmqvist, B.O. Roos, U. Ryde, V. Veryazov, P.-O. Widmark, M. Cossi, B. Schimmelpfennig, P. Neogrady and L. Seijo, Comput. Mater. Sci., 28 (2003) 222.

9 (a) H.-J. Werner and P.J. Knowles, J. Chem. Phys., 89 (1988) 5803; (b) P.J. Knowles and H.-J. Werner, Chem. Phys. Lett., 145 (1988) 514; (c) MOLPRO is a package of *ab initio* programs written by H.-J. Werner and P.J. Knowles, with contributions from R.D. Amos, A. Bernhardsson, A. Berning, P. Celani, D.L. Cooper, M.J.O. Deegan, A.J. Dobbyn, F. Eckert, C. Hampel, G. Hetzer, T. Korona, R. Lindh, A.W. Lloyd, S.J. McNicholas, F.R. Mandy, W. Meyer, M.E. Mura, A. Nicklass, P. Palmieri, R. Pitzer, G. Rauhut, M. Schutz, H. Stoll, A.J. Stone, R. Tarroni, T. Thorsteinsson.

10 T.P. Straatsma, E. Aprà, T.L. Windus, E.J. Bylaska, W. de Jong, S. Hirata, M. Valiev, M.T. Hackler, L. Pollack, R.J. Harrison, M. Dupuis, D.M.A. Smith, J.J. Nieplocha, V. Tipparaju, M. Krishnan, A.A. Auer, E. Brown, G. Cisneros, G.I. Fann, H. Fruchtl, J. Garza, K. Hirao, R. Kendall, J. Nichols, K. Tsemekhman, K. Wolinski, J. Anchell, D. Bernholdt, P. Borowski, T. Clark, D. Clerc, H. Dachsel, M. Deegan, K. Dyall, D. Elwood, E. Glendening, M. Gutowski, A. Hess, J. Jaffe, B. Johnson, J. Ju, R. Kobayashi, R. Kutteh, Z. Lin, R. Littlefield, X. Long, B. Meng, T. Nakajima, S. Niu, M. Rosing, G. Sandrone, M. Stave, H. Taylor, G. Thomas, J. van Lenthe, A. Wong and Z. Zhang, NWChem, A Computational Chemistry Package for Parallel Computers, Version 4.6, Pacific Northwest National Laboratory, Richland, Washington 99352-0999, USA, 2004.

11 P. Pulay, J. Baker and K. Wolinski, http://www.pqs-chem.com.

12 T.D. Crawford, C.D. Sherrill, E.F. Valeev, J.T. Fermann, R.A. King, M.L. Leininger, S.T. Brown, C.L. Janssen, E.T. Seidl, J.P. Kenny and W.D. Allen, PSI 3.2, 2003.

13 Q. Chem. J. Kong, C.A. White, A.I. Krylov, D. Sherrill, R.D. Adamson, T.R. Furlani, M.S. Lee, A.M. Lee, S.R. Gwaltney, T.R. Adams, C. Ochsenfeld, A.T.B. Gilbert, G.S. Kedziora, V.A. Rassolov, D.R. Maurice, N. Nair, Y. Shao, N.A. Besley, P.E. Maslen, J.P. Dombrowski, H. Daschel, W. Zhang, P.P. Korambath, J. Baker, E.F. Byrd, T. van Voorhis, M. Oumi, S. Hirata, C.-P. Hsu, N. Ishikawa, J. Florian, A. Warshel, B.G. Johnson, P.M.W. Gill, M. Head-Gordon and J.A. Pople, J. Comput. Chem., 21 (2000) 1532.

14 See http://www.wavefun.com.

15 S. Brode, H. Horn, M. Ehrig, D. Moldrup, J.E. Rice and R. Ahlrichs, J. Comput. Chem., 14 (1993) 1142.

16 T. Yanai, H. Nakano, T. Nakajima, T. Tsuneda, S. Hirata, Y. Kawashima, Y. Nakao, M. Kamiya, H. Sekino and K. Hirao, Computational Science—International Conference on Computational Science 2003, Lecture Notes in Computer Science (Springer), 2003, pp. 84–95.

17 (a) Y.C. Zheng and J. Almlöf, Chem. Phys. Lett., 214 (1993) 397; (b) K.R. Glaesemann and M.S. Gordon, J. Chem. Phys., 108 (1998) 9959.

18 P.-O. Löwdin, Phys. Rev., 101 (1956) 1730.

19 J. Ivanic and K. Ruedenberg, Theor. Chem. Acc., 106 (2001) 339.

20 M.W. Schmidt and M.S. Gordon, Ann. Rev. Phys. Chem., 49 (1998) 233.

21 M.S. Gordon, J. Phys. Chem., 100 (1996) 3974.

22 G.M. Chaban and M.S. Gordon, J. Phys. Chem. A, 103 (1999) 185.

23 P. Bandyopadhyay, T.J. Dudley and M.S. Gordon, in preparation.

24 Y. Jung, Y. Akinaga, K.D. Jordan and M.S. Gordon, Theor. Chem. Acc., 109 (2003) 268.

25 C.H. Choi, D.-J. Liu, J.W. Evans and M.S. Gordon, J. Am. Chem. Soc., 124 (2002) 8730.

26 G. Chaban, M.W. Schmidt and M.S. Gordon, Theor. Chem. Acc., 97 (1997) 88.

27 J. Ivanic, J. Chem. Phys., 119 (2003) 9364.

28 J. Olsen, B.O. Roos, P. Jørgensen and H.J.A. Jensen, J. Chem. Phys., 89 (1998) 2185.

29 H. Nakano and K. Hirao, Chem. Phys. Lett., 317 (2000) 90.

30 (a) L. Bytautas and K. Ruedenberg, Mol. Phys., 100 (2002) 757; (b) J. Ivanic and K. Ruedenberg, Theor. Chem. Acc., 107 (2002) 220.

31 J. Ivanic and K. Ruedenberg, J. Comput. Chem., 24 (2003) 1250.

32 L. Bytautas and K. Ruedenberg, J. Chem. Phys., 121 (2004) 10852, 10905, 10919.

33 K. Hirao, Chem. Phys. Lett., 190 (1992) 374.

34 H. Nakano, J. Chem. Phys., 99 (1993) 7983.

35 J. Rintelman, I. Adamovic, S. Varganov, M.S. Gordon, J. Chem. Phys., 122 (2005) 044105.

36 M.S. Gordon, M.W. Schmidt, G.M. Chaban, K.R. Glaesemann, W.J. Stevens and C. Gonzalez, J. Chem. Phys., 110 (1999) 4199.

37 (a) P. Piecuch, K. Kowalski, I.S.O. Pimienta and M. McGuire, Int. Rev. Phys. Chem., 21 (2002) 527; (b) P. Piecuch, S.A. Kucharski, K. Kowalski and M. Musial, Comput. Phys. Commun., 149 (2002) 71.

38 (a) K. Raghavachari and J.A. Pople, Int. J. Quantum Chem., 20 (1981) 1067; (b) G.H.F. Diercksen, B.O. Roos and A.J. Sadlej, J. Chem. Phys., 59 (1981) 29; (c) J.E. Rice and R.D. Amos, Chem. Phys. Lett., 122 (1985) 583; (d) K.B. Wiberg, C.M. Hadad, T.J. LePage, C.M. Breneman and M.J. Frisch, J. Phys. Chem., 96 (1992) 671; (e) A.P.L. Rendell, G.B. Backskay, N.S. Hush and N.C. Handy, J. Chem. Phys., 87 (1987) 5976.

39 K. Kowalski and P. Piecuch, in preparation.

40 C.M. Aikens, S.P. Webb, R.L. Bell, G.D. Fletcher, M.W. Schmidt and M.S. Gordon, Theor. Chem. Acc., 110 (2003) 233.

41 W.J. Lauderdale, J.F. Stanton, J. Gauss, J.D. Watts and R.D. Bartlett, Chem. Phys. Lett., 187 (1994) 21.

42 P.J. Knowles, J.S. Andrews, R.D. Amos, N.C. Handy and J.A. Pople, Chem. Phys. Lett., 186 (1991) 130.

43 (a) D. Jayatilaka and T.J. Lee, Chem. Phys. Lett., 199 (1992) 211; (b) T.J. Lee and D. Jayatilaka, Chem. Phys. Lett., 201 (1993) 1; (c) D. Jayatilaka and T.J. Lee, J. Chem. Phys., 98 (1993) 9734; (d) G.D. Fletcher, M.S. Gordon and R.S. Bell, Theor. Chem. Acc., 107 (2002) 57.

44 S. Webb, unpublished results.

45 K. Kowalski and P. Piecuch, J. Chem. Phys., 120 (2004) 1715.

46 D.G. Fedorov, S. Koseki, M.W. Schmidt and M.S. Gordon, Int. Rev. Phys. Chem., 22 (2003) 551.

47 D. Fedorov, M.W. Schmidt, S. Koseki and M.S. Gordon, in: K. Hirao, Y. Ishikawa (Eds.), Recent advances in relativistic molecular theory, Vol. 5, World Scientific, Singapore, 2004, pp. 107–136.

48 (a) S. Koseki, N. Matsunaga, M.W. Schmidt and M.S. Gordon, J. Phys. Chem., 99 (1995) 12764; (b) S. Koseki, M.W. Schmidt and M.S. Gordon, J. Phys. Chem., 102 (1998) 1043; (c) S. Koseki, D.G. Fedorov, M.W. Schmidt and M.S. Gordon, J. Phys. Chem. A, 105 (2001) 8262; (d) S. Koseki, Y. Ishihara, H. Umeda, D.G. Fedorov and M.S. Gordon, J. Phys. Chem., 106 (2002) 785; (e) S. Koseki, Y. Ishihara, D.G. Fedorov, M.W. Schmidt and M.S. Gordon, J. Phys. Chem. A, 104 (2004) 4707.

49 (a) D.G. Fedorov and M.S. Gordon, J. Chem. Phys., 112 (2000) 5611; (b) D.G. Fedorov, M.S. Gordon, M.R. Hoffmann, K.G. Dyall, Low-lying potential energy surfaces, ACS Symposium Series 828, Oxford Press, Oxford (2002).

50 M. Krauss and W.J. Stevens, Ann. Rev. Phys. Chem., 35 (1985) 357.

51 (a) M. Douglas and N.M. Kroll, Ann. Phys., 82 (1974) 89; (b) T. Nakajima and K. Hirao, Chem. Phys. Lett., 329 (2000) 511; (c) T. Nakajima and K. Hirao, J. Chem. Phys., 113 (2000) 7286; (d) W.A. DeJong, R.J. Harrison and D.A. Dixon, J. Chem. Phys., 114 (2001) 48; (e) T. Nakajima and K. Hirao, J. Chem. Phys., 119 (2003) 4105.

52 G.M. Chaban, J.O. Jung and R.B. Gerber, J. Chem. Phys., 111 (1999) 1823.

53 (a) K. Yagi, T. Taketsugu, K. Hirao and M.S. Gordon, J. Chem. Phys., 113 (2000) 1005; (b) K. Yagi, K. Hirao, T. Taketsugu, M.W. Schmidt and M.S. Gordon, J. Chem. Phys., 121 (2004) 1383.

54 N. Matsunaga, G.M. Chaban and R.B. Gerber, J. Chem. Phys., 117 (2002) 3541.

55 D.G. Truhlar and B.C. Garrett, Acc. Chem. Res., 13 (1980) 440.

56. R.P.A. Bettens and M.A. Collins, J. Chem. Phys., 108 (1998) 2424, and references cited therein.

57 http://comp.chem.umn.edu/gamessplus.
58 H.M. Netzloff, M.S. Gordon and M.A. Collins, J. Chem. Phys., submitted for publication.
59 (a) T. Taketsugu and M.S. Gordon, J. Phys. Chem., 99 (1995) 8467; (b) M.S. Gordon, G. Chaban and T. Taketsugu, J. Phys. Chem., 100 (1996) 11512.
60 (a) D.K. Hoffman, R.S. Nord and K. Ruedenberg, Theor. Chim. Acta, 69 (1986) 265; (b) J. Sun and K. Ruedenberg, J. Chem. Phys., 98 (1993) 9707.
61 K. Bondensgaard and F. Jensen, J. Chem. Phys., 104 (1996) 8025.
62 Y. Yamaguchi, M. Frisch, J. Gaw, H.F. Schaefer and J.S. Binkley, J. Chem. Phys., 84 (1986) 2262.
63 A. Komornicki and J.W. McIver, J. Chem. Phys., 70 (1979) 2016.
64 P. Korambath and H.A. Kurtz, in: S.P. Karna, A.T. Yeates (Eds.), Nonlinear optical materials, ACS Symposium Series 628, 1996, ACS, Washington, DC, pp. 133–144.
65 H.A. Kurtz, J.J.P. Stewart and K.M. Dieter, J. Comput. Chem., 11 (1990) 82.
66 M.A. Freitag, B. Hillman, A. Agrawal and M.S. Gordon, J. Chem. Phys., 120 (2004) 1197.
67 M.A. Spackman, J. Comput. Chem., 17 (1996) 1.
68 J.H. Jensen and M.S. Gordon, J. Phys. Chem., 99 (1995) 8091.
69 (a) A.J. Stone, Chem. Phys. Lett., 83 (1981) 233; (b) A.J. Stone and M. Alderton, Mol. Phys., 56 (1985) 1047.
70 W. Chen and M.S. Gordon, J. Phys. Chem., 100 (1996) 14316.
71 J. Nieplocha, R.J. Harrison and R.J. Littlefield, Proceedings of Supercomputing '94, (1994) 340.
72 G.D. Fletcher, M.W. Schmidt, B.M. Bode and M.S. Gordon, Comput. Phys. Commun., 128 (2000) 190.
73 R.M. Olson, M.W. Schmidt, M.S. Gordon and A.P. Rendell, Proceedings of Supercomputing 2003, in press.
74 G.D. Fletcher, M.W. Schmidt and M.S. Gordon, Adv. Chem. Phys., 110 (1999) 267.
75 T. Kudo and M.S. Gordon, J. Phys. Chem. A, 105 (2001) 11276.
76 C.M. Aikens and M.S. Gordon, J. Phys. Chem. A, 108 (2004) 3103.
77 H. Umeda, S. Koseki, U. Nagashima and M.W. Schmidt, J. Comput. Chem., 22 (2001) 1243.
78 T.L. Windus, M.W. Schmidt and M.S. Gordon, Theor. Chim. Acta, 89 (1994) 77.
79 G.D. Fletcher, to be published.
80 T.J. Dudley, M.W. Schmidt and M.S. Gordon, J. Chem. Phys., submitted for publication.
81 Z. Gan, Y. Alexeev, R.A. Kendall and M.S. Gordon, J. Comput. Chem. Phys., 119 (2003) 47.
82 D.G. Fedorov, R.M. Olson, K. Kitaura, M.S. Gordon and S. Kosaki, J. Comput Chem., 25 (2004) 872.
83 (a) C.H. Choi, J. Ivanic, M.S. Gordon and K. Ruedenberg, J. Chem. Phys., 111 (1999) 8825; (b) C.H. Choi, K. Ruedenberg and M.S. Gordon, J. Comput. Chem., 22 (2001) 1484; (c) C.H. Choi, J. Chem. Phys., 120 (2004) 3535.
84 (a) D.G. Fedorov, R.M. Olson, K. Kitaura, M.S. Gordon and S. Koseki, J. Comput. Chem., 25 (2004) 872; (b) D.G. Fedorov and K. Kitaura, J. Chem. Phys., 120 (2004) 6832; (c) D.G. Fedorov and K. Kitaura, Chem. Phys. Lett., 389 (2004) 129; (d) D.G. Fedorov and K. Kitaura, J. Chem. Phys., 121 (2004) 2483.
85 (a) C. White and M. Head-Gordon, J. Chem. Phys., 101 (1994) 6593; (b) C. White and M. Head-Gordon, Chem. Phys. Lett., 257 (1996) 647; (c) C. White and M. Head-Gordon, J. Chem. Phys., 105 (1996) 5061.
86 (a) M.C. Strain, G.E. Scuseria and M.J. Frisch, Science, 271 (1996) 5245; (b) J.C. Burant, M.C. Strain, G.E. Scuseria and M.J. Frisch, Chem. Phys. Lett., 248 (1996) 43; (c) J.C. Burant, M. Strain, G.E. Scuseria and M.J. Frisch, Chem. Phys. Lett., 258 (1996) 45; (d) G.E. Scuseria, J. Phys. Chem. A, 103 (1999) 4782.
87 (a) M. Challacombe, E. Schwegler and J. Almlof, J. Chem. Phys., 104 (1996) 4685; (b) E. Schwegler and M. Challacombe, J. Chem. Phys., 105 (1996) 2726; (c) M. Challacombe, J. Chem. Phys., 113 (2000) 10037; (d) J.A. Board, J.W. Causey, J.F. Leathrum, A. Windemuth and K. Schulten, Chem. Phys. Lett., 198 (1992) 89; (e) J.M. Perez-Jorda and W. Wang, J. Chem. Phys., 104 (1996) 8003; (f) H.G. Petersen, D. Soelvason, J.W. Perram and E.R. Smith, J. Chem. Phys., 101 (1994) 8870.
88 M.S. Gordon, M.A. Freitag, P. Bandyopadhyay, J.H. Jensen, V. Kairys and W.J. Stevens, J. Phys. Chem. A, 105 (2001) 293.
89 I. Adamovic, M.A. Freitag and M.S. Gordon, J. Chem. Phys., 118 (2003) 6725.
90 J. Song and M.S. Gordon, in preparation.
91 H.M. Netzloff and M.S. Gordon, J. Chem. Phys., 121 (2004) 2711.

92 I. Adamovic and M.S. Gordon, Mol. Phys., in press.
93 V. Kairys and J.H. Jensen, J. Phys. Chem. A, 104 (2000) 6656.
94 P.N. Day, R. Pachter, M.S. Gordon and G.M. Merrill, J. Chem. Phys., 112 (2000) 2063.
95 H.M. Netzloff and M.S. Gordon, to be published.
96 H.M. Netzloff and M.S. Gordon, J. Comput. Chem., 25 (2004) 1926.
97 J. Shoemaker, L.W. Burggraf and M.S. Gordon, J. Chem. Phys., 112 (2000) 2994.
98 P. Ren and J.W. Ponder, J. Phys. Chem. B, 107 (2003) 5933, see also, http://dasher.wustl.edu/tinker.
99 (a) C.H. Choi and M.S. Gordon, J. Am. Chem. Soc., 121 (1999) 11311; (b) Y. Jung, C.H. Choi and M.S. Gordon, J. Phys. Chem., 105 (2001) 4039; (c) C.H. Choi and M.S. Gordon, in: Z. Rappaport, Y. Apeloig (Eds.), The chemistry of organic silicon compounds Vol. 3, Wiley Chichester (2001) 821; (d) C.H. Choi and M.S. Gordon, J. Am. Chem. Soc., 124 (2002) 8730; (e) C.H. Choi and M.S. Gordon, in: L.A. Curtiss, M.S. Gordon (Eds.), Computational materials chemistry: Methods and applications, Kluwer Academic, Dordrecht, (2004) 125–190; (f) H.S. Lee, C.H. Choi and M.S. Gordon, J. Am. Chem. Soc., submitted for publication.
100 D. Zorn and M.S. Gordon, in preparation.
101 (a) P. Zapol, L.A. Curtiss, H. Tamura and M.S. Gordon, in: L.A. Curtiss, M.S. Gordon (Eds.), Computational materials chemistry: Methods and applications, Kluwer Academic, Dordrecht, 2004, pp. 266–307; (b) H. Tamura and M.S. Gordon, in preparation.
102 H. Tamura and M.S. Gordon, J. Chem. Phys., 119 (2003) 10318.
103 J. Tomasi and M. Persico, Chem. Rev., 94 (1994) 2027.
104 E. Cances, B. Mennucci and J. Tomasi, J. Chem. Phys., 107 (1997) 3032.
105 E. Cances and B. Mennucci, J. Chem. Phys., 109 (1998) 249.
106 B. Mennucci, C. Amovilli and J. Tomasi, Chem. Phys. Lett., 286 (1998) 221.
107 (a) V. Barone and M. Cossi, J. Phys. Chem. A, 102 (1998) 1995; (b) M. Cossi, N. Rega, G. Scalmani and V. Barone, J. Comput. Chem., 24 (2003) 669.
108 H. Li, C.S. Pomelli and J.H. Jensen, Theor. Chim. Acta, 109 (2003) 71.
109 H. Li and J.H. Jensen, J. Comput. Chem., 25 (2004) 1449.
110 (a) P. Bandyopadhyay and M.S. Gordon, J. Chem. Phys., 113 (2000) 1104; (b) P. Bandyopadhyay, B. Mennucci, J. Tomasi and M.S. Gordon, J. Chem. Phys., 116 (2002) 5023.
111 M.M. Karelson, A.R. Katritzky and M.C. Zerner, Int. J. Quantum Chem., S20 (1986) 521.
112 K. Baldridge and A. Klamt, J. Chem. Phys., 106 (1997) 6622.
113 B.M. Bode and M.S. Gordon, J. Mol. Graphics, 16 (1999) 133.

Theory and Applications of Computational Chemistry: The First Forty Years
Edited by C. Dykstra et al.

CHAPTER 42

How and why coupled-cluster theory became the pre-eminent method in an **ab initio** *quantum chemistry*

Rodney J. Bartlett

Quantum Theory Project Departments of Chemistry and Physics, University of Florida, Gainesville, FL 32611, USA

Abstract

The correlation problem has been a focal point of quantum chemistry for more than the 40 years covered by this volume. Today, coupled-cluster methods typically offer its best numerical solution for molecules of $\lesssim 20$ atoms in terms of rigor and systematic convergence to the right answer. Coupled-cluster (CC) theory is not variational, but, more importantly, it is size-extensive making energy differences meaningful and applications to extended systems like polymers and solids possible. CC introduces higher excitations much more effectively and at less expense than analogous configuration interaction methods. As a non-variational method, analytical gradients in CC theory require new developments compared to other methods. These new developments emphasize the role of the CC functional that underlies the CC theory for energies, forces, density matrices, and properties of ground and equation-of-motion (EOM-CC) excited states. The latter also permits facile applications for single, doubly, etc. ionized and electron-attached states. The evolution of the critical ideas that lead to coupled-cluster theory and its extensive applications in chemistry are enumerated, in an objective, but personal, first-hand account.

42.1 INTRODUCTION

In this book reviewing 40 years of quantum chemistry, I thought it would be interesting to present my personal view on the critical ideas and developments that enabled coupled-cluster (CC) theory to evolve into the reference method for the electronic structure theory of molecules, lying in between the smallest and largest systems. Beyond establishing that fact—which I would argue most recognize today—there are lessons in its development that will repeat themselves with the other methods we have, or will likely develop; over the next 40 years. Furthermore, even in CC theory we have many new developments just being explored, and others soon to be presented, that will further greatly enhance the scope of its applications. In the interest of some brevity, I will limit myself to single reference CC methods for ground states and EOM-CC excited states and associated properties. I discuss various multi-reference CC methods elsewhere [1], and I would have little new to add to that. I will also not dwell upon numerical results, as there are better places for documenting that information. See the reviews [2–5], the papers in the special benchmark issue [6] and the extensive survey of comparative numerical results as a function of correlation and basis set presented in the book by Helgaker et al. [7]. An introduction to the theory with many numerical illustrations is presented in the tutorial [8].

42.2 ORIGINS: $\exp(T_2)|0\rangle$

The development starts with Rayleigh–Schrödinger perturbation theory (RSPT) where the wavefunction and the energy can be written conveniently as

$$\Psi = \Phi_0 + \sum_{k=1}^{\infty} \Psi^k \tag{1}$$

$$\Psi^k = [(E_0 - H_0)^{-1} Q(V - E^1)]^k |\Phi_0\rangle - \sum_{l=2}^{k} E^l (E_0 - H_0)^{-1} Q |\Psi^{k-l}\rangle \tag{2}$$

$$E = \langle \Phi_0 | V | \Psi \rangle = E_0 + \sum_{k=1}^{\infty} E^k \tag{3}$$

$$E^k = \langle \Phi_0 | V | \Psi^{k-1} \rangle \tag{4}$$

assuming intermediate normalization, $\langle \Phi_0 | \Psi \rangle = 1$, $Q = 1 - |\Phi_0\rangle\langle\Phi_0|$, and keeps the resolvent from being singular. If we somewhat naively look at this expression for the energies in RSPT, we see a possible problem. We know the energy has to be an *extensive* property, $E = NE_{\text{unit}} = N(E_0 + E^1 + E^2 + \cdots)_{\text{unit}}$ so for a non-interacting set of molecules, for any energy beyond first order, we have a potentially non-vanishing contribution of E^l to E^k from the renormalization terms. This suggests that we might have terms in the energy that would scale as N^2, N^3 and higher; but this cannot be since each energy order must be extensive itself. For a finite system, this fact might be hidden in the numerical results, but in the infinite limit this cannot be the case. Also in any finite order, it should not be. Brueckner recognized this of necessity for the square well interaction potentials then being considered in nuclei, where he would get a sum of infinities. So he showed that through fourth order, such terms from the renormalization terms in RSPT had to cancel with other terms from the lead part of the perturbed wavefunction [9]. Goldstone [10] extended this observation by proving this was true to all orders of RSPT using diagrammatic techniques. (A less naive development necessarily pays attention to the N dependence of just the conjoint (EPV) terms which have an index in common that remain, and the disjoint ones that cancel, as discussed in our book [11]), but this would still lead us to the same place after substantial algebra).

Hence, Brueckner–Goldstone *many-body perturbation theory* (MBPT) makes it possible to write RSPT in the form,

$$\Psi = \Phi_0 + \sum_{k=1}^{\infty} [R_0 H]^k |\Phi_0\rangle_{\mathrm{L}} \tag{5}$$

$$E = \langle \Phi_0 | H | \Psi \rangle = \sum_{k=0}^{\infty} \langle \Phi_0 | H [R_0 H]^k | \Phi_0 \rangle_{\mathrm{L}} \tag{6}$$

where the L indicates a restriction to linked diagrams, and R_0 is the resolvent operator from above. Φ_0 is some independent particle reference function, like Hartree–Fock, or Kohn–Sham DFT, or a Brueckner determinant, etc. There is no restriction on this choice, but, of course, it is simpler if the orbitals composing the determinant and all excitations from it are eigenfunctions of a one-article operator. Then $H_0 = \sum_i h^{\mathrm{eff}}(i)$, with $h^{\mathrm{eff}}(1)\varphi_{\mathrm{p}} = \varepsilon_{\mathrm{p}}\varphi_{\mathrm{p}}$, the associated one-particle effective Hamiltonian that corresponds to the reference independent particle determinant. Usually this will be a Hermitian operator, but in some cases like Brueckner, that might not be the case. That, too, is not really a problem [12].

MBPT is a dramatic improvement over RSPT. By simply realizing that all the renormalization terms disappear with a proper counting of the lead terms, we have a number of benefits:

- I MBPT is *size-extensive* [13], *guaranteeing* proper scaling with the number of particles or units in the system.
- II MBPT is applicable to polymers, solids, the electron gas, and nuclei.
- III MBPT eliminates the large (unlinked diagram) errors in molecular calculations that beset CI methods, while greatly reducing the level of computational complexity for a given accuracy.
- IV MBPT has the same simple perturbative structure as Brillouin–Wigner perturbation theory, as subject to linked diagrams, $\Psi^k = (R_0 H)^k |\Phi_0\rangle_{\mathrm{L}}$, $E^k = \langle \Phi_0 | H | \Psi^{k-1} \rangle_{\mathrm{L}}$, $E^{2k+1} = \langle \Psi^k | V - E^1 | \Psi^k \rangle_{\mathrm{L}}$, and $E^{2k} = \langle \Psi^k | E_0 - H_0 | \Psi^k \rangle_{\mathrm{L}}$.

So what is wrong with it? In this form it is still perturbative, implicitly assuming that the correlation perturbation is relatively small. For many problems we want more flexiblity than offered by perturbation theory. This leads to non-perturbative approaches where various categories of terms in MBPT are summed to all orders. One such method would be to make denominator shifts, so a denominator in perturbation theory like $\varepsilon_i - \varepsilon_a$ could be replaced by $\varepsilon_i - \varepsilon_a - \langle ai||ai\rangle$. By adding the anti-symmetrized integral, we have $(\varepsilon_i - \varepsilon_a - \langle ai||ai\rangle)^{-1} = (\varepsilon_i - \varepsilon_a)^{-1} + \langle ai||ai\rangle(\varepsilon_i - \varepsilon_a)^{-2} + \langle ai||ai\rangle(\varepsilon_i - \varepsilon_a)^{-3} + \cdots$; where we would sum all orders of new terms that depend upon $\langle ai||ai\rangle$, by the simple expedient of shifting its value by the two-electron integral when summing over the indices, a,i, as we evaluate E^2, E^3; etc. Such denominator shifts were extensively used in MBPT by Kelly [14], for example; see also Ref. [15].

However, a far more pervasive way to remove the perturbative nature of MBPT is offered by coupled-cluster theory, and this takes us to its origin. The usual expression for the linked diagram theorem in the time-dependent development is as a logarithm. Inverting the procedure, Hubbard [16] seems to have been the first to recognize that the linked diagram wavefunction above can be most conveniently written as

$$\Psi = \exp(\hat{T})|\Phi_0\rangle \tag{7}$$

where the operator, $\exp(\hat{T})$, would sum all the linked diagram units that constitute the wavefunction. Consequently, the energy must be

$$E = \langle\Phi_0|H\exp(\hat{T})|\Phi_0\rangle \tag{8}$$

For those familiar with wave-reaction operator perturbation theory [17] $\exp(\hat{T})$ is a realization of the wave-operator, Ω, $\Psi = \Omega\Phi_0$. As $\exp(\hat{T}) = 1 + \hat{T} + \hat{T}^2/2 + \hat{T}^3/3! + \cdots$ we have the *linked* wavefunction separated into *connected* terms, $\hat{T}$, and *disconnected ones*, like all the products of $\hat{T}$s.

The next step was taken by Coester and Kümmel [18,19] who proposed using this ansatz, then called $\exp(\hat{S})$, as a way to develop wavefunctions for nuclei. Though proposing the ansatz, they did not provide any equations for its determination until after that was done by Ĉíẑek, and which arises from a somewhat different route.

That route toward CC theory built upon the work of Sinanoglu [20,21] and Nesbet [22]. These authors conceived of electron correlation as being mostly pair-like. Obviously, most of the electron correlation in molecules had to arise from electrons of opposite spin in the same spatial orbital. So a logical way to address electron correlation was to attempt to account for such interactions, first. This suggests an independent electron pair approximation (IEPA). One way to see this is from Nesbet's formula, which gives the correlation energy from simply projecting the Schroedinger equation, as

$$E_{\text{corr}} = \langle\Phi_0|H|\Psi_{\text{FCI}}\rangle = \sum\langle\Phi_0|H|\Phi_{ij}^{ab}\rangle C_{ij}^{ab} = \sum_{i<j,a<b}\langle ij||ab\rangle C_{ij}^{ab} \tag{9}$$

where Φ_0 is the HF determinant. Via the Brillouin theorem there are no single excitations contributing to the energy. This is manifestly 'pairlike' in that if the CI coefficients for double excitations were known, we could resolve the correlation energy into 'pair' correlations. Of course, C_{ij}^{ab} depends upon all other excitations including singles, triples, etc. However, the pair correlations that arise from double excitations are the most important terms as they account for the second and third orders of the correlation perturbation theory.

An alternative route toward such a theory is offered by MBPT. We know that MBPT(2) with a HF reference gives

$$E^2 = \sum_{i<j,a<b}|\langle ij||ab\rangle|^2/(\varepsilon_i + \varepsilon_j - \varepsilon_a - \varepsilon_b) = \sum_{i<j}\varepsilon_{ij} \tag{10}$$

a sum of pair energies. Since these quantities are not invariant to orbital rotations, their individual values might not mean too much, but their sum is well defined. Here, we could also add some denominator shifts, like $\langle ab||ab\rangle$, $\langle ij||ij\rangle$ and $\langle ai||ai\rangle$, $\langle bi||bi\rangle$, $\langle aj||aj\rangle$,

$\langle bj||bj\rangle$ to try to achieve some improved approximations, while raising further issues with lack of invariance to orbital rotations. If all the above are included in a shift, this is often called Epstein–Nesbet perturbation theory [23], since in CI-based perturbation theory that shifted denominator is $\langle \Phi_0|H|\Phi_0\rangle - \langle \Phi_{ij}^{ab}|H|\Phi_{ij}^{ab}\rangle$.Once we go to MBPT(3), we also would have terms that depend upon the triad, i, j, k of occupied orbitals, so we are forced to extract 'pairlike' expressions from summing over k, for example. Since such third-order terms also contribute to C_{ij}^{ab}, we have the same situation in CI. Sinanoglu used such perturbative expressions to develop kinds of variational equations for independent pair interactions [20,21].

The next critical step was taken by Jiri Ĉíẑek [24,25], who recognized the limitations of IEPAs and introduced *coupled-pair* many-electron theory (CPMET). Today we call this coupled-cluster doubles (CCD) as $T = T_2$ [13,26]. These coupled-pairs properly include all the non-pair interactions while retaining a pair-like correlation structure. They also alleviate the failure of IEPAs and later, variants like coupled-electron pair theory (CEPA) [27,28], that are beset with a failure to be invariant to orbital transformations. CEPA, however, was an attempt to gain some of the advantages of CC theory within a CI computational structure, and still arises in various contexts today [29–31]. As CCD is properly invariant to any transformation among just the occupied or unoccupied space, localization transformations, e.g. change nothing in the CCD solutions for the energy and density matrices.

The CCD wavefunction,

$$\Psi_{\mathrm{CCD}} = \exp(T_2)|\Phi_0\rangle \tag{11}$$

$$\Psi_{\mathrm{CCD}} = \left(1 + T_2 + \frac{1}{2}T_2^2 + \frac{1}{3!}T_2^3 + \cdots\right)|\Phi_0\rangle \tag{12}$$

introduces all double excitations, then products of double excitations as part of the CI quadruple, hextuple, octuple, etc., excitations until the excitation level exceeds the number of electrons in the system. However, even though the exponential is *not* truncated (an essential feature of CC methods) its projection naturally terminates. Hence, projection onto the reference function provides the energy

$$E_{\mathrm{CCD}} = \langle \Phi_0|H\exp(T_2)|\Phi_0\rangle = \langle \Phi_0|HT_2|\Phi_0\rangle \tag{13}$$

and projection only on the double excitations provides the CCD equations,

$$0 = \langle \Phi_{ij}^{ab}|(H - E_{\mathrm{CCD}})\exp(T_2)|\Phi_0\rangle \tag{14}$$

Since H has only one and two-electron terms, the CCD equations have to naturally truncate after T_2^2. From a CI perspective...

$$\frac{1}{2}T_2^2|0\rangle = \frac{1}{4}\sum_{i<j<k<l,a<b,c<d} \Phi_{ijkl}^{abcd}(t_{ij}^{ab}t_{kl}^{cd} + 17 \text{ permutations})$$

[32,33] so these are quadruple excitations, of which there would be $\sim n^4N^4$ if we persisted in using CI technology; but as their coefficients are composed of products of double excitation amplitudes, the CC equations lead, instead, to only $\sim n^2N^2$ amplitudes

(coefficients) and a $\sim n^2N^4$ computational procedure (a quadruple CI would require a $\sim n^4N^6$ computational step!). Also, since

$$\frac{1}{4}\sum_{klcd}\langle{}^{ab}_{ij}|H|^{abcd}_{ijkl}\rangle t^{ab}_{ij}t^{cd}_{kl} = \left(\frac{1}{4}\sum_{klcd}\langle kl||cd\rangle t^{cd}_{kl}\right)t^{ab}_{ij} = E_{\text{CCD}}t^{ab}_{ij},$$

the energy is eliminated from the CC equations *from the beginning* making the results for the energy and the wavefunction consist of only 'linked' diagrams and to be 'size-extensive' [13]. In CI these *unlinked terms* would be the numerically largest part of the quadruple excitations, so their elimination immediately gets us much closer to the right answer. Physically, what are being included are the simultaneous two-particle interactions rather than the 'true' four-particle interactions, which would be introduced by T_4. The former would logically be expected to be the most significant for a Hamiltonian that only has two particle operators, and it does account for all terms through the fourth order of perturbation theory. T_4 arises in fifth order.

The iterative solution of the CCD equations gives

$$E_{\text{CCD}} = E_0 + E^1 + E^2 + E^3 + E^4(DQ) + \cdots \tag{15}$$

showing that the method provides an infinite-order summation of many linked diagrams in MBPT; all through third order (HF reference), and those that arise from disconnected T_2 terms that in CI language would be part of quadruples (all through fourth order), hextuples, etc. The introduction of this form of wavefunction dramatically changed quantum chemistry.

In Ĉíẑek's original paper [24], which derives from an even earlier dissertation, he reports results for semi-empirical Hamiltonians like PPP, but also even a partly *ab initio* result for N_2. However, his use of second-quantized based, diagrammatic techniques to derive the CC equations was unfamiliar to most quantum chemists (see [34]), likely delaying the appreciation of the CC method, although for simple cases like CCD, conventional Slater rule matrix evaluation can be applied [35] Also, explicit rules for diagrams were given and could have been used to derive more complicated CC equations.

The next significant contribution was a paper by Paldus, Ĉíẑek, and Shavitt [36]. These authors reported CCD results for a minimum Slater basis set BH_3 molecule, and compared them to full CI. Although the very small basis (there were only 91 distinct integrals *without* D_{3h} symmetry!) could hardly be said to provide a realistic assessment of electron correlation, the comparison to full CI was unambiguous. Equally important, for the first time some numerical consideration was made of other clusters in CC theory, such as T_3 as part of an extended (ECPMET) method, and due to the quasi-degeneracy one might expect in BH_3, T_3 might be a bit larger. Again diagrammatic methods were used in the derivations.

At the same time as the prior paper [36], we [37–41] were writing several papers about the application of MBPT to molecules. Although Kelly had attempted some one-center applications to water, e.g. where he could use his atomic-based MBPT methods; without having basis functions on every atom in the molecule, this form of MBPT could not be realistically applied to molecules. Hence, we combined MBPT with standard techniques for molecular integrals and SCF calculations to overcome this limitation. At the time we were focused on MBPT and did not know the detailed connection with coupled-cluster theory, but we were using the same type of diagrammatic derivations as in CC theory. In two papers [42,43] we talk about an infinite-order sum of double excitations diagrams, D-MBPT(∞), and in the latter even introduce an operator (diagram) that sums these terms to all orders, but did not realize that that was linearized, LCCD, until soon thereafter. Hence, we had LCCD programmed, and only lacked the contribution from T_2^2 to realize CCD. LCCD is also known as CEPA(0) [27,28]. LCCD is related to CID, differing by neglecting the residual effects of $(T_2)^2$ *after the unlinked diagram cancellation.*

A year or so after our initial MBPT studies, Pople's group joined in this MBPT effort, but chose to call their approach Møller–Plesset (MP) perturbation theory [44,45], and they scruptulously avoided any use of second-quantization or diagrammatic techniques in its implementation. This was much preferred by some! In 1978 we [13] and Pople et al. [46] had independently evaluated the quadruple excitation terms in fourth-order MBPT. John and I met on the eve of the American Conference for Theoretical Chemistry to compare results, and fortunately for our talks, had the same values! At this point, having T_2^2 in our equations, we reported in back-to-back papers in the Proceedings of the ACTC [13,46] the general purpose implementation of CCD and explored it for a few problems. This also offered the first general purpose (as opposed to the few full CIs), numerical measure of quadruple excitations in correlated methods.

42.3 HIGHER EXCITATIONS IN CC THEORY: $\exp(T_1 + T_2 + T_3 + \cdots)|0\rangle$

CC theory is inherently better than an equivalent level of CI because it eliminates unlinked diagrams and as a consequence, is size-extensive [13]. It is also inherently better than an equivalent level of MBPT because it is not limited to finite-orders, or potential difficulties encountered in the convergence of perturbation theory. It is well known, e.g. that ordinary MP perturbation theory does not converge for the electron gas, and this has also been emphasized recently for molecules [47], though resummations (including CC theory) work fine [48]. But, the operable word is 'equivalent level'. For CI, that meant at least single and double excitations, and frequently some more, perhaps even from a multi-reference space. MBPT had been done with single excitations in fourth-order SDQ-MBPT(4) in the above two papers [13,46].

Adding the single excitations from T_1 introduces the CCSD wavefunction,

$$\Psi_{\text{CCSD}} = \exp(T_1 + T_2)|\Phi_0\rangle = \exp(T_1)\exp(T_2)|\Phi_0\rangle \tag{16}$$

This, of course, means we have all CI single excitations, plus the products of single and double excitations like $T_1T_2|\Phi_0\rangle$, which in CI language introduce some triple excitation effects ($\hat{C}_3 = T_3 + T_1T_2 + T_1^3/3!$), as well as additional quadruple excitations like $\frac{1}{2}T_1^2T_2|\Phi_0\rangle$ and $\frac{1}{4!}T_1^4|\Phi_0\rangle$. With a HF reference, single excitations will not contribute to the energy until fourth order, so, generally, their numerical effect is not going to be all that important. However, if we want to use a non-HF reference function—which will become very important in giving CC theory its wide range of applicability—then effects of single excitations arise in first order, just like double excitations. Furthermore, for non-HF cases, these single-excitation terms can be very large. A second problem was the treatment of one-electron properties (or density matrices in general), where we know that single excitations are essential to a correct description. A third element that is also quite significant, is that any single determinant $|\Phi_0\rangle = \exp(T_1)|\Theta_0\rangle$, where Θ_0 is another single determinant [49–51]. So unlike CI methods, CC theory at the CCSD level and beyond, has some capacity to generate its own orbitals that are appropriate to the problem *from any starting set*. The orbitals so generated can be extracted from the CCSD solution by analysis of T_1, *a posteriori*. So CCSD and beyond are insensitive to occupied and virtual rotations, besides being *invariant* to occupied–occupied and virtual–virtual rotations.

To make such situations accessible, CCSD had to be developed. Initially, this looked formidable as CCD required only nine diagrams in its equations, but CCSD required 45 (*see* Ref. [52] *for an interesting anecdote*). George Purvis and I derived the CCSD equations and implemented them in 1982 [53] for closed and UHF open-shells from our many-body viewpoint, which is vastly easier than attempting to derive all equations from Slater's rules for determinantal matrix elements! George Purvis used his KOMMUTE program to check our diagram generation. We also had the advantage of the many-body approach providing natural intermediates, although our first set was hardly the best [53]. Now, it is clear that all CC equations can be reduced to an effective linearized form for CCSDTQ [54] and beyond, using intermediates based upon $\exp(-T)H\exp(T) = \bar{H}$.

A partial alternative to CCSD that provides an analogous singles effect within a different computational strategy was to rotate the orbitals to define Brueckner orbitals that have the property that $T_1 = 0$. This was the route of Dykstra [55]. Making $T_1 = 0$ via orbital rotation only requires four of the $45 - 9 = 36$ diagrams in CCSD to actually be evaluated, making it simpler in that respect than CCSD, but it requires repeated orbital iterations and transformations until convergence is reached.

Later closed shell CCSD programs were written by my Slovakian collaborators [56] in 1985, and then by Schaefer's group [57], after Andy Scheiner spent some time with us to learn how we did it. The third was a program of Lee and Rice [58], then in Handy's group. Now several groups were doing CC theory.

The next critical element in the development of CC theory was to incorporate the connected triple excitations, T_3. Since even CCD puts in the dominant quadruple excitation effects, and CCSD some of the disconnected triple excitations effects, the only term left in fourth-order MBPT comes from T_3, and the triples will be much more important to CC theory than to CI, since CIs unlinked diagrams have a very large role that can only be alleviated by putting in quadruple excitations (see Fig. 42.1). Triples had been explored in the ECPMET discussed above. Kvasnicka et al., Pople et al., Guest and Wilson, Urban et al., and ourselves had included triples in fourth-order MBPT = MP4 [59–64], but no attempt had been made to introduce them into general purpose CC methods. In 1984 we wrote a paper detailing the triple excitation equations in CC theory and reported results for CCSDT-1 [65], which meant the lead contribution of triples was included on top of CCSD. This also made it possible to treat triple excitations 'on-the-fly' in the sense that we never required storage of the $\sim n^3N^3$ T_3 amplitudes.

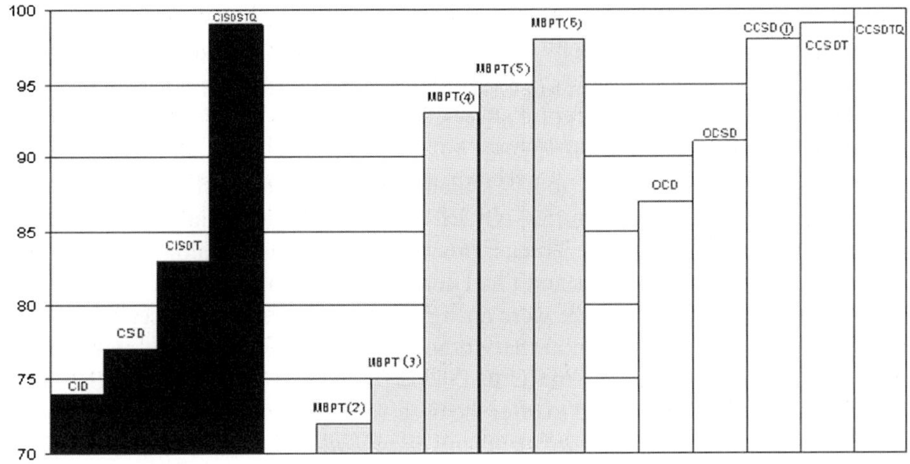

Fig. 42.1. Comparison of CI, MBPT, and CC results with full CI. Results based on DZP basis for BH, HF, and H_2O at Re $1.5R_e$, $2.0R_e$.

When we solved the iterative equations for CCSDT-1 it became evident that the first iterative approximation to triples,

$$t_{ijk}^{abc[2]} = (\varepsilon_i + \varepsilon_j + \varepsilon_k - \varepsilon_a - \varepsilon_b - \varepsilon_c)^{-1} \times \left(\frac{1}{2} \sum_d P(a/bc|k/ij) t_{ij}^{ad} \langle bc||dk\rangle - \frac{1}{2} \sum_l P(i/jk|c/ab) t_{il}^{ad} \langle lc||jk\rangle \right)$$

was a good one. $\{(P(a/bc|k/ij) = 1 - P_{ab} - P_{ac} - P_{ki} - P_{kj} + P_{ab}P_{ki} + P_{ac}P_{ki} + P_{ab}P_{kj} + P_{ac}P_{kj}\}$ [65], and from its connection with MBPT(4), it suggested that we introduce a non-iterative approximation for triples which would save significantly on the two, iterative, $\sim n^3N^4 \sim n^7$ steps. This approximation was nothing more than evaluating $E_4^{[T]} = \sum |t_{ijk}^{abc[2]}|^2 (\varepsilon_i + \varepsilon_j + \varepsilon_k - \varepsilon_a - \varepsilon_b - \varepsilon_c)$ [66]. This provides an energy correct through fourth order in MBPT and a wavefunction correct through second order when using a HF reference function. We called the method CCSD + T(CCSD) to emphasize that the triples were obtained from converged CCSD amplitudes. More commonly today, the designation CCSD [T] is used. We of course, had the effects of singles in our CCSDT-1 program, so could have easily added their effect, including the first (fifth-order contribution) of singles to this equation, but did not do so because the method was only meant to be a correct fourth-order approximation, and selecting one of many fifth-order terms did not seem to be relevant at the time [67] (see also [68]).

Instead, that step was left to Pople, Head-Gordon, and Raghavachari [69,70] who, by virtue of looking at the problem from the viewpoint of what they called 'quadratic CI', meant to be a simplified CC method (but see [71,72]) considered that like in CISD, if the doubles were there, the singles should be too. This added the numerically small, but significant term $E_5^{(S)} = \sum_{i,a} t_i^{a[2]*} t_i^{a[3]} (\varepsilon_i - \varepsilon_a)$, where $t_i^{a[3]} = \frac{1}{4} \sum \langle kl || cd \rangle t_{ikl}^{acd[2]}$ and $t_i^{a[2]} = \frac{1}{2} [\sum_{j,c,d} \langle aj || cd \rangle t_{ij}^{cd[1]} - \sum_{k,j,d} \langle kj || id \rangle t_{kj}^{ad[1]}]$ which together with $E_4^{[T]}$ defines CCSD(T). (On average, CCSD is 7.33 mH from full CI for BH, HF, H_2O, and SiH_2 at R_e, $1.5R_e$, and $2.0R_e$ in a DZP basis; while CCSD [T]'s error is 1.57 mH and CCSD(T)'s 1.19 mH. So the extra term is ~0.4 mH (~0.6%) of the total correction). The main reason for its significance is that it works contrary to $E_4^{[T]}$ which can overshoot in difficult cases. Using amplitudes from our CCSD program [53] (QCISD has few of the 45 diagrams of CCSD) and a former student of mine, Gary Trucks, this term was evaluated along with $E_4^{[T]}$ by Raghavachari et al. [70]. John Watts and I had done the same thing, but while we looked at potential energy curves where there were no significant differences between CCSD [T] and CCSD(T), they, instead, looked at the vibrational frequencies of O_3, a problem we had previously identified as a failure of CCSD [T] [73], where CCSD(T) was much better. Today, CCSD(T) is often considered to offer the best, widely applicable approximation in quantum chemistry [74] *for molecular problems near equilibrium.* For bond breaking, the iterative CCSDT-1 is better, reflecting the advantages of infinite-order *versus* finite-order perturbative approximations. It is also interesting, that later studies of ozone's frequencies, using larger basis sets, quadruple excitations, and multi-reference methods, actually show that those obtained from CCSD(T) are not very good [1], but they were better than what we had at the time, which spurred the applications of CCSD(T).

The final word on CCSD(T) was its generalization to non-HF reference functions [75]. This required adding non-HF diagrams and making a semi-canonical transformation of the orbitals. Despite its finite-order perturbative character, in this way we formulated the problem to be invariant to occupied–occupied and virtual–virtual transformations, just like CC theory itself is. This makes it possible to use ROHF, (quasi) QRHF, Brueckner, Kohn–Sham, or natural orbitals, or in fact nearly any reference function and to have analytical gradients for any of these in ACES II [76]. Several numerical results have been presented [75]. One potentially improved fourth-order approximation to the generalized CCSD(T) is the $\mathbf{\Lambda}$-based functional analogue, CCSD(T)$_{\mathbf{\Lambda}}$ (see Section 41.4 on the CC functional) which has been suggested [77,78], looks encouraging, and deserves detailed assessment.

The full CCSDT was first presented in 1987 [79], before CCSD(T). There is a very small numerical mistake in this paper, caused by the "real" world intervening in the affairs of scientists. Jozef Noga came from Czechoslovakia to work with me for several short visits. The visits were short because 'iron-curtain politics' did not allow him to be out of Czechoslovakia except for brief periods (the same was true for my Polish collaborators.). In those days, it was not unusual for us to work virtually all night to finish things. Jozef had written a CCSDT program (99 diagrams) in a *tour-de-force* during one of those short visits, and we were putting the finishing touches on it the morning his plane was scheduled to leave. At that time of night, Jozef thought he saw a previously overlooked simplification in his program, which had already been proven to be numerically correct by the several independent checks we can make by evaluating diagrams in different ways, checking extensivity, etc. To gain the speed, however, he made the simplification at the last minute, and we reported results *after* this change instead of before, which introduced a slight error. With the help of Gus Scuseria and

Fritz Schaefer, who wrote an independent CCSDT program later [80], this was corrected. The open-shell version of CCSDT was written by Watts [81].

Armed with CCSDT, an $\sim n^8$ method, we naturally pursued CCSDTQ (job security?), and Stan Kucharski wrote the first CCSDTQ program [54]. Now we are up to 180 diagrams and an $\sim n^{10}$ procedure. The convergence of the CC series to full CI is impressive. See Figure 42.1 comparing CC, MBPT, and single reference CI for the old, tried and true full CI examples we used over the years as reference values [82–84].

Problems like bond breaking, multiple important configurations like in ozone, multi-open-shell systems, etc, would appear to require at least some additional excitations in a reference space to obtain a correct description. This is the usual multi-reference argument. At the same time, if we are close enough to the full CI, we know that any single reference function will do. At least at the CCSDTQ level, having quadruple excitations out of a single reference would seem to be largely equivalent to double excitations out of a second, doubly excited reference function [85]. However, the $\sim n^{10}$ dependence is forbidding. Various non-iterative inclusions of quadruples have been considered [86], but the one I like best is what we call 'factorized' quadruple amplitudes, Q_f [87]. It has its origin in recognizing that in the fifth-order MBPT *energy*, we can use the factorization theorem to simplify the expression to one that only requires an $\sim n^6$ evaluation of T_4 contributions. Unlike the energy, we cannot do the same for the T_4 amplitudes, but in some of those all night sessions from the past when working on our expectation value (XCC [86,88]) method, Kucharski and I found that we could 'force' the factorization in the amplitudes, too, with only a very small error in the final results [87,89]. When we do this we simply modify the T_2 equations by a single term, $\frac{1}{2}\langle_{ij}^{ab}|T_2^{\dagger[1]}HT_2^2/2|\Phi_0\rangle$, where $T_2^{[1]}$ indicates its first-order value. This gives us the remarkable result that we introduce the most significant part of T_4 *without even having to consider the T_4 equation*! Adding in the T_3 equation, this term becomes $\frac{1}{2}\langle_{ij}^{ab}|T_2^{\dagger[1]}H(T_2^2/2+T_3|\Phi_0\rangle$ making its computational dependence $\sim n^7$. We have studied CCSDT(Q_f) and related approximations like CCSD(TQ_f) which is only $\sim n^7$ like CCSD(T), but the dependence on the virtuals is $\sim n^5$, instead of the $\sim N^4$ of CCSD(T).

Gwaltney and Head-Gordon introduced CCSD(2) which uses the Q_f approximation above for the quadruples, and, in fact, is equivalent to that of Hirata et al. [48], except the latter authors used the full Q approximation and Gwaltney and Head-Gordon separate left- and right-hand orbital equations that arise from the non-Hermitian operator structure, familiar for excited states from EOM-CC. A comparison of several possible iterative and non-iterative approximations for simple potential curves that include T_4 have been presented [90], including attempts at renormalization [91]. Additional non-iterative corrections through fifth [77] and sixth order [92] have been investigated numerically.

Besides some numerical investigation of higher order clusters accomplished by exploiting a full CI program [93] we continued to push the general purpose single reference CC theory through to pentuples, CCSDTQP [94], plus some approximations to it. Already at the quadruple excitation level we recognized that if we hoped to get the (harmonic) vibrational frequency of N_2 correct to $\sim 1\ \mathrm{cm}^{-1}$ would require pentuple excitations [89], so we added them [95]. The role of pentuples has since been verified by others [96].

The work of Kállay and Surjan [97] and Olsen [98] along with that of Hirata [99], is indicative of a very important development in CC theory. These scientists used automated implementations and Hirata actually got a computer to write efficient, error-free CC programs routinely. The tensor contraction engine approach of Hirata [99], creates CC and MBPT programs for ground, and EOM-CC programs for excited states routinely, and in parallel, building upon the structure of NWChem [100]. Kállay and Gauss use automated, string-based evaluations [101], but have also added analytical gradients to their programs, while not yet focusing on its parallel implementation. These developments are a testament to the place that CC theory has assumed. It is viewed as the best, rigorous, predictive tool available to quantum chemists, that is applicable to the largest number of problems for $\lesssim 20$ atoms. Hence, automatic generation of these very complicated programs was warranted. These developments, of course, remove any need for extending the sequence of methods CCSD, CCSD(T), CCSDT, CCSDT(Q_f), CCSDTQ, CCSDTQP, etc. that involved so much work over about 25 years [3]. But given the paradigm of converging results that has been established in the field—MBPT(2) < CCD < CCSD < CCSD(T) < CCSDT(Q_f) < CCSDTQ < CCSDTQP < Full CI—coupled to sequences of converging basis sets; automatic program implementation for using these methods should enable their extensive application for years to come.

And what about the future of CC theory? With these developments, much more effort can be devoted to multi-reference CC, Fock space CC, XCC, UCC, variational CC, methods like those discussed below for large molecules, and perhaps the ultimate current method, R12-CC [102]—where besides the correlation problem, we have the best current solution to the basis set problem—or a wealth of other methods that do not fit into the basic structure of the CC functional, discussed next, as that is the basis for the automated generation.

42.4 ANALYTICAL GRADIENTS AND THE CC FUNCTIONAL: $E = \langle 0|(1+\Lambda)\bar{H}|0\rangle$, $E^k = \langle 0|(1+\Lambda)\bar{H}^k|0\rangle$.

If there is one essential lesson in the development of CC theory, it is that no quantum chemical method can survive for long without an associated capability for analytical gradients ($F = \mathrm{d}E(\mathbf{R})/\mathrm{d}\mathbf{R}$). Chemistry means atoms and molecules interacting, forming, and breaking bonds. By obtaining the forces on the atoms we can determine molecular structure, vibrational spectra, and transition states. The large number of degrees of freedom ($\sim 3N$), for N atoms, means that without exceptional symmetry, few polyatomic molecules can be studied without an analytical gradient method, and as we break bonds, even high-symmetry buys little. Hence, the importance of analytical force evaluation, while simultaneously solving the Schroedinger equation—and in about the same amount of time—can hardly be overestimated. It was not known how to do this for CC theory until 1984 [103,104].

Forces seem to be superficially simple since the Hellman–Feynman theorem tells us that for some atomic displacement, $\mathrm{d}X_k$,

$$\mathrm{d}E(\mathbf{R})/\mathrm{d}X_k|_0\langle\Psi|\Psi\rangle = \langle\Psi|(\partial\hat{H}/\partial X_k)|_0|\Psi\rangle$$

and this is true provided our quantum chemical calculations use a *complete AO, MO*, and *configuration* basis set. But, of course, in practice we use finite atomic basis sets attached to atoms and truncate our configuration basis. This means we have to also include the non-vanishing contributions from the non-Hellman–Feynman (NHF) terms

$$\langle \mathrm{d}\Psi/\mathrm{d}X_k|\hat{H} - E|\Psi\rangle + \langle\Psi|\hat{H} - E|\mathrm{d}\Psi/\mathrm{d}X_k\rangle$$

To do so, we would need a knowledge of the *first-order perturbed wavefunction* with respect to all atomic displacements. However, there are $\sim 3N$ such perturbed wavefunctions for a molecule with N atoms. Does this mean we have to do $\sim 3N$ calculations in addition to the energy to search a PES? Not necessarily.

Using the chain rule, we can rewrite the NHF expression in terms of the variation of the energy functional $\delta E\{\Psi\}$ as,

$$\begin{aligned}\mathrm{d}(\delta E)/\mathrm{d}X_k &= \mathrm{d}/\mathrm{d}X_k[\langle\delta\Psi|\hat{H} - E|\Psi\rangle + \langle\Psi|\hat{H} - E|\delta\Psi\rangle] \\ &= 2\mathrm{d}/\mathrm{d}X_k[(\partial\delta E/\partial_\chi(\mathbf{R}))\mathrm{d}_\chi(\mathbf{R}) + (\partial\delta E/\partial\mathbf{c})\mathrm{d}\mathbf{c} + (\partial\delta E/\partial\mathbf{C})\mathrm{d}\mathbf{C}]\end{aligned}$$

Hence, if we are to make $\delta E = 0$, we can avoid these terms. But to do so we have to have E optimum with respect to the location of the atomic basis functions, $\chi(\mathbf{R})$; the MO coefficients, $\mathbf{c}(\mathbf{R})$; and the CI coefficients, $\mathbf{C}(\mathbf{R})$. The first cannot be satisfied unless the atomic orbital basis set is 'floated' off the atomic centers to an optimum location [105], while the second requires optimum MO coefficients, and the third optimum CI coefficients. In practice, we will introduce atomic orbital derivatives explicitly, so the AOs can follow their atoms. Now focusing only on the MO and CI coefficients, in SCF we have optimum MOs and no CI term. In MCSCF, both terms would vanish, while in CI, the MO derivatives would remain, but the CI coefficients contribution would vanish. In the non-variational coupled-cluster theory, *neither will vanish and this means that CC theory forces us into some new considerations for analytical forces.*

When either of these quantities are not optimum, we potentially still need to consider $\sim 3N$ such perturbed wavefunctions. Adding in the AO derivatives we replace $\mathrm{d}H/\mathrm{d}X_k$ by $H^{(k)}$ with $E^{(k)}$ the energy derivative. Looking at the form of the NHF terms, from perturbation theory we know the first-order wavefunction has to satisfy the inhomogeneous equation,

$$(E_0 - H_0)\mathrm{d}\Psi/\mathrm{d}X_k = (H^{(k)} - E^{(k)})\Psi^{(0)}$$

where the $\Psi^{(0)} = \Psi(0)$, the wavefunction at the starting geometry. Subject to the usual intermediate normalization, we can rewrite this equation as

$$\mathrm{d}\Psi/\mathrm{d}X_k = R_0(H^{(k)} - E^{(k)})\Psi^{(0)} = R_0H^{(k)}\Psi^{(0)}$$

in terms of the resolvent operator, $R_0 = (E_0 - H_0)^{-1}Q$. With minor manipulations, we can also write $\mathbf{R}_0 = |\mathbf{h}\rangle\langle\mathbf{h}|E_0 - H_0|\mathbf{h}\rangle^{-1}\langle\mathbf{h}| = |\mathbf{h}\rangle R_0\langle\mathbf{h}|$ where we have chosen to replace Q by its representation ('inner projection' [17]) in terms of functions orthogonal to the reference function, $\Psi = \Psi^{(0)}$. In this resolvent form, it is easy to see that

$$\langle\mathrm{d}\Psi/\mathrm{d}X_k|\hat{H} - E|\Psi\rangle + \langle\Psi|\hat{H} - E|\mathrm{d}\Psi/\mathrm{d}X_k\rangle = 2\langle\Psi|\hat{H}R_0H^{(k)}|\Psi\rangle$$

Consequently, if we introduce the auxiliary vector, $\mathbf{\Lambda} = \langle \Psi|\hat{H}|\mathbf{h}\rangle \mathbf{R}_0$, this then gives us $2\mathbf{\Lambda}\langle \mathbf{h}|H^{(k)}|\Psi\rangle$ for the NHF terms.

Note what has been accomplished. In this form, $\mathbf{\Lambda}$ is independent of the perturbation operator so we now compute *one* $\mathbf{\Lambda}$ and use it for *all degrees of freedom*, leaving only the $\langle \mathbf{h}|H^{(k)}|\Psi\rangle$ term to be evaluated for each degree of freedom, which can then be dotted into $\mathbf{\Lambda}$.

This procedure is a manifestation of the interchange theorem familiar from double perturbation theory [106,107] to emphasize that one could choose to do a calculation of a quantity like $\langle \Psi|VR_0W|\Psi\rangle$ by evaluating $R_0W|\Psi\rangle$ or $(R_0V|\Psi\rangle)^\dagger$ first, depending upon which operator lent itself to the greatest simplification. In the particular case of analytical gradients with $\sim 3N$ degrees of freedom, this is a major, indeed, essential, simplification. (For the MO relaxation part of the problem, the same interchange theorem simplifies the coupled-perturbed Hartree–Fock solutions as would occur with CI gradients where the CI coefficients, themselves, are optimum [108]). But our concern was the use of the interchange theorem for the *correlation* problem [103] as the cluster amplitudes do not come from a variational solution.

With the above, we can now move to CC theory where the energy is

$$E_{\mathrm{CC}} = \langle \Phi_0|\hat{H}|\Psi_{\mathrm{CC}}\rangle = \langle \Phi_0|\hat{H}\exp(T)|\Phi_0\rangle = \langle \Phi_0|\exp(-T)\hat{H}\exp(T)|\Phi_0\rangle = \langle \Phi_0|\bar{H}|\Phi_0\rangle.$$

In order to keep track of all terms simply, it is much more convenient to consider our derivative Hamiltonian to be written in second quantization, $H^k = \sum_{p,q}\langle p|h|q\rangle^k p^\dagger q + \frac{1}{4}\sum_{p,q,r,s}\langle pq||rs\rangle^k p^\dagger q^\dagger sr$ ensuring that by virtue of the total derivative of the matrix elements, H^k will properly include all appropriate AO *and* MO derivative terms. (To make the small distinction from the general, first-quantized approach above, we use k instead of (k) in superscript.) Then straightforward differentiation gives

$$E^k_{\mathrm{CC}} = \langle \Phi_0|\bar{H}^k|\Phi_0\rangle + \langle \Phi_0|[\bar{H}, T^k]|\Phi_0\rangle$$

where $\bar{H}^k = \exp(-T)\mathrm{d}H^k/\mathrm{d}X_k\exp(T)$, while the $\mathrm{d}T/\mathrm{d}X_k = T^k$ is in the second term. Since in CC theory the first-order perturbed wavefunction is $\mathrm{d}\Psi/\mathrm{d}X_k = \mathrm{d}T/\mathrm{d}X_k\exp(T)|\Phi_0\rangle = T^k\Psi_{\mathrm{CC}}$, we know from the CC amplitude equations that

$$\langle \mathbf{h}|\bar{H}^k + [\bar{H}, T^k]|\Phi_0\rangle = 0 \tag{17}$$

Liberal use of the resolution of the identity $1 = P + Q$, where $Q\bar{H}P = 0$ are the CC equations, and QT^kP is the only non-vanishing part of the perturbed T operator [3] then gives

$$\langle \mathbf{h}|T^k|\Phi_0\rangle = \langle \mathbf{h}|(E_{\mathrm{CC}} - \bar{H})|\mathbf{h}\rangle^{-1}\langle \mathbf{h}|\bar{H}^k|\Phi_0\rangle \tag{18}$$

Applying the interchange theorem just as we did above, now leads us to

$$E^k_{\mathrm{CC}} = \langle \Phi_0|\bar{H}^k|\Phi_0\rangle + \langle \Phi_0|\bar{H}|\mathbf{h}\rangle \mathbf{R}_0\langle \mathbf{h}|\bar{H}^k|\Phi_0\rangle$$

Finally, choosing to define $\mathbf{\Lambda} = \langle \Phi_0|\bar{H}|\mathbf{h}\rangle \mathbf{R}_0$, we have

$$E^k_{\mathrm{CC}} = \langle \Phi_0|\bar{H}^k|\Phi_0\rangle + \mathbf{\Lambda}\bar{\mathbf{H}}^k \tag{19}$$

Though guided to this form by simply asking what the energy derivative is in CC theory, we can reverse the process (that is use the 'integral Hellman–Feynman theorem') and observe that we can obtain this result by introducing the fundamental CC functional [3,103,104, 109] from the beginning. In operator form,

$$E_{\mathrm{CC}} = \langle \Phi_0 | (1 + \mathbf{\Lambda}) \bar{H} | \Phi_0 \rangle$$

This is an important quantity. Our first development of it followed the above, derivative argument [103,104,109]. This simply means we use first-order perturbation theory with the CC reference as the unperturbed problem to get the first-order energy correction—the gradient. It was also suggested from a different field as a generalization of CC theory [110, 111], but in chemistry we consider energies and their derivatives (properties) to be synonymous as above [111]. Finally, it can also be deduced as a consequence of a Lagrangian multiplier constraint [112].

Regardless, the role of this functional in CC theory is paramount. Varying $\delta\mathbf{\Lambda} = \partial/\partial\lambda_{ij\ldots}^{ab\ldots}$, we obtain

$$Q\bar{H}P = 0 \tag{20}$$

the usual CC equations, while varying w.r.t $\delta T = \partial/\partial t_{ij\ldots}^{ab\ldots}$ we obtain the $\mathbf{\Lambda}$ equations,

$$P\mathbf{\Lambda}Q\bar{H}Q + P\bar{H}Q - E_{\mathrm{CC}}P\mathbf{\Lambda}Q = 0 \tag{21}$$

$$P[\mathbf{\Lambda}, \bar{H}]Q + P\bar{H}Q + P\bar{H}Q\mathbf{\Lambda}Q = 0 \tag{22}$$

The second form uses the commutator to formally eliminate the energy. Before we do that, though, it is apparent that the $\mathbf{\Lambda}$ equations are CI-like, as they are energy dependent and are linear. They also retain disconnected terms in their expressions [11,109], since unlike T, $\mathbf{\Lambda}$ is not composed of solely connected terms.

Because these two stationary conditions are satisfied, the CC functional—unlike the CC energy itself—is stationary. Hence, knowing both T and $\mathbf{\Lambda}$, we can immediately write all energy derivatives simply from a generalized Hellman–Feynman formula [109],

$$E_{\mathrm{CC}}^{k} = \langle \Phi_0 | (1 + \mathbf{\Lambda}) \bar{H}^{k} | \Phi_0 \rangle \tag{23}$$

This also allows us to define the CC 'response' and 'relaxed' generalizations of the usual density matrices [109] making property evaluation in CC theory routine. In particular, we now have for gradients

$$E_{\mathrm{CC}}^{k} = \sum_{p,q} D_{pq}\left(h_{pq}^{k} + \sum_{m} \langle pm || qm \rangle^{k} \right) + \Gamma_{pqrs} \langle pq || rs \rangle^{k} + \sum_{pq} I_{pq} S_{pq}^{k} \tag{24}$$

where we used the 'relaxed' density matrix D_{pq} defined elsewhere [109,113] and discussed below. The derivative integrals are the AO integrals rotated into the MO basis. The I_{pq} is also defined elsewhere [109,113]. For ordinary properties, we have the 'response' density matrices

$$\gamma_{pq} = \langle 0 | (1 + \mathbf{\Lambda}) \exp(-T) p^{\dagger} q \exp(T) | 0 \rangle \tag{25}$$

$$\Gamma_{pqrs} = \langle 0 | (1 + \mathbf{\Lambda}) \exp(-T) p^{\dagger} q^{\dagger} s r \exp(T) | 0 \rangle \tag{26}$$

The difference in γ_{pq} and D_{pq} is the incorporation of orbital relaxation for the reference determinant, so that the results for a dipole moment computed with the 'relaxed' density matrix gives precisely the same results as if it were done by differentiating the CC energy in a finite-field calculation,

$$\vec{\mu}_{\rm el} = \sum_{pq} D_{pq}\langle p|\vec{r}|q\rangle \tag{27}$$

In other words, the HF orbitals are relaxed to accommodate the dipole perturbation, and then the CC calculation of the energy carries this orbital change into the correlation problem. By doing all this analytically, we can incorporate the expressions for the orbital relaxation effect prior to specifying the actual operator, leaving that to be added as h^k_{pq} in Eq. (24) above.

The generalization of the interchange theorem [103] to the correlation problem is what makes CC analytical gradient theory viable, and, indeed, routine today. Also, the introduction of the *response* and the *relaxed* density matrices provides the non-variational CC generalizations of density matrix theory that makes it almost as easy to evaluate a property as with a normal expectation value. They are actually more general, *since they apply to any energy expression whether or not it derives from a wavefunction! This is essential, e.g. for* CCSD(T). The difference is that we require a solution for both T and $\boldsymbol{\Lambda}$ if we want to use untruncated expressions for properties, as is absolutely necessary to define proper critical points. It is certainly true that

$$\vec{\mu}_{\rm el} = \langle 0|\exp(T^\dagger)\vec{r}\exp(T)|0\rangle / \langle 0|\exp(T^\dagger)\exp(T)|0\rangle \tag{28}$$

$$= \langle 0|[\exp(T^\dagger)\vec{r}\exp(T)]_C|0\rangle \tag{29}$$

where the second-form observes that the denominator can be removed in all orders by insisting that the numerator be limited to just connected terms [24,25]. But this form will never terminate. The first form will terminate in the numerator and in the denominator when T exceeds the number of electrons in the system, but not once the denominator is divided into the numerator. Short of that, we have to force some truncation. For CCSD based on HF, this form will be correct through second-order perturbation theory, while CCSDT-1 is correct through third order, for example [114]. Hence, quite reasonable approximations for ordinary (non-gradient) expectation values can be obtained without solving the $\boldsymbol{\Lambda}$ equations.

Such expressions for the energy and associated amplitudes have been considered in the expectation value [88] and unitary CC variants [115]. In fact, UCC is stationary [115], meaning that its solutions do not require use of the interchange theorem for the correlation part of the problem, but would still do so for the MO variation.

Today, analytical gradients for CC methods have been accomplished for CCD [116], and by using the above relaxed density formulation, closed-shell CCSD [117] and CCSDT-1 [118]. Also for CCSD(T) [119,120]. General, open-shell, symmetry-specific CCSD [113], and CCSD(T) have been presented and are widely used in the ACES II program system [121,122]. Now analytical derivatives for the full CCSDT are available [123]. Kállay, Gauss, and Szalay have now added the gradients for higher levels of CC with their automated procedures [124].

Gradients have also been implemented for EOM-CC excited and ionized states [125,126] discussed in Section 41.5, and the spin-flip EOM-CC variant [127].

Gradients provide structures and their numerical second derivatives characterize critical points while providing vibrational frequencies. However, there are still great advantages if the second derivatives can be determined analytically, too. This unique development was accomplished by Gauss and Stanton [128], who implemented analytical Hessians for CC methods from CCSD to CCSD(T) and the full CCSDT. Open-shell (UHF) second derivatives for CC methods were presented by Szalay et al. [129]. Now this has been done automatically for even higher levels of theory with Kállay and Gauss' automated approach [130].

42.5 EXCITED STATES: $\bar{H}R_k = R_k\omega_k$

A natural way to introduce equations for excited states into a quantum chemical approach is to consider stimulating the molecule by a time-varying electric field to which the molecule can respond by excitation, and derive solutions from the time-dependent Schroedinger equation. Analysis then leads to equations for the excitation energies and properties of the excited state eigensolutions like transition moments. In particular, such an approach, after a Fourier transformation from time to frequency, will yield the dynamic polarizability whose spectral expansion is

$$\alpha(\omega) = \sum_k \{|\langle\Psi_0|\mathbf{r}|\Psi_k\rangle|^2[1/(\omega + E_0 - E_k) + 1/(E_0 - E_k - \omega)]\}$$

When $\omega = E_k - E_0 = \omega_k$, we obtain the excitation energy and the residue of the dynamic polarizability at the pole is the dipole strength. Hence, by using the coupled-cluster approach to obtain a solution of the time-dependent Schroedinger equation, one can immediately obtain equations that will yield excitation energies and associated properties. When this is done for Hartree–Fock theory, we make the transition to time-dependent HF(TDHF), and we derive the TDHF, or random-phase approximation (RPA) for excitation energies and associated properties. We can do exactly the same thing by insisting upon a CC solution of the time-dependent Schroedinger equation. This important step was taken by Monkhorst [131] to define a coupled-cluster linear response (CCLR) approach. Other pertinent papers from this time include those of Mukherjee et al. [132] and those by Nakatsuji and Hirao [50,133]. The latter were developing hybrid CI and CC methods for excited states (SAC-CI), that are similar to CCLR and EOM-CC, but they are not equivalent.

In the same volume of the Sanibel proceedings as the Monkhorst paper was an attempt at a different, time-independent approach to excited states in CC theory by Harris [134], that used the equation-of-motion (EOM) formalism of D.J. Rowe [135] (see also [136, 137]). Harris' approach did not lead to viable equations because he proposed to write an excited state as an exponential, $\Psi_k = \exp(\hat{S})\exp(\hat{T})|0\rangle$, where like $\hat{T}$, $\hat{S}$ was an excitation operator, which just redefines $\hat{T}$. (For some special cases such an operator can be used [138], to obtain excited states, but see also the general solutions of the non-linear equations of CC theory [139].) However, the EOM development simply uses a linear operator instead of $\exp(\hat{S})$ [140–144], and it provides exactly the same eigenvalues (not all properties) as the time-dependent linear response CC approach. It also ties directly to the gradient method, to the CC functional, and provides an alternative route to higher order properties, as discussed below. That is, the resolvent in Eq. (18) corresponds to the matrix for EOM-CC eigenstates. It also offers an alternative, and quite appealing, mathematical structure; that extends the time-dependent approach and suggests different, highly useful, approximations that differ from CCLR. Hence, EOM-CC offers an independent, unified treatment of a wealth of different phenomena.

The EOM philosophy is very simple, remarkedly flexible, and transparent in a few lines. Simply consider two stationary state Schroedinger solutions,

$$H\Psi_k = E_k\Psi_k \tag{30}$$

$$H\Psi_0 = E_0\Psi_0 \tag{31}$$

and make the ansatz

$$\Psi_k = \Omega_k\Psi_0 \tag{32}$$

where Ω_k is an operator that creates the kth state from the reference (not necessarily ground) state, Ψ_0. Then, after multiplication from the left by Ω_k; we can subtract the reference state solution from that for the kth state, to obtain

$$[H, \Omega_k]\Psi_0 = \omega_k\Omega_k\Psi_0 \tag{33}$$

$\omega_k = E_k - E_0$. That is called the EOM from its analogy with the Heisenberg equation of motion, that relates a commutator to a time-derivative. We immediately gain the benefit that the excitation energy itself is the eigenvalue, rather than having to compute the difference between two large numbers. Conceptually, the EOM approach attempts to cancel the common elements in the two wavefunctions, focusing on the essential differences. Finally, from the viewpoint of the second-quantized operators involved in the commutator, the resultant operator is of lower rank, making it easier to evaluate it accurately than the two operators separately [145].

Hence, we can introduce a $|\Psi_0\rangle = \exp(T)|0\rangle$ and a suitable form for Ω_k and we have a set of equations for a CC approach to excited states. The primary constraint on Ω_k is that it be able to provide the exact result in the limit. Hence, if we choose a CI-like operator,

$$\Omega_k = r_0 + \sum_{a,i} r_i^a a^\dagger i + \sum_{a<b,i<j} r_{ij}^{ab} a^\dagger i b^\dagger j + \cdots \tag{34}$$

which means a constant plus all single, double, triple, etc., excitations, with coefficients to be determined. Then, obviously, this operator will generate the full CI solution for the kth state, even when working on a single determinant reference. For this choice of Ω_k, the distinction between working on $\exp(T)|0\rangle$ and just $|0\rangle$, is unimportant because $[T, \Omega_k] = 0$. Such a linear operator is the most convenient choice to make in EOM-CC (though not the only one, as a combination of excitation and de-excitation operators could be used, for example). It ties directly to our property approach discussed above, and assumes a very convenient mathematical form. In particular, writing the commutator as a connected expression, EOM-CC becomes

$$[\exp(-T)H\exp(T)\hat{R}_k]_C|0\rangle = \omega_k\hat{R}_k|0\rangle \tag{35}$$

$$\bar{\mathbf{H}}\mathbf{R}_k = \mathbf{R}_k\omega_k \tag{36}$$

where the eigenvector is written as $\mathbf{R}_k$ (not to be confused with the resolvent) to emphasize that it is a right-hand eigenvector to a non-Hermitian matrix, $\bar{\mathbf{H}} = \langle \mathbf{h}\exp(-T)H\exp(T)|\mathbf{h}\rangle$. $\mathbf{h}$ represents the selection of excitations, like single and double excitations in k. In other words, this is a CI problem for excited states built upon a CC ground state. Because of its non-Hermitian character, $\bar{\mathbf{H}}$ also has a left-hand eigenvector, $\mathbf{L}_k\bar{\mathbf{H}} = \mathbf{L}_k\omega_k$, with the same eigenvalue, but $\bar{\mathbf{H}}$ and $\mathbf{L}$ do not have to be connected. We choose the normalization that $\langle 0|\hat{L}_k\hat{R}_l|0\rangle = \delta_{kl}$ [143]. In fact, the very existence of the *stationary-state eigenvectors* is what distinguishes the EOM-CC approach from CCLR, which like most time-dependent approaches does not have eigenvectors *per se*. For energies either eigenvector will suffice, but for properties you need both. EOM-CC provides the excited state one-matrix as $\gamma_{pq}^k = \langle 0|\hat{L}_k\exp(-T)p^\dagger q \times \exp(T)\hat{R}_k|0\rangle$.

The beautiful connection between the ground state problem and the excited state one becomes apparent when we realize that in the CC functional, $L_0 = 1 + \Lambda$ and $R_0 = 1$, as the first pair of eigenfunctions to $\bar{H}$. Then, just as above, we have the elements of the ground state density matrix as $\langle 0|\hat{L}_0\exp(-T)p^\dagger q\exp(T)|0\rangle$. We also have the resolution of the identity, $1 = \sum_k |R_k\rangle\langle L_k|$. As long as Ω_k is a linear operator, the equations are well defined. It would be nice if we could use an exponential operator to vest excited states with the same flexibility as CC does for ground states, but doing it in a straightforward way like $\Omega_k = \exp(\hat{S})$ leads to a non-determinable set of equations. More general exponential ansatz are possible from the Fock space multi-reference CC method of Mukeherjee et al. [146], that are closely related to EOM-CC [147].

Logically, we use the same set of single and double excitations in $\mathbf{h}$ and T, and that defines EOM-CCSD. However, from the EOM viewpoint, that is not really necessary as we have two potentially different approximations. This issue has been studied numerically [148]. From the time-dependent, linear response viewpoint [112,131], the ground state and the excited states which pertain to its response are tied together *via* the differentiation of the ground state, which requires that the choice of excitations be the same for both; that is the standard choice.

A few years ago we extended EOM-CCSD to the full EOM-CCSDT method and made some fairly large basis set (~90 function) calculations, based upon the full triple excitation. CCSDT ground state, and the inclusion of all triple excitations in $\mathbf{h}$ [149]. At the same time others [150] reported a study of H_8. These authors also looked at simple potential energy curves with EOM-CCSDT and its active orbital modification, EOM-CCSDT [151,152]. Now, by virtue of their automated procedures, [153] EOM-CC can be taken to any level. Hirata has similarly done EOM-CCSDTQ [154], and Kállay EOM-CCSDTQP.

For the very flexible EOM structure, we can equally well consider the operator, Ω_k, to correspond to a change in particle number. That is, instead of just exciting an electron by promotion of an occupied orbital into an unoccupied one, we can equally well take the electron all the way out by assuming the excitation into orbital a is actually into the continuum. Since the continuum orbital will have no overlap with any of the square integrable bound state orbitals left, the net result is that the operator $a^\dagger i \rightarrow i^\dagger$, will annihilate an electron from the orbital ϕ_i, so we are describing an ionized state, i.e. IP-EOM-CC. In this case our **h** consists of all determinants that correspond to one less $(n - 1)$ electrons. Similarly, we can take an electron from the continuum and add it to the molecule, causing $a^\dagger i \rightarrow i^\dagger$ which corresponds to a state that has an extra electron, i.e. an electron-attached state, i.e. EA-EOM-CC. **h** consists of all determinants that correspond to $n + 1$ electrons. Obviously, by construction we have a seamless transition in EOM-CC from a Rydberg series of electronic excitations to the ionized states that occur at the continuum. We also have as other solutions of the IP-EOM-CC problem the shake-up states, where ionization is accompanied by excitation. Obvious generalizations to double ionization and double electron attachment have been made [155]. In the interest of efficient numerical implementation, it is important that for proper, fully factorized EOM-CC applications, that we write separate programs for the IP and EA sectors [156,157] rather than simply taking the electronic-excitation (EE-EOM-CC) program and forcing the continuum argument above in its implementation, although that will provide the right answers [158].

EOM-CC makes many multi-reference problems accessible, *since it is multi-reference for the target state*. Since we are diagonalizing a matrix, all determinants in the eigenstate have, whatever weight they require, instead of having to grow to the appropriate value through some basically perturbative means, as in the single reference CC theory. For example, an open-shell singlet excited state is no problem for EOM-CC, since its eigenstate will contain the two equally weighted determinants automatically, just as would CI. However, if the molecule had an open-shell singlet ground state, the EOM-CC recourse would be to have another excited state that could be obtained from single-reference CC theory, and then use EOM-CC to de-excite to the ground state. Alternatively, to treat an open-shell singlet with a single-determinant reference it would be difficult to get the determinants to be properly weighted, when one determinant would be the reference and the other, part of the orthogonal complement. Its coefficient would very slowly grow to that value depending upon the procedure for solving the non-linear CC equations. Most general procedures are based upon some accelerated Jacobi iterations which are like perturbation theory, but for a case like this the perturbation is not small, causing convergence difficulties; and recommending that a two-determinant (multi-reference) CC approach be used instead [159–161].

Regarding symmetry, if the reference CC state is for a closed shell, all EOM states derived from it are spin-eigenstates. If the reference CC state is for a high-spin situation, then we might use either a UHF, ROHF, or QRHF reference, and the resultant CC solution would be very close to a spin eigenfunction, though it is not one. In a clever approach, Szalay and Gauss obtain spin eigenfunctions for the EOM target state with such an open-shell reference [162] by simultaneously considering the CC equations for $\hat{S}^2$.

Krylov and her co-workers have taken the de-excitation EOM to its logical conclusion in what they call their spin-ip (SF-EOM-CC) method. EOM-CC readily lends itself to treating high-spin open shells with ROHF, UHF, or other references, as illustrated for O_2 [142]. Hence, it is straightforward to obtain a high-spin single reference solution and then excite and de-excite to lower spin and higher spin states as required [163]. De-exciting to lower spin states is particularly attractive for studies of complicated radicals. Their method does not yield spin-eigenfunctions, but they would be close.

A large number of numerical results now exist for EOM-CC. A few general comments can be made without trying to be too quantitative. It is apparent that EOM-CCSD does very well for excited states dominated by single excitations, being particularly accurate for Rydberg excitations if the basis set is adequate. Its errors are slightly worse for valence-excited states, and can be entirely wrong for states dominated by double excitations. The obvious explanation is that double excitations provide most of the essential correlation effects for states dominated by singles, so EOM-CCSD is good for such states; but when a state is dominated by doubles, we need at least triple excitations. Then in finer detail, we also see that various approximations to triples like EOM-CCSDT-1, EOM-CCSDT-3 [164], CC3 [165], etc., that improve upon the results for singly excited states, still fail compared to the full CCSDT for dominate doubly excited ones. Now, we even know what happens for some EOM-CCSDTQ results [148,154].

As one would expect, if we describe ionization potentials with IP-EOM-CCSD, we will have similar behavior. We should do quite well for most principal ionizations where the eigenstate is dominated by single excitations, meaning linear combinations of $i^\dagger|0\rangle$ determinants, with $i^\dagger a^\dagger j|0\rangle$ playing most of the correlation and relaxation role; but when the latter 'shake-up' is dominant, then we would logically need $i^\dagger a^\dagger j b^\dagger k|\rangle 0$ in our space to do as well for the shake-up eigenvalues. The latter requires IP-EOM-CCSDT [156], and the complementary EA-EOM-CCSDT method [157] and higher [166]

Just as we required analytical gradients for ground state CC results, one can obtain analytical gradients for EOM-CC excited states, too. This important development was made by Stanton and Gauss [125], offering the first analytical gradient capability for excited states beyond MCSCF and some CIs. With this tool, and its extensions to second-derivatives [167], realistic non-adiabatic spectral simulations have been obtained [168], making it possible to actually argue spectroscopy with the spectroscopists; rather than focus only on idealized vertical excitation energies or vertical, principal ionization potentials. This represents the culmination of much work based on CC theory directed toward spectroscopy; and is a major accomplishment for the field [169,170].

For second-order properties derived from the perturbation, $O^{(1)}$, we have a particularly attractive method that uses the EOM eigenvectors to represent the perturbed wavefunctions. Since the right- and left-hand eigenvectors form a complete set, we know we can write the perturbed wavefunctions in the form,

$$|\Psi^1\rangle = \sum_{k=1} \hat{R}_k|0\rangle c_k^{(1)} \tag{37}$$

$$\langle\Psi^1| = \sum_{k=1} \langle 0|\hat{L}_k d_k^{(1)} \tag{38}$$

This makes the second-order energy,

$$E^2 = \langle 0|(1+\mathbf{\Lambda})(\exp(-T)O^1\exp(T) - E^1)|\Psi^1\rangle \tag{39}$$

$$= \langle 0|L_0(\bar{O}^1 - E^1)R_0(\bar{O}^1 - E^1)|0\rangle \tag{40}$$

$$= \sum_{k=1} \langle 0|L_0(\bar{O}^1 - E^1)|R_k\rangle\langle L_k|(\bar{O}^1 - E^1)|0\rangle/\omega_k \tag{41}$$

$$= \langle 0|(1+\mathbf{\Lambda})(\bar{O}^1 - E^1)|\mathbf{h}\rangle\mathbf{C}^1 \tag{42}$$

The critical element is that the resolvent operator, $R_0 = |\mathbf{R}\rangle\langle\mathbf{L}|E_{CC} - \bar{H}|\mathbf{R}\rangle^{-1}\langle\mathbf{L}| \Rightarrow |\mathbf{h}\rangle\langle\mathbf{h}|E_{CC} - \bar{H}|\mathbf{h}\rangle^{-1}\langle\mathbf{h}|$, showing that it arises from the EOM-CC $\bar{\mathbf{H}}$ eigenfunctions in the configuration space $|\mathbf{h}\rangle$. Furthermore, we solve the linear equation $C^1 = (\mathbf{E}_0 1 - \bar{\mathbf{H}})\bar{\mathbf{O}}^1$ to avoid matrix inversion [144].

Perera et al. have made extensive use of this expression in calculations of NMR coupling constants [171,172], as well as for static and dynamic polarizabilities and with Rozycko, hyperpolarizabilities [173,174]. In the treatment of higher order properties, distinctions between EOM-CC and CCLR appear. These have been discussed [174,175], where the point is made that we do not have *one* completely satisfactory treatment of higher order properties, but two alternative approximations with different characteristics. The EOM-CC approach which uses the (formally) complete set of eigenfunctions for expanding all perturbed wavefunctions as above, leads to the usual sum-over-state expressions, though evaluated as in closed form above, has proven itself in many such studies [171,173]. (The differences with CCLR derive from the degree of completeness which is only guaranteed in the untruncated CC limit.)

The use of EOM-CC as a 'target state' method, rather than one for spectroscopy, as in the gradient work above, also leads us to alternative approaches for 'multi-reference' problems as in the open-shell singlet case, but for more difficult examples. One illustration is the treatment of the ground state of ozone, with its well-known multi-configuration structure. In essence, we would like to treat ozone with a reference space of four determinants, a^2, b^2, $a\bar{b}$ and $b\bar{a}$.

The last two only contribute for unsy mmetric geometries. A straightforward multi-reference application would include them all. In single reference CC, we would only have the option to use one as the reference, with all others being part of the orthogonal complement, making it necessary for their coefficients to grow into appropriate values through the solution of the CC equations; and sometimes this is difficult because higher connected cluster operators than those included might be essential. In double ionization potential (DIP-EOM-CC), we can choose to solve for the CC solution for the hypothetical O_3^{-2} anion, with its reference determinant a^2b^2 and then use the operators $\{i^\dagger j^\dagger, i^\dagger j^\dagger b^\dagger k, \ldots\}$ to produce doubly ionized states, a^0b^2, a^2b^0, $a^1\bar{b}^1$, $b^1\bar{a}^1$. In this way, we treat the quasi-degenerate a and b orbitals equivalently in the target state, as their configurations can assume any value for their respective coefficients, yet we retain, operationally, the convenient single reference structure of CC calculations with its unambiguous ease of application. Lacking an easily applied multi-reference CC theory, this is often a preferable route to the solution to a problem. Its success depends, also, on the insensitivity of CCSD and beyond to orbital rotations. Here, we are using O_3^{-2}

orbitals, which you might expect to be far away from those for ozone, itself. If our method has the invariance of full CI, that is not an issue, but in its truncated form, it will not necessarily achieve the same degree of orbital insensitivity as for ground state CC, primarily because the target states are from CI, rather than exponential type [176]. This can affect the accuracy of some approximations [177].

I alluded to the fact that it would be nice to have an exponential ansatz for the target states in EOM-CC, rather than a linear one. This can be done by introducing the Fock space multi-reference method originally developed by Lindgren [178] and Mukeherjee [146]. This necessitates the introduction of a valence universal wave-operator that requires a hierarchical solution using the ground state (0,0) solution, then the ionized sector (0,1) and electron attached, (1,0) ones, and then the excited states from the (1,1) sector, etc. For the ionized and electron-attached states, there is no difference for principal ionization potentials and their electron affinities than in the corresponding (IP and EA) EOM-CC methods, since the exponential operator in that case degenerates to a linear one. However, for excited states there are a few differences in the two approximations. Comparisons of FS-CCSDT and EOM-CCSDT are presented elsewhere [179]. Only the Fock space approach results in a fully size-extensive, linked diagram structure, as CCLR or EOM-CC depend upon a CI approximation for the excited state, and that cannot be fully linked. However, it will have the size-intensive [165] property for excited states of AB that can be decomposed to those for $A + B$. But only the Fock space and STEOM-CC method, discussed below, has the property of going smoothly to $A^+ + B^-$ [180–182].

The similarity transformed EOM, STEOM-CC [183], approaches the problem somewhat differently, but it also provides an exponential ansatz for excited states, namely $\Psi_k = \exp(\hat{S})\exp(\hat{T})|0\rangle$, where $\exp(\hat{S})$ has a different meaning than before. The method decouples the contributions of higher cluster operators from the lower ones, by using the results for the (1,0) and (0,1) results to define the second similarity transformation, $\hat{S}$, leaving the excited states to be obtained now from a problem of the dimension of a CI singles (CIS) calculation. This method is a kind of exact CIS for the excited states of molecules, at least those dominated by single excitations. It is very attractive for large-scale application as in our work for free-base porphine [184,185]. Extensions by Nooijen and Lotrich have been made for doubly excited states [186].

Though the numerical results will not usually be dramatically different between SAC-CI and EOM-CC, when the former's approximations are made very carefully—as they share similar low-order perturbation approximations—the methods differ dramatically in formal ways and in their computational implementations [187].

42.6 DEVELOPMENTS FOR LARGE MOLECULES AND POLYMERS

As discussed above, the wide applicability and predictive nature of CC methods have established a plateau in the field and led to automated program generation for the standard sequence of approximations of CC and EOM-CC methods. This alleviates any need for further development of the basic CC functional structure, at least for molecules treated in a conventional way. However, one profitable route for further development is to focus on the *other* elements in CC calculations; like localization techniques and removing basis set redundancy to attack the highly non-linear scaling ($>\sim n^6$) of the standard CC methods.

The work of Schütz and Werner [189,190] in the first category has had some notable successes. These authors have demonstrated linear scaling at the CC level, while reporting largely CCSD and CCSD(T) results with >1000 basis functions. The basic structure derives from that of Saebo and Pulay [191], who identify beforehand regions of a large molecule that suit a localized description, and then enforce this by working in a projected AO basis set instead of the usual virtual MO space, with localized MOs for the occupied orbitals. However, much of the benefit depends upon using MBPT(2) for weakly interacting electron pairs, only using CC theory for the relatively small remainder. Also the need for a global reference calculation, often the rate-determining step since MBPT(2) is now so fast in their approach, forces the use of fairly small basis sets. This approach is typical of 'threshold'-based methods where pre-analysis is essential, but with care, can be done successfully and even provide reasonable analytical forces despite the innate discontinuities of the PES due to the thresholds, which is often a failing of such methods.

The alternative of exclusively using the AO basis operationally in CC calculations, as the most local basis possible (avoiding any kind of local MO description for the occupied orbitals), has also been considered [192,193], but has not yet been shown to pay dividends. In fact, unpublished work on polymers [194,195] suggests that AO methods actually do not compare very well with the standard delocal, Bloch periodic symmetry treatment. By the way CCSD has now been applied with periodic symmetry to simple polymers at the CCSD level [196,197].

Our contribution to the linear scaling effort [51,206] has focused on what we call the natural linear scaling CC (NLSCC) method which goes back to local bond–orbital ideas of Malrieu et al. [199,200], and was considered in an *ab initio* CC long ago [201,202]. It also has similarities to Yang's Divide and Conquer approach [203,204]. NLSCC differs from the above in several ways: (1) It never requires a global solution for the large molecule as do Shuetz and Werner, e.g. but only that for the components constituting the large molecule. (2) It exclusively works in a localized, orthogonal MO basis, using the natural localized MOs of Weinhold [205]. This means the rate determining steps are $\sim n^2N^4$ for CCSD, just like normal CCSD, while if we use the AO basis completely, we have to contend with $\sim (n+N)^6$, or for just the virtual space, $\sim n^2(n+N)^4$. The greater sparseness of the AO basis has to offset the poorer scaling to offer competitive results. (3) The integrals and the CC amplitudes tend to fall off as r^{-3} using NLMOs, which is sufficient to fragment the large molecule into overlapping, local units for application. These are molecular units whose bonds are terminated by hydrogens, plus a boundary neighborhood. (4) There is exceptional transferability of CC amplitudes so that a C–C bond looks pretty much the same in terms of its associated correlated amplitudes regardless of what it is bonded to. In other words, we see much of the well-known transferability in chemistry in NLSCC.

In an earlier version of this approach, we exploited the transferability to provide results for the correlation energy and density matrix of diamond, without any treatment for the extended system [198]. Unlike most of the other attempts at linear scaling, which tend to only really be realized in small basis sets, the rate determining step in the CC calculation is for the largest 'molecular' unit, so one can use as large a basis for each such unit as molecular CC approaches will allow. (5) The method can describe bond-breaking and delocalization as long as they are contained within one of the computational units, and if the long-range forces are still small enough by virtue of the r^{-3} fall off. We would not do graphite! (6) There is no need to use periodic boundary conditions making amorphous materials, and 'real' as opposed to idealized polymers, accessible. A recent paper compares the NLSCC approach to that of Schuetz and Werner for their polyglycine example [206]. There is still work to be done on forces, dispersion, and long-range effects to develop a truly adequate approach.

The other development that, I believe, will pay considerable dividends in CC theory is to pay more attention to removing the linear dependency in our quantum chemical descriptions. We all know that the underlying Gaussian bases we use in quantum chemistry offer a great approximation for a lot of things, but they also carry along much excess baggage. Linear dependency at the HF-SCF level is well known, but it also exists in the two-electron integral matrix [207], and in turn, in the CC amplitudes. One manifestation is the observed fact that simply replacing the HF-SCF virtual space by a better chosen set related to it by a rectangular transformation that reduces its dimension, can be done at little cost in the final, correlated result, if the chosen rectangular transformation is a good one! Many such attempts have been made over the years, primarily for CI methods, where natural-orbital iterations have been extensively used [208]. We proposed an optimized virtual orbital space (OVOS) method once upon a time [209], using the MBPT(2) Hylleraas functional to provide optimum second-order energy for a given transformation, and it worked very well. Many other attempts by others are reviewed in that paper. But even simpler than OVOS is the well-known use of frozen natural orbitals in this context [210], which provided the initial guess for OVOS, and was almost as good. If we can reduce the dimension of the virtual space by 50%, we gain at least $\sim 2^4$ in a CC calculation. We are doing this currently with great success for energies and analytical gradients [211]. This gain bypasses any need for a localized treatment of the underlying problem, but combining the two might offer further advantages.

More sophisticated approaches that have the same primary advantage of offering a seamless treatment of large molecules in large basis sets solely based upon doing reasonable mathematics, and without exploiting an *a priori* localized structure, are built either upon the singular value decomposition approach [212,213] or the Cholesky decomposition [207,214]. The former applies to any matrix, while the latter requires a positive-definite one. In the first case, we can replace the matrix by an expansion in vectors weighted by their singular values, which measures their importance. Then we can make a contraction of the usual MO-indexed amplitudes, $t_{ij\ldots}^{ab\ldots}$ by a contracted set, t_X^Y, as determined, in principle, by their singular values. (In practice, we have to obtain the weight factors from some simpler, related problem like MBPT(2).) Now the effective dimension of the CC problem is greatly reduced, again dramatically diminishing the high scaling of the unmodified calculation. We call this *compressed coupled-cluster*. Impressive results have also been obtained by the Cholesky route [214]. Canonizing these techniques into widely used programs, including those that are automatically generated, is an important step toward the future, wide-ranging applications of these highly accurate CC methods.

42.7 ACKNOWLEDGEMENTS

Our work has been accomplished through the efforts of *many* excellent co-workers, many of whose names are in the lists of references as my co-authors, and also for the significant contributions that they have since made. Through their efforts, CC theory became far more than we likely would have envisioned. I also thank my senior colleagues Jiri Ĉíẑek, Joe Paldus, and Isaiah Shavitt, who have been highly supportive of our work from the beginning. I especially appreciate Shi Shavitt's exceptional efforts on behalf of our book, which will be available in 2006, and will provide the detailed analysis of the systematic treatment of the CC equations and diagrams that we have always used, albeit sometimes to the chagrin of our readers, as there was no place to read about all the details (Ref. [67] offers a nice summary). Now there will be. We have benefited through the years from exceptional support which we still enjoy, from the AFOSR (FA9550-04-1-0119), who had the foresight to back the development of CC theory into the method it has become; and also ARO (MURI, No. AA-5-72732-B1), that supports some of our current efforts. I also want to thank Dr. Ajith Perera, my long-term collaborator who helped me to get this chapter ready for publication, not to mention his many other contributions; and Dr. Monika Musial, for her help with the text and her expert handling of high-level CC theory which leads to much of our recent work. This chapter also benefited from written comments of many of my prior co-workers which found their way into the text. Finally, please let me acknowledge that much of our work paralleled that of the late John Pople. Though I never wrote a paper with John, nor had the pleasure of working directly with him; I greatly valued our many interactions over the years, which were invariably stimulating. We will all miss him.

42.8 REFERENCES

1 R.J. Bartlett, Int. J. Mol. Sci., 3 (2003) 579.
2 R.J. Bartlett, J. Phys. Chem., 89 (1989) 93, Feature Article.
3 R.J. Bartlett, in: D.R. Yarkony (Ed.), Modern electronic structure theory, Vol. 2, World Scientific, Singapore, 1995, pp. 1047–1131.
4 J. Paldus, Methods in computational molecular physics, NATO ASI, 1991.
5 M. Urban, I. Cernusak, V. Kello and J. Noga, in: S. Wilson (Ed.), Methods in computational chemistry, Vol. I, Plenum Press, New York, 1987, p. 117.
6 Special Issue on Benchmark Results in Quantum Chemistry, J. Mol. Struct. (THEOCHEM) 400, (1997), Editor, C. Dykstra, This issue offers a wide selection of what quantum chemists consider benchmark results to be, and the use of CC theory dominates.
7 T. Helgaker, P. Jørgensen and J. Olsen, Molecular electronic-structure theory, Wiley, New York, 2000, pp. 817–883.
8 R.J. Bartlett and J.F. Stanton, in: K.B. Lipkowitz and D.B. Boyd (Eds.), Rev. Comput. Chem., 5, VCH, 1994, pp. 65–169.
9 K.A. Brueckner, Phys. Rev., 97 (1955) 1353.
10 J. Goldstone, Proc. R. Soc. London A, 239 (1957) 267.
11 I. Shavitt and R.J. Bartlett, Many-body methods in quantum chemistry: Many-body perturbation theory and coupled-cluster theory, Cambridge Press, Cambridge, MA, 2006, To be published.
12 A. Beste and R.J. Bartlett, J. Chem. Phys., 120 (2004) 8395.
13 R.J. Bartlett and G.D. Purvis, III, Int. J. Quantum Chem., 14 (1978) 561.

14 H.P. Kelly, Adv. Chem. Phys., 14 (1969) 129.
15 R.J. Bartlett and D.M. Silver, J. Chem. Phys., 84 (1976) 4578.
16 J. Hubbard, Proc. R. Soc. London A, 240 (1957) 539.
17 P.O. Löwdin, Int. J. Quantum Chem., 2 (1968) 867.
18 F. Coester, Nucl. Phys., 1 (1958) 421.
19 F. Coester and H. Kümmel, Nucl. Phys., 17 (1960) 477.
20 O. Sinanoglu, J. Chem. Phys., 76 (1962) 706.
21 O. Sinanoglu, Adv. Chem. Phys., 5 (1964) 35.
22 R.K. Nesbet, Adv. Chem. Phys., 14 (1969) 1.
23 R.K. Nesbet, Proc. R. Soc. (London) A, 230 (1955) 312.
24 J. Ĉíẑek, J. Chem. Phys., 45 (1966) 4256.
25 J. Ĉíẑek, Adv. Chem. Phys., 14 (1969) 35.
26 R.J. Bartlett and G.D. Purvis, III, Phys. Scr., 21 (1980) 225.
27 W. Meyer, Int. J. Quantum Chem. Symp., (1971) 55.
28 W. Meyer, J. Chem. Phys., 58 (1973) 10.
29 J.-L. Heully and J.-P. Malrieu, Chem. Phys. Lett., 199 (1992) 545.
30 P. Botschwina, in: J.P. Maier (Ed.), Ion and cluster ion spectroscopy and structure, Elsevier, Amsterdam, 1989.
31 P.G. Szalay, Towards state-specific formulation of multireference coupled-cluster theory: Coupled electron pair approximations (CEPA) leading to multireference configuration interaction (MR-CI) type equations, in: R.J. Bartlett (Ed.), Modern ideas in coupled-cluster methods, World Scientific, Singapore, 1997, pp. 81–123.
32 A.C. Hurley, Electron correlation in small molecules, Academic Press, New York, 1976.
33 J. Paldus and L. Sroubkova, Int. J. Quantum Chem., 3 (1969) 149.
34 R.J. Bartlett, Theor. Chim. Acta, 103 (2000) 273.
35 J. Ĉíẑek and J. Paldus, Int. J. Quantum Chem., 5 (1971) 359.
36 J. Paldus, J. Ĉíẑek and I. Shavitt, Phys. Rev. A, 5 (1972) 50.
37 R.J. Bartlett and D.M. Silver, Phys. Rev. A, 10 (1974) 1927.
38 R.J. Bartlett and D.M. Silver, Chem. Phys. Lett., 29 (1974) 199.
39 R.J. Bartlett and D.M. Silver, Int. J. Quantum Chem. Symp., 8 (1974) 271.
40 D.M. Silver and R.J. Bartlett, Phys. Rev., A13 (1976) 1.
41 R.J. Bartlett and D.M. Silver, J. Chem. Phys., 62 (1975) 3258.
42 R.J. Bartlett, D.M. Silver, in: J.-L. Calais, O. Goscinski, J. Linderberg and Y. Öhrn (Eds.), Quantum science, Plenum Press, New York, 1976.
43 R.J. Bartlett and I. Shavitt, Chem. Phys. Lett., 50 (1977) 190, Erratum: Chem. Phys. Lett., 57 (1978) 157.
44 J.S. Binkley and J.A. Pople, Int. J. Quantum Chem., 9 (1975) 229.
45 J.A. Pople, J.S. Binkley and R. Seeger, Int. J. Quantum Chem. Symp., 10 (1976) 1.
46 J.A. Pople, R. Krishnan, H.B. Schlegel and J.S. Binkley, Int. J. Quantum Chem., 14 (1978) 545.
47 H. Larsen, A. Halkier, J. Olsen and P. Jørgensen, J. Chem. Phys., 112 (2000) 1107.
48 S. Hirata, M. Nooijen, I. Grabowski and R.J. Bartlett, J. Chem. Phys., 114 (2001) 3919.
49 D.J. Thouless, The quantum mechanics of many body systems, Academic Press, New York, 1961, p. 121.
50 H. Nakatsuji and K. Hirao, J. Chem. Phys., 68 (1978) 14279.
51 N. Flocke and R.J. Bartlett, Chem. Phys. Lett., 367 (2003) 80.
52 H.F. Schaefer, III, Private communication. At the St Simons meeting, April 23–25, 2004; Fritz Schaefer tells the story that he and John Pople were in Australia discussing the prospects for CCSD, when John remarked that he thought it would be almost impossible to do, when our paper appeared.
53 G.D. Purvis, III and R.J. Bartlett, J. Chem. Phys., 76 (1982) 1910.
54 S.A. Kucharski and R.J. Bartlett, J. Chem. Phys., 97 (1992) 4282.
55 R.A. Chiles and C.E. Dykstra, J. Chem. Phys., 74 (1981) 4544.
56 J. Noga, V. Kello and M. Urban, COMENIUS/2, Technical Report, 1985 Comenius, University, Bratislava. Numerical results first appeared in M. Urban, S.J. Cole, and R.J. Bartlett, J. Chem. Phys., 83 (1985) 4041.
57 G.E. Scuseria, A.C. Sheiner, T.J. Lee, J.E. Rice and H.F. Schaefer, III, J. Chem. Phys., 86 (1987) 2881.

58 T.J. Lee and J.E. Rice, Chem. Phys. Lett., 150 (1988) 406.
59 R. Krishnan, M.J. Frisch and J.A. Pople, J. Chem. Phys., 72 (1980) 4244.
60 M.J. Frisch, R. Krishnan and J.A. Pople, Chem. Phys. Lett., 75 (1980) 66.
61 R.J. Bartlett, H. Sekino and G.D. Purvis, III, Chem. Phys. Lett., 98 (1983) 66.
62 M.F. Guest and S. Wilson, Chem. Phys. Lett., 72 (1980) 49.
63 V. Kvasnicka, V. Laurinc and S. Biskupic, Chem. Phys. Lett., 67 (1979) 81.
64 M. Urban, J. Noga and V. Kello, Theor. Chim. Acta., 62 (1983) 549.
65 Y.S. Lee, S.A. Kucharski and R.J. Bartlett, J. Chem. Phys., 81 (1984) 5906.
66 M. Urban, J. Noga, S.J. Cole and R.J. Bartlett, J. Chem. Phys., 83 (1985) 4041.
67 S.A. Kucharski and R.J. Bartlett, Adv. Quantum Chem., 18 (1986) 281.
68 R.J. Bartlett, J.D. Watts, S.A. Kucharski and J. Noga, Chem. Phys. Lett., 165 (1990) 513.
69 J.A. Pople, M. Head-Gordon and K. Raghavachari, Chem. Phys. Lett., 87 (1987) 5968.
70 K. Ragavachari, G.W. Trucks, J.A. Pople and M. Head-Gordon, Chem. Phys. Lett., 157 (1989) 479.
71 J. Paldus, J. Ĉíẑek and B. Jeziorski, J. Chem. Phys., 90 (1989) 4356.
72 G.E. Scuseria and H.F. Schaefer, III, J. Chem. Phys., 90 (1989) 3700.
73 J.F. Stanton, W.N. Lipscomb, D.H. Magers and R.J. Bartlett, J. Chem. Phys., 90 (1989) 1077.
74 T.H. Dunning, Jr., J. Phys. Chem. A, 104 (2000) 9062.
75 J.D. Watts, J. Gauss and R.J. Bartlett, J. Chem. Phys., 98 (1993) 8718.
76 J.F. Stanton, J. Gauss, J.D. Watts, M. Nooijen, N. Oliphant, S.A. Perera, P.G. Szalay, W.J. Lauderdale, S.A. Kucharski, S.R. Gwaltney, S. Beck, A. Balkova, D.E. Bernholdt, K.K. Baeck, P. Różyczko, H. Sekino, C. Hober, J. Pittner and R.J. Bartlett. ACES II program is a product of the Quantum Theory Project, University of Florida, Integral packages included are VMOL (J. Almlöf, P. Taylor); VPROPS (P.R. Taylor); A modified version of ABACUS integral derivative package (T.U. Helgaker, H.J.Aa. Jensen, J. Olsen, P. Jørgensen, P.R. Taylor).
77 S.A. Kucharski and R.J. Bartlett, J. Chem. Phys., 108 (1998) 5243.
78 T.D. Crawford and J.F. Stanton, Int. J. Quantum Chem., 601 (1998).
79 J. Noga and R.J. Bartlett, J. Chem. Phys., 86 (1987) 7041, Erratum: J. Chem. Phys., 89 (1988) 3401.
80 G.E. Scuseria and H.F. Schaefer, III, Chem. Phys. Lett., 152 (1988) 382.
81 J.D. Watts and R.J. Bartlett, J. Chem. Phys., 93 (1990) 6104.
82 C.W. Bauschlicher, Jr., S.R. Langhoff, P.R. Taylor and H. Partridge, Chem. Phys. Lett., 126 (1986) 436.
83 C.W. Bauschlicher, Jr. and P.R. Taylor, J. Chem. Phys., 85 (1986) 2779.
84 C.W. Bauschlicher, Jr., S.R. Langhoff, P.R. Taylor, N.C. Handy and P.J. Knowles, J. Chem. Phys., 85 (1986) 1469.
85 N. Oliphant and L. Adamowicz, J. Chem. Phys., 96 (1992) 3739.
86 R.J. Bartlett, S.A. Kucharski and J. Noga, Chem. Phys. Lett., 155 (1989) 133.
87 S.A. Kucharski and R.J. Bartlett, J. Chem. Phys., 108 (1998) 9221.
88 S.A. Kucharski and R.J. Bartlett, Chem. Phys. Lett., 158 (1989) 550.
89 S.A. Kucharski, J.D. Watts and R.J. Bartlett, Chem. Phys. Lett., 302 (1999) 295.
90 M. Musial and R.J. Bartlett, J. Chem. Phys., (2005) in press.
91 K. Kowalski and P. Piecuch, J. Chem. Phys., 113 (2000) 18.
92 S.A. Kucharski and R.J. Bartlett, J. Chem. Phys., 108 (1998) 5255.
93 S. Hirata, M. Nooijen and R.J. Bartlett, Chem. Phys. Lett., 321 (2000) 216.
94 M. Musial, S.A. Kucharski and R.J. Bartlett, J. Chem. Phys., 116 (2002) 4382.
95 M. Musial, S. Kucharski and R.J. Bartlett, J. Mol. Struct. (THEOCHEM), 547 (2001) 269.
96 T.A. Ruden, T. Helgaker, P. Jørgensen and J. Olsen, J. Chem. Phys., 121 (2004) 5847.
97 M. Kállay and P. Surjan, J. Chem. Phys., 115 (2001) 2945.
98 J. Olsen, J. Chem. Phys., 113 (2000) 7140.
99 S. Hirata, J. Phys. Chem. A, 107 (2003) 10154.
100 E. Apra, T.L. Windus, T.P. Straatsma, M. Dupuis, E.J. Bylaska, W. de Jong, S. Hirata, D.M.A. Smith, M.T. Hackler, L. Pollack, R.J. Harrison, J. Nieplocha, V. Tipparaju, M. Krishnan, A.A. Auer, E. Brown, G. Cisneros, G.I. Fann, H. Fruchtl, J. Garza, K. Hirao, R. Kendall, J.A. Nichols, K. Tsemekhman, M. Valiev, K. Wolinski, J. Anchell, D. Bernholdt, P. Borowski, T. Clark, D. Clerc, H. Dachsel, M. Deegan, K. Dyall, D. Elwood, E. Glendening, M. Gutowski, A. Hess, J. Jaffe, B. Johnson, J. Ju,

R. Kobayashi, R. Kutteh, Z. Lin, R. Littlefield, X. Long, B. Meng, T. Nakajima, S. Niu, M. Rosing, G. Sandrone, M. Stave, H. Taylor, G. Thomas, J. van Lenthe, A. Wong and Z. Zhang, NWChem, A computational chemistry package for parallel computers, version 4.6, Pacific Northwest National Laboratory, Richland, WA, USA, 2004.
101 M. Kállay and J. Gauss, J. Chem. Phys., 120 (2004) 6841.
102 J. Noga, W. Kutzelnigg and W. Klopper, Chem. Phys. Lett., 199 (1992) 497.
103 L. Adamowicz, W.D. Laidig and R.J. Bartlett, Int. J. Quantum Chem., 18 (1984) 245.
104 R.J. Bartlett, Anaytical evaluation of gradients in coupled-cluster and many-body perturbation theory, in: P. Jørgensen and J. Simons (Eds.), Geometrical derivatives of energy surfaces and molecular properties, Reidel, Dordrecht, The Netherlands, 1986, pp. 35–61.
105 A.C. Hurley, J. Chem. Phys., 47 (1967) 1275.
106 A. Dalgarno and A.L. Stewart, Proc. R. Soc. London, Series A, 247 (1958) 245.
107 J. Hirschfelder, W. Byers-Brown and S. Epstein, Adv. Quantum Chem., 1 (1964) 255.
108 N.C. Handy and H.F. Schaefer, III, J. Chem. Phys., 81 (1984) 5031.
109 E.A. Salter, G.W. Trucks and R.J. Bartlett, J. Chem. Phys., 90 (1989) 1752.
110 J.S. Arponen, Ann. Phys., 151 (1983) 311.
111 P.G. Szalay, M. Nooijen and R.J. Bartlett, J. Chem. Phys., 103 (1995) 281.
112 H. Koch, H.J.Aa. Jensen, P. Jørgensen and T. Helgaker, J. Chem. Phys., 93 (1990) 3333.
113 J. Gauss, J.F. Stanton and R.J. Bartlett, J. Chem. Phys., 95 (1991) 2639.
114 E.A. Salter, H. Sekino and R.J. Bartlett, J. Chem. Phys., 87 (1987) 502.
115 R.J. Bartlett, S.A. Kucharski, J. Noga, J.D. Watts and G.W. Trucks, in: U. Kaldor (Ed.), Some consideration of alternative ansatz in coupled-cluster theory, Lecture notes in chemistry, Vol. 52, Springer, Heidelberg, 1989, pp. 125–149.
116 G. Fitzgerald, R. Harrison, W.D. Laidig and R.J. Bartlett, J. Chem. Phys., 85 (1986) 5143.
117 A.C. Scheiner, G.E. Scuseria, J.E. Rice, T.J. Lee and H.F. Schaefer, III, J. Chem. Phys., 87 (1987) 5361.
118 G.E. Scuseria and H.F. Schaefer, III, Chem. Phys. Lett., 146 (1988) 23.
119 G.E. Scuseria, J. Chem. Phys., 94 (1991) 442.
120 T.J. Lee and A.P. Rendell, J. Chem. Phys., 94 (1991) 6229.
121 J.D. Watts, J. Gauss and R.J. Bartlett, Chem. Phys. Lett., 200 (1992) 1.
122 J.D. Watts, J. Gauss and R.J. Bartlett, J. Chem. Phys., 98 (1993) 8718.
123 J. Gauss and J.F. Stanton, J. Chem. Phys., 116 (2002) 4773.
124 M. Kállay, J. Gauss and P. Szalay, J. Chem. Phys., 119 (2003) 2991.
125 J.F. Stanton and J. Gauss, J. Chem. Phys., 101 (1994) 8938.
126 M. Kállay and J. Gauss, J. Chem. Phys., 121 (2004) 9257.
127 S.V. Levehenko, T. Wang and A.I. Krylov, J. Chem. Phys., (2005) in press.
128 J. Gauss and J.F. Stanton, J. Chem. Phys., 116 (2002) 4773.
129 P.G. Szalay, J. Gauss and J.F. Stanton, Theor. Chem. Acc., 100 (1993) 5.
130 M. Kállay and J. Gauss, J. Chem. Phys., 120 (2004) 6841.
131 H.J. Monkhorst, Int. J. Quantum Chem. Symp., 11 (1977) 421.
132 S. Ghosh, D. Mukherjee and S.N. Bhattacharyya, Mol. Phys., 43 (1981) 173.
133 H. Nakatsuji, K. Ohta and K. Hirao, J. Chem. Phys., 75 (1986) 409.
134 F.A. Harris, Int. J. Quantum Chem. Symp., 11 (1977) 403.
135 D.J. Rowe, Rev. Mod. Phys., 40 (1968) 153.
136 T.H. Dunning and V. McKoy, J. Chem. Phys., 47 (1967) 1735.
137 J. Simons and W.D. Smith, J. Chem. Phys., 58 (1973) 4899.
138 L. Adamowicz and R.J. Bartlett, Int. J. Quantum Chem., 19 (1986) 217.
139 K. Jankowski and K. Kowalski, J. Chem. Phys., 110 (1999) 9345.
140 H. Sekino and R.J. Bartlett, Int. J. Quantum Chem. Symp., 18 (1984) 255.
141 J. Geertsen, M. Rittby and R.J. Bartlett, Chem. Phys. Lett., 164 (1989) 57.
142 D. Comeau and R.J. Bartlett, Chem. Phys. Lett., 207 (1993) 414.
143 J.F. Stanton and R.J. Bartlett, J. Chem. Phys., 98 (1993) 7029.
144 J.F. Stanton and R.J. Bartlett, J. Chem. Phys., 99 (1993) 5178.
145 J. Simons, this volume.

146 D. Mukeherjee and S. Pal, Adv. Quantum Chem., 20 (1989) 292.
147 M. Musial and R.J. Bartlett, J. Chem. Phys., 121 (2004) 1670.
148 S. Hirata, M. Nooijen and R.J. Bartlett, Chem. Phys. Lett., 326 (2000) 255.
149 S.A. Kucharski, M. Wloch, M. Musial and R.J. Bartlett, J. Chem. Phys., 115 (2001) 8263.
150 K. Kowalski and P. Piecuch, J. Chem. Phys., 115 (2001) 643.
151 K. Kowalski and P. Piecuch, Chem. Phys. Lett., 347 (2001) 237.
152 P. Piecuch, K. Kowalski, I.S. Pimienta and M.J. McGuire, Int. Rev. Phys. Chem., 21 (2002) 527–655.
153 M. Kállay and P.R. Surjan, J. Chem. Phys., 113 (2000) 1359.
154 S. Hirata, J. Chem. Phys., 121 (2004) 51.
155 M. Nooijen and R.J. Bartlett, J. Chem. Phys., 102 (1995) 3629.
156 M. Musial, S.A. Kucharski and R.J. Bartlett, J. Chem. Phys., 118 (2003) 1128.
157 M. Musial and R.J. Bartlett, J. Chem. Phys., 119 (2003) 1901.
158 J.F. Stanton and J. Gauss, J. Chem. Phys., 111 (1999) 8785.
159 A. Balková and R.J. Bartlett, J. Chem. Phys., 99 (1993) 7907.
160 A. Balková and R.J. Bartlett, Chem. Phys. Lett., 193 (1992) 364.
161 P.G. Szalay and R.J. Bartlett, J. Chem. Phys., 103 (1995) 281.
162 P.G. Szalay and J. Gauss, J. Chem. Phys., 112 (2000) 4027.
163 S.V. Levchenko and A.I. Krylov, J. Chem. Phys., 120 (2004) 175.
164 J.D. Watts and R.J. Bartlett, Chem. Phys. Lett., 258 (1996) 581.
165 H. Koch, O. Christiansen, P. Jørgensen, A.M.S. deMeras and T. Helgaker, J. Chem. Phys., 106 (1997) 1808.
166 S. Hirata, M. Nooijen and R.J. Bartlett, Chem. Phys. Lett., 328 (2000) 459.
167 J.F. Stanton and J. Gauss, J. Chem. Phys., 103 (1995) 8931.
168 J.F. Stanton and J. Gauss, J. Chem. Phys., 110 (1999) 6079.
169 M. Doscher, H. Köppel and P.G. Szalay, J. Chem. Phys., 117 (2002) 2645.
170 G. Monninger, M. Förderer, P. Gurtler, S. Kalhofer, S. Petersen, L. Nemes, P.G. Szalay and W. Krätschmer, J. Chem. Phys., 106 (2002) 5779.
171 S.A. Perera, M. Nooijen and R.J. Bartlett, J. Chem. Phys., 104 (1996) 3290.
172 S.A. Perera and R.J. Bartlett, J. Am. Chem. Soc., 118 (1996) 7849.
173 P. Rozyczko, S.A. Perera, M. Nooijen and R.J. Bartlett, J. Chem. Phys., 107 (1997) 6736.
174 P. Rozyczko and R.J. Bartlett, J. Chem. Phys., 107 (1997) 10823.
175 H. Sekino and R.J. Bartlett, Adv. Quantum Chem., 35 (1999) 149.
176 A.D. Yau, S.A. Perera and R.J. Bartlett, Mol. Phys., 100 (2002) 835.
177 M. Tobita, S.A. Perera, M. Musial, R.J. Bartlett, M. Nooijen and J.S. Lee, J. Chem. Phys., 119 (2003) 10731.
178 I. Lindgren, Int. J. Quantum Chem. Symp., 12 (1978) 33.
179 M. Musial and R.J. Bartlett, J. Chem. Phys., 121 (2004) 1670.
180 L. Meissner and R.J. Bartlett, J. Chem. Phys., 102 (1995) 7490.
181 D. Mukherjee, Chem. Phys. Lett., 125 (1986) 207.
182 D. Mukherjee, Int. J. Quantum Chem., 20 (1986) 409.
183 M. Nooijen and R.J. Bartlett, J. Chem. Phys., 107 (1997) 6812.
184 M. Nooijen and R.J. Bartlett, J. Chem. Phys., 106 (1997) 6449.
185 S.R. Gwaltney, M. Nooijen and R.J. Bartlett, J. Chem. Phys., 111 (1999) 58.
186 M. Nooijen and V. Lotrich, J. Chem. Phys., 113 (2000) 494.
187 The fundamental differences between EOM-CC and SAC-CI come from the approximations in the latter and from their CI computational procedure. In CC theory the exponential is never truncated including terms like T_1^4, T_2, T_2^2. SAC-CI normally uses only the latter and then to define approximate CI quadruple excitation coefficients as they basically use CI technology to do all calculations. This means they would have up to $\sim n^4N^4$ quadruple coefficients. This obviously forces severe configuration truncation. To the contrary, the mathematical structure of CC theory necessarily leads to factorized equations, making CCSD and EOM-CCSD properly consist of $\sim n^2N^2$ coefficients. No such factorization is built into SAC-CI. Because of their CI computational procedure, SAC-CI need not recognize the non-Hermitian left- and right-hand eigenvector structure of EOM-CC that leads naturally to excited state density matrices,

transition densities, and especially analytical gradients. Instead, they enforce the orthogonality/non-interacting property of excited states as if they were CI type, despite their neglect of what should be a non-truncated exponential operator that constitutes an essential part of the wavefunction. So to say SAC-CI is just EOM-CC with a few approximations [188] is a misrepresentation. More importantly, it is almost impossible to duplicate SAC-CI results since the configuration space is severely truncated and exactly how is hard to say, plus truncating the cluster operator itself even prohibits the underlying CC result to correspond to the correct value. In a rigorous, systematically converging *ab initio* method, there is no room for these approximations for excited states. On the other hand, because of their approximations SAC-CI can be applied to large, complicated systems, while EOM-CC has not yet embarked upon making simplifying approximations for large molecule applications.

188 H. Nakatsuji, SAC-CI method: theoretical aspects and some recent topics, in: J. Leszczynski (Ed.), Computational Chemistry (Review of Current Trends), Vol. 2, World Scientific, Singapore, 1997.
189 M. Schütz and H.J. Werner, J. Chem. Phys., 111 (1999) 5691.
190 M. Schütz and H.J. Werner, Chem. Phys. Lett., 318 (2000) 370.
191 S. Saebo and P. Pulay, Chem. Phys. Lett., 117 (1985) 37.
192 G.E. Scuseria and P.Y. Ayala, J. Chem. Phys., 111 (1999) 8330.
193 P.Y. Ayala and G.E. Scuseria, J. Comput. Chem., 21 (2000) 1524.
194 R. Podeszwa, M. Tobita, S. Hirata and R.J. Bartlett, To be published.
195 M. Tobita, PhD. Unpublished dissertation, University of Florida, 2001.
196 S. Hirata, I. Grabowski, M. Tobita and R.J. Bartlett, Chem. Phys. Lett., 345 (2001) 475.
197 S. Hirata, R. Podeszwa, M. Tobita and R.J. Bartlett, J. Chem. Phys., 120 (2004) 2581.
198 N. Flocke and R.J. Bartlett, J. Chem. Phys., 118 (2003) 5326.
199 J.P. Malrieu, P. Claverie and S. Diner, Theor. Chim. Acta, 13 (1969) 13.
200 S. Diner, J.P. Malrieu and P. Claverie, Theor. Chim. Acta, 13 (1969) 1.
201 W.D. Laidig, G.D. Purvis and R.J. Bartlett, Chem. Phys. Lett., 97 (1983) 209.
202 W.D. Laidig, G.D. Purvis, III and R.J. Bartlett, J. Phys. Chem., 89 (1985) 2161.
203 W. Yang and J.M. Perez-Jorda, Linear scaling methods for electronic structure calculations, in: P.v.R. Schleyer (Ed.), Encyclopedia of computational chemistry, Wiley, New York, 1998.
204 W. Yang, J. Mol. Struct. (THEOCHEM), 255 (1992) 461.
205 F. Weinhold, Natural bond orbital methods, in: P.v.R. Schleyer, N.L. Allinger, T. Clark, J. Gasteiger, P.A. Kollman, H.F. Schaefer, III and P.R. Schreiner (Eds.), Encyclopedia of computational chemistry, Vol. 3, Wiley, Chichester, UK, 1998.
206 N. Flocke and R.J. Bartlett, J. Chem. Phys., 121 (2004) 10935.
207 N.H.F. Beebe and J. Linderberg, Int. J. Quantum Chem., 12 (1977) 683.
208 C.F. Bender and E.R. Davidson, J. Phys. Chem., 70 (1966) 2675.
209 L. Adamowicz and R.J. Bartlett, J. Chem. Phys., 86 (1987) 6314.
210 C. Sosa, J. Geertsen and R.J. Bartlett, Chem. Phys. Lett., 159 (1989) 148.
211 A. Taube and R.J. Bartlett, Coll. Czech Chem. Comm., in press.
212 T. Kinoshita, O. Hino and R.J. Bartlett, J. Chem. Phys., 119 (2003) 7756.
213 O. Hino, T. Kinoshita and R.J. Bartlett, J. Chem. Phys., 121 (2004) 1206.
214 H. Koch, A.S. de Meras and T.B. Pedersen, J. Chem. Phys., 118 (2003) 9481.

Biographical sketches of contributors

Jean-Marie André received B.Sc. (1965) and Ph.D. (1968) degrees from the Catholic University of Louvain (UCL). After a postdoctoral stay with Enrico Clementi, he became Professor of Chemistry at the "Facultés Universitaires Notre-Dame de la Paix" in Namur where he established the theoretical chemistry laboratory with a primary interest in the field of the electronic properties of polymers. J.M. André has been the Director of this laboratory since 1974. He has written about 300 scientific papers and 7 books, and he is a member of the Royal Academy of Belgium, the International Academy of Quantum Molecular Science, and the European Academy of Science.

Jean-Marie André

Rod Bartlett obtained his Ph.D. at the Florida Quantum Theory Project (QTP) under the direction of P.O. Löwdin and Yngve Öhrn. He went from perturbation theory to pioneering MBPT and coupled-cluster theory for the correlation problem. Today he is the Graduate Research Professor of QTP, inheriting the mantle from his teachers, P.O. Löwdin and John Slater. His work ranges from the quantum theory of non-linear optics to NMR coupling constants to MBPT/CC methods for polymers to predicting the existence of polynitrogen molecules and others some of which have now been found.

Rodney J. Bartlett

Charles Bauschlicher received his Ph.D. from the University of California at Berkeley in 1976; his thesis advisor was Professor H.F. Schaefer. Charles was a postdoctoral fellow with Professor Shavitt at Battelle Memorial Institute. In 1977 he joined ICASE at NASA's Langley Research Center. In 1980 he joined NASA Ames Research Center, becoming a civil servant in 1982. His recent research interests have been astrochemistry and nanotechnology.

Charles W. Bauschlicher, Jr.

After receiving a B.A. degree from Harvard in 1964, Wes Borden spent a year on a Fulbright Fellowship, studying theoretical chemistry with H.C. Longuet-Higgins in the other Cambridge. Professor Borden completed his Ph.D. degree at Harvard in 1968, under the direction of E.J. Corey, and then served 5 years on the Harvard Faculty, before moving to the University of Washington. There he began to collaborate with Ernest Davidson, from whom Professor Borden learned that *ab initio* calculations provided a powerful tool for understanding chemistry. After 31 years at UW, in 2004 Professor Borden accepted a Welch Chair of Chemistry at the University of North Texas.

Weston Thatcher Borden

Joel Bowman received his Ph.D. in chemistry from Caltech in 1975 working under the direction of Aron Kuppermann. After rising through the academic ranks at Illinois Institute of Technology, he moved to Emory University in 1986 and became Samuel Candler Dobbs Professor in 1990. He has been a Visiting Professor at the University of Chicago and Yale University, a JILA Fellow and is a Fellow of the American Physical Society. He is currently serving his second term as Chair of the Chemistry department at Emory. He helped establish the Cherry L. Emerson Center for Scientific Computation at Emory and was its first acting director in 1991. Currently, most of his group's computational chemistry research is done on "Multinode" a 100-cpu Beowulf cluster.

Joel M. Bowman

Brina (Beth) Brauer received her M.S. degree under the supervision of K.J. Kaufmann, and Ph.D. under the supervision of G.B. Schuster, from the University of Illinois at Urbana-Champaign. Her graduate work primarily focused on measurement of intersystem-crossing and reaction rates of aryl-containing carbenes. She moved to Israel 9 years ago and became interested in spectroscopic calculations of biological molecules while working in the laboratory of M.S. de Vries. She has been working in the group of R.B. Gerber for the past 3 years.

B. Brauer

Robert Bukowski received his Ph.D. in chemistry in 1996 at the University of Warsaw, Poland, where he worked with Bogumil Jeziorski. In 1997 he joined the group of Krzysztof Szalewicz at the University of Delaware. His research focuses on the theory of intermolecular interactions and its applications in condensed phase simulations.

Robert Bukowski

Petr Cársky received his Ph.D. in chemistry in 1968 working at Czechoslovak Academy of Sciences under the supervision of Rudolf Zahradník. Except for a single year (in 1989) he had a position at the J. Heyrovský Institute of Physical Chemistry (Academy of Sciences), where he is now Director of the institute. The beginning of his career falls into the era of semiempirical calculations but since the early 1970s he has been active in the field of *ab initio* computational chemistry.

Petr Cársky

Stuart Carter received his Ph.D. in theoretical chemistry in 1966, working under the direction of J.N. Murrell at the Universities of Sheffield and Sussex. After leaving academia to work as a computer programmer with International Computers Ltd, he returned to continue his research in 1976, working again with J.N. Murrell, and also with I.M. Mills at Reading. In 1981, he started a collaboration on rovibrational variational theory with N.C. Handy at Cambridge, which continues to this day. He has had other positions in the meantime, most notably with P. Rosmus in Frankfurt, W. Meyer in Kaiserslautern, R.J. Buenker in Wuppertal and P. Palmieri in Bologna. In 1996, he started a collaboration with J.M. Bowman at Emory, and it is this work that has resulted in the development of the variational code 'Multimode'.

Stuart Carter

Galina Chaban came to the field of computational chemistry in 1983 while being a student at the Moscow Institute of Fine Chemical Technology, Russia. She performed her first studies under the guidance of Dr Nina Klimenko. Then she continued as a research scientist at the Institute of New Chemical Problems in the research group of Professor Oleg Charkin. After immigrating to the USA in 1992, she performed her graduate studies in the group of Professor Mark Gordon at Iowa State University and received her Ph.D. in 1997. In 1998–2000, she worked as a postdoctoral researcher at the Hebrew University of Jerusalem and at the University of California Irvine with Professor Benny Gerber. At the present time, she holds a position of Computational Chemist at the NASA Ames Research Center.

G.M. Chaban

Sudip Chattopadhyay received his M.Sc. degree in chemistry in 1993 from the University of Burdwan, India. He is a gold-medallist in both his graduate and post-graduate examinations for topping the merit lists. In 1995 he joined the group of Professor Debashis Mukherjee and received his Ph.D. degree in 2001 for his work on molecular electronic structure theory. He joined the faculty of the Department of Chemistry, Bengal Engineering and Science University, India in 2004. His research interest includes the development of many-body electronic structure theories and their application to problems of broad chemical interest.

Sudip Chattopadhyay

Richard A. Christie was born in 1975 in Aberdeen, Scotland. He graduated with a first class B.Sc. (Honors) degree in chemistry in 1997 from the Robert Gordon University (Aberdeen, Scotland). In 2004 he earned his Ph.D. in chemistry at the University of Pittsburgh under the guidance of Kenneth D. Jordan. His research has focused on theoretical studies of hydrogen-bonded clusters.

R.A. Christie

Giovanni Ciccotti

Giovanni Ciccotti received his Laurea in Physics in 1967 from the University of Roma "La Sapienza". In 1971 he joined the Science Faculty of the University of Lecce, in 1973 that of Camerino and in 1977 moved back to the Science Faculty of the University of Roma "La Sapienza", at first as Associate Professor and since 1990 as Professor of "Structure of Matter". His activity in Molecular Dynamics started in Paris, while at CECAM (Centre Europeen de Calcul Atomique et Moleculaire) in 1974 and his first paper in MD was in 1975. In 1999 he won the Berni J. Alder CECAM prize and from 2004 he is Fellow of IOP. His first publication, in 1969, was a theoretical paper on elementary particles.

Enrico Clementi

Enrico Clementi received his doctoral degree in chemistry in 1956 from the University of Pavia, Italy. After postdoctoral experiences with Nobel Giulio Natta, Mikael Kasha, Kennet S. Pitzer and Nobel Robert S. Mulliken, since 1961 he directed and performed pioneering research at the IBM Research Division (San Jose, California), the Istituto Donegani (Novara, Italy), the IBM Data System Division (Poughkeepsie and Kingston, NY), the Centro Ricerche e Sviluppo Studi Superiori Sardegna (Cagliari, Italy) and the University L. Pasteur (Strasbourg, France). His work on development, programming and applications on the atomic structure of atoms and molecules, with and without relativistic correction, on correlation energy, on Monte Carlo and Molecular Dynamics for biological systems, on micro-dynamics of liquid flows and on architecture, both hardware and system software, for an experimental parallel computer, is documented in about 450 papers and several books.

Larry Curtiss received his Ph.D. in physical chemistry in 1973 from Carnegie-Mellon University working with Professor John A. Pople. He then spent 2 years as a Battelle Institute Fellow at Battelle Memorial Institute. In 1976 he joined Argonne National Laboratory where he is currently a Senior Scientist in the Chemistry and Materials Science Divisions. His research interests have included methods development and applications of quantum chemical methods to a wide range of problems including hydrogen bonding, catalysis and properties of solid-state materials. He received the University of Chicago Distinguished Performance Award in 1995 and was selected as an AAAS Fellow in 1997. He is the author of over 260 papers and is an ISI highly cited chemist for 1981–1999.

Larry A. Curtiss

Sanghamitra Das received her M.Sc. degree in physics in 2001 from the Calcutta University, India. In 2002, she joined the group of Debashis Mukherjee as a Ph.D. student.

Sanghamitra Das

Erik Deumens' work in computational chemistry has focused on two projects: the development of a theory for the description of the dynamics of electrons and nuclei in molecular reactions as fully coupled degrees of freedom, and the implementation of this theory in computer software. The second project is the building of infrastructure for computation at QTP (Quantum Theory Project, Gainesville, FL), including building clusters, software libraries and courses.

Erik Deumens

Clifford Dykstra received his Ph.D. in chemistry in 1976 working under the direction of H.F. Schaefer at the University of California at Berkeley. He then joined the faculty of the University of Illinois at Urbana-Champaign, becoming Professor of Chemistry in 1988. Two years later, he moved to the chemistry department at Indiana University-Purdue University Indianapolis (IUPUI) where he is now Chancellor's Professor of Chemistry. His first attempt at computational work in chemistry was in 1972 on an IBM 1800. His first TACC publication, an electronic structure CI calculation with Schaefer, appeared in 1975.

Clifford E. Dykstra

Masahiro Ehara received his Ph.D. (engineering) in 1993 under the direction of Professor H. Nakatsuji at Kyoto University. He then joined the Institute for Fundamental Chemistry and Department of Theoretical Chemistry of Heidelberg University as a postdoctoral fellow. In 1995, he moved to Nakatsuji's group at the Department of Synthetic Chemistry and Biological Chemistry in Kyoto University.

M. Ehara

Marcus Elstner studied physics in Munich and Berlin. He received his Ph.D. in physics in 1998 under the direction of Sandor Suhai from the German Cancer Research Center in Heidelberg and Thomas Frauenheim (University of Paderborn). He moved to Paderborn University after a postdoctoral fellowship with E. Kaxiras (Department of Physics) at Harvard from 1999 to 2000. Since December 2002 he is an Assistant Professor (Junior professor) in the Department of Physics at Paderborn University.

Marcus Elstner

Gernot Frenking studied chemistry at the Universities of Aachen, Kyoto and Berlin where he received his Ph.D. in 1979. After obtaining his Habilitation in Theoretical Organic Chemistry in 1984 he moved to the United States where he spent 1 year as a visiting scientist in the group of Professor Henry F. Schaefer III at the University of California at Berkeley. He then worked as a Research Scientist at SRI International in Menlo Park, CA. In 1989 he returned to Germany and became an Associate Professor (C3) for Computational Chemistry at the Philipps-Universität Marburg. In 1998 he was appointed Full Professor (C4) for Theoretical Chemistry. The focus of his research interests has always been the diversity of molecular structures and the understanding of the nature of the chemical bonding which comes from the analysis of quantum chemical calculations. This motivation is beautifully expressed in the famous statement of Charles Coulson: "*Give us insight not numbers*"

Gernot Frenking

Bruce Garrett received a Ph.D. in chemistry in 1977 from the University of California, Berkeley with W.H. Miller. He was a postdoctoral research specialist at the University of Minnesota with D.G. Truhlar (1977–1979) before joining the scientific staff at Battelle Columbus Laboratories. He co-founded Chemical Dynamics Corporation, a contract research organization, where he conducted basic research from 1980 to 1989. He is currently Laboratory Fellow and Associate Director for Molecular Interactions & Transformation in the Chemical Sciences Division at Pacific Northwest National Laboratory. His first computational studies in 1972 involved kinetic Monte Carlo simulations with D.L. Bunker as an Undergraduate Research Assistant at the University of California, Irvine, and resulted in his first TACC publication in 1974.

Bruce C. Garrett

Robert Benny Gerber obtained his B.Sc. in chemistry from the Hebrew University of Jerusalem in 1965. He did his doctoral research with C.A. Coulson at Oxford University in 1968, and was a Postdoctoral Research Fellow with Martin Karplus at Harvard University. Since 1976 he has been on the faculty of the Hebrew University of Jerusalem, where he holds the Saerre K. and Louis P. Fiedler Chair in Chemistry. Since 1990 he is also Professor of Chemistry at the University of California at Irvine. Gerber's current research interests include: vibrational spectroscopy calculations for large molecules; new molecules of the noble gas elements and their formation dynamics; photochemical reactions in low-temperature matrices and clusters; and dynamics of atmospherically important processes.

R.B. Gerber

Born and raised in the New York City area, Mark Gordon received his B.S., Ph.D. and postdoctoral education at Rensselaer Polytechnic Institute, Carnegie-Mellon University (with J.A. Pople) and Iowa State University (with K. Ruedenberg). At North Dakota State University, he rose to Distinguished Professor and Department Chair. At Iowa State University, he is Distinguished Professor and Director of the Applied Mathematical Sciences Program in the Ames Laboratory USDOE. He has been the Chair of the Theoretical Chemistry Subdivision of the American Chemical Society, and the Secretary-Treasurer of its Physical Chemistry Division. He is a Fellow of the American Physical Society, a Fulbright Senior Scholar and was recently elected to the International Academy of Quantum Molecular Science.

Mark S. Gordon

Annick Goursot received her Ph.D. in physical chemistry in 1972 at the University of Marseille, France. After a 2-year postdoc at the University of Montreal, she spent 10 years in the School of Chemistry of Mulhouse. In 1985, she joined the School of Chemistry of Montpellier, where she initiated computational chemistry calculations for functional materials.

A. Goursot

Professor Hall was born in Pennsylvania and graduated from Juniata College with a B.S. degree in chemistry in 1966. After completing his Ph.D. degree with Richard Fenske at the University of Wisconsin in 1971, he accepted an AEI fellowship to study *ab initio* quantum chemistry with Ian Hillier at the University of Manchester (UK). He joined the faculty of Texas A&M University in 1975, rose through the ranks and served as Head of the Department from 1986 to 1994. He currently directs the Laboratory for Molecular Simulation and serves as Executive Associate Dean for the College of Science. In 2004 he was named Davidson Professor of Science. His research interests are primarily directed toward understanding chemical structures and reactions through the application of state-of-the-art quantum calculations.

Michael B. Hall

Nicholas Handy's complete academic career has been at the University of Cambridge, from undergraduate (1960) to Professor and Fellow of the Royal Society (1990). He started research with S.F. Boys, and he has made a number of contributions in quantum chemistry (gradient theory, Møller–Plesset theory, full configuration interaction), in density functional theory (including its promotion and the development of new exchange-correlation functionals), and in theoretical spectroscopy (in particular the variational approach for rovibrational energy levels and consequent spectra). He is a Fellow of St Catharine's College, Cambridge, where he teaches mathematics.

Nicholas C. Handy

Encai Hao received his Ph.D. at Jilin University in China, and then did postdoctoral studies at Emory University (with Tim Lian) and at Northwestern University (with Joe Hupp and George Schatz). He is currently working at Nanofilm, Inc.

Encai Hao

Upendra Harbola is currently a Postdoctoral Fellow in University of California at Irvine. He received his M.Sc. in 1996 from Kumaon University (India) and his Ph.D. in 2003 from Jawaharlal Nehru University (India) where he worked with Professor Shankar P. Das on glass transition phenomena in binary liquids. Presently his research interests include the development of theoretical tools to study the equilibrium and non-equilibrium response functions for optical and transport properties of many electron systems.

Upendra Harbola

Bill Hase received his Ph.D. in chemistry in 1970, working in the research area of experimental physical chemistry under the direction of John W. Simons at New Mexico State University. His research included studies of the methylene singlet–triplet energy gap and of the unimolecular decomposition of vibrationally excited alkane and alkylsilane molecules prepared by chemical activation. His career as a computational chemist began during his postdoctoral work with Don Bunker at the University of California, Irvine. In 1973 he joined the Chemistry Department at Wayne State University, where he remained until 2004, when he assumed the Robert A. Welch Chair in Chemistry at Texas Tech University. He remembers that his first computational chemistry classical trajectory computer program was written in assembly language and run on a PDP-10.

William L. Hase

Jun-ya Hasegawa received his Ph.D. (engineering) in 1998 under the direction of Professor H. Nakatsuji at Kyoto University. He then joined the Department of Theoretical Chemistry of Lund University as a postdoctor. In 1999, he moved back to Nakatsuji's group at Department of Synthetic Chemistry and Biological Chemistry in Kyoto University.

J. Hasegawa

Philippe C. Hiberty studied theory at the University of Paris-Sud with W.J. Hehre, completed his Ph.D. under the supervision of L. Salem, and got a research position at the CNRS. In 1979 he started his postdoctoral research with J.I. Brauman at Stanford University and then with H.F. Schaefer III at Berkeley. He went back to Orsay to join the Laboratoire de Chimie Théorique, where he developed a research program based on valence bond theory. He became Directeur de Recherche in 1986, and in addition, he now teaches Quantum Chemistry at the Ecole Polytechnique in Palaiseau. His research interests are, among others, in the application of quantum chemistry and valence bond theory to fundamental concepts of organic chemistry

Philippe C. Hiberty

Kimihiko Hirao obtained his Ph.D. in 1974 from Kyoto University. He held a post-doctoral position at the University of Alberta (Professor Sigeru Huzinaga) in 1974–1975 and worked with Professor Roy McWeeny at Sheffield University in 1978–1979. He became Professor at Nagoya University in 1988 and moved to The University of Tokyo in 1993. He is the author of more than 250 scientific articles in theoretical chemistry.

Kimihiko Hirao

Hrant P. Hratchian holds a degree from Eastern Michigan University where he completed honors research work with M.C. Milletti. Currently, he is a Graduate Student at Wayne State University under the direction of H.B. Schlegel. His research interests include the development of new methods for reaction path following, transition state optimization and the application of electronic structure theory to organometallic and inorganic chemistry.

Hrant P. Hratchian

Ivan Hubac obtained his Ph.D. in 1971 (applied mathematics) from University of Waterloo, Canada. His supervisors were Professors J. Koutecky, J. Cizek and J. Paldus. After his return to Czechoslovakia from Canada in 1971 he worked at the Department of Mathematics, Chemical faculty, Slovak Technical University, Bratislava. At present he works as Professor of Physics at Faculty of Mathematics, Physics and Informatics, Comenius University, Bratislava, Slovakia. His first quantum chemistry paper appeared in 1967.

Ivan Hubac

Dr Stephan Irle obtained his Ph.D. from the University of Vienna in Austria 1997 under the supervision of Professor Hans Lischka. He then joined the group of Professor Keiji Morokuma in Atlanta as a postdoctoral research associate for 1 year. In 1998 he accepted the faculty-equivalent position of Associate Scientist and Systems Manager at the Cherry L. Emerson Center for Scientific Computation.

Stephan Irle

Bogumil Jeziorski received his M.S. degree in chemistry from the University of Warsaw in 1969. He conducted his graduate work also in Warsaw under the supervision of W. Kolos. After a postdoctoral position at the University of Utah, he was a research associate at the University of Florida and a Visiting Professor at the University of Waterloo, University of Delaware and University of Nijmegen. Since 1990 he has been a Professor of Chemistry at the University of Warsaw. His research has been mainly on the coupled-cluster theory of electronic correlation and on the perturbation theory of intermolecular forces. His other research interests include chemical effects in nuclear beta decay, theory of muonic molecules and relativistic and radiative effects in molecules.

Bogumil Jeziorski

Ken Jordan received his Ph.D. in physical chemistry in 1974 under the direction of Bob Silbey at MIT. He then joined the Department of Engineering and Applied Science, Yale University, as a J.W. Gibbs Instructor, being promoted to Assistant Professor in 1976. In 1978 Professor Jordan moved to the Chemistry Department at the University of Pittsburgh where he is now Professor and Director of the Center for Molecular and Materials Simulations. His interest in the application of computers to chemical problems stems from his graduate student days. Professor Jordan's recent research has focused on the properties of hydrogen-bonded clusters, modeling chemical reactions on surfaces, electron-induced chemistry and the development of new methods for Monte Carlo simulations.

K.D. Jordan

Raymond Kapral carried out his doctoral studies on molecular quantum mechanics at Princeton University and received his Ph.D. in 1967. He pursued his postdoctoral work at the Massachusetts Institute of Technology where his interests switched to non-equilibrium statistical mechanics. He then took a faculty position at the University of Toronto in 1969 where he is currently Professor of Chemistry. He is a Fellow of the Royal Society of Canada and received the Palladium Medal of the Chemical Institute of Canada in 2003. European collaborations have not only provided an opportunity to mix quantum and classical mechanics but also to mix science with good food and wine.

Raymond Kapral

Kwang S. Kim was born in Seoul in 1950. He obtained his Ph.D. degree in chemistry from the University of California in Berkeley, where his mentor was Professor Henry F. Schaefer III. After spending a few years as an IBM postdoctoral fellow with Professor Enrico Clementi and as a Research Assistant Professor at Rutgers University, he joined the Pohang University of Science and Technology in Pohang, Korea. He is currently a Professor, and the Director of the National Creative Research Initiative Center for Superfunctional Materials. His research interests include investigations of intermolecular interactions, clusters, molecular recognition, nanomaterials and molecular devices. He was a recipient of the Korea Science Prize in the year 2004.

Kwang S. Kim

Andreas M. Köster was born in Steinhude, Germany, in 1964. He earned his Ph.D. degree in theoretical chemistry with Professor Karl Jug at the Universtät Hannover in 1992 with a work about chemical reactivity. A short research stay after the Ph.D. with Professor Roman Nalewajski in Krakow, Poland, introduced him to density functional theory. At the beginning of 1993 he started his postdoctoral research with Dennis Salahub in Montréal. Since then he has been involved in the technical implementation and development of the density functional code deMon. In 1995, he returned to Germany to start a Habilitation, and built a new density functional method with auxiliary functions. In 1999, he moved to CINVESTAV in Mexico to take up a titular Professor position in Theoretical Chemistry.

M. Köster

Han Myoung Lee was born in 1969, in Iksan, Korea. After obtaining his undergraduate, master's and Ph. D. degrees from WonKwang University, he joined Professor Kim's group in Pohang in 1997. During his stay in Pohang, he had been involved in detailed theoretical investigations of aqueous and metal clusters. He has also worked on nanoclusters and nanoelectronic/nanophotonic devices. He is currently a Research Assistant Professor at the Center for Superfunctional Materials and a Visiting Scientist at the Pacific Northwest National Laboratory in Richmond, U.S.A.

Han Myoung Lee

Matthias Lein received his Ph.D. in chemistry in 2004 under the supervision of Gernot Frenking at the Fachbereich Chemie der Universität Marburg in Germany. He currently works as a Marsden Postdoctoral Research Fellow with Peter Schwerdtfeger at Massey University in Auckland, New Zealand.

Matthias Lein

Uttam Sinha Mahapatra received his M.Sc. degree in physics in 1992 from the University of Kalyani, India and Ph.D. degree in 2000 for his work on molecular electronic structure with Debashis Mukherjee. In 2001, he joined Lorenz Cederbaum's group at the Theoretische Chemie, University of Heidelberg, Germany with a fellowship of the Alexander von Humboldt Foundation. Presently, he is a faculty of the Department of Physics, Darjeeling Government College, India. His research interests include development of theories of structure and dynamics of atomic and molecular relativistic and non-relativistic many, body theories and their application to chemically interesting systems and study of weak interactions in atoms.

Uttam Sinha Mahapatra

J. Andrew McCammon holds the Joseph E. Mayer Chair of Theoretical Chemistry at the University of California, San Diego (UCSD), and is an Investigator of the Howard Hughes Medical Institute. He received his Ph.D. in chemical physics in 1976 from Harvard University, where he worked with John Deutch on biological applications of statistical mechanics and hydrodynamics. In 1976–1978, he was a Research Fellow at Harvard, where he developed the computer simulation approach to protein dynamics in collaboration with Martin Karplus. He was an Assistant Professor and then M.D. Anderson Professor (1981–1994) at the University of Houston before moving to UCSD. He recalls with pleasure the first views of atomic dynamics in a protein molecule, generated slowly on a pen plotter during his postdoctoral work.

J. Andrew McCammon

Yifat Miller received her M.Sc. in chemistry in 2002 working under the supervision of Professor R.B. Gerber at the Hebrew University of Jerusalem, Israel, and is currently a Ph.D. student in the same group. Her fields of research are: vibrational spectroscopy and reaction dynamics of atmospherically relevant molecules.

Y. Miller

Keiji Morokuma received his Ph.D. in chemistry in 1963 working under the direction of Kenichi Fukui at Kyoto University. After postdoctoral work with Martin Karplus at Columbia and Harvard, he joined the faculty of the University of Rochester in 1967. In 1977, he moved to the Institute for Molecular Science at Okazaki, Japan to head the theoretical department and computer center. In 1993 he joined the faculty of Emory University, Atlanta, GA, where he is currently William Henry Emerson Professor and Director of Cherry L. Emerson Center for Scientific Computation. He is presently President of the International Academy of Quantum Molecular Science.

Keiji Morokuma

Shaul Mukamel, the Chancellor Professor of Chemistry at the University of California, Irvine, received his Ph.D. in 1976 from Tel Aviv University, followed by postdoctoral appointments at MIT and the University of California at Berkeley and faculty positions at Rice Univeristy, the Weizmann Institute and at the University of Rochester. He has been the recipient of the Sloan, Dreyfus, Guggenheim, Alexander von Humboldt Senior Scientist and the Lippincott awards. He is a Fellow of American Physical Society and of the Optical Society of America. His research interests in theoretical chemical physics and biophysics include: developing a Liouville-space quasiparticle approach to femtosecond spectroscopy and to many, body theory of electronic and vibrational excitations of molecules, molecular aggregates, nanostructures and semiconductors; designing optical and infrared pulse sequences for probing structure and folding dynamics of proteins by multidimensional coherent spectroscopies, non-linear X-ray and single molecule spectroscopy; photon statistics; electron transfer and energy funelling in photosynthetic complexes and dendrimers. He is the author of the textbook, Principles of Nonlinear Optical Spectroscopy (Oxford University Press), 1995.

Shaul Mukamel

Debashis Mukherjee is a Professor of Physical Chemistry and the Director of the Indian Association for the Cultivation of Science, Calcutta, India. He has been one of the earliest developers of a class of multi-reference coupled cluster theories and also of the coupled cluster based linear response theory. Other contributions by him are in the resolution of the size-extensivity problem for multi-reference theories using an incomplete model space and in the size-extensive intermediate Hamiltonian formalism. His research interests focus on the development and applications of non-relativistic and relativistic theories of many-body molecular electronic structure and theoretical spectroscopy, quantum many-body dynamics and statistical field theory of many-body systems. He is a member of the International Academy of the Quantum Molecular Science, a Fellow of the Third World Academy of Science, the Indian National Science Academy and the Indian Academy of Sciences. He is the recipient of the Shantiswarup Bhatnagar Prize of the Council of Scientific and Industrial Research of the Government of India.

Debashis Mukherjee

Shigeru Nagase received his Ph.D. degree in chemistry in 1975 under the direction of Professor T. Fueno from Osaka University in Japan. After he did postdoctoral work at the University of Rochester (Professor K. Morokuma) and at the Ohio State University (Professor C. W. Kern), he joined the faculty of Yokohama National University as Associate Professor in 1980, and he became Professor in 1991. In 1995 he moved to Tokyo Metropolitan University. Since April 2001, he has been a Professor of Theoretical Chemistry at the Institute for Molecular Science. His current interests include the theoretical and computational studies of large molecular systems.

Shigeru Nagase

Takahito Nakajima received his Ph.D. in chemistry in 1997 working under the direction of Hiroshi Nakatsuji at Kyoto University. He then joined the Department of Applied Chemistry at the University of Tokyo in 1999. He is an Associate Professor at the Department of Applied Chemistry, the University of Tokyo. He is also a researcher of PRESTO, Japan Science and Technology Corporation (JST). His current research interests include the developments of the relativistic molecular theory and the quantum chemical approach towards large-scale calculations.

Takahito Nakajima

Haruyuki Nakano received his Ph.D. in chemistry in 1993 working under the direction of Professor Shigeki Kato at Kyoto University. He then joined the faculty of the University of Tokyo. In 2003, he moved to Kyushu University at Fukuoka, where he is now Professor of Theoretical Chemistry.

Haruyuki Nakano

Hiroshi Nakatsuji received his Ph.D. (Engineering) in 1971 at Kyoto University. Since 1990, he has been a Professor at Department of Synthetic Chemistry and Biological Chemistry in Kyoto University. He is a member of International Academy of Quantum Molecular Science. He is also a Director of Fukui Institute for Fundamental Chemistry of Kyoto University.

H. Nakatsuji

Yngve Öhrn earned his Doctor of Philosophy degree in 1966 working in the Quantum Chemistry Group at Uppsala University, Sweden under the direction of Per-Olov Löwin. The numerical work at that time was attempted on an ALWAC III E computer. The same year he joined the faculty of the Departments of Chemistry and Physics at the University of Florida, where he has been since then as a Member and Director (1983–1998) of the QTP, an Institute for Theory and Computation in Molecular and Materials Sciences.

Yngve Öhrn

Massimo Olivucci, Ph.D., is Professor of Organic Chemistry at the University of Sienna, Italy. He took his Ph.D. in 1989 with F. Bernardi and from 1989 to 1991 he carried out research work with M.A. Robb at King's College London. As documented by over 130 research papers his research focuses on the investigation of the photochemistry of organic and biological chromophores using computational tools. In 1999 he was awarded the Premio Nazionale "FEDERCHIMICA". He was one of the finalists of the 2001 edition of the EU Descartes Prize and in 2004, he was awarded the "Premio alla Ricerca" of the Organic Chemistry Division of the Italian Chemical Society.

Massimo Olivucci

Dola Pahari received her M.Sc. degree in chemistry in 2000 from the Indian Institute of Technology, Kanpur, India. She is doing Ph.D. on the development and applications of molecular electronic structure theory under the supervision of Debashis Mukherjee. She is interested in developing spin-adapted state-specific many-body theories.

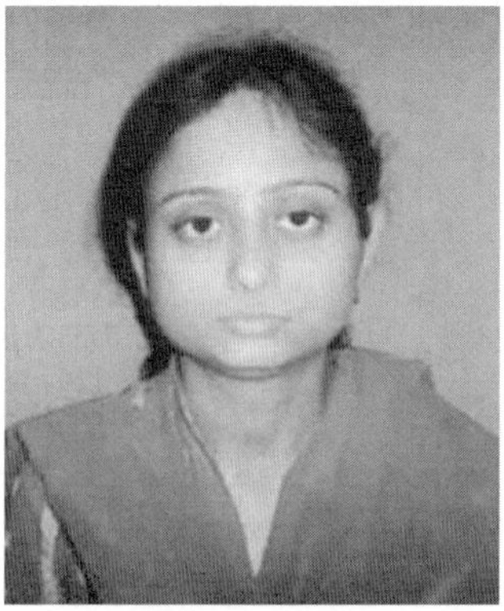

Dola Pahari

Josef (Joe) Paldus defended his M.Sc. Thesis in 1958 at the Charles University in Prague, supervised by V. Hanus and J. Koutecky, and the latter also supervised his Ph.D. Thesis, defended in 1961 at the Czechoslovak Academy of Sciences. He did his postdoctoral studies with D.A. Ramsay at the National Research Council in Ottawa. After emigrating to Canada in 1968 he joined the Department of Applied Mathematics of the University of Waterloo and later also its Chemistry Department and Guelph-Waterloo Center for Graduate Work in Chemistry. Since his obligatory retirement in 2001 he continues his research as a Distinguished Professor Emeritus. He became a Fellow of the Royal Society of Canada in 1983, a Member of the International Academy of Quantum Molecular Sciences a year later and, most recently, a Fellow of the Fields Institute for Research in Mathematical Sciences.

Josef Paldus

Jiri Pittner was born in 1968 in Plzen, Czech Republic. He received his master's degree in physical chemistry in 1991 from the Charles University Prague and his Ph.D. in chemistry in 1996 from the Humboldt University Berlin under the direction of V. Bonacic-Koutecky. In 1997 he joined the J. Heyrovsky Institute of Physical Chemistry (Academy of Sciences of the Czech Republic).

Jiri Pittner

Professor Rabitz's research interests lie at the interface of chemistry, physics and engineering, with principal areas of focus including molecular dynamics, biophysical chemistry, chemical kinetics and optical interactions with matter. An overriding theme throughout his research is the emphasis on molecular scale systems analysis. Professor Rabitz has over 635 publications in the general area of chemical physics. He has been pursuing research in the control of quantum systems since 1986.

Herschel Rabitz

Krishnan Raghavachari received his Ph.D. in 1981 from Carnegie-Mellon University working with Professor John A. Pople on the development and applications of electron correlation techniques in Computational Quantum Chemistry. He then joined Bell Laboratories in Murray Hill, New Jersey, as a Research Scientist. He received the Distinguished Researcher award at Bell Laboratories in 1987. He joined Indiana University as a Professor of Chemistry in 2002. His work covers a broad spectrum of problems ranging from chemical bonding in small clusters to computational investigations of semiconductor and nanoscale materials. He is the author of over 250 papers in Chemistry, Physics and Materials Science. He was elected as a Fellow of the American Physical Society in 2001. He is among the 50 most cited authors in chemistry in the comprehensive period from 1981 to 1997.

Krishnan Raghavachari

Alessandra Ricca received her Ph.D. in chemistry in 1993 from the University of Geneva, Switzerland. Alessandra was a postdoctoral fellow at NASA Ames Research Center with Charles Bauschlicher and a postdoctoral fellow at Stanford University with Charles Musgrave. In 1998 she joined the Computational Chemistry group at NASA Ames Research Center as a Research Scientist and she is now member of the Computational Nanotechnology group at NASA Ames. Her research interests included transition metal chemistry, spectroscopy, medicinal chemistry, thermochemistry and astrochemistry. Her research efforts are currently devoted to nanotechnology.

Alessandra Ricca

Björn O. Roos received his Ph.D. in theoretical physics in 1968 at the University of Stockholm. He became Professor of Theoretical Chemistry at the University of Lund in 1983 and formally retired in June 2002. He is, however, still active as a project leader of the MOLCAS group. He is a member of the Royal Swedish Academy of Sciences and served as a member of the Nobel Committee for chemistry 1986–2000. Roos research has specialized around the description of electron correlation in molecular systems. He has published more than 300 scientific articles. Also, since 1989 Roos has been the leader of the European Summer school in Quantum Chemistry (ESQC).

Björn O. Roos

Dennis Salahub grew up, more or less, in Edmonton, Alberta, Canada. Serafin Fraga was his third-year quantum chemistry professor and that got it started. Fraga suggested Camille Sandorfy at the U de Montreal for grad studies — a superb suggestion — for starting to learn about science, and the world. A few postdocs, 23 years at the UdeM, three at NRC (Director General of the Steacie Institute for Molecular Sciences) in Ottawa, and the last two at the University of Calgary (Vice-President, Research & International), etc. — the common threads? the joy of science, theoretical chemistry, DFT and its applications, the deMon Developers, etc.

D.R. Salahub

George Schatz received his Ph.D. in chemistry in 1976 working for Aron Kuppermann at Caltech. He was a postdoc with John Ross at MIT 1975–6, and moved to Northwestern University in 1976, where he is now Morrison Professor of Chemistry. He has worked both in gas phase and materials theory, including dynamics calculations, electronic structure studies and classical electrodynamics.

George C. Schatz

Steve Scheiner received his Ph.D. in 1976, using theoretical methods to dissect the factors that influence enzyme activity, under the direction of William Lipscomb at Harvard University. He then spent 2 years at Ohio State University as a Weizmann Fellow with William Kern, considering ways in which hydrogen bonds influence biomolecular structure and function. He joined the faculty of Southern Illinois University, Carbondale in 1978, where his research focused on H-bonds and proton transfers. He moved to Utah State University in 2000. He reluctantly admits that he recalls all too vividly punching what seemed like thousands of computer cards for his early calculations.

Steve Scheiner

Reinhard Schinke received his Ph.D. in physics in 1976 at the University of Kaiserslautern, Germany, working in the field of molecular collision theory. In 1980, after 1 year at the IBM research laboratory in San Jose, California, as a postdoc, he entered the department of molecular interactions at the Max-Planck Institute for Fluid Dynamics in Goettingen where he has remained since. His research switched from collisions to the area of photodissociation and more recently to unimolecular reactions. Currently he studies the recombination of ozone with particular emphasis on a dynamical explanation of the pronounced isotope effect, which has been observed both in the atmosphere and in the laboratory. Throughout his scientific career, he has tried to understand experimental observations on the basis of accurate potential energy surfaces and exact dynamics calculations.

Reinhard Schinke

H. Bernhard Schlegel received degrees from the University of Waterloo and Queen's University, Canada (Ph.D., 1979, with S. Wolfe). After postdoctoral studies at Princeton University with K. Mislow and L.C. Allen and at Carnegie-Mellon with J.A. Pople, he joined the Merck, Sharp, and Dohme Research Labs. Since 1980, he has been Professor of Chemistry at Wayne State University. His current research interests include the development of methods for exploring potential energy surfaces for chemical reactions and applying computational quantum chemistry to the study of problems in physical, organic, inorganic and materials chemistry.

H. Bernhard Schlegel

Mike Schmidt received his Ph.D. under the direction of Klaus Ruedenberg from Iowa State University in 1982. He joined Mark Gordon's group directly after graduate school, and has been involved in developments for the GAMESS program ever since. Since 1992 he has been employed by the Department of Chemistry at Iowa State University, except for a year's appointment at Tokyo Metropolitan University. His research interests are MCSCF wavefunctions and their properties, and main group inorganic chemistry.

Michael W. Schmidt

Gustavo Scuseria received his Ph.D. in physics in 1983 working under the direction of Ruben H. Contreras at the University of Buenos Aires. He came to the University of California at Berkeley for postdoctoral work with Fritz Schaefer. In 1989, he joined the chemistry department at Rice University where he is now the Robert A. Welch Professor of Chemistry. He recalls compiling his first FORTRAN programs from boxes of cards on IBM 360 machines and the joyful advent of the VM/CMS operating system supporting 2 MB of virtual memory back in the early 1970s, while still a student. His first TACC publication on vibrational corrections to calculated nuclear spin—spin coupling constants appeared in 1979.

Gustavo E. Scuseria

Sason Shaik received his Ph.D., with Nicholaos D. Epiotis, from the University of Washington. He spent a post doctoral year with Roald Hoffmann at Cornell. In 1980 he started his first academic position at Ben-Gurion University. In 1992 he moved to the Hebrew University, where he is currently the Director of the Lise Meitner Center for Computational Quantum Chemistry. His research interests are in the use of quantum chemistry, and in particular of valence bond theory, to develop paradigms which can pattern data and lead to the generation of new problems. He still recalls the strong sense of creativity while doing the research described in Ref. [5] of his chapter. He started his P450 research in 1998, and has been fascinated ever since!

Sason Shaik

Jack Simons earned his Ph.D. in theoretical chemistry at the University of Wisconsin, Madison in 1970, working with John Harriman as an NSF Fellow. He held an NSF Postdoctoral Fellowship at MIT from 1970 to 1971, working with John Deutch. After that, he joined the faculty at the University of Utah in 1971 where he was appointed to the Henry Eyring Chair in 1989. He has published 300 papers, many dealing with negative ions, and three textbooks on theoretical chemistry, and he supports a web page (http://simons.hec.utah.edu) on this subject. Jack and his wife, Peg, are avid backpackers and skiers who enjoy living in Utah.

Jack Simons

Adalgisa Sinicropi received her degree in chemistry (Laurea in Chimica) cum Laude at the University of Siena in 1999. She received her Ph.D. in chemistry in 2002 submitting a thesis entitled "The Mechanism of Light Energy Wastage and Exploitation in $^1n,\pi^*$ Chromophores". Within her research project she collaborates intensively with M. Olivucci on the application of high-level quantum chemical methods to the mechanistic investigation of photochemical reactions and the study of different aspects of the stereochemistry of organic photoreactions. She has gained extensive experience in computational chemistry that she is now applying to the study of photobiological systems.

Adalgisa Sinicropi

Zdenek Slanina, native of Bohemia, received his M.A. degree from Charles University of Prague and his Ph.D. degree from the former Czechoslovak Academy of Sciences. Since the demise of the Academy, he has been sailing the suddenly accessible high seas, researching and teaching abroad at institutions like Max-Planck Institute for Chemistry, University of Arizona, National Chung-Cheng University, Toyohashi University of Technology and others, pursuing both cluster science and Pacific-rim cultures.

Zdenek Slanina

Viktor Staroverov received his undergraduate education in Minsk, Belarus. In 1997, he completed an M.S. program in chemistry at Brock University under the direction of Stuart M. Rothstein. He received his Ph.D. in chemistry in 2001 from Indiana University working under the supervision of Ernest R. Davidson. In the same year, he joined Gustavo Scuseria's research group at Rice University as a postdoctoral fellow. While in college, he was on track to become a synthetic organic chemist, but found his true calling in theory and computation after several encounters with theoretical chemistry. The earliest of these formative experiences was a course project on isomer enumeration.

Viktor N. Staroverov

Krzysztof Szalewicz received his M.S. and Ph.D. degrees in chemistry from the University of Warsaw in 1973 and 1977, respectively, under the supervision of Wlodzimierz Kolos. He held various permanent and visiting positions at the University of Warsaw, University of Florida, University of Cologne and University of Uppsala. In 1988 he came to the University of Delaware where he is now a Professor of Physics and Chemistry. He was a Visiting Fellow at JILA, Harvard, and Princeton and is a Fellow of the APS. His first computational work included feeding paper tapes with Algol codes into a GIER computer. His research interests have been in many-body theory of electronic structure including the use of explicitly correlated bases, in perturbation theory of intermolecular forces, simulations of condensed phase, chemical effects in beta decay and properties of exotic molecules (containing muons, antiprotons, etc.).

Krzysztof Szalewicz

P. Tarakeshwar was born in Kharagpur, India. After obtaining his Ph.D. in 1994 from the Indian Institute of Technology in Kanpur, he worked as a Scientist in drug design in a pharmaceutical industry in India. In 1997, he joined Professor Kim's group in Pohang wherein his research was focused on computational investigations of intermolecular interactions, molecular clusters containing π systems and nanomaterial design. He is currently working as an Assistant Professor in Computational Sciences at the Korea Institute for Advanced Study in Seoul.

P. Tarakeshwar

Ajit J. Thakkar

Ajit Thakkar, born in Poona, India in 1950, left home at 17 to explore the West. A circuitous route led him to Queen's University in Kingston, Ontario. A summer job programming calculations of virial coefficients and transport cross-sections using FORTRAN IV, dreadful JCL, and punched cards on an IBM 360/50 drew him to computational chemistry. In 1976, he completed a Ph.D. in theoretical chemistry guided by Vedene Smith and influenced by Robert Parr. His faculty career began at the University of Waterloo and, since 1984, continued at the idyllic Fredericton campus of the University of New Brunswick. He is now a University Research Professor, and author of more than 200 articles on molecular properties, electron densities and intermolecular forces.

Walter Thiel

Walter Thiel received his Ph.D. degree in 1973 at the University of Marburg. He spent two postdoctoral years with M.J.S. Dewar at the University of Texas at Austin and then returned to Marburg for his Habilitation. He was Professor at the University of Wuppertal from 1983 to 1992, Visiting Professor at the University of California at Berkeley in 1987, and Full Professor of Chemistry at the University of Zurich from 1992 to 1999. He is now Director at the Max-Planck-Institut fuer Kohlenforschung in Muelheim an der Ruhr. Since 2001 he is affiliated with the University of Duesseldorf as an Honorary Professor. He received the Heisenberg Fellowship in 1982, the Alfried-Krupp Award in 1988 and the WATOC Schroedinger Medal in 2002.

Donald G. Truhlar was born in Chicago in 1944. He received a B.A. in chemistry from St Mary's College of Minnesota and a Ph.D. in chemistry from Caltech. He joined the faculty of the University of Minnesota in 1969, and currently he is Lloyd H. Reyerson Professor of Chemistry and Director of Minnesota Supercomputing Institute. His research interests include theoretical and computational chemical dynamics, chemical structure, statistical mechanics, biochemistry, chemical physics, scientific computation and nanoscience. His honors include an Alfred P. Sloan Fellowship, APS and AAAS Fellowship, an NSF Creativity Award, the George Taylor/Institute of Technology Alumni Society Distinguished Service Award, the ACS Award for Computers in Chemical and Pharmaceutical Research, the Minnesota Award and the NAS Award for Scientific Reviewing.

Donald G. Truhlar

Takao Tsuneda has developed exchange-correlation functionals and correction schemes of density functional theory for calculations of large-scale molecules. He studied under Professor K. Hirao's supervision at the University of Tokyo, and received his Ph.D. in 1997. After the postdoctoral research, he worked as Assistant Professor since 1999. Recently, he took up his new post, Associate Professor, at the University of Tokyo in June, 2004.

Takao Tsuneda

Ad van der Avoird

Ad van der Avoird studied chemical engineering at the Technical University in Eindhoven, The Netherlands, from 1959 to 1964. From 1964 to 1967 he worked at the Battelle Institute in Geneva, Switzerland, and from 1967 to 1971 at the Unilever Research Laboratory in Vlaardingen, where in 1968 he became Head of the Molecular Physics section. In 1968 he obtained his Ph.D. degree at the Technical University in Eindhoven and in the same year he became Part-time Professor at the University of Nijmegen. In 1971 he became Full Professor of Theoretical Chemistry in Nijmegen. Since 1979 he is a member of the Netherlands Royal Academy of Arts and Sciences (KNAW) and since 1997 a member of the International Academy of Quantum Molecular Science.

A. Vela

Alberto Vela was born in Mexico City. He received his B.E. in the Faculty of Chemistry-UNAM. He obtained his Ph.D. in science in 1988 in the Department of Chemistry at UAM-Iztapalapa under the supervision of J.L. Gazquez. In 1993 he was a Visiting Researcher, for almost 2 years, in Professor Dennis R. Salahub's group where he started his journey with the program deMon. From 1983 to 1997 he was Full Professor of the Department of Chemistry at UAM-I where he participated in the creation of the Theoretical Physical Chemistry group. After 14 years in UAM he moved to Cinvestav where he started the Theoretical Chemistry group in 1997. In this later institution he is Full Professor of the Department of Chemistry.

Jacques Weber received a Ph.D. in physics (1969) from the University of Geneva (Switzerland). He was then a Research Fellow at the Quantum Theory Project of the University of Florida (Gainesville, FL) and at the IBM Research Laboratory (San Jose, CA). In 1976, he came back as a Lecturer to the Chemistry Department of the University of Geneva. He was appointed Full Professor of Computational Chemistry in 1989, occupying the first chair of that type created in Switzerland. His long experience in computational chemistry extends from the use of a card puncher for a CNDO program code to the development of methodologies and large-scale applications in density functional theory.

J. Weber

Charles Edwin Webster was born in Florida and graduated from the University of West Florida with a B.S. degree in chemistry/biochemistry in 1995. He received his Ph.D. in chemistry in 1999 at the University of Florida where he worked with Russell S. Drago, Michael C. Zerner, Michael J. Scott and C. Russell Bowers. He then joined Michael B. Hall's research group at Texas A&M University and is currently an Assistant Research Scientist. His research focuses on using theoretical and computational chemistry to research and answer questions in a variety of areas, including biological enzyme catalysis, catalytic and stoichiometric mechanisms of bond activation and functionalization of organic molecules by organometallic transition metal complexes, and the elucidation of structure and bonding of various compounds of interest.

Charles Edwin Webster

M.-H. Whangbo received his Ph.D. in chemistry in 1974 working under the direction of V. H. Smith, Jr. and S. Wolfe at Queen's University, Canada. After postdoctoral studies with S. Wolfe at Queen's University and R. Hoffmann at Cornell University, he joined the faculty of North Carolina State University in 1978. Throughout his scientific career, he has explored qualitative structure–property relationships in discrete molecules, extended solids and surfaces on the basis of formal theory developments and electronic structure calculations.

Myung-Hwan Whangbo

Paul E.S. Wormer was born in Amsterdam, The Netherlands. He studied chemical engineering at the Technical University in Delft, where he got his masters degree (cum laude). He got a Ph.D. degree (cum laude) at the University of Nijmegen in theoretical chemistry (supervisor A. van der Avoird). Currently he is Associate Professor at the Radboud University (Nijmegen, The Netherlands). Several times he has been Visiting Research Professor at the University of Waterloo (Canada) where he collaborated with J. Paldus. P.E.S. Wormer is the (co)author of *ca.* 150 peer-reviewed research papers.

Paul E.S. Wormer

LinLin Zhao did her undergraduate work at Sichuan University in China, and is now a Graduate Student at Northwestern University working with Professor George Schatz. Her work has included studies of electrodynamics and quantum chemistry, including the use of electronic structure theory to calculate optical properties of molecules in the presence of metal clusters.

LinLin Zhao

Guishan Zheng received his Master's Degree in polymer science and physics in 2001 working under the direction of Professor Xue Gi in Nanjing University, China. After that, he joined Professor Keiji Morokuma's group in Emory University in 2001 as a Ph.D. student. Currently, his research interest is mainly in nanoscale chemistry and the development of density functional based tight binding method.

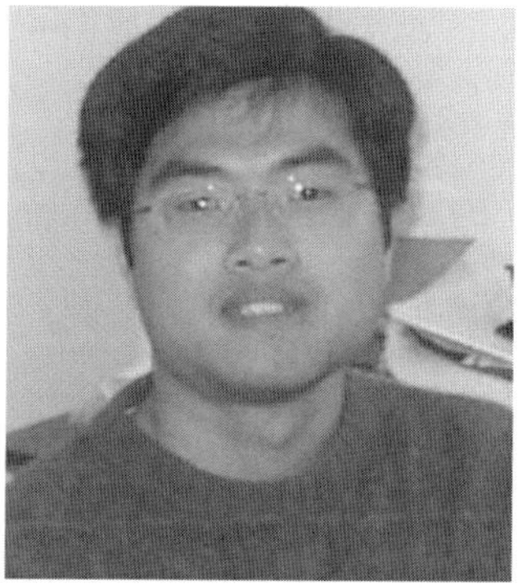

Guishan Zheng

Shengli Zou received B.S. and M.S. degrees in physical chemistry from Shandong University in China and a Ph.D. degree in physical chemistry from Emory University (with Professor Joel M. Bowman) in 2003. He has been a postdoctoral fellow in Professor George C. Schatz's lab at Northwestern University since December 2002. His research interests include the optical properties of nanoparticles and nanoparticle arrays and their applications in biological sensing. He is also interested in the self-assembly of biomolecules.

Shengli Zou

Index

absorption bands 48–9
absorption spectrum 1111–16, 1120–3
ACES/ACES2 467, 474, 1167, 1200, 1206
acetaldehyde 414
acetylene 308–9, 347–54, 499, 1132–3
ACGTO–DFT 1088
acidity 833, 836, 839–40, 843
actinides 743–4
activated rate processes 430–1
activation
  barriers 788, 803, 1040
  energy 72–5, 80–4, 465, 863, 1039–41
  enthalpy 477–8, 864–7, 871
active orbitals 513–19, 582–93, 601–6, 729–30, 740–50
active space 740–4
  see also complete active space
ADF 1086
adiabatic
  approximation 17, 74
  connection 377–8, 530, 675
  excitation energy 1130–1
  ionization 808
  potentials 74, 808
  reaction theory 68, 74, 120
  adsorbates 49, 56, 59, 78, 1100
adsorption 49, 821
AIDS 44–5
alkali fullerides 983–7
alkali metal clusters 315–16, 319–20, 748, 937, 948–9
alkanes 54, 575, 645, 845–6
alkenes 347–54, 836–7
alkyl radicals 645–6
alkylidenes 342
alkylidynes 342
alkynes 347–54, 836–7
all-valence-electron methods 576
AllChem project 1088, 1089
allowed reactions 646–7, 652, 735, 860, 864, 894
allyl radicals 748, 751–6, 860–1
alternant molecular orbitals (AMO) 120, 129
aluminum 314–22, 332–42, 401–2
AM1 166, 416, 791, 847
  Cope rearrangements 862–3
  isomeric fullerenes 893
  quantum chemical molecular dynamics 877–80
  semiempirical quantum-chemistry 561–2, 566–70, 573–6
AMBER 275–6, 278, 280, 283, 286
amino acids 179, 285, 835, 837, 851
amino group conformation 1108–9
2’-amino-4-ethynylphenyl-4’-ethynylphenyl-5’-nitro-1’-benzenethiolate 818–19
ammonia 216–17
  anharmonic vibrational spectroscopy 173, 187
  CH···O hydrogen bond 834–5, 837
  energy decomposition 311–13
  molecular conductance 820, 823–9
  molecular system simulation 101–2
  polyatomic molecule vibrations 259–60
ammonium cations 980–1
AMO *see* alternant molecular orbitals

AMPAC program 893
analytes 53
analytic derivatives 200, 570–1, 1082, 1178, 1206
analytical energy derivatives 132–40, 198–203, 214, 1175–6, 1203–5
analytical gradients 1176, 1185, 1191, 1200–8, 1211–15
Angle Resolved UPS (ARUPS) 1018
ANHAR 253
anharmonicity 78–9, 81–2
  coupling 135, 171, 186–7
  oscillators 135
  perturbation theory 168–73
  scaling factors 167
  vibrational frequencies 178, 181, 183–4, 187
  vibrational spectroscopy 165–90
  vibrational wavefunctions 170, 174
aniline 1025, 1029, 1106, 1111–12, 1137
anions 460, 971–6, 980–1, 1106–11
anisole proton affinities 808–9
anisotropy 53–4
annealing 886, 968, 1183
annihilation operators 123, 375, 391, 448–9, 453–5
Anonymous Parentage for the Inactive (API) excitations 582, 587, 593–9, 611–12, 614–30
anti-H-bonds 832, 838, 844
antibonding
  CH···O hydrogen bond 847–8
  electron-electron repulsion 773–83
  energy decomposition 323–5, 336
  multiconfigurational quantum chemistry 730–5, 749–55
antiferromagnetic spin exchange 776–7
antimony 302, 360–6
antisymmetrization 298–9, 927–8, 1048–52, 1061–2
API *see* Anonymous Parentage for the Inactive
approximate exchange functionals 681–2, 708–14
APW *see* augmented plane waves
aqueous clusters 967–73
argon
  crystals 933, 944–5
  fluids 433–4
  liquid-vapor equilibria 945
  trimers 932–7, 943–51
argon-hydrogen 932–7, 946–7, 1062
argon-hydrogen fluoride trimers 932–7, 946–7
argon-nitric oxide trimers 948
argon-oxide trimers 948–51
aromaticity 878–9, 979
arsenic 360–6
artificial intelligence 107
ARUPS *see* Angle Resolved UPS
asymptotic
  approximation 938
  corrections 530–1, 688–91
  expansion 491–2, 931, 939, 1089
  series 940
ATM *see* Axilrod-Teller-Muto potential
atomic
  charges 570, 1177
  ions 741
  orbitals 117–18, 458, 749–50, 1147–8
atomization energies 538–40, 698, 707–12, 793–4, 800
atomization heat 895
attraction 291, 293–4, 1048
Au–S bonding 816, 819
Au(111) surfaces 814–15, 817
augmented plane waves (APW) 1079
autocorrelation functions 81, 431
auxiliary functions 1081, 1082, 1085–9
avoided crossings 210–12, 459–60, 583, 617, 630, 637–62
Axilrod-Teller-Muto potential (ATM) 923–4, 931, 943–5, 949
azeotrope formation 833

ß-sheets 837, 851
ß-strands 851
B3LYP
  anharmonic vibrational spectroscopy 179–85, 189
  carbon nanostructure self-assembly 879–80
  CH···O hydrogen bond 841
  Cope rearrangements 864–6, 868–71
  deMon codes 1085, 1087
  DFT 530–40
  electrodynamics 61–2
  endohedral metallofullerenes 901–6
  exchange-correlation 699, 709–11, 715
  Fenske–Hall molecular orbitals 1145–7, 1150, 1154–6, 1161–3

G2 and G3 theories 788–91, 798–801, 806–8
GAMESS 1168, 1182
isomeric fullerenes 898–906
molecular conductance 816, 826–7
momentum 501–2
quantum chemical molecular dynamics 879–80
relativistic theory 547, 552
backdonation 326–40, 347–9, 353–63
bacterial photosynthetic reaction centers 1099, 1101, 1121–2, 1124–5
bacteriorhodopsin 276–7, 279–85
Baker–Campbell–Hausdorff (BCH) 23, 474, 596, 598
band gaps 713, 985
band structure 1016–41
conduction 1022–5
electron transfer 1033–41
non-linear optics 1025–33
photoelectron spectra 1016–20
semiconduction 1022–5
barium 314–22
barriers
activation 788, 803, 1040
heights 517–18
transition metal dimers 539
valence bonds 635–52
von Barth–Hedin scheme 677, 685, 697–8, 709–10, 1082
basis functions 199–200, 510, 534, 545
DFT 1081–3, 1089–90
equations of motion 458–60
exchange correlation 726, 746–7, 754
G2 and G3 theories 796, 800
semiempirical methods 569–70
basis set superposition error (BSSE) 534, 823, 1049–51
basis sets
anharmonic vibrational spectroscopy 189
bond functions 296
coupled-clusters 117
G2 and G3 theories 791–2, 795
Gaussian 3, 494, 754, 1083
Gaussian functions 754, 1083
Gaussian-orbital 60, 1083
midbond functions 929
molecular system simulations 93–7, 109
momentum 494–8
optimized 93
polarization function 95, 296, 814
SBKJC 61, 62
water hexamer energetics 1001
bathorhodopsin 282–5
BCH *see* Baker–Campbell–Hausdorff
BD *see* Brueckner doubles
BDE *see* bond dissociation enthalpy
Becke exchange functionals 296, 539, 715–16
see also B3LYP
Bell–Evans–Polanyi (BEP) principle 636
benzenes 476–7, 525–6, 814–19
benzoquinone 987–9, 1106, 1108–11
benzynes 476–7
BEP *see* Bell–Evans–Polanyi
Bergman reaction 477–8
beryllium 305–7, 314–22, 478, 612, 615–19
beta sheets and strands 837, 851
bifurcation 242
binding affinity 41–5
binding energy
carbon nanotubes 822–7
CH···O hydrogen bonds 840–3
core-electron 1125–6
energy decomposition 293
exchange-correlation 690
G2 & G3 theories 796
many-body forces 948
molecular conductance 820, 822–7
molecular system simulations 109
momentum 491
Monte Carlo simulations 997–8, 1001–4
oxygen 822–3
temperature dependence 1003–4
water hexamers 998, 1003–4
biological sensors 53–4
biomimetic molecular switches 285–7
biomolecular simulations 431
biomolecules
CH···O hydrogen bonds 834–5
molecular dynamics 425, 431, 435
semiempirical quantum-chemistry 572–6
biphenyl oligomers 1034
bipolarons 1025, 1038
bis(benzene)chromium 360–6
bismuth bonding 302–3
BLB *see* Extended Brillouin's
Bloch equations 136–7, 469–74
Brillouin–Wigner 469–70, 472

Bloch's theorem 1013, 1027
blue moon ensemble method 431
blue shift 280–2, 831, 837–9, 843–50, 970
BLYP 539, 715–16
BO *see* Born–Oppenheimer
Bofill's updating schemes 205
Bohr model 117
Boltzmann factors 69, 893
Boltzmann-dependent activation 1039
BOMD 1090
bond activation reactions 646–7
bond contraction 838, 844–8, 850
bond directional principle 497
bond dissociation energies 808–9
bond dissociation enthalpy (BDE) 861
bond energy 535–6, 845, 895
  decomposition 291–5, 340, 344, 362
  G2 and G3 theories 808–9
  multiconfigurational quantum chemistry 733, 756, 759
  valence bonds 645–6
bond exchange 658–9
bond lengths 551–2
  Brillouin–Wigner perturbation 467–8
  CH···O hydrogen 836–9, 845, 850
  coordinate systems 201
  Cope rearrangements 859, 864, 866–7, 870–1
  diastatine 552
  energy decomposition 296–315, 328, 334, 337, 340, 349–53
  exchange-correlation functionals 698, 715
  gold 547
  non-covalent binding affinity 43
  photobiology 283–4
  polymer chains 1021–2, 1030
  QM/MM optimizations 214
  SiC3 521
  transition metal dimers 539
  unimolecular reaction rates 406
bond strength 645, 833, 882–3, 969
  energy decomposition 307–11, 329, 338, 356, 366
bonding analysis 295, 320, 338, 347, 367
bonding models 291–5, 314–15, 322, 347, 362
bonding orbitals 733
borane-fluoro bonding 296–302
Born–Huang products 32–3
Born–Oppenheimer (BO) approximation 196, 1030, 1086, 1174
  coupled clusters 117
  intermolecular forces 1066–8
  momentum 483, 490–1
  polymer chains 1030–1
  potential energy surfaces 19
  simulations 415, 1090
  time-dependent molecular theory 17
  unimolecular reaction rates 415
boron bonding 305–9, 311–22, 332–42
boron fluoride (BF) 297–302, 311
boron nitride 907
boron-hydrogen 612–13, 624–7
boron-nitrogen 311–14
boron-phosphorous 311–14
boson operators 378–80, 394–5
BOVB *see* breathing orbital valence bonds
bovine visual pigments 269, 276–7, 279–85
BP *see* Breit–Pauli
bracketing 220–4
Brandow diagrams 123
breathing orbital valence bonds (BOVB) 642–3, 653–4
Breit–Pauli (BP) approximations 548
Breit–Pauli (BP) spin-orbit operators 1174–5
bridges 640–2
Brillouin zones 1013–15
  energy bands 1013–14
Brillouin–Wigner
  Bloch equations 469–70, 472
  coupled-clusters 465–79, 1193
  perturbation 465–79
  resolvent 465, 472–5
broken-symmetry (BS) states 777
bromine 155, 303–5, 526–9
Brønsted parameters 641–2
Brueckner doubles (BD) 794
BS *see* broken-symmetry
BSSE *see* basis set superposition error
Buckminsterfullerenes 444, 875–8, 885–7, 892–8, 983–4
  see also fullerenes
bulk dielectric constants 50–1
bulk water 969

C-conditions 138–9, 473–6
C–H bond dissociation 808, 861

C–H stretching frequencies 831, 837–9, 843–50
C32, C35, C70 885–6, 892, 895–8
C_{60} 444, 788, 875–8, 885–7, 892–8, 980–4
Ca@C72 898, 901–3
$CaFeO_3$ 779–80
calcium 314–22, 779–80, 898, 901–5, 1068–9
calix-4-hydroquinone (CHQ) 984–7
calix[4]quinone-hydroquinone (CQHQ) 985–6
canonical coherent states 23–4
Canonical Van Vleck Perturbation Theory (CVPT) 169
capping processes 875–7, 880–7
Car–Parrinello
  DFT deMon codes 1085, 1086
  unimolecular reaction rates 415
Car–Parrinello molecular dynamics (CMPD) 107, 433, 881, 926
carbenes 342–7, 646–7
carbide diatomics 478
carbocylic conjugated systems 981–2
carbohydrate crystals 837
^{13}C NMR 902–3, 905
carbon
  bonding 305–10, 314–22, 332–47, 406–7
  carbon nanotubes (CNT) 574, 813–14, 820–9, 875–87, 982–3
  carbon–carbon bonds 38, 878
    Cope rearrangements 859–60, 871
    energy decomposition 305–6, 309
    lengths 406–7, 755, 871
    polymer chains 1017, 1021
    strengths 882–3
  clusters 883–7, 892, 893–6
  energetics 893–6
  hyperfine splitting constants 1135
  isomers 875–87
  monophosphides 314–22, 332–42
  nanochemistry 878
  nanomaterials 983–7
  nanostructures self-assembly 6, 875–87
  nanotubes 574, 813–14, 820–9, 875–87, 982–3
  nucleophilic substitution 654–6
  vapor 896
  see also CH···O hydrogen bonds; fullerenes
carbon monoxide
  energy decomposition 296–302, 326–32, 357–60
  excited states equilibrium geometries 1129–32
  Fenske–Hall molecular orbitals 1143, 1151–63
  helium interaction 1068–9
  molecular dynamics 427
  transition metal bonding 326–32
carbonic difluoride 789
carbonyl complex bonding 326–32
carboxy-terminated alkane thioate 54
carbyne complex bonding 342–7
Cartesian coordinates 1056, 1083–5
  momentum 428–9, 483–4, 493
  polyatomic molecules 261, 264
  potential energy surfaces 201–2, 209–14, 228, 231, 241
  time-dependent molecular theory 9–14, 25, 33–6
CAS *see* complete active space
Casimir–Polder integrals 1057
CASPT *see* complete active space perturbation theory
CASPT2 *see* second order perturbation theory
CASSCF *see* complete active space self-consistent fields
CASSI *see* complete active space state interaction
CASVB *see* complete active space valence bonds
catalytic mechanisms 982–3
catalyzed bond activation 646–7
cations 494, 972–3, 977–81
CBS methods 787
CC *see* coupled-clusters
CC-LRT *see* coupled-cluster linear response theory
CCD *see* coupled-cluster doubles
CCSD
  ab initio quantum chemistry 1197–202, 1206–7, 1209, 1211–12, 1214
  anharmonic vibrational spectroscopy 179
  Brillouin–Wigner perturbation theory 465–9, 471–9
  CH···O hydrogen bond 840–1
  coupled-cluster theory 127, 129–34, 136–8
  G2 and G3 theories 786, 791, 794, 799, 808

GAMESS 1170, 1172
intermolecular forces 1063–4
many-body forces 928–9, 935–7, 943, 947
molecular conductance 823, 826
molecular spectroscopy 1130–1
multireference perturbation 520–3
size-consistent state-specificity 583–5, 599, 617
structure stability 521–3
valence bond diagrams 643
water trimers 947–8
CCSD(T)
anharmonic vibrational spectroscopy 179
Brillouin-Wigner perturbation 465–9, 474, 477–8
CH···O hydrogen bond 840–1
coupled-cluster theory 129–34, 137, 1197–202, 1206–14
G2 and G3 theories 786, 791, 794, 799, 808
GAMESS 1170–4, 1180, 1185
intermolecular forces 1063–6
many-body forces 935–7, 943, 947
molecular conductance 823, 826
molecular spectroscopy 1130–1
size-consistent state-specificity 583–4
structure stability 521–3
valence bond diagrams 643
CEBE *see* core-electron binding energy
cellular automata 107
centroid path integral dynamics 435
CEPA *see* coupled electron pair approximation
Ceperley–Alder data 684–5
cesium bonding 314–22
cesium xenon systems 1101, 1112–15
CF *see* Coulson–Fischer
CG-DMS *see* conjugate gradient density matrix search
chair-like transition states 859–61
chameleonic transition states 859–72
charge conductivity 382–4
charge density 130, 298, 483, 658, 977
fitting 1081–2
Hilbert space 387–9
intermolecular forces 1060, 1079
matrices 728, 778
polymer chains 1038
charge distribution 282, 1070–1
DFT 1177, 1181
energy decomposition 296–8, 327, 333, 336–7
many-body forces 932, 941
semiempirical quantum-chemistry 563, 575
charge transfer (CT)
aniline 1108–9
CH···O hydrogen bonds 843–4
excitation energies 534–8
polymer chains 1020–5, 1034
states 81, 575, 657, 1020, 1111, 1173
valence bond diagrams 649–50
charged solitons 1023–5
CHARMM 275–6
chemical accuracy 508–9, 785–7, 1085
chemical bonds 527–8
breaking 155
Brillouin–Wigner perturbation 466
configuration interaction 773–4
coupled-cluster theory 118, 133–6
electronic structure changes 730–4
energy decomposition 291–367
fullerenes 877
main-group compounds 291, 294–326
molecular spectroscopy 1125, 1127
molecular system simulations 90, 108, 111
momentum 497
multiconfigurational quantum chemistry 726, 730–3, 743, 749, 756
photonic reagents 149, 155
polymer chains 1035, 1038
transition metal compounds 291, 294–5, 326–66
valence bond diagrams 635
chemical enhancement 59
chemical reactions
allowed 646–7, 652, 735, 860, 864, 894
forbidden 517, 637, 641, 646–7, 652, 735
radicals 187–8, 644–6, 654, 748, 751–6
chemical reactivity 94, 275, 466, 635–65, 1022
chemical sensors 53–4, 820–9, 965, 980–3
chemical shifts
NMR 570–4, 843–5, 1101, 1115–20
transition metal complexes 354, 1101, 1119, 1137

chemisorption 78, 823–4, 1081, 1084
chirality 875, 880, 893, 907
chlorine
  CH···O hydrogen bonds 835
  energy decomposition 303–5, 311–14, 354–7
  fluorides 809
  hydrogen exchange 526–9
  photonic reagents 155
1-chloro 2-fluorethane 212–13
chloroform 832
CH···O hydrogen bonds 831–52
  C–H stretching frequencies 831, 837–9, 843–50
  early thinking 832–7
  frequency shifts 831, 837–9, 843–50
CHQ *see* calix-4-hydroquinone
chromium 733–5, 740–2
  chromium-hydrogen 741–2
  dimers 539, 741–2
  energy decomposition 347–60
  polymer chains 1034–5
  tetraoxide 1116–18
chromophores 1099, 1101, 1121–3
chromyl chloride 1115–16
CI *see* configuration interaction; conical intersections
clamped-nuclei 919–21
classical coherent states 23–4
classical electrodynamics 47–58
classical electrostatic attraction 293–4
classical mechanics 68–72
classical molecular dynamics 428–32
Clebsch–Gordon coupled product 1057
closed operators 609–10
closed-shell
  atoms 919, 948, 1047–72
  complexes 950
  configuration 136, 627
  molecules 92, 100, 564, 726, 744, 1048–9
  states 582
  wave functions 730–1
clusters
  alkali metal 315–16, 319–20, 748, 937, 948–9
  amplitudes 592–7, 605
  equations of motion 455–7
  hydrogen-bonded 171, 182–5
  interaction energy 942–8
  many-body forces 919–58
  open-shell 948–51
  operators 455–7, 592–7
  rare-gases 921, 924, 945–6
  size-consistent state-specificity 592–7
  surface enhanced Raman Spectra 59–63
  van der Waals 176, 185
  see also coupled-clusters
CMPD *see* Car–Parrinello molecular dynamics
CMS *see* complete model space
CNC polyatomic molecules 1132–3
CNDO *see* complete neglect of diatomic differential overlap
CNT *see* carbon nanotubes
coarse grained methods/models 435–7, 1175, 1179–80
cobalt ions 1034, 1035
coherence/coherent states 21–32, 37, 159–60
collagen 837
Colle–Salvetti formula 701–4, 1082
collective hydrodynamics 427
collision energy 36
collision-induced absorption 1099–1101, 1112–15
collision-induced infrared spectra 1058–9
COLUMBUS 760
combination mode transitions 186–7
commutators 376–8, 394, 447–8, 453–5, 592–3
complete active space (CAS)
  Brillouin–Wigner perturbation 475–6
  intermolecular forces 1065–6
  multiconfigurational quantum chemistry 725, 739, 744–8
  perturbation 512–17
  size-consistent state-specificity 583, 586–93, 603, 616, 625–9
  valence bonds 524–6
  wave functions 725, 744
complete active space perturbation theory (CASPT) 586–7, 1065–6
complete active space self-consistent fields (CASSCF)
  Cope rearrangements 862–5
  coupled-cluster theory 119
  electronic structure 508
  GAMESS 1168–70, 1179, 1184
  intermolecular forces 1065–6

many-body perturbation 511–14, 522–3
molecular spectroscopy 1101, 1133
molecular system simulations 101
multiconfigurational quantum chemistry 725, 729–30, 739–48, 754–7, 760–1
multireference perturbation 508–14
photobiology 273–6, 278–83, 286–7
size-consistent state-specificity 584, 613, 616, 625, 628
state interactions 748, 757
valence bonds 523–9
water molecules 729–30
wave functions 739–44
complete active space state interaction (CASSI) method 748, 757
complete active space valence bonds (CASVB) 523–9
complete intermolecular potential surfaces 1047–72
complete model space (CMS) 136–8, 471–8, 587–93, 606–10
complete multi-configurational SCF 101
complete neglect of diatomic differential overlap (CNDO) 61, 560–3, 568, 834, 1016
complex spectra elucidation 186
complex spin unrestricted Thoulles determinant 33
complexes
acetylene 342–7, 349, 353–4
carbene 342–7
carbonyl 326–32
carbyne 342–7
dihydrogen 357–60
diyl 332–42, 358
ethylene 349, 353–4
Fischer 342–7
octahedral 322, 329
phosphane 313–14, 347–57, 360
Schrock 342–7
compressed coupled-clusters 1215
Compton
profiles 489, 492–5, 499–500
scattering 491–3
computational electrodynamics 47–64
discrete dipole approximations 47, 50, 60
finite difference time domain 47, 50, 61
multiple multipole method 50
computer experiments 437, 1018–19, 1022
concerted mechanisms 646, 862–3, 871
condensed phases 77–83, 153, 430, 919–58
conductance/conduction 48–53, 373–95, 813–29, 937, 1020–5
configuration interaction (CI)
ab initio quantum chemistry 1195–6, 1203
anharmonic vibrational spectroscopy 170–1, 180
coupled-clusters 118–20, 129, 137–8, 1195–6, 1203
direct 739
electron-electron repulsion 773–4
expansions 739
full 611, 614–15, 727, 1168–72, 1174
GAMESS 1168–74
intermolecular forces 1051, 1055, 1066
large-scale 132, 479, 571, 739, 761, 1213
molecular system simulations 98–9
perturbation selection 587
polyatomic molecules 254–63
quadratic 611, 614–15, 727, 1168–72, 1174
singles 61, 200, 561, 1173–4
size-consistent state-specificity 587
valence bond diagrams 642–3
vibrational 254–63
see also multi-reference-configuration interaction; Symmetry-Adapted Cluster–CI
configuration mixing diagrams 637–8, 652–8
configuration state functions (CSF) 515–17, 524, 1172
multiconfigurational quantum chemistry 727, 734, 739–45
size-consistent state-specificity 586–93, 598, 602, 613–16, 619
configurational heat capacity 1000, 1004
configurations, cyclic 934–5
conformational changes 42–5, 285–7, 419, 431, 1018–19
conical intersections (CI) 210–12, 270–2, 275–6, 282–3, 659–62, 742, 1176–8
conjugate gradient density matrix search (CG-DMS) 572
conjugated organic molecules 1101–2, 1106–12
excitation 1101, 1106–12
ionization 1101, 1106–12
spectra 1101, 1106–12
conjugated polymers 1034

connected cluster theorem 121–3
connected diagrams 121–2, 473–5
connectivity 598–9
constrained ensembles 430
constraint satisfaction 681, 688–99
continuum solvent models 1184
continuum treatments 43–4
control fields 150–62
control landscapes 157–8
coordinates
 anharmonic vibrational spectroscopy 170–9
 molecular Hamiltonians 11–16
 potential energy surfaces 201–2, 209–31
 scaling 679–80
 variational transition states 68–83
coordination shells 954–6
Cope rearrangements 859–72
 AMI calculations 862–3
 bond making/breaking 860–4, 871
 calculations 859–72
 dynamic electron correlation 863–5
 interallylic distances 860, 863, 865–7, 869, 871
 kinetic isotope effects 862, 865, 871
 MINDO/3 calculations 862–3
 substituent effects 859, 865–71
copper bonding 347–54
copper tetramers 60–1
copper–pyridine complex 60–3
core states 1100, 1125–8
core-electron binding energy (CEBE) 1125–6
core-ionization spectra 1125
core-valence correlation 795
correction schemes
 asymptotic corrections 530–1, 688–91
 DFT 507–8, 529–40
 electron correlation 1134
 exchange-correlation 690–1, 697
 G2 and G3 theories 787, 791–2, 795–804
 intermolecular forces 1066
 orthogonalization 566–7
correlation
 ab initio quantum chemistry 1204, 1206
 correlation-corrected VSCF 171–2, 174–80, 181–5
 coupled-clusters 130, 1204, 1206
 DFT 507, 669–717
 electrons 5, 118, 124, 730, 736–8, 1134
 energy
  density-gradient expansions 687
  molecular system simulations 94–9, 106–7
  multiconfigurational quantum chemistry 736–8
  scaling 803–7
 G3 theory 795–6
 holes *see* exchange-correlation holes
 lengths 701, 704
 local density approximations 684–6
 potentials 501, 703, 714–15, 761
  see also exchange-correlation
 wavefunctions 701–3, 1168
corresponding states law 1048
Coulomb
 forces 427–8
 holes 99, 736, 738
 integrals 1089
 interactions 78, 493, 563, 712–14, 1047, 1054, 1181
 plasmas 135
 repulsion 779–80
 screening 712–14
 terms 1148
Coulombic Hamiltonian 12–17
Coulombic systems 679
Coulson–Fischer (CF) orbitals 731–3
counterintuitive orbitals 770–2
counterions 280–2
counterpoise method 534, 833–41, 922, 1087
 corrections 833–5, 932, 1050, 1054
coupled electron pair approximation (CEPA)
 ab initio quantum chemistry 1195–7
 size-consistent state-specificity 582–3, 587–9, 592–3, 602–5, 610–15, 618–30
coupled-cluster doubles (CCD)
 ab initio quantum chemistry 1195–202, 1206
 intermolecular forces 1053, 1063
 theory 119, 125–34, 1195–202, 1206
coupled-cluster linear response theory (CC-LRT) 1101, 1207, 1212
coupled-clusters (CC)
 amplitudes 1063
  Brillouin-Wigner perturbation 472–6
  equations of motion methods 456–60
  internal 138

size-consistent state-specificity 586–605
theory 125–39, 1214–15
ansatz 174–5, 468–9, 472, 1063, 1194, 1209, 1213
Brillouin–Wigner perturbation 465–79, 1193
expansion 124, 131–3, 466–8, 585, 1100–2
extended 135
externally corrected 137, 468–9
functionals 1205–7
G2 and G3 theories 786–7, 791, 794
GAMESS 1172–3
intermolecular forces 1063–4
linear 125, 1197
many-body perturbation 119–20, 122–8, 133, 1193–7
multi-reference 135–9, 465–79
state specific 581–631
state universal 468, 474, 585–6
state-selective 137
valence universal 136, 1213
origins 115–40
quantum chemistry 1191–216
SAC–CI method 1101
single-reference 466–77, 582, 602–3, 1192, 1201–2, 1210–12
size-consistent state-specificity 581–631
theory 115–40, 1191–216
coupled-pair many-electron theory (CPMET) 131–2, 1195
coupled-perturbed (CP)-MOD equations 1105
coupling constants 129–37, 509, 673, 753, 1087–8, 1212
coupling matrix elements 241–2
covalent bonds 293–366, 525–9, 644–6, 652–3, 731, 837
covariance functions 1000, 1005
Cp_2Ni_2 1143, 1151–6
CPHF coefficients 1104
CPMET *see* coupled-pair many-electron theory
CR-CCSD(T) 134, 1173
R-CCSD(T) 1066
creation operators 30, 123, 375, 391, 446–9, 453–5
critical phenomena 135, 427
cross sections 36–7
crossing point heights 642–3
crossing resonance energy 643
crystal-vapor interfaces 78
crystals 78
argon 933, 944–5
cohesion energy 943–5, 1059
rare-gas 923–4, 944–6
vacancies 945–6
CSF *see* configuration state functions
CT *see* charge transfer
current, density 387–9, 678
current-voltage (I–V) curves 813–19, 827–8, 1111
curve crossing 637
curvilinear coordinates 76
cusps 736–7, 1134–6
CVPT *see* Canonical Van Vleck Perturbation Theory
CVT rate constants 74–5
cyclic configurations 934–5
cyclic polyenes theory 130
cycloadditions 271, 635, 646–7, 882
cyclobutadiene 478, 659
cyclobutane 271
cyclohexane-1,4-diyl 212, 860–9
cyclophane systems 988–9

DALTON package 760
damped gradient corrections 690–1
damping constants 60
Datacraft 6024 minicomputers 4
Davidson correction 1066
DC *see* Dirac–Coulomb
DCB *see* Dirac–Coulomb–Breit
DCD *see* Dewar–Chatt–Duncanson
DDA *see* discrete dipole approximation
DDI *see* distributed data interface
defect transport 1038
degeneracy
anharmonic vibrational spectroscopy 179–80
chair Cope rearrangements 859–61
molecular system simulations 98–9, 101
multiconfigurational quantum chemistry 734–8
multireference coupled-clusters 465
open-shell monomers 1065–6
perturbation 179–80, 769–70
SAC–CI method 1105
degrees of freedom 14–17, 37–8, 196–7, 1202–7

deMon codes 1079–92
denitrogenation 273–4
density
  difference 298–9, 497
  functional approximation 677–9
  Laplacians 702–3
  momentum 487–93
  number density 484, 490–1, 495
  reciprocity 495–6
  shifts 847–50
density functional theory (DFT)
  ab initio methods 507–8, 529–40
  anharmonic vibrational spectroscopy 182
  B3LYP 530–40, 1085, 1087
  carbon nanostructure self-assembly 879–82
  conceptual 1085, 1146
  constrained-search 1082
  conventional correction schemes 529–32
  Cope rearrangements 859–72
  correction schemes 507–8, 529–40
  deMon codes 1079–92
  design strategies 680–2
  designing functional molecular systems 966–7
  exchange-correlation 669–717
  Fenske–Hall molecular orbitals 1143–7, 1150–1, 1161–3
  force fields 182
  GAMESS 1168
  intermolecular forces 1064–5
  isomeric fullerenes 900
  Jacob's ladder 678
  local-scaling transformation 671
  long-range corrections 529, 532–40
  molecular system simulations 95–103, 108–9
  momentum density 501–2
  motivation for 671–3
  open-ended single-walled carbon nanotubes 880–2
  orbital-dependent 531, 674, 678, 716
  orbital-free 502, 671
  photobiology 272
  potential energy surfaces 197–8
  pyridine–copper tetramer 61–3
  quantal 671
  quantum mechanics/molecular dynamics 434
  spin-polarized 677–9, 1083, 1134
  symmetry-adapted perturbation 928, 952, 954–7, 1064–5
  tight-binding 814, 875–9, 1088–9
  water hexamers 997
  see also Kohn–Sham DFT
density gradient expansions (DGE) 680–1, 686–8, 692–3, 695–9
  second-order coefficient 686
density matrices
  equations of motion 448
  exchange-correlation holes 704–5
  expansions 678, 704–5
  Kohn–Sham 501, 677–8, 682
  momentum 487–9
  natural orbitals 727–30
  non-equilibrium Green Functions 376–7
  one-electron 494, 677
  reduced 487–9, 670–2, 678, 716, 726–8, 736
density of states (DOS) 397–409, 460, 820–9, 1016
derivatives *see individual entries*
design equations 151
designing functional molecular systems 963–89
  clusters 963–80
  flippers 965, 987–9
  ionophores 965, 980–3
  nanomaterials 963–7, 983–7
  nanotubes 963–7, 983–7
  receptors 965, 980–3
  sensors 965, 980–3
desolvation 41
deterministic canonical ensembles 429
deterministic functions 29–32
Dewar–Chatt–Duncanson (DCD) model 326, 333
DFT *see* density functional theory
DGAUSS 1085, 1086
DGE *see* density gradient expansions
DHF *see* Dirac–Hartree–Fock
diabatic curves 639, 643–5, 649
diabatic states 1068
diagrams
  Brandow 123
  connected 121–2, 473–5
  disconnected 121–2, 473–5
  exclusion-principle-violating 130, 366, 582, 587, 603, 927, 949

Feynman-like 121, 124, 433–4
Goldstone 122–7, 1054, 1058
Hugenholtz 122–3
ladder 124–5
linked 121–2, 1066, 1193–8, 1213
resulting 123, 127
ring 124–5
unlinked 121–2, 1066, 1193–8
vacuum 121–2
valence bonds 223–4, 635–65
diastatine 552
diatomic molecules
active space selection 741–2
bonding 296–302
energy decomposition 296–302
first-row 118
many-body forces 924–5
time-dependent molecular theory 17
diazomethyl radical 273
dibromoethane 809
dicarbon 1129–32
dielectric constants 50–1, 53–5, 58–9
dielectric environments 47, 978
Diels–Alder allowed reactions 229, 240–1, 646–7
different orbitals for different spin (DOD) 731
diffuse functions 791–2, 795, 814
diffusion Monte Carlo (DMC) simulations 172–3, 262–3, 948, 953–4
Diffusion Quantum Monte Carlo (DQMC) algorithms 175–6
difluormethane 841–5
difluoromethylene singlet-triplet splitting 513–14
dihalogen bonding 303–5
dihydrogen bonding 296, 297–302, 357–60
dimension formula 118
Paldus–Weyl 118
dimers
CH···O hydrogen bonds 846–7
four-component relativistic molecular theory 547
intermolecular forces 1065–6
long-range corrections 534, 536–40
many-body forces 925, 928–30
metal nanoparticles 58–9
dimethyl oxalate 832
dinitrogen 83, 293–303, 305–6, 308–9, 458–60
diphenyl substituent effects 865–70
dipnicogen bonding 297, 302–5
dipole moments 930
anharmonic infrared intensities 178
cesium xenon systems 1113–14
coupled-cluster functionals 1205–6
intermolecular forces 1048
lithium-hydrogen molecules 628–9
dipoles 971–2, 987
CH···O hydrogen bonds 846–7
discrete approximations 47, 50, 60
electrodynamics 51–2, 59–61
many-body forces 923–4, 947–9, 953
molecular system simulations 95–6
polymer chains 1015, 1027
resonance 51–2
triple terms 923–4
Dirac equations 541
Dirac transformations 486
Dirac–Coulomb (DC) Hamiltonian 541–2
Dirac–Coulomb–Breit (DCB) Hamiltonian 541–2
Dirac–Hartree–Fock (DHF) method 542–3
Dirac–Kohn–Sham (DKS) method 542–3
diradicals 860–72
direct configuration interaction 739
direct dynamic simulations 415–19
direct inversions 203, 207–9
direct nonadiabatic dynamics 9–39
direct reaction dynamics 559, 575
Direct Simulation Monte Carlo (DSMC) method 436
disconnected diagrams 121–2, 473–5
discrete dipole approximations (DDA) 47, 50, 60
dispersion
coefficients 1051
energy 921–47, 1054–69, 1182
interactions 820, 921, 930, 976–81, 1003, 1056
intermolecular forces 1047–72
many-body forces 920–4, 931–2
non-additive interaction energy 938–9
second-order Møller–Plesset energy 1069–72
Dissipative Particle Dynamics (DPD) 436–7
dissipative structures 887
dissociation
constants 8327

energies 193, 467, 552, 808–9
enthalpy 861
hydrogen bonds 612–13, 619–27
intermolecular forces 1065–6
unimolecular reaction rates 403–5, 517–18
distributed data interface (DDI) 1177–9
distributed multipole analysis (DMA) 1181
2,11-dithio[4,4]metametahydroquinocyclophane (MHQC) 988
2,11-dithio[4,4]metametaquinocyclophane (MQC) 988
4,4'-dithioldiphenylacetylene 817–19
2,7-dithiolpyrene 817–19
dividing surfaces 68–72, 75–6, 80–3
diyl complexes 332–42, 358
DK *see* Douglas–Kroll
DKS *see* Dirac–Kohn–Sham
DMA *see* distributed multipole analysis
DMC *see* diffusion Monte Carlo
DMOL 1086
DNA 102–7, 269, 477, 574, 658
DOD *see* different orbitals for different spin
donor-acceptors 311–14, 979
doping processes 1020–5
double donor 937, 969
double precision 5
double-well potentials 1023–5
double-zeta methods 1012
doublet stability 130
doublets, state 345–7, 645, 757–9, 1100–3, 1131–5
Douglas–Kroll (DK) method 549–53
Douglas–Kroll–Hess Hamiltonians 747–8
DPD *see* Dissipative Particle Dynamics
DQMC *see* Diffusion Quantum Monte Carlo
drug delivery 865, 984, 989
drug design 574
drug discovery 41, 44–5
DSMC *see* Direct Simulation Monte Carlo
dyes 1021
dynamic correlation 117–18, 725, 736–8, 744–6, 863–5
dynamic depolarization 51
dynamic electron correlation 643, 736–8, 755, 863–5, 871, 971
dynamic polarizability 755, 939, 1101, 1207, 1212
dynamic reaction paths 237–8
dynamic rendering 37–8
dynamical, time-dependent molecular theory 9–39
dynamical variables 32, 34
Dyson equations 373–82, 388–9, 393–4, 500–1
Dyson orbitals 493–500
DZP basis sets 625

E-CCM *see* extended coupled-cluster method
EA *see* electron affinity
EBK *see* Einstein–Brillouim–Keller
EDA *see* energy decomposition analysis
effective core potentials 61, 563, 569, 901, 1088–9
effective elastic band theory 225–6
effective fragment potential 1177, 1181–4
effective Hamiltonians 583, 584, 746, 766–72
effective radius 51–2, 57
EFISH *see* Electric Field Induced Second Harmonic Generation
EGCI *see* Exponentially Generated CI
eigenmode-following algorithm 996
Einstein–Brillouim–Keller (EBK) quantization 408
elastic band theory 224–6
Electric Field Induced Second Harmonic Generation (EFISH) 1026
electric fields 1026–7, 1207
CH···O hydrogen bond 838, 846–7
lasers 150–1
local 55–9
size-consistent state-specificity 627–8
electrochemical switching 986
electrocyclic reactions 292, 646–7, 652
electrodynamics 47–64
electromagnetic fields 47–64
electron affinity (EA)
coupled-clusters 1211
equations of motion 443–61, 1211
G2 and G3 theories 793–4, 797, 801–2, 805–7
valence bond diagrams 648–9
electron cloud perturbations 840
electron correlation 5, 118, 124, 730, 736–8, 1134
dynamic 643, 736–8, 755, 863–5, 871, 971
effects 118, 500, 565, 755, 935–6, 996–7
energy 599, 792
intramolecular 927–32, 1058

intramonomer 928, 932–3, 1059–61
many electron 118, 124
nondynamic 273, 508, 736–8
static 273, 508, 736–8
electron deficient π-system clusters 979
electron delocalization 664, 774, 781, 1027, 1029
electron density 842–3, 847–50
electron gas 120, 124–8
electron lattice models 135
electron lone pair 335
electron momentum
density 487–502
spectroscopy 493, 1107
electron nuclear dynamics (END) 32–7
electron operators 378–84, 389–92
electron paramagnetic resonance (EPR) 753
electron repulsion integrals 544–5
electron-localized states 774, 781
electron-spin resonance (ESR) 1099–100, 1102, 1133–6
electronegativity 768, 831–9, 849
electronic band structure 774–9
non-spin-polarized 774–5
spin-polarized 774–5, 777–9
electronic degrees of freedom 14–17, 37–8, 196–7
electronic delocalization 658–9
electronic energy 11–12, 565–6
electronic excitation 509, 575–6, 1099–100
electronic Hamiltonian mapping 775–7
electronic polarizability 573, 1028–9, 1031
electronic properties 842–3, 1026–33
electronic reorganization 645
electronic Schrödinger equation 15–17
electronic self energy 391–2
electronic spectroscopy 740–3, 748, 751–6, 833
electronic state energy 756–7
electronic state potentials 748, 756–60
electronically excited states 559, 571, 575–6, 1040
electrons
aqueous clusters 972–3
conduction electrons 48–53, 937
electron-electron repulsion 772–81
mean free paths 51–3
multiconfigurational quantum chemistry 725, 730, 736–8
transfer 373–95, 1033–41
transport 373–95, 1033–41
tunneling 373–95
electrophiles 647–50, 979
electrostatic interactions
attraction 291, 293–4, 1048
bonding 329, 336, 340, 350, 363, 824
CH···O hydrogen bonds 840, 843–4
effects 281, 834, 846
energy 653–5, 930, 976–8, 1054
energy decomposition 291–367
forces 298, 356, 920
intermolecular forces 1053–4
many-body forces 920, 931
embedded clusters 78
empirical
correction factors 792
fits 681, 706–8
potentials 176, 188–9, 434, 921, 926–31, 945–6
EN *see* Epstein–Nesbet
END *see* electron nuclear dynamics
endohedrals 901–7, 983–7
energetics
carbon clusters 893–6
many-body forces 919–58
neutral water clusters 968–9
polymer chains 1013–42
water hexamers 996–7, 1000–5
energy
bands 1013–42, 1079
barriers 1037–9
derivatives 1175–6
diastatine 552
difference plots 619–27
energy decomposition analysis (EDA) 291–367
main-group compound bonding 291, 294–326
transition metal bonding 291, 294–5, 326–66
fluctuations 995, 1001, 1005
G2 and G3 theories 785–810
gaps 643–6, 648–9, 1013–42
Gaussian wave packets 28–9
landscapes 996
partitioning 291, 295
quantum virial theorem 292–3
storage 283–5

three-body interactions 922–38, 944–5, 948–51
enhancement factors 61–3, 698–9
ensemble, canonical 70, 397–429, 954, 999
ensemble-averaged VTST 82–3
enthalpy
of activation 477–8, 864–7, 871
aqueous clusters 968–9
bond dissociation 861
endohedral metallofullerenes 903–6
entropy interplay 891–2, 898
of formation 788–9, 793–4, 797, 802, 805–9
entropy 891–2, 896, 898, 968–9
enzyme catalytic mechanisms 982–3
EOM *see* equations of motion
EOM-CC *see* equations of motion coupled clusters
EPR *see* electron paramagnetic resonance
Epstein–Nesbet (EN)
partition 582, 600–2, 612, 615, 625–7
perturbation 1195
EPV *see* exclusion-principle-violation
equations of motion coupled clusters (EOM-CC) 455–7, 518–20, 1109, 1127
ab initio quantum chemistry 1191–2, 1202, 1207–13
acetylene 1132–3
GAMESS 1174
SAC–CI method 1101
theory 134
equations of motion (EOM)
electron affinity 443–61
ionization potentials 443–61
metastable anion states 457–60
Møller–Plesser approximations 450–3
non-equilibrium Green Functions 379–82
equilibrium
constants 41, 895–6
equilibrium-solvation paths (ESP) 80
geometry 752–6, 789–90, 798–9, 1099–100, 1128–33
internuclear distances 1130–1
equilibrium properties 433
error analysis 1001–2
ES *see* Euler stabilization
ESCA *see* X-ray induced photoelectron spectroscopy
ESHG *see* Electric Field Induced Second Harmonic Generation
ESP *see* equilibrium-solvation paths
ESR *see* electron-spin resonance
ethane 305
ethyl radical dissociation 403–5
ethylene 309, 347–54, 536–8
1-thiol-4-ethynylphenyl-4'-ethynylphenyl-1'-benzenethiolate 817–19
2'-amino-4-ethynylphenyl-4'-ethynylphenyl-5'-nitro-1'-benzenethiolate 818, 819
Euler angles 26
Euler integration 232–4
Euler stabilization (ES) 233–4
Euler–Lagrange equations 32
evolving width Gaussian wave packets 25–9
Ewald summation 428
Ex-EGCI *see* Excited Exponentially Generated CI
Ex-MEG *see* Excited Mixed Exponentially Generated
exact exchange functionals 530–1, 637, 669–717, 1145–9
exchange
energy 682–4, 686–7
exact exchange 530–1, 637, 669–717, 1145–9
functionals
approximate 681, 708–14
DFT 529–40, 669–717
energy decomposition 296
Fenske–Hall molecular orbitals 1148–9
long-range corrections 529–40
holes 672, 681, 692–4, 699–705
hydrogenic atoms 700
LDA 694, 705, 713
PBE 699–701
Taylor expansion 699–705
TPSS 699–701
interactions 920, 932
non-additivity 921–32
potentials 501, 669, 689–91, 698, 713, 1079
repulsion 291–367, 534, 565–7, 773, 840, 843, 844
three-body interactions 922–38
two-body interactions 922–36
exchange-correlation
analytic properties 679–80
approximate exchange 681, 708–14

constraint satisfaction 681, 688–99
conventional 534, 674, 690
density-gradient expansions 680–1, 686–8, 692–3
development progress 669–717
empirical fits 681, 706–8
energy 675–7, 1082
exact exchange 681, 708–14
functionals 669–717
generalized gradient approximation 706, 708–16
Hartree–Fock 707–8, 714, 715
Hohenberg–Kohn principle 670, 673, 677
holes 672, 681, 692–4, 699–705
normalization 692–4
spherically averaged 700
hybrid exchange 681, 708–14
implementation 714–16
kinetic energy density 677–9, 683–4, 696–7
Kohn–Sham DFT 669–717
LDA 680, 682–6, 697–8
LSDA 678–9, 683, 685, 715–16
many-body forces 920, 932
meta GGA 678, 681, 708–16
mixing exchange functionals 681, 708–14
performance 714–16
potentials 553, 674, 677, 714–17, 1064–5, 1081, 1086–9
asymptotic behavior 688–93, 700–3
Van Leeuwen-Baerends model 690
spin-polarizations 678–9, 683–5
excitation
dipolar 51
energy
allyl radicals 752–6
formaldehyde 518–21
long-range corrections 534–8
multiconfigurational quantum chemistry 742
operators 584
spectra 1101, 1106–12, 1115–19
Excited Exponentially Generated CI (Ex-EGCI) 1105
Excited Mixed Exponentially Generated (Ex-MEG) 1105
excited states
coupled-clusters 1192–202, 1207–13
equilibrium geometries 1099–100, 1102, 1128–33
free-base phthalocyanine 1101, 1122–3
GAMESS 1173–5
multi-electron processes 1130–1
SAC–CI method 1099–102, 1105–12, 1115–19, 1122–3, 1128–33
exclusion-principle-violation (EPV) 130, 366, 582–7, 603–5, 927, 949
exit-channel dynamics 417–19
exohedral fullerenes 983–7
exponential ansatz 472, 1063, 1209, 1213
exponential cluster expansion 124–8
Exponentially Generated CI (EGCI) 1105
Extended Brillouin's (BLB) theorem 738–9
extended coupled-cluster method (E-CCM) 135
extended CPMET 131
extended Douglas–Kroll transformations 550–3
extended Hückel tight binding 766
extended Koopmans' theorem 454, 461
extended molecules 814–16
external orbitals 513
externally corrected coupled-cluster methods 468–9
extinction spectra 47–54

fast multipole method 546, 728
FC *see* Franck–Condon
FCI *see* full configuration interaction
FDTD *see* finite difference time domains
FEM *see* floating encapsulate model
Fenske–Hall molecular orbitals 1143–63
Fermi contact hyperfine splitting constants 1099–100, 1133–6
Fermi energy 49, 59, 976, 986
Fermi operators 394–5
Fermi superoperators 378
Fermi vacuum 468
ferrocene 360–6, 1143–6
ferromagnetic spin exchange 776–7
FETs 820
Feynman diagrams 121, 124
Feynman path integrals 433–4
FHF species 653–4
field operators 22, 25, 30–1
field-effect transistors (FETs) 820

fine-grained parallelism 1179
finite difference time domains (FDTD) 47, 50, 61
finite dimension standard valence bond theory 117
finite histograms 995
finite size effects 53
finite temperature
  properties 995–1006
  string method 431
first-order electrical properties 627–9
first-order reaction path following 232–4
first-order reduced density matrices 487–9, 502, 726, 727
first-principles calculation 165–90
first-principles simulations 923–6
first-row diatomic molecules 118
Fischer-type complexes 342–7
fitting functions 937
five-membered rings 1106–8
flexible molecules 429
flexible-monomer potentials 923
flippers 965, 987–9
floating encapsulate model (FEM) 906
fluid dynamics 78–83, 105–8, 426–8, 436
fluorescence 47, 49, 269–70, 274, 276–9
fluorine
  energy decomposition 303–7, 354–7
  hydrogen exchange 526–9
  lead monofluoride 748, 756–60
  molecular system simulations 95
  multireference coupled-clusters 467, 472–3
  unimolecular reaction rates 417–19
FMO *see* fragment molecular orbital
FO *see* fragment orbitals
Fock evaluations 241
Fock matrices 565, 1016
Fock operators 586, 601–2, 615, 745
Fock orbitals 511
Fock space 469
forbidden reactions 517, 637, 641, 646–7, 652, 735
force fields 180–3, 223, 709, 877–8, 920–1
forces
  coupled-clusters 1202–7
  induction 920–2, 931–2, 938–42, 1053–4
  intermolecular 919–58, 1047–72
  see also many-body forces
foreign state intermediates 656–8
form factors 483–4, 488–9, 493–5
formaldehyde
  CH···O hydrogen bonds 833–4, 840–1, 844–5
  inner-shell satellite spectra 1126–7
  unimolecular reactions 412, 517–18
  valence excitation energies 518–21
formic acid dimers 846–7
formyl radicals 401, 402
FORS *see* fully optimized reaction space
FORTRAN 2, 3
four-body effects 947–8, 957
four-component relativistic molecular theory 542–7
Fourier transforms 488–9, 494
Fourier–Hankel method 493
fragment molecular orbital (FMO) 1180–1
fragment orbitals (FO) 650–2
Franck–Condon (FC) factor 1090
Franck–Condon (FC) points 270–2
free energy
  of activation 80
  aqueous clusters 968–9
  barriers 1038–9
  of complexation 41–5
  non-covalent binding affinity 41–5
free-base phthalocyanine 1101, 1122–3
free-base tetrazaporphin 1122–3
freezing-point diagrams 833
frequency shifting 831, 837–9, 843–50, 969–71
frequency-dependent polarizability 60–3, 1051, 1057–61, 1182–4
frontier orbitals 292, 311, 329, 641, 1151–3, 1156
FSGO valence bands 1017
full configuration interaction (FCI) 611, 614–15, 727, 1168–72, 1174
fullerenes 875–87, 897
  endohedral 983–7
  energetics 893–6
  formation 877–9
  giant 875, 883–7
  isomer relative stability 892–3, 896
  metallofullerenes 891–2, 901–7
  self-assembly 875–87
  thermodynamics 893–6
fully optimized reaction space (FORS) 739, 1168–9

functional molecular system design 963–89
furan 808–9, 1106–8

G2 theory 785–94, 808
G3 theory 785–9, 794–810
G3S theory 803–7
G3X theory 799–803
gallium bonding 314–22, 332–42
GAMESS *see* General Atomic and Molecular Electronic Structure System
gas-phase 68–77, 492–3, 897–8
gauge invariance 716
gauge-including atomic orbitals (GIAO) 570
Gaussian
- exchange-correlation 671, 685–6
- functions 754, 1083
- G2 and G3 theories 785–810
- GAUSSIAN package 4, 659, 671, 685–6
- Gaussian type orbitals (GTO) 1058–9, 1080, 1081
- Gaussian80 *ab initio* package 834
- local density approximations 685–6
- program package 834, 1101
- spinors 543–4
- valence bond diagrams 659
- wave packets 21–9

GDIIS 203, 207–9
GEA *see* gradient expansion approximation
Gellman-law expressions 377–8
General Atomic and Molecular Electronic Structure System (GAMESS)
- electronic structure 1167–85
- pyridine–copper tetramer 61–3
- quantum mechanics 1167–77
- quantum mechanics/molecular mechanics 1181–4
- scalable electronic structure theory 1168, 1177–81

general model space (GMC) 138–9, 630
general-multiconfiguration space SCF (GMC-SCF) 516–23
general-*R* method 920–1, 929–30, 942, 1128–33
generalized ensembles 428
generalized Fock operator 586, 602, 615, 745
generalized gradient approximation (GGA)
- asymptotic corrections 691
- correlation wave functions 702–3
- DFT 529
- DFT deMon codes 1084, 1085, 1087
- empirical fits 706
- exchange-correlation 681, 691, 694, 702–3, 706, 708–16
- formation energies 1085
- Kohn–Sham DFT 678
- long-range corrections 533
- meta 678, 681, 708–16
- mixing exchange functionals 681, 708–14
- systematic constraint satisfaction 696–7

generalized Langevin equation 81
germanium bonding 310, 314–22, 342–7
GF *see* Greens function
GFP *see* green fluorescent proteins
GGA *see* generalized gradient approximation
giant fullerenes 883–7
GIAO *see* gauge-including atomic orbitals
Gibbs energy/free energy 893, 895, 903
global hybrid exchange functionals 708–11, 713
glycine energy conformers 181–2
glycine-water complexes 184–5
GMC *see* general model space
GMC-SCF *see* general-multiconfiguration space SCF
gold 47–64, 347–54, 547, 814–19, 974–6
Goldberg polyhedra 897
Goldstone diagrams 121–7, 1054–6, 1058
gradient expansion approximation (GEA) 680–1, 686–8, 692–3, 1084, 1087
gradients
- analytical gradients 1202–7
- conjugate density matrices 572
- coupled-cluster theory 132–40
- potential energy surfaces 198–200
- reduced gradient 681, 708–14

graphical control/analysis 6
graphical representations 104–5
Graphical Unitary Group Approach (GUGA) 739
graphite interaction 823–7
green fluorescent proteins (GFP) 269–70, 274, 276–9
Greens function (GF) 373–95, 453–4
grid generating functions 1089
grid methods 175
Grote–Hynes theory 81–2

ground state
allyl radicals 752–3
G3S theory 803
hydrogen fluoride 95–6
lead monofluoride 748, 756–60
momentum 494–5
photobiology 270–2
pyridine–copper tetramer 60–3
group sum selected operators (GSUM) 1103, 1105
group-13 diyl complexes 332–42
group-14 nonpolar bonding 310
growing string approach 227
GSUM *see* group sum selected operators
GTO *see* Gaussian type orbitals
GUGA *see* Graphical Unitary Group Approach

H_4 model 612–15
hafnium bonding 326–32
half sandwich complexes 320–1
half-and-half hybrids 709, 1085, 1087
halogens 794
Hamiltonian encoding technique 160
Hamiltonians
ab initio vibration SCF 177
condensed phase reactions 79
effective 583, 584, 746, 766–72
Heisenberg 775–7
Hubbard 129
minimal electron nuclear dynamics 33
models 129, 411
multireference coupled-clusters 471–2
non-equilibrium current-carrying states 382–4
non-Hermitian 136, 457, 746, 1209
non-relativistic 783
Pariser–Parr–Pople 129, 560
photonic reagent control 151–2, 157
polyatomic molecules 252–6, 260–5
relativistic 548–9, 783
semiempirical 120
size-consistent state-specificity 583, 592, 601–2
spin 135, 775–7
spin-free 123
time-dependent molecular theory 12–17
transformed 591
two-component relativistic theory 548, 550–3
unimolecular reaction rates 405
Watson 171, 252
zeroth-order 601–2, 745–6
Hamilton's equations 29
hard sphere interactions 427
hardware development 105–8
harmonic
approximations 74, 252–4
frequency 552
oscillators 22–9, 165–6, 252–3, 893, 1026, 1031
Hartree–Fock (HF)
argon-hydrogen fluoride trimers 946–7
average potential 106
coupled-cluster theory 106–7, 117–18, 129–30
eigenvalue functions 1080
electron nuclear dynamics 36
energy 104, 707–8, 931, 946–7, 1000
exchange-correlation 707–8, 714, 715
G2 theory 789
GAMESS 1168, 1172, 1180–1
generalized 101, 120
interaction energy 931, 946–7
many-body forces 931, 946–7
molecular system simulations 95–104, 109–11
momentum 496–7
multiconfigurational quantum chemistry 726, 730–1
orbitals 1206
polymer chains 1012–13
restricted 109–11, 789, 864–5
semiempirical quantum-chemistry 563
stability 129–30
unrestricted 109–11, 732–3, 789
water hexamers 1000
wave functions 730–1
Hartree–Fock–Roothan (HFR) calculations 1144–7, 1149–51, 1154–5
Hartree–Fock–Slater (HFS) approximation 1080
heat
capacity 995, 1000, 1003–6
of formation 793–4, 861, 894–5
of reaction 477–8
of vaporization 895
Heck reaction 646

Heisenberg
Green Functions 378–82, 385–7
Hamiltonians 775–7
Heitler–London (HL) functions 90–2, 109–11, 116–17, 644, 652–3, 731–2
helices 851
helium 1051, 1062, 1068–9
superfluid 946
trimers 932–7, 942–3
helium-hydrogen fluoride 1062
Hellman–Feynman theorem 199–200, 675, 1202–3, 1205
Hermite–Gaussian auxiliary functions 1089
Hessians
anharmonic vibrational spectroscopy 166
coupled-cluster theory 132–40
GAMESS 1178
minimization 215–17
momentum 498–9
potential energy surfaces 198–200, 215–17, 219–20, 230, 240
reaction paths 240
transition states 219–20, 230
heterofullerenes 907
heteroleptic diyl complexes 332–42
hexacarbonyl bonding 327–9
1,5-hexadiene 859–61, 867–8
hexafluorides 322–6
HF *see* Hartree–Fock
HFR *see* Hartree–Fock–Roothan
HFS *see* Hartree–Fock–Slater
high-density scaling 695–6
high-level programming languages 2
high-resolution rotational spectroscopy 947
high-spin (HS) states 777, 948–9
high-temperature
carbon chemistry 876
quantum chemical molecular dynamics 875–87
small fragment chemistry 877
higher level corrections (HLC) 787, 791–2, 795–804
higher-order integrators 236–7
highest occupied molecular orbitals (HOMO) 977, 1150, 1154–63, 1172–3
Hilbert space 375–82, 385–9, 469–70, 583
Hinshelwood–Lindemann model 400
HIV protease 44–5
HL *see* Heitler–London
HLC *see* higher level corrections
Hohenberg–Kohn principle 670, 673, 677, 1081
holes
Coulomb holes 99, 736
exchange 672, 681, 692–4, 699–705, 713
holonomic constraints 429, 430
HOMO *see* highest occupied molecular orbitals
homoleptic diyl complexes 332–42
hopping integral 776–7
host–guest complexes 41
hot spots 56
HS *see* high-spin
Hubbard Hamiltonian 129
Hubbard–Lieb model 135
Hückel-type semiempirical quantum-chemistry 560
Hudson Valley 105–8
Hugenholtz diagrams 122–3
hybrid functionals 530–2, 708–14, 715–17
Coulomb-attenuated 714
global 711–13, 717
half-and-half 709, 1085, 1087
local 681, 712
screened 533, 713
hybrid methods 226–7, 572–3
hybridization 540, 834–9, 883, 974–6
hydration 973
hydrides 357–60, 401–2, 1068–9
hydrocarbons 794
hydrocarboxyl radicals 187–8
hydrodynamic flows 436
hydrogen
abstraction 645–6
atoms 77–8, 485–6, 494–5, 954
bonded clusters 171, 182–5
bonded complexes 185, 698–9
bonding 78, 567–9, 937–8, 966–80, 984, 998
bulk water 969
energy decomposition 296–302, 305–14, 342–7, 354–60
exchange 526–9, 660–1
ground-state momentum 494–5
hydrogen-carbon bond lengths 406–7
hydrogen-chloride 947–8
hydrogen-fluoride 95–6
hydrogen-peroxide 342–7

hyperfine splitting constants 1135
molecules 77, 116, 478, 730–4, 737
multiconfigurational quantum chemistry 730–4
see also CH···O hydrogen bonds
hydroperoxy radicals 401, 402
p-hydroxybenzilideneimidazolone 274, 276–7
Hyleraas correlation 93
hyper-Rayleigh scattering 49
hyperconjugation 307, 847
hypercoordinated species 653–4
hyperfine splitting constants 1099–100, 1102, 1133–6
hyperpolarizability 941–2, 1026–33, 1212
hypochlorous acid 405–6, 411–13

I–V *see* current-voltage
IBM
card 2–3
computers 1177–8
Kingston 91, 104–8
molecular package (IBMOL) 100–2, 106, 1012
Poughkeepsie 104–5
Research Laboratory 97–8
San Jose 97–105
IBMOL 100–2, 106, 1012
IEPA *see* independent electron pair approximation
IET *see* inelastic electron tunneling
IM *see* intermediate Hamiltonians
imidogen helium interaction 1068–9
IMK *see* Ishida, Morokuma and Komornicki
importance sampling 428
improper blue-shifting 848
impulse approximation 491–3
IMS *see* incomplete model space
inactive excitations 593–9
inactive orbitals 725
incomplete model space (IMS) 137–8, 469, 473–6, 581–9, 605–11
independent electron pair approximation (IEPA) 1194–5
independent-particle models (IPM) 117–18
indium 314–22, 332–42
INDO *see* intermediate neglect of diatomic differential overlap
induced-fit effects 45
induction 920–2, 931–2, 1053–4
dispersion 928, 946–7, 967, 1047
energy 926, 930–1, 935, 938–42, 977, 1057
interaction 938, 956, 979–80, 1072
inelastic electron tunneling (IET) 373–4
infrared (IR)
intensity 178, 181–5, 1176–7
spectroscopy 189, 263, 831–3, 1062
inherent structures 1002
inner-core states 1100, 1125–8
inner-shell
ionization 1099–100, 1102, 1125–8
satellite spectra 1125, 1126–7
instability 129, 208
instantaneous interaction energy 291–3
integral approximations 560–70
CNDO 560–3, 568, 834, 1016
INDO 560–3, 567, 569
NNDO 560–3, 565–7, 569–70
ZDO 1148
integrals *see individual types*
intensive energy 445, 455, 461
inter-isomeric equilibrium 906
inter-isomeric separation energy 892
interaction energy 291–367, 766, 919–58, 1000–1
interallylic distances 860, 863, 865–7, 869, 871
interatomic pair correlation 1051
interchange theorem 1204, 1206
intermediate Hamiltonians (IM) 585
intermediate neglect of diatomic differential overlap (INDO) 560–3, 567, 569
intermediate normalization 138, 475, 510, 581, 607
intermolecular forces/interactions 919–58, 982–3, 1047–72
intermolecular perturbation 768
intermonomer separations 920–1, 929–30, 1128–33
internal coordinates 201–2, 213–15, 228–9
internally folded density 489
interpolation 220–4
intramolecular correlation 927–32, 1058
intramolecular dynamics 416–17
Intramolecular Vibration Energy Redistribution (IVR) 167–8
intramonomer correlation 928, 932–3, 1059–61
intrinsic barriers 641–2

intrinsic non-Rice–Ramsperger–Kassel–Marcus kinetics 403–5, 408, 417–19
intrinsic reaction coordinates (IRC) 417–19
intruders
  coupled-cluster theory 136–7, 139, 472–6
  multireference methods 472–6, 584, 597–8, 608
  size-consistent state-specificity 584, 597–8, 608
inversion symmetries 490
iodine 303–5, 526–9
ionic
  bonding 293
  intermediate curves 652–6
  structures 526–7, 652–6, 749
ionization energy 1145–6
ionization potentials (IP)
  coupled-clusters 1211
  equations of motion 443–61, 1211
  G2 and G3 theories 793–4, 797, 802, 805–9
  valence bond diagrams 649–50
ionization spectra 1099–102, 1106–12, 1125–8
ionophores 965, 980–3
ions, active space 741
IP *see* ionization potentials
IPM *see* independent-particle models
IPR *see* isolated pentagon rule
IR *see* infrared
IRC *see* intrinsic reaction coordinates
iridium bonding 326–32
iron 332–42, 347–54, 779–80, 1035
irreducible representation 21, 27, 32, 295, 319–24, 1160
Ishida, Morokuma and Komornicki (IMK) algorithms 233–4
isoelectronic cations 494
isoelectronic species bonding 296–302
isolated pentagon rule (IPR) 892, 894–5, 898–9, 901–3, 905
isomer relative stability 892–3
isomeric endofullerenes 891–907
isomeric fullerenes 891–907
isomers
  book 1006
  cages 875, 883–7, 891–907, 969–73, 996–7, 1006
  prism 969, 996–7, 1002, 1006
  ring 996–7
isotope effects 67, 77, 574, 862, 865, 871
iterative subspaces 203, 207–9
IVE efficiency 419
IVR *see* Intramolecular Vibration Energy Redistribution

Jacobian factor 43
Jacob's ladder 678
Jahn–Teller effect 659, 1020–1, 1160
jellium 59
Jeziorski–Monkhorst ansatz 469, 472–5, 582, 585–6
Jordon transformations 486

Keldysh Green functions 385–7
Keldysh loops 373–82
Ketosteroid Isomerase (KSI) 982–3
kinetic balance 543–4
kinetic control 906–7
kinetic effects 865
kinetic energy 13–16, 677–9, 683–4, 696–7
  density 531, 678, 683, 696–8, 702
kinetic isotope effects 67, 77, 574, 862, 865, 871
Kohn–Sham
  density matrices 501, 677–8, 682
  DFT 669–717, 1064, 1081, 1083
  equations 529, 674–5, 691
  exchange-correlation 674–5
  methods 542–3, 673–5, 677, 1083, 1087
  operators 1146
  orbitals 501–2, 674–8, 682, 690, 696–8, 714
  static isotropic polarizability 540
  theorem 540
  wave functions 674
KOMMUTE 1198
Koopmans' theorem 144–5, 450, 454, 493, 726
  extended 454, 461
Kramers theory 81, 100, 757, 1048
krypton crystals 945–6
Kubo formulas 431–2

La@C82 892, 901, 905–6
laboratory of molecular structure and spectra (LMSS), Chicago 92–5
ladder diagrams 124–5
Lagrange multipliers 133, 429

Lagrangians 18–19, 32–4, 210–12
Langevin equations
  generalized 81
  model 435–6
Langreth–Mehl (LM) functionals 695
lanthanides 741, 743–4
lanthanum 892, 901, 905–6
Laplacian of the density 702–3
large molecule anharmonic vibrational spectroscopy 165–90
lasers 149–62
lattice Boltzmann method 436
lattice gauge theory 135
lattice quantum field theories 135
lattice-gas automation 436
LCAO *see* linear combination of atomic orbitals
LCGTO *see* linear combination of Gaussian type orbitals
LDA *see* local density approximation
lead
  bonding 310, 314–22, 342–7
  fluoride 725, 741, 747–8, 756–60
  oxide 741–2
leapfrog transformation 897
Lebedev grids 1083, 1088
LEDs 1033, 1040–1
Lee–Yang–Parr correlation functional (LYP) 539, 715–16, 1085, 1165
  see also B3LYP
Leffler–Hammond postulate 636
Lennard–Jones fluids 427
Lennard–Jones parameters 433–4
Levenberg–Marquard algorithm 1086
Lewis acids/bases 979
Libby's hypothesis 1035–6
Lie groups 21, 23, 25, 27–9, 32
Lieb–Oxford bounds 1084
ligand binding 44
ligand selectivity 851
light-emitting diodes (LEDs) 1033, 1040–1
linear combination of atomic orbitals (LCAO)
  coupled-cluster theory 117
  DFT deMon codes 1081, 1083–4, 1091
  polymer chains 1011–12, 1015
linear combination of Gaussian type orbitals (LCGTO) 1083–6
linear polyenes 129
linear response theory 431
linear scaling 571–2, 574, 1180–1, 1214–15
linear synchronous transit (LST) 221–3
linked cluster theorem 122
linked diagrams 121–2, 1066, 1193–8, 1213
Liouville equation 435
Liouville space pathways (LSP) 373–82, 385–9, 393–4
Lipkin–Meshkov–Glick model 135
Lippmann–Schwinger equation 470–1
liquid argon 946
liquid helium meniscus 1048
liquid-vapour equilibrium 832–3
liquid water 104, 427–8, 936–7, 954–8, 1001–3
lithium bonding 305–7, 314–22
lithium-hydrogen 612–13, 619–24, 627–9
lithium trimers 948–9
LM *see* Langreth–Mehl
LMO *see* localized molecular orbitals
LMP2 *see* localized second-order Møller–Plesset perturbation
LMSS *see* laboratory of molecular structure and spectra
local [tau] approximation (LTA) 683–4, 698
local density approximation (LDA)
  asymptotic corrections 690
  DFT 530, 1081, 1083–4, 1087
  DFT deMon codes 1081, 1083–4, 1087
  exchange-correlation 680, 682–6, 697–8
  intermolecular forces 1064–5
  long-range corrections 529
  Perdew–Wang scheme 685, 693, 707, 709, 715, 1084
  Perdew–Zunger scheme 531, 679, 684
  systematic constraint satisfaction 697–8
  Vosko–Wilk–Nusair 530, 684–5
local electric fields 55–9
local hybrids 711–12
local spin-density approximations (LSDA)
  Barth–Hedin scheme 677, 685, 697–8, 709–10, 1082
  DFT 678, 679, 1084
  empirical fits 707
  exchange-correlation 678–9, 683, 685, 715–16
  hybrid exchange functionals 709
  Kohn–Sham DFT 678, 679
  Vosko–Wilk–Nusair 530, 684–5

localized molecular orbitals (LMO)
GAMESS 1170–1, 1181
malonaldehyde 1128–9
semiempirical quantum-chemistry 572
valence bonds 523
localized orbitals 523, 726, 1071, 1170–1, 1176–7
localized second-order Møller–Plesset perturbation (LMP2) 998–1006
localized surface plasmon resonance (LSPR) 51, 54, 60
London forces 1048
lone pairs 517
aqueous clusters 971
CH···O hydrogen bonds 848–50
energy decomposition 299–303, 313–14
Fenske–Hall molecular orbitals 1157
multiconfigurational quantum chemistry 748
long time tail 427
long-range corrections/interactions 529, 532–40, 713–14
low-energy isomers 996
LSDA *see* local spin-density approximations
LSP *see* Liouville space pathways
LSPR *see* localized surface plasmon resonance
LST *see* linear synchronous transit
LTA *see* local tau approximation
luminescent jellyfish 269–70, 274, 276–9
LUMO
energy decomposition 311, 313–14, 329
Fenske–Hall molecular orbitals 1150, 1157–8
GAMESS 1172–3
weakly bound clusters 97
LYP *see* Lee–Yang–Parr correlation functional

magnesium 314–22, 499
magnetism 47–64, 776–7, 780–1, 983–7
insulating states 775
main frame computers 1053–9
main-group compounds 291, 294–326
malonaldehyde 1128–9
adiabatic excitation energies 1128–9
optimized structures 1128–9
manganese 1143, 1151–6
manganese tetraoxide 1116–18
manganese-nickel interactions 1153–4
many-body decomposition 998–9
many-body expansions, convergence 920–6, 937, 947–9, 957
many-body forces 919–58
GAMESS 1172–3
trimers 922–3, 932–42, 943–57
water 925, 928–42, 947–8, 951–7
many-body perturbation theory (MBPT) 119–28, 133, 508–23, 1172–3, 1193–7
many-electron Hamiltonian 550–3
Marcus Electron Transfer theory 1033–41
Marcus equations 636, 641
Markov chain 428
mass polarization 13
matrices
charge density 728
conjugate density 572
coupling elements 241–2
Fock 565, 1016
polymer chains 1016
reduced density 672, 677–9, 727–30
response density 1205–6
see also density matrices
matrix element integration 263–4
Matsububara imaginary time 374–5
maximum overlap orbitals 129
Maxwell's equations/theory 48, 51
MBPT *see* many-body perturbation theory
MC *see* Monte Carlo
MC QDPT *see* multi-configuration quasi-degenerate perturbation theory
MCSCF *see* multi-configurational self-consistent fields
MD *see* molecular dynamics
mean absolute deviations 793–4, 797–807
mean field methods 434
medicinal chemistry 574
MEG *see* mixed exponentially generated
melting transition 997
memory 5–6
MEP *see* minimum energy paths
mesoscopic dynamics 435–7
meta generalized gradient approximation 678, 681, 708–16
metal arene complex bonding 360–6
metal nanoparticles 47–64
metal surfaces 813–29
metallacyclic compounds 347–8

metallic clusters 774–5, 974–6
metallic state 774–6
metallocenes 314–22, 360–6
metallofullerenes 891–2, 901–7
metals, polymer chains 1034–5
metastable anions 457–60
methane (CH_4) 258–60, 835–6, 840–5
methanimine 101–2
methanol 182–4, 261–2, 808, 840–1
method of moments coupled-clusters 469
methoxy radical dissociation energy 808
methyl alcohols 417–19
methyl diazenyl diradical 273
methyl fluoride 841–844–5
methyl–carbonyl bonds 155
N-methylactamide (NMA) 182–3
methylene amidogen 257–8
methylene bonding 342–7
methylene singlet-triplet splitting 513–14
methylidynes 360–6
methylphenol proton affinities 808–9
N-methylthioacetamide (NMTAA) 287
Metropolis Monte Carlo simulations 426, 428, 1006
MHQC *see* 2,11-dithio[4,4]metametaquinocyclophane
midbond functions 929
Mie theory 51
Mile-Stoning 431
MINDO 893, 894–5
MINDO/3 calculations 862–3
minicomputers 4–6
minima, potential energy surfaces 195–218
minimal electron nuclear dynamics 32–7
minimum energy paths (MEP) 71–6, 219, 230–41, 270–2
minimum orbital deformation (MOD) 1103–5, 1128–9
mixed exponentially generated (MEG) 1105
mixing exchange functionals 681, 708–14
MM *see* molecular mechanics; MULTIMODE
MMP *see* multiple multipole method
MNDO 561–70
MO *see* molecular orbitals
MOD *see* minimum orbital deformation
mode transitions 186–7
model core potentials 1084, 1089, 1185
model holes 699–701
model pair correlation functions 703–4
model space 583–630
  complete 136–8, 471–8, 582, 587–93, 606–10, 629
  general 138–9, 630
  incomplete 137–8, 469, 473–6, 581–9, 605–11
Modern Techniques in Computational Chemistry (MOTECC) 106
MOLCAS software 754, 757
MOLDEN 1087
molecular currents 373–95
molecular devices 269, 285, 374, 814, 963–89
molecular dynamics (MD) 425–38, 875–87
  ab initio methods 434
  carbon nanostructure self-assembly 875–87
  classical 428–32
  coarse grained dynamics 435–7
  controls 149–62
  fullerene formation 877
  mesoscopic dynamics 435–7
  non-covalent binding affinity 44–5
  photonic reagent control 149–62
  quantum mechanics 432–5, 875–87
  simulations 104–5, 875–87, 1090
molecular electronics 813–29
molecular excited state geometries 1099–100, 1102, 1128–33
molecular flippers 965, 987–9
molecular Hamiltonians 12–17
molecular integral algorithms 1083
molecular junctions 373–95
molecular mechanics (MM) 197–8, 223–4, 1181–4
molecular orbitals (MO)
  correlation diagrams 315–17, 323, 361–4, 366
  coupled-cluster theory 117
  exact relationships 766–7
  fragment 1180–1
  highest occupied 977, 1150, 1154–63, 1172–3
  molecular system simulations 90–2
  multiconfigurational quantum chemistry 725–6, 727–30
  orbital interaction 766–72
  polymer chains 1012–13
  self-consistent fields 560–2, 570–1, 576, 1143, 1144
  semiempirical quantum-chemistry 560–2

valence bonds 636, 643
weakly bound clusters 977
see also localized molecular orbitals
molecular recognition 41–2, 836, 964
molecular sensors 820–7, 828–9
molecular spectroscopy 1099–137
molecular structure and spectra 92–5
molecular system simulations 89–111
molecular theory, dynamical, time-dependent 9–39
molecular wires 374, 382–4
molecule-electromagnetic field interactions 47–64
molecule-lead self energies 389–92
molecule–metal systems 60–3
MOLEKEL 1087
Møller–Plesset (MP)
Brillouin-Wigner perturbation 470
equations of motion 450–3
G2 and G3 theories 785–7, 790–9, 801–3, 805–9
partitions 600, 602, 612, 625, 627
Møller–Plesset second-order perturbation theory (MP2) 744, 820, 825–6, 997, 1172–3
Møller–Plesset third-order perturbation theory (MP3) 785, 796–7, 801–3, 805–7, 1063
MOLPRO software 760
molybdenum bonding 347–60
molybdenum tetraoxide 1116–19
moments 469, 1058–9
see also dipole moments
momentum 483–502
density 487–93, 499–502
space wave functions 484–6, 495–6
mono-determinant valence bond wave functions 664–5
mono-hydrated hydroxide ions 262–3
monosubstituted benzenes 1108–9
monovalent atom bond exchange 658–9
Monte Carlo (MC) simulations 995–1006
ab initio 999
diffusion 172–3, 262–3, 948, 953
direct simulation method 436
fullerene formation 877
Metropolis 426, 428, 1006
molecular dynamics 425–6, 428, 430, 433
molecular systems 104–5
umbrella sampling 430, 574
water 953–7, 995–1006
$MoO_{4-n}S_n^{2-}$ (n=0-4) 1115–19
MOPAC 893
Morokuma decomposition/partitioning 843, 844–5
$MoSeO_4^{2-}$ 1115–19
MOTECC *see Modern Techniques in Computational Chemistry*
MP *see* Møller–Plesset
MP2 *see* Møller–Plesset second-order perturbation theory
MP3 *see* Møller–Plesset third-order perturbation theory
MQC *see* 2,11-dithio[4,4]metametaquinocyclophane
MR *see* multi-reference
MR CC *see* multi-reference coupled-clusters
MR-CI *see* multi-reference-configuration interaction
MRMP *see* multi-reference Møller–Plesset perturbation
MS-Xα *see* multiple scattering-Xα
MT *see* multi-dimensional tunneling
Müller–Brown surfaces 239
Mulliken populations 1148–9
multi-configuration quasi-degenerate perturbation theory (MC QDPT) 508–23
multi-configuration-based approximations 454–5, 1066
multi-configurational quantum chemistry 725–61
active orbitals 725, 727, 729–30, 740–4
CASPT2 725, 744–8, 754–7, 761
CASSCF 725, 729–30, 739–48, 754–7, 760–1
degeneracy 734–8
dynamic correlation 744–6
hydrogen molecules 730–4
multiconfigurational wave functions 738–44
near degeneracy 734–8
relativistic regimes/effects 747–8, 756–60
second-order perturbation 744–6
wave functions 738–44
multi-configurational self-consistent fields (MCSCF)
electronic structure 508, 1169–72, 1178, 1184

GAMESS 1169–72, 1178, 1184
intermolecular forces 1051–2
molecular system simulations 95–103, 109–11
wave functions 738–9
multi-dimensional tunneling (MT) 68, 82–3
approximation 74–5, 82
large-curvature tunneling 74–5
optimization 74–6, 82–3
reaction path curvature 74
small-curvature tunneling 74
multi-electron processes 1128–32
multi-particle collision model 436
multi-reference coupled-clusters (MR CC) 135–9, 465–79
multi-reference Møller–Plesset perturbation (MRMP) 508–23
multi-reference (MR)
equation-of-motion 1210
GAMESS 1172, 1174
perturbation 508–23
SAC–CI method 1100, 1105
second-order perturbation 1172
size-consistent state-specificity 581–631
multi-reference-configuration interaction (MR-CI)
coupled-clusters 118–19
GAMESS 1172, 1174
intermolecular forces 1066
momentum density 500–1
SAC–CI method 1100
multi-state CASPT2 746
MULTIMODE (MM) 251, 254–63
multiple bonds 303, 305, 308–10, 1105, 1173
multiple multipole method (MMP) 50
multiple nonpolar bonding 308–9
multiple scattering-Xα (MS-Xα) method 1080, 1081–83
multiple time step integrators 430
multiplicative scaling 803–7
multipole expansion 929, 944, 1015, 1053–61, 1070
multipole moments 133, 931–3, 938–42, 1057, 1070–2
dipoles 930, 1113–14
induced 931, 938, 940–1
permanent 931, 938, 940, 1057
quadrupole 824
N-electron systems 29–32
N-methylthioacetamide (NMTAA) 287
n-mode coupling 255–60
nanodevice design 963–7, 983–7
nanomaterials 963–7, 978–9, 983–7
nanoparticles 47–64, 574–5
nanotechnology 813–29
nanotubes 820–9, 875–87, 963–7, 983–7
natural bond orbitals 842, 848
natural linear scaling CC (NLSCC) 1214
natural orbitals (NO) 726–30, 733–4, 749–51
coupled-cluster theory 118–19
equations of motion 454
Navier-Stokes equations 89, 436–7
n-body decomposition 998–9
near degeneracy 98–9, 101, 734–8
near-equilibrium trimers 932–7
NEB *see* nudged elastic band
Neél states 777
negative differential resistance 817
NEGF *see* non-equilibrium Green Functions
neglect of diatomic differential overlap (NNDO) 560–3, 565–7, 569–70
neighbor lists 427
NEMD *see* non-equilibrium molecular dynamics
Nesbet's formula 1194, 1195
neutral radical/soliton effect 1022–3
neutral water clusters 968–9
neutron diffraction 835, 837
Newton methods 203–7
nickel 332–42, 1143, 1151–6
nickel tetracarbonyl, ($Ni(CO)_4$) 340–1, 1099–101, 1120–1, 1137
nitric acid 186–7
nitrile-water complexes 184–5
nitrogen
energy decomposition 296, 297–309, 311–14, 360–6
equations of motion 458–9
helium interaction 1068–9
hyperfine splitting constants 1135
momentum density 499–500
nitrogen dioxide 410, 412, 414, 820
nitrogen–copper distances 61–3
nitrosyl hydride 401, 402
NLSCC *see* natural linear scaling CC
NMA *see* *N*-methylactamide

NMR *see* nuclear magnetic resonance
NMTAA *see* *N*-methylthioacetamide
NNDO *see* neglect of diatomic differential overlap
NO *see* natural orbitals
Nobel Prize in Chemistry 1033–4
noble gases 1062
noble metal clusters 974
non-additivity 791–2, 795, 919–58
  forces 920–6, 930–3, 937–8, 944, 954–5
  potentials 920–1, 925, 942–6
non-adiabatic coupling 9–16, 36–7, 919
non-adiabatic methods/theory 9–39, 1173–5
non-canonical transformations 429
non-covalent binding/bonding affinity 41–5
non-covalent interactions *see* intermolecular forces
non-degenerate perturbation 768–9
non-equilibrium dynamics 887
non-equilibrium Green Functions (NEGF) 373–95
non-equilibrium molecular dynamics (NEMD) 432
non-equilibrium solvation 81–2
non-equivalent orbital interactions 778–80
non-Hermitian Hamiltonians 136, 457, 746, 1209
non-IPR 898–901, 907
non-linear excitations 1025–33
non-linear optics 47–9, 963–5, 986, 1025–33
non-local DFT levels 296
non-Newtonian equations 429
non-polar bonding 305–10
non-Rice–Ramsperger–Kassel–Marcus (non-RRKM) kinetics 400–20
normal coordinates 252
normal modes 251–2, 264–5
normal versus counterintuitive orbital interaction 770–2
normalization 692–4
  Bloch 136
  intermediate 138, 475, 510, 581, 607
nuclear curvature 1030, 1032
nuclear degrees of freedom 14–17
nuclear energy derivatives 1175–6
nuclear magnetic resonance (NMR) 902–3, 905, 1212
  chemical shifts 570–4, 843–5, 1101, 1115–20
  transition metal complexes 1101–2, 1115–20
nuclear matter 124, 125–8, 135
nucleic acids 104, 574, 834–7, 851
nucleophiles 647–50, 654–7, 661
nudged elastic band (NEB) theory 226
number density 484, 490–1, 495
number operator 591
numerical integration 93, 232–4, 415, 714, 1083–9
numerical performance 611–29
numerical stability 1088
NuMol 1085

O–H stretch frequencies 969–71
occupation numbers 726–9, 733–4, 750–1
occupation restricted multiple active spaces (ORMAS) 1169
octahedral complexes 322, 329
octal row bonding 305–7
OLEDs 1033, 1040–1
oligopeptide 286–7
OM1, OM2 & OM3 semiempirical methods 567
OMT *see* optimized multidimensional tunneling
one-dimensional reciprocal space 1014
one-electron
  Hamiltonians 766–72
  momentum density 487–8, 490
  theory 774–5
ONIOM method 820–3, 829
open-ended single-walled carbon nanotubes 875–7, 880–7
open-shell systems
  anharmonic vibrational spectroscopy 187–8
  clusters 948–51
  intermolecular forces 1065–6, 1068–9
  non-equivalent orbitals 778–80
  size-consistent state-specificity 584
  three-body interaction energies 948–51
operators
  forms 1205–7
  rank 448–9
  see also individual operators
optical properties 1025–33
  extinction 47, 49–50–4
  hyper-Rayleigh scattering 49

non-linear 47–9, 963–5, 986, 1025–33
scattering 47–54, 60, 77, 491–3, 500
surface enhanced Raman spectroscopy 49, 59–63
optimal control 151–60
optimized effective potential 677
optimized multidimensional tunneling (OMT) 74–5, 83
Optimized Valence Configuration (OVC) 738
optimized virtual orbital space (OVOS) 1215
orbital degeneracy 948–51
orbital expansion coefficients 101
orbital interaction
concepts 765–74, 778–80
counterintuitive 770–2
effective Hamiltonians 766–72
electron-electron repulsion 772–4, 778–80
energy decomposition 291–367
kinetic energy density 677–9
non-equivalent 778–80
normal 770–2
perturbation 767–70
reduced density matrices 677–9
two-electron two orbital 781
orbital invariance 582, 586–9, 603–5, 630, 741–2
orbital mixing 765–6, 772, 782–3
orbital occupation 765–6, 772, 774–5
orbital ordering 780–1
orbital relaxation 298, 332, 928, 1126, 1206
orbital selection rules 651–2
orbital symmetry 641
orbitals
active 725, 727, 729–30, 740–4
atomic 117–18, 458, 749–50, 1147–8
Brueckner 794
Dyson 493–500
external 513
fragment 650–2, 1180–1
inactive 725
localized 523, 726, 1071, 1170–1, 1176–7
maximum overlap 129
natural bond 842, 848
orthogonal 524
Slater 93, 95, 296, 1081
see also molecular orbitals; natural orbitals
ordinary differential equations 232
Organic Electroluminescent Diodes (OLEDs) 1033, 1040–1
organic molecules 742, 748, 751–6
organic nanotubes 965
ORMAS *see* occupation restricted multiple active spaces
orthogonal orbitals 524
orthogonalization corrections 566–7
oscillators
field operators 25
harmonic 22–9, 165–6, 252–3, 893, 1031
strengths 534–7, 755, 1048, 1106–7, 1120–2
osmium 326–32, 347–54, 1116–18
OVC *see* Optimized Valence Configuration
overlap integrals 117, 766–9, 776–7
many-body forces 927, 932, 953
molecular systems simulations 93
semiempirical quantum-chemistry 563–6
overlap matrices 1016
overtones 186–7
OVOS *see* optimized virtual orbital space
oxidative addition reactions 646–7
oxohydrocarbons 837
oxygen
energy decomposition 305–7
hyperfine splitting constants 1135
multireference coupled clusters 471–2
oxygen–oxygen partial RDF 1002–3
sensors 820–5
see also CH···O hydrogen bonds
ozone, O_2 748, 749–51

π bonds 300–14, 324–6, 336–45, 356, 647, 657
p space 484–6, 487–8, 495–6
π-based materials 981–2
π-body decomposition 998–1002
π-conjugated organic molecules 1101–2, 1106–12
π-conjugational stabilization 878–9
π-electron densities 987–9
π-electron semiempirical quantum-chemistry 560
π-π interactions 979–80
π-radicals 777–8
π-system clusters 976–80
PA *see* proton affinity
pair density 672
short-range behavior 700
spherically averaged 700

pair distribution functions 672, 703–4
pair natural orbital methods 1055
pair-wise coupling 180
Pairwise Distance Directed Gaussians (PDDG) 568–9
pairwise-additivity 919–58
  interactions 171, 176, 919–58
  potentials 920, 924–6, 946, 950–7, 1060–2
Paldus–Weyl dimension formula 118
palladium 332–42, 1157–63
parallel computing 1177–80
  architectures 5, 103
parameterization 559, 563–76, 684–6, 695–7, 707–10
  general-purpose 559, 568–9, 576
  special-purpose 559, 569–70, 576
Pariser–Parr–Pople (PPP) method 129, 560
parity violation 135
particles
  cubes 51–2, 56–7
  cylinder 51–2, 56–7
  molecular dynamics 436–7
  oblate 51–3
  particle mesh method 428
  prisms 51–2, 56–8, 969–70, 996–7
  prolate 51–2
  pyramid 51–2, 305, 311, 1111–12
  shape/size effects 47–59
  spherical 48–53, 56–9
  spheroids 51–3, 170
  tetrahedron 52–7
  triangles 50, 52–3, 58
  truncated tetrahedron 55–6
partitioning 291, 295
  partition functions 70–9, 433, 893, 906
    anharmonicity 73, 78–9, 1175
    internal coordinates 73, 76
Partly Separable VSCF (PS-VSCF) 179–80
path optimization 224–7
Pauli repulsion 291–367, 840, 843, 844, 1053–4
PCM *see* Polarizable Continuum Models
PDDG *see* Pairwise Distance Directed Gaussians
peapods 891
peer-reviewed papers 835–6
Peierls distortion 1022
pentagon junctions 894, 898–900
pentamers 954, 957, 968–9
peptides 987
peptidomimetic polycyclic structures 285–7
Perdew–Wang scheme 685, 693, 707, 709, 715, 1084
Perdew–Zunger scheme 531, 679, 684
Perdew's Jacob's ladder 678
performance, size-consistent state-specificity 611–29
pericyclic reactions 270, 860, 864, 871
periodic band structures 1014
perturbation
  degeneracy 179–80, 769–70
  electronegativity 768
  geometry 768
  intermolecular forces/interactions 768, 927–8, 1051–3
  non-degenerate 768–9
  selection 587, 1103–5, 1128–34, 1137
perturbation theory (PT)
  ab initio vibration SCF 177
  anharmonic vibrational spectroscopy 168–9, 179–80
  Brillouin–Wigner 465–79
  CH···O hydrogen bonds 840
  concepts 765–70
  coupled-cluster theory 119–22, 137–8
  many-body perturbation 119–28, 133, 508–23, 1172–3, 1193–7
  multi-reference 465–79, 508–23
  non-equilibrium Green Functions 381–2
  nonadditive forces 930–2
  orbital interaction 767–70
  polyatomic molecules 252–4
  polymer chains 1022
  Rayleigh–Schrödinger 510–11, 1052–3, 1061–2, 1192–3
  size-consistent state-specificity 587, 599–602, 617–18, 625–7
  time-dependent 120–1
  time-independent 120–1, 124
  see also Møller–Plesset
PES *see* potential energy surfaces
PFI-ZEKE *see* pulsed-field ionization zero-electron-kinetic energy
phase space 405–8
phenyl substituent effects 865–70
phenyl-1,5-hexadiene 861–2, 865–70
phenyl–carbonyl bonds 155
1-phenylcyclohexane-1,4-diyl diradicals 869

phonon operators 378–84, 389–92
phosphane complex bonding 354–7
phosphine bonding 357–60
phosphinidene bonding 332–42
phosphorus 302, 311–14, 357–66
photoactive proteins 269–88
photobiology 269–88
ab initio quantum chemistry 272–6
Aequorea victoria 269–70, 274, 276–9
biomimetic molecular switches 285–7
bovine visual pigments 269, 276–7, 279–85
green fluorescent proteins 269–70, 274, 276–9
luminescent jellyfish 269–70, 274, 276–9
quantum chemistry 272–6
Rhodopsin 269, 276–7, 279–85
photochemistry 269–88
cycloaddition 271–2
funnels 270–1, 660
nickel tetracarbonyl 1099–101, 1120–1
pericyclic reactions 270
reaction paths 274–5
reactivity 659–63
SAC–CI method 1099–101, 1120–1
switching/switches 285–6, 986
valence bond diagrams 659–63
photoelectron spectra 1016–20, 1107–8
photoexcitation 270, 281, 661
photofragmentation reactions 1101, 1120–1
photoisomerization path 277, 282–4
photon energy
exploitation 270, 276–7
wastage 269–70, 274, 276–9
photonic reagents 149–62
photoproducts 270–2, 660
photoprotection/photostability 269
photoreceptors 285
photorhodopsin, photoRh 282–3
phthalocyaninines 1099, 1121–5
physisorption 822–3
Pinchas effect 838
pion-nucleon systems 135
Pipek–Mezey localization 1128
planar unsaturated organic molecules 752–6
plasmon
excitation 49–64
frequencies 49–64
lineshape 52–3
localized surface plasmon resonance 51, 54, 60
oscillations 48–9
resonance 49–64
platinum bonding 332–42
PLEDs 1040–1
PM3 methods
carbon nanostructure self-assembly 879–80
CH···O hydrogen bonds 847
isomeric fullerenes 893
MNDO semiempirical quantum-chemistry 566–76
semiempirical methods 559, 561
unimolecular reaction rates 416
vibrational spectroscopy 166, 179
PMF *see* potential of mean force
Poincaré surfaces 405–6
Poisson brackets 20
Poisson–Boltzmann equation 44
polarizability
coupled-clusters 133, 1212
dynamic 755, 939, 1101, 1207, 1212
electronic 573, 1028–9, 1031
frequency-dependent 60–1, 1051, 1057–61, 1176, 1182–4
intermolecular forces 1060–2
lithium-hydrogen molecules 628–9
long-range corrections 540
pyridine–copper tetramer 60–3
vibrational 1031
zero frequency 61
Polarizable Continuum Models (PCM) 1184
polarizable potentials 926, 953
polarization
CH···O hydrogen bonds 843, 844
functions 791–2, 795, 800–1, 814
many-body forces 920, 926, 931, 940–2, 956
model 921, 953–7, 1006
molecular Hamiltonians 13
multipole expanded dispersion 1057–8
non-additive interaction energy 940–2
polymer chains 1026–33, 1038
spin-polarization 678–9, 683–5, 777–8, 1134
water 956

polarons 1025
polyacenes 1039
polyacetylenes 1022–3, 1027–32
polyanilines 1025
Pólya's enumeration theorem 897
polyatomic molecules 165–90, 251–65
polyatomic monomers 925
polyenes 37, 129–30, 576, 1021, 1039
polyethene chains 1027–8
polyethylene 1013, 1015–20
polyhedr Voronoi 1086
polyhedra 893, 896–7
polymer chains 1011–42
  conduction 1020–5
  electron transfer 1033–41
  electronic properties 1026–33
  energy bands/gaps 1013–42
  non-linear excitations 1025–33
  non-linear optics 1025–33
  photoelectron spectra 1016–20
  semiconduction 1020–5
polymer coupled-clusters 1213–15
Polymer Light-Emitting Diodes (PLEDs) 1040–1
polymethime chains 1030–2
polymethine dyes 1021
polymethineimine 1030
POLYMODE 253, 254–5
POLYMOL 1012
polynomial expansions 168–9
polyparaphenylenes 1020, 1025, 1034, 1040
polypeptides 285, 963, 965, 987
polyphosphazenes 1030
polyphosphinoboranes 1030
polypropylenes 1019–20
polypyrrole 1020, 1025
polysilacetylenes 1030
polysilane chains 1027–8
polythiophene 1020, 1025
"pop-out" processes 886, 887
Pople and Walmsley misfit 1024
porphyrins 1101, 1121–5
positron annihilation 493
POSMOL 968
post-Hartree–Fock approaches 118, 128, 966
post-transition state dynamics 416–17
potassium 314–22, 980–1
potential energy
  coupled-clusters 118–19, 135–6
  curves 414–17, 467, 534, 1103–5, 1113, 1120
  fullerene formation 878–9
  Hartree–Fock functions 109–11
  time-dependent molecular theory 16
potential energy surfaces (PES)
  anharmonic vibrational spectroscopy 165, 166
  beryllium insertion 612, 615–19
  boron-hydrogen bond-dissociation 612–13, 624–7
  Cope rearrangements 863–5
  coupled-cluster theory 118–19, 133, 135–6
  electron transfer 1033–41
  G3S theory 803
  intermolecular forces 1058
  lithium-hydrogen 612–13, 619–24, 627–9
  many-body perturbation 509
  minima 195–218
  polyatomic molecules 261–2
  reaction pathways 195–202, 230–43
  SAC–CI method 1103
  size-consistent state-specificity 611, 612–29
  time-dependent molecular theory 11–39
  transition states 195–202, 218–30
  unimolecular reaction rates 409–13
  variational transition state theory 67, 80
potential of mean force (PMF) 80
potentials
  ab initio 104, 167–8, 173–89, 195–243, 951
  curves 118–19, 135–6, 748, 756–60
  empirical 176, 188–9, 434, 921, 926–31, 945–6
  fitting 174
  functions 433–4
  n-mode representation 255–60
  pairwise-additivity 920, 924–6, 946, 950–7, 1060–2
  polynomial expansions 168–9
  quartic force fields 180
  representations 175–6, 255–60
  triple dipoles 923–4
Powell-symmetric-Broyden (PSB) update 220
PPP *see* Pariser–Parr–Pople
predictor-corrector reaction paths 235–6, 242–3

pressure 412–13, 833, 896
product wave functions 15–17
projected frequencies 241–2
projected functions 1052–3
projection operations 472
promotion energy 639, 643, 649–50
promotion gaps 639–48, 658
propagation, Gaussian wave packets 28–9
propagators 120, 430, 443, 453–4, 470–1, 1060–4
protein–ligand interactions 851
proteins 42–5, 104–7, 431
  green fluorescent 269–70, 274, 276–9
proton affinity (PA) 794, 797, 802, 805–9
proton donors 831–9
proton transfer 653–4
protonated Schiff bases 274–6, 279, 286
protonation states 44
PS *see* pseudospectral
PS-VSCF *see* Partly Separable VSCF
PSB *see* Powell-symmetric-Broyden
pseudospectral (PS) approach 545–7
PT *see* perturbation theory
pulse shaper 149–62
pulsed-field ionization zero-electron-kinetic energy (PFI-ZEKE) 1090–1
punch cards 2–3
push–pull polyene pairs 1029
pyramid particles 51–2, 305, 311, 1111–12
pyridine–copper tetramers 60–3

Q projections 591–8, 607–8
QCAS *see* quasi-complete active space
QCASSCF *see* quasi-complete active space self-consistent fields
QCI *see* quadratic configuration interaction
QCPE *see* Quantum Chemistry Program Exchange
QDPT *see* quasi-degenerate perturbation theory
QET *see* quasi-equilibrium theory
QFF *see* quartic force fields
QM *see* quantum mechanics
QM/MM *see* quantum mechanics/molecular mechanics
QST *see* quadratic synchronous transit
quadratic configuration interaction (QCI) 786–7, 791–6, 799–801, 804, 1200
quadratic synchronous transit (QST) 222
quadrupole moments 824, 827
Quantum Chemistry Program Exchange (QCPE) 3–4
quantum classical Liouville equation 435
quantum coherence 159–60
quantum control mechanism 156–62
quantum dynamic time-dependency 11–12
quantum field theory 120–1, 125–6
  lattice 135
quantum fluid mechanics 135
quantum mechanics/molecular mechanics (QM/MM)
  GAMESS 1181–4
  minimization 203, 209–10
  photobiology 275–6, 286
  potential energy surfaces 198, 203, 209–10
  semiempirical quantum-chemistry 572–4, 576
quantum phenomena controls 149–62
quantum Rice–Ramsperger–Kassel–Marcus kinetics 413–15
quantum spin 135
quantum vibrational motion of polyatomic molecules 251–65
quantum virial theorem 292–3
quartic force fields (QFF) 180
quasi-classical electrostatic interaction 291–367
quasi-complete active space (QCAS) 588, 610
quasi-complete active space self-consistent fields (QCASSCF) 514–18
quasi-degeneracy 465, 508–23, 588–9
quasi-degenerate perturbation theory (QDPT) 508–23
quasi-equilibrium theory (QET) 398–420
quasi-fullerenes 898
quasi-Newton methods 204, 206–7
quinone moiety 987–9
quinones 1108–11

r space 486–8, 494–6
R-CCSD(T) 1066
radial distribution functions 1002–3
radicals 187–8, 644–6, 654, 748, 751–6
Raman
  reflection 63
  spectra/intensity 47, 49, 55–64, 944, 1087, 1176–7
random-phase approximation (RPA) 695, 1207

randomly oriented carbon molecules 883–7
rank
operators 585
reduction 449
rare-gas clusters 921, 924, 945–6
rare-gas crystals 923–4, 944–6
rare–gas molecules 185–6
RAS *see* restricted active space
rate constants 68, 74–5, 79, 82–3
rate-equilibrium relationships 636, 640–2
rational function optimization (RFO) 205–6
Rayleigh–Schrödinger perturbation theory (RSPT) 510–11, 1052–3, 1061–2, 1192–3
RCCSD 1066
reaction
coordinates 75–6, 80–3, 431
paths 79, 195–202, 230–43, 260–5
see also individual reactions
reaction path Hamiltonians (RPH) 230–1, 239, 241, 260–4, 265
reaction-diffusion dynamics 436
reactive empirical bond-orders (REBO) 877–8, 882–3, 886–7
reactive trajectories 69–72
real time non-equilibrium Green Functions 373–95
REBO *see* reactive empirical bond-orders
receptors 965, 980–3
reciprocal form factor 489, 493–5
red shift 831, 837–9, 843–50, 969–71
reduced absorption coefficients 1112–15
reduced density matrices 487–9, 502, 672, 677–9, 726, 727–30
reduced gradient following (RGF) 221
reduced resolvent 940
redundant internal coordinates 213–15, 228–9
reference states 30–2
reference wavefunctions 514–17
refractive index 52–5, 833
rehybridization 838–9, 847
relative free energy 42–3
relative stability 892–3, 896
relativistic
corrections 106
Hamiltonians 548, 550–3
methods 1173–5
molecular theory 507, 540–53
regimes/effects 747–8, 756–60
relativistic scheme by eliminating small components (RESC) 548–9
relaxation energy 1039
reorganization energy 1038, 1039
repulsion 291–367, 840–4, 1053–4
Coulomb 779–80
effects 773–4
integrals 544–5
RESC *see* relativistic scheme by eliminating small components
resolution of the identity 24, 32, 1082, 1168, 1204–9
resolvent identity 475
resonance
energy 409–13, 652
integrals 129–30, 294, 566–7, 644, 651
states 409–13
theory 117
widths 409–13
response density matrices 1205–6
response functions 133, 374, 445, 455–8
restricted active space (RAS) 588, 761
restricted Hartree–Fock (RHF) 109–11, 789, 864–5
resulting diagrams 123, 127
RFO *see* rational function optimization
RGF *see* reduced gradient following
Rh *see* Rhodopsin
rhenium bonding 326–32
RHF *see* restricted Hartree–Fock
Rhodopseudomonas viridis 575–6, 1121, 1124–5
Rhodopsin (Rh) 269, 276–7, 279–85
Rice–Ramsperger–Kassel–Marcus (RRKM) theories 397–420
ridge following methods 223
rigid environments 78
rigid-monomers 1058, 1065
approximation 925
potentials 923–5
rigid-rotor and harmonic-oscillator (RRHO) approximation 893
RIMP2 999
ring diagrams 124–5
Ritz variational principle 118, 125
Roothan–Hall equations 1080
rotational constants 552
rotational spectroscopy 172, 265, 835, 925, 947, 995

rovibrational states/levels 37–8, 409, 627, 944–6, 952, 1065
RPA *see* random-phase approximation
RPH *see* reaction path Hamiltonians
RRHO *see* rigid-rotor and harmonic-oscillator
RRKM *see* Rice–Ramsperger–Kassel–Marcus
RSPT *see* Rayleigh–Schrödinger perturbation theory
rubidium bonding 314–22
Runge–Kutta second-order methods 234
ruthenium 347–54, 1116–18, 1157–63
Rydberg excitations 714, 1107, 1126, 1211
Rydberg states 534, 742, 1100

SAC *see* scaling all correlation
SAC–CI *see* Symmetry-Adapted Cluster–Configuration Interaction
saddle methods 79, 226–7
SAM1 893
sandwich compounds 320–1, 360–6
SAPT *see* symmetry-adapted perturbation theory
satellite spectra 1107–8, 1125–7
saturated pressure 896
SBKJC basis sets 61, 62
scaling
  anharmonic vibrational spectroscopy 176–7
  electronic structure theory 1168, 1177–81
  exchange-correlation 679–80
  scale factors 803–7
  scaled quantum mechanics (SCM) 167
  scaling all correlation (SAC) 803–4
  systematic constraint satisfaction 695–6
scanning tunneling microscopy (STM) 373–4
scattering 470–1, 1080–3
  optical 47–54, 60, 77, 491–3, 500
SCC *see* self-correlation correction
SCF *see* self-consistent fields
SCF MO *see* self-consistent field molecular orbitals
Schellman motif 851
Schrock-type complexes 342–7
Schrödinger eigenvalues 11, 15
Schrödinger equation 15–17, 117, 196–7, 443–6, 483–5
SCM *see* scaled quantum mechanics (SCM)
screened hybrids 712–14
seam minimization 210–12
second harmonic generation (SHG) 1026
second order DGE 687–8, 692–3
second order electrical properties 627–9
second order perturbation theory (CASPT2) 518–20, 1065, 1133
  Cope rearrangements 864–5, 871
  Møller–Plesset (MP2) 744, 820, 825–6, 997, 1172–3
  multiconfigurational quantum chemistry 725, 744–8, 754–6, 757, 760–1
  photobiology 273–6, 278, 280–3, 286–7
  size-consistent state-specificity 586–7
second order reaction path following 234–6
second virial coefficients 833, 926, 952, 1049, 1062
selenium 1118–19
self-consistent field molecular orbitals (SCF MO) 93, 560–2, 570–1, 576, 1051
self-consistent fields (SCF)
  convergence 572, 1085–6, 1088–9, 1171
  coupled-clusters 117
  DFT deMon codes 1085, 1088
  Fenske–Hall molecular orbitals 1143–4
  GAMESS 1169–72, 1178
  intermolecular forces 1051, 1053–4
  molecular system simulations 95–103, 109–11
  polymer chains 1015–16
  semiempirical quantum-chemistry 560–2, 570–1, 576
  vibrational 167, 169–72, 174–85, 252–63
self-correlation correction (SCC) 697
self-energies 381–2, 383–4, 389–92
self-exchange charge transfer 1034–5
self-interaction correction (SIC) 531
self-interaction energy 531, 636, 711
self-interaction error 711, 1082
semibullvalene 662–3
semiclassical methods 173
semiconduction 1020–5
semiempirical techniques 559–77
  applications 573–6
  CNDO 61, 560–3, 568, 834, 1016
  INDO 560–3, 567, 569
  MINDO 893, 894–5
  MINDO/3 calculations 862–3

MNDO 561–70
OM1, OM2 & OM3 567
polymer chains 1016
SINDO1 561, 569, 575
see also AM1; PM3 methods
sensors 49, 53–4, 965, 980–3
SERS *see* surface enhanced Raman spectra
SHAKE algorithm 429–30
shake-up effects 134, 1108, 1126, 1210–11
shaped ultra-fast laser pulses 149–62
shell-model 118, 128, 587, 591, 601, 611–12
SHG *see* second harmonic generation
shock compression 435
SIC *see* self-interaction correction
sidewall current flow 813
sigma bonding 778, 783
silicon 654–6
silicon bonding 310, 314–22, 342–7
silicon-containing carbon clusters 521–3
silver 47–64, 347–54, 974–6
SIMOMM 1183
simulated annealing 886–7, 1183
simulations
1930–1960 90–2
carbon nanostructure self-assembly 875–87
DFT 95–103, 108–9
fluid dynamics 105–8
Hartree–Fock 93, 95–104, 106, 109–11
large molecular systems 89–111
MCSCF 95–103, 109–11
self-consistent fields 93, 95–103, 109–11
statistical simulations 105–8
SINDO1 561, 569, 575
single bonds 305–8, 310, 612, 749, 756
single nonpolar bonds 305–7
single reference coupled-clusters (SR CC) 466–8, 474, 582–3
single-reference many-body perturbation theory (SR-MBPT) 509
single-walled carbon nanotubes (SWCNT) 820–3, 875–87
singlet instability 129–30
singlet states 344–5, 376, 773–5, 789
singlet-triplet splitting 513–14
size extensivity 472–6, 1191
iterative corrections 473, 1201
a posteriori correction 469, 476–9, 585
size-consistent state-specific multi-reference (SSMR) methods 581–631
API excitations 582, 587, 593–9, 611–12, 614–15, 617
CAS 583, 586–93, 603, 616
CEPA-like approximations 588–9, 602–5, 612–13, 618–24, 626–7
CMS 587–8, 589–93, 607
Epstein–Nesbet partition 582, 600–2, 612, 615, 625, 627
incomplete model space 581, 584, 606–11
Møller–Plesset partitions 600, 602, 612, 625, 627
performance 611–29
perturbation 587, 599–602, 617–18, 625–7
results 611–29
size-up/size-down mechanisms 875, 887
Slater-type orbitals (STO) 93, 95, 296, 1081
Slater's exchange term 1079–80
SM *see* stabilization methods; supermolecular
smoothed particle hydrodynamics (SPH) 437
S_N2 reaction 412, 648, 651, 656–7
SO *see* spin-orbitals
SOC 757
sodium bonding 314–22
sodium trimers 948–9
solid–fluid phase transition 426
solid-state optoelectronics 135
solid-state reactions theory 77–83
solitons 1022–5
solvated ions 104–5
solvation 42–4, 81–2, 104–5, 973
equilibrium 80–1
nonequilibrium 81–2
separable equilibrium 80
solvents 77–82, 980–3, 1035–9
GAMESS 1181–5
molecular dynamics development 429–35
non-covalent binding affinity 41–4
plasmon lineshape 52–3
SOS *see* sum-over-states
space wave functions 484–6, 495–6
space-symmetry broken Hartree–Fock instabilities 130
SPARTAN 893, 1167
spawning methods 435
species identification 185–6
specific reaction parameters 416, 570, 574–5

spectral shifts 845, 923
spectroscopic constants 624, 625, 759
spectroscopy
  absorption spectra 1101, 1111–16, 1120–5
  anharmonic vibrational 165–90
  electronic 742, 748, 751–6
  excitation spectrum 1099, 1100–2, 1106–7, 1111–12, 1116, 1120–4
  infrared spectra 189, 263, 831–3, 1062
  momentum density determination 493
  Raman spectra 47, 59–64, 944, 1087, 1176–7
  rotational spectroscopy 172, 835, 925, 947, 995
  satellite 1107–8, 1125–7
  time domain/resolved 189, 282–4
  variational transition state theory 77
  vibrational 165–90, 1125–8
  X-ray photoelectron 1016–20, 1125
SPH *see* smoothed particle hydrodynamics
spherical particles 48–53, 56–9
spheroid particles 51–3, 170
spin
  algebras 123
  endohedral fullerenes 983–7
  momentum density 489
  multiplicity 137, 533, 591, 1101–5, 1129
spin-adaptation 581, 590–2, 605, 619
spin-compensation 684–5
spin-density 728, 753
spin-dependence 123
spin-DFT 677–8, 1082
spin-exchange 776–7
spin-flop EOMCC method 469
spin-functions 1210–11
spin-Hamiltonian mapping 775–7
spin–orbit
  corrections 795
  coupling 758, 782–3, 1174–5
  effect 552, 756
  interactions 794, 951
spin-orbitals (SO) 29–32, 726, 728, 731, 758
spin-pairing 644
spin-polarization 678–9, 683–5, 777–8, 1134
spin-scaling 680, 689
spinors 543–4
spiral algorithm 897
spontaneous density fluctuations 374–5, 376–7
SR CC *see* single reference coupled-clusters
SR-MBPT *see* single-reference many-body perturbation theory
S_{RN}^2 mechanism 656–8
$S_{RN}2^c$ mechanism 656–8
SRPES *see* synchrotron radiation photoelectron spectra
SS *see* state-selective/specific
SSH *see* Su, Schrieffer and Heeger
SSMR *see* size-consistent state-specific multi-reference
stability groups 27
stabilization methods (SM) 458–9
standard free energy of complexation 41–5
state correlation diagrams *see* valence bond state correlation
state universal (SU) methods 136–9, 583
state-averaged CASSCF 512
state-selective/specific (SS) methods 137
state-specificity 137, 411, 581–631
state–state processes 36–7
static electron correlation 736–8
static polarizability 540, 1027–33
stationary states 9–16, 20, 432, 1208–9
statistical correlation 1001
statistical mechanics 398, 425–32, 793, 901, 1081
statistical simulations 105–8
step-wise mechanisms 413, 475, 638, 646, 652–3, 656–8
stereo conjugated polymers 1027
stereochemical predictions 650–2
steric repulsion 840, 843, 844
STM *see* scanning tunneling microscopy
STO *see* Slater-type orbitals
stochastic rotation 428, 436
Stokes frequency 56
Stone–Wales transformation 882, 886, 897
stretching frequencies 831, 837–9, 843–50, 969–71
string methods 224, 227, 431
strontium bonding 314–22
SU *see* state universal
Su, Schrieffer and Heeger (SSH) theory 1023–4

substituent effects 865–71
competitive 866, 869–71
cooperative 865–9
Cope rearrangement 865–70
phenyl 859, 865–6, 871
sulfur 322–6, 1118–19
sum-over-states (SOS) 61, 1212
superconductivity 907
superfluid helium 946
supermolecular (SM) methods 966–7, 1049–56, 1069–72
superoperators 373–95
supramolecular structures 41, 1042
surfaces
chemistry 1167, 1183
hopping 434–5
intersections 210–12
selection rules 61
semiempirical quantum-chemistry 575
surface enhanced Raman spectra (SERS) 49, 55–63
surrounding medium effects 52–3
Suzuki coupling 646
SWCNT *see* single-walled carbon nanotubes
switching/switches, photochemistry 285–6, 986
symmetrized perturbation theory 1051
symmetry dilemma 120, 129–30
Symmetry-Adapted Cluster–Configuration Interaction (SAC–CI) 1099–137
analytical energy gradients 1102, 1128–9, 1137
bacterial photosynthetic reaction centers 1099, 1101, 1124–5
collision-induced absorption 1099–101, 1112–15
equilibrium geometries 1099–100, 1102, 1128–33
excited states 1099–102, 1105–12, 1115–19, 1122–3, 1128–33
general-*R* method 920–1, 929–30, 942, 1128, 1129–33
hyperfine splitting constants 1099–100, 1102, 1133–6
ionization spectra 1099–102, 1106–12, 1125–8
method 1102–6
photochemical reactions 1099–101, 1120–1, 1124–5
porphyrins 1101, 1121–5
potential curves 1103–5, 1113, 1120
SD-R method 1102–3, 1132–6
spectra 1107–8, 1112–16, 1122, 1127
transition metals 1099, 1101, 1115–19, 1120–1
van der Waals complexes 1099, 1101, 1112–15
symmetry-adapted perturbation theory (SAPT)
DFT 928, 954–7, 1064–5
intermolecular forces 1060–5
many-body forces 919–58
SAPT-5s 935, 951–7
synchrotron radiation photoelectron spectra (SRPES) 1107–8
systematic constraint satisfaction 695–9

tantalum bonding 326–32
tau-dependent functionals 698
Taylor series 234
TD-DFT *see* time-dependent density functional theory
TDCHF 1060–1
TDHF *see* time-dependent Hartree–Fock
TDVP 24
temperature dependence 135, 412–13, 1003–4
tetrahedra 52–7, 649, 897, 937, 954–5
tetramethylenethane 476
tetraoxo-metal complexes 1116–18
tetraphenyl substituent effects 865–70
thallium bonding 314–22, 332–42
thermal unimolecular dissociation 398
thermochemistry 785–810
thermodynamics
aqueous clusters 968–9
behavior 995–7
carbon clusters 893–6
cycle analysis 42–3
driving forces 861–2
equilibrium 883, 886, 893, 907
intermolecular forces 1058–9
non-covalent binding affinity 42–3
water hexamers 997, 1001
THG *see* third harmonic generation
1-thiol-4-ethynylphenyl-4'-ethynylphenyl-1'-benzenethiolate 817–9
thiophene 1106–8
third harmonic generation (THG) 1026
third row elements 807–8

third-order induction 938, 939–42
third-order integrators 236–7
third-order non-linear optical effects 1032–3
Thomas–Fermi model 669, 680, 702, 717, 1080–2
Thomas–Fermi–Dirac model 669, 1080
Thouless determinant 33
three-body
  forces 943–8
  interaction energy 922–38, 944–5, 948–51
  potentials 953–4, 957
tight binding
  deMon DFT 1088
  high-temperature quantum chemistry 875–9
  molecular conductance 814
  nanomaterials 983
  one-electron Hamiltonian 766, 776
  polymer chains 1024
time evolution 24, 25–9
time-dependent density functional theory (TD-DFT) 274–5, 530–2, 534–7, 1101
time-dependent Hartree–Fock (TDHF) methods 1052, 1207
time-dependent molecular theory 9–39
  coherent states 21–32
  Coulombic Hamiltonian 12–17
  determinantal coherent states 29–32
  direct nonadiabatic dynamics 9–39
  dynamic rendering 37–8
  electron nuclear dynamics 32–7
  Gaussian wave packets 21–9
  Hamiltonians 12–17
  minimal electron nuclear dynamics 32–7
  potential energy surfaces 11–39
  quantum mechanics 18–21
  time-dependent variational principle 18–21
time-dependent perturbations 120–1
time-dependent Schrödinger equation 10–11
time-dependent variational principle 18–21
time-domain experiments 189, 282–4, 399
time-independent perturbation 120–1, 124
time-independent Schrödinger equation 196–7
time-resolved spectroscopy 282–4
time's direction 128
tin bonding 310, 314–22, 342–7
TIP4P water potentials 996, 1002, 1005–6
titanium: sapphire lasers 153
toluene 155
topography 497–8
transition metals
  active space selection 741–3
  catalysed bond activation 646–7
  chemical bonds 291, 294–5, 326–66
  designing functional molecular systems 976
  dimers 534, 538–40
  excitation spectra 1101, 1115–19
  Fenske–Hall molecular orbitals 1143, 1144, 1150–63
  NMR chemical shifts 1101, 1115–19
  orbital ordering 781
  photofragmentation 1101, 1120–1
  SAC–CI method 1099, 1101, 1115–19, 1120–1
transition probability 157–8
transition states (TS)
  assumption 68–72, 76–7, 80–1
  barrier heights 517–18
  Cope rearrangements 859–72
  dividing surfaces 67–81
  fundamental dynamical 68, 70, 80–1
  hydrogen exchange 527–9
  photobiology 270
  potential energy surfaces 195–202, 218–30
  unimolecular reaction rates 413–14
  variational 67–84
transition structures 272, 478, 517
  aromatic 860, 862, 864
  centauric 860–2
  chameleonic 859–60
  Cope rearrangement 860–71
  stabilization 870
transmission coefficients 82–3
transport
  defects 1038
  many-body forces 924
  molecular dynamics 431–4
  molecular wires 374
  properties 431, 434, 924
  rigid environments 78
  solitons 37, 1038
tri-ruthenium dodecacarbonyl 1157–63
triatomic molecules 397–420, 741–2
tricarbonyl-manganate 1143, 1151–6
trideuterium oxide 259–60, 262–3
trimers 922–3, 932–42, 943–57
trimethylamine 101–2

triphenyl substituent effects 865–70
triple dipole terms 923–4
triplet states
electron-electron repulsion 774–7
energy decomposition 345, 363
molecular spectroscopy 1130
multiconfigurational quantum chemistry 732–3
perturbation 476
photobiology 287
valence bonds 645
TRM *see* trust radius method
Trotter factorization 430
truncated tetrahedron particles 55–9
trust radius method (TRM) 205–6
TS *see* transition states
Tully's fewest stitches algorithm 434–5
tumor cell lines 477
tungsten 322–32, 342–60
tunneling
electrons 373–95
multi-dimensional 68, 74–5, 82–3
splittings 259–63, 968, 1062
variational transition states 68, 74–5, 82–3
vibration-rotation 947–8, 952, 953–4
twin excitation/transition states 659–63
two-body
cluster expansions 469
interaction energies 932–7
potentials 951–2
two-component relativistic molecular theory 548–53

ubiquinones 1108–9
UEG *see* uniform electron gas
UFF *see* Universal Force Fields
UGA *see* unitary group approach
UHF *see* unrestricted Hartree–Fock
umbrella sampling 430–1
unexpanded dispersion 1054–6
unfilled vacancy molecules 748, 749–51
uniform electron gas (UEG) 670, 679, 682, 684
uniform scaling 679–80
unimolecular dissociation 403–5, 517–18
unimolecular rate constant 400
unimolecular reaction rates 397–420
direct dynamic simulations 415–19
lifetime distribution 400–3
resonance states 409–13
steps 413–15
unitary group approach (UGA) 126
Universal Force Fields (UFF) 820
unlinked diagrams 121–2, 1066, 1193–8
unrestricted Hartree–Fock (UHF) 109–11, 732–3, 789
unsaturated hydrocarbons 794
unsaturated organic molecules 752–6
unsubstituted methane 835–6
updating schemes 204–5, 208–9, 215, 219–20
uracil 832
uranium oxide 741–2

v-representability problems 1082
vacuum
diagrams 121–2
Fermi 468
state 30
valence bond configuration interaction (VBCI) 642–3
valence bond configuration mixing diagrams (VBCMD) 637–8, 652–8
valence bond state correlation diagrams (VBSCD)
applications 642–52
chemical reactivity 637–52
electronic delocalization 658–9
photochemical reactivity 659–63
twin-state concept 659–63
valence bonds (VB) 635–65
CAS 523–9
coupled-cluster theory 117
diagrams 223–4, 635–65
molecular system simulations 91, 101
structures 523, 527–9, 637–43, 652–4, 662–4, 1055
transition states 223–4
wave functions 664–5
valence double zeta (VDZ) 814, 816, 999, 1001
Valence Effective Hamiltonian (VEH) 1016
valence excitation energies 518–21, 755
valence excited state energy 755
valence ionization spectra 1107–8, 1125
valence polarization functions 800–1
valence properties, polymer chains 1016–17
valence states 1100

valence-Rydberg mixing 746, 1106
valence-universal (VU) methods 136–7, 583
van der Waals
bonding 533–4
clusters 176, 185
collision-induced absorption 1101, 1112–14
complexes 1099, 1101, 1112–15
forces 919, 927–8, 1048
interactions 109, 690, 714, 927, 974
law 1048
molecules 1048, 1064–72
van Vleck perturbation 169, 509, 512
vapor pressure 833
variational principle 12, 18, 32
ab initio vibrational spectroscopy 175
exchange-correlation 674
molecular spectroscopy 1102
multiconfigurational quantum chemistry 733
Ritz 118, 125
variational theory (VT) 137–8, 252–4
variational transition state theory (VTST) 67–84
canonical 69–74, 81–3
condensed phase reactions 77–83
free energy of activation 72, 80–3
gas phase reactions 68–77
quantized dynamical bottlenecks 77, 90
quantized generalized transition state 73–4
rate constants 67–74, 77–83
reaction paths 230–1, 239, 241
solid-state reactions 77–83
thermodynamic formulation 72, 80
variable reaction coordinates 76
VAX 11/780 minicomputer 5
VB *see* valence bonds
VBCI *see* valence bond configuration interaction
VBCMD *see* valence bond configuration mixing diagrams
VBSCD *see* valence bond state correlation diagrams
VCI *see* vibrational configuration interaction
VDZ *see* valence double zeta
vector computing architectures 5
VEH *see* Valence Effective Hamiltonian
velocity 237, 427
verifications, transition states 230
Verlet algorithm 226, 427
vertical charge transfer 649–50
vertical electron affinity 443–4, 648–9
vertical excitation energy 754–6
vertical ionization potentials 443–4, 649–50
vertices 121–3, 128, 898
interaction vertices 122–3, 128
supervertices 123
vibration
polyatomic molecules 251–65
polymer chains 1018–19, 1030–3
vibration-rotation-tunneling (VRT) 947–8, 952, 953–4
vibrational analysis 251–2
vibrational configuration interaction (VCI) 254–63
vibrational energy levels 413–14
vibrational frequencies 72–3, 1175–7
vibrational generalized transition state partition 72–3
vibrational methods
diffusion quantum Monte Carlo 172–3
grid methods 172
harmonic approximation 167–75, 182, 186–7, 252–4, 1030
perturbation theory 168–9
reduced-dimensionality 172
semiclassical methods 173
vibrational modes 60–1, 72–6, 165–80, 262
bending 73, 661–2, 970–1
normal coordinates 251–7, 261–4
normal modes 166–85, 241, 251–64
separability 175
stretching modes/stretching 181–7, 262, 837–40, 968–71
torsion modes/torsions 172–5, 185–7, 260–4
vibrational polarizability 1031
vibrational self-consistent fields (VSCF) 167, 169–72, 174–85, 252–63
correlation-corrected 171, 177–8, 254
DPT2 179
vibrational spectra/spectroscopy 165–90, 1125–8
ab initio 173–87
vibrational transitions 187
combination transitions 186–7
fundamental transitions 167, 187
overtone transitions 186–7
vibrational zero point energy 84, 893, 997

vibrationally adiabatic potentials 74
virial
  coefficients 944–5, 1058–9
    second 833, 926, 952, 1049, 1062
    third 944–5
  theorem 292–3, 300
virtual states 253–4
viscosity 833, 1049
visual pigments 276
Vosko–Wilk–Nusair (VWN) correlation 530, 684–5
VRT *see* vibration-rotation-tunneling
VSCF *see* vibrational self-consistent fields
VT *see* variational theory
VTST *see* variational transition state theory
VU *see* valence-universal
VWN *see* Vosko–Wilk–Nusair

water
  CH···O hydrogen bonds 833–4, 840–5
  clusters
    energetics 968–9
    finite temperature properties 995–1006
    many-body forces 925, 928–42, 947–8, 951–7
    Monte Carlo simulations 995–1006
  dimers 925, 928–30
  hexamers 995–1006
  liquid 104, 427–8, 936–7, 954–8, 1001–3
  many-body forces 925, 928–42, 947–8, 951–7
  molecular dynamics 427–8
  molecules 105, 171, 729, 925–38, 969–86
  natural orbitals 729–30
  non-additive interaction energy 932
  potentials 996, 1002, 1005–6
  three-body potentials 953–4, 957
  trimers 932–42, 947–8, 951–7
  vibrational spectra 1127–8
Watson Hamiltonians 171, 252–4
wave functions
  coupled-clusters 127–8, 138–9, 470–2
  equations of motion 446–7
  exchange-correlation holes 701–3
  GAMESS 1167–85
  intermolecular forces 1052–3
  minimal electron nuclear dynamics 32–7
  multiconfigurational quantum chemistry 725–61
  multireference coupled-clusters 470–2
  valence bonds 664–5
wave packets 21–9
weakly bound clusters 976–80
weight factors 127–8
weighted-density approximation 682
Weizsäcker kinetic energy density 531, 696–7
Wely groups 23
Wick's theorem 394–5
width evolution 25–9
Woodward–Hoffmann reactions 646–7
workstations 5–6, 1059
World War II 1079–80

X-ray photoelectron spectroscopy (XPS) 1016–20, 1125
Xα-Scattered Waves (Xα-SW) 1080, 1083–4
xenon hexafluoride 322–6
XPS *see* X-ray photoelectron spectroscopy

Z-effect 98
Z-matrix coordinates *see* internal coordinates
ZAPT 1178
ZDO *see* zero differential overlap
ZEKE *see* zero-electron-kinetic energy
zero differential overlap (ZDO) 1148
zero frequency polarizability 61
zero-electron-kinetic energy (ZEKE) 1090–1
zero-momentum Hessians 498
zero-point energy (ZPE)
  aqueous clusters 968–9
  energy decomposition 297
  G2 and G3 theories 789–90, 794, 798–800, 803–4
  variational transition state theory 72
  vibrational configuration interaction 260
zeroth-order approximation 920, 927–8
zeroth-order Hamiltonians 601–2, 745–6
zeroth-order operators 1055–6
zeroth-order regular approximation (ZORA) 548
zeroth-order wave functions 1052–3
Ziegler-Natta reaction 105
zinc bonding 314–22
ZINDO 898
ZORA *see* zeroth-order regular approximation
ZPE *see* zero-point energy